HANDBOOK OF
NANOPHYSICS

Handbook of Nanophysics

HANDBOOK OF NANOPHYSICS

Nanomedicine and Nanorobotics

Edited by

Klaus D. Sattler

CRC Press
Taylor & Francis Group
Boca Raton London New York

CRC Press is an imprint of the
Taylor & Francis Group, an **informa** business

CRC Press
Taylor & Francis Group
6000 Broken Sound Parkway NW, Suite 300
Boca Raton, FL 33487-2742

First issued in paperback 2019

© 2011 by Taylor & Francis Group, LLC
CRC Press is an imprint of Taylor & Francis Group, an Informa business

No claim to original U.S. Government works

ISBN-13: 978-1-4200-7546-5 (hbk)
ISBN-13: 978-0-367-38362-6 (pbk)

Library of Congress Cataloging-in-Publication Data

Handbook of nanophysics. Nanomedicine and nanorobotics / editor, Klaus D. Sattler.
      p. cm.
   Includes bibliographical references and index.
   ISBN 978-1-4200-7546-5 (alk. paper)
   1. Nanomedicine. 2. Nanoelectromechanical systems. I. Sattler, Klaus D. II. Title: Nanomedicine and nanorobotics.

R857.N34N353 2009
610.28'4--dc22
                                                                              2009038060

Visit the Taylor & Francis Web site at
http://www.taylorandfrancis.com

and the CRC Press Web site at
http://www.crcpress.com

Contents

PART I Nano-Bio Interfacing

PART II Nanotoxicology

PART III Clinical Significance of Nanosystems

PART IV Medical Imaging

PART V Drug Delivery

Preface

The *Handbook of Nanophysics* is the first comprehensive reference to consider both fundamental and applied aspects of nanophysics. As a unique feature of this work, we requested contributions to be submitted in a tutorial style, which means that state-of-the-art scientific content is enriched with fundamental equations and illustrations in order to facilitate wider access to the material. In this way, the handbook should be of value to a broad readership, from scientifically interested general readers to students and professionals in materials science, solid-state physics, electrical engineering, mechanical engineering, computer science, chemistry, pharmaceutical science, biotechnology, molecular biology, biomedicine, metallurgy, and environmental engineering.

What Is Nanophysics?

Modern physical methods whose fundamentals are developed in physics laboratories have become critically important in nanoscience. Nanophysics brings together multiple disciplines, using theoretical and experimental methods to determine the physical properties of materials in the nanoscale size range (measured by millionths of a millimeter). Interesting properties include the structural, electronic, optical, and thermal behavior of nanomaterials; electrical and thermal conductivity; the forces between nanoscale objects; and the transition between classical and quantum behavior. Nanophysics has now become an independent branch of physics, simultaneously expanding into many new areas and playing a vital role in fields that were once the domain of engineering, chemical, or life sciences.

This handbook was initiated based on the idea that breakthroughs in nanotechnology require a firm grounding in the principles of nanophysics. It is intended to fulfill a dual purpose. On the one hand, it is designed to give an introduction to established fundamentals in the field of nanophysics. On the other hand, it leads the reader to the most significant recent developments in research. It provides a broad and in-depth coverage of the physics of nanoscale materials and applications. In each chapter, the aim is to offer a didactic treatment of the physics underlying the applications alongside detailed experimental results, rather than focusing on particular applications themselves.

The handbook also encourages communication across borders, aiming to connect scientists with disparate interests to begin interdisciplinary projects and incorporate the theory and methodology of other fields into their work. It is intended for readers from diverse backgrounds, from math and physics to chemistry, biology, and engineering.

The introduction to each chapter should be comprehensible to general readers. However, further reading may require familiarity with basic classical, atomic, and quantum physics. For students, there is no getting around the mathematical background necessary to learn nanophysics. You should know calculus, how to solve ordinary and partial differential equations, and have some exposure to matrices/linear algebra, complex variables, and vectors.

External Review

All chapters were extensively peer reviewed by senior scientists working in nanophysics and related areas of nanoscience. Specialists reviewed the scientific content and nonspecialists ensured that the contributions were at an appropriate technical level. For example, a physicist may have been asked to review a chapter on a biological application and a biochemist to review one on nanoelectronics.

Organization

The *Handbook of Nanophysics* consists of seven books. Chapters in the first four books (*Principles and Methods, Clusters and Fullerenes, Nanoparticles and Quantum Dots,* and *Nanotubes and Nanowires*) describe theory and methods as well as the fundamental physics of nanoscale materials and structures. Although some topics may appear somewhat specialized, they have been included given their potential to lead to better technologies. The last three books (*Functional Nanomaterials, Nanoelectronics and Nanophotonics,* and *Nanomedicine and Nanorobotics*) deal with the technological applications of nanophysics. The chapters are written by authors from various fields of nanoscience in order to encourage new ideas for future fundamental research.

After the first book, which covers the general principles of theory and measurements of nanoscale systems, the organization roughly follows the historical development of nanoscience. *Cluster* scientists pioneered the field in the 1980s, followed by extensive work on *fullerenes, nanoparticles,* and *quantum dots* in the 1990s. Research on *nanotubes* and *nanowires* intensified in subsequent years. After much basic research, the interest in applications such as the *functions of nanomaterials* has grown. Many bottom-up and top-down techniques for nanomaterial and nanostructure generation were developed and made possible the development of *nanoelectronics* and *nanophotonics.* In recent years, real applications for *nanomedicine* and *nanorobotics* have been discovered.

Acknowledgments

Many people have contributed to this book. I would like to thank the authors whose research results and ideas are presented here. I am indebted to them for many fruitful and stimulating discussions. I would also like to thank individuals and publishers who have allowed the reproduction of their figures. For their critical reading, suggestions, and constructive criticism, I thank the referees. Many people have shared their expertise and have commented on the manuscript at various stages. I consider myself very fortunate to have been supported by Luna Han, senior editor of the Taylor & Francis Group, in the setup and progress of this work. I am also grateful to Jessica Vakili, Jill Jurgensen, Joette Lynch, and Glenon Butler for their patience and skill with handling technical issues related to publication. Finally, I would like to thank the many unnamed editorial and production staff members of Taylor & Francis for their expert work.

Klaus D. Sattler
Honolulu, Hawaii

Editor

Klaus D. Sattler pursued his undergraduate and master's courses at the University of Karlsruhe in Germany. He received his PhD under the guidance of Professors G. Busch and H.C. Siegmann at the Swiss Federal Institute of Technology (ETH) in Zurich, where he was among the first to study spin-polarized photo-electron emission. In 1976, he began a group for atomic cluster research at the University of Konstanz in Germany, where he built the first source for atomic clusters and led his team to pioneering discoveries such as "magic numbers" and "Coulomb explosion." He was at the University of California, Berkeley, for three years as a Heisenberg Fellow, where he initiated the first studies of atomic clusters on surfaces with a scanning tunneling microscope.

Dr. Sattler accepted a position as professor of physics at the University of Hawaii, Honolulu, in 1988. There, he initiated a research group for nanophysics, which, using scanning probe microscopy, obtained the first atomic-scale images of carbon nanotubes directly confirming the graphene network. In 1994, his group produced the first carbon nanocones. He has also studied the formation of polycyclic aromatic hydrocarbons (PAHs) and nanoparticles in hydrocarbon flames in collaboration with ETH Zurich. Other research has involved the nanopatterning of nanoparticle films, charge density waves on rotated graphene sheets, band gap studies of quantum dots, and graphene foldings. His current work focuses on novel nanomaterials and solar photocatalysis with nanoparticles for the purification of water.

Among his many accomplishments, Dr. Sattler was awarded the prestigious Walter Schottky Prize from the German Physical Society in 1983. At the University of Hawaii, he teaches courses in general physics, solid-state physics, and quantum mechanics.

In his private time, he has worked as a musical director at an avant-garde theater in Zurich, composed music for theatrical plays, and conducted several critically acclaimed musicals. He has also studied the philosophy of Vedanta. He loves to play the piano (classical, rock, and jazz) and enjoys spending time at the ocean, and with his family.

Contributors

Sarbari Acharya
Laboratory of Nanomedicine
Institute of Life Sciences
Bhubaneswar, India

Hashim Uddin Ahmed
Division of Surgery and Interventional
 Science
University College London
London, United Kingdom

Maqsood Ahmed
Centre for Nanotechnology, Biomaterial
 and Tissue Engineering
Division of Surgery and Interventional
 Science
University College London
London, United Kingdom

Javed Ally
Department of Mechanical Engineering
University of Alberta
Edmonton, Alberta, Canada

Ronen Almog
Department of Physical Electronics
School of Electrical Engineering
Faculty of Engineering
Tel Aviv University
Tel Aviv, Israel

Alidad Amirfazli
Department of Mechanical Engineering
University of Alberta
Edmonton, Alberta, Canada

Liliana Arrachea
Departamento de Física Juan José
 Giambiagi
Facultad de Ciencias Exactas y Naturales
Universidad de Buenos Aires
Buenos Aires, Argentina

María Arroyo-Hernández
Instituto de Microelectrónica
 de Madrid
Centro Nacional de Microelectrónica
Consejo Superior de Investigaciones
 Científicas
Madrid, Spain

Manit Arya
Division of Surgery and Interventional
 Science
University College London
London, United Kingdom

Hassan M. E. Azzazy
Department of Chemistry

and

Yousef Jameel Science and Technology
 Research Center
The American University in Cairo
Cairo, Egypt

Biancamaria Baroli
Dipartimento Farmaco Chimico
 Tecnologico
Università di Cagliari
Cagliari, Italy

Stefano Bellucci
Laboratori Nazionali di Frascati
Istituto Nazionale di Fisica Nucleare
Frascati, Italy

Martin Benoit
Physics department and Center
 for Nanoscience
Ludwig-Maximilians-Universität
 München
Munich, Germany

Erem Bilensoy
Department of Pharmaceutical
 Technology
Faculty of Pharmacy
Hacettepe University
Ankara, Turkey

Matthew S. P. Boyles
School of Life Sciences
Edinburgh Napier University
Edinburgh, United Kingdom

Eike Brunner
Fachrichtung Chemie und
 Lebensmittelchemie
Technische Universität Dresden
Dresden, Germany

Montserrat Calleja
Instituto de Microelectrónica
 de Madrid
Centro Nacional de Microelectrónica
Consejo Superior de Investigaciones
 Científicas
Madrid, Spain

Kevin J. Chalut
Department of Physics
University of Cambridge
Cambridge, United Kingdom

Martin J. D. Clift
School of Life Sciences
Edinburgh Napier University
Edinburgh, United Kingdom

and

Division of Histology
Institute of Anatomy
University of Bern
Bern, Switzerland

David P. Cormode
Translational and Molecular Imaging
    Institute
Mount Sinai School of Medicine
New York, New York

Christian Dahmen
Division Microrobotics and Control
    Engineering
Department of Computing Science
Carl von Ossietzky Universität
    Oldenburg
Oldenburg, Germany

Ramiz Daniel
Department of Physical Electronics
School of Electrical Engineering
Faculty of Engineering
Tel Aviv University
Tel Aviv, Israel

Fahima Dilnawaz
Laboratory of Nanomedicine
Institute of Life Sciences
Bhubaneswar, India

Lixin Dong
Nanorobotic Systems Lab
Department of Electrical and Computer
    Engineering
Michigan State University
East Lansing, Michigan

Hermann Ehrlich
Fachrichtung Chemie und
    Lebensmittelchemie
Technische Universität Dresden
Dresden, Germany

Volkmar Eichhorn
Division Microrobotics and Control
    Engineering
Department of Computing Science
Carl von Ossietzky Universität
    Oldenburg
Oldenburg, Germany

Ivan H. El-Sayed
Department of Otolaryngology–Head
    and Neck Surgery
Comprehensive Cancer Center
University of California at
    San Francisco
San Francisco, California

Mostafa A. El-Sayed
Laser Dynamics Laboratory
School of Chemistry and Biochemistry
Georgia Institute of Technology
Atlanta, Georgia

Mark Emberton
Division of Surgery and Interventional
    Science
University College London
London, United Kingdom

Maaike Everts
School of Medicine
University of Alabama at Birmingham
Birmingham, Alabama

Tarek M. Fahmy
Department of Biomedical Engineering
Yale University
New Haven, Connecticut

Christian Falconi
Department of Electronic Engineering
University of Tor Vergata

and

Institute of Acoustics "O.M. Corbino"
    Italian National Research Council
Rome, Italy

Sergej Fatikow
Division Microrobotics and Control
    Engineering
Department of Computing Science
Carl von Ossietzky Universität
    Oldenburg
Oldenburg, Germany

Zahi A. Fayad
Translational and Molecular Imaging
    Institute
Mount Sinai School of Medicine
New York, New York

Teresa F. Fernandes
School of Life Sciences
Edinburgh Napier University
Edinburgh, United Kingdom

Antoine Ferreira
Institut PRISME
ENSI Bourges
Bourges, France

Matthieu Fruchard
Institut PRISME
University of Orleans
Orleans, France

Edward P. Furlani
Institute for Lasers, Photonics and
    Biophotonics
University at Buffalo (SUNY)
The State University of New York
Buffalo, New York

Birgit Gaiser
School of Life Sciences
Edinburgh Napier University
Edinburgh, United Kingdom

Christoph Gerber
National Competence Center for
    Research in Nanoscale Science
Institute of Physics of the University
    of Basel
Basel, Switzerland

Lyndon Gommersall
Department of Urology
Division of Medical Sciences
University of Birmingham
Birmingham, United Kingdom

Armin Grunwald
Institut für Technikfolgenabschätzung
    und Systemanalyse
Karlsruhe Institute of Technology (KIT)
Karlsruhe, Germany

Saskia Hagemann
Division Microrobotics and Control
    Engineering
Department of Computing Science
Carl von Ossietzky Universität
    Oldenburg
Oldenburg, Germany

Michael R. Hamblin
Wellman Center for Photomedicine
Massachusetts General Hospital

and

Department of Dermatology
Harvard Medical School
Boston, Massachusetts

and

Harvard-MIT Division of Health
    Sciences and Technology
Massachusetts Institute of Technology
Cambridge, Massachusetts

Stacey L. Harper
Department of Environmental
    and Molecular Toxicology
Oregon State University

and

Oregon Nanoscience and
    Microtechnologies Institute
Corvallis, Oregon

Xiaohua Huang
Laser Dynamics Laboratory
School of Chemistry and Biochemistry
Georgia Institute of Technology
Atlanta, Georgia

Ying-Ying Huang
Wellman Center for Photomedicine
Massachusetts General Hospital

and

Department of Dermatology
Harvard Medical School
Boston, Massachusetts

and

Aesthetic and Plastic Center
Guangxi Medical University
Nanning, Guangxi, People's Republic
    of China

Morana Jaganjac
Laboratory for Oxidative Stress
Department of Molecular Medicine
Rudjer Boskovic Institute
Zagreb, Croatia

Daniel Jasper
Division Microrobotics and Control
    Engineering
Department of Computing Science
Carl von Ossietzky Universität
    Oldenburg
Oldenburg, Germany

Jonathan C. G. Jeynes
Faculty of Electronics and Physical
    Sciences
Advanced Technology Institute
University of Surrey
Guildford, United Kingdom

Paige L. Johnson
Department of Chemistry and
    Biochemistry
The University of Tulsa
Tulsa, Oklahoma

Helinor J. Johnston
School of Life Sciences
Edinburgh Napier University
Edinburgh, United Kingdom

and

Chemicals and Nanotechnologies Division
Department for Environment, Food, and
    Rural Affairs
London, United Kingdom

Martin Kammer
Fachrichtung Chemie und
    Lebensmittelchemie
Technische Universität Dresden
Dresden, Germany

Raymond Kapral
Department of Chemistry
University of Toronto
Toronto, Ontario, Canada

Pu-Chun Ke
Department of Physics and
    Astronomy

and

Center for Optical Materials Science
    and Engineering Technologies
Clemson University
Clemson, South Carolina

Mo Keshtgar
Centre for Nanotechnology, Biomaterials
    and Tissue Engineering
Division of Surgery and Interventional
    Science
University College London

and

Breast Unit
Royal Free Hampstead NHS Trust
    Hospital
London, United Kingdom

Priscila M. Kosaka
Instituto de Microelectrónica
    de Madrid
Centro Nacional de Microelectrónica
Consejo Superior de Investigaciones
    Científicas
Madrid, Spain

Karina Kulangara
Department of Biomedical Engineering
Duke University
Durham, North Carolina

Hans Peter Lang
National Competence Center for
    Research in Nanoscale Science
Institute of Physics of the University
    of Basel
Basel, Switzerland

Deborah Leckband
Department of Chemistry and Chemical
    and Biomolecular Engineering
University of Illinois
Urbana, Illinois

Kam W. Leong
Department of Biomedical Engineering
Duke University
Durham, North Carolina

Sijie Lin
Department of Physics and
    Astronomy

and

Center for Optical Materials Science
    and Engineering Technologies
Clemson University
Clemson, South Carolina

Hong Luo
Department of Genetics and
    Biochemistry
Clemson University
Clemson, South Carolina

Sylvain Martel
NanoRobotics Laboratory
Department of Computer and Software
    Engineering
Institute of Biomedical Engineering
École Polytechnique de Montréal
Montreal, Quebec, Canada

Constantinos Mavroidis
Bionano Robotics Laboratory
Department of Mechanical and
    Industrial Engineering
Northeastern University
Boston, Massachusetts

Johnjoe McFadden
Faculty of Health and Medical Sciences
University of Surrey
Guildford, United Kingdom

Carlos Medina
School of Pharmacy and Pharmaceutical
 Sciences
Trinity College Dublin
Dublin, Ireland

Amit Meller
Department of Biomedical Engineering
 and Physics
Boston University
Boston, Massachusetts

Johann Mertens
Instituto de Microelectrónica
 de Madrid
Centro Nacional de Microelectrónica
Consejo Superior de Investigaciones
 Científicas
Madrid, Spain

Ranjita Misra
Laboratory of Nanomedicine
Institute of Life Sciences
Bhubaneswar, India

Michael Moskalets
Department of Metal and Semiconductor
 Physics
Kharkiv Polytechnic Institute
National Technical University
Kharkiv, Ukraine

Pawel Mroz
Wellman Center for Photomedicine
Massachusetts General Hospital

and

Department of Dermatology
Harvard Medical School
Boston, Massachusetts

Willem J. M. Mulder
Translational and Molecular Imaging
 Institute
Mount Sinai School of Medicine
New York, New York

Bradley J. Nelson
Institute of Robotics and Intelligent Systems
Eidgenössische Technische Hochschule
 Zürich
Zurich, Switzerland

Klaas Nicolay
Department of Biomedical Engineering
Eindhoven University of Technology
Eindhoven, the Netherlands

Cagdas D. Onal
NanoRobotics Laboratory
Department of Mechanical Engineering
Carnegie Mellon University
Pittsburgh, Pennsylvania

Onur Ozcan
NanoRobotics Laboratory
Department of Mechanical Engineering
Carnegie Mellon University
Pittsburgh, Pennsylvania

Jason Park
Department of Biomedical Engineering
Yale University
New Haven, Connecticut

Marija Poljak-Blazi
Laboratory for Oxidative Stress
Department of Molecular Medicine
Rudjer Boskovic Institute
Zagreb, Croatia

Rachela Popovtzer
Engineering School
Bar-Ilan University
Ramat-Gan, Israel

Arthur Rabner
Department of Physical Electronics
School of Electrical Engineering
Faculty of Engineering
Tel Aviv University
Tel Aviv, Israel

Manfred Radmacher
Institut für Biophysik
Universität Bremen
Bremen, Germany

Marek W. Radomski
School of Pharmacy and Pharmaceutical
 Sciences
Trinity College Dublin
Dublin, Ireland

Sarah H. Radwan
Department of Chemistry

and

Yousef Jameel Science and Technology
 Research Center
The American University in Cairo
Cairo, Egypt

Apparao M. Rao
Department of Physics and
 Astronomy

and

Center for Optical Materials Science
 and Engineering Technologies
Clemson University
Clemson, South Carolina

Jason Reppert
Department of Physics and
 Astronomy

and

Center for Optical Materials Science
 and Engineering Technologies
Clemson University
Clemson, South Carolina

Sarwat B. Rizvi
Centre for Nanotechnology, Biomaterials
 and Tissue Engineering
Division of Surgery and Interventional
 Science
University College London
London, United Kingdom

Sanjeeb Kumar Sahoo
Laboratory of Nanomedicine
Institute of Life Sciences
Bhubaneswar, India

Vaibhav Saini
Screening Technologies Branch
Developmental Therapeutics Program
National Cancer Institute at Frederick
Frederick, Maryland

Vanesa Sanz-Beltran
Faculty of Health and Medical Sciences
University of Surrey
Guildford, United Kingdom

Ram Sasisekharan
Harvard-MIT Division of Health
    Sciences and Technology

and

Koch Institute for Integrative Cancer
    Research

and

Department of Biological Engineering
Massachusetts Institute of Technology
Cambridge, Massachusetts

Marc Schneider
Pharmazeutische Nanotechnologie
Universität des Saarlandes
Saarbrücken, Germany

Alexander Marcus Seifalian
Centre for Nanotechnology, Biomaterials
    and Tissue Engineering
Division of Surgery and Interventional
    Science
University College London

and

Royal Free Hampstead NHS Trust
    Hospital
London, United Kingdom

Yosi Shacham-Diamand
Department of Physical Electronics
School of Electrical Engineering
Faculty of Engineering
Tel Aviv University
Tel Aviv, Israel

and

Department of Applied Chemistry
Waseda University
Tokyo, Japan

Iqbal S. Shergill
Department of Urology
Harold Wood Hospital
London, United Kingdom

S. R. P. Silva
Nano-Electronics Centre
Advanced Technology Institute
University of Surrey
Guildford, United Kingdom

Michael T. Simonich
Department of Environmental
    and Molecular Toxicology
Oregon State University

and

Oregon Nanoscience and
    Microtechnologies Institute
Corvallis, Oregon

Abhalaxmi Singh
Laboratory of Nanomedicine
Institute of Life Sciences
Bhubaneswar, India

Metin Sitti
NanoRobotics Laboratory
Department of Mechanical Engineering

and

Robotics Institute
Carnegie Mellon University
Pittsburgh, Pennsylvania

Gautam V. Soni
Department of Biomedical Engineering
    and Physics
Boston University
Boston, Massachusetts

Venkataramanan Soundararajan
Harvard-MIT Division of Health
    Sciences and Technology

and

Koch Institute for Integrative Cancer
    Research

and

Department of Biological Engineering
Massachusetts Institute of Technology
Cambridge, Massachusetts

Robert W. Stark
Center for Nanoscience

and

Department of Earth and Environmental
    Sciences
Ludwig-Maximilians-Universität
    München
Munich, Germany

Christian Stolle
Division Microrobotics and Control
    Engineering
Department of Computing Science
Carl von Ossietzky Universität
    Oldenburg
Oldenburg, Germany

Frank Stracke
Fraunhofer Institut Biomedizinische
    Technik
Sankt Ingbert, Germany

Gustav J. Strijkers
Department of Biomedical Engineering
Eindhoven University of Technology
Eindhoven, the Netherlands

Javier Tamayo
Instituto de Microelectrónica
    de Madrid
Centro Nacional de Microelectrónica
Consejo Superior de Investigaciones
    Científicas
Madrid, Spain

Robert L. Tanguay
Department of Environmental
    and Molecular Toxicology
Oregon State University

and

Oregon Nanoscience and
    Microtechnologies Institute
Corvallis, Oregon

Yu-Guo Tao
Department of Chemistry
University of Toronto
Toronto, Ontario, Canada

Dale Teeters
Department of Chemistry and
    Biochemistry
The University of Tulsa
Tulsa, Oklahoma

Vladimir P. Torchilin
Department of Pharmaceutical Sciences
Center for Pharmaceutical Biotechnology
    and Nanomedicine
Northeastern University
Boston, Massachusetts

Crystal Y. Usenko
Biological Science
Baylor University
Waco, Texas

Panagiotis Vartholomeos
Control Systems Laboratory
Department of Mechanical Engineering
National Technical University
    of Athens

and

Zenon Automation Technologies
Athens, Greece

Tuan Vo-Dinh
Fitzpatrick Institute for Photonics

and

Department of Biomedical Engineering
    and Chemistry
Duke University
Durham, North Carolina

Meni Wanunu
Department of Biomedical Engineering
    and Physics
Boston University
Boston, Massachusetts

Thomas J. Webster
Division of Engineering

and

Department of Orthopedics
Brown University
Providence, Rhode Island

Michael Weigel-Jech
Division Microrobotics and Control
    Engineering
Department of Computing Science
Carl von Ossietzky Universität
    Oldenburg
Oldenburg, Germany

Tim Wharton
Lynntech Inc.
College Station, Texas

Thomas Wich
Division Microrobotics and Control
    Engineering
Department of Computing Science
Carl von Ossietzky Universität
    Oldenburg
Oldenburg, Germany

Neven Zarkovic
Laboratory for Oxidative Stress
Department of Molecular Medicine
Rudjer Boskovic Institute
Zagreb, Croatia

Lijie Zhang
Division of Engineering

and

Department of Orthopedics
Brown University
Providence, Rhode Island

Yan Zhang
Fitzpatrick Institute for Photonics

and

Department of Biomedical Engineering
    and Chemistry
Duke University
Durham, North Carolina

I

Nano–Bio Interfacing

1

Quantum Dots: Basics to Biological Applications

Sarwat B. Rizvi
University College London

Mo Keshtgar
University College London
Royal Free Hampstead
NHS Trust Hospital

Alexander Marcus
Seifalian
University College London
Royal Free Hampstead
NHS Trust Hospital

1.1 Introduction

A nanometer (nm) is one billionth, or 10^{-9}, of a meter and the most fundamental processes of living matter occur at this scale. Nanotechnology is therefore the manipulation of matter in dimensions <100 nm. The term nanotechnology was defined by Professor Nario Taniguchi in 1974: "Nanotechnology mainly consists of the processing of, separation, consolidation, and deformation of materials by one atom or one molecule." Nanotechnology, therefore, deals with the control of matter on the atomic and molecular scale. The whole concept revolves around the fact that at the nanoscale, materials show very different physical and chemical properties compared to those that they exhibit on a macroscale, hence enabling unique applications. Nanoparticles have found application in many areas including advanced materials, electronics, magnetics and optoelectronics, biomedicine, pharmaceuticals, cosmetics, energy, and catalytic and environmental detection and monitoring [1]. Nanoscience and nanotechnology lie at the interface between physics, chemistry, engineering, and, most importantly, biology.

Nanotechnology aims to produce nanoscale functional structures that carry out procedures with atomic precision. Biological molecules and organic structures can be symbolized to machines described by nanotechnology. For example, the cellular organelles function in a precise manner to maintain the normal cellular physiology through the production of various multifunctional molecules. Not only are the cellular machines highly efficient but they are also precisely regulated by data stored in the DNA, a single DNA molecule being approximately 2 nm in size. The application of nanotechnology to biology therefore seems obvious as the sciences included interact for the development of nanobiotechnology.

The most important application of nanobiotechnology is nanomedicine. This may be defined as the application of nanotechnology in the diagnosis, monitoring, and control of disease at a molecular level using engineered nanodevices and nanostructures. It implies that materials and devices are designed to interact with the body at molecular scales with a high degree of specificity. This can be translated into targeting cellular and tissue-specific applications to achieve maximal therapeutic efficiency with minimum side effects. Hence, the concept of personalized medicine, i.e., the prescription of therapeutics best suited for the individual. This aims for an early, presymptomatic diagnosis coupled with highly effective targeted therapy avoiding generalized side effects. Considering the advances in nanotechnology, the concepts of nanodiagnosis, nanotherapy, and nanorobotics performing nanosurgery at the subcellular level will revolutionize the future of nanomedicine.

Nanoparticles can be classified on the basis of the type of material into

- Metallic, e.g., gold and magnetic nanoparticles
- Semiconductor, e.g., quantum dots (QDs)
- Polymeric, e.g., carbon-based nanoparticles such as fullerenes and nanotubes

Of the many multifunctional nanoparticles that can target, diagnose, and treat disease, semiconductor nanocrystals or QDs have been successfully used as fluorescent tags in many biomedical fields. This chapter is an introduction to QDs, and it looks at their basic physics, possible biological and clinical applications, as well as the limitations that need to be overcome for widespread clinical use.

1.2 Quantum Dots

Fluorophores are substances that absorb photons of light at a particular wavelength and then reemit these at a different wavelength. QDs are fluorescent semiconductor nanocrystals with unique photophysical properties. They are composed of materials from the elements in the periodic groups of II–VI, III–V, or VI, for example, cadmium telluride (Cd from group II and Te from group VI) and indium phosphamide (In from group III and P from group V). Being nanocrystals, they range in size from 2 to 10 nm (10–50 atoms) in diameter. At these small sizes, the properties differ from those in the bulk form of semiconductor materials. Due to the quantum confinement effects of electrons and holes within the core of the nanocrystal, QDs are highly photostable, with broad absorption, narrow and symmetric emission spectra, slow excited state decay rates, and broad absorption cross sections. Their emission color depends on their size, chemical composition, and the surface chemistry and can be tuned from the ultraviolet to the visible and near infrared wavelengths. Their unique optical properties have overcome the shortcomings of the traditional dyes and fluorescent proteins and they promise to be the next generation fluorophores.

1.2.1 Structure of a Quantum Dot

The general structure of a QD comprises an inorganic core semiconductor material, e.g., CdTe or CdSe, and an inorganic shell of a different band gap semiconductor material, e.g., ZnS. This is further coated by an aqueous organic coating to which biomolecules can be conjugated. The choice of the shell and coating is important as the shell stabilizes the nanocrystal core and also alters the photophysical properties, while the coating confers properties that allow its allocation to various applications like determining solubility in aqueous media and providing reactive groups for binding to biological molecules.

A bare nanocrystal core cannot be used as it is highly reactive and toxic resulting in a very unstable structure that is prone to photochemical degradation [2]. Also, the core crystalline structure has surface irregularities that lead to emission irregularities like blinking. Capping the core with a semiconductor material of a higher band gap, e.g., ZnS, not only increases the stability and quantum yield but also passivates the toxicity of the core by shielding reactive Cd^{2+} and Te^{2-} ions from being exposed to photo-oxidative environments, e.g., exposure to air [2]. However, ZnS coating is not sufficient to stabilize the core in biological solutions and therefore a further aqueous coating

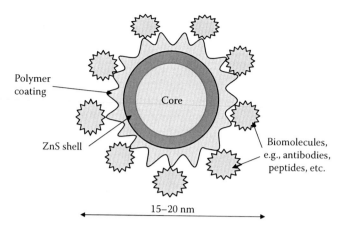

FIGURE 1.1 Structure of a QD.

is required to ensure solubility in biological media. QDs have been coated with a shell of functionalized silica, phospholipid micelles, or linkers like mercaptoacetic acid, mercaptoundecanoic acid, dihydrolipoic acid, or amphiphilic polymers like modified polyacrylic acid to render them soluble in aqueous media [3]. The aqueous coating can then be tagged with various biomolecules of interest, e.g., antibodies, nucleic acids, etc., and different methods of bioconjugation have been described (Figure 1.1).

1.2.2 Basic Physics

To understand the basic physics of the QDs, it is important to review the atomic structure of semiconductor materials. In the bulk form, the electrons exist in a range of energy levels, described as continuous as these energy levels are very close to each other. There are two types of energy levels separated by a region called the band gap. Only two electrons can occupy an energy level at a time. Electrons occupying energy levels below the band gap are described as being in the valence band and those occupying the energy levels above the band gap are described as being in the conduction band (Figure 1.2).

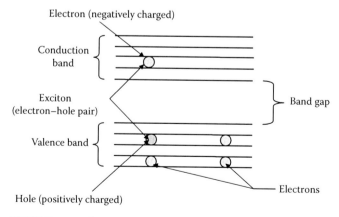

FIGURE 1.2 Schematic diagram of bands, band gap, and electron-hole pairs.

For an electron to jump from the valence to the conduction band, it must acquire energy to cross the band gap. In the bulk form of the semiconductor material, most electrons lack this energy and therefore occupy the valence band. Applying a stimulus, such as heat or voltage, can stimulate the electrons to jump to the conduction band across the forbidden band gap.

As the valence electron jumps across the band gap, it leaves behind a positively charged "hole." The raised electron and hole can then be taken as a pair and this is called an "exciton." The average physical separation between the electron and the hole is referred to as the *Exciton Bohr Radius* and this distance is different for each bulk material. The band gap of a bulk material of a specific composition is fixed and so is the minimal amount of radiation that it must absorb to raise its electron across the band gap. This is because the energy levels in the bulk material are continuous.

The valence electron stays in the conduction band only temporarily before falling back to their corresponding valence position. The electromagnetic radiation emitted as the electron falls back is of a different wavelength from the stimulus required to raise the electron to the conduction band (Figure 1.3). As the band gap of the bulk material is fixed, the energy transmitted as the electron falls back to its valence level in the bulk is also fixed leading to fixed emission frequencies.

As the size of the semiconductor material approaches the size of its Exciton Bohr Radius, its properties seize to resemble the bulk material and the nanocrystal is called a quantum dot (QD). At this size, the energy levels can no longer be continuous and they must be treated as discrete, i.e., there is a finite distance between the different energy levels. This concept of discrete energy levels is called *quantum confinement*.

Secondary to the context of discrete energy levels, the smallest change in the size of the dot has the effect of altering the size of the band gap and hence the band gap energy. As the band gap determines the emission frequency of the QD, it is possible to control the color output by altering the size of the QD with extreme precision. Changing the surface chemistry can also change the band gap energy. This is an extremely valuable property for biological and biomedical applications.

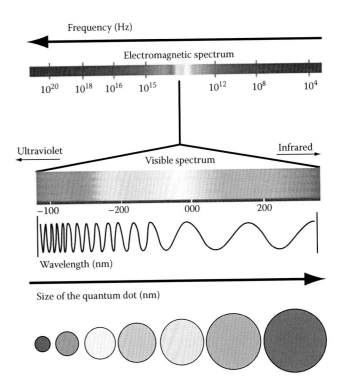

FIGURE 1.4 **(See color insert following page V-2.)** Size tunable emission of the QD.

Another unique optical feature of QDs is a tunable absorption and emission pattern. The emission of QDs can be tuned all the way from the UV to the near infrared of the spectrum such that smaller dots emit in the blue range and the larger dots in the red and near infrared region (Figure 1.4). The bulk semiconductor material displays a uniform absorption spectrum, while the absorption and emission spectra of QDs change as the size of the dot increases.

A QD will not absorb light that has a wavelength longer than the first excitation peak also referred to as *absorption onset* and this depends on the size and composition of the dot. The peak emission wavelength has a slightly longer wavelength than the absorption onset and this energy separation is referred to as the *Stoke's shift*.

Another interesting and useful property of QDs is that the peak emission wavelength is independent of the wavelength of excitation light. This means that variable-sized QDs with different absorption onsets can be excited by a single wavelength of light, as long as this wavelength is shorter than their peak excitation wavelength. This property finds application in multiplexed imaging where a number of different-sized QDs with discrete emission peaks and hence different colors can be excited by a single wavelength of light.

1.3 Methods of QD Synthesis

These can be broadly classified into two types, i.e., organic and aqueous synthesis. Another method of aqueous synthesis using a microwave-assisted technique has also been described [7].

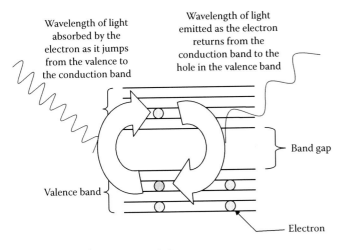

FIGURE 1.3 Fluorescence emission.

1.3.1 Organic Synthesis

Currently most available QDs have been synthesized using the organic synthesis method. First described by the Bawendi group, this method used dimethyl cadmium as the Cd precursor, and trioctyl phosphine (TOP) and trioctyl phosphine oxide (TOPO) as capping reagents [4]. However, dimethyl cadmium is extremely toxic, pyrophoric, expensive, unstable at room temperature and explosive at high temperatures. Another group [5,6] of researchers then pioneered the technique further using CdO as the Cd precursor and this method was simpler and more reproducible than the initial method and hence could be scaled up for industrial application.

However, the QDs produced by the organic phase are insoluble and therefore not applicable to biological systems. A number of methods to solubilize QDs in aqueous media have been developed with good success.

1.3.2 Aqueous Synthesis

This method produces water-soluble QDs and employs a reaction between Cd salts and NaHTe or NaHSe using thiol compounds as capping reagents. Compared to the organic synthesis where high annealing temperatures are used, QDs synthesized in the aqueous phase do not have good crystallinity. Furthermore, they have low quantum yields (QY) [4] and full width half maximum (FWHM), and long reaction time, making preparation a time-consuming and tedious process. On the other hand, aqueous synthesis is a simpler, inexpensive, and reproducible method that can easily be scaled up. Recently various groups have improved the QY of water-soluble QDs by optimizing the synthetic methods and post-synthetic treatment, e.g., by illuminating under room light for 20 days [4].

The microwave-assisted synthesis is a new method described by the Ren group [7] and is based on microwave irradiation with controllable temperatures. This allows a rapid production of size-tunable QDs from green to the near IR. Production of high-quality QDs has been demonstrated using this method in 5–45 min, based on the reaction between Cd^{2+} and NaHTe solution [4]. This method showed significant advantages over the traditional aqueous synthesis like reduced toxicity, good reproducibility, inexpensive, excellent water solubility, stability, and biological compatibility [7]. Also, the QDs produced had the same QY as those generated in the organic phase.

1.4 Surface Modification for Application in Biological Systems

QDs produced by organic synthesis are not water soluble and therefore cannot be applied to biological systems. Their hydrophobic surface capping molecules need to be replaced by hydrophilic ones to improve stability in biological media. A number of techniques of solubilization have been described. These include ligand exchange, silanization, and phase transfer methods.

The ligand exchange method is the simplest and includes the exchange of the hydrophobic surfactant molecules with bifunctional molecules, which are hydrophilic on one side and hydrophobic on the other, to bind to the ZnS shell on the QD [8]. Most often thiols (–SH) are used to bind to the ZnS and carboxyl (–COOH) groups are used as hydrophilic ends. The resulting QDs are soluble in both aqueous and polar solvents.

Surface silanization involves the growth of a silica shell around the nanocrystal. As silica shells are highly cross-linked, they are very stable. However, the drawback is that the process is laborious and the shell may be hydrolyzed [9].

The phase transfer method uses amphiphilic [10,11] polymers to coat the QD surface. The hydrophobic alkyl chains of the polymer interdigitate with the alkyl groups on the QD surface while the hydrophilic groups point outwards to attain water solubility. However, coating with a polymer may leave the overall diameter of the QD and this may pose a limitation in biological applications [12]. Various reports of coating using phospholipid micelles, dithioretol, organic dendrons, and oligomeric ligands are present in literature [13–16].

1.4.1 Bioconjugation

QDs are unique in the feature that they can be conjugated to biological molecules without alteration of the function of these molecules and while retaining their unique photophysical properties. Various methods of bioconjugation have been described including adsorption, electrostatic interaction, mercapto exchange, and covalent linkage.

Using these methods QDs have been conjugated to a number of biological molecules including avidin, biotin, oligonucleotides, peptides, antibodies, DNA, and albumin [8].

1.4.1.1 Adsorption

Molecules like oligonucleotide and various serum albumins [17] can be adsorbed on the surface of the water-soluble QDs. The adsorption is nonspecific and depends on pH, temperature, ionic strength, and surface charge of the molecule [8].

1.4.1.2 Electrostatic Interaction

It has been demonstrated that by electrostatic interaction, proteins engineered with positively charged domains can interact with the negative charges on the QD surface [8]. These conjugates not only have greater photoluminescence but are also more stable than nonconjugated dots.

1.4.1.3 Mercaptoexchange

Many biological molecules have a thiol group that can be tagged on to the surface of a QD by a mercapto exchange process [8]. However, the resulting bond between thiol and Zn is not only weak but also dynamic and this may lead to precipitation of the biomolecules in solution as they easily detach from the QD surface.

1.4.1.4 Covalent Linkage

Functional groups on the QD surface like primary amine, carboxylic acids, and thiols can form a covalent bond with similar groups present on biomolecules. This is a more stable linkage and involves the use of cross-linker molecules [8].

1.4.1.5 Avidin–Biotin Interaction

This is the commonest method of bioconjugation, based on the high-affinity interaction between avidin and biotin. Avidin is attached to antibodies and biotin can be covalently bound to the surface of QDs and vice versa. Many commercially available biomolecules are labeled with avidin or biotin, which makes the use of this method of bioconjugation extremely convenient [4].

engineered fluorescent proteins. A comparison of their properties is given in Table 1.1.

Apart from the advantages there are certain limitations to the QD application. These are based on properties like blinking and photobrightening. Blinking occurs when the QD rapidly alternates between an emitting and nonemitting state. This may cause problems in single-molecule imaging or tracking. Photobrightening, on the other hand, is the increase in the intensity of fluorescence of the QD upon excitation. Although this may be advantageous in certain cases, it poses a problem in fluorescence quantization studies [8]. Both blinking and photobrightening result from the mobile charges present on the surface of the QDs. A significant amount of research is underway to evaluate [19] and overcome these limitations [20,21].

1.5 Why Use Quantum Dots over Traditional Fluorophores?

Fluorescent dyes and proteins like fluorescein, alexaflour488, and rhodamine6G have long been used for monitoring biological events. Fluorescence imaging technology is well established and has been successfully used in research and medicine for biological imaging and clinical diagnostics. QDs were introduced to biological imaging in 1998 [18], and since then have demonstrated a great potential to overcome the limitations of traditional organic dyes and genetically

1.6 Applications in Biology and Biomedicine

The application of QDs in biology and biomedicine relies on their use as fluorescent biological labels for cellular and molecular imaging. Their large surface area allows binding to target and therapeutic molecules and this feature combined with their nanosize can be used to diagnose and treat disease at the cellular and molecular level. Although QDs have been used for a broad range of imaging applications, their vast potential for biomedical applications remains unexplored. Here, we discuss

TABLE 1.1 Comparison between Optical Properties of Traditional Organic Dyes and QDs for Biological Application

Properties	Organic Dyes	Quantum Dots	Advantage of QDs
Excitation spectrum	Narrow	Broad	Organic dyes can only be excited by the light of a specific wavelength due to the narrow excitation spectrum versus QDs, which may be excited by lights of a range of wavelengths, allowing multicolor QDs to be excited by a single wavelength of light.
Emission spectra	Broad and asymmetrical	Narrow and symmetrical	The broad emission spectra of conventional probes may overlap and this limits the number of fluorescent probes that can be tagged to biomolecules for simultaneous imaging in a single experiment. QDs have narrow emission spectra, which can be controlled by altering the size, composition, and surface coatings of the dots. Hence, multiple QDs emitting different colors can be excited by a single wavelength of light making them ideal for multiplexed imaging.
Photobleaching threshold	Low	High	Organic dyes bleach within a few minutes on exposure to light, whereas QDs are extremely photostable due to their inorganic core, which is resistant to metabolic degradation, and can maintain high brightness even after undergoing repeated cycles of excitation and fluorescence for hours. Hence they can be used for long-term monitoring and cell-tracking studies.
Decay lifetime	Fast (<5 ns)	Slow (30–100 ns)	The fluorescence lifetime of QDs is considerably longer than typical organic dyes that decay within a few nanoseconds. This is valuable in overcoming the autofluorescence of background tissues hence improving signal-to-noise ratio.
Quantum yield	Low	High	QDs have higher quantum yields, large absorbance cross section and large saturation intensity than organic fluorophores in aqueous environments making them much brighter probes for in vivo studies and continuous racking experiments over extended periods of time.
Absorbance cross section	Low	High	
Saturation intensity	Low	High	

the application of QDs for in vitro and in vivo imaging as well as their potential clinical applications like use in sentinel lymph node biopsy and photodynamic therapy.

1.6.1 In Vitro Imaging

This includes fixed and live cell imaging. Fixed cell imaging finds application in molecular biology for labeling of cellular proteins and structures, immunohistochemical detection of various molecules of interest using primary and secondary antibodies, and fluorescent in situ hybridization (FISH) technique for DNA and RNA mapping.

1.6.1.1 Fixed Cell Imaging

1.6.1.1.1 Immunofluorescent Labeling of Cellular Proteins and Immunohistochemical Detection in Fixed Cells

The key to the progress in molecular cell biology and medical research is an understanding of the protein and gene expression in cells. Application of fluorescence microscopy for this purpose requires the fixation of target molecules in their natural distribution by making the cell wall penetrable by fluorescent probes. This requires chemical fixation followed by detergent treatment that renders the cell wall permeable to the labeling antibody. The advantages of QDs over organic dyes have already been described above. However, as QDs are larger in size (~10–15 nm) compared to organic dyes (~5 nm), they require different methods of cell fixation and permeabilization for optimum fixation [22].

QD application for fixed cell imaging has widely been explored. QD-Ab conjugates accurately identified membrane-bound protein p-gp (glycoprotein) in fixed MCF7r breast adenocarcinoma cells [4]. The distribution of p-gp was displayed by confocal reconstruction of 3-D imaging and this showed that Q-Ab conjugate labeling was highly sensitive and photostable compared to organic dyes like FITC, alexaFluor488, and R-phycoerythrin.

Immunohistochemical detection of various molecules of interest in human and animal fixed cells is another application of QDs in molecular biology. This involves the use of labeled primary antibodies that will specifically bind to the antigen of interest followed by the application of a secondary antibody. Secondary antibodies are conjugated to organic fluorophores or enzymes that catalyze the deposition of fluorescent substrates at or near the site of the primary antibody, demonstrating that enzyme-based amplification can greatly increase the sensitivity of immunohistochemical detection [23].

QD conjugates can be used for several immunohistochemical applications and have various advantages over organic fluorophores, including their increased photoluminescence, photostability, broad excitation, and narrow emission spectra allowing multitarget labeling. Multilabeling QD protocols for an extremely sensitive immunohistochemical detection have been described [23].

Using QD conjugates, actin and microtubule fibers in the cytoplasm [9] and various nuclear antigens of fixed cancer cells have been stained [24].

Indirect immunofluorescence has been used to identify Her2, microtubules, and nuclear antigens in fixed cancer cells by first incubating fixed cells with a primary Ab, then a biotinylated secondary Ab, and finally with QD-labeled streptavidin. Apart from single-target labeling, double labeling of nuclear antigens and Her2/microtubules with two different QDs has also been demonstrated. The QDs were shown to be several fold more photostable than the organic dye Alexafluor 488 [24].

1.6.1.1.2 Fluorescence In Situ Hybridization

FISH is a powerful molecular technique with a wide range of applications in molecular biology, molecular genetics, and clinical diagnostics. It uses fluorescently labeled DNA probes for gene mapping and identification of chromosomal abnormalities and is an extremely sensitive technique that can provide diagnostic and prognostic results for particular chromosomal disorders. It utilizes the concept of detection of a fluorescent signal at the site of hybridization of a fluorescent dye labeled probe with its homologous chromosomal target. The drawbacks of using rapidly photobleaching and multicolor organic dyes with problems of spectral overlap can be overcome with QDs with their high photostability and discrete spectra allowing multiplexed imaging. Xiao et al. used a QD-FISH probe to analyze human metaphase chromosomes and found that compared to organic dyes like Texas-Red and FITC, QDs were more photostable and significantly brighter making them a more stable and quantitative mode of FISH for research and clinical applications [25]. QDs are also likely to probe single DNA molecules and their interaction with proteins allowing the study of dynamic processes. This should find application in rapid gene mapping and DNA–protein interactions [4].

1.6.1.2 Live Cell Imaging

One of the biggest challenges of cell biology is to explore the biological processing of various cellular events using biomolecules in live cells. This would aid the understanding of dynamic cellular and molecular interactions in vitro and in vivo and monitor them over prolonged periods at high resolution. QDs can be used to label live cells for long-term tracking experiments and they form ideal fluorophores for this purpose [26].

For cell tracking, QDs can be liganded to the extracellular surface or be loaded inside the cells. This can be used for both the in vitro and in vivo labeling of live cells. A number of techniques to label the inside of cells have been suggested. Cells can be labeled by incubating them with QDs that are nonspecifically endocytosed [27,28] or they may be bound to peptides on the cell surface that are specifically endocytosed [29]. Alternatively, they can be introduced into the cell via microinjection [13,30], which though laborious, is the only technique that ensures uniform cytoplasmic distribution.

By far, peptide-mediated intracellular delivery of QDs has been mostly used. This approach is based on the fact that protein transduction domains (PTD) have been used for the passive delivery of drugs across the cell membranes as well as the blood–brain barrier. A number of biomolecules including proteins, oligonucleotides, liposomes, and magnetic nanoparticles have been delivered into cells using PTDs [31]. It has been shown that QDs can be delivered into cells using similar techniques. QDs have been coupled to PTDs via a streptavidin-biotin link [32], covalently [33], by electrostatic adsorption, or adsorption to synthetic PTDs like pep-1 [34]. Of all the approaches for intracellular delivery, coupling PTDs to QDs via streptavidin-biotin is most easily performed [31].

1.6.1.2.1 Single QD Tracking of Membrane Receptors and Signaling Pathways

Various biological processes and pathways like chemotaxis, synaptic regulation, or signal transduction rely on the transmembrane receptors and signaling pathways. QDs with their prolonged photostability, higher photoluminescence, discrete emission spectra, and high signal-to-noise ratio are the ideal fluorophores for single-molecule tracking. Also their nanosize order and allowance for multiplexed imaging gives them significant advantage of the traditional fluorophores to target, detect, and track the dynamics of biological processes for prolonged periods of time, at a truly molecular level. Using single-QD tracking, the dynamics of individual GABA receptors in the axonal growth of spinal neurons has been demonstrated [35].

Fluorophores have been used to study signaling pathways within and between cells, and QDs have been shown to be ideal probes for this field. The dynamics of glycine receptors in neuronal membranes using QD probes has been demonstrated [36]. QDs were used to track individual glycine receptors (GlyRs) and analyze their lateral dynamics in neuronal membranes in living cells over periods of time ranging from milliseconds to minutes. QD labeling enabled imaging for 20 min as compared to the organic dye Cy3, which allowed imaging for only 5 s. Tracking individual dots allowed characterization of multiple diffusion domains showing that QDs are ideal probes for single-molecule studies in live cells.

1.6.1.2.2 Cell Motility Assays

Tumor cell motility determines the ability of cells to migrate and metastasize. QDs can be used to demonstrate the metastatic potential of cancer cells and a number of assays have been developed to distinguish between invasive and noninvasive cancer cell lines [37]. Gu et al. have demonstrated a two dimensional in vitro cell motility assay based on the phagokinetic uptake of QDs by cells as they move across a homogenous layer of QDs leaving behind a fluorescent free tail. The ratio of the trail area to the cell area distinguishes between invasive and noninvasive tumor cells [38,39].

1.6.1.2.3 Fluorescence Resonance Energy Transfer

Fluorescence resonance energy transfer is a process in which energy is transferred from an excited donor to an acceptor particle leading to a reduction in the donor's emission and excited state lifetime and an increase in the acceptor's emission intensity. This happens whenever the distance between the donor and acceptor is smaller than the critical radius known as the Forster radius [2]. FRET is sensitive to the distance between the donor and the acceptor and is used to measure changes in distances rather than absolute distances. It is therefore suitable to study biomolecule conformation, dynamics, and interactions, e.g., monitoring protein conformational changes, protein interactions, and assaying enzyme activity [2]. Using organic dyes for FRET poses the problems of early photobleaching and significant emission overlap between donor and acceptor and QDs provide an excellent alternative. Apart from the above, QD-based FRET technology has found application in monitoring other processes like DNA replication and telomerization for a fast and sensitive DNA detection and DNA array analyses [40]. One potential limitation of using QDs for FRET in biological systems is the size of the nanocrystals, as capping with the ZnS shell not only increases the size but also decreases the distance from the core, hence decreasing FRET efficiency. Also single-molecule experiments can be complicated by QD blinking [8].

1.6.2 In Vivo Imaging

QDs would be ideal probes for in vivo imaging owing to their excellent photoluminescence and photostability under prolonged exposure to laser illumination. Advancement in nanocrystal synthesis, surface coating, and bioconjugation has significantly enhanced their application for in vivo imaging and tracking. Among surface coatings, polyethylene glycol (PEG) has been shown to enhance stability, circulation time, and minimize nonspecific deposition [18]. However, the ideal coating that will completely nullify the toxicity of QDs without altering its unique photophysical properties remains to be discovered.

Recent applications of live animal imaging using QD fluorescence have used mutiphoton microscopy [20,41] and near infrared imaging [42]. QDs have a two-photon absorption cross section several times greater than organic dyes and this property makes them more efficient at probing thick tissue specimens by multiphoton microscopy [8]. Also longer wavelengths extending into the near infrared can be used to excite the QDs allowing near infrared imaging to view structures deep within the biological tissues with minimum photobleaching and photodamage [8]. An interesting application of NIR imaging of QDs is in the sentinel lymph node (SLN) biopsy to aid major cancer surgery.

1.6.2.1 Sentinel Lymph Node Biopsy Using NIR Imaging

Sentinel lymph node biopsy is a means of ultra-staging cancer metastasis and is now the standard of care in breast cancer surgery. It is based on targeting the first draining lymph node of a lymphatic basin at the cancer site to determine the extent of disease spread. Absence of metastasis in the SLN means that the disease is limited and extensive surgery can be avoided. Current tracers for SLN biopsy include the blue dye and radioisotope. However, these have various limitations that can be overcome by the use of QDs that emit in the near infrared range (>700 nm).

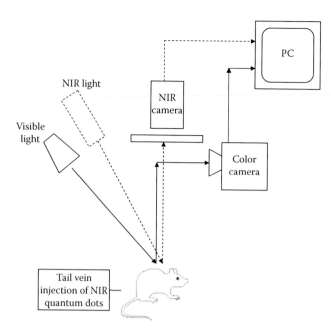

FIGURE 1.5 Schematic diagram of a near infrared imaging system for deep tissue imaging. The near infrared laser light penetrates deep tissues with minimal scatter and excites the QDs that emit in the NIR range. The fluorescence from these dots is detected by the NIR camera, which is basically a CCD camera with a near infrared emission filter in front of the lens. It, therefore, allows only NIR fluorescence to be picked up. These images are superimposed with images from the color camera on a PC to anatomically locate the position of the QDs.

The main problem with live animal imaging is to overcome the background tissue autofluorescence. Near infrared (NIR) imaging can overcome this problem based on the concept that normal tissue chromophores do not absorb or scatter light in the near infrared range. NIR light can therefore penetrate deeper tissues without being scattered and is ideal for imaging in real time. NIR quantum dots (NIRQDs) have successfully been used to demonstrate in vivo sentinel lymph node (SLN) biopsy in mice and pigs [14]. A subdermal injection of pico-molar quantities of NIR QDs entered the lymphatics and the fluorescence could be traced to the SLN in real time by a surgeon using an NIR imaging system (Figure 1.5). This allowed an accurate and sensitive localization and biopsy of the SLN with minimal tissue dissection. This technology promises to overcome the various limitations of the current tracers used for SLN biopsy.

1.6.2.2 Cancer Studies

In vivo cancer targeting by QDs bound to tumor-specific antibodies has also been reported [32]. This may find application as an ultrasensitive tool for early cancer diagnosis. Antibody specific to prostate cancer cell marker PSMA was conjugated to QDs and injected into mice transplanted with human prostate cancer. This accurately localized the tumor, which was clearly imaged in vivo. Their bright luminescence and long lifetime allowed a more accurate and sensitive imaging compared to green fluorescent

protein (GFP). Another group demonstrated the use of QDs to detect early cervical cancer [43]. QDs conjugated to anti-EGFR antibodies were able to detect increased expression of EGFR levels, which correspond to early change from cervical dysplasia to cancer. This can have a huge impact on the early diagnosis and treatment of cancer.

1.6.3 Drug Delivery

Drugs can be targeted to tumors by an enhanced permeability and retention (EPR) effect and this concept has been applied to anticancer agents [3]. QD probes can target and accumulate in tumors by both the EPR effect and the recognition of cancer cell surface biomarkers [32]. Chemotherapeutic agents bound to QD probes, which will recognize and bind to cancer cells, might offer a new strategy for molecular cancer therapy by avoiding systemic toxicity. QDs are one of the many nanoscale platforms being developed as novel drug delivery systems [44]. They have emerged as ideal candidates promising tools for this purpose as they can target specific sites at a molecular level and have unique photophysical properties [45].

1.6.4 Photodynamic Therapy

Photodynamic therapy (PDT) is a an accepted form of cancer therapy for various cancers like skin, breast, mouth, lung, esophageal, and bladder cancer [4]. The current photosensitizers suffer from drawbacks like poor solubility, low quantum yield, poor chemical and photostability, and inability to avoid tissue autofluorescence as the absorption maxima is not in the NIR range. QDs could be used to sensitize the PDT photosensitizer through a FRET mechanism or interact directly with molecular oxygen to generate a reactive oxygen species leading to apoptosis of the cancer cells [46]. The application of QDs as photosensitizers and their mechanism of action have been discussed in detail elsewhere [47]. With their unique optical properties, QDs may become a class of novel photosensitizers in PDT.

1.6.5 Pathogen and Toxin Detection

QD conjugates have been used for the study of various microorganisms and toxin detection. The use of conjugated CdSe QDs for strain and metabolism-specific microbial labeling in a wide variety of bacteria and fungi via the use of surface receptor molecules on the organisms has been demonstrated [48]. Several different pathogens have been targeted including *Giardia lamblia*, *E. coli*, *Listeria monocytogenes*, *Salmonella typhi*, and *Cryptosporidium* [2]. Using this concept of multiplexed imaging, four different QDs were able to detect four toxins including cholera toxin, shiga-like toxin 1, ricin, and staphylococcus enterotoxin B simultaneously from a single well [49]. This QD-based technology will significantly improve the sensitivity of assays for pathogen and toxin detection.

1.7 Limitations

The major limitation to the clinical application of QD technology is the inherent toxicity of the nanocrystal core, which is composed of toxic heavy metals like cadmium, tellurium, or selenium. These heavy metals mainly target the liver, kidney, or spleen, and there are reports to suggest that QDs accumulate in these organs. Although several groups have demonstrated that ZnS shell and capping with hydrophilic coats show no obvious toxicity of CdSe crystals under normal experimental conditions, yet release of free Cd and Se under oxidative environments like exposure to UV light or air causes cytotoxicity, which has also been shown [18]. The latter occurs when the coating is unstable and breaks down to release the toxic core ions into the surrounding. Apart from the composition of the core, the size of the nanocrystal can determine the toxicity through effects on subcellular localization [50].

However, this does not mean that QDs cannot be used in vivo, as parameters like synthesis, processing, and surface coating can minimize the release of free Cd^{2+}. QDs have been shown to exert no toxicity in low concentrations when coated with sheep serum albumin, mercaptoundecanoic acid, mercaptopropionic acid, and PEG-silane coating [18]. However, an ideal coating, which will passivate the toxicity of the nanocrystal core, be biostable, be biocompatible, not alter the photophysical properties of QDs and still keep the size of the crystal small enough for biological application and excretion by the kidney, remains to be developed.

Apart from cytotoxicity, not much is known regarding the immunogenicity, metabolism, degradation, and excretion of the nanocrystals. There have been reports suggesting that nanocrystals accumulate in the liver, kidney, and spleen but whether they are ultimately cleared from the body remains to be explored. A considerable amount of research is required to resolve these issues before QDs can be used for diagnostic and therapeutic purposes in humans.

1.8 Future Advances

The QD technology and its application in biomedicine is still in its infancy. They are promising tools for cancer diagnosis and therapy at the molecular level. Through a powerful and sensitive detection of biomarkers like tumor-specific antigens, conjugated QDs can be used for early diagnosis of cancer. Also a rapid, accurate, nontoxic drug delivery mechanism that will achieve specific therapy without generalized side effects could be achieved through the use of QDs. Near infrared QDs may replace the current tracers for SLN biopsy cancer surgery making the diagnosis of lymph node metastasis a truly minimally invasive procedure, and hence revolutionizing cancer surgery. The most promising aspect of QD application is their use as photosensitizers for PDT. This application is unique as it utilizes their inherent toxicity via the generation of free radicals to target cancer cells. However, there remain many unexplored areas in QD technology that still need to be addressed before their vast potential can be recognized and safely applied in the practice of nanomedicine.

References

1. West JL, Halas NJ. Applications of nanotechnology to biotechnology commentary. *Curr Opin Biotechnol* 2000 Apr; 11(2):215–217.
2. Jamieson T, Bakhshi R, Petrova D, Pocock R, Imani M, Seifalian AM. Biological applications of quantum dots. *Biomaterials* 2007 Nov; 28(31):4717–4732.
3. Iga AM, Robertson JH, Winslet MC, Seifalian AM. Clinical potential of quantum dots. *J Biomed Biotechnol* 2007; 2007(10):76087.
4. Weng J, Ren J. Luminescent quantum dots: A very attractive and promising tool in biomedicine. *Curr Med Chem* 2006; 13(8):897–909.
5. Qu L, Peng X. Control of photoluminescence properties of CdSe nanocrystals in growth. *J Am Chem Soc* 2002 Mar 6; 124(9):2049–2055.
6. Peng ZA, Peng X. Formation of high-quality CdTe, CdSe, and CdS nanocrystals using CdO as precursor. *J Am Chem Soc* 2001 Jan 10; 123(1):183–184.
7. Li L, Qian H, Ren J. Rapid synthesis of highly luminescent CdTe nanocrystals in the aqueous phase by microwave irradiation with controllable temperature. *Chem Commun (Camb)* 2005 Jan 28; 4:528–530.
8. Alivisatos AP, Gu W, Larabell C. Quantum dots as cellular probes. *Annu Rev Biomed Eng* 2005; 7:55–76.
9. Bruchez M, Jr., Moronne M, Gin P, Weiss S, Alivisatos AP. Semiconductor nanocrystals as fluorescent biological labels. *Science* 1998 Sep 25; 281(5385):2013–2016.
10. Wang XS, Dykstra TE, Salvador MR, Manners I, Scholes GD, Winnik MA. Surface passivation of luminescent colloidal quantum dots with poly(dimethylaminoethyl methacrylate) through a ligand exchange process. *J Am Chem Soc* 2004 Jun 30; 126(25):7784–7785.
11. Nann T. Phase-transfer of CdSe@ZnS quantum dots using amphiphilic hyperbranched polyethylenimine. *Chem Commun (Camb)* 2005 Apr 7; 13:1735–1736.
12. Uyeda HT, Medintz IL, Jaiswal JK, Simon SM, Mattoussi H. Synthesis of compact multidentate ligands to prepare stable hydrophilic quantum dot fluorophores. *J Am Chem Soc* 2005 Mar 23; 127(11):3870–3878.
13. Dubertret B, Skourides P, Norris DJ, Noireaux V, Brivanlou AH, Libchaber A. In vivo imaging of quantum dots encapsulated in phospholipid micelles. *Science* 2002 Nov 29; 298(5599):1759–1762.
14. Kim S, Lim YT, Soltesz EG, De Grand AM, Lee J, Nakayama A et al. Near-infrared fluorescent type II quantum dots for sentinel lymph node mapping. *Nat Biotechnol* 2004 Jan; 22(1):93–97.
15. Pathak S, Choi SK, Arnheim N, Thompson ME. Hydroxylated quantum dots as luminescent probes for in situ hybridization. *J Am Chem Soc* 2001 May 2; 123(17):4103–1404.
16. Wang YA, Li JJ, Chen H, Peng X. Stabilization of inorganic nanocrystals by organic dendrons. *J Am Chem Soc* 2002 Mar 13; 124(10):2293–2298.

17. Lakowicz JR, Gryczynski I, Gryczynski Z, Nowaczyk K, Murphy CJ. Time-resolved spectral observations of cadmium-enriched cadmium sulfide nanoparticles and the effects of DNA oligomer binding. *Anal Biochem* 2000 Apr 10; 280(1):128–136.

18. Fu A, Gu W, Larabell C, Alivisatos AP. Semiconductor nanocrystals for biological imaging. *Curr Opin Neurobiol* 2005 Oct; 15(5):568–575.

19. Yao J, Larson DR, Vishwasrao HD, Zipfel WR, Webb WW. Blinking and nonradiant dark fraction of water-soluble quantum dots in aqueous solution. *Proc Natl Acad Sci U S A* 2005 Oct 4; 102(40):14284–14289.

20. Larson DR, Zipfel WR, Williams RM, Clark SW, Bruchez MP, Wise FW et al. Water-soluble quantum dots for multi-photon fluorescence imaging in vivo. *Science* 2003 May 30; 300(5624):1434–1436.

21. Hohng S, Ha T. Near-complete suppression of quantum dot blinking in ambient conditions. *J Am Chem Soc* 2004 Feb 11; 126(5):1324–1325.

22. Ornberg RL, Liu H. Immunofluorescent labeling of proteins in cultured cells with quantum dot secondary antibody conjugates. *Methods Mol Biol* 2007; 374:3–10.

23. Akhtar RS, Latham CB, Siniscalco D, Fuccio C, Roth KA. Immunohistochemical detection with quantum dots. *Methods Mol Biol* 2007; 374:11–28.

24. Wu X, Liu H, Liu J, Haley KN, Treadway JA, Larson JP et al. Immunofluorescent labeling of cancer marker Her2 and other cellular targets with semiconductor quantum dots. *Nat Biotechnol* 2003 Jan; 21(1):41–46.

25. Xiao Y, Barker PE. Semiconductor nanocrystal probes for human metaphase chromosomes. *Nucleic Acids Res* 2004; 32(3):e28.

26. Jaiswal JK, Simon SM. Potentials and pitfalls of fluorescent quantum dots for biological imaging. *Trends Cell Biol* 2004 Sep; 14(9):497–504.

27. Hoshino A, Hanaki K, Suzuki K, Yamamoto K. Applications of T-lymphoma labeled with fluorescent quantum dots to cell tracing markers in mouse body. *Biochem Biophys Res Commun* 2004 Jan 30; 314(1):46–53.

28. Hanaki K, Momo A, Oku T, Komoto A, Maenosono S, Yamaguchi Y et al. Semiconductor quantum dot/albumin complex is a long-life and highly photostable endosome marker. *Biochem Biophys Res Commun* 2003 Mar 14; 302(3):496–501.

29. Chan WC, Nie S. Quantum dot bioconjugates for ultra-sensitive nonisotopic detection. *Science* 1998 Sep 25; 281(5385):2016–2018.

30. Jaiswal JK, Mattoussi H, Mauro JM, Simon SM. Long-term multiple color imaging of live cells using quantum dot bioconjugates. *Nat Biotechnol* 2003 Jan; 21(1):47–51.

31. Lagerholm BC. Peptide-mediated intracellular delivery of quantum dots. *Methods Mol Biol* 2007; 374:105–112.

32. Gao X, Cui Y, Levenson RM, Chung LW, Nie S. In vivo cancer targeting and imaging with semiconductor quantum dots. *Nat Biotechnol* 2004 Aug; 22(8):969–976.

33. Hoshino A, Fujioka K, Oku T, Nakamura S, Suga M, Yamaguchi Y et al. Quantum dots targeted to the assigned organelle in living cells. *Microbiol Immunol* 2004; 48(12):985–994.

34. Mattheakis LC, Dias JM, Choi YJ, Gong J, Bruchez MP, Liu J et al. Optical coding of mammalian cells using semiconductor quantum dots. *Anal Biochem* 2004 Apr 15; 327(2):200–208.

35. Bouzigues C, Levi S, Triller A, Dahan M. Single quantum dot tracking of membrane receptors. *Methods Mol Biol* 2007; 374:81–91.

36. Dahan M, Levi S, Luccardini C, Rostaing P, Riveau B, Triller A. Diffusion dynamics of glycine receptors revealed by single-quantum dot tracking. *Science* 2003 Oct 17; 302(5644):442–445.

37. Pellegrino T, Parak WJ, Boudreau R, Le Gros MA, Gerion D, Alivisatos AP et al. Quantum dot-based cell motility assay. *Differentiation* 2003 Dec; 71(9–10):542–548.

38. Gu W, Pellegrino T, Parak WJ, Boudreau R, Le Gros MA, Alivisatos AP et al. Measuring cell motility using quantum dot probes. *Methods Mol Biol* 2007; 374:125–131.

39. Gu W, Pellegrino T, Parak WJ, Boudreau R, Le Gros MA, Gerion D et al. Quantum-dot-based cell motility assay. *Sci STKE* 2005 Jun 28; 2005(290):l5.

40. Patolsky F, Gill R, Weizmann Y, Mokari T, Banin U, Willner I. Lighting-up the dynamics of telomerization and DNA replication by CdSe-ZnS quantum dots. *J Am Chem Soc* 2003 Nov 19; 125(46):13918–13919.

41. Voura EB, Jaiswal JK, Mattoussi H, Simon SM. Tracking metastatic tumor cell extravasation with quantum dot nanocrystals and fluorescence emission-scanning microscopy. *Nat Med* 2004 Sep; 10(9):993–998.

42. Frangioni JV. In vivo near-infrared fluorescence imaging. *Curr Opin Chem Biol* 2003 Oct; 7(5):626–634.

43. Nida DL, Rahman MS, Carlson KD, Richards-Kortum R, Follen M. Fluorescent nanocrystals for use in early cervical cancer detection. *Gynecol Oncol* 2005 Dec; 99(3 Suppl 1):S89–S94.

44. Cuenca AG, Jiang H, Hochwald SN, Delano M, Cance WG, Grobmyer SR. Emerging implications of nanotechnology on cancer diagnostics and therapeutics. *Cancer* 2006 Aug 1; 107(3):459–466.

45. Hild WA, Breunig M, Goepferich A. Quantum dots—Nano-sized probes for the exploration of cellular and intracellular targeting. *Eur J Pharm Biopharm* 2008 Feb; 68(2):153–168.

46. Samia AC, Chen X, Burda C. Semiconductor quantum dots for photodynamic therapy. *J Am Chem Soc* 2003 Dec 24; 125(51):15736–15737.

47. Samia AC, Dayal S, Burda C. Quantum dot-based energy transfer: Perspectives and potential for applications in photodynamic therapy. *Photochem Photobiol* 2006 May; 82(3):617–625.

48. Kloepfer JA, Mielke RE, Wong MS, Nealson KH, Stucky G, Nadeau JL. Quantum dots as strain- and metabolism-specific microbiological labels. *Appl Environ Microbiol* 2003 Jul; 69(7):4205–4213.

49. Goldman ER, Clapp AR, Anderson GP, Uyeda HT, Mauro JM, Medintz IL et al. Multiplexed toxin analysis using four colors of quantum dot fluororeagents. *Anal Chem* 2004 Feb 1; 76(3):684–688.

50. Lovric J, Bazzi HS, Cuie Y, Fortin GR, Winnik FM, Maysinger D. Differences in subcellular distribution and toxicity of green and red emitting CdTe quantum dots. *J Mol Med* 2005 May; 83(5):377–385.

2
Viral Biology and Nanotechnology

Vaibhav Saini
National Cancer Institute at Frederick

Maaike Everts
University of Alabama at Birmingham

2.1 Introduction

Nanotechnology has been applied to develop advanced products in a variety of industries, such as information technology, cosmetics, and clothing. Similarly, it has great potential for revolutionizing the biomedical industry. For example, it has been envisioned that a nanoscale multifunctional robot—a nanobot—can be developed, capable of searching damaged tissues or cells in the body and repairing them. Although truly mechanical nanobots are still merely hypothetical at this time; it should be noted that viruses operate at a nanoscale; thus, viruses represent naturally occurring "nanobots," which can be manipulated to perform several functions for biomedical applications. For example, viruses can identify cells in the body that display a target receptor, infect these specific cells, and modulate the cellular machinery to drive the formation of viral progeny. Each of these steps in the viral life cycle can be manipulated to perform functions required of a nanobot: identify target cells and manipulate them to change their behavior. This paradigm is most apparent in the role that viruses play as gene delivery vectors for cancer gene therapy. Examples of viruses that have gene therapy vector applications include adenovirus (Ad), adeno-associated virus (AAV), and human immunodeficiency virus (HIV), among others (Saini et al. 2007). Due to their different biology, the use of these vectors is dependent upon the desired outcome. For example, AAV vectors have a carrying capacity for therapeutic genes of <5 kb as compared to ~36 kb for Ad vectors (Saini et al. 2007); thus, AAV vectors are suitable for the incorporation of smaller therapeutic genes

than Ad vectors. Similarly, AAV vector-delivered genes have a prolonged expression spread out over many generations as compared to transient expression achieved using Ad vectors. Thus, AAV vectors are suitable for applications requiring long-term expression of the therapeutic gene, such as potential therapy for cystic fibrosis, whereas Ad vectors are suitable for applications requiring short-term expression of the therapeutic gene, such as tumor therapy (Saini et al. 2007).

Besides this intrinsic role of viruses in nanotechnology, which is outside the scope of this review, in recent years, inorganic nanoparticles (NPs) have been coupled to a variety of these viral vectors. These multifunctional NP-labeled gene therapy viral vectors can be utilized for simultaneous targeting, imaging, and therapy of diseases.

In addition to their role in gene therapy, viruses are currently being studied for a variety of other nanoscale applications, including nanofabrication, where viruses serve as templates to construct nanowires and organic–inorganic batteries. Furthermore, nanotechnology has been used for the detection and elimination of viral pathogens to prevent and eliminate viral diseases. For example, sensitive and cost-effective nanoscale viral antigen detection assays have been developed. Also, nanofiltration techniques have been improved to remove viruses from blood-derived products, thereby greatly reducing the risk of viral transmission through these biomedical products.

Herein, we will describe these applications of nanotechnology in combination with viruses, followed by future expectations and hurdles that must be overcome before these systems can be translated to clinical use.

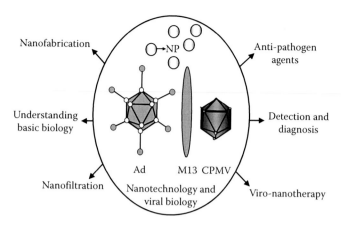

FIGURE 2.1   Applications of nanotechnology and viral biology.

2.2 Background

Viruses have long been recognized as pathogens that lie on the boundary between living and nonliving entities. The viral genome is a fascinating subject, as it can be composed of either single- or double-stranded molecule(s) of either RNA or DNA. Thus, viruses are amenable for genetic manipulation. The realization of this opportunity to genetically engineer viruses has led to the development of viral vectors, containing targeting, imaging, and therapeutic genes, which have been utilized for gene therapy of a variety of diseases, including cancer (Saini et al. 2007). In this regard, it has been recognized that novel therapeutic systems are necessary to achieve cancer eradication. To this end, advancement in nanotechnology harbors well for the development of novel cancer therapies that can be combined with gene therapy. In the following text, we will present examples of the applications of nanotechnology and viral biology in a variety of fields, such as electronics, basic biology, and cancer therapy (Figure 2.1).

A few terms that have been used in this article are defined below:

Gene therapy: A branch of science that deals with achieving therapy by delivering genetic material (DNA or RNA) to target cells.

Viro-nano therapy: A combination of viral biology and nanotechnology to achieve therapy of diseases.

Multifunctional nanoscale agent: A nanodimensional device capable of multiple functions, such as targeting, imaging, and therapy.

2.3 Assembly of Nanoparticles and Viruses into Multicomponent Systems

In order to utilize inorganic NP and organic viral characteristics simultaneously, it is beneficial to combine these two distinct constituents into a single multicomponent system. To construct this system, NPs need to be coupled to viruses or viral vectors, which can be accomplished in two different ways:

1. NPs can be attached to the outer surface of the viruses
2. NPs can be packaged inside the viruses

2.3.1 NPs Coupled to the Outer Surface of Viruses

To attach NPs to the outer surface of the viruses or viral vectors, a variety of natural and engineered chemistries have been exploited. With regard to naturally occurring chemistry, an example is the Chilo iridescent virus (CIV), which has inherent chemical functionality for gold nanoparticle (AuNP) deposition. To achieve this deposition, the naturally occurring repulsive electrostatic forces between AuNPs themselves can be countered by adding electrolytes to the reaction mixture. This allows AuNPs to "seed" the CIV surface. Following the formation of these AuNP seeds, the deposition of Au ions on this biotemplate results in Au shells of varying thickness. Thus, naturally, chemically reactive viruses are amenable for the fabrication of metallodielectric structures that provide cores with a narrower size distribution and smaller diameters (below 80 nm) than currently used silica. These structures can potentially be used for surface plasmon sensing, surface-enhanced Raman spectroscopy, enhancement of the nonlinear optical response of organic materials, and even cancer therapy (Radloff et al. 2005).

With respect to engineered chemistry, it should be realized that viruses harbor the genetic material encoding their own structure, and are thus amenable for engineering desirable chemical functionalities into the capsid using standard molecular biology techniques. In fact, viral vectors tend to have a fairly high plasticity with respect to the incorporation of "binding tags" in their capsid proteins. In this regard, bacteriophages, plant viruses, as well as animal viruses have all been modified appropriately for coupling metal NPs to them.

2.3.1.1 Bacteriophage

M13 bacteriophage has a rodlike filamentous shape (~880 nm long, 6.6 nm wide) and carries a single-stranded DNA. It is a monodisperse biotemplate with an anisotropic shape that can be genetically altered to generate ordered nanocomposite structures. In this regard, two different capsid proteins, a minor (pIII, 5–7 copies) and a major (pVIII, 2700 copies) coat protein, have been engineered to express affinity tags for NPs. For example, to accomplish this, a peptide capable of specific recognition for ZnS was identified through screening a phage library. This peptide was expressed in pIII, which resulted in ordering of ZnS quantum dots (QDs) on the virus (Figure 2.2) (Lee et al. 2002). More complex structures, as explained in another section below, were generated utilizing similar approaches. Besides MS1, another bacteriophage, MS2, was modified to surface-express cysteine residues, and could thus potentially be used for coupling NPs to this sulfhydryl moiety (Peabody 2003). The use of cysteine residues for coupling is beneficial, as covalent coupling chemistry results in permanent attachment of NPs to the viruses.

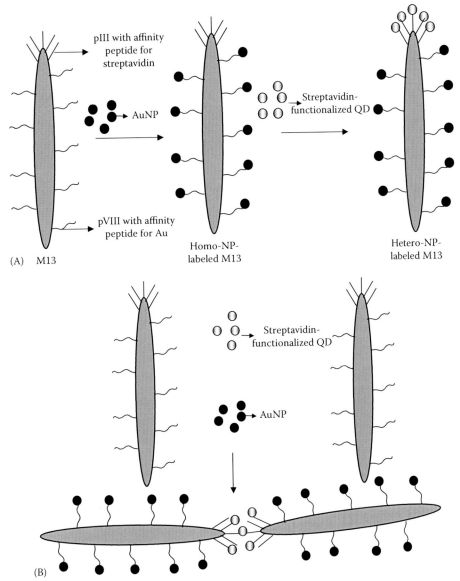

FIGURE 2.2 (A) M13 bacteriophage can be genetically engineered to express different NP-binding affinity peptides on its surface. (B) Hetero-NP-based complex structures can be fabricated when more than one genetically engineered M13 bacteriophage binds to an NP.

2.3.1.2 Plant Viruses

Cysteine functionality has also been displayed on the plant virus cowpea mosaic virus (CPMV) for generating nanowires and virus-QD or virus-carbon nanotube hybrids (Portney et al. 2005). CPMV has an icosahedral shape, is 30 nm in diameter, is composed of 60 copies of two different types of protein subunits, has well-characterized physical and biological properties, and has remarkable stability. In addition, 0.5–1.5 g of virus can be routinely produced from a kilogram of the infected leaves of the black-eyed pea plant (Wang et al. 2002). These useful properties have advanced the utilization of CPMV for nanotechnology purposes.

Turnip yellow mosaic virus (TYMV), another plant virus, has also been considered as a nanobioscaffold. It has a diameter of ~28 nm, carries single-stranded RNA, and is constructed from

180 chemically identical protein subunits that trimerize into 60 asymmetric units, loosely assembled in a $T = 3$ icosahedral symmetry. It can be produced in gram quantities from the infected leaves of either the turnip or Chinese cabbage. As compared to other plant viruses, TYMV has unique advantages for nanotechnology applications. For example, TYMV empty capsids can be isolated naturally from the host plant or generated artificially by treating under high pressure, an alkaline environment, or repeated freeze–thaw processes. In addition, these isolation techniques can form a hole in the capsid, thereby providing a possible route for incorporation of materials, such as NPs or drugs, and raise the possibility of interior modification of the capsid. Also, TYMV has remarkable stability. Thus, it would be advantageous to invest in its development as a biotemplate. As an example of what has already been done, the amine groups on TYMV were

labeled with Fluorescein (FL) *N*-hydroxysuccinimidyl (NHS) ester or *N,N,N′,N′*-tetramethylrhodamine (TMR). The NHS ester and the carboxyl groups were labeled with fluorescein amine under activation with 1-(3-dimethylaminopropyl-3-ethylcarbodiimide) hydrochloride (EDC) and sulfo-NHS (Barnhill et al. 2007). This fluorescently labeled vector can now be utilized for imaging, besides being potentially useful for coupling metal NPs with appropriate functional groups for coupling to amine and carboxyl groups on the virus surface.

2.3.1.3 Animal Viruses

NP-labeled phages and plant viruses have greatly aided in the development and demonstration of the principles of construction of novel organic–inorganic nanobiosystems. However, for translational biomedical purposes, capitalizing on recent advances in gene therapy applications, coupling NPs to human viral-vector-based systems would be optimal. In this regard, due to their well-studied biology and documented clinical trial history, Adenoviral (Ad) vectors are promising candidates for developing NP-labeled vector system for medical purposes. On this basis, our group has previously coupled sulfo-*N*-hydroxy-succinimide-functionalized AuNPs to amine groups present in the capsid of Ad vectors (Everts et al. 2006). This type of coupling is covalent but nonspecific and unfortunately resulted in abrogation of NP-labeled Ad vector infectivity and targeting at higher NP:virus ratios. This may be due to uncontrolled and undesirable coupling of NPs to Ad capsid proteins, such as the fiber knob, that are crucial for Ad vector infectivity and targeting. To circumvent this problem, we explored another type of coupling strategy that is noncovalent and specific. Toward this end, we have utilized Ad vectors expressing a 6-histidine amino acid tag in a major capsid protein, hexon, which were specifically coupled to Nickel (II) Nitrilotriacetic acid (Ni-NTA)-functionalized AuNPs. This coupling strategy dramatically reduced the negative effects on infectivity and completely removed the negative effects on targeting of specifically NP-labeled Ad vector as compared to the nonspecifically labeled virus (Saini et al. 2008).

2.3.2 NPs Packaged Inside Viruses

Besides coupling NPs to the outer surface of viral vectors, NPs can be packaged inside viruses by exploiting the molecular interactions that are necessary for the assembly of mature viral progeny. For example, a red clover necrotic mosaic virus (RCNMV) capsid has been assembled around nanoparticles of Au, $CoFe_2O_4$, and CdSe, utilizing the molecular knowledge of viral capsid assembly. RCNMV has an icosahedral capsid that is composed of 180 identical protein subunits. It has an outer diameter of 36 nm and an inner cavity of ~17 nm. Its genome consists of two single-stranded RNAs: RNA-1 and RNA-2. For the assembly of RCNMV capsid, a 20-nucleotide hairpin structure in RNA-2 hybridizes with RNA-1 to form the origin of assembly (OAS). The OAS initiates capsid assembly by selectively recruiting and orienting capsid protein (CP) subunits. To package NPs inside RCNMV capsid, an OAS mimic oligonucleotide

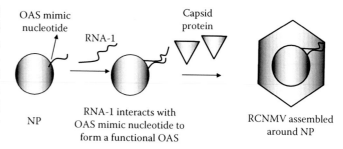

FIGURE 2.3 RCNMV assembly around an NP.

was attached to the NPs, and subsequently RNA-1 was added to generate a synthetic OAS that directed the assembly of RCNMV capsid protein (Figure 2.3) (Loo et al. 2007). Similarly, based on an understanding of capsid assembly, AuNPs, QDs, and magnetic NPs have been incorporated into brome mosaic virus (BMV) (Dragnea et al. 2003, Dixit et al. 2006, Huang et al. 2007). BMV is an icosahedral virus. Its capsid is composed of 180 identical proteins, which form pentameric or hexameric subunits, and has an outside diameter of 28 nm and the inner core diameter of ~18 nm. It has a multipartite genome composed of four single-strand RNA molecules. For BMV virion assembly, the initial interactions occur between the positively charged amino-termini of the coat protein and the negatively charged RNA. These initial interactions are followed by weaker protein–protein interactions between capsid subunits. To incorporate NPs inside the BMV capsid, the surface of the NPs was functionalized to render it negatively charged. These negatively charged NPs interacted with the positively charged termini of coat proteins and resulted in the assembly of BMV capsid around the NPs (Dragnea et al. 2003, Dixit et al. 2006, Huang et al. 2007). Though a variety of NPs have been packaged inside viruses with simple capsids, such as RCNMV and BMV that are composed of one monomer, it is difficult to achieve this internal packaging with complex viruses, such as Ad, whose capsids are composed of many different types of proteins. However, as detailed above, NPs have been coupled to the outside of a variety of viruses, including Ad. It must be noted that while it is relatively easy to couple NPs to the outside of viruses, it can potentially alter the surface properties of the virus and can hamper the interaction of viruses with their target cells.

As described above, a variety of NPs have been coupled to a variety of viral vectors, either nonspecifically or specifically, and either on the surface or inside the virus. In the following sections, we will highlight the current and future applications of these NP-labeled viral vector systems.

2.4 Applications of Nanotechnology and Viral Biology

2.4.1 Nanofabrication

To construct multifunctional nanoscale devices, utilization of either top-to-bottom or bottom-to-top production approaches have been explored. In a top-to-bottom nanofabrication strategy, parts of a macroscale object are chipped away to produce a desired nanoscale object. A subtle and sophisticated alternative is the bottom-to-top fabrication strategy, which is based upon directed self-assembly of nanoscale units into a multicomponent nanosized system. Toward this end, studies into the self-assembly of a nanoscale virus from nanoscale viral components can provide insights into the crucial events required for successful self-assembly of a multicomponent nano-unit. Moreover, natural viral self-assembly has been exploited to construct organic–inorganic systems of moderate sophistication, such as nanowires for electronic applications.

An example of nanoscale self-assembly, as mentioned previously, is provided by the M13 phage. It can be modified to express peptide tags with affinity for different types of NPs, resulting in the assembly of magnetic and semiconducting nanowires (Mao et al. 2004). Moreover, by genetic incorporation of a variety of affinity peptides in a single phage, it is possible to construct simple homo- as well as complex hetero-NP arrays. To elaborate further on this concept, different affinity peptides have genetically been expressed in two different proteins in the M13 capsid: an Au-binding peptide on pVIII and a streptavidin-binding peptide on pIII. These viral vectors were utilized to construct one-dimensional 5 nm AuNP homo-NP arrays on pVIII, which were advanced into continuous metallic nanowires of ~40 nm diameter after incubation for 5 min for electroless deposition (Huang et al. 2005). Furthermore, since this phage also displayed another affinity peptide, it was used to construct complex hetero-NP arrays as well. To this end, streptavidin-functionalized-CdSe QDs were coupled to pIII and AuNPs to pVIII in M13, which resulted in hetero-NP arrays (Huang et al. 2005). Further complexity was imparted to these hetero-nanoparticle arrays when more than one phage bound to the same streptavidin-coated nanoparticle and formed linear wire-dot-wire-like constructs (Figure 2.2). These complex nanowires can be utilized to route electrical carriers providing interconnection for nanodimensional quantum devices (Huang et al. 2005).

In addition to research into the construction of complex nanowires, nanoelectronic avenues for their application have also been investigated. For example, M13-based nanowires have been utilized to construct the anode of a lithium (Li) ion battery. For this, an M13 biotemplate was used to form Au-Cobalt oxide nanowires that were subsequently assembled on polyelectrolyte multilayers, which served as a platform for integrating M13 nanowires into thin and flexible organic–inorganic Li ion batteries (Nam et al. 2006).

Besides nanowires and battery electrodes, viruses can also be utilized to construct long-range three-dimensional (3D) ordered nanoscale solids. In this regard, viruses have been demonstrated to form crystals, a property that can be harnessed for fabrication of 3D ordered solids. Besides viruses, materials such as block copolymers and colloidal crystals can be employed for construction of 3D ordered nanostructures; however, these materials can offer only a limited number of architectures. On the other hand, viruses can provide a broad range of porous and highly organized architectures and symmetries. To illustrate

this point, large nanoscopic cavities and channels, occupying ~50% of the total volume of the body centered cubic crystals of cowpea mosaic virus (CPMV), were utilized for the production of uniquely regular nanocomposites of palladium and platinum. These nanocomposites, which are monolithic 3D-structured solids having nanometer details, may be useful as sensors and in x-ray optical systems (Falkner et al. 2005).

Viruses, such as tobacco mosaic virus (TMV), have also been utilized to provide scaffolds for assembly of oriented high surface area nanomaterials. The rod-shaped TMV is 300 nm in length, 18 nm in diameter, has a hollow inner channel of 4 nm diameter, and is composed of 2130 identical protein subunits that self-assemble in a helix around a single strand of genomic virus RNA. It is stable over a wide range of temperatures and pH values. Thus, it has the properties to serve as a good potential biotemplate. For example, a TMV variant was genetically engineered to display cysteine residues, and utilizing a thiol–Au interaction this TMV was self-assembled onto gold-patterned surfaces in a vertically oriented fashion. This resulted in a more than 10-fold increase in surface area. Following the vertical assembly, uniform metal coatings up to 40 nm in thickness were produced by electroless deposition of nickel and cobalt ions. This virus-templated electrode was tested in a nickel-zinc battery system and was observed to have more than double the total electrode capacity as compared to a non-TMV mutant-templated electrode (Royston et al. 2008). These high surface area viral-templated nanoscale materials have a wide range of applications, such as electrodes, catalyst supports, thermal barriers, sensor arrays, and energy storage devices.

In the applications described above, viral biotemplates have mainly been explored as inactive scaffolds that can be modified for directed assembly of NPs. However, viral biotemplates may actually improve the performance of the assembled nanoelectrical components. In this regard, it has been demonstrated that the electron mobility in silicon nanotubes assembled around a TMV was improved by a factor of four as compared to empty silicon nanotubes (Fonoberov and Balandin 2005). Thus, organic–inorganic systems can enhance the electron transport properties of the semiconductor nanotubes, thereby improving the performance of nanocircuits assembled from them.

Thus, as noted in above examples, viral vectors can serve as efficient biotemplates for assembly and construction of simple nanowires as well as complex electrodes, long-range 3D-ordered solids, and may even improve the electrical properties of nanoelectronic devices assembled from them.

2.4.2 Understanding Basic Biology

The progress in the fields of both virology and nanotechnology can serve as a catalyst to advance our understanding of basic biological processes, such as viral biology and intracellular trafficking. In this regard, as mentioned above, BMV capsids have been assembled around AuNP, QD, and magnetic NP cores (Dragnea et al. 2003, Dixit et al. 2006, Huang et al. 2007). These NP-encapsulated viruses can be utilized to study single viruses in a biological milieu. This would generate detailed information on viral infection, intracellular trafficking, drug delivery, and even imaging of tumors, using tumor-targeted viral vectors in real time in a physiologically relevant setup. This understanding would greatly aid in the development of antiviral drugs targeted to disrupt particular steps critical for viral life cycle, and may lead to a better control and/or treatment of virus-mediated diseases, such as herpes and AIDS. However, this hope is tempered with caution as technical hurdles may limit our ability to incorporate NPs inside complex viruses such as Ad, HSV, and HIV.

Nanotechnology and viral biology have also been harnessed to study long-known, but not completely understood basic cellular processes such as endocytosis, which results in endocytic vesicles of ≤100 nm diameter. To study the effect of size variability of the endocytic guest particle in the subviral size region (≤50 nm) on the process of endocytosis, a saccharide-coated QD, resulting in a QD-conjugated 15 nm diameter sugar ball, was used as an endosome marker. This sugar ball was compared for endocytic capability to Glycoviruses (~50 nm), which are composed of a plasmid DNA enclosed in an amphiphilic glycocluster. This study demonstrated the size-dependence of endocytosis, with the optimal size determined to be 50 nm and anything on either side of it less efficient as potential endocytic material. Therefore, when designing an artificial drug, gene, or molecular probe delivery system, it should be kept in mind that its size should be around that of viruses of about 50 nm (Osaki et al. 2004). This study thus exemplifies how advances in nanotechnology favor advancement of other scientific fields, such as viral and cell biology.

2.4.3 Antipathogen Agents

Nanotechnology has also been explored for developing improved antiviral and antibacterial agents. In this regard, a novel nanoparticle carrier has been developed to enhance the efficacy of a DNA vaccine against HIV. Due to their stability and affordability, DNA vaccines offer a practical alternative for countering HIV infection or progression of AIDS in the developing countries. However, factors such as cellular barriers and nuclease digestion might have been responsible for the low immunogenicity observed for DNA vaccines tested in the nonhuman primate models for HIV. To overcome this limitation, it has been proposed to protect the DNA vaccine by incorporating it inside a shell. Toward this end, a DNA vaccine encoding the envelope protein of HIV-2 (*env*) was condensed inside a novel polycationic adjuvant formulation. This formulation formed nanoparticles in solution, and encapsulating the DNA in this adjuvant resulted in enhanced protein expression. The nanoparticle-encapsulated DNA vaccine resulted in a boosting of the systemic antibody response as compared to naked DNA, when administered intradermally in BALB/c mice. Thus, nanoparticles may serve as adjuvants to increase the effectiveness of DNA vaccines (Locher et al. 2003).

Nanoscale viral vectors have also been utilized to develop targeted antibacterial agents. In this regard, there have been recent reports of resurgence of bacterial resistance to antibiotics, which mandates the use of increased antibiotic concentrations. However, because of toxicity limitations on the maximum dosage of antibiotics that can be administered to patients, it would be beneficial to develop targeted antibiotics. To achieve this, antibody-targeted bacteriophages coupled to the antibiotic chloramphenicol have been developed. This was achieved by coupling chloramphenicol to the aminoglycoside neomycin. One molecule of neomycin has six primary amines amenable for coupling: one of these amines was coupled to one chloramphenicol molecule and the others were available for coupling through EDC chemistry to the carboxyl residues in the phage coat. This antibiotic-coated antibody-targeted phage completely inhibited the growth of both gram-positive (*Staphylococcus aureus*, *Streptococcus pyogenes*) and gram-negative bacteria (*Escherichia coli*). Moreover, the phage-targeted antibiotic was observed to have an improvement in potency by a factor of 20,000 as compared to the free antibiotic (Yacoby et al. 2007).

In addition to phage, a plant virus, cowpea chlorotic mottle virus (CCMV), has also been targeted to *S. aureus* using a biofilm format. CCMV has a diameter of 28 nm, an iscosahedral shape, and an RNA genome. It infects the cowpea plant and has been well studied. Antibody-targeted CCMV conjugated to an imaging agent (Gadolinium) was shown to bind to *S. aureus* in a biofilm. Thus, it could be developed further for possible applications, such as imaging and clearance of the infection sites (Suci et al. 2007). Thus, looking at the provided examples above, it is clear that advances in nanotechnology will advance the development of novel and more effective multifunctional antiviral and antibacterial agents.

2.4.4 Nanofiltration

Viruses are disease-causing pathogens, and therefore a variety of strategies have been developed to either prevent, contain, or treat viral infections in patients. Although vaccination and treatment have their undeniable merits, prevention of viral exposure is an effective way to eliminate viral-based diseases. In addition to encountering viruses *via* infected individuals, it is possible to get infected by the administration of contaminated biomedical plasma products or even *via* drinking water. To eliminate the risk of viral contaminants in plasma products, such as immunoglobulins and coagulation factors, a variety of approaches have been developed that are based on heat inactivation, chemical inactivation, or irradiation (Ng and Dobkin 1985, Preuss et al. 1997, Savage et al. 1998, Scheidler et al. 1998, Omar and Kempf 2002). However, in addition to inactivating the viral contaminants, the use of these methods decreases the yield of the therapeutic proteins by ~20%. Nanofiltration has been developed to circumvent this problem. It has been used to remove both enveloped (HIV) and nonenveloped viruses (hepatitis A virus [HAV], parvovirus B19) (Burnouf and Radosevich 2003). In this regard, due to their small size, it has been difficult to remove nonenveloped viruses using this technique. To resolve this problem, it has been proposed to increase the size of the nonenveloped viruses so that they cannot pass though the filter. To achieve this increase in size, antibodies can be bound to virus. For example, bovine parvovirus (BPV) and bovine enterovirus (BEV) were coated with IgG and successfully eliminated using 20 and 50 nm filters (Omar and Kempf 2002). Besides utilizing antibodies to increase the size of the viral vectors, harmless chemical compounds such as amino acids like glycine can be utilized to aggregate the viruses. In this regard, parvovirus B19, which is between 20 and 25 nm in size, was shown to be aggregated by 0.3 M glycine and retained by a 35 nm filter (Yokoyama et al. 2004). This demonstrates the feasibility of utilizing nanofiltration to remove small viral pathogens such as HAV and parvovirus B19 that are abundantly present in plasma pools, thereby improving the quality of plasma-derived products tremendously (Omar and Kempf 2002). However, flow rates are likely to be problematic for processing of plasma by this technique, which could be improved by using higher pressure to force the plasma through the filters, provided filter integrity is not compromised. In this regard, nanotechnology has yielded materials with improved strength that could be tested for their utility in fabricating filters of higher durability.

Besides plasma products, nanofiltration has also been employed to decontaminate surface water and groundwater. In this regard, protozoans (Giardia and Cryptosporidium) as well as viruses (bacteriophages MS-2 and PRD-1, coliphage Q beta, poliomyelitis virus vaccine, and model viruses Q beta and T4) have been removed from water at a variety of places around the world. This raises the possibility of utilizing nanofiltration for the production of drinking water (Van der Bruggen and Vandecasteele 2003). The above examples underscore the potential of nanofiltration as a nondestructive method for the removal of viruses from important biological fluids.

2.4.5 Detection and Diagnosis

As mentioned earlier, viruses are known to cause diseases. An important aspect of preventing or treating these diseases is to achieve reliable and quick detection of the viral pathogen. In this regard, nanotechnology has helped in the development of methods for improved detection of viruses. To this end, viral detection methods are based on recognition of either viral antigens (immunoassays) or viral genomic sequences (PCR-based methods). However, conventional immunoassays have the disadvantage of low sensitivity, and PCR-based methods suffer from easy contamination and cumbersome processing. To overcome these issues, NPs have been used to develop faster, sensitive, and less labor-intensive assays for detection of viruses. An example of NP-based immunoassay for direct viral detection is the sandwich ELISA assay for Ad detection (Valanne et al. 2005). In this assay, microtiter wells are coated with an antihexon monoclonal antibody, with hexon being the major protein component of the Ad capsid. These antibody-coated wells are subsequently exposed to either purified Ads or nasopharyngeal patient samples. This is followed by incubation with monoclonal anti-hexon

antibodies that have been covalently attached to 107 nm diameter highly fluorescent europium(III)-chelate-doped nanoparticles, which can readily be detected. An individual NP can bind multiple antibodies (theoretically ~159), and thus increases the avidity for the hexon protein tremendously. In addition, the specific fluorescence of the NPs used in this assay is high. This combination of increased avidity for the viral antigen and high fluorescence of the NPs results in a 10–1000-fold improved detection as compared to traditional immunofluorometric assays (Valanne et al. 2005).

Besides improved immunoassays, NPs have also been used to develop viral gene detection based assays. For example, 8–15 nm diameter AuNPs coupled to oligonucleotide detection probes have been used for visual gene chip format-based detection of hepatitis B virus (HBV) and hepatitis C virus (HCV) in serum samples of patients (Wang et al. 2003). This assay is based on a novel method for DNA detection, which utilizes sandwich hybridization and AuNP enhancement techniques using silver (Taton et al. 2000). As no specialized equipment is required for recording signals, this visual detection assay is less expensive and less time consuming than the existing fluorophore-based methods (Wang et al. 2003). This underscores the possibilities for improved pathogen detection due to the advances in nanotechnology.

Besides NPs, higher-order nanofabricated materials, such as nanowires, have also been used for direct, real-time electrical detection of viruses (Patolsky et al. 2004). These nanowires have been attached to receptors that can bind biological pathogens, thus configuring the nanowires as field-effect transistors. The natural interaction of the nanowire-attached receptor with the pathogen can be measured as a change in conductance. For example, antibodies against hemagglutinnin of the influenza A virus have been conjugated to silicon nanowires. The subsequent conductance recordings are characteristic for influenza A virus but not for paramyxovirus or Ad, thereby highlighting the selectivity of this approach (Patolsky et al. 2004).

The above examples point to the immense power of nanotechnology for improved detection and diagnosis of pathogens such as viruses.

2.4.6 Viro-Nano Therapy

The combination of nanotechnology and nanoscale viral vectors for treatment of diseases is known as viro-nano therapy (Saini et al. 2006). It is an active area of research and has potential to provide novel and powerful treatments for diseases such as cancer. In this regard, a combinatorial therapeutic regimen has been observed to be more efficient for tumor treatment. For example, gene therapeutic viral vectors have been combined with chemotherapy, immunotherapy, and radiotherapy for the treatment of tumors. More recently, hyperthermia combined with dendritic-cell-mediated immunotherapy has demonstrated antitumor effects (Guo et al. 2007). Also, in a clinical study, Ad-based gene therapy along with microwave-induced hyperthermia has been shown to reduce the tumor burden (Zhang et al. 2005). Thus, addition of nanotechnology modalities to the already

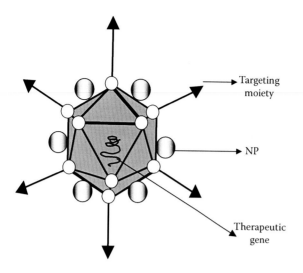

Targeting
moiety

NP

Therapeutic
gene

FIGURE 2.4 Envisioned multifunctional nanoscale agent for therapy.

incorporated therapeutic modalities in viral vectors can greatly enhance the treatment of tumors (Figure 2.4). For example, AuNPs, QDs, and magnetic NPs have recently been shown to be useful for imaging, drug delivery, and hyperthermic tumor cell ablation (O'Neal et al. 2004, Alexiou et al. 2006, Paciotti et al. 2006, Shashkov et al. 2008). Thus, NP-labeled viral vectors can be developed for simultaneous targeting, imaging, and nanotechnology- and gene-therapy-based treatment of primary and metastatic cancer. For example, Ad-, AAV-, and HIV-based viral vectors have been targeted to tumors using antibodies, ligands, and receptors, either noncovalently coupled to the outer surface of the vector or genetically incorporated into the viral genome. Moreover, a variety of imaging modalities, such as fluorescent proteins (green fluorescent protein [GFP] and red fluorescent protein [RFP]), and HSV-TK have been genetically incorporated into the viral vectors. Lastly, tumor suppressor genes, such as p53, and suicide genes have been genetically inserted into the viral genome to achieve gene therapy. Thus, a viral vector that has been targeted on tumors expresses a variety of imaging modalities and contains a therapeutic gene, and when coupled to NPs can potentially be utilized for multifunctional cancer therapy (Saini et al. 2007).

Besides cancer, a combination of nanotechnology and viral-vector-mediated gene therapy could also be beneficial for therapeutic areas such as wound healing. NPs have been utilized in the past for tissue welding, which can be defined as the suturing of incisions. To achieve this, gold nanoshells are added in a solder of albumin, and laser heat is then used to set the solder in and around the incision (Gobin et al. 2005). In the future, nanoshell- or nanorod-mediated wound healing could be further enhanced by Ad vector-mediated delivery of genes encoding therapeutic proteins such as fibromodulin, which promote the process of wound healing (Stoff et al. 2007).

Thus, our multifunctional nanotechnology and viral-vector-based nanotherapy system has the potential to yield novel therapies for a multitude of biomedical conditions endangering human health. With the development of new technologies and a better understanding of the molecular pathways that underlie complex processes such as aging and wound healing, the combined nanotechnology and viral-vector-based gene therapy can be expanded to formulate better treatments for these conditions.

2.5 Critical Discussion

The examples outlined above highlight the importance of the combination of the fields of nanotechnology and viral biology for a symbiotic advancement of these fields individually, as well as in the development of combinatorial multifunctional nanoscale agents for biomedical purposes. While a variety of methods have been developed for coupling NPs to viruses, it would be worthwhile to develop a method to couple even higher amount of NPs, thereby increasing the payload capacity of the viruses. In addition, after perfecting the use of inorganic–organic NP–virus systems for electrical purposes, it may become possible to further miniaturize devices, such as computers, thereby decreasing the cost of production, and increasing portability, accessibility, and information processing speed. This can, in turn, revolutionize fields ranging from online ticket booking systems to biomedical information exchange. However, before realizing these quantum applications, we need to improve upon the higher-order assembly of NP-labeled viruses into complex 3D structures.

A huge challenge for successful cancer therapy is early diagnosis, complete eradication, and no relapse. However, despite significant improvements in cancer therapy, there is no therapy that can guarantee complete tumor ablation and zero chance of relapse. Thus, the envisioned multifunctional NP-labeled viral-vector-based cancer therapy device has the potential to become highly significant for cancer therapy. But, there are as yet unanswered questions about the practical feasibility for using such a device in clinic. A variety of questions pertaining to biodistribution and immunological consequences need to be studied, as outlined in another section of this chapter. Thus, though nanotechnology and viral biology have provided a thrust for interdisciplinary research to develop multifunctional nanodevices, more research is needed to fully realize the potential of this combination.

2.6 Summary

Herein, we have presented the developments pertaining to various aspects of combining nanotechnology and viral biology. In this regard, we have discussed methods for either coupling NPs to the surface of viruses or packaging NPs on the inside of viruses. Subsequently, we have outlined a variety of applications of nanotechnology and viral biology in nanofabrication, understanding basic biology, developing sophisticated antipathogen agents and nanofiltration methods, and finally detecting and diagnosing pathogens and the achievement of tumor therapy. We have also highlighted a few of the critical issues that need to be resolved before nanotechnology and viral biology combination technology can be applied at a large scale.

2.7 Future Perspective

As evidenced by the multitude of research publications on the combination of nanotechnology with virology, interest is high in this particular intersection of two important scientific fields. The applications for cancer detection and treatment in particular have gained a lot of attention from both scientists and the general public. However, from a clinical perspective, the immune response against the NP-labeled viral vectors will need to be carefully considered before these vectors can be utilized in a translational setting. The immune response clears the viral vectors as well as the cells infected with the vectors, resulting in diminished therapeutic efficacy. To achieve immune evasion, viral vectors can be coated with polyethylene glycol (PEG) molecules (Mok et al. 2005), which is a well-studied strategy to protect a variety of nanoparticles such as quantum dots (QDs) and liposomes from the immune response (Li et al. 2002, Ryman-Rasmussen et al. 2007).

Besides the immune response, the toxicity of the NP-labeled viral vector system needs to be studied as well. With regard to QDs, their toxicity is due to highly toxic metals constituents such as Cd. It has been shown that modifications in surface coating can alter QD's cellular interaction and toxicity. However, once QDs enter inside cells, the surface coating is disrupted in the endosomes resulting in leaching of toxic metals and consequently in cell death. Thus, efforts are underway to replace the toxic metal constituents of QDs with nontoxic novel nanotechnology-derived materials to enhance their clinical utility (Hardman 2006, Byrne et al. 2007). However, it should be realized that toxicity is not only determined by the material composition of the main body of the nanoparticle, but also by the surface composition. For example, it has been reported that the toxicity of AuNPs is dependent on their interaction with the cell membrane, which can be modulated based on the coating of these nanoparticles (Goodman et al. 2004).

In addition to the immune response and toxicity, nanoparticle biodistribution should be carefully analyzed. In this regard, a recent study showed that size and charge are important determinants of nanoparticle biodistribution and excretion in vivo (Balogh et al. 2007). Detailed future studies for the resolution of these issues will be critical for the clinical translation of the NP-labeled viral vectors.

Acknowledgments

We would like to thank Dr. Chris Brazel, University of Alabama, Tuscaloosa, AL, for a critical reading of the manuscript.

References

Alexiou, C., Jurgons, R., Seliger, C., and Iro, H. (2006) Medical applications of magnetic nanoparticles. *J Nanosci Nanotechnol* 6: 2762–2768.

Balogh, L., Nigavekar, S. S., Nair, B. M. et al. (2007) Significant effect of size on the in vivo biodistribution of gold composite nanodevices in mouse tumor models. *Nanomedicine* 3: 281–296.

Barnhill, H. N., Reuther, R., Ferguson, P. L., Dreher, T., and Wang, Q. (2007) Turnip yellow mosaic virus as a chemoaddressable bionanoparticle. *Bioconjug Chem* 18: 852–859.

Burnouf, T. and Radosevich, M. (2003) Nanofiltration of plasma-derived biopharmaceutical products. *Haemophilia* 9: 24–37.

Byrne, S. J., Williams, Y., Davies, A. et al. (2007) "Jelly dots": Synthesis and cytotoxicity studies of CdTe quantum dot-gelatin nanocomposites. *Small* 3: 1152–1156.

Dixit, S. K., Goicochea, N. L., Daniel, M. C. et al. (2006) Quantum dot encapsulation in viral capsids. *Nano Lett* 6: 1993–1999.

Dragnea, B., Chen, C., Kwak, E. S., Stein, B., and Kao, C. C. (2003) Gold nanoparticles as spectroscopic enhancers for in vitro studies on single viruses. *J Am Chem Soc* 125: 6374–6375.

Everts, M., Saini, V., Leddon, J. L. et al. (2006) Covalently linked Au nanoparticles to a viral vector: Potential for combined photothermal and gene cancer therapy. *Nano Lett* 6: 587–591.

Falkner, J. C., Turner, M. E., Bosworth, J. K. et al. (2005) Virus crystals as nanocomposite scaffolds. *J Am Chem Soc* 127: 5274–5275.

Fonoberov, V. A. and Balandin, A. A. (2005) Phonon confinement effects in hybrid virus-inorganic nanotubes for nanoelectronic applications. *Nano Lett* 5: 1920–1923.

Gobin, A. M., O'Neal, D. P., Watkins, D. M. et al. (2005) Near infrared laser-tissue welding using nanoshells as an exogenous absorber. *Lasers Surg Med* 37: 123–129.

Goodman, C. M., McCusker, C. D., Yilmaz, T., and Rotello, V. M. (2004) Toxicity of gold nanoparticles functionalized with cationic and anionic side chains. *Bioconjug Chem* 15: 897–900.

Guo, J., Zhu, J., Sheng, X. et al. (2007) Intratumoral injection of dendritic cells in combination with local hyperthermia induces systemic antitumor effect in patients with advanced melanoma. *Int J Cancer* 120: 2418–2425.

Hardman, R. (2006) A toxicologic review of quantum dots: Toxicity depends on physicochemical and environmental factors. *Environ Health Perspect* 114: 165–172.

Huang, Y., Chiang, C. Y., Lee, S. K. et al. (2005) Programmable assembly of nanoarchitectures using genetically engineered viruses. *Nano Lett* 5: 1429–1434.

Huang, X., Bronstein, L. M., Retrum, J. et al. (2007) Self-assembled virus-like particles with magnetic cores. *Nano Lett* 7: 2407–2416.

Lee, S. W., Mao, C., Flynn, C. E., and Belcher, A. M. (2002) Ordering of quantum dots using genetically engineered viruses. *Science* 296: 892–895.

Li, W. M., Mayer, L. D., and Bally, M. B. (2002) Prevention of antibody-mediated elimination of ligand-targeted liposomes by using poly(ethylene glycol)-modified lipids. *J Pharmacol Exp Ther* 300: 976–983.

Locher, C. P., Putnam, D., Langer, R. et al. (2003) Enhancement of a human immunodeficiency virus env DNA vaccine using a novel polycationic nanoparticle formulation. *Immunol Lett* 90: 67–70.

Loo, L., Guenther, R. H., Lommel, S. A., and Franzen, S. (2007) Encapsidation of nanoparticles by red clover necrotic mosaic virus. *J Am Chem Soc* 129: 11111–11117.

Mao, C., Solis, D. J., Reiss, B. D. et al. (2004) Virus-based toolkit for the directed synthesis of magnetic and semiconducting nanowires. *Science* 303: 213–217.

Mok, H., Palmer, D. J., Ng, P., and Barry, M. A. (2005) Evaluation of polyethylene glycol modification of first-generation and helper-dependent adenoviral vectors to reduce innate immune responses. *Mol Ther* 11: 66–79.

Nam, K. T., Kim, D. W., Yoo, P. J. et al. (2006) Virus-enabled synthesis and assembly of nanowires for lithium ion battery electrodes. *Science* 312: 885–888.

Ng, P. K. and Dobkin, M. B. (1985) Pasteurization of antihemophilic factor and model virus inactivation studies. *Thromb Res* 39: 439–447.

O'Neal, D. P., Hirsch, L. R., Halas, N. J., Payne, J. D., and West, J. L. (2004) Photo-thermal tumor ablation in mice using near infrared-absorbing nanoparticles. *Cancer Lett* 209: 171–176.

Omar, A. and Kempf, C. (2002) Removal of neutralized model parvoviruses and enteroviruses in human IgG solutions by nanofiltration. *Transfusion* 42: 1005–1010.

Osaki, F., Kanamori, T., Sando, S., Sera, T., and Aoyama, Y. (2004) A quantum dot conjugated sugar ball and its cellular uptake. On the size effects of endocytosis in the subviral region. *J Am Chem Soc* 126: 6520–6521.

Paciotti, G., Kingston, D., and Tamarkin, L. (2006) Colloidal gold nanoparticles: A novel nanoparticle platform for developing multifunctional tumor-targeted drug delivery vectors. *Drug Develop Res* 67: 47–54.

Patolsky, F., Zheng, G., Hayden, O. et al. (2004) Electrical detection of single viruses. *Proc Natl Acad Sci U S A* 101: 14017–14022.

Peabody, D. S. (2003) A viral platform for chemical modification and multivalent display. *J Nanobiotechnol* 1: 5.

Portney, N. G., Singh, K., Chaudhary, S. et al. (2005) Organic and inorganic nanoparticle hybrids. *Langmuir* 21: 2098–2103.

Preuss, T., Kamstrup, S., Kyvsgaard, N. C. et al. (1997) Comparison of two different methods for inactivation of viruses in serum. *Clin Diagn Lab Immunol* 4: 504–508.

Radloff, C., Vaia, R. A., Brunton, J., Bouwer, G. T., and Ward, V. K. (2005) Metal nanoshell assembly on a virus bioscaffold. *Nano Lett* 5: 1187–1191.

Royston, E., Ghosh, A., Kofinas, P., Harris, M. T., and Culver, J. N. (2008) Self-assembly of virus-structured high surface area nanomaterials and their application as battery electrodes. *Langmuir* 24: 906–912.

Ryman-Rasmussen, J. P., Riviere, J. E., and Monteiro-Riviere, N. A. (2007) Surface coatings determine cytotoxicity and irritation potential of quantum dot nanoparticles in epidermal keratinocytes. *J Invest Dermatol* 127: 143–153.

Saini, V., Zharov, V. P., Brazel, C. S. et al. (2006) Combination of viral biology and nanotechnology: New applications in nanomedicine. *Nanomedicine* 2: 200–206.

Saini, V., Roth, J. C., Pereboeva, L., and Everts, M. (2007) Importance of viruses and cells in cancer gene therapy. *Adv Gene Mol Cell Ther* 1: 30–43.

Saini, V., Martyshkin, D. V., Mirov, S. B. et al. (2008) An adenoviral platform for selective self-assembly and targeted delivery of nanoparticles. *Small* 4: 262–269.

Savage, M., Torres, J., Franks, L., Masecar, B., and Hotta, J. (1998) Determination of adequate moisture content for efficient dry-heat viral inactivation in lyophilized factor VIII by loss on drying and by near infrared spectroscopy. *Biologicals* 26: 119–124.

Scheidler, A., Rokos, K., Reuter, T., Ebermann, R., and Pauli, G. (1998) Inactivation of viruses by beta-propiolactone in human cryo poor plasma and IgG concentrates. *Biologicals* 26: 135–144.

Shashkov, E. V., Everts, M., Galanzha, E. I., and Zharov, V. P. (2008) Quantum dots as multimodal photoacoustic and photothermal contrast agents. *Nano Lett* 8: 3953–3958.

Stoff, A., Rivera, A. A., Mathis, J. M. et al. (2007) Effect of adenoviral mediated overexpression of fibromodulin on human dermal fibroblasts and scar formation in full-thickness incisional wounds. *J Mol Med* 85: 481–496.

Suci, P. A., Berglund, D. L., Liepold, L. et al. (2007) High-density targeting of a viral multifunctional nanoplatform to a pathogenic, biofilm-forming bacterium. *Chem Biol* 14: 387–398.

Taton, T. A., Mirkin, C. A., and Letsinger, R. L. (2000) Scanometric DNA array detection with nanoparticle probes. *Science* 289: 1757–1760.

Valanne, A., Huopalahti, S., Soukka, T. et al. (2005) A sensitive adenovirus immunoassay as a model for using nanoparticle label technology in virus diagnostics. *J Clin Virol* 33: 217–223.

Van der Bruggen, B. and Vandecasteele, C. (2003) Removal of pollutants from surface water and groundwater by nanofiltration: Overview of possible applications in the drinking water industry. *Environ Pollut* 122: 435–445.

Wang, Q., Lin, T., Tang, L., Johnson, J. E., and Finn, M. G. (2002) Icosahedral virus particles as addressable nanoscale building blocks. *Angew Chem Int Ed Engl* 41: 459–462.

Wang, Y. F., Pang, D. W., Zhang, Z. L. et al. (2003) Visual gene diagnosis of HBV and HCV based on nanoparticle probe amplification and silver staining enhancement. *J Med Virol* 70: 205–211.

Yacoby, I., Bar, H., and Benhar, I. (2007) Targeted drug-carrying bacteriophages as antibacterial nanomedicines. *Antimicrob Agents Chemother* 51: 2156–2163.

Yokoyama, T., Murai, K., Murozuka, T. et al. (2004) Removal of small non-enveloped viruses by nanofiltration. *Vox Sang* 86: 225–229.

Zhang, S., Xu, G., Liu, C. et al. (2005) Clinical study of recombinant adenovirus-p53 (Adp53) combined with hyperthermia in advanced cancer (a report of 15 cases). *Int J Hyperthermia* 21: 631–636.

3

Nano-Bio Interfacing with Living Cell Biochips

Yosi Shacham-Diamand*
Tel Aviv University
Waseda University

Ronen Almog
Tel Aviv University

Ramiz Daniel
Tel Aviv University

Arthur Rabner
Tel Aviv University

Rachela Popovtzer
Bar-Ilan University

3.1 Introduction

Biochips are micro-fabricated platforms that use either biological elements for detection, sensing, and monitoring, or use components interacting with biological material, for example, DNA or protein biochips that use biochemical components as part of the detection mechanism along with the support of electronics and electro-optical systems. Such chips may include also other micro-system technologies such as micro-fluidics or micro-electro-mechanical-actuators. Another example is a biochip that detects biological components, such as glucose or urea, using physical, electrical, or chemical effects.

A subset of the biochip-technology family is the whole-cell biochips (WCBC) technology. These chips are platforms that integrate cells as part of their operation. Until recently, the leading application for such technology was for environmental monitoring such as monitoring drinking water safety and checking for pollution hazards (Belkin 2003, Belkin et al. 1997, Mitchell and Gu 2004, Nivens et al. 2004, Hyung-Lee et al. 2005, Popovtzer et al. 2005, Elad et al. 2008, Li et al. 2008). These applications include the development of the biological part, the micro-system technology, and the integration, including cell storage and sampling methods (Bjerketorpet et al. 2006, Dejene et al. 2006). Based on the success of these applications, more applications emerged, such as cells monitoring for cancer (Popovtzer et al. 2006, 2007, 2008). We expect that more applications will emerge

for whole-cell biochips in the near future, as their unique capability will be recognized and their technology will mature.

One key feature of the whole-cell biochip technology is that they interface the biology in a way that explores the cell response as a system (Daniel et al. 2008). Therefore, they are characterized by nonspecific functional sensing unlike most other sensors today that are target specific. The target for most common specific sensors can be a known molecule or a physical condition (i.e., pH, temperature of specific gas, etc.). The whole-cell biochip senses the functional response of the cell; thus, the information they provide is based on the cell system behavior. For example, whole-cell biochips that integrate microbes for water toxicity (English et al. 2006) answer the question "Is the water safe?" and not "Is there a specific toxin in the water?"

In this chapter, we present a simple model of the basic mechanisms of the whole-cell biosensor (WCBS) technology that uses living cells, integrated on a micro-fabricated platform, as the sensor elements. These chips can be classified according to (a) their mode of use and (b) mode of sensing. The mode of use defines two classes of whole-cell biosensors: (1) long-term use whole-cell biosensors and (2) short-term use whole-cell biosensors. The "mode of use" classification is based on the way the cells are handled prior to integration and their integration mode (Bjerketorpet et al. 2006, Dejene et al. 2006). Here are some details of the two classes of use:

1. Whole-cell biosensors for long-term usage—where cells are stored for long term, activated for their application,

---

* The Bernard L. Schwartz Chair for Nano Scale Information Technology.

and are used for various time lengths, either for one time usage or continuous sensing that may take days, weeks, or even months. These sensors are truly biosensors in the classical sense where biology is used for functional detection; that is, the cell's functional response to external excitation is an intrinsic part in the detection process and the biology is part of the detection mechanism that also uses electronics and photonic circuits. Integrating living cells on a chip for long-term usage is not an easy task since both technologies use different materials and processes that may seriously interrupt one another. For example, processing chips requires aggressive cleaning and etching procedures and elevated temperature far beyond the limits of biology while the biological materials may include alkali metal contamination that is hazardous to semiconductor devices. Therefore, most whole-cell biochip technologies today first process the micro-system, then deposit the biology, and finally apply the sample. Also they are currently limited to prokaryotes with specific storage and handling procedures (Bjerketorpet et al. 2006, Dejene et al. 2006). It may be extended to eukaryotes, most likely yeast; however, it is a great challenge to integrate more complex prokaryote cells on simple biochips that are poorly equipped platforms for cell maintenance with limited supply of all the needed cell support. For this purpose, we need to develop an "incubator on a chip" that integrates all the "life support" for the cells.

2. Whole-cell biosensors for short-term use—in this type, the cells are placed shortly before the analysis is done. In this case, the issue of storage, handling, and care of the cells are separated from those of the biochips. The cells can be harvested shortly before application or stored for a long time under optimal conditions in larger and better-equipped facilities.

The mode of sensing is based on the way the information generated by the cells is converted and translated to electronic information. The mode of sensing classification is more complicated than the "mode of use" classification. In this chapter, we will try to define a unified method for that purpose.

A variety of biological assays has been devised for sensing, including colorimetric, fluorescent, bioluminescent, and electrochemical detection (English et al. 2006, Popovtzer et al. 2006, Daniel et al. 2008). The key issue is how to detect the cell response providing the desired information without affecting the cells' viability and metabolism. Note, that we define whole-cell biosensor as a system that includes whole, alive, and functional cells, integrated with photonics and electronics subsystems. We do not consider here methods that break the cell membrane for "postmortem" interrogation of the cell response; we focus on systems that integrate live and functioning cells and explore their response.

Since typical enzymes and other large molecules do not penetrate the cell membrane, we need to use methods that "interrogate" living cells without compromising their viability. Here is a short list of such methods:

1. Mediated response using biochemical agents with smaller molecular weight as messengers (Popovtzer et al. 2006, 2007, 2008)
2. Find ways to penetrate the membrane using small-enough probes, for example, carbon nano-tubes, that will not cause damage (English et al. 2006)
3. Interface the membrane in a way that will allow the desired signals to be transferred from the cell to the external world (e.g., deposit metal dots on the membrane (Dagan-Moscovich et al. 2007, Ben-Yoav and Freeman 2008, Vernick et al. 2008))
4. Interrogate molecules that are expressed at the membrane
5. Use remote methods such as
   a. Optical sensing detecting photo- and bioluminescence effects (Belkin et al. 1998, Belkin 2003)
   b. Electrical methods (such as impedance spectroscopy) that are sensitive to the changes of the dielectric constant inside the cells due to the cell response (Ron et al. 2008)

Another family of whole-cell biochips that is not discussed in detail in this chapter is using multielectrode arrays for neural electrical sensing. This unique application deserves special treatment. The coupling can be either with conducting electrodes, either in close proximity or penetrating, or using field effect devices (Fromherz 2008). So far, a coupling of living cells with electrochemical insulator semiconductor (EIS)-based field-effect devices has been utilized for recording the spontaneous or triggered action potential of some electrogenic cells as well as for cell-acidification detection only (Poghossian and Schöning 2007).

There are many cell-on-chip papers dealing with functional response. A sample of these papers appears in the references (Simpson et al. 2001, Choa 2004, Dalzel et al. 2002, Belkin 2003, Sagi et al. 2003, Mitchell and Gu 2004, Nivens et al. 2004, Sørensen et al. 2006, Poghossian and Schöning 2007, Polyak and Marks 2007). In this chapter, we focus on whole-cell integration where the sensing is due to the expression of proteins by the cells due to some external interaction. We will model the specific sensing equations, in the various cases, in a way that is similar to the modeling of electronic systems. We model the cell operations using a set of state equations with input and output. This approach will utilize the cell as another building block in the system allowing complete modeling using conventional engineering concepts (Daniel et al. 2008). This approach may be useful in the future for other applications such as using cells for biocomputing (Vera et al. 2007).

3.2 System Description

The schematic block diagram of a whole-cell biochip is shown in Figure 3.1.

The input can be any signal that causes protein expression that can be detected later. Whole-cell biosensors can operate either as sensors to some external excitation (e.g., toxicity, heat)

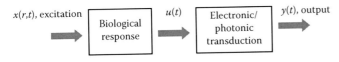

FIGURE 3.1 Schematic block diagram of a whole-cell biochip (r is position, t is time).

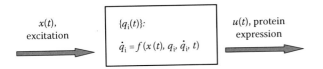

FIGURE 3.2 Schematic diagram of the cell sensors (q_i are the state variables, t is time).

or as differential sensors monitoring changes in the pattern of behavior of cells.

The biological response can be modeled using a set of state equations linking a set of internal variables $\{q_i(t)\}$ that are related to the protein expression. We assume that a small number of rate-limiting equations determine the biological response function, thus simplifying the system. Typically, the signal generation mechanism is nonlinear and is the result of the simultaneous response of 10^3–10^7 cells integrated on the microchip. We assume that each cell generates the signal independently. We do not treat here collective phenomena, such as "quorum sensing" and we assume that all the cells respond similarly. This is probably a very simple assumption and it helps to understand the basic phenomena. A more complex behavior of the cell-on-chips should require a more complex model (Figure 3.2).

The generating function f depends on the following fundamental issues:

1. The output type—there are few fundamental options:
   a. Electrochemical output—in this case, we are looking for a single protein that is generated in response to the excitation. That protein reacts with the substrate, generating products diffusing toward the anode where they are oxidized. Typically, these sensors operate in an amperometric mode where a fixed bias is applied and the current is measured against time.
   b. Optical output—here we have two possible generating functions:
      i. Photoluminescence—where the signal is proportional to the concentration of a photoluminescent protein that is generated in response to the excitation.
      ii. Bioluminescence—in this case, the signal is the result of the reaction between few enzymes that are generated in response to the excitation. The situation here can be rather complicated since the generation of the proteins involves some genetic processes that are either direct, that is, the generation is proportional to the induction, or indirect, that is, the response of a built-in inhibition mechanism is affected by the cell response (Belkin 2003, Polyak and Marks 2007).

   c. Electrical response—probing the electrical properties of the cells, for example, the generation of proteins that can be detected by impedance spectroscopy (Ron et al. 2008). In this case, we assume that the signal is proportional to the concentration of the proteins; therefore, it is, in principle, similar to case a.

Next, we present simple models for optical and electrochemical whole-cell biosensors on chips.

3.3 Model

We assume two parts to the model:

1. The protein generation
2. The signal generation in response to the generated proteins

The first part, the protein response, is modeled assuming that the expression rate depends on the excitation. The generated signal depends on the concentration of the expressed protein. That dependence may follow three basic functions as follows.

3.3.1 Case A: The Signal Is Generated by a Single Product

In this case, we assume that the signal output, Y, is proportional to the concentration of the generated product in response to the excitation:

$$Y = f(C_{pr}) \approx a \cdot C_{pr} + b \tag{3.1}$$

where
  C_{pr} is the product concentration
  a and b are parameters

In this case, we may assume that the output is generated via a first-order reaction. For example, light generation from green fluorescent proteins where the emitted light is proportional to the concentration of the proteins plus an additional background light that is generated from molecules that already exist in the sample under test.

3.3.2 Case B: There Is More Than One Relevant Product

In this case, the cells generate protein and substrate that co-interact to generate the signal. Assume a two-component system:

$$Y = f(C_{pr,1}, C_{pr,2}) \tag{3.2}$$

where $C_{pr,1}$ and $C_{pr,2}$ are the product concentrations. In this case, we assume that there are few precursors generated as a response to the external excitation. These precursors can be enzymes and substrates that interact and the result is the signal, for example, emitted light in the case of bioluminescence. The function depends on the input as well as the output. The possibility of having an internal feedback mechanism may complicate the solution and additionally (the basic mechanism may be nonlinear) require special care in the noise calculation.

3.3.3 Case C: The Signal Is Generated by Interaction between an Enzyme and an Added Substrate

In this case, the signal is generated due to the interaction by the products of the cell's response and an external substrate added to the solution:

$$Y = f(P) \approx a_1 \cdot P + b_1 \qquad (3.3)$$

where

P is the by-product concentration
a_1 and b_1 are constants

In this case, the reaction continuously generates by-products and the signal is the result of the detection of those by-products. The enzyme (E) reacts with the substrate (S) generating an intermediate ES complex that reacts to form the by-product P and the enzyme at a rate of

$$\frac{dP}{dt} \propto ES \qquad (3.4)$$

A full analysis of this expression will be presented in Section 3.4.3.

Therefore, there are two coupled systems: (a) the reactions that determine the rate of generation of the by-products and (b) the reaction that generates the signal from the by-products.

Note that in all of these equations the assumed concentrations are the effective values as the system is not necessarily at equilibrium. For example, there is transport of species from excitation-sensing sites. We assume that since the system is very small, the transport is fast enough and its characteristic time constants are much lower than the cell response rates. However, for larger systems, if needed, the transport effect can be added to the solutions that are presented here.

In the following parts of this chapter, we bring three examples demonstrating whole-cell biochips with different signal-generating equations. The first is a genetically modified *E. coli* on-chip system where green fluorescent proteins are generated in response to the excitation by toxic material. The second is also an *E. coli* on-chip system with a bioluminescent response to toxicant excitation. The last example is an *E. coli* on-chip system with an electrochemical response.

3.4 Examples

In the following section, we bring three examples demonstrating the application of the method that is described in this chapter. The examples are based on three previously published works:

a. Case A and case B: photoluminescence and bioluminescent whole-cell biochip respectively, used for acute toxicity monitoring of water (Jan Roelof van der Meer 2004, D'Souza 2001, Simpson et al. 2001, Choa 2002, Dalzel et al. 2002, Belkin 2003, Sagi et al. 2003, Mitchell and Gu 2004, Nivens et al. 2004, Sørensen et al. 2006, Poghossian and Schöning 2007, Magrisso et al. 2008).

b. Case C: Electrochemical whole-cell biochips used for sensing water toxicity and monitoring cancer cells (Popovtzer et al. 2006, 2007, 2008).

3.4.1 Example Case A: Photoluminescent Whole-Cell Biochip

In this case, the *E. coli* expresses green fluorescent proteins (GFP) in response to the acute toxicity in its environment. We assume that the signal output, for example, light, is directly proportional to the concentration of the fluorescent protein (C_{GFP}). However, the GFP concentration depends on its generation rate by the microbes. This process depends on the type of the reporters, promoters, toxicant, and ambient conditions. As an example to this system, we present here the response of genetically engineered *E. coli* to the induction by nalidixic acid in water (Figure 3.3). The overall response is given in Figure 3.4.

The protein is generated in response to the toxic effect and its generation rate is proportional to the responding promoter concentration that triggers the reporters expressing GFP (Daniel et al. 2008):

$$\frac{dC_{\text{GFP}}}{dt} = \alpha \cdot P_r \qquad (3.5)$$

where P_r is the responding activated promoter's concentration. This variable depends on the bacteria concentration and the concentration of the chemical that is responsible to the

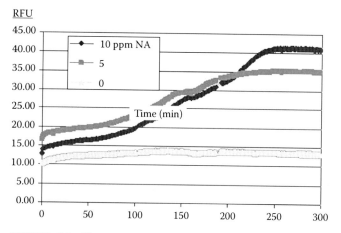

FIGURE 3.3 Fluorescence emission from genetically engineered *E. coli* that were induced by nalidixic acid (NA) (Rabner et al. 2006). The chip integrated 1.5×10^5 microbes per micro-well. The microbes were engineered at the lab of Prof. S. Belkin at the Hebrew University in Jerusalem.

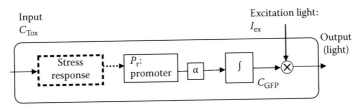

FIGURE 3.4 Schematic drawing of the fluorescent response of the microbes [13] to toxic material in water.

induction. In some cases, we may assume a simple first-order approximation:

$$\frac{dP_r}{dt} = \mu \cdot G_0 \cdot C_{Tox} \tag{3.6}$$

where

μ is the rate constant
G_0 is the concentration of the bacteria (M)
C_{Tox} is the concentration of the toxin that induces the overall response

This model is correct only at the initial stages for a short time. At longer time, the toxicity can turn on all the available promoters and the rate of their generation drops to zero. However, the low concentration regime is the one that is of interest for toxicity sensors, thus this model (Equation 3.6) is relevant to those sensors. This model seems to fit various inducers, such as nalidixic acid, under certain concentrations. When the concentration becomes too high, above the range of 2–5 mg/L for nalidixic acid, we start to see a decline in the overall response, and for higher concentration, above 10 mg/L the microbes' response starts to decline with the increasing concentration of the inducer. This is probably due to the toxic effect on other systems of the cell.

The output Y, the light intensity is proportional to the GFP concentration, C_{GFP} and the excitation, I_{ex}:

$$Y \triangleq I_{out} = \eta \cdot C_{GFP} \cdot I_{ex} \tag{3.7}$$

Thus the complete set of equations can be rewritten using an internal variable set $\{q_i\}$, input $x(t)$, and output $y(t)$ where $x = C_{Tox}$, $q_1 = P_r$, $q_2 = C_{GFP}$, and $y = I_{out}$.

3.4.2 Example Case B: Bioluminescent Whole-Cell Biochip

The exact modeling of such a system is under investigation and the first model appears in Popovtzer et al. (2005). The typical response of such a system appears in the bioluminescent microbe's response to external inductions (Daniel et al. 2008). An example of such a response of the genetically engineered *E. coli* bacteria is shown in Figure 3.5.

The biophysical system in the whole-cell biosensor is reduced to Figure 3.6.

To build the system's state equations, the following variables are defined:

P is the concentration of the fatty acid
S is the concentration of the long-chain aliphatic aldehyde that acts as the substrate
E_L and E_P are the concentrations of the luciferase and reductase enzyme complex, respectively

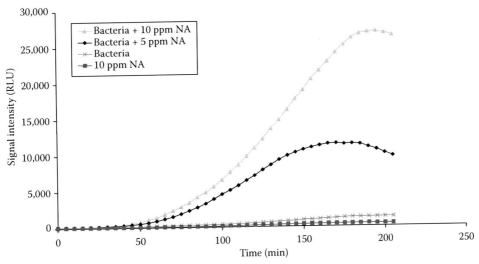

FIGURE 3.5 Bioluminescence vs. time from the genetically engineered bacteria as a function of time for nalidixic acid induction at various concentrations (*E. coli*, recA promoter).

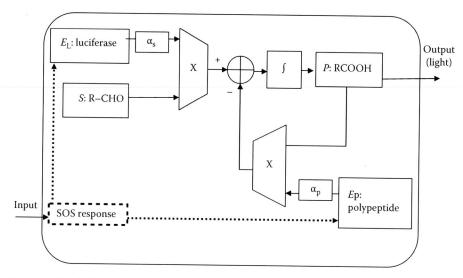

FIGURE 3.6 System diagram of the bioluminescent whole-cell on a chip system.

The output, that is, light, is proportional to P, the fatty acid concentration. P depends on the substrate concentration at $t = 0$, $S(0)$ and the enzymes concentration as

$$\frac{dP}{dt} = -\frac{dS}{dt} = \alpha_s \cdot E_L \cdot S - \alpha_p \cdot E_P \cdot P \tag{3.8}$$

where

α_s is the rate constant in which the luciferase combines to the substrate, in M/min

α_p is the rate constant in which the reductase enzyme complex combines with the substrate

In this case, when the initial substrate concentration is $S(0)$ and any substrate molecule is converted to the product, we may assume that in any time after the reaction starts, $S(t) + P(t) \sim S(0)$. Therefore, we can rewrite the previous expression:

$$\frac{dP}{dt} = \alpha_s \cdot S(0) \cdot E_L - (\alpha_s \cdot E_L + \alpha_p \cdot E_P) \cdot P \tag{3.9}$$

The enzymes (luciferase and reductase) concentration can be determined by the following rate equations:

$$\frac{dE_L}{dt} = \alpha_L \cdot P_r \tag{3.10}$$

$$\frac{dE_P}{dt} = \alpha_L \cdot P_r - \frac{E_P}{\tau_p} \tag{3.11}$$

where P_r is the promoter products' concentration (Daniel et al. 2008). This variable depends on the bacteria concentration and the concentration of the chemical that is responsible to the induction. A zero-order approximation is based on the assumption that the generation of the prompters' products is proportional to the inducer concentration:

$$\frac{dP_r}{dt} = \mu \cdot G_0 \cdot C_{Tox} \tag{3.12}$$

where

μ is the rate constant

G_0 is the concentration of the bacteria (M)

C_{Tox} is the concentration of the toxic material

In this case, C_{Tox} is the input and P is the output. This assumption may be acceptable for a low concentration of the inducer. At a higher concentration it will affect the cell in a way that generally slows down the promoters' generation rate. At high enough concentration, the promoter products' generation rate will decrease as the inducer concentration rate will increase. For the sake of simplicity, we limit ourselves to the condition of low inducer concentration that is usually the useful regime of toxicity sensing. However, in future modeling, medium and high inducer concentration effects should be included.

All the other variables, S, E_P, E_L, and P_r are internal state variables that are a function of C_{NA}, and the initial conditions. Thus the complete set of equations can be rewritten using an internal variable set $\{q_i\}$, input $x(t)$ and output $y(t)$ where $x = C_{Tox}$, $q_1 = P_r$, $q_2 = E_P$, $q_3 = E_L$, and $x = C_{Tox}$. The initial conditions are determined by the specific problem. Usually the internal state and the output are zero before the excitation. However, there might be a situation where they are not equal to 0.

3.4.3 Example Case C: Bioelectrochemical Whole-Cell Biochip

A simple example of a mathematical model for the whole-cell biosensor with electrochemical detection is presented in Popovtzer et al. (2006). In that system, the electrochemical signal was generated by the oxidation of the by-products of the reaction between an external substrate (i.e., *p*-aminophenyl-β-D-galactopyranoside

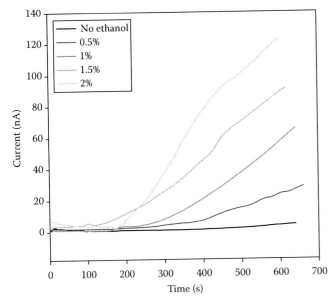

FIGURE 3.7 Amperometric response curves for online monitoring of ethanol using the electrochemical silicon chip. The recombinant *E. coli* containing a promoterless *lacZ* gene fused to the promoter *grpE* exposed to 0.5%–2% concentration of ethanol. The bacteria cultures with the substrate PAPG and the ethanol were placed into the 100 nL volume electrochemical cells on the chip immediately after the ethanol addition (~1 min) and were measured using the amperometric technique at 220 mV. (From Popovtzer, R., PhD thesis, Tel Aviv University, Tel Aviv, Israel, 2007.)

[PAPG]) and a protein (i.e., β-galactosidase) that is expressed in the cell either in response to an external excitation or due to another internal mechanism. Such an electrochemical signal appears in Figure 3.7.

There are few models for the cell response to the toxic material. One option is given as

$$P_r = \mu \cdot G_0 \cdot C_{Tox} \tag{3.13}$$

The generation of the enzyme is assumed to be proportional to the promoter concentration:

$$\frac{dE_T}{dt} = \alpha \cdot P_r \tag{3.14}$$

E_T is the total enzyme concentration in the system. Assuming a linear relation between the promoter's concentration and the toxic material concentration the total enzyme concentration is given as

$$\frac{dE_T}{dt} = k_0 \cdot C_{Tox} \tag{3.15}$$

The enzyme can appear in its free state, E, or in its captured state, ES, there for the total concentration is given by the following equation:

$$E_T = E + ES \tag{3.16}$$

The enzyme–substrate indication was modeled by the Michaelis–Menten equation (Michaelis and Menten 1913). The substrate–enzyme interaction is given as

$$E + S \xrightarrow{k_1} ES \underset{k_2}{\overset{k_3}{\rightleftharpoons}} P + E \tag{3.17}$$

where
   E is the free enzyme concentration
   S is the substrate concentration
   ES is the enzyme–substrate complex concentration
   P is the product concentration

Whole-cell biosensors are not in equilibrium since they respond to an external excitation and there is a functional response increasing the total concentration of the enzyme. However, we will assume that the total enzyme generation rate, dE_T/dt, as given by expression (3.15) to be much slower than the rate at which the Michaelis–Menten equation (3.17) reaches its equilibrium.

In this case, the rate of generation of the ES complex which is given by

$$\frac{d(ES)}{dt} = k_1 \cdot E \cdot S - k_2 \cdot ES - k_3 \cdot ES \tag{3.18}$$

can be presented as

$$\begin{aligned}
\frac{d(ES)}{dt} &= k_1 \cdot (E_T - ES) \cdot S - k_2 \cdot ES - k_3 \cdot ES \\
&= k_1 \cdot E_T \cdot S - (k_1 \cdot S + k_2 + k_3) \cdot ES \\
&= k_1 \cdot E_T \cdot S - \frac{ES}{\tau}
\end{aligned} \tag{3.19}$$

where τ is the effective time constant that describes the characteristic time in which ES reaches its quasi-equilibrium:

$$\tau = \frac{1}{(k_1 \cdot S + k_2 + k_3)} \tag{3.20}$$

The assumption that ES reaches quasi equilibrium is similar to the assumption for equilibrium in the classical treatment of the Michaelis–Menten equation

$$k_1 \cdot E_T \cdot S \approx \frac{ES}{\tau} \quad \rightarrow \quad ES = E_T \cdot \frac{k_1 \cdot S}{k_1 \cdot S + k_2 + k_3} \tag{3.21}$$

For relatively large concentration of substrate $[ES] \sim E_T$ (Michaelis and Menten 1913). This can be controlled since the substrate is externally added to the chip.

The rate of production of the product P is

$$\frac{dP}{dt} = k_3 \cdot ES = k_3 \cdot E_T \cdot \frac{k_1 \cdot S}{k_1 \cdot S + k_2 + k_3} \tag{3.22}$$

The overall reaction is shown in Figure 3.8. In this diagram, E_T is the total enzyme generated by the cell response, ES is the enzyme concentration due to the ES complex decomposition, τ is the effective time constant which is a function of the Michaelis–Menten

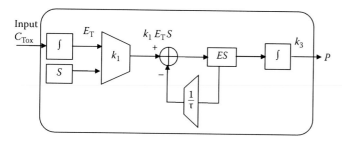

FIGURE 3.8 Schematic diagram of an electrochemical whole cell biochip.

parameters and P is the by-product concentration. The state equations for this system can be rearranged so that x is the input C_{Tox}, $q_1 = E_T$, $q_2 = ES$, and y, the system's electrical output signal is proportional to P.

The intrinsic output variable here is the by-product concentration P. However, the real readout is at the electrode that operates at positive bias to oxidize the by-products. The by-products diffuse through the medium toward the electrode and there is a collection efficiency in which it is described by a parameter k_4.

3.5 Summary and Conclusions

In this chapter, we present a simple model for the intrinsic signal of a whole-cell biochip using a relatively simple set of equations. Using an electrical equivalent for the Michaelis–Menten equation was already presented (Kopelman 1986, 1988, Grima and Schnell 2006). We adapt a rigorous mathematical model linking the biology to the electronics circuits. There are few variations depending on the specific application. They all define the output, y, as a function of internal variables, $\{q_i\}$ that depends on the input x. This approach will simplify the calculation of the system response, that is, signal and noise. This will allow the modeling of system variables such as

a. Minimum detectable signal (MDS)
b. Signal-to-noise ratio (SNR)

Note that the intrinsic signal is fed onto an electrical signal that amplifies the system, but also contributes some noise. There is also the issue of the bandwidth of the system operation. For linear systems, it can be easily determined from the state equations. It would be more complicated for nonlinear detection mechanisms.

The intrinsic signal that depends on the enzymatic activity in the cells is classically described using the Michaelis–Menten model. We used this model and assumed constant reaction constants, k_1, k_2, and k_3. However, evidence shows that in living cells those parameters may slowly vary with time (Kopelman 1988).

$$k \propto t^{1-p} \tag{3.23}$$

where p is a nondimensional index that quantifies the deviations from the classical law of mass action (Michaelis and Menten

1913). This should be included in further refining of the model, especially, in systems where the cells are integrated and used for relatively long period, for example, in water toxicity sensor under constant water flow. The definition of what is long and short periods are still to be determined.

Another issue is that the Michaelis–Menten model assumes a large ensemble of enzyme molecules allowing the setting up of the classical differential equations. However, the question is whether the Michaelis–Menten model holds in the cellular environment where the absolute value of the enzyme molecules is not large. It can be shown that the Michaelis–Menten still holds even for a single enzyme molecule situation although there are some unique features to the signal statistics in that case (Wu 2000, Kopelman 1986, 1988).

Another issue that should be included in a full model is the products' transport in the cell container. This is not important in sensors where the total emission luminescence is almost independent of the by-products' distribution, assuming uniform microbes' distribution. It should be taken into consideration when the microbes are fixed in a solid matrix and their by-products diffuse into the liquid medium. In sensors where the by-products diffuse toward the sensing device—electrodes in the case of electrochemical sensing—the transport may play an important role (Popovtzer et al. 2007) and should be included in the model.

Acknowledgments

The authors thank Prof. Shimshon Belkin from the Hebrew University at Jerusalem with whom we have collaborated on various whole-cell biochip projects in the last 7 years. His research achievements on the genetically engineered *E. coli* for the detection of water toxicity are the base of this work.

References

Belkin, S. 2003. Microbial whole-cell sensing systems of environmental pollutants. *Current Opinion in Microbiology* 6: 206–212.

Belkin, S. et al. 1997. A panel of stress-responsive luminous bacteria for the detection of selected classes of toxicants. *Water Research* 31: 3009–3016.

Belkin, S. et al. 1998. Monitoring subtoxic environmental hazards by stress-responsive luminous bacteria. *Environmental Toxicology and Water Quality* 11: 179–185.

Ben-Yoav, H. and Freeman, A. 2008. Enzymatically attenuated in situ release of silver ions to combat bacterial biofilms: A feasibility study. *Journal of Drug Delivery Science and Technology* 18: 25–29.

Bjerketorpet, J. et al. 2006. Advances in preservation methods: Keeping biosensor microorganisms alive and active. *Current Opinion in Biotechnology* 17: 43–49.

Cho, J.-C. 2004. A novel continuous toxicity test system using a luminously modified freshwater bacterium. *Biosensors and Bioelectronics* 20: 338–344.

Dagan-Moscovich, H. et al. 2007. Nanowiring of the catalytic site of novel molecular enzyme-metal hybrids to electrodes. *Journal of Physical Chemistry C* 111: 5766–5769.

Dalzell, D. J. B. et al. 2002. A comparison of five rapid direct toxicity assessment methods to determine toxicity of pollutants to activated sludge. *Chemosphere* 47: 535–545.

Daniel, R. et al. 2008. Modeling and measurement of a whole-cell bioluminescent biosensor based on a single photon avalanche diode. *Biosensors and Bioelectronics* 24: 882–887.

Dejene, A. et al. 2006. Freeze-drying of sol–gel encapsulated recombinant bioluminescent *E. coli* by using lyo-protectants. *Sensors and Actuators B* 113: 768–773

D'Souza, S. F. 2001. Microbial biosensors. *Biosensors and Bioelectronics* 16: 337–353.

Elad, T. et al. 2008. Microbial whole-cell arrays. *Microbial Biotechnology* 1(2): 137–148.

English, B. P. et al. 2006. Ever-fluctuating single enzyme molecules: Michaelis-Menten equation revisited. *Nature Chemical Biology* 2: 878–894.

Fromherz, P. 2008. Joining microelectronics and microionics: Nerve cells and brain tissue on semiconductor chips. *Solid-State Electronics* 52: 1364–1373.

Grima, R. and Schnell, S. 2006. A systematic investigation of the rate laws valid in intracellular environments. *Biophysical Chemistry* 124: 1–10.

Hyung-Lee, J. et al. 2005. A cell array biosensor for environmental toxicity analysis. *Biosensors and Bioelectronics* 21: 500–507.

Kopelman, R. 1986. Rate-processes on fractals: Theory, simulations, and experiments. *Journal of Statistical Physics* 42: 185–200.

Kopelman, R. 1988. Fractal reaction kinetics. *Science* 241: 1620–1626.

Li, Y.-F. et al. 2008. Construction and comparison of fluorescence and bioluminescence bacterial biosensors for the detection of bio-available toluene and related compounds. *Environmental Pollution* 152: 123–129.

Magrisso, S. et al. 2008. Microbial reporters of metal bioavailability. *Microbial Biotechnology* 1(4): 320–330.

Michaelis, L. and Menten, M. L. 1913. Die kinetik der invertinwirkung. *Biochemische Zeitschrift* 49: 333–369.

Mitchell, R. J. and Gu, M. B. 2004. An *Escherichia coli* biosensor capable of detecting both genotoxic and oxidative damage. *Applied Microbiology and Biotechnology* 64: 46–52.

Nivens, D. E. et al. 2004. Bioluminescent bioreporter integrated circuits: Potentially small, rugged and inexpensive whole-cell biosensors for remote environmental monitoring. *Journal of Applied Microbiology* 96: 33–46.

Poghossian, A. and Schöning, M. J. 2007. Chemical and biological field-effect sensors for liquids—A status report. *Handbook of Biosensors and Biochips*. John Wiley & Sons, Weinheim, Germany.

Polyak, B. and Marks, R. S. 2007. Bioluminescent whole-cell optical fiber sensors. *The Handbook of Biosensors and Biochips*, R. S. Marks et al. (eds.). John Wiley & Sons, London, U.K.

Popovtzer, R. 2007. PhD thesis, Tel Aviv University, Tel Aviv, Israel.

Popovtzer, R. et al. 2005. Novel integrated electrochemical nano-biochip for toxicity detection in water. *Nano Letters* 5: 1023–1027.

Popovtzer, R. et al. 2006. Electrochemical detection of biological reactions using a novel nano-bio-chip array. *Sensors and Actuators B: Chemical* 119: 664–672.

Popovtzer, R. et al. 2007. Mathematical model of whole cell based bio-chip: An electrochemical biosensor for water toxicity detection. *Journal of Electroanalytical Chemistry* 602: 17–23.

Popovtzer, R. et al. 2008. Electrochemical lab on a chip for high-throughput analysis of anticancer drugs efficiency. *Nanomedicine: Nanotechnology, Biology, and Medicine* 4: 121–126.

Rabner, A. et al. 2006. Whole cell luminescence biosensor-based lab-on-chip integrated system for water toxicity analysis. *Proceedings of the SPIE-The International Society for Optical Engineering* 6112: 33–42.

Roelof van der Meer, J. et al. 2004. Illuminating the detection chain of bacterial bioreporters. *Environmental Microbiology* 6(10): 1005–1020.

Ron, A. et al. 2008. Cell-based screening for membranal and cytoplasmatic markers using dielectric spectroscopy. *Biophysical Chemistry* 135: 59–68.

Sagi, E. et al. 2003. Fluorescence and bioluminescence reporter functions in genetically modified bacterial sensor strains. *Sensors and Actuators B* 90: 2–8.

Simpson, M. L. et al. 2001. Whole-cell biocomputing. *Trends in Biotechnology* 19(8): 317–323.

Sørensen, S. J. et al. 2006. Making bio-sense of toxicity: New developments in whole-cell biosensors. *Current Opinion in Biotechnology* 17: 11–16.

Vera, J. et al. 2007. Power-law models of signal transduction pathways. *Cellular Signaling* 19: 1531–1541.

Vernick, S. et al. 2008. Directed metallization of single enzyme molecules with preserved enzymatic activity. *IEEE Transactions on Nanotechnology* 8: 95–99.

Wu, G. 2000. Another integrated form of the Michaelis-Menten equation, its analogy to electrical circuit model and implications for active transporters. *Medical Hypotheses* 54(5): 748–749.

4

Micro- and Nanomechanical Biosensors

María Arroyo-Hernández
Instituto de Microelectrónica de Madrid

Priscila M. Kosaka
Instituto de Microelectrónica de Madrid

Johann Mertens
Instituto de Microelectrónica de Madrid

Montserrat Calleja
Instituto de Microelectrónica de Madrid

Javier Tamayo
Instituto de Microelectrónica de Madrid

4.1 Introduction

Cantilevers were initially developed for atomic force microscopy (AFM), where they were demonstrated to be sensitive enough to allow imaging of surfaces with an atomic resolution. When cantilevers are used as probes for AFM, they need an integrated tip at the free end. This sharp tip allows detecting changes in the topography while scanning the surface with sub-nanometer scale sensitivity in the vertical direction and with high lateral resolution.

Recently, micro- and nano-cantilevers have been employed as sensors using physical principles that are similar to those found in AFM. When cantilevers are used as sensors, the tip is not necessary. Cantilevers can be made in different shapes, sizes, and materials and they can be arranged in arrays with large numbers of elements; some examples are presented in Figure 4.1. These cantilevers can be operated in the so-called static, surface stress, or DC-mode, where an asymmetric molecular binding on the cantilever's top or bottom surface causes an unbalanced surface stress resulting in a measurable deflection up or down. In the dynamic or AC-mode, the adsorbates lead to changes of mass and mechanical properties of the vibrating

system that are translated into a change in the resonant frequency (Bietsch et al. 2004; Ramos et al. 2006, 2007; Waggoner and Craighead 2007).

Research on biosensors is a rapidly progressing and an interdisciplinary field, bringing together physicists and engineers for developing the hardware parts of sensors, chemists to modify sensor surfaces and to synthesize bioreceptor layers, and biologists or doctors interested in specific biological samples or biotechnological processes (Fritz 2008). The application areas include basic research in the life science, health care and medical diagnostics, drug discovery, and environmental monitoring.

Existing biological detection, such as fluorescence microscopy and enzyme-linked immunosorbent assays (ELISAs), are powerful research and clinical tools, but require time-consuming and expensive procedures involving the labeling of samples with a fluorescent or radioactive tag before the analysis, thereby limiting their use in clinical settings where rapid detection of the biological targets in small sample volume is highly desirable.

Cantilever biosensors are emerging as an interesting technique for real time, fast, and direct detection. The development of cantilever biosensors is focused on (1) maximizing the sensitivity, (2) reducing sample volume, (3) reducing false positives and false negatives, (4) detecting molecular species from a single molecule up to high concentrations in real-time without influencing the sample, and (5) creating systems for parallel detection of any

---

M. Arroyo-Hernández and P.M. Kosaka contributed equally to this work.

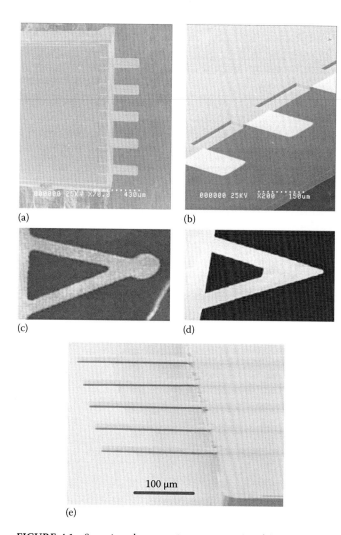

FIGURE 4.1 Scanning electron microscopy images of (a) an array of commercially available SiO$_2$ cantilevers, (b) a section of an array of SiO$_2$ cantilevers, (c) SiO$_2$ cantilever (From Tang, Y. et al., *Sens. Actuators*, 97, 109, 2004. With permission.), (d) silicon cantilever (From Tang, Y. et al., *Sens. Actuators*, 97, 109, 2004. With permission.), and (e) SU-8 cantilevers for high sensitivity.

number of biomolecules of interest (Waggoner and Craighead 2007). In addition, this technique should be cheap, small, portable, and usable for untrained personnel.

This chapter reports the development of micro- and nanomechanical biosensors, starting with the theoretical background and detection techniques. In addition, it covers aspects related to the fabrication and functionalization of the cantilevers and the main applications of this new and promising technology.

4.2 Fundamentals

Nanomechanical biosensors are based on the changes produced by molecular interactions at the nanoscale in micro- and nanosystems. These changes can be both in the position (static, surface stress, or DC-mode) and in the movement (dynamic or AC-mode). The fundamentals of each mode are explained in the following sections.

4.2.1 Static, Surface Stress, or DC-Mode

When molecules are attached to one side of a cantilever, the molecular interactions induce a deformation of the cantilever due to an asymmetrical stress. This deformation, typically measured in terms of the free-end deflection, can be directly inferred from the gradient in the mechanical stress, when the gravity is neglected and no other external forces are applied. The relation between surface stress and the deformation can be calculated from the Stoney's equation:

$$\frac{1}{r} = 6\frac{1-\upsilon}{Ed^2}(\Delta\sigma_t - \Delta\sigma_b) \tag{4.1}$$

where

  r is the curvature radius
  υ is the Poisson coefficient
  E is the Young's modulus
  d is the thickness of the plate
  $(\Delta\sigma_t - \Delta\sigma_b)$ is the differential surface stress between opposite sides

This equation was proposed in 1908 by G. Gerald Stoney (Stoney 1909) for the bending of a thin plate metallized on one side by electrodeposition (Figure 4.2). When Equation 4.1 is applied to a cantilever, a decrease in r means that the deflection at the free end increases.

The cantilever deflection can be toward or against the non-modified side. In the first case, the sign of the stress is negative and referred to as compressive, while in the second case, the sign is positive and referred to as tensile.

The relation between the displacement of the free end of a cantilever (z) and the surface stress can be derived from Stoney's

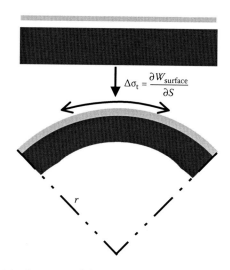

FIGURE 4.2 Depiction of the bending with the radius of curvature r of a thin plate metallized on the top side. In this case, the differential surface stress is negative as the plate bends downward, that is, the top layer expands with respect to the bottom one.

equation, obtaining the following expression for a cantilever with length L (Butt 1996):

$$z \cong 3\frac{(1-\upsilon)L^2}{Ed^2}(\Delta\sigma_t - \Delta\sigma_b) \qquad (4.2)$$

Equations 4.1 and 4.2 show that the deflection signal, and thus the response to the molecular immobilization and/or recognition, can be increased by decreasing the ratio d/L or the stiffness of the cantilever material (AE/L; where A is the area of the cantilever). In the fist case, the strategy is the fabrication of thinner and longer cantilevers. In the second case, the strategy is the investigation of materials with a low Young's modulus. Over the last years, polymers like polystyrene (McFarland et al. 2004; McFarland and Colton 2005), polypropylene (McFarland and Colton 2005), nylon (McFarland and Colton 2005), polyimide (Wang et al. 2003) and SU-8 (Calleja et al. 2005, 2006; Mouaziz et al. 2006; Nordström et al. 2008) have emerged as alternative materials for fabricating microcantilevers due to their lower E than silicon and silicon nitride. Among polymers, SU-8 has been shown to be the best candidate material. SU-8 is a negative epoxy-based photo-resistant material with a low absorption in the near UV range spectrum. SU-8 is cross-linked and because of this is inert to various chemicals (solvents and acids) used in microfabrication. Figure 4.1e shows an array of SU-8 microcantilever. Nevertheless, silicon and silicon nitride are still mainly used for cantilever fabrication due to the well-established fabrication techniques and the possibility of integration with microelectronics.

4.2.2 Dynamic or AC-Mode

4.2.2.1 Natural Frequency of a Cantilever Beam

The dynamic operation mode is devoted to the measurement of the resonance of the cantilever. The transversal vibration of the beam $z(x,t)$ can be described using the following partial differential equation:

$$EI\frac{\partial^4 z(x,t)}{\partial x^4} + \rho S\frac{\partial^2 z(x,t)}{\partial t^2} = 0 \qquad (4.3)$$

where

E represents Young's modulus
I is the moment of inertia
ρ is the cantilever density
S is the cross-section area of the beam

For a rectangular cantilever of length L, width W, and thickness d, the cross-section area and the moment of inertia are $S = Wd$ and $I = Wd^3/12$, respectively. The boundary conditions for a singly clamped beam are $z(0,t) = (\partial z(0,t))/\partial t = 0$ and

$$\frac{\partial^2 z(0,t)}{\partial t^2} = \frac{\partial^3 z(0,t)}{\partial t^3} = 0 \qquad (4.4)$$

Hence, the resonance-cantilever frequency beam is given by

$$f_n = \frac{l_n^2 d}{2\pi L^2}\sqrt{\frac{E}{12\rho}} \qquad (4.5)$$

where

n is the longitudinal mode number
$l_0 = 3.52$ for the fundamental mode

The cantilever resonance frequency, f_0, can also be calculated by using the usual harmonic oscillator formula $f_0 = 1/2\pi\sqrt{k/m}$, where the cantilever mass and spring constant are, respectively, $m = \rho LWd$ and $k = EWd^2/12L^3$.

4.2.2.2 Frequency Shift

4.2.2.2.1 Effect of Temperature

The thermal sensitivity is defined by $S_T = \partial f/f\partial T$ and depends on the value $\partial E/\partial T$ that accounts for the dependence of the material's Young's modulus on temperature. The thermal sensitivity can then be expressed as

$$S_T = \frac{\alpha}{2} + \frac{\partial E}{2E\partial T} \qquad (4.6)$$

where α is coefficient of thermal expansion of the material.

4.2.2.2.2 Effect of Surrounding Medium

In many usual conditions, the resonant frequency is changed by the mass bounded to the cantilever surface (Chen et al. 1995). The mass measurements are handicapped by the presence of damping forces and inertial masses when the experiment is performed in the natural environment of the molecules of interest.

4.2.2.2.3 In Liquid Environment

The frequency shift caused by the surrounding fluid results primarily from a change in the effective mass when the cantilevers vibrate (Sader 1998). The dynamics of a cantilever displacement $z = z(x,t)$ in the beam approximation is described by the Euler–Bernoulli equation (Chon et al. 2000):

$$EI\frac{\partial^4 z(x,t)}{\partial x^4} + \rho S\frac{\partial^2 z(x,t)}{\partial t^2} = F_h(x,t) \qquad (4.7)$$

where $F_h(x,t)$ is the distributed loading applied to the beam due to the response of the liquid to the beam oscillation.

And the resonance frequency of the cantilever beam f is given by

$$f = f_0\left(1 + \frac{\pi\rho_f W}{4\rho d}\right)^{-\frac{1}{2}} \qquad (4.8)$$

where

f_0 is the resonance frequency in vacuum
ρ_f is the density of the fluid

4.2.2.2.4 In Gas Environment

We can consider a vibrating beam subjected to a drag force due to the interaction between the beam and the surrounding medium. The drag force can be split into two components, the first, C_1 representing the viscous damping that leads to energy dissipation; and the second term, C_2 proportional to the acceleration, is the inertial force. The transversal vibration of the beam $z(x,t)$ can be described using the following partial differential equation:

$$EI \frac{\partial^4 z(x,t)}{\partial x^4} + C_1 \frac{\partial z(x,t)}{\partial t} + (\rho S + C_2) \frac{\partial^2 z(x,t)}{\partial z^2} = 0 \qquad (4.9)$$

When the effect of pressure is analyzed, it is useful to define the flow regimes, which are distinguished by numerical values assigned to the Knudsen number (k_n) (Mertens et al. 2003),

$$k_n = \frac{\tau}{W} = \frac{1}{D\gamma W} \qquad (4.10)$$

where

τ is the mean free path of gas molecules
W is the width of the gas layer in motion (usually taken as the cantilever width)
D is the gas number density
γ is the collision cross section

Three different flow regimes can be discerned: the free molecular regime for $k_n > 10$, the transition regime for $10 > k_n > 0.01$, and the viscous regime for $k_n < 0.01$.

In the free molecular regime, the fluid, considered as a rarefied gas, slips with respect to the cantilever surface. The damping is hence proportional to the cantilever velocity,

$$f = f_0 \sqrt{1 - \frac{1}{2Q^2}} \qquad (4.11)$$

For the viscous flow regime, the gas properties are mainly governed by molecule–molecule collisions, giving rise to a continuum in the gas properties. The cantilever acceleration is thereby the major parameter in the damping, inducing an increase in the effective mass of the cantilever

$$f = f_0 \left(1 - \frac{\pi W M}{24 \rho d R T} p - \frac{3}{8 \rho d} \sqrt{\frac{\pi M \mu}{R T}} \sqrt{\frac{p}{f_0}} \right) \qquad (4.12)$$

where

M is the molecular mass of the gas
μ is the gas viscosity
R is the constant for a perfect gas

4.2.2.2.5 Molecular Adsorption

When cantilevers are used in the dynamic mode, the resonance shift upon material adsorption depends on the position of the adsorbate along the cantilever. Moreover, as the size of the cantilever approaches the dimensions of the adsorbed material, the change in the local moment distribution along the cantilever length plays a very important role that in some conditions can overpass the response due to the added mass (Gupta et al. 2006; Tamayo et al. 2006).

We can use a theoretical model based on the Euler–Bernoulli equation for a beam with both mass and flexural rigidity local increase due to the deposited material. The differential equation of the vibration is then given by

$$\frac{\partial^2}{\partial x^2} G(x) \frac{\partial^2 z(x,t)}{\partial x^2} + (\rho S + \lambda(x)) \frac{\partial^2 z(x,t)}{\partial z^2} = 0 \qquad (4.13)$$

where

z is the cantilever transverse displacement
x is the longitudinal coordinate
t is the time
ρ is the cantilever mass density
S is the cross-section area
$\lambda(x)$ is the adsorbed mass per unit length
$G(x)$ is the flexural rigidity of the cantilever

This equation cannot be analytically solved in a general situation. The Rayleigh's method calculates the resonant frequency of a given vibration mode by performing an energy–work balance during a vibration cycle (Ramos et al. 2006). The mean value of the beam flexural work per oscillation cycle is given by

$$U = \frac{1}{4} \int_0^L G^F(x) \left[\frac{\partial^2 w(x,t)}{\partial x^2} \right]^2 dx \qquad (4.14)$$

Whereas the mean kinetic energy per vibration cycle (K) is given by

$$K = \frac{1}{4} \int_0^L (\rho_c S_c + \rho_a S_a(x)) \left[\frac{\partial^2 w(x,t)}{\partial t^2} \right]^2 dx \qquad (4.15)$$

the indexes c and a are related to the cantilever and to the adsorbed molecules, respectively.

The flexural vibration can be supposed as an harmonic oscillation, therefore, it can be written as $w(x,t) = \psi_n(x) \cos(\omega_n t + \beta)$ where ψ_n is the shape of the nth-mode flexural vibration of the unloaded cantilever, ω_n is the nth-mode angular eigenfrequency of the loaded cantilever and β is an arbitrary phase angle. Therefore, the second time derivative is

$$\frac{\partial^2 w(x,t)}{\partial t^2} = \omega_n^2 w(x,t) \qquad (4.16)$$

The flexural eigenfrequencies are then obtained by equalling the mean potential and kinetic energies per oscillation cycle:

$$\omega_n = \left(\frac{\int_0^L G^F(x)\left(\frac{d^2\psi_n(x)}{dx^2}\right)dx}{\rho_c W T_c \int_0^L \left(1+\frac{\rho_a}{\rho_c}\frac{T_a(x)}{T_c}\right)\psi_n^2(x)dx} \right)^{1/2} \quad (4.17)$$

The flexural vibration mode shape for the unloaded cantilever is given by

$$\psi_n(x) = A_n \left[\begin{array}{l} \sin K_n^F x - \sinh K_n^F x \\ + \dfrac{(\sin K_n^F L + \sinh K_n^F L)(\cosh K_n^F x - \cos K_n^F x)}{\cos K_n^F L + \cosh K_n^F L} \end{array} \right] \quad (4.18)$$

where A_n is an arbitrary value of the oscillation amplitude for the flexural vibrations. The flexural eigenvalues satisfies the equation $1 + \cos(K_n^F L)\cosh(K_n^F L) = 0$, which give $K_n^F = 1.8751, 4.6941$, and 7.8548.

In addition, the intrinsic sensitivity of the measurements can be enhanced by using higher resonance frequencies. These frequencies can be reached by shrinking the structures or by using higher resonant modes. Therefore, an increased sensitivity in the mass detection by improving the nanofabrication techniques is predictable (Gupta 2004a; Ekinci et al. 2004; Ilic et al. 2005).

4.3 Detection Techniques

The sensitivity of a biosensor is related to the resolution of measurement that is closely related with the detection techniques used. The techniques for the detection of cantilever deflection and resonance frequency shift can be either optical or electrical (Lavrik et al. 2004). The optical methods include optical beam deflection and interferometry and the electrical methods include piezoresistance, piezoelectric, and capacitance variation.

4.3.1 Optical Detection Techniques

The optical detection techniques can be divided in optical beam deflection and interferometry.

4.3.1.1 Optical Beam Deflection

Optical beam deflection (Figure 4.3a) is widely used due to its easy implementation and good resolution. This technique can measure deflections as small as 0.1 nm. It involves focusing a laser beam at the free end of a cantilever and detecting the reflected beam by a position-sensitive detector (PSD). This method can be used both in air and in liquid with the only requirement that the refractive index of the medium does not change during the experiment.

4.3.1.2 Interferometry

Another family of optical detection techniques is based in the use of interferometers, which are based on the interference between two waves: one reflected from a mirror and another reflected from the cantilever (Figure 4.3b). Using white light interferometry, deflections smaller than 2 nm can be resolved with a lateral resolution of 2 μm (Helm et al. 2005). Wehrneister et al. have developed a Fabry–Perot-based interferometer that allows measuring the bending of the cantilever and the changes in the refractive index simultaneously (Wehrneister et al. 2007). They use a laser as an internal reference to measure absolute deflections. By illuminating using a fiber instead of a laser, Azak et al. have achieved a submicron spot focused on the cantilever that allows measuring nanometer-scale cantilevers (Azark et al. 2007).

4.3.1.3 Arrays of Cantilevers and *Ex Situ* Measurements

The Stoney's equation (Equation 4.1) is only valid if the surface stress changes uniformly along the cantilever length. For that reason, in the last years, some improvements have been performed in order to measure the whole profile and arrays of the cantilevers. One example is the system developed by Jeon et al., consisting of a set of eight light-emitting diodes in equidistant positions along the cantilever and a single position-sensitive detector (Jeon and Thundat 2004). Another approach that

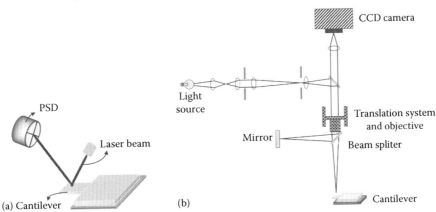

FIGURE 4.3 Schematic illustration of (a) optical beam deflection and (b) interferometry detection methods.

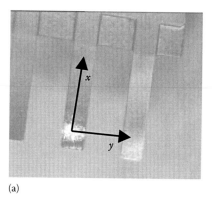
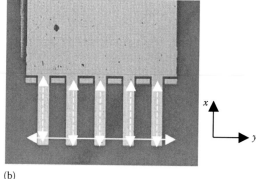

(a)   (b)

FIGURE 4.4 Depiction of the system of measurement of the whole profile of (a) a cantilever and (b) an arrays of cantilevers.

requires only one laser is the system developed by Mertens et al. (2005). In this system, the laser beam was mounted on two magnetic actuators that perform a scan along the longitudinal and the transversal axis of the cantilevers (Figure 4.4). This system allows obtaining a three-dimensional image of the deformation of the cantilevers together with the simultaneous measurement of several cantilevers in the same array. Besides, it allows performing *ex situ* immobilization and recognition assays, instead of real time measurements. For that purpose, the flat area of the chip is used as the reference and the complications of real-time measurements are avoided. In real-time measurements, the cantilevers are placed in a liquid cell and a constant flux containing the analytes is passed through it. Therefore, this configuration is prone to the formation of bubbles during the liquid injection and changes of the temperature or ionic strength of the different liquids that are to be injected resulting in a deflection of the cantilever due to the environmental changes.

4.3.1.4 Bimetallic Effect

As the optical detection techniques are based on the measurement of a light reflected from the cantilever, they are usually coated with a thin metal layer to improve their reflectivity and, thus the sensitivity. Typically, the metallic layer is made of gold with a chromium interlayer to improve the adherence, as this is very useful in the funtionalization of the self-assembled monolayers (explained in detail in Section 4.5). Nevertheless, metal-coated cantilevers are extremely sensitive thermometers due to the difference in the thermal expansion coefficient between the two materials. This difference produces a deflection of about 20–100 nm/K for the standard cantilever dimensions (Barnes et al. 1994). The contribution of this effect to the cantilever deflection has to be taken into account when sensing in the static mode.

4.3.2 Electrical Detection Techniques

Electrical methods are based on some electrical properties of the materials that change when variations in the stress take place. Thus, piezoresistive detection is based on the change

of the electrical resistance of a material due to an applied mechanical stress (Tortonese 1993). Doped single crystal silicon and doped polysilicon are widely used for the fabrication of piezoresistive cantilevers, which typically have two identical beams with piezoresistors integrated in the clamping region. The measurement of the piezoresistivity variation is made using a DC-biased Wheatstone bridge. Piezoelectricity is based on the generation of an electric potential in response to an applied mechanical stress in materials such as ZnO (DeVoe and Pisano 1997). Capacitance methods measure the capacitance between a cantilever and a fixed conductor substrate. The changes in the deflection produce changes in the distance between them that in turn produce a capacitance variation.

The disadvantage of these methods is that they require electrical connections to the cantilever that are more difficult to implement as the size of the cantilevers is reduced. In addition, piezoresistance results in heat dissipation and thermal drift due to current flow through the cantilever and the capacitive read-out is limited for the variations of the dielectric constant of the medium—a limitation enhanced when measured in liquids.

4.4 Fabrication

Silicon is the standard substrate material in integrated circuit (IC) technology. The fabrication of cantilevers uses the processes applied in IC technology combined with micromachining steps.

In general, fabrication of cantilevers is based on two well-established techniques: (1) *bulk micromachining* and (2) *surface micromachining*, schematically represented in Figure 4.5a and b, respectively. The principal difference between these techniques is the sacrificial layer that releases the devices from the substrate when removed. In the case of bulk micromachining significant amounts of the sacrificial layer (bulk silicon wafer, for example) is selectively removed from the substrate (Kovacs et al. 1998). It is usually applied to create devices with three-dimensional (3D) architecture or suspended structures (Lavrik et al. 2004). Surface micromachining is characterized by the fabrication of micromechanical structures from

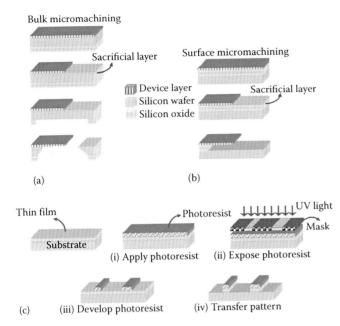

FIGURE 4.5 Schematic representation of (a) bulk micromachining, (b) surface micromachining process steps to produce suspended devices, and (c) photolithography process sequence for structuring a thin-film layer.

the deposited thin films. The original substrate remains intact and it is used as the device base.

Typically, fabrication of cantilevers by bulk micromachining or surface micromachining comprises three basic technological processes: thin film deposition, patterning, and etching that define thickness, lateral sizes, and the surrounding of the cantilever, respectively (Lavrik et al. 2004).

4.4.1 Silicon Cantilevers

In this section, the fabrication processes of the silicon and silicon nitride cantilevers will be explained by detailing the different steps of the process.

4.4.1.1 Thin Film Deposition

The most common thin-film deposition methods are (1) *chemical vapor deposition* (CVD), performed at low pressure (LPCVD), atmospheric pressure (APCVD) or plasma-enhanced (PECVD) and (2) *physical vapor deposition* (PVD), such as sputtering and evaporation (Hierlemann et al. 2003).

4.4.1.2 Patterning

The cantilever shapes can be defined by patterning through a mask—that was previously designed with a computer-assisted design program (CAD) and fabricated by electron-beam lithography—onto a certain material using *photolithography*. In the photolithographic process (Figure 4.5c), a photoresist layer is spin-coated onto the material to be patterned. Next, the photoresist layer is exposed to UV light through the mask. Depending on whether a positive or negative photoresist was used, the exposed

or unexposed photoresist areas are removed during the resist development process (Hierlemann et al. 2003). The remaining photoresist acts as a protective barrier during the following step. After the etching, the remaining photoresist is removed and the next layer can be deposited or patterned.

4.4.1.3 Etching

The etching technique allows the removal of the unwanted regions of the substrate. There are two different categories of etching processes: (1) *wet etching* using liquid chemicals and (2) *dry etching* using gas-phase chemistry. The etching reactions rely on the oxidation of silicon to form compounds that can be physically removed from the substrate. *Wet etching* is usually isotropic and provides a better etch selectivity for the material to be etched in comparison to the other accompanying materials. Highly reactive species, such as acids and bases, are used in wet etching. The most common isotropic silicon etching is called *HNA*, a mixture of hydrofluoric acid (HF), nitric acid (HNO_3), and acetic acid (CH_3COOH) (Kovacs et al. 1998). A simple description of the reaction is that the HNO_3 in the solution oxidizes the silicon, while the fluoride ions from the HF etch the oxidized compound.

$$18HF + 4HNO_3 + 3Si \rightarrow 3H_2SiF_{6(aq)} + 4NO + 8H_2O \quad (4.19)$$

The acetic acid is used to prevent the dissociation of HNO_3 into NO_3^- or NO_2^-, allowing the formation of the species responsible for the oxidation of silicon (Kovacs et al. 1998).

$$N_2O_4 \leftrightarrow 2NO_2 \quad (4.20)$$

Dry etching is generally anisotropic, resulting in a better pattern transfer. A family of fluorine-containing compounds like interhalogens (BrF_3 and ClF_3) and noble gas fluorides (XeF_2) are employed for the dry etching process. As an example, the reaction for dry etching with XeF_2 gas flow is

$$2XeF_2 + Si \rightarrow 2Xe + SiF_4 \quad (4.21)$$

In this reaction, the XeF_2 molecules are physiosorbed on the silicon surface and dissociated to release the volatile xenon atoms, while the fluorine atoms remain to react with the silicon to form the volatile SiF_4 (Williams and Muller 1996). Other reaction examples are very well described in a review written by Williams and Muller (1996).

4.4.2 Polymer Cantilevers

Silicon microfabrication has many advantages, such as (Gupta and Akin 2004): (1) precise control of dimensions, (2) miniaturization of devices, (3) fabrication of an array of devices with very close physical parameter values, (4) batch fabrication leading to a decrease in production cost, and (5) possibility of integration of various functional devices on the same platform. However, with the aim to increase the cantilever sensitivity (see Section 4.2.1), polymers have been also proposed as an alternative material for the production of cantilevers with high sensitivity (Figure 4.1e). The most applied polymer in cantilever fabrication is the SU-8, which is a polymer that can be spin-coated to thicknesses ranging from 1 μm up to 1 mm (Lorenz et al. 1997; Lorentz et al. 1998; Genolet et al. 1999; Lee et al. 2003; Calleja et al. 2003, 2005, 2006; Mouaziz et al. 2006; Nordström et al. 2008) and allows the formation of microstructures using lithography and molding techniques to form quasi-three dimensional devices (Mouaziz et al. 2006). The fact that the SU-8 has about 50 times smaller Young's modulus than silicon has proven useful to provide the cantilevers an enhanced sensitivity (Calleja et al. 2006).

4.5 Functionalization

Functionalizing a cantilever is a critical preparation step because when it is coated with a receptor layer (or linker molecule) that reacts specifically with a particular substance, the sensor signals can be assigned to the substance detection. The receptor layer deposited on the cantilever surface should be (1) *thin*, to avoid changes in the mechanical properties of the cantilever, (2) *uniform*, to generate a uniform stress, (3) *compact*, to avoid interactions with the underneath solid substrate, (4) *stable* against changes in buffer and temperature, and (5) *strongly bound*, that is, the detector molecules should be tightly attached to the cantilever surface and preserve their original recognition specificity, but should have enough flexibility to freely interact with their specific molecule or radical in the environment (Fritz 2008). In order to reuse the sensors several times, the receptor's activity level should not decrease significantly after subsequent recognition assays.

In the dynamic mode or AC-mode (see Section 4.2.2), where the mass change on the cantilever is monitored by changes in the resonance frequency, the cantilevers can be coated on both sides. However, in the static or DC-mode (see Section 4.2.1) the

cantilever functionalization must be performed on one cantilever side and the opposite side should prevent any adsorption of the sample molecules. Many approaches can be used to immobilize the molecular recognition agents to the cantilever surface, depending upon the final application. The most employed methods are silanization of silicon surfaces, self assembled monolayers (SAM) on gold, and deposition of thin polymer films. These are schematically illustrated in Figure 4.6a through c, respectively.

4.5.1 Silanization

Silanization is the modification of surfaces with hydroxyl groups (OH) such as mica, glass, and metal oxide, with organo-silicon molecules. The silanization methods for silicon can also be applied to silicon nitride (Diao et al. 2005), a very common material used in cantilever fabrication.

$$Si-OH + X-SiP_2'P \rightarrow Si-O-SiP_2'P + HX \qquad (4.22)$$

This reaction is generally referred to as organosilanization and is the most common reaction of oxide surfaces (Pesek and Matyska 1997). In this kind of reaction, P' is usually a small organic group such as methyl: X is a reactive group such as (methoxy and ethoxy) chloride. The second organic group P is the terminal group, which is replaced with different functional groups like amino, carboxyl, or thiol groups, and provides the desired properties for the modified surface, like hydrophobicity, hydrophilicity, ion-exchange, etc. (Pesek and Matyska 1997). The commercially available Si wafers usually have a layer of native oxide (~1–1.5 nm thickness), which is formed when exposed to ambient atmosphere. On this layer, water molecules avidly adsorb producing silanol groups (Si–OH), with a high density of silanol groups on the surface (Rye et al. 1997). These silanol groups are used as connection sites for silanization reaction, and the attachment between the terminal group P and the

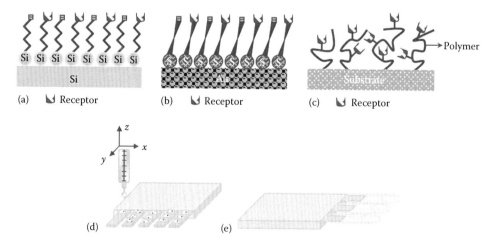

FIGURE 4.6 Schematic illustration for the immobilization of biomolecular receptors on the surface of a cantilever: (a) silanization of silicon surfaces, (b) self assembled monolayers (SAM) on gold, and (c) deposition of thin polymer films. Scheme techniques to immobilize different specific biomolecules on individual cantilevers of a cantilever array: (d) inkjet printing of individual droplets, and (e) incubation in thin glass capillaries or pipette tips.

surface is through a silicon–oxygen–silicon carbon linkage. This type of bonding is thermodynamically and hydrolytically stable. The simplicity of the reaction and the high stability of the product make this esterification process a customary modification route. This is just one example of silanization reaction; descriptions of other methods can be found in the literature (Pesek and Matyska 1997; Pavlovic et al. 2002; Diao et al. 2005; Arroyo-Hernandez et al. 2006, 2008). Bacteria, viruses (Ilic et al. 2004; Zhang and Ji 2004) and, in few cases, proteins (Diao et al. 2005) have been immobilized by the activation methods based on organosilanization.

4.5.2 Self-Assembled Monolayers

Self-assembled monolayers (SAM) are ordered organic assemblies that are spontaneously formed by the adsorption of a surfactant with a specific affinity of its head group to a substrate. The adsorbates organize spontaneously into crystalline or semicrystalline structures (Ulman 1996; Schreiber 2004). The SAM layers are constituted of a headgroup that specifically interacts with the substrate surface and defines the individual SAM system, a chain or backbone and an endgroup that determines the chemical or physical properties of the functionalized surface.

The chemisorption of organosulfur compounds on gold surfaces has attracted much attention as a method for preparing well-organized organic surfaces since 1983 when Allara and Nuzzo (Nuzzo and Allara 1983) showed the formation of alkenethiol monolayers on gold substrates. Spectroscopic, especially, reflection-absorption infrared spectroscopy (Nuzzo et al. 1990; Laibinis et al. 1991) and physical–chemical studies, like contact angle measurements (Bain 1989; Laibinis et al. 1991), have shown that the adsorption of long-chain alkanethiols of the general structure $HS(CH_2)_mX$ ($m \geq 5$, $X = CH_3$, $CHCH_2$, $CONH_2$, $COOH$, NH_2, $N(CH_3)_2$, Cl, Br, etc.) onto gold substrates results in the formation of densely packed films on gold. The reaction may be considered an oxidative addition of the S–H bond to the gold surface, followed by a reductive elimination of the hydrogen (Ulman 1996).

$$X(CH_2)_m SH + Au_n^0 \rightarrow X(CH_2)_m S^- Au^+ \cdot Au_n^0 + \tfrac{1}{2} H_2 \quad (4.23)$$

The bonding of the thiolate group to the gold surface is very strong (homolytic bond strength is around 40 kcal/mol (Ulman 1996)).

The use of thiolated molecules on gold coated cantilevers is the most common functionalization method for DNA (Fritz et al. 2000; Mukhopadhyay et al. 2005a; Stachowiak 2006), antibodies (Wu 2001b; Dutta et al. 2003) and protein (Moulin et al. 1999; Veiseh et al. 2002; Kim et al. 2007).

4.5.3 Polymer Coating

An alternative approach to gold coating and organosilanization is the coating of cantilevers with polymers with properties matching the needs of the ending use (Bergese et al. 2007; Goddard and Hotchkiss 2007). As an example we can mention the work of Bergese et al. (2007) where they coated a microcantilever with a thin film based on *N,N*-dimethylacrylamide to covalently bind amino-modified DNA.

It is important to have in mind that the choice of the link molecule is not trivial. If special care is not taken, the receptor molecules will nonspecifically adsorb on the cantilever surface and will block the specific binding sites. This is also addressed by implementing a blocking chemistry. One example is the use of poly(ethylene glycol) (PEG) as a passive layer, preventing molecular adsorption (Veiseh et al. 2002; Schreiber 2004; Lan et al. 2005). PEG is a nontoxic, non-immunogenic, and non-antigenic polymer that resists protein adhesion. PEG is know to decrease the attractive forces between the solid surfaces and the proteins because of its highly hydrated polymer chains, steric stabilization forces, as well avs, chain mobility (Lan et al. 2005). In addition, its structure prevents the PEG from binding with acidic or basic amino acid chains. It is therefore thermodynamically unfavorable for a protein to bind to the surface beneath a PEG brush (Goddard and Hotchkiss 2007). Proteins like bovine serum albumin (BSA) and casein are also used as blocking agents.

4.5.4 Cantilever Arrays

Immobilizing different specific biomolecules on individual cantilevers of a cantilever array is not an easy task. They can be functionalized by incubation in individual thin glass capillaries or pipette tips (Figure 4.6d) (Fritz 2008) or by inkjet printing (Figure 4.6e) to coat cantilever arrays (Bietsch et al. 2004). Special attention must be given for a homogeneous coating at the clamping region, where the flexible cantilever beam is connected to the bulk silicon. The deflection of the cantilever free end is much more influenced by the bending in the clamping area than by the bending close to the free end of the cantilever (Fritz 2008).

4.6 Applications

4.6.1 Nucleic Acids Biosensors

The detection of specific DNA sequences is of great importance in genome analysis, clinical diagnosis, or drug discovery, among others. For that purpose, the development of a technique that allows measuring a high number of samples with low cost, using low sample amount, and with high specificity and selectivity is required.

In 2000, Fritz et al. demonstrated the potential of nanomechanical systems for DNA biosensing by the detection of the hybridization of 12-mer oligonucleotide and by the discrimination of single base mismatches (Fritz et al. 2000). Since then, many studies have been made in order to improve these results. McKendry et al. were able to simultaneously detect different DNA sequences with nanomolar resolution and even relate the nanomechanical bending with the DNA concentration in the solution (McKendry et al. 2002). They also proved that the biosensor could be reused up to 10 times without losing efficiency.

The detection of hybridization requires a reference cantilever to measure the differential surface stress from the subtraction of the non-specific signal contribution, that is, reactions occurring on the underside of the cantilever, liquid injection spikes, changes in the refractive index, and temperature variations due to the bimetallic effect.

Although the origin of the nanomechanical response for hybridization is not clear, many studies have focused on that problem and different mechanisms have been proposed. In the beginning, Wu et al. proposed that electrostatic interaction and changes in the entropy configuration of molecules could be responsible for the cantilever bending (Wu et al. 2001b). Nevertheless, subsequent works attributed the origin of the response to different mechanisms. McKendry et al. proposed that the origin of cantilever bending after hybridization is mainly due to steric hindrance interactions and discarded the electrostatic interactions as hybridization experiments using increasing ionic strength did not show a relevant nanomechanical response (McKendry et al. 2002). Hagan et al. proposed that the origin of the bending is due to the hydration forces based on the theoretical predictions using an empirical potential derived from independent experiments (Hagan et al. 2002). Liu et al. suggested a flexoelectric origin based on theoretical calculations. The

model considered the cantilever-biomolecule layer as an asymmetric membrane and uses the relation between the polyelectrolytes theory, spontaneous curvature, and electric potential (Liu et al. 2003).

Recently, a new concept of biosensors based on the nano-confinement of water molecules has been developed (Mertens et al. 2008). In this case, the cantilevers are sensitized with a highly compacted DNA monolayer with intermolecular distances between 0.5 and 0.8 nm. The DNA molecules form a nanochannel net that allows the confinement of 2–3 water molecules that produce a sharp tensile stress rise, in the order of 40–70 mN (Figure 4.7) due to the hydrogen bond interaction. This particular feature disappears after hybridization with complementary DNA that provides with a clear fingerprint of the hybridization. This biosensing principle reaches femtomolar sensitivity of the complementary DNA with high selectivity in mixtures (1 nM c-DNA in 1 µM NC-DNA) and allows the detection of single base mismatches.

Besides the genomic applications, the nanomechanical cantilevers have also been used in fundamental research. Among those, the DNA immobilization and the formation of self-assembled monolayers (Bammerlin et al. 2007), the DNA melting (Biswal et al. 2006), or thermodynamics of DNA hybridization (McKendry et al. 2002) are ones that can be highlighted.

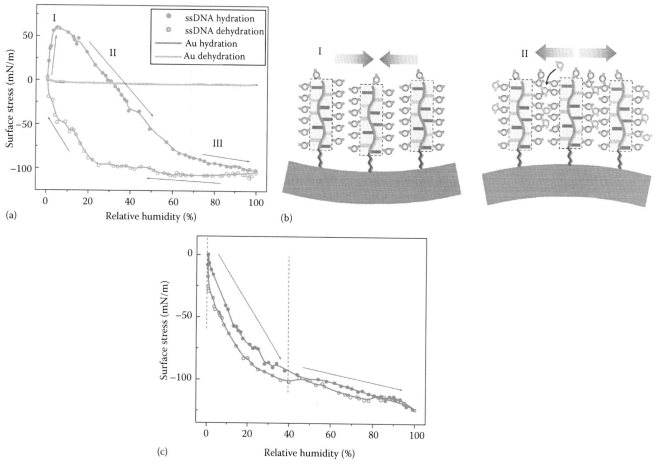

FIGURE 4.7 Hydration dependence of the surface stress of the densely packed self-assembled ssDNA monolayers. (a) Surface-stress variation during a hydration and dehydration cycle for a gold-coated silicon cantilever sensitized with a thiol-modified 16-mer ssDNA probe, (b) schematics of the ssDNA monolayer in stage I and stage II, and (c) effect of hybridization on the surface stress *versus* relative humidity relationship.

Another application of nanomechanical biosensors is the detection of messenger RNAs (mRNAs), which are single stranded RNA molecules that regulate gene expression. The interest is in both genomics and drug administration as they can change the gene expression and the protein abundance. Zhan et al. have demonstrated the utility of nanomechanical cantilever for the early detection of gene 1–8U, which is a potential marker for cancer progression or viral infections. They were able to detect the presence of 1–8U in few minutes with a sensitivity in the picomolar range and to the base mismatches (Zhang et al. 2006).

4.6.2 Protein Biosensors

Microcantilever biosensors can be used to investigate the binding or conformational change of proteins and the characteristics of protein-ligand interactions.

The mechanical response of the microcantilever can be measured as the deflection of the beam caused by the surface stress during the protein-ligand binding onto one side of the microcantilever. The high sensitivity of this technique permits detecting 11.3 mg cystamine dihydrochloride/mL, 0.04 mg glutaraldehyde/mL and 0.5 mg streptavidin/mL solutions (Ji et al. 2008). The technique has also been used to detect two forms of prostate-specific antigen (PSA) over a wide range of concentrations from 0.2 to 60 g/mL in a background of human serum albumin (HSA) and human plasminogen (HP) at 1 mg/mL, making this a clinically relevant diagnostic technique for prostate cancer (Wu et al. 2001a). The ligand-binding domain of the human oestrogen receptor (ERα-LBD) and ERα-LBD can also be detected using the conformation-specific peptides α/βl at protein concentrations of 2.5–20 nM (Mukhopadhyay et al. 2005b). Other techniques use composite self-excited PZT-glass cantilever (4 mm in length and 2 mm wide). These millimeter-sized beams have the sensitivity to measure in real-time protein-protein binding, and the binding rate constant, with nanogram mass-change sensitivity. Protein concentrations of 0.1 and 1.0 mg/mL were successfully detected for the binding of anti-rabbit IgG (biotin conjugated) to rabbit IgG immobilized cantilever and the subsequent binding of captavidin (Campbell and Mutharasan 2005).

Dynamic mode can also be used to measure the proteins' properties. The resonant frequencies of cantilever beams were observed to either increase or decrease upon attachment of the protein layers. This is due to the dependence of the cantilever on both the mechanical properties, mainly stiffness constant, mass upon adsorption, or the conformational change of proteins on nanoscale cantilevers (Gupta et al. 2006).

4.6.3 Molecular Motor Biosensors

Nanoscale actuators, capable of converting chemical or electrical energy into mechanical motion, are needed for a wide range of applications. They can be organized to perform complex mechanical tasks beginning at the nanometer scale but expressed in the macroscopic world. The direct integration of a dynamic biomolecular motor with cantilever arrays provides a way forward to a generation of "smart" bioinspired mechanical devices based on molecular concepts. This novel nanomechanical transduction mechanism integrates "bottom up" molecular design with "top down" microfabrication and does not require applied loads or fluorescent tags. Motility is driven by the ability directly to convert energy into mechanical work, when stimulated by light, electricity, or chemical reagents.

Translating biochemical energy into micromechanical work, in-plane surface stress forces induce micromechanical-bending motion of the cantilever beam. Shu et al. (2005) have investigated the force exerted by surface-tethered DNA motors on microfabricated cantilevers. The reversible DNA nanomachine can be triggered by pH change. The addition of protons causes conformational change from an open to a close state. This induces compressive surface stress that can be measured through cantilever motion. Under acidic conditions (pH 5.0), the base pair interactions between a protonated and an unprotonated cytosine residue form the four-stranded DNA structure, called the i-motif, defined as the *closed* form of the machine. In contrast, at pH > 6.5 the cytosine bases become nonprotonated and can hybridize with the complementary strand, forming the *open* form of the motor.

A synthetic muscle device can be constructed by combining a scalable responsive gel comprising a robust, self-assembled, nanostructured block copolymer at the surface of a cantilever (Howse et al. 2006). The gels composed of the Poly(methylmethacrylate)-Poly(methacrylic acid) (PMMA-PMAA) domains deform reversibly in response to a pH stimulus with a volume change of 3. This chemical motor generates a peak power of 20 mW/kg. It is completely scalable, as the mechanism of operation is the serial addition of molecular shape changes, and can provide reciprocating motion of over 6 orders of magnitude in length scale, from nanometers to millimeters.

A molecular machine-based actuator that displays reversible bending through the cycled addition of the aqueous oxidant and reductant solutions can also be created (Huang et al. 2004). Cantilever bending is driven by mechanical contraction and extension of the inter-ring distance in the surface-bound, bistable, redox-controllable rotaxane molecules.

The regulation of cell-traction forces and cell viscoelasticity plays an essential role in cell division, cell motility, and cell–cell or cell–tissue adhesion, as well as in critical operations, such as wound healing, immune response, or cancer. The microcantilever technique is able to determine some of the mechanical properties of the cells, such as the active and passive responses to controlled stress or strain (Micoulet et al. 2005). Cell elongation can be observed as a reaction against a constant load or the cell force is measured as a response to constant deformation. Passive viscoelastic deformation and active cell response can be discriminated.

4.6.4 Pathogen Biosensors

The need for a fast, ultrasensitive, and inexpensive technique for an effective biological detection at low concentration is becoming necessary for the early detection of harmful organisms like bacteria cells and viruses. The applications include food safety, environmental monitoring, early clinical diagnosis, and national security issues.

Ilic et al. (2000) detected *Escherichia coli* (*E. coli*) bacteria using an array of resonant cantilevers 100 μm long. They could detect 16 *E. coli* cells in air using ambient thermal noise to excite resonances that correspond to a mass of ~6 pg. Using cantilevers coated with antibodies specific to the bacteria they observed a linear dependence of the frequency shift with the number of *E. coli* cells bounded to the antibody layer. In further work with the *E. coli* (Ilic et al. 2001) they used cantilevers 15 μm long and could measure the frequency shift due to the immobilization of a single bacteria cell. They measured the mass of a single *E. coli* as 665 fg. The enhancement in sensitivity can be attributed to the reduced oscillator mass.

A rapid biosensor for active bacterial growth using an oscillating cantilever coated by a nutritive layer was suggested by Gfeller et al. (2005a,b). This sensor was able to detect in real-time the active growth of an *E. coli* cell within 1 h, which is much faster than any conventional culturing method that takes at least 24 h. In addition, these devices have proved to be able to detect antibiotic resistance in less than 2 h (Gfeller et al. 2005b).

Ramos et al. (2006, 2008) have showed that the sign and magnitude of the resonant frequency change depends on the position and extent of the droplets of the bacterial cells on the cantilever. They have observed that the adsorption of bacteria on a resonant cantilever can produce a negative or a positive resonant frequency shift as shown in Figure 4.8. Based on a theoretical one-dimensional mode proposed by the authors, the results have been attributed to the stiffness of bacteria that shifts the resonance to higher frequencies and to the added mass that shifts the resonance to a negative frequency. On the one hand, the resonant response is dominated by the added mass when the bacteria adsorbs onto regions of high-vibration amplitude like that of the cantilever free end. On the other hand, the stiffness effect dominates when the adsorption is near the clamping, a region of small vibrations.

In addition, cantilevers have been employed to detect the mass of a single virus particle. In 2004, Gupta et al. were the first to develop an array of silicon cantilever beams driven by thermal noise and ambient noise as a microresonator sensor to detect the mass of individual virus particles (Gupta et al. 2004b). The relationship between the decrease in the resonant frequency versus the effective number of virus particles observed in the cantilever beam was linear, proving the validity of the measurements. They measured an average dry mass of 9.5 fg for a single *Vaccinia virus* particle. In 2006, Jonhson et al., used cantilever beams driven by a PZT piezoelectric ceramic as a resonating sensor to measure the mass of virus particles (Jonhson et al. 2006). Using a cantilever with dimensions of 6 μm × 4 μm, they measured the average mass of a single *Vaccinia virus* particle to be 7.9 ± 4.6 fg, which is in the expected range of 5–10 fg. These results have shown that the cantilevered structures can be very useful components for the detection of airborne virus particles.

Resonating mechanical cantilevers were used to test the ability to specifically bind and detect small numbers of virus particles captured from the liquid environment. Ilic et al. (2004)

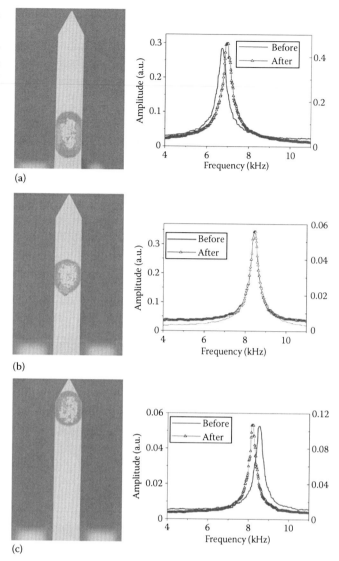

FIGURE 4.8 Optical micrographs of the three silicon cantilevers in which the *E. coli* cells were deposited by ink jet at three different positions along the cantilever length and frequency spectra of the Brownian motion of the these cantilevers before and after the bacteria deposition. The center of the *E. coli* spot is separated from the clamping at (a) 73 μm, (b) 200 μm, and (c) 390 μm, approximately.

have used arrays of chemically functionalized, surface micromachined polycrystalline silicon cantilevers to measure binding events from various concentrations of baculovirus in a buffer solution. Biomolecular binding of the baculovirus to the antibody-treated regions of the cantilever sensor alters the total mass of the mechanical oscillator, changing its natural resonant frequency. At the lowest virus concentration used by the authors and considering a weight of a single baculovirus as ~1.5 fg, the minimum number of virus bounded to the cantilevers that were detected was estimated as 6, suggesting that these oscillators, when operated in the regime where the mechanical quality factor is about 10^4, are capable to detect the binding of a single baculovirus. In addition, by taking the frequency spectra before and after virus binding, the devices could distinguish between the various solutions of different concentrations of baculovirus.

Cantilevers modified by the feline *Coronavirus* (FIP) type I anti-viral antiserum have been developed to detect FIP type I virus in solution, these devices were able to detect a viral concentration as low as 0.1 μg/mL (Velanki and Ji 2006). It was observed that when the FIP type I virus positive samples were injected into the fluid cell where the cantilever was held, the cantilever bent upon recognition of the FIP type I virus by the antiserum on the surface of the cantilever. When a sample that did not contain FIP type I virus was used no bending of the cantilever was observed. FIP is known to be highly prevalent in the cat population and causes a deadly disease called *feline infectious peritonitis* among cats. In addition, FIP is a virus whose structure is very similar to a *Coronavirus* that causes a viral respiratory disease in humans called *severe acute respiratory syndrome* (SARS). This study provides the way for the development of cantilever sensors for human-associated SARS *Coronavirus* that could detect the virus in a fast and single step experiment. The use of cantilevers with the aim of developing portable and low-cost devices for the detection of other viruses like human immunodeficiency virus type I (HIV-1) (Lam et al. 2006) and hepatitis C virus (HCV) helicase (Hwang et al. 2007) have been reported. Nevertheless, to make this label-free detection approach more powerful and sensitive, a variety of improvements must be made, including the use of cantilever arrays, and a tighter control over temperature during incubation and measurements.

4.7 Conclusions

Micro- and nanomechanical biosensors promise to be a revolutionary technology for the next generation of biological sensors that will probably have an impact in the fields of genomics and proteomics. Their major advantages are, wide field of application, versatility (they can be operated in air, vacuum, and liquid environment), can be micro-fabricated at low cost, and need only small sample volumes for their operation. Some problems like their lack of specificity that can be overcome by a sophisticated surface functionalization, and the lack of a detailed theoretical description for all experimental cases still delay a general use of cantilever biosensors, but researchers are currently addressing these issues. The most charming fact in the world of micro- and nanomechanical biosensors is that the ultrathin layers of molecules are able to deflect solid devices.

References

Abu-Lail, N.I.; Kaholek, M.; LaMattina, B.; Clark, R.L.; Zauscher, S. 2006. Micro-cantilevers with end-grafted stimulus-responsive polymer brushes for actuation and sensing. *Sensors and Actuators, B: Chemical* 114: 371–378.

Arroyo-Hernández, M.; Perez-Rigueiro, J.; Martinez-Duart, J.M. 2006. Formation of amine functionalized films by chemical vapour deposition. *Materials Science and Engineering C* 26: 938–941.

Arroyo-Hernández, M.; Perez-Rigueiro, J.; Conde, A. et al. 2008. Characterization of biofunctional thin films deposited by activated vapor silanization. *Journal of Materials Research* 23: 1931–1939.

Azark, O.N.; Shagam, D.M.; Karabacak, K.L. et al. 2007. Nanomechanical displacement detection using fiber-optic interferometry. *Applied Physics Letters* 91: 093112/1–093112/3.

Bain, D.C.; Troughton; E.B.; Tao, Y.T. et al. 1989. Formation of monolayer films by the spontaneous assembly of organic thiols from solution onto gold. *Journal of the American Chemical Society* 111: 321–335.

Bammerlin, M.; Köser, J.; Battiston, F.M.; Hubler, U. 2007. Nanomechanical cantilever bio-sensors for time-resolved detection of DNA and surface layer formation. *Nanotechnology Conference and Trade Show—NSTI Nanotech, Technical Proceedings* vol. 2, Boston, MA, 469–472.

Barnes, R.; Stephenson, R.J.; Woodburn, C.N. et al. 1994. A femtojoule calorimeter using micromechanical sensors. *Review of Scientific Instruments* 65: 3793–3798.

Bergese, P.; Bontempi, E.; Chiari, M.; Colombi, P.; Damin, F.; Depero, L.E.; Oliviero, G.; Pirri, G.; Zucca, M. 2007. Investigation of a biofunctional polymeric coating deposited onto silicon microcantilevers. *Applied Surface Science* 253: 4226–4231.

Bietsch, A.; Zhang, J.; Hegner, M.; Lang, H.P.; Gerber, C. 2004. Rapid functionalization of cantilever array sensors by inkjet printing. *Nanotechnology* 15: 873–880.

Biswal, S.L.; Raorane, D.; Chaiken, A.; Birecki, H.; Majumdar, A. 2006. Nanomechanical detection of DNA melting on microcantilever surfaces. *Analytical Chemistry* 78: 7104–7109.

Butt, H.J. 1996. A sensitive method to measure changes in the surface stress of solids. *Journal of Colloidal Interface Science* 180: 251–260.

Calleja, M.; Tamayo, J.; Johansson, A. et al. 2003. Polymeric cantilever arrays for biosensing applications. *Sensor Letters* 1: 20–24.

Calleja, M.; Nordström, M.; Álvarez, M.; Tamayo, J.; Lechuga, L.M.; Boisen, A. 2005. Highly sensitive polymer-based cantilever-sensors for DNA detection. *Ultramicroscopy* 105: 215–222.

Calleja, M.; Tamayo, J.; Nordström, M.; Boisen, A. 2006. Low-noise polymeric nanomechanical biosensors. *Applied Physics Letters* 88: 113901.

Campbell, G.A.; Mutharasan, R. 2005. Detection and quantification of proteins using self-excited PZT-glass millimeter-sized cantilevers. *Biosensors and Bioelectronics* 21: 597–607.

Chen, G.Y.; Thundat, T.; Wachter, E.A.; Warmack, R.J. 1995. Adsorption-induced surface stress and its effects on resonance frequency of microcantilevers. *Journal of Applied Physics* 77: 3618–3622.

Chon, J.W.M.; Mulvaney, P.; Sader, J.E. 2000. Experimental validation of theoretical models for the frequency response of atomic force microscope cantilever beams immersed in fluids. *Journal of Applied Physics* 87: 3978–3988.

DeVoe, D.L.; Pisano, A.P. 1997. Modeling and optimal design of piezoelectric cantilever microactuators. *Journal of Micromechanical Systems* 6: 266–270.

Diao, J.; Ren, D.; Engstrom, J.R.; Lee, K.H. 2005. A surface modification strategy on silicon nitride for developing biosensors. *Analytical Biochemistry* 343: 322–328.

Dutta, P.; Tipple, C.A.; Lavrik, N.V.; Datskos, P.G.; Hofstetter, H.; Hofstetter, O.; Sepaniak, M.J. 2003. Enantioselectivity sensor based on antibody-mediated nanomechanics. *Analytical Chemistry* 75: 2342–2348.

Ekinci, K.L.; Huang, X.M.H.; Roukes, M.L. 2004 Ultrasensitive nanoelectromechanical mass detection. *Applied Physics Letters* 84: 4469–4471.

Fritz, J. 2008. Cantilever biosensors. *Analyst* 133: 855–863.

Fritz, J.; Baller, M.K.; Lang, H.P.; Rothuizen, H.; Vettiger, P.; Meyer, E.; Güntherodt, H.J.; Gerber, C.; Gimzewski, J.K. 2000. Translating biomolecular recognition into nanomechanics. *Science* 288: 316–318.

Genolet, G.; Brugger, J.; Despont, M. et al. 1999. Soft, entirely photoplastic probes for scanning force microscopy. *Review of Scientific Instruments* 70: 2398–2401.

Gfeller, K.Y.; Nugaeva, N.; Hegner, M. 2005a. Micromechanical oscillators as rapid biosensor for the detection of active growth of *Escherichia coli*. *Biosensors and Bioelectronics* 21: 528–533.

Gfeller, K.Y.; Nugaeva, N.; Hegner, M. 2005b. Rapid biosensor for detection of antibiotic-selective growth of *Escherichia coli*. *Applied and Environmental Microbiology* 71: 2626–2631.

Goddard, J.M.; Hotchkiss, J.H. 2007. Polymer surface modification for the attachment of bioactive compounds. *Progress in Polymer Science* 32: 698–725.

Gupta, A.; Akin, D. 2004a. Detection of bacterial cells and antibodies using surface micromachined thin silicon cantilever resonators. *Journal of Vacuum Science and Technology B* 22: 2785–2791.

Gupta, A.; Akin, D.; Bashir, R. 2004b. Single virus particle mass detection using microresonators with nanoscale thickness. *Applied Physics Letters* 84: 1976–1978.

Gupta, A.; Akin, D.; Bashir, R. 2005. Mechanical effects of attaching protein layers on nanoscale-thick cantilever beams for resonant detection of virus particles. *Proceedings of the IEEE International Conference on Micro Electro Mechanical Systems (MEMS)*, Piscataway, NJ, vol. 22, pp. 746–749.

Gupta, A.K.; Nair, P.R.; Akin, D. et al. 2006. Anomalous resonance in a nanomechanical biosensor. *Proceedings of the National Academy of Sciences of the United States of America* 103: 13362–13367.

Hagan, M.F.; Majumdar, A.; Chakraborty, A.K. 2002. Nanomechanical forces generated by surface grafted DNA. *The Journal of Physical Chemistry B* 106: 10163–10173.

Helm, M.; Servant, J.J.; Saurenbach, F.; Berger, R. 2005. Read-out of micromechanical cantilever sensors by phase shifting interferometry. *Applied Physics Letters* 87: 064101/1–064101/3.

Hierlemann, A.; Brand, O.; Hagleinter, C.; Baltes, H. 2003. Microfabrication techniques for chemical/biosensors. *Proceedings of the IEEE* 91: 839–863.

Howse, J.R.; Topham, P.; Crook, C.J.; Gleeson, A.J.; Bras, W.; Jones, R.A.L.; Ryan, A.J. 2006. Reciprocating power generation in a chemically driven synthetic muscle. *Nano Letters* 6: 73–77.

Huang, T.J. 2007. Towards artificial molecular motor-based electroactive/photoactive biomimetic muscles. *Proceedings of the Society of Photo-Optical Instrumentation Engineers (SPIE)* 6524: H5240.

Huang, T.J.; Brough, B.; Ho, C.-M. et al. 2004. A nanomechanical device based on linear molecular motors. *Applied Physics Letters* 85: 5391–5393.

Hwang, K.S.; Lee, S.-M.; Eom, K. et al. 2007. Nanomechanical microcantilever operated in vibration modes with use of RNA aptamer as receptor molecules for label-free detection of HCV helicase. *Biosensors and Bioelectronics* 23: 459–465.

Ilic, B.; Czaplewski, D.; Zalalutdinov, M. et al. 2000. Mechanical resonant immunospecific biological detector. *Applied Physics Letters* 77: 450–452.

Ilic, B.; Czaplewski, D.; Zalalutdinov, M. et al. 2001. Single cell detection with micromechanical oscillators. *Journal of Vacuum Science and Technology* 19: 2825–2828.

Ilic, B.; Yang, Y.; Craighead, H.G. 2004. Virus detection using nanoelectromechanical devices. *Applied Physics Letters* 85: 2604–2606.

Ilic, B.; Yang, Y.; Aubin, K.; Reichenbach, R.; Krylov, S.; Craighead, H.G. 2005. Enumeration of DNA molecules bound to a nanomechanical oscillator. *Nano Letters* 5: 925–929.

Jeon, S.; Thundat, T. 2004. Instant curvature measurement for microcantilever sensors. *Applied Physics Letters* 85: 1083–1084.

Ji, H.-F.; Gao, H.; Buchapudi, K.R.; Yang, X.; Xua, X.; Schulteb, M.K. 2008. Microcantilever biosensors based on conformational change of proteins. *The Analyst* 133: 434–443.

Jonhson, L.; Gupta, A.K.; Ghafoor, A.; Akin, D.; Bashir, R. 2006. Characterization of vaccinia virus particles using microscale silicon cantilever resonators and atomic force microscopy. *Sensors and Actuators B* 115: 189–197.

Kim, D.J.; Weeks, B.L.; Hope-Weeks, L.J. 2007. Effect of surface conjugation chemistry on the sensitivity of microcantilever sensors. *Scanning* 29: 245–248.

Kovacs, G.T.A.; Maluf, N.I.; Petersen, K.R. 1998. Bulk micromachining of silicon. *Proceedings of the IEEE* 86: 1536–1551.

Laibinis, P.E.; Whitesides, G.M.; Allara, D.L.; Tao, Y.T.; Parikh, A.N.; Nuzzo, R.G. 1991. Comparison of the structures and wetting properties of self-assembled monolayers of normal-alkanethiols on the coinage metal-surfaces, Cu, Ag, Au. *Journal of the American Chemical Society* 113: 7152–7167.

Lam, Y.; Abu-Lail, N.I.; Alam, M.S.; Zauscher, S. 2006. Using microcantilever deflection to detect HIV-1 envelope glycoprotein gp120. *Nanomedicine: Nanotechnology, Biology and Medicine* 2: 222–229.

Lan, S.; Veiseh, M.; Zhang, M. 2005. Surface modification of silicon and gold-patterned silicon surfaces for improved biocompatibility and cell patterning selectivity. *Biosensors and Bioelectronics* 20: 1697–1708.

DeVoe, D.L.; Pisano, A.P. 1997. Modeling and optimal design of piezoelectric cantilever microactuators. *Journal of Micromechanical Systems* 6: 266–270.

Diao, J.; Ren, D.; Engstrom, J.R.; Lee, K.H. 2005. A surface modification strategy on silicon nitride for developing biosensors. *Analytical Biochemistry* 343: 322–328.

Dutta, P.; Tipple, C.A.; Lavrik, N.V.; Datskos, P.G.; Hofstetter, H.; Hofstetter, O.; Sepaniak, M.J. 2003. Enantioselectivity sensor based on antibody-mediated nanomechanics. *Analytical Chemistry* 75: 2342–2348.

Ekinci, K.L.; Huang, X.M.H.; Roukes, M.L. 2004 Ultrasensitive nanoelectromechanical mass detection. *Applied Physics Letters* 84: 4469–4471.

Fritz, J. 2008. Cantilever biosensors. *Analyst* 133: 855–863.

Fritz, J.; Baller, M.K.; Lang, H.P.; Rothuizen, H.; Vettiger, P.; Meyer, E.; Güntherodt, H.J.; Gerber, C.; Gimzewski, J.K. 2000. Translating biomolecular recognition into nanomechanics. *Science* 288: 316–318.

Genolet, G.; Brugger, J.; Despont, M. et al. 1999. Soft, entirely photoplastic probes for scanning force microscopy. *Review of Scientific Instruments* 70: 2398–2401.

Gfeller, K.Y.; Nugaeva, N.; Hegner, M. 2005a. Micromechanical oscillators as rapid biosensor for the detection of active growth of *Escherichia coli*. *Biosensors and Bioelectronics* 21: 528–533.

Gfeller, K.Y.; Nugaeva, N.; Hegner, M. 2005b. Rapid biosensor for detection of antibiotic-selective growth of *Escherichia coli*. *Applied and Environmental Microbiology* 71: 2626–2631.

Goddard, J.M.; Hotchkiss, J.H. 2007. Polymer surface modification for the attachment of bioactive compounds. *Progress in Polymer Science* 32: 698–725.

Gupta, A.; Akin, D. 2004a. Detection of bacterial cells and antibodies using surface micromachined thin silicon cantilever resonators. *Journal of Vacuum Science and Technology B* 22: 2785–2791.

Gupta, A.; Akin, D.; Bashir, R. 2004b. Single virus particle mass detection using microresonators with nanoscale thickness. *Applied Physics Letters* 84: 1976–1978.

Gupta, A.; Akin, D.; Bashir, R. 2005. Mechanical effects of attaching protein layers on nanoscale-thick cantilever beams for resonant detection of virus particles. *Proceedings of the IEEE International Conference on Micro Electro Mechanical Systems (MEMS)*, Piscataway, NJ, vol. 22, pp. 746–749.

Gupta, A.K.; Nair, P.R.; Akin, D. et al. 2006. Anomalous resonance in a nanomechanical biosensor. *Proceedings of the National Academy of Sciences of the United States of America* 103: 13362–13367.

Hagan, M.F.; Majumdar, A.; Chakraborty, A.K. 2002. Nanomechanical forces generated by surface grafted DNA. *The Journal of Physical Chemistry B* 106: 10163–10173.

Helm, M.; Servant, J.J.; Saurenbach, F.; Berger, R. 2005. Read-out of micromechanical cantilever sensors by phase shifting interferometry. *Applied Physics Letters* 87: 064101/1–064101/3.

Hierlemann, A.; Brand, O.; Hagleinter, C.; Baltes, H. 2003. Microfabrication techniques for chemical/biosensors. *Proceedings of the IEEE* 91: 839–863.

Howse, J.R.; Topham, P.; Crook, C.J.; Gleeson, A.J.; Bras, W.; Jones, R.A.L.; Ryan, A.J. 2006. Reciprocating power generation in a chemically driven synthetic muscle. *Nano Letters* 6: 73–77.

Huang, T.J. 2007. Towards artificial molecular motor-based electroactive/photoactive biomimetic muscles. *Proceedings of the Society of Photo-Optical Instrumentation Engineers (SPIE)* 6524: H5240.

Huang, T.J.; Brough, B.; Ho, C.-M. et al. 2004. A nanomechanical device based on linear molecular motors. *Applied Physics Letters* 85: 5391–5393.

Hwang, K.S.; Lee, S.-M.; Eom, K. et al. 2007. Nanomechanical microcantilever operated in vibration modes with use of RNA aptamer as receptor molecules for label-free detection of HCV helicase. *Biosensors and Bioelectronics* 23: 459–465.

Ilic, B.; Czaplewski, D.; Zalalutdinov, M. et al. 2000. Mechanical resonant immunospecific biological detector. *Applied Physics Letters* 77: 450–452.

Ilic, B.; Czaplewski, D.; Zalalutdinov, M. et al. 2001. Single cell detection with micromechanical oscillators. *Journal of Vacuum Science and Technology* 19: 2825–2828.

Ilic, B.; Yang, Y.; Craighead, H.G. 2004. Virus detection using nanoelectromechanical devices. *Applied Physics Letters* 85: 2604–2606.

Ilic, B.; Yang, Y.; Aubin, K.; Reichenbach, R.; Krylov, S.; Craighead, H.G. 2005. Enumeration of DNA molecules bound to a nanomechanical oscillator. *Nano Letters* 5: 925–929.

Jeon, S.; Thundat, T. 2004. Instant curvature measurement for microcantilever sensors. *Applied Physics Letters* 85: 1083–1084.

Ji, H.-F.; Gao, H.; Buchapudi, K.R.; Yang, X.; Xua, X.; Schulteb, M.K. 2008. Microcantilever biosensors based on conformational change of proteins. *The Analyst* 133: 434–443.

Jonhson, L.; Gupta, A.K.; Ghafoor, A.; Akin, D.; Bashir, R. 2006. Characterization of vaccinia virus particles using microscale silicon cantilever resonators and atomic force microscopy. *Sensors and Actuators B* 115: 189–197.

Kim, D.J.; Weeks, B.L.; Hope-Weeks, L.J. 2007. Effect of surface conjugation chemistry on the sensitivity of microcantilever sensors. *Scanning* 29: 245–248.

Kovacs, G.T.A.; Maluf, N.I.; Petersen, K.R. 1998. Bulk micromachining of silicon. *Proceedings of the IEEE* 86: 1536–1551.

Laibinis, P.E.; Whitesides, G.M.; Allara, D.L.; Tao, Y.T.; Parikh, A.N.; Nuzzo, R.G. 1991. Comparison of the structures and wetting properties of self-assembled monolayers of normal-alkanethiols on the coinage metal-surfaces, Cu, Ag, Au. *Journal of the American Chemical Society* 113: 7152–7167.

Lam, Y.; Abu-Lail, N.I.; Alam, M.S.; Zauscher, S. 2006. Using microcantilever deflection to detect HIV-1 envelope glycoprotein gp120. *Nanomedicine: Nanotechnology, Biology and Medicine* 2: 222–229.

Lan, S.; Veiseh, M.; Zhang, M. 2005. Surface modification of silicon and gold-patterned silicon surfaces for improved biocompatibility and cell patterning selectivity. *Biosensors and Bioelectronics* 20: 1697–1708.

DeVoe, D.L.; Pisano, A.P. 1997. Modeling and optimal design of piezoelectric cantilever microactuators. *Journal of Micromechanical Systems* 6: 266–270.

Diao, J.; Ren, D.; Engstrom, J.R.; Lee, K.H. 2005. A surface modification strategy on silicon nitride for developing biosensors. *Analytical Biochemistry* 343: 322–328.

Dutta, P.; Tipple, C.A.; Lavrik, N.V.; Datskos, P.G.; Hofstetter, H.; Hofstetter, O.; Sepaniak, M.J. 2003. Enantioselectivity sensor based on antibody-mediated nanomechanics. *Analytical Chemistry* 75: 2342–2348.

Ekinci, K.L.; Huang, X.M.H.; Roukes, M.L. 2004 Ultrasensitive nanoelectromechanical mass detection. *Applied Physics Letters* 84: 4469–4471.

Fritz, J. 2008. Cantilever biosensors. *Analyst* 133: 855–863.

Fritz, J.; Baller, M.K.; Lang, H.P.; Rothuizen, H.; Vettiger, P.; Meyer, E.; Güntherodt, H.J.; Gerber, C.; Gimzewski, J.K. 2000. Translating biomolecular recognition into nanomechanics. *Science* 288: 316–318.

Genolet, G.; Brugger, J.; Despont, M. et al. 1999. Soft, entirely photoplastic probes for scanning force microscopy. *Review of Scientific Instruments* 70: 2398–2401.

Gfeller, K.Y.; Nugaeva, N.; Hegner, M. 2005a. Micromechanical oscillators as rapid biosensor for the detection of active growth of *Escherichia coli*. *Biosensors and Bioelectronics* 21: 528–533.

Gfeller, K.Y.; Nugaeva, N.; Hegner, M. 2005b. Rapid biosensor for detection of antibiotic-selective growth of *Escherichia coli*. *Applied and Environmental Microbiology* 71: 2626–2631.

Goddard, J.M.; Hotchkiss, J.H. 2007. Polymer surface modification for the attachment of bioactive compounds. *Progress in Polymer Science* 32: 698–725.

Gupta, A.; Akin, D. 2004a. Detection of bacterial cells and antibodies using surface micromachined thin silicon cantilever resonators. *Journal of Vacuum Science and Technology B* 22: 2785–2791.

Gupta, A.; Akin, D.; Bashir, R. 2004b. Single virus particle mass detection using microresonators with nanoscale thickness. *Applied Physics Letters* 84: 1976–1978.

Gupta, A.; Akin, D.; Bashir, R. 2005. Mechanical effects of attaching protein layers on nanoscale-thick cantilever beams for resonant detection of virus particles. *Proceedings of the IEEE International Conference on Micro Electro Mechanical Systems (MEMS)*, Piscataway, NJ, vol. 22, pp. 746–749.

Gupta, A.K.; Nair, P.R.; Akin, D. et al. 2006. Anomalous resonance in a nanomechanical biosensor. *Proceedings of the National Academy of Sciences of the United States of America* 103: 13362–13367.

Hagan, M.F.; Majumdar, A.; Chakraborty, A.K. 2002. Nanomechanical forces generated by surface grafted DNA. *The Journal of Physical Chemistry B* 106: 10163–10173.

Helm, M.; Servant, J.J.; Saurenbach, F.; Berger, R. 2005. Read-out of micromechanical cantilever sensors by phase shifting interferometry. *Applied Physics Letters* 87: 064101/1–064101/3.

Hierlemann, A.; Brand, O.; Hagleiter, C.; Baltes, H. 2003. Microfabrication techniques for chemical/biosensors. *Proceedings of the IEEE* 91: 839–863.

Howse, J.R.; Topham, P.; Crook, C.J.; Gleeson, A.J.; Bras, W.; Jones, R.A.L.; Ryan, A.J. 2006. Reciprocating power generation in a chemically driven synthetic muscle. *Nano Letters* 6: 73–77.

Huang, T.J. 2007. Towards artificial molecular motor-based electroactive/photoactive biomimetic muscles. *Proceedings of the Society of Photo-Optical Instrumentation Engineers (SPIE)* 6524: H5240.

Huang, T.J.; Brough, B.; Ho, C.-M. et al. 2004. A nanomechanical device based on linear molecular motors. *Applied Physics Letters* 85: 5391–5393.

Hwang, K.S.; Lee, S.-M.; Eom, K. et al. 2007. Nanomechanical microcantilever operated in vibration modes with use of RNA aptamer as receptor molecules for label-free detection of HCV helicase. *Biosensors and Bioelectronics* 23: 459–465.

Ilic, B.; Czaplewski, D.; Zalalutdinov, M. et al. 2000. Mechanical resonant immunospecific biological detector. *Applied Physics Letters* 77: 450–452.

Ilic, B.; Czaplewski, D.; Zalalutdinov, M. et al. 2001. Single cell detection with micromechanical oscillators. *Journal of Vacuum Science and Technology* 19: 2825–2828.

Ilic, B.; Yang, Y.; Craighead, H.G. 2004. Virus detection using nanoelectromechanical devices. *Applied Physics Letters* 85: 2604–2606.

Ilic, B.; Yang, Y.; Aubin, K.; Reichenbach, R.; Krylov, S.; Craighead, H.G. 2005. Enumeration of DNA molecules bound to a nanomechanical oscillator. *Nano Letters* 5: 925–929.

Jeon, S.; Thundat, T. 2004. Instant curvature measurement for microcantilever sensors. *Applied Physics Letters* 85: 1083–1084.

Ji, H.-F.; Gao, H.; Buchapudi, K.R.; Yang, X.; Xua, X.; Schulteb, M.K. 2008. Microcantilever biosensors based on conformational change of proteins. *The Analyst* 133: 434–443.

Jonhson, L.; Gupta, A.K.; Ghafoor, A.; Akin, D.; Bashir, R. 2006. Characterization of vaccinia virus particles using microscale silicon cantilever resonators and atomic force microscopy. *Sensors and Actuators B* 115: 189–197.

Kim, D.J.; Weeks, B.L.; Hope-Weeks, L.J. 2007. Effect of surface conjugation chemistry on the sensitivity of microcantilever sensors. *Scanning* 29: 245–248.

Kovacs, G.T.A.; Maluf, N.I.; Petersen, K.R. 1998. Bulk micromachining of silicon. *Proceedings of the IEEE* 86: 1536–1551.

Laibinis, P.E.; Whitesides, G.M.; Allara, D.L.; Tao, Y.T.; Parikh, A.N.; Nuzzo, R.G. 1991. Comparison of the structures and wetting properties of self-assembled monolayers of normal-alkanethiols on the coinage metal-surfaces, Cu, Ag, Au. *Journal of the American Chemical Society* 113: 7152–7167.

Lam, Y.; Abu-Lail, N.I.; Alam, M.S.; Zauscher, S. 2006. Using microcantilever deflection to detect HIV-1 envelope glycoprotein gp120. *Nanomedicine: Nanotechnology, Biology and Medicine* 2: 222–229.

Lan, S.; Veiseh, M.; Zhang, M. 2005. Surface modification of silicon and gold-patterned silicon surfaces for improved biocompatibility and cell patterning selectivity. *Biosensors and Bioelectronics* 20: 1697–1708.

Lavrik, N.V.; Sepaniak, M.J.; Datskos, P.G. 2004. Cantilever transducers as a platform for chemical and biological sensors. *Review of Scientific Instruments* 75: 2229–2253.

Lee, J.; Shen, H.; Kim, S. et al. 2003. Fabrication of atomic force microscopy probe with low spring constant using SU-8 photoresist. *Japanese Journal of Applied Physics Part 2—Letters* 42: L1171–L1174.

Liu, F.; Zhang, Y.; Ou-Yang, Z. 2003. Flexoelectric origin of nanomechanic deflection in DNA-microcantilever system. *Biosensors and Bioelectronics* 18: 655–660.

Lorenz, H.; Despont, M.; Fahrni, N. et al. 1997. SU-8 : A low cost negative resist for MEMS. *Journal of Micromechanics and Microengineering* 7: 121–124.

Lorentz, H.; Despont, M.; Fahrni, N. et al. 1998. High-aspect-ratio, ultrathick, negative-tone near-UV photoresist and its applications for MEMS. *Sensors and Actuators* 64: 33–39.

McFarland, A.W.; Colton, J.S. 2005. Chemical sensing with micromolded plastic microcantilevers. *Journal of Micro-Electromechanical Systems* 14: 1375–1385.

McFarland, A.W.; Poggi, M.A.; Bottomley, L.A.; Colton, J.S. 2004. Production and characterization of polymer microcantilevers. *Review of Scientific Instruments* 75: 2756–2758.

McKendry, R.; Zhang, J.; Arntz, Y. et al. 2002. Multiple label-free biodetection and quantitative DNA-binding assays on a nanomechanical cantilever array. *Proceedings of the National Academy of Sciences of the United States of America* 99: 9783–9788.

Mertens, J.; Finot, E.; Thundat, T. et al. 2003. Effects of temperature and pressure on microcantilever resonance response. *Ultramicroscopy* 97: 119–126.

Mertens, J.; Álvarez, M.; Tamayo, J. 2005. Real-time profile of microcantilevers for sensing application. *Applied Physics Letters* 87: 234102.

Mertens, J.; Rogero, C.; Calleja, M. et al. 2008. Label-free detection of DNA hybridization based on hydration-induced tension in nucleic acids. *Nature Nanotechnology* 3: 301–307.

Micoulet, A.; Spatz, J.P.; Ott, A. 2005. Mechanical response analysis and power generation by single-cell stretching. *ChemPhysChem* 6: 663–670.

Mouaziz, S.; Boero, G.; Popovic, R.S.; Brugger, J. 2006. Polymer-based cantilevers with integrated electrodes. *Journal of Microelectromechanical Systems* 15: 890–895.

Moulin, A.M.; O'Shea, S.J.; Badley, R.A.; Doyle, P.; Welland, M.E. 1999. Measuring surface-induced conformational changes in proteins. *Langmuir* 15: 8776–8779.

Mukhopadhyay, R.; Lorentzen, M.; Kjems, J.; Besenbacher, F. 2005a. Nanomechanical sensing of DNA sequences using piezoresistive cantilevers. *Langmuir* 21: 8400–8408.

Mukhopadhyay, R.; Sumbayev, V.V.; Lorentzen, M.; Kjems, J.; Andreasen, P.A.; Besenbacher, F. 2005b. Cantilever sensor for nanomechanical detection of specific protein conformations. *Nano Letters* 5: 2385–2388.

Nordström, M.; Keller, S.; Lillemose, M. et al. 2008. SU-8 cantilevers for bio/chemical sensing; fabrication, characterisation and development of novel read-out methods. *Sensors* 8: 1595–1612.

Nuzzo, R.G.; Allara, D.L. 1983. Adsorption of bifunctional organic disulfides on gold surfaces. *Journal of the American Chemical Society* 105: 4481–4483.

Nuzzo, R.G.; Dubois, L.H.; Allara, D.L. 1990. Fundamental-studies of microscopic wetting on organic-surfaces. 1. Formation and structural characterization of a self-consistent series of polyfunctional organic monolayers. *Journal of the American Chemical Society* 112: 558–569.

Pavlovic, E.; Quist, A.P.; Gelius, U.; Oscarsson, S. 2002. Surface functionalization of silicon oxide at room temperature and atmospheric pressure. *Journal of Colloid and Interface Science* 254: 200–203.

Pesek, J.J.; Matyska, M.T. 1997. Methods for the modification and characterization of oxide surfaces. *Interface Science* 5: 103–117.

Ramos, D.; Tamayo, J.; Mertens, J.; Calleja, M.; Zaballos, A. 2006. Origin of the response of nanomechanical resonators to bacteria adsorption. *Journal of Applied Physics* 100: 106105/1–16105/3.

Ramos, D.; Calleja, M.; Mertens, J.; Zaballos, Á.; Tamayo, J. 2007. Measurement of the mass and rigidity of adsorbates on a microcantilever sensor. *Sensors* 7: 1834–1845.

Ramos, D.; Tamayo, J.; Mertens, J.; Calleja, M.; Villanueva, L.G.; Zaballos, A. 2008. Detection of bacteria based on the thermomechanical noise of a nanomechanical resonator: Origin of the response and detection limits. *Nanotechnology* 19: 035503/1–035503/9.

Rye, R.R.; Nelson, G.C.; Dugger, M.T. 1997. Mechanistic aspects of alkylchlorosilane coupling reactions. *Langmuir* 13: 2965–2972.

Sader, J.E. 1998. Frequency response of cantilever beams immersed in viscous fluids with applications to the atomic force microscope. *Journal of Applied Physics* 84: 64–76.

Schreiber, F. 2004. Self-assembled monolayers: From 'simple' model systems to biofunctionalized interfaces. *Journal of Physics: Condensed Matter* 16: R881–R900.

Shu, W.; Liu, D.; Watari, M. et al. 2005. DNA molecular motor driven micromechanical cantilever arrays. *Journal of the American Chemical Society* 127: 17054–17060.

Shu, W.; Laue, E.D.; Seshia, A.A. 2007. Investigation of biotin-streptavidin binding interactions using microcantilever sensors. *Biosensors and Bioelectronics* 22: 2003–2009.

Stachowiak, J.C.; Yue, M.; Castelino, K.; Chakraborty, A.; Majumdar, A. 2006. Chemomechanics of surface stresses induced by DNA hybridization. *Langmuir* 22: 263–268.

Stoney, G.G. 1909. The tension of metallic films deposited by electrolysis. *Proceedings of the Royal Society of London A Matter* 82: 172–175.

Tamayo, J.; Ramos, D.; Mertens, J.; Calleja, M. 2006. Effect of the adsorbate stiffness on the resonance response of microcantilever sensors. *Applied Physics Letters* 89: 224104/1–224104/3.

Tang, J.; Fang, J.; Yan, X.; Ji, H.-F. 2004. Fabrication and characterization of SiO_2 microcantilever for microsensor application. *Sensors and Actuatiors* B 97: 109–113.

Tortonese, M.; Barret, R.C.; Quate, C.F. 1993. Atomic resolution with an atomic force microscope using piezoresistive detection. *Applied Physics Letters* 62: 834–836.

Ulman, A. 1996. Formation and structure of self-assembled monolayers. *Chemical Reviews* 96: 1533–1554.

Veiseh, M.; Zareie, M.H.; Zhang, M. 2002. Highly selective protein patterning on gold-silicon substrates for biosensor applications. *Langmuir* 18: 6671–6678.

Velanki, S.; Ji, H.-F. 2006. Detection of feline coronavirus using microcantilever sensors. *Measurement Science and Technology* 17: 2964–2968.

Waggoner, P.S.; Craighead, H.G. 2007. Micro- and nanomechanical sensor for environmental, chemical and biological detection. *Lab on a Chip* 7: 1238–1255.

Wang, X.; Ryu, K.S.; Bullen, D.A. et al. 2003. Scanning probe contact printing. *Langmuir* 19: 8951–8955.

Wehrmeister, J.; Fuß, A.; Saurenbach, F. et al. 2007. Readout of micromechanical cantilever sensors arrays by Fabry-Perot interferometry. *Review of Scientific Instruments* 78: 104105/1–104105/8.

Williams, K.R.; Muller, R.S. 1996. Etch rates for micromachining processing. *Journal of Microelectromechanical Systems* 5: 256–269.

Wu, G.; Datar, R.H.; Hansen, K.M.; Thundat, T.; Cote, R.J.; Majumdar, A. 2001a. Bioassay of prostate-specific antigen (PSA) using microcantilevers. *Nature Biotechnology* 19: 856–860.

Wu, G.; Ji, H.; Hansen, K. et al. 2001b. Origin of nanomechanical cantilever motion generated from biomolecular interactions. *Proceedings of the National Academy of Sciences of the United States of America* 98: 1560–1564.

Zhang, J.; Ji, H.F. 2004. An anti *E-coli* O157: H7 antibody-immobilized microcantilever for the detection of *Escherichia coli* (*E-coli*). *Analytical Sciences* 20: 585–587.

Zhang, J.; Lang, H.P.; Huber, F. et al. 2006. Rapid and label-free nanomechanical detection of biomarker transcripts in human RNA. *Nature Nanotechnology* 1: 214–220.

5

Enzymatic Nanolithography

Manfred Radmacher
Universität Bremen

5.1 Introduction to Enzymes

Enzymes are a special class of proteins that catalyze chemical reactions. In a cell, virtually all chemical reactions do not occur spontaneously because they could otherwise not be controlled by the cell. Thus, in most cases, cells employ enzymes for controlling chemical reactions. There are two reasons for chemical reactions not to occur spontaneously in a cell: either they are energetically not favorable (i.e., the reaction requires energy) or a large activation energy barrier has to be overcome because it slows down the kinetics. In both cases, enzymes can help to catalyze these reactions.

The simplest enzymes are the hydrolases that cleave a molecule into two fragments. The unmodified molecule is called the substrate, the two fragments are called products. One example of the hydrolases is the phosphatases that remove a phosphate group from a protein. The reverse enzyme, called kinase, adds a phosphate to a protein. The two chemical reactions can be denoted as

$$\text{Protein-PO}_4 + \text{Phosphatase} \Rightarrow \text{PO}_4 + \text{Protein} + \text{Phosphatase}$$

$$\text{PO}_4 + \text{Protein} + \text{Kinase} \Rightarrow \text{Protein-PO}_4 + \text{Kinase}$$

Take Note: These "chemical" equations should not be read literally, but more conceptually. For example, the phosphate group in the above equation will be dissolved in water, that is, it will be charged, or it will be transferred to another molecule, a coenzyme. However, these specific details are not important at this point, and can therefore be neglected for now.

So, at the end of the reaction, the enzyme molecule is unmodified and can undergo the same reaction cycle over again.

Thus, its function is equivalent to the function of a catalyzer. Generally, the enzyme molecule will transiently bind the substrate molecule(s), forming a complex of enzyme and substrate, and then perform the chemical modification and release the product(s). In short, this can be written as

$$E + S \Rightarrow ES \Rightarrow E + P$$

As mentioned above, an enzyme may need two (or more) different substrate molecules, for example, kinases will use a protein molecule and a phosphate group as the substrate, or an enzyme may produce two (or more) product species, for example, a phosphatase will produce a protein plus a phosphate group that has been cleaved off the protein. Nevertheless, the general idea outlined here holds in any case (Figure 5.1).

One key feature of enzymes is that they are highly specific. They will bind their substrates with very high efficiency; thus, in most cases they will catalyze only one particular chemical reaction.

In the simplest model—the Michaelis–Menten model—the kinetics of the enzymes, that is, the rate at which the product is produced, can be derived directly. The above reaction scheme can be described by four rate constants that correspond to the on- and off-rate of the formation of the enzyme–substrate complex, and the on- and off-rate of the formation of product from the enzyme substrate complex:

$$E + S \underset{k2}{\overset{k1}{\Leftrightarrow}} ES \underset{k4}{\overset{k3}{\Leftrightarrow}} E + P$$

If the concentration of P is assumed as very low, then the rate constant $k4$ can be neglected. In a biological context, this assumption is always valid, since an enzyme is only active and

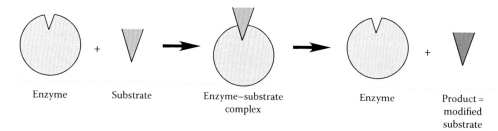

FIGURE 5.1 Generalized reaction scheme of an enzyme. The enzyme will bind a substrate molecule to form an enzyme–substrate complex. The substrate will be modified to yield a product molecule that will be released by the enzyme.

produces the product if the cell needs the product. In enzymatic nanolithography, the same argument holds.

A simple derivation, which can be found in the appendix, yields the following production rate of the product P:

$$V = \frac{\partial P}{\partial t} = v_{max} * \frac{S}{k_m + S} * E_t$$

where

S and P are the concentrations of substrate and product, respectively

E_t is the total concentration of the enzyme that corresponds to the sum of the concentrations of the free enzyme and the enzyme substrate complex

This also corresponds to the total amount of the enzyme added. v_{max} and k_m are parameters characterizing the efficiency of the enzyme, called the turnover number and Michaelis constant, respectively (Figure 5.2).

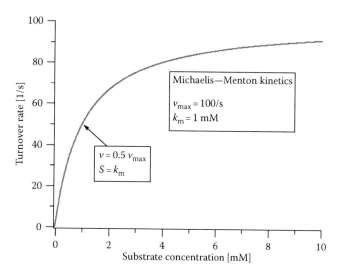

FIGURE 5.2 Production rate of an enzyme described by the Michaelis–Menton kinetics. The enzyme is characterized by two quantities: the maximum turnover number that is achieved at very large substrate concentrations. Typical values for enzymes can range between a few reactions per second up to a few reactions per microsecond. The Michaelis constant is that substrate concentration at which the half-maximum turnover rate is measured. Typical values here are in the mM-regime, in many cases smaller.

By plotting the production rate as a function of the substrate concentration, the meaning of these two parameters can be rationalized easily. The production rate increases with increasing substrate concentration, but will saturate at some value that is defined by the enzyme itself. This value is the time needed for one substrate molecule to bind to the enzyme, and be modified and finally released as product. Thus, the "maximum turnover number" corresponds to the maximum number of enzymatic reactions per enzyme molecule per second. The Michaelis constant is the substrate concentration where the production rate is half of the maximum turnover number, and thus is a measure for the specificity of the enzyme. Typical values for the turnover number will be 100 per second (per enzyme molecule) with a range of 1 per second of up to 600,000 per second for the fastest enzyme. Typical values for the Michaelis constant will be on the order of mM, but might even be in the micromolar regime.

5.2 Techniques for Local Chemical Modifications

The simplest scheme for enzymatic nanolithography is the following: the enzyme molecules are deposited at the area of interest on a surface and then they chemically modify this area of the surface. An implementation of this scheme is by applying DNAse by a pipette to a local area of a surface coated with DNA that will then be digested locally by the DNAse (Hyun et al., 2004). An alternative is to use the product of an enzyme, which is locally produced, to either chemically modify the surface in a second step, or just to adsorb it onto the surface (as has been used with enzyme alkaline phosphatase that produces an insoluble product [Riemenschneider and Radmacher, 2005]).

In these two examples, the local activity is achieved since the enzyme molecules were only present locally, either because they have been applied locally by a pipette or they have been immobilized to an atomic force microscope (AFM) tip that can be positioned at the area of interest.

An alternative is to activate the enzyme molecules locally that could then be present in the entire bulk phase or on the entire surface. Here several possible mechanisms are conceivable, for example, local photo-activation or activation by electrical potentials, which has not been demonstrated yet. Amazingly, mechanically "activating" the substrate by creating local defects in a suitable surface, a phospholipid layer, has been demonstrated to locally hydrolyze this surface (Grandbois et al., 1998).

To summarize, we organize the different schemes in a table:

Modification process	Direct chemical modification by the enzyme	Hyun et al. (2004)
	Indirect chemical modification by the product	
	Adsorption of product to the surface	Riemenschneider and Radmacher (2005)
Localization process	By micro pipette	Hyun et al. (2004)
	By AFM tip	Riemenschneider and Radmacher (2005)
	By dip pen technology	
Local activation process	Photo activation	Not demonstrated yet!
	Electrical surface potential	Not demonstrated yet!
	Mechanical force	Grandbois et al. (1998)

5.3 Demonstrations of Nanolithography

The demonstration of enzymatic nanolithography has been achieved by the dip pen technology, micropipettes, mechanical force exerted by AFM, and the immobilization of the enzyme at the AFM tips. These are discussed below.

5.3.1 Dip Pen Technology

Dip pen technology was demonstrated first in 2002 (Lee et al., 2002) and is widely used nowadays. A solution, for instance, of silanes or alkane-thiols is applied to an AFM tip, such that it is wetted by this liquid. When scanning this tip over a suitable surface in air (gold in the case of alkane-thiols, or glass in the case of silanes), the liquid will be deposited and can bind covalently to the surface. Dip pen technology works fine with simple molecules such as silanes or alkane-thiols. The activity of the enzymes while depositing the enzyme molecules directly by dip pen technology is an issue. Since dip pen nanolithography (DPN) is done in air, there is some risk of denaturing the enzyme molecules with the consequence of decreasing their activity. Consequently, most applications of DPN in combination with the enzymes create a structured surface that is then modified by the enzymes in the solution. One example of this indirect approach is the *in situ* polymerization of caffeic acid on a suitable substrate (Xu et al., 2005): a gold surface, coated with amino terminated alkane-thiols is first modified by DPN to exhibit the carboxyl groups locally. In the presence of horseradish-peroxidase, hydrogen peroxide, and caffeic acid, a polymer is formed in these modified areas. However, strictly speaking, this is not what is meant by enzymatic nanolithography, because the enzyme's activity only modifies a prestructured surface. Li et al. (2008) reported another example, where patterns of enzymes were created by dip pen technology. They structured a gold substrate by dip pen technology such that it was biotinylated locally, and then vesicles with the enzyme

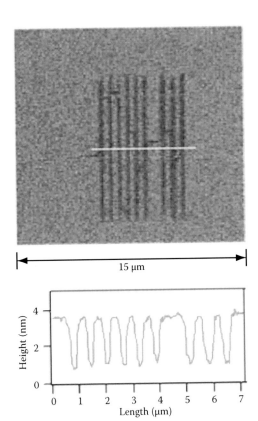

FIGURE 5.3 The enzyme DNAse has been applied by the dip pen technology on a sample where oligonucleotides have been covalently linked to a gold support. The DNAse digests the oligonucleotides creating trenches of 3 nm depth and roughly 200 nm width in the DNA film. (From Hyun, J. et al., *J. Am. Chem. Soc.*, 126, 4470, 2004. With permission.)

ATPase were linked to the structured surface. Again, this is not enzymatic nanolithography as interpreted in this chapter.

Hyun et al. have actually used dip pen technology for direct enzymatic nanolithography, that is, they deposited the enzyme directly by DPN (Hyun et al., 2004). They prepared a sample with the DNA covalently linked to a gold surface. Then they deposited the enzyme DNAse by DPN locally, which cleaves the DNA strands by dip pen technology. An AFM topographical image of trenches of approximately 3 nm depth and 200 nm width showed that the DNA was digested by the enzyme's activity (Figure 5.3).

5.3.2 Micropipettes

An alternative way to deposit enzymes locally on a surface without the possible risk of drying the enzyme solution, as it may occur in DPN, is by the use of micropipettes. Ionescu et al. have used micropipettes with an inner diameter of 100 nm to deposit a protease solution locally (Ionescu et al., 2003, 2005). The enzymes were deposited on a protein sample (BSA, bovine serum albumin physisorbed on glass slides) to locally hydrolyze the proteins on the surfaces. Thus, it was possible to etch the channels in the protein layer of the surface to a depth of 10–20 nm and a width of 150 nm. They used protease trypsin as it

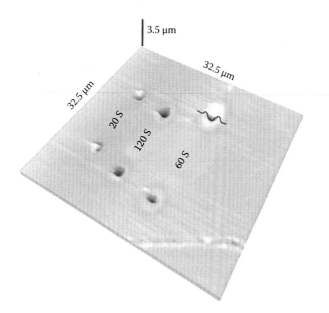

FIGURE 5.4 A micropipette was used to apply a trypsin solution locally that etches holes in a film of BSA adsorbed on glass. (From Ionescu, R.E. et al., *Nano Lett.*, 3, 1639, 2003. With permission.)

is very abundant, and cleaves many different proteins. However, in future, it is possible to use specific proteases to remove specific proteins from the surface and thus create structures that are more complicated (Figure 5.4).

5.3.3 Mechanical Force

By applying a local force, soft samples can be deformed or restructured locally. This has been applied to soft polymeric samples in which by scratching with elevated forces or by vibrating the tip, features like trenches or holes can be created (Wendel et al., 1996). It is even possible to structure covalently attached monomolecular films of alkane-thiol on gold with elevated forces (Wadu-Mesthrige et al., 2001). Grandbois et al. (1998) have applied this concept of mechanically modifying surfaces in a very interesting implementation of enzymatic nanolithography. A lipid film was transferred on a silanized glass slide using the Langmuir–Blodgett technique. The phospholipase hydrolyzes the phospholipids such that single-chained hydrocarbons result that then go into solution. Since the active site, where the enzyme cleaves the phospholipid molecule, is buried deep in the dense lipid layer, the phospholipase can only act at locations of defects in the lipid layer that are usually not present in the lipid monolayer. So, the AFM tip is used to locally destroy the ordering in lipid films, creating defects, and thus making the film susceptible for the action of the enzyme (Figure 5.5).

5.3.4 Enzyme-Coated AFM Tips

Coating AFM tips covalently with enzymes and using them for structuring surfaces by the activity of these enzymes is a very promising route. Here, the active area is given by the sharpness

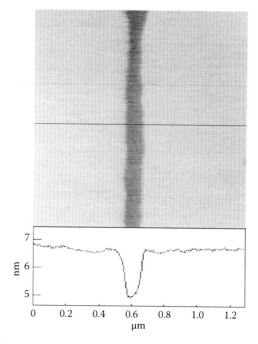

FIGURE 5.5 When scanning at elevated forces along one line of a phospholipid monolayer sample the enzyme phospholipase, which is in solution, can degrade the lipid molecules locally, which consequently go into solution. This allows creating lines with a width of around 100 nm. (From Grandbois, M. et al., *Biophys. J.*, 74, 2398, 1998. With permission.)

of AFM tips, where typical radii of curvature are between 20 and 40 nm. Since the enzyme molecules are covalently attached to the AFM tip, the number of active molecules, that is, those that can reach the surface can be minimized further by using very short spacers for attaching the enzyme to the tip. Thus, this technique promises higher spatial resolution than DPN or micropipettes, where surface tension forces will determine the final resolution.

Enzyme-coated tips have been used in several ways for modifying surfaces. The first application used the enzyme alkaline phosphatase that was covalently linked to an AFM tip (Riemenschneider and Radmacher, 2005). In this chapter, the substrate BCIP that is in solution was modified by the phosphatase to yield the insoluble product NBT that precipitates. In this setup, it is essential that only the very apex of the tip be coated with the enzyme molecules. Otherwise, the precipitating product will be produced over a large area, thus no discernible structures can be written. This is achieved by using the following trick: the enzyme is covalently attached to the protein streptavidin that exhibits four binding sites for the small organic molecule, biotin. In a first step, a glass slide is biotinylated and the streptavidin-phosphatase complex is adsorbed on the glass slide. Due to geometrical reasons, it can only bind to one or two biotin molecules that still have unoccupied binding sites for the biotin. Then, the sample is raster scanned with a biotinylated AFM tip that can bind to the free biotin moieties of streptavidin molecules on the surface. When continuing to scan, all biotin–streptavidin bonds are loaded and either those between the streptavidin and the support or those between the streptavidin and the tip break. In any case, some streptavidin-enzyme conjugates are transferred to the tip so that it is finally coated with enzyme molecules. Since the phosphatase-streptavidin complex is very small (~5–8 nm) and only a short linker is used to bind the biotin to the AFM tip, only the very apex of the AFM tip is coated. Then we transfer this AFM tip to another sample and add BCIP to the buffer. The enzyme molecules on the tip will produce NBT that precipitate on the support. If the tip is in the vicinity of the sample or even in contact with the sample, small precipitates with a typical size of 100 nm can be formed (Figure 5.6).

In a more recent work, enzyme molecules immobilized to an AFM tip were used to modify a suitable surface directly. Nakanmura et al. (2007) used V8 protease that specifically cleaves peptides immobilized to a support. In this study the peptide to be digested by the protease was labeled by two different fluorescent dyes, such that one dye stayed on the support after the cleavage by the protease while the other went into solution. Thus, the unmodified surface fluorescence energy transfer (FRET) between the two dyes could be observed, whereas after cleavage no FRET was observed. In this study, no spatial resolution of the patterns created was reported (Figure 5.7).

A very interesting work was reported recently. Luo et al. (2009) used the enzyme horseradish peroxidase to induce the polymerization of aniline locally. Hydrogen-peroxidase uses hydrogen peroxide as the substrate to produce oxygen radicals. In this case, the oxygen reacts with aniline to form polyaniline that precipitates on the support. Thus, small areas of the polymer with a typical size of 200 nm in width and a few nanometers in height are deposited. Here, it was essential to functionalize only the very apex of the AFM tip with the enzyme, as in the work of Riemenschneider et al. Consequently, the same procedure as above was implemented: a conjugate of the enzyme with streptavidin was immobilized to a biotinylated support that was then scanned with a biotinylated AFM tip to transfer the enzyme molecules to the very apex of the AFM tip. This functionalized AFM tip was then used to allow the polymerization and deposition of poly-aniline locally.

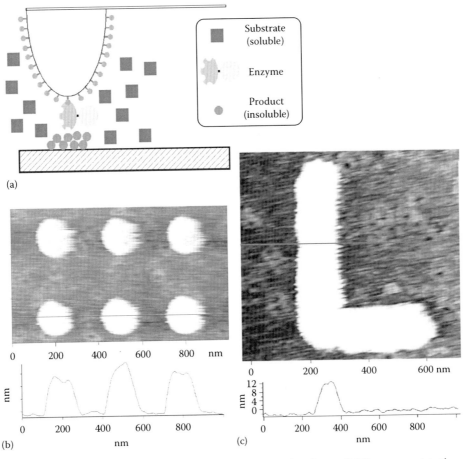

FIGURE 5.6 The enzyme alkaline phosphatase was immobilized to an AFM tip. The substrate BCIP was present in the surrounding medium. The product of the enzymatic reaction will form together with the NBT, a water insoluble complex, which will precipitate on the sample (a). Surface modification was done by two procedures: the tip was resting for 20 s at one spot, and then moved rapidly to the next spot to deposit the next dot (b). Alternatively, the tip was slowly moved with a velocity of 10 nm/s across the sample to form a continuous line (c). Imaging was done after the enzymatic reaction with the same tip in the tapping mode. (From Riemenschneider, L. and Radmacher, M., *Nano Lett.*, 5, 1643, 2005. With permission.)

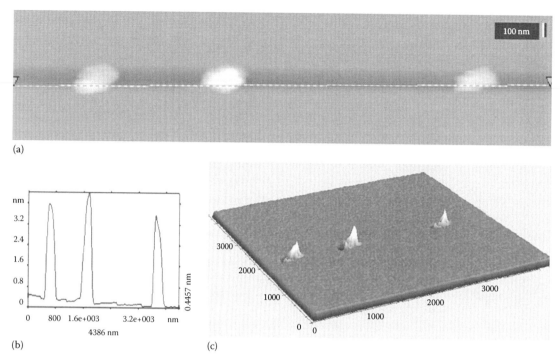

FIGURE 5.7 Luo et al. used horseradish peroxidase that induces the local polymerization of the conducting polymer polyaniline. (From Luo, X. et al., *Chem. Biol.*, 15, 1591, 2009. With permission.)

5.3.5 Immobilization of Enzymes

Several schemes can be used for the immobilization of enzymes on the tip of an AFM. The simplest scheme uses conjugates of the enzyme with streptavidin that will bind strongly to the biotinylated AFM tips, as has been used by Riemenschneider and Radmacher (2005). However, this simple scheme is only possible if a conjugate of the enzyme with the streptavidin is available. A more general approach will apply schemes developed for immobilizing proteins by force spectroscopy in AFM (Hinterdorfer and Dufrêne, 2006). Here, for instance, NTA-terminated alkane-thiols can be used to bind histidine-tagged proteins, or a heterobifunctional polyethylene glycol crosslinker can be used to react with amines on the protein surface.

5.3.6 Activity of Immobilized Enzymes

The activity of enzymes is often reduced when adsorbing them onto surfaces. The degree of the reduction of activity will depend on the surface and the way in which the enzyme is immobilized on the surface. For example, it is conceivable that when physisorbing enzymes directly, the activity will be greatly reduced. However, even in the case of physisorbtion, the activity of the lysozyme was measured directly by the AFM (Radmacher et al., 1994). The activity may be comparable to the situation in bulk when long spacers are used to covalently link the enzymes to a surface. An example of the application of this is the development of protein micro-arrays for diverse biotechnical applications (Merkel et al., 2005).

5.4 Outlook and Perspectives

The future prospects of enzymatic nanolithography depend on three aspects:

- What are the smallest features that can possibly be written?
- What are the classes of surfaces that can be modified by the enzymes?
- What is the ultimate writing speed that can be expected?

5.4.1 What Are the Smallest Features That Can Possibly Be Written?

In terms of feature size, we have to discuss the intrinsic physics of the writing process that delimits the feature size achievable. DPN and micropipettes rely on the surface tension forces, because either a droplet of the liquid (micropipettes) or a neck of a wetting liquid has to be formed between the tip and the surface (DPN). The reported feature sizes created by DPN or micropipettes are between 100 and 200 nm. There is probably some room for improvement. However, I guess that the droplets of the liquid in air with a diameter much smaller than 50 nm can hardly be formed. Therefore, there will be a limit to these techniques, which will probably be somewhat larger than molecular dimensions.

When using an AFM tip for enzymatic nanolithography, the feature size will depend on the geometry of the AFM tip. Since typical tip radii are on the order of 10–50 nm, feature sizes of this order can be expected. Actually, if the spacer used for linking the enzyme to the tip is smaller than its radius of curvature, then even smaller feature sizes can be expected. So, feature sizes on the order of 5–10 nm can be expected with regular AFM tips. When using sharper, possibly ultrasharp tips like carbon nanotubes, feature sizes could further be minimized. The important point is that the limitation is in the geometry of the available tips only and not in some physical principle as surface tension in the other methods. Therefore, I believe that the highest potential in terms of resolution in enzymatic nanolithography is in immobilizing enzymes to AFM tips. However, this argument only holds if the enzyme molecules on the tip directly modify chemical groups on the surface. However, this is yet to be demonstrated. In the enzymatic nanolithography implementations done so far, the product of the enzyme was used to form a precipitate or to stimulate the polymerization of a precipitate. In these cases, diffusion is an issue and results in larger feature sizes.

5.4.2 What Are the Classes of Surfaces That Can Be Modified by Enzymatic Nanolithography?

Before answering this question, we have to answer the following question first: What is the purpose of modifying surfaces locally by enzymes? One answer is that we want to chemically structure a surface such that it can be further functionalized so that active molecules can be bound to specific areas of this surface. Therefore, in structuring surfaces by enzymes, we are looking for surfaces that are—after the modification by enzymes—chemically functional. As an example, in a sample where the DNA is bound to it we can now structure this surface locally by an enzyme such that part of the DNA is cleaved away (e.g., with a DNAse), or locally a second strand of DNA is added to the existing ones (e.g., using a ligase). Now, the surface is chemically different in different locations, and can be further functionalized by adding a complementary strand of the DNA, for instance, which will only bind to the modified areas.

Therefore, obviously, the nucleotides bound to the surfaces and the enzymes modifying the DNA are one interesting class of surfaces that are to be considered. One important advantage of DNAs is that they can be synthesized with a well-defined sequence. Consequently, several examples of enzymatic nanolithography presented here, have employed the enzyme DNAse.

There are two other classes of biological macromolecules, which may be useful in enzymatic nanolithography: proteins and polysaccharides. In several examples discussed above, proteins were used as substrate to proteases that hydrolyze them. Applications where proteins are modified by the enzyme to add some extra functionality to the proteins are even more interesting. This could be achieved by using transferases, for instance, which transfer small chemical groups to proteins.

The group of polysaccharides may be the most promising of all, because polysaccharides are synthesized by the cell with the help of enzymes by chemically adding new sugar groups to pre-existing molecules. Thus, there are a large number of enzymes in nature, which can chemically modify polysaccharides. One advantage of polysaccharides is their stability, for instance, cellulose and chitin are among the most stable biological molecules available. This makes polysaccharides a very interesting option for possible applications, even when the knowledge on their biochemistry is not as advanced as in the case of nucleotides or proteins.

It is hard to predict at this point which enzymes and substrate molecules will be useful in future applications of enzymatic nanolithography. One of the reasons is the sheer number of different enzymes. The enzyme database BRENDA (http://www.brenda-enzymes.org/) lists, as of February 2009, 4905 different enzymes, out of which 282 use the RNA, 106 use the DNA, and 145 use the saccharides as a substrate, and most of them are transferases that will transfer small chemical groups to proteins or other biological macromolecules.

5.4.3 What Is the Ultimate Writing Speed That Can Be Achieved?

If we want to write small features, let us say a linear feature of 1 nm size, in a sequential fashion by scanning an AFM tip along a line with enzyme molecules linked to the tip, the number of enzyme molecules involved will naturally be very small. Let us assume, just for the sake of argument, that we want to digest a protein surface (e.g., coated with BSA) by a typical protease. If we assume a diameter of 3 nm for a BSA molecule and a maximum turn over number of 1000 per second (trypsin can have turnover numbers of up to 600,000, see http://www.brenda-enzymes.org/), we will get a writing speed of 3000 nm per second, which is very impressive. Even if we take into account that the turnover number of immobilized enzymes is reduced, a writing speed of 1 μm/s is conceivable.

5.5 Summary

We have discussed several implementations of enzymatic nanolithography that allow modifying suitable surfaces locally with the help of enzyme molecules. This technique holds large promises, because nanostructuring techniques will eventually arrive at a molecular scale, where any surface modification will be, essentially, chemical in its nature. Thus, employing enzymes, which are the catalysts of nature, makes a lot of sense. A key issue is localizing the enzyme's function, which means in most cases applying the enzyme locally to a surface area of interest. Either this has been done by depositing the enzyme locally, for example, by micropipettes or by the dip pen technology or by immobilizing enzyme molecules to the tip of an AFM. The current state of the art is "writing" the feature on the scale of 100 nm, where, especially, the AFM technology holds the promise to push the performance down to the nanometer, that is, the molecular scale.

Appendix 5.A: Derivation of Michaelis–Menton Kinetics

We look at the following simplified reaction scheme:

$$E + S \underset{k2}{\overset{k1}{\Longleftrightarrow}} ES \underset{k4}{\overset{k3}{\Longleftrightarrow}} E + P \qquad (5.A.1)$$

An enzyme molecule E and a substrate molecule S form a complex ES that dissociates by releasing the product P.

$k1$ and $k3$ are the forward rates of the respective chemical reaction, whereas $k2$ and $k3$ are the backward rates.

We can formulate the following equation for the rates of the chemical reactions:

$$\frac{\partial P}{\partial t} = k3 * ES - k4 * E * P \qquad (5.A.2a)$$

$$\frac{\partial ES}{\partial t} = k1 * E * S - k2 * ES - k3 * ES + k4 * E * P \qquad (5.A.2b)$$

$$\frac{\partial E}{\partial t} = k2 * ES - k1 * E * S \qquad (5.A.2c)$$

$$\frac{\partial S}{\partial t} = k2 * ES - k1 * E * S \qquad (5.A.2d)$$

Here, the capital symbols E, S, P, and ES stand for the concentration of the respective species.

We assume that P is small, since only when the product is needed, the enzyme will be activated. With $P \approx 0$, Equations 5.A.2a and b can be simplified:

$$\frac{\partial P}{\partial t} = k3 * ES \qquad (5.A.3a)$$

$$\frac{\partial ES}{\partial t} = k1 * E * S - (k2 + k3) * ES \qquad (5.A.3b)$$

Since we are interested in the rate after the onset of the chemical reaction, we look at the steady state case:

$$\frac{\partial ES}{\partial t} = 0 \qquad (5.A.4)$$

This simplifies Equation 5.A.3b:

$$0 = k1 * E * S - (k2 + k3) * ES \qquad (5.A.5)$$

$$(k2 + k2) * ES = k1 * E * S \qquad (5.A.6)$$

$$ES = \frac{k1}{k2 + k3} * E * S \qquad (5.A.7)$$

We define the Michaelis constant k_m as

$$k_m = \frac{k2 + k3}{k1} \qquad (5.A.8)$$

which simplifies Equation 5.A.7

$$ES = \frac{1}{k_m} * E * S \qquad (5.A.9)$$

In an experiment, we do not control or measure the concentration E or ES, but we know the total amount of the enzyme added, which may be present as the free enzyme or bound as the ES complex. We define the total enzyme concentration E_t as

$$E_t = E + ES \qquad (5.A.10)$$

$$E = E_t - ES \qquad (5.A.11)$$

We combine Equations 5.A.11 and 5.A.9 to yield

$$ES = \frac{1}{k_m} * (E_t - ES) * S \qquad (5.A.12)$$

We can solve Equation 5.A.12 for ES:

$$ES = \frac{1}{k_m} * E_t * S - \frac{1}{k_m} * ES * S \qquad ES + \frac{1}{k_m} * ES * S = \frac{1}{k_m} * E_t * S$$

$$ES * \left(1 + \frac{S}{k_m}\right) = \frac{S}{k_m} * E_t \qquad ES = \frac{S}{k_m * \left(1 + \dfrac{S}{k_m}\right)} * E_t$$

$$ES = \frac{S}{k_m * \left(\dfrac{k_m}{k_m} + \dfrac{S}{k_m}\right)} * E_t \qquad ES = \frac{S}{k_m * \dfrac{k_m + S}{k_m}} * E_t$$

$$ES = \frac{S}{k_m + S} * E_t$$

$$(5.A.13)$$

The result of Equation 5.A.13 together with Equation 5.A.3a gives us the expression for the production rate v of a simple enzyme following Michaelis–Menten kinetics:

$$V = \frac{\partial P}{\partial t} = k3 * ES = k3 * \frac{S}{k_m + S} * E_t = v_{max} * \frac{S}{k_m + S} * E_t \qquad (5.A.14)$$

Two parameters describe the kinetics of an enzyme: v_{max}, the maximum production or turnover rate, and k_m, the Michaelis constant that corresponds to the substrate concentration where the enzyme operates at half-maximum velocity.

References

Grandbois, M., Clausen-Schaumann, H., and Gaub, H. E. (1998) Atomic force microscope imaging of phospholipid bilayer degradation by phsospholipase A2. *Biophys. J.*, 74, 2398–2404.

Hinterdorfer, P. and Dufrêne, Y. F. (2006) Detection and localization of single molecular recognition events using atomic force microscopy. *Nat. Methods*, 3, 347–355.

Hyun, J., Kim, J., Craig, S. L., and Chilkoti, A. (2004) Enzymatic nanolithography of a self-assembled oligonucleotide monolayer on gold. *J. Am. Chem. Soc.*, 126, 4470–4471.

Ionescu, R. E., Marks, R. S., and Gheber, L. A. (2003) Nanolithography using protease etching of protein surfaces. *Nano Lett.*, 3, 1639–1642.

Ionescu, R. E., Marks, R. S., and Gheber, L. A. (2005) Manufacturing of nanochannels with controlled dimensions using protease nanolithography. *Nano Lett.*, 5, 821–827.

Lee, K.-B., Park, S.-J., Mirkin, C. A., Smith, J. C., and Mrksich, M. (2002) Protein nanoarrays generated by dip.pen nanolithography. *Science*, 295, 1702–1705.

Li, Z., Liu, X., and Zhang, Z. (2008) Preparation of F0F1-ATPase nanoarray by dip-pen nanolithography and its application as biosensors. *IEEE Trans. Nanobiosci.*, 7, 194–199.

Luo, X., Pedrosa, V. A., and Wang, J. (2009) Enzymatic nanolithography of polyaniline nanopatterns using peroxidase-modified AFM tips. *Chem. Biol.*, 15, 5191–5194.

Merkel, J. S., Michaud, G. A., Salcius, M., Schweitzer, B., and Predki, P. F. (2005) Functional protein microarrays: just how functional are they? *Curr. Opin. Biotechnol.*, 16, 447–452.

Nakamura, C., Miyamtot, C., Obatay, I., Takeda, S., Nakamura, C., Miyamoto, C., Nakamura, N., Kageshima, M., Tokumoto, H., Miyake, J., Yabuta, M., and Miyake, J. (2007) Enzymatic nanolithography of FRET peptide layer using V8 protease-immobilized AFM probe. *Biosens. Bioelectron.*, 22, 2308–2314.

Radmacher, M., Fritz, M., Hansma, H. G., and Hansma, P. K. (1994) Direct observation of enzyme activity with the atomic force microscope. *Science*, 265, 1577–1579.

Riemenschneider, L. and Radmacher, M. (2005) Enzyme assisted nanolithography. *Nano Lett.*, 5, 1643–1646.

Wadu-Mesthrige, K., Amro, N. A., Garno, J. C., Xu, S., and Liu, G.-Y. (2001) Fabrication of nanometer-sized protein patterns using atomic force microscopy and selective immobilization. *Biophys. J.*, 80, 1891–1899.

Wendel, M., Irmer, B., Cortes, J., Lorenz, H., Kotthaus, J. P., and Lorke, A. (1996) Nanolithography with an atomic force microscope. *Superlattices Microstruct.*, 21, 1–8.

Xu, P., Uyama, H., Whitten, J. E., Lobayashi, S., and Kaplan, D. L. (2005) Peroxidase-catalyzed in situ polymerization of surface orientated caffeic acid. *J. Am. Chem. Soc.*, 127, 11745–11753.

6

Biomimetic Synthesis of Nanostructures Inspired by Biomineralization

Eike Brunner
Technische Universität Dresden

Hermann Ehrlich
Technische Universität Dresden

Martin Kammer
Technische Universität Dresden

6.1 Introduction

Nanostructures can be found almost everywhere in nature. A recent review article is therefore, entitled "Live as a Nanoscale Phenomenon" (Mann 2008). On one hand, the understanding of the underlying sophisticated biochemical and biophysical mechanisms is of central importance for fundamental biological research. On the other hand, this knowledge turns out to be extremely useful in materials science: Biomimetic* or bioinspired synthesis approaches appear to be very advantageous. In contrast to common synthesis methods, natural nanostructures are formed under favorable, gentle conditions, that is, at ambient temperature, near neutral pH, etc. Biomimetic approaches promise to be environment-friendly and energy-conserving. Furthermore, natural nanostructures are usually unmatched by synthetic materials with respect to their delicacy, precision, and regularity. In summary, the idea of biomimetic materials synthesis is the exploitation of principles observed in nature in order to solve synthesis problems, for example, in nanoscience. Biomineralization turns out to be a particularly important source of inspiration for nanomaterials synthesis. Therefore, this chapter starts with a brief description of our present knowledge concerning the principles of biomineralization. In the light of these principles, a number of important developments in biomimetic nanomaterials synthesis are illustrated with selected examples, such as the synthesis of silica-based as well as calcium-based nanostructures and hybrid materials. It should be noted that there is an increasingly huge number of publications in this rapidly growing field. Due to the limited space and tutorial style of this handbook, a selection had to be made in order to highlight basic principles and major trends. For more comprehensive reviews, the interested reader is directed to recent monographs (Bar-Cohen 2005, Nalwa 2005, Ruiz-Hitzky et al. 2008) and review articles (see, e.g., Sarikaya et al. 2003, Zhang 2003, Vincent et al. 2006, Mann 2008, Weiner 2008).

6.2 Biominerals as a Source of Inspiration for Nanoscience

Biomineralization, that is, the formation and patterning of inorganic materials by living organisms is a common phenomenon in nature and an important source of inspiration for the biomimetic synthesis of nanostructured inorganic materials (Lowenstam and Weiner 1989, Mann 2001, Baeuerlein et al. 2007). Biomineralization encompasses the formation of silica-based compounds, of calcium-based materials such as calcium carbonate or hydroxyapatite, of iron oxides, and others.

Biominerals can be classified following the same strategy as non-biogenic minerals according to their composition based on the anionic constituents. Biominerals are present in many of the 78 mineral classes known today (Skinner 2005). There is an overwhelming variety of biomineralizing organisms: The skeleton of vertebrates is a result of biomineralization processes.

---

* The term "biomimetic" was coined by Otto Schmitt (Schmitt 1969) and describes the transfer of concepts or principles from biology to technology.

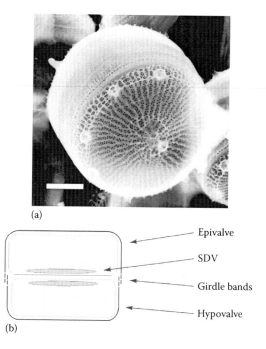

(a)

(b)

            — Epivalve

            — SDV

            — Girdle bands

            — Hypovalve

FIGURE 6.1 Diatom biosilica: micro- and nanopatterning of glass. (a) Scanning electron micrograph of the extracted siliceous cell wall of *Thalassiosira pseudonana*. Scale bar: 1μm. (b) Scheme showing a diatom cell during cell division at the stage where new valves are formed in the SDV. (Reproduced from Gröger, C. et al., *J. Struct. Biol.*, 161, 55, 2008a. With permission.)

Furthermore, approximately 200,000 diatom species (Mann and Droop 1996), 128,000 species of mollusks, 800 coral species, 5,000 sponge species including 500 glass sponges, 700 species of calcareous green, red, and brown algae, and more than 300 species of deep-sea benthic foraminifera exist (Ehrlich et al. 2008a).

Often, the shape and patterning of biominerals is very regular and, as in the case of diatoms, of amazing beauty (see Figure 6.1). Many biominerals exhibit extraordinarily interesting properties from the materials science point of view as well as with respect to biomedical applications. As there is such an overwhelming variety of different biomineralization processes, the discussion must be limited to a few important examples in the following sections.

6.2.1 Biosilica: Diatoms and Glass Sponges as Model Organisms

Diatoms are particularly interesting model systems for the study of biomineralization phenomena (Round et al. 1990). They are eucaryotic algae surrounded by a silica-based cell wall that is constructed like a petri dish (see Figure 6.1). Diatom cell walls are famous for being intricately and ornately structured on the micro- and nanometer scale. Apart from silica, the cell walls also contain organic material, which is embedded in or tightly bound to the silica (Sumper and Brunner 2006, 2008. Sumper and Lehmann 2006) thus making up an extremely interesting composite material of extraordinarily high mechanical strength (Hamm et al. 2003) and fascinating optical properties. Diatoms are assumed to act as photonic crystal slabs (Fuhrmann et al.

2004). Therefore, potential nanotechnological applications of diatom cell walls have recently found increasing interest (Gordon et al. 2008, Hildebrand 2008, Kröger and Poulsen 2008).

Mann (1993) suggested the "molecular tectonics" principle as a general route for biomineralization events: The initial step is *supramolecular pre-organization*, that is, the self-assembly of superstructures such as compartments (e.g., lipid vesicles and protein cages), and extended networks (e.g., protein-polysaccharide networks). As a second step, *interfacial molecular recognition* takes place that can give rise to the nucleation of inorganic particles. The pre-organized supramolecular structures then serve as templates. As the final stage, *cellular processing* takes place involving a variety of genetically controlled cellular processes that determine the unusual shapes and patterns observed in many biominerals. The biosynthesis of diatom cell walls, in principle, follows this route: It takes place during cell division in a highly specialized compartment, the so-called silica deposition vesicle (SDV, see Figure 6.1, Drum and Pankratz 1964). Special biomolecules are involved in cell wall biosynthesis and become at least partly embedded into the siliceous cell wall. In the past years, this entrapped organic material was carefully analyzed. So far, three different classes of such biomolecules have been identified: (1) the silaffins, highly post-translationally modified peptides/proteins of zwitterionic character (Kröger et al. 1999, 2002, Sumper et al. 2007); (2) long-chain polyamines (positively charged in solution, Kröger et al. 2000, Sumper et al. 2005, Sumper and Lehmann 2006); and (3) the highly acidic silacidins (negatively charged in solution, Wenzl et al. 2008). The silaffins were found to be capable of self-assembling into supramolecular aggregates due to their zwitterionic character. The same is true for the long-chain polyamines (Sumper et al. 2003), provided a properly chosen counterion such as orthophosphate, pyrophosphate, or silacidin is present. Both, silaffins as well as long-chain polyamines *in vitro* induce rapid silica precipitation from silicic acid containing solutions. It is therefore assumed that the aforementioned biomolecules exert two functions in cell wall biogenesis: They induce/accelerate silica precipitation and are involved in pattern formation as self-assembling, structure-directing templates (Sumper 2002).

Other frequently studied model organisms are sponges (Porifera), primitive animals: Glass sponges (Hexactinellida, Figure 6.2) originate from the Cambrian period about 600 million years ago. The spicules of certain species can act as biological fiber optics (Aizenberg et al. 2004). The siliceous spicules of silica sponges exhibit an inner axial channel containing a protein filament (Garrone 1978) that is surrounded by silica layers. About 95% of the organic material in this filament consists of special proteins called silicateins as has been shown for demosponges (Shimizu et al. 1998). Silicateins enzymatically act in the biosilification process of the spicules (Shimizu et al. 1998) and are capable of self-assembling (Murr and Morse 2005). They were shown to be phosphorylated (Müller et al. 2005) as observed previously for silaffins from diatoms (Sumper 2002). Phosphorylation is necessary for the ability of silicatein molecules to self-assemble into filaments (Müller et al. 2007).

(a)   (b)

FIGURE 6.2 Glass sponge spicules: silica with unique mechanical and optical properties. *Hyalonema affine* (a) is a typical representative of glass sponges (Hexactinellida: Porifera). This sponge possesses highly flexible silica-collagen based anchoring spicules (b).

In addition to silicateins, Matsunaga et al. (2007) have discovered long-chain polyamines in the spicules of the marine sponge *Axinyssa aculeata* that are capable of self-assembling and inducing silica precipitation *in vitro*. Their structure is similar to the polyamines found in diatoms. However, the sponge-polyamines contain sulfate counterions. Furthermore, the high molecular weight protein collagen was found within the spicules of some representatives of Hexactinellida sponges (Ehrlich and Worch 2007), whereas other hexactinellids exhibit the aminopolysaccharide chitin (Ehrlich et al. 2008b) within the framework skeleton as well as separate spicules.

6.2.2 Calcium-Based Biominerals

About 3.6% of the earth's crust is constituted by calcium (Cameron 1990). The moderately soluble calcium is dissolved in natural waters and occurs in the extracellular fluids of animals and humans. Calcium carbonate ($CaCO_3 \times H_2O$) as well as calcium phosphate ($Ca_3(PO_4)_2$) are formed by living organisms. In general, calcium carbonate may exhibit six different polymorphs: amorphous calcium carbonate (ACC), monohydrocalcite, calcium carbonate hexahydrate (ikaite), vaterite, aragonite, and calcite (Addadi et al. 2003). The calcium phosphate phases found in hard tissues include calcium phosphate dihydrate ($CaHPO_4 \times 2H_2O$, DCPD), octacalcium phosphate ($Ca_8H_2(PO_4)_6 \times 5H_2O$, OCP), tricalcium phosphate ($\text{\ss}-Ca_3(PO_4)_2$, ß-TCP), and hydroxyapatite, ($Ca_5(PO_4)_3$ OH, HAP).

These minerals are the structural basis for the strength of the skeleton, of vertebrates, of eggshells, as well as for the hardness of teeth and claws (Cameron 1990). So far, the biochemical and biophysical mechanisms governing the synthesis of the highly ordered calcium carbonate-based nanocomposites found, for example, in coccoliths (see Figure 6.3) remain a challenge for

10 μm

FIGURE 6.3 Tailoring calcite: Coccolith biocrystals. SEM image of *Coccolithus pelagicus*. (Courtesy of Mineralogical Society of America, Chantilly, VA.)

fundamental research. Nanoparticles or nanocrystals and their precursors may play a major role in biomineralization both *in vivo* and *in vitro*. Instead of conventional nucleation and crystal growth, the assembly of nanocrystallites into superstructures seems to be answer to the question of how nature produces the aforementioned materials (Navrotsky 2004). Cölfen and Antonietti (2005) explained the structure of calcium carbonate biominerals and biomimetically synthesized calcium carbonate materials by a nonclassical crystallization process resulting in so-called mesocrystals. Mesocrystals can be considered as assemblies of crystallographically oriented nanocrystals. They are highly crystalline and behave similarly to "classical" single crystals in diffraction experiments. In contrast to classical single crystals, however, mesocrystals are porous due to the presence of intercalated organic material. The first step in the formation of a mesocrystal is classical nucleation and nanocrystal growth. These nanocrystals are temporarily stabilized by organic additives. The stabilized nanocrystals then self-assemble into a mesocrystal and fuse. However, at least part of the stabilizing organic additives becomes intercalated into the material in contrast to a classical single crystal. The mesocrystal concept turned out to be very successful: It could meanwhile be extended to a variety of materials and structures including "bridged" nanocrystals, nanocrystal superlattices, "sponge" crystals, and porous single crystals. For a recent review see Zhou and O'Brien (2008).

The biogenic calcium phosphate-based minerals found in teeth and bones have superior mechanical properties due to their complex architecture (Gajjeraman et al. 2007). Bone is mainly composed of non-stoichiometric hydrohyapatite (HAP) and collagen. The hierarchical structure of bone can be described as follows (Weiner and Wagner 1998): Nanoscopic platelets of HAP are oriented and aligned within self-assembled collagen fibrils. Lamellae are formed by the parallel arrangement of collagen fibrils. These lamellae are concentrically arranged around blood vessels forming osteons. Finally, the osteons are either packed densely into compact bone or comprise a trabecular network of microporous bone, referred to as spongy or cancellous bone. The shape of the crystallites found in bone is either plate or spherical or cylindrical. Characteristic dimensions for the platelets are 2–4 nm thickness, 50 nm length, and 25 nm width. Note that the length and width are highly variable (Wang et al. 2006). The crystallites form layers that are separated by four layers of triple-helical collagen molecules. This nanoscale composite material is the structural basis for the mechanical properties of bone.

6.3 Biomimetic Nanomaterials Synthesis

6.3.1 Self-Assembly as a Central Principle

Supramolecular pre-organization is the first step and a central principle in biomineralization events (see Section 6.2). Biomimetic materials synthesis approaches, therefore, often mimic the biomolecular self-assembly processes observed in nature. Although the term "self-assembly" is nowadays used and defined in very different ways, most authors agree on the following definition (Lehn 1990, 1995, 2002, Whitesides and Boncheva 2002, Whitesides and Grzybowski 2002, Ariga et al. 2008).

Self-assembly is a process involving pre-existing components, that is, separate or distinct parts of a disordered structure. These components are self-assembled resulting in the formation of an ordered structure or pattern without any external guidance in a reversible manner. The product of the self-assembly process should be stable at thermodynamic equilibrium. Two types of self-assembly are distinguished by their physical properties: static and dynamic self-assembly. Static self-assembly is present if the system is in global or at least local equilibrium and does not dissipate energy. Dynamic self-assembly occurs in systems where the structure-directing processes dissipate energy.

Self-assembly processes occur at various scales starting from molecules ending with galaxies. Molecular self-assembly is of special interest in chemistry (Lehn 2002). Its principles are used as the basis for nano- and mesoscale structure generation in materials science (Whitesides and Grzybowski 2002a). In molecular self-assembly, intramolecular and intermolecular self-assembly processes must be distinguished: A prominent biological example for intramolecular self-assembly is protein folding, that is, the spontaneous formation of globular structures by polypeptide chains under certain chemical conditions. The spontaneous formation of multi-protein complexes like ribosomes as well as of cell membranes from various components are important biological examples for intermolecular self-assembly.

Molecular self-assembly processes are driven by the fine-tuned interplay of various interactions. The result is determined by the information encoded in different molecular properties: molecular shape, surface properties, charge distribution, polarizability, hydrophobicity, etc. (Lehn 2002, Whitesides and Boncheva 2002). These interactions are usually non-covalent, apart from a few exceptions, for example, disulfide bridges in proteins. The most important interactions for biomineralization are hydrogen bonds, electrostatic, and hydrophobic interactions. Hydrophobicity, that is, the exclusion of nonpolar molecules or parts of molecules from water, is especially well-known to be a major effect for self-assembly in biological systems. Self-assembling biomolecules are capable of forming a variety of different superstructures that can be used as templates for the synthesis of inorganic materials or in order to arrange nanoparticles into two- or three-dimensional arrays. Therefore, biomolecular self-assembly processes turned out to be extremely inspiring for biomimetic materials synthesis as can be seen in the following sections. It should, however, be emphasized that molecular self-assembly processes are usually driven by the interplay of various interactions. Minor changes of the aforementioned molecular properties may result in completely changed morphologies. In other words, the structures are determined by a variety of critical parameters that complicate reliable predictions. The rapidly growing amount of—more or less—empirical observations which are summarized in the following chapters may provide the basis for theoretical rationalization; a challenging future topic.

6.3.2 Biomimetic Synthesis of Silica-Based Materials

6.3.2.1 Nanospheres and Related Materials

Biomolecules such as silaffins and long-chain polyamines self-assemble *in vitro* thus forming spherical aggregates that can be used to form spherical silica nanoparticles by the addition of silicic acid or other precursor compounds (see Section 6.2). Numerous synthetic polypeptides/poly(amino acids) as well as polyamines have meanwhile been employed in order to synthesize silica nanospheres (for reviews see, e.g., Sumper and Brunner 2006, Behrens et al. 2007, Brunner and Lutz 2007, Gröger et al. 2008b). Polyallylamine (PAA) is often used for such experiments since it is commercially available and inexpensive. The ability of PAA to accelerate silica polycondensation was shown by Mizutani et al. (1998). Rapid* silica precipitation is only induced if microscopic phase separation occurs, that is, if a microemulsion is formed; a process which can be induced by the addition of properly chosen counterions such as phosphate or pyrophosphate to the PAA solution in complete analogy to the observations made for long-chain polyamines extracted from diatoms (Brunner et al. 2004, Lutz et al. 2005). Experiments carried out on PAA/phosphate assemblies indicate that the aggregation probably results in the formation of a hydrogen-bonded network stabilized by balanced electrostatic

---

* Rapid means within minutes or at least tens of minutes.

interactions (Brunner et al. 2004, Lutz et al. 2005). The size of the PAA/phosphate assemblies in such a phase-separated PAA solution is strictly controlled by the amount of phosphate added to the solution as well as the pH (see Figure 6.4). The droplet size, in turn, dictates the size of the silica nanoparticles, precipitating if monosilicic acid is added. It is therefore assumed, that the monosilicic acid dissolves in the PAA/phosphate aggregates thereby forming a "liquid precipitate" which hardens by silica formation. Interestingly, silica precipitates of completely

different, elongated morphology with spherical holes result if a so-called polyamine-stabilized silica sol is used as a silica precursor instead of a monosilicic acid (Sumper 2004). McKenna et al. (2004) demonstrated for various polypeptides that the silicification of the spherical aggregates formed in solution results in a redistribution of the template molecules within the assemblies.

Patwardhan and Clarson (2002a, 2003) and Patwardhan et al. (2002b,c) studied the silica precipitation behavior of PAA and various other compounds using tetramethoxysilane (TMOS) as the precursor compound. The size of the precipitated spherical silica particles depends on the PAA concentration as well as the molar mass (Patwardhan and Clarson 2002a). Apart from various other parameters, such as the reaction and precursor pre-hydrolysis time or the concentration of TMOS, the authors also analyzed the influence of an externally applied shear during the experiment (Patwardhan et al. 2002a). Elongated or fiber-like structures were then obtained. Other long-chain polyamines apart from PAA, for example, polyethyleneimine (PEI, Patwardhan and Clarson 2002b) as well as amine-terminated dendrimers (Knecht and Wright 2004, Knecht et al. 2005) were used in corresponding biomimetic synthesis experiments. Methylation of the polyamines results in an increase of the silica particle size (Behrens et al. 2007). This is probably due to the increasingly hydrophobic character of methylated polyamines, which in turn, facilitates their aggregation.

Another bioinspired synthesis approach is based on the incorporation of PEI and monosilicic acid into spherical reverse micelles (Bauer et al. 2007). Silica morphologies such, as spherical hollow silica spheres and others could be obtained by this method.

The discovery of silicatein in sponges (see Section 6.2) has inspired a multitude of nanomaterials synthesis approaches that were recently reviewed and summarized (Brutchey and Morse 2008). For example, Cha et al. (2000) used a synthetic block copolypeptide for the biomimetic synthesis of silica spheres as well as well-defined columns of amorphous silica.

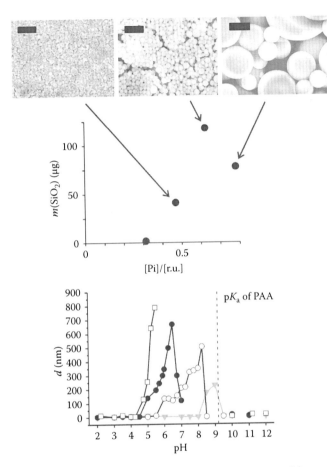

FIGURE 6.4 Silica nanospheres of controlled size. Top: Silica precipitated from aqueous PAA/silicic acid solutions as a function of phosphate concentration. Top: SEM images of the precipitates. The black bar shown in the images defines a length of $2\,\mu m$. For the samples shown here, the following average particle diameters could be determined: left: 170 nm (70 nm); middle: 290 nm (150 nm); right: 2400 nm (1100 nm). The standard deviations of the particle diameters are given in parentheses. The average particle diameter of the precipitated silica nanospheres is correlated with the phosphate concentration. Middle: Amount of precipitated silica as a function of phosphate concentration. (From Brunner, E. et al., *Phys. Chem. Chem. Phys.*, 6, 854, 2004. With permission.) Bottom: Diameter, d, of poly(allylamine) aggregates determined by dynamic light scattering as a function of pH for four different phosphate concentrations: 70 mM (open squares), 55 mM (black circles), 40 mM (open circles), 25 mM (gray triangles). PAA concentration: 1 mM. Lines are drawn to guide the eye. (Reproduced from Lutz, K. et al., *Phys. Chem. Chem. Phys.*, 7, 2812, 2005. With permission.)

6.3.2.2 Nanotubes

Silica nanotubes are another type of nanostructure that could meanwhile be synthesized biomimetically based on the observation that certain molecules self-assemble into corresponding templating superstructures. Special chiral phospholipids are capable of forming tubules that can be used as templates for the deposition of silica (Baral and Schoen 1993) or other inorganic phases (Archibald and Mann 1993). Further templating superstructures used for the synthesis of silica nanotubes are, for example, tobacco mosaic virus (TMV) capsules (Shenton et al. 1999) or peptides self-assembling into nanotubular structures. Lanreotide—a therapeutic octapeptide—self-assembles into monodisperse nanotubes of 24.4 nm diameter and ca. 1.8 nm wall thickness (Valéry et al. 2003). These nanotubes can be used as templating superstructures for the synthesis of unique, double-walled silica nanotubes via the synergy between dynamic

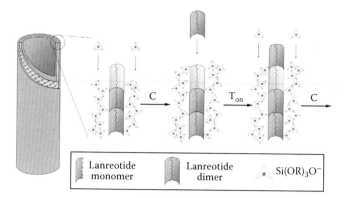

FIGURE 6.5 Double walled silica nanotubes. Dynamic template model for the growth mechanism of peptide-silica fibers. The cationic lanreotide nanotube surface catalyzes the silica condensation by electrostatic forces (step C) and the anionic silica deposit promotes the lanreotide assembly (step T_{on}) by a synergetic neutralization of the system. (Reprinted from Pouget, E. et al., *Nat. Mater.*, 6, 434, 2007. With permission.)

template self-assembly and mineralization (see Figure 6.5, Pouget et al. 2007). Silica tubes could also be synthesized within the pore channels of polycarbonate membranes thus mimicking the influence of spatial confinement usually involved in biomineralization processes (Gautier et al. 2006).

6.3.2.3 Other Structures

A variety of other silica-based nanostructures could meanwhile be produced biomimetically apart from nanospheres and nanotubes. Helical mesoporous silica fibers were synthesized using a ternary surfactant mixture as template (Lin et al. 2007). Nanocomposite silica-polyamine films could be made by a reactive layer-by-layer deposition (Laugel et al. 2007). The so-called direct ink writing technology enables the robotic deposition of polyamine-rich ink in complex 3D patterns closely resembling the patterns found in diatom cell walls (Xu et al. 2006). These patterns can be silicified by immersion in silicic acid-containing solutions that allow to mimic diatom cell walls surprisingly well. Biomimetic silica micropatterning could also be achieved

by the silicification of poly(2-(dimethylamino)ethyl methacrylate) (pDMAEMA) patterns. These patterns were manufactured by the microcontact printing of DMAEMA and a subsequent surface-initiated polymerization (Kim et al. 2005). Silica-carbon nanofibers were biomimetically synthesized which may be useful for electrochemical biosensor systems (Vamvakaki et al. 2008). Regular patterns of silica nanospheres for holographic applications could be produced using the so-called R5-peptide as a template (Brott et al. 2001). The amino acid sequence of this peptide is derived from the sequence of silaffin 1A (Kröger et al. 1999). Periodically arranged silica nanospheres of 12–23 nm size were obtained from lysine-containing tetraethyl orthosilicate (TEOS) solutions (Yokoi et al. 2006). Furthermore, silica nanoparticles could be grown in so-called phospholipid onion phases (El Rassy et al. 2005). Silica nanocomposites were also made from a chimeric spider silk protein (Wong Po Foo et al. 2006).

Inspired by the observation of collagen and chitin in several Hexatinellida sponges, the synthesis of silica-collagen and silica-chitin composites was suggested (Ehrlich et al. 2006, 2008b) as a promising strategy. The structures shown in Figure 6.6 could be obtained by adding silicic acid solutions obtained by the hydrolysis of TMOS or TEOS to homogeneous collagen suspensions. Quantitative studies suggest a direct interaction of silicic acid molecules with the primary amine groups of the collagen. This can be explained by electrostatic interactions between the negatively charged silica species and the positively charged amine groups of the collagen at neutral pH (Heinemann et al. 2007a). Silicification of collagen using the sol-gel technique facilitates the formation of silica-collagen composite materials that originally occur as hydrogels. Drying in a humidified atmosphere resulted in compact monolithic xerogels of considerable mechanical strength, whereas control samples synthesized without collagen cracked under the same conditions (Heinemann et al. 2007b). Compression strength tests showed that up to about 100 MPa could be applied to the composites containing 70% silica and 30% fibrillar collagen. Cell culture experiments demonstrated that the silica-collagen hybrid materials exhibit a proper biocompatibility supporting the adhesion, proliferation, and differentiation of human mesenchymal stem cells and human monocytes.

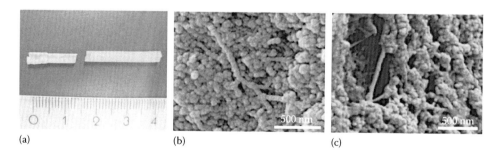

(a)   (b)   (c)

FIGURE 6.6 Silicified collagen. (a) Collagen-silica composites produced by *in vitro* precipitation of silica from a silicic acid solution using collagen fibrils isolated from the spicules of the glass sponge *Hyalonema sieboldi*. (b) SEM image of the *in vitro*-synthesized collagen-silica composite: Silica nanoparticles are deposited on the collagen fibrils. (c) SEM image of a partially demineralized *H. sieboldii* spicule reveals a similar morphology as observed for the *in vitro*-synthesized composite. (From Ehrlich, H. and Worch, H., Collagen: A huge matrix in glass sponge flexible spicules of the meter-long hyalonema seibddi, in *Handbook of Biomineralization*, Vol. 1, Baeuerlein, E. (ed.), Wiley-VCH, Weinheim, Germany, 2007, pp. 23–41. With permission.)

6.3.3 Biomimetic Synthesis of Calcium-Based Materials

6.3.3.1 Calcium Carbonate Materials

In the past years, numerous efforts were made in order to biomimetically synthesize CaCO₃-based materials (see, e.g., Cölfen 2003, Cölfen and Mann 2003, Cölfen et al. 2005, Meldrum and Cölfen 2008). Various morphologies could be produced by template-controlled crystallization. Examples are: single calcite crystals of complex morphology synthesized using a polymer template, complex shapes generated from liquid precursors, as well as numerous nanoparticle superstructures (Dousi et al. 2003). The synthesis of such superstructures requires controlled assembly of the building units (nanoparticles) which strongly depends on the used additives. Mesoscopic structure formation could, for example, be induced by surfactant-mediated aggregation (Li and Mann 2002). Furthermore, water-soluble polymer additives can be used to temporarily stabilize nanoparticle building blocks, which then aggregate in a controlled fashion (Cölfen and Mann 2003). The following templates could be used meanwhile, for the synthesis of biomimetic calcium carbonate structures: self-assembled monolayers/Langmuir monolayers, lipid bilayer stacks, microemulsions, organic hydrogels, vesicles, and functionalized micropatterned surfaces (Kato et al. 2002, Cölfen 2003, Estroff et al. 2004, Viravaidya et al. 2004).

A very impressive example for biomimetically synthesized nano-and microstructured calcite is demonstrated in Figure 6.7

(Mukkamala and Powell 2004). Calcite "microtrumpets" were synthesized *in vitro* from a solution containing 1,3-diamino-2-hydroxypropane-*N,N,N′,N′*-tetraacetic acid. These trumpet-like structures are formed after several hours of crystallization. They closely resemble the structure of coccoliths such as *Discosphaera tubifera* (Figure 6.7e). As can be also seen from Figure 6.7, the formation of larger frameworks or superstructures from smaller crystallites is an important principle in calcium biomineralization and biomimetics. This example, along with others (Mukkamala et al. 2006) shows that the overall morphology of *in vitro* synthesized calcium carbonate compounds can be controlled surprisingly well although the shape and structure of natural biominerals is usually determined by rather complex molecules (proteins, polysaccharides) and genetically controlled mechanisms.

6.3.3.2 Calcium Phosphate Materials

Recent studies have demonstrated the important role of calcium phosphate nanoparticles in the formation of hard tissues in nature as well as for the synthesis of novel biomaterials (Cai and Tang 2008). Since the composite nanostructure of bone (see Section 6.2) must be responsible for its mechanical properties, numerous attempts to synthesize HAP/collagen composites have meanwhile been reported (see, e.g., Kikuchi et al. 2004). Various synthesis routes were proposed: wet-chemical techniques such as the direct precipitation from aqueous solutions, electrochemical depositions,

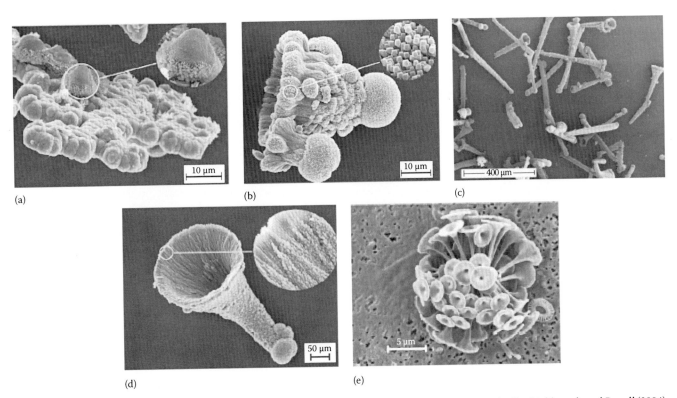

(a) (b) (c)

(d) (e)

FIGURE 6.7 Calcite microtrumpets: Crystal tectonics in action. SEM images of calcite formation as described by Mukkamala and Powell (2004): (a) after 6 h; (b) after 12 h; (c) and (d) after 24 h reaction time. (e) For comparison: The coccolith *D. tubifera*. (Reproduced from Mukkamala, S.B. and Powell, A.K., *Chem. Commun.*, 8, 918, 2004. With permission.)

sol-gel procedures, mechano-chemical as well as hydrother-mal synthesis (Ethirajan et al. 2008, Zhang and Lu 2008). Other concepts suggested the synthesis of hybrid calcium phosphate nanoparticles using organic additives (Ganesan and Epple 2008) or templates such as surfactants, liposomes, block copolymers, self-associated nanogels, supramolecular hydrogels, emulsions, and microemulsions (Schwiertz et al. 2008, Ethirajan et al. 2008). So far, however, the obtained products exhibit neither bone-like nanostructures and mechanical properties nor the desired bone-like behavior as bone substitutes (Kikuchi et al. 2004). Recently, HAP-reinforced polymer biocomposites were suggested as robust synthetic bone substitutes (Roeder et al. 2008).

Other calcium phosphate phases apart from HAP are also involved in bone mineralization. Termine and Posner (1967) proposed that amorphous calcium phosphate (ACP) may be a precursor phase in bone formation. It should then be possible to fill small spaces (gaps) within the collagen fibrils by introducing ACP and its subsequent transformation into HAP. This so-called transient precursor strategy, that is, the initial deposition of a less ordered mineral and its subsequent transformation into a crystalline phase is often found in nature, for example, among invertebrates (Weiner 2006).

The function of collagen in biomineralization is not only determined by its structure but also by various modifications as well as by its interactions with molecules such as non-collagenous proteins (NCPs) and possibly mono- or oligo-meric sugars (see, for example, Ehrlich et al. 2005, George and Veis 2008, and references therein). Usually, NCPs are highly charged since they are rich in aspartate and glutamic acid as well as in phosphoserine (Olszta 2007). Despite their relatively low concentrations, NCPs play an important role in the biom-ineralization process, for example, by controlling the nucle-ation of apatite (George and Veis 2008). During secondary bone formation, the organization of the crystallites is directed by the collagen fibril matrix. The resulting *intrafibrillar* crys-tallites are extremely small (only a few unit cells thick) and are stabilized by the surrounding organic matrix. Such nanocrys-tals may also form at the surface and between collagen fibers. They are then referred to as *interfibrillar* crystals (Olszta et al. 2007). Recently, it was reported that the mineral-organic interphase in bone is lined by a polysaccharide—chondroitin sulfate (ChS, Wise et al. 2007) consisting of repeated disac-charide units: *N*-acetylgalactosamine sulfate and glucuronic acid containing a carboxylate group. Previous investigations indicated that ChS is present at the early stages of bone and tooth formation and is considered to be involved in regulating mineral deposition and crystal morphology during osteogen-esis (Jiang et al. 2005). The self-assembly of HAP nanocrys-tals on chondroitin-sulfate was also studied *in vitro* (Rhee and Tanaka 2002).

In summary, negatively charged carboxylate groups are sug-gested as being capable of binding calcium ions and of induc-ing the nucleation of the related biominerals in nature. It is hypothesized that the carboxyl groups constitute the organic-mineral interface in numerous biominerals (Gilbert et al.

2005). In other words, the carboxyl group may be the molecu-lar "glue" connecting the organic template with the inorganic mineral phase.

Glucuronic acid is present in various proteoglycans (deco-rin, versican, biglycan, syndecan) and glycosaminoglycans (hyaluronic acid, chondroitin sulfates, dermatan sulfate, hepa-rin, keratin sulfate) and it continues to attract research interest with respect to biomineralization and biomimetic nanomate-rials synthesis. Ehrlich et al. (2008c,d) showed that the biomi-metic carboxymethylation of collagen caused by the reaction with glucuronic acid facilitates the mineralization of collagen. SEM images of unmodified collagen fibrils (Figure 6.8a) exhibit surfaces without any crystalline formations even after 48 and 72 h of mineralization. In contrast, carboxymethyllysine (CML)-collagen fibrils facilitate the formation of crystallites just after 24 h (Figure 6.8b). Crystals with a morphology characteristic for octacalcium phosphate grow perpendicular to the fiber axis of the collagen fibril. Figure 6.8c and d confirms the formation of nano- and microcrystalline OCP layers at the surface of CML-collagen fibrils (Ehrlich et al. 2008d). Their orientation is dic-tated by the orientation of the carboxymethyllysine residues. First, calcium ions bind to the carboxylate groups of the CML residues. Subsequently, phosphate ions bind to the calcium ions. In other words, CML residues are the nucleation sites for OCP crystal formation. These crystals grow spontaneously with a preferential orientation due to the chemical interactions with the pre-organized carboxylate groups of the CML-based colla-gen template (Figure 6.8c and d).

6.3.3.3 Fluorapatite-Gelatine-Nanocomposites

Gelatine is formed by the irreversible hydrolysis of collagen. It is a promising candidate for template-based syntheses due to its intrinsic biocompatibility, ability to interact with HAP surfaces, availability, and low cost (Ethirajan et al. 2008). The use of gela-tine as a matrix for biomimetic mineralization was studied under various conditions (Kniep and Simon 2006). Recently, Ethirajan et al. (2008) suggested a novel strategy to synthesize HAP inside crosslinked gelatine nanoparticles serving as defined nanore-actors for the crystal growth, by constituting a confined envi-ronment. The formation of HAP inside the particles follows Ostwald's rule of stages: First, an amorphous phase is formed which is interesting by itself since it has potential uses as a resorbable bone substitute. Subsequently, this composite mate-rial transforms into the thermodynamically stable HAP via an OCP intermediate.

The biomimetic system apatite-gelatine was also success-fully exploited in a so-called double-diffusion arrangement. Starting from oppositely arranged reservoirs containing aqueous solutions of calcium and phosphate/fluoride, respec-tively, the ions migrate into the central, gelatine-containing compartment (Kniep and Busch 1996, Simon et al. 2004). The obtained fluorapatite-gelatine nanocomposites exhibit frac-tal a morphology as described by Kniep and Simon (2008). Growth of the composites starts with an elongated hexagonal prism, which develops into the so-called first dumb-bell state

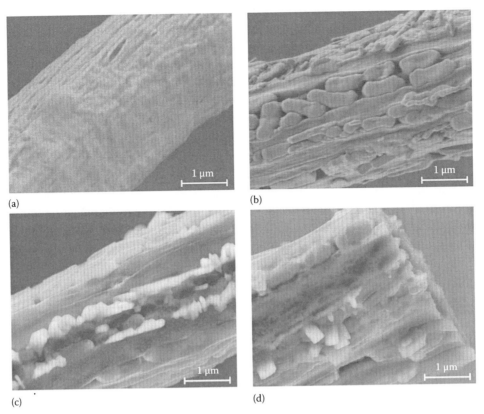

FIGURE 6.8 Octacalciumphosphate nanocrystal growth on modified collagen fibrils. SEM images of collagen fibrils in a calcium phosphate mineralization experiment. No inorganic formations are observed on the non-modified collagen fibrils after 24 h of mineralization (a). In contrast, crystalline structures intercalated between the nanofibrils occur at the surface of carboxymethylated collagen fibrils after 24 h (b). After a 48 h growth, the characteristic crystal morphology of OCP could be observed (c,d). (Reproduced from Ehrlich, H. et al., *Int. J. Biol. Macromol.*, 44, 51, 2008c. With permission.)

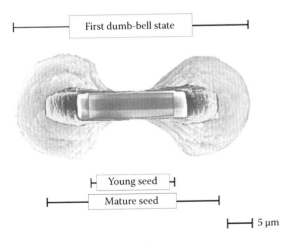

FIGURE 6.9 Fluoroapatite-gelatine composites. Superimposition of SEM images of the different growth stages of (fractal) fluorapatite–gelatine nanocomposite aggregates. (Reproduced from Kniep, R. and Simon, P., *Angew. Chem. Int. Ed.*, 47, 1405, 2008. With permission.)

(Figure 6.9). Scanning electron microscopy images of different growth states are superimposed: the central "young" seed grows into a "mature" one (Figure 6.9) and subsequently splits into the first dumb-bell state (Figure 6.9). The structure of the young seeds has already been investigated by high-resolution TEM (HRTEM) methods. These results revealed that the inner architecture of the young seeds is built by a parallel stacking of elongated nanocomposite subunits oriented with their long [001] axes parallel to the seed. Each nanocomposite subunit grows around a central protein triple helix (Kniep and Simon 2008). The obtained composite (Figure 6.10) can be considered to be a mesocrystal (see Section 6.2), that is, an arrangement of aligned individual nanocrystals with a common crystallographic orientation giving rise to scattering properties similar to those of single crystals. As can been seen in Figure 6.10, the protein molecules within the nanocomposite are aligned parallel to the long axis of the young seed, which is the crystallographic c axis. The polar triple helices exhibit opposite charges at their ends resulting in a macroscopic electric dipole moment of the nanocomposite (Simon et al. 2006).

Finally, it should be noted that the development of nanostructured calcium phosphate based materials (nanoceramics) as well as hybrid materials with hierarchically organized porosity caused a conceptual change: the regeneration of hard tissue is now the preferred goal rather than its replacement (Vallet-Regi et al. 2008).

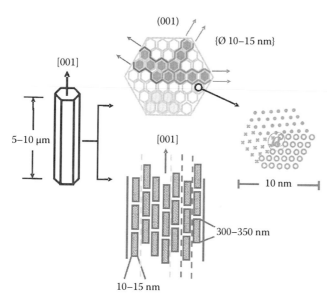

FIGURE 6.10 "Hidden" hierarchy of microfibrils in fluoroapatite-gelatine composites. Schematic description of the nanocomposite inner architecture of a young fluorapatite–gelatine seed derived from HRTEM investigations. Left: Elongated hexagonal-prismatic habit of the seed. Middle bottom: parallel-rod stacking of (self-similar) elongated nano-subunits along [001] as well as "accidental" variations in rod clustering by mechanical treatment (dashed lines). Middle top: Nanostructure in (001) with grain boundaries representative of a hexagonal material, as well as variations in accidental clustering by breaking and preferred fracture directions (arrows). Right: Nanomosaic structure about 10 nm in diameter nucleated by (around) a central gelatine macromolecule (@). (Reproduced from Kniep, R. and Simon, P., *Angew. Chem. Int. Ed.*, 47, 1405, 2008. With permission.)

6.3.4 Nanoparticle Assemblies, Nanowires, and Others

The use of nanoscale effects often requires the organization of nanoparticles in two- or three-dimensional assemblies that can be produced biomimetically. Nanoparticle vesicles spontaneously self-assemble from homopolymer polyamine solutions containing water-soluble, citrate stabilized CdSe/CdS quantum dots (Cha et al. 2003). It is proposed that this is due to the formation of charge-stabilized hydrogen bonds between the positively charged polyamines and the negatively charged citrate molecules in analogy to the polyamine/phosphate assemblies described in Section 6.3.2. The ability of certain peptides, so-called inorganic binding peptides of binding to inorganic surfaces such as noble metals (Au, Ag, Pt, Pd), semiconductors (e.g., Cu_2O, ZnO, GaAs, ZnS), ionic crystals (e.g., $CaCO_3$, Cr_2O_3, Fe_2O_3) and others can be exploited in order to produce various nanoparticle assemblies such as nanoparticle surface coatings (for a review, see, e.g., Sarikaya et al. 2003).

The biomimetic synthesis of nanowires is of special interest with respect to electronic applications. Gold nanowires could be synthesized by coating of self-assembled peptide nanowires formed with a histidine-rich peptide (Djalali et al. 2002). Silver nanowires were made by the reduction of ionic silver within self-assembled nanotubes formed by a peptide with Alzheimer's β-amyloid diphenylalanine structural motif (Reches and Gazit 2003). A method for the controlled patterning of self-assembled peptide nanotubes could also be introduced which allows an arrangement of the nanotubes in a defined manner (Reches and Gazit 2006). Conducting nanowires could be manufactured by the deposition of gold and silver at the surface of self-assembled amyloid fibers (Scheibel et al. 2003). DNA can also be used as a template for the fabrication of metallic nanowires (for a review, see Mertig and Pompe 2005). Bimetallic Ag-Au nanowires were synthesized by the metallization of artificial DNA duplexes (Fischler et al. 2007).

Apart from the aforementioned silica- and calcium-based materials that are inspired to a large extent by the observations made on biomineralization phenomena, other nanomaterials could also be synthesized biomimetically. One example is the synthesis of titania particles induced by protamine—an unstructured cationic protein—in solution from a water-stable titanium precursor compound (Jiang et al. 2008).

6.4 Summary

In summary, it can be stated that biomineralization research continues to impact on biomimetic or bio-inspired synthesis approaches. Here, we have discussed a few important examples for biomineralization phenomena (Chapter 2) as well as their consequences for nanomaterials sciences: Biosilica, for example, from diatoms and sponges is a composite material consisting of amorphous silica and special biomolecules. Calcium-based biominerals are often composites made up of small calcium-based crystallites and certain biomolecules. The special biomolecules involved in the process of biomineral formation fulfill different functions, for example, by acting as structure-directing templates as well as by enhancing the nucleation or precipitation of the mineral phase. Therefore, the biomolecules found in the described biominerals have meanwhile inspired a considerable number of biomimetic synthesis experiments as illustrated in Section 6.3. Such biomimetic approaches meanwhile enable the synthesis of various nanostructures such as nanospheres, nanotubes, nanowires as well as nanoparticle arrays. Molecular self-assembly is of central importance: Self-assembled superstructures formed, for example, by biomolecules such as proteins, lipids, and polyamines can be exploited successfully as templating superstructures. Ongoing research dealing with the complicated biochemical and biophysical processes underlying the formation of nanostructures in living organisms will hopefully further inspire biomimetic synthesis approaches and lead to even more sophisticated materials in the future.

Acknowledgments

The authors wish to thank Prof. Annie K. Powell (Karlsruhe) and Dr. Paul Simon (Dresden) for fruitful discussions as well as for providing them with figures. Thanks are further due to Renate Schulze (Dresden) for proofreading the manuscript.

References

Addadi, L., Raz, S., Weiner, S. 2003. Taking advantage of disorder: Amorphous calcium carbonate and its roles in biomineralization. *Adv. Mater.* 15: 959–970.

Aizenberg, J., Sundar, V. C., Yablon, A. W. et al. 2004. Biological glass fibers: Correlation between optical and structural properties. *Proc. Natl. Acad. Sci. USA* 1001: 3358–3363.

Archibald, D., Mann, S. 1993. Template mineralization of self-assembled anisotropic lipid microstructures. *Nature* 354: 430–433.

Ariga, K., Hill, J. P., Lee, M. P. et al. 2008. Challenges and breakthroughs in recent research on self-assembly. *Sci. Technol. Adv. Mater.* 9: 14–109.

Baeuerlein, E., Behrens, P, Epple, M. 2007. *Handbook of Biomineralization.* Vols. 1–3. Weinheim, Germany: Wiley-VCH.

Baral, S., Schoen, P. 1993. Silica-deposited phospholipid tubules as a precursor to hollow submicron-diameter silica cylinders. *Chem. Mater.* 5: 145–147.

Bar-Cohen, Y. 2005. *Biomimetics: Biologically Inspired Technologies.* Boca Raton, FL: CRC Press.

Bauer, C. A., Robinson, D. B., Simmons, B. A. 2007. Silica formation in confined environments via bioinspired polyamine catalysis at near-neutral pH. *Small* 3: 58–62.

Behrens, P., Jahns, M., Menzel, H. 2007. The polyamine silica system: A biomimetic model for the biomineralization of silica. In *Handbook of Biomineralization: Biomimetic and Bioinspired Chemistry*, Vol. 2, eds. P. Behrens, E. Bäuerlein, pp. 3–18. Weinheim, Germany: Wiley-VCH.

Brott, L. L., Naik, R. R., Pikas, D. J. et al. 2001. Ultrafast holographic nanopatterning of biocatalytically formed silica. *Nature* 413: 291–293.

Brunner, E., Lutz, K. 2007. Solid-state NMR in biomimetic silica formation and silica biomineralization. In *Handbook of Biomineralization: Biomimetic and Bioinspired Chemistry*, Vol. 2, eds. P. Behrens, E. Bäuerlein, pp. 19–38. Weinheim, Germany: Wiley-VCH.

Brunner, E., Lutz, K., Sumper, M. 2004. Biomimetic synthesis of silica nanospheres depends on the aggregation and phase separation of polyamines in aqueous solution. *Phys. Chem. Chem. Phys.* 6: 854–857.

Brutchey, R. L., Morse, D. E. 2008. Silicatein and the translation of its molecular mechanism of biosilicification into low temperature nanomaterial synthesis. *Chem. Rev.* 108: 4915–4934.

Cameron, J. N. 1990. Unusual aspects of calcium metabolism in aquatic animals. *Annu. Rev. Physiol.* 52: 77–95.

Cai J., Tang, R. 2008. Calcium phosphate nanoparticles in biomineralization and biomaterials. *J. Mater. Chem.* 18: 3775–3787.

Cha, J. N., Stucky, G. D., Morse, D. E. et al. 2000. Biomimetic synthesis of ordered silica structures mediated by block copolypeptides. *Nature* 403: 289–292.

Cha, J. N., Birkedal, H., Euliss, L. E. et al. 2003. Spontaneous formation of nanoparticle vesicles from homopolymer polyelectrolytes. *J. Am. Chem. Soc.* 125: 8285–8289.

Cölfen, H. 2003. Precipitation of carbonates: Recent progress in controlled production of complex shapes. *Curr. Opin. Colloid Interface Sci.* 8: 23–31.

Cölfen, H., Antonietti, M. 2005. Mesocrystals: Inorganic superstructures made by highly parallel crystallization and controlled alignment. *Angew. Chem. Int. Ed.* 44: 5576–5591.

Cölfen, H., Mann, S. 2003. Higher-order organization by mesoscale self-assembly and transformation of hybrid nanostructures. *Angew. Chem. Int. Ed.* 42: 2350–2365.

Djalali, R., Chen, Y.-f., Matsui, H. 2002. Au nanowire fabrication from sequenced histidine-rich peptide. *J. Am. Chem. Soc.* 124: 13660–13661.

Dousi, E., Kallitsis, J., Chrissanthopoulos, A. et al. 2003. Calcite overgrowth on carboxylated polymers. *J. Cryst. Growth* 253: 496–503.

Drum, R. W., Pankratz, H. S. 1964. Post mitotic fine structure of *Gomphonema parvulum. J. Ultrastruct. Res.* 10: 217–223.

Ehrlich, H., Worch, H. 2007. Collagen: A huge matrix in glass sponge flexible spicules of the meter-long hyalonema sieboldi. In *Handbook of Biomineralization*, Vol. 1, ed. E. Baeuerlein, pp. 23–41. Weinheim, Germany: Wiley-VCH.

Ehrlich, H., Douglas, T., Scharnweber, D. et al. 2005. Hydroxyapatite crystal growth on modified collagen I-templates in a model dual membrane diffusion system. *Z. Anorg. Allg. Chem.* 631: 1825–1830.

Ehrlich, H., Heinemann, S., Hanke, T. et al. 2006. Hybrid materials from a silicate-treated collagen matrix, methods for the production thereof and the use thereof. WO 2008/023025, PCT/EP2007/058694.

Ehrlich, H., Koutsoukos, P. G., Demadis, K. D., Pokrovsky, O. S. 2008a. Principles of demineralization: Modern strategies for isolation of organic frameworks. Part I. Common definitions and history. *Micron* 39: 1062–1091.

Ehrlich, H., Janussen, D., Simon, P. et al. 2008b. Nanostructural organization of naturally occurring composites. Part II. Silica-chitin-based biocomposites. *J. Nanomat.* (available online at doi:10.1155/2008/670235).

Ehrlich, H., Hanke, T., Frolov, A. et al. 2008c. Modification of collagen *in vitro* with respect to formation of N^{ε}-carboxymethyllysine. *Int. J. Biol. Macromol.* 44: 51–56.

Ehrlich, H., Hanke, T., Born, R. et al. 2008d. Mineralization of biomimetically carboxymethylated collagen fibrils in a model dual membrane diffusion system. *J. Membr. Sci.* 326: 254–259.

El Rassy, H., Belamie, E., Livage, J. et al. 2005. Onion phases as biomimetic confined media for silica nanoparticle growth. *Langmuir* 21: 8584–8587.

Estroff, L., Addadi, L., Weiner, S. et al. 2004. An organic hydrogel as a matrix for the growth of calcite crystals. *Org. Biomol. Chem.* 2: 137–141.

Ethirajan, A., Ziener, U., Chuvilin, A. et al. 2008. Biomimetic hydroxyapatite crystallization in gelatine nanoparticles synthesized using a miniemulsion process. *Adv. Funct. Mater.* 18: 2221–2227.

Fischler, M., Simon, U., Nir, H. et al. 2007. Formation of bimetallic Ag-Au nanowires by metallization of artificial DNA duplexes. *Small* 3: 1049–1055.

Fuhrmann, T., Landwehr, S., El Rharbi-Kucki, M. et al. 2004. Diatoms as living photonic crystals. *Appl. Phys. B* 78: 257–260.

Gajjeraman, S., Narayanan, K., Hao, J. et al. 2007. Matrix macromolecules in hard tissues control the nucleation and hierarchical assembly of hydroxyapatite. *J. Biol. Chem.* 282: 1193–1204.

Ganesan, K., Epple, M. 2008. Calcium phosphate nanoparticles as nuclei for the preparation of colloidal calcium phytate. *New J. Chem.* 32: 1326–1330.

Garrone, R. 1978. Phylogenesis of connective tissue. Morphological aspects and biosynthesis of sponge intercellular matrix. In *Frontiers of Matrix Biology*, Vol. 5, ed. L. R. Créteil, pp. 108–158. Basel, Germany: Karger-Verlag.

Gautier, C., Lopez, P., Hemadi, M. et al. 2006. Biomimetic groth of silica tubes in confined media. *Langmuir* 22: 9092–9095.

George, A., Veis, A. 2008. Phosphorylated proteins and control over apatite nucleation, crystal growth, and inhibition. *Chem. Rev.* 108: 4670–4693.

Gilbert, P. U. P. A., Albrecht, M., Frazer, B. H. 2005. The organicmineral interface in biominerals. *Rev. Mineral. Geochem.* 59: 157–185.

Gordon, R., Losic, D., Tiffany, M. A. et al. 2008. The glass menagerie: Diatoms for novel applications in nanotechnology. *Trends Biotechnol.* 27: 116–127.

Gröger, C., Sumper, M., Brunner, E. 2008a. Silicon uptake and metabolism of the marine diatom *Thalassiosira pseudonana*: Solid-state ^{29}Si NMR and fluorescence microscopic studies. *J. Struct. Biol.* 161: 55–63.

Gröger, C., Lutz, K., Brunner, E. 2008b. Biomolecular self-assembly and its relevance in silica biomineralization. *Cell Biochem. Biophys.* 50: 23–39.

Hamm, C. E., Merkel, R., Springer, O. et al. 2003. Architecture and material properties of diatom shells provide effective mechanical protection. *Nature* 241: 841–843.

Heinemann, S., Ehrlich, H., Knieb, C. et al. 2007a. Biomimetically inspired hybrid materials based on silicified collagen. *Int. J. Mater. Res.* 98: 603–608.

Heinemann, S., Heinemann, C., Ehrlich, H., et al. 2007b. A novel biomimetic hybrid material made of silicified collagen: Perspectives for bone replacement. *Adv. Eng. Mater.* 9: 1061–1068.

Henriksen, K., Stipp, S. L. S., Young, J. R. et al. 2003. Tailoring calcite: Nanoscale AFM of coccolith biocrystals. *Am. Miner.* 88: 2040–2044.

Hildebrand, M. 2008. Diatoms, biomineralization processes, and genomics. *Chem. Rev.* 108: 4855–4874.

Jiang, H., Liu, X. Y., Zhang, G. et al. 2005. Kinetics and template nucleation of self-assembled hydroxyapatite nanocrystallites by chondroitin sulfate. *J. Biol. Chem.* 280: 42061–42066.

Jiang, Y., Yang, D., Zhang, L. et al. 2008. Biomimetic synthesis of titania nanoparticles induced by protamine. *Dalton Trans.* 4165–4171.

Kato, T., Sugawara, A., Hosoda, N. 2002. Calcium carbonate-organic hybrid materials. *Adv. Mater.* 14: 869–877.

Kikuchi, M., Ikoma, T., Itoh, S. et al. 2004. Biomimetic synthesis of bone-like nanocomposites using the self-organization mechanism of hydroxyapatite and collagen. *Compos. Sci. Technol.* 64: 819–825.

Kim, D. J., Lee, K.-B., Lee, T. G. et al. 2005. Biomimetic micropatterning of silica by surface-initiated polymerization and microcontact printing. *Small* 1: 992–996.

Knecht, M. R., Wright, D. W. 2004. Amine-terminated dendrimers as biomimetic templates for silica nanosphere formation. *Langmuir* 20: 4728–4732.

Knecht, M. R., Sewell, S. L., Wright, D. W. 2005. Size control of dendrimer-templated silica. *Langmuir* 21: 2058–2061.

Kniep, R., Busch, S. 1996. Biomimetic growth and self-assembly of fluorapatite aggregates by diffusion into denatured collagen matrices. *Angew. Chem. Int. Ed.* 35: 2624–2623.

Kniep, R., Simon, P. 2006. Fluorapatite-gelatine-nanocomposites: Self-organized morphogenesis, real structure and relations to natural hard materials. *Top. Curr. Chem.* 270: 73–125.

Kniep, R., Simon, P. 2008. "Hidden" hierarchy of microfibrils within 3D-periodic fluorapatite-gelatine nanocomposites: Development of complexity and form in a biomimetic system. *Angew. Chem. Int. Ed.* 47: 1405–1409.

Kröger, N., Poulsen, N. 2008. Diatoms—From cell wall biogenesis to nanotechnology. *Annu. Rev. Genet.* 42: 83–107.

Kröger, N., Deutzmann, R., Sumper, M. 1999. Polycationic peptides from diatom biosilica that direct silica nanosphere formation. *Science* 286: 1129–1132.

Kröger, N., Deutzmann, R., Bergsdorf, C. et al. 2000. Species-specific polyamines from diatoms control silica morphology. *Proc. Natl. Acad. Sci. USA* 97: 14133–14138.

Kröger, N., Lorenz, S., Brunner, E. et al. 2002. Self-assembly of highly phosphorylated silaffins and their function in biosilica morphogenesis. *Science* 298: 584–586.

Laugel, N., Hemmerlé, J., Porcel, C. et al. 2007. Nanocomposite silica/polyamine films prepared by a reactive layer-by-layer deposition. *Langmuir* 23: 3706–3711.

Lehn, J.-M. 1990. Perspectives in supramolecular chemistry—From molecular recognition towards molecular information processing and self-organization. *Angew. Chem. Int. Ed.* 29: 1304–1319.

Lehn, J.-M. 1995. *Supramolecular Chemistry: Concepts and Perspectives*. Weinheim, Germany: Wiley VCH.

Lehn, J.-M. 2002. Toward self-organization and complex matter. *Science* 295: 2400–2403.

Li, M., Mann, S. 2002. Emergent nanostructures: Water-induced mesoscale transformation of surfactant-stabilized amorphous calcium carbonate nanoparticles in reverse microemulsions. *Adv. Funct. Mater.* 12: 773–779.

Lin, G.-L., Tsai, Y.-H., Lin, H.-P. et al. 2007. Synthesis of mesoporous silica helical fibers using a catanionic-neutral ternary surfactant in a highly dilute silica solution: Biomimetic silicification. *Langmuir* 23: 4115–4119.

Lowenstam, H. A., Weiner, S. 1989. *On Biomineralization*. Oxford, NY: Oxford University Press.

Lutz, K., Gröger, C., Sumper, M. et al. 2005. Biomimetic silica formation: Analysis of the phosphate-induced self-assembly of polyamines. *Phys. Chem. Chem. Phys.* 7: 2812–2815.

Mann, D. G., Droop, S. J. M. 1996. 3. Biodiversity, biogeography and conservation of diatoms. *Hydrobiologia* 336: 19–32.

Mann, S. 1993. Molecular tectonics in biomineralization and biomimetic materials chemistry. *Nature* 365: 499–500.

Mann, S. 2001. *Biomineralization: Principles and Concepts Ion Bioinorganic Materials Chemistry*. Oxford, U.K.: Oxford University Press.

Mann, S. 2008. Life as a nanoscale phenomenon. *Angew. Chem. Int. Ed.* 47: 5306–5320.

Matsunaga, S., Sakai, R., Jimbo, M. et al. 2007. Long-chain polyamines (LCPAs) from marine sponge: Possible implication in spicule formation. *ChemBioChem* 8: 1729–1735.

McKenna, B. J., Birkedal, H., Bartl, M. H. et al. 2004. Micrometer-sized spherical assemblies of polypeptides and small molecules by acid-base chemistry. *Angew. Chem. Int. Ed.* 43: 5652–5655.

Meldrum, F. C., Cölfen, H. 2008. Controlling mineral morphologies and structures in biological and synthetic systems. *Chem. Rev.* 108: 4332–4432.

Mertig, M., Pompe, W. 2005. Biomimetic fabrication of DNA-based metallic nanowires and networks. In *Nanobiotechnology*, eds. C. M. Niemeyer, C. Mirkin, pp. 256–272. Weinheim, Germany: Wiley-VCH.

Mizutani, T., Nagase, H., Fujiwara, N. et al. 1998. Silicic acid polymerization catalyzed by amines and polyamines. *Bull. Chem. Soc. Jpn.* 71: 2017–2022.

Müller, W. E. G., Rothenberger, M., Boreiko, A. et al. 2005. Formation of siliceous spicules in the marine demosponge *Suberites domuncula. Cell Tissue Res.* 321: 285–297.

Müller, W. E. G., Boreiko, A., Schloßmacher, U. et al. 2007. Fractal-related assembly of the axial filament in the demosponge *Suberites domuncula*: Relevance to biomineralization and the formation of biogenic silica. *Biomaterials* 28: 4501–4511.

Mukkamala, S. B., Powell, A. K. 2004. Biomimetic assembly of calcite microtrumpets: Crystal tectonics in action. *Chem. Commun.* 8: 918–919.

Mukkamala, S. B., Anson, C. E., Powell, A. K. 2006. Modelling calcium carbonate biomineralization processes. *J. Inorg. Biochem.* 100: 1128–1138.

Murr, M. M., Morse, D. E. 2005. Fractal intermediates in the self-assembly of silicatein filaments. *Proc. Natl. Acad. Sci. USA* 102: 11657–11662.

Nalwa, H. S. 2005. *Handbook of Nanostructured Biomaterials and Their Applications in Nanobiotechnology*. Los Angeles, CA: American Scientific Publishers.

Navrotsky, A. 2004. Energetic clues to pathways to biomineralization: Precursors, clusters, and nanoparticles. *Proc. Natl. Acad. Sci. USA* 101: 12096–12101.

Olszta, M. J., Cheng, X., Jee, S. S. et al., 2007. Bone structure and formation: A new perspective. *Mater. Sci. Eng.* R58: 77–116.

Patwardhan, S. V., Clarson, S. J. 2002a. Silicification and biosilicification, Part 4. Effect of template size on the formation of silica. *J. Inorg. Organomet. Polym.* 12: 109–116.

Patwardhan, S. V., Clarson, S. J. 2002b. Silicification and biosilicification, Part 3. The role of synthetic polymers and peptides at neutral pH. *Silicon Chem.* 1: 207–214.

Patwardhan, S. V., Clarson, S. J. 2003. Silicification and biosilicification, Part 5. An investigation of the silica structures formed at weakly acidic pH and neutral pH as facilitated by cationically charged macromolecules. *Mater. Sci. Eng. C* 23: 495–499.

Patwardhan, S. V., Mukherjee, N., Clarson, S. J. 2002a. Formation of fiber-like amorphous silica structures by externally applied shear. *J. Inorg. Organomet. Polym.* 11: 117–121.

Patwardhan, S. V., Mukherjee, N., Clarson, S. J. 2002b. Effect of process parameters on the polymer synthesis of silica at neutral pH. *Silicon Chem.* 1: 47–55.

Pouget, E., Dujardin, E., Cavalier, A. et al. 2007. Hierarchical architectures by synergy between dynamical template self-assembly and biomineralization. *Nat. Mater.* 6: 434–439.

Reches, M., Gazit, E. 2003. Casting metal nanowires within discrete self-assembled peptide nanotubes. *Science* 300: 625–627.

Reches, M., Gazit, E. 2006. Controlled patterning of aligned self-assembled peptide nanotubes. *Nat. Nanotechnol.* 1: 195–200.

Rhee, S.-H., Tanaka, J. 2002. Self-assembly phenomenon of hydroxyapatite nanocrystals on chondroitin sulphate. *J. Mater. Sci.: Mater. Med.* 13: 597–600.

Roeder, R. K., Converse, G. L., Kane, R. J. et al. 2008. Hydroxyapatite-reinforced polymer biocomposites for synthetic bone substitutes. *JOM* 60(3): 38–45.

Round, F., Crawford, R., Mann, D. 1990. *The Diatoms*. Cambridge, U.K.: Cambridge University Press.

Ruiz-Hitzky, E., Ariga, K., Lvov, Y. M. 2008. *Bio-Inorganic Hybrid Nanomaterials*. Weinheim, Germany: Wiley-VCH.

Sarikaya, M., Tamerler, C., Jen, A. K.-Y. et al. 2003. Molecular biomimetics: Nanotechnology through biology. *Nat. Mater.* 2: 577–585.

Scheibel, T., Parthasarathy, R., Sawicki, G. et al. 2003. Conducting nanowires built by controlled self-assembly of amyloid fibers and selective metal deposition. *Proc. Natl. Acad. Sci. USA* 100: 4527–4532.

Schmitt, O. 1969. Some interesting and useful biomimetic transforms. In: *Third International Biophysics Congress*, Boston, MA: p. 297.

Schwiertz, J., Meyer-Zaika, W., Ruiz-Gonzalez, L. et al. 2008. Calcium phosphate nanoparticles as templates for nanocapsules prepared by the layer-by-layer technique. *J. Mater. Chem.* 18: 3831–3834.

Shenton, W., Douglas, T., Young, M. et al. 1999. Inorganic-organic nanotube composites from template mineralization of tobacco mosaic virus. *Adv. Mater.* 11: 253–256.

Shimizu, K., Cha, J., Stucky, G. D. et al. 1998. Silicatein α: Cathepsin L-like protein in sponge biosilica. *Proc. Natl. Acad. Sci. USA* 95: 6234–6238.

Simon, P., Göbel, C., Carrillo-Cabrera, W. et al. 2004. Fluorapatit-gelatine-composite: Biomimetic morphogenesis and real structure. *Z. Anorg. Allg. Chem.* 630: 1760–1766.

Simon, P., Zahn, D., Lichte, H. et al. 2006. Intrinsic electric dipole fields and the induction of hierarchical form developments in fluorapatite-gelatine nanocomposites: A general principle for morphogenesis of biominerals? *Angew. Chem. Int. Ed.* 45: 1911–1915.

Skinner, H. C. W. 2005. Biominerals. *Mineral. Mag.* 69(5): 621–641.

Sumper, M. 2002. A phase separation model for the nanopatterning of diatom biosilica. *Science* 295: 2430–2433.

Sumper, M. 2004. Biomimetic patterning of silica by long-chain polyamines. *Angew. Chem. Int. Ed.* 43: 2251–2254.

Sumper, M., Brunner, E. 2006. Learning from diatoms: Nature's tools for the production of nanostructured silica. *Adv. Funct. Mater.* 16: 17–26.

Sumper, M., Brunner, E. 2008. Silica biomineralisation in diatoms: The model organism *Thalassiosira pseudonana*. *ChemBioChem* 9: 1187–1194.

Sumper, M., Lehmann, G. 2006. Silica pattern formation in diatoms: Species-specific polyamine biosynthesis. *ChemBioChem* 7: 1419–1427.

Sumper, M., Lorenz, S., Brunner, E. 2003. Biomimetic control of size in the polyamine-directed formation of silica nano-spheres. *Angew. Chem. Int. Ed.* 42: 5192–5195.

Sumper, M., Brunner, E., Lehmann, G. 2005. Biomineralization in diatoms: Characterization of novel polyamines associated with silica. *FEBS Lett.* 579: 3765–3769.

Sumper, M., Hett, R., Lehmann, G., Wenzl, S. 2007. A code for lysine modifications of a silica biomineralizing silaffin protein. *Angew. Chem. Int. Ed.* 46: 8405–8408.

Termine, J. D., Posner, A. S. 1967. Amorphous/crystalline inter-relationships in bone mineral. *Calcif. Tiss. Res.* 1: 8–23.

Valéry, C., Paternostre, M., Robert, B. et al. 2003. Biomimetic organization: Octapeptide self-assembly into nanotubes of viral capsid-like dimension. *Proc. Natl. Acad. Sci. USA* 100: 10258–10262.

Vallet-Regi, M., Daniel, A. Arcos Navarette. 2008. *Biomimetic Nanoceramics in Clinical Use*. Cambridge, U.K.: RSC Publishing.

Vamvakaki, V., Hatzimarinaki, M., Chaniotakis, N. 2008. Biomimetically synthesized silica-carbon nanofiber archi-tectures for the development of highly stable electrochemi-cal biosensor systems. *Anal. Chem.* 80: 5970–5975.

Vincent, J. F. V., Bogatyreva, O. A., Bogatyrev, N. R. et al. 2006. Biomimetics: Its practice and theory. *J. R. Soc. Interface* 3: 471–482.

Viravaidya, C., Li, M., Mann, S. 2004. Microemulsion-based syn-thesis of stacked calcium carbonate (calcite) superstruc-tures. *Chem. Commun.* 19: 2182–2183.

Wang, L., Nancollas, G., Henneman, Z. J. et al. 2006. Nanosized particles in bone and dissolution intensivity of bone min-eral. *Biointerphases* 1: 106–111.

Weiner, S. 2006. Transient precursor strategy in mineral forma-tion of bone. *Bone* 39: 431–433.

Weiner, S. 2008. Biomineralization: A structural perspective. *J. Struct. Biol.* 163: 229–234.

Weiner, S., Wagner, H. D. 1998. The material bone: Structure mechanical function relations. *Ann. Rev. Mater. Sci.* 28: 271–298.

Wenzl, S., Hett, R., Richthammer, P. et al. 2008. Silacidins: Highly acidic phosphopeptides from diatom shells assist in silica precipitation in vitro. *Angew. Chem. Int. Ed.* 120: 1729–1732.

Whitesides, G. M., Boncheva, M. 2002. Beyond molecules: Self-assembly of mesoscopic and macroscopic components. *Proc. Natl. Acad. Sci. USA* 99: 4769–4774.

Whitesides, G. M., Grzybowski, B. 2002. Self-assembly at all scales. *Science* 295: 2418–2422.

Wise, S., Maltsev, S., Davies, M. E. et al. 2007. The mineral-organic interface in bone is lined by polysaccharide. *Chem. Mat.* 19: 5055–5057.

Wong Po Foo, C., Patwardhan, S. V., Belton, D. J. et al. 2006. Novel nanocomposites from spider silk-silica fusion (chimeric) proteins. *Proc. Natl. Acad. Sci. USA* 103: 9428–9433.

Xu, M., Gratson, G. M., Duoss, E. B. et al. 2006. Biomimetic silici-fication of 3D polyamine-rich scaffolds assembled by direct ink writing. *Soft Matter* 2: 205–209.

Yokoi, T., Sakamoto, Y., Terasaki, O. et al. 2006. Periodic arrange-ment of silica nanospheres assisted by amino acids. *J. Am. Chem. Soc.* 128: 13664–13665.

Zhang, S. 2003. Fabrication of novel biomaterials through molec-ular self-assembly. *Nat. Biotechnol.* 21: 1171–1178.

Zhang, Y., Lu, J. 2008. A mild and efficient biomimetic synthesis of rodlike hydroxyapatite particles with a high aspect ratio using polyvinylpyrrolidone as capping agent. *Cryst. Growth Des.* 8: 2101–2107.

Zhou, L., O'Brien, P. 2008. Mesocrystals: A new class of solid materials. *Small* 4: 1566–1574.

Nanotubes for Biotechnology

Jonathan C. G. Jeynes
University of Surrey

Vanesa Sanz-Beltran
University of Surrey

Johnjoe McFadden
University of Surrey

S. R. P. Silva
University of Surrey

7.1 Introduction

Currently, a range of nanoparticles are being investigated for their application to life sciences, including gold nanoparticles and quantum dots. But, this chapter will primarily discuss only nanotubes. The definition of a nanotube is not fixed, although most researchers generally accept that nanotubes have a diameter from 1 to 100 nm and can be anything from 100 nm to a few microns in length (Martin and Kohli, 2003; Kohli and Martin, 2005).

Nanotubes often have unique properties owing to their nanometer size, which the same material in bulk does not possess (e.g., difference in melting point, electronic and optical properties). They also have many advantages over spherical nanoparticles, for example, a larger surface area with which to attach a range of drug and targeting modalities. Moreover, there are a range of unique possibilities that nanotubes allow, for example, filling the inside with a drug cargo while making the outside biocompatible. Furthermore, nanotubes allow possibilities that other nanoparticles cannot offer, for example, tubular membranes that can sort out proteins into different sizes, or sense different analytes in solution.

Figure 7.1 illustrates the types of applications nanotubes have in biotechnology. These include drug and gene delivery to cancer cells and tumors, scaffolds for tissue growth, antimicrobials, therapeutics, biosensing/diagnostics, and biological tools such as molecular sieving or separation. This chapter focuses mainly on these applications. The important and still controversial topic of nanotube toxicity is also briefly considered.

7.1.1 Types of Nanotubes

There are a range of nanotubes made out of different materials, including inorganic (Feldman et al., 1995; Tenne, 2006)

carbon (Iijima, 1991), organosilicon polymers (Linsky et al., 1971), self-assembling peptide sequences (Ghadiri et al., 1993; Fernandez-Lopez et al., 2001; Scanlon and Aggeli, 2008), proteins and porins, DNA (Yin et al., 2008), lipids (Yager and Schoen, 1984; Price and Patchan, 1991; Schnur, 1993; Goldstein et al., 2001; Selinger et al., 2001) and template-synthesized nanotubes (Kohli and Martin, 2005), and a number of these reviews shall be summarized in the next section.

Much research at the biotechnology nanotube interface is conducted on carbon nanotubes because of their intriguing electronic and physical properties (Dai, 2002). Carbon nanotubes (CNT) are hollow cylinders consisting of single or multiple sheets of graphene wrapped into a cylinder. They are grown either as single-walled tubes with diameters between 1 and 10 nm, or as multi-walled tubes, when they can be up to 100 nm depending on how many walls they have. Figure 7.2 shows a single-walled nanotube that has been coupled to the protein ferritin by a linker molecule (the coupling is explained in the following Section 7.1.2).

Another type of fascinating nanotube is made from self-assembling peptides. They consist of an even number of alternating D- and L-amino acids, and are illustrated in Figure 7.3. They self-assemble into tubes by spontaneously forming hydrogen bonds, and in turn, the tubes align to form parallel arrays (Ghadiri et al., 1993). The number of amino acids can determine the size of the tube; for example, a 12 amino acid ring gives a nanotube diameter of 1.3 nm (Khazanovich et al., 1994). DNA is similarly governed to produce nanotubes, with different sequences of base pairs giving different sizes of tubes (Yin et al., 2008). Lipids also self-assemble in solution, with hydrophilic and hydrophobic forces forcing microtubule-like structures with variable lengths and diameters (Schnur, 1993). These lipid-based tubes have micellar phases, where the hydrophobic

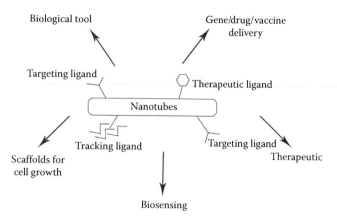

FIGURE 7.1 Schematic for the uses that nanotubes have to biotechnology after being functionalized with the appropriate ligand.

part of the molecule interacts with water, while the hydrophobic part interacts with other lipids.

7.1.2 Making Nanotubes Biocompatible

Making nanotubes compatible with a biological environment is one of the first challenges for any application to biotechnology, as many nanotubes are insoluble in water, and are naturally inert. Much research has been devoted to "functionalizing" nanotubes, which means making them chemically active. Of course, using nanotubes that are made from water-soluble molecules like peptides or DNA does circumvent this problem.

Functionalizing nanotubes can take many forms, including treating them with acids or bases, producing chemical groups like carboxylic (COOH) acids (Liu et al., 1998; Jeynes et al., 2008),

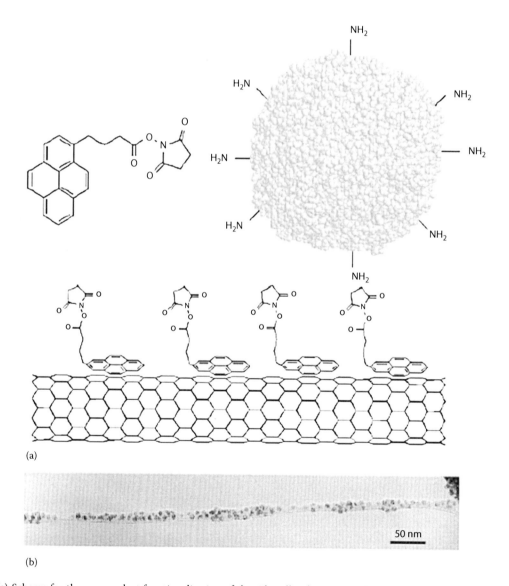

FIGURE 7.2 (a) Scheme for the noncovalent functionalization of the sidewalls of nanotubes for protein immobilization. (b) A TEM image showing ferritin immobilized on a suspended nanotube by using the scheme in (a). (Reproduced from Dai, H., *Acc. Chem. Res.*, 35, 1035, 2002. With permission.)

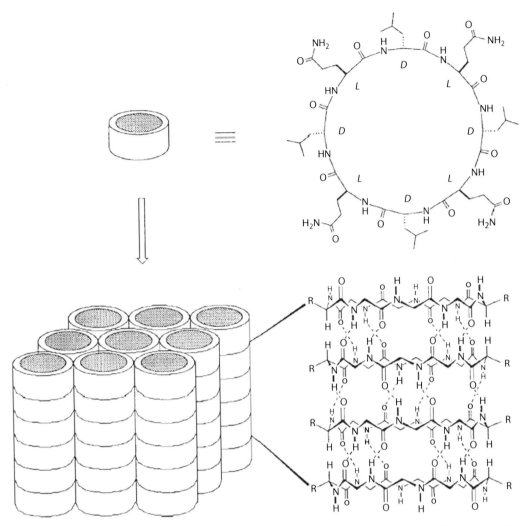

FIGURE 7.3 Cyclic peptide structures with alternating D- and L-amino acids adopting flat ring-shaped conformations and, depending on the peptide sequence and the conditions employed, assembling into ordered parallel arrays of solid-state nanotubes. The D- and L-(dextrorotary and levorotary) amino acids are optical isomers. The illustration emphasizes the antiparallel ring stacking and the presence of extensive intersubunit hydrogen-bonding interactions (for clarity most side chains are omitted). (Reproduced from Hartgerink, J.D. et al., *J. Am. Chem. Soc.*, 118, 43, 1996. With permission.)

encasing them in a surfactant (O'Connell et al., 2002), or coating them with biomolecules like DNA and proteins (Zheng et al., 2003; Jeynes et al., 2006). In Figure 7.2, a non-covalent form of functionalization is illustrated where the hydrophobic benzene rings of the linker molecule are attracted to the hydrophobic CNT and held there with overlapping π-orbitals. The succinimidyl moiety on the linker molecule can then react to the amine (NH$_2$) groups of the ferritin protein via standard coupling chemistry.

may also prove useful by either discouraging or encouraging cell growth; and there are other completely different applications ranging from novel sensing devices to new tools for molecular manipulations. This is because nanotubes are a completely new type of material that allows for new types of devices and processes to be conceived. This is a very exciting field and the reader should not be surprised if other applications emerge that are not included in this already wide-ranging list.

7.2 Application of Nanotubes

Nanotubes have been investigated to discover their compatibility and effect on mammalian cells in tissue culture (*in vitro*) as well as experiments in mice, rat and rabbit models (*in vivo*). Nanotubes may prove useful in carrying drug (or DNA/RNA) cargoes across cell membranes; or alternatively as therapeutics by utilizing their unique optical and physical properties. They

7.2.1 Drug Delivery

One major problem in cancer therapy is getting the relevant drug specifically to the cancer cell without the drug affecting healthy cells in the body. The use of nanoparticles as a means of carrying drugs into targeted cells is being researched extensively.

Carbon nanotubes are particularly good candidates for drug delivery because they enter cells naturally (Lacerda et al., 2007).

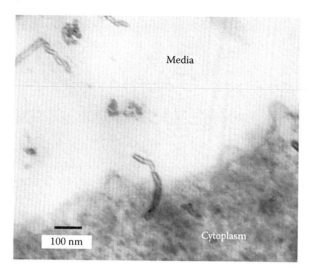

FIGURE 7.4 CNT penetrating through the cell membrane. (Reproduced from Pantarotto, D. et al., *Angew. Chem. Int. Ed.*, 43, 5242, 2004b. With permission.)

Figure 7.4 shows a nanotube as it penetrates through the cell membrane of a human cell while Figure 7.5 shows CNTs in cells using the characteristic CNT Raman spectroscopy signal. This property of cell penetration has been capitalized on by chemically modifying nanotubes to attach drugs, fluorescent trackers, and antibodies. We shall discuss a few examples in detail.

There is some debate over the exact uptake mechanism of nanotubes, with some proposing endocytosis (where the cell takes up particles through an active process) (Kam and Dai, 2005; Kam et al., 2006), and others proposing mechanical injection (Pantarotto et al., 2004a; Kostarelos et al., 2007). The most likely explanation is that tubes prepared in different ways are taken up in different manners according to their sizes and chemical functionalities.

A good example of drug uptake is the use of amphotericin B, which is an antifungal agent but can be toxic to mammalian cells due to poor water solubility and aggregation. This drug was covalently linked to MWNTs along with a fluorescent marker for tracking. The CNTs were taken up into mammalian cells without any toxicity while at the same time they retained potent antifungal activity (Wu et al., 2005). Similarly, anticancer drugs like methotrexate and *cis*-platin are shown to be more effective when attached to CNTs (Pastorin et al., 2006; Feazell et al., 2007).

7.2.2 Gene Delivery

Another relatively new innovation is short interfering strands of RNA (siRNA), which can switch off genes that have mutagenic consequences (Shen, 2008). Similarly, the use of DNA to vaccinate against a variety of diseases is being intensively investigated (Lu et al., 2008). However, the uptake of DNA/RNA into cells is low because it is not transported across the membrane. Ordinarily in tissue culture, cells are either electroporated or treated with lipofectin to create pores in the membrane through which DNA plasmids can pass. Attaching DNA/RNA to nanotubes is an effective method of getting the nucleic acids into cells because of their cell-penetrating properties. Getting DNA into the cells of animals is much more problematic as the circulatory system quickly clears injected molecules from the body. An efficient delivery vector is therefore needed, especially as other vector approaches such as viruses can be dangerous. This was proved when a patient died after being administered a virus vaccine in 1999 (Somia and Verma, 2000; Zallen, 2000).

Much research has focused on using nanotubes as vectors, and for a good review of carbon nanotubes see Lacerda et al. (2008). Plasmids expressing marker proteins (e.g., green fluorescent protein) have been attached to CNTs and transfected into cells giving similar or better levels of expression compared to conventional techniques using liposomes (Pantarotto et al., 2004b; Singh et al., 2005). Gold nanowires have also been used to transfect cells with DNA (Kuo et al., 2008) while peptide nanotubes have been used to transport single-stranded DNA into cells (Yan et al., 2007).

A good example is siRNA attached to CNTs, which prevented the transcription of green fluorescence protein once

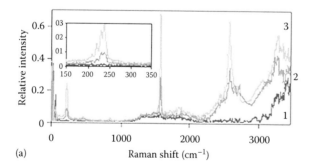

(a)

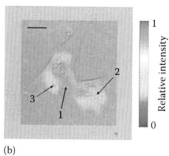

(b)

FIGURE 7.5 Spectra and corresponding Raman intensity area maps of live 3T3 fibroblast and myoblast stem cells. (a) Combined Raman and fluorescence spectra of live murine 3T3 cells incubated with DNA-suspended SWNTs. The three spectra correspond to locations on the area map. Absence of a $267\,cm^{-1}$ Raman peak suggests minimal SWNT–SWNT contact (inset). (b) Area map of Raman radial breathing mode (between 200 and $300\,cm^{-1}$) intensity of nanotubes in live 3T3 cells after 48 h in culture overlaid onto an optical micrograph of the same region. (Reproduced from Heller, D.A. et al., *Adv. Mater.*, 17, 2793, 2005. With permission.)

they were internalized into cells (Kam et al., 2005a). Similarly, the delivery of siRNA to human T cells by CNTs silenced the expression of the HIV-specific cell-surface receptor, so these cells were less susceptible to the virus (Liu et al., 2007b). The effect was far enhanced compared to controls because of the CNTs.

7.2.3 Use of Nanotubes as Therapeutic Agents

Nanoparticles and nanotubes have physical and optical properties that are utilized so that the particle itself becomes a therapeutic agent. One good example is the use of Near Infra Red (NIR) radiation to heat particles once they are in cancer cells. Normal cells and indeed, animal tissue are relatively transparent to NIR radiation whereas nanotubes or nanoparticles can be strong NIR absorbers. Cells which have encapsulated nanoparticles heat up when exposed to NIR radiation and die, a process called hyperthermia or photothermal cancer treatment. For a good overview of this application using gold nanoparticles see Jain et al. (2007).

Cancer cells that internalize bio-functionalized CNTs can be selectively killed by exposure to NIR radiation (using an 800 nm laser) (Kam et al., 2005b). Figure 7.6 shows CNTs solubilized in water by a phospholipid functionalized with folate. These functionalized CNTs enter cancer cells which naturally over-express folate receptors. Very short pulses of the laser did not kill cells with internalized CNTs, but prolonged exposure (2 min) caused cell death. This was seen immediately after exposure, demonstrated by changes to the morphology of adherent cells from stretch to rounded, with an eventual detachment from the substrate. Similar work was demonstrated by using antibodies attached to CNTs which targeted specific cancer cell types (Chakravarty et al., 2008).

Presumably, local heating of the cell is sufficient to cause disruption to its structural integrity, leading to rapid cell death. Indeed, researchers using light microscopy, showed bubbles forming around clusters of gold nanoparticles in human leukemia cells as they were heated by a laser (Lapotko et al., 2006). As the bubble expanded, it mechanically disrupted the cell membrane resulting in cell death. Other work using gold nanorods showed that cell death was increased 10-fold when the rods were attached to the membrane rather than internalized in the cell, as microbubbles resulting from cavitation caused membrane rupture (Tong et al., 2007).

NIR irradiation of nanotubes opens the route for targeting tumors in cancer patients treated with nanotubes. Indeed, gold nanoshells injected into tumors in mice absorb NIR radiation and can shrink tumors (Hirsch et al., 2003).

Moreover, it has also been shown that the absorbance of a Radio Frequency (RF) field by CNTs also results in the emission of heat. It is mainly the metallic CNTs which absorb the RF as metallic materials absorb long wavelengths. In tissue culture experiments, cells incubated with CNTs and then exposed to RF were killed. Similarly, tumors induced in rabbits shrunk after they were injected with CNTs and then exposed to RF (Gannon et al., 2007).

Nanotubes have many other types of therapeutic properties. For example, peptide nanotubes have shown antibiotic behavior, selectively killing both gram negative and positive bacteria compared to mammalian cells. It is thought that the peptides lodge themselves in the walls of the bacteria, causing rapid cell death (Fernandez-Lopez et al., 2001).

In terms of therapeutic applications *in vivo*, CNTs functionalized with tumor markers can target tumor cells when injected intravenously into mice (Liu et al., 2007a). Indeed, CNTs can deliver a vaccine to the foot and mouth disease virus and elicit a strong anti-peptide antibody in mice (Pantarotto et al., 2003). Moreover, out of a range of nanoparticles, CNTs offered the best delivery of the drug erythropoietin (EPO) when injected into the intra-small intestines of rats (Venkatesan et al., 2005). A follow up study showed that short CNTs were far more effective delivery vehicles than long CNTs, presumably because they were more efficiently distributed in the body and taken up by cells (Ito et al., 2007).

7.2.4 Using Nanotubes as Molecular Tools or Scaffolds

Nanotubes can be used as tools with which to manipulate or sort biological molecules as well as being used as scaffolds on which to grow and monitor mammalian cells.

Membranes containing gold nanotubes with diameters of 1 nm can separate small molecules according to size (Jirage et al., 1997) while membranes containing larger holes and functionalized with antibodies can separate proteins (Yu et al., 2001). Indeed, membrane nanotubes can be used as ion-based sensors. By measuring the transmembrane ion current between two solutions separated by a nanotube membrane, the flow of analytes between the solutions can be electrically measured. These devices can measure analyte concentration, and can even be switched on or off through the presence of a particular molecule or drug that effectively blocks the membrane-to-solution flow (Kang and Martin, 2001; Martin et al., 2001; Steinle et al., 2002). Perhaps the most elegant ion channel nanotube is that found in nature itself; on the inside of tubular proteins such as α-hemolysin. This protein spans a membrane so that when analytes (from metal ions to small molecules) pass through the pore in the middle, a drop in the membrane potential gives an accurate measure of the concentration or identity of analyte (Sanchez-Quesada et al., 2000; Ashkenasy et al., 2005).

Nanotubes can also enhance the growth of cells or strengthen a tissue scaffold matrix (Harrison and Atala, 2007). For example, carbon nanotubes have been used to strengthen polymers like chitosan, which are biodegradable materials used to stimulate growth of tissue encouraging, for example, the regrowth of bone (Wang et al., 2005b; Boccaccini et al., 2007; Misra et al., 2007; Shi et al., 2007). Carbon nanotube substrates can also increase the proliferation of osteoblast (bone) cells (Supronowicz et al., 2002; Zanello et al., 2006) as well as stimulate greater electrical

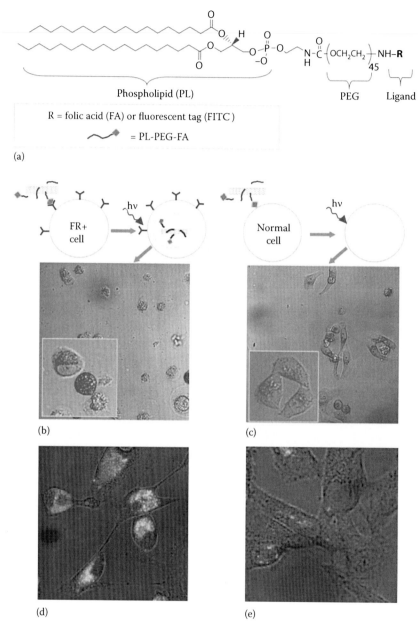

(a)

(b)

(c)

(d)

(e)

FIGURE 7.6 Selective targeting and killing of cancer cells. (a) Chemical structure of PL-PEG-FA and PL-PEG-FITC synthesized by conjugating PL-PEG-NH$_2$ with FA or FITC, respectively, for solubilizing individual SWNTs. (PEG = poly ethylene glycol, FITC = fluorescein isothiocyanate). (b) (Upper) Schematic of selective internalization of PL-PEG-FA-SWNTs into folate over expressing (FR+) cells via receptor binding and then NIR 808-nm laser radiation. (Lower) Image showing death of FR+ cells with rounded cell morphology after the process in Upper (808-nm laser radiation at 1.4 W cm^2 for 2 min). (Inset) Higher magnification image shows details of the killed cells. (c) (Upper) Schematic of no internalization of PL-PEG-FA-SWNTs into normal cells without available FRs. (Lower) Image showing normal cells with no internalized SWNTs that are unharmed by the same laser radiation condition as in (b). (Inset) Higher magnification image shows a live normal cell in stretched shape. (d) Confocal image of FR+ cells after incubation in a solution of SWNTs with two cargoes (PL-PEG-FA and PL-PEG-FITC). The strong green FITC fluorescence inside cells confirms the SWNT uptake with FA and FITC cargoes. (e) The same as d for normal cells without abundant FRs on cell surfaces. There is little green fluorescence inside cells, confirming little uptake of SWNTs with FA and FITC cargoes. (Magnifications: 20×). (Reproduced from Kam, N.W.S. et al., *Proc. Natl. Acad. Sci. U.S.A.*, 102, 11600, 2005b. With permission.)

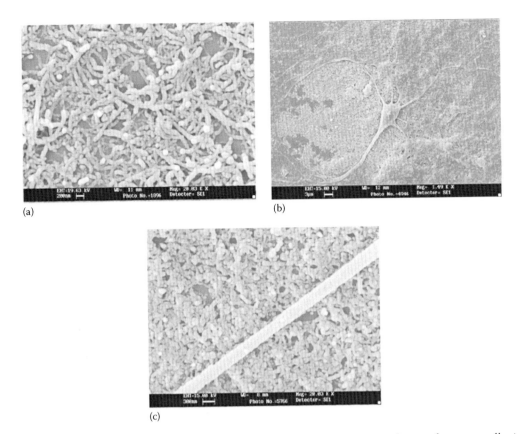

(a)

(b)

(c)

FIGURE 7.7 Purified multiwalled carbon nanotubes (MWNT) layered on glass are permissive substrates for neuron adhesion and survival. (a) Micrographs taken by the scanning electron microscope showing the retention on glass of MWNT films after an 8-day test in culturing conditions. (b) Neonatal hippocampal neuron growing on dispersed MWNT after 8 days in culture. The surface structure, composed of films of MWNT and peptide-free glass, allows neuron adhesion. Dendrites and axons extend across MWNT, glia cells, and glass. The relationship between dendrite and MWNT is very clear in the image in (c), were a neurite is traveling in close contact to carbon nanotubes (From Zanello, L.P. et al., *Nano Lett.* 6, 562, 2006. With permission.)

activity in neurones (Lovat et al., 2005). Figure 7.7 shows osteoblast cells proliferating on a CNT substrate. Titanium nanotubes can increase blood-clotting time and are being investigated for potential uses in bandages (Roy et al., 2007).

7.2.5 Biosensing and Diagnostics Using Nanotubes

Nanotubes have particular potential for biosensing as they can be readily coupled to biomolecules owing to their size and chemistry. Moreover, many types of nanotubes have conductivity and electrical properties that allow sensitive detection of biomolecules in real time. Furthermore, miniaturization coupled with microfluidic devices, allows many analytes to be monitored simultaneously. This opens the possibility of bedside devices which could rapidly diagnose disease rather than having to send samples to a laboratory.

Nanotubes can detect electrochemical signals from biomolecules such as the redox reaction that occurs when glucose is broken down by the enzyme glucose oxidase. The advantage of nanotubes over other detection systems is that electrons resulting from a reaction can be coupled directly to the nanotube, rather than a mediator molecule, increasing sensitivity. Moreover, nanotubes can be deposited or grown between electrodes in a field effect transistor (FET) configuration allowing real time

detection of molecules without the need for biochemical labeling to amplify signal. Carbon nanotubes are particularly appealing for their use as biosensors because of their natural conductance properties. A good review of this area is by Kim et al. (2007).

Figure 7.8 illustrates the two basic biosensing capabilities that nanotubes possess; transistor-based and electrochemical-based biosensing. On the left-hand side of Figure 7.8, the basic configuration of FET device-based biosensors is shown. Here, a semiconducting material is connected to the source and the drain electrode through which current is injected and collected, respectively, while the conductance is measured through the gate electrode.

Much success has been achieved using silicon nanotubes FET devices. These are good biosensors because they have a high surface to volume ratio, and can be doped p or n type using conventional techniques (Patolsky et al., 2006b). Indeed, real time detection of a variety of molecules can be achieved using silicon nanowire FET devices, including DNA hybridization (Hahm and Lieber, 2004), multiplex cancer detection (Zheng et al., 2005), various other small molecule–protein interactions (Patolsky et al., 2006c), chemical species (Cui et al., 2001), as well as single virus detection (Patolsky et al., 2004b). In another innovative experiment, nerve cells grown across a series of FETs had the action potential tracked down the axion (Patolsky et al., 2006a).

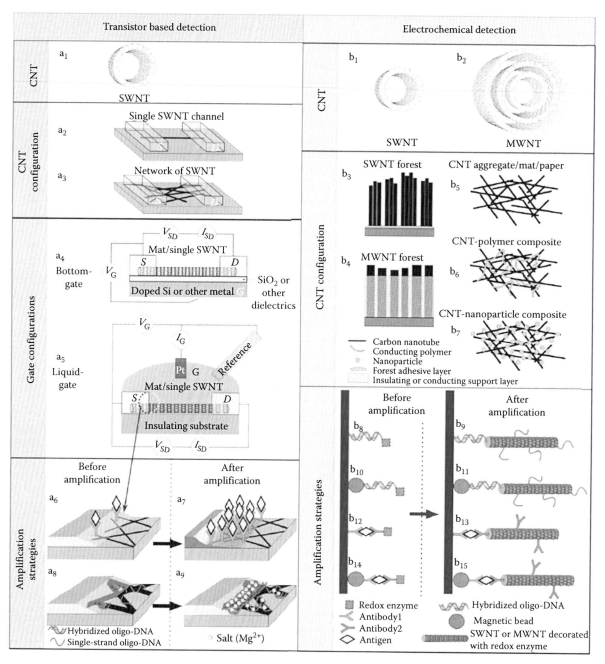

FIGURE 7.8 Schematic representation of biomolecular sensing using carbon nanotubes in various device configuration and signal amplification strategies. (Reproduced from Kim, S.N. et al., *Adv. Mater.*, 19, 3214, 2007. With permission.)

A review of this topic can be found in Patolsky et al. (2006c). Figure 7.9 shows the electrical detection and optical images of single viruses as they adsorb and then desorb away from the silicon nanowire FET.

Carbon nanotubes (CNTs) are also used as FET biosensors, detecting a variety of protein, and antibody binding events (Chen et al., 2003), DNA hybridization (Star et al., 2006; Tang et al., 2006), and thrombin detection using aptamers (a short single-stranded piece of DNA that binds proteins) (So et al., 2005). The mechanism of electrical sensing differs though from

silicon FET devices in that much of the sensing area comes from the Schottky barrier interface between the electrode and the CNT (Chen et al., 2004), rather than along the length of the nanotube. The Schottky barrier was utilized to construct a CNT device with electrodes made by angled-deposition to give maximum contact area between the electrodes and the CNTs (Byon and Choi, 2006).

On the right-hand side of Figure 7.8, common approaches to create electrochemical biosensors (Katz and Willner, 2004; Gooding, 2005) are shown. These differ from FETs in their

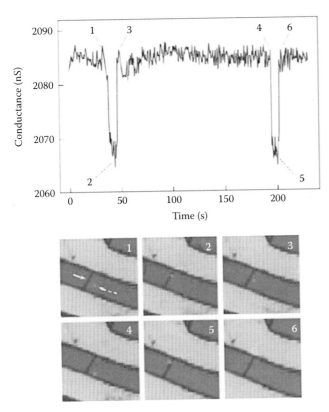

FIGURE 7.9 Conductance (upper) and optical (lower) data recorded simultaneously vs. time for a single silicon nanowire device after the introduction of influenza A solution. Combined bright-field and fluorescence images correspond to time points 1–6 indicated in the conductance data; virus appears as a dot in the images. The solid arrow in image 1 highlights the position of the nanowire device, and the dashed arrow indicates the position of a single virus. (Reproduced from Patolsky, F. et al., *Proc. Natl. Acad. Sci. U.S.A.*, 101, 14017, 2004b. With permission.)

sensing mechanism. They can be made by depositing CNTs in a mat making an electrode where reversible cyclic voltammetry is monitored. The first report of this approach was to follow the redox of cytochrome *c* and blue copper protein azurin adsorbed onto nanotubes (Davis et al., 1997). Another approach is to grow forests of nanotubes, attach biomolecules covalently and monitor electrochemical reactions. A good example of this is glucose oxidase (GOx) where the redox cofactor (flavin adenine dinucleotide, FAD) is embedded deep in the enzyme tertiary structure. In conventional biosensing devices, mediators are used to exchange electrons with a flat electrode, but by using a CNT forest, the electron transfer rate is greatly enhanced by coupling GOx or FAD directly to nanotubes (Patolsky et al., 2004a; Withey et al., 2006). Similar approaches can detect DNA hybridization (Katz and Willner, 2004; Wang, 2005) or antigens (Warsinke et al., 2000) using a variety of electrochemically active labels such as alkaline phosphatase or horse radish peroxidise.

CNTs can also sense the electrical activity of brainwaves. By using a forest of vertically aligned CNTs attached to an electrode substrate, the electrophysiology of the brain can be monitored directly without using conventional conducting gels (Ruffini et al., 2008).

7.3 Toxicity of Nanotubes

There is debate over whether or not nanotubes are in fact toxic to cells. Much of the research has been based on CNTs, which are suspected of having possible toxic effects due to their insolubility and physical properties. There are many reports *in vivo* and *in vitro* with mixed findings.

Exploring toxicity in human cell culture, one team found that after a number of days growth with carbon material, cell viability was less than with the controls. A more pronounced effect was found with larger carbon-based materials in this order; carbon black (most toxic), carbon fibers and then MWNTs (Magrez et al., 2006). Moreover, when the MWNTs had functional groups, they became more toxic, lending support to the hypothesis that larger carbon materials are more toxic because of the oxidative species (functional groups) that cover the surface. Other researchers found toxicity with human epidermal (skin) cells in a dose-dependent manner (Shvedova et al., 2003). However, a number of other groups have looked into the cytotoxicity of CNTs *in vitro* and have concluded that they are not toxic when using them to delivery therapeutic agents (see discussion in Sections 7.2.1 and 7.2.2). It has also been noted that the specific type of toxicity test performed can influence the result as CNTs can react with some cytotoxicity assay reagents (Worle-Knirsch et al., 2006).

There have been a number of studies performed *in vivo* (for comprehensive reviews see Donaldson et al., 2006; Lacerda et al., 2006; Lam et al., 2006). One study showed that lung lesions and granulomas (microscopic nodules) were produced in rat's lungs after they had been instilled (a process similar to injection) with 99% pure SWNT, that is cleared of all metal contaminants (Warheit et al., 2004), and other studies with MWNT on rats showed similar results (Muller et al., 2005; Shvedova et al., 2005). Indeed, 0.1 or 0.5 mg per mouse of SWNT material produced by different methods and instilled in the intratracheal space, produced similar granulomas in the lung tissue (Lam et al., 2004; Shvedova et al., 2005). Indeed, pulmonary exposure to CNTs also has a negative effect on other parts of the body, in particular the heart (Li et al., 2007). The reasons for apparent CNT toxicity have included their oxidative potential, and their fibrous nature as they agglomerate into large bundles due to van der Waals forces. Indeed, one study showed that very long MWNT have asbestos like qualities (Poland et al., 2008). Figure 7.10 shows a macrophage failing to ingest such a MWNT. Interestingly, smaller tubes are easily ingested so the toxicity of CNTs is very likely to be dependent on the particular size range and production process (i.e., type of contaminates). It is often assumed that the toxicity of nanoparticles is dependent on the overall surface area and surface chemistry. However, this is not always true as the toxicity of TiO_2 nanorods are not dependent on the size or the surface of the nanoparticle. They can be less toxic than quartz when used as a positive control to elicit inflammation and granulomas in rats' lungs (Warheit et al., 2006). So the whole story of nanotube toxicity is likely to be complicated with exceptions to the rule.

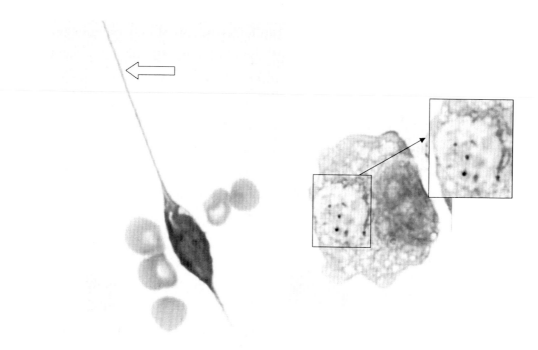

FIGURE 7.10 Effect of fiber length on phagocytosis by peritoneal macrophages. Like long-fiber amosite (asbestos), long MWNTs also lead to frustrated phagocytosis as shown by the open arrow (E-erythrocytes). In contrast, short MWNTs can be readily phagocytosed (see inset). (From Poland, C.A. et al., *Nat. Nanotechnol.*, 3, 423, 2008. With permission.)

There have been a number of studies *in vivo* which have assessed the clearance of CNTs from the body of rats or mice, and looked at the possible therapeutic potential. One report showed that Iodine (^{125}I)-labeled SWNT, which was injected into mice in a variety of places (e.g., stomach, interperitoneal), was cleared from the body rapidly with no tissue damage or distress reported (Wang et al., 2004). This is corroborated by another report that shows that the half-life of the CNTs in the body is about 30 min (Singh et al., 2006). Other reports show PEGylated (see Figure 7.6 for an example of a PEGylated CNT) CNTs have a longer blood circulation time than pristine CNTs and greater tumor uptake (Yang et al., 2007, 2008). Other workers used 2-photon luminescence to visualize injected gold nanorods in the ear blood vessels of mice, showing the extent of circulation round the body (Wang et al., 2005a).

7.4 Conclusions and Future Perspectives

Nanotube technology has come a long way in terms of synthesis, preparation and application to biotechnology. We have seen that there are many areas of biotechnology research where nanotubes play a vital role, including biosensing, drug delivery, therapeutics, molecule separation, and tissue scaffolding. It is clear that without nanotubes, this field would be severely limited in scope and many avenues of research would be closed.

In terms of which types of nanotubes are most suitable for biotechnology, it seems that there are opportunities to use different types for different applications. However, it does seem that much research has focused on carbon nanotubes, perhaps because of the intense interest this new type of material generated in the late 1990s. These nanotubes are truly remarkable, and as we have seen in the chapter, have a broad scope of applications. It is likely that in the next decade we shall see them entering the arena of healthcare, particularly in drug delivery and biosensing.

Perhaps, one area that could be capitalized on more is that of filling nanotubes with molecules or drugs. The potential to fill and then cap nanotubes, while designing the cap to fall off when it enters a target cell and then to deliver of the drug, is an intriguing possibility that could offer very interesting results.

Other nanotubes have much potential, particularly peptide nanotubes. These have the inherent quality of being modeled by a computer so that the desired chemical groups can be positioned on the outside and on the inside of the tube. This area of research is in its infancy but there is much potential to produce designer tubes for specific functions (i.e., binding of a particular drug).

Before this occurs though, more research needs to be performed on the toxicity of nanotubes as there have been conflicting reports. It seems that the preparation of nanotubes is very important with the size, the length and the surface chemistry being vital factors in determining toxicity levels. It is not clear which is the safest nanotube to use, whether it is gold, carbon or an organic type. Any material however inert it may seem, can take on a different nature in nanometer dimensions and can have unexpected detrimental effects on cells.

Clearly, in any healthcare setting, the preparation of nanotube suspensions would have to be standardized, a task which is no small feat. Moreover, the cost of production would have to be reduced as most nanotube types are produced in small quantities (mg per batch), and scalability of production to industrial quantities would have to be investigated.

Overall, nanotubes are a fascinating and novel class of materials which have improved areas of biotechnology. Nanotubes will almost certainly have an integral part to play in healthcare and biotechnology within the coming decade and the future of this area of research is very exciting.

References

Ashkenasy, N., Sanchez-Quesada, J., Bayley, H., and Ghadiri, M. R. (2005) Recognizing a single base in an individual DNA strand: A step toward DNA sequencing in nanopores. *Angewandte Chemie-International Edition*, 44, 1401–1404.

Boccaccini, A. R., Chicatun, F., Cho, J., Bretcanu, O., Roether, J. A., Novak, S., and Chen, Q. (2007) Carbon nanotube coatings on Bioglass-based tissue engineering scaffolds. *Advanced Functional Materials*, 17, 2815–2822.

Byon, H. R. and Choi, H. C. (2006) Network single-walled carbon nanotube-field effect transistors (SWNT-FETs) with increased Schottky contact area for highly sensitive biosensor applications. *Journal of the American Chemical Society*, 128, 2188–2189.

Chakravarty, P., Marches, R., Zimmerman, N. S., Swafford, A. D. E., Bajaj, P., Musselman, I. H., Pantano, P., Draper, R. K., and Vitetta, E. S. (2008) Thermal ablation of tumor cells with anti body-functionalized single-walled carbon nanotubes. *Proceedings of the National Academy of Sciences of the United States of America*, 105, 8697–8702.

Chen, R. J., Bangsaruntip, S., Drouvalakis, K. A., Kam, N. W. S., Shim, M., Li, Y. M., Kim, W., Utz, P. J., and Dai, H. J. (2003) Noncovalent functionalization of carbon nanotubes for highly specific electronic biosensors. *Proceedings of the National Academy of Sciences of the United States of America*, 100, 4984–4989.

Chen, R. J., Choi, H. C., Bangsaruntip, S., Yenilmez, E., Tang, X. W., Wang, Q., Chang, Y. L., and Dai, H. J. (2004) An investigation of the mechanisms of electronic sensing of protein adsorption on carbon nanotube devices. *Journal of the American Chemical Society*, 126, 1563–1568.

Cui, Y., Wei, Q. Q., Park, H. K., and Lieber, C. M. (2001) Nanowire nanosensors for highly sensitive and selective detection of biological and chemical species. *Science*, 293, 1289–1292.

Dai, H. (2002) Carbon nanotubes: Synthesis, integration and properties. *Accounts of Chemical Research*, 35, 1035–1044.

Davis, J. J., Coles, R. J., and Hill, H. A. O. (1997) Protein electrochemistry at carbon nanotube electrodes. *Journal of Electroanalytical Chemistry*, 440, 279–282.

Donaldson, K., Aitken, R., Tran, L., Stone, V., Duffin, R., Forrest, G., and Alexander, A. (2006) Carbon nanotubes: A review of their properties in relation to pulmonary toxicology and workplace safety. *Toxicological Sciences*, 92, 5–22.

Feazell, R. P., Nakayama-Ratchford, N., Dai, H., and Lippard, S. J. (2007) Soluble single-walled carbon nanotubes as longboat delivery systems for platinum(IV) anticancer drug design. *Journal of the American Chemical Society*, 129, 8438–8439.

Feldman, Y., Wasserman, E., Srolovitz, D. J., and Tenne, R. (1995) High-rate, gas-phase growth of Mos2 nested inorganic fullerenes and nanotubes. *Science*, 267, 222–225.

Fernandez-Lopez, S., Kim, H. S., Choi, E. C., Delgado, M., Granja, J. R., Khasanov, A., Kraehenbuehl, K. et al. (2001) Antibacterial agents based on the cyclic D,L-alpha-peptide architecture. *Nature*, 412, 452–455.

Gannon, C. J., Cherukuri, P., Yakobson, B. I., Cognet, L., Kanzius, J. S., Kittrell, C., Weisman, R. B. et al. (2007) Carbon nanotube-enhanced thermal destruction of cancer cells in a noninvasive radiofrequency field. *Cancer*, 110, 2654–2665.

Ghadiri, M. R., Granja, J. R., Milligan, R. A., Mcree, D. E., and Khazanovich, N. (1993) Self-assembling organic nanotubes based on a cyclic peptide architecture. *Nature*, 366, 324–327.

Goldstein, A. S., Amory, J. K., Martin, S. M., Vernon, C., Matsumoto, A., and Yager, P. (2001) Testosterone delivery using glutamide-based complex high axial ratio microstructures. *Bioorganic and Medicinal Chemistry*, 9, 2819–2825.

Gooding, J. J. (2005) Nanostructuring electrodes with carbon nanotubes: A review on electrochemistry and applications for sensing. *Electrochimica Acta*, 50, 3049–3060.

Hahm, J. and Lieber, C. M. (2004) Direct ultrasensitive electrical detection of DNA and DNA sequence variations using nanowire nanosensors. *Nano Letters*, 4, 51–54.

Harrison, B. S. and Atala, A. (2007) Carbon nanotube applications for tissue engineering. *Biomaterials*, 28, 344–353.

Hartgerink, J. D., Granja, J. R., Milligan, R. A., and Ghadiri, M. R. (1996) Self-assembling peptide nanotubes. *Journal of the American Chemical Society*, 118, 43–50.

Heller, D. A., Baik, S., Eurell, T. E., and Strano, M. S. (2005) Single-walled carbon nanotube spectroscopy in live cells: Towards long-term labels and optical sensors. *Advanced Materials*, 17, 2793–2799.

Hirsch, L. R., Stafford, R. J., Bankson, J. A., Sershen, S. R., Rivera, B., Price, R. E., Hazle, J. D., Halas, N. J., and West, J. L. (2003) Nanoshell-mediated near-infrared thermal therapy of tumors under magnetic resonance guidance. *Proceedings of the National Academy of Sciences of the United States of America*, 100, 13549–13554.

Iijima, S. (1991) Helical microtubules of graphitic carbon. *Nature*, 354, 56–58.

Ito, Y., Venkatesan, N., Hirako, N., Sugioka, N., and Takada, K. (2007) Effect of fiber length of carbon nanotubes on the absorption of erythropoietin from rat small intestine. *International Journal of Pharmaceutics*, 337, 357–360.

Jain, P. K., El-Sayed, I. H., and El-Sayed, M. A. (2007) Au nanoparticles target cancer. *Nano Today*, 2, 18–29.

Jeynes, J. C. G., Jeynes, C., Kirkby, K. J., Rummeli, A., and Silva, S. R. P. (2008) RBS/EBS/PIXE measurement of single-walled carbon nanotube modification by nitric acid purification treatment. *Nuclear Instruments and Methods in Physics Research Section B-Beam Interactions with Materials and Atoms*, 266, 1569–1573.

Jeynes, J. C. G., Mendoza, E., Chow, D. C. S., Watts, P. C. R., Mcfadden, J., and Silva, S. R. P. (2006) Generation of chemically unmodified pure single-walled carbon nanotubes by solubilizing with RNA and treatment with ribonuclease A. *Advanced Materials*, 18, 1598–1602.

Jirage, K. B., Hulteen, J. C., and Martin, C. R. (1997) Nanotubule-based molecular-filtration membranes. *Science*, 278, 655–658.

Kam, N. W. S. and Dai, H. J. (2005) Carbon nanotubes as intracellular protein transporters: Generality and biological functionality. *Journal of the American Chemical Society*, 127, 6021–6026.

Kam, N. W. S., Liu, Z., and Dai, H. J. (2005a) Functionalization of carbon nanotubes via cleavable disulfide bonds for efficient intracellular delivery of siRNA and potent gene silencing. *Journal of the American Chemical Society*, 127, 12492–12493.

Kam, N. W. S., Liu, Z. A., and Dai, H. J. (2006) Carbon nanotubes as intracellular transporters for proteins and DNA: An investigation of the uptake mechanism and pathway. *Angewandte Chemie-International Edition*, 45, 577–581.

Kam, N. W. S., O'connell, M., Wisdom, J. A., and Dai, H. J. (2005b) Carbon nanotubes as multifunctional biological transporters and near-infrared agents for selective cancer cell destruction. *Proceedings of the National Academy of Sciences of the United States of America*, 102, 11600–11605.

Kang, M. S. and Martin, C. R. (2001) Investigations of potential-dependent fluxes of ionic permeates in gold nanotubule membranes prepared via the template method. *Langmuir*, 17, 2753–2759.

Katz, E. and Willner, I. (2004) Biomolecule-functionalized carbon nanotubes: Applications in nanobioelectronics. *Chemphyschem*, 5, 1085–1104.

Khazanovich, N., Granja, J. R., Mcree, D. E., Milligan, R. A., and Ghadiri, M. R. (1994) Nanoscale tubular ensembles with specified internal diameters—Design of a self-assembled nanotube with a 13-Angstrom pore. *Journal of the American Chemical Society*, 116, 6011–6012.

Kim, S. N., Rusling, J. F., and Papadimitrakopoulos, F. (2007) Carbon nanotubes for electronic and electrochemical detection of biomolecules. *Advanced Materials*, 19, 3214–3228.

Kohli, P. and Martin, C. R. (2005) Template-synthesized nanotubes for biotechnology and biomedical applications. *Journal of Drug Delivery Science and Technology*, 15, 49–57.

Kostarelos, K., Lacerda, L., Pastorin, G., Wu, W., Wieckowski, S., Luangsivilay, J., Godefroy, S. et al. (2007) Cellular uptake of functionalized carbon nanotubes is independent of functional group and cell type. *Nature Nanotechnology*, 2, 108–113.

Kuo, C. W., Lai, J. J., Wei, K. H., and Chen, P. (2008) Surface modified gold nanowires for mammalian cell transfection. *Nanotechnology*, 19, 7.

Lacerda, L., Bianco, A., Prato, M., and Kostarelos, K. (2006) Carbon nanotubes as nanomedicines: From toxicology to pharmacology. *Advanced Drug Delivery Reviews*, 58, 1460–1470.

Lacerda, L., Bianco, A., Prato, M., and Kostarelos, K. (2008) Carbon nanotube cell translocation and delivery of nucleic acids in vitro and in vivo. *Journal of Materials Chemistry*, 18, 17–22.

Lacerda, L., Raffa, S., Prato, M., Bianco, A., and Kostarelos, K. (2007) Cell-penetrating CNTs for delivery of therapeutics. *Nano Today*, 2, 38–43.

Lam, C. W., James, J. T., Mccluskey, R., and Hunter, R. L. (2004) Pulmonary toxicity of single-wall carbon nanotubes in mice 7 and 90 days after intratracheal instillation. *Toxicological Sciences*, 77, 126–134.

Lam, C. W., James, J. T., Mccluskey, R., Arepalli, S., and Hunter, R. L. (2006) A review of carbon nanotube toxicity and assessment of potential occupational and environmental health risks. *Critical Reviews in Toxicology*, 36, 189–217.

Lapotko, D. O., Lukianova, E., and Oraevsky, A. A. (2006) Selective laser nano-thermolysis of human leukemia cells with microbubbles generated around clusters of gold nanoparticles. *Lasers in Surgery and Medicine*, 38, 631–642.

Li, Z., Hulderman, T., Salmen, R., Chapman, R., Leonard, S. S., Young, S. H., Shvedova, A., Luster, M. I., and Simeonova, P. P. (2007) Cardiovascular effects of pulmonary exposure to single-wall carbon nanotubes. *Environmental Health Perspectives*, 115, 377–382.

Linsky, J. P., Paul, T. R., and Kenney, M. E. (1971) Planar organosilicon polymers. *Journal of Polymer Science Part a-2: Polymer Physics*, 9, 143–160.

Liu, J., Rinzler, A. G., Dai, H. J., Hafner, J. H., Bradley, R. K., Boul, P. J., Lu, A. et al. (1998) Fullerene pipes. *Science*, 280, 1253–1256.

Liu, Z., Cai, W. B., He, L. N., Nakayama, N., Chen, K., Sun, X. M., Chen, X. Y., and Dai, H. J. (2007a) In vivo biodistribution and highly efficient tumour targeting of carbon nanotubes in mice. *Nature Nanotechnology*, 2, 47–52.

Liu, Z., Winters, M., Holodniy, M., and Dai, H. J. (2007b) siRNA delivery into human T cells and primary cells with carbon-nanotube transporters. *Angewandte Chemie-International Edition*, 46, 2023–2027.

Lovat, V., Pantarotto, D., Lagostena, L., Cacciari, B., Grandolfo, M., Righi, M., Spalluto, G., Prato, M., and Ballerini, L. (2005) Carbon nanotube substrates boost neuronal electrical signaling. *Nano Letters*, 5, 1107–1110.

Lu, S., Wang, S. X., and Grimes-Serrano, J. M. (2008) Current progress of DNA vaccine studies in humans. *Expert Review of Vaccines*, 7, 175–191.

Magrez, A., Kasas, S., Salicio, V., Pasquier, N., Seo, J. W., Celio, M., Catsicas, S., Schwaller, B., and Forro, L. (2006) Cellular toxicity of carbon-based nanomaterials. *Nano Letters*, 6, 1121–1125.

Martin, C. R. and Kohli, P. (2003) The emerging field of nanotube biotechnology. *Nature Reviews Drug Discovery*, 2, 29–37.

Martin, C. R., Nishizawa, M., Jirage, K., Kang, M. S., and Lee, S. B. (2001) Controlling ion-transport selectivity in gold nanotubule membranes. *Advanced Materials*, 13, 1351–1362.

Misra, S. K., Watts, P. C. P., Valappil, S. P., Silva, S. R. P., Roy, I., and Boccaccini, A. R. (2007) Poly(3-hydroxybutyrate)/ Bioglass (R) composite films containing carbon nanotubes. *Nanotechnology*, 18, 075701.

Muller, J., Huaux, F., Moreau, N., Misson, P., Heilier, J. F., Delos, M., Arras, M., Fonseca, A., Nagy, J. B., and Lison, D. (2005) Respiratory toxicity of multi-wall carbon nanotubes. *Toxicology and Applied Pharmacology*, 207, 221–231.

O'connell, M. J., Bachilo, S. M., Huffman, C. B., Moore, V. C., Strano, M. S., Haroz, E. H., Rialon, K. L. et al. (2002) Band gap fluorescence from individual single-walled carbon nanotubes. *Science*, 297, 593–596.

Pantarotto, D., Briand, J. P., Prato, M., and Bianco, A. (2004a) Translocation of bioactive peptides across cell membranes by carbon nanotubes. *Chemical Communications*, 1, 16–17.

Pantarotto, D., Partidos, C. D., Hoebeke, J., Brown, F., Kramer, E., Briand, J. P., Muller, S., Prato, M., and Bianco, A. (2003) Immunization with peptide-functionalized carbon nanotubes enhances virus-specific neutralizing antibody responses. *Chemistry and Biology*, 10, 961–966.

Pantarotto, D., Singh, R., Mccarthy, D., Erhardt, M., Briand, J. P., Prato, M., Kostarelos, K., and Bianco, A. (2004b) Functionalized carbon nanotubes for plasmid DNA gene delivery. *Angewandte Chemie-International Edition*, 43, 5242–5246.

Pastorin, G., Wu, W., Wieckowski, S., Briand, J. P., Kostarelos, K., Prato, M., and Bianco, A. (2006) Double functionalisation of carbon nanotubes for multimodal drug delivery. *Chemical Communications*, 11, 1182–1184.

Patolsky, F., Timko, B. P., Yu, G. H., Fang, Y., Greytak, A. B., Zheng, G. F., and Lieber, C. M. (2006a) Detection, stimulation, and inhibition of neuronal signals with high-density nanowire transistor arrays. *Science*, 313, 1100–1104.

Patolsky, F., Weizmann, Y., and Willner, I. (2004a) Long-range electrical contacting of redox enzymes by SWCNT connectors. *Angewandte Chemie-International Edition*, 43, 2113–2117.

Patolsky, F., Zheng, G. F., Hayden, O., Lakadamyali, M., Zhuang, X. W., and Lieber, C. M. (2004b) Electrical detection of single viruses. *Proceedings of the National Academy of Sciences of the United States of America*, 101, 14017–14022.

Patolsky, F., Zheng, G. F., and Lieber, C. M. (2006b) Fabrication of silicon nanowire devices for ultrasensitive, label-free, real-time detection of biological and chemical species. *Nature Protocols*, 1, 1711–1724.

Patolsky, F., Zheng, G. F., and Lieber, C. M. (2006c) Nanowire-based biosensors. *Analytical Chemistry*, 78, 4260–4269.

Poland, C. A., Duffin, R., Kinloch, I., Maynard, A., Wallace, W. A. H., Seaton, A., Stone, V., Brown, S., Macnee, W., and Donaldson, K. (2008) Carbon nanotubes introduced into the abdominal cavity of mice show asbestos-like pathogenicity in a pilot study. *Nature Nanotechnology*, 3, 423–428.

Price, R. and Patchan, M. (1991) Controlled release from cylindrical microstructures. *Journal of Microencapsulation*, 8, 301–306.

Roy, S. C., Paulose, M., and Grimes, C. A. (2007) The effect of TiO$_2$ nanotubes in the enhancement of blood clotting for the control of hemorrhage. *Biomaterials*, 28, 4667–4672.

Ruffini, G., Dunne, S., Fuentemilla, L., Grau, C., Farres, E., Marco-Pallares, J., Watts, P. C. P., and Silva, S. R. P. (2008) First human trials of a dry electrophysiology sensor using a carbon nanotube array interface. *Sensors and Actuators A: Physical*, 144, 275–279.

Sanchez-Quesada, J., Ghadiri, M. R., Bayley, H., and Braha, O. (2000) Cyclic peptides as molecular adapters for a pore-forming protein. *Journal of the American Chemical Society*, 122, 11757–11766.

Scanlon, S. and Aggeli, A. (2008) Self-assembling peptide nanotubes. *Nano Today*, 3, 22–30.

Schnur, J. M. (1993) Lipid Tubules—A paradigm for molecularly engineered structures. *Science*, 262, 1669–1676.

Selinger, J. V., Spector, M. S., and Schnur, J. M. (2001) Theory of self-assembled tubules and helical ribbons. *Journal of Physical Chemistry B*, 105, 7157–7169.

Shen, Y. (2008) Advances in the development of siRNA-based therapeutics for cancer. *Idrugs*, 11, 572–578.

Shi, X. F., Sitharaman, B., Pham, Q. P., Liang, F., Wu, K., Billups, W. E., Wilson, L. J., and Mikos, A. G. (2007) Fabrication of porous ultra-short single-walled carbon nanotube nanocomposite scaffolds for bone tissue engineering. *Biomaterials*, 28, 4078–4090.

Shvedova, A. A., Castranova, V., Kisin, E. R., Schwegler-Berry, D., Murray, A. R., Gandelsman, V. Z., Maynard, A., and Baron, P. (2003) Exposure to carbon nanotube material: Assessment of nanotube cytotoxicity using human keratinocyte cells. *Journal of Toxicology and Environmental Health—Part A*, 66, 1909–1926.

Shvedova, A. A., Kisin, E. R., Mercer, R., Murray, A. R., Johnson, V. J., Potapovich, A. I., Tyurina, Y. Y. et al. (2005) Unusual inflammatory and fibrogenic pulmonary responses to single-walled carbon nanotubes in mice. *American Journal of Physiology-Lung Cellular and Molecular Physiology*, 289, L698–L708.

Singh, R., Pantarotto, D., Mccarthy, D., Chaloin, O., Hoebeke, J., Partidos, C. D., Briand, J. P., Prato, M., Bianco, A., and Kostarelos, K. (2005) Binding and condensation of plasmid DNA onto functionalized carbon nanotubes: Toward the construction of nanotube-based gene delivery vectors. *Journal of the American Chemical Society*, 127, 4388–4396.

Singh, R., Pantarotto, D., Lacerda, L., Pastorin, G., Klumpp, C., Prato, M., Bianco, A., and Kostarelos, K. (2006) Tissue biodistribution and blood clearance rates of intravenously administered carbon nanotube radiotracers. *Proceedings of the National Academy of Sciences of the United States of America*, 103, 3357–3362.

So, H. M., Won, K., Kim, Y. H., Kim, B. K., Ryu, B. H., Na, P. S., Kim, H., and Lee, J. O. (2005) Single-walled carbon nanotube biosensors using aptamers as molecular recognition elements. *Journal of the American Chemical Society*, 127, 11906–11907.

Somia, N. and Verma, I. M. (2000) Gene therapy: Trials and tribulations. *Nature Reviews Genetics*, 1, 91–99.

Star, A., Tu, E., Niemann, J., Gabriel, J. C. P., Joiner, C. S., and Valcke, C. (2006) Label-free detection of DNA hybridization using carbon nanotube network field-effect transistors. *Proceedings of the National Academy of Sciences of the United States of America*, 103, 921–926.

Steinle, E. D., Mitchell, D. T., Wirtz, M., Lee, S. B., Young, V. Y. and Martin, C. R. (2002) Ion channel mimetic micropore and nanotube membrane sensors. *Analytical Chemistry*, 74, 2416–2422.

Supronowicz, P. R., Ajayan, P. M., Ullmann, K. R., Arulanandam, B. P., Metzger, D. W., and Bizios, R. (2002) Novel current-conducting composite substrates for exposing osteoblasts to alternating current stimulation. *Journal of Biomedical Materials Research*, 59, 499–506.

Tang, X. W., Bansaruntip, S., Nakayama, N., Yenilmez, E., Chang, Y. L., and Wang, Q. (2006) Carbon nanotube DNA sensor and sensing mechanism. *Nano Letters*, 6, 1632–1636.

Tenne, R. (2006) Inorganic nanotubes and fullerene-like nanoparticles. *Nature Nanotechnology*, 1, 103–111.

Tong, L., Zhao, Y., Huff, T. B., Hansen, M. N., Wei, A., and Cheng, J. X. (2007) Gold nanorods mediate tumor cell death by compromising membrane integrity. *Advanced Materials*, 19, 3136–3141.

Venkatesan, N., Yoshimitsu, J., Ito, Y., Shibata, N., and Takada, K. (2005) Liquid filled nanoparticles as a drug delivery tool for protein therapeutics. *Biomaterials*, 26, 7154–7163.

Wang, J. (2005) Nanomaterial-based electrochemical biosensors. *Analyst*, 130, 421–426.

Wang, H. F., Huff, T. B., Zweifel, D. A., He, W., Low, P. S., Wei, A., and Cheng, J. X. (2005a) In vitro and in vivo two-photon luminescence imaging of single gold nanorods. *Proceedings of the National Academy of Sciences of the United States of America*, 102, 15752–15756.

Wang, S. F., Shen, L., Zhang, W. D., and Tong, Y. J. (2005b) Preparation and mechanical properties of chitosan/carbon nanotubes composites. *Biomacromolecules*, 6, 3067–3072.

Wang, H. F., Wang, J., Deng, X. Y., Sun, H. F., Shi, Z. J., Gu, Z. N., Liu, Y. F., and Zhao, Y. L. (2004) Biodistribution of carbon single-wall carbon nanotubes in mice. *Journal of Nanoscience and Nanotechnology*, 4, 1019–1024.

Warheit, D. B., Laurence, B. R., Reed, K. L., Roach, D. H., Reynolds, G. A. M., and Webb, T. R. (2004) Comparative pulmonary toxicity assessment of single-wall carbon nanotubes in rats. *Toxicological Sciences*, 77, 117–125.

Warheit, D. B., Webb, T. R., Sayes, C. M., Colvin, V. L., and Reed, K. L. (2006) Pulmonary instillation studies with nanoscale TiO₂ rods and dots in rats: Toxicity is not dependent upon particle size and surface area. *Toxicological Sciences*, 91, 227–236.

Warsinke, A., Benkert, A., and Scheller, F. W. (2000) Electrochemical immunoassays. *Fresenius Journal of Analytical Chemistry*, 366, 622–634.

Withey, G. D., Lazareck, A. D., Tzolov, M. B., Yin, A., Aich, P., Yeh, J. I., and Xu, J. M. (2006) Ultra-high redox enzyme signal transduction using highly ordered carbon nanotube array electrodes. *Biosensors and Bioelectronics*, 21, 1560–1565.

Worle-Knirsch, J. M., Pulskamp, K., and Krug, H. F. (2006) Oops they did it again! Carbon nanotubes hoax scientists in viability assays. *Nano Letters*, 6, 1261–1268.

Wu, W., Wieckowski, S., Pastorin, G., Benincasa, M., Klumpp, C., Briand, J. P., Gennaro, R., Prato, M., and Bianco, A. (2005) Targeted delivery of amphotericin B to cells by using functionalized carbon nanotubes. *Angewandte Chemie-International Edition*, 44, 6358–6362.

Yager, P. and Schoen, P. E. (1984) Formation of tubules by a polymerizable surfactant. *Molecular Crystals and Liquid Crystals*, 106, 371–381.

Yan, X. H., He, Q., Wang, K. W., Duan, L., Cui, Y., and Li, J. B. (2007) Transition of cationic dipeptide nanotubes into vesicles and oligonucleotide delivery. *Angewandte Chemie-International Edition*, 46, 2431–2434.

Yang, S. T., Fernando, K. A. S., Liu, J. H., Wang, J., Sun, H. F., Liu, Y. F., Chen, M. et al. (2008) Covalently PEGylated carbon nanotubes with stealth character in vivo. *Small*, 4, 940–944.

Yang, S. T., Guo, W., Lin, Y., Deng, X. Y., Wang, H. F., Sun, H. F., Liu, Y. F. et al. (2007) Biodistribution of pristine single-walled carbon nanotubes in vivo. *Journal of Physical Chemistry C*, 111, 17761–17764.

Yin, P., Hariadi, R. F., Sahu, S., Choi, H. M. T., Park, S. H., Labean, T. H., and Reif, J. H. (2008) Programming DNA tube circumferences. *Science*, 321, 824–826.

Yu, S. F., Lee, S. B., Kang, M., and Martin, C. R. (2001) Size-based protein separations in poly(ethylene glycol)-derivatized gold nanotubule membranes. *Nano Letters*, 1, 495–498.

Zallen, D. T. (2000) US gene therapy in crisis. *Trends in Genetics*, 16, 272–275.

Zanello, L. P., Zhao, B., Hu, H., and Haddon, R. C. (2006) Bone cell proliferation on carbon nanotubes. *Nano Letters*, 6, 562–567.

Zheng, M., Jagota, A., Semke, E. D., Diner, B. A., Mclean, R. S., Lustig, S. R., Richardson, R. E., and Tassi, N. G. (2003) DNA-assisted dispersion and separation of carbon nanotubes. *Nature Materials*, 2, 338–342.

Zheng, G. F., Patolsky, F., Cui, Y., Wang, W. U., and Lieber, C. M. (2005) Multiplexed electrical detection of cancer markers with nanowire sensor arrays. *Nature Biotechnology*, 23, 1294–1301.

8

Nanoscale Forces in Protein Recognition and Adhesion

Deborah Leckband
*University of Illinois
at Urbana-Illinois*

8.1 Introduction

Adhesion is essential in biology. Cell adhesion is required to maintain the structural organization of all multicellular organisms across all anatomical length scales. Cells also transduce mechanical signals and respond by regulating adhesion, motility, and differentiation. Other adhesive interactions between cells are central to immunity. Pathogens also exploit adhesive interactions with cells to infect hosts. Determining the molecular mechanisms underlying these processes is central to understanding the fundamental basis of a host of biological processes that underlie both the organization of healthy tissues and diseases such as cancer.

Force probe techniques are ideal tools for investigating the molecular forces, strength of adhesion, and molecular mechanisms in biological adhesion. Different nanoforce probes, such as the atomic force microscope, can quantify the strengths of single molecular bonds. However, while the forces to rupture single molecular bonds reveal a wealth of information regarding individual molecular linkages, biological adhesion frequently involves the collective behavior of tens to thousands of proteins. It is therefore important to explore not only the mechanical strengths of single protein linkages, but to also determine how populations of bonds govern biological interactions.

Furthermore, while adhesion is a central parameter defining many biological interactions, adhesion measurements alone are insufficient to determine the relationships between molecular architectures and their nanomechanical functions. Importantly, both the range and magnitude of forces are often functionally relevant, particularly in complex environments such as the cell surface. Many adhesion proteins as well as other glycoproteins extend large distances from the cell membrane. Often the range of protein interactions, facilitated by these large structures, is considered critical to their biological function.

The surface force apparatus (SFA) quantifies molecular forces between extended surfaces as a function of their separation distance. The unique advantage of this instrument is that it quantifies molecular forces over nanometer distances. With regard to biological adhesion, the SFA also measures the adhesion energy per area due to molecular, e.g., protein interactions between extended surfaces such as biological membranes. The latter configuration is arguably more relevant to cell adhesion than single molecule studies. Two distinct differences between the SFA and the other molecular force probes are the following: (1) it measures the absolute separations between two surfaces to within 0.1 nm and (2) it quantifies the interaction energies between macroscopic surfaces. It is sufficiently sensitive to be able to probe interactions with energies on the order of the thermal energy k_BT (2.48 kJ/mol or 0.59 kcal/mol at room temperature).

This chapter describes the use of the surface force apparatus in nanoscale measurements of protein adhesion. The principles of surface force apparatus measurements are described, and two test cases are cited to illustrate how these measurements can elucidate unique information relating molecular structures to the mechanical functions of bio-adhesion molecules.

8.2 Surface Force Apparatus— Measurement Principles

The surface force apparatus (SFA) is unique among molecular force probes in that it quantifies the magnitudes and distance dependence of molecular forces between two macroscopic, curved surfaces as a function of the intersurface separation. This instrument differs from other force probe techniques, including the atomic force microscope, in two important ways. First, it directly determines the separation distance between the interacting surfaces over distances from <1 nm to several hundred nanometers, with ±0.1 nm resolution. Second, it quantifies the integrated force, or energy, between macroscopic surfaces.

8.2.1 Interferometric Determination of Absolution Surface Separations of Absolute Surface Separations

With the SFA, the absolute separation between two interacting surfaces is determined directly within ±0.1 nm by multiple beam interferometry (Born and Wolf, 1980; Israelachvili, 1973; Israelachvili and Adams, 1978). The samples in the instrument are supported on the surfaces of cleaved, atomically flat mica sheets that are fixed to the surfaces of two, crossed hemicylindrical, macroscopic silica lenses (Figure 8.1a). The top surfaces of the hemi-cylindrical lenses are polished with radii of curvature of 1–2 cm.

The surfaces of the mica adjacent to the silica lenses are coated with reflective 50 nm silver mirrors, so that the region between the silver mirrors constitutes the resonant cavity of a Fabry–Perot interferometer (Born and Wolf, 1980; Tolansky, 1951). Light transmitted by the interferometer comprises interference fringes of equal chromatic order (FECO), whose wavelengths are determined by the thicknesses and refractive indices of the different films between the two reflective silver surfaces (Figures 8.1b and 8.2) (Born and Wolf, 1980; Israelachvili, 1973; Tolansky, 1951). The transmitted wavelengths, therefore, shift as one changes the gap distance, D, between the two surfaces (Figure 8.1c and d).

In SFA measurements, the absolute distance between the opposing surfaces, that is, the thickness of the intervening layer, is determined from the measured wavelengths of light transmitted through the interferometer. In measurements between supported lipid bilayers or between protein monolayers interacting across a thin water film, the interferometer comprises five layers (Figure 8.2): two mica sheets, two bilayer/protein films, and the aqueous gap between them. Their respective thicknesses are Y, Z, and D, and μ_1, μ_2, and μ_3 are the corresponding refractive indices of each of the layers. One could use a multilayer matrix approach described in textbooks to determine D, if Y, Z, and the refractive indices are known (Born and Wolf, 1980). However, an analytical expression for the wavelengths transmitted through a five-layer interferometer showed that the wavelength shifts, $\Delta\lambda$, are related to the changes in the separation ΔD by (see Figure 8.2) (Israelachvili, 1973),

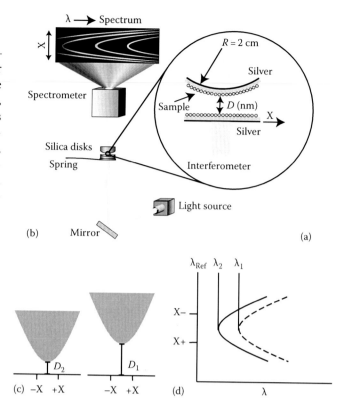

(b)

(a)

FIGURE 8.1 The surface force apparatus. (a) Samples in the SFA are supported on two hemicylindrical, transparent silica lenses oriented at right angles to each other. The equivalent geometry is a sphere interacting with a flat plate. (b) The samples with the reflecting silver mirrors form the resonant cavity of a Fabry–Perot interferometer. White light transmitted by the interferometer comprises a series of interference fringes, which are separated with an imaging spectrometer. The curvature of the fringes corresponds to the curvature of the contact region between the samples. (c) As the distance between the curved surfaces shifts from D_1 to the smaller distance D_2. (d) The wavelengths of the interference fringes shift from λ_1 to the shorter wavelength λ_2.

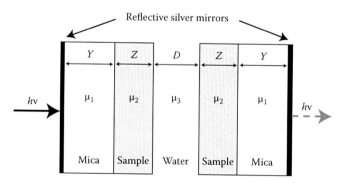

FIGURE 8.2 Schematic of the multilayer films within the resonant cavity of the interferometer of the apparatus. In measurements between the protein layers, the five films would be mica, bilayer with supported proteins, and water. The thicknesses of the films are, respectively, Y, Z, and D. The corresponding refractive indices are μ_1, μ_2, and μ_3.

$$\Delta D = \frac{n\Delta\lambda}{2\mu_1} - 2Z \qquad (8.1)$$

for odd order fringes (*n* odd) and

$$\Delta D\mu_3^2 = \frac{n\Delta\lambda}{2\mu_1} - 2Z\mu_2^2 \quad \text{for even order fringes (}n\text{ even)} \qquad (8.2)$$

One thus quantifies ΔD from the measured values of $\Delta\lambda$. For a given sample composition in the interferometer, it is also possible to determine D directly, by solving the reflectivity equations using either the multilayer matrix method or the analytical expression.

With Fabry–Perot interferometers, the use of a white light source (continuous spectrum) enables the quantification of continuous changes in the surface separation. Therefore, if the separation between the disks changes from D_1 to D_2 as in Figure 8.1c, then the transmitted wavelengths will shift from λ_1 to λ_2 (Figure 8.1d). If $D_2 < D_1$, then the fringes shift to shorter wavelengths, as in Figure 8.1d. This ability to directly quantify the surface separation differs from other force probes, which only record the relative movement of a cantilever or bead rather than the absolute probe-sample distance (Leckband and Israelachvili, 2001).

The ± 0.1 nm distance resolution achieved is due, in part, to the sharpness of the interference fringes. This results from the high coefficient of finesse of the interferometer that is due to the high reflectivity (~ 0.98) of the silver mirrors (Born and Wolf, 1980). Sharp interference fringes are essential for the high precision wavelength measurements needed for the nanoscale distance determinations. The standard error in wavelength measurements is typically ± 0.02 nm. With this resolution in $\Delta\lambda$ and the dependence of the transmitted wavelengths on D, one readily determines the absolute surface separation D within ± 0.1 nm.

The interferometer also images the shape of the intersurface contact region (Figure 8.1b and c) (Tolansky and Omar, 1952). Because of the curvature of the silica lenses, the intersurface distance increases with the distance from the center of the contact (Figure 8.1a) (Israelachvili, 1973; Tadmor et al., 2003). As a result, the transmitted wavelength increases with the radial distance from the center. The interference patterns projected onto an imaging spectrometer (Figure 8.1b) reveal the local geometry and the dimensions of the contact area (Israelachvili and Adams, 1978; Israelachvili and McGuiggan, 1990; Tolansky and Omar, 1952).

This imaging capability has at least two important consequences. First, one can directly quantify the local curvature at the point of contact between the surfaces. This is important because the geometry scales the magnitude of the intersurface force (Hunter, 1989; Israelachvili, 1992a). This is addressed in greater detail below (Section 8.2.2). Second, one can unambiguously establish whether changes in the cantilever position are due to a decrease in D or to deformations in soft materials between the surfaces (Leckband and Israelachvili, 2001). In several examples, this imaging interferometry was used to study biological materials. In one particular example, this approach enabled the visualization of the hemi-fusion between lipid bilayers in real time (Helm and Israelachvili, 1993; Helm et al., 1989; Helm et al., 1992; Israelachvili, 1992a; Leckband and Israelachvili, 2001; Leckband et al., 1993).

Another advantage of the optical technique of the SFA is that it enables determinations of the refractive indices of nanometer thick films confined between the two surfaces. In Equation 8.1, changes in the odd order fringes only depend on the refractive index of the mica μ_1. By contrast, the even order fringes depend on the refractive indices of all three layers: mica, biological samples on the mica, and the intervening water layer. Because the refractive indices of mica and water are known, one can determine the refractive indices of the biological samples, μ_2, from the wavelength shifts of the odd and even order fringes, together with Equations 8.1 and 8.2. Furthermore, it is possible to estimate the protein coverage from the measured refractive index relative to a densely packed protein film, similar to ellipsometry.

8.2.2 SFA Measures the Energy Per Area between Two Surfaces

A consequence of the sample geometry and large radius of curvature is that the SFA quantifies the energy per area (rather than the force) versus distance between two equivalent flat plates (Israelachvili, 1992b; Israelachvili and Adams, 1978; Leckband and Israelachvili, 2001). The SFA quantifies the total force between two crossed, macroscopic hemi-cylinders. However, this geometry effectively integrates the force law between the two surfaces to give the interaction energy per unit area between equivalent flat plates. To appreciate this, consider the differential area dA at a distance Z between the tip of the sphere and a flat surface (Figure 8.3). A sphere interacting with a flat plate is geometrically equivalent to two crossed cylinders, if the radius of the sphere $R = (R_1 R_2)^{1/2}$ where R_1 and R_2 are the radii of the cylinders (Israelachvili, 1992b). In SFA measurements, the geometric average radius is ~ 1 cm. With this macroscopic radius, one can assume that the surfaces are locally flat on the nanometer scale of molecular forces. The differential force per area between locally flat patches on the two surfaces at a separation Z is $dF(Z)/dA = f_m(Z)\rho$ where $f_m(Z)$ is the force per bond at Z and ρ is the number of bonds per area. When moving radially out from the center, the surface separation increases by δ. The differential force per area changes correspondingly by $dF_D(Z + \delta)/dA = f_m(Z + \delta)\rho = f_{flat}(Z + \delta)$ (Figure 8.3a), where f_{flat} is the force per area between two locally flat surface elements, dA (cf. Figure 8.3b). The total force between the sphere and the opposing plate is the sum of all interactions between these small surface elements: $F_c^T(Z) \approx \sum_{\delta=0}^{\delta=\infty} f_{flat}(Z + \delta)$. To integrate

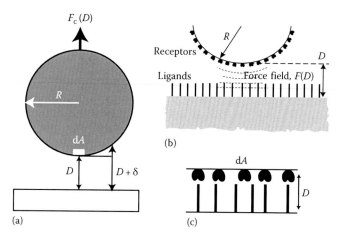

FIGURE 8.3 Illustration of the Derjaguin approximation and the interactions between the molecules on opposing macroscopic, curved surfaces. (a) The net force between the sphere and flat surface $F_c(D)$. Here R is the radius of curvature of the sphere, dA is the differential surface element, and D is the minimum distance between the sphere and the plate. (b) Illustration of discrete molecular interactions between a sphere and a flat plate. The distance between the molecules varies radially from the center of contact. (c) On a molecular length scale, the differential surface element dA on the sphere (b) is locally flat on the scale of the molecules.

the expression on the right from $\delta = 0$ to $\delta = \infty$ requires a change of variables to account for the geometry (Hunter, 1989; Israelachvili, 1992b). This introduces a factor of $2\pi R$ where R is the radius of the sphere (Israelachvili, 1992b). Integration gives

$$F_c^T(D) = 2\pi R \int_{Z=D}^{Z=\infty} f_{flat}(Z)dZ = 2\pi R E_{flat}(D)$$ where $F_c^T(D)$ is the

total force between the sphere and plate at a separation distance D, and E_f is the energy per area between two equivalent flat plates at separation D (Hunter, 1989; Israelachvili, 1992b). This relationship shows that the force between a macroscopic sphere and a plate (or between two crossed cylinders) is directly proportional to the energy per area between equivalent flat plates. This is the well-known Derjaguin approximation that applies when $R \gg D$ (Hunter, 1989; Israelachvili, 1992b).

An important consequence of this relationship is that it demonstrates that the curvature only affects the magnitude of the force between crossed-cylinders through the $2\pi R$ prefactor: $F_c(D) = 2\pi R E_f(D)$. The radius of curvature does not distort the intersurface potential. There is no dependence on R of the qualitative features of the force versus distance profile. Between two crossed cylinders, as in SFA measurements, R is the geometric average radius of the hemicylinders, $R = (R_1 R_2)^{1/2} \sim 1$ cm (Hunter, 1989; Israelachvili, 1992b). Since the range of molecular forces D is less than 0.2 μm, the Derjaguin approximation applies. This relationship has been tested and validated in over 60 years of research.

The Derjaguin approximation is based on the assumption that the surfaces are locally flat on the molecular length scale. This is easily demonstrated. Consider opposing proteins and receptors, for example, on a sphere and opposing flat plate. At different

radial distances, the molecules are at different relative distances (Figure 8.3b). Using the cord theorem, one can show that, if the radius is ~1 cm, then moving radially from the center by two microns causes a shift in the relative positions of the opposing proteins (or δ in Figure 8.3a) of only 0.2 nm. This is much smaller than the size of the proteins or their range of interaction. This validates the assumption.

The relationship between the normalized force F_c/R and the energy per area between planar surfaces E_f applies generally, even when the molecules interact through complicated oscillatory force laws with multiple repulsive maxima and attractive minima. Numerical calculations using surface element integration confirmed this. For example, Sivasankar et al. (2001) calculated the normalized force F_c/R between two curved surfaces coated with molecules that interact through an oscillatory force law. The calculated net force between the curved surfaces as a function of the distance $F_c^T(D)$ agreed quantitatively with the calculated energy between two equivalent plates, when scaled by $2\pi R$. Sivasankar et al. (2001) also showed that the probe curvature only distorts the normalized force curves when $R < 1$ μm. Because $R \sim 1$ cm in a typical SFA experiment, the radius scales the amplitude of the normalized force but it does not affect the shape of the force–distance curve.

8.2.3 Force Sensitivity in SFA Measurements

The SFA can quantify molecular interactions with energies ~1 $k_B T$ or 0.6 kcal/mol at room temperature. The normalized force sensitivity $\Delta F/R$ is ±0.1 mN/m or 0.1 mJ/m², as determined from the deflection of a sensitive leaf spring that supports the lower disk. The force sensitivity ΔF of the leaf springs used in SFA measurements is ~1 nN. This is lower than the ~1 pN sensitivity of cantilevers used in AFM measurements of single bond strengths. However, when measuring forces over large distances between *surfaces*, the measurement sensitivity depends on the normalized force $\Delta F/R$ (Leckband and Israelachvili, 2001). If we compare the sensitivities of the two approaches in which the probe radii are 1 cm and 10 nm, then the force sensitivity of the method using the smaller probe would have to be $R_1/R_2 = 10^{-2}$ m/10^{-8} m $= 10^6$ greater than that using the larger probe. In order to measure the same force with an AFM probe and the SFA, for example, the force resolution would have to be $\Delta F_2 = 10^{-9}$ N $\times 10^{-6} = 10^{-15}$ N. This is lower than $\Delta F \sim \pm 1$ pN that is typically achieved with an AFM. Because of the resolution of both the normalized force and the absolute distance measurements, the SFA is currently the most sensitive technique for quantifying normalized force–distance profiles between surfaces over large distances.

8.2.4 Adhesion Energies

To quantify adhesion energies, one determines the work to separate two surfaces (Hunter, 1989; Israelachvili, 1992b; Johnson et al., 1971). Between macroscopic, curved surfaces, the normalized pull-off force F_{po}/R is directly proportional to the adhesion energy per area E_A between equivalent flat plates. However, the exact relationship depends on whether the solids deform when in contact (Israelachvili, 1992b). In many SFA measurements, the mica surfaces and epoxy used to fix them to the silica deform and flatten under the influence of the intersurface forces and/or the external load. If the surfaces deform, then the Johnson–Kendall–Roberts theory for the adhesion between deformable surfaces relates the pull-off force to the adhesion energy per area by $E_A = F_{po}/1.5\pi R$ (Johnson et al., 1971). However, in the limit of small deformations, the adhesion energy is better described by the Derjaguin–Müller–Toporov theory, in which $E_A = F_{po}/2\pi R$ (Israelachvili, 1992b).

The pull-off force is proportional to the average adhesion energy per area E_A, but the magnitude depends on the density of adhesion proteins on the surfaces. The adhesion energies are therefore normalized by the protein or ligand surface density Γ, to account for this. We normalize the data by the more dilute molecule, since this is the limiting reagent determining the adhesion. The average adhesion energy per bond E_b is then estimated, by normalizing the adhesion energy by the protein coverage Γ. Taking into account the Boltzmann distribution between bound and free states, the estimated, average bond energy is $E_b = E_A(1 + \exp(-E_b/k_B T))/\Gamma$. This is an average over a large population of molecules. It assumes an average protein distribution such that there is no accumulation in the contact region or at the perimeter. Furthermore, any inactive protein in the population will lower the specific activity (adhesion/moles protein). For these reasons, the thus estimated bond energies are lower bounds.

The comparison with equilibrium bond energies raises an important question concerning whether the adhesion energy reflects the Gibbs free energy of the bonds or the activation energy for unbinding. The former is only possible if the system is at equilibrium during the detachment process. By contrast, single molecule bond rupture is an inherently non-equilibrium measurement, and the rupture force scales with the logarithm of the dissociation rate (Balsera et al., 1997; Dudko et al., 2006; Evans and Ritchie, 1997; Hummer and Szabo, 2003).

There are important similarities and differences between single bond rupture and adhesion between surfaces. In both cases, the external, tensile force lowers the activation energy for unbinding and accelerates the rate of de-adhesion (Evans and Ritchie, 1997; Hummer and Szabo, 2003). The force to rupture a single bond within 1 ms, for example, depends on the activation energy for unbinding and on the pulling rate. In single molecule studies, the bonds rarely reform, so the system is far from equilibrium. Between surfaces, however, the bound and unbound states can achieve dynamic equilibrium (Li and Leckband, 2006; Vijayendran et al., 1998). During detachment, the surviving bonds hold the surfaces together, and broken bonds can reform on the measurement time scale. How then does adhesion scale with bond properties in this scenario?

To address this question, we conducted Brownian dynamics simulations of the forced detachment of parallel plates bridged by multiple bonds in parallel. The simulations done at different separation rates showed that the adhesion scales with the Gibbs free energy of the bonds when the separation velocity is near-equilibrium. Under these conditions, the system is thermally equilibrated during the detachment process, and the adhesion is independent of the pulling rate. Far from equilibrium, however, the adhesion increases roughly with the logarithm bond dissociation rate. The excess detachment force, relative to the equilibrium value, is due to an increasing number of kinetically trapped bonds at pull-off. In this non-equilibrium regime, adhesion scales with the logarithm of the bond dissociation rate, or with the activation energy for unbinding.

The critical pulling rate v_c defines the cross over between the equilibrium and non-equilibrium pulling regimes. This parameter is defined by the distance from the ground state to the transition state (or the bond length L_b) divided by the intrinsic lifetime of the unstressed bond: $v_c = L_b/\tau_0$ (Li and Leckband, 2006). If the time to pull the bond to the transition state is much less than the intrinsic lifetime, then the bonds will be thermally equilibrated during the separation. This parameter v_c could be used to define measurement conditions. However, relaxation times and bond lengths are not always known, so that the equilibrium, pulling-rate-independent regime is established empirically.

With the SFA, the adhesion measurements are conducted over several minutes. To identify the rate-independent, equilibrium regime, measurements are typically conducted at decreasing separation speeds, until the measured adhesion is rate-independent. It is generally assumed that the adhesion energy reflects the equilibrium bond energy when the measured values are pulling-rate-independent, as suggested by the Brownian dynamics simulations (Li and Leckband, 2006).

8.2.5 Sample Requirements and Strategies

Because the SFA measurements record the integrated force over large areas (~300 μm^2) the molecules on the surfaces should be uniformly distributed. Achieving this can be challenging, but one of the more straightforward approaches for protein studies is to use supported lipid bilayers and engineered proteins with epitope tags (Figure 8.4). In order to immobilize proteins directly to supported bilayers, there are several commercially available lipid analogs with reactive headgroups that are capable of forming strong physical or covalent bonds with different amino acids. For example, maleimide-functionalized lipid headgroups covalently bind cysteines (Yeung et al., 1999), whereas nitrilo-tri-acetic acid (NTA) headgroups chelate oligohistidine tags in the presence of nickel (Sivasankar et al., 1999). If reactive amino acids or epitope tags are uniquely positioned on the proteins, then the surface attachment via these sites will orient the proteins on the membrane. For example, a unique cysteine engineered into the sequence of cytochrome b5 was used to control the immobilized orientation of cytochrome b5 on supported

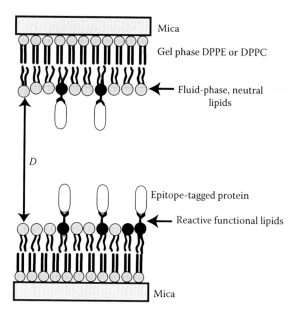

FIGURE 8.4 Supported lipid bilayers used to immobilize oriented protein monolayers in surface force apparatus studies. Asymmetric bilayers, formed by Langmuir–Blodgett deposition, consist of the gel phase DPPE (or DPPC) and an outer leaflet containing a mixture of neutral lipids and lipids with modified head groups. The neutral lipids may be in the gel phase or the fluid phase at the experimental temperature, in order to yield immobile or mobile lipids/proteins, respectively. The proteins attach to the functionalized lipids in the outer monolayer. Normalized forces are measured as a function of the distance D between the bilayer surfaces. (Adapted from Leckband, D., Surface force apparatus measurements of molecular forces in biological adhesion, in *Handbook of Molecular Force Spectroscopy*, Noy, A. (ed.), Springer, Heidelberg, Germany, 2007, pp. 1–27.)

lipid membranes (Yeung et al., 1999). Proteins could also specifically dock to ligands that are covalently bound to lipid head groups. The best-known example is streptavidin, which binds to biotinylated lipid head groups. The resulting, oriented streptavidin monolayers can also form two-dimensional crystals on lipid films (Calvert and Leckband, 1997; Darst et al., 1991).

Alternatively, one can immobilize the epitope-tagged or the covalently labeled proteins via an intermediate protein monolayer that is bound to the membrane. Proteins biotinylated at discrete sites will bind to and orient on streptavidin monolayers (Leckband et al., 1995b; Yeung and Leckband, 1998; Yeung et al., 1999). An alternative is to use oligohistidine-tagged or cysteine-tagged fragments of protein A. Protein A binds the Fc domain of immunoglobulin G (IgG). The histidine-tagged protein A will assemble on NTA-head-groups of functionalized lipids in the membranes to provide oriented docking sites for proteins engineered with Fc domains (Johnson et al., 2003). Engineered glycosyl–phospho–inositol (GPI) anchors can also be used to orient proteins on lipid membranes (Perez et al., 2005).

The best results are obtained with asymmetric supported-bilayers prepared by the Langmuir–Blodgett deposition on the

mica sheets. Cleaved mica sheets are atomically flat, so that the surface roughness does not affect either the distance or the force measurements. In these supported bilayers, the first layer is di-phosphatidyl ethanolamine or di-phosphatidyl choline in the gel phase (Figure 8.4). Fluorescence recovery after photobleaching measurements showed that the diffusion coefficients of the fluid lipids supported on the densely packed, gel phase lipids are similar to the lipid bilayers spread on glass (Leckband, unpublished observations). By contrast, if the first layer adjacent to the mica is deposited from the lipids in the fluid phase, the resulting layer is amorphous. This causes defects in the outermost monolayer and reduces the lipid diffusivity (Calvert and Leckband, 1997). As a cautionary note, lipid bilayers formed by the Langmuir–Blodgett deposition often contain pinhole defects. The proper choice of lipid and the rate of the Langmuir–Blodgett transfer onto the solid support can minimize the size and density of these defects (Bassereau and Pincet, 1997).

To control the protein mobility and the protein density on the membrane, the reactive lipid analogs can be mixed with neutral lipids that are either in the fluid or gel phase at the experimental temperature. The protein anchors are mobile in the former case, but not in the latter. As long as the two lipid components are miscible, the lipid analogs with their attached proteins will distribute uniformly on the bilayer. The exception is when the proteins aggregate or form two-dimensional crystals on the lipid films, as in the case of streptavidin (Calvert and Leckband, 1997).

The lateral mobility of membrane-anchored proteins is important in biological adhesion measurements. In SFA measurements, receptors and ligands on opposing membranes are not initially in register, and may need to diffuse laterally to engage binding targets on the opposite surface. Without lateral mobility, the measured adhesion energy can drop several-fold relative to the adhesion between fluid membranes at the same protein densities (Leckband et al., 1994).

8.3 Case Studies

Surface force investigations of proteins include streptavidin (Leckband et al., 1994; Leckband et al., 1995a), cytochrome *c*, cytochrome b5 (Yeung and Leckband, 1998; Yeung et al., 1999), antibodies (Leckband et al., 1995b), cadherins (Prakasam et al., 2006a,b; Sivasankar et al., 1999, 2001; Zhu et al., 2003), the neural cell adhesion molecule (Johnson et al., 2004, 2005), SNAREpin proteins (Li et al., 2007), and the immune proteins CD2, CD58, and CD48 (Bayas et al., 2007; Zhu et al., 2002). These studies each exploited the distance resolution of the SFA, in order to quantify the interaction forces, adhesion energies, and the dimensions of protein complexes. This section describes two examples that illustrate the unique insights into structure–function relationships that can be obtained from surface force apparatus measurements of proteins.

8.3.1 CD2 Family of Cell Adhesion Proteins in Immunity

8.3.1.1 Structure and Function of CD2 and Its Ligands

Cellular immunity results from interactions between thymus cells (T-cells) and antigen-presenting-cells (APCs) that present foreign antigens on their surfaces. Several proteins are involved in the formation of this important intercellular interaction. In particular, binding between the Major Histocompatibility Complex (MHC) on the APCs and the T-Cell Receptor (TCR) on T-cells triggers an immune response. The formation of these intercellular junctions and the association of the MHC and TCR are facilitated by auxiliary proteins such as CD2 and CD58 (Davis and van der Merwe, 1996a,b).

CD2 is expressed on T-cells. Its ligand CD58 is expressed on APCs. In addition to CD58, CD2 has other ligands, all of which are members of the "CD2 protein family" (Davis and van der Merwe, 1996a). The structure of CD2 was determined by both x-ray crystallography and by NMR (Bodian et al., 1994; Davis and van der Merwe, 1996b; Jones et al., 1992). The structures of the extracellular domain of CD2 and of its ligand human CD58 (Ikemizu et al., 1999) and rat CD48 (Evans et al., 2006) were also determined. Members of the CD2 protein family have a similar overall architecture (Figure 8.5a), although their chemical compositions differ (Davis and van der Merwe, 1996a). The extracellular regions of proteins in this family consist of two, tandemly-linked immunoglobulin (Ig)-type domains (D1 and D2) that are bound to cell membranes by hydrophobic anchors (Figure 8.5a). The end-to-end lengths of the extracellular regions are ~7.5 nm. Based on the structure of the complex between the outer D1 domains of CD2 and human CD58 (Wang et al., 1999), and on biochemical data, the proteins are postulated to bridge cell membranes in a head-to-head configuration as shown in Figure 8.5b (Davis and vanderMerwe, 1996b; Evans et al., 2006; Ikemizu et al., 1999; McAlister et al., 1996; Wang et al., 1999). As described in Section 8.3.1.2, surface force apparatus measurements tested this model and quantified the dimensions of the protein–protein complexes.

In the proposed head-to-head configuration, the CD2–CD58 complex is predicted to span a membrane–membrane gap of ~13.5 nm (Davis and van der Merwe, 1996b). It has also been postulated that this distance is functionally important because it closely matches the membrane spacing spanned by the MHC-TCR complex (Davis and van der Merwe, 1996b, Davis et al., 1998). CD2 may therefore function as both an adhesive protein and a scaffold to control the inter-membrane spacing, in order to facilitate binding between the MHC and the TCR on adjacent cells. Consistent with this hypothesis, CD2 molecules engineered with an additional immunoglobulin (Ig) domain spacer impede T-cell activation (Davis et al., 2003).

8.3.1.2 Dimensions of the CD2–CD58 Complex from Force–Distance Profiles

Surface force apparatus measurements between the oriented, immobilized extracellular fragments of CD2 and CD58 tested the protein-binding model and quantified the adhesion energy (Bayas et al., 2007; Zhu et al., 2002). These measurements used the extracellular regions of the proteins bound directly to the membrane via oligohistidine tags. The extracellular domains of both proteins were engineered with oligohistidine tails at the

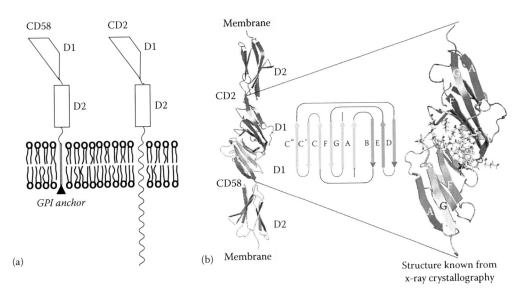

FIGURE 8.5 CD2–CD58 complex. (a) General architecture of the CD2 protein family. The proteins are anchored to the membrane via hydrophobic tails. The extracellular region consists of two domains D1 and D2. This figure shows two different hydrophobic anchors observed in this family. (b) Proposed head-to-head binding alignment between the CD2 and its ligands. The structure of the complex between the outer D1 domains is shown on the right side. (Modified from Bayas, M.V. et al., *Biophys. J.*, 84, 2223, 2003.)

C-terminus of the second domain (D2). This ensured their selective immobilization and proper orientation on NTA–DLGE lipid layers (cf. Figure 8.4). The SFA measurements then quantified the normalized force as a function of the distance D between the membrane surfaces.

Figure 8.6 shows a normalized force–distance profile measured between the CD2 and the CD58 monolayers. Force–distance curves were measured during approach (decreasing D) and separation (increasing D). During approach, the onset of the repulsion ($F/R > 0$) at $D < 16$ nm is due to steric and osmotic repulsion between the two protein layers. The proteins are each 7.5 nm in length. This behavior therefore agrees with the expected range of repulsion between end-on oriented proteins. Upon surface separation, protein–protein binding caused the normalized force to drop below zero. At the minimum in the curve at 15.3 ± 0.5 nm the CD2–CD58 bonds failed, and the surfaces jumped out of contact (out arrow). The minimum in the curve corresponds to the maximum gradient in the intersurface potential, or the maximum attractive force. The mechanical instability that results in the jump-out occurs when the gradient of the potential exceeds the spring constant. The difference between the final resting position of the surfaces after pull-off ($F = 0$) and the position of the adhesive minimum gives the spring deflection at adhesive failure. The product of the spring constant and the jump distance gives the normalized pull-off force.

The end-to-end dimension of the CD2–CD58 complex under tension determines the membrane separation at pull-off. This distance also includes a ~1.0 nm contribution from the anchoring NTA tethers on each of the opposing membranes. Taking the tethers into account, the end-to-end dimension of the complex is $(15.3 \pm 0.5) - (2 \times 1.0) = 13.3 \pm 0.5$ nm. This value agrees quantitatively with the predicted dimensions of the complex based on the crystallographic dimensions of the individual proteins, and on the head-to-head configuration of the complex formed between the outer D2 domains of CD2 and CD58 (Davis et al., 1998, 2003; van der Merwe and Davis, 2003).

The resolution of the distance measurements enabled the experimental validation of the predicted dimensions of the CD2–CD58 complex. These results further confirmed that the CD2–CD58 complex is structurally matched to the size of the TCR–MHC complex that the intercellular adhesive junctions must accommodate. The results supported both the adhesive and the proposed scaffolding roles of these proteins.

8.3.1.3 CD2-Mediated Adhesion Energies and the Role of Ion Pair Interactions

From the pull-off force indicated in Figure 8.6, the adhesion energy per area between the two protein monolayers was 0.38 ± 0.1 mJ/m². Normalizing this by the lower CD58 density of 9.5×10^3 proteins/μm², as described above, gives an average, estimated CD2–CD58 bond energy of $8.6 \pm 0.3\ k_B T$. Here, k_B is the Boltzmann constant and T is the absolute temperature.

One of the questions addressed with the SFA concerned the role of charged amino acid side chains in stabilizing the protein–protein bond under force. Both the structure in Figure 8.5b and site directed mutagenesis studies suggested that electrostatic interactions between charged side chains at the CD2–CD58 interface contribute to the stability of the complex. Steered molecular dynamics simulations of the forced rupture of the CD2–CD58 complex identified key amino acids in CD2 that appear to form major load-bearing contacts with CD58 (Bayas et al., 2003). The following amino acids in CD2 were then targeted for substitution by neutral alanine, in order to test whether the simulations predict their relative contributions to the adhesion strength: namely, aspartate 31, lysine 41, lysine 51, and lysine 91 (D31, K41, K51, and K91). The simulations predicted that the relative contributions of these amino acids to adhesion follow the order K51 > D31 > K91 > K41. Subsequent surface force measurements with the CD2 mutants D31A, K41A, K51A, and K91A in which the amino acid was changed to alanine quantified the loss of charged groups in CD2–CD58 adhesion.

Force profiles measured between these CD2 mutants and wild type CD58 were nearly identical to that in Figure 8.6, except

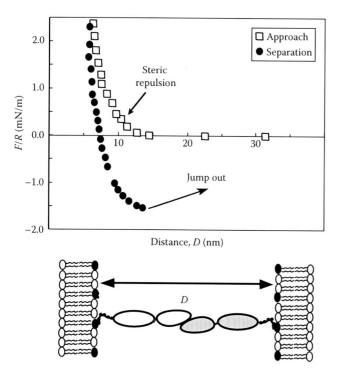

FIGURE 8.6 Normalized force versus distance profile between monolayers of human the CD2 and CD58. The open triangles show the advancing force profile measured during decreasing distance, D, and the filled circles show the receding force profile measured during separation (increasing D). At the minimum in the curve (maximum attractive force), the bonds rupture and the sample surfaces pull out of contact (out arrow). (Adapted from Bayas, M.V. et al., *J. Biol. Chem.*, 282, 5589, 2007.)

TABLE 8.1 Importance of Electrostatic Interactions for CD2–CD58 Complex Stability

CD2 Variant	Simulations	Adhesion Energy, k_BT
Wild type	—	8.6 ± 0.5
K41A	Not critical	7.4 ± 0.5
K91A	Modest effect	6.8 ± 0.5
D31A	Critical	5.6 ± 0.6
K51A	Critical	5.0 ± 0.6

that the adhesion energies differed (Bayas et al., 2007). The proteins adhered at the same membrane distances, within experimental error. Mutation-dependent changes in the bond energies are summarized in Table 8.1. The trend in bond strengths predicted by the simulations suggests the relative importance of charged groups stabilizing the protein–protein interface. The predicted trend agrees qualitatively with the experimentally measured, relative decreases in the average bond energy following the corresponding elimination of the charged side chain in CD2. This result experimentally validates the simulation results. In so doing, the measurements also confirm that complementary charges at the protein–protein interface indeed confer mechanical stability to the CD2–CD58 bond.

In summary, these force measurements verified the dimensions of the CD2–CD58 complex, and confirmed the predicted orientation and size relative to the TCR–MHC complex. They also quantified the impact of the specific amino acid side chains on the magnitude of the protein–protein adhesion, and verified the relative importance of charged side chains for the mechanical stability of the CD2–CD58 bond.

8.3.2 Cadherin-Mediated Cell Adhesion: Characterizing Complex Binding Mechanisms by Direct Force Measurements

8.3.2.1 Cadherins in Biology

Cadherins consist of a family of calcium-dependent cell surface proteins that mediate adhesion between cells in all solid tissues. Cadherins are essential for a host of physiological functions, including the formation and organization of tissue architecture (Gumbiner, 1996, 2005). They bridge adjacent cells by binding to similar cadherins on the opposite cell surface.

The structures of classical cadherins comprise the extracellular region that embeds the adhesive function, a transmembrane domain, and a cytoplasmic domain (Figure 8.7a) (Yap et al., 1997). The extracellular segment folds into five cadherin-type extracellular (EC) domains that are numbered 1–5 from the N-terminal domain (EC1–5) (Figure 8.7a and b). Structural studies of cadherin fragments and of cadherin junctions in tissues (Boggon et al., 2002; He et al., 2003) suggest that cadherins adhere through a simple mechanism that only involves interactions between the opposing terminal extracellular domains (EC1) (Figure 8.7c). This simple postulated mechanism predicts several properties of cadherin junctions that are readily tested by a variety of quantitative biophysical approaches, including surface force apparatus measurements. An important question concerns the identity of the minimal functional segment of the protein and the functions of the different EC domains.

Despite the availability of static structures, it is essential to test models with alternative biophysical approaches because structures

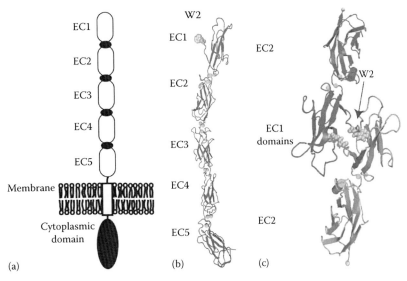

(a)   (b)   (c)

FIGURE 8.7 Classical cadherin structure and model of homophilic binding. (a) The classical cadherins are transmembrane proteins with a cytoplasmic domain, a transmembrane region, and an extracellular region. The extracellular region folds into five structurally homologous cadherin extracellular (EC) domains number 1–5 from the N-terminus. (b) Crystal structure of the extracellular region of Xenopus C-cadherin. (c) Binding between the N-terminal domains of the C-cadherin extracellular region in which the Trp[2] (W2) residues from opposing cadherins dock into the hydrophobic pocket of the opposed protein. (Reproduced from Leckband, D., *Cell Mol. Bioeng.*, 1, 312, 2008. With permission.)

are only one form of physical evidence for molecular mechanisms. One cannot see bond energies, assembly kinetics, bond strengths, or adhesion energies. In cases where multiple protein–protein interactions may contribute to adhesion, a single structure represents only one of several possible, relevant states. Conformational changes, multiple binding states, binding/unbinding kinetics, and cooperativity can further elaborate adhesion mechanisms. Except for the simplest binding interactions, validating adhesion models requires multiple approaches in order to establish molecular mechanisms and characterize protein function. Proposed models are thus substantiated, not only by static structures, but also by quantitative, approaches such as described here.

8.3.2.2 Detection of a Multistage Cadherin-Binding Mechanism

Surface force measurements directly tested the simple binding model for cadherins proposed based on the x-ray structures of the extracellular domain (cf. Figure 8.7c). Measurements were conducted with cadherins directly bound to membranes via C-terminal oligohistidine tags as well as with Fc-tagged cadherin dimers immobilized to protein A monolayers. The studies described here used recombinant cadherin ectodomains (EC1–5) fused to the Fc fragment of a human IgG antibody (EC1–5-Fc) (Figure 8.8). The EC1–5-Fc was immobilized on immobilized, oriented protein A fragments on the supported planar bilayers. The protein A and Fc-tag contribute 2 and 4.3 nm, respectively, to the steric thickness of each cadherin monolayer. Figure 8.8 illustrates the sample architecture. The fully extended length of the Fc-tagged ectodomain is ~27 nm. The distances refer to the distance between the membranes.

The protein samples were built up by first immobilizing an engineered fragment of protein A with a C-terminal oligohistidine tail. The latter protein binds and orients on supported bilayers in which the outer leaflet contains fluid phase NTA-functionalized lipids (Figure 8.8). The protein A then captures Fc-tagged cadherin via the Fc domain, and orients the cadherins on the bilayer.

Figure 8.9a shows the normalized force versus distance profiles between the cadherin monolayers. Forces were measured during approach (decreasing D) and separation (increasing D). During approach, the oriented cadherin monolayers repel at $D < 55$ nm, due to steric and osmotic repulsion between the molecular layers. During separation, we measured adhesion at three distinct membrane gap distances.

These different adhesive interactions were identified by controlling the membrane separation distance prior to detachment. The principle of these measurements is illustrated in Figure 8.9b. Initially, the surfaces are brought to a separation distance $D1$ and then separated. During surface separation, one would measure adhesion, if the proteins formed an adhesive bond at gap distances $D > D1$. Sampling a range of distances during the approach and separation would detect the force maxima and minima at different membrane separations, if they exist. In the case of the cadherin interactions, if 45 nm < $D1$ < 53 nm, the cadherins adhered at 53.1 ± 0.8 nm. However, at membrane

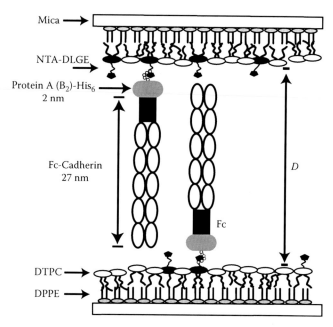

FIGURE 8.8 Schematic of the molecular architecture of the immobilized Fc-tagged cadherin ectodomains used in surface force measurements. The cadherin Fc dimer is immobilized on opposite planar lipid bilayers. The first layer is diphosphatidyl ethanolamine (DPPE). The outer layer contains lipids with nitrilo-triacetic acid (NTA) headgroups. A hexahistidine-tagged protein A fragment binds to the NTA head group, and the protein A in turn captures the Fc-tagged cadherin extracellular domain fragment. In some cases, the NTA-DLGE lipids are mixed with neutral di-tritanoly phosphatidyl choline (DTPC). The distance D is the separation between the membrane surfaces. The measured thickness of the immobilized protein A fragment is 2 nm, and the extended end-to-end length of the Fc-tagged cadherin fragment is ~22.5 nm. (Reproduced from Leckband, D., *Cell Mol. Bioeng.*, 1, 312, 2008. With permission.)

distances 39 nm < $D2$ < 45 nm, the surfaces adhered at 44.5 ± 1.0 nm. Similarly, if $D3$ < 38 nm, the proteins adhere at 38.2 ± 0.6 nm. Three adhesive interactions were thus quantified at distinctly different membrane gap distances (Figure 8.9b). The force data demonstrated that the outermost bond is the weakest, and the innermost bond is the strongest.

This same approach as illustrated in Figure 8.9b was used to quantify oscillatory force curves with multiple maxima and minima at different intersurface spacings. Israelachvili and Pashley (Israelachvili and Pashley, 1983) thus identified up to eight water layers confined in the thin gap between two mica surfaces in water. Other examples detected multilayers of cytochrome c adjacent to mica surfaces, as well as multiple polymer layers in melts confined between the mica sheets (Horn and Israelachvili, 1988; Kekicheff et al., 1990). In the latter cases, the periodicity of oscillations of maxima and minima in the force–distance curves corresponded exactly to the theoretically predicted dimensions of the successive molecular layers (Attard and Parker, 1992; Frink and vanSwol, 2000).

In the force data in Figure 8.9, adhesion at 53 nm is consistent with binding between the outermost EC1 domains. Adhesion

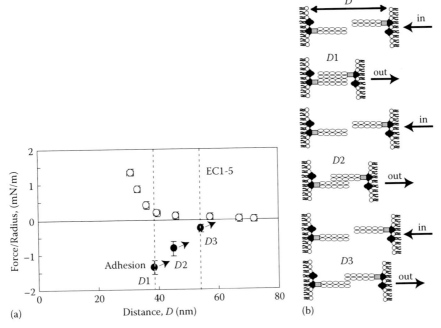

FIGURE 8.9 Normalized force-distance profiles between the oriented C-cadherin monolayers. (a) Force distance profile between the oriented cadherin monolayers. Normalized forces were measured during approach (white circles) and separation (black circles). Upon approach, the steric repulsion results in a positive force ($F/R > 0$) at $D < 55$ nm as seen in the force–distance curve (white circles) in (a). If the proteins adhere, then the net force becomes attractive ($F/R < 0$) on separation. At the minimum in the curve (black circles), the bonds yield and the protein layers pull out of contact (right pointing arrows). Adhesive minima were measured at membrane separations $D = 38.2 \pm 0.6$ nm, 44.5 ± 1.0 nm, and 53.1 ± 0.8 nm. (b) Illustration of the method used to detect the three bonds in (a). After bringing the proteins in to a particular separation distance (left pointing arrows), the surfaces were separated (right pointing arrows). If the minimum separation distance was D1 < 38 nm, then the proteins adhered at 38 nm. When 39 nm < D2 < 45 nm, the proteins bound at 44 nm. When 47 nm < D3 < 54 nm, adhesion was at 53.1 ± 0.8 nm. All distances 35 nm < D < 54 nm were thus tested. (Reproduced from Leckband, D., *Cell Mol. Bioeng.*, 1, 312, 2008. With permission.)

at the other distances indicates that the proteins form discrete bonds that involve other domains and span different membrane gaps. This multi-stage binding mechanism was demonstrated with three different classical cadherins that have different chemical compositions but exhibit the same overall structure (Prakasam et al., 2006b, Leckband and Prakasam, 2006).

8.3.2.3 Binding Dynamics: Cadherins Spontaneously Pull Membranes Together

Time dependent surface force measurements showed that cadherins initially bind via the outermost domains at ~54 nm, but then pull the membranes in to a gap distance of 38 nm. The interferometric technique of the SFA enables the direct real-time observation of dynamic changes in inter-surface forces or molecular orientations that cause changes in the surface separation distances. This is visualized by monitoring real-time changes in the wavelengths of the interference fringes, and by interfacing the imaging spectrometer (cf. Figure 8.1b) with a video camera. Previous real-time dynamic measurements of membrane fusion, for example, recorded the initiation and spreading of a hemi-fused region between two opposed lipid bilayers (Helm et al., 1989).

Similarly, dynamic measurements demonstrated that the opposed cadherins spontaneously pull the membranes together from an initial distance of ~53 nm, where their N-terminal

domains first contact, to a final resting position of 38 nm. This process occurs spontaneously without the application of an external force. This finding is significant because it indicates that the final equilibrium cadherin complex spans an intermembrane distance of 38 nm. After accounting for the length of the Fc-tags and the protein A fragment, this distance is consistent with membrane gap distances in EM images of desmosomal cadherin junctions in skin (He et al., 2003).

This spontaneous process is illustrated in Figure 8.10, which shows three successive snapshots from the recorded movement of the fringes. The images were obtained at the start of the movement, during the movement, and at the final resting position. At initial contact between the outer EC1 domains, the vertical fiduciary line was positioned to the left of the rightmost fringe in the doublet (2 closely spaced fringes). Within ~2 min, the fringe moved leftward (decreasing λ) and finally came to rest just to the left of the vertical line. The total decrease in the wavelength $\Delta\lambda$ of the circled fringe corresponds to a 13 nm decrease in the relative membrane separation. The instrument is sufficiently stable that the surfaces drift by <1 nm during the same period (~2 min). This process is cadherin-dependent. These results illustrate how the dynamics of molecular interactions that result in changes in the equilibrium surface separation, can be followed in real-time with the optical technique of the SFA.

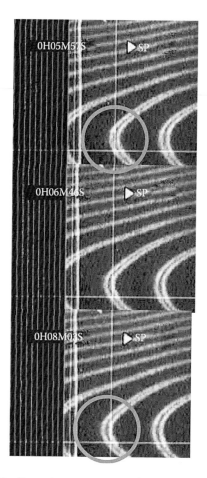

FIGURE 8.10 Dynamic measurements of the spontaneous assembly of cadherin junctions. The three images are snapshots from a real-time video recording of shift in the wavelengths of the interference fringes as the cadherins spontaneously pull the two membranes from a separation of ~52 nm to a final resting position at 39 nm. The vertical line to the left of the rightmost fringe in the circled doublet is the reference. The upper panel shows the fringes after the surfaces were placed at a separation distance of ~52 nm, and the leftmost fringe in the doublet is to the right of the vertical reference line. Over the course of ~2 min, the fringes spontaneously move leftward, and finally halt when the fringe is to the left of the vertical line. The final resting position is at a separation distance of 39 nm. (Reproduced from Leckband, D., *Cell Mol. Bioeng.*, 1, 312, 2008. With permission.)

8.3.2.4 Interpreting Force–Distance Profiles between Cadherin Monolayers

To interpret these force–distance profiles and the positions of the adhesive minima in terms of the protein structure, one compares the distances of pull-off to the protein dimensions. Here we consider the extended length of the cadherin ectodomain, accounting for the dimensions of the anchoring protein layers (Fc-tag, protein A fragment, and NTA head group). In the case of the cadherin ectodomain, although protein adopted a bent structure in the crystal (Boggon et al., 2002), steered molecular dynamics simulations indicate that it behaves like a semi-flexible rod (Sotomayor and Schulten, 2008). In the simulations, when

the ectodomains were subjected to very small, physiological forces of ~50 pN, the bent cadherin structure readily straightened to adopt an end-to-end length of 22 nm (Sotomayor and Schulten, 2008). This straightening should similarly occur under the small tensile forces exerted in the force probe measurements, and was used to interpret the force–distance curves.

The position of the greatest adhesion relative to the length of the ectodomain under low tension indicates that the ectodomains could fully interdigitate at a membrane separation of 38 nm (Figure 8.9b, *D1*) (Zhu et al., 2003). However, it is important to point out that the proteins could also adopt other configurations consistent with the measured end-to-end length (see Section 8.3.2.4). It is not always possible to distinguish between different protein configurations based on the pull-off distance alone.

Force measurements with domain deletion mutants were used to identify the structural regions required for the three, different adhesive states in Figure 8.9a. In this case, the removal of critical EC domains would ablate any bonds that require the deleted protein region. The impact of the mutations is readily observed with the SFA from changes in the occurrence and magnitude of the spatially separated adhesive states. Several different EC domain deletion mutants identified different structural regions required for each of the adhesive states in Figure 8.9a. These include the fragments EC 1245 and EC 12, which lack domains EC3 and EC3–5, respectively (see Table 8.2). Four additional EC deletion mutants were investigated (Zhu et al., 2003). The main results are summarized in Table 8.2 and are described here.

First, all fragments that contain extracellular (EC) domains 1–3 formed three, spatially separated bonds with the same order of relative adhesion strength as exhibited by the full ectodomain, EC1–5. The spatial separation between the three bound states was quantitatively the same for the different fragments (Table 8.2), within experimental error. The positions of the minima changed because the deletion mutants have different (shorter) overall lengths. This approach localized the multistate binding mechanism to the outer three domains.

Second, removing EC3 (third domain) eliminated the strongest bonds, so that fragments lacking EC3 only bound weakly

TABLE 8.2 Intermembrane Distances at Cadherin Bond Failure

Protein Fragment	First Bond (Å)	Second Bond (Å)	Third Bond (Å)
EC1–5	38.2 ± 0.6	44.5 ± 1.0	53.1 ± 0.8
EC1–3	21.1 ± 0.9	26.9 ± 0.2	34.9 ± 0.5
	(Δ = 17.1 nm)	(Δ = 17.6)	(Δ = 18.2)
EC1–2	—	—	27.1 ± 0.9
			(Δ = 26.0)
EC345	38.6 ± 0.4	—	—
EC1245	—	—	42.4 ± 0.8
			(Δ = 10.7)

Notes: The values in the parenthesis are the changes in the binding distances between the mutants (identical proteins on both membranes), relative to the intermembrane distances at which the full-length cadherin ectodomains adhere.

via the outermost EC1 domain. Additionally, the EC12 fragment formed a single bond with EC12 and also with the full ectodomain (EC1–5). Third, EC345 adhered to both the full-length ectodomain and a second EC345 fragment at 38 ± 0.5 nm and 39 ± 0.5 nm, respectively. The adhesion energy of this fragment was low, but the binding was at the same distance as the strongest bond formed between the full ectodomains, that is, ~38 nm.

Taken together, these force measurements with structural mutants confirm that cadherins form multiple bound states that require at least two EC domains. The EC1 domains adhere relatively weakly at the largest membrane separations, consistent with the model in Figure 8.1c. The cadherins also form two other bonds with greater adhesion energies that require EC3.

8.3.2.5 Model Testing Using Complementary Force Measurement Techniques

The surface force measurements provide unique details regarding the mechanisms by which proteins may mediate cell adhesion. The spatial details cannot be obtained by other current force probe measurements. In addition, one obtains the adhesion energies rather than the (non-equilibrium) tensile strengths (rupture forces) of the protein bonds. The structural interpretation of the force measurements, especially with complex proteins such as cadherins, generates molecular models of protein-mediated adhesion. These models are typically tested with other, complementary approaches. Importantly, the biological relevance of such findings also needs to be demonstrated with live cells.

Independent measurements of the kinetics of cadherin-mediated binding between pairs of live cells confirmed that the initial adhesion occurs by a two-stage process that requires cadherin domains EC3 and EC1 (Chien et al., 2008). Initial, fast binding between EC1 domains is followed by a slower transition to a second binding state that requires EC3. The cadherin structural requirements for the biphasic kinetics agreed qualitatively with the surface force measurements. Moreover, the time-dependent behavior is analogous to the dynamic force data illustrated in Figure 8.10. These cell studies further confirmed that nanoscale protein interactions measured with the surface force apparatus indeed control initial cell–cell binding.

Single bond rupture studies using atomic force microscopy (AFM) investigated the number and strengths of bonds formed between cadherin ectodomains. The results of these measurements further supported both the surface force measurements and the cell binding kinetics. The AFM measurements confirmed that cadherins form multiple bound states. Again, the EC1 domains form weak bonds with fast kinetics. The full-length protein forms the weak, EC1-dependent bonds, but they also form additional stronger bonds with slow kinetics. Studies of the time-dependent distribution of the weak and strong cadherin bonds also demonstrated that cadherins first form the weak, EC1 bonds, but then transition to a second state that is characterized by higher rupture forces and slower kinetics. The results of the latter time-dependent shift parallel both the dynamic surface force measurements and the cell binding kinetics.

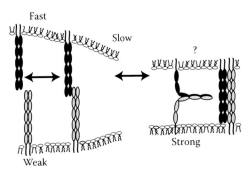

FIGURE 8.11 Hypothetical cadherin binding mechanism. In this hypothetical mechanism, cadherins rapidly bind via their outer EC1 domains, but then slowly transition to a second adhesive state characterized by slow kinetics, stronger, higher probability binding, and a smaller inter-membrane distance. The configurations on the right are compatible with the majority of the structural and biophysical studies, including those described in this chapter.

The importance of these force measurements, and the surface force apparatus results in particular, is that they clearly demonstrate that the simple model (Figure 8.7c) based on structural data alone is insufficient to account for the complexity of cadherin binding. Taken together, complementary force measurement approaches provide compelling evidence for a model in which the proteins rapidly bind via their EC1 domains on initial contact to form a weak bond. They subsequently transition to a more stable configuration that is characterized by stronger adhesion and slower kinetics. The structural basis for the first state is supported by crystallography. However, there is currently no structure of other adhesive configurations. Different protein configurations (Figure 8.11) are compatible with both the positions of the adhesive minima in SFA measurements and the electron density maps of cell–cell junctions in tissues, but further studies are needed to elucidate the complete binding mechanism.

Additional studies with other biophysical, structural and genetic tools should reveal the detailed molecular mechanism of cadherin junction assembly. This case study demonstrates the capacity of SFA measurements to reveal unique, fundamental aspects of the molecular level interactions, as well as the value of complementing such studies with other force probe measurements, structural studies, and genetic manipulation.

8.4 Summary

In conclusion, this review highlights the important capabilities of the surface force apparatus for investigations of biological adhesion. The forces and distances obtained in these measurements provide unique information regarding the molecular mechanisms and the forces governing adhesion in biology. The specific examples of the CD2 and cadherin described in this chapter clearly demonstrate that the nanoscale structures of these multi-domain adhesion proteins impact both the magnitude and the range of interactions between cell membranes. The SFA is a powerful instrument, which complements other biophysical approaches, including single molecule AFM studies.

Acknowledgments

This work was supported by NIH GM51338, by NSF CBET 0853705, and by the Reid T. Milner Professorship to DEL.

Abbreviations

AFM	Atomic force microscopy
APC	Antigen presenting cell
EC	Cadherin extracellular domain
EC1-5	Cadherin extracellular domains one through five
MHC	Major histocompatibility complex
SFA	Surface force apparatus
TCR	T-cell receptor

References

Attard, P. and Parker, J.L. 1992. Oscillatory solvation forces: A comparison of theory and experiment. *J. Phys. Chem.* 92: 5086–5093.

Balsera, M., Stepaniants, S., Izrailev, S., Oono, Y., and Schulten, K. 1997. Reconstructing potential energy functions from simulated force-induced unbinding processes. *Biophys. J.* 73: 1281–1287.

Bassereau, P. and Pincet, F. 1997. Quantitative analysis of holes in supported bilayers providing the adsorption energy of surfactants on solid substrate. *Langmuir* 13: 7003–7007.

Bayas, M.V., Schulten, K., and Leckband, D. 2003. Forced detachment of the CD2–CD58 complex. *Biophys. J.* 84: 2223–2233.

Bayas, M.V., Kearney, A., Avramovic, A., van der Merwe, P.A., and Leckband, D.E. 2007. Impact of salt bridges on the equilibrium binding and adhesion of human CD2 and CD58. *J. Biol. Chem.* 282: 5589–5596.

Bodian, D.L., Jones, E.Y., Stuart, D.I., and Davis, S.J. 1994. Crystal structure of the extracellular region of the human cell adhesion molecule CD2 at 2.5 A resolution. *Structure* 2: 755–766.

Boggon, T., Murray, J., Chappuis-Flament, S., Wong, E., Gumbiner, B.M., and Shapiro, L. 2002. C-Cadherin ectodomain structure and implications for cell adhesion mechanisms. *Science* 296: 1308–1313.

Born, M. and Wolf, E. 1980. *Principles of Optics*. Oxford, NY: Pergamon.

Calvert, T.L. and Leckband, D. 1997. Two-dimensional protein crystallization at solid-liquid interfaces. *Langmuir* 13: 6737–6745.

Chien, Y.-H., Jiang, N., Li, F., Zhang, F., Zhu, C., and Leckband, D. 2008. Two-stage cadherin kinetics require multiple extracellular domains but not the cytoplasmic region. *J. Biol. Chem.* 283: 28454–28563.

Darst, S.A., Ahlers, M., Meller, P.H., Kubalek, E.W., Blankenburg, R., Ribi, H.O., Ringsdorf, H., and Kornberg, R.D. 1991. Two-dimensional crystals of streptavidin on biotinylated lipid layers and their interactions with biotinylated macromolecules. *Biophys. J.* 59: 387–396.

Davis, S.J. and van der Merwe, P.A. 1996a. CD2-An exception to the immunoglobulin superfamily concept. *Science* 273: 1241–1242.

Davis, S.J. and van der Merwe, P.A. 1996b. The structure and ligand interactions of CD2: Implications for T-cell function. *Immunol. Today* 17: 177–187.

Davis, S.J., Ikemizu, S., Wild, M.K., and van der Merwe, P.A. 1998. CD2 and the nature of protein interactions mediating cell-cell recognition. *Immunol. Rev.* 163: 217–236.

Davis, S.J., Ikemizu, S., Evans, E.J., Fugger, L., Bakker, T.R., and van der Merwe, P.A. 2003. The nature of molecular recognition by T cells. *Nat. Immunol.* 4: 217–224.

Dudko, O.K., Hummer, G., and Szabo, A. 2006. Intrinsic rates and activation free energies from single-molecule pulling experiments. *Phys. Rev. Lett.* 96: 108101.

Evans, E. and Ritchie, K. 1997. Dynamic strength of molecular adhesion bonds. *Biophys. J.* 72: 1541–1555.

Evans, E.J., Castro, M.A., O'Brien, R., Kearney, A., Walsh, H., Sparks, L.M., Tucknott, M.G., Davies, E.A., Carmo, A.M., van der Merwe, P.A. et al. 2006. Crystal structure and binding properties of the CD2 and CD244 (2B4)-binding protein, CD48. *J. Biol. Chem.* 281: 29309–29320.

Frink, L.J. and vanSwol, F. 2000. A common theoretical basis for surface forces apparatus, osmotic stress, and beam bending measurements of surface forces. *Colloid Surf. A: Physicochem. Eng. Aspects* 162: 25–36.

Gumbiner, B.M. 1996. Cell adhesion: The molecular basis of tissue architecture and morphogenesis. *Cell* 84: 345–357.

Gumbiner, B.M. 2005. Regulation of cadherin-mediated adhesion in morphogenesis. *Nat. Rev. Mol. Cell. Biol.* 6: 622–634.

He, W., Cowin, P., and Stokes, D.L. 2003. Untangling desmosomal knots with electron tomography. *Science* 302: 109–113.

Helm, C.A. and Israelachvili, J.N. 1993. Forces between phospholipid bilayers and relationship to membrane fusion. *Methods Enzymol.* 220: 130–143.

Helm, C.A., Israelachvili, J.N., and McGuiggan, P.M. 1989. Molecular mechanisms and forces involved in the adhesion and fusion of amphiphilic bilayers. *Science* 246: 919–922.

Helm, C.A., Israelachvili, J.N., and McGuiggan, P.M. 1992. Role of hydrophobic forces in bilayer adhesion and fusion. *Biochemistry* 31: 1794–1805.

Horn, R.G. and Israelachvili, J.N. 1988. Molecular organization and viscosity of a thin film of molten polymer between two surfaces as probed by force measurements. *Macromolecules* 21: 2836–2841.

Hummer, G. and Szabo, A. 2003. Kinetics from nonequilibrium single-molecule pulling experiments. *Biophys. J.* 85: 5–15.

Hunter, R. 1989. *Foundations of Colloid Science*. Oxford, NY: Oxford University Press.

Ikemizu, S., Sparks, L.M., van der Merwe, P.A., Harlos, K., Stuart, D.I., Jones, E.Y. and Davis, S.J. 1999. Crystal structure of the CD2-binding domain of CD58 (lymphocyte function-associated antigen 3) at 1.8-A resolution. *Proc. Natl. Acad. Sci. U.S.A* 96: 4289–4294.

Israelachvili, J. 1973. Thin film studies using multiple-beam interferometry. *J. Colloid Interface Sci.* 44: 259–272.

Israelachvili, J. 1992a. Adhesion forces between surfaces in liquids and condensable vapours. *Surf. Sci. Rep.* 14: 110–159.

Israelachvili, J. 1992b. *Intermolecular and Surface Forces.* New York: Academic Press.

Israelachvili, J.N. and Adams, G.E. 1978. Measurement of forces between two mica surfaces in aqueous electrolyte solutions in the range 0-100 nm. *J. Chem. Soc. Faraday Trans. I* 75: 975–1001.

Israelachvili, J. and McGuiggan, P. 1990. Adhesion and short-range forces between surfaces: New apparatus for surface force measurements. *J. Mater. Res.* 5: 2223–2231.

Israelachvili, J.N. and Pashley, R.M. 1983. Molecular layering of water at surfaces and origin of repulsive hydration forces. *Nature* 306: 249–250.

Johnson, K.L., Kendall, K., and Roberts, A.D. 1971. Surface energy and the contact of elastic solids. *Proc. R. Soc. Lond. A* 324: 301–313.

Johnson, C.P., Jensen, I.E., Prakasam, A., Vijayendran, R., and Leckband, D. 2003. Engineered protein a for the orientational control of immobilized proteins. *Bioconjug. Chem.* 14: 974–978.

Johnson, C.P., Fujimoto, I., Perrin-Tricaud, C., Rutishauser, U., and Leckband, D. 2004. Mechanism of homophilic adhesion by the neural cell adhesion molecule: Use of multiple domains and flexibility. *Proc. Natl. Acad. Sci. U.S.A* 101: 6963–6968.

Johnson, C.P., Fujimoto, I., Rutishauser, U., and Leckband, D.E. 2005. Direct evidence that neural cell adhesion molecule (NCAM) polysialylation increases intermembrane repulsion and abrogates adhesion. *J. Biol. Chem.* 280: 137–145.

Jones, E.Y., Davis, S.J., Williams, A.F., Harlos, K., and Stuart, D.I. 1992. Crystal structure at 2.8 Å resolution of a soluble form of the cell adhesion molecule CD2. *Nature* 360: 232–239.

Kekicheff, P., Ducker, W.A., Ninham, B.W., and Pilen, M.P. 1990. Multilayer adsorption of cytochrome c on mica around isoelectric pH. *Langmuir* 6: 1704–1708.

Leckband, D. 2007. Surface force apparatus measurements of molecular forces in biological adhesion. In *Handbook of Molecular Force Spectroscopy*, (ed. A. Noy), pp. 1–27. Heidelberg, Germany: Springer.

Leckband, D. 2008. From single molecules to living cells: Nanomechanics of cell adhesion. *Cell. Mol. Bioeng.* 1: 312–326.

Leckband, D. and Israelachvili, J. 2001. Intermolecular forces in biology. *Q. Rev. Biophys.* 34: 105–267.

Leckband, D. and Prakasam, A. 2006. Mechanism and dynamics of cadherin adhesion. *Annu. Rev. Biomed. Eng.* 8: 259–287.

Leckband, D.E., Helm, C.A., and Israelachvili, J. 1993. Role of calcium in the adhesion and fusion of bilayers. *Biochemistry* 32: 1127–1140.

Leckband, D., Schmitt, F.-J., Israelachvili, J., and Knoll, W. 1994. Direct force measurements of specific and nonspecific protein interactions. *Biochemistry* 33: 4611–4624.

Leckband, D., Müller, W., Schmitt, F.-J., and Ringsdorf, H. 1995a. Molecular mechanisms determining the strength of receptor-mediated intermembrane adhesion. *Biophys. J.* 69: 1162–1169.

Leckband, D.E., Kuhl, T.L., Wang, H.K., Müller, W., and Ringsdorf, H. 1995b. 4-4-20 Anti-fluorescyl IgG Fab' recognition of membrane bound hapten: Direct evidence for the role of protein and interfacial structure. *Biochemistry* 34: 11467–11478.

Li, F. and Leckband, D. 2006. Dynamic strength of molecularly bonded surfaces. *J. Chem. Phys.* 125: 194702.

Li, F., Pincet, F., Perez, E., Eng, W.S., Melia, T.J., Rothman, J.E., and Tareste, D. 2007. Energetics and dynamics of SNAREpin folding across lipid bilayers. *Nat. Struct. Mol. Biol.* 14: 890–896.

McAlister, M.S.B., Mott, H.R., van der Merwe, P.A., Campbell, I.D., Davis, S.J., and Driscoll, P.C. 1996. NMR analysis of interacting soluble forms of the cell-cell recognition molecules CD2 and CD48. *Biochemistry* 35: 5982–5991.

Perez, T.D., Nelson, W.J., Boxer, S.G., and Kam, L. 2005. E-cadherin tethered to micropatterned supported lipid bilayers as a model for cell adhesion. *Langmuir* 21: 11963–11968.

Prakasam, A., Chien, Y.H., Maruthamuthu, V., and Leckband, D.E. 2006a. Calcium site mutations in cadherin: Impact on adhesion and evidence of cooperativity. *Biochemistry* 45: 6930–6939.

Prakasam, A.K., Maruthamuthu, V., and Leckband, D.E. 2006b. Similarities between heterophilic and homophilic cadherin adhesion. *Proc. Natl. Acad. Sci. U.S.A* 103: 15434–15439.

Sivasankar, S., Brieher, W., Lavrik, N., Gumbiner, B., and Leckband, D. 1999. Direct molecular force measurements of multiple adhesive interactions between cadherin ectodomains. *Proc. Natl. Acad. Sci. U.S.A* 96: 11820–11824.

Sivasankar, S., Gumbiner, B.M., and Leckband, D. 2001. Direct measurements of multiple adhesive alignments and unbinding trajectories between cadherin extracellular domains. *Biophys. J.* 80: 1758–1768.

Sotomayor, M. and Schulten, K. 2008. The allosteric role of the Ca^{2+} switch in adhesion and elasticity of C-cadherin. *Biophys. J.* 94: 4621–4633.

Tadmor, R., Chen, N., and Israelachvili, J.N. 2003. Thickness and refractive index measurements using multiple beam interference fringes (FECO). *J. Colloid Interface Sci.* 264: 548–553.

Tolansky, S. 1951. Applications of multiple-beam interferometry. *Nature* 167: 815–816.

Tolansky, S. and Omar, M. 1952. Evaluation of small radii of curvature using the light-profile microscope. *Nature* 170: 758–759.

van der Merwe, P.A. and Davis, S.J. 2003. Molecular interactions mediating T cell antigen recognition. *Annu. Rev. Immunol.* 21: 659–684.

Vijayendran, R., Hammer, D., and Leckband, D. 1998. Simulations of the adhesion between molecularly bonded surfaces in direct force measurements. *J. Chem. Phys.* 108: 1162–1169.

Wang, J.H., Smolyar, A., Tan, K., Liu, J.H., Kim, M., Sun, Z.Y., Wagner, G., and Reinherz, E.L. 1999. Structure of a heterophilic adhesion complex between the human CD2 and CD58 (LFA-3) counter receptors. *Cell* 97: 791–803.

Yap, A., Brieher, W.M., Ruschy, M., and Gumbiner, B.M. 1997. Lateral clustering of the adhesive ectodomain: A fundamental determinant of cadherin function. *Curr. Biol.* 7: 308–315.

Yeung, C. and Leckband, D. 1997. Molecular level characterization of microenvironmental influences on the properties of immobilized proteins. *Langmuir* 13: 6746–6754.

Yeung, C., Purves, T., Kloss, A.A., Kuhl, T.L., Sligar, S., and Leckband, D. 1999. Cytochrome c recognition of immobilized, orientational variants of cytochrome b5: Direct force and equilibrium binding measurements. *Langmuir* 15: 6829–6836.

Zhu, B., Davies, E.A., van der Merwe, A., and Leckband, D. 2002. Direct measurements of heterotypic adhesion between the cell adhesion proteins CD2 and CD48. *Biochemistry* 42: 12163–12170.

Zhu, B., Chappuis-Flament, S., Wong, E., Jensen, I.E., Gumbiner, B.M. and Leckband, D. 2003. Functional analysis of the structural basis of homophilic cadherin adhesion. *Biophys. J.* 84: 4033–4042.

9

Force Spectroscopy on Cells

Martin Benoit
*Ludwig-Maximilians-
Universität München*

9.1 Introduction

In the human body, a vast variety of different molecules is circulating, diffusing, and interacting. Cell membranes contain many proteins that are responsible for interacting and communicating with their environment. Some examples are adhesion molecules for cell anchorage or locomotion, membrane pores for molecular exchange, and other receptor molecules. The receptor molecules specifically screen for information mediated by ligand-molecules (e.g., hormones) that match the binding site. Some cellular receptor molecules act like "noses," programmed to transduce the event of a bound ligand through the membrane into the cell by a conformational change (e.g., G-protein-coupled receptors). A change in conformation is a synonym for a mechanical deformation of the molecule. This often triggers a molecular reaction inside the cell by a signaling cascade. A few ligand molecules (e.g., chemokines) might change the behavior of the whole cell, for example, the directed motion along a concentration gradient. For cellular motion, reversible adhesion and force are required. In order to react adequately to external stimuli, a cell can utilize different concepts for tuning its adhesion by directly strengthening the adhesion of an adhesion molecule (affinity), by increasing the number of available adhesion molecules (avidity), or by altering the properties of the cellular anchor of the adhesion molecule. Additionally, the cell can distribute an external load to its adhesion molecules either in a sequential manner (pealing off from the adhesion site one bond after the other at low forces) or in parallel to the grouped adhesion molecules (clustered weak bonds that share the load in parallel and add it up to a very high force). To resolve the concepts behind cellular adhesion, techniques with a single molecular resolution as well as techniques that reveal multi-molecular arrangements are required. Force spectroscopy is a technique that measures forces within or between individual molecules. Performing such experiments on living cells at the level of single molecules not only reveals the strength of a molecular bond but also adhesion strategies and mechanical reactions of cells related to external forces. Like the receptor molecules, the adhesion molecules also scrutinize their environment for specific ligands, but they rather aim at mediating motility or anchorage of the cell. Some adhesion molecules also change their conformation if a ligand has bound and trigger molecular reactions inside the cell. With this concept, the cell can sense the mechanical and chemical properties of the environment it adheres to and consequently react to it. In the following chapter, the technique of force spectroscopy is applied to address questions about cell adhesion forces. How strong is the cell's adherence? What is the maximum force with which a cell can adhere? How does the cell regulate the adhesion strength? What force binds a single adhesion molecule to its ligand?

First, we detail the measurement of the cellular adhesion forces by using the atomic force microscope (AFM).

9.2 AFM and Force Spectroscopy

Like every scanning microscope, the AFM also consists of three basic units: a sensor (interacting with a surface very locally), a scanner (positioning either the probe or the sensor with respect to the sensor or the probe, respectively), and a detector (collecting and converting the signal from the sensor). The force microscope is a mechanical instrument with a resolution to image atoms. Therefore, it is commonly called the atomic force microscope (Binnig et al., 1986). In contrast to the electron microscope (EM), the AFM can even operate in a liquid environment, gaining access to functional biomolecules and living cells. The central sensor of the instrument is a sharp tip on a cantilever that probes the sample mechanically. If the tip, mounted to a small beam that acts like a spring, experiences a force the beam is deflected.

In a small range, this deflection of the beam is proportional to the exerted force and is characterized by a spring constant k according to Hooke's law. A laser reflected from the backside of the cantilever spring reports the deflection Δz to a segmented photodetector with a sub-nanometer resolution. The force on the tip $\Delta F = k^*\Delta z$ can be detected at a few pico Newton resolution. In general, a piezo actuator positions and scans the sample in x–y–z direction at sub-nanometer accuracy. A typical setup of an AFM is shown in Figure 9.1.

The measured signal from the segmented photodetector coordinated with the three-dimensional position of the piezo yields a high-resolution image of the sample surface. Single molecules (Figure 9.3) and even single atoms are visualized with this technique (Hansma et al., 1988; Oesterhelt et al., 2000).

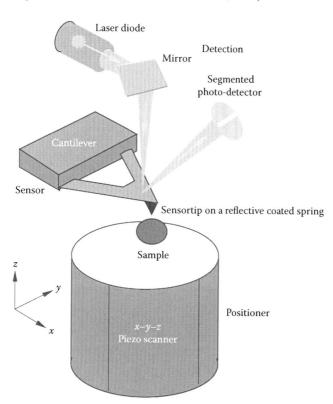

FIGURE 9.1 The AFM consists of three elements: The positioner, a voltage sensitive piezo crystal, coordinates the interaction of the sample and the cantilever tip with sub-nanometer precision. The force sensor is the micro fabricated cantilever with a small tip mounted to a reflective coated beam (here triangular). The movement of the reflected laser in the photo detector reports the deflection of this beam, due to forces acting on the tip.

For the improved force spectroscope designs, the scanning of the sample in the x–y direction is often disabled, in order to optimize the resolution in the z direction ($\Delta F = k^*\Delta z$). During the approach, in a typical force experiment, the repulsive force trace gives viscoelastic information about the sample. After a time of contact at a certain repulsive force, this force trace is reversed and even stretches the sample if adhesion occurs. Therefore, the sample has to be immobilized to the substrate. For receptor–ligand interaction measurements, spacer molecules are typically inserted into the system to avoid nonspecific interactions of the sample with the solid surfaces of the substrate and the tip (Figure 9.2). Such a nonspecific interaction is the first peak measured in the force graph of Figure 9.3 that was performed with a bare tip. Force spectroscopy experiments improve if molecular handles covalently link the spacer to a specific atom of the investigated molecule instead of "grabbing" the molecule nonspecifically with a bare tip. In the next examples of single molecule force experiments, both methods are utilized. The nonspecific method requires no tip functionalization, but the few "good force traces" have to be selected from the nonspecific. If the length of the molecule is known, a good selection criterion is the measured length of the molecule during the stretching calculated from an elastic model (see page 9-6).

9.2.1 Manipulating the Nano-World (Angstrøms, Pico Newtons, and Molecular Handles)

In the following example, the force spectroscopy is combined with high-resolution imaging of the AFM. A plain cantilever at high resolution imaged the bacteriorhodopsin (BR)—the light driven proton pump in the extracellular purple membrane of *Halobacterium salinaris*. BR consists of seven transmembrane alpha helices (helix "a" to "g" in Figure 9.3) forming a pore for protons. Three of these pores form a ring-like structure, as visualized in Figure 9.3. A nonspecific force spectroscopy experiment on this membrane led to the extraction of an individual BR molecule from this purple membrane and showed a typical unfolding pattern that perfectly correlates with the amino acid sequences of the pairs of alpha helices and their lengths.

Molecular handles cloned to the BR at specific amino acids in the loops between the helices or to the extracellular terminus of helix "a" provided certain access points of the molecule directly by a complementary receptor on the cantilever tip (Muller et al., 2002). However, the quality of an AFM image is reduced by such a functionalization of the cantilever. Here, we see two concepts of single molecule force experiments: nonspecific adhesion to a bare cantilever versus specific adhesion to a functionalized cantilever. Both concepts have complementary advantages and disadvantages.

The second concept, with defined spacer and specific handles, is highly reproducible and has an increased success rate. In particular, it has been validated for the generation of specific unfolding signatures of individual molecules (Carrion-Vazquez et al., 2000; Clausen-Schaumann et al., 2000; Rief et al., 2000a; Dietz and Rief, 2004). The use of linker molecules reduces the

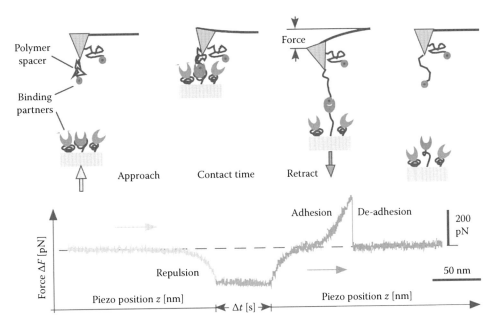

FIGURE 9.2 Sketch of single molecule force experiment with the recorded force trace below. The force sensor approaches at zero force until the repulsive interaction with the surface deflects the cantilever upward. This is represented by a negative repulsion force in the graph below. After a defined contact time the piezo motion is reversed and the repulsive force at the sensor tip is released. An adhesive interaction due to a receptor–ligand interaction deflects the force sensor downward. This is represented by the positive force characteristic for the spacer molecule (here PEG). Finally, the applied force overcomes the strength of the receptor ligand bond (here biotin–avidin) and the measured force drops instantaneously to zero.

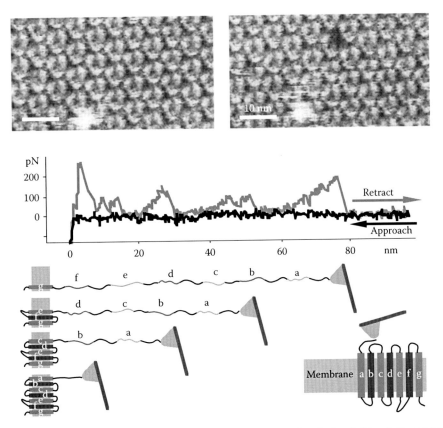

FIGURE 9.3 AFM images of individual bacteriorhodopsin molecules in a bacteria surface, before (left) and after (right) a single molecule force experiment. The peaks of the force curve between the two images are represented by the scheme below indicating (except the first peak from nonspecific interaction) the unfolding of alpha helices in pairs until the final extraction of helix "g" from the membrane. (From Oesterhelt, F. et al., *Science*, 288, 143, 2000. With permission.)

nonspecific interaction of the molecule with the surfaces (e.g., the bare cantilever tip).

The first concept is easy in preparation and has good imaging properties. To decide whether the measured force trace was specific or originated from some other nonspecific interaction, one hint is provided by the image lacking one molecule after the force experiment. Other hints are the characteristic force signature of the investigated molecule (should it be known already) and the constraints for the length of the molecule. According to this example, each experiment with a length that does not roughly match the amino acid length of the BR, indicates that a different molecule did attach to the tip (too long), or the BR molecule was "fished" somewhere else other than the terminus of the helix "a" (too short).

The use of polymeric molecular constructs that include an unknown molecule in the middle of a sequence of known molecules guarantees that the unknown force signature is included in the force trace if more than half of the known molecules were recognized in that force trace (Dietz and Rief, 2004; Kufer et al., 2005; Puchner et al., 2008b).

9.2.2 Equilibrium Thermodynamics versus Forced Unbinding

Viruses and bacteria developed molecules that specifically bind to the molecules on their target cell. In return, immune cells continuously screen the body for foreign molecules and produce antibodies that specifically bind to any molecule that does not belong to the organism. Pharmacology and pharmaceutical industry have characterized such receptor–ligand interactions for quite some time by their dissociation constant K_D to screen for more effective ligands, antibodies, or inhibitors with better K_D values, to develop better and more specific medication.

Force spectroscopy has opened up new avenues for measuring molecular interaction strengths. What is the mechanical force that holds an individual ligand in its receptor? In the illustration (Figure 9.4), this will be the force needed to pull the bond out of the potential valley of the binding energy, ΔG, and to lift it over the hill of the activation energy, ΔG_{on}. Theoretical work based on classical thermodynamics was offered by Bell (1978) and translated to force measurements by Evan (1998).

In thermodynamic equilibrium, we can measure the dissociation constant K_D of two substances a and b forming a complex ab. K_D is the fraction of the concentrations of the unbound ligands [a] and [b], and the complex [ab]:

$$K_D = \frac{[a][b]}{[ab]} = \frac{k_{off}^0}{k_{on}^0} \tag{9.1}$$

K_D also resembles the fraction of the kinetics k_{on} (how fast does a ligand bind at a given concentration?) and k_{off} (how fast is a ligand released?).

The binding energy is composed by

$$\Delta G = \Delta G_{off} - \Delta G_{on} \tag{9.2}$$

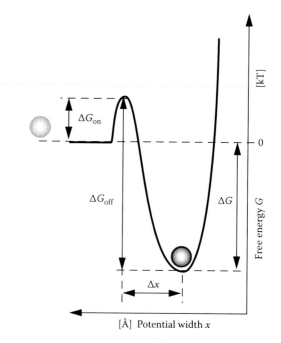

FIGURE 9.4 A symbolic receptor–ligand complex is represented by the sphere in the minimum of the potential valley. The energy barrier, ΔG_{off}, consisting of the Gibbs free energy, ΔG, and the activation energy, ΔG_{on}, has to be overcome to separate the receptor–ligand complex.

For 1 mol of molecules, the free energy without an external force is

$$\Delta G = \Delta H - T\Delta S = -N_A k_B T \ln K_D \tag{9.3}$$

k_B is the Boltzmann constant. Thus, for a single bond we can formulate

$$K_D = e^{\frac{\Delta G}{k_B T}} \tag{9.4}$$

With force spectroscopy, we study the forces needed to separate the ligand from its receptor.

While the values gained from the thermodynamic equilibrium experiments do not change within a certain concentration limit, the unbinding force of a biotin–avidin bond, initially measured at 160 pN in 1996, turned out to range from 0 pN to more than 300 pN. Unfortunately, we will see that there is no "typical force" of a molecular bond but the force depends on the experimental setup and the history of the force application. A force sensor with a lower spring constant can measure lower forces. We will see that a bond generally opens at higher forces the lower the ambient temperature T is. An unbinding experiment done with faster separation velocities tends to result in higher forces than a slow experiment. What then is the sense in measuring such a force at all? It seems like a famous dilemma from quantum mechanics: The design of the experiment anticipates the result. The value of the wave function is created at the moment of measurement (particle or wave, momentum or location, moment or energy). A key to this "dilemma of the force spectroscopy" is the thermal energy (Brownian motion) that randomly influences various packages of energy in the order of k_BT on every molecular bond.

9.2.2.1 Gedankenexperiment

If we consider an experiment without any external force, we will find a certain fraction (K_D) of molecules dissociated even without an external force! At higher temperatures, this fraction will be even bigger until the thermal energy (k_BT) far exceeds the energy of the molecular bond (ΔG) and almost no bond is stable anymore (Equation 9.4). Increasing the ambient temperature T in Figure 9.4 is like flooding the potential valley with thermal energy in a way that the complex "floats" on the increasing energy levels until it exceeds the energy barrier. This is the melting temperature of a molecular complex. In the absence of an external force, the fraction of dissociated bonds is solely generated by the stochastic impact of the thermal energy of the ambient temperature. If we apply a force at this temperature to that molecular bond, we support the thermal energy in dissociating the bond by adding mechanical energy and by rectifying the thermal motion of the complex: Once it is unbound, it stays unbound. If we apply the force very slowly, the probability for the bond to dissociate at forces close to zero due to the Brownian motion is still quite high. If we pull faster, we reduce the time to await the thermally driven unbinding. This results in a reduced unbinding probability and a reduced contribution to the bond dissociation of the thermal energy. Accordingly, the fraction of the additional required mechanical energy to overcome the potential barrier finally increases. As a consequence, we measure a statistically higher unbinding force according to the increasing remaining mechanical energy the faster we load the force to

the bond. This behavior is represented by the concept of the loading rate (how fast is the force loaded to the bond?) as seen in Figure 9.5 (Merkel et al., 1999).

The Van't Hoff–Arrhenius equation under external force considers the separation of a single bond with a rate k_{off}:

$$k_{off} = v \cdot e^{-\frac{\Delta G_{off} - F\Delta x}{k_BT}} = k_{off}^0 \cdot e^{\frac{F\Delta x}{k_BT}} \qquad (9.5)$$

where

v is a factor that can be viewed as a frequency that probes the potential landscape of the bond

k_{off}^0 is the equilibrium off-rate in the absence of force

From this equation, we can see that the energy landscape depicted in Figure 9.11 becomes tilted by the subtraction of $F^*\Delta x$. The stronger the force, the more the potential is shifted. Deeper inner barriers become the maximum of the potential, the more the potential is shifted. The force experiment then probes the next potential valley behind the new highest barrier. After some calculations to be followed by literature (Walton et al., 2008), we come to the key equation for the most probable force F to be measured:

$$F(f) = \frac{k_BT}{\Delta x} \cdot \ln\left(f \cdot \frac{\Delta x}{k_{off} \cdot k_BT} \right) = \frac{\Delta G_{off}}{\Delta x} \cdot \ln\left(f \cdot \frac{\Delta x}{v \cdot k_BT} \right) \qquad (9.6)$$

F logarithmically depends on the loading rate $f = dF/dt$. The Brownian motion affecting the bond results in a stochastic deviation around this force value. Thus, we call it the most probable rupture force. A well-conducted force spectroscopy experiment collects a statistically significant amount of

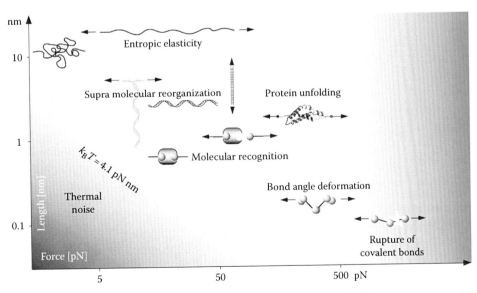

FIGURE 9.5 The potential width (Δx) versus force (F) graphic opens an energy window (white) for force measurements. Molecular bonds weaker than k_BT (located in the gray area below the diagonal line) are not stable at room temperature ($T = 293\,\text{K}$). It is not possible to exert forces along a molecule that are stronger than the weakest covalent bond without disintegrating the molecule (gray area on the right hand side).

unbinding force measurements to determine this most probable unbinding force from the force distribution histogram (see Figure 9.12). From Equation 9.6, we can now understand the behavior of the forces in the overview graph of Figure 9.10: The most probable measured force increases with the logarithm of the loading rate and decreases with the temperature T (particularly since $\Delta G = \Delta H - T\Delta S$). Now we also understand that early MD-simulations on the biotin–avidin complex pulled with the velocities of nm/ps probed potentials that are far beyond the thermal energy range. Such deep potentials were probed force-spectroscopically by Michel Grandbois et al. for the first time in 1999 (Grandbois et al., 1999).

The AFM can be used for single molecule force experiments on biomolecules only in a window (Figure 9.5) between the force resolution limit of a few piconewton, a barrier of the thermal energy $k_{B}T$, and the strength of the weakest covalent bond in the spanned molecule chain. Below the AFM force resolution limit down to the Femto–Newton level, the optical and magnetic traps still can probe long ranged molecular interaction (Rief et al., 2000b; Kruithof et al., 2008). We know very well about that restriction due to the thermal energy, $k_{B}T$, at the experiments' temperature, since a complex with a binding energy $\Delta G \leq k_{B}T$ is statistically open and not stable enough to be probed by force spectroscopy.

9.2.2.2 Single Molecule Force Measurement versus Ensemble Measurement of K_{D}

From a mechanistic point of view, the thermal energy represented by the Brownian motion of the molecules steadily kicks against the complex from various directions and with various velocities and momentums. This stochastically forces the bond to open from time to time. With a force spectroscopic experiment, we only pull in one single direction. The slower we pull the more probable the thermal energy helps to open the bond. All these considerations are based on the Bell–Evans model (Bell, 1978; Evans, 1998). In a force spectroscopic experiment, we detect a single molecular force in cooperation with the thermal energy. We need statistically relevant numbers of single molecule measurements to identify the most probable unbinding force, and with the help of the Bell–Evans model, we can explore the potential landscape of the bond. The correlation to equilibrium thermodynamics with the help of the Jarzynski theorem (Collin et al., 2005) is sometimes complicated. The forces of many molecular interactions have been calculated back to the natural off-rates k_{off} measured by techniques such as calorimetry or SPR. Sometimes, the differences are still significant (Dettmann et al., 2000; Morfill et al., 2007) because force spectroscopy sets an unbinding direction that might not be used by equilibrium measurements, as we will see later (page 9-7). SPR and calorimetry also use theories, models, and parameters for their machines to recalculate the off-rate, the on-rate, and the dissociation constant, K_{D}, respectively. Another basic difference is that SPR and calorimetry do ensemble measurements, while force-spectroscopy averages single measurements of individual molecules.

Nevertheless, force spectroscopy directly and uniquely allows access to the most probable adhesion force, F, the potential width, Δx, and the unbinding energy, ΔG_{off}. This is complementary information to the equilibrium constant, K_{D}, and the off-rate, k_{off} (Morfill et al., 2007). For each reliable force experiment, the temperature and all of the viscoelastic constraints of each experimental setup have to be taken into account. The experimental setup includes the hard substrate over the molecular anchor (spacer included), the complex itself, the second anchor (spacer), and the cantilever with a known spring constant (see Figure 9.1). In a common first order approximation, an effective spring constant is calculated by the instruments' spring constant multiplied by the pulling velocity. The common approximation of the loading rate as the spring constant multiplied by the sensor velocity only holds for a rigid and stiff connection of the binding partners to the substrate and the force sensor. Any molecular spacer in series to the bond with elastic properties in the range of the force sensor or softer will change the loading rate in a sometimes nonlinear manner (Rief et al., 1997b). This is numerically adjustable if a model can represent the mechanical behavior of the spacer. With such a model of the setup's viscoelastic properties, the characteristics of the bond (potential width, Δx, and depth, ΔG) can be evaluated precisely (e.g., WLC-model as shown later).

9.3 Mechanical Properties of Single Molecules

According to the second law of thermodynamics, polymers like DNA or proteins maximize their entropy by relaxing into the energetically lowest conformation (random coil or even specifically folded proteins). Applying an external force to the ends of such a polymer increases the end-to-end distance, x, to an energetically less favored conformation with decreased entropy. This less-favored conformation generates a restoring force F due to the reduced entropy.

9.3.1 Entropic Polymer Elasticity (the Worm-Like Chain Model)

A simple model for such a molecule is the freely jointed chain (FJC) model, consisting of N segments of the length d^* (Figure 9.6).

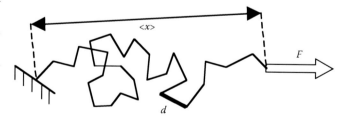

FIGURE 9.6 Sketch of the FJC model: Force F, end-to-end distance $\langle x \rangle$ and length of each chain segment d.

---

* A DNA single strand consists of phosphate groups with a connecting sugar ring ($d = 5.7\,\text{Å}$), containing a base.

From thermodynamics, we can derive the average end-to-end distance <*x*> under an external Force *F* as a partial derivative of *G* with respect to *F* at a constant temperature *T*:

$$<x> = -\left(\frac{\partial G}{\partial F}\right)_T = -k_B T \frac{\partial \ln(Z(F))}{\partial F}$$

$$= Nd\left(\coth\left(\frac{Fd}{k_B T}\right) - \frac{k_B T}{Fd}\right) = NdL\left(\frac{Fd}{k_B T}\right) \quad (9.7)$$

G is the free energy of the polymer conformation
Z is the partition function
L is the Langevin function

In order to apply this function to a measured force curve of a stretched polymer the inverse Langevin function describes the average force <*F*> that is required to extend the end-to-end distance *x*:

$$<F(x)> = \frac{k_B T}{d} L^{-1}\left(\frac{x}{Nd}\right) \quad (9.8)$$

Assuming a semi flexible chain with a bending stiffness *B*, the characteristic length *p* can be defined as persistence length: $p = B/k_B T$. For forces $F \gg k_B T/p$, an approximation known as worm-like chain (WLC) model holds:

$$F(x) = \frac{k_B T}{4p}\left(4\frac{x}{L} + \frac{1}{(1-(x/L))^2} - 1\right) \quad (9.9)$$

$L = Nd$ is the overall length of the polymer backbone. Other approximations for polymers with different properties are the freely rotating chain (FRC) and the freely jointed springs (FJS) model (Hugel et al., 2005). The result of fitting a polymer model to a force distance trace gives the typical persistence length* and the overall length of the polymer's backbone, respectively (Rief et al., 1997a; Oesterhelt et al., 2000).

9.3.2 Single Molecule Experiments

Many force measurements of the elastic properties of single molecules have been performed (Butt et al., 2005). As mentioned before, the design (temperature, loading rate, spring constant etc.) of such an experiment is important to gain useful results. In the following section, we will focus on the reaction pathway of the separating receptor–ligand complex and the influence of the position of the "molecular handles" on the measured force.

An external force might define an "artificial" reaction pathway to a complex (arrow in Figure 9.7) over an activation barrier invisible for calorimetric measurements following along the "natural" pathway. This barrier might appear only in force measurements at increasing loading rates. At slow loading rates close

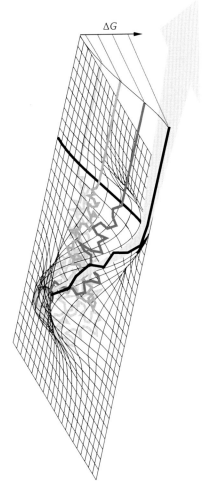

FIGURE 9.7 Sketch of a complex bound in a potential valley with an activation barrier (Δ*G*). Unbinding trajectories with increasing loading rates light to black become increasingly restricted to the reaction coordinate (big arrow) over the barrier by the external force.

to the natural off-rate (light gray trajectory), this barrier might stay undetected, whereas at higher loading rates it is measured (darker trajectories).[†] A reaction coordinate perpendicular to the sketched one Figure 9.7 would not explore this barrier even at high loading rates.

9.3.2.1 Hands-on Molecules (DNA)

Measurements of the adhesion force between two individual DNA strands show us that the relation between the binding energy, Δ*G*, and measured force, *F(f)*, is not trivial, but nevertheless easy to understand.

In Figure 9.8, force spectroscopic experiments stretching a double stranded DNA molecule are shown at different velocities.

Whereas the force curves are similar until the end of the force plateau at all velocities, a hump appears at increased velocities at the

---

* Does not necessarily correlate to the physical length of the polymer segment.

† An additional reason for the higher forces in the MD simulations at high loading rates is the appearance of such barriers that otherwise are circumvented at normal velocities.

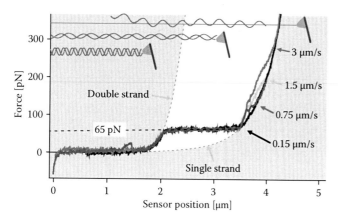

FIGURE 9.8 Four force experiments stretching a 2 μm long DNA double strand at different velocities (3, 1.5, 0.75, and 0.15 μm/s). A sketch of the DNA molecule fixed between the substrate and the force-sensing cantilever tip depicts the three stretching regimes. First regime: the double stranded DNA is stretched (along the dotted gray borderline) up to the second regime. The second regime (white background) links to the third regime via a force plateau at 65 pN (dotted black line). The third regime stretches a single DNA strand (along the dotted gray borderline). The dotted gray borderlines are WLC-fits to the force curve.

end of the plateau. Corresponding to Figure 9.7, an activation barrier becomes visible here that might be related to the friction that occurs while separating and unwinding the two strands at high velocities.

In the first regime, the two strands are wound up well and are stably connected to a double strand by the Watson–Crick base pairs until the force increases up to 65 pN. Now, the DNA undergoes structural changes and the strands start to separate until the end of the force plateau. This resembles a phase transition like a melting process to single stranded DNA. This pronounced plateau is a transition in thermodynamic equilibrium; thus the force does not depend on velocity.

If we consider the same experiment (as in Figure 9.8) with the force sensor attached to the blue strand, while the red strand stays attached to the substrate (shear mode), the experiment would terminate at the end of the 65 pN plateau: As soon as the two strands are separated, the connection to the substrate is lost, the force drops to zero and no single strand is stretched further.

Here, we understand that the mount for the handles connecting the molecules to the force spectroscopy experiment determines the reaction pathway. DNA is an ideal molecule for explaining the effect of the molecular handles in force spectroscopy, as three kinds of elucidating force experiments are possible:

The single strand stretch mode, where the handles are only connected to one strand at opposite ends (Figure 9.8).

The shear mode, where the handles are connected to either strand at opposite ends (Figure 9.9 left).

The zipper mode, where the handles are connected to either strand at the same end, as presented in the next example (see Figure 9.9 right).

In Figure 9.9, the DNA-oligomer sequence of the 30 base-pair DNA strands is identical in both modes. Thus, the binding energy, ΔG, is also identical, but the measured forces, even though pulled at the same velocity, are different. This example again shows: when opening a complex, the measured force is not correlated to the binding energy in a trivial relation. However, the forced unbinding experiments reveal new insights into the molecular structure of a bond. As expected, the length of the DNA-oligomer tunes the unbinding force (Figure 9.9 left). In the shear mode, a longer DNA-oligomer leads to an increased unbinding force but not higher than the 65 pN force plateau (BS-transition). In the shear geometry, the base pairs cooperatively resist the increasing force by equally distributing the overall force among each other. The shear potential is deep but the length is just a few Ångström. In the zipper mode, the full force is loaded sequentially to a single base

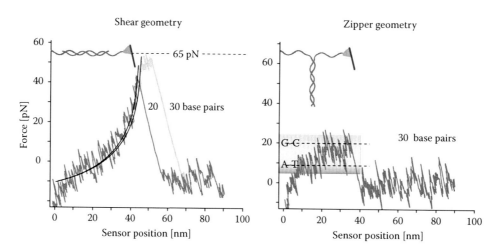

FIGURE 9.9 Unbinding force curves of DNA oligomers in the shear mode (dark curve 20 and gray curve 30 base pairs) and in the zipper mode (right curve 30, base pairs). Black lines represent WLC fits to the force curves. In the zipper mode, the force balances between 9 pN (A-T base pairing force) and 20 pN (G-C base pairing force). (From Rief, M. et al. *Nat. Struct. Biol.*, 6, 346, 1999; Courtesy of Zimmermann, J. and Kufer, S.)

pair, one by one. The resulting forces are dominated by the base pairing forces of AT = 9 pN and GC = 20 pN.* The zipper potential is shallower, but its length directly corresponds to the length of the DNA. Looking at this example, we easily understand why the result of a force spectroscopy experiment strongly depends on its design (where are the handles fixed?). In terms of force spectroscopy as an alternative technique to determine ΔG this is bad news, but it can be useful to understand nano structures, molecular geometry, and mechanics (Dietz and Rief, 2006; Puchner et al., 2008a). DNA provides three force standards: in the shear mode, a long DNA will clamp the maximum force at 65 pN, in the zipper mode a DNA that is purely composed from AT base pairs clamps the force at 9 pN, and if it is purely composed from GC base pairs at 20 pN. A sequence of hierarchically designed DNA force experiments incorporating one single DNA 20-mer single strand on the cantilever tip that matches in shear geometry to a transporter strand immobilized with a remaining sequence in shear geometry to a substrate yields a molecular cut and paste apparatus† (Kufer et al., 2008).

9.3.2.2 Receptor–Ligand Interaction Forces of the Biotin–Avidin Complex

Receptor–ligand interactions and intermolecular recognition are essential for life. Typically, biomolecules interact in a noncovalent manner. They are meant to separate again so that they can be reused again.‡

The first non-covalent interaction that was studied with force spectroscopy, the avidin–biotin interaction, is also one of the strongest. Biotin, a small peptide (Vitamin H) meanwhile became commonly used as a strong non-covalent standard linker (together with the tetrameric protein avidin or streptavidin) for biomolecular investigations (labeling, functionalization, or immobilization).

In less than 10 years of its invention, the AFM (Binnig et al., 1986) turned out to be a tool to obtain high-resolution images of atoms, molecules, and living cells (Radmacher et al., 1992), and pioneering work on intermolecular force measurements was done. The biotin–avidin bond was the first to be studied by AFM force-spectroscopy (Florin et al., 1994). Consequently, the separation of these interacting molecules was the first to be calculated by MD simulations initialized by Grubmüller et al. (1996). At that time, molecular dynamic simulations of a few picoseconds were extremely time consuming and the experimental timescales of milliseconds were still hard to address.

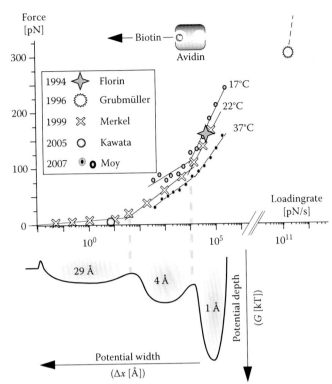

FIGURE 9.10 Investigations on the biotin–avidin interaction force plotted against the logarithm of the force-loading rate pioneered by AFM force-spectroscopy (Florin) and by early molecular dynamic simulation (Grubmüller, dotted line connects to further points out of the graphics range). Thorough studies confined the theory of this molecular bond by bio-force sensors (Merkel), by optical trap (Kawata), and by AFM (Moy, who also revealed a temperature dependency). Line fits indicate distinct regimes of the intermolecular bond: A schematic drawing of a possible potential landscape representing these regimes of the biotin–avidin interaction derived from Schulten et al. is sketched below the graph and outlined in Figure 9.11.

The most coherent picture of the biotin–avidin interaction is presented by Merkel et al. (1999), Rico and Moy (2007), and Isralewitz et al. (2001). The resulting potential landscape of the biotin–avidin interaction is schematically indicated in Figure 9.10. The very shallow and long ranged part (29 Å) of the potential is only accessible at slow loading rates (below 40 pN/s) and with very soft spring constants (<1 pN/nm) of BFS or OT.§ (A list of the spring constants from the experiments in Figure 9.10 is given in pN/nm [instrument]: Florin 39 [AFM], Grubmüller 2800 [MDS],¶ Merkel 0.1–3 [BFS], Moy 10 [AFM], and Kawata 0.045 [OT].)

---

* Unfortunately, the thermal energy of the Brownian motion exceeds the signal of the individual separating base pairs and defeats reading the DNA sequence with this technique.

† The free single-stranded end of the 50-mer transporter DNA sticking to a substrate in the 20-mer zipper geometry is picked up by the stronger force of the 20-mer shear geometry on the cantilever tip. The transporter DNA finally is transferred to a matching single strand 30-mer on another substrate in a force experiment between the stronger 30-mer shear geometry on the substrate and the 20-mer shear geometry on the cantilever.

‡ This does not hold for the so-called suicide-couplers; they covalently bind. Once bound, they will never open again (Kufer et al. 2005).

§ The lower force of the OT measurement might be due to a local heating above room temperature by the laser trap.

¶ The molecular dynamic simulations are performed at very high velocities in a completely different time frame (picoseconds instead of milliseconds). The mechanical properties of the molecular bond are different and the forces quite high and even up to nN, as indicated by the dashed line in such high velocities are inaccessible for the force measurements so far due to the hydrodynamic effects.

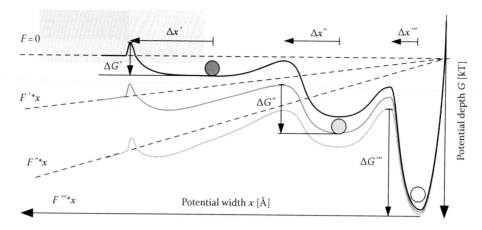

FIGURE 9.11 The three valleys of the biotin–avidin potential landscape in thermodynamic equilibrium without external force are sketched in black. Applying an external force tilts the energy landscape by a factor $F*x$ as sketched in grey. The shallow $\Delta x' = 29$ Å wide, part of the potential with an energy barrier height of $\Delta G_{off}'$ is probed in the force window $0 \leq F \leq F'$. An inner energy barrier $\Delta G_{off}''$ becomes "visible" behind the lowered barrier $\Delta G_{off}'$ in the force window $F' \leq F \leq F''$ that tilts the potential further. Now, the part of the potential with the width of $\Delta x'' = 4$ Å is probed behind that barrier. Further tilting the potential landscape, the energy barrier $\Delta G_{off}'''$ finally surmounts $\Delta G_{off}''$ and the last part of the potential landscape $\Delta x''' = 1$ Å is probed at forces $F \geq F''$.

The intermediate potential valley (4 Å) is represented by the steeper slope in the plot and the deepest and narrowest part of the potential (1 Å) is represented by the steepest slope to the right-end at loading rates above 10 nN/s.

Literally, this loading rate dependent behavior of the force regimes can be visualized (see Figure 9.11) by a potential tilt ($F*x$) through the applied force. This picture is directly derived from the Van't Hoff–Arrhenius equation (Equation 9.6) under external force.

From a physicist's point of view, it is favorable to use spacer molecules that are well represented by models (WLC, FJC, etc.) for studying molecular interactions by force spectroscopy. Here it is possible to exactly recalculate the contributions of the incorporated molecules from the data. However, does this represent the "natural" situation of the investigated complex? The next example shows a complex of interest that is immobilized, as naturally produced, on a bacterium. Unfortunately, this natural spacer (the P-pilus) has a complex force-extension characteristic and cannot be described by a basic thermodynamic model!

9.3.2.3 The PAP-G–Galabiose Bond: Single Molecule Force Spectroscopy on a Cell

In this example of the force measurement between the galabiose and the PAP-G molecule, we study a system that was designed naturally for an external loading force. The P-pilus is a structure of PAP-units that are non-covalently linked to each other like a row of matching pieces of a puzzle. The PAP-units additionally are stacked together, forming a stiff helical rod. At the very tip of the P-pilus, the adhesion molecule PAP-G specifically binds to the galabiose on the surface of an epithelial cell of the urinary tract. The question of how to correctly apply the handles to the receptor ligand complex is solved by itself, because we perform the experiment as close to the real situation as possible.

The galabiose is covalently linked to the PEG spacer on the AFM cantilever at the same side-group where it is *in vivo* linked to the glycocalix of a cell membrane. The PAP-G unit remains linked to the *Escherichia coli* bacterium as it is assembled on a P-pilus. A typical force-extension plot of a P-pilus is shown in Figure 9.12 with a highly reproducible pattern for all velocities: a plateau at 27 pN, followed by a steeper shoulder and then a WLC-like behavior. With a numerical reproduction of this curve, the load to the bond during an experiment can be simulated in a Monte-Carlo calculation. Together with the Bell–Evans model, a certain set of parameters for the potential width $\Delta x = 7 \pm 1.5$ Å and an off-rate $= 8*10^{-4}$/s resembles the measured force distribution correctly (Figure 9.12). As we would expect from the previous consideration, the unbinding probability at 27 pN is increased because the bond is probed a few seconds in the long unstacking plateau at 27 pN. The most probable rupture force at 48 pN matches the force range of the transition 2. In the force range above 100 pN, the PAP-units of P-pilus would start to disintegrate irreversibly (Lugmaier et al., 2008). With this mechanism, the *E. coli* bacterium tries to stick to a target cell. In case of shear flow, the *E. coli* might be flushed away because the connection to the galabiose would be loaded and each bond would open up one by one if the force would exceed 48 pN.* However, thanks to the P-pilus, the bond is loaded only at 27 pN for a while and all the neighboring bonds on the other pili are also loaded in parallel at 27 pN. It is the P-pilus' strategy to parallelize each of the PAP-G bonds, so that the overall force is not 48 pN in a sequence (like in the unzipping of DNA) but continuously N times 27 pN (where N is the number of the bound pili). If the end of the plateau is reached, the force increases slowly but steadily. The range of the

---

* *E. coli* might use a stronger bond with a stronger force, but then it would be hard to reopen the bond and to reuse it at another binding site or the pilus might irreversibly disintegrate.

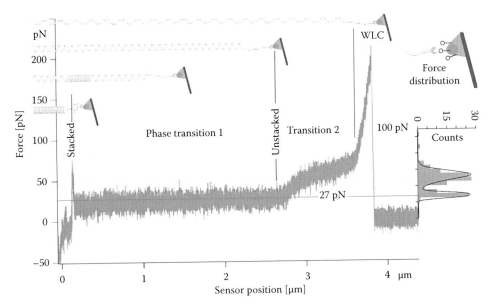

FIGURE 9.12 Force extension trace of a P-pilus and the force distribution between PAP-G and galabiose. A sketch of the unstacking pilus extended between the substrate (*E. coli*) and the galabiose decorated cantilever tip depicts four stretching regimes until the PAP-G unit releases the galabiose: first stretching the fully stacked pilus (left of the "stacked" line). Second coexistence of stacked and unstacked (phase transition 1) until the third phase: fully unstacked pilus undergoes a second transition (transition 2). The fourth phase behaves like a worm-like chain (WLC). The force distribution (histogram to the right of the unbinding forces) shows two peaks at 27 and 48 pN fitted by a Monte-Carlo simulation (black line).

transition 2 might be enough to add up the further increased force from some other P-pili reaching the transition 2. Since the P-pilus is a very "expensive" structure for the *E. coli* bacterium, the galabiose bond is released before forces above 100 pN could damage the integrity of the pilus. The PAP-G-hook is pulled back and sticks out on the stiff stacked pilus, waiting to bind to a next galabiose molecule.

Analogous to the *E. coli* experiment, the next chapter presents approaches for conducting force measurements between interacting molecules in their natural environment—the membrane of a living cell.

9.4 Force Spectroscopy on Living Cells

With some effort, the receptor and adhesion molecules of cells can be isolated and purified for single molecule force experiments. Some molecules cannot become covalently linked to the force sensor. Membrane-anchored molecules, in particular, often turn out to be insoluble in water. Immobilizing such (trans-) membrane proteins in artificial lipid membranes (supported bilayers) on a substrate is an alternative to gain access to membrane proteins with force spectroscopy (Dewa et al., 2006).

As an alternative, in this section, force measurements between single molecules and the living cells as well as force measurements between living cells at single molecular resolution will be described. The interacting molecules are provided in the "correct conformation" by the cell and the force is coupled into the molecules by the "natural handles." This opens up a broad field

of study of cellular mechanisms and strategies that tune the adhesion strength of a cell at the molecular level (Parot et al., 2007; Helenius et al., 2008; Ludwig et al., 2008; Schmitz and Gottschalk, 2008; Selhuber-Unkel et al., 2008). The cell adhesion molecules of the integrin family are prominent examples for cellular adhesion regulators (see page 9-14). As the cell, reacting to its environment, might change its adhesion, a cell adhesion experiment can even serve as a reporter for the function of pharmaceutics (drugs, hormones, and chemokines) that trigger an intracellular reaction with an impact on adhesion (Schmitz and Gottschalk, 2008). Most mammalian cells are optimized to metabolize at 37°C. To conduct an AFM experiment at temperatures unequal to the ambient room temperature attracts drift effects to the temperature sensitive force sensor. Unfortunately, the advantages of studying the molecular interactions between single molecules located in cellular membranes are counterbalanced by the fact that there will never be a perfect model of the very complex cell that serves as a "spacer" between the substrate and the molecule or the molecule and the cantilever, respectively. Each cell might react differently due to the cell cycle, the last feeding period, the temperature changes during preparation, and the exerted force. The cell as "spacer" contains several billions of molecules. The complex interplay of all of them renders the mechanical characteristics of the cell, that is, the linker between the adhesion molecules and the substrate, and the force sensor, respectively.

From solid-state physics, we can borrow classic models about viscoelastic bodies and approximate them to the mechanical answer of a cell in a force experiment.

9.4.1 Cell Mechanics (Theory)

In a simplified way, the major players in cell elasticity are the cytoskeleton (with its actin filaments, microtubules, and intermediate filaments), the cytosol, and the cellular membrane. The cellular membrane separates the intracellular from the extracellular space and incorporates the adhesion molecules. Some cells have smooth surfaces, others expose "fingers" (filopodia or microvilli) protruding the membrane by actin filaments (Alberts et al., 2002). In a typical force experiment, the relaxed membrane easily adapts to the shape of the indenting cantilever. The membrane supporting cytoskeleton starts to exert a pronounced elastic counter force to the further indenting cantilever tip compared to the weak contribution of the compliant lipid membrane.

On a millisecond timescale, the elastic property of the cell is predominant, but already in the second timescale, viscous and plastic deformations of the cellular components occur. To describe the cellular reaction to the indenting cantilever, general concepts of modeling viscoelastic properties of matter from solid-state physics are used. Here, we focus on a elasticity model by Heinrich Hertz that was developed originally for two interacting elastic spheres [Hertz model (Hertz et al., 1882)] and on some basic arrangements of dashpots and springs (Kelvin-, Maxwell-, and Voigt-model) representing viscoelasticity models (Fung, 1993).

In a very simplified way, one can interpret the force interaction between two cells as the interaction of two elastic spheres (see Figure 9.13). By a few approximations, the general Hertz-model can be reduced to the following equations:

Equation 9.10 describes a compliant spherical object (radius: R, elastic modulus E and the Poisson ratio: v) indented at a distance z by a rigid plane (= a rigid sphere with an infinite radius):

$$F_{sphere} = \frac{4}{3}\sqrt{R} \cdot \frac{E}{1-v^2} \cdot \sqrt{z^3} \qquad (9.10)$$

For symmetric reasons, the same Equation 9.10 is valid for a rigid spherical object (radius: R) indenting an elastic object with a plane surface by a distance z. A second equation describes a rigid conical object (semi-angle of the cone: α) indenting a plane compliant object by the distance z (this holds for a typical cantilever with a conical tip indenting a cell, that at the tips' scale for small indentations approximately has a plane shaped surface):

$$F_{cone} = \frac{2}{\pi}\tan(\alpha) \cdot \frac{E}{1-v^2} \cdot z^2 \qquad (9.11)$$

In most of the cases, the Poisson ratio v can be set to 0.5 that represents a homogenous and incompressible elastic medium.

The Force F is proportional to the 3/2 power of the indentation z in Equation 9.10 and to the square of the indentation z in Equation 9.11.

From this static Hertz model we can derive the Young's modulus E of a cell from each force curve when approaching the cell with the force sensor (Radmacher, 2002; Wojcikiewicz et al., 2004; Lamontagne et al., 2008).

The viscoelasticity of a cell can be evaluated better by adding viscous elements to the description. An ideal viscous element creates a force that is proportional to the applied velocity with inverse orientation. In a typical experiment, the viscous drag force is constant because the force sensor travels with a constant velocity (changing sign from approach to retract after the contact with the cell).

Figure 9.14 depicts force-versus-distance plots of virtual AFM force experiments at constant velocity on idealized models composed from dashpots and springs. An ideal Hooke's spring would travel from zero with a constantly increasing force, whereas an ideal dashpot would jump with the velocity to a force according to the viscosity at this velocity and stay constant at

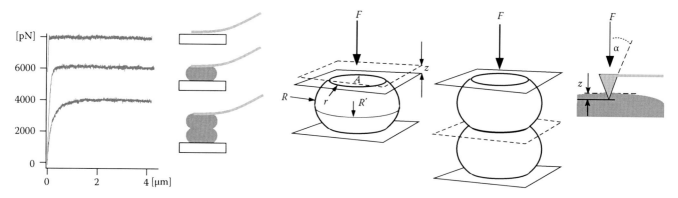

FIGURE 9.13 The elastic modulus E can be derived from the analysis of the indentation force curves, using the Hertz model. The indentation force curves are derived from a plain cantilever indenting a rigid plane substrate (1), a tipless cantilever indenting a red blood cell (2) and a red blood cell immobilized on a tipless cantilever indenting another red blood cell (3). In the middle, the geometrical schematics that are used in the Hertz-model are shown: the indenting force F, the indentation depth z and the radius of the sphere R. (contact surface A, contact radius r and extended radius R' are not needed in the simplified Hertz-model shown here). To the right the geometry for the conical indenter is sketched. (Modified from Benoit, M. and Gaub, H.E., *Cells Tissues Organs*, 172, 174, 2002.)

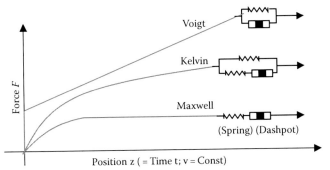

FIGURE 9.14 Force distance traces of the elastic bodies (as depicted from above: Voigt-model, Kelvin-model, and Maxwell-model) reacting to an elongation constantly increasing with time.

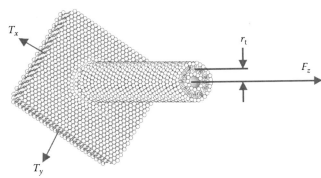

FIGURE 9.15 A schematic drawing of a membrane tether with the radius, r_t, pulled from lipid bilayer. The membrane tension is represented by T_x and T_y and the force F_z pulling the membrane tether and its length l_t is measured by the AFM.

this force. The Voigt, Kelvin,[*] and Maxwell model are combinations of springs and dashpots and are a first choice for describing the viscoelastic properties of cells.

In some typical cell-adhesion force curves shown in Figures 9.20, 9.21, and 9.26, elucidating the huge variety of viscoelastic responses of the cellular "spacer" signatures of the presented models might already be identified. To get a clearer picture of the mechanics behind these traces we need to give more attention to the cellular spacer.

9.4.1.1 Tether

A typical pattern in which cells perform in combination with adhesion, are lipid membrane tubes called "tethers" (Waugh and Hochmuth, 1987). Tethers are pulled out from the membrane around a membrane-anchored adhesion spot when the cantilever is retracted while the adhesion spot is still intact (Figure 9.15).

A membrane tether forms as a lipid bilayer tube with typical diameters of 10–200 nm, counterbalancing the energies of the membrane's tension and the membrane's curvature (Raucher and Sheetz, 1999; Marcus and Hochmuth, 2002; Sun et al., 2005; Harmandaris and Deserno, 2006). The diameter also depends on the lipid and protein composition of the membrane, on the ambient temperature, and on the amount of molecules (e.g., actin filaments) that are pulled within the tube. Neuronal cells tend to pull long tethers of up to millimeters until they detach![†] (Hochmuth et al., 1996).

A typical membrane tension T_{cell} for cells is usually about 15 fN/nm (Discher et al., 1998) that induces an equilibrium tether force F_t in a tether of the radius r_t (r_t is defined in the center between the two lipid bilayers)

$$F_t = 2\pi \cdot r_t \cdot T_{cell} \tag{9.12}$$

The stiffness of the lipid membrane does not favor a narrow curvature and as a consequence of the interplay between the membrane stiffness, B ($1.8*10^{-19}$J)[‡] (Hwang and Waugh, 1997), the membrane tension, T_{cell}, and the axial tether force, F_t, a formula for the typical tether radius, r_t, was found (Waugh and Hochmuth, 1987):

$$r_t = 2\pi \frac{B}{F_t} \tag{9.13}$$

By combining Equations 9.12 and 9.13 the tether force of a lipid membrane tether, depending on the membrane tension T_{cell} is calculated:

$$F_t = 2\pi\sqrt{B \cdot T_{cell}} \tag{9.14}$$

The typical physiological membrane tension (T_{cell} = 15 fN/nm) would result in a typical steady-state axial tether force F_t of 9 pN at a radius, r_t, of 110 nm. When a tether is pulled at constant velocity, a steady flow of lipids into the growing tether is recruited from the cell membrane. An additional (constant) force adds up proportional to the pulling velocity and is composed of the friction and the viscosity at the tether's foot. Consequently, the axial tether force, F_t, at the tip is higher than the one calculated from the tension in the cell membrane. At membrane (hyper) tensions of T_{cell}^{max} between 2 and 20 pN/nm, a lipid membrane would disintegrate (Evans et al., 2003). Equation 9.14 calculates the maximum force, F_t^{max} that a cellular membrane tether could withstand between 120 and 300 pN. Even though the bilayer might withstand tether forces above 300 pN, the tether radius would come below 3 nm and the bilayer would collapse. The bending rigidity and the maximum membrane tension largely depend on the lipid composition and the ambient temperature (Sackmann, 1995; Seifert and Lipowsky, 1995; Heimburg, 2007).

---

[*] The Voigt and the Maxwell model can be represented by the Kelvin model if either the parallel spring is set equal to zero (Maxwell) or the spring in series to the dashpot is set indefinitely stiff (Voigt).

[†] This usually excludes neuronal cells from the force experiments, because no AFM piezo will travel millimeters at a high resolution!

[‡] The transition from fluid to solid phase of lipids in the membrane (induced by temperature, electrical potentials, and lipid mixtures) markedly can change the mechanical properties of the cellular membrane.

Tethers pulled from cells can vary in diameter and viscoelastic behavior. If actin bundles or membrane proteins are pulled within the tether or if the membrane tension, T_{cell}, is very low, the tether radius, r_t, can be increased to some hundred nanometers. If the tether radius exceeds 100 nm and several actin filaments are included, the distinction between the tether and the microvillus blurs.

9.4.1.2 Tether Model

Usually cells have a large reservoir of membrane to keep the membrane tension constant. Cells can refill lipids into the membrane from the caveolae and by exocytosis. (Sens and Turner, 2006). In a force experiment at constant velocity, we recognize a membrane tether by the almost constant force plateau in the force trace. A constant load is exerted to the bond(s) attaching the tether to the force sensor, while the small tube is pulled from the cellular membrane at constant velocity (see Figure 9.16). In an extreme case where the bond does not open until the membrane reservoir of the cell is used up completely, this constant force would increase (as indicated by the dashed line in Figure 9.16). Neurons in particular have a "never ending" membrane reservoir. In a first approximation, tethers behave like viscous elements due to the hampered viscous flow of the cell membrane through the foot of the tether. Tethers keep the force exerted to the bond constant according to the pulling velocity and thence the loading rate of the bond close to zero (= force clamp analogous to the BS-transition plateau of DNA in Figure 9.8). Compared to a steadily increasing force of an elongating spring, such a tether at constant force can maximize the adhesion energy of a bond that is strong enough to stand this constant tether force. Compared to a Hooke's spring, the energy (force times distance) increases quadratically with the pulled distance and the force linearly up to the level that breaks the bond. The tether keeps the force below that bond breaking level, while the energy increases only linearly but for a very long distance (like a releasing fishing line). As discussed above, *E. coli* and gram-negative bacteria lacking a lipid membrane as an outer layer had to develop the very complicated pilus structure to mimic the benefit of that viscous behavior of membrane tethers (Lugmaier et al., 2008).

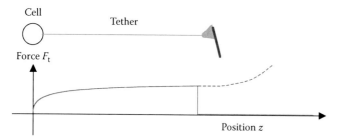

FIGURE 9.16 Tether schematics of a cantilever pulling a single tether from a cell and a typical force trace of a membrane tether. If a membrane tether is pulled from a small cell (platelet or red blood cell) with a limited membrane reservoir, the membrane tension would increase due to the limited surface area to cover the volume of the cell and the increased need for of membrane for the pulled tether. This would lead to an increased force as indicated by the dotted line.

Equipped with these simplified viscoelastic models, we will later see that from cell-adhesion force-spectroscopic experiments interesting results and conclusions can be drawn about molecular anchoring. First, we will look at the main players that all these efforts on modeling the cellular spacer were made for—the cell adhesion molecules.

9.4.2 Molecular Concepts in Cell Adhesion

Molecular biologists maintain a large and increasing library of known adhesion molecules and their molecular data and these entities are available for scientists all around the world via the Internet.

Unfortunately, some molecules, namely, the cell adhesion molecules, very often lose their natural behavior when they are extracted from the cellular membrane and transferred into experiments. In their natural environment, the cellular membranes are fixed either by a lipid anchor or by one or more hydrophobic transmembrane regions stabilized in the membrane. In a force experiment, basic lipid anchors can withstand forces of about 20 ± 10 pN (at a separation velocity of 5 μm/s). The strength of a lipid anchor depends on the lipid layer composition (Evans, 1998). Transmembrane anchors hold stronger forces (Oesterhelt et al., 2000) and additionally allow further attachment to the cytoskeleton inside the cell. Such transmembrane anchors can utilize conformational changes of the connected adhesion molecule in the extracellular space: induced by an adhesion at the extracellular region, a reaction at the intracellular region is triggered by a change in conformation (outside in signaling) (Pierres et al., 2007). The adhesive site of the molecule can also be tuned from inside the cell via the transmembrane anchor to the extracellular space (inside out signaling).

9.4.2.1 Adhesion Molecules

Adhesion molecules are divided into families and subclasses. In Figure 9.17 a few molecules and their reaction partners are depicted as examples.

Adhesion molecules can adhere to their partners in different ways classified as followed.

Homophilic: only molecules of the same kind can adhere to each other (csA and cadherins)

Heterophilic: the interacting partners are molecules of different kinds ($\alpha_4\beta_1 <\to$ V-CAM-1)

Mediated: the interacting adhesion molecules need a third molecule to mediate adhesion ($\alpha_{2b}\beta_3$-fibrinogen-$\alpha_{2b}\beta_3$).

Some molecules can form multimers to enhance the adhesion or the selectivity with an analogous classification: homo multimerization (clustering) or hetero multimerization (integrins always occur as hetero dimers, the formation of lipid rafts is also seen as hetero multimerization)

Cell-to-cell and cell-to-surface adhesion involves a variety of different adhesion systems. Tight junctions form impermeable barriers within an epithelial cell layer. Adhesion belts (adherens junctions) and desmosomes (formed by cadherins and other proteins) serve to link to actin, intermediate filaments, and other

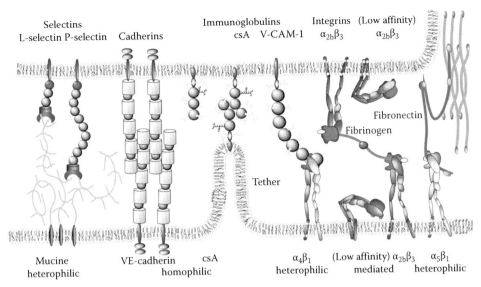

FIGURE 9.17 Survey of the most important mammalian cell adhesion molecules: Selectins, cadherins, members of the immunoglobulin super-family (csA and V-Cam-1) and integrins. Cell adhesion molecules interact, homophilic, heterophilic, or use linker molecules (fibrinogen) to mediate adhesion between the adhesion molecules. The selectins specifically bind to glyco-proteins (mucins) cadherins form dimers and interact homophilically with another cadherin dimer. In a molecular cluster, they can multimerize to a very strong adhesion disk. Contact-site A (csA) and the vascular cell adhesion molecule (V-CAM-1) are representing the huge super family of the immunoglobulins. Integrins also form dimers like the cadherines, but heterodimers. They consist of an α- and a β-monomer and can change their affinity by changing their conformation (low affinity of $\alpha_{2b}\beta_3$). Fibronectin is a constituent of the extracellular matrix. The integrin recognizes the RGD peptide sequence of the fibronectin, of V-CAM-1, of fibrinogen and of all the other molecules specifically binding the integrins.

components of the cytoskeleton. Gap junctions formed by connexins allow ion exchange between cells. Selectins and integrins form specific connections to specific extracellular binding sites and mediate cellular recognition and targeted locomotion.

Integrins are interesting adhesive membrane receptors, as they are known to be individually tuned by the cell in adhesion strength (low, medium, and high affinity). They are therefore widely utilized by cells to mediate adhesion, cell sorting, and targeted migration. All integrins require divalent anions (e.g., Ca^{2+}, Mg^{2+}) for binding if they recognize a peptide sequence RGD (*Arg*inin-*Gly*cin-*Asp*aragin) in the binding partner. By shifting the concentrations of divalent anions in the medium, integrins artificially can be switched between affinity states: no divalent anions = low affinity, presence of Ca^{2+} = intermediate affinity, and presence of Mg^{2+} replacing Ca^{2+} = high affinity. Integrins are hetero dimeric transmembrane proteins: presently 19 α-monomers and 8 β-monomers are known in the human body. The RGD binding domain is located in the α-monomer. The combination of an α-monomer and a β-monomer so far results in 24 different known integrins specific to certain binding partners (e.g., $\alpha_5\beta_1$ specifically binds to fibronectin, $\alpha_6\beta_1$ to laminin, $\alpha_1\beta_2$ to I-CAM, $\alpha_4\beta_1$ to V-CAM...)

Most of the integrins at physiological salt concentrations are in an intermediate or low affinity state diffusing in the membrane. They either change to an adhesive conformation when diffusing into an activating part of the membrane where cytoskeleton-related molecules connect to the integrins' cytoplasmic part (inside out signaling). Integrins also change their conformation when sensing an external binding partner or an external force (outside in signaling). For example, the integrin's β_3 subunit binds via cytoplasmic talin to actin that is a constituent of the cytoskeleton. On the other hand, phosphorylation of the β_3 cytoplasmic tail by the cell prevents binding to talin. Additionally, the integrin changes into a conformation that releases a bound ligand from the extracellular integrin binding site (inside out signaling). The full functionality of integrins is definitely maintained only in an appropriate lipid membrane. Furthermore, the tuning of the integrins' affinity by the cell can be utilized as a reporter for adhesion-relevant processes inside the cell. Obviously, these molecules and their dependency on intracellular processes can only be investigated properly in a cellular membrane of a living cell.

9.4.2.2 Adhesion Strategies

Different strategies are possible for a cell to control its adhesion strength:

Avidity: How many binding competent molecules are available in the membrane to be accessed by the binding partner?

Affinity: How strong is the bond? Affinity is described, commonly, by the dissociation constant, K_D, or the off-rate. The unbinding force and the molecular bond potential (width Δx and depth ΔG) also represent the affinity in force experiments.

Anchoring: How is the adhesion molecule linked to the cell? Molecules freely diffusing in the membrane might reach the adhesion site faster than molecules restricted by a connection to the cytoskeleton. However, a pure lipid anchor holds approximately 20 pN only, whereas a transmembrane anchor with up to approximately 100 pN is much stronger. The strongest group of anchors connects the intracellular domain of the adhesion molecule to the actin, tubulin or other filaments of the cytoskeleton with up to the nano-Newton range. The anchorage of an adhesion molecule defines the mechanical environment ("spacer") of the adhesion molecule. It controls the lateral motility in the membrane and the loading rate of the applied force to the adhesion site.

Clustering: This is a strategic combination of avidity, affinity, and anchoring that increases the affinity of an adhesion spot with a strong anchor to the cytoskeleton by multimerizing the binding competent adhesion molecules (desmosomes, gap junctions, tight junctions, and focal adhesion). Such a cluster can withstand forces of approximately 30 nN. Clusters contain a self-healing mechanism through the rebinding of broken bonds. Dissociating bonds are not pulled apart due to the neighboring molecules keeping the split partners in close proximity. Such a force can either rip a cell or the cluster apart. The molecular clustering resembles the unbinding experiment with the shear DNA: the adhesion strength of each single molecule adds up to one big de-adhesion event instead of opening (pealing off) one individual bond after the other (as in the zipper mode in Figure 9.9) (Besser and Safran, 2006; Erdmann and Schwarz, 2006). The cell has a wide variety of possibilities via the anchoring to influence the adhesion. An adhesion molecule anchored in an actin-rich protrusion (microvillus) exposed on its tip has a higher probability to probe any object coming close to the cell than the one on the retracted membrane parts would have. As already mentioned, a molecule purely sitting in the membrane is limited in its binding force to the strength of the membrane anchor, even though the affinity of the binding site might be much stronger. It will be either ripped out of the membrane upon the stronger adhesion or form a membrane tether. The cell can tune the level of the tethers' force plateau by the lipid composition and lipid rafts.*

9.4.3 Cell Adhesion (Force Measurements)

Cells are the smallest units of life. As individuals (amoeba), they adapted very well to the environment during evolution and now they have to adequately react to several actual environmental changes in order to survive. In multicellular organisms, cells started to become specialists for certain tasks (immune cells, neurons, endothelial cells, etc.) to better adapt to environmental changes or even change the environment altogether (to a better one?). Communication between the cells, sorting, migration, homing, and many other functions have to be maintained in the multicellular organism. Therefore, some cellular reactions are universal, while some are only present in heart muscle cells, inner ear cells, red blood cells, and so on. Thus,

for cell adhesion experiments there is no universal protocol. In the worst case, a new protocol has to be elaborated for each cell type. Nevertheless, a few basic principles will be described in the following examples.

9.4.3.1 Force Sensor Preparations for Cell Adhesion Measurements

The problem of mounting the molecular handles correctly is solved by the cell, but now we have to face a general problem of cellular force spectroscopic experiments: how to immobilize cells?

Very recent and universal methods are based on the aspiration of the cell on a small hole in a substrate by a tiny pressure gradient (Pamir et al., 2008) or in a system of micro-fluidic channels (Ryu et al., 2008). Latest cantilever designs might even allow the aspiration of the cell to the cantilever in the near future (Godin et al., 2007). However, presently one has to immobilize a cell on the cantilever by an adhesive coating (Benoit, 2002). In Figure 9.18, the possible designs of cell adhesion force measurements are summarized.

In a classical AFM experiment, a bare cantilever is used for imaging a cell. This already allows the detection of the nonspecific adhesion to the material of the cantilever tip (e.g., Si (SiO), SiN, etc.). In general, soft cantilevers with a spring constant of less than 10 pN/nm and blunt (at least non-sharpened) tips, pyramidally shaped with radii larger than 20 nm, are recommended in connection with cells. Cells survive uncontrolled indentations of soft cantilevers more often than of hard cantilevers with sharpened tips. The AFM imaging resolution on soft and rough samples is reduced anyway, so a non-sharpened tip is not a regression. Presently the best cantilever marked for this purpose is the MLCT cantilever by Park Scientific (now distributed by Veeco Instruments). The spring constant is said to be 10 pN/nm, but due to fabrication uncertainties, this number might vary by more than 100%. Therefore, it is important to independently determine the spring constant of each cantilever. To probe more specific interactions, the cantilever can be functionalized with an adhesion molecule. At this point, one should be aware of the fact that plain coating of the cantilever in the solvated molecule leads to some major obstacles concerning force measurements:

- Denaturation of the molecule on the surface of the tip
- Bad orientation of the molecule on the surface of the tip
- Harvesting of the molecule from the surface of the tip by the cell during force measurements
- Nonspecific interaction of the cell with uncoated areas of the tip or denatured molecules on the tip

A proper functionalization of the cantilever covalently links the molecule of interest via a non-adhesive spacer (e.g., PEG) to the tip (page 9-10)

9.4.3.1.1 Single Cell on the Cantilever

For a "functionalization" of the cantilever with cells, the pure coating recipe is sufficient, since now the cell will be the surface with which the sample interacts. Depending on the cell type,

---

* Controlled temperature is a crucial prerequisite for reasonable cell adhesion measurements: The viscosity of the cellular membrane extensively depends on the temperature. Drastic changes in viscosity take place at the transition temperature from the liquid to the solid phase of the lipid bilayer (Heimburg, 2007).

Force Spectroscopy on Cells

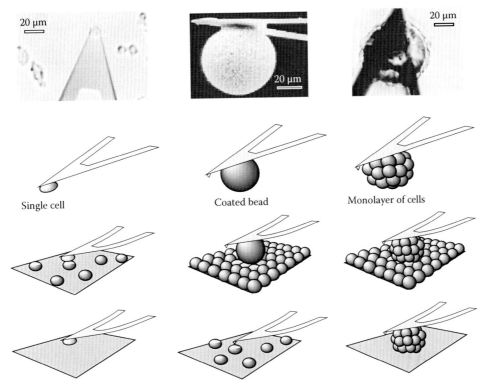

FIGURE 9.18 The left column of images: light microscopic image of a *D. discoideum* cell immobilized on a tipless cantilever. Other cells on the substrate are below the focus plane. Schematics below: cell functionalized cantilever, force measurements of single cells, and force measurement of a single cell to substrates. Middle column of images: REM micrograph of a sepharose bead glued to a cantilever. Schematics below: force sensor with microbead, force measurements between the bead and a cell layer. For the sake of completeness of the force measurement of a cantilever tip on single cells. Last column: light microscopy image of bone cells grown on a glass bead on a cantilever (focused at the center of the bead). Schematics below: cell layer functionalized bead, force measurements between two cell layers, and force measurement between cell layers with substrates.

polylisins, concavalin A, or fibronectin are among the good candidates for precoating the cantilever. With such a precoated (preferentially tipless) cantilever, a cell waiting on a weak adhering surface (e.g., BSA coated Petri dish) can easily be "fished" by a cantilever that is gently brought into contact with the cell for seconds. With tiny tweezers, the tip might be removed from the cantilever mechanically if no tipless cantilever is available.[*] The freshly attached cell might adhere more firmly to the cantilever after 1–10 min[†] before the "cell functionalized" force sensor is used for force experiments on a substrate (page 9-18), on another cell (Puech et al., 2006), and (page 9-21).

9.4.3.1.2 Sphere on the Cantilever

Adhesion measurements on confluent cell layers with a tipless cantilever often results in a badly defined contact area. One possible way of improvement is to glue a sphere (radius 5–50 μm) to the cantilever. Sepharose, Agarose, Latex, or glass beads can

be glued to the cantilever by a drop (less than a pico liter) of two-component epoxy. Such a sphere functionalized with adhesion molecules can be useful to probe different cell layers in cell culture dishes. (page 9-20)

9.4.3.1.3 Cell Layer on the Cantilever

Finally, such a coated bead on the cantilever can be cultivated in a cell culture after gently injecting a couple of cells onto the bead on the cantilever.[‡] After a few days, a cell layer grows on the sphere (see Figure 9.18 top right).

This configuration can be probed on surfaces or other cell layers (page 9-23).

The complexity of the cell adhesion experiments increases with the number of contributing cells and molecules. Starting with interactions between a single cell and a defined functionalized surface (e.g., cantilever tip or glass slide), followed by interactions between two individual cells, then between a cell layer and functionalized surfaces and finally multicellular interactions between cell layers.

---

[*] In this case, the spring constant of the triangular cantilever can be halved by pinching of the second leg in the same way as the tip.

[†] Do not lift the cantilever out of the liquid unless you want to remove the attached cell. Each detachment of the cell passivates the cantilever surface with remaining cellular membrane patches.

[‡] As for all preparation steps the cantilever lays downside up in order to access the tip and to prevent damages to it.

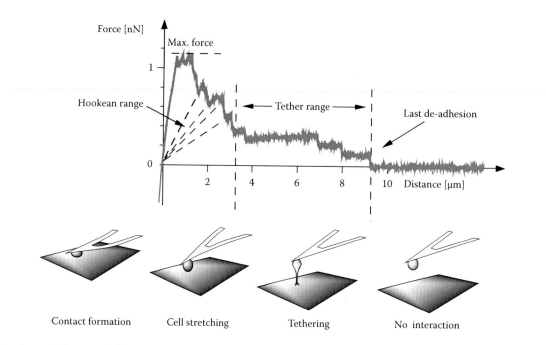

FIGURE 9.19 A typical force graph from a single-melanoma cell-adhesion experiment after a 1 min contact at 200 pN to a V-CAM-1 functionalized substrate at 37°C. The steep initial increase of the force reflects the predominantly elastic stretching of the whole cell. Around the maximum force, an increased number of intensive unbinding events take place. In the descending shoulder single de-adhesion steps become discernible with a force loading slope that is close to a Hookean behavior (indicated by dashed lines to the origin). Then in the tethering region, the slope is close to zero until the cell fully detaches from the substrate. The scheme below the force trace illustrates the situation of the cell in the force experiment. (Modified from Benoit, M. and Gaub, H.E., *Cells Tissues Organs*, 172, 174, 2002.)

9.4.3.2 Single Cell-to-Surface Measurements

The adhesion force measurement of a single cell to a substrate is the most prevalent experiment, as it involves the complexity of only one cell. Either a functionalized surface of the cantilever tip on a cell or a cell-functionalized cantilever on a functionalized substrate realizes this configuration. For example, a cantilever functionalized with the lectin (*Helix Pomatia*) that specifically recognizes red blood cells of the blood group A, repeatedly measured adhesion forces, while scanning a sample of mixed red blood cells of group A and O (Grandbois et al., 2000). With this label-free technique, individual red blood cells of group A were localized on this sample. In the MAC-mode such images of cells are obtained very fast at high resolution (Schindler et al., 2000).

To test the adhesion behavior of the same cell on different substrates, the second configuration with a single cell on the cantilever (Figure 9.19) is particularly useful. From the complex force traces of a cell-adhesion experiment, scientists usually extract the following numbers (Franz et al., 2007):

The *initial slope* is the approximately linear first increase of the adhesive force (Figure 9.19). It represents the elastic elements of the cell (Figure 9.20).

The *maximum adhesion force* indicates the highest force in the force plot. This is a rough first approach to quantify the adhesion strength.

The *slope prior to a de-adhesion event* is a hint for the actual cellular "spacer." A slope close to zero indicates a tether, whereas

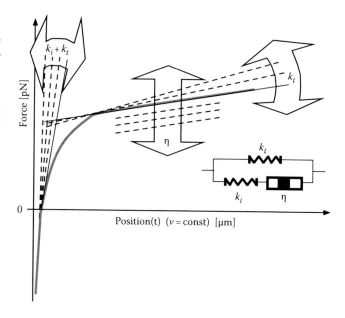

FIGURE 9.20 How elements of the Kelvin-model tune a force graph of a tether at constant pulling velocity.

a steep slope results from a strong spacer. If the extrapolation of the slope comes close to the origin* of the force distance plot, the spacer behaves like a Hookean spring.

---

* Defined as the intersection between the force-zero line with the first increase of the force plot.

The *distance from the original cell surface* (origin) of a de-adhesion event allows an upper estimate for the force loaded to the bond opened at this de-adhesion event.

The *force step size* of a de-adhesion event allows for a lower estimate of the actual unbinding force. Solely the very last de-adhesion event is an exact measure of the unbinding force, the de-adhesion events before might have been higher but appear lower due to the still existing force connections mediated by possibly non-independent cellular components between the surface and the cantilever.

The *area* spanned by the force trace above the zero line has the dimension of energy. It rather reflects the energy dissipated by the separation of the cell from the surface, than an "adhesion energy" of the molecular bonds of the tested surface.

The *adhesion rate* is not determined by a single force measurement. It is the fraction of force curves with adhesion events from a whole set of at least 50 force curves (e.g., all force measurements with contact times of 2 s on a substrate without adhesion molecules).

The *bond formation rate* is determined by the number of recognized adhesion events per force curve. From the measurement in Figure 9.19 this number is not determined. It might be registered as ">10." In analogy to the adhesion rate there is "another" bond formation rate that refers to all detectable single bonds of all force curves of the whole set of measurements.

With a leukocyte (Jurkat) on the cantilever, Moy et al. thoroughly investigated the interaction between the integrin LFA-1 ($\alpha_1\beta_2$) and a cell adhesion molecule ICAM-1 or ICAM-2 on a substrate (Wojcikiewicz et al., 2006).

In this intense study a large number of data were collected at 25°C (room temperature) to analyze the binding potentials of the integrin-I-CAM complexes with the Bell–Evans model at different loading rates (determined by the slope prior to the de-adhesion event). The recalculation from the measured most probable de-adhesion forces into the characteristics of the bond potential ΔG, Δx, and the off-rates matches very well to the picture of a receptor–ligand interaction. Two affinity states of the integrin that can be switched artificially by replacing the Ca^{2+} ions in the binding pocket of the integrin by Mg^{2+} were resolved. The adhesion force, in the presence of the Mg^{2+} increased by 25 pN for the I-CAM-1 and by 10 pN for I-CAM-2 bonding to the Mg^{2+} activated integrin. The adhesion strength determined by the area under the force (dissipated de-adhesion energy) was used to quantify the adhesion strength. I-CAM-1 showed a stronger de-adhesion energy than ICAM-2. A stimulation by PMA to induce molecular clustering of the integrin resulted in a pronounced increase (fourfold for I-CAM-1 and threefold for I-CAM-2) of de-adhesion energy.

Another integrin VLA-4 ($\alpha_4\beta_1$) also present in this leukocyte cell specifically recognizes the vascular cell adhesion molecule V-CAM-1. By cell adhesion measurements, the specific interaction between the integrin ($\alpha_4\beta_1$) and the adhesion molecule V-CAM-1 immobilized on a substrate (50–100/μm^2) was studied at short contact times of some milliseconds for two reasons

(Schmitz et al., 2008). One reason was to resemble the natural situation for the leukocyte cell that has to react very fast to a signal on the surface to establish adhesion. The other was to study the initiation of this adhesion on the single molecule level but not after an acquisition of several adhesion molecules leading to strong forces.

This study did emphasize the anchorage of the integrin ($\alpha_4\beta_1$) in the cellular membrane. For this reason, the individual force distance traces (as shown in Figure 9.21) have been thoroughly analyzed.

The trained eye recognizes the typical signature of membrane tethers from these curves (in particular in curve 5 of Figure 9.21). By selecting such ideal tether curves, the viscoelastic parameters of the membrane could be characterized utilizing the Kelvin-model (Figure 9.14) and fitting it to the traces (Figure 9.20): $k_i = 0.26$ pN/nm, $k_t = 1.6$ fN/nm and $\eta = 6$ fNs/nm.

As all integrins, the $\alpha_4\beta_1$ integrin has an increased affinity too, if Ca^{2+} is replaced by Mg^{2+}. However, the most probable adhesion force did not significantly change (from 26 to 25 pN with Mg^{2+}). This is because the membrane tethers act like force clamps. At 3 $\mu m/s$ pulling velocity and 37°C, the tethers pulled from the Jurkat cells keep the receptor ligand complex at the constant force of 26 pN. For thermodynamic reasons, the integrins in the presence of Ca^{2+} are stochastically distributed in low affinity and high affinity states. The replacement of Ca^{2+} by Mg^{2+} shifts

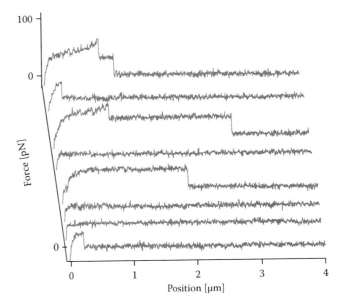

FIGURE 9.21 Subsequent force curves of Jurkat cells on a cantilever probed on a V-Cam-1-coated surface after contacts of 100 ms at 50 pN. Characteristic measures are the maximum force, the number of de-adhesion events per curve, the position and the step height of a de-adhesion event, or the area ("energy") spanned between the curve and the zero force line. The typical signature of a tether (the almost constant force plateau) ideally is represented in curve 5 (from above) in curve 3 is the longest tether a short one in the first curve overlaid by another adhesive feature. Curves 2 and 8 are too short to discern whether a tether was the origin of this trace. The curves 4, 6, and 7 are counted as non-adhesive events in the measure of the adhesion rate.

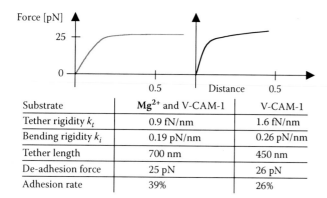

Substrate	Mg^{2+} and V-CAM-1	V-CAM-1
Tether rigidity k_t	0.9 fN/nm	1.6 fN/nm
Bending rigidity k_i	0.19 pN/nm	0.26 pN/nm
Tether length	700 nm	450 nm
De-adhesion force	25 pN	26 pN
Adhesion rate	39%	26%

FIGURE 9.22 Scheme of the "typical" force traces from force measurements between VLA-4 ($\alpha_4\beta_1$ integrin) and V-CAM-1, V-CAM-1 with Ca^{2+} replaced by **Mg^{2+}** (left). The differences in the force slopes are subtle: tethers are shorter, stiffer in the initial slope, and slightly steeper in the force plateau without Mg^{2+}.

this distribution toward a higher fraction of integrins with high affinity (Figure 9.22).

Nevertheless, the bond formation rate (fivefold), the adhesion rate, and the tether length significantly (twofold) are increased in the presence of Mg^{2+}. This indicates that the avidity of high affinity integrins has been increased. The viscous element η did not change. It remains unclear whether the tether length is increased, because Mg^{2+} alters the stiffness of the membrane or because the integrins are distributed into membrane areas of different stiffness. There is a likelihood that integrins in stiff areas (lipid rafts) show higher affinity in untreated cells. Adding Mg^{2+} activates all integrins including those in the soft areas.

Mg^{2+} artificially increases the affinity of an integrin, but the measured force does not indicate a stronger binding force. Here we clearly see how the cellular spacer dictates the measured, most probable, adhesion force by the tether plateau.* SDF-1 is a reporter for an inflammation in the tissue that is presented on the surface of blood vessel cells in the neighborhood of the inflammation. The chemokine SDF-1 is known to stimulate the G-protein-coupled CXCR-4 receptor of the Jurkat cells. The aim of the lymphocyte is to instantaneously react on that SDF-1 signal on the vascular endothelium with strong adhesion before the blood stream pushes it out of the region of inflammation. The tethering constantly slows down the lymphocyte, to allow for scrutinizing the vessel walls for SDF-1 molecules. The lymphocyte instantaneously stops and firmly adheres to subsequently leave the blood system for defeating the injury in the tissue.

In the age of nanotechnology, a very sophisticated experiment became possible. A non-adhesive surface was nano-patterned with hexagonally arranged adhesion spots at defined distances in the range of a few tens of nanometers. This nanofabricated surface was used to investigate the spatial requirements of receptor arrangements in molecular adhesion clusters (focal

adhesion spots). The experiments revealed that cell adhesion, proliferation, and differentiation strongly depend on the nanoscale arrangement of adhesion ligands and in particular that the spacing between the single nanometer-sized adhesion spots is the key player for defining the cell's fate (Arnold et al., 2004). Adhesion force measurements show that such nanostructures influence cell adhesion strength and adhesion cluster formation during the first 5 min of adhesion itself (Selhuber-Unkel et al., 2008). Therefore, a cell immobilized on the cantilever stays in contact with such a nano-patterned adhesion surface with different spacings between adhesion sites for several seconds. While retracting the cell after the time of contact, from the force traces pronounced adhesion peaks up to some nN are identified. In Figure 9.23, the de-adhesion forces are summarized for the different spaced nano-patterns and contact times (Selhuber et al., 2006). By simultaneous fluorescence microscopy, these peaks were correlated to fluorescing focal adhesion sites on the substrate established by the cell at the time of contact (Selhuber-Unkel et al., submitted).

9.4.3.3 Spheres on Cells

Another example to defeat the cellular malfunction in the tissue is by the selective uptake of therapeutics in small vehicles by the targeted cells in an organism. This targeted uptake of specially designed vehicles, in particular, loaded with DNA by cells is a strong aim in medical research. The molecular composition of the external surface of such a vehicle shall determine the type of the target cell. In an adhesion force experiment, that mimics the small vehicle by a sphere of 5 μm radius immobilized to an AFM cantilever, the initial binding force of such a vehicle to a

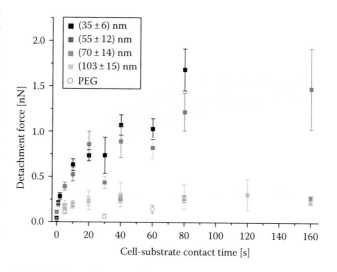

FIGURE 9.23 The critical spacing for the cooperative arrangement of adhesion molecules in focal contact is between 55 and 70 nm. Cell de-adhesion forces after contacts up to 160 min on substrates with different spaced adhesion sites (black and dark gray 55 nm and smaller; medium gray and light gray 70 nm and larger). Contacts on a plain PEG surface do not show significant adhesion. (From Selhuber-Unkel, C. et al., *Biophys. J.*, 98, 543, 2010. With permission.)

---

* The force might be even slightly decreased by the less rigid membrane environment of the majority of integrins.

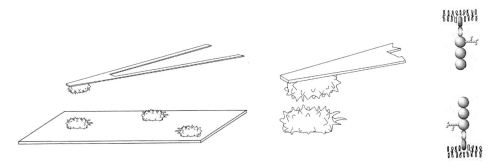

FIGURE 9.24 The *Dictyostelium* cell on the cantilever is brought in contact with another *Dictyostelium* cell as gently and short that only single csA molecules will interact with each other (hypothetical structure of a contact site A molecule on the right).

cell was quantified. Two different surfaces, positively (NH_2) and negatively (COOH) charged spheres, have been investigated on two different cell lines (MCF-10A and MDA-MB-4355) at 37°C in the nutrient medium (Munoz Javier et al., 2006). Since cells merely show a negative net charge by the sugar groups of the glycocalix, the results expectedly showed a higher adhesion rate of the positively charged spheres. The most probable "molecular"* adhesion force is increased from 20 pN (negative) to 25 pN for the positively charged spheres after a contact of 1 ms at 50 pN. Because of the increase in the adhesion rate from 20% (negative) to almost 80%, the net charge of the MCF-10A cell line appeared much more negatively charged than the MDA-MB-4355 cell line with a moderate increase from 30% (negative) to almost 50%.

9.4.3.4 Single Cell-to-Cell Measurements

Dictyostelium is a single cell organism (amoeba), which has been studied by bio-scientists for many years (Bozzaro et al., 2004; Jin and Hereld, 2006). It can change from a single cell organism into a multicellular organism (slug) by switching the active genes in the nucleus. For this purpose, *Dictyostelium* cells meet by the hot spot of a chemokine signal (cAMP) sent out from each switched cell. The switched cells start to produce a Ca^{2+} independent adhesion molecule csA (contact site A) in the extracellular membrane. In cell-to-cell adhesion measurements between *Dictyostelium discoideum* cells the adhesion force of the homophilic interaction between two individual csA molecules was to be investigated (Figure 9.24). This experiment now involves two unknown spacers into the force measurement. Luckily, the cells approximately behave identically and the sets of adhesion molecules in the cell membrane are known. Nevertheless, how is it possible to discern from the force signal, which pair of adhesion molecules from the orchestra of integrins and other adhesion molecules present on these cells interacts? *Dictyostelium* is a robust organism that even stands the removal of environmental Ca^{2+} for some hours. As we know, Ca^{2+} and Mg^{2+} are essential divalent anions to mediate the integrin adhesion. Luckily, all integrins and adhesion molecules of the *Dictyostelium* refuse to contribute to the adhesion measurements after the removal of these anions.

The adhesion rate dropped to 3% (nonspecific interaction) after contacts of 0.1 s at 50 pN. By the adhesion force measurement without divalent anions, a most probable de-adhesion force between 20 and 23 pN was detected for the homophilic interaction between individual csA molecules at loading rates between 3 and 9 pN/s. After prolonged contacts—1 and 2 s—the force histograms showed pronounced force peaks at multiples of 23 pN (Benoit et al., 2000).

The csA molecule is known to be anchored solely in the external lipid layer of the membrane (ceramide anchor). A genetically modified mutant of *Dictyostelium* with a transmembrane anchor was used to test whether the bond between two csA molecules brakes or whether the molecule is extracted from the lipid bilayer. Since the force measurements showed no significant change between the two species the anchor is believed to be at least as strong as the molecular interaction and thus the csA is extracted from the membrane in less than 50% of the adhesion events (Benoit et al., 2000).

In a very challenging study (Panorchan et al., 2006a,b), the homophilic interaction forces between individual cadherin molecules (VE-, N- and E-cadherin) were measured (Figure 9.25). Despite the fact that the measurements have been performed between living cells, the force traces in this study do not show tethers but instead show WLC-like load to the bond. On the one hand, probably due to the strong spreading of the cells on the substrate, they might prevent tether formation by a high membrane tension. On the other hand, the cadherins (the E-cadherin in particular) might well be connected to the cytoskeleton by catenin complexes. By applying the Bell–Evans formalism, even an inner barrier of the E-cadherins' potential landscape was identified. The weakest de-adhesion forces were found in N-cadherin 17–30 pN, followed by the VE-cadherin interaction between 32 and 50 pN, and finally E-cadherin de-adhesion forces between 29 and 73 pN were measured at loading rates between 50 and 5000 pN/s.

9.4.3.5 Cell Layer-to-Surface Measurements

The presented examples of cell-adhesion force measurements aimed to measure initial adhesion or fast molecular processes at the level of single molecules. These measurements are important and they are practical. In contrast, long-term adhesion processes

---

* There is no statement on which molecules are responsible for the positive or the negative adhesion.

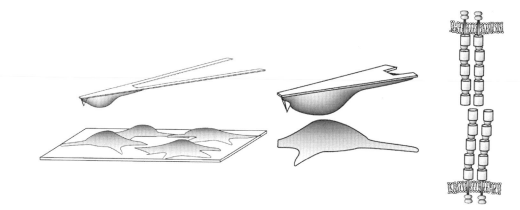

FIGURE 9.25 The homophilic interaction between individual cadherin dimers was measured by call adhesion force spectroscopy between HUVEC cells strongly adhering to the substrate and the cantilever. Schematics of the experiment and of the interacting cadherin dimers.

are difficult to measure in a force experiment. Cells that seem to like a surface for the first few seconds might decide to push it away after having explored it for an hour. How will the adhesion forces involved in the artificial bones and implants develop? With a bone cell layer on a cantilever potential, implant surfaces were probed to find out the durablilty, for the acceptance of implants by the adhering cells (Benoit and Gaub, 2002). The cell-to-surface contacts can be prolonged to several minutes, maybe up to an hour, but then the drift becomes a limiting factor and, force spectroscopy is not applicable for this purpose.

Another step into understanding the complexity of the cell adhesion experiments is made by the force measurements with cell layers. The number of cells interacting with the surface is unknown and the adhesion molecules contributing in parallel to the force trace are high. A typical force graph of a fibronectin-coated sphere mounted to the cantilever after a contact of 20 min at 5 nN on a layer of confluent RL cells is shown in Figure 9.26. After the contact, an approximated Hookean stretching of the cell layer (left arrow) takes place until by an increasing number of dissociating bonds and the progressed disentangling of membrane and cytoskeleton the maximum force is reached. The measured maximum adhesion forces are up to three orders of magnitudes higher than in a single molecule experiment. The large maximum adhesion force of 20 nN is the sum of several hundred or thousands of single molecules each contributing with their weak adhesion forces. Some of these contributing molecules

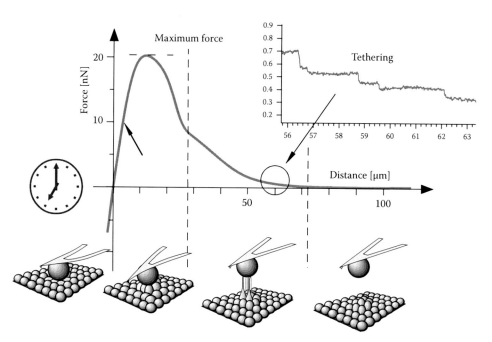

FIGURE 9.26 A typical force graph from an adhesion experiment between a fibronectin-functionalized sphere on a cell layer of JAR cells after a 20 min contact at 5 nN. The steep initial increase of the force reflects the predominantly elastic stretching of the participating cells. The individual unbinding events are not resolved in the scaling of this figure. Zooming in the descending shoulder of the tethering region unravels de-adhesion steps of individual membrane tethers. The scheme below the force trace illustrates the situation of the cells in the force experiment.

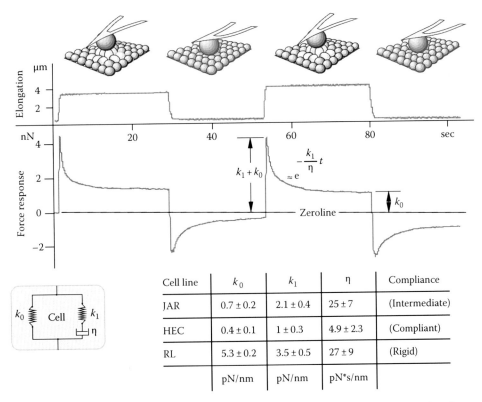

FIGURE 9.27 Schematics of a viscoelastic force experiment on a cell layer of a JAR, HEC, or RL cell line. Below the rectangular elongation pattern with plateau steps of 25 s and the measured force response of the cell layer. After the step jump that elongates the two parallel springs of the Kelvin model the dashpot relaxes logarithmically until k_1 is relaxed and k_0 keeps the tension constant. (Modified from Benoit, M. and Gaub, H.E., *Cells Tissues Organs*, 172, 174, 2002.)

Cell line	k_0	k_1	η	Compliance
JAR	0.7 ± 0.2	2.1 ± 0.4	25 ± 7	(Intermediate)
HEC	0.4 ± 0.1	1 ± 0.3	4.9 ± 2.3	(Compliant)
RL	5.3 ± 0.2	3.5 ± 0.5	27 ± 9	(Rigid)
	pN/nm	pN/nm	pN*s/nm	

are still resolved as individual de-adhesion events when zooming into the tethering region in the force traces' descending shoulder. An adhesive interaction length of several tens of micrometers is rarely seen in single molecule force measurements.

Such a configuration is perfectly suited to investigate the viscoelastic properties of a cell layer probed by a force load following a rectangular step function. With the fibronectin functionalized sphere the load can be indenting or pulling (Figure 9.27). If the force steps stay far below the maximum adhesion force the transition to the tethering region will not be reached within the duration of the force plateau until the back step. A typical force trace is represented in Figure 9.27. Modeling the cell layer with the Kelvin model, the force response is reproduced. The values of $k_1(=k_i)$, $k_0(=k_t)$, and η ($=\eta$) were determined from the force step experiment for different cell lines as listed in Figure 9.27.

Compared to a single tether the values of $\mu 1$, $\mu 0$, and $\eta 1$ are larger by orders of magnitudes.

9.4.3.6 Cell Layers on Cell Layers

The highest level of complexity is reached by experiments between two cell layers. The surface geometry of the two cell layers is not defined already and can roughly be estimated by the indentation force, the radius of the sphere, and the elasticity of the interacting cells. The viscoelastic behavior of the two interacting cell layers can only be estimated from additional experiments (e.g., Figure 9.27). The advantages of this experimental design are that the cells can polarize, establish intercellular communication and other epithelial cell habits close to their natural behavior. The typical cluster of adhesion molecules as in tight junctions, gap junctions, or focal adhesions are established within the confluent cell layers. While the two cell layers are brought into contact, they might start to communicate and establish complex adhesion patterns (Pierres et al., 2007). Thousands of adhesion molecules and receptors are contributing to the measured de-adhesion forces in the range of several nN. The organization of a strong molecular adhesion cluster is known to last several minutes (Kawakami et al., 2001). Recognition, signaling, transport, and diffusion processes determine the time of the molecular composition at the adjacent cell membranes.

With fluorescence microscopy, molecular arrangements in focal adhesion spots can be visualized. Electron micrographs resolve these multi-molecular adhesion structures in fixed cells. The direct correlation of these images to the force is complemented by measurements (Selhuber et al., 2006) and theory (Schwarz et al., 2006). The following cell adhesion measurements between cell layers were conducted after contact times of up to 40 min at contact forces of 5 nN* and separation velocities of up to 7 µm/s (Thie et al., 1998).

---

* This relatively high force is necessary to ensure a stable cell contact for such a long time despite the drift effects.

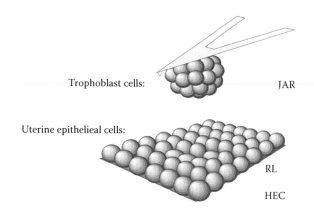

Trophoblast cells:　　　　　　　　　　　　　　　　JAR

Uterine epithelieal cells:

　　　　　　　　　　　　　　　　　　　　　　　　RL

　　　　　　　　　　　　　　　　　　　　　HEC

FIGURE 9.28 The JAR-cell line grown to fluency on the cantilever is brought into contact with either the confluent cell layer of RL cells or HEC cells for several minutes. The RL cell layer is known to firmly arrest the JAR sphere, whereas the HEC cell layer does not.

These pioneering force measurements from the last century are the first AFM adhesion force investigations involving living cell layers. The experimental setup comes as close as possible to the native situation in the human body: JAR-cells from a trophoblast cell line, grown on a sphere resembling the natural trophoblast structure in size, shape, and cellular arrangement of the apical region. RL95-2 cells or HEC-1-A cells, from uterine cell lines cultured in a petri dish resemble the uterine epithelial layer. Both cell layers where held in contact for several minutes at 5 nN with the trophoblast cell layer on the sphere until separation (as shown in Figure 9.28).

How long would it take to firmly arrest the trophoblast layer on either cell layer? Comparing the shorter contacts of 1 or 10 min to HEC or RL cell layers indicates a stronger adhesion to the HEC cells with respect to the maximum adhesion forces. Regarding the area under the curve representing the dissipated adhesion energy, after 10 min an enhanced energy dissipation is present for contacts between the RL and the JAR cell layers. After 20 min, the RL cells firmly connect their adhesion molecules into clusters that are strongly connected with the cytoskeleton in the complex interplay with the JAR cell layer. This was not evident from the maximum adhesion force, but by the adhesion pattern recorded while retracting the cellular trophoblast sphere on the AFM cantilever (see Figure 9.29) from the RL-cell layer and the HEC-cell layer, respectively. The adhesion strength can be quantified by the dissipated adhesion energy (area spanned by the force trace). In a view through the light microscope, after such a strong de-adhesion event at least one of the cell layers appeared to be severely damaged.

This crude method of investigating cellular adhesion is a relatively small step toward the origin of life and birth. These experiments about the homing of the dividing embryonic cells inside

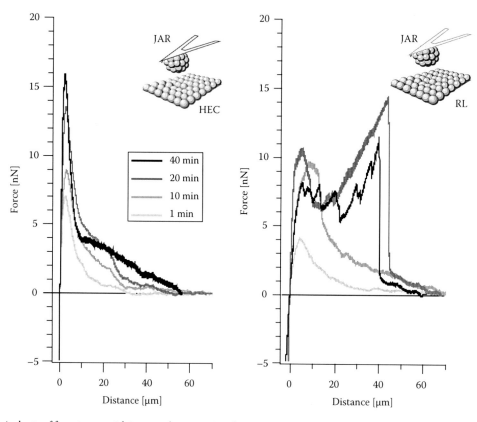

FIGURE 9.29 Typical sets of force traces with increased contact time from 1 to 40 min. Traces from the adhesion force measurements between a trophoblast cell layer (JAR) and a endometrium cell layer (HEC) on the left side, traces between a trophoblast cell layer and another endometrium cell layer (RL) on the right. (Modified from Thie, M. et al., *Hum. Reprod.*, 13, 3211, 1998.)

the trophoblast sphere in the female uterus show how tiny the steps of science are and how complex the mystery of life might be. On the other hand, this crude method neglects all the single molecules in the process in order to quantify a general phenomenon of cell adhesion, that is, molecular clustering. A zoom into the descending shoulders of the force traces allows for the counting of each single tether and measuring its step height that pools at 75 ± 15 pN.

The descending shoulders of the force curves, at higher force resolution, again uncover the individual detachment events of single integrins on membrane tethers (see Figure 9.26). Whether this adhesion pattern is a feature of the integrins $\alpha_4\beta_1$ and $\alpha_4\beta_6$ forming desmosomal structures and focal adhesion spots (Kawakami et al., 2001) or interactions contributed by other adhesion molecules remains a question.

Two general adhesion concepts appear from the experiment:

- The pealing off between the two cell layers, molecule by molecule, tether by tether, in analogy to one by one base pair in the DNA zipper configuration.
- The equally distributed force to many weak molecules in parallel as in the DNA in shear geometry.

Here the molecular clustering sums up a pronounced de-adhesion rupture event of 15 nN. From the force measurements of single integrins ($\alpha_4\beta_1$) a typical force appears to be between 20 and 60 pN at loading rates of 10–100 pN/s. Approximately 100–1000 molecular integrin bonds contribute to the measured clusters. They all have to be connected to the cytoskeleton, otherwise the cluster would separate from the cell forming a tether that we know will finally break at a force of about 300 pN since such a strong cluster of molecules would not release the tether again.

9.5 Conclusions

Experimentalists can directly measure mechanical properties of cells, cell layers and cellular membrane by force spectroscopy. Based on the parameters extracted from these measurements the complexity of the cell is described approximately by simplified viscoelasticity models. This is important for understanding cellular functions and the role of the cell in the background of the adhesion signal in an adhesion force measurement. The cell has a passive contribution as a spacer and as an agent for embedding the adhesion molecule, and it plays an active role when tuning the adhesion molecule or its molecular anchor. On a longer timescale, the formation of adhesion clusters is an important contribution of the active cell. From the incomplete collection of adhesion force experiments with the atomic force microscope ranging from single molecules to multi-molecular adhesion clusters, basic adhesion strategies of cells are unscrambled from the presented examples of force traces.

The application of the loading rate concept by Bell and Evans, that was developed to mechanically determine the adhesion strength of receptor–ligand interactions, has to be adapted to the mechanical properties of the cell, if possible. The membrane

tether is of particular interest in this context for two reasons. It probes the adhesion complex in a force clamp at constant force and small loading rate on the one hand; on the other hand, the force trace contains subtle information about the membrane embedding of the adhesion molecule. So far, the molecular interactions have been studied from the chemical and thermodynamical point of view. Indeed, the molecular interactions of biomolecules on the scale of nanometers are mechanistic. These molecules are little machines and mechanical tools, plugs, and suspensions that can be described by classical mechanics. The cell-adhesion force measurements allow for this aspect and strongly contribute to the understanding of intracellular processes. Cell-adhesion force measurements even enable direct label free access to intracellular processes that imply changes in adhesion or viscoelastic behavior. For this reason, force measurements on the single molecular level gain interest not only for nano-bio-physics but also for molecular biology and medical sciences.

While classical physics aims to find "laws" that are valid for as many classes as possible, biology classically specifies the differences between classes, subclasses, and individuals of subclasses even down to the base pairs of their DNA. Physicists say: "a cell…" while biologists say, "we took an endothelial cell from the upper third of the dorsal endometrium in the early S2-phase of a 12 days old male…" For a cell-adhesion measurement with molecular resolution, the biophysicist has a dilemma. The information from the individual adhesion molecule is embedded in the concert of all the participating molecules of the cell including the membrane and the cytoskeleton. It is impossible to distinguish the contribution of each of them from the measured force signal.* The individual behavior of the investigated cell not only distorts a measured force trace, but furthermore it may not be determinable from which adhesion molecule a detected force signal originated from, because there are many classes of adhesion molecules, in different affinity states, present on the cellular surface.

With the knowledge of the typical adhesion patterns presented here, even complex force traces can be analyzed if a known signature of a certain molecule or a typical multi-molecular arrangement is identified in the trace. In some cases, the contribution of certain adhesion molecules can be ruled out just by luck or smart experimental design, by exclusion, blocking experiments, or by genetic knock out manipulation and silencing of a cell line. Subtracting the identified contributions of the "unwanted" molecules from measured force distributions often enhances the specific signal. Cells are organized to manage environmental strokes and often induce new uncertainties through unforeseen reactions, while the scientist tries to exclude molecules from the force experiment. Cells often have a back up for an eliminated molecule. Therefore, adhesion force experiments, conducted by

---

* Nuclear physicists have built enormous detectors to measure almost each particle from a collision experiment of just two elemental particles. Will biophysicists have to build similar detectors to measure the adhesion of just two cells?

an interdisciplinary team with members from chemical, medical, biological, and physical sciences benefit from the merge of knowledge, scientific strategies, and points of view to adequately reflect the complexity of each individual cell.

9.6 Outlook

Planar patch clamp technology turned out to be not only a perfect platform for AFM measurements on non-adherent cells, but also an extension toward simultaneous electrophysiological measurements, an application that is extremely attractive for pharmacological research. Here, for example, the mechanical signal from the cell can be correlated in time with the activity of membrane pores (Pamir et al., 2008).

Force spectroscopy not only characterizes antibodies with respect to their interaction force with their specific ligand, but also might identify diseases that are caused by the malfunction of cellular adhesion and optimize related medication.

Force measurements with the AFM sequentially measure, one interaction after the other at high resolution in force and space in a very time-consuming manner. A recent concept, that still has a high resolution in force but measures billions of interactions within a few seconds is the molecular force balance (Blank et al., 2004; Albrecht et al., 2008). This force balance might be a potential technique for high throughput force measurements on cells in the near future.

Acknowledgement

Many thanks to Prof. Dr. Hermann Gaub, for his great support and powerful lab space.

Abbreviations

A	Adenine (DNA base pair of thymine)
AFM	Atomic force microscope (sometimes, scanning force microscope SFM)
BR	Bacteriorhodopsin (optical driven proton pump of *H. salinaris*)
BFS	Bio-force sensor (red blood cell utilized as force sensor in a pipette)
BS-transition	Transition from the naturally B-formed DNA to the S-formed (S = stretched)
BSA	Bovine serum albumin
cAMP	Cyclic adenosine mono phosphate (chemokine)
C	Cytosine (DNA base pair of guanine)
csA	Contact site A
Ca^{2+}	Calcium divalent cation
CeNS	Center of Nano Science
CXCR	Receptor for cytosines of the CXC family (CXCR-4 is specific for SDF-1)
DNA	Deoxyribonucleic acid
E	Young's modulus (elastic modulus)
E-cadherin	Epithelial cadherin (cell adhesion molecule)
E. coli	*Escherichia coli* (bacterium)
EM	Electron microscope (SEM, scanning electron microscope)
FJC	Freely jointed chain (polymer model)
FRC	Freely rotating chain (polymer model)
G	Guanine (DNA base pair of cytosine)
GFP	Green fluorescent protein
HEC-1-A	Human uterine epithelial cell line
I-CAM	Inter cellular adhesion molecule
JAR	Human trophoblast-like cell line
K_D	Dissociation constant
k_B	Boltzmann constant
k_{off}	Unbinding rate
$k_B T$	Thermal energy equivalent of two degrees of freedom at a certain temperature
MCF-10A	Human mammary gland epithelial cell line
MD (MDS)	Molecular dynamics simulation
MDA-MB-4355	Human mammary gland epithelial cell line
Mg^{2+}	Magnesium divalent cation
MLCT	Description of industrial cantilevers (micro lever for contact and tapping mode)
N-cadherin	Neuronal cadherin (cell adhesion molecule)
nm	Nano meter
OT	Optical trap or optical tweezers (trapped bead in a laser focus as force sensor)
PAP-G	Last unit of the P-pilus consisting of several PAP-units that binds to galabiose
PN	Pico Newton
PMA	Parametoxyamphetamine (influences intracellular cell signaling)
RGD	(*Arg*inin-*Gly*cin-*Asp*aragin) integrin specific peptide sequence
RL95-2	Human uterine epithelial cell line
SDF-1	Stromal derived factor-1
Si	Silicon
SiO	Silicon oxide
SiN	Silicon nitride
T	Thymine (DNA Base pair of adenine)
V-CAM	Vascular cell adhesion molecule
VE-cadherin	Vascular endothelial cadherin (cell adhesion molecule)
VLA-4	Very late antigen-4
WLC	Worm-like chain (polymer model)
ΔG	Gibbs free energy
v	Poisson ratio (compressibility)

References

Alberts, B., Johnson, A., Lewis, J., Raff, M., Roberts, K., and Walter, P. (2002) *Molecular Biology of the Cell*, Garland Science, New York.

Albrecht, C. H., Neuert, G., Lugmaier, R. A., and Gaub, H. E. (2008) Molecular force balance measurements reveal that dsDNA unbinds under force in rate dependent pathways. *Biophys J*, 94, 4766–4774.

Arnold, M., Cavalcanti-Adam, E. A., Glass, R., Blummel, J., Eck, W., Kantlehner, M., Kessler, H., and Spatz, J. P. (2004) Activation of integrin function by nanopatterned adhesive interfaces. *Chemphyschem*, 5, 383–388.

Bell, G. I. (1978) Models for the specific adhesion of cells to cells. *Science*, 200, 618–627.

Benoit, M. (2002) Cell adhesion measured by force spectroscopy on living cells. *Methods Cell Biol*, 68, 91–114.

Benoit, M. and Gaub, H. E. (2002) Measuring cell adhesion forces with the atomic force microscope at the molecular level. *Cells Tissues Organs*, 172, 174–189.

Benoit, M., Gabriel, D., Gerisch, G., and Gaub, H. E. (2000) Discrete interactions in cell adhesion measured by single-molecule force spectroscopy. *Nat Cell Biol*, 2, 313–317.

Besser, A. and Safran, S. A. (2006) Force-induced adsorption and anisotropic growth of focal adhesions. *Biophys J*, 90, 3469–3484.

Binnig, G., Quate, C. F., and Gerber, C. (1986) Atomic force microscope. *Phys Rev Lett*, 56, 930–933.

Blank, K., Lankenau, A., Mai, T., Schiffmann, S., Gilbert, I., Hirler, S., Albrecht, C., Benoit, M., Gaub, H. E., and Clausen-Schaumann, H. (2004) Double-chip protein arrays: Force-based multiplex sandwich immunoassays with increased specificity. *Anal Bioanal Chem*, 379, 974–981.

Bozzaro, S., Fisher, P. R., Loomis, W., Satir, P., and Segall, J. E. (2004) Guenther Gerisch and *Dictyostelium*, the microbial model for ameboid motility and multicellular morphogenesis. *Trends Cell Biol*, 14, 585–588.

Butt, H.-J. R., Cappella, B., and Kappl, M. (2005) Force measurements with the atomic force microscope: Technique, interpretation and applications. *Surf Sci Rep*, 59, 1.

Carrion-Vazquez, M., Oberhauser, A. F., Fisher, T. E., Marszalek, P. E., Li, H., and Fernandez, J. M. (2000) Mechanical design of proteins studied by single-molecule force spectroscopy and protein engineering. *Prog Biophys Mol Biol*, 74, 63–91.

Clausen-Schaumann, H., Seitz, M., Krautbauer, R., and Gaub, H. E. (2000) Force spectroscopy with single bio-molecules. *Curr Opin Chem Biol*, 4, 524–530.

Collin, D., Ritort, F., Jarzynski, C., Smith, S. B., Tinoco, I. Jr., and Bustamante, C. (2005) Verification of the Crooks fluctuation theorem and recovery of RNA folding free energies. *Nature*, 437, 231–234.

Dettmann, W., Grandbois, M., Andre, S., Benoit, M., Wehle, A. K., Kaltner, H., Gabius, H. J., and Gaub, H. E. (2000) Differences in zero-force and force-driven kinetics of ligand dissociation from beta-galactoside-specific proteins (plant and animal lectins, immunoglobulin G) monitored by plasmon resonance and dynamic single molecule force microscopy. *Arch Biochem Biophys*, 383, 157–170.

Dewa, T., Sugiura, R., Suemori, Y., Sugimoto, M., Takeuchi, T., Hiro, A., Iida, K., Gardiner, A. T., Cogdell, R. J., and Nango, M. (2006) Lateral organization of a membrane protein in a supported binary lipid domain: Direct observation of the organization of bacterial light-harvesting complex 2 by total internal reflection fluorescence microscopy. *Langmuir*, 22, 5412–5418.

Dietz, H. and Rief, M. (2004) Exploring the energy landscape of GFP by single-molecule mechanical experiments. *Proc Natl Acad Sci USA*, 101, 16192–16197.

Dietz, H. and Rief, M. (2006) Protein structure by mechanical triangulation. *Proc Natl Acad Sci USA*, 103, 1244–1247.

Discher, D. E., Boal, D. H., and Boey, S. K. (1998) Simulations of the erythrocyte cytoskeleton at large deformation. II. Micropipette aspiration. *Biophys J*, 75, 1584–1597.

Erdmann, T. and Schwarz, U. S. (2006) Bistability of cell-matrix adhesions resulting from nonlinear receptor–ligand dynamics. *Biophys J*, 91, L60–L62.

Evans, E. (1998) Energy landscapes of biomolecular adhesion and receptor anchoring at interfaces explored with dynamic force spectroscopy. *Faraday Discuss*, 111, 1–16.

Evans, E., Heinrich, V., Ludwig, F., and Rawicz, W. (2003) Dynamic tension spectroscopy and strength of biomembranes. *Biophys J*, 85, 2342–2350.

Florin, E. L., Moy, V. T., and Gaub, H. E. (1994) Adhesion forces between individual ligand-receptor pairs. *Science*, 264, 415–417.

Franz, C. M., Taubenberger, A., Puech, P. H., and Muller, D. J. (2007) Studying integrin-mediated cell adhesion at the single-molecule level using AFM force spectroscopy. *Sci Stke*, 2007, l5.

Fung, Y. C. (1993) *Biomechanics—Mechanical Properties of Living Tissues*, Springer, New York.

Godin, M., Bryan, A. K., Burg, T., Babcock, K., and Manalis, S. R. (2007) Measuring the mass, density and size of particles and cells using a suspended microchannel resonator. *Appl Phys Lett*, 91, 123121.

Grandbois, M., Beyer, M., Rief, M., Clausen-Schaumann, H., and Gaub, H. E. (1999) How strong is a covalent bond? *Science*, 283, 1727–1730.

Grandbois, M., Dettmann, W., Benoit, M., and Gaub, H. E. (2000) Affinity imaging of red blood cells using an atomic force microscope. *J Histochem Cytochem*, 48, 719–724.

Grubmuller, H., Heymann, B., and Tavan, P. (1996) Ligand binding: Molecular mechanics calculation of the streptavidin-biotin rupture force. *Science*, 271, 997–999.

Hansma, P. K., Elings, V. B., Marti, O., and Bracker, C. E. (1988) Scanning tunneling microscopy and atomic force microscopy: Application to biology and technology. *Science*, 242, 209–216.

Harmandaris, V. A. and Deserno, M. (2006) A novel method for measuring the bending rigidity of model lipid membranes by simulating tethers. *J Chem Phys*, 125, 204905.

Heimburg, T. (2007) Front matter. *Thermal Biophysics of Membranes*.

Helenius, J., Heisenberg, C. P., Gaub, H. E., and Muller, D. J. (2008) Single-cell force spectroscopy. *J Cell Sci*, 121, 1785–1791.

Hertz, H. (1882) Über die Berührung fester elastischer Körper. *Reine Angewandte Mathematik*, 92, 156–171.

Hochmuth, F. M., Shao, J. Y., Dai, J., and Sheetz, M. P. (1996) Deformation and flow of membrane into tethers extracted from neuronal growth cones. *Biophys J*, 70, 358–369.

Hugel, T., Rief, M., Seitz, M., Gaub, H. E., and Netz, R. R. (2005) Highly stretched single polymers: Atomic-force-microscope experiments versus ab-initio theory. *Phys Rev Lett*, 94, 048301.

Hwang, W. C. and Waugh, R. E. (1997) Energy of dissociation of lipid bilayer from the membrane skeleton of red blood cells. *Biophys J*, 72, 2669–2678.

Isralewitz, B., Gao, M., and Schulten, K. (2001) Steered molecular dynamics and mechanical functions of proteins. *Curr Opin Struct Biol*, 11, 224–230.

Jin, T. and Hereld, D. (2006) Moving toward understanding eukaryotic chemotaxis. *Eur J Cell Biol*, 85, 905–913.

Kawakami, K., Tatsumi, H., and Sokabe, M. (2001) Dynamics of integrin clustering at focal contacts of endothelial cells studied by multimode imaging microscopy. *J Cell Sci*, 114, 3125–3135.

Kruithof, M., Chien, F., de Jager, M., and van Noort, J. (2008) Subpiconewton dynamic force spectroscopy using magnetic tweezers. *Biophys J*, 94, 2343–2348.

Kufer, S. K., Dietz, H., Albrecht, C., Blank, K., Kardinal, A., Rief, M., and Gaub, H. E. (2005) Covalent immobilization of recombinant fusion proteins with hAGT for single molecule force spectroscopy. *Eur Biophys J*, 35, 72–78.

Kufer, S. K., Puchner, E. M., Gumpp, H., Liedl, T., and Gaub, H. E. (2008) Single-molecule cut-and-paste surface assembly. *Science*, 319, 594–596.

Lamontagne, C. A., Cuerrier, C. M., and Grandbois, M. (2008) AFM as a tool to probe and manipulate cellular processes. *Pflugers Arch*, 456, 61–70.

Ludwig, T., Kirmse, R., Poole, K., and Schwarz, U. S. (2008) Probing cellular microenvironments and tissue remodeling by atomic force microscopy. *Pflugers Arch*, 456, 29–49.

Lugmaier, R. A., Schedin, S., Kuhner, F., and Benoit, M. (2008) Dynamic restacking of *Escherichia coli* P-pili. *Eur Biophys J*, 37, 111–120.

Marcus, W. D. and Hochmuth, R. M. (2002) Experimental studies of membrane tethers formed from human neutrophils. *Ann Biomed Eng*, 30, 1273–1280.

Merkel, R., Nassoy, P., Leung, A., Ritchie, K., and Evans, E. (1999) Energy landscapes of receptor–ligand bonds explored with dynamic force spectroscopy. *Nature*, 397, 50–53.

Morfill, J., Blank, K., Zahnd, C., Luginbuhl, B., Kuhner, F., Gottschalk, K. E., Pluckthun, A., and Gaub, H. E. (2007) Affinity-matured recombinant antibody fragments analyzed by single-molecule force spectroscopy. *Biophys J*, 93, 3583–3590.

Muller, D. J., Kessler, M., Oesterhelt, F., Moller, C., Oesterhelt, D., and Gaub, H. (2002) Stability of bacteriorhodopsin alpha-helices and loops analyzed by single-molecule force spectroscopy. *Biophys J*, 83, 3578–3588.

Munoz Javier, A., Kreft, O., Piera Alberola, A., Kirchner, C., Zebli, B., Susha, A. S., Horn, E., Kempter, S., Skirtach, A. G., Rogach, A. L., Radler, J., Sukhorukov, G. B., Benoit, M., and Parak, W. J. (2006) Combined atomic force microscopy and optical microscopy measurements as a method to investigate particle uptake by cells. *Small*, 2, 394–400.

Oesterhelt, F., Oesterhelt, D., Pfeiffer, M., Engel, A., Gaub, H. E., and Muller, D. J. (2000) Unfolding pathways of individual bacteriorhodopsins. *Science*, 288, 143–146.

Pamir, E., George, M., Fertig, N., and Benoit, M. (2008) Planar patch-clamp force microscopy on living cells. *Ultramicroscopy*, 108, 552–557.

Panorchan, P., George, J. P., and Wirtz, D. (2006a) Probing inter-cellular interactions between vascular endothelial cadherin pairs at single-molecule resolution and in living cells. *J Mol Biol*, 358, 665–674.

Panorchan, P., Thompson, M. S., Davis, K. J., Tseng, Y., Konstantopoulos, K., and Wirtz, D. (2006b) Single-molecule analysis of cadherin-mediated cell-cell adhesion. *J Cell Sci*, 119, 66–74.

Parot, P., Dufrene, Y. F., Hinterdorfer, P., Le Grimellec, C., Navajas, D., Pellequer, J. L., and Scheuring, S. (2007) Past, present and future of atomic force microscopy in life sciences and medicine. *J Mol Recognit*, 20, 418–431.

Pierres, A., Prakasam, A., Touchard, D., Benoliel, A. M., Bongrand, P., and Leckband, D. (2007) Dissecting subsecond cadherin bound states reveals an efficient way for cells to achieve ultrafast probing of their environment. *FEBS Lett*, 581, 1841–1846.

Puchner, E. M., Alexandrovich, A., Kho, A. L., Hensen, U., Schafer, L. V., Brandmeier, B., Grater, F., Grubmuller, H., Gaub, H. E., and Gautel, M. (2008a) Mechanoenzymatics of titin kinase. *Proc Natl Acad Sci USA*, 105, 13385–13390.

Puchner, E. M., Franzen, G., Gautel, M., and Gaub, H. E. (2008b) Comparing proteins by their unfolding pattern. *Biophys J*, 95, 426–434.

Puech, P. H., Poole, K., Knebel, D., and Muller, D. J. (2006) A new technical approach to quantify cell-cell adhesion forces by AFM. *Ultramicroscopy*, 106, 637–644.

Radmacher, M. (2002) Measuring the elastic properties of living cells by the atomic force microscope. *Methods Cell Biol*, 68, 67–90.

Radmacher, M., Tillamnn, R. W., Fritz, M., and Gaub, H. E. (1992) From molecules to cells: Imaging soft samples with the atomic force microscope. *Science*, 257, 1900–1905.

Raucher, D. and Sheetz, M. P. (1999) Characteristics of a membrane reservoir buffering membrane tension. *Biophys J*, 77, 1992–2002.

Rico, F. and Moy, V. T. (2007) Energy landscape roughness of the streptavidin-biotin interaction. *J Mol Recognit*, 20, 495–501.

Rief, M., Clausen-Schaumann, H., and Gaub, H. E. (1999) Sequence-dependent mechanics of single DNA molecules. *Nat Struct Biol*, 6, 346–349.

Rief, M., Gautel, M., Oesterhelt, F., Fernandez, J. M., and Gaub, H. E. (1997a) Reversible unfolding of individual titin immuno-globulin domains by AFM. *Science*, 276, 1109–1112.

Rief, M., Oesterhelt, F., Heymann, B., and Gaub, H. E. (1997b) Single molecule force spectroscopy on polysaccharides by atomic force microscopy. *Science*, 275, 1295–1297.

Rief, M., Gautel, M., and Gaub, H. E. (2000a) Unfolding forces of titin and fibronectin domains directly measured by AFM. *Adv Exp Med Biol*, 481, 129–136; discussion 137–141.

Rief, M., Rock, R. S., Mehta, A. D., Mooseker, M. S., Cheney, R. E., and Spudich, J. A. (2000b) Myosin-V stepping kinetics: A molecular model for processivity. *Proc Natl Acad Sci USA*, 97, 9482–9486.

Ryu, W., Huang, Z., Sun Park, J., Moseley, J., Grossman, A. R., Fasching, R. J., and Prinz, F. B. (2008) Open micro-fluidic system for atomic force microscopy-guided in situ electrochemical probing of a single cell. *Lab Chip*, 8, 1460–1467.

Sackmann, E. (1995) Physical basis of self-organization and function of membranes: Physics of vesicles. In Lipowsky, R. and Sackmann, E. (Eds.) *Structure and Dynamics of Membranes*, Elsevier, Amsterdam, the Netherlands.

Schindler, H., Badt, D., Hinterdorfer, P., Kienberger, F., Raab, A., Wielert-Badt, S., and Pastushenko, V. (2000) Optimal sensitivity for molecular recognition MAC-mode AFM. *Ultramicroscopy*, 82, 227–235.

Schmitz, J. and Gottschalk, K. E. (2008) Mechanical regulation of cell adhesion. *Soft Matter*, 4(7), 1373–1387.

Schmitz, J., Benoit, M., and Gottschalk, K. E. (2008) The viscoelasticity of membrane tethers and its importance for cell adhesion. *Biophys J*, 95, 1448–1459.

Schwarz, U. S., Erdmann, T., and Bischofs, I. B. (2006) Focal adhesions as mechanosensors: The two-spring model. *Biosystems*, 83, 225–232.

Seifert, U. and Lipowsky, R. (1995) The morphology of vesicles. In Lipowsky, R. and Sackmann, E. (Eds.) *Structure and Dynamics of Membranes*, Elsevier, Amsterdam, the Netherlands.

Selhuber, C., Blummel, J., Czerwinski, F., and Spatz, J. P. (2006) Tuning surface energies with nanopatterned substrates. *Nano Lett*, 6, 267–270.

Selhuber-Unkel, C., Lopez-Garcia, M., Kessler, H., and Spatz, J. P. (2008) Cooperativity in adhesion cluster formation during initial cell adhesion. *Biophys J*, 95, 5424–5431.

Selhuber-Unkel, C., Erdmann, T., López-García, M., Kessler, H., Schwarz, U. S., and Spatz, J. P. (2010) Cell adhesion strength is controlled by intermolecular spacing of adhesion receptors. *Biophys J*, 98, 543–551.

Sens, P. and Turner, M. S. (2006) Budded membrane microdomains as tension regulators. *Phys Rev E Stat Nonlin Soft Matter Phys*, 73, 031918.

Sun, M., Graham, J. S., Hegedus, B., Marga, F., Zhang, Y., Forgacs, G., and Grandbois, M. (2005) Multiple membrane tethers probed by atomic force microscopy. *Biophys J*, 89, 4320–4329.

Thie, M., Rospel, R., Dettmann, W., Benoit, M., Ludwig, M., Gaub, H. E., and Denker, H. W. (1998) Interactions between trophoblast and uterine epithelium: Monitoring of adhesive forces. *Hum Reprod*, 13, 3211–3219.

Walton, E. B., Lee, S., and Van Vliet, K. J. (2008) Extending Bell's model: How force transducer stiffness alters measured unbinding forces and kinetics of molecular complexes. *Biophys J*, 94, 2621–2630.

Waugh, R. E. and Hochmuth, R. M. (1987) Mechanical equilibrium of thick, hollow, liquid membrane cylinders. *Biophys J*, 52, 391–400.

Wojcikiewicz, E. P., Zhang, X., and Moy, V. T. (2004) Force and compliance measurements on living cells using atomic force microscopy (AFM). *Biol Proc Online*, 6, 1–9.

Wojcikiewicz, E. P., Abdulreda, M. H., Zhang, X., and Moy, V. T. (2006) Force spectroscopy of LFA-1 and its ligands, ICAM-1 and ICAM-2. *Biomacromolecules*, 7, 3188–3195.

10

Nanoscale Magnetic Biotransport

Edward P. Furlani
University at Buffalo (SUNY)

10.1 Introduction

Magnetic micro/nanoparticles with biofunctional coatings are finding increasing use in fields such as microbiology, biomedicine, and biotechnology, where they are used to label, transport, and separate biomaterials, and to deliver therapeutic drugs to a target tissue (Safarik and Safarikova 2002; Berry and Curtis 2003; Pankhurst et al. 2003; Pedro et al. 2003; Arrueboa et al. 2007; Majewski and Thierry 2007). These particles, as shown in Figure 10.1, are well suited for bioapplications for several reasons. First, they are nontoxic and are well tolerated by living organisms. Indeed, magnetic (magnetite) nanoparticles occur naturally in a diverse number of species and organisms including humans, dolphins, homing pigeons, bees, and magnetotactic bacteria (Safarik and Safarikova 2002). Second, they can be synthesized in sizes that range from a few nanometers to tens of nanometers with a narrow size distribution, which makes them ideal for probing and manipulating micro/nanoscale bioparticles and biosystems. Third, magnetic nanoparticles can be custom tailored with appropriate surface treatments to enhance biocompatibility and enable biofunctional coating with affinity biomolecules for highly specific binding with a desired biomaterial. Fourth, magnetic nanoparticles exhibit superparamagnetic behavior, i.e., they are easily magnetized by an applied magnetic field, but revert to an unmagnetized state once the field is removed. Thus, they experience a magnetic force when subjected to a local field gradient, and they can be used to separate or immobilize magnetically labeled biomaterials from a carrier fluid using an external magnetic field. Significantly, the relatively low permeability of an aqueous carrier fluid enables efficient coupling between an applied field and magnetically labeled biomaterial. Furthermore, the low intrinsic magnetic susceptibility of most biomaterials provides substantial contrast between labeled and unlabeled material, which enables a high degree of selectivity and detection. Magnetic labeling has advantages over conventional fluorescence and chemiluminescence-based biolabels. For example, small samples of magnetically labeled material can be detected using ultrasensitive ferromagnetic "spin valve" sensors, which can be integrated into microfluidic-based diagnostic systems.

To summarize, the small size, large surface-to-volume ratio, biofunctionality, and superparamagnetic behavior of magnetic nanoparticles make them well suited for probing and manipulating bioparticles and biosystems such as proteins (5–50 nm), viruses (20–450 nm), genes (2 nm wide and 10–100 nm long), or whole cells (10–100 μm) (Pankhurst et al. 2003) (Figure 10.2). In this regard, it is important to note that submicron- or micron-sized magnetic particles are also widely used for bioapplications (Figure 10.1b). However, these are not formed from homogeneous magnetic material; rather, they usually consist of magnetic nanoparticles encapsulated within a nonmagnetic matrix, thereby forming a composite microparticle. The matrix material is typically an inert organic polymer, silica, or ceramic, and functions not only to contain the particles but also to protect them from degradation due to environmental factors such as oxidation. Magnetic microparticles can be coated with appropriate molecules to enable a desired functionality such as biotargeting and drug delivery, as described below.

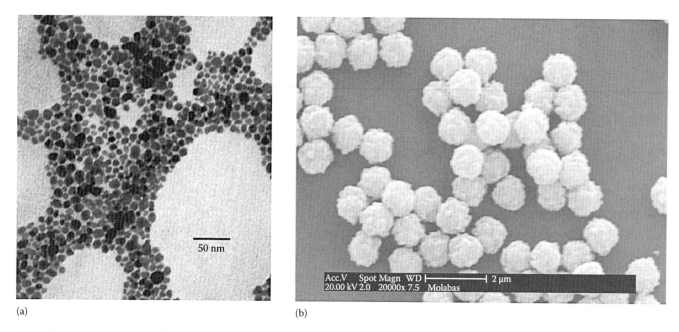

(a)   (b)

FIGURE 10.1   Magnetic particles: (a) TEM of Fe_3O_4 nanoparticles without size selection and (b) polymeric microparticles with embedded magnetic nanoparticles. (Courtesy of Dynal Biotech, Oslo, Norway.)

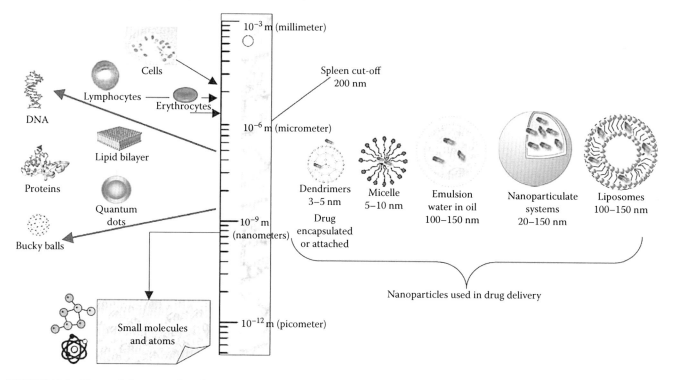

FIGURE 10.2   Nanoscale biomaterials and nanoparticle systems for drug delivery. (Adapted from Arrueboa, M. et al., *Nanotoday*, 2(3), 22, 2007. With permission.)

Magnetic particles have additional significant advantages and uses that are not directly related to biotransport. Specifically, they can be designed to absorb energy at a resonant frequency from a time-varying magnetic field, which enables their use for therapeutic hyperthermia of tumors. In RF hyperthermia, specifically, magnetic nanoparticles are directed to the malignant tissue and then irradiated with an AC magnetic field of sufficient magnitude and duration to heat the tissue to 42°C for 30 min or more, which is sufficient to destroy the tissue (Moroz et al. 2002). Studies demonstrate that RF hyperthermia could be used as an alternate or an adjuvant to other cancer therapies (Yanase et al. 1998; Gupta and Gupta 2005; Hergt and Duzt 2007). Magnetic nanoparticles are also used for bioimaging, both optically, using surface-bound fluorophores for biophotonic

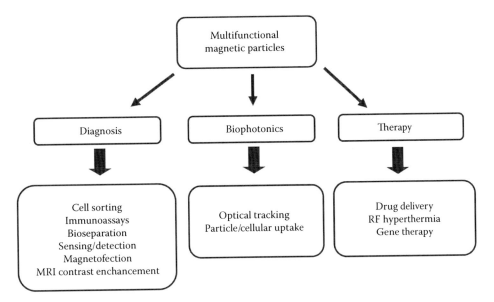

FIGURE 10.3 Bioapplications of magnetic particles.

applications (Levy et al. 2002; Prasad 2003, 2004; Sahoo et al. 2005), and magnetically, where they serve as contrast agents for enhanced MRI. Common bioapplications of magnetic particles are listed in Figure 10.3.

In this chapter, we study the transport of biofunctional magnetic nanoparticles in a carrier fluid under the influence of a magnetic field. We begin with a review of key concepts of magnetism, including the classification of magnetic materials. We then summarize the properties of magnetic nanoparticles. This is followed by a detailed discussion of the physics and equations governing magnetic particle transport in a viscous medium. We discuss two different transport models: a classical Newtonian model for predicting the motion of individual particles, and a drift-diffusion model for predicting the behavior of a concentration of nanoparticles, which accounts for the effects of Brownian motion. Next, we review specific biotransport applications including magnetic bioseparation, drug delivery, and magnetofection. We demonstrate the transport models via application to these processes. We conclude the chapter with an outlook for future trends in this field.

10.2 Fundamentals of Magnetism

Magnetism is a fundamental phenomenon of nature, and the interest in it dates back to antiquity. Magnetism is mediated by a magnetic field, which arises from the motion of charged matter, e.g., the orbital motion of electrons around an atomic nucleus, or the flow of electrons in an electric current. A magnetic field is also generated by the quantum-mechanical phenomenon of spin. The fundamental element in magnetism is the magnetic dipole. This can be thought of as a pair of closely spaced magnetic poles, or equivalently as a small current loop. A magnetic dipole has a dipole moment, **m**, and the magnetization, **M**, of a material is a measure of the net magnetic dipole moment per unit volume:

$$\mathbf{M} = \lim_{\Delta V \to 0} \frac{\sum_i \mathbf{m}_i}{\Delta V}, \qquad (10.1)$$

where $\sum_i \mathbf{m}_i$ is a vector sum of the dipole moments contained in the elemental volume, ΔV.

10.2.1 Units of Magnetism

Three different systems of units are commonly used in the study of magnetism. These are the CGS or Gaussian system, and two MKS or SI systems that are referred to as the Kennelly and Sommerfeld conventions, respectively (Furlani 2001). Throughout this chapter, we use SI units in the Sommerfeld convention wherein the magnetic flux Φ is in weber (Wb), the flux density **B** is in tesla (T) or Wb/m^2, and both the field strength **H** and magnetization **M** are in A/m. The units for the SI (Sommerfeld) and CGS systems are as follows:

Symbol	Description	SI	CGS
H	Magnetic field strength	A/m	Oe
B	Flux density	Tesla	Gauss
M	Magnetization	A/m	emu/cm^3
Φ	Flux	Weber	Maxwells

The conversion factors for these systems are as follows:

$$1\,Oe = 1000/4\pi\ A/m$$
$$1\,Gauss = 10^{-4}\ T$$
$$1\,emu/cm^3 = 1000\ A/m$$
$$1\,Maxwell = 10^{-8}\ Wb$$

In the Sommerfeld convention, the dipole moment, **m**, is measured in $A \cdot m^2$. In the Kennelly and CGS systems, it is measured in $Wb \cdot m$ and emu, respectively ($1\,emu = 4\pi \times 10^{-10}\ Wb \cdot m$). If a magnetic dipole is subjected to an applied **B** field, it acquires energy $E = -\mathbf{m} \cdot \mathbf{B}$, and experiences a torque $\mathbf{T} = \mathbf{m} \times \mathbf{B}$. The force on a magnetic dipole is discussed in a separate section below.

In the MKS system, the fields are related by the constitutive relation

$$\mathbf{B} = \mu_0 \left(\mathbf{H} + \mathbf{M} \right), \tag{10.2}$$

where $\mu_0 = 4\pi \times 10^{-7}$ T·m/A is the permeability of free space. In linear, homogeneous, and isotropic magnetic media, both $\mathbf{B}$ and $\mathbf{M}$ are proportional to $\mathbf{H}$. Specifically, $\mathbf{B} = \mu \mathbf{H}$ and $\mathbf{M} = \chi_m \mathbf{H}$, where μ and χ_m are the permeability and susceptibility of the material, respectively. These coefficients are related to one another, i.e., $\mu = \mu_0(\chi_m + 1)$ or $\chi_m = \mu/\mu_0 - 1$. The constitutive relations need to be modified for nonlinear, inhomogeneous, or anisotropic materials. A material is magnetically nonlinear if μ depends on $\mathbf{H}$, otherwise it is linear. For nonlinear materials, $\mathbf{B} = \mu(H)\mathbf{H}$ and $\mathbf{M} = \chi_m(H)\mathbf{H}$.

10.2.2 Magnetic Materials

Magnetic materials fall into one of the following categories: diamagnetic, paramagnetic, ferromagnetic, antiferromagnetic, and ferrimagnetic (Furlani 2001). Diamagnetic materials have no net atomic or molecular magnetic moment. When these materials are subjected to an applied field, atomic currents are generated that give rise to a bulk magnetization that opposes the field. Bismuth is an example of a diamagnetic material. Paramagnetic materials have a net magnetic moment at the atomic level due to unfilled electronic shells, but the coupling between neighboring moments is weak. These moments tend to align with an applied field, but the degree of alignment decreases at higher temperatures due to randomizing thermal effects. Ferromagnetic materials have a net magnetic moment at the atomic level, but unlike paramagnetic materials, there is a strong coupling between neighboring moments. This coupling gives rise to a spontaneous alignment of the moments over macroscopic regions called domains. The domains undergo further alignment when the material is subjected to an applied field. Lastly, antiferromagnetic and ferrimagnetic materials have oriented atomic moments with neighboring moments antiparallel to one another. In antiferromagnetic materials, the neighboring moments are equal, and there is no net magnetic moment. In ferrimagnetic materials, the neighboring moments are unequal, and there is a net magnetic moment. This is due to a difference in either the number of moments in each direction, or the size of the moments in alternating directions, or both.

10.3 Magnetic Nanoparticles

Magnetic nanoparticles typically range from 1 to 100 nm in diameter. However, larger particles, several hundred nanometers in diameter or even micron-sized particles, can be fabricated by encapsulating magnetic nanoparticles in organic (e.g., polymeric) or inorganic materials, as shown in Figure 10.1b. Magnetic nanoparticles are finding increasing use in a diverse range of applications including high-density data storage, magnetic nanocomposites for electronic components, high-strength nanostructured permanent magnet materials, ultrasensitive magnetic sensors, novel ferrofluids, and magnetic inks. Currently, the most prominent applications of magnetic nanoparticles are in the fields of bioscience and biomedicine. In this section, we briefly review the synthesis and properties of these particles.

10.3.1 Synthesis of Nanoparticles

Methods for synthesizing magnetic nanoparticles have evolved over several decades, and new methods continue to be developed and refined. There are two basic approaches to nanoparticle synthesis: bottom-up and top-down. In a bottom-up approach, elemental building blocks such as atoms, molecules, or clusters are assembled into nanoparticles. This approach relies on the energetics of the process to guide the assembly. Examples of bottom-up chemical methods include coprecipitation, sonochemical reactions, sol-gel synthesis, microemulsions, hydrothermal reactions, and hydrolysis and thermolysis of precursors (Sahoo et al. 2004, 2005, 2008; Willard et al. 2004; Lu et al. 2007). The top-down approach involves the reduction of larger-scale matter to desired nanoscale structures, and is generally subtractive in nature. Top-down methods include photolithography, mechanical machining/polishing, laser beam and electron beam processing, and electrochemical removal.

The most commonly used methods for preparing magnetic nanoparticles involve some form of bottom-up chemical approach. Such methods are routinely used to prepare particles of different materials including oxides such as magnetite Fe_3O_4 and maghemite γ-Fe_2O_3; pure metals such as Fe, Ni, and Co; ferrites of the form $MeO \cdot Fe_2O_3$, where Me = Mg, Zn, Mn, Ni, Co, ...); and alloys such as $CoPt_3$ and FePt (Willard et al. 2004; Lu et al. 2007). It should be noted that the most widely used magnetic nanoparticles are the oxides, Fe_3O_4 and γ-Fe_2O_3.

10.3.2 Properties of Magnetic Nanoparticles

The properties of a magnetic nanoparticle are typically very different from those of the constituent bulk material. They can be strong functions of the physical attributes of the particle including its size, shape, chemical composition, crystal structure, and morphology, as well as myriad surface effects. The magnetic response of noninteracting nanoparticles depends on their physical attributes as well as environmental factors such as the ambient temperature, the applied field, and the manner in which the temperature and field are varied. On the other hand, the magnetic behavior of an assembly of interacting particles depends not only on the particle attributes and environmental factors, but also on the spacing between particles and the distribution of particle size.

The behavior of noninteracting magnetic particles depends on their underlying magnetic state, which is size dependent.

We consider four different states: multidomain (MD), pseudo-single domain (PSD), single domain (SD), and superparamagnetic (SP), which are ordered relative to the particle size to which they apply (i.e., larger to smaller).

10.3.2.1 Multidomain Particles

The multidomain state occurs in bulk materials. Specifically, a macroscopic specimen consists of a multitude of domains, wherein regions of uniform magnetization are separated by domain walls. Domains form in such a way so as to minimize the total energy of a specimen. The variables governing domain formation include the exchange and magnetocrystalline energies, as well as the saturation magnetization. These, in turn, depend on material composition and temperature. Thus, different materials have different domain state size dependence, and the domain state of a given specimen will vary with temperature. Domains vary in size, shape, and orientation, and the bulk magnetization of a specimen is defined by the collective behavior of all its domains. Field-induced magnetization reversal in multidomain particles occurs via domain wall nucleation and motion.

Since the energy required for these processes is relatively low, multidomain particles tend to have a relatively low remanent magnetization M_r and coercivity H_c (Figure 10.4a). A typical magnetization curve (M vs. H) for a multidomain specimen is shown in Figure 10.4b, which shows the saturation magnetization M_s and saturating field H_s.

As the size of a multidomain specimen decreases, it transitions to a pseudo-single domain state, with properties and behavior that are intermediate between the MD and SD states. While the magnetic behavior of PSD particles is not fully understood, they are thought to have relatively few domains, and tend to have MD-like low coercivity, but SD-like high remanence. The PSD state occurs in Fe_3O_4 (magnetite) particles in the size range between 0.1 and 20 μm (Figure 10.4a).

10.3.2.2 Single Domain Particles

As the size of a magnetic specimen continues to decrease, there is a critical volume below which the energetics favor a single domain state. The transition occurs when the magnetostatic energy E_{ms} of the particle becomes comparable to the energy E_{wall} required to

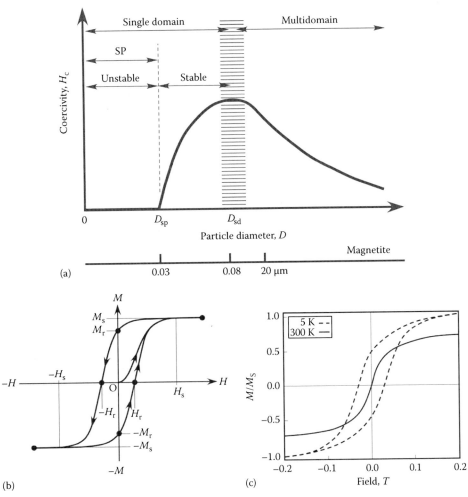

FIGURE 10.4 Properties of magnetic particles: (a) coercivity vs. particle diameter, (b) magnetization curve for multidomain particles, and (c) magnetization curves for superparamagnetic particles (superparamagnetic at room temperature 300 K, but not below the blocking temperature 5 K).

form a domain wall within the particle. The critical diameter D_{sd} for this transition for a spherical particle depends on the material, and is typically in the range of a few tens of nanometers: e.g., $D_{sd} = 15$, 55, and 128 nm for Fe, Ni, and Fe_3O_4, respectively. In single domain particles, below the characteristic Curie temperature of the material, the individual spins are ferromagnetically coupled and uniformly aligned, and the particle becomes spontaneously magnetized with a relatively large magnetic moment, typically 10^3–$10^5 \mu_B$, where $\mu_B = 9.274 \times 10^{-24}$ J/T is the Bohr magneton. In the absence of an applied field, the magnetization of the particle lies along a preferred direction (easy axis), which is defined by magnetic anisotropies that can be due to many factors including the particle's shape, chemical composition, crystalline structure and residual strain, as well as various surface effects. The magnetization will remain in this state as long as the anisotropy energy $E_a = K_{eff}V$ is large relative to the thermal energy kT, where K_{eff} and V are the effective anisotropy constant and particle volume, respectively, $k = 1.38 \times 10^{-23}$ m²·kg/s²·K is Boltzmann's constant, and T is the temperature in Kelvin. Field-induced magnetization reversal in such particles occurs via spin rotation against the anisotropy "force" (there are no domain walls to move). Consequently, they exhibit a larger coercivity than multidomain materials because it is energetically more difficult to overcome the anisotropy energy than it is to nucleate or move a domain wall. The magnetization curve takes the form of a square hysteresis loop, and thus single domain particles are magnetically hard, having both high coercivity and high remanence (Figure 10.4a).

10.3.2.3 Superparamagnetic Particles

As the volume of a single domain particle decreases, its anisotropy energy eventually reduces to a level where it is comparable to the thermal energy. At this point, the particle has a diameter D_{sp} and the magnetization can spontaneously flip from one easy direction to another, even when there is no applied field. The particle is said to be in a superparamagnetic (SP) state (Figures 10.4a and c, $T = 300$ K curve). The thermally induced frequency f of switching between the different easy axes directions is given by

$$f = f_0 \exp\left(-\frac{E_b}{k_BT}\right), \tag{10.3}$$

where

  f_0 is an attempt frequency, which is typically taken to be 10^9 Hz
  E_b is the total energy barrier to switching ($E_b = E_a$ if there is no applied field)

If an ensemble of superparamagnetic particles is subjected to changes in the applied field or temperature, the magnetization will approach an equilibrium value in a period $\tau = 1/f$, which is called the relaxation time, first derived by Neel, i.e.,

$$\tau = \tau_0 \exp\left(\frac{K_{eff}V}{k_BT}\right), \tag{10.4}$$

where $\tau_0 = 10^{-9}$ s. If τ is shorter than the time required to measure the magnetization of the nanoparticles, the nanoparticles are said to be in a superparamagnetic state. On the other hand, if τ is longer than the measurement time, the nanoparticles are said to be "blocked." Below the blocking temperature, the particles possess coercivity and remanent magnetization, and the magnetization as a function of temperature and magnetic field is similar to those of bulk ferromagnetic materials below the Curie temperature, as shown in Figure 10.4c ($T = 5$ K curve). The temperature that defines the transition between the blocked and unblocked states is called the blocking temperature T_B, if $T < T_B$ the particle is blocked, whereas if $T > T_B$ it is unblocked. Note that the blocking temperature depends on several factors including the particle size, the effective anisotropy, the measurement time, and the applied field.

The magnetic response of an ensemble of identical noninteracting superparamagnetic particles is essentially the same as that of a paramagnet except that the magnetic moment is not that of a single atom, but rather of a single domain particle that contains 10^5 atoms. Thus, the susceptibility is much higher than that of a common paramagnet, and hence the term superparamagnetic. The magnetization of an ensemble of paramagnetic particles is given by

$$M = n\mu L\left(\frac{\mu H_a}{kT}\right), \tag{10.5}$$

where

  $L(y) = \coth(y) - 1/y$ is the Langevin function
  n is the number of particles per unit volume ($n\mu$ is the saturation magnetization of the assembly)
  $\mu = M_sV$ is the magnetic moment per particle, which is obviously a strong function of the particle size

The magnetization is zero, if there is no applied field, the coercivity and remanent magnetization are zero, and there is a very high field saturation, as shown in Figure 10.4c ($T = 300$ K).

10.4 Magnetic Particle Transport

Magnetophoresis involves the manipulation of colloidal magnetic particles in a viscous medium using an applied magnetic field. The motion of a magnetic particle in a carrier fluid is governed by several factors including (a) the magnetic force, (b) viscous drag, (c) particle–fluid interactions (perturbations to the flow field), (e) gravity/buoyancy, (g) thermal kinetics (Brownian motion), and (h) interparticle effects including (1) magnetic dipole interactions, (2) electric double-layer interactions, and (3) van der Waals force. In this section, we review the basic equations governing magnetophoretic particle transport. To simplify the analysis, we consider particles in low concentration and we neglect particle/fluid interactions and interparticle effects.

In many applications, particle diffusion due to Brownian motion can also be neglected, which further simplifies the analysis. This is true when the applied (e.g., magnetic) forces on the particle dominate the force it experiences due to random collisions with the fluid molecules. However, for a given application, it is not always clear whether or not Brownian motion can be ignored, as the applied force can vary substantially throughout the region of interest. Nevertheless, the effects of Brownian motion are usually ignored for particles that are greater than a few tens of nanometers in diameter, as discussed below.

In this section, we present two different models for predicting magnetic particle transport in a carrier fluid, each governing a different regime of transport. In the first model, we neglect Brownian motion and use classical Newtonian physics to predict the motion of individual particles, submicron-sized or larger. In the second model, we account for Brownian motion by solving a drift-diffusion equation for the behavior of a concentration of noninteracting magnetic nanoparticles.

10.4.1 Newtonian Particle Transport

We consider the motion of a spherical magnetic particle in a viscous carrier fluid under the influence of an applied field. We restrict our attention to slow flow regimes where the magnetic and viscous drag forces dominate, and we neglect Brownian motion. The particle has a density ρ_p, radius R_p, volume $V_p = \frac{4}{3}\pi R_p^3$, and mass $m_p = \rho_p V_p$. We use classical Newtonian dynamics to study particle motion (Furlani 2006):

$$m_p \frac{d\mathbf{u}_p}{dt} = \mathbf{F}_m + \mathbf{F}_f + \mathbf{F}_g, \tag{10.6}$$

where

    $\mathbf{u}_p$ is the velocity of the particle
    $\mathbf{F}_m$, $\mathbf{F}_f$, and $\mathbf{F}_g$ are the magnetic, fluidic, and gravitational forces, respectively

The inertial term $m_p(d\mathbf{u}_p/dt)$ is often ignored for submicron particles as their mass is negligible. When this is the case, the particle motion is based on a simple force balance:

$$\mathbf{F}_m + \mathbf{F}_f + \mathbf{F}_g = 0. \tag{10.7}$$

We can solve this for the instantaneous particle velocity $\mathbf{u}_p$ given an expression for the fluidic force. Specifically, the fluidic force is obtained using Stokes' law for the viscous drag on a sphere as described below:

$$\mathbf{F}_f = -6\pi\eta R_{hyd,p}(\mathbf{u}_p - \mathbf{u}_f), \tag{10.8}$$

where

    $R_{hyd,p}$ is the effective hydrodynamic radius of the particle
    η and $\mathbf{u}_f$ are the viscosity and velocity of the fluid, respectively

It is important to note that $R_{hyd,p}$ is greater than the physical radius of the particle as surface-bound materials contribute to

the viscous drag. We substitute Equation 10.8 into Equation 10.7 and obtain $\mathbf{u}_p = \mathbf{u}_f + \gamma(\mathbf{F}_m + \mathbf{F}_g)$ or

$$\frac{d\mathbf{x}_p}{dt} = \mathbf{u}_f + \gamma(\mathbf{F}_m + \mathbf{F}_g), \tag{10.9}$$

where $\gamma = 1/(6\pi\eta R_{hyd,p})$ is the mobility of the particle. We use either Equation 10.6 or 10.9 to predict the motion of a particle depending on the significance of the particle inertia. We estimate this as follows. Consider Equation 10.6 in one dimension with no gravitational force:

$$m_p \frac{du_p}{dt} = 6\pi\mu R_p(u_f - u_p). \tag{10.10}$$

This equation has a solution of the form (with $u_p(0) = 0$):

$$u_p(t) = u_f\left(1 - e^{-\frac{t}{\tau}}\right), \tag{10.11}$$

where $\tau = (m_p/(6\pi\mu R_p)) = ((2\rho_p R_p^2)/9\mu)$ is a time constant for the particle to obtain its terminal velocity, i.e., the local velocity of the fluid. For Fe_3O_4 particles in water with a radius between 1 and 100 nm, τ ranges from 1.11 ps to 11.1 ns, respectively. Thus, for many applications the nanoparticles acquire their terminal velocity on a time scale that is much shorter than the overall transport time through a system. When this is the case, we ignore particle inertia and solve Equation 10.9 for the motion of the particle.

10.4.2 Drift-Diffusion Transport

As noted above, Equation 10.6 does not take into account Brownian motion, which can influence particle motion when the particle diameter D_p is sufficiently small. We estimate this diameter using the following criterion (Gerber et al. 1983):

$$|\mathbf{F}|D_p \leq kT, \tag{10.12}$$

where $|\mathbf{F}|$ is the magnitude of the applied force acting on the particle. This condition implies that Brownian motion needs to be taken into account when the energy exerted by the applied force in moving the particle a distance equal to its diameter is less than or comparable to thermal energy kT. In order to apply Equation 10.12, one needs to estimate $|\mathbf{F}|$. If a field source is specified, and the magnetic force is the dominant force, then one can estimate $|\mathbf{F}|$ for a given particle over the region of interest. Gerber et al. have studied the capture of Fe_3O_4 particles in water using a single magnetic wire, and have estimated the critical particle diameter for their application to be $D_{c,p} \equiv kT/|\mathbf{F}| = 40$ nm, i.e., $|\mathbf{F}| = 0.1$ pN (Gerber et al. 1983). For particles with a diameter below $D_{c,p}$ (which is application dependent) one solves a drift-diffusion equation for the particle volume concentration c, rather than the Newtonian equation for the trajectory of a single particle. Specifically, c is governed by the following equation (Gerber et al. 1983; Fletcher 1991; Furlani and Ng 2008),

$$\frac{\partial c}{\partial t} + \nabla \cdot \mathbf{J} = 0, \qquad (10.13)$$

where $\mathbf{J} = \mathbf{J}_D + \mathbf{J}_A$ is the total flux of particles, which includes a contribution $\mathbf{J}_D = -D\nabla c$ due to diffusion, and a contribution $\mathbf{J}_A = \mathbf{U}c$ due to the drift of the particles under the influence of applied forces. The diffusion coefficient D is given by the Nernst–Einstein relation $D = \gamma kT$, where $\gamma = 1/(6\pi\eta R_{hyd,p})$ is the mobility of the particle. The drift velocity $\mathbf{U}$ in $\mathbf{J}_A$ is obtained from Equation 10.6 in the limit of negligible inertia ($m_p(d\mathbf{u}_p/dt) \to 0$), i.e., by setting $\mathbf{F}_m + \mathbf{F}_f + \mathbf{F}_g = 0$. Specifically, from Equation 10.9 we find that $\mathbf{U} = \gamma\mathbf{F}$, where $\mathbf{F} = (\mathbf{u}_f/\gamma) + \mathbf{F}_m + \mathbf{F}_g$. Note that if the Stokes' drag is the only force, then $\mathbf{U} = \mathbf{u}_f$.

10.4.3 Magnetic Force

We model the magnetic force using an "effective" dipole moment approach wherein the magnetized particle is replaced by an "equivalent" point dipole (Furlani and Ng 2006). The magnetic force on the dipole (and hence on the particle) is given by

$$\mathbf{F}_m = \mu_f(\mathbf{m}_{p,eff} \cdot \nabla)\mathbf{H}_a, \qquad (10.14)$$

where

  μ_f is the permeability of the transport fluid
  $\mathbf{m}_{p,eff}$ is the "effective" dipole moment of the particle
  $\mathbf{H}_a$ is the applied magnetic field intensity at the center of the particle, where the equivalent point dipole is located

The dipole approximation has been used for decades to compute the force on submicron magnetic particles. The validation of this approximation has recently been confirmed via particle trajectory measurements in a microfluidic system (Sinha et al. 2007).

  In order to determine the magnetic force, one needs to know the field distribution and the magnetic response (i.e., M vs. H) of the particle. The magnetic field distribution can be determined once the field source and system materials are specified. In many bioapplications, the materials (e.g., carrier fluid, fluidic system, and biomaterial) are nonmagnetic and the field can be obtained in an analytical form. Otherwise, it can be determined using numerical techniques such as the finite element method.

  In order to determine the effective moment $\mathbf{m}_{p,eff}$, we need a model for the magnetization of the particle. A magnetization model that takes into account self-demagnetization and magnetic saturation has been developed and is briefly summarized here (Furlani 2006; Furlani and Sahoo 2006; Furlani and Ng 2006). Consider a magnetic nanoparticle with a radius R_p and volume V_p in the presence of an applied magnetic field $\mathbf{H}_a$. Assume

that particle is uniformly magnetized, and that the magnetization is a linear function of the field intensity up to saturation, at which point it remains constant at a value M_s. Specifically, below saturation,

$$\mathbf{M}_p = \chi_p \mathbf{H}_{in}, \qquad (10.15)$$

where

  $\chi_p = \mu_p/\mu_0 - 1$ is the susceptibility of the particle
  μ_p is its permeability

The field inside the particle is a superposition of the applied field and the internal demagnetization field, $H_{in} = H_a - H_{demag}$. H_{demag} is the self-demagnetization field that accounts for the magnetization of the particle, i.e., its magnetic "surface charge." It is well known that $H_{demag} = M/3$ for a uniformly magnetized spherical particle with magnetization M in free space (Furlani 2001).

  If the particle is suspended in a magnetically linear fluid of permeability μ_f, the force on it in an applied field $\mathbf{H}_a$ is (Furlani and Ng 2006)

$$\mathbf{F}_m = \mu_f V_p \frac{3(\chi_p - \chi_f)}{\left[(\chi_p - \chi_f) + 3(\chi_f + 1)\right]}(\mathbf{H}_a \cdot \nabla)\mathbf{H}_a, \qquad (10.16)$$

where χ_f is the susceptibility of the carrier fluid. For most bioapplications $|\chi_f| \ll 1(\mu_f \approx \mu_0)$, in which case Equation 10.16 reduces to

$$\mathbf{F}_m = \mu_0 V_p \frac{3(\chi_p - \chi_f)}{(\chi_p - \chi_f) + 3}(\mathbf{H}_a \cdot \nabla)\mathbf{H}_a. \qquad (10.17)$$

Under this assumption we also find that

$$\mathbf{M}_p = \frac{3(\chi_p - \chi_f)}{(\chi_p - \chi_f) + 3}\mathbf{H}_a. \qquad (10.18)$$

Equation 10.18 applies below saturation. However, when the particle is saturated, $\mathbf{M}_p = \mathbf{M}_{sp}$. We account for both conditions by expressing the magnetization in terms of the applied field as follows:

$$\mathbf{M}_p = f(H_a)\mathbf{H}_a, \qquad (10.19)$$

where

$$f(H_a) = \begin{cases} \dfrac{3(\chi_p - \chi_f)}{(\chi_p - \chi_f) + 3} & H_a < \left(\dfrac{(\chi_p - \chi_f) + 3}{3(\chi_p - \chi_f)}\right)M_{sp} \\[4mm] \dfrac{M_{sp}}{H_a} & H_a \geq \left(\dfrac{(\chi_p - \chi_f) + 3}{3(\chi_p - \chi_f)}\right)M_{sp} \end{cases} \qquad (10.20)$$

and $H_a = |\mathbf{H}_a|$. The dipole moment is $\mathbf{m}_{p,eff} = V_p\mathbf{M}_p$ or $\mathbf{m}_{p,eff} = V_p f(H_a)\mathbf{H}_a$.

It is instructive to evaluate Equation 10.14 using the expression for $\mathbf{m}_{p,eff}$. Consider a two-dimensional system in which the field source is assumed to be of infinite extent in the z-direction, and therefore the z-components of the magnetic field and force are zero. Under this assumption, we obtain

$$\mathbf{H}_a(x, y) = H_{ax}(x, y)\hat{\mathbf{x}} + H_{ay}(x, y)\hat{\mathbf{y}}, \qquad (10.21)$$

and

$$\mathbf{F}_m(x, y) = F_{mx}(x, y)\hat{\mathbf{x}} + F_{my}(x, y)\hat{\mathbf{y}}, \qquad (10.22)$$

where

$$F_{mx}(x, y) = \mu_0 V_p f(H_a)\left[H_{ax}(x, y)\frac{\partial H_{ax}(x, y)}{\partial x} + H_{ay}(x, y)\frac{\partial H_{ax}(x, y)}{\partial y}\right], \qquad (10.23)$$

and

$$F_{my}(x, y) = \mu_0 V_p f(H_a)\left[H_{ax}(x, y)\frac{\partial H_{ay}(x, y)}{\partial x} + H_{ay}(x, y)\frac{\partial H_{ay}(x, y)}{\partial y}\right]. \qquad (10.24)$$

We can evaluate $\mathbf{F}_m(x, y)$ given an expression for the field distribution $\mathbf{H}_a(x, y)$. Analytical expressions for the field distributions of common magnetic sources (i.e., cylindrical and rectangular rare-earth magnets, and rectangular current carrying conductors) can be found in the literature (Gerber et al. 1983; Fletcher 1991; Furlani 2001, 2006; Furlani and Sahoo 2006; Furlani and Ng 2006, 2008; Furlani et al. 2007).

10.4.4 Fluidic Force

The fluidic force is predicted using a modified form of Stokes' law for the viscous drag on a sphere (Liu et al. 2007), i.e.,

$$\mathbf{F}_f = -6\pi\eta R_{hyd,p} f_{D_p}(\mathbf{u}_p - \mathbf{u}_f)\pi, \qquad (10.25)$$

where

$R_{hyd,p}$ is the hydrodynamic radius
η and $\mathbf{u}_f$ are the viscosity and velocity of the fluid, respectively

The term f_{D_p} is a drag coefficient that accounts for the influence of a solid wall in the vicinity of the moving particle, i.e.,

$$f_{D_p} = \left[1 - \frac{9}{16}\left(\frac{D_p}{D_p + 2z}\right) + \frac{1}{8}\left(\frac{D_p}{D_p + 2z}\right)^3 - \frac{45}{256}\left(\frac{D_p}{D_p + 2z}\right)^4 - \frac{1}{16}\left(\frac{D_p}{D_p + 2z}\right)^5\right], \qquad (10.26)$$

where

D_p is the diameter of the particle
z is the distance between the bottom of the particle and the wall

Note that far from the wall, $f_{D_p} = 1$ and Equation 10.25 reduces to the usual Stokes' drag formula, which strictly applies to a single isolated particle in an infinite uniform flow field. However, for most applications the flow field is not uniform but rather varies throughout the fluidic system (e.g., laminar flow through a microchannel). Nevertheless, the particle diameter is typically much smaller than the dimensions of the fluidic system, and thus the fluid velocity is usually relatively constant across the particle. Thus, we use Equation 10.25 to estimate the viscous drag force on a particle at a given time, using the particle velocity at that time and the fluid velocity at the position of the particle at that time.

10.4.5 Gravitational Force

The gravitational force takes into account buoyancy and is given by

$$F_g = -V_p(\rho_p - \rho_f)g, \qquad (10.27)$$

where

$F_g = |\mathbf{F}_g|$ is the magnitude of the force
ρ_p and ρ_f are the densities of the particle and fluid, respectively
$g = 9.8\,\text{m/s}^2$ is the acceleration due to gravity

The force is often ignored when analyzing the motion of submicron particles, as it is usually much weaker than the magnetic force (Furlani 2006; Furlani and Sahoo 2006). However, it is instructive to evaluate this force for a one micron ($R_p = 0.5\,\mu\text{m}$) Fe_3O_4 particle in water ($\rho_p = 5000\,\text{kg/m}^3$, $\rho_f = 1000\,\text{kg/m}^3$). We evaluate the force due to the particle mass F_m and buoyancy F_b separately, $F_m = \rho_p \frac{4}{3}\pi R_p^3 g$ and $F_b = \rho_f \frac{4}{3}\pi R_p^3 g$, and obtain $F_m = 2.618 \times 10^{-2}\,\text{pN}$, and $F_b = 0.523 \times 10^{-2}\,\text{pN}$, which are typically an order of magnitude smaller than the applied magnetic force.

10.5 Bioapplications

As noted in the introduction, the use of magnetic particles for bioapplications is proliferating. Current applications include magnetic drug and gene delivery, RF hyperthermia (Moroz et al. 2002), cell sorting, (Zborowski and Chalmers 2007), bioassays (Osaka et al. 2006; Majewski and Thierry 2007), magnetofection (Schillinger et al. 2005), magnetic twisting cytometry (Stamenovic and Wang 2000), MRI (Moroz et al. 2002), biosensing, and bioseparation (Safarik and Safarikova 2004) (Figure 10.3). In this section, we discuss the biofunctionalization of magnetic particles and the applications in which the particles are used primarily to transport, immobilize, or separate biomaterial.

10.5.1 Biofunctionalization

Magnetic micro/nanoparticles need to be functionalized (treated with appropriate surface coatings) to be of use in bioapplications (Figure 10.5). For such applications, there are several issues to consider: colloidal stability, biocompatibility, and biotargeting (Berry and Curtis 2003; Lu et al. 2007; Majewski and Thierry 2007). Colloidal stability implies that the particles do not aggregate in a carrier fluid. Nanoparticles in a colloidal dispersion undergo Brownian motion and frequently collide with one another. The dispersion will not be stable if the particles aggregate upon collision. Attractive interactions such as van der Waals, electrostatic, and magnetic dipole-dipole forces can cause irreversible aggregation. These forces need to be countered if the colloid is to remain stable. To this end, the particles are surface-modified to create electrostatic or steric repulsive forces. Electrostatic stabilization involves coating the nanoparticles with materials that produce an electrical double layer on the surface. The nanoparticles are stabilized by mutual electrostatic repulsion, which is substantial because of their large area-to-volume ratio. Steric stabilization is achieved by coating the nanoparticles with surfactants "surface acting agents" or polymers. Surfactants are typically organic compounds that are amphiphilic, i.e., they contain both hydrophilic groups (heads) pointing outward, and hydrophobic groups (tails) pointing inward. Surfactants for magnetic nanoparticles typically contain functional groups such as carboxylic acids, phosphates, and sulfates. Common polymeric stabilization coatings include dextran, polyethylene glycol (PEG), and polyvinyl alcohol (PVA), which also aid in biocompatibility (Majewski and Thierry 2007).

Biocompatibility of a particle implies that it will function within a complex biosystem with minimal undesired interactions (e.g., nonspecific adsorption events) that would otherwise degrade its functionality, e.g., ability to recognize or bind to a target biomaterial. For example, for *in vivo* applications such as drug targeting, magnetic particles move through the vascular system and must be treated with an appropriate coating to resist opsonization (Arrueboa et al. 2007). In opsonization, opsonins (plasma proteins including antibodies) bind to the particles and attract phagocytes, which remove particles from the vascular system via phagocytosis with subsequent accumulation in organs of the reticuloendothelial system (RES) such as the spleen and liver. The amount and types of plasma proteins that bind to the particles is determined by their physicochemical surface characteristics. Commonly used biocompatible coatings include dextran, PEG, and PVA, as previously noted. Inorganic materials such as gold and silica can also be used as biocompatible coating materials.

Biotargeting functionality is added to magnetic particles by coating their surface with appropriate biotargeting agents such as antibodies, proteins, and charged molecules (Figure 10.5). For example, magnetic nanoparticles with specific surface-bound antibodies can bind to target cells that have the antibody receptors (Pamme and Wilhelmb 2006; Zborowski and Chalmers 2007). Magnetic particles coated with immunospecific agents have been successfully bound to red blood cells, lung cancer cells, bacteria, and urological cancer cells, among others.

10.5.2 Bioseparation and Bioassays

The ability to separate, sort, or immobilize biomaterials such as proteins, enzymes, nucleic acids (e.g., RNA and DNA), or whole cells in their native environment is fundamental to the detection and analysis of such entities (Molday et al. 1997; Safarik and Safarikova 2002, 2004; Pankhurst et al. 2003; Majewski and Thierry 2007). Magnetic separation or immobilization of a biomaterial involves the use of magnetic micro/nanoparticles with

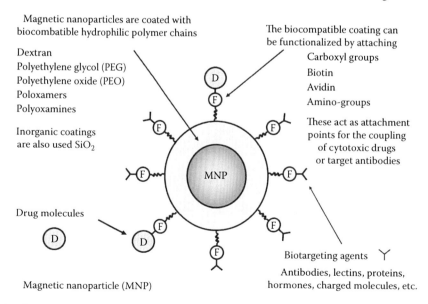

Magnetic nanoparticles are coated with biocombatible hydrophilic polymer chains

Dextran
Polyethylene glycol (PEG)
Polyethylene oxide (PEO)
Poloxamers
Polyoxamines

Inorganic coatings are also used SiO_2

Drug molecules

Magnetic nanoparticle (MNP)

The biocompatible coating can be functionalized by attaching
Carboxyl groups
Biotin
Avidin
Amino-groups

These act as attachment points for the coupling of cytotoxic drugs or target antibodies

Biotargeting agents
Antibodies, lectins, proteins, hormones, charged molecules, etc.

FIGURE 10.5 Biofunctional magnetic nanoparticle.

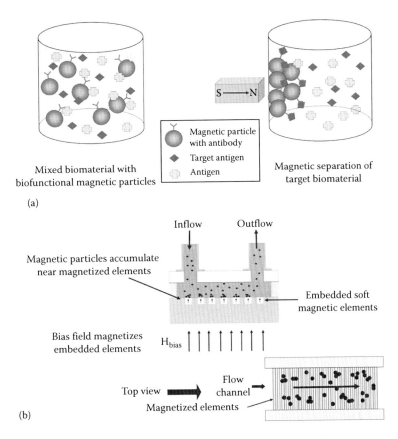

FIGURE 10.6   Magnetic bioseparation: (a) illustration of bioseparation process and (b) magnetophoretic microsystem for bioseparation.

surface-immobilized affinity ligands that are selected to bind with the target biomaterial. Magnetic particles can be used to separate the target biomaterial from a low concentration sample using an applied field (Figure 10.6a). Once separated, the biomaterial can be released into an appropriate medium in a sufficiently high concentration to enable a desired analysis by removing the field. Magnetic separation is used for sample preparation in various applications including high-throughput genome isolation for sequencing or polymerase chain reactions to carry out genotyping (Hawkins et al. 1997). The separation of nucleic acids from their native complex mixtures is required to enable processes such as sequencing, amplification, hybridization, and detection. Magnetic particles have been used for various immunoassays including fluorescence or enzyme immunoassays (Yazdankhah et al. 1999; Schuster et al. 2000).

Magnetic bioseparation is usually implemented using either a direct or indirect approach (Safarýk and Safarýkova 1997). In the more common direct approach, the magnetic particles are functionalized with specific ligands that will bind with the target biomaterial. These particles are mixed with a solution containing the biomaterial, and the mixture is allowed to incubate until a sufficient amount of the target biomaterial binds to the particles. The magnetically labeled biomaterial is then separated from the solution using a magnetic separation system, and then re-released in higher concentration in an appropriate medium for further processing.

In the indirect approach, the target biomaterial is first incubated in solution with an affinity ligand (primary antibody), which is added in free form. After a sufficient amount of biomaterial binds to the primary antibody, magnetic particles with surface-bound secondary antibodies (antibodies against the primary antibodies) are introduced, and the mixture is allowed to incubate until a sufficient amount of the target biomaterial becomes magnetically labeled. The labeled material is then magnetically separated and re-released in higher concentration for further processing.

In conventional magnetic separation systems rare-earth magnets or electromagnets are used to produce a nonuniform field distribution throughout the separation region. When magnetic particles enter this region they experience a force that moves them toward areas of high field gradient where they can be captured as shown in Figure 10.6a. The particles have a high susceptibility and acquire a dipole moment in an external field, but quickly relax back to an unmagnetized state once the field is removed. Thus, when the field is removed, the separated particles disperse into a desired solution for further processing.

Conventional magnetic separation systems have drawbacks in that they tend to be relatively large, costly, and complex, requiring significant energy to operate. Moreover, the accurate manipulation of microscopic particles in small sample volumes is awkward and time consuming in such systems, and the ability to precisely monitor the separation process is limited.

However, advances in microsystem technology have led to the development of novel integrated magnetic bioseparation microsystems that are energy efficient and ideal for the analysis and monitoring of small samples (Choi et al. 2000, 2001; Gijs 2004; Han and Frazier 2004, 2005; Pamme 2006; Hardt and Friedhelm 2007).

An example of a microsystem for magnetic bioseparation is shown in Figures 10.6b, 10.7, and 10.8. This system consists of an array of integrated soft-magnetic elements embedded in a nonmagnetic substrate beneath a microfluidic channel (Bu et al. 2008). A bias field is used to magnetize the elements, which produce a field gradient that imparts a force to magnetic particles within the microchannel. This system has been studied using a Newtonian transport model (Furlani 2006, Furlani and Sahoo 2006). The equations of motion for a magnetic nanoparticle traveling through the microchannel can be written in component form as follows:

$$m_\mathrm{p} \frac{du_{\mathrm{p},x}}{dt} = F_{\mathrm{m}x}(x,y) - 6\pi\eta R_\mathrm{p}\left[u_{\mathrm{p},x} - \frac{3\bar{u}_\mathrm{f}}{2}\left[1 - \left(\frac{y - (h + h_\mathrm{c} + t_\mathrm{b})}{h_\mathrm{c}}\right)^2\right]\right],$$

(10.28)

$$m_\mathrm{p} \frac{du_{\mathrm{p},y}}{dt} = F_{\mathrm{m}y}(x,y) - 6\pi\eta R_\mathrm{p} u_{\mathrm{p},y},$$

(10.29)

$$u_{\mathrm{p},x}(t) = \frac{dx_p}{dt}, \quad u_{\mathrm{p},y}(t) = \frac{dy_p}{dt},$$

(10.30)

where

- h and h_c are the half heights of the magnet and fluidic channel, respectively
- t_b is the thickness of the base of the channel (i.e., the distance from the top of the magnetic elements to the lower edge of the fluid)

Equations 10.28 through 10.30 constitute a coupled system of first-order ordinary differential equations (ODEs) that are solved subject to initial conditions for $x(0)$, $y(0)$, $u_{\mathrm{p},x}(0)$, and $u_{\mathrm{p},y}(0)$. These equations can be solved numerically using various techniques such as the Runge-Kutta method.

We demonstrate the use of Equations 10.28 through 10.30 via application to a microsystem with a fluid channel that is 200 μm high, 1 mm wide, and 10 mm long ($h_\mathrm{c} = 100\,\mu\mathrm{m}$, $w_\mathrm{c} = 500\,\mu\mathrm{m}$ in Figure 10.8 III). The fluid is nonmagnetic ($\chi_\mathrm{f} = 0$), and has a viscosity and density equal to that of water, $\eta = 0.001\,\mathrm{N\cdot s/m^2}$, and $\rho_\mathrm{f} = 1000\,\mathrm{kg/m^3}$. There are 10 permalloy (78% Ni 22% Fe) elements embedded immediately beneath the microchannel (Furlani 2001). Each element is 100 μm high, and 100 μm wide ($h = 50\,\mu\mathrm{m}$, $w = 50\,\mu\mathrm{m}$), and they are spaced 50 μm apart (edge to edge). The bias field is 0.25 T, which is provided by a single NdFeB magnet positioned beneath the microsystem. We assume that this field is sufficient to magnetize the permalloy elements to saturation, $M_\mathrm{es} = 8.6 \times 10^5\,\mathrm{A/m}$. We consider the transport of Fe_3O_4 particles that have the following properties: $R_\mathrm{p} = 250\,\mathrm{nm}$, $\rho_\mathrm{p} = 5000\,\mathrm{kg/m^3}$, and $M_\mathrm{sp} = 4.78 \times 10^5\,\mathrm{A/m}$. We adopt a magnetization model for the particles that is consistent with Equation 10.20 when $\chi_\mathrm{p} \gg 1$, i.e.,

$$f(H_\mathrm{a}) = \begin{cases} 3 & H_\mathrm{a} < M_\mathrm{sp}/3 \\ M_\mathrm{sp}/H_\mathrm{a} & H_\mathrm{a} \geq M_\mathrm{sp}/3 \end{cases}.$$

(10.31)

We study the behavior of a particle as it moves through the microsystem. We assume that the particle enters the microchannel to the left of the first element at $x(0) = -10w$, and compute its trajectory as a function of its initial height above the magnetized elements: $\Delta y = 10, 20, \ldots, 140\,\mu\mathrm{m}$ (i.e., initial heights of $y(0) = 60, 70, \ldots, 190\,\mu\mathrm{m}$). The average fluid velocity is $\bar{u}_\mathrm{f} = 10\,\mathrm{mm/s}$, and the particle enters the channel with this velocity, $u_\mathrm{p}(0) = 10\,\mathrm{mm/s}$. The computed particle trajectories are shown in Figure 10.9a. It is easy to identify each trajectory with its entry height

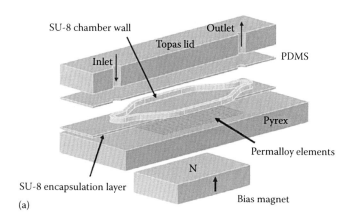

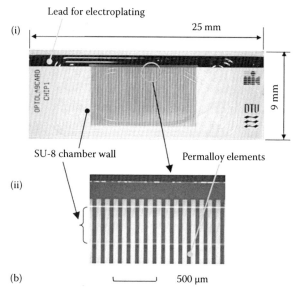

FIGURE 10.7 Passive magnetophoretic microsystem: (a) exploded view of the microsystem and (b) image of chip with electroplated permalloy elements and SU-8 fluidic chamber wall. (Adapted from Bu, M. et al., *Sens. Actuators A*, 145–146, 430, 2008. With permission.)

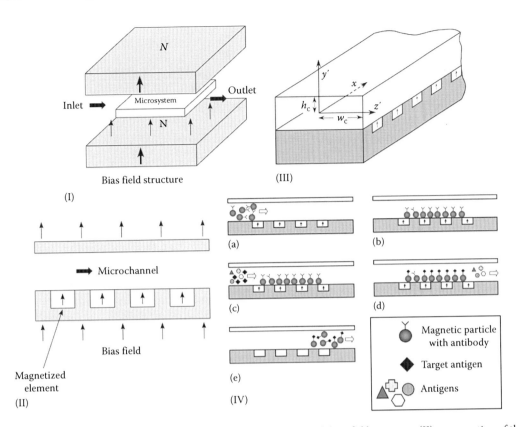

FIGURE 10.8 Magnetophoretic microsystem for bioseparation: (I) microsystem with bias field structure, (II) cross-section of the microsystem showing the microchannel, (III) geometry and reference frame for the microfluidic channel, and (IV) cross-section of microsystem illustrating bioseparation sequence: (a) magnetic particles with surface-bound antibodies enter the microchannel with the bias field applied and the elements magnetized, (b) magnetized elements capture the particles, (c) target antigens are introduced into the microchannel, (d) target antigens become immobilized on captured magnetic particles, (e) the bias field is removed and the magnetic elements revert to an unmagnetized state releasing the separated material for further processing.

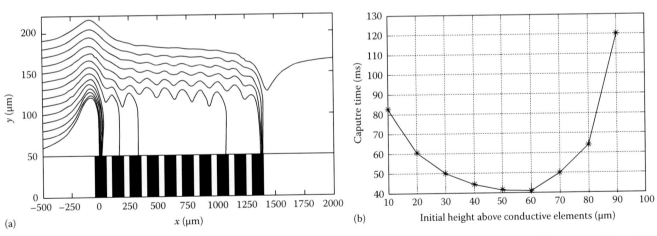

FIGURE 10.9 Bioseparation microsystem analysis: (a) Fe_3O_4 particle trajectory vs. entry height above the magnetized elements and (b) particle capture time vs. entry height above the magnetized elements.

as this is the y-intercept for the plot. According to the analysis, particles that enter the microchannel 0–130 μm above the magnetized elements (i.e., $y = 50$–180 μm) will be captured, but those entering at a greater height will pass through the system. The capture time (the time it takes for a particle to reach the bottom of the microchannel where it is held in place) is plotted as a function of the entry height in Figure 10.9b. This plot shows that particles entering at heights 0–90 μm above the elements will be captured within 130 ms. Notice that the capture time is minimum for particles that enter the channel 50–60 μm above the elements. Particles that enter at lower heights ($\Delta y = 0, \ldots, 40$ μm) experience a substantial vertical repulsive (upward) force prior to reaching the first element, which tends to extend their travel distance thereby increasing their capture time. Particles that enter at higher heights ($\Delta y = 70, \ldots, 140\,\mu m$) bypass the first element, and therefore have extended trajectories and longer capture times.

The capture efficiency of this system can be estimated as follows: If we assume that the particles are uniformly distributed height-wise as they enter the microchannel, then the percentage of particles captured will be equal to the maximum predicted capture entry height 130 μm, divided by the total height of the microchannel, which is 200 μm. Based on this analysis, approximately 65% of the particles that enter the microsystem will be captured. The transport model Equations 10.28 through 10.30 can be used to optimize the capture efficiency of the system.

Magnetic nanoparticles are also finding increasing use for cell-sorting applications, especially at the microscale (Safarik and Safariková 2004; Zborowski and Chalmers 2007). Immunomagnetic cell sorting involves the use of surface markers in the form of magnetic nanoparticles conjugated to antibodies, which are designed to bind to specific antibody receptors on target cells (magnetic labeling). The magnetically labeled cells are sorted using an applied field, which couples to the cellular-bound magnetic nanoparticles, thereby imparting a force to the cell enabling its controlled movement. Pamme and Wilhelmb (2006) have fabricated and characterized a continuous flow, microfluidic cell sorter. They have also applied a Newtonian particle transport model to analyze its performance. It should be noted that red blood cells have a sufficiently high magnetic susceptibility that they can be directly manipulated using a magnetic field, without the need for magnetic labeling (Zborowski et al. 2003). Particle transport models have been successfully applied to immunomagnetic cell sorting, and magnetic manipulation of blood cells. Numerous systems have been analyzed using these methods including quadrupole cell sorters (Sun et al. 1998; Zborowski et al. 1999, 2003; Zborowski and Chalmers 2007), continuous blood cell separators (Takayasu et al. 1982, 2000), and microfluidic-based blood cell separators (Han and Frazier 2004, 2005; Furlani 2007).

10.5.3 Magnetic Drug Targeting

Conventional cancer treatments include surgery, chemotherapy, radiation, immunotherapy, and combinations thereof. The success of these therapies can be limited by various factors including inaccessibility of the tumor, surgical risk, deleterious side effects due to systemic distribution of toxic pharmaceuticals, and collateral irradiation of healthy tissue. An alternate and relatively new cancer treatment involves magnetic drug targeting wherein magnetic particles carrying anticancer agents are directed to malignant tissue using an applied field. This therapy can improve the effectiveness of the treatment while reducing undesired side effects. The growing application of magnetic targeting is due to rapid progress in the development of functionalized magnetic micro/nanoparticles that are designed to target a

specific tissue, and effect local chemo-, radio- and gene therapy at a tumor site (Davis 1997; Berry and Curtis 2003; Pankhurst et al. 2003; Marcucci and Lefoulon 2004; Ferrari 2005; Dobson 2006a; Arrueboa et al. 2007).

In magnetic drug targeting, the therapeutic agent can be either encapsulated into a magnetic micro/nanosphere or conjugated on its surface, as shown in Figure 10.2 (Cinteza et al. 2006; Arrueboa et al. 2007; Zheng et al. 2007). Magnetic particles with bound drug molecules are injected into the vascular system upstream from the malignant tissue (Figure 10.10a). They can be immobilized at the tumor site using a local magnetic field gradient produced by an external field source. Particle accumulation at the tumor is often augmented by magnetic agglomeration, and the efficiency of the accumulation depends on various physiological parameters including particle size, surface characteristics, field strength, blood flow rate, etc.

Drug targeting can be passive or active. In passive targeting, therapeutic particles "leak" out of the vascular system and accumulate at the tumor due to the enhanced permeability of the vasculature that is found near developing tumors. Specifically, most solid tumors exhibit a vascular pore cut-off size from 380 to 780 nm (Arrueboa et al. 2007). Tumor-limited lymphatic drainage adds to this process, and the combined effects result in selective accumulation of the particles in tumor tissue, which is known as enhanced permeation and retention.

In active drug targeting, magnetic particles are functionalized with molecules that are capable of selectively recognizing their target tissue and binding to it (Figure 10.5). Target recognition can occur at different levels, from whole organs/tissue to receptors on specific cells. Thus, therapeutic nanoparticles can provide treatment at both the organ/tissue and cellular level. In the latter case, they can internalize and effect therapy within a cell via processes such as endocytoses or phagocytosis. Cellular targeting can be based on epitopes or receptors that are over- or exclusively expressed in tumor cells, or other over-expressed species such as low molecular weight ligands (folic acid, thiamine, sugars), proteins (transferrin, antibodies, lectins), peptides (RGD, LHRD), polysaccharides (hyaluronic acid), polyunsaturated fatty acids, and DNA. The most common type of targeting molecules are antibodies, lectins, proteins, hormones, charged molecules, and low molecular weight ligands such as folate (Sudimack and Lee 2000; Arrueboa et al. 2007).

Upon achieving a sufficient particle concentration at a tumor, drug molecules can be released from their carrier particles by changing physiological conditions such as pH, osmolality, or temperature, or by enzymatic activity (Berry and Curtis 2003). Since the therapeutic agents are localized to regions of diseased tissue, higher dosages can be applied, which enables more effective treatment. This is in contrast to less-selective conventional chemotherapy wherein a toxic drug is distributed systemically throughout the body, potentially harming healthy tissue.

Magnetic drug targeting has been studied using surface tumors (Alexiou et al. 2000; Berry and Curtis 2003), and small animal models including rabbits (Alexiou et al. 2000), swine (Goodwin et al. 1999, 2001), and rats (Pulfer and Gallo 1999;

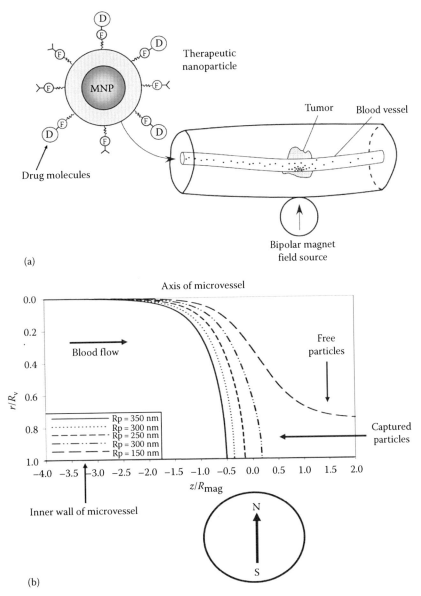

FIGURE 10.10 Magnetic drug targeting: (a) noninvasive drug targeting in a microvessel and (b) trajectories of Fe_3O_4 nanoparticles in a microvessel (cross-section of bias magnet shown for reference).

Pulfer et al. 1999). The clinical trials have produced encouraging results that range from the permanent remission of squamous cell carcinoma in New Zealand White Rabbits (Alexiou et al. 2000), to the effective treatment of breast cancer in humans (Lubbe et al. 1996, 2001).

In addition to the clinical trials, theoretical research has been conducted to study the magnetic drug targeting process. Specifically, Newtonian and drift-diffusion transport models have been used to determine the viability of noninvasive therapy, and to design magnetic fields for clinical applications (Chen et al. 2004, 2005; Ritter et al. 2004; Aviles et al. 2005; Yellen et al. 2005; Furlani and Ng 2006; Furlani and Furlani 2007). It is important to note that these studies are based on very simplistic models of blood flow (e.g., laminar fluid flow through a rigid blood vessel) and, consequently, they provide only a first-order estimate of the

magnetic targeting process. In reality, blood vessels are complex flexible structures, and whole blood is a complex fluid consisting of a suspension of red and white blood cells, and platelets in plasma. A rigorous analysis of particle transport through the vascular system needs to take these complexities into account and is beyond the scope of our discussion. Nevertheless, the transport models presented here enable estimates of particle trapping as a function of key variables including particle size, blood flow rate, location and strength of the field source, and the magnitude of the field gradient at the tumor site.

It is instructive to apply a transport model to a specific system. To this end, we consider a drug targeting system in which the particles are captured at the tumor site by a rare-earth cylindrical magnet positioned outside the body, as shown in Figure 10.10a (Furlani and Ng 2006; Furlani and Furlani 2007).

The magnet is vertically magnetized through its cross-section, and produces a nonuniform field distribution in nearby tissue. It is of infinite extent into the page (*y* direction), with a magnetization that is orthogonal to the blood flow. The magnetic force that it produces is based on an analytical expression for the field distribution within the microvessel, combined with a linear magnetization model for the magnetic response of particles as described by Equation 10.20. The fluidic analysis is based on the assumption of laminar blood flow through a cylindrical microvessel, and the effect of the blood cells is taken into account via use of an effective bulk viscosity η that depends on the hematocrit, which is the percentage by volume of packed red blood cells in a given sample of blood. The Newtonian transport equations for particle motion in the *x*–*z* plane the (*z*-axis is along the length of the microvessel) can be written in component form as

$$m_p \frac{du_{px}}{dt} = -\mu_0 V_p f(H_a) M_s^2 R_{mag}^4 \frac{(x+d)}{2((x+d)^2 + z^2)^3} - 6\pi\eta R_p u_{px},$$

(10.32)

and

$$m_p \frac{du_{pz}}{dt} = -\mu_0 V_p f(H_a) M_s^2 R_{mag}^4 \frac{z}{2((x+d)^2 + z^2)^3}$$
$$- 6\pi\eta R_p \left[u_{pz} - 2\bar{u}_f \left(1 - \frac{x^2}{R_v^2} \right) \right],$$

(10.33)

where R_{mag} is the radius of the magnet.

To demonstrate the theory, we choose a microvessel with a radius $R_v = 75\,\mu m$, and an average flow velocity $\bar{u}_f = 15\,mm/s$. The magnet is 4 cm in diameter ($R_{mag} = 2.0\,cm$) with a magnetization $M_s = 1 \times 10^6\,A/m$. For the calculation of effective viscosity, we assume a hematocrit of 45% (Furlani and Ng 2006). The initial conditions for the particles are as follows. They start on the axis of the microvessel, far enough upstream so that the magnetic force is initially negligible. They have an initial velocity equal to the axial flow velocity (i.e., $u_x(0) = 0$, $u_y(0) = 0$, and $u_z(0) = 2\bar{u}_f = 30\,mm/s$). We predict the trajectories of five different sized Fe_3O_4 particles: $R_p = 150, 200, 250, 300,$ and $350\,nm$, for a magnet-to-vessel distance $d = 2.5\,cm$ (i.e., the center of the microvessel is 0.5 cm from the surface of the magnetic). In these plots, the radial position $r = |x|$ of the particle is normalized with respect to the microvessel radius R_v, and the axial position z is normalized with respect to the magnet radius R_{mag}. The predicted trajectories, along with the magnetic geometry, are shown in Figure 10.10b. This plot shows that all the particles except the smallest ($R_p = 150\,nm$) are captured by the magnet. The predictions take only a few minutes to complete, and the model can be used to estimate capture efficiency as a function of key variables including the particle size and magnetic properties, and the flow rate, etc. To date, various studies have shown that noninvasive magnetic targeting can be achieved using submicron Fe_3O_4 particles, when the target tissue is within several centimeters from the surface of the body.

10.5.4 Magnetofection

Magnetofection is a relatively new process, and interest in it is growing rapidly. In magnetofection, magnetic carrier particles with surface-bound gene vectors are magnetically attracted toward target cells for transfection. Transfection broadly refers to the process of introducing nucleic acids into cells by nonviral methods (Schillinger et al. 2005). In a typical *in vitro* magnetofection system, the target cells are located at the bottom of a fluidic chamber and a rare-earth magnet beneath the chamber provides a magnetic force that attracts the biofunctional particles toward the cells, as shown in Figure 10.11 (Schillinger et al. 2005; Dobson 2006a,b). Magnetofection has distinct advantages over traditional transfection methods. Specifically, high transfection rates are obtained with significantly low vector doses, and the process time is dramatically reduced from hours to minutes (Plank et al. 2003). However, despite the advantages and growing use of magnetofection, relatively few authors have studied particle transport and accumulation for this process.

We apply the drift-diffusion transport model to analyze the performance of a conventional magnetofection system (Furlani and Ng 2008). We study the accumulation of Fe_3O_4 nanoparticles at the base of a cylindrical fluidic chamber positioned above a rare-earth NdFeB magnet. We assume that the chamber has a radius $R_c = 2\,mm$ and length $L_c = 3\,mm$, and that it is positioned 1 mm above the magnet. The magnet has a radius $R_m = 2.5\,mm$ and length $L_m = 5\,mm$ and is magnetized to saturation, $M_s = 8 \times 10^5\,A/m$ ($B_r = 1\,T$). The chamber and magnet dimensions are representative of a 96 well plate magnetofection system. We further assume that the fluid in the chamber is nonmagnetic ($\chi^f = 0$) with a viscosity and density equal to that of water, $\eta = 0.001\,N \cdot s/m^2$, and $\rho_f = 1000\,kg/m^3$. Fe_3O_4 nanoparticles have a density $\rho_p = 5000\,kg/m^3$ and a saturation magnetization $M_{sp} = 4.78 \times 10^5\,A/m$. We also assume that the hydrodynamic radius of a particle is the same as its physical radius, $R_p = R_{p,hyd}$. We adopt a magnetization model as in Equation 10.31.

First, we compute the magnetic force along a series of horizontal lines $0 \le r \le 2R_m$ corresponding to different heights $z = 1, 1.5, 2, 2.5,$ and $3\,mm$ above the magnet. We choose a particle radius $R_p = 100\,nm$ and evaluate the force using closed-form expressions (Furlani and Ng 2008). The predicted force values are compared with corresponding numerical data obtained using finite element analysis (FEA) in Figure 10.12. The magnitude of the force varies across the chamber and decreases with distance from the magnet. The maximum force on the particle is sub nano-Newton, but substantially greater forces can be achieved using larger particles as the force scales with R_p^3. The axial force F_{mz} attracts particles downward toward the surface of the magnet. Farther from magnet F_{mz} is strongest along the *z*-axis (centerline of the magnet). However, closer to the magnet it peaks off-axis, which implies that the particles will have a pronounced accumulation in an annulus at the bottom of chamber. The radial force F_{mr} peaks (in a negative sense) above the radial edge of the magnet and acts to move the particles radially inward toward the *z*-axis, away from the edge. Thus, particle accumulation is expected to

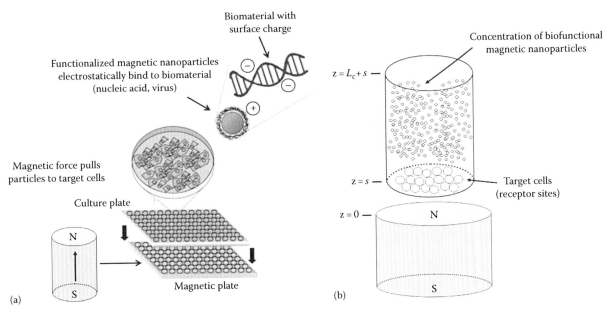

FIGURE 10.11 Magnetofection: (a) culture plate with an array of cell cultures positioned above a magnetic plate with an array of cylindrical rare-earth magnets and (b) a single cell culture with nanoparticles with surface-bound gene vectors.

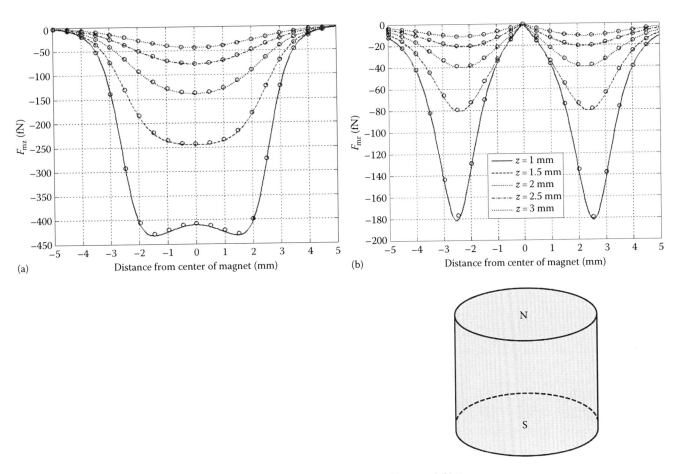

FIGURE 10.12 Magnetic force above a cylindrical rare-earth magnet (o = FEA): (a) F_{mz} and (b) F_{mr}.

be focused toward the center of the chamber and minimal near the edge of the magnet.

Next, we study particle transport. The 2D drift-diffusion Equation 10.13 takes the form

$$\frac{\partial c}{\partial t} = \frac{1}{r}\frac{\partial}{\partial r}\left(r\left(D\frac{\partial c}{\partial r} - U^r c\right)\right) + \frac{\partial}{\partial z}\left(D\frac{\partial c}{\partial z} - U^z c\right), \quad (10.34)$$

where U^r and U^z are drift velocities in r and z directions. The computational domain is $0 \le r \le R_c$ and $z_{c,b} \le z \le z_{c,t}$ where R_c is the radius of chamber, and $z_{c,b}$ and $z_{c,t}$ denote the locations of the bottom and top of the chamber, respectively. We solve Equation 10.34 numerically using the finite-volume method (FVM), and use an upwind scheme to obtain the particle flux (Furlani and Ng 2006). We apply an initial condition in which there is a uniform particle volume concentration throughout the chamber $c(z,0) = c_0$. We impose zero-flux Neumann boundary conditions at the top ($z = z_{z,t}$) and outer edge ($r = R_c$) of the chamber, and a Dirichlet condition $c(r, z_{c,b}, t) = 0$ at the base. The latter condition mimics the magnetofection process wherein nanoparticles that reach the base of the container are removed from the computation as it is assumed that they bind with receptor sites on target cells and therefore no longer influence particle transport. It is assumed that there are a sufficient number of receptors to accommodate all of the particles in the chamber. We compute particle accumulation by summing the number of particles that reach the base of the container during each time step.

We fix the particle radius $R_p = 50\,\text{nm}$ and compute particle accumulation as a function of the separation s between the magnet and the chamber: $s = 0.5, 1.0, 1.5,$ and $2.0\,\text{mm}$. We compare these results with diffusion-limited accumulation, i.e., with no applied field (Figure 10.13a). We find that the rate of magnetically induced accumulation is in orders of magnitude faster than

that of diffusion-limited (no field) accumulation, which is consistent with experimental data shown in Figure 10.13b (Plank et al. 2002). Furthermore, the rate of accumulation increases substantially with decreasing separation. The transport model enables rapid parametric analysis of particle accumulation, which is useful for optimizing novel magnetofection systems.

10.6 Conclusions and Future Prospects

In this chapter, we have provided an overview of biofunctional magnetic micro/nanoparticles and their use as transport agents in bioapplications. We have summarized the basic properties of these particles and their functionalization for bioapplications. We have also discussed the physics governing particle motion in a viscous medium under the influence of an applied field (magnetophoresis). We have presented two different transport models that describe the magnetophoretic behavior of magnetic particles at two different scales: a Newtonian transport model for larger (submicron-micron) particles, and a drift-diffusion transport model for smaller (nanoscale) particles that takes into account Brownian motion. We have discussed specific bioapplications including bioseparation, magnetic drug targeting, and magnetofection. We have demonstrated the use of the transport models via application to these processes.

As for the future, the use of magnetic particles in bioapplications is in its infancy, and will undoubtedly grow significantly in the coming years. Methods for synthesizing, functionalizing, and sensing these particles have advanced rapidly, giving rise to a proliferation of new applications. The small size, large surface-to-volume ratio, inherent multifunctional nature (magnetic manipulation and RF absorption), and unique superparamagnetic behavior of magnetic nanoparticles makes them well suited for bioapplications, both *in vitro* and *in vivo*.

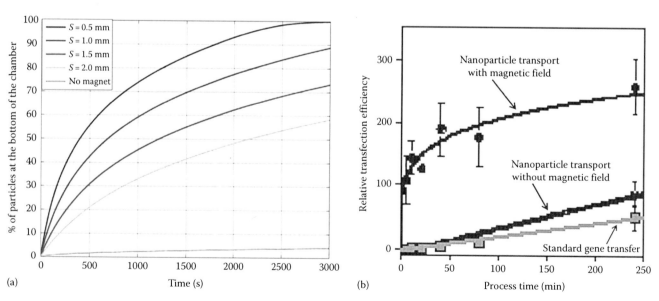

FIGURE 10.13 Nanoparticle accumulation: (a) % of particles at the base of the chamber vs. magnet-to-chamber spacing s and (b) measured relative transfection efficiency, with and without an applied field. (Adapted from Plank, C. et al., *Eur. Cells Mater.*, 3(Suppl. 2), 79, 2002. With permission.)

The most significant advances in future applications of magnetic particles will likely involve the use of microfluidic systems in which magnetic particles will transport or immobilize a biomaterial to facilitate its detection or analysis. The advantages of microfluidic-based analytical systems are numerous (Nguyen and Wereley 2006). Specifically, they function with very small sample sizes, and are capable of rapid processing and a high level of automation with high reliability and low cost. Currently, much research is devoted to the development of multifunctional "Micro-Total Analysis Systems" (μTAS) that are capable of preprocessing and separating biosamples, mixing them with desired reagents, and selectively subjecting them to specific testing and detection (Gijs 2004; Pamme 2006; Hardt and Friedhelm 2007). Once developed, such systems will revolutionize medical diagnostics and treatment, with far-reaching benefits to society as a whole.

Acknowledgments

I would like to thank Dr. Atem Pliss, Dr. Yudhisthira Sahoo, and Prof. Mark Swihart for their insight and technical discussions during the preparation of this chapter.

References

Alexiou, C., Arnold, W., Klein, R.J. et al. 2000. Locoregional cancer treatment with magnetic drug targeting. *Cancer Res.* **60**(23): 6641–6648.

Arrueboa, M., Fernández-Pachecoa, R., Ibarraa, R.M. et al. 2007. Magnetic nanoparticles for drug delivery. *Nanotoday* **2**(3): 22–32.

Aviles, M.O., Ebner, A.D., Chen, H.T. et al. 2005. Theoretical analysis of a transdermal ferromagnetic implant for retention of magnetic drug carrier particles. *J. Magn. Magn. Mater.* **293**: 605–615.

Berry, C.C. and Curtis, A.S.G. 2003. Fictionalization of magnetic nanoparticles for applications in biomedicine. *J. Phys. D: Appl. Phys.* **36**: R198–R206.

Bu, M., Christensen, T.B., Smistrup, K., Wolff, A., and Hansen, M.F. 2008. Characterization of a microfluidic magnetic bead separator for high-throughput applications. *Sens. Actuators A* **145–146**: 430–436.

Chen, H.T., Ebner, A.D., Kaminski, M.D. et al. 2004. Analysis of magnetic drug carrier particle capture by a magnetizable intravascular stent: 1: Parametric study with single wire correlation. *J. Magn. Magn. Mater.* **284**: 181–194.

Chen, H.T., Ebner, A.D., Kaminski, M.D. et al. 2005. Analysis of magnetic drug carrier particle capture by a magnetizable intravascular stent-2: Parametric study with multi-wire two-dimensional model. *J. Magn. Magn. Mater.* **293**: 616–632.

Choi, J.-W., Ahn, C.H., Bhansali, S. et al. 2000. A new magnetic bead-based, filterless bio-separator with planar electromagnet surfaces for integrated bio-detection systems. *Sens. Actuators* **68**: 34–39.

Choi, J.-W., Liakopoulos, T.M., and Ahn, C.H. 2001. An on-chip magnetic bead separator using spiral electromagnets with semi-encapsulated permalloy. *Biosens. Bioelectron.* **16**: 409–416.

Cinteza, L.O., Ohulchanskyy, T.Y., Sahoo, Y. et al. 2006. Diacyllipid micelle-based nanocarrier for magnetically guided delivery of drugs in photodynamic therapy. *Mol. Pharm.* **3**(4): 415–423.

Davis, S.S. 1997. Biomedical applications of nanotechnology-implications for drug targeting and gene therapy. *Trends Biotechnol.* **15**: 217–223.

Dobson, J. 2006a. Magnetic micro- and nano-particle-based targeting for drug and gene delivery. *Nanomedicine* **1**(1): 31–37.

Dobson, J. 2006b. Gene therapy progress and prospects: Magnetic nanoparticle-based gene delivery. *Gene Ther.* **13**: 283–287.

Ferrari, M. 2005. Cancer nanotechnology: Opportunities and challenges. *Nat. Rev. Cancer* **5**(3): 161–171.

Fletcher, D. 1991. Fine particle high gradient magnetic entrapment. *IEEE Trans. Magn.* **27**: 3655–3677.

Furlani, E.P. 2001. *Permanent Magnet and Electromechanical Devices; Materials, Analysis and Applications*. New York: Academic Press.

Furlani, E.P. 2006. Analysis of particle transport in a magnetophoretic microsystem. *J. Appl. Phys.* **99**: 024912:1–11.

Furlani, E.P. 2007. Magnetophoretic separation of blood cells at the microscale. *J. Phys. D: Appl. Phys.* **40**: 1313–1319.

Furlani, E.J. and Furlani, E.P. 2007. A model for predicting magnetic targeting of multifunctional particles in the microvasculature. *J. Magn. Magn. Mater.* **312**(1): 187–193.

Furlani, E.P. and Ng, K.C. 2006. Analytical model of magnetic nanoparticle capture in the microvasculature. *Phys. Rev. E* **73**(6): 061919:1–10.

Furlani, E.P. and Ng, K.C. 2008. Nanoscale magnetic biotransport with application to magnetofection. *Phys. Rev. E* **77**: 061914:1–8.

Furlani, E.P. and Sahoo Y. 2006. Analytical model for the magnetic field and force in a magnetophoretic microsystem. *J. Phys. D: Appl. Phys.* **39**: 1724–1732.

Furlani, E.P., Sahoo, Y., Ng, K.C. et al. 2007. A model for predicting magnetic particle capture in a microfluidic bioseparator. *Biomed. Microdev.* **9**(4): 451–63.

Gijs, M.A.M. 2004. Magnetic bead handling on-chip: New opportunities for analytical applications. *Microfluid Nanofluid* **1**: 22–40.

Gerber, R., Takayasum, M., and Friedlander F.J. 1983. Generalization of HGMS theory: The capture of ultrafine particles. *IEEE Trans. Magn.* **19**(5): 2115–2117.

Goodwin, S.C., Peterson, C., Hob, C. et al. 1999. Targeting and retention of magnetic targeted carriers (MTCs) enhancing intra-arterial chemotherapy. *J. Magn. Magn. Mater.* **194**: 132–139.

Goodwin, S.C., Bittner, C.A., Peterson, C.L. et al. 2001. Single-dose toxicity study of hepatic intra-arterial infusion of doxorubicin coupled to a novel magnetically targeted drug carrier *Toxicol. Sci.* **60**: 177–183.

Gupta, A.K. and Gupta M. 2005. Synthesis and surface engineering of iron oxide nanoparticles for biomedical applications. *Biomaterials* **26**: 3995–4021.

Hafeli, U., Schutt, W., Teller, J., and Zborowski, M. (eds.) 1997. *Scientific and Clinical Applications of Magnetic Carriers.* New York: Plenum Press.

Han, K.H. and Frazier, A.B. 2004. Continuous magnetophoretic separation of blood cells in microdevice format. *J. Appl. Phys.* **96**: 5797–5802.

Han, K.H. and Frazier, A.B. 2005. Diamagnetic capture mode magnetophoretic microseparator for blood cells. *J. Micromech. Syst.* **14**(6): 422–1431.

Hardt, S. and Friedhelm, S. (eds.) 2007. *Microfluidic Technologies for Miniaturized Analysis Systems.* New York: Springer.

Hawkins, T.L., McKernan, K.J., Jacotot, L.B. et al. 1997. DNA sequencing: A magnetic attraction to high-throughput genomics. *Science* **276**: 1887–1889.

Hergt, R. and Duzt, S. 2007. Magnetic particle hyperthermia-biophysical limitations of a visionary tumor therapy. *J. Magn. Magn. Mater.* **311**: 187–192.

Kim, C.K. and Lim, S.J. 2002. Recent progress in drug delivery systems for anticancer agents. *Arch. Pharm. Res.* **25**: 229–239.

Levy, L., Sahoo, Y., Kim, K.-S. et al. 2002. Nanochemistry: Synthesis and characterization of multifunctional nanoclinics for biological applications. *Chem. Mater.* **14**: 3715–3721.

Liu, C., Lagae, L., Wirix-Speetjens, R. et al. 2007. On-chip separation of magnetic particles with different magnetophoretic mobilities. *J. Appl. Phys.* **101**: 024913:1–4.

Lu, A.-H., Salabas, E.L., and Schüth, F. 2007. Magnetic nanoparticles: Synthesis, protection, functionalization, and application. *Angew. Chem. Int. Ed.* **46**: 1222–1244.

Lubbe, A.S., Bergemann, C., Riess, H. et al. 1996. Clinical experiences with magnetic drug targeting: A phase I study with 4′-epidoxorubicin in 14 patients with advanced solid tumors. *Cancer Res.* **56**: 4686–4693.

Lubbe, A.S., Alexiou, C., and Bergemann, C. 2001. Clinical applications of magnetic drug targeting. *J. Surg. Res.* **95**(2): 200–206.

Mah, C., Zolotukhin, I., Fraites, T.J. et al. 2000. Microsphere-mediated delivery of recombinant AAV vectors in vitro and in vivo. *Mol. Ther.* **1**: S239.

Mah, C., Fraites, T.J., Zolotukhin, I. et al. 2002. Improved method of recombinant AAV2 delivery for systemic targeted gene therapy. *Mol. Ther.* **6**: 106–112.

Majewski, P. and Thierry, B. 2007. Functionalized magnetite nanoparticles—Synthesis, properties, and bio-applications. *Crit. Rev. Solid State Mater. Sci.* **32**: 203–215.

Marcucci, F. and Lefoulon, F. 2004. Active targeting with particulate drug carriers in tumor therapy: Fundamentals and recent progress. *Drug Disc. Today* **9**(5): 219–228.

Molday, R.S., Yen, S.P.S., and Rembaum, A. 1997. Application of magnetic microspheres in labeling and separation of cells. *Nature* **268**: 437–438.

Moroz, P., Jones, S.K., and Gray, B.N. 2002. Magnetically mediated hyperthermia: Current status and future directions. *Int. J. Hyperthermia* **18**: 267–84.

Nguyen, N.-T. and Wereley, S.T. 2006. *Fundamentals and Applications of Microfluidics.* Boston, MA: Artech House.

Osaka, T., Matsunaga, T., Nakanishi, T. et al. 2006. Synthesis of magnetic nanoparticles and their application to bioassays. *Anal. Bioanal. Chem.* **384**: 593–600.

Pamme, N. 2006. Magnetism and microfluidics. *Lab Chip* **6**: 24–38.

Pamme, N. and Wilhelmb, C. 2006. Continuous sorting of magnetic cells via on-chip free-flow magnetophoresis. *Lab Chip* **6**: 974–980.

Pankhurst, Q.A., Connolly, J., Jones, S.K. et al. 2003. Applications of magnetic nanoparticles in biomedicine. *J. Phys. D: Appl. Phys.* **36**: R167–R181.

Pedro, T., Morales, M.P., Veintemillas-Verdaguer, S. et al. 2003. The preparation of magnetic nanoparticles for applications in biomedicine. *J. Phys. D: Appl. Phys.* **36**: R182–R197.

Plank, C., Schillinger, U., Scherer, F. et al. 2003. The magnetofection method: Using magnetic force to enhance gene therapy. *Biol. Chem.* **384**: 737–747.

Plank, C., Scherer, F., Schillinger, U. et al. 2002. Magnetofection: Enhancing and targeting gene delivery by magnetic force. *Eur. Cells Mater.* **3**(Suppl. 2): 79–80.

Pulfer, S.K. and Gallo, J.M. 1999. Enhanced brain tumor selectivity of cationic magnetic polysaccharide microspheres. *J. Drug Target.* **6**: 215–228.

Pulfer, S.K., Ciccotto, S.L., and Gallo, J.M. 1999. Distribution of small magnetic particles in brain tumor-bearing rats. *J. Neuro-Oncol.* **41**: 99–105.

Prasad, P.N. 2003. *Introduction to Biophotonics.* Hoboken, NJ: John Wiley & Sons.

Prasad, P.N. 2004. *Nanophotonics.* Hoboken, NJ: John Wiley & Sons.

Ritter, J.A., Ebner, A.D., Daniel, K.D. et al. 2004. Application of high gradient magnetic separation principles to magnetic drug targeting. *J. Magn. Magn. Mater.* **280**: 184–201.

Safarik, I. and Safarikova, M. 2002. Magnetic nanoparticles and biosciences. *Monatsh. Chem.* **133**: 737–759.

Safarik, I. and Safarikova, M. 2004. Magnetic techniques for the isolation and purification of proteins and peptides. *Biomagn. Res. Technol.* **2**(7): 1–17.

Safarýk, I. and Safarýkova, M. 1997. *Scientific and Clinical Applications of Magnetic Carriers*, Hafeli, U., Schutt, W., Teller, J., and Zborowski, M. (eds.). New York: Plenum Press.

Sahoo, Y., Furlani, E.P., and Bergey, E.J. 2008. Magnetic nanoparticles: Structure and bioapplications. In *Encyclopedia of Biomaterials and Biomedical Engineering*, Bowlin, G.L. and Wnek, G. (eds.). New York: Taylor & Francis.

Sahoo, Y., Cheon, M., Wang, S. et al. 2004. Field-directed self-assembly of magnetic nanoparticles. *J. Phys. Chem. B* **108**(11): 3380–3383.

Sahoo, Y., Goodarzi, A., Swihart, M.T. et al. 2005. Aqueous ferrofluid of magnetite nanoparticles: Fluorescence labeling and magnetophoretic control. *J. Phys. Chem. B.* **109**: 3879–3885.

Sahoo, Y., He, Y., Swihart M.T. et al. 2005. An aerosol mediated magnetic colloid: Study of nickel nanoparticles. *J. Appl. Phys.* **98**(5): 054308:1–6.

Schillinger, U., Brilla, T., Rudolph, C. et al. 2005. Review—Advances in magnetofection-magnetically guided nucleic acid delivery. *J. Magn. Magn. Mater.* **293**: 501–508.

Schuster, M., Wasserbauer, E., Ortner, C. et al. 2000. Short cut of protein purification by integration of cell-disrupture and affinity extraction. *Bioseparation* **9**: 59–67.

Sinha, A., Ganguly, R., De, A.K. et al. 2007. Single magnetic particle dynamics in a microchannel. *Phys. Fluids* **19**: 117102:1–5.

Stamenovic, D. and Wang, N. 2000. Cellular response to mechanical stress: Invited review: Engineering approaches to cytoskeletal mechanics. *J. Appl. Physiol.* **89**: 2085–2090.

Sudimack, J. and Lee, R.J. 2000. Targeted drug delivery via the folate receptor. *Adv. Drug Deliv. Rev.* **41**: 147–162.

Sun, L., Zborowski, M., Moore, L.R. et al. 1998. Continuous, flow-through immunomagnetic cell sorting in a quadrupole field. *Cytometry A* **33**: 469–475.

Takayasu, M., Duske, N., Ash, S.R. et al. 1982. HGMS studies of blood–cell behavior in plasma. *IEEE Trans. Magn.* **18**(6): 1520–1522.

Takayasu, M., Kelland, D.R., and Minervini, J.V. 2000. Continuous magnetic separation of blood components from whole blood. *IEEE Trans. Appl. Supercond.* **10**: 927–930.

Willard, M.A., Kurihara, L.K., Carpenter, E.E. et al. 2004. Chemically prepared magnetic nanoparticles. *Int. Mater. Rev.* **49**(3–4): 125–170.

Yanase, M., Shinkai, M., Honds, H. et al. 1998. Antitumor immunity induction by intracellular hyperthermia using magnetite cationic liposomes. *Jpn. J. Cancer Res.* **89**: 775–782.

Yazdankhah, S.P., Slverd, L., Simonsen, S. et al. 1999. Development and evaluation of an immunomagnetic separation-ELISA for the detection of *Staphylococcus aureus* thermostable nuclease in composite milk. *Vet. Microbiol.* **67**: 113–125.

Yellen, B.B., Forbesb, Z.G., Halversona, D.S. et al. 2005. Targeted drug delivery to magnetic implants. *J. Magn. Magn. Mater.* **293**: 647–654.

Zborowski, M. and Chalmers, J.J. 2007. *Magnetic Cell Separation*, Vol. 32 (Laboratory Techniques in Biochemistry and Molecular Biology). New York: Elsevier Science.

Zborowski, M., Sun, L., Moore, L.R. et al. 1999. Continuous cell separation using novel magnetic quadrupole flow sorter. *J. Magn. Magn. Mater.* **194**: 224–230.

Zborowski, M., Ostera, G.R., Moore, L.R. et al. 2003. Red blood cell magnetophoresis. *Biophys. J.* **84**: 2638–2645.

Zheng, Q., Ohulchanskyy, T.Y., Sahoo, Y. et al. 2007. Water-dispersible polymeric structure co-encapsulating a novel hexa-peri-hexabenzocoronene core containing chromophore with enhanced two-photon absorption and magnetic nanoparticles for magnetically guided two-photon cellular imaging. *J. Phys. Chem. C*, **111**(45): 16846–16851.

Nanomechanical Sensors for Biochemistry and Medicine

Hans Peter Lang
Institute of Physics of the
University of Basel

Christoph Gerber
Institute of Physics of the
University of Basel

11.1 Introduction

The purpose of biosensors in biochemistry and medicine is to provide a highly sensitive and selective method for reliable, rapid, and preferably continuous monitoring of certain chemicals and chemical processes in a biochemical and a physiological environment. In most cases, specific biochemical interactions between various biological ligands are utilized for the sensing of binding events. The sensor's input is a physical property or an interaction of a physical, chemical, or biochemical nature. This input is processed via the sensor's transduction element into a recordable signal.

Many different ways of transduction exist, such as electrochemical, electromechanical, electroacoustical, photoelectric, electromagnetic, magnetic, electrostatic, thermoelectric, electric, and mechanical transductions. In a biosensor, a biological transduction element is combined with a physicochemical detection element. The sensitive biological part may consist of biological materials such as tissues, microorganisms, organelles, cell receptors, enzymes, proteins, peptides, polysaccharides, antibodies, or nucleic acids. The sensor element responds with a specific signal, such as changes in electric potential, electrical current, conductance or impedance, the intensity and the phase of electromagnetic radiation, and also in mass, temperature, viscosity, strain, or stress.

Transduction methods comprise among others (1) optical processes involving spectroscopy (absorption, fluorescence, phosphorescence, Raman) and refraction; (2) electrochemical processes like electrolysis and voltammetry; (3) mass detection via resonance frequency shifts in quartz crystal microbalances, surface acoustic wave devices, or microfabricated resonant structures; (4) array techniques, e.g., charge-coupled device camera readout of fluorescence-labeled spotted arrays; and (5) nanomechanical cantilevers, which are the focus of this article.

Biosensors have a major application potential in daily life, e.g., for bacteria detection to avoid contamination of food and for the detection of life-threatening bio-agents. While the main drivers are the health care and the food industry, authenticity issues (genuine products, detection of genetically modified food) are also of importance.

The total biochip market size is projected to grow to about $4.9 billion in 2012 with an annual growth rate of 12.3% (Fuji-Keizai 2008). The key requirements for biosensors are high selectivity, cost effectiveness speed, and reliability. Their main applications lie in the fields of medicine, health, diagnostics, in the food/beverage sectors, cosmetics, perfumes, and in safety and security issues (terrorism prevention). Biosensor strategies include bioinformatics, i.e., a multiple probe/measurement approach including statistical evaluation of acquired data.

Biosensors are most economically produced by batch microfabrication in large numbers and may include novel smart materials, functional coatings, or nanoparticles. Food contamination still poses a common problem, even in the most developed countries, and foodborne diseases (e.g., campylobacteriosis and salmonellosis) have reached epidemic proportions in several countries. Emerging health issues, such as contamination from acrylamide or dioxins, rotten meat, avian flu, bovine spongiform encephalopathy (BSE), and genetically modified organisms (GMOs) are creating additional concerns among both the public and the decision makers (WHO 2006).

11.2 Microcantilever Array Sensors

11.2.1 Sensor Concept

In this chapter, we focus on nanomechanical microcantilever sensor arrays for the detection of biochemical processes. Microcantilevers have been used for many years in atomic force microscopy (AFM), a technique pioneered by the IBM Zurich Research Laboratory in Switzerland (Binnig et al. 1986). AFM employs a microcantilever with a sharp tip to image a nonconductive surface with a lateral and a vertical resolution on the atomic scale. For the application as a sensor, neither the tip nor the surface are needed: the sensor response is generated by processes taking place on the surface of the microcantilever. These processes involve the adsorption of molecules on the microcantilever's surface, resulting in the bending of the cantilever beam via the generation of interface stress and strain. The specificity of the detection process is considerably enhanced if the specific probe receptor molecules are attached to one of the surfaces of the microcantilever to bind the target molecules.

Here, an array of eight microcantilevers is used (Figure 11.1a). Typically, the upper surface of a microcantilever is coated with a thin layer of a material that exhibits an affinity to molecules in the environment. This surface is referred to as the "functionalized" surface of the microcantilever. The other surface of the microcantilever (typically the lower surface) may be left uncoated or is coated with a passivation layer, which is either inert or does not show significant affinity to the molecules to be detected (Figure 11.1b). To functionalize a microcantilever surface, a metal layer is often deposited. Metal surfaces, such as gold, may be used to covalently bind a monolayer representing the chemical surface sensitive to the target molecules. Frequently, a monolayer of thiol molecules, offering well-defined surface chemistry, is formed on the gold surface, providing a template for subsequent molecule adsorption. The gold layer also often serves as a reflection layer in case the cantilever bending is read out optically.

11.2.2 Compressive and Tensile Surface Stress

If molecules are adsorbed on the upper (functionalized) surface, then a downward bending of the microcantilever will result due to the formation of surface stress. The surface stress is called "compressive," because the adsorbed layer of molecules (e.g., a monolayer of alkylthiols) produces a downward bending of the microcantilever away from its functionalized side. In case of the opposite situation, i.e., if the microcantilever bends upward, one would speak of "tensile stress." If both the upper and the lower surface of the microcantilever are subjected to surface stress changes, then the situation is much more complex, because a predominant compressive stress formation on the lower microcantilever surface will appear as tensile stress on the upper surface. Therefore, it is extremely important to properly passivate the lower surface in such a way that, ideally, no stress-generating adsorption processes take place on the lower surface of the microcantilever.

Various strategies can be used to passivate the lower surface of microcantilevers. For biochemical systems, the application of a thin layer of 2-[methoxy-poly(ethyleneoxy)propyl]trimethoxysilane will create a pegylated surface that is almost inert toward the adsorption of biological layers. Only the actual experiment will show whether the passivation layer was really efficient, for, as such, passivated cantilevers will not show a substantial bending response upon exposure to an analyte.

11.2.3 Differential Stress Measurements

Single microcantilevers may bend due to effects other than the formation of surface stress during the adsorption of molecules. The major influences for such bending are a thermal drift, or an interaction with the environment, especially if the microcantilever is operated in a liquid. Furthermore, a non-specific physisorption of molecules on the cantilever surface or a nonspecific binding to the receptor molecules during measurements may contribute to the drift.

To exclude such influences, the simultaneous measurement of reference microcantilevers aligned in the same array as the sensing microcantilevers is crucial (Lang et al. 1998). The difference in responses from the reference and the sensor microcantilevers yields the net bending signal, and even small sensor signals can be extracted from large microcantilever deflections without being dominated by undesired effects. When only single

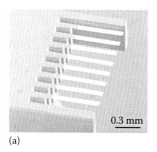
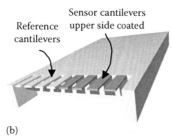

(a)   (b)

FIGURE 11.1 (a) Scanning electron microscopy image of a silicon microcantilever array. (b) Schematic drawing showing the upper functionalized and the lower passivated surfaces of four sensor cantilevers (darker color, right part of the array) and four passivated reference cantilevers (left part of the array). The thick solid side bars are on the one hand for mechanical protection of the cantilevers, on the other hand they represent a solid reference surface, e.g., for a reference baseline.

microcantilevers are used, no thermal-drift compensation is possible. To obtain useful data under these circumstances, both microcantilever surfaces have to be chemically well defined. One of the surfaces, typically the lower one, should be passivated; otherwise, the microcantilever response will be convoluted with undesired effects originating from uncontrolled reactions taking place on the lower surface.

With a pair of microcantilevers, reliable measurements are obtained. One of them is used as the sensor microcantilever (coated typically on the upper side with a molecule layer that shows affinity to the molecules to be detected), whereas the other microcantilever serves as the reference. The sensor microcantilever should be coated with a passivation layer on the upper surface so as not to exhibit affinity to the molecules to be detected. Thermal drifts are canceled out of difference responses, i.e., the difference in the deflections of the sensor and the reference microcantilevers are taken (differential measurements). Alternatively, both microcantilevers are used as sensors (sensor layers on the upper surfaces), and the lower surface is passivated.

The use of an array of microcantilevers is recommended in which some cantilevers are used either as sensor or as reference microcantilevers so that multiple difference signals can be evaluated simultaneously. The thermal drift is canceled out since one surface of all microcantilevers, typically the lower one, is left uncoated or coated with the same passivation layer.

11.3 Modes of Operation

11.3.1 Static Mode

The gradual bending of a microcantilever, as a result of a progressing molecular coverage, is referred to as an operation in the "static mode" (Figure 11.2a). Various environments are possible, such as vacuum, ambient environment, and liquids. In a gaseous environment, molecules adsorb on the functionalized sensing surface and form a molecular layer, provided there is affinity for the molecules to adhere to the surface.

Polymer sensing layers show a partial sensitivity because molecules from the environment diffuse into the polymer layer at different rates, mainly depending on the size and the solubility of the molecules in the polymer layer. By selecting polymers out of a wide range of hydrophilic/hydrophobic ligands, the chemical affinity of the surface can be influenced, because

different polymers vary in diffusion suitability for polar/unpolar molecules. Thus, for a detection in the gas phase, the polymers can be chosen according to the detection problem, i.e., what the applications demand. Typical chemicals that can be detected are volatile organic compounds (VOCs), such as solvent vapors.

A static-mode operation in liquids, however, usually requires rather specific sensing layers based on molecular recognition, like in DNA hybridization or in antigen–antibody recognition.

11.3.2 Dynamic Mode

Information on the mass change and the amount of molecules adsorbed on the microcantilever surface can be obtained by oscillating the microcantilever at its eigenfrequency (Figure 11.2b). However, the surface coverage is basically not known and molecules on the surface might be exchanged with molecules from the environment in a dynamic equilibrium.

Tracking the eigenfrequency of the microcantilever during mass adsorption or desorption is done to obtain information about these processes. The eigenfrequency is identical to the resonance frequency of an oscillating microcantilever if its elastic properties remain unchanged during the molecule adsorption/desorption process and if the damping effects are negligible. This mode of operation is called the "dynamic mode." The microcantilever is used as a microbalance, as with mass addition on the cantilever surface, the cantilever's eigenfrequency will shift to a lower value. The mass change on a rectangular cantilever is calculated (Thundat et al. 1994) according to

$$\Delta m = \left(\frac{k}{4\pi^2} \times 0.24 \right) \times \left(\frac{1}{f_1^2} - \frac{1}{f_0^2} \right), \qquad (11.1)$$

where
f_0 is the eigenfrequency before the mass change occurs
f_1 is the eigenfrequency after the mass change has happened

The cantilever spring constant k is calculated according to

$$k = \frac{Ewt^3}{4l^3}, \qquad (11.2)$$

where
E is Young's modulus ($E_{Si} = 1.3 \times 10^{11}$ N/m^2 for Si(100))
w, t, and l are the width, the thickness, and the length of the cantilever

Dynamic mode operation in a liquid environment poses a challenge because of the strong damping of the cantilever oscillation due to the high viscosity of the surrounding media. This results in a low quality factor Q of the oscillation, and the resonance frequency shift is difficult to track with high resolution. The quality factor is defined as

$$Q = \frac{2\Delta f}{f_0}. \qquad (11.3)$$

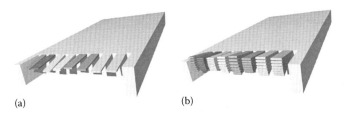

(a) (b)

FIGURE 11.2 (a) Static mode of operation. The individual cantilevers are bent down to a certain extent, depending on the magnitude of surface stress formed during adsorption of a molecular layer. (b) Dynamic mode of operation. The magnitude of oscillation might vary for each individual cantilever.

In air, a frequency resolution Δf of below 1 Hz is easily achieved, in contrast to a liquid environment, where resolution values of about 20 Hz are already considered very good. With damping or changes in the elastic properties of the cantilever during the experiment, e.g., a stiffening or softening of the spring constant by adsorption of a molecule layer, the measured resonance frequency will not be exactly the same as the eigenfrequency, and the mass derived from the frequency shift will be inaccurate.

Unlike in ultrahigh vacuum conditions (Ilic et al. 2004, Ekinci and Roukes 2005), where the resonance frequency is equal to the eigenfrequency, the two terms eigenfrequency and resonance frequency should be carefully distinguished for operation in a strong damping environment, as described in literature (Braun et al. 2005).

11.3.3 Further Modes of Operation

Microcantilevers coated with metal layers are also prone to thermal effects because thermal expansion differences in the cantilever and the coating layer will also contribute to bending when the temperature is varied. This effect is used in another mode of operation referred to as the "heat mode." There, cantilever bending occurs because of differing thermal expansion coefficients in the sensor layer and in cantilever materials (Gimzewski et al. 1994).

Heat changes are either caused by external influences, such as a change in temperature, and occur directly on the surface by exothermal, e.g., catalytic reactions, or are due to the material properties of a sample attached to the apex of the cantilever. The latter technique is known as micromechanical calorimetry. The sensitivity of the cantilever heat mode is in orders of magnitude higher than that of the traditional calorimetric methods performed on milligram samples, as it only requires nanogram amounts of sample and achieves nanojoules (Berger et al. 1996) to picojoules (Bachels and Schäfer 1999, Bachels et al. 1999) sensitivity. Static, dynamic, and heat measurement modes have established cantilevers as versatile tools to perform experiments in nanoscale science with very small amounts of material.

Mass-change determination can be combined with varying environment temperature conditions to obtain a method introduced in the literature as "micromechanical thermogravimetry" (Berger et al. 1998). The sample under investigation is mounted directly onto the cantilever. Its mass should not exceed several hundred nanograms. In case of adsorption, desorption, or decomposition processes, mass changes in the picogram range can be observed in real time by tracking the resonance-frequency shift.

In photon-absorbing materials, a fraction of energy is converted into heat. This photothermal heating can be measured as a function of the light wavelength to provide optical absorption data of the material. The interaction of light with a bimetallic cantilever creates heat on the cantilever surface, resulting in a bending of the cantilever (Barnes et al. 1994). Such bimetallic-cantilever devices are capable of detecting heat flows due to an optical heating power as low as 100 pW, being two orders of magnitude better than in conventional photothermal spectroscopy. Recently, this technique has been applied for reliable and quick detection of explosives (Krause et al. 2008, Van Neste et al. 2008).

11.4 Microcantilever Functionalization

For reliable operation of microcantilever sensors, it is essential that the surfaces of the cantilevers are coated in a reproducible and robust manner to provide suitable receptor surfaces for the analyte molecules to be detected. Such coatings should be specific, homogeneous, stable and reproducible. Microcantilever sensors might be designed to be either reusable or for single use only.

For static mode measurements, one side of the cantilever should be passivated to block undesired, unspecific adsorption. Often, the microcantilever's upper side, which will be referred to as the sensor side, is coated with a 20 nm thick layer of gold to provide a platform for the binding of the receptor molecules, for example, by means of thiol chemistry, whereas the lower side is passivated using silane chemistry to provide an inert surface such as polyethylene glycol silane. Silanization is performed first on the silicon microcantilever. Subsequently, a gold layer is deposited on the top side of the microcantilever, leaving the lower side unchanged. It is very important that the method for microcantilever coating is fast, reproducible, reliable, and allows one or both cantilever surfaces to be coated separately. Various ways are reported to coat a microcantilever with functional molecular layers. Here, two different strategies are highlighted.

11.4.1 Coating in Microcapillary Arrays

It is essential that every microcantilever in an array can be coated separately with a functional layer. This requirement can be achieved by confining each cantilever in a dimension-matched microcapillary filled with the liquid containing the molecules to be deposited on the microcantilever (Figure 11.3). Therefore, the microcantilevers of the array are inserted into disposable glass microcapillaries filled with liquid containing the probe molecules. The outer diameter of the glass capillaries is 240 μm so that they can be easily and neatly placed next to each other to accommodate the pitch of the cantilevers in the array (250 μm). Their inner diameter is 180 μm, allowing sufficient room to

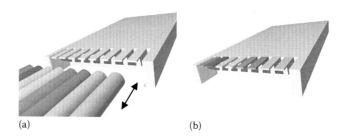

FIGURE 11.3 Functionalization of a microcantilever array in dimension-matched glass microcapillaries filled with a solution of probe molecules. (a) Before insertion and (b) after incubation.

insert the cantilevers (width: 100 μm) safely. This method has been successfully applied for the deposition of a variety of materials onto cantilevers, such as self-assembled monolayers (Fritz et al. 2000a), thiol-functionalized single-stranded DNA oligonucleotides (Fritz et al. 2000b, McKendry et al. 2002, Zhang et al, 2006), and proteins (Arntz et al. 2003, Backmann et al. 2005). Incubation of the microcantilever array in the microcapillaries takes from a few seconds (the self-assembly of alkanethiol monolayers) to several tens of minutes (coating with protein solutions). The microcapillary functionalization unit may be placed in an environment of saturated vapor of the solvent used for the probe molecules to avoid drying out of the solutions.

11.4.2 Coating Using an Inkjet Spotter

The disadvantage of coating in microcapillary arrays is that manual alignment of the microcantilever array and the functionalization tool is required, and therefore the technique is not suitable for coating large numbers of cantilever arrays. Moreover, the upper and the lower surfaces of the microcantilevers are exposed to the same solution containing the probe molecules. For ligands that bind covalently, e.g., by gold-thiol coupling, only the upper surface will be coated, provided the gold layer is only on the upper surface of the microcantilever. For coating with polymer layers, microcapillary arrays are not suitable, because both surfaces of the microcantilever would be coated with polymers. This would be inappropriate for static mode measurements, where an asymmetry between the upper and lower surface is required.

A method suitable for coating many cantilever sensor arrays in a rapid and reliable way is inkjet spotting (Bietsch et al. 2004a,b), see Figure 11.4. An *x–y–z* positioning system allows a fine nozzle (typical capillary diameter: 70 μm) to be positioned with an accuracy of approx. 10 μm over a cantilever. Individual droplets (diameter: 60–80 μm, volume 0.1–0.3 nL) can be dispensed individually by means of a piezo-driven ejection system in the inkjet nozzle. When the droplets are spotted with a pitch smaller than 0.1 mm, they merge and form continuous films. By adjusting the number of droplets deposited on cantilevers, the resulting film thickness can be controlled precisely. The inkjet-spotting technique allows a cantilever to be coated within

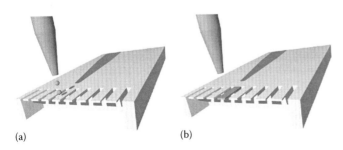

(a)   (b)

FIGURE 11.4 Functionalization of a microcantilever array using an inkjet spotting nozzle. (a) Individual droplets are ejected from the nozzle onto the upper surface of the microcantilever. (b) An individual microcantilever has been coated with a film of probe molecules.

seconds and yields very homogeneous, reproducibly deposited films of well-controlled thickness.

The successful coating of self-assembled alkanethiol monolayers, polymer solutions, self-assembled DNA single-stranded oligonucleotides (Bietsch et al. 2004b), and protein layers has been demonstrated. In conclusion, inkjet spotting has turned out to be a very efficient and versatile method for functionalization that can even be used to coat arbitrarily shaped sensors reproducibly and reliably (Lange et al. 2002, Savran et al. 2003).

11.5 Experimental Setup

11.5.1 Measurement Setup for a Liquid Environment

In general, a measurement set-up for cantilever arrays consists of four major parts, see Figure 11.5: (1) the measurement chamber containing the cantilever array; (2) an optical or electrical system to detect the cantilever deflection (e.g., laser sources, collimation lenses and a position-sensitive detector [PSD], or piezoresistors and Wheatstone-bridge detection electronics); (3) electronics to amplify, process, and acquire the signals from the detector; and (4) a gas- or liquid-handling system to inject samples reproducibly into the measurement chamber and purge the chamber. The cantilever sensor array is located in an analyte chamber with a volume of 3–90 μL, which has inlet and outlet ports for gases or liquids. The cantilever deflection is determined by means of an array of eight vertical-cavity surface-emitting lasers (VCSELs) arranged at a linear pitch of 250 μm that emit at a wavelength of 760 nm into a narrow cone of 5°–10°. The light of each VCSEL is collimated and focused onto the apex of the corresponding cantilever by a pair of achromatic doublet lenses, 12.5 mm in diameter. This size has to be selected in such a way that all eight laser beams pass through the lens close to its center to minimize scattering, chromatic, and spherical aberration artifacts. The light is then reflected off the gold-coated surface of the cantilever and hits the surface of a PSD. PSDs are light-sensitive photo-potentiometer-like devices that produce photocurrents at two opposing electrodes. The magnitude of the photocurrents depends linearly on the distance of the impinging light spot from the electrodes. Thus, the position of an incident light beam can easily be determined with micrometer precision. The photocurrents are transformed into voltages and amplified in a preamplifier. As only one PSD is used, the eight lasers cannot stay switched on simultaneously. Therefore, a time-multiplexing procedure is used to switch the lasers on and off sequentially at typical intervals of 10–100 ms. The resulting deflection signal is digitized and stored together with time information on a personal computer (PC), which also controls the multiplexing of the VCSELs as well as the switching of the valves for the liquid handling system. The measurement set-up for liquids (Figure 11.5) consists of a poly-etheretherketone (PEEK) liquid cell, which contains the cantilever array and is sealed by a viton O-ring and a glass plate. The VCSELs and the PSD are mounted on a metal frame around the liquid cell.

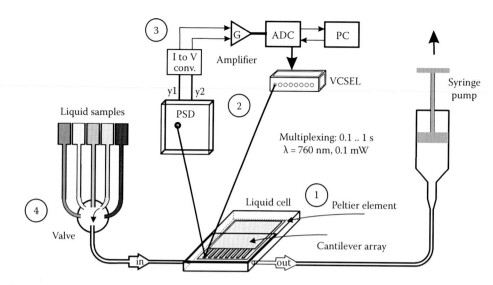

FIGURE 11.5 Schematic drawing of the measurement setup: (1) measurement chamber with microcantilever array, (2) deflection readout system (optical beam deflection), (3) amplification electronics, and (4) liquid-handling system: the liquid is pulled from individual reservoirs through the measurement chamber using a motorized syringe.

After preprocessing the position of the deflected light beam in a current-to-voltage converter and amplifier stage, the signal is digitized in an analog-to-digital converter and stored on a PC. The liquid cell is equipped with inlet and outlet ports for liquids. They are connected via a 0.18-mm-inner-diameter Teflon tubing to individual thermally equilibrated glass containers, in which the biochemical liquids are stored. A six-position valve allows the inlet to the liquid chamber to be connected to each of the liquid-sample containers separately. The liquids are pulled through the liquid chamber by means of a syringe pump connected to the outlet of the chamber. A Peltier element is situated very close to the liquid-containing volume of the chamber to allow temperature regulation within the chamber. The entire experimental set-up is housed in a temperature-controlled box regulated with an accuracy of 0.01 K to the target temperature.

11.5.2 Application I: Patient's Breath Characterization

The first application discussed here is an experiment in a gaseous environment (Baller et al. 2000). The experimental setup is basically the same as the one in liquids, except for the fact that exhaled air collected from a patient is pushed by a syringe pump through the measurement chamber. Before using modern diagnosis tools, medical doctors examined the patient's breath to detect diseases, since certain diseases can be recognized by an examination of exhaled air. Examples of such illnesses are the following: (1) Diabetes mellitus (type II diabetes), a severe, chronic form of diabetes caused by insufficient production of insulin and resulting in abnormal metabolism of carbohydrates, fats, and proteins. This disease involves the presence of acetone in the patient's breath. (2) Uremia, a toxic condition resulting from kidney disease in which there is a retention of waste products in the bloodstream. These waste products are normally

excreted in the urine. A compound found in a patient's breath associated with uremia is dimethylamine.

Breath samples of two patients suffering from diabetes mellitus and uremia were taken and stored in medical plastic bags for exhaled air samples. For a comparison, breath samples from healthy persons were also investigated for reference. For each measurement, 10 mL of exhaled air was removed from the medical plastic bag under temperature-controlled conditions, and injected into the microcantilever array measurement chamber. Each cantilever is coated with a different polymer and responds in its own characteristic way to the breath sample during the exposure time of 6 min because the rate of diffusion of the substances in exhaled air is different for each polymer. Also, during the purging process of the chamber (cleaning with dry nitrogen for 8 min from a second syringe), the desorption characteristics are unique to each polymer. All flow rates were set to 1 mL gas per minute. The microcantilever deflections were found to be very reproducible for samples from the same patient, but dissimilar for sick and healthy persons (Figure 11.6a through d). The amount of data was reduced by extracting the deflections of the eight microcantilevers at three different points in time (at 320, 420, and 520 s after start of the measurement, cf. vertical lines in Figure 11.6) during exposure to the exhaled air sample, and at four different points during the purging process of the measurement chamber with dry nitrogen gas (at 620, 720, 820, and 920 s). The reduced data set consisted of cantilever deflections of eight cantilevers at seven different points in time, i.e., 8 × 7 = 56 cantilever deflection values, which characterize one measurement in a 56 dimensional space. The mathematical method of principal component analysis (PCA) projects this 56 dimensional information into 2 dimensions, whereby the largest differences between measurements are determined in a least-square fit procedure. The two axis of a two-dimensional PCA plot are referred to as the principal components. The PCA reveals the most dominant

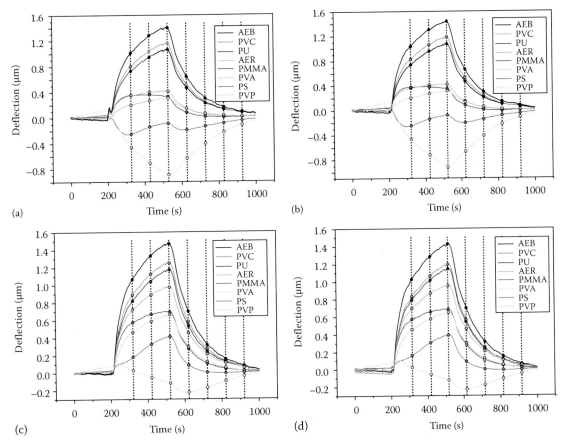

FIGURE 11.6 (a and b) Two independent measurements taken from a healthy person (cantilever deflection traces acquired during the injection of a breath sample into the measurement chamber). (c and d) *Ditto*, but for a patient suffering from uremia. Polymer coatings: AEB = araldite epoxy resin type B, PVC = polyvinyl chloride, PU = polyurethane, AER = araldite epoxy resin type R, PMMA = polymethylmethacrylate, PVA = polyvinyl alcohol, PS = polystyrene, PVP = polyvinylpyridine. (Data courtesy of Daniel Schmid, University Hospital Basel, Basel, Switzerland.)

deviations in the responses for the different patients' breath samples in measured data. Clear clustering of breath measurements of healthy persons and of patients with acetone breath (diabetes) and uremia is observed in Figure 11.7. The symbols in the PCA

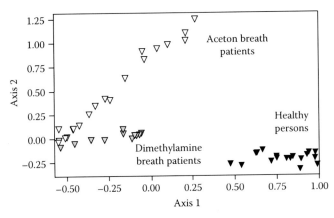

FIGURE 11.7 Principal component analysis plot revealing clearly separated groups of triangles (each symbol is a breath measurement) that allows a clear distinction between healthy persons from patients with acetone or dimethylamine in their breath. (Data courtesy of Daniel Schmid, University Hospital Basel, Basel, Switzerland.)

plot (Figure 11.7) indicate the individual measurements. Three different clusters of points are observed, allowing a distinction between healthy persons, acetone breath patients and dimethylamine breath patients. We conclude that the microcantilever technique allows a fast and a noninvasive detection of diseases in patients' breath samples (Schmid et al. 2008).

11.5.3 Application II: DNA Hybridization Sensing

This example demonstrates the capability of cantilever array sensors to detect biochemical reactions. Each cantilever is functionalized with a specific biochemical probe receptor, sensitive for detection of the corresponding target molecule. The main advantage of cantilever array sensors is that measurements of differences in the responses of sensor and reference cantilevers can be evaluated. Measuring the deflection of only one cantilever will yield misleading results that might give rise to an incorrect interpretation of the cantilever-deflection trace. Therefore, at least one of the cantilevers (the sensor cantilever) is coated with a sensitive layer that exhibits an affinity to the molecules to be detected, whereas other cantilevers are

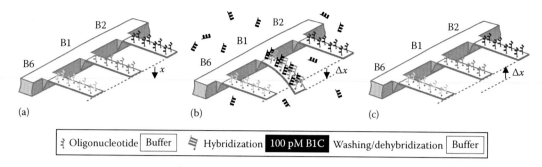

FIGURE 11.8   (a) Schematic drawing of a cantilever sensor array coated with single stranded DNA oligonucleotides. (b) Hybridization with the matching DNA sequence bends the second cantilever due to the formation of surface stress. (c) The initial state is restored after purging with a dehybridization agent.

coated with a molecular layer that does not show an affinity to them (reference cantilevers).

The biochemical system to be investigated here involves a DNA hybridization experiment in liquid using a thiolated 12-mer oligonucleotide sequence from the Bio B biotin synthetase gene (EMBL accession number: J04423). Three surface-bound probes were selected, Bio B1 (5′-SH-C_6-ACA TTG TCG CAA-3′, C_6 is a spacer), Bio B2 (5′-SH-C_6-TGC TGT TTG AAG-3′), and Bio B6 (5′-SH-C_6-TCA GGA ACG CCT-3′), which were immobilized by thiol binding onto the gold-coated upper surface of a cantilever in an array (Figure 11.8a).

Please note that the sequences are selected in length in such a way that stress generation is expected to occur close to the cantilever surface. With much longer sequences, the experiment would not necessarily work because the stress would be generated too far away from the surface (Alvarez et al. 2004).

The target complements called Bio B1C, B2C, and B6C are diluted in a 5× sodium saline citrate (ssc) buffer at 100 pM concentration. Upon injection of the matching sequence to Bio B1, i.e., Bio B1C, the sensor cantilever coated with Bio B1 is expected to bend, whereas the reference cantilever coated with Bio B2 as well as that coated with Bio B6 will not bend (Figure 11.8b). After thorough rinsing with an unbinding agent, the cantilever coated with Bio B1 will bend back to its initial position (Figure 11.8c). The bending is due to the formation of surface stress during the hybridization process because of steric crowding, because a double-stranded DNA requires more space than a single-stranded DNA.

The actual experiment proceeds as follows (Figure 11.9): First, the liquid cell with the functionalized cantilever array is filled with an ssc buffer. After a stable deflection base line has been achieved, the ssc buffer is injected after 4 min for 3 min. All cantilevers deflect, but once the injection is over, a stable baseline is reached again. At 18 min, the target Bio B1C is injected, which is supposed to hybridize with the Bio B1 probe, but not with the Bio B2 or the Bio B6 probe. All cantilevers deflect, but the deflection magnitude of the Bio B1-coated cantilever is much larger than those of the Bio B2- and the Bio B6-coated cantilever. Finally, at 37 min, the ssc buffer is injected again and a stable baseline is reached. From the deflection data shown in Figure 11.9a, it seems that no conclusive result can be obtained from individual cantilever responses only, as both the sensor and the reference cantilevers bend. However,

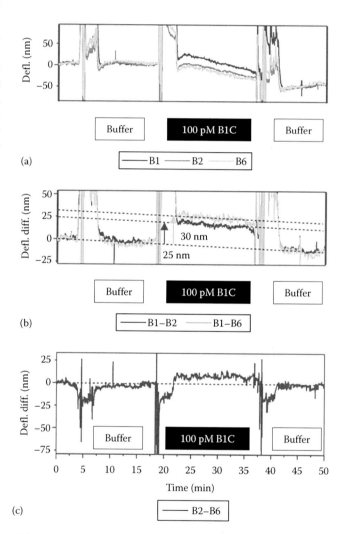

FIGURE 11.9   (a) Deflection traces of sensor (functionalized with DNA oligonucleotide sequence Bio B1) and reference cantilevers (functionalized with DNA oligonucleotide sequences Bio B2 and Bio B6, respectively). (b) Differences B1–B2 (signal: 25 nm) and B1–B6 (signal: 30 nm) of the bending responses of the sensor cantilever B1 and the reference cantilevers B2 and B6. (c) Difference in responses of the two reference probes B2–B6 (signal: <5 nm). (Data courtesy of Jiayun Zhang, University of Basel, Basel, Switzerland.)

a clear deflection signal is observed when calculating the difference in deflection responses from probes Bio B1 (sensor) and reference Bio B2 (Figure 11.9b), or the difference in deflection responses from probes Bio B1 and reference Bio B6. The differential deflection magnitudes obtained are 25 nm (B1–B2) or 30 nm (B1–B6), respectively. The difference in deflection responses between two reference cantilevers yields no signal or only a very small signal that can be attributed to an unspecific binding of B1C to one of the reference probes, supposedly to B2, as the difference B2–B6 yields a small positive signal of less than 5 nm, see Figure 11.9c. We conclude that it is absolutely mandatory to use at least two cantilevers in an experiment, a reference cantilever and a sensor cantilever, to be able to cancel out undesired artifacts such as thermal drift or unspecific adsorption.

11.6 Applications in a Biochemical Environment

The following sections give an overview on the research performed with microcantilevers in the field of biochemistry and medicine. The examples given only represent a selection of some of the publications in this field in the last few years, and are not meant to be comprehensive.

11.6.1 pH Sensing

Control of pH is often important in biochemical reactions. Hence, this section concerns the measurement of pH using microcantilevers by measuring their deflection as a function of pH. Microcantilevers coated with self-assembled monolayers of mercaptohexadecanoic acid (MHA, hydrophilic) and hexadecanethiol (HDT, hydrophobic) bend due to the presence of hydrogen ions, as interfacial stress develops depending on pH values and ionic strength (Fritz et al. 2000a). At a low pH, MHA is protonated, whereas at a high pH, MHA is deprotonated. SiO_2 and silicon nitride microcantilevers were also found to exhibit a deflection dependency with pH when coated with 4-aminobutyltriethoxysilane, 11-mercaptoundecanoic acid, and Au/Al-coated over a pH range 2–12. Aminosilane-modified SiO_2/Au cantilevers performed robustly over the pH range 2–8 yielding 49 nm deflection/pH unit, while Si_3N_4/Au cantilevers performed well at the pH 2–6 and 8–12, producing a 30 nm deflection/pH unit (Ji et al. 2001a,b). Microcantilevers with poly(methacrylic acid) (PMAA) and poly(ethylene glycol) dimethacrylate coating were found to be sensitive to pH changes (Bashir et al. 2002). Hydrogel coatings were also found to be sensitive to pH (Zhang et al. 2004a). The dependence of the micromechanical responses to different ionic strength and ion species present in the aqueous environment is discussed in detail (Watari et al. 2007), highlighting the critical role of counter- and co-ions on surface stress.

11.6.2 Ion Sensing

Detection of ions using microcantilevers requires receptor molecules on their surface to be able to recognize ions selectively in solution. Coupling of the ions to the receptor sites involves conformational changes of the receptor and also a generation of interfacial stress that is transduced to the microcantilever, which, in turn, responds by bending. Using microcantilevers coated with a self-assembled monolayer of triethyl-12-mercaptododecylammonium bromide on gold CrO_4^{2-} ions are detected at a concentration of 10^{-9} M. Other anions, such as Cl^-, Br^-, CO_3^{2-}, HCO_3^-, and SO_4^{2-} do not deflect such modified cantilevers significantly (Ji et al. 2001b). Hg^{2+} has been measured at a concentration of 10^{-11} M using a microcantilever coated with gold. Almost no affinity to other cations exists, such as K^+, Na^+, Pb^{2+}, Zn^{2+}, Ni^{2+}, Cd^{2+}, Cu^{2+}, and Ca^{2+} (Xu et al. 2002). Adsorption characteristics of Ca^{2+} ions as a function of concentration in an aqueous $CaCl_2$ solution was investigated in the static and the dynamic mode (Cherian et al. 2002). Microcantilevers functionalized with the metal-binding protein, AgNt84-6, are able to detect heavy metal ions like Hg^{2+} and Zn^{2+}, but are insensitive to Mn^{2+} (Cherian et al. 2003). Hydrogels containing benzo-18-crown-6 have been used to modify microcantilevers for measurements of the concentration of Pb^{2+} in aqueous solutions (Liu and Ji 2004). Using different thiolated ligands as self-assembled monolayers (SAMs) functionalized on silicon microcantilevers coated with gold allows the detection of Cs^+, Co^{2+}, and Fe^{3+} (Dutta et al. 2005). In an electrochemical application, a gold coated microcantilever is utilized as the working electrode to detect Cr(VI) (Tian et al. 2005). Others use 11-undecenyltriethylammonium bromide (Boiadjiev et al. 2005) or sol–gel layers (Carrington et al. 2006) for detection of Cr(VI). Based on the EDTA–Cd(II) complex and its binding capability to bovine serum albumine (BSA), an antibody-based Cd(II) sensor using microcantilevers is presented (Velanki et al. 2007).

11.6.3 Glucose

Living cells use glucose as a source of energy. Chemically, glucose is a monosaccharide or simple sugar, also known as blood sugar. Detection of glucose concentrations is of outmost importance also to determine the medical condition of a patient. Glucose sensing via microcantilevers is achieved by coating the cantilevers with the enzyme glucose oxidase on gold (Subramanian et al. 2002) or via polyethyleneimine (PEI) conjugation (Yan et al. 2004). Glucose concentrations between 0.2 and 20 mM could be detected (Pei et al. 2004). In another study, a detection range between 2 and 50 mM is reported for glucose. No signal is observed for fructose, mannose, and galactose (Yan et al. 2005).

11.6.4 Hydrogen Peroxide (H_2O_2)

Hydrogen peroxidase provides oxygen in enzymatically controlled reactions. Hydrogen peroxide is detected at the nM level using multilayer modified microcantilevers functionalized through a layer-by-layer nanoassembly technique via intercalation of the enzyme horseradish peroxidase. The magnitudes of bending were found to be proportional to the concentrations of hydrogen peroxide (Yan et al. 2006a).

11.6.5 DNA, RNA

Observation of DNA hybridization using microcantilevers provides valuable information on the similarity of genetic sequences, whereby changes in single nucleotides are detectable. The microcantilever technique does not use an additional polymerase chain reaction (PCR) amplification and is label-free. Specific DNA hybridization detection was observed via surface stress changes related to the transduction of receptor-ligand binding into a direct nanomechanical response of microfabricated cantilevers without the need for external labeling or amplification. The differential deflection of the cantilevers was found to provide a true molecular recognition signal despite the large responses of individual cantilevers. The hybridization of complementary oligonucleotides shows that a single base mismatch between two 12-mer oligonucleotides is clearly detectable (Fritz et al. 2000b). The findings were confirmed or modeled by several groups (Hansen et al. 2001, Hagan et al. 2002). Hybridization in a complex nonspecific background was observed in a complement concentration range between 75 nM and 2 μM (McKendry et al. 2002) following the Langmuir model kinetics (Marie et al. 2002). Enzymatic processes were directly performed on a microcantilever functionalized with DNA incorporating a Hind III restriction endonuclease site, followed by digestion with Hind III to produce DNA comprising a single-stranded end on the cantilever surface. Ligase was used to couple a second DNA molecule with a compatible end to the DNA on the cantilever (Stevenson et al. 2002). Using a gold nanoparticle–labeled DNA, microcantilevers have been used to detect DNA strands with a specific sequence in the dynamic mode, whereby a concentration of 23 pM could still be detected, as well as, a single basepair mismatch (Su et al. 2003). Whereas the adsorption of a thiol functionalized single-stranded DNA is easily observed, hybridization cannot be observed if long hydrocarbon spacer molecules between a single strand DNA and a thiol anchor are used (Alvarez et al. 2004). A very high sensitivity is obtained by creating localized binding sites with gold nanodots. Consecutive selective bonding of double-stranded DNA molecules through a thiol linker allows the detection of a single 1587 basepair DNA molecule (Ilic et al. 2005). DNA hybridization is also observed using piezoresistive cantilevers (Marie et al. 2002, Gunter et al. 2004). A different technique to read out the microcantilever deflections in an array is reported (Alvarez and Tamayo 2005). There, the optical beam deflection technique is combined with the scanning of a laser beam illuminating the cantilevers of an array sequentially. DNA hybridization is also reported using polymer SU-8 cantilevers (Calleja et al. 2005). Mukhopadhyay et al. report 20 nM hybridization sensitivity using piezoresistive cantilevers and DNA sequences with an overhang extension distal to the surface (Mukhopadhyay et al. 2005a) A larger array comprising 20 microcantilevers is described in Lechuga et al., 2006. Moreover, the authors present integration of the array with microfluidics. Surface stress changes in response to thermal dehybridization, or melting, is reported (Biswal et al. 2006). The dependence of salt concentration and hybridization efficiency is discussed in detail (Stachowiak et al. 2006). Two different DNA-binding proteins, the transcription factors SP1 and NF-kappa B are investigated (Huber et al. 2006). Phase transition and stability issues of DNA are discussed in Biswal et al. (2007). A differential gene expression of the gene 1-8U, a potential marker for cancer progression or viral infections, has been observed in a complex background. The measurements provide results within minutes at the picomolar level without target amplification, and are sensitive to base mismatches (Zhang et al. 2006).

11.6.6 Proteins and Peptides

Proteins are larger organic molecules composed of amino acids arranged in a linear chain and connected by peptide bonds between the carboxyl and amino groups of adjacent amino acid residues. Proteins are involved in all vital metabolic processes in cells, providing, e.g., mechanical functions in muscle cells or, in the case of enzymes, catalyze biochemical processes. Further importance lies in cell signaling, immune responses, and cell adhesion processes, as well as digestion. Since proteins often also have a secondary and tertiary structure, i.e., they are folded in a complex way, it is essential to investigate protein interaction under conditions as close as possible to their native environment in the cell or organism. This requirement is fulfilled by using adequate buffers, pH, and temperature conditions. Microfabricated cantilevers were utilized to detect the adsorption of low-density lipoproteins and their oxidized form on heparin, and to detect the adsorption of bovine serum albumine and immunoglobuline G (IgG) (Moulin et al. 2000). The activity, stability, lifetime and re-usability of monoclonal antibodies to myoglobin covalently immobilized onto microfabricated cantilever surfaces was investigated (Grogan et al. 2002). Using piezoresistive microcantilevers, the interaction of the anti-bovine serum albumin (a-BSA) with the bovine serum albumin (BSA) was studied (Kooser et al. 2003). Continuous label-free detection of two cardiac biomarker proteins (creatin kinase and myoglobin) was demonstrated using an array of microfabricated cantilevers functionalized with covalently anchored anti-creatin kinase and anti-myoglobin antibodies (Arntz et al. 2003). Label-free protein detection was reported using a microcantilever functionalized with DNA aptamers receptors for Taq DNA polymerase (Savran et al. 2004). A label-free detection of the C-reactive protein (CRP) using a resonant frequency shift in piezoresistive cantilevers was described (Lee et al. 2004), utilizing the specific binding characteristics of the CRP antigen to its antibody, which was immobilized with Calixcrown SAMs on Au. Receptors on microcantilevers for serotonin, but insensitive to its biological precursor with a similar structure tryptophan were described (Zhang et al. 2004b). Using single-chain fragment antibodies instead of complete antibodies allowed a lowering of the limit of detection to concentrations of about 1 nM (Backmann et al. 2005). Wee et al. (2005) reported the detection of the prostate-specific antigen (PSA) and the C-reactive protein. The detection of the human oestrogen receptor in the free and the oestradiol-bound conformation could be distinguished (Mukhopadhyay et al. 2005b). The Ca^{2+} binding protein calmodulin changed its conformation in the presence or absence of Ca^{2+} resulting in a microcantilver deflection change (Yan et al. 2006b). No effect was observed upon exposure to K^+ and Mg^{2+}. The detection of the activated cyclic adenosine monophosphate (cyclic AMP)-dependent protein kinase was performed in the dynamic mode employing a peptide derived from the heat-stable protein kinase inhibitor (Kwon et al. 2007). The detection of streptavidin at a 1–10 nM concentration was reported using biotin-coated cantilevers (Shu et al. 2007). Using glutathione-*S*-transferase (GST) for the detection of GST antibodies, a sensitivity of 40 nM was obtained (Dauksaite et al. 2007). A two-dimensional multiplexed real-time, label-free antibody-antigen binding assay by optically detecting nanoscale motions of two-dimensional arrays of microcantilever beams was presented (Yue et al. 2008). The PSA was detected at 1 ng/mL using antibodies covalently bound to one surface of the cantilevers. Conformational changes in membrane protein patches of bacteriorhodopsin proteoliposomes were observed with microcantilevers through a prosthetic retinal removal, i.e., bleaching (Braun et al. 2006). Using an analog of the myc-tag decapeptide, binding of anti-myc-tag antibodies was reported (Kim et al. 2003).

11.6.7 Lipid Bilayers, Liposomes, Cells

Larger biochemical arrangements of molecules include lipid bilayers in biological membranes or whole cells, which can also be examined using microcantilevers. Cantilever array sensors can sense the formation by vesicle fusion of supported phospholipid bilayers of 1,2-dioleoyl-*sn*-glycero-3-phosphocholine (DOPC) on their surface and can monitor changes in the mechanical properties of lipid bilayers (Pera and Fritz 2007). Liposomes were detected based on their interaction with the protein C2A, which recognized the phosphatidylserine exposed on the surface of the liposome (Hyun et al. 2006). Individual *Escherichia coli* (*E. coli*) O157:H7 cell-antibody binding events using microcantilevers operated in the dynamic mode were reported (Ilic et al. 2001). The contractile force of self-organized cardiomyocytes was measured on biocompatible poly(dimethylsiloxane) cantilevers, representing a microscale cell-driven motor system (Park et al. 2005). Resonating cantilevers were used to detect individual phospholipid vesicle adsorption in liquid. A resonance frequency shift corresponding to an added mass of 450 pg has been measured (Ghatnekar-Nilsson et al. 2005).

11.6.8 Spores, Bacteria, and Viruses

Even larger biological entities include fungal spores, whole bacteria, and viruses. Micromechanical cantilever arrays have been used for a quantitative detection of the vital fungal spores of *Aspergillus niger* and *Saccharomyces cerevisiae*. The specific adsorption and growth on concanavalin A, fibronectin or immunoglobulin G cantilever surfaces was investigated. Maximum spore immobilization, germination, and mycelium growth was observed on the immunoglobulin G functionalized cantilever

surfaces, as measured from shifts in resonance frequency within a few hours, being much faster than standard petri dish cultivation (Nugaeva et al. 2005). Short peptide ligands can be used to efficiently capture Bacillus subtilis (a simulant of Bacillus anthracis) spores in liquids. Fifth-mode resonant frequency measurements were performed before and after dipping microcantilever arrays into a static B. subtilis solution showing a substantial decrease in frequency for binding-peptide-coated microcantilevers as compared to that for control peptide cantilevers (Dhayal et al. 2006). A new approach for investigating antibiotic reaction mechanisms that could speed up the development of new antibiotics has been reported recently (Ndieyira et al. 2008) using microcantilever arrays to explore the mechanisms of antibiotic interactions with mucopeptides—components of bacterial cell walls—down to a sensitivity of 10 nM, and at clinically relevant concentrations in blood serum.

11.6.9 Medical

Diseases can often be identified or characterized by the presence of certain specific biochemical molecules. If receptor ligands exist for these target molecules, then these molecules are likely to be detected by receptor sites attached to a microcantilever, provided the binding events are transduced into a nanomechanical response, i.e., bending of the microcantilever. A bioassay of the PSA using microcantilevers has been presented (Wu et al. 2001), covering a wide range of concentrations from 0.2 ng/mL to 60 μg/mL in a background of human serum albumin (HSA). Detection has been confirmed by another group using microcantilevers in the resonant mode (Hwang et al. 2004, Lee et al. 2005). The feasibility of detecting severe acute respiratory syndrome associated coronavirus (SARS-CoV) using microcantilever technology was studied in a publication (Velanki and Ji 2006) by showing that the feline coronavirus (FIP) type I virus can be detected by a microcantilever modified by a feline coronavirus (FIP) type I anti-viral antiserum. A method for quantification of a prostate cancer biomarker in urine without sample preparation using monoclonal antibodies was described (Maraldo et al. 2007).

11.7 Outlook

Cantilever-sensor array techniques have turned out to be a very powerful and highly sensitive tool to study physisorption and chemisorption processes, as well as to determine material-specific properties such as heat transfer during phase transitions. Experiments in liquids have provided new insights into such complex biochemical reactions as the hybridization of DNA or molecular recognition in antibody–antigen systems or proteomics.

Future developments must go toward technological applications, in particular, to find new ways to characterize real-world samples such as clinical samples. The development of medical diagnosis tools requires an improvement of the sensitivity of a large number of genetic tests to be performed with small amounts of single donor-blood or body-fluid samples at low cost.

From a scientific point of view, the challenge lies in optimizing cantilever sensors to improve their sensitivity to the ultimate limit: the detection of individual molecules.

Several fundamentally new concepts in microcantilever sensing are available in recent literature, which could help to achieve these goals: the issue of a low-quality factor of resonating microcantilevers in liquid has been elegantly solved by fabrication of a hollow cantilever that can be filled with biochemical liquids. Confining the fluid to the inside of a hollow cantilever also allows a direct integration with conventional microfluidic systems, and significantly increases sensitivity by eliminating high damping and viscous drag (Burg and Manalis 2003) Biochemical selectivity can be enhanced by using enantioselective receptors (Dutta et al. 2003). Other shapes for micromechanical sensors like microspirals could be advantageous for biochemical detection (Ji et al. 2006). Miniaturization of microcantilevers into "true" nanometric dimensions, by using nanowires (Cui et al. 2001), single wall carbon nanotubes (Singh et al. 2007), or graphene sheets (Sakhaee-Pour et al. 2008) will further increase sensitivity.

Acknowledgments

We thank the European Union FP 6 Network of Excellence FRONTIERS for support. The work described here is funded partially by the National Center of Competence in Research in Nanoscience (Basel, Switzerland), the Swiss National Science Foundation, and the Commission for Technology and Innovation (Bern, Switzerland).

References

Alvarez, M and Tamayo, J. 2005. Optical sequential readout of microcantilever arrays for biological detection. *Sens. Actuators B: Chem.* 106: 687–690.

Alvarez, M, Carrascosa, LG, Moreno, M et al. 2004. Nanomechanics of the formation of DNA self-assembled monolayers and hybridization on microcantilevers. *Langmuir* 20: 9663–9668.

Arntz, Y, Seelig, JD, Lang, HP et al. 2003. Label-free protein assay based on a nanomechanical cantilever array. *Nanotechnology* 14: 86–90.

Bachels, T and Schäfer, R. 1999. Formation enthalpies of Sn clusters: A calorimetric investigation. *Chem. Phys. Lett.* 300: 177–182.

Bachels, T, Tiefenbacher, F, and Schäfer, R. 1999. Condensation of isolated metal clusters studied with a calorimeter. *J. Chem. Phys.* 110: 10008–10015.

Backmann, N, Zahnd, C, Huber, F et al. 2005. A label-free immunosensor array using single-chain antibody fragments. *Proc. Natl. Acad. Sci. U.S.A.* 102: 14587–14592.

Baller, MK, Lang, HP, Fritz, J et al. 2000. A cantilever array based artificial nose. *Ultramicroscopy* 81: 1–9.

Barnes, JR, Stephenson, RJ, Welland, ME, Gerber, C, and Gimzewski, JK. 1994. Photothermal spectroscopy with femtojoule sensitivity using a micromechanical device. *Nature* 372: 79–81.

Bashir, R, Hilt, JZ, Elibol, O, Gupta A, and Peppas, NA. 2002. Micromechanical cantilever as an ultrasensitive pH microsensor. *Appl. Phys. Lett.* 81: 3091–3093.

Berger, R, Gerber, C, Gimzewski, JK, Meyer, E, and Guntherodt, HJ. 1996. Thermal analysis using a micromechanical calorimeter. *Appl. Phys. Lett.* 69: 40–42.

Berger, R, Lang, HP, and Gerber, C. 1998. Micromechanical thermogravimetry. *Chem. Phys. Lett.* 294: 363–369.

Bietsch, A, Hegner, M, Lang, HP, and Gerber, C. 2004a. Inkjet deposition of alkanethiolate monolayers and DNA oligonucleotides on gold: Evaluation of spot uniformity by wet etching. *Langmuir* 20: 5119–5122.

Bietsch, A, Zhang, J, Hegner, M, Lang, HP, and Gerber, C. 2004b. Rapid functionalization of cantilever array sensors by inkjet printing. *Nanotechnology* 15: 873–880.

Binnig, G, Quate, CF, and Gerber, C. 1986. Atomic force microscope. *Phys. Rev. Lett.* 56: 930–933.

Biswal, SL, Raorane, D, Chaiken, A, Birecki, H, and Majumdar, A. 2006. Nanomechanical detection of DNA melting on microcantilever surfaces. *Anal. Chem.* 78: 7104–7109.

Biswal, SL, Raorane, D, Chaiken, A, and Majumdar, A. 2007. Using a microcantilever array for detecting phase transitions and stability of DNA. *Clin. Lab. Med.* 27: 163–171.

Boiadjiev, VI, Brown, GM, Pinnaduwage, LA, Goretzki, G, Bonnesen, PV, and Thundat, T. 2005. Photochemical hydrosilylation of 11-undecenyltriethylammonium bromide with hydrogen-terminated Si surfaces for the development of robust microcantilever sensors for Cr(VI). *Langmuir* 21: 1139–1142.

Braun, T, Backmann, N, Vögtli, M et al. 2006. Conformational change of bacteriorhodopsin quantitatively monitored by microcantilever sensors. *Biophys. J.* 90: 2970–2977.

Braun, T, Barwich, V, Ghatkesar, MK et al. 2005. Micromechanical mass sensors for biomolecular detection in a physiological environment. *Phys. Rev. E* 72: 031907.

Burg, TP and Manalis, SR. 2003. Suspended microchannel resonators for biomolecular detection. *Appl. Phys. Lett.* 83: 2698–2700.

Calleja, M, Nordstrom, M, Alvarez, M, Tamayo, J, Lechuga, LM, and Boisen, A. 2005. Highly sensitive polymer-based cantilever-sensors for DNA detection. *Ultramicroscopy* 105: 215–222.

Carrington, NA, Yong, L, and Xue, ZL. 2006. Electrochemical deposition of sol-gel films for enhanced chromium(VI) determination in aqueous solutions. *Anal. Chim. Acta* 572: 17–24.

Cherian, S, Gupta, RK, Mullin, BC, and Thundat, T. 2003. Detection of heavy metal ions using protein-functionalized microcantilever sensors. *Biosens. Bioelectron.* 19: 411–416.

Cherian, S, Metha, A, and Thundat, T. 2002. Investigating the mechanical effects of adsorption of Ca^{2+} ions on a silicon nitride microcantilever surface. *Langmuir* 18: 6935–6939.

Cui, Y, Wei, Q, Park, H, and Lieber, CM. 2001. Nanowire nanosensors for highly selective detection of biological and chemical species. *Science* 293: 1289–1292.

Dauksaite, V, Lorentzen, M, Besenbacher, F, and Kjems, J. 2007. Antibody-based protein detection using piezoresistive cantilever arrays. *Nanotechnology* 18: 125503.

Dhayal, B, Henne, WA, Doorneweerd, DD, Reifenberger, RG, and Low, PS. 2006. Detection of Bacillus subtilis spores using peptide-functionalized cantilever arrays. *J. Am. Chem. Soc.* 128: 3716–3721.

Dutta, P, Chapman, PJ, Datskos, PG, and Spaniak, MJ. 2005. Characterization of ligand-functionalized microcantilevers for metal ion sensing. *Anal. Chem.* 77: 6601–6608.

Dutta, P, Tipple, C, Lavrik, N, and Datskos, P. 2003. Enantioselective sensors based on antibody-mediated nanomechanics. *Anal. Chem.* 75: 2342–2348.

Ekinci, KL and Roukes, ML. 2005. Nanoelectromechanical systems. *Rev. Sci. Instrum.* 76: 061101.

Fritz, J, Baller, MK, Lang, HP et al. 2000a. Stress at the solid-liquid interface of self-assembled monolayers on gold investigated with a nanomechanical sensor. *Langmuir* 16: 9694–9696.

Fritz, J, Baller, MK, Lang, HP et al. 2000b. Translating biomolecular recognition into nanomechanics. *Science* 288: 316–318.

Fuji-Keizai 2008. *Biochip Trends for Drug R&D and Diagnostics—Companies, Equipment, Consumables, Software, Services and World Market.* Fuji-Keizai USA, Inc., San Jose, CA. http://www.researchandmarkets.com/reports/599371/

Ghatnekar-Nilsson, S, Lindahl, J, Dahlin, A et al. 2005. Phospholipid vesicle adsorption measured in situ with resonating cantilevers in a liquid cell. *Nanotechnology* 16: 1512–1516.

Gimzewski, JK, Gerber, C, Meyer, E, and Schlittler, RR. 1994. Observation of a chemical reaction using a micromechanical sensor. *Chem. Phys. Lett.* 217: 589–594.

Grogan, C, Raiteri, R, O'Connor, GM et al. 2002. Characterisation of an antibody coated microcantilever as a potential immunobased biosensor. *Biosens. Bioelectron.* 17: 201–207.

Gunter, RL, Zhine, R, Delinger, WG, Manygoats, K, Kooser, A, and Porter, TL. 2004. Investigation of DNA sensing using piezoresistive microcantilever probes. *IEEE Sens. J.* 4: 430–433.

Hagan, MF, Majumdar, A, and Chakraborty, AK. 2002. Nanomechanical forces generated by surface grafted DNA. *J. Phys. Chem. B* 106: 10163–10173.

Hansen, KM, Ji, HF, Wu, GH et al. 2001. Cantilever-based optical deflection assay for discrimination of DNA single-nucleotide mismatches. *Anal. Chem.* 73: 1567–1571.

Huber, F, Hegner, M, Gerber, C, Guntherodt, HJ, and Lang, HP. 2006. Label free analysis of transcription factors using microcantilever arrays. *Biosens. Bioelectron.* 21: 1599–1605.

Hwang, KS, Lee, JH, Park, J, Yoon, DS, Park, JH, and Kim, TS. 2004. In-situ quantitative analysis of a prostate-specific antigen (PSA) using a nanomechanical PZT cantilever. *Lab. Chip* 4: 547–552.

Hyun, SJ, Kim, HS, Kim, YJ, and Jung, HI. 2006. Mechanical detection of liposomes using piezoresistive cantilever. *Sens. Actuators B: Chem.* 117: 415–419.

Ilic, B, Craighead, HG, Krylov, S et al. 2004. Attogram detection using nanoelectromechanical oscillators. *J. Appl. Phys.* 95: 3694–3703.

Ilic, B, Czaplewski, D, Zalalutdinov, M et al. 2001. Single cell detection with micromechanical oscillators. *J. Vac. Sci. Technol. B* 19: 2825–2828.

Ilic, B, Yang, Y, Aubin, K, Reichenbach, R, Krylov, S, and Craighead, HG. 2005. Enumeration of DNA molecules bound to a nanomechanical oscillator. *Nano Lett.* 5: 925–929.

Ji, HF, Hansen, KM, Hu, Z, and Thundat, T. 2001a. Detection of pH variation using modified microcantilever sensors. *Sens. Actuators B: Chem.* 72: 233–238.

Ji, HF, Lu, YQ, Du, HW, Xu, XH, and Thundat, T. 2006. Spiral springs and microspiral springs for chemical and biological sensing. *Appl. Phys. Lett.* 88: 063504.

Ji, HF, Thundat, T, Dabestani, R, Brown, GM, Britt, PF, and Bonnesen, PV. 2001b. Ultrasensitive detection of CrO_4^{2-} using a microcantilever sensor. *Anal. Chem.* 73: 1572–1576.

Kim, BH, Mader, O, Weimar, U, Brock, R, and Kern, DP. 2003. Detection of antibody peptide interaction using microcantilevers as surface stress sensors. *J. Vac. Sci. Technol. B* 21: 1472–1475.

Kooser, A, Manygoats, K, Eastman, MP, and Porter, TL. 2003. Investigation of the antigen antibody reaction between anti-bovine serum albumin (a-BSA) and bovine serum albumin (BSA) using piezoresistive microcantilever based sensors. *Biosens. Bioelectron.* 19: 503–508.

Krause, AR, Van Neste, C, Senesac, L, Thundat, T, and Finot, E. 2008. Trace explosive detection using photothermal deflection spectroscopy. *J. Appl. Phys.* 103: 094906.

Kwon, HS, Han, KC, Hwang, KS et al. 2007. Development of a peptide inhibitor-based cantilever sensor assay for cyclic adenosine monophosphate-dependent protein kinase. *Anal. Chim. Acta* 585: 344–349.

Lang, HP, Berger, R, Andreoli, C et al. 1998. Sequential position readout from arrays of micromechanical cantilever sensors. *Appl. Phys. Lett.* 72: 383–385.

Lange, D, Hagleitner, C, Hierlemann, A, Brand, O, and Baltes, H. 2002. Complementary metal oxide semiconductor cantilever arrays on a single chip: Mass-sensitive detection of volatile organic compounds. *Anal. Chem.* 74: 3084–3095.

Lechuga, LM, Tamayo, J, Alvarez, M et al. 2006. A highly sensitive microsystem based on nanomechanical biosensors for genomics applications. *Sens. Actuators B: Chem.* 118: 2–10.

Lee, JH, Hwang, KS, Park, J, Yoon, KH, Yoon, DS, and Kim, TS. 2005. Immunoassay of prostate-specific antigen (PSA) using resonant frequency shift of piezoelectric nanomechanical microcantilever. *Biosens. Bioelectron.* 20: 2157–2162.

Lee, JH, Yoon, KH, Hwang, KS, Park, J, Ahn, S, and Kim, TS. 2004. Label free novel electrical detection using micromachined PZT monolithic thin film cantilever for the detection of C-reactive protein. *Biosens. Bioelectron.* 20: 269–275.

Liu, K and Ji, HF. 2004. Detection of Pb^{2+} using a hydrogel swelling microcantilever sensor, *Anal. Sci.* 20: 9–11.

Maraldo, D, Garcia, FU, and Mutharasan, R. 2007. Method for quantification of a prostate cancer biomarker in urine without sample preparation. *Anal. Chem.* 79: 7683–7690.

Marie, R, Jensenius, H, Thaysen, J, Christensen, CB, and Boisen, A. 2002. Adsorption kinetics and mechanical properties of thiol-modified DNA-oligos on gold investigated by microcantilever sensors. *Ultramicroscopy* 91: 29–36.

McKendry, R, Zhang, J, Arntz, Y et al. 2002. Multiple label-free biodetection and quantitative DNA-binding assays on a nanomechanical cantilever array. *Proc. Natl. Acad. Sci. U.S.A.* 99: 9783–9787.

Moulin, AM, O'Shea, SJ, and Welland, ME. 2000. Microcantilever-based biosensors. *Ultramicroscopy* 82: 23–31.

Mukhopadhyay, R, Lorentzen, M, Kjems, J, and Besenbacher, F. 2005a. Nanomechanical sensing of DNA sequences using piezoresistive cantilevers. *Langmuir* 21: 8400–8408.

Mukhopadhyay, R, Sumbayev, VV, Lorentzen, M, Kjems, J, Andreasen, PA, and Besenbacher, F. 2005b. Cantilever sensor for nanomechanical detection of specific protein conformations. *Nano Lett.* 5: 2385–2388.

Ndieyira, JW, Watari, M, Barrera, AD et al. 2008. Nanomechanical detection of antibiotic—Mucopeptide binding in a model for superbug drug resistance, *Nat. Nanotechnol.* 3: 691–696.

Nugaeva, N, Gfeller, KY, Backmann, N, Lang, HP, Duggelin, M, and Hegner, M. 2005. Micromechanical cantilever array sensors for selective fungal immobilization and fast growth detection. *Biosens. Bioelectron.* 21: 849–856.

Park, J, Ryu, R, Choi, SK et al. 2005. Real-time measurement of the contractile forces of self-organized cardiomyocytes on hybrid biopolymer microcantilevers. *Anal. Chem.* 77: 6571–6580.

Pei, JH, Tian, F, and Thundat, T. 2004. Glucose biosensor based on the microcantilever. *Anal. Chem.* 76: 292–297.

Pera, I and Fritz, J. 2007. Sensing lipid bilayer formation and expansion with a microfabricated cantilever array. *Langmuir* 23: 1543–1547.

Sakhaee-Pour, A, Ahmadian, MT, and Vafai, A. 2008. Applications of single-layered graphene sheets as mass sensors and atomistic dust detectors. *Solid State Commun.* 145: 168–172.

Savran, CA, Burg, TP, Fritz; J, and Manalis, SR. 2003. Microfabricated mechanical biosensor with inherently differential readout. *Appl. Phys. Lett.* 83: 1659–1661.

Savran, CA, Knudsen, SM, Ellington, AD, and Manalis, SR. 2004. Micromechanical detection of proteins using aptamer-based receptor molecules. *Anal. Chem.* 76: 3194–3198.

Schmid, D, Lang, HP, Marsch, S, Gerber, C, and Hunziker, P. 2008. Diagnosing disease by nanomechanical olfactory sensors—System design and clinical validation. *Eur. J. Nanomed.* 1: 44–47.

Shu, W, Laue, ED, and Seshia, AA. 2007. Investigation of biotin-streptavidin binding interactions using microcantilever sensors. *Biosens. Bioelectron.* 22: 2003–2009.

Singh, G, Rice, P, and Mahajan, RL. 2007. Fabrication and mechanical characterization of a force sensor based on an individual carbon nanotube, *Nanotechnology* 18: 475501.

Stachowiak, JC, Yue, M, Castelino, K, Chakraborty, A, and Majumdar, A. 2006. Chemomechanics of surface stresses induced by DNA hybridization. *Langmuir* 22: 263–268.

Stevenson, KA, Mehta, A, Sachenko, P, Hansen, KM, and Thundat, T. 2002. Nanomechanical effect of enzymatic manipulation of DNA on microcantilever surfaces. *Langmuir* 18: 8732–8736.

Su, M, Li, S, and Dravid VP. 2003. Microcantilever resonance-based DNA detection with nanoparticle probes. *Appl. Phys. Lett.* 82: 3562–3564.

Subramanian, A, Oden, PI, Kennel, SJ et al. 2002. Glucose biosensing using an enzyme-coated microcantilever. *Appl. Phys. Lett.* 81: 385–387.

Thundat, T, Warmack, RJ, Chen, GY, and Allison, DP. 1994. Thermal and ambient-induced deflections of scanning force microscope cantilevers. *Appl. Phys. Lett.* 64: 2894–2896.

Tian, F, Boiadjiev, VI, Pinnaduwage, LA, Brown, GM, and Thundat, T. 2005. Selective detection of Cr(VI) using a microcantilever electrode coated with a self-assembled monolayer. *J. Vac. Sci. Technol. A* 23: 1022–1028.

Van Neste, CW, Senesac, LR, Yi, D, and Thundat, T. 2008. Standoff detection of explosive residues using photothermal microcantilevers. *Appl. Phys. Lett.* 92: 134102.

Velanki, S and Ji, H.-F. 2006. Detection of feline coronavirus using microcantilever sensors. *Meas. Sci. Technol.* 17: 2964–2968.

Velanki, S, Kelly, S, Thundat, T, Blake, DA, and Ji, HF. 2007. Detection of Cd(II) using antibody-modified microcantilever sensors. *Ultramicroscopy* 107: 1123–1128.

Watari, M, Galbraith, J, Lang, HP et al. 2007. Investigating the molecular mechanisms of in-plane mechanochemistry on cantilever arrays. *J. Am. Chem. Soc.* 129: 601–609.

Wee, KW, Kang, GY, Park, J et al. 2005. Novel electrical detection of label-free disease marker proteins using piezoresistive self-sensing micro-cantilevers. *Biosens. Bioelectron.* 20: 1932–1938.

WHO 2006. World Health Organization, Report: Food and health in Europe: A new basis for action (Nov. 2006). Eds., Aileen Robertson, Cristina Tirado, Tim Lobstein, Marco Jermini, Cecile Knai, Jørgen H. Jensen, Anna Ferro-Luzzi, and W.P.T. James. WHO Regional Publications, European Series, No. 96, Albany, New York. http://www.euro.who.int/InformationSources/Publications/Catalogue/20040130_8

Wu, G, Datar, RH, Hansen, KM, Thundat, T, Cote, RJ, and Majumdar, A. 2001. Bioassay of prostate-specific antigen (PSA) using microcantilevers. *Nat. Biotechnol.* 19: 856–860.

Xu, XH, Thundat, TG, Brown, GM, Ji, HF. 2002. Detection of Hg^{2+} using microcantilever sensors. *Anal. Chem.* 74: 3611–3615.

Yan, X, Hill, K, Gao, H, and Ji, HF. 2006b. Surface stress changes induced by the conformational change of proteins. *Langmuir* 22: 11241–11244.

Yan, XD, Ji, HF, and Lvov, Y. 2004. Modification of microcantilevers using layer-by-layer nanoassembly film for glucose measurement. *Chem. Phys. Lett.* 396: 34–37.

Yan, XD, Shi, XL, Hill, K, and Ji, HF. 2006a. Microcantilevers modified by horseradish peroxidase intercalated nano-assembly for hydrogen peroxide detection. *Anal. Sci.* 22: 205–208.

Yan, XD, Xu, XHK, and Ji, HF. 2005. Glucose oxidase multilayer modified microcantilevers for glucose measurement. *Anal. Chem.* 77: 6197–6204.

Yue, M, Stachowiak, JC, Lin, H, Datar, R, Cote, R, and Majumdar, A. 2008. Label-free protein recognition two-dimensional array using nanomechanical sensors. *Nano Lett.* 8: 520–524.

Zhang, YF, Ji, HF, Snow, D, Sterling, R, and Brown, GM. 2004a. A pH sensor based on a microcantilever coated with intelligent hydrogel. *Instrum. Sci. Technol.* 32: 361–369.

Zhang, J, Lang, HP, Huber, F et al. 2006. Rapid and label-free nanomechanical detection of biomarker transcripts in human RNA. *Nat. Nanotechnol.* 1: 214–220.

Zhang, YF, Venkatachalan, SP, Xu, H et al. 2004b. Micromechanical measurement of membrane receptor binding for label-free drug discovery. *Biosens. Bioelectron.* 19: 1473–1478.

12

Analyzing Individual Biomolecules Using Nanopores

Meni Wanunu
Boston University

Gautam V. Soni
Boston University

Amit Meller
Boston University

12.1 Introduction

Emerging from the broad diversity of nanoscale-sensing platforms, nanopores are single-molecule (SM) sensors in a class of their own, capable of detection, analysis, and the manipulation of single molecules with high throughput. Much like gel electrophoresis, where the native electric charge of biomolecules is used to move the molecule in the direction of an external electrical field, in the nanopore technique, electrical force is used to thread biomolecules through a narrow nanopore constriction. Nanopores offer an unparalleled set of advantages over other single-molecule sensing methodologies: (1) Biomolecules can be detected without chemical or radioactive labeling, thus enabling the development of extremely low-cost and rapid diagnostic tools. (2) High analyte sensitivity can be achieved, as nanomolar concentrations are routinely sufficient. In conjunction with advances in nanofluidics, nanopore sensors integrated with sub-µL cells are extremely well suited for the amplification-free detection of trace samples. (3) Nanopores are a unique force apparatus for the analysis of biomolecular interactions, in that the immobilization of the molecules onto surfaces or beads is not required. (4) Finally, sensing using nanopores does not require making or breaking chemical bonds, and thus the technique is nondestructive. Despite the great utility of this technique for current and future applications, its underlying principles are simple, borrowing from the resistive sensing principles developed by Coulter in the 1950s (Coulter, 1953). In this chapter, we provide a detailed overview of the experimental aspects of the nanopore technique. We highlight its short-term and long-term prospects, and describe the progress made thus far with natural and synthetic pores for single-biomolecule detection and analysis. We finally discuss future possible directions of nanopore sensing, and the potential impacts of the technique on biophysics and biomedical research.

12.1.1 Nanopores: The Resistive Sensing Technique

A nanopore is a nanoscale aperture through a membrane impermeable to ions. Typical nanopore experiments are conducted by immersing both sides of the membrane in an aqueous solution of monovalent electrolyte (e.g., KCl), and applying voltage across the membrane using two electrodes. The applied bias results in the transport of ions through the nanopore, providing a baseline "open pore" current signal (see Figure 12.1a), which scales with the ion mobility, pore geometry, and bias according to

$$i \approx \left[(n\mu)_{K^+} + (n\mu)_{Cl^+} \right] \frac{d^2}{l} V \qquad (12.1)$$

where

n and μ are the ion densities and the mobilities, respectively
d is the pore diameter
l is its length
V is the applied voltage

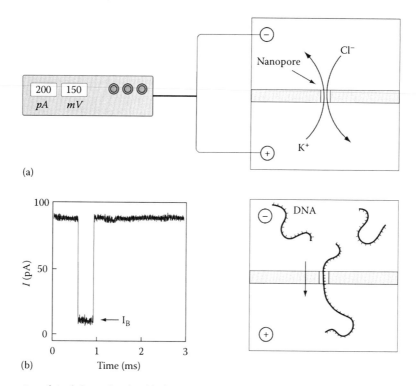

(a)

(b)

FIGURE 12.1   Nanopore sensing of single biomolecules. (a) The measured signal is the ion current of an electrolyte through the nanopore when voltage is applied using a high-gain amplifier. (b) Biopolymer analysis (in this case DNA) is achieved by analyzing the current blockade events, which correlate with the biomolecular properties.

When the current through a single nanopore is recorded, information about the properties of molecules in the solution can be retrieved. For example, when a biomolecule stochastically enters the pore, the ion current is impeded by the lower effective pore volume (as biomolecules have much smaller mobilities than ions). This causes a sudden attenuation of the residual electrolyte current from its open pore value. When the molecule exits the pore, the pore current is restored to its open pore value (see Figure 12.1b). These current variations can be readily picked up using high-gain current amplifiers, which measure down to picoamperes of current at high temporal resolutions (~10 μs). Obviously, high salt concentrations improve the signal by providing an ion current sufficiently above the noise level.

Once a charged biomolecule enters the pore, the electric field electrophoretically drives it through the pore with a force proportional to its charge and the applied field. The size, concentration, and charge of biomolecules can be measured by analyzing the depth, rate, and duration of these residual current signals, respectively. While the electric force on charged biomolecules facilitates detection, uncharged species can also be detected as they enter the pore either stochastically or assisted by a convective force (e.g., hydrostatic pressure). Since nanopore sensing does not rely on any specific property of the biomolecule, the scope of nanopore sensing is extremely broad.

12.1.2 Nanopores: Emerging Technology for DNA Sequencing

DNA is a uniformly charged biopolymer that can be driven through nanopores by an applied field. Over a decade ago, it was demonstrated that single-stranded nucleic acid (ssNA) polymers could be driven through a lipid-embedded α-hemolysin (α-HL) protein channel by the application of a transmembrane voltage (Kasianowicz et al., 1996). Since the channel dimensions (1.5 nm) are only slightly larger than the ssNA cross section (~1.2–1.3 nm) (Saenger, 1988), polymer translocation was forced to occur in a single-file manner (see Figure 12.1b). Later studies using NA homopolymers revealed a striking feature: RNA diblock copolymers (e.g., poly(A_nC_n) displayed bi-level current events, where each level corresponded to the current level of the homopolymer type (Akeson et al., 1999)). Moreover, similar experiments with homopurine polymers displayed different current levels than homopyrimidines (Meller et al., 2000). This result suggested that nanopores are sensitive to the local biopolymer structure, the information conveniently read by analysis of the residual electrolyte current signal. In essence, the residual electrolyte current through the nanopore "reports" local biopolymer properties, at resolutions determined at best by the pore length itself (for α-HL, this is ~6 nm, or 12 bases). The effective resolution is a complex quantity that also depends on the translocation speed of the polymer, the measurement bandwidth, and the signal-to-noise ratio of the measurement.

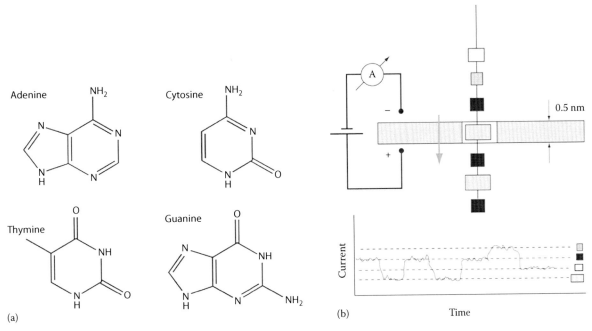

FIGURE 12.2 (a) The chemical structure of the four nucleotides. (b) Top: Schematic of the electronic measurement of ion current through the pore as different nucleobases enter the atomically thin nanopore. Bottom: Hypothesized contrast in the signal for the four nucleobases.

An immediate offshoot of this idea was to ask: Assuming individual bases can be detected, can the DNA sequence of a translocating polymer be decoded (i.e., as shown schematically in Figure 12.2b)? Can a DNA sequence be artificially "read," just as RNA polymerase molecules do when they travel along a single DNA strand during transcription? Despite this appealing idea, one might imagine that the complexity of unraveling a DNA sequence using a simple hole is outstanding. Before we list the enormous difficulties associated with nanopore sequencing, we must emphasize the potential role of nanopores for DNA sequencing. Although current DNA sequencing technologies have optimized the Sanger method to provide highly efficient and rapid sequencing, the costs associated with sequencing a complete genome are unlikely to decrease below the level required for personalized genomics. This has motivated the development of widely different methods for sequencing a complete genome for a fraction of its current cost (Branton et al., 2008). Nanopores are outstanding candidates for this task, provided that a physical mechanism for the identification of the nucleobases identity is implemented, since they allow processing of extremely long reads and a re-reading of the same molecule to reduce the error rate. Thus, nanopore sequencing techniques are expected to yield the high-throughput and cost reduction needed for next-generation DNA-sequencing methods—expected to complete the sequencing of a mammal-sized genome for under $1000.

We now outline the main challenges for ion-current-based DNA sequencing. As illustrated in Figure 12.2, the theoretical spatial resolution of a nanopore sensor is ideally determined by its resistive volume, and defined by its diameter and depth. Therefore, resolving individual bases require nanopores of depths comparable to a single base. The challenge then is to create a *stable*,

atomically thin pore that does not degrade over time, particularly when a biopolymer is continually threaded through it. Assuming that an ideal nanopore is available, a prerequisite for current-based DNA sequencing is that each of the four DNA bases provide a distinguishable ion-current level, and that the differences between these signals be greater than the sensor's electrical noise. This is inherently difficult, considering the fact that both purines and both pyrimidines have similar molecular weights, and are thus expected to produce similar ion-blockade "footprints."

Next, the translocation speed of the polymer to be read must be slow enough to allow sufficient current sampling from each translocating base, given the measurement time resolution. The average translocation time per nucleotide in the ssNA translocation experiment through α-HL is ~1 μs, during which only ~60 ions pass through the pore (at room temperature, 1 M KCl, and 100 mV bias, the blocked ion-current level is ~10 pA). Since this signal is comparable to the electrical noise in the system, measured at a full-bandwidth of 100 kHz, the practical discrimination of the bases relies on minimizing the noise to its theoretical limits. Coupled to the reduced noise, it is necessary to slow down translocation speeds by at least an order of magnitude without degrading the ion-current signal, in order to unambiguously "call" the identity of each base.

Each of these challenges is a grand obstacle and to date, no pore has been able to resolve individual bases in a polynucleotide. However, the quest for sequencing DNA molecules using a nanopore is far from abandoned, as several alternative approaches and workarounds are currently in development (Branton et al., 2008). A promising sequencing approach, which does not rely on the fast dynamics of voltage-driven translocation, involves enzymatically regulated kinetics. In this approach, a processive DNA exonuclease is fused to an engineered α-HL pore, and sequentially each nucleotide in the polymer is digested in the pore proximity. Each cleaved nucleotide is then sequentially sampled by the pore, where each of the four bases provides a distinguishable current level. Since the nucleotide processing time determines the sequencing rate, up to tens of milliseconds are available for sampling each base. Although the activity of the fused exonuclease to α-HL has yet to be demonstrated, this technique remains promising since all four nucleobases have already been resolved, as they exhibit different residual ion-current values in an α-HL mutant adapted with a molecular recognition element (see Figure 12.3 in Section 12.2).

Undoubtedly, designing a multifunctional protein complex is a main avenue toward rapid DNA sequencing, as well as an enormous scientific feat. However, techniques that rely on lipid bilayers are often deemed impractical due to the fragility of the bilayer and the difficulties in inserting only one protein channel. Since the new millennium, alternative nanopore-based approaches for DNA sequencing have been developed, primarily relying on recent developments in synthetic nanopore fabrication. Various methods for fabricating synthetic nanopores in different materials have been developed, allowing additional sensing modalities to be integrated with the electrolyte current signal (e.g., optical or electronic detection). Of the different substrate types, ultrathin inorganic solid-state membranes are gaining utility as the membrane of choice, since pores can be accurately controlled in three dimensions by sub-nanometer resolution imaging using the transmission electron microscope (TEM). The merits of solid-state nanopores include integration capabilities with other sensing schemes, physical and chemical robustness, and the availability of methods to control the chemical interface for more sophisticated sensing (Gyurcsanyi, 2008). Apart from the potential for single-molecule DNA sequencing, a broader range of SM studies of biopolymers can be carried out, like the analysis of double-stranded DNA (dsDNA), proteins, antibodies, as well as different biomolecular complexes (Dekker, 2007; Healy et al., 2007; Wanunu and Meller, 2008).

12.1.3 The Future of Nanopores: Current and Prospective Applications

From both the scientific and commercial perspectives, nanopore research is thriving. Recent experimental and theoretical works have converged to adequately describe the complex physics involved in biopolymer transport through biological pores. Nanopore force spectroscopy was developed to investigate interactions between biomolecules at the single-molecule level (Bates et al., 2003). Synthetic nanopores are gradually taking center stage in the nanopore community, evidenced by a rising number of demonstrations of single-biomolecule detection. Studies of dsDNA translocation through solid-state nanopores have revealed substantial differences in the transport mechanism as compared to ssDNA through α-HL, attributed to differences in the polymer stiffness and the relative nanopore dimensions (Storm et al., 2005a; Wanunu et al., 2008). Solid-state pores have shown promise for detecting Watson–Crick mismatches in a DNA duplex (Mcnally et al., 2008), analyzing protein molecules (Han

et al., 2006; Fologea et al., 2007), and detecting antigen–antibody immunocomplexes in solution (Han et al., 2008). Today, several commercial establishments have emerged in order to harvest the huge potential of nanopore technology for biomolecular sensing and DNA sequencing by developing novel nanopore sensors.

Nanopores are also at the interface of several disciplines, from the materials science of nanopore fabrication, to the physics of polymer transport, to the chemistry of nanopore interface modification, and the biology of biomolecules in the confined environment of the pore. In this chapter, we illustrate some basic concepts and features of nanopores, provide a survey of important works in the field, as well as some foresight into the future of nanopores. The rest of this chapter is organized as follows: In Section 12.2, we provide the reader with a survey of selected types of engineered nanopores, both organic and inorganic. In Section 12.3, we discuss in detail the fundamental process of biopolymer translocation through small pores, as addressed both theoretically and experimentally. In Section 12.4, we highlight how force is used to probe biomolecular interactions using nanopores. Section 12.5 concludes this chapter with a discussion of some of the present and future challenges associated with nanopores, as well as our view of the promising methods currently being developed for tackling these problems.

12.2 Nanopore Types and Properties

12.2.1 Bioengineered Protein Pores

Bayley and coworkers have been both pioneers and innovators, for the elucidation of the crystal structure of heptameric α-HL and for the utilization of engineered α-HL mutants for sensing small molecules. For example, α-HL was engineered to include a divalent metal ion-binding site, which showed distinct current blockade patterns for each type of ion, used for studying ion binding kinetics (Kasianowicz et al., 1999; Braha et al., 2000). Organic analytes, such as TNT, (Guan et al., 2005) inositol 1,4,5-trisphosphate (Cheley et al., 2002), and others (Gu et al., 1999; Braha et al., 2005) were detected using engineered α-HL mutants and/or transient adapters. Engineered α-HL mutants were found to efficiently incorporate a chiral molecular adapter (β-cyclodextrin, or β-CD) when inserted from its *trans* side (Kang et al., 2006). This was used to discriminate among the enantiomers of chiral drugs. The addition of (*R*-) or (*S*-) ibuprofen to the solution resulted in distinct ion-current levels, presumably corresponding to distinct diastereomeric complexes with the host β-CD adapter. This rapid enantiomeric detection method was also used for sensing the racemization kinetics of another drug, thalidomide, at μM levels. Most notably, the discrimination of ribonucleoside and deoxyribonucleoside monophosphates was recently demonstrated using a singly mutated α-HL equipped with an amino-modified β-CD adapter (see Figure 12.3a). When residue 113 (methionine) of α-HL was replaced with arginine, aminated-β-CD molecules were found to reside in the pore for sufficient periods to allow the observation of complexation with different nucleoside phosphates. As seen in Figure 12.3b, the amplitude of these complexes is different for each

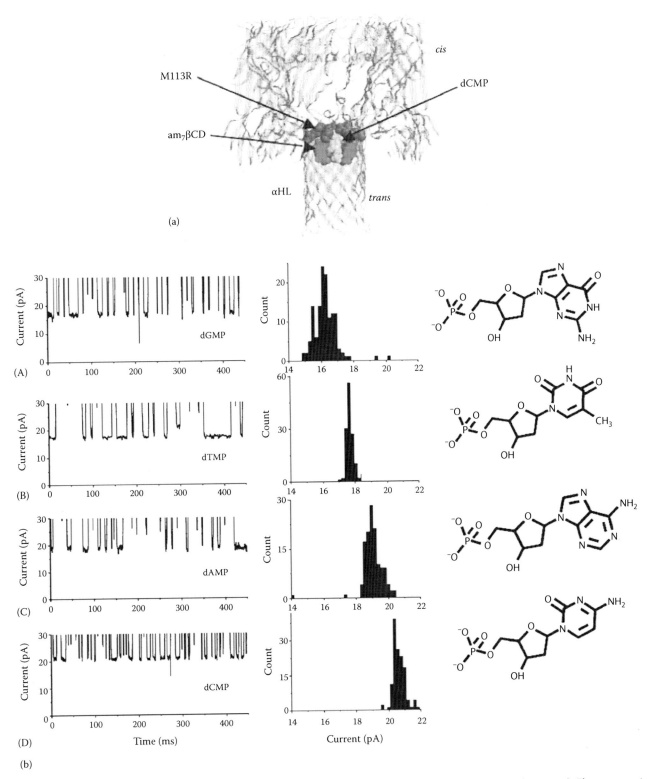

FIGURE 12.3   (a) Model of the heptameric α-HL pore (7AHL), in which Met-113 has been substituted with Arg (see arrow). The cross section of an am₇βCD adapter is shown in the pore (dark gray), and an incorporated dCMP is shown in light gray. (b) Normalized current levels for each dNMP is shown from single (M113R)₇ pores at +130 mV with 40 μM am₇βCD. dNMP (5 μM) was added to the cis chamber. (A) dGMP; (B) dTMP; (C) dAMP; (D) dCMP. Event histograms based on 133 events for each dNMP are shown at center, with the structure of the dNMP to the right. (Reproduced from Astier, Y. et al., *J. Am. Chem. Soc.*, 128, 1705, 2006. With permission.)

one of the bases. This distinction among the bases forms a basis for a controlled digestion/nanopore-sequencing method.

While the above works show the versatility of protein channels, several challenges require attention before a viable nanopore-sensing platform can be developed: (1) the fragility of the bilayer presented difficulties in running an experiment, (2) the challenge of inserting a single protein channel (important for current-based measurements) is frustrating, and (3) integration of protein channel sensors with other sensing schemes is difficult, since protein channels are not fixed in space, but rather freely diffuse throughout the bilayer. In addition to these inherent difficulties with membrane-embedded channels, α-HL itself has fixed dimensions, which allow only certain molecules to be detected. While other single protein channels have been inserted into bilayers for purposes of molecular detection, most natural channels display self-gating (i.e., fluctuations of the current in the absence of analyte), therefore presenting difficulties in sensing molecules. The above challenges have initiated a quest for more robust low-noise synthetic pores of controllable dimensions and surface properties. We provide here a survey of selected nanopore types that have been used in key applications of single-molecule analysis.

12.2.2 Nanopores in Glass/Quartz

Karhanek and coworkers described the use of nanopipettes made of pulled capillaries for use as stochastic sensors for DNA translocation (Karhanek et al., 2005). In this system, a glass (or quartz) capillary is drawn using a precision puller to a fine pipette (see Figure 12.4a). An attractive feature of this technique is that nanopipettes can be made rather cheaply using a straightforward technique. The mean orifice size is estimated by measuring ion current through the pipettes. Scanning electron microscopy (SEM) can give information about the shape of the pipette orifice, although the resolution of any imaging method is insufficient to determine the exact orifice size, especially for nanopores of molecular dimensions (i.e., sub-10 nm). A major drawback of this technique is that the orifice size is not accurately controlled at the nm level. However, small orifice sizes have been achieved (e.g., 6 nm), with the size estimated from electrochemical measurements (Shao and Mirkin, 1997). White et al. reported the fabrication of nanopipette electrodes down to 15 nm diameter by sealing a glass capillary over a sharp Pt wire (see Figure 12.4b), followed by an etching of the glass and Pt (Zhang et al., 2004; Wang et al., 2006). An advantage of the glass pipette as a substrate is that the glass substrate exhibits low capacitance, which translates to comparable noise at high bandwidth as protein channels in lipid bilayers.

12.2.3 Nanopores in Plastic Films

Using the track-etch method (Fleischer et al., 1972), polymer films (e.g., polycarbonate, polyimide, polyethylene terephthalate) can be irradiated with heavy nuclei to produce tracks in the film. By placing a mask on the polymer and a sensitive ion

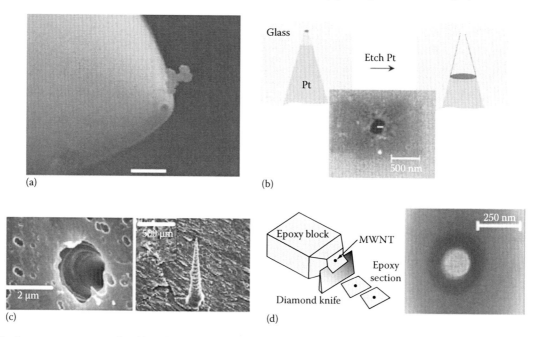

(a)   (b)

(c)   (d)

FIGURE 12.4 Some nanopore examples. (a) High-resolution scanning electron microscope (SEM) image of a nanopipette tip, showing a steep cone angle up to the pore with a diameter 52.2 nm (Scale bar: 200 nm) (Reproduced from Karhanek, M. et al., *Nano Lett.*, 5, 403, 2005. With permission). (b) Fabrication step for a nanopore electrode, and the SEM image of the pore that resulted from etching the same electrode (scale bar drawn across the nanopore: 102 nm) (Reproduced from Zhang, B. et al., *Anal. Chem.*, 76, 6229, 2004. With permission). (c) Left: SEM image of the wide side of a track-etched polycarbonate membrane (scale bar: 2 μm). Right: SEM image of a gold nanowire plated into conical nanopores during the etching process (scale bar: 5 μm) (Reproduced from Harrell, C.C. et al., *Nano Lett.*, 2, 194, 2006. With permission). (d) Scheme of nanopore fabrication by slicing the epoxy orthogonally to the embedded multi-wall carbon nanotube (MWNT), as well as an SEM image of the slice (dark region corresponds to the MWNT). (Reproduced from Ito T. et al., *Acc. Chem. Res.*, 37, 937, 2004. With permission.)

detector behind it, ion discharge can be feedback-controlled in such a way that only a single ion is transmitted through the film. Following irradiation, the polymer film is chemically etched in a bath until a pore is formed. The etching then ceases by placing a stop bath on the other side of the film (see Figure 12.4c). Using this technique, the pore size is estimated by measuring ion current through the pores in solution. Control over the pore shape (e.g., from conical to cylindrical) (Harrell et al., 2006), their size, and their chemical functionality (e.g., choice of polymer, modification) (Siwy et al., 2004, 2005; Vlassiouk and Siwy, 2007; Ali et al., 2008), renders such pores versatile platforms for SM studies. SM detection of DNA, (Schiedt et al., 2005) proteins, (Heins et al., 2005a) and other macromolecules (Heins et al., 2005b) have been demonstrated using these pores. More recently, a method based on local heating induced by an Ar⁺ laser was used to fabricate nanopores of sub-10 nm dimensions in Apiezon Wax films (Wu et al., 2006). Sizing of the nanopores in each case was made by fitting the pore conductance to the expected ion flux based on the ion mobilities and the voltage. However, since the pore dimensions in these cases cannot be directly observed at the nanoscale, sizing is often done by measuring changes in the relative conductance upon translocation of single DNA molecules. Since the DNA moves much slower than ions in the pore, the magnitude of pore blockage during DNA entry should indicate the fractional current corresponding to DNA entry (this approximation is valid at high ionic strengths only, as surface charge dominates ionic transport at a low ionic strength).

12.2.4 Nanopores from Wire Templates

Crooks and coworkers have used multiwall carbon nanotubes (MWNTs) as templates for the fabrication of sub-100 nm nanopores (Ito et al., 2003a,b). In this method, epoxy is polymerized over a stretched MWNT, and then the mold is sectioned using a microtome (see Figure 12.4d). Sohn and Saleh used a PDMS-based microfabrication algorithm to produce 300 nm channels, used for sensing DNA molecules (Saleh and Sohn, 2003). Alternatively, Yang and coworkers fabricated GaN nanotubes and used them as channels for the resistive sensing of single DNA molecules (Goldberger et al., 2006). Using a ZnO nanowire as an etchable inner template, the GaN tube size can be controlled by controlling the ZnO thickness (down to 30 nm i.d.).

12.2.5 Nanopores in Thin Solid-State Membranes

Possibly, the most attractive method for the fabrication of nearly two-dimensional nanoscale pores, involves a precision top-bottom sculpturing of nanopores in thin silicon nitride or silicon dioxide membranes. Solid-state materials are attractive substrates for fabricating nanostructures since they can be processed with sub-nanometer precision, and integrated with multiple detection elements. The fabrication of nanopores in silicon-supported membranes was introduced by Li and coworkers using ion-beam sculpting (Li et al., 2001). In

this technique, an energetic Ar⁺ ion beam is used to sputter a low-stress silicon nitride (SiN) membrane, which contains a bowl-shaped cavity. The sputtering thins the membrane, eventually resulting in the formation of a nanopore, as detected by the leakage of Ar⁺-ions to the other side of the membrane. Nanopore sizes down to 1.5 nm have been fabricated using this method, as determined by TEM.

Storm and coworkers have substituted the ion-beam method with a simpler procedure, in which a focused electron beam (e-beam) of a commercial TEM was used (Storm et al., 2003, 2005a). Using this technique, a large (100 nm) microfabricated pore (the pore was fabricated from a silicon-on-insulator [SOI] hole coated with SiO₂) was irradiated with an intense 300 kV TEM beam, which resulted in the slow shrinking of the pore by the fluidization of the SiO₂ layer. The direct use of the TEM for shaping the nanopores allows both the fabrication and the analysis of the nanopore shape in real time with sub-nanometer TEM imaging precision, as shown in Figure 12.5. The drilling of nanopores through SiO₂/SiN membranes using the TEM was

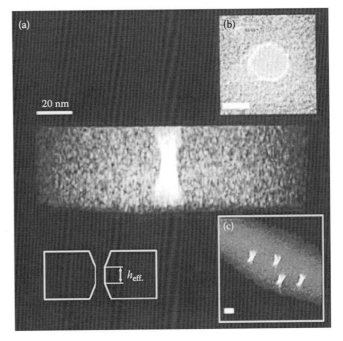

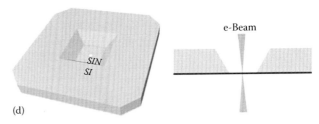

FIGURE 12.5 Side-view of (a) 3D reconstructed TEM tomogram of an 8 nm pore in a 50 nm membrane. (b) Top view of the same pore. (c) Tomogram of a 2 × 2 nanopore array. (d) Schematic of the silicon chip with a silicon nitride window (left) and an electron beam drilling of a nanopore (right). (Reproduced from Kim, M.J. et al., *Adv Mater.*, 18, 3149, 2006. With permission.)

later reported (Keyser et al., 2005). In this method, the e-beam of a field-emission TEM is used to locally ablate the membrane contents, resulting in a hole on the order of the focused beam dimensions (3–5 nm). Then, the e-beam intensity is adjusted to either shrink or enlarge the pore size. Using this method, the pore size can be tuned with a sub-nanometer resolution, and pores in the range of 1–20 nm can be reproducibly made. Recently, a significant reduction in the nanopore formation time enabled the production of highly uniform nanopores and nanopore arrays (Kim et al., 2006). Focused ion-beams (FIB) have also been used to create nanopores in oxide/nitride membranes, which can be further shrunk by either localized deposition (Nilsson et al., 2006) or fluidization via an e-beam (Lo et al., 2006). While the FIB method has advantages of rapid large-scale fabrication, the pore dimensions are not as well controlled as with the TEM-based method.

The striking features of nanopores in ultrathin solid-state membranes are the following: (1) they can be fabricated in under a minute; (2) their dimensions can be directly measured with a sub-nanometer resolution using the TEM, TEM-tomography, as well as electron energy loss spectroscopy (EELS); (3) they exhibit chemical and physical robustness; (4) They can be chemically modified with organic (Wanunu and Meller, 2007) or inorganic (Chen et al., 2004) films to change their interfacial properties; and (5) arrays can be fabricated for the parallel detection of single molecules. Therefore, while other reviews discuss in more detail other synthetic nanopore types, we chose to primarily focus the remainder of this chapter on solid-state nanopores.

12.2.6 Nanopore Properties

Of the large variety of synthetic pores studied to date, surely one nanopore system is better suited for detecting a given set of molecules than another. So, which qualities of a nanopore are desirable, and which are undesirable? To answer this, we identify

three categories: nanopore geometry, interfacial nanopore properties, and dielectric properties of the membrane. Pore geometry and its interfacial properties directly affect its conductance. These effects are discussed first. The dielectric properties of the membrane mainly affect its capacitance, and therefore is a main contributor to the nanopore electrical noise. These effects are discussed in Section 12.2.7.

The nanopore dimensions are crucial in determining the signal contrast expected from molecular entry. As a first-order approximation, the change in electrolyte current between a vacant pore and an occupied pore is proportional to the changes in relative pore volume, where the pore volume is defined by the volume in which most of the electrolyte resistance falls. For a cylindrical pore of diameter d and height h, this volume is approximately $\pi d^2 h$. Since the pore conductivity should scale as $G \propto d^2/h$, the amplitude of the signal from a molecule with cross section a is maximum for an infinitely thin pore with a similar diameter as the molecule (i.e., $h \to 0$ and $d \to a$, respectively).

At the nanoscale, however, synthetic nanopores are seldom perfect cylinders. The structure of solid-state nanopores is relatively easy to study, since they are stable under the normal imaging doses of the TEM. Recent 3D tomographic imaging of solid-state nanopores fabricated in SiN using an e-beam reveals a truncated double-cone structure, with a half angle of 30° ± 2° and an effective nanopore length of h_{eff}, where h is the SiN membrane thickness (see inset to Figure 12.7). Moreover, preliminary evidence shows that the pore shape can be controlled by the irradiation parameters, such as the focus and the intensity of the beam (Storm et al., 2003; Kim et al., 2007a; Wu et al., 2008). For example, Wu and coworkers have recently performed a spatial elemental analysis of the oxygen, the nitrogen, and the silicon atoms around a forming pore using EELS. The results, shown in Figure 12.6, reveal that the profile of the high-intensity electron beam during expansion of a pore determines its overall shape: A defocused beam results in a thinner pore than a focused beam,

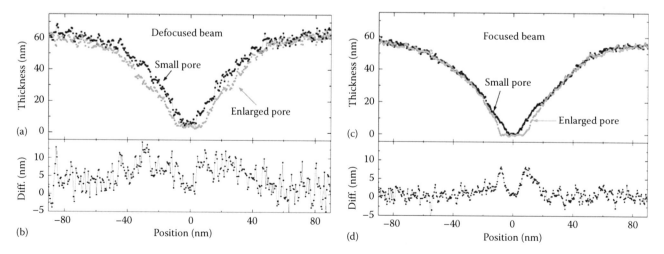

FIGURE 12.6 Thickness variation profiles around nanopores before and after enlargement with a (a) 20 nm defocused beam and a (c) 10 nm focused beam. Changes in the nanopore geometry can be seen from the difference profiles of the thicknesses before and after nanopore enlargements (b) and (d). (Reproduced from Wu, M.-Y., et al., *Nano Lett.*, 9, 479, 2008. With permission.)

presumably due to the fluidization of the surrounding matrix around the pore using a defocused beam.

Ion conductivity measurements of SiN nanopores in the size range 3–15 nm was used to validate the pore size: the nanopores typically yielded ohmic (linear and symmetrical) *I–V* curves, which indicate a symmetrical pore structure, with the conductivity *G* extracted from the slope of the *I–V* curve. The expected conductivity can be calculated by integrating over the typical hourglass pore geometry found from 3D tomography, as described by the following equation:

$$G = \frac{\pi d^2}{4}\sigma\left(\frac{\delta\tan\alpha + 1}{h + h_{\text{eff}}\delta\tan\alpha}\right) \tag{12.2}$$

where

$\sigma = (\mu_K + \mu_{Cl})n_{KCl}e$ is the specific conductance at a number density n_{KCl}, equal to 15.04 and 3.01 for 1.0 and 0.2 M KCl, respectively (μ_K and μ_{Cl} are the electrophoretic mobilities of K⁺ and Cl⁻, respectively, and *e* is the elementary charge unit)

$\delta = (h - h_{\text{eff}})/d$, h_{eff} is the width of the cylindrical region in the nanopore

α is the cone half angle

Figure 12.7 depicts the dependence of *G* on the pore diameter (as determined by TEM). The solid lines are fits to Equation 12.2, taking into account the pore structure, as determined by TEM 3D tomography, and parameterized by the geometrical factors α and h_{eff}.

The interfacial properties of the nanopore can greatly affect single-molecule detection, as they determine the capture rate,

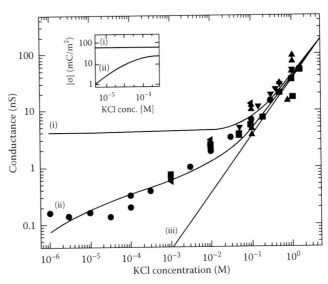

FIGURE 12.8 Conductance values of 10 individual pores (10 ± 2 nm) measured for KCl concentrations from 1 μM to 1 M. Each individual pore has its own symbol. The black upright triangles represent individual pores that were used at a single salt concentration. The lines labeled (i), (ii), and (iii) show the results of calculations as predicted by bulk behavior, a model for constant surface charge, and a model for a variable surface charge, respectively (see Smeets et al., 2006a). The inset shows the values of the surface-charge density vs. salt concentrations on a log–log scale, for both the constant surface charge model ((i), σ = 60 mC m⁻²) and the variable-charge model (ii). (Reproduced from Smeets, R.M.M., et al., *Nano Lett.*, 6, 89, 2006a. With permission.)

the dwell times of molecules in the pore, as well as the dependence of the molecular signal on the ionic strength of the electrolyte. For example, the surface charge of the nanopore can alter the speed at which the molecule translocates through a membrane (Kim et al., 2007b). Surface charge effects were recently studied by investigation of salt dependence on the pore conductivity and the current during DNA translocation (Ho et al., 2005; Smeets et al., 2006a). Smeets and coworkers found that the surface charge of 10 nm diameter pores fabricated in SiO₂ membranes dominates the pore conductivity at a low salt concentration, leading to a nonlinear dependence of *G* on the salt concentration at KCl concentrations below 100 mM (see Figure 12.8).

12.2.7 Electrical Noise in Solid-State Nanopores

The practical limits and bandwidth of nanopore detection is determined by the electrical noise of the system. The system is affected by both low-frequency ("1/*f*") and high-frequency noise, which determine the system's temporal resolution, and its long-term stability, respectively. In analyzing the various factors contributing to noise, it is constructive to differentiate between noise generated internally by the nanopore device (due to fluctuations in pore-conductance, the membrane capacitance, etc.) and the instrumental noise (headstage, amplifier, etc.). We note

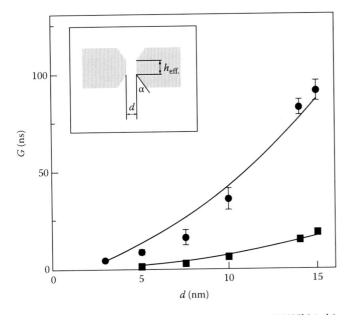

FIGURE 12.7 Conductance of solid-state nanopores in 1 M KCl (circle) and 0.2 M KCl (squares) as a function of the pore size. The solid line is a fit to Equation 12.2. (Reproduced from Kim, M.J. et al., *Adv Mater.*, 18, 3149, 2006. With permission.)

that the low-frequency noise is "resistive" in nature and can be readily characterized using standard instrumentation (in other words, the instrument noise is much lower and therefore fully decoupled from the nanopore noise). However, high-frequency noise is "capacitive" in nature and more complex to resolve, since the decoupling of the nanopore noise from the amplifier noise is harder to model. Moreover, high-gain analog current amplifiers are bandwidth-limited (i.e., Uram et al. (2008) measured an upper limit of ~55 kHz for the popular Axopatch 200B), and their step response function would depend on the coupled capacitance of the system. Thus, key to low-noise nanopore measurements is to minimize the membrane capacitance.

The ion mobility fluctuation theory developed by Hooge (1969) can generally be used to model the open pore current noise by considering two main contributions to its power spectral density (PSD): A Johnson noise term in the high-frequency regime ($S_I = 4\pi^2 c^2 f^2 S_V$), and a "flicker" noise term ($S_I/I^2) = (\alpha/N_c f)$ in low-frequency regime (see Figure 12.9). In these expressions, f is the frequency, c is the pore capacitance, S_V is the voltage noise power spectral density, α is Hooge's parameter and N_c is the number of charge carriers. In the high-frequency regime, pore capacitance and other parameters like pore resistance, electrode resistance, as well as the dielectric loss parameter for the non-ideal pore capacitance can be measured by analyzing the step response of the current–voltage relationship (Smeets et al., 2008). In the low-frequency regime, the current power spectral density (S_I) is measured at a given averaged open pore current (I) and the number of charge carriers (N_c) can be changed (i.e., by changing the salt concentration of the electrolyte) to obtain an estimate of the ratio of Hooge's parameter to the number of charge carriers. This ratio was found to decrease with the salt concentration as the conductance of the pore increased. Combining the two noise models, the noise behavior of the

solid-state nanopores can be satisfactorily fit for the entire frequency range (see Figure 12.9).

This simple model of nanopore noise is quite intuitive and has been widely used for modeling ion-flow fluctuations through channels. The square dependence of S_I on the system's capacitance at high frequencies dictates that an effective way to reduce noise is to reduce the capacitance. One way to achieve a lower capacitance is to reduce the solution contact area of silicon by "painting" the silicon chip with an insulating material, (Wanunu and Meller, 2008) resulting in a typical reduction of the RMS noise by a factor of 5. Another method for lowering the capacitive noise involves designing a thick dielectric layer (e.g., SiO_2) between the membrane and the Si support during the chip fabrication process. In Figure 12.10, we compare the power spectral density (PSD) plots for 4 nm pores (at +300 mV applied voltage, 21°C, 100 kHz low-pass filter) drilled in $10 \times 10 \mu m^2$ SiN windows fabricated in chips that either contain (gray) or do not contain (black) a 1 μm thick thermal SiO_2 layer between the SiN membrane and the Si wafer. It is clear from the plots that the thermal oxide displays excellent insulating properties, which reduce the high-frequency noise dramatically (>1 kHz). This noise reduction allows measurements at maximum amplifier bandwidths, even for pore/analyte systems in which the signal amplitude is ~100 pA.

While the noise model presented above is an extremely simple approximation, it ignores some common "real" sources of additional noise, which may be surface charge fluctuations in the pore area (Chen et al., 2004), nano-bubble formation (Smeets

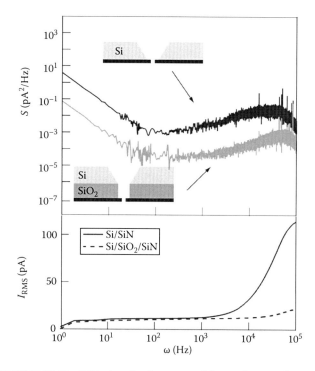

FIGURE 12.10 PSD plots for 4 nm pores fabricated in similar SiN membranes, either with (black) or without (gray) an intermediate, 1 μm thick thermal SiO_2 layer (pictures are not to scale; gray trace shifted down by 100 in *y*-axis for clarity). Bottom: integrated PSD plots, showing the effect of capacitance on the high frequency noise.

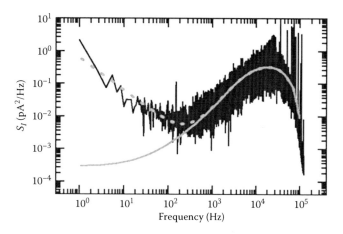

FIGURE 12.9 Current power spectral density (PSD) of a 15.6 nm diameter nanopore (black) and the calculated power spectrum (gray solid line). The red dotted line results from an addition of the measured low-frequency noise to the calculated values. (Reproduced from Smeets, R.M.M. et al., *Proc. Natl. Acad. Sci. USA*, 105, 417, 2008. With permission.)

et al., 2006b), or temporal current fluctuations due to surface contamination. A full description of nanopore noise which takes into account these and other factors is yet to be derived. From a practical standpoint, it is well accepted that cleaned hydrophilic pores are essential to make high quality low-noise measurements with nanopores. Pores that display higher resistance than predicted by their geometrical model (Kim et al., 2006; Smeets et al., 2006a) due to unknown impurity in the pore vicinity (e.g., residual carbon deposited during the e-beam pore fabrication) generally exhibit a higher noise level, particularly evident in the low-frequency regime. Different methods to achieve clean hydrophilic pores have been tested, involving O_2 plasma cleaning (Storm et al., 2005a) and/or piranha cleaning followed by pore wetting using low-viscosity solvents (e.g., methanol) (Wanunu and Meller, 2007, 2008). Piranha cleaning was found to be highly effective for repetitively obtaining similar pore characteristics, enabling a high volume of nanopore measurements to be carried out, including multiple measurements from the same nanopore (Wanunu et al., 2008).

12.3 DNA Translocation Dynamics

While in the previous section we discussed some of the methods used to fabricate nanopores, and nanopore characteristics, in terms of their open pore current, here we focus on the most fundamental process associated with nanopore sensing, namely, the translocation of biopolymers through the pore. Generally speaking, we split biopolymer translocation into two main steps: (a) capture, involving one of the biopolymer ends "finding" the pore and (b) sliding, in which the biopolymer moves through the pore, under the influence of the electrical force. For each successful threading of a biopolymer, there are multiple unsuccessful events, or "collisions" (Meller, 2003; Wanunu et al., 2008). However, to date, many of the details associated with the collision and capture processes have remained obscure. The main difficulty has been the lack of direct current observation of the capture process that typically results in extremely brief and small amplitude events as compared with the complete threading events. The relative size of the nanopore to the biopolymer cross section has been shown to affect the fraction of collisions to full events (Wanunu et al., 2008).

In contrast, much more is known about the sliding process of biopolymers in the nanopore, which is the main focus of this section. We split our discussion in two: (1) The large pores regime ($d > a$), where the electrical driving forces are primarily balanced by frictional drag forces outside the pore, resulting in fast translocation times. In particular, the typical timescales obtained in these experiments are orders of magnitude shorter than the biopolymer's own relaxation time (i.e., Zimm time), and the process is said to be highly in nonequilibrium. (2) The small pores regime ($d \approx a$), where the electrical driving force is primarily balanced by biopolymer interactions with the pore walls and the work required to confine the biopolymer in the nanopore. The resulting dynamics yield much slower translocation timescales that are comparable or longer than the biopolymers'

Zimm time. The physics underlying the biopolymer sliding in these two regimes are drastically different, and will be discussed separately.

The two main experimental parameters measured during the translocation process were shown in Figure 12.1. First, the total duration of the event is t_D, which corresponds to the time required to slide the whole polymer through the pore. Second, the fractional amplitude of the event is defined as the ratio of the blocked current to the open current, i.e., $I_B = i_b/i_O$. This quantity is a measure of the molecular properties, such as the molecular size and charge.

12.3.1 The Balance between Frictional Drag and Interactions

Biological systems at room temperature obey the low Reynolds number limit where inertial forces can be neglected when compared to viscous forces. Thus, the system is over-damped and drag forces are proportional to the velocity with which the biopolymer is pulled through the viscous media. In large pores (where the pore diameter is much larger than the DNA cross section) the interactions of the biopolymer with the pore walls are small, and drag forces outside the pore dominate the sliding process. In contrast, small pores (where d is only slightly larger than the biopolymer's cross section), are more complicated. When biopolymer/pore interactions are primarily limited to the pore itself, these forces dominate the process leading to a linear scaling of the translocation time with biopolymer length. A good example for this behavior is the translocation of ssDNA through the α-HL pore (Meller, 2003). However, this behavior is not universal: when the DNA strongly interacts with the pore as well as the membrane, super-linear dependencies of the translocation time on biopolymer length are observed. This behavior is typical for small solid-state nanopore in SiN membranes. In this case, the hydrodynamic drag of the biopolymer inside or outside the pore can be neglected in comparison with the friction-induced interactions. We split our discussion into these two limits.

12.3.1.1 Large Solid-State Pores

Neglecting pore/polymer interactions, Storm et al. (2005b) have studied the translocation of dsDNA (~2 nm) through ~10 nm solid-state pores. They have considered two types of frictional drag: (1) Drag inside the pore, which scales as $F_P = 2\pi\eta v l_{pore} a/(d - a)$, where η is the fluid viscosity, v is the translocation velocity, l_{pore} and d are, respectively, the pore length and the diameter, and a is the biopolymer's cross section. (2) Drag outside the pore. For example, when the biopolymer coil structure is modified as more and more biopolymer sections are threaded, a frictional drag experienced by the polymer segments outside the pore is also modified. In the simplest case, if the biopolymer can be assumed to form a Gaussian coil with a radius $R_g(t)$ at each step of the sliding process, the shrinking coil will be subject to friction: $F_c = 6\pi\eta R_g \, dR_g/dt$, as shown in Figure 12.11.

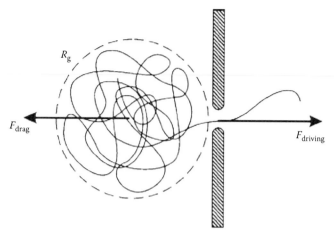

FIGURE 12.11 The balance between the two dominating forces determines the dynamics of the translocation of a DNA molecule through a nanopore: A driving force that locally pulls the DNA through the pore and a viscous drag force that acts on the entire DNA blob. At time t, the DNA on the left side of the pore has a radius of gyration of R_g, indicated by the dashed line. (Reproduced from Storm, A.J. et al., *Nano Lett.*, 5, 1193, 2005b. With permission.)

In this case, since F_c depends on L (through R_g), the translocation time will display a super-linear dependence on the polymer length, namely: $t_T \sim L^{2\nu}$ where $\nu \approx 0.59$ is the Flory exponent for a polymer in 3D. This simplified prediction agreed well with the measured results. We note, however, that the scaling argument leading to these estimates cannot be used to obtain the absolute magnitude of the translocation time, and requires that the process be in equilibrium at any stage. However, the translocation times obtained in the experiments were much shorter than the polymer's equilibrium times, suggesting that the process is not at equilibrium, and implying that more elaborated biopolymer structures may be formed, altering the frictional forces.

12.3.2 Theoretical Studies of DNA–Pore Interactions

While in large nanopores the translocation dynamics of DNA molecules are in the fast regime, i.e., much faster than the Zimm time, in the small nanopore regime (i.e., $a \leq d \leq 2a$) the DNA is more confined, and therefore interacts with the pore walls. A fair number of different modes of DNA translocation have been modeled theoretically. Here, we will only summarize a few of these studies, which explicitly considered polymer–pore interactions. We note how, in this slow-sliding regime, entropic contributions to the biopolymer configurations play a role, especially in experiments using dsDNA.

Focusing only on the nanopore–polymer interactions, Lubensky and Nelson (1999) modeled a voltage-driven translocation as a first-passage problem (with diffusion and drift components) to find the probability distribution $\psi(t)$ for a polymer of length L to exit from a pore, as seen in Figure 12.12. These predictions can be experimentally observed: In Figure 12.12b, we show dwell-time distributions for poly(dC)$_{100}$ (dark) and poly(dA)$_{100}$ (light) translocation through α-HL pore (Meller et al., 2000). While the shape of the poly(dC)$_{100}$ distribution is nearly Gaussian, the bulkier adenine residues in poly(dA)$_{100}$ interact more extensively with α-HL, resulting in a distribution with a broader tail.

To investigate the effect of polymer–pore interactions on the width of the distribution tail, Luo et al. (2007) have used Langevin dynamics to model a 2D polymer translocation through a nanopore. In this paper, the DNA chains were represented as bead-spring chains held together by a Lennard-Jones (LJ) potential, with additional constraints accounting for the polymer's elasticity and for excluded volume interactions. Polymer–pore interactions were modeled by placing a LJ potential between the pore and the beads. In Figure 12.13, we show numerically obtained translocation distributions for

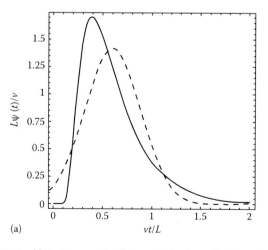

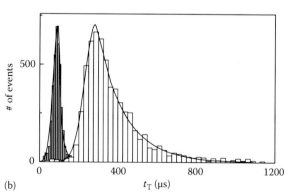

(a)

(b)

FIGURE 12.12 (a) First-passage dwell-time distribution $\psi(t)$ (dimensioned by L/ν) for a polymer of length L (time axis is re-dimensioned by ν/L) translocating through a pore (solid line), and a Gaussian distribution (dashed line) with the same mean and variance for comparison. (Reproduced from Lubensky, D.K. and Nelson, D.R., *Biophys. J.*, 77, 1824, 1999. With permission.) (b) Measured dwell-time distributions for poly(dC)$_{100}$ (dark) and poly(dA)$_{100}$ (light) through α-HL, showing the influence of DNA/pore interactions on the distribution shape. (Reproduced from Meller, A. et al., *Proc. Natl. Acad. Sci. USA*, 97, 1079, 2000. With permission.)

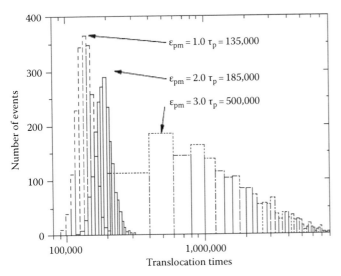

FIGURE 12.13 Simulated dwell-time distributions for a polymer translocating through a pore (solid line), at different values of polymer–pore interactions (ε_{pm}). (Reproduced from Luo, K. et al., *Phys. Rev. Lett.*, 99, 148102, 2007. With permission.)

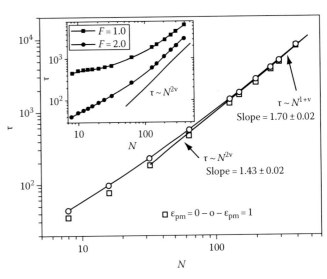

FIGURE 12.14 Dwell-time distributions for a polymer translocating through a pore (solid line), at different values of polymer–pore interactions (ε_{pm}). Inset shows results for $\varepsilon_{pm} = 3$. (Reproduced from Luo, K. et al., *Phys. Rev. Lett.*, 99, 148102, 2007. With permission.)

the same polymer as a function of the pore–monomer interaction parameter ε_{pm}. As seen from the distributions, increasing interactions result in increased peak dwell times (τ_p), but even more strikingly, increase in the breadth of the dwell-time distribution. Increasing pore–polymer interactions reduce the effective driving force of the DNA in the pore, which in turn increases its dwell time. However, the increasing number of possible conformations the polymer can assume, as it resides in the pore for a longer time due to reduced driving force, results in broad distributions. In fact, for $\varepsilon = 3.0$, while the most probable dwell time is $\tau_p = 500,000$, the mean dwell time is much above this value. This timescale can be estimated by fitting the tail of the dwell-time distribution, i.e., $P(t, t > \tau_p)$ to a mono-exponentially decaying function $P(t) = A \exp(-t/\tau)$, where, $\tau_{mean} \approx \tau$.

An interesting problem in the physics of polymer translocation is concerned with how average dwell times of a polymer scale with its length. This problem is particularly important for DNA translocation, as it predicts the spacing (on a time parameter) between different DNA length, analogous to the relationship between DNA length and the distance between observed bands in an electrophoresis gel experiment. While the simple diffusion-drift model of Nelson and Lubensky predict a linear dependence with polymer length (as experimentally determined for ssDNA through α-HL), a number of theoretical works have predicted a nonlinear dependence. Kantor and Kardar (2004) have found that based on the polymer's entropic free energy, forced translocation should exhibit a power law scaling on the polymer length, i.e., $\tau \propto l^\alpha$, where a lower limit for α is 1.5 for a polymer in 3D. Moreover, this exponent is strongly affected by the way the force is applied on the polymer: pulling the polymer from its end yields a power-law dependence with $\alpha \sim 2$, while unforced motion is expected to yield a power law of 2.2. An interesting

finding in these simulations is that this effective power law is somewhat dependent on the polymer length, increasing mildly for longer polymers.

Luo and coworkers have found a similar increase in length dependence for longer polymers. In Figure 12.14, the dimensionless translocation time τ is plotted against the polymer length N under different applied forces and pore–polymer interactions for a polymer in 2D. The scaling predicted for the polymer is related to the polymer's Flory exponent ν, transitioning from a dependence on $N^{2\nu}$ for short polymers to $N^{1+\nu}$ for long polymers. In 3D, the values of these power laws are 1.2 and 1.6, respectively.

Vocks et al. (2008) have numerically modeled the polymer undergoing Rouse dynamics (i.e., in the absence of hydrodynamic interactions), and have found a different power-law dependence. Remarkably, the polymer's stretch/relax motion in the vicinity of the pore influences the length-dependent dynamics, leading to a power law scaling of $(1 + 2\nu)/(1 + \nu)$, which is 1.37 in 3D. This is explained as follows: when the polymer is in the pore, the electric field pulls the polymer chain to generate a tension imbalance. After a monomer leaves the pore, the chain takes a finite time to relax, and therefore, there is a backward tension release that needs to take place in the pore direction. With an increasing polymer length, this effect adds to the slowing down the DNA, resulting in an effective power law scaling.

In summary, several theoretical works have tackled the problem of polymer translocation through small pores. Numerical simulations show that the dwell-time distributions are toward longer timescales and broader when interactions between the polymer and the pore increase (as experimentally seen for ssDNA homopolymers of purines and pyrimidines). In all cases, although the exact power law exponent of dwell times with polymer length is still debated (see Table 12.1), it is quite clear that the length–time relationship is nonlinear. Moreover, since each theoretical work has made some approximations, a comparison

TABLE 12.1 Power Law Scaling of DNA Translocation Dynamics Obtained in Models and Simulations

Author	Exponent	Model Details
Nelson et al.	1	Analytical model
Kantor et al.	1.2	Polymer numerical simulation
Luo et al.	1.2 → 1.59	Fluctuating bond model
Cacciuto et al.	1.2	Hard sphere MC
Milchev et al.	1.65	Bead-spring MC
Dubbeldam et al.	1.5	Scaling arguments, fractional calculus and MC
Strom et al.	1.2	Scaling arguments
Vocks et al.	1.37	Rouse dynamics MC

with detailed experimental studies is crucial for the delineation of the governing factors in DNA translocation.

12.3.3 Translocation Dynamics of dsDNA through Small Pores

To illustrate the transition between the fast translocations to the slow translocation process, Wanunu et al. (2008) have recently compared the DNA translocation properties of two pore sizes, 4 and 8 nm. Figure 12.15 displays an event scatter plot of the blocked current vs. dwell time measured for an 8000 bp dsDNA sample under a 300 mV bias at room temperature, for the two pore sizes, as indicated. As expected, the fraction of the unblocked current is much smaller at the larger pore as compared with the smaller pore (~0.9 vs. ~0.5, respectively). Most striking is a shift of nearly two orders of magnitude in the typical

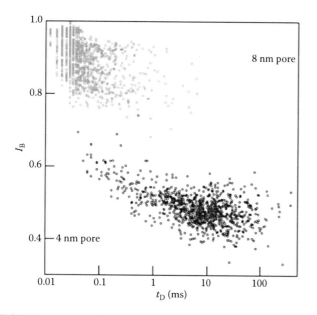

FIGURE 12.15 Semilog I_B vs. t_D scatter plots measured for 8000 bp DNA at the indicated nanopore diameters ($V = 300$ mV, $T = 21.0°C \pm 0.1°C$). Two salient features emerge upon decreasing the pore size: (1), a decrease in I_B (from 0.9 to 0.5); and (2), a drastic increase in t_D of nearly two orders of magnitude. (Reproduced from Wanunu, M. et al., *Biophys. J.*, 95, 4716, 2008. With permission.)

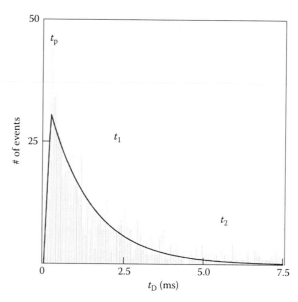

FIGURE 12.16 Dwell-time distribution for translocations of a 6000 bp DNA fragment (for a 4 nm pore at 21°C, 300 mV applied voltage). The solid curve line is an exponential fit to the data. (Reproduced from Wanunu, M. et al., *Biophys. J.*, 95, 4716, 2008. With permission.)

dwell times, from ~0.1 to ~10 ms, in going from the 8 to the 4 nm pores. This result cannot be explained by an increased drag in the nanopore, which could only account for a threefold increase in dwell time when going from the 8 to the 4 nm pore, suggesting that a more elaborated polymer/pore/membrane interaction is taking place.

To characterize the translocation process, a statistical analysis of thousands of DNA translocation events is necessary. A typical dwell-time distribution, measured for a 6000 bp fragment through a 4 nm pore is displayed in Figure 12.16. The shape of the distribution resembles the first-passage time distribution shown in Figure 12.12: the probability of translocation is close to zero at the early times and it sharply increases, reaching a clear peak at =200 µs. Unlike the theoretical model, the experimental distribution exhibits a broad decay at the longer times. A systematic study of different DNA lengths revealed that the distributions clearly follow bi-exponential distributions, with two timescales (t_1 and t_2). The solid line illustrates the typical features of the translocation distribution, with a sharp increase followed by an exponential decay with a timescale $t_1 = 1.4$ ms. Bi-exponential fit is not shown here for clarity, despite the fact that the population is more prevalent for a DNA length greater than 3500 bp (see reference Wanunu et al., 2008 for further discussion of t_2).

The effect of pore size on the dwell-time distributions was further studied at a higher resolution by preparing a set of nanopores in the range of ~2.7–5 nm, and measuring the typical dwell times as explained above. As seen in Figure 12.17, a decrease of over an order of magnitude in dwell time was observed when the pore diameter was increased by a mere ~2.3 nm. This increase is much larger than the expected increase based on the viscous drag alone, illustrating the importance of DNA–pore interactions in

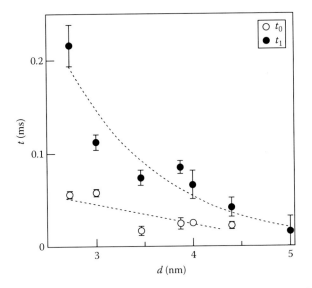

FIGURE 12.17 Plots of the collision timescale (t_0, *open circles*) and the translocation timescale (t_1, *solid circles*) for 400 bp DNA as a function of nanopore diameters (*d*) in the range of 2.7–5 nm. The lines are guides to the eye. (Reproduced from Wanunu, M. et al., *Biophys. J.*, 95, 4716, 2008. With permission.)

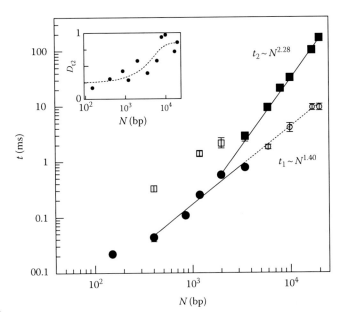

FIGURE 12.18 Log–log plot of DNA translocation timescales as a function of DNA length (*N*) measured using a 4 nm pore. Bi-exponential fits to the dwell-time distributions yield two time scales: t_1 (circles) and t_2 (squares) that follow power laws with $\alpha_1 = 1.40$ and $\alpha_2 = 2.28$, respectively. The inset displays the dependence of the relative weight (the fraction of events in the t_2 population vs. the t_1 population) as a function of DNA length. (Reproduced from Wanunu, M. et al., *Biophys. J.*, 95, 4716, 2008. With permission.)

determining the biopolymer's dynamics. Notably, the collision timescale t_0 is only mildly affected by the pore size.

As explained above, the dwell-time distributions show two distinct timescales (t_1 and t_2), obtained by fitting bi-exponential functions to the distributions. Notably, at the limits of either short or long DNA molecules, the dwell-time distributions are mostly mono-exponential, with t_1 dominating the short molecules, and t_2 the longer molecules. Intermediate DNA lengths require the use of two exponentials. Figure 12.18 displays a summary of the characteristic translocation times of 11 different DNA lengths, from ~150 to 20,000 bp, measured using a 4 nm pore. For a DNA molecule shorter than ~3500 bp the prominent timescale (t_1) follows a power-law dependence on *N*, with a power of 1.40, a striking agreement with the computer simulations displayed in Figure 12.14. This result can be attributed to the DNA–pore interactions. However, at the longer polymer lengths where t_2 dominates, a much stronger power-law dependence emerges: $t_2 \sim N^{2.3}$. While a number of alternative explanations for this stronger length dependence can be made, it has been attributed to the possible, additional interactions of the long DNA coil, with the silicon nitride membrane or with itself.

To better quantify the strength of these interactions, Wanunu and coworkers measured the temperature dependence of the translocation process for different DNA lengths, revealing the strength of binding interactions of the DNA molecules with the pore and/or the membrane. Arrhenius plots for the timescales for different lengths showed binding strengths of $12.0 \pm 0.5 \, k_B T$, while for t_2 timescales, increasing Arrhenius slopes were found for increasing DNA lengths (from $18 \, k_B T$ to $48 \, k_B T$). The increasing Arrhenius slopes correspond to a greater extent of interactions with the pore, suggesting an interaction-based mechanism for the observed slowing down.

While the exact nature of these interactions remains to date a subject of current and future studies, it is well known that the SiN membranes are negatively charged, and that even a small surface charge can result in sufficient DNA adsorption on the pore and membrane surfaces. The coexistence of two striking experimental results, namely, a power-law dependence of the translocation dwell time vs. polymer length, and the strong dependence of this process on pore size, indicates that both interactions, outside as well as inside the pore, should be considered. It is unclear, however, as to which is the predominant one, and if the velocity of the DNA sliding is uniform. Clearly, more experiments and theoretical studies must be carried out before the underlying physics can be unraveled. However, from a practical standpoint, the longer timescales obtained with small pores are desirable, as they improve the temporal resolution of the nanopore technique, as well as the method's sensitivity to small variations in the local cross section of the biomolecule.

12.4 Nanopores for Biomolecular Analyses

The response of a biomolecule to an applied force can reveal its mechanical properties, its interactions with other biomolecules, as well as its self-interactions. Single-molecule techniques, like atomic force microscopy (AFM) and optical/magnetic tweezers, are the most direct methods of exerting and measuring forces on biomolecules, and thus have been applied to a broad variety of biological systems, ranging from nucleic acids to enzymes to motor proteins. A more recent addition to these methods is the nanopore-based force spectroscopy. The nanopore technique utilizes the native electrical charge of a biomolecule to exert force when it is threaded through a nanoscale constriction. Importantly, this method does not require the formation of a physical attachment between the biomolecules and the pore. Thus, a higher throughput can be achieved, as compared with methods that require surface attachments. Additionally, in the nanopore method, the force is not applied on the entire molecule at once. Rather, the pore constriction itself exerts a local shear force only on those parts of the biomolecular complex that form contact with the pore. This permits, at least in principle, the analysis of complex biomolecular structures (Gerland et al., 2004).

The ability to apply force on individually charged biopolymers opens up a wide range of possibilities for monitoring molecular properties at high precision. We first describe here, studies that concern the magnitude of force applied on individual DNA molecules inside a solid-state nanopore. These studies directly reveal the effective charge of the DNA inside the nanopore per unit length. Unzipping of dsDNA can be achieved by pulling a single-stranded DNA overhang (connected to a dsDNA stem) through a sub-2 nm pore, smaller than the dsDNA diameter. We discuss the use of solid-state nanopores to unzip DNA hairpin structures and measure their kinetics. Finally, we review Nanopore Force Spectroscopy (NFS), a method used to apply time-varying forces on biomolecular complexes and to probe their response. As an example of this method, we consider the interactions of an enzyme (Exonuclease I) with an ssDNA molecule in a α-HL nanopore.

12.4.1 Estimation of Static Forces in Nanopores

In the previous section, we discussed the forces affecting the translocation *dynamics* of biopolymers through nanopores. We noted that the electrical driving force is balanced by the drag exerted on the biopolymer coil, as well as the interactions with the pore and the membrane (Figure 12.19). The translocation process is extremely fast, and under most conditions the biopolymer does not relax to its equilibrium state, and the process is thus said to be out of thermal equilibrium. In this section, we are interested in another type of measurement, in which the biopolymer is held *statically* in the pore, by connecting one of its ends to a soft spring, while the electrical field pulls at its other end. The advantage of these measurements is that they can be used to directly quantify the forces exerted on the DNA at rest and, in particular, to evaluate the effective charge of DNA threaded in the pore.

The electrical force exerted on a charged biopolymer residing in the pore (specifically a DNA molecule) is given by $F_{el} = (q_{eff}/a)\Delta V$, where ΔV is the voltage applied across the membrane, and q_{eff}/a is the effective charge per unit length of the DNA molecule, and a is the basepair height. Considering the bare DNA charge of 1 electron charge per base ($q = 2\,e^-$/bp with $a = 0.34$ nm), a 100 mV potential across the pore would result in a force of ~100 pN on a dsDNA molecule. This estimation ignores the charge condensation that reduces the "effective" charge to

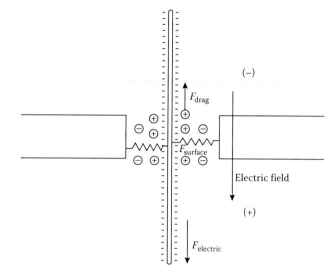

FIGURE 12.19 Schematic of dominant forces experienced by DNA inside a nanopore. Dynamics of the DNA strand inside the pore is controlled by a complex interplay of these forces.

a fraction of the bare charge, thus resulting in a much smaller force value. Bulk measurements using gel electrophoresis provide a wide range of values of q_{eff} from 0.12 e^-/bp (Smith and Bendich, 1990; Gurrieri et al., 1999) to 1.0 e^-/bp (Schellman, 1977). Moreover, bulk estimations may not be applicable to the nanopore geometry, as the effective charge in small nanopores could be further reduced due to an increased ion condensation resulting from the electrostatic properties of the pore (Hu and Shklovskii, 2008). It is therefore essential to directly probe the effective charge of biopolymers in nanopores.

Recently, direct measurements of the electrical force exerted on a dsDNA molecule threaded through a solid-state nanopore was performed by coupling an optical tweezers apparatus to a nanopore setup (Keyser et al., 2006; Trepagnier et al., 2007). Polystyrene beads, attached to one end of λ-DNA, trapped in a force-calibrated optical tweezers were brought close to the membrane until the free end of the DNA entered the pore due to the applied voltage across the pore. The electrical force (F_{el}) pulling the DNA to the other side of the membrane was countered by the force applied on the trapped bead. When the forces are balanced, the DNA motion is frozen and the optical tweezer is used to quantify the forces applied on the stretched DNA, at rest.

Keyser and coworkers measured the stall force using relatively large nanopores (~6–11 nm), at ionic strengths ranging from 0.02 to 1 M KCl to be 0.24 pN mV^{-1} (e.g., 24 pN at 100 mV). The static force applied on the DNA was found to be independent of pore size (in the range of 6–11 nm) and on the buffer salt conditions (0.02–1 M). It was thus concluded, that under the conditions used, q_{eff} was independent of pore size and ionic strength, with a value of ~0.5 e^-/bp—remarkably close to Manning's counterion condensation around DNA in bulk (Manning, 1978).

A refined picture of the static forces applied on DNA in nanopores was recently developed by Luan et al. who employed

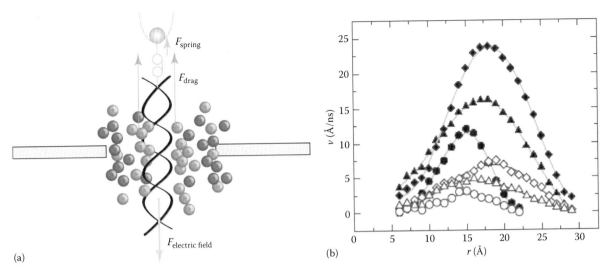

FIGURE 12.20 MD simulation of DNA in a nanopore under applied electric field. (a) A short DNA duplex is held stationary in the nanopore by "connecting" one of its ends to a soft spring, while its other end is pulled by the electric field. Additionally, drag force (due to electroosmotic flow) is present. Ion distribution of K^+ (filled circles) and the Cl^- ions (open circles) around the DNA is schematically shown. (b) Water flow profile in three types of nanopores as a function of distance from the center of DNA. Circles: 2.25 nm radius pore, triangles: 3 nm radius pore with rough surface, and rhombus: 3 nm radius pore with smooth walls. Open and close symbols are simulations done at external electric field values of 0.125 and 0.5 V/6.4 nm, respectively. (Reproduced from Luan, B. and Aksimentiev, A., *Phys. Rev. E Stat. Nonlin. Soft Matter Phys.*, 78, 021912, 2008. With permission.)

molecular dynamics (MD) simulations to study this problem (Luan and Aksimentiev, 2008). The main features of the MD studies are schematically shown in Figure 12.20a. In bulk electrolyte solution, as well as in nanopores, motion and forces of DNA are modeled by considering 20 bp DNA [poly(dA)$_{20}$-poly(dT)$_{20}$] in 0.1 M KCl solution under an applied electric field. The electrical force applied on the DNA was balanced by connecting the DNA to a weak harmonic spring with a given spring constant, simulating the optical tweezers. Bead displacement was used to estimate the force, revealing a linear dependence of the force with the applied electric field, and proportionality constant of 0.25 e$^-$.

Interestingly, upon further investigation Luan and coworkers found that although the counterion distribution around DNA is independent of the solution viscosity, the stall force itself is viscosity-dependent, in an apparent contradiction to the classical Manning theory. It was noted that the redistribution of ions around the negatively charged DNA surface results in an unbalanced total electrolyte charge, which is driven by the electric field, generating an electroosmotic flow of water near the DNA. The additional drag applied on the DNA due to this flow is directed in an opposite direction to the electrical field, thus reducing the effective force sensed by the tweezers when the DNA is at rest. In other words, there are two contributions to the force acting on the stalled DNA: (1) the electrophoretic force, which can be related to the DNA's electrophoretic mobility, under the applied electric field E ($v = \mu E$), and (2) a hydrodynamic force due to the electroosmotic flow ($F_{drag} = \xi v$, where ξ is Stokes' coefficient of DNA moving with velocity v in solution). The ratio of the simulated stall force to $\xi\mu E$ was found to be 1, regardless of the solution viscosity or the value of E.

Moreover, MD simulations using different nanopore diameters and a corrugated surface revealed that while the ion distribution around the DNA depends on nanopore size, the stall force barely does. At the same time, the stall force was found to be much higher in nanopores with rough surfaces. Thus, the frictional force between the electrolytes flow and the pore surface affects the electro-osmotic flow, and thereby the hydrodynamic drag applied to the DNA (Figure 12.20b). Overall, the stall force could be described by: $F_{stall} = \mu\xi E$, (or $F = q_{eff}E$ where $q_{eff} = \xi\mu$) satisfying both the linear electric field dependence on the force as well as its dependence on viscosity. Since both ξ and μ are directly related to the biopolymer dynamics, they can be accurately measured using an active control technique (Bates et al., 2003), circumventing the need for the more complicated optical tweezers based method. Yet, the combination of the optical tweezers and the nanopore can be used to perform a controlled "flossing" of a single molecule of dsDNA repeatedly through a nanopore (Trepagnier et al., 2007), and can potentially be applied to probe the spatial location of proteins bound on DNA, as well as be used as a force probe to unbind biological complexes.

12.4.2 DNA Unzipping

The unzipping of double-stranded nucleic acids is a fundamental process in molecular biology. It is involved in a wide range of cellular processes, such as: DNA replication, DNA transcription, translation initiation, and RNA interference. The forces and the timescales associated with the breakage of the bonds stabilizing the secondary and the tertiary structures of nucleic acids can now be studied at the single molecule level, revealing information masked by ensemble averaging, including short-lived intermediate states and multi-step kinetic processes. In particular, nanopores can be used to directly apply and measure unzipping forces on individual DNA and RNA molecules, eliminating the need for molecular linkers, surface immobilization of the molecules, and the global application of force.

The protein pore α-HL is ideal for nucleic acid unzipping studies. Its heptameric structure, composed of cap and stem portions, has been shown to be highly stable even under high temperatures (Jung et al., 2006). The cap portion of α-HL, which is usually assembled on the "*cis*" side of the membrane, contains a wide vestibule-like mouth, which can accommodate double-stranded nucleic acids. The stem portion, which spans the phospholipid membrane, is a nearly cylindrical water-filled channel with an inner diameter ranging from 1.4 to 1.8 nm. It, therefore, geometrically permits the passage of single-stranded DNA or RNA molecules, but blocks the translocation of double-stranded nucleic acids.

Sauer-Budge et al. (2003) have demonstrated that a DNA duplex molecule (composed of a 100-mer oligo hybridized to a complementary 50-mer oligo, such that a 50-mer 3′ single-stranded overhang is formed), must be unzipped when the 3′ overhang is threaded through the pore. Quantitative PCR analysis showed that while both the 50-mer and 100-mer oligos were present in the *cis* chamber, only the 100-mer oligo was found in the *trans* chamber, after the detection of hundreds of nanopore blockade events. The most plausible explanation for this observation was that unzipping had occurred at the pore, leaving all 50-mer oligos in the *cis* chamber while the 100-mer oligos passed through the pore to the *trans* side. Furthermore, the distribution of nanopore blockade durations (or dwell times) displayed a characteristic mean time of ~435 ms, orders

of magnitude longer than the timescale associated with the translocation of single-stranded DNAs of comparable length. Introducing a 6-base mismatch in the duplex region of the hybridized sample resulted in a shortening of the characteristic timescale by more than a factor of two.

While these static-voltage experiments provided evidence that the α-HL pore can be used to unzip DNA duplexes, these measurements were limited to a narrow range of voltages (120–180 mV), making it difficult to precisely determine the kinetic properties of the process. Mathé and coworkers extended the DNA unzipping studies to hairpin molecules, employing Dynamic Voltage Control (DVC) (Mathe et al., 2004, 2006). They systematically probed the distributions of the unzipping times of three DNA hairpin molecules composed of duplex regions (10, 9, or 7 bp) and a 3′ poly(dA)50 ssDNA overhang, over a wide range of both voltage (30–150 mV) and temperatures. Figure 12.21 displays an example of a typical unzipping event using DVC, and the corresponding distribution of ~1500 unzipping events used to determine the characteristic unzipping time, t_U. These measurements revealed that a single-base mismatch (in the 10 bp hairpin) strongly shifts the unzipping timescale toward shorter times and can therefore be easily detected. They demonstrated that single-nucleotide mutations can be readily detected in both unlabeled and unmodified DNA hairpins and in DNA hybrids.

Although biological nanopores have been model systems for studying DNA unzipping kinetics, the fragility of the lipid membrane and the diffusion of channels in the membrane are major drawbacks toward prospective biotechnological applications. However, as discussed in Section 12.2, progress in nanopore fabrication methodology has allowed researchers to

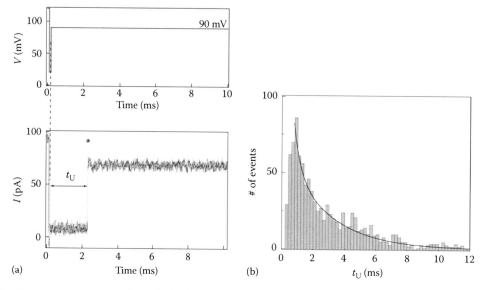

FIGURE 12.21 Bond rupture using a step in the applied voltage (or the force) when a molecule is threaded through the pore. (a) Top panel represents the applied voltage wave in each and every event. An abrupt drop in the measured ion current (lower panel) triggers the voltage wave generation. After the molecule is threaded inside the pore, the voltage is raised to a fixed level (90 mV in this case) and the current is monitored until an upward transition, signaling the bond rupture, is observed. The delay time from the moment that the step is applied to the rupture time (marked with a star) is the unzipping time, t_U. (b) A distribution of t_U measured for >1000 events as shown on the left. The distribution display a clear peak at ~1 ms and a decaying exponential tail with average time ~2.7 ms.

fabricate nanopores at the 1–2 nm scale, suitable for DNA unzipping studies, at an extreme pH, temperature and voltage conditions. Here, we discuss the recent progress in this area, which utilizes sub-2 nm solid-state nanopores fabricated in thin SiN membranes.

Initial attempts at understanding the unzipping of dsDNA through small solid-state nanopores utilized the quantitative PCR of the sample solution, rather than the direct observation of individual unzipping events (Zhao et al., 2008). Working with very small pores (pore size ~1–2.3 nm) and high bias voltages (0.5–4 V), current transients associated with DNA interaction with the pore were observed. However, the molecular details in these measurements were masked due to high noise and low bandwidth. Therefore, it was not possible to assign specific dynamics features corresponding to the process of DNA translocation or unzipping to the measured large current transients. The PCR data, nevertheless, showed that a voltage threshold is required for translocation of the hairpin molecules through the pore, which varied as a function of pore diameter and the DNA duplex structure of the DNA. Because a lower threshold was obtained for smaller pores (1–1.5 nm) as compared with the larger pores (1.5–2.3 nm) it was concluded that the DNA unzipped in the smaller pore case was stretched and translocated (without unzipping) in the latter case.

An unequivocal experimental proof of DNA hybrids and DNA hairpin unzipping was more recently provided by Mcnally et al. (2008). High-bandwidth nanopore experiments (~75 kHz) allowed a sufficient resolution to distinguish between ssDNA molecules (dA$_{120}$), blunt end dsDNA molecule (24 bp), and hybrid DNA (HP50) of 50 bp dsDNA with a flanking poly(dA)$_{50}$, interacting with a sub-2 nm solid-state nanopore. These three molecules yielded well-distinguished signals (Figure 12.22), which clearly illustrated, for the first time, the ability to resolve individual DNA unzipping events using a solid-state pore. The introduction of a two-base mismatch in the HP50 hybrid resulted in a clear shift in the characteristic dwell time to shorter times, as expected if the molecules are unzipped. This experiment also rejected the possibility of hybrid DNA translocating through the pore without unzipping or escaping from the pore against the voltage gradient.

Mcnally and coworkers also studied the temperature dependence of the unzipping process in solid-state pores. Using temperatures in the range of 0°C–21°C and a variety of different hairpin duplex lengths (N = 5, 10, 16, and 24), the dependence of the unzipping timescale on temperature was measured. At first approximation, it was expected that the unzipping time would follow the first-order Arrhenius kinetics: $t_U = A\exp((\Delta G - qV)/k_B T)$, which is an activation energy barrier, and represents the reduction in the energy due to the electrical voltage and DNA charge. For DNA hairpins longer than 10 bp, it was found that to a good approximation t_U depends exponentially on T^{-1}, in accordance with the above equation. For the shorter duplexes (5 or 10 bp), a weaker dependence on temperature was found, which is expected for smaller energy barriers, if the applied voltage is sufficient to decrease the barrier below ~$k_B T$.

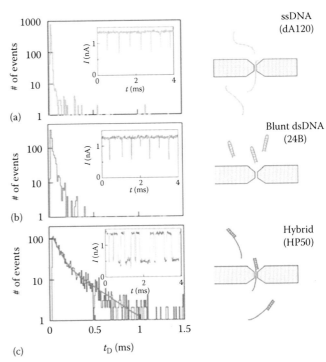

FIGURE 12.22 Dwell-time distributions for three DNA samples through 2 nm pores. (a) An ssDNA molecule consisting of 120-mer poly deoxyadenine molecule. (b) Blunt dsDNA, consisting of a 24 bp duplex region and a 6-base loop. (c) A hybrid single-strand/duplex molecule, consisting of a 100-mer strand hybridized to a fully complementary 50-mer at one end. While the dwell times for ssDNA and blunt dsDNA are short (<21 µs), noticeably longer dwell times are observed for the hybrid molecule. The complete dwell-time distribution for the hybrid molecule can be modeled as a biexponential function with decay timescales $t_0 \sim 40$ µs and $t_1 \sim 240$ µs, corresponding to a collision and translocation/unzipping process, respectively. Typical events for the three samples are shown as insets to the dwell-time distributions. (Reproduced from McNally, B. et al., *Nano Lett.*, 8, 3418, 2008. With permission.)

In addition, Mcnally and coworkers showed that a nanopore can distinguish between two 24 bp DNA hairpins, where 2 bp mismatches were introduced either at the center or at the end of the duplex. The end-mismatch duplex yielded similar unzipping kinetics as a 22 bp DNA. In contrast, the center mismatch, each equivalent to the ~11 bases case, displayed a much faster kinetics, in accordance with the temperature studies of short duplexes. In summary, these results showed (1) Arrhenius kinetics of DNA unzipping as measured by nanopore based experiments, and (2) introduction of a two-base mismatch altered the binding energy landscape of the duplex in a position-sensitive manner.

These studies unambiguously demonstrate that solid-state nanopores can be used to unzip short DNA duplexes and that the unzipping timescale is the rate-limiting step for DNA translocation through a sub-2 nm pore, in an analogous way to the biological pores studied earlier. Due to differences in the local charge and the DNA–pore surface interactions between the biological and the solid-state nanopores, the forces and the dwell times of unzipping are slightly different but the underlying physical description of the unzipping process was proven to be identical. With advancement in the nanopore fabrication technology, a more consistent picture of the underlying processes will emerge. These results represent an important milestone towards the development of an enzyme-free, ultra-fast DNA sequencing method, involving simultaneous sequential DNA unzipping with an optical readout of the fluorescent probes (Soni and Meller, 2007) as well as a high throughput sequence variability detection using nanopores.

12.4.3 Nanopore Force Spectroscopy

An important example of exploiting the merits of nanopores as a single molecule tool is the force spectroscopy measurements. Similar to conventional dynamic force spectroscopy (Evans, 2001), Nanopore Force Spectroscopy (NFS) probes the energy landscape of a bound two species system along the force-induced unbinding pathway, revealing single molecule biochemical information on molecular dissociation. This method has been successfully used in a variety of biophysical problems of DNA unzipping (Mathe et al., 2004, 2006) and DNA–protein interaction measurements (Hornblower et al., 2007). To obtain meaningful statistics with a single molecule approach, a large number of measurements needed to be performed which is challenging for conventional force probe methods (like optical tweezers and AFM) as individual molecules are addressed and measured one at a time, a slow and tedious process. NFS on the other hand allows a rapid determination of association and dissociation rate measurements without surface immobilization or biomolecule modifications.

Hornblower and coworkers have recently presented a general method to apply local electrical force to individual ssDNA–protein complexes (Hornblower et al., 2007). They illustrated the approach by studying the binding kinetics of an enzyme, Exonuclease I, to ssDNA. Exonuclease I binds to 3′ end of ssDNA and catalyzes its cleavage in the presence of Mg^{2+}. Catalysis can be prevented if the 3′ is not phosphorylated or in the absence of Mg^{2+}, forming stable complexes. Due to the large size of the enzyme, the complex cannot translocate through the pore, but rather shifts the characteristic dwell time of the lone oligonucleotide to much longer timescales when the enzyme dissociates from the oligo under the influence of the electrical force exerted on the oligo. Counting the number of events with a longer event duration (DNA–protein bound system), compared with the number of shorter events (free DNA) at the experimental concentration of ssDNA and exoI protein, provides a direct means to estimate their equilibrium binding constant (K_D). The dissociation and the association constant of the system could be measured, self consistently, by applying NFS to the system. In these measurements, the applied voltage was ramped automatically upon the capture of the complexes in the nanopore, at a constant rate, until the rupture of the ssDNA–Exo I complexes was observed (stars in Figure 12.23a). The distributions of the voltages associated with the rupture voltages were constructed from hundreds of rupturing events (Figure 12.23b), and the experiment was repeated for different voltage ramps. The end result is a curve describing the critical rupture voltage (the peaks of the distributions, as shown in Figure 12.23b) as a function of the voltage ramp (Figure 12.23c). The NFS curve clearly displays two regimes: (1) At very small loading rates (or ramps) the critical voltage does not depend on the ramp and can be approximated by $V_C \sim V_\beta (\log(K_D^{-1}))^{-1}$ with $V_\beta = k_B T/q_{eff}$. (2) At the large ramps

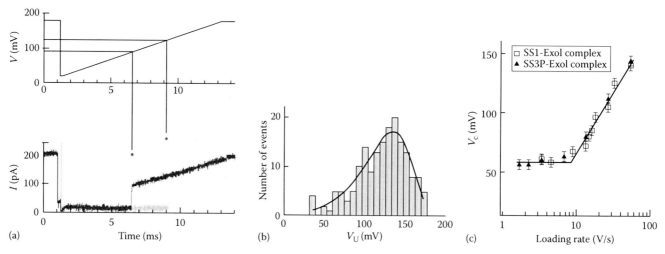

FIGURE 12.23 NFS of DNA–protein complexes. (a) Current and voltage profiles during two typical events in a nanopore voltage-ramp force spectroscopy experiment. Plotted are the time-dependent voltages applied to each individual complex (top), and the measured pore current (bottom). Dissociation of the complexes results in an abrupt rise of the current (asterisks) to the open-pore level. Dashed lines are shown to indicate the voltage where dissociation occurred for each of the two events (VU). (b) A typical distribution of the dissociation voltage, measured at a loading rate of 30 V s⁻¹. (c) Plots of log loading rate vs. VC for SS3P-ExoI and SS1-ExoI complexes. The VC for each loading rate was determined to be the maximum of the histogram of at least 500 dissociation events, such as those shown in Figure 12.3b. Error bars represent the width of the bins in each histogram. (Reproduced from Hornblower, B. et al., *Nat. Methods* 4, 315, 2007. With permission.)

limit it is predicted that the critical voltage will grow with the logarithm of the ramp value: $V_C \sim V_\beta \log(r/V_\beta k_d)$ where r is the ramp value. This curve can thus be used to determine the kinetic binding/unbinding constants of the ssDNA–ExoI interaction, and in general any ssDNA–protein system. Hornblower and coworkers obtained $k_d = 37 \pm 6\,\mathrm{s}^{-1}$ and $k_a = 10^8\,\mathrm{s}^{-1}\,\mathrm{M}^{-1}$, consistent with the previous bulk studies for DNA–ExoI interaction. Thus, the two-regime curve for the V_C vs. log rate obtained by NFS, revealed the kinetics of DNA–ExoI binding that would otherwise be difficult to detect, and that would usually require labeling of the compounds with fluorescent tags (e.g., for anisotropy measurements) demonstrating the power of this method for the analysis of nucleic acid and protein interactions. In these experiments, Hornblower and coworkers show a general method that can be applied to other DNA–protein complexes using protein or solid-state nanopores.

12.5 Conclusions and Summary

In this chapter, we discussed the concept of electrical resistive sensing in nanopores, and described some of the fundamental physical principles relevant to nanopore detection, as well as some of the outstanding challenges in this field. After surveying a number of nanopore types, in Section 12.2, we discussed the factors governing noise in nanopores, since they dictate the temporal resolution of the method and our ability to discriminate small differences in analyte size or charge. Understanding noise factors in solid-state nanopore is an ongoing effort, which directly impacts on virtually any experiment. It ties into the development of nanopore coating methodologies, and novel fabrication techniques, which are active areas of research. In Section 12.3, we discussed the basic phenomena of driven biopolymer translocation through nanopores. Clearly a combination of entropic forces and interactions with the pore walls and the membrane must be considered, balancing the external electrical force, driving the process, and leading to the non-linear dynamics observed in experiments. Progress in this area is rapid, and more experiments as well as theoretical models and computer simulation, help to uncover the major factors associated with this process. In Section 12.4, we considered the force balance acting on a "stalled" biopolymer (dsDNA) in the nanopore. These experiments allow researchers to better estimate the effective charge of the confined DNA bases in the nanopore. Finally, in Section 12.5, we provided a few example studies where the nanopore was employed to investigate a biomolecular problem, such as DNA unzipping kinetics or DNA/enzyme interactions. This is a small sample of a growing area of single-molecule studies of biophysical systems using nanopores.

The particular features of the nanopore method, i.e., its ability to probe label-free, individual biopolymers, opens up vast possibilities in genomics, proteomics, drug discovery, forensics, as well as biophysics, molecular biology, and other areas. Here, we have focused our discussion in order to highlight the basic physical concepts that underlie nanopore detection. Although more research must be performed in order to clearly quantify and understand the factors leading to the observed phenomena, such as DNA dynamics, it is worthwhile indicating that a decade of extensive research in this field has exposed the main physical rules that must be applied to elucidate the experimental observations. Undoubtedly, in the next few years, researchers will build on this knowledge to realize more sophisticated and versatile nanopore sensors, which display much higher signal/noise characteristics, and a superior temporal and a spatial resolution. This field will continue to be fueled by practical applications—the development of ultrafast and low cost methods for molecular diagnostics, and specifically DNA sequencing and genotyping, due to some of the unique features of the method, as highlighted in Section 12.1. Some of the most exciting developments will involve the articulation of the pores with transverse electrical probes or other modalities, such as optical detection, tailoring the nanopore sensors to future biomolecular detection tasks, with unprecedented accuracy, speed, and low cost.

References

Akeson, M., Branton, D., Kasianowicz, J. J., Brandin, E., and Deamer, D. W. (1999) Microsecond time-scale discrimination among polycytidylic acid, polyadenylic acid, and polyuridylic acid as homopolymers or as segments within single RNA molecules. *Biophys J*, 77, 3227–3233.

Ali, M., Schiedt, B., Healy, K., Neumann, R., and Ensinger, A. (2008) Modifying the surface charge of single track-etched conical nanopores in polyimide. *Nanotechnology*, 19, 85713.

Astier, Y., Braha, O., and Bayley, H. (2006) Toward single molecule DNA sequencing: Direct identification of ribonucleoside and deoxyribonucleoside 5 monophosphates by using an engineered protein nanopore equipped with a molecular adapter. *J Am Chem Soc*, 128, 1705–1710.

Bates, M., Burns, M., and Meller, A. (2003) Dynamics of DNA molecules in a membrane channel probed by active control techniques. *Biophys J*, 84, 2366–2372.

Braha, O., Gu, L. Q., Zhou, L. et al. (2000) Simultaneous stochastic sensing of divalent metal ions. *Nat Biotechnol*, 18, 1005–1007.

Braha, O., Webb, J., Gu, L. Q., Kim, K., and Bayley, H. (2005) Carriers versus adapters in stochastic sensing. *Chemphyschem*, 6, 889–892.

Branton, D., Deamer, D. W., Marziali, A. et al. (2008) The potential and challenges of nanopore sequencing. *Nat Biotechnol*, 26, 1146–1153.

Cheley, S., Gu, L. Q., and Bayley, H. (2002) Stochastic sensing of nanomolar inositol 1,4,5-trisphosphate with an engineered pore. *Chem Biol*, 9, 829–838.

Chen, P., Mitsui, T., Farmer, D. B. et al. (2004) Atomic layer deposition to fine-tune the surface properties and diameters of fabricated nanopores. *Nano Lett*, 4, 1333–1337.

Coulter, W. H. (1953). U.S. Patent 2,656,508.

Dekker, C. (2007) Solid-state nanopores. *Nat Nanotechnol*, 2, 209–215.

Evans, E. (2001) Probing the relation between force—Lifetime—And chemistry in single molecular bonds. *Annu Rev Biophys Biomol Struct*, 30, 105–128.

Fleischer, R. L., Alter, H. W., Furman, S. C., Price, P. B., and Walker, R. M. (1972) Particle track etching. *Science*, 178, 255–263.

Fologea, D., Ledden, B., Mcnabb, D. S., and Li, J. L. (2007) Electrical characterization of protein molecules by a solid-state nanopore. *Appl Phys Lett*, 91.

Gerland, U., Bundschuh, R., and Hwa, T. (2004) Translocation of structured polynucleotides through nanopores. *Phys Biol*, 1, 19–26.

Goldberger, J., Fan, R., and Yang, P. (2006) Inorganic nanotubes: A novel platform for nanofluidics. *Acc Chem Res*, 39, 239–248.

Gu, L. Q., Braha, O., Conlan, S., Cheley, S., and Bayley, H. (1999) Stochastic sensing of organic analytes by a pore-forming protein containing a molecular adapter. *Nature*, 398, 686–690.

Guan, X., Gu, L. Q., Cheley, S., Braha, O., and Bayley, H. (2005) Stochastic sensing of TNT with a genetically engineered pore. *Chembiochem*, 6, 1875–1881.

Gurrieri, S., Smith, S. B., and Bustamante, C. (1999) Trapping of megabase-sized DNA molecules during agarose gel electrophoresis. *Proc Natl Acad Sci USA*, 96, 453–458.

Gyurcsanyi, R. E. (2008) Chemically-modified nanopores for sensing. *TrAc Trends Anal Chem*, 27, 627–639.

Han, A., Creus, M., Schurmann, G. et al. (2008) Label-free detection of single protein molecules and protein-protein interactions using synthetic nanopores. *Anal Chem*, 80, 4651–4658.

Han, A. P., Schurmann, G., Mondin, G. et al. (2006) Sensing protein molecules using nanofabricated pores. *Appl Phys Lett*, 88, 093901.

Harrell, C. C., Siwy, Z. S., and Martin, C. R. (2006) Conical nanopore membranes: Controlling the nanopore shape. *Small*, 2, 194–198.

Healy, K., Schiedt, B., and Morrison, A. P. (2007) Solid-state nanopore technologies for nanopore-based DNA analysis. *Nanomedicine*, 2, 875–897.

Heins, E. A., Baker, L. A., Siwy, Z. S., Mota, M. O., and Martin, C. R. (2005a) Effect of crown ether on ion currents through synthetic membranes containing a single conically shaped nanopore. *J Phys Chem B*, 109, 18400–18407.

Heins, E. A., Siwy, Z. S., Baker, L. A., and Martin, C. R. (2005b) Detecting single porphyrin molecules in a conically shaped synthetic nanopore. *Nano Lett*, 5, 1824–1829.

Ho, C., Qiao, R., Heng, J. B. et al. (2005) Electrolytic transport through a synthetic nanometer-diameter pore. *Proc Natl Acad Sci USA*, 102, 10445–10450.

Hooge, F. (1969) 1/f noise is no surface effect. *Phys Lett A*, 29, 139.

Hornblower, B., Coombs, A., Whitaker, R. D. et al. (2007) Single-molecule analysis of DNA-protein complexes using nanopores. *Nat Methods*, 4, 315–317.

Hu, T. and Shklovskii, B. I. (2008) Theory of DNA translocation through narrow ion channels and nanopores with charged walls. *Phys Rev E Stat Nonlin Soft Matter Phys*, 78, 032901.

Ito, T., Sun, L., and Crooks, R. M. (2003a) Observation of DNA transport through a single carbon nanotube channel using fluorescence microscopy. *Chem Commun (Camb)*, 7, 1482–1483.

Ito, T., Sun, L., and Crooks, R. M. (2003b) Simultaneous determination of the size and surface charge of individual nanoparticles using a carbon nanotube-based Coulter counter. *Anal Chem*, 75, 2399–2406.

Ito, T., Lun, S., Henriquez, R. R., and Crooks, R. M. (2004) A carbon nanotube-based Coulter nanoparticle counter. *Acc Chem Res*, 37, 937–945.

Jung, Y., Bayley, H., and Movileanu, L. (2006) Temperature-responsive protein pores. *J Am Chem Soc*, 128, 15332–15340.

Kang, X. F., Cheley, S., Guan, X., and Bayley, H. (2006) Stochastic detection of enantiomers. *J Am Chem Soc*, 128, 10684–10685.

Kantor, Y. and Kardar, M. (2004) Anomalous dynamics of forced translocation. *Phys Rev E*, 69, 21114.

Karhanek, M., Kemp, J. T., Pourmand, N., Davis, R. W., and Webb, C. D. (2005) Single DNA molecule detection using nanopipettes and nanoparticles. *Nano Lett*, 5, 403–407.

Kasianowicz, J. J., Brandin, E., Branton, D., and Deamer, D. W. (1996) Characterization of individual polynucleotide molecules using a membrane channel. *Proc Natl Acad Sci USA*, 93, 13770–13773.

Kasianowicz, J. J., Burden, D. L., Han, L. C., Cheley, S., and Bayley, H. (1999) Genetically engineered metal ion binding sites on the outside of a Channel's transmembrane beta-barrel. *Biophys J*, 76, 837–845.

Keyser, U. F., Koeleman, B. N., Dorp, S. V. et al. (2006) Direct force measurements on DNA in a solid-state nanopore. *Nat Phys*, 2, 473–477.

Keyser, U. F., Krapf, D., Koeleman, B. N. et al. (2005) Nanopore tomography of a laser focus. *Nano Lett*, 5, 2253–2256.

Kim, M. J., Mcnally, B., Murata, K., and Meller, A. (2007a) Characteristics of solid-state nanometre pores fabricated using a transmission electron microscope. *Nanotechnology*, 18, 205302.

Kim, Y. R., Min, J., Lee, I. H. et al. (2007b) Nanopore sensor for fast label-free detection of short double-stranded DNAs. *Biosens Bioelectron*, 22, 2926–2931.

Kim, M. J., Wanunu, M., Bell, D. C., and Meller, A. (2006) Rapid fabrication of uniformly sized nanopores and nanopore arrays for parallel DNA analysis. *Adv Mater*, 18, 3149–3153.

Li, J., Stein, D., Mcmullan, C. et al. (2001) Ion-beam sculpting at nanometre length scales. *Nature*, 412, 166–169.

Lo, C., Thomas, A., and Alexey, B. (2006) Fabrication of symmetric sub-5 nm nanopores using focused ion and electron beams. *Nanotechnology*, 17, 3264.

Luan, B. and Aksimentiev, A. (2008) Electro-osmotic screening of the DNA charge in a nanopore. *Phys Rev E Stat Nonlin Soft Matter Phys*, 78, 021912.

Lubensky, D. K. and Nelson, D. R. (1999) Driven polymer translocation through a narrow pore. *Biophys J*, 77, 1824–1838.

Luo, K., Ala-Nissila, T., Ying, S. C., and Bhattacharya, A. (2007) Influence of polymer-pore interactions on translocation. *Phys Rev Lett*, 99, 148102.

Manning, G. S. (1978) The molecular theory of polyelectrolyte solutions with applications to the electrostatic properties of polynucleotides. *Q Rev Biophys*, 11, 179–246.

Mathe, J., Arinstein, A., Rabin, Y., and Meller, A. (2006) Equilibrium and irreversible unzipping of DNA in a nanopore. *Europhys Lett*, 73, 128–134.

Mathe, J., Visram, H., Viasnoff, V., Rabin, Y., and Meller, A. (2004) Nanopore unzipping of individual DNA hairpin molecules. *Biophys J*, 87, 3205–3212.

Mcnally, B., Wanunu, M., and Meller, A. (2008) Electromechanical unzipping of individual DNA molecules using synthetic sub-2 nm pores. *Nano Lett*, 8, 3418–3422.

Meller, A. (2003) Dynamics of polynucleotide transport through nanometre-scale pores. *J Phys Condens Matter*, 15, R581–R607.

Meller, A., Nivon, L., Brandin, E., Golovchenko, J., and Branton, D. (2000) Rapid nanopore discrimination between single polynucleotide molecules. *Proc Natl Acad Sci USA*, 97, 1079–1084.

Nilsson, J., Lee, J. R., Ratto, T. V., and Létant, S. E. (2006) Localized functionalization of single nanopores. *Adv Mater*, 18, 427–431.

Saenger, W. (1988) *Principles of Nucleic Acid Structure*, New York, Springer-Verlag.

Saleh, O. A. and Sohn, L. L. (2003) Direct detection of antibody-antigen binding using an on-chip artificial pore. *Proc Natl Acad Sci USA*, 100, 820–824.

Sauer-Budge, A. F., Nyamwanda, J. A., Lubensky, D. K., and Branton, D. (2003) Unzipping kinetics of double-stranded DNA in a nanopore. *Phys Rev Lett*, 90, 238101.

Schellman, J. A. (1977) Electrical double layer, zeta potential, and electrophoretic charge of double-stranded DNA. *Biopolymers*, 16, 1415–1434.

Schiedt, B., Healy, K., Morrison, A. P., Neumann, R., and Siwy, Z. S. (2005) Transport of ions and biomolecules through single asymmetric nanopores in polymer films. *Nucl Instrum Methods Phys Res Sect B Beam Interact Mater Atoms*, 236, 109–116.

Shao, Y. and Mirkin, M. V. (1997) Fast kinetic measurements with nanometer-sized pipets. Transfer of potassium ion from water into dichloroethane facilitated by dibenzo-18-crown-6. *J Am Chem Soc*, 119, 8103–8104.

Siwy, Z., Heins, E., Harrell, C. C., Kohli, P., and Martin, C. R. (2004) Conical-nanotube ion-current rectifiers: The role of surface charge. *J Am Chem Soc*, 126, 10850–10851.

Siwy, Z., Trofin, L., Kohli, P. et al. (2005) Protein biosensors based on biofunctionalized conical gold nanotubes. *J Am Chem Soc*, 127, 5000–5001.

Smeets, R. M., Keyser, U. F., Krapf, D. et al. (2006a) Salt dependence of ion transport and DNA translocation through solid-state nanopores. *Nano Lett*, 6, 89–95.

Smeets, R. M., Keyser, U. F., Wu, M. Y., Dekker, N. H., and Dekker, C. (2006b) Nanobubbles in solid-state nanopores. *Phys Rev Lett*, 97, 088101.

Smeets, R. M. M., Keyser, U. F., Dekker, N. H., and Dekker, C. (2008) Noise in solid-state nanopores. *Proc Natl Acad Sci USA*, 105, 417–421.

Smith, S. B. and Bendich, A. J. (1990) Electrophoretic charge density and persistence length of DNA as measured by fluorescence microscopy. *Biopolymers*, 29, 1167–1173.

Soni, G. V. and Meller, A. (2007) Progress toward ultrafast DNA sequencing using solid-state nanopores. *Clin Chem*, 53, 1996–2001.

Storm, A. J., Chen, J. H., Ling, X. S., Zandbergen, H. W., and Dekker, C. (2003) Fabrication of solid-state nanopores with single-nanometre precision. *Nat Mater*, 2, 537–540.

Storm, A. J., Chen, J. H., Zandbergen, H. W., and Dekker, C. (2005a) Translocation of double-strand DNA through a silicon oxide nanopore. *Phys Rev E Stat Nonlin Soft Matter Phys*, 71, 051903.

Storm, A. J., Storm, C., Chen, J. et al. (2005b) Fast DNA translocation through a solid-state nanopore. *Nano Lett*, 5, 1193–1197.

Trepagnier, E. H., Radenovic, A., Sivak, D., Geissler, P., and Liphardt, J. (2007) Controlling DNA capture and propagation through artificial nanopores. *Nano Lett*, 7, 2824–2830.

Uram, J. D., Ke, K., and Mayer, M. (2008) Noise and bandwidth of current recordings from submicrometer pores and nanopores. *ACS Nano*, 2, 857–872.

Vlassiouk, I. and Siwy, Z. S. (2007) Nanofluidic diode. *Nano Lett*, 7, 552–556.

Vocks, H., Panja, D., Barkema, G. T., and Ball, R. C. (2008) Pore-blockade times for field-driven polymer translocation. *J Phys Condens Matter*, 20, 095224.

Wang, G., Zhang, B., Wayment, J. R., Harris, J. M., and White, H. S. (2006) Electrostatic-gated transport in chemically modified glass nanopore electrodes. *J Am Chem Soc*, 128, 7679–7686.

Wanunu, M. and Meller, A. (2007) Chemically modified solid-state nanopores. *Nano Lett*, 7, 1580–1585.

Wanunu, M. and Meller, A. (2008) Single molecule analysis of nucleic acids and DNA-protein interactions using nanopores. In Selvin, P. and Ha, T. (Eds.) *Laboratory Manual on Single Molecules*. Cold Spring Harbor, NY, Cold Spring Harbor Press.

Wanunu, M., Sutin, J., Mcnally, B., Chow, A., and Meller, A. (2008) DNA translocation governed by interactions with solid-state nanopores. *Biophys J*, 95, 4716–4725.

Wu, S., Park, S. R., and Ling, X. S. (2006) Lithography-free formation of nanopores in plastic membranes using laser heating. *Nano Lett*, 6, 2571–2576.

Wu, M.-Y., Smeets, R. M. M., Zandbergen, M. et al. (2008) Control of shape and material composition of solid-state nanopores. *Nano Lett*, 9, 479–484.

Zhang, B., Zhang, Y., and White, H. S. (2004) The nanopore electrode. *Anal Chem*, 76, 6229–6238.

Zhao, Q., Comer, J., Dimitrov, V. et al. (2008) Stretching and unzipping nucleic acid hairpins using a synthetic nanopore. *Nucleic Acids Res*, 36, 1532–1541.

II

Nanotoxicology

13

Chances and Risks of Nanotechnology

Armin Grunwald
*Institut für
Technikfolgenabschätzung
und Systemanalyse*

13.1 Introduction and Overview

The public perception of nanotechnology was initially characterized completely by the expected positive qualities and consequences. In the meantime, however, a debate about the risks of nanotechnology has developed. Questions have been, and are still being, asked about the toxicity of nanoparticles; debates have begun about the regulation of nanomaterials; and ethical analyses have been conducted. The public debate about nanotechnology has been spurred on by position papers published by nongovernmental organizations (NGOs) (e.g., ETC Group 2003). Organizations and corporations maintain working groups on the social aspects of nanotechnology, such as the group called Responsible Production and Use of Nanomaterials, of the German Union of the Chemical Industry (Verband der Chemischen Industrie). Ministries and other government bodies hold workshops and public discussions on the risks of nanotechnology. International organizations such as the European Commission (Nordmann 2004) and UNESCO (ten Have 2007) have commissioned expert groups to prepare position papers (Renn and Roco 2006).

The early ethical studies on nanotechnology (about 2003) focused above all on the *need* to have ethics in and for nanotechnology (Baird et al. 2004). The ethically relevant aspects of nanotechnology that they named are an indication of a groping and very insecure (or phrased positively, open) approach to this relatively new field of science and technology. The concept "nanoethics" was quickly coined in this connection. In the meantime, a number of anthologies have been published about the social issues of nanotechnology in general and about ethical issues in

particular (e.g., Schummer and Baird 2006, Allhoff et al. 2007; Banse et al. 2008), and the first international journal *NanoEthics* has been founded. Nanoethics is now established as a new field of applied ethics, with ties especially to the ethics of technology and risks, to bioethics, and to medical ethics.

There are problems of definition, however, with the *object* of ethical reflection about nanotechnology, namely, the field of nanotechnology itself (Decker 2006; Schmid and Decker 2003; Schmid et al. 2006, Chap. 2). So far, nanotechnology is less technology than science, since it primarily takes place in a laboratory as basic science, which is the reason that people often speak of the nanosciences. The applications of nanotechnology that have already made it to the market, such as sun creams, automobile tires, and specially treated surfaces, hardly appear sufficient to constitute a challenge to ethics. The features of nanotechnology that are conceptually and methodologically fascinating as an object of ethics lie somewhere else. An ethical challenge is posed by the expectations and fears, the hopes and concerns, and the promises and the anxieties that accompany or are tied to nanotechnology. The objects of ethical reflection on nanotechnology go far beyond aspects of *current* nanotechnology and include in particular the *futures* that, metaphorically speaking, nanotechnology promises or threatens.

Another conceptual difficulty is the meaning of the word "ethical" itself. The word itself is not clearly defined in general usage and is often equated with "moral," or the two are mixed in speaking of "ethical, moral issues." A recently published anthology entitled *Nanoethics* (Allhoff et al. 2007) contains despite its title, for example, hardly any chapters on the *ethics* of nanotechnology, but rather chapters on many other socially relevant issues surrounding nanotechnology. Following the philosophical tradition (Rip and Swierstra 2007), we attach great importance to maintaining a strict conceptual distinction between ethics and morality. "Morality" is understood to refer to habitual and customary behavior, values, virtues, rules of behavior, and norms that are actually acknowledged by individuals, groups, or society as a whole. Morals form the totality of the normative orientations governing action and decision making that are recognized and observed in a certain context. "Ethics" refers, in contrast, to the systematic and theoretical reflection on this morality. Such reflection is only required when the acknowledged morality does not provide unambiguous guidance in a specific situation, for example, if there is a conflict between different moral precepts, or if the problem to be solved is so new that the established moral precepts are inappropriate for providing a solution (a situation that occurs frequently with new forms of technology). In the following, we refer to such a situation as "normative uncertainty." Ethical reflection serves to overcome normative uncertainty. Normative uncertainty can exist in conflicts, ambiguities, or uncertainty about the moral precept that is regarded "correct" or appropriate in an individual case. In the absence of conflicts between interests, values, norms, or rights there is no ethical problem.

Ethical issues about nanotechnology thus only arise if the existing precepts guiding action (e.g., values, norms, or laws) are challenged by advances in nanotechnology, if they prove inadequate, or if they lead to conflicts with regard to handling the new technologies and their consequences (e.g., their risks). Against this background, the goal of this chapter is to present and structure the state of ethical analyses of nanotechnology and, above all, to identify the issues on which there is a need for ethical reflection.

First, we present the very diverse motivations that led to the initial ethical analyses of nanotechnology (Section 13.2). This is followed by a systematic overview of the thematic areas in which progress in nanotechnology is leading to ethical concerns in certain fields of application (Section 13.3). Then there is a detailed study of the field of nanobiotechnology and of the discussion of risk being conducted there (Section 13.4). Finally, we provide a perspective on how ethics in nanotechnology can be expected to develop (Section 13.5).

13.2 Background: Motivations of Nanoethics

The first comments on the necessity of confronting the social issues, including ethical ones, surrounding nanotechnology stem from the years following the announcement of the American National Nanotechnology Initiative (NNI 1999). The predominant patterns of argument that were raised regarding the urgent need of ethical reflection were, in summary: "Without an attention to ethics, it would not be possible to ensure efficient

and harmonious development, to cooperate between people and organizations, to make the best investment choices, to prevent harm to other people, and to diminish undesirable economic implications" (Roco 2007, p. xi). We examine these motivations more closely in the following since they form an essential setting for an interpretation of ethical reflection.

13.2.1 Exhausting Potentials for Innovation

At the beginning of the debate about the risks of nanotechnology it was noted that nanotechnology and the ethical reflection on it developed at two very different speeds (Mnyusiwalla et al. 2003). One train of thought went as follows. While nanotechnology developed rapidly, there was hardly any interest in ethical issues. For example, not a single project of this nature received any financial support in the United States in 2003, while work was being conducted in hundreds of purely technical projects: "As a result, the ethics of nanotechnology have not really been addressed even when funds have been specifically allocated for doing so. In 2001 the U.S. National Nanotechnology Initiative allocated between $16 and $28 million for the study of its societal implications but only about half that budget was used" (Ball 2003). This concern, which we refer to as one over a "policy of innovation," is that the grave and growing gap between rapid advances in nanotechnology and its inadequate ethical appraisal could lead to a development in which the expected advantages and opportunities promised by nanotechnology might not be achieved, for example, because of public rejection: "We believe that there is danger of derailing NT [nanotechnology] if serious study of NT's ethical, environmental, economic, legal and social implications [...] does not reach the speed of progress in the science" (Mnyusiwalla et al. 2003, p. R9). The study of the ethical issues and of the consequences of nanotechnology is thus necessary to be able to introduce innovations in modern societies. Its absence might otherwise result in the threat of public resistance, which would hinder both progress and the social utilization of the expected advantages and benefits of science and technology: "The only way to avoid such a moratorium [following ETC Group 2003, A.G.] is to immediately close the gap between the science and ethics of NT. [...] Either the ethics of NT will catch up or the science will slow down" (Mnyusiwalla et al. 2003, p. R12).

The authors suggest measures for helping ethics in this chase, based on the experience gained from the ethical research that accompanied the human genome project. This experience includes the appropriate support of ethical, legal, and social implication activities (ELSI), the motivation of next-generation scientists in these fields (building capacity), an interdisciplinary and transdisciplinary networking instead of ethical reflection in an ivory tower, and the participation of developing countries and the public. In view of the possibility that ethical reflection might have a negative impact on innovation, the demand "mind the gap" appears to be clever for a policy of innovation: the earlier the possible ethical problems from nanotechnology innovations are recognized, the better it might be possible to avoid them constructively.

13.2.2 Potential for Large Social Transformation

Nanotechnology's presumed depth of penetration and its potential for social transformation are often given as reasons for the necessity of studying it from an ethical perspective. In this vein, Allhoff et al. (2007, see the preface) assert that "Nanotechnology will eventually impact every area of our world." In view of the revolutionary potential that many attribute to nanotechnology, the ethical, legal, and social implications have been examined by commissions and study groups for years (e.g., Coffrin and MacDonald 2004; Nanoforum 2004) and demands have been raised, especially from politicians, that nanotechnology be developed in a responsible manner. Behind these demands is the perception that the presumed substantial influence of nanotechnology on future development is tied to a high degree of uncertainty: "Tremendous transformative potential comes with tremendous anxieties" (Nordmann 2004, p. 4). The revolutionary force of nanotechnology means that each existing normative framework has to fail when confronting these new developments: "In this regard, the advocate could maintain that issues are transformative or revolutionary in some particular way and that, whatever other ethical frameworks we have already developed, those frameworks will be ill-equipped to deal with the force that nanotechnology represents" (Allhoff 2007, p. 198).

On the one hand, the reason given for the need to do an ethical study of the expected transformative potential appears logical heuristically: if the transformative potential of nanotechnology is great and if likewise the uncertainties about the forms it might take and consequences are also great, then it is very probable that normative uncertainty and conflicts would manifest themselves in a pluralistic society. On the other hand, however, this argument does not provide any specific reasons where the ethical challenges lie or how we can search for them. It only states that if we were to look, then it is quite probable that we would find them.

13.2.3 Unintended Side Effects

The fact that science and technology often have unintended consequences belongs to the most frequent causes of ethical reflection. Among the *unintended consequences* of technology are, for example, the risks of developments in science and technology to society and the environment that take the form of catastrophic accidents or of creeping changes. Classic questions that are asked are: Which risks can or should be accepted in view of the hoped positive consequences? How should risk-opportunity analyses and comparative risk evaluations be carried out? When should the precautionary principle be applied in view of our insufficient knowledge (Harremoes et al. 2002)? And how are we to deal with

situations in which it is unclear whether there is any risk at all (Wiedemann and Schütz 2008)?

Even before the ETC Group issued its call for a moratorium (ETC Group 2003), toxicologists had drawn attention to unanswered questions in connection with nanomaterials, especially nanoparticles (Colvin 2003). The background for this was past experience, namely, that the use of new materials, especially chemicals, can have negative consequences for the environment and our health. The history of asbestos, in particular, is a warning (Gee and Greenberg 2002). Initially this led to demands for empirical research into the consequences, often referred to as environment-safety-health (or ESH) studies. These demands were subsequently put in the context of ELSI studies.

The normative background of this argumentation consists primarily of the perception of the responsibility of science as this has developed in view of the unintended consequences in previous cases. Normative uncertainty arises as soon as classic risk management is no longer possible because of insufficient knowledge, or in other words when questions are raised as to the applicability and possible consequences of the precautionary principle (see Section 13.3.1). There are substantial differences between the demands for a moratorium, for increased research into consequences, and for utilizing the potential of nanotechnology as quickly as possible, some of the causes of which are different normative prejudices.

13.2.4 Apocalyptic Fears

One of the factors present when demands for an ethical council were first raised were fears of a possible apocalyptic side of nanotechnology (Joy 2000). Moor and Weckert (2004), for example, tied an ethics of nanotechnology to the expectations that nanotechnology "would include [...] how to minimize the risk of runaway robots." Jean-Pierre Dupuy put the apocalyptic dimension of Joy's fears at the center of his conceptualization of an ethics of nanotechnology (Dupuy and Grinbaum 2004; Dupuy 2005). He argues that, in as much as nanotechnological dreams make use of the principles of self-organization and thus could release self-organizing nanomachines into the world, they would ultimately and inevitably lead to an absolute catastrophe (Dupuy and Grinbaum 2004, in continuation and radicalization of the thoughts of Bill Joy). A focus on apocalyptic scenarios has been an acknowledged pattern of argumentation in studies considering the ethics of technology following Hans Jonas' "imperative of responsibility" (Jonas 1984) and is certainly part of the spectrum of ethical considerations of nanotechnology. It overlooks, however, features of nanotechnology that might serve ethics as the possible starting point for ethical considerations that are below the threshold of apocalyptic threats and, thus, all developments that are already in a technical or nearly technical state of development (such as the potential risks from nanoparticles). Moreover, another risk in restricting oneself to this manner of argumentation is that ethical reflections will not be perceived or taken seriously by those who reject them as nonsense or at least as pure speculation visions of the future in which these apocalyptic fears are based on technology. On the other hand, the fact that apocalyptic dangers are made a topic signals the presence of normative uncertainty and thus the necessity of ethical reflection.

13.3 Ethical Questions for Nanotechnology: The State of the Art

In the last few years, an informal "canon" has crystallized for charting the ethical questions surrounding nanotechnology (e.g., Khushf 2004; Grunwald 2005; Ach and Siep 2006; Allhoff et al. 2007). The state of the art of reflection on these issues is presented in the following section.

13.3.1 Health Issues and Nanomedicine

Nanotechnology affects or may affect human health via two channels: nanotech materials and products could, on the one hand, lead to health problems and hazards (see Section 13.3.1.1). On the other, many hopes are related with the emerging field of nanomedicines (Section 13.3.1.2).

13.3.1.1 Nanoparticles and Human Health

Nanotechnology is used to create new chemical substances or to study known chemical substances as particles that have changed in form and size in order to introduce these substances into the economy. By either generally adding nanoparticles or attaching them to key targets, it is possible to improve the properties of the materials or sometimes even to create new ones, such as for treating surfaces, in cosmetics, or in sun creams, but also in the particularly sensitive area of groceries.

The injection of such substances into the economy can have consequences for one's health. Nanomaterials can be unintentionally released at different positions and can enter the human body in different ways (e.g., by inhalation), where they can possibly lead to undesired reactions. This raises the fundamental question of the acceptability of the possible risks, of possible necessary precautions, of a comparison of the expected advantages and the possible risks, and of limits that take the form of environmental or health standards. In accordance with the different phases in the life cycle of technology, it is necessary to distinguish the production (e.g., for job security), use (e.g., for consumer protection), and disposal (e.g., the long-term storage of the materials).

Despite the intensive toxicological study, the state of our knowledge about how nanoparticles spread and about their toxic impact on the environment and health is very incomplete. Because of the specific difficulties of following nanoparticles in the human body and because of their diversity, even the development of appropriate measurement techniques is an extremely complex and time-consuming task. By means of animal experiments we have discovered that certain nanoparticles

can be biologically active. The state of toxicological research (e.g., Schmid et al. 2006, Chap. 5) does not permit us, however, to characterize the risks of nanomaterials well qualitatively or even to record them quantitatively.

For this reason, classic risk management, which works with quantitative risks and determines the limits of acceptability, does not work. "The tools developed in that discipline cannot be used when so little is known about the possible dangers that no meaningful probability assessments are possible" (Hansson 2006, S. 316). This is an example of an "unclear risk" (Wiedemann and Schütz 2008), namely, a risk whose magnitude cannot be indicated, for which the nature and extent of the damage are unknown, and where it is even unknown whether a danger exists *at all*: "Much of the public discussion about nanotechnology concerns possible risks associated with the future development of that technology. It would therefore seem natural to turn to the established discipline for analyzing technological risks, namely, risk analysis, for guidance about nanotechnology. It turns out, however, that risk analysis does not have much to contribute here" (Hansson 2006, p. 315). Precautionary considerations—such as are foreseen for environmental issues within the context of European framework legislation (von Schomberg 2005)—must instead be continued until toxicological research makes conventional risk management possible (Haum et al. 2004). This by no means indicates that restrictive measures or even a moratorium have to be automatically introduced. On the contrary, the point is to systematically and continuously evaluate our growing knowledge of toxicology in order to be able to draw very rapid consequences in case alarming results are discovered (Grunwald 2008).

13.3.1.2 Nanomedicine

According to an early and widespread definition, which is admittedly unnecessarily stilted, nanomedicine is "(1) the comprehensive monitoring, control, construction, repair, defence, and improvement of human biological systems, working from the molecular level, using engineered nanodevices and nanostructures; (2) the science and technology of diagnosing, treating, and preventing disease and traumatic injury, of relieving pain, and of preserving and improving human health, using molecular tools and molecular knowledge of the human body; (3) the employment of molecular machine systems to address medical problems, using molecular knowledge to maintain and improve human health at the molecular scale" (Freitas 1999, p. 418).

The use of nanotechnology is expected, according to more or less realistic scenarios, to produce improvements in medical diagnoses and therapies. With the aid of diagnostic instruments based on nanotechnology, illnesses or dispositions for illnesses could be recognized earlier than is now possible: "Although many of the ideas developed in nanomedicine might seem to be in the realm of science fiction, only a few more steps are needed to make them come true, so the 'time-to-market' of these technologies will not be as long as it seems today. Nanotechnology will soon allow many diseases to be monitored, diagnosed, and treated in a minimally invasive way and it thus holds great promise of improving health and prolonging life. Whereas molecular or personalized medicine will bring better diagnosis and prevention of disease, nanomedicine might very well be the next breakthrough in the treatment of disease" (Kralj and Pavelic 2003, p. 1012).

With regard to therapy, there is a prospect that treatments might be developed with the help of nanotechnology that are dedicated and free of side effects. The widespread application of dosage systems at the nanoparticle level could lead to very precise drug delivery in the human body and thus to substantial advances in drug treatments and to a drastic reduction in side effects. The principle of hyperthermia using magnetic fluids is to inject nanoparticles containing iron into a tumor and to cause them to vibrate by means of magnetic alternating fields, which could heat and kill the tumor tissue without damaging the surrounding tissue (Paschen et al. 2004, p. 225ff.). Furthermore, the biocompatibility of artificial implants can be improved by means of nanotechnology (Freitas 2003).

The potential of these developments is considerable, even though for now this is only a potential, and some of the conditions of its realization are uncertain. Consideration of the ethical consequences nevertheless also has to include the possible risks: "There is an urgent need to improve the understanding of toxicological implications of nanomedicines in relation to the specific nanoscale properties currently being studied, in particular in relation to their proposed clinical use by susceptible patients. In addition, due consideration should be given to the potential environmental impact and there should be a safety assessment of all manufacturing processes. Risk-benefit assessment is needed in respect of both acute and chronic effects of nanomedicines in potentially predisposed patients—especially in relation to target disease. A shift from risk assessment to pro-active risk management is considered essential at the earliest stage of the discovery, and the development of new nanomedicines" (ESF 2005, p. 9).

There is probably no field of science that is so used to dealing with risks in such a matter of fact manner and that is so practiced in doing so as medicine and pharmaceutics. This makes it prima facie unlikely that direct medical applications of nanotechnology will raise conceptionally new ethical questions. The application of nanotechnology in medicine comprises another step in medical progress, but does not pose a new moral challenge in the medical applications that are of interest in this context. The purpose of the normative idea of healing remains to develop approaches to improving the possibility of healing while reducing the side effects (e.g., drug delivery) and maintaining or further developing established ethical standards in dealing with risks in research, diagnosis, and therapy.

13.3.2 Environment and Sustainable Development

The application of nanotechnology is expected to produce a significant relaxation of the burden on the environment: a saving of material resources, the reduction in the mass of by-products that are a burden on the environment, an improved efficiency in transforming energy, a reduction in energy consumption, and the removal of pollutants from the environment (Fleischer and Grunwald 2008). Initial studies appear to confirm the tendency of these expectations, yet they also point out the limits (e.g., Fiedeler 2007). A number of studies on precisely the issue of the sustainability of nanotechnology have been published in the meantime (e.g., JCP 2008).

But all the potential positive contributions to sustainable development may come at a price. The unintended release of nanomaterials into the environment can lead to previously unknown effects. Synthetic nanoparticles are foreign bodies in the biosphere. They can reach the environment by means of emissions during production, during the everyday use of products, or during disposal. It is, for example, possible for titanium dioxide particles in sun creams to be released during bathing and to reach the sea. The nanoparticles in sprays remain intentionally reactive for a long time and can even be transported long distances and diffusely distributed by air, or accumulate at certain locations. Issues such as mobility, responsiveness, persistence, pulmonary penetration, and solubility have to be taken into consideration to evaluate the potential spread of nanoparticles (Colvin 2003). The ambivalence of technology with respect to sustainable development applies to nanotechnology, too (Fleischer and Grunwald 2008). The production, use, and disposal of products containing nanomaterials may lead to their appearance in air, water, soil, or even organisms. Nanoparticles could eventually be transported as aerosols over great distances and be distributed diffusely. Despite many research initiatives throughout the world, only little is known about the potential environmental and health impacts of nanomaterials. This situation applies also and above all for substances which do not occur in the natural environment, such as fullerenes or nanotubes. The challenge of acting under circumstances with high uncertainties, but with the nanoproducts already at the marketplace, is the heart of the ethical challenges by nanoparticles.

Questions of eco- or human toxicity of nanoparticles, on nanomaterial flow, on the behavior of nanoparticles in spreading throughout various environmental media, on their rate of degradation or agglomeration, and their consequences for the various conceivable targets are, however, not ethical questions. In these cases, empirical-scientific disciplines, such as human toxicology, eco-toxicology, or environmental chemistry, are competent. They are to provide the knowledge basis for practical consequences for working with nanoparticles and for disseminating products based on them. However, as the debate on environmental standards of chemicals or radiation has shown, the results of empirical research do not determine how society should react. Safety and environmental standards—in our case for dealing with nanoparticles—are to be based on sound knowledge but cannot logically be derived from that knowledge. In addition, normative standards, for example, concerning the intended level of protection, the level of public risk acceptance, and other societal and value-laden issues enter the field. Because of this situation it is not surprising that frequently conflicts about the *acceptability* of risks occur (Grunwald 2008). Therefore, the field of determining the acceptability and the tolerability of environmental risks of nanoparticles is an ethically relevant issue.

Decisive for a comprehensive assessment of nanotechnology or of the corresponding products from an environmental point of view is that the entire course of the products lifetime is taken into consideration. This extends from the primary storage sites to the transportation and the manufacturing processes to the product's use, ending finally with its disposal (Fleischer and Grunwald 2002). In many areas, however, nanotechnology is still in an early phase of its development, so that the data about its life cycle that would be needed for life-cycle assessment are far from being available. Empirical research on the persistence, long-term behavior, and whereabouts of nanoparticles in the environment as well as on their respective consequences would be necessary to enable us to act responsibly in accordance with ethical criteria of long-term responsibility and with regard to our use of, and contact with, our natural environment.

13.3.3 Privacy and Surveillance Issues

From the very beginning, the threat posed by new surveillance and supervision technology to one's privacy has been viewed as one of the ethical features of nanotechnology that needed to be taken into consideration (e.g., Moor and Weckert 2004; Nanoforum 2004). Nanotechnology provides a number of opportunities for personal data to be collected, stored, and distributed on a large scale. It is conceivable that miniaturization might lead to the development of sensor and storage technology that drastically increases the possibility of collecting data from victims without them noticing it (Moor and Weckert 2004; van den Hoven 2007). Such miniaturization, in connection with the limited human capacity to see, might make nanotechnology nearly invisible. Furthermore, the combination of miniaturization and networking of the surveillance equipment could substantially obstruct the present control options and regulations to protect personal data or even make them completely obsolete. In the military sphere, espionage will have new opportunities (Altmann 2006). Passive surveillance of humans might in the more distant future be supplemented by active surveillance (that the targets themselves might not notice) if, for example, it proved possible to construct a direct technical access to the nervous system or the brain. Some of these scenarios are not only considered realistic but even as certain developments: "But what is not speculation is that with the advent of nanotechnology invasions of privacy and unjustified control of others will increase" (Moor and Weckert 2004, p. 306).

Health services are particularly sensitive with regard to the private sphere. The development of small units for preparing diagnoses—"lab on a chip" (see van Merkerk 2007)—can make it possible to prepare comprehensive personalized diagnoses and prognoses on the basis of a person's health data. The demands placed on data privacy and thus on the protection of one's privacy therefore must be high. The "lab on a chip" technology cannot only facilitate individual diagnoses, but also permit fast and cheap mass screenings. The rapid decoding of the genetic

disposition of individuals can move into the reach of normal hospital activity or of nonclinical service providers. Anyone could undergo an examination for a genetic disposition to certain illnesses or could be pressed by an employer or insurance company to do so. In this way, individuals could be exposed to social pressure that would limit their capacity to act freely.

All of these questions about privacy, surveillance, and data protection are not specific to nanotechnology. Surveillance technology has attained a substantial state of the art even without nanotechnology, which itself raises questions about the survival of individual privacy (e.g., EPTA 2006). So-called smart tags are already making use of radio frequency identification (RFID) technology in access control, in ticketing such as in public mass transportation, and in logistics. Currently they have a size of several tenths of a millimeter in three dimensions, which means that with the human eye they are hardly noticeable. Further miniaturization will permit a further reduction in size and the addition of more functions. This will be possible even without a contribution by nanotechnology, but nanotechnology will certainly support and accelerate such developments.

The discussions of ethical issues in information science, the life sciences, and medicine have long included ethically relevant questions such as about a right to knowledge or ignorance, a personal right to certain data, and a right to privacy. They have also encompassed discussions about data protection and about possible undesired social eigendynamics as a consequence of a drastic extension of genetic and other tests. The developments in nanotechnology can accelerate or facilitate the achievement of technical opportunities, and they can extend the technical limitations that currently exist and thus increase the urgency of possibly problematic consequences. In this field, however, they do not lead to *new* ethical questions, even though the handling of this issue is doubtlessly relevant.

13.3.4 Distributional Justice

Problems of a fair distribution arise in principle in every field of technical innovation. Since scientific and technical progress requires substantial investment, it usually takes place where the largest economic and personnel resources are already present. Technical progress frequently exacerbates an unfair distribution that may already be present, even if only as a tendency. All the research, development, and production based on nanotechnology require capacities that in practice can only be provided by highly developed countries. Because this field of technology of its very nature affects a cross-section of society, these countries have to satisfy a variety of different instrumental and organizational prerequisites. Today and in the foreseeable future, only a few countries satisfy these conditions, such as the United States, Europe, Japan, China, Russia, and several other countries on the verge of joining them.

It is true that some have expressed high expectations that nanotechnology might benefit precisely the developing countries, such as if nanotechnology would make it possible to pursue a cheap mass production of products needed in these countries (Mnyusiwalla et al. 2003). Currently there is no tendency apparent that the technology gap between rich and poor could be diminished as a result of nanotechnology, even if at least a number of developing countries have invested heavily in research on nanotechnology (Schummer 2007). Optimistic hopes that nanotechnology will provide a rapid solution to the great developmental problems must be relativized: "The selected issues discussed in this chapter allow drawing mostly pessimistic conclusions on the impact of nanotechnology on developing countries" (Schummer 2007, p. 303). At any rate, these hopes will not come true as a result of nanotechnology as such, but require political regulative measures that take the forces of civil society into account (Invernizzi 2008). On the other hand, it should not be forgotten that nanotechnology might be tied to substantial economic problems in some developing countries. If, for example, cotton were replaced on a large scale by nanofibers or the rubber in automobile tires by nanoparticles, then the economic development of the countries that depend on the export of these natural resources would be endangered.

The problems posed by an unfairly distributed access to new technology frequently appear particularly clearly in medicine, where inequality often means unfairness: "The elements outlined above lead us to think that without stringent regulation by the state it is difficult for the important advances in nanomedicine to reach all levels of society and ease the load on the poor, especially in developing countries" (Foladori 2008, p. 211). One of the early demands was for developing countries to be involved in the examination of the ethical aspects of nanotechnology (Mnyusiwalla et al. 2003), in order not to exclude large portions of the world's population from the benefits of the expected potentials and, in consequence, to make the victims of discrimination.

The subject of a fair distribution is thus very relevant in connection with nanotechnology. On the other hand, the problems of fair distribution belong to the ethical issues of modern technology as a matter of *principle*. These problems are *also* relevant to nanotechnology, but are not specific to it.

13.3.5 Military

The military applications of new technology are frequently at the focus of ethical considerations, at least of those expressed in public, in the work of critical scientists (e.g., Altmann 2006) and in the position papers of NGOs. The use of nanotechnology in arms opens the door to improved weapons, innovative materials, and new areas of application. One option is the development of a nanoscale powder for use in propellants and explosives in order to increase the energy efficiency and speed of the explosion. Nanoscale electronic, sensory, and electromagnetic components could improve the performance of our capacity to steer and control military vehicles and make them more robust. This could further strengthen the current trend to unmanned and autonomous systems in the air, at sea, and in space. There are numerous potential applications in military intelligence that are based on the use of nanotechnological components for sensors, sensory

systems, and sensory networks. Even the field of weapons and munitions will be influenced very immediately by the improved sensory capacities and the improved computer power and storage capacity that nanotechnology influences. Developments in nanotechnology will presumably have substantial consequences on military personal, including at the level of personal equipment (keyword: "Soldier as a System"). In the forefront is the effort to equip soldiers with additional functionality without substantially increasing the weight of the equipment (Moore 2007). All of these developments can be viewed as an increased efficiency compared to previous military technology.

A politically and ethically relevant secondary consequence of these developments could be proliferation, if the assumption is made that the customary moral standards for the application of technology continue to apply in war. Not all countries observe the standards equally well, and some dictatorships could be tempted to use the new technologies internally (such as for surveillance). These more efficient weapons could even fall into the hands of terrorists. The smallness of nanotechnology is occasionally given as a reason for concern that nanotechnological techniques could lead to the construction of substantially smaller bombs and explosives, which would be considerably easier for terrorists to build and use. Another secondary affect that some fear is a new arms spiral, which would be driven solely by the speculative concern that potential opponents could arm themselves with nanotechnological weapons.

In view of these secondary effects, it is not to be expected that dramatically new ethical questions caused by nanotechnological advances in military technology will be raised. The reason is that the types of secondary effects and the related concerns are all known to us from previous developments in military technology. If really new ethical questions are raised in this field, then they are only in connection with developments in military technology that are rather speculative and visionary. Lines of development that would certainly challenge the existing moral orientations would be the creation of entirely new types of weapons. There are no limits on speculation of this nature in connection with visions about nanotechnology. Members of this repertoire include biological weapons manufactured by means of artificial or technically modified viruses and swarms of nanorobots. An emphatic reservation regarding such speculation is appropriate here; it is possible that such ethical reflections might be diagnosed as premature (Keiper 2007).

13.3.6 Human Enhancement

In combination with biotechnology and medicine, nanotechnology offers perspectives that go beyond the traditional medical tasks of healing illnesses, such as "improving" the human body, or radically transforming, or redesigning it. Although the examples of medical applications of nanotechnology (Section 13.3.6) are still within a, so to speak, traditional framework because the goal was to heal deviations from a healthy state, i.e., the classical goal of medicine, "enhancement" goes a category beyond the healing.

Human enhancement can refer to the extension of the physical and psychic capacities of people. For example, new or extended sensory functions could be implemented, such as through the extension of the electromagnetic spectrum that an eye is capable of perceiving. A simple cochlear or retinal implantation is today already possible. Advances in nanoinformatics could lead to improvements so that such implants approximate the dimensions of natural systems and their performance, or that they are later given new functions. It is also possible for new interfaces between man and machine to be created by means of a direct coupling of machines in IT systems to the human brain ("brain chip"). Another focus of the debate about human enhancements is the lengthening of the human lifetime, even to visions of abolishing death.

By extending these lines of development into the speculative, it would be possible to study topics such as the increased integration of technology in the human body, the increasing convergence of man and technology, and the creation of "cyborgs" as technically extended humans or technology that is given human traits. Although these perspectives are today speculative, they raise anthropological questions as to our image of man and to the relationship between man and technology (Jotterand 2008). This raises, at the same time, the question of the degree to which humans *may*, *should*, or *want* to go on transforming the human body. Initially, ethical analysis has to confront the semantic and hermeneutic problems that are tied to the concepts of healing, doping, and improving since the latter are factors that play a determining role regarding each of the relevant normative parameters (Grunwald 2008, Chap. 9; Jotterand 2008).

The practical relevance of such ethical questions may at first glance appear minimal considering the possible "technical improvement" of humans by measures that depend strongly on nanotechnology. This assessment is contradicted by two factors. First, the vision of a technical improvement of humans is presented completely seriously. Research projects are planned in this vein, and milestones to achieving this goal are defined, with nanotechnology playing the role of an "enabling technology." The idea of a convergence of nanotechnology, biotechnology, information science, and the cognitive or neurosciences, which is presented at the level of atoms, and which is supposed to become possible by means of the technical improvement of humans (Roco and Bainbridge 2002), has already influenced research funding. Second, technical improvements are by no means so dramatically new. On a rudimentary scale, at least, they are established, as is shown by the examples of cosmetic surgery as a technical improvement of body features that are perceived as being unaesthetic, or of the way in which at least some of the drugs used for improving performance (e.g., Ritalin) are already being employed in the framework of what is called everyday doping (Farah et al. 2004). It is not difficult to make the prognosis that the options and practice of technical improvements of humans will increase. A demand for them is probable given our experience with the shortcomings and failures of humans. There can be no doubt that there is a need for ethical reflection

on this issue in view of the normative uncertainty and potential for moral conflict that is linked to it.

Previous answers from ethics go in different directions. Liberal eugenics could draw the conclusion that the selection of the goals of interventions to change human features should be left to the individual preferences of market participants (Habermas 2001). Deontological ethics would above all ask about the possible violation of the rights of those affected. Religious moralists would emphasize both the fact that human self-understanding is traditional and deeply tied to civilization and that man is a temporally finite being. The representatives of social group could focus on the potential loss of previous options for making life successful, such as the option of humans to live, as a "normal" human or as a handicapped being, in a society of technically equipped humans (Wolbring 2006): Others could fear to be forced to adjust when the start is made with technical improvements. It is apparent that ethical questions are linked with this topic to a degree that has hardly been present in debates about medical ethics or anthropology.

13.4 Focus: Responsible Nanobiotechnology

The goal of nanobiotechnology is to influence and form living systems by using the means of nanotechnology, up to ultimately creating "artificial" life in the field of synthetic biology (Brown 2004). In view of the even larger depth of this human intervention in nature, the question as to whether such developments are responsible becomes more drastic (Dupuy 2007).

13.4.1 Scientific Responsibility

Due to the ability of living systems to reproduce, modify, and organize themselves, the risk situation in nanobiotechnology and in synthetic biology is clearly different from that of abiotic synthetic nanoparticles (Grunwald 2008). The very early stage of development in nanobiotechnology and synthetic biology and the distance from the market, however, prevent any discussion in concrete terms of the risks or even of unclear risks resulting from products or processes. There are no specific synthetic living beings or their parts under risk observation in a manner similar to titanium dioxide or silver oxide nanoparticles in marketed products. Just the opposite: the subject of the ethical debate is the *process of research itself*. Thus, the focus here is on questions of the responsibility of the scientists and disciplines involved, of the accountability of certain areas of research, and of the relationship between the self-regulation and self-obligation of science and state regulation.

It is thus not surprising that the well-known conference of Asilomar (1975) is repeatedly cited as a model for future steps in the field of nanobiotechnology (e.g., Boldt and Müller 2008). That conference took place in a situation in which a global spirit of optimism regarding genetic engineering could be observed while at the same time the first signs of public criticism and demands for state regulation could be heard. The outcome of the conference was that genetic engineers committed themselves to taking responsibility and caution. Interpretations of the conference are controversial. On the one hand, it is praised as a positive example of science proactively assuming responsibility; on the other hand, it mainly served the purpose of preempting state regulation so that genetic engineers could carry on conducting their research with as little interference as possible (de Vriend 2006).

At the second global conference on synthetic biology (2006), there was an attempt to follow up these actions at Asilomar and a corresponding declaration was passed (Maurer et al. 2006). This, however, only refers to the possible *military* use of synthetic biology and puts up a set of self-obligations for possibilities of this kind (http://syntheticbiology.org/SB2Declaration.html):

> First, we support the organization of an open working group that will undertake the coordinated development of improved software tools that can be used to check DNA synthesis orders for DNA sequences encoding hazardous biological systems; we expect that such software tools will be made freely available.
>
> Second, we support the adoption of best-practice sequence checking technology, including customer and order validation, by all commercial DNA synthesis companies; we encourage individuals and organizations to avoid patronizing companies that do not systematically check their DNA synthesis orders.
>
> Third, we support ongoing and future discussions within international science and engineering research communities for the purpose of developing creative solutions and frameworks that directly address challenges arising from the ongoing advances in biological technology, in particular, challenges to biological security and biological justice.
>
> Fourth, we support ongoing and future discussions with all stakeholders for the purpose of developing and analyzing inclusive governance options, including self-governance, that can be considered by policymakers and others such that the development and application of biological technology remains overwhelmingly constructive.

Self-obligations in science have, however, also come in for criticism. On the occasion of the same conference on synthetic biology, 35 nongovernmental organizations (including ETC Group, Greenpeace, Third World Network) wrote a joint letter on the subject, from which a long passage can be cited here in which they refer specifically to this restriction to a possible military aspect of synthetic biology (ETC 2006; also mentioned in de Vriend 2006, p. 49; Wolbring 2006):

> Moreover, the social, economic, ethical, environmental, and human rights concerns that arise from the field of synthetic biology go far beyond deterring bioterrorists and "evildoers." Issues of ownership (including intellectual property), direction and control of the science, technology,

processes and products must also be thoroughly considered. Society—especially social movements and marginalized peoples—must be fully engaged in designing and directing dialogue on the governance of synthetic biology. Because of the potential power and scope of this field, discussions and decisions concerning these technologies must take place in an accessible way (including physically accessible) at local, national and global levels. In the absence of effective regulation it is understandable that scientists are seeking to establish best practices but the real solution is for them to join with society to demand broad public oversight and governmental action to ensure social wellbeing.

This letter is characterized overall by

- Mistrust of scientific self-regulation and self-obligations. This is undemocratic as scientists should not be allowed to decide such far-reaching questions affecting their own activities.
- Demands for a broad investigation of the social consequences of synthetic biology instead of a restriction to abuse scenarios, e.g., by terrorists.
- Emphasis on the necessity for including social groups in dialogs about the agenda of research and the handling of possible social consequences.

The normative uncertainty or the conflict that is symbolized by this letter and the activities at the second conference on synthetic biology concern the *distribution of responsibilities* for the further research process. What influence do scientists, the public, the state, or other social actors or areas have on the further course of events in nanobiotechnology and synthetic biology in particular? What role, indeed, should a "policy of knowledge" (Stehr 2004) play, in which it would be necessary to decide which knowledge is desirable to acquire and which should rather be prevented? And to determine how responsibility and accountability should correspondingly be distributed?

The question is *which* responsibility *should* be attributed to scientists in the fields of nanobiotechnology and synthetic biology. This is often demanded in such a way that scientists are supposed to reflect on the consequences of their actions in a manner that constitutes a complete assessment of the technology. This is often done with the implicit hope that if scientists assessed the results of their own actions comprehensively, they would make judgments in a responsible manner and act accordingly, and that negative and unintended consequences could be avoided completely or to a large extent. These expectations are, however, doomed to failure. This is because technology assessment in general and in nanobiotechnology and in synthetic biology in particular are characterized by

- The need for *social* consultation (on state sponsorship of research, on government policy toward science and technology, and on regulating the context of technical development by means of legislation, judicial decisions, or economic measures)

- The necessity of a systemic perspective on the extent of the consequences
- The problem of anticipating and *evaluating an uncertain future* (systemic feedback, long-distance effects in both spatial and temporal terms, the lack of social causal relationships)
- The *evaluation* problem (how to evaluate the consequences of technology with regard to desirability or tolerability in view of the plural nature of society and the diversity of opinions)

Both *inter- and transdisciplinary processes* are necessary to be able to at least approximate a solution. Neither individual scientists nor disciplines such as synthetic biology, nor even philosophy, can address these questions alone with any prospect of success. Scientists in nanobiotechnology and synthetic biology are experts in their fields, and not in the possible social consequences of their actions or for the question of the acceptability of uncertain risks and dealing with them. Thus, a warning should go out against unreflected optimism in terms of responsibility, regardless of whether biologists commit themselves to these expectations or society tries to commit them to it.

When it comes to attributing responsibility, a broader approach is thus necessary, which does justice to the realities of a society with an extensive division of labor, citizens' claims for democratic participation, and the specific circumstances in the sciences. The democratic public should be considered first and foremost here. A deliberation about the agenda of science, specifically here of synthetic biology, that is conducted in a democratic matter is one of the demands for a transparent relationship between science, politics, and the public (Habermas 1968).

Taking seriously demands for participation by a democratic public as well as by decision-making processes that are politically legitimized, however, does not lead to releasing nanobiotechnology and synthetic biology from all responsibility. It is rather the case that this field is justifiably expected to provide transparent information to the public. This is particularly true in cases of potentially worrying developments that might lead society to not only initiate processes of ethical reflection or technology assessment in order to analyze and evaluate the problem systematically, but also generally when it is a question of shaping the scientific agenda. This specific responsibility in the area of early provision of information lies in the fact that scientists possess particular cognitive competence in their own area and are the first to have certain information. This responsibility also includes participation in interdisciplinary and social dialogs and in political counsel. Science, including synthetic biology, is part of society and not something external to society. What is thus desired of science is that it reflects on its role in society and actively accepts this role in its many aspects.

A professional code of ethics or declarations of commitment can certainly be useful and valuable in many situations: "A code of ethics and standards should emerge for biological engineering

as it has done for other engineering disciplines" (Church 2005, p. 423). But they replace neither ethical reflection nor shaping of democratic opinion. This is particularly true in the face of the new options for technically modified or newly created life forms and the higher costs of preventive care incurred by these, which have become the focus of attention due to developments in synthetic biology. In the future, it will be a case of operationalizing the principle of preventive care, for instance through appropriate containment strategies following the example of developments in genetic engineering which at each step of the way broaden the knowledge base with systematic investigations of the consequences and in this way transform the initially "unclear" risk to a risk in the sense of classical risk management (cf. Grunwald 2008, Chap. 7.2). To do this, it requires responsible scientists, an interested democratic public that expects clarification and demands its own participation, and measures at state level which certainly do not have to be immediately turned into regulations but can be expressed, e.g., as a systematic observation of further developments in nanobiotechnology and in synthetic biology in the framework of technology assessment, accompanied by philosophical interpretation of the risks and opportunities arising in each case with reference to the normative frameworks involved and through ethical reflection.

13.4.2 Chances and Risks

It is still too early to describe the chances and risks of developments in nanobiotechnology in concrete terms. Both exist probably most clearly in the areas of environment science and health: "It [synthetic biology] has potential benefits, such as the development of low-cost drugs or the production of chemicals and energy by engineered bacteria. But it also carries risks: manufactured bioweapons and dangerous organisms could be created, deliberatively or by accident" (Church 2005, p. 423).

13.4.2.1 Areas of Application

The areas of application of nanobiotechnology are in the field of medicine and more generally in the life sciences. Aspects from medicine that deserve emphasis (see Paschen et al. 2004 for detailed information) are new diagnostic means such as biosensors and the possibility of permanently monitoring the status of someone's health, the procedure for precisely transporting active substances and targeting their deposition (drug delivery), and new biocompatible materials and surfaces (e.g., Freitas 2003). The following areas of application can be distinguished in synthetic biology:

- Nanomanufacturing and nanostructuring using biobased methods. The goal is the use of the principle of self-organization of molecular units to create more complex structures in order to achieve certain functions (biomineralization).
- Production of the desired chemicals by means of bacteria that are manufactured or modified in a dedicated manner (plan drug factories, de Vriend 2006, p. 30).

- The technical utilization of functional biomolecules and hybrid systems (e.g., of biomolecular motors and actuators) in technical systems or in combination with nonbiological components.
- The realization of functions at the interface between biological and technical materials and systems (e.g., for neuroimplants or prostheses).

These areas of application and the corresponding functional combinations make it possible for biosensors and biomembranes, for instance, to be employed in environmental technology, and the use of photoenergetic processes can provide biological support for photovoltaics. For example, it would evidently be economically and ecologically significant to be technically able to copy photosynthesis in synthetic biology. The scenarios for similarly far-reaching applications can also be found in biomedicine.

One particular point that will need a great deal of attention concerns the far-reaching hopes placed in the technical use of processes of self-organization, independent of the concrete fields of application. The utilization of phenomena of self-organization, including the possibility of replicating technologically created or modified "living things," is an intentional part of the program and leads to self-organization becoming increasingly significant: "The paradigm of *complex, self-organizing systems* envisioned by von Neumann is stepping ahead at an accelerated pace, both in science and in technology.... We are taking more and more control of living materials and their capacity for self-organization and we use them to perform mechanical functions" (Dupuy and Grinbaum 2004, p. 292; this leads the authors to express serious concerns about the risks, see Section 13.4.2.2).

There are also economic reasons for the fascination of self-organization. If we could succeed in making materials organize themselves in such a manner that technically desired functions or properties would arise, this might be substantially less expensive than if these properties had to be created by humans, such as by means of a nanotechnological top-down procedure in which atoms are given a specific arrangement. The goal would thus be to replace construction processes that are designed laboriously by manual action by processes that run by themselves. Human intervention would thus be moved to a different level. The builder would become the controller of a nature that carries out the construction: "The self-assembling properties of biological systems, such as DNA molecules, can be used to control the organisation of species such as carbon Nanotubes, which may ultimately lead to the ability to 'grow' parts of an integrated circuit rather than having to rely upon expensive top-down techniques" (de Vriend 2006, p. 19).

13.4.2.2 Risks

The primary fears of risks connected with nanobiotechnology and synthetic biology concern not only the possible, unintended, negative consequences for our health and environment, but also the intended utilization of this technological potential for novel biological weapons. "In the case of synthetic biology, specific risks in need of close scrutiny and monitoring are uncontrolled self-replication and spreading of organisms outside the lab, and deliberate misuse by terrorist groups or individuals or by 'biodesigner-hackers'" (Boldt and Müller 2008, p. 387).

A well-known scenario from the debate about organisms modified by genetic technology concerns a product of synthetic biology, e.g., an artificial or modified virus, that might escape from a laboratory and cause considerable risks to the environment or health, and there being no way of recapturing it. These are risks that involve biological safety (see de Vriend 2006). Beyond posing an immediate danger to living species and individuals, the genetic pool of certain species might be contaminated as a result of genetic transfer, which would thus lead to ongoing and irreversible modifications.

A major cause for the expectation of such potential risks is that synthetic biology could create living things that are alien to the natural biosphere, and we do not possess any evolutionary experience for dealing with these in ecosystems and organisms (according to de Vriend 2006). The consequences of the release of such partially or completely invented living things would thus be impossible to calculate. One accurate response to this is that the probability is only small that such artificial living things could survive and then cause damage in the natural world because of their lack of adaptation into natural processes. But even a small probability is a finite one, which is why problems of this nature do require careful observation. In analogy to the well-known "gray goo" as a possible consequence of our losing control of self-replicating nanorobots (Joy 2000), a "green goo" was described as the scenario of such a catastrophe: "[t]he green goo scenario, somewhat more likely, if only because it is more easily in reach, suggests that a DNA-based artificial organism might escape from the lab and cause enormous environmental damage" (Edwards 2006, p. 202f).

This scenario shows the ambivalence of referring to the potential of self-organization. If self-organization is above all used to refer to capacities that are considered positive and whose technical use promises many, especially economic advantages, the capacities of living systems for self-organization and for self-replication also pose a special type of potential threat. As the debate about the gray goo scenario in nanotechnology showed, positive visions can quickly turn into negative ones (Schmid et al. 2006, Chap. 5.4). The technical utilization and organization of living processes lead us to a fundamental and inherent risk, namely, that we can lose control of living systems in a much more dramatic manner than we can of technical systems because they pass on their properties and can multiply.

Following the debate about the risks of genetically modified organisms, the precautionary principle was developed and implemented in European law for dealing with the challenges posed to biosafety (see on this Grunwald 2008, Chap. 7). Within the framework of a gradual step-by-step procedure and a commitment at every step to carefully examine the consequences of that step, a layered containment strategy was established. Accordingly, the affected research initially takes place under high-security conditions, subsequently in "normal" laboratories, then in controlled open experiments, and finally, currently on application, in controlled plantings that keep a minimum distance from fields used for nongenetically modified agriculture. This has led to a successive reduction in our initially complete ignorance about the consequences of genetically modified organisms in our natural environment.

The essence of another argument is that synthetic biology, by means of intentionally creating new or technically modified living objects, intervenes to a considerable degree in the course of natural evolution. While natural evolution progresses only in small steps, and large modifications only take place over extremely long periods of time, man is now endeavoring to take control of evolution in just a few years or decades by creating artificial life: "Ponder for a moment the incredible hubris of the entire endeavor of bionanotechnology. The natural environment has taken billions of years to perfect the machinery running our bodies and the bodies of all other living things. And in a single generation we usurp this knowledge and press it to our own use" (Goodsell 2004, S. 309). In this sense, the massive *acceleration* of natural development by means of nanobiotechnological procedures is a special challenge that is being viewed even in synthetic biology as a risk factor. Once in circulation, artificial cells—even if they are based on knowledge gained from natural cells—will not have had millions of years of evolution behind them, but possibly only a few years of experiments in a laboratory. The new construction of cells or the reprogramming of viruses must be placed under special observation.

Finally, the complexity of the processes in molecular biology per se poses a gateway for potential risks, as has already been shown in the first experiments with forms of gene therapy: "A more concrete rationalization of this distrust is the feeling that unpredictable consequences can follow from rearranging a complex system that is not fully understood. The recent history of gene therapy offers us a cautionary tale. To introduce genes into patients, it has been necessary to use what is essentially a piece of nanotechnology—an adapted virus is used to introduce the new genetic material into the cell. But fatalities have occurred due to unexpected interactions of the viral vector with the cell" (Jones 2004, p. 214 ff.).

All of the patterns of argumentation reviewed above refer solely to the possible *unintended* consequences of synthetic biology or its applications. Yet we must consider not only the possibility of the unintended release of synthetic living objects that cannot be subsequently retrieved, but also the possibility of the *intentional* construction of novel biological weapons on the basis of newly constructed or modified cells ("biosecurity," de Vriend 2006, p. 54). The products or techniques of synthetic biology, or perhaps only the knowledge produced by it, might be misused for military purposes in government weapons, programs, or by terrorists. With regard to concrete possibilities, our fantasies have more or less free rein, and in view of the fact that hardly anything is known about military programs of this nature, there is a danger of chasing after conspiracy theories. One consequence of this state of ignorance is that no purely imagined details should be made known, such as those regarding intentionally reprogrammed viruses. On the contrary, the issue should be the ethical aspects of the *possibility* of such developments. It should be

taken into consideration in this respect that high-tech techniques must be applied in order to create synthetic biological weapons: "Contrary to popular belief, however, a biological weapon is not merely an infectious agent but a complex system consisting of (1) a supply of pathogen [...]; (2) a complex formulation of chemical additives that is mixed to stabilize it and preserve its infectivity and virulence during storage; (3) a container to store and transport the formulated agent and (4) an efficient dispersal mechanism" (Tucker and Zilinskas 2006, p. 39).

This would mean that it is by no means an easy matter to produce and use such biological weapons. Since the scientific and logistic effort required would be considerable, such a development is rather improbable in the area of terrorism. Nonetheless scenarios have been created even for this area in which synthetic biological weapons could be created by, for example, a fanatic lone operator, an expert in synthetic biology who could misuse his specialist knowledge to develop such living objects out of personal hate and to use them against the object of his or her hate, or by a biohacker who, analogous to computer hackers, could construct a damaging virus just to demonstrate that it was technically possible or to rouse public awareness (de Vriend 2006). And there is also the military field, which in some countries has almost limitless resources at its disposal and hardly any logistic problems. In this sense and despite all the uncertainty and improbability involved, the conscious creation of harmful synthetic or technically modified living objects must indeed be considered a potential risk of synthetic biology.

13.4.3 Nanotechnology in the Realm of the Living

The aim of nanobiotechnology is to understand the biological functional units from a molecular perspective and to create in a controlled manner the functional components of living systems on a nanoscale by utilizing technical materials and electronic and chemical interfaces (Schmid et al. 2006, Chap. 3.3). Nanobiological knowledge can be used *to create new* functionality in living systems by modifying natural biomolecules, by modifying cellular design, or by designing artificial cells. The traditional, scientifically formed self-understanding of biology in the sense of *comprehending* vital processes is redefined in synthetic biology (Woese 2004, Ball 2005) as a *reinvention* of nature, as the creation of artificial life on the basis of our knowledge about "natural" life. In this way, biology becomes a technical science (de Vriend 2006) in which "[...] the preexisting nanoscale devices and structures of the cell can be adapted to suit technological goals" (Ball 2005, p. R1).

Synthetic biology differentiates between an approach that uses *artificial* molecules in order to reproduce biotic systems and an approach that uses the elements of "classical" biology and combines them to form new systems that function in a "nonnatural" manner (Benner and Sismour 2005). The thought of creating artificial life (AL) or a technically modified life that is partially equipped with new functions is the force behind this: "how far can it [life] be reshaped to accommodate unfamiliar materials, circumstances and tasks?" (Ball 2005, p. R3). Examples of these efforts range from the design of artificial proteins to the creation of virus imitations or the reprogramming of viruses, up to attempts to program cells to perform desired functions (Ball 2005, Benner and Sismur 2005, pp. 534–540).

The starting point of synthetic biology is to model biotic units as complex technical relationships and to break them into simpler technical ones (deconstructing life according to de Vriend 2006). While this is still, so to speak, a form of *analytic* biology, it becomes a *synthetic* one when the knowledge about individual processes of life that has been obtained from technical modeling and the corresponding experiments is combined and utilized in a manner to be able to achieve as a dedicated result certain "useful functions" (de Vriend 2006). "Seen from the perspective of synthetic biology, nature is a blank space to be filled with whatever we wish" (Boldt and Müller 2008, p. 388). Cells are interpreted as being machines that consist of components. The language and approach of mechanical engineering are transferred to living systems.

Decisive for scientific and technological progress is the *combination* of knowledge about molecular biology and genetic techniques with the new opportunities offered by nanotechnology. The prerequisite for a precise design of artificial cells would be an adequately complete understanding of all the necessary subcellular processes and interactions. Our current state of knowledge is still far from this. Much of the research and development in nanobiotechnology serve to provide such knowledge (Schmid et al. 2006, Chap. 3.3).

13.5 Conclusions: Future Perspectives of Nanoethics

The ethics of nanotechnology have developed within a period of just a few years, together with other activities regarding the investigation and evaluation of their social consequences. However, it was (and to some extent still is) the case that there was some uncertainty as to whether nanotechnology created *new* ethical questions, whether it created ethical questions at all, and if it did, what would be the appropriate scientific and systematic framework for adequately dealing with these questions and providing answers to them (Allhoff 2007; Nordmann 2007a; Grunwald 2008). Today, it is no longer doubted that nanotechnology involves *relevant* ethical questions; the question of whether these challenges are new or to what extent they are new is only of secondary importance in the face of the practical necessity of ethical reflection.

However, it has transpired that there will probably not be any autonomous "nanoethics" as a new subfield within applied ethics equivalent to, e.g., medical ethics, information ethics or bioethics, as was expected until recently (Lin and Allhoff 2007). The reason for this is, on the one hand, that the ethical questions are not bound up with nanotechnology per se but always with its specific applications. The ethical questions relating to specific applications, in turn, already have their own place in applied ethics, for instance specifically in medical ethics, information ethics, or bioethics. Nanotechnology leads to partly new ethical questions in these fields through new application options, but does not add an entirely new field to the range of applied ethics.

On the other hand, the concept of nanotechnology, which would have to endow a new subfield of this kind with the specifics, will itself probably disintegrate in the future. In the development of nanotechnology, it is notable that the term "nanotechnology" was used almost exclusively in the singular until around 2004, but since then the term "nanotechnologies", in the plural, has had increasingly wider repercussions (prominently, e.g., in Nordmann et al. 2006). The term "nanosciences" is also usually used in the plural form. There is much emphasis on the term "nanotechnology" as merely a blanket covering many research and development works that are very heterogeneous in detail. The term "nanotechnology" has had a great impact: it has gained public perception, created political awareness and a readiness to provide sponsorship, instilled a certain fascination and for quite some time cast a favorable light on technical progress. One probable development is that the prefix "nano" will remain and may even become more widespread. We will continue to talk of nanomaterials, nanobiotechnology, nanomedicine, or nanoelectronics, but probably there will be less talk of nanotechnology as an entity as technical and scientific differentiation progresses.

Ethical reflection will follow this trend. And here there will probably be a "disentanglement" (Nordmann 2007b) of the different parts of "nanoethics," which will return "home" to the established subfields of applied ethics. The right place to conduct professional ethical reflection on nanotechnology is where the central ethical questions in the treatment of normative uncertainties are discussed, where the relevant traditions exist, and the necessary competence is present. This is precisely the case in the subfields of applied ethics. Disentanglement here means that questions of nanomedicine are dealt with in medical ethics, questions of data protection and privacy in information ethics, questions of dealing with nanoparticles in risk ethics, and questions of technically modified life in bioethics. "Nanoethics" will not become a new branch of applied ethics. However, there will be a need for ethicists to demonstrate their readiness to openly address ethical aspects of nanotechnology (or –technologies) across the borders of classical subdivisions of ethics and conduct a dialog with scientists in the fields of natural and engineering sciences.

References

Ach, J. S. and Siep, L. (eds.). 2006: *Nano-Bio-Ethics. Ethical and Social Dimensions of Nanobiotechnology*. Berlin, Germany: Lit-Verlag.

Allhoff, F. 2007. On the autonomy and justification of nanoethics. *NanoEthics* 1: 185–210.

Allhoff, F., Lin, P., Moor, J., and Weckert, J. (eds.). 2007. *Nanoethics. The Ethical and Social Implications of Nanotechnology*. Hoboken, NJ: Wiley.

Altmann, J. 2006. *Military Nanotechnology: Potential Application and Preventive Arms Control*. London, U.K.: Routledge.

Baird, D., Nordmann, A., and Schummer, J. (eds.). 2004. *Discovering the Nanoscale*. Amsterdam, the Netherlands: IOS Press.

Ball, P. 2003. *Nanoethics and the Purpose of New Technologies*. Online available at: <http://www.whitebottom.com/philipball/docs/Nanoethics.doc > (accessed February 24, 2008).

Ball, P. 2005. Synthetic biology for nanotechnology. *Nanotechnology* 16: R1–R8.

Banse, G., Grunwald, A., Hronszky, I., and Nelson, G. (eds.). 2008. *Assessing Societal Implications of Converging Technological Development*. Berlin, Germany: Edition Sigma.

Benner, S. A. and Sismour, A. M. 2005. Synthetic biology. *Nature Reviews/Genetics* 6: 533–543.

Boldt, J. and Müller, O. 2008. Newtons of the leaves of grass. *Nature Biotechnology* 26: 387–389.

Brown, C. 2004. BioBricks to help reverse-engineer life. *EE Times* June 11, 2004.

Church, G. 2005. Let us go forth and safely multiply. *Nature* 438: 423.

Coffrin, T. and MacDonald, C. 2004. Ethical and social issues in nanotechnology. Annotated bibliography. Online available at: <http://www.ethicsweb.ca/nanotechnology/bibliography.html > (accessed September 3, 2008).

Colvin, V. 2003. Responsible nanotechnology: Looking beyond the good news. Centre for biological and environmental nanotechnology at Rice University. Online available at: <http://www.eurekalert.org/> (accessed May 2, 2008).

de Vriend, H. 2006. *Constructing Life. Early Social Reflections on the Emerging Field of Synthetic Biology*. The Hague, the Netherlands: Rathenau Institute.

Decker, M. 2006. Eine definition von nanotechnologie: Erster schritt für ein interdisziplinäres nanotechnology assessment. In *Nanotechnologien im Kontext*, eds. A. Nordmann, J. Schummer, and A. Schwarz, pp. 33–48. Berlin, Germany: Akademischen Verlagsgesellschaft.

Dupuy, J.-P. 2005. The philosophical foundations of nanoethics. Arguments for a method. *Lecture at the Nanoethics Conference*, University of South Carolina, Columbia, SC, March 2–5, 2005.

Dupuy, J.-P. 2007. Complexity and uncertainty: A prudential approach to nanotechnology. In *Nanoethics. The Ethical and Social Implications of Nanotechnology*, eds. F. Allhoff, P. Lin, J. Moor, and J. Weckert, pp. 119–132. Hoboken, NJ: Wiley.

Dupuy, J.-P. and Grinbaum, A. 2004. Living with uncertainty: Toward the ongoing normative assessment of nanotechnology. *Techné* 8: 4–25. Wieder abgedruckt in *Nanotechnology Challenges: Implications for Philosophy, Ethics and Society*, eds. I. Schummer and D. Baird, 2006, pp. 287–314. Singapore: World Scientific Publishing.

Edwards, S. A. 2006. *The NanoTech Pioneers: Where Are They Taking Us?* Weinheim, Germany: Wiley-VCH.

EPTA—European Parliamentary Technology Assessment Network. 2006. *ICT and Privacy in Europe*. Online available at: < http://www.eptanetwork.org > (accessed September 3, 2008).

ESF—European Science Foundation. 2005. *Nanomedicine. An ESF Forward Look*. Online available at: <http://www.esf.org/publications/forward-looks.html> (accessed September 3, 2008).

ETC Group. 2003. *From Genomes to Atoms. The Big Down*. Atomtech: Technologies converging at the nano-scale. Online available at: <http://www.etcgroup.org> (accessed September 3, 2008).

ETC-Group. 2006. *Extreme genetic Engineering*. An introduction to synthetic biology. Online available at: < http://www.etcgroup.org > (accessed September 3, 2008).

Farah, M. J., Illes, J., Cook-Deegan, R. et al. 2004. Neurocognitive enhancement: What can we do and what should we do? *Nature Reviews: Neuroscience* 5: 421–425.

Fiedeler, U. 2007. *The Role of Nanotechnology in Chemical Substitution*. Brüssel, Belgium: European Parliament.

Fleischer, T. and Grunwald, A. 2002. Technikgestaltung für mehr Nachhaltigkeit–Anforderungen an die Technik folgenabschätzung. In *Technikgestaltung für eine nachhaltige Entwicklung. Von der Konzeption Zur Umsetzung*, eds. Grunwald, A., pp. 95–146, Berlin: Edition Sigma.

Fleischer, T. and Grunwald, A. 2008. Making nanotechnology developments sustainable. A role for technology assessment? *Journal of Cleaner Production* 16 (8–9): 889–898.

Foladori, G. 2008. Converging technologies and the poor. The case of nanomedicine and nanobiotechnology. In *Assessing Societal Implications of Converging Technological Development*, eds. G. Banse, A. Grunwald, I. Hronszky, and G. Nelson, pp. 193–216. Berlin, Germany: Edition Sigma.

Freitas, R. A. Jr. 1999. *Nanomedicine, Volume I: Basic Capabilities*. Georgetown, TX: Landes Bioscience.

Freitas, R. A. Jr. 2003. *Nanomedicine, Vol. IIA: Biocompatibility*. Georgetown, TX: Landes Bioscience.

Gee, D. and Greenberg, M. 2002. Asbestos: From 'magic' to malevolent mineral. In *The Precautionary Principle in the 20th century. Late Lessons from Early Warnings*, eds. P. Harremoes, D. Gee, M. MacGarvin et al., pp. 49–63. London, U.K.: Earthscan.

Goodsell, D. S. 2004. *Bionanotechnology. Lessons from Nature*. Hoboken, NJ: Wiley-Liss.

Grunwald, A. 2005. Nanotechnology—A new field of ethical inquiry? *Science and Engineering Ethics* 11: 187–201.

Grunwald, A. 2008. *Auf dem Weg in eine nanotechnologische Zukunft. Philosophisch-ethische Fragen*. Freiburg, Baden-Württemberg, Germany: Karl Alber.

Habermas, J. 1968. Verwissenschaftlichte Politik und öffentliche Meinung. In *Technik und Wissenschaft als Ideologie*, ed. J. Habermas, pp. 120–145. Frankfurt, Germany: Suhrkamp Verlag.

Habermas, J. 2001. *Die Zukunft der Natur des Menschen*. Frankfurt, Germany: Suhrkamp Verlag.

Hansson, S. O. 2006. Great uncertainty about small things. In *Nanotechnology Challenges—Implications for Philosophy, Ethics and Society*, eds. J. Schummer and D. Baird, pp. 315–325. Singapore: World Scientific Publishing.

Harremoes, P., Gee, D., MacGarvin, M. et al. (eds.). 2002. *The Precautionary Principle in the 20th Century. Late Lessons from Early Warnings*. London, U.K.: Earthscan.

Haum, R., Petschow, U., Steinfeldt, M., and von Gleich, A. 2004. *Nanotechnology and Regulation within the Framework of the Precautionary Principle*. Berlin, Germany: Institut für ökologische Wirtschaftsforschung.

Invernizzi, N. 2008. Nanotechnology for developing countries. Asking the wrong question. In *Assessing Societal Implications of Converging Technological Development*, eds. G. Banse, A. Grunwald, I. Hronszky, and G. Nelson, pp. 229–239. Berlin, Germany: Edition Sigma.

Jonas, H. 1984. *The Imperative of Responsibility*. London, U.K.: Routledge (initial version: Jonas, H. 1979. *Das Prinzip Verantwortung. Versuch einer Ethik für die technologische Zivilisation*. Frankfurt, Germany: Suhrkamp.)

Jones, R. A. L. 2004. *Soft Machines. Nanotechnology and Life*. Oxford, U.K.: University Press.

Jotterand, F. 2008. Beyond therapy and enhancement: The alteration of human nature. *Nanoethics* 2: 15–23.

Joy, B. 2000. Why the future doesn't need us. In *Wired*, April 2000. Wieder abgedruckt in *Nanoethics. The Ethical and Social Implications of Nanotechnology*, eds. F. Allhoff, P. Lin, J. Moor, and J. Weckert, pp. 17–30. Hoboken, NJ: Wiley.

Keiper, A. 2007. Nanoethics as a discipline? *The New Atlantis. A Journal of Technology & Science* 16(Spring): 55–67.

Khushf, G. 2004. The ethics of nanotechnology—visions and values for a new generation of science and engineering. In *Emerging Technologies and Ethical Issues in Engineering*, ed. National Academy of Engineering, pp. 29–55. Washington, DC: National Academy of Engineering.

Kralj, M. and Pavelic, K. 2003. Medicine on a small scale. How molecular medicine can benefit from self-assembled and nanostructured materials. *EMBO Reports* 4: 1008–1012.

Lin, P. and Allhoff, F. 2007. Nanoscience and nanoethics: Defining the disciplines. In *Nanoethics. The Ethical and Social Implications of Nanotechnology*, eds. F. Allhoff, P. Lin, J. Moor, and J. Weckert, pp. 3–16. Hoboken, NJ: Wiley.

Maurer, S. M., Lucas, K. V., and Terrel, S. 2006. *From Understanding to Action: Community Based Options for Improving Safety and Security in Synthetic Biology*. Berkeley, CA: University of California.

Mnyusiwalla, A., Daar, A. S., and Singer, P. A. 2003. Mind the gap. Science and ethics in nanotechnology. *Nanotechnology* 14: R9–R13.

Moor, J. and Weckert, J. 2004. Nanoethics: Assessing the nanoscale from an ethical point of view. In *Discovering the Nanoscale*, eds. D. Baird, A. Nordmann, and J. Schummer, pp. 301–310. Amsterdam, the Netherlands: IOS Press.

Moore, D. 2007. Nanotechnology and the military. In *Nanoethics. The Ethical and Social Implications of Nanotechnology*, eds. F. Allhoff, P. Lin, J. Moor, and J. Weckert, pp. 267–275. Hoboken, NJ: Wiley.

Nanoforum (ed.). 2004. *Nanotechnology. Benefits, Risks, Ethical, Legal, and Social Aspects of Nanotechnology*. Report online available at: < www.nanoforum.org > (accessed September 8, 2008).

NNI—National Nanotechnology Initiative. 1999. *National Nanotechnology Initiative*. Washington, DC.

Nordmann, A. 2004. *Converging Technologies—Shaping the Future of European Societies, High Level Expert Group "Foresighting the New Technology Wave"*. Brussels, Belgium: European Commission.

Nordmann, A. 2007a. If and then: A critique of speculative NanoEthics. *Nanoethics* 1: 31–46.

Nordmann, A. 2007b. Entflechtung—Ansätze zum ethisch-gesellschaftlichen Umgang mit der Nanotechnologie. In *Nano—Chancen und Risiken aktueller Technologien*, eds. A. Gazsó, S. Greßler, and F. Schiemer, pp. 215–229. Berlin, Germany: Springer.

Nordmann, A., Schummer, J., and Schwarz, A. (eds.). 2006. *Nanotechnologien im Kontext. Philosophische, ethische und gesellschaftliche Perspektiven*. Berlin, Germany: Akademische Verlagsanstalt.

Paschen, H., Coenen, C., Fleischer, T. et al. 2004. *Nanotechnologie. Forschung und Anwendungen*. Berlin, Germany: Springer.

Renn, O. and Roco, M. C. 2006. The risk governance of nanotechnology. *Journal of Nanoparticle Research* 8: 153–191.

Rip, A. and Swierstra, T. (2007): Nano-ethics as NEST-ethics: Patterns of moral argumentation about new and emerging science and technology. *NanoEthics* 1: 3–20.

Roco, M. C. 2007. Foreword: Ethical choices in nanotechnology development. In *Nanoethics. The Ethical and Social Implications of Nanotechnology*, eds. F. Allhoff, P. Lin, J. Moor, and J. Weckert, 5f. Hoboken, NJ: Wiley.

Roco, M. C. and Bainbridge, W. S. (eds.). 2002. *Converging Technologies for Improving Human Performance*. Arlington, VA: National Science Foundation.

Schmid, G. and Decker, M. (eds.). 2003. *Small Dimensions and Material Properties. A Definition of Nanotechnology*. Europäische Akademie Graue Reihe, Bad Neuenahr-Ahrweiler, Germany.

Schmid, G., Ernst, H., Grünwald, W. et al. 2006. *Nanotechnology—Perspectives and Assessment*. Berlin, Germany: Springer.

Schummer, J. 2007. Impact of nanotechnologies on developing countries. In *Nanoethics. The Ethical and Social Implications of Nanotechnology*, eds. F. Allhoff, P. Lin, J. Moor, and J. Weckert, pp. 291–307. Hoboken, NJ: Wiley.

Schummer, J. and Baird, D. (eds.). 2006. *Nanotechnology Challenges—Implications for Philosophy, Ethics and Society*. Singapore: World Scientific Publishing.

Stehr, N. 2004. *The Governance of Knowledge*. London, U.K.: Transaction Publishers.

ten Have, H. (ed.). 2007. *Nanotechnologies, Ethics and Politics*. Paris, France: UNESCO.

Tucker, J. B. and Zilinskas, R. A. (2006). The promise and perils of synthetic biology. *The New Atlantis* 12: 25–45.

van den Hoven, J. 2007. Nanotechnology and privacy: Instructive case of RFID. In *Nanoethics. The Ethical and Social Implications of Nanotechnology*, eds. F. Allhoff, P. Lin, J. Moor, and J. Weckert, pp. 253–266. Hoboken, NJ: Wiley.

van Merkerk, R. (ed.). 2007. *Intervening in Emerging Technologies—A CTA of Lab-on-a-Chip Technology*. Utrecht, the Netherlands: Koninklijk Nederlands.

von Schomberg, R. 2005. The precautionary principle and its normative challenges. In *The Precautionary Principle and Public Policy Decision Making*, eds. E. Fisher, J. Jones, and R. von Schomberg, pp. 141–165, Cheltenham, U.K.: Edward Elgar Press.

Wiedemann, P. and Schütz, H. (eds.). 2008. *The Role of Evidence in Risk Characterization. Making Sense of Conflicting Data.* Weinheim, Germany: Wiley-VCH.

Woese, C. R. 2004. A new biology for a new century. *Microbiology and Molecular Biology Reviews* 68(2): 173–186.

Wolbring, G. 2006. *The Triangle of Enhancement Medicine, Disabled People, and the Concept of Health: A New Challenge for HTA, Health Research, and Health Policy.* Online available at: www.ihe.ca/documents/hta/HTA-FR23.pdf

Human and Natural Environment Effects of Nanomaterials

Birgit Gaiser
Edinburgh Napier University

Martin J. D. Clift
Edinburgh Napier University
University of Bern

Helinor J. Johnston
Edinburgh Napier University

Matthew S. P. Boyles
Edinburgh Napier University

Teresa F. Fernandes
Edinburgh Napier University

14.1 Introduction

Nanotechnology can be defined as the manipulation, precision placement, measurement, modeling, or manufacture of materials at the nanometer (nm) scale (Donaldson et al. 2001). The field of nanotechnology is anticipated to provide lighter, stronger, smaller, and more efficient and durable products such as stain-free clothing, as well as exploitation for environmental remediation (Reijnders 2006). The attraction of producing and exploiting nanomaterials is a consequence of the fact that the properties of nano-sized materials are expected (and have also been demonstrated) to be strikingly different from bulk forms of the same material (Service 2004). To put the size of nanomaterials into perspective, 1 nm, or 1 billionth of a meter (10^{-9}), is the diameter of 10 hydrogen atoms, 4 nm is the size of a single protein molecule, 1000 nm is equal to the size of a pollen grain, and 100,000 nm can be equated to the average width of one human hair (Whitesides 2003). It is recognized that the definition of a "nanomaterial" (NM), defined as "a material that has one or more external dimensions in the nanoscale, or which is nanostructured" (nanomaterials can exhibit properties that differ from those of the same material that do not have nanoscale features) (BSI 2007, SCENIHR 2007b, ISO 2008), is different from a "nanoparticle" (NP) defined as a material with three external dimensions in the nanoscale) (BSI 2007, SCENIHR 2007b).

The nanotechnology industry is a rapidly developing field, with the amount of money invested into its research and development, as well as the economic value of the industry increasing continuously every year (Lux Research 2006). Concomitant with this is the level of nanotechnology-related products that are manufactured and distributed for use in a wide and diverse range of consumer, industrial, and technological applications, such as clothing, cosmetics, computer technology, household paints, medicine, and sporting equipment (Maynard 2006). Due to the constant influx of nanotechnology-related products in these areas, it is essential that rigorous assessment of their risk to both humans and the environment is conducted in order to implement a set of stringent safety guidelines for their production, use, and disposal, so that the promise surrounding the industry can be realized (Hoet et al. 2004, Maynard et al. 2007).

In recent years, the importance of risk assessment in association with nanotechnology has been widely recognized by the scientific community, governments, and regulating bodies. This was emphasized in 2004 in a report published by the Royal Society and Royal Academy of Engineers investigating the possible opportunities and uncertainties of nanotechnology (Dowling et al. 2004). It stated that although nanotechnology does have the potential to provide highly efficient products for a wide and diverse range of applications, it is imperative that an understanding of the possible health and occupational risks of the diverse array of different materials intended for use is attained urgently. It was further highlighted by this report that both biological systems and the environment should be considered in relation to the potential hazardous effects of the manufacturing, use, and disposal of nanomaterials (Dowling et al. 2004). These suggestions were further emphasized in recent reports by the Department for Environment, Food and Rural

Affairs (DEFRA) (NRCG 2005, 2007), the Scientific Committee on Emerging and Newly Identified Health Risks (SCENIHR 2006, 2007a, 2009), and the Council for Science and Technology (CST, 2007). In addition to such reports, a large number of conferences, workshops, journals, books, and interest groups have been established that are dedicated to the risk assessment of nanomaterials. Within such interest groups, news on risk assessment of nanomaterials are quickly communicated within the research community on Web sites that are dedicated to the safety of nanomaterials (such as www.safenano.org) in recognition of the fact that communication between different groups for the benefit of quicker and better nanomaterial risk assessment is paramount.

It is intended that this chapter will provide an introduction into the fundamental principles of toxicology, as well as the specific and novel challenges presented by nanomaterials in relation to their risk assessment. This will be achieved via the illustration of how the specific properties of nanomaterials can influence their toxicity, and why there is a concern surrounding their exposure as compared to larger particles of the same material (Service 2004).

14.2 Background

14.2.1 The Theory of Toxicology

Toxicology is the study of the adverse effects of substances on living organisms (Timbrell 1999). One of the fundamental principles of toxicology is that the "risk" posed by a substance is a function of its potential to cause harm, or "hazard," and the amount of substance a biological system is "exposed" to. This is defined by the following principle:

$$\text{Risk} = \text{Hazard} \times \text{Exposure}$$

It is necessary to consider exposure levels of substances, as according to Paracelsus, all materials are toxic if exposure occurs in sufficient quantities (Timbrell 1999); specifically, "All things are poison and nothing is without poison, only the dose permits something not to be poisonous."

These fundamental toxicology principles imply that substances with a low hazard generally pose a low risk. If, however, there is a high-enough exposure to these low hazard/substances, then these can be harmful, and even fatally toxic. An extreme example of this is the potential for fatal poisoning via overconsumption of water, also known as reaching a state of "hyperhydration." Alternatively, the risk from substances with a high hazard/high risk can be significantly reduced by imposing restrictions on their exposure.

14.2.2 Nanotoxicology: A Short Historical Perspective

Concern surrounding the exposure of humans to nanomaterials derives from their small size, and emanates from two independent findings that separately recognized that as particle size decreases, toxicity generally increases.

First, Ferin et al. (1992) recognized that titanium dioxide (TiO_2) particle size was fundamental in driving its toxicity. In this study, rats were exposed, via inhalation, to TiO_2 (21 and 250 nm in diameter) for 12 weeks, and examination of the consequences of TiO_2 exposure were evaluated over a 64 week postexposure period. Alternatively, rats were administered TiO_2 of various sizes (12, 21, 230, and 250 nm in diameter) via a single intratracheal instillation and toxicological investigations were made 24 h postexposure. It was demonstrated that the smaller-sized TiO_2 (<100 nm) elicited a greater acute pulmonary inflammatory response (characterized by neutrophil infiltration) following exposure compared to the larger-sized TiO_2. The smaller-sized TiO_2 particles (<100 nm diameter) were also found to remain within the lung longer (501 days) than the larger TiO_2 particles (174 days), highlighting that the clearance of smaller particles from the lung was slower. The prolonged retention of smaller TiO_2 particles in the lung was suggested to derive from the finding that they were able to translocate to the pulmonary interstitium more efficiently than the larger TiO_2 particles. This was facilitated by the fact that smaller particles were not efficiently taken up by alveolar macrophages, which thereby allowed for their interaction with alveolar type-1 epithelial cells. In addition, it was found that an increased mass dose (translating to an increased number of particles and decreased particle size) promoted the movement of particles within the pulmonary system. It was also observed that the number of particles present, particle size, delivered dose, and the delivered dose rate also had an effect on the translocation process. The study therefore highlighted that particle size was fundamental to their toxicity, and that particles <100 nm were particularly hazardous.

Second, epidemiological studies conducted in the 1990s, found a positive correlation between the level of particulate air pollution and increased morbidity and mortality rates. Adverse health effects were manifested predominantly in susceptible individuals who had preexisting pulmonary or cardiovascular disease (Dockery et al. 1993, Schwartz 1994, Pope and Dockery 1999). The detrimental health outcomes of particulate air pollution were suggested to be driven by its PM_{10} content. PM_{10} is defined as particulate matter collected through a size selective inlet that has a 10 μm cut off and 50% efficiency. The composition of PM_{10} is complex and includes carbon particulates, transition metals, endotoxins, windblown dusts, and pollen (Wilson et al. 2002). All of the components are thought to contribute to the observed toxicity (Stone 2000, Donaldson and Stone 2003). However, the ultrafine component of PM_{10} has been held principally accountable for eliciting the adverse health effects that were highlighted in epidemiological studies. This led to the generation of the "ultrafine hypothesis" (Seaton et al. 1995), whereby ultrafine (uf) particles (<100 nm) were considered to be particularly hazardous to health. It is also relevant that the findings were in agreement with those of Ferin et al. (1992), whereby smaller particles had a greater toxic potential. Peters et al. (1997), who reported the association between respiratory ill health and the number of ambient ultrafine particles inhaled, also supported the ultrafine hypothesis.

A number of investigators have subsequently demonstrated that particle size is fundamental to particle toxicity. In a study conducted by Li et al. (1997), the pulmonary toxicity of PM_{10} was examined within *in vivo* (6 h postexposure) and *in vitro* (24 h postexposure) settings. Following intratracheal instillation of PM_{10} in rats, an increase in inflammation (characterized by a neutrophil influx into the lungs) and cytotoxicity (indicated by increased levels of lactate dehydrogenase [LDH]) was measured in the bronchoalveolar lavage (BAL) fluid. An increase in BAL protein can be indicative of increased epithelial permeability. Depletion in the antioxidant glutathione within the BAL fluid was also observed, implying that oxidative stress development was a component of the pulmonary response to PM_{10}. *In vitro*, BAL leucocytes isolated from PM_{10} exposed or control rats were cultured and an increase in tumor necrosis factor-alpha (TNF-α) production was witnessed. Furthermore, A549 cells exposed to PM_{10} were observed to have increased permeability; thus corroborating the *in vivo* findings. The *in vitro* and *in vivo* findings, therefore, are indicative of the pro-inflammatory and oxidative potential of PM_{10} in addition to illustrating their ability to promote epithelial cell permeability. The response exhibited by the ultrafine carbon black (ufCB) or CB *in vivo* was also considered and compared to that exhibited by PM_{10}. It was demonstrated that the inflammatory response stimulated by ufCB was greater in magnitude compared to that of CB and PM_{10}. The findings of this study therefore emphasize the size dependency of particle toxicity and also provide an insight into the mechanisms that drive the observed toxicity.

A comparative analysis of CB-(260 nm) and the ufCB-(14 nm) related pulmonary toxicity, following the exposure of rats, was assessed by Li et al. (1999). The pro-inflammatory effect of ufCB was greater than that of CB, when instilled at an equal mass dose of 125 μg, thereby confirming that ultrafine particles cause more inflammation than their larger counterparts, and is thought to be driven by their larger surface area. Specifically, a 50% increase in neutrophils was observed in rats exposed to ufCB compared to CB particles, 6 h postexposure. Rats instilled with ufCB also exhibited an increase in BAL TNF-α, LDH release, as well as a rise in total BAL protein, which were absent in rats instilled with CB (Li et al. 1999). However, it was shown that both particle sizes caused a decrease in glutathione content in the lung, although the highest levels of depletion were observed in rats treated with ufCB, again illustrating the pro-oxidative behavior of ultrafine particles. In conclusion, Li et al. (1999) suggested that the superior toxicity of ufCB, in comparison to CB, was related to the size and surface area of the particle type.

These findings were supported by Brown et al. (2001) who examined the effects of the intratracheal instillation of 64, 202, and 535 nm diameter polystyrene microspheres in rats. It was observed that 64 nm particles induced an inflammatory response that was typified by the influx of neutrophils, LDH, and protein release into the BAL fluid, which were more pronounced than those exhibited by an equal mass dose of 202 and 535 nm-sized polystyrene microspheres. Further analysis also showed that the level of inflammation in the rat lung was related to the surface area of the particles; accordingly, the greater the particle surface area, the greater the inflammatory response observed. These findings were subsequently supported by Duffin et al. (2002), who demonstrated that the pulmonary inflammatory response initiated by a number of low toxicity particles (TiO_2 and CB) in rats was driven by their surface area

so that particles <100 nm in diameter were more toxic than their larger counterparts.

Stoeger et al. (2006) also demonstrated the specific surface area of nanomaterials was related to their inflammogenic potential within mice. Mice were instilled with 5, 20, and 50 μg (ranging in their surface area from 30 to 800 m^2/g) of a series of different particles, including flame-soot particles, spark-generated ultrafine carbon particles (ufCP), and diesel exhaust particles (DEP). The inflammatory response was characterized by the increased levels of LDH, total protein, TNF-α, interleukin (IL)-1β, and macrophage inflammatory protein-2 (MIP-2) in the BAL fluid that was associated with the infiltration of polymorphonuclear (PMN) leukocytes into the lung. ufCP were more potent than the other particles tested. It is also relevant that the inflammatory response was not detectable at doses ≤20 cm^2. It was concluded that although the surface area of particles indicates their ability to cause inflammation, as exemplified by the fact that ultrafine particles cause a heightened inflammatory response compared to larger particles, a threshold exists (Stoeger et al. 2006).

The findings of the studies discussed therefore support the hypothesis that nanomaterials have the potential to elicit toxicity that is driven by their pro-inflammatory and pro-oxidant nature that may drive adverse health effects within humans. In addition, although these studies are limited to only a few specific particle types (namely, CB, polystyrene microspheres, and TiO_2), it is suggested that particle size, particle mass dose, surface area, and surface reactivity are relevant in driving the toxicity of nanomaterials. The findings therefore provide concern surrounding exposure to nanomaterials, due to their similarity in size to the ultrafine particles used, while also providing insight into the experimental approaches that should be considered with highest priority when considering the toxicity of nanomaterials. However, nanomaterials are a diverse group of materials that vary with regard to a variety of properties, including their shape structure, surface chemistry, and composition, which are thought to contribute to their potential toxicity (Oberdorster et al. 2005a, 2007). Therefore making generalizations about nanomaterials may be difficult to achieve, given that their fate and effects are not solely driven by their small size.

It is relevant, therefore, that the perceived risks associated with nanomaterials have emanated from their small size due to knowledge on the behavior of ultrafine particles. Nanomaterials, however, are defined as "a material that has one or more external dimensions in the nanoscale" (BSI 2007, SCENIHR 2007b, ISO 2008). Conversely, ultrafine particles are defined as having all dimensions under 100 nm in diameter (BSI 2007, SCENIHR 2007b, ISO 2008). Therefore, although all ultrafine particles can be described as nanomaterials (or nanoparticles) not all nanomaterials can be described as ultrafine particles; consequently, the extrapolation of the findings should be approached with caution. This is also supported by the knowledge that there is already evidence that ultrafine particles, of a variety of compositions and forms, do not always behave similarly, indicating that their ability to cause toxicity is not fully comparable (SCENIHR 2007a, 2009). Accordingly, Dick et al. (2003) investigated four different ultrafine particle types (CB, cobalt, nickel, and TiO$_2$) to determine which attributes of particles influenced their toxicity *in vitro* and *in vivo*. It was observed that the inflammatory response elicited by the different particle types was not comparable, and that the toxicity of the different particle types studied could be ranked, from highest to lowest, in the following order: carbon = cobalt > nickel > titanium dioxide; therefore, attributes other than their similarity in size were responsible for driving the observed toxicity. Furthermore, Xia et al. (2006) compared the toxicity of ultrafine ambient particles, polystyrene microspheres (positively and negatively charged) and a range of engineered nanomaterials [CB, TiO$_2$, fullerol (C$_{60}$(OH)$_{24}$), etc.] to RAW 264.7 macrophages *in vitro*. These authors reported that the ambient ultrafine particles and positively charged polystyrene nanospheres induced the highest levels of toxicity, evidenced by their capability of inducing reactive oxygen species (ROS) production, glutathione depletion, and organelle damage. The results indicate that the composition and charge of the nanomaterial influences their toxicity. This study, however, used only one cell type, and it is important to highlight that the other cell types may show different sensitivities to the nanomaterial panel.

14.2.2.1 Definition of Nanotoxicology

Further to the discussion in this section, nanotoxicology can therefore be described as a multidisciplinary science including material science, chemistry, physics, and medicine (Donaldson and Tran 2004). Research in this area focuses on fully understanding the relationship between properties such as particle size, surface area and reactivity, dose, and composition in determining the toxicity of nanomaterials. Nanotoxicology focuses on the assessment of effects of nanomaterials, but instrumental in this process is the development and implementation of specific protocols for the investigation of the toxicity of nanomaterials.

The following sections will illustrate the specific challenges facing nanotoxicology research, as well as providing examples of how physical, chemical, and toxicokinetic properties of specific nanomaterials affect their risk.

14.2.3 Exposure Routes of Nanomaterials

The availability and toxicity of a substance to an organism are determined not only by the dose an entire organism is exposed to, but via a set of properties referred to as "toxicokinetics," which describe uptake, transport, metabolism, sequestration to different compartments, and finally elimination of a substance from the organism. These parameters are essential since the toxicity of the substances is dependent on which organs or cell types are exposed, which form the substance is in, for example, bound to serum protein, aggregated, dissolved, oxidized, and how long the substance remains at the site of exposure.

Routes by which substances can enter the human body include ingestion, inhalation, injection, and permeation through the skin (Figure 14.1). Therefore, the potential for nanomaterials to

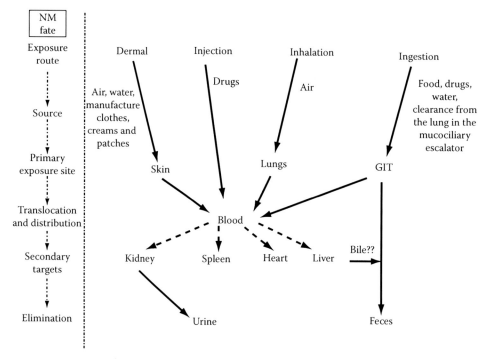

FIGURE 14.1 Routes of exposure and potential fate(s) of nanomaterials within the human body. The wide-ranging products that are expected to contain NPs allow exposure via a variety of routes. A primary target of NP toxicity is considered if toxicity has the potential to occur at the site of exposure. If toxicity occurs as a consequence of the translocation of NPs to sites that are distal to that of the portal of entry then the organ is considered a secondary target (with some, but not all examples outlined). The ability of NPs to translocate to distal sites is dependent on the barrier function and clearance mechanisms evident at the exposure site (GIT = gastrointestinal tract). (Adapted from Oberdorster, G. et al., *Particle Fibre Toxicol.*, 2, 8, 2005b.)

exhibit toxicity at their site of entry (namely, the lungs, skin, and gut) is of great relevance. This is of importance as studies that have evaluated the toxicity of ultrafine particles (when investigating the components responsible for driving PM_{10} toxicity) concentrated on the impact of particulate exposure within specific target organs. The respiratory system was considered to be the primary target for particle toxicity due to the fact that exposure to PM_{10} occurred via inhalation and, as reported, following epidemiological investigation that ultrafine particles can cause respiratory ill health (Peters et al. 1997). The impact of ultrafine particles on the cardiovascular system was also investigated as mortality associated with exposure to PM_{10} was also associated with a reduction in cardiovascular function (Schwartz 1997). However, the realization that the nanomaterial localization is not restricted to their portal of entry has become evident, as they can become distributed to organs that are distal to their site of exposure, so that the nanomaterial toxicity can be potentially exerted at a number of secondary targets (Oberdorster et al. 2005a). Therefore, toxicity at a number of targets (including the liver, brain, spleen, and kidneys) is worthy of consideration, when assessing the toxicity of nanomaterials on human health. For example, the liver is considered to be a preferential site of NP exposure subsequent to exposure via a variety of routes (Ogawara et al. 1999, Nemmar et al. 2001, Takenaka et al. 2001, Oberdorster et al. 2002). Thus, it is imperative that the ability for nanomaterials to have detrimental consequences for normal liver function be investigated.

The analyses of the effects of nanomaterials on the skin has widely suggested that the optimal opportunity for the uptake of nanomaterials by the epidermis is when the outer skin cells are broken, such as when the skin is sunburnt or diseased (Corachan 1988). Mortensen et al. (2008) reported on the skin penetration of carboxylated quantum dots (QDs) in SKH-1 mice. These authors observed qualitatively higher levels of penetration in the ultraviolet radiation (UVR) exposed mice when compared to the non-UVR exposed individuals. Additionally, it has also been hypothesized by Tinkle et al. (2003) that when broken skin is flexed it is more sensitive to penetration by nanomaterials, such as TiO_2, as it forms a more permeable environment, allowing the translocation of the nanomaterials to the lymph nodes and then subsequently into the blood circulation.

Currently, research into the effects of nanomaterial exposure via injection is limited, and is concentrated within medicinal applications or in determining the distribution of nanomaterials subsequent to exposure. The aim of injecting nanomaterials is to target specific organs, tissues, and cells, and thus the nanomaterials used will be coated with such coatings as polymers in order to negate their identification by the immune system, specifically, the phagocytic cells (such as macrophages). It is essential that research be performed in this area, so that the advantages posed by nanomedicine (Section 14.3.3) are gained.

It is also foreseeable, and therefore of concern, that the release of nanomaterials into the environment could increase the concentration of atmospheric particles (Borm et al. 2006), which has obvious health implications, when considering the epidemiological findings of PM_{10} toxicity, in addition to the potential damage to the environment.

In order to fully understand the effects of nanomaterials on both human health and their impact on the environment, it is essential that increased research be performed in each of these exposure areas in order to fulfill the potential advantages of nanotechnology.

14.2.4 Determinants of Nanomaterial Toxicity

In addition to the material that composes nanomaterials, a number of additional properties can influence their uptake, distribution, and toxicity.

14.2.4.1 Size and Surface Area

The size of nanomaterials is especially relevant within the field of nanotechnology, due to the fact that it is believed that it is within nano dimensions (1–100 nm) that materials are most likely to have altered properties (Dowling et al. 2004). The size drives the interest surrounding the exploitation of nanomaterials but also contributes to the concern regarding their toxicity (Oberdorster et al. 2005a). One of the key facts in nanomaterial behavior and potential toxicity is the increase in the surface area that is associated with their small size when compared to their larger counterparts. Therefore, as the material size decreases, the surface area per unit mass increases, which equates to a greater proportion of its atoms/molecules being displayed on the surface of the material (Nel et al. 2006). In many studies, the toxicity of nanomaterials was best correlated to their surface area, and not when exposure was expressed as mass dose (Oberdorster 2001, Duffin et al. 2002). "Quantum effects" are also important in dictating the reactivity of nanomaterials due to the size range within which this field is constrained (Dowling et al. 2004) so that properties that are not previously encountered in larger forms of the same material are thus expressed. Specifically, the small size of nanomaterials results in electron confinement and restricted bond angles that translate into the emergence of unique physicochemical properties that not only make the NPs useful for exploitation in new applications and products, but also potentially determine their biological reactivity and toxicity (Stone and Kinloch 2007). The size dependent toxicity has been repeatedly demonstrated *in vitro* and *in vivo* leading to the conclusion that, in general, the toxicity is likely to increase as the material size decreases (see, for example, Li et al. 1999).

14.2.4.2 Shape

Engineered nanomaterials are produced in a variety of defined shapes, compared to naturally occurring particles, such as rods, tubes, and/or rings (Aitken et al. 2006). The shape of the nanomaterial can influence its characteristics, such as its bioavailability, its surface area, and its ability to persist in the organism (biopersistence) (Oberdorster et al. 2005a). For example, a high "aspect ratio" of length to width of fibrous materials (such as asbestos fibers or carbon nanotubes (CNTs)) can inhibit material uptake by phagocytosing cells, as well as their removal from the site of exposure and injury.

14.2.4.3 Surface Reactivity and Chemistry

In addition to the larger total surface area described in Section 14.2.4.1, the percentage of atoms a nanomaterial exposes on its surface increases significantly with a decreasing diameter (Oberdorster et al. 2005a, Stone and Kinloch 2007). This is particularly important for nanomaterials that (1) consist of reactive materials such as quartz or metals, (2) contain reactive contaminants such as transition metals or organic molecules on their surface (for example, PM_{10}), or (3) are modified to bear reactive groups on their surface.

Surface chemistry can be intentionally altered to influence the transport and uptake of nanomaterials by cells, as well as to target nanomaterials to specific organs or cell types when used for medical applications (Pison et al. 2006). Common surface coatings of nanomaterials include amino groups (positively charged) or carboxyl groups (negatively charged). These materials will react in different ways and with different structures (Oberdorster et al. 2005a). In addition, changes to the surface coating of nanomaterials may significantly alter their size (such as QDs—for example, see Hardman et al. (2006)). Surface modifications can also prevent the dissolution of materials, which otherwise may dissolve.

14.2.4.4 Aggregation and Adsorption

Materials in suspension, as prepared for toxicological studies, often tend to aggregate (Wick et al. 2007). This can significantly alter their properties, such as diameter, surface area, and number of available "monodispersed" nanomaterials. The characterization of the aggregation state of nanomaterials in different studies is integral to the interpretation of results, as the same nanomaterial may exist in very different aggregation states depending on the medium used for suspension (Foucaud et al. 2007).

Achieving a monodispersed suspension is possible via the use of specific dispersants (including proteins, solvents, detergents, and natural dispersants, such as humic or fulvic acids in experiments involving aquatic organisms), stirring and sonication (Wick et al. 2007). Care must be taken to choose specific dispersants that do not, individually influence the testing system used (see for example, Monteiro-Riviere et al. 2005b or Klaine et al. 2008).

Adsorption describes the binding of substances to the surface of a nanomaterial. In nanotoxicology, adsorption is an important issue when drawing conclusions on the effects of nanomaterials on a biological system. This is relevant, as both the reagents used in toxicological tests (Belyanskaya et al. 2007), as well as the products of toxic reactions (such as cytokines or LDH; Clift et al. 2008) can be adsorbed onto the surface of particles and subsequently provide misleading results. Adsorption is also potentially important for nanomaterials used in nanomedicine. For example, it is known that particles can change their properties by acquiring a protein coating within the target organism and thus not perform their targeted function (Pison et al. 2006).

14.2.5 Adverse Effects of Nanomaterials

A number of findings have been repeatedly encountered when assessing the toxicity of nanomaterials and so, a common response to exposure may be apparent. Enhanced ROS production within cells, mediated by nanomaterial exposure, is thought to stimulate an inflammatory response, with exposure ultimately culminating in cell death.

14.2.5.1 Oxidative Stress and Inflammation

ROS are unstable molecules that are highly reactive. Oxidative stress occurs when there is an imbalance between oxidants and antioxidants within cells, so that oxidant presence is favored due to the excessive production of oxidants and the depletion of antioxidants (MacNee and Rahman 2001), and is thought to be a hallmark of nanomaterial exposure. The oxidation of cellular structures such as lipids, proteins, and DNA stimulates their modification or degradation to compromise their normal function so that cell death ensues (Ryter et al. 2007). For this reason, it is therefore imperative that cells are able to overcome any increase in the ROS production to sustain normal cell function. ROS generation is therefore able to stimulate a protective response within cells to limit their propensity for damage. This includes the participation of antioxidants (which aspire to remove ROS from the cell), the implementation of repair mechanisms (to restore oxidized molecules to their original state), and preventative measures that are due to the oxidant-mediated expression of genes, the products of which have antioxidant capabilities (Ryter et al. 2007). These defenses, however, may be overwhelmed by nanomaterial exposure, with oxidative stress implicated with evidence of cellular antioxidant depletion (Stone et al. 1998), increased production of the ROS (Park et al. 2008), increase in oxidized cellular components (Sayes et al. 2005) or prevention of nanomaterial toxicity through pretreatment with antioxidants (Shvedova et al. 2007), within animal models or cells in exposed to nanomaterials.

Additionally, as previously discussed in Section 14.2.2, nanomaterials appear to exhibit responses that are pro-inflammatory in nature. The response is generally characterized by the infiltration of neutrophils *in vivo*, and an increased production of pro-inflammatory mediators (including TNFα and IL-8) *in vitro* and *in vivo*. The inflammatory response is thought to be stimulated by a rise in intracellular ROS that stimulates the activation of transcription factors, such as NF-κβ, that in turn promote the expression of pro-inflammatory genes, and has been experimentally proven (see for example Manna et al. 2005).

14.2.5.2 Cytotoxicity

Cytotoxicity describes the potential of a substance to elicit cell death. Cell death can be described as an irreversible loss of cell function and structure that predominantly occurs via apoptosis or necrosis (Patel and Gores 1995). Determining the ability of nanomaterials to impact on cell viability can be identified by detecting the number of live (viability assays) or dead cells (cytotoxicity testing) subsequent to the exposure of cells. Consequently, oxidative and inflammatory responses, associated with nanomaterial exposure may culminate in cell death if the protective responses fail to overcome the damaging responses initiated. This, however, is not always dose-dependent. Determinants in the cytotoxicity of a particular nanomaterial also include the period of exposure as well as the specific cells/organs/tissues that are exposed. A reduction in cell viability is therefore common when evaluating the toxicity of nanomaterials, and can be used as a basis for comparing the toxicity of different nanomaterials.

14.3 Recent Developments in Nanotoxicology: Carbon-Based Nanomaterials

Although there is a plethora of nanomaterials made of different substances and having a range of different characteristics, such as metal oxides, this chapter will specifically focus on carbon-based nanomaterials.

14.3.1 High Aspect Ratio Nanomaterials

As with other nanomaterials, high aspect ratio nanomaterials (HARN) have been found to have a huge array of applications and, with prolific invested interest, are promising to be extremely beneficial. The classified group, HARN, in itself comprises of a collection of different types including nanowires, nanorods, nanobelts, and nanosprings (Oberdorster et al. 2007); and are determined by high aspect ratio dimensions of a length-to-width aspect ratio of greater than 3:1. As introduced in the previous section, one member of HARN that has received a particularly high level of interest are CNTs. The remainder of this discussion therefore, will largely focus on CNTs, with a brief outline of their characteristics, the toxicological implications of CNTs following exposure, as well as their potential applications.

14.3.1.1 Carbon Nanotubes

The proposed use of CNTs is in both the medical and industrial fields, although CNTs have been present in the atmosphere for thousands of years, with examples of crystalline and polycrystalline aggregated CNTs found within a 10,000 year-old ice core (Esquivel and Murr 2004). CNTs have since accumulated in the environment because of the combustion of substances such as propane and natural gases through the use of gas stoves or hot water heaters. CNTs are also emitted from gas power plants and industrial-sized gas utilities (Bang et al. 2004, Murr et al. 2004) and are deposited through vehicle combustion systems and break-lining degradation (Murr and Bang 2003).

After the realization of intentionally specific production of CNTs, reported by Iijima (1991) over a decade ago, the interest in CNTs' production, utilization, and subsequent toxicological investigation has soared at a staggering rate.

CNTs (Figure 14.2) are produced as either single-walled carbon nanotubes (SWCNTs) with a single, continuous graphite tube, termed graphene, or multi-walled carbon nanotubes (MWCNTs) consisting of any number of graphene tubes held within each other on a common axis. They can have lengths of up to micrometers and widths of hundreds to only a few nanometers (Popov 2004, Wei et al. 2007). The production of MWCNTs is seen to be much higher than SWCNTs (Mitchell et al. 2007) and is now estimated to reach many hundreds of metric tons each year (Eklund et al. 2007).

14.3.1.1.1 Biological Interactions and Toxicological Implications of HARN

As with other nanomaterials, the properties that make HARN attractive to many industries are consequently the same properties that raise concerns regarding their potential toxicity: highly reactive surfaces and components, poor solubility, and a high aspect ratio of sometimes over 1000. This generates a great concern toward their biocompatibility and reactivity (Mitchell et al. 2007). The behavior toward biological systems and the relative toxicity of CNTs is dependent on a number of factors, possibly the most important being exposure route, material concentration, and sample purity (Yehia et al. 2007). With all this in mind, and

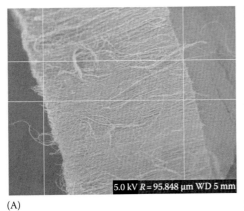
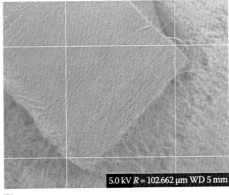

(A)   (B)

FIGURE 14.2 (A) Scanning electron microscopy image (SEM) of MWCNT carpet growth, grown by chemical vapor deposition using an iron catalyst on a quartz substrate, with average length of 96 μm (B) SEM image of MWCNT carpet growth, grown using an iron catalyst on a quartz substrate, with average length of 103 μm (grid identifies MWCNT length).

with the large-scale industrial production of CNTs, both current and planned, it is proposed that the main exposure route would be through inhalation. Therefore, the remainder of this section will concentrate on the pulmonary toxicity of CNTs, the evaluation of particle components (impurities), and redox potential.

14.3.1.1.2 Health Effects

14.3.1.1.2.1 Deposition and Clearance

The deposition and clearance of CNTs from biological systems is a contentious subject and will obviously vary considerably based on the exposure route and the material size, shape, electrostatic behavior, and functionalization. The deposition of aerosolized CNTs throughout the pulmonary system is a subject that in itself causes much controversy as it may be governed by their electrostatic nature, resulting in the accumulation of fibers into agglomerates; the size of which may not allow inhalation into areas of the lung that other nanomaterials can reach (Muller et al. 2005). This makes inhalation studies problematic, with respiratory investigations being done using instillation or aspiration techniques. Using these methods it has been shown that 80% of a MWCNTs sample can remain within a rat lung for at least 60 days, this concentration is reduced to 36% with physical grinding (resulting in shorter tubes) prior to exposure (Muller et al. 2005), indicating the importance of length. Dispersal is another key factor, with well dispersed samples of both MWCNTs and SWCNTs displaying enhanced distribution within lung tissue (Muller et al. 2005, Mercer et al. 2008), while agglomerates formed from a less dispersed CNTs sample will remain localized (Muller et al. 2005, Shvedova et al. 2005), and will predominately not reach the smaller airways (Muller et al. 2005). This increased dispersal is often associated with a higher toxicity, but it has been shown by Murr et al. (2005) and Shvedova et al. (2005) that this is not always the case, as the agglomerated samples can in fact result in a higher toxicity.

As mentioned, the electrostatic nature of CNTs will influence their inhalation but, as stated by Shvedova et al. (2005), it is plausible that the inhalation of CNTs could take place, particularly through increased production or environmental exposure; and although problematic, the inhalation of MWCNTs has been investigated. It was found that although alveolar macrophages laden with CNTs and particle deposition (both dispersed and agglomerated) were dose dependently increased throughout the lungs, there was no further evidence of pulmonary toxicity (Mitchell et al. 2007). However, systemically, a reduced and inadequate immune response was found, through the measure of T-cell activation and proliferation, in mice exposed to MWCNTs via inhalation, and may indicate an avoidance of the defense mechanisms of the respiratory system (Mitchell et al. 2007). CNTs instillation and aspiration studies have also indicated a systemic influence or translocation of CNTs from the lungs, when SWCNTs exposed mice presented signs of oxidative stress and DNA damage in the aortic tissue and a tendency for increased atherosclerosis (under the right conditions), coinciding with the oxidative stress evident in the lungs (Li et al. 2007). It is not, however, clear if this influence in the cardiovascular system is a direct result of the exposure to SWCNTs, or if the influence is an indirect result of pulmonary toxicity. However, the clearance of MWCNTs was assessed by Deng et al. (2007); after intratracheal instillation a large proportion of MWCNTs remain in the lung for only 1 day, and are very quickly removed, with no evidence of transfer into the blood (Figure 14.3). This clearance is likely due to the ability of alveolar macrophages to attach and readily phagocytose CNTs (Pulskamp et al. 2007a). However, this raises a question of fiber length and biopersistence and introduces the "fiber paradigm." The maximum length for successful phagocytosis by alveolar macrophages is believed to be 20 μm (Dörger et al. 2001). Therefore, fibers such as CNTs, are believed to be particularly toxic when they have a length of over 20 μm, a width of less than 3 μm, and are durable and biopersistent (Donaldson and Tran 2004). This results in phagocytic cells attempting and failing full phagocytosis, ultimately undergoing frustrated phagocytosis that leads to prolonged deposition, biopersistence, and the induction of oxidative stress, acute and chronic inflammation, enhanced cell proliferation (Ye et al. 1999, Dörger et al. 2001, Brown et al. 2007, Poland et al. 2008), and formation of foreign body giant cells (FBGC) (Poland et al. 2008).

Other routes of exposure have been investigated, often, however, using functionalized CNTs (fCNTs). Intravenous injection of taurine (a naturally occurring amino acid derivative) fMWCNTs are

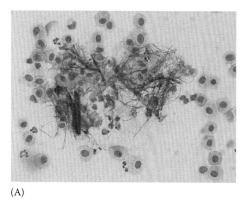
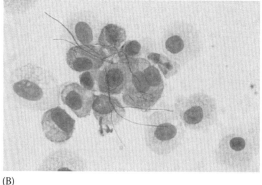

(A)   (B)

FIGURE 14.3 Light microscopy images of bronchioalveolar lavage (BAL) cells treated with MWCNTs (62.5 μg/animal by instillation, 18 h exposure before lavage). Images shown are at (A) ×40 (B) ×100 magnification.

found to quickly transfer to the liver, heart, lung, and spleen, and remain unchanged in the liver for at least 28 days (Deng et al. 2007). Intraperitoneal injection of SWCNTs, functionalized for solubility and radiotracing, are also found to extensively distribute throughout the body, predominantly depositing in the stomach, kidneys, and bone, and remaining at relatively high levels within the bone for 11 days (Wang et al. 2004). Direct stomach intubation has been shown to cause no entry of CNTs into the blood, but they entered the intestinal tract and were excreted via the feces (Deng et al. 2007). The distribution and clearance of CNTs has also been examined by Singh et al. (2006), who found that the intravenous administration of the CNTs, again functionalized, resulted in distribution throughout many organs and tissues within 30 min, irrespective of surface charge with no organ specific tendencies (Singh et al. 2006). However, they were subsequently and quickly excreted through the urinary tract, with the excretion pattern relatively unchanged by surface charge (Singh et al. 2006). This relatively safe behavior of CNTs shown by Deng et al. (2007), Singh et al. (2006), and Wang et al. (2004) may or may not be attributed to functionalization. However, this provides some answer to the durability of CNTs and again raises the question of biopersistence as transmission electron microscopy images of tissue and urine samples show the CNTs to be unchanged and intact (Singh et al. 2006, Deng et al. 2007).

14.3.1.1.3 CNTs' Toxicity

In vivo, CNTs are found to increase granuloma formation, alveolar wall thickening (Lam et al. 2004, Shvedova et al. 2005, 2007, Mercer et al. 2008), and fibrosis (Muller et al. 2005, Shvedova et al. 2005, 2007); with observed increased cellular proliferation in response to SWCNTs and significant increase in the profibrotic growth factors: platelet-derived growth factor −B and −C (Mangum et al. 2006). Additional cell injury has been observed by an increase in LDH, protein (Muller et al. 2005, Shvedova et al. 2005), γ-glutamyl transferase (GGT), oxidative stress and alveolar type II (AT-II) cells (indicating alveolar type I (AT-I) cell death) (Shvedova et al. 2005). An increase in granuloma formation (and enhanced inflammatory response) was observed following injection of particularly high aspect ratio CNTs into the peritoneal cavity, especially, when compared to negative controls and shorter CNTs samples, further elucidating the "fiber paradigm," but also when compared to a long asbestos sample (long fiber amosite) (Poland et al. 2008), raising great concerns for the toxicity and increased production of CNTs.

Dose-dependent deterioration of lung function and lung defense is also observed in response to CNTs, with an increased expiratory time, and decreased clearance of infectious agents, such as bacteria (Shvedova et al. 2005); with the ability of the CNTs to appropriate proteins key in pulmonary defense (Salvador-Morales et al. 2007, Shvedova et al. 2007). Two proteins found within pulmonary surfactant (SP-A and SP-D) play a key role in the pulmonary defense against allergens and microorganisms (Salvador-Morales et al. 2007), through the induction of calcium-signaling-dependent phagocytosis (Ohmer-Schrock et al. 1995). Salvador-Morales et al. (2007) reported that calcium-ion-dependent binding occurs between functionalized (with ketones, aldehydes, or carboxyl

groups) CNTs and these two surfactant proteins. This highlights two possible outcomes: firstly, the proteins are not available for defense against pathogen infection; and secondly, it again raises the importance of fiber length—the role of SP-A and SP-D in pulmonary defense involves the enhancement of phagocytosis. Therefore, there is probably the encouragement of alveolar macrophages to attempt to phagocytose impossible tube lengths (Salvador-Morales et al. 2007), with the similar outcome of frustrated phagocytosis, biopersistence, and enhanced and maintained inflammation. In terms of an inflammatory response, an influx of various immune cells is found (Muller et al. 2005, Shvedova et al. 2005, Mercer et al. 2008, Poland et al. 2008) in response to CNTs, with an accompanying rise in pro-inflammatory cytokines (Muller et al. 2005, Shvedova et al. 2005, Poland et al. 2008).

In vitro SWCNTs are shown to cause the release of high levels of TGF-β1 by macrophages (Shvedova et al. 2005) indicating the need for tissue regeneration, and both MWCNTs and SWCNTs are shown to cause a marked increase in the pro-inflammatory cytokine TNF-α when exposed to lipopolysaccharide (LPS) activated macrophages (Pulskamp et al. 2007a). Little cytokine release (TNF-α, IL-1β, IL-12, IL-10, IL-8) is observed in response to SWCNTs alone (Murr et al. 2005, Shvedova et al. 2005, Pulskamp et al. 2007a,), while MWCNTs are seen to increase TNF-α production, TNF-α mRNA expression, and LDH release, when well dispersed (Muller et al. 2005), but not when aggregated (Murr et al. 2005). At realistic doses CNTs have not been shown to cause significant increases in cell death, either by apoptosis or necrosis in alveolar macrophages (Pulskamp et al. 2007a,b); but are found to reduce mitochondrial membrane potential (Pulskamp et al. 2007a).

14.3.1.1.3.1 Oxidative Stress and Toxic Components of CNTs

The ability for CNTs to cause oxidative stress is an important aspect of their toxicity and is therefore well documented, although the literature is not always complementary.

CNTs have been shown to directly and indirectly influence the redox state of a cell and cause oxidative stress. The presence of ROS, and subsequent oxidative stress, is clearly evident in response to SWCNTs with increased measurements of lipid peroxidation, extracellular superoxide dismutase cleavage, decreased antioxidants glutathione and ascorbic acid (Shvedova et al. 2007), and conversion of superoxide radicals to hydroxyl radicals (Kagan et al. 2006); with the ability of CNTs to cause oxidative burst and direct generation of ROS clearly demonstrated by Pulskamp et al. (2007b). However, these data, alongside data collected by Kagan et al. (2006), raise the question of toxicity influencing factors found within CNT samples, including the content of iron and amorphous carbons. Iron is known to be a particularly toxic agent with a high redox potential (Papanikolaou and Pantopoulos 2005), the presence of which in the amphibole group of asbestos is believed to partly contribute to their toxicity (Shukla et al. 2003), as it is in CNT samples containing iron remnants from the production process (Oberdorster et al. 2005a,b). It was shown that purified (reduced metal content) CNT samples lack the ability to initiate an oxidative burst or produce ROS (Pulskamp et al. 2007b) and that non-purified samples with a high metal content can not only cause oxidative burst but do so in only 10 min, and dose dependently after 24 h (Pulskamp et al. 2007b), resulting in a greater ability in generating hydroxyl radicals in activated macrophages (Kagan et al. 2006). Interestingly, CNT samples with reduced levels of amorphous carbon cause a relatively delayed increase in ROS irrespective of iron content (Pulskamp et al. 2007b). The work by Pulskamp et al. (2007a) substantiate this behavior, wherein pristine SWCNTs and MWCNTs were found to induce ROS production in alveolar macrophages, and SWCNTs with a reduced metal content resulted in no ROS increase compared to controls.

The specificity of ROS production is also dependent on CNT' composition. Peroxynitrite ($ONOO^-$) can be generated in samples with a high metal content as well as a rapid and sustained generation of the superoxide anion (O_2^-), in response to CNT samples with high metal and amorphous carbon content, but is limited to 24 h postexposure when amorphous carbon is removed (Pulskamp et al. 2007b).

However, in direct conflict, *in vitro* investigation has found no evidence of oxidative stress through the measure of oxidative stress marker, HO-1 expression (Pulskamp et al. 2007b). Similar to Kagan et al. (2006), Shvedova et al. (2005) found no evidence of SWCNTs exposed macrophages initiating oxidative burst or production of nitric oxide (NO) or attempting to actively phagocytose the fibers. It was also shown by Pulskamp et al. (2007a) through the measure of nitrite concentrations that NO is not produced in response to SWCNTs or MWCNTs. Additionally it was shown that LPS induced formation of NO is dampened with the inclusion of MWCNTs. This is believed to be due to the binding of CNTs to the protein inducible nitric oxide synthase (iNOS) (Pulskamp et al. 2007a). This again raises questions to the hosts' ability to react to inhaled pathogens with previous exposure to the CNTs.

As previously mentioned the iron content of asbestos and CNTs is implicated in their toxicity. As a common production method of CNTs (catalytic chemical vapor deposition) uses iron as a catalyst, there is often a high level of iron in a CNT sample. It is therefore important to assess the influence this component has on CNT toxicity. Although some believe the iron contained in a sample of CNT is not readily available, and will therefore not, in itself, cause duress upon CNTs-exposed biological systems, this is often assumed using observations of the location of iron using electron microscopy. The difference in effect between the purified and non-purified CNT samples has been shown by Kagan et al. (2006) and Pulskamp et al. (2007b). More importantly, it has been shown by using measurements of iron mobilization and redox activity that although most iron remains enclosed and inaccessible within the CNTs, a small quantity is released into suspension fluids, which is sufficient to cause redox reactions, subsequently causing single strand breaks in DNA (Guo et al. 2007).

14.3.1.1.3.2 Other Exposure Routes for CNTs
A large proportion of this review has concentrated on the respiratory toxicity of CNTs, as this is the most likely common exposure method, although other routes of exposure include dermal, intestinal, and intravenous.

Human epidermal keratinocytes (HEK) exposed to CNTs are found to readily take them up, where they appear free in the cytoplasm and held within vacuoles. Although CNTs were not found within the cells' nuclei, they were found in close proximity and possibly scoring the nuclear membrane (Monteiro-Riviere et al. 2005a). CNTs are found to induce free radical production and cause oxidative stress when exposed to HEK (Shvedova et al. 2003), resulting in morphological changes (Shvedova et al. 2003) and reduced viability (Shvedova et al. 2003, Monteiro-Riviere et al. 2005a) with dose and time dependent increase in a pro-inflammatory reaction (Monteiro-Riviere et al. 2005a). When human dermal fibroblasts were exposed to purified (iron removed) SWCNTs, reduced cell survival and cell adhesion were observed, more so than when treated with other carbon nanomaterials, such as MWCNTs, active carbon, CB or carbon graphite (all purified) (Tian et al. 2006). A reduced expression of adhesion and cell cycle–related proteins were observed upon SWCNTs' exposure, in comparison to untreated cells, and an increase in cell death in response to a purified SWCNT sample that was not observed in the sample containing its catalytic iron content (Tian et al. 2006).

When administered intravenously CNTs are likely to be functionalized, which will change their behavior and interactions considerably, often reducing toxicity, as discussed in Section 14.3.1.1.2.1. However, toxicological research into this exposure route has shown that relatively pure SWCNTs can cause a reduction in the cell number of human peripheral blood lymphocytes, not through apoptosis or necrosis, but through reduced metabolic activity (Zeni et al. 2008). While MWCNTs have been shown to reduce the viability of human T-lymphocytes, significantly more so when samples are oxidized, compared to pristine samples. This decrease in viable cells was found to be dose and time dependent in both MWCNT samples and was through the initiation of apoptosis (Bottini et al. 2006), causing programmed cell death. With functionalization, the lack of response upon fCNTs' intravenous administration reported by Deng et al. (2007) and Singh et al. (2006), is likely to relate to the type of functionalization. For example, ammonium functionalized SWCNTs were found to elicit no activation or cytotoxicity of primary T-cells, B-cells, or macrophages, while polyethylene glycol (PEG) functionalized SWCNTs (although they did not cause significant cytotoxicity or T-cell and B-cell activation) were readily taken up by primary immune cells and caused the activation of macrophages, which was not seen in response to PEG alone (Dumortier et al. 2006).

The research into CNTs' toxicity to date has at times clearly demonstrated the ability of CNTs to follow not only the intrinsic toxicity and pathogenicity of other nanomaterials, but also that of asbestos, in particular circumstances. It is therefore vital that as the production and utilization of CNTs increase, the investigations into CNT toxicity continues, as there is a clear need to fill the gaps and elucidate the contradictions found within the literature. This applies particularly to those relating to sample purity, in respect to metal components and carbon structures found within CNT samples to the clarification of the extent of respiratory deposition upon actual CNTs inhalation, to the likely exposure levels to be expected during large-scale production, to the degree of biopersistence (through tube length and durability), and the translocation of CNTs within and from the respiratory system. All of these points need to be considered and clarified to aid the progression of safe production and utilization of CNTs.

14.3.2 Additional Carbon-Based Nanomaterials

14.3.2.1 Ultrafine Carbon Black

As previously stated, the ultrafine component of PM_{10} was identified as a constituent of particulate air pollution that mediated the toxicity evident in epidemiological studies. In order to assess the mechanisms driving the toxicity of particulate air pollution, ufCB was often used as a representative ultrafine particle to model the composition of PM_{10} in studies that aimed to characterize the consequences of PM_{10} exposure. The ufCB was deemed a particularly relevant "model" particle to use as it comprises of up to 50% of the total mass of PM_{10} (Donaldson and Stone 2003).

Therefore, a number of studies were conducted using ufCB (see for example Li et al. 1999) to assess the toxic potential of ultrafine particles within lung models. There is a vast array of background information available that indicates that ufCB is more potent at inducing inflammation in the lung than fine CB (Li et al. 1999, Brown et al. 2000, Wilson et al. 2002), with the toxicity driven by the larger surface area of ufCB (Stoeger et al. 2006). *In vitro*, ufCB has also been shown to elicit a pro-inflammatory response, characterized by the expression of the cytokines TNF-α in macrophages (Brown et al. 2004) and IL-8 by epithelial cells (Monteillier et al. 2007). In macrophage models, the up-regulation of TNF-α has been shown to be controlled by a complex series of events that ultimately results in the activation of mediators of gene transcription due to rises in the development of intracellular calcium and oxidative stress (Brown et al. 2004). Consequently, due to the known toxicity of ufCB, it is now considered to be a benchmark nanomaterial, often used in the comparison of effects of exposure to other engineered nanomaterials. It is also necessary to consider ufCB toxicity itself since it has a number of commercial applications. For example, it is used as a toner in photocopiers and printers and is also used by the rubber industry in the manufacture of tires (Reijnders 2006).

14.3.2.2 Carbon Fullerenes (C_{60})

Carbon fullerenes (buckminsterfullerene or "buckyballs," C_{60}), are composed exclusively from carbon atoms (60 in total) that are arranged in a spherical, football-like, cage structure, and have a diameter of about 1 nm (Whatmore 2006). Fullerene production can occur naturally, as they can be released from combustion processes, such as forest fires (Powell and Kanarek 2006) or be intentionally produced for a specific purpose. It is known that the water solubility of fullerenes is low, and for this reason, a number of techniques have been employed to improve

their solubility, suspension, and reduce their tendency to aggregate. This is achieved through the surface modification of fullerene molecules, by preparation using solvents or through their extended stirring in water (Dhawan et al. 2006). Therefore, the implications of the use of these processes have been a focus of a number of studies using fullerenes.

There is evidence of C_{60} toxicity. Sayes et al. (2005) found that C_{60} exerted cytotoxicity through lipid peroxidation (and resultant membrane damage) that is reliant on the development of oxidative stress. It was also observed that the administration of the antioxidant ascorbic acid could prevent the development of oxidative damage and therefore the appearance of a C_{60} oxidant-mediated toxicity. Kamat et al. (2000) also demonstrated that C_{60} could exhibit toxicity through the generation of ROS and lipid peroxidation within rat liver microsomes. These findings correlate to those demonstrating that ultrafine particulate–mediated toxicity is associated with oxidative stress. Furthermore, it has been suggested that the surface modification of C_{60} impacts on its toxicity, specifically that "pure" C_{60} is more toxic than hydroxylated $C_{60}(OH)_{24}$ *in vitro* (Sayes et al. 2004), and that this could aid in the design of nanomaterials to render them less toxic.

Contrary to findings that have demonstrated that C_{60} induces oxidative stress to mediate its toxicity, there is evidence that C_{60} can act as a free radical scavenger, and thus have antioxidant properties (Gharbi et al. 2005). This was illustrated by the fact that C_{60} administration can protect the liver from carbon tetrafluoride-mediated liver damage due to its free radical scavenging activity. The antioxidant potential of C_{60} is thought to be dependent on its dispersion, and thus aggregates of C_{60} are thought not to exhibit this antioxidant property (Gharbi et al. 2005). For example, Gharbi et al. (2005) provided evidence that C_{60} pretreatment was able to protect against the toxicity of carbon tetrachloride within rats. For this reason, fullerenes may be exploited as therapeutic agents, within the treatment of oxidative driven diseases.

14.3.3 Technical Applications of Carbon-Based Nanomaterials

The advantageous properties of CNTs, including incredible strength (Salvetat-Delmotte and Rubio 2002), high electrical current carrying ability (Wei et al. 2007), and easy functionalization (Wei et al. 2007), allow an extensive range of applications, in many industries. The impact may be most noticeable in medicine, including support structures for tissue regeneration (Popat et al. 2007, Abarrategi et al. 2008), vessels for drug delivery systems (Wu et al. 2005, Popat et al. 2007), or delivery plasmid DNA in gene therapy (Pantarotto et al. 2004, Liu et al. 2005), and as diagnostic and imaging tools (Wu et al. 2005). The benefits in using these materials are evident as the interactions of fCNTs with cells and biological systems often appear innocuous. Nevertheless, with this rise in use there is obviously an increase in exposure, and the possibility of biological complications. It is therefore vital that the risk of environmental and occupational exposure is properly assessed. In addition, due to the structure of C_{60} nanomaterials they, similarly to CNTs, have also been identified as possible drug delivery systems for nanomedicine.

14.3.4 Nanomedicine

Nanomedicine is an emerging sub-discipline of nanotechnology that intends to engineer innovative medicines to encompass the many areas of drug delivery, drug design, imaging, therapeutics, and diagnostics at the nm scale (Sahoo and Lahasetwar 2003, Salata 2004, Ferrari 2005, Jain 2005, Moghimi et al. 2005, Caruthers et al. 2007).

The European Science Foundation (ESF) describes nanomedicine as the science and technology of diagnosing, treating, and preventing disease and traumatic injury, of relieving pain, and preserving and improving human health, using molecular tools and knowledge of the human body (ESF 2005). It further defines nanomedicine as the science and technology of nm size scale complex systems, consisting of at least two components, one of which being the active ingredient (ESF 2005). The report states, however, that the nm size scale complexes can range from 1 to 1000 nm in size, and therefore by definition not all nanomedicines are nanomaterials, following the standard definition (BSI 2007, SCENIHR 2007b, ISO 2008).

The use of nanomaterials in medicine, however, due to their unique size-dependent properties, is widely agreed to be advantageous in the production of nanopharmaceuticals (Silva 2004, Thrall 2004, Moghimi et al. 2005, Pison et al. 2006, Caruthers et al. 2007). Comparable to all engineered nanomaterials, nanomedicines can comprise of either homogeneous, heterogeneous, or multi-functional structures, consisting of a core with either a surrounding surface (shell) material, or attached macromolecular branches and can be further constituted by manipulations to exhibit a range of various surface coatings/charges (Pison et al. 2006, Caruthers et al. 2007). Examples of nanomaterials intended for use in medicine include, colloidal gold NPs, ferromagnetic beads, such as super paramagnetic iron-oxide nanoparticles (SPIONs), both SWCNTs and MWCNTs, C_{60} fullerenes, micelles, dendrimers, and QDs (Svenson and Tomalia 2005, Azzazy et al. 2006, 2007, Lacerda et al. 2006, Mornet et al. 2006, Sahoo et al. 2007).

These specific nanomaterials, as well as many more, are being developed for their application in specific areas of clinical medicine, such as diagnostics and bioimaging (Penn et al. 2003, Alivisatos 2004, Sharma et al. 2006), polymer therapeutics (West and Halas 2003, Chellat et al. 2005), drug and gene delivery systems (Moghimi 1995, Reynolds et al. 2003, Emerich and Thanos 2006), antibody technologies (Sahoo et al. 2007), polymer-drug conjugates (Jagur-Grodzinski 1999), polymer-protein, and antibody conjugates (Pison et al. 2006), as well as tissue engineering and repair (Kawasaki and Player 2005, Azzazy et al. 2006).

Although the ability to manufacture and manipulate nanomaterials to suit a specific purpose for medical application is well documented (Moghimi et al. 2005, Pison et al. 2006, Caruthers et al. 2007), there are major concerns as to the effects of these

nanomedicines once administered within the human body (ESF 2005). These include the effectiveness of gene and protein delivery, the biocompatibility of nanomedicines and their ability to overcome biological barriers, as well as the most advantageous routes of administration (Azzazy et al. 2006, Pison et al. 2006).

One particular aspect concerning the use of nanomaterials as nanomedicines is their possible toxic effects and potential to cause adverse health effects following exposure (ESF 2005, Medina et al. 2007). The ESF have stated that an urgent understanding of the toxicological implications of nanomedicines is essential prior to their proposed clinical use (ESF 2005). The toxicity of many of the nanomaterials intended for use in medical applications, such as CNTs, SPIONs, and dendrimers, however, is not well documented. This is mainly due to the ability to manipulate the surface of these nanomaterials, forming many different surface coatings and surface charges, which have been previously suggested as possible parameters of toxicity and adverse health effects following exposure to engineered nanomaterials (Oberdorster et al. 2005b). It is imperative, therefore, that an understanding of the potential deleterious effects of the wide range of nanomaterials to be used in nanomedicine, taking into consideration the many different routes of exposure that these NPs could be administered (e.g., via injection, orally, and dermal), is attained in order to determine their (1) effectiveness, (2) sustainable use, and (3) potential hazardous effects, prior to their application within a clinical setting.

Although, as previously highlighted, there is a large and diverse array of NPs intended for use in nanomedicine, the remainder of this discussion will specifically focus on QDs.

14.3.4.1 Quantum Dots

QDs are a type of nanoparticle intended for use in medical applications (Pison et al 2006). These highly fluorescent, semiconductor nanomaterials range from 1 to 25 nm in diameter, but can be altered to up to 100 nm by a variety of surface modifications, often with polymers of varying chain lengths (Hardman 2006).

QDs are spherical with a heterogeneous structure consisting of a core material, such as cadmium telluride (CdTe) or cadmium selenide (CdSe). This core is often coated with a shell, for example zinc sulphide (ZnS). The shell can then be covered with particular chemical surface coatings such as neutral organic (hydrophobic), amino or amine (positively charged and hydrophilic), or carboxyl (negatively charged and hydrophilic) moieties. A further surface coating of polymers such as PEG can also be attached to QDs.

Due to their size-dependent fluorescence spectra, narrow emission, and high photostability, QDs have been postulated to be good alternative bio-imaging tools *in vivo* and *in vitro* (Jaiswal et al. 2003, Ness et al. 2003, Lidke et al. 2004, Nisman et al. 2004, Gao et al. 2005, Giepmans et al. 2005, Mason et al. 2005, Pinaud et al. 2006). Their ability to emit light in the infrared spectrum that can penetrate tissue, combined with the possibility of targeting specific cell types through coating materials, carries the potential to make them an advantageous and non-invasive diagnostic tool for visualizing structures such as tumors (Wu et al.

2003, Gao et al. 2004, Kim et al. 2004, Michalet et al. 2005), as well as the tracing of cancer cells *in vivo* (Hoshino et al. 2004a). It is these characteristics, however, together with the component material of QDs that have resulted in the concerns raised in regards to their potential toxicity.

However, knowledge about the toxicity of QDs, as well as the cellular responses and processes following exposure to these nanomaterials are limited. Furthermore, as some of the metals present in the heterogeneous composition of these nanomaterials, such as Cd have been reported to be highly cytotoxic (IPCS 1992), it is imperative that an understanding of the potential toxic and adverse health effects, as well as the cellular interactions of QD is attained prior to any form of medical use in humans.

QDs come in various sizes, core types, with or without shell, and with or without additional coatings. Therefore, any toxicity observed could be related to either of these properties, in addition to dose, exposure time, and cell type or organism exposed. It is not surprising that some of the results found to date are conflicting or warranting further investigation.

14.3.4.2 QD Toxicity

An early study by Derfus et al. (2004) on the possible adverse health effects of QDs assessed the toxic potential of "naked" QDs (QDs with no shell layer) with a core material of CdSe on hepatocyte cells *in vitro*. Derfus et al. (2004) reported a concentration-dependent toxicity, related to the release of free ions of the toxic heavy metal cadmium (Cd^{2+}) from the core of the QD following its oxidation. The hypothesis that the release of Cd^{2+} from the core would be reduced by the addition of a shell was not confirmed, since studies have shown Cd^{2+} to be released from QDs both with and without a shell layer (Derfus et al. 2004, Kloepfer et al. 2005). It has even been suggested that the shell might decrease the stability of the QD and subsequently cause a greater toxic response (Michalet et al. 2005, Clift et al. 2008). Further research into how effective the shell layer can be in reducing Cd-related toxicity, as well as what shell material would be most suitable for QDs is required at this stage.

It was further suggested by Derfus et al. (2004) that the addition of chemical surface coatings to QD with a shell layer could possibly reduce QD toxicity. This was refuted in a subsequent investigation by Hoshino et al. (2004b), who reported that the type of QD surface chemistry is an indicator of toxicity rather than the core material, as suggested by Derfus et al. (2004). It was found that QDs with a negatively charged, carboxylated chemical surface coating were less toxic compared to those with a positively charged, amine coating. It was also highlighted that the amine coated QD were more susceptible to a reduction in stability over time, in comparison to the carboxylated QD that were observed to maintain structure. These findings were further supported by Shiohara et al. (2004) who found positively charged CdSe/ZnS QD to reduce in stability over a 6 h period with no apparent adverse effects on hepatocyte cell viability *in vivo*. It was further observed by Shiohara et al. (2004) that CdSe/ZnS QD were highly toxic irrespective of coating and induced cell death even at low concentrations.

Further studies found QD toxicity to increase concomitantly with a decrease in size (Lovric et al. 2005a). The size of QDs was also found to be specific to the localization of the QD inside the cell. Another interesting finding from Lovric et al. (2005a) was related to cell death. It was observed that cell death was more prominent within the smaller-sized QDs. Pretreatment of QDs with antioxidants, however, was found to negate any form of cell death to occur. It has also been reported by Lovric et al. (2005b) that "naked" CdTe QDs inhibit cell function, increase production of ROS, and subsequently induce cell death via apoptosis. Direct comparisons, however, between the various findings of Lovric et al. (2005a,b) cannot be made due to the difference in the surface coating on the QD examined.

14.3.4.3 Fate of QDs

From the studies mentioned in the previous section, it has been further highlighted that the intracellular localization of the QDs might be important for possible toxic and adverse cellular responses. The actual cellular process related to the uptake of QDs is limited. As described in the study by Hoshino et al. (2004b), carboxylated, negatively charged QDs were taken up via a form of endocytosis, and subsequently observed to be located in endosome-like vesicles within EL-4 murine tumor cells. Studies on the effects of CdSe/ZnS QD on eukaryotic cells, have also found avidin-conjugated and organic CdSe/ZnS QD to be taken up into endosomes (Jaiswal et al. 2003, Derfus et al. 2004), positively charged CdSe/ZnS QD into lysosomes (Hanaki et al. 2003), as well as CdSe/ZnS QD covered with a micelle/polymer coating into organelles (Dubertret et al. 2002). In relation to these findings, further investigation of CdSe/ZnS QD with an amphiphilic micelle coating in mice has reported that the coating on QD influences the mechanism of uptake, as well as the location of the NP (Larson et al. 2003). Studies into the uptake of QDs, however, have predominantly examined CdSe/ZnS QD, and have not investigated the numerous other QDs available. Further research is required therefore to fully understand the cellular processes and fate of the many different types of QDs.

In summary, the toxic effects, as well as the cellular processes, of QDs are not fully understood. The lack of knowledge on this nanomaterial type can be attributed to the plethora of different QDs available, as well as testing systems and protocols. In addition, since each form of QD has a very specific set of physiochemical properties, it is difficult to relate the different types of QDs to one another, and therefore to determine a scientific opinion on the adverse health effects of QDs. It is therefore imperative that increased research is focused on QDs in order to attain a definitive understanding of the toxic potential of these nanomaterials and their role within the cell, prior to any human exposure in an occupational setting or for medical use.

14.3.5 Ecotoxicology of Manufactured Nanomaterials

It is generally accepted that the environmental fate of nanomaterials, as well as their effects on environmental species, depends on the specific type and composition of the nanomaterial

(Klaine et al. 2008). Currently, it is not possible to attribute precise environmental effects to specific nanomaterials due to the severe lack of knowledge in this area (SCENIHR 2009).

Despite this, however, investigation into the fate and effects of nanomaterials on the environment has continuously increased since the first study in 2004 (Oberdorster 2004). In 2008, an influx of original papers (e.g., see Klaine et al. (2008) and SCENIHR (2009), for reviews) published in the field helped to elucidate the fate and effects of nanomaterials on the environment, however, it is clear that much work is still required in this field. Previously, research into the effects of nanomaterials on the environment, as well as individual, non-human, target species has been limited to a small number of test species and nanomaterials. Mainly, aquatic organisms have been studied, predominantly freshwater invertebrates and fish. Additionally, considerable work has also focused upon the effects of nanomaterials on microorganisms, although at this moment in time little is known as to the effects of nanomaterials on primary producers, as well as organisms living in sediment and soil systems.

The increased production and use of nanotechnology related products will result in greater amounts of nanomaterials reaching the environment, due to their inevitable disposal via waste (USA EPA 2007). In addition to these, and in a potentially large number of cases, nanomaterials will actually be incorporated into the remediation of contaminated environments, such as nanoscale zero valent iron (Fe^0) NPs for groundwater remediation (Henn and Waddill 2006), therefore making the release of nanomaterials into the environment intentional. Waste treatment systems, landfill sites, and the combustion of products containing nanomaterials are examples and means of how nanomaterials may end up in the environment (USA EPA 2007). Although it is most likely that the nanomaterials will be present within the environment as modified forms of their primary counterparts (Ju-Nam and Lead 2008), it is essential that further research is conducted in these areas in order to fully understand the environmental impact of nanomaterials and nanotechnology.

14.3.5.1 Fate of Nanomaterials in the Environment

The fate and behavior of nanomaterials also depends on the characteristics of the receiving environment (such as waste water and soil) and not only on the physical and/or chemical characteristics of the nanomaterial itself (Chen and Elimelech 2008, Saleh et al. 2008). Following release into the environment, nanomaterials may undergo a number of transformations such as dissolution, speciation, biotic or abiotic transformation (degradation), aggregation and/or agglomeration or settlement (Boxall et al. 2007a, Klaine et al. 2008). In addition, the nanomaterials are also likely to associate with other dissolved, colloidal, and particulate matter already present within the environment. The degree to which these processes occur will depend on the nanomaterial composition and the receiving environment. Nevertheless, the aggregation and/or agglomeration of the nanomaterials may lead to their settlement and subsequent association with sediments and soils (Baalousha et al. 2008, Klaine et al. 2008, Navarro et al. 2008a). It is likely, therefore, that organisms, such as worms or

crustaceans, living within sedimentary systems will experience high levels of exposure to nanomaterials via ingestion and respiration, and via their skin, exoskeleton, cuticle, or other body surfaces (Fernandes et al. 2007, Handy et al. 2008a). Furthermore, these species play a pivotal role in food chains, often providing the link between deposit feeders and higher trophic level species. It is therefore of key importance that substrate dwelling organisms are subjected to more detailed studies assessing uptake, accumulation, and effects (Klaine et al. 2008).

Currently, it is not known at what concentrations and quantities of nanomaterials exist in both aquatic and sedimentary systems (Boxall et al. 2007b, Mueller and Nowack 2008, SCENIHR 2009). Due to a lack of scientific studies and standardized methodologies, estimates for the quantity and concentration of nanomaterials in the environment have only been derived from scenarios based on predicted uses of products containing nanomaterials and not from actual environmental measurements (Boxall et al. 2007b, Mueller and Nowack 2008). It is important, that an evaluation of exposure levels be conducted so that an adequate assessment of potential environmental effects can take place. This evaluation must consider that nanomaterials might exist in higher levels in some areas due to their tendency to aggregate/agglomerate and potentially to adsorb to, or associate with, organic matter. Due to these issues, therefore, methods needed to measure nanomaterials in the environment are currently under development (Christian et al. 2008, Hassellöv et al. 2008, Tiede et al. 2008a, 2009).

14.3.5.2 Effects of Nanomaterials on the Environment

Nanotoxicology studies using cell cultures and animal models indicate that nanomaterials can cause oxidative stress, potentially inducing inflammation and subsequent disease states such as atherosclerosis and chronic obstructive pulmonary disease (COPD) (Ferin et al. 1992, Li et al. 1999, Brown et al. 2001, Oberdorster et al. 2005a, 2007). It is well known that the physical and chemical characteristics contribute, partly, to the biological activity of nanomaterials (SCENIHR 2009). Such characteristics can include size, surface area, shape, composition, charge, crystal structure, and solubility.

Upon entry to the environment (Figure 14.4), the nanomaterials might stay within a specific environmental compartment (air, soil, water, or sediment) or may translocate across boundaries. This translocation may occur in both directions, depending on the type of nanomaterial and the physical and chemical conditions of the receiving media. Organisms will have a role in the fate of these nanomaterials, directly or indirectly, via either their exudates or other such like organic material. Organisms may take up nanomaterials from the environment via respiratory surfaces, roots, leaves, skin, exoskeleton, or other body surfaces, as well as via their food. Within aquatic systems, nanomaterials may be taken up directly from materials in suspension or from sedimentary systems.

The entry processes of nanomaterials into microbial organisms can occur via a number of mechanisms, including diffusion (passive uptake), specific or non-specific active uptake, or even following cell membrane damage (Klaine et al. 2008). Microbial

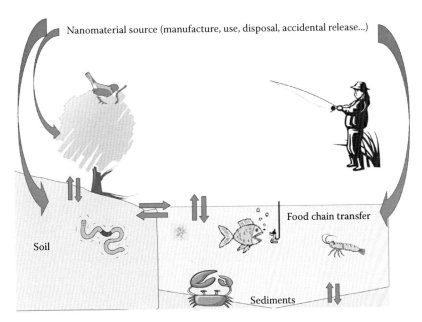

FIGURE 14.4 Upon entry to the environment, nanomaterials might stay within a specific environmental compartment (air, soil, water, or sediment) or may translocate across boundaries (short arrows). This translocation may occur in both directions, depending on the type of nanomaterial and the physical and chemical conditions of the receiving media. Organisms will have a role in the fate of these nanomaterials, directly or indirectly, via either their exudates or other such like organic material. Organisms may take up nanomaterials from the environment via respiratory surfaces, roots, leaves, skin, exoskeleton, or via their food. Within aquatic systems, nanomaterials may be taken up directly from materials in suspension or from sedimentary systems.

responses following exposure to and their subsequent entry of nanomaterials into the organism, include DNA damage, interference with energy metabolism, and the respiratory system, as well as both oxidative damage and ROS production (Sondi and Salopek-Sondi 2004, Lyon et al. 2006, Wiesner et al. 2006, Li et al. 2008, Lyon and Alvarez 2008, Klaine et al. 2008).

Silver, at the nanoscale, is used in a wide range of products as an antimicrobial agent, for example in wound dressings (Chen and Schluesener 2008). It is not clear, however, whether the antimicrobial effects are simply due to the dissolved silver (silver ions) or the effects observed are related to the size of the silver material (Luoma 2008). It has been suggested, however, that silver NPs may accumulate in the membrane of *Escherichia coli* (*E. coli*) bacteria, causing the cell walls to pit, thus increasing cell permeability and potentially inducing cell death via necrosis (Sondi and Salopek-Sondi 2004). In addition to silver, TiO_2-coated hollow glass beads have also been shown to inhibit the photosynthetic activity of cyanobacteria and diatoms, suggesting that these nanomaterials have a potentially useful application in preventing excessive algal growth (Kim and Lee 2005). Despite increased research being performed in relation to bacteria, there is still a severe lack of information relative to the effects of nanomaterials on soil species. A number of studies have investigated the effects of nanomaterials on the emergence and growth of plant species, specifically assessing the effects of alumina nanoparticles (Yang and Watts 2005, Murashov 2006), SWCNTs (Cañas et al. 2008) and copper NPs (Lee et al. 2008). In the studies of Yang and Watts (2005), as well as Murashov (2006), although negative effects were reported, it was unclear

if these effects were due to the nanoparticulate state of the aluminum, or if they were related to a soluble fraction of aluminum ions. Additionally, Cañas et al. (2008) have indicated that SWCNTs can affect the root elongation of crop species, while Lee et al. (2008) have shown the reduction in the growth of seedlings following exposure to copper NPs. The findings of Lee et al. (2008) are similar to those of Lin and Xing (2007), who reported that the copper NPs reduce seed germination and root growth due to the accumulation of copper NPs in the root cells. Investigation into additional soil species has shown TiO_2 NPs to reduce significantly the enzymatic activity of terrestrial isopods (*Porcellio scaber*) following exposure via ingestion. Despite these findings, however, the survival and growth of *Porcellio scaber* were not affected (Jemec et al. 2008). In addition, it has recently been reported that the reproductive ability of earthworms (*Eisenia veneta*) is significantly affected following the consumption of double-walled CNTs via their food (Scott-Forsmand et al. 2008).

In relation to aquatic species and microbes, assessment of the effects of nanomaterials has primarily focused upon primary producers [such as the freshwater microalgae *Pseudokirchneriella subcapitata* (Franklin et al. 2007, van Hoeke et al. 2008), *Desmodesmus subspicatus*, (Hund-Rinke and Simon 2006, Navarro et al. 2008a,b) and the marine macroalga *Fucus serratus* (Nielsen et al. 2008)], invertebrates [predominantly *Daphnia* species have been used, although other crustaceans, such as copepods and isopods have also been investigated (Hund-Rinke and Simon 2006, Lovern and Klaper 2006, Oberdorster et al. 2006, Templeton et al. 2006,

Zhu et al. 2006, Fernandes et al. 2007, Roberts et al. 2007, Jemec et al. 2008, Rosenkranz et al. 2009)], and fish [including rainbow trout (*Oncorhynchus mykiss*), zebra fish (*Danio rerio*), largemouth bass (*Micropterus salmoides*), fathead minnow (*Pimephales promelas*), and Japanese medaka (*Oryzias latipes*) (Oberdorster 2004, Griffitt et al. 2007, 2008, Federici et al. 2007, Lee et al. 2007, Smith et al. 2007, Zhang et al. 2007)]. In the majority of the studies referenced above, the authors have used specific endpoints such as mortality, growth, feeding, and reproduction as assessment tools relative to the effects of nanomaterials on each specific organism. Additionally, most of the studies have used specific biomarkers, such as assessment of oxidative stress assessed via TBARS or lipid peroxidation (either within specific cells, upon a specific organ, or within an entire organism), genetic damage, damage to specific cell organelles (such as the mitochondria), as well as the nucleus (Henry et al. 2007, Smith et al. 2007, Gagné et al. 2008, Handy et al. 2008a,b).

Those studies which have focused upon the effects of nanomaterials (such as TiO_2, CB, and polystyrene-latex NPs) on aquatic crustaceans (including *Daphnia magna*, *Artemia salina*, *Hyallela azteca*, and *Amphiascus tenuiremis*) indicate that nanomaterials can be ingested resulting in their accumulation within the gastrointestinal tract (Oberdorster et al. 2006, Fernandes et al. 2007, Roberts et al. 2007, Templeton et al. 2006, Rosenkranz et al. 2009). It has also been shown that the nanomaterials can adhere to the exoskeleton surfaces of the exposed organisms, suggesting that potentially multiple routes of exposure and absorption occur simultaneously. With regard to the *Daphnia magna*, Rosenkranz et al. (2009) observed that the fluorescent polystyrene NPs (20 nm) rapidly entered the fat-storing droplets of neonate *Daphnia magna* (Figure 14.5).

Investigations into fish species have reported a range of effects that include dose-dependent increases in ventilation rate, gill pathologies (edema, altered mucocytes, and hyperplasia), and mucus secretion (nanomaterial precipitation in the gill mucus) following exposure to nanomaterials (Handy et al. 2008a). Oxidative stress has also been observed to occur

in the gills and intestine, as well as in the brain of fish species (Handy et al. 2008a). Pathologies observed in the brain of rainbow trout (*Oncorhynchus mykiss*) included possible aneurysms or swellings on the ventral surface of the cerebellum (Smith et al. 2007). It was subsequently suggested by Smith et al. (2007) that SWCNTs are a respiratory toxicant in trout and that the observed cellular pathologies suggest cell cycle defects, neurotoxicity, and unspecified blood borne factors that could possibly mediate systemic pathologies. NPs, specifically polystyrene NPs, have also been reported to accumulate in both the eggs and embryos of the reported accumulations of polystyrene nanoparticles in eggs and embryos of the medaka *Oryzias antipes* (Kashiwada 2006), thus suggesting that NPs could pass through the membrane of the gills and/or intestine and enter the circulatory system. Despite these studies, little is known and understood regarding the fate of nanomaterials in the vast array of environmental species.

Although, as previously highlighted, the exposure of environmental species to nanomaterials is predominantly via the waste from nanotechnology, including the manufacturing of nanomaterials and use of nanomaterials (e.g., use of nanomaterials by humans in cosmetics, weathering, leachate, or washing of products containing nanomaterials and their subsequent disposal through the waste stream), species may also be exposed to nanomaterials via their food. The effects of nanomaterial exposure via the food chain, however, are not well studied or understood. Recently, Holbrook et al. (2008) reported that QDs could be transferred across different levels, from ciliated protozoans (*Tetrahymena pyriformis*) to rotifers (*Brachionus calyciflorus*). Although transfer was observed, that is, predators acquired QDs from their prey, bioconcentration (accumulation from surrounding environment) in the ciliates was limited and no biomagnification (enrichment across levels) in the rotifers was detected. Bouldin et al. (2008) also reported the transfer of the QDs from dosed algae (*Pseudokirchneriella subcapitata*) to the freshwater crustacean *Ceriodaphnia dubia*. These studies support the previous research by Fortner et al. (2005), who observed C_{60} fullerene NPs to accumulate in microbial cells and subsequently in worms, due to their consumption of the microbes.

14.3.5.3 Current State of the Field and Basis for Future Research

It is clear from the previous discussion that there are a number of areas that require immediate attention so that adequate hazard evaluation and environmental risk assessment can take place.

One area receiving increasing discussion and debate is the specific methodology used to assess the impact of nanomaterials on the environment in a laboratory setting. As previously discussed, nanomaterials (such as CNTs and fullerenes) are usually dispersed with the aid of a number of chemicals in order to maintain a stable and well-dispersed suspension. The use of mechanical or chemical means to suspend nanomaterials, however, may lead to changes in the properties and behavior of the test material (Brant et al. 2005, SCENIHR 2009). Despite this,

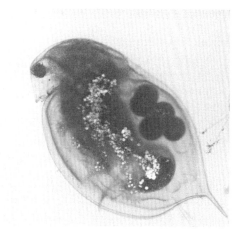

FIGURE 14.5 Confocal microscopy image of *Daphnia magna* exposed to 20 nm fluorescent carboxylated polystyrene beads (2.6 μg/L for 2 h).

it is unclear how, and if, these chemical dispersants influence the outcome of the laboratory-based studies. Tetrahydrofuran (THF) has been shown, however, to significantly alter the effects of nanomaterials in laboratory-based environmental studies, causing an observed enhancement of nanomaterial toxicity, such as reduced survival in larval zebrafish and other test species (Henry et al. 2007, Klaine et al. 2008). It is accepted and understood, however, that nanomaterials will encounter such chemicals that are used to stabilize and disperse nanomaterials in a laboratory setting. Despite this, it is essential that the effects observed within the laboratory be attributed to a direct effect of the nanomaterial and not resultant of either the material or an experimental artifact (Brant et al. 2005). Therefore, the role of naturally occurring chemicals, such as humic and fulvic acids, are now being assessed in relation to their use as dispersal agents for nanomaterials, as well as their consequent role in the bioavailability of nanomaterials (Gao et al. 2009). Currently, it is not fully understood in which conditions these naturally occurring acids may increase the dispersal of nanomaterials, and thus, either potentially enhancing or reducing the bioavailability of nanomaterials. In addition, it is widely agreed that it is imperative that standardized protocols for nanoecotoxicology research are established (SCENIHR 2009). It is essential that these experimental approaches consider appropriate positive and negative controls, including comparisons of nano-sized materials to their equivalent, larger, material counterparts. In addition, characterization of exposures via appropriate methods should be carried out and chemical analyses undertaken, as possible (Klaine et al. 2008).

Another aspect of "nanoecotoxicology" that raises increasing concern and debate is the ability to determine the precise number and amount of manufactured nanomaterials that are currently present and that enter the environment (via a variety of sources) daily. At this moment there are no monitoring methods that would allow the direct environmental assessment of these aspects (Tiede et al. 2009), nor is there any quantitative knowledge on the rates of release of nanomaterials into the environment. Preliminary research that has investigated these concerns however, has reported that depending on the nanomaterial, it is likely that any observed effects may partially result from its dissolution. Currently, it is not understood to what extent the effects observed for nanomaterials can be attributed to the dissolved form or to the nanomaterial itself, as it has been suggested that the nanomaterial effect reported could be a consequence of both the fraction and size of the nanomaterial (Franklin et al. 2007, Navarro et al. 2008b). Additionally, Lin and Xing (2008) have also suggested that phytotoxicity (reduced seed germination and root growth) observed following exposure of a range of plant species (radish, rape, ryegrass, lettuce, corn, and cucumber) to ZnO nanoparticles was unlikely to be completely due to dissolved Zn. Similar findings were also reported by Griffit et al. (2008) who drew a similar conclusion with respect to nanocopper in their studies on zebrafish (*Danio rerio*). These authors reported that the observed mortality, morphological effects, and global gene expression patterns in the gill could not be solely attributed to the dissolved fraction. It is essential, therefore, that further research is focused upon these aspects, as well as all the other aspects discussed in this section in order to help determine the actual effects that nanotechnology and nanomaterials might pose to the environment.

In conclusion, despite the recent advances in this area, there is still a severe lack of knowledge relating to the fate and effects of nanomaterials in the environment. In order to gain a thorough understanding of the interactions of nanomaterials with the environment and species living therein, it is essential that future studies are conducted upon a range of taxa, from different environmental strata, with diverse feeding modes and at different levels in the food chain, enabling the evaluation of a wide range of different and potential modes of nanomaterial uptake. In addition, short and long-term assessments of a variety of toxicological endpoints, and investigation of the fate of nanomaterial within whole organisms as well as key tissues and cells of species that are potentially threatened by nanomaterial exposure should be performed. Multispecies studies should also be undertaken so that different media, as well as different modes of uptake and trophic levels, are assessed to evaluate their relevance in terms of exposure. Finally, investigation into the diet of all organisms, as well as any possible interactions with other environmental contaminants should be conducted in order to determine the role of nanomaterials as carriers, and assess their uptake and translocation within whole organisms, and their tissues and cells.

14.4 Summary and Future Perspectives for Nanotoxicology

In summary, studies of the toxicological properties of nanomaterials have produced some very interesting and important results. The field is still very young, however, and is developing in parallel to new types, variations, and applications of nanomaterials. It is hoped, therefore, that researchers, regulating bodies and manufacturers of nano-products assess and understand the potential risk of these new materials during, or preferably, prior to their mass production, and not, as has been the case with a number of substances used in new technologies, in hindsight.

As the number of nanomaterials and their applications continues to rise, especially with developments in nanomedicine that will result in the direct application of nanomaterials into the human body, nanotoxicology will continue to be an important discipline. New classes of nanomaterials are constantly being developed, and challenges include characterization of nanomaterial properties such as size, surface area, and dispersion in different environments, as well as visualization in suspensions, organisms, tissues, and cells. Another challenge is to hypothesize and support new paradigms for the toxicity of nanomaterials, such as the fiber paradigm, and a set of reference materials in order to make predictions regarding the fate and effects of "new" products from these reference materials. The alternative, a complete characterization of nanomaterial properties and effects in a large number of species, tissues, cell cultures, and environments, for each nanomaterial of any type and size, is time- and cost-intensive, and thus not economically viable or possible. In addition, further challenges also include the modeling of specific exposure scenarios, modeling of material uptake and the proposal of viable alternatives of *in vivo* testing where appropriate.

Despite these many challenges, astonishing progress has been made in the last few years regarding new methodologies for analyzing, characterizing, and visualizing nanomaterials together with a significant improvement in the quality of nanotoxicological research due to the ability of research groups to join efforts, particularly, in the area of nomenclature and methodology development. All of these have resulted in a more coordinated approach being developed that will provide the foundations for the scientific assessment of the risk associated with nanoscience and nanotechnology.

Acknowledgments

The authors would like to thank staff and students at the Edinburgh Napier University who have contributed with data, information, and images to this work, as well as supervision, guidance, and technical support. We would like to acknowledge particularly the following: Vicki Stone, David Brown, Philipp Rosenkranz, Lesley Young, Marina Mocogni, John Kinross, and Nick Christofi, for their contribution to this work.

References

Abarrategi, A., Gutiérrez, M. C., Moreno-Vicente, C., Hortigüela, M. J., Ramos, V., López-Lacomba, J. L., Ferrer, M. L., del Monte, F. 2008. Multiwall carbon nanotube scaffolds for tissue engineering purposes. *Biomaterials*, 29(1), 94–102.

Aitken, R. J., Chaudry, M. Q., Boxall, A. B. A., Hull, M. 2006. Manufacture and use of nanomaterials: current status in the UK and global trends. *Occupational Medicine*, 56, 300–306.

Alivisatos, P. 2004. The use of nanocrystals in biological detection. *Nature Biotechnology*, 22(1), 47–52.

Azzazy, H. M. E., Mansour, M. M. H., Kazmierczak, S. C. 2006. Nanodiagnostics: A new frontier for clinical laboratory medicine. *Clinical Chemistry*, 52(7), 1238–1246.

Azzazy, H. M. E., Mansour, M. M. H., Kazmierczak, S. C. 2007. From diagnostics to therapy: Prospects of quantum dots. *Clinical Biochemistry*, 40, 917–927.

Baalousha, M., Manciulea, A., Cumberland S., Kendall, K., Lead, J. R. 2008. Aggregation and surface properties of iron nanoparticles: Influence of pH and natural organic matter. *Environmental Toxicology and Chemistry*, 27, 1875–1882.

Bang, J. J., Guerrero, P. A., Lopez, D. A., Murr, L. E., Esquivel, E. V. 2004. Carbon nanotubes and other fullerene nanocrystals in domestic propane and natural gas combustion streams. *Journal of Nanoscience and Nanotechnology*, 4(7), 716–718.

Belyanskaya, L., Manser, P., Spohn, P., Bruinink, A., Wick, P. 2007. The reliability and limits of the MTT reduction assay for carbon nanotubes–cell interaction. *Carbon*, 45, 2643–2648.

Borm, P. J. A., Robbins, D., Haubold, S., Kuhlbusch, T., Fissan, H., Donaldson, K., Schins, R. et al. 2006. The potential risks of nanomaterials: A review carried out for ECETOC. *Particle and Fibre Toxicology*, 3, 11–35. http://www.particleandfibretoxicology.com/content/3/1/11 [consulted April 2009].

Bottini, M., Bruckner, S., Nika, K., Bottini, N., Bellucci, S., Magrini, A., Bergamaschi, A., Mustelin, T. 2006. Multiwalled carbon nanotubes induce T lymphocyte apoptosis. *Toxicology Letters*, 160(2), 121–126.

Bouldin, J. L., Ingle, T. M., Sengupta, A., Alexander, R., Hanningan, R. E., Buchanan, R. A. 2008. Aqueous toxicity and food chain transfer of quantum dots in freshwater algae and *Ceriodaphnia dubia*. *Environmental Toxicology and Chemistry*, 27(9), 1958–1963.

Boxall, A. B. A., Tiede, K., Chaudhry, Q. 2007a. Engineered nanomaterials in soils and water: How do they behave and could they pose a risk to human health. *Nanomedicine*, 2(6), 919–927.

Boxall, A. B. A., Chaudhry, Q., Sinclair, C., Jones, A., Aitken, R., Jefferson, B., Watts, C. 2007b. Current and future predicted environmental exposure to engineered nanoparticles. DEFRA Report.

Brant, J., Nelevoanet, H., Hotze, M., Wiesner, M. 2005. Comparison of electrokinetic properties of colloidal fullerenes (n-C_{60}) formed using two procedures. *Environmental Science and Technology*, 39, 6343–6351.

British Standards Institution (BSI) 2007. Publicly Available Specification (PAS) 136; Terminology for nanomaterials; First published December 2007.

Brown, D. M., Donaldson, K., Borm, P. J., Schins, R. P., Dehnhardt, M., Gilmour, P., Jimenez, L. A., Stone, V. 2004. Calcium and ROS-mediated activation of transcription factors and TNF-alpha cytokine gene expression in macrophages exposed to ultrafine particles. *American Journal of Physiology–Lung Cellular and Molecular Physiology*, 286(2), L344–L353.

Brown, D. M., Kinloch, I. A., Bangert, U., Windle, A. H., Walter, D. M., Walker, G. S., Scotchford, C. A., Donaldson, K., Stone, V. 2007. An in vitro study of the potential of carbon nanotubes and nanofibres to induce inflammatory mediators and frustrated phagocytosis. *Carbon*, 45(9), 1743–1756.

Brown, D. M., Wilson, M. R., MacNee, W., Stone, V., Donaldson, K. 2001. Size-dependent proinflammatory effects of ultrafine polystyrene particles: A role for surface area and oxidative stress in the enhanced activity of ultrafines. *Toxicology and Applied Pharmacology*, 175, 191–199.

Brown, D. M., Stone, V., Findlay, P., MacNee, W., and Donaldson, K. 2000. Increased inflammation and intracellular calcium caused by ultrafine carbon black is independent of transition metals or other soluble components. *Occupational and Environmental Medicine*, 57(10), 685–691.

Cañas, J. E., Long, M., Nations, S., Vadan, R., Dai, L., Luo, M., Ambikapathi, R., Lee, E. R., Olszyk, D. 2008. Effects of functionalized and nonfunctionalized single walled carbon nanotubes on root elongation of selected crop species. *Environmental Toxicology and Chemistry*, 27(9), 1922–1931.

Caruthers, S. D., Wickline, S. A., Lanza, G. M. 2007. Nanotechnological applications in medicine. *Current Opinion in Biotechnology*, 18, 26–30.

Chellat, F., Merhi, Y., Moreau, A., Yahia, L. H. 2005. Therapeutic potential of nanoparticulate systems for macrophage targeting. *Biomaterials*, 26, 7260–7275.

Chen, K. L., Elimelech, M. 2008. Interaction of fullerene (C_{60}) nanoparticles with humic acid and alginate coated silica surfaces: Measurements, mechanisms, and environmental implications. *Environmental Sciences and Technology*, 42, 7607–7614.

Chen, X., Schluesener, H. J. 2008. Nanosilver: A nanoproduct in medical application. *Toxicology Letters*, 176, 1–12.

Christian, P., Von der Kammer, F., Baalousha, M., Hofmann, Th. 2008. Nanoparticles: Structure, properties, preparation and behaviour in environmental media. *Ecotoxicology*, 17, 326–343.

Clift, M. J. D., Brown, D. M., Rothen-Ruthishauser, B., Duffin, R., Donaldson, K., Proudfoot, L., Guy, K., Stone, V. 2008. Comparing a panel of commercially available quantum dots and polystyrene nanoparticles with differing surface characteristics; an analysis of their uptake and toxicity in a murine macrophage cell line. *Toxicology and Applied Pharmacology*, 232, 418–427.

Corachan, M., Tura, J. M., Campo, E., Soley, M., Traveria, A. 1988. Podoconiosis in Aequatorial Guinea. Report of two cases from different geological environments. *Tropical and Geographical Medicine*, 40(4), 359–364.

Deng, X., Jia, G., Wang, H., Sun, H., Wang, X., Yang, S., Wang, T., Liu, Y. 2007. Translocation and fate of multi-walled carbon nanotubes in vivo. *Carbon*, 45(7), 1419–1424.

Derfus, A. M., Chan, W. C. W., Bhatia, S. N. 2004. Probing the cytotoxicity of semiconductor quantum dots. *Nano Letters*, 4(1), 11–18.

Dhawan, A., Taurozzi, J. S., Pandey, A. K., Shan, W., Miller, S. M., Hashsham, S. A., Tarabara, V. V. 2006. Stable colloidal dispersions of C60 fullerenes in water: evidence for genotoxicity. *Environmental and Science Technology*, 40(23), 7394–7401.

Dockery, D. W. 1993. Epidemiologic study design for investigating respiratory health effects of complex air pollution mixtures. *Environmental Health Perspectives*, 101(S4), 187–191.

Dick, C. A., Brown, D. M., Donaldson, K., Stone, V. 2003. The role of free radicals in the toxic and inflammatory effects of four different ultrafine particle types. *Inhalation Toxicology*, 15, 39–52.

Donaldson, K., Stone, V. 2003. Current hypotheses on the mechanisms of toxicity of ultrafine particles. *Ann 1st Super Sanita*, 39, 405–410.

Donaldson, K., Stone, V., Clouter, A., Renwick, L. and MacNee, W. 2001. Ultrafine particles. *Occupational and Environmental Medicine*, 58(3), 211–216.

Donaldson, K., Tran, C. L. 2004. An introduction to the short-term toxicology of respirable industrial fibres. *Mutation Research/Fundamental and Molecular Mechanisms of Mutagenesis*, 553(1–2), 5–9.

Dowling, A., Clift, R., Grobert, N., Hutton, D., Oliver, R., O'Neill, O., Pethica, J., Pidgeon, N., Porritt, J., Ryan, J., Seaton, A., Tendler, S., Welland, M., Whatmore, R. 2004. *Nanoscience and Nanotechnologies: Opportunities and Uncertainties*. London, U.K.: Royal Society and Royal Academy of Engineering. http://www.nanotec.org.uk/report/Nano%20 report%202004%20fin.pdf

Dörger, M., Münzing, S., Allmeling, A.-M., Messmer, K., Krombach, F. 2001. Differential responses of rat alveolar and peritoneal macrophages to man-made vitreous fibers in vitro. *Environmental Research*, 85(3), 207–214.

Dubertret, B., Skourides, P., Norris, D. J., Noireaux, V., Brivanlou, A. H., Libchaber, A. 2002. In vivo imaging quantum dots encapsulated in phospholipids micelles. *Science*, 289, 1759–1762.

Duffin, R., Tran, C. L., Clouter, A., Brown, D. M., MacNee, W., Stone, V., Donaldson, K. 2002. Reactivity in the acute pulmonary inflammatory response to particles. *The Annals of Occupational Hygiene*, 46(S1), 242–245.

Dumortier, H., Lacotte, S., Pastorin, G., Marega, R., Wu, W., Bonifazi, D., Briand, J. P., Prato, M., Muller, S., Bianco, A. 2006. Functionalized carbon nanotubes are non-cytotoxic and preserve the functionality of primary immune cells. *Nano Letters*, 6(7), 1522–1528.

Eklund, P., Ajayan, P., Blackmon, R., Hart, A. J., Kong, J., Pradhan, B., Rao, A., Rinzler, A. 2007. International assessment of research and development of carbon nanotube manufacturing and application. WTEC panel report. http://www.dtic.mil/cgi-bin/GetTRDoc?AD=ADA472146 &Location=U2&doc=GetTRDoc.pdf

Emerich, D. F., Thanos, C. G. 2006. The pinpoint promise of nanoparticle-based drug delivery and molecular diagnosis. *Biomolecular Engineering*, 23, 171–184.

Esquivel, E. V., Murr, L. E. 2004. A TEM analysis of nanoparticulates in a Polar ice core. *Materials Characterization*, 52(1), 15–25.

European Science Foundation (ESF). 2005. Nanomedicine; An ESF—European Medical Research Councils (EMRC) Forward Look Report.

Federici, G., Shaw, B. J., Handy, R. D. 2007. Toxicity of titanium dioxide to rainbow trout (*Oncorhynchus mykiss*): Gill injury, oxidative stress, and other physiological effects. *Aquatic Toxicology*, 84, 415–430.

Ferin, J., Oberdorster, G., Penney, D. P. 1992. Pulmonary retention of ultrafine and fine particles in rats. *American Journal of Respiratory Cell and Molecular Biology*, 6, 535–542.

Fernandes, T. F., Christofi, N., Stone, V. 2007. The environmental implications of nanomaterials. In: *Nanotoxicology: Characterization, Dosing and Health Effects* (eds. N. A. Monteiro-Riviere and C. L. Tran). Taylor & Francis/CRC Press, Boca Raton, FL.

Ferrari, M., 2005. Nanovector therapeutics. *Current Opinion in Chemical Biology*, 9, 343–346.

Foucaud, L., Wilson, M. R., Brown, D. M., Stone, V. 2007. Measurement of reactive species production by nanoparticles prepared in biologically relevant media. *Toxicology Letters*, 174(1–3), 1–9.

Fortner, J. D., Lyon, D. Y., Sayes, C. M., Boyd, A. M., Falkner, J. C., Hotze, E. M., Alemany, L. B., Tao, Y. J., Guo, W., Ausman, K. D., Colvin, V. L., Hughes, J. B. 2005. C_{60} in water: Nanocrystal formation and microbial response. *Environmental Science and Technology*, 39, 4307–4316.

Franklin, N., Rogers, N., Apte, S., Batley, G., Gadd, G., Casey, P. 2007. Comparative toxicity of nanoparticulate ZnO, Bulk ZnO, and $ZnCl_2$ to a freshwater microalga (*Pseudokirchneriella subcapitata*): The importance of particle solubility. *Environmental Science and Technoloyg*, 41, 8484–8490.

Gagné, F., Auclair, J., Turcotte, P., Fournier, M., Gagnona, C., Sauvé, S., Blaise, C. 2008. Ecotoxicity of CdTe quantum dots to freshwater mussels: Impacts on immune system, oxidative stress and genotoxicity. *Aquatic Toxicology*, 86, 333–340.

Gao, X., Cui, Y, Levenson, R. M., Chung, L. W. K., Nie, S. 2004. *In vivo* cancer targeting and imaging with semiconductor quantum dots. *Nature Biotechnology*, 22(8), 969–976.

Gao, X., Yang, L., Petro, J. A., Marshall, F. F., Simons, J. W., Nie, S. 2005. *In vivo* molecular and cellular imaging with quantum dots. *Current Opinion in Biotechnology*, 16, 63–72.

Gao, J., Youn, S., Hovsepyan, A., Llaneza, V. L., Wang, Y., Bitton, G., Bonzongo, J.-C. J. 2009. Dispersion and toxicity of selected manufactured nanomaterials in natural river water samples: Effects of water chemical composition. *Environmental Science and Technology*, 43, 3322–3328.

Gharbi, N., Pressac, M., Hadchouel, M., Szwarc, H., Wilson, S. R., Moussa, F. 2005. [60] fullerene is a powerful antioxidant in vivo with no acute or subacute toxicity. *Nano Letters*, 5(12), 2578–2585.

Giepmans, B. N. G., Deerinck, T. J., Smarr, B. L., Jones, Y. Z., Ellisman, M. H. 2005. Correlated light and electron microscopic imaging of multiple endogenous proteins using quantum dots. *Nature Methods*, 2(10), 743–749.

Griffitt, R. J., Luo, J., Gao, J., Bonzongo, J. -C. 2008. Effects of particle composition and species on toxicity of metallic nanomaterials in aquatic organisms. *Environmental Toxicology and Chemistry*, 27(9), 1972–1978.

Griffitt, R. J., Weil, R., Hyndman, K. A., Denslow, N. D., Powers, K., Taylor, D., Barber, D. S. 2007. Exposure to copper nanoparticles causes gill injury and acute lethality in zebrafish (*Danio rerio*). *Environmental Science and Technology*, 41, 8178–8186.

Guo, L., Morris, D. G., Liu, X., Vaslet, C., Hurt, R. H., Kane, A. B. 2007. Iron bioavailability and redox activity in diverse carbon nanotube samples. *Chemistry of Materials*, 19(14), 3472–3478.

Hanaki, K. I., Momo, A., Oku, T., Komoto, A., Maenoson, S., Yamaguchi, Y. 2003. Semiconductor quantum dot albumin complex is a long-life and highly photostable endosome marker. *Biochemical and Biophysical Research Communications*, 302, 496–501.

Handy, R. D., Henry, T. B., Scown, T. M., Johnston, B. D., Tyler, C. R. 2008a. Manufactured nanoparticles: Their uptake and effects on fish—A mechanistic analysis. *Ecotoxicology*, 17, 396–409.

Handy, R. D., von der Kammer, F., Lead, J. R., Hassellöv, M., Owen, R., Crane, M. 2008b. The ecotoxicology and chemistry of manufactured nanoparticles. *Ecotoxicology*, 17, 287–314.

Hardman, R. 2006. A toxicologic review of quantum dots: Toxicity depends on physiochemical and environmental factors. *Environmental Health Perspectives*, 114(2), 165–172.

Hassellöv, M., Readman, J. W., Ranville, J. F., Tiedje, K. 2008. Nanoparticle analysis and characterization methlogies in environmental risk assessment of engineered nanoparticles. *Ecotoxicology*, 17, 344–361.

Henn, K. W., Waddill, D. W. 2006. Utilization of nanoscale zero-valent iron for source remediation—A case study. *Remediation*, Spring 2006, 57–77.

Henry, T. B., Menn, F.-M., Fleming, J. T., Wilgus, J., Compton, R. N., Sayler, G. S. 2007. Attributing effects of aqueous C_{60} nano-aggregates to tetrahydrofuran decomposition products in larval zebrafish by assessment of gene expression. *Environmental Health Perspectives*, 115(7), 1059–1065.

Hoet, P. H. M., Bruske-Hohlfeld, I., Salata, O. V. 2004. Nanoparticles–known and unknown health risks. *Journal of Nanobiotechnology*, 2, 12.

Holbrook, R. D., Murphy, K. E., Morrow, J. B., Cole, K. D. 2008. Trophic transfer of nanoparticles in a simplified invertebrate food web. *Nature Nanotechnology*, 3, 352–355.

Hoshino, A., Fujioka, K., Oku, T., Masakazu, S., Sasaki, Y. F., Ohta, T., Yasuhara, M., Suzuki, K., Yamamoto, K. 2004b. Physicochemical properties and cellular toxicity of nanocrystal quantum dots depend on their surface modification. *Nano Letters*, 4(11), 2163–2169.

Hoshino, A., Hanaki, K., Suzuki, K., Yamamoto, K. 2004a. Applications of T-lymphoma labelled with fluorescent quantum dots to cell tracing markers in mouse body. *Biochemical and Biophysical Research Communications*, 314, 46–53.

Hund-Rinke K., Simon M. 2006. Ecotoxic effects of photocatalytic active nanoparticles (TiO_2) on algae and daphnids. *Environmental Sciences and Pollution Research*, 13, 225–232.

Iijima, S. 1991. Helical microtubules of graphitic carbon. *Nature*, 354(6348), 56–58.

International Organization for Standardization (ISO) 2008. Technical Specification (ISO/TS) 27687:2008; Nanotechnologies—Terminology and definitions for nano-objects—Nanoparticle, nanofibre and nanoplate; First published 2008-08-15.

International Programme on Chemical Safety (IPCS) 1992. Cadmium, Environmental Health Criteria 134. http://www.inchem.org/documents/ehc/ehc/ehc134.htm [Consulted April 2009].

Jagur-Grodzinski, J. 1999. Biomedical application of functional polymers. *Reactive and Functional Polymers*, 39, 99–138.

Jain, K. K. 2005. The role of nanobiotechnology in drug discovery. *Drug Discovery Today*, 10(21), 1435–1442.

Jaiswal, J. K., Mattoussi, H., Mauro, J. M., Simon, S. M. 2003. Long-tem multiple colour imaging of live cells using quantum dot bioconjugates. *Nature Biotechnology*, 21, 47–51.

Jemec, A., Drobne, D., Remskar, M., Sepcic, K., Tisler T. 2008. Effects of ingested nano-sized titanium dioxide on terrestrial Isopods (*Porcellio scaber*). *Environmental Toxicology and Chemistry*, 27, 1904–1914.

Ju-Nam, Y., Lead, J. R. 2008. Manufactured nanoparticles: An overview of their chemistry, interactions and potential environmental implications. *The Science of the Total Environment*, 400(1–3), 396–414.

Kagan, V. E., Tyurina, Y. Y., Tyurin, V. A., Konduru, N. V., Potapovich, A. I., Osipov, A. N., Kisin, E. R. et al. 2006. Direct and indirect effects of single walled carbon nanotubes on RAW 264.7 macrophages: Role of iron. *Toxicology Letters*, 165(1), 88–100.

Kamat, J. P., Devasagayam, T. P., Priyadarsini, K. I., Mohan, H. 2000. Reactive oxygen species mediated membrane damage induced by fullerene derivatives and its possible biological implications. *Toxicology*, 155(1–3), 55–61.

Kashiwada, S. 2006. Distribution of nanoparticles in the see-through Medaka (*Oryzias latipes*). *Environmental Health Perspectives*, 114(11), 1697–1702.

Kawasaki, E. S., Player, A. 2005. Nanotechnology, nanomedicine, and the development of new, effect therapies for cancer. *Nanomedicine: Nanotechnology, Biology, and Medicine*, 1, 101–109.

Kim, S.-C., Lee, D.-K. 2005. Preparation of TiO$_2$-coated hollow glass beads and their application to the control of algal growth in eutrophic water. *Microchemical Journal*, 80, 227–232.

Kim, S., Lim, Y. T., Soltesz, E. G., De Grand, A. M., Lee, J., Nakayama, A., Parker, J. A. et al. 2004. Near-infrared fluorescent type II quantum dots for sentinel lymph node mapping. *Nature Biotechnology*, 22(1), 93–97.

Klaine, S. J., Avarez, P. J. J., Batley, G. E., Fernandes, T. F., Handy, R. D., Lyon, D. Y., Mahendra, S., McLaughlin, M. J., Lead, J. R. 2008. Nanomaterials in the environment: Behavior, fate, bioavailability, and effects. *Environmental Toxicology and Chemistry*, 27, 1825–1851.

Kloepfer, J. A., Mielke, R. E., Nadeau, J. L. 2005. Uptake of CdSe and CdSe/ZnS quantum dots into bacteria via purine-dependent mechanisms. *Applied and Environmental Microbiology*, 71(5), 2548–2557.

Lacerda, L., Bianco, A., Prato, M., Kostarelos, K. 2006. Carbon nanotubes as nanomedicines: From toxicology to pharmacology. *Advanced Drug Delivery Reviews*, 58, 1460–1470.

Lam, C. W., James, J. T., McCluskey, R., Hunter, R. L. 2004. Pulmonary toxicity of single-wall carbon nanotubes in mice 7 and 90 days after intratracheal instillation. *Toxicological Sciences*, 77(1), 126–134.

Larson, D. R., Zipfel, W. R., Williams, R. M., Clark, S. W., Bruchez, M. P., Wise, F. W. 2003. Water-soluble quantum dots for multiphoton fluorescence imaging *in vivo*. *Science*, 300, 1434–1436.

Lee, K. L., Nallathamby, P. D., Browning, L. M., Osgood, C. J., Xu, X.-H. N. 2007. In vivo imaging of transport and biocompatibility of single silver nanoparticles in early development of zebrafish embryos. *ACS Nano*, 1(2), 133–143.

Lee, W. M., An, Y. J., Yoon, H., Kweon, H. S. 2008. Toxicity and bioavailability of copper nanoparticles to the terrestrial plants mung bean (*Phaseolus radiatus*) and wheat (*Triticum aestivum*): Plant agar test for water-insoluble nanoparticles. *Environmental Toxicology and Chemistry*, 27, 1915–1921.

Li, X. Y., Brown, D., Smith, S., MacNee, W., Donaldson, K. 1999. Short-term inflammatory responses following intratracheal instillation of fine and ultrafine carbon black in rats. *Inhalation Toxicology*, 11, 709–731.

Li, X. Y., Gilmour, P. S., Donaldson, K., MacNee, W. 1997. In vivo and in vitro proinflammatory effects of particulate air pollution (PM10). *Environmental Health Perspectives*, 105(S5), 1279–1283.

Li, Z., Hulderman, T., Salmen, R., Chapman, R., Leonard, S. S., Young, S. H., Shvedova, A., Luster, M. I., Simeonova, P. P. 2007. Cardiovascular effects of pulmonary exposure to single-wall carbon nanotubes. *Environmental Health Perspectives*, 115(3), 377–382.

Li, D., Lyon, D. Y., Li, Q., Alvarez, P. J. J. 2008. Effect of soil sorption and aquatic natural organic matter on the antibacterial activity of a fullerene water suspension. *Environmental Toxicology and Chemistry*, 27, 1888–1894.

Lidke, D. S., Nagy, P., Heintzmann, R., Arndt-Jovin, D. J., Post, J. N., Grecco, H. E., Jares-Erijman, E. A., Jovin, T. M. 2004. Quantum dot ligands provide new insights into erbB/HER receptor-mediated signal transduction. *Nature Biotechnology*, 22(2), 198–203.

Lin, D., Xing, B. 2007. Phytotoxicity of nanoparticles: Inhibition of seed germination and root growth. *Environmental Pollution*, 150(2), 243–250.

Lin, D., Xing, B. 2008. Root uptake and phytotoxicity of ZnO nanoparticles. *Environmental Science and Technology*, 42, 5580–5585.

Liu, Y., Wu, D. C., Zhang, W. D., Jiang, X., He, C. B., Chung, T. S., Goh, S. H., Leong, K. W. 2005. Polyethylenimine-grafted multiwalled carbon nanotubes for secure non-covalent immobilization and efficient delivery of DNA. *Angewandte Chemie (International Edition in English)*, 44(30), 4782–4785.

Lovern, S. B., Klaper R. 2006. *Daphnia magna* mortality when exposed to titanium dioxide and fullerene (C$_{60}$) nanoparticles. *Environmental Toxicology and Chemistry*, 25, 1132–1137.

Lovric, J., Bazzi, H. S., Cuie, Y., Fortin, G. R. A., Winnik, F. M., Maysinger, D. 2005a. Differences in subcellular distribution and toxicity of green and red emitting CdTe quantum dots. *Journal of Molecular Medicine*, 83, 377–385.

Lovric, J., Cho, S. J., Winnik, F. M., Maysinger, D. 2005b. Unmodified cadmium telluride quantum dots induce reactive oxygen species formation leading to multiple organelle damage and cell death. *Chemistry and Biology*, 12, 1–8.

Luoma, S. 2008. Silver nanotechnologies and the environment: Old problems or new challenges? Project on Emerging Technologies (PEN) 15. Pew Charitable Trusts, Woodrow Wilson International Center for Scholars.

Lux Research Interactive. 2006. The Nanotech Report, 4th Edition Presentation. June 19 2006.

Lyon, D. Y., Alvarez, P. J. J. 2008. Fullerene water suspension (nC60) exerts antibacterial effects via ROS-independent protein oxidation. *Environmental Science and Technology*, 42, 8127–8132.

Lyon, D. Y., Adams, L. K., Falkner, J. C., Alvarez, P. J. J. 2006. Antibacterial activity of fullerene water suspensions: Effects of preparation method and particle size. *Environmental Science and Technology*, 40, 4360–4366.

MacNee, W., Rahman, I. 2001. Is oxidative stress central to the pathogenesis of chronic obstructive pulmonary disease? *Trends in Molecular Medicine*, 7, 55–62.

Mangum, J., Turpin, E., Antao-Menezes, A., Cesta, M., Bermudez, E., Bonner, J. 2006. Single-walled carbon nanotube (SWCNT)-induced interstitial fibrosis in the lungs of rats is associated with increased levels of PDGF mRNA and the formation of unique intercellular carbon structures that bridge alveolar macrophages in situ. *Particle and Fibre Toxicology*, 3(1), 15.

Manna, S. K., Sarkar, S., Barr, J., Wise, K., Barrera, E. V., Jejelowo, O., Rice-Ficht, A. C., Ramesh, G. T. 2005. Single-walled carbon nanotube induces oxidative stress and activates nuclear transcription factor-KB in human keratinocytes. *Nano Letters*, 5, 1676–1684.

Mason, J. N., Farmer, H., Tomlinson, I. D., Schwartz, J. W., Savchenko, V., DeFelice, L. J., Rosenthal, S. J., Blakely, R. D. 2005. Novel fluorescence based approaches for the study of biogenic amine transporter localization, activity and regulation, *Journal of Neuroscience Methods*, 143, 3–25.

Maynard, A. D. 2006. Nanotechnology: assessing the risks. *Nanotoday*, 1(2), 22–33.

Maynard, A. D., Coull, B. A., Gryparis, A., Schwartz, J. 2007. Mortality risk associated with short-term exposure to traffic particles and sulfates. *Environmental Health Perspectives*, 115(5), 751–755.

Medina, C., Santos-Martinez, M. J., Radomski, A., Corrigan, O. I., Radomski, M. W. 2007. Nanoparticles: Pharmacological and toxicological significance. *British Journal of Pharmacology*, 150, 552–558.

Mercer, R. R., Scabilloni, J., Wang, L., Kisin, E., Murray, A. R., Schwegler-Berry, D., Shvedova, A. A., Castranova, V. 2008. Alteration of deposition pattern and pulmonary response as a result of improved dispersion of aspirated single-walled carbon nanotubes in a mouse model. *American Journal of Physiology. Lung Cellular and Molecular Physiology*, 294(1), L87–L97.

Michalet, X., Pinaud, F. F., Bentolila, L. A., Tsay, J. M., Doose, S., Li, J. J., Sundaresan, G., Wu, A. M., Gambhir, S. S., Weiss, S. 2005. Quantum dots for live cells, in vivo imaging, and diagnostics. *Science*, 307, 538–544.

Mitchell, L. A., Gao, J., Vander Wal, R., Gigliotti, A., Burchiel, S. W., McDonald, J. D. 2007. Pulmonary and systemic immune response to inhaled multiwalled carbon nanotubes. *Toxicological Sciences*, 100(1), 203–214.

Moghimi, S. M. 1995. Introduction: Targeting of drugs and delivery systems. *Advanced Drug Delivery Reviews*, 17, 1–3.

Moghimi, S. M., Hunter, A. C., Murray, J. C. 2005. Nanomedicine: Current status and future prospects. *The FASEB Journal*, 19, 311–330.

Monteiller, C., Tran, L., MacNee, W., Faux, S., Jones, A., Miller, B., Donaldson, K. 2007. The pro-inflammatory effects of low-toxicity low-solubility particles, nanoparticles and fine particles, on epithelial cells in vitro: the role of surface area. *Occupational and Environmental Medicine*, 64(9), 609–615.

Monteiro-Riviere, N. A., Inman, A. O., Wang, Y. Y., Nemanich, R. J. 2005b. Surfactant effects on carbon nanotube interactions with human keratinocytes. *Nanomedicine*, 1, 293–299.

Monteiro-Riviere, N. A., Nemanich, R. J., Inman, A. O., Wang, Y. Y., Riviere, J. E. 2005a. Multi-walled carbon nanotube interactions with human epidermal keratinocytes. *Toxicology Letters*, 155(3), 377–384.

Mornet, S., Vasseur, S., Grasset, F., Veverka, P., Goglio, G., Demourgues, A., Portier, J., Pollert, E., Duguet, E. 2006. Magnetic nanoparticle design for medical applications. *Progress in Solid State Chemistry*, 34, 237–247.

Mortensen, L. J., Oberdorster, G., Pentland, A. P., DeLouise, L. A. 2008. In vivo skin penetration of quantum dot nanoparticles in the murine model: The effect of UVR. *Nano Letters*, 8, 2779–2787.

Mueller, N., Nowack, B. 2008. Exposure modeling of engineered nanoparticles in the environment. *Environmental Science and Technology*, 42, 4447–4453.

Muller, J., Huaux, F., Moreau, N., Misson, P., Heilier, J.-F., Delos, M., Arras, M., Fonseca, A., Nagy, J. B., Lison, D. 2005. Respiratory toxicity of multi-wall carbon nanotubes. *Toxicology and Applied Pharmacology*, 207(3), 221–231.

Murashov, V. 2006. Letter to the editor: Comments on particle surface characteristics May, L., Watts, D.J., *Toxicology Letters*, 2005, 158, 122–132. *Toxicology Letters*, 164, 185–187.

Murr, L. E., Bang, J. J. 2003. Electron microscope comparisons of fine and ultra-fine carbonaceous and non-carbonaceous, airborne particulates. *Atmospheric Environment*, 37(34), 4795–4806.

Murr, L. E., Garza, K. M., Soto, K. F., Carrasco, A., Powell, T. G., Ramirez, D. A., Guerrero, P. A., Lopez, D. A., Venzor, J., III 2005. Cytotoxicity assessment of some carbon nanotubes and related carbon nanoparticle aggregates and the implications for anthropogenic carbon nanotube aggregates in the environment. *International Journal on Environmental Research and Public Health*, 2(1), 31–42.

Murr, L. E., Soto, K., Esquivel, E., Bang, J., Guerrero, P., Lopez, D., Ramirez, D. 2004. Carbon nanotubes and other fullerene-related nanocrystals in the environment: A TEM study. *JOM Journal of the Minerals, Metals and Materials Society*, 56(6), 28–31.

Navarro, E., Baun, A., Behra, R., Hartmann, N. B., Filser, J., Miao, A.-J., Quigg, A., Santschi, P. H., Sigg, L. 2008a. Environmental behavior and ecotoxicity of engineered nanoparticles to algae, plants, and fungi. *Ecotoxicology*, 17, 372–386.

Navarro, E., Piccapietra, F., Wagner, B., Marconi, F., Kaegi, R., Odzak, N., Sigg, L., Tabehra, A. 2008b. Toxicity of silver nanoparticles to *Chlamydomonas reinhardtii*. *Environmental Science and Technology*, 42(23), 8959–8964.

Nel, A., Xia, T., Mädler, L., Li, N. 2006. Toxic potential of materials at the nanolevel. *Science*, 311(5761), 622–627.

Nemmar, A., Vanbilloen, H., Hoylaerts, M. F., Hoet, P. H., Verbruggen, A., Nemery, B. 2001. Passage of intratracheally instilled ultrafine particles from the lung into the systemic circulation in hamster. *American Journal of Respiratory and Critical Care Medicine*, 164(9), 1665–1668.

Ness, J. M., Akhtar, R. S., Latham, C. B., Roth, K. A. 2003. Combined tyramide signal amplification and quantum dots for sensitive and photostable immunofluorescence detection. *The Journal of Histochemistry and Cytochemistry*, 51(8), 981–987.

Nielsen, H. D., Berry, L. S., Stone, V., Fernandes, T. F. 2008. Interactions between carbon black nanoparticles and the brown algae *Fucus serratus*: inhibition of fertilization and zygotic development. *Nanotoxicology*, 2(2), 88–97.

Nisman, R., Dellaire, G., Ren, Y., Li, R., Bazett-Jones, D. P. 2004. Application of quantum dots as probes of correlative fluorescence, conventional, and energy-filtered transmission electron microscopy. *The Journal of Histochemistry and Cytochemistry*, 52(1), 13–18.

NRCG (2005). Characterising the potential risks posed by engineered nanoparticles–A first UK Government Research Report. Defra, London.

NRCG (2007). Characterising the potential risks posed by engineered nanoparticles–A second UK Government Research Report. Defra, London. http://www.defra.gov.uk/environment/quality/nanotech/documents/nanoparticles-riskreport07.pdf

Oberdorster, E. 2004. Manufactured nanomaterials (Fullerenes, C60) induce oxidative stress in the brain of juvenile largemouth bass. *Environmental Health Perspectives*, 112, 1058–1062.

Oberdorster, G. 2001. Pulmonary effects of inhaled ultrafine particles. *International Archives of Occupational and Environmental Health*, 74, 1–8.

Oberdorster, G., Maynard, A., Donaldson, K., Castranova, V., Fitzpatrick, J., Ausman, K., Carter, J. et al. 2005b. Principles for characterizing the potential human health effects from exposure to nanomaterials: Elements of a screening strategy. *Particle and Fibre Toxicology*, 2, 8.

Oberdorster, G., Oberdorster, E., Oberdorster, J. 2005a. Nanotoxicology: An emerging discipline evolving from studies of ultrafine particles. *Environmental Health Perspectives*, 113, 823–839.

Oberdorster, G., Sharp, Z., Atudorei, V., Elder, A., Gelein, R., Lunts, A., Kreyling, W., Cox, C. 2002. Extrapulmonary translocation of ultrafine carbon particles following whole-body inhalation exposure of rats. *Journal of Toxicology and Environmental Health*, 65, 1531–1543.Oberdorster, G., Stone, V., Donaldson, K. 2007. Toxicology of nanoparticles: A historical perspective. *Nanotoxicology*, 1(1), 2–25.

Oberdorster, E., Zhu, S., Blickley, T. M., McClellan-Green, P., Haasch, M. L. 2006. Ecotoxicology of carbon-based engineered nanoparticles: Effects of fullerene (C60) on aquatic organisms. *Carbon*, 44, 1112–1120.

Ogawara, K., Yoshida, M., Furumoto, K., Takakura, Y., Hashida, M., Higaki, K., Kimura, T. 1999. Uptake by hepatocytes and biliary excretion of intravenously administered polystyrene microspheres in rats. *Journal of Drug Targeting*, 7(3), 213–221.

Ohmer-Schrock, D., Schlatterer, C., Plattner, H., Schlepper-Schafer, J. 1995. Lung surfactant protein A (SP-A) activates a phosphoinositide/calcium signaling pathway in alveolar macrophages. *Journal of Cell Science*, 108(12), 3695–3702.

Pantarotto, D., Singh, R., McCarthy, D., Erhardt, M., Briand, J. P., Prato, M., Kostarelos, K., Bianco, A. 2004. Functionalized carbon nanotubes for plasmid DNA gene delivery. *Angewandte Chemie (International Edition in English)*, 43(39), 5242–5246.

Papanikolaou, G., Pantopoulos, K. 2005. Iron metabolism and toxicity. *Toxicology and Applied Pharmacology*, 202(2), 199–211.

Park, E. J., Choi, J., Park, Y. K., Park, K. 2008. Oxidative stress induced by cerium oxide nanoparticles in cultured BEAS-2B cells. *Toxicology*, 245, 90–100.

Patel, T., Gores, G. J. 1995. Apoptosis and hepatobiliary disease. *Hepatology*, 21, 1725–1741.

Penn, S. G., He, L., Natan, M. J. 2003. Nanoparticles for bioanalysis. *Current Opinion in Chemical Biology*, 7, 609–615.

Peters, A., Wichmann, H. E., Tuch, T., Heinrich, J., Heyder, J. 1997. Respiratory effects are associated with the number of ultrafine particles. *American Journal of Respiratory and Critical Care Medicine*, 155(4), 1376–1383.

Petersen, E. J., Huang, Q., Weber, W. J. 2008. Bioaccumulation of radio-labeled carbon nanotubes by *Eisenia foetida*. *Environmental Science and Technology*, 42, 3090–3095.

Pinaud, F., Michalet, X., Bentolila, L. A., Tsay, J. M., Doose, S., Li, J. J., Iyer, G., Weiss, S. 2006. Advances in fluorescence imaging with quantum do bio-probes. *Biomaterials*, 27, 1679–1687.

Pison, U., Welte, T., Giersig, M., Groneberg, D. A. 2006. Nanomedicine for respiratory diseases, *European Journal of Pharmacology*, 533, 341–350.

Poland, C. A., Duffin, R., Kinloch, I., Maynard, A., Wallace, W. A. H., Seaton, A., Stone, V., Brown, S., MacNee, W., Donaldson, K. 2008. Carbon nanotubes introduced into the abdominal cavity of mice show asbestos-like pathogenicity in a pilot study. *Nature Nanotechnology*, 3(7), 423–428.

Popat, K. C., Eltgroth, M., LaTempa, T. J., Grimes, C. A., Desai, T. A. 2007. Decreased Staphylococcus epidermis adhesion and increased osteoblast functionality on antibiotic-loaded titania nanotubes. *Biomaterials*, 28(32), 4880–4888.

Pope, C. A., Dockery, D. W. 1999. Epidemiology of particle effects. *Air Pollution and Health*, 31, 674–705.

Popov, V. N. 2004. Carbon nanotubes: Properties and application. *Materials Science and Engineering: R: Reports*, 43(3), 61–102.

Powell, M. C., Kanarek, M. S. 2006. Nanomaterial health effects—part 1: background and current knowledge. *Wisconsin Medical Journal*, 105(2), 16–20.

Pulskamp, K., Diabaté, S., Krug, H. F. 2007a. Carbon nanotubes show no sign of acute toxicity but induce intracellular reactive oxygen species in dependence on contaminants. *Toxicology Letters*, 168(1), 58–74.

Pulskamp, K., Wörle-Knirsch, J. M., Hennrich, F., Kern, K., Krug, H. F. 2007b. Human lung epithelial cells show biphasic oxidative burst after single-walled carbon nanotube contact. *Carbon*, 45(11), 2241–2249.

Reijnders, L. 2006. Cleaner technology and hazard reduction of manufactured nanoparticles. *Journal of Cleaner Production*, 14(2), 124–135.

Reynolds, A. R., Moghimi, S. M., Hodivala-Dilke, K. 2003. Nanoparticle-mediated gene delivery to tumour neovasculature. *Trends in Molecular Medicine*, 9(1), 2–4.

Roberts, A., Mount, A. S., Seda, B., Souther, J., Qiao, R., Lin, S., Ke, P. C., Rao, A. M., Klaine, S. J. 2007. In vivo biomodification of lipid-coated carbon nanotubes by *Daphnia magna*. *Environmental Science and Technology*, 41, 3025–3029.

Rosenkranz, P., Stone, V., Chaudry, Q., Fernandes, T. F. 2009. A comparison of nanoparticle and fine particle uptake by *Daphnia magna*. *Environmental Toxicology and Chemistry*, 28(10), 2142–2149.

Ryter, S. W., Kim, H. P., Hoetzel, A., Park, J. W., Nakahira, K., Wang, X., Choi, A. M. 2007. Mechanisms of cell death in oxidative stress. *Antioxidants and Redox Signalling*, 9, 49–89.

Sahoo, S. K., Labhasetwar, V. 2003. Nanotech approaches to drug delivery and imaging. *Drug Discovery Today*, 8(24), 1112–1120.

Sahoo, S. K., Parveen, S., Panda, J. J. 2007. The present and future of nanotechnology in human health care. *Nanomedicine: Nanotechnology, Biology, and Medicine*, 3, 20–31.

Salata, O. V. 2004. Applications of nanoparticles in biology and medicine. *Journal of Nanobiotechnology*, 2(3). http://www.jnanobiotechnology.com/content/2/1/3 [accessed April 2009].

Saleh, N. B., Pfefferle, L. D., Elimelech, M. 2008. Aggregation kinetics of multiwalled carbon nanotubes in aquatic systems: Measurements and environmental implications. *Environmental Science and Technology*, 42, 7963–7969.

Salvador-Morales, C., Townsend, P., Flahaut, E., Vénien-Bryan, C., Vlandas, A., Green, M. L. H., Sim, R. B. 2007. Binding of pulmonary surfactant proteins to carbon nanotubes; potential for damage to lung immune defense mechanisms. *Carbon*, 45(3), 607–617.

Salvetat-Delmotte, J.-P., Rubio, A. 2002. Mechanical properties of carbon nanotubes: A fiber digest for beginners. *Carbon*, 40(10), 1729–1734.

Sayes, C. M., Gobin, A. M., Ausman, K. D., Mendez, J., West, J. L., Colvin, V. L. 2005. Nano-C_{60} cytotoxicity is due to lipid peroxidation, *Biomaterials*, 26, 7587–7595.

Schwartz, J. 1994. Air pollution and daily mortality: a review and meta analysis. *Environmental Research*, 64, 36–52.

Schwartz, J. 1997. Air pollution and hospital admissions for cardiovascular disease in Tucson. *Epidemiology*, 8(4), 371–377.

Scientific Committee on Emerging and Newly Identified Health Risks (SCENIHR). 2006. Modified opinion (after public consultation) on the appropriateness of existing methodologies to assess the potential risks associated with engineered and adventitious products of nanotechnologies. March 10, 2006. European Commission Health and Consumer Protection Directorate-General, Brussels, Belguim.

Scientific Committee on Emerging and Newly Identified Health Risks (SCENIHR). 2007a. Opinion on the appropriateness of the risk assessment methodology in accordance with the Technical Guidance Documents for new and existing substances. June 21–22, 2007. European Commission Health and Consumer Protection Directorate-General, Brussels, Belguim.

Scientific Committee on Emerging and Newly Identified Health Risks (SCENIHR). 2007b. Opinion on the scientific aspects of the existing and proposed definitions relating to products of nanoscience and nanotechnologies, November 29, 2007. European Commission Health and Consumer Protection Directorate-General, Brussels, Belguim.

Scientific Committee on Emerging and Newly Identified Health Risks (SCENIHR). 2009. Risk Assessment of Products of Nanotechnologies. January 19, 2009. European Commission, Brussels, Belgium. European Commission Health and Consumer Protection Directorate-General, Brussels, Belguim.

Scott-Forsmand, J. J., Krogh, P. H., Scaefer, M., Johansen, A. 2008. The toxicity testing of double-walled nanotubes-contaminated food to *Eisenia veneta* earthworms. *Ecotoxicology and Environmental Safety*, 71, 616–619.

Seaton, A., MacNee, W., Donaldson, K., Godden, D. and Stone, V. 1995. Particulate air pollution and acute health effects. *Lancet*, 345(8943), 176–178.

Service, R. F. 2004. Nanotechnology grows up. *Science*, 304, 1732–1734.

Sharma, P., Brown, S., Walter, G., Santra, S., Moudgil, B. 2006. Nanoparticles for bioimaging. *Advances in Colloid and Interface Science*, 123–126, 471–485.

Shiohara, A., Hoshino, A., Hanaki, K., Suzuki, K., Yamamoto, K. 2004. On the cyto-toxicity caused by quantum dots. *Microbiology and Immunology*, 48(9), 669–675.

Shukla, A., Ramos-Nino, M., Mossman, B. 2003. Cell signaling and transcription factor activation by asbestos in lung injury and disease. *The International Journal of Biochemistry and Cell Biology*, 35(8), 1198–1209.

Shvedova, A. A., Castranova, V., Kisin, E. R., Schwegler-Berry, D., Murray, A. R., Gandelsman, V. Z., Maynard, A., Baron, P. 2003. Exposure to carbon nanotube material: Assessment of nanotube cytotoxicity using human keratinocyte cells. *Journal of Toxicology Environmental Health Part A*, 66(20), 1909–1926.

Shvedova, A. A., Kisin, E. R., Mercer, R., Murray, A. R., Johnson, V. J., Potapovich, A. I., Tyurina, Y. Y. et al. 2005. Unusual inflammatory and fibrogenic pulmonary responses to single-walled carbon nanotubes in mice. *American Journal of Physiology. Lung Cellular and Molecular Physiology*, 289(5), L698–L708.

Shvedova, A. A., Kisin, E. R., Murray, A. R., Gorelik, O., Arepalli, S., Castranova, V., Young, S.-H. et al. 2007. Vitamin E deficiency enhances pulmonary inflammatory response and oxidative stress induced by single-walled carbon nanotubes in C57BL/6 mice. *Toxicology and Applied Pharmacology*, 221(3), 339–348.

Silva, G. A. 2004. Introduction to nanotechnology and its applications to medicine. *Surgical Neurology*, 61, 216–220.

Singh, R., Pantarotto, D., Lacerda, L., Pastorin, G., Klumpp, C. D., Prato, M., Bianco, A., Kostarelos, K. 2006. Tissue biodistribution and blood clearance rates of intravenously administered carbon nanotube radiotracers. *Proceedings of the National Academy of Sciences of the United States of America*, 103(9), 3357–3362.

Smith, C., Shaw, B., Handy, R. 2007. Toxicity of single walled carbon nanotubes on rainbow trout, (*Oncorhyncos mykiss*): Respiratory toxicity, organ pathologies, and other physiological effects. *Aquatic Toxicology*, 82, 94–109.

Sondi, I., Salopek-Sondi, B. 2004. Silver nanoparticles as antimicrobial agent: A case study on *E. coli* as a model for Gram-negative bacteria. *Journal of Colloid Interface Science*, 275, 177–182.

Stoeger, T., Reinhard, C., Takenaka, S., Schroeppel, A., Karg, E., Ritter, B., Heyder, J., Schulz, H. 2006. Instillation of six different ultrafine carbon particles indicates a surface area threshold dose for acute lung inflammation in mice. *Environmental Health Perspectives*, 114(3), 328–333.

Stone, V. 2000. Environmental air pollution. *American Journal of Respiratory and Critical Care Medicine*, 162, S44–S47.

Stone, V., Johnson, G. D., Wilton, J. C., Coleman, R., Chipman, J. K. 1994. Effect of oxidative stress and disruption of Ca^{2+} homeostasis on hepatocyte canalicular function in vitro. *Biochemical Pharmacology*, 47:625–632.

Stone, V., Kinloch, I. A. 2007. Nanoparticle interaction with biological systems and subsequent activation of intracellular signaling mechanisms. In: Monteiro-Riviere NA, Tran CL, editors. Nanotoxicology. New York: CRC Press.

Stone, V., Shaw, J., Brown, D. M., MacNee, W., Faux, S. P., Donaldson, K. 1998. The role of oxidative stress in the prolonged inhibitory effect of ultrafine carbon black on epithelial cell function. Toxicology in Vitro, 12, 649–659.

Svenson, S., Tomalia, D. A. 2005. Dendrimers in biomedical applications—Reflections on the field. *Advanced Drug Delivery Reviews*, 57, 2106–2129.

Takenaka, S., Karg, E., Roth, C., Schulz, H., Ziesenis, A., Heinzmann, U., Schramel, P., Heyder, J. 2001. Pulmonary and systemic distribution of inhaled ultrafine silver particles in rats. *Environmental Health Perspectives*, 109(S4), 547–551.

Templeton, R., Ferguson, P. L., Washburn, K., Scrivens, W., Chandler, G. T. 2006. Life-cycle effects of single walled carbon nanotubes (SWCNTs) on an estuarine meiobenthic copepod. *Environmental Science and Technology*, 40, 7387–7393.

Thrall, J. H., 2004. Nanotechnology and medicine. *Radiology*, 230, 315–318.

Tian, F., Cui, D., Schwarz, H., Estrada, G. G., Kobayashi, H. 2006. Cytotoxicity of single-wall carbon nanotubes on human fibroblasts. *Toxicology In Vitro*, 20(7), 1202–1212.

Tiede, K., Boxall, A. B. A., Tear, S. P., Lewis, J., David, H., Hassellov, M. 2008a. Detection and characterisation of engineered nanoparticles in food and the environment. *Food Additives and Contaminants*, 25(7), 795–821.

Tiede, K., Hassellöv, M., Breitbarthc, E., Chaudhry, Q., Boxall, A. B. A. 2009. Considerations for environmental fate and ecotoxicity testing to support environmental risk assessments for engineered nanoparticles. *Journal of Chromatography A*, 1216(3), 503–509.

Timbrell, J. 1999. *Principles of Biochemical Toxicology*. Taylor & Francis/CRC Press, Boca Raton, FL.

Tinkle, S. S., Antonini, J. M., Rich, B. A., Roberts, J. R., Salmen, R., DePree, K., Adkins, E. J. 2003. Skin as a route of exposure and sensitization in chronic beryllium disease. *Environmental Health Perspectives*, 111(9), 1202–1208.

USA EPA 2007. Nanotechnology White Paper. Prepared for the U.S. Environmental Protection Agency by members of the Nanotechnology Workgroup, a group of EPA's Science Policy Council. EPA 100/B-07/001. February 2007.

Van Hoecke, K., de Schamphelaere, K. A. C., van der Meeren, P., Lucas, S. P., Janssen, C. R. 2008. Ecotoxicity of silica nanoparticles to the green alga *Pseudokirchneriella subcapitata*: importance of surface area. *Environmental Toxicology and Chemistry*, 27(9), 1948–1957.

Wang, H., Wang, J., Deng, X., Sun, H., Shi, Z., Gu, Z., Liu, Y., Zhao, Y. 2004. Biodistribution of carbon single-wall carbon nanotubes in mice. *Journal of Nanoscience and Nanotechnology*, 4(8), 1019–1024.

Warheit, D. B., Driscoll, K. E., Oberdorster, G., Walker, C., Kuschner, M., Hesterberg, T. W. 1995. Contemporary issues in fiber toxicology. *Fundamental and Applied Toxicology*, 25, 171–183.

Wei, W., Sethuraman, A., Jin, C., Monteiro-Riviere, N. A., Narayan, R. J. 2007. Biological properties of carbon nanotubes. *Journal of Nanoscience and Nanotechnology*, 7(4–5), 1284–1297.

West, J. L., Halas, N. J. 2003. Engineered nanomaterials for biophotonics applications: Improving sensing, imaging and therapeutics. *Annual Review of Biomedical Engineering*, 5, 285–292.

Whatmore, R. W. 2006. Nanotechnology—what is it? Should we be worried? *Occupational Medicine*, 56, 295–299.

Whitesides, G. M. 2003. The 'right' size in nanobiotechnology. *Nature Biotechnology*, 21(10), 1161–1165.

Wick, P., Manser, P., Limbach, L. K., Dettlaff-Weglikowska, U., Krumreich, F., Roht, S., Stark, W. J., Bruinink, A. 2007. The degree and kind of agglomeration affect carbon nanotube cytotoxicity. *Toxicology Letters*, 168, 121–131.

Wiesner, M. R., Lowry, G. V., Alvarez, P., Dionysiou, D., Biswas, P. 2006. Assessing the risks of manufactured nanomaterials. *Environmental Science and Technology*, 40(14), 4336–4345.

Wilson, M. R., Lightbody, J. H., Donaldson, K., Sales, J., and Stone, V. 2002. Interactions between ultrafine particles and transition metals in vivo and in vitro. *Toxicology and Applied Pharmacology*, 184(3), 172–179.

Wu, X., Liu, H., Liu, J., Haley, K. N., Treadway, J. A., Larson, J. P., Ge, N., Peale, F., Bruchez, M. P. 2003. Immunofluorescent labelling of cancer marker Her2 and other cellular targets with semiconductor quantum dots. *Nature Biotechnology*, 21, 41–46.

Wu, W., Wieckowski, S., Pastorin, G., Benincasa, M., Klumpp, C., Briand, J. P., Gennaro, R., Prato, M., Bianco, A. 2005. Targeted delivery of amphotericin B to cells by using functionalized carbon nanotubes. *Angewandte Chemie (International Edition in English)*, 44(39), 6358–6362.

Xia, T., Kovochick, M., Brant, J., Hotze, M., Sempf, J., Oberley, T., Sioutas, C., Yeh, J. I., Wiesner, M. R., Nel, A. E. 2006. Comparison of the abilities of ambient and manufactured nanoparticles to induce cellular toxicity according to the oxidative stress paradigm. *Nano Letters*, 6, 1794–1807.

Yang, L., Watts, D. J. 2005. Particle surface characteristics may play an important role in phytotoxicity of alumina nanoparticles. *Toxicology Letters*, 158, 122–132.

Ye, J., Shi, X., Jones, W., Rojanasakul, Y., Cheng, N., Schwegler-Berry, D., Baron, P., Deye, G. J., Li, C., Castranova, V. 1999. Critical role of glass fiber length in TNF-alpha production and transcription factor activation in macrophages. *American Journal of Physiology*, 276(3 Pt 1), L426–L434.

Yehia, H., Draper, R., Mikoryak, C., Walker, E., Bajaj, P., Musselman, I., Daigrepont, M., Dieckmann, G., Pantano, P. 2007. Single-walled carbon nanotube interactions with HeLa cells. *Journal of Nanobiotechnology*, 5(1), 8.

Zeni, O., Palumbo, R., Bernini, R., Zeni, L., Sarti, M., Scarfi, M. R. 2008. Cytotoxicity investigation on cultured human blood cells treated with single-walled carbon nanotubes. *Sensors*, 8, 488–499.

Zhang, X., Sun, H., Zhang, Z., Niu, Q., Chen, Y., Crittenden, J. C. 2007. Enhanced bioaccumulation of cadmium in carp in the presence of titanium dioxide nanoparticles. *Chemosphere*, 67, 160–166.

Zhu, S., Oberdorster, E., Haasch, M. L. 2006. Toxicity of an engineered nanoparticle (fullerene, C(60)) in two aquatic species, *Daphnia* and fathead minnow. *Marine Environmental Research*, 62(Suppl. 1), S5–S9.

15

Toxicology, Diagnostics, and Therapy Functions of Nanomaterials

Stefano Bellucci
Istituto Nazionale di Fisica Nucleare

15.1 Introduction

The purpose of nanotechnology is not merely creating useful or functional materials and devices by manipulating matter at the nanometer length scale but, most importantly, exploiting the novel properties of materials that arise owing to their nanoscale. Simply meeting the length scale criterion of 1–100 nm is not really nanotechnology, rather it is a necessary condition; the corresponding sufficient condition consists in taking advantage of the novel (physical, chemical, mechanical, electrical, optical, magnetic, etc.) properties that result solely because of going from the bulk to the nanoscale.

Since 2001, when the United States announced a National Nanotechnology Initiative (NNI) aimed at creating a dedicated program to explore nanotechnology (see http://www.nano.gov/), many other countries, including the EU (see, e.g., http://www.euronanoforum2007.eu/), Japan (see www.nanonet.go.jp), and China (see, e.g., http://www.sipac.gov.cn), followed up with their own nanotechnology research programs. It is important to recall the special character of nanotechnology as both a pervasive and enabling technology, with potential impact in all sectors of the economy: electronics, computing, data storage, materials and manufacturing, health and medicine, energy, transportation, environment, national security, space exploration, and others. In this chapter, we focus on nanotechnology applications in the biomedical sector. We discuss the use of nanoparticles for the therapy of different kinds of tumors. Then we illustrate the use of magnetic nanoparticles in medical diagnostics, as well as for achieving drug or radiation delivery.

Carbon nanotubes (CNTs) are an example of a carbon-based nanomaterial [1] that has won enormous popularity in nanotechnology for its unique properties and applications [2]. CNTs have physicochemical properties that are highly desirable for use within the commercial, environmental, and medical sectors. Moreover, CNTs yield a promising material for the assembly of nanodevices, based upon new CNT–composite materials. Here, we will review the tunable synthesis of multi-walled CNT–silica nanoparticle composite materials, to realize a bionanotechnological platform for versatile uses, including providing an interface between living cells and biosensor arrays. Supramolecular nanostructures that are discussed here also have a potential interest as electron–acceptor complexes and for their properties as fluorescent nanocomposites for applications ranging from biosensors to electronics.

The inclusion of CNTs to improve the quality and performance of many widely used products, as well as potentially in medicine, is likely to increase dramatically the occupational and public exposure to CNT-based nanomaterials in the near future. Hence, it is of utmost importance to explore the yet-almost unknown issue of the toxicity of this new material. There is a certain amount of work published about CNT toxicity at present in the literature. The reader can see, for example, the results of Ref. [3]. In addition, for a partial list of several "in vitro" and "in vivo" studies, one can see Ref. [4].

In this chapter, we compare the toxicity of pristine and oxidized multi-walled CNTs on human T cells and find that the latter are more toxic and induce massive loss of cell viability through programmed cell death at doses of 400 μg/mL, which corresponds to approximately 10 million CNTs per cell. Pristine, hydrophobic CNTs were less toxic and a 10-fold lower concentration of either CNT type was not nearly as toxic. Our results suggest that CNTs indeed can be very toxic at sufficiently high concentrations and that careful toxicity studies need to be undertaken, particularly in conjunction with nanomedical applications of CNTs.

15.2 Nanotechnology for Tumor Therapy

Over the past three and a half decades, since the beginning of the U.S. National Cancer Initiative in 1971, there have been major advances in the diagnosis and treatment of cancer. However, the severe toll cancer continues to impose on our society represents one of the major healthcare concerns of our nations. One out of every two men, and one out of every three women will be confronted with a cancer diagnosis in their lifetime. Conventional treatments currently rely heavily upon radiation and chemotherapy, which are extremely invasive and painstakingly plagued by very serious side effects. Nanotechnology yields the hope for new methods for a noninvasive therapy, capable of minimizing side effects. One of the promising approaches consists in the targeted destruction of cancerous cells using localized heating.

The use of thermal cancer therapies is beneficial in many respects over the conventional tumor removal by surgery. Indeed, normally, most thermal approaches have a very small degree of invasiveness; they are relatively simple to perform and may enable physicians to treat tumors embedded in vital regions where surgical removal is unfeasible. Ideally, the activating energy to heat the tumor would be targeted on the embedded tumor with minimal effect on surrounding healthy tissue. Unfortunately, conventional heating techniques such as focused ultrasound, microwaves, and laser light do not discriminate between tumors and surrounding healthy tissues. Thus, success has been modest and, typically, treatments result in some damage to surrounding tissue.

Recent work suggests that nanostructures designed to attach to cancerous cells may provide a powerful tool for producing highly localized energy absorption at the sites of cancerous cells. Indeed, work since 2003 at La Charité Hospital in Berlin (see http://www.germanyinfo.org/relaunch/info/publications/week/2003/030613/misc2.html) with scientists at the F. Schiller-Universität Jena show that magnetic nanoparticles interstitially injected directly into the tumor, and heated with radio-frequency radiation [5], can destroy cancer cells in a human brain tumor and are also believed to enhance the effects of subsequent radiation therapy. Nanoparticles localize on the tumor due to a special biomolecularly modified outer layer and leave the surrounding healthy tissue with a minimum damage.

In this way, it was proven that iron-oxide nanoparticles, with diameters 10,000 times smaller than that of a human hair, can be introduced inside cancer cells and then treated in such a way as to produce a significant damage to tumor cells in order to fight a particularly aggressive form of brain cancer called glioblastoma, although the method can be employed to treat other forms of the disease as well. The procedure involves coating the iron-oxide nanoparticles with an organic substance, such as the sugar glucose, before injecting them into the tumor. Cancer cells, having a fast metabolism and correspondingly high energy needs, are much more eager to eat up the sugar-coated nanoparticles, in comparison with healthy cells that appear minimally or not at all affected.

In this selective procedure, the magnetic field is responsible for the heating up of the nanoparticles in the cancerous tissue, reaching temperatures of up to 45°C with the result of destroying many of the tumor cells or at least weakening them to such an extent that conventional methods, such as radiation or chemotherapy, can then more easily and effectively get rid of them. The treatment, known as magnetic fluid hyperthermia, was successfully used to prolong the life of laboratory rats that were implanted with malignant brain tumors. Rats receiving nanotherapy lived four times as long as rats receiving no treatment. Then the therapy was given to 15 patients suffering from *Glioblastoma multiforme*, the most common primary brain tumor and the most aggressive form of brain cancer (with a 6–12 months life expectancy prognosis in humans).

The treatment can be particularly attractive to doctors working with tumors in the brain since the nanoparticles can be targeted on the cancerous tissue, so that the therapy could be suitable for curing tumors that lie outside the reach of conventional surgical treatment, such as those situated deep in the brain or in regions that are responsible for essential tasks like speech or motor functions. However, there is also a very strong feeling that the nanoparticles in the brain can cause serious problems due to the axonal transport and the translocation of nanoparticles (see, e.g., [3,6]).

In principle, the hyperthermia therapy is not limited to just various types of brain cancer. Since breast tumors do not lie in the immediate vicinity of essential organs, one can hope to apply the treatment, heating the cancerous tissue up to yet higher temperatures in order to get a very effective cure of breast tumor, which may even be combined with parallel treatments relying upon conventional radiation therapy and chemotherapy.

However, one should bear in mind a caveat: keeping the amount of metal injected into the body to stay under a certain level is the best way to keep the danger of "nanopoisoning" at a relatively low level. After all, it should be remembered that nanoparticles are already used routinely in magnetic resonance (see Merbach and Toth [7]) therapy for the diagnosis of liver tumors.

After the therapy, nanoparticles do not have to be removed and are slowly metabolized. Since, so far, no harmful side effects from thermotherapy with magnetic nanoparticles could be observed, neither on animals nor on human beings, in 2004 nanoparticles have started to be applied for treating human prostate carcinomas at the Clinic for Urology, Charité—University Medicine, Berlin, Germany.

Moving to countries outside Europe, in order to complete the survey, we observe that in Japan [8], work at Nagoya University with magnetite cationic liposomes (MCLs) combined with heat-shock proteins has shown great potential in cancer treatment as well. Using MCLs, one locally generates heat in a tumor by placing test mice in an alternating magnetic field and does not cause the body temperature of the test animal to rise. After injection of MCLs and application of a magnetic field, the tumor and body temperature differed by 6°C. The combined treatment strongly inhibited tumor growth over a 30 day period and complete regression of tumors was observed in 20% of the mice.

Finally, in the United States, researchers at Rice University recently reported work on mice in which gold-coated nanoparticles treated to attach to cancerous cells were heated using infrared radiation. Sources of infrared radiation can be tuned to transmit at a narrow band of electromagnetic frequencies. Additionally, the "nanoshells" ' size can be changed to absorb a particular infrared radiation frequency. Hence, one can choose a frequency of the infrared radiation that couples with the gold-coated nanoparticles, while at the same time does not couple to the tissue of the body, thus enabling the selective destruction of cancerous cells and tumors [9]. The results of a preliminary experiment with mice treated with the nanoshells-infrared radiation therapy have proven to be very encouraging.

In conclusion, we can say that progress in nanotherapy, obtained worldwide by several groups using independent techniques, shows the global interest in nanoscience and its potential application to innovative medical technologies. This justifies the expectation that nanotechnology will soon yield a powerful tool for treating cancer.

15.3 Nanotechnology for Diagnostics and Drug Delivery*

Magnetic nanoparticles have been used as markers in biomedical diagnostics. The aggregation behavior of magnetic nanoparticle suspensions has been investigated by magnetorelaxometry [1]. In fact, due to the fact that bound and unbound nanoparticles have different magnetic relaxation times, biochemical-binding reactions can be detected by means of a SQUID-high resolution measurement technique (Figure 15.1). Magnetic relaxation immunoassays were realized by means of this technique. In addition, *in vivo* applications of magneto-relaxometry seem possible, for example, in cancer diagnostics.

_____
* The source for the material contained in this section is Strem Chemicals; for their nanochemistry items and applications see, e.g., Ref. [10].

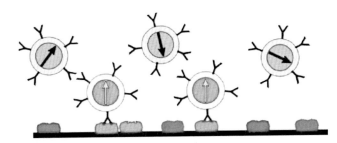

FIGURE 15.1 Immobilization of magnetic nanoparticles by antibody-antigen coupling (Contact: Dietmar.Eberbeck@ptb.de).

However, the use of magnetic nanoparticles is not limited to the above-mentioned applications, but can also be extended for achieving drug or radiation delivery. In particular, small magnetic particles can be engineered to carry therapeutic chemicals or radiation for tumor control. Because they are magnetic, the particles can be guided by an external magnetic field and can be forced to move with or against the flow of blood in veins or arteries or held in a fixed position once they have been conveyed to a target organ, and possibly retrieved when treatment has ended. Such work is in progress at the Argonne National Laboratory and the University of Chicago.

One is, of course, interested in using biosensors and biolabels to understand living cells. Nanotechnology has the potential to increase our ability to understand the fundamental working of living cells. Many potential applications for nanomaterials as biosensors and biolabels are under investigation. They have found use in cellular studies, enhanced spectroscopic techniques, biochips, and protein and enzyme analysis. Fluorescent nanoparticles can be used for cell labeling and magnetic nanoparticles may be utilized as sensors. Multicolor labeling of both fixed and living cells with fluorescent nanoparticles conjugated with biological ligands that specifically bind against certain cellular targets enables the recording of diffusion pathways in receptor cells.

The uptake of nanoparticles into the vesicular compartments around the nucleus of cells can be used to label the cells so that their pathway and fate can be followed. The nanoparticles exhibit reduced photobleaching as compared to traditional dyes and are passed on to daughter cells during cell division, therefore allowing for much longer-term observation. Magnetic nanoparticles can also act as sensors for assessing how external stresses affect changes in intracellular biochemistry and gene expression.

We can ask ourselves, how can nanotechnology improve medical diagnostics? Naturally, the early detection of a disease remains a primary goal of the medical community. Nanotechnology holds great promises for enabling the achievement of this goal. Nanoparticles in particular have exhibited tremendous potential for detecting fragments of viruses, precancerous cells, disease markers, and indicators of radiation damage. Biomolecule-coated ultra-small superparamagnetic iron oxide (USPIO) particles injected into the blood stream recognize target molecular markers present inside cells and induce a specific signal for detection by magnetic resonance imaging (MRI). This technology may allow for the detection of individual cancer cells months or years earlier than traditional diagnostic tools, which require the presence of hundreds of cancer cells.

In this respect, we can also address the issue of how bio-barcode amplification assays (BCA) can make use of nanoparticles in disease detection. A nanoparticle-based BCA utilizes gold nanoparticles and magnetic microparticles attached to large numbers of DNA strands and antibodies for a specific disease marker. The marker binds to the nano- and microparticles forming a complex that is separated from the sample using a magnetic field. Heating the complexes releases the DNA barcodes, which emit an amplified signal due to their large numbers. This BCA technology has been applied to the detection of markers for Alzheimer's disease and is being investigated for numerous others.

There are also stimulating suggestions that nanotechnology can improve targeted drug delivery. Targeted drug delivery systems can convey drugs more effectively and/or more conveniently, increase patient compliance, extend the product life cycle, provide product differentiation, and reduce health care costs. Drug delivery systems that rely on nanomaterials also allow for the targeted delivery of compounds characterized by low oral bioavailability due to poor water solubility, permeability, and/or instability and provide for longer sustained and controlled release profiles. These technologies can increase the potency of traditional small-molecule drugs in addition to potentially providing a mechanism for treating previously incurable diseases.

There are many other applications for nanomaterials in the medical and pharmaceutical sector, which we only have the possibility, owing to limitations in the length of these notes, to merely list. Areas currently under investigation include gene therapy, antibacterial/antimicrobial agents for burn and wound dressings, repair of damaged retinas, artificial tissues, prosthetics, enhancing signals for MRI examinations, and as radio frequency controlled switching of complex biochemical processes.

15.4 Carbon Nanotubes

The development of nanomaterials is currently underway in laboratories worldwide for medical and biotechnological applications including gene delivery [12], drug delivery [13], enzyme immobilization [14], and biosensing [15]. The most commonly used materials are gold [16], silica, and semiconductors. Silica nanoparticles have been widely used for biosensing and catalytic applications due to their large surface area-to-volume ratio, straightforward manufacture, and the compatibility of silica chemistry with the covalent coupling of biomolecules [17]. A key challenge in nanotechnology is the more precise control of nanoparticle assembly for the engineering of particles with the desired physical and chemical properties. Much research is currently focused on CNT as a promising material for the assembly of nanodevices, based upon new CNT–composite materials, such as CNT with a thin surface cover [18] or CNT bound to nanoparticles [19], in order to tailor their properties for specific applications.

In this section, reviewing the results reported in [20], we present the tunable synthesis of multi-walled CNT–silica-nanoparticle composite materials. Instead of coupling prefabricated silica nanobeads to CNT, we chose to grow the silica nanobeads directly onto functionalized multi-walled CNT by reaction of tetraethyl- or tetramethyl-orthosilicate (TEOS or TMOS) with a functionalized CNT precursor, prepared by coupling aminopropyltriethoxysilane (APTEOS) to a functionalized multi-walled CNT through a carboxamide bond, using a water-in-oil microemulsion to strictly control the nanobead size.

Several nested and straight cylindrical graphene sheets form the body of the ideal multi-walled CNT. In reality, nanotubes usually appear curved and have topological defects. Under strong oxidizing conditions (conc. HNO$_3$), nanotubes can be cut into shorter and straighter pieces having carboxylic acid groups at both their tips and at imperfections on their walls [21]. We oxidized multi-walled CNT with outer diameters of approximately 20–40 nm and lengths of 5–10 mm (NanoLab, Inc., Newton, MA) (Figure 15.2a) by refluxing in conc. HNO$_3$ for

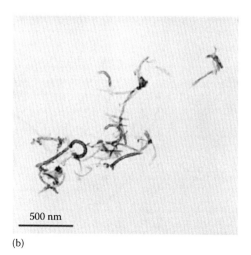

| (a) | (b) |

FIGURE 15.2 Multi-walled CNTs before (a) and after (b) oxidation in nitric acid.

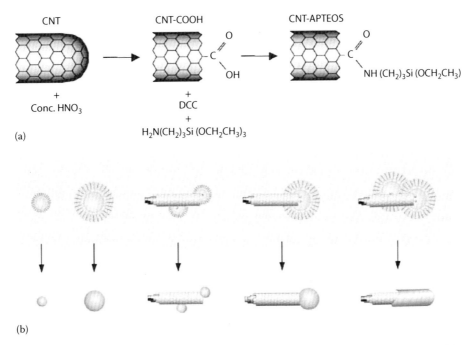

FIGURE 15.3 Scheme for preparing the CNT–nanoparticle composite. (a) Oxidation and preparation of the CNT–APTEOS precursor. (b) Formation of silica nanobeads in reverse micelles in a water-in-oil microemulsion. Inclusion of CNT–APTEOS nucleates the formation of nanobeads on the covalently linked propyltriethoxysilane groups (dots inside the micelles) by reaction with TEOS or TMOS.

6 h, followed by several washes with distilled water. The oxidized CNT (CNT-COOH) were shorter and straighter (Figure 15.2b). Their carboxylic acid groups greatly facilitated their dispersion in aqueous solutions, as well as their further functionalization (Figure 15.3a). A detailed description of the procedure for generating the activated CNT precursor (CNT-APTEOS) to the composite from CNT-COOH by activation of its carboxylic acid groups can be found in [20].

Using the procedure described in [20], we obtained new CNT–nanocomposites consisting of CNT with covalently attached silica nanobeads (Figure 15.4). Non-oxidized CNT (with negligible COOH content) did not support any composite formation (not shown). The inverse microemulsion system resulted in nanobeads covalently linked to the CNT only at locations functionalized with triethoxy-silane groups, while the bare graphitic wall of the pristine CNT did not associate with reverse micelles. Transmission electron microscopic (TEM) images revealed morphologies indicative of different nanobead diameters. Small nanoparticles were found to decorate the walls and ends of the CNT prepared using TMOS as precursor (Figure 15.4a through c). In many cases, small nanoparticle aggregates were observed to be associated with the CNT (Figure 15.4c), as expected for the high density of functional groups on the CNT. Under the conditions used for the synthesis of larger nanoparticles, individual nanobeads either decorated the CNT (Figure 15.4d through f) or had a uniform silica coating around the entire CNT (Figure 15.4g and h). We also observed some functionalized CNT that appeared to have silica within their tubes (Figure 15.4i). The internal presence of silica was not observed with the non-treated nanotubes. Further work is in progress to better understand the filling mechanism.

In summary, we covalently coated CNTs with silica nanoparticles of different sizes. Perhaps, the most valuable feature of our work [20] is that the architecture of the obtained assemblies can be largely controlled by varying the conditions in the synthesis. Thus, the length of CNT is regulated by the oxidation time (Figure 15.2) and the size of the nanobeads by using microemulsion conditions that yield micelles of a particular size. Indeed, silica nanobeads were prepared in a water-in-oil microemulsion system in which the water droplets served as nanoreactors [22]. The size of the final nanospheres was mainly regulated by the dimension of the water droplets, and, therefore, by the molar ratio of water to surfactant (w). Smaller nanobeads were prepared by reducing w. Furthermore, the dimension of the final product can be controlled by varying the molar ratio of water to precursor (h), the molar ratio of precursor to catalyst (n), by choosing the reactivity of the precursor, and the reaction time and temperature. The values of the variable parameters (w, h, n) used can be found in Table 1 of [20].

Because the chemical properties of the silica surface are particularly versatile and silica can be doped with fluorescent [23], magnetic [24], or biological macromolecules [25], nanostructures with a wide range of morphologies suitable for different applications can be obtained. We anticipate that further refinement of our water-in-oil microemulsion approach for creating novel nanostructures combined with procedures for isolating discrete products will allow us to combine different nanostructures into higher order assemblies that could be useful for a variety of applications, including yielding an interface between living cells and biosensor arrays.

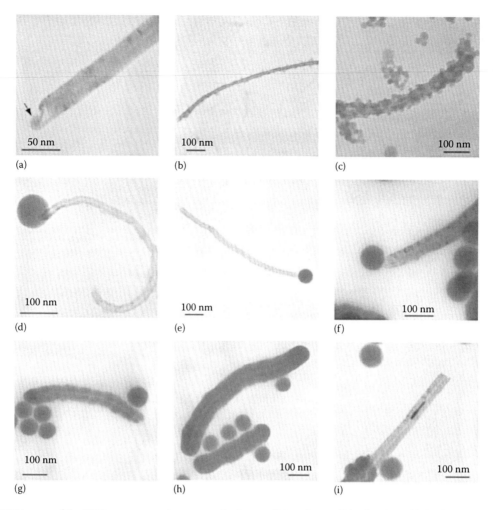

FIGURE 15.4   TEM images of the CNT–nanocomposites prepared using conditions for small (a–c) or large (d–i) silica nanobeads. The arrow in panel (a) indicates a nanobead at the tip of the CNT. The arrow in panel (i) indicates a polymerized silica inside a CNT.

15.5 Supramolecular Nanostructures

In [26] we constructed and characterized supramolecular nano-structures consisting of ruthenium-complex luminophores, which were directly grafted onto short oxidized single-walled CNTs or physically entrapped in silica nanobeads that were covalently linked to short oxidized single-walled CNTs or hydrophobically adsorbed onto full-length multi-walled CNTs. These structures were evaluated as potential electron–acceptor complexes also for their properties as fluorescent nanocomposites for use in biosensors.

The carboxylic acid groups of oxidized SWCNT that originated from the nitric acid-oxidation were covalently tethered to the ruthenium-complexes (Figure 15.5a) or luminophore-doped silica nanobeads (Figure 15.5b), whereas the full-length MWCNT had the ruthenium complex-doped silica nanobeads introduced onto their surfaces by hydrophobic adsorption *via* π–π interactions to maintain the intact CNT π-electronic structure.

The absorbance spectrum of $Ru(ap)(bpy)_2$ in dimethylforma-mide (DMF) exhibited a narrow peak at approximately 289 nm,

which we attribute to a ligand-to-ligand $\pi \rightarrow \pi^*$ transition, and by two broad bands at approximately 375 and 460 nm, which we attribute to $d_{Ru} \rightarrow \pi^*$ ligand singlet metal-to-ligand charge-transfer transitions (Figure 15.6a). On excitation at 460 nm, the steady state emission of $Ru(ap)(bpy)_2$ in DMF revealed an emission band centered at 606 nm (Figure 15.6b).

The absorbance spectrum of $Ru(ap)(bpy)_2$-decorated SWCNT dissolved in DMF was broad and slightly blue-shifted compared to that from $Ru(ap)(bpy)_2$ in DMF to confirm that luminophores decorated the nanotube surface (Figure 15.6a). The spectroscopic contribution of $Ru(ap)(bpy)_2$ grafted onto SWCNT was calculated by subtracting from the absorbance spectrum of the nanostructure that of oxidized SWCNT (dissolved in DMF), with matching absorption at 900 nm, because the spectroscopic contribution of the luminophore was absent for that wavelength. The peaks in the calculated absorbance spectrum of $Ru(ap)(bpy)_2$ grafted on SWCNT were centered at the same wavelengths and slightly broader compared to those of $Ru(ap)(bpy)_2$ in DMF. On excitation at 460 nm, the steady-state emission spectrum of $Ru(ap)(bpy)_2$-decorated SWCNT in

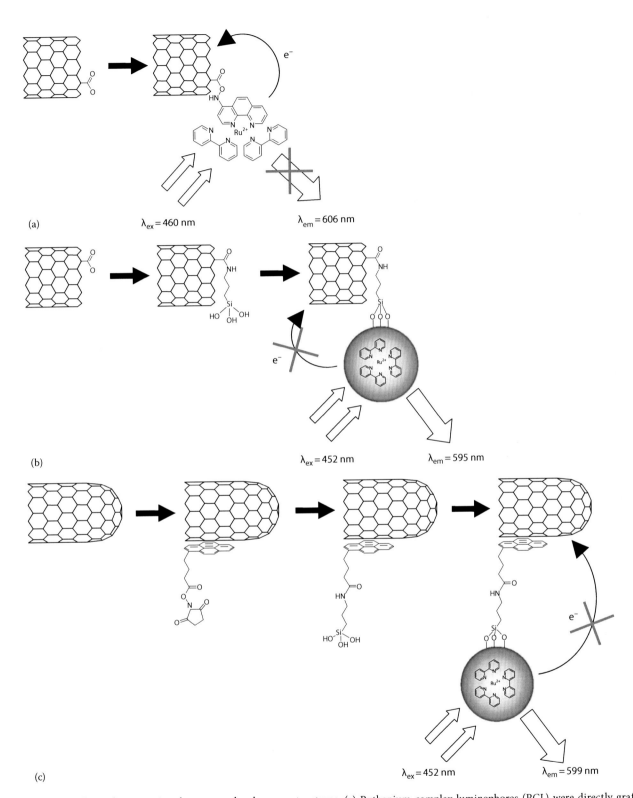

FIGURE 15.5 Scheme for preparing the supramolecular nanostructures. (a) Ruthenium-complex luminophores (RCL) were directly grafted onto oxidized SWCNT to form a supramolecular donor–acceptor nanostructure. (b) RCL-doped silica nanobeads were covalently linked to short oxidized SWCNT to form a supramolecular fluorescent nanostructure. (c) RCL-doped silica nanobeads were hydrophobically adsorbed onto full-length MWCNT via π–π interactions to form a supramolecular fluorescent nanostructure with CNT having the π-electronic structure intact.

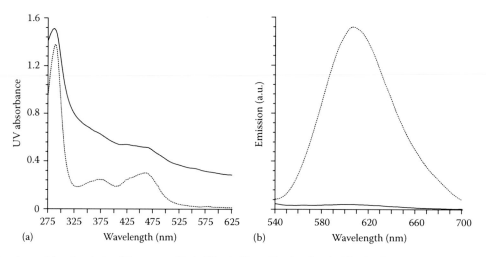

FIGURE 15.6   Absorbance (a) and emission (b) spectra of Ru(ap)(bpy)2 (dotted line) and Ru(ap)(bpy)2-decorated SWCNT (solid line) dispersions in DMF. For the emission spectra we matched the absorptions, at the 460 nm excitation wavelength, of free Ru(ap)(bpy)2 and Ru(ap)(bpy)2 grafted onto SWCNT, calculated by subtracting from the absorbance spectrum of the nanostructure that of oxidized SWCNT.

DMF revealed a strongly quenched (>98%) 606 nm photoluminescent peak compared to that of Ru(ap)(bpy)$_2$ in DMF (Figure 15.6b). Emission spectra were collected after having matched the absorptions, at the 460 nm excitation wavelength, of free Ru(ap)(bpy)$_2$ and Ru(ap)(bpy)$_2$ grafted onto SWCNT. Excitation at 375 nm showed similar quenching to further confirm the formation of the linked supramolecular electron donor–acceptor complexes between the metallo-organic luminophores, which acted as electron-transfer agents, and the CNTs, which acted as electron acceptors. Addition of free luminophore to the nanocomposite dispersions increased their emission intensities. These increases in emission suggest that, at the low concentrations used, dynamic (collisional) quenching was not responsible for the observed fluorescence quenching, which could only be caused by photo-induced charge injection from the metal-to-ligand charge-transfer (both singlet and triplet) excited states of the luminophore into the conduction band of the SWCNT.

TEM images of fluorescent Silica Nanobeads (fSNB) showed uniform diameter (13 ± 1 nm) silica nanobeads. Absorbance spectra of both free and fSNB-encapsulated Ru(bpy) in ethanol (EtOH) were characterized by a narrow peak at approximately 290 nm and by a broad plateau at approximately 450 nm, which we attribute to a ligand-to-ligand $\pi \rightarrow \pi^*$ transition and a $d_{Ru} \rightarrow \pi^*$ ligand singlet metal-to-ligand charge-transfer transition, respectively. The peaks in the spectrum of the fSNB in EtOH were slightly red-shifted and broader compared to those of free Ru(bpy) (Figure 15.7a). In steady state experiments, excitation of fSNB in EtOH at 452 nm produced an emission band that was enhanced, red-shifted and broader compared to that of Ru(bpy) (Figure 15.7b). The emission spectra were collected after having matched the absorptions, at the 452 nm excitation wavelength, of free Ru(bpy) and fSNB. These results suggest that the silica bead network may have affected the electrostatic environment surrounding the entrapped luminophores and

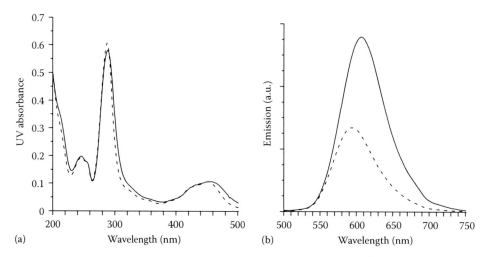

FIGURE 15.7   Absorbance (a) and emission (b) spectra of Ru(bpy) (dotted line) and fSNB (solid line) dispersions in EtOH. For the emission spectra, we matched the absorptions, at the 452 nm excitation wavelength, of free Ru(bpy) and fSNB.

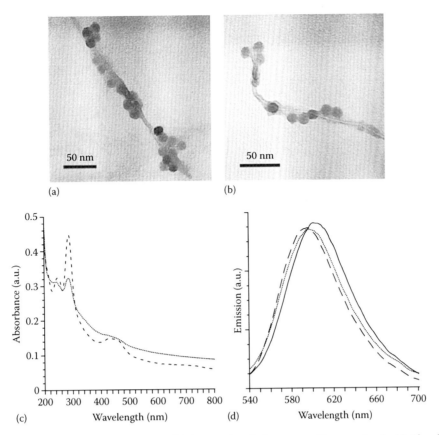

FIGURE 15.8 TEM images of SWCNT (a) and MWCNT (b) decorated with fluorescent silica nanobeads. (c) Absorbance spectra of fSNB-decorated SWCNT (dashed line) and fSNB-decorated MWCNT (dotted line) dispersions in EtOH. (d) Emission spectra of fSNB (solid line), fSNB-decorated SWCNT (dashed line) and fSNB-decorated MWCNT (dotted line) dispersions in EtOH. For the emission spectra we matched the absorption values, at the 452 nm excitation wavelength, of free fSNB and fSNB grafted onto CNT, calculated by subtracting from the absorbance spectrum of the nanocomposite that of silylated CNT.

that encapsulation protected Ru(bpy) from any dynamic self-quenching caused by collisional encounters.

The absorbance spectra of both fSNB-decorated SWCNT (Figure 15.8a) and MWCNT (Figure 15.8b) dissolved in EtOH showed broad peaks at approximately the same wavelengths as those from fSNB to confirm that fluorescent nanoparticles decorated the nanotube surfaces (Figure 15.8c). The spectroscopic contribution of fSNB grafted onto the CNT surface was calculated by subtracting from the absorbance spectrum of the nanocomposite that of silylated CNT with matching absorption at 900 nm, as the spectroscopic contribution of free fSNB was absent for that wavelength. The peaks in the calculated absorbance spectra of fSNB grafted onto both (both SW and MW) CNT were slightly red-shifted compared to those of free fSNB in EtOH. On excitation at 452 nm, the steady-state emission spectrum of fSNB-decorated CNT exhibited slightly quenched (<5%), narrower, and blue-shifted emission peaks compared to that of free fSNB (Figure 15.8d). The emission spectra were collected after having matched the absorptions, at the 452 nm excitation wavelength, of free fSNB and fSNB grafted onto CNT. The observed slight quenching could be addressed to the error in the calculation of fSNB grafted onto CNT and/or to the electron transfer from the luminophores close to the fSNB silica surface

to the CNT. Therefore, the silica host was able to avoid the quenching of the fluorescence due to the charge injection from the metal-to-ligand charge-transfer excited states of the luminophore into the conduction band of the quencher (CNT). This leads to the realization of a supramolecular fluorescent nanostructure useful for a large variety of applications ranging from biosensors to electronics, especially, in case of use of pristine full-length CNT that are characterized by intact π-electronic structure.

In summary, we synthesized in [26] three supramolecular nanostructures based on CNT and ruthenium-complex luminophores. The first nanostructure consisted of short oxidized SWCNT covalently decorated by ruthenium-complexes that act as light-harvesting antennae by donating their excited-state electrons to the SWCNT. This nanocomposite represents an excellent donor–acceptor complex, which may be particularly useful for the construction of photovoltaic devices based on metallo-organic luminophores. The second and the third nanostructures consisted of metallo-organic luminophore-doped silica nanobeads covalently linked to short oxidized SWCNT or hydrophobically adsorbed onto full-length MWCNT. In these nanocomposites, the silica network prevented the fluorescence quenching because excited-state electrons could not be readily donated to the CNT conduction band. Because the physical and chemical properties of the silica nanobeads are so versatile, and the π-electronic structure of the CNT can be kept intact by using a non-destructive modification of the nanotube structure, we consider these nanocomposites to have a promise for a variety of applications ranging from the biosensors to electronics.

15.6 Cellular Toxicity of Carbon Nanotubes

Not enough is yet known about the toxicity of CNTs that exist in many different forms and can be chemically modified and/or functionalized with biomolecules. Pristine single-walled CNTs are extremely hydrophobic tubes of hexagonal carbon (graphene) with diameters as small as 0.4 nm and lengths up to micrometers. Multi-walled CNTs consist of several concentric graphene tubes and diameters of up to 100 nm. These pristine CNTs are chemically inert and insoluble in aqueous media and are therefore of little use in biological or medical applications. Due to the hydrophobicity and tendency to aggregate, they are harmful to living cells in culture [27].

As we have seen in one of the above sections, for many applications, CNTs are oxidized in strong acid to create hydroxyl and carboxyl groups [28], particularly in their ends, to which biomolecules or other nanomaterials can be coupled [20]. These oxidized CNTs are much more readily dispersed in aqueous solutions and have been coupled to oligonucleotides, proteins, or peptides. Indeed, CNTs have been used as vehicles to deliver macromolecules that are not able pass through the cellular membrane by themselves into cells [29].

The toxicity of CNTs is very specific for certain types of nanotubes and specific tissues and cells under very specific conditions and is not yet known, and in particular the toxicity of oxidized CNTs. Therefore, we compared in [30] these two types of CNTs (i.e., pristine and oxidized ones) in a number of functional assays with human T lymphocytes, which would be among the first exposed cell types upon intravenous administration of CNTs in therapeutic and diagnostic nanodevices.

We found that, especially for high concentration (>1 ng/cell), carbon black (CB) is less toxic than pristine CNTs, therefore suggesting the relevance of the structure and topology (CB is amorphous) on the evaluation of the toxicity of a carbonaceous nanomaterial. Moreover, we found that oxidized CNTs are more toxic than pristine CNTs for both the analyzed concentrations, although they are considered better suited for biological applications. This may well be because they are better dispersed in aqueous solution and therefore reach a higher concentration of free CNTs at similar weight per volume values. We calculated that the less toxic amount of 40 μg/mL of CNTs is equal to an order of magnitude of 10^6 individual CNTs per cell in our experiments, based on an average length of 1 μm and a diameter of 40 nm, giving an average molecular mass of 5×10^9 Da.

While our results in [30] do not imply that CNTs should be abandoned for biological or medical purposes, our study sets an upper limit for the concentrations of CNTs that can be used. We recommend that CNTs be used a much less than 1 ng/cell and that cell viability and wellbeing be followed carefully with all new forms of CNTs and CNT-containing nanodevices.

It is likely that CNT toxicity will depend on many other factors than concentration, including their physical form, their diameter, their length, and the nature of attached molecules or nanomaterials. However, our conclusion applies, strictly speaking, only for T lymphocytes *in vitro* under the specific culturing conditions, for the time periods considered, with these specific types and preparations of nanotubes. Therefore, it is important to make it clear that this upper limit only pertains to our *in vitro* model and not to any *in vivo* animal or human situation; and may also not apply to other types of non-phagocytic cells.

15.7 Separation of Fluorescent Material from Single Wall Carbon Nanotubes

For biotechnological uses [31], a high level of purity is required to avoid undesired toxic effects from impurities. Contaminants in SWNT can be classified as carbonaceous (amorphous carbon and graphitic nanoparticles) and metallic (typically transition metal catalysts). It is well documented that nickel, which in combination with yttrium is used as a catalyst in the production of arc-discharged nanotubes, is cytotoxic [32]. Common SWNT purification methods based on oxidation (nitric acid and/or air) have the potential disadvantage of modifying the CNT by introducing functional groups and defects. Other less rigorous purification techniques rely upon filtration, centrifugation, and chromatography. Recently, electrophoresis of nitric acid-treated arc-discharged SWNT was used to separate tubular carbon from fluorescent nanoparticles [33].

As we reported in [34], fluorescent nanoparticles were isolated from both pristine and nitric acid-oxidized commercially available CNTs that had been produced by an electric arc method. The pristine and oxidized CNT-derived fluorescent nanoparticles exhibited a molecular-weight-dependent photoluminescence in the violet-blue and blue to yellowish-green ranges, respectively. The molecular weight dependency of the photoluminescence was strongly related to the specific supplier.

We analyzed the composition and morphology of the fluorescent nanoparticles derived from pristine and oxidized nanotubes from one supplier. We found that the isolated fluorescent materials were mainly composed of calcium and zinc. Moreover, the pristine CNT-derived fluorescent nanoparticles were hydrophobic and had a narrow distribution of maximal lateral dimension. In contrast, the oxidized CNT-derived fluorescent nanoparticles were superficially oxidized and/or coated by a thin carbon layer, had the ability to aggregate when dispersed in water, and exhibited a broader distribution of maximal lateral dimension.

The first sample we treated, purified CNT (pCNT), was composed of as-prepared nanotubes (AP-SWNT) purified by air oxidation, and the second one, oxidized CNT (oxCNT), was obtained by nitric acid oxidation of AP-SWNT. Both samples had a carbon content in the 80–90 wt.% range and approximately 10 wt.% nickel/yttrium catalyst (4:1). Samples of graphite, CB, pCNT, and oxCNT samples were dispersed in aqueous sodium dodecyl sulfate (SDS, 1 wt.%) surfactant using an ultrasonic bath.

Our spectral results reported in [34] may be explained by the presence of fluorescent particles (FP) with variable dimensions and chemistries in the NT samples. The predominant FP in pCNT had a mean mw below 30 kDa and exhibited mainly violet-blue photoluminescence and excitation at 315 nm. In contrast, in the fourth fraction (30–100 kDa) these spectral properties were weaker. All four fractions exhibited 450 and 485 nm photoluminescence on 365 nm excitation. The FP in the pCNT sample may originate from a nickel and yttrium-containing catalyst that had been covered by a few thin layers of metal oxide and/or carbide during the synthetic process, as was reported by Martinez et al. [35].

Our results suggest that the presence of the FP contaminants in these samples as aggregates or occlusions on NT walls led to the quenching of their photoluminescence. Sonication in the presence of the surfactant SDS was able to enrich the solution in micelle-embedded FP. To support this argument, Martinez et al. found that the material encasing the catalyst accompanying nitric acid-oxidized NT masked the metal core from detection by surface-sensitive techniques such as x-ray photoelectron spectroscopy. Their study may also explain the results of Xu et al., who, despite using energy-dispersive x-ray spectroscopy, were unable to discern any metal in the fluorescent electrophoresis fractions from oxidized SWNT [33].

In our study [34], the FP from the oxCNT exhibited more size-dependence in their photoluminescence that ranged from greenish-blue to orange, than those from the pCNT. The oxCNT FP were also more hydrophilic, probably because their carbon shells became carboxylated on oxidation, as was reported by Xu et al. In addition, our spectra suggest the presence of a less-abundant FP fraction that had a mean mw below 3 kDa and 425 nm emission peaks on excitation at 315 nm. Based on comparing these spectral properties with those of the FP from pCNT, FP from the oxCNT also contained fluorescent components derived from the pCNT.

The above findings led us to design a new SWNT purification method. Each dispersed NT sample was ultracentrifuged (as described above). The pellet was subjected to three additional rounds of dispersion in vehicle, sonication for 5 min, and ultracentrifugation. A spectrum indicative of few residual FP was exhibited by oxCNT-in-water. The spectra of the pCNT-in-SDS and oxCNT-in-SDS samples resembled that of graphite to demonstrate the absence of fluorescent contaminants. These results suggest that, in addition to amplifying FP photoluminescence, the surfactant facilitated FP removal. In summary, we now have a simple route consisting of surfactant-assisted dispersal followed by ultracentrifugation for removing FP contaminants from both pristine and nitric acid-treated SWNT.

In summary, we isolated, fractionated by molecular weight, and characterized FP from pCNT and oxCNT received from several suppliers. These FPs were responsible for the photoluminescence of electric arc-produced CNT in the visible range and were likely composed of impurities that were present in the graphite rods used for the production of the CNT. Spectroscopic analysis of the samples revealed some common supplier-independent features, specifically that the FP derived from the pCNT exhibited a violet-blue photoluminescence, whereas the FP derived from the oxCNT exhibited photoluminescence ranging from blue to yellowish-green.

In contrast, the molecular weight dependency for both the pristine and oxidized CNT-derived fractions was strongly related to the specific supplier. This can be explained by differing fabrication processes leading to different physical and chemical aggregation of the impurities present in the graphite rod. We recorded HRTEM images and EDX analysis of the FP isolated from the CNT (Carbon Solutions, Inc., Riverside, CA)-derived molecular weight fractions. The FP derived from the pCNT exhibited a narrow range in width, whereas the FP derived from the oxCNT were larger, had a broader width range, and formed hydrophilic aggregates in water.

Moreover, EDX analysis of the fractions from the oxCNT-in-water supernate suggested that their FP were superficially oxidized and/or coated by a thin carbon layer. We should stress that this section may explain why many nanoparticle preparations are so much more toxic in some studies than in others. The presence of these contaminants in CNT preparations may significantly complicate toxicity experiments, yet biologists and bio-medical scientists are often unaware of their occurrence.

15.8 Conclusions and Outlook

The field of nanoscience has been witnessing a rapid growth in the last decade. Recently, the attention of the community of nanoscientists has increasingly been focusing on technological applications. Nanotechnology has been emerging as an enabling technology, with high potential impact on virtually all fields of human activity (industrial, health-related, biomedical, environmental, economy, politics, etc.), yielding high expectations for a solution to the main needs of society, although having to address open issues with respect to its sustainability and compatibility.

The fields of application of the research in nanoscience include aerospace, defense, national security, electronics, biology, and medicine. There has been a significant progress in the understanding achieved in recent years, both from the theoretical and experimental points of view, along with a strong interest to assess the current state of the art of this fast-growing field. This interest has, at the same time, stimulated research collaboration among the different stakeholders in the area of nanoscience and the corresponding technological applications, prompting possibly the organization and presentation of joint projects in the near future involving both industry and public research.

In this chapter, we focused in particular on the biological and medical fields and described current and possible future developments in nanotechnological applications in such areas. Nanostructured, composite materials for drug delivery, biosensors, diagnostics, and tumor therapy were reviewed here as examples. CNTs were discussed as a primary example of emerging nanomaterials for many of the above-mentioned applications.

In addition, we wish to mention recent works aimed to modify the surfaces of superparamagnetic iron oxide nanoparticles (SPION) in order to reduce the cytotoxicity and enhance the cellular uptake of the nanoparticles [36]. A quantitative evaluation of prospective targeted drug or gene delivery, allowed by magnetically transported SPION, can be found in [37]. Last but not least, *in vitro* and *in vivo* membrane transport studies of targeted delivery have been quantitatively compared through the characterization of the magnetically induced mobility of SPION [38].

The consequences of CNT exposure have been much less studied in airway epithelial cells (AEC) than in macrophagic cells, although the former represent the first body barrier for inhaled particles and the contact between nanomaterials and these cells may be prolonged. One of the most important characteristics of AEC is the maintenance of tight junctional complexes that allow strictly polarized secretory functions, prevent the entry of pathogens and chemicals, and may participate in signal transduction that regulate gene expression [39]. Several airway cell lines, such as CaLu-3, a cell line derived from a human lung adenocarcinoma, maintain this property. These cells have the capability, once cultured on permeable filters, of forming very tight monolayers with trans-epithelial electrical resistance (TEER) values >1000 Ω/cm^2.

Using CaLu-3 cells, we carried out experiments [40] with commercial SWCNT and MWCNT, presenting a low level of metal contamination although synthesized through the carbon vapor deposition (CVD) method, or with a mixture of MW/SWCNT synthesized with the arc discharge method by the present author's group for nanotechnology at INFN-LNF, totally metal-free. CB (Printex 90™ 14 nm diameter, Degussa, Italy) was used as a control, amorphous probe. All the NM were used at doses ranging from 5 to 100 μg/mL. The carbon nanomaterials, used at the highest nominal dose [100 μg/mL], had very different effects on TEER [40]. While laboratory-grade nanotubes and CB were completely ineffective, MWCNTs produced a progressive decrease in TEER, SWCNTs produced a late and smaller, yet significant, decrease.

When carbon nanomaterials were added during the establishing of tight functional complexes, as assessed from the progressive increase of TEER during the growth of CaLu-3 monolayers, their effect was again strictly dependent on the type of material used. For instance, MWCNTs had a profound inhibitory effect and substantially suppressed the formation of a high-resistance epithelium. The suppression of the barrier function was associated with an increase in the paracellular (i.e., between cells and not through the cells) permeability of CaLu-3 monolayers to mannitol, without substantial changes in cell viability [40].

Since the maintenance of the barrier efficiency is a very important factor for the prevention of respiratory diseases and the determination of the biological effects of exposure to xenobiotics, it is conceivable that MWCNT-induced changes may have important functional consequences, such as the enhanced sensitivity to pathogens and the increased likelihood of extrapulmonary translocation from the airways to other body compartments [39,40]. The observed changes are independent on the mass concentration of the material and could be attributed to the high number of structural irregularities that could greatly increase their chemical reactivity.

Another important issue when considering the health hazards related to the exposure to CNTs is related to the mutagenesis [41]. In addition, the toxicity of water-soluble MWCNTs has been discussed [42]. The small size and high surface area of these nanoparticles seem to be related to their biological activities [43]. Definitely, precautions in their manipulation need to be taken [44].

Acknowledgment

The author acknowledges members of his group for nanotechnology at INFN-LNF and collaborators for their work on the developments discussed in this chapter. This research has been funded in part by the Italian Ministry for University and Research MIUR, within the project PRIN "Interaction of novel nanoparticulate materials with biological systems: testing models for human health risk assessment" and by the Italian Ministry of Health and ISPESL within the project "Metodologie innovative per la valutazione del rischio da esposizione occupazionale a nanomateriali."

References

1. S. Iijima. *Nature*, 1991, 354, 56.

2. M. S. Dresselhaus, G. Dresselhaus, and P. Avouris (Eds.). 2001. *Carbon Nanotubes: Synthesis, Structure, Properties and Applications*. Springer, Berlin, Germany.

3. A. Nel, T. Xia, L. Mädler, and N. Li. *Science*, 2006, 311(5761), 622–627; K. Donaldson, V. Stone, A. Clouter, L. Renwick, and W. MacNee. *Occup. Environ. Med.*, 2001, 58, 211–216; M. P. Holsapple, N. A. Monteiro-Riviere, N. J. Walker, and K. V. Thomas. *Toxicol. Sci.*, 2005, 88, 12–17; C. W. Lam, J. T. James, R. McCluskey, S. Arepalli, R.L. Hunter. *Crit. Rev. Toxicol.*, 2006, 36, 189–217; A. D. Maynard, P. A. Baron, M. Foley, A. A. Shvedova, E. R. Kisin, and V. Castranova. *J. Toxicol. Environ. Health* A, 2004, 67(1), 87–107; C. W. Lam, J. T. James, R. McCluskey, and R. L. Hunter. *Toxicol. Sci.*, 2004, 77, 126–134; C. Jia, S. Batterman, C. Godwin, and G. Hatzivasilis. *Environ. Sci. Technol.*, 2005, 27, 636–651.

4. K. Donaldson, R. Aitken, L. Tran, V. Stone, R. Duffin, G. Forrest, and A. Alexander. *Toxicol. Sci.* 2006; 92(1), 5–22.

5. R. Hergt, R. Hiergeist, J. Hilger, W. A. Kaiser, Y. Lapatnikov, S. Margel, and U. Richter. *J. Magn. Magn. Mater.*, 2004, 270, 345.

6. G. Oberdörster, Z. Sharp, V. Aludorei, A. Elder, R. Gelein, W. Kreyling et al. *Inhal. Toxicol.*, 2004, 46, 437–445.

7. A. E. Merbach and E. Toth (Eds.). 2001. *The Chemistry of Contrast Agents in Medical Magnetic Resonance Imaging.* John Wiley & Sons, Chichester, U.K.

8. A. Ito, F. Matsuoka, H. Honda, and T. Kobayashi. *Cancer Immunol. Immunother.*, 2004, 53(1), 26.

9. D. P. O'Neal, L. R. Hirsch, N. J. Halas, J. D. Payne, and J. L. West. *Cancer Lett.*, 2004, 209, 171.

10. H. Bönnemann, W. Brijoux, R. Brinkmann, M. Feyer, W. Hofstadt, G. Khelashvili, N. Matoussevitch, and K. Nagabhushana. Nanostructured transition metals. *Strem Chem.*, 2008, XXI(1), 1–18.

11. D. Eberbeck, F. Wiekhorst, U. Steinhoff, and L. Trahms. *J. Phys.: Condens. Matter*, 2006, 18, S2829–S2846.

12. D. Luo, E. Han, N. Belcheva, and W. M. Saltzman. *J. Control. Release*, 2004, 95, 333; A. K. Salem, P. C. Searson, and K. W. Leong. *Nat. Mater.*, 2003, 2, 668.

13. G. F. Paciotti, L. Myer, D. Weinreich, D. Goia, N. Pavel, R. E. McLaughlin, and L. Tamarkin. *Drug Deliv.*, 2004, 11, 169; K. S. Soppimath, T. M. Aminabhavi, A. R. Kulkarni, and W. E. Rudzinski. *J. Control. Release*, 2001, 70, 1.

14. P. Nednoor, M. Capaccio, V. G. Gavalas, M. S. Meier, J. E. Anthony, and L. G. Bachas. *Bioconjug. Chem.*, 2004, 15, 12; T. Konno, J. Watanabe, and K. Ishihara. *Biomacromolecules*, 2004, 5, 342.

15. X. L. Luo, J. J. Xu, W. Zhao, and H. Y. Chen. *Biosens. Bioelectron.*, 2004, 19, 1295; S. Hrapovic, Y. Liu, K. B. Male, and J. H. Luong. *Anal. Chem.*, 2004, 76, 1083.

16. M. C. Daniel and D. Astruc. *Chem. Rev.*, 2004, 104, 293.

17. W. Tan, K. Wang, X. He, X. J. Zhao, T. Drake, L. Wang, and R. P. Bagwe. *Med. Res. Rev.*, 2004, 24, 621; S. Santra, P. Zhang, K. Wang, R. Tapec, and W. Tan, *Anal. Chem.*, 2001, 73, 4988; X.X. He, K. Wang, W. Tan, B. Liu, X. Lin, C. He, D. Li, S. Huang, and J. Li. *J. Am. Chem. Soc.*, 2003, 125, 7168.

18. T. Seeger, Ph. Redlich, N. Grobert, M. Terrones, D. R. M. Walton, H. W. Kroto, and M. Rühle. *Chem. Phys. Lett.*, 2001, 339, 41; E. Whitsitt, and A. R. Barron, *Nano Lett.*, 2003, 3, 775.

19. H. Kim and W. Sigmund. *Appl. Phys. Lett.*, 2002, 81, 2085; J. M. Haremza, M. A. Hahn, and T. D. Krauss. *Nano Lett.*, 2002, 2, 1253; S. Ravindran, S. Chaudhary, B. Colburn, M. Ozkan, and C. S. Ozkan. *Nano Lett.*, 2003, 3, 447; S. Lee and W. Sigmund. *Chem. Commun.*, 2003, 6, 780; J. Sun, L. Gao, and M. Iwasa. *Chem. Commun.*, 2004, 7, 832.

20. M. Bottini, L. Tautz, H. Huynh, E. Monosov, N. Bottini, M. I. Dawson, S. Bellucci, and T. Mustelin. *Chem. Commun.*, 2005, 6, 758.

21. J. Liu, A. G. Rinzler, H. Dai, J. H. Hafner, R. K. Bradley, P. J. Boul, A. Lu et al. *Science*, 1998, 280, 1253.

22. J. Esquena, Th. F. Tadros, K. Kostarelos, and C. Solans. *Langmuir*, 1997, 13, 6400; F. J. Arriagada, and K. Osseo-Asare. *J. Colloid Interface Sci.*, 1999, 211, 210.

23. R. P. Bagwe, C. Yang, L. R. Hilliard, and W. Tan. *Langmuir*, 2004, 20, 8336.

24. H. H. Yang, S. Q. Zhang, X. L. Chen, Z. X. Zhuang, J. G. Xu, and X. R. Wang. *Anal. Chem.*, 2004, 76, 1316.

25. G. Fiandaca, E. Vitrano, and A. Cupane. *Biopolymers*, 2004, 74, 55.

26. M. Bottini, A. Magrini, A. Di Venere, S. Bellucci, M. I. Dawson, N. Rosato, A. Bergamaschi, and T. Mustelin. *J. Nanosci. Nanotechnol.*, 2006, 6, 1381.

27. D. Cui, F. Tian, C. S. Ozkan, M. Wang, and H. Gao. *Toxicol. Lett.*, 2005, 155, 73; N. A. Monteiro-Riviere, R. J. Nemanich, A. O. Inman, Y. Y. Wang, and J. E. Riviere. *Toxicol. Lett.*, 2005, 155, 377.

28. J. Liu, A. G. Rinzler, H. Dai, J. H. Hafner, R. K. Bradley, P. J. Boul, A. Lu, T. Iverson, K. Shelimov, C. B. Huffman, F. Rodriguez-Macias, Y. S. Shon, T. R. Lee, D. T. Colbert, and R. E. Smalley. *Science*, 1998, 280, 1253.

29. D. Pantarotto, J. P. Briand, M. Prato, and A. Bianco. *Chem. Commun.*, 2004, 16; N. W. Shi Kam, T. C. Jessop, P. A. Wender, and H. Dai. *J. Am. Chem. Soc.*, 2004, 126, 6850.

30. M. Bottini, S. Bruckner, K. Nika, N. Bottini, S. Bellucci, A. Magrini, A. Bergamaschi, and T. Mustelin. *Toxicol. Lett.*, 2006, 160, 121.

31. R. H. Baughman, A. A. Zakhidov, and W. A. de Heer. *Science*, 2002, 297, 787.

32. M. D. Pulido and A. R. Parrish. *Mutat. Res.*, 2003, 533, 227.

33. X. Xu, R. Ray, Y. Gu, H. J. Ploehn, L. Gearheart, K. Raker, and W. A. Scrivens, *J. Am. Chem. Soc.*, 2004, 126, 12736.

34. M. Bottini, C. Balasubramanian, M. I. Dawson, A. Bergamaschi, S. Bellucci, and T. Mustelin, *J. Phys. Chem. B Condens. Matter Mater. Surf. Interfaces Biophys.*, 2006, 110, 831.

35. M. T. Martinez, M. A. Callejas, A. M. Benito, W. K. Maser, M. Cochet, J. M. Andres, J. Schreiber, O. Chauvet, and J. L. Fierro. *Chem. Commun.*, 2002, 7(9), 1000.

36. A. J. Gupta and M. Gupta. *Biomaterials*, May 2005, 26(13), 1565–1573.

37. F. G. Mondalek, Y. Y. Zhang, B. Kropp, R. D. Kopke, X. Ge, R. L. Jackson, and K. J. Dormer. *J. Nanobiotechnol.*, 2006, 4, 4.

38. A. L. Barnes, R. A. Wassel, F. Mondalek, K. Chen, K. J. Dormer, and R. D. Kopke. *Biomagn. Res. Technol.*, 2007, 5, 1.

39. E. Bergamaschi, O. Bussolati, A. Magrini, M. Bottini, L. Migliore, S. Bellucci, I. Iavicoli, and A. Bergamaschi. *Int. J. Immunopathol. Pharmacol.* 2006, 19(4), 3–10.

40. B. M. Rotoli, O. Bussolati, M. G. Bianchi, A. Barilli, C. Balasubramanian, S. Bellucci, and E. Bergamaschi. *Toxicol. Lett.*, 2008, 178, 95–102.

41. A. Di Sotto, M. Chiaretti, G. A. Carru, S. Bellucci, and G. Mazzanti. *Toxicol. Lett.*, 2009, 184(3), 192–197.

42. M. Chiaretti, G. Mazzanti, S. Bosco, S. Bellucci, A. Cucina, F. Le Foche, G. A. Carru, S. Mastrangelo, A. Di Sotto, R. Masciangelo, A. M. Chiaretti, C. Balasubramanian, G. De Bellis, F. Micciulla, N. Porta, G. Deriu, and A. Tiberia. *J. Phys.: Condens. Matter*, 2008, 20, 474203.

43. M. De Nicola, S. Bellucci, E. Traversa, G. De Bellis, F. Micciulla, and L. Ghibelli. *J. Phys.: Condens. Matter*, 2008, 20, 474204.

44. M. De Nicola, D. Mirabile Gattia, S. Bellucci, G. De Bellis, F. Micciulla, R. Pastore, A. Tiberia, C. Cerella, M. D'Alessio, M. Vittori Antisari, R. Marazzi, E. Traversa, A. Magrini, A. Bergamaschi, and L. Ghibelli. *J. Phys.: Condens. Matter*, 2007, 19, 395013.

Cross references

Toxicology of nanomaterials

Nanomaterial standards for efficacy and toxicity assessment

Human health implications of nanomaterial exposure

Tunneling nanotubes for the exchange of components between animal cells

16

Cell Oxidative Stress: Risk of Metal Nanoparticles

Marija Poljak-Blazi
Rudjer Boskovic Institute

Morana Jaganjac
Rudjer Boskovic Institute

Neven Zarkovic
Rudjer Boskovic Institute

16.1 Introduction

Nanoparticles are of great scientific interest as a bridge between bulk materials and sophisticated constructs that can be useful in various fields of material and life sciences. A bulk material should have constant physical properties regardless of its size, but at the nanoscale this is often not the case. The properties of materials change as their size approaches the nanoscale and as the percentage of atoms at the surface of a material becomes significant. Interesting and sometimes even unexpected properties of nanoparticles are partly due to the aspects of the surface of the material dominating the properties in lieu of the bulk properties. In contrast to many efforts aimed at exploiting the desirable properties of nanoparticles for medicine, there are limited attempts at evaluating the potentially undesirable effects of these particles when administered for medical purposes. Nanoparticles are likely to be the cornerstones of innovative nanomedical devices to be used for drug discovery and delivery, discovery of biomarkers, and molecular diagnostics.

16.2 Nanotechnology

The growing use of nanotechnology in high-tech industries is likely to become an essential way for modern humans to be exposed to engineered nanoparticles. The list of the most common

applications of nanoparticles includes energy, electronics, consumer goods, environmental catalysts and filters, cosmetics (e.g., using ZnO and TiO_2), and medicine. From a medical point of view, nanoparticles are attractive as alternative drug and vaccine delivery mechanisms (e.g., inhalation, oral in place of injection), tissue engineering technology such as for bone growth promotion, cancer therapies, biocompatible coatings for cardiovascular implants, bio-labeling for the detection of biomarkers (e.g., using Au), carriers for drugs with low water solubility, fungicides (e.g., using ZnO), magnetic resonance imaging (MRI) contrast agents (e.g., using super-paramagnetic iron oxide), new dental composites, etc.

Nanotechnology is being applied in medical sciences to try and create personalized medicine relying on biomarker-based proteomics and genomic technologies, while nanoparticles can be used for qualitative or quantitative *in vivo* or *ex vivo* diagnosis by concentrating, amplifying, and protecting a biomarker from degradation to provide a more sensitive analysis (Geho et al. 2004).

Finally, it must be mentioned that the novel and rapidly developing field of lipidomics will certainly become important for nanosciences since biomembranes, such as the cell membrane that surrounds the liquid cell body maintaining the integrity of the cell, gain their physical and biological features due to complex lipid–protein structures and interactions. The transport of biologically relevant substances through cellular and tissue membranes requires complex structure/function relationships, which eventually allow only biocompatible substances to be efficient in living systems. From this point of view, minute membranous structures composed of fat and proteins might be considered as natural nanoparticles that act as cellular compartments regulating the essential biochemical process of the living cell, such as cell respiration on the level of mitochondria or lysosomal enzymatic activities.

16.3 Nanoparticles

16.3.1 Definition of Nanoparticles

Nanoparticles are tiny materials, usually referred to as particles with a size of up to 100 nm. Nanoparticles exhibit completely new physicochemical properties different to bulk materials of the same composition or improved properties based on specific characteristics (size, distribution, morphology, phase, etc.). Such properties make them very attractive for commercial and medical development. Nanoparticles can be made of a wide range of materials, the most common being metal oxide ceramics, nonoxide ceramics, silicates and metals, and others, e.g., those based on polymer materials. Nanoparticles present several different forms (flakes, spheres, dendritic shapes, cubes, etc.; see Figure 16.1).

While metal and metal oxide nanoparticles in use are typically spherical, silicate nanoparticles have a flaky shape with two of their dimensions in the range of 100–1000 nm. They are generally designed and manufactured with physical properties tailored to meet the needs of the specific application for which they will be used.

16.4 Metal Nanoparticles

Metal nanoparticles are highly oxidizable; therefore, their stabilization by suitable passive surface layer is normally necessary. A promising technique could be the preparation of core/shell nanoparticles by various gas phase synthesis methods such as arc-discharge, etc. The *passivation* treatment should be completed before exposing the nanoparticles to ambient air. The coating of the nanoparticles by various hydrophilic/phobic substances is another important issue, which is, according to the opinion of experts, in a rather developed stage. Metallic nanoparticles such as iron oxide nanoparticles (15–60 nm) generally comprise a class of super-paramagnetic agents that can be coated with dextran, phospholipids, or other compounds to inhibit aggregation and enhance stability. The particles are used as passive or active targeting agents (Gupta and Gupta 2005). Gold shell nanoparticles, other metal-based agents, are a category of spherical nanoparticles consisting of a dielectric core covered by a thin metallic shell, which is typically gold. These particles possess highly favorable optical and chemical properties for biomedical imaging and therapeutic applications (Figure 16.2; Zivkovic et al. 2005b, Hirsch et al. 2006).

Metal particles can be relatively easily functionalized to drive self-assembly (e.g., structured materials) or for binding to substrates (e.g., biological sensors, controlled drug delivery mechanisms, etc.). Metal nanoparticles can be used for toxic, i.e., biocidal applications. Stable silver nanoparticles can be prepared in a form suitable for effective biocide water-born polymeric coatings and are incorporated into antibacterial ceramics for bathrooms fixtures, wound dressings (bandages), or even sport socks. Increased protective properties of coatings can also be

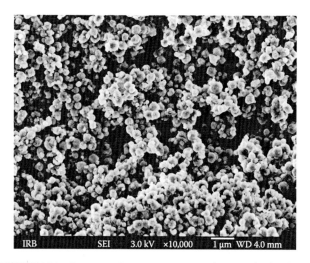

FIGURE 16.1 Scanning electron microscopy photograph of zeolite A nano-crystals obtained by hydrothermal treatment of clear aluminosilicate solutions (dimensions in the range of 100–500 nm). (Courtesy of Dr. Boris Subotic, Rudjer Boskovic Institute, Zagreb, Croatia.)

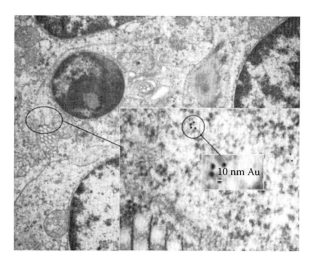

FIGURE 16.2 Gold nanoparticles used for detection of ROS-induced cellular lipid peroxidation product 4-hydroxynonenal (HNE)-protein adducts formed during oxidative stress of activated phagocytic leukocytes (transmission electron microscopy photograph). Gold nanoparticles are attached to the monoclonaly antibody which recognizes specifically HNE-protein adducts.

achieved by adding metal nanoparticles (e.g., Zn, Pb, and Mn). Doping polymer composites with complex oxide nanoparticles dramatically increases the efficiency of the composites during exposure to strongly aggressive media. Due to their chemical properties, especially catalytic, metal oxides are those metal composites showing the most interesting bioreactive potential.

Reactivity can be considered to be the most relevant aspect for catalysis and related applications (sensors, etc.). Combining the reactivity and catalytic activity hits some important application fields, such as fuels (and fuel additives), fuel cells, and explosives. The catalysis is enhanced by a high surface area-to-volume ratio and potential homogenous distribution of nanoparticles. This can lead to a reduction in the required amounts of the commonly used platinum group precious metals (widely used in fuel cells and in catalytic converters) and the opening up of new possibilities for less-commonly used metals such as gold, which only becomes an effective catalyst in cases when it is in a nanoparticle form.

Transparency is often also essential, as in the case of photo-catalytic self-cleaning windows and can even be a key property on its own, as in well-known sunscreen applications. Rare earth oxides are sensitive to air moisture and other contaminants. The chemical reactivity can therefore strongly influence the surface properties, in particular the light emission from the surface. Besides, the high surface-to-volume ratio enhances this effect and makes the nanoscale crucial for taking advantage of the high surface-to-volume ratio.

On the other hand, a high surface-to-volume ratio makes nanoparticles reactive and exposes them to environmental influences; in particular, that can be important for metal nanoparticles if exposed to the reactive oxygen species.

16.5 Reactive Oxygen Species

During the last 20 years, oxygen-free radicals and related highly reactive substances known as the reactive oxygen species (ROS), which are generated from oxygen molecules by all aerobic cells, have been of great interest in experimental and clinical medicine (Halliwell and Gutteridge 1989).

ROS comprise oxygen radicals, such as superoxide anion ($O_2^{\cdot-}$) and hydroxyl radical ($HO^{\cdot}$), as well as many other reactive oxygen nonradical species (which do not contain in their formula sign ($\cdot$) which indicates unpaired electron in radicals) such as hydrogen peroxide (H_2O_2), singlet oxygen ($^1O_2$), hypochlorous acid (HOCl), and ozone (O_3). These chemical species can be produced from endogenous and exogenous sources. Potential endogenous sources include inflammation, mitochondria, cytochrome P450 metabolism, microsomes, and peroxisomes (Valko et al. 2006). Environmental agents (e.g., cigarette smoke, carcinogens) are well-known exogenous sources of ROS that generate chair reactions of the ROS production in cells.

While oxygen-free radicals contain unpaired electrons and therefore react in microseconds within a short distance of nanometers with any molecule nearby, ROS, such as hydrogen peroxide, are often involved in complex biochemical reactions that have important biological purposes.

We consider oxygen-free radicals under the term ROS while the ROS do not need to have unpaired electrons. Nevertheless, they are a highly reactive chemical species and are often produced in excess thus causing the process of oxidative stress. ROS are continuously formed in organisms during physiological and pathological metabolic processes by oxygen reduction or excitation (Feher et al. 1987, Halliwell 1995).

They are physiologically generated in small quantities during metabolic processes (Braughler and Hall 1989, Hall and Braughler 1989) or during irradiation (e.g., x-rays and UV light). The ROS are important by-products of mitochondrial (cellular) respiration that implies catalyzed electron transport reactions (Valko et al. 2006). They are also formed during metal-catalyzed reactions (Valko et al. 2006), as well as during a process of "oxidative burst" that occurs in case of inflammation mediated by phagocytic leukocytes and serves as a crucial mechanism of defense against potentially harmful microbes, cancer cells, or particles including nanoparticles (Figure 16.3; Zivkovic et al. 2005a, Zivkovic et al. 2007, Jaganjac et al. 2008).

Under the term phagocytic leukocytes, we consider white blood cells that can destroy potentially dangerous agents by ingestion and digestion (phagocytosis). During phagocytosis inside of the phagocytes, specific membranous structures are formed denoting phagosomes that digest potentially dangerous microbes or particles by production of ROS and digestive enzymes.

Therefore, phagocytes undergo desirable oxidative stress (excessive production of ROS), which is localized inside of the phagosomes, while the spread of oxidative stress is prevented by cellular membranous structures and intracellular antioxidants such as enzymes superoxide dismutases (SOD), catalase, and glutathione peroxidase/glutathione reductase.

However, oxidative bursts also include extracellular processes, i.e., during oxidative bursts of leukocytes, ROS are being produced due to the leukocyte activation even outside of the leukocytes. In this case, of particular importance is enzyme myeloperoxidase (MPO), which generates hypochloric acid (HOCl) that can trigger a further chain reaction of oxidative stress. Since phagocytes are able to migrate out of the blood vessels aiming to destroy potentially harmful agents almost anywhere in the body, it is easy to imagine that the application of metal nanoparticles (or any other particles foreign to the body) that could be interpreted by the organism as potential danger would cause a general inflammatory response targeting these particles by oxidative bursts wherever these particles are present.

ROS are present also as "pollutants" in the atmosphere and are spontaneously generated from oxygen even at a high rate of 1%–3%. Therefore, it is not surprising that an average person has to metabolize up to 2 kg of ROS every year (Halliwell and Gutteridge 1989).

ROS play a dual role in biological systems. In small amounts, they are involved mostly in desirable physiological responses.

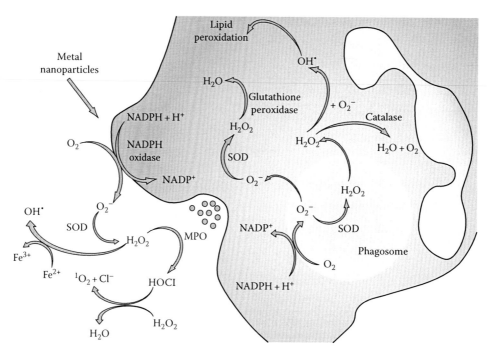

FIGURE 16.3 "Oxidative burst"—complex process of ROS production by phagocytic leukocytes (phagocytes—polymorphonuclear leukocytes and macrophages) during inflammation.

Namely, ROS at low concentrations may function as physiological mediators of cellular responses to various stimuli (Schreck and Baeuerle 1991). ROS generated by phagocytes act as regulators of other immune reactive cells (Hellstrand et al. 1994) and of apoptotic processes (Lundqvist-Gustafsson and Bengtsson 1999). Recently, we have reported that granulocytes play an important role in the defense against cancer in the early stages of tumor development, while their oxidative burst is an essential mechanism of the anticancer response, which might even lead to tumor regression (Jaganjac et al. 2008).

In contrast, overproduction of ROS is cytotoxic and damages macromolecules (DNA, proteins, sugars, and lipids) (Esterbauer et al. 1991). Excessive ROS may act in a chain reaction of the vicious circle of self-generated overproduction and additionally result in the cellular secretion of biologically important molecules such as chemotactic factors, growth factors, and proteolytic enzymes, causing cell and tissue destruction and altering the expression of signaling proteins and genes. The harmful effects of ROS are opposed by the antioxidant defense systems (nonenzymatic antioxidants in addition to antioxidant enzymes) (Halliwell 1996). However, oxidative damage accumulates during life and is assumed to play a key role in the development of cancer, neurodegenerative disorders (Parkinson's disease, Alzheimer's disease), arthritis, and in many other age-dependent diseases, including aging (Halliwell and Gutteridge 1989). When antioxidant defense mechanisms are decreased or if overproduction of radicals occurs, cells are unable to maintain this oxidative steady state (denoted otherwise as oxidative homeostasis).

The loss of oxidative homeostasis may result in damage to the affected cells and to the surrounding tissue due to miss-activation of signaling pathways, inflammatory cytokine production, altered gene expression, and other cellular alterations leading to their malignant transformation, degeneration, and even death.

Toyokuni et al. (1995) have discussed the significance of persistent oxidative stress in cancer. Namely, persistent oxidative stress and ROS production may constantly activate transcription factors and induce the expression of proto-oncogenes, it may also induce DNA (genomic instability), activate specific antioxidant systems in cancer cells, or specifically damage certain protease inhibitors thus facilitating tumor invasion and metastasis.

Furthermore, ROS are involved in the cellular response to toxins and various injuries, in the induction of cell growth (proliferation), and in the function of the number of cellular regulatory signaling systems (Valko et al. 2006). ROS themselves are known to act as cellular signal transduction messengers and as activators of signal transduction pathways, which are necessary to keep the cells functional and allow them to "communicate" with other, even remote cells, as in the case of hormonal or cytokine activities (Figure 16.4; Valko et al. 2006).

16.6 Oxidative Stress

Oxidative stress is a balance shift of oxido-reductive reactions to oxidation, resulting in the excessive production of ROS (Sies 1991; Figure 16.5). During inflammation, phagocytes in particular those denoted as granulocytes due to their high amounts of intracellular granule (ROS producing membranous micro/nanoparticles containing digestive enzymes), release high amounts of ROS by the activation of xanthin oxidase, NADPH oxidase, and phospholipase by the leakage of activated oxygen from mitochondria during oxidative phosphorylation, etc. (Pincemail 1995).

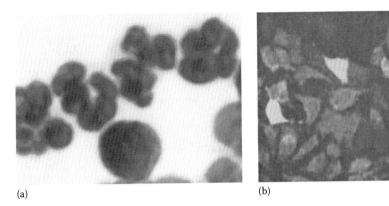

(a)   (b)

FIGURE 16.4 **(See color insert following page V-2.)** Intracellular ROS detection. (a) Superoxide anion oxygen free radical (O_2^-) in functional granulocyte leukocytes (Nitro blue tetrasolium method). (b) Hydrogen peroxide (H_2O_2, ROS, but not oxygen free radical) in human uterine cervical carcinoma cell line (HeLa cells) (DCFH-DA fluorescence method).

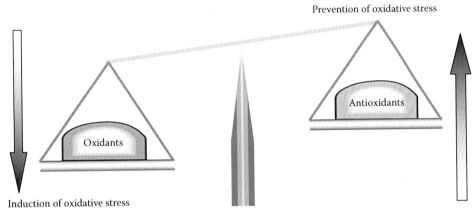

FIGURE 16.5 Oxidative stress, a balance shift of oxido-reductive reactions to oxidation.

Hydroxyl radical (HO·) is probably the most harmful ROS because of its high reactivity and instability and it is the only ROS that has enough energy to cause lipid peroxidation (Porter 1984, Darley-Usmar et al. 1995).

During aggressive oxidative stress, ROS damage membranes of the cells oxidizing their polyunsaturated fatty acids (PUFA) thus causing a chain reaction of lipid peroxidation, which ends in the destruction of biomembranes, consequently killing cells (Figure 16.6).

The final products of lipid peroxidation are reactive aldehydes such as 4-hydroxyalkenals and other similar α,β-unsaturated aldehydes that also act as inflammatory mediators and factors

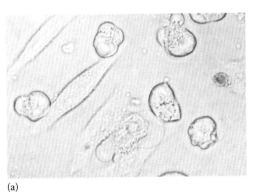 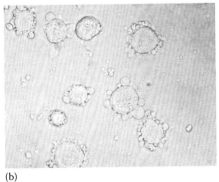

(a)   (b)

FIGURE 16.6 Lipid peroxidation and the cellular decay (Human carcinoma HeLa cells—magnification 650×). In the initial stage of oxidative stress the cells "behave individually" due to their differential capacities to tolerate ROS (a), hence cell membrane is intact in some cells maintaining their structural integrity, while in other cells lipid peroxidation causes formation of the membrane "blebs" (balloons developed due to the destruction of the lipids upon ROS attack). In advanced stage (b) there are no intact cells, some were already destroyed leaving only the "cellular debris" (minute cellular fragments), while blebs dominate in the membranes of the other cells.

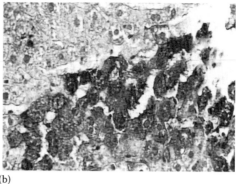

(a)   (b)

FIGURE 16.7   Immunochemical detection of the HNE-protein conjugates in cells (dark gray color). (a) activated granulocytes—lipid peroxidation is present mostly in the vicinity of the leukocytes membranes. (b) liver tissue after needle injury (biopsy)—lipid peroxidation affects the cells stressed by mechanical (needle puncture) stress, leaving the rest of the cells free of damage. In both cases monoclonal antibodies specific for the HNE-protein adducts were used to detect the lipid peroxidation product HNE bond to the cellular proteins, while the monoclonal antibody reaction with the HNE-protein adducts was visualized by the use of DAB staining giving dark gray color in contrast to the hematoxylin staining of the (parts of) cells free of HNE (light gray color).

of carcinogenesis (Smith and Marnett 1991, Yamamoto 1992, Uchida 2003, Zarkovic et al. 2006).

A great number of these aldehydes have been isolated from biological samples where they may promote and reinforce cell damage induced by oxidative stress (Liu et al. 1999, 2000, Uchida et al. 1999, Zarkovic et al. 1999, Kumagai et al. 2000, Poli and Schaur 2000). Lipid-derived aldehydes are more stable than ROS and therefore can diffuse to targets far from the initial oxidative injury. Among these aldehydes, of particular biochemical and biomedical relevance is 4-hydroxy-2-nonenal (HNE), denoted as a "second toxic messenger of free radicals" that is nowadays considered to be a major bioactive marker of lipid peroxidation (see Figure 16.7; Esterbauer et al. 1991, Zollner et al. 1991, Zarkovic et al. 1999, Zarkovic 2003).

16.7 Hierarchical Cellular Oxidative Stress Model—Oxidative Homeostasis

Recently, Li et al. proposed the hierarchical (Figure 16.8) cellular oxidative stress model (Li et al. 2008), assuming that oxidative stress implies a cellular stress response, which activates a number of the redox-sensitive signaling cascades (Li et al. 2003). This concept is in its essence based on the older concept of persistent oxidative stress in cancer (Toyokuni et al. 1995) on one hand, and on the other it deals with the cellular aspects of the broader concept of oxidative homeostasis relevant for various diseases and aging (Wildburger et al. 2009).

At low concentrations, ROS may function as physiological mediators of cellular responses to various stimuli (Schreck and Baeuerle 1991). ROS generated by phagocytes act as regulators of other immune reactive cells (Hellstrand et al. 1994) and of apoptotic processes (Lundqvist-Gustafsson and Bengtsson 1999). Recently, we have reported that granulocytes play an important role in the defense against cancer in the early stages

of tumor development, while their oxidative burst is an essential mechanism of the anticancer response, which might even lead to tumor regression (Jaganjac et al. 2008). That is in agreement with the fact that overproduction of ROS is cytotoxic and damages macromolecules (Esterbauer et al. 1991). Persistent ROS production may activate transcription factors and induce the expression of proto-oncogenes, it may also affect DNA (genomic instability), modulate antioxidant systems in cancer cells, or specifically damage certain protease inhibitors thus facilitating tumor invasion and metastasis.

When cells are exposed to mild oxidative stress, the transcription factor denoted "nuclear factor erythroid 2-related factor 2" (Nrf2) is induced. Nrf2 is known to regulate transcriptional activation of various antioxidant and detoxification enzymes (e.g., NADPH oxidase, catalase, superoxide dismutase, etc.; Li et al. 2008). Nanoparticle-induced oxidative stress may cause negative changes of this protective pathway (Li et al. 2003). A further increase in ROS production to the severe oxidative stress level can result in proinflammatory and cytotoxic effects. In these signaling pathways, redox-sensitive mitogen activated (MAP) kinases and NF-κB cascades are responsible for the proinflammatory effects by inducing the release of cytokines, chemokines, and adhesion molecules. Even higher oxidative stress levels are cytotoxic and involve mitochondrial perturbation, i.e., the mitochondrial leakage that turns a well-controlled respiratory chain of ROS transport within the mitochondria into the spread or ROS out from the mitochondria into the cellular cytoplasm, followed by the release of factors that induce programmed cell death—apoptosis.

It should be mentioned that the concept of persistent oxidative stress in cancer assumes not only that in malignant cells ROS production at subtoxic level causes malignant transformation but also enhances the intense growth of cancer, which is otherwise often treated by therapies that induce further production of ROS (radiotherapy, some chemotherapies, etc.). Thus, the aim of therapeutically induced oxidative stress is to gain levels of ROS production high enough to kill cancer cells.

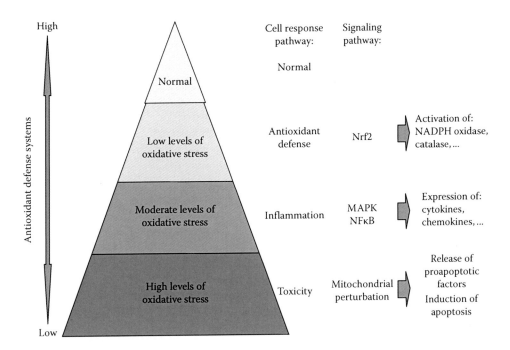

FIGURE 16.8 The involvement of different oxidative stress levels on cellular stress response.

On the other hand, the concept of oxidative homeostasis assumes that persistent oxidative stress can be physiological and can increase the defense against severe oxidative stress. However, even such oxidative stress could eventually increase the level of cellular damage that will lead to degeneration and "programmed cell death" (apoptosis), which are the causes and consequences of persistent oxidative stress and are associated with aging.

Finally, it should be said that high amounts of ROS that exceed the capacity of antioxidants and other defense mechanisms to inactivate their toxicity will lead to sudden cellular death (as in the case of membrane destruction upon lipid peroxidation) known as necrosis, while necrosis will cause inflammatory response and allow the further spread of oxidative stress.

Accordingly, we can expect that nanoparticles, in particular those composed of metal oxides that are able to generate ROS in the living system, could have not only desirable effects, but also undesirable side effects including (co)carcinogenic effects, proinflammatory effects, and degeneration of the affected cells and tissues, depending on their dose, structure, and the sensitivity of the cells to the metal-triggered oxidative stress.

16.8 Nanoparticles-Mediated ROS Generation

Industrial emissions, ground water contaminations, sun creams, functionalized cloths, cosmetics, and many others are potential sources of nanoparticle exposure. Excessive exposure to certain nanoparticles can cause inflammation, cellular damage, and cancer (Poljak-Blazi et al. 2009). Of particular interest are metal

nanoparticles because some metals (e.g., iron) that are necessary for the normal function of living organisms and could be supplied to the organism only from the outside (usually by food) bare the risk of toxicity due to their pro-oxidant capacities. Therefore, nanoparticles can cause the generation of ROS and damage the cells or influence cell signaling.

Nanoparticle-mediated free radical formation can be achieved through several mechanisms. Fenton-type reactions and Haber–Weiss-type reactions are major mechanisms for metal nanoparticle-mediated free radical generation (Figure 16.9). Accordingly, co-administration of hydrogen peroxide (H_2O_2) with metals induced the generation of ROS. However, since hydrogen peroxide is ubiquitous, the administration of pro-oxidant metal particles will always bear a risk of increased ROS production and consequential oxidative stress. Therefore, constructing metal nanoparticles that could be readily used by living cells must be done in a safe way preventing uncontrolled oxidative stress. On the other hand, if such particles could induce desirable oxidative stress, as in the case of the host defense against microbes or cancer, they might be of very high biomedical value.

In Fenton-type reactions, a transition metal ion reacts with H_2O_2 generating an oxidized metal ion and OH$^\bullet$ radical. There are many metals that can be involved in Fenton-type reactions, the most common are iron and copper but the efficiencies at which they produce OH$^\bullet$ radicals significantly varies between them (Valko et al. 2006).

In Haber–Weiss-type reactions, an oxidized metal ion reduced by $O_2^{\bullet-}$ reacts with H_2O_2 to generate OH$^\bullet$ radicals. Activated phagocytes produce large amounts of $O_2^{\bullet-}$ (Bréchard and Tschirhart 2008), but its conversion to H_2O_2 is too slow to

Fenton-type reaction:

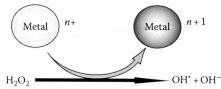

Haber-Weiss-type reactions:

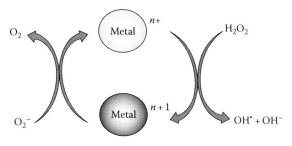

FIGURE 16.9 Fenton-type reaction and Haber-Weiss-type reactions. In the Fenton-type reaction reactive metal particles catalyze decay of hydrogen peroxide into the most reactive hydroxyl radical and hydroxyl ion, while in Haber-Weiss type or reaction superoxide radical is also involved to be transformed into "normal" oxygen molecule.

be physiologically important. When the appropriate metal ion is present, it acts as a catalyst in Haber–Weiss reactions (Aust et al. 1993).

Metal-nanoparticles, such as chromium, cobalt, and vanadium, can be involved both in the Fenton-type reactions and in the Haber–Weiss-type reactions (Valko et al. 2006).

In addition to the above mentioned mechanisms, metal ions can also generate free radicals (Vidrio et al. 2008) or induce cell signaling pathways that react with cellular molecules (Valko et al. 2006). Numerous metal-nanoparticles can be reduced by enzymes (e.g., glutathione reductase) to generate intermediates in favor of further ROS development. Furthermore, some metal-nanoparticles activate intercellular radical-producing systems thus promoting ROS formation (Smith et al. 2001).

16.9 Nanoparticle-Induced Oxidative Stress and Consequential Diseases

High levels of nanoparticles can lead to skin, bladder, liver, lung, and respiratory tract cancers as well as type II diabetes mellitus, cardiovascular diseases, and asthma attacks (Shi et al. 2004). Arsenic (As), beryllium (Be), cadmium (Cd), chromium (Cr), and nickel (Ni) are just some of the known metallic carcinogens for humans (Shi et al. 2004). Furthermore, pro-oxidants such as iron (Fe) and copper (Cu) may catalyze lipid peroxidation thus inducing the chain reaction of ROS generation leading eventually to fat decomposition into reactive aldehydes, as previously mentioned, that could promote the further spread of toxicity

resulting in tissue damage and an altered cellular stress response and growth control.

It must be stated that several studies have shown that various nanoparticles induce ROS production through different mechanisms causing oxidative stress in cells. However, the exact mechanism of ROS formation remains uncertain for the majority of nanoparticles. An example is the case of copper, a metal that is an essential component of several endogenous antioxidant enzymes, in particular cytoplasmatic Cu/Zn dependent superoxide dismutase (known also as SOD1). However, although copper should be considered as an important metal for antioxidant defense, because it acts as a pro-oxidant metal known also as the most potent inducer of lipid peroxidation, copper nanoparticles could also have negative effects for the cells. The depletion of copper has been recently shown to inhibit angiogenesis in a wide variety of cancer cells and xenograft systems (Finney et al. 2009). Therefore, it has yet to be elucidated why high levels of copper often found in the vicinity of cancer cells facilitate their proliferation and tumor progression (Wu et al. 2004, Turski and Thiele 2009).

Altered antioxidant defense levels, altered redox enzyme activity, and macromolecular damage are some of the features of oxidative stress, which may have a crucial role in carcinogenesis (Shi et al. 2004). Thus, the depletion in the tripeptide glutathione (GSH), which acts as the most potent free-radical scavenger, is considered a marker of oxidative stress (Spear and Aust 1995). While GSH is the most important in preserving cellular redox status, free radicals induced by nanoparticles can cause GSH oxidation (Fenoglio et al. 2008) resulting in decreased GSH levels. Furthermore, this may result in attenuated and altered cellular antioxidant defenses creating preconditions for progressive oxidative stress caused by nanoparticles and cell-generated ROS.

Catalase, glutathione peroxidase, and superoxide dismutase are three major enzymes that metabolizes ROS, while nanoparticle-induced cell oxidative stress causes changes in their activity (Shi et al. 2004). Nanoparticles are also capable of affecting other redox-sensitive enzymes, such as thioredoxin reductase (NADPH-dependent flavoenzyme) that plays an important role in the cellular response to oxidative stress (Shi et al. 2004) and its inhibition is associated with the induction of apoptosis (Marzano et al. 2007, Cox et al. 2008).

Nanoparticle-induced cell oxidative stress can lead to various types of DNA damage, such as DNA strand breaks, DNA oxidation, DNA protein crosslinks, and alkali-labile sites (Kawanishi et al. 2002, Shi et al. 2004). One of the major products of ROS-induced DNA damage is 8-Hydroxyl-2′-deoxyguanosine (8-OHdG). 8-OHdG is considered to be a biomarker of oxidative stress to DNA. Exposure to nanoparticles was found to lead to increased levels of 8-OHdG in human cells (Eblin et al. 2006) and in animal models (Inoue et al. 2006) thus proving that some nanoparticles can cause aggressive oxidative stress resulting in mutagenic and carcinogenic DNA damage.

16.10 Cellular Signaling Affected by Metal Nanoparticles

Metal nanoparticles are capable of inducing oxidative stress by disturbing the natural oxidation/reduction balance in cells, as already mentioned. The resulting oxidative stress may also affect levels and functions of numerous redox-sensitive receptors and genes (e.g., growth factor receptors and transcription factors) and/or induce cell death causing apoptosis (programmed cell death). Furthermore, the processing of the signals that cells get from the other cells, even from the distance (e.g., signals carried over hormones or growth factors—cytokines) known under the term "cell signaling pathways" influenced by metal-nanoparticles may mimic extracellular ligands (e.g., insulin) or physical conditions such as hypoxia.

Allen and Tresini (2000) have given a tabular summary of many of the redox effects on gene expression and signaling pathways as well as the differences that may occur between different cell models that are currently known to exist. The most thoroughly examined are MAP kinase and NF-κB signal transduction pathways, and many of the redox effects on the cells are mediated either directly or indirectly through these pathways.

Growth factor receptors are tyrosine kinases that play a key role in the transmission of information from outside the cell into the cytoplasm and the nucleus. This is achieved through mitogen-activated protein kinase (MAPK) signaling. Activated growth factor receptor often results in ROS production. A variety of growth factor receptors (e.g., epidermal growth factor [EGF] and platelet derived growth factor [PDGF]) and the insulin receptor have been reported to be affected by carcinogenic metal nanoparticles (Nowak et al. 2002, Drevs et al. 2003, Zhang et al. 2003). Some metal nanoparticles (Ar, Be, Co, and Ni) are capable of inducing growth factor receptors (e.g., EGF, PDGF, and vascular endothelial growth factor [VEGF]) and their respective downstream genes (e.g., ras, MAPK kinase, ERK kinase). The disregulation of growth factor receptors plays important roles in cell growth (proliferation), cancer spread (metastases development), formation of new blood vessels (angiogenesis), and regulation of the cellular calcium levels important for the activity of numerous enzymes crucial for cellular metabolism.

It was recently reported that iron chelation also alters the pattern of ROS-induced phosphorylation of stress-activated protein kinases SAPK/JNK and p38 MAPK. Thus, the ROS-induced increases in cellular-free iron contribute to signaling events triggered during the oxidative stress response (Deb et al. in press). Cellular defense against the toxicity of ROS is the enhancement of detoxifying enzymes. The regulation of many detoxifying enzymes is mediated by the antioxidant response element (ARE) that is located in the promoted region of related genes. In eukaryotes, there are only a few transcription factors known to be activated by ROS. One of them is Nrf2. Normally, Nrf2 is present in the cytoplasm as an inactive Keap1-Nrf2 complex, but after direct attack by electrophiles or ROS or indirect actions such as phosphorylation by PKC, Nrf2 is liberated from Keap1

repression, translocates into the nucleus, and binds to the ARE sequence to initiate gene expression (Nguyen et al. 2000).

Some other kinases, namely, Src kinases and Janus kinase, which are non-receptor protein kinases are also activated by ROS (Esposito et al. 2003, Marrero et al. 2005). Accordingly, even low concentrations of metal-nanoparticles were reported to induce lymphocyte apoptosis through the ROS-induced activation of the Src family of tyrosine kinases (Balamurugan et al. 2002). Activated Src binds to cell membranes and initiates several signaling pathways (e.g., NF-κB).

The nuclear transcription factors NF-κB and AP-1 are considered to be major stress response transcription factors. NF-κB and AP-1 regulate the expression of many cytotoxic and pro-inflammatory genes (Shi et al. 2004). Many metal nanoparticles (e.g., arsenic, chromium, nickel, lead, and iron) have been reported to activate AP-1. Recently, we have reported that ingestion of iron polymaltose stimulated the activation of p65, p50, and RelB subunits of NF-κB in the peritoneal macrophages. The results showed time-dependent immunomodulatory effects of iron polymaltose indicating that these effects might be achieved via an induction of the intracellular signaling for NF-κB activation in peritoneal phagocytes (macrophages) and p65 subunit in spleen cells (Poljak-Blazi et al. 2009).

While it was already reported 15 years ago that hydroxyl radicals generated by metal particles have an important role in NF-κB activation (Ye et al. 1995), recent findings showing that nanoparticles may enhance the DNA binding activity of the above mentioned transcription factors, while antioxidants may diminish this action (Shi et al. 2004) suggest the high importance that metal nanoparticles could have in the regulation of the redox signaling and biomedical effects of ROS.

In favor of this statement are the findings on the effects of metal nanoparticles in the best known tumor suppressor gene. The tumor-suppressor gene p53 is involved in apoptosis, cell cycle control, and control of DNA repair and plays a guarding role in maintaining genome integrity and the accuracy of chromosome segregation. Hence, p53 is considered to be one of the oxidative stress response factors. Metal nanoparticles increase the expression of the p53 gene (Xu et al. 2008) and p53 downstream proteins (e.g., p21). Furthermore, iron affects p53 transcription and expression, while iron alone induces p53 mutations (Poljak-Blazi et al. 2000).

Metal-nanoparticles, UV radiation, mitogenic stimuli, as well as ROS themselves can also activate another oncogene denoted *ras* (Valko et al. 2006). Ras gene products are membrane-bound G proteins and their main function is to regulate cell growth and oppose apoptotic effects. Thus, the same nanoparticle might consequently affect various, even opposite signaling pathways, in any case influencing the growth of the cells and eventually causing their altered behavior or death.

The best-characterized direct targets of ROS are protein tyrosine phosphatases (PTPs). Cysteine residues of PTPs are very reactive and susceptible to oxidative damage by free radicals, ROS such as hydrogen peroxide, and other oxidants causing reversible inactivation.

Furthermore, this plays an important role in redox control and cell signaling (Valko et al. 2006).

Hypoxia-inducible factors (HIFs) are transcription factors that respond to changes in available oxygen in the cellular environment (e.g., to decreases in oxygen) (Smith et al. 2008). The HIF activity levels in cells influence tumorigenesis and angiogenesis. It was described that metal nanoparticles can mimic hypoxia (Hilliard et al. 2003), thus inducing HIF-1 activity (Gao et al. 2002a) for which PI3K (phosphoinositide 3-kinase)/Akt signaling is required (Gao et al. 2002b).

Finally, it should be stressed again that programmed cell death, apoptosis, is strongly influenced by ROS. Apoptosis is an important part of normal cell development and is controlled by cell signal transduction following the cell cycle arrest (Strasser et al. 2000). Hydrogen peroxide, produced by metal nanoparticles, may induce apoptosis through ROS-mediated reactions (Leonard et al. 2004). Metal nanoparticles, like iron, may also induce apoptosis through p53-dependent or p53-independent (Poljak-Blazi et al. 2000) mechanisms, while major lipid peroxidation product HNE is considered to be a crucial pro-apoptotic factor in various types of cells (Zarkovic 2003).

16.11 Natural Zeolites

In this paragraph, a case of zeolites (micro or nanoparticles) will be presented in particular from the point of view of desirable bioactivities of certain zeolites related to their influence on oxidative stress and the cellular metabolism of ROS.

Natural zeolites are volcanic minerals with unique characteristics. Their chemical structure classifies them as hydrated aluminosilicates comprised of hydrogen, oxygen, aluminum, and silicon arranged in an interconnecting lattice structure. Their unique structure makes zeolites different from other aluminosilicates, due to their ability to absorb gas and water and exchange ions (Mumpton 1999).

Since many biochemical processes are closely related to ion exchange, adsorption, and catalysis, it is believed that natural and synthetic zeolites could make significant contributions to medicine and the pharmaceutical industry. For instance, since the 1990s, zeolites have been used as a contrast medium in diagnostics (Young et al. 1995), as anti-diarrheal drugs (Rodrigues-Fuentes et al. 1997), as antibacterial and antifungal drugs (Maeda and Nose 1999), as supporters or transporters for enzymes and antibodies (Xing et al. 2000.), etc. They are also used to promote the healing of cuts and wounds and are very effective as glucose adsorbents, so diabetes mellitus patients could use them (Concepcion-Rosabal et al. 1997). They reversibly bind small molecules such as oxygen or nitric oxide, possess size and shape selectivity, possibly mimic metallo-enzymes (Bedioui 1995), and have immunomodulatory activity (Ueki et al. 1994, Lim et al. 1997, Pavelić et al. 2002).

Natural clinoptilolite. A fine powder of natural clinoptilolites from Slovakia was obtained by tribomechanical micronization (MZ; micronized zeolite). Particle size distribution curves of the MZ were taken by a Mastersize XLB (Malvern) laser light-

scattering particle size analyzer. A particle size analysis of the clinoptilolite showed that the maximum frequency of particles appeared at $1\,\mu m$. Tribomechanically treated natural clinoptilolite contained approximately 80 wt% clinoptilolite. The remaining 20% consisted of silica, montmorollonite, and mordenite zeolites. The chemical composition of clinoptilolite was SiO_2 70.06%, Al_2O_3 12.32%, Fe_2O_3 1.48%, CaO 3.42% MgO 0.96%, TiO_2 0.71%, P_2O_5 0.05%, MnO 0.02%, Na_2O 0.68%, K_2O 2.38%, SO_3 0.17%, H_2O 7.3%. The humidity at 105°C was max. 6%, pH 6.9–7.1, specific mass $2.39\,g/cm^3$, specific area 360–390 m^2/g, and NH_4^+ substitution capacity 8500 mg NH_4^+/kg (Pavelić et al. 2002).

Mice and dogs suffering from various tumor types who were treated with tribo-mechanically processed zeolite (MZ) had an improvement in their overall health status, prolongation of life span, and a decrease of tumor size in some cases. In addition, toxicology studies on mice and rats demonstrated that the same treatment did not have any negative effects. *In vitro* tissue culture studies showed that zeolite-inhibited protein kinase B (c-Akt), induced the expression of $p21^{WAF1/CIP1}$ and $p27^{KIP1}$ tumor suppressor proteins and blocked cell proliferation in several cancer cell lines (Pavelić et al. 2001, 2002).

After i.p. application of such zeolites, the number of peritoneal macrophages was increased as well as their production of superoxide radical. Opposite to that, nitrite oxide (NO) generation was totally abolished. At the same time, translocation of p65 subunits of NFκB in splenic cells was observed.

In healthy mice fed zeolite for 28 days, the concentration of lipid-bound sialic acid in serum was increased, but lipid peroxidation in the liver was decreased indicating either direct or more likely indirect antioxidant effects of zeolite. In parallel, lymphocytes from the lymph nodes of these mice were activated and therefore expressed a significantly higher allogeneic graft-*versus*-host (GVH) reaction than cells of control mice. The immunomodulating effects of zeolite were further supported by finding the translocation of p65 (NFκB subunit) to the nucleus of spleen cells was observed after i.p. application of zeolite (Pavelić et al. 2002).

An important factor for such bioactivities of zeolite might be its silicate structure, because it is known that the exposure of alveolar macrophages to silicate particles leads to the activation of mitogen-activated protein kinases (MAPK), protein kinase C, and stress-activated protein kinases (SAPK) (Lim et al. 1997).

Furthermore, important transcription factors such as AP-1 and NFκB are also activated by silicates and zeolites as is the enhanced expression of pro-inflammatory cytokines such as IL-1α, IL-6, or TNF-α (Simeonova et al. 1997) that might affect the growth of cancer and the tumor/host relationship.

We summarize an example of the immunomodulating and anticancer effects of MZ *in vivo* (see Figure 16.10). MZ causes local inflammation at the place of application, e.g., peritoneum, resembling other metal particles. Macrophages and neutrophils are attracted and activated, as shown by increased ROS (O_2^-) production. Activated macrophages produce TNF-α that, together with some other stimulants (e.g., other cytokines, ROS, or changed

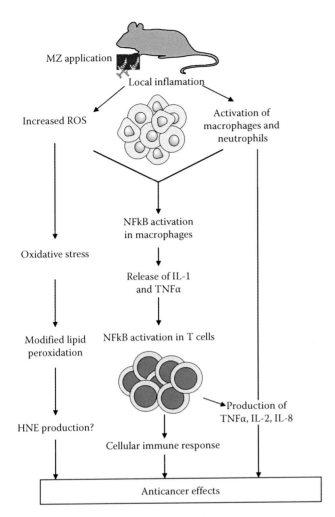

FIGURE 16.10 Possible pathways of tribo-mechanically processed zeolite (MZ) induced stimulation of cellular immune response and anticancer activities.

calcium concentration), stimulate splenic T-cells. Since products of the genes that are regulated by NFκB also cause its activation, this type of positive regulatory loop may amplify and perpetuate the local inflammatory response.

Our hypothesis is that MZ acts the same way after *per os* administration, affecting intestinal mucosa and macrophages. The results of experiments with local allogeneic graft *versus* host reactions, as well as strong reductions of immunogenic melanoma B16 lung metastasis, support this hypothesis. This is in agreement with accumulating evidence that MZ could play an important role in modifying the immune system as well as with the report that silica, silicates, and aluminosilicates act as nonspecific immunomodulators similar to superantigens.

MZ inhibited the proliferation of several cell lines after incubation with (0.5–50 mg/mL) pre-treated medium or after the addition directly to the culture medium. The most significant results were observed in the activity of Akt protein that was strongly inhibited after MZ treatment of the cancer cells, which resulted in growth inhibition and an increase in the apoptosis of cancer cells, but only in the presence of serum.

The absorption of bioactive serum components, at least *in vitro*, could be one of the mechanisms of the MZ action. In favor of this assumption are findings of an increased lipid peroxidation product HNE-protein conjugate formation in the presence of MZ. HNE has a high affinity of binding to proteins, so it can bind major serum protein albumin, which can thus attenuate its toxicity. HNE can also act as a bifunctional growth regulating factor (cytotoxic or growth stimulatory), probably by interfering with the activity of the humoral growth factors (Zarkovic et al. 1993, 1999, Zarkovic 2003).

Cancer and inflammation can be considered to be interfering processes that share two common pathophysiological mechanisms: cytokine network and oxidative stress. Both of these processes involve lipid peroxidation and are linked by tumor stroma. TNF-α and TGF-β are among the cytokines that are likely to be mediators of this complex network system that interfere with lipid peroxidation and consequently with the biological effects of HNE (known as a second messenger of free radicals; Esterbauer et al. 1991, Zarkovic et al. 1999).

Treatment with MZ reduced the lipid peroxidation level in tumor-bearing mice and significantly increased the SOD content in the liver of healthy mice fed with MZ supplementation for 3 weeks. Macrophages of the treated mice were activated since they generated a nine times higher amount of O_2^- than those of control mice. This could explain the toxic (at higher doses) and inflammatory (at lower doses) effects of MZ administered in the peritoneum.

Malignant and normal stromal cells respond differently to the antioxidative effects of MZ (Figure 16.11), which means that MZ could have antioxidative and antitumor effects at the same time (Zarkovic et al. 2003).

Thus, it seems that the MZ particles could be used as nutraceuticals (food supplements) during cytostatic therapy. The potential benefits of such therapies, particularly if novel currently designed nanoparticles of synthetic zeolite (Figure 16.1) are to be used, have yet to be studied.

16.12 Nanotoxicology

Nanoparticles may exert toxic effects (Figure 16.11), therefore, the undesirable effects of nanoparticles are studied in a new research field denoted nanotoxicology (Donaldson et al. 2004, Service 2004, Seaton and Donaldson 2005).

Metal-induced toxicity is very well reported in the literature. One of the major mechanisms behind metal toxicity has been attributed to oxidative stress. A growing amount of data provide evidence that metals are capable of interacting with nuclear proteins and DNA causing the oxidative deterioration of biological macromolecules (Leonard et al. 2004). There is a vast of literature regarding metal nanoparticle-induced cell oxidative stress, e.g., Cd (Liu et al. in press), Pb (Khan et al. 2008), Au (Jia et al. 2009), Al (Tripathi et al. 2009), Ar and Hg (Flora et al. 2008), Ag (Arora et al. 2008), and many others.

Nanoparticles are widely used in consumer products and can reach the body through inhalation, ingestion, skin contact, and

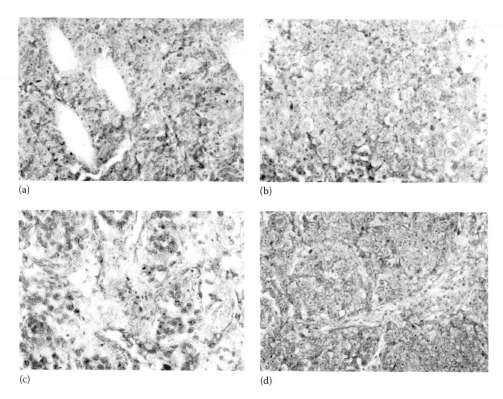

(a)

(b)

(c)

(d)

FIGURE 16.11 Immunohistochemistry for HNE-protein adducts in murine W256 carcinoma tissue. (a) Untreated-tumor composed of tumor cells which do not express spontaneous presence of HNE (major lipid peroxidation product) and are therefore lightly gray stained, while white color represents the traces of the muscles destroyed by invading carcinoma in the hind limb of rat. (b) W256 carcinoma growing in the hind limb of rat treated by per oral application of MZ. Slight immunopositivity (light to dark color) detected by specific monoclonal antibody against HNE-protein adducts can be observed in some cancer cells. (c) W256 carcinoma treated by well known chemotherapeutic drug doxorubicin (Adriamycin), which induces production of HNE (darker cells in the less compact, i.e., destroyed cancer tissue). (d) Tumor of the rat treated both by doxorubicin and MZ, showing immunopositivity for the HNE-protein adducts in malignant cells (clusters of darker gray colored cells), while non-malignant stromal cells were spared from unspecific oxidative stress induced by the drug (parts of tissue presented in central and right part of photo as less dense tissue of light gray color).

injection. When nanoparticles are deposited in organs, they can modify the physiochemical properties of living matter at the nanolevel thus causing adverse biological reactions (Oberdörster et al. 2005a,b). On the other hand, nanoparticles are also used in diagnostics and in biomedicine for the development of potential new drugs. Therefore, the biomedical aspects of nanosciences should be studied together with research on the toxicological and environmental side effects of nanoparticles.

The urban pollution with nanoparticles deriving from combustion sources (e.g., industry, vehicles) contribute to respiratory and cardiovascular morbidity and mortality (Pope 2001, Peters and Pope 2002, Brook 2004). The toxicity of nanoparticles will also depend on their metabolic turnover, i.e., whether they are persistent or cleared from the different organs and whether the host can raise an effective response to sequester or dispose of the particles. Therefore, there is a need for the careful consideration of the benefits and side effects of the use of nanoparticles in medicine (pharmacology and therapeutics) and the potential toxicity of the nanoparticles.

An example of the advantage of using nanoparticles *versus* microparticles is given in Figure 16.12. Iron acting as a pro-oxidant achieves pro-apoptotic effects in particular if used in

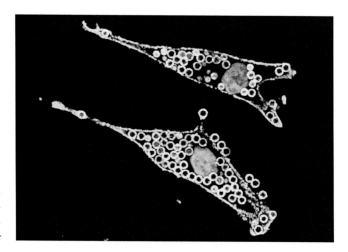

FIGURE 16.12 (See color insert following page V-2.) Metal nanoparticle toxicity: Bone marrow-derived stem cells labeled with large iron oxide particles (2.5 μm) undergo apoptosis (Stained for Annexin V—apoptotic marker). (Courtesy of Dr. Kishore Bhakoo Stem Cell Imaging Group, MRC Clinical Sciences Centre, Faculty of Medicine, Imperial College London, London, U.K., http://sci.csc. mrc.ac.uk.)

the form of bigger particles (2.5 μm), while if the same metal substance is applied in smaller form as nanoparticles it does not induce apoptosis (cell death).

On the other hand, as mentioned in the case of zeolites, even the particles that are not absorbed could gain biomedical effects, particularly if they influence oxidative homeostasis and/or cellular metabolic or signaling pathways. Therefore, the toxicity of metal nanoparticles will depend on their bioavailability, metabolic turn over, and their biological activities.

16.13 Impact of Nanoparticles on Vital Systems

The most common way of nanoparticles entering into the body is through the respiratory system. High amounts of nanoparticles are released in the environment since nanotechnology became available for different industries (Lam et al. 2004). Atmospheric pollutants are rich in nanoparticles (approximately 10^7 particles/cm³ of air that are less than 300 nm in diameter) that may cause lung injury via oxidative stress leading to the activation of different transcription factors with upregulation of proinflammatory protein synthesis (Schins et al. 2000, Medina et al. 2007).

These undesirable effects of the pollutant nanoparticles encouraged a novel therapeutic approach for lung cancer that consists of aerosol therapy using nanoparticles as a drug carrier system (Kim et al. 2004, Tseng et al. 2008). Thus, the drug is efficiently and noninvasively delivered to the lung where it can directly exert its effects before it is degraded or metabolized.

However, these challenging biomedical approaches still need to be intensively developed (Medina et al. 2007). Namely, there are data suggesting that the engineered nanoparticles may contribute to pulmonary morbidity by eliciting proinflammatory effects in the lung and/or by acting as an adjuvant for allergic inflammation. It is possible, therefore, that through the elicitation of an oxidative stress mechanism, engineered nanoparticles may contribute to proinflammatory disease processes in the lung. There is no evidence at this stage, however, that engineered nanoparticles may contribute to any known human pulmonary disease. It should be stressed that asbestos has a micro- and not a nanofiber form, so the well-known toxic effects of these microfibers should not be used as an example of nanoparticle toxicity (Brayner 2008). However, asbestos fibers induce an inflammatory response based on persistent oxidative stress, and from this point of view any particles able to induce such a stress response acting as an irritant for the tissue carries a risk of various pathologies based on oxidative stress and chronic inflammation. Therefore, understanding the link between particle-induced oxidative stress and inflammation provides us with a toxicological paradigm necessary to base the toxicity screening of engineered nanoparticles (Li et al. 2008).

Since nanoparticles could escape normal phagocytic defenses and gain access to the systemic blood stream, they could afterwards be redistributed on the level of the entire organism

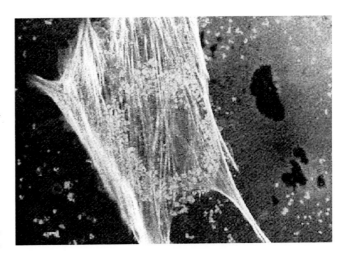

FIGURE 16.13 (See color insert following page V-2.) Bone marrow-derived stem cells labeled with dextran-coated iron oxide nanoparticles conjugated with Tat-FITC. Tat is used as a transfection agent and FITC enables histological verification. (Courtesy of Dr. Kishore Bhakoo Stem Cell Imaging Group, MRC Clinical Sciences Centre, Faculty of Medicine, Imperial College London, London, U.K., http://sci.csc.mrc.ac.uk.)

(Figure 16.13; Medina et al. 2007). Furthermore, nanoparticles can be taken up by the nerve endings of the olfactory bulb and translocated to the central nervous system (CNS). It should be mentioned that the lipid peroxidation caused by excessive oxygen-free radicals and related ROS may lead to an altered function of the blood–brain barrier caused by HNE binding to the cellular proteins in the endothelial and astroglial cells (Schlag et al. 1997, Mertsch et al. 2001). The increased permeability of the blood–brain barrier may further enhance the possibility of nanoparticles to penetrate CNS expressing chronic and acute neurotoxicity for some metal nanoparticles.

Nanoparticles can also be absorbed directly from food, water, drugs, and cosmetics. When nanoparticles enter the skin and gastrointestinal tract, they can induce oxidative stress and pro-inflammatory responses (Medina et al. 2007). Nanoparticles can directly cause the induction of proinflammatory responses, inhibition of cell growth, and reduction of endothelial nitric oxide synthase (Yamawaki and Iwai 2006) as well as the inhibition of cell function and the induction of apoptosis (Cui et al. 2005).

Therefore, it should be said that not only metal nanoparticles but the other nanoparticles as well could have multiple effects on a living system depending on their structural/functional relationship and cellular oxidative stress response. A very good example of this is the finding of lipid peroxidation as a major toxicity principle for the cytotoxicity of C60 nanoparticles (Sayesa et al. 2005), while the major lipid peroxidation product HNE could be used for the functionalization of carbon nanotubes to promote neuronal cell growth (Mattson et al. 2000). Eventually, the use of nanoparticles will lead to the biomedical use of antioxidant substances such as melatonin (Schaffazick et al. 2005), allowing targeted antioxidant effects to protect cells from the toxicity of oxidative stress.

16.14 Conclusion

While nanotechnologies could be considered to be the future of the modern medical industry, the health risks of environmental nanoparticles should be minimized by the optimization of their structure/function relationship that will allow a minimal risk of undesirable oxidative stress. Complementary to that, nanoparticles, in particular metal nanoparticles, are expected to be used in the future as efficient pro-oxidant therapies against specific target cells such as cancer or aggressive microbes and parasites.

References

Allen, R.G. and Tresini, M. 2000. Oxidative stress and gene regulation. *Free Radic Biol Med* 28(3): 463–499.

Arora, S., Jain, J., Rajwade, J.M., and Paknikar, K.M. 2008. Cellular responses induced by silver nanoparticles: In vitro studies. *Toxicol Lett* 179(2): 93–100.

Aust, S.D., Chignell, C.F., Bray, T.M., Kalyanaraman, B., and Mason, R.P. 1993. Free radicals in toxicology. *Toxicol Appl Pharmacol* 120: 168–178.

Balamurugan, K., Rajaram, R., Ramasami, T., and Narayanan, S. 2002. Chromium(III)-induced apoptosis of lymphocytes: Death decision by ROS and Src-family tyrosine kinases. *Free Radic Biol Med* 33: 1622–1640.

Bedioui, F. 1995. Zeolite-encapsulated and clay-intercalated metal porphyrin phthalocyanine and schiff-base complexes as models for biomimetic oxidation catalysts: An overview. *Coord Chem Rev* 144: 39–68.

Braughler, J.M. and Hall, E.D. 1989. Central nervous system trauma and stroke. I. Biochemical considerations for oxygen radical formation and lipid peroxidation. *Free Radic Biol Med* 6: 289–301.

Brayner, R. 2008. The toxicological impact of nanoparticles. *Nano Today* 3: 48–55.

Bréchard, S. and Tschirhart, E.J. 2008. Regulation of superoxide production in neutrophils: Role of calcium influx. *J Leukoc Biol* 84: 1223–1237.

Brook, R.D., Franklin, B., Cascio, W., Hong, Y.L., Howard, G., Lipsett, M., Luepker, R., Mittleman, M., Samet, J., Smith S.C., and Tager, I. 2004. Air pollution and cardiovascular disease—A statement for healthcare professionals from the expert panel on population and prevention science of the American Heart Association. *Circulation*. 109(21) 2655–2671.

Concepcion-Rosabal, B., Rodrigues-Fuentes, G., and Simon-Carballo, R. 1997. Development and featuring of the zeolitic active principle FZ: A glucose adsorbent. *Zeolites* 19: 47–50.

Cox, A.G., Brown, K.K., Arner, E.S., and Hampton, M.B. 2008. The thioredoxin reductase inhibitor auranofin triggers apoptosis through a Bax/Bak-dependent process that involves peroxiredoxin 3 oxidation. *Biochem Pharmacol* 76: 1097–1109.

Cui, D., Tian, F., Ozkan, C.S., Wang, M., and Gao, H. 2005. Effect of single wall carbon nanotubes on human HEK293 cells. *Toxicol Lett* 155: 73–85.

Darley-Usmar, V.M., Mason, R.P., Chamulitrat, W., Hogg, N., and Kalyanarman, B., Lipid peroxidation and cardiovascular disease. In *Immunopharmacology of Free Radical Species*, D. Blake and P.G. Winyard (eds.), Academic Press, London, U.K., 1995, pp. 23–37.

Deb, S., Johnson, E.E., Robalinho-Teixeira, R.L., and Wessling-Resnick, M. 2009. Modulation of intracellular iron levels by oxidative stress implicates a novel role for iron in signal transduction. *Biometals* 22(5): 855–862.

Donaldson, K., Donaldson, K., Stone, V., Tran, C.L., Kreyling, W., and Borm, P.J.A. 2004. Nanotoxicology. *Occup Environ Med* 61: 727–728.

Drevs, J., Medinger M., Schmidt-Gersbach, C., Weber, R., and Unge, C. 2003. Receptor tyrosine kinases: The main targets for new anticancer therapy. *Curr Drug Targets* 4: 113–121.

Eblin, K.E., Bowen, M.E., Cromey, D.W., et al. 2006. Arsenite and monomethylarsonous acid generate oxidative stress response in human bladder cell culture. *Toxicol Appl Pharmacol* 217: 7–14.

Esposito, F., Chirico, G., Montesano Gesualdi, N., et al. 2003. Protein kinase B activation by reactive oxygen species is independent of tyrosine kinase receptor phosphorylation and requires SRC activity. *J Biol Chem* 278: 20828–20834.

Esterbauer, H., Schaur, R.J., and Zollner, H. 1991. Chemistry and biochemistry of 4-hydroxynonenal, malonaldehyde and related aldehydes. *Free Radic Biol Med* 11: 81–128.

Feher, J., Csomos, G., and Vereckei, A. 1987. Physiological free radical reactions. In *Free Radical Reactions in Medicine*, G. Schlag and H. Redl (eds.), Springer-Verlag, Berlin, Germany, pp. 18–39.

Fenoglio, I., Corazzari, I., Francia, C., Bodoardo, S., and Fubini, B. 2008. The oxidation of glutathione by cobalt/tungsten carbide contributes to hard metal-induced oxidative stress. *Free Radic Res* 42: 437–745.

Finney, L., Vogt, S., Fukai, T., and Glesne, D. 2009. Copper and angiogenesis: Unravelling a relationship key to cancer progression. *Clin Exp Pharmacol Physiol* 36(1): 88–94.

Flora, S.J., Mittal, M., and Mehta, A. 2008. Heavy metal induced oxidative stress and its possible reversal by chelation therapy. *Indian J Med Res* 128(4): 501–523.

Gao, N., Ding, M., Zheng, J.Z., et al. 2002b. Vanadate induced expression of hypoxiainducible factor 1 and vascular endothelial growth factor through phosphatidylinositol 3-kinase/Akt pathway and reactive oxygen species. *J Biol Chem* 277: 31963–31971.

Gao, N., Jiang, B.H., Leonard, S.S., et al. 2002a. p38 signalling-mediated hypoxiainducible factor 1alpha and vascular endothelial growth factor induction by Cr(VI) in DU145 human prostate carcinoma cells. *J Biol Chem* 277: 45041–45048.

Geho, D.H., Lahar, N., Ferrari, M., Petricoin, E.F., and Liotta, L.A. 2004. Opportunities for nanotechnology-based innovation in tissue proteomics. *Biomed Microdevices* 6: 231–239.

Gupta, A.K. and Gupta, M. 2005. Synthesis and surface engineering of iron oxide nanoparticles for biomedical applications. *Biomaterials* 26: 3995–4021.

Hall, E.D. and Braughler, J.M. 1989. Central nervous system trauma and stroke. II. Physiological and pharmacological evidence for involvement of oxygen radicals and lipid peroxidation. *Free Radic Biol Med* 6: 303–313.

Halliwell, B. 1995. The biological significance of oxygen-derived species. In *Active Oxygen in Biochemistry*, J. S. Valentine, C. S. Foote, A. Greenberg, and J. F. Liebman (eds.), Blackie Academic Professional, London, U.K., pp. 313–335.

Halliwell, B. 1996. Antioxidants in human health and disease. *Annu Rev Nutr* 16: 33–50.

Halliwell, B. and Gutteridge, J.M.-C. *Free Radicals in Biology and Medicine*, Clarendon Press, Oxford, U.K., 1989.

Hellstrand, K., Asea, A., Dahlgren, C., and Hermodsson, S. 1994. Histaminergic regulation of NK cells. Role of monocyte-derived reactive oxygen metabolites. *J Immunol* 153(11): 4940–4947.

Hilliard, G., Ferguson, T., and Millhorn, D.E. 2003. Cobalt inhibits the interaction between hypoxia-inducible factor-alpha and von Hippel–Lindau protein by direct binding to hypoxiainducible factor-alpha. *J Biol Chem* 278: 15911–15916.

Hirsch, L.R., Gobin, A.M., and Lowery, A.R. 2006. Metal nanoshells. *Ann Biomed Eng* 34: 15–22.

Inoue, K., Takano, H., Yanagisawa, R., et al. 2006. Effects of airway exposure to nanoparticles on lung inflammation induced by bacterial endotoxin in mice. *Environ Health Perspect* 114: 1325–1330.

Jaganjac, M., Poljak-Blazi, M., Zarkovic, K., Schaur, R.J., and Zarkovic, N. 2008. The involvement of granulocytes in spontaneous regression of Walker 256 carcinoma. *Cancer Lett* 260: 180–186.

Jia, H.Y., Liu, Y., Zhang, X.J., Han, L., Du, L.B., Tian, Q., and Xu, Y.C. 2009. Potential oxidative stress of gold nanoparticles by induced-NO releasing in serum. *J Am Chem Soc* 131(1): 40–41.

Kawanishi, S., Hiraku, Y., Murata, M., and Oikawa, S. 2002. The role of metals in site-specific DNA damage with reference to carcinogenesis. *Free Radic Biol Med* 32: 822–832.

Khan, D.A., Qayyum, S., Saleem, S., and Khan, F.A. 2008. Lead-induced oxidative stress adversely affects health of the occupational workers. *Toxicol Ind Health* 24(9): 611–618.

Kim, H.W., Park, I.K., Cho, C.S., et al. 2004. Aerosol delivery of glucosylated polyethylenimine/phosphatase and tensin homologue deleted on chromosome 10 complex suppresses Akt downstream pathways in the lung of K-ras null mice. *Cancer Res* 64: 7971–7976.

Kumagai, T., Kawamoto, Y., Nakamura, Y., et al. 2000. 4-Hydroxy-2-nonenal, the end product of lipid peroxidation, is a specific inducer of cyclooxygenase-2 gene expression. *Biochem Biophys Res Commun* 273: 437–441.

Lam, C.W., James, J.T., McCluskey, R., and Hunter, R.L. 2004. Pulmonary toxicity of single-wall carbon nanotubes in mice 7 and 90 days after intratracheal instillation. *Toxicol Sci* 77: 126–134.

Leonard, S.S., Harris, G.K., and Shi, X. 2004. Metal-induced oxidative stress and signal transduction. *Free Radic Biol Med* 37: 1921–1942.

Li, N., Hao, M., Phalen, R.F., Hinds, W.C., and Nel, A.E. 2003. Particulate air pollutants and asthma. A paradigm for the role of oxidative stress in PM-induced adverse health effects. *Clin Immunol* 109: 250–265.

Li, N., Xia, T., and Nel, A.E. 2008. The role of oxidative stress in ambient particulate matter-induced lung diseases and its implications in the toxicity of engineered nanoparticles. *Free Radic Biol Med* 44: 1689–1699.

Lim, Y., Kim, S.H., Kim, K.A., Oh, M.W., and Lee, K.H. 1997. Involvement of protein kinase C, phospholipase C, and protein tyrosine kinase pathways in oxygen radical generation by asbestos-stimulated alveolar macrophages. *Environ Health Perspect* 105(Suppl. 5): 1325–1327.

Liu, W., Akhand, A.A., Kato, M., et al. 1999. 4-Hydroxynonenal triggers an epidermal growth factor receptor-linked signal pathway for growth inhibition. *J Cell Sci* 112: 2409–2417.

Liu, W., Kato, M., Akhand, A.A., et al. 2000. 4-Hydroxynonenal induces a cellular redox status-related activation of the caspase cascade for apoptotic cell death. *J Cell Sci* 113: 635–641.

Liu, J., Qu, W., and Kadiiska, M.B. 2009. Role of oxidative stress in cadmium toxicity and carcinogenesis. *Toxicol Appl Pharm* 238(3 Special Issue SI): 209–214.

Lundqvist-Gustafsson, H. and Bengtsson, T. 1999. Activation of the granule pool of the NADPH oxidase accelerates apoptosis in human neutrophils. *J Leukoc Biol* 65(2): 196–204.

Maeda, T. and Nose, Y. 1999. A new antibacterial agent: Antibacterial zeolite. *Artif Organs* 23: 129–130.

Marrero, M.B., Fulton, D., Stepp, D., and Stern, D.M. 2005. Angiotensin II-induced signaling pathways in diabetes. *Curr Diabetes Rev* 1: 197–202.

Marzano, C., Gandin, V., Folda, A., Scutari, G., Bindoli, A., and Rigobello, M.P. 2007. Inhibition of thioredoxin reductase by auranofin induces apoptosis in cisplatin-resistant human ovarian cancer cells. *Free Radic Biol Med* 42: 872–881.

Mattson, M.P., Haddon, R.C., and Rao, A.M. 2000 Molecular functionalization of carbon nanotubes and use as substrates for neuronal growth. *J Mol Neurosci* 14: 175–182.

Medina, C., Santos-Martinez, M.J., Radomski, A., Corrigan, O.I., and Radomski, M.W. 2007. Nanoparticles: Pharmacological and toxicological significance. *Br J Pharmacol* 150: 552–558.

Mertsch, K., Blasig, I., and Grune, T. 2001. 4-Hydroxynonenal impairs the permeability of an in vitro rat blood–brain barrier. *Neurosci Lett* 314: 135–138.

Mumpton, F.A. 1999. La roca magica: Uses of natural zeolites in agriculture and industry. *Proc Natl Acad Sci USA* 96: 3463–3470.

Nowak, A.K., Lake, R.A., Kindler, H.L., and Robinson, B.W. 2002. New approaches for mesothelioma: Biologics, vaccines, gene therapy, and other novel agents. *Semin Oncol* 29: 82–96.

Nguyen, T., Huang, H.C., and Pickett, C.B. 2000. Transcriptional regulation of the antioxidant response element. Activation by Nrf2 and repression by MafK. *J Biol Chem* 275(20): 15466–15473.

Oberdörster, G., Maynard, A., Donaldson, K., et al. 2005a. Principles for characterizing the potential human health effects from exposure to nanomaterials: Elements of a screening strategy. *Part Fibre Toxicol* 2: 8.

Oberdörster, G., Oberdörster, E., and Oberdörster, J. 2005b. Nanotoxicology: An emerging discipline evolving from studies of ultrafine particles. *Environ Health Perspect* 113: 823–839.

Pavelić, K., Hadžija, M., Bedrica, L.J., et al. 2001. Mechanically treated natural clinoptilolite zeolite—New adjuvant agent in anticancer therapy. *J Mol Med* 78: 708–720.

Pavelić, K., Katić, M., Šverko, V., et al. 2002. Immunostimulatory effect of natural clinoptilolite as a possible mechanism of its antimetastatic ability. *J Cancer Res Clin Oncol* 128: 37–44.

Peters, A. and Pope, C.A. III (2002). Cardiopulmonary mortality and air pollution. *Lancet* 360: 1184–1185.

Pincemail, J. 1995. Free radicals and antioxidants in human disease. In *Analysis of Free Radicals in Biological Systems*, A. E. Favier, J. Cadet, B. Kalyanaraman, M. Fontecave, and J.-L. Pierre (eds.), Birkhäuser Verlag, Basel, Switzerland, pp. 83–89.

Poli, G. and Schaur, R.J. 2000. 4-Hydroxynonenal in the pathomechanisms of oxidative stress. *IUBMB Life* 50: 315–321.

Poljak-Blazi, M., Jaganjac, M., Mustapic, M., Pivac, N., and Muck-Seler, D. 2009. Acute immunomodulatory effects of iron polyisomaltosate in rats. *Immunobiology* 214: 121–128.

Poljak-Blazi, M., Kralj, M., Hadzija, M.P., Zarkovic, N., Zarkovic, K., and Waeg, G. 2000. Involvement of lipid peroxidation, oncogene expression and induction of apoptosis in the antitumorous activity of ferric-sorbitol-citrate. *Cancer Biother Radiopharm* 15: 285–293.

Pope, C.A. 2001. Particulate air pollution, C-reactive protein, and cardiac risk. *Eur Heart J* 22: 1149–1150.

Porter, N.A. 1984. Chemistry of lipid peroxidation. *Methods Enzymol* 105: 273–282.

Rodrigues-Fuentes, G., Barrios, M.A., Iraizoz, A., Perdomo, I., and Cedre, B. 1997. Enterex-anti-diarrheic drug based on purified natural clinoptilolite. *Zeolite* 19: 441–448.

Sayesa, C., Gobinb, A., Ausmanc, K., et al. 2005 Nano-C60 cytotoxicity is due to lipid peroxidation. *Biomaterials* 26: 7587–7595.

Schaffazick, S.R., Pohlmann, A.R., de Cordova, C.A., Creczynski-Pasac, T.B., and Guterres, S.S. 2005. Protective properties of melatonin-loaded nanoparticles against lipid peroxidation. *Int J Pharm* 289: 209–213.

Schins, R.P.F., McAlinden, A., MacNee, W., et al. 2000. Persistent depletion of I kappa B alpha and interleukin-8 expression in human pulmonary epithelial cells exposed to quartz particles. *Toxicol Appl Pharmacol* 167: 107–117.

Schlag, G., Zarkovic, K., Redl, H., Zarkovic, N., and Waeg, G. 1997. Brain damage secondary to hemorrhagic shock in baboons. In *Shock, SepSis and Organ Failure, 5th Wiggers Bernard Conference*, G. Schlag, H. Redl, D.L. Traber (eds.), Springer-Verlag, Heidelberg, Germany, pp. 3–17.

Schreck, R. and Baeuerle, P.A. 1991. A role for oxygen radicals as second messengers. *Trends Cell Biol* 1(2–3): 39–42.

Seaton, A. and Donaldson, K. 2005. Nanoscience, nanotoxicology, and the need to think small. *Lancet* 365: 923–924.

Service, R.F. 2004. Nanotoxicology: Nanotechnology grows up. *Science* 304: 1732–1734.

Shi, H., Hudson, L.G., and Liu, K.J. 2004. Oxidative stress and apoptosis in metal ion-induced carcinogenesis. *Free Radic Biol Med* 37: 582–593.

Sies, H. 1991. *Oxidative Stress Oxidants and Antioxidants*, H. Sies (ed.), Academic Press, London, U.K., pp. 15–22.

Simeonova, P., Torium, W., Kommineni, C., et al. 1997. Molecular regulation of IL-6 activation by asbestos in lung epithelial cell-role of reactive oxygen species. *J Immunol* 159: 3921–3928.

Smith, W.L. and Marnett, L.J. 1991. Prostaglandin endoperoxide synthase: Structure and catalysis. *Biochim Biophys Acta* 1083: 1–17.

Smith, K.R., Klei, L.R., and Barchowsky, A. 2001. Arsenite stimulates plasma membrane NADPH oxidase in vascular endothelial cells. *Am J Physiol Lung Cell Mol Physiol* 280: L442–L449.

Smith, T.G., Robbins, P.A., and Ratcliffe, P.J. 2008. The human side of hypoxia-inducible factor. *Br J Haematol* 141: 325–334.

Spear, N. and Aust, S.D. 1995. Effects of glutathione on Fenton reagent-dependent radical production and DNA oxidation. *Arch Biochem Biophys* 324: 111–116.

Strasser, A., O'Connor, L., and Dixit, V.M. 2000. Apoptosis signaling. *Annu Rev Biochem* 69: 217–245.

Toyokuni, S., Okamoto, K., Yodoi, J., and Hiai, H. 1995. Persistent oxidative stress in cancer. *FEBS Lett* 358: 1–3.

Tripathi, S., Mahdi, A.A., Nawab, A., Chander, R., Hasan, M., Siddiqui, M.S., Mahdi, F., Mitra, K., and Bajpai, V.K. 2009. Influence of age on aluminum induced lipid peroxidation and neurolipofuscin in frontal cortex of rat brain: A behavioral, biochemical and ultrastructural study. *Brain Res* 1253:107–116.

Tseng, C.L., Wu, S.Y., Wang, W.H., et al. 2008. Targeting efficiency and biodistribution of biotinylated-EGF-conjugated gelatin nanoparticles administered via aerosol delivery in nude mice with lung cancer. *Biomaterials* 29: 3014–3022.

Turski, M.L. and Thiele, D.J. 2009. New roles for copper metabolism in cell proliferation, signaling, and disease. *J Biol Chem* 284: 717–721.

Uchida, K. 2003. 4-Hydroxy-2-nonenal: A product and mediator of oxidative stress. *Prog Lipid Res* 42: 318–343.

Uchida, K., Shiraishi, M., Naito, Y., Torii, Y., Nakamura, Y., and Osawa, T. 1999. Activation of stress signaling pathways by the end product of lipid peroxidation. 4-Hydroxy-2-nonenal is a potential inducer of intracellular peroxide production. *J Biol Chem* 274: 2234–2242.

Ueki, A., Yamguchi, M., Ueki, H., et al. 1994. Polyclonal human T cell activation by silicate in vitro. *Immunology* 82: 332–335.

Valko, M., Rhodes. C.J., Moncol, J., Izakovic, M., and Mazur, M. 2006. Free radicals, metals and antioxidants in oxidative stress-induced cancer. *Chem Biol Interact* 160: 1–40.

Vidrio, E., Jung, H., and Anastasio, C. 2008. Generation of hydroxyl radicals from dissolved transition metals in surrogate lung fluid solutions. *Atmos Environ* 42: 4369–4379.

Wildburger, R., Mrakovcic, L., Stroser, M., Andrisic, L., Borovic Sunjic, S., Zarkovic, K., and Zarkovic, M. 2009. Lipid peroxidation and age-associated diseases—Cause or consequence? *Turkiye Klinikleri* 29: 7–11.

Wu, T., Sempos, C.T., Freudenheim, J.L., Muti, P., and Smit, E. 2004. Serum iron, copper and zinc concentrations and risk of cancer mortality in U.S. adults. *Ann Epidemiol* 14: 195–201.

Xing, G.W., Li, X.W., Tian, G.L., and Ye, Y.H. 2000. Enzymatic peptide synthesis in organic solvent with different zeolites as immobilization matrixes. *Tetrahedron* 56: 3517–3522.

Xu, J., Lian, L.J., Wu, C., Wang, X.F., Fu, W.Y., and Xu, L.H. 2008. Lead induces oxidative stress, DNA damage and alteration of p53, Bax and Bcl-2 expressions in mice. *Food Chem Toxicol* 46: 1488–1494.

Yamamoto, S. 1992. Mammalian lipoxygenases: Molecular structures and functions. *Biochim Biophys Acta* 1128: 117–131.

Yamawaki, H. and Iwai, N. 2006. Mechanisms underlying nanosized airpollution-mediated progression of atherosclerosis: Carbon black causes cytotoxic injury/inflammation and inhibits cell growth in vascular endothelial cells. *Circ J* 70: 129–140.

Ye, J., Zhang, X., Young, H.A., Mao, Y., and Shi, X. 1995. Chromium(VI)-induced nuclear factor-kappa B activation in intact cells via free radical reactions. *Carcinogenesis* 16: 2401–2405.

Young, S.W., Qing, F., and Rubin, D., et al. 1995. Gadolinum zeolite as an oral contrast agent for magnetic resonance imaging. *J Magn Reson Imaging* 5: 499–508.

Zarkovic, N. 2003. 4-Hydroxynonenal as a bioactive marker of pathophysiological processes. *Mol Asp Med* 24: 281–291.

Zarkovic, N., Ilic, Z., Jurin, M., Schaur, R.J., Puhl, H., and Esterbauer, H. 1993. Stimulation of HeLa cell growth by physiological concentrations of 4-hydroxynonenal. *Cell Biochem Funct* 11: 279–286.

Zarkovic, K., Uchida, K., Kolenc, D., Hlupic, L., and Zarkovic, N. 2006. Tissue distribution of lipid peroxidation product acrolein in human colon carcinogenesis. *Free Radic Res* 40: 543–552.

Zarkovic, N., Zarkovic, K., Schaur, R.J., et al. 1999. 4-Hydroxynonenal as a second messenger of free radicals and growth modifying factor. *Life Sci* 65: 1901–1904.

Zarkovic, N., Zarkovic, K., Kralj, M., et al. 2003. Anticancer and antioxidative effects of micronized zeolite clinoptilolite. *Anticancer Res* 23: 1589–1596.

Zhang, P., Gao, W.Y., Turner, S., and Ducatman, B.S. 2003. Gleevec (STI-571) inhibits lung cancer cell growth (A549) and potentiates the cisplatin effect in vitro. *Mol Cancer* 2: 1.

Zivkovic, M., Poljak-Blazi, M., Egger, G., Borovic-Sunjic, S., Schaur, R.J., and Zarkovic, N. 2005a. Oxidative burst and anticancer activities of rat neutrophils. *Biofactors* 24: 305–312.

Zivkovic, M., Poljak-Blazi, M., Zarkovic, K., Mihaljevic, D., Schaur, R.J., and Zarkovic, N. 2007. Oxidative burst of neutrophils against melanoma B16-F10. *Cancer Lett* 232: 100–108.

Zivkovic, M., Zarkovic, K., Skrinjar, L., et al. 2005b. A new method for detection of HNE-histidine conjugates in rat inflammatory cells. *Croat Chem Acta* 78: 91–98.

Zollner, H., Schaur, R.J., and Esterbauer, H. 1991. Biological activities of 4-hydroxyalkenals. In *Oxidative Stress: Oxidants and Antioxidants*, H. Sies (ed.), Academic Press, London, U.K., pp. 319–336.

Fullerene C$_{60}$ Toxicology

Crystal Y. Usenko
Baylor University

Stacey L. Harper
Oregon State University
Oregon Nanoscience and
Microtechnologies Institute

Michael T. Simonich
Oregon State University
Oregon Nanoscience and
Microtechnologies Institute

Robert L. Tanguay
Oregon State University
Oregon Nanoscience and
Microtechnologies Institute

17.1 Carbon Fullerenes

Fullerenes were discovered in 1985 by Robert Curl, Richard Smalley, and Harold Kroto who later received the Nobel Prize for their discovery [1]. Today, carbon fullerenes are one of the most commonly used and most highly studied nanomaterial classes. Despite a wealth of information on the biological interactions of fullerenes, the principles that govern those interactions are largely unknown. This chapter presents information on the state-of-the-science in fullerene research, presenting new fullerene technologies and the implications of such technologies for living systems. We start with a general introduction to carbon fullerenes and methods for assessing their biological activity and toxic potential, and then discuss what is known about how and why carbon fullerenes interact with biological processes. The chapter concludes with a new paradigm to ensure the safety and benefit of future nanotechnologies.

17.1.1 What Are Carbon Fullerenes?

Fullerene C$_{60}$

Fullerenes comprise a family of molecules made entirely of carbon cages forming hollow spheres, ellipsoids, cylindrical tubes, or planes. The two most referenced carbon fullerenes are C$_{60}$ and carbon nanotubes. C$_{60}$ consists of 60 carbon atoms arranged in a cage-like structure resembling a soccer ball, with 20 hexagons and 12 pentagons [1]. They are often referred to as "buckyballs" or "buckminsterfullerene," named after the architect Richard Buckminster Fuller who designed geodesic domes. Buckyballs are approximately 7 Å in diameter and chemically electronegative in the pristine state. Alterations in the surface chemistry of C$_{60}$ can change the physicochemical properties, imparting tremendous industrial and consumer product value. These same properties potentially affect how C$_{60}$ derivatives interact in biological systems. For example, the simple addition of hydroxyl groups to the C$_{60}$ surface increases its water solubility.

17.1.2 Where Do Fullerenes Come From?

Fullerenes can be produced from the natural and anthropogenic combustion of carbon-based material. C$_{60}$ has been identified in a 10,000-year-old ice core as evidence for its natural formation. Industrially, C$_{60}$ is produced at high temperatures when an arc is formed between two graphite rods in a helium gas chamber and solvent is extracted. This method, the Kratschmer technique, results in ~40% product (C$_{60}$), while the remainder is comprised of other forms of fullerenes and soot [2,3]. Unlike carbon nanotubes and many other nanomaterials, metals are not used during synthesis, and thus, are not considered potential contaminants of C$_{60}$ [4].

17.1.3 What Are the Potential Uses of C_{60}?

Numerous applications have been proposed for C_{60} due to the unusual properties they exhibit. They are thermally stable, requiring temperatures >1000°C to break the primary particles apart. The functional stability of C_{60} provides superior strength and durability to new building and industrial materials. Combining C_{60} with alkali metals such as K, Cs, or Na has yielded new superconducting materials. Applications of C_{60} as lubricant, rocket fuel, drug delivery systems, solar cells, and energy storage have also been explored. Although these applications are not yet in production, applications of C_{60} in cosmetics, sunscreens, and lotions are already in the marketplace and widely touted as powerful antioxidants.

17.1.4 Are There Risks to Human Health from Fullerenes?

$$\boxed{\text{Risk} = \text{Exposure} \times \text{Hazard}}$$

For a nanomaterial to pose a risk to human or environmental health, there must be the potential for exposure, and the nanomaterial must interact with biological systems to produce an adverse biological response. Limited information is currently available on the likelihood of occupational or incidental exposure to fullerenes. However, exposures will obviously increase as more applications of this technology enter the marketplace. According to the material safety data sheets (MSDS) provided with the commercial purchase of C_{60}, it is considered equivalent to carbon black and few precautions are recommended for working with C_{60}. While C_{60} and carbon black are elementally identical, studies suggest an inverse correlation between the size and biological activity of nanomaterials. Smaller nanomaterials have a larger surface-to-mass ratio providing more area for interaction with cells and proteins. Despite this fact, knowledge of adequate safety precautions for fullerene use is not commonly available.

17.2 General Principles of Toxicology

Toxicology is the study of the adverse effects of chemicals on living organisms. Four pharmacokinetic parameters determine the degree of chemical–biological interaction when assessing toxicity: absorption, distribution, metabolism, and excretion. The route of exposure influences the magnitude and rate chemical absorption into the bloodstream and its distribution in the body. The chemical accumulates in the target organ(s) where it is metabolized. Poor up absorption is characterized by rapid excretion.

17.2.1 How Are Biological Effects of Exposure Determined?

Numerous model systems have been used to evaluate biological impacts and they fall into one of two main categories: *in vitro*

(cell culture) or *in vivo* (whole organism). *In vitro* methods are commonly used for their lower cost relative to animal models, and rapid assay times. *In vitro* studies can also be used to elucidate a mechanism of action. By removing all other physiological processes, and the many associated variables, specific receptor-mediated responses may be rapidly determined in a cell monoculture.

In vitro models also carry significant limitations. First, cell-to-cell signaling is disrupted, so processes that would normally respond to an exposure of the organism, i.e., homeostatic feedback mechanisms, are lacking. Specific cell types may have differential sensitivities to chemical exposure. Therefore, no response *in vitro* to a test chemical may simply be a function of the cell type employed. Cell culture studies cannot be used to assess the impacts of chemical exposure on development, which is the life stage of greatest sensitivity to chemical exposure. Finally, cell culture models are routinely exposed via the culture medium, circumventing real-world exposure routes and body distribution. Thus, a negative effect of an exposure on cultured cells may be registered regardless of whether that cell type would be targeted *in vivo*.

In vivo models are essential for assessing multiple routes of exposure. The magnitude and direction of a biological response to an inhalation exposure may differ considerably from the response to ingestion. *In vivo* models are also essential for determining chemical distribution and organ targeting. The use of *in vivo* models provides a far superior indication of the potential effects of a chemical exposure in humans.

There are some limitations to *in vivo* models in addition to the cost and length of studies. All data must still be related to human health. Because model species range from insects to mammals, the relevance of toxicological results to human safety is not directly translatable. Homology across species is essential for predicting the response of one species based on data from another. An unavoidable fact is that *in vivo* testing is often expensive and time consuming, particularly with mammalian models. Striking the right balance between cost, model relevance to humans, and the speed with which studies can be conducted is the most efficient and effective means of toxicological investigation. While the mouse and rat models remain the gold standards for human relevance, the zebrafish strikes perhaps the best cost-to-benefit ratio with the added attractiveness of being readily amenable to high throughput chemical genetics screens.

It is often necessary to employ both *in vitro* and *in vivo* testing to fully understand chemical interactions with a biological system. Testing can be set up using a tiered approach that starts with cell culture evaluations. If the results demonstrate an adverse response, researchers then use *in vivo* studies to further characterize the toxicity. Europe's REACH program utilizes such a tiered testing approach based on the production amount. Under this testing strategy, chemicals/nanomaterials produced in larger volumes will be evaluated using more stringent testing procedures.

17.3 Considerations for Nanotoxicology

Nanotoxicology is an emerging subdiscipline devoted to investigating the biological consequences of nanomaterial exposure. Nanomaterials present several new challenges for assessing toxicity. For instance, toxicological prediction for traditional chemistries is at least partly based on standard properties such as pK_a, solubility, and log P. Log P is a partition coefficient of a chemical between two solvents, such as water and octanol, and thus is a good indicator of solubility in biological fluids. Typically, more lipophilic chemicals will be sequestered in fat stores, while hydrophilic ones will be more readily excreted in the urine. The pK_a often determines the uptake of a chemical in different pH environments. Thus, it is possible to predict some likely chemical behaviors in a biological system based solely on structural similarity. This is not yet the case for nanomaterials. The novelty and diversity of nanomaterials combined with a severely limited dataset of structure–activity relationships mean that science cannot yet predict the likely chemical behaviors based on physicochemical properties such as particle core composition, size, shape, charge, surface chemistry, surface area, agglomeration state, and redox potential. It is instead necessary for chemical engineers, materials scientists, and chemists to work with biologists and toxicologists in researching the biological and environmental impacts of nanomaterials.

17.3.1 Why Is There Disagreement about the Biological Effects of Nanomaterials?

Currently, the primary literature is flooded with contradictory reports about the benefits versus hazards of nanomaterials. This is due in part to a lack of physicochemical data, and to the fact that nanomaterials are chemically altered by their immediate environment. Plasticity over time in the number of ligands or surface groups attached to the nanoparticle surface may result in profound differences in toxic potential. Moreover, variations in nanomaterial synthesis techniques can yield multiple congeners within a preparation. In addition to the differences in the number of surface functional groups, nanomaterials may be present as agglomerates, groups of particles held together by noncovalent interactions. Fullerenes, in particular, form large agglomerations in most bio-compatible solvents. Agglomeration results in a distribution of nanoparticle sizes, the largest of which may precipitate from solution. The size, free energy of formation, and stability of agglomerates will directly affect the bio-availability and toxicity of such nanomaterials.

Solvent choice also complicates nanotoxicology comparisons. In particular, tetrahydrofuran (THF) has been used for more than half of the studies of pristine C$_{60}$. While it is ideally evaporated prior to exposure, trace amounts of the solvent have been linked to the adverse biological responses previously reported [5]. In an attempt to obtain an environmentally relevant exposure solution, C$_{60}$ has been stirred in water in direct sunlight for several months, forming water soluble colloids. This method, however, does not evaluate the alteration to the physicochemical properties of the outer shell of the colloid, which will in turn alter the bioavailability and toxicity of the fullerenes.

Since the potential for exposure is a principal component of any analysis of chemical risk, it is important to determine the primary uses, the routes of exposure, and the potential for biological interaction. For example, since fullerenes are used in cosmetics, it is necessary to evaluate the capacity for dermal uptake and photoexcitability. In the environment, fullerenes could be transported via the atmosphere or through aquatic systems. While pristine C$_{60}$ will quickly precipitate from solution and bind to sediment and dissolved organic matter, becoming essentially inert, functionalized C$_{60}$ remains in the water column where it is more readily transported and more readily enters the food chain [6]. Precise characterization of nanomaterial preparations under study is critical for precise estimations of the respective risk.

17.3.2 How Are Nanomaterial Features Characterized?

Current state-of-the-art techniques for characterizing nanomaterials include transmission electron microscopy (TEM), scanning electron microscopy (SEM), atomic force microscopy (AFM), ultra violet-visible spectral analysis (UV-Vis), and nuclear magnetic resonance spectral analysis (NMR). TEM and SEM are used to directly image nanomaterial particles for shape, size, and distribution in the dry state; however, they do not provide the hydrodynamic radius or interactive space of the materials, nor do they account for surface functionalization. AFM provides the atomic force as a visual representation of the topography of materials in the dry state, but is cost- and labor-intensive. UV-Vis can be used to qualitatively assess nanomaterial size, but is limited to materials that exhibit plasmonic effects that change with size. NMR is often used to assess the presence of free ligands, unwanted byproducts, and excess reactants. It provides a spectral analysis for impurities, but does not directly identify impurities. These instrumental analytical approaches are not sufficient for the analysis of complex nanomaterial mixtures. New analytical techniques for nanomaterials are being developed but lag behind the pace of nanomaterial innovation.

17.4 Uptake and Biodistribution of Fullerenes

The route of exposure affects the ultimate biodistribution and toxicity of a chemical or a nanomaterial. Nanomaterials can enter the body through breathing, skin or eye contact, nasal uptake into the olfactory bulb, ingestion, or injection. Fullerenes have been evaluated for multiple potential routes of exposure.

Routes of exposure

Inhalation/breathing–Nanomaterials may be inhaled into the body through the nose or mouth

Dermal/skin contact–Nanomaterials may be absorbed through the skin into the body

Eye contact–Nanomaterials may be absorbed through the eyes

Neuronal uptake/olfactory bulb–Nanomaterials may be taken up directly into exposed neurons of the olfactory bulb in the nasal cavity

Ingestion/swallowing–Nanomaterials may be ingested through the mouth and taken up into the body from the gastrointestinal tract

Injection–Nanomaterials may be injected into the body

17.4.1 How Do Fullerenes Get Inside the Biological Systems?

Inhalation of fullerenes is possible because they are produced in common combustion processes and transported through the atmosphere. Studies have shown that C_{60} adsorbs onto airborne polycyclic aromatic hydrocarbon (PAH) particulates [7]. Furthermore, fullerenes are currently used in some cosmetics, which increases the potential for dermal exposure and uptake. C_{60} and $C_{60}(OH)_x$ were absorbed through the epidermal layers and demonstrated rapid uptake into the dermal cells [8,9]. Large C_{60} agglomerations have been shown to pass through cellular membranes [8,10]. Together, these examples suggest that, at a minimum, the use of personal protective equipment (lab coat, gloves, chemical hood, and/or particulate respirator) is warranted when handling fullerenes in a concentrated form.

The rate and consequences of fullerene dermal uptake, while likely dependent on the functionalization chemistry, have not been adequately characterized. The phenomenon is important, not only from a safety standpoint, but also as a potential drug delivery technology. Likewise, the inhalation of fullerenes is concomitantly a potential hazard and a novel drug delivery technology. The first study to address the hazard was recently made in the rat model [11]. Unmodified C_{60} nano-(2.22 mg/m³, 55 nm diameter) and micro-(2.35 mg/m³, 0.93 μm diameter) particles were inhaled by rats for 3 h per day for 10 consecutive days after which hematology, serum chemistry, bronchoalveolar lavage fluid parameters, and lung particle burden were analyzed. Only minimal changes relative to control animals were detected in these toxicological endpoints [11]. While encouraging, these results may differ for longer duration inhalation studies and/or

in other models such as mice and hamsters, which each display very different responses to TiO_2 microparticle exposure [12].

17.4.2 Where Do Fullerenes Go Inside the Biological Systems?

Once inside the body, fullerene uptake and metabolism will determine the biological consequences of exposure. Fullerene pharmacokinetics, however, remain a virtual unknown. Several studies have evaluated the biodistribution of functional group-radiolabeled fullerenes. Ingested radiolabeled, water soluble C_{60} was rapidly eliminated via the feces in rats, while intravenous administration led to rapid tissue distribution of C_{60} to the liver and across the blood brain barrier [13]. After several weeks, fullerenes had accumulated in the hair and skeletal muscle, with only 2% eliminated [13], suggesting the possibility of chronic toxicity. C_{60} and hydroxylated C_{60} injected into mice also concentrated in the liver and spleen [14,15]. The cellular uptake of ^{14}C-doped C_{60} was rapidly incorporated into lipid bilayers [16,17]. Labeled compound studies, however, bear the caveat that the label may be cleaved giving false indication of the chemical's fate. Better methods are not yet available for the pharmacokinetic study of highly functionalized C_{60s}.

Other studies have shown that C_{60} is internalized by monocyte macrophages and deposited in secondary lysosomes, the nuclear membrane, and inside the nucleus of the cell [18]. Numerous studies have shown that water soluble, photoexcited fullerenes can oxidatively damage DNA [19–21]. This fullerene property, while of some relevance to human safety, obviously has major implications for advancing photo-chemotherapy.

17.5 Biological Impact of Fullerene Exposure

A nanomaterial is biologically active if it alters normal cellular processes. By determining the mechanisms of fullerene toxicity, it should be possible to design nanomaterials that are biologically benign. Different assays in multiple models indicate oxidative stress as the primary mechanism of fullerene toxicity. Induction of oxidative stress is a common mechanism of toxicity for numerous chemicals. It is caused by the accumulation of reactive oxygen species (ROS) in excess of a cell's ability to detoxify the reactive intermediates, or to readily repair the resulting damage. Biological systems undergo oxidative stress by exposure to an oxidizing agent, by metabolic generation of ROS, or by inhibition of antioxidant pathway(s) [22].

17.5.1 What Are the Biological Consequences of Oxidative Stress?

Oxidative stress can cause enzyme inhibition, protein and/or DNA damage, alterations in gene expression, lipid peroxidation, and cell death. During lipid peroxidation, ROS reacts with lipids to form a variety of peroxide products that increase cellular permeability and have cytotoxic effects at high concentrations. This latter effect can alter cellular proliferation and, at lower concentrations, prevent nuclear factor-κB activation *in vitro*, a potential method of blocking chronic inflammatory disease and cancer proliferation [23,24].

Aldehyde products of lipid peroxidation may elicit neurodegenerative disease and chronic inflammation. Since this endpoint can be detected using *in vitro* methods and colorimetric assays, several C$_{60}$ studies have tested for lipid peroxidation as an indicator of oxidative stress [25–27]. The most noted study was conducted by Oberdorster et al. who detected lipid peroxidation in the brains of largemouth bass exposed to C$_{60}$ [27]. Likewise, these assays were used to demonstrate the ability of modified C$_{60}$ to attenuate lipid peroxidation, as in the case of Chen et al. who found that polyhydroxylated C$_{60}$ decreased lipid peroxidation in lung cells that were exposed to an oxidative agent [28].

Photoexcitation of functionalized fullerenes is known to cleave DNA more efficiently than the photoexcitation of pristine C$_{60}$ [29], and water-solubilized C$_{60}$ is known to damage DNA more effectively than C$_{60}$ solubilized in less polar solvents. It is important to note that no study has demonstrated a C$_{60}$-induced formation of tumors or cancer.

17.5.2 How Could C$_{60}$ Induce Oxidative Stress?

Fullerenes have the dual ability to scavenge free radicals and to generate them [30–33]. This dual function depends on surface functionalization and the environment. In cell culture, pristine C$_{60}$ was synergistic with oxidative stress-inducing agents and elicited toxicity via lipid peroxidation [26,31]. *In vivo*, pristine C$_{60}$ induced oxidative stress and lipid peroxidation in fish [27,34]. In contrast, carboxyl and amino acid functionalizations of C$_{60}$ are less toxic and are powerful antioxidants [35,36].

Due to its unique spherical structure and large size, C$_{60}$ has the ability to accept up to 6 electrons [16]. Upon UV or visible photoexcitation, the resulting singlet C$_{60}$ reacts with O$_2$ to form singlet oxygen (^{1}O$_2$) [37,38]. The amount of radiation necessary to raise the fullerene to the triplet state varies with surface functionalization [39]. In general, more functionalization yields a less oxidative fullerene species [39].

When photoexcited, both C$_{60}$ and hydroxylated C$_{60}$ can induce membrane damage in rat hepatic and tumor microsomes by generation of ROS in a time- and concentration-dependent manner [25,40]. Although this feature of fullerenes would be undesirable for healthy living cells, it is applicable for the treatment of certain tumors [40]. Conversely, surface modifications, such as carboxylation, impart ROS scavenging ability to the fullerene [31,41,42]. In fact, carboxylated C$_{60}$ has been patented as an antioxidant supplement to increase lifespan [43].

17.5.3 What Do Cell Culture Studies Tell Us about Fullerene Exposure?

Over half of fullerene toxicology studies have been conducted *in vitro* and over half of those described cytotoxic effects.

In studies where there was a cytoprotective effect, the fullerene had been functionalized with amino acids or other antioxidants.

In comparing cell culture studies, there is some variation in responses depending on the cell type. In a study by Sayes et al. that compared three different cell types, the concentration at which 50% of the cells died (LC$_{50}$) ranged from 2 to 50 parts per billion (ppb) [26]. The study demonstrated that cell death was due to membrane disruption and that co-exposure with antioxidants inhibited cell death. The proposed mechanism of action was demonstrated at the cellular level at low exposure concentrations. Additionally, cell culture studies have found water-stirred C$_{60}$ to cause DNA damage at concentrations as low as 2.2 ppb [20]. In a study investigating immune-activation by amino acid functionalized C$_{60}$ in human embryonic kidney (HEK) cells, fullerenes were found to significantly increase cytokine IL-8 at a concentration of 40 ppm [44].

While there is uncertainty regarding the cytotoxic effects of C$_{60}$, *in vitro* evidence suggests that functionalized C$_{60}$ can have beneficial properties that warrant further investigation. Since C$_{60}$ is able to pass through the lipid membrane of a cell and interact with the nucleus, fullerenes could serve as efficient drug delivery systems. For example, adding folate to the surface might allow specific targeting to cancer cells that overproduce folate receptors. It is also essential to know the interaction of the functional groups within the cell, and whether or not they are cleaved from the fullerene. Cleavage of the functional group, resulting in pristine C$_{60}$, could have negative impacts on the cell.

17.5.4 What Do Whole Animal Studies Tell Us about Fullerene Exposure?

Few studies have been conducted on the metabolism of fullerenes. One study using functionalized C$_{60}$ found that fullerenes and their derivatives remain unchanged in the liver, which is the main site of metabolism [16]. Hydroxylated fullerenes, in particular, suppressed the typical cytochrome P450 response to xenobiotic exposure [45]. It is unclear, however, if other functionalized fullerenes are metabolically activated. To date, only one study has demonstrated increased cytochrome P450 levels following exposure to C$_{60}$. In this study, CYP2K1-like protein was upregulated following exposure to 0.5 ppm water-stirred C$_{60}$ [46]. Conversely, in a paper published by the same group, CYP expression was not increased, even at C$_{60}$ concentrations greater than 2 ppm [47]. Extensive investigation is still needed for an understanding of fullerene metabolism.

Nearly half of the published fullerene toxicology has been completed *in vivo*, but few have evaluated the histology of exposed organs. One study found increased lipid peroxidation in the brains of fish exposed to C$_{60}$ via water [27], but these effects may have been caused by the solvent THF. A majority of the *in vivo* studies have concentrated on the effects at the organismal level. Primary testing parameters include mortality, growth, physiologic malformations, or behavioral abnormalities. Few studies have found C$_{60}$ to induce mortality, with the exception of embryonic zebrafish, which are highly sensitive to exposure during development. Only a few studies have noted responses such as pain and weight loss [13,48].

All of the *in vivo* studies that have been conducted employed high concentrations of C_{60} (ppb-ppm) to elicit a significant response such as stunted growth or mortality. These high-dose experiments may be complicated by solvent co-toxicity. For example, fullerenes prepared through the THF method resulted in an LC_{50} of 0.8 ppm in fish, whereas the LC_{50} of water-stirred C_{60} was greater than 35 ppm [47]. Experiments testing the acute and chronic exposure to low-dose C_{60} are still lacking.

The strong potential for inhalation exposure to fullerenes during manufacture, consumer product use, and from air pollution necessitates our understanding of the hazard. In a study by Baker et al., C_{60} had a longer half-life in the lung (26 days) than microcarbon [11]. Furthermore, it induced higher levels of protein. Despite the use of pristine C_{60}, adverse effects were not detected in any of the organs assessed. Hydroxylated C_{60} inhalation exposure was quickly taken up in the lung and had relatively slow clearance [49]. The impact on the tissue (whether beneficial or adverse) was not evaluated, but indicates that the material is given sufficient opportunity to interact with the cells. Intratracheal instillation of C_{60} or $C_{60}(OH)_{24}$ induced only transient inflammatory and cell injury effects at 1 day post exposure with no adverse lung tissue effects observed at 3 months postinstillation from exposures to the highest dose of the two types of fullerenes. Determining if there is a health risk of inhalation exposure is particularly important for fullerene factory workers and those researching fullerenes. The unique concerns for other routes of exposure also apply to inhalation exposure. For example, does the size of the agglomeration influence toxicity? Studies to date do not indicate an elevated risk to fullerene inhalation exposure, however, a study by Fujitani et al. does demonstrate that C_{60} of different sizes of agglomerations are present in the air of a fullerene factory [51]. Furthermore, the types of agglomerations present are influenced by activity [51].

17.6 A New Paradigm to Ensure Safety and Benefits of Nanotechnologies

Rapid growth of the nanotechnology industry will be accompanied by the increased exposure of humans and the environment to nanomaterials. Evaluations of the safety of these materials lag behind the rapid marketplace entry of nanomaterial-containing consumer products. Thorough and timely evaluation of nanomaterial safety will ensure lasting public trust in the nanotechnology industry as well as avoid unnecessary health crises and the accompanying litigation damages. Moreover, good toxicology done "up front" will provide the nanotechnology industry with information needed to direct the development of safe nanomaterial products [52–55]. This will only happen if product development concerns and product safety concerns substantially overlap. Nanotechnology is here to stay and virtually the entire periodic table is exploitable for developing nanotechnology. The diversity of nanomaterials makes it impractical to utilize current toxicological paradigms to evaluate every new material. As a case study, fullerenes demonstrate technical challenges in assessing exposure risk from nanomaterials. Science is on the cusp of designer nanomaterials with medicinal and industrial potentials that might never be realized with traditional materials. The full human benefit of these materials will only be realized if safety is integral to their design.

References

1. Kroto, H.W., Heath, J.R., Obrien, S.C., Curl, R.F., and Smalley, R.E. 1985. C-60—Buckminsterfullerene. *Nature*, 318(6042): 162–163.

2. Kratschmer, W., Lamb, L.D., Fostiropoulos, K., and Huffman, D.R. 1990. Solid C-60—A new form of carbon. *Nature*, 347(6291): 354–358.

3. Haufler, R.E., Conceicao, J., Chibante, L.P.F., Chai, Y., Byrne, N.E., Flanagan, S., Haley, M.M. et al. 1990. Efficient production of C60 (Buckminsterfullerene), C60H36, and the solvated buckide ion. *Journal of Physical Chemistry*, 94(24): 8634–8636.

4. Hu, J.T., Odom, T.W., and Lieber, C.M. 1999. Chemistry and physics in one dimension: Synthesis and properties of nanowires and nanotubes. *Accounts of Chemical Research*, 32(5): 435–445.

5. Henry, T.B., Menn, F.M., Fleming, J.T., Wilgus, J., Compton, R.N., and Sayler, G.S. 2007. Attributing effects of aqueous C-60 nano-aggregates to tetrahydrofuran decomposition products in larval zebrafish by assessment of gene expression. *Environmental Health Perspectives*, 115(7): 1059–1065.

6. Lecoanet, H.F., Bottero, J.Y., and Wiesner, M.R. 2004. Laboratory assessment of the mobility of nanomaterials in porous media. *Environmental Science and Technology*, 38(19): 5164–5169.

7. Yang, K., Zhu, L.Z., and Xing, B.S. 2006. Adsorption of polycyclic aromatic hydrocarbons by carbon nanomaterials. *Environmental Science and Technology*, 40(6): 1855–1861.

8. Monteiro-Riviere, N.A. and Inman, A.O. 2006. Challenges for assessing carbon nanomaterial toxicity to the skin. *Carbon*, 44(6): 1070–1078.

9. Xia, X.R., Monteiro-Riviere, N.A., and Riviere, J.E. 2006. Trace analysis of fullerenes in biological samples by simplified liquid-liquid extraction and high-performance liquid chromatography. *Journal of Chromatography A*, 1129(2): 216–222.

10. Levi, N., Hantgan, R.R., Lively, M.O., Carroll, D.L., and Prasad, G.L. 2006. C60-Fullerenes: Detection of intracellular photoluminescence and lack of cytotoxic effects. *Journal of Nanobiotechnology*, 4: 14–25.

11. Baker, G.L., Gupta, A., Clark, M.L., Valenzuela, B.R., Staska, L.M., Harbo, S.J., Pierce, J.T., and Dill, J.A. 2008. Inhalation toxicity and lung toxicokinetics of C-60 fullerene nanoparticles and microparticles. *Toxicological Sciences*, 101(1): 122–131.

12. Bermudez, E., Mangum, J.B., Wong, B.A., Asgharian, B., Hext, P.M., Warheit, D.B., and Everitt, J.I. 2004. Pulmonary responses of mice, rats, and hamsters to subchronic inhalation of ultrafine titanium dioxide particles. *Toxicological Sciences*, 77(2): 347–357.

13. Yamago, S., Tokuyama, H., Nakamura, E., Kikuchi, K., Kananishi, S., Sueki, K., Nakahara, H., Enomoto, S., and Ambe, F. 1995. In-vivo biological behavior of a water-miscible fullerene - C-14 labeling, absorption, distribution, excretion and acute toxicity. *Chemistry and Biology*, 2(6): 385–389.

14. Moussa, F., Pressac, M., Genin, E., Roux, S., Trivin, F., Rassat, A., Ceolin, R., and Szwarc, H. 1997. Quantitative analysis of C-60 fullerene in blood and tissues by high-performance liquid chromatography with photodiode-array and mass spectrometric detection. *Journal of Chromatography B-Analytical Technologies in the Biomedical and Life Sciences*, 696(1): 153–159.

15. Qingnuan, L., yan, X., Xiaodong, Z., Ruili, L., qieqie, D., Xiaoguang, S., Shaoliang, C., and Wenxin, L. 2002. Preparation of (99m)Tc-C(60)(OH)(x) and its biodistribution studies. *Nuclear Medicine and Biology*, 29(6): 707–710.

16. Jensen, A.W., Wilson, S.R., and Schuster, D.I. 1996. Biological applications of fullerenes. *Bioorganic and Medicinal Chemistry*, 4(6): 767–779.

17. Wang, I.C., Tai, L.A., Lee, D.D., Kanakamma, P.P., Shen, C.K.F., Luh, T.Y., Cheng, C.H., and Hwang, K.C. 1999. C-60 and water-soluble fullerene derivatives as antioxidants against radical-initiated lipid peroxidation. *Journal of Medicinal Chemistry*, 42(22): 4614–4620.

18. Porter, A.E., Muller, K., Skepper, J., Midgley, P., and Welland, M. 2006. Uptake of C-60 by human monocyte macrophages, its localization and implications for toxicity: Studied by high resolution electron microscopy and electron tomography. *Acta Biomaterialia*, 2(4): 409–419.

19. Bernstein, R., Prat, F., and Foote, C.S. 1999. On the mechanism of DNA cleavage by fullerenes investigated in model systems: Electron transfer from guanosine and 8-oxoguanosine derivatives to C-60. *Journal of the American Chemical Society*, 121(2): 464–465.

20. Dhawan, A., Taurozzi, J.S., Pandey, A.K., Shan, W.Q., Miller, S.M., Hashsham, S.A., and Tarabara, V.V. 2006. Stable colloidal dispersions of C-60 fullerenes in water: Evidence for genotoxicity. *Environmental Science and Technology*, 40(23): 7394–7401.

21. Zhao, X.C., Striolo, A., and Cummings, P.T. 2005. C-60 binds to and deforms nucleotides. *Biophysical Journal*, 89(6): 3856–3862.

22. Anderson, M.E. 1998. Glutathione: An overview of biosynthesis and modulation. *Chemico-Biological Interactions*, 111–112: 1–14.

23. Garg, A. and Aggarwal, B.B. 2002. Nuclear transcription factor-kappa B as a target for cancer drug development. *Leukemia*, 16(6): 1053–1068.

24. Sethi, G., Sung, B., and Aggarwal, B.B. 2008. Nuclear factor-kB activation: From bench to bedside. *Experimental Biology and Medicine*, 233(1): 21–31.

25. Kamat, J.P., Devasagayam, T.P.A., Priyadarsini, K.I., Mohan, H., and Mittal, J.P. 1998. Oxidative damage induced by the fullerene C-60 on photosensitization in rat liver microsomes. *Chemico-Biological Interactions*, 114(3): 145–159.

26. Sayes, C.M., Gobin, A.M., Ausman, K.D., Mendez, J., West, J.L., and Colvin, V.L. 2005. Nano-C60 cytotoxicity is due to lipid peroxidation. *Biomaterials*, 26(36): 7587–7595.

27. Oberdorster, E. 2004. Manufactured nanomaterials (Fullerenes, C-60) induce oxidative stress in the brain of juvenile largemouth bass. *Environmental Health Perspectives*, 112(10): 1058–1062.

28. Chen, Y.W., Hwang, K.C., Yen, C.C., and Lai, Y.L. 2004. Fullerene derivatives protect against oxidative stress in RAW 264.7 cells and ischemia-reperfused lungs. *American Journal of Physiology-Regulatory Integrative and Comparative Physiology*, 287(1): R21–R26.

29. Nakamura, E. and Isobe, H. 2003. Functionalized fullerenes in water. The first 10 years of their chemistry, biology, and nanoscience. *Accounts of Chemical Research*, 36(11): 807–815.

30. Yamakoshi, Y., Umezawa, N., Ryu, A., Arakane, K., Miyata, N., Goda, Y., Masumizu, T., and Nagano, T. 2003. Active oxygen species generated from photoexcited fullerene (C-60) as potential medicines: O-2(-center dot) versus O-1(2). *Journal of the American Chemical Society*, 125(42): 12803–12809.

31. Isakovic, A., Markovic, Z., Todorovic-Markovic, B., Nikolic, N., Vranjes-Djuric, S., Mirkovic, M., Dramicanin, M. et al. 2006. Distinct cytotoxic mechanisms of pristine versus hydroxylated fullerene. *Toxicological Sciences*, 91(1): 173–183.

32. Mori, T., Takada, H., Ito, S., Matsubayashi, K., Miwa, N., and Sawaguchi, T. 2006. Preclinical studies on safety of fullerene upon acute oral administration and evaluation for no mutagenesis. *Toxicology*, 225(1): 48–54.

33. Dugan, L.L., Gabrielsen, J.K., Yu, S.P., Lin, T.S., and Choi, D.W. 1996. Buckminsterfullerenol free radical scavengers reduce excitotoxic and apoptotic death of cultured cortical neurons. *Neurobiology of Disease*, 3(2): 129–135.

34. Usenko, C.Y., Harper, S.L., and Tanguay, R.L. 2008. Fullerene C-60 exposure elicits an oxidative stress response in embryonic zebrafish. *Toxicology and Applied Pharmacology*, 229(1): 44–55.

35. Gharbi, N., Pressac, M., Hadchouel, M., Szwarc, H., Wilson, S.R., and Moussa, F. 2005. [60]Fullerene is a powerful antioxidant in vivo with no acute or subacute toxicity. *Nano Letters*, 5(12): 2578–2585.

36. Hu, Z., Guan, W.C., Wang, W., Huang, L.Z., Tang, X.Y., Xu, H., Zhu, Z., Xie, X.Z., and Xing, H.P. 2008. Synthesis of amphiphilic amino acid C-60 derivatives and their protective effect on hydrogen peroxide-induced apoptosis in rat pheochromocytoma cells. *Carbon*, 46(1): 99–109.

37. Markovic, Z. and Trajkovic, V. 2008. Biomedical potential of the reactive oxygen species generation and quenching by fullerenes (C-60). *Biomaterials*, 29(26): 3561–3573.

38. Pickering, K.D. and Wiesner, M.R. 2005. Fullerol-sensitized production of reactive oxygen species in aqueous solution. *Environmental Science and Technology*, 39: 1359–1365.

39. Prat, F., Stackow, R., Bernstein, R., Qian, W.Y., Rubin, Y., and Foote, C.S. 1999. Triplet-state properties and singlet oxygen generation in a homologous series of functionalized fullerene derivatives. *Journal of Physical Chemistry A*, 103(36): 7230–7235.

40. Kamat, J.P., Devasagayam, T.P.A., Priyadarsini, K.I., and Mohan, H. 2000. Reactive oxygen species mediated membrane damage induced by fullerene derivatives and its possible biological implications. *Toxicology*, 155(1–3): 55–61.

41. Dugan, L.L., Lovett, E.G., Quick, K.L., Lotharius, J., Lin, T.T., and O'Malley, K.L. 2001. Fullerene-based antioxidants and neurodegenerative disorders. *Parkinsonism and Related Disorders*, 7(3): 243–246.

42. Bogdanovic, G., Kojic, V., Dordevic, A., Canadanovic-Brunet, J., Vojinovic-Miloradov, M., and Baltic, V.V. 2004. Modulating activity of fullerol C-60(OH)(22) on doxorubicin-induced cytotoxicity. *Toxicology in Vitro*, 18(5): 629–637.

43. Ali, S.S., Hardt, J.L., Quick, K.L., Kim-Hans, J.S., Erlanger, B.F., Huang, T.T., Epstein, C.J., and Dugan, L.L. 2004. A biologically effective fullerene (C_{60}) derivative with superoxide dismutase mimetic properties. *Free Radical Biology and Medicine*, 37(8): 1191–1202.

44. Rouse, J.G., Yang, J.Z., Barron, A.R., and Monteiro-Riviere, N.A. 2006. Fullerene-based amino acid nanoparticle interactions with human epidermal keratinocytes. *Toxicology in Vitro*, 20(8): 1313–1320.

45. Ueng, T.H., Kang, J.J., Wang, H.W., Cheng, Y.W., and Chiang, L.Y. 1997. Suppression of microsomal cytochrome P450-dependent monooxygenases and mitochondrial oxidative phosphorylation by fullerenol, a polyhydroxylated fullerene C-60. *Toxicology Letters*, 93(1): 29–37.

46. Zhu, S.Q., Oberdorster, E., and Haasch, M.L. 2006. Toxicity of an engineered nanoparticle (fullerene, C-60) in two aquatic species, Daphnia and fathead minnow. *Marine Environmental Research*, 62: S5–S9.

47. Oberdorster, E., Zhu, S.Q., Blickley, T.M., McClellan-Green, P., and Haasch, M.L. 2006. Ecotoxicology of carbon-based engineered nanoparticles: Effects of fullerene (C-60) on aquatic organisms. *Carbon*, 44(6): 1112–1120.

48. Lin, A.M.Y., Fang, S.F., Lin, S.Z., Chou, C.K., Luh, T.Y., and Ho, L.T. 2002. Local carboxyfullerene protects cortical infarction in rat brain. *Neuroscience Research*, 43(4): 317–321.

49. Xu, J.Y., Li, Q.N., Li, J.G., Ran, T.C., Wu, S.W., Song, W.M., Chen, S.L., and Li, W.X. 2007. Biodistribution of Tc-99m-C-60(OH)(x) in Sprague-Dawley rats after intratracheal instillation. *Carbon*, 45(9): 1865–1870.

50. Sayes, C.M., Marchione, A.A., Reed, K.L., and Warheit, D.B. 2007. Comparative pulmonary toxicity assessments of C-60 water suspensions in rats: Few differences in fullerene toxicity in vivo in contrast to in vitro profiles. *Nano Letters*, 7(8): 2399–2406.

51. Fujitani, Y., Kobayashi, T., Arashidani, K., Kunugita, N., and Suemura, K. 2008. Measurement of the physical properties of aerosols in a fullerene factory for inhalation exposure assessment. *Journal of Occupational and Environmental Hygiene*, 5(6): 380–389.

52. Hurt, R.H., Monthioux, M., and Kane, A. 2006. Toxicology of carbon nanomaterials: Status, trends, and perspectives on the special issue. *Carbon*, 44(6): 1028–1033.

53. Guzman, K.A.D., Taylor, M.R., and Banfield, J.F. 2006. Environmental risks of nanotechnology: National nanotechnology initiative funding, 2000–2004. *Environmental Science and Technology*, 40(5): 1401–1407.

54. Harper, S.L., Dahl, J.L., Maddux, B.L.S., Tanguay, R.L., and Hutchison, J.E. 2008. Proactively designing nanomaterials to enhance performance and minimize hazard. *International Journal of Nanotechnology*, 5(1): 124–142.

55. Hoet, P.H., Bruske-Hohlfeld, I., and Salata, O.V. 2004. Nanoparticles—Known and unknown health risks. *Journal of Nanobiotechnology*, 2(1): 12.

III

Clinical Significance of Nanosystems

Pharmacological Significance of Nanoparticles

Carlos Medina
Trinity College Dublin

Marek W. Radomski
Trinity College Dublin

18.1 Introduction

18.1.1 Nanotechnology Applied to Medicine

18.1.1.1 Concepts and Historical Perspectives

The unique properties of nanosized matter and nanoparticles (1–100 nm) have made them very attractive for infinite applications in cosmetics, electronics, aerospace, and the computer industry.

Medicine has also been fascinated by the exciting prospects of nanotechnology and now *nanomedicine* is a rapidly growing area of medical research. That fascination may have started in the 1960s when Isaac Asimov, in his book, *Fantastic Voyage*, reduced five people to a microscopic fraction of their original size and sent them in a miniaturized atomic submarine through a dying man's carotid artery to treat an intractable blood clot in his brain. Since then, researchers have been investigating if nanoparticles and nanodevices could be used in medicine and patients suffering from cancer, and infectious, neurological, degenerative, or cardiovascular diseases are likely to benefit from this research (Lanza et al., 2006; Caruthers et al., 2007; Singh et al., 2007; Zuo et al., 2007; Lammers et al., 2008).

Although the association between nanotechnology and medicine is becoming increasingly apparent and a new hybrid discipline called nanomedicine is gaining acceptance, an accurate definition of the term and scope of the new science is still under debate.

From a European perspective, the European Science Foundation (ESF) defines nanomedicine as "the comprehensive monitoring, repair, and improvement of all human biological systems, working from the molecular level using engineered devices and nanostructures to achieve medical benefit. This definition covers the use of analytical tools to give a better understanding of the molecular basis of a disease, as well as the design of nano-sized therapeutic and drug delivery systems that deliver more effective therapies." In addition, ESF defined five main disciplines of nanomedicine: analytical tools; nanoimaging; nanomaterials and nanodevices; novel therapeutics and drug delivery systems; and clinical, regulatory, and toxicological issues.

From an American perspective, the National Institutes of Health (NIH) defines nanomedicine as "an offshoot of nanotechnology, which refers to highly specific medical intervention at the molecular scale for curing disease or repairing damaged tissues, such as bone, muscle, or nerve."

Yet another perspective is given in the nanomedicine glossary in "Nanotechnology Now" that describes nanomedicine as "(1) the comprehensive monitoring, control, construction, repair, defence, and improvement of all human biological systems, working from the molecular level, using engineered nanodevices and nanostructures; (2) the science and technology of diagnosing, treating, and preventing disease and traumatic injury, of relieving pain, and of preserving and improving human health, using molecular tools and molecular knowledge of the human body; (3) the employment of molecular machine systems to address medical problems, using molecular knowledge to maintain and improve human health at the molecular scale."

Therefore, for the purpose of this chapter, we will broadly consider nanomedicine as the application of nanotechnology to the prevention, diagnosis, and treatment of different diseases in the human body. In this context, "nanopharmacology" will be defined as a nanomedical discipline concerned with drug characterization and development.

18.1.1.2 General Applications

There are several applications of nanotechnology in medicine, and some of them are well established; however, others are still at the experimental stage.

It is worth emphasizing that nanometer-sized particles are in the same range of dimensions as antibodies, membrane receptors, nucleic acids, and proteins. These characteristics and the possibility of modulating their physicochemical properties make nanoparticles powerful tools for imaging, diagnosis, pharmacology, and therapeutics (Zhang et al., 2007). Already, the range of nanomedical applications is diverse, comprising chemotherapeutic agents, pacemakers, hearing aids, glucose monitoring, and drug delivery systems (Medina et al., 2007; Jain, 2008).

Nanoparticles are currently being tested for *molecular imaging* to get a more accurate diagnosis of various pathologic processes with high-quality images. Currently, imaging diagnosis is not only limited to a simple description of anatomic structures, but can also involve molecular imaging of cellular signaling. The physicochemical characteristics of the nanoparticles allow the redirection and the concentration of the marker at the site of interest. In fact, nanoparticle-based imaging contrast agents have been shown to improve the sensitivity and specificity of ultrasound, fluorescence, nuclear, computed tomography, and magnetic resonance imaging (Schmitz et al., 2001; Lanza and Wickline, 2003; Wickline and Lanza, 2003). Patients suffering from cancer and atherosclerosis are likely to benefit from these new applications, as they image cancer cells or atherosclerotic plaque before the disease process manifests itself with clinical symptoms.

For *diagnostic applications*, nanoparticles allow the selective and sensitive detection of disease molecular markers by amplifying and protecting a biomarker from degradation, to provide more sensitive analysis (Nam et al., 2003; Geho et al., 2004). Initial studies with magnetic nanoparticle probes coated with antibodies and single "bar code" DNA fragments are able to amplify signals of small abundant biomolecules. For instance, it has been shown that nanoparticle-based bio–bar-code is able to detect free prostate-specific antigen (PSA) at low attomolar concentrations, facilitating the detection of prostate cancer at a very early stage (Nam et al., 2003). This finding could change dramatically the approach to diagnose and treat this disease as it could improve cancer screening and diagnosis, and, therefore, curative treatment could be performed early on during the course of malignancy.

As *therapeutic delivery systems*, nanoparticles allow the targeted delivery and controlled release of different drugs to the site of action, while minimizing possible side effects that result from the systemic administration of medicines (Fonseca et al., 2002; Bhadra et al., 2003). Currently, drug delivery is the most advanced application of nanotechnology in pharmacology and therapeutics and also in the entire medical field. Indeed,

nanoparticle-based drug delivery systems are gaining application in the pharmaceutical industry since nanoparticle-based drugs may have improved pharmacokinetic properties, such as higher solubility and better biodistribution, compared to small molecule drugs.

In this chapter, we will primarily focus on recent advancements of nanotechnology pertinent to pharmacology and therapeutics. Some of these concepts have been recently reviewed by our group (Medina et al., 2007).

18.1.1.3 Multifunctional Nanoparticles

Initially, nanoparticles designed for medical applications were engineered to provide a single function (monofunctional nanoparticle), for instance, a liposome that is only used as a nanoparticle delivery system. However, multifunctional nanoparticles have now been synthesized. These multifunctional nanoparticles combine several functionalities in a single stable construct. For example, a nanoparticle can be used as a nanoparticle-based delivery system but at the same time can be modified with an imaging agent to monitor the drug transport process (Figure 18.1). Yang et al. have used multifunctional magneto-polymeric nanohybrids composed of magnetic nanocrystals and anticancer drugs encapsulated by an amphiphilic block copolymer (Yang et al., 2007). In a different approach, multifunctional nanoparticles have been also used for in vivo imaging and siRNA delivery and silencing in cancer (Farokhzad et al., 2006).

However, despite the promising features of multifunctional nanoparticles, more studies need to be carried out to elucidate the bioavailability and potential increased toxicity of hybrid nanoparticles (Sanvicens and Marco, 2008).

18.1.2 Personalized Medicine and Nanomedicine

Personalized medicine may be defined as individualized or individual-based-therapy, which allows the prescription of

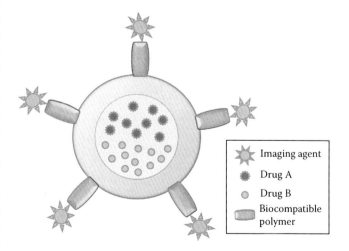

FIGURE 18.1 Multifunctional nanoparticles can combine nanoparticles for imaging (such as quantum dots), a stabilizer polymer to ensure biocompatibility and therapeutic compounds (drugs A and B).

precise treatments best suited for a single patient (Jain, 2002), thus matching the right drug to the right patient.

Medicine has always attempted to be personalized. Historically, the patient–doctor relationship always had aspirations to heal each person. Recently, the use of genomic tools to study the influence of several genetic factors on drug action and metabolism (pharmacogenomics) holds promise to advance personalized medicine (Meyer and Zanger, 1997). The cornerstone of pharmacogenomics is the recognition that individuals may differ in their genetic structure (genetic polymorphisms) and that these differences may modulate the individual patient response to standard treatment. This concept has already been considered in drug discovery and development as well as in the clinical use of drugs. The main goal of such personalized medicine would be to design the appropriate treatment for a patient according to his or her genotype. But even if genomic markers are discovered to pinpoint patients at high risk of suffering from side effects or high probability of nonresponse to standard treatment, those should be treated with alternative drugs that in many cases are not as efficient as they should be. In addition, ethical issues are involved in the development of personalized medicine in the area of genetic testing.

How would nanotechnology assist in solving problems encountered by the field of personalized medicine? Given the inherent nanoscale functions of the biological components of living cells and the fact that pathology often affects subcellular processes, nanoparticles have the potential of working at this level to correct the underlying molecular pathology. For example, nontherapeutic response to standard treatment could be due to genetic polymorphism, leading to the under-expression of cellular systems required for drug action, such as transporters or receptors. Nanosized particles can deliver drugs across different barriers through small capillaries into individual cells and their subcellular structures, fulfilling the role of personalized medicine (Jain, 2005a,b).

18.1.3 Nanotoxicology

Like most new technologies, including all emerging medicines and medical devices, there is a rising debate concerning the possible toxic effects derived from the use of nanoparticles. They can act on living cells at the nanolevel resulting not only in biologically desirable effects, but also in undesirable effects. The main characteristics of the nanoparticles are their small size and increased percent surface molecules, and these can affect not only the physicochemical properties of the material but also the biological reactivity such as uptake and interactions with biological tissues. Therefore, nanoparticles *per se* when used for imaging or drug delivery could affect the biological function at the cellular and protein levels. This is of particular relevance to nonbiodegradable nanoparticles, which accumulate and persist for longer times in the body exerting side effects (Kipen and Laskin, 2005). To study the potential adverse effects of particles at nanoscale in depth, a new science, that of *nanotoxicology*, has emerged. Nanotoxicology can be defined as the study of the interactions of nanoparticles with biological systems with an emphasis on establishing the relationship, if any, between the physical and chemical properties of nanoparticles and the induction of toxicological responses (Service, 2004, 2005). The relationship between the human exposure to nanorods of asbestos, biopersistence of this material, and the development of global epidemic of lung mesothelioma in the 1960s is a classical example of the toxicity of nanomaterials (Donaldson and Tran, 2004). Carbon nanotubes, a group of nanostructures with a very useful profile for nanomedical applications, are a more recent example of potential pulmonary (Li et al., 2007b) and vascular (Radomski et al., 2005) toxicity of nanomaterials (Figure 18.2a and b). Therefore, development of novel nanoparticles for medicine must proceed in tandem with assessment of any toxicological and environmental side effects of these particles. In fact, according to ESF, basic research in the areas of materials science and device fabrication as well as individual and environmental

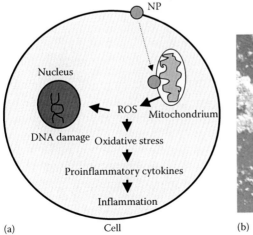

(a) Cell (b) Platelet aggregates

FIGURE 18.2 Toxicity induced by nanoparticles (NPs): (a) Nanoparticles, when entering the cells, may induce oxidative stress with the generation of reactive oxygen species (ROS), leading to an inflammatory response and DNA damage. (b) Nanoparticles may also induce platelet aggregation.

safety and toxicology must be pursued using multidisciplinary approach with careful consideration given to clinical, ethical, and societal issues.

18.2 Pharmacokinetics, Pharmacodynamics, and Nanoparticles

Pharmacokinetics is the study of the processes by which a drug is absorbed, distributed, metabolized, and eliminated by the body. In other words, it is the study of what the body does to a drug. In contrast, pharmacodynamics is the study of what drugs do to the body.

Traditionally, drugs have been produced by exploring small molecules and their structures to produce pharmacologically active materials. These molecules are designed to have a useful combination of lipophilic and hydrophilic properties so that they can distribute all over the body quite quickly via the bloodstream and by diffusion through different tissues and cells. However, the use of these pharmacological agents is frequently limited by pharmacodynamic problems such as low efficacy (the drug is not active enough to exert expected effects) or the lack of selectivity (the drug acts on multiple molecular targets). In addition, many drugs have poor pharmacokinetic profiles including poor solubility or low bioavailability. Finally, some drugs can be recognized as "nonself" by the immune system (high immunogenicity), which then provokes immune reactions against the medicine. Nanoparticles have the potential to improve the poor pharmacodynamic or pharmacokinetic profile of drugs. For instance, when a free drug is introduced into the body, it can be rapidly taken up by phagocytic cells, metabolized by the liver, and excreted by the kidneys. However, the use of materials in nanoscale may facilitate great modifications of pharmacokinetic properties of drugs such as solubility and diffusivity. Furthermore, nanoformulations of drugs can protect active compounds from the attack of the immune surveillance system (the reticuloendothelial system [RES]), thus protecting the drug from premature inactivation during its transport. Finally, nanoformulations show decreased rate of liver degradation and metabolism as well as reduced renal clearance. All these effects can result in prolonged blood circulation with an enhanced tissue accumulation and therapeutic effect.

The following sections provide a more comprehensive discussion related to the use of nanoparticles to improve pharmacokinetics and pharmacodynamics of classical drugs.

18.2.1 Solubility

Approximately 40% of different pharmacological agents are very difficult to formulate due to the lack of solubility (Liversidge and Cundy, 1995; Merisko-Liversidge and Liversidge, 2008). If a drug must penetrate a biological membrane, the molecule generally must possess some hydrophobic or lipophilic properties to be absorbed properly. Poorly soluble drugs generally have

low dissolution velocity that can result in low, variable absorption and a poor therapeutic response. But in many cases, classical strategies of pharmacological development to deal with this issue have not been really successful and finally even a very valuable drug is rejected early on in its development. Based on the understanding that the rate of drug dissolution is the primary driving force behind improved pharmacokinetic properties, drugs can be formulated as nanometer-sized particles resulting in an increased surface area due to reduction in particle size. Therefore, nanoparticle formulations may overcome this solubility problem as dissolution rate can be substantially increased (Merisko-Liversidge et al., 2003). However, this increase in surface energy can cause the nanometer-sized drug particles to spontaneously aggregate, therefore stabilizers are undoubtedly needed to disperse the formulation into discrete particles. One clear example is nanocrystalline dispersions made particularly for oral delivery that comprise drug, water, and stabilizers. Stabilizers that are usually used to aid the dispersion of particles are either polymers and/or surfactants. Polymers act as the primary stabilizers whereas surfactants are used as the secondary stabilizers. To be effective, the stabilizers must wet the drug crystals and provide a barrier to prevent agglomeration. In the absence of the appropriate stabilizer, the high surface energy of nanoparticles would tend to agglomerate or aggregate the drug crystals.

18.2.2 Nanoparticle Distribution

18.2.2.1 Barriers to Nanoparticle Distribution

Biological compartments act as potential barriers to the passage of nanoparticles, and there are a few of them in our body to be considered, such as the epithelium, blood, the immune system including the network of lymphatic vessels and nodes and the RES, and finally the extravasation of the nanoparticles to site-specific targeting through the extracellular matrix (ECM).

18.2.2.1.1 Epithelium and Blood

The epithelium can be considered as the first biological barrier to nanoparticle distribution, following environmental exposure, inhalation, or ingestion. Organ-specific epithelium includes lung, gut, and skin epithelia. There is some evidence that the smaller the particle diameter is, the faster they could diffuse through the epithelium (Szentkuti, 1997). Indeed, it has been found that inhaled (99m)Tc-labeled carbon particles (<100 nm) pass to the blood circulation within 1 min following exposure in humans (Nemmar et al., 2002). For therapeutics, many nanoparticles can be administered by parenteral injection, therefore, this first biological barrier could be easily avoided. In this case, the first compartment encountered by injectable nanoparticles is blood. Nanoparticles in biological fluids such as blood and plasma associate with a range of biopolymers, especially proteins, organized into the "protein corona" (Lundqvist et al., 2008). Interestingly, the nanoparticle size and surface properties determine the composition of the protein corona.

18.2.2.1.2 The Immune System

The network of blood and lymphatic vessels provides natural routes for the distribution of nanoparticles to the target tissues; however, at the same time when nanoparticles enter the bloodstream, in addition to encountering plasma proteins, they face immune cells. In general, particulate materials can be rapidly cleared out from the body either by monocytes or leukocytes in the blood stream or by the liver and spleen macrophages (RES) (Moghimi et al., 2001). Opsonization, which is the surface deposition of blood opsonic factors such as immunoglobulins and complementary factors, facilitates the binding of nanoparticles to phagocytic cells, particularly macrophages and leukocytes (Figure 18.3). Therefore, the amount of nanoparticles available to reach the desired site would be decreased, reducing the efficacy of the drug. Hence, evasion of nanoparticulate binding to phagocytic cells could be accomplished to a certain extent by interference with opsonization. For instance, it has been described that nanoparticles covered by uncharged liposomes have a longer circulation time in rats compared to anionic liposomes (Senior and Gregoriadis, 1982). This could be explained by the fact that uncharged liposomes are poor activators of the complement system and consequently with the process of complement opsonization (Devine and Bradley, 1998). In addition, polyethylene glycol (PEG) polymer may be conjugated onto the surface of the nanoparticles, which is a relatively inert hydrophilic polymer that provides good steric hindrance for preventing protein binding. Indeed, it has been previously shown that PEG may provide increased biodistribution, prolonged systemic circulation lifetime, and reduced liver uptake in rabbits by interfering with the opsonization process (Vandorpe et al., 1997).

18.2.2.1.3 Extravasation

The next barrier of nanoparticle distribution is the vascular endothelium. The endothelial tissue is the layer of cells that covers the interior of the blood vessels, preventing most particles from exiting the bloodstream. However, there are basically two mechanisms by which nanoparticles may extravasate to the ECM: (1) paracellular route, although the presence of tight junctions between endothelial cells (gaps are extremely small, approximately 2 nm in diameter) will prevent most nanoparticles to exit the circulation; (2) most likely, nanoparticles can pass through the endothelium by a cellular route called transcytosis or vesicle transport, a mechanism for transcellular transport in which an endothelial cell encloses extracellular material in an invagination of the cell membrane to form a vesicle, then moves the vesicle across the cell to eject the material through the opposite cell membrane by the reverse process (Oh et al., 2007). This mechanism is size-dependent, therefore, the smaller the molecule is the easier it passes through. This process involves caveolae that are caveolin-coated, omega-shaped plasmalemmal invaginations 60–70 nm in diameter particularly abundant in endothelial tissue. In addition to these two physiological mechanisms, under some pathological conditions such as inflammation or cancer, the endothelial tissue can become leaky, allowing nanoparticles to exit the bloodstream more easily, leading to accumulation of nanoparticles in tissues. In cancer, the defective vascular architecture, created due to the rapid vascularization necessary to serve fast-growing tissues, in conjunction with poor lymphatic drainage allows a phenomenon known as the "enhanced permeability and retention" (EPR) effect (Brannon-Peppas and Blanchette, 2004). By exploring all these effects, including EPR effect, nanomedicine could improve targeting drug molecules to the site of action in cancer or other pathologies (Figure 18.4).

Once out of the circulation, the next barrier to nanoparticle distribution is the ECM. Cells are embedded in a connective tissue matrix made up of collagen, hyaluronic acid, and proteoglycans among others (Schuppan and Riecken, 1990) impairing particles to move along. However, the nature of the ECM is that some water channels are present and may permit the passage of different particles. In fact, it has been previously shown that nanoparticles can translocate locally to lymphatic nodes when they are injected subcutaneously, a process which involves skin lymphatic vessels (Kim et al., 2004).

18.2.2.1.4 Cellular Uptake

One of the main goals in nanomedicine is the precise drug release into highly specified target. This objective can be achieved by the small size of nanoparticles, which can penetrate into the cells. However, nanoparticles should recognize the site of interest for specific targeting once they have avoided all the physiological barriers as described above. Therefore, it is essential to design strategies that enable nanoparticles to recognize their target cells and also enable nanoparticles to enter the cells and then gain access to specific organelles.

Surface functionalization of nanoparticles is one of those design strategies that can offer several possibilities for drug targeting in terms of cellular binding, uptake, and intracellular transport (Figure 18.5).

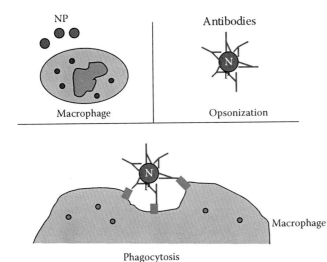

FIGURE 18.3 Opsonization and phagocytosis of nanoparticles (NPs) by macrophages. Antibody binding to NPs can opsonize the material and facilitate its uptake and destruction by phagocytic cells.

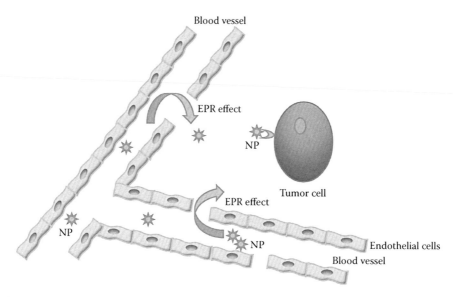

FIGURE 18.4 EPR effect. Nanoparticles (NPs) can be designed to utilize the EPR to exit the blood vessels in the tumor and target surface receptors on the tumor cells.

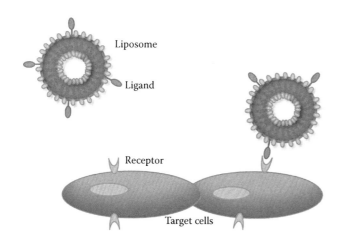

FIGURE 18.5 Functionalization of nanoparticles. For instance, a liposome can be functionalized conjugating appropriate ligands for specific receptors that are overexpressed in inflammatory or tumor cells.

Entry into cells is basically governed by biological mechanisms of endocytosis. The routes of endocytosis mostly in operation are the receptor-mediated routes. These routes require recognition of some ligand by a specific biological receptor. Site-specific drug delivery is therefore viable because as part of the pathological phenotype of diseases, such as in cancer or inflammation, cells often express several surface receptors or antigens that are not expressed by their normal counterparts (Bakowsky et al., 2008). Indeed, conjugating appropriate ligands for specific receptors should result in a higher drug concentration at the tissue of interest, thus increasing the efficacy of the drug action (Bies et al., 2004; Bakowsky et al., 2008).

Particulates usually end intracellularly in endosomes or lysosomes followed by particle degradation. Therefore, once nanoparticles have been taken up by the target cells, it becomes

critical that the carrier can protect successfully the molecule from degradation by intracellular endolysosomal systems resulting in their efficient intracellular delivery (Torchilin et al., 2001). In addition, for drugs such as dexamethasone, receptors of which are cytoplasmic, it may be important to retain the drug in the cytoplasmic compartment to enhance its therapeutic efficacy (Adcock, 2000). With surface functionalization, nanoparticles can be delivered to either the cellular cytoplasm or organelles. For instance, it has been found that changes in the surface charge of nanoparticles (from anionic to cationic) can facilitate a rapid endolysosomal escape into the cellular cytosol, avoiding particle degradation and enhancing their therapeutic effect (Panyam et al., 2002).

Finally, when they have exerted their functions, nanoparticles should be degraded without any side effect. One approach to do this would be to cover nanoparticles with biodegradable materials, such as polymers or liposomes, which do not accumulate in the body and possibly are risk-free (Sapra et al., 2005).

18.2.3 Interactions with Blood Cells

We have already mentioned the interactions of the nanoparticles with the immune system, but in the circulation too, there are other cells or cell elements such as red blood cells and platelets that are likely to interact with nanoparticles.

Hemolysis is the damage to red blood cells or erythrocytes leading to the release of hemoglobin to the surrounded tissue. In some circumstances, nanoparticles could induce a mild hemolysis leading to the adsorption of hemoglobin by nanoparticles, facilitating the interaction with phagocytic cells and clearance by RES. In few occasions, severe hemolysis can lead to anemia, which in some cases can be life-threatening. Basically, two mechanisms for drug-mediated hemolysis have been suggested, non-immune mediated (direct nanoparticle–erythrocyte interactions) and immune mediated (drug-specific antibody). To minimize these effects, modifications of the nanoparticle surface should be considered. For instance, it has been suggested that cationic chains of water-soluble fullerenes can induce significant hemolysis whereas neutral or anionic moieties do not exert any effect on erythrocytes (Bosi et al., 2004). Accordingly, it has been shown that nanoparticles covered by PEG induce less damage to erythrocytes (Kim et al., 2005).

Thrombogenicity is the ability of some materials to interact with platelets, leading to platelet activation and aggregation. It has previously been found that mixed carbon nanoparticles and nanotubes, both MWCNT and SWCNT, are able to induce platelet aggregation in vitro (Figure 18.2b) and, in addition, accelerate the rate of vascular thrombosis in rat carotid artery (Radomski et al., 2005). Furthermore, it has been shown that nanoparticles can directly induce cytotoxic morphological changes in human umbilical vein endothelial cells, induction of proinflammatory responses, inhibition of cell growth, and reduction of endothelial nitric oxide synthase (Yamawaki and Iwai, 2006). No studies evaluating the effects of nanoparticle size and charge have been conducted so far. However, it seems again that decreasing particle surface charge by coating nanoparticles with PEG or designing biodegradable nanoparticles may decrease platelet activation and aggregation (Koziara et al., 2005; Li et al., 2007a).

18.3 Therapeutic Applications of Nanoparticles

18.3.1 Drug Discovery and Development

Technological achievements in nanobiotechnology are being applied to improve mainly diagnostics and drug delivery. However, now many researchers are using nanomedicine and nanoparticles to improve drug discovery and development. One of these applications has been used in proteomics. It is known that protein identification is a crucial step for target identification and validation in the early stages of drug discovery. However, most protocols to date are not able to identify less abundant proteins that can be only studied at nanoscale protein analysis (Jain, 2005b). For instance, single-walled carbon nanotubes have been explored as a platform for investigating surface–protein and protein–protein binding and developing highly specific electronic biomolecule detectors (Chen et al., 2003).

Another application would be the use of lipoparticle, which enables integral membrane proteins to be solubilized while retaining their intact structural conformation. Retaining the native structural conformation of membrane-bound receptors is vital during assay development for optimization. This approach was developed initially to study ligands binding to membrane proteins in HIV (Hoffman et al., 2000).

In addition, some nanoparticles could be potential drugs for the future, for example, fullerenes that are novel carbon allotrope with a polygonal structure made up exclusively by 60 carbon atoms. These nanoparticles are characterized by having numerous surfaces of attachment that can be functionalized

for tissue binding (Bosi et al., 2003). Interestingly, there is some evidence that fullerenes bind and inactivate circulating and intracellular free radicals, preventing cell injury and cell death. Furthermore, fullerenes may exert their antioxidant properties by catalyzing superoxide into hydrogen peroxide and oxygen. Indeed, it has been previously shown that fullerene C60 derivative exerts superoxide dismutase (SOD) mimetic properties (Ali et al., 2004). Therefore, fullerenes have the potential to be used as treatment in diseases associated with excessive generation of oxygen-derived reactive species.

18.3.2 Drug Delivery

As previously mentioned, drug delivery is the major application of nanotechnology in medicine nowadays. The efficacy of different drugs such as chemotherapeutical agents is often limited by dose-dependent side effects. Indeed, anticancer drugs, which usually have a large volume of distribution, are toxic to both normal and cancer cells. Therefore, one of the main challenges in drug delivery in general and cancer chemotherapy in particular is precise controlled drug release into highly specified target. This objective can be achieved by miniaturizing the delivery systems to become much smaller than their targets. Controlled drug delivery involves the association of a drug with a carrier system, thereby allowing modulation of the pharmacokinetic properties and biodistribution of the drug. For all drugs, including gene therapy constructs, an ideal delivery vehicle should have the following characteristics: be biocompatible, biodegradable, nonimmunogenic, nontoxic; able to carry a variety of types of molecular agents without changing its own or their chemical constitution; and able to release the drug in a controlled manner. Out of the nanoparticles available to date, liposomes and polymers are the most used nanoparticles as drug delivery systems worldwide (Table 18.1 and Figure 18.6), since these compounds are biodegradable, they do not accumulate in the body and are possibly risk-free (Sapra et al., 2005).

Liposomes are nanoparticles comprising lipid bilayer membranes surrounding an aqueous interior. The amphiphilic molecules used for the preparation of these compounds have similarities with biological membranes and have been used for

TABLE 18.1 Applications of Nanoparticles in Medicine

Type of Nanoparticle	Application	Indication
Liposomes	Drug delivery	Cancer
		Vaccine
		Infections
Polymers	Drug delivery	Hepatitis
Dendrimers	Therapeutics	Cancer
		Inflammation
		HIV/AIDS
Carbon nanotubes	Imaging	Atomic force microscopy
Quantum dots	Imaging	Labelling reagents
		Western blot
		Flow cytometry

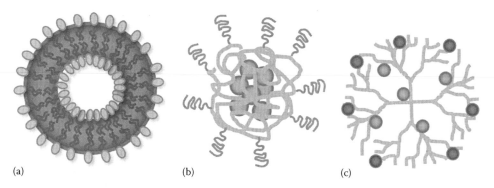

(a)   (b)   (c)

FIGURE 18.6   Schematic illustration of different nanoparticles: (a) liposome, (b) polymeric nanoparticle, and (c) dendrimer.

improving the efficacy and safety of different drugs (Hofheinz et al., 2005). The active compound can be located either in the aqueous spaces, if it is water-soluble, or in the lipid membrane, if it is lipid-soluble. In addition, liposomes protect the encapsulated drugs from undesired effects of external conditions and also can be functionalized with specific ligands that can target specific cells. However, there are some limitations with the use of liposomes as delivery systems. Liposomes are usually rapidly cleared out from the body by phagocytic cells, resulting in a poor therapeutic index. Recently, a new generation of liposomes called "stealth liposomes" has been developed to overcome this problem. Usually, this new strategy involves formulating long-circulating liposomes by coating the liposome surface with biocompatible polymers such as PEG. The polymer layer provides a protective shell over the liposome surface and suppresses liposome recognition by opsonins, conferring the ability to evade the interception by the immune system, and therefore, a longer half-life (Moghimi and Szebeni, 2003).

Polymers, such as polysaccharide chitosan nanoparticles, have been used for some time now as drug delivery systems (Agnihotri et al., 2004). Recently, water-soluble polymer hybrid constructs have been developed. These are polymer–protein conjugates or polymer–drug conjugates. Polymer conjugation to proteins reduces immunogenicity, prolongs plasma half-life, and enhances protein stability and solubility. Polymer–drug conjugation promotes cancer targeting through the enhanced permeability and retention effect and, at the cellular level following endocytic capture, allows lysosomotropic drug delivery (Lee, 2006).

Besides drug-encapsulated liposomes and polymer–drug conjugates that represent practically 80% of the total nano delivery systems, other nanoparticle platforms such as nano-emulsions (Kumar et al., 2008; Tagne et al., 2008) and dendrimers (Caminade et al., 2008; Svenson and Chauhan, 2008) have also shown therapeutic potential. *Emulsions* comprise oil in water type mixtures that are stabilized with surfactants to maintain size and shape. The lipophilic material can be dissolved in an organic solvent that is emulsified in an aqueous phase. *Dendrimers* refer to synthetic, three-dimensional, core–shell structures that can be synthesized for a wide range of applications.

18.3.3 How Do Nanoparticles Recognize the Target Cells?

Targeting drugs and controlled drug delivery to their sites of action is still a major challenge in pharmaceutical research trying to improve the therapeutic index of drugs by increasing their localization to specific organs, tissues, or cells and by decreasing their potential toxic side effects at normal sensitive sites. With the use of nanotechnology, targeting drug molecules to the site of action is becoming a reality. There are different approaches to achieve this important objective:

18.3.3.1 Functionalization of Nanoparticles Conjugating Appropriate Ligands for Specific Receptors on Target Cells

As mentioned in Section 18.2.2.1.4, in some pathological conditions such as inflammation and cancer, cells may express different receptors and proteins that are not normally expressed by healthy cells. Therefore, specific ligands attached to the surface of the nanoparticles can selectively recognize and bind to different receptors or molecules on target cells, thus increasing the concentration of the drug in the required site. Proof-of-principle experiments have shown that it is possible to attach targeting ligands to the nanoparticle surface (antibodies, peptides, or small molecules), and thus prepared constructs can successfully deliver drugs to the required site (Sapra and Allen, 2002; Kocbek et al., 2007). In general, the main issue with nanoparticle conjugation is to cover the nanoparticle surface with the sufficient amount of ligands to achieve efficient binding and recognition, but at the same time too many ligands may trigger off recognition and activation of immune system, leading to construct clearance. Therefore, small molecules have been considered with great attention for the design of nanoparticles with specific recognition properties such as folate-conjugated liposomes as high affinity receptors for folic acid are greatly enriched on different cancer cells (Lee and Low, 1994).

18.3.3.2 Accumulation of Nanoparticles in the Required Site

The enhanced permeability and retention phenomenon occurring in cancer can be used for selective drug targeting using nanoparticles. The classical example of the exploitation of EPR effect for cancer nanotherapeutics is the use of liposomal vincristine (an anticancer drug that interferes with the formation of mitotic spindle and inhibits cell proliferation) for the treatment of non-Hodgkin lymphoma (Sarris et al., 2000). It is known that the high neurotoxicity of vincristine necessitates the reduction of the dose during treatment and this has been associated with lower remission and survival rates in patients with Hodgkin's disease. It has been found that vincristine encapsulated in liposomes persists longer in the circulation (prolonged half-life) and reaches higher concentrations in cancer tissue and lymph nodes than in neural tissue (Kanter et al., 1994). The increased concentrations of drug in cancer tissue and relatively low concentrations in healthy tissues detected with liposome vincristine correlate with reduced neurotoxicity in experimental animals (Boman et al., 1994) and humans (Gelmon et al., 1999). The beneficial effects of drug carrier have been ascribed to the ability of liposomes to selectively extravasate in the leaky vasculature present in the rapidly growing cancers (Yuan et al., 1994).

Another approach to increase the accumulation of drug-loaded nanoparticles is the development of pH-sensitive nanoparticles that are stable at physiological pH (pH 7.4), but undergo destabilization under acidic conditions, thus leading to the release of their contents. The concept of pH-sensitive nanoparticles emerged from the observation that some pathological tissues, such as cancer and inflamed tissues, exhibit an acidic environment as compared to normal tissues (Torchilin et al., 1993). In fact, pH-sensitive liposomes have been already developed (Simões et al., 2004). However, there are some limitations for their use in vivo: first, in cancer, the highest acidity levels are found in the core (center) of malignant growth, which is usually avascular (contains little or no blood vessels), thus with the very limited drug access; second, the pH in cancer tissues is greatly heterogeneous with no correlation between perivascular pH and nearest vessel blood flow (Helmlinger et al., 1997); third, the lowest pH that is found in cancer is usually around 6.5, which makes the design of pH-sensitive nanoparticles in a range of 0.9 unit of pH very challenging.

18.3.4 Main Clinical Applications

18.3.4.1 Cancer

Even when new molecules for cancer treatment are discovered, there are many limitations for their clinical use, such as drug resistance at cancer level and toxicity due to the fact that anticancer agents act also on healthy cells. Therefore, the use of nanotechnology in oncology to overcome all these issues is promising. Hence, it is crucial to develop new strategies for targeted delivery of anticancer agents to cancer tissues. These strategies should be guided by the combined knowledge of cancer pathophysiology and the pharmacokinetic and pharmacodynamic profiles of nanoparticles in vivo.

18.3.4.1.1 Non-Surface Modified Nanoparticles

Non-surface-modified nanoparticles are not coated with any material to prevent protein binding, opsonization, recognition,

and rapid clearance from the blood stream by phagocytic cells of RES (passive targeting). The liver is essential for drug metabolism and excretion as many phagocytic cells of RES are localized in this organ. Nanoparticles, when injected intravenously, are cleared predominantly by the liver macrophages called Kupffer cells (Moghimi et al., 2001). The liver retention of noncoated nanoparticles could be beneficial for the treatment of some types of cancers where RES is a plausible target, for example, in the primary liver cancer or for the treatment of liver metastasis that occurs frequently in cancers of various origins. It is proposed that after the drug-containing nanoparticles are taken up by Kupffer cells, nanoparticles become leaky and release their content. Although promising results were shown in animal models with the use of doxorubicin-loaded nanoparticles (Chiannilkulchai et al., 1990), no benefits have been reported in humans (Valle et al., 2005).

18.3.4.1.2 Surface-Modified Nanoparticles

If the RES cells were not the intended therapeutic target, nanoparticles should be designed to avoid interactions with the RES system. One of the approaches involves formulating long-circulating nanoparticles by coating their surface with biocompatible polymers such as PEG. These "Stealth Nanoparticles" are "invisible" to phagocytic cells that do not phagocytize particles, leading to prolonged half-life of drug–nanoparticle constructs. The mechanism by which the drug is delivered to cancer from nanoparticles seems to be related to the EPR effect. The clinically available pegylated liposomal doxorubicin for the treatment of different cancers, including ovarian cancer (Gordon et al., 2000) and T-cell lymphomas (Pulini et al., 2007), is the prime example of "stealth" technology. In this formulation, doxorubicin is packaged in a liposome made of a bilayer sphere of lipids with an outer coating of PEG (Green and Rose, 2006). The clinical experience is consistent with superior therapeutic and safety profiles of liposome doxorubicin when compared with standard preparation.

Recently, a lot of efforts have been devoted to develop nanoparticles that are able to recognize the target cells and selectively deliver the drug to cancer tissues. This aim can be achieved, as explained before, by functionalization of nanoparticles. Among the different approaches of active targeting, liposomes using an antibody as a targeting ligand and a lipid vesicle as a carrier for drugs have attracted much attention. Also, folate-conjugated liposomes have been studied for active targeting. The rationale of this approach has been well established in a variety of in vitro methods (Lee and Low, 1994; Maruyama et al., 1999), however, only a limited number of preclinical studies report successful targeting in vivo (Mamot et al., 2005; Kennel et al., 2008).

18.3.4.2 Infections

Since the discovery of penicillin by Alexander Fleming in 1928, much progress has been made in the treatment of different infectious diseases. However, currently there are new global challenges such as HIV infection, viral, bacterial, and parasite resistance, which are still responsible for millions of deaths worldwide. Therefore, the need for new therapeutic strategies is pressing.

18.3.4.2.1 Active Treatment

The macrophage is a specialized host defense cell involved in the body defense system against various infectious diseases. Although most microorganisms are killed by these cells, many pathogenic organisms have developed ways for resisting macrophage destruction. Hidden against the cytodestruction and lysis within macrophage subcellular compartment, lysosome pathogens such as *Leishmania, Listeria monocytogenes*, and *Mycobacterium tuberculosis* may persist for weeks or months, silently fueling infection. Therefore, passive targeting of nanoparticles with encapsulated antimicrobial agents to infected macrophages is a valid strategy for the treatment of some types of infectious diseases. Degradation of the nanoparticles by lysosomes will lead to the release of the antimicrobial agents into the endolysosomal system or macrophage cytoplasm. The very high local concentrations of antimicrobial drugs are then lethal to pathogens. Indeed, different liposomal formulations have been developed for the treatment of human leishmaniasis, particularly liposomal amphotericin B. This compound has been used successfully for the treatment of visceral leishmaniasis in adults (Laguna et al., 2003) and children (di Martino et al., 1997). Preclinical data are also available on the treatment of listeriosis in animals with antibiotic-containing nanoparticles (Fattal et al., 1991; Jaakohuhta et al., 2007). In addition, some animal studies have also shown that antituberculosis agent–loaded nanoparticles are efficacious, resulting in bacterial clearance from infected organs (Ahmad et al., 2008; Verma et al., 2008).

One of the main challenges in the management of infectious diseases is the treatment of viral infections. Viruses are small infectious agents that do not have a metabolic machinery of their own; therefore, they have to use the metabolic processes of the host cell they infect to replicate. Despite the research made in this field so far, one of the main problems is the drug resistance due to the viral mutations. Nanotechnology has been applied to the treatment of viral infections, providing promising therapeutic approaches. The treatment of viral hepatitis is one of the most challenging global therapeutic tasks. Recently, biotechnologicals such as interferon alpha showed some therapeutic effectiveness in viral hepatitis. However, due to the drug resistance and a relatively short half-life of interferon alpha, many patients do not respond to the treatment (Papatheodoridis and Cholongitas, 2004). Therefore, interferon alpha has been now formulated as pegylated liposome preparation. This has resulted in sustained-release delivery system of interferon alpha with promising therapeutic effects (Vyas et al., 2006).

Another huge therapeutic and global problem represents infection with HIV retrovirus. The recently developed antiretroviral drugs may prolong lives yet the HIV virus keeps on generating more resistant strains. More recently, antiretroviral drug–loaded nanoparticles have been developed for the treatment of HIV infection due to the fact that nanoparticles offer protection to antiretroviral drugs from viral nuclease attack. Many studies have been performed in vitro systems with great success (Bender et al., 1996; Zimmer, 1999; Lambert et al., 2001); however, all these data remain to be confirmed in vivo.

18.3.4.2.2 Vaccines

The creation of vaccines is one of medicine's most important accomplishments. In fact, many infectious diseases that caused millions of deaths worldwide are now under control. However, there are not vaccines available for all infectious diseases, such as HIV and viral hepatitis C. In many cases, this is due to the fact that the antigen itself is only very weakly immunogenic and evokes only weak immune responses; therefore, an adjuvant (immune response amplifier) is needed to intensify the body immune response.

Vaccines formulated using nanotechnology were first developed in 1970 using nanoparticles as encapsulation method to ensure protection of viral antigens against degradation by human peptidases prior to their uptake by macrophages. Nowadays, nanoparticles can be designed in such a sophisticated way that they can mimic empty virus envelopes or capsids, thus intensifying the host immune response. Despite the research in this field, only one nanosized vaccine adjuvant has been approved for human use so far, MF59 (Ott et al., 1995). MF59 is an oil-in-water emulsion composed of <250 nm droplets and has been shown to be a potent stimulator of cellular and humoral responses to subunit antigens in both animal models and clinical studies (Dupuis et al., 1998; Peek et al., 2008). Other adjuvants generated using nanotechnology are nanoparticles produced by combining a protein antigen, cholesterol, phospholipid, and the saponin adjuvant Quil A (ISCOM). A randomized, double-blind clinical study in young adults comparing the immune responses induced by influenza ISCOM vaccines and conventional vaccines found that the ISCOM vaccines resulted in more rapid rises of antibodies against the vaccine strains (Rimmelzwaan et al., 2000).

18.3.4.3 Cardiovascular Diseases

Perhaps one of the most active areas of cardiovascular research of immediate clinical significance is the pursuit to identify, quantify, and treat vulnerable and unstable plaque, the so-called atherosclerotic plaque. Atherosclerosis is the inflammatory process of vasculature characterized by the accumulation of lipid/debris/apoptotic macrophages within the vessel wall (atherosclerotic plaques). The plaques when ruptured attract thrombus (blood clot) formation, limiting blood and oxygen flow to such important organs as heart (coronary atherosclerosis) or brain (cerebral atherosclerosis). For some time it has been recognized that thrombosis associated with plaque rupture is the principal cause of acute coronary syndromes (angina and myocardial infarction) and strokes. Site-targeted nanoparticles offer the opportunity for local drug delivery in combination with molecular imaging as it has been demonstrated with the use of perfluorocarbon nanoparticles for the treatment of atherosclerotic plaque (Lanza et al., 2006). In addition, antirestenotic (preventing reocclusion of recanalized vessels) drugs containing nanoparticles could be used to prevent restenosis as they could be delivered directly to the vessel wall (Lanza et al., 2000). Furthermore, some approaches have been made to treat thrombi formation by nanomedicine. In fact, it has been previously shown that liquid perfluorocarbon nanoparticle emulsions can be targeted to thrombi *in vitro* and *in vivo* (Lanza et al., 1996).

The selective drug delivery to intravascular sites of interest presents a unique opportunity to target clot-dissolving therapeutics to cerebral sites of thromboembolism while decreasing the risk of hemorrhagic complications (as enhanced clot dissolution will encourage blood loss, bleeding) and increasing the effectiveness of thrombolytic therapy.

18.3.4.4 Central Nervous System Disorders

The delivery of drugs in the brain is one of the main challenges in drug delivery and therapeutics. The blood brain barrier (BBB), the biological surface dividing the blood compartment from neurons in the brain, is very well known to be the best gatekeeper towards external agents. This has been an important issue for the treatment of different central nervous system (CNS) disorders, particularly in neurodegenerative conditions such as Alzheimer's and Parkinson's diseases. However, nanosized materials can cross the BBB as previously has been suggested in different studies and could be used for the treatment of CNS disorders (Fechter et al., 2002; Oberdörster, 2004). One of the key issues in neurodegenerative diseases is oxidative stress. In fact, it has been found that increased oxidative stress and free radical production contributes to brain damage. Therefore, it would be desirable to reduce CNS disorder–associated oxidative stress. For instance, it has been found that cerium oxide nanoparticles are indeed a valid therapeutic strategy as they are endowed with long-lasting antioxidant properties (Singh et al., 2007). However, other nanoparticles may enhance oxidative stress and free radical generation in the brain (Oberdörster, 2004). Furthermore, some particles have been implicated in the pathogenesis of Parkinson's disease (Olanow, 2004). Therefore, further studies on toxicological assessments of ceria are needed before they can be used in clinical trials.

18.4 Conclusions

The fascination of medicine with nanoparticles has already resulted in the huge stimulation of research on the diagnostic, imaging, and therapeutic potential of tiny particles. It remains to be seen if this fascination will progress to tangible medical benefits in developing novel and viable diagnostic and therapeutic strategies.

Acknowledgments

This work was supported by PI grant from Science Foundation Ireland to MWR. Carlos Medina is a Science Foundation Ireland Stokes Lecturer.

References

Adcock IM (2000) Molecular mechanisms of glucocorticosteroid actions. *Pulmonary Pharmacology & Therapeutics* 13:115–126.

Agnihotri SA, Mallikarjuna NN, and Aminabhavi TM (2004) Recent advances on chitosan-based micro- and nanoparticles in drug delivery. *Journal of Controlled Release* 100:5–28.

Ahmad Z, Pandey R, Sharma S, and Khuller GK (2008) Novel chemotherapy for tuberculosis: Chemotherapeutic potential of econazole- and moxifloxacin-loaded PLG nanoparticles. *International Journal of Antimicrobial Agents* 31:142–146.

Ali SS, Hardt JI, Quick KL, Sook Kim-Han J, Erlanger BF, Huang T-T, Epstein CJ, and Dugan LL (2004) A biologically effective fullerene (C60) derivative with superoxide dismutase mimetic properties. *Free Radical Biology and Medicine* 37:1191–1202.

Bakowsky H, Richter T, Kneuer C, Hoekstra D, Rothe U, Bendas G, Ehrhardt C, and Bakowsky U (2008) Adhesion characteristics and stability assessment of lectin-modified liposomes for site-specific drug delivery. *Biochimica et Biophysica Acta (BBA)—Biomembranes* 1778:242–249.

Bender AR, von Briesen H, Kreuter J, Duncan IB, and Rubsamen-Waigmann H (1996) Efficiency of nanoparticles as a carrier system for antiviral agents in human immunodeficiency virus-infected human monocytes/macrophages in vitro. *Antimicrobial Agents and Chemotherapy* 40:1467–1471.

Bhadra D, Bhadra S, Jain S, and Jain NK (2003) A PEGylated dendritic nanoparticulate carrier of fluorouracil. *International Journal of Pharmaceutics* 257:111–124.

Bies C, Lehr C-M, and Woodley JF (2004) Lectin-mediated drug targeting: History and applications. *Advanced Drug Delivery Reviews* 56:425–435.

Boman NL, Masin D, Mayer LD, Cullis PR, and Bally MB (1994) Liposomal vincristine which exhibits increased drug retention and increased circulation longevity cures mice bearing P388 tumors. *Cancer Research* 54:2830–2833.

Bosi S, Da Ros T, Spalluto G, and Prato M (2003) Fullerene derivatives: An attractive tool for biological applications. *European Journal of Medicinal Chemistry* 38:913–923.

Bosi S, Feruglio L, Da Ros T, Spalluto G, Gregoretti B, Terdoslavich M, Decorti G, Passamonti S, Moro S, and Prato M (2004) Hemolytic effects of water-soluble fullerene derivatives. *Journal of Medicinal Chemistry* 47:6711–6715.

Brannon-Peppas L and Blanchette JO (2004) Nanoparticle and targeted systems for cancer therapy. *Advanced Drug Delivery Reviews* 56:1649–1659.

Caminade AM, Turrin CO, and Majoral JP (2008) Dendrimers and DNA: Combinations of two special topologies for nanomaterials and biology. *Chemistry—A European Journal* 14:7422–7432.

Caruthers SD, Wickline SA, and Lanza GM (2007) Nanotechnological applications in medicine. *Current Opinion in Biotechnology* 18:26–30.

Chen RJ, Bangsaruntip S, Drouvalakis KA, Wong Shi Kam N, Shim M, Li Y, Kim W, Utz PJ, and Dai H (2003) Noncovalent functionalization of carbon nanotubes for highly specific electronic biosensors. *Proceedings of the National Academy of Sciences of the United States of America* 100:4984–4989.

Chiannilkulchai N, Ammoury N, Caillou B, Devissaguet JP, and Couvreur P (1990) Hepatic tissue distribution of doxorubicin-loaded nanoparticles after i.v. administration in reticulosarcoma M 5076 metastasis-bearing mice. *Cancer Chemotherapy and Pharmacology* 26:122–126.

Devine DV and Bradley AJ (1998) The complement system in liposome clearance: Can complement deposition be inhibited? *Advanced Drug Delivery Reviews* 32:19–39.

di Martino L, Davidson RN, Giacchino R, Scotti S, Raimondi F, Castagnola E, Tasso L et al. (1997) Treatment of visceral leishmaniasis in children with liposomal amphotericin B. *Journal of Pediatric* 131:271–277.

Donaldson K and Tran CL (2004) An introduction to the short-term toxicology of respirable industrial fibres. *Mutation Research/Fundamental and Molecular Mechanisms of Mutagenesis* 553:5–9.

Dupuis M, Murphy TJ, Higgins D, Ugozzoli M, van Nest G, Ott G, and McDonald DM (1998) Dendritic cells internalize vaccine adjuvant after intramuscular injection. *Cellular Immunology* 186:18–27.

Farokhzad OC, Cheng J, Teply BA, Sherifi I, Jon S, Kantoff PW, Richie JP, and Langer R (2006) Targeted nanoparticle-aptamer bioconjugates for cancer chemotherapy in vivo. *Proceedings of the National Academy of Sciences of the United States of America* 103:6315–6320.

Fattal E, Rojas J, Youssef M, Couvreur P, and Andremont A (1991) Liposome-entrapped ampicillin in the treatment of experimental murine listeriosis and salmonellosis. *Antimicrobial Agents and Chemotherapy* 35:770–772.

Fechter LD, Johnson DL, and Lynch RA (2002) The relationship of particle size to olfactory nerve uptake of a non-soluble form of manganese into brain. *Neurotoxicology* 23:177–183.

Fonseca C, Simoes S, and Gaspar R (2002) Paclitaxel-loaded PLGA nanoparticles: Preparation, physicochemical characterization and in vitro anti-tumoral activity. *Journal of Controlled Release* 83:273–286.

Geho DH, Lahar N, Ferrari M, Petricoin EF, and Liotta LA (2004) Opportunities for nanotechnology-based innovation in tissue proteomics. *Biomed Microdevices* 6:231–239.

Gelmon KA, Tolcher A, Diab AR, Bally MB, Embree L, Hudon N, Dedhar C, Ayers D, Eisen A, Melosky B, Burge C, Logan P, and Mayer LD (1999) Phase I study of liposomal vincristine. *Journal of Clinical Oncology* 17:697–705.

Gordon AN, Granai CO, Rose PG, Hainsworth J, Lopez A, Weissman C, Rosales R, and Sharpington T (2000) Phase II study of liposomal doxorubicin in platinum- and paclitaxel-refractory epithelial ovarian cancer. *Journal of Clinical Oncology* 18:3093–3100.

Green AE and Rose PG (2006) Pegylated liposomal doxorubicin in ovarian cancer. *International Journal of Nanomedicine* 1:229–239.

Helmlinger G, Yuan F, Dellian M, and Jain RK (1997) Interstitial pH and pO2 gradients in solid tumors in vivo: High-resolution measurements reveal a lack of correlation. *Nature Medicine* 3:177–182.

Hoffman TL, Canziani G, Jia L, Rucker J, and Doms RW (2000) A biosensor assay for studying ligand-membrane receptor interactions: Binding of antibodies and HIV-1 Env to chemokine receptors. *Proceedings of the National Academy of Sciences of the United States of America* 97:11215–11220.

Hofheinz RD, Gnad-Vogt SU, Beyer U, and Hochhaus A (2005) Liposomal encapsulated anti-cancer drugs. *Anticancer Drugs* 16:691–707.

Jaakohuhta S, Härmä H, Tuomola M, and Lövgren T (2007) Sensitive *Listeria* spp. immunoassay based on europium(III) nanoparticulate labels using time-resolved fluorescence. *International Journal of Food Microbiology* 114:288–294.

Jain KK (2002) Personalized medicine. *Current Opinion in Molecular Therapeutics* 4:548–558.

Jain KK (2005a) Role of nanobiotechnology in developing personalized medicine for cancer. *Technology in Cancer Research & Treatment* 4:645–650.

Jain KK (2005b) The role of nanobiotechnology in drug discovery. *Drug Discovery Today* 10:1435–1442.

Jain KK (2008) Nanomedicine: Application of nanobiotechnology in medical practice. *Medical Principles and Practice* 17:89–101.

Kanter PM, Klaich GM, Bullard GA, King JM, Bally MB, and Mayer LD (1994) Liposome encapsulated vincristine: Preclinical toxicologic and pharmacologic comparison with free vincristine and empty liposomes in mice, rats and dogs. *Anticancer Drugs* 5:579–590.

Kennel SJ, Woodward JD, Rondinone AJ, Wall J, Huang Y, and Mirzadeh S (2008) The fate of MAb-targeted Cd125mTe/ZnS nanoparticles in vivo. *Nuclear Medicine and Biology* 35:501–514.

Kim D, El-Shall H, Dennis D, and Morey T (2005) Interaction of PLGA nanoparticles with human blood constituents. *Colloids and Surfaces B: Biointerfaces* 40:83–91.

Kim S, Lim YT, Soltesz EG, De Grand AM, Lee J, Nakayama A, Parker JA et al. (2004) Near-infrared fluorescent type II quantum dots for sentinel lymph node mapping. *Nature Biotechnology* 22:93–97.

Kipen HM and Laskin DL (2005) Smaller is not always better: Nanotechnology yields nanotoxicology. *American Journal of Physiology. Lung Cellular and Molecular Physiology* 289:L696–L697.

Kocbek P, Obermajer N, Cegnar M, Kos J, and Kristl J (2007) Targeting cancer cells using PLGA nanoparticles surface modified with monoclonal antibody. *Journal of Controlled Release* 120:18–26.

Koziara JM, Oh JJ, Akers WS, Ferraris SP, and Mumper RJ (2005) Blood compatibility of cetyl alcohol/polysorbate-based nanoparticles. *Pharmaceutical Research* 22:1821–1828.

Kumar M, Misra A, Mishra AK, Mishra P, and Pathak K (2008) Mucoadhesive nanoemulsion-based intranasal drug delivery system of olanzapine for brain targeting. *Journal of Drug Targeting* 16:806–814.

Laguna F, Videla S, Jimenez-Mejias ME, Sirera G, Torre-Cisneros J, Ribera E, Prados D et al. (2003) Amphotericin B lipid complex versus meglumine antimoniate in the treatment of visceral leishmaniasis in patients infected with HIV: A randomized pilot study. *The Journal of Antimicrobial Chemotherapy* 52:464–468.

Lambert G, Fattal E, and Couvreur P (2001) Nanoparticulate systems for the delivery of antisense oligonucleotides. *Advanced Drug Delivery Reviews* 47:99–112.

Lammers T, Hennink WE, and Storm G (2008) Tumour-targeted nanomedicines: Principles and practice. *British Journal of Cancer* 99:392–397.

Lanza GM and Wickline SA (2003) Targeted ultrasonic contrast agents for molecular imaging and therapy. *Current Problems in Cardiology* 28:625–653.

Lanza GM, Abendschein DR, Hall CS, Scott MJ, Scherrer DE, Houseman A, Miller JG, and Wickline SA (2000) In vivo molecular imaging of stretch-induced tissue factor in carotid arteries with ligand-targeted nanoparticles. *Journal of the American Society of Echocardiography* 13:608–614.

Lanza GM, Wallace KD, Scott MJ, Cacheris WP, Abendschein DR, Christy DH, Sharkey AM, Miller JG, Gaffney PJ, and Wickline SA (1996) A novel site-targeted ultrasonic contrast agent with broad biomedical application. *Circulation* 94:3334–3340.

Lanza G, Winter P, Cyrus T, Caruthers S, Marsh J, Hughes M, and Wickline S (2006) Nanomedicine opportunities in cardiology. *Annals of the New York Academy of Sciences* 1080:451–465.

Lee LJ (2006) Polymer nano-engineering for biomedical applications. *Annals of Biomedical Engineering* 34:75–88.

Lee RJ and Low PS (1994) Delivery of liposomes into cultured KB cells via folate receptor-mediated endocytosis. *The Journal of Biological Chemistry* 269:3198–3204.

Li Z, Hulderman T, Salmen R, Chapman R, Leonard SS, Young SH, Shvedova A, Luster MI, and Simeonova PP (2007b) Cardiovascular effects of pulmonary exposure to single-wall carbon nanotubes. *Environmental Health Perspectives* 115:377–382.

Li X, Radomski A, Corrigan OI, and Radomski MW (2007a) The effects of poly(lactide-co-glycolide) (PLGA) and chitosan based nanoparticles on human platelet aggregation, in *Pharmaceutical Sciences World Congress*, April 22–25, 2007, Amsterdam, the Netherlands.

Liversidge GG and Cundy KC (1995) Particle size reduction for improvement of oral bioavailability of hydrophobic drugs: I. Absolute oral bioavailability of nanocrystalline danazol in beagle dogs. *International Journal of Pharmaceutics* 125:91–97.

Lundqvist M, Stigler J, Elia G, Lynch I, Cedervall T, and Dawson KA (2008) Nanoparticle size and surface properties determine the protein corona with possible implications for biological impacts. *Proceedings of the National Academy of Sciences* 105:14265–14270.

Mamot C, Drummond DC, Noble CO, Kallab V, Guo Z, Hong K, Kirpotin DB, and Park JW (2005) Epidermal growth factor receptor-targeted immunoliposomes significantly enhance the efficacy of multiple anticancer drugs in vivo. *Cancer Research* 65:11631–11638.

Maruyama K, Ishida O, Takizawa T, and Moribe K (1999) Possibility of active targeting to tumor tissues with liposomes. *Advanced Drug Delivery Reviews* 40:89–102.

Medina C, Santos-Martinez MJ, Radomski A, Corrigan OI, and Radomski MW (2007) Nanoparticles: Pharmacological and toxicological significance. *British Journal of Pharmacology* 150:552–558.

Merisko-Liversidge E, Liversidge GG, and Cooper ER (2003) Nanosizing: A formulation approach for poorly-water-soluble compounds. *European Journal of Pharmaceutical Sciences* 18:113–120.

Merisko-Liversidge EM and Liversidge GG (2008) Drug nanoparticles: Formulating poorly water-soluble compounds. *Toxicologic Pathology* 36:43–48.

Meyer UA and Zanger UM (1997) Molecular mechanisms of genetic polymorphisms of drug metabolism. *Annual Review of Pharmacology and Toxicology* 37:269–296.

Moghimi SM and Szebeni J (2003) Stealth liposomes and long circulating nanoparticles: Critical issues in pharmacokinetics, opsonization and protein-binding properties. *Progress in Lipid Research* 42:463–478.

Moghimi SM, Hunter AC, and Murray JC (2001) Long-circulating and target-specific nanoparticles: Theory to practice. *Pharmacological Reviews* 53:283–318.

Nam J-M, Thaxton CS, and Mirkin CA (2003) Nanoparticle-based bio-bar codes for the ultrasensitive detection of proteins. *Science* 301:1884–1886.

Nemmar A, Hoet PHM, Vanquickenborne B, Dinsdale D, Thomeer M, Hoylaerts MF, Vanbilloen H, Mortelmans L, and Nemery B (2002) Passage of inhaled particles into the blood circulation in humans. *Circulation* 105:411–414.

Oberdörster E (2004) Manufactured nanomaterials (Fullerenes, C60) induce oxidative stress in the brain of juvenile largemouth bass. *Environmental Health Perspectives* 112:1058–1062.

Oh P, Borgstrom P, Witkiewicz H, Li Y, Borgstrom BJ, Chrastina A, Iwata K, Zinn KR, Baldwin R, Testa JE, and Schnitzer JE (2007) Live dynamic imaging of caveolae pumping targeted antibody rapidly and specifically across endothelium in the lung. *Nature Biotechnology* 25:327–337.

Olanow CW (2004) Manganese-induced parkinsonism and Parkinson's disease. *Annals of the New York Academy of Sciences* 1012:209–223.

Ott G, Barchfeld GL, Chernoff D, Radhakrishnan R, van Hoogevest P, and Van Nest G (1995) MF59. Design and evaluation of a safe and potent adjuvant for human vaccines. *Pharmaceutical Biotechnology* 6:277–296.

Panyam J, Zhou W-Z, Prabha S, Sahoo SK, and Labhasetwar V (2002) Rapid endo-lysosomal escape of poly(DL-lactide-co-glycolide) nanoparticles: Implications for drug and gene delivery. *FASEB Journal* 16:1217–1226.

Papatheodoridis GV and Cholongitas E (2004) Chronic hepatitis C and no response to antiviral therapy: Potential current and future therapeutic options. *Journal of Viral Hepatitis* 11:287–296.

Peek LJ, Middaugh CR, and Berkland C (2008) Nanotechnology in vaccine delivery. *Advanced Drug Delivery Reviews* 60:915–928.

Pulini S, Rupoli S, Goteri G, Pimpinelli N, Alterini R, Tassetti A, Scortechini AR et al. (2007) Pegylated liposomal doxorubicin in the treatment of primary cutaneous T-cell lymphomas. *Haematologica* 92:686–689.

Radomski A, Jurasz P, Alonso-Escolano D, Drews M, Morandi M, Malinski T, and Radomski MW (2005) Nanoparticle-induced platelet aggregation and vascular thrombosis. *British Journal of Pharmacology* 146:882–893.

Rimmelzwaan GF, Nieuwkoop N, Brandenburg A, Sutter G, Beyer WEP, Maher D, Bates J, and Osterhaus ADME (2000) A randomized, double blind study in young healthy adults comparing cell mediated and humoral immune responses induced by influenza ISCOM(TM) vaccines and conventional vaccines. *Vaccine* 19:1180–1187.

Sanvicens N and Marco MP (2008) Multifunctional nanoparticles—Properties and prospects for their use in human medicine. *Trends in Biotechnology* 26:425–433.

Sapra P and Allen TM (2002) Internalizing antibodies are necessary for improved therapeutic efficacy of antibody-targeted liposomal drugs. *Cancer Research* 62:7190–7194.

Sapra P, Tyagi P, and Allen TM (2005) Ligand-targeted liposomes for cancer treatment. *Current Drug Delivery* 2:369–381.

Sarris AH, Hagemeister F, Romaguera J, Rodriguez MA, McLaughlin P, Tsimberidou AM, Medeiros LJ, Samuels B, Pate O, Oholendt M, Kantarjian H, Burge C, and Cabanillas F (2000) Liposomal vincristine in relapsed non-Hodgkin's lymphomas: Early results of an ongoing phase II trial. *Annals of Oncology* 11:69–72.

Schmitz SA, Winterhalter S, Schiffler S, Gust R, Wagner S, Kresse M, Coupland SE, Semmler W, and Wolf K-J (2001) USPIO-enhanced direct MR imaging of thrombus: Preclinical evaluation in rabbits. *Radiology* 221:237–243.

Schuppan D and Riecken EO (1990) Molecules of the extracellular matrix: Potential role of collagens and glycoproteins in intestinal adaptation. *Digestion* 46:2.

Senior J and Gregoriadis G (1982) Is half-life of circulating small unilamellar liposomes determined by changes in their permeability?. *FEBS Letter* 145:109–114.

Service RF (2004) Nanotoxicology: Nanotechnology grows up. *Science* 304:1732–1734.

Service RF (2005) Nanotechnology: Calls rise for more research on toxicology of nanomaterials. *Science* 310:1609.

Simões S, Moreira JN, Fonseca C, Düzgünes N, and Pedroso de Lima MC (2004) On the formulation of pH-sensitive liposomes with long circulation times. *Advanced Drug Delivery Reviews* 56:947–965.

Singh N, Cohen CA, and Rzigalinski BA (2007) Treatment of neurodegenerative disorders with radical nanomedicine. *Annals of the New York Academy of Sciences* 1122:219–230.

Svenson S and Chauhan AS (2008) Dendrimers for enhanced drug solubilization. *Nanomedicine* 3:679–702.

Szentkuti L (1997) Light microscopical observations on luminally administered dyes, dextrans, nanospheres and microspheres in the pre-epithelial mucus gel layer of the rat distal colon. *Journal of Controlled Release* 46:233–242.

Tagne J-B, Kakumanu S, and Nicolosi RJ (2008) Nanoemulsion preparations of the anticancer drug dacarbazine significantly increase its efficacy in a xenograft mouse melanoma model. *Molecular Pharmaceutics* 5:1055–1063.

Torchilin VP, Rammohan R, Weissig V, and Levchenko TS (2001) TAT peptide on the surface of liposomes affords their efficient intracellular delivery even at low temperature and in the presence of metabolic inhibitors. *Proceedings of the National Academy of Sciences of the United States of America* 98:8786–8791.

Torchilin VP, Zhou F, and Huang L (1993) pH-Sensitive liposomes. *Journal of Liposome Research* 3:201–255.

Valle JW, Dangoor A, Beech J, Sherlock DJ, Lee SM, Scarffe JH, Swindell R, and Ranson M (2005) Treatment of inoperable hepatocellular carcinoma with pegylated liposomal doxorubicin (PLD): Results of a phase II study. *British Journal of Cancer* 92:628–630.

Vandorpe J, Schacht E, Dunn S, Hawley A, Stolnik S, Davis SS, Garnett MC, Davies MC, and Illum L (1997) Long circulating biodegradable poly(phosphazene) nanoparticles surface modified with poly(phosphazene)-poly(ethylene oxide) copolymer. *Biomaterials* 18:1147–1152.

Verma RK, Kaur J, Kumar K, Yadav AB, and Misra A (2008) Intracellular time course, pharmacokinetics, and biodistribution of Isoniazid and Rifabutin following pulmonary delivery of inhalable microparticles to mice. *Antimicrobial Agents and Chemotherapy* 52:3195–3201.

Vyas SP, Rawat M, Rawat A, Mahor S, and Gupta PN (2006) Pegylated protein encapsulated multivesicular liposomes: A novel approach for sustained release of interferon Î±. *Drug Development and Industrial Pharmacy* 32:699–707.

Wickline SA and Lanza GM (2003) Nanotechnology for molecular imaging and targeted therapy. *Circulation* 107:1092–1095.

Yamawaki H and Iwai N (2006) Mechanisms underlying nanosized air-pollution-mediated progression of atherosclerosis: carbon black causes cytotoxic injury/inflammation and inhibits cell growth in vascular endothelial cells. *Circulation Journal* 70:129–140.

Yang J, Lee CH, Ko HJ, Suh JS, Yoon HG, Lee K, Huh YM, and Haam S (2007) Multifunctional magneto-polymeric nanohybrids for targeted detection and synergistic therapeutic effects on breast cancer13. *Angewandte Chemie International Edition* 46:8836–8839.

Yuan F, Leunig M, Huang SK, Berk DA, Papahadjopoulos D, and Jain RK (1994) Mirovascular permeability and interstitial penetration of sterically stabilized (stealth) liposomes in a human tumor xenograft. *Cancer Research* 54:3352–3356.

Zhang L, Gu FX, Chan JM, Wang AZ, Langer RS, and Farokhzad OC (2007) Nanoparticles in medicine: Therapeutic applications and developments. *Clinical Pharmacology and Therapeutics* 83:761–769.

Zimmer A (1999) Antisense oligonucleotide delivery with polyhexylcyanoacrylate nanoparticles as carriers. *Methods* 18:286–295.

Zuo L, Wei W, Morris M, Wei J, Gorbounov M, and Wei C (2007) New technology and clinical applications of nanomedicine. *Medical Clinics of North America* 91:845–862.

19

Organs from Nanomaterials

Maqsood Ahmed
University College London

Alexander Marcus
Seifalian
University College London
Royal Free Hampstead
NHS Trust Hospital

19.1 Introduction

Considerable progress has been made in medicine and healthcare over the past century resulting in an increase in life expectancy in many countries, generating an aging population. This has led to a sharp increase in age-related health problems. Chronic degenerative diseases, such as myocardial infarction, stroke, arthritis, and atherosclerosis, are associated with tissue degeneration and can result in organ dysfunction. Over 8 million surgical procedures are performed annually to treat tissue or organ failure in the United States alone, with over 80,000 still awaiting transplantation (Langer and Vacanti, 1993). The number of patients awaiting organs always exceeds that of donors with the average waiting time being 5 years in the United States, resulting in 40% of patients dying before a donor is found (UNOS, 2002). Therefore, in order to overcome these limitations, the concept of tissue engineering has emerged.

Tissue engineering is a relatively new branch of medicine that aims to replace, reconstruct, or support organ function by either culturing cells harvested from the patient or donor onto a suitable synthetic material, which is then implanted in the patient's body where tissue regeneration is required, with the scaffold gradually degenerating. Alternatively, the scaffold is directly implanted into the body, stimulating the surrounding tissue to mature and proliferate on the template itself, resulting in *in vivo* tissue regeneration. Research into the synthetic materials, termed biomaterials, has become a key component of tissue engineering. The first generation of biomaterials were sought to be bioinert, eliciting a minimal immune response from the host. Their aim was to achieve good functional properties, such as mechanical strength and durability, adequately matching those of the replaced tissue. The next generation of biomaterials aims to enhance cellular attachment and functions such as growth,

differentiation, and morphogenesis with the scaffold containing chemical and structural information, controlling tissue formation, analogous to cell–cell communication. By controlling the properties of biomaterials at the nanoscale, these scaffolds are becoming a distinct possibility. This chapter will aim to review the various applications of nanomaterials in a tissue engineering context for the development of complex organs including liver and cardiovascular implants.

19.2 Nanomaterials and Nanocomposites

Nanotechnology is a new field of science concerned with the control of matter on an atomic and molecular scale. It generally involves developing materials, devices, or structures 100 nm or smaller. The basic premise of the field is that bulk properties of materials made from nanosized structures vary considerably from the original, resulting in materials with unique physical, mechanical, and chemical properties. Nanocomposites should be clearly differentiated from nanocrystalline and nanophased materials, which refer to single phases in the nanometer range. Nanocomposite refers to composites of more than one independent solid phase, where at least one dimension is in the nanometer range. These solid phases may be amorphous, semicrystalline, or crystalline, or combinations thereof. When the components of a composite are mixed, phase separation normally occurs, resulting in microcomposites (Kannan et al., 2005). The fillers in nanocomposites, however, are able to intercalate between the layers of the matrix, or disperse uniformly even further within the matrix, forming exfoliating nanocomposites and maximizing the surface area for component interaction (Figure 19.1).

Nanocomposites can have unique properties; the nanoscale size of the special hybrids affords the molecular-level control

Phase separated   Intercalated   Exfoliated

FIGURE 19.1 Types of synthesized nanocomposites.

of polymer dynamics, surface bulk properties, and biological functions combining the characteristics of the parent constituents into a single material. The extraordinary versatility of nanocomposites springs from the large selection of biopolymers and fillers available. Existing biopolymers include, but are not limited to, polysaccharides, aliphatic polyesters, polypeptides and proteins, and polynucleic acids, whereas fillers include clays, hydroxyapatite, single- and multi-walled nanotubes, carbon nanofibers, and metal nanoparticles (Hule and Pochan, 2007; Madbouly et al., 2007).

Nanocomposites are prevalent throughout nature in shells, bone, and teeth, which are composed of alternating calcium carbonate ($CaCO_3$) and nanoscale aragonite asperities and, therefore, are 3000 times stronger than monolithic $CaCO_3$ crystals (Zaremba et al., 1996; Gao et al., 2003). This sharp increase in mechanical strength of nanocomposites over their parent constituents is thought to be based on the arrangement of the nanofillers in the soft polymer matrix. While the nanofillers bestow tensile strength to the composite, the interfacial shear attributable to the lubricating proteins allows for efficient load transfer due to the significantly enhanced surface area/volume ratio of the composite (Tiwari et al., 2004). The Griffin criterion states that in composites with fillers below a critical length, the strength of the composite, even if cracked, is virtually equivalent to a solid crystal. This holds true for nanocomposites and explains the relative immunity of nanocomposites to fracture (Gao et al., 2003).

19.3 Surface Nanomorphology

The properties of these nanocomposite materials do not simply depend on the properties of their individual parents but also on their morphology and interfacial characteristics. The influence of surface features or topography on cellular adhesion, growth, migration, and orientation has long been recognized (Curtis and Varde, 1964; Curtis and Wilkinson, 1997, 1999). Mechanoforces, originating from small-scale forces, transmitted directly or indirectly through cell interactions with surface micro and nano-topography, have been implicated in cell signaling (Curtis et al., 2006). The extracellular matrix (ECM) is a natural scaffold that maintains tissue architecture and encourages physiological cell growth into a three-dimensional configuration while providing binding sites for cell adhesion molecules (Dillow and

Tirrell, 1998). The ECM is not a smooth structure; it is covered with grooves, ridges, pits and pores, and fibrillar networks composed mostly of collagen and elastin fibers with diameters ranging from 10 to 300 nm (Wight, 1996). The nanostructure of the ECM is not completely random; there is evidence to suggest that there is a certain degree of order with mesh-like lattices found in skin while orthogonal lattices are found in the cornea; meanwhile tendons, ligaments, and muscles all have parallel aligned fibrils (Kjaer, 2004).

The nanopatterning of tissue engineering scaffolds for organ development is becoming more prominent. The improvements in nanofabricating techniques are advancing the search for a "smart" scaffold, which mimics the nanostructured morphology of the ECM. Techniques such as chemical vapor deposition, photolithography, reactive ion etching, and three-dimensional printing have all proved to be useful but it is a process known as electrospinning that is generating some promising results (Ashammakhi et al., 2008; Nair et al., 2004). Electrospinning is a highly versatile, inexpensive method that produces nanofibers of both natural and synthetic polymers by electrically charging a suspended droplet of polymer melt or solution (Keun Kwon et al., 2005; Shields et al., 2004; Zong et al., 2005). This technique allows for the control of thickness, composition, and porosity of nanofiber meshes with a relatively simple experimental setup. Polymer meshes of fibers, with diameters comparable to the native ECM, can be produced. The high porosity and surface area of nanofibers allows for favorable cell interactions, which is promising for tissue engineering applications.

While these nanomaterials and nanocomposites are still at an early stage of development, this rapidly growing field has attracted the attention of many research groups leading to them having a wide range of potential biomedical applications, in diverse areas, ranging from novel sensing technologies and surface modifications to implant technology summarized in Table 19.1. In tissue engineering, the most significant application of nanomaterials is at the biomaterial–cell interface where it has found widespread applications in reducing the athrombogeneity of scaffolds, increasing cellular attachment, and controlling differentiation while inferring an increase in the mechanical strength of the scaffold.

19.4 Organ Development

19.4.1 Cardiovascular Applications

Vascular disease is the leading cause of death in the Western world (United States Government, 2007). Tissue engineering vascular grafts that can treat fatal arterial occlusions has become a major topic of research. The ideal graft must fulfill several requirements: it must resist narrowing of the lumen by intimal thickening, have comparable biomechanical properties, and possess thromboresistant properties. The endothelium is the thin layer of cells that line the interior surface of blood vessels, maintaining vessel integrity with various dynamic mechanisms preventing intimal hyperplasia (IH) and thrombosis (Bunting et al.,

TABLE 19.1 Some Examples of Nanomaterials and Their Biomedical Applications

Nanomaterial	Size	Application	Function	References
Microstructured hard metal coated with diamond nanolayers	—	Surgical blades	Low physical adhesion to materials or tissues, chemical, and biological inertness, low friction coefficient decreasing penetration force necessary	Vincent and Doting (1989)
Carbon nanotubes	Diameter 30–50 nm, length several μm	Catheters for minimally invasive surgery, vascular implants	Reduced thrombogenicity, improved mechanical properties, i.e., better recovery to original shape, very good handling characteristics, and high resistance to fracture	Gilmore et al. (2007), Hu et al. (2004)
Diamond-like carbon (DLC)	Film thickness: 5–310 nm	Implant coatings for orthopedics, dentistry, and vascular system	Ultrahigh hardness, improved toughness, low friction, good adhesion to titanium alloys, and promising biocompatibility characteristics. Improved hemocompatibility increased EC adhesion	Sheeja et al. (2004), Roy and Lee (2007), Roy et al. (2009), Kobayashi et al. (2005)
Nanocrystalline silver	5–10 nm	Wound dressing, cancer therapy	Effective antibacterial and fungicidal activity, adequate debridement, fluid exudation from the wound	Wright et al. (1999), Wu et al. (2008)
POSS	5 nm	Drug delivery, dental application, cardiovascular and biomedical devices, and tissue engineering	Anti-thrombogenic; improved EC adhesion, biodurability, oxidation resistance, biocompatibility; reduction in flammability, and oxygen permeability	Alobaid et al. (2006), Kannan et al. (2005, 2006a,c), McCusker et al. (2005)
Nano-hydroxyapatite composites with a variety of polymers such as PVA, PET, and natural biopolymers like chitosan	5–50 nm	Bone fixation biomaterial structures, tissue engineering	Improved mechanical, physicochemical, biocompatibility properties. Cell adhesion. Proliferation and spreading morphology all improved. Life time of orthopedic implants could increase to upward of 40 years	Dimitrievska et al. (2008), Nayar et al. (2008)
Polyurethane with gold nanoparticles embedded	4.5–5 nm	Biomedical devices, imaging, and tissue engineering	Better cellular proliferation, lower platelet activation, and reduced bacterial adhesion demonstrated for the PU nanocomposite with 43.5 or 65 ppm of Au than the pure	Copland et al. (2004), Hsu et al. (2008)
Titanium dioxide nanoparticles	50–90 nm	Dental and orthopedic implants	Excellent biocompatibility and bioactivity, enhanced growth of apatite	Cui et al. (2005a), Liu et al. (2005)
Fumed silica nanoparticles with NO donors	5–300 nm	Intravascular sensors, extracorporeal circuit tubings, catheters, and vascular grafts	NO release resulting in reduced platelet activation, adhesion, smooth muscle cell proliferation, bacterial cell adhesion leading to more biocompatible medical devices	Frost and Meyerhoff (2004), Frost et al. (2005), Zhang et al. (2003)

Notes: PVC, poly vinyl chloride; PET, poly ethylene terephthalate.

1977). *In vitro* studies of endothelialising grafts with cultured endothelial cells (ECs) have shown that a confluent endothelium can improve the long-term patency of the graft by preventing thrombogenic complications (Seifalian et al., 2009; Thomas et al., 2003). However, the process of extracting and culturing ECs for *in vitro* endothelialization of grafts is labor intensive requiring a great deal of time and expertise making it expensive and restricted to a few specialized clinics. Therefore, materials that induce *in situ* endothelialization of grafts, without IH and thrombosis formation, are highly desirable (Figure 19.2). The use of nanocomposite materials for graft constructions allows the control of surface morphology for cellular adhesion and growth while maintaining comparable biomechanical properties. Some of the fillers and polymers used in cardiovascular tissue engineering are summarized in Table 19.2.

Polyurethane has been investigated for several decades for developing biomedical devices with a wide range of applications, especially in the cardiovascular field (Tiwari et al., 2002).

Physicochemical   Biofunctionalization

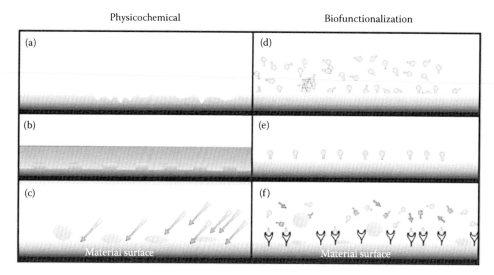

FIGURE 19.2 Examples of various physical: (a) topographical-surface roughening; (b) ordered patterning, chemical: (c) chemical modification, and biofunctionalization techniques: (d) passive coating; (e) covalently linked; (f) peptide linker to enhance *in situ* endothelialization.

TABLE 19.2 Some Examples of Nanoparticles Used in Cardiovascular Devices

Nanoparticle	Size	Function	Comment	References
Carbon nanotubes	Diameter 30–50 nm, length several μm	Improved hemocompatibility	Can apply electric fields—change cell behavior such as viability, orientation, differentiation	Endo et al. (2005), Gilmore et al. (2007), Meng et al. (2005)
Diamond-like carbon (DLC)	Film thickness: 5–310 nm	Improved hemocompatibility increased EC adhesion	Can be doped with numerous elements: Si, N, F to improve properties	Hasebe et al. (2006), Roy et al. (2009)
POSS	5 nm	Anti-thrombogenic, improved EC adhesion, biodurability, biocompatibility	POSS-modified poly carbonate urethane implanted subcutaneously into back of healthy adult sheep for 36 months—no evidence of inflammatory layer or capsule formation	Kannan et al. (2006a,b, 2005), Raghunath et al. (2009), Tiwari et al. (2002)
Bisphosphonate	—	Physical and biomechanical properties comparable to unmodified PUs, calcification resistant	5 month long study with bisphosphonate-modified sheep valve single replacement—no sign of calcium of phosphorus build up	Alferiev et al. (2003)
Nanofibers	Diameter: 100–800 nm, Porosity: 60%–70%	Improved cell spreading, viability and attachment, suitable mechanical properties for regulation of normal cell function in vascular tissue engineering	Nanofibers can be synthesized with a wide range of polymers at various diameters and can be coated with various ECM proteins	Nair et al. (2004), Venugopal et al. (2005)

However, the use of this material for cardiovascular prostheses is hampered by calcification, thrombosis, tearing, and biodegradation. To overcome these problems, a nanocomposite using polyhedral oligomeric silsesquioxane poly(carbonate-urea)urethane (POSS-PCU) has been developed (Kannan et al., 2005). POSS-PCU nanocomposite improves cell adhesion characteristics. As these nanocages occupy minimal volumes within the polymer, relatively greater surface areas of PU are available, which allows for improved endothelialization (Kannan et al., 2006a). Upon adhering, the ECs have good proliferating characteristics and are capable of forming a confluent monolayer. Cell

adhesion and proliferation on the polymer was excellent both for ECs and stem cells (Alobaid et al., 2006). In addition to this, the POSS-PCU nanocomposite was found to be anti-thrombogenic; POSS-PCU has an antiplatelet effect by both repelling their surface adsorption and lowering the binding strength of platelets to the polymer (Kannan et al., 2006b). Calcification is one of the primary causes of heart valve failure. The replacement of a diseased human heart valve is now commonplace, with approximately 275,000 valve replacements done each year and while the exact mechanism of valve calcification is unknown; minor modifications of the biomaterial surface has yielded promising results. The bisphosphonation of the polyurethane backbone, through bromoalkylating the hard segment, allows the localized delivery of bisphophonate, which inhibits the calcification of a polyurethane subdermal implant without impeding growth, development, or function (Alferiev et al., 2003). A bisphosphonate-modified heart valve was placed in a sheep model for 5 months and showed no sign on calcification, inflammation, or host rejection around the site of implantation.

Since their discovery in 1991, carbon nanotubes (CNTs) have found numerous applications in engineering and biomedicine. By incorporating CNTs into poly(styrene-isobutylene-styrene) (SIBS), a triblock copolymer extensively used as a coating for medical devices (Pinchuk et al., 2008), L-929 mouse fibroblast cells were shown to grow exponentially while platelet activation and red blood cell disruption were markedly reduced (Gilmore et al., 2007; Meng et al., 2005). Furthermore, it has been shown that increasing the electrical conductivity of an implantable material results in a change in cell behavior, such as cell viability and migration, and decreases the foreign body response to implanted devices thus making them more biocompatible (Seal et al., 2001; Zhao et al., 2004). Therefore, the CNT/SIBs coating opens up the possibility of controlling cell behavior through the intrinsic conductivity of the material and the ability to apply electric fields to the coatings. The major concern of using CNTs is their cytotoxicity profile (Cui et al., 2005b; Muller et al., 2005). However, the biocompatibility of CNTs with human umbilical vein endothelial cells and neuronal cells has been demonstrated suggesting that there is a potential use for CNTs in biomedical applications (Flahaut et al., 2006; Hu et al., 2004).

Fumed silica (FS) nanoparticles dispersed as inorganic polymeric fillers in polyurethanes and silicon rubber matrixes have also been utilized to fabricate or coat a range of cardiovascular implants (Zhang et al., 2003). FS nanoparticles, ranging in size from 5 to 300 nm, can be functionalized with nitrosthiols and diazeniumdiolates forming potent nitric oxide (NO) donors upon exposure to copper (II) ions, ascorbate, or visible light irradiation (Frost and Meyerhoff, 2004; Frost et al., 2005). NO has been shown to be a powerful inhibitor of platelet activation and adhesion, bacterial cell adhesion, and smooth muscle cell proliferation associated with restenosis and neointimal hyperplasia (Baek et al., 2002; Chaux et al., 1998; Mellion et al., 1981; Nablo et al., 2001; Radomski et al., 1987). Therefore, by anchoring effective NO donors to FS particles, the biocompatibility of many biomedical devices can be enhanced improving their performance considerably. The NO-releasing characteristics are dependent on the specific structure of the alkyl amine attached to the silica surface. However, the major limitations of these NO-releasing nanocomposites is the finite nature of the released NO calling into question there ability to coat long-term implants. Controlling the release rates of NO to achieve a constant flux is also proving to be a challenge (Figure 19.3).

Endothelial cell adhesion and growth has also been demonstrated on materials which mimic the natural nanostructural characteristics of the arterial wall. Biomaterials with nanoscale surface features have been developed and shown to improve cell adhesion, growth, and function (Miller et al., 2004, 2005, 2007; Pezzatini et al., 2008). Poly(lactic-*co*-glycolic acid) (PLGA) was treated with sodium hydroxide (NaOH) and cast onto silastic molds resulting in random and uncontrollable nanoscaled surface features (Miller et al., 2004). Both vascular smooth muscle cell (SMC) and EC densities were improved on treated PLGA surfaces possibly due to an increase in the adsorption of fibronectin and vitronectin, key proteins for mediating cell density on nanostructured PLGA (Miller et al., 2007). The increase in fibronectin and vitronectin adhesion is most likely due to the polymer surface becoming more hydrophilic following NaOH treatment—increased surface hydrophilicity has been shown to promote fibronectin adsorption (van Kooten et al., 2004). In addition to increased surface hydrophilicity, altering

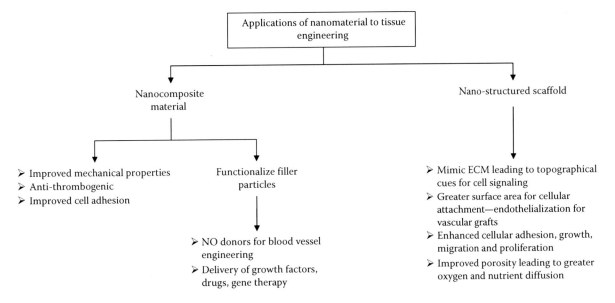

FIGURE 19.3 Overview of the applications of nanomaterials in organ development.

the surface roughness to a nanoscale topography increases the surface charge density, which can also greatly effect the adsorption of proteins (Gessner et al., 2002). Cellular function was also shown to improve; cells grown on nanostructured surfaces were observed to have very long filopodia protruding from the cell body allowing the cell to scout the surrounding area and interact with the nanometer structures (Chung et al., 2003). Furthermore, an increase in matrix metalloproteinases (MMPs), enzymes linked to cell movement and adhesion to substrata, was observed from the supernatant of EC cultured on nanostructured surfaces (Pezzatini et al., 2008).

Nanofibers, produced by electrospinning, are among the most promising scaffolds due to their versatility and their many desirable properties: biocompatibility, ECM-like architecture, high porosity, adjustable mechanical and biodegradable properties, surface modification capabilities, and the ability to load drugs or genes. Aligned or random nanofibers, with a diameter in the range of 3 nm to several micrometers, can be developed (Zhang et al., 2005). By aligning the nanofibers, greater mechanical strength and modulus of nanofiber can be achieved and some degree of control over direction of cell growth can be administered (Xu et al., 2004). By controlling fiber density and thus porosity, control over the diffusion of nutrients and waste products can be maintained (Ishii et al., 2005; Thorvaldsson et al., 2008). Both SMCs and ECs have been shown to attach, grow, and maintain function on nanofibers made from a variety of polymers including polycaprolactone, polyethylene terephthalate, and poly(L-lactid-*co*-ε-caprolactone) (He et al., 2005a,b, 2006; Ma et al., 2005; Venugopal et al., 2005). Electrospinning can be further utilized to overcome the inherent problem of obtaining cell infiltration into the pores of the scaffold by concurrently electrospraying smooth muscle cells while electrospinning the polymer (Stankus et al., 2007).

19.4.2 Bioartificial Liver

Up to 2 million people are reported to be suffering from chronic liver disease in the United Kingdom and deaths from chronic liver disease have increased by eight times in men aged 35–44 years, and seven times in women, over the past three decades (British Liver Trust, 2009). The liver transplant list in the United Kingdom between April 1, 2006, and March 31, 2007, included a total of 1200 patients, of which only 54 received transplantation while 8 died (Nhs Blood and Transplant, 2008, 2009). Thus, it is easy to imagine the impact of an alternative therapy for liver disease by means of implantable or bioartificial liver tissue engineered replacements.

The main focus of attention for hepatic tissue engineering so far has been to recreate various aspects of liver structure which are deemed as important in a perfusion setting: an immediate supply of oxygen and nutrients and the existence of oxygen/nutrient gradients, presence of bile ductules and the recognition of hepatocyte cellular junctions, and polarity and basal membrane contact dependence in the native liver (Gerlach et al., 2008). Isolated hepatocyte tissue culture results in hepatocytes,

which lose their specific function in a matter of hours (Cui et al., 2005a). Therefore, suitable substrates to which hepatocytes can be attached or anchored to must be developed; which can enhance the initial number, survival, and growth of transplanted cells with the key challenge being to maximize cellular function and survival. The ideal scaffold for liver tissue engineering should have the following features:

1. Must be biocompatible
2. High porosity of more than 90%–95%
3. Controllable pore size
4. Larger pores (200–400 μm)
5. Cells must be within 200–300 μm of perfusion medium or blood supply

A suitable substrate combined with cell transplantation can offer many other advantages such as (1) providing sufficient volume for transplantation, (2) the potential to improve efficiency by optimizing shape and composition of the matrices and/or by growth factors and ECMs to the polymeric scaffold, (3) is easily portable and easily handled, (4) potentially preservable, and (5) unlike cell injection, cell transplantation into polymeric scaffolds could be reversible (Bruns et al., 2005; Fiegel et al., 2008; Sellaro et al., 2007; Zavan et al., 2005).

Recently, inverted colloidal crystal (ICC) scaffolds were developed, which allow the control of elasticity, porosity, and pore size (Kotov et al., 2004; Liu et al., 2007; Shanbhag et al., 2005a,b). Colloid crystals are hexagonally packed lattices of microspheres. ICCs are similarly organized structures; however, the microspheres are replaced with a cavity and the interstitial spaces are filled with a polymer of choice resulting in a honeycomb resembling scaffold (Kotov et al., 2004). These biodegradable and highly elastic scaffolds can be synthesized in a wide range of sizes and porosity, which make them an attractive option for liver tissue scaffolds. By varying the microsphere diameter, some control over cellular interactions and migration can also be exercised (Shanbhag et al., 2005a,b).

Upon conventional layered culture, hepatocytes tend to flatten and spread, leading to a disruption of cell cytoskeleton and a loss of cell polarity (Arterburn et al., 1995). By forming a multilayer of hepatocytes between layers of ECM—a sandwich culture—an improvement in the cell polarity and an increase in the density of cell–cell junctions are observed (Moghe et al., 1996). However, an ECM-based sandwich culture is limited in its applications due to the complex composition of the ECM with batch-to-batch variations, mass transfer barriers hampering nutrient exchange and uncontrollable coating of the ECM double layers (Langer and Tirrell, 2004). A synthetic sandwich culture scaffold was synthesized which was able to improve mass transfer while maintaining similar polarity as the ECM-based culture (Du et al., 2008). Furthermore, cell–cell interactions and functionality were also seen to improve. The synthetic sandwich culture consisted of a bottom layer of galactosylated polyethylene terephthalate (PET) followed by a monolayer of hepatocytes and a top support of a GRGDS-modified porous PET track-etched membrane (GRGDS = glycine, arginine, glycine, asparagine, serine).

A multilayered nanofilm for hepatocellular culture was developed using the versatile layer by layer (LBL) method. LBL nanofilms are promising as they are easy to manufacture and their chemical, mechanical, and biofunctional properties can be readily manipulated (Kidambi et al., 2004, 2007; Richert et al., 2004). LBL nanofilms are formed by the adsorption of positively and negatively charged polyelectrolytes, which can then be biofunctionalized through the adsorption of proteins and biomolecules capable of transmitting signals to contracting cells (Jessel et al., 2006; Wittmer et al., 2007). The mechanical properties of these LBL nanofilms can be well controlled through the choice of polymer, solution conditions (pH and ionic strength), number of layers, and post-formation chemical cross-linking; which is a useful attribute as hepatocytes are well known to be highly sensitive to various substrate properties (Semler and Moghe, 2001; Semler et al., 2000, 2005).

Another method of maintaining a high level of cellular function *in vivo* is to form spheroids—spherical, multicellular aggregates—which resemble tissue and contain a high degree of cell–cell contacts (Khalil et al., 2001; Landry et al., 1985). These spheroids have been shown to maintain viability for extended periods of time and support liver-specific functions including albumin production, urea synthesis, and cytochrome P450 activity (Landry et al., 1985; Lin et al., 1995). Spheroids have been seen to form when hepatocytes are cultured on nanofiber scaffolds made from galactosylated poly(ε-caprolactone-*co*-ethyl ethylene phosphate) (Chua et al., 2005). The use of a nanofiber mesh allows the hepatocyte to aggregate around and within the mesh enhancing the overall cell–substrate interaction. While spheroids are a vast improvement on monolayer and sandwich culture, nutrient supply is still by diffusion; therefore they still suffer from a lack of supply of oxygen and nutrients if the spheroids are greater than 100 μm (Glicklis et al., 2004). They lack the close cell proximity to a sinusoid that could continuously replenish nutrients and oxygen.

While considerable progress has been made in the search of a bioartificial liver, with some examples shown in Table 19.3, there is still a limited understanding of the long-term *in vitro* and *in vivo* characterization of porous 3D scaffolds. Specifically, the long-term effect of the incorporation of inorganic bioactive phases on the degradation and ion release kinetics of these highly porous systems. In this regard, the development of appropriate characterization techniques coupled with predictive analytical models is mandatory in order to be able to comprehensively assess the degradation of these systems with respect to pore structures. This also includes research directed at assessing the suitability of bioactive composite scaffolds for enhancing the angiogenesis and vascularization of tissue/scaffold constructs (Boccaccini and Gough, 2007; Glicklis et al., 2004).

TABLE 19.3 Some Examples of Liver Tissue Engineering Experiments Using Nanomaterial Scaffolds

Nanomaterial Scaffold	Porosity/Pore Size	Biodegradable	Cell Type	Comment	References
Sodium silicate inverted colloidal crystal	Variable pore diameter of 10, 75, and 160 μm	No	Human HEP G2	Wide variety of pore sizes and porosity can be synthesized	Kotov et al. (2004), Liu et al. (2007), Shanbhag et al. (2005a,b)
Galactosylated poly(ε-caprolactone-*co*-ethyl ethylenephosphate) film and nanofiber	Nanofiber with diameter 760 nm	Yes	Rat hepatocytes	Hepatocytes cultured on nanofiber, as opposed to a film, engulf nanofiber and form spheroids	Chua et al. (2005)
GRGDS-polyethylene terephthalate (PET)(top support) and galactosylated PET(bottom support)	Density: 3×10^7 pores/cm² pore diameter: 800 nm	No	Rat hepatocytes	Improved cell–cell interaction and differentiation over 14 day culture than collagen sandwich culture	Du et al. (2008)
Poly(ester amide) with nanofabricated collagen	—	No	Rat hepatocytes	Nanotopography of collagen improves adhesion of cells	Bettinger et al. (2008)
Braided PET and polysulfone hollow nanofibers	Variable pore size: 160–320 μm	No	Rat hepatocytes	Able to control pore size by choice of braiding angle	Hoque et al. (2007)
Organic–inorganic hybrid scaffold from polydimethylsiloxane and tetraethoxysilane (PDMS-TEOS)	Pore size 150–212 μm	No	Human HEP G2	Much improved albumin secretion and cell clustering	Kataoka et al. (2005)
Poly-lactic-*co*-glycolic acid (PLGA) coated with collagen/gelatine/oxygen plasma	Varied depending on the combination (50–500 μm)	Yes	Rat hepatocytes	Considerably improved cell density in perfusion conditions rather than static	Hasirci et al. (2001)
Poly (DL-lactic acid) substratum (PLA)	—	Yes	Rat hepatocytes	Rapid self-organization of 3D spheroids. Spheroids formed exhibit hepatocyte-specific functionality (CYP-450 activity and albumin secretion) after almost 2 months in static culture	Riccalton-Banks et al. (2003)

19.4.3 Bone Regeneration

Bone disease presents a serious clinical problem. Osteoporosis affects an estimated 75 million people in Europe, United States, and Japan. In the year 2000, there were an estimated 9 million new osteoporotic fractures, of which 1.6 million were at the hip, 1.7 million were at the forearm, and 1.4 million were clinical vertebral fractures. Europe and the Americas accounted for 51% of all these fractures, while most of the remainder occurred in the Western Pacific region and Southeast Asia. This emphasizes the need to develop treatments to target better bone health. Ideally, the solution to most of these problems would be whole orthopedic implants replacing the diseased bone. Conventional implants made of titanium and its associated alloys have shown that current limitations include donor-site morbidity, potential risks of rejection and infection, and high cycle fatigue failure due to loading over many years (Johnell and Kanis, 2006; Teoh, 2000). When considering load-bearing orthopedic implants, the following must be taken into account: (1) Osseointegration—the ability of the implant to integrate with surrounding bone so as to ensure that the implant does not loosen. (2) Biocompatibility—nontoxic and not cause any inflammatory or immunological response. (3) High corrosion and wear resistance—ensure longevity of implant in patient without the need to revise surgery. (4) Mechanical properties—suitable hardness, tensile strength, modulus, and elongation (Geetha et al., 2009).

Bone tissue engineering strategies aim to regenerate natural bone tissue at the site of a severe bone defect by using a biomaterial scaffold on which the appropriate cells and/or growth factors are encouraged to grow. For the scaffold to be effective, it must be biocompatible and have suitable mechanical properties and degradability. The degradation behavior of the polymer scaffold can be altered with the introduction of nanoparticles. Degradation of conventional PLGA scaffolds produces products which lowers the local pH and induces an inflammatory reaction, damaging bone cell health at the implant site. Titania nanoparticles dispersed into a PLGA scaffold has been shown to decrease the harmful changes in pH for PLGA degradation, hence improving the biocompatibility of the implant by reducing the inflammatory response (Liu et al., 2006).

A wide variety of nanoparticles have been used as fillers in polymer matrixes to improve the effectiveness of the scaffold in bone tissue engineering applications. Titanium dioxide (TiO_2) nanoparticles have been shown to be able to mimic the nanometer surface topography and roughness found in osseous tissue; poly(D, L lactic acid) (PDLLA)-based composite films filled with different TiO_2 nanoparticle concentrations showed an improved homogeneous distribution of non-stoichiometric hydroxyapatite (HA) nanocrystals (Gerhardt et al., 2007). A hybrid composite polymer consisting of fumed silica nanoparticles and 2-hydroxyethylmethacrylate (HEMA) showed improved mechanical properties and enhanced cell adhesion when assayed with primary cultures of human osteoblasts (OB), which might lead to a more efficient implant fabrication process (Schiraldi et al., 2004). Functionalized nanoparticles can also be embedded into the polymer scaffolds allowing us to deliver specific substances, which can influence cellular growth. A study combining dexamethasone-loaded carboxymethylchitosan/poly(amidoamine) dendrimer (Dex-loaded CMCht/PAMAM) nanoparticles with both HA and SPCL scaffolds showed that we are able to directly influence cellular fate with the delivery of dexamethasone inside the cells (Oliveira et al., 2009). CNTs are well known to give added strength to the polymeric scaffold; however, it was also shown that by embedding single-walled carbon nanotubes (SWCNTs) into poly propylene fumarate (PPF), the density of bone tissue ingrowth was substantially denser. After a 12 week implantation in a rabbit model, the SWCNT-modified PPF showed a threefold greater bone tissue ingrowth compared to the controls (Sitharaman et al., 2008).

Another focus of bone tissue engineering is the use of nanofibers to construct a three-dimensional scaffold on which osteogenic cells might be grafted on. Both bioactive ceramics and polymers have been developed and analyzed for use as tissue engineering scaffolds. Osteogenesis is able to occur as bioactive ceramics, which are chemically similar to natural bone, are able to provide a bony contact to form bonds with the host bone (Hench and Wilson, 1984). Chitosan has also been extensively used in bone tissue engineering as it can be molded into porous structures which allow osteoconduction (Di Martino et al., 2005). The limitations associated with the independent use of chitosan can be overcome by introducing HA to mimic the natural structure of bone as a nanocomposite of minerals and proteins. It was demonstrated that there were more pronounced cytoplasmic extensions and formation of extracellular matrix when mouse pre-osteoblasts were grafted on these modified chitosan-nHA scaffolds (Di Martino et al., 2005; Thein-Han and Misra, 2009). Further fabrication of biomimetic HA/chitosan-gelatin network composites in the form of 3D-porous scaffolds shows improved adhesion, expression, and proliferation of rat calvaria osteoblasts (Zhao et al., 2002). Self-assembled nanofibers and nanotubes have also been used for bone tissue engineering purposes. A recent experiment has shown that a peptide-amphiphile (PA) with cell-adhesive ligand RGD, capable of self-assembly into a nanostructured fibrous scaffold, is able to better direct the mineralization of HA to form a composite material where it mimics the natural alignment of collagen fibrils and HA crystals in bone (Hartgerink et al., 2001). The fabrication of a biodegradable nano-HA/polyphosphazene microsphere 3-D scaffold which offers cytocompatibilty and suitable mechanical properties has also demonstrated good osteoblast cell adhesion and alkaline phosphatase expression (Nukavarapu et al., 2008). Electrospraying of HA nanoparticles on electrospun poly(L-lactic acid)-*co*-poly(ε-caprolactone) and gelatin nanofibers (PLACL/gel) also shows enhanced human fetal osteoblast cells (hFOB) proliferation and enzyme activity, suggesting that the nanoparticles create a better osteophilic environment for growth and mineralization of osteoblasts (Gupta et al., 2009).

One challenge of bone tissue engineering is to improve the bonding of an orthopedic implant with juxtaposed bone. An important criterion for an orthopedic implant, as mentioned above, is the formation of adequate osseointegration between the biomaterial implant and bone tissue. It has been shown that it is possible to have enhanced resorption activity by osteoclast-like cells cultured on nanophase ceramics (Webster et al., 2001). One common implant failure mechanism is due to the extensive formation of scar tissue around the implant. Nanomaterials have the potential to overcome this limitation. The most common bone cement material used clinically today is poly(methyl methacrylate), (PMMA) (Gupta et al., 2009; Li et al., 1995). The addition of magnesium oxide (MgO) nanoparticles into the PMMA has been shown to drastically decrease the temperature during PMMA solidification reducing the damaging effects the exothermic reaction on surrounding bone thus improving the degree of osseointegration (Ricker et al., 2008). Barium sulfate nanoparticles, replacing barium sulfate microparticles, have also been introduced in PMMA cements and have shown a twofold increase in the fatigue life of the commercial acrylic bone cement (PMMA), improving the longevity of the implant (Gomoll et al., 2008).

19.4.4 Nanotoxicology

With the ever-increasing use of nanoparticles and nanomaterials, the adverse health effects have come under the spotlight. Nanotechnology allows us to exploit its unique properties of high surface reactivity and ability to cross cell membranes; however, the very same properties could have severe side effects, which if left unchecked, could have drastic implications on the environment and human health and safety (Günter Oberdörster and Oberdörster, 2005; Nel et al., 2006; Vallyathan and Gwinn, 2006).

There are numerous mechanisms with which nanomaterials and nanoparticles can influence cells including interacting with the surface lipid membrane, cell receptors, and entering the cell via passive or active processes. Once in the cell, the critical component of toxicity of the nanomaterial is its oxidative potential (Nel et al., 2006; Vallyathan and Gwinn, 2006). *In vitro* studies using a wide range of cells (lung, liver, skin) have shown that increased oxidative stress reduces the cells antioxidant defense mechanisms and can lead to DNA damage (Gurr et al., 2005; Sagai et al., 1993; Sayes et al., 2006b). Lipid peroxidation and increased membrane permeability have also been associated with nanomaterials such as fullerenes (Sayes et al., 2005). However, its also been demonstrated that functionalizing certain nanomaterials such as fullerenes or CNTs can drastically alter its toxicology profile, as can its fiber aspect length/diameter ratio (Dumortier et al., 2006; Magrez et al., 2006; Sayes et al., 2006a). Therefore, it is clear that the interaction of nanosized materials with biological systems is not a simple, straightforward process that can be readily predicted.

Regulatory agencies have a considerable role to play in recognizing that nanosized substances differ significantly from conventional-sized substances and thus while the toxicology of the base material may be well defined, the nanosized form may behave markedly differently. Detailed investigations should be conducted to evaluate toxicology profiles of nanosized materials resulting in some governmental regulations for the use of nanotechnology with efforts such as the International Council on Nanotechnology (ICON 2004) being encouraged and supported.

19.5 Summary

With the continued shortage of organs for transplantation, tissue engineering provides a novel solution. The use of nanomaterials in tissue engineering and organ development has quickly gathered pace and has become an intensely researched field. Designing athrombogenic surfaces for vascular grafts and nanostructured, nanopatterned biomimicking scaffolds are becoming well established leading to improved cell attachment, growth, differentiation, and morphogenesis. Technological advances, such as electrospinning, have led to rapid biofabrication and tissue assembly replacing the expensive and time-consuming bioreactor-based methods. However, there are still numerous challenges to overcome before tissue engineered organs become a reality. Suitable cell sources need to be identified, the rules governing cell growth and differentiation need to be understood and suitable scaffolds, which are capable of providing the necessary oxygen and nutrients to densely packed cells in whole organs, need to be developed. To overcome these problems, a multidisciplinary approach needs to be taken with life scientists working hand in hand with engineers, material scientists, and mathematicians.

References

Alferiev, I., Stachelek, S. J., Lu, Z., Fu, A. L., Sellaro, T. L., Connolly, J. M., Bianco, R. W., Sacks, M. S., and Levy, R. J. 2003, Prevention of polyurethane valve cusp calcification with covalently attached bisphosphonate diethylamino moieties, *J. Biomed. Mater. Res. A*, 66(2), 385–395.

Alobaid, N., Salacinski, H. J., Sales, K. M., Ramesh, B., Kannan, R. Y., Hamilton, G., and Seifalian, A. M. 2006, Nanocomposite containing bioactive peptides promote endothelialisation by circulating progenitor cells: An in vitro evaluation, *Eur. J. Vasc. Endovasc. Surg.*, 32(1), 76–83.

Arterburn, L. M., Zurlo, J., Yager, J. D., Overton, R. M., and Heifetz, A. H. 1995, A morphological study of differentiated hepatocytes in vitro, *Hepatology*, 22(1), 175–187.

Ashammakhi, N., Ndreu, A., Nikkola, L., Wimpenny, I., and Yang, Y. 2008, Advancing tissue engineering by using electrospun nanofibers, *Regen. Med.*, 3(4), 547–574.

Baek, S. H., Hrabie, J. A., Keefer, L. K., Hou, D., Fineberg, N., Rhoades, R., and March, K. L. 2002, Augmentation of intrapericardial nitric oxide level by a prolonged-release nitric oxide donor reduces luminal narrowing after porcine coronary angioplasty, *Circulation*, 105(23), 2779–2784.

Bettinger, C. J., Kulig, K. M., Vacanti, J. P., Langer, R., and Borenstein, J. T. 2008, Nanofabricated Collagen-Inspired Synthetic Elastomers for Primary Rat Hepatocyte Culture, *Tissue Eng Part A*, 15(6), 1321–1329.

Boccaccini, A. R. and Gough, J. E. 2007, *Tissue Engineering Using Ceramics and Polymers*. Cambridge, U.K.: Woodhead Publishing and Maney Publishing on behalf of the Institute of Materials, Minerals & Mining.

British Liver Trust. 2009, Facts about liver disease. http://www.britishlivertrust.org.uk/home/media-centre/facts-about-liver-disease.aspx

Bruns, H., Kneser, U., Holzhuter, S., Roth, B., Kluth, J., Kaufmann, P. M., Kluth, D., and Fiegel, H. C. 2005, Injectable liver: A novel approach using fibrin gel as a matrix for culture and intrahepatic transplantation of hepatocytes, *Tissue Eng.*, 11(11–12), 1718–1726.

Bunting, S., Moncada, S., and Vane, J. R. 1977, Antithrombotic properties of vascular endothelium, *The Lancet*, 310, 1075–1076.

Chaux, A., Ruan, X. M., Fishbein, M. C., Ouyang, Y., Kaul, S., Pass, J. A., and Matloff, J. M. 1998, Perivascular delivery of a nitric oxide donor inhibits neointimal hyperplasia in vein grafts implanted in the arterial circulation, *J. Thorac. Cardiovasc. Surg.*, 115(3), 604–612.

Chua, K. N., Lim, W. S., Zhang, P., Lu, H., Wen, J., Ramakrishna, S., Leong, K. W., and Mao, H. Q. 2005, Stable immobilization of rat hepatocyte spheroids on galactosylated nanofiber scaffold, *Biomaterials*, 26, 2537–2547.

Chung, T. W., Liu, D. Z., Wang, S. Y., and Wang, S. S. 2003, Enhancement of the growth of human endothelial cells by surface roughness at nanometer scale, *Biomaterials*, 24, 4655–4661.

Copland, J. A., Eghtedari, M., Popov, V. L., Kotov, N., Mamedova, N., Motamedi, M., and Oraevsky, A. A. 2004, Bioconjugated gold nanoparticles as a molecular based contrast agent: implications for imaging of deep tumors using optoacoustic tomography, *Mol Imaging Biol*, 6, 341–349.

Cui, C., Liu, H., Li, Y., Sun, J., Wang, R., Liu, S., and Lindsay Greer, A. 2005a, Fabrication and biocompatibility of nano-TiO$_2$/titanium alloys biomaterials, *Mater. Lett.*, 59(24–25), 3144–3148.

Cui, D., Tian, F., Ozkan, C. S., Wang, M., and Gao, H. 2005b, Effect of single wall carbon nanotubes on human HEK293 cells, *Toxicol. Lett.*, 155(1), 73–85.

Curtis, A. S. and Varde, M. 1964, Control of cell behavior: Topological factors, *J. Natl. Cancer Inst.*, 33, 15–26.

Curtis, A. and Wilkinson, C. 1997, Topographical control of cells, *Biomaterials*, 18(24), 1573–1583.

Curtis, A. and Wilkinson, C. 1999, New depths in cell behaviour: Reactions of cells to nanotopography, *Biochem. Soc. Symp.*, 65, 15–26.

Curtis, A. S., Dalby, M., and Gadegaard, N. 2006, Cell signaling arising from nanotopography: Implications for nanomedical devices, *Nanomedicine*, 1(1), 67–72.

Di Martino, A., Sittinger, M., and Risbud, M. V. 2005, Chitosan: A versatile biopolymer for orthopaedic tissue-engineering, *Biomaterials*, 26(30), 5983–5990.

Dillow A. K. and Tirrell M. 1998, Targeted cellular adhesion at biomaterial interfaces, *Curr. Opin. Solid State and Mater. Sci.*, 3(3), 252–259.

Dimitrievska, S., Petit, A., Ajji, A., Bureau, M. N., and Yahia, L. 2008, Biocompatibility of novel polymer-apatite nanocomposite fibers, *J Biomed Mater Res A*, vol. 84(1), 44–53.

Du, Y., Han, R., Wen, F., Ng San San, S., Xia, L., Wohland, T., Leo, H. L., and Yu, H. 2008, Synthetic sandwich culture of 3D hepatocyte monolayer, *Biomaterials*, 29(3), 290–301.

Dumortier, H., Lacotte, S., Pastorin, G., Marega, R., Wu, W., Bonifazi, D., Briand, J. P., Prato, M., Muller, S., and Bianco, A. 2006, Functionalized carbon nanotubes are non-cytotoxic and preserve the functionality of primary immune cells, *Nano Lett.*, 6(7), 1522–1528.

Christransen, C., Lindsay, R., Melton, L. J., Genant, H. Kanis, J., Glfier, C. C., Delmas, P. D., Riggs, B. L., Burckhardt, P., and Martin, T. J. 1997, Who are candidates for prevention and treatment for osteoporosis? *Osteoporos. Int.*, 7(1), 1–6.

Endo, M., Koyama, S., Matsuda, Y., Hayashi, T., and Kim, Y. A. 2005, Thrombogenicity and blood coagulation of a micro-catheter prepared from carbon nanotube-nylon-based composite, *Nano Lett.*, 5(1), 101–105.

Fiegel, H. C., Kaufmann, P. M., Bruns, H., Kluth, D., Horch, R. E., Vacanti, J. P., and Kneser, U. 2008, Hepatic tissue engineering: From transplantation to customized cell-based liver directed therapies from the laboratory, *J. Cell Mol. Med.*, 12(1), 56–66.

Flahaut, E., Durrieu, M. C., Remy-Zolghadri, M., Bareille, R., and Baquey, C. 2006, Investigation of the cytotoxicity of CCVD carbon nanotubes towards human umbilical vein endothelial cells, *Carbon*, 44(6), 1093–1099.

Frost, M. C. and Meyerhoff, M. E. 2004, Controlled photoinitiated release of nitric oxide from polymer films containing S-nitroso-N-acetyl-DL-penicillamine derivatized fumed silica filler, *J. Am. Chem. Soc.*, 126(5), 1348–1349.

Frost, M. C., Reynolds, M. M., and Meyerhoff, M. E. 2005, Polymers incorporating nitric oxide releasing/generating substances for improved biocompatibility of blood-contacting medical devices, *Biomaterials*, 26(14), 1685–1693.

Gao, H., Ji, B., Jäger, I. L., Arzt, E., and Fratzl, P. 2003, Materials become insensitive to flaws at nanoscale: Lessons from nature, *Proc. Natl. Acad. Sci. U.S.A.*, 100(10), 5597–5600.

Geetha, M., Singh, A. K., Asokamani, R., and Gogia, A. K. 2009, Ti based biomaterials, the ultimate choice for orthopaedic implants—A review, *Prog. Mater. Sci.*, 54(3), 397–425.

Gerhardt, L. C., Jell, G., and Boccaccini, A. 2007, Titanium dioxide (TiO$_2$) nanoparticles filled poly(d,l lactid acid) (PDLLA) matrix composites for bone tissue engineering, *J. Mater. Sci.: Mater. Med.*, 18(7), 1287–1298.

Gerlach, J. C., Zeilinger, K., and Patzer II, J. F. 2008, Bioartificial liver systems: Why, what, whither?, *Reg. Med.*, 3(4), 575.

Gessner, A., Lieske, A., Paulke, B. R., and Müller, R. H. 2002, Influence of surface charge density on protein adsorption on polymeric nanoparticles: Analysis by two-dimensional electrophoresis, *Eur. J. Pharm. Biopharm.*, 54(2), 165–170.

Gilmore, K. J., Moulton, S. E., and Wallace, G. G. 2007, Incorporation of carbon nanotubes into the biomedical polymer poly(styrene-[beta]-isobutylene-[beta]-styrene), *Carbon*, 45(2), 402–410.

Glicklis, R., Merchuk, J. C., and Cohen, S. 2004, Modeling mass transfer in hepatocyte spheroids via cell viability, spheroid size, and hepatocellular functions, *Biotechnol. Bioeng.*, 86(6), 672–680.

Gomoll, A. H., Fitz, W., Scott, R. D., Thornhill, T. S., and Bellare, A. 2008, Nanoparticulate fillers improve the mechanical strength of bone cement, *Acta Orthopaedica*, 79(3), 421–427.

Günter Oberdörster, E. O. and Oberdörster, J. 2005, Nanotoxicology: An emerging discipline evolving from studies of ultrafine particles, *Environ. Health Perspect.*, 113(7), 823–839.

Gupta, D., Venugopal, J., Mitra, S., Giri Dev, V. R., and Ramakrishna, S. 2009, Nanostructured biocomposite substrates by electrospinning and electrospraying for the mineralization of osteoblasts, *Biomaterials*, 30(11), 2085–2094.

Gurr, J. R., Wang, A. S. S., Chen, C. H., and Jan, K. Y. 2005, Ultrafine titanium dioxide particles in the absence of photoactivation can induce oxidative damage to human bronchial epithelial cells, *Toxicology*, 213(1–2), 66–73.

Hartgerink, J. D., Beniash, E., and Stupp, S. I. 2001, Self-assembly and mineralization of peptide-amphiphile nanofibers, *Science*, 294(5547), 1684–1688.

Hasebe, T., Shimada, A., Suzuki, T., Matsuoka, Y., Saito, T., Yohena, S., Kamijo, A., Shiraga, N., Higuchi, M., Kimura, K., Yoshimura, H., and Kuribayashi, S. 2006, Fluorinated diamond-like carbon as antithrombogenic coating for blood-contacting devices, *J. Biomed. Mater. Res. A*, 76(1), 86–94.

Hasirci, V., Berthiaume, F., Bondre, S. P., Gresser, J. D., Trantolo, D. J., Toner, M., and Wise, D. L. 2001, Expression of Liver-Specific Functions by Rat Hepatocytes Seeded in Treated Poly(Lactic-co-Glycolic) Acid Biodegradable Foams, *Tissue Eng*, 7(4), 385.

He, W., Ma, Z., Yong, T., Teo, W. E., and Ramakrishna, S. 2005a, Fabrication of collagen-coated biodegradable polymer nanofiber mesh and its potential for endothelial cells growth, *Biomaterials*, 26(36), 7606–7615.

He, W., Yong, T., Teo, W. E., Ma, Z., and Ramakrishna, S. 2005b, Fabrication and endothelialization of collagen-blended biodegradable polymer nanofibers: Potential vascular graft for blood vessel tissue engineering, *Tissue Eng.*, 11(9–10), 1574.

He, W., Yong, T., Ma, Z. W., Inai, R., Teo, W. E., and Ramakrishna, S. 2006, Biodegradable polymer nanofiber mesh to maintain functions of endothelial cells, *Tissue Eng.*, 12(9), 2457.

Hench, L. L. and Wilson, J. 1984, Surface-active biomaterials, *Science*, 226(4675), 630–636.

Hoque, M. E., Mao, H. Q., and Ramakrishna, S. 2007, Hybrid braided 3-D scaffold for bioartificial liver assist devices, *J Biomater Sci Polym Ed*, 18(1), 45–58.

Hsu, S. H., Tang, C. M., and Tseng, H. J. 2008, Gold Nanoparticles Induce Surface Morphological Transformation in Polyurethane and Affect the Cellular Response, *Biomacromolecules*, 9(1), 241–248.

Hu, H., Ni, Y., Montana, V., Haddon, R. C., and Parpura, V. 2004, Chemically functionalized carbon nanotubes as substrates for neuronal growth, *Nano Lett.*, 4(3), 507–511.

Hule, R. A. and Pochan, D. 2007, Polymer nanocomposites for biomedical applications, *MRS Bull.*, 32(354), 358.

Ishii, O., Shin, M., Sueda, T., and Vacanti, J. P. 2005, In vitro tissue engineering of a cardiac graft using a degradable scaffold with an extracellular matrix-like topography, *J. Thorac. Cardiovasc. Surg.*, 130(5), 1358–1363.

Jessel, N., Oulad-Abdelghani, M., Meyer, F., Lavalle, P., Haîkel, Y., Schaaf, P., and Voegel, J. C. 2006, Multiple and time-scheduled in situ DNA delivery mediated by Î²-cyclodextrin embedded in a polyelectrolyte multilayer, *Proc. Natl. Acad. Sci.*, 103(23), 8618–8621.

Johnell, O. and Kanis, J. A. 2006, An estimate of the worldwide prevalence and disability associated with osteoporotic fractures, *Osteoporos. Int.*, 17(12), 1726–1733.

Kannan, R. Y., Salacinski, H. J., De Groot, J., Clatworthy, I., Bozec, L., Horton, M., Butler, P. E., and Seifalian, A. M. 2006, The Antithrombogenic Potential of a Polyhedral Oligomeric Silsesquioxane (POSS) Nanocomposite, *Biomacromolecules*, 7(1), 215–223.

Kannan, R. Y., Salacinski, H. J., Butler, P. E., and Seifalian, A. M. 2005, Polyhedral oligomeric silsesquioxane nanocomposites: The next generation material for biomedical applications, *Acc. Chem. Res.*, 38(11), 879–884.

Kannan, R., Salacinski, H., Sales, K., Butler, P., and Seifalian, A. 2006a, The endothelization of polyhedral oligomeric silsesquioxane nanocomposites, *Cell Biochem. Biophys.*, 45(2), 129–136.

Kannan, R. Y., Salacinski, H. J., De Groot, J., Clatworthy, I., Bozec, L., Horton, M., Butler, P. E., and Seifalian, A. M. 2006, The antithrombogenic potential of a polyhedral oligomeric silsesquioxane (POSS) nanocomposite, *Biomacromolecules*, 7(1), 215–223.

Kataoka, K., Nagao, Y., Nukui, T., Akiyama, I., Tsuru, K., Hayakawa, S., Osaka, A., and Huh, N. H. 2005, An organic-inorganic hybrid scaffold for the culture of HepG2 cells in a bioreactor, *Biomaterials*, 26(15), 2509–2516.

Keun Kwon, I., Kidoaki, S., and Matsuda, T. 2005, Electrospun nano- to microfiber fabrics made of biodegradable copolyesters: Structural characteristics, mechanical properties and cell adhesion potential, *Biomaterials*, 26(18), 3929–3939.

Khalil, M., Shariat-Panahi, A., Tootle, R., Ryder, T., McCloskey, P., Roberts, E., Hodgson, H., and Selden, C. 2001, Human hepatocyte cell lines proliferating as cohesive spheroid colonies in alginate markedly upregulate both synthetic and detoxificatory liver function, *J. Hepatol.*, 34(1), 68–77.

Kidambi, S., Lee, I., and Chan, C. 2004, Controlling primary hepatocyte adhesion and spreading on protein-free polyelectrolyte multilayer films, *J. Am. Chem. Soc.*, 126(50), 16286–16287.

Kidambi, S., Sheng, L., Yarmush, M. L., Toner, M., Lee, I., and Chan, C. 2007, Patterned co-culture of primary hepatocytes and fibroblasts using polyelectrolyte multilayer templates, *Macromol. Biosci.*, 7(3), 344–353.

Kjaer, M. 2004, Role of extracellular matrix in adaptation of tendon and skeletal muscle to mechanical loading, *Physiol. Rev.*, 84(2), 649–698.

Kobayashi, S., Ohgoe, Y., Ozeki, K., Sato, K., Sumiya, T., Hirakuri, K. K., and Aoki, H. 2005, Diamond-like carbon coatings on orthodontic archwires, *Diamond & Related Materials*, vol. 14, pp. 1094–1097.

Kotov, N. A., Liu, Y., Wang, S., Cumming, C., Eghtedari, M., Vargas, G., Motamedi, M., Nichols, J., and Cortiella, J. 2004, Inverted colloidal crystals as three-dimensional cell scaffolds, *Langmuir*, 20(19), 7887–7892.

Landry, J., Bernier, D., Ouellet, C., Goyette, R., and Marceau, N. 1985, Spheroidal aggregate culture of rat liver cells: Histotypic reorganization, biomatrix deposition, and maintenance of functional activities, *J. Cell Biol.*, 101(3), 914–923.

Langer, R. and Tirrell, D. A. 2004, Designing materials for biology and medicine, *Nature*, 428(6982), 487–492.

Langer, R. and Vacanti, J. P. 1993, Tissue engineering, *Science*, 260(5110), 920–926.

Li, J., Fartash, B., and Hermansson, L. 1995, Hydroxyapatite-alumina composites and bone-bonding, *Biomaterials*, 16(5), 417–422.

Lin, K. H., Maeda, S., and Saito, T. 1995, Long-term maintenance of liver-specific functions in three-dimensional culture of adult rat hepatocytes with a porous gelatin sponge support, *Biotechnol. Appl. Biochem.*, 21(Pt 1), 19–27.

Liu, H., Slamovich, E. B., and Webster, T. J. 2006, Less harmful acidic degradation of poly(lacticco-glycolic acid) bone tissue engineering scaffolds through titania nanoparticle addition, *Int. J. Nanomed.*, 1(4), 541–545.

Liu, Y., Wang, S., Krouse, J., Kotov, N. A., Eghtedari, M., Vargas, G., and Motamedi, M. 2007, Rapid aqueous photo-polymerization route to polymer and polymer-composite hydrogel 3D inverted colloidal crystal scaffolds, *J. Biomed. Mater. Res. A*, 83(1), 1–9.

Liu, X., Zhao, X., Fu, R. K. Y., Ho, J. P. Y., Ding, C., and Chu, P. K. 2005, Plasma-treated nanostructured TiO2 surface supporting biomimetic growth of apatite, *Biomaterials*, 26(31), 6143–6150.

Ma, Z., Kotaki, M., Yong, T., He, W., and Ramakrishna, S. 2005, Surface engineering of electrospun polyethylene terephthalate (PET) nanofibers towards development of a new material for blood vessel engineering, *Biomaterials*, 26(15), 2527–2536.

Madbouly, S. A., Otaigbe, J. U., Nanda, A. K., and Wicks, D. A. 2007, Rheological behavior of POSS/polyurethaneâ Urea nanocomposite films prepared by homogeneous solution polymerization in aqueous dispersions, *Macromolecules*, 40(14), 4982–4991.

Magrez, A., Kasas, S., Salicio, V., Pasquier, N., Seo, J. W., Celio, M., Catsicas, S., Schwaller, B., and Forro, L. 2006, Cellular toxicity of carbon-based nanomaterials, *Nano Lett.*, 6(6), 1121–1125.

McCusker, C., Carroll, J. B., and Rotello, V. M. 2005, Cationic polyhedral oligomeric silsesquioxane (POSS) units as carriers for drug delivery processes, *Chem Commun (Camb)*, 8, 996–998.

Mellion, B. T., Ignarro, L. J., Ohlstein, E. H., Pontecorvo, E. G., Hyman, A. L., and Kadowitz, P. J. 1981, Evidence for the inhibitory role of guanosine 3′, 5′-monophosphate in ADP-induced human platelet aggregation in the presence of nitric oxide and related vasodilators, *Blood*, 57(5), 946–955.

Meng, J., Kong, H., Xu, H. Y., Song, L., Wang, C. Y., and Xie, S. S. 2005, Improving the blood compatibility of polyurethane using carbon nanotubes as fillers and its implications to cardiovascular surgery, *J. Biomed. Mater. Res. A*, 74(2), 208–214.

Miller, D. C., Thapa, A., Haberstroh, K. M., and Webster, T. J. 2004, Endothelial and vascular smooth muscle cell function on poly(lactic-co-glycolic acid) with nano-structured surface features, *Biomaterials*, 25(1), 53–61.

Miller, D. C., Haberstroh, K. M., and Webster, T. J. 2005, Mechanism(s) of increased vascular cell adhesion on nano-structured poly(lactic-co-glycolic acid) films, *J. Biomed. Mater. Res. A*, 73(4), 476–484.

Miller, D. C., Haberstroh, K. M., and Webster, T. J. 2007, PLGA nanometer surface features manipulate fibronectin interactions for improved vascular cell adhesion, *J. Biomed. Mater. Res. A*, 81(3), 678–684.

Moghe, P. V., Berthiaume, F., Ezzell, R. M., Toner, M., Tompkins, R. G., and Yarmush, M. L. 1996, Culture matrix configuration and composition in the maintenance of hepatocyte polarity and function, *Biomaterials*, 17(3), 373–385.

Muller, J., Huaux, F., Moreau, N., Misson, P., Heilier, J. F., Delos, M., Arras, M., Fonseca, A., Nagy, J. B., and Lison, D. 2005, Respiratory toxicity of multi-wall carbon nanotubes, *Toxicol. Appl. Pharmacol.*, 207(3), 221–231.

Nablo, B. J., Chen, T. Y., and Schoenfisch, M. H. 2001, Sol-gel derived nitric-oxide releasing materials that reduce bacterial adhesion, *J. Am. Chem. Soc.*, 123(39), 9712–9713.

Nair, L. S., Bhattacharyya, S., and Laurencin, C. T. 2004, Development of novel tissue engineering scaffolds via electrospinning, *Expert. Opin. Biol. Ther.*, 4(5), 659–668.

Nayar, S., Pramanick, A. K., Sharma, B. K., Das, G., Ravi, K. B., and Sinha, A. 2008, Biomimetically synthesized polymer-hydroxyapatite sheet like nano-composite, *J Mater Sci Mater Med*, 19(1), 301–304.

Nel, A., Xia, T., Madler, L., and Li, N. 2006, Toxic potential of materials at the nanolevel, *Science*, 311(5761), 622–627.

Nhs Blood and Transplant 2008. UK transplant activity report 2007–2008: Liver activity. https://www.uktransplant.org.uk/ukt/statistics/transplantactivityreport=transplantactivityreport:jsp. 2009.

Nukavarapu, S. P., Kumbar, S. G., Brown, J. L., Krogman, N. R., Weikel, A. L., Hindenlang, M. D., Nair, L. S., Allcock, H. R., and Laurencin, C. T. 2008, Polyphosphazene/nano-hydroxyapatite composite microsphere scaffolds for bone tissue engineering, *Biomacromolecules*, 9(7), 1818–1825.

Oliveira, J. M., Sousa, R. A., Kotobuki, N., Tadokoro, M., Hirose, M., Mano, J. F., Reis, R. L., and Ohgushi, H. 2009, The osteogenic differentiation of rat bone marrow stromal cells cultured with dexamethasone-loaded carboxymethylchitosan/poly(amidoamine) dendrimer nanoparticles, *Biomaterials*, 30(5), 804–813.

Pezzatini, S., Morbidelli, L., Gristina, R., Favia, P., and Ziche, M. 2008, A nanoscale fluorocarbon coating on PET surfaces improves the adhesion and growth of cultured coronary endothelial cells, *Nanotechnology*, 19(27), 275101.

Pinchuk, L., Wilson, G. J., Barry, J. J., Schoephoerster, R. T., Parel, J. M., and Kennedy, J. P. 2008, Medical applications of poly(styrene-block-isobutylene-block-styrene) ("SIBS"), *Biomaterials*, 29(4), 448–460.

Radomski, M. W., Palmer, R. M., and Moncada, S. 1987, The anti-aggregating properties of vascular endothelium: Interactions between prostacyclin and nitric oxide, *Br. J. Pharmacol.*, 92(3), 639–646.

Raghunath, J., Zhang, H., Edirisinghe, M. J., Darbyshire, A., Butler, P. E., and Seifalian, A. M. 2009, A new biodegradable nanocomposite based on polyhedral oligomeric silsesquioxane nanocages: Cytocompatibility and investigation into electrohydrodynamic jet fabrication techniques for tissue-engineered scaffolds, *Biotechnol. Appl. Biochem.*, 52(Pt 1), 1–8.

Riccalton-Banks, L., Liew, C., Bhandari, R., Fry, J., and Shakesheff, K. 2003, Long-term culture of functional liver tissue: three-dimensional coculture of primary hepatocytes and stellate cells, *Tissue Eng*, 9(3), 401–410.

Richert, L., Boulmedais, F., Lavalle, P., Mutterer, J., Ferreux, E., Decher, G., Schaaf, P., Voegel, J. C., and Picart, C. 2004, Improvement of stability and cell adhesion properties of polyelectrolyte multilayer films by chemical cross-linking, *Biomacromolecules*, 5(2), 284–294.

Ricker, A., Liu-Snyder, P., and Webster, T. J. 2008, The influence of nano MgO and BaSO$_4$ particle size additives on properties of PMMA bone cement, *Int. J. Nanomed.*, 3(1), 125–132.

Roy, R. K. and Lee, K. R. 2007, Biomedical applications of diamond-like carbon coatings: A review, *J Biomed Mater Res B Appl Biomater*, vol. 83B(1), 72–84.

Roy, R. K., Choi, H. W., Yi, J. W., Moon, M. W., Lee, K. R., Han, D. K., Shin, J. H., Kamijo, A., and Hasebe, T. 2009, Hemocompatibility of surface-modified, silicon-incorporated, diamond-like carbon films, *Acta Biomater.*, 5(1), 249–256.

Sagai, M., Saito, H., Ichinose, T., Kodama, M., and Mori, Y. 1993, Biological effects of diesel exhaust particles. I. In vitro production of superoxide and in vivo toxicity in mouse, *Free Radic. Biol. Med.*, 14(1), 37–47.

Sayes, C. M., Gobin, A. M., Ausman, K. D., Mendez, J., West, J. L., and Colvin, V. L. 2005, Nano-C60 cytotoxicity is due to lipid peroxidation, *Biomaterials*, 26(36), 7587–7595.

Sayes, C. M., Liang, F., Hudson, J. L., Mendez, J., Guo, W., Beach, J. M., Moore, V. C., Doyle, C. D., West, J. L., Billups, W. E., Ausman, K. D., and Colvin, V. L. 2006a, Functionalization density dependence of single-walled carbon nanotubes cytotoxicity in vitro, *Toxicol. Lett.*, 161(2), 135–142.

Sayes, C. M., Wahi, R., Kurian, P. A., Liu, Y., West, J. L., Ausman, K. D., Warheit, D. B., and Colvin, V. L. 2006b, Correlating nanoscale titania structure with toxicity: A cytotoxicity and inflammatory response study with human dermal fibroblasts and human lung epithelial cells, *Toxicol. Sci.*, 92(1), 174–185.

Schiraldi, C., D'Agostino, A., Oliva, A., Flamma, F., De Rosa, A., Apicella, A., Aversa, R., and De Rosa, M. 2004, Development of hybrid materials based on hydroxyethylmethacrylate as supports for improving cell adhesion and proliferation, *Biomaterials*, 25(17), 3645–3653.

Seal, B. L., Otero, T. C., and Panitch, A. 2001, Polymeric biomaterials for tissue and organ regeneration, *Mater. Sci. Eng.: R: Rep.*, 34(4–5), 147–230.

Seifalian, A. M., Tiwari, A., Rashid, T., Salacinski, H., and Hamilton, G. 2009, Letter to the editor, *Artif. Organs*, 26(2), 209–210.

Sellaro, T. L., Ravindra, A. K., Stolz, D. B., and Badylak, S. F. 2007, Maintenance of hepatic sinusoidal endothelial cell phenotype in vitro using organ-specific extracellular matrix scaffolds, *Tissue Eng.*, 13(9), 2301–2310.

Semler, E. J. and Moghe, P. V. 2001, Engineering hepatocyte functional fate through growth factor dynamics: The role of cell morphologic priming, *Biotechnol. Bioeng.*, 75(5), 510–520.

Semler, E. J., Ranucci, C. S., and Moghe, P. V. 2000, Mechanochemical manipulation of hepatocyte aggregation can selectively induce or repress liver-specific function, *Biotechnol. Bioeng.*, 69(4), 359–369.

Semler, E. J., Lancin, P. A., Dasgupta, A., and Moghe, P. V. 2005, Engineering hepatocellular morphogenesis and function via ligand-presenting hydrogels with graded mechanical compliance, *Biotechnol. Bioeng.*, 89(3), 296–307.

Shanbhag, S., Wang, S., and Kotov, N. A. 2005a, Cell distribution profiles in three-dimensional scaffolds with inverted-colloidal-crystal geometry: Modeling and experimental investigations, *Small*, 1(12), 1208–1214.

Shanbhag, S., Woo Lee, J., and Kotov, N. 2005b, Diffusion in three-dimensionally ordered scaffolds with inverted colloidal crystal geometry, *Biomaterials*, 26(27), 5581–5585.

Sheeja, D., Tay, D. B. K., and Nung, L. N. 2004, Feasibility of diamond-like carbon coatings for orthopaedic applications, *Diamond & Related Materials*, 13(1), 184–190.

Shields, K. J., Beckman, M. J., Bowlin, G. L., and Wayne, J. S. 2004, Mechanical properties and cellular proliferation of electrospun collagen type II, *Tissue Eng.*, 10(9–10), 1510–1517.

Sitharaman, B., Shi, X., Walboomers, X. F., Liao, H., Cuijpers, V., Wilson, L. J., Mikos, A. G., and Jansen, J. A. 2008, In vivo biocompatibility of ultra-short single-walled carbon nanotube/biodegradable polymer nanocomposites for bone tissue engineering, *Bone*, 43(2), 362–370.

Stankus, J. J., Soletti, L., Fujimoto, K., Hong, Y., Vorp, D. A., and Wagner, W. R. 2007, Fabrication of cell microintegrated blood vessel constructs through electrohydrodynamic atomization, *Biomaterials*, 28(17), 2738–2746.

Teoh, S. H. 2000, Fatigue of biomaterials: A review, *Int. J. Fatigue*, 22(10), 825–837.

Thein-Han, W. W. and Misra, R. D. 2009, Biomimetic chitosan-nanohydroxyapatite composite scaffolds for bone tissue engineering, *Acta Biomater.*, 5(4), 1182–1197.

Thomas, A. C., Campbell, G. R., and Campbell, J. H. 2003, Advances in vascular tissue engineering, *Cardiovasc. Pathol.*, 12(5), 271–276.

Thorvaldsson, A., Stenhamre, H., Gatenholm, P., and Walkenstrom, P. 2008, Electrospinning of highly porous scaffolds for cartilage regeneration, *Biomacromolecules*, 9(3), 1044–1049.

Tiwari, A., Pandey, K. N., Mathur, G. N., and Nema, S. K. 2004, Revealing the concept of polyarylsilaneimide quasi nano composite formation correlation with macroscopic properties and electrical parameters, *Mat Res Innovat*, 8(2), 103–114.

Tiwari, A., Salacinski, H., Seifalian, A. M., and Hamilton, G. 2002, New prostheses for use in bypass grafts with special emphasis on polyurethanes, *Cardiovasc. Surg.*, 10(3), 191–197.

Tiwari, A., Pandey, K. N., Mathur, G. N., and Nema, S. K. 2004, Revealing the concept of polyarylsilaneimide quasi nano composite formation correlation with macroscopic properties and electrical parameters, *Mater. Res. Innovat.*, 8(2), 103–114.

United States Government. chronic diseases overview. http://www.cdc.gov/nccdphp/overviewtext:htm. 2007.

UNOS. 2002 OPTN/SRTR Annual Report. 2002. Unpublished work.

Vallyathan, V. and Gwinn, M. R. 2006, Nanoparticles: Health effects—pros and cons, *Env. Health Perspect.*, 114(12), 1818–1825.

van Kooten, T. G., Spijker, H. T., and Busscher, H. J. 2004, Plasma-treated polystyrene surfaces: Model surfaces for studying cell-biomaterial interactions, *Biomaterials*, 25(10), 1735–1747.

Venugopal, J., Zhang, Y. Z., and Ramakrishna, S. 2005, Fabrication of modified and functionalized polycaprolactone nanofibre scaffolds for vascular tissue engineering, *Nanotechnology*, 16(10), 2138–2142.

Vincent, J. G. and Doting, J. 1989, From flintstone to diamond blade: a new multifunctional instrument for use in coronary surgery, *Eur J Cardiothorac Surg*, 3(4), 373–375.

Webster, T. J., Ergun, C., Doremus, R. H., Siegel, R. W., and Bizios, R. 2001, Enhanced osteoclast-like cell functions on nanophase ceramics, *Biomaterials*, 22(11), 1327–1333.

Wight, T. N., 1996, The vascular extracellular matrix, in *Atherosclerosis and Coronary Artery Disease*, Ross, R., Topol, E. J., Fuster, V. ed., Lippincott-Raven, Philadelphia, PA, pp. 421–440.

Wittmer, C. R., Phelps, J. A., Saltzman, W. M., and Van Tassel, P. R. 2007, Fibronectin terminated multilayer films: Protein adsorption and cell attachment studies, *Biomaterials*, 28(5), 851–860.

Wright, J. B., Lam, K., Hansen, D., and Burrell, R. E. 1999, Efficacy of topical silver against fungal burn wound pathogens, *Am J Infect Control*, 27(4), 344–350.

Wu, Q., Cao, H., Luan, Q., Zhang, J., Wang, Z., Warner, J. H., and Watt, A. A. 2008, Biomolecule-assisted synthesis of water-soluble silver nanoparticles and their biomedical applications, *Inorg Chem*, 47(13), 5882–5888.

Xu, C. Y., Inai, R., Kotaki, M., and Ramakrishna, S. 2004, Aligned biodegradable nanofibrous structure: A potential scaffold for blood vessel engineering, *Biomaterials*, 25(5), 877–886.

Zaremba, C. M., Belcher, A. M., Fritz, M., Li, Y., Mann, S., Hansma, P. K., Morse, D. E., Speck, J. S., and Stucky, G. D. 1996, Critical transitions in the biofabrication of abalone shells and flat pearls, *Chem. Mater.*, 8(3), 679–690.

Zavan, B., Brun, P., Vindigni, V., Amadori, A., Habeler, W., Pontisso, P., Montemurro, D., Abatangelo, G., and Cortivo, R. 2005, Extracellular matrix-enriched polymeric scaffolds as a substrate for hepatocyte cultures: In vitro and in vivo studies, *Biomaterials*, 26(34), 7038–7045.

Zhang, H., Annich, G. M., Miskulin, J., Stankiewicz, K., Osterholzer, K., Merz, S. I., Bartlett, R. H., and Meyerhoff, M. E. 2003, Nitric oxide-releasing fumed silica particles: Synthesis, characterization, and biomedical application, *J. Am. Chem. Soc.*, 125(17), 5015–5024.

Zhang, Y., Lim, C. T., Ramakrishna, S., and Huang, Z. M. 2005, Recent development of polymer nanofibers for biomedical and biotechnological applications, *J. Mater. Sci. Mater. Med.*, 16(10), 933–946.

Zhao, M., Bai, H., Wang, E., Forrester, J. V., and McCaig, C. D. 2004, Electrical stimulation directly induces pre-angiogenic responses in vascular endothelial cells by signaling through VEGF receptors, *J. Cell Sci.*, 117(Pt 3), 397–405.

Zhao, F., Yin, Y., Lu, W. W., Leong, J. C., Zhang, W., Zhang, J., Zhang, M., and Yao, K. 2002, Preparation and histological evaluation of biomimetic three-dimensional hydroxyapatite/chitosan-gelatin network composite scaffolds, *Biomaterials*, 23(15), 3227–3234.

Zong, X., Bien, H., Chung, C. Y., Yin, L., Fang, D., Hsiao, B. S., Chu, B., and Entcheva, E. 2005, Electrospun fine-textured scaffolds for heart tissue constructs, *Biomaterials*, 26(26), 5330–5338.

20

Nanotechnology for Implants

Lijie Zhang
Brown University

Thomas J. Webster
Brown University

20.1 Introduction: Nanotechnology and Nanomaterials—Biomimetic Tools for Implants

In 1959, Nobel award winner Richard Feynman first proposed the seminal idea of nanotechnology by suggesting the development of molecular machines. Ever since, the scientific community has investigated the role that nanotechnology can play in every aspect of society. The intrigue of nanotechnology comes from the ability to control material properties by assembling such materials at the nanoscale. The tunable material properties that nanotechnology can provide were stated in Norio Taniguchi's paper in 1974 where the term "nanotechnology" was first used in a scientific publication (Taniguchi 1974). Nanotechnology has achieved tremendous progress in the past several decades. Recently, nanomaterials, which are materials with basic structural units, grains, particles, fibers, or other constituent components smaller than 100 nm in at least one dimension (Siegel and Fougere 1995), have evoked a great amount of attention for improving disease prevention, diagnosis, and treatment.

The intrigue in nanomaterial research for regenerative medicine is easy to see and is widespread. For example, from a material property point of view, nanomaterials can be made of metals, ceramics, polymers, organic materials, and composites thereof, just like conventional or micron-structured materials. Nanomaterials include nanoparticles, nanoclusters, nanocrystals, nanotubes, nanofibers, nanowires, nanorods, nanofilms, etc. To date, numerous top-down and bottom-up nanofabrication technologies, such as electrospinning, phase separation, self-assembly processes, thin-film deposition, chemical vapor deposition, chemical etching, nano-imprinting, photolithography, electron beam or nanosphere lithographies (Freeman et al. 2008), are available for the design of nanomaterials with ordered or random nanotopographies (Figure 20.1). Nanomaterials can also be grown or self-assembled into nanotubes/nanofibers, which can even more accurately stimulate the dimensions of natural entities, such as collagen fibers. After decreasing the material size into the nanoscale, dramatically increased surface area, surface roughness, or surface area-to-volume ratios of nanomaterials may lead to a higher surface reactivity with many superior physiochemical properties (i.e., mechanical, electrical, optical, catalytic, and magnetic properties) (Fahlman 2007). Therefore, nanomaterials with such excellent properties have been extensively investigated in a wide range of biomedical applications, in particular regenerative medicine.

With the striking increase in the world's population, there are enormous demands each year for various biomedical implants to repair diseased or lost tissues. However, conventional tissue replacements (such as autografts and allografts) have a variety of problems that cannot satisfy high-performance demands necessary for today's patient. Consequently, tissue engineering (or regenerative medicine) emerged initially defined by Robert Langer and Joseph Vacanti as "an interdisciplinary field that applies the principles of engineering and life sciences toward the development of biological substitutes that restore, maintain, or improve tissue function" (Langer and Vacanti 1993). However, it is clear today that materials used in a wide range of tissue engineering applications still require improvement. Since natural tissues or organs have features that are nanometer in dimension and cells directly interact with (and create) nanostructured extracellular matrices (ECMs), the biomimetic features and excellent physiochemical properties of nanomaterials play a key role in stimulating cell growth as well as guiding tissue regeneration.

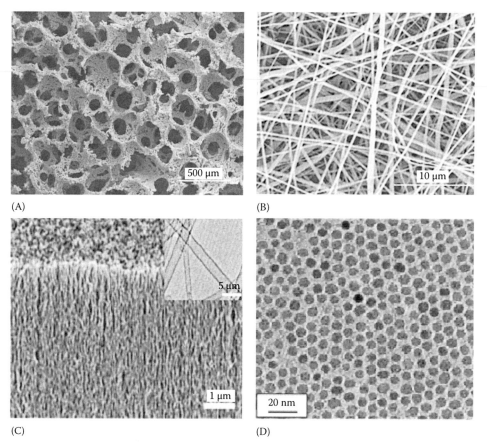

(A)   (B)

(C)   (D)

FIGURE 20.1   (A) Scanning electron microscopy (SEM) image of poly(L-lactic acid) (PLLA) nanofibrous scaffold with interconnected spherical macropores created by a phase-separation technique. (From Chen, V.J. and Ma, P.X., *Biomaterials*, 25, 2065, 2004.) (B) Electrospun polycaprolactone/hydroxyapatite/gelatin (PCL/HA/Gel, 1:1:2) nanofibers which significantly improved osteoblast functions for bone tissue engineering applications. (From Venugopal, J.R. et al., *Artif. Organs*, 32, 388, 2008.) (C) A densely aligned single-wall carbon nanotube (SWCNT) forest grown with novel water-assisted chemical vapor deposition in 10 min. (From Sun, S. et al., *J. Am. Chem. Soc.*, 126, 273, 2004.) (D) Transmission electron microscopy (TEM) image of monodispersed magnetic Fe_3O_4 nanoparticles (6 nm) deposited from their hexane dispersion and dried at room temperature.

Even though it was a field in its infancy a decade ago, currently, numerous researchers fabricate cytocompatible biomimetic nanomaterial scaffolds encapsulating cells (such as stem cells, chondrocytes, and osteoblasts) for tissue engineering applications. In this chapter, we will focus on the recent in vitro and, more importantly, in vivo progress toward the use of nanomaterials for bone, cartilage, vascular, neural, and bladder tissue engineering applications. As the next frontier in nanotechnology research, toxicity concerns of nanomaterials and nanoparticles during manufacturing and/or implantation will be covered as well.

20.2 Unique Surface Properties of Nanomaterials

It is important to note that while this chapter discusses how nanotechnology is increasing tissue growth, it first emphasizes why nanomaterials should even be considered for regenerative medicine. First, protein interactions are essential for mediating cell functions on implants. Applications in which a firm understanding

of protein–surface interactions are needed include biomaterials, tissue engineering materials, implants, bioseparations, biosensors, biochips, food science, and aquatics (to only name a few). The first event that occurs when an implant is placed in the body is that water molecules adhere to the surface (in different ways for different surfaces) followed by the formation of a layer of proteins from bodily fluid. Next, cells of the body explore the adsorbed proteins and may subsequently attach to specific binding sites exposed in these proteins. The types of cells that recognize the adsorbed proteins and attach to the implant surface, both initially and chronically, determine the fate of the implant. Will the implant successfully integrate into the juxtaposed tissues or will the body wall it off to exclude it from interacting as much as possible?

Properties of the implant surface control these initial water–surface interactions, which subsequently regulate which proteins will adsorb, how many proteins will adsorb, and what their conformations will be once they adsorb. For these reasons, it is important to design biomaterials with surfaces that optimize protein adsorption for controlling desirable cell responses pertinent to specific implant functionality.

The properties of a material at its surface, especially chemistry, roughness, and energy are what elicit water molecule layers to adhere and subsequently mediate the adsorption of different types of proteins. Surfaces in an aqueous environment elicit the adsorption of ordered water layers at the interface with bulk water. These water layers are capable of excluding solutes and can be up to 100 μm thick with some hydrophilic surfaces (Webster 2001). Clearly, the nature of the surface influences the water interface and subsequent protein adsorption. Differences in the monolayer of proteins adsorbed to material surfaces are based on such properties as surface roughness, energy, and chemistry, as will be expanded in the next sections. The atomic structure and bonding interactions can vary considerably even within a material that appears to have the same chemical composition. An example of this is when considering grain size variations that are related to the overall surface roughness. Increased surface roughness can also serve to promote surface energy; thus, many surface properties are intertwined and changing one will change others. Lastly, the energy at the surface of a material controls its wettability, which many consider to be the most important aspect to promote cell adhesion.

Moreover, the topography of nanomaterials has been used to control protein interactions. Until recently, the size of the features on the surface of a biomaterial was not considered to be an important design feature to control protein interactions. However, one important example of how topography can influence protein interactions involves nanomaterials. Materials with micron-sized features and up (such as those commonly used in current implants) have significantly different material properties than those with nanoscale features. In fact, the National Science Foundation's definition for nanomaterials states that they not only have smaller constituent sizes (such as grains, particles, and fibers), but also have significantly altered properties compared to conventional materials. Conventional materials have more atoms in bulk and less at the surface, while the reverse is true for nanophase materials. Nanomaterials also have increased defect sites at their surfaces (the highly reactive regions of materials) and therefore greater opportunities for binding to proteins and enzymes (Webster 2001). This gives nanophase materials increased catalytic activity, which can be very economical for numerous enzymatic processes.

Other enhanced properties of nanomaterials include superior optical, mechanical, and electrical properties. An example of mechanical property changes when transforming from conventional to nanometer grain sizes is that the smaller grain sizes present in nanophase ceramics slide over one another more easily than in a conventional material; this makes nanophase ceramics less brittle and more ductile when compared to conventional ceramics. It is now also clear that the unique, biologically inspired surface roughness of nanomaterials promotes protein interactions that are important to increase the regeneration of bone, cartilage, vascular, bladder, and nervous system tissues. Nanomaterials are biologically inspired materials since tissues in our bodies have nanoscale constituent dimensions due to the presence of proteins and hydroxyapatite (as in the case for bone).

Another property that influences protein interactions is surface energy. For example, some studies have shown that hydrophilic, surfaces tend to manipulate fibronectin conformation to have more RGD (Arginine-Glycine-Aspartic acid, a well-known cell-binding sequence)-binding sites available to cells than charged and hydrophobic surfaces (Webster 2001). This trend also holds true for subsequent cell-binding interactions. The hydrophilic and neutral surfaces obtain the best cell-binding interactions followed by charged and then hydrophobic surfaces. Contact angle analysis can be used to calculate the surface energy of a sample. Since proteins are charged molecules, a surface that is also charged will influence their interactions. Thus, the unique surface features (i.e., topography) and energy (i.e., wettability) of nanomaterials makes them excellent candidates for various tissue growth applications.

20.3 Nanomaterials for Improved Tissue Engineering Applications

20.3.1 Promise of Nanomaterials for Bone and Cartilage Tissue Regeneration

Today, various bone fractures, osteoarthritis, osteoporosis, and bone cancer represent common and significant clinical problems. The National Center for Health Statistics (NCHS) reported that bone fractures numbered 1039,000 in 2004 in the United States. In addition, around 118,700 patients (home health care) had osteoarthritis and associated disorders in 2000. The American Academy of Orthopedic Surgeons also reported that in just a 4 year period, there was an 83.72% increase in the number of hip replacements performed from nearly 258,000 procedures in 2000 to 474,000 procedures in 2004 (AAOS 2004). Such traumatic bone and cartilage damage happens frequently each year. A similar trend has been documented for other industrialized countries as well. However, traditional implant materials may only last 10–15 years on average and implant failures originating from implant loosening, inflammation, infection, osteolysis, and wear debris frequently occur. It is clearly urgent to develop a new generation of cytocompatible bone and cartilage substitutes to regenerate bone/cartilage tissue at defect sites, which will last the lifetime of the patient.

Importantly, when examining nature, using nanotechnology for regenerative medicine becomes obvious. For example, bone is an assembled nanocomposite that consists of a protein-based soft hydrogel template (i.e., collagen, non-collagenous proteins [laminin, fibronectin, vitronectin], and water) and hard inorganic components (hydroxyapatite, HA, $Ca_{10}(PO_4)_6(OH)_2$) (Webster 2001, Zhang et al. 2009a) (Figure 20.2A). Specifically, 70% of the bone matrix is composed of nanocrystalline HA, which is typically 20–80 nm long and 2–5 nm thick (Kaplan et al. 1994). Other protein components in the bone ECM are also nanometer in dimension. This self-assembled nanostructured ECM in bone closely surrounds and controls mesenchymal stem cell, osteoblast (bone-forming cell), osteoclast (bone-resorbing cells), and fibroblast functions. Moreover, cartilage is a low regenerative

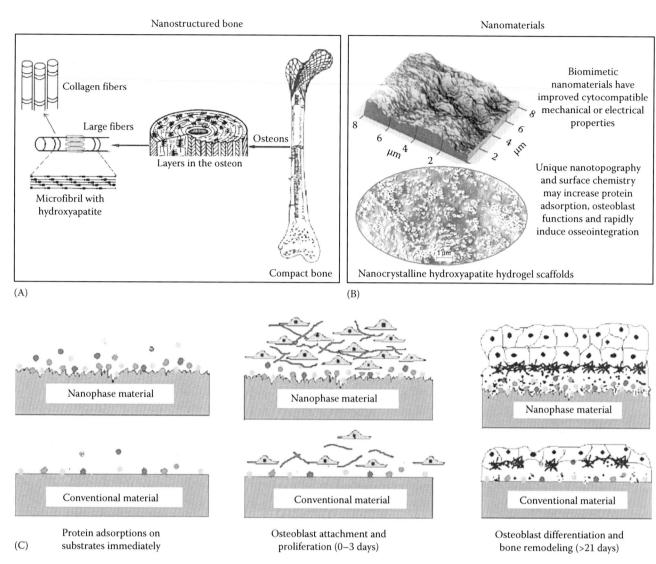

FIGURE 20.2 The biomimetic advantages of nanomaterials. (A) The nanostructured hierarchal self-assembly of bone. (B) Nanophase titanium (top, the atomic force microscopy image) and nanocrystalline HA/HRN hydrogel scaffold (bottom, the SEM image). (C) Schematic illustration of the mechanism by which nanomaterials may be superior to conventional materials for bone regeneration. The bioactive surfaces of nanomaterials mimic those of natural bones to promote greater amounts of initial protein adsorption and efficiently stimulate more new bone formation compared to conventional materials.

tissue composed of a small percentage of chondrocytes and a dense nanostructured ECM rich in collagen fibers, proteoglycans, and elastin fibers. The limited regenerative properties of cartilage originate from an inhibited chondrocyte mobility in the dense ECM, absence of progenitor cells, and vascular networks necessary for efficient cartilage tissue repair (Vasita and Katti 2006). The design of novel nanomaterials that possess not only excellent mechanical properties but that are also biomimetic in terms of their nanostructure (Figure 20.2B) has become quite popular in order to improve osteoblast and chondrocyte functions.

Furthermore, in addition to the dimensional similarity to bone/cartilage tissue, nanomaterials also exhibit unique surface properties (such as surface topography, surface chemistry, surface wettability, and surface energy) due to their significantly increased surface area and roughness compared to conventional or micron-structured materials. As is known, material surface properties mediate specific protein (such as fibronectin, vitronectin, and laminin) adsorption and bioactivity before cells adhere on implants, further regulating cell behavior and dictating tissue regeneration (Webster 2001). Furthermore, an important criterion for designing orthopedic implant materials is the formation of sufficient osseointegration between synthetic materials and bone tissue. Studies have demonstrated that nanostructured materials with cell-favorable surface properties may promote greater amounts of specific protein interactions to more efficiently stimulate new bone growth compared to conventional materials (Degasne et al. 1999, Webster et al. 2000, 2001a) (Figure 20.2C). This may be one of the underlying mechanisms why nanomaterials are superior to conventional materials for tissue growth. Thus, by controlling surface properties, various nanophase ceramic, polymer, metal, and composite scaffolds have been designed for bone/cartilage tissue engineering applications.

Nanophase ceramics, especially nano-hydroxyapatite (HA, a native component of bone), are popular bone substitutes, coatings, and other filler materials due to their documented ability to promote mineralization. The nanometer grain sizes and high surface fraction of grain boundaries in nanoceramics increase osteoblast functions (such as adhesion, proliferation, and differentiation). For example, some in vitro studies demonstrated that nanophase HA (67 nm grain size) significantly enhanced osteoblast adhesion and strikingly inhibited competitive fibroblast adhesion compared to conventional, 179 nm grain size HA, after just 4 h of culture (Webster et al. 2000). Researchers believe they know why. Specifically, researchers have elucidated the highest adsorption of vitronectin (a protein well known to promote osteoblast adhesion) on nanophase ceramics, which may explain the subsequent enhanced osteoblast adhesion on these materials (Webster et al. 2000). In addition, enhanced osteoclast-like cell functions (such as the synthesis of tartrate-resistant acid phosphatase (TRAP) and the formation of resorption pits) have also been observed on nano HA compared to conventional HA (Webster et al. 2001b). In a recent study, Nukavarapu et al. fabricated a biodegradable nano-hydroxyapatite/polyphosphazene microsphere 3-D

scaffold, which had suitable mechanical properties (compressive moduli of 46–81 MPa) and cytocompatibility properties for bone tissue engineering applications (Nukavarapu et al. 2008a). It should not be surprising that nanostructured composites have similar mechanical properties to bone since bone itself is a nanostructured composite.

Importantly, such results have not been limited to in vitro studies. In fact, in vivo (specifically, rat) studies also demonstrated that nanocrystalline HA accelerated new bone formation on tantalum scaffolds when used as an osteoconductive coating compared to uncoated or conventional micron-sized HA-coated tantalum (Sato 2006). Histological examination (Figure 20.3) revealed that nanocrystalline HA coatings promoted greater amounts of new bone growth in the rat calvaria than uncoated or conventional HA-coated tantalum after 6 weeks of implantation. Similar tendencies have been reported for other nanoceramics including alumina, zinc oxide, and titania, thus providing strong evidence that it may not matter what implant chemistry is fabricated to have nanometer surface features to promote bone growth. For example, osteoblast adhesion increased by 146% and 200% on nanophase zinc oxide (23 nm) and titania (32 nm) compared to microphase zinc oxide (4.9 μm) and titania (4.1 μm), respectively (Colon et al. 2006). Furthermore, nanophase zinc oxide, nanophase titania, and nanofiber alumina enhanced collagen synthesis, alkaline phosphatase activity, and calcium mineral deposition

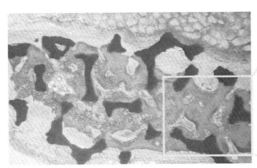
Uncoated tantalum

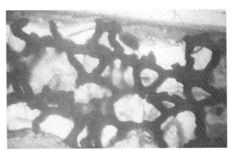
Conventional HA coated tantalum

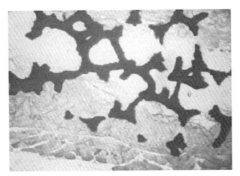
Nanocrystalline HA coated tantalum

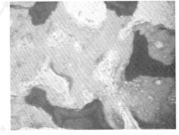
Nanocrystalline HA coated tantalum

FIGURE 20.3 **(See color insert following page V-2.)** Histology of rat calvaria after 6 weeks of implantation of uncoated tantalum, conventional HA-coated tantalum and nanocrystalline HA-coated tantalum. Greater amounts of new bone formation occur in the rat calvaria when implanting nanocrystalline HA-coated tantalum than uncoated and conventional HA-coated tantalum. Red represents new bone and blue represents collagen. (Adapted from Sato, M. Nanophase hydroxyapatite coatings for dental and orthopedic applications, PhD thesis, Purdue University, West Lafayette, IN, 2006.)

by osteoblasts compared to conventional equivalents (Colon et al. 2006, Webster et al. 2005).

Because collagen in bone and cartilage is a triple helix self-assembled into nanofibers 300 nm in length and 1.5 nm in diameter, many recent efforts have been dedicated to exploring the influence that novel biomimetic nanofibrous or nanotubular scaffolds have on regenerative medicine by following a bottom-up self-assembly process. Specifically, Hartgerink et al. reported that a peptide-amphiphile (PA) with the cell-adhesive ligand RGD (Arg-Gly-Asp) self-assembled into supramolecular nanofibers (Figure 20.4A and B) (Hartgerink et al. 2001). Importantly, by directly nucleating and aligning HA on the long axis of a nanofiber, a new nanofiber composite was designed with the same self-assembly pattern as collagen and HA crystals in bone. Moreover, Hosseinkhani et al. investigated mesenchymal stem cell (MSC) behaviors on self-assembled PA nanofiber scaffolds (Hosseinkhani et al. 2006). Significantly enhanced osteogenic differentiation of MSC occurred in the 3-D PA scaffold compared to 2-D static tissue culture. RGD-modified PA nanofibers promoted the maximum amount of alkaline phosphatase activity and osteocalcin content by osteoblasts.

Promise has also been demonstrated for other novel nanostructured self-assembled chemistries. For example, osteogenic helical rosette nanotubes obtained through the self-assembly of DNA base pairs (Guanine∧Cytosine) in aqueous solutions (Figure 20.4C) have been reported for bone tissue engineering applications. They have tailorable amino acid and peptide side chains (such as lysine, RGD, and KRSR (Lys-Arg-Ser-Arg, which selectively promotes osteoblast adhesion and inhibits fibroblast adhesion)) and are excellent mineralization templates

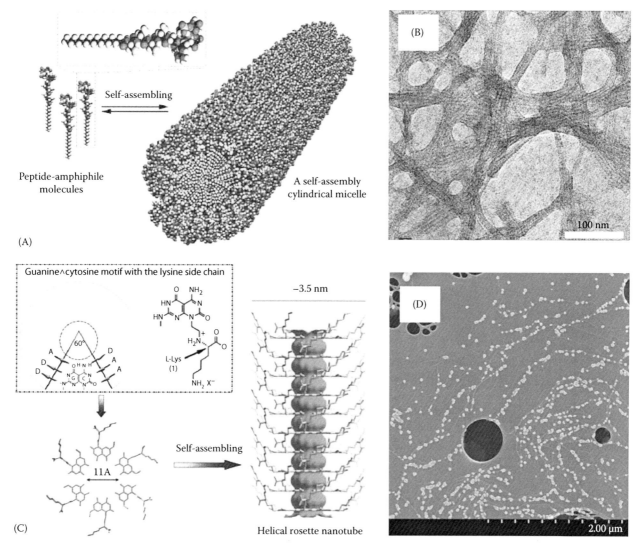

FIGURE 20.4 Self-assembled nanofibers and nanotubes for bone/cartilage tissue engineering applications. (A) Schematic illustration of the self-assembly process of peptide-amphiphiles functionalized with RGD to form a nanofiber 7.6 ± 1 nm in diameter. (Adapted from Hartgerink, J.D. et al., *Science*, 294, 1684, 2001.) (B) TEM image of the above self-assembled nanofibers. (C) Schematic illustration of the self-assembly process of the Guanine∧Cytosine DNA base pairs forming helical rosette nanotubes (HRNs). (D) SEM images of biomimetic nano HA aligned with HRNs on a porous carbon TEM grid.

to assemble a biomimetic nanotube/HA structure (Figure 20.4D). Furthermore, significantly improved osteoblast adhesion has been observed on helical rosette nanotubes regardless of whether they are incorporated into hydrogels or coated on titanium (compared to untreated controls) (Zhang et al. 2008a,b). Cartilage tissue engineering has also benefited from nanostructured self-assembled chemistries. Specifically, Kisiday et al. designed a self-assembling peptide (the peptide KLD-12, Lys-Leu-Asp) hydrogel for cartilage repair (Kisiday et al. 2002). The chondrocyte-encapsulated scaffold supported chondrocyte differentiation and promoted the synthesis of a cartilage-like ECM (rich in proteoglycans and type II collagen) in 3-D cell culture after 4 weeks, thus showing promise for cartilage tissue engineering. In summary, by this self-assembly process, one can create a biologically inspired 3-D scaffold with self-assembled biomimetic features more suitable for reconstructing 3-D bone and cartilage.

In addition, due to their superior cytocompatible, mechanical, and electrical properties, carbon nanotubes/nanofibers (CNTs/CNFs) are ideal scaffold candidates for bone tissue engineering applications (Zhang et al. 2009b). In a recent study, compared to larger fibers, 60 nm diameter CNFs significantly increased osteoblast adhesion and concurrently decreased competitive cell (fibroblast, smooth muscle cell, etc.) adhesion in order to stimulate sufficient osseointegration (Price et al. 2003). Other research efforts have also demonstrated that CNTs are suitable to promote osteoblast functions (Zanello et al. 2006). Recently, an in vivo study of ultra-short single-wall carbon nanotube (SWCNT) polymer nanocomposites after implanting them into rabbit femoral condyles and subcutaneous pockets for up to 12 weeks was reported (Sitharaman et al. 2008). The nanocomposites exhibited favorable hard and soft tissue responses after 4 and 12 weeks. They induced a 300% greater bone volume than all other experimental groups at 4 weeks and 200% greater bone growth at defect sites than control polymers without CNTs after 12 weeks. Besides, CNT/CNF reinforced polymer nanocomposites have also demonstrated excellent electrical conductivity for tissue regeneration. For instance, using biodegradable polylactic acid (PLA)/CNT composites as an example, an 80/20% (w/w) PLA/CNT composite exhibited ideal electrical conductivity for bone growth while PLA was an insulator and not appropriate for electrically stimulating bone growth. Specifically, the PLA/CNT composite promoted a 46% increase in osteoblast proliferation and a 300% increase in calcium content after electrical stimulation for 2 and 21 days compared to PLA alone, respectively (Supronowicz et al. 2002). These studies indicated that the CNTs/CNFs and their composites can serve as osteogenic scaffolds with good cytocompatibility properties, reinforced mechanical properties, and improved electrical conductivity to effectively enhance bone tissue growth.

As mentioned above, synthetic and natural polymers (e.g., polyglycolic acid (PGA), poly(lactic-co-glycolic acid) (PLGA), poly(l-lactic acid) (PLLA), PLA, gelatin, collagen, and chitosan) are excellent candidates for bone/cartilage tissue engineering applications due to their biodegradability and ease of fabrication.

Nanoporous or nanofibrous polymer matrices can be fabricated via electrospinning, phase separation, particulate leaching, chemical etching, and 3-D printing techniques. For cartilage applications, there has been great interest incorporating chondrocytes or progenitor cells (such as stem cells) into the 3-D polymer or composite scaffolds during electrospinning (Fecek et al. 2008, Li et al. 2005, 2008). For example, in vitro chondrogenesis of MSCs was investigated in an electrospun poly(ε-caprolactone) (PCL) nanofibrous scaffold (Li et al. 2005). The differentiation of the stem cells into chondrocytes in the nanofibrous scaffold was comparable to an established cell pellet culture. However, the easily fabricated and modified nanofibers possessed much better mechanical properties to overcome the disadvantages of using cell pellets and, thus, were presented as ideal candidates for stem cell transplantation during clinical cartilage repair. Because the infiltration of cells is usually inhibited by small pore sizes of electrospun polymer nanofibers leading to uneven cell distributions in the scaffold, this recent study improved chondrocyte seeding technology and obtained a homogeneous cell–PLLA nanofiber composite (Figure 20.5) (Li et al. 2008). The results showed that chondrocytes were uniformly present throughout the entire cell-nanofiber composite and the scaffold in a dynamic bioreactor developed into a smooth cartilage-like tissue with more total collagen and improved mechanical properties relative to that obtained in static culture. Moreover, significantly increased chondrocyte functions (specifically, adhesion, proliferation, and matrix synthesis) on 3-D nanostructured PLGA created via chemical etching were observed (Park et al. 2005). It is important to note that the controllable biodegradable properties of these PLGA, PLLA, and PLA polymers make them suitable for bone tissue engineering and orthopedic applications. Table 20.1 shows the degradation time and the mechanical properties of some common polymers in detail. The biodegradable polymer scaffolds hold the potential to provide suitable mechanical support during bone tissue growth and gradually degrade in vivo into nontoxic natural metabolites after bone healing (Athanasiou et al. 1996, Balasundaram and Webster 2007).

In bone tissue engineering, there are a large number of studies that report the promise of biomimetic 3-D nanostructured polymer scaffolds, which encapsulate stem cells and/or osteoblasts. For instance, Venugopal and colleagues electrospun a fibrous nanocomposite of PCL/HA/Gel (Gel is gelatin) at a ratio of 1:1:2 (Figure 20.1B). The results demonstrated that osteoblast proliferation, alkaline phosphatase activity, and mineralization were highest on the highly flexible PCL/HA/Gel nanocomposite when compared to other PCL nanofibrous scaffolds (Venugopal et al. 2008). Recently, a novel polymer/calcium phosphate composite was developed for bone tissue engineering applications. These nanofibrous fibrin-based composites promoted osteoblast alkaline phosphatase activity as well as osteoblast marker gene (mRNA) expression to support bone maturation both in vitro and in vivo in a mouse calvarial defect model (Osathanon et al. 2008).

Last but not the least, nanophase metals have been extensively investigated for orthopedic applications due to their

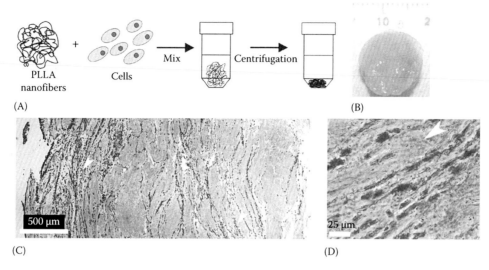

(A)   (B)

(C)   (D)

FIGURE 20.5 **(See color insert following page V-2.)** (A) Schematic illustrating an efficient cell seeding method into a cell-nanofiber composite for cartilage tissue engineering applications. (B) Image of a shiny cartilage-like tissue from the cell-nanofiber composite after 42 days of culture. (C) Low-magnification histology showing a well-dispersed chondrocyte distribution throughout the nanofiber scaffold after 1 day of cell culture (the cross section). (D) High-magnification histology showing distinct cell populations among the nanofibers. Arrows point to chondrocytes dispersed among nanofibers. (Adapted from Li, W.J. et al., *Tissue Eng. Part A*, 14, 639, 2008.)

TABLE 20.1 The Degradation Time and the Mechanical Properties of Common Polymers

Polymer	Approximate Degradation Time (Month)	Tensile Modulus (GPa)	Tensile Strength (MPa)
PLLA	>36	3	60–70
PGA	6–12	5–7	60–80
PGA-co-TMC	12–15	2.4	60
PDLLA	12–15	2	40–50
PDLLA/GA (85:15)	6–12	2	40–50

Source: Balasundaram, G. and Webster, T.J., *Macromol. Biosci.*, 7, 635, 2007.

higher surface roughness, energy, and presence of more particle boundaries at the surface compared with conventional micron metals. Specifically, Webster et al. provided the first evidence that nanophase Ti, Ti6Al4V, and CoCrMo significantly enhanced osteoblast adhesion compared to respective conventional metals (Webster and Ejiofor 2004). In addition, linearly nano-featured patterned Ti can be created via electron beam evaporation (Puckett et al. 2008). This study revealed that the nanoregion of the patterned Ti induced greater osteoblast adhesion than the micron-rough regions and also controlled osteoblast morphology and alignment. Moreover, an electrochemical method known as anodization is a well-established nano-surface modification technique used to fabricate highly porous TiO_2 nanotube layers on Ti. Through the anodization of Ti in dilute hydrofluoric acid (HF) electrolyte solutions, nanotubes with diameters around 100 nm and lengths around 500 nm can be implanted into the TiO_2 layers of Ti. Greatly improved osteoblast functions on nanotubular anodized Ti compared to unanodized Ti in vitro has been reported (Yao et al. 2005). Moreover, increased

chondrocyte adhesion was also observed on anodized nanotubular Ti compared to unanodized Ti in a recent study, thus, suggesting the possibility of promoting cartilage growth on anodized Ti (Burns et al. 2008).

20.3.2 Promise of Nanomaterials for Vascular Tissue Regeneration

Due to the increasing presence of vascular diseases (such as atherosclerosis), vascular grafts of greater efficacy to replace damaged blood vessels are needed. For example, the American Heart Association reports that coronary heart disease mostly caused by atherosclerosis has led to 451,326 deaths in 2004 and is the single leading cause of death in the United States today (AHA 2004). In addition, peripheral arterial disease related to blood vessels outside of the heart and brain affects about 8 million Americans. Over 500,000 coronary and periphery bypass surgeries were performed in the United States in 2005. Since vascular tissue is a layered structure possessing numerous nanostructured features (i.e., due to the presence of collagen and elastin in the vascular ECM), nanomaterials have shown much promise to improve vascular cell (specifically, endothelial and smooth muscle cells) functions to inhibit thrombosis and severe inflammation.

For example, Choudhary et al. reported that vascular cell adhesion and proliferation were greatly improved on nanostructured Ti compared to conventional Ti (Figure 20.6) (Choudhary et al. 2007). Interestingly, greater endothelial cell adhesion, total elastin, and collagen synthesis were observed on nanostructured Ti than respective vascular smooth muscle cell functions after 5 days in culture; such results are important as they suggest an increased probability of endothelialization compared to current vascular material problems of overgrown smooth muscle cells

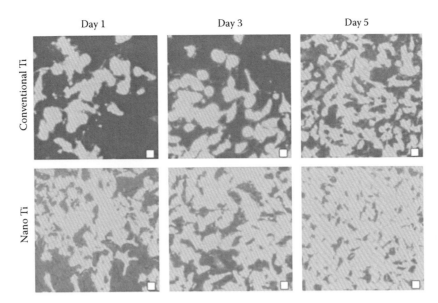

FIGURE 20.6 Fluorescence microscopy images of greatly increased endothelial cell proliferation on nanostructured Ti compared to conventional Ti. Scale bar is 10 μm. (Adapted from Choudhary, S. et al., *Tissue Eng. Part A*, 13, 1421, 2007.)

into the vascular lumen. It was speculated that the increased nano-roughness and particle boundaries on nanostructured Ti contributed to the observed favorable endothelial cell functions. In addition, Miller et al. created biodegradable PLGA vascular grafts with nanometer surface features through chemical etching in NaOH and through a cast-mold technique (Miller et al. 2004a, 2005, 2007). Results demonstrated that both those created through chemical etching and polymer cast-mold techniques possessed random nanometer structures on PLGA, which promoted endothelial or vascular smooth muscle cell proliferation compared to the conventional PLGA (Miller et al. 2004a). A further study provided evidence that nanostructured PLGA promoted more fibronectin and vitronectin adsorption from serum than conventional PLGA, thus, leading to the greater vascular cell responses in the nanostructured PLGA (Miller et al. 2005). In order to elucidate specific nanometer surface features that promoted vascular cell responses, 500, 200, and 100 nm polystyrene spheres were used to cast PLGA (Miller et al. 2007).

Results demonstrated that the PLGA with 200 nm structures promoted vascular cell responses and greater fibronectin interconnectivity compared to smooth PLGA and PLGA with 500 nm surface features (Figure 20.7).

Such results have been translated into the design of 3-D polymer scaffolds as several random and aligned 3-D nanofiber scaffolds have been fabricated for vascular applications. For example, Lee and colleagues fabricated and evaluated a variety of electrospun collagen, elastin, and synthetic polymers (such as PLLA, PLGA, and PCL) nanofiber scaffolds for vascular graft applications (Lee et al. 2007). These scaffolds have tailorable mechanical properties and exceptional cytocompatibility properties for vascular applications. Specifically, extensive smooth muscle cell infiltration was observed in the collagen/elastin/PLLA scaffold after 21 days of culture. By electrospinning on a rotating disk collector, Xu et al. fabricated an aligned PLLA-CL (75:25) nanofibrous scaffold, which mimicked the oriented fibril structure in the medial layer of an artery (Xu et al. 2004). Not only did

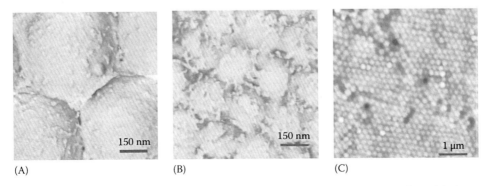

(A)   (B)   (C)

FIGURE 20.7 Atomic force microscopy images of fibronectin (5 μg/mL)-coated PLGA cast nanosphere surfaces. (A) Phase images of fibronectin adsorbed on PLGA with 500 nm surface features showed no interconnectivity between proteins. (B) Phase images of fibronectin adsorbed on PLGA with 200 nm surface features showed significant interconnectivity between fibronectin. (C) PLGA with 200 nm surface features only. (Adapted from Miller, D.C. et al., *J. Biomed. Mater. Res.*, 81A, 678, 2007.)

coronary artery smooth muscle cells favorably interact with that scaffold, but cells also oriented along the fiber, further emulating the natural environment. In addition to the electrospinning method, self-assembled peptides have been formulated into scaffolds to mimic the vascular basement membrane showing excellent cytocompatibility properties for vascular tissue repair. Genové et al. functionalized three peptide sequences from two basement membrane proteins (specifically, laminin and collagen IV) onto a self-assembled peptide scaffold (Genové et al. 2005). These tailorable self-assembled scaffolds enhanced endothelialization and improved nitric oxide release and laminin as well as collagen IV deposition by the endothelial cell monolayer. These results indicated the promise of biomimetic nanoscaffolds for improving vascular tissue engineering applications and when coupled with the aforementioned promise of nanomaterials for orthopedic applications, suggests a possible widespread use of nanomaterials for numerous tissue engineering applications.

20.3.3 Promise of Nanomaterials for Neural Tissue Regeneration

In addition to aiding in orthopedic and vascular tissue regeneration, nanomaterials are also helping to heal damaged nerves. In particular, nervous system injuries, diseases, and disorders occur far too frequently. In the United States, there are about 250,000–400,000 patients suffering from a spinal cord injury each year (Travis Roy Foundation Web site). Although various cell therapies and implants have been investigated, repairing damaged nerves and achieving full functional recovery are still challenging considering the complexity of the nervous system. For example, nearly 50,000 patients die among the average 1.4 million Americans that sustain traumatic brain injuries each year (Langlois et al. 2004). Generally, the nervous system can be divided into two main parts: the central nervous system (CNS) (including the brain and the spinal cord) and the peripheral nervous system (PNS) (including the spinal and autonomic nerves). These two systems have two different repair procedures after injury (Figure 20.8) (Bahr and Bonhoeffer 1994, Huang and Huang 2006, Zhang et al. 2005). For the PNS, the damaged axons usually regenerate and recover via proliferating Schwann cells, phagocytosing myelin by macrophages or monocytes, forming bands of Bünger by the bundling of Schwann cells, and sprouting axons in the distal segment (Evans 2001). However, it is difficult to re-extend and reinnervate axons to recover functions in the CNS because the CNS is absent of Schwann cells to promote axonal growth. More importantly, due to the influence of astrocytes, meningeal cells, and oligodendrocytes, the thick glial scar tissue typically formed around today's neural biomaterials will prevent proximal axon growth and inhibit neuron regeneration (Zhang et al. 2005). For these reasons, CNS injuries may cause severe functional damage and are much more difficult to repair than PNS injuries.

The ideal materials for neural tissue engineering applications should have excellent cytocompatible, mechanical, and electrical properties. Without good cytocompatibility properties,

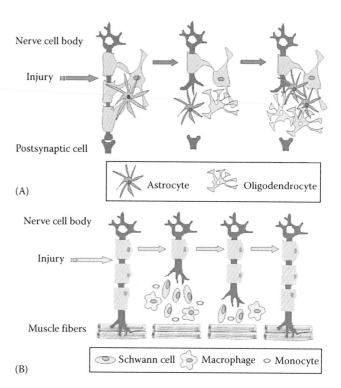

FIGURE 20.8 Schematic graphs of injured nerve regeneration in the central and peripheral nervous systems. (A) Central nervous system recovery process with glial scar tissue formation and (B) peripheral nervous system recovery process involving the activity of Schwann cells, macrophages, and monocytes. (Adapted from Bahr, M. and Bonhoeffer, F., *Trends Neurosci.*, 17, 473, 1994; Zhang, L. et al., Carbon nanotubes and nanofibers for tissue engineering applications, in *The Area of 'Carbon,'* (ed.) C. Liu, Research Signpost (Ed.), Trivandrum, India, 2009b, in press.)

materials may fail to improve neuron growth and at the same time may elicit severe inflammation or infection. Without sufficient mechanical properties, the scaffold may not last long enough to serve as a structure to physically support neural tissue regeneration. In addition, superior electrical properties of scaffolds are required to help stimulate and control neuron behavior under electrical stimulation, thereby more effectively guiding neural tissue repair. To date, various natural and synthetic materials have been adopted as nerve grafts to repair severely damaged nerves by bridging nerve gaps and guiding neuron outgrowth. However, there are still many shortcomings for these neural biomaterials: for autografts, it is usually difficult to collect sufficient donor nerves from patients and it is possible donor site nerve functions may be impaired (Terzis et al. 1997), and for allografts, inflammation, rejection, and transmission of diseases may frequently occur leading to implant failure (Zalewski and Gulati 1981). Other traditional biomaterials (such as silicon probes used in neuroprosthetic devices and polymers used as nerve conduits) used for neural tissue repair have been limited by the extensive formation of glial scar tissue around materials as well as poor mechanical and electrical properties not optimal for nerve regrowth. Nanotechnology provides a wide platform to develop novel and improved neural tissue engineering materials and therapy including designing nanofiber/nanotube scaffolds with exceptional cytocompatibility and conductivity properties to boost neuron activities. Nanomaterials have also been used to encapsulate various neural stem cells and Schwann cells into biomimetic nanoscaffolds to enhance nerve repair.

For example, work by Ramakrishna's research group has led to the fabrication of various nanofibrous PLLA or PCL scaffolds via electrospinning and phase separation; such scaffolds have demonstrated excellent cytocompatibility properties for neural tissue engineering applications (Koh et al. 2008, Prabhakaran et al. 2008, Yang et al. 2004). Recently, they incorporated laminin (a neurite-promoting ECM protein) into electrospun PLLA nanofibers in order to create a biomimetic scaffold for peripheral nerve repair (Koh et al. 2008). The results showed that neurite outgrowth improved on lamnin-PLLA scaffolds through facile blended electrospinning. In another recent report, electrospun PCL/chitosan nanofiber scaffolds exhibited improved mechanical properties compared to chitosan (Prabhakaran et al. 2008). A 48% increase in Schwann cell proliferation was observed on these PCL/chitosan scaffolds compared to PCL scaffolds alone after 8 days of culture. In addition, Zhang and colleagues also reported favorable neural cell responses on the self-assembled peptide nanofiber scaffold (called SAPNS). Holmes et al. reported that the self-assembled peptide scaffold supported neuronal cell functions, neurite outgrowth, and functional synapse formation among neurons (Holmes et al. 2000). Furthermore, Ellis-Behnke et al. investigated SAPNS for in vivo axon regeneration in the CNS (Ellis-Behnke et al. 2006). The SAPNS aided in CNS regeneration to help axonal growth, even "knitting" the brain tissue together and successfully improving functional recovery.

Due to the fact that carbon nanotubes/fibers have excellent electrical conductivity, strong mechanical properties, and have similar nanoscale dimensions to neurites, they have been used to guide axon regeneration and improve neural activity as biomimetic scaffolds at neural tissue injury sites. In particular, Mattson et al. found for the first time that neurons grew on multiwalled carbon nanotubes (MWCNTs) (Mattson et al. 2000). They observed over a 200% increase in total neurite length and nearly a 300% increase in the number of branches and neurites on MWCNTs coated with 4-hydroxynonenal compared to uncoated MWCNTs. Hu et al. revealed that different surface charges of MWCNTs, obtained through chemical functionalization, resulted in different neurite outgrowth patterns (such as neurite length, branching, and the number of growth cones) (Hu et al. 2004). They demonstrated that positively charged MWCNTs significantly increased the number of growth cones and neurite branches compared to negatively charged MWCNTs, thus, controlling neural growth. Furthermore, Lovat et al. demonstrated that purified MWCNTs potentially boosted electrical signal transfer of neuronal networks (Figure 20.9A and B) (Lovat et al. 2005). Moreover, highly ordered CNT/CNF matrices or freestanding nanotube films have been fabricated for neural tissue engineering applications (Firkowska et al. 2006, Gheith et al. 2005). For instance, Gheith et al. investigated the

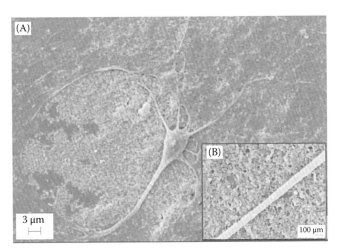

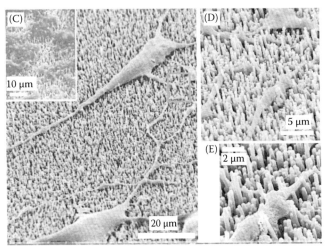

FIGURE 20.9 SEM images of neural cell adhesion on carbon nanotube/fiber substrates. (A) Neonatal hippocampal neurons adherent on purified MWCNT glass substrates with extended neurites after 8 days; inset image (B) shows a single neurite in close contact to CNTs. (Adapted from Lovat, V. et al., *Nano Lett.*, 5, 1107, 2005. With permission.) (C), (D), and (E) PC12 neural cells grown freestanding on vertically aligned CNFs coated with polypyrrole at different magnifications. (Adapted from Nguyen-Vu, T.D.B. et al., *IEEE Trans. Biomed. Eng.*, 54, 1121, 2007. With permission.)

biocompatibility of a freestanding positively charged SWCNT/polymer thin-film membrane prepared by layer-by-layer assembly (Gheith et al. 2005). They observed that 94%–98% of neurons were viable on the SWCNT/polymer films after a 10 day incubation. The SWCNT/polymer films favorably induced cell differentiation, guided neuron extension, and directed more elaborate branches than controls.

In order to inhibit extensive astrocyte functions that result in the formation of glial scar tissue, different weight ratios of high surface energy CNFs were incorporated into polymers and the result showed for the first time that astrocyte adhesion can be effectively inhibited by using CNF/polymer composites (McKenzie et al. 2004). In addition, decreased astrocyte proliferation was observed on nanostructured CNFs, thus, leading to decreased glial scar tissue formation on such materials. On the other hand, Nguyen-Vu et al. fabricated a vertically aligned CNF nanoelectrode array by coating a thin film of electronic conductive polymer (such as polypyrrole) for neural implants (Nguyen-Vu et al. 2007). The vertical CNF arrays had more open and mechanically robust 3-D structures as well as better electrical conductivity which contributed to forming an intimate neural–electrical interface between cells and nanofibers (Figure 20.9C and E). Gabay et al. developed a novel method to fabricate islands of CNT on substrates. Neurons preferably attached on the CNT islands and further extended their neurites to form interconnected neural networks according to pre-designed patterns (Gabay et al. 2005). In this manner, the CNTs/CNFs and their composites are promising scaffold candidates for injured neural tissue repair.

Studies have also provided evidence that individual CNTs/CNFs may be useful in treating neurological damage when combined with stem cells. Stem cells have the potential to differentiate and self-renew into controllable, desirable cell types: that is, neural stem cells in the CNS can differentiate into neurons and astrocytes (Reynolds and Weiss 1992). Thus, many efforts have focused on impregnating multipotential stem cells into CNTs/CNFs and other nanoscaffolds, which can be directly transplanted into injury sites and assist neural tissue recovery. However, a challenging problem has been how to effectively deliver and selectively differentiate stem cells into favorable neuronal cell types at injury sites in order to regenerate desirable tissue. Although the underlying mechanisms triggering differentiation of stem cells are not totally clear, accumulated evidence has indicated that novel biomimetic nanomaterials may contribute to selective stem cell differentiation (without the use of growth factors) (Jan and Kotov 2007, Lee et al. 2009). For example, Lee et al. injected CNFs impregnated with stem cells into stroke-damaged neural tissue in rat brains and found extensive neural stem cell differentiation with little glial scar tissue formation in vivo (Lee et al. 2009). After 1 and 3 weeks of animal implantation, histological sections showed that neural stem cells favorably differentiated into neurons (Figure 20.10A and B) and little to no glial scar tissue (Figure 20.10C and D) formed around CNFs compared to controls (only implanting stem cells without CNFs or implanting CNFs without cells). In addition, Jan et al. successfully differentiated mouse embryonic neural stem cells including neurospheres and single cells into neurons on layer-by-layer assembled SWCNT/polyelectrolyte composites (Jan and Kotov 2007). Specifically, the layer-by-layer SWCNT composites promoted slightly more neurons and fewer astrocytes on substrates during a 7 day culture period than poly-L-ornithine (a common substrate for neural stem cell studies). Clearly, CNTs/CNFs played an important role in effectively delivering stem

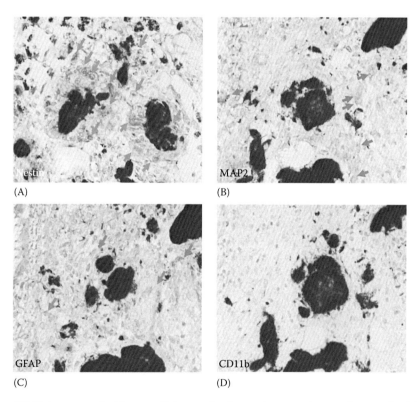

(A)   (B)

(C)   (D)

FIGURE 20.10 Histology of CNFs impregnated with stem cells into stroke-damaged rat neural tissue after 3 weeks. (A) and (B), Numerous active neuroprogenitor cells and fully differentiated neurons (brown stained cells, marked by nestin and MAP2, respectively) around CNFs. (C) and (D) Few glial cells interacting with CNFs led to little or no glial scar tissue formation. GFAP is a marker for astrocytes; CD11b is a marker for activated microglia cells. Black areas in the images are CNFs. Scale bar is 25 μm. (Adapted from Lee, J.E. et al., *Int. J. Nanomed.*, 2009, in press.)

cells into injured sites and promoted stem cells to differentiate into favorable neurons to repair damaged neural tissues.

20.3.4 Promise of Nanomaterials for Bladder Tissue Regeneration

Nanomaterials have also been used in soft tissues such as the bladder. As the sixth most common cancer in the United States, urinary bladder cancer affects over 53,200 Americans and leads to 12,200 deaths annually (AUA 2008). Although standard treatments such as surgery to remove bladder tumors followed by radiation, chemotherapy, and immunotherapy have improved, various complications (such as systemic infections, flu-like symptoms, and cancer recurrence) with these procedures are still too commonly reported. Sometimes, radical cystectomy by removing parts or even the entire bladder is needed. However, such a drastic approach requires the implantation of a bladder tissue replacement to quickly recover bladder functions. As emerging bladder tissue engineering materials, nanomaterials provide a promising approach to more efficiently improve bladder tissue regeneration for some of the same reasons mentioned above (biologically inspired roughness, increased surface energy, selective protein adsorption, etc.). In particular, Harrington and colleagues have coated a series of branched or linear self-assembling PA nanofibers with cell-adhesive RGDS (Arg-Gly-Asp-Ser) on traditional PGA scaffolds (Harrington et al. 2006). Human bladder smooth muscle cell densities on the branched PA/PGA nanocomposite were greater than on the uncoated PGA after 17 days of culture. In a recent review, they mentioned a model of encapsulation of bladder smooth muscle cells and urothelial cells into a PA/PGA gel with specific growth factors (Figure 20.11) (Harrington et al. 2008). Due to their ability to mimic the oriented nanostructured bladder ECM, electrospun polymer nanofibers have been used in bladder tissue engineering. Baker et al. showed that bladder smooth muscle cells were aligned on oriented electrospun polystyrene

scaffolds similar to the native bladder tissue (Baker et al. 2006). This study also demonstrated that argon plasma-treated electrospun polystyrene nanofibers significantly improved smooth muscle cell attachment. Fibrinogen has also been electrospun into a scaffold for urinary tract tissue regeneration (McManus et al. 2007). This study demonstrated that human bladder smooth muscle cells rapidly migrated into, proliferated onto, and remodeled the 3-D fibrinogen scaffold.

Other nanostructured polymers with superior biocompatibility properties have been widely investigated by Haberstroh and colleagues for bladder tissue regeneration applications (Pattison et al. 2005, 2007, Thapa et al. 2003). For instance, this research group used nanotextured PLGA and poly(ether urethane) (PU) films to successfully enhance bladder smooth muscle cell functions (Thapa et al. 2003). Through chemical etching technologies, PLGA and PU were transformed from their native nano-smooth surface features into those possessing a high degree of nano-roughness. This study revealed that nano-roughness played a critical role in promoting bladder smooth muscle cell proliferation once the influence of surface chemistry change was eliminated (through cast-mold techniques using the chemical-treated polymer as the cast). Recently, Pattison et al. also demonstrated that nanostructured PLGA and PU 3-D scaffolds prepared by solvent casting and salt leaching methods significantly enhanced bladder smooth muscle cell functions and ECM protein synthesis compared to conventional nano-smooth polymers in vitro (Pattison et al. 2007). Furthermore, preliminary in vivo studies have provided evidence that nanostructured polymer scaffolds form little to no calcium oxalate stones (stone formation is a common problem during bladder replacements) in augmented rat bladders. Although there are many unknowns for the use of nanomaterials in bladder tissue engineering applications, it is undoubtedly a promising future research direction to utilize these biomimetic nanomaterials with progenitor cells to regenerate bladder tissue in resected bladder cancerous tissue locations.

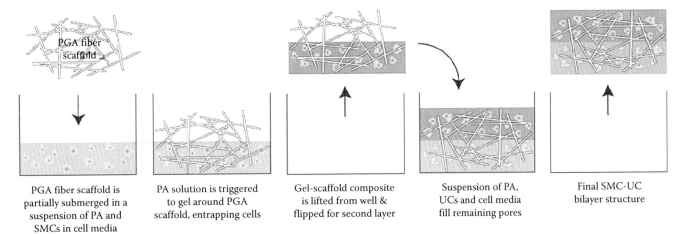

FIGURE 20.11 Schematic illustration of the bilayer smooth muscle cell/urothelial cell (SMC-UC) encapsulation in a PA/PGA gel. (Adapted from Harrington, D.A. et al., *World J. Urol.*, 26, 315, 2008. With permission.)

20.4 Potential Risks of Nanomaterials toward Human Health

As described, nanotechnology has achieved tremendous progress in a relatively short time period in medical applications. Of course, this means that nanomaterials have entered widespread industrial production. For instance, nano ceramics are commercially available as new bone grafts or as implant coating materials (i.e., nano HA paste—Ostim® from Obernburg, Germany and nano-beta-tricalcium phosphate—Vitoss from Orthovita, Malvern, Pennsylvania) (Wagner et al. 2006). However, it is important to note that the research on nanomaterials for tissue engineering applications is still at its infancy and, most importantly, the influence of nanomaterials on human health and the environment is not well understood. In particular, toxic responses to nanoparticles generated from the degradation of implanted nanomaterials via wear debris from artificial joints with nanofeatures and heavy metals (iron, nickel, and cobalt catalysts) remaining in CNTs have all been reported. Many reports on the cellular uptake of nanoparticles in the lungs, immune system, as well as other organs have been published (Gutwein et al. 2004, Hoet et al. 2004, Lam et al. 2004). Nanoparticle uptake by endothelial cells, alveolar macrophages, pulmonary or intestinal epithelium, nerve cells, etc., has been reviewed and, thus, may possess a problem for this field if not thoroughly understood before widespread use (Hoet et al. 2004). Specifically, the viability of osteoblasts in vitro when cultured in the presence of nano alumina and titania particles for 6 h was investigated (Gutwein and Webster 2004). This study demonstrated that ceramic nanoparticles were safer to osteoblasts than conventional, micron-sized, ceramic particles. In contrast, in an in vivo study, it has been showed that CNTs were more toxic than carbon black in the lungs, which may be a serious occupational health hazard in chronic inhalation exposures (Lam et al. 2004). Sometimes nanoparticle interactions with biomolecules in vivo or their aggregation states may change their toxicity to humans. But the often contradictory results of current studies are clearly not enough to provide the final answer concerning nanomaterial toxicity. In-depth investigations of nanomaterials on human health and the environment are necessary to fully realize whether nanoparticles should be used in biomedical applications.

20.5 Conclusions

To date, there has been an exponential increase in studies using nanotechnology for tissue engineering applications. To be concise, this chapter only covered the recent progress using nanomaterials for bone, cartilage, vascular, neural, and bladder tissue regeneration. Many additional recent reviews of nanotechnology applications for the specific regeneration of tissues can be found (Christenson et al. 2007, Harrington et al. 2008, Liu-Snyder and Webster 2006, Miller et al. 2004b, Nukavarapu et al. 2008b, Seidlits et al. 2008, Vasita and Katti 2006, Webster and Ahn 2007, Zhang et al. 2008c). Nanotechnology approaches for the regeneration of other types of tissues (such as the muscle, skin, kidney, liver, pancreas, and the immune system) have also been reviewed (Khademhosseini et al. 2008, Nukavarapu et al. 2008b). Although many challenges may lie ahead, synthetic nanomaterials can mimic properties of the natural ECM of all tissues and, thus, show great potential for improving numerous tissue engineering applications. Particularly, due to their excellent cytocompatibility properties, research interest has been evoked to use nanomaterials as the next generation of tissue repair materials. In the future, the underlying mechanisms of the in vivo interactions between nanomaterials and cells at the molecular level will significantly advance this field.

References

American Academy of Orthopedic Surgeons (AAOS). 2004. URL: http://www.aaos.org/Research/stats/patientstats.asp.

American Heart Association (AHA). 2004. URL: http://www.americanheart.org/.

American Urological Association (AUA). 2008. URL: http://www.urologyhealth.org/adult/index.cfm?cat=03&topic=37.

Athanasiou, K. A., Niederauer, G. G., and Agrawal, C. M. 1996. Sterilization, toxicity, biocompatibility and clinical applications of polylactic acid/polyglycolic acid copolymers. *Biomaterials* 17: 93–102.

Bahr, M. and Bonhoeffer, F. 1994. Perspectives on axonal regeneration in the mammalian CNS. *Trends Neurosci.* 17: 473–479.

Baker, S. C., Atkin, N., Gunning, P. A. et al. 2006. Characterisation of electrospun polystyrene scaffolds for three-dimensional in vitro biological studies. *Biomaterials* 27: 3136–3146.

Balasundaram, G. and Webster, T. J. 2007. An overview of nanopolymers for orthopedic applications. *Macromol. Biosci.* 7: 635–642.

Burns, K., Yao, C., and Webster, T. J. 2008. Increased chondrocyte adhesion on nanotubular anodized titanium. *J. Biomed. Mater. Res. A* 88: 561–568.

Chen, V. J. and Ma, P. X. 2004. Nano-fibrous poly(-lactic acid) scaffolds with interconnected spherical macropores. *Biomaterials* 25: 2065–2073.

Choudhary, S., Haberstroh, K. M., and Webster, T. J. 2007. Enhanced functions of vascular cells on nanostructured Ti for improved stent applications. *Tissue Eng. Part A* 13: 1421–1430.

Christenson, E. M., Anseth, K. S., van den Beucken, J. J. et al. 2007. Nanobiomaterial applications in orthopedics. *J. Orthop. Res.* 25: 11–22.

Colon, G., Ward, B. C., and Webster, T. J. 2006. Increased osteoblast and decreased *Staphylococcus epidermidis* functions on nanophase ZnO and TiO$_2$. *J. Biomed. Mater. Res. A* 78: 595–604.

Degasne, I., Basle, M. F., Demais, V. et al. 1999. Effects of roughness, fibronectin and vitronectin on attachment, spreading, and proliferation of human osteoblast-like cells (Saos-2) on titanium surfaces. *Calcified Tissue Int.* 64: 499–507.

Ellis-Behnke, R. G., Liang, Y. X., You, S. W. et.al. 2006. Nano neuro knitting: Peptide nanofiber scaffold for brain repair and axon regeneration with functional return of vision. *PNAS* 103: 5054–5059.

Evans, G. R. D. 2001. Peripheral nerve injury: A review and approach to tissue engineered constructs. *Anat. Rec.* 263: 396–404.

Fahlman, B. D. 2007. *Materials Chemistry*. Dordrecht, the Netherlands: Springer.

Fecek, C., Yao, D., Kaçorri, A. et al. 2008. Chondrogenic derivatives of embryonic stem cells seeded into 3D polycaprolactone scaffolds generated cartilage tissue in vivo. *Tissue Eng. Part A* 14: 1403–1413.

Firkowska, I., Olek, M., Pazos-Peréz, N., Rojas-Chapana, J., and Giersig, M. 2006. Highly ordered MWNT-based matrixes: Topography at the nanoscale conceived for tissue engineering. *Langmuir* 22: 5427–5434.

Freeman, J. W., Wright, L. D., Laurencin, C. T., and Bhattacharyya, S. 2008. Nanofabrication techniques. In *Biomedical Nanostructures*, eds. K. E. Gonsalves, C. R. Halberstadt, C. T. Laurencin, and L. S. Nair, 3–24. Hoboken, NJ: John Wiley & Sons, Inc.

Gabay, T., Jakobs, E., Ben-Jacob, E., and Hanein, Y. 2005. Engineered self-organization of neural networks using carbon nanotube clusters. *Physica A* 350: 611–621.

Genové, E., Shen, C., Zhang, S., and Semino, C. E. 2005. The effect of functionalized self-assembling peptide scaffolds on human aortic endothelial cell function. *Biomaterials* 26: 3341–3351.

Gheith, M. K., Sinani, V. A., Wicksted, J. P., Matts, R. L., and Kotov, N. A. 2005. Single-walled carbon nanotube polyelectrolyte multilayers and freestanding films as a biocompatible platform for neuroprosthetic implants. *Adv. Mater.* 17: 2663–2670.

Gutwein, L. G. and Webster, T. J. 2004. Increased viable osteoblast density in the presence of nanophase compared to conventional alumina and titania particles. *Biomaterials* 25: 4175–4183.

Harrington, D. A., Cheng, E. Y., Guler M. O. et al. 2006. Branched peptide-amphiphiles as self-assembling coatings for tissue engineering scaffolds. *J. Biomed. Mater. Res.* 78A: 157–167.

Harrington, D. A., Sharma, A. K., Erickson, B. A., and Cheng, E. Y. 2008. Bladder tissue engineering through nanotechnology. *World J. Urol.* 26:315–322.

Hartgerink, J. D., Beniash, E., and Stupp, S. I. 2001. Self-assembly and mineralization of peptide-amphiphile nanofibers. *Science* 294: 1684–1688.

Hata, K., Futaba, D. N., Mizuno, K., Namai, T.,Yumura, M., and Iijima, S. 2004. Water-assisted highly efficient synthesis of impurity-free single-walled carbon nanotubes. *Science* 306: 1362–1364.

Hoet, P. H., Bruske-Hohlfeld, I., and Salata, O. V. 2004. Nanoparticles—Known and unknown health risks. *J. Nanobiotechnol.* 2: 12.

Holmes, T. C., de Lacalle, S., Su, X., Liu, G., Rich, A., and Zhang, S. 2000. Extensive neurite outgrowth and active synapse formation on self-assembling peptide scaffolds. *PNAS* 97: 6728–6733.

Hosseinkhani, H., Hosseinkhani, M., Tian, F., Kobayashi, H., and Tabata, Y. 2006. Osteogenic differentiation of mesenchymal stem cells in self-assembled peptide-amphiphile nanofibers. *Biomaterials* 27: 4079–4086.

Hu, H., Ni, Y. C., Montana, V., Haddon, R. C., and Parpura, V. 2004. Chemically functionalized carbon nanotubes as substrates for neuronal growth. *Nano Lett.* 4: 507–511.

Huang, Y. C. and Huang, Y. Y. 2006. Biomaterials and strategies for nerve regeneration. *Artif. Organs* 30: 514–522.

Jan, E. and Kotov, N. A. 2007. Successful differentiation of mouse neural stem cells on layer-by-layer assembled single-walled carbon nanotube composite. *Nano Lett.* 7: 1123–1128.

Kaplan, F. S., Hayes, W. C., Keaveny, T. M., Boskey, A., Einhorn, T. A., and Iannotti, J. P. 1994. Form and function of bone. In *Orthopedic Basic Science*, ed. S. R. Simon, 127–185. Rosemont, IL: American Academy of Orthopaedic Surgeons.

Khademhosseini, A., Borenstein, J., Toner, M., and Takayama, S. 2008. *Micro- and Nanoengineering of the Cell Microenvironment: Technologies and Applications*. Norwood, MA: Artech House.

Kisiday, J., Jin, M., Kurz, B. et al. 2002. Self-assembling peptide hydrogel fosters chondrocyte extracellular matrix production and cell division: Implications for cartilage tissue repair. *Proc. Natl. Acad. Sci.* 99: 9996–10001.

Koh, H. S., Yong, T., Chan, C. K., and Ramakrishna, S. 2008. Enhancement of neurite outgrowth using nanostructured scaffolds coupled with laminin. *Biomaterials* 29: 3574–3582.

Lam, C. W., James, J. T., McCluskey, R., and Hunter, R. L. 2004. Pulmonary toxicity of single-wall carbon nanotubes in mice 7 and 90 days after intratracheal instillation. *Toxicol. Sci.* 77: 126–134.

Langer, R. and Vacanti, J. P. 1993. Tissue engineering. *Science* 260: 920–926.

Langlois, J. A., Rutland-Brown, W., and Thomas, K. E. 2004. *Traumatic Brain Injury in the United States: Emergency Department Visits, Hospitalizations, and Deaths, Centers for Disease Control and Prevention*. Atlanta, GA: National Center for Injury Prevention and Control.

Li, W. J., Jiang, Y. J., and Tuan, R. S. 2008. Cell-nanofiber-based cartilage tissue engineering using improved cell seeding, growth factor, and bioreactor technologies. *Tissue Eng. Part A* 14: 639–648.

Li, W. J., Tuli, R., Okafor, C. et al. 2005. A three-dimensional nanofibrous scaffold for cartilage tissue engineering using human mesenchymal stem cells. *Biomaterials* 26: 599–609.

Lee, J. E., Kim, J. H., Kim, J. Y., Kang, D., and Webster, T. J. 2009. Repair of stroke induced neural tissue damage through implantation of carbon nanofibers impregnated with stem cells. *Int. J. Nanomed.* (in press).

Lee, S. J., Yoo, J. J., Lim, G. J., Atala, A., and Stitzel, J. 2007. In vitro evaluation of electrospun nanofiber scaffolds for vascular graft application. *J. Biomed. Mater. Res.* 83A: 999–1008.

Liu-Snyder, P. and Webster, T. J. 2006. Designing drug-delivery systems for the nervous system using nanotechnology: Opportunities and challenges. *Expert Rev. Med. Devices* 3: 683–687.

Lovat, V., Pantarotto, D., Lagostena, L. et al. 2005. Carbon nanotube substrates boost neuronal electrical signaling. *Nano Lett.* 5: 1107–1110.

Mattson, M. P., Haddon, R. C., and Rao, A. M. 2000. Molecular functionalization of carbon nanotubes and use as substrates for neuronal growth. *J. Mol. Neurosci.* 14: 175–182.

McKenzie, J. L., Waid, M. C., Shi, R., and Webster, T. J. 2004. Decreased functions of astrocytes on carbon nanofiber materials. *Biomaterials* 25: 1309–1317.

McManus, M., Boland, E., Sell, S. et al. 2007. Electrospun nanofibre fibrinogen for urinary tract tissue reconstruction. *Biomed. Mater.* 2: 257–262.

Miller, D. C., Haberstroh, K. M., and Webster, T. J. 2005. Mechanism(s) of increased vascular cell adhesion on nanostructured poly(lactic-*co*-glycolic acid) films. *J. Biomed. Mater. Res.* 73A: 476–484.

Miller, D. C., Haberstroh, K. M., and Webster, T. J. 2007. PLGA nanometer surface features manipulate fibronectin interactions for improved vascular cell adhesion. *J. Biomed. Mater. Res.* 81A: 678–684.

Miller, D. C., Thapa, A., Haberstroh, K. M., and Webster, T. J. 2004a. Endothelial and vascular smooth muscle cell function on poly(lactic-co-glycolic acid) with nano-structured surface features. *Biomaterials* 25: 53–61.

Miller, D. C., Webster, T. J., and Haberstroh, K. M. 2004b. Technological advances in nanoscale biomaterials: The future of synthetic vascular graft design. *Expert Rev. Med. Devices* 1: 259–268.

Nguyen-Vu, T. D. B., Chen, H., Cassell, A. M., Andrews, R. J., Meyyappan, M., and Li, J. 2007. Vertically aligned carbon nanofiber architecture as a multifunctional 3-D neural electrical interface. *IEEE Trans. Biomed. Eng.* 54: 1121–1128.

Nukavarapu, S. P., Kumbar, S. G., Brown, J. L. et al. 2008a. Polyphosphazene/nano-hydroxyapatite composite microsphere scaffolds for bone tissue engineering. *Biomacromolecules* 9: 1818–1825.

Nukavarapu, S. P., Kumbar, S. G., Nair, L. S., and Laurencin, C. T. 2008b. Nanostructures for tissue engineering/regenerative medicine. In *Biomedical Nanostructures*, eds. K. E. Gonsalves, C. R. Halberstadt, C. T. Laurencin, and L. S. Nair, 377–407. Hoboken, NJ: John Wiley & Sons, Inc.

Osathanon, T., Linnes, M. L., Rajachar, R. M., Ratner, B. D., Somerman, M. J., and Giachelli, C. M. 2008. Microporous nanofibrous fibrin-based scaffolds for bone tissue engineering. *Biomaterials* 29: 4091–4099.

Park, G. E., Pattison, M. A., Park, K., and Webster, T. J. 2005. Accelerated chondrocyte functions on NaOH-treated PLGA scaffolds. *Biomaterials* 26: 3075–3082.

Pattison, M. A., Wurster, S., Webster, T. J., and Haberstroh, K. M. 2005. Three-dimensional, nano-structured PLGA scaffolds for bladder tissue replacement applications. *Biomaterials* 26: 2491–2500.

Pattison, M., Webster, T. J., Leslie, J., Kaefer, M., and Haberstroh, K. M. 2007. Evaluating the in vitro and in vivo efficacy of nano-structured polymers for bladder tissue replacement applications. *Macromol. Biosci.* 7: 690–700.

Prabhakaran, M. P., Venugopal, J. R., Chyan, T. T. et al. 2008. Electrospun biocomposite nanofibrous scaffolds for neural tissue engineering. *Tissue Eng. Part A* 14: 1787–1797.

Price, R. L., Waid, M. C., Haberstroh, K. M., and Webster, T. J. 2003. Selective bone cell adhesion on formulations containing carbon nanofibers. *Biomaterials* 24: 1877–1887.

Puckett, S., Pareta, R., and Webster, T. J. 2008. Nano rough micron patterned titanium for directing osteoblast morphology and adhesion. *Int. J. Nanomed.* 3: 229–241.

Reynolds, B. A. and Weiss, S. 1992. Generation of neurons and astrocytes from isolated cells of the adult mammalian central nervous system. *Science* 255: 1707–1710.

Sato, M. 2006. Nanophase hydroxyapatite coatings for dental and orthopedic applications. PhD thesis, Purdue University, West Lafayette, IN.

Seidlits, S. K., Lee, J. Y., and Schmidt, C. E. 2008. Nanostructured scaffolds for neural applications. *Nanomedicine* 3: 183–199.

Siegel, R. W. and Fougere, G. E. 1995. Mechanical properties of nanophase metals. *Nanostruct. Mater.* 6: 205–216.

Sitharaman, B., Shi, X., Walboomers, X. F. et al. 2008. In vivo biocompatibility of ultra-short single-walled carbon nanotube/biodegradable polymer nanocomposites for bone tissue engineering. *Bone* 43: 362–370.

Sun, S., Zeng, H., Robinson, D. B. et al. 2004. Monodisperse MFe_2O_4 (M = Fe, Co, Mn) nanoparticles. *J. Am. Chem. Soc.* 126: 273–279.

Supronowicz, P. R., Ajayan, P. M., Ullmann, K. R., Arulanandam, B. P., Metzger, D. W., and Bizios, R. 2002. Novel current-conducting composite substrates for exposing osteoblasts to alternating current stimulation. *J. Biomed. Mater. Res.* 59: 499–506.

Taniguchi, N. 1974. On the basic concept of 'NanoTechnology.' *Proceedings of International Conference on Precision Engineering (ICPE)*, Tokyo, Japan, 18–23.

Terzis, J. K., Sun, D. D., and Thanos, P. K. 1997. Historical and basic science review: Past, present and future of nerve repair. *J. Reconstr. Microsurg.* 13: 215–225.

Thapa, A., Miller, D. C., Webster, T. J., and Haberstroh, K. M. 2003. Nano-structured polymers enhance bladder smooth muscle cell function. *Biomaterials* 24: 2915–2926.

Travis Roy foundation. URL: http://www.travisroyfoundation. org/pages/resources-stats.htm.

Vasita, R. and Katti, D. S. 2006. Nanofibers and their applications in tissue engineering. *Int. J. Nanomed.* 1: 15–30.

Venugopal, J. R., Low, S., Choon, A. T., Kumar, A. B., and Ramakrishna, S. 2008. Nanobioengineered electrospun composite nanofibers and osteoblasts for bone regeneration. *Artif. Organs* 32: 388–397.

Wagner, V., Dullaart, A., Bock, A. K., and Zweck, A. 2006. The emerging nanomedicine landscape. *Nat. Biotechnol.* 24: 1211–1217.

Webster, T. J. 2001. Nanophase ceramics: The future orthopedic and dental implant material. In *Advances in Chemical Engineering*, ed. J. Y. Ying, 125–166. New York: Academic press.

Webster, T. J. and Ahn, E. S. 2007. Nanostructured biomaterials for tissue engineering bone. *Adv. Biochem. Eng. Biotechnol.* 103: 275–308.

Webster, T. J. and Ejiofor, J. U. 2004. Increased osteoblast adhesion on nanophase metals: Ti, Ti$_6$Al$_4$V, and CoCrMo. *Biomaterials* 25: 4731–4739.

Webster, T. J., Hellenmeyer, E. L., and Price, R. L. 2005. Increased osteoblast functions on theta + delta nanofiber alumina. *Biomaterials* 26: 953–960.

Webster, T. J., Ergun, C., Doremus, R. H., Siegel, R. W., and Bizios, R. 2000. Specific proteins mediate enhanced osteoblast adhesion on nanophase ceramics. *J. Biomed. Mater. Res.* 51: 475–483.

Webster, T. J., Schadler, L. S., Siegel, R. W., and Bizios, R. 2001a. Mechanisms of enhanced osteoblast adhesion on nanophase alumina involve vitronectin. *Tissue Eng.* 7: 291–301.

Webster, T. J., Ergun, C., Doremus, R. H., Seigel, R. W., and Bizios, R. 2001b. Enhanced osteoclast-like functions on nanophase ceramics. *Biomaterials* 22: 1327–1333.

Xu, C. Y., Inai, R., Kotaki, M., and Ramakrishna, S. 2004. Aligned biodegradable nanofibrous structure: A potential scaffold for blood vessel engineering. *Biomaterials* 25: 877–886.

Yang, F., Murugan, R., Ramakrishna, S., Wang, X., Ma, Y. X., and Wang, S. 2004. Fabrication of nano-structured porous PLLA scaffold intended for nerve tissue engineering. *Biomaterials* 25: 1891–1900.

Yao, C., Perla, V., McKenzie, J. L., Slamovich, E. B., and Webster, T. J. 2005. Anodized Ti and Ti$_6$Al$_4$V possessing nanometer surface features enhances osteoblast adhesion. *J. Biomed. Nanotechnol.* 1: 68–73.

Zalewski, A. A. and Gulati, A. K. 1981. Rejection of nerve allografts after cessation of immunosuppression with cyclosporin A. *Transplantation* 31: 88–89.

Zanello, L. P., Zhao, B., Hu, H., and Haddon, R. C. 2006. Bone cell proliferation on carbon nanotubes. *Nano Lett.* 6: 562–567.

Zhang, N., Yan, H., and Wen, X. 2005. Tissue-engineering approaches for axonal guidance. *Brain Res. Rev.* 49: 48–64.

Zhang, L., Ramsaywack, S., Fenniri, H., and Webster, T. J. 2008a. Enhanced osteoblast adhesion on self-assembled nanostructured hydrogel scaffolds. *Tissue Eng. Part A* 14: 1353–1364.

Zhang, L., Chen, Y. P., Rodriguez, J., Fenniri, H., and Webster, T. J. 2008b. Biomimetic helical rosette nanotubes and nanocrystalline hydroxyapatite coatings on titanium for improving orthopedic implants. *Int. J. Nanomed.* 3: 323–334.

Zhang, L., Sirivisoot, S., Balasundaram, G., and Webster, T. J. 2008c. Nanoengineering for bone tissue engineering. In *Micro- and Nanoengineering of the Cell Microenvironment: Technologies and Applications*, eds. A. Khademhosseini, J. Borenstein, M. Toner, and S. Takayama, 431–460. Norwood, MA: Artech house.

Zhang, L., Sirivisoot, S., Balasundaram, G., and Webster, T. J. 2009a. Nanomaterials for improved orthopedic and bone tissue engineering applications. In *Advanced Biomaterials: Fundamentals, Processing and Applications*, eds. B. Basu, D. Katti, and A. Kumar, Hoboken, NJ: John Wiley & Sons, Inc., (in press).

Zhang, L., Ercan, B., and Webster, T. J. 2009b. Carbon nanotubes and nanofibers for tissue engineering applications. In *The Area of 'Carbon,'* ed. C. Liu, Trivandrum, India: Research Signpost, (Ed.), (in press).

21

Nanotechnology for the Urologist

Hashim Uddin Ahmed
University College London

Lyndon Gommersall
University of Birmingham

Iqbal S. Shergill
Harold Wood Hospital

Manit Arya
University College London

Mark Emberton
University College London

21.1 Introduction

This chapter reviews important aspects of nanotechnology research and clinical applications in the field of urology. There is particular emphasis on key preclinical and clinical studies to provide a working understanding of recent and potential applications in the diagnosis, treatment, and long-term management of urological patients. It is widely expected that nanotechnology and nanomedicine will have a significant impact on urological research and clinical practice allowing urologists to intervene at the cellular and molecular level. With structured and safe implementation, nanotechnologies have the potential to revolutionize urological practice (Gommersall et al. 2007).

21.2 Background

21.2.1 Revision of Basic Nanomedicine

Nanotechnology is the study, design, creation, synthesis, manipulation, and application of materials, devices, and systems through the control of matter at the nanometer scale. Nanotechnology involves the utilization of man-made products no larger than 1–1000 nm (i.e., a few atoms to smaller than a single cell). A dictionary definition (Nano nano-pref. (1) Extremely small nanoid. (2) One-billionth [10^{-9} m] nanometer) elucidates the scale of this field and allows us to define that nanoscale particles are in the 10^{-9} m dimension range, consistent with the magnitude of most synthetic nanoparticles to date. For a real perspective, the width of a DNA molecule is 2.5 nm; cell membranes are 6–10 nm thick; and most proteins are between 5 and 20 nm in diameter. Therefore, most conventional molecular research is already proceeding in nanoscale dimensions. Nanomaterials can be grouped into a number of distinct entities. These include nanovectors, nanotubes, and nanosensors for targeted drug delivery; nanowires and nanocantilever arrays for early detection of precancerous and malignant lesions; and nanopores for DNA sequencing (BECON 2006).

21.2.2 Is Nanotechnology Relevant to Urology?

Urology is in a key position to benefit from nanotechnology. The combination of functional abnormalities (e.g., hypercontractile or hypocontractile bladder, benign prostatic hyperplasia), reconstructive problems (e.g., bladder reconstruction, urinary sphincters), and cancer (e.g., early detection and imaging, staging, minimally invasive treatments) all lend themselves to the development of these novel technologies. Currently, the three main areas of integration of synthetic nanotechnology with potential availability to urologists are either for the delivery of pharmaceuticals, as an adjunct to conventional imaging, minimally invasive therapies, and tissue engineering. Most of this research focuses around cancer. A summary of the current nanotechnology applications in urology is included in Table 21.1 (Humes 2000).

21.3 State of the Art

21.3.1 Nanotechnology in Imaging and Diagnosis

Nanotechnology has the opportunity to enhance the detection of pathology in a specific and sensitive manner. The current mainstay of medical diagnosis with traditional circulating contrast media utilize broadly untargeted approaches. These techniques

TABLE 21.1 Overview of the Use of Nanotechnology in Urology

Pathology	Nanotechnology
Prostate-specific antigen	PSA has been detected at clinically significant concentrations with microcantilevers as well as a PSA blotting paper nanotest
Ultrasound imaging	Low-density lipid nanoparticles have been used to enhance ultrasound imaging
Prostate cancer staging	Ultrasmall supermagnetic iron oxides (USPIOs) and MRI to enhance lymph node staging in prostate cancer
Prostate cancer Penile cancer Bladder cancer	Peritumor injection of 70 MBq (99m) Tc-nanocolloid has been shown to enhance detection of sentinel lymph nodes
Prostate cancer	Targeted bioconjugate nanoparticle to deliver docetaxel/paclitaxel
Bladder cancer	Aluminum phthalocyanine tetrasulfonate, a known photosensitizer, encapsulated in transferrin-conjugated liposomes to treat bladder cancer
Hormone refractory prostate cancer	Experimental use of liposomal doxyrubicin
Metastatic prostate cancer	The LHRH analogue Leuprorelin is conjugated within polylactic acid co-glycolic acid liposome microspheres immunotargeted nanoshells
Tissue engineering	Cell adhesion and growth of human bladder smooth muscle cells on porous, nano-dimensional polymeric 3D poly(lactic-co-glycolic acid) scaffolds
	Electrospun fibrinogen nanofiber matrix for tissue reconstruction
Implantable kidney	Nanoporous silicon filtration membranes utilize nanofabrication methods for the potential treatment of patients with acute renal failure
Enseal™ device (SurgRx, Palo Alto, California)	This device utilizes millions of nanoparticles embedded into the instrument to regulate the temperature of the bipolar diathermy

Source: Alexis, F. et al., *Urol. Oncol.*, 26(1), 74, Jan–Feb 2008. With permission.

are well established and use contrast media to enhance the differentiation of structures. With enhanced specificity, markers of early stage disease in malignancy could localize tumors before they progress and become incurable by conventional methods. This targeted approach could then utilize the same systems to deliver therapeutics at high concentrations to the tumor, limiting side effects and damage to the individual.

21.3.1.1 Molecular Diagnostics

Nanotechnology has the potential to revolutionize molecular diagnostics to elucidate the key molecular defects for a particular disease. This will require high-throughput detection devices that require nanogram quantities of analytics and reagents. With the current cost of molecular diagnostics escalating, it may be that increased use will inevitably expose these sophisticated tests to the economies of scale and decrease costs, making such testing available at the bedside.

Urological cancers such as prostate and renal malignancies are exemplified by the early detection of small tumors of unknown

long-term clinical significance. In other words, the use of routine abdominal ultrasound has led to the incidental detection of small renal tumors, which may or may not pose a risk to life expectancy. Similarly, the incremental and at times controversial use of prostate-specific antigen (PSA) has resulted in a greater proportion of low-risk cancers occurring in younger men. The incidence of prostate cancer is rising in Europe and the United States, with an incidence of 30,000 per year in the United Kingdom and 250,000 per year in the United States. It is generally accepted that the rise in incidence is due to use of the blood serum test, PSA, as a screening test. When men present with a raised PSA or other risk parameters for harboring prostate cancer, they are advised to undergo a transrectal ultrasound guided prostate biopsy. Thirty thousand men have a prostate biopsy in the United Kingdom and over 1 million biopsies are carried out in the United States.

21.3.1.2 New Nanotechnology Diagnostic Tests

Using microcantilevers, which can be thought of as flexible beams resembling a row of diving boards, that can be coated with molecules capable of binding biomarkers, the quantification of PSA at clinically significant concentrations was reported (Wu et al. 2001). The use of a PSA "blotting paper" nanotest assay was reported, and there was a good correlation between this novel test and conventional PSA testing (Azzouzi et al. 2002). However, the reliability decreased inversely with the PSA value. Subsequently, a novel reagent consisting of gold nanoparticles and using biolyte-selective surface-enhanced Raman scattering responses detected free PSA levels of 1 ng/mL in human sera. Further work showed that after the recrystallization of the bacterial cell-surface layer fusion protein on gold chips precoated with thiolated secondary cell-wall polymer, a monomolecular protein lattice could be exploited as a sensing layer in surface plasmon resonance biochips to detect PSA (Pleschberger et al. 2004). More recently, Shulga et al. (2008) developed a new spectrophotometric method using covalently attached capture antibody labeled with alkaline phosphatase for the detection of free PSA. Briman et al. (2007) described the production and use of a novel electronic device architecture for the quantitative detection and measurement of PSA.

Telomerase activity, a marker of limitless replicative potential, is often elevated in malignancy. Grimm et al. (2004) have developed nanoparticles that switch on their magnetic state by annealing with telomerase-synthesized TTAGGG sequences. This can then be detected by benchtop magnetic resonance relaxometers. They comment that these nanoparticles are biocompatible and may be able to detect molecular lesions in vivo.

21.3.1.3 New Nanotechnology Imaging Tests

Overall, men undergoing systematic transrectal ultrasound guided biopsy of 6–12 cores of prostatic tissue have approximately 1 in 4 probability of being diagnosed with prostate cancer. Of these, about half are diagnosed with low-risk disease. It therefore follows that the majority of men presenting with an abnormal PSA do not have prostate cancer or have insignificant disease that need not be treated. However, PSA is false positive-prone

(7 out of 10 men in this category will still not have prostate cancer) and false negative-prone (2.5 out of 10 men with prostate cancer have no elevation in PSA) (Thompson et al. 2004, 2008). In addition, once diagnosed, it is also generally accepted that not all men with localized prostate cancer need treatment and a significant proportion with low-to-intermediate risk disease could be under surveillance (Bill-Axelson et al. 2005, Cooperberg et al. 2004). In the setting of prostate cancer, any new imaging test could be used as a triage diagnostic test (Bossuyt et al. 2006). In addition to the diagnostic inaccuracy inherent in transrectal prostate biopsy, there is also a significant risk of infection/sepsis, haematuria, haematospermia, pain/discomfort, dysuria, and urinary retention. Equally important, from a public health perspective, if these men could be designated as low risk, not undergo a biopsy and discharged from further follow-up, valuable healthcare resources could be saved. If an imaging tool could reliably prevent these men from undergoing diagnosis, this would represent a significant breakthrough. Further, in those with a diagnosis of significant prostate cancer, by using accurate imaging prior to biopsy, accurate risk stratification of cancer (volume, grade) is possible by ensuring the proper sampling of a suspicious lesion.

New diagnostic agents used in medical imaging are now being developed, which can be considered as nanotechnologies. Magnetic resonance imaging (MRI) has been used clinically for targeted imaging with gadolinium-based (Oyewumi et al. 2004) and iron-oxide (Harisinghani et al. 2003)-based nanoparticles for the enhancement of conventional diagnostic modalities. This latter imaging tool has allowed the development of techniques to accurately stage disease in a number of cancers. In addition, low-density lipid nanoparticles have been used to enhance ultrasound imaging (May et al. 2002). For each current imaging discipline, it could be possible to develop nanoparticles that can provide enhanced detection and subsequently targeted therapeutic capabilities (Sullivan and Ferrari 2004).

MRI became useful for constructing a high-resolution image of internal structures when relaxation time—the time it takes for the protons to emit their signal—was taken into consideration. There are two components to this relaxation, which have different time frames and that can be detected, T1 and T2. Different types of tissues will exhibit different T1 and T2 values. Fat has a high signal and will appear bright on a T1-weighted (T1W) image. Water has low signal appearing dark on a T1W scan. However, tumors are not easily characterized and therefore the use of contrast agents may aid in early detection and precise staging. A paramagnetic contrast agent, such as a gadolinium compound, can be administered, so that pre-contrast T1W and post-contrast images can be obtained for greater diagnostic utility. The staging of urological cancers is also fraught with difficulties, as there is debate as to whether lymph nodes should be routinely removed during extirpative surgery. Accurate lymph node staging in urologic cancer is important in deciding clinical prognosis and optimizing treatment. The gold standard for ascertaining lymph node status is radical lymphadenectomy. This carries risks of surgical complications such as lymphocele

and peripheral oedema as well as operative complications and inaccuracy of sampling error. Imaging techniques such as computed tomography, positron emission tomography, and MRI are good at detecting positive lymph nodes but have limitations on specificity and sensitivity because they rely on inaccurate factors such as size and morphology of the lymph node (Feldman et al. 2008).

Preoperative imaging tests utilizing nanoparticle contrast-enhanced images for identifying cancer positive lymph nodes may be of benefit. MRI compounds known as ultrasmall supermagnetic iron oxides (USPIOs) or monocrystalline iron oxide nanoparticles (MIONS) are currently under investigation (Shen et al. 1993). The small size of these nanoparticles allows them to cross capillary walls and traverse into lymph nodes in one of two ways: first, by directly crossing into capillaries from venules to the medullary sinuses of lymph nodes and second, by extravasating across permeable capillary vessels using nonspecific transcytosis, and then into lymph vessels that take them to the respective regional nodes. Once in the lymph node, the nanoparticles are phagocytosed by macrophages of the reticuloendothelial system (RES), causing nanoparticle accumulation. Magnetic nanoparticle accumulation in benign nodes causes a drop in T2W signal. However, lymph nodes affected by cancer cells do not have functioning RES macrophages and therefore have a high signal on T2W scans. When a lymph node is completely replaced by metastatic cells, it shows no contrast uptake and stays bright. Equally, when lymph nodes are only partly affected with metastatic cells, specific quantitative criteria are needed to assess the images. In prostate cancer, this property has been shown to increase the sensitivity and specificity of the detection of otherwise occult lymph node metastasis. In a seminal paper by Harisinghani et al. (2003), the technique of the preoperative intravenous injection of USPIOs with MRI 24 h later identified 33 of 80 prostate cancer patients with positive lymph nodes after radical retropubic prostatectomy. They reported that 100% of the patients with positive lymph nodes would have been detected preoperatively with the USPIO nanotechnology (i.e., 100% sensitivity) compared to only 45.4% with current size criteria for MRI alone. Interestingly, the technique also had a specificity of 95.7% on a patient-by-patient basis (Figure 21.1). Deserno et al. (2004) showed similar results in 58 patients with bladder cancer using the same technique. However, this study only carried out surgical lymphadenectomy in 46 while the rest had image-guided biopsy of nodes that were larger than 8 mm. MR lymphangiography using these nanoparticles detected positive lymph nodes with a sensitivity and a specificity of 96% and 95%, respectively. This is compared with the sensitivity of 76% and the specificity of 97% in conventional MRI in the same patient group. The same group (Tabatabaei et al. 2005) showed that nanoparticle-enhanced MR lymphangiography in a small group of seven patients with penile squamous cell carcinoma was good with sensitivity and specificity for the detection of metastatic deposits in individual lymph nodes of 100% and 97%, respectively. Further, in testicular cancer, in a group of 18 men who had a radical orchidectomy for germ cell tumor, MR lymphangiography detected nodal metastases with a sensitivity of 88.2% and a specificity of 92% (Harisinghani et al.

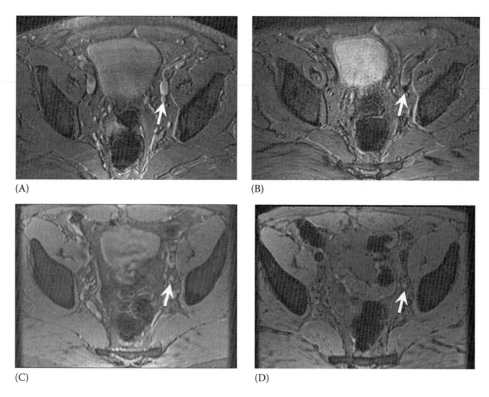

FIGURE 21.1 MR lymphography using nanoparticles. (A) Precontrast image demonstrating lymph node in patient with prostate cancer (arrow). (B) Post contrast. The node is now homogeneously dark demonstrating that it is benign (verified at pathology) (C) Precontrast image in another patient with prostate cancer. Small lymph node (arrow). (D) Post contrast. Lymph node retains high signal demonstrating likely malignant metastatic lymph node (verified at pathology). (From Feldman, A.S. et al., *Urol. Oncol.*, 26(1), 65, 2008. With permission.)

2005). The results from this trial are promising, but are clearly a preliminary evaluation of MR lymphangiography in germ cell tumor limited by the fact that histological evaluation was on the whole with needle biopsy. It is important that large multicenter prospective trials are conducted to confirm the results of these trials outside of expert centers and compared to other novel imaging modalities such as PET-CT. In addition, the need to perform both pre-contrast and post-contrast images 24 h apart may be a significant limitation, although using post-contrast images alone may be sufficient. Overall, contrast MR lymphangiography may be cost effective (Harisinghani et al. 2006, Hovels et al. 2004).

Second-generation superparamagnetic compounds of cross-linked iron oxide have also been developed (Schellenberger et al. 2002, Wunderbaldinger et al. 2002). Such dendritic nanoparticles have multiple arms that can complex with targeting moieties, such as antibodies or enzyme substrates, permitting both magnetic and optical properties. The sentinel lymph node concept revolves around the idea that there are one or two lymph nodes that an organ drains to initially and if this lymph node is positive then it is likely that lymph node spread to the other nodes in the chain is likely and a full lymphadenectomy warranted. Conversely, if the sentinel lymph node is negative, then a full lymphadenectomy is unlikely to be of therapeutic benefit and could potentially be avoided. The concept has gained most popularity in penile and breast cancers (Ahmed et al. 2006). In Switzerland, Studer's group have taken another approach using

nanocolloids for the descriptive mapping of sentinel lymph node anatomy using fusion imaging between single photon emission computed tomography (SPECT) and CT following intraprostatic injection of technetium-99m nanocolloid (Warncke et al. 2007). They investigated 10 patients with organ-confined prostate cancer and injected technetium-99m nanocolloid into each lobe under ultrasound guidance. Scintigraphy was performed 1 h later and fusion images with CT were scored blindly by an experienced radiologist and nuclear medicine physician. An average of 10 lymph nodes were detected per patient (range 2–19) and the location of positive lymph nodes was highly variable. They concluded that the lymphatic drainage of the prostate appears to be more extensive than previously described in the literature. Pre-sacral and common iliac regions seem to be primary rather than secondary drainage sites. Para-aortic and inguinal lymph nodes have not been considered as primary drainage sites previously. This suggests that significant portions of sentinel lymph nodes are not resected with routine extended or radical pelvic lymphadenectomy. In addition, a similar approach has been used in penile cancer (Kroon et al. 2005) and in bladder cancer by Liedberg et al. (2006) with 70 MBq (99m)Tc-nanocolloid. Of the 32 pathological lymph node positive cases at cystectomy, 26 (81%) were detected by nanotechnology, resulting in a 19% false positive rate (Wu et al. 2001).

Semiconductor nanocrystals known as quantum dots have been increasingly utilized as biological imaging and labeling probes because of their unique optical properties, including broad absorption with narrow photoluminescence spectra, high quantum yield, low photobleaching, and resistance to chemical degradation (Kaji et al. 2007). Although hydrophobic and toxic properties have until recently limited their use in vivo, by altering the surface structure and limiting the use of toxic semiconductors such as selenium and cadmium, clinical potential may become a reality. In addition, the surface modification of QDs with antibodies, aptamers, peptides, or small molecules that bind to antigens present on the target cells or tissues has resulted in the development of sensitive and specific targeted imaging and diagnostic modalities for in vitro and in vivo applications. Bagalkot et al. (2007) recently used a functionalized surface of fluorescent QDs with the A10 RNA aptamer, which recognizes the extracellular domain of the prostate specific membrane antigen (PSMA). They developed a targeted quantum dot imaging system capable of differential uptake and imaging of prostate cancer cells that express the PSMA protein. In addition, they intercalated doxurubicin, an antineoplastic anthracycline agent, into the double-stranded stem of the A10 RNA aptamer. Furthermore, several groups have successfully imaged angiogenesis with MRI in animal models using nanoparticles by targeting the αvβ3-integrin extracellular matrix protein and this may have implications for cancer imaging research in future (Winter et al. 2003).

21.3.2 Cancer Treatment Using Nanotechnology

Treatment options for two important urological cancers—small renal tumors and early localized prostate cancer—vary between extremes with surveillance and close monitoring on the one hand and radical extirpative surgery on the other. The latter carries certainty of cancer control but high risk of side effects and impact on quality of life, while the former carries certainty over lack of side effects, but the burden of anxiety and surveillance, which comes from leaving a cancer untreated (Ahmed et al. 2007, Silverman et al. 2008). A middle ground in which only the cancer is treated leaving normal tissue unaffected and therefore side effects are at a minimum is the holy grail of cancer research. Nanotechnology enables the use of powerful cytotoxic drug specific to cancer cells without damaging effects to normal structures such as the neurovascular bundles or external sphincter that surround the prostate or normal renal parenchyma, for example. The problem is in designing nanoparticles specific to cancer cells, which can be taken up by these cells alone so that the cytotoxic effect is delivered to just those cells over a protracted period in order to get adequate cell kill (Gommersall et al. 2008, Haley and Frenkel 2008).

In addition, the treatment of the other end of the spectrum of cancer—metastatic disease—currently centers on chemotherapeutic agents. Although cancer cells are inherently more vulnerable than normal cells to the effect of chemotherapy agents, the drugs are not selective and can cause injury to normal tissues. Indeed, the toxicity subjected to normal cells impacts and limits

dose and frequency of chemotherapeutic regimens leading to a high proportion of failures. In order to achieve the goal of selective cancer cell kill, novel carriers linked to cancer-specific targets at the molecular level (cells, vasculature, and extracellular stroma) are necessary (Haley and Frenkel 2008).

There are a number of temperature-based technologies currently being evaluated for the treatment of localized prostate and kidney cancer. Thermal ablative technologies offer a minimally invasive option to conventional surgery by delivering a localized lethal temperature to malignant tissue while minimizing damage to surrounding normal tissue. Cryotherapy (Babaian et al. 2008, Pareek and Nakada 2005) and high-intensity-focused ultrasound (HIFU) (Illing and Chapman 2007) are being used in the treatment of localized prostate carcinoma while radio frequency ablation, along with cryotherapy, have been recently used for treating small renal masses (McDougal et al. 2005, Mouraviev et al. 2007). These technologies also face the challenges of surgery, in that precision is required in targeting tumors in order to avoid the toxicity of treatment that comes from damage to adjacent structures.

Nanotechnology may clearly have advantages in cell specificity, small size, and guided tumor toxicity (Table 21.2). Two delivery platforms can be utilized to target with nanoparticles

TABLE 21.2 Nanoparticle Platforms for Drug Delivery

Nanoparticle	Material	Structure
Polymeric	Poly(lactide-co-glycolide) Poly(lactide) Poly(caprolactone) Poly(orthoester) Poly(anhydride) Poly(beta-aminoester)	60–250 nm
Lipsome	Doxil®/Caelyx®: PEG-DSPE:HSPC/cholesterol (5:56:39) DaunoXome®: DSPC/cholesterol (2:1)	20–250 nm
Dendrimer branched	Poly(amidoamine) Poly(ethylenimine) Poly(peptide)	5–200 nm

Source: Alexis, F. et al., *Urol. Oncol.,* 26(1), 74, Jan–Feb 2008.

Note: Polymeric nanoparticles have hydrophilic, hydrophobic, and therapeutic drugs incorporated into the polymer. Liposomes have hydrophilic and hydrophobic agents inside the hollow liposome and within the hydrophobic membrane, respectively. Dendrimers are branched molecules that have therapeutic or diagnostic agents within them.

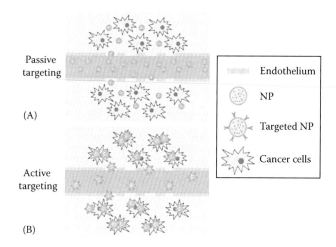

FIGURE 21.2 Passive and active targeting by nanoparticles. (A) Passive targeting. Nanoparticles extravasate into tumor due to local leaky microvasculature. (B) Active targeting. Concentration of nanoparticles by binding of specific ligands to antigens on cell surface. (From Alexis, F. et al., *Urol. Oncol.*, 26(1), 74, Jan–Feb 2008. With permission.)

(Alexis et al. 2008). Active targeting relies on ligand receptor interactions (Yezhelyev et al. 2006) (Loo et al. 2005), while passive targeting takes advantage of enhanced tumor capillary permeability and subsequent preferential retention of intravenously administered particles (Maeda et al. 2003) (Figure 21.2).

One factor leading to the lack of progress in this matter has been preventing or avoiding degradation of the nanoparticles in vivo. In general, the particle size is constructed for the nanoparticle to pass through fenestrations of the leaky cancer endothelium. This results in the preferential concentration of the nanoparticle at tumor sites alone. Particle size also determines if the circulating nanoparticle is opsonized and cleared by the RES. The RES is predominantly distributed in the liver, lungs, spleen, and bone marrow. To such uptake, the surface of the nanoparticle has to be modified. One approach is to alter the nanoparticle's composition. For example, pegylation involves the addition of polyethylene glycol on the surface. This technology has given rise to the "STEALTH" particle, which is not detected by the macrophage (Storm et al. 1995). Another way to reduce opsonization is to use hydrophilic surfaces, which involves the incorporation of hydrophilic polymers that coat the nanoparticle and that can repel plasma proteins. This allows the nanoparticle to escape macrophage-directed opsonization and is deemed the "cloud" effect (Jeon et al. 1991).

Liposomal delivery systems have now overcome the RES (Gabizon and Papahadjopoulos 1992). Liposomes have been used for many years for the treatment of metastatic prostate cancer. The decapeptide leuprorelin has been conjugated within polylactic acid co-glycolic acid liposome microspheres as monthly and three monthly injections in the treatment of metastatic prostate cancer (Okada and Toguchi 1995). Nanotechnologies have utilized key differences in physiological parameters of the malignant environment. Liposomal delivery of chemotherapy has been utilized in the overexpression of fenestrations in cancer neovasculature to increase drug concentration at tumor sites (Langer 1998). This strategy has been used in the treatment of Kaposi's sarcoma for over a decade and more recently in breast cancer (Park 2002).

Liposomal drug delivery can be enhanced in a number of ways including local hyperthermia with thermostable liposome (Maruyama et al. 1993), photosensitizers (Derycke et al. 2004), and ultrasound activation (May et al. 2002). Derycke and de Witte (2002) investigated the use of aluminium phthalocyanine tetrasulfonate, a known photosensitizer, encapsulated in transferrin-conjugated liposomes. Transferrin is overexpressed in bladder cancer cells due to the increased requirement for iron. They reported that photodynamic therapy for conjugated liposomes was more effective compared to the unconjugated controls (Nakanishi et al. 2003). Liposomes have also been used for gene therapy targeting of renal cell carcinoma. Nakanishi et al. (2003) have targeted renal cell carcinoma cell lines and fresh ex vivo renal cell carcinoma tissue with cationic multilamellar liposomes containing the human interferon ß gene (IAB-1). They report significant cytotoxic effects with the addition of the IAB-1 gene in this setting. This approach mirrors many other research protocols that are trying to realize effective transport of oligonucleotides or plasmid DNA through the malignant cell membrane. This nonviral gene therapy approach has been trialed by other researchers in other malignancies (Santhakumaran et al. 2004).

As the folate receptor is often overexpressed in a multitude of cancers, folate receptor-mediated drug delivery is therefore an attractive option to target cancer cells rather than benign folate-receptor negative cells, and the effectiveness of folate-linked, lipid-based nanoparticles as a vector for DNA transfection and for suicide gene therapy to treat prostate cancer, has been reported (Hattori and Maitani 2004). Further progress has been made in manufacturing nanoparticles with an oleic acid shell that can be loaded with hydrophobic drugs for the sustained release of chemotherapy. In vitro, these approaches have shown dose-dependent antiproliferative effects compared to controls. This vehicle can also be labeled with antibodies for use in imaging (Jain et al. 2005). In prostate cancer cell lines, polypropylenimine dendrimers have been used to deliver c-myc triplex-forming oligonucleotides to inhibit transcription of this oncogene in vitro (Santhakumaran LM et al. 2004). These 130–280 nm nanoparticles caused a 65% decrease in c-myc expression, making them useful candidates for gene transfer in malignancy. Finally, in vitro targeting of synthesized antibody against prostate-specific membrane antigen (a well-characterized antigen expressed on the surface of prostate cancer cells) with conjugated dendrimer nanoparticles has also been shown to be a suitable platform for targeted molecule delivery into appropriate antigen-expressing cells (Thomas et al. 2004). This approach could target cancer cells specifically and spare noncancerous normal cells.

Kam et al. (2005) used near-infrared light at 700–1100 nm to optically stimulate single-walled carbon nanotubes (SWNTs). The transparency of biological tissues to this wavelength of light and the strong absorption of SWNTs in this spectrum make SWNTs uniquely suited to the delivery of olignucleotides. In addition, near-infrared targeting can cause localized heating of the SWNTs to cause cell destruction in vitro. Folate labeling has further enabled the internalization of nanotubes into folate receptor-expressing cancer cells rather than the folate receptor-negative benign tissue. These SWNTs have been shown to have limited cytotoxicity when used in vitro.

Gold nanoshells (GNS) can be designed to have a diameter ranging from 1 nm to greater than 100 nm. One group has developed a 110 nm diameter nanoshell with a dielectric silica core and a 10 nm thick gold shell designed to specifically optimize optical scattering and absorption properties (Hirsch et al. 2003). This particle can absorb intense near infrared (NIR) energy. By using this property and the local delivery of NIR light, it may be possible to activate and heat these gold nanoshells in a precise manner to achieve specific ablation of malignant tissue. GNS exert their ablative effect only when stimulated by NIR light, which excites electrons in the outer gold shell. When these return to their relaxed state, heat is emitted. NIR-activated nanoshell tumor ablation was first tested in an in vitro human breast cancer cell line (Hirsch et al. 2003). Addition of a nanoshell suspension to cultured SK-BR-3 cells demonstrated no cytotoxic effect, but laser initialization showed a zone of cell death limited to the area of the laser alone. Another group have shown that this technique can be used to ablate two human prostate cancer cell lines (PC-3 and C4-2) (Stern and Cadeddu 2008) (Stern et al. in press) (Figure 21.3). Another group (Everts et al. 2006) has shown active targeting using gold nanoparticles delivered by an adenovirus vector; this is the first time such a technique has been demonstrated. This work is yet to undergo further work in urological cancers.

Biodegradable polymeric nanoparticles have been the subject of great interest for drug delivery applications. Nanoparticles can build up in malignant tissue after systemic administration, as the biodistribution is determined by physical and biochemical properties (particle size, nature of the polymer and drug, surface biochemical properties) (Avgoustakis 2004). Cheng et al. (2007) have recently shown that the surface functionalization of nanoparticles with the A10 PSMA aptamer significantly enhances delivery of nanoparticles to tumors compared to equivalent nanoparticles lacking the A10 PSMA aptamer (Figure 21.4). This is the first report of a tumor-specific targeting of a nanoparticle–aptamer system in vivo and may result in the development effective therapies for disseminated prostate cancer. This work was furthered by the same group using docetaxel (Dtxl)-encapsulated nanoparticles formulated with biocompatible and biodegradable poly(D,L-lactic-co-glycolic acid)-block-poly(ethylene glycol) (PLGA-*b*-PEG) copolymer and surface functionalized with the A10 RNA aptamers that recognize the extracellular domain of the PSMA molecule. They showed that these docetaxel-encapsulated nanoparticle–aptamer

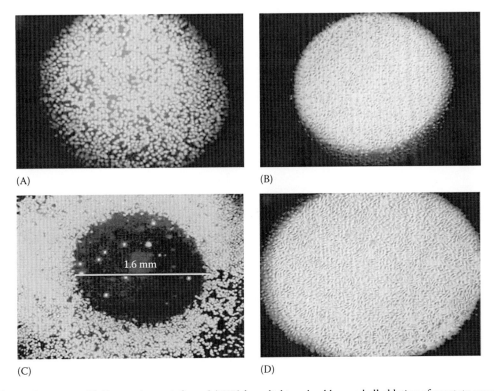

(A)

(B)

1.6 mm

(C)

(D)

FIGURE 21.3 (Magnification 100X). Targeted near infra red (NIR) laser light and gold nanoshell ablation of prostate cancer tissue cultures. (A) and (B) Control PC-3 cells with NIR application without gold nanoshells. (C) PC-3 cells with gold nanoshells and NIR laser application. Zone of nonviability of cells (calcein stain) corresponds to NIR spot of light only. (D) PC-3 cells from (C) demonstrating morphology of cells is intact despite lack of viability. (From Stern, J.M. and Cadeddu, J.A., *Urol. Oncol.*, 26(1), 93, 2008. With permission.)

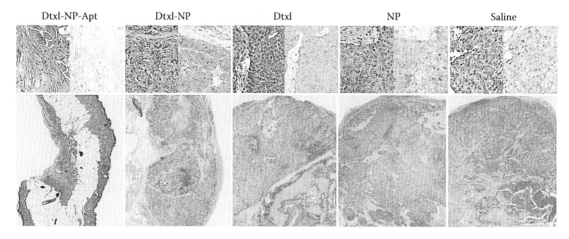

FIGURE 21.4 PLGA-*b*-PEG–COOH nanoparticle synthesis and aptamer conjugation. Docetaxel encapsulated into nanoparticle. (From Cheng, J. et al., *Biomaterials*, 28(5), 869, 2007. With permission.)

bioconjugates (Dtxl-NP-Apt) bind to the PSMA protein expressed on the surface of LNCaP prostate epithelial cells and get taken up by these cells resulting in significantly enhanced in vitro cellular toxicity compared with non-targeted nanoparticles that lack the PSMA aptamer (Farokhzad et al. 2006) (Figure 21.5). Docetaxel, when used systemically, can prolong the survival of patients with hormone-resistant prostate cancer (Petrylak et al. 2004, Tannock et al. 2004). So controlled release of docetaxel targeted to prostate cancer cells may result in enhanced cytotoxicity and antitumor efficacy with fewer systemic side effects.

This could in turn make it a potential therapeutic modality for the management of localized prostate cancer.

Other models are also in development. Transferrin receptor-targeted drug delivery system was developed in a murine model of prostate cancer. Nanoparticles of encapsulated paclitaxel with surface-conjugated transferrin deliver higher paclitaxel doses into the tumor over a sustained time period compared with conventional paclitaxel delivery. The nanoparticle avoids use of the paclitaxel vehicle, Cremophor- EL, and its related toxicities, while at the same can target prostate cancer cells with

FIGURE 21.5 (See color insert following page V-2.) Effect of targeted bioconjugate nanoparticles. Histological slides demonstrating excised tumors from mice treated with saline, pegylated PLGA nanoparticle without docetaxel, docetaxel alone, docetaxel-encapsulated nanoparticle and docetaxel-encapsulated nanoparticle–aptamer bioconjugates. The docetaxel–nanoparticle–aptamer confirmed absence of residual cancer. All others showed variable PSMA staining showing degrees of tumor viability. (From Farokhzad, O.C. et al., *Proc. Natl. Acad. Sci. USA*, 103(16), 6315, 2006. With permission.)

up-regulated transferrin receptors (Sahoo et al. 2004). A pegy-lated lipid-based nanoparticle with a conjugated folate ligand has been developed for targeted drug delivery of a suicide gene to prostate cancer xenografts from LNCaP and PC-3 cell lines (Hattori and Maitani 2005). This nanoparticle delivers a Herpes simplex virus thymidine kinase gene that phosphorylates a pro-drug ganciclovir to a toxic triphosphate, which in turn blocks cellular DNA synthesis. The particle binding appears to be to the extracellular domain of the PSMA receptor and enters the cell by endocytosis. Such concepts may also be useful for benign pros-tatic hyperplasia. Peng et al. (2007) recently used a degradable, poly(beta-amino ester) polymer, poly(butane-diol-diacrylate-co-amino-pentanol)(C32), to develop a nanoparticle system to deliver a diphtheria toxin suicide gene (DT-A) driven by a pros-tate-specific promoter to cells. These C32/DT-A nanoparticles were directly injected into the normal prostate as well as pros-tate tumors in mice. Almost half the normal prostates showed a significant size reduction as a result of apoptosis. Injection with naked DT-A-encoding DNA had little effect. Significant apopto-sis was also seen in C32/DT-A-injected prostate cancer lesions. Importantly, the surrounding tissue damage was avoided (Peng et al. 2007).

A promising technique that has reached clinical application is magnetic nanoparticle thermotherapy. This is a novel minimally invasive approach, developed for interstitial thermal therapy, which uses a fluid containing biocompatible superparamagnetic nanoparticles (magnetic fluid) injected directly into superficial or deep-seated tumors. This is then selectively heated in an alter-nating magnetic field. Thermal energy is released to the surround-ing medium as a result of physical processes that differ according to the size of the magnetic material used and the strength of the applied magnetic field. Due to their aminosilane-type coat-ing, the nanoparticles (MFL AS, MagForce Nanotechnologies AG, Berlin, Germany) are taken up intracellularly by differen-tial endocytosis (Jordan et al. 1996). Moreover, selective uptake into prostate cancer cells has been shown in vitro and offers the perspective of tumor-cell-selective hyperthermia (Jordan et al. 2001). Animal studies on mouse mammary carcinoma, glioblas-toma, and prostate cancer have demonstrated the feasibility and efficacy of this heating method, as well as a very low clearance rate of the nanoparticles from tumors, allowing for serial heat treatments following a single magnetic fluid injection. In a pros-tate tumor model, thermoablative temperatures up to 70°C were achieved (Johannsen et al. 2005a) and it has now been extended to human phase I trials (Johannsen et al. 2005a,b). The feasibility and efficacy of combined thermal therapy using magnetic nano-particles and irradiation has also been shown in a rat model of prostate cancer (Johannsen et al. 2006). Johanssen et al. (2007b) investigated the use of magnetic nanoparticles to heat prostatic tissue in an alternating-current magnetic-field applicator in a number of patients with recurrent prostate cancer after previ-ous radiotherapy (Figure 21.6). Invasive thermometry was used to monitor heating to 39.4°C–48.5°C. This novel interstitial treatment was applied for six weekly hyperthermia sessions of 60 min duration. Several interesting concepts can be extrapolated

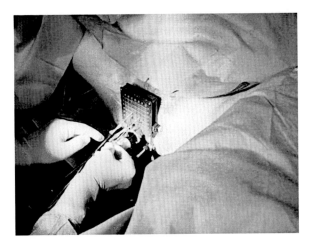

FIGURE 21.6 Magnetic nanoparticle thermotherapy. Magnetic nanoparticle suspension injected transperineally into prostate using a brachytherapy grid and ultrasound setting to guide the placement of the needle. (From Johannsen, M. et al., *Int. J. Hyperthermia*, 21, 637, 2005b. With permission.)

from this investigation. Injected nanoparticles remained within the prostate for the 6-week treatment duration and could be detected on CT to ensure adequate prostate distribution with rectal and urethral sparing. The concept of combining mild hyperthermia with low dose rate irradiation has been proposed for some time as being promising, but has never been realized clinically (Kampinga and Dikomey 2001, Armour and Raaphorst 2004). To this end, a phase II study in previously untreated patients with clinically localized prostate cancer and interme-diate risk criteria has been started at the same institution. In this trial, patients receive LDR brachytherapy (125-iodine seeds, prescription dose 145 Gy) combined with magnetic nanoparticle thermotherapy.

21.3.3 Tissue Engineering and Nanotechnology

The development of tissue engineered urological structures would revolutionize the treatment of a diverse range of uro-logical problems including extirpative surgery for muscle invasive bladder cancer, functional disorders of the bladder, and lower urinary tract as well as tissue damage occurring secondary to trauma. The research is in its infancy, but pro-grammes are developing (Pattison et al. 2005, Staack et al. 2005, Ma et al. 2005).

When patients with spina bifida, spinal cord injury, or other bladder insult develop high urinary storage pressure, hydro-nephrosis, and upper tract injury, it may be necessary to carry out reconstruction of the bladder in order to reduce these cor-ollaries of ineffectual bladder function (Gurocak et al. 2007, Harrington et al. 2008). The use of a polymer Wlms [PLGA, PCL, polyurethane] treated with sodium hydroxide to degrade the polymer surface has been reported recently (Pattison et al. 2005). The degradation creates changes on the surface of the polymer with nanoscale features that can enhance bladder smooth muscle cell adhesion (Thapa et al. 2003). Electrospun

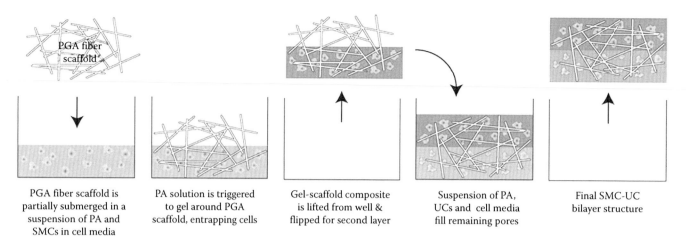

PGA fiber scaffold is partially submerged in a suspension of PA and SMCs in cell media

PA solution is triggered to gel around PGA scaffold, entrapping cells

Gel-scaffold composite is lifted from well & flipped for second layer

Suspension of PA, UCs and cell media fill remaining pores

Final SMC-UC bilayer structure

FIGURE 21.7 Representation of manufacture of PA-PGA scaffold for potential use in bladder tissue engineering. (From Harrington, D.A. et al., *World J. Urol.*, 26(4), 315, 2008. With permission.)

fibrinogen nanofiber matrix technology has also been investigated (Pham et al. 2006). In a recent application to bladder tissue engineering, Baker et al. (2006) reported on bladder smooth muscle cell attachment and growth on polystyrene electrospun scaffolds. Other examples of electrospun scaffolds for bladder regeneration have also been reported using fibrinogen (McManus et al. 2007) and cellulose acetate (Han and Gouma 2006). The potential for all of these systems lies in the researcher's control over fiber alignment and diameter. While conventional microfiber polymer scaffolds are not as easy to influence, nanofiber scaffolds may offer greater flexibility over the cytoskeletal arrangement of adherent cells, and possibly allow greater influence over phenotype expression. One group has explored another approach to forming bioscaffolds with nanoscale features. They looked at self-assembling peptide-amphiphile (PA) systems (Silva et al. 2004, Hartgerink et al. 2002). PAs are designed for self-assembly through the covalent attachment of hydrophobic and highly charged hydrophilic segments. Within the peptide segment, amino acid residues are chosen which promote alpha-sheet formation among the individual molecules. The terminal end of each PA can include a biologically relevant epitope that can signal cells or bind other molecules. The resulting supramolecular structures are 6–8 nm diameter nanofibers, which are hundreds of nanometers long. These nanofibers interact to form a self-supporting aqueous gel (Figure 21.7). As one example of this work, researchers have embedded human bladder smooth muscle cells and urothelial cells in a PA-scaffold composite with basic fibroblast growth factor, which has been shown to stimulate smooth cell proliferation and extracellular matrix production (Beqaj et al. 2005, Imamura et al. 2007). After 3 weeks of in vivo incubation in a subcutaneous nude rat model, it was found that the embedded human cells were retained and composed most of the scaffold cellular content. In addition, a separate and distinct urothelial cell layer could be maintained in the bilayer system. These PA systems demonstrate promise for localized placement of cells within a scaffold, and directed growth factor delivery during

regeneration. Despite continual progress in advancing our ability to regenerate bladder tissue, the field is yet to produce a complete solution that offers both the ability to store and expel urine in a safe and controllable manner.

21.3.4 Role of Nanotechnology in Surgery

Advances are constantly made in the development of a diverse range of surgical instruments. Robotic and robot-assisted surgeries are also propagating throughout the globe. Specific to urology, the Enseal™ device (SurgRx, Palo Alto, California) has been used to seal the dorsal venous complex at laparoscopic radical retropubic prostatectomy. This device utilizes millions of nanoparticles embedded into the instrument to regulate the temperature of the bipolar diathermy. In theory, this minimizes collateral thermal spread and tissue damage with effective sealing of vessels up to 7 mm in size. Evidence in abstract form has reported its safe use in sealing the dorsal venous complex during robotic-assisted laparoscopic prostatectomy (Lee et al. 2005). This report suggests that the Enseal device is effective, but they concede that the device takes longer to use. Its use could have implications for decreasing urethral, sphincter, and nerve damage, which could improve continence and decrease the incidence of erectile dysfunction after prostatectomy. The potential for nanosurgery in future urological practice is appealing. Nanotweezers could have a place in vasectomy reversal or varicocoele repair while "nano-urobots" could have a place in cystoscopy, ureteroscopy, and fulguration of urological tumors.

21.4 Critical Discussion

Fundamental to the success of nanotechnology is its perceived safety by the public. Many concerns have been aired concerning the use of manufactured nanoparticles. The impact on human health has been assessed in the United Kingdom as part of a document published by the Office of Science and Technology. This independent article produced by the Royal Society and Royal

Academy of Engineering represents an exhaustive discussion on the potential exposure to nanoparticles. It raises concerns that nanoparticles, due to their size and ability to pass across cellular membranes, represent a potential biohazard. They recommend that more research is directed at the toxicology of nanoparticles. The report compares nanotechnology development with asbestos fibers, which caused widespread health and safety concerns. Regulation to prevent misuse will be paramount to enable maximum benefit with minimal risk for these powerful cutting-edge technologies.

With stringent attention to safety, nanotechnology could have a widespread beneficial impact on many aspects of the delivery of urological care. Quite naturally, the main emphasis on nanotechnology research in the field of urology has centered on the use nanoparticles to detect, localize, specifically target and ablate cancer. This research will rely heavily on cross-disciplinary work that aims to discover potential genetic and protein targets that are cancer specific so that imaging contrast agents and drug delivery models can be further developed and applied in clinical trials for the benefit of patients.

21.5 Future Perspectives

The safety and integrity of using nanoparticles for therapeutic work clearly needs to be established. This will depend on the discovery and design of specific ligands that will allow targeting of specific tissues such as malignant cells without the nanoparticle being discovered and degraded or changed by the host immune system. In addition, the delivery of cytotoxic agents needs to be localized and controlled as well as sustained so that accurate and complete cell kill can occur. Nanotechnologists will need to continue their work hand in hand with molecular biologists so that advances in the molecular and genetic profiling of tissue can aid the development of targeted nanoparticles. Exciting in vitro and in vivo applications of nanotechnology have especially been reported in the diagnosis and treatment of cancer, but at least at present, there is a paucity of clinical studies to verify its potential use. As urological cancer has the greatest potential to benefit from nanotechnology, urologists should be encouraged to actively participate with basic science researchers working in this field. Commencement of phase 1 and phase 2 trials incorporating nanotechnology to evaluate novel diagnostic and therapeutic approaches for the benefit of patients should be encouraged.

21.6 Summary

A major aim of medicine has long been the early and accurate diagnosis of clinical conditions, providing an efficient treatment without secondary effects. With the emergence of nanotechnology, the achievement of this goal seems closer than ever. To this end, the development of novel materials and devices operating at the nanoscale range provides new and powerful tools for imaging, diagnosis, and therapy.

References

Ahmed HU, Arya M, Minhas S (2006). Dynamic sentinel lymph node biopsy in penile cancer. *Expert Rev Anticancer Ther* 6(7):963–967.

Ahmed HU, Pendse D, Illing R, Allen C, van der Meulen JH, Emberton M (2007). Will focal therapy become a standard of care for men with localized prostate cancer? *Nat Clin Pract Oncol* 4(11):632–642.

Alexis F, Rhee JW, Richie JP, Radovic-Moreno AF, Langer R, Farokhzad OC (2008). New frontiers in nanotechnology for cancer treatment. *Urol Oncol* 26(1):74–85.

Armour EP, Raaphorst GP (2004). Long duration mild temperature hyperthermia and brachytherapy. 30. *Int J Hyperthermia* 20(2):175–189.

Avgoustakis K (2004). Pegylated poly(lactide) and poly(lactide–co–glycolide) nanoparticles: Preparation, properties and possible application in drug delivery. *Curr Drug Deliv* 1:321–333.

Azzouzi R, Cormie L, Valeri A et al. (2002). Relevance of PSA nanotest on capillary blood in organized mass screening of prostate cancer. 1308. American Urological Association Annual Meeting, May 25–30, 2002, Orlando, FL.

Babaian RJ, Donnelly B, Bahn D et al. (2008). Best practice statement on cryosurgery for the treatment of localized prostate cancer. *J Urol* 180(5):1993–2004.

Bagalkot V, Zhang L, Levy-Nissenbaum E et al. (2007). Quantum dot-aptamer conjugates for synchronous cancer imaging, therapy, and sensing of drug delivery based on bi-fluorescence resonance energy transfer. *Nano Lett* 7(10):3065–3070.

Baker SC, Atkin N, Gunning PA et al. (2006). Characterisation of electrospun polystyrene scaffolds for three-dimensional in vitro biological studies. *Biomaterials* 27:3136–3146.

BECON Nanoscience and Nanotechnology Symposium Report (2006). National Institutes of Health Bioengineering Consortium, June 2000.

Beqaj SH, Donovan JL, Lu DB et al. (2005). Role of basic fibroblast growth factor in the neuropathic bladder phenotype. *J Urol* 174:1699–1703.

Bill-Axelson A, Holmberg L, Ruutu M et al.; Scandinavian Prostate Cancer Group Study No. 4 (2005). Radical prostatectomy versus watchful waiting in early prostate cancer. *N Engl J Med* 352(19):1977–1984.

Bossuyt PM, Irwig L, Craig J, Glasziou P (2006). Comparative accuracy: Assessing new tests against existing diagnostic pathways. *BMJ* 332:1089–1092.

Briman M, Artukovic E, Zhang L, Chia D, Goodglick L, Gruner G (2007). Direct electronic detection of prostate-specific antigen in serum. *Small* 3: 758–762.

Cheng J, Teply BA, Sherifi I et al. (2007). Formulation of functionalized PLGA-PEG nanoparticles for in vivo targeted drug delivery. *Biomaterials* 28(5):869–876.

Cooperberg MR, Lubeck DP, Meng MV, Mehta SS, Carroll PR (2004). The changing face of low-risk prostate cancer: Trends in clinical presentation and primary management. *J Clin Oncol* 22(11):2141–2149.

Derycke AS, de Witte PA (2002). Transferrin-mediated targeting of hypericin embedded in sterically stabilized PEG-liposomes. *Int J Oncol* 20(1):181–187.

Derycke AS, Kamuhabwa A, Gijsens A et al. (2004). Transferrin-conjugated liposome targeting of photosensitizer AlPcS4 to rat bladder carcinoma cells. *J Natl Cancer Inst* 96(21):1620–1630.

Deserno WM, Harisinghani MG, Taupitz M et al. (2004). Urinary bladder cancer: Preoperative nodal staging with ferumoxtran-10-enhanced MR imaging. *Radiology* 233:449–456.

Everts M, Saini V, Leddon JL et al. (2006). Covalently linked Au nanoparticles to a viral vector: Potential for combined photothermal and gene cancer therapy. *Nano Lett* 6:587.

Farokhzad OC, Cheng J, Teply BA et al. (2006). Targeted nanoparticle-aptamer bioconjugates for cancer chemotherapy in vivo. *Proc Natl Acad Sci USA* 103(16):6315–6320.

Feldman AS, McDougal WS, Harisinghani MG (2008). The potential of nanoparticle-enhanced imaging. *Urol Oncol* 26(1):65–73.

Gabizon A, Papahadjopoulos D (1992). The role of surface charge and hydrophilic groups on liposome clearance in vivo. *Biochim Biophys Acta* 1103(1):94–100.

Gommersall L, Shergill IS, Ahmed HU et al. (2007). Nanotechnology and its relevance to the urologist. *Eur Urol* 52(2):368–375.

Gommersall L, Shergill IS, Ahmed HU, Arya M, Grange P, Gill IS (2008). Nanotechnology in the management of prostate cancer. *BJU Int* 102(11):1493–1495.

Grimm J, Perez JM, Josephson L, Weissleder R (2004). Novel nanosensors for rapid analysis of telomerase activity. *Cancer Res* 64(2):639–643.

Gurocak S, Nuininga J, Ure I, DeGier RPE, Tan MO, Feitz W (2007). Bladder augmentation: Review of the literature and recent advances. *Indian J Urol* 23:452–457.

Haley B, Frenkel E (2008). Nanoparticles for drug delivery in cancer treatment. *Urol Oncol* 26(1):57–64.

Han D, Gouma P-I (2006). Electrospun bioscaffolds that mimic the topology of extracellular matrix. *Nanomed Nanotechnol Biol Med* 2:37–41.

Harisinghani MG, Barentsz J, Hahn PF et al. (2003). Noninvasive detection of clinically occult lymph-node metastases in prostate cancer. *N Engl J Med* 348(25):2491–2499.

Harisinghani MG, Saksena MA, Hahn PF et al. (2006). Ferumoxtran-10- enhanced MR lymphangiography: Does contrast-enhanced imaging alone suffice for accurate lymph node characterization? *AJR* 186:144–148.

Harisinghani MG, Saksena M, Ross R et al. (2005). A pilot study of lymphotrophic nanoparticle-enhanced magnetic resonance imaging technique in early stage testicular cancer: A new method for noninvasive lymph node evaluation. *Urology* 66:1066–1071.

Harrington DA, Sharma AK, Erickson BA, Cheng EY (2008). Bladder tissue engineering through nanotechnology. *World J Urol* 26(4):315–322.

Hartgerink JD, Beniash E, Stupp SI (2002). Peptide-amphiphile nanofibers: A versatile scaffold for the preparation of self-assembling materials. *Proc Natl Acad Sci USA* 99:5133–5138.

Hattori Y, Maitani Y (2004). Enhanced in vitro DNA transfection efficiency by novel folate-linked nanoparticles in human prostate cancer and oral cancer. *J Control Release* 97:173–183.

Hattori Y, Maitani Y (2005). Folate-linked lipid-based nanoparticle for targeted gene delivery. *Curr Drug Deliv* 2:243–252.

Hirsch LR, Stafford RJ, Bankson JA et al. (2003). Nanoshell-mediated nearinfrared thermal therapy of tumors under magnetic resonance guidance. *Proc Natl Acad Sci USA* 100:13549.

Hovels AM, Heesakkers RA, Adang EM et al. (2004). Cost analysis of staging methods for lymph nodes in patients with prostate cancer: MRI with a lymph node-specific contrast agent compared to pelvic lymph node dissection or CT. *Eur Radiol* 14:1707–1712.

Humes HD (2000). Bioartificial kidney for full renal replacement therapy. *Semin Nephrol* 20(1):71–82.

Illing R, Chapman A (2007). The clinical applications of high intensity focused ultrasound in the prostate. *Int J Hyperthermia* 23(2):183–191.

Imamura M, Kanematsu A, Yamamoto S et al. (2007). Basic fibroblast growth factor modulates proliferation and collagen expression in urinary bladder smooth muscle cells. *Am J Physiol Renal Physiol* 293:F1007–F1017.

Jain TK, Morales MA, Sahoo SK, Leslie-Pelecky DL, Labhasetwar V (2005). Iron oxide nanoparticles for sustained delivery of anticancer agents. *Mol Pharm* 2:194–205.

Jeon SI, Lee JH, Andrade JD et al. (1991). Protein-surface interactions in the presence of polyethylene oxide. I. Simplified theory. *J Colloid Interface Sci* 142:149–158.

Johannsen M, Gneveckow U, Thiesen B et al. (2007a). Thermotherapy of prostate cancer using magnetic nanoparticles: Feasibility, imaging, and three-dimensional temperature distribution. *Eur Urol* 52(6):1653–1661.

Johannsen M, Gneveckow U, Taymoorian K et al. (2007b). Morbidity and quality of life during thermotherapy using magnetic nanoparticles in locally recurrent prostate cancer: Results of a prospective phase I trial. *Int J Hyperthermia* 23:315–323.

Johannsen M, Thiesen B, Gneveckow U et al. (2006). Thermotherapy using magnetic nanoparticles combined with external radiation in an orthotopic rat model of prostate cancer. *Prostate* 66(1):97–104.

Johannsen M, Thiesen B, Jordan A et al. (2005a). Magnetic fluid hyperthermia (MFH) reduces prostate cancer growth in the orthotopic Dunning R3327 rat model. *Prostate* 64(3):283–292.

Johannsen M, Gneveckow U, Eckelt L et al. (2005b). Clinical hyperthermia of prostate cancer using magnetic nanoparticles: Presentation of a new interstitial technique. *Int J Hyperthermia* 21:637–647.

Jordan A, Scholz R, Maier-Hauff K et al. (2001). Presentation of a new magnetic field therapy system for the treatment of human solid tumors with magnetic fluid hyperthermia. *J Magn Magn Mater* 225:118–126.

Jordan A, Wust P, Scholz R et al. (1996). Cellular uptake of magnetic fluid particles and their effects in AC magnetic fields on human adenocarcinoma cells in vitro. *Int J Hyperthermia* 12(6):705–722.

Kaji N, Tokeshi M, Baba Y (2007). Quantum dots for single bio-molecule imaging. *Anal Sci* 23(1):21–24.

Kam NW, O'Connell M, Wisdom JA, Dai H (2005). Carbon nanotubes as multifunctional biological transporters and near-infrared agents for selective cancer cell destruction. *Proc Natl Acad Sci USA* 102:11600–11605.

Kampinga HH, Dikomey E (2001). Hyperthermic radiosensitization: Mode of action and clinical relevance. *Int J Radiat Biol* 77(4):399–408.

Kroon BK, Horenblas S, Meinhardt W et al. (2005). Dynamic sentinel node biopsy in penile carcinoma: Evaluation of 10 years experience. *Eur Urol* 47(5):601–606.

Langer R (1998). Drug delivery and targeting. *Nature* 392(6679 Suppl):5–10.

Lee D, Lee JT, Sheperd D, Abrahams H (2005). Preliminary use of the Enseal™ system for sealing of the dorsal venous complex during robotic assisted laparoscopic prostatectomy. Abstract 1186. American Urology Association Annual Meeting, 2005, San Antonio, TX.

Liedberg F, Chebil G, Davidsson T, Gudjonsson S, Mansson W (2006). Intraoperative sentinel node detection improves nodal staging in invasive bladder cancer. *J Urol* 175(1):84–88.

Loo C, Lowery A, Halas N et al. (2005). Immunotargeted nanoshells for integrated cancer imaging and therapy. *Nano Lett* 5:709.

Ma Z, Kotaki M, Inai R, Ramakrishna S (2005). Potential of nanofiber matrix as tissue-engineering scaffolds. *Tissue Eng* 11(1–2):101–109.

Maeda H, Fang J, Inutsuka T et al. (2003). Vascular permeability enhancement in solid tumor: Various factors, mechanisms involved and its implications. *Int Immunopharmacol* 3:319.

Maruyama K, Unezaki S, Takahashi N, Iwatsuru M (1993). Enhanced delivery of doxorubicin to tumor by long-circulating thermosensitive liposomes and local hyperthermia. *Biochim Biophys Acta* 1149(2):209–216.

May DJ, Allen JS, Ferrara KW (2002). Dynamics and fragmentation of thick-shelled microbubbles. *IEEE Trans Ultrason Ferroelectr Freq Control* 49(10):1400–1410.

McDougal WS, Gervais DA, McGovern FJ et al. (2005). Long-term followup of patients with renal cell carcinoma treated with radio frequency ablation with curative intent. *J Urol* 174:61.

McManus M, Boland E, Sell S et al. (2007). Electrospun nanofibre Wbrinogen for urinary tract tissue reconstruction. *Biomed Mater* 2:257–262.

Mouraviev V, Joniau S, Van Poppel H, Polascik TJ (2007). Current status of minimally invasive ablative techniques in the treatment of small renal tumours. *Eur Urol* 51(2):328–336.

Nakanishi H, Mizutani Y, Kawauchi A et al. (2003). Significant antitumoral activity of cationic multilamellar liposomes containing human IFN-beta gene against human renal cell carcinoma. *Clin Cancer Res* 9(3):1129–1135.

Okada H, Toguchi H (1995). Biodegradable microspheres in drug delivery. *Crit Rev Ther Drug Carrier Syst* 12(1):1–99.

Oyewumi MO, Yokel RA, Jay M, Coakley T, Mumper RJ (2004). Comparison of cell uptake, biodistribution and tumor retention of folate-coated and PEG-coated gadolinium nanoparticles in tumor-bearing mice. *J Control Release* 95(3):613–626.

Pareek G, Nakada SY (2005). The current role of cryotherapy for renal and prostate tumors. *Urol Oncol* 23:361.

Park JW (2002). Liposome-based drug delivery in breast cancer treatment. *Breast Cancer Res.* 4(3):95–99.

Pattison MA, Wurster S, Webster TJ, Haberstroh KM (2005). Three-dimensional, nano-structured PLGA scaffolds for bladder tissue replacement applications. *Biomaterials* 26:2491–2500.

Peng W, Anderson DG, Bao Y, Padera RF Jr, Langer R, Sawicki JA (2007). Nanoparticulate delivery of suicide DNA to murine prostate and prostate tumors. *Prostate* 67(8):855–862.

Petrylak DP, Tangen CM, Hussain MH et al. (2004). Docetaxel and estramustine compared with mitoxantrone and prednisone for advanced refractory prostate cancer. *N Engl J Med* 351:1513–1520.

Pham QP, Sharma U, Mikos A (2006). Electrospinning of polymeric nanofibers for tissue engineering applications: A review. *Tissue Eng* 12:1197–1211.

Pleschberger M, Saerens D, Weigert S et al. (2004). An S-layer heavy chain camel antibody fusion protein for generation of a nanopatterned sensing layer to detect the prostate-specific antigen by surface plasmon resonance technology. *Bioconjug Chem* 15:664–671.

Sahoo SK, Ma W, Labhasetwar V (2004). Efficacy of transferrin-conjugated paclitaxel-loaded nanoparticles in a murine model of prostate cancer. *Int J Cancer* 112:335–340.

Santhakumaran LM, Thomas T, Thomas TJ (2004). Enhanced cellular uptake of a triplex forming oligonucleotide by nanoparticles formation in the presence of polypropylenimine dendrimers. *Nucleic Acids Res* 32:2102–2112.

Schellenberger EA, Hogemann D, Josephson L, Weissleder R (2002). Annexin V-CLIO: A nanoparticle for detecting apoptosis by MRI. *Acad Radiol* 9(Suppl 2):S310–S311.

Shen T, Weissleder R, Papisov M, Bogdanov A, Brady TJ (1993). Monocrystalline iron oxide nanocompounds (MION): Physicochemical properties. *Magn Reson Med* 29(5):599–604.

Shulga OV, Zhou D, Demchenko AV, Stine KJ (2008). Detection of free prostate specific antigen (fPSA) on a nanoporous gold platform. *Analyst* 133:319–322.

Silva G, Czeisler C, Neice K et al. (2004). Selective differentiation of neural progenitor cells by high-epitope density nanofibers. *Science* 303:1352–1357.

Silverman SG, Israel GM, Herts BR, Richie JP (2008). Management of the incidental renal mass. *Radiology* 249(1):16–31.

Staack A, Hayward SW, Baskin LS, Cunha GR (2005). Molecular, cellular and developmental biology of urothelium as a basis of bladder regeneration. *Differentiation* 73(4):121–133.

Stern JM, Cadeddu JA (2008). Emerging use of nanoparticles for the therapeutic ablation of urologic malignancies. *Urol Oncol* 26(1):93–96.

Stern JM, Stanfield J, Lotan YA, Park S, Hsieh JT, and Cadeddu JA (2007) Efficacy of laser-activated gold nanoshells to ablate prostate cancer cells in vitro. *J Endourol* 21(8): 939–943.

Storm G, Belliot T, Daemen D et al. (1995). Surface modification of nanoparticles to oppose uptake by the mononuclear phagocytic system. *Adv Drug Deliv Rev* 17:31–48.

Sullivan DC, Ferrari M (2004). Nanotechnology and tumor imaging: seizing an opportunity. *Mol Imaging* 3(4):364–369.

Tabatabaei S, Harisinghani M, McDougal WS (2005). Regional lymph node staging using lymphotrophic nanoparticle enhanced magnetic resonance imaging with ferumoxtran-10 in patients with penile cancer. *J Urol* 174:923–927.

Tannock IF, de Wit R, Berry WR et al. (2004). Docetaxel plus prednisone or mitoxantrone plus prednisone for advanced prostate cancer. *N Engl J Med* 351:1502–1512.

Thapa A, Webster TJ, Haberstroh KM (2003). Polymers with nanodimensional surface features enhance bladder smooth muscle cell adhesion. *J Biomed Mater Res A* 67:1374–1383.

Thomas TP, Patri AK, Myc A et al. (2004). In vitro targeting of synthesized antibodyconjugated dendrimer nanoparticles. *Biomacromolecules* 5:2269–2274.

Thompson I, Pauler D, Goodman P et al. (2004). Prevalence of prostate cancer among men with a prostate-specific antigen level < or = 4.0 ng per milliliter. *N Engl J Med* 350(22):2239–2246.

Thompson IM, Tangen CM, Kristal AR (2008). Prostate-specific antigen: A misused and maligned prostate cancer biomarker. *J Natl Cancer Inst* 100:1487–1488.

Warncke SH, Mattei A, Fuechsel FG, Z'Brun S, Krause T, Studer UE (2007). Detection rate and operating time required for gamma probe-guided sentinel lymph node resection after injection of technetium-99m nanocolloid into the prostate with and without preoperative imaging. *Eur Urol* 52(1):126–132.

Winter PM, Caruthers SD, Kassner A et al. (2003). Molecular imaging of angiogenesis in nascent Vx-2 rabbit tumors using a novel alpha(nu)beta3-targeted nanoparticle and 1.5 tesla magnetic resonance imaging. *Cancer Res* 63(18):5838–5843.

Wu G, Datar RH, Hansen KM, Thundat T, Cote RJ, Majumdar A (2001). Bioassay of prostate-specific antigen (PSA) using microcantilevers. *Nat Biotechnol* 19:856–860.

Wunderbaldinger P, Josephson L, Weissleder R (2002). Crosslinked iron oxides (CLIO): A new platform for the development of targeted MR contrast agents. *Acad Radiol* 9(Suppl 2):S304–S306.

Yezhelyev MV, Gao X, Xing Y et al. (2006). Emerging use of nanoparticles in diagnosis and treatment of breast cancer. *Lancet Oncol* 7:657.

IV

Medical Imaging

22

Quantum Dots for Nanomedicine

Sarah H. Radwan
The American University in Cairo

Hassan M. E. Azzazy
The American University in Cairo

22.1 Introduction

Quantum dots (QDs) are inorganic fluorophores that possess unique optical properties such as broad excitation spectra, narrow emission spectra, high photostability, high quantum yields, and long fluorescence lifetimes. These properties make them more attractive than current fluorophores for biomedical applications. The emission wavelengths of QDs can be controlled by varying their sizes and compositions allowing the simultaneous detection of multiple targets. In addition, their nano-size gives them distinctive properties that differ from bulk materials such as a large surface area-to-volume ratio, chemically tailorable surfaces, and increased solubility enabling their use for *in vivo* imaging, *in vitro* diagnostics, drug delivery, and photodynamic therapy (PDT).

22.2 Structure of QDs

QDs are semiconductor nanocrystals composed of a core surrounded by a shell, and have diameters that range from 2 to 9.5 nm (Michalet et al., 2005). The size of QDs relative to different biological entities is shown in Figure 22.1. Semiconductors have a valence band that is completely filled with electrons followed by a band gap and then an empty conduction band. When a semiconductor is struck by a photon, with energy higher than the band gap energy, an electron is excited to the conduction band leaving a hole (of opposite electric charge behind). The electron and its hole are attracted to each other by Coulomb force and together are referred to as an exciton.

Excitons in a QD are confined in three dimensions of space and since the size of the QDs is in the same order as the Bohr exciton radius (the distance between an electron and its hole in an exciton), this leads to the quantum confinement effect. This effect causes the valence and conduction bands within QDs to turn into quantized energy levels with energy values directly related to the QD size. This effect is also responsible for the unique optical properties of QDs (Michalet et al., 2005; Azzazy et al., 2007; Jamieson et al., 2007).

22.2.1 Core

The core of a QD is made of a semiconductor material. The core is either composed of atoms of groups II–VI such as CdSe, CdS, or CdTe; or groups III–V such as InP, or InAs; or groups IV–VI such as PbSe (Michalet et al., 2005; Liu, 2006; Jamieson et al., 2007). The cores are highly reactive due to the large surface area-to-volume ratio, which makes them unstable (Jamieson et al., 2007). Surface defects in the nanocrystals can act as traps causing intermittent fluorescence (switching between fluorescent and nonfluorescent states) or the so-called blinking, which is due to the trapping and untrapping of the electron or hole. This phenomenon of blinking decreases the quantum yield, which is defined by the following equation (Michalet et al., 2005; Jamieson et al., 2007):

$$\text{Quantum yield} = \frac{\text{\# of photons emitted}}{\text{\# of photons absorbed}}$$

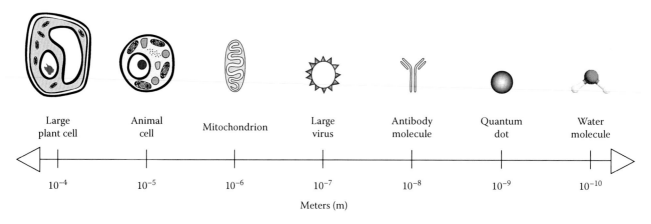

FIGURE 22.1 The relative sizes of different biological entities in relation to a QD.

22.2.2 Shell

The shell is made of another semiconductor material that has a larger band gap than the core material such as ZnS or CdS. The larger band gap energy of the shell confines excitation and emission to the core (Bruchez et al., 1998). In addition, the shell also reduces blinking and protects the surface atoms of the highly reactive core from oxidation, photochemical degradation, and other chemical reactions that may lead to a decrease in fluorescence intensity, i.e., quenching. Therefore, the shell increases quantum yields, which could reach up to 90%, as well as increases photostability (Michalet et al., 2005; Jamieson et al., 2007).

22.3 Solubilization and Functionalization

QDs are insoluble in water since they are synthesized in hydrophobic surfactants such as trioctylphosphine oxide (TOPO) and hexadecylamine (Qu and Peng, 2002). They can be synthesized by different methods (please refer to other volumes in this series for details). There are several methods to make QDs soluble. The simplest method is to exchange the hydrophobic surface ligand, TOPO, with hydrophilic thiolated molecules such as mercaptoacetic acid (MAA), mercaptopropionic acid (MPA), dithiothreitol (DTT), dihydrolipoic acid (DHLA), or with oligomeric phosphines, dendrons, or peptides (Michalet et al., 2005; Liu, 2006). The second method is to encapsulate the QD in phospholipid micelles, polymer beads or shells, or to add a layer of amphiphilic block copolymers, amphiphilic polysaccharides, or silica shells (Michalet et al., 2005). The addition of a silica shell, which is also known as surface silanization, makes the QDs very stable since the silica shells are highly cross-linked (Bruchez et al., 1998).

By modifying their surface chemistries, QDs can be used in different applications. Different targeting moieties could be attached to the surface such as streptavidin, avidin, antibodies, oligonucleotides, receptor ligands, or peptides (Figure 22.2). Furthermore, the ability to attach multiple moieties to a single QD makes QDs multifunctional and efficient in multiplexing assays. Also, attaching multiple ligands to a single QD increases avidity for the target cell and makes QDs superior to

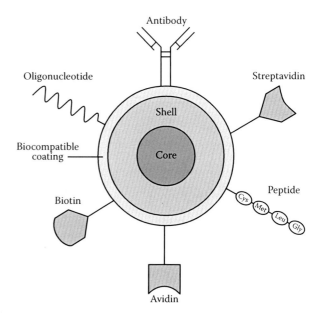

FIGURE 22.2 Schematic diagram of the structure of a QD with different targeting moieties attached to the surface.

conventional fluorophores in imaging techniques in which each fluorophore is usually conjugated to only one ligand and detection becomes dependent on target concentration (Frangioni, 2003; McNeil, 2005; Delehanty et al., in press).

22.4 Optical Properties of QDs

When QDs are excited, they fluoresce. Fluorescence occurs when the excited electron in the QD goes back to the ground state emitting a photon with a longer wavelength than the one absorbed (Stoke's shift). Size, composition, and shape all play key roles in determining the wavelength of the emitted light. The larger the QD, the longer the wavelength of emitted light. This is because as the size of the QD increases, more energy levels exist and are closer to each other, and therefore the electron will need less energy to be excited and in turn the emitted light will be of lower energy and a longer wavelength. QDs can emit light of wavelengths ranging from ultraviolet (UV) to infrared (IR)

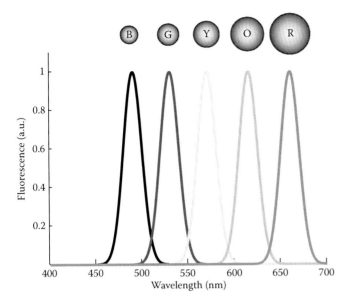

FIGURE 22.3 The relationship between size of QDs and their emission wavelengths. As QD size increases, the wavelength of emitted light increases. (B, blue; G, green; Y, yellow; O, orange; R, red).

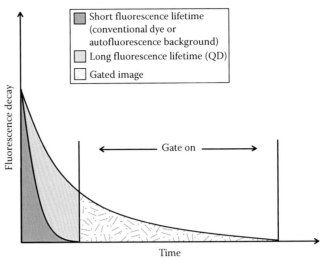

FIGURE 22.4 The fluorescence decay of a long-lived species and that of a shorter-lived one. Time-gated imaging allows the detection of photons arriving within a specified time window. The shorter-lived species or background fluorescence are discarded.

(Jamieson et al., 2007). The effect of varying the size of QDs on their emission wavelengths is represented in Figure 22.3. QDs also possess other unique optical properties such as increased brightness, resistance to photobleaching, and a high quantum yield. They also have a long fluorescence lifetime (a measure of the time a fluorophore spends in the excited state) that is utilized in time-gated imaging (Box 22.1). QDs are ideal for multiplexed time-gated detection since they emit light after most of the background autofluorescence is exhausted. However, they still emit light fast enough to retain the high photon turnover rate required to achieve high sensitivity. Time gating was used to create images of specific mouse fibroblasts termed 3T3 cells (Dahan et al., 2001). Only photons arriving in a specified time window (35–65 ns), after excitation by a laser pulse, are detected thus decreasing background noise, increasing signed-to-background ratio (GBR), and enhancing sensitivity. On the other hand, conventional organic fluorophores have a short lifetime fluorescence and emit light within a few nanoseconds upon excitation, which overlaps with the background autofluorescence; therefore, they

are not practical for time-gated imaging. Table 22.1 lists all the optical properties of QDs in comparison to conventional fluorophores.

22.5 Intracellular Delivery

It is important to mention the different methods of intracellular delivery before discussing the biomedical applications of QDs since the effective delivery of QDs is essential in many *in vitro* as well as *in vivo* applications such as imaging, cellular labeling, drug delivery, and PDT. There are mainly two methods by which QDs can be delivered into cells and tissues: endocytosis and physical methods. The easiest way to deliver QDs intracellularly is to take advantage of the normal intracellular transport processes such as endocytosis, which is also the least damaging delivery method.

Endocytosis is a mechanism by which materials are transported into the cell. Cells engulf the material with the plasma membrane which then fuses and pinches off forming an intracellular vesicle. This occurs with most water-soluble material since they cannot cross the hydrophobic cell membrane by diffusion

---

Box 22.1   Time-Gated Detection

Time-gated detection enables the separation of the signal of a long-lived species, such as a QD, from the signal of a shorter-lived species or from the autofluorescence background.

This is done by detecting only the photons that arrive during a specified time window after excitation by the laser pulse.

Therefore, to decrease background noise, photons hitting in the first few nanoseconds are disregarded thus increasing the SBR and sensitivity (Figure 22.4).

TABLE 22.1 A Comparison between the Optical Properties of QDs and Those of Conventional Fluorophores

Optical Properties	Conventional Fluorophores	QDs
Width of excitation spectrum	Narrow, excitation by light of a specific wavelength depending on the dye used	Broad, excitation by light of a broad range of wavelengths
Width of emission spectrum	Broad, red tailed	Narrow and symmetric (~25–30 nm full width at half maximum)
Photostability	Unstable, photobleaches in a few minutes on exposure to light	Stable, resistant to photobleaching over long periods of time (minutes to hours)
Fluorescence lifetime	Short, <5 ns	Long, ~30–100 ns
Quantum yield	Low, less than 15% in aqueous environments	Very high, 50%–60% in aqueous environments up to 90% in organic solvents
Effective Stoke's shift	Small	Large
Molar extinction coefficient[a]	Small	Large (the shorter the excitation wavelengths, the higher the extinction coefficient)
Two-photon action cross sections[b]	Low, 10,000–20,000 GM units at 800 nm	Very high, 2–3 orders of magnitude larger than conventional dyes, up to 47,000 GM units

[a] A measure of how strongly a substance absorbs light. It is used to convert units of absorbance into units of molar concentration and is determined by measuring the absorbance at a reference wavelength for 1 M of the target solution in a cuvette with a path length of 1 cm.

[b] A measure of brightness for imaging and is measured in Goeppert-Mayer (GM) units.

(Samia et al., 2006). QDs capped with DHLA entered cells successfully by endocytosis. The hydrophilic carboxylated ligands facilitated the attachment of QDs to cell surfaces through electrostatic interactions; however, this method results in the intracellular delivery of QDs to any cell and therefore it is nonspecific (Jaiswal et al., 2003).

To increase specificity as well as efficiency, receptor-mediated endocytosis could be used, where the QDs are labeled with receptor ligands to specifically target cell surface receptors (Medintz et al., 2008). QDs conjugated to transferrin have been shown to undergo receptor-mediated endocytosis by mammalian cells (Michalet et al., 2005). Targeting the integrin receptors is also possible by conjugating QDs to peptides containing the arginine–glycine–aspartate (RGD) motif. Epidermal growth factor (EGF) receptors could also be targeted by different methods. One of the methods makes use of the high binding affinity of biotin for streptavidin (Box 22.2) in which biotinylated EGF protein is conjugated to streptavidin-coated QDs. EGF receptors are attractive targets since they are overexpressed on the surfaces of many cancer cells (Delehanty et al., in press). Folate receptors are also overexpressed in cancer cells and can be targeted by conjugating QDs to folic acid (Misra, 2008). QDs conjugated to antibodies could also target cell surface receptors. For example, anti-type I insulin-like growth factor receptor (IGF1R) antibody conjugated to QDs were used to detect and measure IGF1R levels in breast cancer cells (Zhang et al., in press).

Assisted endocytosis is a third type of endocytosis that is receptor-independent. The use of certain biomolecules such as protein transduction domains (PTDs), which are short peptide sequences, facilitates the uptake by adhering to cellular surfaces resulting in endocytosis. The use of PTDs could be the most efficient method for the intracellular delivery of QDs *in vivo* (Delehanty et al., in press). Examples of PTDs are the HIV Tat peptide and Pep-1 (available commercially as Chariot™) (Mattheakis et al., 2004; Kovar et al., 2007). Antibodies could be

used to bind cell surface proteins on specific cells that undergo endocytosis. Pancreatic cancer cells were labeled and imaged by using QDs conjugated to an anti-Claudin-4 antibody to target the Claudin-4 membrane protein, which is overexpressed on the surface of these cells (Qian et al., 2007). Endocytosis of QDs into cells could also be assisted by encapsulating QDs into micelles or by using delivery vehicles such as liposomes and dendrimers. Liposomes are vesicles surrounded by a phospholipid bilayer, whereas dendrimers are repeatedly branched polymers having highly symmetrical structures. They are both used to deliver DNA and drugs into cells. Liposomes allow the intracellular delivery of hydrophilic QDs by entrapping them in the aqueous vesicle, while hydrophobic QDs are delivered into cells by entrapping them in the phospholipid bilayer. In addition, by choosing appropriate lipids, liposomes could fuse with cellular membranes allowing the delivery of hydrophobic QDs to the cell membrane (Delehanty et al., in press). Cationic liposomes were shown to have a relatively higher delivery efficiency than dendrimers (Derfus et al., 2004). Another way to facilitate endocytosis is to encapsulate QDs in cholesterol-bearing pullulan (CHP) modified with amino groups (CHPNH$_2$). A pullulan is a water-soluble polysaccharide consisting of maltotriose units linked through α-1,6-glucosidic bonds. The CHPNH$_2$-QDs complexes had a higher delivery efficiency than cationic liposomes (Hasegawa et al., 2005). Polyethyleneimine (PEI) fused with polyethylene glycol (PEG) could encapsulate QDs and facilitate their endocytosis and their release into the cytoplasm. PEI has the ability to disrupt the acidic endosomes by attracting protons. This method is more efficient than other amphiphilic polymers since the QD-conjugates are smaller in size, which is necessary for intracellular access to organelles. They are also more stable in acidic environments (Duan and Nie, 2007).

The physical methods that could be used for intracellular delivery are electroporation, microinjection, the use of protein microbubbles, particle bombardment, or scrape-loading (Samia

BOX 22.2   STRONG BINDING AFFINITY OF NATURAL BIOMOLECULES

1. Biotin-(Strept) Avidin System

Biotin ($C_{10}H_{16}N_2O_3S$), also known as vitamin H or B_7, is a small molecule (MW 244.3 Da) found in little amounts in all living cells. It is necessary for cell growth and the metabolism of fats and amino acids. It has a very strong binding affinity to the tetrameric glycoprotein avidin, which is found in high levels in egg whites. This protein consists of four identical subunits. Each subunit can bind to one molecule of biotin and therefore avidin can bind up to four molecules of biotin simultaneously with high specificity. The dissociation constant (K_d) of this complex is 10^{-15} M making it the strongest known non-covalent biological interaction between a protein and a ligand. Biotin also binds strongly to streptavidin, which is purified from the bacterium *Streptomyces avidinii* and is also produced recombinantly. It is a tetrameric protein with a similar dissociation constant. However, streptavidin has lower nonspecific binding than avidin (Diamandis and Christopoulos, 1991).

Biotin

2. Enzyme-Substrate

Most of the enzymes are proteins that act as biological catalysts increasing the rate of metabolic reactions in living cells. The site on the enzyme where the substrate binds is called the active site and such binding requires structural complementarity that reflects the high specificity of the enzyme. The enzyme binds to its substrate in a manner resembling a lock and key (the old model) and chemically alters the substrate into a product. The unchanged enzyme is then free to interact with another substrate molecule.

3. Antibody–Antigen

This form of interaction is similar to an enzyme-substrate interaction; however, the interaction does not lead to an irreversible chemical change in the antibody or the antigen. Antibodies, also known as immunoglobulins, are glycoproteins that are found in the human serum and other secreted body fluids such as milk, tears, and saliva. They are used by the immune system to recognize foreign objects such as pathogens that enter the body. They are formed of two heavy chains and two light chains and bind to their target antigens through various noncovalent interactions. The antibody binds through its variable domains of the light and heavy chains and a high degree of structural complementarity is required for interaction to occur making it highly specific. The dissociation constant varies for different antibody-antigen complexes. In some cases, cross reactivity could occur where an antibody reacts with an unrelated antigen with an identical or similar epitope (binding site of the antigen).

et al., 2006; Medintz et al., 2008). However, these methods are disruptive and could damage the cell membrane (Medintz et al., 2008). In electroporation, electric fields are used to make the cell membrane temporarily permeable by creating pores through which QDs can enter the cell (Derfus et al., 2004). The microinjection technique involves the direct injection of QDs into the cytoplasm. QDs encapsulated in phospholipid micelles were microinjected into *Xenopus* embryos for *in vivo* imaging (Dubertret et al., 2002). PEG-coated QDs, functionalized with the appropriate localization sequence peptides, were microinjected into the cytoplasm of HeLa cells (the cell line used in research originally obtained from cervical cancer cells) to target the mitochondria and nucleus (Derfus et al., 2004). However, microinjection is time consuming since it is based on serial injections and therefore the number of cells that can be labeled is limited (Medintz et al., 2008). Microbubbles are gas-filled bubbles surrounded by a protein, polymer, or a lipid shell, typically, a lipid monolayer. The microbubbles are doped with QDs and burst upon sonication freeing the QDs and allowing their entry into the cell (Samia et al., 2006). Particle bombardment was used to transport DNA-coated gold microparticles into cells. A high-voltage electric discharge device was used to accelerate the particles to high velocity to allow their entry into cells (Yang et al., 1990). Similarly, this technique could be applied to the intracellular delivery of QDs (Samia et al., 2006). In scrape-loading, the cells are adhered onto the surface of a culture dish and by scraping them off the dish, holes are produced in the plasma membrane through which QDs can enter (Partridge et al., 1996). This method is not widely used because it results in extreme damage to the cultured cells (Delehanty et al., in press). The pros and cons of the main delivery methods are listed in Table 22.2. Scrape-loading, particle bombardment, and

TABLE 22.2 The Advantages and Disadvantages of Different Intracellular Delivery Methods for QDs

Method of Intracellular Delivery of QDs		Advantages	Disadvantages
Endocytosis	Non-specific endocytosis	Simple (no functionalization required)	Nonspecific uptake
		Non-invasive	Endosomal trapping
		Could be used *in vivo*	Used only for whole-cell labeling
	Receptor-medicated endocytosis	Simple	Endosomal trapping
		Non-invasive	Used only for whole cell-labeling
		Increase in specificity and efficiency	
		Could be used *in vivo*	
	Assisted endocytosis	Simple	Possible aggregate formation (e.g., in case of using liposomes)
		Non-invasive	
		High efficiency	Toxicity of some of the assisting molecules
		Could be used *in vivo*	
		Possible endosomal escape and cytoplasmic labeling	
Physical methods	Electroporation	Endosomal escape and cytoplasmic labeling of a large population of cells	Invasive
			High rate of cell death
		Could be used *in vivo*	Aggregate formation
	Microinjection	No aggregate formation	Invasive
		Endosomal escape and cytoplasmic labeling	Serial technique (one cell at a time)
		Labeling of intracellular structures and organelles using localization peptide sequences conjugated to QD	Time consuming
			Low volume analysis
			Not practical for *in vivo* applications
		High efficiency	Devices are expensive

microbubbles are not mentioned in the table since they have not been widely used.

22.6 Biomedical Applications

22.6.1 Fluorescent Cell Labeling

Fluorescent cell labeling is used to visualize whole cells or subcellular structures by labeling them with fluorophores. Labeling whole cells is useful for the detection of specific cell populations or pathogens, cell tracking, and cell lineage experiments while the simultaneous labeling of subcellular structures is useful for understanding intracellular processes (Michalet et al., 2005; Medintz et al., 2008). The conventional fluorophores typically used are organic dyes, fluorescent proteins, or lanthanide chelates (Jamieson et al., 2007).

22.6.1.1 Advantages of QDs in Fluorescent Cell Labeling

The two main challenges facing conventional fluorophores are long-term monitoring (since they suffer from photobleaching) and simultaneous multicolor labeling of different structures. Simultaneous multicolor labeling using conventional fluorophores will require excitation by light of more than one wavelength since they have narrow absorption spectra. Furthermore, their broad emission spectra can result in spectral overlap between the different dyes (Jamieson et al., 2007; Medintz et al., 2008).

QDs are stable and can resist photobleaching allowing for the long-term monitoring of labeled substances. They have broad absorption spectra and narrow and symmetric emission spectra,

which enables the simultaneous multicolor labeling of cells and organelles (Medintz et al., 2008).

22.6.1.2 Delivery and Labeling Methods

Whole cell labeling could be carried out by microinjection, electroporation, or simply endocytosis of QDs since it is the simplest delivery method (Michalet et al., 2005). The labeling of cell surface proteins is carried out by the functionalization of QDs with different ligands including streptavidin, avidin, antibodies, receptor ligands, or peptides. The most common method for labeling cell surface proteins is the three-layer approach that involves using a primary antibody specific to the target, a biotinylated secondary antibody, and streptavidin-coated QDs. A two-layer approach could also be used in which a primary antibody specific to the target is recognized by a secondary antibody conjugated to a QD. As for cytoplasmic or nuclear target labeling, the challenge in this case is to prevent the QDs from getting trapped in the endocytic pathway, which would prevent them from reaching their intracellular target. Therefore, endocytosis would not be a recommended delivery method. Microinjection of QDs, conjugated to the appropriate localization peptides, would be the method of choice (Michalet et al., 2005).

QDs have been used to label intracellular structures and proteins such as microtubules, actin, mitochondria, and nuclear antigens as well as cell surface proteins such as cancer markers like Her2, mortalin, and *p*-glycoprotein. QDs have also been used for labeling cell surface receptors such as glycine receptors, integrin, erb/HER receptors, AMPA receptors, $GABA_c$ receptors, and TrkA receptors (for reviews, see Michalet et al., 2005 and Jamieson et al., 2007).

22.6.2 Förster Resonance Energy Transfer Analysis

Examples of FRET-based molecular probes are molecular beacons and Taqman® probes. Molecular beacons are formed from a fluorophore (donor) and a quencher molecule (acceptor) conjugated to the ends of a single-stranded oligonucleotide sequence that has a hairpin structure. In this state, the fluorescence of the donor fluorophore is quenched as a result of FRET to the acceptor, which is in close proximity to the donor. When the complementary target DNA is present, it binds to the molecular beacon, causing the hairpin structure to open. This increases the distance between the two molecules causing the fluorophore to restore its fluorescence. Taqman probes are based on the same concept; however, the fluorophore and the quencher are conjugated to the ends of a linear oligonucleotide and in this state, the fluorophore is quenched. These probes are used in real-time polymerase chain reaction (PCR) to monitor DNA amplification. During PCR, the Taqman probes anneal to the target DNA. During the extension of the primers and the replication of the target DNA, the Taqman probes are cleaved by the 5′ exonuclease activity of *Taq* DNA polymerase and therefore fluorescence of the donor is restored (Didenko, 2001). There are several problems associated with FRET-based molecular probes including background fluorescence of the unbound probes, which can mask the fluorescence target signals, spectral crosstalk or overlap between donor and acceptor emissions and the direct excitation of the acceptor (Zhang et al., 2005).

QDs offer high detection specificity with minimum crosstalk and without excitation of the acceptor. The unbound probes of QDs produce a very low level of background fluorescence in FRET applications. QDs can also amplify the target signal by binding to multiple targets at a time, which increases the energy transfer efficiency; therefore, smaller quantities of the target can be detected with high sensitivity and without the need for target amplification. They can also be used in multiplexing assays detecting different targets simultaneously (Zhang et al., 2005). In FRET applications, QDs are typically used as energy donors since they are not good energy acceptors (Clapp et al., 2005).

22.6.2.1 QD-FRET Applications

A prototype immunoassay using QD-FRET has been designed in which bovine serum albumin (BSA) antigen was conjugated to QDs emitting red light, while anti-BSA antibody was conjugated to QDs emitting green light. The antigen–antibody binding affinity (Box 22.2) was assessed using an enzyme-linked immunosorbent assay (ELISA). The formation of the immune complex decreases the emission of the green QDs and increases the emission of the red QDs due to FRET. When the immune complex was exposed to unlabeled BSA antigen, the emission of the green QDs was restored (Wang et al., 2002).

A nanosensor for maltose sensing has been designed in which QDs were conjugated to the maltose binding protein (MBP), which can bind maltose or an analog sugar–dye complex that acts as a nonemissive quencher molecule (Figure 22.5). When the QD is bound to the analog, fluorescence is quenched due to FRET to the analog. However, upon adding maltose to the solution, the analog is replaced by maltose (since MBP has a higher binding affinity for maltose) and the fluorescence is restored, which increases with increasing maltose concentration. In this experiment, FRET efficiency was enhanced by coating each QD with an average of 10 MBP molecules (Medintz et al., 2003).

To measure the activity of proteases, a QD was conjugated to multiple peptide sequences, each conjugated to a quencher.

Box 22.3   Förster Resonance Energy Transfer

Förster resonance energy transfer (FRET) is also known as fluorescence resonance energy transfer. It is the non-radiative transfer of excitation energy from an excited donor to an acceptor (through dipole–dipole interaction) within a certain distance known as the Förster radius. The Förster radius is the distance between the donor and acceptor at which energy transfer efficiency is 50% and is typically less than 10 nm (Didenko, 2001). The following equation shows the relationship between energy transfer efficiency (E), the donor–acceptor separation distance (r), and the Förster radius (R_o):

$$E = \frac{R_o^6}{R_o^6 + r^6} \qquad \text{Yeh et al. (2005)}$$

Another important condition in FRET is that the emission spectrum of the donor must overlap with the absorption spectrum of the acceptor. FRET decreases the fluorescence intensity and the fluorescence lifetime of the donor and increases them in the acceptor (Didenko, 2001). FRET can measure changes in distance between the donor and the acceptor and therefore could be used for measuring protein conformational changes, monitoring protein interactions, and observing enzymatic activity as well as in genetic applications such as the study of telomerization and DNA replication, hybridization, and cleavage (Jamieson et al., 2007; Medintz et al., 2008).

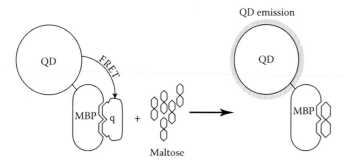

FIGURE 22.5 Schematic diagram of a QD-MBP nanosensor. MBP can bind to either maltose or to a sugar analog which acts as a quencher. When the quencher is bound to MBP, the QD is quenched as a result of FRET. Upon addition of maltose, the quencher is displaced and fluorescence of QD is restored.

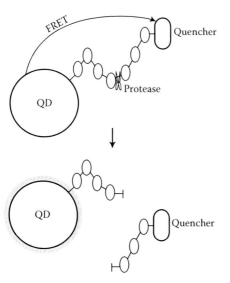

FIGURE 22.6 Schematic diagram of a QD-peptide nanosensor. QD attached to a quencher via a peptide sequence which contains cleavage sites for a specific protease. The QD is quenched as a result of FRET but on addition of the protease, the peptide is cleaved and fluorescence of the QD is restored.

The peptide sequence contains cleavage sites recognized by a protease. In the absence of the protease, the QD is quenched as a result of FRET to the quencher, but when the protease is present, the peptide sequence is cleaved and the QD fluoresces, signaling the presence of an active enzyme (Figure 22.6). The detection of protease activity of different enzymes such as caspase-1, thrombin, collagenase, and chymotrypsin was achieved by changing the peptide sequence. A different scenario was implemented for the detection of caspase-1 in which Cy3 fluorophore was conjugated to the peptide instead of the quencher and therefore, upon cleavage, Cy3 emissions decreased while QD emissions increased (Medintz et al., 2006). Based on the same principle, the increased activity of extracellular matrix metalloproteinases in cancerous cells was used to differentiate between healthy and cancerous breast cell cultures. The QD was conjugated to tetra-peptides, each labeled with a dye. Upon the addition of the QD-FRET probes to the breast cancer cultures, the peptide sequences were cleaved leading to increased QD fluorescence intensity and a reduction of the intensity of the dyes (Shi et al., 2006). In a different study, the proteolytic activity of trypsin as well as the inhibition efficiency of trypsin inhibitors were monitored. However, it was difficult to determine the inhibition efficiency of large trypsin inhibitors such as protein molecules. This might have been due to the displacement of the dye-labeled peptides from the QDs by the protein molecules causing an increase in the fluorescence intensity of the QDs even in the absence of trypsin (Shi et al., 2007).

In genetic applications, molecular beacons employing QDs as donor fluorophores can be used for the detection of nucleic acids and point mutations using the FRET phenomenon. In 2004, the first QD-labeled molecular beacon (Figure 22.7) was designed, which enabled long-term analysis permitted by the high photostability of QDs (Kim et al., 2004a).

22.6.2.2 Challenges Facing QD-FRET

There are four challenges facing QD-FRET. The first challenge is that the outer coatings of QDs increase their size and this in turn increases the distance between the energy acceptor and the QD core, which can sometimes decrease the efficiency of FRET.

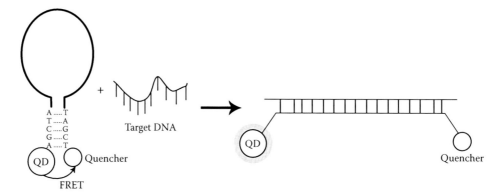

FIGURE 22.7 Schematic diagram of a QD-conjugated molecular beacon. The QD is quenched at first as a result of FRET to the quencher which is in close proximity. On addition of the target DNA sequence, the molecular beacon opens to allow hybridization to the target. This increases the distance between the QD and the quencher causing the QD to restore its fluorescence.

However, this problem was partially solved by using a two-step FRET approach in which a QD was conjugated to a Cy3-labeled MBP, which in turn was conjugated to the Cy3.5-labeled sugar analog allowing FRET from the QD to Cy3.5 via a Cy3 bridge. Upon adding maltose, the analog is replaced and the Cy3 emissions increase. Despite the fact that the overall efficiency is reduced, the use of the two-step FRET was crucial to produce a functional sensor (Medintz et al., 2003). The second challenge is the steric hindrance resulting from the conjugation of multiple energy acceptors to a single QD. When using QD-FRET for the detection of proteases, the conjugation of multiple peptide sequences (with bound energy acceptors) to each QD decreases the accessibility of proteases to the conjugated peptides. The third challenge is that the environmental conditions, such as the ionic strength and pH, can also affect fluorescence intensity (Jamieson et al., 2007). Finally, the fourth challenge is that the displacement of dye-labeled peptides from the surfaces of QDs by larger protein molecules in solution has also been reported (Shi et al., 2007).

22.6.3 *In Vivo* Fluorescence Imaging

In vivo imaging can be limited by tissue absorbance, autofluorescence, scattering, limited penetration depth, and the inability of long-term monitoring (Box 22.4). The criteria that increase effectiveness of *in vivo* fluorescence imaging are listed in Box 22.4. QDs meet most of these criteria making them better image contrast agents than the conventional organic dyes. QDs are bright, resistant to photobleaching, and can be made water soluble. They have large extinction coefficients, long fluorescence lifetimes, and high quantum yields. Their size-tunable emissions permit the production of QDs that absorb and fluoresce in the NIR region, allowing deep tissue penetration and imaging with minimum scattering and autofluorescence (Frangioni, 2003; Michalet et al., 2005). A second method used to achieve deep tissue penetration is the two-photon excitation confocal microscopy in which two NIR photons are absorbed sequentially by a QD. This phenomenon is called upconversion luminescence since the emitted photon has a higher energy than the excitation photon. The higher energy of the emitted photon is compensated by the absorption of two photons achieving deeper tissue penetration with lower excitation energy, which is not possible using conventional fluorophores (Chen, 2008). It should be noted that QDs can even exhibit multiphoton (three or more) excitation because of their broad absorption spectra and very high two-photon action cross sections (Grecco et al., 2004).

22.6.3.1 Self-Illuminating QDs and Signal Amplification

Self-illuminating QDs improve *in vivo* imaging and are based on the phenomenon of bioluminescence resonance energy transfer (BRET). QDs are conjugated to a bioluminescent molecule such as the enzyme luciferase, which acts as the energy donor. Upon the addition of its substrate (coelenterazine), luciferase is activated and emits a photon that excites the bound QD (acceptor) causing its fluorescence. Therefore, BRET is similar to FRET, but the donor energy in BRET results from the chemical reaction catalyzed by luciferase rather than the absorption of excitation

---

BOX 22.4 *IN VIVO* FLUORESCENCE IMAGING

In vivo fluorescence imaging is the creation of images of live human or animal tissues or organs by using non-invasive photon penetration. This is carried out by first targeting fluorophores to the specific tissues then using light to excite these fluorophores. When the fluorophores relax to the ground state, they fluoresce, emitting light of a longer wavelength. The emitted light can then be detected using a high sensitivity charge-coupled device (CCD) camera or another optical detector (Kovar et al., 2007).

Sometimes components inside tissues, such as water, melanin, lipids, or hemoglobin, absorb the excitation photon. The unlabeled tissue could also emit this light leading to a phenomenon called "tissue autofluorescence." Absorbance and tissue autofluorescence can limit the SBR by masking the signals produced from labeled substances. Most tissues exhibit autofluorescence when visible light is used, therefore near-infrared (NIR; 700–1000 nm) light is used instead to decrease absorbance and autofluorescence and hence increase SBR (Frangioni, 2003).

Another event that could occur during imaging is the scattering of light (Frangioni, 2003). Scattering as well as tissue absorbance decrease the penetration depth. As the wavelength of light decreases, scattering and absorbance increase, which results in penetration depths of only a few millimeters. Since tissue absorbance is lower in the NIR region, light can penetrate to depths of several centimeters (Frangioni, 2003; Kovar et al., 2007). Therefore, to limit absorbance, autofluorescence, and scattering, NIR light is used for excitation instead of visible light, thereby increasing SBR and penetration depth.

The fluorophore-related criteria that increase the effectiveness of *in vivo* imaging are water solubility, excitation and emission wavelengths in the NIR region, high photostability, long fluorescence lifetime, a large extinction coefficient, high quantum yield, low non-specific binding, low *in vivo* toxicity, and rapid blood clearance of the unbound fluorophores. The tissue-related criteria are minimum tissue absorbance, autofluorescence, and scattering (Frangioni, 2003; Kovar et al., 2007).

light. In imaging, this leads to signal amplification and removes the need for excitation light. This in turn decreases the tissue absorbance and autofluorescence resulting in low background noise and a high SBR (Frangioni, 2006; So et al., 2006; Jamieson et al., 2007). Alternatively, signal amplification could be achieved by receptor-mediated endocytosis by which QDs accumulate inside the cells (Frangioni, 2003; Kovar et al., 2007).

22.6.3.2 *In Vivo* Imaging Applications Using QDs

QDs were injected intravenously into mice to image their blood vessels using two-photon excitation confocal microscopy. The QDs had two-photon action cross sections up to 47,000 GM units, which is much larger than that of conventional fluorophores (Larson et al., 2003). QDs with four different surface coatings were tested for imaging in mice. Coating QDs with PEG decreased their accumulation in the liver and bone marrow, and the long-term experiments showed that QDs remained fluorescent for nearly 4 months *in vivo* (Ballou et al., 2004). An intradermal injection of NIR-QDs was carried out in live mice and pigs. The QDs migrated to sentinel lymph nodes that were imaged at a depth of 1 cm with almost no background noise (Kim et al., 2004b). This proves the efficiency of NIR-QDs for *in vivo* imaging and the possibility of using them for surgical guidance.

22.6.3.3 Challenges Facing QDs for *In Vivo* Imaging

An effective contrast agent should bind and be retained by the target for a sufficient time to allow imaging, whereas its unbound fraction should be rapidly cleared from the blood circulation. The challenges that face QDs as contrast agents are their toxicity, owing to their composition and rapid uptake (before binding to their targets) by the reticuloendothelial system (RES), which decreases their blood concentration. Coating with PEG could overcome the rapid clearance of QDs by RES since it decreases the chances of opsonization (coating of foreign material that enters the cell with antibodies or other components of the immune system that facilitates its elimination) (McNeil, 2005; Jamieson et al., 2007). On the other hand, PEG coating will increase the size of QDs preventing the clearance of the unbound fraction resulting in a high background and a low SBR (Frangioni, 2003; Jamieson et al., 2007; Kovar et al., 2007).

Another barrier for the use of QDs for *in vivo* imaging is the increased size of the functionalized QDs, which could reach up to 100 nm (Hardman, 2006). This limits tissue penetration and may induce unfavorable immune responses by the host. In addition, the binding of host antibodies to QDs may render them ineffective. Large functionalized QDs will stay in circulation longer and could eventually be taken up by the liver, which may prevent imaging of organs close to the liver. It should be noted that the liver is particularly sensitive to cadmium toxicity, which would be an issue if QDs made of cadmium are used (Jamieson et al., 2007; Kovar et al., 2007).

22.6.4 Diagnostic Applications

22.6.4.1 Molecular Assays

22.6.4.1.1 Nucleic Acids

Conventional fluorophores suffer from photobleaching and cannot be used in multiplexing assays as their emission spectra overlap. QDs, on the other hand, have long-term photostability and can be used in multiplexing assays, owing to their broad absorption spectra and narrow emission spectra. They can also be excited by a single wavelength. These properties make QDs more efficient probes than conventional fluorophores for fluorescence *in situ* hybridization (FISH) applications (Box 22.5) (Bentolila and Weiss, 2006). QDs have been used in FISH for the detection of human metaphase chromosomes as well as the detection of the ERBB2/HER2/ neu gene (Xiao and Barker, 2004). QD-FISH probes were also used in the simultaneous detection of different genetic sequences. In addition, QDs were used in FISH to detect four different mRNA transcripts in neurons of mouse midbrain sections (Chan et al., 2005). QDs were also used in RNA interference studies by monitoring small interference RNA (siRNA) delivery and assessing the degree of gene silencing (Chen et al., 2005; Derfus et al., 2007).

DNA microarrays allow high-throughput gene expression profiling (by measuring the amount of cDNA derived from mRNA) as well as genotyping and mutation detection. Typical DNA microarrays contain immobilized DNA probes that are complementary to the target DNA and are labeled with fluorophores to allow detection (for further details, see Joos et al., 2003; Chaudhuri, 2005). DNA-QD probes were used in DNA

BOX 22.5   FLUORESCENCE *IN SITU* HYBRIDIZATION

FISH is a technique used to optically detect the location of oligonucleotide targets such as DNA or mRNA in tissue sections. This is carried out by using fluorophore-labeled oligonucleotide probes that hybridize to the target. The fluorescence emitted by the bound fluorophore can then be detected using fluorescence microscopy. This technique is used in the clinical detection of chromosomal abnormalities related to genetic disorders and in the determination of gene copy number in tumors with gene amplification as well as in viral detection (Xiao and Barker, 2004; Bentolila and Weiss, 2006).

microarrays for the detection of synthetic oligonucleotides corresponding to genomic sequences in hepatitis B and C viruses. Capture strands immobilized on the microarray hybridize to the viruses. Probe strands conjugated to QDs then hybridize to the capture strands allowing simultaneous detection of both viruses (Gerion et al., 2003).

22.6.4.1.2 *Single Nucleotide Polymorphisms*

Single nucleotide polymorphisms (SNPs) are the most common form of genetic variation in which a single nucleotide, at a given site in the human genome, differs among individuals. If the sequence of a certain gene from two individuals differs in one SNP, this will give rise to two alleles of the gene. It is estimated that there are more than 10 million SNPs in the entire human genome and SNPs located within certain genes are associated with disease susceptibility in humans (Hirschhorn and Daly, 2005).

A prototype assay for multiplexed DNA detection was developed using QDs of different sizes embedded into polymeric beads (Han et al., 2001). The beads are conjugated to oligonucleotide probes specific to the target DNA, which is either directly labeled with a fluorophore or with biotin for binding to fluorescently labeled avidin. By using single-bead spectroscopy analysis, the presence and abundance of each target is detected by the signal produced by the fluorophore. Additionally, the QDs of different sizes inside each microbead emit different colors, and increasing the number of a particular QD increases the intensity of its emitted color. Therefore, using different ratios of QDs inside each microbead generates a unique emission spectrum for each target thus identifying its sequence. Theoretically speaking, if 6 different colors and 10 intensity levels for each color are used, the color-coding of one million different biomolecules could be achieved. The single-color encoded beads are highly uniform and reproducible resulting in bead identification accuracies of almost 100%. Following this model, Xu et al. (2003) developed a new method for SNP genotyping. Two different QDs, each at three different intensities (achieved by varying the number of each QD), were embedded into microbeads resulting in nine different microbeads. The microbeads were conjugated to oligonucleotide probes targeting different SNPs. In this system, biotinylated primers were used to identify each SNP using multiplex PCR. The biotinylated amplicons were then incubated with the microbeads, and hybridization took place only when there was a perfect match. Streptavidin-Cy5 was then added, which bound to the biotinylated amplicons. A flow cytometer was then used to detect the target signal (Cy5), which indicates the presence and abundance of each SNP, and the coding signals of the QDs inside the microbead, which identifies the specific SNP (Figure 22.8). This technique was accurate and sensitive producing the same results as Taqman in-house assays for 940 genotypes. Both of the above methods are homogeneous methods that do not require a separation step prior to detection.

22.6.4.1.3 *Genetic Mutations*

Using DNA microarrays, QDs conjugated to oligonucleotides were used for the detection of SNPs and single-base deletions in the human p53 gene (Gerion et al., 2003). The p53 gene is a tumor suppressor gene that has been found to be mutated in more than 50% of human tumors (Denissenko et al., 1997). Another method for SNP detection is the QD-based electrochemical coding in which four different QDs (ZnS, CdS, PbS, and CuS) are conjugated to the four mononucleotides, A, C, G, and T, respectively (Liu et al., 2005). The QD-mononucleotide conjugates are then added sequentially to a solution containing the eight different possible SNPs, which are captured on magnetic beads. Each QD-mononucleotide conjugate binds to its complementary base in addition to previously bound conjugates. For example, when C-CdS is added, it binds to its complementary base, G, at any SNP site. After washing, G-PbS is then added, which binds to C at any SNP site as well as the C-CdS already bound. The resulting assemblies can be magnetically separated and transferred to an electrochemical cell where the QDs are then dissolved and detected electrochemically. Unique voltammograms are created containing up to four characteristic peaks (corresponding to the four different metals that the QDs are made of) revealing the identity of each mismatch in only one electrochemical run (Figure 22.9). The detection of known two-base mutations in a single DNA target was also possible using electrochemical coding. Another method for the detection of SNPs using QDs was developed using an oligonucleotide ligation assay (Figure 22.10) and a two-color fluorescence coincidence analysis (Yeh et al., 2006). A biotinylated capture probe and a fluorophore-labeled reporter probe hybridize to a perfect match target and are then ligated by a DNA ligase forming a fluorescent ligation product (FLP), which is then separated from the template DNA target by heat. The FLPs can then bind to streptavidin-coated QDs forming QD-FLP nanoassemblies that are excited by a single wavelength producing simultaneous emissions that are detected as coincident signals. In the presence of a mismatch, the two probes are not ligated and no QD-FLP nanoassemblies are formed, and in turn, no coincident signals are produced. This assay was highly sensitive with the ability to detect zeptomolar (10^{-21} M) target concentrations with an allele discrimination selectivity factor of more than 10^5.

22.6.4.2 Immunoassays

A sandwich potentiometric immunoassay was designed using CdSe QDs. The assay was carried out using 96 well plates where the antigen is sandwiched between two antibodies with the secondary antibody conjugated to CdSe QDs. With the addition of hydrogen peroxide, the QDs are oxidized and Cd^{2+} ions are released, which are then detected using Cd^{2+}-ion-selective electrodes. The detection limit was found to be <10 fmol in 150 µL sample wells (Thurer et al., 2007).

Another sensitive immunoassay was used to detect the tumor marker total prostate specific antigen using a sandwich assay but was carried out on a carbon substrate surface. In this method, the secondary antibodies were biotinylated and then incubated with streptavidin-coated QDs. Fluorescence imaging was then used to detect the QDs. The detection limit in undiluted human serum samples was found to be 0.25 ng/mL (Kerman et al., 2007).

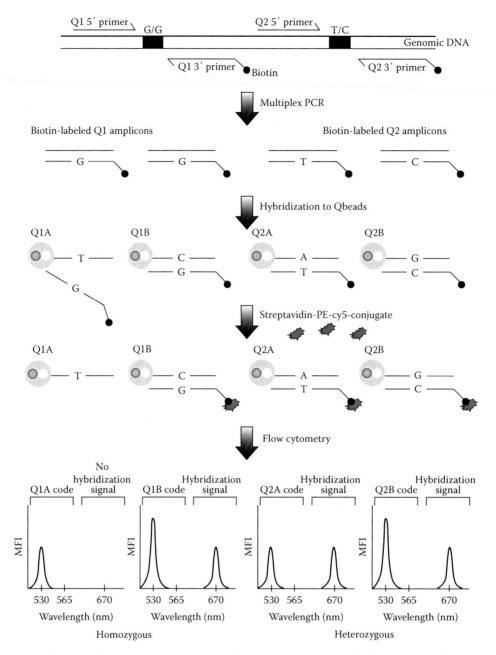

FIGURE 22.8 Schematic diagram of SNP genotyping using QDs embedded into microbeads. Multiplex PCR using biotinylated primers is used to amplify each SNP from genomic samples. The biotinylated amplicons hybridize to microbeads only when there is a perfect match. Streptavidin-Cy5 is then added which binds to the biotinylated amplicons. Flow cytometric analysis detects the target signal (Cy5), indicating the presence and abundance of each SNP genotype, and the coding signals of the QDs inside the microbead, which identifies the specific SNP. (Reprinted with Xu, H. et al., *Nucl. Acids Res.*, 31, e43, 2003. With permission.)

An example of a multiplex fluoroimmunoassay for the detection of four different toxins simultaneously was developed in which QDs of different sizes (colors) were conjugated to the appropriate antibodies and excited using a single wavelength. The toxins detected were shiga-like toxin 1, ricin, cholera toxin, and staphylococcal enterotoxin B with detection limits of 300, 30, 10, and 3 ng/mL, respectively (Goldman et al., 2004).

QD-based fluoroimmunoassays were also used for the simultaneous detection of the foodborne pathogenic bacteria,

Escherichia coli 0157:H7 and *Salmonella typhimurium*. Target bacterial cells were separated from samples using magnetic beads coated with the specific antibodies. QDs of different sizes, conjugated to the appropriate antibodies, reacted with the bead-cell complexes. The intensity of the fluorescence emission of the final complexes was measured for quantitative detection of the two species of bacteria. The fluorescence intensity as a function of cell number was found for both species and the detection limit was 10^4 colony forming units/mL (Yang and Li, 2006). Other

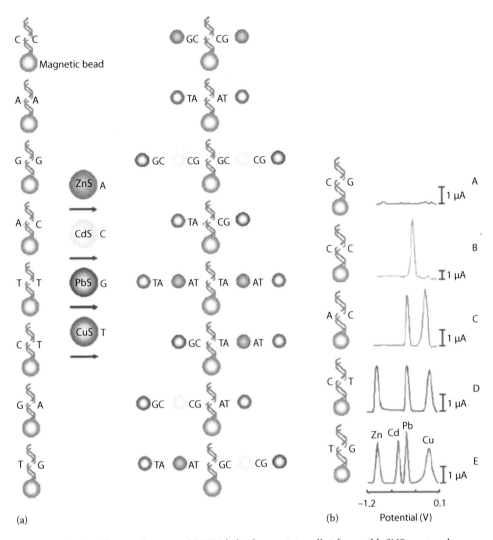

FIGURE 22.9 Electrochemical coding for SNP detection. (a) DNA hybrids containing all eight possible SNPs captured on magnetic beads followed by the sequential addition of the four different QDs, each attached to a different mononucleotide and the resulting assemblies; (b) Voltammograms for some chosen mismatches. (A) Control (complementary DNA), no signal detected. (B) C-C SNP, only one peak corresponding to Pb with twofold higher intensity than other SNPs. (C) A-C SNP, two peaks for Cu and Pb. (D) C-T SNP, three peaks of Pb, Zn, and Cu. (E) T-G SNP, four peaks of Zn, Cd, Pb, and Cu. (Reprinted with Liu, G. et al., *J. Am. Chem. Soc.*, 127, 38, 2005. With permission.)

species of bacteria were also detected using QD-based immunoassays such as *Cryptosporidium parvum*, *Giardia lamblia*, and *Listeria monocytogenes* (Lee et al., 2004; Zhu et al., 2004; Tully et al., 2006).

Protein microarrays are used for protein expression profiling, identification of protein–protein interactions, and in diagnostics to measure the amount of proteins in biological samples. A protein microarray is a miniaturized assay in which captured protein molecules such as antibodies are immobilized on a solid surface and bind to their target proteins present in the solution. After washing, the target protein can be detected by different methods. For example, in an antibody microarray, fluorescently labeled detector antibodies are added that bind to the target antigen forming a sandwich and producing a fluorescent signal that is proportional to the concentration of the antigen (for a review on protein microarrays, see Sydor and

Nock, 2003). QDs could be conjugated to detector antibodies instead of conventional fluorophores. A QD-protein microarray was developed for the detection of cancer markers in biological specimens (serum, plasma, and other body fluids) at picomolar (10^{-12} M) concentration. First, two different models of QD probes were tested. In one assay, detector antibodies specific to a selected target were directly conjugated to QDs. In a different assay, streptavidin-coated QDs were conjugated to biotinylated detector antibodies specific to the target molecule. The fluorescence intensity signals produced by the streptavidin QD-biotinylated antibody probes were 30 times higher than the signals produced by the QD-antibody probes resulting in a more sensitive assay. Using the more sensitive QD probe in a series of multiplexing experiments, it was possible to detect six different cytokines including tumor necrosis factor alpha and interleukins (Zajac et al., 2007).

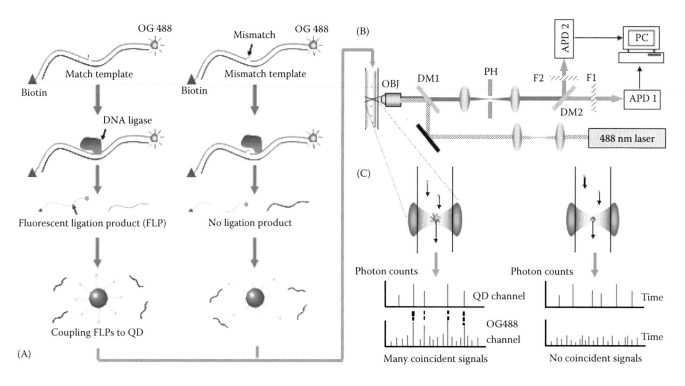

FIGURE 22.10 Schematic diagram of a homogeneous oligonucleotide ligation assay for the detection of SNPs using QD-mediated two-color coincidence analysis. (A) A biotinalated capture probe and a fluorophore-labeled reporter probe hybridize to the target and are ligated by a DNA ligase only in the presence of a perfect match forming an FLP. The FLPs can then bind to streptavidin-coated QDs forming QD-FLP nanoassemblies (left). In the presence of a mismatch, the two probes are not ligated and therefore no FLPs are formed (right). (B) The QD-FLP nanoassemblies pass through a small detection volume of a confocal fluorescence detection system and are detected using a single wavelength excitation. (C) The QD-FLPs produce simultaneous emissions that are detected as coincident signals (left). In case of a mismatch, no coincident signals are produced (right). (Reprinted with Yeh, H.C. et al., *Nucl. Acids Res.*, 34, e35, 2006. With permission.)

22.6.4.2.1 Challenges Facing QDs in Immunoassays

There are commercially available QD conjugation kits allowing the conjugation of different targeting moieties to QDs. However, there is no control over the ratio of the conjugated moieties, their orientation, or avidity, which can make assay development and optimization difficult. There is also a limit to the spectral resolution of the signals obtained from multiplexed immunoassays. Only the emission spectra separated by 15 nm in their intensity maxima can be resolved (Jamieson et al., 2007). Additionally, antibody cross-reactivity is an issue for any immunoassay, however, this issue is not related to the use of QDs (Medintz et al., 2008).

22.6.4.3 Single Molecule Detection

Single molecule detection (SMD) allows detection of targets without the need for their amplification. In the case of nucleic acids, this eliminates the need for PCR and thus decreases errors due to contamination. SMD also allows quantitative analysis with high sensitivity and specificity (Agrawal et al., 2006). Conventional fluorophores are not bright enough and suffer from photobleaching and crosstalk between their emissions. Their narrow absorption spectra requires the use of two laser excitation beams, which introduces the difficulty of par-focality, i.e., the cofocusing of the two laser beams (Agrawal et al., 2006).

QDs offer improved brightness and photostability, size tunable narrow emission spectra, which results in improved SBR, and minimum crosstalk. Their broad absorption spectra allow excitation with a single light source (Agrawal et al., 2006).

Several systems for SMD using QDs were developed for the detection of nucleic acids or proteins. Agrawal et al. (2006) designed a homogeneous real-time assay for the SMD of nucleic acids, proteins, or intact viruses in a microfluidic channel without the need for target amplification or a separation step. This was achieved using confocal microscopy and two-color fluorescence coincidence analysis. In this method, two different QDs (red and green) are used to bind to the same target at two different sites forming a sandwich complex, thus leading to enhanced sensitivity and specificity of detection (Figure 22.11). QDs can be excited by a single wavelength, which eliminates the need for two excitation laser beams that are difficult to cofocus into the small detection volume of a few femtoliters (10^{-15} L). The red and green emitted photons are detected simultaneously using two separate detectors on two channels and their arrival times are analyzed. If an unbound green QD passes through the detection volume, a photon burst on the green channel will be detected, while in the case of a red QD, a photon burst on the red channel will be detected. However, if the sandwich complex passes, fluorescent bursts will be detected in the green and red channels

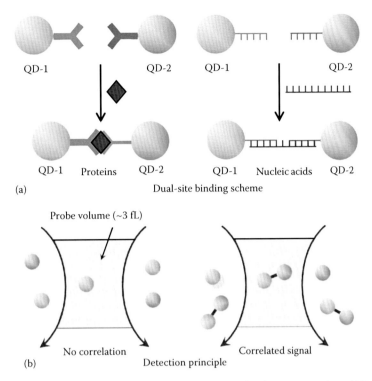

FIGURE 22.11 Schematic diagram of the binding method in a single-molecule sandwich assay using QDs. (a) Two QDs of different sizes (colors) are used to bind to the same target at two different sites. The target could be a protein or a nucleic acid; (b) the femtoliter detection volume containing unbound QD probes (left) and bound QD sandwich pairs (right). (Reprinted with Agrawal, A. et al., *Anal. Chem.*, 78, 1061, 2006. With permission.)

simultaneously as well as the coincidence channel since the photons will arrive at the same time. In the rare event that two unbound QDs pass simultaneously, fluorescent bursts will be detected in both the red and green channels but will not give strong coincident signals. Figure 22.12 shows real-time photon burst data obtained for the four different scenarios.

A QD-FRET homogeneous assay for the detection of DNA was also designed (Zhang et al., 2005). The DNA nanosensor consists of a streptavidin-coated QD, biotinylated capture probes, and fluorophore(Cy5)-labeled reporter probes. The target DNA hybridizes with the reporter probe from one end and the capture probe from the other. The streptavidin-coated QDs then bind to the biotinylated capture probes. Each QD acts as a target concentrator by binding to multiple DNA targets. This amplifies the signal and increases the efficiency of FRET. When the QD is excited, it acts as the FRET donor transferring energy to the acceptor (Cy5). Using confocal fluorescence spectroscopy, a fluorescent burst can be detected whenever a single nanosensor assembly passes through the detection volume. The fluorescent bursts of the donor and acceptor are detected by different detectors and the presence of the target is indicated by the simultaneous detection of bursts from both acceptor and donor. However, if the target is absent, only fluorescent bursts from the donor are detected. This system allows SMD without the need for target amplification and with a 100-fold higher sensitivity than other DNA detection probes such as molecular beacons. Blinking could be a possible limitation to the use of QDs in such assays;

it should be noted that, although the shell reduces blinking, it does not eliminate it completely (Michalet et al., 2005). Coating the QD surface with thiol moieties or using QDs in free suspension might help solve this problem (Alivisatos et al., 2005).

22.6.4.4 Western Blot

Conventional Western blot analysis is used for the detection of proteins using gel electrophoresis to separate native proteins by their three-dimensional structures or denatured proteins according to the size of their polypeptide chains. The proteins are then transferred to a membrane, either nitrocellulose or polyvinylidene difluoride, where they are detected. There are different methods for the detection of the blotted proteins, however, the most common one is by using primary antibodies specific to the target protein and enzyme-conjugated secondary antibodies. Upon the addition of the appropriate substrate, a colored product is formed that stains the immobilized protein band. QDs can be used instead of enzymes to detect picogram levels of proteins. The unique optical properties of QDs especially their narrow emission enables multiplexing, easy image acquisition, and quantification. In addition, the QD-labeled blots can be stored at 4°C–8°C, with minimal loss of signal, for later imaging (Ornberg et al., 2005). QD-based Western blots were used to detect tracer proteins, such as telomeric repeat binding factors and its interacting nuclear protein 2, directly in cell lysates, which is not possible using standard Western blot technology. In conventional Western blotting, the preliminary steps of immunoprecipitation

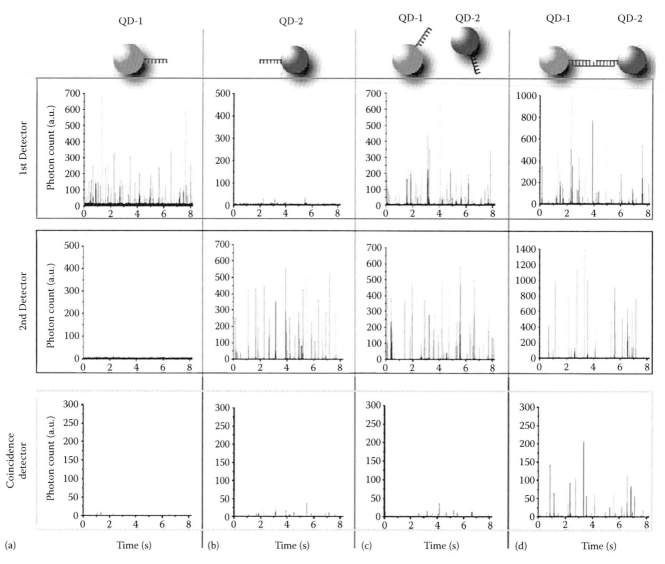

FIGURE 22.12   Real-time photon bursts data recorded from the first detector (detects emission of QD-1), the second detector (detects emission of QD-2), and the coincidence detector. (a) In the presence of QD-1 probes, signals are obtained only from the first detector; (b) in the presence of QD-2 probes, signals are obtained only from the second detector; (c) in the presence of both unbound QD probes, signals are obtained from both detectors but not from the coincidence detector; (d) in the presence of the sandwich pairs, signals are obtained from all three detectors. (Reprinted with Agrawal, A. et al., *Anal. Chem.*, 78, 1061, 2006. With permission.)

and concentration have to be carried out for the detection of tracer proteins. The QD-based Western blot analysis, however, does not require these steps making the procedure faster and less labor-intensive (Bakalova et al., 2005).

22.6.4.5 Immunohistochemistry

Immunohistochemistry is used to label proteins in tissue sections using antibodies that are either conjugated to enzymes or fluorophores. Conventional fluorophores suffer from photobleaching while QDs have high levels of brightness and are photostable and therefore can be used instead of conventional fluorophores. An immunohistochemical protocol was developed combining the sensitivity of enzyme signal amplification with the photostability of QDs for the detection of intracellular

antigens in mouse and rat brain tissue sections (Ness et al., 2003). QDs can also be used in quantitative multiplex immunohistochemistry. QDs of different sizes conjugated to biotinylated primary antibodies enabled the simultaneous staining of three co-localized proteins in tonsillar tissue. The use of spectral imaging enables the quantification of the fluorescence signal produced by QDs as well as percentage colocalization, facilitating standardized, high throughput use in clinical trial studies (Sweeney et al., 2008).

Growth hormone, and prolactin proteins and their mRNAs were also detected using immunohistochemistry in combination with *in situ* hybridization. The detection of each protein with its mRNA was carried out on different slides. For the detection of the proteins, immunohistochemical staining was carried out using primary antibodies specific to the target proteins and then adding QDs conjugated to secondary antibodies. For the detection of mRNAs, biotinylated oligonucleotide probes were allowed to hybridize to the target mRNAs. QDs of a different size coated with streptavidin were then added and attached to the biotinylated probes. This technique enabled the simultaneous visualization of each protein and its mRNA in three dimensions (Matsuno et al., 2005).

22.6.5 Therapy

22.6.5.1 Photodynamic Therapy

22.6.5.1.1 *Photophysical Processes and Competing Reactions of the Excited Photosensitizer*

On absorbing light, the photosensitizer gets excited and an electron moves from the ground state (S_0) to one of the excited singlet states (S_n). The photosensitizer then relaxes to the first excited singlet state (S_1), a process known as internal conversion or nonradiative relaxation. When the photosensitizer relaxes all the way back to the ground state (S_0) emitting light in the process, this is referred to as fluorescence ($S_1 \rightarrow S_0$). It could also undergo nonradiative relaxation. A third possibility is that an intersystem crossover could occur in which the photosensitizer moves from the first excited singlet state (S_1) to the first excited triplet state (T_1) (Josefsen and Boyle, 2008).

When the photosensitizer is in the T_1 state, several competing reactions could take place. First, the photosensitizer could go back to S_0 emitting light, which is referred to as phosphorescence ($T_1 \rightarrow S_0$) (Josefsen and Boyle, 2008). Alternatively, the photosensitizer could undergo nonradiative relaxation. However, in the presence of oxygen, the photosensitizer could collide with oxygen and become deactivated, bringing it back to S_0. An electron or energy transfer from the photosensitizer to oxygen could also occur. The energy transfer causes the formation of singlet oxygen ($^1O_2$), which is the main cytotoxic agent in PDT. This reaction is also known as a type II reaction (De Rosa and Crutchley, 2002). All the competing reactions are represented in Figure 22.13 and a simplified Jablonski energy diagram is shown in Figure 22.14, which illustrates all the

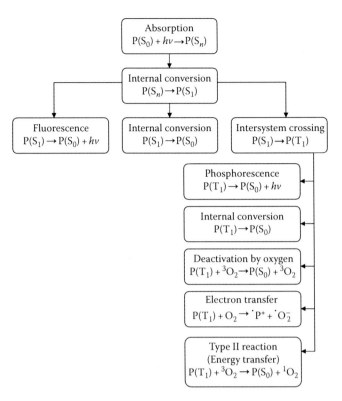

FIGURE 22.13 Competing processes that can take place in the excited photosensitizer.

photophysical processes of the excited photosensitizer and the generation of singlet oxygen.

22.6.5.1.2 *Photochemical Reactions and Formation of Reactive Oxygen Species*

The photochemical reactions in PDT are responsible for the formation of singlet oxygen as well as the highly reactive oxygen species (ROS) that can cause cell damage and death. When the photosensitizer is in T_1, two photochemical reactions could occur. In type I reactions, the photosensitizer interacts with cellular substrates. Type I reactions are divided into two subtypes. In type I(i) reactions, an electron is transferred from the substrate to the

---

BOX 22.6   PHOTODYNAMIC THERAPY

PDT is a technique used in the treatment of different types of cancer. It utilizes a photosensitizing drug, light of a specific wavelength, and molecular oxygen. The photosensitizing drug is injected intravenously into the patient's bloodstream where it enters all the cells but is retained only by the tumor cells (due to leaky vasculature and a lack of lymphatics). Light of a specific wavelength is then focused on the tumor area. The photosensitizer uses this light to excite molecular oxygen ($^3O_2$) into its singlet state ($^1O_2$), which is the main cytotoxic agent in PDT. The advantages of PDT are that it is noninvasive, localized, inexpensive, and efficient. Some of the conventional photosensitizers, such as Photofrin®, have been approved by the U.S. Food and Drug Administration for clinical application (Chen, 2008; Josefsen and Boyle, 2008).

Photosensitizer + Light + $^3O_2 \rightarrow {}^1O_2$ (cytotoxic agent that leads to tumor destruction)

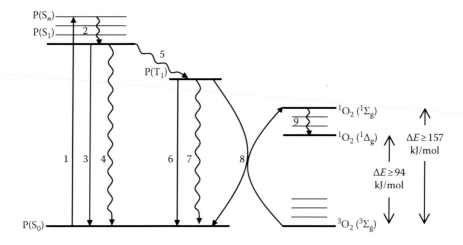

FIGURE 22.14 Simplified Jablonski energy diagram of photosensitizer and singlet oxygen generation in photodynamic therapy.

1 = Absorption   6 = Phosphorescence
2 = Internal conversion   7 = Internal conversion
3 = Fluorescence   8 = Energy transfer through collision with
4 = Internal conversion $^3O_2$ (Type II reaction)
5 = Intersystem crossing   9 = Internal conversion

$^3O_2$ ($^3\Sigma_g$) = molecular oxygen, spin aligned.
$^1O_2$ ($^1\Sigma_g$) = singlet oxygen, spin paired in two different antibonding orbitals.
$^1O_2$ ($^1\Delta_g$) = singlet oxygen, spin paired in the same antibonding orbital.

photosensitizer, which can then interact with oxygen forming a superoxide anion radical. This radical is very reactive and can form other ROS through a series of reactions that can cause cell death (Josefsen and Boyle, 2008). Type I(ii) reactions occur when the excited photosensitizer gains a hydrogen atom from substrates forming free radicals, which can then form ROS including peroxides, which can also cause cell death (Josefsen and Boyle, 2008).

In type II reactions that occur due to collision with oxygen, the photosensitizer goes back to S_0 losing energy, which is used to convert oxygen from the ground triplet state ($^3O_2$) to either of the two excited singlet states ($^1O_2$). The first form of singlet oxygen ($^1\Sigma_g$) has higher energy since it is formed by inverting the spin of one of the outermost electrons, violating Hund's rule, which states that in half-filled orbitals of identical energies, each electron occupies an empty orbital and they all must have

the same spin. This form rapidly loses energy and converts to the more stable singlet form ($^1\Delta_g$) in which both electrons are paired in one of the antibonding orbitals. Figure 22.15 shows the molecular orbital diagrams of the three forms of oxygen. The resulting singlet oxygen is still very reactive and can also cause cell death. Reactions type I and II (Box 22.7) can occur simultaneously, however, type II reactions are the predominant ones in PDT (Josefsen and Boyle, 2008).

22.6.5.1.3 QDs in PDT

Most of the conventional photosensitizers such as organic dyes (e.g., rose bengal, methylene blue, and eosin) and tetrapyrroles (e.g., porphyrins, chlorins, bacteriochlorins, and phthalocyanines [Figure 22.16]) have several disadvantages including a low extinction coefficient that requires the administration of large

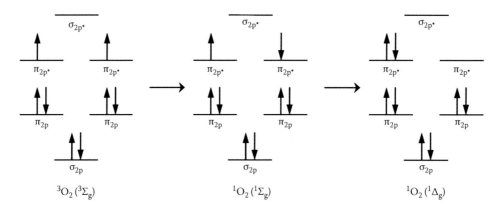

FIGURE 22.15 Molecular orbital diagrams of molecular and singlet oxygen. $^3O_2$ ($^3\Sigma_g$) = molecular oxygen, spin aligned. $^1O_2$ ($^1\Sigma_g$) = singlet oxygen, spin paired in two different antibonding orbitals. $^1O_2$ ($^1\Delta_g$) = singlet oxygen, spin paired in the same antibonding orbital.

BOX 22.7   PHOTOCHEMICAL REACTIONS AND FORMATION OF ROS

Type I(i) reactions

$$P^* + S \rightarrow {}^\bullet P^- + {}^\bullet S^+$$

$${}^\bullet P^- + O_2 \rightarrow P + {}^\bullet O_2^-$$

$$2{}^\bullet O_2^- + 2H^+ \rightarrow O_2 + H_2O_2$$

$${}^\bullet O_2^- + Fe^{3+} \rightarrow O_2 + Fe^{2+}$$

$$Fe^{2+} + H_2O_2 \rightarrow Fe^{3+} + {}^\bullet OH + OH^-$$

$${}^\bullet O_2^- + {}^\bullet OH \rightarrow {}^1O_2 + OH^-$$

Type I(ii) reactions

$$P^* + RH \rightarrow PH^\bullet + R^\bullet$$

$$R^\bullet + O_2 \rightarrow RO_2{}^\bullet$$

$$RO_2{}^\bullet + RH \rightarrow RO_2H + R^\bullet$$

Type II reactions

$$P^* + {}^3O_2 \rightarrow P + {}^1O_2$$

P: photosensitizer
S: substrate
P^*: excited photosensitizer
R: alkyl group

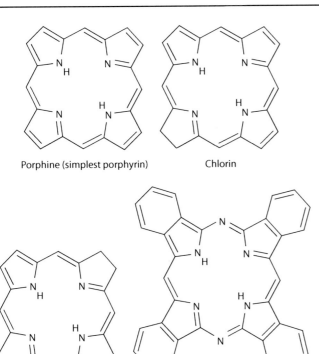

Porphine (simplest porphyrin)   Chlorin

Bacteriochlorin   Phthalocyanine

FIGURE 22.16 The conventional photosensitizers.

doses for PDT to be effective. They also absorb wavelengths shorter than 700 nm and suffer from photobleaching as well as poor target selectivity. Additionally, some of the conventional photosensitizers accumulate in the skin causing patients to become photosensitive and prone to sunburn for a period lasting

6–8 weeks after the termination of treatment (Chen, 2008). The properties of an efficient photosensitizer are listed in Box 22.8 and QDs possess most of these properties. In addition to the triplet state (see Section 22.6.5.1.1), QDs also have a dark exciton state and several surface states with long fluorescence lifetimes. These long-lived states provide sufficient time for the interaction of the excited QD with molecular oxygen or cellular substrates and thus increase the efficiency of PDT (Chen, 2008). Furthermore, QDs are nontoxic in the absence of light yet are cytotoxic upon exposure to UV radiation, and appropriate molecules such as antibodies can be attached to their surfaces to ensure specific tumor targeting (Bakalova et al., 2004).

There are several concerns, however, over the use of QDs in PDT. The thick coating layer of QDs could decrease the energy transfer between the QD and oxygen and thus decrease its photosensitizing ability and in turn decrease the singlet oxygen quantum yield. Their overall photostability in biological environments, toxicity, as well as clearance from the body are all issues that have to be critically addressed before their full use in PDT (Bakalova et al., 2004; Samia et al., 2006). The use of QDs alone as photosensitizers could be inefficient since the singlet oxygen quantum yield is very low as was reported by Samia et al. (2003) who used CdSe QDs as photosensitizers to generate singlet oxygen in toluene. The singlet oxygen quantum yield was around 5%, which is very low compared to the 40%–60% yield of conventional photosensitizers. This problem, however, could be solved by the continuous exposure to UV/visible light irradiation, which may increase the singlet oxygen quantum yield. On the other hand, the use of QD-photosensitizer conjugates has been proposed as a better system, as detailed in the following section.

Box 22.8   Properties of an Efficient Photosensitizer

- Constant composition (Bakalova et al., 2004; Josefsen and Boyle, 2008)
- Water solubility to avoid aggregation (Bakalova et al., 2004)
- High quantum yield of singlet oxygen ($\Phi_\Delta \geq 0.5$), which was shown to be higher at shorter wavelengths (Chen, 2008; Josefsen and Boyle, 2008)
- High quantum yield of the excited triplet state ($\Phi_T \geq 0.5$) (Bakalova et al., 2004; Josefsen and Boyle, 2008)
- High efficiency of energy transfer (φ_{en}) (Samia et al., 2006)

  The following equation shows the relationship between the previous three parameters:

  $$\Phi_\Delta = \Phi_T \varphi_{en}$$

  $$= \Phi_T \left(\frac{k_{en} \left[{}^3O_2 \right]}{k_r + k_{nr} + k_q \left[{}^3O_2 \right]} \right) \quad \text{DeRosa et al. (2002)}$$

  where
  - k_{en} is the rate constant of energy transfer
  - k_r is the rate constant of the radiative relaxation of the T_1 state
  - k_{nr} is the rate constant of the nonradiative relaxation of the T_1 state
  - k_q is the sum of rate constants of the quenching of the T_1 state by triplet oxygen through deactivation, electron transfer, and energy transfer

- Triplet-state energy, which is the difference between the ground singlet state (S_0) and first excited triplet state (T_1), should be ≥ 94 kJ/mol, the energy needed to convert molecular oxygen into singlet oxygen ($1 \Delta g$) (Bakalova et al., 2004)
- Long triplet state lifetime ($>1 \mu s$) to allow for the interaction of the excited photosensitizer with molecular oxygen or cellular substrates (Chen, 2008; Josefsen and Boyle, 2008)
- High photostability and resistance to photobleaching (Bakalova et al., 2004)
- High extinction coefficient (Chen, 2008)
- Maximum absorption in the NIR region to prevent tissue absorbance and to allow deep tissue penetration (Chen, 2008)
- Selective targeting to tumor cells only and not healthy cells (Chen, 2008)
- Lack of toxicity in the absence of light (Bakalova et al., 2004)
- Rapid body clearance to avoid side effects (Bakalova et al., 2004)
- Simple synthesis strategy and available and inexpensive starting materials (Bakalova et al., 2004; Josefsen and Boyle, 2008)

PROPERTIES OF AN EFFICIENT ENHANCER

- Water solubility
- Lack of toxicity
- Matching emission spectrum of enhancer to the absorption spectrum of photosensitizer
- High extinction coefficient
- High emission quantum yield
- Easy conjugation to photosensitizer

22.6.5.1.4 QD-Photosensitizer Conjugates

QDs conjugated to a conventional photosensitizer are used to enhance its excitation through FRET (Bakalova et al., 2004; Chen, 2008). This method seems to be more efficient in generating singlet oxygen than the use of QDs alone as photosensitizers. The properties of a good enhancer are listed in Box 22.8. For QDs to be used as enhancers, the emission spectrum of QD must match the absorption spectrum of the photosensitizer and this is easily achieved since QDs have size-tunable emission spectra (Chen, 2008). Samia et al. used QDs conjugated to a conventional photosensitizer (the silicon phthalocyanine-based compound, Pc4), which showed an impressive 77% FRET efficiency. However, the wavelength used for excitation of the QDs (488 nm) cannot penetrate the tissue (Samia et al., 2003; Chen, 2008).

More recently, two photosensitizers, rose bengal and chlorin e6, were covalently attached to peptide coated-QDs to obtain high singlet oxygen quantum yields (Tsay et al., 2007). PDT is achieved by the excitation of the QDs leading to the activation of the photosensitizers through FRET. There are two methods that enable the simultaneous use of QDs for imaging as well as PDT. The first method is by limiting the FRET efficiency between the photosensitizer and the QD, and only using one excitation wavelength for the QD. The high absorption cross sections of QDs enable the transfer of sufficient energy needed to activate the photosensitizers for PDT with minimal loss of the fluorescence needed for imaging. The second method is by using two excitation wavelengths for directly exciting the photosensitizer and the QD. In this case, the QD and the

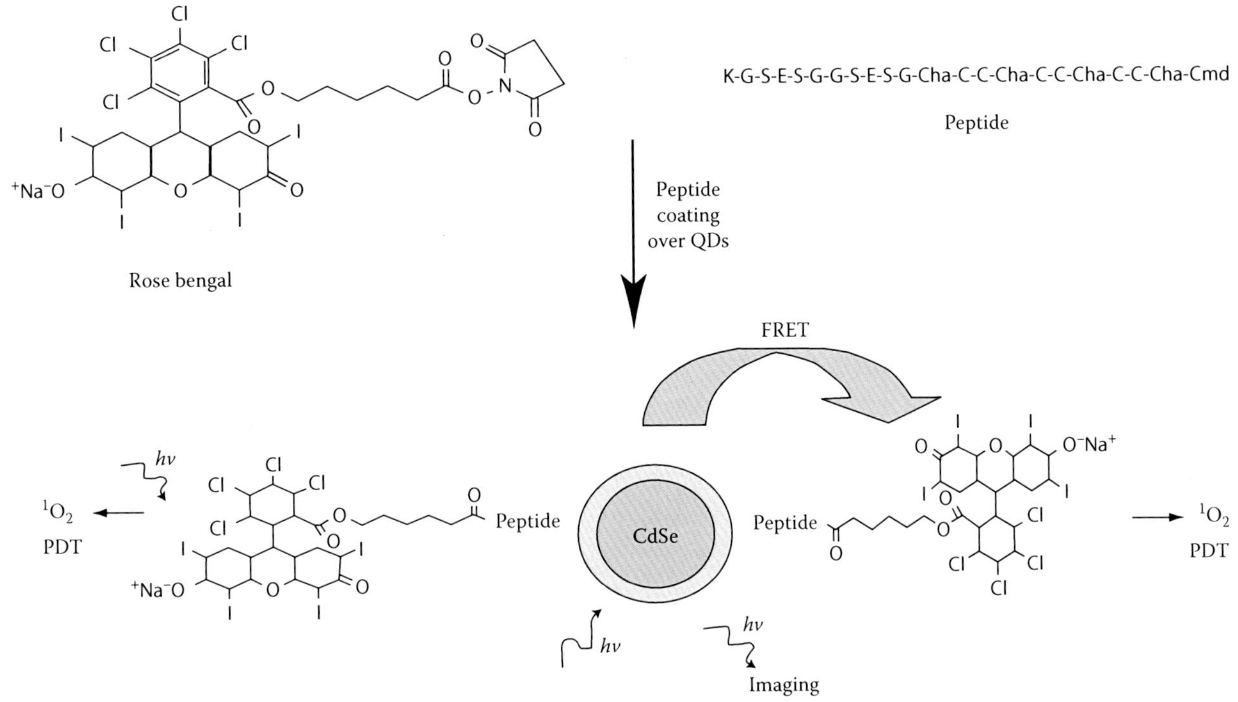

FIGURE 22.17 The simultaneous use of QD-photosensitizer conjugates in photodynamic therapy as well as imaging. (Reprinted with Tsay, J.M. et al., *J. Am. Chem. Soc.*, 129, 6865, 2007. With permission.)

photosensitizer are spectrally unmatched, i.e., the emission spectrum of the QD does not match the absorption spectrum of the photosensitizer and therefore FRET will not occur. The two methods are shown in Figure 22.17 in the case of rose bengal.

22.6.5.1.5 QDs in Combined Radiation and Photodynamic Therapies

As mentioned in Box 22.8, the singlet oxygen yield increases when shorter wavelengths are used to excite the photosensitizers. However, at short wavelengths, deep tissue penetration is not possible and this is why PDT is used only to treat tumors in accessible locations such as skin, head, and neck cancers and is ineffective in the treatment of deep-lying tumors. To solve this problem, x-ray luminescent nanoparticles such as QDs in conjugation with conventional photosensitizers can be used in which the QDs are irradiated with x-rays causing them to emit light, which activates the photosensitizer. In this case, 100% of the produced light is delivered to the photosensitizer, whereas, in traditional PDT, only a small percent of the light can penetrate into the tissue. X-rays are known for their ability to penetrate deep into tissue, together with the fact that they have short wavelengths, allows not only deep tissue penetration but the generation of a high quantum yield of singlet oxygen as well with no need for external light. This combined therapy is effective for the treatment of deep tumors, is inexpensive, and compared to conventional radiation therapy, lower doses of localized radiation can be used,

which reduces damage to the healthy cells (Chen and Zhang, 2006; Chen, 2008).

22.6.5.1.6 The Intracellular Delivery of QDs

For intracellular delivery, QDs or QD-conjugates could be encapsulated in delivery vehicles such as liposomes. However, this delivery vehicle could decrease the quantum yield of singlet oxygen (the most significant factor in PDT). This is because the singlet oxygen produced would first have to diffuse out of the carrier system and it may not reach the target because of its short range of action. Another problem encountered with the use of delivery vehicles is their size. If the size is not small enough, this could decrease the absorption of the excitation light, which in turn could decrease the singlet oxygen quantum yield (Josefsen and Boyle, 2008).

Another approach for the delivery of QDs that eliminates the need for carrier systems is the use of targeting moieties such as antibodies, yet as mentioned above, the surface coatings should not be too thick or else this will also decrease singlet oxygen quantum yield (Bakalova et al., 2004). Folic acid was used for delivering a nanoparticle ($LaF_3:Tb^{3+}$)-porphyrin conjugate to target tumor cells expressing folate receptors. Folic acid was shown to have no effect on the singlet oxygen quantum yield making it a good targeting moiety in PDT (Liu et al., 2008).

22.6.5.2 Tumor Targeting and Drug Delivery

Active targeting is carried out by conjugating the QDs to antibodies or cell surface receptor ligands that target tumor cells.

After they reach their target cell, QDs are then internalized by endocytosis (McNeil, 2005). Active targeting of tumor cells expressing folate receptors or EGF receptors has been achieved by conjugating QDs to folic acid or EGF protein, respectively. Folic acid is a better targeting moiety than antibodies due to its small size, lack of immunogenicity, and simple conjugation (Chen and Zhang, 2006).

Passive targeting utilizes the fact that tumor tissues are surrounded by a leaky vasculature and a diminished number of lymphatics and this allows the unlabeled QDs to escape the leaky vasculature and accumulate inside the tumor site surrounding the cancer cells rather than targeting them directly. This phenomenon is referred to as the enhanced permeation and retention effect. As mentioned previously, uncoated QDs are taken up by the RES, which can prevent them from reaching their intended target. To avoid this from happening, the QDs are coated with hydrophilic molecules such as PEG, which increases their circulating time and allows them to escape the RES (McNeil, 2005).

A tumor targeting QD could also act as a drug delivery vehicle. QDs could be conjugated to therapeutic drugs (Delehanty et al., in press). For example, in PDT, the photosensitizing drug could be conjugated to QDs for delivery into cancer cells. In this case, the tumor-targeting QDs act as a delivery vehicle as well as FRET donors (Bakalova et al., 2004). QDs could also aid other drug delivery systems in controlling the release of drugs. For example, thio-modified CdS QDs were used as chemically removable caps to keep neurotransmitters and drug molecules inside a mesoporous silica nanosphere-based system. Disulfide bond-reducing molecules, such as DTT and β-mercaptoethanol, are then used to trigger the release of the drugs (Lai et al., 2003).

22.7 Toxicity

The possible routes of exposure to QDs are inhalation, ingestion, or absorption through the skin. Inhalation of QDs could be problematic since QDs are easily endocytosed by cells and can reside in cells for long periods of time ranging from weeks to months. The risks of exposure through the skin and ingestion are unknown. In addition, little is known about their metabolism or excretion (Hardman, 2006). Assessing the true toxicity of QDs is a difficult matter since not all QDs are alike, and variable factors such as their size, charge, concentration, choice of core and shell material, biocompatible coating, as well as dosage and exposure time can affect their toxicity. However, generally speaking, the toxicity of QDs arises from two main factors: the toxic heavy metals from which the core is made that can leak upon oxidation by air or exposure to UV radiation, and the ROS formed upon the excitation of QDs (Medintz et al., 2008). The addition of the shell can protect the core from oxidation by air but cannot protect it from UV radiation, and further coating with larger molecules such as BSA can only partially protect the QD core from photo-oxidation and hence decrease the leakage but cannot stop it completely (Alivisatos et al., 2005; Hardman, 2006). This puts limits to their use *in vivo* due to the

possibility of intracellular instability and degradation. Heavy metals such as Cd can cross the blood–brain barrier, build up in adipose tissue with a half-life of more than 10 years, and distribute in all tissues with particular toxicity to the liver and kidneys. Cd is also a probable carcinogen (Hardman, 2006). The size of functionalized QDs also plays a role in their toxicity. Functionalized QDs larger than 10 nm are too big for renal filtration (Frangioni, 2003; Medintz et al., 2008). QDs are also cytotoxic and at high concentrations affect cell growth and function, as was seen by their effect on *Xenopus* embryo development (Dubertret et al., 2002). QD hydrophilic coatings used for solubilization such as MPA and MAA are also mildly cytotoxic (Jamieson et al., 2007).

22.8 Conclusion

QDs have substantial utility for numerous biomedical applications. These semiconductor nanocrystals are characterized by high quantum yields, large extinction coefficients, remarkable photostability, and broad absorption spectra along with narrow size-tunable photoluminescent emission spectra. Unlike conventional organic fluorophores, QDs of different sizes can be simultaneously excited by a single wavelength far removed from their respective emissions and emit intense and stable light in all colors of the rainbow. This makes them suitable for applications that require the simultaneous detection of multiple biomarkers. Their long fluorescence lifetime and high photostability make them useful in applications such as time-gated imaging, long-term monitoring of labeled substances, as well as SMD. QDs can be made biocompatible by several strategies including surface silanization and coating with a polymer shell, thus enabling their utilization in biological systems. Conjugation of QDs to a variety of biomolecules, such as oligonucleotides and proteins including antibodies adapts them for specific target recognition. Their nano-size allows them to penetrate cell membranes and interact with cellular molecules. Although further work is necessary to thoroughly assess the safety aspects of QDs and fully optimize their use for biomedical applications, QDs hold the potential to propel medicine toward new frontiers.

Abbreviations

BRET	Bioluminescence resonance energy transfer
BSA	Bovine serum albumin
CCD	Charge-coupled device
CHP	Cholesterol-bearing pullulan
$CHPNH_2$	Cholesterol-bearing pullulan modified with amino groups
DHLA	Dihydrolipoic acid
DTT	Dithiothreitol
EGF	Epidermal growth factor
ELISA	Enzyme-linked immunosorbent assay
FISH	Fluorescence *in situ* hybridization
FLP	Fluorescent ligation product
FRET	Förster (Fluorescence) resonance energy transfer

IGF1R	Type I insulin-like growth factor receptor
IR	Infrared
MAA	Mercaptoacetic acid
MBP	Maltose-binding protein
MPA	Mercaptopropionic acid
NIR	Near-infrared
PCR	Polymerase chain reaction
PDT	Photodynamic therapy
PEG	Polyethylene glycol
PEI	Polyethyleneimine
PTD	Protein transduction domains
QD	Quantum dot
RES	Reticuloendothelial system
ROS	Reactive oxygen species
SBR	Signal-to-background ratio
siRNA	Small interference RNA
SMD	Single molecule detection
SNP	Single nucleotide polymorphism
TOPO	Trioctylphosphine oxide
UV	Ultraviolet

Acknowledgments

We would like to thank Sandy Barsoum from the Center for Learning and Teaching for professional assistance with Figures 22.1 through 22.4 and Bishoy Hanna from the Biology Department for his assistance with Figures 22.5 through 22.7.

References

Agrawal, A., C. Zhang, T. Byassee, R. A. Tripp, and S. Nie. 2006. Counting single native biomolecules and intact viruses with color-coded nanoparticles. *Analytical Chemistry* 78:1061–1070.

Alivisatos, A. P., W. Gu, and C. Larabell. 2005. Quantum dots as cellular probes. *Annual Review of Biomedical Engineering* 7:55–76.

Azzazy, H. M., M. M. Mansour, and S. C. Kazmierczak. 2007. From diagnostics to therapy: Prospects of quantum dots. *Clinical Biochemistry* 40:917–927.

Bakalova, R., H. Ohba, Z. Zhelev, M. Ishikawa, and Y. Baba. 2004. Quantum dots as photosensitizers? *Nature Biotechnology* 22:1360–1361.

Bakalova, R., Z. Zhelev, H. Ohba, and Y. Baba. 2005. Quantum dot-based western blot technology for ultrasensitive detection of tracer proteins. *Journal of the American Chemical Society* 127:9328–9329.

Ballou, B., B. C. Lagerholm, L. A. Ernst, M. P. Bruchez, and A. S. Waggoner. 2004. Noninvasive imaging of quantum dots in mice. *Bioconjugate Chemistry* 15:79–86.

Bentolila, L. A. and S. Weiss. 2006. Single-step multicolor fluorescence in situ hybridization using semiconductor quantum dot-DNA conjugates. *Cell Biochemistry and Biophysics* 45:59–70.

Bruchez, M. Jr., M. Moronne, P. Gin, S. Weiss, and A. P. Alivisatos. 1998. Semiconductor nanocrystals as fluorescent biological labels. *Science* 281:2013–2016.

Chaudhuri, J. D. 2005. Genes arrayed out for you: The amazing world of microarrays. *Medical Science Monitor: International Medical Journal of Experimental and Clinical Research* 11:RA52–RA62.

Chan, P., T. Yuen, F. Ruf, J. Gonzalez-Maeso, and S. C. Sealfon. 2005. Method for multiplex cellular detection of mRNAs using quantum dot fluorescent in situ hybridization. *Nucleic Acids Research* 33:e161.

Chen, W. 2008. Nanoparticle fluorescence based technology for biological applications. *Journal of Nanoscience and Nanotechnology* 8:1019–1051.

Chen, W. and J. Zhang. 2006. Using nanoparticles to enable simultaneous radiation and photodynamic therapies for cancer treatment. *Journal of Nanoscience and Nanotechnology* 6:1159–1166.

Chen, A. A., A. M. Derfus, S. R. Khetani, and S. N. Bhatia. 2005. Quantum dots to monitor RNAi delivery and improve gene silencing. *Nucleic Acids Research* 33:e190.

Clapp, A. R., I. L. Medintz, B. R. Fisher, G. P. Anderson, and H. Mattoussi. 2005. Can luminescent quantum dots be efficient energy acceptors with organic dye donors? *Journal of the American Chemical Society* 127:1242–1250.

Dahan, M., T. Laurence, F. Pinaud et al. 2001. Time-gated biological imaging by use of colloidal quantum dots. *Optics Letters* 26:825–827.

De Rosa M. C. and R. J. Crutchley. 2002. Photosensitized singlet oxygen and its applications. *Coordination Chemistry Reviews* 233–234:351–371.

Delehanty, J. B., H. Mattoussi, and I. L. Medintz. 2009. Delivering quantum dots into cells: Strategies, progress and remaining issues. *Analytical and Bioanalytical Chemistry* 393:1091–1105.

Denissenko, M. F., J. X. Chen, M. S. Tang, and G. P. Pfeifer. 1997. Cytosine methylation determines hot spots of DNA damage in the human P53 gene. *Proceedings of the National Academy of Sciences of the United States of America* 94:3893–3898.

Derfus, A. M., W. C. W. Chan, and S. N. Bhatia. 2004. Intracellular delivery of quantum dots for live cell labeling and organelle tracking. *Advanced Materials* 16:961–966.

Derfus, A. M., A. A. Chen, D. H. Min, E. Ruoslahti, and S. N. Bhatia. 2007. Targeted quantum dot conjugates for siRNA delivery. *Bioconjugate Chemistry* 18:1391–1396.

Diamandis, E. P. and T. K. Christopoulos. 1991. The biotin-(strept) avidin system: Principles and applications in biotechnology. *Clinical Chemistry* 37:625–636.

Didenko, V. V. 2001. DNA probes using fluorescence resonance energy transfer (FRET): Designs and applications. *BioTechniques* 31:1106–1121.

Duan, H. and S. Nie. 2007. Cell-penetrating quantum dots based on multivalent and endosome-disrupting surface coatings. *Journal of the American Chemical Society* 129:3333–3338.

Dubertret, B., P. Skourides, D. J. Norris, V. Noireaux, A. H. Brivanlou, and A. Libchaber. 2002. In vivo imaging of quantum dots encapsulated in phospholipid micelles. *Science (New York, N.Y.)* 298:1759–1762.

Frangioni, J. V. 2003. In vivo near-infrared fluorescence imaging. *Current Opinion in Chemical Biology* 7:626–634.

Frangioni, J. V. 2006. Self-illuminating quantum dots light the way. *Nature Biotechnology* 24:326–328.

Gerion, D., F. Chen, B. Kannan et al. 2003. Room-temperature single-nucleotide polymorphism and multiallele DNA detection using fluorescent nanocrystals and microarrays. *Analytical Chemistry* 75:4766–4772.

Goldman, E. R., A. R. Clapp, G. P. Anderson et al. 2004. Multiplexed toxin analysis using four colors of quantum dot fluororeagents. *Analytical Chemistry* 76:684–688.

Grecco, H. E., K. A. Lidke, R. Heintzmann et al. 2004. Ensemble and single particle photophysical properties (two-photon excitation, anisotropy, FRET, lifetime, spectral conversion) of commercial quantum dots in solution and in live cells. *Microscopy Research and Technique* 65:169–179.

Han, M., X. Gao, J. Z. Su, and S. Nie. 2001. Quantum-dot-tagged microbeads for multiplexed optical coding of biomolecules. *Nature Biotechnology* 19:631–635.

Hardman, R. 2006. A toxicologic review of quantum dots: Toxicity depends on physicochemical and environmental factors. *Environmental Health Perspectives* 114:165–172.

Hasegawa, U., S. M. Nomura, S. C. Kaul, T. Hirano, and K. Akiyoshi. 2005. Nanogel-quantum dot hybrid nanoparticles for live cell imaging. *Biochemical and Biophysical Research Communications* 331:917–921.

Hirschhorn, J. N. and M. J. Daly. 2005. Genome-wide association studies for common diseases and complex traits. *Nature Reviews Genetics* 6:95–108.

Jaiswal, J. K., H. Mattoussi, J. M. Mauro, and S. M. Simon. 2003. Long-term multiple color imaging of live cells using quantum dot bioconjugates. *Nature Biotechnology* 21:47–51.

Jamieson, T., R. Bakhshi, D. Petrova, R. Pocock, M. Imani, and A. M. Seifalian. 2007. Biological applications of quantum dots. *Biomaterials* 28:4717–4732.

Joos, L., E. Eryuksel, and M. H. Brutsche. 2003. Functional genomics and gene microarrays—The use in research and clinical medicine. *Swiss Medical Weekly: Official Journal of the Swiss Society of Infectious Diseases, the Swiss Society of Internal Medicine, the Swiss Society of Pneumology* 133:31–38.

Josefsen, L. B. and R. W. Boyle. 2008. Photodynamic therapy and the development of metal-based photosensitisers. *Metal-Based Drugs* 2008:276109.

Kerman, K., T. Endob, M. Tsukamotoa, M. Chikaea, Y. Takamuraa, and E. Tamiyaa. 2007. Quantum dot-based immunosensor for the detection of prostate-specific antigen using fluorescence microscopy. *Talanta* 71:1494–1499.

Kim, J. H., D. Morikis, and M. Ozkan. 2004a. Adaptation of inorganic quantum dots for stable molecular beacons. *Sensors and Actuators B* 102:315–319.

Kim, S., Y. T. Lim, E. G. Soltesz et al. 2004b. Near-infrared fluorescent type II quantum dots for sentinel lymph node mapping. *Nature Biotechnology* 22:93–97.

Kovar, J. L., M. A. Simpson, A. Schutz-Geschwender, and D. M. Olive. 2007. A systematic approach to the development of fluorescent contrast agents for optical imaging of mouse cancer models. *Analytical Biochemistry* 367:1–12.

Lai, C. Y., B. G. Trewyn, D. M. Jeftinija et al. 2003. A mesoporous silica nanosphere-based carrier system with chemically removable CdS nanoparticle caps for stimuli-responsive controlled release of neurotransmitters and drug molecules. *Journal of the American Chemical Society* 125:4451–4459.

Larson, D. R., W. R. Zipfel, R. M. Williams et al. 2003. Water-soluble quantum dots for multiphoton fluorescence imaging in vivo. *Science (New York, N.Y.)* 300:1434–1436.

Lee, L. Y., S. L. Ong, J. Y. Hu et al. 2004. Use of semiconductor quantum dots for photostable immunofluorescence labeling of *Cryptosporidium parvum*. *Applied and Environmental Microbiology* 70:5732–5736.

Liu, G., T. M. Lee, and J. Wang. 2005. Nanocrystal-based bioelectronic coding of single nucleotide polymorphisms. *Journal of the American Chemical Society* 127:38–39.

Liu, W. T. 2006. Nanoparticles and their biological and environmental applications. *Journal of Bioscience and Bioengineering* 102:1–7.

Liu, Y., W. Chen, S. P. Wang, and A. G. Joly. 2008. Investigation of water-soluble x-ray luminescence nanoparticles for photodynamic activation. *Applied Physics Letters* 92: 0439011–0439013.

Matsuno, A., J. Itoh, S. Takekoshi, T. Nagashima, and R. Y. Osamura. 2005. Three-dimensional imaging of the intracellular localization of growth hormone and prolactin and their mRNA using nanocrystal (Quantum dot) and confocal laser scanning microscopy techniques. *The Journal of Histochemistry and Cytochemistry: Official Journal of the Histochemistry Society* 53:833–838.

Mattheakis, L. C., J. M. Dias, Y. J. Choi et al. 2004. Optical coding of mammalian cells using semiconductor quantum dots. *Analytical Biochemistry* 327:200–208.

McNeil, S. E. 2005. Nanotechnology for the biologist. *Journal of Leukocyte Biology* 78:585–594.

Medintz, I. L., A. R. Clapp, H. Mattoussi, E. R. Goldman, B. Fisher, and J. M. Mauro. 2003. Self-assembled nanoscale biosensors based on quantum dot FRET donors. *Nature Materials* 2:630–638.

Medintz, I. L., A. R. Clapp, F. M. Brunel et al. 2006. Proteolytic activity monitored by fluorescence resonance energy transfer through quantum-dot-peptide conjugates. *Nature Materials* 5:581–589.

Medintz, I. L., H. Mattoussi, and A. R. Clapp. 2008. Potential clinical applications of quantum dots. *International Journal of Nanomedicine* 3:151–167.

Michalet, X., F. F. Pinaud, L. A. Bentolila et al. 2005. Quantum dots for live cells, in vivo imaging, and diagnostics. *Science (New York, N.Y.)* 307:538–544.

Misra, R. D. 2008. Quantum dots for tumor-targeted drug delivery and cell imaging. *Nanomedicine (London, England)* 3:271–274.

Ness, J. M., R. S. Akhtar, C. B. Latham, and K. A. Roth. 2003. Combined tyramide signal amplification and quantum dots for sensitive and photostable immunofluorescence detection. *The Journal of Histochemistry and Cytochemistry: Official Journal of the Histochemistry Society* 51:981–987.

Ornberg, R. L., T. F. Harper, and H. Liu. 2005. Western blot analysis with quantum dot fluorescence technology: A sensitive and quantitative method for multiplexed proteomics. *Nature Methods* 2:79–81.

Partridge, M., A. Vincent, P. Matthews, J. Puma, D. Stein, and J. Summerton. 1996. A simple method for delivering morpholino antisense oligos into the cytoplasm of cells. *Antisense & Nucleic Acid Drug Development* 6:169–175.

Qian, J., K. T. Yong, I. Roy et al. 2007. Imaging pancreatic cancer using surface-functionalized quantum dots. *The Journal of Physical Chemistry B* 111:6969–6972.

Qu, L. and X. Peng. 2002. Control of photoluminescence properties of CdSe nanocrystals in growth. *Journal of the American Chemical Society* 124:2049–2055.

Samia, A. C., X. Chen, and C. Burda. 2003. Semiconductor quantum dots for photodynamic therapy. *Journal of the American Chemical Society* 125:15736–15737.

Samia, A. C., S. Dayal, and C. Burda. 2006. Quantum dot-based energy transfer: Perspectives and potential for applications in photodynamic therapy. *Photochemistry and Photobiology* 82:617–625.

Shi, L., V. De Paoli, N. Rosenzweig, and Z. Rosenzweig. 2006. Synthesis and application of quantum dots FRET-based protease sensors. *Journal of the American Chemical Society* 128:10378–10379.

Shi, L., N. Rosenzweig, and Z. Rosenzweig. 2007. Luminescent quantum dots fluorescence resonance energy transfer-based probes for enzymatic activity and enzyme inhibitors. *Analytical Chemistry* 79:208–214.

So, M. K., C. Xu, A. M. Loening, S. S. Gambhir, and J. Rao. 2006. Self-illuminating quantum dot conjugates for in vivo imaging. *Nature Biotechnology* 24:339–343.

Sweeney, E., T. H. Ward, N. Gray et al. 2008. Quantitative multiplexed quantum dot immunohistochemistry. *Biochemical and Biophysical Research Communications* 374:181–186.

Sydor, J. R. and S. Nock. 2003. Protein expression profiling arrays: Tools for the multiplexed high-throughput analysis of proteins. *Proteome Science* 1:3.

Thurer, R., T. Vigassy, M. Hirayama, J. Wang, E. Bakker, and E. Pretsch. 2007. Potentiometric immunoassay with quantum dot labels. *Analytical Chemistry* 79:5107–5110.

Tsay, J. M., M. Trzoss, L. Shi et al. 2007. Singlet oxygen production by peptide-coated quantum dot-photosensitizer conjugates. *Journal of the American Chemical Society* 129:6865–6871.

Tully, E., S. Hearty, P. Leonard, and R. O'Kennedy. 2006. The development of rapid fluorescence-based immunoassays, using quantum dot-labelled antibodies for the detection of Listeria monocytogenes cell surface proteins. *International Journal of Biological Macromolecules* 39:127–134.

Wang, S., N. Mamedova, N. A. Kotov, W. Chen, and J. Studer. 2002. Antigen/antibody immunocomplex from CdTe nanoparticle bioconjugates. *Nano Letters* 2:817–822.

Xiao, Y. and P. E. Barker. 2004. Semiconductor nanocrystal probes for human metaphase chromosomes. *Nucleic Acids Research* 32:e28.

Xu, H., M. Y. Sha, E. Y. Wong et al. 2003. Multiplexed SNP genotyping using the Qbead system: A quantum dot-encoded microsphere-based assay. *Nucleic Acids Research* 31:e43.

Yang, L. and Y. Li. 2006. Simultaneous detection of *Escherichia coli* O157:H7 and *Salmonella typhimurium* using quantum dots as fluorescence labels. *The Analyst* 131:394–401.

Yang, N. S., J. Burkholder, B. Roberts, B. Martinell, and D. McCabe. 1990. In vivo and in vitro gene transfer to mammalian somatic cells by particle bombardment. *Proceedings of the National Academy of Sciences of the United States of America* 87:9568–9572.

Yeh, H. C., S. Y. Chao, Y. P. Ho, and T. H. Wang. 2005. Single-molecule detection and probe strategies for rapid and ultrasensitive genomic detection. *Current Pharmaceutical Biotechnology* 6:453–461.

Yeh, H. C., Y. P. Ho, Ie. M. Shih, and T. H. Wang. 2006. Homogeneous point mutation detection by quantum dot-mediated two-color fluorescence coincidence analysis. *Nucleic Acids Research* 34:e35.

Zajac, A., D. Song, W. Qian, and T. Zhukov. 2007. Protein microarrays and quantum dot probes for early cancer detection. *Colloids and Surfaces B: Biointerfaces* 58:309–314.

Zhang, C. Y., H. C. Yeh, M. T. Kuroki, and T. H. Wang. 2005. Single-quantum-dot-based DNA nanosensor. *Nature Materials* 4:826–831.

Zhang, H., D. Sachdev, C. Wang, A. Hubel, M. Gaillard-Kelly, and D. Yee. 2009. Detection and downregulation of type I IGF receptor expression by antibody-conjugated quantum dots in breast cancer cells. *Breast Cancer Research and Treatment* 114:277–285.

Zhu, L., S. Ang, and W. T. Liu. 2004. Quantum dots as a novel immunofluorescent detection system for *Cryptosporidium parvum* and *Giardia lamblia*. *Applied and Environmental Microbiology* 70:597–598.

Relaxivity of Nanoparticles for Magnetic Resonance Imaging

Gustav J. Strijkers
*Eindhoven University
of Technology*

Klaas Nicolay
*Eindhoven University
of Technology*

23.1 Introduction

23.1.1 General

Magnetic resonance imaging (MRI) is one of the leading non-invasive imaging techniques in clinical diagnostics as well as in research settings. The success of MRI partly results from its ability to produce detailed anatomical images of living subjects with resolutions down to typically 1 mm for clinical scanners and even lower than 50 μm with high-field MRI scanners. Even more important is that, apart from the anatomical information, MRI is also able to provide metabolic and physiological parameters, such as the perfusion of tissues and heart function. Contrast agents are frequently used in MRI examinations, for example, to enhance the local contrast in pathological tissue or for imaging of the vasculature (Strijkers et al., 2007).

The use of nanoparticles as an MRI contrast agent has seen a tremendous increase over the last decade (Mulder et al., 2007a). In research labs all over the world, smart contrast agents are under development on the basis of magnetic nanoparticles intended for molecular imaging purposes. Molecular imaging sets out to noninvasively depict cellular and molecular processes in living tissues (Weissleder and Mahmood, 2001). Such information could provide a molecular fingerprint of disease, which can be used for diagnosis, aid in treatment decisions, or monitor the effects of therapy. The integration of nanoparticles with therapeutic drugs will offer the possibility to image the local delivery of the drug (Sun et al., 2008).

Commonly, the contrast in MR images is generated on the basis of local variations in proton density and the T_1 and $T_2^{(*)}$ relaxation times of tissues. This chapter will focus on MRI contrast agents that act by locally lowering T_1 and $T_2^{(*)}$ thereby increasing the contrast with the surrounding tissue. Other mechanisms could contribute to the contrast as well, e.g., the transfer of magnetization between molecules (magnetization transfer [MT] and the chemical exchange of saturated magnetization [CEST]) or the tissue-dependent self-diffusion of water (diffusion contrast). However, these will not be discussed here in detail, apart from where diffusion and exchange play a role in T_1 and $T_2^{(*)}$ relaxation effects. Roughly, we can divide the relaxation time lowering contrast agents in two classes.

The first class comprises the superparamagnetic nanoparticle contrast agents, usually iron-oxide based, which exert a strong decrease in the $T_2^{(*)}$ relaxation time (see Section 23.3 for a detailed description of superparamagnetism). Figure 23.1A shows a schematic drawing of an iron-oxide nanoparticle. In order to prevent aggregation and to make the particles water soluble and applicable for intravascular administration, the iron-oxide core is generally coated with a biocompatible shell consisting of dextran, citrate monomers, or phospholipids (Weissleder et al., 1990; van Tilborg et al., 2006; Schellenberger et al., 2008). Nanoparticles

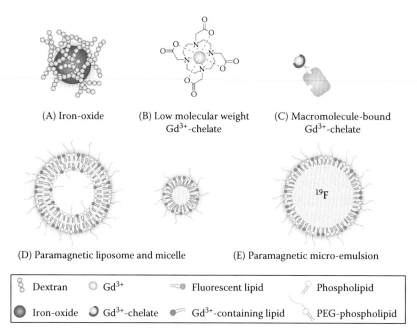

(A) Iron-oxide   (B) Low molecular weight Gd^{3+}-chelate   (C) Macromolecule-bound Gd^{3+}-chelate

(D) Paramagnetic liposome and micelle   (E) Paramagnetic micro-emulsion

| ⧂ Dextran | ◯ Gd^{3+} | ⊸ Fluorescent lipid | ⫼ Phospholipid |
| ● Iron-oxide | ◎ Gd^{3+}-chelate | ⊸ Gd^{3+}-containing lipid | ⫷ PEG-phospholipid |

FIGURE 23.1 Exemplar MRI contrast agents. (A) Iron-oxide nanoparticle. To make the agent biocompatible the iron-oxide core is encapsulated in a water soluble shell, usually dextran. (B) Gd(DOTA)$^-$, a low molecular weight Gd^{3+}-based contrast agent. Gd^{3+} is coordinated by electronic interactions with eight moieties of the chelate, leaving one coordination site accessible for the surrounding water. (C) Protein-bound Gd^{3+}-chelate. The protein could be used for targeting purposes or to prolong the blood circulation half life. The increased mass of the construct leads to improved relaxation properties. (D) Paramagnetic lipid-based nanoparticle contrast agents. Liposomes and micelles can be equipped with hundreds to thousands of amphiphilic Gd^{3+}-chelates resulting in very potent contrast agents. The increased rotational correlation time of the aggregate leads to improved relaxation properties at clinical field strengths. The nanoparticles additionally could be equipped with fluorescent lipids for fluorescence imaging purposes (PEG = polyethylene glycol). (E) Micro-emulsion, a perfluorocarbon core surrounded by a monolayer of lipids. Similar as to the liposomes and micelles, the micro-emulsion can contain amphiphilic Gd^{3+}-chelates and fluorescent lipids for MRI and fluorescence imaging purposes.

based on iron-oxide have been exploited extensively as MRI contrast agents during the last two decades. The nanoparticles with a typical core diameter in the order of 4–50 nm are referred to as superparamagnetic iron-oxide (SPIO) particles, ultrasmall superparamagnetic iron-oxide (USPIO) particles, very small superparamagnetic iron-oxide particles (VSOP), monocrystalline iron-oxide (MION) particles, or cross-linked iron-oxide (CLIO) particles depending on their size, crystalline structure, coating, and higher-order organization (Weissleder et al., 1990; Wang et al., 2001; Laurent et al., 2008). Micron-sized particles of iron-oxide (MPIO) with a size of up to approximately 6 μm were applied as targeted contrast agent and to visualize individual cells (Hinds et al., 2003; Shapiro et al., 2005).

Figure 23.2 exemplifies an application of targeted iron-oxide nanoparticles in imaging acute brain inflammation (McAteer et al., 2007). In this study, monoclonal antibodies to VCAM-1 and P-selectin were conjugated to 1 μm-sized MPIOs and injected in mice in which the vasculature of the left striatum was activated by the injection of interleukin-1β as a model for acute inflammation. Figure 23.2A shows MR images of the brains with specific areas of low signal intensity in the left side, which reflect specific the binding of the targeted MPIOs to the activated brain endothelium. Signal voids were only observed for specific particles (VCAM-MPIO and VCAM/P-selectin-MPIO), whereas

injection with unspecific particles (IgG-MPIO) did not result in signal decrease. Specific binding could be blocked by a pretreatment of the uncoupled VCAM-1 antibody. Note that the left and right hemispheres are equally intense in the absence of specific binding by the MPIOs. Figure 23.2B is a three-dimensional reconstruction of the signal voids, demonstrating that the VCAM-MPIO binding delineates the architecture of the cerebral vasculature, which is absent for a pretreatment with VCAM-1 antibody (Figure 23.2C). This study illustrates the potent contrast effect exerted by the micron-sized iron-oxide nanoparticles and the new opportunities for their application in the detection and quantification of pathology that is not visible on traditional anatomical images.

The second major class of contrast agents is composed of the paramagnetic metal ion–containing agents (see Section 23.3 for a detailed description of paramagnetism). The metal ion serves to induce a substantial decrease in T_1 without causing substantial line broadening. The most commonly used metal ion that satisfies these criteria is the lanthanide Gd^{3+}, but also Mn^{2+} and Fe^{3+} are effective agents because of their large paramagnetic moments. Free metal ions are not well tolerated by living tissue and therefore the ion is generally coordinated with a protective chelate to form a nontoxic complex. Figure 23.1B displays a Gd^{3+} complex with DOTA (tetraazacyclododecanetetraacetic acid).

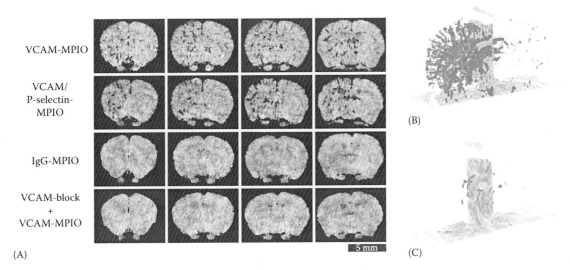

FIGURE 23.2 Targeted iron-oxide nanoparticles detect acute brain inflammation. (A) T_2^*-weighted gradient echo images of mice brains in which the vasculature of the left striatum was activated by injection of interleukin-1β as a model for acute inflammation. Each row displays four images of the same brain. Mice were injected with 1 μm-sized MPIOs. MPIOs were conjugated with monoclonal antibodies to (top row) VCAM-1 and (second row) VCAM-1 as well as P-selectin. Injection with (third row) IgG-MPIO as a negative control did not result in significant signal decrease. Specific binding could be blocked by a pretreatment of uncoupled VCAM-1 antibody (bottom row). (B) Three-dimensional reconstruction of VCAM-MPIO binding delineating the vasculature of the left brain hemisphere. (C) Pretreatment with uncoupled VCAM-1 antibody abolished binding of the VCAM-MPIO. (Reproduced from McAteer, M.A., *Nat. Med.*, 13(10), 1253, 2007. With permission.)

The Gd^{3+} ion has nine coordination sites of which eight are coordinated with the DOTA assuring the stability of the complex. The remaining site allows for the coordination with a water molecule, which is necessary to effectively induce the T_1 lowering effect (see Section 23.4). Most commercially available Gd^{3+}-based paramagnetic contrast agents are low molecular weight poly(aminocarboxylate) complexes of which $Gd(DTPA)^{2-}$ (Magnevist), $Gd(DOTA)^-$ (Dotarem), Gd(DTPA-BMA) (Omniscan), and GdHPDO3A (Prohance) are among the best known.

A considerable drawback of the Gd^{3+}-based agents is that a relatively high concentration of approximately 0.1 mM is needed to obtain sufficient contrast, which precludes the detection of sparse molecular processes in molecular imaging applications. To solve this problem, in recent years, a wide variety of macromolecular and nanoparticle paramagnetic contrast agents have been developed with the goal of increasing the particle detection sensitivity. This can be done by optimizing the Gd^{3+}-complex to increase its intrinsic T_1 lowering capacity (Aime et al., 2004) as well as by increasing the Gd^{3+} payload within a single nanoparticle to amplify the local Gd^{3+} concentration in the tissue. Examples of such approaches are shown in Figure 23.1C through E. Figure 23.1C displays a Gd^{3+}-complex conjugated to a macromolecule (antibody, protein). The larger molecular weight, slower thermal translation, and rotation, as compared to the low molecular weight chelate alone, leads to an increased T_1 lowering effect (as will be explained in Section 23.4). For further improvement, the targeting ligand could be decorated with multiple Gd^{3+}-chelates. If such an approach does not suffice, nanoparticles have the potential to boost the particle detection sensitivity orders of magnitude. Nanoparticles

can be equipped with thousands to hundreds of thousands of Gd^{3+} entities per particle, increasing the detection sensitivity accordingly, leading to detection limits in the (sub)micromolar range. Examples of such nanoparticles that have been developed and successfully applied in molecular MRI studies are liposomes, micelles (Figure 23.1D), and micro-emulsions (Figure 23.1E). These are examples of lipid-based nanoparticles. Lipids are amphiphilic molecules; they contain both a hydrophobic (nonpolar) and a hydrophilic (polar) domain. In an aqueous environment, the dualistic character induces self-association into a variety of aggregates, among which are liposomes, micelles, and emulsions (Mulder et al., 2006). The particles can be designed to contain several kinds of lipids, such as Gd^{3+}-containing lipids for MR detection and fluorescent lipids for fluorescent microscopy. A monolayer of lipids can also be used to enclose a hydrophobic perfluorocarbon core (micro-emulsion) in which case ^{19}F MRI could additionally be used to detect the nanoparticles (Ahrens et al., 2005). Micelles typically have a diameter between 10 and about 50 nm, while liposomes and micro-emulsions in general range in size from 80 to 400 nm.

Figure 23.3 exemplifies the application of targeted Gd^{3+}-containing liposomes in a mouse tumor model to aid in the detection of newly formed (angiogenic) blood vessels in the tumor (Mulder et al., 2007b). Figure 23.3A shows the pre-injection T_1-weighted MR image of the mouse with a large subcutaneous tumor indicated with the arrows. Figure 23.3B displays the MR image after a liposomal contrast agent was injected. The liposomal agent was conjugated with multiple RGD-peptide moieties, which bind to $\alpha_v\beta_3$-integrin expressed on the activated

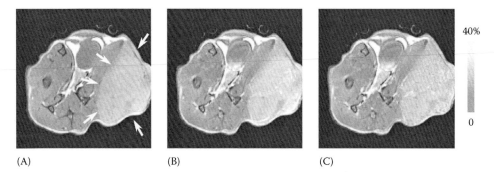

FIGURE 23.3 RGD-conjugated paramagnetic liposomes used for MRI of tumor angiogenesis. (A) Pre-injection T_1-weighted MR image of a mouse with a large subcutaneous B16F10 melanoma tumor, indicated with the yellow arrows. (B) T_1-weighted MR image after the liposomal contrast agent was injected. (C) Angiogenic activity could be quantified by pixel-by-pixel comparison of pre- and post-contrast images. Pixels that are significantly enhanced are color-coded with their percentile signal enhancement according to the scale on the right. (Reproduced from Mulder, W.J., *FASEB J.*, 21(2), 2007b, 378. With permission.)

endothelium in angiogenic blood vessels. A signal increase in the tumor was observed in areas predominantly located at the rim of the tumor, where the highest density of angiogenic blood vessels was expected. The angiogenic activity could be quantified with a pixel-by-pixel comparison of the pre- and post-contrast images, color-coding only significantly enhanced pixels (Figure 23.3C). This study illustrates the novel applications of powerful Gd^{3+}-based contrast agents in the characterization of disease processes at the sub-voxel scale.

23.1.2 Organization of This Chapter

This chapter is intended to give an overview and to provide a physical understanding of the spin–lattice (T_1) and spin–spin ($T_2^{(*)}$) relaxation mechanisms of superparamagnetic and paramagnetic nanoparticle contrast agents. To that aim, the basic T_1 and T_2 relaxation theory and the parameters determining contrast in the MR images are discussed in Section 23.2. Next, Section 23.3 deals with the T_2 and T_2^* relaxation induced by the superparamagnetic iron-oxide nanoparticles. Section 23.4 discusses the T_1 and T_2 relaxation of low molecular weight and nanoparticle Gd^{3+}-based contrast agents and Section 23.5 concludes this chapter with some perspectives.

23.2 Relaxation

23.2.1 Magnetic Resonance Imaging

MRI is a method that exploits the equilibrium net nuclear magnetization that is produced by the tiny imbalance in spin-up and spin-down hydrogen nuclei when water molecules are placed in a magnetic field. The equilibrium magnetization can be excited by an additional externally applied magnetic field. This magnetic field is called a radio-frequency (RF) pulse; it is short (pulse) and the RF-pulse should have a frequency that matches the precession frequency of the hydrogen nuclear magnetic moment (Larmor frequency). The application of an RF-pulse will result in the rotation of the macroscopic nuclear magnetization vector. The amount of rotation will depend on the strength

and duration of the RF-pulse; 90° and 180° pulses are commonly used. MR signals are subsequently recorded as induction voltages, i.e., free-induction-decay signals or spin-echoes. In a typical MRI pulse sequence, RF-pulses are suitably combined with magnetic field gradients to produce many echoes, which are subsequently utilized to reconstruct the MR image (Haacke et al., 1999; Vlaardingerbroek and Den Boer, 1999). The RF-pulses bring the magnetization into an excited state and there are forces driving the magnetization back to equilibrium, processes that are known as relaxation. Most of the contrast agents generate a contrast by changing the relaxation times of the surrounding water hydrogen nuclei. Therefore, before focusing on the specific contrast agents, the basic relaxation mechanisms (Cowan, 1997) and how these can be exploited to generate contrast in the MR images will be explained.

23.2.2 Spin-Lattice Relaxation

After excitation of the nuclear magnetization by RF-pulses, the spin-lattice or longitudinal relaxation mechanism is responsible for returning the magnetization M_z back to the Boltzmann equilibrium M_0 with time t, according to

$$M_z = M_0(1 - e^{-t/T_1}). \tag{23.1}$$

The spin-lattice relaxation is characterized by the time constant T_1 and therefore also referred to as T_1-relaxation. Spins in the high-energy state, anti-parallel to the applied magnetic field, can make transitions to the lower energy state, parallel to the magnetic field, by either spontaneous or stimulated emission of energy. The probability for spontaneous emission is extremely low as it depends on the third power of the Larmor frequency and therefore does not significantly contribute to T_1. The spin-lattice relaxation is thus dominated by stimulated emission. The stimulation is supplied by magnetic fields fluctuating at the Larmor frequency, which are present due to continuous random molecular motions (translation and rotation) of the molecules. In order to qualitatively understand T_1, we therefore have to understand the nature of the interaction as well as the frequency distribution of the molecular motions. In this section, the T_1-relaxation of the hydrogen nuclei in water will be discussed in detail. The additional effects of contrast agents will be discussed in Sections 23.3 and 23.4.

The origin of the stimulated transitions lies in the dipole–dipole interactions. Each nuclear moment can be considered a magnetic dipole moment, which produces a magnetic field in its surroundings. A water hydrogen nucleus placed in the MRI magnet therefore does not only experience the static magnetic field $\vec{B}_0$ but also the magnetic field $\vec{B}_d$ that originates from the magnetic moments of the other hydrogen nucleus on the same water molecule and the hydrogen nuclei of surrounding water molecules. The total magnetic field experienced by a certain hydrogen nucleus therefore amounts to

$$\vec{B}_t = \vec{B}_0 + \vec{B}_d, \tag{23.2}$$

where the total dipole field is given by

$$\vec{B}_d = \sum_i \vec{b}_i = \frac{\mu_0}{4\pi} \sum_i \left(\frac{\vec{m}}{\vec{r}_i^3} - \frac{(\vec{m} \cdot \vec{r}_i)\vec{r}_i}{\vec{r}_i^5} \right), \tag{23.3}$$

with

$\vec{b}_i$ the dipole field produced by a single moment

$\vec{m}$ the dipole moment at distance $\vec{r}_i$

i runs over all the neighboring moments

Note that for hydrogen nuclei at a distance of 0.2 nm $|\vec{b}_i| \sim 2 \times 10^{-4}$ T, which means that the dipolar field is about four orders of magnitude smaller than $\vec{B}_0$ and therefore can be treated as a small disturbance of the main magnetic field. Figure 23.4A provides a classical view of a number of water molecules in a liquid state. A hydrogen nucleus is not only under the influence of the main magnetic field $\vec{B}_0$, but also interacts with the dipole fields of the neighboring hydrogen nuclei. From this picture, it is clear that not only the magnitude of $\vec{B}_d$ plays a role in the dipolar interaction but also the relative orientation of molecules, since the total magnetic field is the vector sum of the applied magnetic field and additional dipole fields. Furthermore, molecular movement, such as rotation and diffusion (Figure 23.4B) modulate the magnitude of the interaction with time resulting from time-dependent changes in orientations and distances. Interactions between nuclear spins in the same molecule are most important since their distance $\vec{r}_i$ is small and fixed. Therefore, the interaction is strong and only modulated by the relative orientation of the moments with respect to the

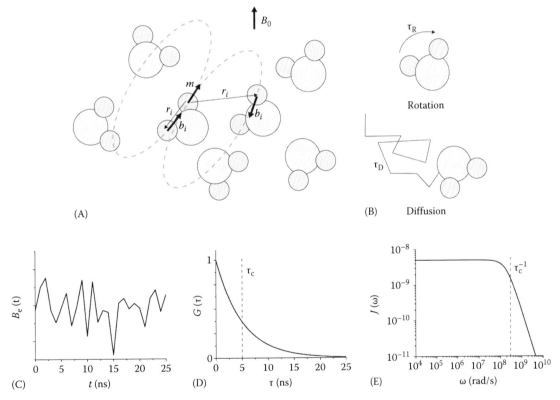

FIGURE 23.4 Classical view of dipole–dipole interactions between neighboring protons in water and rotational and diffusional autocorrelation functions. (A) The nuclear magnetic moment m of a proton produces a dipolar field b_i in its surrounding. (B) Rotational and diffusional molecular movements contribute to the longitudinal and transversal relaxation. (C) The effective magnetic field $B_e(t)$ experienced by the water proton spins is a time-fluctuating random function. (D) The autocorrelation function (Equation 23.5) for rotational and diffusional motion is well described by an exponentially with time decaying function. The corresponding correlation time $\tau_c = 5$ ns, indicated by the dashed line. (E) Lorentzian spectral density function (Equation 23.7) for $\tau_c = 5$ ns as function of frequency.

externally applied magnetic field. Interactions between nuclear spins of neighboring molecules are less efficient because these are modulated in both distance as well as orientation and the dipole field decays with the third power of the distance (Equation 23.3).

In order to understand the role of molecular motions, we revert to a classical view where we consider each water molecule independently, experiencing an average time-dependent magnetic dipole field produced by the neighboring hydrogen nuclei. A complete quantum-mechanical description of the problem would include a statistical many-body analysis. Nevertheless, for a basic understanding of the features of relaxation, a classical approach suffices. Under this assumption, a scalar magnetic field $B_e(t)$ is defined, which represents the effective time-varying local magnetic field that is experienced by the specific nucleus under consideration. Details on molecular motion and relative orientations are contained within the time-dependence of $B_e(t)$. The rotational and diffusional movements (Brownian motion) are random (Figure 23.4B), at least on timescales larger than the time it takes for a water molecule to rotate about its own axis or the time it takes to diffuse a few molecular distances. Under this assumption, $B_e(t)$ is a random fluctuating function, as drawn schematically in Figure 23.4C. In other words, there is no correlated movement for large timescales. For shorter timescales, the water molecules have not moved much and thus there is a high degree of correlation with the previous timepoints. To capture the timescale at which correlations exists, one can calculate the so-called autocorrelation function

$$G(\tau) = \langle B_e(t)B_e(t+\tau) \rangle, \tag{23.4}$$

which describes the mean time variation of the fluctuations of $B_e(t)$. For long times τ, $G(\tau)$ must go to zero when no large timescale correlations are present. For shorter timescales, correlations and thus $G(\tau)$ are high. The autocorrelation function $G(\tau)$ for rotational and diffusional motion in general is well described by an exponential decaying function

$$G(\tau) = G(0)e^{-\tau/\tau_c}, \tag{23.5}$$

as shown in Figure 23.4D for $\tau_c = 5\,\mathrm{ns}$ and $G(0) = 1$. Here, τ_c is the correlation time that defines the timescale for which correlations exist. Rotation (τ_R) and diffusion (τ_D) in general possess different correlation times, which means that $G(\tau)$ may display multi-exponential decay. The frequency distribution of the time variations in the effective field and the autocorrelation function are related via the Fourier-transform

$$J(\omega, \tau_c) = \int_{-\infty}^{\infty} G(\tau)e^{-i\omega\tau}d\tau, \tag{23.6}$$

which is called the spectral density function. For the autocorrelation function given by Equation 23.5, the spectral density function normalized for $J(0, \tau_c) = \tau_c$ reduces to

$$J(\omega, \tau_c) = \frac{\tau_c}{1 + \omega^2\tau_c^2}, \tag{23.7}$$

which is the well-known Lorentzian distribution of angular frequencies.

Now that the strength of the interaction (dipole–dipole) and the frequency distribution of the fluctuations of the dipole field (spectral density function) are known, the remaining link in understanding the T_1 relaxation is to consider the transition probabilities of two interacting nuclear moments. For this we consider a spin system of two equal spins in four possible configurations: $|\uparrow\uparrow\rangle$, $|\uparrow\downarrow\rangle$, $|\downarrow\uparrow\rangle$, or $|\downarrow\downarrow\rangle$, which are the eigenstates of the Zeeman Hamiltonian. In order to contribute to the T_1 relaxation mechanism, either one (e.g., from $|\uparrow\downarrow\rangle$ to $|\uparrow\uparrow\rangle$) or two ($|\downarrow\downarrow\rangle$ to $|\uparrow\uparrow\rangle$) spins have to change from anti-parallel to parallel orientation, requiring emission of energy at the proton Larmor frequency ω_1 or twice the Larmor frequency $2\omega_1$, respectively. Therefore, T_1 relaxation in liquids requires modulations of the dipolar field, i.e., rotation and diffusion of molecules, at the Larmor frequency or twice the Larmor frequency. By a full quantum-mechanical evaluation of the dipolar interaction Hamiltonian of two equal nuclear spins, it can be shown that

$$\frac{1}{T_1} = \frac{3\gamma_1^4\hbar^2\mu_0^2}{160\pi^2 r^6}\left[J(\omega_1, \tau_c) + 4J(2\omega_1, \tau_c)\right], \tag{23.8}$$

where

γ_1 is the proton gyromagnetic ratio
$\omega_1 = \gamma_1 B_0$ the proton Larmor frequency
r is the distance between the two nuclear moments

The latter equation finally integrates both the strength of the dipolar interaction as well as its dependence on the spectral distribution function.

23.2.3 Spin–Spin Relaxation

After excitation of the nuclear magnetization by a 90° RF-pulse, the spins are considered to be in a coherent state, precessing about the main magnetic field with the Larmor frequency. Due to spin–spin relaxation, the transversal magnetization decays with time t, according to

$$M_{xy} = M_0 e^{-t/T_2}, \tag{23.9}$$

in which M_0 represents the magnetization in the transversal plane directly after excitation. The spin–spin or transversal relaxation is characterized by the time constant T_2 and is therefore also referred to as T_2-relaxation.

For a basic understanding of the spin–spin relaxation, we again refer to Figure 23.4A and the classical view in which each individual nuclear moment is considered separately, experiencing a modulated dipole field produced by its surroundings. Individual magnetic moments will experience a different total magnetic field $B_e(t)$ due to the modulation of the dipole fields and therefore will precess at a slightly different resonance frequency. Consequently, spins that were precessing coherently right after excitation will rapidly dephase with respect to each other, leading to a fast decrease in M_{xy}. Again, for a full understanding of the transversal relaxation we have to consider the transition probabilities of two interacting spins characterized by the eigenstates $|\uparrow\uparrow\rangle$, $|\uparrow\downarrow\rangle$, $|\downarrow\uparrow\rangle$, and $|\downarrow\downarrow\rangle$. In contrast to the spin–lattice relaxation, T_2-relaxation does not exclusively require a stimulated transition with the absorption of energy. Therefore, not only transitions for which one or two spins change orientation contribute to T_2, but also the transitions that keep the total energy equal, that is $|\uparrow\downarrow\rangle$ to $|\downarrow\uparrow\rangle$ and $|\downarrow\uparrow\rangle$ to $|\uparrow\downarrow\rangle$. As a result, the transversal relaxation is not only dependent on the spectral density function at ω_I and $2\omega_I$, but also at $\omega = 0$. A full quantum mechanical evaluation yields

$$\frac{1}{T_2} = \frac{3\gamma_I^4\hbar^2\mu_0^2}{320\pi^2 r^6}\left[3J(0,\tau_c) + 5J(\omega_I,\tau_c) + 2J(2\omega_I,\tau_c)\right]. \quad (23.10)$$

23.2.4 T_2^*-Relaxation

Because T_2-relaxation is also dependent on field variations at zero frequency, any external field variation in space and time will contribute to dephasing in the transversal plane. Such field variations may not only be caused by the dipolar fields of neighboring nuclear spins (microscopic origin), but may also originate from macroscopic field inhomogeneities such as those caused by the imperfect shimming of the magnet or regional susceptibility differences between tissues or near tissue borders (macroscopic origin). The dipolar contributions are considered to be an intrinsic property of the tissue, whereas macroscopic field inhomogeneities also have an instrumental origin. The accelerated transversal relaxation is referred to as T_2^* (pronounce T_2-star), according to

$$\frac{1}{T_2^*} = \frac{1}{T_2} + \frac{1}{T_2'}, \quad (23.11)$$

with T_2 and T_2' the dipolar contributions and macroscopic contributions, respectively. T_2 and T_2^* can be distinguished using the appropriate pulse sequences. The spin-echo sequence is T_2-weighted since the 180° pulse reverses the dephasing of spins due to macroscopic field inhomogeneities. The gradient-echo sequence on the other hand is T_2^*-weighted because both dipolar and macroscopic fields contribute to dephasing. The distinction between T_2 and T_2^* however is not always straightforward and absolute. Superparamagnetic contrast agents produce local magnetic field variations causing dephasing, which depending on the timescale and type of the pulse sequence may or may not be reversible (see Section 23.3).

23.2.5 Relaxation in the Rotating Frame

Dipolar magnetic field fluctuations of large macromolecules in tissues are characterized by high τ_c, and therefore only weakly contribute to the T_1-relaxation at typical MRI field strengths (1.5–7 T), because here the $J(\omega, \tau_c)$ cutoff frequency τ_c^{-1} lies far below the Larmor frequency and consequently the $J(\omega_I, \tau_c)$ and $J(2\omega_I, \tau_c)$ contributions are quite small. Nevertheless, the macromolecular content and composition may contain important information on tissue status and pathological conditions. A spin-locking technique allows for studying the relaxation at low frequencies without reverting to low-field measurements with less signal-to-noise.

The principle of a spin-locking experiment is as follows. Usually, the nuclear magnetization after a 90° RF-pulse will decay to zero in the transversal plane due to spin–spin relaxation. However, if a transverse magnetic field of magnitude B_{SL} rotating with the Larmor frequency will be applied after the RF-pulse, the transversal magnetization will not relax to zero but will start to precess around B_{SL} with angular frequency $\omega_{SL} = \gamma_I B_{SL}$. The reason is that the frequency of the spin-lock field B_{SL} equals the Larmor frequency, which means that B_{SL} is a static field in the rotating frame of reference. This situation can therefore be compared to the precession of the magnetization around the main magnetic field B_0 as is observed in the laboratory frame. The transverse magnetization relaxes not to zero, but to the value appropriate to the spin-lock field B_{SL} and because of its analogy with T_1 relaxation, the time constant of relaxation is referred to as $T_{1\rho}$. It can be shown that

$$\frac{1}{T_{1\rho}} = \frac{3\gamma_I^4\hbar^2\mu_0^2}{160\pi^2 r^6}\left[\frac{3}{2}J(2\omega_{SL},\tau_c) + \frac{5}{2}J(\omega_I,\tau_c) + J(2\omega_I,\tau_c)\right], \quad (23.12)$$

where $J(\omega_I)$ and $J(2\omega_I)$ are small and therefore $T_{1\rho}$ is primarily sensitive to the spectral distribution function at the low spin-lock frequency. The most basic way to measure $T_{1\rho}$ is to record a free-induction-decay following the spin-lock field. The signal intensity will follow the function

$$M_{xy} = M_0 e^{-\tau_{SL}/T_{1\rho}}, \quad (23.13)$$

in which τ_{SL} is the duration of the spin-lock pulse. The spin-lock frequency may be varied in order to probe the spectral density function distribution.

23.2.6 Using Relaxation to Generate Contrast in MR Images

Figure 23.5A displays the T_1 and T_2 relaxation times at 1.5 and 7 T as a function of τ_c in the range between 0.01 and 1000 ns. The graph was calculated using Equations 23.8 and 23.10. Biological tissue is rather complicated; it is composed of many cellular and extracellular compartments in which water behaves diversely with different correlation times. Furthermore macromolecules, proteins, DNA, iron deposits, *etc.* are sources of time varying dipole fields with a broad range of correlation times. Nevertheless, the quantitative T_1 and T_2 found in biological tissues are quite well described by the basic Equations 23.8 and 23.10. Basically,

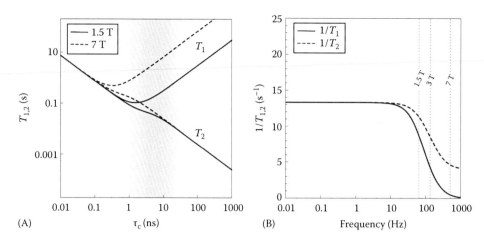

FIGURE 23.5 Correlation time and frequency dependence of the T_1 and T_2 relaxation times. (A) T_1 and T_2 as function of the correlation time τ_c for magnetic field strengths of 1.5 and 7 T. The curves were calculated using Equations 23.8 and 23.10. The shaded area indicates the approximate range of correlation times that correspond to the relaxation times found in biological tissues. (B) Inverse relaxation times as function of Larmor frequency. Frequencies corresponding to relevant clinical magnetic field strengths are indicated with vertical dashed lines.

in tissues T_1 at 1.5, T ranges between 300 ms for fat to typically 900 ms for brain gray matter. Cerebrospinal fluid (CSF) can have a very long T_1 up to 4 s. The spin–lattice relaxation times in tissues increase at higher field strengths. The spin–spin relaxation times (T_1) typically range between 40 ms in the liver caused by a high iron content to about 100 ms in brain gray matter. The spin–spin relaxation (T_2) is less dependent on magnetic field strength. Indeed, such T_1 and T_2 can be found in Figure 23.5A for correlation times in the range indicated by the shaded area. Figure 23.5B shows the frequency dependence of the inverse relaxation times for $\tau_c = 5$ ns, showing that T_1 increases significantly with frequency. T_2 increases to a lesser extent or could even stay constant with the frequency, depending on the correlation time τ_c. Frequencies corresponding to relevant clinical magnetic field strengths are indicated with vertical dashed lines.

Because spin–lattice and spin–spin relaxation times differ and vary between tissues, they provide the main source of contrast in MR images. Contrast can be generated by tuning the appropriate time delays in the MRI sequence. For a basic understanding, we will use the example of the spin-echo and the gradient-echo sequence. The spin-echo sequence consists of a 90° RF-pulse followed by a 180° RF-pulse producing an induction signal at TE, the echo time. The repetition time TR is the time between successive repetitions of the sequence. The amplitude of the spin-echo induction voltage is dependent on T_1, T_2, and the sequence parameters according to

$$V(t) = k\rho(1 - e^{-TR/T_1})e^{-TE/T_2}, \qquad (23.14)$$

where

    ρ is the proton density

    k is an instrumental proportionality factor, which includes among others the sensitivity of the RF-coil and amplification factors

Figure 23.6A (progressive saturation) displays the behavior of $V(t)$ as a function of TR for a short and long T_1. This behavior

is called progressive saturation, as it is a result of a progressive decrease in longitudinal magnetization by successive RF-pulses. Two tissues with different longitudinal relaxation times will produce a different spin-echo signal depending on the choice of TR and hence will obtain a different contrast in the resulting MR image. Another common way to obtain a T_1 contrast is to apply a 180° inversion RF-pulse preparation before a particular imaging sequence, in which case the longitudinal magnetization will follow an exponential recovery according to $1 - 2e^{-TI/T_1}$, with TI the inversion time that can be tuned to obtain the largest difference between short and long T_1 relaxation times as shown schematically in Figure 23.6B (inversion recovery). Figure 23.6C (spin-echo) shows the spin-echo sequence and its dependence on TE. Dephasing as a consequence of macroscopic field inhomogeneities in the first half of TE is reversed during the second half of TE by the rephasing nature of the 180° RF-pulse. Therefore, the spin-echo sequence is predominantly dependent on the microscopic dipolar fields characterized by T_2. As shown in the figure, by the appropriate choice of TE, the resulting spin-echo amplitude for tissues with a different T_2 will be different resulting in a contrast in the MR image.

The gradient-echo sequence uses a set of opposite magnetic gradient fields to dephase and rephase spins and produce an echo. The echo induction voltage amplitude of the so-called RF-spoiled gradient-spoiled gradient-echo sequence behaves according to

$$V(t) = k\rho \frac{(1 - e^{-TR/T_1})\sin\alpha}{1 - \cos\alpha\, e^{-TR/T_1}} e^{-TE/T_2^*}, \qquad (23.15)$$

in which α is the flip-angle of the RF-pulses. Although this looks more complicated, it reduces to the spin-echo for $\alpha = 90°$, apart from the fact that the gradient-echo sequence is dependent on T_2^* instead of T_2 because the rephasing 180° pulse is missing. Similar to the spin-echo sequence, a contrast can be obtained by

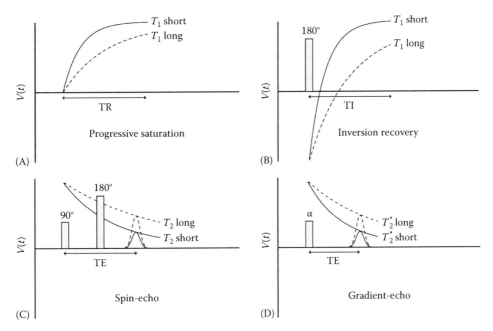

FIGURE 23.6 Generation of contrast in MR images exploiting differences in T_1, T_2, and T_2^*. Solid and dashed lines indicate the time evolution of the echo induction voltage $V(t)$ as function of the relevant delays in the MR sequence for different T_1 and T_2 relaxation times. (A) Progressive saturation using a T_1-weighted spin-echo sequence with short repetition time TR. Tissues with short T_1 will produce a higher echo signal and consequently will appear bright on the resulting MR image (positive contrast). (B) Inversion recovery experiment using a 180° preparation pulse. Contrast between tissues can be tuned positive or negative by variation of the inversion time TI. (C) T_2-weighted spin-echo sequence with long echo time TE. Tissues with short T_2 will produce a lower echo signal and thus appear darker in the MR image (negative contrast). (D) The gradient echo sequence is T_2^*-weighted, resulting in negative contrast for tissues with short T_2^*.

choosing an appropriate TE with respect to T_2^*, resulting from differences in echo amplitude as shown schematically in Figure 23.6D (gradient-echo). There are many more MRI sequences and preparation schemes that display more complicated dependence on T_1, T_2, T_2^*, or $T_{1\rho}$. Nevertheless, the fact that a contrast can be obtained by considering the appropriate sequence timing parameters with respect to the relaxation times is something they all have in common.

23.2.7 Relaxivity

The superparamagnetic and paramagnetic contrast agents discussed in the next sections act by shortening the tissue water relaxation times. The relaxation time in the presence of the contrast agent can be written as

$$\frac{1}{T_i} = \frac{1}{T_i^0} + \frac{1}{T_i^{CA}}, \quad \text{with } i \in \{1, 2\}, \qquad (23.16)$$

where

T_i^0 is the intrinsic relaxation time of the tissue
T_i^{CA} is the contribution of the contrast agent

The potency of a contrast agent to shorten the relaxation times T_1 and T_2 for a given contrast agent concentration [CA] is called

relaxivity r_1 and r_2, respectively, expressed in units $\text{mM}^{-1}\,\text{s}^{-1}$. The contrast agent contribution to the relaxation can therefore be rewritten as

$$\frac{1}{T_i^{CA}} = r_i[\text{CA}], \quad \text{with } i \in \{1, 2\}. \qquad (23.17)$$

Relaxivity r_i (lower case) is to be distinguished from the relaxation rate R_i (upper case), defined as

$$R_i = \frac{1}{T_i}, \quad \text{with } i \in \{1, 2\}. \qquad (23.18)$$

Although the relaxivity is a contrast agent specific parameter, it also depends on the solvent, spatial distribution, and aggregation state, which could be tissue dependent *in vivo*. The linearity of $1/T_i$ with [CA] is therefore not observed under all circumstances (for specific examples, see Sections 23.3.3 and 23.4.4). An alternative way of classifying contrast agents is based on their specific r_2/r_1 ratio. For r_2/r_1 in the range of approximately 1–5, a contrast agent is suitable for generating a positive contrast using T_1-weighted sequences. When r_2/r_1 is larger than about 5, the contrast agent is best visualized using T_2- or T_2^*-weighted imaging.

23.3 Superparamagnetic Nanoparticles

23.3.1 Magnetic Properties of Iron-Oxide Nanoparticles

23.3.1.1 Magnetism

Magnetism fundamentally arises from the movement of electrical charges. Consequently, magnetic fields and forces play an important role whenever moving and orbiting charges are found, such as for the orbital and spin moments of elementary particles (e.g., electrons, protons) for the nuclear and electron shells of atoms as well as for electrical currents moving through wires. To describe their basic magnetic properties, materials are historically classified on their behavior when subjected to an externally applied magnetic field. *Diamagnetic* molecules and materials consist of atoms with paired electrons, such that no net magnetization exists. Nevertheless, when put in a magnetic field, these systems are weakly magnetized proportional to the magnitude of the applied field but in the opposite direction (Figure 23.7A). Diamagnetism is a result of the change in orbital motion of the electrons, which tends to oppose the external magnetic field (Lenz's law). The magnetic susceptibility χ_m of diamagnetic materials is low (of the order of 10^{-5}) and negative. *Paramagnetic* systems consist of atoms that possess a net magnetic moment due to the presence of unpaired electrons. The magnetic moments, however, do not or only very weakly interact, such that there is no cooperative behavior. A paramagnetic system can be magnetized by an external magnetic field with a magnetic susceptibility that is positive and much larger than for the diamagnetic systems. Up to moderate fields, the induced magnetization is linear with applied field strengths (Figure 23.7B).

Many of the magnetic nanoparticles used as MRI contrast agents consist of *ferromagnetic* materials. A ferromagnetic system consists of atoms with a magnetic moment, which additionally have a strong and positive interaction (Figure 23.7C). Below the so-called Curie temperature T_c, the magnetic moments will cooperatively align in parallel, which leads to magnetization even in the absence of an external field. The magnetization increases rapidly with increasing field strengths and saturates at a higher magnetic field. Apart from ferromagnetic ordering, there are also materials for which the ordering is in an antiparallel fashion. If all the magnetic moments have equal magnitude, the net magnetization is zero and such material is classified as an *antiferromagnetic* system (Figure 23.7D). If the material consists of atoms with unequal magnetic moments, a net magnetization will be present and the system is called *ferrimagnetic* (Figure 23.7E) giving rise to a similar magnetization behavior as the ferromagnetic materials.

23.3.1.2 Magnetism of Colloidal Iron-Oxide Nanoparticles

There are many stoichiometric iron-oxides and iron-oxyhydroxides, of which magnetite (Fe_3O_4), maghemite (γ-Fe_2O_3), goethite (α-FeOOH), and hematite (α-Fe_2O_3) are among the better known ones. Most iron-oxide nanoparticles for MRI applications consist of magnetite or maghemite (Corot et al., 2006; Laurent et al., 2008). Magnetite has a crystallographic inverse spinel structure (Figure 23.8), which can be written as $Fe^{3+}_A(Fe^{3+}, Fe^{2+})_BO_4$, where Fe^{2+} occupies the tetrahedral A-sites (surrounded by four oxygen atoms) and Fe^{3+} occupies the octahedral B-sites (surrounded by six oxygen atoms). Magnetite is a ferrimagnet with a saturation magnetization M_{sat} of 510 kA/m. Maghemite can be considered to be a Fe^{2+}-vacant magnetite, with a structure written as $(Fe^{3+}_8)_A(Fe^{3+}_{40/3}[]_{8/3})_BO_{32}$, where [] represents a vacancy. Maghemite has a saturation magnetization

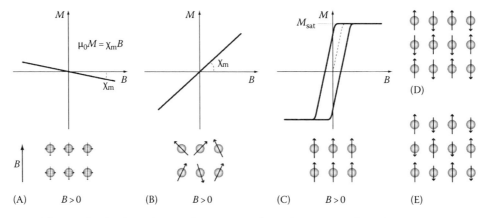

FIGURE 23.7 Magnetic behavior of molecules and materials in an externally applied magnetic field. The curves in (A–C) represent the induced magnetization *M* (unit A/m) as function of the applied magnetic field *B*. (A) A diamagnetic material is weakly magnetized with opposite orientation as a consequence of orbital interactions. (B) A paramagnetic material possesses non-interacting magnetic moments due to unpaired electrons which are magnetized in the direction of the applied magnetic field. Up to moderate fields the induced magnetization is linear with the applied field strength. (C) Ferromagnetic materials possess strongly positively interacting unpaired electrons, which leads to a large magnetization even in the absence of an external field. (D) In antiferromagnetic materials the ordering is in an anti-parallel fashion, resulting in a zero net magnetization. (E) In ferrimagnetic materials the magnetic moments are also ordered anti-parallel. However, they are unequal in magnitude resulting in a nonzero net magnetization.

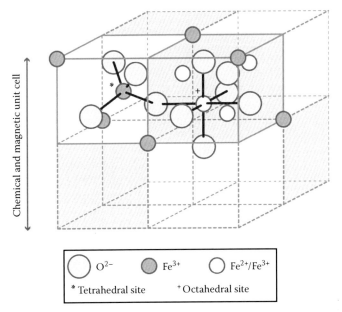

Chemical and magnetic unit cell

○ O^{2-} ● Fe^{3+} ○ Fe^{2+}/Fe^{3+}

* Tetrahedral site   + Octahedral site

FIGURE 23.8 Chemical and magnetic unit cell of the inverse spinel iron-oxide crystallographic structure. The unit cell can be divided into eight octants. There are two types of octants containing iron atoms at tetrahedral A-sites (*) and octahedral B-sites (+), indicated in the figure by white and gray octants, respectively. In each of the octants the oxygen ions occupy the same location.

of 380 kA/m. As a comparison, the saturation magnetization of Fe is 1751 kA/m, which would make ferromagnetic iron metal nanoparticles a much more effective MRI contrast agent. However, in aqueous solution, elemental iron nanoparticles will rapidly oxidize into nonmagnetic iron-oxides and iron-oxyhydroxides.

In bulk form, magnetite and maghemite consist of many magnetic domains, which in the zero field are oriented in a fashion that minimizes the total magnetostatic, magnetocrystalline, and magnetostrictive energy, as shown schematically in Figure 23.9A. As a result, the net magnetization of the material may be zero. When subjected to an applied field, the domains orient parallel to the field and the material becomes magnetized (Figure 23.9B). Iron-oxide nanoparticles used as the MRI contrast agent, however, are generally much smaller than an individual magnetic domain, which ideally results in a nanoparticle that acts as a single-domain fully magnetized sphere. The orientation of the magnetization with respect to the crystalline structure of the nanoparticle is determined by the magnetic anisotropy. The bulk magnetocrystalline anisotropy energy of Fe_3O_4 is cubic and in first approximation given by $E_c = K_c V (\alpha_1^2 \alpha_2^2 + \alpha_2^2 \alpha_3^2 + \alpha_1^2 \alpha_3^2)$, with $K_c = -1.1 \times 10^4$ J/m³. Here α_1, α_2, and α_3 denote the direction cosines of the magnetization vector relative to the cubic axes and V the volume of the crystal. The preferred axis of the magnetization is called the easy axis because this orientation minimizes the anisotropy energy. For bulk Fe_3O_4, this is the <1 1 1> direction, which is shown schematically in Figure 23.9C. Because in practice the magnetic nanoparticles are not always perfect spheres and may more resemble ellipsoids, the anisotropy energy is usually dominated by an extra unidirectional shape anisotropy energy $E_s = K_s V \varphi^2$, with $K_s < 0$ and φ the direction cosine with the long axis of the ellipsoid. Additionally, surface roughness, impurities, and crystal defects may pin the magnetization in certain directions, while dipole interactions between particles can lead to other preferred orientations.

When the magnetic particles in a colloidal suspension are exposed to an external magnetic field, the magnetization of individual particles will rotate and align with respect to the

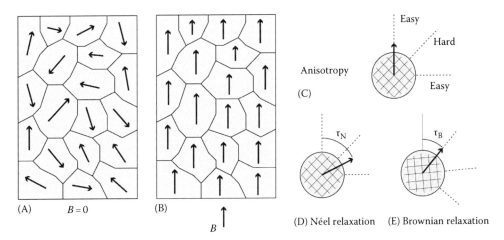

FIGURE 23.9 Bulk ferromagnetic material and superparamagnetic iron-oxide nanoparticles. (A) In bulk form in zero magnetic field, ferromagnetic materials consist of many domains which minimize the total magnetic and magnetocrystalline energy, resulting in zero net magnetization. (B) When subjected to a magnetic field, the domains orient parallel to the field and the bulk material becomes magnetized. (C) Iron-oxide nanoparticles are generally smaller than typical magnetic domains and therefore act as a single-domain fully magnetized sphere. The magnetocrystalline anisotropy of Fe_3O_4 is cubic with an easy axis of magnetization in the <1 1 1> direction, indicated by the square grid. Often the particles have an ellipsoid shape leading to an extra unidirectional anisotropy that dominates the cubic one. (D) Néel relaxation describes the reorientation of the magnetization within the particle, characterized by the correlation time τ_N. (E) Brownian relaxation involves rotation of particle and magnetization as a whole, characterized by the correlation time τ_B.

magnetic field. This is governed by two rotation modes (Néel, 1949; Brown, 1963). The first one is called Néel relaxation and describes the rotation of the magnetization within the particle (Figure 23.9D). Note that here the term relaxation refers to the reorientation of the dipole moment of the nanoparticle and not to the T_2-relaxation. For a full reversal of the magnetization, e.g., from antiparallel to parallel with respect to the applied magnetic field, the field energy has to overcome the magnetic anisotropy energy. The rotation correlation time involved in Néel relaxation under the assumption that $(E_c + E_s) \ll k_b T$ is

$$\tau_N = 10^{-9} e^{(E_c + E_s)/k_b T}, \tag{23.19}$$

which describes the Arrhenius kinetics of the magnetization to cross the magnetic anisotropy energy maxima (Fannin and Charles, 1989). The other mode of rotation is Brownian relaxation, which involves the rotation of the particle and magnetization as a whole (Figure 23.9E). The correlation time involved in Brownian motion is

$$\tau_B = \frac{3V\eta}{k_b T}, \tag{23.20}$$

with

V the volume of the nanoparticle
η the viscosity of the carrier fluid

The total correlation time is the sum of the two contributions

$$\frac{1}{\tau_c} = \frac{1}{\tau_N} + \frac{1}{\tau_B}. \tag{23.21}$$

For particles larger than about 7 nm, the Brownian relaxation mechanism dominates the Néel contribution, since the latter one increases exponentially with volume (Rosensweig, 2002).

Under these circumstances, the peculiar situation occurs that although fundamentally the nanoparticle is ferrimagnetic, the magnetization is nevertheless capable of rotating freely in the viscous fluid under the influence of an applied magnetic field and thermal fluctuations. The magnetization curve will therefore more closely resemble the paramagnetic curve of Figure 23.7B than the ferromagnetic curve of Figure 23.7C. The magnetization behaves like a paramagnet, however, with large magnetic moments. This unusual magnetization behavior therefore has been named *superparamagnetism* (Livingston and Bean, 1959). The magnetization curve will follow the classic Langevin function

$$M = M_{sat}[\coth(x) - x^{-1}] \quad \text{with} \tag{23.22}$$

$$x = \frac{M_{sat} V B}{k_b T}, \tag{23.23}$$

which is linear with applied field B for small B as shown in Figure 23.7B.

23.3.2 T_2-Relaxation Induced by Iron-Oxide Nanoparticles

When considering the effect of the magnetic iron-oxide nanoparticle on the T_2-relaxation time of surrounding water protons, we have to identify the relevant correlation times, similar to those in Section 23.2.1 for water protons only. First of all, the magnetic moment of the superparamagnetic particles fluctuates with the Néel correlation time τ_N. Secondly, there is the Brownian motion of the water and the nanoparticle. Assuming that the Brownian diffusional motion of water is the more rapid one, we can define a characteristic diffusion correlation time for water to diffuse (in three dimensions) in the neighborhood of the nanoparticle as

$$\tau_D = R^2/D, \tag{23.24}$$

with

R the radius of the nanoparticle
D the diffusion coefficient of water

This is the characteristic time it takes for water to diffuse a distance equal to the particle diameter. While for the case of pure water the dipole interaction of neighboring hydrogen nuclear spins could be neglected, this is certainly not the case for the large nanoparticles that possess a very large magnetic moment with respect to the nuclear moment of protons. The T_2-relaxation is therefore primarily governed by the so-called *outer sphere relaxation*, which describes the dephasing of the hydrogen nuclear spins diffusing in the inhomogeneous field created by the iron-oxide nanoparticles; the situation is shown schematically in Figure 23.10A.

In order to estimate the magnitude of the inhomogeneous magnetic field surrounding the nanoparticles, we need to consider the z-component of the dipole field

$$\Delta B_z = \frac{1}{3} \Delta\chi_m B_0 R^3 \frac{2z^2 - x^2 - y^2}{(x^2 + y^2 + z^2)^{5/2}} \tag{23.25}$$

Here, $\Delta\chi_m$ is the magnetic susceptibility difference between the nanoparticle and the water. For $x = R$, $y = 0$, and $z = 0$, Equation 23.25 reduces to

$$\Delta B_z = -\frac{1}{3} \Delta\chi_m B_0, \tag{23.26}$$

which is the magnetic field at the equator of the nanoparticle. The latter is a good estimate for the magnetic field differences experienced by a water proton diffusing in the neighborhood of the nanoparticle. ΔB_z is associated with a shift in the proton Larmor frequency

$$\Delta\omega = \gamma_I \Delta B_z. \tag{23.27}$$

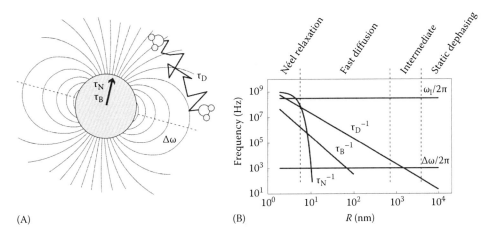

FIGURE 23.10 Transversal relaxation induced by an iron-oxide nanoparticle. (A) T_2-relaxation of water hydrogen nuclei is caused by diffusion in the fluctuating inhomogeneous magnetic dipole field produced by the nanoparticle. (B) Frequencies, corresponding to correlation times and magnetic fields, as function of iron-oxide particle radius R. Different regimes of T_2-relaxation are indicated with vertical dashed lines. The following parameters were used for this calculation: $B_0 = 7\,\mathrm{T}$, $T = 293\,\mathrm{K}$, $D = 2.2 \times 10^{-3}\,\mathrm{mm^2/s}$, $\eta = 8.9 \times 10^{-4}\,\mathrm{Pa\cdot s}$, and $\Delta\chi_m = 1 \times 10^{-5}$.

The shift in Larmor frequency $\Delta\omega$ and the Larmor frequency ω_l are the characteristic frequencies that determine the timescale of the problem and have to be compared with the inverse of the relevant correlation times (τ_N, τ_B, τ_D). Based on this comparison, the nanoparticle size-dependence of the T_2-relaxation rate can be roughly divided into four regimes as shown in Figure 23.10B: (1) the Néel relaxation regime, (2) the fast diffusion regime, (3) the intermediate regime, and (5) the static dephasing regime. Note that these regimes do not have absolute boundaries and that they are often further subdivided in literature (Yung, 2003; Laurent et al., 2008).

23.3.2.1 Néel Relaxation Regime

For the smallest iron-oxide nanoparticles (below $R \approx 7\,\mathrm{nm}$), the relaxation is dominated by the correlation time $\tau^{-1} = \tau_N^{-1} + \tau_D^{-1}$, which in this regime is of the same order of magnitude as the Larmor frequency ω_l. The diffusion of water is very fast as compared to nanoparticle dimensions and the system is therefore in a motional narrowing regime, which means that the spatial variations of the magnetic field are averaged out by the fast motion of the water protons. Consequently, T_2-relaxation is dominated by diffusional and Néel fluctuations at the Larmor frequency, leading to a quite moderate relaxivity where r_2 is only moderately higher than r_1. The r_2/r_1 ratio is low (~1–3) making these contrast agents quite effective T_1-relaxation agents enabling the generation of positive contrast in T_1-weighted MR images (Taboada et al., 2007). The calculation of the r_1 and r_2 relaxivity of such small iron-oxide nanoparticles is generally based on the classical outer sphere theory (Section 23.4.3). However, calculations like this remain challenging due to difficulties in correctly accounting for the nanoparticle magnetic anisotropy (Roch et al., 1999).

23.3.2.2 Fast Diffusion Regime

Iron-oxide particles with a radius between about 7 nm and a few hundred nanometers are in the fast diffusion regime. Here, the

inverse Néel relaxation correlation time τ_N^{-1} is negligibly small and the inverse diffusional correlation time τ_D^{-1} is much smaller than the Larmor frequency making Néel and Larmor contributions ineffective. The T_2-relaxation is therefore governed by the fast diffusion of the water hydrogen nucleus through the inhomogeneous magnetic field of the nanoparticle. Relaxation is still in a motional narrowing regime since τ_D^{-1} is much larger than $\Delta\omega$, meaning that the diffusion averages out the effect of a single magnetic nanoparticle. For estimating the magnitude of T_2, we can refer to the expression for the motional narrowing of high-resolution NMR line-widths, which can be derived using the Bloch equation (Cowan, 1997)

$$\frac{1}{T_2} = \left\langle \omega_{loc}^2 \right\rangle \tau_D. \tag{23.28}$$

Using $\omega_{loc} = \Delta\omega$, f is the effective volume fraction of nanoparticles in the solution, and a equals a scaling factor we obtain for the fast diffusion regime

$$\frac{1}{T_2^{CA}} = af \left\langle \Delta\omega^2 \right\rangle \tau_D. \tag{23.29}$$

A rigorous treatment of the outer sphere contributions yields $a = 16/135$ for perfect motional narrowing (Gillis and Koenig, 1987). The factor a will be larger for large nanoparticles and low concentrations when motional averaging is less effective. The $\Delta\omega^2$ dependence can also be recognized in the equations for outer sphere relaxation induced by paramagnetic agents, as will be explained in Section 23.4.3. As the effective volume fraction f is proportional to the concentration of nanoparticles [CA] in the solvent, Equation 23.29 is consistent with Equation 23.17. Equation 23.29 also holds for T_2^*, since the diffusional correlation time could be shorter than the typical echo-time in the MRI experiment, in which case, relaxation cannot be effectively reversed by a 180° RF-pulse.

23.3.2.3 Intermediate Regime

Particles with a radius between a few hundred nanometers and several micrometers are in the intermediate regime characterized by τ_D^{-1} of the same order as $\Delta\omega$. The motional narrowing approximation fails in this regime but at the same time the hydrogen nuclear spins cannot be considered static on the timescale $\Delta\omega$. This represents a difficult case for which no general analytical equations have been formulated yet. In this regime, a distinct difference in T_2 and T_2^* can be observed and measured using spin-echo and gradient-echo sequences, respectively. With increasing echo-time, the 180° RF-pulse of the spin-echo sequence increasingly fails to effectively reverse static dephasing contributions.

23.3.2.4 Static Dephasing Regime

For the largest particles (typically larger than several micrometers in diameter), the inverse diffusional correlation time τ_D^{-1} is much smaller than $\Delta\omega$, which means that water protons are quasi static. In this regime, diffusional translation has little effect on the relaxation. The magnitude of the transverse relaxation can be estimated from the simple consideration that $1/T_2$ is proportional to dephasing as a consequence of the spread in Larmor frequencies $\Delta\omega$. Therefore,

$$\frac{1}{T_2^{\mathrm{CA}}} = af\Delta\omega, \qquad (23.30)$$

with

  f the volume fraction of the nanoparticles
  a a geometric factor, which equals $2\pi/(3\sqrt{3})$ for randomly placed stationary magnetic nanoparticles (Brown, 1961)

For the spin-echo, this means that the nanoparticle does not contribute to T_2-relaxation since dephasing during the first half of the echo-time is perfectly rephased during the second half, similar to macroscopic field inhomogeneities. Of course this represents an extreme case when diffusional motion is completely absent. In practice, there will always be some contribution of diffusional motion to the observed relaxation rate. For the gradient-echo sequence in the static dephasing regime, Equation 23.30 represents T_2^* decay.

23.3.3 Changing Relaxivity: Responsive Agents

The dependence of the transversal relaxivity on the size of the iron-oxide nanoparticle represents an interesting opportunity to design contrast agents of which the relaxivity changes upon activation by biological or chemical processes in a cellular environment. The basic idea is that activation induces the aggregation of the iron-oxide nanoparticles, which amplifies the transversal relaxivity of the so formed clusters. We will start from relatively small nanoparticles for which the fast diffusion regime applies (Equation 23.29). Consider a spherical cluster with radius R_c composed of N iron-oxide particles. The effective magnetization

of the cluster and the resulting $\Delta\omega$ is proportional to the number of iron-oxide particles within the cluster volume and therefore

$$\Delta\omega \propto \frac{N}{R_c^3}. \qquad (23.31)$$

Since the iron-oxide nanoparticle aggregate is not necessarily tightly packed, the relation between the number of particles within a cluster and the cluster size is through a fractal dimension

$$N = R_c^{\mathrm{fd}}, \qquad (23.32)$$

where fd is the fractal dimension. For the limiting case of fully packed spherical aggregates, fd = 3; whereas for more loosely packed aggregates fd ≈ 2. The diffusional correlation time depends on the square of the cluster radius, thus

$$\tau_D \propto R_c^2,$$

and the relative volume fraction f of the clusters changes according to

$$f \propto \frac{R_c^3}{N}. \qquad (23.33)$$

Substituting Equations 23.31 through 23.33 in Equation 23.29 yields an expression for the dependence of the transversal relaxation with a cluster radius

$$\frac{1}{T_2^{\mathrm{CA}}} \propto R_c^{\mathrm{fd}-1}. \qquad (23.34)$$

In the limit fd = 3, this reduces again to a square dependence with a cluster diameter, while for more loosely packed clusters, e.g., fd ≈ 2, a linear dependence should be observed.

The above approach of manipulating the transversal relaxation rate upon clustering has been exploited in the design of nanoparticle contrast agents that sense a certain biological or chemical activity (Perez et al., 2003; Harris et al., 2006; Shapiro et al., 2006; Taktak et al., 2007; Hong et al., 2008). A model system to study transversal relaxation changes induced by clustering was presented by Taktak et al. (2007). In this study, 16 nm radius amino-cross-linked dextran-coated iron-oxide (CLIO) nanoparticles were conjugated with up to 72 biotin molecules per nanoparticle. Up to 4 biotin molecules can bind with high affinity to the avidin protein. Upon addition to the aqueous CLIO solution, the avidin induced clustering of the nanoparticles. Concentration of the avidin was varied to yield varying cluster dimensions with a radius up to $R_c \approx 75$ nm. The experiment is depicted schematically in Figure 23.11A. The transversal and longitudinal relaxation rates as a function of the resulting cluster radius (at a constant iron concentration) are shown in

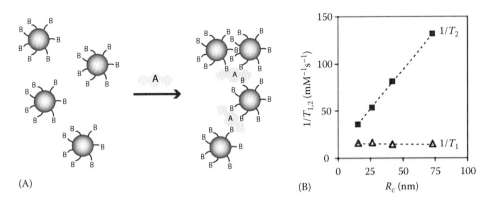

FIGURE 23.11 Changes in transverse relaxation rate upon clustering of iron-oxide nanoparticles. (A) Schematic drawing of the experiment. Iron-oxide nanoparticles conjugated with biotin (B = biotin) will cluster upon addition of avidin (A = avidin). (B) The transverse relaxation rate increases linearly with cluster radius R_c, while the longitudinal relaxation rate remains constant. (Reproduced from Taktak, S., *Anal. Chem.*, 79(23), 8863, 2007. With permission.)

Figure 23.11B. The transversal relaxation rate increased linearly with a cluster radius, indicative of a loosely packed aggregate. The longitudinal relaxation rate remained constant with size, which suggests that water access to the nanoparticle was not limited by the clustering (see also next section).

23.4 Paramagnetic Agents

23.4.1 Paramagnetic Relaxation

The physics of relaxation induced by paramagnetic elements in solutions will be explained using the example of the Gd^{3+}-based contrast agents, which are the most commonly used contrast agents for MRI. Gadolinium is a rare-earth metal with an atomic mass of 157.25 u, an atomic radius of 0.254 nm, and a neutral electron configuration $[Xe]4f^7 5d^1 6s^2$. Gadolinium forms Gd^{3+} ions with the electron configuration $[Xe]4f^7$. The 4f electron shell is half filled with parallel spin orientation according to Hund's Rules, resulting in the spin quantum number $S = 7/2$, which provides a large paramagnetic moment. The presence of such large fluctuating paramagnetic moments in a solution has strong effects on the nuclear spin relaxation of the solvent (water) nuclei.

A Gd^{3+} ion coordinated to a protective chelate (e.g., DTPA, DOTA) is schematically drawn in Figure 23.12A. The chelate leaves space for one water molecule to directly coordinate with Gd^{3+} in the so-called inner sphere (coordination number $q = 1$). Three contributions to the nuclear spin relaxation of the surrounding water hydrogen nuclei can be distinguished. The first one is *inner sphere relaxation* (is), which results from the contact and dipolar interactions of water hydrogen nuclei with the Gd^{3+} ion in the primary coordination sphere. The second contribution is the *outer sphere relaxation* (os), which describes the relaxation of the hydrogen nuclear spins diffusing in the inhomogeneous fluctuating field created by the Gd^{3+} magnetic moment. This contribution resembles the outer sphere relaxation discussed for the iron-oxide nanoparticles, although for Gd^{3+} different relevant length scales and correlation times have to be considered. A third contribution arises from water linked with hydrogen-bonds to the chelate in the second coordination sphere. The interaction

strengths of this *second sphere relaxation* (2s) contribution are poorly understood and are therefore not considered separately from the outer sphere relaxation. The total relaxivity of a paramagnetic contrast agent is therefore given by

$$r_i = r_i^{is} + r_i^{os}, \quad \text{with } i \in \{1, 2\}. \tag{23.35}$$

The superscripts is and os refer to inner sphere and outer sphere relaxation, respectively. The next sections will discuss a quantitative approach to understanding the inner sphere and outer sphere contributions to the longitudinal and transversal relaxation in the presence of a paramagnetic contrast agent.

23.4.2 Inner Sphere Relaxation

Inner sphere relaxation originates from the exchange of water molecules between the primary coordination site and the bulk solvent, characterized by the exchange correlation time τ_m. The longitudinal inner sphere relaxivity is given by

$$r_1^{is} = \frac{q}{[H_2O]} \frac{1}{T_{1m} + \tau_m}, \tag{23.36}$$

whereas the transversal relaxivity is

$$r_2^{is} = \frac{q}{[H_2O]\tau_m} \left[\frac{T_{2m}^{-1}(\tau_m^{-1} + T_{2m}^{-1}) + \Delta\omega_m^2}{(\tau_m^{-1} + T_{2m}^{-1})^2 + \Delta\omega_m^2} \right] \tag{23.37}$$

where

q is the number of simultaneously coordinated water molecules ($q = 1$ for $Gd(DTPA)^{2-}$ and $Gd(DOTA)^-$)

$[H_2O]$ is the concentration of water in the solvent (assumed 55.6 M for pure water)

$\Delta\omega_m$ is the chemical shift induced by the paramagnetic ion

T_{1m} and T_{2m} are the longitudinal and transversal relaxation times of a water proton coordinated to the Gd^{3+} paramagnetic center in the absence of exchange. Equation 23.36 shows that if water

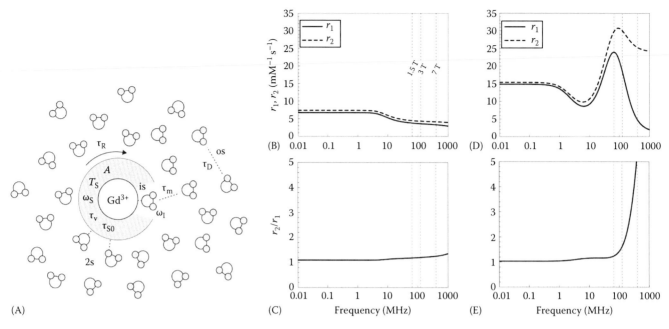

FIGURE 23.12 Longitudinal and transversal relaxation of Gd^{3+}-based contrast agent. (A) Schematic drawing of Gd^{3+} encapsulated in a chelate with one coordination site for direct water access. Relevant parameters are indicated in the picture. Relaxation is caused by inner sphere (is) relaxation and exchange of water in the first coordination sphere, as well as outer sphere (os) relaxation of water diffusing in the time fluctuating magnetic dipole field of the Gd^{3+} ion. Second sphere (2s) relaxation of water hydrogen-bonded to the chelate is generally not separately considered from outer sphere relaxation. (B) Calculated frequency dependence of r_1 and r_2 for a low molecular weight Gd^{3+}-chelate. The following parameters were used for this calculation (Caravan et al., 1999): $q = 1$, $r = 0.313$ nm, $d = 0.35$ nm, $D = 2.2 \times 10^{-3}$ mm^2/s, $\tau_m = 77$ ns, $\tau_R = 58$ ps, $\tau_v = 25$ ps, $\tau_{S0} = 72$ ps, and $A = 0$. (C) Ratio r_2/r_1 as function of frequency for low-molecular weight Gd^{3+}-chelate. (D) Calculated frequency dependence of r_1 and r_2 for a Gd^{3+}-based nanoparticle. For this calculation $\tau_R = 1.32$ ns, corresponding to the longer rotational correlation time found in protein-bound Gd^{3+}-chelates and Gd^{3+}-based nanoparticle. (E) Corresponding ratio r_2/r_1 as function of frequency.

exchange is sufficiently fast, T_{1m} determines the longitudinal relaxivity. The value of T_{1m} and T_{2m} are given by the so-called Solomon–Bloembergen–Morgan equations (Solomon, 1955; Bloembergen, 1957; Bloembergen and Morgan, 1961) and are composed of a dipole–dipole (dd) interaction term and a scalar (sc) interaction term

$$\frac{1}{T_{im}} = \frac{1}{T_i^{dd}} + \frac{1}{T_i^{sc}}, \quad \text{with } i \in \{1, 2\}. \quad (23.38)$$

The dipolar term arises from dipole–dipole interactions through space, while the scalar interaction arises from contact interactions through the chemical bond. The dipolar longitudinal and transversal relaxation are given by

$$\frac{1}{T_1^{dd}} = \frac{2}{15}\left(\frac{\mu_0}{4\pi}\right)^2 \gamma_I^2 \gamma_S^2 \hbar^2 S(S+1) r^{-6} \left[3J(\omega_I, \tau_{c1}) + 7J(\omega_S, \tau_{c2})\right]$$
$$(23.39)$$

and

$$\frac{1}{T_2^{dd}} = \frac{1}{15}\left(\frac{\mu_0}{4\pi}\right)^2 \gamma_I^2 \gamma_S^2 \hbar^2 S(S+1) r^{-6}$$
$$\times \left[4J(0, \tau_{c1}) + 3J(\omega_I, \tau_{c1}) + 13J(\omega_S, \tau_{c2})\right] \quad (23.40)$$

with γ_I and γ_S the proton and electron gyromagnetic ratios, ω_I and ω_S the proton and electron Larmor frequency, S the electron spin quantum number, and r the proton-paramagnetic ion distance. The scalar contributions are

$$\frac{1}{T_1^{sc}} = \frac{2}{3}S(S+1)\left(\frac{A}{\hbar}\right)^2 J(\omega_S, \tau_{e2}) \quad (23.41)$$

and

$$\frac{1}{T_2^{sc}} = \frac{1}{3}S(S+1)\left(\frac{A}{\hbar}\right)^2 \left[J(0, \tau_{e1}) + J(\omega_S, \tau_{e2})\right]. \quad (23.42)$$

The strength of the scalar interaction is governed by the electron-nuclear hyperfine coupling constant A. Dipolar and scalar relaxation are modulated by the rotational correlation time of the Gd^{3+}-chelate τ_R, the water exchange correlation time τ_m, as well as the longitudinal and transversal electronic relaxation times of the Gd^{3+} ion, T_{S1} and T_{S2}. The overall resulting correlation times relevant to the dipolar and scalar relaxation terms are

$$\frac{1}{\tau_{ci}} = \frac{1}{T_{Si}} + \frac{1}{\tau_m} + \frac{1}{\tau_R}, \quad \text{with } i \in \{1, 2\} \quad (23.43)$$

and

$$\frac{1}{\tau_{ei}} = \frac{1}{T_{Si}} + \frac{1}{\tau_m}, \quad \text{with } i \in \{1, 2\}. \qquad (23.44)$$

The electronic longitudinal and transversal relaxation times of the Gd^{3+} ion are also frequency dependent with

$$\frac{1}{T_{S1}} = \frac{1}{5\tau_{s0}\tau_v}\left[J(\omega_S, \tau_v) + 4J(2\omega_S, \tau_v)\right] \qquad (23.45)$$

and

$$\frac{1}{T_{S2}} = \frac{1}{10\tau_{s0}\tau_v}\left[3J(0, \tau_v) + 5J(\omega_S, \tau_v) + 2J(2\omega_S, \tau_v)\right]. \qquad (23.46)$$

Fluctuations of the electronic moment due to collisions and molecular vibrations of the Gd^{3+}-chelate complex govern the electronic relaxation, characterized by the correlation time τ_v and the electronic relaxation time at zero field τ_{s0}. Note the similarity of Equations 23.45 and 23.46 with Equations 23.8 and 23.10.

23.4.3 Outer Sphere Relaxation

Outer sphere relaxation arises from the complex dynamics of the solvent molecules and the Gd^{3+}-chelate diffusing and rotating in each other's surroundings, as well as the electronic fluctuations of the Gd^{3+} ions (Freed, 1978; Kruk et al., 2001a,b; Helm, 2006). Outer sphere relaxivities are given by

$$r_1^{os} = \frac{32\pi}{405}\left(\frac{\mu_0}{4\pi}\right)^2 \gamma_I^2\gamma_S^2\hbar^2 S(S+1)N_A d^{-3}\tau_D$$
$$\times\left[3H(\omega_I) + 7H(\omega_S)\right] \qquad (23.47)$$

and

$$r_2^{os} = \frac{16\pi}{405}\left(\frac{\mu_0}{4\pi}\right)^2 \gamma_I^2\gamma_S^2\hbar^2 S(S+1)N_A d^{-3}\tau_D\left[4 + 3H(\omega_I) + 13H(\omega_S)\right], \qquad (23.48)$$

with

$$H(\omega) = Re\left[\frac{1 + \frac{1}{4}(i\omega\tau_D + \tau_D/T_{S1})^{\frac{1}{2}}}{1 + (i\omega\tau_D + \tau_D/T_{S1})^{\frac{1}{2}} + \frac{4}{9}(i\omega\tau_D + \tau_D/T_{S1}) + \frac{1}{9}(i\omega\tau_D + \tau_D/T_{S1})^{\frac{3}{2}}}\right] \qquad (23.49)$$

where

N_A is Avogadro's number
d is the distance of the closest approach of solvent water molecules to the Gd^{3+} ion

The numerical prefactor in Equations 23.47 and 23.48 differ somewhat between different available models. A rigorous derivation of these equations would be beyond the scope of this chapter. Nevertheless, some basic physical features can be recognized. The relaxation rate depends on the square of the magnitude of the nuclear and electronic moments. This can also be recognized in Equation 23.29 for the relaxation rate induced by iron-oxide nanoparticles in the fast diffusion regime, which is dependent on $\Delta\omega^2$ related to the square of the value of the iron-oxide magnetic moment. The relaxivity falls off with distance to the power of 3, which means that the relaxation is only effective at a few molecular distances, in contrast to the iron-oxide nanoparticles, which due to their larger size and magnetic moment are able to induce relaxation at diffusional length scales.

23.4.4 Combined Equations

The measurement of the relaxivity as function of frequency (magnetic field), named nuclear magnetic relaxation dispersion (NMRD), is a powerful experimental technique to characterize contrast agents and to learn which parameters determine or limit the value of the relaxivity. Due to experimental constraints, commonly only the frequency dependence of r_1 is measured. The combined inner and outer sphere relaxivity equations have a large number of unknown parameters (correlation times: τ_m, τ_D, τ_R, τ_v, τ_{s0}; distances: r, d; hyperfine coupling constant A), which makes a direct fit of the NMRD curve with so many free parameters a difficult task. Therefore, usually a number of parameters are determined using alternative methods. For example, r and d may be estimated from molecular calculations and the diffusion correlation time may be estimated from the diffusion constant. The exchange correlation time τ_m can be separately determined through the temperature dependence of the transverse ^{17}O NMR relaxation time (Aime et al., 1998). The hyperfine coupling constant is small and in general only contributes significantly at frequencies below approximately 10 MHz, much lower than clinical field strengths.

Figure 23.12B and C show example calculations of the frequency dependence (NMRD) of the longitudinal and transversal relaxivity using the combined equations for inner sphere and outer sphere relaxivity. Figure 23.12B represents a typical calculation for clinically approved low molecular weight contrast agents, such as $Gd(DTPA)^{2-}$. The longitudinal and transversal relaxivity as a function of frequency are rather featureless curves with a low frequency plateau, decreasing at frequencies above approximately 10 MHz. Note that the frequency axis of the NMRD curve has a logarithmic scale. The clinically relevant field strengths 1.5, 3, and 7 T are indicated with dashed vertical lines. The ratio r_2/r_1, plotted in Figure 23.12C, is about 1.1 increasing somewhat at higher frequencies. The low r_2/r_1 ratio makes these contrast agents suitable for generating a positive contrast in T_1-weighted images, also at higher field strengths, although the absolute value of r_1 is somewhat lower and thus less favorable at higher fields.

For the calculation in Figure 23.12D, the rotational correlation time has been increased by about a factor of 23. This represents the calculation for a macromolecular or nanoparticle contrast agent, e.g., a liposomal agent, in which the Gd^{3+}-containing lipids in the membrane are slowed down in their movements. As a result, there is a dramatic effect on the shape of NMRD curves. The values of longitudinal and transversal relaxivity at low frequencies are increased by more than a factor of 2. At higher frequencies, the longitudinal relaxivity has a characteristic maximum at about 1.5 T after which it decreases sharply at higher field strengths. The transversal relaxivity has a maximum between 1.5 and 3 T, after which it only moderately decreases at higher fields. As a consequence, the r_2/r_1 ratio, plotted in Figure 23.12E, starts to increase at field strengths above 1.5 T, making the agent less and less suitable for imaging with T_1-weighted sequences. The calculated nanoparticle relaxivity is defined per mM of Gd^{3+}, which means that the overall relaxivity of a nanoparticle contrast agent containing thousands of Gd^{3+}-chelates is accordingly higher.

The above calculations represent a simple example of how the theory of the longitudinal and transversal relaxivity of the Gd^{3+}-based agents can be used to understand and possibly optimize the relaxivity of nanoparticle agents. Parameters that could be optimized are the rotational correlation time (via the mass, size, or composition of the particle), the exchange and electronic correlation times, as well as the coordination number (via the choice of chelate) (Yang et al., 2008).

23.4.5 Exchange and Compartmentalization

As mentioned before, Equation 23.36 implies that for optimal longitudinal relaxivity the exchange correlation time τ_m should be negligibly small compared to T_{1m}, because the exchange of coordinated water with the bulk has to be fast enough not to limit the propagation of the relaxation effect. This is generally the case for the low molecular weight contrast agents. However, for paramagnetic liposomes with Gd^{3+}-containing lipids in the liposomal membrane, it has been observed that water access to the internal leaflet of the membrane bilayer posed a limiting factor to the relaxivity (Strijkers et al., 2005).

The effect of water exchange also plays a role on a larger spatial scale. The theory of the paramagnetic relaxation outlined in Sections 23.4.1 through 23.4.4 implicitly assumed that the relaxation effect induced by the paramagnetic center of either the first, second, or outer coordination sphere origin has an effect on all bulk water molecules. In other words, the Gd^{3+}-chelates are assumed to be distributed homogeneously in the solvent and mixing is fast enough to spatially affect the complete voxel in the time frame of the NMR experiment. This is not an issue when studying the contrast agents in aqueous solution in test tubes or for contrast agents distributed in the vasculature. However, the application of contrast agents in tissues or in a cellular environment involves spatially separated water-containing compartments, such as the vasculature, the extracellular environment, the cell cytoplasm, and subcellular compartments. These are physically separated by cell membranes that could act as barriers preventing the efficient mixing of water molecules. The observed relaxation times are therefore a voxel-averaged contribution from the different compartments that may include the effects of equilibrium inter-compartmental water exchange.

Recently, a number of studies have reported on the reduced relaxivity of cell-internalized contrast agents (Kobayashi et al., 2001; Billotey et al., 2003; Geninatti Crich et al., 2006; Simon et al., 2006; Terreno et al., 2006; Brekke et al., 2007a,b; Kok et al., 2009). All the studies noted that the observed longitudinal relaxation time induced by the cell-internalized agent was considerably higher than could be anticipated on the basis of the contrast agent concentration in relation to the contrast agent relaxivity determined from separate measurements in an aqueous solution. The effect was coined *relaxivity quenching* and attributed to limited water exchange consequent to contrast agent entrapment in cytoplasmic or subcellular compartments.

To illustrate the effects of compartmentalization on the T_1-relaxation, a model voxel is considered that contains two compartments denoted by A and B, as drawn schematically in Figure 23.13A. Compartment B, with volume V_B, is assumed to be small as compared with compartment A, with volume V_A. The contrast agent will be distributed in compartment B, such that

$$\frac{1}{T_{1B}} = \frac{1}{T_{1A}} + r_1[CA], \tag{23.50}$$

where T_{1A} and T_{1B} are the longitudinal relaxation times of compartments A and B, respectively, in the absence of exchange. If water does exchange between the two compartments, the overall longitudinal relaxation rate of the voxel will not necessarily be a simple volume-weighted average, but an average that includes the effects of inter-compartmental water exchange, according to (Bloembergen and Morgan, 1961; Luz and Meiboom, 1964; Andrasko and Forsen, 1974; Huster et al., 1997)

$$\frac{1}{T_1} = \frac{1}{T_{1A}} + \frac{V_B}{V_A} \frac{1}{T_{1B} + \tau_{BL}}. \tag{23.51}$$

where, τ_{BL} is the mean residence lifetime (correlation time) of water in compartment B. Figure 23.13B illustrates the effect on the longitudinal relaxation rate as a function of the contrast agent voxel concentration $[CA]_V$. When the exchange is fast ($\tau_{BL} = 0$), the two compartments are well mixed and the longitudinal relaxation rate is linear with the concentration and the slope represents the r_1 of the agent. Upon increasing τ_{BL}, however, one observes the saturation of the longitudinal relaxation with concentration, and the relaxivity, i.e., the slope of the curve, decreases (quenches).

The theory of relaxivity quenching has recently been generalized to include three compartments, e.g., extracellular, cytoplasmic, and vesicular (endosomal, lysosomal) compartments, of arbitrary relative volume fractions (Strijkers et al., 2009). Although the relaxivity quenching in many cases is an undesired

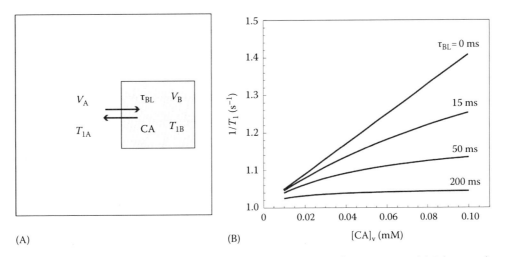

FIGURE 23.13 Quenching of the longitudinal relaxation due to compartmentalization of contrast agent. (A) Schematic drawing of two compartments with volume V_A and V_B. Water exchanges between the two compartments and the mean lifetime of water in compartment B is τ_{BL}. T_{1A} and T_{1B} are the longitudinal relaxation times of water protons in the two compartments. Contrast agent (CA) is assumed to be located in compartment B only. (B) Calculation of the longitudinal relaxation rate as function of contrast agent voxel concentration $[CA]_V$, for varying values of τ_{BL}. Other input parameters were: $V_B/V_A = 0.01$, $T_{1A} = 1$ s, and $r_1 = 4$ mM^{-1} s^{-1}.

effect, it also opens opportunities for studying the compartmentalization with MRI *in vivo* and for designing responsive agents that are released from a quenched compartment upon biological activation.

23.5 Perspectives and Summary

This chapter has focused primarily on the relaxivity of iron-oxide and Gd^{3+}-based MR contrast agents, as these are the most widespread and commonly used agents. There exists a wealth of literature on the design and applications of complexes and nanoparticles endowed with Gd^{3+} ions as contrast agents for *in vivo* MRI. Furthermore, a recent paper by Laurent et al. provides an excellent and complete overview of the new developments in the design, characterization, and application of iron-oxide nanoparticles (Laurent et al., 2008).

Nevertheless, there have been several other types of MRI contrast agents under investigation in recent years that have attracted considerable attention because of favorable relaxation or bioapplicable properties. Dysprosium-based complexes display high transversal relaxation rates due to slow water exchange, making them a suitable alternative to Gd^{3+}-based complexes for application at high field strengths (Vander Elst et al., 2002). Europium (Eu^{2+}) is a lanthanide very similar to Gd^{3+}. It also has seven unpaired electrons with reasonable relaxivity properties. Eu^{2+}-based complexes have been explored as redox responsive agents (Burai et al., 2000). Manganese (Mn^{2+}) is a potent paramagnetic relaxation agent with five unpaired electrons (Koretsky and Silva, 2004). Mn(DPDP) (dipyridoxyl-diphosphate) is a clinically approved contrast agent indicated for intravenous administration in the detection, localization, characterization, and evaluation of liver lesions. In ionic form, Mn^{2+} is used as a Ca^{2+} analog with applications in functional studies of brain connectivity (Saleem et al., 2002) and as viability markers in the heart

(Waghorn et al., 2008). Alternatively, nanoparticles of MnO and Gd$_2$O$_3$ have been prepared, solubilized, and investigated for their relaxation properties (McDonald and Watkin, 2003; Na et al., 2007). Aluminosilicates (zeolites) with a porous structure have been explored as carriers for Gd^{3+} ions and displayed interesting relaxation properties (Csajbok et al., 2005). Promising new high-relaxivity types of contrast agents are the Gd^{3+}-containing fullerenes (gadofullerenes) and carbon nanotubes (Bolskar, 2008). Their relaxivity is thus far not well understood.

A completely different class of contrast agents is based on a magnetization transfer mechanism. Such agents were named chemical exchange saturation transfer (CEST) agents (Ward et al., 2000; Aime et al., 2002; Snoussi et al., 2003; Zhang et al., 2003a,b; Aime et al., 2005a; Sherry and Woods, 2008). The action of a CEST agent is based on a system that contains one or more pools of exchangeable protons with a sharply defined resonance frequency and a chemical shift, well separated from the bulk water peak. Upon selective saturation of the exchangeable protons, this saturated magnetization will be transferred to the bulk water, leading to a drop in the water hydrogen nuclear spin signal intensity. CEST agents are therefore negative contrast agents, similar to the iron-oxides. Many small diamagnetic organic compounds, such as sugars and amino acids, have pools of exchangeable protons and may therefore serve as CEST agents. Certain paramagnetic lanthanide (notably Eu, Dy, Ho, Er, Tm, and Yb) chelates, so-called PARACEST agents, are particularly suited as they possess a large chemical shift allowing for the exploitation of faster exchange rates resulting in a larger CEST effect. Aime and coworkers have developed a liposome with a shift reagent entrapped within the liposomal cavity (Aime et al., 2005b; Terreno et al., 2007, 2008a). Because of the shift reagent, water protons in the liposomal compartment have a resonance frequency different from the bulk water protons. Selective saturation of these protons leads to a large CEST effect. The shift of the resonance frequency can be tuned, which allows for the application of several CEST agents simultaneously (Aime et al., 2005a; Terreno et al., 2008b).

In summary, this chapter provided a basic understanding of the T_1 and T_2 relaxation induced by paramagnetic Gd^{3+}-based and superparamagnetic iron-oxide nanoparticle contrast agents. The relaxation theory of the classical paramagnetic and superparamagnetic contrast agents is well developed. Nevertheless, challenges remain as contrast agents become more and more complex, both in composition as well as in structure. A good example is the gadofullerenes with high relaxivity values, which are still rather poorly understood. Furthermore, with the *in vivo* application of sophisticated contrast agents, the complex interplay between the contrast agent and the local physiological environment has to be accounted for. Examples were given of how the clustering of nanoparticles and compartmentalization could lead to changed relaxivity.

Acknowledgments

Luce Vander Elst, Robert Müller, and Enzo Terreno are acknowledged for valuable discussions and Bram Coolen, Sjoerd Hak, and Erik Sanders are acknowledged for carefully editing the manuscript. The authors' work in the field of MRI contrast agents is funded in part by the European Commission FP6-projects DiMI (project number LSHB-CT-2005-512146) and MediTrans (project number NMP4-CT-2006-026668), as well as by the BSIK program entitled Molecular Imaging of Ischemic Heart Disease (project number BSIK03033). Parts of these studies are performed in the framework of the European Cooperation in the field of Scientific and Technical Research (COST) D38 Action Metal-Based Systems for Molecular Imaging Applications.

Glossary

Amino acid	An organic compound containing an amino group (NH_2), a carboxylic acid group (COOH), and various side groups. Amino acids link together to form onproteins or function as metabolic intermediates.
Angiogenesis	A physiological process involving the growth of new blood vessels from preexisting vessels.
Antibody	A protein used by the body's immune system to recognize harmful substances.
$\alpha_v\beta_3$-Integrin	A receptor expressed on the cell surface of angiogenic blood vessels.
Avidin	A protein that can bind up to four molecules of biotin simultaneously with a high degree of affinity and specificity.
Biotin	Vitamin H or B7, with the chemical formula $C_{10}H_{16}N_2O_3S$. Biotin binds with high affinity to the protein avidin.
Cell membrane	A lipid bilayer separating the inside and the outside of cells.
Cerebrum	Part of the vertebrate brain.
Cerebrospinal fluid	Bodily fluid that occupies cavities within and around the brain.
Cytoplasm	Part of the cell enclosed by the cell membrane.
Dextran	Long-chain polymer of glucose.
DNA	Deoxyribonucleic acid. A long polymer containing the genetic information of living organisms and some viruses.
Endosome	Intracellular membrane-enclosed compartment.
Endothelium	Layer of cells that line the interior surface of blood vessels.
Extracellular	Outside the cell.
Gray matter	Brain tissue containing neural cell bodies.
Interleukin-1β	Protein excreted by macrophages and involved in inflammatory processes.
Intracellular	Inside the cell.
Lysosome	Intracellular membrane-enclosed compartment containing digestive enzymes.
Macromolecule	A large molecule in the context of biochemistry.
Pathology	Related to disease processes.
Peptide	Short polymer of amino acids.
Perfusion	Process of blood delivery via a capillary bed to biological tissue.
Protein	Organic compound made of amino acids.
P-selectin	Protein found on the endothelium involved in the recruitment of white blood cells.
Striatum	Part of the vertebrate brain.
Subcutaneous tumor	Cancerous tissue present under the skin. To study tumor development in a controlled condition, cancerous cells are often implanted under the skin in mice.
Vasculature	Arrangement of blood vessels in the body or body part.
VCAM-1	Protein found on the endothelium involved in the recruitment of cells of the immune system.

References

Ahrens, E. T., R. Flores, H. Xu, and P. A. Morel. 2005. In vivo imaging platform for tracking immunotherapeutic cells. *Nat Biotechnol* 23(8):983–987.

Aime, S., A. Barge, D. Delli Castelli, F. Fedeli, A. Mortillaro, F. U. Nielsen, and E. Terreno. 2002. Paramagnetic lanthanide(III) complexes as pH-sensitive chemical exchange saturation transfer (CEST) contrast agents for MRI applications. *Magn Reson Med* 47(4):639–648.

Aime, S., M. Botta, M. Fasano, and E. Terreno. 1998. Lanthanide(III) chelates for NMR biomedical applications. *Chem Soc Rev* 27:19.

Aime, S., L. Calabi, C. Cavallotti, E. Gianolio, G. B. Giovenzana, P. Losi, A. Maiocchi, G. Palmisano, and M. Sisti. 2004. [Gd-AAZTA]: A new structural entry for an improved generation of MRI contrast agents. *Inorg Chem* 43(24):7588–7590.

Aime, S., C. Carrera, D. Delli Castelli, S. Geninatti Crich, and E. Terreno. 2005a. Tunable imaging of cells labeled with MRI-PARACEST agents. *Angew Chem Int Ed Engl* 44(12):1813–1815.

Aime, S., D. Delli Castelli, and E. Terreno. 2005b. Highly sensitive MRI chemical exchange saturation transfer agents using liposomes. *Angew Chem Int Ed Engl* 44(34):5513–5515.

Andrasko, J. and S. Forsen. 1974. NMR-study of rapid water diffusion across lipid bilayers in dipalmitoyl lecithin vesicles. *Biochem Biophys Res Commun* 60(2):813–819.

Billotey, C., C. Wilhelm, M. Devaud, J. C. Bacri, J. Bittoun, and F. Gazeau. 2003. Cell internalization of anionic maghemite nanoparticles: Quantitative effect on magnetic resonance imaging. *Magn Reson Med* 49(4):646–654.

Bloembergen, N. 1957. Proton relaxation times in paramagnetic solutions. *J Chem Phys* 27(2):572–573.

Bloembergen, N. and L. O. Morgan. 1961. Proton relaxation times in paramagnetic solutions. Effects of electron spin relaxation. *J Chem Phys* 34(3):842.

Bolskar, R. D. 2008. Gadofullerene MRI contrast agents. *Nanomedicine* 3(2):201–213.

Brekke, C., S. C. Morgan, A. S. Lowe, T. J. Meade, J. Price, S. C. Williams, and M. Modo. 2007a. The in vitro effects of a bimodal contrast agent on cellular functions and relaxometry. *NMR Biomed* 20(2):77–89.

Brekke, C., S. C. Williams, J. Price, F. Thorsen, and M. Modo. 2007b. Cellular multiparametric MRI of neural stem cell therapy in a rat glioma model. *Neuroimage* 37(3):769–782.

Brown, R. 1961. Distribution of fields from randomly placed dipoles: Free-precession signal decay as result of magnetic grains. *Phys Rev* 121:1379–1382.

Brown, W. F. 1963. Thermal fluctuations of a single domain particle. *J Appl Phys* 34(4):1319.

Burai, L., E. Toth, S. Seibig, R. Scopelliti, and A. E. Merbach. 2000. Solution and solid-state characterization of Eu(II) chelates: A possible route towards redox responsive MRI contrast agents. *Chemistry* 6(20):3761–3770.

Caravan, P., J. J. Ellison, T. J. McMurry, and R. B. Lauffer. 1999. Gadolinium(III) chelates as MRI contrast agents: Structure, dynamics, and applications. *Chem Rev* 99(9):2293–2352.

Corot, C., P. Robert, J. M. Idee, and M. Port. 2006. Recent advances in iron oxide nanocrystal technology for medical imaging. *Adv Drug Deliv Rev* 58(14):1471–1504.

Cowan, B. 1997. *Nuclear Magnetic Resonance and Relaxation.* Cambridge, U.K.: Cambridge University Press.

Csajbok, E., I. Banyai, L. Vander Elst, R. N. Muller, W. Zhou, and J. A. Peters. 2005. Gadolinium(III)-loaded nanoparticulate zeolites as potential high-field MRI contrast agents: Relationship between structure and relaxivity. *Chemistry* 11(16):4799–4807.

Fannin, P. C. and S. W. Charles. 1989. The study of a ferrofluid exhibiting both Brownian and Neel relaxation. *J Phys D Appl Phys* 22(1):187–191.

Freed, J.H. 1978. Dynamic effects of pair correlation-functions on spin relaxation by translational diffusion in liquids. II. Finite jumps and independent T1 processes. *J Chem Phys* 68(9):4034–4037.

Geninatti Crich, S., C. Cabella, A. Barge, S. Belfiore, C. Ghirelli, L. Lattuada, S. Lanzardo, A. et al. 2006. In vitro and in vivo magnetic resonance detection of tumor cells by targeting glutamine transporters with Gd-based probes. *J Med Chem* 49(16):4926–4936.

Gillis, P. and S. H. Koenig. 1987. Transverse relaxation of solvent protons induced by magnetized spheres: Application to ferritin, erythrocytes, and magnetite. *Magn Reson Med* 5(4):323–345.

Haacke, E.M., R.W. Brown, M.R. Thompson, and R. Venkatesan. 1999. *Magnetic Resonance Imaging, Physical Principles and Sequence Design.* New York: John Wiley & Sons.

Harris, T. J., G. von Maltzahn, A. M. Derfus, E. Ruoslahti, and S. N. Bhatia. 2006. Proteolytic actuation of nanoparticle self-assembly. *Angew Chem Int Ed Engl* 45(19):3161–3165.

Helm, L. 2006. Relaxivity in paramagnetic systems: Theory and mechanisms. *Prog Nucl Magn Reson Spectrosc* 49(1):45–64.

Hinds, K. A., J. M. Hill, E. M. Shapiro, M. O. Laukkanen, A. C. Silva, C. A. Combs, T. R. Varney, R. S. Balaban, A. P. Koretsky, and C. E. Dunbar. 2003. Highly efficient endosomal labeling of progenitor and stem cells with large magnetic particles allows magnetic resonance imaging of single cells. *Blood* 102(3):867–872.

Hong, R., M. J. Cima, R. Weissleder, and L. Josephson. 2008. Magnetic microparticle aggregation for viscosity determination by MR. *Magn Reson Med* 59(3):515–520.

Huster, D., A. J. Jin, K. Arnold, and K. Gawrisch. 1997. Water permeability of polyunsaturated lipid membranes measured by O-17 NMR. *Biophys J* 73(2):855–864.

Kobayashi, H., S. Kawamoto, T. Saga, N. Sato, T. Ishimori, J. Konishi, K. Ono, K. Togashi, and M. W. Brechbiel. 2001. Avidin-dendrimer-(1B4M-Gd)(254): A tumor-targeting therapeutic agent for gadolinium neutron capture therapy of intraperitoneal disseminated tumor which can be monitored by MRI. *Bioconjugate Chem* 12(4):587–593.

Kok, M. B., S. Hak, W. J. Mulder, D. van der Schaft, G. J. Strijkers, and K. Nicolay. 2009. Cellular compartmentalization of internalized paramagnetic liposomes strongly influences both T1 and T2 relaxivity. *Magn Reson Med* 61(5):1022–1032.

Koretsky, A. P. and A. C. Silva. 2004. Manganese-enhanced magnetic resonance imaging (MEMRI). *NMR Biomed* 17(8):527–531.

Kruk, D., T. Nilsson, and J. Kowalewski. 2001a. Nuclear spin relaxation in paramagnetic systems with zero-field splitting and arbitrary electron spin. *Phys Chem Chem Phys* 3(22):4907–4917.

Kruk, D., T. Nilsson, and J. Kowalewski. 2001b. Outer-sphere nuclear spin relaxation in paramagnetic systems: A low-field theory. *Mol Phys* 99(17):1435–1445.

Laurent, S., D. Forge, M. Port, A. Roch, C. Robic, L. Vander Elst, and R. N. Muller. 2008. Magnetic iron oxide nanoparticles: Synthesis, stabilization, vectorization, physicochemical characterizations, and biological applications. *Chem Rev* 108(6):2064–2110.

Livingston, J. and C. Bean. 1959. Anisotropy of superparamagnetic particles as measured by torque and resonance. *J Appl Phys* 30(4):318s.

Luz, Z. and S. Meiboom. 1964. Proton relaxation in dilute solutions of cobalt(2) + nickel(2) ions in methanol + rate of methanol exchange of solvation sphere. *J Chem Phys* 40(9):2686–2692.

McAteer, M. A., N. R. Sibson, C. von Zur Muhlen, J. E. Schneider, A. S. Lowe, N. Warrick, K. M. Channon, D. C. Anthony, and R. P. Choudhury. 2007. In vivo magnetic resonance imaging of acute brain inflammation using microparticles of iron oxide. *Nat Med* 13(10):1253–1258.

McDonald, M. A. and K. L. Watkin. 2003. Small particulate gadolinium oxide and gadolinium oxide albumin microspheres as multimodal contrast and therapeutic agents. *Invest Radiol* 38(6):305–310.

Mulder, W. J., A. W. Griffioen, G. J. Strijkers, D. P. Cormode, K. Nicolay, and Z. A. Fayad. 2007a. Magnetic and fluorescent nanoparticles for multimodality imaging. *Nanomedicine* 2(3):307–324.

Mulder, W. J., G. J. Strijkers, G. A. van Tilborg, A. W. Griffioen, and K. Nicolay. 2006. Lipid-based nanoparticles for contrast-enhanced MRI and molecular imaging. *NMR Biomed* 19(1):142–164.

Mulder, W. J., D. W. van der Schaft, P. A. Hautvast, G. J. Strijkers, G. A. Koning, G. Storm, K. H. Mayo, A. W. Griffioen, and K. Nicolay. 2007b. Early in vivo assessment of angiostatic therapy efficacy by molecular MRI. *FASEB J* 21(2):378–383.

Na, H. B., J. H. Lee, K. An, Y. I. Park, M. Park, I. S. Lee, D. H. Nam et al. 2007. Development of a T1 contrast agent for magnetic resonance imaging using MnO nanoparticles. *Angew Chem Int Ed Engl* 46(28):5397–5401.

Néel, L. 1949. Theorie du trainage magnetigue des ferromagnetigues en grains fins avec applications aux terres cuites, *Ann Geophys* 5:99–136.

Perez, J. M., F. J. Simeone, Y. Saeki, L. Josephson, and R. Weissleder. 2003. Viral-induced self-assembly of magnetic nanoparticles allows the detection of viral particles in biological media. *J Am Chem Soc* 125(34):10192–10193.

Roch, A., R. N. Muller, and P. Gillis. 1999. Theory of proton relaxation induced by superparamagnetic particles. *J Chem Phys* 110(11):5403–5411.

Rosensweig, R. E. 2002. Heating magnetic fluid with alternating magnetic field. *J Magn Magn Mater* 252:370–374.

Saleem, K. S., J. M. Pauls, M. Augath, T. Trinath, B. A. Prause, T. Hashikawa, and N. K. Logothetis. 2002. Magnetic resonance imaging of neuronal connections in the macaque monkey. *Neuron* 34(5):685–700.

Schellenberger, E., J. Schnorr, C. Reutelingsperger, L. Ungethum, W. Meyer, M. Taupitz, and B. Hamm. 2008. Linking proteins with anionic nanoparticles via protamine: Ultrasmall protein-coupled probes for magnetic resonance imaging of apoptosis. *Small* 4(2):225–230.

Shapiro, M. G., T. Atanasijevic, H. Faas, G. G. Westmeyer, and A. Jasanoff. 2006. Dynamic imaging with MRI contrast agents: Quantitative considerations. *Magn Reson Imaging* 24(4):449–462.

Shapiro, E. M., S. Skrtic, and A. P. Koretsky. 2005. Sizing it up: Cellular MRI using micron-sized iron oxide particles. *Magn Reson Med* 53(2):329–338.

Sherry, A. D. and M. Woods. 2008. Chemical exchange saturation transfer contrast agents for magnetic resonance imaging. *Annu Rev Biomed Eng* 10:391–411.

Simon, G. H., J. Bauer, O. Saborovski, Y. Fu, C. Corot, M. F. Wendland, and H. E. Daldrup-Link. 2006. T1 and T2 relaxivity of intracellular and extracellular USPIO at 1.5T and 3T clinical MR scanning. *Eur Radiol* 16(3):738–745.

Snoussi, K., J. W. Bulte, M. Gueron, and P. C. van Zijl. 2003. Sensitive CEST agents based on nucleic acid imino proton exchange: Detection of poly(rU) and of a dendrimer-poly(rU) model for nucleic acid delivery and pharmacology. *Magn Reson Med* 49(6):998–1005.

Solomon, I. 1955. Relaxation processes in a system of 2 spins. *Phys Rev* 99(2):559–565.

Strijkers, G. J., S. Hak, M. B. Kok, C. S. Springer, and K. Nicolay. 2009. A three-compartment T1-relaxation model for intracellular paramagnetic contrast agents. *Magn Reson Med* 61(5):1049–1058.

Strijkers, G. J., W. J. Mulder, R. B. van Heeswijk, P. M. Frederik, P. Bomans, P. C. Magusin, and K. Nicolay. 2005. Relaxivity of liposomal paramagnetic MRI contrast agents. *Magn Reson Mater Phys* 18(4):186–192.

Strijkers, G. J., W. J. Mulder, G. A. van Tilborg, and K. Nicolay. 2007. MRI contrast agents: Current status and future perspectives. *Anticancer Agents Med Chem* 7(3):291–305.

Sun, C., J. S. Lee, and M. Zhang. 2008. Magnetic nanoparticles in MR imaging and drug delivery. *Adv Drug Deliv Rev* 60(11):1252–1265.

Taboada, E., E. Rodriguez, A. Roig, J. Oro, A. Roch, and R. N. Muller. 2007. Relaxometric and magnetic characterization of ultrasmall iron oxide nanoparticles with high magnetization. Evaluation as potential T1 magnetic resonance imaging contrast agents for molecular imaging. *Langmuir* 23(8):4583–4588.

Taktak, S., D. Sosnovik, M. J. Cima, R. Weissleder, and L. Josephson. 2007. Multiparameter magnetic relaxation switch assays. *Anal Chem* 79(23):8863–8869.

Terreno, E., A. Barge, L. Beltrami, G. Cravotto, D. D. Castelli, F. Fedeli, B. Jebasingh, and S. Aime. 2008a. Highly shifted LIPOCEST agents based on the encapsulation of neutral polynuclear paramagnetic shift reagents. *Chem Commun (Camb)* 5:600–602.

Terreno, E., C. Cabella, C. Carrera, D. Delli Castelli, R. Mazzon, S. Rollet, J. Stancanello, M. Visigalli, and S. Aime. 2007. From spherical to osmotically shrunken paramagnetic liposomes: An improved generation of LIPOCEST MRI agents with highly shifted water protons. *Angew Chem Int Ed Engl* 46(6):966–968.

Terreno, E., D. D. Castelli, L. Milone, S. Rollet, J. Stancanello, E. Violante, and S. Aime. 2008b. First ex-vivo MRI co-localization of two LIPOCEST agents. *Contrast Media Mol Imaging* 3(1):38–43.

Terreno, E., S. Geninatti Crich, S. Belfiore, L. Biancone, C. Cabella, G. Esposito, A. D. Manazza, and S. Aime. 2006. Effect of the intracellular localization of a Gd-based imaging probe on the relaxation enhancement of water protons. *Magn Reson Med* 55(3):491–497.

van Tilborg, G. A., W. J. Mulder, P. T. Chin, G. Storm, C. P. Reutelingsperger, K. Nicolay, and G. J. Strijkers. 2006. Annexin A5-conjugated quantum dots with a paramagnetic lipidic coating for the multimodal detection of apoptotic cells. *Bioconjugate Chem* 17(4):865–868.

Vander Elst, L., A. Roch, P. Gillis, S. Laurent, F. Botteman, J. W. Bulte, and R. N. Muller. 2002. Dy-DTPA derivatives as relaxation agents for very high field MRI: The beneficial effect of slow water exchange on the transverse relaxivities. *Magn Reson Med* 47(6):1121–1130.

Vlaardingerbroek, M.T. and J.A. Den Boer. 1999. *Magnetic Resonance Imaging*. Berlin, Heidelberg, Germany: Springer-Verlag.

Waghorn, B., T. Edwards, Y. Yang, K. H. Chuang, N. Yanasak, and T. C. Hu. 2008. Monitoring dynamic alterations in calcium homeostasis by T (1)-weighted and T (1)-mapping cardiac manganese-enhanced MRI in a murine myocardial infarction model. *NMR Biomed* 21(10):1102–1111.

Wang, Y. X., S. M. Hussain, and G. P. Krestin. 2001. Super-paramagnetic iron oxide contrast agents: Physicochemical characteristics and applications in MR imaging. *Eur Radiol* 11(11):2319–2331.

Ward, K. M., A. H. Aletras, and R. S. Balaban. 2000. A new class of contrast agents for MRI based on proton chemical exchange dependent saturation transfer (CEST). *J Magn Reson* 143(1):79–87.

Weissleder, R. and U. Mahmood. 2001. Molecular imaging. *Radiology* 219(2):316–333.

Weissleder, R., P. Reimer, A. S. Lee, J. Wittenberg, and T. J. Brady. 1990. MR receptor imaging: Ultrasmall iron oxide particles targeted to asialoglycoprotein receptors. *AJR Am J Roentgenol* 155(6):1161–1167.

Yang, J. J., J. Yang, L. Wei, O. Zurkiya, W. Yang, S. Li, J. Zou et al. 2008. Rational design of protein-based MRI contrast agents. *J Am Chem Soc* 130(29):9260–9267.

Yung, K. T. 2003. Empirical models of transverse relaxation for spherical magnetic perturbers. *Magn Reson Imaging* 21(5):451–463.

Zhang, S., M. Merritt, D. E. Woessner, R. E. Lenkinski, and A. D. Sherry. 2003a. PARACEST agents: Modulating MRI contrast via water proton exchange. *Acc Chem Res* 36(10):783–790.

Zhang, S., R. Trokowski, and A. D. Sherry. 2003b. A paramagnetic CEST agent for imaging glucose by MRI. *J Am Chem Soc* 125(50):15288–15289.

Nanoparticle Contrast Agents for Medical Imaging

David P. Cormode
Translational and Molecular Imaging Institute

Willem J. M. Mulder
Translational and Molecular Imaging Institute

Zahi A. Fayad
Translational and Molecular Imaging Institute

24.1 Introduction

Medical imaging has undergone significant development over the past three to four decades. New imaging modalities have been introduced, the quality of images has improved greatly, and the variety of information that can be derived from these systems has increased, allowing excellent diagnosis of disease and monitoring of therapy. For example, magnetic resonance imaging (MRI) is used to visualize tumors, computed tomography (CT) can delineate the coronary arteries, ultrasound is used in prenatal checkups, and positron emission tomography (PET) is used as a readout for glucose metabolism. The information derived from such imaging is due to contrast in the images. The contrast may be manipulated by adjusting the imaging parameters in some techniques, such as MRI, but the introduction of exogenous agents to provide additional contrast has been very useful in all methods of imaging. In some techniques like PET or single photon emission tomography (SPECT), contrast agents are a prerequisite.

Research in contrast agents for medical imaging is currently heavily focused on nanoparticles (Corot et al., 2006; Manchester and Singh, 2006; Azzazy et al., 2007; Cormode et al., 2007; Huang et al., 2007c; Mulder et al., 2007; Wickline et al., 2007). This interest is due in particular to the possibilities that nanoparticles allow in molecular imaging, the field where techniques that permit the detection of biological processes on the cellular level are developed (Weissleder and Mahmood, 2001; Mahmood and Josephson, 2005). The advantages of nanoparticles in this area include excellent contrast due to their nanoparticulate nature, e.g., the fluorescence of quantum dots (Medintz et al., 2005), the ease of integration of multiple properties such as multiple types of contrast (Nahrendorf et al., 2008), extended circulation times (Klibanov et al., 1990, Allen et al., 1991), and high payloads (Schmieder et al., 2005). These advantages have led to some extremely impressive nanoparticle platforms that have been reported, such as agents that can be detected by two or more imaging modalities (Nahrendorf et al., 2008), agents that also deliver therapeutics (Winter et al., 2006), agents that can detect particular types of cells (Amirbekian et al., 2007), or agents that allow entirely new imaging systems to be developed (Durr et al., 2007; Huang et al., 2007b).

Nanotechnology is frequently described as the study of materials that are 100 nm or less in one or more dimensions (Midgley et al., 2007), and the recent explosion of interest in this field is due to the unusual properties that occur in such materials. Within the range of materials that come under the heading of nanomedicine, however, many exceed the 100 nm limit in every proportion (Lanza et al., 2006). This is because some of the advantageous properties of nanoparticles for therapeutic or medical-imaging applications are not lost by exceeding that limit, for example, high payload or extended circulation times. Therefore, the term nanoparticles in this chapter will include particles considerably larger than this 100 nm limit.

Before discussing the different classes of nanoparticles in use or in development as contrast agents today, we will describe the topic of molecular imaging, explaining why nanoparticles are attractive for this technology. Then we will give a brief outline of the physics of the main types of medical-imaging techniques in use in the clinic and in research today, i.e., MRI, PET, CT, fluorescence, and ultrasound. We will discuss the pros and cons of these techniques and the advantages of the nanoparticles that are exploited as contrast agents by each. Finally, we will set out different classes of nanoparticulate contrast agents (i.e., iron oxides, quantum dots, micelles, liposomes, emulsions, lipoproteins, gold nanoparticles, silica, viruses, and other novel agents), describing each and giving examples of their use.

24.2 Molecular Imaging

Molecular imaging has been described as the "*in vivo* characterization and measurement of biologic processes at the cellular and molecular level" (Weissleder and Mahmood, 2001). In any disease, certain types of cells, cell receptors, or compounds excreted by cells are overexpressed or underexpressed when compared to

normal tissue. The aim of molecular imaging is to produce contrast agents that are targeted to one of the overexpressed species and, when administered, will preferentially accumulate in the diseased tissue. In addition to allowing earlier diagnosis, the extent of accumulation may indicate the stage of the disease, the severity of the case, the response to treatment, or the fundamental information about the expression of that type of cell, cell receptor, or compound in that disease. Therefore, molecular imaging is an important new field in medicine that will allow enormous improvements in our understanding and treatment of disease and has shown to be extremely valuable in drug development.

In the current state of the art of nanoparticle contrast agents, contrast-generating material, fluorescence, targeting, and other functionalities are often combined. Figure 24.1A depicts a generalized representation of a nanoparticle for use as a molecular-imaging contrast agent. Contrast may be located within the core

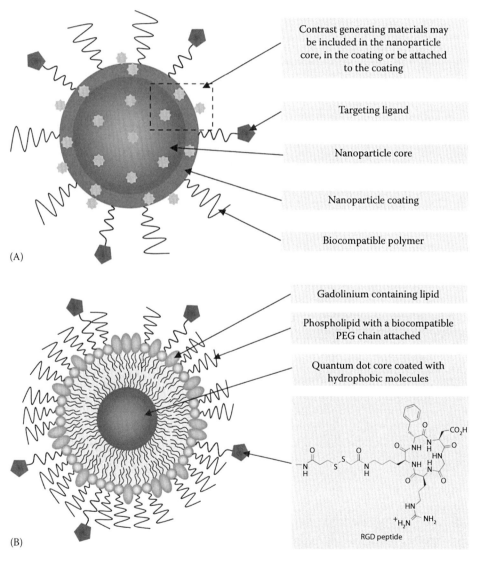

Contrast generating materials may be included in the nanoparticle core, in the coating or be attached to the coating

Targeting ligand

Nanoparticle core

Nanoparticle coating

Biocompatible polymer

(A)

Gadolinium containing lipid

Phospholipid with a biocompatible PEG chain attached

Quantum dot core coated with hydrophobic molecules

RGD peptide

(B)

FIGURE 24.1 (A) A generalized depiction of a nanoparticle for use as a molecular imaging contrast agent. (B) A paramagnetic, quantum dot-based contrast agent targeted to the $\alpha_v\beta_3$ integrin (found in angiogenic tissue) via the RGD peptide, as reported in Mulder et al. (2006). (Adapted from Cormode, D.P. et al., *Arterioscler. Thromb. Vasc. Biol.*, 29, 992, 2009. With permission).

of the particle, within the shell, or attached to the surface. This flexibility is extremely convenient as it allows multiple types of contrast to be included. For example, a nanoparticle has been reported that possesses a paramagnetic iron oxide core for MRI and the fluorophore FITC attached to the surface (Kelly et al., 2005). The common types of contrast-generating materials used are explained in Section 24.3. In addition to contrast-generating materials, the particle may have a payload of therapeutics. The combination of imaging and drug delivery is known as "theranostics" and is a powerful emerging paradigm (Del Vecchio et al., 2007). Targeting ligands may be antibodies, antibody fragments, proteins, peptides, peptomimetics, aptamers, sugars, or small molecules. Some nanoparticles used for molecular imaging rely on passive targeting such as the enhanced permeability and retention (EPR) effect where nanoparticles accumulate in tissues that have a leaky vasculature such as tumors (Wu et al., 1993; Kwon and Kataoka, 1995).

To further illustrate the structure of nanoparticle-based contrast agents for molecular imaging, an example of a recently reported quantum dot agent (Mulder et al., 2006) is depicted in Figure 24.1B. The core of the particle is a quantum dot, which provides contrast for fluorescence. The quantum dot is coated with hydrophobic molecules and is encapsulated in a mixed layer of amphiphiles. The advantage of this one-step encapsulation strategy is that three desired properties are simultaneously incorporated into the amphiphile layer: gadolinium for MRI contrast, PEG (polyethyleneglycol) chains that prolong the circulation half-life of the particle (Klibanov et al., 1990), and a targeting ligand. Angiogenesis is the process of forming new blood vessels, which occurs extensively in cancer (Zetter, 2008) and in atherosclerosis (Choudhury et al., 2004). This particle is targeted by the cyclic RGD peptide that binds to the $\alpha_v\beta_3$-integrin (Zitzmann et al., 2002), which is overexpressed in the angiogenic tissue (Avraamides et al., 2008). The authors incubated this agent with human umbilical vein endothelial cells, which express the $\alpha_v\beta_3$-integrin, and much higher contrast agent uptake was observed via both fluorescent imaging and MR imaging of pellets of these cells as compared to cells incubated with similar untargeted particles.

24.3 The Physics of Medical Imaging

In this section, we will briefly outline the physical principles of the main medical-imaging techniques currently used in clinical medicine: magnetic resonance imaging, positron emission tomography, computed tomography, and ultrasound, as well as the most promising fluorescence systems in development. We will discuss the advantages and disadvantages of each, and how nanoparticle contrast agents are being applied to them.

24.3.1 Magnetic Resonance Imaging

Strijkers et al. discuss the physics associated with MRI in detail in an accompanying chapter in this volume; however, we will briefly discuss the physical principles involved. The nuclei of atoms possess many different properties, among which is spin (*S*). The spin of many nuclei is zero; however, others possess non-zero spin values, such as 1/2 in the case of protons (the hydrogen nucleus). MRI usually probes hydrogen nuclei due to their very high concentration in tissue, but other nuclei are occasionally investigated, such as fluorine, which allows the direct visualization of fluorine-based contrast agents (Morawski et al., 2004; Ahrens et al., 2005). Most elements possess an isotope that has a nonzero spin value, but the natural occurrence of such isotopes is often very low, so the imaging time required to gain good signal-to-noise ratios is excessively long.

A nucleus with a spin number of zero has only one spin state, but a nucleus with a nonzero spin number can exist in multiple spin states denoted by m_s where the following different values are possible: $m_s = S, S - 1, S - 2..., -S$. For example, for the hydrogen nucleus $S = \frac{1}{2}$ and so $m_s = \frac{1}{2}$ or $-\frac{1}{2}$. In the absence of a magnetic field, the different spin states have equal energy; however, when a magnetic field is applied, the states split in energy, in proportion to the strength of the magnetic field ($\Delta E = (h/2\pi)\gamma B_0$, where γ is the gyromagnetic ratio of the nucleus). In order to produce a reasonably large difference in energies (yet still five orders of magnitude lower than room temperature thermal energy) between the states, enormous magnetic fields are required. The fields used in clinical medicine are 0.5–3 T, while in experimental settings, fields as high as 17.6 T are used (Neuberger et al., 2007; Van De Ven et al., 2007). In comparison, the earth's magnetic field varies from 30 to 60 μT, depending on the location.

In MRI, hydrogen nuclei are excited from the lower-energy state to the higher-energy state using radiowaves. The nuclei then relax to the lower-energy state by emitting radiowaves. The rate of this relaxation is highly dependent on the environment of the proton and is described by the parameters T_1 and T_2 (longitudinal and transverse relaxation, respectively). Due to the slightly differing conditions that exist around protons in different tissues and the judicious choice of imaging sequences, tissues can be distinguished by the rate of relaxation (signal received). The shorter the T_1 value of a tissue (e.g., using what is known as a T_1 weighted scan), the faster the protons relax from the excited to the ground state, the greater the signal received, and the brighter that tissue looks in an MRI image. Conversely, when T_2 weighted scans are used the shorter the T_2 value of a tissue the *darker* that tissue appears.

The place of contrast agents in this imaging technique is to decrease T_1 or T_2 and thus produce additional contrast. This is done primarily by metal ions with large numbers of unpaired electrons. These ions cause small, local inhomogeneities in magnetic fields, which lead the protons to relax more swiftly. The main ions used to reduce T_1 are Gd^{3+} and Mn^{2+}, whereas superparamagnetic iron oxide nanoparticles are used to shorten T_2 (Gupta et al., 2005; Bottrill et al., 2006; Na et al., 2007). In addition, chemical shift saturation transfer (CEST) agents have recently begun to be trialed as contrast media for MRI (Woods et al., 2006). T_1 or T_2 are inversely proportional to the concentration of the contrast agent, as described by the following equation:

$$\frac{1}{T_{n(\text{obs})}} = \frac{1}{T_{n(\text{inh})}} + r_n[\text{CA}]$$

where

r_n is the relaxivity of the contrast agent

[CA] is the concentration

$T_{n(\text{inh})}$ is the T_n of the substance without a contrast agent

$T_{n(\text{obs})}$ is the observed T_n in the presence of the contrast agent

$n = 1$ or 2

Assuming an r_1 value of $5\,\text{mM}^{-1}\text{s}^{-1}$, a $T_{1(\text{inh})}$ of 4 s, and a minimum detectable change of T_n of 20%, a contrast agent concentration of $10\,\mu\text{M}$ is necessary to produce appreciable contrast in an image. Increasing relaxivity makes the contrast agent more effective and has been approached in three major ways. First, by including more paramagnetic metal ions per contrast agent. This has been widely achieved though the use of nanoparticles, which can have as many as 100,000 gadolinium ions incorporated (Winter et al., 2003). The ability to deliver a high payload of contrast agent is one of the main strengths of nanoparticles. Second, reducing the tumbling rate of the metal ion may increase the r_1. Nanoparticles tumble less swiftly than small molecules and therefore chelated metals at the nanoparticle surface have a higher r_1. Third, in the case of gadolinium-based contrast agents, major efforts have centered around producing stable chelates that allow the coordination of the gadolinium ion with greater numbers of water molecules (Raymond and Pierre, 2005). The more water molecules have access to the gadolinium ion, the more protons can be relaxed in a given time, and hence greater contrast can be produced.

MRI provides excellent images of the soft tissues of the body with high spatial resolution and does not expose the patient to ionizing radiation. It has the disadvantages that the scans may be long in duration, the equipment is very expensive, and some patients cannot be scanned due to metallic implants, claustrophobia, or an inability to remain still. Also, this technique is relatively insensitive to contrast agents; however, developments in nanoparticle contrast agents mean that this problem is being overcome.

24.3.2 Positron Emission Tomography

This modality (imaging technique) relies entirely on the use of contrast agents that contain an unstable nucleus that is prone to positron emission. The only USFDA-approved agent for PET is an analogue of glucose, 2-fluoro-2-deoxyglucose (FDG), where the fluorine is the F-18 isotope, but other agents, based around this and other radioactive isotopes, are also used in experimental settings. The radiolabeled agent is injected into the patient an hour or so before imaging. The unstable nucleus in the agent decays into a more stable form via emission of a positron, which is annihilated by collision with an electron, resulting in the production of a pair of photons that travel in opposite directions. The patient is encircled by a ring of detectors; when two photons are detected on opposite sides of the ring simultaneously this registers as a signal. Over time many positron emissions are detected and a

picture is built up from the distribution of the agent within the patient. The distribution of FDG is a good reflection of glucose uptake and phosphorylation by cells in the body. For example, in atherosclerosis (the buildup of cholesterol in arteries) FDG is known to accrue in areas of inflammation and may be used in the assessment of this disease (Rudd et al., 2002).

This modality has the advantage that it is very sensitive to contrast agents. Concentrations as low as 10^{-12} M may be detected, and thus, using positron-emitting nuclei coupled to targeting ligands, this technique can be used to detect very sparse disease markers or diseases at a very early stage. The very high sensitivity is due to the very low natural level of positron emitters; however, the concomitant disadvantage of this technique is that very little anatomical information can be derived. To overcome this problem, PET-imaging devices are often combined with another imaging modality that gives anatomical information, such as CT or recently MRI (Catana et al., 2008). PET also has poor spatial resolution because positrons travel a significant distance (ca. 2–3 mm) prior to annihilation (Ter-Pogossian, 1985). The last major issue with this technique is that patients are exposed to ionizing radiation.

Due to the very high sensitivity of PET, the prime driver of research using nanoparticles in MRI, the ability to deliver high payloads of contrast does not apply. Recently, however, some of the other advantages of nanoparticles have been exploited for this technique, primarily the ease of incorporating multiple sources of contrast such as positron emission for PET and fluorescence (Rossin et al., 2005). In addition, one of the flaws of small-molecule contrast agents is that the validation of their cellular distribution and eventual fate is difficult (Nahrendorf et al., 2008). Such validation using nanoparticles can be achieved using confocal microscopy or histology.

24.3.3 Computed Tomography

Computed tomography images the body using x-rays. The patient is situated in a donut-shaped apparatus that contains a mobile x-ray source and several rows of detectors (Figure 24.2).

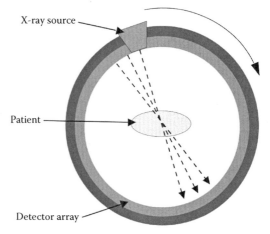

FIGURE 24.2 Schematic diagram of a computed tomography imaging system.

The x-rays are attenuated via photoelectric or Compton interactions with the nuclei of the atoms that make up the tissues of the patient, and the detectors record the transmitted radiation. Computational analysis of the results from the detectors allows images of the patient to be formed. Due to the differing x-ray attenuation properties of different organs excellent images of soft tissues may be created. This technique is very insensitive to contrast; very high concentrations of contrast agents are required. Contrast agents are based on elements that attenuate x-rays very strongly, i.e., heavy elements such as iodine or gold.

Contrast agents for CT that are currently licensed for use are small molecules that include several iodine atoms. These agents are applied to image the vasculature. The main advantage of these agents is that (despite the enormous doses used) they are cleared from the body rapidly via the kidneys. The advantages nanoparticles have to offer as contrast agents are the ability to deliver the huge amounts of contrast required and the ease of incorporation of multiple properties. Nanoparticles have, in fact, a long history of use as contrast agents in x-ray-reliant imaging techniques (Yu and Watson, 1999). Thorotrast, for example, was widely used from the 1920s to the 1950s and was a dextran-coated colloidal (3–10 nm) suspension of ThO_2. Its use was discontinued due to safety concerns surrounding the radioactivity of thorium. Molecular imaging of macrophages in atherosclerosis was recently achieved using CT, where the contrast agent was an iodine-based nanoparticle (Hyafil et al., 2007). This was especially impressive as CT had previously been thought too insensitive to contrast agents for molecular imaging to be accomplished (Weissleder and Mahmood, 2001).

Aside from the disadvantage of being insensitive to contrast agents, this technique exposes the patient to ionizing radiation. However, the equipment is simple to operate compared to, for example, MRI. In addition, there is no restriction on the types of patients that may be imaged using this technique, and the scans take only seconds to complete.

24.3.4 Ultrasound

Ultrasound exploits the reflection of high-frequency sound waves (2–18 MHz, far above the range of human hearing) to make images. The equipment used for ultrasound imaging is quite different from that used in the other imaging modalities, being quite compact, and portable, cart-based systems are available. As opposed to placing the patient in the center of a large, ring-shaped device, a hand-held transducer, connected to a pulse generator and a signal amplifier, is placed on the skin of the patient. Special gels are applied to the skin of the patient to ensure good conduction of sound. In the transducer is a piezoelectric crystal, a substance which mechanically deforms upon the application of an electric current. The pulse generator sends a series of electrical pulses to the transducer, which results in rapid vibration and consequently the production of ultrasound from the piezoelectric crystal. Small amounts of the emitted sound are reflected from areas where there are changes in density in the body, which includes boundaries between organs, but

also within organs where there are variations in structure. Some of these reflections or echoes are received by the transducer and converted back into electric current. The signals received in this manner contain two types of information. First, the distance from the transducer of the structure that caused the echo can be inferred from the time the echo is received after the pulse was emitted. Second, the type of interface that produced the echo can be determined from the amplitude of the signal received. These signals are processed into an image of the subject, which is normally a slice going into the patient, although it is possible to generate 3D images. Ultrasound is frequently used to create the images of fetuses, with which the general public are familiar.

Liposomes and emulsions have been reported as contrast agents for ultrasound (Lanza et al., 1996; Huang et al., 2002), but currently most researchers utilize microbubbles (Villanueva and Wagner, 2008). These microbubbles are composed of inert gases such as perfluorocarbons or nitrogen covered in a shell made of polymers, lipids, or albumin. Two microbubble formulations are USFDA-approved for use in patients. Ultrasound incident on microbubbles causes the gas-filled interior to compress and oscillate, which produces an echo that can be distinguished from the signals from tissue. Molecular imaging may be performed using microbubbles when targeting ligands such as antibodies (Villanueva et al., 1998), peptides (Weller et al., 2005), sugars (Villanueva et al., 2007), and so forth, are attached.

Ultrasound can produce "live" imaging, with image refresh rates of 30 frames per second (Klibanov, 2006), has good spatial resolution, and the equipment is very cheap. Furthermore, the boundaries between phases, e.g., between tissue and bone or between tissue and the lungs can be very well defined. The disadvantage that goes hand-in-hand with this strength is that structures beyond such a boundary, such as the brain, cannot be imaged. Additionally, the depth of tissue penetration of sound is low, which means that structures that lie deep within the body are difficult to image, especially for obese patients.

24.3.5 Fluorescence

There are many techniques that exploit fluorescence, including confocal microscopy, fluorescence microscopy, fluorescence imaging, fluorescence tomography, and two-photon fluorescence. All these techniques are important in the field of molecular imaging, especially confocal microscopy, which, by detecting fluorophores incorporated into contrast agents and through use of immunofluorescent staining, allows the cellular fate of agents postinjection to be determined (Mulder et al., 2007). Although indocyanine green, a fluorescent dye, is licensed for use in patients as a diagnostic for cardiac function, no fluorescence-based imaging technique is used in the clinic yet. However, prototype fluorescence-imaging devices for breast cancer are being tested (Intes, 2005) and will soon be widely available, with devices for other applications to follow shortly. In addition, all the aforementioned techniques are very important in the development of new contrast agents and drugs and in the research of diseases (Sosnovik and Weissleder, 2005).

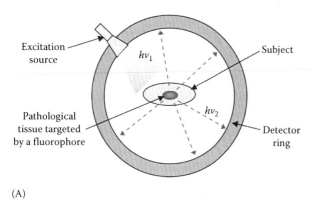

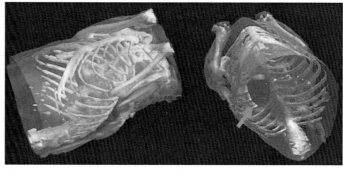

(A)   (B)

FIGURE 24.3 (A) Schematic depiction of fluorescence tomography equipment. (B) Different projections of reconstructions of combined CT and fluorescence tomography images of a mouse bearing a tumor expressing GFP. The tumor is shown in red and highlighted by the arrows. (Reprinted from Ntziachristos, V., *Annu. Rev. Biomed. Eng.*, 8, 1, 2006. With permission.)

A full explanation of the physics of all these fluorescence-based techniques is beyond the scope of this chapter, and the reader is directed to other sources for that information (Ntziachristos et al., 2003; Ntziachristos, 2006). The physical basis of the fluorescence of a molecule or particle relies upon there being an above-the-ground (lowest energy) electronic state to which the molecule can be excited by an incident photon interacting with an electron. Once excited to this higher-energy state, the molecule can relax via a variety of pathways to the ground state. Useful fluorophores relax via radiative pathways with high efficiency, thus producing fluorescent light.

For the application providing fluorescent contrast in imaging diseased tissue, the fluorescence must pass through intervening tissue and be received outside the body. The wavelength window at which the body absorbs light least is the near infrared (NIR) and is approximately from 650 to 900 nm (Sosnovik and Weissleder, 2005). Good fluorophores for these applications emit in the near infrared. While conventional fluorophores (i.e., small, organic molecules) such as NBD or Cy5.5 (Levi et al., 2007) are very often used in fluorescence-based molecular imaging, quantum dots are widely used as they have a number of significant advantages such as wide excitation windows; narrow, tunable emission spectra; high resistance to photobleaching; and high fluorescence efficiency.

To image a subject using fluorescence tomography, first a fluorophore contrast agent is administered, then the subject is placed in a chamber from which light is excluded. The subject is irradiated with light of a wavelength that excites the fluorophore that is now distributed through the body (Figure 24.3A). A ring of detectors that surround the subject picks up the light that the fluorophores emit. Determination of the distribution of contrast in the subject requires a complex calculation involving the location of the excitation source, the attenuation of the light of the tissue, and the scattering of the light of the tissue; this calculation results in a 3D depiction of the distribution of the contrast fluorophore. As this technique does not provide anatomical information, MR or CT images are sometimes co-acquired, upon which the fluorescence results can be superimposed. For example, Zacharakis et al. combined CT and fluorescence tomography images of a mouse bearing a tumor that expressed green fluorescent protein (GFP), as shown in Figure 24.3B (Zacharakis et al., 2005; Ntziachristos, 2006). The protein is the source of the fluorescent contrast in this case; no additional contrast agent is applied.

The main advantages of fluorescence tomography are the high sensitivity to contrast, speed of imaging, low equipment cost, and the ability to detect multiple fluorophores simultaneously, which gives the potential to perform diagnoses for multiple markers. The major limitations are a low spatial resolution and the low tissue penetration depth of light. The latter limitation means that only small animals or relatively superficial (within centimeters of the skin) diseases such as breast cancer or pediatric disorders can currently be imaged, although technical improvements may allow investigation of deeper lying organs.

24.4 Classes of Nanoparticle Contrast Agents

24.4.1 Iron Oxides

Iron oxide nanoparticles are widely used to provide contrast in MR imaging (Pankhurst et al., 2003; Bulte and Kraitchman, 2004; Corot et al., 2006; Mccarthy et al., 2007) and have other applications in nanomedicine such as thermal therapy and drug delivery (Duguet et al., 2006). They are usually synthesized via a co-precipitation of Fe^{2+} and Fe^{3+} under basic conditions to form superparamagnetic Fe_3O_4 (magnetite) nanoparticles (Gupta and Gupta, 2005). A variety of biocompatible materials such as PEG (Zhang et al., 2002), dextran (Berry et al., 2003), polyvinylpyrrolidone (Liu et al., 2007), polyacrylic acid (Yoshida et al., 2007), or chitosan (Li et al., 2006) have been used as coatings. The conditions of synthesis are usually tuned to produce particles that consist of a single domain, are superparamagnetic (i.e., they are magnetized only when an external field is applied) and are smaller than 15 nm. These nanoparticles align with the magnetic field, then flip away from alignment at a rate whose time constant (τ) is given by the equation:

$$\tau = \tau_0 \exp\left(\frac{KV}{k_B T}\right)$$

where

τ_0 is property of the particles
K is the anisotropy constant
V is the crystal volume
k_B is the Boltzmann constant
T is the temperature

Small particles give a small volume and hence KV is comparable to $k_B T$, the time constant is short and a rapidly fluctuating magnetic field is produced around the particles. Iron oxide nanoparticles produce negative contrast in MRI (darkening of the image) due to a reduction in T_2 caused by the local inhomogeneities in the magnetic field that the particles create, according to the following equation:

$$\frac{1}{T_2^*} = \frac{1}{T_2} + \gamma \frac{\Delta B_0}{2}$$

where

ΔB_0 is the change in the field caused by the iron oxide
γ is the gyromagnetic ratio

Iron oxide has the advantage that it is detected with greater sensitivity than gadolinium-based agents. On the other hand, gadolinium produces positive contrast (brightening in the image), whereas iron oxide produces negative contrast (darkening in the image). Positive contrast in an image can readily be ascribed to the accumulation of the contrast agent, whereas there are many sources of negative contrast making it hard to be certain that image darkening is due to iron oxide gathering in the tissue. The development of positive contrast sequences for iron oxide may ameliorate this issue in the future (Briley-Saebo et al., 2007).

Feridex® and Resovist® are iron-oxide-based agents that have been licensed for clinical use (Corot et al., 2006). Feridex is coated with dextran, whereas Resovist has a carboxydextran coating and both are large (60–180 nm) aggregates of the coating into which multiple small (ca. 5 nm) iron cores are embedded (Wang et al., 2001). These agents may be used for liver tumor diagnosis (Reimer and Tombach, 1998). These particles are simple in design, composed of only iron oxide cores and a biocompatible coating. Macrophages, however, take up a similar, smaller agent (15–30 nm), Sinerem, due to their affinity for the dextran coating of this particle. This phenomenon was exploited by Reuhm et al. to image the level of macrophage expression in the aorta of a rabbit model of atherosclerosis (Ruehm et al., 2001).

In order to direct iron oxides to other targets, Weissleder and coworkers have proposed cross-linked iron oxide (CLIO), where the dextran coating of the iron oxide particle is cross-linked by epichlorohydrin and ammonia (Wunderbaldinger et al., 2002). This process forms a particle that exposes many amine groups

that may be utilized for modification with targeting ligands and/or other functionalities such as fluorophores. Targets such as VCAM-1, E-selectin, apoptotic cells, and others have been successfully imaged *in vivo* using this platform (Schellenberger et al., 2004; Funovics et al., 2005; Kelly et al., 2005). Recently, a highly multifunctional particle based on this CLIO technology has been reported where the agent may be detected by MRI, PET, and optical techniques (Nahrendorf et al., 2008). The MRI contrast comes from the iron oxide core, while ^{64}Cu, attached to the particle through chelates at the surface, gives PET contrast and a NIR fluorophore is also attached to the surface to allow detection by fluorescence-based imaging. Macrophages still have an affinity for this modified dextran-coated particle and the authors were able to image the expression of this cell type in the aorta of the apoE-KO mouse model of atherosclerosis using the three techniques (Figure 24.4). A significant drop in the T_2 of the aortic root was observed when MR images pre- and 24 h postinjection were compared (Figure 24.4A and B, color scale indicates T_2, note change from blue to green in inset images). In the PET-CT images, substantial PET signal was observed around the aortic arch in apoE-KO mice, but not in wild-type mice (Figure 24.4C and D, PET signal is red-orange, the aorta is colored blue). This is because macrophages are not present in the aorta of the latter non-atherosclerotic mice. NIRF imaging of excised aortas also indicated the presence of the particle (Figure 24.4E). Other strategies for targeting iron oxides have been adopted, for example, attachment of proteins to lipids surrounding hydrophobic

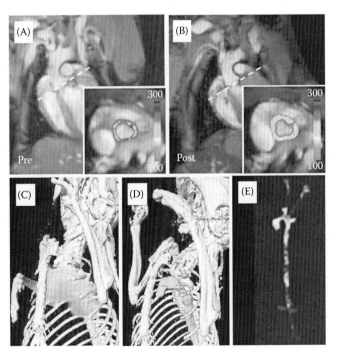

FIGURE 24.4 (See color insert following page V-2.) Trimodal imaging of macrophage expression: T_2-weighted MR imaging of the aortic root of apoE-KO mice (A) pre- and (B) 24 h postinjection with the agent; PET-CT image of (C) an apoE-KO mouse and (D) a wild-type mouse 24 h postinjection. (E) A NIRF image of an excised aorta from an apoE-KO mouse 24 h postinjection. (From Nahrendorf, M. et al., *Circulation*, 117, 379, 2008. With permission.)

iron oxide particles (Van Tilborg et al., 2006), oxidative conjugation of antibodies to dextran-coated iron oxides (Weissleder et al., 1992), or through silane chemistry (Zhang and Zhang, 2005).

Iron oxides have been widely used in cell tracking, where the iron particles are internalized by cells, and thus the cells may be tracked by MRI as they migrate through tissues (Bulte et al., 2002). There are many diseases for which treatment with cells cultured outside the body is attractive, and the motivation for cell tracking is to ascertain the fate of the administered cells—the treatment will only be effective if the cells gather in the correct sites. Due to the high resolution of MRI and the biocompatibility and the strong contrast induced by iron oxides, the MRI–iron oxide combination is a popular method for cell tracking. Targeted agents have been used to label cells (Bulte et al., 1999), but it has been found that untargeted agents such as Feridex can be successful for the purpose as well (Franklin et al., 1999). Often a polycationic transfection agent is used to speed uptake, such as lipofectamine (Hoehn et al., 2002), but they have not been used in all cases: passive uptake (Franklin et al., 1999) or techniques like magnetoelectroporation also work (Walczak et al., 2005), depending on the cell type. These methods have been very successful, but care must be taken that the loading procedure does not affect the biochemistry of the cells (Dodd et al., 2001).

An excellent example of a cell tracking study was reported by De Vries and colleagues where the application of Feridex-labeled dendritic cells to human melanoma patients was investigated (De Vries et al., 2005). Tumor antigen-loaded dendritic cells can be used to produce an immune response against cancer, but are efficacious only in a minority of patients (Figdor et al., 2004). The authors proposed that the reason for the minority response to treatment was due to insufficient delivery of the cells to the target organs, the lymph nodes. The cells were cultured in the presence of Feridex and tested to ensure adequate uptake of iron and unperturbed function of the cells. Having passed these tests, the cells were administered to lymph nodes of the patients under ultrasound guidance. As can be seen in Figure 24.5A through C, it was possible to visualize the cells postinjection and track their subsequent migration. Histology on the lymph nodes after removal of these tissues confirmed the accumulation of the dendritic cells as observed by MRI (Figure 24.5D). In 50% of the cases, however, it was found that the cells were not accurately injected into the lymph nodes, but into nearby tissue, which may explain the low efficacy of this form of cell therapy.

24.4.2 Quantum Dots

Quantum dots have a number of advantages over small-molecule fluorophores (Figure 24.6): wide wavelength windows in which they may be excited (Leatherdale et al., 2002),

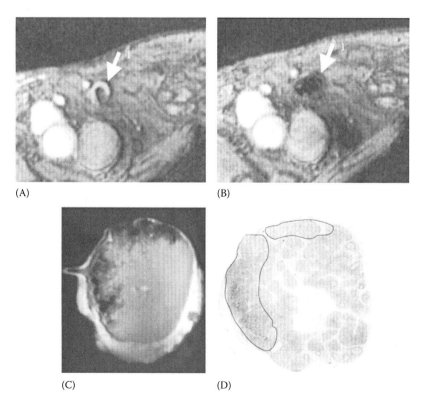

(A)   (B)

(C)   (D)

FIGURE 24.5 De Vries et al. used MRI scans of Feridex loaded dendritic cells to track the migration of this cell type. (A) MR image of a lymph node (arrow) pre and (B) postinjection with Feridex loaded cells. (C) A high resolution MR image of an excised lymph node whose dark areas are caused by infiltration of iron oxide loaded cells, as confirmed by histology (D) where iron appears blue. (Adapted from De Vries, I.J.M. et al., *Nat. Biotechnol.*, 23, 1407, 2005. With permission.)

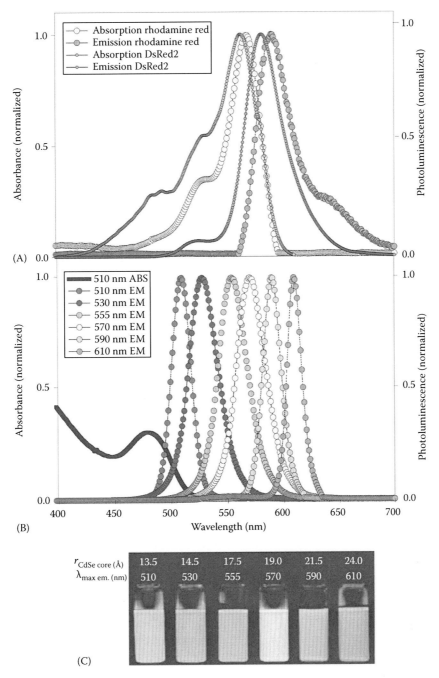

FIGURE 24.6 **(See color insert following page V-2.)** Comparison of the emission and absorption spectra of (A) Rhodamine red and DsRed2 and (B) quantum dots whose core size varies from 13.5 to 24 Å. A photograph of the quantum dot solutions appears in (C). (Reproduced from Medintz, I.L. et al., *Nat. Mater.*, 4, 435, 2005. With permission.)

narrow, tunable emission spectra as shown in Figure 24.6B (Bruchez et al., 1998), a high resistance to photobleaching (Sukhanova et al., 2004), and they fluoresce with high efficiency (Medintz et al., 2005). The narrow and tunable emission spectra means that several different types of quantum dots can be simultaneously detected, distinguished by their differing emission wavelengths (Kobayashi et al., 2007). The reader is referred to the accompanying chapters in this book for information on the semiconductor quantum confinement effects that produce these advantageous properties.

To create quantum dots with optimal fluorescence properties, it is necessary to synthesize them under hydrophobic conditions, resulting in hydrophobically capped particles. For use as contrast agents, it is necessary to transfer the quantum dots to aqueous conditions, which was initially achieved by exchanging the capping molecules for ligands that are bifunctional—i.e., possess a group that can bind to the particle surface and another group that is hydrophilic, thus rendering the particle water-soluble. An example of such a capping molecule is mercaptoacetic acid (Chan and Nie, 1998). This process, however, tends to alter or adversely impact the fluorescent properties of the quantum dots and the current favored solubilization technique is to enrobe these nanoparticles in amphiphilic polymers or lipids (Dubertret et al., 2002; Wu et al., 2003), as highlighted in Figure 24.1B.

Quantum dots have applications in cell tracking via fluorescence techniques (Derfus et al., 2004), which is useful for investigating cell therapies, as described in Section 24.4.1. Furthermore, quantum dots can be conjugated with targeting ligands such as antibodies or peptides, making them very useful for immunofluorescence and molecular imaging in small animals (Xing et al., 2007). In these applications, the fluorescence occurs via a source exciting a quantum dot, which emits light. This is not the only fluorescence situation in which quantum dots can be exploited. Quantum dots can also be involved in fluorescence resonance energy transfer (FRET), where the fluorescence of the quantum dot is absorbed by a nearby organic dye and the quantum dot fluorescence can longer be detected. This can be exploited for detecting substrates such as maltose, as reported by Mauro and coworkers (Medintz et al. 2003). In this study, maltose-binding protein (MBP) was attached to quantum dots and incubated with cyclodextrin that had been modified with QSY9 dye. This cyclodextrin molecule bound to the MBP and the attached dye absorbed the fluorescence from the quantum dots. When maltose was added to this system, the cyclodextrin was displaced from the MBP and hence the fluorescence of the quantum dot was restored. Furthermore, when enzymes that catalyze bioluminescent reactions, such as luciferase, are attached to quantum dots they may be used for fluorescence imaging with no external light source. This is because the light produced by the luciferase reaction excites the quantum dot, in a process known as bioluminescence resonance energy transfer (BRET). This has been shown to be practical *in vivo*, by injecting these quantum dots into mice (So et al., 2006).

24.4.3 Micelles

When amphiphiles or lipids self-assemble in solution, a variety of nanoparticulate structures can be formed, such as micelles, liposomes, or other more exotic structures like threads (Moschetta et al., 2002). Micelles are small (5–50 nm) and the lipids that compose them are arranged so that their hydrophilic headgroups face water, forming the exterior of the particle, while their hydrophobic tails form the core of the particle. Liposomes are far larger particles, generally greater than 100 nm in diameter. The lipids that are being used determine the type of particle produced. Lipids with a high bending modulus form micelles, whereas lipids with a low bending modulus form liposomes (Sandstrom et al., 2007). Micelles can be made functional by including specially modified

lipids or by carrying a hydrophobic cargo in the core. For the micelle-based particle highlighted in Section 24.2, gadolinium chelating and peptide-based targeting lipids were included in the formulation while quantum dots were carried in the particle cores.

Much effort is put into research of long-circulating contrast materials for CT, as the currently USFDA-approved agents only allow a very short time window for imaging. Torchilin and coworkers have reported a micelle formed from block copolymers, one part of which is composed of hydrophilic ethylene glycol subunits, the other part is hydrophobic and contains triiodobenzene derivatives (Trubetskoy et al., 1997). This particle was used to visualize the blood pool in rats as much as 3 h postinjection (Torchilin et al., 1999). Furthermore, some groups have focused on designing micelles that provide contrast for imaging modalities such as PET or SPECT by including radioactive metals (Torchilin, 2002). The design of micelle contrast agents advanced by the inclusion of targeting moieties, for example, an Italian group reported a micelle formed from a mixture of gadolinium-labeled lipids and CCK peptides that make the nanoparticle specific for tumors that form in the lung, colon, and pancreas (Accardo et al., 2004). Recently we have reported a micelle that includes gadolinium to yield contrast for MRI, a fluorophore and an antibody targeted to oxidized LDL, a substance that is influential in the progression of atherosclerosis (Briley-Saebo et al., 2008). When injected into atherosclerotic mice, this agent was observed by MRI to accumulate in their diseased arteries, via an increase in signal intensity of over 125%. Fluorescence investigations revealed the accumulation of these micelles in the macrophage cells of the arteries and the mechanism of particle uptake into macrophages *in vitro*.

24.4.4 Liposomes

Liposomes were first discovered in the 1960s and are hence one of the most established nanoplatforms (Bangham and Horne, 1964). They consist of lipid bilayers formed into spheres, so that they possess an aqueous internal compartment (Figure 24.7A). This internal compartment is frequently exploited to carry water-soluble drugs or contrast agents to prevent degradation of these cargoes or to yield long circulation times. For example, a long-circulating contrast agent for computed tomography was created by incorporating high concentrations of short-circulating USFDA-approved iodinated molecules into liposomes (Mukundan et al., 2006). The authors illustrated the long-circulating nature of these iodine-loaded liposomes by acquiring CT images of mice injected with this agent after 120 min, where good delineation of the blood vessels can still be observed (Figure 24.7B).

Liposomes can be used as MRI contrast agents if either their aqueous core is loaded with gadolinium chelates (Tilcock et al., 1989) or the phospholipid bilayer is modified to contain gadolinium (Kabalka et al., 1991). When the gadolinium is included in the phospholipid layer at least some of the gadolinium is

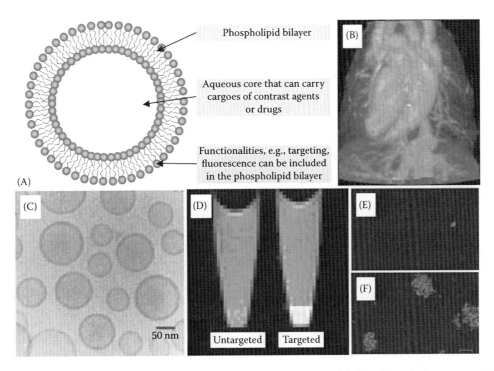

FIGURE 24.7 (A) Schematic depiction of the cross-section of a liposome. (B) Delineation of the blood vessels of a mouse in CT imaging by application of a liposome encapsulating an iodinated molecule. (C) Cryo-TEM of a liposome solution confirming the structure of the particles. (D) MR image of apoptotic cells incubated with paramagnetic and fluorescent liposomes that are (left) untargeted or (right) targeted to apoptosis. Confocal microscopy images of apoptotic cells incubated with the untargeted (E) and targeted (F) liposomes. (From Mukundan, S. et al., *Am. J. Roentgenol.*, 186, 300, 2006; Van Tilborg, G.A.F. et al., *Bioconjugate Chem.*, 17, 741, 2006. With permission.)

exposed to the bulk water, whereas gadolinium in the core can only interact with water contained within the liposome. This can lead to higher relaxivities for the former type of particle, as relaxivity depends on as much water as possible interacting with the gadolinium ions. A liposomal contrast agent has been reported where the gadolinium is incorporated in the phospholipid layer along with a fluorophore (Mulder et al., 2004; Van Tilborg et al., 2006). Furthermore, the particle was coated in PEG, yielding a long circulation half-life, and the protein annexin A5 was attached at the end of some of the PEG chains. Annexin A5 binds phosphatidylserine, a marker for apoptosis, which is an important process in a number of diseases such as cancer and atherosclerosis. The liposomal contrast agent was characterized using cryo-TEM, a technique where the structure in solution of fluxional particles such as liposomes can be imaged due to flash-freezing during sample preparation. The results of this cryo-TEM can be seen in Figure 24.7C, where cross-sections of the liposomes can clearly be seen as dark circles. Preferential uptake of this particle, as compared to untargeted liposomes, by apoptotic Jurkat cells was seen via MRI and confocal microscopy (Figure 24.7D through F).

Liposomes have applications as contrast agents for ultrasound, also. Their contrast-inducing properties for ultrasound may be due to pockets of gas trapped in between their phospholipid bilayers (Huang et al., 2002) and they have been largely superseded by microbubbles for this application (Villanueva and Wagner, 2008). Lastly, liposomes have been used as contrast agents for PET,

for example via inclusion of an 18F-labeled lipid in the particle (Marik et al., 2007).

24.4.5 Emulsions

Emulsions are tiny droplets of hydrophobic liquid suspended in water by means of amphiphilic coatings (Figure 24.8A). As with other amphiphile-based particles, emulsions may be imbued with multifarious properties such as target specificity, fluorescence, or paramagnetism by using specially modified lipids in their synthesis. One of the advantages of emulsions is that their typically large sizes (>150 nm) allow them to carry high payloads (Schmieder et al., 2005). Various types of payload can be included in the hydrophobic core of emulsions, most notably perfluorocarbons, which allow "hotspot" imaging of the location of particles via ^{19}F MRI. Ahrens et al. exploited this possibility by using perfluoropolyether-based emulsions that contained the fluorophore to label dendritic cells. The labeling was shown to be successful by confocal microscopy of these cells, where green coloration in the images indicated particle uptake (Figure 24.8B). The cells were tracked postinjection into mice using ^{19}F MR images (red-yellow coloration) overlaid onto images acquired using conventional ^{1}H MRI (grayscale), as shown in Figure 24.8C through E (Ahrens et al., 2005).

Emulsions have been exploited as contrast agents for most medical-imaging technologies. They have been applied for

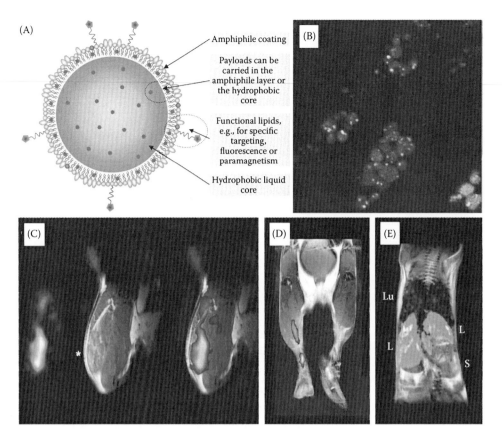

FIGURE 24.8 (**See color insert following page V-2.**) Emulsions used as contrast agents. (A) Schematic depiction of emulsions. (B) Ahrens et al. used perfluoropolyether emulsions that contained the fluorophore DiI to label dendritic cells. Confocal microscopy indicated a high level of particle uptake, as evidenced by the fluorescence of these cells. (C–E) ^{19}F MRI (red-yellow coloration) combined with conventional MRI (grayscale) allows tracking of these cells *in vivo*. (Adapted from Cormode, D.P. et al., *Arterioscler. Thromb. Vasc. Biol.*, 29, 992, 2009; Ahrens, E.T. et al., *Nat. Biotechnol.*, 23, 983, 2005. With permission.)

conventional MRI by including gadolinium-labeled lipids in the amphiphilic coating, while emulsions that contain bromine in their cores have been used for CT (Winter et al., 2005). Furthermore, emulsions may be used to provide contrast for ultrasound due to their echogenic properties (Lanza et al., 1996).

24.4.6 Lipoproteins

Lipoproteins are naturally occurring nanoparticles. They are composed of a hydrophobic core of triglycerides and cholesterol esters covered with a phospholipid monolayer, into which an amphiphatic protein is embedded (Figure 24.9A). The main types of lipoprotein are high-density lipoprotein (HDL), low-density lipoprotein (LDL), very-low-density lipoprotein (VLDL), and chylomicrons. There are substantial differences in the size of these particles: HDL is normally in the 7–13 nm range, LDL 22–28 nm, and VLDL upwards of 30 nm (Nichols et al., 1986). Furthermore, the primary protein constituent of HDL is apoA-I, whereas that of LDL and VLDL is apoB. Lipoproteins transport cholesterol through the body and are involved in a number of other processes such as regulation of

inflammation (Lusis, 2000). LDL is linked to the genesis and progression of atherosclerosis, while elevated levels of HDL have been shown to reduce or reverse the progression of atherosclerosis (Shah, 2006).

Lipoproteins have a number of advantages over other nanoscale molecular-imaging platforms. First, as endogenous nanocarriers, lipoproteins are naturally biocompatible, biodegradable, and nonimmunogenic, which are important limitations for many nanosystems. Second, lipoproteins possess excellent size-control through their apolipoprotein components. Third, lipoproteins, due to their small size and physical properties, penetrate easily into tissues, which is key for certain diagnostic applications such as atherosclerosis imaging (Frias et al., 2006). Lipoproteins are well suited to be modified into molecular-imaging contrast agents (Cormode et al., 2007). The phospholipids may be exchanged for ones that are paramagnetic, radioactive, or fluorescent (Shaish et al., 2001; Frias et al., 2006), while the apolipoprotein may be used to direct the particle to its natural targets, such as macrophages in atherosclerosis. Interestingly, the amine groups from exposed lysine residues may be used to attach targeting species in order to redirect the particles to different targets (Zheng et al., 2005). In addition, it is possible that the core

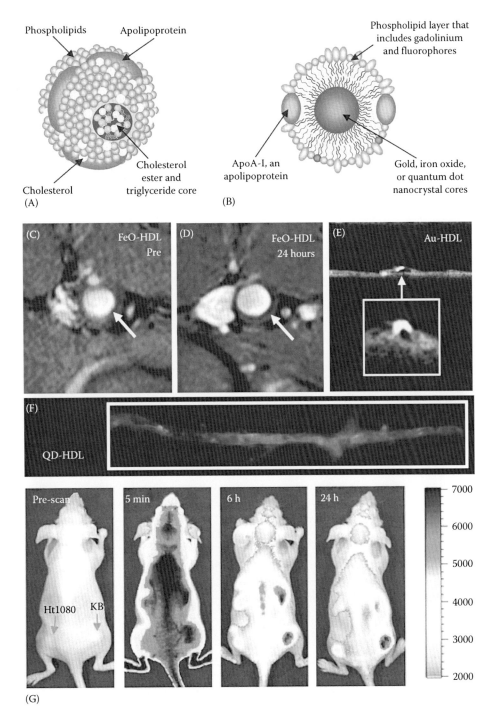

FIGURE 24.9 **(See color insert following page V-2.)** (A) A schematic depiction of the structure of lipoproteins. (B) A schematic depiction of nanocrystal core HDL. MR images of the aorta of an apoE-KO mouse pre- (C) and 24h postinjection (D) with FeO-HDL. (E) Micro CT image of the excised aorta of an apoE-KO mouse injected with Au-HDL. (F) Fluorescence image of the aorta of an apoE-KO mouse injected with QD-HDL. (G) *In vivo* fluorescence images of a mouse bearing a KB and a HT1080 tumor on its right and left flank, respectively. The agent became preferentially concentrated in the tumor that overexpressed folate receptors. (Adapted from Cormode, D.P. et al., *Nano Lett.*, 8, 3715, 2008c; Chen, J. et al., *J. Am. Chem. Soc.*, 129, 5798, 2007. With permission.)

may be substituted with hydrophobic contrast-generating materials (Cormode et al., 2008c). Asides from the use of these agents as molecular-imaging contrast agents, imaging the interactions or distribution of lipoproteins *in vivo* is also attractive as there is a great need to further understand their biochemistry.

Fayad and coworkers favor the use of HDL for modification as a contrast agent and have reconstituted HDL with gadolinium and fluorophore-labeled phospholipids included in the synthesis feedstocks (Frias et al., 2004). This paramagnetic particle was applied in atherosclerotic mice in order to visualize cholesterol-rich plaques using MRI (Frias et al., 2006). Furthermore, we have reported a fully synthetic version of this contrast agent, where the apoA-I was replaced with 37pA, a peptide that forms effective HDL-mimics (Cormode et al., 2008a). Recently, we reported HDL whose core was modified to include diagnostically active nanocrystals, i.e., gold for CT, iron oxide for MRI, or quantum dots for fluorescence (Figure 24.9B). The iron oxide core HDL (FeO-HDL) produced darkening in the aorta of an atherosclerotic mouse when pre- and 24 h postinjection images are compared (Figure 24.9C and D). Gold-core HDL (Au-HDL) was shown to be taken up in the aortas of atherosclerotic mice via micro CT imaging (Figure 24.9E), and quantum dot core HDL (QD-HDL) produced excellent fluorescence in these aortas (Figure 24.9F).

Zheng et al. have developed an approach where LDL is labeled with amphiphilic fluorophores via incubation (Zheng et al., 2005; Chen et al., 2007). In addition, they targeted the particle to folic acid receptor overexpressing tumors by covalently attaching folate acid residues to apoB. Binding of these particles to KB tumor cells was initially shown using confocal microscopy (Zheng et al., 2005). Subsequently, these particles were applied *in vivo* using mice bearing two tumors in their flanks (Figure 24.9G) of which one KB tumor overexpressed the folate receptor and the contralateral HT1080 tumor did not (Chen et al., 2007). Five minutes after injection the agent was distributed throughout the mouse, but after 24 h the agent had localized preferentially in the KB tumor indicating that the folate targeting was successful. In addition, both this group and Aime's group have also reported gadolinium modified LDL that has been used to image cancer cells that overexpress the LDL receptor using MRI (Corbin et al., 2006; Crich et al., 2007).

24.4.7 Gold Nanoparticles

There has been massive interest in gold nanoparticles over the past 15 years due to their unique optical and electrochemical properties (Davis and Beer, 2005) and their applications in catalysis (Crooks et al., 2001), sensors (Haick, 2007; Cormode et al., 2008b), biodiagnostics (Rosi and Mirkin, 2005), disease treatment (Hainfeld et al., 2004; Paciotti et al., 2004; Pissuwan et al., 2007), medical imaging (Huang et al., 2007c; Kim et al., 2007), and materials (Hegmann et al., 2007; Klajn et al., 2007). Gold nanoparticles may be synthesized with excellent control of core size (Hiramatsu and Osterloh, 2004), unusual core morphologies such as rods (Durr et al., 2007) or shells (Skrabalak et al., 2007), and many different biologically compatible coatings such as polyvinylpyrrolidone (Yang et al., 2006) or PEG (Otsuka et al., 2003). There have been fewer medical-imaging studies using gold nanoparticles than using iron oxides, but the understanding of gold nanoparticle synthesis and surface chemistry is much more advanced and they may be exploited in many ways for imaging. Therefore, an increase in the number of reports that exploit gold nanoparticles as contrast agents is expected.

The properties of gold nanoparticles used for medical-imaging purposes are the strong x-ray attenuation of this element, an array of optical properties or as a scaffold for other contrast-inducing species such as gadolinium (Debouttière et al., 2006). Hainfeld et al. were the first to report the use of gold particles as an x-ray contrast agent (Hainfeld et al., 2006). They applied commercially sourced gold nanoparticles to a rat and were able to visualize the vasculature in x-ray images of the hindquarters (Figure 24.10A). Jeong and coworkers synthesized PEG-coated gold nanoparticles, which they used to image the heart and great vessels of a Sprague–Dawley rat (Figure 24.10B). Furthermore, this agent can be used to delineate tumors in the liver (Kim et al., 2007).

The father and son El-Sayed team is one of the major proponents of harnessing the optical properties of gold nanoparticles for contrast. They have exploited the intense light scattering of gold nanoparticles sized 30–100 nm (Orendorff et al., 2006) to distinguish human oral squamous carcinoma cells from noncancerous and other types of cancer by targeting the nanoparticles with an EGFR antibody (El-Sayed et al., 2005). The advantage of this technique is that the only equipment required is an ordinary microscope in its dark field setting. Nie and coworkers were able to detect tumors located in the flank of mice by taking advantage of the surface-enhanced Raman scattering of systemically injected EGFR-targeted gold nanoparticles (Qian et al., 2008). As a consequence of their very high emission under two-photon luminescence conditions due to excitation of their longitudinal surface plasmon and the possibility to be excited and imaged in the NIR window, gold nanorods have been proposed as contrast agents for this type of fluorescence imaging (Wang et al., 2005). Indeed, gold nanorods such as those pictured in Figure 24.10C have been used to identify skin cancer cells in tissue phantoms (Durr et al., 2007). A typical image from these experiments is shown in Figure 24.10D. Lastly, the optical properties of gold nanoparticles are suitable for the generation of contrast in reflectance imaging, an approach which has been performed using oligonucleotide-functionalized particles (Nitin et al., 2007).

24.4.8 Silica

Silica nanoparticles benefit from syntheses that can be tuned to create very monodisperse particle populations ranging in average diameter from less than 50 nm (Osseoasare and Arriagada, 1990) to 1000 nm (Stober et al., 1968). There are two interesting features of silica nanoparticle synthesis that make them particularly attractive for use as contrast agents. First, silica can be synthesized in the presence of cetyltrimethylammonium bromide, which results in mesoporous (honeycombed) particles (Trewyn et al., 2007). The pores in this silica can be loaded with contrast-generating material, creating effective contrast delivery vehicles (Tsai et al., 2007). Second, silica readily nucleates and grows on other nanoparticles, resulting in a silica shell of tunable thickness around the original nanoparticle. This has now been achieved with gold, iron oxide, quantum dots, and other types of core (Lu et al., 2002; Graf et al., 2003; Koole et al., 2008).

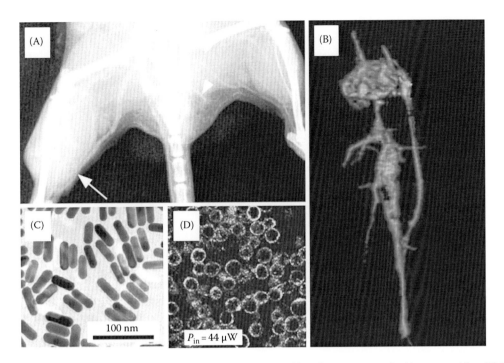

FIGURE 24.10 (A) X-ray image where the blood vessels of a rat are delineated by administration of gold nanoparticles. (B) CT angiogram of the heart and great vessels of a rat obtained by Kim et al. postinjection of PEG-coated gold nanoparticles. (C) TEM image of gold nanorods which were targeted with an EGFR antibody and (D) used to image cancer cells using two-photon luminescence imaging. (From Hainfeld, J.F. et al., *Br. J. Radiol.*, 79, 248, 2006; Kim, D. et al., *J. Am. Chem. Soc.*, 129, 7661, 2007; Durr, N.J. et al., *Nano Lett.*, 7, 941, 2007. With permission.)

The silica shell protects the diagnostically active cores from degradation and allows the size of the particle to be easily increased to the desired size.

Consequently, multiple silica-based contrast agents have been reported and some have been applied *in vivo* (Gerion et al., 2007; Taylor et al., 2008). We have recently published a paper where we included a quantum dot in the core of a silica nanoparticle and compared the biocompatibility of the particles in mice when bare or with a lipid coating (Van Schooneveld et al., 2008). We included gadolinium chelates and PEG in the lipid coating to allow MR visualization and to provide a long circulation half-life. The quantum dot allowed the determination of the blood half-life of the particles and of the particle tissue distribution via fluorescent imaging of blood samples and of tissue sections. The lipid coating resulted in greatly improved biocompatibility over the bare particles, with the latter forming aggregates in the lungs that blocked the blood vessels and caused necrosis in the liver. Surface modification strategies such as lipid coating are therefore recommended to make silica-based contrast agents biocompatible.

24.4.9 Viruses

Viruses are widely used in gene therapy (St George, 2003) and, as an extension of this concept, have recently been proposed as vectors for the delivery of contrast agents for medical imaging. Viruses are robust and uniform assemblies of protein subunits that include nucleic acids in their core (Manchester and Singh, 2006). Several groups have reported viruses that have

gadolinium included as MRI contrast agents. In the first such report, the cowpea chlorotic mottle virus was incubated with Gd^{3+} ions, which bound to metal-binding sites present in the proteins that make up this virus (Allen et al., 2005). Later reports have focused on first attaching a metal chelator to the surface of the virus, to which gadolinium is subsequently bound. This results in stronger and better defined binding of gadolinium to the virus and a r_1 of 14 mM^{-1}s^{-1} (Anderson et al., 2006). Recently Datta et al. reported a virus appended with novel chelates that allow faster water exchange and hence r_1 values of between 30.7 and 41.6 mM^{-1}s^{-1} (Datta et al., 2008).

It is also possible to modify viruses to act as contrast agents, by substituting the RNA core of the virus with diagnostically active nanocrystals such as gold, iron oxide, or quantum dots (Dixit et al., 2006; Huang et al., 2007a). When such nanocrystals possess a negative surface charge (e.g., when coated with carboxylic acid group) and are added to a solution of virus protein subunits, a switch in the solution conditions leads the protein to self-assemble around the nanocrystal (Chen et al., 2006). Cheon and coworkers have reported iron-oxide-labeled viruses which they showed to be taken up in cells by MR imaging (Huh et al., 2007).

24.4.10 Other Nanoparticulate Contrast Agents

A large number of other nanoparticle systems have been adapted for use as contrast agents, such as fullerenes, carbon nanotubes, zeolites, gadolinium oxides, and titania (Sitharaman et al., 2005; Bridot et al., 2007; Endres et al., 2007; Pereira et al., 2008; Shu et al., 2008). Fullerenes may be synthesized to encapsulate gadolinium, to act as MR contrast agents. Shu et al. found that such gadofullerenes, when organically modified on their surfaces, formed 30 nm aggregates (Shu et al., 2008). Antibodies to GFP were attached to these nanoparticles and their recognition properties for GFP probed by fluorescence techniques. Zeolites were adapted as MR contrast agents by replacing Al^{3+} ions in their formulation with various lanthanide elements (Pereira et al., 2008). While the effect of these doped zeolites on T_1 was found to be minor, their effect on T_2 was substantial, rendering them effective as T_2-contrast agents. The ineffectiveness for T_1 reduction was ascribed to poor access of water to the lanthanide ions for these zeolites.

Single-walled carbon nanotubes (SWNTs) are fluorescent in the near infrared (O'Connell et al., 2002). This property has been exploited by Cherukuri et al. to examine the pharmacokinetics of SWNTs in rabbits (Cherukuri et al., 2006). SWNTs coated with the surfactant Pluronic F108 were injected into rabbits and sequential blood draws were taken from the rabbits. The fluorescence from the blood allowed the circulation half-life of the SWNTs to be calculated to be 1 h. Alternatively, carbon nanotubes may be adapted to act as contrast agents by including contrast-generating materials, such as gadolinium. Sitharaman et al. included gadolinium in carbon nanotubes by first cutting them into short sections via fluorination and pyrolysis followed by sonication in $GdCl_3$ solution to create a contrast agent for MRI (Sitharaman et al., 2005).

Gadolinium oxide nanoparticles that were coated with silica, which incorporated a fluorophore (Bridot et al., 2007), have been reported. These nanoparticles were further coated with PEG chains and their contrast-inducing properties and biocompatibility were investigated. It was found that the MR contrast induced by gadolinium oxide nanoparticles 2.2–3.8 nm in diameter was about twice that produced by commercially available gadolinium chelates. The biodistribution of these nanoparticles was investigated via exploitation of their fluorescent properties and it was found that the majority of the dose was excreted via the bladder. Endres et al. attached gadolinium chelates known as DO3A to the surface of titanium particles to which DNA was absorbed (Endres et al., 2007). These particles were shown to be taken up in PC12 cells via x-ray fluorescence and MRI, thereby providing indications of successful DNA delivery to the cells.

24.5 Critical Discussion

The technology exists to adapt many types of nanoparticles to be used as contrast agents for medical imaging. Nanoparticulate contrast agents exhibit significant advantages over small-molecule-based agents. For example, long-circulating, nanoparticle-based contrast agents are available for CT. Additionally, nanoparticles, such as quantum dots, can be engineered to provide significantly superior contrast. Targeted nanoparticles for molecular imaging are allowing the elucidation of disease processes and the effect of drugs upon these processes, providing enormously valuable information. The particles discussed allow

a wide variety of diseases to be investigated using most of the available imaging techniques.

A major difficulty that faces the field is that, while a tremendous number of nanoparticulate contrast agents have been synthesized and reported, only a limited subset of them are likely to have sufficiently low toxicity profiles to be effective in animal imaging and even fewer for imaging in patients. Nanoparticles will be subject to the regulatory hurdles that apply to all medicinal products, but a key difficulty for nanoparticles is that many of them will not be excreted quickly via the urine, as is the case for small-molecule drugs and contrast agents, but will accumulate in the liver (Choi et al., 2007). How they will then be processed and degraded by biological systems is unclear. Toxicology for nanomedicine is a field that needs to vigorously advance over the next few years for the translation of these very useful particles to clinical use (Shaw et al., 2008).

24.6 Future Perspective

We expect to see some agents standardized and developed into products that will be used in animals for the investigation of disease processes and in the field of drug discovery and development. Furthermore, a more limited number of agents will be licensed for use in patients, allowing better diagnoses and readouts for response to therapy. On the nanotechnology side, we expect to see (1) further advances in nanoparticle synthesis that produce more effective contrast agents, (2) new uses for nanoparticles as contrast agents, such as the recent application of gold particles as optical contrast agents (Durr et al., 2007), and (3) new imaging techniques developed for which nanoparticles will be excellent contrast agents. The field of nanotoxicology will emerge and a large number of scientists will become skilled in this area, allowing the identification of low toxicity materials from which safe nanoparticles can be constructed for use *in vivo*.

References

Accardo, A., Tesauro, D., Roscigno, P. et al. (2004) Physicochemical properties of mixed micellar aggregates containing CCK peptides and Gd complexes designed as tumor specific contrast agents in MRI. *J. Am. Chem. Soc.*, 126, 3097–3107.

Ahrens, E. T., Flores, R., Xu, H., and Morel, P. A. (2005) In vivo imaging platform for tracking immunotherapeutic cells. *Nat. Biotechnol.*, 23, 983–987.

Allen, M., Bulte, J. W. M., Liepold, L. et al. (2005) Paramagnetic viral nanoparticles as potential high-relaxivity magnetic resonance contrast agents. *Magn. Reson. Med.*, 54, 807–812.

Allen, T. M., Hansen, C., Martin, F., Redemann, C., and Yau-Young, A. (1991) Liposomes containing synthetic lipid derivatives of poly(ethylene glycol) show prolonged circulation half-lives in vivo. *Biochim. Biophys. Acta* 1066, 29–36.

Amirbekian, V., Lipinski, M. J., Briley-Saebo, K. C. et al. (2007) Detecting and assessing macrophages in vivo to evaluate atherosclerosis noninvasively using molecular MRI. *Proc. Natl. Acad. Sci. U.S.A.*, 104, 961–966.

Anderson, E. A., Isaacman, S., Peadbody, D. S. et al. (2006) Viral nanoparticles donning a paramagnetic coat: Conjugation of MRI contrast agents to the MS2 capsid. *Nano Lett.*, 6, 1160–1164.

Avraamides, C. J., Garmy-Susini, B., and Varner, J. A. (2008) Integrins in angiogenesis and lymphangiogenesis. *Nat. Rev. Cancer*, 8, 604–617.

Azzazy, H. M. E., Mansour, M. M. H., and Kazmierczak, S. C. (2007) From diagnostics to therapy: Prospects of quantum dots. *Clin. Biochem.*, 40, 917–927.

Bangham, A. D. and Horne, R. W. (1964) Negative staining of phospholipids and their structural modification by surface active agents as observed in electron microscope. *J. Mol. Biol.*, 8, 660.

Berry, C. C., Wells, S., Charles, S., and Curtis, A. S. G. (2003) Dextran and albumin derivatised iron oxide nanoparticles: Influence on fibroblasts in vitro. *Biomaterials*, 24, 4551–4557.

Bottrill, M., Kwok, L., and Long, N. J. (2006) Lanthanides in magnetic resonance imaging. *Chem. Soc. Rev.*, 35, 557–571.

Bridot, J.-L., Faure, A.-C., Laurent, S. et al. (2007) Hybrid gadolinium oxide nanoparticles: Multimodal contrast agents for in vivo imaging. *J. Am. Chem. Soc.*, 129, 5076–5084.

Briley-Saebo, K. C., Mulder, W. J. M., Mani, V. et al. (2007) Magnetic resonance imaging of vulnerable atherosclerotic plaques: Current imaging strategies and molecular imaging probes. *J. Magn. Reson. Imaging*, 26, 460–479.

Briley-Saebo, K. C., Shaw, P. X., Mulder, W. J. M. et al. (2008) Targeted molecular probes for imaging atherosclerotic lesions with magnetic resonance using antibodies that recognize oxidation-specific epitopes. *Circulation*, 117, 3206–3215.

Bruchez, M., Jr., Moronne, M., Gin, P., Weiss, S., and Alivisatos, A. P. (1998) Semiconductor nanocrystals as fluorescent biological labels. *Science*, 281, 2013–2016.

Bulte, J. W. M. and Kraitchman, D. L. (2004) Iron oxide MR contrast agents for molecular and cellular imaging. *NMR Biomed.*, 17, 484–499.

Bulte, J. W. M., Duncan, I. D., and Frank, J. A. (2002) In vivo magnetic resonance tracking of magnetically labeled cells after transplantation. *J. Cereb. Blood Flow Metab.*, 22, 899–907.

Bulte, J. W. M., Zhang, S. C., Van Gelderen, P. et al. (1999) Neurotransplantation of magnetically labeled oligodendrocyte progenitors: Magnetic resonance tracking of cell migration and myelination. *Proc. Natl. Acad. Sci. U.S.A.*, 96, 15256–15261.

Catana, C., Procissi, D., Wu, Y. et al. (2008) Simultaneous in vivo positron emission tomography and magnetic resonance imaging. *Proc. Natl. Acad. Sci. U.S.A.*, 105, 3705–3710.

Chan, W. C. W. and Nie, S. M. (1998) Quantum dot bioconjugates for ultrasensitive nonisotopic detection. *Science*, 281, 2016–2018.

Chen, J., Corbin, I. R., Li, H. et al. (2007) Ligand conjugated low-density lipoprotein nanoparticles for enhanced optical cancer imaging in vivo. *J. Am. Chem. Soc.*, 129, 5798–5799.

Chen, C., Daniel, M.-C., Quinkert, Z. T. et al. (2006) Nanoparticle-templated assembly of viral protein cages. *Nano Lett.*, 6, 611–615.

Cherukuri, P., Gannon, C. J., Leeuw, T. K. et al. (2006) Mammalian pharmacokinetics of carbon nanotubes using intrinsic near-infrared fluorescence. *Proc. Natl. Acad. Sci. U.S.A.*, 103, 18882–18886.

Choi, H. S., Liu, W., Misra, P. et al. (2007) Renal clearance of quantum dots. *Nat. Biotechnol.*, 25, 1165–1170.

Choudhury, R. P., Fuster, V., and Fayad, Z. A. (2004) Molecular, cellular and functional imaging of atherothrombosis. *Nat. Rev. Drug Discov.*, 3, 913–925.

Corbin, I. R., Li, H., Chen, J. et al. (2006) Low-density lipoprotein nanoparticles as magnetic resonance imaging contrast agents. *Neoplasia*, 8, 488–498.

Cormode, D. P., Briley-Saebo, K. C., Mulder, W. J. M. et al. (2008a) An apoA-I mimetic peptide HDL-based MRI contrast agent for atherosclerotic plaque composition detection. *Small*, 4, 1437–1444.

Cormode, D. P., Davis, J. J., and Beer, P. D. (2008b) Anion sensing porphyrin functionalized nanoparticles. *J. Inorg. Organomet. Polym.*, 18, 32–40.

Cormode, D. P., Mulder, W. J. M., Fisher, E. A., and Fayad, Z. A. (2007) Modified lipoproteins as contrast agents for molecular imaging. *Future Lipidol.*, 2, 587–590.

Cormode, D. P., Skajaa, T., Van Schooneveld, M. M. et al. (2008c) Nanocrystal core high-density lipoproteins: A multimodal molecular imaging contrast agent platform. *Nano Lett.*, 8, 3715–3723.

Cormode, D. P., Skajaa, T., Fayad, Z. A., and Mulder, W. J. M. (2009) Nanotechnology in medical imaging: Probe design and applications. *Arterioscler. Thromb. Vasc. Biol.*, 29, 992–1000.

Corot, C., Robert, P., Idee, J.-M., and Port, M. (2006) Recent advances in iron oxide nanocrystal technology for medical imaging. *Adv. Drug Deliv. Rev.*, 58, 1471–1504.

Crich, S. G., Lanzardo, S., Alberti, D. et al. (2007) Magnetic resonance imaging detection of tumor cells by targeting low-density lipoprotein receptors with Gd-loaded low-density lipoprotein particles. *Neoplasia*, 9, 1046–1056.

Crooks, R. M., Zhao, M., Sun, L., Chechik, V., and Yueng, L. K. (2001) Dendrimer-encapsulated metal nanoparticle: Synthesis, characterization, and applications to catalysis. *Acc. Chem. Res.*, 34, 181–190.

Datta, A., Hooker, J. M., Botta, M. et al. (2008) High relaxivity gadolinium hydroxypyridonate viral capsid conjugates: Nanosized MRI contrast agents. *J. Am. Chem. Soc.*, 130, 2546–2552.

Davis, J. J. and Beer, P. D. (2005) Nanoparticles: Generation, surface functionalization, and ion sensing. In Schwarz, J. A., Contescu, C. I., and Putyera, K. (Eds.) *Encyclopedia of Nanoscience and Nanotechnology*. New York, Marcel Dekker.

De Vries, I. J. M., Lesterhuis, W. J., Barentsz, J. O. et al. (2005) Magnetic resonance tracking of dendritic cells in melanoma patients for monitoring of cellular therapy. *Nat. Biotechnol.*, 23, 1407–1413.

Deboutitière, P.-J., Roux, S., Vocanson, F. et al. (2006) Design of gold nanoparticles for magnetic resonance imaging. *Adv. Funct. Mater.*, 16, 2330–2339.

Del Vecchio, S., Zanneti, A., Fonti, R., Pace, L., and Salvatore, M. (2007) Nuclear imaging in cancer theranostics. *Q. J. Nucl. Med. Mol. Im.*, 51, 152–163.

Derfus, A. M., Chan, W. C. W., and Bhatia, S. N. (2004) Intracellular delivery of quantum dots for live cell labeling and organelle tracking. *Adv. Mater.*, 16, 961–966.

Dixit, S. K., Goicochea, N. L., Daniel, M.-C. et al. (2006) Quantum dot encapsulation in viral capsids. *Nano Lett.*, 6, 1993–1999.

Dodd, C. H., Hsu, H. C., Chu, W. J. et al. (2001) Normal T-cell response and in vivo magnetic resonance imaging of T cells loaded with HIV transactivator-peptide-derived superparamagnetic nanoparticles. *J. Immunol. Methods*, 256, 89–105.

Dubertret, B., Skourides, P., Norris, D. J. et al. (2002) In vivo imaging of quantum dots encapsulated in phospholipid micelles. *Science*, 298, 1759–1762.

Duguet, E., Vasseur, S., Mornet, S., and Devoisselle, J.-M. (2006) Magnetic nanoparticles and their applications in medicine. *Nanomedicine*, 1, 157–168.

Durr, N. J., Larson, T., Smith, D. K. et al. (2007) Two-photon luminescence imaging of cancer cells using molecularly targeted gold nanorods. *Nano Lett.*, 7, 941–945.

El-Sayed, I. H., Huang, X., and El-Sayed, M. A. (2005) Surface plasmon resonance scattering and absorption of anti-EGFR antibody conjugated gold nanoparticles in cancer diagnostics: Applications in oral cancer. *Nano Lett.*, 5, 829–834.

Endres, P. J., Paunesku, T., Vogt, S., Meade, T. J., and Woloschak, G. E. (2007) DNA-TiO$_2$ nanoconjugates labeled with magnetic resonance contrast agents. *J. Am. Chem. Soc.*, 129, 15760–15761.

Figdor, C. G., De Vries, I. J. M., Lesterhuis, W. J., and Melief, C. J. M. (2004) Dendritic cell immunotherapy: Mapping the way. *Nat. Med.*, 10, 475–480.

Franklin, R. J. M., Blaschuk, K. L., Bearchell, M. C. et al. (1999) Magnetic resonance imaging of transplanted oligodendrocyte precursors in the rat brain. *Neuroreport*, 10, 3961–3965.

Frias, J. C., Ma, Y., Williams, K. J., Fayad, Z. A., and Fisher, E. A. (2006) Properties of a versatile nanoparticle platform contrast agent to image and characterize atherosclerotic plaques by magnetic resonance imaging. *Nano Lett.*, 6, 2220–2224.

Frias, J. C., Williams, K. J., Fisher, E. A., and Fayad, Z. A. (2004) Recombinant HDL-like nanoparticles: A specific contrast agent for MRI of atherosclerotic plaques. *J. Am. Chem. Soc.*, 126, 16316–16317.

Funovics, M., Montet, X., Reynolds, F., Weissleder, R., and Josephson, L. (2005) Nanoparticles for the optical imaging of tumor E-selectin. *Neoplasia*, 7, 904–911.

Gerion, D., Herberg, J., Bok, R. et al. (2007) Paramagnetic silica-coated nanocrystals as an advanced MRI contrast agent. *J. Phys. Chem. C*, 111, 12542–12551.

Graf, C., Vossen, D. L. J., Imhof, A., and Van Blaaderen, A. (2003) A general method to coat colloidal particles with silica. *Langmuir*, 19, 6693–6700.

Gupta, A. K. and Gupta, M. (2005) Synthesis and surface engineering of iron oxide nanoparticles for biomedical applications. *Biomaterials*, 26, 3995–4021.

Haick, H. (2007) Chemical sensors based on molecularly modified metallic nanoparticles. *J. Phys. D: Appl. Phys.*, 40, 7173–7186.

Hainfeld, J. F., Slatkin, D. N., and Smilowitz, H. M. (2004) The use of gold nanoparticles to enhance radiotherapy in mice. *Phys. Med. Biol.*, 49, N309–N315.

Hainfeld, J. F., Slatkin, D. N., Focella, T. M., and Smilowitz, H. M. (2006) Gold nanoparticles: A new x-ray contrast agent. *Br. J. Radiol.*, 79, 248–253.

Hegmann, T., Qi, H., and Marx, V. M. (2007) Nanoparticles in liquid crystals: Synthesis, self-assembly, defect formation and potential applications. *J. Inorg. Organomet. Polym.*, 17, 483–508.

Hiramatsu, H. and Osterloh, F. E. (2004) A simple large-scale synthesis of nearly monodisperse gold and silver nanoparticles with adjustable sizes and with exchangeable surfactants. *Chem. Mater.*, 16, 2509–2511.

Hoehn, M., Kustermann, E., Blunk, J. et al. (2002) Monitoring of implanted stem cell migration in vivo: A highly resolved in vivo magnetic resonance imaging investigation of experimental stroke in rat. *Proc. Natl. Acad. Sci. U.S.A.*, 99, 16267–16272.

Huang, X., Bronstein, L. M., Retrum, J. et al. (2007a) Self-assembled virus-like particles with magnetic cores. *Nano Lett.*, 7, 2407–2416.

Huang, S. L., Hamilton, A. J., Pozharski, E. et al. (2002) Physical correlates of the ultrasonic reflectivity of lipid dispersions suitable as diagnostic contrast agents. *Ultrasound Med. Biol.*, 28, 339–348.

Huang, X., El-Sayed, I. H., Qian, W., and El-Sayed, M. A. (2007b) Cancer cells assemble and align gold nanorods conjugated to antibodies to produce highly enhanced, sharp and polarized surface Raman spectra: A potential cancer diagnostic marker. *Nano Lett.*, 7, 1591–1597.

Huang, X., Jain, P. K., El-Sayed, I. H., and El-Sayed, M. (2007c) Gold nanoparticles: Interesting optical properties and recent applications in cancer diagnostics and therapy. *Nanomedicine*, 2, 681–693.

Huh, Y. M., Lee, E. S., Lee, J. H. et al. (2007) Hybrid nanoparticles for magnetic resonance imaging of target-specific viral gene delivery. *Adv. Mater.*, 19, 3109–3112.

Hyafil, F., Cornily, J. C., Feig, J. E. et al. (2007) Noninvasive detection of macrophages using a nanoparticulate contrast agent for computed tomography. *Nat. Med.*, 13, 636–641.

Intes, X. (2005) Time-domain optical mammography SoftScan: Initial results. *Acad. Radiol.*, 12, 934–947.

Kabalka, G. W., Davis, M. A., Holmberg, E., Maruyama, K., and Huang, L. (1991) Gadolinium-labeled liposomes containing amphiphilic Gd-DTPA derivatives of varying chain-length—Targeted MRI contrast enhancement agents for the liver. *Magn. Reson. Imaging*, 9, 373–377.

Kelly, K. A., Allport, J. R., Tsourkas, A. et al. (2005) Detection of vascular adhesion molecule-1 expression using a novel multimodal nanoparticle. *Circ. Res.*, 96, 327–336.

Kim, D., Park, S., Lee, J. H., Jeong, Y. Y., and Jon, S. (2007) Antibiofouling polymer-coated gold nanoparticles as a contrast agent for in vivo x-ray computed tomography imaging. *J. Am. Chem. Soc.*, 129, 7661–7665.

Klajn, R., Bishop, K. J. M., Fialkowski, M. et al. (2007) Plastic and moldable metals by self-assembly of sticky nanoparticle aggregates. *Science*, 316, 261–264.

Klibanov, A. L. (2006) Microbubble contrast agents targeted ultrasound imaging and ultrasound-assisted drug-delivery applications. *Invest. Radiol.*, 41, 354–362.

Klibanov, A. L., Maruyama, K., Torchilin, V. P., and Huang, L. (1990) Amphipatic polyethyleneglycols effectively prolong the circulation time of liposomes. *FEBS Lett.*, 268, 235–238.

Kobayashi, H., Hama, Y., Koyama, Y. et al. (2007) Simultaneous multicolor imaging of five different lymphatic basins using quantum dots. *Nano Lett.*, 7, 1711–1716.

Koole, R., Van Schooneveld, M. M., Hilhorst, J. et al. (2008) On the incorporation mechanism of hydrophobic quantum dots in silica spheres by a reverse microemulsion method. *Chem. Mater.*, 20, 2503–2512.

Kwon, G. S. and Kataoka, K. (1995) Block copolymer micelles as long-circulating drug vehicles. *Adv. Drug Deliv. Rev.*, 16, 295–309.

Lanza, G. M., Wallace, K. D., Scott, M. J. et al. (1996) A novel site-targeted ultrasonic contrast agent with broad biomedical application. *Circulation*, 94, 3334–3340.

Lanza, G. M., Winter, P. M., Caruthers, S. D. et al. (2006) Nanomedicine opportunities for cardiovascular disease with perfluorocarbon nanoparticles. *Nanomedicine*, 1, 321–329.

Leatherdale, C. A., Woo, W.-K., Mikulec, F. V., and Bawendi, M. G. (2002) On the absorption cross section of CdSe nanocrystal quantum dots. *J. Phys. Chem. B*, 116, 7619–7622.

Levi, J., Cheng, Z., Gheysens, O. et al. (2007) Fluorescent fructose derivatives for imaging breast cancer cells. *Bioconjugate Chem.*, 18, 628–634.

Li, B. Q., Jia, D. C., Zhou, Y., Hu, Q. L., and Cai, W. (2006) In situ hybridization to chitosan/magnetite nanocomposite induced by the magnetic field. *J. Magnetism Magn. Mater.*, 306, 223–227.

Liu, H. L., Ko, S. P., Wu, J. H. et al. (2007) One-pot polyol synthesis of monosize PVP-coated sub-5 nm Fe_3O_4 nanoparticles for biomedical applications. *J. Magn. Magn. Mater.*, 310, E815–E817.

Lu, Y., Yin, Y., Mayers, B. T., and Xia, Y. (2002) Modifying the surface properties of superparamagnetic iron oxide nanoparticles through a sol-gel approach. *Nano Lett.*, 2, 183–186.

Lusis, A. J. (2000) Atherosclerosis. *Nature*, 407, 233–241.

Mahmood, U. and Josephson, L. (2005) Molecular MR imaging probes. *Proc. IEEE*, 93, 800–808.

Manchester, M. and Singh, P. (2006) Virus-based nanoparticles (VNPs): Platform technologies for diagnostic imaging. *Adv. Drug Deliv. Rev.*, 58, 1505–1522.

Marik, J., Tartis, M. S., Zhang, H. et al. (2007) Long-circulating liposomes radiolabeled with [18F]fluorodipalmitin ([18F]FDP). *Nucl. Med. Biol.*, 34, 165–171.

Mccarthy, J. R., Kelly, K. A., Sun, E. Y., and Weissleder, R. (2007) Targeted delivery of multifunctional magnetic nanoparticles. *Nanomedicine*, 2, 153–167.

Medintz, I. L., Clapp, A. R., Mattoussi, H. et al. (2003) Self-assembled nanoscale biosensors based on quantum dot FRET donors. *Nat. Mater.*, 2, 630–638.

Medintz, I. L., Uyeda, H. T., Goldman, E. R., and Mattoussi, H. (2005) Quantum dot bioconjugates for imaging, labelling and sensing. *Nat. Mater.*, 4, 435–446.

Midgley, P. A., Ward, E. P. W., Hungria, A. B., and Thomas, J. M. (2007) Nanotomography in the chemical, biological and materials sciences. *Chem. Soc. Rev.*, 36, 1477–1494.

Morawski, A. M., Winter, P. M., Yu, X. et al. (2004) Quantitative "Magnetic Resonance Immunohistochemistry" with ligand-targeted 19F nanoparticles. *Magn. Reson. Med.*, 52, 1255–1262.

Moschetta, A., Frederik, P. M., Portincasa, P., Vanberge-Henegouwen, G. P., and Van Erpecum, K. J. (2002) Incorporation of cholesterol in sphingomyelin-phosphatidylcholine vesicles has profound effects on detergent-induced phase transitions. *J. Lipid Res.*, 43, 1046–1053.

Mukundan, S., Ghaghada, K. B., Badea, C. T. et al. (2006) A liposomal nanoscale contrast agent for preclinical CT in mice. *Am. J. Roentgenol.*, 186, 300–307.

Mulder, W. J. M., Griffioen, A. W., Strijkers, G. J. et al. (2007) Magnetic and fluorescent nanoparticles for multimodality imaging. *Nanomedicine*, 2, 307–324.

Mulder, W. J. M., Koole, R., Brandwijk, R. J. et al. (2006) Quantum dots with a paramagnetic coating as a bimodal molecular imaging probe. *Nano Lett.*, 6, 1–6.

Mulder, W. J. M., Strijkers, G. J., Griffioen, A. W. et al. (2004) A liposomal system for contrast-enhanced magnetic resonance imaging of molecular targets. *Bioconjugate Chem.*, 15, 799–806.

Na, H. B., Lee, J. H., An, K. et al. (2007) Development of a T1 contrast agent for magnetic resonance imaging using MnO nanoparticles. *Angew. Chem. Int. Ed.*, 46, 5397–5401.

Nahrendorf, M., Zhang, H., Hembrador, S. et al. (2008) Nanoparticle PET-CT imaging of macrophages in inflammatory atherosclerosis. *Circulation*, 117, 379–387.

Neuberger, T., Gulani, V., and Webb, A. (2007) Sodium renal imaging in mice at high magnetic fields. *Magn. Reson. Med.*, 58, 1067–1071.

Nichols, A. V., Krauss, R. M., and Musliner, T. A. (1986) Nondenaturing polyacrylamide gradient gel electrophoresis. *Method Enzymol.*, 128, 417–431.

Nitin, N., Javier, D. J., and Richards-Kortum, R. (2007) Oligonucleotide-coated metallic nanoparticles as a flexible platform for molecular imaging agents. *Bioconjugate Chem.*, 18, 2090–2096.

Ntziachristos, V. (2006) Fluorescence molecular imaging. *Annu. Rev. Biomed. Eng.*, 8, 1–33.

Ntziachristos, V., Bremer, C., and Weissleder, R. (2003) Fluorescence imaging with near-infrared light: New technological advances that enable in vivo molecular imaging. *Eur. Radiol.*, 13, 195–208.

O'connell, M. J., Bachilo, S. M., Huffman, C. B. et al. (2002) Band gap fluorescence from individual single-walled carbon nanotubes. *Science*, 297, 593–596.

Orendorff, C. J., Sau, T. K., and Murphy, C. J. (2006) Shape-dependent plasmon-resonant gold nanoparticles. *Small*, 2, 636–639.

Osseoasare, K. and Arriagada, F. J. (1990) *Colloid Surf.*, 50, 321–339.

Otsuka, H., Nagasaki, Y., and Kataoka, K. (2003) PEGylated nanoparticles for biological and pharmaceutical applications. *Adv. Drug Deliv.*, 55, 403–419.

Paciotti, G. F., Meyer, L., Weinreich, D. et al. (2004) Colloidal gold: A novel nanoparticle vector for tumor directed drug delivery. *Drug Deliv.*, 11, 169–183.

Pankhurst, Q. A., Connolly, J., Jones, S. K., and Dobson, J. (2003) Applications of magnetic nanoparticles in biomedicine. *J. Phys. D: Appl. Phys.*, 36, R167–181.

Pereira, G. A., Ananias, D., Rocha, J. et al. (2008) NMR relaxivity of Ln^{3+}-based zeolite-type materials. *J. Mater. Chem.*, 15, 3832–3837.

Pissuwan, D., Valenzuela, S. M., Miller, C. M., and Cortie, M. B. (2007) A golden bullet? Selective targeting of toxoplasma gondii tachyzoites using antibody functionalized gold nanorods. *Nano Lett.*, 7, 3808–3812.

Qian, X., Peng, X.-H., Ansari, D. O. et al. (2008) In vivo tumor targeting and spectroscopic detection with surface-enhanced Raman nanoparticle tags. *Nat. Biotechnol.*, 26, 83–90.

Raymond, K. N. and Pierre, V. C. (2005) Next generation, high relaxivity gadolinium MR1 agents. *Bioconjugate Chem.*, 16, 3–8.

Reimer, P. and Tombach, B. (1998) Hepatic MRI with SPIO: Detection and characterization of focal liver lesions. *Eur. Radiol.*, 8, 1198–1204.

Rosi, N. L. and Mirkin, C. A. (2005) Nanostructures in biodiagnostics. *Chem. Rev.*, 105, 1547–1562.

Rossin, R., Pan, D., Qi, K. et al. (2005) ^{64}Cu-labeled folate-conjugated shell cross-linked nanoparticles for tumor imaging and radiotherapy: Synthesis, radiolabeling, and biologic evaluation. *J. Nucl. Med.*, 46, 1210–1218.

Rudd, J. H. F., Warburton, E. A., Fryer, T. D. et al. (2002) Imaging atherosclerotic plaque inflammation with [F-18]-fluorodeoxyglucose positron emission tomography. *Circulation*, 105, 2708–2711.

Ruehm, S. G., Corot, C., Vogt, P., Kolb, S., and Debatin, J. F. (2001) Magnetic resonance imaging of atherosclerotic plaque with ultrasmall superparamagnetic particles of iron oxide in hyperlipidemic rabbits. *Circulation*, 103, 415–422.

Sandstrom, M. C., Johansson, E., and Edwards, K. (2007) Structure of mixed micelles formed in PEG-lipid/lipid dispersions. *Langmuir*, 23, 4192–4198.

Schellenberger, E. A., Sosnovik, D., Weissleder, R., and Josephson, L. (2004) Magneto/optical annexin V, a multimodal protein. *Bioconjugate Chem.*, 15, 1062–1067.

Schmieder, A. H., Winter, P. M., Caruthers, S. D. et al. (2005) Molecular MR imaging of melanoma angiogenesis with alpha(nu)beta(3)-targeted paramagnetic nanoparticles. *Magn. Reson. Med.*, 53, 621–627.

Shah, P. K. (2006) HDL/apoA-I infusion for atherosclerosis management: An emerging therapeutic paradigm. *Future Lipidol.*, 1, 55–64.

Shaish, A., Keren, G., Chouraqui, P., Levkovitz, H., and Harats, D. (2001) Imaging of aortic atherosclerotic lesions by 125I-LDL, 125I-Oxidized-LDL, 125I-HDL and 125I-BSA. *Pathobiology*, 69, 225–229.

Shaw, S. Y., Westly, E. C., Pittet, M. J. et al. (2008) Perturbational profiling of nanomaterial biologic activity. *Proc. Natl. Acad. Sci. U.S.A.*, 105, 7387–7392.

Shu, C.-Y., Ma, X.-T., Zhang, J.-F. et al. (2008) Conjugation of a water-soluble gadolinium endohedral fulleride with an antibody as a magnetic resonance imaging contrast agent. *Bioconjugate Chem.*, 19, 651–655.

Sitharaman, B., Kissell, K. R., Hartman, K. B. et al. (2005) Superparamagnetic gadonanotubes are high-performance MRI contrast agents. *Chem. Commun.*, 3, 3915–3917.

Skrabalak, S. E., Au, L., Lu, X., Li, X., and Xia, Y. (2007) Gold nanocages for cancer detection and treatment. *Nanomedicine*, 2, 657–668.

So, M. K., Xu, C. J., Loening, A. M., Gambhir, S. S., and Rao, J. H. (2006) Self-illuminating quantum dot conjugates for in vivo imaging. *Nat. Biotechnol.*, 24, 339–343.

Sosnovik, D. and Weissleder, R. (2005) Magnetic resonance and fluorescence based molecular imaging technologies. *Prog. Drug Res.*, 62, 86–114.

St George, J. A. (2003) Gene therapy progress and prospects: Adenoviral vectors. *Gene Ther.*, 10, 1135–1141.

Stober, W., Fink, A., and Bohn, E. (1968) Controlled growth of monodisperse silica spheres in micron size range. *J. Colloid Interface Sci.*, 26, 62–69.

Sukhanova, A., Devy, M., Venteo, L. et al. (2004) Biocompatible fluorescent nanocrystals for immunolabeling of membrane proteins and cells. *Anal. Biochem.*, 324, 60–67.

Taylor, K. M. L., Kim, J. S., Rieter, W. J. et al. (2008) Mesoporous silica nanospheres as highly efficient MRI contrast agents. *J. Am. Chem. Soc.*, 130, 2154–2155.

Ter-Pogossian, M. M. (1985) Positron emission tomography instrumentation. In Reivich, M. and Alavi, A. (Eds.) *Positron Emission Tomography*. New York, Alan R. Liss Inc.

Tilcock, C., Unger, E., Cullis, P., and Macdougall, P. (1989) Liposomal Gd-DTPA—Preparation and characterization of relaxivity. *Radiology*, 171, 77–80.

Torchilin, V. P. (2002) PEG-based micelles as carriers of contrast agents for different imaging modalities. *Adv. Drug Deliv. Rev.*, 54, 235–252.

Torchilin, V. P., Frank-Kamenetsky, M. D., and Wolf, G. L. (1999) CT visualization of blood pool in rats by using long-circulating, iodine-containing micelles. *Acad. Radiol.*, 6, 61–65.

Trewyn, B. G., Slowing, I., Giri, S., Chen, H.-T., and Lin, V. S. Y. (2007) Synthesis and functionalization of a mesoporous silica nanoparticle based on the sol-gel process and applications in controlled release. *Acc. Chem. Res.*, 40, 846–853.

Trubetskoy, V. S., Gazelle, G. S., Wolf, G. L., and Torchilin, V. P. (1997) Block-copolymer of polyethylene glycol and polylysine as a carrier of organic iodine: Design of long-circulating particulate contrast medium for x-ray computed tomography. *J. Drug Target.*, 4, 381–388.

Tsai, C. -P., Hung, Y., Chou, Y.-H. et al. (2007) High-contrast paramagnetic fluorescent mesoporous silica nanorods as a multifunctional cell-imaging probe. *Small*, 4, 186–191.

Van De Ven, R. C. G., Hogers, B., Van Den Maagdenberg, A. et al. (2007) T-1 relaxation in in vivo mouse brain at ultra-high field. *Magn. Reson. Med.*, 58, 390–395.

Van Schooneveld, M. M., Vucic, E., Koole, R. et al. (2008) Improved biocompatibility and pharmacokinetics of silica nanoparticles by means of a lipid coating: A multimodality investigation. *Nano Lett.*, 8, 2517–2525.

Van Tilborg, G. A. F., Mulder, W. J. M., Deckers, N. et al. (2006) Annexin A5-functionalized bimodal lipid-based contrast agents for the detection of apoptosis. *Bioconjugate Chem.*, 17, 741–749.

Villanueva, F. S., Jankowski, R. J., Klibanov, S. et al. (1998) Microbubbles targeted to intercellular adhesion molecule-1 bind to activated coronary artery endothelial cells. *Circulation*, 98, 1–5.

Villanueva, F. S., Lu, E. X., Bowry, S. et al. (2007) Myocardial ischemic memory imaging with molecular echocardiography. *Circulation*, 115, 345–352.

Villanueva, F. S. and Wagner, W. R. (2008) Ultrasound molecular imaging of cardiovascular disease. *Nat. Clin. Pract. Cardiovasc. Med.*, 5, S26–S32.

Walczak, P., Kedziorek, D. A., Gilad, A. A., Lin, S., and Bulte, J. W. M. (2005) Instant MR labeling of stem cells using magnetoelectroporation. *Magn. Reson. Med.*, 54, 769–774.

Wang, H. F., Huff, T. B., Zweifel, D. A. et al. (2005) In vitro and in vivo two-photon luminescence imaging of single gold nanorods. *Proc. Natl. Acad. Sci. U.S.A.*, 102, 15752–15756.

Wang, Y. X. J., Hussain, S. M., and Krestin, G. P. (2001) Superparamagnetic iron oxide contrast agents: Physicochemical characteristics and applications in MR imaging. *Eur. Radiol.*, 11, 2319–2331.

Weissleder, R. and Mahmood, U. (2001) Molecular imaging. *Radiology*, 219, 316–333.

Weissleder, R., Lee, A. S., Khaw, B. A., Shen, T., and Brady, T. J. (1992) Antimyosin-labeled monocrystalline iron-oxide allows detection of myocardial infarct—MR antibody imaging. *Radiology*, 182, 381–385.

Weller, G. E. R., Wong, M. K. K., Modzelewski, R. A. et al. (2005) Ultrasonic imaging of tumor angiogenesis using contrast microbubbles targeted via the tumor-binding peptide arginine-arginine-leucine. *Cancer Res.*, 65, 533–539.

Wickline, S. A., Neubauer, A. M., Winter, P. M., Caruthers, S. D., and Lanza, G. M. (2007) Molecular imaging and therapy of atherosclerosis with targeted nanoparticles. *J. Magn. Reson. Imaging*, 25, 667–680.

Winter, P. M., Morawski, A. M., Caruthers, S. D. et al. (2003) Molecular imaging of angiogenesis in early-stage atherosclerosis with alphavbeta3-integrin-targeted nanoparticles. *Circulation*, 108, 2270–2274.

Winter, P. M., Neubauer, A. M., Caruthers, S. D. et al. (2006) Endothelial alpha(v)beta(3) integrin-targeted fumagillin nanoparticles inhibit angiogenesis in atherosclerosis. *Arterioscler. Thromb. Vasc. Biol.*, 26, 2103–2109.

Winter, P. M., Shukla, H. P., Caruthers, S. D. et al. (2005) Molecular imaging of human thrombus with computed tomography. *Acad. Radiol.*, 12, S9–13.

Woods, M., Woessner, D. E., and Sherry, A. D. (2006) Paramagnetic lanthanide complexes as PARACEST agents for medical imaging. *Chem. Soc. Rev.*, 35, 500–511.

Wu, N. Z., Da, D., Rudoll, T. L. et al. (1993) Increasing microvascular permeability contributes to preferential accumulation of stealth liposomes in tumor tissue. *Cancer Res.*, 53, 3765–3770.

Wu, X. Y., Liu, H. J., Qi, L. J. et al. (2003) Immunofluorescent labeling of cancer marker Her2 and other cellular targets with semiconductor quantum dots. *Nat. Biotechnol.*, 21, 41–46.

Wunderbaldinger, P., Josephson, L., and Weissleder, R. (2002) Crosslinked iron oxides (CLIO): A new platform for the development of targeted MR contrast agents. *Acad. Radiol.*, 9, S304–306.

Xing, Y., Chaudry, Q., Shen, C. et al. (2007) Bioconjugated quantum dots for multiplexed and quantitative immunohistochemistry. *Nat. Protoc.*, 2, 1152–1165.

Yang, Y., Wang, W., Li, J., Mu, J., and Rong, H. (2006) Manipulating the solubility of gold nanoparticles reversibly and preparation of water-soluble sphere nanostructure through micellar-like solubilization. *J. Phys. Chem. B*, 110, 16867–16873.

Yoshida, M., Roh, K. H., and Lahann, J. (2007) Short-term biocompatibility of biphasic nanocolloids with potential use as anisotropic imaging probes. *Biomaterials*, 28, 2446–2456.

Yu, S. B. and Watson, A. D. (1999) Metal-based x-ray contrast media. *Chem. Rev.*, 99, 2353–2377.

Zacharakis, G., Kambara, H., Shih, H. et al. (2005) Volumetric tomography of fluorescent proteins through small animals in vivo. *Proc. Natl. Acad. Sci. U.S.A.*, 102, 18252–18257.

Zetter, B. R. (2008) The scientific contributions of M. Judah Folkman to cancer research. *Nat. Rev. Cancer*, 8, 647–654.

Zhang, Y. and Zhang, J. (2005) Surface modification of monodisperse magnetite nanoparticles for improved intracellular uptake to breast cancer cells. *J. Colloid Interface Sci.*, 283, 352–357.

Zhang, Y., Kohler, N., and Zhang, M. (2002) Surface modification of superparamagnetic magnetite nanoparticles and their intracellular uptake. *Biomaterials*, 23, 1553–1561.

Zheng, G., Chen, J., Li, H., and Glickson, J. D. (2005) Rerouting lipoprotein nanoparticles to selected alternate receptors for the targeted delivery of cancer diagnostic and therapeutic agents. *Proc. Natl. Acad. Sci. U.S.A.*, 102, 17757–17762.

Zitzmann, S., Ehemann, V., and Schwab, M. (2002) Arginine-glycine-aspartic acid (RGD)-peptide binds to both tumor and tumor-endothelial cells in vivo. *Cancer Res.*, 62, 5139–5143.

Optical Nanosensors for Medicine and Health Effect Studies

Tuan Vo-Dinh
Duke University

Yan Zhang
Duke University

25.1 Introduction

Nanosensors provide important tools for monitoring biotargets and molecular signaling processes within single living cells, thus providing the critical information that could be critical to biomedical research and clinical applications. Fiberoptic nanosensors can be fabricated to have nanoscale dimensions, which make them suitable for sensing intracellular/intercellular physiological and biological parameters in submicron environments. A large variety of fiberoptic chemical sensors and biosensors have been developed in our laboratory for environmental and biochemical monitoring (Vo-Dinh et al., 1987; Tromberg et al., 1988; Alarie et al., 1990a; Alarie et al., 1990b; Alarie and Vo-Dinh, 1991; Bowyer et al., 1991; Vo-Dinh et al., 1991; Vo-Dinh et al., 1993; Alarie and Vo-Dinh, 1996). Tapered fibers with submicron distal diameters between 20 and 500 nm have been developed for near-field scanning optical microscopy (NSOM) (Pohl, 1991; Betzig and Chichester, 1993). NSOM was also used to achieve sub-wavelength 100 nm spatial resolution in Raman detection (Deckert et al., 1998; Zeisel et al., 1998). Chemical nanosensors were developed for monitoring calcium and nitric oxide, among other physicochemicals in single cells (Tan et al., 1992a; Tan et al., 1992b). Vo-Dinh and coworkers have developed nanobiosensors with antibody probes to detect biochemical targets inside living single cells (Cullum et al., 2000a; Cullum and Vo-Dinh, 2000b; Vo-Dinh et al., 2000a; Vo-Dinh and Cullum, 2000b; Vo-Dinh et al., 2000c; Vo-Dinh et al., 2001; Kasili et al., 2002; Kasili et al., 2004a; Kasili and Vo-Dinh, 2004b; Song et al., 2004; Kasili and Vo-Dinh, 2005; Vo-Dinh and Kasili, 2005; Vo-Dinh et al., 2006;

Vo-Dinh, 2008). This chapter presents an overview of the principle, development, and applications of fiberoptic nanosensors for medicine and health-effect studies.

Optical fibers having nanoscale tips were developed initially to serve as scanning probes in near-field optical microscopes (Pohl, 1991). The development of one such probe capable of obtaining measurements with a spatial sub-wavelength resolution was reported (Betzig and Chichester, 1993). Due to its high spatial resolution (sub-wavelength), near-field microscopy has received great interest and has been used in many applications (Pohl, 1991). For example, near-field surface-enhanced Raman spectroscopy (NF-SERS) has been used for the measurement of single-dye and dye-labeled DNA molecules with a resolution of 100 nm (Deckert et al., 1998; Zeisel et al., 1998). In this application, DNA strands labeled with the dye brilliant cresyl blue (BCB) were spotted on a SERS-active substrate that was prepared by evaporation of silver on a nanoparticle-coated substrate. The silver-coated nanostructured substrates are capable of inducing the SERS effect, which could enhance the Raman signal of the adsorbate molecules up to 10^8 times (Vo-Dinh, 1998). NF-SERS spectra were collected by illuminating the sample using the nanoprobe and detecting the SERS signals using a spectrometer equipped with a charge-coupled device (CCD). Raster scanning the fiber nanoprobe over the sample produced a two-dimensional SERS image of the DNA on the surface of the substrate with sub-wavelength spatial resolution. Near-field optical microscopy promises to be an area of growing research that could potentially provide an imaging tool for monitoring individual cells and even biological molecules. Single-molecule detection and imaging schemes using nanofibers could open new possibilities in the investigation of the complex biochemical reactions and pathways in biological and cellular systems leading to important applications in medicine and health-effect studies.

25.2 Fundamentals of Biosensors

A unique application of optical fibers is their use in biosensors. A biosensor generally consists of a measurement system that utilizes a probe with biological recognition element, often called a bioreceptor, and a transducer (Vo-Dinh et al., 1993; Vo-Dinh and Cullum, 2000b). Two basic operating principles of a biosensor include (1) "biological recognition" and (2) "sensing." A biosensor can be generally defined as a device that consists of two basic components connected in series: (1) a biological recognition system, often called a bioreceptor, and (2) a transducer. The recognition system is aimed at providing the sensor with a high degree of selectivity for the analyte to be measured. The interaction between the analyte and the bioreceptor is aimed at producing an effect measured by the transducer, which converts the information into a measurable effect, such as an electrical signal.

Biosensors can use various transduction methods: (1) optical detection methods, (2) electrochemical detection methods, and (3) mass-based detection methods. Other detection methods include voltaic and magnetic methods. New types of transducers are constantly being developed for use in biosensors. In this chapter, special emphasis is placed on optical transducing principles. Figure 25.1 illustrates the conceptual principle of the biosensing process.

Bioreceptors are responsible for binding the analyte of interest to the sensor for the measurement. Bioreceptors may take many forms and can generally be classified into five different major categories: (1) antibody/antigen, (2) enzymes, (3) nucleic acids/DNA, (4) cellular structures/cells, and (5) biomimetic. The following section discusses the example of antibody probes. Antigen–antibody (Ag–Ab) binding reaction, which is a key mechanism by which the immune system detects and eliminates foreign matter, provides the basis for specificity of immunoassays. Antibodies, which are complex biomolecules made up of hundreds of individual amino acids, are produced by immune system cells when such cells are exposed to substances or molecules that are called antigens. The antibodies called forth following antigen exposure have recognition/binding sites for specific molecular structures (or substructures) of the antigen. The process in which antigen and antibody interact is analogous to a lock and key fit, in which specific configurations of a unique key enable it to open a lock. Due to this three-dimensional shape fitting, and the diversity inherent in individual antibody makeup, it is possible to find an antibody that can recognize and bind to any one of the large variety of molecular shapes. This unique property of antibodies is the key to their usefulness in antibody-base biosensor, often referred to as immunosensors. One can then use such antibodies as specific probes to recognize and bind to an analyte of interest that is present, even in extremely small amounts, within a large number of other chemical substances. Advances

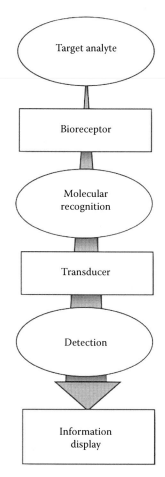

FIGURE 25.1 Operating principle of biosensor systems.

in spectrochemical instrumentation, laser miniaturization, biotechnology, and fiberoptics research have provided opportunities for novel approaches in the development of sensors for the detection of chemicals and biological materials of environmental and biomedical interest. Since the first development of a remote fiberoptic immunosensor for *in situ* detection of the chemical carcinogen benzo[a]pyrene (Vo-Dinh et al., 1987), antibodies have become common bioreceptors used in biosensors today.

25.3 Design and Fabrication of Optical Nanosensors

25.3.1 Fabrication of Fiberoptic Nanoprobes

This section discusses the protocols and instrumental systems involved in the fabrication of fiberoptic nanoprobes. Two methods are generally used for preparing the nanofiber tips. The so-called heat and pull method is the most commonly used. This method consists of local heating of a glass fiber using a laser or a filament and subsequently pulling the fiber apart. The shape of the nanofiber tips obtained depends on controllable experimental parameters such as the temperature and the timing of the procedure. The second method, often referred to as the "Turner's method," involves chemical etching of glass fibers. In a variation of the standard etching scheme, the taper is formed inside the polymer cladding of the glass fibers. The description of these fabrication methods is given in the following section.

The fabrication of nanosensors requires techniques capable of making reproducible optical fibers with submicron-size diameter core. As the laser-pulling process is a time-dependent heating effect, laser power, timing of pulling, velocity setting, and pulling force all contribute to the taper shape and tip size. Since transmission efficiency is highly dependent on the taper shape, it is crucial to control the tip shape in the fabrication of high-quality nanoprobes. In our lab, nanoprobes were pulled from a large-core optical fiber using a micropipette puller (Sutter Instruments P-2000). Figure 25.2 illustrates the experimental procedures for the fabrication of nanofibers using the heat and pull procedure (Vo-Dinh et al., 2000a). A scanning election microscopy (SEM) photograph of one of the fiber probes fabricated for studies is shown in Figure 25.3. The distal end of the nanofiber is approximately 40 nm.

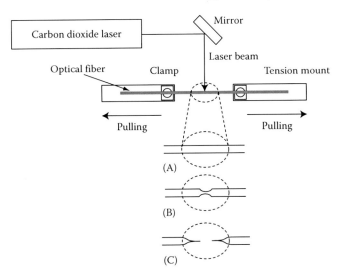

FIGURE 25.2 The "heat and pull" method for the fabrication of nanofibers. (A)–(C) represent the course of pull in time. In (C) a nanotip has been formed.

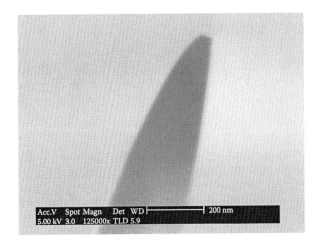

FIGURE 25.3 Scanning electron photograph of an uncoated nanofiber. The size of the fiber tip diameter is approximately 40 nm.

Chemical etching using HF is the basis of the second method for fabricating optical nanofibers. There are two variations of the HF etching method, one method involving the use of a mixture of HF acid and organic solvent, known as Turner etching (Turner, 1984), and the second using only HF, known as tube etching (Hoffmann et al., 1995; Lambelet et al., 1998; Stockle et al., 1999). In the Turner method, a fiber is placed in the meniscus between the HF and the organic overlayer and over time, a small tip is formed, with a smooth, large-angled taper. This large taper angle provides much more light at the tip of the fiber, which in turn greatly increases the sensitivity of the nanosensors. The reproducibility of the Turner method is strongly affected by environmental parameters, such as temperature and vibration, because of the dual chemical nature of the etching process. To avoid this problem, a variation of the etching method was developed, which involves a tube-etching procedure. In this procedure, an optical fiber with a silica core and an organic cladding material is placed in an HF solution. The HF slowly dissolves the silica core producing a fiber with a large taper angle and nanometer-sized tip. The HF begins first to dissolve the fiber's silica core, while not affecting the organic cladding material. This unaffected cladding creates localized convective currents in the HF solution, which causes a tip to be formed. After some time, more of the silica core is dissolved until it emerges above the surface of the HF solution. At this juncture, the HF is drawn up the cladding walls via capillary action and runs down the silica core to produce a nanometer-sized tip. By varying the time of HF exposure and the depth to which the fiber is submerged in the HF solution, one can control the size of the fiber tip and the angle of the taper. Once the tip has been formed, the protruding cladding can be removed either with a suitable organic solvent or by simply burning it off. Nanotips fabricated using etching procedures, which can be designed to have sharp tips, have been used in NSOM studies to detect SERS-labeled DNA molecules on solid substrates at the sub-wavelength spatial resolution (Deckert et al., 1998; Zeisel et al., 1998).

To prevent light leakage of the excitation light on the tapered side of the fiber, the sidewall of the tapered end is then coated with a thin layer of metal, such as silver, aluminum, or gold (75–100 nm thickness) using a thermal evaporation metal-coating device. Such a coating system is illustrated in Figure 25.4. The fiber probe is attached on a rotating motor inside a thermal evaporation chamber (Vo-Dinh et al., 1993; Cullum and Vo-Dinh, 2000b; Vo-Dinh et al., 2000a). The fiber is pointing away from the evaporation source with an angle of approximately 25°. While the probe is rotated, the metal is allowed to evaporate onto the tapered side of the fiber tip to form a thin coating. The tapered end is coated with ~75–100 nm of metal in a Quorum Technologies E6700 Bench Top Vacuum Evaporator. With the metal coating, the size of the probe tip is approximately 250–300 nm (Figure 25.5). Since the fiber tip is pointed away from the metal source, it remains free from any metal coating. A nanoaperture was formed on the distal end of the fiber tip for optical excitation and subsequent binding with bioreceptors. The size of the nanoaperture is related to the angle between fiber

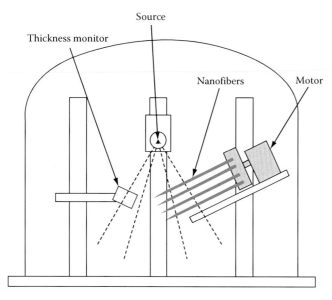

FIGURE 25.4 Instrumental setup for coating the nanofiber tips with metals.

axis and evaporation direction. For example, if the angle is less than 20°, most of the fibers are fully covered with metals and no aperture is visible using SEM. On the contrary, a larger area of the distal end of the fiber tip will be exposed if the angle is higher than 30°. The optimal angle of inclination can be determined through the characterization of nanoapertures under SEM.

Metallic coating process is an important step in nanoprobe fabrication. Silver has been used in nanoprobe fabrication (Vo-Dinh et al., 2000a). It has high reflectivity in the visible and IR range and is very stable in aqueous solutions as long as oxidizing agents or complexing agents are not present. But silver layer tarnish under ordinary atmospheric conditions and do not have a high reflectance below 400 nm. It is desirable to use the nanoprobe right after evaporation. Gold thin film was also demonstrated to be a very stable coating under environmental conditions, whereas it does not have high reflectance in the visible range. An interface layer such as Cr is required to increase adhesion between gold and fiber silica surface. Aluminum is the most desirable material to use because it has the highest extinction coefficient of all metals. Aluminum adheres to fibers more firmly than silver or gold so that no interface layer is required

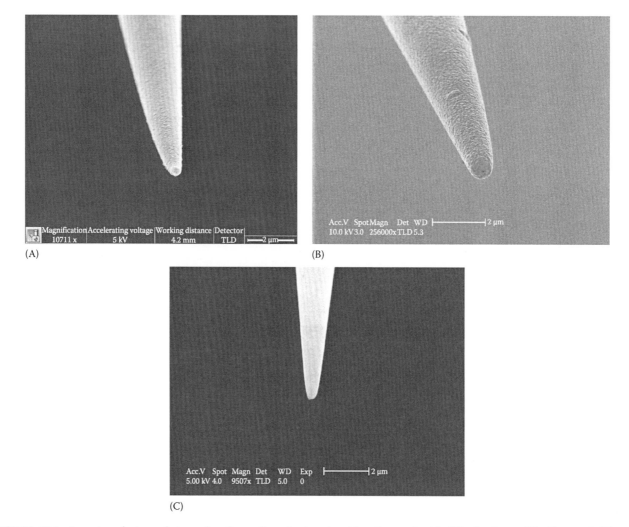

FIGURE 25.5 Scanning electron photographs of nanofibers having the sidewall coated with (A) aluminum, (B) silver, and (C) gold.

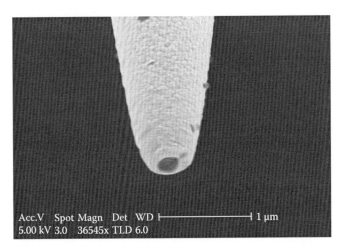

FIGURE 25.6 FIB-etched nanoprobe with aperture diameter of 200 nm.

and general cleaning does not affect the coating. The aluminum is inert toward corrosive agents since a protective oxide layer forms upon contact with the air. However, it is difficult to evaporate aluminum as a thin film while maintaining smooth films with small grain size (Holland, 1956). Grainy films contribute to the high background in near-field sensing. Higher coating rates minimize the grain size to ensure better production efficiency.

Traditional manufacturing processes still limit the quality of metal-coated fiber probes. Optical throughput of pulled nanoprobes is limited by the sharp taper angle. Chemical etched tips have higher throughput; however, they do not have a flat distal end as the laser-pulled ones, as it is difficult to form well-defined nanoapertures in shadow evaporation. Moreover, shadow evaporation often leads to either complete or irregular coated tip. Grainy structures of metal thin film increase the distance between the aperture and the sample, which reduces the resolution and intensity. It is also easy to form pin holes at the taper region that causes leaking. The aperture deviates from ideal circular shape because of grains. A quantitative analysis of probe transmission efficiency becomes difficult. Focused ion beam (FIB) fabrication of nanostructures has been applied on optical fibers for chemical sensing (Dhawan et al., 2008). FIB milling for nanostructure formation allows precise control of size and shape in nanometer accuracy. We have applied FIB for nanoprobe fabrication. By milling with a focused ion beam, an aperture with controllable shape and diameter was achieved (Figure 25.6). The angle of evaporation is not necessary in FIB, therefore reducing the chance of pin-hole formation. A clean aperture free from grains also facilitates the subsequent functionalization for biosensing. FIB processing is a promising technique in nanoprobe fabrication in addition to laser pulling and chemical etching.

25.3.2 Immobilization of Bioreceptors

The next step in the preparation of the nanobiosensor probes involves the covalent immobilization of bioreceptor molecules onto the fiber tip. For antibody binding, several strategies can be used to retain the antibody at the sensing probe. Antibodies can be immobilized onto the nanofiber probes by using a chemical immobilization method. Silane modification techniques eliminate the nonspecific binding potential of silica for biomolecules. The modification of silica surface provides sites for coupling affinity ligands through covalent derivatization with a functional silane containing a functional group. For example, reaction of silica with 3-aminopropyltriethoxysilane (APTS) under the appropriate conditions coats the surface with primary amino groups for conjugation with electrophilic groups. The selection of appropriate silane functional group for surface modification provides a broad range of properties for the subsequent coupling of biomolecules. In our study, the fiber is derivatized in 10% GOPS in H_2O (v/v) at 90°C for 3 h. The pH of the mixture is maintained below 3 with concentrated HCl (1 M). After derivatization, the fiber is washed in ethanol and dried overnight in a vacuum oven at 105°C. The fiber is then coated with silver as described previously. The derivatized fiber is activated in a solution of 100 mg/mL 1,1′ carbonyldiimidazole (CDI) in acetonitrile for 20 min followed by rinsing with acetonitrile and then phosphate-buffered saline (PBS). The fiber tip is then incubated in a 1.2 mg/mL antibody solution (PBS solvent) for 4 days at 4°C and then stored overnight in PBS to hydrolyze any unreacted sites. The fibers are then stored at 4°C with the antibody immobilized tips stored in PBS. This procedure has been shown to maintain over 95% antibody activity (Vo-Dinh et al., 2000a).

25.4 Experimental Method and Instrumentation

25.4.1 Experimental Procedure

A unique capability of nanosensors is the possibility to monitor living single cells in vivo (Cullum and Vo-Dinh, 2000b; Vo-Dinh et al., 2000a; Vo-Dinh and Cullum, 2000b; Vo-Dinh et al., 2001). This section describes the experimental procedures for growing cell cultures for analysis using the nanosensors. Cell cultures were grown in a water-jacketed cell culture incubator at 37°C in an atmosphere of 5% CO_2 in air. Clone 9 cells, a rat liver epithelial cell line, were grown in Ham's F-12 medium (Gibco) supplemented with 10% fetal bovine serum and an additional 1 mM glutamine (Gibco). In preparation for an experiment, 1×10^5 cells in 5 mL of medium were seeded into standard dishes (Corning Costar Corp.). The growth of the cells was monitored daily by microscopic observation and, when the cells reached a state of confluence of 50%–60%, the analyte solution was added and left in contact with the cells for 18 h (i.e., overnight). This procedure is designed to incubate the cells with the analyte molecules for subsequent monitoring using the nanosensors. The growth conditions were chosen so that the cells would be in the log-phase growth during the chemical treatment, but would not be so close to confluence that a confluent monolayer would form by the termination of the chemical exposure. The analyte solution was prepared as a 1 mM stock solution in reagent grade methanol and further diluted in reagent grade ethanol (95%) prior to addition to the cells. Following chemical treatment, the medium containing the analyte was aspirated and replaced with standard growth medium, prior to the nanoprobe procedure.

Measurements with nanosensors were then performed using the following protocols. A culture dish of cells was placed on the pre-warmed microscope stage, and the nanoprobe, mounted on the micropipette holder, was moved into position (i.e., in the same plane of the cells), using bright field microscopic illumination, so that the tip was outside the cell to be probed. The total magnification was usually 400x. Under no room light and no microscope illumination, the laser shutter was opened to illuminate the optical fiber for excitation of the analyte molecules bound to the antibodies at the fiber tip. Usually, if the silver coating on the nanoprobe was appropriate, no light leaked out of the sidewall of the tapered fiber. Only a faint glow of laser excitation at the tip could be observed on the nanoprobe. A reading was first taken with the nanoprobe outside the cell and the laser shutter closed. The nanoprobe was then moved into the cell, inside the cell membrane, and extended into the cellular compartments of interest. The laser was again opened, and readings were then taken and recorded as a function of time during which the nanoprobe was inside the cell.

25.4.2 Instrumentation

The optical measurement system used for monitoring single cells using the nanosensors is schematically illustrated in Figure 25.7 (Cullum and Vo-Dinh, 2000b; Vo-Dinh et al., 2000a; Vo-Dinh and Cullum, 2000b; Vo-Dinh et al., 2001). Laser excitation light, either with the 325 nm line of a HeCd laser (Omnichrome, 8 mW laser power) or the 488 nm line of an argon-ion laser (Coherent, 10 mW), was focused onto a 600 μm delivery fiber, which was connected to the nanofiber through an SMA connector. The nanofiber was secured to a micromanipulator on the microscope. The experimental setup used to probe single cells was adapted for this purpose from a standard micromanipulation/microinjection apparatus. A Nikon Diaphot 300 inverted microscope (Nikon, Inc., Melville, New York) with Diaphot 300/Diaphot 200 Incubator, to maintain the cell cultures at ~37°C on the microscope stage, was used for these experiments. The micromanipulation equipment used consisted of MN-2 (Narishige Co., LTD, Tokyo, Japan) Narishige three-dimensional manipulators for coarse adjustment, and Narishige MMW-23 three-dimensional hydraulic micromanipulators for final movements. The optical fiber nanoprobe was mounted on a micropipette holder (World Precision Instruments, Inc., Sarasota, Florida). The fluorescence emitted from the cells was collected by the microscope objective and passed through an appropriate longpass dichroic mirror to eliminate the laser excitation scatter light. The fluorescence beam was then focused onto a photomultiplier tube (PMT) for detection. The output from the PMT was passed through a picoammeter and recorded on a strip chart recorder or a personal computer (PC) for further data treatment. To record the fluorescence of analyte molecules binding to antibodies at the fiber tip, a Hamamatsu PMT detector assembly (HC125–2) was mounted in the front port of the Diaphot 300 microscope, and fluorescence was collected via this optical path (80% of available light at the focal plane can be collected through the front port). A charge-coupled device (CCD) mounted onto another port of the microscope could be used to record images of the nanosensor monitoring single cells.

25.5 Applications

25.5.1 Intracellular Fluorescence Measurements Using Nanosensors

The small size of fiberoptic nanosensors allows for measurements in submicron environments and for probing individual chemical species in specific locations throughout a cell. Previously, such measurements were limited to the field of fluorescence microscopy, where a fluorescent dye was inserted into a cell and allowed to diffuse throughout. Depending upon the particular fluorescent dye that was chosen, changes in the fluorescence properties of the dye could then be monitored, in an imaging modality, as the dye came in contact with the analyte of interest. Since this technique relies on imaging the fluorescent dye, it requires the homogenous dispersion of the dye through the various locations in the cell, which is limited by intracellular conditions (i.e., pH, etc.) or often does not even occur due to compartmentalization by the cell. Fiberoptics nanobiosensors, therefore, could offer significant improvements over such methods to the problems associated with cellular diffusion. Optical nanofiber probes without antibody probes have been fabricated and used to monitor fluorescence emission from chemical species inside living single cells (Vo-Dinh et al., 2000c). In this study, mouse epithelial cells were incubated with a fluorescent dye by incubating the cells in the dye solution and allowing membrane permeabilization to take place. Another procedure for loading cells with fluorophores involved the method called "scrape loading." In the scrape loading procedure, a portion of the cell monolayer was removed by mechanical means, and cells along the boundary of this "scrape" were transiently permeabilized, allowing the dye to enter these cells. The dye was subsequently washed away, and only permeabilized cells retained the dye molecules, as they were not internalized by cells with intact membranes.

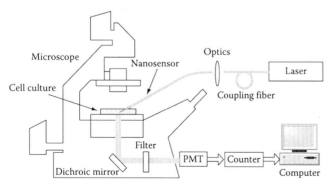

FIGURE 25.7 Instrumental system for fluorescence measurements of single cells using nanosensors.

Following incubation, the fluorescence signal of fluorescent dye molecules in single cells was detected using the optical nanofibers. Micromanipulators were used to move the optical fiber into contact with the cell membrane of a cell to be monitored. The fiber tip was then gently inserted just inside the cell membrane for fluorescence measurements. An argon-ion laser was used and passed through the optical nanofiber and used for excitation. The dye molecules inside the cell were excited, and their fluorescence emission was collected and detected using the microscope instrumental system described previously. Background measurements were performed with cells that were not loaded with the fluorophores. Fluorescence signals were successfully detected inside the fluorophore-loaded cells and not inside non-loaded cells. As another control, the optical fiber probe was then moved to an area of the specimen where there were no cells, and the laser light was again passed down the fiber for excitation. No visible fluorescence (at the emission wavelength) was detected for this control measurement, thus demonstrating the successful detection of the fluorescent dye molecules inside single cells. These results demonstrated the capability of optical nanofibers for measurements of fluorophores in intracellular environments of single living cells.

25.5.2 Monitoring Carcinogenic Benzo[a] pyrene and Related Biomarkers in Single Living Cells

Monitoring the transport of environmental carcinogens such as polycyclic aromatic compounds (e.g., benzo[a]pyrene) into cells is essential for understanding the health effects of these species. Nanosensors having antibody-based probes used to measure fluorescent targets inside a single cell have been demonstrated (Cullum and Vo-Dinh, 2000b; Vo-Dinh et al., 2000a; Vo-Dinh and Cullum, 2000b; Vo-Dinh et al., 2001). As the sizes of individual cells are generally small (1–10 μm), the success of intracellular investigations depends on several factors, including the sensitivity of the measurement system, the selectivity of the probe, and the small size of the nanofiber probes. In a previous work (Vo-Dinh et al., 2000a), the antibody probe was targeted against benzopyrene tetrol (BPT), an important biological compound, which was used as a biomarker of human exposure to the carcinogen benzo[a]pyrene (BaP), a polycyclic aromatic hydrocarbon of great environmental and toxicological interest because of its mutagenic/carcinogenic properties and its ubiquitous presence in the environment. Benzo[a]pyrene has been identified as a chemical carcinogen in laboratory animal studies (Moreno-Bondi et al., 2000). A photograph of an antibody-based nanosensor inside a single cell is shown in Figure 25.8. The small size of the probe allowed manipulation of the nanosensor at specific locations within the cells. The cells were first incubated with BPT prior to measurements using the experimental procedures described previously. Interrogation of single cells for the presence of BPT was then carried out using antibody nanoprobes for excitation and a photometric system for fluorescence signal detection.

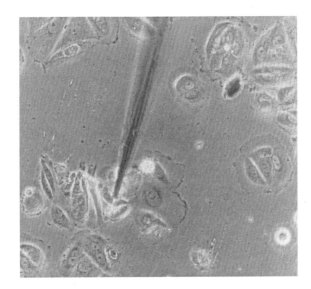

FIGURE 25.8 Photograph of single-cell sensing using the nanosensor system. (The small size of the probe allowed manipulation of the nanoprobe at specific locations within a single cell). (Taken from Vo-Dinh, T. et al., *Nat. Biotechnol.*, 18, 764, 2000a. With permission.)

Nanobiosensors for BPT were used for the measurement of intracellular concentrations of BPT in the cytoplasm of two different cell lines: (1) human mammary carcinoma cells and (2) rat liver epithelial cells, following treatment of the culturing media with an excess of BPT. Figure 25.9 shows a digital image of the nanosensor actually being inserted into a single human mammary carcinoma cell. The measurements were performed on rat liver epithelial cells (Clone 9) used as the model cell system. The cells had been previously incubated with BPT molecules prior to measurements. The results demonstrated the possibility of *in situ* measurements of BPT inside a single cell.

In this study, the nanosensors employed single-use bioprobes because the probes were used to obtain only one measurement at a specific time and could not be reused due to the strong

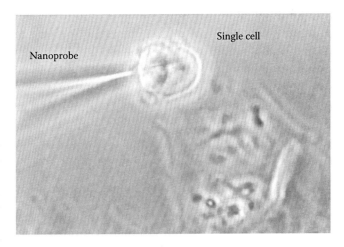

FIGURE 25.9 Digital image of a nanosensor inserted into a single human mammary carcinoma cell.

association constant of the antibody–antigen binding process. The antibody probes, however, could be regenerated using ultrasound methods. Our laboratory has successfully developed a method using ultrasound to noninvasively release antigen molecules from the antibodies, and, therefore, to regenerate antibody-based biosensors (Moreno-Bondi et al., 2000). The results of the measurements with antibody against breast cancer antigen illustrate the effectiveness and potential of the regenerable immunosensor. A 65% removal of the antigens bound to the monoclonal antibodies immobilized on the fiber surface is attained after ultrasound regeneration. The ultrasound regeneration scheme is a nondestructive approach that has a great potential to be applied to nanosensors. The results demonstrate the effectiveness of this innovative ultrasound-based approach for biosensor regeneration, i.e., releasing the antigen from the antibody probe. We have performed calibration measurements of solutions containing different BPT concentrations ranging from 1.56×10^{-10} M to 1.56×10^{-8} M in order to obtain a quantitative estimation of the amount of BPT molecules detected. By plotting the increase in fluorescence from one concentration to the next versus the concentration of BPT and fitting these data with an exponential function in order to simulate a saturated condition, a concentration of $(9.6 \pm 0.2) \times 10^{-11}$ M has been determined for BPT in the individual cell investigated (Cullum and Vo-Dinh, 2000b; Vo-Dinh et al., 2000a; Vo-Dinh and Cullum, 2000b; Vo-Dinh et al., 2000c; Vo-Dinh et al., 2001).

In another study, nanosensors have been developed for *in situ* measurements of the carcinogen BaP (Kasili et al., 2002).

Detection of BaP transport inside single cells is of great biomedical interest since it can serve as a means for monitoring BaP exposure, which can lead to DNA damage (Vo-Dinh, 1989). In order to perform these measurements, it is necessary to use antibodies targeted to BaP. The fluorescent BaP molecules are bound by interaction with the immobilized antibody receptor, forming a receptor–ligand complex. Following laser excitation of this complex, a fluorescence response from BaP provides a basis for the quantification of BaP concentration in the cell being monitored. The fluorescence signal generated allows for a high sensitivity of detection. The intracellular measurements of BaP depend on the reaction times involved. The reaction time established in this study for antibody–BaP complexing was 5 min. This was used as a standard time to enable calibration from fiber to fiber. Additionally, the nanosensors were calibrated using standard analytical procedures using measurements of known concentration of reference solutions.

25.5.3 Detection of Apoptosis in Cells Treated with Anticancer Drugs

During the last few years, there has been increasing interest in developing nanotools for evaluating the effects of anticancer drugs, such as the onset of apoptosis caused by these drugs in living cells. The cell death process known as apoptosis is executed in a highly organized fashion, indicating the presence of well-defined molecular pathways. Caspase activation is a hallmark of

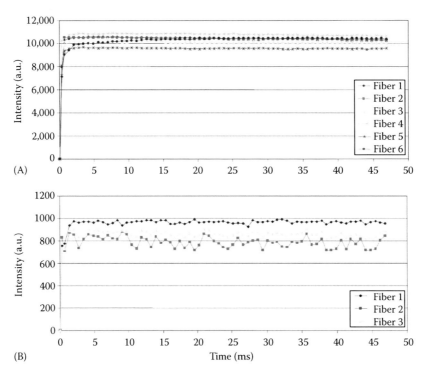

FIGURE 25.10   (A) Detection of caspase-9 activity performed with nanosensors inserted into single live MCF-7 cells in experimental group. (B) Background-corrected intracellular measurements of treated control group of MCF-7 cells. (From Kasili, P.M. et al., *J. Am. Chem. Soc.*, 126, 2799, 2004a. With permission.)

apoptosis, and probably one of the earlier markers that signals the apoptotic cascade (Wolf and Green, 1999; Hengartner, 2001; Ricci et al., 2003). These cysteine proteases are activated during apoptosis in a self-amplifying cascade. A variety of experimental evidence suggests that caspase activation is essential for the apoptotic process to take place, although not all cell death is dependent upon caspase activation. Caspases have an essential role both in the initial signaling events of apoptosis and in the downstream processes that produce the various hallmark signs of apoptosis. Activation of caspases like caspases 2, 8, 9, and 10 leads to proteolytic activation of "downstream" caspases such as 3, 6, and 7.

Nanobiosensors have been used to monitor the onset of the mitochondrial pathway of apoptosis in a single living cell by detecting enzymatic activities of caspase-9 (Kasili et al., 2004a). Minimally invasive analysis of single live MCF-7 cell for caspase-9 activity was demonstrated using the optical nanobiosensor that employed a modification of an immunochemical assay format for the immobilization of nonfluorescent enzyme substrate, leucine-glutamicacid-histidine-asparticacid-7-amino-4-methyl coumarin (LEHD-AMC). LEHD-AMC covalently attached on the tip of an optical nanobiosensor was cleaved during apoptosis by caspase-9 generating free AMC. An evanescent field was used to excite cleaved AMC and the resulting fluorescence signal was detected. By quantitatively monitoring the changes in fluorescence signals, caspase-9 activity within a single living MCF-7 cell was detected. Photodynamic therapy (PDT) protocols employing δ-aminolevulinic acid (ALA) were used to induce apoptosis in MCF-7 cells. The substrate LEHD-AMC was cleaved by caspase-9 and the released AMC molecules were excited and emitted a fluorescence signal. By comparing the fluorescence from an apoptotic cell and an uninduced control, we detected and identified caspase-9 activity. By comparing the fluorescence signals from apoptotic cells induced by photodynamic treatment and nonapoptotic cells, we successfully detected caspase-9 activity, which indicates the onset of apoptosis in the cells.

The results show that the fluorescence signals obtained from the cells that were both incubated with ALA and photoactivated by laser excitation (Figure 25.10A) were much higher than the signal obtained from control groups, i.e., without laser activation (Figure 25.10B) (Kasili et al., 2004a). The presence and detection of cleaved AMC in a single live MCF-7 cell as a result of these experiments reflect the presence of caspase-9 activity and the occurrence of apoptosis. These results indicate that apoptosis can be monitored in vivo using optical nanobiosensors in a single living cell.

Acknowledgments

The author acknowledges the contribution of G.D. Griffin, J.P. Alarie, B.M. Cullum, and P. Kasili. This research is sponsored by the National Institutes of Health (1R01ES014774 and R01-EB006201).

References

Alarie, J. P. and T. Vo-Dinh 1991. A fiberoptic cyclodextrin-based sensor. *Talanta* 38: 529–534.

Alarie, J. P. and T. Vo-Dinh 1996. Antibody-based submicron biosensor for benzo[a]pyrene DNA adduct. *Polycyclic Aromat. Comp.* 8: 45–52.

Alarie, J. P., J. R. Bowyer, M. J. Sepaniak et al. 1990a. Fluorescence monitoring of a benzo[a]pyrene metabolite using a regenerable immunochemical-based fiberoptic sensor. *Anal. Chim. Acta* 236: 237–244.

Alarie, J. P., M. J. Sepaniak, and T. Vo-Dinh 1990b. Evaluation of antibody immobilization techniques for fiber optic-based fluoroimmunosensing. *Anal. Chim. Acta* 229: 169–176.

Betzig, E. and R. J. Chichester 1993. Single molecules observed by near-field scanning optical microscopy. *Science* 262: 1422–1425.

Bowyer, J. R., J. P. Alarie, M. J. Sepaniak et al. 1991. Construction and evaluation of a regenerable fluoroimmunochemical-based fiber optic biosensor. *Analyst* 116: 117–122.

Cullum, B. M., G. D. Griffin, G. H. Miller et al. 2000a. Intracellular measurements in mammary carcinoma cells using fiberoptic nanosensors. *Anal. Biochem.* 277: 25–32.

Cullum, B. M. and T. Vo-Dinh. 2000b. The development of optical nanosensors for biological measurements. *Trends Biotechnol.* 18: 388–393.

Deckert, V., D. Zeisel, R. Zenobi et al. 1998. Near-field surface enhanced Raman imaging of dye-labeled DNA with 100-nm resolution. *Anal. Chem.* 70: 2646–2650.

Dhawan, A., J. F. Muth, D. N. Leonard et al. 2008. Focused in beam fabrication of metallic nanostructures on end faces of optical fibers for chemical sensing applications. *J. Vac. Sci. Technol. B: Microelectron. Nanometer Struct.* 26: 2168–2173.

Hengartner, M. O. 2001. Apoptosis—DNA destroyers. *Nature* 412: 27–29.

Hoffmann, P., B. Dutoit, and R. P. Salathe. 1995. Comparison of mechanically drawn and protection layer chemically etched optical fiber tips. *Ultramicroscopy* 61: 165–170.

Holland, L. 1956. *Vacuum Deposition of Thin Films*. New York: Wiley.

Kasili, P. M., J. M. Song, and T. Vo-Dinh. 2004a. Optical sensor for the detection of caspase-9 activity in a single cell. *J. Am. Chem. Soc.* 126: 2799–2806.

Kasili, P. M. and T. Vo-Dinh. 2004b. Detection of polycyclic aromatic compounds in single living cells using optical nanoprobes. *Polycyclic Aromat. Comp.* 24: 221–235.

Kasili, P. M. and T. Vo-Dinh. 2005. Optical nanobiosensor for monitoring an apoptotic signaling process in a single living cell following photodynamic therapy. *J. Nanosci. Nanotechnol.* 5: 2057–2062.

Kasili, R. M., B. M. Cullum, G. D. Griffin et al. 2002. Nanosensor for in vivo measurement of the carcinogen benzo[a]pyrene in a single cell. *J. Nanosci. Nanotechnol.* 2: 653–658.

Lambelet, P., A. Sayah, M. Pfeffer et al. 1998. Chemically etched fiber tips for near-field optical microscopy: A process for smoother tips. *Appl. Opt.* 37: 7289–7292.

Moreno-Bondi, M. C., J. Mobley, J. P. Alarie et al. 2000. Antibody-based biosensor for breast cancer with ultrasonic regeneration. *J. Biomed. Opt.* 5: 350–354.

Pohl, D. W. 1991. Scanning near-field optical microscopy (SNOM). In *Advances in Optical and Electron Microscopy*, eds. C. J. R. Sheppard and T. Mulevy, 243–312. London, U.K.: Academic.

Ricci, J. E., R. A. Gottlieb, and D. R. Green. 2003. Caspase-mediated loss of mitochondrial function and generation of reactive oxygen species during apoptosis. *J. Cell Biol.* 160: 65–75.

Song, J. M., P. M. Kasili, G. D. Griffin et al. 2004. Detection of cytochrome c in a single cell using an optical nanobiosensor. *Anal. Chem.* 76: 2591–2594.

Stockle, R., C. Fokas, V. Deckert et al. 1999. High-quality near-field optical probes by tube etching. *Appl. Phys. Lett.* 75: 160–162.

Tan, W. H., Z. Y. Shi, and R. Kopelman. 1992a. Development of submicron chemical fiber optic sensors. *Anal. Chem.* 64: 2985–2990.

Tan, W. H., Z. Y. Shi, S. Smith et al. 1992b. Submicrometer intracellular chemical optical fiber sensors. *Science* 258: 778–781.

Tromberg, B. J., M. J. Sepaniak, J. P. Alarie et al. 1988. Development of antibody-based fiber-optic sensors for detection of a benzo[a]pyrene metabolite. *Anal. Chem.* 60: 1901–1908.

Turner, D. R. 1984. Patent No. 4,469,554.

Vo-Dinh, T. 1989. *Chemical Analysis of Polycyclic Aromatic Compounds*. New York: Wiley.

Vo-Dinh, T. 1998. Surface-enhanced Raman spectroscopy using metallic nanostructures. *TrAC, Trends Anal. Chem.* 17: 557–582.

Vo-Dinh, T. 2008. Nanosensing at the single cell level. *Spectrochim. Acta, Part B* 63: 95–103.

Vo-Dinh, T., J. P. Alarie, B. M. Cullum et al. 2000a. Antibody-based nanoprobe for measurement of a fluorescent analyte in a single cell. *Nat. Biotechnol.* 18: 764–767.

Vo-Dinh, T. and B. Cullum. 2000b. Biosensors and biochips: Advances in biological and medical diagnostics. *Fresenius' J. Anal. Chem.* 366: 540–551.

Vo-Dinh, T., B. M. Cullum, and D. L. Stokes. 2001. Nanosensors and biochips: Frontiers in biomolecular diagnostics. *Sens. Actuators, B* 74: 2–11.

Vo-Dinh, T., G. D. Griffin, J. P. Alarie et al. 2000c. Development of nanosensors and bioprobes. *J. Nanoparticle Res.* 2: 17–27.

Vo-Dinh, T., G. D. Griffin, and M. J. Sepaniak. 1991. Fiberoptics immunosensors. In *Fiber Optic Chemical Sensors and Biosensors*, ed. O. S. Wolfbeis, 217–257. Boca Raton, FL: CRC Press.

Vo-Dinh, T. and P. Kasili. 2005. Fiber-optic nanosensors for single-cell monitoring. *Anal. Bioanal. Chem.* 382: 918–925.

Vo-Dinh, T., P. Kasili, and M. Wabuyele. 2006. Nanoprobes and nanobiosensors for monitoring and imaging individual living cells. *Nanomedicine* 2: 22–30.

Vo-Dinh, T., M. J. Sepaniak, G. D. Griffin et al. 1993. Immunosensors: Principles and applications. *Immunomethods* 3: 85–92.

Vo-Dinh, T., B. J. Tromberg, G. D. Griffin et al. 1987. Antibody-based fiberoptics biosensor for the carcinogen benzo(a)pyrene. *Appl. Spectrosc.* 41: 735–738.

Wolf, B. B. and D. R. Green 1999. Suicidal tendencies: Apoptotic cell death by caspase family proteinases. *J. Biol. Chem.* 274: 20049–20052.

Zeisel, D., V. Deckert, R. Zenobi et al. 1998. Near-field surface-enhanced Raman spectroscopy of dye molecules adsorbed on silver island films. *Chem. Phys. Lett.* 283: 381–385.

V

Drug Delivery

Multifunctional Pharmaceutical Nanocarriers

Vladimir P. Torchilin
Northeastern University

26.1 Brief Introduction

Various pharmaceutical nanocarriers, such as nanospheres, nanocapsules, liposomes, micelles, polymeric nanoparticles, dendrimers, nanotubes, cell ghosts, lipoproteins, and some others, are widely used for experimental (and even clinical) delivery of therapeutic and diagnostic agents (Gregoriadis 1988, Müller 1991, Rolland 1993, Torchilin 2006a,b, 2008). The surface modification of these carriers is often used to control their properties in a desirable fashion and make them simultaneously perform several different functions. The most important results of such modification(s) include

- Increased longevity and stability in circulation
- Changed biodistribution
- Targeting effect
- Stimuli(pH or temperature)-sensitivity
- Contrast properties

Frequent surface modifiers (used separately or simulatenously) include

- Soluble synthetic polymers (to achieve carrier longevity)
- Specific ligands, such as antibodies, peptides, folate, transferrin, sugar moieties (to achieve the targeting effect)
- pH- or temperature-sensitive copolymers (to impart stimuli-sensitivity)
- Chelating compounds, such as EDTA, DTPA, or deferoxamine (to add a diagnostic/contrast moiety onto a drug carrier)

Evidently, different modifiers can be combined on the surface of the same nanoparticular drug carrier providing it with a combination of useful properties (e.g., longevity and targetability, targetability and stimuli-sensitivity, or targetability and contrast properties).

The paradigm of using nanoparticulate pharmaceutical carriers to enhance the *in vivo* efficiency of many drugs, anticancer drugs first of all, has established itself over the past decade both in pharmaceutical research and clinical settings and does not need any additional justification. Now, within the frame of this concept, it is important to develop multifunctional stimuli-responsive nanocarriers, i.e., nanocarriers that, depending on the particular requirements, can possess a combination of the following abilities:

- Circulate long
- Target the site of the disease via both nonspecific and/or specific mechanisms, such as the enhanced permeability and retention (EPR) effect and ligand-mediated recognition
- Respond to local stimuli characteristic of the pathological site by, for example, releasing an entrapped drug or deleting a protective coating under the slightly acidic conditions inside tumors facilitating thus the contact between drug-loaded nanocarriers and cancer cells
- Provide an enhanced intracellular delivery of an entrapped drug

Additionally, these carriers can be supplied with contrast moieties to follow their real-time biodistribution and target accumulation. Some other, more exotic properties can be added to the list, such as magnetic sensitivity. Certain examples of multifunctional matrices for oral and tumoral delivery have already been reviewed in the articles by Bernkop-Schnurch and Walker (2001) and van Vlerken and Amiji (2006).

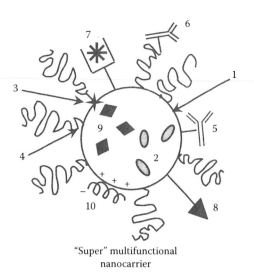

"Super" multifunctional
nanocarrier

FIGURE 26.1 Schematic structure of multifunctional "super" nano-carrier. The nanocarrier (1) is loaded with a drug (2) and can be additionally modified (separately, simultaneously, or in any combination) with a nondetachable (3) or detachable (4) protective polymeric coat for longevity, targeting ligand attached to the nanocarrier surface (5) or to the distal tips of protective polymer chains (6), contrast moeity (7) for visualization, and cell-penetrating peptide (8) for intracellular delivery. The system can also carry superparamagnetic particles (9) for visualization and positively charged sites for complexing DNA (10).

To be able to behave this way, a drug carrier should simultaneously carry on its surface various moieties capable of functioning in a certain orchestrated order. The schematic structure of such pharmaceutical carriers is shown in Figure 26.1. However, systems like this still represent a challenge. (Strictly speaking, the term "multifunctionality" may also be applicable to pharmaceutical carriers simultaneously loaded with more than one drug type, but such systems are out of our consideration here.)

Here, we will also briefly desribe two of the popular pharmaceutical nanocarriers that will be frequently used as examples in further paragraphs—liposomes (mainly for the delivery of water-soluble drugs) and micelles (for the delivery of poorly soluble drugs) (see Figure 26.2). Liposomes are artificial phospholipid vesicles with sizes varying from 50 to 1000 nm and even more, which can be loaded with a variety of water-soluble drugs (into their inner aqueous compartment) and sometimes even with water-insoluble drugs (into the hydrophobic compartment of the phospholipid bilayer). Liposomes have been considered to be promising drug carriers for well over two decades (Lasic 1993, Torchilin 2005). They are biologically inert and completely biocompatible, they cause practically no toxic or antigenic reactions, and drugs included into liposomes are protected from the destructive action of the external media. The association of drugs with carriers, such as liposomes, has pronounced effects on the pharmacokinetic profile of the drug resulting in delayed drug absorption, restricted drug biodistribution, decreased volume of drug biodistribution, delayed drug clearance, and retarded drug metabolism (Allen et al. 1995b). Plain liposomes are rapidly eliminated from the blood and are captured by the

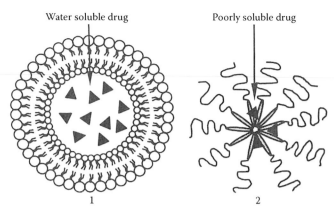

FIGURE 26.2 Schematic structures of the liposome (1) and micelle (2), and their loading with a drug. In liposomes water soluble drug is entrapped into the aqueous interior, while water insoluble (hydrophobic drug) into the liposome membrane. In micelles, hydrophobic drug is incorporated into the micelle core.

cells of the reticulo-endothelial system (RES), primarily in the liver and spleen, as a result of the rapid opsonization of the liposomes. Most liposomes are internalized by phagocytic cells via endocytosis and destined to lysosomes for degradation (Senior 1987). The use of targeted liposomes, i.e., liposomes selectively accumulating inside the affected organ or tissue, may increase the efficacy of the liposomal drug and decrease the loss of liposomes and their contents in RES. To obtain targeted liposomes, many protocols have been developed to bind corresponding targeting moieties, including antibodies, to the liposome surface without affecting the liposome integrity and antibody properties (Lasic 1993, Torchilin 1985). However, the approach with immunoliposomes may nevertheless be limited because of the short lifetime of the liposomes and immunoliposomes in circulation (Senior 1987). The majority of antibody-modified liposomes still accumulate in the liver, which hinders their significant accumulation in target tissues, particularly those with a diminished blood supply (ischemic or necrotic areas) and/or those with a low concentration of a target antigen. Dramatically better accumulation can be achieved if the circulation time of liposomes could be extended leading to the increased total quantity of immunoliposomes passing through the target and increasing their interactions with target antigens. This is why long-circulated (usually coated with polyethylene glycol [PEG], i.e., PEGylated) liposomes have attracted so much attention over the last decade (Lasic and Martin 1995). It was demonstrated that unique properties of long-circulating and targeted liposomes could be combined in one preparation (Torchilin et al. 1996), where antibodies or other specific binding molecules have been attached to the water-exposed tips of PEG chains (Torchilin et al. 2001a), see further.

Micelles, including polymeric micelles (Torchilin 2007, Trubetskoy and Torchilin (1995,1996)), represent another promising type of pharmaceutical carrier. Micelles are colloidal dispersions and have a particle size within the 5–100 nm range. An important property of micelles is their ability to increase the solubility and bioavailability of poorly soluble pharmaceuticals.

The use of certain special amphiphilic molecules as micelle building blocks can also introduce the property of micelle extended blood half-life upon intravenous administration. Because of their small size (5–50 nm), micelles demonstrate a spontaneous penetration into the interstitium in the body compartments with a leaky vasculature (tumors and infarcts) by the EPR effect; a form of selective delivery termed as "passive targeting" (Maeda et al. 2000). It has been repeatedly shown that micelle-incorporated anticancer drugs, such as adriamycin (Kwon and Kataoka 1995) better accumulate in tumors than in nontarget tissues, thus minimizing the undesired drug toxicity toward normal tissue. Diffusion and accumulation parameters for drug carriers in tumors have recently been shown to be strongly dependent on the cutoff size of the tumor blood vessel wall, and the cutoff size varies for different tumors (Yuan et al. 1995). Specific ligands (such as antibodies and/or certain sugar moieties) can be attached to the water-exposed termini of hydrophilic blocks (Torchilin et al. 2001a). In the case of targeted micelles, a local release of a free drug from micelles in the target organ should lead to the increased efficacy of the drug, while the stability of the micelles en route to the target organ or tissue should contribute better drug solubility and toxicity reduction due to less interaction with the nontarget organs.

26.2 Longevity: Basic Property of Phamaceutical Carriers

There is a frequent need to add the longevity in the blood to other properties of nanoparticulate drug delivery systems, and long-circulating pharmaceuticals and pharmaceutical carriers currently represent an important and still growing area of biomedical research (see Cohen and Bernstein 1996, Lasic and Martin 1995, Moghimi and Szebeni 2003, Torchilin 1998). There are quite a few important reasons for making long-circulating drugs and drug carriers:

- To maintain a required level of a pharmaceutical agent (both therapeutic and diagnostic ones) in the blood for extended time intervals. Long-circulating diagnostic agents are of primary importance for blood pool imaging, which helps in evaluating the current state of blood flow and discovering its irregularities caused by pathological lesions. Blood substitutes represent another important area for the use of long-circulating pharmaceuticals, when artificial oxigen carriers should present in the circulation long enough (Winslow et al. 1996).
- To allow long-circulating drug-containing microparticulates or large macromolecular aggregates to slowly accumulate (so-called passive targeting or accumulation via an impaired filtration mechanism; see Maeda 2001, Maeda et al. 2000) in pathological sites with affected and leaky vasculatures (such as tumors, inflammations, and infarcted areas) and to improve or enhance drug delivery in those areas (Figure 26.3; Gabizon 1995, Maeda 2001, Maeda et al. 2000).

- To achieve a better targeting effect for targeted (specific ligand-modified) drugs and drug carriers allowing for more time for their interaction with the target (Torchilin 1996) due to larger number of passages of targeted pharmaceuticals through the target.

The chemical modification of drugs and drug carriers with certain synthetic polymers is the most frequent way to add the *in vivo* longevity to other functions of drugs and drug carriers. Hydrophilic polymers have been shown to protect individual molecules and solid particulates from interaction with different solutes. This phenomenon relates to the stability of various aqueous dispersions (Molyneux 1984), and within the pharmaceutical field it helps to protect drugs and drug carriers from undesirable interactions with biological milieu components. The term "steric stabilization" has been introduced to describe the phenomenon of polymer-mediated protection (Napper 1983). The most popular and successful method used to obtain long-circulating, biologically stable nanoparticles is their coating with certain hydrophilic and flexible polymers, primarily with PEG, as was first suggested for liposomes in the articles by Allen et al. (1991), Klibanov et al. (1990), Maruyama et al. (1991), Papahadjopoulos et al. (1991), and Senior et al. (1991). On the biological level, coating nanoparticles with PEG sterically hinders the interactions of blood components with their surface and reduces the binding of plasma proteins with PEG particles as was demonstrated for liposomes in the articles by Allen (1994), Chonn et al. (1991, 1992), Lasic et al. (1991), Senior et al. (1991), and Woodle (1993). This prevents drug carrier interaction with opsonins and slows down their fast capture by RES (Senior 1987). The mechanisms for preventing opsonization by PEG include the shielding of the surface charged, increased surface hydrophilicity (Gabizon and Papahadjopoulos 1992), enhanced repulsive interaction between polymer-coated nanocarriers, the blood component (Needham et al. 1992), and the formation of the polymeric layer over the particle surface, which is impermeable for other solutes even

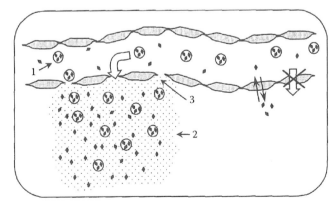

FIGURE 26.3 The schematics of the enhanced permeability and retention (EPR) effect. Continuous vasculature does not allow for the extravasation of large molecules and small particles (1) in normal tissues; however, such extravasation and accumulation in the interstitium become possible in pathological areas (tumors, infarcts, imflammations) (2) with the compromised (leaky) vasculature (3).

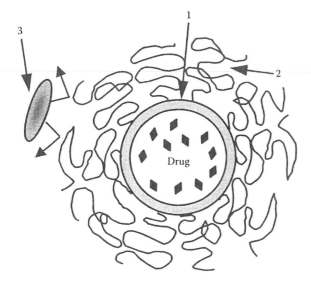

FIGURE 26.4 Schematics of steric protection of nanoparticles (1) with surface-grafted polymer, such as PEG (2), which prevents opsonins (3) from attaching to the surface of the particle and makes it long-circulating.

at relatively low polymer concentrations (Figure 26.4; Gabizon and Papahadjopoulos 1992, Torchilin et al. 1994a).

Theoretical considerations have been experimentally confirmed by studying the efficacy of liposome-incorporated fluorescent marker quenching by macromolecular quencher from the solution, depending on the polymer presence on the liposomal membrane (Torchilin et al. 1994a). It was shown that, for example, the quenching of the liposome incorporated phospholipid derivative of fluorescein with anti-fluorescein antibody was sharply inhibited in the presence of even relatively small quantities of PEG (below 1 mol%) on the liposome. Since the whole quenching process is limited only by macromolecular quencher diffusion from the solution to the liposome surface, it is evident that the presence of PEG on the surface creates pronounced diffusional hindrances for this process (Torchilin et al. 1994a).

Although quite a few polymers have been tried as steric protectors for nanoparticular drug carriers (Torchilin and Trubetskoy 1995b), which will be discussed further, the majority of research on long-circulating drugs and drug carriers was performed with the use of PEG as a sterically protecting polymer because of the following very attractive combination of the properties of PEG:

- Excellent solubility of PEG in aqueous solutions and its ability to bind to a lot of water molecules
- High flexibility of its polymer chain
- Very low toxicity, immunogenicity, and antigenicity
- Lack of accumulation in the RES cells
- Minimal influence on specific biological properties of modified pharmaceuticals (Pang 1993, Powell 1980, Yamaoka et al. 1994, Zalipsky 1995)

It is also important that PEG is not biodegradable and subsequently does not form any toxic metabolites. On the other hand,

PEG molecules with a molecular weight below 40 kDa are readily excretable from the body via the kidneys. From the practical point of view, PEG is easily commercially available in a variety of molecular weights. PEGs, which are normally used for the modification of drugs and drug carriers, have a molecular weight from 1,000 to 20,000 Da. Single-terminus reactive (semitelehelic) PEG derivatives are often used for the modification of pharmacologically important substances without the formation of cross-linked aggregates and heterogenic products. Currently, there exist many chemical approaches to synthesize activated derivatives of PEG and to couple these derivatives with a variety of drugs and drug carriers. An extensive review of these methods and their applicability for solving various problems in the drug delivery area was conducted by several authors (Torchilin 2002, Veronese 2001, Zalipsky 1995).

Despite the well-developed chemistry of PEG coupling, the search for alternative sterically protecting polymers is quite active. This might be explained by both the patent situation around PEG and its derivatives and the hope to get even better control over the properties of the modified drugs and drug carriers. The understanding of the mechanism underlying the stabilizing effect of PEG onto liposomes, permitted to suggest a whole number of other polymers which, together with PEG, can be used as effective steric protectors for various nanocarriers (Torchilin 1996, Torchilin et al. 1995b). These polymers should be biocompatible, soluble, hydrophilic, and have a highly flexible main chain. Synthetic polymers of the vinyl series, such as poly(acryl amide) (PAA), poly(vinyl pyrrolidone) (PVP), and some others may serve as the most evident examples of other potentially protective polymers. It was shown that amphiphilic derivatives of PAA and PVP can be synthesized by the radical polymerization of the corresponding monomers in the presence of a hydrophobic growing chain terminator, and the products obtained provided effective protection to liposomes *in vivo* (Chonn et al. 1992, Lasic et al. 1991, Torchilin et al. 2001a). Other amphiphilic polymers with well soluble and flexible hydrophilic moiety, such as amphiphilic poly(acryloyl morpholine) (PAcM), have also been successfully used as liposome steric protectors (Torchilin and Trubetskoy 1995b, Torchilin et al. 1995b). Amphiphilic PAcM and PVP were also prepared by synthesizing carboxyl-terminated polymers (PAcM-COOH, PVP-COOH) (Monfardini et al. 1995, Ranucci et al. 1994, Sartore et al. 1994), which were then modified with PE.

Biodistribution and blood clearance experiments with polymer-modified 111-In-labeled liposomes in CD-1 mice clearly demonstrated that amphiphilic palmitoyl(Pal)- or PE-containing derivatives of PAcM, PAA, and PVP provide effective protection to liposomes *in vivo* similar to PEG-PE, which agrees well with theoretical considerations and other experiments (Torchilin and Papisov 1994, Torchilin et al. 1993a, 1994a,c, 1995b). Half-clearance times for PVP-Pal-, PAA-Pal-, and PEG-liposomes with a 2.5% mol content of protective polymer are ca. 45, 80, and 80 min, respectively. For PVP-Pal-, PAA-Pal-, and PEG-liposomes with a 6.5% mol content of protective polymer, the half-clearance times are approximately 120, 140, and 170 min, respectively; whereas the half-clearance time for "plain" liposomes of the same size is only about 10–15 min. The extent of protective activity for different polymers toward liposomes *in vivo* depends on the length of the hydrophobic anchor, the polymer molecular weight, and the structure and quantity of the protecting polymer on the liposome surface (Torchilin and Trubetskoy 1995b, Torchilin et al. 1994a). The protective activity of polymers with a short hydrophobic moiety or with higher MW of the hydrophilic block decreases. From the thermodynamic point of view, a short hydrophobic group is unable to keep a large hydrophilic polymeric block on the carrier (liposome) surface: the energy of the polymeric chain motion in the solution might be higher than the energy of this group interaction with the hydrophobic surface or with phospholipid surroundings within the liposomal membrane, and such polymers might be relatively easily removed. A similar result can be observed when the size of the hydrophilic block is too large.

Liposomes containing 5 mol% of distearoyl-PE covalently linked to poly(2-methyl-2-oxazoline) or poly(2-ethyl-2-oxazoline) also exhibit extended blood circulation time and decreased uptake by the liver and spleen (Woodle et al. 1994). A similar observation was made by Maruyama et al. (1994) with phosphatidyl polyglycerols. Recently, the prolonged circulation time of doxorubicin-loaded liposomes in rats was observed when the liposome surface was modified with polyvinyl alcohol (Takeuchi et al. 1999).

The most important biological consequence of nanocarrier modification with protecting polymers is a sharp increase in its circulation time and a decrease in their RES (liver) accumulation (Klibanov et al. 1990, Lasic and Martin 1995, Torchilin et al. 1994a). From a clinical point of view, it is extremely important that various long-circulating liposomes of a relatively small size (100–200 nm) were shown to effectively accumulate in many tumors via the "impaired filtration" mechanism (Gabizon 1995, Gabizon and Papahadjopoulos 1988, Maeda 2001, Maeda et al. 2000). As a result, PEG-coated and other long-circulating liposomes were prepared containing a variety of anticancer agents, such as doxorubicine, arabinofuranosylcytosine, adriamycin, and vincristin (Allen et al. 1992, Boman et al. 1994, Gabizon et al. 1994, Huang et al. 1994). The biggest success was achieved with PEG-liposome-incorporated doxorubicine, which has already demonstrated very good clinical results (Ewer et al. 2004, Gabizon 1995, Rose 2005). An analysis of the pharmacokinetics of long-circulating nanocarriers (using PEG-liposomes) was performed by Allen et al. (1995a). In general, the association of drugs with nanocarriers has pronounced effects on pharmacokinetics: delayed drug absorption, restricted drug biodistribution, decreased volume of drug biodistribution, delayed drug clearance, and retarded drug metabolism (Hwang 1987). All these effects are determined by the hindered interstitial penetration of a drug and lesser drug accessibility for the biological milieu because of entrapment into the drug carrier. The presence of a protective polymer on the carrier surface changes all these parameters still further (Klibanov et al. 1990, Senior et al. 1991). Thus, while "plain" liposomes have nonlinear, saturable kinetics,

long-circulating liposomes demonstrate dose-independent, nonsaturable, and log-linear kinetics (Allen and Hansen 1991, Huang et al. 1992, Mayhew et al. 1992). All pharmacokinetic effects depend on the route of liposome administration and their size and composition, and are always less expressed for sterically protected PEG-carriers (Allen et al. 1989, Liu et al. 1991, 1992, Maruyama et al. 1992).

The current research on PEG-liposomes focuses on new methods of attaching PEG onto the liposome surface in a removable fashion in order to facilitate the liposome capture by the cell after the PEG-liposomes accumulate within the target site via the EPR effect (Maeda et al. 2001), and PEG coating is detached under the action of local pathological conditions (pH decrease in tumors). Thus, the spontaneous incorporation of PEG-lipid conjugates into the liposome membrane from PEG-lipid micelles was shown to be very effective and did not disturb the vesicles (Sou et al. 2000). The relative importance of the PEG chain length and lipid anchor length for the stability of the surface modification with PEG was investigated by Sadzuka et al. (2003). New detachable PEG conjugates are described in the article by Zalipsky et al. (1999b), where the detachment process is based on the mild thiolysis of the dithiobenzylurethane linkage between the PEG and amino-containing substrate (such as PE). Serum stable, long-circulating PEGylated pH-sensitive liposomes were also prepared using, on the same liposome, the combination of PEG and a pH-sensitive terminally alkylated copolymer of *N*-isopropylacrylamide and methacrylic (Roux et al. 2004). The attachment of the pH-sensitive polymer to the surface of liposomes might facilitate liposome destabilization and drug release in compartments with decreased pH values, such as tumors or intracellular organelles, endosomes.

Contrary to liposomes, the surface modification of hydrophobic particles can be performed by physical adsorption of a polymer on a particle surface, or by chemical grafting of polymer chains onto a particle. Possible examples of the first case include the absorption of a series of polyethylene oxide and polypropylene oxide copolymers (Pluronic/Tetronic™ or Poloxamer/Poloxamine™ surfactants) on the surface of polystyrene latex particles via the hydrophobic interaction mechanism. Interestingly, the absorption of Poloxamer-type copolymers takes place only on the solid particles with a clearly hydrophobic surface: no interaction is detected, for example, between such copolymers and the liposome surface (Moghimi et al. 1991). There have been numerous studies on the blood clearance and biodistribution of these particles upon their coating with surfactants, including the use of surfactants for particle protection from the uptake by the reticulo-endothelial system upon intravenous injection (Illum and Davis 1983) and enhanced delivery to lymph nodes after subcutaneous administration (Moghimi et al. 1994).

Hydrophilic linear polymers with a terminal lipid or fatty acyl group, such as PEG-PE or PAA-Pal (Klibanov et al. 1990, Torchilin et al. 1994b, 2001b) have also been tried as potential steric protectors for nanoparticular carriers. To investigate the behavior of such polymers on the surface of nanoparticulates,

Polybeads™ polystyrene latex particles with a diameter of 100 nm were used. PAA-Pal (MW 12,000) and PEG-PE (MW 5,000) have been used to coat the surface of latex particles. The particle diameter increase upon polymer adsorption was found to be ~5 nm for PEG-PE and ~20 nm for PAA-Pal, which proves the formation of a protective polymer layer over the surface of latex particle. The surface of polystyrene nanoparticles was also modified with a hydrophobically modified dextran and dextran-PEG copolymer to find that the stability of the dispersion can be controlled by both the type of surface-adsorbed polymer and its density on the surface (de Sousa Delgado et al. 2000).

Another important type of amphipathic polymer includes amphiphilic copolymers in which in the aqueous surrounding the hydrophobic block itself is able to form a solid phase (core, particle), while the hydrophilic part remains as a surface-exposed "protective cloud." An example of such a structure is the block-copolymer of PEG and polylactide-glycolide (PEG-PLAGA) (Gref et al. 1994, 1995, Krause et al. 1985). Using the PLAGA-PEG copolymer, one can prepare long-circulating particles with an insoluble (solid) PLAGA core and water-soluble PEG shell covalently linked to the core (Gref et al. 1994, 1995). To study the biological behavior of PEG-PLAGA nanospheres, they were labeled by the incorporation of hydrophobic ^{111}In-diethylene teriamine pentaacetic acid stearylamine into the PLAGA core during the solvent evaporation procedure. Blood clearance and biodistribution experiments in BALB/c mice have demonstrated that the protective effect of PEG in this system depends on the content of its block. Clearance and liver accumulation patterns reveal that the higher the content of the PEG block, the slower the clearance and the better protection from the liver uptake. Similar effects on the longevity and biodistribution of microparticular drug carriers might be achieved by direct chemical attachment of the protective polyethylene oxide chains onto the surface of preformed particles (Harper et al. 1991).

The surface of PLAGA microspheres was also modified by the adsorption of the polylysine-PEG copolymer, which resulted in a dramatic decrease of plasma protein adsorption on modified nanoparticles (Muller et al. 2003). Similarly, when the surface of polycyanoacrylate particles was modified with PEG, it also resulted in their increased longevity in circulation allowing for their diffusion into the brain tissue (Calvo et al. 2001, Peracchia et al. 1999). Drug (fluorouracil)-containing dendrimer nanoparticles surface-modified with PEG demonstrated better drug retention and less hemolytic activity (Bhadra et al. 2003). Comparative studies on the modification of the surface of colloidal polycaprolactone carriers with PEG or chitosan demonstrated that PEG coating accelerates the transport of nanoparticles across the epithelium of the ocular mucosa, while chitosan coating favors the retention of modified nanoparticles in the superficial layers of the epithelium (De Campos et al. 2003). This observation provides an opportunity to control certain biological properties of nanocarriers by properly chosen surface modification.

Recently, a lot of studies have been performed with the surface modification of superparamagnetic nanoparticles, which are now considered to be promising agents for drug delivery into regional lymph nodes and for diagnostic imaging purposes (Gupta and Gupta 2005). Similar to other nanocarriers, PEG-modified magnetite nanospheres demonstrated an increased colloidal stability and improved localization in lymph nodes (Illum et al. 2001). Cell culture experiments have confirmed that surface PEGylation changes the interaction of modified iron oxide particles with fibroblasts (Gupta and Curtis 2004). Grafting PEG onto the surface of gold particles via mercaptosilanes expectedly resulted in decreased protein adsorption onto modified particles and less platelet adhesion (Zhang et al. 2001).

Similar to liposomes, protecting the particle surface with hydrophilic flexible polymeric chains results in a substantial decline in phagocytosis by liver macrophages and a subsequent prolongation of the circulation time. Upon intravenous administration, hydrophobic particles of sub-micron size get opsonized with macrophage-recognizable but not yet identified serum proteins (Patel and Moghimi 1990). The absorption of the above copolymers leads not only to the decrease of particle uptake by resident macrophages in the liver but, after coating with some specific copolymers, can redirect the injected nanoparticles to other organs (Porter et al. 1992). For example, the coating of 60 nm polystyrene latex with Poloxamer 407 results in increased particle accumulation in bone marrow. The same group has demonstrated that an analogous procedure also helps substantially to alter the biodistribution of subcutaneously injected nanospheres. Coating of 60 nm diameter polystyrene nanospheres with certain Poloxamer/Poloxamine copolymers results in their increased accumulation in regional lymph nodes. The optimal length of the copolymer polyoxyethylene block for this particular purpose has been found to be 5–15 oxyethylene units. Noncoated particles normally stay at the injection site while particles coated with longer polyoxyethylene-containing copolymers are not retained in the nodes and eventually appear in the systemic circulation (Moghimi et al. 1994). Surface modification of polystyrene latexes with PEG was also successfully applied to make long-circulating particles and study their penetration into tumors (Hobbs et al. 1998, Monsky et al. 1999, Yuan et al. 1995).

Certain differences exist in the biological behavior of hydrophobic nanoparticulates and liposomes coated with the same polymers. Thus, the results of biodistribution studies in mice demonstrated that PEG-PE-coated Polybeads (polystyrene) particles, as one can expect, stayed in circulation for a long time ($t_{1/2} = 4\,\text{h}$). However, unlike PAA-coated liposomes (Torchilin et al. 1994b), PAA-coated nanospheres cleared from the blood as fast as noncoated particles with a similar pattern of liver accumulation. At the same time, spleen accumulation of PAA-coated particles was substantially reduced compared with noncoated ones. Different biological effects of the same amphiphilic polymer coating on the behavior of different particulates *in vivo* have also been reported previously. It was shown that Poloxamer 407 could protect from RES uptake latex nanospheres, but not the liposomes of similar size (Moghimi et al. 1991). This difference in the *in vivo* behavior has been explained by the different orientations of poly(acryl amide) chains on the particle surface compared with the surface of the liposome, which results in a different degree of carrier protection from absorbing plasma proteins. In the article by Allen (1994), the way the polymer attached to the surface was pointed out as the major difference in liposomes with poloxamers or PEG-PE. With poloxamers, the hydrophobic region lies perpendicularly to the acyl chain region of the bilayer, whereas for PEG-PE the hydrophobic anchor is parallel to the acyl chains. As a result, the thickness of the coating is considerably less for poloxamers on liposomes than for poloxamers on nanoparticles (Jamshaid et al. 1988, Woodle et al. 1992). For liposomes, PEG-PE works better than poloxamer F-108 (Woodle et al. 1992). Still, the behavior of a hydrophilic moiety in the surrounding aqueous media remains pretty much the same in all cases.

One more interesting case is the modification of cell surface with PEG. One of the main reasons behind such modification is the decrease of the immunological recognition of corresponding cells, for example during cellular transplantation (such as a blood transfusion). The surface of red blood cells was covalently modified with methoxy-PEG, which significantly diminished the immunologic recognition of surface antigens on these cells (Murad et al. 1999). Other blood cells as well as some tissue cells have also been surface-modified with PEG derivatives and an effective immunocamouflage was observed both *in vitro* and *in vivo* (Scott and Chen 2004, Scott and Murad 1998). A nice review on cell surface modification can be found in the article by Kellam et al. (2003).

26.3 How to Combine Targeting and Longevity

Further development of the concept of pharmaceutical nanocarriers involves the attempt to add the property of the specific target recognition to the carrier's ability to circulate long, i.e., simultaneously attach both the protecting polymer and the targeting moiety on the surface of the nanocarrier, such as liposome (Abra et al. 2002, Blume et al. 1993, Torchilin et al. 1992). The targeting of drug carriers with the aid of ligands specific to cell surface-characteristic structures allows for selective drug delivery to those cells. To obtain "simple" targeted nanocarriers, a variety of methods have been developed to attach corresponding vectors (antibodies, peptides, sugar moieties, folate, and other ligands) to the carrier surface. Thus, for example, numerous methods for antibody coupling to liposomes had been reviewed for the first time long ago (Torchilin 1984, 1985). These methods are relatively simple and allow for binding sufficient numbers of antibody molecules to a liposome surface without compromising the liposome integrity or antibody affinity and specificity. One of the routine methods for antibody coupling to liposomes includes protein covalent binding to a reactive group on the liposome membrane. On the other hand, the chemical modification of a nonmembrane hydrophilic protein (such as an antibody) with a hydrophobic reagent (such as a phospholipid residue) increases the affinity of the modified protein toward liposomes.

Protein binding via a hydrophobic anchor is firm: ΔG of the transfer of a single –CH2 group from water into an organic phase is 0.7 kcal/mol, and the equivalent number of anchoring –CH2 groups in a hydrophobic "tail" can reach dozens. It was shown that hydrophobic groups can be easily incorporated into proteins by treating them with long-chain fatty acid chlorides (Torchilin et al. 1980), activated phospholipids (e.g., oxidized phosphatidyl inositol) (Torchilin et al. 1982), or N-glutaryl phosphatidyl ethanolamine (Weissig et al. 1986). Some other methods of antibody "hydrophobization" are also available; see the review in the article by Torchilin (1985).

Modification with specific antibodies was also successfully applied to nonliposomal nanocarriers. Thus, nanoparticles made of gelatin and human serum albumin were modified with the HER2 receptor-specific antibody trastuzumab via the avidin-biotin system (Wartlick et al. 2004). These surface-modified nanoparticles were effectively endocytosed by HER2-overexpressing cells. The anti-CD3 antibody attached to gelatin particles via the same avidin-biotin system in order to enhance the interaction of these particles with lymphocytes (Balthasar et al. 2005). Two different antibodies specific for CD14 and prostate-specific membrane antigen were used to modify the surface of dendrimer nanoparticles (Thomas et al. 2004), which acquired the ability to specifically bind to the cells overexpressing corresponding antigens.

There are evident reasons for the design of ligand-coated long-circulating drug carriers:

- A ligand (an antibody, another protein, peptide, or carbohydrate) attached to the carrier surface may increase the rate of its elimination from the blood and uptake in the liver and the spleen, see for example, Klibanov (1998), and the presence of the sterically protecting polymer may compensate for this effect.
- Longevity of the specific ligand-bearing nanocarrier may allow for its successful accumulation even in targets with diminished blood flow or with a low concentration of the surface antigen.

Early experiments attempting to combine longevity and targetability in one preparation have been performed with liposomes by a simple co-incorporation of an antibody and PEG into the membrane of the same liposome. Although, under certain circumstances (careful selection of a protective polymer/targeting moiety ratio), such an approach can work, the protective polymer, however, may create steric hindrances for target recognition with the targeting moiety (Benhar et al. 1994, Khaw et al. 1991, Torchilin et al. 1992). The possible negative consequences of combining specific ligands with protecting polymers may include the following:

- A ligand (an antibody, another protein, peptide, or carbohydrate) attached to the carrier surface may increase the rate of its uptake in the liver and the spleen, despite the presence of a PEG brush or another "long-circulating" molecule on the carrier surface; see, for example, the article by Klibanov (1998).

- Ligand-coated drug carriers might result in the development of an unwanted immune response by the patient against the ligand or other carrier components. The extent of anti-liposome antibody development, for example, will depend on the character of the ligand (a small peptide or Fv fragment should be much less immunogenic than a complete foreign IgG molecule) and the liposome composition (Benhar et al. 1994, Harding et al. 1997, Park et al. 2001).

- The amount of the ligand attached to the carrier may be critical to ensure successful binding with the target while maintaining the extended circulation of the carrier. Thus, the use of drug carriers with a lower surface density of the ligand may be allowed to extend the carrier circulation time and to improve the overall *in vivo* targeting efficacy to smaller targets with limited blood flow. Such carriers, however, may not be the best to bind to the same target *in vitro*, when compared with the ones fully coated with ligand.

In an attempt to combine longevity and targetability in one preparation, early experiments have been performed with liposomes by simple co-immobilization of an antibody and PEG on the surface of the same liposome. The protective polymer, however, may create steric hindrances for target recognition with the targeting moiety, which can be overcome by careful selection of a protective polymer/targeting moiety ratio (Benhar et al. 1994, Torchilin et al. 1992). To confirm that under certain circumstances long-circulating PEG-coated liposomes (or other nanoparticulate drug carriers) can be made targeted by the co-incorporation of an antibody onto their surface, *in vivo* studies have been performed with PEG-liposomes additionally containing anti-cardiac myosin antibody (Torchilin et al. 1992) capable of specific recognition of ischemic or/and necrotic cardiomyocytes. Such double-modified liposomes specifically targeted the region of ischemically compromised myocardium in rabbits and dogs with experimental myocardial infarction (Khaw et al. 1991, Torchilin et al. 1992).

To achieve better selective targeting of PEG-coated liposomes, it is advantageous to attach the targeting ligand to the particles not directly, but via a PEG spacer arm, so that the ligand is extended outside of the dense PEG brush excluding steric hindrances for its binding to the target receptors. So, the natural idea has arisen to couple potential ligands to the far (distal) end of the liposome-grafted polymeric chain. All experimental data available so far describe ligand attachment to the distal ends of PEG chains. Currently, various advanced technologies are used, and the targeting moiety is usually attached above the protecting polymer layer by coupling it with the distal water-exposed terminus of activated liposome-grafted polymer molecules (Blume et al. 1993, Torchilin et al. 2001a). Since PEG-lipid conjugates used for liposome preparation are derived from commercial methoxy-PEG (mPEG) and carry only nonreactive methoxy terminal groups, several attempts have been made to functionalize PEG-lipid conjugates. For this purpose, a number of derivatives

of end-group functionalized lipopolymers of general formula X-PEG-PE (Zalipsky 1995, Zalipsky et al. 1998), where X represents a reactive functional group-containing moiety, while PEG-PE represents a urethane-linked conjugate of PE and PEG, were introduced. Most of the end-group functionalized PEG-lipids were synthesized from a few heterobifunctional PEG derivatives containing hydroxyl and carboxyl or amino groups. Typically the hydroxyl end-group of PEG was utilized to form a urethane attachment with the hydrophobic lipid anchor, PE, while the amino or carboxyl groups were utilized for conjugation or further functionalization reactions. To further simplify the coupling procedure and to make it applicable for single-step binding of a large variety of amino group-containing ligands (including antibodies, proteins, and small molecules) to the distal end of nanocarrier-attached polymeric chains without the use of any potentially toxic compounds, a new amphiphilic PEG derivative, *p*-nitrophenylcarbonyl-PEG-PE (*p*NP-PEG-PE) was introduced (Torchilin et al. 2000, 2001a, 2003b). *p*NP-PEG-PE readily incorporates adsorbs on hydrophobic nanoparticles or incorporates into liposomes and micelles via its phospholipid residue, and easily binds any amino group-containing compound via its water-exposed *p*NP group forming a stable and nontoxic urethane (carbamate) bond. The reaction between the *p*NP group and the ligand amino group proceeds easily and quantitatively at a pH level around 8.0, while excessive free *p*NP groups are easily eliminated by spontaneous hydrolysis.

Three general strategies have been employed to assemble ligand-bearing long-circulating PEGylated nanocarriers, first of all, liposomes (Zalipsky et al. 1997, 1998). The first approach involves the modification of preformed liposomes and thus does not require the complex and labor-intensive preparation of pure ligand-PEG-lipids. Since the approach involves conjugation after the formation of liposomes, some of the reactive end groups might remain on the outer surface creating a possibility of crosslinking through the multiple attachments to a single protein (ligand) molecule. According to the second approach, pure ligand-PEG-lipid conjugate is mixed with other liposomal matrix-forming components, e.g., lecithin and cholesterol, and then made into unilamellar vesicles (DeFrees et al. 1996, Gabizon et al. 1999, Wong et al. 1997, Zalipsky et al. 1997). Three-component conjugates containing various ligands: vitamins (Gabizon et al. 1999, Wong et al. 1997), peptides, and saccharides (DeFrees et al. 1996, Zalipsky et al. 1997) were synthesized by reacting the appropriate X-PEG-DSPE with suitably functionalized ligands. In some instances, the attachment of the ligand to PEG was performed first, followed by a lipid ligation to the other end group of the polymer chain (Zalipsky et al. 1999a). In the third approach, it was demonstrated that mPEG-DSPE can be inserted into preformed liposomes achieving similar external surface densities and *in vivo* performance as PEGylated liposomes prepared by lipid mixing/extrusion approach (Yoshioka 1991). A similar insertion strategy was recently applied to several ligand-PEG-DSPE conjugates (Zalipsky et al. 1997). The insertion methodology allows for the achievement of the same external surface densities of both PEG-tethered ligands and mPEG chains as by the lipid mixing/extrusion process (Zalipsky et al. 1997). Perhaps, most importantly, the insertion approach overcomes the drawbacks of both of the methods discussed earlier. As long as the inserted ligand-lipopolymer was purified and properly characterized, the resulting ligand-bearing liposome is free of any extraneous reactive functional groups, and all the ligand moieties are positioned only on the outer leaflet of the liposomal membrane.

Although various monoclonal antibodies have been shown to deliver liposomes and other nanocarriers to many targets, the search for the optimization of the properties of immunoliposomes and long-circulating immunoliposomes still continues. The majority of research relates to cancer targeting, which utilizes a variety of antibodies. Internalizing antibodies are required to achieve an improved therapeutic efficacy of antibody-targeted liposomal drugs as was shown using B-lymphoma cells and internalizable epitopes (CD19) as an example (Sapra and Allen 2002). An interesting concept was developed to target HER2-overexpressing tumors using anti-HER2 liposomes (Park et al. 2001). The antibody CC52 against rat colon adenocarcinoma CC531 attached to PEGylated liposomes provided specific accumulation of liposomes in a rat model of metastatic CC531 (Kamps et al. 2000).

A nice illustration of how the addition of the targeting function onto a long-circulating drug-loaded nanocarrier can dramatically enhance the activity of the drug was obtained by us, when we demonstrated that the nucleosome-specific monoclonal antibody capable of recognition of various tumor cells via the tumor cell surface-bound nucleosomes improved Doxil® (doxorubicin in long-circulating PEGylated liposomes) targeting to tumor cells and increased its cytotoxicity (Lukyanov et al. 2004a) both *in vitro* and *in vivo* (Figure 26.5). GD2-targeted immunoliposomes with a novel antitumoral drug, fenretinide, inducing apoptosis in neuroblastoma and melanoma cell lines, demonstrated strong anti-neuroblastoma

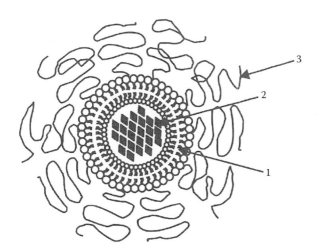

FIGURE 26.5 Schematic structure of Doxil® particle—~100 nm liposome (1) loaded with doxorubicin (2) and coated with PEG (3). To make it immunotargeted, specific antibodies can be attached to some of PEG tips.

activity both *in vitro* and *in vivo* in mice (Raffaghello et al. 2003). scFv antibody-modified liposomes were used to target cytotoxic drugs to biological targets, such as ED-B fibronectin (Marty and Schwendener 2005). The combination of immunoliposome and endosome-disruptive peptide improves the cytosolic delivery of liposomal drug, increases cytotoxicity, and provides a new approach to constructing targeted liposomal systems as shown with the diphtheria toxin A chain incorporated together with pH-dependent fusogenic peptide diINF-7 into liposomes specific towards ovarian carcinoma (Mastrobattista et al. 2002).

The surface modification with antibodies was also applied to target some other pharmaceutical nanocarriers, in particular cancer-targeted (see Brannon-Peppas and Blanchette 2004 for review). Nanoparticles made of poly(lactic acid) were surface-modified with PEG and with anti-transferrin receptor monoclonal antibody to produce PEGylated immunoparticles with the size of about 120 nm and containing ~65 bound antibody molecules per single particle (Olivier et al. 2002). Mammalian cells (NIH3T3, 32D, Ba/F3, hybridoma 9E10) were surface-modified with distal terminus-activated oleyl-PEG, and various proteins (streptavidin, EGFP, and antibody) were successfully attached to the activated PEG termini (Kato et al. 2004) producing a potentially interesting drug delivery system.

A similar combination of longevity and targetability can be also achieved by using some other specific ligands attached to long-circulating preparations. Thus, since transferrin (Tf) receptors (TfR) are overexpressed on the surface of many tumor cells, antibodies against TfR as well as Tf itself are among popular ligands for targeting various nanoparticular drug carriers including liposomes to tumors and inside tumor cells (Hatakeyama et al. 2004). Recent studies involve the coupling of Tf to PEG on PEGylated liposomes in order to combine longevity and targetability for drug delivery into solid tumors (Ishida et al. 2001). A similar approach was applied to deliver agents into tumors for photodynamic therapy including hypericin (Derycke and De Witte 2002, Gijsens et al. 2002) and for intracellular delivery of cisplatin into gastric cancer (Iinuma et al. 2002). Interestingly, the increase in the expression of the TfR was also discovered in post-ischemic cerebral endothelium, which was used to deliver Tf-modified PEG-liposomes to a post-ischemic brain in rats (Omori et al. 2003). Tf (Joshee et al. 2002), as well as anti-TfR antibodies (Tan et al. 2003, Xu et al. 2002), was also used to facilitate gene delivery into cells by cationic liposomes. Tf-mediated liposome delivery was also successfully used for brain targeting. Immunoliposomes with the OX26 monoclonal antibody to the rat TfR were found to concentrate on brain microvascular endothelium (Huwyler et al. 1996).

Targeting tumors with folate-modified nanocarriers also represents a popular approach, since folate receptor (FR) expression is frequently overexpressed in many tumor cells. After early studies demonstrated the possibility of the delivery of macromolecules (Leamon and Low 2001) and then liposomes (Lee and Low 1994) into living cells utilizing FR endocytosis, which could bypass multidrug resistance, the interest in folate-targeted drug delivery by liposomes grew rapidly (see important reviews by Gabizon et al. 2004 and Lu and Low 2002a). Liposomal daunorubicin (Ni et al. 2002) as well as doxorubicin (Pan et al. 2003b) were delivered into various tumor cells via folate receptors and demonstrated increased cytotoxicity. Folate-targeted liposomes have been suggested as delivery vehicles for boron neutron capture therapy (Stephenson et al. 2003) and used also for targeting tumors with haptens for tumor immunotherapy (Lu and Low 2002b). Various biodegradable nanoparticles have also been surface-modified with folate for enhanced tumor targeting. Folate was attached to the surface of cyanoacrylate-based nanoparticles via activated PEG blocks (Stella et al. 2000). Shell cross-linked nanoparticles were also prepared with surface-attached folate residues (Pan et al. 2003a). Similarly, PEG-polycaprolactone-based particles were surface-modified with folate and, after being loaded with paclitaxel, demonstrated increased cytotoxicity (Park et al. 2005). Superparamagnetic magnetite nanoparticles were modified with folate (with or without PEG spacer) and demonstrated better uptake by cancer cells, which can be used for both diagnostic (magnetic resonance imaging agents) and therapeutic purposes (Choi et al. 2004, Zhang et al. 2002). The combination of liposome pH-sensitivity and specific ligand targeting for cytosolic drug delivery utilizing decreased endosomal pH values was described for both folate and Tf-targeted liposomes (Kakudo et al. 2004, Shi et al. 2002, Turk et al. 2002).

Several other combinations of longevity and targeting have been described. Thus, vasoactive intestinal peptide (VIP) was attached to PEG-liposomes with radionuclides to target them to VIP-receptors of the tumor, which resulted in an enhanced breast cancer inhibition in rats (Dagar et al. 2003). PEG-liposomes were targeted by RGD peptides to integrins of tumor vasculature and, being loaded with doxorubicin, demonstrated an increased efficiency against C26 colon carcinoma in murine model (Schiffelers et al. 2003). RGD-peptide was also used to prepare liposomes targeted to integrins on activated platelets for specific cardiovascular targeting (Lestini et al. 2002). Angiogenic homing peptide was used to achieve targeted delivery to vascular endothelium of drug-loaded liposomes in experimental treatment of tumors in mice (Asai et al. 2002). Hyaluronan-modified long-circulating liposomes loaded with mitomycin C are active against tumors overexpressing hyaluronan receptors (Peer and Margalit 2004).

A strategy was developed to bind sugar residues to liposomal MPB-PE (Hansen et al. 1995, Zalipsky et al. 1996). A triantennary galactose-terminated cholesterol derivative was prepared by coupling tris (galactosyloxymethyl)-aminomethane to cholesterol, using glycyl and succinyl as intermediate hydrophilic spacer moieties (Kempen et al. 1984). The attachment of galactose residues was used to target liposomal drugs to the liver for therapy of liver tumors or metastases (Matsuda et al. 2001). The ability of galactosylated liposomes to concentrate in parenchymal cells was applied for gene delivery in these cells; see Hashida et al. (2001) for a review.

26.4 The Function of Stimuli-Sensitivity

An additional function can be added to long-circulating PEGylated pharmaceutical carriers, which allows for the detachment of protecting polymer (PEG) chains under the action of certain local stimuli characteristic of pathological areas, such as a decreased pH value or an increased temperature usually noted for inflamed and neoplastic areas. The matter is that the stability of PEGylated nanocarriers may not always be favorable for drug delivery. In particular, if drug-containing nanocarriers accumulate inside the tumor, they may be unable to easily release the drug to kill the tumor cells. Likewise, if the carrier has to be taken up by a cell via an endocytic pathway, the presence of the PEG coat on its surface may preclude the contents from escaping the endosome and being delivered in the cytoplasm. In order to solve these problems, for example, in the case of long-circulating liposomes, the chemistry was developed to detach PEG from the lipid anchor in the desired conditions. Labile linkage that would degrade only in the acidic conditions characteristic of the endocytic vacuole or the acidotic tumor mass can be based on the diorto esters (Guo and Szoka 2001), vinyl esters (Boomer and Thompson 1999), cystein-cleavable lipopolymers (Zalipsky et al. 1999b), double esters, and hydrazones that are quite stable at a pH level around 7.5 but hydrolyzed relatively fast at pH values of 6 and below (Guo and Szoka 2001, Kratz et al. 1999, Zhang et al. 2004). Polymeric components with pH-sensitive (pH-cleavable) bonds are used to produce stimuli-responsive drug delivery systems that are stable in the circulation or in normal tissues, however, degrade and release the entrapped drugs in body areas or cell compartments with lowered pH, such as tumors, infarcts, inflammation zones or cell cytoplasm or endosomes (Roux et al. 2002a, 2004, Simoes et al. 2004). Since in "acidic" sites the pH drops from the normal physiological value of 7.4 to pH 6 and below, chemical bonds used so far to prepare the acidic pH-sensitive carriers have included vinyl esters, double esters, and hydrazones that are reasonably stable at pH values around 7.5 but hydrolyzed relatively fast at pH values of 6 and below (Guo and Szoka 2001, Kratz et al. 1999, Zhang et al. 2004). By now, a variety of liposomes (Leroux et al. 2001, Roux et al. 2002b) and micelles (Lee et al. 2003a,b, Sudimack et al. 2002) have been described that include the components with the above-mentioned bonds as well as a variety of drug conjugates capable of releasing such drugs as adriamycin (Jones et al. 2003), paclitaxel (Suzawa et al. 2002), doxorubicin (Potineni et al. 2003), and DNA (Cheung et al. 2001, Venugopalan et al. 2002, Yoo et al. 2002) in acidic cell compartments (endosomes) and pathological body areas under acidosis. New detachable PEG conjugates are also described in Zalipsky et al. (1999b), where the detachment process is based on the mild thiolysis of the dithiobenzylurethane linkage between PEG and an amino-containing substrate (such as PE). Serum-stable, long-circulating PEGylated pH-sensitive liposomes were also prepared using the combination of PEG and a pH-sensitive terminally alkylated copolymer of *N*-isopropylacrylamide and methacrylic (Roux et al. 2004) on the same liposome, since the attachment of the pH-sensitive polymer to the surface of the liposomes might facilitate liposome destabilization and drug release in compartments with decreased pH values. The combination of liposome pH-sensitivity and specific ligand targeting for cytosolic drug delivery utilizing decreased endosomal pH values was described for folate- and Tf-targeted liposomes (Kakudo et al. 2004, Shi et al. 2002, Turk et al. 2002).

Dendrimeric systems derived from diaminobutane poly (propylene imine) with surface-attached PEG and loaded with various drugs demonstrated acid-sensitivity and were capable of releasing incorporated drugs when titrated with acids followed by the addition of sodium chloride solution (Paleos et al. 2004).

The stimuli-sensitivity of PEG coats can also allow for the preparation of more sophisticated multifunctional drug delivery systems with certain "hidden" functions, which, under normal circumstances, are "shielded" by the protective PEG coat, however become exposed after PEG detaches. Thus, a nanoparticular drug delivery system (DDS) should be able (a) to accumulate in the required organ or tissue and then (b) penetrate inside the target cells delivering its load (drug or DNA). Organ or tissue (tumor, infarct) accumulation could be achieved by the passive targeting via the EPR effect or by the antibody-mediated active targeting, while the intracellular delivery could be mediated by certain internalizable ligands (folate, transferrin) or by cell-penetrating peptides (CPPs, such as TAT or polyArg). When in the blood, the cell-penetrating function should be temporarily inactivated (shielded) to prevent a nonspecific drug delivery into nontarget cells; however, already being inside the target, the delivery system should lose its protective coat, expose the cell-penetrating function, and provide intracellular drug delivery (see the schematic pattern in Figure 26.6) (Sawant et al. 2005). To be able to behave this way, DDS should simultaneously carry on its surface various moieties capable of functioning in a certain orchestrated way. In other words, ideal nanoparticulate DDS should possess the ability to switch on and switch off certain functions under the action of local stimuli characteristic of the target pathological zone. Another important requirement is that different properties of the multifunctional nanocarrier are coordinated in an optimal fashion. Thus, for example, if the system is to be constructed that can provide the combination of the longevity allowing for the target accumulation via the EPR effect and specific cell binding allowing for its internalization by target cells, two requirements have to be met. First, the half-life of the carrier in the circulation should be long to fit EPR effect requirements and, second, the internalization of the carrier within the target cells should proceed fast not to allow for carrier degradation and drug loss in the interstitial space. However, systems like this still are not common.

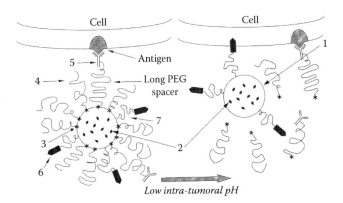

FIGURE 26.6 "Smart" stimuli-sensitive nanocarriers combining longevity, targetability, and cell-penetrating function. Nanocarrier (1) loaded with a drug (2) is grafted using pH-sensitive bonds (3) with sterically protective polymer chains (4), some of which carry targeting ligand, such as antibody (5) on distal tips. Cell-penetrating function (6) is attached to the surface of the nanocarrier via short spacer arm (7). Under normal conditions (in the blood, pH 7.4), the cell-penetrating function is shielded by the protecting polymer chains. However, in acidified pathological areas (such as tumors or infarcts), protective polymer splits away; cell-penetrating function becomes exposed and facilitates the intracellular delivery of the nanocarrier and its drug/DNA load.

26.5 Approaches to Intracellular Delivery of Nanocarriers

Many pharmaceutical agents, including various large molecules (proteins, enzymes, antibodies) and even drug-loaded pharmaceutical nanocarriers, need to be delivered intracellularly to exert their therapeutic action inside cytoplasm or onto a nucleus or other specific organelles, such as lysosomes, mitochondria, or endoplasmic reticulum (Figure 26.7). This group includes preparations for gene and antisense therapy, which have to reach cell nuclei; pro-apoptotic drugs, which target mitochondria; lysosomal enzymes, which have to reach the lysosomal compartment; and some others. The intracellular transport of different biologically active molecules is one of the key problems in drug delivery in general. In addition, the introcytoplasmic drug delivery in cancer treatment may overcome such important obstacles in anticancer chemotherapy as multidrug resistance. However, the lipophilic nature of the biological membranes restricts the direct intracellular delivery of such compounds. The cell membrane prevents big molecules such as peptides, proteins, and DNA from spontaneously entering cells unless there is an active transport mechanism as in the case of some short peptides. Under certain circumstances, these molecules or even small particles can be taken from the extracellular space into cells by the receptor-mediated endocytosis (Varga et al. 2000). The problem, however, is that any molecule/particle entering the cell via the endocytic pathway becomes entrapped into the endosome and eventually ends in the lysosome, where active degradation processes under the action of the lysosomal enzymes take place. As a result, only a small fraction of unaffected substance appears

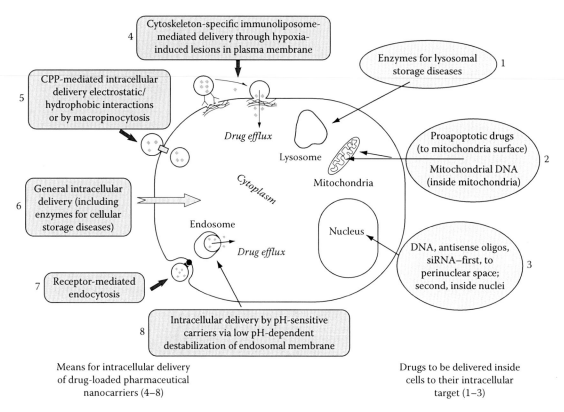

FIGURE 26.7   Intracellular delivery of drugs and DNA. When we need it and how it can be achieved.

in the cell cytoplasm. The methods like microinjection or electroporation used for the delivery of membrane-impermeable molecules in cell experiments are invasive in nature and could damage the cellular membrane (Arnheiter and Haller 1988, Chakrabarti et al. 1989). Much more efficient are the noninvasive methods, such as the use of pH-sensitive carriers including pH-sensitive liposomes (Straubinger et al. 1985) (which under the low pH inside endosomes destabilize the endosomal membrane liberating the entrapped drug into the cytoplasm) and cell-pentrating molecules.

Another problem is that even after being safely delivered into the cell cytoplasm, drugs still have to find their way to specific organelles (nuclei, lysosomes, mitochondria), where they are expected to utilize their therapeutic potential. This is especially important in the case of gene delivery. Gene therapy is now considered by many as a major contender in the future treatment protocols for various diseases, including cancer (Roth et al. 1998) and AIDS (Lisziewicz et al. 1995). Though some early initial trails were encouraging, the latest results have raised certain concerns (Aalto-Setala and Vuorio 1997). Viral vectors for DNA delivery suffer from nonspecificity and inherent risks of virus-induced complications. Nonviral delivery systems, first of all, cationic lipids/liposomes (Farhood et al. 1995), also have certain drawbacks, such as the same nonspecificity and cytotoxic reactions (Filion and Phillips 1997, Scheule et al. 1997), though new cationic lipid derivatives with decreased toxicity are currently under development (Tang and Hughes 1999). Still, the traditional routes of internalization of

DNA carriers by endocytosis or pinocytosis with subsequent degradation of the delivered DNA by lysosomal nucleases strongly limit the efficacy of transfection (Xu and Szoka 1996). From this point of view, the development of a new method that can deliver genetic constructs directly into the cytoplasm of the target cells would be highly desirable.

The use of cationic lipids and cationic polymers as transfection vectors for the efficient intracellular delivery of DNA was suggested in 1987 (Wu and Wu 1987, Xu and Szoka 1996). Currently, this is a huge and well-developed field reflected in numerous reviews and monographs (see one of the recent reviews in Elouahabi and Ruysschaert 2005). Complexes between cationic lipids (such as Lipofectin®, an equimolar mixture of N-[1-(2,3-dioleyloxy)propyl]-N,N,N-trimethylammonium chloride—DOTMA and dioleoyl phosphatidyl ethanolamine—DOPE) and DNA (lipoplexes) and complexes between cationic polymers, such as polyethyleneimine (PEI) (Kunath et al. 2003) and DNA (polyplexes) are formed because of strong electrostatic interactions between the positively charged carrier and negatively charged DNA. A slight net positive charged of already formed lipoplexes and polyplexes is believed to facilitate their interaction with negatively charged cells and improve transfection efficiency (Sakurai et al. 2000). Endocytosis (including the receptor-mediated endocytosis) was repeatedly confirmed as the main mechanism of lipoplex/polyplex internalization by cells (Ogris et al. 2001). Of special importance is the fact that despite endocytosis-mediated uptake lipolexes and polyplexes, DNA does not end in lysosomes but releases in the cytoplasm due to the destabilization of the endosomal membrane provoked by the lipid or polymeric component of the complexes. In particular, lipoplexes fuse with the endosomal membrane when they contain a fusogenic lipid, lDOPE (Felgner et al. 1994), which easily undergoes the transition from the bilayer to the hexagonal phase facilitating the fusion (Koltover et al. 1998). In the case of polyplexes, which cannot directly destabilize the endosomal membrane, the mechanism of DNA escape from the endosomes is associated with the ability of polymers, such as PEI, to strongly protonate under the acidic pH inside the endosome and create a charge gradient eventually provoking a water influx and endosomal swelling and disintegration (Boussif et al. 1995). In both cases, however, DNA-containing complexes when released into the cytosol dissociate allowing for nuclear entry of free DNA. The nuclear translocation of the plasmid DNA is relatively inefficient because of the barrier function of the nuclear membrane and the small size of nuclear pores (~25 nm); in addition, DNA degrades rather rapidly under the action of cytoplasmic nucleases (Pollard et al. 2001). It was estimated that only 0.1% of plasmids undergo nuclear translocation from the cytosol (Pollard et al. 1998). The attachment of nuclear localization sequences to plasmid DNA may significantly enhance its nuclear translocation and transfection efficiency (Branden et al. 1999). New approaches in using multifunctional carriers for DNA delivery include the application of bimetallic nanorods that can simultaneously bind compacted DNA plasmid and targeting ligands in a spatially defined manner (Salem et al. 2003).

Different methods of liposomal content delivery into the cytoplasm have been elaborated by adding the pH-sensitivity function to liposomal preparations (Torchilin et al. 1993b). According to one of these methods, the liposome is made of pH-sensitive components and, after being endocytosed in the intact form, it fuses with the endovacuolar membrane under the action of the lowered pH inside the endosome and destabilizes it, releasing its content into the cytoplasm (Torchilin et al. 1993b). Thus, namely, endosomes become the gates from the outside into the cell cytoplasm (Sheff 2004). This approach was reviewed many times in various publications (in 2004, the endosomal escape by pH-sensitive drug delivery systems was specifically discussed in a special issue of *Advanced Drug Delivery Reviews* #56, J. C. Leroux, ed.). Cellular drug delivery mediated by pH-sensitive liposomes is not a simple intracellular leakage from the lipid vesicle since the drug has to also cross the endosomal membrane (Asokan and Cho 2003). It is usually assumed that inside the endosome the low pH and some other factors destabilize the liposomal membrane, which, in turn, interacts with the endosomal membrane provoking its secondary destabilization and drug release into the cytoplasm. The presence of fusogenic lipids in the liposome composition, such as unsaturated DOPE, with their ability to easily adopt an inverted hexagonal phase, is usually required to render pH-sensitivity to liposomes (Shalaev and Steponkus 1999). Although multifunctional long-circulating PEGylated DOPE-containing pH-sensitive liposomes have a decreased pH-sensitivity, they still effectively deliver their contents into the cytoplasm (see a good review in the article by Simoes et al. 2004). Antisense oligonucleotides were delivered into cells by anionic pH-sensitive PE-containing liposomes, which are stable in the blood; however, undergo phase transition at acidic endosomal pH and facilitate oligo release into cell cytoplasm (see the recent review in Fattal et al. 2004). New pH-sensitive liposomal additives were recently described including oleyl alcohol (Sudimack et al. 2002) and pH-sensitive morpholine lipids (mono-stearoyl derivatives of morpholine) (Asokan and Cho 2003). Serum stable, long-circulating PEGylated pH-sensitive liposomes were also prepared using, on the same liposome, the combination of PEG and a pH-sensitive terminally alkylated copolymer of *N*-isopropylacrylamide and methacrylic (Roux et al. 2004). The combination of liposome pH-sensitivity and specific ligand targeting for cytosolic drug delivery utilizing decreased endosomal pH values was described for both folate and Tf-targeted liposomes (Kakudo et al. 2004, Shi et al. 2002, Turk et al. 2002). The additional modification of pH-sensitive liposomes with an antibody results in pH-sensitive immunoliposomes. The advantages of antibody-bearing pH-sensitive liposomes include cytoplasmic delivery, targetability, and facilitated uptake (i.e., improved intracellular availability) via the receptor-mediated endocytosis. The successful application of pH-sensitive immunoliposomes has been demonstrated in the delivery of a variety of molecules including fluorescent dyes, antitumor drugs, proteins, and DNA (Geisert et al. 1995).

Micelles (polymeric micelles) can also demonstrate pH-sensitivity and the ability to escape from endosomes. Thus, micelles prepared from PEG-poly(aspartate hydrazone adriamycin) easily release an active drug at lowered pH values typical for endosomes and facilitate its cytoplasmic delivery and toxicity against cancer cells (Bae et al. 2005). Alternatively, micelles for the intracellular delivery of antisense oligonucleotides (ODN) were prepared from ODN-PEG conjugates complexed with a cationic fusogenic peptide, KALA, and provided much higher intracellular delivery of the ODN that could be achieved with free ODN (Jeong et al. 2003). One could also enhance an intracellular delivery of drug-loaded micelles by adding lipid components used in membrane-destabilizing Lipofectin® to their composition. Thus, PEG-lipid micelles, for example, carry a net negative charge (Lukyanov et al. 2004b), which might hinder their internalization by cells. On the other hand, it is known that the net positive charge usually enhances the uptake of various nanoparticles by cells, and after the endocytosis the drug/DNA-loaded particles could escape from the endosomes and enter a cell's cytoplasm through the disruptive interaction of the cationic lipid with endosomal membranes (Hafez et al. 2001). The compensation of this negative charge by the addition of positively charged lipids to PEG-PE micelles could improve the uptake by cancer cells of drug-loaded mixed PEG-PE/positively charged lipid micelles. It is also possible that after the enhanced endocytosis, such micelles could escape from the endosomes and enter the cytoplasm of cancer cells. With this in mind, an attempt was made to increase an intracellular delivery and, thus, the anticancer activity of the micellar paclitaxel by preparing paclitaxel-containing micelles from the mixture of PEG-PE and Lipofectin®

lipids (LL) (Wang et al. 2005). When studying the cellular uptake of various fluorescently labeled micelles in adherent BT-20 cells, it was found that while both "plain" PEG-PE micelles and PEG-PE/LL micelles were endocytosed by BT-20 cells as confirmed by the presence of fluorescent endosomes in cells after 2 h of co-incubation with fluorescently labeled micelles, in the case of PEG-PE/LL micelles, however, endosomes became partially disrupted and their content was released into the cell cytosol. The addition of LL facilitating the intracellular uptake and cytoplasmic release of paclitaxel-containing PEG-PE/LL micelles resulted in a substantially increased cell death compared with that under the action of free paclitaxel or paclitaxel delivered using noncationic LL-free PEG-PE micelles. In A2780 cancer cells, the IC50 values of free paclitaxel, paclitaxel in PEG-PE micelles, and paclitaxel in PEG-PE/LL micelles were 22.5, 5.8, and 1.2 μM, respectively. In BT-20 cancer cells, the IC50 values of the same preparations were 24.3, 9.5, and 6.4 μM, respectively (Wang et al. 2005). Multifunctional polymeric micelles capable of pH-dependent dissociation and drug release when loaded with doxorubicin and supplemented with biotin as cancer cell-interacting ligands were also described in the article by Lee et al. (2005).

In addition to membrane-destabilizing lipid components, there exists a large family of membrane-destabilizing anionic polymers that can also enhance the endosomal escape of various drugs and biomacromolecules (Yessine and Leroux 2004). This family includes various carboxylated polymers, copolymers of acrylic and methacrylic acids, copolymers of maleic acid, polymers, and copolymers of *N*-isopropylacrylamide (NIPAM). Copolymers of NIPAM demonstrate a lower critical solution (solubility/insolubility switch) at physiological temperatures and when precipitate, destabilize the biomembranes with which they are interacting (Chen and Hoffman 1995). Such polymers can be attached to the surface of drug/DNA-loaded liposomes or polymeric micelles allowing for endosomal destabilization and cytoplasmic escape.

A new approach in intracellular drug delivery has recently emerged, which allows form bypassing lysosomes and is based on the modification of drugs and drug carriers (including the multifunctional ones) with certain proteins and peptides demonstrating a unique ability to penetrate into cells ("protein transduction" phenomenon). This function can be added on top of the longevity and targetability of the pharmaceutical drug-loaded nanocarriers.

It was demonstrated that the trans-activating transcriptional activator (TAT) protein from HIV-1 enters various cells when added to the surrounding media (Frankel and Pabo 1988). The same is true about several other cell-penetrating proteins and peptides (CPPs). All CPPs are divided into two classes: the first class consists of amphipathic helical peptides, such as transportan and model amphipathic peptide (MAP), where lysine (Lys) is the main contributor to the positive charge, while the second class includes arginine(Arg)-rich peptides, such as TAT (47–57) and Antp or penetratin (Hallbrink et al. 2001). TAT is a transcription activating factor with 86 amino acids and contains a cysteine-rich region important for metal-linked dimerization

in vitro (Frankel et al. 1988a,b) and a highly basic region (with two lysines and six arginines in nine residues) involved in nuclear and nucleolar localization (Ruben et al. 1989, Siomi et al. 1990) and RNA binding (Weeks et al. 1990). TAT peptide (TATp) includes residues 37–72 and contains a region with residues 38–49 that can adopt an α-helical structure with amphipathic characteristics and a cluster of basic amino acids 49–57 i.e., minimal protein transduction domain (PTD) (Loret et al. 1991). The protein transduction sequence for TAT includes residues 47–57 (11-mer; Tyr-Gly-Arg-Lys-Lys-Arg-Arg-Gln-Arg-Arg-Arg) (Schwarze et al. 2000). Homeoproteins are a class of transcription factors that bind DNA through a specific sequence of 60 amino acids, called the homeodomain. The homeodomain of Antennapedia, a Drosophilia homeoprotein (Antp), was internalized by mammalian nerve cells and accumulated in their nuclei modifying the morphology of the neurons (Joliot et al. 1991). The third helix of the antennapedia homeodomain is involved in the translocation process. The minimal PTD of Antp, called penetratin, is the 16-mer peptide (43–58 residues) present in the third helix of the homeodomain (Derossi et al. 1994). Other CPPs that can be used for the modification of nanocarriers include VP22, which is a major structural component of HSV-1 possessing a remarkable property of transport between cells (Elliott and O'Hare 1997); transportan, a 27 amino acid-long chimeric CPP containing the peptide sequence from the amino terminus of the neuropeptide galanin, which is bound via the lysine residue with the membrane-interacting wasp venom peptide, mastoparan (Pooga et al. 1998); and 18-mer amphipathic model peptide with the sequence KLALKLALKALKAALKLA (Oehlke et al. 1998), which translocates plasma membranes of mast cells and endothelial cells by both energy-dependent and energy-independent mechanisms and is capable of transporting different cargoes. In terms of the cellular uptake and cargo delivery kinetics, MAP has the fastest uptake, followed by transportan, TATp (48–60), and penetratin. Similarly, MAP has the highest cargo delivery efficiency, followed by transportan, TATp (48–60), and penetratin. The membrane disturbing potential of these peptides was proportional to the hydrophobic moment of the peptide (Hallbrink et al. 2001).

The recent data assume more than one mechanism for CPPs and CPP-mediated intracellular delivery of various molecules and particles. The TAT-mediated intracellular delivery of large molecules and nanoparticles was proved to proceed via the energy-dependent macropinocytosis with subsequent enhanced escape from endosomes into the cell cytoplasm (Wadia et al. 2004), while individual CPPs or CPP-conjugated small molecules penetrate cells via electrostatic interactions and hydrogen bonding and do not seem to depend on the energy (Rothbard et al. 2004). Since traversal through cellular membranes represents a major barrier for the efficient delivery of macromolecules into cells, CPPs, whatever their mechanism of action is, may serve to transport various drugs and even drug-loaded pharmaceutical carriers into mammalian cells *in vitro* and *in vivo*.

It was first shown in 1999 (Josephson et al. 1999) that CPPs could internalize nanosized particles into the cells. Superparamagnetic iron oxide (SPIO) particles conjugated with TAT peptide and fluorescein isothiocyanate were taken up quickly by T cells, B cells, and macrophages followed by the migration of the conjugate primarily to the cytoplasm, which could be tracked readily by MRI (Kaufman et al. 2003). Biocompatible dextran-coated SPIO nanoparticles derivatized with TAT (48–57) were internalized into lymphocytes over 100-fold more efficiently than nonmodified particles. TAT PTD conjugated to shell cross-linked nanoparticles demonstrated the transduction into cells (Liu et al. 2001). Such particles, with a mean size of 41 nm and carrying an average of 6.7 TAT molecules per single particle, were localized in cytoplasmic and nuclear compartments showing the potential for the MRI of cell trafficking and/or magnetic separation of *in vivo* homed cells. Studies showed that TAT peptide enhanced nanoparticle clearance from the vascular compartment, and nanoparticles were present throughout the liver, rather than only in the cells adjacent to the vascular compartment (endothelial cells/Kupffer cells). Further characterization on the number of TATp molecules required for an efficient delivery of magnetic nanoparticles revealed that higher numbers of TAT peptide molecules (above 10 per single SPIO particle) enhanced the intracellular accumulation of such particles with the 100-fold increase in cell labeling efficiency, decreasing the detection thresholds of labeled cells for *in vivo* tracking (Zhao et al. 2002). The homing of T cells to the spleen was monitored *in vivo* using MRI after loading T cells with TATp-derivatized SPIO nanoparticles, without interfering with the normal T cell response (Dodd et al. 2001). Gold nanoparticles modified with nuclear localization peptides showed localization in the cytoplasm and/or nuclei, depending on the peptide sequence and cell line. Conjugates of gold nanoparticles modified with 9-mer TAT PTD did not enter the nuclei of NIH-3T3 or HepG2 cells, and the cellular uptake of the conjugate was temperature-dependent, suggesting an endosomal pathway of uptake (Tkachenko et al. 2004). The combination of longevity, magnetic properties, and the ability to penetrate inside cells results in pharmaceutical nanopreparations with new and unique properties including contrast properties allowing for MR visualization of cells taking up such particles.

It was also demonstrated that even relatively large particles, such as liposomes, could be delivered into various cells by multiple TATp or other CPP molecules attached to their surface. The translocation of TATp-liposomes into cells required the direct interaction of the liposomal TATp with the cell surface (Levchenko et al. 2003, Torchilin et al. 2001c). Complexes of TATp-liposomes with a plasmid (plasmid pEGFP-N1 encoding for the Green Fluorescence Protein, [GFP]) were used for successful *in vitro* transfection of various tumor cells and normal cells as well as for the *in vivo* transfection of tumor cells in mice bearing Lewis lung carcinoma (Torchilin et al. 2003a) (the combination of positive charge for DNA complexation and cell-penetrating functions). The frozen sections of tumors excised from experimental animals after 72 h demonstrated intensive green fluorescence (Torchilin and Levchenko 2003, Torchilin et al. 2003a). It was also shown that both penetratin and TAT peptide enhanced the translocation efficiency of liposomes in proportion to the number of peptide molecules attached to the liposome surface.

The uptake was peptide- and cell-type dependent and as few as five peptide molecules per liposome could enhance the intracellular delivery of liposomes (Tseng et al. 2002). Antp (43–58) and TAT (47–57) peptides when coupled to small unilamellar liposomes (~100 nm) showed a higher uptake in tumor cells and dendritic cells than unmodified control liposomes. The uptake was time- and concentration-dependent and at least 100 PTD molecules per small unilamellar liposomes were necessary for the efficient translocation into cells. The uptake was inhibited by the pre-incubation of liposomes with heparin, confirming the role of heparan sulfate proteoglycans in PTD-mediated uptake. Antp-liposomes were considered as a carrier system for an enhanced cell-specific delivery of liposome-entrapped molecules (Marty et al. 2004). Thus, the transduction function could have great potential to deliver a wide range of large molecules and small particles both *in vitro* and *in vivo*.

26.6 Reporter Groups

To make it possible to use pharmaceutical nanocarriers for diagnostic/imaging purposes as well as to allow for their real-time biodistribution and target accumulation, the contrast reporter moieties can be added to multifunctional nanocarriers. Currently used medical imaging modalities include the following:

- Gamma-scintigraphy (based on the application of gamma-emitting radioactive materials)
- Magnetic resonance (MR; a phenomenon involving the transition between different energy levels of atomic nuclei under the action of radiofrequency signal)
- Computed tomography (CT; the modality utilizing ionizing radiation—x-rays—with the aid of computers to acquire cross-images of the body and three-dimensional images of areas of interest)
- Ultrasonography (this modality utilizes the irradiation with ultrasound and is based on the different passage rate of ultrasound through different tissues)

Whatever imaging modality is used, medical diagnostic imaging requires that the sufficient intensity of a corresponding signal from an area of interest be achieved in order to differentiate this area from surrounding tissues. Unfortunately, nonenhanced imaging techniques are useful only when relatively large tissue areas are involved in the pathological process. To solve a problem and to achieve a sufficient attenuation in the case of small lesions, contrast agents are used to absorb certain types of signals (irradiation) much stronger than the surrounding tissues. The contrast agents are specific for each imaging modality, and as a result of their accumulation in certain sites of interest, those sites may be easily visualized when the appropriate imaging modality is applied (Torchilin et al. 1995c).

Among nanocarriers successfully supplemented with contrast agents, liposomes and micelles draw special attention because of their easily controlled properties and good pharmacological characteristics. Liposomes may incorporate contrast agents in both internal water compartments and membranes. Two very general approaches are the most often used and efficient to prepare liposomes for gamma- and MR-imaging. First, metal is chelated into a soluble chelate (such as DTPA) and than included into the water interior of a liposome (Tilcock et al. 1989). Alternatively, DTPA or a similar chelating compound may be chemically derivatized by the incorporation of a hydrophobic group, which can anchor the chelating moiety onto the liposome surface during or after liposome preparation (Kabalka et al. 1991a). Different chelators and different hydrophobic anchors were tried for the preparation of ^{111}In, ^{99m}Tc, Mn-, and Gd-liposomes (Glogard et al. 2002, Grant et al. 1989, Kabalka et al. 1991b, Phillips and Goins 1995, Schwendener et al. 1990, Tilcock 1993, Torchilin 1997, Torchilin and Trubetskoy 1995a). In the case of MR imaging, for a better MR signal, all reporter atoms should be freely exposed for interaction with water. Membranotropic chelating agents—such as DTPA-stearylamine (DTPA-SA) (Schwendener et al. 1990) or DTPA-phosphatidyl ethanolamine (DTPA-PE) (Kabalka et al. 1991a)—consist of the polar head containing a chelated paramagnetic atom and the lipid moiety that anchors the metal-chelate complex in the liposome membrane. This approach has been shown to be far more superior in terms of the relaxivity of the final preparation when compared with liposome-encapsulated paramagnetic ions (Barsky et al. 1992, Putz et al. 1994, Unger et al. 1990) due to the decrease in the rotational correlation times of the paramagnetic moiety rigidly connected with relatively large particles. Liposomes with membrane-bound paramagnetic ions also demonstrate the reduced risk of the leakage of potentially toxic metals in the body.

The amphiphilic chelating probes (paramagnetic Gd-DTPA-PE and radioactive ^{111}In-DTPA-SA) were also incorporated into PEG(5 kDa)-PE micelles and used *in vivo* for MR and scintigraphy imaging. In micelles, the lipid part of the molecule can be anchored in the micelle's hydrophobic core while a more hydrophilic chelate is localized on the hydrophilic shell of the micelle. The main feature that makes PEG-lipid micelles attractive for diagnostic imaging applications is their size. Due to the lipid bilayer curvature limitations, it is not possible to prepare liposomes that are smaller than a certain minimal diameter (usually, 70–100 nm) (Enoch and Strittmatter 1979). For some diagnostic applications, the administration of diagnostic particulates of a significantly smaller size is required. One such situation arises, for example, during percutaneous lymphography as it has been shown with polystyrene nanospheres that particulate uptake in the primary lymph node after subcutaneous injection is drastically increased with decreasing particle size especially when nanospheres with a diameter below 100 nm are used (Davis et al. 1993).

To still further increase the liposome load with diagnostic moieties, amphiphilic polychelating polymers (PAP) were synthesized consisting of the main chain with multiple side chelating groups capable of firm binding many reporter metal atoms and a hydrophobic terminal group allowing for polymer adsorption onto hydrophobic nanoparticles or incorporation into the hydrophobic domains of liposomes or micelles (Torchilin 2000); Figure 26.8. Such surface modifications of the nanocarriers

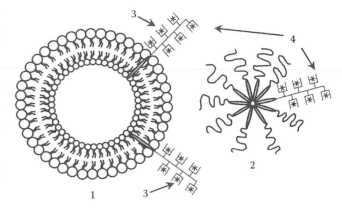

FIGURE 26.8 Loading of liposome (1) or micelle (2) with reporter metal (3), such as Gd for MRI or with 111-In for Gamma-scintigraphy, via the nanocarrier-incorporated polychelating amphiphilic polymer, PAP (4).

allow for a sharp increase in the number of bound reporter metal atoms per particle and image signal intensity. In the case of MR, metal atoms chelated into polymer side groups are directly exposed to the water environment that enhances the relaxivity of the paramagnetic ions and leads to the corresponding enhancement of the vesicle contrast properties (Torchilin 1997, 1999, Trubetskoy and Torchilin 1994).

An interesting example of the application of PAP-nanoparticles for actual *in vivo* imaging is the MRI of lymphatic system components with Gd-loaded nanocarriers. Liposomes and micelles have been studied as delivery vehicles to the lymphatic (Torchilin et al. 1995c, Trubetskoy et al. 1996). It has been shown that radioactively labeled small negatively charged liposomes are the most efficient in targeting rat regional lymph nodes after the subcutaneous administration (Patel et al. 1984). The optimal diameter of liposomes that localize in the lymph nodes after the peritoneal administration in rats is approximately 200 nm (Hirano and Hunt 1985). Liposomes loaded with chelated paramagnetic ions (mostly Gd, Dy, Mn, Fe) have been demonstrated to be useful as MRI contrast agents mostly for the visualization of the macrophage-rich tissues such as organs of the reticuloendothelial system (Schwendener et al. 1990, Unger et al. 1989). In experimental rabbits, transverse scans obtained after subcutaneous administration of the suspension of Gd-PAP-liposomes into the right forepaw demonstrated that the axillary and subscapular lymph nodes can be seen on the scan taken only 5 min postinjection (Trubetskoy and Torchilin 1994). The approach was additionally proven by the fast and informative diagnostic visualization of VX$_2$ human sarcoma in a rabbit popliteal lymph node when, with 200 nm Gd-PAP-liposomes, the tumor was clearly seen 10 min postinjection. Particles (micelles) formed by self-assembled amphiphilic polymers (such as PEG-PE) can also be loaded with amphiphilic polychelates carrying diagnostically important metal ions such as 111-In and Gd. Upon subcutaneous administration, micelles penetrate the lymphatics and effectively visualize its elements in corresponding imaging modalities. Micelles mostly stay within lymph fluid rather than accumulating in the nodal macrophages (because of the protective effect of surface PEG fragments) and rapidly move via the lymphatic pathway. They can serve as fast and efficient lymphangiographic agents for scintigraphy or MRI.

The overall performance of Gd-PAP-liposomes or -micelles for target visualization might be further improved by the additional incorporation of amphiphilic PEG onto the liposome membrane or micelle surface, which can be explained by increased $1/T_1$ values of PEG-Gd-nanoparticles because of the presence of an increased amount of PEG-associated water protons in the close vicinity of chelated Gd ions located on the liposomal membrane (Torchilin et al. 1995a, Trubetskoy et al. 1995). A multifunctional approach certainly assists here, since in addition to the enhanced relaxivity, the coating of the liposome surface with PEG polymer can help in avoiding the contrast agent uptake at the site of injection by resident phagocytic cells. Both PAP-bearing liposomes and micelles additionally containing PEG on their surface can also serve as long-circulating contrast agents for the blood pool gamma- or MR-imaging.

26.7 Some Chemistry Involved

Preparing multifunctional nanocarriers with controlled properties requires the conjugation of proteins, peptides, polymers, cell-penetrating moieties, reporter groups, and other functional ligands to the carrier surface (although, in certain cases, the functional components may be loaded inside the nanocarrier or distributed within the nanocarrier structure; thus, for example, fine ferromagnetic particles can be loaded inside liposomes or polymeric nanoparticles to make them magnetic). This attachment can proceed noncovalently via the hydrophobic adsorption of certain intrinsic or specially inserted hydrophobic groups in the ligands to be attached onto or into the surface of the nanocarrier. Thus, amphiphilic polymers or hydrophobically modified proteins can adsorb on the hydrophobic surface of polystyrene nanoparticles (Yuan et al. 1995) or incorporate into the phospholipid membrane of liposomes (Torchilin 1998) or the hydrophobic core of micelles (Torchilin 2001). More frequently, the attachment is performed chemically, via the interaction of reactive groups generated on the carrier surface and certain groups in the molecule to be attached. In the case of liposomes, the most popular drug delivery system and convenient example of techniques used, the conjugation methodology, is based on three main reactions that are quite efficient and selective:

- Reaction between activated carboxyl groups and amino groups yielding an amide bond
- Reaction between pyridyldithiols and thiols yielding disulfide bonds
- Reaction between maleimide derivatives and thiols yielding thioether bonds (Torchilin and Klibanov 1993)

Some other approaches also exist, for example yielding the carbamate bond via the reaction of the *p*-nitropheylcarbonyl groups introduced onto the surface of nanocarriers with an amino group of various ligands (Torchilin et al. 2001b). The detailed

review of numerous coupling procedures and used protocols for attaching the whole variety of surface modifiers to drug carriers can be found in Klibanov et al. (2003) and Torchilin and Weissig (2003).

It was shown, for example, that carboxylic groups of immunoglobulins can be activated by water-soluble carbodiimide; activated protein then can be bound to free amino-group-containing surfaces, such as phosphatidyl ethanolamine(PE)-containing liposomes (Dunnick et al. 1975). For further ligand attachment, corresponding reactive groups on the surface of the nanocarriers can be premodified with the aid of heterobifunctional cross-linking reagents, such as popular *N*-succinimidyl-3(2-pyridyldithio) propionate (SPDP) used to synthesize a PE derivative further used for the coupling to SH-containing proteins (Leserman et al. 1980). Another possibility is to rely on the reaction of the thiol groups on a ligand (protein) with the maleimide-carrying surfaces (phospholipid molecules, in case of liposomes). This approach (Martin and Papahadjopoulos 1982) is now one of the most widely used in research and practical applications. Different commercially available maleimide reagents can be used for the preparation of maleimide-carrying phospholipids in a simple single-step procedure. Various high and low molecular weight compounds have been attached to liposomes by using pyridyldithiopropionyl-PE or maleimide reagents (Klibanov et al. 2003, Torchilin et al. 2003c). The application of free thiol groups located on immunoglobulin Fab fragments is also attractive. It is believed that these SH groups are positioned far from the antigen-binding sites, enabling the nanocarrier-bound antibody fragments to retain their specific interaction with antigens.

Some ligands carry carbohydrate residues, which can be easily oxidized to yield aldehyde groups that can react with surface aminogroups with the formation of the Schiff bases (Heath et al. 1980). Nanocarriers (such as liposomes) containing surface-exposed carboxylic groups were used for the attachment of different ligands (Kung and Redemann 1986). In the case of liposomes, such liposomes can be prepared by various techniques and activated with water-soluble carbodiimide directly prior to the addition of ligand. The same chemical reactions can be used to attach nonmodified proteins and peptides to various nanocarriers, including pre-formed liposomes, containing membrane-incorporated reactive lipid derivatives, such as *N*-glutaryl-PE or glutaryl-cardiolipin (Bogdanov et al. 1988, Weissig and Gregoriadis 1992, Weissig et al. 1990). The use of a four-tailed hydrophobic cardiolipin derivative instead of a two-tailed PE derivative allows for a decrease in the number of amino groups involved in the conjugation reaction at the same degree of hydrophobicity. This results in better activity preservation by the hydrophobized and liposome-attached protein (Needham et al. 1992, Weissig et al. 1986). Some current methods for attaching various (mainly, targeting) ligands to nanocarriers are reviewed in the article by Nobs et al. (2004).

Some special methods are designed to attach various sterically protective polymers to the surface of nanocarriers (see Section 26.2 and the first paragraph of this section). Thus, for example, to make PEG capable of incorporation into the liposomal membrane, the reactive derivative of hydrophilic PEG is single terminus-modified with hydrophobic moiety (usually, the residue of PE or long chain fatty acid is attached to PEG-hydroxysuccinimide ester) (Klibanov et al. 1990, 1991). In the majority of protocols, PEG-PE is used, which must be added to the lipid mixture prior to the liposome formation. Alternatively, a suggestion of synthesizing single end-reactive derivatives of PEG able to be coupled with certain reactive groups (such as maleimide) on the surface of already prepared liposomes, referred to as the post-coating method, was made (Maruyama et al. 1995). Currently, numerous studies on the preparation and properties of polymer-modified liposomes exist that have been well reviewed in several important books (Gregoriadis 1993, Lasic and Barenholz 1996, Lasic and Martin 1995). The spontaneous incorporation of PEG-lipid conjugates into the liposome membrane from PEG-lipid micelles was also shown to be very effective and did not disturb the vesicles (Sou et al. 2000).

References

Aalto-Setala, K. and Vuorio, E. 1997. Gene therapy of single-gene disorders: Preface to the special section. *Ann Med* 29: 549–551.

Abra, R. M. et al. 2002. The next generation of liposome delivery systems: Recent experience with tumor-targeted, sterically-stabilized immunoliposomes and active-loading gradients. *J Liposome Res* 12: 1–3.

Allen, T. M. 1994. The use of glycolipids and hydrophilic polymers in avoiding rapid uptake of liposomes by the mononuclear phagocyte system. *Adv Drug Deliv Rev* 13: 285–309.

Allen, T. M. and Hansen, C. 1991. Pharmacokinetics of stealth versus conventional liposomes: Effect of dose. *Biochim Biophys Acta* 1068: 133–141.

Allen, T. M. et al. 1989. Liposomes with prolonged circulation times: Factors affecting uptake by reticuloendothelial and other tissues. *Biochim Biophys Acta* 981: 27–35.

Allen, T. M. et al. 1991. Liposomes containing synthetic lipid derivatives of poly(ethylene glycol) show prolonged circulation half-lives in vivo. *Biochim Biophys Acta* 1066: 29–36.

Allen, T. M. et al. 1992. Stealth liposomes: An improved sustained release system for 1-beta-D-arabinofuranosylcytosine. *Cancer Res* 52: 2431–2439.

Allen, T. M. et al. 1995a. Pharmacokinetics of long-circulating liposomes. *Adv Drug Deliv Rev* 16: 267–284.

Allen, T. M. et al. 1995b. Pharmacokinetics and anti-tumor activity of vincristine encapsulated in sterically stabilized liposomes. *Int J Cancer* 62: 199–204.

Arnheiter, H. and Haller, O. 1988. Antiviral sTATe against influenza virus neutralized by microinjection of antibodies to interferon-induced Mx proteins. *EMBO J* 7: 1315–1320.

Asai, T. et al. 2002. Anti-neovascular therapy by liposomal DPP-CNDAC targeted to angiogenic vessels. *FEBS Lett* 520: 167–170.

Asokan, A. and Cho, M. J. 2003. Cytosolic delivery of macromolecules. II. Mechanistic studies with pH-sensitive morpholine lipids. *Biochim Biophys Acta* 1611: 151–160.

Bae, Y. et al. 2005. Preparation and biological characterization of polymeric micelle drug carriers with intracellular pH-triggered drug release property: Tumor permeability, controlled subcellular drug distribution, and enhanced in vivo antitumor efficacy. *Bioconjug Chem* 16: 122–130.

Balthasar, S. et al. 2005. Preparation and characterisation of antibody modified gelatin nanoparticles as drug carrier system for uptake in lymphocytes. *Biomaterials* 26: 2723–2732.

Barsky, D. et al. 1992. Theory of paramagnetic contrast agents in liposome systems. *Magn Reson Med* 24: 1–13.

Benhar, I. et al. 1994. Rapid humanization of the Fv of monoclonal antibody B3 by using framework exchange of the recombinant immunotoxin B3(Fv)-PE38. *Proc Natl Acad Sci U S A* 91: 12051–12055.

Bernkop-Schnurch, A. and Walker, G. 2001. Multifunctional matrices for oral peptide delivery. *Crit Rev Ther Drug Carrier Syst* 18: 459–501.

Bhadra, D. et al. 2003. A PEGylated dendritic nanoparticulate carrier of fluorouracil. *Int J Pharm* 257: 111–124.

Blume, G. et al. 1993. Specific targeting with poly(ethylene glycol)-modified liposomes: Coupling of homing devices to the ends of the polymeric chains combines effective target binding with long circulation times. *Biochim Biophys Acta* 1149: 180–184.

Bogdanov, A. A., Jr. et al. 1988. Protein immobilization on the surface of liposomes via carbodiimide activation in the presence of N-hydroxysulfosuccinimide. *FEBS Lett* 231: 381–384.

Boman, N. L. et al. 1994. Liposomal vincristine which exhibits increased drug retention and increased circulation longevity cures mice bearing P388 tumors. *Cancer Res* 54: 2830–2833.

Boomer, J. A. and Thompson, D. H. 1999. Synthesis of acid-labile diplasmenyl lipids for drug and gene delivery applications. *Chem Phys Lipids* 99: 145–153.

Boussif, O. et al. 1995. A versatile vector for gene and oligonucleotide transfer into cells in culture and in vivo: Polyethylenimine. *Proc Natl Acad Sci U S A* 92: 7297–7301.

Branden, L. J. et al. 1999. A peptide nucleic acid-nuclear localization signal fusion that mediates nuclear transport of DNA. *Nat Biotechnol* 17: 784–787.

Brannon-Peppas, L. and Blanchette, J. O. 2004. Nanoparticle and targeted systems for cancer therapy. *Adv Drug Deliv Rev* 56: 1649–1659.

Calvo, P. et al. 2001. Long-circulating PEGylated polycyanoacrylate nanoparticles as new drug carrier for brain delivery. *Pharm Res* 18: 1157–1166.

Chakrabarti, R. et al. 1989. Transfer of monoclonal antibodies into mammalian cells by electroporation. *J Biol Chem* 264: 15494–15500.

Chen, G. and Hoffman, A. S. 1995. Graft copolymers that exhibit temperature-induced phase transitions over a wide range of pH. *Nature* 373: 49–52.

Cheung, C. Y. et al. 2001. A pH-sensitive polymer that enhances cationic lipid-mediated gene transfer. *Bioconjug Chem* 12: 906–910.

Choi, H. et al. 2004. Iron oxide nanoparticles as magnetic resonance contrast agent for tumor imaging via folate receptor-targeted delivery. *Acad Radiol* 11: 996–1004.

Chonn, A. et al. 1991. Separation of large unilamellar liposomes from blood components by a spin column procedure: Towards identifying plasma proteins which mediate liposome clearance in vivo. *Biochim Biophys Acta* 1070: 215–222.

Chonn, A. et al. 1992. Association of blood proteins with large unilamellar liposomes in vivo. Relation to circulation lifetimes. *J Biol Chem* 267: 18759–18765.

Cohen, S. and Bernstein, H., Eds. 1996. *Microparticulate Systems for the Delivery of Proteins and Vaccines.* Marcel Dekker, New York.

Dagar, S. et al. 2003. VIP grafted sterically stabilized liposomes for targeted imaging of breast cancer: In vivo studies. *J Control Release* 91: 123–133.

Davis, S. S. et al. 1993. Microspheres for targeting drugs to specific body sites. *J Control Release* 24: 157–163.

De Campos, A. M. et al. 2003. The effect of a PEG versus a chitosan coating on the interaction of drug colloidal carriers with the ocular mucosa. *Eur J Pharm Sci* 20: 73–81.

de Sousa Delgado, A. et al. 2000. Surface modification of polystyrene nanoparticles using dextrans and dextran-POE copolymers: Polymer adsorption and colloidal characterization. *J Biomater Sci Polym Ed* 11: 1395–1410.

DeFrees, S. A. et al. 1996. Sialyl Lewis x liposomes as a multivalent ligand and inhibitor of E-selectinmediated cellular adhesion. *J Am Chem Soc* 118: 6101–6104.

Derossi, D. et al. 1994. The third helix of the Antennapedia homeodomain translocates through biological membranes. *J Biol Chem* 269: 10444–10450.

Derycke, A. S. and De Witte, P. A. 2002. Transferrin-mediated targeting of hypericin embedded in sterically stabilized PEG-liposomes. *Int J Oncol* 20: 181–187.

Dodd, C. H. et al. 2001. Normal T-cell response and in vivo magnetic resonance imaging of T cells loaded with HIV transactivator-peptide-derived superparamagnetic nanoparticles. *J Immunol Methods* 256: 89–105.

Dunnick, J. K. et al. 1975. Vesicle interactions with polyamino acids and antibody: In vitro and in vivo studies. *J Nucl Med* 16: 483–487.

Elliott, G. and O'Hare, P. 1997. Intercellular trafficking and protein delivery by a herpesvirus structural protein. *Cell* 88: 223–233.

Elouahabi, A. and Ruysschaert, J. M. 2005. Formation and intracellular trafficking of lipoplexes and polyplexes. *Mol Ther* 11: 336–347.

Enoch, H. G. and Strittmatter, P. 1979. Formation and properties of 1000-A-diameter, single-bilayer phospholipid vesicles. *Proc Natl Acad Sci U S A* 76: 145–149.

Ewer, M. S. et al. 2004. Cardiac safety of liposomal anthracyclines. *Semin Oncol* 31: 161–181.

Farhood, H. et al. 1995. The role of dioleoyl phosphatidylethanolamine in cationic liposome mediated gene transfer. *Biochim Biophys Acta* 1235: 289–295.

Fattal, E. et al. 2004. "Smart" delivery of antisense oligonucleotides by anionic pH-sensitive liposomes. *Adv Drug Deliv Rev* 56: 931–946.

Felgner, J. H. et al. 1994. Enhanced gene delivery and mechanism studies with a novel series of cationic lipid formulations. *J Biol Chem* 269: 2550–2561.

Filion, M. C. and Phillips, N. C. 1997. Toxicity and immunomodulatory activity of liposomal vectors formulated with cationic lipids toward immune effector cells. *Biochim Biophys Acta* 1329: 345–356.

Frankel, A. D. and Pabo, C. O. 1988. Cellular uptake of the tat protein from human immunodeficiency virus. *Cell* 55: 1189–1193.

Frankel, A. D. et al. 1988a. Tat protein from human immunodeficiency virus forms a metal-linked dimer. *Science* 240: 70–73.

Frankel, A. D. et al. 1988b. Dimerization of the tat protein from human immunodeficiency virus: A cysteine-rich peptide mimics the normal metal-linked dimer interface. *Proc Natl Acad Sci U S A* 85: 6297–6300.

Gabizon, A. and Papahadjopoulos, D. 1988. Liposome formulations with prolonged circulation time in blood and enhanced uptake by tumors. *Proc Natl Acad Sci U S A* 85: 6949–6953.

Gabizon, A. and Papahadjopoulos, D. 1992. The role of surface charge and hydrophilic groups on liposome clearance in vivo. *Biochim Biophys Acta* 1103: 94–100.

Gabizon, A. et al. 1994. Prolonged circulation time and enhanced accumulation in malignant exudates of doxorubicin encapsulated in polyethylene-glycol coated liposomes. *Cancer Res* 54: 987–992.

Gabizon, A. et al. 1999. Targeting folate receptor with folate linked to extremities of poly(ethylene glycol)-grafted liposomes: In vitro studies. *Bioconjug Chem* 10: 289–298.

Gabizon, A. et al. 2004. Tumor cell targeting of liposome-entrapped drugs with phospholipid-anchored folic acid-PEG conjugates. *Adv Drug Deliv Rev* 56: 1177–1192.

Gabizon, A. A. 1995. Liposome circulation time and tumor targeting: Implications for cancer chemotherapy. *Adv Drug Deliv Rev* 16: 285–294.

Geisert, E. E., Jr. et al. 1995. Transfecting neurons and glia in the rat using pH-sensitive immunoliposomes. *Neurosci Lett* 184: 40–43.

Gijsens, A. et al. 2002. Targeting of the photocytotoxic compound AlPcS4 to Hela cells by transferrin conjugated PEG-liposomes. *Int J Cancer* 101: 78–85.

Glogard, C. et al. 2002. Liposomes as carriers of amphiphilic gadolinium chelates: The effect of membrane composition on incorporation efficacy and in vitro relaxivity. *Int J Pharm* 233: 131–140.

Grant, C. W. et al. 1989. A liposomal MRI contrast agent: Phosphatidylethanolamine-DTPA. *Magn Reson Med* 11: 236–243.

Gref, R. et al. 1994. Biodegradable long-circulating polymeric nanospheres. *Science* 263: 1600–1603.

Gref, R. et al. 1995. The controlled intravenous delivery of drugs using PEG-coated sterically stabilized nanospheres. *Adv Drug Deliv Rev* 16: 215–233.

Gregoriadis, G., Ed. 1988. *Liposomes as Drug Carriers: Recent Trends and Progress.* John Wiley & Sons, New York.

Gregoriadis, G. 1993. *Liposome Technology*, 2nd edn. CRC Press, Boca Raton, FL.

Guo, X. and Szoka, F. C., Jr. 2001. Steric stabilization of fusogenic liposomes by a low-pH sensitive PEG-diortho ester-lipid conjugate. *Bioconj Chem* 12: 291–300.

Gupta, A. K. and Curtis, A. S. 2004. Surface modified superparamagnetic nanoparticles for drug delivery: Interaction studies with human fibroblasts in culture. *J Mater Sci Mater Med* 15: 493–496.

Gupta, A. K. and Gupta, M. 2005. Synthesis and surface engineering of iron oxide nanoparticles for biomedical applications. *Biomaterials* 26: 3995–4021.

Hafez, I. M. et al. 2001. On the mechanism whereby cationic lipids promote intracellular delivery of polynucleic acids. *Gene Ther* 8: 1188–1196.

Hallbrink, M. et al. 2001. Cargo delivery kinetics of cell-penetrating peptides. *Biochim Biophys Acta* 1515: 101–109.

Hansen, C. B. et al. 1995. Attachment of antibodies to sterically stabilized liposomes: Evaluation, comparison and optimization of coupling procedures. *Biochim Biophys Acta* 1239: 133–144.

Harding, J. A. et al. 1997. Immunogenicity and pharmacokinetic attributes of poly(ethylene glycol)-grafted immunoliposomes. *Biochim Biophys Acta* 1327: 181–192.

Harper, G. R. et al. 1991. Steric stabilization of microspheres with grafted polyethylene oxide reduces phagocytosis by rat Kupffer cells in vitro. *Biomaterials* 12: 695–700.

Hashida, M. et al. 2001. Cell-specific delivery of genes with glycosylated carriers. *Adv Drug Deliv Rev* 52: 187–196.

Hatakeyama, H. et al. 2004. Factors governing the in vivo tissue uptake of transferrin-coupled polyethylene glycol liposomes in vivo. *Int J Pharm* 281: 25–33.

Heath, T. D. et al. 1980. Covalent attachment of horseradish peroxidase to the outer surface of liposomes. *Biochim Biophys Acta* 599: 42–62.

Hirano, K. and Hunt, C. A. 1985. Lymphatic transport of liposome-encapsulated agents: Effects of liposome size following intraperitoneal administration. *J Pharm Sci* 74: 915–921.

Hobbs, S. K. et al. 1998. Regulation of transport pathways in tumor vessels: Role of tumor type and microenvironment. *Proc Natl Acad Sci U S A* 95: 4607–4612.

Huang, S. K. et al. 1992. Pharmacokinetics and therapeutics of sterically stabilized liposomes in mice bearing C-26 colon carcinoma. *Cancer Res* 52: 6774–6781.

Huang, S. K. et al. 1994. Liposomes and hyperthermia in mice: Increased tumor uptake and therapeutic efficacy of doxorubicin in sterically stabilized liposomes. *Cancer Res* 54: 2186–2191.

Huwyler, J. et al. 1996. Brain drug delivery of small molecules using immunoliposomes. *Proc Natl Acad Sci U S A* 93: 14164–14169.

Hwang, K. J. 1987. Liposome pharamacokinetics. In: *Liposomes: From Biophysics to Therapeutics.* Ed. M. J. Ostro, Dekker, New York, pp. 109–156.

Iinuma, H. et al. 2002. Intracellular targeting therapy of cisplatin-encapsulated transferrin-polyethylene glycol liposome on peritoneal dissemination of gastric cancer. *Int J Cancer* 99: 130–137.

Illum, L. et al. 2001. Development of systems for targeting the regional lymph nodes for diagnostic imaging: In vivo behaviour of colloidal PEG-coated magnetite nanospheres in the rat following interstitial administration. *Pharm Res* 18: 640–645.

Illum, S. L. and Davis, S. S. 1983. Effect of the nonionic surfactant poloxamer 338 on the fate and deposition of polystyrene microspheres following intravenous administration. *J Pharm Sci* 72: 1086–1089.

Ishida, O. et al. 2001. Liposomes bearing polyethyleneglycol-coupled transferrin with intracellular targeting property to the solid tumors in vivo. *Pharm Res* 18: 1042–1048.

Jamshaid, M. et al. 1988. Poloxamer sorption on liposomes: Comparison with polystyrene latex and influence on solute efflux. *Int. J. Pharm.* 48: 125–131.

Jeong, J. H. et al. 2003. Novel intracellular delivery system of antisense oligonucleotide by self-assembled hybrid micelles composed of DNA/PEG conjugate and cationic fusogenic peptide. *Bioconjug Chem* 14: 473–479.

Joliot, A. et al. 1991. Antennapedia homeobox peptide regulates neural morphogenesis. *Proc Natl Acad Sci U S A* 88: 1864–1868.

Jones, M. C. et al. 2003. pH-sensitive unimolecular polymeric micelles: Synthesis of a novel drug carrier. *Bioconjug Chem* 14: 774–781.

Josephson, L. et al. 1999. High-efficiency intracellular magnetic labeling with novel superparamagnetic-Tat peptide conjugates. *Bioconjug Chem* 10: 186–191.

Joshee, N. et al. 2002. Transferrin-facilitated lipofection gene delivery strategy: Characterization of the transfection complexes and intracellular trafficking. *Hum Gene Ther* 13: 1991–2004.

Kabalka, G. W. et al. 1991a. Gadolinium-labeled liposomes containing amphiphilic Gd-DTPA derivatives of varying chain length: Targeted MRI contrast enhancement agents for the liver. *Magn Reson Imaging* 9: 373–377.

Kabalka, G. W. et al. 1991b. Gadolinium-labeled liposomes containing various amphiphilic Gd-DTPA derivatives: Targeted MRI contrast enhancement agents for the liver. *Magn Reson Med* 19: 406–415.

Kakudo, T. et al. 2004. Transferrin-modified liposomes equipped with a pH-sensitive fusogenic peptide: An artificial viral-like delivery system. *Biochemistry* 43: 5618–5628.

Kamps, J. A. et al. 2000. Uptake of long-circulating immunoliposomes, directed against colon adenocarcinoma cells, by liver metastases of colon cancer. *J Drug Target* 8: 235–245.

Kato, K. et al. 2004. Rapid protein anchoring into the membranes of Mammalian cells using oleyl chain and poly(ethylene glycol) derivatives. *Biotechnol Prog* 20: 897–904.

Kaufman, C. L. et al. 2003. Superparamagnetic iron oxide particles transactivator protein-fluorescein isothiocyanate particle labeling for in vivo magnetic resonance imaging detection of cell migration: Uptake and durability. *Transplantation* 76: 1043–1046.

Kellam, B. et al. 2003. Chemical modification of mammalian cell surfaces. *Chem Soc Rev* 32: 327–337.

Kempen, H. J. et al. 1984. A water-soluble cholesteryl-containing trisgalactoside: Synthesis, properties, and use in directing lipid-containing particles to the liver. *J Med Chem* 27: 1306–1312.

Khaw, B. A. et al. 1991. Gamma imaging with negatively charge-modified monoclonal antibody: Modification with synthetic polymers. *J Nucl Med* 32: 1742–1751.

Klibanov, A. L. 1998. Antibody-mediated targeting of PEG-coated liposomes. In: *Long Circulating Liposomes: Old Drugs, New Therapeutics*. Eds. M. C. Woodle and G. Storm, Springer, Berlin, Germany, p. 269.

Klibanov, A. L. et al. 1990. Amphipathic polyethyleneglycols effectively prolong the circulation time of liposomes. *FEBS Lett* 268: 235–237.

Klibanov, A. L. et al. 1991. Activity of amphipathic poly(ethylene glycol) 5000 to prolong the circulation time of liposomes depends on the liposome size and is unfavorable for immunoliposome binding to target. *Biochim Biophys Acta* 1062: 142–148.

Klibanov, A. L. et al. 2003. Long-circulating sterically protected liposomes. In: *Liposomes: A Practical Approach*, 2nd edn. eds. V. P. Torchilin and V. Weissig, Oxford University Press, Oxford, U.K.; New York, pp. 231–265.

Koltover, I. et al. 1998. An inverted hexagonal phase of cationic liposome-DNA complexes related to DNA release and delivery. *Science* 281: 78–81.

Kratz, F. et al. 1999. Drug-polymer conjugates containing acid-cleavable bonds. *Crit Rev Ther Drug Carrier Syst* 16: 245–288.

Krause, H. J. et al. 1985. Polylactic acid nanoparticles, a colloidal drug delivery system for lipophilic drugs. *Int. J. Pharm.* 27: 145–155.

Kunath, K. et al. 2003. Low-molecular-weight polyethylenimine as a non-viral vector for DNA delivery: Comparison of physicochemical properties, transfection efficiency and in vivo distribution with high-molecular-weight polyethylenimine. *J Control Release* 89: 113–125.

Kung, V. T. and Redemann, C. T. 1986. Synthesis of carboxyacyl derivatives of phosphatidylethanolamine and use as an efficient method for conjugation of protein to liposomes. *Biochim Biophys Acta* 862: 435–439.

Kwon, G. S. and Kataoka, K. 1995. Block copolymer micelles as long-circulating drug vehicles. *Adv Drug Deliv Rev* 16: 295–309.

Lasic, D. D. 1993. *Liposomes: From Physics to Applications*. Elsevier, Amsterdam, the Netherlands; New York.

Lasic, D. D. and Barenholz, Y. 1996. *Handbook of Nonmedical Applications of Liposomes*. CRC Press, Boca Raton, FL.

Lasic, D. D. and Martin, F. J., Eds. 1995. *Stealth Liposomes*. CRC Press, Boca Raton, FL.

Lasic, D. D. et al. 1991. Sterically stabilized liposomes: A hypothesis on the molecular origin of the extended circulation times. *Biochim Biophys Acta* 1070: 187–192.

Leamon, C. P. and Low, P. S. 2001. Folate-mediated targeting: From diagnostics to drug and gene delivery. *Drug Discov Today* 6: 44–51.

Lee, E. S. et al. 2003a. Polymeric micelle for tumor pH and folate-mediated targeting. *J Control Release* 91: 103–113.

Lee, E. S. et al. 2003b. Poly(L-histidine)-PEG block copolymer micelles and pH-induced destabilization. *J Control Release* 90: 363–374.

Lee, E. S. et al. 2005. Super pH-sensitive multifunctional polymeric micelle. *Nano Lett* 5: 325–329.

Lee, R. J. and Low, P. S. 1994. Delivery of liposomes into cultured KB cells via folate receptor-mediated endocytosis. *J Biol Chem* 269: 3198–3204.

Leroux, J. et al. 2001. N-isopropylacrylamide copolymers for the preparation of pH-sensitive liposomes and polymeric micelles. *J Control Release* 72: 71–84.

Leserman, L. D. et al. 1980. Targeting to cells of fluorescent liposomes covalently coupled with monoclonal antibody or protein A. *Nature* 288: 602–604.

Lestini, B. J. et al. 2002. Surface modification of liposomes for selective cell targeting in cardiovascular drug delivery. *J Control Release* 78: 235–247.

Levchenko, T. S. et al. 2003. Tat peptide-mediated intracellular delivery of liposomes. *Methods Enzymol* 372: 339–349.

Lisziewicz, J. et al. 1995. Antitat gene therapy: A candidate for late-stage AIDS patients. *Gene Ther* 2: 218–222.

Liu, D. et al. 1991. Large liposomes containing ganglioside GM1 accumulate effectively in spleen. *Biochim Biophys Acta* 1066: 159–165.

Liu, D. et al. 1992. Role of liposome size and RES blockade in controlling biodistribution and tumor uptake of GM1-containing liposomes. *Biochim Biophys Acta* 1104: 95–101.

Liu, J. et al. 2001. Nanostructured materials designed for cell binding and transduction. *Biomacromolecules* 2: 362–368.

Loret, E. P. et al. 1991. Activating region of HIV-1 Tat protein: Vacuum UV circular dichroism and energy minimization. *Biochemistry* 30: 6013–6023.

Lu, Y. and Low, P. S. 2002a. Folate-mediated delivery of macromolecular anticancer therapeutic agents. *Adv Drug Deliv Rev* 54: 675–693.

Lu, Y. and Low, P. S. 2002b. Folate targeting of haptens to cancer cell surfaces mediates immunotherapy of syngeneic murine tumors. *Cancer Immunol Immunother* 51: 153–162.

Lukyanov, A. N. et al. 2004a. Tumor-targeted liposomes: Doxorubicin-loaded long-circulating liposomes modified with anti-cancer antibody. *J Control Release* 100: 135–144.

Lukyanov, A. N. et al. 2004b. Increased accumulation of PEG-PE micelles in the area of experimental myocardial infarction in rabbits. *J Control Release* 94: 187–193.

Maeda, H. 2001. The enhanced permeability and retention (EPR) effect in tumor vasculature: The key role of tumor-selective macromolecular drug targeting. *Adv Enzyme Regul* 41: 189–207.

Maeda, H. et al. 2000. Tumor vascular permeability and the EPR effect in macromolecular therapeutics: A review. *J Control Release* 65: 271–284.

Maeda, H. et al. 2001. Mechanism of tumor-targeted delivery of macromolecular drugs, including the EPR effect in solid tumor and clinical overview of the prototype polymeric drug SMANCS. *J Control Release* 74: 47–61.

Martin, F. J. and Papahadjopoulos, D. 1982. Irreversible coupling of immunoglobulin fragments to preformed vesicles. An improved method for liposome targeting. *J Biol Chem* 257: 286–288.

Marty, C. and Schwendener, R. A. 2005. Cytotoxic tumor targeting with scFv antibody-modified liposomes. *Methods Mol Med* 109: 389–402.

Marty, C. et al. 2004. Enhanced heparan sulfate proteoglycan-mediated uptake of cell-penetrating peptide-modified liposomes. *Cell Mol Life Sci* 61: 1785–1794.

Maruyama, K. et al. 1991. Effect of molecular weight in amphipathic polyethyleneglycol on prolonging the circulation time of large unilamellar liposomes. *Chem Pharm Bull (Tokyo)* 39: 1620–1622.

Maruyama, K. et al. 1992. Prolonged circulation time in vivo of large unilamellar liposomes composed of distearoyl phosphatidylcholine and cholesterol containing amphipathic poly(ethylene glycol). *Biochim Biophys Acta* 1128: 44–49.

Maruyama, K. et al. 1994. Phosphatidyl polyglycerols prolong liposome circulation in vivo. *Int J Pharm* 111: 103–107.

Maruyama, K. et al. 1995. Targetability of novel immunoliposomes modified with amphipathic poly(ethylene glycol)s conjugated at their distal terminals to monoclonal antibodies. *Biochim Biophys Acta* 1234: 74–80.

Mastrobattista, E. et al. 2002. Functional characterization of an endosome-disruptive peptide and its application in cytosolic delivery of immunoliposome-entrapped proteins. *J Biol Chem* 277: 27135–27143.

Matsuda, I. et al. 2001. Antimetastatic effect of hepatotropic liposomal adriamycin on human metastatic liver tumors. *Surg Today* 31: 414–420.

Mayhew, E. G. et al. 1992. Pharmacokinetics and antitumor activity of epirubicin encapsulated in long-circulating liposomes incorporating a polyethylene glycol-derivatized phospholipid. *Int J Cancer* 51: 302–309.

Moghimi, S. M. et al. 1991. The effect of poloxamer-407 on liposome stability and targeting to bone marrow: Comparison with polystyrene microspheres. *Int J Pharm* 68: 121–126.

Moghimi, S. M. et al. 1994. Surface engineered nanospheres with enhanced drainage into lymphatics and uptake by macrophages of the regional lymph nodes. *FEBS Lett* 344: 25–30.

Moghimi, S. M. and Szebeni, J. 2003. Stealth liposomes and long circulating nanoparticles: Critical issues in pharmacokinetics, opsonization and protein-binding properties. *Prog Lipid Res* 42: 463–478.

Molyneux, P. 1984. *Water-Soluble Synthetic Polymers: Properties and Behavior.* CRC Press, Boca Raton, FL.

Monfardini, C. et al. 1995. A branched monomethoxypoly(ethylene glycol) for protein modification. *Bioconjug Chem* 6: 62–69.

Monsky, W. L. et al. 1999. Augmentation of transvascular transport of macromolecules and nanoparticles in tumors using vascular endothelial growth factor. *Cancer Res* 59: 4129–4135.

Muller, M. et al. 2003. Surface modification of PLGA microspheres. *J Biomed Mater Res A* 66: 55–61.

Müller, R. H. 1991. *Colloidal Carriers for Controlled Drug Delivery and Targeting: Modification, Characterization, and In Vivo Distribution.* Wissenschaftliche Verlagsgesellschaft, CRC Press, Stuttgart, Germany; Boca Raton, FL.

Murad, K. L. et al. 1999. Structural and functional consequences of antigenic modulation of red blood cells with methoxypoly(ethylene glycol). *Blood* 93: 2121–2127.

Napper, D. H. 1983. *Polymeric Stabilization of Colloidal Dispersions.* Academic Press, London, U.K.; New York.

Needham, D. et al. 1992. Repulsive interactions and mechanical stability of polymer-grafted lipid membranes. *Biochim Biophys Acta* 1108: 40–48.

Ni, S. et al. 2002. Folate receptor targeted delivery of liposomal daunorubicin into tumor cells. *Anticancer Res* 22: 2131–2135.

Nobs, L. et al. 2004. Current methods for attaching targeting ligands to liposomes and nanoparticles. *J Pharm Sci* 93: 1980–1992.

Oehlke, J. et al. 1998. Cellular uptake of an alpha-helical amphipathic model peptide with the potential to deliver polar compounds into the cell interior non-endocytically. *Biochim Biophys Acta* 1414: 127–139.

Ogris, M. et al. 2001. DNA/polyethylenimine transfection particles: Influence of ligands, polymer size, and PEGylation on internalization and gene expression. *AAPS PharmSci* 3: E21.

Olivier, J. C. et al. 2002. Synthesis of pegylated immunonanoparticles. *Pharm Res* 19: 1137–1143.

Omori, N. et al. 2003. Targeting of post-ischemic cerebral endothelium in rat by liposomes bearing polyethylene glycol-coupled transferrin. *Neurol Res* 25: 275–279.

Paleos, C. M. et al. 2004. Acid- and salt-triggered multifunctional poly(propylene imine) dendrimer as a prospective drug delivery system. *Biomacromolecules* 5: 524–529.

Pan, D. et al. 2003a. Folic acid-conjugated nanostructured materials designed for cancer cell targeting. *Chem Commun (Camb)* (19): 2400–2401.

Pan, X. Q. et al. 2003b. Antitumor activity of folate receptor-targeted liposomal doxorubicin in a KB oral carcinoma murine xenograft model. *Pharm Res* 20: 417–422.

Pang, S. N. J. 1993. Final report on the safety assessment of Polyethylene Glycols (PEGs) -6, -8, -32, -75, -150, -14M, -20M. *J Am Coll Toxicol* 12: 429–457.

Papahadjopoulos, D. et al. 1991. Sterically stabilized liposomes: Improvements in pharmacokinetics and antitumor therapeutic efficacy. *Proc Natl Acad Sci U S A* 88: 11460–11464.

Park, E. K. et al. 2005. Preparation and characterization of methoxy poly(ethylene glycol)/poly(epsilon-caprolactone) amphiphilic block copolymeric nanospheres for tumor-specific folate-mediated targeting of anticancer drugs. *Biomaterials* 26: 1053–1061.

Park, J. W. et al. 2001. Tumor targeting using anti-her2 immuno-liposomes. *J Control Release* 74: 95–113.

Patel, H. M. and Moghimi, S. 1990. Tissue specific opsonins and phagocytosis of liposomes. In: *Targeting of Drugs 2: Optimization Strategies*, Vol. 199. Eds. G. Gregoriadis, A. C. Allison, and G. Poste, Plenum Press, New York, p. 87.

Patel, H. M. et al. 1984. Assessment of the potential uses of liposomes for lymphoscintigraphy and lymphatic drug delivery. Failure of 99m-technetium marker to represent intact liposomes in lymph nodes. *Biochim Biophys Acta* 801: 76–86.

Peer, D. and Margalit, R. 2004. Loading mitomycin C inside long circulating hyaluronan targeted nano-liposomes increases its antitumor activity in three mice tumor models. *Int J Cancer* 108: 780–789.

Peracchia, M. T. et al. 1999. Stealth PEGylated polycyanoacrylate nanoparticles for intravenous administration and splenic targeting. *J Control Release* 60: 121–128.

Phillips, W. T. and Goins, B. 1995. Targeted delivery of imaging agents by liposomes. In: *Handbook of Targeted Delivery of Imaging Agents*. Ed. V. P. Torchilin, CRC Press, Boca Raton, FL, pp. 149–173.

Pollard, H. et al. 1998. Polyethylenimine but not cationic lipids promotes transgene delivery to the nucleus in mammalian cells. *J Biol Chem* 273: 7507–7511.

Pollard, H. et al. 2001. Ca2 + -sensitive cytosolic nucleases prevent efficient delivery to the nucleus of injected plasmids. *J Gene Med* 3: 153–164.

Pooga, M. et al. 1998. Cell penetration by transportan. *FASEB J* 12: 67–77.

Porter, C. J. et al. 1992. The polyoxyethylene/polyoxypropylene block co-polymer poloxamer-407 selectively redirects intravenously injected microspheres to sinusoidal endothelial cells of rabbit bone marrow. *FEBS Lett* 305: 62–66.

Potineni, A. et al. 2003. Poly(ethylene oxide)-modified poly(beta-amino ester) nanoparticles as a pH-sensitive biodegradable system for paclitaxel delivery. *J Control Release* 86: 223–234.

Powell, G. M. 1980. Polyethylene glycol. In: *Handbook of Water-Soluble Gums and Resins*. Ed. R. L. Davidson, McGraw-Hill, New York, pp. 1–31.

Putz, B. et al. 1994. Mechanisms of liposomal contrast agents in magnetic resonance imaging. *J Liposome Res* 4: 771–808.

Raffaghello, L. et al. 2003. Immunoliposomal fenretinide: A novel antitumoral drug for human neuroblastoma. *Cancer Lett* 197: 151–155.

Ranucci, E. et al. 1994. Synthesis and molecular weight characterization of low molecular weight end-functionalized poly(4-acryloymorpholine). *Macromol Chem Phys* 195: 3469–3479.

Rolland, A. 1993. *Pharmaceutical Particulate Carriers: Therapeutic Applications Carriers*. Marcel Dekker, New York.

Rose, P. G. 2005. Pegylated liposomal doxorubicin: Optimizing the dosing schedule in ovarian cancer. *Oncologist* 10: 205–214.

Roth, J. A. et al. 1998. Gene therapy for non-small cell lung cancer: A preliminary report of a phase I trial of adenoviral p53 gene replacement. *Semin Oncol* 25: 33–37.

Rothbard, J. B. et al. 2004. Role of membrane potential and hydrogen bonding in the mechanism of translocation of guanidinium-rich peptides into cells. *J Am Chem Soc* 126: 9506–9507.

Roux, E. et al. 2002a. Polymer based pH-sensitive carriers as a means to improve the cytoplasmic delivery of drugs. *Int J Pharm* 242: 25–36.

Roux, E. et al. 2002b. Steric stabilization of liposomes by pH-responsive N-isopropylacrylamide copolymer. *J Pharm Sci* 91: 1795–1802.

Roux, E. et al. 2004. Serum-stable and long-circulating, PEGylated, pH-sensitive liposomes. *J Control Release* 94: 447–451.

Ruben, S. et al. 1989. Structural and functional characterization of human immunodeficiency virus tat protein. *J Virol* 63: 1–8.

Sadzuka, Y. et al. 2003. Effect of polyethyleneglycol (PEG) chain on cell uptake of PEG-modified liposomes. *J Liposome Res* 13: 157–172.

Sakurai, F. et al. 2000. Effect of DNA/liposome mixing ratio on the physicochemical characteristics, cellular uptake and intracellular trafficking of plasmid DNA/cationic liposome complexes and subsequent gene expression. *J Control Release* 66: 255–269.

Salem, A. K. et al. 2003. Multifunctional nanorods for gene delivery. *Nat Mater* 2: 668–671.

Sapra, P. and Allen, T. M. 2002. Internalizing antibodies are necessary for improved therapeutic efficacy of antibody-targeted liposomal drugs. *Cancer Res* 62: 7190–7194.

Sartore, L. et al. 1994. Low molecular weight end-functionalized poly(N-vinylpyrrolidone) for the modifications of polypeptide aminogroups. *J Bioact Compact Polym* 9: 411–427.

Sawant, R. M. et al. (2005) Creating multifunctional drug delivery systems: Micelles made of pH-responsive amohiphilic polymers with hidden biotin moiety. In: *32nd International Symposium on Controlled Release*. Controlled Release Society, Inc., Miami, pp. 406.

Scheule, R. K. et al. 1997. Basis of pulmonary toxicity associated with cationic lipid-mediated gene transfer to the mammalian lung. *Hum Gene Ther* 8: 689–707.

Schiffelers, R. M. et al. 2003. Anti-tumor efficacy of tumor vasculature-targeted liposomal doxorubicin. *J Control Release* 91: 115–122.

Schwarze, S. R. et al. 2000. Protein transduction: Unrestricted delivery into all cells? *Trends Cell Biol* 10: 290–295.

Schwendener, R. A. et al. 1990. A pharmacokinetic and MRI study of unilamellar gadolinium-, manganese-, and iron-DTPA-stearate liposomes as organ-specific contrast agents. *Invest Radiol* 25: 922–932.

Scott, M. D. and Chen, A. M. 2004. Beyond the red cell: Pegylation of other blood cells and tissues. *Transfus Clin Biol* 11: 40–46.

Scott, M. D. and Murad, K. L. 1998. Cellular camouflage: Fooling the immune system with polymers. *Curr Pharm Des* 4: 423–438.

Senior, J. et al. 1991. Influence of surface hydrophilicity of liposomes on their interaction with plasma protein and clearance from the circulation: Studies with poly(ethylene glycol)-coated vesicles. *Biochim Biophys Acta* 1062: 77–82.

Senior, J. H. 1987. Fate and behavior of liposomes in vivo: A review of controlling factors. *Crit Rev Ther Drug Carrier Syst* 3: 123–193.

Shalaev, E. Y. and Steponkus, P. L. 1999. Phase diagram of 1,2-dioleoylphosphatidylethanolamine (DOPE):water system at subzero temperatures and at low water contents. *Biochim Biophys Acta* 1419: 229–247.

Sheff, D. 2004. Endosomes as a route for drug delivery in the real world. *Adv Drug Deliv Rev* 56: 927–930.

Shi, G. et al. 2002. Efficient intracellular drug and gene delivery using folate receptor-targeted pH-sensitive liposomes composed of cationic/anionic lipid combinations. *J Control Release* 80: 309–319.

Simoes, S. et al. 2004. On the formulation of pH-sensitive liposomes with long circulation times. *Adv Drug Deliv Rev* 56: 947–965.

Siomi, H. et al. 1990. Effects of a highly basic region of human immunodeficiency virus TAT protein on nucleolar localization. *J Virol* 64: 1803–1807.

Sou, K. et al. 2000. Poly(ethylene glycol)-modification of the phospholipid vesicles by using the spontaneous incorporation of poly(ethylene glycol)-lipid into the vesicles. *Bioconjug Chem* 11: 372–379.

Stella, B. et al. 2000. Design of folic acid-conjugated nanoparticles for drug targeting. *J Pharm Sci* 89: 1452–1464.

Stephenson, S. M. et al. 2003. Folate receptor-targeted liposomes as possible delivery vehicles for boron neutron capture therapy. *Anticancer Res* 23: 3341–3345.

Straubinger, R. M. et al. 1985. pH-sensitive liposomes mediate cytoplasmic delivery of encapsulated macromolecules. *FEBS Lett* 179: 148–154.

Sudimack, J. J. et al. 2002. A novel pH-sensitive liposome formulation containing oleyl alcohol. *Biochim Biophys Acta* 1564: 31–37.

Suzawa, T. et al. 2002. Enhanced tumor cell selectivity of adriamycin-monoclonal antibody conjugate via a poly(ethylene glycol)-based cleavable linker. *J Control Release* 79: 229–242.

Takeuchi, H. et al. 1999. Prolonged circulation time of doxorubicin-loaded liposomes coated with a modified polyvinyl alcohol after intravenous injection in rats. *Eur J Pharm Biopharm* 48: 123–129.

Tan, P. H. et al. 2003. Antibody targeted gene transfer to endothelium. *J Gene Med* 5: 311–323.

Tang, F. and Hughes, J. A. 1999. Use of dithiodiglycolic acid as a tether for cationic lipids decreases the cytotoxicity and increases transgene expression of plasmid DNA in vitro. *Bioconjug Chem* 10: 791–796.

Thomas, T. P. et al. 2004. In vitro targeting of synthesized antibody-conjugated dendrimer nanoparticles. *Biomacromolecules* 5: 2269–2274.

Tilcock, C. 1993. Liposomal paramagnetic magnetic resonance contrast agents. In: *Liposome Technology*, 2nd edn, Vol. 2. Ed. G. Gregoriadis, CRC Press, Boca Raton, FL, pp. 65–87.

Tilcock, C. et al. 1989. Liposomal Gd-DTPA: Preparation and characterization of relaxivity. *Radiology* 171: 77–80.

Tkachenko, A. G. et al. 2004. Cellular trajectories of peptide-modified gold particle complexes: Comparison of nuclear localization signals and peptide transduction domains. *Bioconjug Chem* 15: 482–490.

Torchilin, V. P. 1984. Immobilization of specific proteins on liposome surface: Systems for drug targeting. In: *Liposome Technology*, Vol. 3. Ed. G. Gregoriadis, CRC Press Boca Raton, FL, pp. 75–94.

Torchilin, V. P. 1985. Liposomes as targetable drug carriers. *Crit Rev Ther Drug Carrier Syst* 2: 65–115.

Torchilin, V. P. 1996. How do polymers prolong circulation times of liposomes. *J Liposome Res* 9: 99–116.

Torchilin, V. P. 1997. Surface-modified liposomes in gamma- and MR-imaging. *Adv Drug Deliv Rev* 24: 301–313.

Torchilin, V. P. 1998. Polymer-coated long-circulating microparticulate pharmaceuticals. *J Microencapsul* 15: 1–19.

Torchilin, V. P. 1999. Novel polymers in microparticulate diagnostic agents. *Chemtech* 29: 27–34.

Torchilin, V. P. 2000. Polymeric contrast agents for medical imaging. *Curr Pharm Biotechnol* 1: 183–215.

Torchilin, V. P. 2001. Structure and design of polymeric surfactant-based drug delivery systems. *J Control Release* 73: 137–172.

Torchilin, V. P. 2002. Strategies and means for drug targeting: An overview. In: *Biomedical Aspects of Drug Targeting*. Eds. V. Muzykantov and V. P. Torchilin, Kluwer Academic Publishers, Boston, MA, pp. 3–26.

Torchilin, V. P. 2005. Recent advances with liposomes as pharmaceutical carriers. *Nat Rev Drug Discov* 4: 145–160.

Torchilin, V. P., Ed. 2006a. *Delivery of Protein and Peptide Drugs in Cancer*. Imperial College Press, London, U.K.

Torchilin, V. P., Ed. 2006b. *Nanoparticulates as Drug Carriers*. Imperial College Press, London, U.K.

Torchilin, V. P. 2007. Micellar nanocarriers: Pharmaceutical perspectives. *Pharm Res* 24: 1–16.

Torchilin, V. P., Ed. 2008. *Multifunctional Pharmaceutical Nanocarriers*. Springer, New York.

Torchilin, V. P. and Klibanov, A. L. 1993. Coupling and labeling of phospholipids. In: *Phospholipids Handbook*. Ed. G. Cevc, Marcel Dekker Inc., New York, pp. 293–321.

Torchilin, V. P. and Levchenko, T. S. 2003. TAT-liposomes: A novel intracellular drug carrier. *Curr Protein Pept Sci* 4: 133–140.

Torchilin, V. P. and Papisov, M. I. 1994. Why do polyethylene glycol-coated liposomes circulate so long? *J Liposome Res* 4: 725–739.

Torchilin, V. P. and Trubetskoy, V. S. 1995a. In vivo visualizing of organs and tissues with liposomes. *J Liposome Res* 5: 795–812.

Torchilin, V. P. and Trubetskoy, V. S. 1995b. Which polymers can make nanoparticulate drug carriers long-circulating? *Adv Drug Deliv Rev* 16: 141–155.

Torchilin, V. P. and Weissig, V., Eds. 2003. *Liposomes: A Practical Approach*, 2nd edn. Oxford University Press, Oxford, U.K.; New York.

Torchilin, V. P. et al. 1980. Incorporation of hydrophilic protein modified with hydrophobic agent into liposome membrane. *Biochim Biophys Acta* 602: 511–521.

Torchilin, V. P. et al. 1982. Phosphatidyl inositol may serve as the hydrophobic anchor for immobilization of proteins on liposome surface. *FEBS Lett* 138: 117–120.

Torchilin, V. P. et al. 1992. Targeted accumulation of polyethylene glycol-coated immunoliposomes in infarcted rabbit myocardium. *FASEB J* 6: 2716–2719.

Torchilin, V. P. et al. 1993a. Polymer-coated immunoliposomes for delivery of pharmaceuticals: Targeting and biological stability. In: *20th International Symposium on Controlled Release of Bioactive Materials*. Controlled Release Society, Washington, DC, pp. 194–195.

Torchilin, V. P. et al. 1993b. pH-sensitive liposomes. *J Liposome Res* 3: 201–255.

Torchilin, V. P. et al. 1994a. Poly(ethylene glycol) on the liposome surface: On the mechanism of polymer-coated liposome longevity. *Biochim Biophys Acta* 1195: 11–20.

Torchilin, V. P. et al. 1994b. Amphiphilic vinyl polymers effectively prolong liposome circulation time in vivo. *Biochim Biophys Acta* 1195: 181–184.

Torchilin, V. P. et al. 1994c. Targeted delivery of diagnostic agents by surface-modified liposomes. *J Controlled Release* 28: 45–58.

Torchilin, V. P. et al. 1995a. PEG-modified liposomes for gamma- and magentic resonance imaging. In: *Stealth Liposomes*. Eds. D. D. Lasic and F. J. Martin, CRC Press, Boca Raton, pp. 225–231.

Torchilin, V. P. et al. 1995b. New synthetic amphiphilic polymers for steric protection of liposomes in vivo. *J Pharm Sci* 84: 1049–1053.

Torchilin, V. P. et al. 1995c. Magnetic resonance imaging of lymph nodes with GD-containing liposomes. In: *Handbook of Targeted Delivery of Imaging Agents*. Ed. V. P. Torchilin, CRC Press, Boca Raton, FL, pp. 403–413.

Torchilin, V. P. et al. 1996. Poly(ethylene glycol)-coated anti-cardiac myosin immunoliposomes: Factors influencing targeted accumulation in the infarcted myocardium. *Biochim Biophys Acta* 1279: 75–83.

Torchilin, V. P. et al. 2000 PEG-Immunoliposomes: Attachment of monoclonal antibody to distal ends of PEG chains via p-Nitrophenylcarbonyl groups. In: *27th International Symposium on Controlled Release of Bioactive Materials*, Controlled Release Society, Inc., Paris, France, pp. 217–218.

Torchilin, V. P. et al. 2001a. p-Nitrophenylcarbonyl-PEG-PE-liposomes: Fast and simple attachment of specific ligands, including monoclonal antibodies, to distal ends of PEG chains via p-nitrophenylcarbonyl groups. *Biochim Biophys Acta* 1511: 397–411.

Torchilin, V. P. et al. 2001b. Amphiphilic poly-N-vinylpyrrolidones: Synthesis, properties and liposome surface modification. *Biomaterials* 22: 3035–3044.

Torchilin, V. P. et al. 2001c. TAT peptide on the surface of liposomes affords their efficient intracellular delivery even at low temperature and in the presence of metabolic inhibitors. *Proc Natl Acad Sci U S A* 98: 8786–8791.

Torchilin, V. P. et al. 2003a. Cell transfection in vitro and in vivo with nontoxic TAT peptide-liposome-DNA complexes. *Proc Natl Acad Sci U S A* 100: 1972–1977.

Torchilin, V. P. et al. 2003b. Immunomicelles: Targeted pharmaceutical carriers for poorly soluble drugs. *Proc Natl Acad Sci U S A* 100: 6039–6044.

Torchilin, V. P. et al. 2003c. Surface modifications of liposomes. In: *Liposomes: A Practical Approach*, 2nd edn. Eds. V. P. Torchilin and V. Weissig, Oxford University Press, Oxford, U.K.; New York, pp. 193–229.

Trubetskoy, V. S. and Torchilin, V. P. 1994. New approaches in the chemical design of Gd-containing liposomes for use in magnetic resonance imaging of lymph nodes. *J Liposome Res* 4: 961–980.

Trubetskoy, V. S. and Torchilin, V.P. 1995. Use of polyoxyethylene-lipid conjugates as long-circulating carriers for delivery of therapeutic and diagnostoc agents. *Adv Drug Deliv Rev* 16: 311–320.

Trubetskoy, V. S. and Torchilin, V.P. 1996. Polyethyleneglycol based micelles as carriers of therapeutic and diagnostic agents. *STP Pharma Sci* 6: 79–86.

Trubetskoy, V. S. et al. 1995. Controlled delivery of Gd-containing liposomes to lymph nodes: Surface modification may enhance MRI contrast properties. *Magn Reson Imaging* 13: 31–37.

Trubetskoy, V. S. et al. 1996. Stable polymeric micelles: Lymphnagiographic contrast media for gamma scintigraphy and magnetic resonance imaging. *Acta Radiol* 3: 232–238.

Tseng, Y. L. et al. 2002. Translocation of liposomes into cancer cells by cell-penetrating peptides penetratin and tat: A kinetic and efficacy study. *Mol Pharmacol* 62: 864–872.

Turk, M. J. et al. 2002. Characterization of a novel pH-sensitive peptide that enhances drug release from folate-targeted liposomes at endosomal pHs. *Biochim Biophys Acta* 1559: 56–68.

Unger, E. C. et al. 1989. Hepatic metastases: Liposomal Gd-DTPA-enhanced MR imaging. *Radiology* 171: 81–85.

Unger, E. et al. 1990. Biodistribution and clearance of liposomal gadolinium-DTPA. *Invest Radiol* 25: 638–644.

van Vlerken, L. E. and Amiji, M. M. 2006. Multi-functional polymeric nanoparticles for tumour-targeted drug delivery. *Expert Opin Drug Deliv* 3: 205–216.

Varga, C. M. et al. 2000. Receptor-mediated targeting of gene delivery vectors: Insights from molecular mechanisms for improved vehicle design. *Biotechnol Bioeng* 70: 593–605.

Venugopalan, P. et al. 2002. pH-sensitive liposomes: Mechanism of triggered release to drug and gene delivery prospects. *Pharmazie* 57: 659–671.

Veronese, F. M. 2001. Peptide and protein PEGylation: A review of problems and solutions. *Biomaterials* 22: 405–417.

Wadia, J. S. et al. 2004. Transducible TAT-HA fusogenic peptide enhances escape of TAT-fusion proteins after lipid raft macropinocytosis. *Nat Med* 10: 310–315.

Wang, J. et al. 2005. Polymeric micelles for delivery of poorly soluble drugs: Preparation and anticancer activity in vitro of paclitaxel incorporated into mixed micelles based on poly(ethylene glycol)-lipid conjugate and positively charged lipids. *J Drug Target* 13: 73–80.

Wartlick, H. et al. 2004. Highly specific HER2-mediated cellular uptake of antibody-modified nanoparticles in tumour cells. *J Drug Target* 12: 461–471.

Weeks, K. M. et al. 1990. Fragments of the HIV-1 Tat protein specifically bind TAR RNA. *Science* 249: 1281–1285.

Weissig, V. and Gregoriadis, G. 1992. Coupling of aminogroup bearing ligands to liposomes. In: *Liposome Thechnology*, Vol. 3. Ed. G. Gregoriadis, CRC Press, Boca Raton, FL, pp. 231–248.

Weissig, V. et al. 1986. A new hydrophobic anchor for the attachment of proteins to liposomal membranes. *FEBS Lett* 202: 86–90.

Weissig, V. et al. 1990. Covalent binding of peptides at liposome surfaces. *Die Pharmazie* 45: 849–850.

Winslow, R. M. et al. 1996. Blood substitutes: New challenges. Birkhäuser, Boston, MA.

Wong, J. Y. et al. 1997. Direct measurement of a tethered ligand-receptor interaction potential. *Science* 275: 820–822.

Woodle, M. C. 1993. Surface-modified liposomes: Assessment and characterization for increased stability and prolonged blood circulation. *Chem Phys Lipids* 64: 249–262.

Woodle, M. C. et al. 1992. Liposome leakage and blood circulation: Comparison of absorbed block copolymers with covalent attachment of PEG. *Int J Pharm* 88: 327–334.

Woodle, M. C. et al. 1994. New amphipatic polymer-lipid conjugates forming long-circulating reticuloendothelial system-evading liposomes. *Bioconjug Chem* 5: 493–496.

Wu, G. Y. and Wu, C. H. 1987. Receptor-mediated in vitro gene transformation by a soluble DNA carrier system. *J Biol Chem* 262: 4429–4432.

Xu, L. et al. 2002. Systemic tumor-targeted gene delivery by anti-transferrin receptor scFv-immunoliposomes. *Mol Cancer Ther* 1: 337–346.

Xu, Y. and Szoka, F. C., Jr. 1996. Mechanism of DNA release from cationic liposome/DNA complexes used in cell transfection. *Biochemistry* 35: 5616–5623.

Yamaoka, T. et al. 1994. Distribution and tissue uptake of poly(ethylene glycol) with different molecular weights after intravenous administration to mice. *J Pharm Sci* 83: 601–606.

Yessine, M. A. and Leroux, J. C. 2004. Membrane-destabilizing polyanions: Interaction with lipid bilayers and endosomal escape of biomacromolecules. *Adv Drug Deliv Rev* 56: 999–1021.

Yoo, H. S. et al. 2002. Doxorubicin-conjugated biodegradable polymeric micelles having acid-cleavable linkages. *J Control Release* 82: 17–27.

Yoshioka, H. 1991. Surface modification of haemoglobin-containing liposomes with polyethylene glycol prevents liposome aggregation in blood plasma. *Biomaterials* 12: 861–864.

Yuan, F. et al. 1995. Vascular permeability in a human tumor xenograft: Molecular size dependence and cutoff size. *Cancer Res* 55: 3752–3756.

Zalipsky, S. 1995. Chemistry of polyethylene glycol conjugates with biologically active molecules. *Adv Drug Deliv Rev* 16: 157–182.

Zalipsky, S. et al. 1996. Long-circulating, polyethylene glycol-grafted immunoliposomes. *J Control Release* 39: 153–161.

Zalipsky, S. et al. 1997. Poly(ethylene glycol)-grafted liposomes with oligopeptide or oligosaccharide ligands appended to the termini of the polymer chains. *Bioconjug Chem* 8: 111–118.

Zalipsky, S. et al. 1998. Biologically active ligand-bearing polymer-grafted liposomes. In: *Targeting of Drugs 6: Strategies for Stealth Therapeutic Systems*. Ed. G. Gregoriadis, Plenum Press, New York, pp. 131–139.

Zalipsky, S. et al. 1999a. New chemoenzymatic approach to glycolipopolymers: Practical preparation of functionally active galactose-poly(ethylene glycol)-distearoylphosphatidic acid (Gal-PEG-DSPA) conjugate. *Chem. Commun.* 118: 653–654.

Zalipsky, S. et al. 1999b. New detachable poly(ethylene glycol) conjugates: Cysteine-cleavable lipopolymers regenerating natural phospholipid, diacyl phosphatidylethanolamine. *Bioconjug Chem* 10: 703–707.

Zhang, F. et al. 2001. Modification of gold surface by grafting of poly(ethylene glycol) for reduction in protein adsorption and platelet adhesion. *J Biomater Sci Polym Ed* 12: 515–531.

Zhang, J. X. et al. 2004. Pharmaco attributes of dioleoylphosphatidylethanolamine/cholesteryl-hemisuccinate liposomes containing different types of cleavable lipopolymers. *Pharmacol Res* 49: 185–198.

Zhang, Y. et al. 2002. Surface modification of superparamagnetic magnetite nanoparticles and their intracellular uptake. *Biomaterials* 23: 1553–1561.

Zhao, M. et al. 2002. Differential conjugation of tat peptide to superparamagnetic nanoparticles and its effect on cellular uptake. *Bioconjug Chem* 13: 840–844.

27
Nanotechnology and Drug Delivery

Fahima Dilnawaz
Institute of Life Sciences

Sarbari Acharya
Institute of Life Sciences

Ranjita Misra
Institute of Life Sciences

Abhalaxmi Singh
Institute of Life Sciences

Sanjeeb Kumar Sahoo
Institute of Life Sciences

27.1 Significance of Nanotechnology in Drug Delivery

Nanotechnology is the manipulation of materials or devices at the nanometer scale (one billionth of a meter), that is, at the level of individual atoms and molecules. The word "nano" is derived from the Greek word for dwarf. The timescale of nanotechnology history began with a talk given in 1959 by physicist Richard Feynman, titled "There's plenty of room at the bottom." The wide perspective of nanotechnology is applicable to all spheres of life. Nanotechnology has an integral role in the rapid advancement of the medical sciences. The emerging and growing field of nanotechnology in the area of medical applications can be termed as nanomedicine, which account for more than 50% of all publications and patent filing worldwide (Wagner 2005). Therefore, much attention has been paid for intense research on nanotechnology-mediated drug delivery and drug targeting.

The conventional drug delivery techniques have many pharmacodynamics and pharmacokinetic limitations, such as low efficacy, poor solubility, low bioavailability (the fraction of an administered dose of drug that reaches the systemic circulation), and quick clearance by reticuloendothelial system (RES), a part

of the immune system. The reduced drug efficacy could be due to the instability of drug inside the cell, unavailability due to multiple targeting or chemical properties of delivering molecules, alterations in genetic makeup of cell surface receptors, overexpression of efflux pumps, changes in signaling pathways with the progression of disease, or drug degradation. Thus, to overcome the lacuna of conventional therapy, nanotechnology comes into the picture (Smith and Van de Waterbeemd 1999).

The goal of overcoming the limitations of existing drug delivery methods is mainly achieved by these small-sized particles that can penetrate across different barriers through small capillaries into individual cells. An ideal targeting system for drug delivery should have properties like (a) biocompatibility (having no toxic or injurious side effects on biological systems), biodegradability (ability to break down into nontoxic products by the body enzymes for safe removal) with low antigenicity, (2) maintenance of drug integrity till the target is reached, (3) avoidance of side effects, (4) ability to pass through biological membranes like cell membranes and nuclear membranes, (5) target recognition and association, and (6) controlled drug release. Nanotechnology is relatively new and the full scope of contribution of this technological advance in the field of human health care remains unexplored. Recent advances suggest that nanotechnology will have a profound impact on disease prevention, diagnosis, and treatment (Emerich 2005, Sahoo and Labhasetwar 2003, Sahoo et al. 2007). Nanotechnology is opening new therapeutics for many agents that cannot be used effectively as conventional formulations because of their poor bioavailability. The reformulation of the drug with smaller particles may increase its bioavailability (El-Shabouri 2002). Nanoparticles, liposomes, micelles, and dendrimers are some of the nanovehicles used as delivery agents that can be prepared to entrap, encapsulate, or bind targeting molecules. This improves the solubility, stability, and absorption of several drugs, avoiding the RES, thus protecting the drug from premature inactivation during its transport (Parveen and Sahoo 2006). The biodistribution or distribution of drug-loaded nanocarriers in tissues and pharmacokinetic parameters of the drug in nanocarriers was also significantly improved compared to free drug (Roco 2003, Shaffer 2005). Gene therapy is a novel form of drug delivery that enlists the synthetic machinery of the patient's cells to produce a therapeutic agent. Applications of gene therapy are not limited to rare inherited diseases but extend potentially to common acquired disorders. Genetic therapies for diseases have been blocked for some time by difficulties in safely delivering the therapeutic genetic material to the affected areas of the body. Nanotechnology, using advanced polymers as a delivery mechanism, may revive genetic therapy as a tool for curing diseases. The nanotechnology-based approach used by the researchers has minimal toxic side effects to normal cells. Researchers have developed nanoparticles capable of delivering genes, which could have the potential cure for several genetically transmitted diseases (Miller 2008, Templeton 2002).

Thus, with the pressing need for the development of new medicines, nanotechnology is being applied to diverse fields of medicine such as oncology, cardiovascular therapy, and in the treatment of other chronic diseases. At present, nanotechnology is being used for the management and treatment of the patient's health in the forms of molecular diagnostics (biomarkers) and drug delivery tools. In this chapter, we focus on a few aspects of the significance of nanotechnology in medical sciences.

27.2 Nanotechnology: A Platform for Drug Delivery

For drug delivery and other biomedical applications, different kinds of nanovehicles are fabricated to increase the efficacy. The current "nano" drug delivery systems in the nanometer range comprise of nanovehicles such as liposomes, nanoparticles, polymeric micelles, dendrimers, and nanocrystals, as depicted in Figure 27.1.

27.2.1 Liposomes

Liposomes can be defined as small artificial vesicles of spherical shape that can be produced from natural nontoxic phospholipids and cholesterol. Liposomes were discovered by Alec. D. Bangham in 1961 at the Babraham Institute, Cambridge; he defined these as lipid spheres containing an aqueous core (Bangham et al. 1965). Liposomes have become very versatile

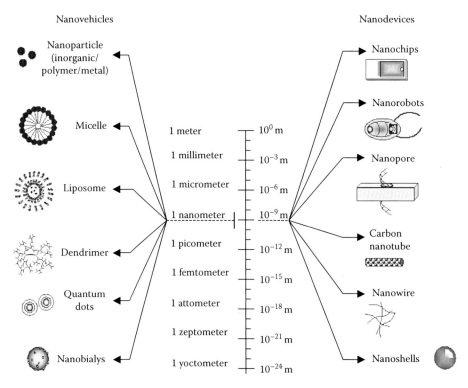

FIGURE 27.1 Types of nanovehicles and nanodevices formed in the nanoscale range.

tools in biology, biochemistry, and medicine and are widely used to treat infectious diseases and to deliver anticancer drugs either by endocytic pathway or by the fusion of liposomal surfaces with the cell membrane. It is the first type of nanovehicle to be used to create therapeutic agents with novel characteristics. Various targeting ligands can be attached to their surface, allowing their surface attachment and accumulation in diseased areas for better treatment of diseases.

Liposomes are broadly classified by their structure as spherical, concentric, multilamellar, and unilamellar. The size, lamellarity (unilamellar and multilamellar), and lipid composition of the bilayers influence many of the important properties like fluidity, permeability, stability, and structure. Thus, the properties of liposomes can be controlled and customized by varying the lipid composition, size, surface charge, and method of preparation to serve specific needs. The properties are also influenced by external parameters like temperature, ionic strength, and the presence of certain molecules nearby. Based on the size and the number of bilayers, liposomes are classified into three classes. Small unilamellar vesicles (SUV) are of 25–50 nm in diameter and are surrounded by a single lipid layer. Large unilamellar vesicles (LUV) are similar to SUV but these are heterogeneous group of vesicles surrounded by a single lipid layer. Multilamellar vesicles (MLV) consist of several lipid layers separated from one another by an aqueous layer of solution (Sahoo and Labhasetwar 2003).

Liposomes deliver the drugs to the target sites by means of both active and passive targeting processes. Unmodified liposomes are rapidly cleared by the RES. To escape from the RES clearance, liposomes are modified to get the long circulating liposomes that are known as "stealth" liposomes (Woodle and Lasic 1992). In stealth liposome formulations, the surface is grafted by polyethylene glycol (PEG) chains or inclusion of phosphatidylinositol and gangliosides to get a hydrophilic surface (Lasic 1998). These are also used for ligand-mediated targeted drug delivery.

Liposomes are extensively studied for encapsulation of drugs. Both water-soluble (hydrophilic) and fat-soluble (hydrophobic) drugs can be entrapped into liposomes. Hydrophilic drugs will be trapped inside the core, while hydrophobic drugs are incorporated within the phospholipid bilayer. The lipid bilayer of the liposome can fuse with other bilayers (e.g., cell membrane), thus delivering the liposome contents into the cells. Liposomes deliver drug by diffusion method or by direct cell fusion. Another strategy for liposome drug delivery is to target endocytosis events. Liposomes of particular size range can be prepared that are able to target natural macrophage phagocytosis. Liposomes can also be decorated with opsonins and ligands to activate endocytosis in different cell types. Examples of liposome-mediated drug delivery are Doxorubicin (Doxil) and Daunorubicin (Daunoxome), which are currently being marketed as liposome delivery systems. These formulations have an extended circulation time by virtue of their small size and sterically stabilized surfaces (Gregoriadis 1995). A series of modifications have been done in liposomal structures to develop liposomes capable of prolonged circulation and enhanced tumor accumulation that

are able to escape from mononuclear phagocytes. As these drug-loaded liposomal systems provide improved pharmacokinetics, reduced toxicities to a number of organ sites, and potentially increased tumor uptake, they are widely used for drug delivery (Allen 1997, Gabizon et al. 1998).

27.2.2 Nanoparticles

Particles of nanometer (10^{-9} m) size are called nanoparticles. Nanoparticles are colloidal particles that range in size from 10 to 1000 nm in diameter (Sahoo et al. 2002a). Nanoparticles can be formulated from both organic and inorganic materials. Organic materials include natural polymers (like chitosan, sodium alginate, agarose, and synthetic-PLGA, PLA, PVA, PBCA, and PCL), proteins (gelatin and albumin), and lipids (triglycerides, fatty acids, and sterols) while inorganic materials include calcium phosphate, silica, and iron oxide (Sahoo and Labhasetwar 2006). Polymeric nanoparticles constitute a versatile drug delivery system that can potentially overcome physiological barriers and target the drugs to specific cells or intracellular compartments either by passive or ligand-mediated targeting approach (Parveen and Sahoo 2008). In these particles, the therapeutic agents are entrapped, adsorbed, or chemically coupled on to the polymeric matrix. They allow a slow and sustained release of drugs from the polymeric matrix for a longer period of time.

Nanoparticles can be prepared using different techniques like solvent emulsification-evaporation method, solvent diffusion method, interfacial polymerization, nanoprecipitation, coacervation, spray drying, spray congealing, supercritical fluid, hot and cold homogenization, and microemulsion method (Mohanraj and Chen 2006). It is necessary to characterize the nanoparticles to establish the understanding and control of nanoparticle synthesis and applications. These can be characterized using various techniques like electron microscopy such as transmission electron microscopy (TEM), scanning electron microscopy (SEM), atomic force microscopy (AFM), dynamic light scattering (DLS), x-ray photoelectron spectroscopy (XPS), x-ray diffractometry (XRD), Fourier transform infrared spectroscopy (FTIR), matrix-assisted laser-desorption time-of-flight mass spectrometry (MALDI-TOF), and ultraviolet-visible spectroscopy.

In nanomedicine, nanoparticles are used for drug delivery as they have many advantages: they increase drug solubilization, protect drug from degradation, and decrease the toxic side effects of drugs. They produce a prolonged release, improve the bioavailability, and modify the pharmacokinetics of the drug. Thus nanoparticles provide a targeted delivery of drug at different cellular or tissue level maintaining appropriate tissue distribution of the drug (Parveen and Sahoo 2006). The main advantage of using nanoparticles for drug delivery applications is their small size by which they can penetrate through smaller capillaries and be taken up by cells and accumulate at the target sites (Desai et al. 1996, Desai et al. 1997, Panyam et al. 2003, Thomas and Klibanov 2003) Recent studies demonstrated the rapid escape of nanoparticles from the endo-lysosomal compartment to the cytoplasmic compartment (Panyam et al. 2002). Other advantages include the use of biodegradable materials for the preparation of nanoparticles, which allow sustained release of therapeutic agents from the polymeric matrix. Also, the surface of nanoparticles can be modified to alter biodistribution of drugs or can be conjugated to a ligand to achieve target-specific drug delivery (Moghimi et al. 2001, Moghimi and Hunter 2001). Various factors such as the polymer material, emulsifiers used for stabilization, polymer composition, method of preparation, and adsorption of some polymers influence the properties of nanoparticles (Sahoo et al. 2002a). Researchers have shown that PLGA nanoparticles were efficiently taken up by dendritic cells *in vitro* (Desai et al. 1997). This helps in selective activation of T cell-mediated immune response. The tetanus toxin C fragments (TTC) conjugated PLGA-PEG-biotin nanoparticles have the potential to serve as drug delivery system that selectively targets neuronal cells *in vitro*. This system may have applications for delivering therapeutics to neurons affected by neurodegenerative diseases and may allow retrospective transport delivery to the central nervous system (Townsend et al. 2007).

Drugs can be targeted either by passive or active mechanisms. Active targeting of drugs is achieved by conjugating the nanoparticles by specific ligand or antibody to target specific cells or tissues (Lamprecht et al. 2001, Thomas and Klibanov 2003). In passive targeting, the drug in nanoparticles reach the target organ through enhanced permeability retention (EPR) effect (Maeda 2001, Sahoo et al. 2002b). As nanoparticle is a colloidal carrier system, these are rapidly opsonized and cleared by the macrophages of RES. The nanoparticles can be protected from RES clearance by simply modifying the size, surface charge, and surface hydrophobicity. Particles less than 100 nm, with hydrophilic surface formed by various hydrophilic surfactants like polyethylene oxide and polyethylene glycol, undergo relatively less opsonization and clearance by RES. Alternatively, infusion of nanoparticle suspension to the accessible target organ or tissue is done using infusion catheters. For lymphatic tumor metastasis, the nanoparticles are subcutaneously injected to target the drug to lymphatic tissues. Nanoparticles are targeted to specific sites by surface modifications, which provide specific biochemical interactions with the receptors expressed on target cells. Sahoo et al. have demonstrated the increased efficacy of paclitaxel-loaded nanoparticles on conjugation with transferrin in a murine model of prostate cancer (Sahoo et al. 2004). They have shown that the transferrin conjugation to nanoparticles enhances therapeutic efficacy of the encapsulated paclitaxel as compared to that with unconjugated nanoparticles. Another important function of nanoparticles is their ability to deliver drugs to the target site crossing several biological barriers like blood–brain barrier (BBB) (Fisher and Ho 2002, Lockman et al. 2002). By coating the nanoparticles with polysorbates, the drug-loaded nanoparticles can be transported across the BBB, allowing brain targeting following an intravenous injection (Kreuter 2001). Poly (butylcyano acrylate) nanoparticles are coated with polysorbates-80 for the delivery of hexapeptide dalargin and

other agents into the brain (Sahoo and Labhasetwar 2003). Thus nanoparticles are emerging as a promising tool for intracellular delivery of most of the anticancer drugs that are practically insoluble (like taxol) and sensitive drugs (such as protein and oligonucleotides) (Brigger et al. 2002).

27.2.3 Dendrimers

Dendrimers are a new class of polymeric materials. These are highly branched, monodisperse macromolecules which were first discovered in the early 1980s by Donald Tomalia and coworkers. These dendrimers comprise a series of well-defined branches around an inner core (Boas and Heegaard 2004). They have attracted significant attention for drug delivery applications due to their nanometer size range, ease of preparation and functionalization, and their ability to display multiple copies of surface groups for biological reorganization processes (Padilla De Jesus et al. 2002, Quintana et al. 2002).

Dendrimers interact with outer molecular environment through their terminal groups. The interior of a dendrimer can be made either hydrophobic or hydrophilic by changing the dendrimer termini. The structure of dendrimer has a great impact on their physical and chemical properties. These can be roughly divided into two categories. One is low molecular weight and the other is high molecular weight species (Klajnert and Bryszewska 2001). Dendrimers can be synthesized either from the central core and work out toward the periphery (divergent synthesis) or in a top-down approach starting from the outermost residues (convergent synthesis) (Sahoo and Labhasetwar 2003). Dendrimers can be used as coating agents to protect or deliver drugs to specific sites in the body or as time-release vehicles for biologically active agents. Water-soluble dendrimers are capable of binding and solubilizing small acidic hydrophobic molecules with antifungal or antibacterial properties which are released upon contact with the target organism. Dendrimers have some unique properties like presence of internal cavities and their globular shape. Thus, the therapeutic agents are encapsulated within the macromolecule interior (Kihara et al. 2002, Padilla De Jesus et al. 2002). Also, drug molecules can be covalently attached to the outer surface groups of dendrimers. A dendrimer can be water soluble when its end-group is hydrophilic, like a carboxyl group. A water-soluble dendrimer with internal hydrophobicity can be designed, which would allow to encapsulate a hydrophobic drug. Dendrimers are of particular interest for cancer applications not only because of their defined and reproducible size, but more importantly, because it is easy to attach a variety of other molecules to the surface. Such molecules could include tumor-targeting agents (including but not restricted to monoclonal antibodies), imaging contrast agents to pinpoint tumors, delivery of drug molecules delivery to the tumor, and reporter molecules that might detect if an anticancer drug is working. For the retention and fast release of the encapsulated drug molecules, stabilizers like polyethylene oxide (PEO) chains are introduced on the periphery of dendrimers which has expanded the scope of dendrimer to be used in cancer therapy

(Liu et al. 2000). As compared to the linear polymers, the higher generation dendrimers occupy a smaller hydrodynamic volume because of their globular structure (Nierengarten et al. 2001); hence they are more useful in drug and gene delivery purposes.

Therapeutic agents can also be attached to a dendrimer to direct the delivery of genes. Dendrimers which act as carriers are called vectors. In gene therapy, these vectors can transfer genes through the cell membrane into the nucleus (Klajnert and Bryszewska 2001). For example, in early studies, polyamidoamine (PAMAM) dendrimers are used for gene delivery applications. Various formulations of dendrimers with different anticancer drugs have been prepared. For example, Duncan and coworkers (Kojima et al. 2000) have prepared conjugates of PAMAM dendrimers with Cisplatin. Cisplatin is a potent anticancer drug but because of nonspecific toxicity and poor water solubility it is not very effective in treatment. But the drug with dendrimer or in carrier system showed increased solubility, decreased systemic toxicity, and selective accumulation in solid tumors. Both hydrophobic drugs and PEO moieties can be attached to the dendrimer periphery by using a careful synthetic strategy with two different chain-end functionalities (Liu and Frechet 1999). Tumor cells can be targeted by using multivalent folic acid conjugate. Recently, in boron neutron-capture method of treatment of cancer, a folate-modified PAMAM dendrimer has been successfully used for the delivery of boron isotopes (^{10}B). In addition to this, PAMAM dendrimers are also used for antimicrobial therapy. For example, PAMAM dendrimer silver complexes show slow release of silver which has antimicrobial activity against various Gram-positive bacteria (Balogh et al. 2001). To date more than 50 families of dendrimers have been developed, each with unique properties for drug/gene delivery.

27.2.4 Nanorobots

Nanorobots are theoretical microscopic devices of nanometer size. These work at atomic, molecular, and cellular level, performing tasks in both the medical and industrial fields. A robot that allows precision interactions with nanoscale objects or can manipulate with nanoscale resolution can be defined as a nanorobot. The technology in which machines or robots of microscopic scale of nanometer size are created is called nanorobotics. Artificial nonbiological nanorobots are yet to be created, hence it remains as a hypothetical concept. They can also be used to identify cancer cells and destroy them. These can be used as an alternative to chemotherapy for cancer treatment (Casal et al. 2003). The radiation treatment kills not only cancer cells but also the healthy human cells, causing hair loss, fatigue, nausea, depression, and other symptoms. But nanorobots would seek out for cancer cells and destroy them, dispelling the disease at the source, leaving healthy cells untouched. A person having a nanorobotic treatment could expect to have no awareness of the molecular device working inside him other than rapid betterment of his health (Cavalcanti and Freitas Jr. 2002).

Scientists report that the exterior of nanorobots can be constructed of carbon atoms in a diamondiod structure (Freitas Jr.

1999). Because of its inert properties, strength, and smooth surface, it will lessen the triggering of the body's immune system. Nanorobots can be manufactured in nanofactories in size of an average desktop printer. Raw materials for making the nanorobots would be nearly cost-free and the process is pollution-free. Nanorobots can be an extremely affordable and highly attractive technology in near future. Nanorobots will respond to acoustic signals. It will receive power or even reprogramming instruction from an external source via sound waves. These nanorobots can be strategically positioned throughout the body. The doctor could not only monitor a patient's progress but also change the instructions of the nanorobots inside the body to progress to another stage of healing. After the completion of the tasks, the nanorobots would be flushed from the body. Nanorobots are applied to reverse the aging process (wrinkles, loss of bone mass, and other age-related conditions are treatable at the cellular level). Nanorobots have a lot of applications such as early diagnosis and targeted drug delivery for cancer, biomedical instrumentation, surgery, pharmacokinetics, and diabetes. Nanorobots may be utilized for attachment to transmigrating inflammatory cells or white blood cells to reach inflamed tissues and assist in their healing process. For the treatment of cancer, nanorobots will be applied in chemotherapy through precise chemical dosage administration. These may be used to deliver anti-HIV drugs. These nanorobots for drug delivery are called "pharmacytes" by Freitas (Freitas Jr. 1999). In the human body, these could be used as ancillary devices for injured organs to process specific chemical reactions. The researchers have analyzed how the nanorobots use different strategies in drug delivery. For example, the nanorobots could employ different sensory capabilities like chemical and temperature sensors as well as random movement. The nanorobots operate in a virtual environment comparing random, thermal, and chemical control techniques. The nanorobot architecture model has nanobioelectronics as the basis for manufacturing integrated system devices with embedded nanobiosensors and actuators, which facilitates its application for medical target identification and drug delivery.

Nanorobots are also good candidates for industrial applications. They are used to remove carbon dioxide from air, repair the ozone hole, scrub the water of pollutants, and restore our ecosystem. They can also be used in the detection of toxic chemicals and the measurement of their concentration in the environment.

27.2.5 Nanochips

Miniaturization of analytical and bioanalytical processes has become an important area of research and development during the past 10 years (Kricka et al. 1994, Van den Berg and Bergveld 1995:).The main advantages of the new devices are integration of multiple steps in complex analytical procedures, diversity of application, sub-microliter consumption of reagents and sample, and portability. Representing the endpoint of miniaturization is the nanochip (nanometer-sized features). The nanochip devices are built at the nanometer scale from individual atoms and molecules (Drexler et al. 1991, Fahy 1993). Micro miniaturization of analytical procedures is having significant impact on diagnostic testing, and enables highly complex clinical testing to be miniaturized and permits testing to move from the central laboratory into non-laboratory settings. Nanotechnology chips with biosensors can mark genes, guide drug discovery, monitor body functioning, and perceive biological and chemical pathogens.

Currently, nanochips are in their developmental stage and their use in practical purposes is still a futuristic goal. However, self-assembling organic molecules of 0.5 mm diameter or 30 mm long lipid tubules or 0.7–0.8 nm diameter cyclic peptide tubes provide the grounds for the development nanochips (Ghadiri et al. 1993). These cyclic peptides were designed with an alternating even number of d- and l- amino acids, which interact through hydrogen bonding into an array of self-assembled nanotubes. One of the first applications was based on their membrane interactions. As the cyclic peptide nanotubes are toxic to bacteria, they were demonstrated to serve as novel antibiotic agents (Fernandez-Lopez et al. 2001). Other potential applications including drug delivery (as these structures can serve as nanocontainers) are in progress, which provides grounds for some optimism for this avenue of development. These miniature devices such as nanorobots and nanochips can carry out integrated diagnosis and therapy by refined and minimally invasive procedures and even nanosurgery, as an alternative to crude surgery. Nanochips are also thought to markedly improve implants and tissue engineering approaches as well. Silicon-based nanochips with itching and self-assembling techniques can be used to deliver drugs on the diseased molecules. Less than one-tenth the size of a dime, nanochips can also detect up to 100 DNA-based disease markers for various diseases simultaneously, including some cancers. The nanochips can also be used to develop new tests designed to generate faster results that would enable doctors to avoid prescribing strong broad-spectrum antibiotics.

A pharma company, Velbionanotech, at Bangalore, India, is designing drugs for various diseases such as heart disease, kidney stones, AIDS, and cancer using short fragments of DNA as a new type of drug. These drugs are assembled in nanochips for delivery in the human body, which are effective in treating the sick/diseased and healing the injured. Researchers are emphasizing on the view that drug development costs could be reduced by using nanochips to test various medications or combinations of chemicals. The tests would use thin and sharp nanoprobes so that they could enter the cells and leave a few molecules of a particular drug behind before exiting, leaving the cell intact and alive. Applications of nanochips in the medical field also include the identification of certain diseases, like cystic fibrosis and scanning of DNA for signs of predispositions of other ailments (Cui et al. 2001). These nanochips are used to deliver anticlotting drugs during cardiac surgery by some researchers (Cui et al. 2001). Nanochips will also be used at some point in the next several years to develop a genetic profile of patients to help doctors determine which drugs will have the least side effects. Nanochips can also function as biosensors and help in detection of proteins in the body that signal presence of a disease.

Certain kinds of DNA that recognize cancer will be attached to these metallic chips and injected into the body. It is like a probe that goes out and seeks cancer cells and attaches to them. Nanochips are also used to detect gene mutations responsible for Mendelian disorders which determine the etiology of complex diseases, including heart disease, diabetes, and neuropsychiatric traits. The Nanogen's NanoChip platforms employ hybridization-based technology, using fluorescent detection and electronic control of the target or probe, to obtain clear genotype signal relative to background, and increased flexibility relative to similar chip-based single nucleotide polymorphism genotyping platforms.

Thus, the synergy between nanochip as a tool and drug delivery process can be achieved very soon. This would help in effective and targeted drug delivery, which is something we can look forward to in the future.

27.3 Nanosize Carriers for Gene Therapy

Gene therapy is the insertion of the genes into an individual cell to treat or prevent hereditary diseases, in which a defective mutant allele is replaced with a functional one. The insertion of gene into the individual cell is mediated by viral vectors such as retro virus, adenovirus, adeno-associated viruses, and popular viral vectors (Ferber 2001). The problem associated with the viral vector is the toxicity, immune and inflammatory responses, gene control, and targeting issue; additionally, there is always a fear of recovering of the virus to cause disease. To overcome this, much interest has been shown for non-viral-mediated gene transfer techniques (Figure 27.2). The advantage of using nonviral vectors is due to their nontoxicity, and less immune reaction, repeated administration at a very low cost. The most widely used nonviral vectors are liposome-mediated cationic polymers and nanoparticles (Johnson-Saliba and Jans 2001, Wagner 2005). The physical properties of nanoparticles, including their morphology, size, charge density, and colloidal stability, are important parameters to determine the overall efficacy of nanoparticles to act as potential nonviral gene delivery vehicles.

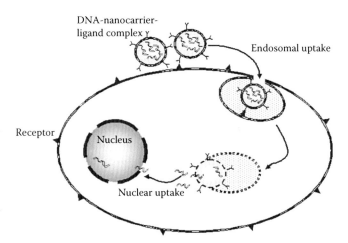

FIGURE 27.2 Schematic diagram of nanocarrier in gene delivery.

27.3.1 Inorganic Nanoparticles for Gene Therapy

Inorganic nanoparticles generally possess versatile properties suitable for cellular delivery, including wide availability, rich functionality, good biocompatibility, and potential capability of targeted delivery. Many inorganic materials, such as calcium phosphate, gold, carbon materials, silicon oxide, iron oxide, and layered double hydroxide (LDH), have been studied. Inorganic nanoparticles show low toxicity and promise for controlled delivery properties, thus presenting a new alternative to viral carriers and cationic carriers. The most widely used inorganic nanoparticle is silica.

27.3.1.1 Mesoporous Silica Nanoparticles

Mesoporous silica nanoparticles (MSNs) were the first synthesized inorganic materials in 1992 by researchers working for the Mobil Corporation. The MSNs have enhanced biocompatibility with high surface areas and pore volumes (Slowing et al. 2007). The functionalization of the MSNs with organic moieties or other nanostructures can bring controlled release and molecular recognition capabilities to these mesoporous materials for drug/gene delivery and sensing applications. MSNs as a novel therapeutic drug delivery agent provide neuroprotection to acrolein-mediated cell death (Cho et al. 2008).

Highly monodisperse organically modified silica (ORMOSIL) and fluorescently labeled silica nanoparticles are being used for drug and gene delivery. ORMOSIL encapsulated by fluorescent dye and surface functionalized by cationic amino group is engineered by the micellar system (Roy et al. 2005). The nanoparticle acting as drug delivery vehicle along with multiple probes provides a promising direction for the targeted therapy and real-time monitoring of the drug action (Roy et al. 2005).

27.3.1.2 Bioceramic Nanoparticles

A mixture of ceramics and mineral oxides such as silica oxide and aluminum oxide efficiently complexed with plasmid DNA forms the bioceramic nanoparticles. This mixture emits far-infrared rays. The ability of bioceramics nanoparticles to act as potential nonviral gene delivery vehicles was well studied. Bioceramic nanoparticles can be synthesized using negatively charged silica (SiO_2), neutrally charged hydroxyapatite (HA), and positively charged zirconia (ZrO_2) nanoparticles as nonviral vectors for efficient *in vivo* gene delivery and targeting to spleen. Various positively charged amino acids (lysine (Lys), histidine (His), arginine (Arg)) and the cationic polymeric amino acids poly-L-lysine (PLL), polyarginine, polyhistidine, and protamine sulfate (PS) are coated for the overall charge modulations. Tan et al. (2007) have compared three ceramic particles that is ($SiO2$, HA, ZrO_2) and found the $PS-SiO_2$ nanoparticles (bioceramic nanoparticle) to be an efficient gene delivery vehicle with targeting specificity to spleen as the spleen has been documented to be one of the most important lymphoid organs involved in the initiation of immune responses (Tan et al. 2007).

27.3.2 Protein Nanoparticles for DNA Delivery

Proteins are large organic molecules that perform highly specific and complex functions in the body, making them an ideal instrument to diagnose and cure disease. Protein particles can be designed in such a way that one can control their size and shape, and preserve the protein's biofunctionality, stability, solubility, and concentration, while minimizing interparticle forces that cause aggregation. There are various proteins that are used for drug and gene delivery process; here we cite a few examples where protein serves as an ideal gene delivery vehicle.

27.3.2.1 Albumin

Albumin is a biodegradable and biocompatible protein. This protein is widely used for microsphere and nanosphere applications because of nonantigenicity and it contributes to the major portion of serum. Human serum albumin (HAS)/bovine serum albumin (BSA) possess many amino and carboxylic groups available for the covalent modification of the drug and protein. Albumin nanoparticles are used for the DNA delivery because DNA–albumin can avoid the opsonization and uptake by the mononuclear phagocytic system (MPS) encountered by the positively charged complexes *in vivo* (Rhaese et al. 2003, Wartlick et al. 2004). Nanoparticles consisting of DNA, human serum albumin (HSA), and polyethyleneimine (PEI), that is, HSA–PEI–DNA nanoparticles were prepared by co-encapsulation of PEI as a lysosomotropic agent at varying nitrogen-to-phosphate (N/P) ratios by the coacervation technique and tested for transfection efficiency in vitro with the aim of generating a nonviral gene delivery vehicle and intravenous administration.

27.3.2.2 Gelatin

Gelatin is a natural protein produced by partial hydrolysis of collagen extracted from various animal sources. Gelatin is used in making nanoparticles because it has a long record of safe use in the human body. Gelatin-based nanoparticles have been prepared to deliver therapeutic genes to human breast cancer tumors implanted in mice (Kommareddy and Amiji 2007). For the enhancement of intracellular delivery potential of gelatin, thiolated gelatin was developed. Thiolated gelatin was formed by the covalent modification of the epsilon-amino groups of gelatin with 2-iminothiolane. The soluble form of extracellular domain (outer region) of vascular endothelial growth factor receptor-1 (VEGF-R1 or sFlt-1) is inserted into plasmid DNA and was encapsulated with poly (ethylene glycol) (PEG)-modified gelatin-based nanoparticles. These nanoparticles were injected into the bloodstream, and 15% of the dose found its way into the tumor, where it produced a protein, that cut off the blood supply to the tumor and makes the tumor a factory for its own destruction. This kind of gene delivery system had been attempted, but with viral vectors, which, although effective, are in many cases toxic to the recipient. PEGylated nonviral vectors are mostly suitable for gene delivery application due to its nontoxicity, no upper limit on the plasmid DNA size, and ease to manufacture. The application of this system can be a versatile method for drug delivery not only for cancer but also for heart diseases, macular degeneration, glaucoma, and diabetes.

27.3.3 Lipid-Based Nanoparticle for Gene Delivery

Lipids are important components of cellular membranes. Lipids are broadly defined as fat-soluble (lipophilic), naturally occurring molecules, such as fats, oils, waxes, cholesterol, sterols, and fat-soluble vitamins (such as vitamins A, D, E, and K), monoglycerides, diglycerides, phospholipids, and others. Novel cationic solid lipid nanoparticles are used as nonviral vectors for gene delivery. Examples of lipid-based nanoparticles serving as gene delivery are seen in human non-small cell lung cancer tumors in mice where these lipid nanoparticles drastically reduce the number and size of the tumors. In cancer treatment, chemotherapy and other different modes of therapy including surgery and radiation are implemented. Now there is a possibility of combined targeted gene therapy. FUSI is a novel tumor-suppressor gene identified in human chromosome. Loss of expression and deficiency of posttranslational modification of FUSI protein have been found in a majority of human lung cancers. For the dual gene therapy, both p53 (well-known tumor suppressor) and FUSI protein were wrapped in the nanoparticles. The FUSI works with p53 to force the lung cancer cells to kill themselves—a process known as apoptosis. These genes do not have much effect on normal tissue or normal cells when they are overexpressed. The FUSI/p53 combination also activates a cell suicide pathway based in the mitochondria cells. The cancer cell lines treated with the gene combination had two to three times more cells killed by apoptosis than either gene nanoparticle had done individually. The nanoparticle delivery system consists of a plasmid gene expression cassette loaded with DNA that encodes either the p53 or the FUSI protein. This is wrapped tightly in a form of cholesterol to protect it from the body's defense mechanisms as DNA cannot be delivered naked in cancer therapy. The positively charged nanoparticles are delivered to negatively charged cancer cell membrane accumulated mainly in the tunoregenic lung cells, where the genes repeatedly express either p53 or FUSI tumor-suppressing proteins (Roth et al. 2004).

27.3.4 Polymeric Nanoparticle as Gene Delivery Agents

A number of different polymers both synthetic and natural have been used in formulating biodegradable nanoparticles. Such polymeric nanoparticles have the advantage of sustaining the release of the encapsulated therapeutic agents over a period of time. There are several studies regarding PLGA nanoparticles or modification of these polymers which are used to deliver plasmid DNA for different therapeutical purposes, delivering therapeutic genes to the diseased cells. Different polymeric nanoparticles can also be used as gene delivery agents. However, recently a polymer termed C32 derived from poly (β-amino ester) was developed; this polymer is capable of delivering genes to cancer cells

more efficiently and with less toxicity than other screened polymers that have been tested. When this was injected into prostate tumors in animals, tumor growth was suppressed or reversed (in 40% of cases), relative to untreated tumors. Moreover, it was also very efficient at delivering its genetic payload to healthy muscle cells. This feature may help safeguard the healthy tissue surrounding tumors, offering a significant improvement over currently available therapies that tend to damage the healthy tissue near the cancer (Zucates et al. 2005).

27.3.4.1 Chitosan

Chitosan, a biodegradable polymer is a good component for oral drug and gene delivery. The properties of chitosan like nontoxicity, ability to prolong residence time in the gastrointestinal tract through mucoadhesion, and enhanced absorption by increasing cellular permeability are the major factors contributing to become a component of oral dosage forms and gene delivery systems. For an effective oral drug administration, chitosan is desirable to combat the challenges owing to the nature of the gastrointestinal tract. The highly acidic pH in the stomach and the presence of enzymes such as pepsin can cause protein degradation (Allemann et al. 1998). Secreted pancreatic enzymes in the membrane and the lumen of the intestine may also cause substantial loss of drug activity (Bernkop-Schnürch and Krajicek 1998). These obstacles lead to poor oral bioavailability for many drugs, proteins, peptides, nucleic acids, vaccines, and gene delivery. Additional barriers for effective DNA delivery include endosomal escape, nuclear localization, transcription, translation, protein processing, and protein secretion into plasma. To overcome these physical and degradative barriers, formulations of the drug or gene into the nanoparticles was developed (Bowman and Leong 2006). Such particles may partially protect the entrapped drug or gene from degradation and improve the cellular uptake through endocytosis. The drug cyclosporine A has also been efficiently encapsulated in chitosan nanoparticles (El-Shabouri 2002). The *in vitro* studies with reporter genes showed that chitosan could be readily formulated into DNA nanoparticles which are able to transfect some cell lines like HEK293 cells and HeLa cells better than other nanoparticles (Kiang et al. 2004).

27.3.4.2 PLGA

Biodegradable polymers like PLGA and PLA are under investigation to deliver genes in cancer therapy. In a study conducted by Prabha and Labhasetwar (2004), antiproliferative activity of wild-type p53 gene (a tumor-suppressor gene) loaded nanoparticles in breast cancer cell line was depicted. Cells transfected with wild-type p53 DNA-loaded nanoparticles demonstrated sustained and significantly greater antiproliferative effect than those treated with naked p53 DNA. He et al. (2004) have formulated thymidin kinase gene-loaded nanoparticles, which showed better expression of DNA in hepatocarcinoma cells. As gene delivery vehicles, PLGA was successfully used to transfect VEGF gene into myocardial cells. Successful transfer of vascular endothelial growth factor (VEGF (165)) into cardiac myocytes was done with plasmid DNA-loaded PLGA nanoparticles for treating ischemic myocardium, a heart disease. The direct gene transfer increased the number of capillaries and improved the cardiac function.

Researchers have functionalized the PLGA nanoparticle surface with a transferrin or linear RGD peptide or both which increased the retinal delivery of nanoparticles and, subsequently, the intraceptor gene expression in retinal vascular endothelial cells, photoreceptor outer segments, and retinal pigment epithelial cells when compared to nonfunctionalized nanoparticles. Choroidal neovascularization (CNV) is the creation of new blood vessels in the choroid layer of the eye. This is a common and leading cause of blindness in adult population (over 50 years old). Engineered transferrin, arginine-glycine-aspartic acid (RGD) peptide or dual-functionalized poly-(lactide-*co*-glycolide) nanoparticles target the choroidal neovascularization (CNV) lesions on intravenous administration. The results of the study show that nanoparticles allow targeted delivery to the neovascular eye but not the control eye on intravenous administration (Singh et al. 2009). Thus, surface-functionalized nanoparticles allow targeted gene delivery to the neovascular eye on intravenous administration thus, inhibiting the progression of CNV.

27.3.5 Magnetic Nanoparticle–Mediated Gene Delivery

Magnetic nanoparticles (MNPs) include iron oxide particles that can be attracted and guided by an external magnetic field for drug and gene delivery. They have promised potential as novel tools for diagnostics and drug delivery. Since magnetic field is generally harmless to living creatures, for transient exposures, MNPs can be concentrated and targeted to specific organs (Alexiou et al. 2000). In addition, magnetic resonance imaging and some magnetic detection systems can detect the strength and distribution of MNPs in living organisms without invasive procedures. Therefore, combining MNPs with existing systems can generate new tools for drug delivery and diagnostics. Functionalized MNPs are usually formulated by the encapsulation of MNP (magnetite) into the biodegradable polymer matrix and used for drug delivery, diagnosis, magnetic resonance imaging (MRI), magnetic cell separation, tissue repair, hyperthermia, and magnetofaction (Hafelli et al. 1997, Johannsen et al. 2006, Radisic et al. 2006, Tiefenauer 2007). The superparamagnetic iron oxide contrast agent has been approved by FDA for MRI. Commercial iron oxide nanoparticles such as Endoderm and Resovit are used as contrast agents in MRI for the diagnosis and exact tissue location determination of the brain (Cengelli et al. 2006, Wang et al. 2003), cardiac infracts, liver lesion, and tumors (Bulte and Kraitchman 2004, Dudeck et al. 2006). The MNPs tend to accumulate at higher levels due to the differences in tissue composition and/or endocytotic uptake process.

The magnetic iron oxide nanoparticles coated with PEI, a polymer may be a potential gene carrier with high transfection efficacy for cancer gene therapy. The nanoparticles were tested as gene vectors with *in vitro* transfection models. Tumor necrosis factor–related apoptosis-inducing ligand (TRAIL) induces apoptosis or cell death in cancer cells. The MNPs were used as gene carrier to transfect TRAIL gene into MCF-7 cells, a breast cancer cell line (Weizhong et al. 2004).

27.3.6 Nanomagnet-Mediated Gene Therapy

In gene therapy, nanomagnets are inserted into cells carrying genes to fight tumors, resulting in many more cells successfully reaching and invading malignant tumors. This technique involves inserting nanomagnets into monocytes, a type of white blood cells used to carry gene therapy and injecting the cells into the bloodstream. A small magnet is placed over the tumor to create a magnetic field, which attracts more monocytes into the tumor. The use of nanoparticles to enhance the uptake of therapeutically armed cells by tumors could create a new era in gene therapy, one in which delivery of the gene therapy vector to the diseased site is much more effective. This new technique could also be used to help deliver therapeutic genes in other diseases like arthritic joints or ischemic heart tissue (Muthana et al. 2008).

27.4 Nanotechnology-Based Devices for Drug Delivery and Sensing

Advances in nanotechnology and nanomedicine have heralded the advent of several innovative nanodevices which are set to revolutionize the field of drug delivery. In recent years, many researchers in nanoscience have used unique, programmable, and molecular recognition properties of DNA to build nanostructures by self-assembly and to realize artificial, machine-like devices (Simmel and Dittmer 2005). These DNA-based nanodevices assume an intermediate position between the biological and synthetic worlds and are capable of drug delivery. They also act as novel biosensors with built-in information-processing capability. These nanodevices can significantly detect proteins, viruses, DNA, and disease biomarkers in any bodily fluid, including tears, urine, and saliva. This could pave the way for faster, more accurate medical diagnostic tests and may ultimately save lives by allowing early disease detection.

27.4.1 Nanotechnology-Based Immunization Devices

A few nanotech-enabled medical products and pharmaceuticals are either in the market or in the pipeline, and right behind them is a vast new research and development infrastructure. Today, nanotechnology brings us closer in curing mankind's deadliest diseases. But before that could happen, science needed to go smaller—smaller than the murderous viruses, small enough to attack tumors, and destroy them from within. The different nanotechnology-based immunodevices include nanoarrays, nanochips, and cantilevers which are helpful in curing diseases starting from polio to the deadly cancer.

The small dimensions of this technology have led to the use of nanoarrays and nanochips as test platforms (Jain 2003). One advantage of this technology is the potential to analyze a sample for an array of infectious agents on a single chip. Applications include the identification of specific strains or serotypes of disease agents, such as specific influenza strains, or the differentiation of diseases caused by different viruses but with similar clinical signs, such as vesicular viral diseases. Scientists from the University of Rochester have created a nanoscale device capable of detecting one quadrillionth of a gram of biological matter, or about the size of certain viruses. In the future, the sensor may be able to detect influenza, severe acute respiratory syndrome (SARS), bird flu, and other viruses. The sensor is a hexagonal array of tiny cavities, each 240 nm in diameter, carved into a very thin slab of silicon using a beam of electrons.

The microfabrication and application of arrays of silicon cantilever beams as microresonator sensors with nanoscale thicknesses were used to detect the mass of individual virus particles. The dimensions of the fabricated cantilever beams were in the range of 4–5 μm in length, 1–2 μm in width, and 20–30 nm in thickness. The virus particles used in the study were vaccinia virus, which is a member of the *Poxviridae* family and forms the basis of the smallpox vaccine. The frequency spectra of the cantilever beams, due to thermal and ambient noise, were measured using a laser Doppler vibrometer under ambient conditions. The change in resonant frequency as a function of the virus particle mass binding on the cantilever beam surface forms the basis of the detection scheme. These devices can be very useful as components of biosensors for the detection of airborne virus particles (Gupta et al. 2004).

Tuberculosis is one of the most resistant and challenging diseases to protect against and the successful results of aerosol delivery using nanoparticle technology offers a potentially new platform for immunization. The novel aerosol version of the common tuberculosis (TB) vaccine, administered directly to the lungs as an oral mist, offers significantly better protection against the disease in experimental animals than a comparable dose of the traditional injected vaccine. In the aerosol vaccine, particles formed at micrometer and nanometer scales and in spherical and elongated shapes appeared to improve dispersal in the mouth. Spray drying of the vaccine is much lower in cost than BCG (the vaccine for TB), easily scalable for manufacturing, and ideal for needle-free use such as via inhalation. New instruments are also being devised to detect existing nanoparticles in the body that indicate certain diseases or conditions. One of them, known as A β-derived diffusible ligand (ADDL), is associated with Alzheimer's disease. Researchers have made a discovery of detecting the disease at nanomolar level by using a nanodevice.

Experts believe that quantum dots (QDs), nanopores, and other devices will become the order of the day and may be available for clinical use in 5–15 years to eradicate different diseases starting from polio to cancer and thus revolutionize the medical field.

27.4.2 DNA-Based Nanodevices

Nanotechnology is of great use for medical diagnosis and various nanoparticles have exhibited tremendous potential for detecting disease markers, precancerous cells, fragment of viruses, and other indicators. Bionanotechnology merges biomaterials with technical materials, often in an unconventional marriage, that is, the participating materials seldom having been observed together. The biomaterials include DNA and RNA which serve as controllable and programmable scaffolds for organizing functional nanomaterial in the design, fabrication, and characterization of nanometer-scale electronic devices and sensors with increased selectivity and sensitivity as compared with many conventional assays that rely on molecular probes (Young and Kiang 2005). The molecular recognition properties of DNA molecules combined with the distinct mechanical properties of single and double strands of DNA can be utilized for the construction of these nanodevices (Simmel and Dittmer 2005). With the help of DNA, it is possible to devise autonomous nanodevices, to grab and release molecules, and also to perform simple information-processing tasks (Simmel and Dittmer 2005). Among the newly developed nanodevices, QDs, nanowires, nanotubes, nanocantilevers, nanopores, and nanoshells are most promising and can be applied for various cancer treatments. It is envisioned that these nanodevices which are hybrids of biologic molecules and synthetic polymers can enter cells and the organelles to interact directly with DNA and proteins and hence help in treatment.

27.4.2.1 DNA-Directed Nanowires

DNA nanowires are one of the intriguing applications of DNA-based nanotechnology. DNA has a favorable dimension, that is, 1–2 nm which makes it an ideal template for the construction of nanometer-scale electronic circuits. These DNA template nanowires have diverse applications especially in the field of electronics where these wires facilitate the reduction of feature size of the integrated circuit. The basic idea behind the making of DNA-directed nanowires was first demonstrated by Braun et al. in 1998 where DNA was utilized as a template to grow conducting silver nanowires using gold electrodes separated by a distance of 12–16 nm (Braun et al. 1998). Nanowires can also be constructed using palladium (Richter et al. 2000), platinum (Ford et al. 2002), gold (Harnack et al. 2002), and copper (Monson and Woolley 2003) using methods similar to Braun et al. (1998). Nanowires are fast becoming a reality with significant applications to the diagnosis, management, and treatment of all urooncological diseases like bladder, prostrate, and kidney cancers also. Other potential applications for nanowires include the early sensing of breast and ovarian malignancies.

The researchers have showed that silicon nanowires (Si-NWs) promise highly sensitive dynamic label-free electrical detection of the gene for cystic fibrosis more efficiently than conventional tests for the disease. Cystic fibrosis (CF) is the most common fatal genetic disease among people of European origin. Scientists in Idaho and Korea are reporting development of a protein coating that may turn these nanowires into a new drug delivery system allowing the use of lower doses of medicine and hence being less harmful to normal cells. Semiconductor materials (Si nanowires) were also used to deliver genetic material which can penetrate both human embryonic kidney cells and mouse embryonic stem cells (Kim 2007). These findings provide a novel way of introducing foreign material into the cell for treatment purposes.

Biological macromolecules such as proteins and nucleic acids are typically charged in aqueous solution and, as such, can be detected readily by nanowire sensors when appropriate receptors are linked to the nanowire active surface. Protein molecules like biotin, single-stranded DNA (ssDNA), and viral solutions like influenza A virus solutions can be easily detected due to increase in conductance of these nanowire sensors. The first example demonstrating the ability of nanowire field effect devices to detect species in liquid solutions was demonstrated in 2001 for the case of hydrogen ion concentration or pH sensing (Cui et al. 2001).

27.4.2.2 DNA Functionalized with Carbon Nanotubes

Emergence of carbon nanotubes has provided a basis for further generation of nanoscale devices. These nanostructures have been used for various purposes like molecular electronics (Tans et al. 1998), hydrogen storage media (Dillon et al. 1997), and scanning probe microscope tips (Wong et al. 1998). They have also been modified for therapeutic interest. Excellent progress has been made in harnessing the potential of carbon nanotubes for several drug delivery and other applications. Functionalized carbon nanotubes have been demonstrated to deliver proteins, nucleic acids, drugs, antibodies, and other therapeutics. Emerging developments in this area are pointing toward the successful utilization of carbon nanotubes for drug delivery. Carbon nanotubes are functionalized to better target specific cells in the body, like the gastrointestinal (GI) tract, solid tumors in various locations, and diseased cells (Pingang et al. 2006). Therefore, this device could be used for applications such as targeted drug/DNA/cell delivery for oral and systemic therapy for cancers of the colon, breast, and ovaries and conveniently modified to treat various other maladies as well. Nanotubes have also been proved as versatile sensors for detecting small organic and inorganic molecules as well as biological molecules of DNA and protein. It has thus dramatically promoted the development of DNA biosensing techniques, especially electrochemical DNA biosensors (Pingang et al. 2006).

As drug delivery agents, nanotubes offer a structural advantage in being extremely thin yet very long and offer larger surface area to graft and regulate the amount of drug required. Chemically functionalized single-walled carbon nanotubes (SWNT) have shown promise in tumor-targeted accumulation in mice and exhibit biocompatibility, excretion via biliary pathway, and little toxicity. Thus, nanotube-based drug delivery is promising for high treatment efficacy and minimum side effects for future cancer therapy with low drug doses.

Nanotubes loaded with gadolinium ions show promise as MRI contrast agents, while those with radioactive iodine or astatine (an element in the same family as iodine) are being developed as cancer-killing therapeutics by researchers at TDA Research, Inc., based in Wheat Ridge. Researchers at Stanford University have succeeded in functionalizing carbon nanotubes to carry anticancer drugs to tumors while avoiding damage to healthy tissues. The nanotubes were also targeted to cancer cells by attaching folic acid to the surface of the nanotubes. Folic acid binds to a folic acid receptor protein found in abundance on the surfaces of many types of cancer cells. Experiments have also been done to deliver drugs such as amphotericin B (an antifungal treatment) and methotrexate (an anticancer drug) using the functionalized carbon nanotubes. The modified nanotubes have also been used to ferry a small peptide into the nuclei of fibroblast cells.

Modifications of carbon nanotubes have lead to more diverse range of applications especially in the field of biological sensors. One such modification is the attachment of DNA molecule onto the nanostructures. These modified nanotubes then can be used for detecting biomolecules in solution. Fabricating nanotube electrode arrays in place of the conventional metal or carbon electrodes provides a thousand times more the number of electrodes in the same small space. As a result, the sensitivity of the sensor and the probability of detection increase drastically. By placing a thousand nanotube probes in the space of one of today's metal electrodes, we can detect DNA sequences from less than a thousand strands. This is sensitive enough to directly measure mRNAs in a drop of blood or a piece of tiny tissue sample and multiple targeting and detection is possible. Moreover it is not time consuming as sample preparation is quite easy and inexpensive as compared to the bulky analytical equipments available in the markets. These nanotubes also help in detecting low concentrations of mercury ions in whole blood, opaque solutions, and living mammalian cells and tissues where the optical sensing is usually poor or ineffective.

Thus due to their excellent electrical conductivity, carbon nanotubes are poised to become the key component of ultrafast, miniaturized diagnostic gear that may soon be able to detect the earliest signs of cancer from a pinprick of blood right in a doctor's office.

27.4.3 Nanopore and DNA Sequencing

The ability to comprehensively sequence any person's genome is the type of quantum leap needed to usher in an age of personalized medicine where a complete knowledge of the DNA sequence of an individual can yield fundamental insight into disease diagnosis, treatment, and possibly prevention by health care providers. To date, the most reliable method of determining a genetic sequence is through the use of the Sanger chain termination chemistry; however, a typical mammalian sequence costs millions of dollars and requires many months for completion because it relies on multiple costly PCR reactions. Besides this, the other limitation of Sanger chain termination chemistry includes slow speed and relatively shorter strands of DNA being sequenced in each gel. Thus alternative methods of sequencing are in high demand.

Nanopore sequencing is one of the most promising technologies being developed as a cheap and fast alternative to the conventional Sanger sequencing method. Nanopore-based methods have come under intense investigation recently due to the promise of inexpensive ultra-high-throughput sequencing of DNA with an entire mammalian sequence being determined at a cost of under $1000. Setting a milestone toward fast DNA sequencing, a solid-state nanopore device detects the difference between single molecules of double- and single-stranded DNA at high speed, with high accuracy, and at extreme pH (Liu and Balasubramanian 2003).

A nanopore is a small pore in an electrically insulating membrane that can be used as a molecular probe or detector. Nanopore sequencing, introduced in the mid-1990s, requires a sensor comparable in size to the DNA molecule itself. The sensor interacts with the individual nucleotides in a DNA chain and distinguishes between them on the basis of chemical, physical, or electrical properties. A voltage applied across the membrane generates an ionic current and pulls the negatively charged DNA molecules through the pore (*Science Daily*, March 21, 2001). As the nucleotides of the DNA migrate across the membrane, they partially block the pore in different ways depending on their size and ionic current through the pore due to passing counter ions is characteristically affected. The physicists used different mathematical calculations and computer modeling of the motions and electrical fluctuations of DNA molecules to determine how to distinguish each of the four different bases (A, G, C, and T) that constitute a strand of DNA.

There are two general methodologies to prepare nanometer-scale pores. The first attempt to produce nanopores were made with a-hemolysin (Akeson et al. 1999). Beyond a-hemolysin, the use of other membrane-bound proteins has been also investigated to seek for various-sized organic pores that might exhibit even better applicability (Rostovtseva et al. 2002, Song et al. 1998, Szabo et al. 1997, 1998). However, a major drawback of this system is its lack of mechanical robustness. Indeed, these biological membranes tend to rupture within a few hours, thus limiting their application in practical sensing devices. However, researchers have now come up with a major breakthrough finding that will aid the reproducible fabrication of robust, synthetic single-nanopore membranes (Wharton et al. 2007). This is mostly done using silica-formulating synthetic nanopores. Recent advances in silica nanotechnology (Ho et al. 2005, Storm et al. 2003) have been exploited to manufacture pores in thin synthetic membranes with sub-nanometer precision. Immersed in an electrolytic solution and under the influence of an electric field, nanopores can be deployed to filter and monitor translocation of DNA and other charged macromolecules.

27.4.4 Nanopore as Disease Biomarker

Biomarkers are specific proteins that can be objectively measured and evaluated as an indicator of normal biological or pathogenic processes as well as pharmacological responses to a therapeutic intervention. They are characteristically produced during a disease. Tests based on biomarkers have been around for more than half a century, but interest in their application for diagnostics and drug discovery as well as development has increased

remarkably since the beginning of the twenty-first century. Biomarkers are useful not only for diagnosis of some diseases but also for understanding the pathomechanism as well as the basis for development of therapeutics. Nanopores that mimic the pores in biological membranes could be used to detect the early stages of diseases like cancer. Present technologies involve detection of molecular changes in cells and tissues using different imaging techniques which is both time consuming as well as expensive. Nanopores promise an early detection since it detects molecular change in a cell even if present at nanomolar range without altering the samples for examination. These attractive features offer new and interesting possibilities for sensing and will most likely result in the development of a new generation of chemical and biological sensors with enhanced sensitivity and selectivity (Patolsky and Lieber 2005).

In the membranes of living cells are tiny holes, or nanopores, that connect the outside world to the inside world of the cell both electrically and chemically. These channels open and close in response to different stimuli like difference in membrane potential, presence of ligands, and mechanical stress (Hille 2001). The researchers aim is that these artificial pores serve as ultrasensitive biosensors to detect proteins that act as disease biomarkers. This new approach entails detecting in the patient's blood a very minute amount of a chemical substance, a biomarker, that indicates that the disease is present, but in its very early stages. These smart abiotic nanopores possess the advantages of biological nanopores resistive pulse sensors but do not suffer from their lacunas and fragilities (Choi et al. 2006).

The principle of operation of these sensors is based on electrochemical conductance variations. The sensors work by detecting changes in the current applied across the membrane when the biomarker is present. When the biomarker binds to the tip of the nanopore or passes through the nanopore, it blocks the current from passing through, leaving a trace of the molecule in the current measurement. In these devices, distinct pore clusters are selectively surface functionalized with specific antibodies that are in turn incorporated into microscale arrays (Choi et al. 2006). Protein markers are routinely detected at nanomolar concentrations resulting from the formation and then binding of antibody–antigen complexes to a metallic substrate.

A variety of approaches are used to prepare abiotic nanopores including focused ion beam etching of silicon nitride and oxide (Li et al. 2003, Storm et al. 2003), soft lithographic techniques (Saleh and Sohn 2003), embedded carbon nanotubes (Ito et al. 2004), femto pulsed laser–based technique on glass and track-etched conical nanopores in polymeric membranes (Harrell et al. 2006, Siwy et al. 2002).

27.5 Nanoscale Materials for Imaging

In medical science, imaging provides the anatomical and physiological information of the human body without the surgical procedure. Imaging gives the most essential information for the modern clinical care. In many aspects, the requirement for imaging is the same as those of drug targeting. Imaging should be quick so that the patient is in a comfortable state and should result in maximizing throughput, whereas the chemotherapy involves the sustained delivery of drugs over a long period of time. For the imaging application, it is essential to control the biodistribution and uptake of pharmaceutical agents. In some cases, it can be a simple vascular agent for intravenous injection, or nonabsorbable agents administered orally. For site-specific targeting, additional vector molecule conjugated to polymers (like antibody, hormone, receptor binding peptide polymer conjugate) are used in order to promote targeted uptake in lesion or tumor tissues. In vivo imaging of these nanoscale systems can be carried out using various types of imaging techniques, including single photon emission computed tomography (SPECT), positron emission tomography (PET), magnetic resonance imaging (MRI), fluorescence microscopy, computed tomography (CT), and ultrasound. Here, we focus on few selected imaging modalities with documented applicability in vivo.

27.5.1 Polymeric Imaging Agents

An important branch of pharmaceutical sciences is aimed at utilizing the polymers and their conjugates for the targeted drug delivery with medical imaging. Few polymer conjugates have proven to be the potential candidates for the imaging agent. The majority of the traditional x-ray contrast agents are dense (for attenuating), inert, nonspecific materials such as barium for oral and iodine for intravenous use. Amphilic polymers and polychelating agents are used for a variety of image contrast applications. Polymer-based iodine complexes for increasing circulation times have been in use since 1995. Methoxy-polyethylene glycol (MPEG) with iodine has been used as a long circulating blood pool contrasting agent for CT imaging (Torchilin 2002).

27.5.2 Liposomes for Imaging

As normal vasculature is not leaky and vasoactive intestinal peptide (VIP) receptors exist only in the extravascular space, the imaging agent could not be distributed to the normal tissues to give high background signal (Dagar et al. 2003). Therefore liposomes have been developed as carriers for a variety of contrast agents and radiopharmaceuticals. Actively targeted 99 m Tcliposomes have been developed for SPECT imaging. Various studies have shown overexpression of a 28 amino acid mammalian neuropeptide, VIP receptors, which exist homogeneously in surgically resected human breast cancer and biopsies (Bertini et al. 2004). Imaging studies showed that sterically stabilized liposomes (SSLs) encapsulating 99 mTc-hexamethylpropyleneamine oxime (HMPAO) with covalently attached VIP ligand accumulated significantly more in breast cancer than normal breast tissues. These data favorably support VIP as an ideal targeting agent to increase *in vivo* specificity and accumulation of liposomes in tumor sites over normal tissues (Koo et al. 2005). Paramagnetic liposomes loaded with gadolinium (Gd) exhibited a threefold increase in relaxivity compared to conventional paramagnetic complexes gadoterate meglumine (Gd-DOTA) and gadolinium-diethylene

triamine pentaacetic acid (Gd-DTPA) (Bertini et al. 2004). The use of liposomes in ultrasound and scintigraphic imaging provides more examples of imaging (Dagar et al. 2003). Palmityl-D-glucuronide (PGlcUA) was effective in increasing the circulation half-life of the liposomes and expression of reporter gene by using the charge-coupled device (CCD) camera and micro-PET, after a single intravenous injection of 1,2-dioleoyl-3-trimethylammonium-propane (DOTAP):cholesterol DNA liposome complexes (Iyer et al. 2002). Therefore this noninvasive, clinically applicable imaging method will greatly aid in the optimization of future gene therapy.

27.5.3 Dendrimer Nanocomposites for Imaging

Smaller dendrimer-based Gd MRI contrast agents (molecular weight 60 kDa) have been developed to overcome the prolonged retention times and toxicity of larger dendrimer and albumin MRI products (Kobayashi et al. 2003). These smaller dendrimer-based MRI agents were more quickly excreted from the kidneys and were able to make visualization of vascular structures better than Gd-DTPA due to less extravasation (Kobayashi and Brechbiel 2004).

27.5.4 Quantum Dots for Cellular Imaging

QDs are semiconductor nanocrystalline aggregates (artificial atoms) of a few hundred atoms. These manifest stable (nonquenching) fluorescent properties at various wavelengths and also offer a new array of coding in gene expression and *in vivo* imaging. They have a future role in imaging because of their small size (2–8 nm). This small size confers unique optical and electronic properties of tunable fluorescent emission with varying particle size or composition. These are the particles whose physical dimensions are smaller than the excitation of the Bohr's radius (Panyam et al. 2003). *In vivo* imaging of QD after injection involves excitation using a spatially broad source of long wavelength and capturing the resulting fluorescence with a sensitive CCD camera. By using the QD-labeled tumor cells, it was possible to follow their paths after the intravenous administration as they extravasated into the lungs (Voura et al. 2004). This can be potentially used to study metastatic tumor extravasation. Zinc sulfide and cadmium selenide (ZnS–CdSe) QD is modified to make them water soluble and they are coated with the targeting peptide sequence which binds to the lungs, blood vessels, and lymphatic tumors, respectively, when administered intravenously to mice (Klostranec and Chan 2006). QDs are also associated with phospholipid micelles and silicon nanospheres to improve solubilization and to reduce the uptake by liver and bone. Ligands and other effective compounds can also be conjugated to QD surfaces for imaging purpose. QDs represent a new tool of significant potential in neuroscience research. QDs are useful for experiments that are limited by the restricted anatomy of neuronal and glial interactions, such as the small size of the synaptic cleft, or between an astrocyte process and a neuron. Because of their extremely small size and optical resolution, they are also well suited for tracking the molecular dynamics of intracellular and/or intercellular molecular processes over long timescales. Studies using QDs in neuroscience illustrate the potential of this technology.

27.5.4.1 QD Micelles

QDs filled in the micelle core forms QD micelles. It provides support to the micelle itself. QD micelles are composed of cadmium selenide/zinc sulfide QDs inside micellar cores made from poly (ethylene glycol)/phosphatidylethanolamine (PEG-PE) and phosphatidylcholine). These QD micelles were stable for months, even in a 1 M salt solution, whereas empty micelles degrade and form aggregates after only a few days. The injected QDs remain within the cells and were nontoxic, inactive, and stable. Probably these could be used to label all embryonic cell types. QDs have superior optical properties which are most promising alternative methods to organic dyes for fluorescence-based bioapplications (proteins, DNA, antibodies). The QD can be conjugated to probes to monitor the cellular functions. Monosized, hydrophobic nanocrystals were synthesized within the hydrophobic cores of micelles which is composed of a mixture of surfactants and phospholipids containing head groups functionalized with polyethylene glycol (–PEG), –COOH, and –NH$_2$ groups (Fan et al. 2005). Biocompatibility was provided by PEG and the other groups were used for further biofunctionalization. The resulting water-soluble metal and semiconductor nanocrystal–micelles preserve the optical properties of the original hydrophobic nanocrystals. These nanocrystals–micelles preserve their original hydrophobic magnetic nanocrystals. Testing in yeast cell suggests the biocompatibility of these micelles. The oscillating magnetic fields can manipulate the magnetic micelles, which kills the live cells, providing opportunity for magnetodynamic therapy without having any side effect (Fan et al. 2005).

27.5.4.2 Aptamer-QD Conjugates

QD aptamer conjugates can be used for the detection of proteins. QDs can be surface functionalized with specific aptamer which are small DNA-RNA oligonucleotide. QDs were surface functionalized with aptamer known to bind thrombin protein (produced during an injury for clotting of blood) to develop thrombin-detecting QD aptamer conjugate. Aptamer-based specific recognition and QD-based fluorescence labeling are important in biosensing. These techniques have been coupled together to construct a new kind of fluorescent QD-labeled aptamer (QD–Apt) nanoprobe by conjugating GBI-10 aptamer to the QD surface. GBI-10 is an ssDNA aptamer for tenascin-C, which distributes on the surface of glioma cells as a dominant extracellular matrix protein. The QD–Apt nanoprobe can recognize tenascin-C on the human glioma cell surface, which will be helpful for the development of new convenient and sensitive *in vitro* diagnostic assays for glioma (Chen et al. 2005).

Aptamer bioconjugates with fluorescent nanocrystals are used for labeling tumor cells. Aptamers that bind to prostate-specific membrane antigen (PSMA) were conjugated to luminescent cadmium selenide (CdSe) and cadmium telluride (CdTe)

nanocrystals for cell-labeling studies. The aptamer–nanocrystal conjugates showed specific targeting of both fixed and live cells that overexpressed PSMA.

27.5.5 Magnetic Nanoparticle for Imaging

An MNP serves as an ideal image agent. Colloidal iron oxide nanoparticle formulated with dextran is clinically used as an MRI contrast agent (Gandon 1991). The covalent coupling of insulin-coated nanoparticles to cell surface receptors prevents internalization and reduces cellular toxicity in vitro (Gupta et al. 2003). Multiple crystals of iron oxide are embedded in the nanoparticular matrices such as polyacrylamide with surface attachment of PEG (Moffat et al. 2003) and solid lipid nanoparticles (Peira et al. 2003). Iron oxide in such polyacrylamide nanoparticles when injected into rats bearing orthotopic 9 L gliomas had significant increase in relaxivity and increase in circulation half-life up to threefold (Moffat et al. 2003).

27.5.6 Nanobialys as a Delivery and Imaging Agent

Nanobialys are ultraminiature, bialys-shaped particles. The name "bialy" is a Polish word meaning bread roll without a hole with different toppings. Nanobialys are smart particles that can deliver both the drug and imaging agents directly into the bloodstream, into tumors or atherosclerotic plaques. These particles are rounded, flat with a dimple at the center with some resemblance to the structure of red blood cells. Nanobialys are important imaging materials developed as a substitute to gadolinium-based contrast agents for MRI scans. Studies have shown that the use of gadolinium as an MRI contrast agent can be harmful with patients having severe kidney disease, diabetes, and related cardiovascular problems. Nanobialys contain manganese instead of gadolinium. Manganese is an element found naturally in the body. The bulk of a nanobialy is a synthetic polymer that can accept a variety of medical, imaging, or targeting components.

Researchers have reported that targeted manganese-carrying nanobialys readily attached themselves to fibrin molecules, which are found in atherosclerotic plaques and blood clots. Laboratory-made clots then glowed brightly in MRI scans. In addition, the manganese in the nanobialys is tied up, so it stays with the particles, making them very safe. This newly developed smart contrast agent can also carry both hydrophilic and hydrophobic drugs. Nanobialys can be used for a wide range of medicinal applications (Pan et al. 2008).

27.6 Summary

Nanotechnology is progressing very fast and is making great impact in the field of therapeutics. Early diagnosis of diseases, problems of multiple drug resistance, and efficient cellular uptake can be overcome by drug therapy integrated with nanoscience. The development of engineered nanoparticles with substantial biomedical significance has posed new opportunities for pharmacology and therapeutics. Nanoparticles capable of carrying soluble and insoluble drugs have been created for delivery of combination therapy for cancer. Besides drug delivery, gene therapy, biosensing, and imaging devices are upcoming examples of nanotechnology being applied to potential improvement and treatment of cardiac diseases, neurological disorders, cancers, and other anomalies. Nanomaterials and nanoparticles are likely to be cornerstones of innovative nanomedical devices, which can be used for drug discovery and delivery and molecular diagnostics and thus the long-term therapeutic improvement of medical science through nanotechnology is to be addressed.

27.7 Future Perspective

Nanotechnology is receiving a lot of attention with the never-seen-before enthusiasm because of its future potential that can literally revolutionize each field in which it is being exploited. In drug delivery, nanotechnology is just beginning to make an impact. Drug delivery is one of the most dynamic and fast-growing sectors of the pharmaceutical industry. Clinically useful drug delivery systems need to deliver a certain amount of drug that can be therapeutically effective, and often over an extended period of time. Such requirements can be met by the nanoscale drug delivery systems manufactured by nanotechnology. Nanomedicine has incredible potential for revolutionizing therapeutics and diagnostics under the premise of developing ingenious nanodevices. Many of the current nanodelivery systems, however, are remnants of conventional drug delivery systems that happen to be in the nanometer range, such as liposomes, polymeric micelles, nanoparticles, dendrimers, nanocrystals, nanochips, and nanoneedles. These are now called nanovehicles and are used in drug delivery. The most promising application of site-specific drug delivery can be done using MNPs. These nanoparticles can carry therapeutic agents on their surface, which can be driven to the target organ under an external magnetic field.

The current methods of preparing nanoparticles are mainly based on double emulsion methods, solvent exchange, and other techniques. The main problems with the current methods are the low drug loading capacity, low loading efficiency, and poor ability to control the size distribution. Utilizing nanotechnologies, one could allow manufacturing of nanoparticles with high loading efficiency and homogeneous particle size.

Nanoscale devices can sense the environment, process information, or convert energy from one form to another. They include nanoscale sensors, which exploit the huge surface area of carbon nanotubes and other nanostructured materials to detect environmental contaminants or biochemicals. Other products of evolutionary nanotechnology are semiconductor nanostructures such as QDs and quantum wells that are being used to build better solid-state lasers. Scientists are also developing even more sophisticated ways of encapsulating molecules and delivering them on demand for targeted drug delivery.

Nanotechnology (nanomaterials and nanoscale devices) for diagnosis, treatment, and monitoring diseases is a fast-developing area of biomedical research and its amalgamation with engineering, pharmaceutical, and medical sciences will become the order of the day in the years to come.

Acknowledgment

SKS would like to thank the Department of Biotechnology, Government of India, for providing the grants no. BT/04(SBIRI)/48/2006-PID and BT/PR7968/MED/14/1206/2006.

References

Akeson, M., Branton, D., Kasianowicz, J. J. et al. 1999. Microsecond time-scale discrimination among polycytidylic acid, polyadenylic acid, and polyuridylic acid as homopolymers or as segments within single RNA molecules. *Biophys J* 77: 3227–3233.

Alexiou, C., Arnold, W., Klein, R. J. et al. 2000. Locoregional cancer treatment with magnetic drug targeting. *Cancer Res* 60: 6641–6648.

Allemann, E., Leroux, J.-C., and Gurny, R. 1998. Polymeric nano- and microparticles for the oral delivery of peptides and peptidomimetics. *Adv Drug Deliv Rev* 34: 171–189.

Allen, T. M. 1997. Liposomes opportunities in drug delivery. *Drugs* 54 Suppl 4: 8–14.

Balogh, L., Swanson, D. R., Tomalia, D. A. et al. 2001. Dendrimer-silver complexes and nanocomposites as antimicrobial agents. *Nano Lett* 1: 18–21.

Bangham, A. D., Standish, M. M., and Watkins, J. C. 1965. Diffusion of univalent ions across the lamellae of swollen phospholipids. *J Mol Biol* 13: 238–252.

Bernkop-Schnürch, A. and Krajicek, M. E. 1998. Mucoadhesive polymers as platforms for peroral peptide delivery and absorption: Synthesis and evaluation of different chitosan-EDTA conjugates. *J Control Release* 50: 215–223.

Bertini, I., Bianchini, F., Calorini, L. et al. 2004. Persistent contrast enhancement by sterically stabilized paramagnetic liposomes in murine melanoma. *Magn Reson Med* 52: 669–672.

Boas, U. and Heegaard, P. M. 2004. Dendrimers in drug research. *Chem Soc Rev* 33: 43–63.

Bowman, K. and Leong, K. W. 2006. Chitosan nanoparticles for oral drug and gene delivery. *Int J Nanomed* 1: 117–128.

Braun, E., Eichen, Y., Sivan, U. et al. 1998. DNA-templated assembly and electrode attachment of a conducting silver wire. *Nature* 391: 775.

Brigger, I., Dubernet, C., and Couvreur, P. 2002. Nanoparticles in cancer therapy and diagnosis. *Adv Drug Deliv Rev* 54: 631–651.

Bulte, J. W. and Kraitchman, D. L. 2004. Monitoring cell therapy using iron oxide MR contrast agents. *Curr Pharm Biotechnol* 5: 567–584.

Casal, A., Hogg, T., and Cavalcanti, A. 2003. Nanorobots ascellular assistants in inflammatory responses. *Proc. IEEE BCATS Biomedical Computation at Stanford 2003 Symposium*, IEEE Computer Society, Stanford, CA.

Cavalcanti, A. and Freitas Jr, R. A. 2002. Autonomous multi-robot sensor-based cooperation for nanomedicine. *Intl J Nonlinear Sci Numerical Simulation* 3: 743–746.

Cengelli, F., Maysinger, D., Tschudi-Monnet, F. et al. 2006. Interaction of functionalized superparamagnetic iron oxide nanoparticles with brain structures. *J Pharmacol Exp Ther* 318: 108–116.

Chen, H. H., Le Visage, C., Qiu, B. et al. 2005. Mr imaging of biodegradable polymeric microparticles: A potential method of monitoring local drug delivery. *Magn Reson Med* 53: 614–620.

Cho, Y., Shi, R., Borgens, R. B. et al. 2008. Functionalized mesoporous silica nanoparticle-based drug delivery system to rescue acrolein-mediated cell death. *Nanomedicine* 3: 507–519.

Choi, Y., Baker, L. A., Hillebrenner, H. et al. 2006. Biosensing with conically shaped nanopores and nanotubes. *Phys Chem Chem Phys* 8: 4976–4988.

Cui, Y., Wei, Q., Park, H. et al. 2001. Nanowire nanosensors for highly sensitive and selective detection of biological and chemical species. *Science* 293: 1289–1292.

Dagar, S., Rubinstein, I., and Onyuksel, H. 2003. Liposomes in ultrasound and gamma scintigraphic imaging. *Methods Enzymol* 373: 198–214.

Desai, M. P., Labhasetwar, V., Amidon, G. L. et al. 1996. Gastrointestinal uptake of biodegradable microparticles: Effect of particle size. *Pharm Res* 13: 1838–1845.

Desai, M. P., Labhasetwar, V., Walter, E. et al. 1997. The mechanism of uptake of biodegradable microparticles in caco-2 cells is size dependent. *Pharm Res* 14: 1568–1573.

Dillon, A. C., Jones, K. M., Bekkedahl, T. A. et al. 1997. Storage of hydrogen in single-wall carbon nanotubes. *Nature* 386: 377.

Drexler, E., Peterson, C., and Perganit, G. 1991. *Unbounding the Future: The Nanotechnology Revolution*, William Morrow and Co Inc., New York, p. 304.

Dudeck, O., Bogusiewicz, K., Pinkernelle, J. et al. 2006. Local arterial infusion of superparamagnetic iron oxide particles in hepatocellular carcinoma: A feasibility and 3.0 T MRI study. *Invest Radiol* 41: 527–535.

El-Shabouri, M. 2002. Positively charged nanoparticles for improving the oral bioavailability of cyclosporine-A. *Int J Pharm* 249: 101–108.

Emerich, D. F. 2005. Nanomedicine-prospective therapeutics and diagnostic applications. *Expert Opin Biol Ther* 5: 1–5.

Fahy, G. M. 1993. Molecular nanotechnology. *Clin Chem* 39: 2011–2016.

Fan, H. Y., Chen, Z., Brinker, C. et al. 2005. Synthesis of organosilane functionalized nanocrystal micelles and their self-assembly. *J Am Chem Soc* 127: 13746–13747.

Ferber, D. 2001. Gene therapy: Safer and virus-free? *Science* 294: 1638–1642.

Fernandez-Lopez, S., Kim, H. S., Choi, E. C. et al. 2001. Antibacterial agents based on the cyclic D,L-alpha-peptide architecture. *Nature* 412: 452–455.

Fisher, R. S. and Ho, J. 2002. Potential new methods for antiepileptic drug delivery. *CNS Drugs* 16: 579–593.

Ford, W. E., Wessels, J., and Harnack, O. 2002. Process for immobilization. Of nucleic acid molecules on a substrate. Europen Patent 1207207.

Freitas Jr., R. A. 1999. Basic capabilities. *Landes Biosci Nanomed* 1:385.

Gabizon, A., Goren, D., Cohen, R. et al. 1998. Development of liposomal anthracyclines: From basics to clinical applications. *J Control Release* 53: 275–279.

Gandon, P. 1991. Non-invasive assessment of hepatic iron stores by MRI. *The Lancet* 363: 357–362.

Ghadiri, M. R., Granja, J. R., Milligan, R. A. et al. 1993. Self-assembling organic nanotubes based on a cyclic peptide architecture. *Nature* 366: 324–327.

Gregoriadis, G. 1995. Engineering liposomes for drug delivery: Progress and problems. *Trends Biotechnol* 13: 527–537.

Gupta, A., Akin, D. and Bashir, R. 2004. Single virus particle mass detection using microresonators with nanoscale thickness. *Appl Phys Lett* 84: 1976.

Gupta, A. K., Berry, C., Gupta, M. et al. 2003. Receptor-mediated targeting of magnetic nanoparticles using insulin as a surface ligand to prevent endocytosis. *IEEE Trans Nanobiosci* 2: 255–261.

Hafelli, U., Schutt, W., Teller, J. et al., 1997. *Scientific and Clinical Applications on Magnetic Carriers*. Plenum, New York, p. 527.

Harnack, O., Ford, W. E., Yasuda, A. et al. 2002. Tris(hydroxymethyl) phosphine capped gold particles template by DNA as nanowire precursors. *Nano Lett* 2: 919–923.

Harrell, C. C., Siwy, Z. S., Martin, C. R. et al. 2006 Conical nanopore membranes: Controlling the nanopore shape. *Small* 2: 194–198.

He, Q., Liu, J., Sun, X. et al. 2004. Preparation and characteristics of DNA nanoparticles targeting to hepatocarcinoma cells. *World J Gastroenterol* 10: 660.

Hille, B. 2001. *Ion Channels of Excitable Membranes*, 3rd edn. Sinauer Associates, Sunderland, MA.

Ho, C., Qiao, R., Heng, J. B. et al. 2005. Electrolytic transport through a synthetic nanometer-diameter pore. *Proc Natl Acad Sci U S A* 102: 10445–10450.

Ito, T., Sun, L., Henriquez, R. R. et al. 2004. A carbon nanotube-based coulter nanoparticle counter. *Acc Chem Res* 37: 937–945.

Iyer, M., Berenji, M., Templeton, N. S. et al. 2002. Noninvasive imaging of cationic lipid-mediated delivery of optical and pet reporter genes in living mice. *Mol Ther* 6: 555–562.

Jain, K. 2003. Nanodiagnostics: Application of nanotechnology in molecular diagnostics. *Expert Rev Mol Diagn* 3: 153–161.

Johannsen, M., Gneveckow, U., Thiesen, B. et al. 2006. Thermotherapy of prostate cancer using magnetic nanoparticles: Feasibility, imaging, and three-dimensional temperature distribution. *Eur Urol* 52: 1653–1662.

Johnson-Saliba, M. and Jans, D. A. 2001. Gene therapy: Optimising DNA delivery to the nucleus. *Curr Drug Targets* 2: 371–399.

Kiang, T., Bright, C., Cheung, C. Y. et al. 2004. Formulation of chitosan-DNA nanoparticles with poly(propyl acrylic acid) enhances gene expression. *Biomater Sci Polym Ed* 15: 1405–21.

Kihara, F., Arima, H., Tsumi, T. et al. 2002. Effects of structure of polyamido amine dendrimer on gene transfer efficiency of the dendrimer conjugate with alpha-cyclodextrin. *Bioconjugate Chem* 13: 1211–1219.

Kim, W. 2007. Interfacing silicon nanowires with mammalian cells. *J Am Chem Soc* 129: 7228–7229.

Klajnert, B. and Bryszewska, M. 2001. Dendrimers: Properties and applications. *Acta Biochim Pol* 48: 199–208.

Klostranec, J. M. W. and Chan, C. W. 2006. Quantum dots in biological and biomedical research: Recent progress and present challenges. *Adv Mater* 18: 1953–1964.

Kobayashi, H. and Brechbiel, M. W. 2004. Dendrimer-based nanosized MRI contrast agents. *Curr Pharm Biotechnol* 5: 539–549.

Kobayashi, H., Kawamoto, S., Jo, S. K. et al. 2003. Macromolecular MRI contrast agents with small dendrimers: Pharmacokinetic differences between sizes and cores. *Bioconjug Chem* 14: 388–394.

Kojima, C., Kono, K., Maruyama, K. et al. 2000. Synthesis of polyamidoamine dendrimers having poly(ethylene glycol) grafts and their ability to encapsulate anticancer drugs. *Bioconjug Chem* 11: 910–917.

Kommareddy, S. and Amiji, M. 2007. Poly(ethylene glycol)-modified thiolated gelatin nanoparticles for glutathione-responsive intracellular DNA delivery. *Nanomedicine* 3: 32–42.

Koo, O. M., Rubinstein, I., and Onyuksel, H. 2005. Role of nanotechnology in targeted drug delivery and imaging: A concise review. *Nanomed: Nanotechnol, Biol Med* 1: 193–212.

Kreuter, J. 2001. Nanoparticulate systems for brain delivery of drugs. *Adv Drug Deliv Rev* 47: 65–81.

Kricka, L. J., Nozaki, O., and Wilding, P. 1994. Micromechanics and nanotechnology: Implications and applications in the clinical laboratory. *J Int Fed Clin Chem* 6: 52–56.

Lamprecht, A., Ubrich, N., Yamamoto, H. et al. 2001. Biodegradable nanoparticles for targeted drug delivery in treatment of inflammatory bowel disease. *J Pharmacol Exp Ther* 299: 775–781.

Lasic, D. D. 1998. Novel applications of liposomes. *Trends Biotechnol* 16: 307–321.

Li, J., Gershow, M., Stein, D. et al. 2003. DNA molecules and configurations in a solid-state nanopore microscope. *Nat Mater* 2: 611–615.

Liu, D. S. and Balasubramanian, S. 2003. A proton-fuelled DNA nanomachine. *Angew Chem Int Ed* 42: 5734.

Liu, M. and Frechet, J. M. 1999. Designing dendrimers for drug delivery. *Pharm Sci Technol Today* 2: 393–401.

Liu, M., Kono, K., and Frechet, J. M. 2000. Water-soluble dendritic unimolecular micelles: Their potential as drug delivery agents. *J Control Release* 65: 121–131.

Lockman, P. R., Mumper, R. J., Khan, M. A. et al. 2002. Nanoparticle technology for drug delivery across the blood-brain barrier. *Drug Dev Ind Pharm* 28: 1–13.

Maeda, H. 2001. The enhanced permeability and retention (EPR) in tumor vasculature: The key role of tumor-selective macromolecular drug targeting. *Adv Enzyme Regul* 41: 189–207.

Miller, A. D. 2008. Towards safe nanoparticle technologies for nucleic acid therapeutics. *Tumori* 94: 234–245.

Moffat, B. A., Reddy, G. R., McConville, P. et al. 2003. A novel polyacrylamide magnetic nanoparticle contrast agent for molecular imaging using MRI. *Mol Imaging* 2: 324–332.

Moghimi, S. M. and Hunter, A. C. 2001. Capture of stealth nanoparticles by the body's defences. *Crit Rev Ther Drug Carrier Syst* 18: 527–550.

Moghimi, S. M., Hunter, A. C., and Murray, J. C. 2001. Long-circulating and target-specific nanoparticles: Theory to practice. *Pharmacol Rev* 53: 283–318.

Mohanraj, V. J. and Chen, Y. 2006. Nanoparticles-a review. *Trop J Pharm Res* 5: 561–573.

Monson, C. F. and Woolley, A. T. 2003. DNA-templated construction of copper nanowires. *Nano Lett* 3: 359–363.

Muthana, M., Scotts, D., Farrow, N. et al. 2008. A novel magnetic approach to enhance the efficacy of cell-based gene therapies. *Gene Ther* 15: 902–910.

Nierengarten, J. F., Eckert, J. F., Rio, Y. et al. 2001. Amphiphilic diblock dendrimers: Synthesis and incorporation in langmuir and langmuir-blodgett films. *J Am Chem Soc* 123: 9743–9748.

Padilla De Jesus, O. L., Ihre, H. R., Gagne, L. et al. 2002. Polyester dendritic systems for drug delivery applications: In vitro and in vivo evaluation. *Bioconjug Chem* 13: 453–461.

Pan, D., Caruthers, S. D., Hu, G. et al. 2008. Ligand-directed nanobialys as theranostic agent for drug delivery and manganese-based magnetic resonance imaging of vascular targets. *J Am Chem Soc* 130: 9186.

Panyam, J., Zhou, W. Z., Prabha, S. et al. 2002. Rapid endolysosomal escape of poly(dl-lactide-co-glycolide) nanoparticles: Implications for drug and gene delivery. *FASEB J* 16: 1217–1226.

Panyam, J., Sahoo, S. K., Prabha, S. et al. 2003. Fluorescence and electron microscopy probes for cellular and tissue uptake of poly(d,l-lactide-co-glycolide) nanoparticles. *Int J Pharm* 262: 1–11.

Parveen, S. and Sahoo, S. K. 2006. Nanomedicine: Clinical applications of polyethylene glycol conjugated proteins and drugs. *Clin Pharmacokinet* 45: 965–988.

Parveen, S. and Sahoo, S. K. 2008. Polymeric nanoparticles for cancer therapy. *J Drug Target* 16: 108–123.

Patolsky, F. and Lieber, C. M. 2005. Nanowire nanosensors. *Mater Today* 8: 20–28.

Peira, E., Marzola, P., Podio, V. et al. 2003. In vitro and in vivo study of solid lipid nanoparticles loaded with superparamagnetic iron oxide. *J Drug Target* 11: 19–24.

Pingang, H., Ying, X. and Yuzhi, F. 2006. Applications of carbon nanotubes in electrochemical DNA biosensors. *Microchim Acta* 152: 175–186.

Prabha, S. and Labhasetwar, V. 2004. Nanoparticle mediated wild-type p53 gene delivery results in sustained antiproliferative activity in breast cancer cells. *Mol Pharm* 1: 211.

Quintana, A., Raczka, E., Piehler, L. et al. 2002. Design and function of a dendrimer-based therapeutic nanodevice targeted to tumor cells through the folate receptor. *Pharm Res* 19: 1310–1316.

Radisic, M., Iyer, R. K., and Murthy, S. M. 2006. Micro- and nanotechnology in cell separation. *Int J Nanomed* 1: 3–14.

Rhaese, S., von Briesen, H., Rubsamen-Waigmann, H. et al. 2003. Human serum albumin-polyethylenimine nanoparticles for gene delivery. *J Control Release* 92: 199–208.

Richter, J., Seidel, R., Kirsch, R. et al. 2000. Nanoscale palladium metallization of DNA. *Adv Mater* 12: 507.

Roco, M. C. 2003. Nanotechnology: Convergence with modern biology and medicine. *Curr Opin Biotechnol* 14: 337–346.

Rostovtseva, T. K., Komarov, A., Bezrukov, S. M. et al. 2002. Dynamics of nucleotides in VDAC channels: Structure-specific noise generation. *Biophys J* 82: 193–205.

Roth, D. M., Lai, N. C., Gao, M. H. et al. 2004. Nitroprusside increases gene transfer associated with intracoronary delivery of adenovirus. *Hum Gene Ther* 15: 989–994.

Roy, I., Ohulchanskyy, T. Y., Bharali, D. J. et al. 2005. Optical tracking of organically modified silica nanoparticles as DNA carriers: A nonviral, nanomedicine approach for gene delivery. *Proc Natl Acad Sci U S A* 102: 279–284.

Sahoo, S. K. and Labhasetwar, V. 2003. Nanotech approaches to drug delivery and imaging. *Drug Discov Today* 8: 1112–1120.

Sahoo, S. K. and Labhasetwar, V., 2006, Biodegradable PLGA/PLA nanoparticles for anticancer therapy, in: *Nanotechnology in Cancer Therapy*, Ed. Amiji, M. M. CRC Press, Boca Raton, FL, pp. 241–248.

Sahoo, S. K., Panyam, J., Prabha, S. et al. 2002a. Residual polyvinyl alcohol associated with poly (d,l-lactide-co-glycolide) nanoparticles affects their physical properties and cellular uptake. *J Control Release* 82: 105–114.

Sahoo, S. K., Sawa, T., Fang, J. et al. 2002b. Pegylated zinc protoporphyrin: A water-soluble heme oxygenase inhibitor with tumor-targeting capacity. *Bioconjug Chem* 13: 1031–1038.

Sahoo, S. K., Ma, W., and Labhasetwar, V. 2004. Efficacy of transferrin-conjugated paclitaxel-loaded nanoparticles in a murine model of prostate cancer. *Int J Cancer* 112: 335–340.

Sahoo, S. K., Parveen, S., and Panda, J. J. 2007. The present and future of nanotechnology in human health care. *Nanomed Nanotechnol, Biol Med* 3: 20–31.

Saleh, O. A. and Sohn, L. L. 2003. An artificial nanopore for molecular sensing. *Nano Lett* 3: 37–38.

Shaffer, C. 2005. Nanomedicine transforms drug delivery. *Drug Discov Today* 10: 1581–1582.

Simmel, F. C. and Dittmer, W. U. 2005. DNA nanodevices. *Small* 1: 284–299.

Singh, S. R., Grossniklaus, H. E., Kang, S. J. et al. 2009. Intravenous transferrin, RGD peptide and dual-targeted nanoparticles enhance anti-VEGF intraceptor gene delivery to laser-induced CNV. *Gene Ther* 16: 645–659.

Siwy, Z. S., Gu, Y. C., Spohr, H. A. et al. 2002. Nanofabricated voltage-gated pore. *Biophys J* 82: 266A.

Slowing, I. I., Trewyn, B. G., Giri, S. V. et al. 2007. Mesoporous silica nanoparticles for drug delivery and biosensing applications. *Adv Funct Mater* 17: 1225–1236.

Smith, D. A. and Van de Waterbeemd, H. 1999. Pharmacokinetics and metabolism in early drug discovery. *Curr Opin Chem Biol* 3: 373–378.

Song, J., Midson, C., Blachly-Dyson, E. et al. 1998. The topology of VDAC as probed by biotin modification. *J Biol Chem* 273: 24406–24413.

Storm, A. J., Chen, J. H., Ling, X. S. et al. 2003. Fabrication of solid-state nanopores with single-nanometre precision. *Nat Mater* 2: 537–540.

Szabo, I., Bàthori, G., Tombola, F. et al. 1997. DNA translocation across planar bilayers containing *Bacillus* subtilis ion channels. *J Biol Chem* 272: 25275–25282.

Szabo, L., Bathori, G., Tombola, F. et al. 1998. Double-stranded DNA can be translocated across a planar membrane containing purified mitochondrial porin. *FASEB J* 12: 495–502.

Tan, K., Cheang, P., Ho, I. A. W. et al. 2007. Nanosized bioceramic particles could function as efficient gene delivery vehicles with target specificity for the spleen. *Gene Ther.*14: 828–835.

Tans, S. J., Verschueren, A. R. M., and Dekker, C. 1998. Room-temperature transistor based on a single carbon nanotube. *Nature* 393: 49.

Templeton, N. S. 2002. Liposomal delivery of nucleic acids in vivo. *DNA Cell Biol* 21: 857–867.

Thomas, M. and Klibanov, A. M. 2003. Conjugation to gold nanoparticles enhances polyethyleneimine's transfer of plasmid DNA into mammalian cells. *Proc Natl Acad Sci U S A* 100: 9138–9143.

Tiefenauer, L. X., 2007, Magnetic nanoparticles as contrast agents for medical diagnosis, in: *Nanotechnology in Biology and Medicine: Methods, Devices, and Applications*, Vol. Section D, Ed. Vo-Dinh, T. CRC Press, Taylor & Francis, Boca Raton, FL, pp. 1–20.

Tomalia, D. A., Baker, H., Dewald, J., Hall, M., Kallos, G., Martin, S., Roeck, J., Ryder, J., and Smith, P. 1985. A new class of polymers: Starburst-Dendritic Macromolecules. *Polymer Journal* 17: 117–132.

Torchilin, V. P. 2002. Peg-based micelles as carriers of contrast agents for different imaging modalities. *Adv Drug Deliv Rev* 54: 235–252.

Townsend, S. A., Evrony, G. D., Gu, F. X. et al. 2007. Tetanus toxin c fragment-conjugated nanoparticles for targeted drug delivery to neurons. *Biomaterials* 28: 5176-5184.

Van den Berg, A. and Bergveld, P. 1995. *Micro Total Analysis Systems*. Kluwer Academic Publishers, Dordrecht, the Netherlands, p. 311.

Voura, E. B., Jaiswal, J. K., Mattoussi, H. et al. 2004. Tracking metastatic tumor cell extravasation with quantum dot nanocrystals and fluorescence emission-scanning microscopy. *Nat Med* 10: 993–998.

Wagner, E. 2005. Polymer nonviral delivery vehicles, in: *Encyclopedia of Diagnostic Genomics and Proteomics*. Marcel Dekker, New York, pp. 1047–1051.

Wang, S. J., Brechbiel, M. and Wiener, E. C. 2003. Characteristics of a new MRI contrast agent prepared from polypropyleneimine dendrimers, generation 2. *Invest Radiol* 38: 662–668.

Wartlick, H., Spankuch-Schmitt, B., Strebhardt, K. et al. 2004. Tumour cell delivery of antisense oligonucleotides by human serum albumin nanoparticles. *J Control Release* 96: 483–495.

Weizhong, W., Chunfang, X., and Hua, W. 2004. Use of PEI-coated magnetic iron oxide nanoparticles as gene vectors. *J Huazhong University of Sci Technol Med Sci* 24: 618–620.

Wharton, J. E., Jin, P., Sexton, L. T. et al. 2007. A method for reproducibly preparing synthetic nanopores for resistive-pulse biosensors. *Small* 3: 1424–1430.

Wong, S. S., Joselevich, E., Woolley, A. T. et al. 1998. Covalently functionalized nanotubes as nanometer probes for chemistry and biology. *Nature* 394: 52.

Woodle, M. C. and Lasic, D. D. 1992. Sterically stabilized liposomes. *Biochim Biophys Acta* 1113: 171–199.

Young, S. and Kiang, C.-H. 2005. *Handbook of Nanostructured Biomaterials and Their Applications in Nanobiotechnology*, American Scientific Publishers, Valencia, CA, Vol. 2. pp. 224–246.

Zucates, G. T., Little, S. R., Anderson, D. G. et al. 2005. Poly(β-amino ester)s for DNA delivery. *Isr J Chem* 45: 477–485.

Targeting Magnetic Particles for Drug Delivery

Javed Ally
University of Alberta

Alidad Amirfazli
University of Alberta

28.1 Introduction

Magnetic targeting has the potential to enhance the effectiveness of drug delivery by providing an external means of guiding drug particles within the body. If magnetic fields could be used to control the motion of drug particles through the body, this would allow site-specific delivery to target locations, improving treatment efficacy and reducing systemic toxicity by increasing drug concentration at the target site and reducing the overall amount of drug administered.

To design magnetic targeting systems for drug delivery, an understanding of magnetic fields, forces, and magnetic particle transport within the body is required. There are four aspects of a magnetic targeting system that must be considered:

- The physiological environment of the magnetic particles
- The magnetic field, which must be designed to produce the desired particle trajectories
- The particles and other magnetic material in the system
- The particle trajectories that result from the interaction of the fluid, magnetic, and other forces together with the system geometry

The concept of magnetic drug targeting was first considered in the 1970s, when polymer-coated magnetic microparticles were first developed [1]. The development of magnetic drug targeting was further aided by the development of neodymium–iron–boron (Nd-Fe-B) permanent magnets in the 1980s. Nd-Fe-B magnets provide an inexpensive means of producing strong magnetic fields and field gradients with a small amount of material by positioning permanent magnets outside of the body. Other magnetic field sources include electromagnets and magnetic resonance imaging (MRI) machines. Superconducting magnets have also been proposed for magnetic targeting. The availability and variety of magnetic particles has also dramatically increased since the 1970s. The main challenge in magnetic targeting is combining these technologies for the effective delivery of drug particles to target sites in the body.

Many studies of magnetic targeting have focused on delivery via the circulatory system [2–7]. Recently, magnetic targeting of inhaled aerosols has emerged as an area of interest [8–14]. It has been demonstrated in animal studies that the administration of chemotherapeutic agents via inhaled aerosol could be effective in treating tumors in the lung [15–17]. However, there is a need to limit exposure of healthy lung and airway tissue to the drug [15]. Magnetic targeting would make inhaled aerosol chemotherapy more effective by concentrating the inhaled drug particles at the target site.

This chapter presents the relevant theoretical background for the four aspects of magnetic targeting and then illustrates their practical implications using examples of the emerging technique of magnetic aerosol drug targeting, based on results from the literature.

28.2 Theoretical Background

28.2.1 Fluid Forces

Typically, magnetic particles are introduced into the body in a fluid flow, for example, blood flow or inhaled air. The transport of the particles is governed by the fluid mechanics of the particle environment. In most cases, viscous forces are dominant, and the Stokes drag expression can be used to determine the forces required to pull the particles from the flow to the target site.

28.2.1.1 Stokes Drag: Fully Immersed Particles

In magnetically targeted drug delivery, particles are normally introduced into a fluid flow in the body. The particles are directed to the target site by forces produced by a magnetic field; these forces are opposed by fluid forces. For micro- and nanoparticles in most blood vessels and airways, the fluid forces near the target site are dominated by the viscous drag of the fluid; the inertia of the particles can usually be ignored. The relative importance of viscous drag relative to inertia is quantified by the Reynolds number Re, which for a particle immersed in a fluid, is given by

$$Re = \frac{\rho V d}{\eta} \tag{28.1}$$

where
- ρ is the fluid density
- V is the characteristic velocity of the flow
- d is the particle diameter
- η is the fluid viscosity

If the Reynolds number for a spherical particle is much less than 1, the drag force can be considered to be purely viscous, in which case the drag force is given by the Stokes flow expression for drag on a fully immersed sphere:

$$\vec{F}_D = 3\pi\eta d(\vec{v}_f - \vec{v}_p) \tag{28.2}$$

According to the Stokes expression in Equation 28.2, the drag force on a particle $\vec{F}_D$ is governed by the particle diameter, fluid viscosity, and the difference between the velocity of the particle, $\vec{v}_p$, and the fluid velocity $\vec{v}_f$.

28.2.1.2 Stokes Drag: Partially Immersed Particles

Unlike particles in the circulatory system that are fully immersed in the blood, particles depositing in the airways will be partially immersed in the liquid lining of the airway surface. In order to determine the fluid forces on these particles that must be overcome by the magnetic field, the case of viscous drag on partially immersed sphere must be considered. In this case, the drag force is given by [18]

$$\vec{F}_D = 3\pi\eta d(\vec{v}_f - \vec{v}_p) f(\theta) \tag{28.3}$$

As for fully immersed spheres, the drag force depends on the liquid viscosity, particle diameter, and velocity difference between

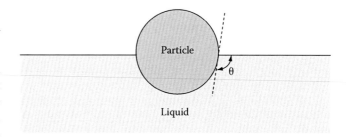

FIGURE 28.1 The contact angle between a solid, spherical particle and air–liquid interface is the angle formed between the interface and the tangent of the particle at the three-phase contact line, where the three phases meet. The contact angle represents the extent of particle immersion, and depends upon the properties of the particle and liquid. The lower the contact angle, the more immersed the particle is in the liquid phase.

the particle and fluid. For a partially immersed particle, there is an additional dependence on the function $f(\theta)$, which depends upon the particle contact angle, θ. The particle contact angle is the angle between a particle's surface and the air–liquid interface (measured through the liquid phase), as shown in Figure 28.1. The particle contact angle depends on the wetting properties of the particle, which can be controlled during particle synthesis depending on the desired particle behavior.

28.2.1.3 Aerodynamic Diameter

In many cases, it is easier to measure the settling velocity of particles in a fluid of known viscosity due to a known force, for example, gravity, than it is to measure the particle size and shape. For this reason, particles are often described in terms of their aerodynamic or hydrodynamic equivalent diameter d_e. The equivalent diameter is determined from Equation 28.2, with the assumption that all other quantities are known and that the Stokes drag assumption is applicable. The equivalent diameter can be thought of as the diameter of a spherical particle that would experience the same drag force in a given flow as an irregular particle. For nonspherical particles, the equivalent diameter will depend on the orientation of the particle relative to the flow.

28.2.1.4 Internal Flows

As seen in Equations 28.2 and 28.3 the fluid velocity plays a role in the drag force on the particles. In an internal flow, such as a blood vessel or airway, the Reynolds number is

$$Re = \frac{\rho V d_h}{\eta} \tag{28.4}$$

where d_h is the hydraulic diameter of the vessel. If $Re < 1$, the flow will be laminar. If $Re > 2300$, the flow will normally be turbulent, and there will be random fluctuations in the velocity of the fluid at various length scales. For intermediate values of Re, there may be turbulence at some length scales and frequencies. Turbulence in a flow will alter the trajectories of particles, and may increase or decrease deposition. The distinction between Equations 28.1

and 28.4 is that Equation 28.1 describes the flow around a particle, whereas Equation 28.4 is concerned with the flow conditions in a blood vessel or airway.

28.2.2 Magnetic Fields

28.2.2.1 Maxwell's Equations

The geometry and strength of the magnetic fields used in drug targeting determine the magnetic force on the drug particles. Magnetic fields are fundamentally related to moving electric charges. As such, for the most general case, the behavior of both magnetic and electric fields, that is, electromagnetic fields, must be considered. Electromagnetic fields behave according to Maxwell's laws, shown below in differential form. These laws govern the possible characteristics of magnetic fields and the response of magnetic materials to magnetic fields. Maxwell's laws have important consequences for determining the types of magnetic fields and magnetic targeting implementations that are possible:

$$\nabla \cdot \vec{E} = \frac{\rho_f}{\varepsilon_0} \tag{28.5}$$

$$\nabla \cdot \vec{B} = 0 \tag{28.6}$$

$$\nabla \times \vec{E} = -\frac{\partial \vec{B}}{\partial t} \tag{28.7}$$

$$\nabla \times \vec{B} = \mu_0 \vec{J} + \mu_0 \varepsilon_0 \frac{\partial \vec{E}}{\partial t} \tag{28.8}$$

Maxwell's laws relate the electric field $\vec{E}$ and magnetic field $\vec{B}$ to one another as well as the electrical charge density ρ_f and current density $\vec{J}$. The constants ε_0 and μ_0 are the permittivity and permeability of free space, respectively. The SI unit of magnetic field strength is the Tesla (T), equivalent to the Weber per square meter. The *cgs* unit of magnetic field is the Gauss (10,000 G = 1 T), and is still found in the literature.

Gauss' law (Equation 28.5) states that the amount of electric charge in an arbitrary volume creates an electric field with a net flux through the surface of the volume. This is because there can be an excess of positive or negative electric charge density ρ_f. In Equations 28.5 through 28.8, there is no similar term for magnetic charges. This is because ultimately magnetism results from the motion of electric charges, as shown in Equation 28.4. In magnetic materials, this occurs due to electron motion at the atomic level. Hence, unlike electric charges magnetic "charges"—north and south poles—are always paired as dipoles. This is reflected in Equation 28.6, which states that the net magnetic flux through a volume is always zero.

The relationship between magnetic fields and moving electric charges is defined by Equations 28.7 and 28.8. Equation 28.7 shows that a magnetic field changing with time will give rise to an electric field. Equation 28.8 states that an electric current or changing electric flux will give rise to a magnetic field—this is the principle behind electromagnets.

28.2.2.2 Magnetostatic Assumption

For the magnetic drug targeting systems considered in this chapter, dynamic and electric effects are typically neglected, that is, only the magnetostatic case is considered. This is an approximation, as electric charges can exist on particles, particularly in aerosols, and dynamic effects are present due to particle motion. Nonetheless, neglecting these effects generally allows sufficient accuracy. As such, Maxwell's equations reduce to

$$\nabla \cdot \vec{B} = 0 \tag{28.9}$$

$$\nabla \times \vec{B} = 0 \tag{28.10}$$

Thus, in the magnetostatic approximation, the magnetic field can be expressed in terms of a scalar potential, φ, where

$$\nabla \varphi = \vec{B} \tag{28.11}$$

$$\nabla^2 \varphi = 0 \tag{28.12}$$

One of the most important implications of Equation 28.12 is that an externally applied magnetic field will always pull on magnetic particles within the body (for all practical purposes, i.e., excluding diamagnetic materials). The particles will always be drawn toward the external magnet, and cannot be pushed away or held at a point away from the magnet's surface by the magnetic field alone, regardless of the magnet type or configuration. This is because the Laplace equation (Equation 28.12) cannot have an internal maximum. This result is proven mathematically for magnetic targeting in two dimensions by Grief and Richardson [4].

Since particles in the body, can only be pulled toward an external magnet, there are several consequences for the design of magnetic targeting systems. To be retained at a point in the body, magnetic particles must be held against the target site. This requires that the magnetic field be oriented correctly, such that the target site is between the source of particles (e.g., blood flow) and the magnet pole. This also means that magnetic targeting is more effective for areas close to the surface of the body, since the maximum force on the particles will occur near the magnet's surface.

28.2.3 Magnetization

Electrons in an atom (bound currents) produce a magnetic moment. In most atoms and molecules, the magnetic moments of the electrons cancel one another producing very weak, if any, magnetic behavior. In magnetic materials, however, they do not. The magnetic properties of these materials depend on the alignment of the atomic and molecular magnetic moments. The alignment of these moments is referred to as magnetization. In most materials, magnetization is produced by an external magnetic field. In permanent magnets, the magnetic moments are aligned without applying any external magnetic field.

28.2.3.1 Magnetic Susceptibility

When describing magnetic materials, it is useful to consider the magnetic field in terms of the magnetic field strength and magnetization. The magnetic field strength is the component of the magnetic field occurring due to free currents, that is, conventional electrical currents through a wire, or due to an externally applied magnet. The magnetic field strength $\vec{H}$ is related to the magnetic field by the permeability of free space:

$$\vec{B} = \mu_0 \vec{H} \tag{28.13}$$

The SI unit of magnetic field strength is the Ampere per meter (A/m); the *cgs* unit is the Oersted (1 Oe = $(1000/4\pi)$ A/m). When a magnetic material is present, there is an additional magnetic field strength component $\vec{M}$, called the magnetization of the material, due to the alignment of the magnetic moments of bound currents in the material:

$$\vec{B} = \mu_0(\vec{H} + \vec{M}) \tag{28.14}$$

The magnetic field strength and magnetization can be related by the magnetic susceptibility χ of a material:

$$\vec{M} = \chi \vec{H} \tag{28.15}$$

The value of the magnetic susceptibility of a material determines its response to an externally applied field. The susceptibility of a material may vary with the magnetic field applied, although for fields typically used in magnetic targeting (up to 0.5 T) it can be considered constant.

28.2.3.2 Magnetic Materials

In paramagnetic materials, the atoms or molecules possess a magnetic moment, but are randomly aligned, resulting in no net magnetization. When a magnetic field is applied, the atoms and molecules align, and produce magnetization. The difference between paramagnetic and ferromagnetic materials is that when the magnetic field is removed, the magnetization of a paramagnetic material disappears. This does not happen instantaneously; the amount of time required is known as the relaxation time of the material.

In ferromagnetic materials, the atoms and molecules have a strong magnetic moment, and form magnetic domains within the material. In the absence of an external field, however, the domains are randomly aligned and there is no net magnetization. When an external field is applied, the domains align, and the material is magnetized. After the external field is removed, the magnetization of a ferromagnetic material persists. Permanent magnets are ferromagnetic materials. Ferromagnetic materials lose their properties and become paramagnetic above a particular temperature known as the Curie temperature, above which the alignment of domains becomes disordered due to the independent magnetization of the material becoming too weak to overcome thermal fluctuations.

The persistent magnetization of ferromagnetic materials means that care must be taken when including them in the design of devices for magnetic targeting or the study of magnetic particles, as they may produce undesired residual magnetic fields. This is particularly important when studying the small-scale behavior of magnetic particles, which may be effected by very weak fields.

If a ferromagnetic material is magnetized and demagnetized at high frequency, heat is dissipated from the hysteresis loss (see below). This feature of ferromagnetic materials is used in magnetic particle hyperthermia, in which magnetic particles are used to destroy tumors by heating them.

Superparamagnetic materials are composite materials containing ferromagnetic nanoparticles, each of which acts as a single domain. The nanoparticles are small enough to be randomly oriented by Brownian motion in the absence of a magnetic field, making the superparamagnetic material behave as a paramagnetic material with the high saturation magnetism and magnetic susceptibility of a ferromagnetic material.

Diamagnetic materials are materials in which the magnetic moments align in the opposite direction of an applied magnetic field, that is, the magnetic susceptibility is negative. This effect is very weak, and can generally be neglected in magnetic targeting. Water, and thus the human body, is weakly diamagnetic. For designing magnetic targeting systems, the body can generally be considered nonmagnetic, with the same magnetic susceptibility as air or vacuum.

28.2.3.3 Demagnetization and Saturation

Consider a magnetized object in an external magnetic field. The magnetic dipoles within the object are aligned with the magnetic field. Within the object, however, the aligned poles also produce a magnetic field in the opposite direction to the applied field. This is called the demagnetizing field. The demagnetizing field will counteract the magnetization of the object due to the external field. The extent of the reduction in magnetization is expressed in terms of the demagnetization factor β [19]:

$$\vec{M} = \frac{\chi}{1 + \beta\chi} \vec{H} \tag{28.16}$$

For a sphere, $\beta = 1/3$ [19]. For a sphere with high magnetic susceptibility, the particle magnetization approaches the limit $\vec{M} = 3\vec{H}$. For small susceptibilities, the demagnetization factor is not significant as shown in Figure 28.2.

Magnetization results from the alignment of the magnetic dipoles in a material with an external magnetic field. Once all of the dipoles are aligned, the material cannot be magnetized further, and reaches its magnetization saturation. Saturation is an important consideration for the design of magnetic targeting systems because it represents a practical upper limit beyond which increasing the magnetic field will not be useful. Saturation magnetization will depend on the particle material. Ferromagnetic materials typically have higher saturation magnetization than paramagnetic materials; superparamagnetic materials offer a

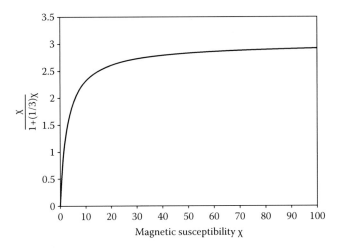

FIGURE 28.2 Magnetization factor of a sphere with increasing magnetic susceptibility.

combination of the high saturation magnetism of ferromagnetic materials and the near-instantaneous demagnetization of paramagnetic materials.

Magnetization properties of materials are typically described by magnetization curves. A magnetization curve is a plot of magnetization against the applied magnetic field; examples for a paramagnetic and ferromagnetic material are given in Figure 28.3. The slope of a magnetization curve is the magnetic susceptibility χ.

In the paramagnetic material, the magnetization increases nearly linearly with the applied field until the material becomes saturated, as shown in Figure 28.3a. Figure 28.3b shows the magnetization curve for a ferromagnetic material. For ferromagnetic materials such as permanent magnets, the initial magnetization may be nonzero. The magnetization increases with the applied magnetic field until saturation. As the applied field is reduced, the magnetization decreases, but follows a different path due to magnetic hysteresis. This occurs because additional energy is required to return the magnetic domains in the material to their original configuration as the applied

field is removed. A non-magnetized ferromagnetic material may thus retain some magnetization after a magnetic field is applied and then removed. The energy lost is converted to heat; this is exploited in magnetic hyperthermia, in which ferromagnetic particles, once at the target site, are used to kill tumors by heating them, using oscillating magnetic fields.

Magnetic nanoparticles are delivered typically collectively, for example, as a composite particle containing nanoparticles in a polymer matrix. In this case, the collective magnetization behavior of the nanoparticles must be considered in order to determine the effective susceptibility of the composite particle. As this may be difficult to do analytically, such particles may be characterized by observing their motion in a fluid of known viscosity and in a known magnetic field. Using the equation of motion for a magnetic particle in a fluid, described in the next section (Equation 28.25), the effective magnetic susceptibility of the particle may be determined empirically.

28.2.4 Methods for Applying Magnetic Fields

28.2.4.1 Permanent Magnets

Magnetic fields are most commonly applied using permanent magnets. Permanent magnets are ferromagnetic materials, that is, they retain their magnetization in the absence of a magnetic field. There are various permanent magnet materials available; the most commonly used being Nd-Fe-B magnets. Typical permanent magnets provide maximum magnetic flux densities up to 2 T with gradients and field gradients of 1–6 T/m [20].

28.2.4.2 Electromagnets

Electromagnets consist of coils of wire with metal pole pieces. Various field geometries can be produced by using different pole piece designs, and the field strength can be varied by varying the current passed through the wire. For magnetic fields strong enough for magnetic targeting, the currents required are large, and cooling of the magnets is required, increasing the size and complexity of the system considerably.

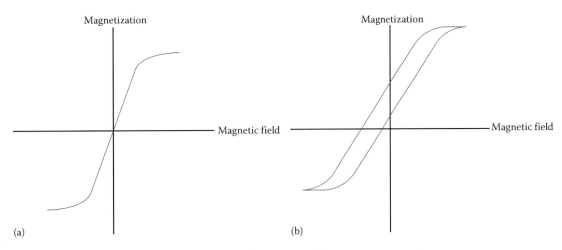

FIGURE 28.3 Magnetization curves for (a) paramagnetic and (b) magnetized ferromagnetic materials.

28.2.4.3 Superconducting Magnets

Superconducting magnets include electromagnets that use superconducting wire and bulk superconducting magnets. Superconducting electromagnets use superconducting wire to dramatically reduce the resistance of magnet coils, allowing much higher currents than copper wire with reduced heating. These systems are complex, however, as the coils must be maintained at cryogenic temperatures (<80 K). Superconducting bulk magnets are similar to permanent magnets, only much stronger, and must also be maintained at cryogenic temperatures.

28.2.4.4 Local Gradients

As shown in the next section, to exert a force on a particle requires both sufficient field strength and gradient. The magnetic field gradient decays rapidly with distance from the magnet. As such, instead of simply increasing the magnetic field strength, the idea of targeting particles by creating local gradients within the target area has been suggested. This is accomplished by implanting a piece of magnetic material such as a stainless steel stent. When a uniform magnetic field is applied, the stent is magnetized, creating local gradients around it and capturing particles. Although invasive, this method overcomes the issue of decaying magnetic gradient with distance from the magnet poles; as such, uniform fields, which can be generated over large distances by extracorporeal magnets, can be used. There have been several studies demonstrating the effectiveness of this method [6,21].

28.2.5 Magnetic Forces

28.2.5.1 Force on a Dipole

It is often useful to approximate a magnetic particle as an infinitesimally small magnetic dipole—two magnetic "charges" with an infinitesimally small distance between them. The magnetic moment $\vec{m}$ of the dipole is then given based on the volume V and magnetic susceptibility of the particle—shown below for a sphere:

$$\vec{m} = \vec{M}V = \frac{4}{3}\pi R^3 \chi \vec{H} \qquad (28.17)$$

The force $\vec{F}_M$ on a spherical magnetic particle due to a magnetic field is given by:

$$\vec{F}_M = \mu_0 \vec{m} \cdot \nabla \vec{H} = \frac{4}{3}\pi\mu_0 R^3 \chi \vec{H}\nabla \vec{H} \qquad (28.18)$$

The variables appearing in Equation 28.18 show the parameters governing the magnetic force on a particle. It is important to note that the magnetic force is proportional to the cube of particle radius. This means magnetic forces will scale up dramatically with increasing particle size, and conversely, very small particles, that is, nanoparticles, will be difficult to target with magnetic fields. This size dependence also highlights the importance of particle aggregation in targeting systems—aggregates of small particles behave as larger particles, and are more strongly affected by magnetic fields.

The magnetic field strength term, $\vec{H}$, stems from the particle magnetization. If the particle is saturated, the term $\vec{H}$ in the expression will, in fact, not increase with any increase in the applied magnetic field. The gradient term $\nabla \vec{H}$ is also important. If there is no field gradient, there will be no force exerted on a magnetic particle. This issue becomes crucial when working away from the pole piece of a magnet, which is a necessity for fields applied external to the body. Far from the pole piece, the field gradient is very small.

Descriptions of magnetic fields used for targeting in the literature vary. In many cases, only the magnetic field strength at the pole face is given; in some cases, the gradient at the pole face is given as well. This may lead to the incorrect impression that stronger or larger magnets produce greater forces on magnetic particles. In fact, when selecting or designing magnets for targeting it is more useful to consider the magnets in terms of the product of the field strength and field gradient, as it appears in Equation 28.18, over distance. Figure 28.4 shows this product for three different cylindrical Nd-Fe-B magnets along their center axes. As seen from Figure 28.4, simply increasing the size of the magnet does not necessarily translate to stronger forces on the particles at a distance. The aspect ratio and shape of the magnet will also determine the field strength and gradient. To provide a useful description of a magnetic field for targeting, the field strength, gradient, and their orientations at the same distance from the pole face at the target site must be specified.

If the demagnetization effect is included in the force equation, the expression in Equation 28.19 is recovered [19]:

$$\vec{F}_M = \frac{4}{3}\pi\mu_0 R^3 \left(\frac{\chi}{1+(1/3)\chi}\right)\vec{H}\nabla\vec{H} \qquad (28.19)$$

From Equation 28.19, it can be seen that as the particle magnetic susceptibility increases, the magnetic force approaches an upper limit, which for a sphere is given by

$$\vec{F}_M = 4\pi\mu_0 R^3 \vec{H}\nabla\vec{H} \qquad (28.20)$$

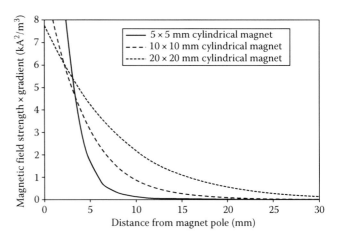

FIGURE 28.4 The product of magnetic field strength and gradient over distance along the center axes of three different Nd-Fe-B permanent magnets.

Equation 28.20 gives the maximum magnetic force that may be exerted on a particle regardless of its magnetic susceptibility. This result shows that there is limited benefit in improving the magnetic susceptibility of particles. The same results can be derived by integrating Gauss' equation over an entire particle.

28.2.5.2 Magnetic Torque on a Particle

In a uniform magnetic field, there will not be a translational force on a magnetic particle. If the particle is not spherical, however, there will be a torque tending to align the long axis of the particle with the magnetic field. For a particle approximated as a magnetic dipole, the magnitude of the magnetic torque T_M is given by [22]

$$T_M = \mu_0 V M H \sin \alpha \tag{28.21}$$

where

V is the particle volume
M is the magnitude of the dipole magnetization
H is the magnitude of the magnetic field
α is the angle between the direction of the magnetic field and the long axis of the particle

Equation 28.21 is valid for axisymmetric particles such as ellipsoids and rods.

28.2.6 Motion of a Magnetic Particle in a Viscous Fluid

In magnetic targeting, particles are usually captured from a viscous fluid, e.g., blood or inhaled air, at a target site. In order to determine the trajectories of depositing particles, an equation of motion for an individual particle can be derived by considering the magnetic, viscous, and gravitational forces on the particle. These forces give the following force balance equation:

$$m \frac{d\vec{v}_p}{dt} = \vec{F}_M + m\vec{g} + \vec{F}_D \tag{28.22}$$

where

m is the particle mass
$\vec{v}_p$ is the particle velocity
t is time
$\vec{F}_M$ is the magnetic force on the particle
$\vec{g}$ is acceleration due to gravity
$\vec{F}_D$ is the drag force on the particle

The drag and magnetic forces are given by Equations 28.2 and 28.18, resulting in the expression

$$\vec{v}_p = \left[\vec{v}_f + \tau\left(\frac{\mu_0\chi}{\rho_p}\vec{H}\nabla\vec{H} + \vec{g}\right)\right](1 - e^{-t/\tau}) + \vec{v}_{p,0}e^{-t/\tau} \tag{28.23}$$

where

$\vec{v}_{p,0}$ is the initial particle velocity
τ is a time constant defined as

$$\tau = \frac{m}{3\pi\mu d} \tag{28.24}$$

Since τ is very large for micro- and nanoparticles, inertial effects can be neglected, giving the final form of the equation of motion of a magnetic particle as

$$\vec{v}_p = \vec{v}_f + \tau\left(\frac{\mu_0\chi}{\rho_p}\vec{H}\nabla\vec{H} + \vec{g}\right) \tag{28.25}$$

Equation 28.25 can be used to determine the trajectory of a magnetic particle in a fluid and known magnetic field by integration. The above analysis is valid only for spheres; for non-spherical particles, magnetic torques must also be considered. Furthermore, such formulation does not account for potential particle–particle interaction (see next section).

The above Lagrangian analysis is based on the motion of an individual particle. The motion of particles can also be predicted with an Eulerian model, using particle concentration, c, instead. In this approach, the particle motion would be determined using the convection equation [4]:

$$\frac{\partial c}{\partial t} + \nabla \cdot (c\vec{v}_p) = 0 \tag{28.26}$$

Diffusion may play an important role in the motion of nanoparticles, particularly, in the bloodstream because red blood cells are comparable in size [4]. An advantage of using particle concentrations is that the effects of diffusion can also be included, by including the diffusive flux as follows [4]:

$$\frac{\partial c}{\partial t} + \nabla \cdot (c\vec{v}_p) = \nabla \cdot (D\nabla c) \tag{28.27}$$

The diffusivity of the particles D depends on the drag coefficient of the particles, and, in blood, the size of the particles relative to red blood cells [4]. Various approximations can be applied to determine the diffusivity [23].

28.2.7 Particle Interactions

The previous analysis neglects interactions between particles. Although the assumption of non-interacting particles may be sufficiently accurate for designing magnetic fields, an understanding of particle interaction is necessary to accurately predict the magnetic particle behavior in the body.

28.2.7.1 Magnetic Particle Interactions

A magnetized particle in a uniform magnetic field creates a local distortion in the field immediately around it, as illustrated in Figure 28.5. This produces a field gradient, which means that the particle exerts a force on other nearby magnetic particles. The magnetic field shown in Figure 28.5 is given by the expression

$$\varphi = -H_0 r\cos\theta + \frac{H_0}{8}d^3\left(\frac{\chi}{\chi+3}\right)\frac{\cos\theta}{r^3} \tag{28.28}$$

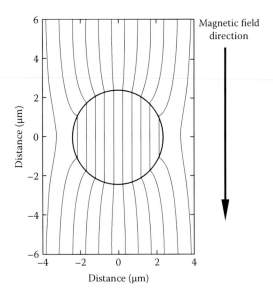

FIGURE 28.5 Magnetic field lines of a uniform magnetic field distorted by a paramagnetic sphere.

When two identical particles are aligned pole-to-pole in a uniform magnetic field, there will be an attractive force between them, given by

$$F_{parallel} = \frac{-3\pi\mu_0^2(\chi/(\chi+3))^2 d^6 H_0^2}{16r^4} \qquad (28.29)$$

If the two particles are side-by-side, however, there will be a repulsive force, given by the expression

$$F_{perpendicular} = \frac{3\pi\mu_0^2(\chi/(\chi+3))^2 d^6 H_0^2}{8r^4} \qquad (28.30)$$

This means particles will tend to attract other particles to their poles, which are aligned to the magnetic field. This attraction is responsible for magnetic particles forming chain-like aggregates when a magnetic field is applied. The chain shape is a result of the attraction at the poles and the repulsive field gradient produced around that particle perpendicular to the magnetic field direction that also pushes particles to the poles. These gradients are responsible for the mutual interactions of magnetic particles that cause aggregates to be formed before and after particle deposition at the target site. Note that the expressions in Equations 28.29 and 28.30 above are approximations; it is assumed that the magnetic moments of the particles are due solely to the uniform magnetic field, and are unaffected by the field distortions caused by the particles themselves.

The presence and formation of particle aggregates is advantageous for magnetic targeting, despite the analytical difficulties it introduces. Aggregates behave as larger particles, and experience larger magnetic forces than individual particles. Since the aggregate shape is complex, however, it is difficult to analytically predict the aggregate drag coefficient and magnetization. One

possibility is to model aggregates as larger ellipsoidal particles, for which expressions for the drag coefficient and magnetization are available. For example, a particle aggregate oriented with a fluid flow and magnetic field along its long axis could be modeled as a prolate spheroid with an aspect ratio γ. In this case, the viscous drag force on the particle is given by the expression [24]

$$\vec{F}_D = 3\pi d \left(\frac{4+\gamma}{5} \right) \eta(\vec{v}_f - \vec{v}_p) \qquad (28.31)$$

The magnetization of the aggregate could also be determined using the dipole approximation, including the appropriate demagnetization factor for a prolate spheroid [25].

$$\vec{M} = \left(\frac{\gamma(\gamma^2-1)^{-1/2}\cosh^{-1}\gamma - 1}{\gamma^2 - 1} \right) \chi\vec{H} \qquad (28.32)$$

Since the aggregate considered is nonspherical, magnetic torque would also have to be considered, if the aggregate were not aligned with the magnetic field and fluid flow. It should be noted that in practice, depending on the drug/particle formulation and delivery strategy, excessively large aggregates might cause problems with the uptake of active ingredients by the tissue at the target site.

28.2.7.2 Hydrodynamic Particle Interactions

Particles will also create local distortions in the velocity field due to viscous drag, as shown in Figure 28.6. If the particles are close together, these distortions will affect neighboring particles. The velocity field, and thus drag force, may also be distorted around particles moving close to a wall, for example, the lining of a blood vessel or airway.

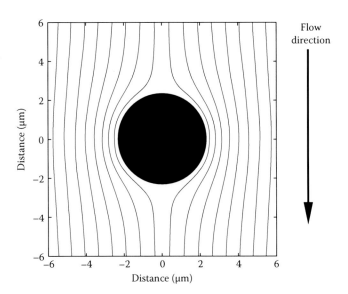

FIGURE 28.6 Streamlines of the velocity field around a spherical particle in Stokes flow from top to bottom.

TABLE 28.1 Stokes Drag Expression Correction Factors for Particle Interactions

Particle Configuration	Expression [26]	
Two particles moving perpendicular to their line of centers	$\lambda = 1 - \dfrac{3}{8}\dfrac{d}{l_p} + \dfrac{9}{64}\dfrac{d^2}{l_p^2} - \dfrac{59}{512}\dfrac{d^3}{l_p^3}$	(28.34)
Two particles moving along their line of centers	$\lambda = 1 - \dfrac{3}{4}\dfrac{d}{l_p} + \dfrac{9}{16}\dfrac{d^2}{l_p^2} - \dfrac{19}{64}\dfrac{d^3}{l_p^3}$	(28.35)
Particle moving parallel to a wall	$\lambda = \left(1 - \dfrac{9}{32}\dfrac{d}{l_p} + \dfrac{1}{64}\dfrac{d^3}{l_p^3}\right)^{-1}$	(28.36)
Particle moving perpendicular to a wall	$\lambda = \left(1 - \dfrac{9}{16}\dfrac{d}{l_p} + \dfrac{1}{16}\dfrac{d^3}{l_p^3}\right)^{-1}$	(28.37)

For fully immersed particles, the effect of these distortions can be accounted for by adding a parameter, λ, to Equation 28.2 [24]:

$$\vec{F}_D = \lambda\left[3\pi\eta d(\vec{v}_f - \vec{v}_p)\right] \qquad (28.33)$$

The value of λ depends on the particle interaction. For two similarly sized spherical particles, λ will depend on the inter-particle separation l_p. The effect of wall interactions will depend on the distance l_w of the particle from the wall. Approximate expressions for λ are given in Table 28.1 for several types of interaction between equal-sized spherical particles [26]. The expressions in Table 28.1 have been truncated to third order accuracy for conciseness; for the full expressions and their derivations, see [26].

28.3 Magnetic Drug Targeting

28.3.1 Applications of Magnetic Targeting

Although a variety of drugs can be coupled to magnetic particles for delivery, chemotherapy stands to benefit most significantly and has been the main focus of magnetic drug targeting research. This is because in most cases that can be effectively treated, the target sites for chemotherapy are localized. In these cases, the goal is to deliver as much drug as possible to the target site and nowhere else in the body, in order to minimize systemic toxicity of the drug. This is particularly true for lung cancer that provides the motivation for studying magnetic targeting of inhaled aerosols.

The harmful side effects associated with chemotherapy arise from the fact that the drugs will attack healthy as well as cancerous tissue. In non-targeted delivery, the drugs circulate throughout the body, and large doses are required to build up an effective concentration in a tumor. By reducing the amount of drug required to deliver enough of the drug to the tumor, the effectiveness of the drug could be increased and the side effects reduced. Reducing the amount of drug could also reduce the cost of chemotherapy. Chemotherapy drugs are among the most expensive manufactured substances, for example, paclitaxel which costs $3000 per gram [27]. Even a slight reduction in the amount required for treatment could lead to significant cost savings.

Various chemotherapy agents, including mitoxantrone [28], doxorubicin [2], epirubicin [2], paclitaxel [2], have been successfully coupled to magnetic carriers. Magnetic targeting of chemotherapy drugs in the circulatory system has been shown to appreciably increase drug concentration in tumors in animal studies using swine, rabbits, and rats [1]. In a study by Alexiou et al., the required dosage of mitoxantrone required to achieve tumor remission in rabbits was reduced by 80% by magnetic targeting [28]. Although the results of these animal studies are encouraging, there has been little conclusive proof of magnetic targeting being useful in human studies to date [29]. One of the reasons for this is the increase in scale from an animal model to that of a human. Scaling of magnetic fields and forces must be considered when looking at the results of animal studies (Table 28.2).

Although chemotherapy has been the main focus of magnetic targeting research, other therapies, such as anti-inflammatories and antibiotics [1], have been suggested as possible applications of magnetic targeting. As in the case of chemotherapy, these treatments would also benefit by reducing systemic exposure to the drugs used. Targeting of radioactive magnetic particles has also been suggested for cancer treatment [30]. Magnetic targeting may also be useful in enhancing diffusion-mediated processes such as gene transfection. This has been demonstrated in *in vitro* studies [12].

Magnetic particles are also used as contrast agents for MRI. Magnetic particles can be concentrated in a region of the body with magnetic field, allowing for enhanced imaging, which can in turn allow delivery of drugs to the target site more precisely, for example, by direct injection.

TABLE 28.2 Chemotherapy Drugs Coupled to Magnetic Particles

Drug	Form	Size	Reference
Doxorubicin	Albumin-coated magnetite	1–2 μm	Hafeli and Chstellain [29]
Doxorubicin	Irregular carbon-coated iron	0.5–5 μm	Hafeli and Chstellain [29]
Dactinomycin	Polycyanoacrylate particles	220 nm	Hafeli and Chstellain [29]
Oxantrazole	Chitosan particles	530 nm	Hafeli and Chstellain [29]
Methotrexate	Solid lipid	450–570 nm	Hafeli and Chstellain [29]
Carminomycine	Ferrocarbon	100 nm	Hafeli and Chstellain [29]
Epirubicin	Ferrofluid	100 nm	Lübbe et al. [2]

28.3.2 Magnetic Particles

Magnetic carriers can take different forms, depending on the drug to be delivered and the delivery method. The forms include ferrofluids, inorganic magnetic particles coated with a drug, magnetoliposomes, core-shell particles, and composite polymer particles.

Magnetic particles generally consist of magnetic material bonded to a drug component. The drug may be directly adsorbed to the magnetic material, bonded to a polymer coating on the magnetic material, or in solution with the liquid component of a ferrofluid. Table 28.1 gives a number of examples of drugs that have been bonded to magnetic particles. Ferrofluids are stable colloidal suspensions of magnetic particles in a liquid containing a surfactant preventing particle aggregation. Ferrofluids generally behave as magnetic liquids, and are the basis for the study of magnetohydrodynamics. Figure 28.7 illustrates these different particle configurations. Although drugs can be bonded directly to magnetic nanoparticles, the magnetic moments of nanoparticles are typically insufficient for manipulation with magnetic fields in the body. As a result, the nanoparticles must be suspended in a polymer matrix of a microparticle or a liquid (ferrofluid). Magnetic nanoparticles can also be coated with a phospholipid bilayer to form magnetoliposomes, which can be tailored for biocompatibility or functionality [31].

Particles have also been developed which can be triggered to release drugs by magnetic fields. Polymer particles containing drugs can be manufactured with magnetite nanoparticles embedded in them that will cause the particles to rupture and release their payload when a magnetic field is applied [30]. Similar mechanisms can also be used to control the diffusion of drugs from particles [30]. Particles that are similarly responsive to oscillating fields have also been developed. This is an advantage because the particles could be guided by static fields and activated by oscillating fields.

28.3.3 Implementations of Magnetic Targeting

28.3.3.1 Deposition in the Circulatory System

In magnetic drug targeting via the circulatory system, magnetic particles are injected into a vein or artery as near as possible to the target site to minimize circulation time. A magnetic field is applied to pull the particles from the blood flow to the target site and hold them there. The magnetic fields are applied using external magnets near the target sites. The magnetic force on the particles due to the magnetic field must overcome the viscous drag force of the blood flow so that the particles are not carried past or away from the target site.

28.3.3.2 Deposition in the Airways

For magnetic targeting in the airways, there is greater distinction between the deposition and retention aspects. For deposition at

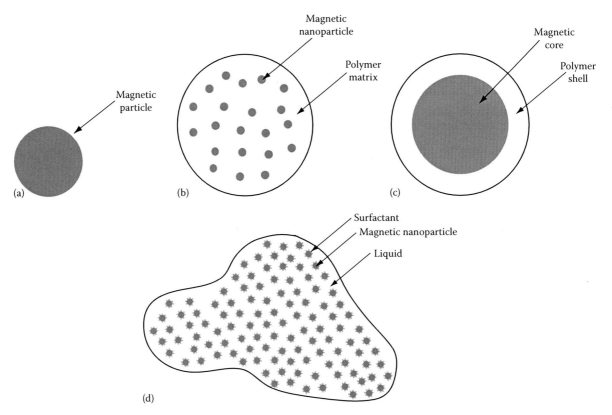

FIGURE 28.7 Various drug delivery particles: (a) a single magnetic particle onto which drugs can be directly adsorbed (b) a composite particle containing magnetic nanoparticles (c) a core-shell type particle with magnetic material encased in a polymer capsule (d) a ferrofluid consisting of magnetic particles suspended in liquid with a surfactant for stability.

the target site, particles would be introduced into the lung via the inhaled aerosol. A magnetic field causes the particles to deposit at the target site from the aerosol flow. This magnetically induced deposition process is somewhat similar to that in the bloodstream, as the ideal trajectories of the particles would be determined by the magnetic field and viscous drag. However, the airflow in the lungs may be affected by turbulence due to the higher Reynolds number, and particle interactions (magnetic, hydrodynamic, and electrostatic) would also be different.

After deposition, particles in the airway must be retained at the target site. The airways are lined with a layer of mucus that is constantly propelled toward the oropharynx. This mucus layer acts as the primary defense of the airways against the inhaled particles. For effective magnetic targeting, the drag force exerted on the particles by the mucus layer, as it carries the particles up the airway must be overcome by the magnetic field. The study of particle retention is complicated by the fact that the particles form aggregates immediately after deposition [9] that are difficult to model theoretically.

28.3.3.3 Gradient Fields

The mechanisms of magnetically targeting drug deposition in the airways and circulatory systems are similar: the magnetic field must produce forces on the particles sufficient to overcome the drag force exerted by the fluid in which the particles are entrained. In order for a field to produce a force on a particle, the field must have a gradient in addition to being sufficiently strong. A uniform magnetic field will not produce any force on a non-magnetized particle.

For drug targeting, magnetic fields are applied by magnets placed outside of the body. Various types of magnets may be used; typical field strengths are summarized in Table 28.3. The most common method is to use permanent magnets. Permanent magnets are a simple, cost-effective means of producing high-strength and high-gradient magnetic fields. However, careful design is required to produce the desired magnetic field geometry. In addition, proper orientation of the magnetic field is necessary in order to direct the particles.

Electromagnets can also be used to produce suitable gradient fields for magnetic targeting, in some cases even exceeding the strength of permanent magnetic materials. Conventional electromagnets typically require liquid cooling to produce such fields. Superconducting magnets can produce fields that far exceed those achievable by permanent magnets [32]. However,

TABLE 28.3 Typical Maximum Field Strengths and Gradients for Various Magnets

Magnet Type	Field Strength (T)	Field Gradient (T/m)
Permanent Nd-Fe-B	2	6
Bulk superconducting magnet	4.5	200
Electromagnet	0.6	120
MRI	3	0.1

such magnets are much more complex and costly. An example of a high-strength and high-gradient electromagnet system is a magnetic resonance imaging (MRI) machine. Another advantage of electromagnet systems is that the magnetic field strength and orientation can be dynamically controlled. Such control would be useful for the dynamic targeting and guidance of magnetic particles, as investigated by Sylvain et al. [33].

28.3.3.4 Magnetic Stents

Magnetic stents are pieces of magnetizable material, for example, stainless steel mesh, implanted in the body to aid in trapping particles from the blood. This approach takes advantage of the fact that magnetic materials will distort a magnetic field. Even in a uniform magnetic field, the magnetized material will have a gradient field around it that will attract the magnetic particles as they pass. This is an advantage, as strong uniform magnetic fields are easier to produce with existing magnet technologies. The disadvantage of magnetic stents is that this approach is invasive, as the material must be implanted. Stents have been shown to be effective in *in vitro* and numerical studies of particle deposition in blood vessels [21].

28.3.3.5 Embolization

Although not a drug delivery technique, magnetic particle embolization has been proposed as a method for treating tumors by blocking the blood supply to the tumor [2,7]. Emboli blocking the blood vessels supplying a tumor can be created by using magnetic fields to form large aggregates in the blood vessels, taking advantage of the tendency of magnetized particles to mutually attract one another. This attraction occurs because, like wires in a magnetic stent, the magnetic material in the particles will distort the magnetic field, producing local gradients. A gradient field would still be required for embolization, however, in order to hold the emboli against the pressure of the blood flow. The feasibility of magnetic particle embolization has been demonstrated using *in vitro* [7] and animal models [2]. The fate of emboli is unclear, however, and must be carefully studied to determine the extent of exposure of the entire body after the magnetic field is removed. A possible solution to the emboli being spread through the body is the use of biodegradable particles with a high rate of degradation, for example, polysaccharide systems.

28.3.3.6 Uniform Magnetic Fields

Uniform magnetic fields, although they do not produce translational forces on non-magnetized particles, may still be useful for targeting, as they do produce magnetic torque and cause particles to aggregate. Uniform fields are considerably easier to produce over large distances than gradient fields. This may be accomplished using several magnet configurations.

Helmholz coils are matched pairs of core-less electromagnet coils with spacing equal to the coil radius, as shown in Figure 28.8. Applying a current in the same direction through both coils produces a uniform magnetic field in the central region between the coils directed along the coil axis. This configuration is useful because the coils are far from the region of interest of

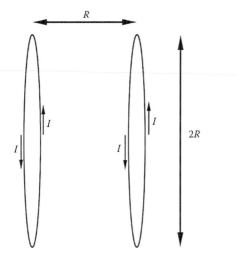

FIGURE 28.8 A Helmholz coil pair. Two electromagnet coils of radius *R* are spaced at a distance *R* apart. A current *I* is passed through both coils in the same direction, producing a region with a uniform magnetic field in between the coils.

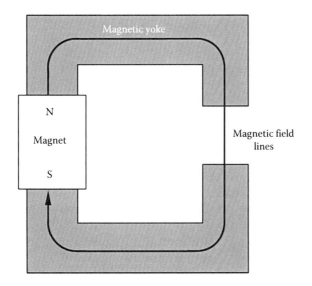

FIGURE 28.9 Schematic of a magnetic circuit. The magnetic field lines from the magnet (permanent or electromagnet) pass through a magnetic yoke across an air gap. The field in the air gap may be nearly uniform, or may be modified by changing the geometry of the poles at either the top or bottom of the air gap.

the magnetic field, allowing ample space for placing specimens and for the cooling system for the coils.

Uniform magnetic fields may also be produced by permanent magnets in magnetic circuits, as illustrated in Figure 28.9. In a magnetic circuit, the magnetic field lines pass through a yoke made of a soft magnetic material (i.e., a material that does not permanently retain magnetization) with a gap. Within the gap, the magnetic field is nearly uniform. The geometry of the magnetic circuit poles on either side of the air gap may also be modified to produce gradient fields [34]. Magnetic circuits may be made with either permanent or electromagnets.

28.4 Magnetic Aerosol Drug Targeting

Chemotherapy for lung cancer by conventional aerosol drug delivery has also shown promise in animal studies [15], however, its effectiveness is limited due to potential damage to healthy airway tissue. This application would clearly benefit from application of magnetic targeting, and in this section, recent approaches will be discussed.

28.4.1 Physiological Environment

The airways can be divided into three regions: extrathoracic (mouth–throat), tracheobronchial (airways), and alveolar (deep lung) [35]. The extrathoracic airways consist of the mouth and throat. The tracheobronchial region consists of the airways that form an inverted tree-like structure of bifurcating tubes, branching off from the trachea, as shown in Figure 28.10. The deep lung consists of the smaller branches from the tracheobronchial "tree" that are covered with alveoli where gas exchange occurs. The geometry of the airways varies between individuals. This poses a challenge for magnetic targeting, as the magnetic field must be aligned with the flow in order to direct inhaled particles to the target site. For the first branches of the airways, this variability is not as significant as it is deeper in the lung. In some applications, only the first branches of the airways may be of interest—for example, most lung cancers begin and develop in the lining cells of the central bronchi, that is, the first branches of the bronchial tree [36,37].

A typical airflow rate for breathing at rest with a mouthpiece is 18 L per minute [35], although this value can vary between individuals. In the trachea and bronchi, turbulence may affect the deposition efficiency. As the airways become smaller, the flow becomes more laminar. The Reynolds number in the airways decreases logarithmically from ~1000 in the trachea to ~10 in the tenth generation of branches [35]. These Reynolds numbers fall

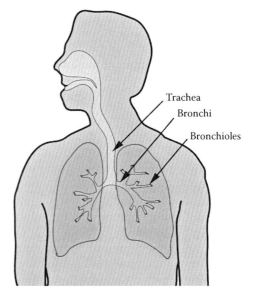

FIGURE 28.10 The large branches of the human airways.

into the intermediate range between laminar and turbulent flow discussed previously, which means that the flows in the airways will be somewhat laminar, although the inertial effects may still be significant. In the lung, these inertial effects occur in the form of a radial "swirling" component of the velocity in the airways due to the branching structure of the bronchial tree [35].

In the airways, there are three types of particle deposition in non-targeted delivery: impaction, sedimentation, and diffusion [35]. Impaction occurs when particles deviate from the flow streamlines due to their inertia, and is primarily responsible for particle deposition at bifurcations in the airways. Sedimentation occurs due to gravitational settling of particles in the airways. Diffusion is the motion of particles due to random molecular motion.

Various deposition models have been developed to predict aerosol deposition in the lung, based on lung geometry, fluid mechanics, and the mechanisms of particle deposition. Deposition is correlated to particle size, although due to individual variability in lung geometry and air flows, there are no general rules governing particle deposition [35]. The particle size referred to here is the equivalent aerodynamic diameter, discussed previously. Large particles will deposit in the upper airways (>10 μm diameter). Small particles (<3 μm diameter) may not deposit at all, and may simply be inhaled and exhaled. For tracheobronchial deposition, 6 μm is ideal for conventional aerosols. A comprehensive overview of inhaled aerosol drug delivery can be found in [35].

The airway surface is protected from inhaled particles like dust, bacteria, and other contaminants by a lining layer of mucus that is propelled by ciliated epithelium beneath and toward the mouth to be removed. Particles depositing in the airways are cleared by this mechanism that must be counteracted by magnetic forces for effective targeted inhaled aerosol drug delivery.

The airway surface lining consists of a layer of thick, viscoelastic mucus that flows on a watery periciliary layer, as shown in Figure 28.11. The cilia on the epithelium beat within the periciliary layer, and their tips engage the mucus layer, propelling it up the airways. The cilia frequencies are synchronized by the hydrodynamics of the cilia beating next to one another. The elasticity of the mucus allows the tips of the cilia to "grab" the mucus layer and push it. The viscoelastic properties of the airway mucus are crucial for the functioning of the mucociliary clearance mechanism. If the mucus is too thick or too thin, the clearance mechanism will not function. The viscoelasticity of the mucus is determined by the bonds between molecules, and can be disrupted by various chemical agents.

The mucus layer in normal tracheobronchial airways is 5–10 μm thick and the periciliary layer is 5–7 μm thick [38]. This thickness of the mucus layer can vary dramatically in response to various pulmonary conditions. In healthy adults, the tracheal mucus velocity is approximately 7–10 mm/min [39]. The mucus velocity is lower in smaller airways, and changes depending upon the health of the airways. The continuity of the mucus layer also varies throughout the airways, as does the number of cilia [40]. For drug delivery purposes in the first generations of the airways, however, it is reasonable to assume particle deposition upon a continuous layer of mucus. The particles upon landing on the mucus layer most likely will be partially immersed, depending on the particle wetting characteristics, and the viscosity and thickness of the mucus layer. For partially immersed particles, the appropriate drag coefficient must be used, that is, the value of $f(\theta)$ in Equation 28.3 must be known. This requires understanding of the wetting properties of the particles used, that is, the particle contact angle with the mucus layer.

As described previously, the drag force on particles deposited at the airway surface depends on the particle contact angle. Schürch et al. have studied the wetting behavior of particles at the air–liquid interfaces of the airways extensively [41]. Their findings have shown that small particles will tend to be highly immersed in the mucus layer, whereas large particles will protrude. This is due to the presence of a layer of surfactant on the mucus that reduces its surface tension. These findings suggest that nanoparticles may be completely engulfed in the mucus layer, but larger particle aggregates will straddle the interface. Since the majority of particles will deposit as aggregates, the latter case is more appropriate to consider.

The surfactant layer on the mucus surface will cause an additional force resisting the motion of a particle along the air–liquid interface of the airways. Figure 28.12 shows a solid particle moving along an interface with surfactant molecules spread at the interface. As the particle moves, it dilutes the surfactant concentration in its wake [42]. This results in a surface tension gradient across the particle, with higher surface tension in the particle wake than in front of the particle. The surface tension gradient produces an additional resisting force opposing the particle motion (this is sometimes called the surface pressure acting on a particle). The extent of this effect depends upon the particle

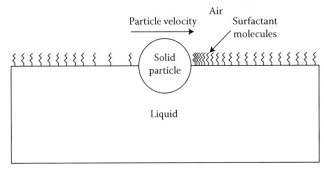

FIGURE 28.12 Surfactant distribution around a particle moving at an air–liquid interface. The imbalance of surfactant in the immediate vicinity of the particle gives rise to a force resisting particle motion, in addition to viscous drag.

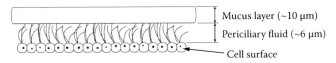

FIGURE 28.11 Schematic of the airway surface lining.

speed, and the diffusivity of the surfactant, as diffusion acts to restore the uniform distribution of the surfactant after it is disrupted by the particle [42].

28.4.2 Magnetic Aerosol Deposition

In magnetically targeted deposition, a magnetic field is used to cause particles to deposit from an aerosol at a target site. As this is a new area, the number of studies on magnetic aerosol particle deposition in the literature is limited, particularly relating to magnetic drug targeting. Three principal works in the field are discussed in this section, each illustrating different approaches to magnetic drug deposition from the aerosol.

Ally et al. [8] were first to study the deposition of magnetic particles in the airways for drug delivery, using *in vitro* and numerical models. Iron particles, 1–3 μm in diameter were deposited from the aerosol in a model airway consisting of a rectangular channel with similar flow rate and dimensions to the human trachea. The Reynolds number of the aerosol flow in the channel in this study was ~600. The Reynolds number for the particles, however, was on the order of 0.05, hence the Stokes expression (Equation 28.2) was used to compute the drag on the particles. Magnetic fields were applied by placing Nd-Fe-B permanent magnets in the channel and by placing the entire apparatus in a magnetic circuit that produced a uniform field perpendicular to the aerosol flow. By placing the permanent magnets within the channel, high field gradients in the region of the aerosol flow were ensured, which is required for a magnetic force to be exerted (Equation 28.18). While such close proximity was shown to produce high particle collection efficiencies, it is not achievable in a human-scale situation with an external magnet, since the magnetic gradient falls off rapidly with distance as illustrated previously in Figure 28.4.

The magnetic fields and gradients produced a translational force on the particles, pulling them to the bottom of the channel. Collection efficiencies of 45% were achieved with a single magnet in the channel; 65% with two magnets and 80% with three magnets placed in series. The diminishing returns of adding magnets to the channel show the importance of the location of the magnet with respect to the aerosol flow. After the initial magnet, the aerosol flow near the subsequent magnets is depleted. Thus, the subsequent magnets are further from the higher concentrations of particles. This difference highlights the importance of distance and proper placement of magnets for targeting.

Ally et al. also developed a Lagrangian model for predicting particle trajectories [8]. Equation 28.25 was integrated to determine the particle trajectories using measured and simulated data for the velocity and magnetic fields in their apparatus. Particle interactions were not considered. For the case of one magnet, the model predicted the aerosol deposition within 8% error. For the two- and three-magnet cases, the deposition was underpredicted by up to 20%. Closer examination of the results shows that in all configurations, the deposition due to the first magnet encountered by the flow was predicted within 8%, with the error increasing for subsequent magnets. This error is likely due to the aggregation occurring between particles that do not deposit, and highlights the importance of inter-particle interactions.

In the case of the uniform field used by Ally et al. [8], only minimal deposition (<5% of the aerosol) was observed. Although this amount was small, it is more than would be expected due to gravitational sedimentation alone, in spite of the fact that the uniform field would not have produced significant translational force on the particles, as the field gradient was nearly zero (Equation 28.8). The deposition observed in this case was likely due to particle aggregation. The particles would distort the uniform field, producing local gradients (Figure 28.5), resulting in inter-particle attractive forces that caused the particles to aggregate. The aggregates would be more likely to deposit due to gravity due to their size. This result also highlights the importance of inter-particle interactions as well as the field gradient.

Martin and Finlay [13,14] tested another strategy for depositing particles that takes advantage of the fact that, although uniform fields do not produce translational forces, they produce torques on nonspherical particles. In these experiments, high aspect ratio magnetic particles were used. When a uniform magnetic field was applied to an aerosol of these particles, the particles aligned with the magnetic field due to the magnetic torque exerted by the field (Equation 28.21). In the small airways of the lung, if the particles are aligned perpendicular to the airway walls by a magnetic field, they will be more likely to deposit due to the particle tips being caught on the airway walls.

Martin and Finlay tested this method of deposition using cromoglicic acid particles coupled to magnetite nanoparticles [13]. The particles were cylindrical and their sizes were 0.47 μm in diameter and 3 μm in length. Deposition in the small airways was studied using an *in vitro* model based on the terminal bronchioles in the lung [14]. The model airways were 0.5 mm in diameter, with parent generations 8 mm long and daughter branches 2 mm long. The model bifurcating airways were assembled with the final apparatus containing 126 parent airways, with a total flow rate of 22 L/min for inhalation.

The magnetic field in Martin and Finlay's work [14] was applied by placing the model airway between the opposed poles of two flat permanent magnets aligned perpendicular to the airflow. This arrangement produced a uniform magnetic field across the model airway. The uniform field would cause particles to align perpendicular to the airflow, producing deposition as their tips caught on the airway walls.

The results in [14] showed that application of the magnetic field produced an average collection efficiency of 3.3%, compared to 1.9% without the magnetic field. These results show that the magnetic field was effective in increasing deposition, although much less than a gradient field. Although gradient fields enable higher deposition, uniform fields can be easily produced over much larger distances, as discussed previously. This approach could benefit from further investigation into how particles align with an airflow and other optimizations. The proposed approach may also be improved by examining whether it can take advantage of particle aggregation.

Dames et al. [11] studied particle deposition in the lungs of mice. The aerosol was produced with a nebulizer and introduced by artificial ventilation. The scale of the flow must be considered here—the mouse airways are much smaller than human airways, thus would have a lower Reynolds number, that is, more laminar airflow. The magnetic field was applied using an electromagnet. The pole piece of the electromagnet was designed to focus the magnetic flux in a small region, producing very high field gradients in order to draw particles toward one lung instead of the other.

The particles used by Dames et al. [11] were superparamagnetic iron oxide nanoparticles (SPIONs) mixed with nebulized water droplets, called "nanomagnetosols." The suspended nanoparticles made the water droplets magnetically responsive, more so than individual particles. This method also allowed drugs to be coupled to either the iron oxide particles or dissolved in the liquid phase, allowing for greater versatility in delivery. Both methods were demonstrated to be effective.

Dames et al. [11] found that deposition in one lung over the other could be increased eight times by application of the magnetic field. In these cases, however, the magnet pole was placed very close to the targeted lung, which was exposed by thoracotomy. When the magnet was applied to an intact mouse, that is, further away from the lung, there was a 2.5 times difference between the two lungs. This clearly illustrates the importance of distance in magnetic targeting. Scaling of this approach to the human body would require careful design of a magnet to produce a sufficient field gradient, as well as consideration of the more turbulent fluid mechanics of the larger human airways.

28.4.3 Overcoming Particle Clearance in the Airways

For effective drug targeting in the airways, retention of particles at a target is as important as deposition. In order for magnetic particles to be retained in the airways, the magnetic force on the particles must overcome the forces from the mucus clearance mechanism. The drag forces were described previously. For particles to be retained there are two possible mechanisms. First, particles may be pulled in direct opposition to the mucus flow. Second, the particles could be pulled into the mucus layer. In the second case, particle clearance may be overcome either by the particles penetrating the mucus layer, or by the particles simply pushing down on the cilia beneath, disrupting their function.

Ally et al. have also studied particle retention at the airway surface using *in vitro* and *ex vivo* animal tissue [9,10] models. In the *in vitro* studies, the airway surface liquid was simulated by a flow of liquid. Particles were deposited on the surface of the liquids, and magnetic fields applied to retain them. The liquids used were water, glycerol, and locust bean gum. These liquids allowed for testing over a range of viscosities. The locust bean gum has been used previously in the literature as a mucus stimulant [43].

The particles used in these experiments were 1–3 μm diameter iron particles. The magnetic fields were applied using various Nd-Fe-B magnets that produced radially symmetric fields. The principal component of the magnetic field was directed into the liquid interface. Particles were deposited at the liquid surface, and then retained with another applied magnetic field.

The higher the liquid viscosity, the stronger the magnetic field required to retain particles in the *in vitro* experiments. Liquid viscosity was found to play a much more significant role than flow rate, which appeared to have little or no effect upon retention. This is in contrast to the relationship suggested by Equation 28.3, which shows the viscosity and liquid speed having similar effects on the drag force. The reason for this discrepancy is that viscosity plays an additional role in particle aggregation. The same magnetic field was used in all of the experiments to deposit particles before a different field was applied for retention. After deposition, before the retention field was applied, the deposited particles were observed to form aggregates. The size of the aggregates formed depended linearly upon the viscosity of the liquid [9]. Since aggregates experience larger magnetic forces than individual particles, when the retention field was applied, the magnetic forces would have been greater in the low viscosity liquids in addition to the drag forces being smaller. Hence, viscosity has a dual effect on particle retention, making it a key parameter. Since the particles were pinned to the air–liquid interface, aggregate formation was driven by the component of the magnetic field parallel to the interface. This component of the field also tended to concentrate the particles.

There is also a component of the magnetic field perpendicular to the interface. The results of Ally et al. [9] showed that this component of the field tends to pull the particles into the interface. This component of the magnetic field could be useful for retention in the airways to hold particles against the airway epithelium. If the particles could be pulled through the air–liquid interface in the airways, the particles could be held in this manner. This approach depends on proper orientation of the magnetic field, since an externally applied magnetic field can only pull particles toward the magnet pole, as discussed previously.

Similar experiments performed by Ally et al. simulated the airway surface using an *ex vivo* animal model. The tissue on a bullfrog (*Rana catesbeiana*) palate is similar to the tissue in the airways, and continues to function after harvesting, if handled and stored properly. This model has been used extensively in the literature to model mucus clearance [44]. Particles were deposited on the palate tissue and observed in a similar manner to the liquid interface experiments. Even with extremely strong magnetic fields, only very large iron oxide particles were retained—because the magnetic force scales with the cube of the particle size, as shown previously in Equation 28.18. This highlights the challenge of retaining particles in the airways in the presence of a functioning mucociliary clearance mechanism with the intrinsic purpose of removing particles. Possible strategies for improving particle retention may include designing magnetic fields with stronger uniform field components to enhance particle aggregation, that is, producing effectively larger particles, or reducing the viscosity of the mucus using a mucolytic. The mucolytic approach, that is, altering the physiological environment, was also tested by Ally et al. using bullfrog palates [9,10]. These tests showed the same correlation between aggregate size and liquid viscosity as seen in the previous *in vitro* study [9]. With mucolytics applied, particles could be retained using reasonably practical magnetic fields. These results show that magnetic particle retention in the airways is feasible, although in addition to using a sufficiently powerful magnet, other considerations such as particle aggregation, field alignment, and the physiological environment are important.

28.5 Challenges for Magnetic Drug Targeting

There are several challenges that must be met for magnetically targeted drug delivery, whether through the circulatory system or the airways. These are

- Distance
- Magnetic field geometry
- Particle size

28.5.1 Distance

The greatest challenge in magnetic targeting is the decrease in magnetic field and gradient with distance. Applications within the human body may require the magnet to be 10 cm or more from the target site. This means the field and gradient will typically be very small. Even if extremely large magnets are used, sharp gradients at such distances are difficult to produce. A "brute force" solution to this challenge seems unlikely—simply increasing the size and power of magnets will not work. Instead, better understanding of the behavior and transport of magnetic particles in the body, that is, particle aggregation and the physiological environment, is required to make optimal use of fields that can be practically produced. Alternatives to high gradient fields, such as uniform fields or locally distorted fields should also be considered for targeting applications.

Great care is required in comparing animal and human studies due to the issue of scale as well. Although in medical science, animal studies are used to validate protocols prior to clinical trials, the scaling of magnetic fields may limit the usefulness of this validation from an engineering perspective. Consideration of the magnetic field and force scaling is necessary to ensure that the effective treatment achieved in animal tests can be replicated on a human scale.

28.5.2 Magnetic Field Geometry

Another challenge is the fact that magnetic fields will always pull particles out of the body—that is, toward the magnet. This means that for particles to be retained, they must be held against a tissue surface. This has significant implications. For example, consider a target site is located in a vessel close to the body's surface, but on the opposite side of the vessel. The magnetic field would have to be applied from the opposite side of the body—far from the apparently conveniently located target site. Possible

solutions to this issue include the use of magnetic stents to create local field gradients within the body, and improving targeting over long distances as discussed above.

28.5.3 Particle Size

The trend toward using nanoparticles in medicine is also a challenge. Magnetic forces scale with particle volume—decreasing particle diameter by half reduces the force on the particle by 88%. As such, smaller particles are only weakly influenced by magnetic fields. For this reason, nanoparticles must typically be delivered collectively, for example, in the form of degradable composite particles or in a ferrofluid. Another possible solution is to take advantage of particle aggregation; this would require further study. Optimization of particle shape to improve aggregation and produce larger magnetic forces may also prove useful.

References

1. Pankhurst, Q.A., Connolly, J., Jones, S.K., and Dobson, J. 2003. Applications of magnetic nanoparticles in biomedicine. *J. Phys. D: Appl. Phys.* 36: R167–R181.
2. Lübbe, A.S., Alexiou, C., and Bergemann, C. 2001. Clinical applications of magnetic drug targeting. *J. Surg. Res.* 95: 200–206.
3. Forbes, Z.G., Yellen, B.B., Barbee, K.A., and Friedman, G. 2003. An approach to targeted drug delivery based on uniform magnetic fields. *IEEE Trans. Magn.* 39: 3372–3377.
4. Grief, A.D. and Richardson, G. 2005. Mathematical modelling of magnetically targeted drug delivery. *J. Magn. Magn. Mater.* 293: 455–463.
5. Goodwin, S., Peterson, C., Hoh, C., and Bittner, C. 1999. Targeting and retention of magnetically targeted carriers (MTCs) enhancing intra-arterial chemotherapy. *J. Magn. Magn. Mater.* 194: 132–139.
6. Chen, H., Ebner, A.D., Rosengart, A.J., Kaminski, M.D., and Ritter, J.A. 2004. Analysis of magnetic drug carrier particle capture by a magnetizable intravascular stent: 1. Parametric study with single wire correlation. *J. Magn. Magn. Mater.* 284: 181–194.
7. Liu, J., Flores, G.A., and Sheng, R. 2001. In-vitro investigation of blood embolization in cancer treatment using magnetorheological fluids. *J. Magn. Magn. Mater.* 225: 209–217.
8. Ally, J., Martin, B., Khamesee, M.B., Roa, W., and Amirfazli, A. 2005. Magnetic targeting of aerosol particles for cancer therapy. *J. Magn. Magn. Mater.* 293: 442–449.
9. Ally, J., Amirfazli, A., and Roa, W. 2006. Factors affecting magnetic retention of particles in the upper airways: An in vitro and ex vivo study. *J. Aerosol Med.* 19: 491–509.
10. Ally, J., Roa, W., and Amirfazli, A. 2008. Use of mucolytics to enhance magnetic particle retention at a model airway surface. *J. Magn. Magn. Mater.* 320: 1834–1843.
11. Dames, P., Gleich, B., Flemmer, A. et al. 2008. Targeted delivery of magnetic aerosol droplets to the lung. *Nature Nanotechnol.* 2: 495–499.

12. Plank, C. 2008. Nanomagnetosols: Magnetism opens up new perspectives for targeted aerosol delivery to the lung. *Trends Biotech*. 26: 59–63.

13. Martin, A.R. and Finlay, W.H. 2008. Alignment of magnetite-loaded high aspect ratio aerosol drug particles with magnetic fields. *Aerosol Sci. Tech*. 42: 295–298.

14. Martin, A.R. and Finlay, W.H. 2008. Enhanced deposition of high aspect ratio aerosols in small airway bifurcations using magnetic field alignment. *Aerosol Sci*. 39: 679–690.

15. Dahl, A.R., Grossi, I.M., Houchens, D.P. et al. 2000. Inhaled isotretinoin (13-cis retinoic acid) is an effective lung cancer chemopreventive agent in A/J mice at low doses: A pilot study. *Clin. Cancer Res*. 6: 3015–3024.

16. Rao, R.D., Markovic, S.N., and Anderson, P.M. 2003. Aerosol therapy for malignancy involving the lungs. *Curr. Cancer Drug Tar*. 3: 239–250.

17. Otterson, G.A., Villalona-Calero, M.A., Sharma, S. et al. 2007. Phase I study of inhaled Doxorubicin for patients with metastatic tumors to the lungs. *Clin. Cancer Res*. 13: 1246–1252.

18. Radoev, B., Nedjalkov, M., and Djakovich, V. 1992. Brownian motion at liquid-gas interfaces. 1. diffusion coefficients of macroparticles at pure interfaces. *Langmuir* 8: 2962–2965.

19. Henjes, K. 1994. The traction force in magnetic separators. *Meas. Sci. Technol*. 5: 1105–1108.

20. Hatch, G.P. and Stelter, R.E. 2001. Magnetic design considerations for devices and particles used for biological high-gradient magnetic separation (HGMS) systems. *J. Magn. Magn. Mater*. 225: 262–276.

21. Aviles, M.O., Ebner, A.D., and Ritter, J.A. 2008. Implant assisted-magnetic drug targeting: Comparison of in vitro experiments with theory. *J. Magn. Magn. Mater*. 320: 2704–2713.

22. Shine, A.D. and Armstrong, R.C. 1987. The rotation of a suspended axisymmetric ellipsoid in a magnetic field. *Rheol. Acta* 26: 152–161.

23. Wang, L.N-H. and Keller, K.H. 1985. Augmented transport of extracellular solutes in concentrated erythrocyte suspensions in Couette flow. *J. Colloid Interface Sci*. 103: 210–225.

24. Panton, R.L. 1996. *Incompressible Flow*, 2nd edn. New York: John Wiley & Sons, Inc.

25. Evetts, J.E. 1992. *Concise Encyclopedia of Magnetic and Superconducting Materials*. Oxford, NY: Pergamon Press.

26. Happel, J. and Brenner, H. 1965. *Low Reynolds Number Hydrodynamics*. London, U.K.: Prentice-Hall, Inc.

27. Suffness, M. 1995. *Taxol: Science and Applications*. London, U.K.: CRC Press.

28. Alexiou, C., Schmid, R.J., Jurgons, R. et al. 2006. Targeting cancer cells: Magnetic nanoparticles as drug carriers. *Eur. Biophys. J*. 35: 446–450.

29. Hafeli, U.O. and Chstellain, M. 2006. Magnetic nanoparticles as drug carriers. In *Nanoparticulates as Drug Carriers*, ed. V.P. Torchilin. London, U.K.: Imperial College Press, pp. 397–418.

30. Hafeli, U.O. 2003. Magnetically modulated therapeutic systems. *Int. J. Pharm*. 277: 19–24.

31. Marcela, G. and Kannan, K.M. 2005. Synthesis of magneto-liposomes with monodisperse iron oxide nanocrystal cores for hyperthermia. *J. Magn. Magn. Mater*. 293: 265–270.

32. Takeda, S., Mishima, F., Fujimoto, S., Izumi, Y., and Nishijima, S. 2007. Development of magnetically targeted drug delivery system using superconducting magnet. *J. Magn. Magn. Mater*. 311: 367–371.

33. Sylvain, M. 21–24 May 2008. Automatic transport of magnetic particles in the blood vessels using a clinical MRI system. *Paper Presented to the Seventh International Conference on the Scientific and Clinical Applications of Magnetic Carriers*, Vancouver, Canada.

34. Khamesee, M.B., Kato, N., Nomura, Y., and Nakamura, T. 2002. Design and control of a microrobotic system using magnetic levitation. *IEEE-ASME Trans. Mechatronics* 7: 1–14.

35. Finlay, W.H. 2001. *The Mechanics of Inhaled Pharmaceutical Aerosols: An Introduction*. London, U.K.: Academic Press.

36. Spencer, H. 1985. *Pathology of the Lung*, 4th edn. Oxford, NY: Pergamon Press.

37. Schlesinger, R.B. and Lippman, M. 1978. Selective particle deposition and bronchogenic carcinoma. *Environ. Res*. 15: 424–431.

38. Sleigh, M.A., Blake, J.R., and Liron, N. 1988. The propulsion of mucus by cilia. *Am. Rev. Respir. Dis*. 137: 726–741.

39. Morgan, L., Pearson, M., de Iongh, R. et al. 2004. Scintigraphic measurement of tracheal mucus velocity in vivo. *Eur. Respir. J*. 23: 518–522.

40. Martin, D.E. and Youtsey, J.W. 1988. *Respiratory Anatomy & Physiology*. St. Louis, MO: The C.V. Mosby Company.

41. Schürch, S., Geiser, M., Lee, M.M., and Gehr, P. 1999. Particles at the airway interfaces of the lung. *Colloids Surf. B*. 15: 339–353.

42. Dmitrov, K., Avramov, M., and Radoev, B. 1994. Brownian motion at liquid-gas interfaces. 3. Effect of insoluble surfactants. *Langmuir* 10: 1596–1599.

43. King, M., Brock, G., and Lundell, C. 1985. Clearance of mucus by simulated cough. *J. Appl. Physiol*. 58: 1776–1782.

44. King, M. 1998. Experimental models for studying mucociliary clearance. *Eur. Respir. J*. 11: 222–228.

29

Biodegradable Nanoparticles for Drug Delivery

Jason Park
Yale University

Tarek M. Fahmy
Yale University

29.1 Introduction: Nanomedicine and Biodegradable Nanoparticles

Advances in the development of nanoscale technologies are beginning to have broader impacts on the understanding, treatment, and prevention of disease through applications that the National Institutes of Health collectively refer to as "nanomedicine" (Moghimi et al. 2005). This nascent field encompasses a vast array of disparate technologies with potential biomedical applications. Some examples of the variety of technologies being studied include functionalized carbon nanotubes, nanofabricated silicon-based sensing devices and DNA nanomachines. In all of these endeavors, researchers are searching for new ways to use nanotechnology to improve health. At the forefront of nanomedicine is a subdivision focused on the rational delivery and targeting of therapeutic or diagnostic compounds via nanoscale particulates, that is, "nanoparticles." The subcellular and nanoscale size ranges of these systems allow for designed escape through fenestrations in epithelial linings, deep penetration into tissues, and efficient uptake by target cells and tissues for the improved delivery of therapeutic or diagnostic compounds (Moghimi et al. 2001, Soppimath et al. 2001, Brannon-Peppas and Blanchette 2004).

This chapter focuses on a discussion of solid biodegradable nanoparticles. These are of particular interest because of their multifaceted advantages. In addition to their nanoscale size, nanoparticles based on biodegradable materials can be utilized to solubilize a concentrated payload of drug, improve drug stability and bioavailability, and extend drug or gene effect through sustained delivery. Moreover, the use of biodegradable, biocompatible materials reduces the risk of unwanted toxicities and adverse effects. Biodegradable nanoparticles have proven to be versatile platforms for the delivery of a large variety of compounds and have been well studied for the delivery of small molecules (Yoo and Park 2001, Mu and Feng 2003, Farokhzad et al. 2006), proteins (Couvreur and Puisieux 1993, Vila et al. 2002), and nucleotides (Cohen et al. 2000, Mao et al. 2001, Hood et al. 2002) as well as small inhibitory RNA (siRNA) (Khan et al. 2004, Schiffelers et al. 2004, Yuan et al. 2006). Notable advances in the design and engineering of these systems have included the modification of the nanoparticle surface to improve stability and circulation throughout the body (Moghimi et al. 2001), biospecific targeting against cellular ligands or extracellular matrix components (Takeuchi et al. 2001, Kim et al. 2005, Farokhzad et al. 2006, Sinha et al. 2006), and the incorporation of diagnostic imaging agents (Fahmy et al. 2007).

29.1.1 Biodegradable Materials for Drug Delivery

The development of biodegradable materials for drug delivery began with the realization that a wide variety of compounds—hydrophilic or hydrophobic, small molecule or protein or gene—could be conjugated to or encapsulated within a matrix of biodegradable polymers. These agents could then be slowly released over time with the degradation of the polymer and/or diffusion out of the matrix (Saltzman 2001). A fundamental advantage of biodegradable nanoparticles over nonbiodegradable formats is the coupling of the controlled release of encapsulated agents with particle degradation, an attractive feature for systemic administration in the body since removal or clearance of the carrier does not complicate treatment. In addition, since encapsulated compounds are made bioavailable only upon release, relatively large doses can be entrapped in the polymeric matrix and sustained delivery achieved for periods lasting up to months (Saltzman 2001). Some of the first commercially approved drug delivery systems utilizing biodegradable materials were implantable synthetic polymeric microparticle formulations designed to elongate the therapeutic effect of peptides such as GnRH (Lupron Depot®) or human growth hormone (Nutropin Depot®) (Okada and Toguchi 1995). These systems sought to better fit the delivery of encapsulated agents to their therapeutic windows by creating a sustained release depot of drug, thereby decreasing the need for frequent and/or bolus dosing of drug (Figure 29.1). Nanoscale biodegradable delivery systems allow more sophisticated means to modulate drug effect through increased control over nanoparticle

distribution and uptake. By exerting control over the release of drug as well as the biodistribution of the particle, it may be possible to greatly improve drug efficacy and decrease toxicity.

Biodegradable nanoparticles have largely been engineered from materials already known to be biocompatible and often commonly used in medicine for other purposes. To enable the efficient encapsulation and delivery of drug, these materials are generally polymeric in nature. The use of polymers for the fabrication of these systems facilitates easy tuning of the drug release profiles through simple changes in the polymer composition or molecular weight. Additionally, these polymers can be designed to degrade through physiological processes (i.e., hydrolysis or enzymatic degradation), leaving biocompatible monomeric by-products. Therefore, one immediate benefit is that of toxicity and safety: at a biochemical level, the materials comprising many biodegradable nanoparticles are relatively harmless. For example, the polyester polylactic acid, a commonly used biomaterial, degrades *in vivo* through hydrolysis, resulting in the release of lactic acid monomers that can be safely eliminated through natural mechanisms (Anderson and Shive 1997). Through a proper understanding of material composition/degradation as well as a detailed investigation of the final biologic and immunogenic effects of nanoparticles, these systems can be engineered to leave little impact beyond the ones intended.

One additional, less obvious benefit is that manufacturing methods for biodegradable nanoparticles have proven to be flexible. Not surprisingly, there exist innumerable, often laboratory-specific methods for manufacturing nanoparticles depending on the preferred biomaterial, compound(s) to be encapsulated and

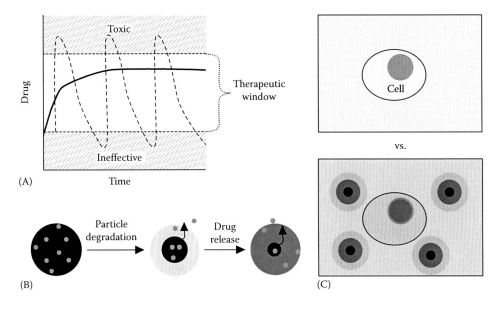

FIGURE 29.1 Benefits of controlled, localized release. The objective of drug delivery is to improve the efficacy and safety of drugs. (A) Most traditional delivery methods (e.g., oral or intravenous) result in tidal drug concentrations with high peak concentrations post-administration followed by clearance and need for frequent re-administration. Sustained release strategies (solid line) seek to maintain therapeutic concentrations from less frequent doses. (B) This profile can be achieved by the release of drug from biodegradable particles. As particles degrade over time, encapsulated drug is released and made bioavailable. (C) Due to limits imposed by systemic toxicities, drug concentration at the target site may be lower than desired. Localization of drug delivery vehicles allows for more concentrated delivery to the target, resulting in improved efficacy and lower toxicities.

the desired properties. Variation of parameters in the choice of the starting material, manufacturing method, and drug encapsulation technique offer nearly infinite possibilities for altering the relevant characteristics of nanoparticles for drug delivery: size, morphology, charge, surface functionality, and drug release profile.

29.2 Particle Fabrication Materials

An ever-growing number of biomaterials, both synthetic and natural, have been used to formulate biodegradable nanoparticles. Some of the most commonly used polymers are discussed here. These can be divided into synthetic polymers (e.g., polylactides/glycolides, polyanhydrides, polyacrylates, and polylactones) and natural polymers (e.g., alginate, chitosan, cellulose, collagen, and albumin). The synthetic polymers are generally advantageous in terms of longer periods of sustained release and better control over degradation/release rates via well-established formulation strategies. However, these methods can be compromised by harsher formulation processes—a consideration for the delivery of labile proteins and other sensitive compounds. Natural polymers provide better biocompatibility and generally milder formulation processes, but are limited in their delivery capacity and time frame, tunability of release, and targeting applications.

29.2.1 Synthetic Polymers

29.2.1.1 Polyesters (PLA/PGA/PLGA/PCL)

The most commonly used and extensively investigated materials for biodegradable nanoparticles have been the hydrophobic polymer polylactic acid (PLA), the hydrophilic polyglycolic acid (PGA), and their copolymers poly(lactide-*co*-glycolide) (PLGA) (Saltzman 2001). As polyesters, these polymers will eventually hydrolyze under physiologic conditions to their biologically compatible monomers (lactic acid and/or glycolic acid), which are then eliminated via natural mechanisms, such as metabolisis through the citric acid cycle (Anderson and Shive 1997, Saltzman 2001). The copolymer PLGA is of special interest because the degradation rate of the polymer and corresponding drug release rate can be modulated in predictable fashion from days to months depending on the ratio of PLA to PGA and the molecular weight of the polymer. In the case of PLGA and many other hydrolyzable polymers, higher ratios of hydrophobic (e.g., lactic acid) segments inhibit penetration of water into the polymer matrix and result in slower degradation while increased ratios of hydrophilic segments increase the rate of degradation and drug release (Anderson and Shive 1997). These polymers have been studied extensively because of their excellent safety and toxicity profiles; they have a long history of use in humans for resorbable sutures, bone implants and screws, contraceptive implants, and as graft material/scaffolds for tissue engineering research (Saltzman 2001). Particles and scaffolds made out of PLA or PLGA were among the first materials to be used clinically for drug delivery purposes, and the delivery of small molecules,

peptides, and genes have been well demonstrated from these materials. Closely related to the polylactides/glycolides is polycaprolactone (PCL), which is also degraded through hydrolysis of ester linkages and has likewise been FDA-approved for use as suture material. Unsurprisingly, given its chemical structure (Figure 29.2) the degradation rates of PCL nanoparticles tend to be even slower than those of pure PLA materials (Uhrich et al. 1999, Saltzman 2001).

29.2.1.2 Polyanhydrides

Polyanhydrides are polymers characterized by the anhydride bonds connecting monomer units that can be aliphatic, unsaturated, or aromatic depending on the desired properties. Polyanhydrides have a long history of use in drug delivery devices as they biodegrade through hydrolysis to release nontoxic dicarboxylic acids that are readily metabolized (Uhrich et al. 1999, Saltzman 2001). The degradation rate of polyanhydrides is primarily controlled through modulation of the relative hydrophobicity of the monomer units, due to the high water instability of the anhydride bond. The Gliadel® wafer, used for the sustained delivery of carmustine for treatment of glioma, is formulated with a polyanhydride consisting of a 20:80 ratio of a hydrophobic carboxyphenoxypropane monomer and a hydrophilic sebacic acid monomer (Saltzman 2001).

29.2.1.3 Polyacrylates

Polyacrylates are another class of biodegradable polymers that have been used in medical applications ranging from tissue adhesives to hydrogels and contact lenses. Indeed, poly(2-hydroxyethyl methacrylate) (pHEMA) is among the most biocompatible of all synthetic polymers, with minimal local tissue reaction after implantation. However, the high water stability of these polymers and minimal *in vivo* degradation compromise their use in drug delivery applications (Uhrich et al. 1999, Saltzman 2001). Polycyanoacrylates are an example of polyacrylates that have been modified to degrade rapidly enough for drug delivery purposes and can be readily formulated into nanoparticles.

29.2.2 Natural Polymers

29.2.2.1 Polysaccharides (Cellulose/Chitin/ Chitosan/Dextran/Alginate)

Polysaccharides are among the most abundant and diverse polymers in nature and have been widely investigated for use in biodegradable nanoparticles. Some of the most commonly used polymers for nanoparticles have been cellulose, chitin, chitosan, dextran, and alginate (Saltzman 2001). Cellulose is a long straight polymer consisting entirely of glucose residues connected by β-(1,4) linkages. A similar polymer, chitin, can be obtained by substitution of the glucose residues with *N*-acetylglucosamine. Another closely related linear polymer is chitosan, which is composed of randomly distributed β-(1,4)-linked D-glucosamine and *N*-acetyl-D-glucosamine. Biodegradable nanoparticles have been made through the cross-linking of these linear polymers,

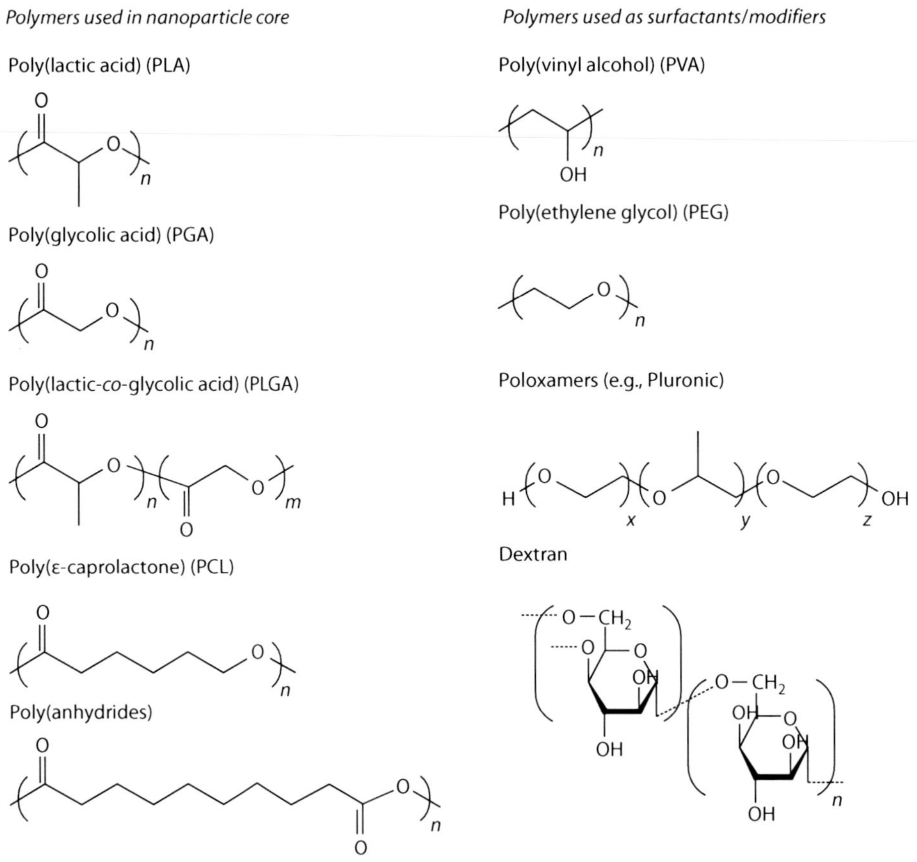

Polymers used in nanoparticle core

Poly(lactic acid) (PLA)

Poly(glycolic acid) (PGA)

Poly(lactic-*co*-glycolic acid) (PLGA)

Poly(ε-caprolactone) (PCL)

Poly(anhydrides)

Polymers used as surfactants/modifiers

Poly(vinyl alcohol) (PVA)

Poly(ethylene glycol) (PEG)

Poloxamers (e.g., Pluronic)

Dextran

FIGURE 29.2 Structures of some commonly used synthetic polymers in biodegradable nanoparticles.

as well as by co-incorporation or even modification of other (e.g., synthetic) polymers through covalent bonds.

Alginate is another linear polysaccharide, consisting of D-mannuronic and L-guluronic acid. Interestingly, alginate forms a gel in the presence of bivalent cations (e.g., Ca^{2+}). This property has been widely utilized to create hydrogels for tissue engineering applications and microparticles for drug delivery. Preparation of nanoscale alginate particles has proven to be more difficult, but there remains significant interest due to the highly hydrophilic and biocompatible nature of alginate. Another natural polysaccharide of interest is dextran. Dextran is composed of glucose residues that are exclusively in α-(1,6) linkages, resulting in a helical conformation. While dextran can been directly formulated into nanoparticles, it has more often been used for the modification of nanoparticle properties, such as engineering of hydrophilic surfaces for enhanced circulation of the particles in the blood pool (Moghimi et al. 2001), or as an excipient or modifier of encapsulated drugs (Mitra et al. 2001).

29.2.2.2 Naturally Occurring Proteins (Albumin/Collagen)

Albumin and collagen are, respectively, the most abundant serum and extracellular matrix proteins found in animals and thus have been studied frequently as biomaterials (Saltzman

2001). Formulation methods have varied, but success has been observed with the adjustment of existing techniques used to create protein powders (e.g., spray-drying and lyophilization). Recombinant human albumin nanoparticles are already FDA-approved as delivery vehicles in the form of Abraxane®, a paclitaxel-loaded albumin nanoparticle formulation. The albumin nanoparticle functions as an excipient/solubilizing factor for the highly hydrophobic chemotherapeutic drug paclitaxel, obviating side effects sometimes observed with traditional excipients such as Cremophor.

29.2.3 Solubilization and Stabilization of Particle Formulations

The manufacture of nanoparticles invariably requires the use of solvents to solubilize the polymers and surfactants to stabilize the nanoparticle structure. The potential for accidental incorporation of these reagents calls for careful consideration of the reagents used and of the procedures for removing excess/unwanted reactants and products. Some of the most commonly used materials are listed here.

Most of the synthetic polymers previously discussed are water-insoluble and are formulated into nanoparticles using emulsion methods, predicating the use of organic solvents to dissolve the polymer and surfactants to stabilize the aqueous emulsion. The residual solvent left in the polymer matrix after nanoparticle manufacture can cause problems ranging from degradation of encapsulated drug and destabilization or deformation of the nanoparticle, to direct toxic biological effects. Solvents commonly used for nanoparticle manufacture include alcohols (e.g., methanol, ethanol, and trifluoroethanol), chlorocarbons (e.g., methylene chloride and chloroform), a variety of polar aprotic solvents (e.g., dimethyl formamide, dimethyl sulfoxide, and dioxane), and other organics such as ethyl acetate. The natural polymers, on the other hand, are generally water-soluble and do not usually require the use of organic solvents. However, their high water solubility makes formulation into nanoscale structures capable of encapsulating drugs more difficult. Cross-linking of the polymer has been a popular method but requires the use of potentially toxic cross-linking reagents. For example, divalent cations can be used to cross-link alginate to form particles, but some of the cations that form the most stable structures (Zn^{2+}, Co^{2+}, and Ni^{2+}) can cause significant toxicities in the concentrations needed for sufficient stabilization of nanoparticles (Pandey and Khuller 2005).

Regardless of the starting material, emulsifiers and stabilizers are often needed to stabilize the initial nanoparticle structure and/or prevent subsequent aggregation and fusion. With many synthetic polymers, amphiphilic molecules are used to stabilize the interface between the hydrophobic organic/polymer and hydrophilic aqueous phases, stabilizing an emulsion containing nano-sized droplets that form into nanoparticles. Common examples include polyvinyl alcohol (PVA) and poloaxmers, a triblock copolymer consisting of a hydrophobic central chain of polyoxypropylene flanked by two hydrophilic polyethylene chains (Figure 29.2). Polyvinyl alcohol (PVA) has been the most commonly used emulsifier for PLGA and similar synthetics, largely because it has performed well and populations of small and relatively uniform particles can be consistently recovered. Other biocompatible and commonly used surfactants/emulsifiers have included polysorbates, dextran, and PVA, or poloaxmers of varying molecular lengths. As with solvents and cross-linking agents, an issue to consider is that although the emulsifier is "washed" off during wash or dialysis steps after nanoparticle formulation, a fraction often remains associated with the polymer surface, potentially affecting the physical and chemical properties of the nanoparticle. Sahoo et al. noted that residual PVA associated with PLGA nanoparticles increased the hydrophilic properties of the nanoparticle surface and directly contributed to decreased cellular uptake (Sahoo et al. 2002).

29.3 Particle Formulation Methods

29.3.1 Polymerization of Monomers

Biodegradable nanoparticles are prepared through one of two main methods: (1) polymerization of monomers with *in situ* nanoparticle formation or (2) dispersion of preformed polymers

in proper solvents and subsequent stabilization after some form of mechanical input. The development of high-grade preformed polymers, growing commercial availability of specialized copolymers, and time and material quality considerations have generally favored the latter method. The primary benefit of *in situ* methods is that all of the materials used in nanoparticle production can be selected to be biocompatible and encapsulated compounds spared from significant exposure to harsh formulation processes such as organic solvents or high-energy sources. Couvreur et al. were able to produce polycyanoacrylate nanoparticles, without any irradiation or initiator, through the mechanically-induced polymerization of methyl or ethyl cyanoacrylate using a vortexer and an acidified aqueous phase (Couvreur et al. 1979).The drug was dissolved in the polymerization medium and encapsulated instantaneously upon nanoparticle formation. Similarly, Hu et al. demonstrated the production of chitosan-polyacrylate nanoparticles via polymerization of acrylic acid in a chitosan solution (Hu et al. 2002).

Nonetheless, a number of polymerization methods require the use of initiators or conditions that may be inherently toxic or damaging to the desired payload (e.g., use of divalent cations, free radical polymerization). Furthermore, encapsulation efficiency can be a significant issue with these methods. The entrapment of drug throughout the polymer matrix requires the initial dispersion of the drug throughout the entire polymerization medium, or loading after nanoparticle formation, and a significant portion of drug can be lost. This chapter focuses on the use of preformed natural and synthetic polymers.

29.3.2 Cross-Linking of Natural Polymers

Naturally occurring polymers can be suitable for drug delivery purposes but by their very nature, which lends them biocompatibility and high water solubility, they tend to be more difficult to formulate into nanoscale carriers. Strategies for the use of naturally occurring polymers include co-incorporation or modification of other polymers via copolymerization, or covalent bonding. The simplest method is to use natural polymers in lieu of, or in conjunction with synthetic polymers in well-established synthetic formulation strategies. Kawashima et al. co-incorporated chitosan with PLGA to modify the surface of PLGA nanoparticles, making them "mucoadhesive" for potential oral delivery applications (Takeuchi et al. 2001). Likewise, many natural polymers can be conjugate to synthetic polymers through covalent bonds, extending beneficial properties to synthetic polymeric nanoparticles. This approach can even extend to direct modification of the drug. Mitra et al. conjugated doxorubicin molecules to dextran in order to improve the encapsulation of this moderately hydrophilic chemotherapeutic into chitosan nanoparticles (Mitra et al. 2001).

Natural polymers can also be used alone in dispersion methods such as coacervation. These tend to be particularly useful if the drug or protein of interest is capable of interaction with the polymer, via electrostatic or other means. In these methods, the polymer and drug are mixed together in an emulsion or in solution and energy added via vortexing or sonication to form stable nanoscale structures that can be recovered via filtration or centrifugation. Mao et al. demonstrated the formation of chitosan–DNA nanoparticles by vortexing together a sodium acetate-buffered solution containing chitosan and a sodium sulfate solution containing DNA. The preservation of DNA activity was demonstrated via efficient transfection using the nanoparticles. These nanoparticles were even modifiable via PEGylation or attachment of peptides or proteins for improved stability, nanoparticle uptake, and transfection (Mao et al. 2001).

Another method is to cross-link the polymer via chemical or physical means, resulting in drug entrapment in a nanoscale hydrophilic gel. For example, although alginate generally forms into larger scale gels, fine adjustment of the relative proportions of Ca^{2+} to alginate in an aqueous solution allows cation-induced rearrangement of alginate molecules into microdomains or microgels that can be recovered as nanoparticles via high-speed centrifugation. The duration required for the formation of enough cross-links to stabilize the nanoparticles is referred to as gelling/curing time, and it varies depending on the drug, cross-linking agent, and polymer. With alginate particles, the gel strength decreases in the presence of metals in the following order: $PB^{2+}> Cu^{2+}> Ba^{2+}> Sr^{2+}> Cd^{2+}> Ca^{2+}> An^{2+}> Co^{2+}> Ni^{2+}$. However, only Sr^{2+}, Ba^{2+}, and Ca^{2+} are considered to be biologically safe cross-linkers for use *in vivo* (Pandey and Khuller 2005). Couvreur et al. demonstrated that alginate nanoparticles could be formed and stabilized using a solution containing calcium and poly-L-lysine. Doxorubicin dissolved in the alginate medium was entrapped and could be delivered from the particles (Rajaonarivony et al. 1993). A significant benefit to this approach is the formation of highly hydrophilic nanoparticles. This may represent a method to reduce opsonization and clearance without the need for additional surface engineering (e.g., PEGylation).

29.3.3 Dispersion of Pre-Formed Polymers into Particles

The most commonly studied polymers for biodegradable nanoparticles have been the ester-linked, hydrolyzable PLA, PGA, PLGA, and PCL (Soppimath et al. 2001). The techniques of choice for forming nanoparticles from these materials have been variations on a method known as solvent evaporation (see Table 29.1). Depending on whether the drug to be encapsulated is soluble in the organic or aqueous phase, these methods are identified as "single" or "double" emulsion methods. They may also be referred to as oil (O) in water (W), or water in oil in water methods (i.e., O/W single emulsion versus W/O/W double emulsion). In these methods, the pre-formed polymer is dissolved in an organic solvent such as dichloromethane. The drug is then either dissolved directly in the solvent (for the O/W method) or dissolved in an aqueous solution (W/O/W method) and dispersed into the organic solution via vortexing or sonication with a tip probe. This mixture is then emulsified through sonication or high-pressure/speed homogenization in an aqueous solution containing a surfactant/emulsifying agent such as PVA. The mechanical energy creates an emulsion which is stabilized by the emulsifying agent. Upon formation of a stable emulsion, the organic solvent is evaporated off through continuous stirring of the solution—this process may sometimes be accelerated through increased temperature/decreased pressure, though it is usually carried out at room temperature. Evaporation of the solvent results in condensation of drug and polymer, and the formation of "hardened" individual nanoparticles. A simplified schematic of this technique is described in Figure 29.3. Hardened nanoparticles can then be collected via filtration or centrifugation, and the excess surfactant and unencapsulated drug washed away with sterile water or media. Due to their susceptibility to hydrolysis, nanoparticles are typically lyophilized and frozen for long-term storage.

TABLE 29.1 Common Nanoparticle Formulation Methods

Technique	Polymers	Advantages	Disadvantages
Solvent evaporation/ (O/W, W/O/W) emulsion	Synthetic, natural	Versatility and ease-of-use. Delivery of wide range of compounds. Highly tunable.	Low encapsulation efficiency and "burst release" of hydrophilic compounds. Exposure of drug to organics and mechanical energy.
Spontaneous emulsion/ nanoprecipitation	Synthetic	Rapid particle formation. High encapsulation efficiency of hydrophobic compounds. Minimal exposure to mechanical energy.	Very low encapsulation efficiency of hydrophilic compounds.
In situ polymerization	Synthetic	Minimal exposure of drug to harsh formulation processes.	Loss/dilution rate of compound. Potential toxicities of reactants/initiators.
Polymer cross-linking	Natural	Formation of highly hydrophilic, biocompatible particles.	Very rapid release of encapsulated compounds. Minimal ability to tune release or modify surface for targeting.
Spray drying	Synthetic, protein	Consistent and highly scalable process.	Difficult to modify process to encapsulate hydrophilic compounds. Difficult to achieve nanoscale particle diameters.
Polymer-drug conjugation	Synthetic, natural	Very high encapsulation rates. Highly tunable release characteristics.	Covalent modification of drug may alter safety/ efficacy profile.

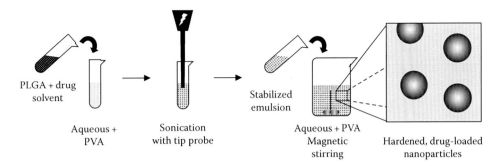

PLGA + drug
solvent

Aqueous +
PVA

Sonication
with tip probe

Stabilized
emulsion

Aqueous + PVA
Magnetic
stirring

Hardened, drug-loaded
nanoparticles

FIGURE 29.3 Solvent emulsion method for nanoparticle manufacture. Relatively simple manufacturing schemes are among the major advantages of polymeric biodegradable nanoparticles. A generic single emulsion (oil in water, O/W) method is shown here. Drug and polymer (PLGA) are dissolved in an organic solvent. This mixture is then added to an aqueous phase consisting of water and a surfactant/emulsifier (PVA). A stabilized emulsion is created by the addition of energy via sonication. Solvent evaporation and nanoparticle hardening is then facilitated by dispersion of the emulsion in a larger aqueous phase with magnetic stirring. Excess drug and surfactants can be removed and nanoparticle collected via centrifugation or filtration.

A modified version of the solvent evaporation method is the "spontaneous emulsion" method. In this method, a water-soluble solvent such as acetone or methanol is added to the polymer-containing organic phase. Upon addition to the aqueous phase, spontaneous diffusion of the water-soluble solvent out of the organic phase creates an interfacial turbulence that leads to nanoparticle formation. By increasing the concentration of the water-soluble solvent, the particle size can be decreased (Soppimath et al. 2001). Niwa et al. demonstrated that the rapid diffusion of acetone from dichloromethane or chloroform to an aqueous phase containing PVA could result in PLGA nanoparticle diameters as small as 200 nm without any additional source of energy (e.g., homogenization/sonication), and that chemotherapeutics with hydrophilic (5-fluorouracil) or hydrophobic (indomethacin) properties could be encapsulated and delivered (Niwa et al. 1993). A similar method is the "nanoprecipitation" method, which is based on the interfacial deposition of polymer following displacement of a solvent (Govender et al. 1999). Here, the polymer and drug are dissolved in a semi-polar solvent such as acetonitrile. The organic phase is then slowly added drop wise to an aqueous phase, which is magnetically stirred to aid in solvent evaporation, and nanoparticles can be recovered via filtration or centrifugation. Govender et al. demonstrated encapsulation and release of a hydrophilic drug (procaine hydrochloride) from 200 nm diameter PLGA particles using this technique (Govender et al. 1999). In general, however, very low encapsulation efficiencies of hydrophilic compounds or proteins are attained when using the spontaneous emulsion or nanoprecipitation methods, making these methods better suited for the encapsulation of hydrophobic compounds (Niwa et al. 1993, Govender et al. 1999).

A recognized limitation with PLGA nanoparticles remains the encapsulation of hydrophilic or water-soluble drugs, due to the rapid migration of drug into the aqueous phase. The result is often low drug loading and poor encapsulation efficiencies. This has been observed as an especially significant drawback with the spontaneous emulsion and nanoprecipitation methods. A separate yet related challenge is the "burst" release of drug. It is common to observe a biphasic release profile, wherein a significant portion of the total encapsulated dose is rapidly released upon rehydration of nanoparticles, followed by a period of more sustained, diffusion-mediated release (Figure 29.4B). The burst effect is thought to be due to the relatively large surface area of nanoparticles and the nonhomogeneous (i.e., surface-preferential) distribution of drug throughout the polymer matrix (Saltzman 2001). Thus, surface-associated drug or drug encapsulated near the surface is rapidly released with little control, while the more homogeneously distributed drug is released by diffusion and degradation of the polymer, resulting in sustained release. This theory is seen in practice with the burst effect frequently observed when delivering hydrophilic drugs and proteins, and higher encapsulation efficiencies and more linear release profiles observed with hydrophobic drugs of comparable molecular weight. Strategies to electrochemically stabilize the drug in the nanoparticle core, or decrease its ionization or solubility in water can improve encapsulation and decrease the extent of the initial burst (Govender et al. 1999). For example, Janes et al. were able to double the encapsulation of doxorubicin in chitosan nanoparticles by complexing the positively charged drug with polyanionic dextran sulfate (Janes et al. 2001).

A more unique strategy to encapsulate drug is to covalently conjugate the drug to a biodegradable polymer and formulate the conjugate into polymeric nanoparticles or micelles. Yoo et al. were able to encapsulate and deliver doxorubicin from PEGylated PLGA nanoparticles by conjugating the hydroxyl group of a PLGA–PEG molecule to the primary amino group of a doxorubicin molecule (Yoo and Park 2001). The clear downside is that chemical modification of the drug is not always possible and may also result in diminished or altered activity. However, a significant advantage over encapsulation of doxorubicin by emulsion methods was observed in that, by better linking release of drug to polymer degradation, the "burst effect" was diminished and a more linear sustained release profile was demonstrated. In addition, compared to a loading of 0.3% (w/w) and encapsulation efficiency of 6.7% for unconjugated doxorubicin,

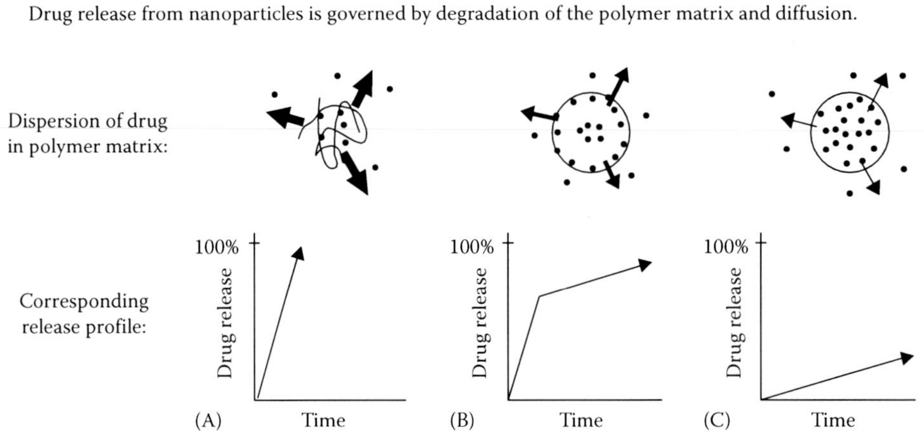

Drug release from nanoparticles is governed by degradation of the polymer matrix and diffusion.

FIGURE 29.4 Different types of drug release from biodegradable nanoparticles. Drug and polymer characteristics influence the release profile. (A) Nanoparticles made from a loosely cross-linked hydrophilic polymer, such as alginate, allow for deep penetration of water and rapid diffusion of drug. (B) Nanoparticles made from slowly degrading synthetic polymers tend to have longer time courses of drug release. Some compounds may be poorly distributed throughout the core of the particle, resulting in biphasic release patterns with a period of rapid release of surface-associated drug, followed by degradation-coupled sustained release of entrapped drug. (C) Hydrophobic compounds tend to be better entrapped in the nanoparticle core and are released in more linear fashion with degradation.

the conjugated form was encapsulated at 10-fold greater levels and with 96.6% efficiency (Yoo and Park 2001). Thus, the direct linking of compounds to polymers may be a viable and necessary strategy for compounds that are difficult to encapsulate with more traditional methods.

Modulation of many of the parameters or process variables in the described nanoparticle manufacturing methods will have significant impact on particle characteristics such as size, morphology, and charge. These variables include concentration and molecular weight of the monomer/polymer, concentration and molecular weight of surfactant/emulsifier, temperature, pH, and whether the emulsion is formed using vortexing, ultrasonication, or high-speed homogenization. A few generalizations can be made: lower energy (e.g., vortexing versus sonication/homogenization) and lower surfactant concentrations will generally result in larger particles; decreased polymer in organic concentration usually results in smaller diameter particles; and increased polymer weight can be expected to increase the degradation time. However, it is impossible to neatly summarize the effects of parameter optimization on final particle characteristics not only due to the large number of variables, but also because the individual properties of the drug and polymer themselves greatly influence these characteristics.

Thus, most studies begin with attempts to encapsulate new compounds via a variety of different methods, followed by parameter optimization. Highly water-soluble drugs usually require a double emulsion method, while water insoluble drugs can be easily encapsulated using single or spontaneous emulsions or nanoprecipitation. Highly labile compounds are better encapsulated using methods that do not result in significant exposure to organics or high energies, and thus may be better suited for delivery using a natural polymer. On the other hand, longer desired periods of release necessitate the use of synthetic polymers.

29.4 Particle Characterization Methods

29.4.1 Size, Morphology, and Electrostatics

The key features that govern nanoparticle interactions in the body are particle size and surface characteristics such as charge, morphology, and surface functional groups (Panyam and Labhasetwar 2003). Many techniques have been developed to evaluate and quantify these properties, and some of the more basic practices are discussed here. Particle diameter is usually measured in one of two ways. The first is by electron microscopy (Figure 29.5). Nanoparticles are commonly dispersed onto a metallic stub and sputter-coated with a conductive material such as gold for the visualization of nanoparticle morphology and size via scanning or transmission electron microscopy. Quantitative analysis of particle size and morphology can be made with the aid of various software packages, including one provided free-of-charge by the NIH (ImageJ).

A second and arguably more important method of measuring particle size is the use of light scattering to determine hydrodynamic diameter. These include Coulter counters with appropriate filters and dynamic light scattering (DLS) machines. Measurement of nanoparticle size via these methods gives an understanding of the true hydrodynamic diameter of nanoparticles, allowing identification of potential issues such as nanoparticle aggregation in aqueous solutions. The effect of surface modifications, such as PEGylation, on particle diameter and aggregation can also be measured. The benefit of using DLS is that many of the machines are also capable of making simultaneous zeta potential (surface charge) measurements. The electrostatic properties of nanoparticles have important implications for their *in vivo* fate—anionic nanoparticles show a high affinity for the cell membrane, and may be more efficiently internalized, while cationic nanoparticles

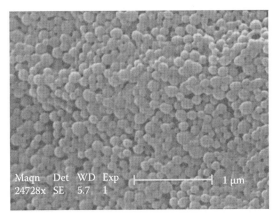 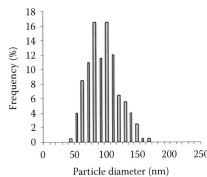

FIGURE 29.5 Nanoparticle imaging.

may attract negatively charged serum proteins and be more readily opsonized (Moghimi et al. 2001).

29.4.2 Drug Encapsulation and Release

The loading capacity and encapsulation efficiency of a nanoparticle formulation method are key metrics for reporting on the methodology. The loading capacity is a measurement of the total encapsulated dose. A higher loading is generally desirable due to the reduced quantity of polymer needed for administration, although the optimal loading is ultimately dependent on factors such as drug release rate, nanoparticle distribution and uptake, and optimal drug concentrations. Encapsulation efficiency is a measure of the efficiency with which a compound is incorporated into nanoparticles. High encapsulation efficiency is always desirable, especially, when considering the high cost of many drugs and therapeutic proteins. Drug loading is usually expressed as mass or units of drug per mass of nanoparticles, and encapsulation efficiency as a percentage.

Drug loading is measured by rapid degradation of the polymeric matrix of the nanoparticle and quantification of the recovered drug, generally via the detection method for the free, unencapsulated drug (e.g., absorbance, fluorescence, and ELISA).

$$\text{Drug loading}$$
$$= \frac{\text{Mass of drug in nanoparticles}}{\text{Mass of nanoparticles}}$$
$$\text{Encapsulation Efficiency}$$
$$= \frac{\text{Measured loading}}{\text{Theoretical max loading}} \times 100$$

This can be achieved by dissolving the nanoparticle in an organic or other solution capable of rapid degradation. PLGA nanoparticle loadings are often measured by degradation of nanoparticles in dimethyl sulfoxide or 1 N NaOH with 1% SDS (often used for protein and peptides). Nanoparticles can also be dissolved in an organic (e.g., dichloromethane) and drug recovered via extraction into a mobile phase solution (e.g., acetonitrile). High performance liquid chromatography (HPLC) and other analytical chemistry methods (e.g., mass spectrometry) are commonly used to quantify drug after separation of the polymer. If extraction from nanoparticles proves difficult, one method is to identify the starting quantity of the drug and subtract the quantity of the unencapsulated drug after particle formulation. Validations of measurement and detection methods are required for an accurate measurement of nanoparticle loading. Addition of a known quantity of free drug to blank (non-drug loaded) nanoparticles and measurement of drug after the extraction protocol is an important control to determine the efficiency of the extraction protocol.

Ultimately, the most important characteristic may be the release profile of the drug, including the bioactivity of the drug after encapsulation and release. Drug entrapped in the polymeric matrix of a biodegradable nanoparticle is released in sustained fashion through (1) diffusion of the drug in the matrix, and (2) degradation of the matrix. *In vitro* measurement of drug release is usually performed under physiologic buffered conditions. Drug loaded (and blank) nanoparticles are suspended in a physiologically relevant buffer such as phosphate buffered saline and incubated at 37°C. An incubated shaker is typically used to agitate the particles and simulate fluid convection. At relevant time points, the particles are then centrifuged and the supernatant removed for the measurement of drug concentration. The buffer is replaced for collection of multiple time points, and the release is generally described as a cumulative profile over the time period of interest. The bioactivity of the drug in this released fraction can be evaluated using a variety of techniques (e.g., ELISA, cell viability/ cytotoxicity studies, etc.) that are specific to the drug.

29.5 Applications: Targeted and Multifunctional Nanoparticles

One of the primary benefits of using biodegradable materials for drug delivery is the sustained delivery of drug achieved by slow and continuous release from the degrading polymer matrix. This enables more infrequent dosing and lower systemic peak concentrations. Among the first drug delivery systems to be used in the clinic were biodegradable implants and microspheres that utilized these properties in the sustained parenteral delivery of peptides. Sustained delivery from a depot can also be utilized to achieve effective localized delivery in situations where the formulation can be accurately administered to the site of interest. For example, with stereotactically implanted devices or microparticles, sustained release from the depot can result in an accumulation of local drug concentrations and minimize the systemic dose and any associated toxicities. An excellent example is the Gliadel® wafer, a biodegradable implant designed to line the cavity and provide sustained release of chemotherapeutic drug after surgery to remove malignant glioma. However, in cases where a systemic treatment approach is necessary, such as metastatic cancers, the deliberate placement of delivery devices is not feasible. A different approach is needed.

The development of nanoscale biodegradable particle systems as drug delivery platforms has enabled means that are more sophisticated in modulating drug effect through targeted delivery to specific cells and tissues.

29.5.1 Passive Targeting (The EPR Effect)

The first challenge in the development of effective nanoparticle drug delivery systems remains the systemic distribution of particles following intravenous administration and interaction with the body's natural clearance mechanisms. Following intravenous injection, drugs are normally distributed through the vascular and lymphatic systems. Nanoparticles, even when made from highly biocompatible materials, are rapidly cleared from the circulation by phagocytosis through a process that is facilitated by the surface deposition of opsonizing factors and the activation of complement proteins (Moghimi et al. 2001). Opsonization and clearance are influenced heavily by the size and surface characteristics of the particles (size, geometry, charge, functional groups, etc.). Compared to smaller particles, nanoparticles over 200 nm in diameter have been found to activate the complement system more efficiently and are cleared more rapidly (Moghimi et al. 2001). Smaller particles, however, have a relatively larger amount of surface area and are therefore more prone to problems related to aggregation. In either case, significant surface charge (especially cationic) and the presence of hydrophobic moieties have been identified as significant factors contributing to aggregation, protein deposition, and subsequent particle clearance.

The problems of particle aggregation, deposition of opsonizing factors and activation of complement factors can be largely obviated by appropriate engineering of the particle surface. The most common and successful strategy to date has been the coating of particles in a layer of hydrophilic polymer such as poly(ethylene glycol) (PEG) to shield charges or hydrophobic residues and reduce protein opsonization. This is sometimes referred to as "steric stabilization" of the particle, and it allows for prolonged systemic circulation. The advantages of prolonged circulation of a drug include increased area under curve (AUC) and extended drug effect, but prolonged circulation of drug-loaded nanoparticles provides an additional benefit. It has been shown that long circulating particles of sufficiently small diameter are capable of extravasating through the fenestrations in inflamed or disrupted vasculature, such as those seen in some solid tumors. These extravasating particles then accumulate in the tissue bed through a phenomenon known as the "enhanced permeation and retention" (EPR) effect (Brigger et al. 2002) (Figure 29.6). In the case of solid tumors with compromised and leaky vasculature, localized drug delivery can thereby be achieved largely as a function of particle size, and is referred to as "passive targeting." Research has shown that the optimal size for these long-circulating particles is between 50 and 200 nm, as particles outside this range are subject to leakage into capillaries or filtration in the kidneys and liver (Moghimi et al. 2001). The passive targeting of tumors has in fact been demonstrated clinically with 100 nm diameter PEGylated liposomes delivering the anticancer drug doxorubicin, and recent research indicates that the same effects may be achieved using biodegradable nanoparticles (Brigger et al. 2002).

29.5.2 Active (Biospecific) Targeting

Recent advances in the engineering of nanoparticles have focused on the active targeting of nanoparticles to specific tissues and cells and their effects once they reach the target site. These include sophisticated methodologies for the biospecific targeting of particles and the engineering of cellular uptake and intracellular delivery. With new discoveries in the biophysical and immunologic markers of disease, more and more tools are becoming available

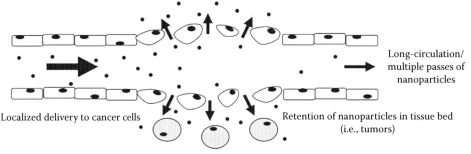

FIGURE 29.6 Passive targeting of nanoparticles via the EPR effect. Shielding of nanoparticles with hydrophilic polymers reduces opsonization and clearance, allowing for prolonged systemic circulation. Nanoparticles of sufficiently small diameter (<200 nm) are then capable of accumulating in tissues with compromised vasculature (e.g., solid tumors) via the "enhanced permeation and retention" (EPR) effect, resulting in targeted delivery of encapsulated drug.

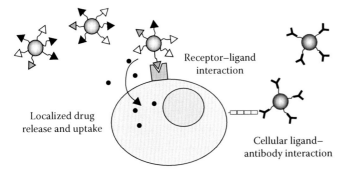

Receptor–ligand interaction

Localized drug release and uptake

Cellular ligand–antibody interaction

FIGURE 29.7 Targeting delivery using cell-specific ligands and antibodies. Nanoparticles can be targeted using ligands specific to an ever-growing library of cellular markers. These ligands include antibodies (Y-shapes) and peptides, oligomers, aptamers, and even small molecules (closed triangles). The challenge lies in the attachment to the particle, correct presentation, and preservation of activity of these ligands.

to help identify and target diseased cells over healthy ones. These tools have included antibodies, proteins, peptide or nucleic acid sequences, and even small molecules that bind to or "recognize" disease-specific ligands such as tumor-associated antigens. Biodegradable nanoparticles have been formulated to display these molecules on their surface to better target the delivery of their internal payload (Figure 29.7). The objective remains improved drug efficacy through the accumulation of nanoparticles at target sites and subsequent sustained and localized delivery. Additionally, although nanoparticle size inherently promotes greater efficiency in uptake by cells, the specific targeting of nanoparticles to internalizing ligands is a strategy shown to dramatically improve intracellular delivery and enhance payload efficacy. There exist endeavors to augment these effects by enhancing endosomal escape of internalized nanoparticles (Panyam and Labhasetwar 2003).

The principal advantage of nanoscale drug delivery systems is the targeted delivery of therapeutics to increase drug efficacy and decrease toxicity. As previously discussed, nanoparticle size is a fundamental means for achieving passive targeting of tumors, via the preferential extravasation and accumulation of particles through endothelial fenestrations in tumor sites. The state of the art has become the engineering of biospecific *active* targeting using antibodies, receptor ligands, and even small molecules and peptides. The clinical advantage of nanoscale carriers over combination immunotherapeutic or targeted drug formulations (e.g., antibody-drug conjugates) is of payload capacity—the specific delivery of large amounts of therapeutic or imaging agents per biorecognition event (Ferrari 2005). Thus, while chemotherapeutics can be linked to monoclonal antibodies via a chemical cross-linker, resulting in the delivery of several drug molecules per antibody binding event (Gu et al. 2007), payloads of an order of magnitude greater can be delivered per the same recognition/binding event by loading drugs into the surface ligand-modified, targeted nanoparticle. The principle strategies used to engineering biospecific targeting of nanoparticles can be described as antibody–antigen, peptide–antigen, or ligand–receptor interactions.

29.5.2.1 Antigen Recognition Targeting (e.g., Antibody and Peptide Targeting)

A wide variety of strategies exists for targeting nanoparticles against cancer. As most tumors develop abnormal neovasculature, one of the primary strategies has been to target tumor vasculature using anti-VEGF antibodies or integrin ($\alpha v \beta$)–targeting ligands (Hood et al. 2002). The Arg-Gly-Asp (RGD) peptide has also been shown to be an effective target for vascularized tumors (Schiffelers et al. 2004). Likewise, many tumors have been found to overexpress various antigens or ligands to which monoclonal antibodies can be developed for immunotargeting of nanoparticles. In fact, there exist a number of monoclonal antibodies already approved for stand-alone or combination immunotherapy (e.g., rituximab for the CD20 ligand on B-cell lymphomas, trastuzumab for the HER2/neu growth receptor in breast cancer) that have been used to target nanoparticles. Nobs et al. reported a 6-fold increase in uptake by presenting rituximab or trastuzumab on the surfaces of PLA particles (Nobs et al. 2006).

29.5.2.2 Ligand-Receptor Targeting (e.g., Targeting the Folate Receptor)

The folate receptor is another targetable ligand highly overexpressed in many breast and ovarian cancers, and has been well studied for the targeting of nanoparticles. The carboxylic group of the vitamin folic acid is covalently conjugated to the polymer or a flexible linker such as bi-functionalized PEG, allowing normal recognition and binding to the folate receptor. Kim et al. were able to modify the anionic surface of PLGA nanoparticles using a cationic polylysine-PEG-folate conjugate, demonstrating increased uptake of labeled nanoparticles via a folate receptor-dependent pathway (Kim et al. 2005). Zhang et al. were able to demonstrate that the surface-specific presentation of folate molecules on paclitaxel-loaded PLGA nanoparticles resulted in 1.5–1.7-fold increase in nanoparticle uptake, which however increased drug effect from 50% killing of cancer cells to 92% efficacy (Zhang et al. 2007).

The ultimate objective of targeting nanoparticles to tumor-specific antigens is to translate improved *in vitro* particle uptake into improved *in vivo* drug delivery and efficacy. Despite enormous efforts devoted to developing targeted nanoparticles, there remain many challenges and limitations. Attachment of antibodies to a biodegradable nanoparticle may improve uptake and specificity to tumors, but may also significantly increase the hydrodynamic diameter and alter the surface properties of the particle, leading to increased clearance or unexpected sites of accumulation. Nonetheless, the early data are promising; Cheng et al. modified the carboxyl terminus of PLGA-PEG-COOH with an RNA aptamer targeting the prostate specific membrane antigen (Farokhzad et al. 2006, Cheng et al. 2007). Nanoparticles made from this conjugate demonstrated increased accumulation of drug in subcutaneous prostate cancer tumors in mice and improved tumor response, a noteworthy *in vivo* demonstration of the potential benefits of targeted nanoparticle-based delivery. The benefits of accurately targeted nanoparticles may even extend beyond those anticipated from improved drug delivery. Goren et al. noted that folate-targeted delivery of doxorubicin-containing liposomes resulted in bypass of multidrug resistance, a phenomenon also observed by Cuvier et al. using polycyanoacrylate nanoparticles (Cuvier et al. 1992, Goren et al. 2000).

29.5.2.3 Internalization of Nanoparticles and Nucleotide Delivery

The bypass of multidrug resistance observed by Cuvier and Goren may be directly related to the improved intracellular delivery of drugs after internalization of nanoparticles. deVerdiere et al. found that sustained and local delivery from particles was not sufficient to reverse resistance by separating particles from cells via a membrane that allowed free diffusion of doxorubicin, but not particles. Thus, particle-cell contact, and most likely uptake, was the critical factor. Biodegradable nanoparticles can be internalized through a number of different mechanisms: pinocytosis, phagocytosis, and endocytosis via clathrin coated pits or caveoli (Moghimi et al. 2001). One application of note for improved intracellular delivery is the sustained intracellular delivery of nucleotides. The advent of new therapeutics, such as biologics and siRNA, are driving a need to find new, more efficient delivery methods. The potential for sustained, nontoxic delivery to the cytoplasm makes biodegradable nanoparticles a prime candidate for delivery of these compounds. The functional delivery of a wide variety of sequence-specific inhibitors, including siRNA, from PLGA micro- and nanoparticles has been demonstrated by Khan et al. (2004) and Yuan et al. (2006). Likewise, Cohen-Sacks et al. were able to use PLGA nanoparticles to deliver antisense platelet-derived growth factor receptor to prevent restenosis *in vivo* (Cohen-Sacks et al. 2002). Thus, biodegradable nanoparticles represent a promising solution for the sustained delivery and prolonged effect of therapeutics based on small interfering RNA (siRNA), antisense oligonucleotides, ribozymes, and DNAzymes.

29.5.3 Multifunctional Nanoparticles

The versatility of biodegradable materials has enabled the development of multifunctional nanoparticles; for example, nanoparticles that utilize multiple ligands to reduce clearance and improve targeting valency and avidity. Another example of this versatility is the co-delivery of multiple agents, a feature that may feature prominently due to the potency of many combination therapeutic regimens. A noteworthy example is the "nanocell" developed by Sengupta et al. capable of temporal release of two different agents (Sengupta et al. 2005). By encapsulating a doxorubicin-PLGA conjugate nanoparticle within a PEGylated lipid envelope containing combretastatin, sequential delivery of the two compounds was achieved. Therefore, delivery of the cytotoxic chemotherapeutic drug occurred *after* the anti-angiogenic agent, elegantly resulting in combination therapy while simultaneously obviating the problem of decreased drug penetration in tumors after anti-angiogenic therapy.

The versatility of biodegradable nanoparticles as targeted delivery platforms has also led to the development of multifunctional platforms that include diagnostic imaging agents (Fahmy et al. 2007). The utility of such a system would be based on the ability to image diseased tissues with simultaneous delivery of therapeutic agents. These systems could be utilized for purposes ranging from a better understanding of drug pharmacokinetics and optimal dosing, to improved monitoring of treatment response. Although several nanoparticle imaging systems exist (e.g., gold nanoparticles, dendrimers, quantum dots, and iron oxide particles), biodegradable nanoparticles offer additional advantages in terms of delivery potential, targetability, and toxicities (Fahmy et al. 2007). As a demonstration of enhanced targetability, chitosan nanoparticles loaded with gadolinium, an MR contrast agent, were found to be internalized through endocytosis in several cancer cell lines at levels of orders of magnitude higher than Magnevist®, the currently available formulation of gadolinium (Brannon-Peppas and Blanchette 2004). Even compared to dendrimers, which have been well studied as drug delivery vehicles, most biodegradable nanoparticle systems offer superior advantages in terms of delivery of a variety of agents and payload capacity per biorecognition event (Ferrari 2005, Fahmy et al. 2007). Thus, multifunctional nanoparticles may represent a novel method to deliver a therapeutic payload while simultaneously providing clinicians with enhanced tools for diagnosis and the monitoring of disease progression and response to treatment.

29.6 Conclusion

The forces driving the development of nanoscale drug delivery systems are tremendous. A growing number of novel drugs—hydrophobic small molecule drugs, biologics, nucleic acids, and siRNA—face significant delivery challenges through traditional oral or injection routes. These therapeutics serve to highlight the importance of research and development of delivery technologies. In addition, novel methods of drug delivery will enable reformulation, improvement, and development of new uses for existing drugs, as well as drugs previously shelved for toxicity or solubility issues. The excellent biocompatibility and safety of starting materials, versatility of encapsulated compounds and delivery profiles, and significant advances in surface engineering and targetability have made biodegradable nanoparticles one of the most promising areas of study in the field of nanomedicine.

References

Anderson, J. M. and Shive, M. S. (1997) Biodegradation and biocompatibility of PLA and PLGA microspheres. *Advanced Drug Delivery Reviews*, 28, 5–24.

Brannon-Peppas, L. and Blanchette, J. O. (2004) Nanoparticle and targeted systems for cancer therapy. *Advanced Drug Delivery Reviews*, 56, 1649–1659.

Brigger, I., Dubernet, C., and Couvreur, P. (2002) Nanoparticles in cancer therapy and diagnosis. *Advanced Drug Delivery Reviews*, 54, 631–651.

Cheng, J., Teply, B. A., Sherifi, I., Sung, J., Luther, G., Gu, F. X., Levy-Nissenbaum, E., Radovic-Moreno, A. F., Langer, R., and Farokhzad, O. C. (2007) Formulation of functionalized PLGA-PEG nanoparticles for in vivo targeted drug delivery. *Biomaterials*, 28, 869–876.

Cohen, H., Levy, R. J., Gao, J., Fishbein, I., Kousaev, V., Sosnowski, S., Slomkowski, S., and Golomb, G. (2000) Sustained delivery and expression of DNA encapsulated in polymeric nanoparticles. *Gene Therapy*, 7, 1896–1905.

Cohen-Sacks, H., Najajreh, Y., Tchaikovski, V., Gao, G., Elazer, V., Dahan, R., Gati, I., Kanaan, M., Waltenberger, J., and Golomb, G. (2002) Novel PDGF beta R antisense encapsulated in polymeric nanospheres for the treatment of restenosis. *Gene Therapy*, 9, 1607–1616.

Couvreur, P. and Puisieux, F. (1993) Nanoparticles and microparticles for the delivery of polypeptides and proteins. *Advanced Drug Delivery Reviews*, 10, 141–162.

Couvreur, P., Kante, B., Roland, M., Guiot, P., Bauduin, P., and Speiser, P. (1979) Polycyanoacrylate nanocapsules as potential lysosomotropic carriers—Preparation, morphological and sorptive properties. *Journal of Pharmacy and Pharmacology*, 31, 331–332.

Cuvier, C., Roblot-Treupel, L., Millot, J. M., Lizard, G., Chevillard, S., Manfait, M., Couvreur, P., and Poupon, M. F. (1992) Doxorubicin-loaded nanospheres bypass tumor cell multidrug resistance. *Biochemical Pharmacology*, 44, 509–517.

Fahmy, T. M., Fong, P. M., Park, J., Constable, T., and Saltzman, W. M. (2007) Nanosystems for simultaneous imaging and drug delivery to T cells. *AAPS Journal*, 9, E171–E180.

Farokhzad, O. C., Cheng, J. J., Teply, B. A., Sherifi, I., Jon, S., Kantoff, P. W., Richie, J. P., and Langer, R. (2006) Targeted nanoparticle-aptamer bioconjugates for cancer chemotherapy in vivo. *Proceedings of the National Academy of Sciences of the United States of America*, 103, 6315–6320.

Ferrari, M. (2005) Cancer nanotechnology: Opportunities and challenges. *Nature Reviews Cancer*, 5, 161–171.

Goren, D., Horowitz, A. T., Tzemach, D., Tarshish, M., Zalipsky, S., and Gabizon, A. (2000) Nuclear delivery of doxorubicin via folate-targeted liposomes with bypass of multidrug-resistance efflux pump. *Clinical Cancer Research*, 6, 1949–1957.

Govender, T., Stolnik, S., Garnett, M. C., Illum, L., and Davis, S. S. (1999) PLGA nanoparticles prepared by nanoprecipitation: Drug loading and release studies of a water soluble drug. *Journal of Controlled Release*, 57, 171–185.

Gu, F. X., Karnik, R., Wang, A. Z., Alexis, F., Levy-Nissenbaum, E., Hong, S., Langer, R. S., and Farokhzad, O. C. (2007) Targeted nanoparticles for cancer therapy. *Nano Today*, 2, 14–21.

Hood, J. D., Bednarski, M., Frausto, R., Guccione, S., Reisfeld, R. A., Xiang, R., and Cheresh, D. A. (2002) Tumor regression by targeted gene delivery to the neovasculature. *Science*, 296, 2404–2407.

Hu, Y., Jiang, X. Q., Ding, Y., Ge, H. X., Yuan, Y. Y., and Yang, C. Z. (2002) Synthesis and characterization of chitosan-poly(acrylic acid) nanoparticles. *Biomaterials*, 23, 3193–3201.

Janes, K. A., Fresneau, M. P., Marazuela, A., Fabra, A., and Alonso, M. J. (2001) Chitosan nanoparticles as delivery systems for doxorubicin. *Journal of Controlled Release*, 73, 255–267.

Khan, A., Benboubetra, M., Sayyed, P. Z., Ng, K. W., Fox, S., Beck, G., Benter, I. F., and Akhtar, S. (2004) Sustained polymeric delivery of gene silencing antisense ODNs, siRNA, DNAzymes and ribozymes: in vitro and in vivo studies. *Journal of Drug Targeting*, 12, 393–404.

Kim, S. H., Jeong, J. H., Chun, K. W., and Park, T. G. (2005) Target-specific cellular uptake of PLGA nanoparticles coated with poly(L-lysine)-poly(ethylene glycol)-folate conjugate. *Langmuir*, 21, 8852–8857.

Mao, H. Q., Roy, K., Troung-Le, V. L., Janes, K. A., Lin, K. Y., Wang, Y., August, J. T., and Leong, K. W. (2001) Chitosan-DNA nanoparticles as gene carriers: Synthesis, characterization and transfection efficiency. *Journal of Controlled Release*, 70, 399–421.

Mitra, S., Gaur, U., Ghosh, P. C., and Maitra, A. N. (2001) Tumour targeted delivery of encapsulated dextran-doxorubicin conjugate using chitosan nanoparticles as carrier. *Journal of Controlled Release*, 74, 317–323.

Moghimi, S. M., Hunter, A. C., and Murray, J. C. (2001) Long-circulating and target-specific nanoparticles: Theory to practice. *Pharmacological Reviews*, 53, 283–318.

Moghimi, S. M., Hunter, A. C., and Murray, J. C. (2005) Nanomedicine: Current status and future prospects. *FASEB Journal*, 19, 311–330.

Mu, L. and Feng, S. S. (2003) A novel controlled release formulation for the anticancer drug paclitaxel (Taxol (R)): PLGA nanoparticles containing vitamin E TPGS. *Journal of Controlled Release*, 86, 33–48.

Niwa, T., Takeuchi, H., Hino, T., Kunou, N., and Kawashima, Y. (1993) Preparations of biodegradable nanospheres of water-soluble and insoluble drugs with D,L-lactide glycolide copolymer by a novel spontaneous emulsification solvent diffusion method, and the drug release behavior. *Journal of Controlled Release*, 25, 89–98.

Nobs, L., Buchegger, F., Gurny, R., and Allemann, E. (2006) Biodegradable nanoparticles for direct or two-step tumor immunotargeting. *Bioconjugate Chemistry*, 17, 139–145.

Okada, H. and Toguchi, H. (1995) Biodegradable microspheres in drug-delivery. *Critical Reviews in Therapeutic Drug Carrier Systems*, 12, 1–99.

Pandey, R. and Khuller, G. K. (2005) Alginate as a drug delivery carrier. In *Handbook of Carbohydrate Engineering*. Yarema, K. J. (ed.), Taylor & Francis Group, Boca Raton, FL. Panyam, J. and Labhasetwar, V. (2003) Biodegradable nanoparticles for drug and gene delivery to cells and tissue. *Advanced Drug Delivery Reviews*, 55, 329–347.

Rajaonarivony, M., Vauthier, C., Couarraze, G., Puisieux, F., and Couvreur, P. (1993) Development of a new drug carrier made from alginate. *Journal of Pharmaceutical Sciences*, 82, 912–917.

Sahoo, S. K., Panyam, J., Prabha, S., and Labhasetwar, V. (2002) Residual polyvinyl alcohol associated with poly (D,L-lactide-co-glycolide) nanoparticles affects their physical properties and cellular uptake. *Journal of Controlled Release*, 82, 105–114.

Saltzman, W. M. (2001) *Drug Delivery*, Oxford, New York.

Schiffelers, R. M., Ansari, A., Xu, J., Zhou, Q., Tang, Q. Q., Storm, G., Molema, G., Lu, P. Y., Scaria, P. V., and Woodle, M. C. (2004) Cancer siRNA therapy by tumor selective delivery with ligand-targeted sterically stabilized nanoparticle. *Nucleic Acids Research*, 32, e149.

Sengupta, S., Eavarone, D., Capila, I., Zhao, G. L., Watson, N., Kiziltepe, T., and Sasisekharan, R. (2005) Temporal targeting of tumour cells and neovasculature with a nanoscale delivery system. *Nature*, 436, 568–572.

Sinha, R., Kim, G. J., Nie, S. M., and Shin, D. M. (2006) Nanotechnology in cancer therapeutics: Bioconjugated nanoparticles for drug delivery. *Molecular Cancer Therapeutics*, 5, 1909–1917.

Soppimath, K. S., Aminabhavi, T. M., Kulkarni, A. R., and Rudzinski, W. E. (2001) Biodegradable polymeric nanoparticles as drug delivery devices. *Journal of Controlled Release*, 70, 1–20.

Takeuchi, H., Yamamoto, H., and Kawashima, Y. (2001) Mucoadhesive nanoparticulate systems for peptide drug delivery. *Advanced Drug Delivery Reviews*, 47, 39–54.

Uhrich, K. E., Cannizzaro, S. M., Langer, R. S., and Shakesheff, K. M. (1999) Polymeric systems for controlled drug release. *Chemical Reviews*, 99, 3181–3198.

Vila, A., Sanchez, A., Tobio, M., Calvo, P., and Alonso, M. J. (2002) Design of biodegradable particles for protein delivery. *Journal of Controlled Release*, 78, 15–24.

Yoo, H. S. and Park, T. G. (2001) Biodegradable polymeric micelles composed of doxorubicin conjugated PLGA-PEG block copolymer. *Journal of Controlled Release*, 70, 63–70.

Yuan, X. D., Li, L., Rathinavelu, A., Hao, J. S., Narasimhan, M., He, M., Heitlage, V., Tam, L., Viqar, S., and Salehi, M. (2006) siRNA drug delivery by biodegradable polymeric nanoparticles. *Journal of Nanoscience and Nanotechnology*, 6, 2821–2828.

Zhang, Z. P., Lee, S. H., and Feng, S. S. (2007) Folate-decorated poly(lactide-co-glycolide)-vitamin E TPGS nanoparticles for targeted drug delivery. *Biomaterials*, 28, 1889–1899.

VI

Response to Nanomaterials

30

Uptake of Carbon-Based Nanoparticles by Mammalian Cells and Plants

Pu-Chun Ke
Clemson University

Sijie Lin
Clemson University

Jason Reppert
Clemson University

Apparao M. Rao
Clemson University

Hong Luo
Clemson University

30.1 Introduction

Over the past decade, efforts have been made toward understanding and predicting the fate of nanomaterials in biological systems (Ke and Qiao, 2007) and, quite recently, in the environment (Colvin, 2003; Nel et al., 2006; Maynard et al., 2006). The motivations for making such efforts are twofold. First, it is desirable to utilize the unique physiochemical properties of nanomaterials for implementing new applications of nanotechnology, primarily within the realms of biosensing and nanomedicine. Second, it has become apparent that the safe development of nanotechnology must be guided by research on the fate of nanoparticles in living systems (Maynard et al., 2006). It is estimated that a few thousand tons of engineered nanomaterials are currently produced each year, and over 600 consumer products on the market are related to or derived from nanotechnology. Conceivably, these engineered nanomaterials will eventually be discharged into the ecological systems comprising water, air, soil, and, most importantly, the dynamic food chains that are intimately related to human health (Mauter and Elimelech, 2008).

The exploration of nanoparticles in living systems has revealed itself as a topic of tremendous complexity, excitement, and, at times, frustration. The more we examine biological responses to nanoparticles, the more divergent is the behavior of nanomaterials from that of bulk materials. One distinct feature of nanomaterials is their versatility in transporting and transforming in the liquid phase, fuelled by their chemical reactions, hydrophobic interactions, electrostatic forces, hydrogen bonding, as well as their pi-stacking with biological and organic molecules and complexes. In this chapter, we attempt to highlight the recent progress made in the investigation of the fate of nanoparticles in mammalian cells and plant species, and to demonstrate that an exploration of the biological applications of nanotechnology must be based on a sound understanding of the fate of nanomaterials in living systems. We wish to convey the idea that describing nanomaterials in living systems is an area that requires a combined effort from materials, chemistry, physics (especially biophysics), biology (especially cell biology, toxicology, immunology), genetics, medicine, as well as environmental science and engineering. The recent establishment of two Centers of Environmental Implications of Nanotechnology (CEIN) by the National Science Foundation highlights the recognition of this urgency by a major government funding agency. Naturally, research addressing the uptake and the fate of nanoparticles in living systems has shown a growing trend of team research, often involving inputs from multiple disciplines and affiliations, as well as participants from multiple countries.

30.2 Background (History and Definitions)

30.2.1 Nanomaterials

It is now commonly acknowledged that materials in which at least one of its three dimensions is less than 100 nm can be categorized as a nanostructured material. With the advent of various physical and chemical synthesis methods, most materials can be readily prepared in their nanostructured forms. Of these, the carbon-based nanomaterials are the most widely studied. They are unique in the sense that they can be prepared in bulk quantities in the form of two-dimensional graphene, one-dimensional carbon nanotubes, or zero-dimensional fullerenes (Figure 30.1). A single sheet of graphite is referred to as graphene, while a carbon nanotube is thought of as a single graphene sheet rolled into a seamless cylinder with typical aspect ratios (length/diameter) exceeding 1000. Depending on the synthesis conditions, carbon nanotubes favor either single-, double-, or multi-walled morphologies. An extremely short length single-walled carbon nanotube resembles the structure of fullerenes in which a group of 60 carbon atoms link covalently into soccer-ball shaped molecules called C_{60}. Besides C_{60}, the fullerene family also comprises of bigger molecules such as C_{70}, C_{76}, etc. (Shoji et al., 2004).

There are three common ways to synthesize carbon nanotubes. The first method is the electric arc discharge in which a catalyst-impregnated graphite electrode is vaporized by an electric arc (~100 A) in an inert atmosphere. The condensed carbon soot deposits on the inner surface of the water-cooled arc chamber (Journet et al., 1997). The second method, commonly referred to as pulsed laser vaporization, uses a pulsed laser beam

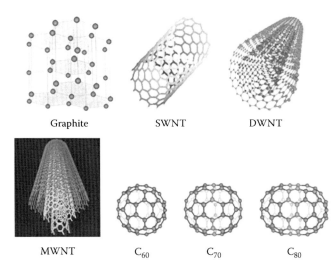

FIGURE 30.1 A single layer (shown shaded) in graphite is called graphene. A seamless cylinder of graphene represents a single-walled carbon nanotube (SWNT). Concentrically aligned SWNTs lead to double- (DWNT) and multi- (MWNT) walled carbon nanotube structures. A covalently bonded cluster of 60 carbon atoms results in a soccer-ball shaped structure called fullerene. The fullerene family also comprises of closed clusters of C_{70}, C_{80}, etc.

(Nd:YAG, 1064 nm) that is focused onto a catalyst-impregnated graphite target. The ablated material from the target generates soot that is collected in a water-cooled cold finger (Thess et al., 1996). Both techniques are used widely to produce SWNTs. As-prepared soot contains 60%–70% SWNTs and the remainder is amorphous carbon and nanoparticles. The third method of synthesis is chemical vapor deposition (CVD). This technique is the most promising in producing bulk quantities of carbon nanotubes. The nanotubes are formed by the pyrolysis of a hydrocarbon source seeded by catalyst particles (typically ferromagnetic particles such as Fe, Co, Ni, etc., are used) in an inert atmosphere and between 700°C and 1200°C. The CVD process is usually the method of choice for preparing MWNTs, although reports have shown that the SWNTs can also be grown (Andrews et al., 1999; Bachilo et al., 2003; Hata et al., 2004).

30.2.2 Uptake of Nanoparticles by Mammalian Cells

Nanoparticles possess much higher specific surface areas than their bulk counterparts (Navarro et al., 2008) and have found applications as acting catalysts for facilitating biochemical reactions and biological processes (Mitchell et al., 2002), as substrates for detecting single antibody binding and gene sequencing (Wang et al., 2004a,b), and as carriers for gene and drug delivery (Ke and Qiao, 2007). In the last case, nanoparticles have mostly been applied to mammalian cells and organisms, with the first *in vitro* CNT gene and protein delivery demonstrated by Dai's group at Stanford (Kam et al., 2004), Prato and Bianco's groups at the University of Trieste and CNRS (Pantarotto et al., 2004), and RNA polymer delivery by our group at Clemson (Lu et al., 2004). The current understanding is that the uptake of CNTs may occur irrespectively of mammalian cell types (Kostarelos et al., 2007). Fullerenes, with C_{60} and C_{70} being the most abundant, have also been examined with respect to their potential in medicinal and sensory applications. Mauzerall et al. reported that C_{70} had the tendency of accumulating in the lipid bilayer (Hwang and Mauzerall, 1993), suggesting the possibility of a photoinduced electron transport across cell membranes mediated by C_{70}. Ashcroft et al. demonstrated the formation of an immunoconjugate using a fullerene C_{60} and a murine anti-gp240 melanoma antibody (Ashcroft et al., 2006). The bioactivity of the conjugate was shown to be preserved for specific drug delivery.

Regarding toxicity, a topic inherently inseparable from research discussions on nanoparticles in living systems, Oberdorster (2004) reported an oxidative stress induced by fullerene C_{60} in the brain of a juvenile large-mouth bass. Sayes et al. (2004) found acute cytotoxicity induced by C_{60} but minimal side effects associated with functionalized fullerenes. Contrary to the above studies, Gharbi et al. (2005) found that fullerene C_{60} was a powerful antioxidant with no acute or subacute *in vivo* toxicity. This study further suggested that the use of organic solvents such as tetrahydrofuran (THF) could be the actual cause of fullerene toxicity reported by the others.

Using molecular dynamics simulations, our group provided a first study on the biophysical mechanism of cytotoxicity (Qiao et al., 2007). In agreement with Saye's observations (Sayes et al., 2004), our atomistic simulations showed that a pristine C_{60} molecule enlarged in its proximity the area per lipid molecule by 15% in a dipalmitoylphosphatidylcholine (DPPC) bilayer, while a functionalized $C_{60}(OH)_{20}$ molecule caused the shrinkage of the top leaflet of the bilayer by 25%. Our recent experimental study (Salonen et al., 2008) further showed that the supramolecular complex of C_{70} and phenolic gallic acid could cause a remarkable physical contraction of up to 25% in HT-29 colon cancer cells. Due to the structural representativeness and the abundance of phenolic acids in tea and red wine, as well as in ecological plant species, this study generalizes our understanding of the fate of nanoparticles in biological systems as well as in the environment (Figure 30.2).

It should be commented that research on cell responses to nanoparticles is nothing but conclusive. The current trend of research in this area has shifted from reporting phenomenological observations to addressing the mechanisms of cellular and tissue biodistribution of nanoparticles, the impact of nanoparticles on genome integrity (genotoxicity), and the ways of minimizing or utilizing cell responses to nanoparticles for gene sequencing, biosensing, gene and drug delivery, and novel and alternative therapeutics.

All the complexities associated with the nanoparticle uptake may arise from this simple fact: most nanoparticles are inherently hydrophobic but can become water soluble through covalent functionalization or noncovalent surface coating (supramolecular assembly). The covalent modification of carbon nanotubes, for example, involves the esterification or the amidation of acid-oxidized nanotubes and the side-wall covalent attachment of functional groups (Sano et al., 2001; Bahr et al., 2001; Sun et al., 2001; Banerjee and Wong, 2002; Pompeo and Resasco, 2002). However, these covalent schemes are often characterized by uncertainties in determining reaction efficacy and by undesirable modifications in the physical and chemical properties of the nanotubes (Pantarotto et al., 2003; Matarredona et al., 2003; Yurekli et al., 2004). In comparison, the noncovalent modifications of the carbon nanotubes employ a nonspecific attachment of proteins (Shim et al., 2002; Chen et al., 2003), linear bio- and synthetic polymers (DNA, RNA, polyvinyl pyrrolidone, polystyrene sulfonate), and surfactants (sodium dodecyl sulfate or SDS, etc.) (O'Connell et al., 2001, 2002; Zheng et al., 2003; Rao et al., 2004; Strano et al., 2004; Wang et al., 2004a,b). Many surfactants, organic solvents, and residues, however, cause cytotoxicity and/or other side effects that limit the biocompatibility of carbon nanotubes. In addition to increased solubility, functionalized carbon nanotubes also invoke minimal immunogenicity, and show no apparent tissue/organ accumulation. Developing alternative schemes is thus crucial for facilitating the full range of biological and biomedical applications for nanomaterials and their derivatives.

It should be commented that observing or detecting the uptake of nanoparticles is experimentally challenging. Because of the small dimension of nanoparticles, they are not traceable by bright-field microscopy unless they accumulate to form agglomerates. To confirm the presence of nanoparticles inside cells, researchers often employ confocal fluorescence microscopy (Lu et al., 2004; Kostarelos et al., 2007). The major

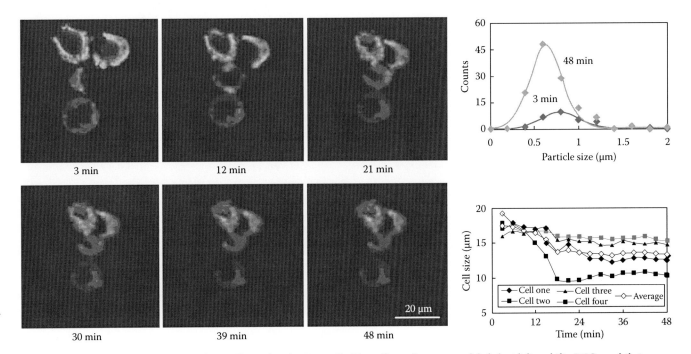

FIGURE 30.2 Real-time interaction of C_{70}-gallic acid and HT-29 cells. The cell membranes were labeled with lipophilic $DiIC_{18}$ and their cross sections appeared as red "rings." Over time, the cells were mechanically contracted due to the mutual interactions between the C_{70}-gallic acid nanoparticles. (From Salonen, E. et al., *Small*, 4, 1986, 2008. With permission.)

advantage of this technique is its sectioning property that discerns the axial locations of the nanoparticles with respect to cell membranes and their internal organelles. However, without fluorescently labeled biomolecules or functional groups, carbon-based nanoparticles exhibit minimal fluorescence or photoluminescence in the visible range and can only be detected in UV and near-infrared. While trespassing upon cell membranes or shuttling within the cellular space, single carbon nanotubes can be traced by following their photoluminescence in near-infrared (Bachilo et al., 2002). This approach requires the use and adaptation of an infrared CCD camera and is headed by Wiseman's group at Rice University (Tsyboulski et al., 2005) and Strano's group, now at MIT (Heller et al., 2005). One obvious advantage of this method is its minimal damage to cells and the high signal-to-noise ratio due to the suppressed cell autofluorescence in near infrared. Alternatively, carbon nanotubes and fullerenes in extra- and intracellular spaces may be detected using the technique of fluorescence energy transfer (Lin et al., 2006), or by tracing the fluorescence of the RNA polymer (Lu et al., 2004) or fullerene-gallic acid (Salonen et al., 2008). Furthermore, molecular dynamics simulations may provide complementary and insightful information on the biophysical mechanisms of the uptake of nanoparticles (Salonen et al., 2008).

30.2.3 Uptake of Nanoparticles by Plants

30.2.3.1 Structures of Plant Cells and Important Plant Species in the Food Chain

A general understanding of the structure of plant cells will help one to understand how carbon-based nanoparticles work as drug delivery or biosensing agents, as well as the potential toxicity due to their penetration into cell structures.

Cells are the basic units of life. All the eukaryotic cells contain the same basic components, that are essentially a set of membrane-enclosed compartments and a network of fibrous polymers called cytoskeleton (Buchanan et al., 2000). Cell membranes are a bilayer of polar lipid molecules and associated proteins, of which the plasma membrane defines the boundary of each cell and helps create and maintain electrochemically distinct environments within and outside the cell. Other membrane-bounded organelles include nucleus, vacuoles, peroxisomes, mitochondria, plastids (in plant and algal cells), as well as the endoplasmic reticulum and the Golgi apparatus.

The lipid bilayer of the cell membranes serves as a general permeability barrier to the diffusion of most polar molecules and a scaffolding of certain proteins. The membrane-spanning proteins within the lipid bilayer function to selectively transport and redistribute ions, small organic molecules and metabolites, or to transmit signals across the plasma membrane and intracellular membranes. They can also function to process lipids, assemble glycoproteins and polysaccharides, as well as to provide mechanical links between cytosolic and cell wall compounds in plant cells.

A cytoskeleton is a network of interconnected filamentous protein polymers composed of three major families of proteins: intermediate filaments, actin, and tubulin. A cytoskeleton runs throughout the cell within the cytosol, anchoring proteins and other macromolecules and supporting organelles during and after their synthesis to provide structural stability to the cytoplasm. It also mediates the spatial organization of components within the cell, and directs the movements of the cell or its contents, as well as participates in processing cellular information and provides directional cues within the cell.

Unlike animal cells, the plant cells do not have centrioles, lysosomes, intermediate filaments, cilia, or flagella. However, they do possess a number of other specialized structures. First, every plant cell is surrounded by a cell wall that is a highly organized composite of many different polysaccharides, proteins, and aromatic substances. These are mainly cellulose and hemicellulose, pectin, and in many cases lignin, and are secreted by the protoplast on the outside of the cell membrane. This contrasts with the chitin-containing fungus cell walls and the peptidoglycan-containing bacterial cell walls. The cell wall largely dictates the shape of a plant cell and its rigidity protects the cell. It is permeable to small soluble proteins and other solutes, and regulates the life cycle of the plant organism. Second, a plant cell has chloroplasts that contain chlorophyll and the biochemical systems for light harvesting and photosynthesis. This makes plants energetically self-supporting and autotrophic. Third, a plant cell has a sap-filled, large central vacuole enclosed by a single layer of unit membrane known as the tonoplast. This single membrane organelle produces turgor pressure against the cell wall for support; it controls the movement of molecules between the cytosol and the sap; stores organic acids, sugars, salts, O_2, CO_2, and pigments; and digests waste proteins and organelles. Fourth, plant cells have specialized cell–cell communication pathways known as plasmodesmata, the microscopic channels of approximately 50–60 nm in diameter at the mid-point that traverse the cell walls and enable transport and communication between them. Plasmodesmata enable direct, regulated, symplastic intercellular transport of substances between cells. Plant cells can use both passive and active transport to move molecules and ions through the passage.

Plants can be broadly categorized into vascular and nonvascular types. Vascular plants have the above-ground shoot system (leaves, buds, stem, flowers, and seeds) and the underground root system (roots, tubers, and rhizomes). They also have evolved specialized tissues, a xylem and a phloem. The xylem is involved in structural support and tends to conduct water and minerals from the roots to the leaves. The phloem cells conduct food—the photosynthetic products produced in leaves to the rest of the plant. Major food crops, such as rice, maize, wheat, etc., are all vascular plants.

30.2.3.2 State of the Art of Plant Uptake of Nanoparticles

Most early studies on nanoparticle uptake focused on animal systems. Very few studies have been conducted to evaluate the potential impact of nanoparticles on the ecological systems, including plants, fish, and wildlife and the impact of nanoparticles entering the food chain on human and animal health.

In early work with aquatic ecosystems, it has been reported that uncoated fullerenes exerted oxidative stress and caused severe lipid peroxidation in fish brain (Oberdorster, 2004). Water flea, *Daphnia magna*, ingested lipid-coated nanotubes through normal feeding and utilized the lipid coating as a food source (Roberts et al., 2007). After the ingestion, the nanotubes were no longer water soluble and accumulated on the external surfaces of the organisms making them less mobile. These results show that nanoparticles can accumulate and affect aquatic animals.

Research on nanoparticle effects on higher plants is limited but a few studies have reported both negative and positive effects of nanoparticles. On the positive side, Lu et al. (2002) reported that, at low concentrations, a mixture of nano-SiO_2 and nano-TiO_2 increased nitrate reductase activity, enhanced the water and the nutrient uptake, stimulated the antioxidant system and hastened germination and the growth of soybean (*Glycine max*). Nano-TiO_2 (at 0.25%) increased seed germination, plant dry weight, chlorophyll production, and the RuBP activity and the rate of photosynthesis of spinach (*Spinacia oleracea*) while concentrations greater than 0.4% were detrimental to plant growth (Zheng et al., 2005). Nano-TiO_2 at 0.25% has been reported to accelerate the rate of O_2 evolution of spinach leaves under illumination and to protect chlorophyll from photooxidation under long-term illumination suggesting that the enhancement of photosynthesis might be related to elevated photochemical reactions of chloroplast (Hong et al., 2005a,b). Nano-TiO_2 at 0.25% increased the activities of superoxide dismutase, catalase, and peroxidase while decreasing the accumulation of reactive oxygen-free radicals (Hong et al., 2005b). Chlorophyll-bound gold and silver nanoparticles have been shown to enhance the production of excited electrons in the photosynthetic complex thus enhancing photosynthesis (Govorov and Carmeli, 2007).

On the contrary, Yang and Watts (2005) reported that uncoated alumina nanoparticles (at 2000 mg/L) reduced the root growth of corn, carrot, cucumber, soybean, and cabbage seedlings while alumina nanoparticles coated with phenanthrene had no effect on root growth. They attributed the protective effect of coated alumina nanoparticles to their ability to scavenge free radicals and prevent oxidative damage. However, it was not clear if the toxicity of uncoated alumina nanoparticles was a direct effect of nanoparticles or the dissolution of aluminum ion.

Nanoparticles can be made from a variety of materials and their toxicity may vary significantly. A recent study by University of Massachusetts researchers evaluated plant (radish, rape, rye, lettuce, corn, and cucumber) toxicity to five types of nanoparticles: aluminum, alumina, multi-walled carbon nanotubes (MWNTs), zinc, and zinc oxide (Lin and Xing, 2007). Their results show that the toxicity can vary among different types of nanoparticles and plant species. At 2000 mg/L, aluminum, alumina, and MWNT suspensions did not affect seed germination but Zn and ZnO suspensions inhibited germination of rye and corn seeds. Suspensions of Zn and ZnO (2000 mg/L) significantly inhibited

the root growth of corn and terminated the root growth of all other five species tested. MWNT suspension had no significant effect on root growth. Alumina suspensions reduced corn root growth but had no effect on other crops. Aluminum suspensions had no effect on cucumber roots but promoted the root growth of radish and rape seedlings and retarded the root growth of rye and lettuce seedlings. The IC_{50} of nano-Zn and nano-ZnO were estimated to be near 50 mg/L. The toxicity to Zn and ZnO nanoparticles was related to the direct effect of nanoparticles on the root surface; not due to the ion dissolution. In a follow-up study, Lin and Xing (2008) reported that ZnO nanoparticles damaged root tip, entered root cells, and inhibited seedling growth, thus reducing the biomass. However, the translocation of ZnO nanoparticles from root to shoots was minimal in their study.

In most of these quoted studies, seeds were incubated in nanoparticle suspensions prior to germination. It is not understood if plant roots can directly take up nanoparticles from the growing media and transport them through the vascular system. Furthermore, the effects of nanoparticles on physiological/metabolic processes and the productivity have not been investigated. In the next section, we present our approach on the uptake of nanoparticles by plants and show how the integration of nanoparticles by plants may affect the physiological processes and the productivity of the plants.

30.3 Uptake of Nanoparticles by Plants

30.3.1 Natural Organic Matter for Nanoparticle Suspension

Natural organic matter (NOM), originating from decomposed plants and animals, is a heterogeneous mixture of organic humic acids, hydrophilic acids, proteins, lipids, amino acids, and hydrocarbons. The collective term of NOM is considered a network with a hydrophobic interior and a hydrophilic exterior. Such an amphiphilic structure of NOM permits its binding with hydrophobic nanoparticles and allows for the transport and the biodistribution of nanomaterials in the environment. As a result, NOM is also regarded a major pollutant carrier in the environment (Baker, 1991).

NOM has recently been found to be effective in suspending MWNTs (Hyuang et al., 2007). To mimic the distribution of nanoparticles in natural water sources, we suspended various types of nanoparticles—SWNTs, MWNTs, C_{60}, and C_{70}, in NOM. As shown in Figure 30.3, other than SWNTs, all types of carbon nanoparticles were well dispersed in the NOM. Our hypothesis is that the morphology of carbon nanostructures dictates their binding with NOM. The binding energy is believed to depend on the interaction between the hydrophobic interiors of the NOM and the surfaces of the carbon nanoparticles, as well as the pi-stacking between the aromatic moieties of NOM and the aromatic rings of the carbon nanoparticles. For fullerenes C_{60} and C_{70}, each of a diameter of less than 1 nm, NOM could act as the guest while fullerene agglomerations act as the hosts (Figure 30.3g), forming supramolecular complexes in aqueous solutions.

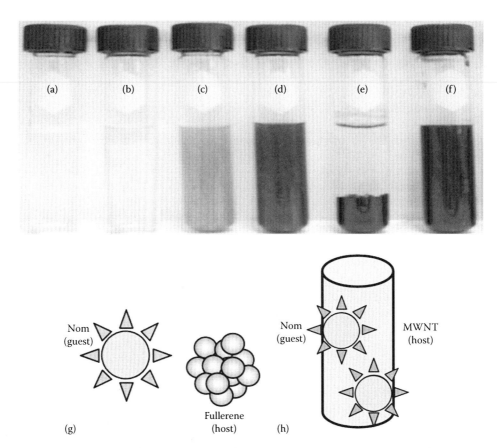

FIGURE 30.3   (a) Milli-Q water; (b) NOM of 0.1 mg/mL; (c) C_{60} of 0.4 mg/mL suspended in NOM; (d) C_{70} of 0.4 mg/mL suspended in NOM; (e) SWNTs of 0.4 mg/mL precipitate out of NOM. Notice NOM is depleted from the solution as well. (f) MWNTs of 0.4 mg/mL, well dispersed in NOM of 0.1 mg/mL. (g) Hypothesized model for fullerenes suspended in NOM. (h) Hypothesized model for MWNTs suspended in NOM.

For larger structures such as MWNTs, their carbon surfaces can be approximated as flat sheets that may be stacked with NOM to elicit solubility. SWNTs, on the other hand, with their small diameter (~1 nm) and large aspect ratio, must first form bundles for NOM to bind to, and the complexes of SWNT-NOM could eventually be pulled out of the solution due to the gravitational force (Figure 30.3e).

NOM in freshwater ecosystems usually assumes a concentration 1~100 mg/L. To mimic the natural ecosystems, we prepared a Nordic NOM solution of 100 mg/L in Milli-Q water and suspended fullerene C_{70} and MWNTs in the NOM. The solubility of nanoparticles in NOM was first estimated as the following. We measured the absorbance a_1 of a nanoparticle suspension using a Biomate 3 immediately after the sample was prepared. After one day of settlement, the absorbance a_2 of the nanoparticle suspensions was measured. The nominal concentration c_1 of the nanoparticle suspension was calculated by dividing the quantity of nanoparticles added to the suspension by the volume of the suspension. The actual concentration of the nanoparticle suspension was derived from this simple formula: $c_2 = c_1 * a_2/a_1$. Typically, 50% or less carbon-based nanoparticles can be stably suspended in the NOM.

The size distribution of the nanoparticle-NOM suspension was measured using a Zetasizer®, a dynamic light scattering device that is effective for characterizing nanoparticle suspensions. Here, "size" refers to the hydrodynamic diameter of the nanoparticle. According to Rayleigh's approximation, the intensity of scattering of a nanoparticle is proportional to the sixth power of its diameter. The sample of NOM in Milli-Q water (100 mg/L) displayed a number of size peaks at, say, 1.11 nm (area percentage: 38.9%), 4.89 nm (17.7%), and 31.21 nm (21.1%). This size distribution indicates that the filtration of NOM may be necessary when examining nanoparticle uptake by plants. Due to the complex origin of the NOM and the dynamic nature of their conformation in the aqueous phase, repeatability of such size characterization is low. Such a low repeatability is further reflected by the fact that nanoparticles suspended in NOM often have rather unstable size distributions.

30.3.2 Plant Culture for Uptake

To study the plant uptake of nanoparticles, we chose to use rice (*Oryza sativa*) as the starting plant materials for our research. Rice is one of the most important crop species and a major staple food, feeding about half of the world's population (Bajaj and Mohanty, 2005). Understanding how rice cells interact with different nanoparticles will greatly facilitate similar studies using other food crop plants, shedding light on the impact of nanoparticles on plant growth, human health, and environmental protection.

Rice seeds were first germinated in a growth chamber using a nutrition medium in the presence of various concentrations of different nanoparticles. The rice-germination buffer is composed of half strength MS basal salts and vitamins (Murashige and Skoog, 1962), and 7.5 g/L sucrose. The pH of the medium was adjusted to 5.7 before autoclaving at 120°C for 20 min. Freshly harvested rice seeds (*O. sativa* L. ssp. *japonica*, cv Taipei 309) of similar size were soaked in an ethanol solution for 30 s and surface sterilized twice in Clorox® bleach with two drops of Tween-20™ (Polysorbate 20) and vigorously shaken for 30 min. After rinsing several times in sterile distilled water, 10 seeds were transferred to each 100 × 15 mm Petri dish that contained different concentrations of C_{70}-NOM and MWNT-NOM for germination at 25 ± 1°C for 2 weeks. Rice seeds that germinated under the same conditions in the presence of NOM, but without nanoparticles were also included in our experiments as controls. To examine the potential impact of chemicals used for seed sterilization on plant nanoparticle uptake, additional control experiments with the seeds washed by distilled water without chemical sterilization were conducted. There was no significant difference in the nanoparticle uptake observed between the two treatments.

The 2 week-old seedlings maintained in the growth chamber with a culture medium were then transplanted into soil in a big pot and grown to maturity in a greenhouse under normal day/light conditions. The standard maintenance procedures, such as watering, fertilization, and disease and pest controls were followed at the various developmental stages of rice growth. For each sample concentration, at least five plants were maintained for analyses. These plants were referred to as the first generation. To investigate the possible generational transmission of nanomaterials, mature seeds from the control plants and the C_{70}-treated plants were harvested (approximately 6 months after germination) and 60 seeds of a similar size for each plant were chosen and sterilized using the same procedure as described above. Ten seeds were planted, one in each Petri dish filled with the rice-germination medium and kept at 25 ± 1°C for 2 weeks. These germinated plants, without further addition of nanomaterials, were termed as the second generation. Here, we refer to the plants grown in the germination buffer as the control, and the plants in NOM suspended in the germination buffer as the negative control. The same amounts of NOM were used for C_{70}-NOM and MWNT-NOM for each selected concentration.

30.3.3 Bright-Field Microscopy of Uptake

Bright-field microscopy, whose resolution is limited at an airy diffraction of ~λ/2, has been shown as an effective tool in describing the uptake of nanoparticles by plants. Nanoparticles suspended in NOM are highly mobile and can readily accumulate to form micron-sized aggregates when encountering each other in or near the vascular system. Over time, these aggregates grow in size due to the hydrophobic interaction between the nanoparticles. When combined with fluorescence microscopy, Fourier transform infrared spectroscopy (FTIR), and electron microscopy, bright-field microscopy offers key information on the biodistribution of nanoparticles in plants. In addition to being the least expensive, bright-field microscopy is also the most straightforward to operate and is the least invasive.

For plants incubated with a fullerene C_{70}-NOM for approximately 1 week, we observed the appearance of black aggregates in the seeds, roots, stems, and leaves (Figure 30.4). The appearance of black aggregates was much less prevalent in plant compartments when incubated with MWNT-NOM. A rather large amount of MWNTs was observed on the root surfaces of the plants (Figure 30.5), possibly due to the large dimension of MWNTs and their high adsorption for root surfaces.

Our bright-field microscopy (Figure 30.6) suggests that the uptake of nanoparticles shared the same vascular system with

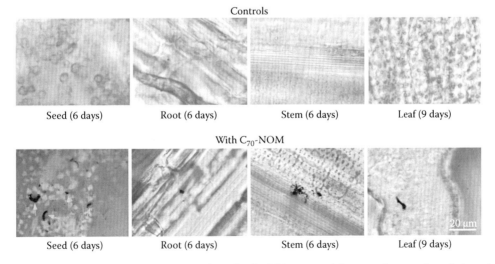

FIGURE 30.4 Uptake of C_{70} by rice plants. The upper row shows bright-field images of the controls approximately 1 week since incubation. Micron-sized black aggregates of nanoparticles are observed in the corresponding bottom row images of plants treated with C_{70}-NOM. The scale bar is 20 μm for all images. (From Lin et al., *Small*, 5, 1128, 2009. With permission.)

FIGURE 30.5 Adsorption of MWNTs (black aggregates) on rice-plant root surfaces.

water and nutrients, and the transportation of the nanoparticles occurred from the plant roots to the stems, the leaves, and the harvested seeds (if there are any), driven by transpiration (i.e., the evaporation of water through plant leaves), passive diffusion, osmotic pressure, capillary forces, etc. This uptake mechanism was proposed based on the observation that fullerene aggregates

were found mostly in or near the vascular system, less in the cortex, and much less in the epidermis of the plants. Also, fullerene aggregates were found to decrease in plant roots and gradually built up in plant stems and leaves over time. One intriguing observation we made was the accumulation of fullerene C_{70} aggregates in the stomata cells of the plant leaves (Figure 30.7). Since stomata cells are pores responsible for gas exchange, this accumulation implies the impact of nanoparticle uptake on plant photosynthesis, as well as development. It should be noted that even for the second-generation plants (generated from the harvested seeds of the plants exposed to nanoparticles in their first 2 weeks, without direct exposure to nanoparticles) the appearance of fullerene nanoparticles was abundant along the vascular system (Figure 30.8).

30.3.4 SEM and TEM of Uptake

In addition to observing the tissue distribution of nanoparticle uptake, as discussed above using bright-field microscopy, we further examined the cellular distribution of nanoparticles using electron microscopy. The principle of electron microscopy lies in constructing images using scattered (scanning electron microscopy or SEM) or refracted (transmission electron microscopy or TEM) electron beams from the sample under study. As a result of the short "wavelength" of electrons, as compared to photons,

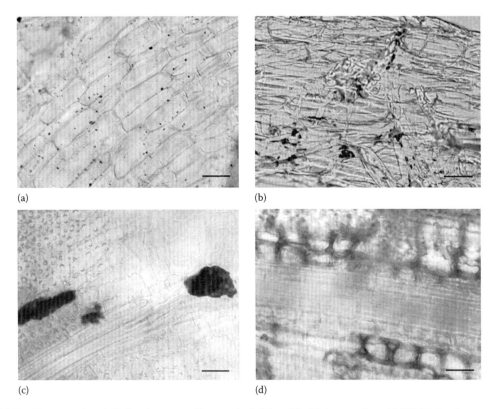

(a)

(b)

(c)

(d)

FIGURE 30.6 (a) Significant uptake of C_{70} by plant roots (C_{70}: nominal 400 mg/L). (b) Aggregation of MWNTs in plant roots (MWNT: nominal 400 mg/L). Little MWNTs are seen inside the root cells. (c) Aggregation of C_{70} in the vascular system and nearby tissues of a leaf tissue (C_{70}: nominal 400 mg/L). The vascular system is partially blocked by the aggregation. (d) Leakage of C_{70} (darker areas) from the vascular system to nearby leaf tissue (C_{70}: nominal 400 mg/L). All scale bars: 20 μm. (From Lin et al., *Small*, 5, 1128, 2009. With permission.)

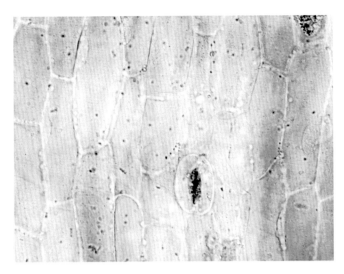

FIGURE 30.7 Bright-field image of a plant leaf. Black aggregates appear at the openings of the stomata cells, hindering water evaporation and gas exchange. No aggregates were found for control plants (data not shown).

FIGURE 30.8 Bright-field image of a 2 week-old second-generation plant leaf. Note the appearance of a black fullerene aggregate in the vascular system.

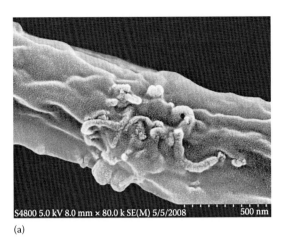

(a)

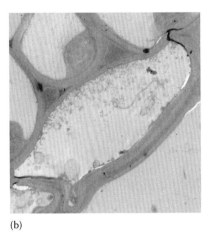

(b)

FIGURE 30.9 (a) Adsorption of MWNTs onto a rice-plant root. (b) Uptake of C_{70} and their aggregation in rice-plant leaf cells.

electron microscopy typically possesses an image resolution of ~1 nm or finer. In other words, electron microscopy can resolve structures hundred times smaller than does bright-field microscopy. However, this superior resolution of electron microscopy comes at a cost: specimens for electron microscopy ought to be dried (or denatured for biological samples) and often coated with metallic nanoparticles to afford electron conductivity. In these experiments, specifically, 10 samples of rice plant roots and leaves were evenly coated with a thin film of platinum (~5 nm). SEM imaging was performed using an FE-SEM microscope operating at 5 kV. For TEM imaging, 10 samples of the roots and the leaves of rice plants were placed in 3.5% Glutaraldehyde, post

fixed in osmium tetroxide, dehydrated in a graded series of ethanol, and embedded in an LR White-embedding media. Samples were polymerized overnight and sectioned using a microtome. Tissues of plant roots and leaves were cut at 60–90 nm. The lattice spacing of C_{70} particles was analyzed by performing a fast Fourier transform (FFT) of the TEM images, using the software "Diffractogram."

Figure 30.9a shows an SEM image of MWNTs coated on the surface of rice plant roots, consistent with our observation in Figure 30.5 that MWNTs were mostly adsorbed onto the root surfaces without uptake. Figure 30.9b shows a TEM image of the appearance of fullerene C_{70} particles in the rice-plant leaf cells. The uptake of the fullerene nanoparticles was evident although the mechanisms involved in the uptake remained unclear.

30.3.5 Monitoring Uptake Using FTIR

Spectroscopic tools such as Raman and FTIR have been used extensively to probe the presence or absence of fullerenes since C_{60}, C_{70}, etc., exhibit unique spectral fingerprints. To confirm that the aggregates were composed of C_{70} or C_{70} derivatives, the Fourier transform (FT) Raman and the infrared (IR) spectra were acquired at room temperature for both the first- and second-generation rice plants. For the FT-Raman spectroscopy measurements, a 1064 nm laser beam was focused to a 1 mm spot size, and the Raman-scattered signal was detected using a liquid nitrogen-cooled germanium detector. A Fourier transform infrared spectrometer equipped with a deuterated triglycine sulfate detector was used to collect the infrared absorption spectrum of the selected samples. The FTIR absorption measurements used ~3 mg of the root, the leaf, or the stem mixed with ~50 mg of KBr powder and pressed into a ~5 mm diameter. The sample chamber was evacuated to eliminate an interfering IR absorption by water vapor and the CO_2 present in the ambient atmosphere.

Typical FT-Raman (lower traces) and IR-spectral (upper traces) fingerprints are presented in Figure 30.10 for C_{70}, control, first-generation seeds and leaves, and second-generation leaves. Clearly, the dominant FT-Raman (indicated by "+") and the FTIR (indicated by "diamonds") features (Eklund et al., 1995) of C_{70} were observed in the first-generation seeds and leaves and in the second-generation leaves, thus confirming the uptake and transmission of C_{70}. After collection of the absorption spectrum, each of the C_{70} peaks was fit to a Lorentzian line shape and the area under the peak (integrated intensity) was calculated via equation: $I = A/\Gamma \pi$, where A was the amplitude, and Γ was the FWHM. This area was then converted into a percent uptake of C_{70} by dividing it by the total area of all the combined samples.

To quantify the dynamics of the C_{70} uptake, a detailed FTIR study was carried out for the roots, stems, and seeds of the first-generation rice plants when the concentration of C_{70} was increased from 20 to 800 mg/L. As shown in Figure 30.11c, C_{70} particles were prevalent in the roots as well as in the stems and leaves of the 2 week-old plants, while the distribution of C_{70} in these plants

showed no significant concentration dependence. The prevalence of C_{70} in plant leaves and roots is also evident in Figures 30.6a and 30.9b. For the mature (6 month-old) plants, however, C_{70} was predominantly present in or near the stems' vascular systems, less in the leaves, and understandably even less in the seeds due to the multiplied uptake rates (green bars). Furthermore, no C_{70} was left in the roots of the mature plants, suggesting a robust transport of nanomaterials from the plant roots to the leaves.

To compare the uptake capacity of plant seeds vs. roots, we germinated two sets of rice seeds in a rice-germination buffer and in a C_{70}-NOM mixed rice-germination buffer (20 mg/L), respectively. Within 3 days, these seeds started germination to first produce shoots and then roots. One week after shooting at the three-leaf stage, the seeds, no longer able to provide sufficient nutrients for the newly germinated plants, detached from the seedlings. At this point, we transferred the seedlings from the rice-germination buffer to be in contact for 1 week with the C_{70} suspensions (20 mg/L), prior to the FTIR study of the roots, the stems and the leaves of these plants. This set of samples was termed as "roots exposed." The other set of samples that had been exposed to nanoparticles from the beginning of the germination was termed as "seeds + roots exposed" and the FTIR study was conducted for the roots, the stems, and the leaves of these plants at the end of the second week. Since shoots usually come out 1–2 days earlier than roots during seed germination, C_{70} taken up by the seeds could first be transported to the shoots (stems and leaves) and then to the roots. This may have led to more accumulation of nanoparticles in the leaves than in the roots (Figure 30.11d, "seeds + roots exposed," dark gray bars). The "roots exposed" samples showed a different trend of nanoparticle translocation possibly because C_{70} first entered the roots and was then transported to the stems and the leaves (Figure 30.11d, "roots exposed," gray bars).

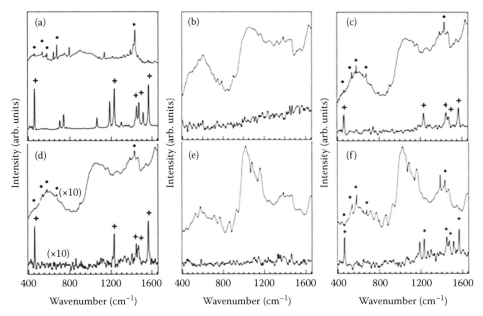

FIGURE 30.10 FT-Raman (lower traces) and FTIR (upper traces) of (a) C_{70}, (b) control leaf, (c) first-generation leaf, (d) second-generation leaf with uptake of C_{70}, (e) control seed, and (f) first-generation seed with uptake of C_{70}. (From Lin et al., *Small*, 5, 1128, 2009. With permission.)

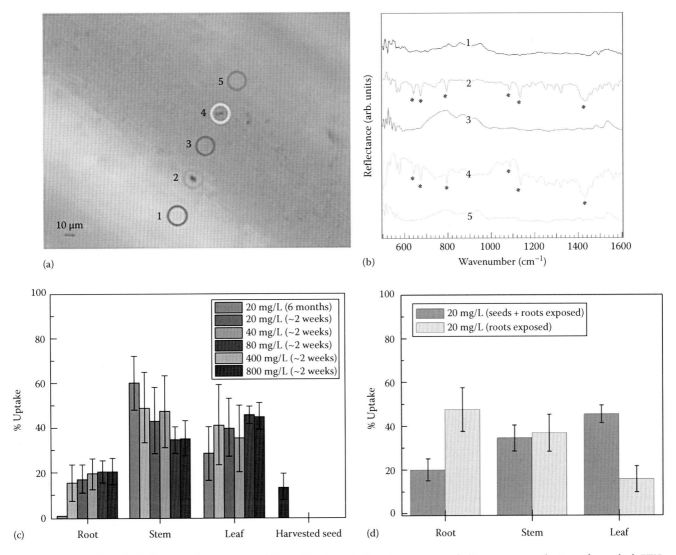

FIGURE 30.11 (a) Bright-field image of a sectioned leaf tissue. The circles indicate the positions of a line-scan across the tissue from which FTIR spectra (panel b) were obtained. (b) Infrared features corresponding to C_{70} are seen in spectra labeled 2 and 4. (c) Uptake dynamics for rice plants of ~2 weeks and 6 months after germination. All samples were initially exposed to C_{70} for 2 weeks through their seeds. Only the 6 month-old plants had harvested seeds. (d) Comparison of uptake dynamics for rice plants exposed through "seeds + roots exposed" (dark gray bars) and "roots exposed" (gray bars) to C_{70} of 20 mg/L. (From Lin et al., *Small*, 5, 1128, 2009. With permission.)

In contrast to C_{70}, the uptake of MWNTs at concentrations of 20–800 mg/L was found to be insignificant, with few black aggregates appearing in the vascular system and almost none in the plant tissues (Figure 30.6b). In our study the minimal uptake of MWNTs was not characterized also due to the relatively weak Raman and non-existent IR signatures for the MWNTs (Reich et al., 2001).

30.3.6 Critical Discussion

The accumulation and the transformation of nanoparticles in plant tissues and cells suggests a plausible mechanism for nanoparticle uptake: a dynamic competition between nanotransport driven by water and nanoparticle convections and the physical hindrances of plant tissues and nanoparticle aggregation.

Individual nanoparticles may enter plant roots through osmotic pressure, capillary forces, pores on cell walls, and intercellular plasmadesmata (Smith, 1978), or via the highly regulated symplastic route. Once in the plant roots and stems, individual C_{70} nanoparticles may share the vascular system with water and nutrients and may be transported via transpiration. Individual C_{70} nanoparticles may also form aggregates or even clog the vascular system (Figure 30.6c) due to hydrophobic interaction or may leak into nearby tissues and cells (Figure 30.6d) via the mechanisms that were discussed above for plant roots. At high concentrations, C_{70} aggregation within the vascular system and in plant tissues and cells is expected to interfere with the nutrients and the water uptake, and hinder plant development. Indeed, flowering of the rice plants incubated with C_{70}-NOM (400 mg/L) was delayed by at least 1 month and their seed-setting rate reduced by 4.6%, compared to the controls or the NOM-fed plants. MWNTs, meanwhile, are larger one-dimensional nanostructures and, unless oriented approximately perpendicular to plant tissues, are less likely to enter plants (Figure 30.9a). Indeed, our bright-field and electron-microscopy imaging (Figure 30.6f inset) showed that MWNTs adsorbed to the plant root surfaces possibly because of the high affinity of the tubes for the epidermis and the waxy casparian strips of the roots. In light of the dynamic-binding process of MWNT-NOM and of the Stokes–Einstein relation (Batchelor, 1973) that entails a slower-diffusion rate for a larger-sized MWNT-NOM than for NOM or C_{70}-NOM, a concentration gradient of NOM can form across the plant roots, which would favor the transport of NOM through plant tissues and the apoplastic pathway (Yeo et al., 1987).

30.4 Future Perspective

The uptake, the translocation and the transmission of nanoparticles in generations of food-crop plants suggest the potential impacts of nanotechnology on the environment and for food safety. The effects of nanoparticles on plant growth and development also raise questions as to what extent genetic and molecular mechanisms may mediate plant responses to nanoparticles exposure, and how to control such responses for mitigating the adverse effects of nanomaterials on plant development and the environment. It is noted from our electron-microscopy imaging (Figure 30.12) that nanoparticles may induce physical damage or chemical oxidation to the plant cell walls or the plasma membranes to facilitate their uptake. Once getting into the cells, nanoparticles may further penetrate into other membrane-bounded compartments, such as the nucleus, the chloroplast and the mitochondria, exerting on the functionality of these organelles. It is very likely that the nanoparticles taken up by plant cells may modify the structure of the plasma membrane, and other inner membrane systems, activate or inactivate

proteins associated with membranes or those in the cytosol. These could lead to a dramatic change in the signal-transduction cascade in plant cells and alternate the expression of a specific subset of genes, and consequently impact on plant development. Future studies employing proteomics and genomics approaches should allow an identification of key target genetic components and regulatory networks being affected by particular nanoparticles, or those that are involved in plant response to nanomaterial exposure. Using transgenic approaches, the expression of these nanoparticle-specific target genes could then be turned off or down-regulated; or turned on or up-regulated in the nanoparticle-treated plants to examine how these genes are involved in the dynamic process of plant uptake of nanoparticles and in mediating plant responses to the nanoparticle invasion. The fundamental knowledge gained from these studies will provide significant insight into the genetic mechanisms governing plant-nanoparticle interaction. This knowledge can be used to develop molecular strategies to genetically engineer important crop species for a modified expression of selected target genes or pathways for positive plant-nanoparticle interaction, thus contributing to agriculture production.

Nanoparticle penetration of the plant cell wall also provides great opportunities to explore the potential of using nanomaterials as vehicles for the delivery of DNA, proteins, nucleotides and chemicals into plant cells as is done routinely in mammalian cells (Bharali et al., 2005; Roy et al., 2005). Early attempts using the mesoporous silica nanoparticle (MSN) system demonstrated DNA and chemical transport into isolated plant cells and intact leaves. However, this still requires a "gene gun"-mediated cell-wall breakage for delivery (Torney et al., 2007). Using a plasmid DNA encoding a reporter gene, the jellyfish green-florescent protein (GFP), we have recently showed successful noninvasive, poly(amidoamine) dendrimer-mediated DNA delivery into plant cells (Pasupathy et al., 2008). Further studies for a better understanding of nanoparticle-plant cell interaction will not

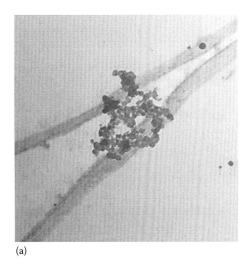

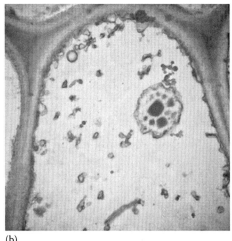

(a)   (b)

FIGURE 30.12   (a) C_{70} aggregates translocating across rice-plant roots cells. (b) C_{70} aggregates translocating across a rice-plant leaf cell, possibly interacting with the cell nucleus.

only significantly enhance plant science research and advance our knowledge in plant biology, but also lead to the development of mature and novel tools for agricultural biotechnology and agricultural production.

30.5 Summary

This chapter reviewed nanoparticle uptake by mammalian cells and plants. We have especially presented the state of the art of nanoparticle uptake by plants. It is our understanding that this chapter will serve as a reference for researchers and the general public who are interested in the perspectives of integrating nanoscience with biology, for the practices of nanomedicine, biosensing, toxicology, and environmental science and engineering. As nanoparticles propagate into the environment or, more specifically, enter the pipeline of the food chain, it has become increasingly crucial to obtain a fundamental and sound understanding of the fate and behavior of nanoparticles in the liquid phase and in living systems. Obtaining such an understanding is nontrivial, mainly due to the complexity of nanoparticles in transforming, transporting, and integrating with their host biological and ecological systems. However, with carefully designed interdisciplinary approaches, as illustrated in this chapter, research in this area can offer guidance to the safe development of nanotechnology and contribute to a sustainable environment that is intimately connected to human health.

Acknowledgments

Ke acknowledges an NSF Career award #CBET-0744040. Luo acknowledges USDA grants BRAG 2007-33522-18489 and CSREES SC-1700315. This is Technical Contribution No. 5565 of the Clemson University Experiment Station.

References

Andrews, R., Jacques, D., Rao, A. M., Derbyshire, D., Qian, D., Fan, X., Dickey, E. C., and Chen, J. 1999. Continuous production of aligned carbon nanotubes: A step closer to commercial realization. *Chem. Phys. Lett.* 303: 467–474.

Ashcroft, J. M., Tsyboulaski, D. A., Hartman, K. B., Zakharian, T. Y., Marks, J. W., Weisman, R. B., Rosenblum, M. G., and Wilson, L. J. 2006. Fullerene (C_{60}) immunoconjugates: Interaction of water-soluble C_{60} derivatives with the murine anti-gp240 melanoma antibody. *Chem. Commun.* 28: 3004–3006.

Bachilo, S., Balzano, L., Herrera, J. E., Pompeo, F., Resasco, D. E., and Weisman, D. E. 2003. Narrow (n,m)-distribution of single-walled carbon nanotubes grown using a solid supported catalyst. *J. Am. Chem. Soc.* 125: 11186–11187.

Bachilo, S. M., Strano, M. S., Kittrell, C., Hauge, R. H., Smalley, R. E., and Weisman, R. B. 2002. Structure-assisted optical spectra of single-walled carbon nanotubes. *Science* 298: 2361–2366.

Bahr, J. L., Mickelson, E. T., Bronikowski, M. J., Smalley, R. E., and Tour, J. M. 2001. Dissolution of small diameter single-wall carbon nanotubes in organic solvents? *Chem. Commun.* 1: 193–194.

Bajaj, S. and Mohanty, A. 2005. Recent advances in rice biotechnology—Towards genetically superior transgenic rice. *Plant Biotechnol. J.* 3: 275–307.

Baker, R. A. 1991. *Organic Substances and Sediments in Water*. Vol. 1–3, Lewis Publishers, Chelsea, MI.

Banerjee, S. and Wong, S. S. 2002. Structural characterization, optical properties, and improved solubility of carbon nanotubes functionalized with Wilkinson's catalyst. *J. Am. Chem. Soc.* 124: 8940–8948.

Batchelor, G. K. 1973. *An Introduction to Fluid Dynamics*. Cambridge University Press, Cambridge, U.K.

Bharali, D. J. et al. 2005. Organically modified silica nanoparticles: A nonviral vector for in vivo gene delivery and expression in the brain. *Proc. Natl. Acad. Sci. USA* 102: 11539–11544.

Buchanan, B. B., Gruissem, W., and Jones R. L. 2000. Biochemistry & molecular biology of plants. American Society of Plant Physiologists, Rockville, MD.

Chen, R. J., Bangsaruntip, S., Drouvalakis, K. A., Shi Kam, N. W., Shim, M., Li, Y., Kim, W., Utz, P. J., and Dai, H. J. 2003. Noncovalent functionalization of carbon nanotubes for highly specific electronic biosensors. *Proc. Natl. Acad. Sci. USA* 100: 4984–4989.

Colvin, V. L. 2003. The potential environment impact of engineered nanomaterials. *Nat. Biotechnol.* 21: 1166–1170.

Eklund, P. C., Rao, A. M., Wang, Y., Zhou, P., Wang, K., Holden, J. M., Dresselhaus, M. S., and Dresselhaus, G. 1995. Optical properties of C_{60} and C_{70} based solid films. *Thin Film Solids* 257: 211.

Gharbi, N., Pressac, M., Hadchouel, M., Szwarc, H., Wilson, S. R., and Moussa, F. 2005. [60]Fullerene is a powerful antioxidant in vivo with no acute or subacute toxicity. *Nano Lett.* 5: 2578–2585.

Govorov, A. O. and Carmeli, I. 2007. Hybrid structures composed of photosynthetic system and metal nanoparticles: Plasmon enhancement effect. *Nano Lett.* 7: 620–625.

Hata, K., Futaba, D., Mizuno, K., Namai, T., Yumura, M., and Iijima, S. 2004. Water-assisted highly efficient synthesis of impurity-free single-walled carbon nanotubes. *Science* 306: 1362–1364.

Heller, D. A., Baik, S., Eurell, T. E., and Strano, M. S. 2005. Single-walled carbon nanotube spectroscopy in live cells: Toward long-term labels and optical sensors. *Adv. Mater.* 17: 2793–2799.

Hong, F. S., Zhou, J., Liu, C., Yang, F., Wu, C., Zheng, L., and Yang, P. 2005a. Effect of nano-TiO_2 on photochemical reaction of chloroplast of spinach. *Biol. Trace Element Res.* 105: 269–279.

Hong, F. S., Yang, F., Liu, C., Gao, Q., Wan, Z. G., Gu, F. G., Wu, C., Ma, Z. N., Zhou, J., and Yang, P. 2005b. Influences of nano-TiO_2 on the chloroplast aging of spinach under light. *Biol. Trace Element Res.* 104: 249–260.

Hwang, K. C. and Mauzerall, D. 1993. Photoinduced electron transport across a lipid bilayer mediated by C_{70}. *Nature* 361: 138–140.

Hyuang, H., Fortner, J. D., Hughes, J. B., and Kim, J. H. 2007. Natural organic matter stabilizes carbon nanotubes in the aqueous phase. *Environ. Sci. Technol.* 41: 179–184.

Journet, C., Maser, W., Bernier, P., Loiseau, A., Chapelle, M., Lefrant, S., Deniard, P., Lee, R., and Fischer, J. 1997. Large-scale production of single-walled carbon nanotubes by the electric-arc technique. *Nature* 388: 756.

Kam, N. W. S., Jessop, T. C., Wender, P. A., and Dai, H. J. 2004. Nanotube molecular transporters: Internalization of carbon nanotube-protein conjugates into mammalian cells. *J. Am. Chem. Soc.* 126: 6850–6851.

Pasupathy, K., Lin, S., Hu, Q., Luo, H., and Ke, P. C. 2008. Direct plant gene delivery with a poly(amidoamine) dendrimer. *Biotechnol. J.* 3: 1078–1082.

Ke, P. C. and Qiao, R. 2007. Carbon nanomaterials in biological systems. *J. Phys. Condens. Matter* 19: 373101(25 pp).

Kostarelos, K., Lacerda, L., Pastorin, G., Wu, W., Wieckowski, S., Luangsivilay, J., Godefroy, S., Pantarotto, D., Briand, J.-P., Muller, S., Prato, M., and Bianco, A. 2007. Cellular uptake of functionalized carbon nanotubes is independent of functional group and cell type. *Nat. Nanotechnol.* 2: 108–113.

Lin, D. and Xing, B. 2007. Phytotoxicity of nanoparticles: Inhibition of seed germination and root growth. *Environ. Pollut.* 150: 243–250.

Lin, D. and Xing, B. 2008. Root uptake and phytotoxicity of ZnO nanoparticles. *Environ. Sci. Technol.* 42: 5580–5585.

Lin, S., Keskar, G., Wu, Y., Wang, X., Mount, A. S., Klaine, S. J., Moore, J. M., Rao, A. M., and Ke, P. C. 2006. Detection of phospholipid-carbon nanotube translocation using fluorescence energy transfer. *Appl. Phys. Lett.* 89: 143118-1-3.

Lin, S., Reppert, J., Hu, Q., Hudson, J. S., Reid, M. L., Ratnikova, T. A., Rao, A. M., Luo, H., and Ke, P. C. 2009. Uptake, translocation, transmission of carbon nanomaterials in rice plants. 5: 1128–1132.

Lu, C. M., Zhang, C. Y., Wen, J. Q., Wu, G. R., and Tao, M. X. 2002. Research of the effect of nanometer materials on germination and growth enhancement of Glycine max and its mechanism. *Soybean Sci.* 21: 168–172.

Lu, Q., Moore, J. M., Huang, G., Mount, A. S., Rao, A. M., Larcom, L. L., and Ke, P. C. 2004. RNA polymer translocation with single-walled carbon nanotubes. *Nano Lett.* 4: 2473–2477.

Matarredona, O., Rhoads, H., Li, Z. R., Harwell, J. H., Balzano, L., and Resasco, D. E. 2003. Dispersion of single-walled carbon nanotubes in aqueous solutions of the anionic surfactant NaDDBS. *J. Phys. Chem. B* 107: 13357–13367.

Mauter, M. S. and Elimelech, M. 2008. Environment applications of carbon-based nanomaterials. *Environ. Sci. Technol.* 42: 5843–5859.

Maynard, A. D., Aitken, R. J., Butz, T., Colvin, V., Donaldson, K., Oberdörster, G., Philbert, M. A., Ryan, J., Seaton, A., Stone, V., Tinkle, S. S., Tran, L., Walker, N. J., and Warheit, D. B. 2006. Safe handling of nanotechnology. *Nature* 444: 267–269.

Mitchell, D. T., Lee, S. B., Trofin, L., Li, N., Nevanen, T. K., Söderlund, H., and Martin, C. R. 2002. Smart nanotubes for bioseparations and biocatalysis. *J. Am. Chem. Soc.* 124: 11864–11865.

Murashige, T. and Skoog, F. 1962. A revised medium for rapid growth and bioassays with tobacco tissue cultures. *Physiol. Plant.* 15: 473–497.

Navarro, E., Baun, A., Behra, R., Hartmann, N. B., Filser, J., Miao, A. J., Quigg, A., Santschi, P.H., and Sigg, L. 2008. Environmental behavior and ecotoxicity of engineered nanoparticles to algae, plants, and fungi. *Ecotoxicology* 17: 372–386.

Nel, A., Xia, T., Madler, L., and Li, N. 2006. Toxic potential of materials at the nanolevel. *Science* 311: 622–627.

Oberdorster, E. 2004. Manufactured nanomaterials (fullerenes, C60) induce oxidative stress in the brain of juvenile large mouth bass. *Environ. Health Perspect.* 112: 1058–1062.

O'Connell, M. J., Boul, P., Ericson, L. M., Huffman, C., Wang, Y., Haroz, E., Kuper, C., Tour, J., Ausman, K. D., and Smalley, R. E. 2001. Reversible water-solubilization of single-walled carbon nanotubes by polymer wrapping. *Chem. Phys. Lett.* 342: 265–271.

O'Connell, M. J., Bachilo, S. M., Huffman, C. B., Moore, V. C., Strano, M. S., Haroz, E. H., Rialon, K. L., Boul, P. J., Noon, W. H., Kittrell, C., Ma, J., Hauge, R. H., Weisman, R. B., and Smalley, R. E. 2002. Band gap fluorescence from individual single-walled carbon nanotubes. *Science* 297: 593–596.

Pantarotto, D., Partidos, C. D., Hoebeke, J., Brown, F., Kramer, E., Briand, J.-P., Muller, S. Prato, M., and Bianco, A. 2003. Immunization with peptide-functionalized carbon nanotubes enhances virus-specific neutralizing antibody responses. *Chem. Biol.* 10: 961–966.

Pantarotto, D., Singh, R., McCarthy, D., Erhardt, M., Braind, J.-P., Prato, M., Kostarelos, K., and Bianco, A. 2004. Functionalized carbon nanotubes for plasmid DNA gene delivery. *Angew. Chem.* 43: 5242–5246.

Pompeo, F. and Resasco, D. E. 2002. Water solubilization of single-walled carbon nanotubes by functionalization with glucosamine. *Nano Lett.* 2: 369–373.

Qiao, R., Roberts, A. P., Mount, A. S., Klaine, S. J., and Ke, P. C. 2007. Translocation of C_{60} and its derivatives across a lipid bilayer. *Nano Lett.* 7: 614–619.

Rao, R., Lee, J., Lu, Q., Keskar, G., Freedman, K. O., Floyd, W. C., Rao, A. M., and Ke, P. C. 2004. Single-molecule fluorescence microscopy and Raman spectroscopy studies of RNA bound carbon nanotubes. *Appl. Phys. Lett.* 85: 4228–4230.

Reich, S., Thomsen, C., Duesberg, G. S., and Roth, S. 2001. Intensities of the Raman-active modes in single and multi-wall nanotubes. *Phys. Rev. B* 63: 041401.

Roberts, A. P., Mount, A. S., Seda, B., Qiao, R., Lin, S., Ke, P. C., Rao, A. M., and Klaine, S. J. 2007. In vivo biomodification of lipid-coated carbon nanotubes by Daphnia magna. *Environ. Sci. Technol.* 41: 3025–3029.

Roy, I. et al. 2005. Optical tracking of organically modified silica nanoparticles as DNA carriers: A nonviral, nanomedicine approach for gene delivery. *Proc. Natl. Acad. Sci. USA* 102: 279–284.

Salonen, E., Lin, S., Reid, M. L., Allegood, M. S., Wang, X., Rao, A. M., Vattulainen, I., and Ke, P. C. 2008. Real-time translocation of fullerene reveals cell contraction. *Small* 4: 1986–1992.

Sano, M., Kamino, A., Okamura, J., and Shinkai, S. 2001. Self-organization of PEO-graft-single-walled carbon nanotubes in solutions and langmuir-blodgett films. *Langmuir* 17: 5125–5128.

Sayes, C. M., Fortner, J. D., Guo, W., Lyon, D., Boyd, A. M., Ausman, K. D., Tao, Y. J., Sitharaman, B., Wilson, L. J., Hughes, J. B., West, J. L., and Colvin, V.L. 2004. The differential cytotoxicity of water-soluble fullerenes. *Nano Lett.* 4: 1881–1887.

Shim, M., Shi Kam, N. W., Chen, R. J., and Dai, H. J. 2002. Functionalization of carbon nanotubes for biocompability and biomolecular recognition. *Nano Lett.* 2: 285–288.

Shoji, Y., Tashiro, K., and Aida, T. 2004. Selective extraction of higher fullerenes using cyclic dimers of zinc porphyrins. *J. Am. Chem. Soc.* 126: 6570–6571.

Smith, H. 1978. *The Molecular Biology of Plant Cells.* University of California Press, Berkeley, CA.

Strano, M. S., Zheng, M., Jagota, A., Onoa, G. B., Heller, D. A., Barone, P. W., and Usrey, M. L. 2004. Understanding the nature of the DNA-assisted separation of single-walled carbon nanotubes using fluorescence and Raman spectroscopy. *Nano Lett.* 4: 543–550.

Sun, Y., Wilson, S. R., and Schuster, D. I. 2001. High dissolution and strong light emission of carbon nanotubes in aromatic amine solvents. *J. Am. Chem. Soc.* 123: 5348–5349.

Thess, A., Lee, R., Nikolaev, P., Dai, H., Petit, P., Robert, J., Xu, C., Lee, Y. H., Kim, S. G., Rinzler, A. G., Colbert, D. T., Scuseria, G., Tománek, D., Fischer, J., and Smalley, R. E. 1996. Crystalline ropes of metallic carbon nanotubes. *Science* 273: 483–487.

Torney, F., Trewyn, B. G., Lin, V. S. Y., and Wang, K. 2007. Mesoporous silica nanoparticles deliver DNA and chemicals into plants. *Nat. Nanotechnol.* 2: 295–300.

Tsyboulski, D. A., Bachilo, S. M., and Weisman, R. B. 2005. Versatile visualization of individual single-walled carbon nanotubes with near-infrared fluorescence microscopy. *Nano Lett.* 5: 975–979.

Wang, J., Liu, G., and Jan, M. R. 2004a. Ultrasensitive electrical biosensing of proteins and DNA: Carbon-nanotube derived amplification of the recognition and transduction events. *J. Am. Chem. Soc.* 126: 3010–3011.

Wang, H., Zhou, W., Ho, D. L., Winey, K. I., Fischer, J. E., Glinka, C. J., and Hobbie, E. K. 2004b. Dispersing single-walled carbon nanotubes with surfactants: A small angle neutron scattering study. *Nano Lett.* 4: 1789–1793.

Yang, L. and Watts, D. 2005. Particle surface characteristics may play an important role in phytotoxicity of alumina nanoparticles. *Toxicol. Lett.* 158: 122–132.

Yeo, A. R., Yeo, M. E., and Flowers, T. J. 1987. The contribution of an apoplastic pathway to sodium uptake by rice roots in saline conditions. *J. Exp. Bot.* 38: 1141–1153.

Yurekli, K., Mitchell, C. A., and Krishnamoorti, R. 2004. Small-angle neutron scattering from surfactant-assisted aqueous dispersions of carbon nanotubes. *J. Am. Chem. Soc.* 126: 9902–9903.

Zheng, L., Hong, F., Lu, S., and Liu, C. 2005. Effect of nano-TiO_2 on strength of naturally aged seeds and growth of spinach. *Biol. Trace Element. Res.* 104: 83–91.

Zheng, M., Jagota, A., Semke, E. D., Diner, B. A., McLean, R. S., Lustig, S. R., Richardson, R. E., and Tassi, N. G. 2003. DNA-assisted dispersion and separation of carbon nanotubes. *Nat. Mater.* 2: 338–242.

Penetration of Metallic Nanomaterials in Skin

Biancamaria Baroli
Università di Cagliari

31.1 Introduction

Our societies are being revolutionized by nanosciences and, as a consequence, many nanotechnological products* are entering the market and becoming more and more familiar. However, the essence of a nanoproduct is not always clear to the general public, which has increasing concerns about their safety.

Therefore, before starting with the main topic of this chapter, it would be worthwhile to first define what a nanotechnological product is. To this end, the best approach is to begin from the definition of nanoscience, which is the "study of phenomena and manipulation of materials at atomic, molecular, and macromolecular scale, where properties differ significantly from those at larger scale" (What are nanoscience and nanotechnologies, *The Royal Society & The Royal Academy of Engineering*, 2008). The most important information in this statement is that materials (e.g., gold) behave differently if they are in bulk or nanometric state (e.g., gold ingot versus gold nanoparticle) because quantum effects dominate at the latter scale. In addition, nanotechnologies are the "design, production, characterization, and application of structure, devices, and systems by controlling shape and size at the nanometre scale" (What are nanoscience and nanotechnologies, *The Royal Society & The Royal Academy of Engineering*, 2008). Therefore, nanotechnological products are those whose dimensions range between the atomic level and 100 nm, are produced using nanotechnologies, and possess properties not observable in the bulk state.

These findings are extremely appealing and frightening at the same time. Appealing because the potential industrial applications of each specific material might increase tremendously, along with the strategies to solve problems. In this regard, even the way of treating and diagnosing illnesses is changing from conventional medical to nanomedical approaches. Nanomedicine focuses on nanoscale interactions between individual cells (or one of their organelle/structure) and individual bioactive molecules. Its goals are (1) optimizing pharmaceutical and/or regenerative therapies for individual patient, and (2) diagnosing diseases using nanotechnological bio-imaging platforms/techniques. On the other hand, potentialities of nanoproducts are also frightening because body defenses might be overcome at the nanometer scale and/or new side effects might develop along with the exploitation of these appealing new properties.

Consequently, this chapter discusses metallic[†] nanomaterial penetration into/through the skin from these two opposite perspectives: potentialities and risks.

31.2 Background (History and Definition)

Skin is the most vast organ of the human body with defensive functions (Elias 2005, Lee et al. 2006). Its complex structure and composition provide physical (Elias 2005, Lee et al. 2006), antimicrobial (Barak et al. 2005, Niyonsaba et al. 2006, Schmid-Wendtner

---

* In this contribution, nanotechnological products and nanoproducts are interchangeably used.

† If not otherwise stated, metallic nanomaterial accounts for carbon, metal, metal oxide, and salt ones. In general, nanomaterials may have different shapes.

and Korting 2006), immunological (Mathers and Larregina 2006, Seiffert and Granstein 2006), and metabolic (Hikima et al. 2005, Bickers and Athar 2006) barriers against the entrance and further diffusion of chemical and biological foreign agents. Therefore, everything that cannot be physically kept outside of the body will be attempted to be neutralized, attacked by the immune system, or degraded by enzymatic reactions. Actually, all these diversified defensive barriers exist because there are indeed molecules that can penetrate (i.e., enter to a certain extent) and permeate (i.e., enter and reach the dermis) the skin.

As well as having barriers, skin is also one of the administration routes used to deliver therapeutics locally and systemically (Hadgraft and Lane 2005, Brown et al. 2006). Local treatments benefit from administration ease, convenience (e.g., limited waste), and patient compliance. In contrast, the transdermal delivery of a systemically active drug has the great advantage of avoiding the hepatic first-pass effect (i.e., therapeutic dose is not metabolized prior to reaching the bloodstream) and being extremely well accepted by patients. However, the ingress of a drug or a drug delivery system (e.g., liposome and transfersomes), as well as any other foreign agent, is hindered by skin protective barriers, which could be sensed as a disadvantage. Nevertheless, numerous formulations, techniques, and devices have successfully been developed to weaken skin barriers and improve absorption (Barry 2004, Benson 2005, 2006, Power 2007, Sivamani et al. 2007).

Among formulations, colloidal systems have received great attention because of their potential to act as transdermal carriers (Cevc 2004, Benson 2006, Elsayed et al. 2007). However, and with few exceptions, it was concluded that those formulations enhancing drug permeation acted as skin fluidizers (see Section 31.2.2.2, under *Physicochemical properties of the vehicle dispersing a penetrating molecule*) rather than carriers. This discussion has recently been reanimated because of sunscreen and nanomaterial toxicological evaluation (Hoet et al. 2004, Oberdörster et al. 2005, Nohynek et al. 2007, Warheit et al. 2007), and found scientists divided into two categories: those who were able to observe nanomaterial penetration into the skin, and those who were not able to. Consequently, one may wonder why some scientists detected no penetration and, most importantly, whether nanomaterial skin penetration may theoretically be possible.

31.2.1 Skin Anatomical Structure

At the macroscopic level, skin is composed of three layers, epidermis, dermis, and hypodermis, of which only the outermost and the innermost layers significantly influence skin penetration, epidermis because of its defensive barriers and dermis because of its capillary anastomoses that create several concentration gradients between epidermis and dermis. Consequently, skin penetration is a thermodynamically driven event, which is hampered by epidermal defensive barriers.

At the microscopic level, it is possible to note that epidermis is a stratified squamous epithelial layer, organized into five strata, and whose most abundant cell phenotype is keratinocyte. These cells suffer a continuum and progressive modification

(i.e., cytomorphosis) that proceeds from the innermost (i.e., stratum basale) to the outermost stratum (i.e., stratum corneum (SC)) of the epidermis. This process keeps the epidermis healthy and defensively competent, forms and maintains the outermost cornified stratum (i.e., SC) of dead cells (i.e., corneocytes), and hence allows us to distinguish between the viable epidermis (VE; thickness: 50–100 μm; Johnson et al. 1997) and the SC (thickness: 10–20 μm; Johnson et al. 1997, Cevc 2004). Actually, the latter may be further classified into compact and disjunct SC based on the integrity of corneodesmosomes (i.e., SC desmosomes, which are cell membrane structures used to hold cells together). Disjunct SC is located more externally because of normal skin exfoliation that ends cytomorphosis' phases (Elias 1983).

SC is generally referred to as the skin barrier. However, this is the resultant of positive cooperation and interaction between SC composition, structure, and supramolecular organization of SC lipidic matrix (Elias and Choi 2005). To summarize, SC cross-sectional organization is frequently described as a wall by the brick-and-mortar model, where bricks are alternated, staggered, and interdigitated corneocytes and the mortar is a lipidic matrix surrounding them (Elias 1983, Nemes and Steinert 1999). More precisely, corneocytes are hexagonal in shape (diameter 40 μm; thickness 0.3–0.8 μm; area: 1100–1200 μm²/cell; Cevc 2004), are arranged in 15 layers, and are spaced from each other by a gap of ca. 75 nm (air-dried conditions; Johnson et al. 1997). Consequently, a tortuous path runs along corneocytes' perimeter. This path is filled with a lipidic matrix, and obstructed by intact and degraded corneodesmosomes (Al-Amoudi et al. 2005). In addition, this "wall" is not uniform since corneocytes arrange themselves in columnar clusters, which have a superficial area of up to 12 dead cells. Within each cluster, corneocytes overlap laterally (15% to ≤30% in humans; Cevc 2004). Each cluster is then separated by each other by boundaries of lower penetration resistance, which may be identified with superficial corrugations (i.e., furrows; Cevc 2004).

As previously mentioned, corneocytes are dead cells, whose former cytoplasm have been filled with keratin and keratin–filaggrin complex. In addition, their former plasma membrane was transformed into a thick proteic structure called cell envelope (Lazo and Downing 1999, Kalinin et al. 2002). Corneocyte composition is involved in preserving SC hydration (through proteolytic processes; Rawlings et al. 1994) and SC enzyme activity (for cytomorphosis completion). The latter is indirectly involved in lipidic matrix arrangement. In fact, corneocyte cell envelope is used as a scaffold to covalently bind, and thus properly align, the first layer of intercorneocyte lipids (i.e., omega-hydroxyceramides). This arrangement forces the other lipids to self-assemble in head–head tail–tail repeating bilayers (cross-sectional observation; Elias 1990, Wertz 1996, Norlén et al. 2007).

Mortar lipids may be classified into cholesterol and its derivatives, ceramides, free fatty acids, and triglycerides (Lampe et al. 1983, Elias 1990, Bouwstra and Ponec 2006, Norlén et al. 2007). Relative percentages differ depending on extraction methods and health of donors (Lampe et al. 1983, Law et al. 1995, Lazo and Downing 1999, Weerheim and Ponec 2001, Kalinin et al. 2002, Farwanah et al. 2005). However, it is worth noting that approximately 80% of these lipids are nonpolar. Nonetheless, their heads are hydrophilic because they contain hydroxyl groups that can establish hydrogen bonds with adjacent lipids or water (Bouwstra and Ponec 2006). Therefore, intercorneocyte lipidic matrix is not uniform either, since it is composed of repeating hydrophobic and hydrophilic areas (Elias 1983, Al-Amoudi et al. 2005, Bouwstra and Ponec 2006, Norlén et al. 2007). In addition, the close observation of a single bilayer (head–tail tail–head) has revealed that it could be 6 or 13 nm long (Bouwstra and Ponec 2006). The latter configuration assures a fluid lipophilic central area of 2.37 nm and two other contiguous areas of 4.57 nm, where fluidity decreases moving toward the lipid heads (Bouwstra and Ponec 2006). Consequently, it can be estimated that the fluid area would, respectively, be ≤6.94 nm or ≤5.42 nm when considering a contribution of 1/2 or 1/3 from the contiguous semifluid areas. In contrast, lipid heads delimit a hydrophilic region (i.e., SC aqueous pores) whose diameter has been estimated to be 2.8 ± 1.3 nm (Tang et al. 2001). Actually, many scientists have attempted to model and extrapolate aqueous pore dimensions from diffusiveness, and it is worth remembering that published dimensions vary from 0.4 to 36 nm (Cevc 2004). According to Cevc (2004), there might be different types of aqueous pores within the skin, and the superficial dimension of the pore may not be maintained inside the aqueous channel (Cevc 2004).

Finally, by shifting our observation from a cross-sectional to a three-dimensional (3D) examination, it is possible to note that lipid bilayers are mostly assembled in orthorhombic lateral packing (Bouwstra and Ponec 2006). This is a compact 3D organization that confers impermeability to SC. Nevertheless, it is possible to also find a less-compact 3D organization, the hexagonal lateral packing, in nonhealthy patients and in the outermost part of disjunct SC. This less-compact packing, respectively, relates to an altered lipid composition, and with sebum penetration into desquamating layers (Lavrijsen et al. 1995, Weerheim and Ponec 2001, Bouwstra and Ponce 2006).

Furthermore, when considering SC chemical composition from a macroscopic perspective, some other information may be found. First, due to the respective lipophilicity and hydrophilicity of SC and VE, a hydrophilic–lipophilic gradient exists within the epidermis, and more precisely between SC surface and the stratum granulosum (i.e., outermost layers of VE). This gradient hampers the ingress of highly lipophilic agents in the VE (Moss et al. 2002). Second, SC lipid composition (i.e., intercorneocyte lipids, sebum lipids, and lipids excreted by sweat glands; see later) creates a nonlinear pH gradient between SC surface (pH: 4.5–5.5; SC acid mantle) and the stratum granulosum (pH: ca. 7) (Elias 2005, Lee et al. 2006, Schmid-Wendtner and Korting 2006). This gradient maintains an optimal environment for SC enzyme activity (for cytomorphosis completion), acts as an antimicrobial barrier, and a penetration barrier, which hinders the ingress of charged agents (Swarbrick et al. 1984, Wagner et al. 2003, Elias 2005, Lee et al. 2006, Schmid-Wendtner and Korting 2006).

The description of the skin that has been given so far pictures a nanoporous, heterogeneous, macroscopically lipophilic, but also partially hydrophilic at the nanoscale, and complex organized "wall." Nevertheless, the provided description is not complete because it misses that of some skin appendages, which grant macroporosity to skin. These appendages are the sweat and sebaceous glands.

Sweat glands are coiled tubular glands that extend from SC to dermis or hypodermis (2–5 mm length) and are involved in thermoregulation and excretion of acids and body wastes (Kierszenbaum 2002). Their serous fluid is an aqueous mixture (median pH 5.3) of organic acids, carbohydrates, amino acids, nitrogenous substances, vitamins, and electrolytes (Patterson et al. 2000, Hirokawa et al. 2007, Meyer et al. 2007). Sweat is excreted at 2–20 nL/min speed from an orifice of 60–80 μm in diameter (Kierszenbaum 2002).

In contrast, sebaceous glands are associated with hair follicles whose infundibulum is used to expel sebum on skin surface (Kierszenbaum 2002) at a speed of 0.1 mg/cm² of skin/h (Meidan et al. 2005). Sebum is a mixture of squalene, waxes, cholesterol derivatives, triglycerides, fatty acids, and cell debris (Stewart 1992). However, some triglycerides are transformed into free fatty acids after their excretion. This modification allows sebum to become a lipophilic fluid with a water-emulsifying potential (Magnusson et al. 2001). Sebum is excreted through an orifice whose dimensions range between 10 and 210 μm in diameter, depending on body site (Otberg et al. 2004). Nonetheless, the hair thickness should be subtracted from the orifice diameter to evaluate the actual area (and volume) of the infundibulum that could be used for penetration or excretion. Pilosebaceous units extend down to the dermis (2–4 mm length), where they receive innervation and blood supply through the dermal papilla (Kierszenbaum 2002). SC follows this invagination for less than 100 μm, after which no physical barrier exists against penetration into VE.

In conclusion, skin anatomy appears to theoretically support the possibility of nanomaterial skin penetration.

31.2.2 Skin Penetration and Diffusion

Although the previous paragraph showed that skin is a macroporous and nanoporous barrier, the presence of orifices on its surface may not be a sufficient reason to conclude that nanomaterial skin penetration will really occur *in vivo*. In this regard, some scientists always say that skin has defended our bodies since the beginning of time keeping out bacteria and viruses. In addition, another typical comment is that we would develop many different diseases and illnesses, if skin were not protecting us.

Nevertheless, by critically analyzing the situation, it is possible to start arguing that bacteria (300–5000 nm) and viruses (18–450 nm) are, on average, too large for skin nanoporosity (<6.9 nm in the intercorneocyte lipids) as recently commented elsewhere (Baroli 2008). In addition, very few scientists take into consideration the fact that skin, as well as materials, could behave differently at the nanoscale. Moreover, there is no reason why skin penetration has to be strictly associated with skin permeation and local and/or systemic diseases. Finally, we must not forget that skin, in everyday life, may also be damaged thus favoring the ingress of foreign agents. In any case, let us proceed with the discussion on whether skin porosity is sufficient to claim penetration.

31.2.2.1 Skin Penetration Pathways

To answer this question we could first analyze whether skin porosity may be associated with penetration pathways. Earlier in this chapter, it was reported that skin macroporosity is provided by gland openings, while skin nanoporosity by the lipophilic (i.e., tail–tail region) and hydrophilic (i.e., head–head region) areas within the intercorneocyte space.

These anatomical features have been extensively studied by pharmaceutical technologists, and it was initially hypothesized that the intercorneocyte space could provide the "intercellular penetration pathway," which could be further distinguished into "apolar intercellular route" and "polar intercellular route." When referring to systemically delivered drugs, these pathways are generally called "transepidermal a-/polar intercellular permeation pathways." Many different studies support the existence of this pathway, and indicate that small (<600 Da), moderate lipophilic (Moss et al. 2002), and uncharged molecules (Swarbrick et al. 1984) are the best candidates to be transdermally delivered using the intercellular apolar route. In contrast, the penetration of water and polar molecules through the polar route was shown to be poor (Potts and Francoeur 1991). For what concerns dimensions, it has recently been estimated that molecules, such as lidocaine and nicotine, whose molecular weights are of ca. 300 Da, have an average dimension of 1 nm (Nohynek et al. 2007).

Consequently, it could be concluded that nanomaterials whose dimensions are below 6.9 nm may penetrate skin, whereas those approaching 1 or 2 nm may also permeate skin, both using the intercellular routes. However, it can also be envisaged that small dimensions of nanomaterials are not sufficient to claim penetration or permeation if additional physicochemical properties are missed. In contrast, dimensions are indeed discriminating for those nanomaterials possessing appropriate physicochemical properties but not appropriate dimensions.

Along with the intercellular route, the existence of a transfollicular route was also hypothesized. However, and despite large superficial openings of sweat glands and hair follicles, the significance of this route has been questioned for a long time. Doubts arose due to the following observations: (1) The openings of sweat glands and hair follicles are, respectively, estimated to be, on average, 0.01% (Scheuplein 1967) and 0.1% (Szabo 1962) of the total skin surface area. (2) The outward movement of sweat and sebum during their excretion may be an obstacle to the inward movement of a penetrating agent. (3) The duct of hair follicle does not always remain open. In fact, when hair does not grow, it is closed by a plug of shedded corneocytes glued together with dry sebum and other cell detritus (Lademann et al. 2005).

Nevertheless, recent studies have clearly shown that (1) hair follicles are nonuniformly sized and distributed in the body (Otberg et al. 2004). For instance, a maximum of 292 follicles/cm² has been detected in the forehead, whereas other body sites account for less than 50 follicles/cm². In addition, by calculating hair follicle density (area of orifices versus total skin area), one finds that this number is 1.28% ± 0.24% in the forehead, and 0.09% ± 0.04% in forearm. Even the volume of the follicular infundibulum varies in different body locations and reaches its maximum in the forehead, where it is 0.19 ± 0.07 mm³ (Otberg et al. 2004). (2) Hair follicles may be found in an active or inactive state. In active hair follicle, sebum is secreted and hair grows. More recently, these states have, respectively, been indicated as open and closed hair follicles (Lademann et al. 2005). It has been shown that molecules and particles penetrate into active hair follicles (Domashenko et al. 2000, Hoffman 2000, Lademann et al. 2001), and that closed hair follicles may be opened by a soft peeling, thus increasing their potential to act as penetration routes (Lademann et al. 2005). (3) Large molecules (Wu et al. 2001, Dokka et al. 2005), particles (Toll et al. 2004, Vogt et al. 2006, Lademann et al. 2007), and small hydrophilic molecules (Hueber et al. 1994, Essa et al. 2002, El Maghraby et al. 2001) preferentially permeate skin through the hair follicle infundibulum. In contrast, only few contributions on penetration through sweat glands have been reported in Medline (Changez et al. 2006a and b).

In summary, it seems that the transfollicular penetration pathway may be used by those agents whose dimensions are below those of gland openings (i.e., 10–210 μm), and which are able to disperse themselves into sweat or sebum.

31.2.2.2 Factors Affecting Skin Permeation

Factors affecting skin permeations are generally classified into (1) location and conditions of the skin at the application site, (2) physicochemical characteristics of the penetrating molecule, and (3) physicochemical properties of the vehicle dissolving/dispersing a penetrating molecule. As one may note, two of them refer to penetrating molecules and not to nanomaterials. Nevertheless, these same factors will indubitably also affect nanomaterial penetration and permeation.

Location and conditions of skin at application site. As one may envisage from the previous discussion, skin integrity and regional variation, density and dimensions of orifices, composition and thickness of sweat-sebum emulsion covering SC, and dimensions of aqueous pores and lipidic fluid paths set the basic conditions affecting the skin absorption of any agent. In this regard, it should be mentioned that, in the same healthy individual, not only appendage density and distribution but also SC thickness varies, reaching its maximum width (100–200 μm) in palms and soles (Kierszenbaum 2002). In addition, SC chemical composition and structure, and thus its barriers' function, vary with age, race, and sex. For instance, it is well known that permeability diminishes with age (Mitsui 1997a).

Moreover, it should not be underestimated that skin may lose its integrity or may suffer alterations due to dermatological (e.g., atopic dermatitis, psoriasis, and ichthyosis) and other pathological conditions (e.g., inflammation, burn, and infections), damage and trauma, extensive use of detergents, or prolonged exposure to air-conditioned nonhumidified environments (Rawlings et al. 1994, Lavrijsen et al. 1995, Suhonen et al. 1999, Kalinin et al. 2002, Farwanah et al. 2005, Bouwstra and Ponec 2006). In addition, exfoliation or abrasion with cosmetics, sand, or similar products may remove some layers of the SC, thus making penetration easier.

Although some of the effects induced by the above-mentioned causes are rather obvious, SC dehydration (Suhonen et al. 1999, Barry 2004) produced by detergents and/or dry air remains one of the subtlest causes of skin penetration risk in modern days. In fact, dehydration is generally associated with skin exfoliation, cracking, and/or fissuring, which create breaches that could be used by foreign agents (e.g., metallic nanomaterials) to penetrate and potentially permeate skin. It is intriguing to note that when the amount of water in SC ranges between 20% and 50% (normal water content is 15%–20%), corneocytes swell, reducing SC compactness and resistance to diffusion (Roberts and Walker 1993, van Hal et al. 1996). Nevertheless, it has been shown that hydration-induced SC modifications moderately increase only the permeation of moderate polar molecules (Wurster and Kramer 1961, Behl et al. 1980, 1981, 1982, 1983).

Finally, another variable affecting skin absorption is temperature. In fact, skin temperature may increase due to irradiation (e.g., sun exposure), fever, or higher vascularization (e.g., blood flow, or increased number of capillaries). Under these circumstances, skin permeability also increases due to intercorneocyte lipids' fluidization and/or improved dermic clearance.

Therefore, it can be deducted that location and conditions of skin at the application site will influence the absorption of any agent, not only molecules, since structure and properties of the skin barrier could be modified. With regard to nanomaterials, it can be further hypothesized that their penetration might be favored in those locations where there is a higher density of hair follicles, and in those areas where skin thickness has been decreased by exfoliation or abrasion. In addition, SC hydration may favor nanomaterial penetration through the transepidermal route, especially into the disjunct SC. Moreover, it has been reported that the SC hydrophilic-lipophilic gradient may also favor the penetration of particulate formulations applied nonocclusively (Cevc and Blume 1992). Consequently, it is reasonable to state that the formulation where nanomaterials are dispersed and the application method will definitely affect the absorption outcomes for any skin condition. In other words, this means that both *in vitro* and *in vivo* investigations should be carefully planned to mimic real exposure and application conditions, and that data belonging to experiments with different settings should not be compared.

Physicochemical characteristics of the penetrating molecule. All the studies performed to ameliorate drug skin permeation revealed that there are several physicochemical parameters that a molecule has to possess to be a good permeant or diffusant. These can be summarized as follows.

1. Generally, molecules are not administered as they are, but are formulated into a pharmaceutical product. Therefore, they will be applied on the skin in the presence of other molecules, which, together with the drug, will interact and/or compete for absorption. Therefore, the first requisite is drug solubility and concentration in its dispersing vehicle. Since permeation is a thermodynamically driven event, and occurs as a result of *n* partitioning and diffusion processes, higher concentration of drugs within the vehicle (i.e., thermodynamic activity) relates with higher amount of drugs transdermally delivered (Moser et al. 2001). Solubility in the vehicle presumes that (a) vehicle facilitates drug transfer from vehicle to skin, or that (b) vehicle and drug concurrently enter into the skin by dissolving in it, thus creating a continuity between the pharmaceutical product and the skin.

   As far as metallic nanomaterials are concerned, one can hypothesize that these same parameters may affect nanomaterial skin absorption as well. This hypothesis is developed assuming that, very rarely, metallic nanomaterials will accidentally or intentionally come into contact with skin in their powder state. Specifically, it is more likely that they will be supplied as a dispersion rather than a solution. It can therefore be concluded that if nanomaterials possess all those properties (see later within this section) granting them skin penetration ability, and if they are kept homogeneously dispersed in the formulation, if they are smaller than skin openings, and if the dispersing vehicle is able to dissolve or to be emulsified in sweat, and/or sebum, and/or sweat-sebum emulsion, and/or lipidic SC pathways, increasing their concentration in the vehicle should increase the probability of recovering a higher amount of them in skin. Moreover, this may be particularly true for those nanomaterials, called nanoclusters (i.e., aggregates of few atoms), whose dimensions are ≤1 nm. In fact, it may be reasonable to hypothesize that nanomaterials sized at, and below, 1 nm should behave similarly to a molecule sized below 300 Da (Nohynek et al. 2007).

2. Following the discussion on molecular weight, it has been shown that good molecular permeants are sized below 600 Da (Barry 2004). However, this finding does not exclude that larger molecules may enter the SC and disorganize it, which may consequently facilitate the intentional or nonintentional ingress of other agents. Therefore, it could be concluded that the dimension of nanomaterials influences (a) their ability to enter the skin, (b) the selection of the penetration route, and (c) the coefficient of diffusion in the dispersing vehicle and in the skin (see later within this section).

3. It should also be highlighted that, in contrast to molecules, the shape of nanomaterials together with their dimensions may additionally influence skin absorption. It should in fact be remembered that carbon and metallic nanomaterials are generally rigid and of various shapes, such as spherical, ellipsoid, triangular, cubic, prism-like, and needle-like (What

are nanoscience and nanotechnologies, *The Royal Society & The Royal Academy of Engineering*, 2008). Therefore, the association of the most commonly used word "nanoparticle" with a "spherical shape" is not straightforward when one refers to nanomaterials. Moreover, if one now imagines the diffusion of such nanoparticles in a narrow capillary or through a porous structure, it is suddenly obvious that shape and orientation will greatly influence this process (influence of shape is discribed in Section 31.3.3.).

4. Another parameter influencing the skin diffusion of molecules is the pKa or pKb of the penetrating molecule and pH of the vehicle. According to Brönsted theory* acids (HA) and bases (B) are, respectively, those neutral or charged molecules that can produce or accept hydrogenions (H$^+$), thus transforming them into their conjugate base (A$^-$) and acid (BH$^+$), as shown below.

$$HA + H_2O \leftrightharpoons H^+ + A^- \quad \text{and} \quad B + H_2O \leftrightharpoons BH^+ + OH^-.$$

The conjugate species (A$^-$ and BH$^+$) of a weak acid (HA) or base (B) are generally strong. These dissociation processes may be quantitatively expressed—at equilibrium—by the following equations:

$$K_a = ([H^+] \times [A^-])/[HA] \quad \text{and} \quad K_b = ([BH^+] \times [OH^-])/[B],$$

where the amount of each species is given in molar concentrations. For weak acids and bases, these equilibrium constants (Ka and Kb) range below 1×10^{-3}, and hence their logarithmic forms are preferentially used:

$$pK_a = -\log K_a \quad \text{and} \quad pK_b = -\log K_b.$$

Therefore, as pK_a and pK_b increase, the strength of the acid or base decreases. Finally, since these equations describe equilibria, it is possible to deduct that vehicle and skin pH influence the degree of dissociation of the ionizable molecule, and therefore the concentration of the un-ionized fraction.

These parameters are important because it was shown that only the un-ionized fraction of the penetrating molecule will be transported to VE (Swarbrick et al. 1984). Coming back to nanomaterials, it should be emphasized that they are commonly formulated to not aggregate. Under this perspective, the use of superficially charged nanomaterials could be one valuable strategy. However, charges generally belong to the coating and not to the core material. Therefore, it is the coating and not the main material that will engage interactions with skin components, thus influencing nanomaterial skin absorption. As regards how metallic nanoparticles are evaluated, one

should note that they could be made of pure metals (e.g., Au, Pt, and Fe) whose oxidation state is zero, oxides (e.g., Fe_2O_3, TiO_2, and ZnO) where atoms are linked to each other by covalent bonds, or extremely low soluble salts (e.g., CdS and CdSe) where atoms are linked to each other by bonds less polar than those of oxides. It should also be highlighted that some pure metals are prone to be oxidized by air, and therefore their core is generally coated with a thin layer of core metal oxides. However, the modification of superficial properties (e.g., by oxidation) may further vary the nanomaterial diffusion coefficient in the dispersing vehicle (see later within this section). It can also be foreseen that in those cases where atoms have a different electronegativity†, dipole nanomaterials could be obtained. These, however, would aggregate. Therefore, it is conceivable to state that coating will prevent aggregation or chemical modification of nanomaterials.

By transposing what was found for molecules to metallic nanomaterials, one will have to consider the pKa or pKb of the coating and the pH of the dispersing medium. Moreover, following the previous discussion, one may also be prompt to conclude that charged nanomaterials might have little skin penetration potential. However, this would be a simplistic answer, because it does not contemplate that (a) the amount of superficial charges may vary upon nanomaterial contact with the skin acid mantle and (b) along nanomaterial skin absorption. In addition, (c) positively or negatively charged surfaces may differently interact with the different components of the skin and routes of penetration. In this regard, it should be remembered that one of the skin barriers is the nonlinear pH gradient between SC surface (pH: 4.5–5.5; SC acid mantle) and the stratum granulosum (pH: ca. 7) (Elias 2005, Lee et al. 2006, Schmid-Wendtner and Korting 2006). This barrier is, however, influenced by other factors (e.g., pathological diseases and skin integrity) and inter-patient variations can be observed *in vivo*. In addition, it has been proved that this gradient is modified when the skin is assayed *in vitro* (Wagner et al. 2003). Consequently, since it is extremely difficult, for now, to show what is happening inside the skin during the diffusion of a charged agent, the real contribution of charged surfaces on nanomaterial skin absorption may remain unrevealed for some other years.

5. Another parameter, that is often set aside, is skin binding and metabolism (Täuber 1989). Skin has different enzymatic systems (e.g., aryl hydrocarbon hydroxylase, 7-ethoxycoumarin-deethylase, epoxide hydrolase, cytochrome-P450-dependent monooxygenase, esterases, aminopeptidases, peptidases, and reductases) that are used for endogenous substance conversion (e.g., hormones, steroids, inflammatory mediators, and lipids) and for exogenous substance eliminations. The metabolic potential of skin is estimated

---

* The Brönsted theory may be found in any textbook of inorganic chemistry.

† A covalent bond becomes ionic at a 50% level when atoms differ in electronegativity by 1.7.

to be 2% of the hepatic one. In addition and for what concerns binding, physical and/or chemical interactions between skin and penetrating agent components may be envisaged.

Therefore, it can easily be expected that binding may avoid further penetration, whereas metabolism may reduce the amount of molecules that can effectively diffuse to the dermis. This statement also applies to metallic nanomaterials. Nevertheless, when evaluating nanomaterial metabolism one should consider that both the coating and the central core of nanomaterials could suffer chemical reactions (e.g., oxidation, hydrolysis, and deamination). These reactions, while not causing a real rupture of the nanomaterial itself, do modify its superficial properties. Consequently, many other properties could be modified as well. Among these, it is possible to enumerate (a) the hydrodynamic radius, or more generally, the nanomaterial size. (b) The superficial properties of nanomaterials, which can affect (c) the potential of skin binding and (d) the potential to self-aggregate or to aggregate when in contact with specific structures/components of the skin. (e) Finally, and as will be discussed later, a change in size and/or superficial properties may influence, and possibly change, the nanomaterial diffusion coefficient in the skin. Consequently, it is possible to conclude that describing the ability or inability of a nanomaterial to be absorbed into/through the skin without considering variables mentioned above would be rather inexact.

Finally, after considering size, shape, charges, concentration, stability against environmental- and skin-induced chemical reactions, and binding potential, two other important parameters will be discussed. These are the partition coefficient and the diffusion coefficient.

6. According to European Community legislation, the partition coefficient (K or log P) is defined "as the ratio at the equilibrium concentration (C_i) of a dissolved substance in a two-phase system consisting of two largely immiscible solvents" (Partition Coefficient, *The Consumer Products Safety & Quality*, 2008):

$$K_{OW} = \frac{C_{n-octanol}}{C_{water}},$$

where K_{OW} in this case is the oil in water partition coefficient.

Generally, this ratio is given in its logarithmic form (log K or log P). It is remembered that K depends on temperature, and that ionizable substances should be tested in their un-ionized forms. Many organic solvents have been used to correlate the K_{OW} with skin absorption, finding that n-octanol and isopropyl myristate versus water or aqueous buffers are the combinations that better mimic the complex lipid/polar nature of the SC (Barry 1983a).

Therefore, this coefficient has been used to indicate the affinity of a molecule to different solvents (e.g., oil in water partition coefficient: K_{OW}) or biological tissues. In skin

absorption, there are two important coefficients of this kind: the SC to vehicle partition coefficient (K_{SC-VH}) and the VE to SC partition coefficient (K_{VE-SC}). By considering skin as a nonuniform membrane composed of different sandwiched layers, one could also indicate the dermis-to-VE and blood-to-dermis partition coefficients. However, the K_{SC-VH} is essential to establish a high initial concentration of the penetrating agent in the first skin layers, and thus a high concentration gradient. In contrast, the second K_{VE-SC} accounts for the hydrophilic-lipophilic gradient between the SC and the VE. In other words, the partition coefficient of a penetrating molecule should be adequate to enable it to leave the dissolving/dispersing vehicle and to cross the different skin layers, thus avoiding its sequestration. In general, very lipophilic molecules will easily partition in SC but will leave it with difficulty, whereas hydrophilic molecules will suffer poor penetration (Moss et al. 2002).

In the case of metallic nanomaterials no partition coefficients have been reported so far. This may be due to the fact that (a) nanomaterials are not generally supplied as dry powders, but as (b) liquid formulations where they are dispersed and not dissolved in their dispersing medium. In addition, (c) surfactants or other molecules within the dispersing medium may affect the results of the assay. Moreover, (d) it should not be forgotten that nanomaterial coating will definitely influence the results, and therefore nanoparticle and coating should be considered as a whole, and different coatings may produce different partition coefficients. Finally, (e) one also has to ponder that nanomaterials might aggregate, which is going to negatively affect the results of this assay. In conclusion, there might be the need for specific and validated protocols to assess this parameter for metallic nanomaterials.

7. As previously commented, another important parameter is the diffusion coefficient (D) of the penetrating molecule in its vehicle and in the skin (Barry 1983b). The first one may be calculated, for homogeneous liquids, using the following equation:

$$D = \frac{kT}{(6\pi\eta r)},$$

where

k is the Boltzmann's constant ($k = R/N$; R: gas constant; N: Avogadro's number)

T is the absolute temperature

η is the viscosity of the vehicle

r is the radius of the diffusing molecule, assuming it to be spherical ($V = 4/3\pi r^3$)

However, if solute and solvent sizes are comparable, D is equal to $kT/(4\pi\eta r)$.

Without entering into mathematical details, it is herein summarized that D is proportional to one over the molecular weight of the examined molecule (i.e., $D \propto 1/MW^n$; $n = 1/3$), and it does not vary much when molecular weights range between 100 and 1000 Da. In contrast, changes in the polarity of small molecules greatly affect their diffusion coefficient. Other factors affecting D are the concentration of, and interaction between, diffusing molecules (e.g., aggregation), the viscosity of the dispersing medium that can also be altered by diffusant concentration (it increases) and temperature (it decreases), the interactions between diffusants and the dispersing medium (e.g., complexation), and the temperature (D increases by rising it).

However, as the dispersing medium becomes more complex than a homogeneous liquid (e.g., gel, emulsion, polymer, and skin), the previous equation is no longer obeyed. Although an extensive mathematical discussion is beyond the scope of this chapter, one can certainly agree that the proposed equation has its utility because it allows us to ponder factors/parameters affecting diffusivity (e.g., temperature, viscosity, aggregation, and complexation). For instance, when considering the diffusion of a potential nonspherical nanomaterial, it should be considered that the term r in the previous equation (or any other equation attempting to model particle skin diffusion) is going to continuously vary depending on nanomaterial spatial orientation. In addition, it has also been shown that in complex membrane (e.g., skin), the dimension and the hydrodynamic volume of the diffusant are more important than its molecular weight.

In conclusion, physicochemical parameters affecting nanomaterial ingress into, and diffusion through, the skin are almost the same as the ones that influence molecule skin absorption, and may be enumerated as follows: dimensions, shape, charge, chemical and physical stability, diffusion coefficient, and partition coefficient.

Physicochemical properties of the vehicle dispersing a penetrating molecule. As was previously mentioned, when a formulation is applied on the skin, all its components are subjected to skin absorption, although to a different extent. It should also be highlighted that almost all formulations are applied nonocclusively in real life. This application method allows ingredients to either evaporate or to be absorbed. Therefore, formulations change their composition during skin penetration, and it can be foreseen that different synergisms/interactions between penetrating agent and/or vehicle ingredients and/or skin components may be established.

It was in fact shown that, for example, the diffusion coefficient of a penetrating molecule in its vehicle is reduced as vehicle viscosity increases, which in turn retards, and possibly avoids, its partitioning into the skin. In addition, lipophilic formulations may compete with SC for penetrating molecule partitioning; extremely lipophilic ones may also retain penetrating molecules. Moreover, as previously mentioned, occlusive formulations may moderately increase skin absorption of specific molecules. Finally, and perhaps most importantly, it should not be underestimated that molecules such as solvents, surfactants, and enhancers entering the vehicle composition may alter, to different extent, the SC barriers thus increasing the absorption of dispersed/dissolved molecules therein (Barry 2004, Benson 2005).

By examining the dispersing media of metallic nanomaterials, one may note that they could be distinguished in buffered aqueous dispersions (e.g., quantum dots and metallic nanomaterials), organic solvent dispersions (e.g., metallic or magnetic nanomaterials), or emulsions (e.g., metal oxides such as TiO_2 and ZnO nanoparticles). Following previous reasoning, each vehicle indubitably has a completely different impact on skin barriers, penetrating routes, and hence absorption. In this regard, it was shown that large oligonucleotides were transfollicularly delivered to the dermis only if they were formulated in an oil in water (O/W) cream (Dokka et al. 2005), and that some iron nanoparticles reached the stratum granulosum most likely because of their surfactant-rich aqueous dispersing medium (Baroli et al. 2007). Moreover, and especially for those vehicle containing surfactants, it could be useful to calculate their hydrophile-lipophile balance (HLB) number, which indicates whether a molecule is more lipophilic or hydrophilic. This number could be used to evaluate the emulsifying potential of investigated formulations, and can be easily obtained using the following equations and tables reported in the cited reference (Mitsui 1997b) and other textbooks on pharmaceutical technology.

HBL number = Σ (group values of hydrophilic groups) + Σ (group values of lipophilic groups) + 7.

In addition, the below equation should preferentially be used for nonionic surfactants:

$$HBL\,number = 7 + 11.7 \times \log\left(M_w/M_o\right),$$

where M_w and M_o are, respectively, the molecular weights of hydrophilic and lipophilic groups. In addition, it is possible to know in advance what type of surfactant is needed to emulsify a mixture of oils. This can easily be calculated by the following equation:

Incognitos HBL number

$$= [\Sigma \,(\text{required HLB}_{oil\,x} \times \text{percentage}_{oil\,x})]/\Sigma\,\text{percentages}_{oil\,x}.$$

In conclusion, physicochemical parameters of the dispersing vehicle affecting nanomaterial ingress into, and diffusion through, the skin are almost the same as those that influence molecule skin absorption, and may be enumerated as follows: lipophilicity, viscosity, presence of permeation enhancer in vehicle composition, ability to be emulsioned/dispersed with/in sebum, and/or sweat, and/or sweat-sebum emulsion, and/or SC lipid matrix.

31.2.2.3 Modeling Skin Permeation

The modeling of molecule permeation through the skin has received a great deal of attention in the last 40 years. However, no equations will be herein reported because they start from the assumption that only the penetrating molecule diffuses into and through the skin, and that composition of the dispersing vehicle is not modified (e.g., evaporation, diffusion of skin sebum, sweat or lipids into the vehicle, and diffusion of vehicle ingredients into the skin) while dissolved/dispersed molecules are absorbed. Actually, assumptions are more numerous, and determine that the mathematical model does not describe the entire scenario. Another reason why no equations will be discussed here is that nobody has shown that the skin could be permeated (i.e., diffusion from SC to dermis) by metallic nanomaterials. To date, only penetration has been clearly demonstrated in some articles.

31.3 Presentation of the State of the Art and Critical Discussion

Studies on skin absorption of metallic nanomaterials are not numerous, and therefore they will be summarized in the following paragraphs. A critical discussion of disclosed results will follow the summary of each study.

31.3.1 Magnetic Nanoparticles

Magnetic nanoparticles, sized 10–2 nm in diameter, have been enthusiastically proposed for (1) cell labeling and targeting, (2) drug delivery and targeting, (3) magnetic resonance imaging, (4) hyperthermia-based tumor therapies, (5) magnetic transfection of cells (Penn et al. 2003, Gupta and Gupta 2005, Neuberger et al. 2005, Wang et al. 2007). However, to date, there is only one article (Baroli et al. 2007) reporting skin penetration of such nanomaterials.

In this chapter (Baroli et al. 2007), the penetration ability of iron-oxide- and iron-based rigid spherical nanoparticles was assessed using human full-thickness skin. In particular, iron oxide nanoparticles were coated with TMAOH and dispersed in an aqueous solution of TMAOH, whereas iron nanoparticles were coated with AOT and dispersed in a AOT-rich aqueous solution. In addition, iron oxide nanoparticles were homogeneous in size and measured 6.9 ± 0.9 nm. In contrast, iron nanoparticles were of different sizes (<100 nm), but 51% had a diameter of 4.9 ± 1.3 nm. Skin was occlusively exposed to nanoparticles for up to 24 h, and then examined using a transmission electron microscope (TEM) and a scanning electron microscope (SEM) equipped with an x-ray micro-analyzer (energy dispersion spectrometry, EDS), which is generally indicated with the acronym EDS-SEM. Results showed that nanoparticles were able to passively penetrate skin through the transepidermal and transfollicular routes. In fact, particles were found in the deepest layers of SC, at the stratum granulosum–SC interface, and in hair follicles. In addition, in some exceptional cases, particles were also found in the VE. This finding was correlated with the surfactant-rich dispersing vehicle. No particles were found in the receptor fluids, as assessed by an inductively coupled plasma-optical emission spectrometer (ICP-OES).

These results are in agreement with previous discussions, and provide new perspectives on nanomaterials absorption into/through the skin. It seems, in fact, that below 10 nm, rigid nanoparticles may also use the transepidermal route for penetration. In addition, these results provide further proof that the dispersing vehicle may favor or hamper particle penetration, as assessed

by a skin barrier integrity evaluation test (i.e., skin impedance). It is in fact conceivable that, due to nanoparticles' tiny dimensions, any modification suffered by the SC lipid matrix may create new penetration paths within the SC, or enlarge those already existing (aqueous pores: 2.8 ± 1.3 nm; previously estimated fluid lipophilic paths: above 2.37 nm and below 6.94 or 5.42 nm).

31.3.2 Titanium Oxides and Zinc-Oxide-Based Nanoparticles

Another class of metallic nanomaterials consists of nanometric titanium oxide and zinc oxide, which are able to scatter UV light (i.e., TiO_2: 14–120 nm; ZnO: 20–30 nm). Therefore, they are extensively used in the cosmetics industry, such as in sunscreens. In such formulations, however, nanoparticle dimensions are generally larger than few nanometers (i.e., TiO_2: 14 nm–several μm; ZnO: 30–200 nm), and particles are also commonly coated with aluminum oxide, silicium dioxide, or silicon oils. A recent review on the health safety of these nanoparticles compared different studies and concluded that titanium oxide and zinc oxide particles do not sufficiently penetrate the skin, and therefore particles may only be found in the outermost layer of SC (Nohynek et al. 2007).

These conclusions are not surprising for rigid particles above 100 nm. In contrast, what is surprising is the absence of particles in hair follicles. However, in absorption studies, it would be limiting to not consider the enhancing or retarding effect of nanoparticle dispersing medium. As was previously commented, it is the entire formulation (i.e., drug or nanomaterial plus dispersing vehicle) that will contribute to drug or nanoparticle absorption or external segregation. Actually, sunscreens should be formulated to stick to the skin (Nohynek 2008), and therefore the presence of permeation enhancers in their composition would seem erratic. Therefore, it is not at all surprising that these particular formulations may retain nanoparticles on skin surfaces. In addition, it can be speculated that these formulations may also form an occlusive film of particles upon water evaporation, and that would further prevent their penetration.

Nevertheless, titanium oxide and zinc oxide penetration studies do not seem to contemplate or imitate skin conditions when sunscreens are applied at the seaside. In this case, skin will most likely be fully hydrated, hot for IR irradiation (actual skin temperature depends on latitude), smoothly peeled (similar to a mild tape-stripping) by the action of saline water and sand, and possibly inflamed (depending of skin phototype); all conditions that can favor penetration of any agent (see previous discussions). Physicochemical properties of nanomaterial coatings are also commonly not reported. Therefore, these studies do not seem to reveal a complete scenario.

31.3.3 Quantum Dots

Quantum dots (QDs) are nanocrystals composed of elements in groups II/VI, III/V, and IV/VI of the periodic table. QDs are semiconductors and are recommended for electrical and optical applications (Walter 2003, Quantum dot *Wikipedia* 2008,

Quantum dots explained *Evident technologies* 2008). QDs are also used as fluorescent probes and labels because their fluorescence does not quench (Monteiro-Riviere, N. A., personal communication, October 30, 2008). Even though commercially available, QDs range below 10 nm, and they are usually coated to increase dispersibility and decrease metal-core toxic leaks. Therefore, their hydrodynamic radius should more correctly be used when evaluating their skin absorption potential. To date, there are a few studies on QD skin passive penetration (Ryman-Rasmussen et al. 2006, Zhang and Monteiro-Riviere 2008, Zhang et al. 2008). In addition, one can also find an article that contemplates the use of low-frequency sonophoresis (Paliwal et al. 2006) to enhance QD skin penetration. However, this study will not be further commented on because QD penetration is not passive.

In the earliest study (Ryman-Rasmussen et al. 2006), neutral, anionic and cationic, spherical and ellipsoid QDs (hydrodynamic radius 14–45 nm) dispersed in borate buffers (pH 8.7 for neutral and cationic QDs, and pH 9.0 for anionic QDs) were left in contact (40 μL, 1 μM) to dermatomed porcine skin for 8 and 24 h. The protocol setup comprises of flow-through diffusion cells, skin from the back, and a blood-mimicking receptor fluid (1.2 mM KH_2PO_4, 32.7 mM $NaHCO_3$, 2.5 mM $CaCl_2$, 4.8 mM KCl, 1.2 mM $MgSO_4$ ċ $7H_2O$, 118 mM NaCl, 1200 mg/L D-glucose, 4.5% BSA, 5 U/mL heparin, 30 μg/mL amikacin, and 12.5 U/mL penicillin G) indicated by the authors as "perfusate." Skin was equilibrated with the perfusate for 30 min prior to starting the experiments. Perfusate was run at a flow rate of 2 mL/h. Perfusate and skin (confocal microscopy) were assayed for fluorescence. Results showed that all types of spherical QDs were able to penetrate the SC, and were found in the epidermis and dermis after 8 h. Neutral and cationic ellipsoid QDs reached only the VE in 8 h. In contrast, 24 h were needed to observe anionic ellipsoid QDs in the VE. No fluorescence was detected in the receptor fluid.

These results show that shape may influence nanomaterial skin absorption, as previously commented. However, it is difficult to state, in the absence of skin barrier integrity evaluation tests (e.g., skin impedance and transepidermal water loss (TEWL)), whether QDs penetrated because of their size* and physicochemical properties, or because of the basicity of the dispersing medium. It is known that basic or acid vehicles should be avoided because they could damage skin. It is common sense to avoid our hands in bleach for more than a few seconds. Actually, the authors commented on this potential criticism of their work, and they stated that no modification was observed under microscope examination. However, barrier integrity tests would have provided information on potential barrier alteration that is not visible under microscope analysis, as commented on in Baroli et al. (2007). It is surprising, in fact, that charged nanomaterials penetrated intact skin (see previous discussion). In addition, it would have been important to assess the isoelectric point of tested QDs to know at

what pH charged QDs would have become neutral. In fact, taking into consideration ellipsoid QDs, where penetration is partially hindered by shape, one may realize that cationic QDs diffused into skin better than anionic QDs. Therefore, one could speculate that cationic QDs could have interacted with acid species within superficial skin layers (pH of porcine skin is similar to human skin, Monteiro-Riviere, N. A., personal communication, October 30, 2008). For example, the basic dispersing medium might have forced skin free fatty acid (FFA) to dissociate, and their conjugate bases might have interacted with cationic QDs. Following this reasoning, one could also hypothesize that the basic dispersing medium might have forced esters of FFAs to hydrolyze, producing FFAs and alcohols. Then, these two species might possibly have partially neutralized the pH of the dispersing medium (FFAs), partially interacted with cationic QDs (FFAs), and acted as permeation enhancers (alcohols). Actually, ester hydrolysis might also support anionic QD penetration at later times. Nevertheless, and although the penetration mechanism was not revealed by the authors, what is extremely important is that QDs penetrated and diffused through the skin. It should not be underestimated that people working in the manufacturing departments of nanotechnology industries will be exposed to any type of nanoparticle-containing formulations. Therefore, from a safety perspective, this work is highly relevant.

In another study (Zhang et al. 2008), the penetration ability of nail-shaped core-shell (i.e., cadmium/selenide core and cadmium sulfide shell) and PEG-coated QDs was assessed. These QDs were produced in the authors' laboratories and were neutrally charged. They were dispersed in an aqueous vehicle (neutral pH) obtained by evaporation under vacuum of a previous chloroform-water dispersion. The radius and hydrodynamic radius of QDs were shown to be, respectively, 8.40×5.78 nm (by TEM) and 39–40 nm (by size-exclusion chromatography). Skin absorption studies were performed using dermatomed (400 μm thickness) porcine back skin, flow-through diffusion cells, and a receptor fluid (same composition as in Ryman-Rasmussen et al. 2006) that was run at a flow rate of 2 mL/h (37°C, pH 7.3–7.5). After equilibrating the skin for 30 min, 40 μL of QD aqueous dispersions (i.e., 1, 2, and 10 μM) was applied on the skin for 24 h. The perfusate was analyzed for fluorescence and cadmium (Cd) recovery using an ICP-OES; skin was observed using a TEM and a confocal microscope.

The results of this study showed that neither fluorescence nor Cd was recovered in the perfusate. In contrast, QDs were recovered in the skin as a function of their concentration. At the lowest concentration, QDs were only found in SC layers, whereas at the highest concentration they were found in SC, and, sometimes, in the upper epidermis and outer root sheath of the hair follicle. In addition, TEM analysis revealed that QDs were localized within the intercorneocyte lipids of the outermost SC layers. The authors compared these data with previous results (Ryman-Rasmussen et al. 2006), and deduced that since no alteration of the skin was visible under microscope observation, the lower penetration of nail-shaped QDs (Zhang et al. 2008) was a consequence of their shape.

---

* Please have in mind that previously given dimensions of skin macro- and nanoporosity refer to human skin.

It is indubitable that shape highly influences the diffusion coefficient of a material, as previously discussed. In addition, one can easily experiment on how differently shaped particles would flow when they are free to fall from a funnel. However, neglecting the contribution of the vehicle on the skin absorption of a dissolved/dispersed agent should carefully be stated in the absence of skin barrier integrity tests' data. In addition, by considering the dimensions of penetration pathways that were discussed earlier, 30–40 nm particles should theoretically be too large* to experience a deep penetration. Consequently, they should theoretically be found only in disjunct SC, which therefore is in agreement with the authors' findings.

In another study (Zhang and Monteiro-Riviere 2008), the authors investigated QD penetration ability as a function of skin conditions and application methods. Two types of QDs were used: they were both negatively charged and dispersed into a 50 mM borate buffer (pH 9.0). However, they differed in shape and hydrodynamic radius (i.e., spherical, 14 nm; ellipsoid, 18 nm). QDs were applied on the skin for 8 and 24 h (40 μL, 1 μM). In this study, authors used back rat skin (Wistar rats, 10 weeks of age) that was (a) preventively (*in vivo*) tape-stripped (10 times) to remove SC, (b) preventively (*in vivo*) abraded with sandpaper (60 times), or (c) left intact. Skin samples subjected to flexion were first fixed to the flexing apparatus (flex skin at 45°, 20 flexes/min, 60 min of flexing), exposed to QDs, and kept hydrated (from the dermis side) by a tissue wipe soaked with saline solution. After flexion, tape-stripping or abrasion, skin samples were clamped into flow-through diffusion cells. Absorption experiments were conducted at 37°C, using a modified Krebs-Ringer bicarbonate buffer as a perfusion solution (flow rate 2 mL/h, pH 7.3–7.5). The perfusate was analyzed for fluorescence and cadmium (Cd) recovery (ICP-OES); skin was observed using a confocal microscope.

The results of this study showed that neither fluorescence nor Cd was recovered in the perfusate. In addition, tape-stripping removed most of the hairs and all SC. Abrasion removed all SC, hairs, and most of the VE. In this case, the authors suggested the removal of the basal membrane (located between VE and dermis) based on QD penetration into dermis of abraded skin. No QDs reached the dermis in those experiments where tape-stripped skin was used. Spherical QDs appeared to penetrate better than ellipsoid ones, which is in perfect agreement with the work of Ryman-Rasmussen et al. (2006). Moreover, tested QDs slightly penetrated intact flexed and nonflexed skin. The authors suggested that poor penetration may be caused by poor contact of the formulation with the skin (because of hairs, although clipped), interspecies differences, and potential charge bindings.

By comparing the work of Zhang and Monteiro-Riviere (2008) with that of Ryman-Rasmussen et al. (2006), it is striking to note that the same QDs differently penetrated the intact skin of two different species (i.e., respectively, rat and pig). In fact, using pig skin, spherical negatively charged QDs were found in epidermis

and dermis already after 8 h, whereas the ellipsoid ones only after 24 h. As the authors commented, hair follicle density is much higher in rats than in pigs. However, in addition to what the authors suggested to explain poor penetration, it could also have been worth pondering whether the amount, composition, and thickness of sebum or sebum-sweat emulsion on rat skin could have influenced it. By assuming that the hypothesis formulated herein for the work of Ryman-Rasmussen et al. (2006) is true, poor penetration observed in the work of Zhang and Monteiro-Riviere (2008) could be explained as follows. The basic dispersing medium interacted with sebum/sebum-sweat emulsion prior to reaching SC layers. Potential damages/modification of SC composition and hence barrier could have been hindered or avoided. Or simply, the basic dispersing medium was not dissolved/emulsioned well in rat sebum/sebum-sweat emulsion. It is also known that the arrangement and overlapping of corneocytes differs in different species, and human skin should preferentially be used in skin absorption studies. Moreover, to hypothesize any penetration mechanism, it would definitively have been useful to have data on skin barrier integrity (Baroli et al. 2007) and on potential modification of superficial skin pH (Wagner et al. 2003).

31.3.4 Carbon Nanotubes and Fullerenes

Carbon nanotubes (CNTs) and fullerenes (C_{60}) are nanomaterials made of carbon and in the shape of multi-walled or single-walled hollow tubes (MWCNTs, SWCNTs) and buckyballs (i.e., hollow sphere), respectively (Carbon nanotube 2008, Fullerene Wikipedia 2008). They belong to the same family of carbon allotropes. An SWCNT may be imagined as a rolled sheet of graphite. In contrast, an MWCNT derives from the coaxial assembly of several SWCNTs (Wei et al. 2007). Both CNTs and C_{60} have been proposed for medical applications.

For what concerns skin absorption, a remarkable paper has recently been published (Rouse et al. 2007). In this study, the authors investigated skin penetration ability of a fullerene-FITC-peptide (3.5 nm) dispersed in PBS using dermatomed porcine skin. Skin was flexed or nonflexed and exposed to nanomaterials for 8 and 24 h. Flow-through diffusion cells were used. The receptor fluid (same composition as in Ryman-Rasmussen et al. 2006) was perfused at a flow rate of 1.75 mL/min (actually, 2 mL/h, as confirmed by authors. Monteiro-Riviere, N. A., personal communication, October 30, 2008). In this study, the authors showed that tested nanomaterial diffused into the skin by passive penetration and as a function of flexing time and exposure.

By comparing the work of Rouse et al. (2007) with the studies on QDs (Ryman-Rasmussen et al. 2006, Zhang and Monteiro-Riviere 2008, Zhang et al. 2008), one realizes that individual fullerenes are much smaller than QDs, and therefore their enhanced penetration is in agreement with the previous theoretical discussion on skin absorption mechanisms. Although the authors reported that studied fullerenes formed aggregates in PBS (40–250 nm) by increasing their concentration, they did not provide the following information: reversibility potential of

---

* Too large even for porcine skin.

aggregates, reversibility potential of aggregates under mechanical stress, percentage of individual fullerenes vs. percentage of aggregates, percentage of individual fullerenes vs. percentage of aggregates under mechanical stress, and potential effects of skin pH on fullerene aggregation. Therefore, even in this case, it is difficult to correlate these findings with specific penetration mechanisms. However, once more, penetration was clearly shown.

31.4 Summary

The aforementioned studies on nanomaterial skin absorption reveal that nanomaterials do penetrate the skin. Although, in general, mechanisms supporting penetration, and in some cases further diffusion into deeper and viable layers of the skin, are not critically addressed or explained, for now, it is possible to draw the following conclusions. Size is certainly the most important factor concerning skin absorption of nanomaterials as commented herein, and shown by Baroli et al. (2007) and Rouse et al. (2007). The effect of charges remains unclear, because it should be evaluated jointly with vehicle effects on skin barriers. In addition, due to interspecies' differences in skin architecture and composition, it should also be recommended to compare studies with identical (or very similar) protocols.

Nonetheless, the smallest nanomaterials do penetrate and, in some cases, they were able to reach the dermis. This finding can be commented on from two main perspectives. Pharmaceutical technologists may enthusiastically start developing drug delivery systems and diagnostics that could be systemically delivered through the skin, or other formulations to treat local diseases. In contrast, toxicologists should certainly consider potential toxicological risks. This latter concern is not hysterical but has its own rationale on the observation that many nanomaterials generate reactive oxygen species when incubated with different cells or when inoculated *in vivo*.* It could be therefore questioned whether these same reactive species could be produced within the SC, where, although no alive cells may be found, many different enzymes regulate the final phases of the cytomorphosis, whose correct completion assures proper skin barriers. This concern is of great importance when one considers that people working in the manufacturing departments of nanotechnology industries could be exposed to any type of nanomaterial-containing formulations.

31.5 Future Perspective

The author of this chapter therefore feels that issues of both perspectives could be solved by taking a multidisciplinary approach to understand mechanisms allowing the ingress and further diffusion of nanomaterials into and through the skin. It might be possible to take advantage of nanotechnological products without risking consumers' health only by understanding mechanisms and the ways to control them.

---

* Too many articles could be cited. Please search for nanoparticles and oxidative stress.

Acknowledgments

Prof. Monteiro-Riviere N. A., Dr. Onnis V., and Dr. Congiu C. are kindly acknowledged for numerous discussions and unfailing interest.

References

Al-Amoudi, A., Dubochet, J., Norlén, L. 2005. Nanostructure of the epidermal extracellular space as observed by cryo-electron microscopy of vitreous sections of human skin. *Journal of Investigative Dermatology* 124: 764–777.

Barak, O., Treat, J. R., James, W. D. 2005. Antimicrobial peptides: Effectors of innate immunity in the skin. *Advances in Dermatology* 21: 357–374.

Baroli, B. 2008. Nanoparticles and skin penetration. Are there any potential toxicological risks? *Journal of Consumer Protection and Food Safety* 3: 1–2.

Baroli, B., Ennas, M. G., Loffredo, F., Isola, M., Pinna, R., Lopez-Quintela, M. A. 2007. Penetration of metallic nanoparticles in human full-thickness skin. *Journal of Investigative Dermatology* 127: 1701–1712.

Barry, B. W. 1983a. Basic principles of diffusion through membranes. In *Dermatological Formulation: Percutaneous Absorption*, pp. 69–72. New York/Basel, Switzerland: Marcel Dekker Inc.

Barry, B. W. 1983b. Basic principles of diffusion through membranes. In *Dermatological Formulation: Percutaneous Absorption*, pp. 75–85. New York/Basel, Switzerland: Marcel Dekker Inc.

Barry, B. W. 2004. Breaching the skin's barrier to drugs. *Nature Biotechnology* 22: 165–167.

Behl, C. R., Flynn, G. L., Kurihara, T. et al. 1980. Hydration and percutaneous absorption (I): Influence of hydration on alkanol permeation through hairless mouse skin. *Journal of Investigative Dermatology* 75: 346–352.

Behl, C. R., Barrett, M., Flynn, G. L. 1981. Hydration and percutaneous absorption (II): Influence of hydration on water and alkanol permeation through Swiss (furry) mouse skin-comparison with hairless mouse skin data. *Journal of Pharmaceutical Sciences* 70: 1212–1225.

Behl, C. R., Barrett, M., Flynn, G. L. et al. 1982. Hydration and percutaneous absorption (III): Influence of stripping and scalding on hydration alterations of the permeability of hairless mouse skin to water and n-alkanols. *Journal of Pharmaceutical Sciences* 71: 229–234.

Behl, C. R., El-Sayed, A. A., Flynn, G. L. 1983. Hydration and percutaneous absorption (IV): Influence of hydration on n-alkanol permeation through rat skin-comparison with hairless and Swiss (furry) mice. *Journal of Pharmaceutical Sciences* 72: 79–82.

Benson, H. A. 2005. Transdermal drug delivery: Penetration enhancement techniques. *Current Drug Delivery* 2: 23–33.

Benson, H. A. 2006. Transfersomes for transdermal drug delivery. *Expert Opinion on Drug Delivery* 3: 727–737.

Bickers, D. R., Athar, M. 2006. Oxidative stress in the pathogenesis of skin disease. *Journal of Investigative Dermatology* 126: 2565–2575.

Bouwstra, J. A., Ponec, M. 2006. The skin barrier in healthy and diseased state. *Biochimica et Biophysica Acta* 1758: 2080–2095.

Brown, M. B., Martin, G. P., Jones, S. A., Akomeah, F. K. 2006. Dermal and transdermal drug delivery systems: Current and future prospects. *Drug Delivery* 13: 175–187.

Carbon Nanotube. 2008. Wikipedia, viewed October 31, 2008, <http://en.wikipedia.org/wiki/Carbon_nanotube>.

Cevc, G. 2004. Lipid vesicles and other colloids as drug carriers on the skin. *Advanced Drug Delivery Reviews* 56: 657–711.

Cevc, G., Blume, G. 1992. Lipid vesicles penetrate into intact skin owing to the transdermal osmotic gradients and hydration force. *Biochimica et Biophysica Acta* 1104: 226–232.

Changez, M., Chander, J., Dinda, A. K. 2006a. Transdermal permeation of tetracaine hydrochloride by lecithin microemulsion: In vivo. *Colloids and Surfaces B, Biointerfaces* 48: 58–66.

Changez, M., Varshney, M., Chander, J., Dinda, A. K. 2006b. Effect of the composition of lecithin/n-propanol/isopropyl myristate/water microemulsions on barrier properties of mice skin for transdermal permeation of tetracaine hydrochloride: In vitro. *Colloids and Surfaces B, Biointerfaces* 50: 18–25.

Dokka, S., Cooper, S. R., Kelly, S., Hardee, G. E., Karrasw, J. G. 2005. Dermal delivery of topically applied oligonucleotides via follicular transport in mouse skin. *Journal of Investigative Dermatology* 124: 971–975.

Domashenko, A., Gupta, S., Cosartelis, G. 2000. Efficient delivery of transgenes to human hair follicle progenitor cells using topical lipoplex. *Nature Biotechnology* 18: 420–423.

El Maghraby, G. M. M., Williams, A. C., Barry, B. W. 2001. Skin hydration and possible shunt route penetration in controlled estradiol delivery from ultradeformable and standard liposomes. *The Journal of Pharmacy and Pharmacology* 53: 1311–1322.

Elias, P. M. 1983. Epidermal lipids, barrier function and desquamation. *Journal of Investigative Dermatology* 80 Suppl: 44s–49s.

Elias, P. M. 1990. The importance of epidermal lipids for the stratum corneum barrier. In *Topical Drug Delivery Formulations*, eds. D. W. Osborne and A. H. Amann, pp. 13–28. New York/Basel, Switzerland: Marcel Dekker Inc.

Elias, P. M. 2005. Stratum corneum defensive functions: An integrated view. *Journal of Investigative Dermatology* 125: 183–200.

Elias, P. M., Choi, E. H. 2005. Interactions among stratum corneum defensive functions. *Experimental Dermatology* 14: 719–726.

Elsayed, M. M., Abdallah, O. Y., Naggar, V. F., Khalafallah, N. M. 2007. Lipid vesicles for skin delivery of drugs: Reviewing three decades of research. *International Journal of Pharmaceutics* 332: 1–16.

Essa, E. A., Bonner, M. C., Barry, B. W. 2002. Human skin sandwich for assessing shunt route penetration during passive and iontophoretic drug and liposome delivery. *The Journal of Pharmacy and Pharmacology* 54: 1481–1490.

Farwanah, H., Raith, K., Neubert, R. H. H., Wohlrab, J. 2005. Ceramide profiles of the uninvolved skin in atopic dermatitis and psoriasis are comparable to those of healthy skin. *Archives of Dermatological Research* 296: 514–521.

Fullerene. 2008. Wikipedia, viewed October 31, 2008, <http://en.wikipedia.org/wiki/Fullerene>.

Gupta, A. K., Gupta, M. 2005. Synthesis and surface engineering of iron oxide nanoparticles for biomedical applications. *Biomaterials* 26: 3995–4021.

Hadgraft, J., Lane, M. E. 2005. Skin permeation: The years of enlightenment. *International Journal of Pharmaceutics* 305: 2–12.

Hikima, T., Tojo, K., Maibach, H. I. 2005. Skin metabolism in transdermal therapeutic systems. *Skin Pharmacology and Physiology* 18: 153–159.

Hirokawa, T., Okamoto, H., Gosyo, Y., Tsuda, T., Timerbaev, A. R. 2007. Simultaneous monitoring of inorganic cations, amines and amino acids in human sweat by capillary electrophoresis. *Analytica Chimica Acta* 581: 83–88.

Hoet, P. H. M., Brüske-Hohlfeld, I., Salata, O. V. 2004. Nanoparticles—Known and unknown health risks. *Journal of Nanobiotechnology* 2: 12.

Hoffman, R. M. 2000. The hair follicle as a gene therapy target. *Nature Biotechnology* 18: 20–21.

Hueber, F., Schaefer, H., Wepierre, J. 1994. Role of transepidermal and transfollicular routes in percutaneous absorption of steroids: In vitro studies on human skin. *Skin Pharmacology* 7: 237–244.

Johnson, M. E., Blankschtein, D., Langer, R. 1997. Evaluation of solute permeation through the stratum corneum: Lateral bilayer diffusion as the primary transport mechanism. *Journal of Pharmaceutical Sciences* 86: 1162–1172.

Kalinin, A. E., Kajava, A. V., Steinert, P. M. 2002. Epithelial barrier function: Assembly and structural features of the cornified cell envelope. *Bioessays* 24: 789–800.

Kierszenbaum, A. L. 2002. Integumentary system. In *Histology and Cell Biology. An Introduction to Pathology*, pp. 299–318. St Louis, MO: Mosby Inc.

Lademann, J., Otberg, N., Richter, H. et al. 2001. Investigation of follicular penetration of topically applied substances. *Skin Pharmacology and Applied Skin Physiology* 14: 17–22.

Lademann, J., Otberg, N., Jacobi, U., Hoffman, R. M., and Blume-Peytavi, U. 2005. Follicular penetration and targeting. *The Journal of Investigative Dermatology. Symposium Proceedings* 10: 301–303.

Lademann, J., Richter, H., Teichmann, A. et al. 2007. Nanoparticles—An efficient carrier for drug delivery into the hair follicles. *European Journal of Pharmaceutics and Biopharmaceutics* 66: 159–164.

Lampe, M. A., Williams, M. L., Elias, P. M. 1983. Human stratum corneum lipids: Characterization and regional variations. *Journal of Lipid Research* 24: 120–130.

Lavrijsen, A. P. M., Bouwstra, J. A., Gooris, G. S., Weerheim, A., Boddé, H. E., Ponec, M. 1995. Reduced skin barrier function parallels abnormal stratum corneum lipid organization in patients with lamellar ichthyosis. *Journal of Investigative Dermatology* 105: 619–624.

Law, S., Wertz, P. W., Swartzendruber, D. C., Squier, C. A. 1995. Regional variation in content, composition and organization of porcine epithelial barrier lipids revealed by thin-layer chromatography and transmission electron microscopy. *Archives of Oral Biology* 40: 1085–1091.

Lazo, N. D., Downing, D. T. 1999. A mixture of alpha-helical and 3(10)-helical conformations for involucrin in the human epidermal corneocyte envelope provides a scaffold for the attachment of both lipids and proteins. *The Journal of Biological Chemistry* 274: 37340–37344.

Lee, S. H., Jeong, S. K., Ahn, S. K. 2006. An update of the defensive barrier function of skin. *Yonsei Medical Journal* 47: 293–306.

Magnusson, B. M., Walters, K. A., Roberts, M. S. 2001. Veterinary drug delivery: Potential for skin penetration enhancement. *Advanced Drug Delivery Reviews* 50: 205–227.

Mathers, A. R., Larregina, A. T. 2006. Professional antigen-presenting cells of the skin. *Immunologic Research* 36: 127–136.

Meidan, V. M., Bonner, M. C., Michniak, B. B. 2005. Transfollicular drug delivery—Is it a reality? *International Journal of Pharmaceutics* 306: 1–14.

Meyer, F., Laitano, O., Bar-Or, O., McDougall, D., Heingenhauser, G. J. 2007. Effect of age and gender on sweat lactate and ammonia concentrations during exercise in the heat. *Brazilian Journal of Medical and Biological Research* 40: 135–143.

Mitsui, T. (ed.). 1997a. Cosmetics and skin. In *New Cosmetic Science*, pp. 38–46. Amsterdam, the Netherlands: Elsevier Science B.V.

Mitsui, T. (ed.). 1997b. Cosmetics and physical chemistry. In *New Cosmetic Science*, pp. 167–169. Amsterdam, the Netherlands: Elsevier Science B.V.

Monteiro-Riviere, N. A. 2008. Personal Communication. October 30.

Moser, K., Kriwet, K., Froehlich, C., Kalia, Y. N., Guy, R. H. 2001. Supersaturation: Enhancement of skin penetration and permeation of a lipophilic drug. *Pharmaceutical Research* 18: 1006–1011.

Moss, G. P., Dearden, J. C., Patel, H., Cronin, M. T. D. 2002. Quantitative structure–permeability relationships (QSPRs) for percutaneous absorption. *Toxicology in Vitro* 16: 299–317.

Nemes, Z., Steinert, P. M. 1999. Bricks and mortar of the epidermal barrier. *Experimental & Molecular Medicine* 31: 5–19.

Neuberger, T., Schöpf, B., Hofmann, H., Hofmann, M., von Rechenberg, B. 2005. Superparamagnetic nanoparticles for biomedical applications: Possibilities and limitations of a new drug delivery system. *Journal of Magnetism and Magnetic Materials* 293: 483–496.

Niyonsaba, F., Nagaoka, I., Ogawa, H. 2006. Human defensins and cathelicidins in the skin: Beyond direct antimicrobial properties. *Critical Reviews in Immunology* 26: 545–576.

Nohynek, G. J. 2008. Safety of nanotechnology in cosmetics and sunscreens. *Proceedings of 7th Conference and Workshop on Biological Barriers and Nanomedicine*, Saarbruecken, Germany, p. 7.

Nohynek, G. J., Lademann, J., Ribaud, C., Roberts, M. S. 2007. Grey goo on the skin? Nanotechnology, cosmetic, and sunscreen safety. *Critical Reviews in Toxicology* 37: 251–277.

Norlén, L., Gil, I. P., Simonsen, A., Descouts, P. 2007. Human stratum corneum lipid organization as observed by atomic force microscopy on Langmuir–Blodgett films. *Journal of Structural Biology* 158: 386–400.

Oberdörster, G., Oberdörster, E., Oberdörster, J. 2005. Nanotoxicology: An emerging discipline evolving from studies of ultrafine particles. *Environmental Health Perspectives* 113: 823–839.

Otberg, N, Richter, H., Schaefer, H., Blume-Peytavi, U., Sterry, W., Lademann, J. 2004. Variations of hair follicle size and distribution in different body sites. *Journal of Investigative Dermatology* 122: 14–19.

Paliwal, S., Menon, G. K., Mitragotri, S. 2006. Low-frequency sonophoresis: Ultrastructural basis for stratum corneum permeability assessed using quantum dots. *Journal of Investigative Dermatology* 126: 1095–1101.

Partition Coefficient. 2008. *Consumer Products Safety & Quality*, viewed October 27, 2008, available at: <http://ecb.jrc.ec.europa.eu/documents/Testing-Methods/ANNEXV/A08web1992.pdf>.

Patterson, M. J., Galloway, S. D., Nimmo, M. A. 2000. Variations in regional sweat composition in normal human males. *Experimental Physiology* 85: 869–875.

Penn, S. G., He, L., Natan, M. J. 2003. Nanoparticles for bioanalysis. *Current Opinion in Chemical Biology* 7: 609–615.

Potts, R. O., and Francoeur, M. L. 1991. The influence of stratum corneum morphology on water permeability. *Journal of Investigative Dermatology* 96: 495–499.

Power, I. 2007. Fentanyl HCl iontophoretic transdermal system (ITS): Clinical application of iontophoretic technology in the management of acute postoperative pain. *British Journal of Anaesthesia* 98: 4–11.

Quantum Dot. 2008. Wikipedia, viewed October 31, 2008, <http://en.wikipedia.org/wiki/Quantum_dot>.

Quantum Dots explained. 2008. Evident technologies, viewed October 31, 2008, <http://www.evidenttech.com/qdot-definition/quantum-dot-introduction.php>.

Rawlings, A. V., Scott, I. R., Harding, C. R., Bowser, P. A. 1994. Stratum corneum moisturization at the molecular level. *Journal of Investigative Dermatology* 103: 731–740.

Roberts, M. S., Walker, M. 1993. Water: The most natural penetration enhancer. In *Pharmaceutical Skin Penetration Enhancement*, eds. K. A. Walters and J. Hadgraft, pp. 1–30. New York: Marcel Dekker, Inc.

Rouse, J. G., Yang, J., Ryman-Rasmussen, J. P., Barron, A. R., Monteiro-Riviere, N. A. 2007. Effects of mechanical flexion on the penetration of fullerene amino acid-derivatized peptide nanoparticles through skin. *Nano Letters* 7: 155–60.

Ryman-Rasmussen, J. P., Riviere, J. E., Monteiro-Riviere, N. A. 2006. Penetration of intact skin by quantum dots with diverse physicochemical properties. *Toxicological Sciences* 91: 159–165.

Scheuplein, R. J. 1967. Mechanism of percutaneous absorption (II): Transient diffusion and the relative importance of various routes of skin penetration. *Journal of Investigative Dermatology* 48: 79–88.

Schmid-Wendtner, M. H., Korting, H. C. 2006. The pH of the skin surface and its impact on the barrier function. *Skin Pharmacology and Physiology* 19: 296–302.

Seiffert, K., Granstein, R. D. 2006. Neuroendocrine regulation of skin dendritic cells. *Annals of the New York Academy of Sciences* 1088: 195–206.

Sivamani, R. K., Liepmann, D., Maibach, H. I. 2007. Microneedles and transdermal applications. *Expert Opinion on Drug Delivery* 4: 19–25.

Stewart, M. E. 1992. Sebaceous gland lipids. *Seminars in Dermatology* 11: 100–105.

Suhonen, T. M., Bouwstra, J. A., Urtti, A. 1999. Chemical enhancement of percutaneous absorption in relation to stratum corneum structural alterations. *Journal of Controlled Release* 59: 149–161.

Swarbrick, J., Lee, G., Brom, J., Gensmantel, N. P. 1984. Drug permeation through human skin II: Permeability of ionizable compounds. *Journal of Pharmaceutical Sciences* 73: 1352–1355.

Szabo, G. 1962. The number of eccrine sweat glands in human skin. *Advances in Biology of Skin* 3: 1–5.

Tang, H., Mitragotri, S., Blankschtein, D., Langer, R. 2001. Theoretical description of transdermal transport of hydrophilic permeants: Application to low frequency sonophoresis. *Journal of Pharmaceutical Sciences* 90: 545–568.

Täuber, U. 1989. Drug metabolism in the skin: Advantages and disadvantages. In *Transdermal Drug Delivery: Developmental Issues and Research Initiatives*, eds. J. Hadgraft and R. H. Guy, pp. 99–112. New York: Marcel Dekker Inc.

The Royal Society and Royal Academy of Engineering on Nanotechnology and Nanoscience. What are nanoscience and nanotechnologies, July 29, 2004. In *Nanoscience and Nanotechnologies: Opportunities and Uncertainties.* The Royal Society and Royal Academy of Engineering on Nanotechnology and Nanoscience, viewed October 31, 2008, <http://www.nanotec.org.uk/finalReport.htm>.

Toll, R., Jacobi, U., Richter, H., Lademann, J., Schaefer, H., Blume-Peytavi, U. 2004. Penetration profile of microspheres in follicular targeting of terminal hair follicles. *Journal of Investigative Dermatology* 123: 168–176.

van Hal, D. A., Jeremiasse, E., Junginger, H. E., Spies, F., Bouwstra, J. A. 1996. Structure of fully hydrated human stratum corneum: A freeze-fracture electron microscopy study. *Journal of Investigative Dermatology* 106: 89–95.

Vogt, A., Combadiere, B., Hadam, S. et al. 2006. 40 nm, but not 750 or 1,500 nm, nanoparticles enter epidermal CD1a+ cells after transcutaneous application on human skin. *Journal of Investigative Dermatology* 126: 1316–1322.

Wagner, H., Kostka, K. H., Lehr, C. M, Schaefer, U. F. 2003. pH profiles in human skin: Influence of two in vitro test systems for drug delivery testing. *European Journal of Pharmaceutics and Biopharmaceutics* 55: 57–65.

Walter, K. 2003. When semiconductors go nano. S&TR, November 2003, pp. 4–10. *Lawrence Livermore National Laboratory*, viewed September 10, 2008, <http://www.llnl.gov/str/November03/pdfs/11_03.1.pdf>.

Wang, M. D., Shin, D. M., Simons, J. W., Nie, S. 2007. Nanotechnology for targeted cancer therapy. *Expert Review of Anticancer Therapy* 7: 833–837.

Warheit, D. B., Borm, P. J., Hennes, C., Lademann, J. 2007. Testing strategies to establish the safety of nanomaterials: Conclusions of an ECETOC workshop. *Inhalation Toxicology* 19: 631–643.

Weerheim, A., Ponec, M. 2001. Determination of stratum corneum lipid profile by tape stripping in combination with highperformance thin-layer chromatography. *Archives of Dermatological Research* 293: 191–199.

Wei, W., Sethuraman, A., Jin, C., Monteiro-Riviere, N. A., Narayan, R. J. 2007. Biological properties of carbon nanotubes. *Journal of Nanoscience and Nanotechnology* 7: 1–14.

Wertz, P. W. 1996. The nature of the epidermal barrier: Biochemical aspects. *Advanced Drug Delivery Reviews* 18: 283–294.

Wu, H., Ramachandran, C., Bielinska, A. U. et al. 2001. Topical transfection using plasmid DNA in a water-in-oil nanoemulsion. *International Journal of Pharmaceutics* 221: 23–34.

Wurster, D. E., Kramer, S. F. 1961. Investigation of some factors influencing percutaneous absorption. *Journal of Pharmaceutical Sciences* 50: 288–293.

Zhang, L. W., Monteiro-Riviere, N. A. 2008. Assessment of quantum dot penetration into intact, tape-stripped, abraded and flexed rat skin. *Skin Pharmacology and Physiology* 21: 166–180.

Zhang, L. W., Yu, W. W., Colvin, V. L., Monteiro-Riviere, N. A. 2008. Biological interactions of quantum dot nanoparticles in skin and in human epidermal keratinocytes. *Toxicology and Applied Pharmacology* 228: 200–211.

32

Nanoparticulate Systems and the Dermal Barrier

Frank Stracke
*Fraunhofer Institut
Biomedizinische Technik*

Marc Schneider
Universität des Saarlandes

32.1 Introduction

Though there have always been nanoscale environmental compounds interacting with the human organism, a significant interest in these interactions did not arise until artificial nanoparticulate compounds began to find wide acceptability (Oberdörster et al. 2005). At present, the rapidly growing number of nanoparticulate systems, with respect to chemistry, size, shape, and formulation, and the expanding areas of application make it essential to investigate the various biological aspects of nanoparticle exposure to the human organism. On the one hand, health hazards are to be identified and assessed; on the other hand, medical and pharmaceutical potentials may be discovered and exploited. The behavior of nanoparticulate substances in biosystems and their physiological effects can neither be extrapolated straightforwardly from the bulk properties nor can they be predicted from the molecular properties of the constituents. So, in general, all the expertise is to be obtained experimentally. The first step of any interaction between an organism and any compound is the uptake of the compound from the environment. Several pathways of absorption exist for human (and most animal) organisms:

1. The oral uptake pathway includes all absorption routes of the digestive tract, i.e., via the stomach or the intestine epithelium.
2. The pulmonary uptake route, i.e., absorption through the epithelia of the lung following inhalation. This is a crucial route for nanoparticle risk assessment.

3. The mucosal uptake pathways combine all the remaining epithelial routes that do not belong to the oral and the pulmonary absorption routes, e.g., vaginal and ocular pathways. If the oral and the nasal mucosae are part of the oral and the pulmonary absorption routes, respectively, is subject to the particular definition.
4. The transdermal or cutaneous absorption route includes all pathways via the skin and its appendages (hair follicles, sweat, and sebaceous glands).

In this chapter, we focus on skin and its barrier as well as its sink functions to artificial nanoparticulate compounds. Since it is an overview on general nanoparticles-skin interactions, no specific physiological responses to particular particle composition are considered, but only effects arising from the particle size, shape, and physical/physicochemical properties. Furthermore, only healthy skin is treated herein. Injured and inflamed skin can be entered even by particles larger than the nanometer scale due to the loss of its barrier function (Oberdörster et al. 2005). This is also the case for chemically induced irritations of the skin, e.g., by nonpolar solvents and strong alkalines.

The chapter starts with an introduction to skin morphology and the composition principles concerning its barrier and absorption properties. Beyond these, numerous other functions are conducted by our biggest organ (temperature and mechanical sensing, immunological, metabolic, and endocrine, etc.), which cannot be treated in the limited scope of this chapter.

The dermal transport of exogenous compounds with respect to their chemical and physical properties is reviewed, and the consequences to expect for the uptake of nanoparticulate materials are discussed. Thereafter, a survey of the techniques to measure, visualize, and investigate dermal penetration and permeation is given. These methods are assessed for their suitability in nanoparticle cutaneous transport studies. After a basic chemical classification of nanoparticulate compounds, the results of various penetration and permeation experiments available in the literature are presented and discussed.

32.2 Basic Aspects of Skin Morphology and Transport

The skin is the most visible and obvious barrier of the body separating the inside from the outside. In this context, the skin is responsible for separating an "aqueous" inner compartment from a dry outer compartment and is therefore highly polarized in its functionality and composition. Its main function is that of water regulation, which is inherently connected with thermoregulation that makes use of a large evaporation enthalpy of water. Furthermore, the skin represents a huge barrier for all kinds of xenomaterials that might be of a noxious nature.

An important aspect to the understanding of the interaction of any kind of material with the skin is skin morphology and its composition that is well adapted to its function. The complex tasks the skin has to fulfill is reflected in its complex composition resulting in several layers with different properties, inter alia concerning the penetration behavior. Overall, three major layers can be distinguished (Figure 32.1):

1. The *hypodermis* (also *subcutis*) or *subcutaneous fatty tissue* (HD)
2. The *dermis* (D)
3. The *epidermis* (ED) which can be further subdivided into the outermost horny layer, the so-called *stratum corneum* (SC) and the *viable epidermis*

32.2.1 Hypodermis, Subcutaneous Fatty Tissue

The subcutaneous fat layer (hypodermis) of several millimeters thickness acts mainly as a heat insulator and a mechanical cushion, and stores readily available high-energy chemicals. It is vascularized and has no chemical barrier function.

32.2.2 Dermis

As for most skin layers, the thickness of the dermis depends on the body site and ranges from 3 to 5 mm. The major task of the dermis is to withstand mechanical stress. The dermis consists of a matrix of connective tissue comprising collagen, elastin, and reticulin, and is interspersed by skin appendages, such as sweat glands, pilosebaceous units, and hair follicles that also penetrate the layer above. Furthermore, nerves, lymphatic and blood vessels are located in this skin layer. Blood vessels are found directly beneath the *stratum basale* of the viable epidermis, supplying nutrients and removing metabolites. Fibroplasts are the main cell type present in the dermis and are responsible for the making and the degradation of the extracellular matrix (ECM) (Tobin 2006). For systemic drug absorption, both the blood system and the lymphatic system are responsible, acting as sinks, and hence keeping the drug concentration in the dermis low.

32.2.3 Epidermis

The human epidermis consists of two very different areas: one composed of cornified corneocytes (CC) that form the outermost layer and the inner part of the epidermis that is the viable epidermis composed of living cells (Figure 32.1b).

The cornified cells of the stratum corneum are embedded into a complex matrix formed by aqueous filaments and lipid bilayers consisting of various classes of lipids, e.g., ceramides, cholesterol, cholesterol esters, free fatty acids, and triglycerides (Wertz Downing 1989, Tobin 2006). The averaged viscosity of this intercellular matrix is on the order of 1 Pa·s, comparable to the viscosity

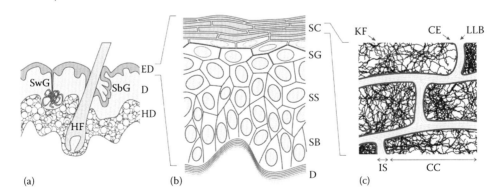

FIGURE 32.1 Human skin morphology: (a) full skin cartoon with epidermis ED, dermis D, and hypodermis HD as well as a sweat gland SwG and a hair follicle HF with a sebaceous gland SbG. (b) Zoom view of the epidermis including the viable epidermis with stratum basale SB, stratum spinosum SS, stratum granulosum SG, and the nonviable, cornified stratum corneum SC, which is the actual dermal barrier. (c) The stratum corneum with the rigid corneocytes CC, dead cells densely packed with keratinous filaments KF in a matrix of connecting proteins and confined by the cornified envelope CE, separated by the intercellular space IS filled with lipid compounds (fatty acids, ceramides, etc.) and aqueous films to form lamellar lipid bilayers LLB. The corneocytes are interconnected via the intercellular space by desmosomes (not shown).

of glycerol. Structurally, this top layer of the skin can be described by the so-called brick-and-mortar model (Elias 1983), which is sketched in Figure 32.1c. The distances between the corneocytes are in the range of ~100 nm whereas the continuous phases only offer a width of ~10 nm (Wang et al. 2006, Naegel et al. 2008b). The stratum corneum with its unique structure is of major significance for the barrier function of the skin, controlling the percutaneous absorption of dermally applied substances and regulating fluid homeostasis. The thickness of the stratum corneum varies depending on the body location and is thicker at stress areas such as the soles of the feet and the palms than in other areas. For most of the body, one can assume a thickness of 10–25 μm. The *stratum corneum* (SC) swells severalfold by hydration what is of relevance for penetration. The SC is formed by the cells and the materials originating at the bottom of the epidermis, the basal layer or *stratum germinativum*. A continuous migration of the cells from the basal layer to the top secures the renewal of the top skin layers within ~28 days (Tobin 2006, Roberts and Walters 2008). This base layer of the viable epidermis is followed by the *stratum spinosum*, the *stratum granulosum*, and the stratum corneum. The movement from the basal layer to the SC through the different states or skin layers is accompanied by a steady differentiation of the cells. While in the beginning the cells contain a nucleus and are held together with desmosomes or hemidesmosomes, they lose the nucleus, get less strongly attached to each other, produce lipids that are stored in Odland bodies or lamellar bodies, and are excreted into the intercellular space to form the lipid matrix (→ mortar) (stratum granulosum), get flattened, and become cornified (corneocytes, → bricks). The cornification takes place due to the production of keratinous filaments that are then strongly interconnected by a protein called filaggrin (*filament aggre*gating prote*in*).

Furthermore, the viable epidermis contains melanocytes that produce melanin for light protection, and Langerhans cells, responsible for the immune response of the skin. Another important aspect is the absence of any vascularization in the viable epidermis. So, to enter the vascular system of an organism, a substance has to permeate the epidermis and to penetrate the dermis, where the peripheral blood vessels are found.

32.2.4 Skin Appendages

Additionally to the skin structures described so far, there are the so-called skin appendages like hair follicles, sweat glands, and sebaceous glands. These structures originate below the skin surface and form channels through the skin to the skin surface. Their overall contribution to the skin surface is as small as 1% but nevertheless they are an important part of the skin for biological functionality. However, we will not focus here on the function and the possible role for penetration of the appendages, which is reviewed elsewhere (Vogt et al. 2005).

32.2.5 Penetration and Permeation

From a review of literature it is often seen that not much care has been taken regarding the correct use of the terms *penetration* and

permeation, although they have different meanings. Therefore, we just want to highlight that penetration describes the process of entry into a body or layer, while permeation describes the passage through the body or the layer into the following structure.

The easiest way to quantitatively measure and report transport over a barrier (i.e., permeation over full skin, epidermis or isolated stratum corneum) of thickness h, is to employ a stationary flux J_{SS} (per area) from an *infinite dose* donor compartment (this means, the donor reservoir concentration does not decrease upon efflux) over the barrier to a sink acceptor compartment (this means, the acceptor reservoir does not concentrate upon influx). Accordingly, Fick's first law of diffusion becomes

$$J(z) = D\frac{\partial c(z)}{\partial z} \Rightarrow J_{SS} = D\frac{c_m}{h} \tag{32.1}$$

where
   D is the diffusion coefficient
   c is the concentration
   z is the normal dimension

The concentration c_m at the barrier surface on the donor compartment side is determined by the partition coefficient K_m between reservoir solvent and barrier material, and the donor compartment concentration c:

$$K_m = \frac{c_m}{c} \tag{32.2}$$

Together, this leads to the well-established equation (Dancik et al. 2008)

$$J_{SS} = K_m D\frac{c}{h} \tag{32.3}$$

In order to have a measure independent of the donor reservoir concentration, the permeability coefficient was introduced:

$$K_P = \frac{J_{SS}}{c} = K_m \frac{D}{h} \tag{32.4}$$

Another parameter, feasible for the comparison of different solutes, is the *maximum flux* J_{max}, observed when the donor compartment is saturated with the solute molecules. This ascertains the highest achievable concentration difference over the barrier, since the partition coefficient is the ratio of solubilities in the respective phases $K_m = S_m/S$. Hence,

$$J_{max} = D\frac{S_m}{h} = K_m D\frac{S}{h} = K_P S \tag{32.5}$$

Since the solubility is not a defined quantity for nanoparticulate colloids, J_{max} is not an appropriate measure for nanoparticle transport (Figure 32.2).

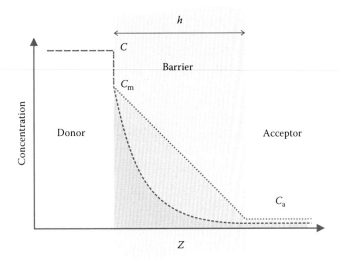

FIGURE 32.2 Scheme of the barrier transport model: The concentration in the barrier at the donor compartment surface, C_m, is determined by the donor reservoir concentration and the partition coefficient, K_m. At an early stage, the concentration gradient varies over the barrier distance (dashed line). After balancing to a constant value (dotted line), steady-state conditions are achieved (assuming negligible acceptor concentrations, c_a).

32.2.6 Pathways for Skin Absorption

Regarding the absorption of any kind of material, two general pathways can be considered: along the skin appendages or through the stratum corneum and the underlying layers. Even though the appendages present only a small portion of the surface they are considered to contribute and might even be addressed for a directed delivery because of the depth they reach. The hair follicles especially seem to be a promising target for nanoparticulate carrier systems (Lademann et al. 2006, 2007). The invasion of a substance into a skin appendage is not an absorption process itself. Compounds inside an appendage are still on the outside of the body by definition. Nevertheless, accumulation in

such structures may lead to faster and more efficient uptake due to altered barrier morphology and enhanced exposure times in the appendages.

Focusing on the absorption across the stratum corneum, two possible pathways are obvious: through the corneocytes (bricks) or along the intercellular spaces along the lipid matrix (mortar). The latter pathway seems to be most suited for penetration, offering higher diffusivity although the pathway is much longer (Figure 32.3). A full understanding of the processes responsible for the penetration process has not yet been obtained. A close interaction of theoretical modeling approaches based on experimentally extracted data is necessary and is in process (e.g., Hansen et al. 2008, Naegel et al. 2008a,b). Generally, the SC is assumed to be the main barrier for absorption. The hydrophilicity of the absorbent is therefore crucial. Overall, the SC can be considered to be a lipoidic compartment, and hence lipophilic molecules can distribute more easily into the SC and penetration is facilitated. Furthermore, the absorbent needs to fit into the intercellular space and needs to move along the lipid phase or the aqueous phase, respectively, restricting the space available, and therefore, the size of the penetrating species influences the absorption behavior. However, the layer-wise buildup of the skin offers an additional barrier, especially for those species that have successfully entered the SC due to their high lipophilicity. For permeation into the underlying viable epidermis, which constitutes a hydrophilic compartment, certain hydrophilicity is necessary (Moghimi et al. 1999). As a consequence, only substances with lipophilicity not too high and not too low are well suited for skin absorption (Figure 32.4).

These two aspects of penetration and permeation, the size of an absorbent as well as its partition coefficient, are expressed in the formula of Potts and Guy who found a phenomenological expression based on molecular properties to describe the absorption (Potts and Guy 1992):

$$\log K_P = \log\left(\frac{D}{h}\right) + f \cdot \log K_{oct} - \beta'' \cdot MW \qquad (32.6)$$

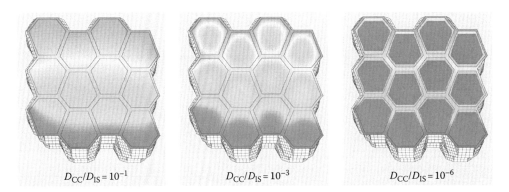

FIGURE 32.3 (**See color insert following page V-2.**) Color-coded simulations of three diffusing compounds in a 3D stratum corneum model, calculated for different diffusivity ratios D_{CC}/D_{IS} of the compounds between corneocytes (bricks, CC) and in the intercellular space (mortar, IS). (Top view; red indicates high concentration). For nanoparticulate compounds, a strictly intercellular route (right case) will be due. (Courtesy of M. Heisig, IWR, Ruprecht-Karls-Universität Heidelberg, Germany; From Feuchter, D. et al., *Comput. Vis. Sci.*, 9(2), 117, 2006. With permission.)

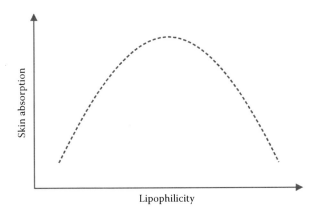

FIGURE 32.4 Relation between absorption and lipophilicity. Optimal absorption conditions are for $1 \leq \log (K_{oct}) \leq 3$ (K_{oct} being the octanol/water-partition coefficient).

Hereby, D is the not-restricted diffusivity of the permeant in the membrane (skin), h represents the length of the diffusion pathway, MW is the molecular weight,* K_{oct} is the octanol/water-partition coefficient,[†] K_p is the permeability coefficient, and f, β'' empirical constants. Several attempts were made to optimize this relatively crude phenomenological expression but only slight improvements were achieved. Most approaches find the molecular weight to be a strong effector and a reliable measure to predict permeabilities and maximum fluxes of solutes through the stratum corneum, as shown by Magnusson et al. (2004).

Empirical findings by Bos and Meinardi (2000), known as the "500 Dalton rule," even indicate a molar-mass exclusion effect for dermal penetration that is not predicted by the quantitative penetration models of Potts and Guy (1992) and Magnusson et al. (2004). Nevertheless, it is shown that the diffusion of high-MW corpuscles is drastically impeded in the stratum corneum. As nanoparticles have molar masses in the order of 10^6 Da and more, no significant transport over an intact dermal barrier is to be expected for human beings.

From another point of view, one can estimate the "best-case" flux of particles over the stratum corneum by the classical diffusion theory. Under the plausible assumption, particle migration occurs exclusively along the intercellular route (see Figure 32.1c), they have to diffuse over a distance of some 100 μm (say $h' = 500$ μm) via the tortuous intercellular space that has a cross sectional area of about $\varepsilon = 1\%$ of the total surface area (Moghimi et al. 1996). Using the coarse simplification of a homogeneous membrane matrix (hence neglecting any size-exclusion effects at pores, adsorption at corneocyte surfaces, and phase transition steps) of diffusivity D and furthermore *infinite dose* conditions on the peripheral surface, one may calculate the *steady-state* flux parameters over the barrier according to Equations 32.3 through 32.5. The diffusivity D for spherical nanoparticles is approximated by the Stokes–Einstein relation and data from Moghimi et al. (1996):

TABLE 32.1 "Best-Case" Transport Parameters of Nanoparticles Over a Typical Stratum Corneum Barrier

r (nm)	K_p (cm·s⁻¹)	J_{SS} (s⁻¹·cm⁻²) with $c = 10^{14}$ (cm⁻³)	C_{max} (cm⁻³)	J_{max} (s⁻¹·cm⁻²)
5	8.9×10^{-11}	8.9×10^3	1.4×10^{18}	1.3×10^8
10	4.5×10^{-11}	4.5×10^3	1.8×10^{17}	7.9×10^6
25	1.8×10^{-11}	1.8×10^3	1.1×10^{16}	2.0×10^5
50	0.9×10^{-11}	0.9×10^{31}	1.4×10^{15}	1.3×10^4

$$J_{SS} = \varepsilon \cdot D \frac{c}{h'} = \varepsilon \cdot \frac{kT}{6\pi\eta r} \cdot \frac{c}{h'} \qquad (32.7)$$

where
  k is the Boltzmann constant
  η is the dynamic viscosity
  r is the particle radius
  h' is the diffusion distance
  ε is the area fraction of intercellular spaces

Obviously, the particle flux drops with the reciprocal of r, the same holds for the mass flux assuming a constant mass concentration c in the donor compartment. A modified maximum-particle flux can be defined by assuming a maximum-nanoparticle density being 74% ("closest packed") of the reciprocal of the particle volume:

$$J_{max} = \varepsilon \cdot \frac{kT}{6\pi\eta r} \cdot \frac{c_{max}}{h'} = 0.74 \cdot \varepsilon \cdot \frac{kT}{6\pi\eta} \cdot \frac{3}{4\pi h'} \cdot \frac{1}{r^4} \qquad (32.8)$$

Thus, the maximum-particle flux decreases with the fourth power of the particle radius. Some exemplifying results of such estimations are given in Table 32.1. In reality, the first particles will not arrive at the inner interface of the barrier until a lag time has elapsed. During this lag time, the steady-state conditions, i.e., the constant gradient over the barrier, develops. This time also increases with the particle radius. Together with the impeding effects due to the inhomogeneous, porous nature of the stratum corneum, as well as the finite dose and the exposure time, the transport of nanoparticulate material into the human skin is strongly limited far below this best-case estimation.

32.3 Experimental Techniques for the Investigation of Cutaneous Absorption

Like in other fields of experimental research, the investigation of matter transport processes into and through the skin does not know a common *silver bullet* technique. The choice of experimental methods and conditions should always be governed by the scientific task to be dealt with, the particular properties of the compounds under investigation, and in the present case of course ethical aspects. Most of the studies, and in particular risk assessment, cannot be performed using human beings.

---

* For most molecules the molecular weight correlates well with the molecular volume.
[†] Accessibility of K_{oct} is better than the value of the membrane/water coefficient K_m.

So the first step in a serious dermal transport study is to choose an appropriate skin model for the respective transport process. This might be excised human skin, animal skin, vital animal models, or artificial cell-culture skin models, all of which have advantages and drawbacks. Excised human skin comes closest to realistic conditions, but is rarely available on time in a genuine state. Mostly it is used after some preparation and frozen storage. Animal skin may be freshly prepared for each experiment, but deviates from human skin in thickness, morphology, and appendage density. According to Brandau and Lippold, skin permeability across the species is in the following descending order: rabbit > rat > guinea-pig > mini-pig > Rhesus monkey > man (Brandau and Lippold 1982). Scott has demonstrated that skin permeability could be related to interspecies differences in skin structures (Scott et al. 1991). For mouse skin (Schätzlein and Cevc 1998) as well as for pig skin (Carrer et al. 2008) lipid-filled channel-like structures are reported. The penetration is directly related to such structures. Furthermore, the lipid composition of the skin depends on the nutrition provided and hence has an impact on the penetration behavior (Monteiro-Riviere et al. 2008). Besides the models taken from living beings, skin models based on cell cultures were developed and artificial membranes, with different lipid coatings, are also used (Netzlaff et al. 2005, Schaefer et al. 2008).

Overall, care has to be taken to assign data from one model to the other as this might be different for each investigated compound. Further on, the choice of a technique for the measurement of the transport is an essential aspect. These techniques strongly vary, depending on whether penetration or permeation, invasive or noninvasive, empirical or mechanistic studies are carried out, and on the requirements of the skin model. Finally, the nature of the penetrating compound must be considered to allow an accurate quantification. In this section, we will introduce the different techniques for dermal transport studies and associate them to the different application fields. Furthermore each method is evaluated in terms of the potentials and obstacles in the application to nanoparticulate compounds.

32.3.1 Franz Diffusion Cell and Saarbruecken Penetration Model

The most common techniques in dermal transport studies expose isolated barrier membranes (full skin or isolated layers) to a concentrated donor solution on the peripheral side in order to measure permeation by time-dependent sampling from a counter compartment on the inner side, or to measure penetration by interrupting the transport and preparing a depth profile of the solute in the barrier membrane. The latter is done by normal sectioning of the skin sample and the subsequent chemical analysis of the sections. If such depth profiles are to be determined to different times, the experiment must be repeated for each profile, since the analysis is consumptive. Both the permeation and penetration measurements yield averaged data over comparatively large areas. Hence, these techniques are feasible for empirical studies, balance out physiological deviations, and

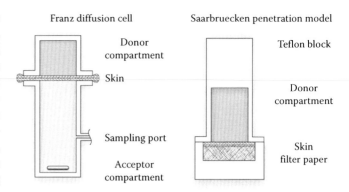

FIGURE 32.5 Schematic depiction of a Franz diffusion cell (FD-C) (left) and Saarbruecken penetration model (SB-M) (right). The Franz diffusion cell is a two-compartment model enabling the application of a drug formulation (top compartment) and the possibility to follow the permeation kinetics into a buffer being in contact with the skin (lower compartment) by taking samples through the sampling port. In contrast, the SB-M offers the possibility to investigate the skin in a less hydrated—more natural—state. However, only the penetration into the skin but not the permeation through the skin can be evaluated.

enable reproducible results, but completely blur the details and physiological pathways of the transport.

These classical methods to investigate (drug) permeation and penetration into skin are often based on the Franz diffusion cell (Franz 1975, Hotchkiss et al. 1990, 1992, Wagner et al. 2000, Larsen et al. 2003, OECD 2004) and on the Saarbruecken penetration model (Wagner et al. 2000) in combination with sectioning methods (cryosectioning and tape stripping [Tsai et al. 1991, Higo et al. 1993, Dreher et al. 1998, 2005, Weigmann et al. 1999]) (Figure 32.5). Both techniques are conducted at $(32 \pm 1)°C$ to mimic the *in vivo* temperature conditions. Both techniques can be applied to homogenously dissolved as well as nanoparticulate colloidal formulations. Only the subsequent chemical analysis has to account for the particular nature of the compound.

32.3.1.1 Franz Diffusion Cell

The Franz diffusion cell offers to utilize different skin models (filters, lipid-soaked filters, and membranes, excised animal, and human skin). However, the skin is always in contact with the buffer solution in the receptor compartment, resulting in a skin in fully hydrated state. This model allows studying permeation kinetics across the skin barrier. Fick's first law of diffusion based on the described conditions

$$J_{SS} = \frac{Q}{A(t - t_{\text{lag}})} = K_m D \frac{\Delta c}{h} \qquad (32.9)$$

facilitates a quantitative treatment of the data. Hereby represents J_{ss} the steady-state flux per unit area A across the barrier, t is the exposure time, Q is the integrated amount of drug transported at time t (mass or number of moles), and t_{lag} represents the early period of exposure, in which steady state conditions are not yet fulfilled. The flux is related to the diffusivity D of the drug across the layer of interest, the concentration difference Δc between the donor and the acceptor compartment (assuming sink conditions, Δc can be taken as the concentration in the donor compartment c, and Equation 32.9 becomes Equation 32.3), and the thickness of the layer. At any time, the permeation study can be interrupted for the preparation of a depth profile by sectioning the barrier.

32.3.1.2 Saarbruecken Penetration Model

In contrast, the Saarbruecken penetration model is a method without acceptor reservoir only relying on a filter paper to guarantee sink conditions. Due to the absence of an acceptor reservoir no permeation studies are possible. However, the absence of excess water results in a more realistic penetration situation. The drug is brought into contact with the skin, and after different incubation times the penetration profiles are determined.

The profiling of the drug penetration in the described and related methods is often based on the so-called tape stripping and the cryosectioning techniques.

32.3.1.3 Tape Stripping

The tape-stripping method is an evaluated and accepted approach for the determination of the drug amount in the stratum corneum. An adhesive tape is pressed with fixed pressure on the skin surface. Removal of the tape disrupts the skin and removes a well-defined layer that can be analyzed after dissolution (GPC, HPLC, UV/vis-spectrometry, radio labeling, etc.). As a result, one can plot the amount of drug per layer against the depth of the layer (Tsai et al. 1991, Weigmann et al. 1999, Surber et al. 2001, Dreher et al. 2005).

32.3.1.4 Cryosectioning

For deeper layers the tape-stripping method is not applicable. The water content of the viable epidermis is too high to effectively use adhesive tape to remove the tissue. Therefore, the tissue is frozen by expanding CO_2 to have a fast freezing procedure that should not influence on the drug distribution (fluid N_2 is another option for such a freezing process). Thereafter, the frozen skin is mounted on a precooled metal bloc in the microtome (kept at $\sim -20°C$). The skin is then moved with micrometer precision (micrometer screw) and cut parallel to the surface. The cuts were pooled and evaluated for the amount of drug.

32.3.2 Attenuated Total Reflection Infrared Spectrometry

If electromagnetic radiation undergoes total internal reflection at an interface toward a medium of lower refractive index, a so-called *evanescent field* is formed in that medium. The evanescent field declines exponentially with distance from the interface, so that all interactions take place within a depth on the order of half the radiation wavelength. A "zig-zag" beam guidance setup allows multiple interactions between an IR beam totally reflected in a high refraction crystal and

the specimen at the crystal surface. In this fashion, vibrational absorption spectra can be recorded from layers of a few microns with high sensitivity. This technique is mostly used with *Fourier transform* infrared spectrometry, called then attenuated total reflection Fourier transform infrared spectrometry (ATR-FTIR).

Tape strips (32.3.1.3) can directly be applied to ATR-FTIR crystal surfaces, making this technique a convenient method for measuring penetration depth profiles (Higo et al. 1993, Pirot et al. 1997). A procedure for quick permeation assessment is to attach the entire barrier (full skin, epidermis, etc.) to the crystal surface and to apply a formulation onto the peripheral skin surface. The ATR-FTIR measure now gives the amount of the investigated compound that permeated the barrier to a given time. Since there is no acceptor compartment, the compound is accumulated near the barrier/crystal interface, hence corrupting steady-state conditions.

32.3.3 Imaging Techniques for Inorganic Material: TEM/SEM

In order to analyze inorganic particulate materials, the high-resolution techniques—transmission electron microscopy (TEM) and scanning electron microscopy (SEM)—are the most spread and used approaches. The materials inherently give a good contrast due to their huge amount of electrons, and are therefore comparably easy to distinguish from biological and organic material. Furthermore, the obtainable resolution exceeds the resolution of light microscopy by a factor of 100, so individual particles can be resolved and imaged.

In general, the electron microscopy methods can be compared with light microscopy. Here we find transmitted-light-(TEM) and reflected-light microscopes (SEM). In addition, the EM methods allow for element identification based on characteristic x-ray radiation and inelastically scattered electron dispersion spectra. Sectioning of the specimen is required for all electron-microscopic studies. Since the experimental effort is high and only small areas can be investigated, this technique is beneficial in detailed mechanistic studies on dermal penetration, but is ineligible for a systematic quantification of cutaneous transport properties (Figure 32.6).

32.3.3.1 Transmission Electron Microscopy

Free electrons are accelerated and focused on the specimen introduced into a vacuum (10^{-7} Torr) chamber. The electrons need to penetrate through the sample and will be detected on the opposite site. Therefore, thin samples (typically 30–300 nm) are necessary to allow electron penetration. Scattering at the structures of the sample are responsible for image formation. The final TEM image is a projection of a 3D sample into a 2D picture. Care has to be taken with respect to the interpretation of the obtained 2D images. Reverse projection of circular structures to spheres in 3D—often found in scientific publications—is only possible if further information on the sample is available.

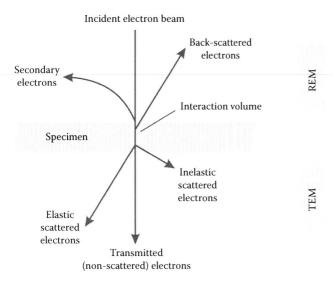

FIGURE 32.6 Sketch highlighting the different interactions of an electron beam with a specimen and the possible microscopic applications. (Courtesy of Center for Microscopy and Image Analysis *ZMB*, Zurich, Switzerland.)

32.3.3.2 Scanning Electron Microscopy

A focused electron beam is moved pointwise across the sample surface (in vacuum). In contrast to TEM, the detection and imaging of the sample is based on the secondary or back-scattered electrons. The thickness of the specimen is therefore not a limiting factor. The information from the sample originates from different regions of the sample: (1) secondary electrons can only leave the sample if generated at the surface or close to it; (2) back-scattered electrons cover a volume of up to 1–$2\,\mu m$ depth.

The specimen needs to be water-free and needs to have a conducting surface. Hence, a considerable sample preparation and a coating of the surface are typically necessary (e.g., thin gold layers).

32.3.4 Noninvasive Imaging Techniques

Noninvasive techniques have in common a capability of "remote" quantitative measurement, and hence the potential of depth profiling or even 3D imaging in closed tissue systems. Nevertheless these methods can also be applied to sectioned specimens.

32.3.4.1 Magnetic Resonance Techniques

Nuclear and electron magnetic resonance techniques are versatile techniques for *in vivo* and *in vitro* investigations for various purposes, due to chemical and dynamic sensitivity as well as the large observation depth. The major drawback of magnetic resonance imaging for studies on skin is the comparatively low resolution. Magnetic resonance microscopy is able to resolve the main dermal structures like epidermis, dermis, and hair follicles, but fails in resolving stratum corneum and the cellular pattern of the epidermis (Richard et al. 1991, Song

et al. 1997, Bittoun et al. 2006). Nevertheless time-dependent distribution profiles of some compounds were measured using a bench-top nuclear magnetic instrument (Backhouse et al. 2004) and of the spin label ^{15}N-PDT by electron paramagnetic resonance (EPR) imaging (He et al. 2001). Though magnetic resonance techniques are not widespread in the field of dermal transport research, these techniques have considerable potential, in particular with respect to particulate compounds (Richardson et al. 2005).

32.3.4.2 Confocal Fluorescence Microscopy

Fluorescence is a process based on light absorption and the subsequent emission of light with a certain wavelength shift (Figure 32.8). This implies that only compounds that (1) absorb the illumination light and (2) convert a part of the absorbed energy back into light are detectable by fluorescence techniques. These particular compounds are called fluorescent dyes, labels, markers, or simply fluorophores. A deliberate fluorescence experiment can beat most other technical approaches by ease, selectivity, and contrast. The major drawback, in particular to penetration studies, is that often compounds under investigation do not show fluorescence. There are two approaches to overcome this obstacle. First, a fluorescent label can be attached to the target compound. This is a feasible way to study the migration characteristics of heavy-weight compounds like macromolecules and particles, i.e., compounds that do not significantly alter physical and chemical properties upon fluorescent labeling. Second, the actual compound under investigation can be replaced by a fluorescent molecule of akin physical and chemical properties, a so-called dummy (molecule). A dummy experiment is a feasible way for fluorescence studies of small molecule compound migration, since a fluorescent label to a small molecule would drastically change almost any molecular parameter relevant to penetration (e.g., molecular weight, charge, shape, and volume, K_{oct}, response to pH alterations).

Having an appropriate fluorescent compound, its penetration progress can be easily explored by repeated imaging of the specimen without the need of drawing samples and disturbing the process. Using fluorescence microscopy, one can reveal the penetration routes in detail. Three-dimensional subcellular resolved imaging is enabled by confocal fluorescence microscopy (and two-photon microscopy, see Section 32.3.4.4), often abbreviated CLSM (confocal laser scanning microscopy). In CLSM, an excitation laser is focused into the sample by a microscope objective. The evoked fluorescent light from the sample is collected by the same objective and separated from reflected laser light by a dichroic mirror (so-called epifluorescence geometry). Then an intermediate image of the focal spot is adjusted to an adapted pinhole in a way that all fluorescent light from out of the focal volume is rejected. All light that passed the pinhole is detected and can be spatially associated to the focal coordinates. The image formation is conducted by either scanning the sample stage or by scanning the focus using scanning mirrors (galvoscanners).

Confocal fluorescence microscopy is frequently used for penetration studies of molecular (Alvarez-Roman et al. 2004b, Grams et al. 2004) and particulate (Alvarez-Roman et al. 2004a,b,c, Ryman-Rasmussen et al. 2006) compounds as well as for particles as vehicle for molecular compounds (Alvarez-Roman et al. 2004c, Shim et al. 2004). Some of these studies are carried out noninvasively, the others in vertical-section samples. All (linear) fluorescence techniques bear the risk of phototoxicity to the specimen and photodecomposition of the fluorophore, a source of artifact in quantitative fluorescence imaging. Due to absorption and scattering of the visible excitation and emission light in the biological specimen, the observation depth is limited. Ultraviolet excitation is completely impeded for these reasons. Single particles can only be resolved, if the mean distance of particles significantly exceeds the diffraction limit of the microscope (Stracke et al. 2006).

32.3.4.3 Confocal Raman Microscopy and CARS

The Raman Effect is based on the inelastic scattering of light by molecules. The scattered photon frequency is reduced by the amount of a molecular vibrational resonance frequency (Figure 32.8). Hence, a complete vibrational spectrum can be obtained by subtracting the wavenumbers of scattered light from the wavenumber of the illuminating laser. If arranged in a confocal setup like described for confocal fluorescence microscopy before, one gets vibrational spectra of a 3D resolved focal volume, and has access to Raman spectral imaging. This is the major advantage of Raman investigations: Noninvasive high-chemical-information 3D imaging without the restriction to fluorescent compounds, fluorescent labeling, or fluorescent dummies. Unfortunately Raman scattering has a very low efficiency (the respective molecular cross section are about 15 orders of magnitude smaller than for fluorescence). This means that very sensitive detection technology, very efficient laser wavelength rejection, and long acquisition times are indispensable, resulting in enormous image formation times and moreover very large raw data files. Due to the high information content of the data, the interpretation of Raman spectral images is sophisticated compared to fluorescence images. Since nearly every molecular and crystalline compound shows Raman scattering, the signal from the diffraction limited focal volume (~1 μm^3) is a superposition of Raman spectra of all contained compounds; a careful linear decomposition of the signal has to be performed for the quantification of the compound under investigation. Since nanoparticles are smaller than the focal volume, this holds true also for nanoparticulate formulations.

Confocal Raman imaging was mainly applied in dermal penetration studies as a tool to obtain depth profiles $I(z)$ instead of entire 3D images $I(x,y,z)$ of major skin compounds, such as water and *natural moisturizing factors* by Puppels et al. (Caspers et al. 2003) as well as phospholipids (Xiao et al. 2005), but up to now no particulate compound penetration was explored using Raman microscopy.

A very recent technology to obtain luminescence vibrational spectra with higher efficiency and hence much faster image formation is *coherent antistokes Raman scattering* microscopy (CARS), which is based on a four-wave mixing process (Figure 32.8). This technology employs an excitation spectroscopy setup with two coincident short-pulsed laser sources, one of them tunable. The emission intensity is measured at a fixed wavelength and plotted against the wave number difference of the coincident lasers to yield a vibrational spectrum. Arranged as a laser scanning setup, CARS microscopy allows 3D diffraction limited resolution images without confocal optics. A review on biomedical applications of CARS microscopy by Evans and Xie (2008) gives a good overview of the basics and an exemplifying penetration image of retinol into mouse skin. To our knowledge, serious penetration studies, either on dissolved or particular compounds, using this novel technique, have not been conducted so far.

32.3.4.4 Two-Photon Microscopy

Two-photon microscopy (TPM, 2PM) is a laser scanning fluorescence technique akin to confocal microscopy. All differences originate from the different excitation mechanism. Instead of the absorption of a single photon, two-photons of half the transition wavelength are absorbed simultaneously. The two-photon absorption process requires enormous light intensities, and depends on the intensity in a squared manner. Such intensities can only be enabled by strongly focusing high peak power pulsed lasers (such as mode-locked Ti:sapphire or fiber lasers in the near infrared, NIR) into the specimen. Light–matter interactions like absorption only occur in a tiny focal volume of less than 1 fL. This strict confinement of absorption has two major advantages: (1) The detected fluorescence can be unambiguously associated to the focal volume. Three-dimensional resolution is hence an inherent feature of two-photon microscopy, effacing the need for a confocal pinhole optic. So, unlike in confocal microscopy, even fluorescence photons scattered in the specimen may be collected for detection. Together with the strongly enhanced penetration depth of NIR radiation into biologic matter, multiphoton microscopy shows a superior observation capability in deep tissue, often only limited by the working distance of the microscope objective. (2) Due to the absence of out-of-focus absorption of NIR radiation, phototoxicity and photobleaching in the specimen are drastically reduced. Furthermore, even UV-absorbing fluorophores like many natural compounds (keratin, NADH, and flavines, see Figure 32.7), otherwise inaccessible, can be addressed using two-photon microscopy.

Obviously, like confocal microscopy, multiphoton imaging has the same restrictions concerning investigated compounds in penetration studies: nonfluorescent compounds have to be labeled or replaced by fluorescent dummies in a deliberate way. A couple of penetration studies were performed in order to reveal detailed dermal penetration pathways of hydrophilic and hydrophobic drug dummies (Yu et al. 2002, 2003) and to visualize the release and uptake of a drug dummy from a fluorescently labeled nanoparticulate polymer carrier (Stracke et al. 2006) (Figure 32.8).

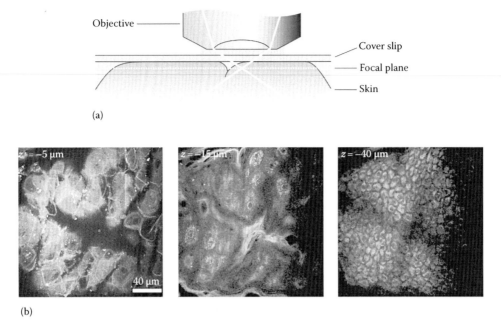

(a)

(b)

FIGURE 32.7 Image formation in two-photon microscopy: the focus of the excitation laser beam is scanned over the focal plane to yield a fluorescence image thereof. Stepwise vertical displacement of the focal plane leads to a 3D reconstruction of the specimen. Two-photon micrographs of focal planes 5, 15, and 40 μm beneath the skin surface are shown, revealing stratum corneum, stratum granulosum, and stratum spinosum morphology. (Courtesy of M. Schwarz, Fraunhofer IBMT, St. Ingbert, Germany.)

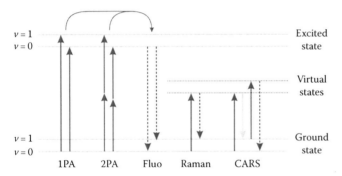

FIGURE 32.8 Energy level (Jablonski) scheme of the optical techniques described herein. 1PA: single photon (=linear) absorption, 2PA: two-photon absorption, fluorescence emission, Raman scattering, and CARS. Dashed lines denote emissions; the gray arrow denotes the tuneable laser in CARS. ν is a vibronic quantum number. Virtual states of molecules are auxiliary constructs, but do not physically exist.

32.4 Penetration and Permeation Studies on Nanoparticulate Material

Recently the interaction of nanoscaled solid carriers with skin has attracted some interest in the area of drug delivery. Applying drugs in pharmaceutical formulations containing nanoparticulate material as drug delivery devices was considered to be a promising approach (Alvarez-Roman et al. 2004a,c, Kohli and Alpar 2004, Toll et al. 2004, Luengo et al. 2006, Ryman-Rasmussen et al. 2006, 2007a,b, Stracke et al. 2006). Safety issues gained in parallel more and more room (Nohynek et al. 2008, Vega-Villa et al. 2008).

In general, for particles, two possible routes through the skin have to be considered: the intercellular route, following the lipid channels between the corneocytes to the deeper skin layers; and the appendage route (hair follicles, sweat glands). Both ways have shown considerable interaction with nanoparticulate formulations (Alvarez-Roman et al. 2004a,b, Toll et al. 2004, Lademann et al. 2007).

In this section, an overview of penetration and permeation studies on nanoparticulate compounds into human skin and its models will be given, with an emphasis on the utilized detection and imaging techniques, skin models, and penetration conditions. It is assessed to what extent obtained results can be transferred to practical scenarios. This overview will be subdivided according to the particles' chemical composition, which concomitantly includes a classification according to the application foci of the studies: risk assessment for metal and mineral nanoparticles, or pharmaceutical applications for organic nanoscale compounds.

32.4.1 Particle Classification

(Arrange in such a way that Figure 32.9 is immediately below this head)

32.4.2 Polymeric Particles

Looking at drug delivery systems, the most promising and mostly applied carrier technology is based on polymer materials – especially on biomaterials that offer an intrinsic biocompatibility and biodegradability. Hence, there are several studies on drug penetration after of polymeric encapsulation that show marked

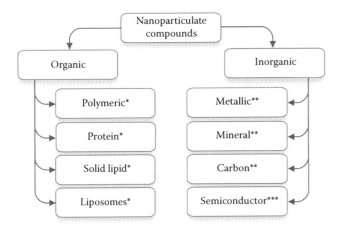

FIGURE 32.9 Classification scheme of nanoparticulate compounds according to their chemical composition. Asterisks denote the major application focus of skin penetration studies: (*) transport studies for pharmaceutical applications, drug delivery, etc., (**) penetration and uptake studies for assessment of health risks, (***) academic purposes.

differences between the conventional and nanoparticulate formulations (Alvarez-Roman et al. 2004a,c, Kohli and Alpar 2004, Shim et al. 2004, Toll et al. 2004, Luengo et al. 2006). A crucial question for the investigation of nanoparticulate drug delivery carriers is the site of drug release from the particles, i.e., does the release occur on the skin surface leaving the carrier particles outside or do the particles penetrate the skin to release the drug within the tissue? However, yet there is only limited information on circumstances allowing dermal penetration of polymeric nanoparticles. FITC-dextran particles of different sizes (up to 4 μm) were investigated under mechanical stress and without stress as a reference experiment (Tinkle et al. 2003). A clear

cutoff with respect to the size could be determined (≤1 μm), but only for skin under mechanical stress. Nevertheless, these data are quite surprising with respect to the size of the penetrating species. Kohli and Alpar also found latex particles (up to 500 nm) (Kohli and Alpar 2004) to penetrate while applying mechanical stress to the barrier. Shim and coworkers applied polymer-based particles of 40 and 130 nm onto the skin of guinea-pigs and detected clear differences between the penetrated drug amounts. However, working with hairless animals did not show any penetration into the skin, clearly indicating that the hair follicles might be an important pathway for skin invasion of particulate materials (Shim et al. 2004).

On the other hand, without applying mechanical stress to the skin, polymer particles $d \sim 300$ nm seemed not to penetrate into human skin within 6 h after application but release a drug dummy for cutaneous absorption. This was observed by two-photon and CLSM (Stracke et al. 2006). The simultaneous investigation in particle migration and load release and uptake was enabled by multimodal fluorescence strategy: the particles were covalently labeled with fluorescein and physically loaded with Texas Red as a drug dummy for release. Furthermore, dermal structures were visualized by keratinous autofluorescence. The excitation wavelengths were chosen to allow to separately investigate each of the fluorescent compounds. The covalently labeled particles could only be observed in the skin furrows and on the skin surface, whereas the released Texas Red significantly penetrated the epidermis (Figure 32.10).

These investigations were carried out on human skin biopsies stored frozen. The frozen storage preserves the barrier properties of skin, but other relevant characteristics are likely altered (e.g., pH). Furthermore, no more cellular autofluorescence from the viable epidermis due to NADH and flavines is left.

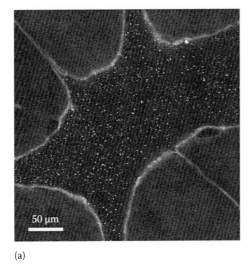

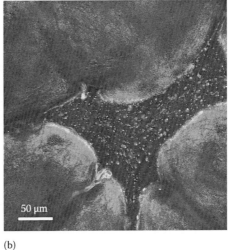

(a) (b)

FIGURE 32.10 (See color insert following page V-2.) PLGA nanoparticles in human skin furrows 5 h after administration. (a) A two-photon micrograph at $z = 15$ μm subsurface depth showing superficial keratin fluorescence and clearly resolved single particles. (b) A pseudocolor overlay of two-photon and confocal images at $z = 28$ μm revealing the release of a drug dummy from the particles and its cutaneous uptake. The two-photon channel (green) shows keratinous layers and single particles, the 488 nm excited confocal channel (blue) addresses the fluorescein-labeled particles solely, and the 543 nm excited confocal channel (red) exclusively shows the dummy compound (Texas Red).

32.4.3 Protein Particles

These kinds of particles, although already commercially available for oral application, have not yet been applied on skin.

32.4.4 Solid Lipid Nanoparticles

In the early 1990s, solid lipid nanoparticles (SLN) were introduced as drug carriers in the pharmaceutical field. In general, SLN are composed of physiological solid lipids manufactured by a high pressure homogenization process. Such systems applied onto skin result in higher drug permeation. However, this enhanced drug absorption is not the result of penetrating particles but of the occlusive effect as a result of surface coverage (Souto et al. 2007). No intact penetration of the SLN is reported to our knowledge.

32.4.5 Liposomes

Liposomes are composed of a closed bilayer of phospholipids offering a hydrophobic compartment (lipid layer) as well as a hydrophilic compartment (inner liposomal space). Their advantage, with respect to pharmaceutical application as drug carrier, is the variety of drugs to be incorporated as well as the biocompatibility, inherently connected with natural phospholipids. Hence, liposomes are the biggest group of nanoparticulate carriers used for application in cosmetics or for therapeutic purposes.

Regarding penetration, ultraflexible liposomal structures are assumed to penetrate successfully (Cevc 1996, Van Den Bergh et al. 1999), whereas conventional liposomes failed (Van Kuijk-Meuwissen et al. 1998a,b).

As the liposomes are bigger in size (≥50 nm) than the skin openings, a driving force large enough to drag the liposomes through the skin is required. Cevc and coauthor proposed a mechanism to explain the observed dermal penetration based on a hydration-gradient-driven transport (Cevc and Gebauer 2003). Thus the different activity of water on the two barrier sides is responsible for aggregate movement, and is increased with surface area and its hydrophilicity. The problem is similar to the situation of a molecule traveling through a narrow pore. The activation energy for a molecule, $\Delta G_m^{\#}$, increases logarithmically with the partition coefficient, K_m, for the molecule (RT: thermal energy)

$$\Delta G_m^{\#} = -RT \cdot \ln K_m. \tag{32.10}$$

The energy needed by an intact vesicle to enter into a pore will depend on the size and the elasticity of the membrane (deformability). This can be accounted for separating the free energy for the elasticity and the free energy to break the vesicle resulting in the free energy of vesicle deformation (Cevc and Gebauer 2003).

Cevc and Gebauer gave an analytical equation incorporating several vesicular properties and found that a low rigidity of a vesicle favors an energetically inexpensive membrane deformation. This sounds quite reasonable. As a driving force the

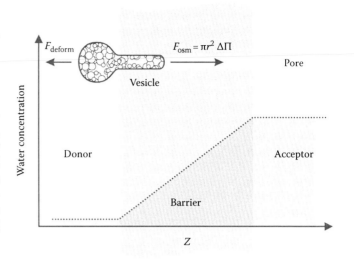

FIGURE 32.11 Scheme of mechanism of ultradeformable vesicle transport through a stratum corneum pore. The vesicle will penetrate into the pore if the force from the osmotic pressure exceeds the deformation resistance of the vesicle.

osmotic pressure difference, allowing permeation if the free energy gain exceeds the work of permeant penetration into the pore is suggested:

$$\Delta\Pi \cdot V_m = \Delta G_m^{\#}. \tag{32.11}$$

The water activity gradient in the stratum corneum is reported to be also responsible for the motion of the liposomes along the pores of the intercellular space.

As a concluding remark, we would suggest a force-based ansatz instead of an argumentation relying on different energetic states to describe this problem (Figure 32.11).

32.4.6 Metallic Particles

Beryllium nanoparticles (up to 1 μm) were found in the stratum corneum using tape stripping as well as in deeper skin layers based on TEM imaging (Tinkle et al. 2003). Maghemite (γ-Fe$_2$O$_3$) and iron core shell particles ($d \leq 20$ nm) were shown to penetrate into the top layer and eventually can reach the viable epidermis of human skin (Baroli et al. 2007). This study was conducted using electron dispersion spectrometry SEM on vertical skin sections. This technique allows chemical identification of iron particles in the tissue. In addition, these particles were also found in the hair follicles. The authors tried to investigate the possible mechanism as well and investigated the barrier properties (resistance), and found that the formulation vehicles (solvent) have already a marked influence on the barrier that cannot be identified with the visualization tools. However, the authors hydrated the skin deliberately before applying the nanoparticulate formulations that might significantly alter the barrier properties as a result of swelling.

32.4.7 Mineral Particles

Mineral particles are of huge relevance due to their production and usage in large amounts. In 2003/2004, approximately 1000 tons of these materials were fabricated for their usage in sun protection agents with sizes between 50 and 500 nm (Nohynek et al. 2008). Sun-screen-grade nanoparticles are composed of titanium dioxide (TiO_2) and zinc oxide (ZnO).

Regarding the passive diffusion of TiO_2, the EU Scientific Committee on Cosmetics and Non-Food Products (SCCNP) (SCCNFP 2000) published a paper based on studies with micro- and nanosized material. Herein they state that these particles remain on the skin surface or on the outer layer of the stratum corneum and do not penetrate into or through the living skin (Tan et al. 1996, Dussert et al. 1997, Lademann et al. 1999). Confirmation was obtained with studies on human, porcine, or murine skin (Pflücker et al. 2001, Schulz et al. 2002, Gamer et al. 2006, Mavon et al. 2007) for particles within a size range between 10 and 100 nm. These data were recently confirmed by the outcome of an European research project (Nanoderm, Gontier et al. 2008). Here, TiO_2-particles are only found in the top layer of the stratum corneum and the openings of the hair follicle (Lekki et al. 2007). Similar results were obtained for ZnO (Gamer et al. 2006, Cross et al. 2007).

32.4.8 Carbon Nanotubes/Fullerenes

Carbon materials are considered as drug delivery systems due to their inert nature and their volume to be filled with drugs (Lacerda et al. 2006). With respect to the interaction of these materials with skin, several studies have been undertaken to examine the toxic impact of these materials on skin cells (keratinocytes) (Monteiro-Riviere et al. 2005, Lam et al. 2006, Rouse et al. 2006). To our knowledge, only one work investigated the penetration behavior of carbon materials, i.e., fullerenes into skin (Rouse et al. 2007). Rouse et al. coupled a peptide with a fullerene and added a fluorescent marker (3.5 nm size over all) to facilitate the observation of the complex. No penetration of the complex in the skin was found as long as the skin was untreated. Simulating the mechanical stress of the skin (such as walking)—flexing of the skin—resulted in penetration of the particulate material into skin. In addition, the penetration was dependent on the time the mechanical stress was applied; flexing needs to go on for at least 60 min to allow penetration. This may indicate that the structure of the skin is altered applying mechanical stress.

32.4.9 Semiconductor Nanocrystals

Semiconductor nanocrystals, also known as quantum dots (QD), are widely used for noninvasive imaging purposes. The QD offer several advantageous properties that offer superior detection and experimental control compared to many other nanoparticulate materials. For all penetration processes, the size of the permeant is considered to be one of the main factors. QD are, without additional surface coating, available in a very small size range,

below 10 nm. Besides the size-dependent fluorescent emission, the QD show typically small size distributions and favorable optical properties. The inherent fluorescence guarantees a label-free (not altering the chemical composition) material (Hild et al. 2008). Furthermore, the powerful and convenient fluorescent methods can be noninvasively applied. Typically, most of the experimental data are based on confocal microscopy, sometimes supported with TEM images and quantitative analysis (Zhang et al. 2008b).

As a result, they are applied manyfold in several areas and are used as well for the understanding of the interaction of nano-sized materials with skin. There are several publications dealing with QD and the interaction with skin. An important aspect is that none of the present publications deals with human but with murine (Chu et al. 2007, Zhang and Monteiro-Riviere 2008a) (mouse, rat) and pig skin (Ryman-Rasmussen et al. 2006, Zhang et al. 2008b). As the different skin types do differ and no information on particle penetration exists care should be taken to generalize the different data.

In a recent work, healthy skin was compared with damaged skin regarding the QD penetration. Skin was damaged applying tape stripping to remove the stratum corneum or a sand paper was used to physically damage the skin and facilitate penetration because of a missing or ruptured main barrier. However, no clear penetration was observed (Zhang and Monteiro-Riviere 2008a). Rat skin without stratum corneum, as well as healthy rat skin, did not show any permeability for the types of QD used, in contrast to the results obtained for pig skin with the same particles (Ryman-Rasmussen et al. 2006). Only the abraded skin showed some penetration according to the authors, although this trend is not obvious (Zhang and Monteiro-Riviere 2008a). Surprisingly, following earlier results, even flexing did not facilitate the penetration of the QD into the skin.

In contrast to these data obtained for rat skin, for porcine and mice strong penetration was reported (Ryman-Rasmussen et al. 2006, Chu et al. 2007). The findings in these two papers are really surprising and unexpected, and with respect to other data on particle penetration the reliability might be questioned. Ryman-Rasmussen et al. applied two types of QD—spherical and ellipsoidal—with different surface coatings (PEG-amine-, carboxy-, and PEG-functionalized) from a borate buffer (pH = 9) onto porcine skin, and found that all particles do penetrate into skin after 24 h. Neither the size nor the surface modification had any influence on the QD absorption. Following the penetration of molecules (Potts–Guy equation) this is really surprising for particulate material. Chu and coworkers applied QD *in vivo* as well as *in vitro* on mouse skin (which differs from human skin morphology, see Section 32.3). With fluorescence methods they found penetration into the skin, and in addition analyzing the animal organs with inductive coupled plasma atomic emission spectrometry (ICP-AES), the QD were detected in the lung, the heart, the liver, and kidneys. Comparable high levels of the QD were kept over 1 week. The presence of particles in these organs has not yet been proved because ICP only detects ionic species, not particulate material.

Just recently, another study identified a further condition destroying/reducing the barrier function of skin: ultraviolet radiation (UVR) (Mortensen et al. 2008). Carboxylated QD applied on mouse skin in a glycerol vehicle showed increased penetration if the skin is exposed to UV radiation (290–400 nm) compared to low level of penetrated QD without illumination. However, the overall penetration was low in both cases. The authors could even demonstrate the possible pathway of the intruding QD along the intercellular space using silver-enhanced TEM imaging.

32.5 Conclusion

The penetration of particulate materials into skin is a very complex issue. Unfortunately, there is no mechanistic concept of particle penetration into skin (and even more none of a global nanoparticle biophysics). Everything in this novel field of skin transport depends on phenomenological descriptions of experimental results, only valid for a particular scenario. Ab initio predictions are therefore not a practical approach to describe the penetration processes. For this reason, up to now only experimental data is available and will be indispensable also in the future. However, the experimental strategies cannot be transferred straightforwardly from the nonparticulate situation, but have to take into account the particulate nature of the analyte and not only the material. Furthermore, practical and ethical considerations limit the freedom of the experimental design. In addition, the choice of the experimental parameters as well as the following interpretation of the obtained data strongly depends on the field of the research and the particular problem investigated. As a consequence, a standardized operation procedure is difficult to establish, but desirable. Overall, this novel field of research requires an interdisciplinary expertise combining appropriate technical, analytical, and physiological knowledge to yield reliable results.

Nevertheless, the growing number of nanoparticle occurrence in manufacturing and application need a better understanding of the relevant mechanisms of nanoparticulate penetration into skin, not least for an adequate risk assessment.

References

Alvarez-Roman, R., Naik, A., Kalia, Y., Guy, R. H., and Fessi, H. 2004a. Skin penetration and distribution of polymeric nanoparticles. *Journal of Controlled Release* 99(1): 53–62.

Alvarez-Roman, R., Naik, A., Kalia, Y. N., Fessi, H., and Guy, R. H. 2004b. Visualization of skin penetration using confocal laser scanning microscopy. *European Journal of Pharmaceutics and Biopharmaceutics* 58(2): 301–316.

Alvarez-Roman, R., Naik, A., Kalia, Y. N., Guy, R. H., and Fessi, H. 2004c. Enhancement of topical delivery from biodegradable nanoparticles. *Pharmaceutical Research* 21(10): 1818–1825.

Backhouse, L., Dias, M., Gorce, J. P., Hadgraft, J., McDonald, P. J., and Wiechers, J. W. 2004. GARField magnetic resonance profiling of the ingress of model skin-care product ingredients into human skin in vitro. *Journal of Pharmaceutical Sciences* 93(9): 2274–2283.

Baroli, B., Ennas, M. G., Loffredo, F., Isola, M., Pinna, R., and Lopez-Quintela, M. A. 2007. Penetration of metallic nanoparticles in human full-thickness skin. *Journal of Investigative Dermatology* 127(7): 1701–1712.

Bittoun, J., Querleux, B., and Darrasse, L. 2006. Advances in MR imaging of the skin. *NMR in Biomedicine* 19(7): 723–730.

Bos, J. D. and Meinardi, M. M. H. M. 2000. The 500 Dalton rule for the skin penetration of chemical compounds and drugs. *Experimental Dermatology* 9(3): 165–169.

Brandau, R. and Lippold, B. 1982. *Dermal and Transdermal Absorption*. Stuttgart, Germany: Wissenschaftliche Verlag Gesellschaft, 257 pp.

Carrer, D. C., Vermehren, C., and Bagatolli, L. A. 2008. Pig skin structure and transdermal delivery of liposomes: A two-photon microscopy study. *Journal of Controlled Release* 132: 12–20.

Caspers, P. J., Lucassen, G. W., and Puppels, G. J. 2003. Combined in vivo confocal Raman spectroscopy and confocal microscopy of human skin. *Biophysical Journal* 85(1): 572–580.

Cevc, G. 1996. Transfersomes, liposomes and other lipid suspensions on the skin: Permeation enhancement, vesicle penetration, and transdermal drug delivery. *Critical Reviews in Therapeutic Drug Carrier Systems* 13(3–4): 257–388.

Cevc, G. and Gebauer, D. 2003. Hydration-driven transport of deformable lipid vesicles through fine pores and the skin barrier. *Biophysical Journal* 84(2 I): 1010–1024.

Chu, M., Wu, Q., Wang, J., Hou, S., Miao, Y., Peng, J., and Sun, Y. 2007. In vitro and in vivo transdermal delivery capacity of quantum dots through mouse skin. *Nanotechnology* 18(45): 455103.

Cross, S. E., Innes, B., Roberts, M. S., Tsuzuki, T., Robertson, T. A., and McCormick, P. 2007. Human skin penetration of sunscreen nanoparticles: In-vitro assessment of a novel micronized zinc oxide formulation. *Skin Pharmacology and Physiology* 20(3): 148–154.

Dancik, Y., Jepps, O. G., and Roberts, M. S. 2008. Physiologically based pharmacokinetics and pharmacodynamics of skin. In: M. S. Roberts and K. A. Walters, editors. *Dermal Absorption and Toxicity Assessment*, 2nd edn. New York/London, U.K.: Informa, pp. 179–207.

Dreher, F., Arens, A., Hostynek, J. J., Mudumba, S., Ademola, J., and Maibach, H. I. 1998. Colorimetric method for quantifying human stratum corneum removed by adhesive-tape-stripping. *Acta Dermato-Venereologica* 78(3): 186–189.

Dreher, F., Modjtahedi, B. S., Modjtahedi, S. P., and Maibach, H. I. 2005. Quantification of stratum corneum removal by adhesive tape stripping by total protein assay in 96-well microplates. *Skin Research and Technology* 11(2): 97–101.

Dussert, A. S., Gooris, E., and Hemmerle, J. 1997. Characterization of the mineral content of a physical sunscreen emulsion and its distribution onto human stratum corneum. *International Journal of Cosmetic Science* 19(3): 119–129.

Elias, P. M. 1983. Epidermal lipids, barrier function, and desquamation. *Journal of Investigative Dermatology* 80: 44–49.

Evans, C. L. and Xie, X. S. 2008. Coherent antistokes Raman scattering microscopy: Chemical imaging for biology and medicine. *Annual Reviews of Analytical Chemistry* 1: 883–890.

Feuchter, D., Heisig, M., and Wittum, G. 2006. A geometry model for the simulation of drug diffusion through the stratum corneum. *Computing and Visualization in Science* 9(2): 117–130.

Franz, T. J. 1975. Percutaneous absorption. On the relevance of in vitro data. *Journal of Investigative Dermatology* 64(3): 190–195.

Gamer, A. O., Leibold, E., and Van Ravenzwaay, B. 2006. The in vitro absorption of microfine zinc oxide and titanium dioxide through porcine skin. *Toxicology In Vitro* 20(3): 301–307.

Gontier, E., Ynsa, M.-D., Bíró, T., Hunyadi, J., Kiss, B., Gáspár, K., Pinheiro, T., Silva, J.-N., Felipe, P., Stachura, J., Dabros, W., Reinert, T., Butz, T., Moretto, Ph., and Surlève-Bazeille, J.-E. 2008. Is there penetration of titania nanoparticles in sunscreens through skin? A comparative electron and ion microscopy study. *Nanotoxicology* 2(4): 218–231.

Grams, Y. Y., Bouwstra, J. A., Whitehead, L., and Cornwell, P. 2004. On-line visualization of dye diffusion in fresh unfixed human skin. *Pharmaceutical Research* 21(5): 851–859.

Hansen, S., Henning, A., Naegel, A., Heisig, M., Wittum, G., Neumann, D., Kostka, K. H., Zbytovska, J., Lehr, C. M., and Schaefer, U. F. 2008. In-silico model of skin penetration based on experimentally determined input parameters. Part I: Experimental determination of partition and diffusion coefficients. *European Journal of Pharmaceutics and Biopharmaceutics* 68(2): 352–367.

He, G., Samouilov, A., Kuppusamy, P., and Zweier, J. L. 2001. In vivo EPR imaging of the distribution and metabolism of nitroxide radicals in human skin. *Journal of Magnetic Resonance* 148(1): 155–164.

Higo, N., Naik, A., Bommannan, D. B., Potts, R. O., and Guy, R. H. 1993. Validation of reflectance infrared spectroscopy as a quantitative method to measure percutaneous absorption *in vivo*. *Pharmaceutical Research* 10: 1500–1506.

Hild, W. A., Breunig, M., and Goepferich, A. 2008. Quantum dots—Nano-sized probes for the exploration of cellular and intracellular targeting. *European Journal of Pharmaceutics and Biopharmaceutics* 68(2): 153–168.

Hotchkiss, S. A., Chidgey, M. A. J., Rose, S., and Caldwell, J. 1990. Percutaneous absorption of benzyl acetate through rat skin in vitro. 1. Validation of an in vitro model against in vivo data. *Food and Chemical Toxicology* 28(6): 443–447.

Hotchkiss, S. A. M., Hewitt, P., Caldwell, J., Chen, W. L., and Rowe, R. R. 1992. Percutaneous absorption of nicotinic acid, phenol, benzoic acid and triclopyr butoxyethyl ester through rat and human skin in vitro: Further validation of an in vitro model by comparison with in vivo data. *Food and Chemical Toxicology* 30(10): 891–899.

Kohli, A. K. and Alpar, H. O. 2004. Potential use of nanoparticles for transcutaneous vaccine delivery: Effect of particle size and charge. *International Journal of Pharmaceutics* 275(1–2): 13–17.

Lacerda, L., Bianco, A., Prato, M., and Kostarelos, K. 2006. Carbon nanotubes as nanomedicines: From toxicology to pharmacology. *Advanced Drug Delivery Reviews* 58(14): 1460–1470.

Lademann, J., Weigmann, H. J., Rickmeyer, C., Barthelmes, H., Schaefer, H., Mueller, G., and Sterry, W. 1999. Penetration of titanium dioxide microparticles in a sunscreen formulation into the horny layer and the follicular orifice. *Skin Pharmacology and Applied Skin Physiology* 12(5): 247–256.

Lademann, J., Richter, H., Schaefer, U. F., Blume-Peytavi, U., Teichmann, A., Otberg, N., and Sterry, W. 2006. Hair follicles—A long-term reservoir for drug delivery. *Skin Pharmacology and Physiology* 19: 232–236.

Lademann, J., Richter, H., Teichmann, A., Otberg, N., Blume-Peytavi, U., Luengo, J., Weiß, B., Schaefer, U. F., Lehr, C. M., Wepf, R. et al. 2007. Nanoparticles—An efficient carrier for drug delivery into the hair follicles. *European Journal of Pharmaceutics and Biopharmaceutics* 66(2): 159–164.

Lam, C. W., James, J. T., McCluskey, R., Arepalli, S., and Hunter, R. L. 2006. A review of carbon nanotube toxicity and assessment of potential occupational and environmental health risks. *Critical Reviews in Toxicology* 36(3): 189–217.

Larsen, R. H., Nielsen, F., Sørensen, J. A., and Nielsen, J. B. 2003. Dermal penetration of fentanyl: Inter- and intraindividual variations. *Pharmacology and Toxicology* 93(5): 244–248.

Lekki, J., Stachura, Z., Dabros, W., Stachura, J., Menzel, F., Reinert, T., Butz, T., Pallon, J., Gontier, E., Ynsa, M. D. et al. 2007. On the follicular pathway of percutaneous uptake of nanoparticles: Ion microscopy and autoradiography studies. *Nuclear Instruments and Methods in Physics Research, Section B: Beam Interactions with Materials and Atoms* 260(1): 174–177.

Luengo, J., Weiss, B., Schneider, M., Ehlers, A., Stracke, F., König, K., Kostka, K.-H., Lehr, C.-M., and Schaefer, U. F. 2006. Influence of nanoencapsulation on human skin transport of flufenamic acid. *Skin Pharmacology and Physiology* 19(4): 190–197.

Magnusson, B. M., Anissimov, Y. G., Cross, S. E., and Roberts, M. S. 2004. Molecular size as the main determinant of solute maximum flux across the skin. *Journal of Investigative Dermatology* 122(4): 993–999.

Mavon, A., Miquel, C., Lejeune, O., Payre, B., and Moretto, P. 2007. In vitro percutaneous absorption and in vivo stratum corneum distribution of an organic and a mineral sunscreen. *Skin Pharmacology and Physiology* 20(1): 10–20.

Moghimi, H. R., Williams, A. C., and Barry, B. W. 1996. A lamellar matrix model for stratum corneum intercellular lipids. II. Effect of geometry of the stratum corneum on permeation of model drugs 5-fluorouracil and oestradiol. *International Journal of Pharmaceutics* 131(2): 117–129.

Moghimi, H. R., Barry, B. W., and Williams, A. C. 1999. Stratum corneum and barrier performance: A model lamellar structural approach. In: R. L. Bronaugh and H. I. Maibach, editors. *Drugs—Cosmetics—Mechanisms—Methodology*. New York/Basel, Switzerland/Hong Kong, China: Marcel Dekker, pp. 515–553.

Monteiro-Riviere, N. A., Nemanich, R. J., Inman, A. O., Wang, Y. Y., and Riviere, J. E. 2005. Multi-walled carbon nanotube interactions with human epidermal keratinocytes. *Toxicology Letters* 155(3): 377–384.

Monteiro-Riviere, N. A., Baynes, R. E., and Riviere, J. E. 2008. Animal skin morphology and dermal absorption. In: M. S. Roberts and K. A. Walters, editors. *Dermal Absorption and Toxicity Assessment*, 2nd edn. New York/London, U.K.: Informa Health Care, pp. 17–35.

Mortensen, L. J., Oberdörster, G., Pentland, A. P., and DeLouise, L. A. 2008. In vivo skin penetration of quantum dot nanoparticles in the murine model: The effect of UVR. *Nano Letters* 8(9): 2779–2787.

Naegel, A., Hansen, S., Neumann, D., Lehr, C. M., Schaefer, U. F., Wittum, G., and Heisig, M. 2008a. Erratum to "In-silico model of skin penetration based on experimentally determined input parameters. Part II: Mathematical modelling of in-vitro diffusion experiments. Identification of critical input parameters" [*European Journal of Pharmaceutics and Biopharmaceutics* 68 (2008) 368–379] (DOI:10.1016/j.ejpb.2007.05.018). *European Journal of Pharmaceutics and Biopharmaceutics* 68(3): 846.

Naegel, A., Hansen, S., Neumann, D., Lehr, C. M., Schaefer, U. F., Wittum, G., and Heisig, M. 2008b. In-silico model of skin penetration based on experimentally determined input parameters. Part II: Mathematical modelling of in-vitro diffusion experiments. Identification of critical input parameters. *European Journal of Pharmaceutics and Biopharmaceutics* 68(2): 368–379.

Netzlaff, F., Lehr, C. -M., Schaefer, U. F., and Wertz, P. W. 2005. The human epidermis models EpiSkin®, SkinEthic® and EpiDerm®: An evaluation of morphology and their suitability for testing phototoxicity, irritancy, corrosivity, and substance transport. *European Journal of Pharmaceutics and Biopharmaceutics* 60(2): 167–178.

Nohynek, G. J., Dufour, E. K., and Roberts, M. S. 2008. Nanotechnology, cosmetics and the skin: Is there a health risk? *Skin Pharmacology and Physiology* 21(3): 136–149.

Oberdörster, G., Oberdörster, E., and Oberdörster, J. 2005. Nanotoxicology: An emerging discipline evolving from studies of ultrafine particles. *Environmental Health Perspectives* 113(7): 823–839.

OECD. 2004. Guidance document for the conduct of skin absorption studies. OECD Series on Testing and Assessment. No 28. Paris, France: Environment Directorate.

Pflücker, F., Wendel, V., Hohenberg, H., Gärtner, E., Will, T., Pfeiffer, S., Wepf, R., and Gers-Barlag, H. 2001. The human stratum corneum layer: An effective barrier against dermal uptake of different forms of topically applied micronised titanium dioxide. *Skin Pharmacology and Applied Skin Physiology* 14(suppl. 1): 92–97.

Pirot, F., Kalia, Y. N., Stinchcomb, A. L., Keating, G., Bunge, A., and Guy, R. H. 1997. Characterization of the permeability barrier of human skin in vivo. *Proceedings of the National Academy of Sciences of the United States of America* 94(4): 1562–1567.

Potts, R. O. and Guy, R. H. 1992. Predicting skin permeability. *Pharmaceutical Research* 9(5): 663–669.

Richard, S., Querleux, B., Bittoun, J., Idy-Peretti, I., Jolivet, O., Cermakova, E., and Leveque, J. L. 1991. In vivo proton relaxation times analysis of the skin layers by magnetic resonance imaging. *Journal of Investigative Dermatology* 97(1): 120–125.

Richardson, J. C., Bowtell, R. W., Mäder, K., and Melia, C. D. 2005. Pharmaceutical applications of magnetic resonance imaging (MRI). *Advanced Drug Delivery Reviews* 57(8): 1191–1209.

Roberts, M. S. and Walters, K. A. 2008. Human skin morphology and dermal absorption. In: M. S. Roberts and K. A. Walters, editors. *Dermal Absorption and Toxicity Assessment*, 2nd edn. New York/London, U.K.: Informa, p. 678.

Rouse, J. G., Yang, J., Barron, A. R., and Monteiro-Riviere, N. A. 2006. Fullerene-based amino acid nanoparticle interactions with human epidermal keratinocytes. *Toxicology In Vitro* 20(8): 1313–1320.

Rouse, J. G., Yang, J., Ryman-Rasmussen, J. P., Barron, A. R., and Monteiro-Riviere, N. A. 2007. Effects of mechanical flexion on the penetration of fullerene amino acid-derivatized peptide nanoparticles through skin. *Nano Letters* 7(1): 155–160.

Ryman-Rasmussen, J. P., Riviere, J. E., and Monteiro-Riviere, N. A. 2006. Penetration of intact skin by quantum dots with diverse physicochemical properties. *Toxicological Sciences* 91(1): 159–165.

Ryman-Rasmussen, J. P., Riviere, J. E., and Monteiro-Riviere, N. A. 2007a. Surface coatings determine cytotoxicity and irritation potential of quantum dot nanoparticles in epidermal keratinocytes. *Journal of Investigative Dermatology* 127(1): 143–153.

Ryman-Rasmussen, J. P., Riviere, J. E., and Monteiro-Riviere, N. A. 2007b. Variables influencing interactions of untargeted quantum dot nanoparticles with skin cells and identification of biochemical modulators. *Nano Letters* 7(5): 1344–1348.

SCCNFP. 2000. *Opinion of the Scientific Committee on Cosmetic Products and Non-Food Products Intended for Consumer Concerning Titanium Dioxide*. Brussels, Belgium: European Commission.

Schaefer, U. F., Hansen, S., Schneider, M., Luengo Contreras, J., and Lehr, C. M. 2008. Models for skin absorption and skin toxicity testing. In: C. Ehrhardt and K.-J. Kim, editors. *Drug Absorption Studies: In Situ, In Vitro, and In Silico Models*. New York: Springer, pp. 3–33.

Schätzlein, A. and Cevc, G. 1998. Non-uniform cellular packing of the stratum corneum and permeability barrier function of intact skin: A high-resolution confocal laser scanning microscopy study using highly deformable vesicles (Transfersomes). *British Journal of Dermatology* 138(4): 583–592.

Schulz, J., Hohenberg, H., Pflücker, F., Gärtner, E., Will, T., Pfeiffer, S., Wepf, R., Wendel, V., Gers-Barlag, H., and Wittern, K. P. 2002. Distribution of sunscreens on skin. *Advanced Drug Delivery Reviews* 54(suppl.): 157–163.

Scott, R. C., Corrigan, M. A., Smith, F., and Mason, H. 1991. The influence of skin structure on permeability: An intersite and interspecies comparison with hydrophilic penetrants. *Journal of Investigative Dermatology* 96(6): 921–925.

Shim, J., Kang, H. S., Park, W. S., Han, S. H., Kim, J., and Chang, I. S. 2004. Transdermal delivery of minoxidil with block copolymer nanoparticles. *Journal of Controlled Release* 97(3): 477–484.

Song, H. K., Wehrli, F. W., and Ma, J. 1997. In vivo MR microscopy of the human skin. *Magnetic Resonance in Medicine* 37(2): 185–191.

Souto, E. B., Almeida, A. J., and Müller, R. H. 2007. Lipid nanoparticles (SLN®, NLC®) for cutaneous drug delivery: Structure, protection and skin effects. *Journal of Biomedical Nanotechnology* 3(4): 317–331.

Stracke, F., Weiss, B., Lehr, C. -M., König, K., Schaefer, U. F., and Schneider, M. 2006. Multiphoton microscopy for the investigation of dermal penetration of nanoparticle-borne drugs. *Journal of Investigative Dermatology* 126(10): 2224–2233.

Surber, C., Schwarb, F. P., and Smith, E. W. 2001. Tape-stripping technique. *Journal of Toxicology—Cutaneous and Ocular Toxicology* 20(4): 461–474.

Tan, M. H., Commens, C. A., Burnett, L., and Snitch, P. J. 1996. A pilot study on the percutaneous absorption of microfine titanium dioxide from sunscreens. *Australasian Journal of Dermatology* 37(4): 185–187.

Tinkle, S. S., Antonini, J. M., Rich, B. A., Roberts, J. R., Salmen, R., DePree, K., and Adkins, E. J. 2003. Skin as a route of exposure and sensitization in chronic beryllium disease. *Environmental Health Perspectives* 111(9): 1202–1208.

Tobin, D. J. 2006. Biochemistry of human skin—Our brain on the outside. *Chemical Society Reviews* 35(1): 52–67.

Toll, R., Jacobi, U., Richter, H., Lademann, J., Schaefer, H., and Blume-Peytavi, U. 2004. Penetration profile of microspheres in follicular targeting of terminal hair follicles. *Journal of Investigative Dermatology* 123(1): 168–176.

Tsai, J. C., Weiner, N. D., Flynn, G. L., and Ferry, J. 1991. Properties of adhesive tapes used for stratum corneum stripping. *International Journal of Pharmaceutics* 72(3): 227–231.

Van Den Bergh, B. A. I., Vroom, J., Gerritsen, H., Junginger, H. E., and Bouwstra, J. A. 1999. Interactions of elastic and rigid vesicles with human skin in vitro: Electron microscopy and two-photon excitation microscopy. *Biochimica et Biophysica Acta—Biomembranes* 1461(1): 155–173.

Van Kuijk-Meuwissen, M. E. M. J., Junginger, H. E., and Bouwstra, J. A. 1998a. Interactions between liposomes and human skin in vitro, a confocal laser scanning microscopy study. *Biochimica et Biophysica Acta—Biomembranes* 1371(1): 31–39.

Van Kuijk-Meuwissen, M. E. M. J., Mougin, L., Junginger, H. E., and Bouwstra, J. A. 1998b. Application of vesicles to rat skin in vivo: A confocal laser scanning microscopy study. *Journal of Controlled Release* 56(1–3): 189–196.

Vega-Villa, K. R., Takemoto, J. K., Yánez, J. A., Remsberg, C. M., Forrest, M. L., and Davies, N. M. 2008. Clinical toxicities of nanocarrier systems. *Advanced Drug Delivery Reviews* 60(8): 929–938.

Vogt, A., Mandt, N., Lademann, J., Schaefer, H., and Blume-Peytavi, U. 2005. Follicular targeting—A promising tool in selective dermatotherapy. *The Journal of Investigative Dermatology Symposium Proceedings/The Society for Investigative Dermatology, Inc. [and] European Society for Dermatological Research* 10(3): 252–255.

Wagner, H., Kostka, K. -H., Lehr, C. -M., and Schaefer, U. F. 2000. Drug distribution in human skin using two different in vitro test systems: Comparison with in vivo data. *Pharmaceutical Research* 17(12): 1475–1481.

Wang, T. F., Kasting, G. B., and Nitsche, J. M. 2006. A multiphase microscopic diffusion model for stratum corneum permeability. I. Formulation, solution, and illustrative results for representative compounds. *Journal of Pharmaceutical Sciences* 95(3): 620–648.

Weigmann, H. -J., Lademann, J., Meffert, H., Schaefer, H., and Sterry, W. 1999. Determination of the horny layer profile by tape stripping in combination with optical spectroscopy in the visible range as a prerequisite to quantify percutaneous absorption. *Skin Pharmacology and Applied Skin Physiology* 12(1–2): 34–45.

Wertz, P. W. and Downing, D. T. 1989. Stratum corneum: Biological and biochemical considerations. In: J. Hadgraft and R. H. Guy, editors. *Transdermal Drug Delivery*. New York/Basel, Switzerland: Marcel Dekker, pp. 1–22.

Xiao, C., Moore, D. J., Rerek, M. E., Flach, C. R., and Mendelsohn, R. 2005. Feasibility of tracking phospholipid permeation into skin using infrared and Raman microscopic imaging. *Journal of Investigative Dermatology* 124(3): 622–632.

Yu, B., Kim, K. H., So, P. T. C., Blankschtein, D., and Langer, R. 2002. Topographic heterogeneity in transdermal transport revealed by high-speed two-photon microscopy: Determination of representative skin sample sizes. *Journal of Investigative Dermatology* 118(6): 1085–1088.

Yu, B., Blankschtein, D., Langer, R., Kim, K. H., and So, P. T. C. 2003. Visualization of oleic acid-induced transdermal diffusion pathways using two-photon fluorescence microscopy. *Journal of Investigative Dermatology* 120(3): 448–455.

Zhang, L. W. and Monteiro-Riviere, N. A. 2008a. Assessment of quantum dot penetration into intact, tape-stripped, abraded and flexed rat skin. *Skin Pharmacology and Physiology* 21(3): 166–180.

Zhang, L. W., Yu, W. W., Colvin, V. L., and Monteiro-Riviere, N. A. 2008b. Biological interactions of quantum dot nanoparticles in skin and in human epidermal keratinocytes. *Toxicology and Applied Pharmacology* 228(2): 200–211.

Cellular Response to Continuous Nanostructures

Kevin J. Chalut
University of Cambridge

Karina Kulangara
Duke University

Kam W. Leong
Duke University

33.1 Introduction

The primary goal of this chapter is to discuss cellular response to topographical cues at the nanoscale. We introduce the reader to the cellular machinery necessary to adhere and interact with substrates, present current fabrication techniques capable of generating nanoscale features on surfaces, and review how cells respond to these topographical cues at the nanoscale.

33.1.1 Cellular Adhesion to Substrates

In analogy to a tent on the ground, most cells have to be anchored to a surface; without anchorage, cells act more as a fluid and these anchorage-dependent cells would be unable to properly execute their function (Ingber, 2003a). Additionally, cells probe the elasticity and topography of their surroundings and use the information they gather to position and orient themselves, which has significant implications for development and organogenesis, tissue maintenance, angiogenesis, and wound healing (Schwarz and Bischofs, 2005).

As early as 1911, Ross G. Harrison first noted the influence of the substratum on cell orientation and migration, reporting the stereotropism of embryonic cells (Harrison, 1911). In his study, he used a spider web as substrate. He concluded that cells need a solid framework to execute their function and that the substratum exerts a stereotropism or physical guidance that dictates cell movement and function.

The initial contact of a cell with the extracellular matrix (ECM) and the substrate is via the integrin receptors and the focal adhesion structures. Integrin receptors are large trans-membrane proteins, which form a heterodimer containing one α- and one β-subunit. They bind with their extracellular domain to ECM proteins primarily in the short amino acid arginine glycine aspartate (RGD) sequence, present in fibronectin, laminin, and vitronectin. The earliest form of integrin-mediated contacts are focal complexes. These small (~500 nm) but highly dynamic point contacts called focal complexes are located at the leading edge of lamellipodia, membrane protrusions, and are associated with motility. When the lamellipodia retract or stop protruding, focal contacts are replaced by focal adhesions (Zaidel-Bar et al., 2003). Anchor proteins such as vinculin, paxillin, talin, α-actinin, and filamin are then sequentially recruited to the adhesion sites, which bind the cytoskeleton (CSK) to the membrane. Larger, more stable streak-like complexes called focal adhesions are characterized by the presence of zyxin and the concomitant assembly of actin bundles. The transition from focal contacts to mature focal adhesions involves cytoskeletal tension driven by cross-bridging interactions of actin and myosin filaments (actomyosin), which applies force at cell–matrix adhesions (Riveline et al., 2001). Subsequently, downstream signaling of proteins in the Rho family of guanosine triphosphate–ases (GTPases) (Riveline et al., 2001; Discher et al., 2005) occurs.

Fibrillar adhesion characterized by elongated fibrils or array of dots are distributed in more central regions of the cell and contain high levels of tensin and little or no phosphotyrosine (Zamir and Geiger, 2001). In addition to their role in cellular adhesion and migration, focal adhesions are active mechanosensors (Schwarz et al., 2006). Focal adhesions elongate in the

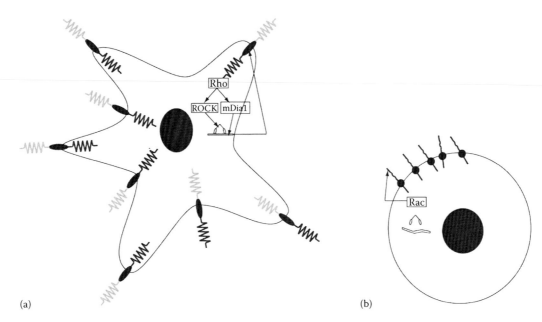

(a)   (b)

FIGURE 33.1 A rigid microenvironment (a) couples to focal complexes and tends to result in upregulation of Rho and downregulation of Rac, while the converse is true in a softer microenvironment (b). Downstream targets of Rho are ROCK, which activates myosin II contractility, and mDia1, which regulates actin polymerization and the growth of microtubules. Increased contractility results in the recruitment of integrins to focal complexes resulting in a transition to focal adhesion. This feedback mechanism may be regulated by microtubules, which are built into the loop. A softer microenvironment results in upregulation of Rac and downregulation of Rho, which results in focal contacts and lamellipodia gathering at the leading edge of the cell, increasing migration and motility.

direction of stress fibers to reflect the direction of the internal force, and are constantly in flux, disassembling and assembling to drive motility and the cell cycle (Ballestrem et al., 2001).

The Rho family of GTPases includes Rho, Rac, and Cdc 42 among others, which regulate the formation of stress fibers, lamellipodia, and filopodia, respectively (Nobes and Hall, 1995). Integrin ligation can activate several pathways. The first downstream pathway is focal adhesion kinase (FAK) that upregulates Rac and downregulates Rho (Wang et al., 2001). Rac is involved in cell spreading and migration as well as cytoskeletal reorganization. Another pathway involves Rho activation, which is generally antagonistic to Rac activation. The transition from focal complex to focal adhesion is characterized by a downregulation of Rac and an upregulation of Rho (Schwarz et al., 2006). The main downstream targets of Rho are Rho-associated kinase (ROCK) and mDia1. ROCK activates myosin II, which regulates contractility in cells, and mDia1 is active in actin polymerization and microtubule (MT) growth (Riveline et al., 2001). Focal adhesions are likely involved in a positive feedback loop in which the activation of ROCK leads to increased contractility, in turn recruiting additional integrins to the focal adhesion, which leads to additional ROCK activation in a feed-forward manner (Figure 33.1) (Schwarz and Bischofs, 2005). There must be a regulatory mechanism that breaks the loop; a good candidate is mDia1, which induces the growth of MTs, known to be a regulatory mechanism for focal adhesions (Krylyshkina et al., 2003).

The stress fibers connecting focal adhesions can be considered as force dipoles exerting lateral forces at the membrane of the cells, and all forces at the membrane exerted on the microenvironment

are localized at the focal adhesion. Additionally, the direction of force is generally in the direction of the elongation of the focal adhesion. The reason for this is as follows. The total free energy includes the pulling energy, the elastic energy of the aggregate, and the free energy of the assembled proteins in the surrounding cytoplasm. Increasing the pulling force also increases the elastic energy, and the stress is relieved by the recruitment of ambient proteins (Shemesh et al., 2005). The thermodynamic nature of cell–matrix adhesions explains their highly dynamic character. Over 100 proteins can join the cell–matrix aggregate, and the assemblies are constantly dissembling and reassembling (Figure 33.2). In what follows, we outline the treatment by Schwarz and Bischofs (Bischofs and Schwarz, 2003; Bischofs et al., 2004) to

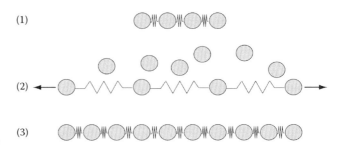

FIGURE 33.2 Self-assembly of cell–matrix contacts upon pulling force. In step 1, the protein aggregate is relaxed. In step 2, a pulling force results in an increase in elastic energy. In step 3, ambient proteins insert into aggregates resulting in an overall stress relaxation. This sequence is responsible for the assembly of focal contacts and the transition to focal adhesions with pulling forces and tension in the CSK/ECM network.

describe the response of cells to the mechanical properties of their substrate through cell–matrix adhesions.

Cell–matrix connections can be described as a field of focal adhesions connected to each other through stress fibers and then to the extracellular environment, which is described by isotropic linear elasticity theory. Under the assumptions of isotropic linear elasticity theory, the extracellular environment is described by its Young's modulus, E, and the Poisson ratio, v : E describes rigidity while v describes the relative weight of compression and shear modes.

The cell actively pulls on its elastic microenvironment, and it is assumed that the amount of work invested in exploring its environment would be minimized. Since the cell–matrix contacts are coupled throughout the actin–CSK in such a way that the forces are balanced, only pairs of opposing forces need be considered. In the language of elasticity theory, this is called an anisotropic force contraction dipole, represented by a tensor $P_{ij} = P n_i n_j$ where P is the dipole strength and n is the orientation. The work required to build up the force dipole at cell position r_n is proportional to the strain of the microenvironment, u_{ij}:

$$\Delta W = P_{ij} u_{ij}^{e}(r_n) \tag{33.1}$$

The optimal cellular organization will minimize ΔW. Since the cell is generally prestressed by actomyosin activity, $P < 0$ (associated with contraction), and therefore, the strain tensor is optimally negative (compressive strain, or a stiffer microenvironment).

If we consider the effect of the external environment on a single focal adhesion to be that of a linear spring, then Equation 33.1 indicates that a focal adhesion has to invest energy $W = F^2/K$ into the spring, where K is the linear spring coefficient, analogous to the Young's modulus E. The force is then minimized by maximizing K, the stiffness of the spring. This analogy helps explain cell spreading: if the elasticity of the microenvironment is anisotropic, the buildup of force will be more efficient in some directions than others, assuming resource allocation is approximately equal at each focal adhesion, and those contacts will outgrow others. An isotropic environment will result in a more symmetric, generally round or stellate, cell morphology.

There are two relevant and interesting boundary conditions we can use to solve Equation 33.1: homogeneous external strain in the medium and a physical boundary to the medium. In the case of homogeneous external strain, the equations of 3D isotropic elasticity give

$$\Delta W = \frac{-Pp}{E}\left[(1+v)\cos^2\theta - v\right] \tag{33.2}$$

where θ is the orientation angle relative to the externally applied stress $p < 0$. The minimal ΔW corresponds to $\theta = 0$, so the cell orients preferentially in the direction of stress (Figure 33.3).

In the case of a physical boundary, one informative case lends itself to exact solution: that of a semi-infinite half plane with free and clamped boundary conditions. Physically, this

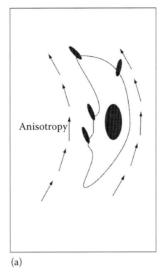
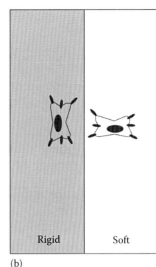

(a)   (b)

FIGURE 33.3 A schematic representing durotaxis—the movement of cells in the direction of increasing rigidity in the microenvironment. (a) The cell follows the distribution of strain on a substrate. (b) The cell orients itself in the direction of greater strain, parallel to a boundary with a softer medium, and perpendicular to a boundary with a more rigid medium.

corresponds to a case of a homogeneous medium with a boundary in which the interface has infinite rigidity for clamped boundary conditions and vanishing rigidity for free boundary conditions (see Figure 33.3). If we consider a force dipole a distance d away from the surface and an angle θ with respect to the surface, the solution to Equation 33.1 for free and clamped boundary conditions is

$$\Delta W = \frac{P^2}{Ed^3}\left(a_v + b_v \cos^2\theta + c_v \cos^4\theta\right) \tag{33.3}$$

The coefficients a, b, and c are functions of v; here we are concerned with them only in the sense that all coefficients are positive for free boundary conditions and negative for clamped boundary conditions. Therefore, the buildup of force is most efficient when $\theta = 0$ for clamped boundary conditions and $\theta = \pi/2$ for free boundary conditions. The interpretation of Equation 33.3 is as follows: cells orient perpendicularly to clamped boundaries and parallel to free boundaries. We conclude from this, similar to the interpretation of Equation 33.2, that the mechanical activity of a cell is more efficient on a rigid surface, and the cell is attracted to the rigid surface, whereas there is an aversive response to a free (nonrigid) surface. This phenomenon is called durotaxis (Lo et al., 2000).

The cell behavior described above should be considered in the context of cell function, which is strongly determined by mechanical stimuli. Certain cells, such as endothelial cells (ECs) in blood capillaries (Davis and Hill, 1999) or osteocytes in bone (Donahue, 2000) function properly only when subjected to a certain amount of strain. It is also important to emphasize that the results above exist on a timescale that corresponds to active behavior (see Section 33.1.4): Cyclic alterations of stretch lead to exactly the opposite results with cells orienting away from the direction of cyclic mechanical forces (Wang and Grood, 2000); cells resist passive deformations.

Experimental evidence bears out these phenomena. It was first reported in 1997 that substrate compliance influences cell behavior, in which morphological changes were observed in ECs and fibroblasts (Pelham and Wang, 1997); later, cellular response to substrate compliance was noticed in other cell lines (Deroanne et al., 2001; Engler et al., 2004). Also, it has been shown that substrate elasticity influences growth and apoptosis of normal, but not cancerous, cells (Wang et al., 2000). Other experimental observations support the above claims. For example, inhibition of myosin activates Rac (Katsumi et al., 2002) (Figure 33.1), thereby inhibiting focal adhesions and increasing cell motility (versus contraction). Additionally, cell-to-cell contacts at a high cell density override cues resulting from substrate compliance (Yeung et al., 2005).

Perhaps the most important result to arise from studies of the mechanical properties of the substrate is that matrix elasticity directs stem cell lineage differentiation (Engler et al., 2006). In this study, Discher's group showed, using undifferentiated mesenchymal stem cells, that neurogenic markers are dominant in a soft microenvironment, myogenic (muscle) markers are dominant in a microenvironment of intermediate rigidity, and osteogenic (bone) markers are dominant in a stiff microenvironment. The rigidity of the microenvironments mimics that of brain, muscle, and bone, respectively. Notably, the inhibition of myosin using blebbistatin blocks all elasticity-directed lineage differentiation.

The above concepts fit into a Lagrangian formulation: the cell behaves in a way that the forces at work in the cell do work most efficiently on the environment. In the case of substrate compliance, the cell works most efficiently on a rigid substrate.

The CSK is linked to these cellular adhesions and plays an important role in orchestrating the cellular response to a given environment. It forms a mechanical network in which the balance of forces influences biochemical signaling pathways and ultimately cell function. In Section 33.2, we will survey results and experimental techniques regarding the role of external forces, internal forces, and topography in cell function.

33.1.2 The Cytoskeleton

The CSK provides structural integrity to cells, but also serves as a structure for molecular transport. Intracellular and extracellular forces acting on the cell are distributed through the CSK. In most eukaryotic cells, the CSK comprises three classes of proteins: microfilaments, MTs, and intermediate filaments (IFs). Microfilaments composed of actin are mostly distributed as bundles adjacent to the cell membrane, supporting projections in the cell membrane such as filopodia, or microspikes (involved in cell sensing) or lamellipodia (involved in cell crawling). Another type of actin is a loosely spaced network of stress fibers that is involved in cell contraction (Janmey et al., 1991). Actin has been localized in the nucleus and has recently been found to play an important role in gene transcription (Pederson and Aebi, 2005). MTs are

hollow cylinders capable of bearing compression (Mizushima-Sugano et al., 1983). According to the tensegrity theory (Ingber, 2003a), discussed in more detail later, they provide a counterbalance to actin stress fibers. MTs propagate outward from the centromere to the cell periphery, thereby providing a surveying device for the center of the cell as well as a plane and mechanism for cell division. Both actin and MTs work in conjunction with highly efficient motor proteins (myosin II for actin, kinesin and dynein for MTs), which are driven by the hydrolysis of ATP to ADP, to provide pre-stress to the cellular matrix in addition to their essential role as mechanisms and regulators of signal transduction (Janmey, 1998). The role of IFs has been far less studied, but is proving to be just as important as that of actin and MTs. IFs have been implicated in distributing tensile forces throughout the cell (Toivola et al., 2005) and have been associated in signal transduction (Kim and Coulombe, 2007). IFs are tough rope-like fibers oriented largely around the nucleus; they are classified into four types expressed in different cell types: type I, or keratin-based proteins; type II, including vimentin and desmin; type III, including neurofilament proteins; and type IV, nuclear lamins. IFs become particularly active when the cell is under stress, and are purported to be the tensile elements of a tensegrity network (Tint et al., 1991). IFs together with actin compose the CSK element that integrates the nucleus into the mechanical network (Helmke et al., 2003); therefore, they may play the role of directly communicating forces to the nucleus.

First, we will provide a framework for understanding the role of the cellular adhesions and CSK by discussing prevailing theories regarding cytoskeletal dynamics. Explanations about cytoskeletal dynamics include, but are not limited to the sol-gel hypothesis in which cytoskeletal polymers undergo a sol-gel transition under stress (Janmey et al., 1990); soft glassy rheology (SGR) in which the gel phase is better described as a soft glassy phase (Fabry et al., 2001); and tensegrity, or tensional integrity, in which the CSK is a network regulated by tensional and compressive elements (Ingber, 1993). The tensegrity theory also serves as a possible explanation of signal transduction by the CSK; percolation theory (Forgacs, 1995), which will not be discussed here, is also a theory about signal transduction within the cytoskeletal network based on the interconnectivity of this network. Discussion of signal transduction in the CSK will be saved for a later section. One caveat to bear in mind is that most work understanding CSK dynamics has been performed *in vitro*; putting the prevailing theories in the context of cell behavior in integrated, hierarchical systems is a work in progress.

33.1.3 Tensegrity and Mechanotransduction

Tensegrity (tensional integrity) networks comprise opposing tension and compression networks that balance each other, thereby providing structural stability (Fuller, 1961). It has been proposed that active and passive cell mechanics are determined largely by principles of tensegrity, with actomyosin acting as the tensional component while MTs act as the compressive component (Ingber, 1993). The IFs serve an important role in structural integrity and can provide structural support in lieu of microfilaments (Charras and Horton, 2002). They appear to serve as "guy wires," and are tethered to the nucleus, previewing the notion that they serve as an important link in the network from the CSK to the nucleus together with the microfilaments (Helmke et al., 2003). The primary tenet of biological tensegrity is that biological materials, including cells, tissues, and organs, are prestressed networks that provide an integrative network in which to translate and respond to mechanical signals (Ingber, 2003a,b). Tensegrity is related to, and inseparable from, mechanotransduction, in which the cytoskeletal network translates mechanical signals into biochemical signals. Tensegrity acknowledges the CSK as the primary regulator of mechanotransduction, and further postulates that mechanotransduction is modulated by cytoskeletal pre-stress, cell shape, and the balance of forces within the cell and ECM.

33.1.4 Viscoelasticity of the Cytoskeleton

Cellular microrheology involves a measurement of stress, or applied force per unit area, to the strain, or response of the cell to the stress. Measurement of these quantities leads to a description of the cell through the elastic modulus (the ratio of stress to strain) and the loss modulus (the ratio of stress to strain rate). Many groups have measured the elastic moduli and loss moduli of cells and a consensus has emerged: cells act in one of two mechanical paradigms, a soft glass or a semiflexible polymer network, depending on the timescale of observation (Pullarkat et al., 2007). The crossover between the two is approximately 10^2 s; with shorter timescales dominated by mechanical behavior (elastic and loss modulus) that scales with frequency with a power-law exponent of 3/4 (see Figure 33.4) (Deng et al., 2006).

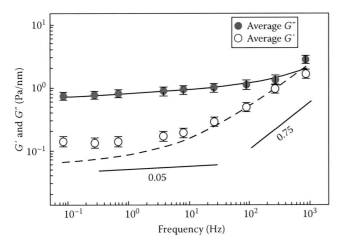

FIGURE 33.4 Elastic modulus (G') and loss modulus (G'') measured against frequency for bovine trachea smooth muscle cells using magnetic twisting cytometry, and fit to a two-term power law. The elastic modulus, which represents the elastic response of the cell to stress, undergoes a clear transition at approximately 10 Hz. Below 10 Hz it shows a very weak power law dependence reminiscent of a soft-glass regime, and above which shows a stronger dependence reminiscent of an elastic polymer network. (From Deng, L.H. et al., *Nat. Mater.*, 5, 636, 2006. With permission.)

The $f^{3/4}$ behavior is governed by dynamics reminiscent of a filamentous polymer network, such as reconstituted F-actin networks (Morse, 1998; Gardel et al., 2003, 2004). In the mechanical paradigm of elastic networks, there is insufficient time available for inelastic structural rearrangements, which means that the mechanical properties of the cell are governed by bending of the CSK filaments driven by thermal fluctuations (Bursac et al., 2005). It is worth noting that this entropy-driven behavior depends on the density of CSK filaments and density of crosslinkers; with sufficiently low density of filaments or cross-linkers, this behavior can become enthalpy-driven, and obey a linear response to stress (Gardel et al., 2004).

Cellular dynamics exist in a different mechanical paradigm at longer timescales (>10–100 Hz), in which nonequilibrium behavior dominates and the generalized Stokes–Einstein relationship breaks down (Bursac et al., 2005). Here, the mechanical properties of cells exhibit a very weak power-law dependence on frequency, f^β, with the exponent $0.1 < \beta < 0.3$ (the values depends on the region of the cell in which the mechanical properties as well as cell type) (Fabry et al., 2001; Desprat et al., 2005). Remarkably, this very weak dependence on timescale holds over many (at least five) decades of frequency. The featureless spectrum over such a wide frequency range indicates that there are no specific molecular mechanisms dominating cellular response, but that cellular behavior and response exists at a higher level of structural organization. It has been postulated that cells obey SGR, in which a material exists in a disordered and metastable state until stressed, at which point the material fluidizes and slowly returns to its original configuration (Deng et al., 2006). Since temperature fluctuations are insufficient to overcome the energies binding elements within a soft glass, SGR also predicts an effective temperature (Fabry et al., 2003). The effective temperature can be thought of as the level of agitation capable of moving the system away from a metastable state; it defines the power-law exponent and indicates the higher level structural organization of the cell. Additionally, effective temperature defines proximity of a cell to solid or fluid state. It was shown recently that effective temperature is equivalent to pre-stress (Trepat et al., 2007), which is a remarkable unification with the tensegrity model. If true, it represents further evidence that cellular dynamics are indeed largely governed by pre-stress.

Cellular function is largely regulated by the balance of forces between the CSK and the ECM and is mediated by cell–matrix contacts. Environmental cues such as matrix rigidity and topography can cause conformational changes in cell–matrix contacts, which influence biochemical signaling as well as the morphology of the cell and the forces within the cell. These various elements exist within feedback loops that largely determine the ways in which the cell interacts with its environment.

33.1.5 Geometry Sensing

Other important characteristics of substrates that influence cellular behavior are substrate curvature and topographical cues. The shape of the environment directs cell function, which is important to understand cellular response to continuous nanostructures. The cellular elements most directly involved with geometry sensing are the lipid bilayers, which separate the contents of the cell from the environment; transmembrane proteins and channels, which typically exist in mechanical equilibrium with the cell membrane; and geometry-sensing proteins such as proteins in the BAR (bin, amphysin, RVS) domain. The BAR domain is a crescent-shaped dimer that is a member of the amphiphysin protein family involved in both membrane dynamics and sensing of substrate curvature (Habermann, 2004).

The lipid bilayer spans a thickness of two lipids, which are amphiphilic molecules comprising a polar head group and a nonpolar fatty acid tail (Alberts, 2008). The hydrophobic fatty acid tails face toward the middle of the bilayer, and they are shielded from water by the polar head group. The lipid bilayer has spontaneous curvature in which forces are balanced: an outward force arises because the inner membrane has a smaller surface area and therefore greater tension than the outer membrane; and an inward force balances the outer force due to the free-energy cost that would be incurred by exposing the fatty acid tails to water in the environment (Zimmerberg and Kozlov, 2006). The forces in the membrane are highly anisotropic and a configurational alteration of a transmembrane protein can upset the mechanical equilibrium in the membrane, potentially causing transmembrane channels to open or changes in the configuration of other transmembrane proteins. Alternatively, the membrane can respond directly to geometry and alter the configuration of the transmembrane proteins and channels.

Cells respond in significant ways to the topography of their environment. Cellular response to geometry should always be considered in the context that cell shape is an important indication, and probably regulator, of cell function. For instance, a more rounded morphology is associated with cell death and spreading is associated with differentiation (Chen et al., 1997). In addition to direct changes in the balance of forces in the membrane including transmembrane proteins, there are several hypotheses that may explain the mechanisms by which cells respond to substrate geometry. Recent studies indicate that BAR-domain proteins serve an important role in endocytosis and may recognize and seek concave surfaces such as nanoposts or fibers (Figure 33.5). Moreover, BAR-domain proteins serve as a binding site for small GTPases including Rac (Habermann, 2004). On the other hand, cell attachment to convex surfaces (such as nanocaved patterns) can induce the opening of mechanosensitive ion channels such as K⁺ (Patel et al., 2001) (Figure 33.6). This opening is caused by the differential tension between the inner and outer layers of the lipid bilayer of the cell membrane.

Different mechanisms underlie the cellular response to convex and concave substrates. The sensing of concave substrate curvature by BAR-domain proteins coupled with its affinity for GTPases such as Rac could explain changes in function and cell motility (Vogel and Sheetz, 2006). The selective opening of membrane channels due to surface indentations is also highly influential on cell function, for instance, activating the TREK-1 channel activity (Chemin et al., 2005).

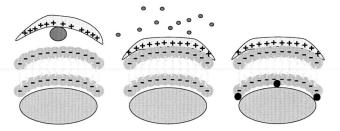

FIGURE 33.5 The BAR (bin, amphysin, RVS) domain is a crescent-shaped dimer with 12 positively charged residues that bind to the negatively charged polar head groups of the lipid membrane. The BAR domain is rigid enough to act as a scaffold for the membrane and produce curvature. Additionally, it acts as a binding site for small G-proteins such as Rac (shown attached to BAR domain in the left panel and then released in the middle panel) that enhance focal complexes (represented in black in the figure) and, therefore, cell spreading and traction forces. The affinity of BAR-domain proteins for convex curvatures enhances recruitment and release of GTPases and may be responsible for increased proliferation and cell spreading observed on concave surfaces.

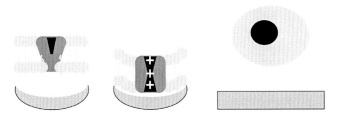

FIGURE 33.6 Adherence of cell to convex surface causes differential tension between inner and outer layer of membrane, potentially opening ion channels and thereby altering cell function. Cell attachment decreases on nanocaved surfaces with respect to a flat substrate.

33.1.6 Fabrication of Nanostructures

Micro- and nanofabrication advances have been fueled by the catalyst and microelectronics industries (Wood, 2007). Nanopatterned surfaces can be produced by a variety of methods. Biological studies require a large patterned sample area to culture enough cells so as to permit molecular biology assays. The inherent variability of a biological sample also necessitates multiple experiments to ascertain reproducibility.

In general, nanomanufacturing requires high resolution and low cost for fabricating devices. It might include the fabrication of an expensive master from which the features can be replicated to a polymeric material at low costs in large quantities in an inexpensive process. Most of the nanofabrication techniques rely on lithographic methods (Wood, 2007).

The most widely used conventional lithography technique is contact mode photolithography (Wood, 2007). For this technique, a transparent polymer or a metal-coated glass slide is generated with the desired features and serves as a mask. The features are then transferred to the photocurable resist sensitive to UV radiation, after which the resist is developed and the features transferred to the substrate. In a biological context, this method is more readily used to fabricate 1–2 μm structures although it can produce features below 400 nm (Falconnet et al., 2006). Electron beam (e-beam) lithography has extremely high resolution capabilities and therefore is often used to produce nano-featured masters. However e-beam fabricated pattern can only be made accessible to biological applications through the use of replication technologies. X-ray lithography can also be used as an alternative to fabricate small feature size patterns (Cerrina and Marrian, 1996; Shipway et al., 2000). Colloidal lithography (Shipway et al., 2000) uses colloidal nanoparticles to produce nanoscale topographies through self-assembly or assisted by chemical patterning (Yi Dong et al., 2006).

Nanopatterns can be transferred to polymers by replica molding (Wood, 2007). A degassed resin is poured onto the master and the replica is separated from the master after curing. Resins commonly used include elastomeric polymer such as polydimethylsiloxane (PDMS). Nanoimprint lithography allows the replication of nanoscale features in thermoplastic polymers such as polymethylmethacrylate, polystyrene, or UV-curable materials such as acrylates and epoxies. For this technique, the master mold is placed in contact with the polymer surface and both are heated and pressed. When the temperature is higher than the glass transition temperature of the polymer, the latter starts flowing and with the applied pressure fills the gaps in the master mold. After cooling, the mold and the replica can be separated.

Nanoscale-sized fibers produced by electrospinning have also attracted significant attention for their use in biological applications (Chew et al., 2006). Fibrous scaffolds display a high surface-to-volume ratio favorable for cell attachment and a highly macroporous structure to facilitate mass transport of nutrients and waste products. Electrospinning can yield continuous fibers from 3 nm to several microns in diameter, encapsulate drugs in the fibers, and be conveniently applied to any polymer soluble in low boiling solvents. It has, therefore, become a popular technique of fabricating nanostructured biomaterials for tissue engineering applications.

The development of techniques such as nanoimprint lithography and electrospinning to fabricate nanoscale topographies composed on biomedical polymers has greatly facilitated studies ranging from mechanistic analysis to understand cellular response to nanotopography to applications of regenerative medicine. Such scaffolds or substrates with nanoscale features are appealing because they may mimic the microenvironment in vivo, where cells are often surrounded by nanoscale features such as collagen fibers in the ECM. A schematic demonstrating some types of nanostructures is given in Figure 33.7.

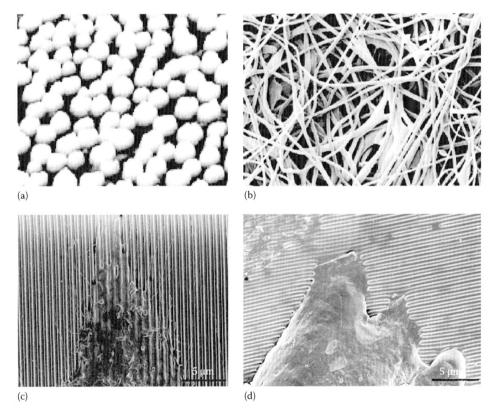

(a)   (b)

(c)   (d)

FIGURE 33.7 Schematic demonstrating several different types of nanostructures. (a) Nanoislands, 13 nm in height and 263 nm in diameter. (From Dalby, M.J. et al., *Exp. Cell Res.*, 276, 1, 2002b. With permission.) (b) SEM micrograph of electrospun fibers. (From Moroni, L. et al., *Biomaterials*, 27, 4911, 2006. With permission.) (c) SEM micrograph of human mesenchymal stem cells cultured on polystyrene surfaces comprising nanopatterned gratings with 350 nm linewidth, 700 nm pitch, and 350 nm depth; and (d) SEM micrograph of human mesenchymal stem cells cultured on PDMS surfaces comprising nanopatterned gratings with 350 nm linewidth, 700 nm pitch, and 350 nm depth. It is noteworthy that there is a visible deformation of the PDMS nanopatterned substrate, while there is no visible deformation of the nanopatterned polystyrene substrate. The PDMS substrate is much more compliant than the polystyrene substrate.

33.2 Cellular Response to Nanotopography

One way to influence cell shape is by using a defined adhesive surface surrounded by a nonadhesive substrate in which cells are constrained to particular orientation and shape on a substrate with adhesive features at the micron scale (Weiss and Garber, 1952). Contact guidance influences gene expression (Curtis and Seehar, 1978), which has been observed in fibroblasts cultured on microgrooves (Dalby et al., 2003b). It has also been shown that the size of adhesive micropatterns regulates whether mesenchymal stem cells differentiate into adipocytes or osteoblasts (McBeath et al., 2004).

33.2.1 Focal Adhesion, Cell Spreading, and Alignment on Nanotopography

Much of the pioneering work in this field has come from Wilkinson and Curtis (Curtis and Seehar, 1978; Curtis and Clark, 1990; Wójciak-Stothard et al. 1995; Curtis and Wilkinson, 1998; Curtis et al. 2001; Wilkinson et al. 2002). Recently, they produced nanoscale islands by blending polystyrene (PS) and poly(4-bromostyrene) (PBrS) and studied the effect of this nanotopography on fibroblasts (Dalby et al., 2002b). The nanoislands, which were 13 nm in height and 263 nm in diameter, increased spreading and proliferation in fibroblasts (see Figure 33.7a). The nanoislands also upregulated genes encoding for G-proteins involved with CSK conformation and the formation and control of filopodia and lamellipodia. As the nanoislands were increased beyond 35 nm, cell adhesion and spreading decreased, and cellular response was decreased in kind (Dalby et al., 2003a, 2004c). Similar results were observed in ECs with decreasing cellular response to nanoislands of 13, 35, and 95 nm, respectively (Dalby et al., 2002a). Dalby et al. (2004d) also studied fibroblast behavior on nanocolumns produced by colloidal lithography on a polymethylmethacrylate (PMMA) substrate; the nanocolumns were 160 nm in height, 100 nm in diameter, and had 230 nm center-to-center spacing. Cell spreading and adhesion were decreased, the CSK was more

diffuse, the number of filopodia increased, and focal adhesions were smaller. Intriguingly, fibroblasts attempted to endocytose the nanocolumns (Dalby et al., 2004a).

Considering their role as a primary mechanosensor that can trigger signaling activity related to gene expression, cell proliferation, and cell survival, focal adhesions are of vital importance in understanding cellular response to continuous nanostructures. The height of nanoislands or nanocolumns is likely to determine if the focal adhesion sites can form a continuous structure on the surface or if they are spatially restricted to the top of the nanoislands or columns. The size and spatial arrangement of the focal adhesion structures will in turn result in increased or reduced cell spreading.

Donahue's group has done recent work investigating the focal adhesion proteins and CSK dynamics in osteoblasts on continuous nanostructures. In the first study, Lim et al. (2007) used poly-(L-lactide)/polystyrene (PLLA/PS) demixed nanoscale pit features of 14, 29, and 45 nm to investigate focal adhesions. Osteoblasts cultured on 14 and 29 nm pits were much more spread than on 45 nm, where they showed a more rounded morphology. At 14 and 29 nm, the cells exhibited higher levels of paxillin (a focal adhesion protein) and slightly higher levels of vinculin (another focal adhesion protein) (Figure 33.8) and expressed much higher levels of actin stress fibers than cells cultured on 45 nm pits. Additionally, the 14 and 29 nm cultures expressed a much higher level of FAK, indicating the involvement of the focal adhesion signaling pathway in causing the osteoblast spreading. These results are in line with studies from Dalby (Dalby et al., 2002a,b, 2003a, 2004c) in which small islands (under 35 nm) increased cellular response including proliferation and adhesion. Also from Donahue's group, Hansen et al. (2007) used PLLA/PS demixed nanoislands ranging between 11 and 38 nm to investigate osteoblast viscoelasticity as a response to nanofeatures using atomic force microscopy. Osteoblasts exhibited an increasing Young's modulus (stiffness) from 11 to 38 nm, and also from PS to glass. The authors note that the surface features of bone are approximately 32 nm, which indicates a possible biological origin of their observations.

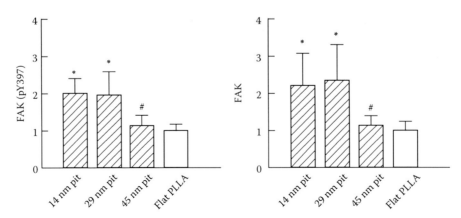

FIGURE 33.8 Expression of focal adhesion kinase (FAK) on nanopitted surfaces of 14, 29, and 45 nm compared to flat PLLA. * indicates statistically significant differences between flat and 45 nm samples. (From Lim, J.Y. et al., *Biomaterials*, 28, 1787, 2007. With permission.)

Recent experiments using chemical nanopatterning techniques confirm the concept that cell-adhesive ligand spacing regulates cell behavior. Arnold et al. produced RGD-coated gold nanodots at spacings of 28–85 nm using micelle nanolithography. Their findings suggest that 58–73 nm spacing in RGD molecules are optimal for osteoblast adhesion, integrin clustering, and activation. Greater spacing leads to reduced osteoblast adhesion, spreading, and integrin-mediated focal contact formation (Arnold et al., 2004). Topography in the nanoscale might restrict integrin clustering to certain areas on the topography. Teixeira et al. (2003) observed alignment of human corneal epithelial on ridges 70 nm wide with 400 nm pitch and 600 nm depth. Aligned axonal outgrowth was observed on 100–400 nm wide grooves, where the nerve cell process is preferentially located on ridge edges rather than in groves (Johansson et al., 2006). Cell adhesion on nanotopography is further exploited in a microfluidics system to enrich human breast cancer cells by adhesion difference (Kwon et al., 2007).

Cells respond differently to convex and concave surfaces: 13 nm high nanoposts increased cell spreading and proliferation (Dalby et al., 2004e) while some nanopitted surfaces decreased adhesion relative to flat substrates (Martines et al., 2004). Dalby et al. (2004b) also investigated convex substrate curvature by manufacturing poly(ε-caprolactone) (PCL) pits with diameters of 35, 75, and 120 nm. As the diameter of the pits increased, the cell spreading decreased and fewer actin stress fibers were visible. Also, as the diameter of the pit increased, the number of filopodia increased, though there was evidence that filopodia were interacting with pits down to 35 nm.

Cells tend to orient along the line of minimal curvature (Dunn and Heath, 1976; Dunn and Brown, 1986). In the sense that cells explore their microenvironment in the most efficient way possible, this can be easily explained: following the line of minimal curvature minimizes the membrane and cytoskeletal deformation and reconfiguration (which requires energy) necessary to explore a curved environment. Furthermore, cells orient along a grooved topography (Dalby et al., 2003b; Yim et al., 2005). This is also explained by cells minimizing the energy needed to explore their environment as they are not adding the cost in energy necessary by crossing boundaries, an insight buttressed by the results of Dunn and Brown (1986), where the alignment is found to be directly proportional to groove width and spacing in the range observed (from 1.65 to 8.96 μm width and 3.0 to 32.0 μm spacing).

Studies by Curtis and Dalby suggest that cells can distinguish symmetries in topographies. Osteoblasts spreading was reduced on ordered nanopits, but increased on more randomly oriented nanopits (Biggs et al., 2007). Furthermore, ordered or disordered topographies influence the fate of mesenchymal stem cells as discussed below (Dalby et al., 2007b).

A series of studies explored the effect of nanotopographical cues on smooth muscle cells (SMCs) and ECs, which are important in tissue engineering for vascular graft and bladder replacement implants. Miller et al. (2004) studied the effect of surface chemistry and nanotopography using two different nanofeatures in the range of 50–100 nm. The first was produced by etching poly(D, L-lactide-*co*-glycolide) (PLGA) with NaOH and the second was poly(dimethylsiloxane) (PDMS) produced by replica molding from a NaOH-etched nanostructured PLGA, followed by casting PLGA on the PDMS mold. While SMCs responded favorably to both substrates, EC adhesion and proliferation was decreased on the NaOH-etched nanostructured surfaces. This led the authors to conclude that ECs were more sensitive to surface chemistry and SMCs were more sensitive to nanotopography. Yim et al. (2005) further studied the effect of nanotopography on SMCs by synthesizing PMMA and PDMS surfaces comprising nanopatterned gratings with 350 nm linewidth, 700 nm pitch, and 350 nm depth. SMCs aligned well with the gratings and showed significant elongation; however, proliferation was decreased. Furthermore, the microtubule-organizing centers (MTOCs), which are associated with cell migration, polarized in a direction toward a wound edge in an *in vitro* wound healing assay when SMCs were cultured on a flat glass substrate. In contrast, the SMCs polarized along the direction of the grating on nanopatterned surfaces in the *in vitro* wound assay. This study, along with the results in Miller et al., indicates that SMCs respond very strongly to nanotopographical cues. As discussed above, nanotopography on a 2D substrate generally decreases cell proliferation as opposed to nanofibers as we will see in Section 33.2.3.

33.2.2 Nanotopography as Immune Response Modulator

An interesting aspect of nanotopography is its potential to modulate immune response to an implant. Andersson et al. (2003) found that nanoscale features influenced cytokine production and morphology in T24 uroepithelial cells. Three types of TiO_2 surfaces were synthesized: flat; grooved (depth 184 nm, ridge and groove width 15 μm); and hemispheric (density of 12/μm², center-to-center spacing of 250 nm, average diameter of 170 nm, and a height of 100 nm). There was no difference in adhesion or proliferation among the three substrates; however, cells on the hemispheric substrate were less spread and more stellate. At the same time, cytokine production was similar on the flat and grooved substrates while it was significantly decreased on the hemispheric substrate. This study represents one of the first in which a specific function, such as cytokine production, is associated with nanotopographical cues of a substrate.

33.2.3 Topographical Cues Derived from Nanofibers

Early work on electrospun nanofibrous scaffolds, designed to deliver nanotopographical cues to cells similar to those provided by ECM, has produced interesting results. Mo et al. (2004) studied the cell response of SMCs and ECs on electrospun poly(L-lactide-*co*-ε-capro-lactone) (PLLA-CL) nanofibers ranging in diameter from 500 nm to 1.2 μm. Both ECs and SMCs showed good attachment and proliferation on the nanofibers. The cells were also observed to migrate, forming a continuous monolayer. Lee et al. (2005) investigated human ligament fibroblasts on aligned and random electrospun polyurethane nanofibers. Cells showed increased proliferation on both random and aligned nanofibers as compared to flat substrates. However, the amount of collagen produced was much higher in aligned versus random nanofibers. When cyclic strain was applied to the nanofibers, the amount of collagen was significantly increased when the cells were aligned perpendicular to strain and less so in the parallel direction. The increase in collagen production is likely related to the organization of CSK along the nanofibers as the CSK can serve as a direction-sensitive strain gauge. Collagen production was not increased in randomly oriented nanofibers when cyclic strain was introduced. These results motivate further studies of creating 3D continuous nanostructures to mimic _in vivo_ microenvironment for cell culture.

Recent studies of cellular response to both 3D and 2D continuous nanostructures, which are summarized in Table 33.1, have focused on mechanisms of nanotopographical influence as well as achieving specific functional results. Rho et al. (2006) investigated the effects of collagen nanofibers on normal human keratinocytes. To create a biomimetic ECM, collagen was electrospun and cross-linked to present a randomly oriented nanofibrous matrix. To further investigate the effect of ECM proteins, type I collagen, fibronectin, and laminin were adsorbed onto the matrix. In a wound healing model in rats, noncoated collagen nanofibers did not promote good cell adhesion as compared to a noncoated PS surface. In contrast, collagen- and laminin-coated nanofibers stimulated significantly better adhesion and spreading as compared to TCPS with type I collagen better than laminin. Interestingly, fibronectin, an ECM glycoprotein that plays an important role in cell adhesion and wound healing, appeared to have no positive effect in this setting. Additionally, collagen nanofibers coated with type I collagen promoted wound healing in the rat model at a much faster rate than the control group. This study highlights the fact that in many cases surface chemistry may be as important as, if not more so than, topography in considering cell–substrate interactions.

TABLE 33.1 Summary of Research on Cellular Response to Nanotopography[a]

Cell Type	Nanofeature	Proliferation/Morphology	Cell Response
Fibroblast	Nanoisland 13, 35, and 95 nm	Increased spreading and proliferation with decreasing height (<35 nm)	Enhanced CSK organization (<13 nm)
Fibroblast	Nanopits 35, 75, and 120 nm	Decreased spreading with increasing diameter	As diameter increased, fewer stress fibers and more filopodia
Smooth muscle cells (SMCs) and endothelial cells (ECs)	NaOH nanofeatures ranging from 50–100 nm	Increased proliferation for SMCs, decreased proliferation for ECs	
SMCs and ECs	Etched nanofeatures ranging from 50–100 nm	Increased proliferation for SMCs and ECs	
SMCs	Grooves with 350 nm linewidth, 700 nm pitch, 350 nm depth	Aligned and elongated, proliferation decreased	SMCs favored nanotopographical cues over cues from wound assay
Epithelial cells	Flat, nanometric grooves, and hemispheric	No change in adhesion or proliferation, less spread on hemispheric	Cytokine production decreased on hemispheric
Fibroblasts	Electrospun (ES) nanofibers, aligned and random	Increased proliferation	Increased ECM (collagen) on aligned versus randomly oriented nanofibers
Keratinocytes	Collagen nanofibers coated with collagen, laminin, or fibronectin	Collagen and laminin-coated fibers promoted significantly better adhesion	Collagen-coated nanofiber promoted wound healing
Mouse embryonic fibroblasts	3D nanofibers coated with collagen (CNF), laminin (LNF), or fibronectin (FNF)	CNF and FNF resulted in fibroblasts with in vivo morphology, less stress fibers	Downregulation in Rac for CNF and FNF coupled with decreased spreading
Osteoblasts	Nanopits 14, 29, and 45 nm	14 and 29 nm more spread, more focal adhesions	Paxilin, Vinculin, and FAK more highly expressed in 14 nm and 29 nm
MCF-7	350 nm nanometric grooves		RACK1 overexpressing cells exhibited more focal adhesions and less contact guidance
Osteoblasts	Nanopits 14, 29, and 45 nm	14 and 29 nm more spread, more focal adhesions	Paxilin, Vinculin, and FAK more highly expressed in 14 nm and 29 nm
Human mesenchymal stem cells (hMSCs)	ES nanofibers diameters ranging from 1 to 270 μm with varying porosity	Adhesion and proliferation varies with fiber diameter, porosity	Response best at 10 μm with nanopores, poor response at higher diameters
hMSCs	350 nm nanometric grooves	CSK and nucleus aligned and elongated along direction of grooves	Induction to neuronal lineage, response depends on size of features
hMSCs	Ordered and disordered nanopits diameter 120 nm and depth 100 nm	Cell spreading increased on disordered features	Induction to osteopathic lineage on disordered features

[a] See text for references.

Continuing with recent studies involving electrospun nanostructures, Ahmed et al. (2007) investigated the response of mouse embryonic fibroblasts (MEFs) on nanofibrillar surfaces composed of electrospun polyamide nanofibers and adsorbed with type I collagen, fibronectin, or laminin-1. The authors investigated morphology, CSK dynamics, and the distribution of myosin II-B on the three nanofibrillar surfaces as well as corresponding 2D surfaces composed of Aclar. Compared to flat coated surfaces, culture of MEFs on fibronectin and collagen I-coated 3D surfaces resulted in an almost complete loss of stress fibers. Instead, there was an abundance of thin actin fibrils aligning along the long axis of the cell to enforce an apical-ventral symmetry normally observed in fibroblasts *in vivo*. There were concurrent changes in the number and organization of focal adhesions and an upregulation in Rac expression (Nur et al., 2006) along with a downregulation of Rho and Cdc42. As is the case when there is an upregulation of Rac, the amount of spreading was greatly reduced. Laminin-1 did not alter fibroblast phenotype to nearly the extent induced by collagen I and fibronectin. This study again highlights the benefit of a biomimetic microenvironment offered by a nanofibrous scaffold with the appropriate topographical and biochemical cues.

33.2.4 Mechanism of Cellular Response to Nanotopography

Understanding of the underlying mechanism of cellular response to a given nanotopography is poor. A recent study examined the signal transduction mechanisms involved in the response of MCF-7 breast cancer cells to 350 nm grooves (Dalby et al., 2008). MCF-7 cells were transfected with either an empty vector (NEO cells) or with RACK1, a signaling protein that plays a central role in cell adhesion and contact guidance. The effect of nanotopography was much more pronounced for the NEO-transfected cells than the RACK1-transfected cells (Figure 33.9). Vinculin staining showed that NEO cells had few focal adhesions on flat and even less on grooved substrates, while RACK1 overexpressing cells showed significantly more

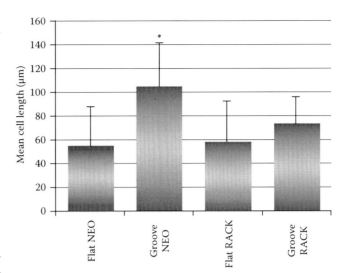

FIGURE 33.9 MCF-7 cells overexpressing RACK1 are compared with the empty vector (NEO) on flat versus 350 nm nanometric grooved substrate. * indicates statistically significant ($p < 0.05$) differences. Reported differences are compared to control (NEO cells on planar surface). (From Dalby, M.J. et al., *Biomaterials*, 29, 282, 2008. With permission.)

focal adhesions either way. When RACK1 was depleted with an antisense vector, the cells elongated even more than NEO cells on the same nanogrooved substrates, demonstrating fewer lamellipodia on flat surfaces and almost none on grooved substrates (Figure 33.10). These findings indicate that RACK1 has an inhibitory effect on contact guidance, and that a reduction in focal contacts and disassembly of actin stress fibers is necessary for contact guidance. Another report states that topography in the nanoscale is imprinted in the cell CSK and this imprinting is dependant on the integrin β3 subunit (Wood et al., 2008). When combined with the observation that Src activity is inhibited in response to alignment on microgrooved substrates (Hamilton and Brunette, 2007), it is becoming increasingly clear that integrins and the downstream effectors Src and RACK1 provide an important regulation pathway in the cellular response to topographical cues.

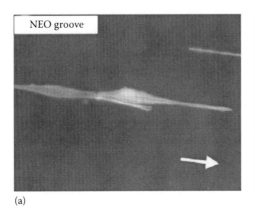
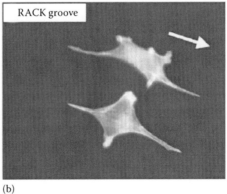

(a) (b)

FIGURE 33.10 Morphological differences between NEO cells (a) and RACK1 overexpressing cells on nanogrooved substrate (b) (white arrow indicates direction of grooves). (From Dalby, M.J. et al., *Biomaterials*, 29, 282, 2008. With permission.)

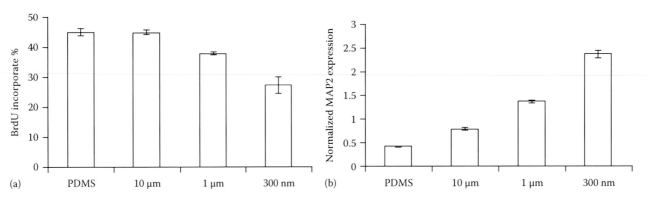

FIGURE 33.11 The proliferation of hMSCs (a) decreases with decreasing feature size while the MAP2 expression, which indicates neuronal induction, increases (b). (From Yim, E.K. et al., *Exp. Cell Res.*, 313, 1820, 2007. With permission.)

33.2.5 Potential of Nanotopography in Directing Stem Cell Differentiation

In terms of response to nanostructures, one of the most interesting cell types with clear interest in tissue and cell engineering is human mesenchymal stem cells, known to have the capacity to differentiate into cells of the stromal lineage including adipogenic (fat), chondrogenic (cartilage), osteoblastic (bone), myoblastic (muscle), and fibroblastic (connective tissue) (Bianco and Robey, 2001; Oreffo et al., 2005). In an early study on the effect of electrospun scaffolds on human mesenchymal stem cells (hMSCs), Moroni et al. (2006) fabricated scaffolds from poly(ethylene oxide terephthalate)-poly(butylene terephthalate) (PEOT/PBT) copolymers (see Figure 33.7b). The authors found that cell attachment and proliferation were highly dependent on fiber diameter ranging from approximately 1 to 270 μm. The cells showed the highest level of adhesion and proliferation when cultured on 10 μm fibers, followed by smaller fibers (1 and 4 μm) and then significantly lower on the larger fibers (21 and 270 μm). The 10 μm fibrous surface also induced significantly better adhesion and proliferation than its nonporous counterpart. It is interesting that fiber curvature on the order of cell size produces the maximal cellular response. The authors noted, however, that the fiber diameter closest to the cell size, 21 μm, showed poor cellular response, and this poor response is probably for different reasons than that observed on 270 μm fibers (which is essentially flat to a cell). It could be that cell motility is enhanced when the fiber diameter is close to the cell dimension (Smilenov et al., 1999), resulting in a lower proliferation rate.

Yim et al. (2007) studied the effect of a PDMS nanograting of linewidth 350 nm on hMSCs (Figure 33.7c and d). The CSK and nucleus of the hMSCs were observed to align and elongate along the direction of the grating. Similar to the earlier report on SMCs (Yim et al., 2005), proliferation of hMSCs was significantly reduced on the nanopatterns as compared to a flat control. At the same time, and most interestingly, there was a significant upregulation of markers such as MAP2, Tuj1, and GFAP that are indicative of induction into a neuronal lineage for hMSCs cultured on nanopatterns versus flat control. Another interesting result is that, as the grating size was reduced from 10 μm to 1 μm to 300 nm, the proliferation decreased and the normalized

MAP2 expression increased (Figure 33.11) indicating that there is a functional trade-off between proliferation and differentiation. Moreover, the upregulation of MAP2 expression was greater on nanograting than retinoic acid on a flat substrate, suggesting that the nanotopographical cue in this case is more pronounced than the biochemical cue. Also, these results indicate that, for differentiation to occur, the cues must be at the nanoscale. This study highlights the potential of applying nanotopographical cue to influence the differentiation of stem cells.

In another study, Dalby et al. (2007b) investigated the response of hMSCs to nantopographical cues using polymer demixing of PS and PBrS to synthesize nanopits of 120 and 100 nm deep. Five nanopitted patterns were developed with an average center-to-center spacing of 300 nm: A square array (SQ); hexagonal array (HEX); a disordered square array with pits displaced randomly by up to 50 or 20 nm on both axes from their position in a true square (DSQ50 and DSQ20, respectively); and pits placed randomly over a 150 μm by 150 μm field (RAND, see Figure 33.7d). The goal of this study was to investigate whether hMSCs could demonstrate osteospecific differentiation. Based on studies of osteoprogenitors and hMSCs, the authors concluded that DSQ50 showed the highest level of osteogenic induction versus planar controls and the other patterns. hMSCs were then cultured on DSQ50 and compared to flat controls with or without dexamethasone (DEX), a corticosteroid that can induce bone formation. Markers associated with osteospecific differentiation were significantly upregulated on DSQ50 samples, although not as much as they were upregulated for DEX samples. This study indicates that disordered nanotopographical cues may be able to induce differentiation; it is interesting that osteospecific differentiation was not observed on RAND or ordered samples.

33.3 Summary

There are two ideas we would like to highlight in the discussion of cellular response to nanostructures: first, the mechanisms and pathways of the induction of force from the environment through the cell; and second the reorganization of the cell machinery that may be responsible for changing cell behavior in response to nanotopographical cues.

There are four elements to consider in summary of the mechanisms of force transduction, which have been reviewed in detail above and in Janmey and Weitz, (2004), Dalby (2005), and Vogel and Sheetz (2006). The first element is the ECM/basement membrane (ECM/BM) that acts as a substrate for the cell. The stiffness and chemical constitution of the ECM/BM dictates in large part how the cells behave. Small changes in the organization or chemical cues in the ECM/BM will be transmitted via the integrins to the cell.

The second consideration is the cell–matrix contacts which act as the mediator of cellular response to external cues from the substrate. Both the number of contacts, their spacing, and the size of the contacts and their turnover rate modulates cell morphology. The number and size of cell–matrix contacts influence cell spreading due to the correlation with formation of stress fibers, which largely regulates adhesion, proliferation (Ahmed et al., 2007), and differentiation (Dalby et al., 2007b). The number and size of cell–matrix contacts also influence the growth of MTs (Kaverina et al., 2002) although it is unclear from the literature whether this is a direct or indirect effect. The size of cell–matrix contacts influences downstream signaling of G-proteins (Discher et al., 2005), which regulate spreading and motility (Ballestrem et al., 2001). Importantly, the formation of focal adhesions modulates tension in the cytoskeletal network, which is an essential indicator and effector of function (Ingber, 2003a).

The third element to consider is the CSK, which is tethered to ECM/BM proteins via cell–matrix connections. The CSK reorganizes in response to cues from the environment. This reorganization can include, among other events, fluidization in response to high-frequency stress (Deng et al., 2006), increased tension in the CSK network in response to cell anchorage and formation of focal adhesions (Discher et al., 2005), or increased turnover in response to motile behavior (Ballestrem et al., 2001). The CSK is a highly inhomogeneous network that transmits forces through the whole cell (Wang and Suo, 2005), and it appears that some level of redundancy is exercised in the cell to ensure that the entire cell responds to a locally applied force (Forgacs, 1995) or reorganization of the ECM/BM. While microfilaments and MTs are associated with cell–matrix contacts and the structural integrity of the cell, IFs form a direct, solid-state connection through which signals propagate to the cell nucleus (Maniotis et al., 1997; Hu et al., 2005).

The fourth element to consider is the cell nucleus, which is stiffer and more viscous than the cytoplasm (Guilak et al., 2000; de Vries et al., 2007), is highly integrated into the structure and mechanical properties of the cell as a whole (Caille et al., 2002), and appears to act as a molecular shock absorber with a strict compression limit (Dahl et al., 2004). The nuclear envelope comprises a stiff lamin network (Goldman et al., 2002) through which the more viscous chromatin can move and into which chromatin can bind. Due to the charged nature of chromatins and the fact that they are mechanically integrated with the CSK through lamins, they are highly sensitive to the osmotic environment (Guilak, 1995) as well as forces transduced from the environment (Dalby et al., 2007a). The cell nucleus, in particular

its mechanical properties and its response to external forces, is extremely important. The viscoelasticity of the cell nucleus has been associated with cancer (Backman et al., 2000), apoptosis (Kerr et al., 1972), and stem cell differentiation (Constantinescu et al., 2006), and the response of the nucleus to environmental stimuli is an active area of research (Cui and Bustamante, 2000; Dahl et al., 2005; Dahl et al., 2006; Chalut et al., 2008).

With the above four elements in mind, we discuss the second idea, which is the reorganization of the probable machinery of changes in cell behavior due to nanotopographical cues. Regarding this machinery, there are several considerations. First, previous work has demonstrated that MTs specifically invade regions of applied force (Kaverina et al., 2002) and applied tension results in a greater number of kinetochore MTs (King and Nicklas, 2000). Though there are limits to this increase, it could potentially enhance the possibility of CSK binding to chromosomes. Second, IFs transmit force directly to chromatin through the lamin network, and IFs reorient in response to CSK reorganization, leading to nuclear distortion and nucleoli rearrangement along the applied axis of the force (Maniotis et al., 1997). Lamins which connect chromatin are a highly elastic network; therefore, mechanical stress will change the organization of lamins and, therefore, the distribution and spatial position of chromatin (Dalby et al., 2007a). Third, the position of chromosomes is highly influential on cell function. The relative position of chromosomes is generally consistent (Heslop-Harrison, 2000) and is likely compartmentalized into discrete territories (Cremer and Cremer, 2001). The location of a gene within a chromosome territory affects its access to mechanisms responsible for specific cellular functions; therefore, changes in position affect genome regulation.

In summary, the hypothesis is as follows (Figure 33.12). Nanotopographical cues can simulate ECM/BM, which in turn affects the formation of cell–matrix contacts and therefore also affects organization of the CSK. The reorganization of the CSK has two important effects: first, MTs invade regions of applied stress increasing the number of kinetochore MTs and affecting cell division and structure; second, formation of stress fibers can reorient the IF network, which is directly linked with centromeres, whose relative position is highly influential on genome regulation. These effects could be responsible for any changes in gene expression presented by cells due to nanotopographical cues.

We would like to present several thoughts regarding future research directions. First, the link between nanotopographical cues and nuclear deformation should be further examined. There is a gap in the research when it comes to mechanism: there is not yet a clear experimentally observed link between nanotopographical cues, alteration of chromatin position, and cell function. Linking nuclear shape to nanotopographical cues could assist in filling that gap. Second, the cell line, and its origin, is important. A certain set of substrate parameters may result in a high level of response for SMCs yet promote poor response in human mesenchymal stem cells. Further, the response of a particular type of cell will vary depending on whether or not it is a primary or established cell line. A third thought, which has been

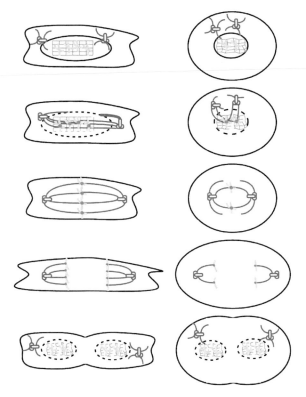

FIGURE 33.12 Depiction of the effect ... under minimal applied stress (right). The microtubule organizing centers are on the periphery of the cells with filamentous extensions, the lamin network is depicted as a grid in the nucleus, the centromeres are X-shaped structures in light gray in the grid, and the kinetochore microtubules are the circle in the middle of the centromeres. Applied stress can increase number of kinetochore microtubules and relative position of centromeres that are part of lamin network as well as microtubule organizing centers. The relative position of chromosomes within a chromosome territory affects genome regulation and therefore cell function. This may, in part, help explain the effect of nanotopographical cues on cell function.

discussed at length above, is that cells respond to, in no particular order, the surface chemistry, surface tension, shape (including the dimensionality), and texture of their substrate. It is clear from the results presented here that cells will respond to these four parameters—surface chemistry, surface tension, shape, and texture—largely according to the parameters they encounter in their *in vivo* environment. For instance, osteoblasts will tend to respond most favorably to stiff environments with nanofeatures similar to the topology of the matrix within the skeleton, which is around 30 nm (Palin et al., 2005). The key to engineering function in cells by manipulation of the substrate will depend on the ability to control these four parameters and understand the ways in which they relate to one another.

33.4 Future Perspectives

Further studies are necessary to address the underlying mechanisms of a cellular response to a given nanotopography. Understanding these mechanisms will lead to the design

of biomaterials that can elicit a given cellular response. This is especially attractive for implants. For instance, a surface topography on an implanted medical device that is able to reduce the inflammatory immune response and subsequent scar formation will lead to an increased life span of the device.

Surfaces with nanotopographical features and nanofibers serve as novel tools to unravel basic biological questions of cellular adhesion, adhesion-mediated cellular signaling, and mechanotransduction. Mechanical factors important for determining a cellular response are nanometer-level spacing of molecules, curvature and rigidity of the substrate, and ECM components. One of these molecules is talin, which has integrin-binding sites along 55 nm (Zhang et al., 2008). Defects in mechanotransduction underlie many diseases such as cancers, immune disorders, genetic malformations, and neuropathies. The tools of nanotechnology and modern cell biology allow us to investigate many of the physical aspects of complex cellular processes at the micrometer and nanometer scales.

References

Ahmed, I., Ponery, A. S., Nur, E. K. A., Kamal, J., Meshel, A. S., Sheetz, M. P., Schindler, M., and Meiners, S. (2007) Morphology, cytoskeletal organization, and myosin dynamics of mouse embryonic fibroblasts cultured on nanofibrillar surfaces. *Mol Cell Biochem*, 301, 241–249.

Alberts, B. (2008) *Molecular Biology of the Cell*, New York, Garland Science.

Andersson, A. S., Backhed, F., Von Euler, A., Richter-Dahlfors, A., Sutherland, D., and Kasemo, B. (2003) Nanoscale features influence epithelial cell morphology and cytokine production. *Biomaterials*, 24, 3427–3436.

Arnold, M., Cavalcanti-Adam, E. A., Glass, R., Blummel, J., Eck, W., Kantlehner, M., Kessler, H., and Spatz, J. P. (2004) Activation of integrin function by nanopatterned adhesive interfaces. *Chemphyschem*, 5, 383–388.

Backman, V., Wallace, M. B., Perelman, L. T., Arendt, J. T., Gurjar, R., Muller, M. G., Zhang, Q. et al. (2000) Detection of preinvasive cancer cells. *Nature*, 406, 35–36.

Ballestrem, C., Hinz, B., Imhof, B. A., and Wehrle-Haller, B. (2001) Marching at the front and dragging behind: Differential alphaVbeta3-integrin turnover regulates focal adhesion behavior. *J Cell Biol*, 155, 1319–1332.

Bianco, P. and Robey, P. G. (2001) Stem cells in tissue engineering. *Nature*, 414, 118–121.

Biggs, M. J., Richards, R. G., Gadegaard, N., Wilkinson, C. D., and Dalby, M. J. (2007) The effects of nanoscale pits on primary human osteoblast adhesion formation and cellular spreading. *J Mater Sci Mater Med*, 18, 399–404.

Bischofs, I. B. and Schwarz, U. S. (2003) Cell organization in soft media due to active mechanosensing. *Proc Natl Acad Sci U S A*, 100, 9274–9279.

Bischofs, I. B., Safran, S. A., and Schwarz, U. S. (2004) Elastic interactions of active cells with soft materials. *Phys Rev E Stat Nonlin Soft Matter Phys*, 69, 021911.

Bursac, P., Lenormand, G., Fabry, B., Oliver, M., Weitz, D. A., Viasnoff, V., Butler, J. P., and Fredberg, J. J. (2005) Cytoskeletal remodelling and slow dynamics in the living cell. *Nat Mater*, 4, 557–561.

Caille, N., Thoumine, O., Tardy, Y., and Meister, J. J. (2002) Contribution of the nucleus to the mechanical properties of endothelial cells. *J Biomech*, 35, 177–187.

Cerrina, F. and Marrian, C. (1996) A path to nanolithography. *MRS Bull*, 21, 56–62.

Chalut, K. J., Chen, S., Finan, J. D., Giacomelli, M. G., Guilak, F., Leong, K. W., and Wax, A. (2008) Label-free, high-throughput measurements of dynamic changes in cell nuclei using angle-resolved low coherence interferometry. *Biophys J*, 94, 4948–4956.

Charras, G. T. and Horton, M. A. (2002) Single cell mechanotransduction and its modulation analyzed by atomic force microscope indentation. *Biophys J*, 82, 2970–2981.

Chemin, J., Patel, A. J., Duprat, F., Lauritzen, I., Lazdunski, M., and Honore, E. (2005) A phospholipid sensor controls mechanogating of the K+ channel TREK-1. *Embo J*, 24, 44–53.

Chen, C. S., Mrksich, M., Huang, S., Whitesides, G. M., and Ingber, D. E. (1997) Geometric control of cell life and death. *Science*, 276, 1425–1428.

Chew, S. Y., Wen, Y., Dzenis, Y., and Leong, K. W. (2006) The role of electrospinning in the emerging field of nanomedicine. *Curr Pharm Des*, 12, 4751–4770.

Constantinescu, D., Gray, H. L., Sammak, P. J., Schatten, G. P., and Csoka, A. B. (2006) Lamin A/C expression is a marker of mouse and human embryonic stem cell differentiation. *Stem Cells*, 24, 177–185.

Cremer, T. and Cremer, C. (2001) Chromosome territories, nuclear architecture and gene regulation in mammalian cells. *Nat Rev Genet*, 2, 292–301.

Cui, Y. and Bustamante, C. (2000) Pulling a single chromatin fiber reveals the forces that maintain its higher-order structure. *Proc Natl Acad Sci U S A*, 97, 127–132.

Curtis, A.S. and Clark, P. (1990) The effects of topographic and mechanical properties of materials on cell behavior. *Crit Rev Biocompat*, 5, 343–363.

Curtis, A. S. and Seehar, G. M. (1978) The control of cell division by tension or diffusion. *Nature*, 274, 52–53.

Curtis, A. S. and Wilkinson, C.D. (1998) Reaction of cells to topography. *J Biomater Sci*, 9, 1313–1329.

Curtis, A.S., Casey, B., Gallagher, J.O., Pasqui, D., Wood, M.A., and Wilkinson, C.D. (2001) Substratum nanotopography and the adhesion of biological cells. Are symmetry or regularity of nanotopography important? *Biophys Chem*, 94, 275–283.

Dahl, K. N., Kahn, S. M., Wilson, K. L., and Discher, D. E. (2004) The nuclear envelope lamina network has elasticity and a compressibility limit suggestive of a molecular shock absorber. *J Cell Sci*, 117, 4779–4786.

Dahl, K. N., Engler, A. J., Pajerowski, J. D., and Discher, D. E. (2005) Power-law rheology of isolated nuclei with deformation mapping of nuclear substructures. *Biophys J*, 89, 2855–2864.

Dahl, K. N., Scaffidi, P., Islam, M. F., Yodh, A. G., Wilson, K. L., and Misteli, T. (2006) Distinct structural and mechanical properties of the nuclear lamina in Hutchinson-Gilford progeria syndrome. *Proc Natl Acad Sci U S A*, 103, 10271–10276.

Dalby, M. J. (2005) Topographically induced direct cell mechanotransduction. *Med Eng Phys*, 27, 730–742.

Dalby, M. J., Riehle, M. O., Johnstone, H., Affrossman, S., and Curtis, A. S. (2002a) In vitro reaction of endothelial cells to polymer demixed nanotopography. *Biomaterials*, 23, 2945–2954.

Dalby, M. J., Yarwood, S. J., Riehle, M. O., Johnstone, H. J., Affrossman, S., and Curtis, A. S. (2002b) Increasing fibroblast response to materials using nanotopography: Morphological and genetic measurements of cell response to 13-nm-high polymer demixed islands. *Exp Cell Res*, 276, 1–9.

Dalby, M. J., Childs, S., Riehle, M. O., Johnstone, H. J., Affrossman, S., and Curtis, A. S. (2003a) Fibroblast reaction to island topography: Changes in cytoskeleton and morphology with time. *Biomaterials*, 24, 927–935.

Dalby, M. J., Riehle, M. O., Yarwood, S. J., Wilkinson, C. D., and Curtis, A. S. (2003b) Nucleus alignment and cell signaling in fibroblasts: Response to a micro-grooved topography. *Exp Cell Res*, 284, 274–282.

Dalby, M. J., Berry, C. C., Riehle, M. O., Sutherland, D. S., Agheli, H., and Curtis, A. S. (2004a) Attempted endocytosis of nano-environment produced by colloidal lithography by human fibroblasts. *Exp Cell Res*, 295, 387–394.

Dalby, M. J., Gadegaard, N., Riehle, M. O., Wilkinson, C. D., and Curtis, A. S. (2004b) Investigating filopodia sensing using arrays of defined nano-pits down to 35 nm diameter in size. *Int J Biochem Cell Biol*, 36, 2005–2015.

Dalby, M. J., Giannaras, D., Riehle, M. O., Gadegaard, N., Affrossman, S., and Curtis, A. S. (2004c) Rapid fibroblast adhesion to 27 nm high polymer demixed nano-topography. *Biomaterials*, 25, 77–83.

Dalby, M. J., Riehle, M. O., Sutherland, D. S., Agheli, H., and Curtis, A. S. (2004d) Changes in fibroblast morphology in response to nano-columns produced by colloidal lithography. *Biomaterials*, 25, 5415–5422.

Dalby, M. J., Riehle, M. O., Sutherland, D. S., Agheli, H., and Curtis, A. S. (2004e) Use of nanotopography to study mechanotransduction in fibroblasts–methods and perspectives. *Eur J Cell Biol*, 83, 159–169.

Dalby, M. J., Biggs, M. J., Gadegaard, N., Kalna, G., Wilkinson, C. D., and Curtis, A. S. (2007a) Nanotopographical stimulation of mechanotransduction and changes in interphase centromere positioning. *J Cell Biochem*, 100, 326–338.

Dalby, M. J., Gadegaard, N., Tare, R., Andar, A., Riehle, M. O., Herzyk, P., Wilkinson, C. D., and Oreffo, R. O. (2007b) The control of human mesenchymal cell differentiation using nanoscale symmetry and disorder. *Nat Mater*, 6, 997–1003.

Dalby, M. J., Hart, A., and Yarwood, S. J. (2008) The effect of the RACK1 signalling protein on the regulation of cell adhesion and cell contact guidance on nanometric grooves. *Biomaterials*, 29, 282–289.

Davis, M. J. and Hill, M. A. (1999) Signaling mechanisms underlying the vascular myogenic response. *Physiol Rev*, 79, 387–423.

De Vries, A. H., Krenn, B. E., Van Driel, R., Subramaniam, V., and Kanger, J. S. (2007) Direct observation of nanomechanical properties of chromatin in living cells. *Nano Lett*, 7, 1424–1427.

Deng, L. H., Trepat, X., Butler, J. P., Millet, E., Morgan, K. G., Weitz, D. A., and Fredberg, J. J. (2006) Fast and slow dynamics of the cytoskeleton. *Nat Mater*, 5, 636–640.

Deroanne, C. F., Lapiere, C. M., and Nusgens, B. V. (2001) In vitro tubulogenesis of endothelial cells by relaxation of the coupling extracellular matrix-cytoskeleton. *Cardiovasc Res*, 49, 647–658.

Desprat, N., Richert, A., Simeon, J., and Asnacios, A. (2005) Creep function of a single living cell. *Biophys J*, 88, 2224–2233.

Discher, D. E., Janmey, P., and Wang, Y. L. (2005) Tissue cells feel and respond to the stiffness of their substrate. *Science*, 310, 1139–1143.

Donahue, H. J. (2000) Gap junctions and biophysical regulation of bone cell differentiation. *Bone*, 26, 417–422.

Dunn, G. A. and Brown, A. F. (1986) Alignment of fibroblasts on grooved surfaces described by a simple geometric transformation. *J Cell Sci*, 83, 313–340.

Dunn, G. A. and Heath, J. P. (1976) A new hypothesis of contact guidance in tissue cells. *Exp Cell Res*, 101, 1–14.

Engler, A., Bacakova, L., Newman, C., Hategan, A., Griffin, M., and Discher, D. (2004) Substrate compliance versus ligand density in cell on gel responses. *Biophys J*, 86, 617–628.

Engler, A. J., Sen, S., Sweeney, H. L., and Discher, D. E. (2006) Matrix elasticity directs stem cell lineage specification. *Cell*, 126, 677–689.

Fabry, B., Maksym, G. N., Butler, J. P., Glogauer, M., Navajas, D., and Fredberg, J. J. (2001) Scaling the microrheology of living cells. *Phys Rev Lett*, 87(14), 148102:1-4.

Fabry, B., Maksym, G. N., Butler, J. P., Glogauer, M., Navajas, D., Taback, N. A., Millet, E. J., and Fredberg, J. J. (2003) Time scale and other invariants of integrative mechanical behavior in living cells. *Phys Rev E*, 68, 041914:1-18.

Falconnet, D., Csucs, G., Grandin, H. M., and Textor, M. (2006) Surface engineering approaches to micropattern surfaces for cell-based assays. *Biomaterials*, 27, 3044–3063.

Forgacs, G. (1995) On the possible role of cytoskeletal filamentous networks in intracellular signaling: An approach based on percolation. *J Cell Sci*, 108, 2131–2143.

Fuller, B. (1961) Tensegrity. *Portfolio Artnews Annual*, 4, 112–127.

Gardel, M. L., Valentine, M. T., Crocker, J. C., Bausch, A. R., and Weitz, D. A. (2003) Microrheology of entangled F-actin solutions. *Phys Rev Lett*, 91, 158302.

Gardel, M. L., Shin, J. H., Mackintosh, F. C., Mahadevan, L., Matsudaira, P. A., and Weitz, D. A. (2004) Scaling of F-actin network rheology to probe single filament elasticity and dynamics. *Phys Rev Lett*, 93, 188102:1-4.

Goldman, R. D., Gruenbaum, Y., Moir, R. D., Shumaker, D. K., and Spann, T. P. (2002) Nuclear lamins: Building blocks of nuclear architecture. *Genes Dev*, 16, 533–547.

Guilak, F. (1995) Compression-induced changes in the shape and volume of the chondrocyte nucleus. *J Biomech*, 28, 1529–1541.

Guilak, F., Tedrow, J. R., and Burgkart, R. (2000) Viscoelastic properties of the cell nucleus. *Biochem Biophys Res Commun*, 269, 781–786.

Habermann, B. (2004) The BAR-domain family of proteins: A case of bending and binding? *EMBO Rep*, 5, 250–255.

Hamilton, D. W. and Brunette, D. M. (2007) The effect of substratum topography on osteoblast adhesion mediated signal transduction and phosphorylation. *Biomaterials*, 28, 1806–1819.

Hansen, J. C., Lim, J. Y., Xu, L. C., Siedlecki, C. A., Mauger, D. T., and Donahue, H. J. (2007) Effect of surface nanoscale topography on elastic modulus of individual osteoblastic cells as determined by atomic force microscopy. *J Biomech*, 40, 2865–2871.

Harrison, R. G. (1911) On the stereotropism of embryonic cells. *Science*, 34, 279–281.

Helmke, B. P., Rosen, A. B., and Davies, P. F. (2003) Mapping mechanical strain of an endogenous cytoskeletal network in living endothelial cells. *Biophys J*, 84, 2691–2699.

Heslop-Harrison, J. S. (2000) Comparative genome organization in plants: From sequence and markers to chromatin and chromosomes. *Plant Cell*, 12, 617–636.

Hu, S., Chen, J., Butler, J. P., and Wang, N. (2005) Prestress mediates force propagation into the nucleus. *Biochem Biophys Res Commun*, 329, 423–428.

Ingber, D. E. (1993) Cellular tensegrity: Defining new rules of biological design that govern the cytoskeleton. *J Cell Sci*, 104, 613–627.

Ingber, D. E. (2003a) Tensegrity I. Cell structure and hierarchical systems biology. *J Cell Sci*, 116, 1157–1173.

Ingber, D. E. (2003b) Tensegrity II. How structural networks influence cellular information processing networks. *J Cell Sci*, 116, 1397–1408.

Janmey, P. A. (1998) The cytoskeleton and cell signaling: Component localization and mechanical coupling. *Physiol Rev*, 78, 763–781.

Janmey, P. A. and Weitz, D. A. (2004) Dealing with mechanics: Mechanisms of force transduction in cells. *Trends Biochem Sci*, 29, 364–370.

Janmey, P. A., Hvidt, S., Lamb, J., and Stossel, T. P. (1990) Resemblance of actin-binding protein actin gels to covalently cross-linked networks. *Nature*, 345, 89–92.

Janmey, P. A., Euteneuer, U., Traub, P., and Schliwa, M. (1991) Viscoelastic properties of vimentin compared with other filamentous biopolymer networks. *J Cell Biol*, 113, 155–160.

Johansson, F., Carlberg, P., Danielsen, N., Montelius, L., and Kanje, M. (2006) Axonal outgrowth on nano-imprinted patterns. *Biomaterials*, 27, 1251–1258.

Katsumi, A., Milanini, J., Kiosses, W. B., Del Pozo, M. A., Kaunas, R., Chien, S., Hahn, K. M., and Schwartz, M. A. (2002) Effects of cell tension on the small GTPase Rac. *J Cell Biol*, 158, 153–164.

Kaverina, I., Krylyshkina, O., Beningo, K., Anderson, K., Wang, Y. L., and Small, J. V. (2002) Tensile stress stimulates microtubule outgrowth in living cells. *J Cell Sci*, 115, 2283–2291.

Kerr, J. F., Wyllie, A. H., and Currie, A. R. (1972) Apoptosis: A basic biological phenomenon with wide-ranging implications in tissue kinetics. *Br J Cancer*, 26, 239–257.

Kim, S. and Coulombe, P. A. (2007) Intermediate filament scaffolds fulfill mechanical, organizational, and signaling functions in the cytoplasm. *Genes Dev*, 21, 1581–1597.

King, J. M. and Nicklas, R. B. (2000) Tension on chromosomes increases the number of kinetochore microtubules but only within limits. *J Cell Sci*, 113 Pt 21, 3815–3823.

Krylyshkina, O., Anderson, K. I., Kaverina, I., Upmann, I., Manstein, D. J., Small, J. V., and Toomre, D. K. (2003) Nanometer targeting of microtubules to focal adhesions. *J Cell Biol*, 161, 853–859.

Kwon, K. W., Choi, S. S., Lee, S. H., Kim, B., Lee, S. N., Park, M. C., Kim, P., Hwang, S. Y., and Suh, K. Y. (2007) Label-free, microfluidic separation and enrichment of human breast cancer cells by adhesion difference. *Lab Chip*, 7, 1461–1468.

Lee, C. H., Shin, H. J., Cho, I. H., Kang, Y. M., Kim, I. A., Park, K. D., and Shin, J. W. (2005) Nanofiber alignment and direction of mechanical strain affect the ECM production of human ACL fibroblast. *Biomaterials*, 26, 1261–1270.

Lim, J. Y., Dreiss, A. D., Zhou, Z., Hansen, J. C., Siedlecki, C. A., Hengstebeck, R. W., Cheng, J., Winograd, N., and Donahue, H. J. (2007) The regulation of integrin-mediated osteoblast focal adhesion and focal adhesion kinase expression by nanoscale topography. *Biomaterials*, 28, 1787–1797.

Lo, C. M., Wang, H. B., Dembo, M., and Wang, Y. L. (2000) Cell movement is guided by the rigidity of the substrate. *Biophys J*, 79, 144–152.

Maniotis, A. J., Chen, C. S., and Ingber, D. E. (1997) Demonstration of mechanical connections between integrins cytoskeletal filaments, and nucleoplasm that stabilize nuclear structure. *Proc Natl Acad Sci U S A*, 94, 849–854.

Martines, E., McGhee, K., Wilkinson, C., and Curtis, A. (2004) A parallel-plate flow chamber to study initial cell adhesion on a nanofeatured surface. *IEEE Trans Nanobiosci*, 3, 90–95.

McBeath, R., Pirone, D. M., Nelson, C. M., Bhadriraju, K., and Chen, C. S. (2004) Cell shape, cytoskeletal tension, and RhoA regulate stem cell lineage commitment. *Dev Cell*, 6, 483–495.

Miller, D. C., Thapa, A., Haberstroh, K. M., and Webster, T. J. (2004) Endothelial and vascular smooth muscle cell function on poly(lactic-co-glycolic acid) with nano-structured surface features. *Biomaterials*, 25, 53–61.

Mizushima-Sugano, J., Maeda, T., and Miki-Noumura, T. (1983) Flexural rigidity of singlet microtubules estimated from statistical analysis of their contour lengths and end-to-end distances. *Biochim Biophys Acta*, 755, 257–262.

Mo, X. M., Xu, C. Y., Kotaki, M., and Ramakrishna, S. (2004) Electrospun P(LLA-CL) nanofiber: A biomimetic extracellular matrix for smooth muscle cell and endothelial cell proliferation. *Biomaterials*, 25, 1883–1890.

Moroni, L., Licht, R., De Boer, J., De Wijn, J. R., and Van Blitterswijk, C. A. (2006) Fiber diameter and texture of electrospun PEOT/PBT scaffolds influence human mesenchymal stem cell proliferation and morphology, and the release of incorporated compounds. *Biomaterials*, 27, 4911–4922.

Morse, D. C. (1998) Viscoelasticity of tightly entangled solutions of semiflexible polymers. *Phys Rev E*, 58, R1237-R1240.

Nobes, C. D. and Hall, A. (1995) Rho, rac, and cdc42 GTPases regulate the assembly of multimolecular focal complexes associated with actin stress fibers, lamellipodia, and filopodia. *Cell*, 81, 53–62.

Nur, E. K. A., Ahmed, I., Kamal, J., Schindler, M., and Meiners, S. (2006) Three-dimensional nanofibrillar surfaces promote self-renewal in mouse embryonic stem cells. *Stem Cells*, 24, 426–433.

Oreffo, R. O., Cooper, C., Mason, C., and Clements, M. (2005) Mesenchymal stem cells: Lineage, plasticity, and skeletal therapeutic potential. *Stem Cell Rev*, 1, 169–178.

Palin, E., Liu, H. N., and Webster, T. J. (2005) Mimicking the nanofeatures of bone increases bone-forming cell adhesion and proliferation. *Nanotechnology*, 16, 1828–1835.

Patel, A. J., Lazdunski, M., and Honore, E. (2001) Lipid and mechano-gated 2P domain K(+) channels. *Curr Opin Cell Biol*, 13, 422–428.

Pederson, T. and Aebi, U. (2005) Nuclear actin extends, with no contraction in sight. *Mol Biol Cell*, 16, 5055–5060.

Pelham, R. J., Jr. and Wang, Y. (1997) Cell locomotion and focal adhesions are regulated by substrate flexibility. *Proc Natl Acad Sci U S A*, 94, 13661–13665.

Pullarkat, P. A., Fernandez, P. A., and Ott, A. (2007) Rheological properties of the eukaryotic cell cytoskeleton. *Phys Rep-Rev Sect Phys Lett*, 449, 29–53.

Rho, K. S., Jeong, L., Lee, G., Seo, B. M., Park, Y. J., Hong, S. D., Roh, S., Cho, J. J., Park, W. H., and Min, B. M. (2006) Electrospinning of collagen nanofibers: Effects on the behavior of normal human keratinocytes and early-stage wound healing. *Biomaterials*, 27, 1452–1461.

Riveline, D., Zamir, E., Balaban, N. Q., Schwarz, U. S., Ishizaki, T., Narumiya, S., Kam, Z., Geiger, B., and Bershadsky, A. D. (2001) Focal contacts as mechanosensors: Externally applied local mechanical force induces growth of focal contacts by an mDia1-dependent and ROCK-independent mechanism. *J Cell Biol*, 153, 1175–1186.

Schwarz, U. S. and Bischofs, I. B. (2005) Physical determinants of cell organization in soft media. *Med Eng Phys*, 27, 763–772.

Schwarz, U. S., Erdmann, T., and Bischofs, I. B. (2006) Focal adhesions as mechanosensors: The two-spring model. *Biosystems*, 83, 225–232.

Shemesh, T., Geiger, B., Bershadsky, A. D., and Kozlov, M. M. (2005) Focal adhesions as mechanosensors: A physical mechanism. *Proc Natl Acad Sci U S A*, 102, 12383–12388.

Shipway, A. N., Katz, E., and Willner, I. (2000) Nanoparticle arrays on surfaces for electronic, optical, and sensor applications. *Chemphyschem*, 1, 18–52.

Smilenov, L. B., Mikhailov, A., Pelham, R. J., Marcantonio, E. E., and Gundersen, G. G. (1999) Focal adhesion motility revealed in stationary fibroblasts. *Science*, 286, 1172–1174.

Teixeira, A. I., Abrams, G. A., Bertics, P. J., Murphy, C. J., and Nealey, P. F. (2003) Epithelial contact guidance on well-defined micro- and nanostructured substrates. *J Cell Sci*, 116, 1881–1892.

Tint, I. S., Hollenbeck, P. J., Verkhovsky, A. B., Surgucheva, I. G., and Bershadsky, A. D. (1991) Evidence that intermediate filament reorganization is induced by ATP-dependent contraction of the actomyosin cortex in permeabilized fibroblasts. *J Cell Sci*, 98 (Pt 3), 375–384.

Toivola, D. M., Tao, G. Z., Habtezion, A., Liao, J., and Omary, M. B. (2005) Cellular integrity plus: Organelle-related and protein-targeting functions of intermediate filaments. *Trends Cell Biol*, 15, 608–617.

Trepat, X., Deng, L. H., An, S. S., Navajas, D., Tschumperlin, D. J., Gerthoffer, W. T., Butler, J. P., and Fredberg, J. J. (2007) Universal physical responses to stretch in the living cell. *Nature*, 447, 592–595.

Vogel, V. and Sheetz, M. (2006) Local force and geometry sensing regulate cell functions. *Nat Rev Mol Cell Biol*, 7, 265–275.

Wang, H. B., Dembo, M. and Wang, Y. L. (2000) Substrate flexibility regulates growth and apoptosis of normal but not transformed cells. *Am J Physiol Cell Physiol*, 279, C1345–C1350.

Wang, H. B., Dembo, M., Hanks, S. K., and Wang, Y. (2001) Focal adhesion kinase is involved in mechanosensing during fibroblast migration. *Proc Natl Acad Sci U S A*, 98, 11295–11300.

Wang, J. H. and Grood, E. S. (2000) The strain magnitude and contact guidance determine orientation response of fibroblasts to cyclic substrate strains. *Connect Tissue Res*, 41, 29–36.

Wang, N. and Suo, Z. G. (2005) Long-distance propagation of forces in a cell. *Biochem Biophys Res Commun*, 328, 1133–1138.

Weiss, P. and Garber, B. (1952) Shape and movement of mesenchyme cells as functions of the physical structure of the medium: Contributions to a quantitative morphology. *Proc Natl Acad Sci U S A*, 38, 264–280.

Wilkinson, C. D. Riehle, M., Wood, M., Gallagher, J., and Curtis A. S. (2002) The use of materials patterned on a nano- and micro-metric scale in cellular engineering. *Mater Sci Eng C*, 19, 263–269.

Wojciak-Stothard, B., Curtis, A.S., McGrath, M., Sommer, I., Wilkinson, C.D., and Monaghan, W. (1995) Role of the cytoskeleton in the reaction of fibroblasts to multiple grooved substrata. *Cell Motility Cytoskel*, 31, 147–158.

Wood, M. A. (2007) Colloidal lithography and current fabrication techniques producing in-plane nanotopography for biological applications. *J R Soc Interface*, 4, 1–17.

Wood, M. A., Bagnaninchi, P., and Dalby, M. J. (2008) The beta integrins and cytoskeletal nanoimprinting. *Exp Cell Res*, 314, 927–935.

Yeung, T., Georges, P. C., Flanagan, L. A., Marg, B., Ortiz, M., Funaki, M., Zahir, N., Ming, W., Weaver, V., and Janmey, P. A. (2005) Effects of substrate stiffness on cell morphology, cytoskeletal structure, and adhesion. *Cell Motil Cytoskeleton*, 60, 24–34.

Yi Dong, K., Kim Min, J., Turner, L., Breuer K., S., and Kim, D.-Y. (2006) Colloid lithography-induced polydimethylsiloxane microstructures and their application to cell patterning. *Biotechnol Lett*, 28, 169–173.

Yim, E. K., Reano, R. M., Pang, S. W., Yee, A. F., Chen, C. S., and Leong, K. W. (2005) Nanopattern-induced changes in morphology and motility of smooth muscle cells. *Biomaterials*, 26, 5405–5413.

Yim, E. K., Pang, S. W., and Leong, K. W. (2007) Synthetic nanostructures inducing differentiation of human mesenchymal stem cells into neuronal lineage. *Exp Cell Res*, 313, 1820–1829.

Zaidel-Bar, R., Ballestrem, C., Kam, Z., and Geiger, B. (2003) Early molecular events in the assembly of matrix adhesions at the leading edge of migrating cells. *J Cell Sci*, 116, 4605–4613.

Zamir, E. and Geiger, B. (2001) Molecular complexity and dynamics of cell-matrix adhesions. *J Cell Sci*, 114, 3583–3590.

Zhang, X., Jiang, G., Cai, Y., Monkley, S. J., Critchley, D. R., and Sheetz, M. P. (2008) Talin depletion reveals independence of initial cell spreading from integrin activation and traction. *Nat Cell Biol*, 10, 1062–1068.

Zimmerberg, J. and Kozlov, M. M. (2006) How proteins produce cellular membrane curvature. *Nat Rev Mol Cell Biol*, 7, 9–19.

VII

Cancer Therapy

34

Nanotechnology for Targeting Cancer

**Venkataramanan
Soundararajan**
*Massachusetts Institute
of Technology*

Ram Sasisekharan
*Massachusetts Institute
of Technology*

34.1 Introduction

There exists a delicate balance between life and death within all biological systems. It is estimated that an average of 60 billion cells undergo programmed cell death or apoptosis per day in the adult human body, counterbalancing cell division (Alberts 1983). This equilibrium facilitates the maintenance of an optimal density of cells in healthy tissue. Cancer or tumor is a disproportion that is pathologically characterized by uncontrollable cell division and apoptosis evasion. Tumors are sustained by reservoirs

of pro-growth signals and additionally have a distinct insensitivity to anti-growth signals (Hanahan and Weinberg 2000). Tumor tissues thrive on angiogenesis, which is the construction of new blood vessels. These hastily erected blood vessels ensure adequate supply of nutrients and oxygen for the greedy cancer cell colonies (Figg and Folkman 2008). Expanding tumors are typified by an obstinate lack of boundary consciousness and they often impinge on neighboring organelles, disrupting their normal function. Cancer cells, like pathogens, are further capable of metastasizing to non-neighboring organs by hitching onto lymphatic or blood vessels. Metastasis, which is characteristic of malignant tumors, permits the rapid colonization of multiple organs with tumor-favoring microenvironments (Harold 2001). Such an irrepressible spread of the disease is usually responsible for the eventual death of the organism.

A number of cancer-causing agents, also known as carcinogens, have been identified (Milman and Weisburger 1994). Sustained exposure to carcinogens causes irreparable genetic mutations, which increases the risk of cancer development. The carcinogenic potential of materials such as asbestos, arsenic, coal tar, heavy metals, tobacco, and dyes as well as environmental risk factors such as exposure to ultraviolet radiation is well characterized. Further, some pathogens have been recognized as cancer-causative agents (Table 34.1). For example, chronic infection with the human papillomavirus (HPV) has been linked with cervical cancer development (Wu 2004). Carcinogenic synergies between multiple agents have also been recorded. For instance, carcinogenic synergy between hepatitis B virus and aflatoxin has been observed to drastically amplify the possibility of liver cancer development (Zur Hausen et al. 2006). Owing to the ubiquitous presence of carcinogens as well as numerous cases of congenital susceptibility to cancer, the prevention of the disease in its totality is extremely challenging.

The detection of cancer at its earliest stages is essential for the effective treatment of the disease. The ability to monitor cellular microenvironments and subcellular compartments with high sensitivity is in turn fundamental to the timely recognition of tumorous metamorphosis. This necessitates the miniaturization of sensing technology. Furthermore, miniaturization is also essential for the design of effective anticancer therapeutic systems because the systemic administration of chemotherapy or anticancer chemicals results in high levels of toxicity to healthy tissues. With the advent of nanotechnology, there has been a surge of attempts to "sense" molecular signatures of cancer at its onset and "target" therapeutics to cancer cells.

TABLE 34.1 Pathogens with Suspected Roles in Promoting Carcinogenesis

Pathogen	Family	Cancer Type
Hepatitis B virus (HBV)	Hepadnaviridae	Liver
Hepatitis C virus (HCV)	Hepadnaviridae	Liver
Human papilloma virus (HPV)	Papillomaviridae	Cervix, anus, vulva, oropharynx
Simian virus 40 (SV40)	Polyomaviridae	Lung
Herpes simplex virus-2 (HSV-2)	Herpesviridae	Cervix
Epstein-Barr virus (EBV)	Herpesviridae	Nasopharynx, blood
Human T-lymphotropic virus (HTLV)	Retroviridae	Blood, bone marrow
Helicobacter pylori	Helicobacteraceae	Stomach, pancreas
Salmonella typhi	Enterobacteriaceae	Gallbladder
Streptococcus bovis	Streptococcaceae	Colorectum
Chlamydia pneumoniae	Chlamydiaceae	Lung
Mycoplasma	Mycoplasmataceae	Stomach, intestine, colon, esophagus

A nanometer is one billionth of a meter and devices that are less than hundreds of nanometers in dimension are known as nanodevices. Nanoscale devices are typically smaller than human cells and are able to achieve cellular entry. Such devices can further readily interact with biological macromolecules owing to their comparable dimensions. Nanotechnology, hence, presents an unprecedented insight into the complex regulatory and signaling network of biomolecular interactions that motivate cancerous transformation of normal cells. Nanotechnology has also provided significant impetus to the development of promising approaches for fighting cancer on various fronts (Figure 34.1).

This chapter describes the full spectrum of nanodevices for molecular diagnostics and therapeutics, specifically highlighting the state-of-the-art nanoscale techniques for targeting cancer. On the diagnostics front, the development of nanoscale systems for the molecular sensing of tumorigenesis fingerprints and for *in vivo* imaging of the tumor microenvironment is discussed. On the therapeutics front, the development of nanoscale platforms for targeted drug delivery, combination chemotherapy, and sustained drug release is described. Nanoscale systems for other purposes such as multidrug resistance reversal and personalized medicine are then discussed with emphasis on the significance of computational nanotechnology in this post-genomic era. The chapter concludes with a summary of opportunities and challenges involving the translation of nanoscale anticancer devices to the clinical realm.

34.2 Nanoscale Molecular Sensing and Early Malignancy Detection

The exquisite sensitivity of nanoscale technology and its ability to monitor subcellular compartments presents clinicians with a paradigm shift in the identification of tumors at elementary stages (Amiji 2006). The design of nanoscale carcinogenesis detection systems is motivated by advancements in molecular sensing, nanoelectromechanical systems (NEMS), nanofluidics, and ultrasensitive imaging technologies. This section showcases these emerging nanoscale opportunities for the timely detection of cancer.

34.2.1 Molecular Combing for Unearthing Genomic Instability

Molecular combing is a prospective high-resolution technology that provides for the linearization and alignment of deoxyribonucleic acid (DNA) molecules thus permitting a thorough analysis of the encoded genetic information. This technique is useful for unearthing mutated genomic domains that may have been missed by conventional sequencing–based screens in cancer patients (Weitzman 2001). Since conventional screenings produce considerable proportion of false negative diagnoses, "cancer combing" is a major step forward toward the development of foolproof procedures for the identification of mutations that arise during the genesis of cancer.

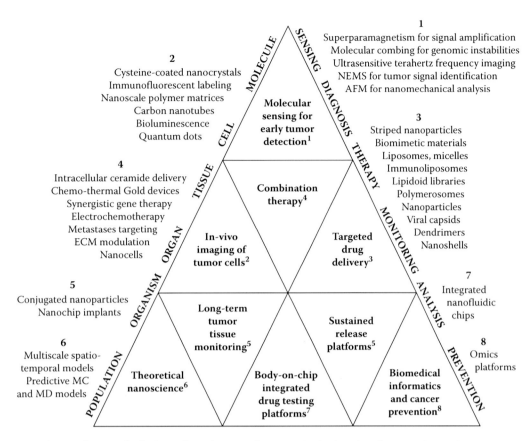

FIGURE 34.1 Compilation of nanoscale platforms for early tumor detection, *in vivo* imaging of tumor microenvironment, targeted drug delivery, sustained release, long-term tumor monitoring, combination chemotherapy, body-on-a-chip, and biomedical informatics.

34.2.2 Nanomechanical Cellular Analysis with Atomic Force Microscopy

The atomic force microscope (AFM) can be used to measure the stiffness and elasticity of materials at nanoscale resolutions with high precision. The AFM has been employed to show that lung, breast, and pancreatic tumor cells are all substantially softer than normal cells (Cross et al. 2007). Such studies enable the rational differentiation of cancer cells from noncancerous cells of the same tissue microenvironment, in spite of their general similarity in shape. These studies also contribute to our understanding of the change in mechanical properties underlying each cellular phenotypic event that constitutes carcinogenesis. However, the translation of these *in vitro* studies to viable *in vivo* platforms would be paramount to the deployment of cellular elasticity sensing platforms in real-time tumor detection applications.

34.2.3 Tumor Signal Sensing with Nanoelectromechanical Systems

NEMS are nanoscale devices that are constructed out of components such as nanocantilevers, nanosieves, nanowires, and nanochannels. These devices are useful for the collection and analysis of biological signals. NEMS can aid in the timely detection of rare molecular signals associated with malignancy (Cui et al. 2001). For example, nanowire sensor arrays

with surface receptors are demonstrated to recognize even femtomolar concentrations of carcinoembryonic antigen and prostate-specific antigen from as few as 10 tumor cells (Zheng et al. 2005). This ability to identify tumor molecular markers at small concentrations holds immense potential for the early detection of malignancy.

34.2.4 Tumor Signal Amplification with Superparamagnetic Nanoparticles

Superparamagnetic materials are generally nonmagnetic materials that are composed of small, randomly oriented ferromagnetic clusters. However, when an external magnetic field is applied, thermal fluctuations aid in orienting the clusters resulting in "switching on" of the materials magnetism (Bean and Livingston 1959). Magnetic resonance imaging or MRI involves scanning materials with externally applied magnetic fields. Superparamagnetic materials are ideal candidates for the amplification of low-amplitude tumor signals owing to their polarization during MRI scans. This property of superparamagnetic nanoparticles has been used for noninvasive illumination of the anatomical contours of brain tumors in the early stages of development (Simberg et al. 2007). Tumor signal amplification with superparamagnetic nanoparticles has also been helpful in locating tumor colonies that may not appear on conventional

MRI scans owing to the associated poor signal-to-noise (SNR) ratio. This high-sensitivity technology is useful for the timely and accurate detection of cancer.

34.2.5 Ultrasensitive Nanoscale Imaging Technology

Conventional optical interferometer systems use resolution-limited, Fourier transform spectroscopy (FTS) to monitor signals from biological samples. Although this provides for faster signal processing and more affordable system design, FTS is unsuitable for nanoscale sensing which by its very nature, demands extremely high resolutions for sensitive scanning. Terahertz spectroscopy systems designed with quantum cascade structures employ far-infrared radiation to extract molecular spectral information with very high resolutions (Ferguson and Zhang 2002). This ultrasensitive imaging technology offers some insight into the native conformations of biomolecules, many of which have collective vibration modes in the terahertz range. One of the applications of this technology is the biomolecular characterization of tumor protein–ligand interactions and this is beneficial for the early detection of cancer (Menikh et al. 2002).

The following section examines tumor-specific nanoscale imaging systems.

34.3 *In Vivo* Imaging of the Tumor Microenvironment with Nanoscale Systems

Nanoscale devices are commonly employed for many medical imaging applications (West et al. 2006). Owing to their smaller dimensions compared to the relatively larger pores on the leaky tumor microvasculature, nanoscale devices preferentially accumulate within tumor tissues. This "auto-targeting" of nanoscale devices enables localization of the imaging agent in tumor microenvironments. Nanoscale imaging provides compelling snapshots of the biological mechanisms governing tumor maturation while also enabling effective monitoring of disease spread and assessment of therapeutic efficacy. This section discusses the use of nanoscale systems for *in vivo* imaging of the tumor microenvironment.

34.3.1 Quantum Dots

Quantum dots (QDs) are nanoparticles less than 10 nm in diameter, possessing unique optical, chemical, and electronic characteristics. These include improved signal intensity, narrow emission spectra, good photo-stability, single-wavelength excitation of multiple colors, size-tunable light emission, resistance to photo-bleaching, and ease of surface functionalization (Bruchez and Hotz 2007). Such distinctive properties render QDs as ideal fluorescent probes for the simultaneous illumination of multiple subcellular compartments with high resolution, appreciable sensitivity, and agreeable color contrast. A variety

of QD probes have been developed for monitoring the tumor microenvironment. These probes are generally composed of an amphiphilic polymer matrix encapsulated within QDs that are surface-functionalized with tumor-targeting ligands. The powerful combination of ultrasensitivity and ultraprecise targeting enables high-resolution imaging of tumor cells and subcellular structures.

34.3.2 PEGylated Quantum Dots

Polymeric QDs have limited *in vivo* lifetimes (~minutes). However, they may be optimized for imaging tumors by ensuring adequate circulation time and sustained fluorescence over longer periods. The conjugation of polyethylene glycol (PEG) molecules to the QD polymer coat produces a manifold increase in their lifetime. The resulting fluorescent signal from PEGylated QDs also lasts for several months after injection into the blood stream. This increased lifetime of PEGylated QDs is due to the presence of a hydration shell around the device that improves its ability to resist pressure (Allen et al. 2002). The hydration sphere in turn is stabilized by a rich n etwork of hydrogen bonds between PEG and water molecules. Furthermore, the excellent structural fit of PEG with the tetrahedral lattice of water molecules enhances the stability of the hydration sphere. Given their greatly increased circulation lifetime, PEGylated QDs are ideal tools for sustained monitoring of tumor cells.

34.3.3 Immunofluorescent Label-Conjugated Quantum Dots

PEGylated QDs can conveniently be conjugated to biomolecules through covalent chemical bonding. These conjugated QDs allow for selective labeling of cell surface receptors, cytoskeleton components, nuclear antigens, and other biomolecules. Immunofluorescent labeling is one technique by which the molecular footprint of tumors may be mapped to aid in the design of appropriate therapeutic systems. For instance, QDs linked with streptavidin and immunoglobin molecules have been used to selectively label the Her2 receptors on the surface of live breast cancer cells (Wu et al. 2003). Such selective immunofluorescent labeling of tumor biomarkers helps to track the progress of the disease and to evaluate the efficacy of therapeutic regimens.

34.3.4 Bioluminescent Quantum Dots

QDs generally require blue light for efficient excitation. However, the visible frequency range has low tissue-penetration capability and is associated with random excitation of endogenous fluorophores resulting in high levels of background noise (Arnone et al. 2000). Bioluminescent QDs can auto-fluoresce and therefore do not call for any external excitation. Bioluminescence is the phenomenon of light production involving the conversion of biochemical energy to light energy in organisms such as fireflies, glow worms, honey mushrooms, gulper eels, coral, and vibrionaceae. Bioluminescence resonance energy transfer (BRET) is a technology that is prevalently used in a variety of biotechnological applications for harnessing bioluminescence from natural and genetically engineered organisms (Shrestha and Deo 2006). BRET involves the transformation of a naturally occurring fluorophore protein such as luciferase into a bioluminescent probe by conjugating it with QDs. Since fluorophores such as luciferase are of fragile structure, bioluminescent QDs are likely to possess constrained fluorescent activity and highly reduced circulation lifetimes. In order to retain the activity of these autofluorescent devices, synthetic biology techniques are employed to design stable variants of the luciferase gene. Furthermore, genetically engineered fluorophores may be tailored to emit light of shorter wavelength than the natural firefly luciferase protein, thereby synergizing better with the absorption spectrum of QDs. Bioluminescent QDs are powerful tools for probing the anatomical contours of living systems provided the *in vivo* resistance to degradation by host immune system is enhanced.

34.3.5 Cysteine-Coated Quantum Dot Nanocrystals

Coating of QD nanocrystals generally causes variations in their optical properties that adversely affect their pharmacokinetics. However, coating of QD nanocrystals with the sulfur-rich cysteine amino acid renders them bright and stable (Liu et al. 2007). The cysteine coating also prevents the undesirable adsorption of random serum proteins onto the QD surface that typically leads to enlargement and distortion. Furthermore, unbound QDs are rapidly cleared via the kidney's filtering mechanisms. This suppresses the otherwise rampant background noise contribution of unbound QDs. Cysteine-coated QD nanocrystals are hence particularly useful for the *in vivo* imaging of microenvironments wherein the tumor signal-to-background noise ratio is generally low.

34.3.6 Luminescent Carbon Nanotubes

Carbon nanotubes (CNTs) are cylindrical carbon allotropes of length-to-diameter ratios greater than a million. These nanostructures are characterized by extraordinary strength, chemical inertness, and thermal conductivity (Reich et al. 2004). Hybrid nanostructures of CNTs conjugated to luminescent QDs are used to selectively illuminate cancer cells. An added advantage of these hybrid nanostructures is the additional cavity volume present within the CNT where substantial payloads of anticancer drug can be stored. The successful translation of these luminescent CNTs to clinical trials would require toxicity assays relating to long-term CNT exposure.

34.3.7 Nanoscale Polymeric Matrices

Polymeric matrices are highly versatile, biocompatible, biodegradable devices that are extensively used for *in vivo* imaging applications and a well known example is poly-lactic-*co*-glycolic acid (LaVan et al. 2003). These devices also offer the advantages of high agent encapsulation efficiency and specific accumulation within tumor sites. Moreover, polymeric matrices can be conveniently surface-conjugated to PEG molecules for increasing their circulation lifetime and functionalized with tumor-targeting molecules for homing the device specifically to tumor sites (Aubin-Tam and Hamad-Schifferli 2008). These characteristics that render polymeric matrices attractive for *in vivo* imaging are also responsible for the extensive use of these devices as targeted drug delivery vehicles.

The following section describes the design of targeted nanoscale drug delivery systems with emphasis on the recently discovered biomaterials and technologies that are being utilized for anticancer applications.

34.4 Targeting the Nanoscale Delivery of Chemotherapy to Tumors

The development of nanoscale systems for targeted delivery of drugs to cancer cells is a major focus of chemotherapy, primarily for the purpose of toxicity reduction. Nanoscale systems that are smaller than 200 nm in diameter effectively navigate through the leaky tumor microvasculature and aggregate selectively within the tumor interstitial space. They are thereafter contained within the tumor microenvironment owing to the dysfunctional lymphatic drainage. Devices that rely primarily on this enhanced permeation and retention (EPR) for delivering drugs to tumors are known as passively targeted or autotargeted systems. Active targeting, on the other hand, implies that targeting ligands such as polysaccharides, antibodies, peptides, nucleic acid aptamers, or other small biomolecules are conjugated to the surface of the nanodevice. Biomolecules are carefully screened to identify potential targeting ligands that bind selectively to receptors that are unique to, or overexpressed in, cancerous cells. A compilation of overexpressed cell surface receptors in various tumor types is presented (Table 34.2). The conjugation of appropriate tumor-specific targeting molecules onto drug-encapsulated nanodevices ensures selective delivery of the agent to cancerous cells and hence minimal toxicity to noncancerous cells (Figure 34.2). Although passive and active targeting are both helpful for directing nanoscale devices to tumor tissues, the latter mode is associated with greater reduction in chemotherapeutic toxicity, specifically to healthy cells in the immediate neighborhood of tumor cells. This section outlines the advancements in targeting the nanoscale delivery of anticancer drugs.

34.4.1 Biodegradable Polymeric Nanoparticles

The duration of therapy for various diseases ranges from a few hours to several months. Application-specific injectable drug delivery devices are being developed across this therapeutic duration spectrum (Chasin and Langer 1990). These devices are typically constituted of polymeric materials that are suitable

TABLE 34.2 Overexpressed Cell Surface Receptors Classified According to Incidence of Cancer Type

Overexpressed Cell Surface Receptor	Cancer Type (Incidence If Available)
Estrogen receptor alpha (ERα)	Breast (70%), ovary
Progesterone receptor (PR)	Breast (64%), uterus
Human epidermal growth factor receptor 2 (HER2)	Pancreas (26%), bladder (44%), Cervix (77%), breast (30%), stomach
Endothelin-A receptor (ET$_A$R)	Breast (45.3%), ovary
Platelet-derived growth factor receptor (PDGFR)	Liver (22.1%), breast (39.2%)
Fibroblast growth factor receptor 2 (FGFR-2)	Sarcoma, breast (5%–10%)
Heparin Sulfate Glycosaminoglycan (HSGAG)	
Neuropilin-1 (NRP-1)	Breast, colon, prostate, pituitary
Androgen receptor (AR)	Prostate (43%)
Insulin-like growth factor-1 receptor (IGF$_1$R)	Cervix
G-Protein coupled receptor (GPCR)	Brain (57%), stomach, prostate (60%)
Cannabinoid-specific receptors (CBR)	Liver (62%)
Chemokine receptor (CXCR4)	Skin
Vascular endothelial growth factor receptor (VEGFR)	Colorectal, ovary, pancreas, breast
Interleukin-13 (IL-13) receptor	Skin, brain

for *in vivo* use, that is, nontoxic, noncarcinogenic, nonmutagenic, nonallergenic, and noninflammatory. Materials satisfying these criteria are classified as biodegradable, bioeliminable or permanent, depending on the mode of their *in vivo* clearance

(Table 34.3). Biodegradable materials are initially in the solid or gel phase and are thereafter broken down into natural metabolites within the body by hydrolytic or enzymatic activity (Hasirci et al. 2001). PLGA is an example of a biodegradable material that has been approved for clinical use by the Food and Drug Administration (FDA). PLGA is completely hydrolyzed into lactic acid and glycolic acid which are components of the Kreb's cycle and are hence naturally metabolized by the body. The characteristic degradation half lives of biodegradable polymers varies from a few minutes to several years based upon the ease of hydrolysis (Table 34.4). Matrices composed of biodegradable polymers degrade by cross-link, side chain, or backbone degradation and the polymer molecular weight plays an important role in determining the average degradation lifetimes (Figure 34.3a). Other characteristics of biodegradable polymers that are known to influence their hydrolysis rate include hydrophobicity, steric effects (local structure, glass transition), microstructure (porosity, phase separation, crystallinity), oligomer solubility, autocatalysis, and pH of the medium (Kumar 2007). Biodegradable polymeric devices are classified as bulk or surface eroding based on the hydrolysis and diffusion rates (von Burkersroda et al. 2002). Bulk eroding polymer matrices undergo uniform, instantaneous wetting and degradation happens throughout the bulk of the matrix, whereas water diffuses much more slowly into surface-eroding polymer matrices and degradation hence happens only at the exposed surface of these matrices (Figure 34.3b). The drug release profile associated with surface-eroding polymer matrices is generally monophasic whereas bulk eroding polymer matrices display more biphasic release profiles with a distinct initial "burst release" phase. During the burst phase,

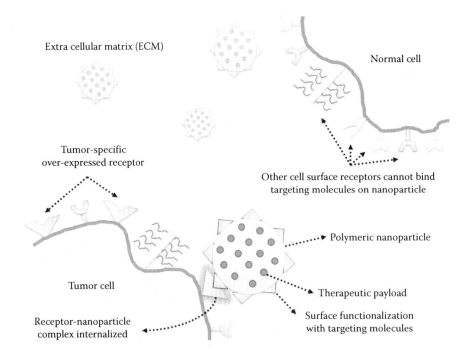

FIGURE 34.2 Active targeting of nanoparticles to tumor-specific overexpressed receptors.

TABLE 34.3 Polymeric Biomaterials Classified According to Mode of *In Vivo* Clearance

Clearance Mode	Polymer	Chemical Structure
Biodegradable	Poly(glycolic acid)	
	Poly(lactic acid)	
	Poly(lactide-*co*-glycolide)	
	Poly(ε-caprolactone)	
	Poly(malic acid)	
	Poly(ethylene carbonate) Poly(propylene carbonate)	R = H/CH$_3$
	Poly(anhydride)	
Bioeliminable	Poly(alkyl cyanoacrylate)	
	Poly(ethylene oxide)	
Permanent/implant	Poly(methyl methacrylate) poly(hydroxyethyl methacrylate)	R = CH$_3$/CH$_2$CH$_2$OH
	Poly(*N*-isopropyl acrylamide)	R = CH(CH$_3$)$_2$

TABLE 34.4 Degradation Lifetimes of Hydrolysis-Susceptible Polymers

Polymer Class	Degradation Half–Life
Polyanhydrides	<10 min
Polyorthoesters	~5 h
Polyesters	~20 months
Polyphosphazenes	~5 years
Polyamides	>50,000 years

bulk eroding nanoparticles release a significant volume of their drug payload and this is undesirable for therapeutic applications (Langer and Folkman 1978). However, drug molecules may be conjugated chemically to polymer molecules to ensure non-burst release.

34.4.2 Liposomes, Micelles, and Polymerosomes

Liposomes are nanoscale spheres composed of one or more bilayer, self-assembled, concentric lipid membranes which can be used to encapsulate various biological molecules. Micelles are self-assembled, spherical lipid monolayers with a hydrophobic core—hydrophilic shell architecture. Polymerosomes are composed of synthetic polymer amphiphiles and are structurally similar to liposomes. Certain characteristics of these lipid-based carriers enable their use for chemotherapeutic drug delivery. These include biodegradability, biocompatibility, drug insulation, ease of surface functionalization, and, encapsulation of

drugs with wide-spectrum physicochemical properties (Peer et al. 2007). Furthermore, some of these carriers can be loaded with multiple drug combinations. For example, the inner cavity of liposomes being aqueous is ideal for encapsulating hydrophilic drugs, while the volume between successive lipid membranes is well suited for housing relatively hydrophobic drugs (Figure 34.4a). While these advantages have resulted in extensive deployment of lipid-based nanoscale carriers in therapeutic applications, the synthesis of these carriers remains cumbersome. This challenge is overcome by the use of lipidoid libraries.

34.4.3 Lipidoid Libraries

The time-intensive synthesis of liposomes is a major limitation to the throughput of this class of delivery system. Lipidoid libraries are designed by the rapid and parallel combination of amino molecules with alkyl-akrylates and alkyl-acrylamides. This multiplexed synthesis is capable of producing more than a 1000 different lipid-like structures in an accelerated manner (Akinc et al. 2008). Lipidoids are a viable platform for the targeted intracellular delivery of short strands of ribonucleic acid (RNA). The targeted approach is necessary because direct introduction of RNA into the bloodstream results in its breakdown by the body's immune system. The RNA may be engineered to block the action of selectively targeted oncogenes, while leaving other cellular mechanisms intact. This exquisite specificity offers a new tool for harnessing the potential of RNA interference (RNAi) which is aimed at suppressing the expression of specific proteins that have been associated with cancer.

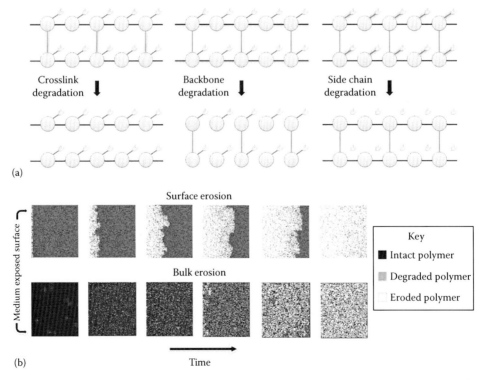

FIGURE 34.3 (a) Mechanisms of polymer degradation; (b) modes of polymeric matrix erosion.

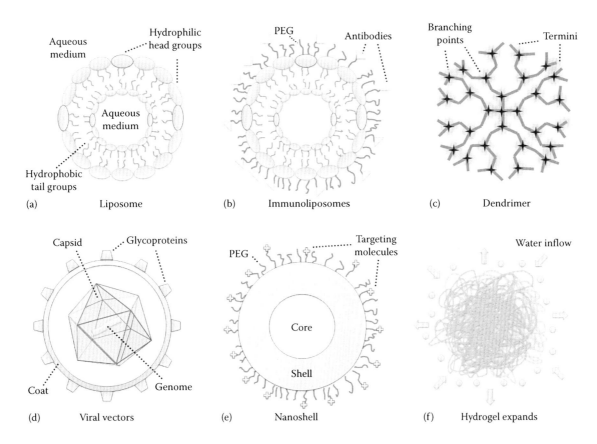

FIGURE 34.4 Structures of common nanoscale drug delivery vehicles—(a) Liposomes; (b) immunoliposomes; (c) dendrimers; (d) viral capsids; (e) nanoshells; and (f) hydrogels.

34.4.4 Immunoliposomes

Antibodies or immunoglobulins are proteins used by the immune system to identify and neutralize bacteria, viruses, and other foreign bodies. The chemical conjugation of monoclonal antibodies to the distal ends of PEGylated liposomes yields immunoliposomes (Figure 34.4b). These are useful both for the intracellular targeting of therapeutic nucleotides aimed at malignant cells as well as for roping in the immune systems arsenal (Kontermann 2006). Moreover, unlike liposomes that are incapable of permeating the blood-brain barrier composed of the brain capillary endothelial wall, immunoliposomes undergo receptor-mediated transcytosis. This has been used to successfully deliver small molecule anticancer drugs such as doxorubicin to brain tumors (Huwyler et al. 1996).

34.4.5 Dendrimers

Dendrimers are synthetic organic macromolecules structured like branched trees (Figure 34.4c). The fabrication of these macromolecules is relatively new to polymer sciences (Tomalia and Frechet 2003). Dendrimers are classified as bioeliminable because they are cleared rapidly from blood by the kidneys owing to their nanoscale (<10 nm) dimensions. They are also water soluble, have well-defined chemical structures, and are completely biocompatible. Dendrimers can also be conveniently conjugated

with appropriate targeting molecules with moderate affinity and high avidity. The term "avidity" refers to multivalent collective binding. High avidity of multi-ligand conjugated dendrimer nanovehicles is responsible for manifold increase in their targeting ability (Hong et al. 2007). The moderate affinity of dendrimers is important because high binding affinity of nanocarriers to their targets decreases the probability of their tissue penetration due to "binding-site barrier" effects. Although dendrimers are well suited for targeted drug delivery applications, their mass production is a challenging task.

34.4.6 Natural and Synthetic Viral Capsids

Capsids are protein shells of viruses consisting of several oligomeric units. These are structured as icosahedrons (20 equilateral triangular faces) or other complex geometries and house the viral genetic matter (Figure 34.4d). Like carbon buckyballs, viral capsids self-assemble into near-spherical geometries with high surface area-to-volume ratios (Wagner and Hewlett 2004). The specificity of biomolecular interactions between glycoproteins on the host cell surface and the virus coat dictates cellular entry, implying that certain cells are intrinsically more susceptible to infection with certain vectors. This is the rationale behind cell-specific targeted gene therapy with viral vectors (Table 34.5). Furthermore, vectors can be tailored for tumor-centric

TABLE 34.5 Tumor Targeting with Viral Capsids and Vectors

Viral Family	Type	Capsid	Target Molecules	Tumor Type
Adenoviridae	DNA	Icosahedral	Urokinase-type plasminogen activator	Skin, prostate
			Matrix metalloproteinases (MMP)	
Herpesviridae	DNA	Icosahedral	Urokinase-type plasminogen activator	Brain
			Matrix metalloproteinases (MMP)	
Poxviridae	DNA	Complex	Urokinase-type plasminogen activator	Liver
Reoviridae	RNA	Icosahedral	Matrix metalloproteinases (MMP)	Skin
Picornaviridae	RNA	Icosahedral	—	—
Rhabdoviridae	RNA	Helical	Matrix metalloproteinases (MMP)	Skin
Paramyxoviridae	RNA	Helical	Signaling lymphocytic activation molecule	Bladder, skin, ovary, brain
			Membrane cofactor protein (MCP)	

applications by appropriately modulating the capsid surface glycoproteins. For instance, the incorporation of tumor-targeting peptides onto the capsids of recombinant adeno-associated human parvovirus transforms this typically broad-range infective vector into a tumor-specific delivery vehicle (Grifman et al. 2001). The advantageous features of viral capsids such as their stable structure, target specificity, and ease of surface functionalization have also prompted the design of synthetic nano-capsids for drug delivery applications (Cattaneo et al. 2008).

34.4.7 Targeted Nanoshells

Nanoshells are nanoscale concentric spheres with a core-shell architecture that provides for the accommodation of multiple drugs as payload (Figure 34.4e). The core-shell dimensions can further be manipulated, allowing for the fine-tuning of key optical properties of the device, such as light absorption at different wavelengths. Nanoshells that are surface functionalized with targeting biological molecules are able to seek out and bind selectively to the surface receptors of cancerous cells. Illumination of these tumor-bound nanoshells with infrared light leads to increase in their temperature and this is useful to selectively destroy tumor cells (Hirsch et al. 2003). Thermal treatment of tumors with IR and targeted nanoshells does not have the toxic side effects of chemotherapy such as fatigue, immune suppression, and hair loss.

34.4.8 Striped Nanoparticles

While gold nanoparticles coated with alternating bands of two different molecular "stripes" are able to enter cells, random coating of these molecules onto nanoparticles does not enable their internalization (Verma et al. 2008). Furthermore, while gold nanoparticles sans the molecular stripes rupture membranes leading to cell death, nanoparticles with the stripes are able to internalize nondestructively. It is proposed that similar material properties of striped nanoparticles and cell surfaces may play a critical role in mediating the former's nondestructive cellular entry (Jackson et al. 2004). Striped nanoparticles are hence ideal for targeting the delivery of anticancer drugs to the cytoplasm of tumor cells.

34.4.9 Biomimetic Smart Material Nanodevices

Materials may be classified as "regular" or "smart" depending on their ability to adapt to external stimuli. The properties of smart materials can be significantly altered in a controlled manner using specific external stimuli to which the material is sensitive. Regular materials, on the other hand, are able to offer only their characteristic and often limited set of properties (Langer and Tirrell 2004). Most regular materials (e.g., steel) are a sharp contrast to biological materials that show definitive adaptability to environmental stimuli (e.g., plants spreading out to maximize the area exposed to sunlight). Biomimetics refers to the engineering of stimulus-responsive smart systems based upon design principles borrowed from the study of biological materials (Campbell 1995). The ability to quantitatively predict the change in properties of smart materials with varying external stimulus is beneficial for the rationale design of responsive materials for biomedical applications. There are numerous examples of nanoscale "smart" anticancer technologies (Table 34.6). However, many of these technologies are associated with safety concerns relating to deployment of smart materials *in vivo*. One exception is stimulus-responsive nanogels that are biodegradable and hence safe for *in vivo* use.

34.4.10 Hydrogels

Hydrogels are water-soluble, cross-linked networks that swell with water inflow (Figure 34.4f). Hydrogels are classified according to the kind of chemical bonding involved. Ease of chemical modification, *in situ* formability and biodegradability are some of the properties that make hydrogels desirable for drug delivery (Khademhosseini et al. 2006). Moreover hydrogels can be readily tailored with recognition sites such as adhesion or collagenase sequences for a variety of biological applications (Mann et al. 2001). Further, the environmental responsiveness of hydrogels to stimuli such as change in pH, ionic strength, or presence of analytes, is very useful for the regulation of network swelling. This in turn controls the kinetics of drug release from hydrogels. For instance, pH-responsive nanoscale hydrogels or nanogels maybe designed to swell specifically within the relatively acidic

TABLE 34.6 Stimulus-Responsive Smart Materials in Nanoscale Antitumor Applications

Material	Stimulus	Variable	Antitumor Applications
Piezoelectric	Force	Voltage	Sensing of tumor cells
Shape memory alloy/polymer	Temperature	Shape	Controlled drug release
Magnetic shape memory alloy	Magnetic field		Anti-inflammatory drug delivery
pH-sensitive polymers	pH	Volume	Targeted drug delivery
Electrorheostatic	Electric field		Controlled drug release
Magnetorheostatic	Magnetic field	Viscosity	Tumor cell separation
Non-Newtonian fluid	Force		
Halochromic	pH		High-contrast imaging
Electrochromic	Electric field		Subcellular imaging
Thermochromic	Temperature	Color	Tumor microenvironment monitoring
Photochromic	Light intensity		
Supramolecular assemblies			Metastasis and angiogenesis detection
Chemically cross-linked hydrogels	Enzymes	Entropy	Agent encapsulation in nanoparticles
Enzyme-responsive surfaces			Combination chemotherapy

environment of cancer cells thereby releasing the drug payload in a targeted and controlled manner (Lee et al. 2008). While hydrogels are useful for delivery of drugs over short durations, their widespread therapeutic use is challenged by difficulties in obtaining sustained release. Some of the other nanoscale systems with sustained drug release capabilities are discussed in the following section.

34.5 Nanoscale Systems for Sustained Drug Release and Tumor Monitoring

Sustained release of agents over several weeks is required for many applications such as monitoring of the tumor microenvironment and anticancer drug delivery (Langer and Folkman 1978). The design of nanoscale systems that are tailored for sustained release applications is outlined in this section.

34.5.1 Drug-Conjugated Polymeric Nanoparticles

The chemical conjugation of drug molecules to polymer fragments prevents rapid dissolution of the drug into the medium. The mechanism governing sustained release from drug-conjugated polymeric nanoparticles is as follows (Figure 34.5a). Upon repeated hydrolytic or enzymatic cleavage of the polymer backbones, the fragments become increasingly smaller in size. The dissolution of the drug–polymer conjugates commences only after the molecular weight of these fragments decreases below a certain threshold that is characteristic of the polymer. The dissolved drug molecules then begin diffusing out of the nanoparticle. An added advantage of drug-conjugated nanoparticles is the absence of the initial burst release phase which, as was discussed earlier, is characteristic of drug-encapsulated polymeric nanoparticles. Burst release in the latter occurs primarily because of the near-instantaneous efflux of medium-exposed drug molecules present in the vicinity of the carrier surface (Figure 34.5a). In the case of drug-conjugated polymeric nanoparticles, the chemical binding of the drug molecules to the polymer backbone prevents instantaneous diffusion of even these medium-exposed drug molecules on the carrier surface. Drug-conjugated polymeric nanoparticle systems have been used for the sustained delivery of chemotherapeutic agents. For example, PLGA-doxorubicin nanoconjugates have been employed to effectively deliver the chemotherapeutic drug doxorubicin to tumors over several weeks (Yoo et al. 2000). Additionally, the polymers constituting the nanoparticles may be surface functionalized with PEG and tumor-targeting molecules, similar to other nanodevices. While the use of drug-conjugated polymeric nanoparticles is associated with these multiple advantages, implants of nanoscale devices help in attaining a more locally targeted delivery in the immediate neighborhood of the implantation site (Grayson et al. 2003).

34.5.2 Nanochip Implants

Polymeric multi-reservoir microchips are designed for temporally controlled local release of multiple agents over several weeks (Grayson et al. 2003). By varying the enclosing "gate" characteristics that determine the kinetics of polymer matrix hydrolysis, unique temporal release profiles are obtained for each drug contained within the chip reservoirs. Polymeric nanochips serve as preprogrammed implants for the sustained and controlled release of combination chemotherapy or for simultaneous delivery of drugs and imaging agents for parallel therapy and monitoring. However, tumor sites are not always conducive for procedures such as implantation and it hence becomes important to develop minimally invasive platforms for sustained combination chemotherapy. The following section examines such technologies that are driving nanoscale combination chemotherapy.

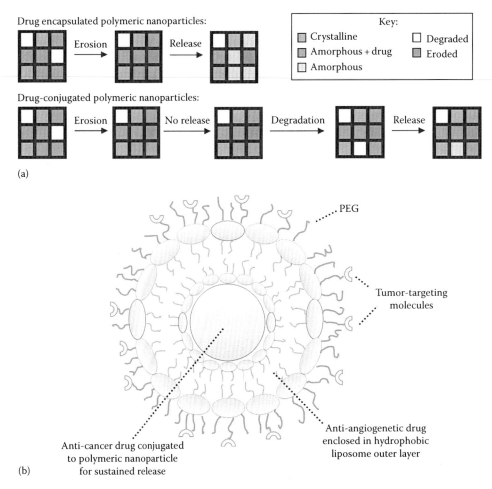

FIGURE 34.5 (See color insert following page V-2.) (a) Mechanistic rationale for sustained release from drug-conjugated polymeric nanoparticles and comparatively faster release from drug-encapsulated polymeric nanoparticles and (b) structure of the nanocell and function of the constituent biomaterials for antiangiogenesis and anticancer combination therapy.

34.6 Nanoscale Combination Therapy: Angiogenesis, Metastasis, and Beyond

Cancer is a complex disease that arises from the dysregulation of several biological networks that are frequently interconnected (Keith et al. 2005). Consequently, the single drug–single target approach of monotherapies is often less effective than combination chemotherapies that can synergistically target multiple processes simultaneously. This is verified by assessments of numerous successful clinical combination therapies for cancer (Zimmerman et al. 2007). Therapeutic synergy involves one or more of the following effects—multiplicative increase of cancer cell death, decreased dosage requirement of each drug, reduced toxicity, and minimized development of drug resistance (Chou 2006). The last three effects are a natural consequence of using multiple drugs because the decreased requirement for each drug alleviates issues of toxicity and drug resistance. A compilation of synergistic interactions amongst some of the commonly administered cancer drugs indicates an abundance of mechanistic synergy that can be harnessed by combination therapies (Table 34.7).

However, the effective administration of combination therapies requires very precise spatiotemporal control in the release of each drug owing to the sensitivity associated with functional rewiring of cellular biochemical signaling networks. This in turn calls for the design of sophisticated nanoscale delivery platforms that can carry multiple payloads and disburse them selectively to the cancerous tissues at appropriate times. This section examines such nanoscale platforms for the administration of cancer combination chemotherapy.

34.6.1 Nanocells

Antiangiogenesis drugs curb the growth of new blood vessels thereby starving the cancer colony of oxygen and nutrients. This limits the expansion of tumor tissues. However, delivery of just antiangiogenesis drugs does not suffice to defeat cancer owing to subsequent systemic hypertoxicity (Folkman 1996). Moreover, the destruction of tumor blood supply by antiangiogenesis drugs cuts off potential routes for the delivery of chemotherapeutic agents (Carmeliet and Jain 2000). The design of the nanocell is based upon the principle that the chemotherapeutic agent could be released from within the tumor after its

TABLE 34.7 Cancer Combination Chemotherapy

	Cap	Cyc	Doc	Dox	Epi	Flu	Gem	Lap	Let	Met	Mit	Pac	Tam	Tra
Cap			■					■				■		▦
Cyc		▦	■	▦	■	■		▦		■		■		■
Doc	■	■	▦	□								▦	□	▦
Dox			□		▦	■						□	■	■
Epi		■			▦	■				■				
Flu		■		■	■	▦				■				
Gem							▦					■		
Lap	■							▦	▦	■		■	■	■
Let		▦						■	▦					
Met		■			■	■			▦	▦	■	■		
Mit										■	▦			
Pac	■	▦		□			■	■		■	▦		□	■
Tam			□	■				■				□	▦	□
Tra		■	▦	■				■			■	□		▦

Note: Synergistic (■), additive (▦), and antagonistic (□) interactions between common cancer drugs. The following three letter abbreviations are used for the drugs: **Cap**ecitabine (*Xeloda*®), **Cyc**lophosphamide (*Cytoxan*®), **Doc**etaxel (*Taxotere*®), **Dox**orubicin (*Adriamycin*®), **Epi**rubicin (*Pharmorubicin*®), **Flu**orouracil (5FU), **Gem**citabine (*Gemzar*®), **Lap**atinib (*Tyverb*®), **Let**rozole (*Femara*®), **Met**hotrexate (MTX), **Mit**oxantrone, **Pac**litaxel (*Taxol*®), **Tam**oxifen (*Nalvodex*®), **Tra**stuzumab (*Herceptin*®).

blood supply is restricted (Sengupta et al. 2005). The nanocell is a dual-compartment device that derives its name from its resemblance to a cell (Figure 34.5b). The device is composed of an outer PEGylated liposomal vesicle enclosing a polymeric nanoparticle in its interior. The antiangiogenesis drug is encapsulated within the lipid layers and the interior polymeric nanoparticle houses the chemotherapeutic agent. Owing to the nanoscale dimensions of this device and the predominance of enhanced permeation and retention (EPR) in the tumor microenvironment as outlined earlier, the nanocells are selectively targeted to tumor tissues. Once the nanocells are lodged within the tumors, the outer layer releases the antiangiogenesis agent thereby cutting off tumor blood supply. The sustained release of the chemotherapeutic agent then commences from within the tumor tissue. Combination chemotherapy with nanocells is more efficient than nanoscale monotherapy and conventional combination chemotherapy in inhibiting tumor growth. The nanocell platform can be further extended to house drugs that target other features of cancer such as drug resistance (Sengupta and Sasisekharan 2007).

34.6.2 Nanoscale Modulation of Tumor Extracellular Matrix

The surface of cancer cells is rich in a plethora of complex polysaccharides such as heparan sulfate glycosaminoglycans (GAGs) that are known to regulate tumorigenesis, tumor progression, neovascularization, and metastasis (Sasisekharan et al. 2002). Furthermore, the extracellular matrix (ECM) is composed of macromolecular networks that restrict liberal transport of materials thereby limiting the efficiency of anti-cancer systems (Goodman et al. 2007). The ECM and cell surface can be modulated with enzymes such as collagenase that degrades collagen fibers or heparanase that degrades heparan sulfate GAGs. These ECM-modulating enzymes play critical roles in several aspects of tumor biology (Liu and Sasisekharan 2005). The combination of anticancer and antiangiogenetic drugs with such enzymes that alter the ECM is hence crucial for maximally synergistic therapeutic benefit. For instance, conjugation of collagenase onto the surface of nanoparticles for site-specific degradation of collagen networks produces manifold increase in nanoparticle mobility and internalization (Goodman et al. 2007).

34.6.3 Chemo-Thermal Combo-Therapy with Targeted Gold Nanoparticles

Targeted thermal therapy generally inhibits expansion of tumor tissues and antitumor drugs are more adept at killing cancerous cells at elevated temperatures (Everts et al. 2006). Gold nanoparticles have the ability to transduce light into heat and this property appears promising for two-pronged chemo-thermal combination therapy. For example, targeted gold nanoparticles can carry drugs as payload and concomitantly transduce light to heat in the tumor microenvironment resulting in more efficient antitumor activity (Visaria et al. 2006).

34.6.4 Combination Gene Therapy for Synergistic Targeting of Multiple Pathways

Monotherapies typically target specific genes involved in specific pathways responsible for motivating cancer. Combination therapies on the other hand target multiple genetic pathways and harness the resulting synergy to produce a multiplicative increase in cancer cell death (Keith et al. 2005). However, the use of nanoscale platforms for controlled delivery is paramount to realizing the full potential of combination gene therapy. For instance, controlled delivery of multiple tumor-suppressor genes with nanoparticle-based systems is more effective than the individual monotherapies in treating cancer (Deng et al. 2007).

34.6.5 Nanoscale Combo-Therapies for Targeting Metastasis and Angiogenesis

Metastasis from cancer colonies are significantly more difficult to detect and treat than the primary tumor itself. However, since newly formed metastatic tumor colonies thrive on the development of fresh blood vessels by angiogenesis, nanoparticle-based systems may be used to target drugs specifically to the sites of maximal blood vessel formation (Murphy et al. 2008). The targeting approach involves sensing the gradients of pro-angiogenetic factors and directing the nanoparticle-based systems toward increasing concentrations of these factors. This integrated approach to treatment of metastasis and angiogenesis with nanoparticle-based systems is more effective than monotherapies in suppressing the spread of malignancy.

34.6.6 Fighting Intractable Tumors with Nanoparticle-Based Combo-Therapies

Cancer cells develop resistance to anticancer agents leading to the formation of intractable tumors. However, pretreatment of intractable tumors with subtherapeutic levels of a potent but highly toxic anticancer agent followed by nanoparticle-based delivery of a second anticancer agent sensitizes tumors cells to the latter (Kano et al. 2007). Nanoscale systems enable the administration of such drug cocktails with precise spatiotemporal targeting.

The following section focuses on nanoscale systems for tumor drug resistance reversal.

34.7 Drug Resistance Reversal in Tumor Cells with Nanotechnology

Multi drug resistance (MDR) is a property of certain pathological bacterial and tumor cells that successfully defy a wide spectrum of potent apoptosis-inflicting drugs (Krishan and Arya 2002). Increased drug efflux, enzymatic drug deactivation, increased membrane permeability, altered drug molecule receptor sites, and creation of compensatory metabolic pathways are some of the mechanisms employed by tumor cells to attain MDR (Figure 34.6a). Nanoparticle-based delivery provides an alternative route

to the internalization of drugs since it utilizes a different set of receptors that are not modified by MDR (Figure 34.6b). This section focuses on nanoscale strategies for MDR reversal.

34.7.1 Direct Intracellular Ceramide Delivery with Polymeric Nanoparticles and Liposomes

Ceramide is the pro-apoptotic mediator that is suppressed by the overexpression of glucosyl ceramide synthase (GCS) enzyme in drug-resistant tumor cells. Intravenous injection of ceramide is not feasible owing to high systemic toxicity, necessitating nanoscale-targeted delivery of ceramide directly to the cytoplasm of cancer cells. Nanoparticles that are co-encapsulated with ceramide and an anticancer drug are effective in reversing MDR and killing tumor cells that are normally insensitive to the anticancer drug. For instance, co-encapsulation of ceramide and paclitaxel in polymeric nanoparticles promotes apoptotic signaling in human ovarian cancer cells with MDR and results in nearly 100-fold increase in their chemotherapeutic sensitization (Devalapally et al. 2008). Nanoliposomal formulations of pro-apoptotic ceramide and the anticancer drug sorafenib are also able to completely eliminate breast and skin cancer cell lines *in vitro* (Tran et al. 2008). While ceramide inhibits MDR, sorafenib attacks angiogenesis and promotes apoptosis. Liposomes and polymeric nanoparticles are hence useful for direct cytoplasmic delivery of MDR reversing agents and pro-apoptotic drugs.

34.7.2 Nanoscale Electrochemotherapy

Electrochemotherapy is a relatively new cancer treatment modality. This form of therapy uses electric pulses to increase the permeability of cancer cell membranes, followed by the delivery of chemotherapeutic agents. For instance, electroporation combined with delivery of bleomycin is effective in increasing the membrane permeability and promoting the internalization of bleomycin molecules into a variety of cancer cell lines *in vitro* (Gothelf et al. 2003). However, electrochemotherapy as an *in vivo* MDR-treatment modality will be practicable only upon adapting the electroporation technology on targeted, safe therapeutic platforms (Wagner 2007). The following section discusses the development of integrated platforms for testing a variety of combination drug cocktails.

34.8 Body-on-a-Chip: Systems Approach to Drug Testing on Integrated Platforms

Microfluidics is the study of fluid behavior at microscales. Microfluidic chips have been designed to host a variety of human cells and mimic different body tissues (Shuler and Xu 2007). This effectively bridges *in vitro* and *in vivo* testing by simulating full-fledged tissue microenvironments. Such integrated body-on-a-chip platforms allow for the targeted screening

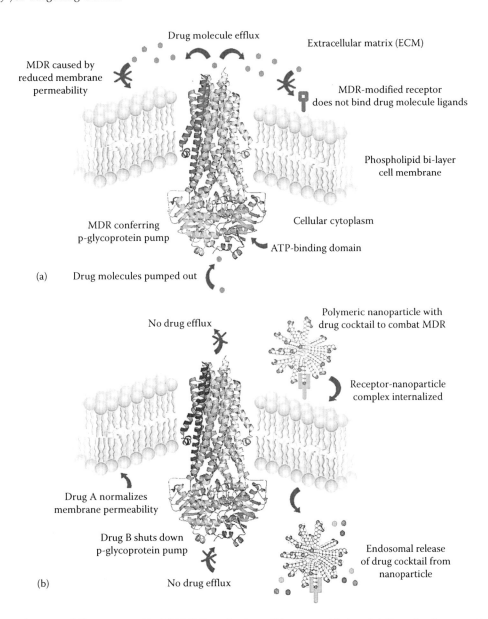

Drug molecule efflux

Extracellular matrix (ECM)

MDR caused by
reduced membrane
permeability

MDR-modified receptor
does not bind drug molecule ligands

Phospholipid bi-layer
cell membrane

MDR conferring
p-glycoprotein pump

Cellular cytoplasm

ATP-binding domain

(a)   Drug molecules pumped out

No drug efflux

Polymeric nanoparticle with
drug cocktail to combat MDR

Receptor-nanoparticle
complex internalized

Drug A normalizes
membrane permeability

Drug B shuts down
p-glycoprotein pump

Endosomal release
of drug cocktail from
nanoparticle

(b)   No drug efflux

FIGURE 34.6 (See color insert following page V-2.) (a) MDR mechanisms; (b) nanoparticle-based delivery for circumvention of MDR.

and selection of the best-suited drugs instead of the conventional shot-in-the-dark approach. These platforms are useful to test the efficacy and toxicity of preclinical, anticancer drugs, thereby saving time and money associated with conventional drug screening processes. Furthermore, this technology holds promise for "personalized medicine" wherein patient-specific information is used to predetermine drug efficacy and side effects prior to actual administration.

34.9 Theoretical and Computational Nanotechnology

Computational nanotechnology refers to the design of models for elucidating the mechanistic underpinnings of nanoscale phenomena. An understanding of theoretical nanosciences is fundamental to designing new nanoscale technologies for targeting

cancer (Rieth and Schommers 2006). This section discusses a few integrated semiempirical models of nanoscale systems for anticancer applications.

34.9.1 Modeling Polymeric Nanoparticle Erosion and Drug Release

The degradation kinetics of polymeric networks within polymer-based nanoscale and microscale delivery systems determines the rate of drug release. Several computational models have been developed to simulate hydrolysis-driven erosion of polymeric matrices (Siepmann and Gopferich 2001). Many of these models employ two-dimensional pixel grids to represent polymeric matrices and associate the pixels with a degradation lifetime that is derived from the characteristic rate constant of hydrolysis for that polymer. Such pixel-based Monte Carlo (MC) simulations have been successful in predicting the mechanistic underpinnings of polymer hydrolysis and have also been used to quantitatively estimate matrix erosion kinetics, porosity changes, microenvironmental pH variations, and drug release rates.

34.9.2 Multiscale Spatiotemporal Modeling of Nanoparticle Penetration in Tumor Spheroids

The tissue-penetration efficiency of nanoparticles increases with the use of collagenase enzyme to modulate the extracellular environment. Semiempirical, multiscale models of nanoparticle penetration into tumor spheroids are able to accurately predict the spatiotemporal distribution profiles of nanoparticles of various dimensions (Goodman et al. 2007). Furthermore, tissue-specific circulation of the agents released from the nanoparticles can also be estimated from these models. Multiscale models provide a platform to bridge molecular-, cellular-, and tissue-level phenomena that prevail in tumor microenvironments. Multiscale models also collectively contribute to our understanding of tumor architectures and how they impinge on the efficiency of nanoparticle-based delivery systems.

34.10 Summary

There is an emerging understanding that a systems approach to targeting cancer will be fundamental to the development of effective diagnostic, therapeutic, and preventive tools (Hartwell et al. 2006). In this post-genomic era, proteomics (Timms 2008), glycomics (Raman et al. 2005) and other "omics" platforms are greatly advancing our knowledge of the molecular and cellular underpinnings of cancer. In order to transform this knowledge into modular, multifunctional, and potent nanoscale devices for fighting cancer, the National Cancer Institute (NCI) has established the "Alliance for nanotechnology in cancer" (http://nano.cancer.gov/). Other countries have similarly embarked on the nano-war against cancer, resulting in a spurt of innovative anticancer approaches. However, seamless integration of anticancer efforts in nanotechnology, bioinformatics, and modern molecular biology for successful clinical application is in its infancy. Cost efficacy, high throughput, standardization in design, development of benchmarks, long-term toxicity studies, and creation of opportunities for synergistic interaction of engineers, biologists, and physicians are some future milestones toward the assimilation of nano-oncology into mainstream clinical biomedicine.

Acknowledgments

This work was supported by the Singapore-MIT Alliance for Research and Technology (SMART). The authors thank Dr. V. Sasisekharan for insightful discussions, David Weingeist, and Luke Robinson for help with compiling information on clinical combination chemotherapies, David Eavarone for helpful discussions on polymeric nanoparticles, Barghavi Govindarajan for careful review of this chapter, and Ada Ziolkowski for assistance with correspondence.

References

Akinc, A. et al. A combinatorial library of lipid-like materials for delivery of RNAi therapeutics. *Nature Biotechnology*, 26 (2008): 561–569.

Alberts, B. *Molecular Biology of the Cell*. Garland Science, Taylor & Francis Group, Oxfordshire, U.K., 1983.

Allen, C. et al. Controlling the physical behavior and biological performance of liposome formulations through use of surface grafted poly(ethylene glycol). *Bioscience Reports*, 22(2) (2002): 225–250.

Amiji, M.M. *Nanotechnology for Cancer Therapy*, Vol. 82. CRC Press, Boca Raton, FL, 2006.

Arnone, D., C. Ciesla, and M. Pepper. Terahertz imaging comes into view. *Physics World*, 4 (April 2000): 35–40.

Aubin-Tam, M.-E. and K. Hamad-Schifferli. Structure and function of nanoparticle-protein conjugates. *Biomedical Materials*, 3 (2008): 1–17.

Bean, C.P. and J.D. Livingston. Superparamagnetism. *Journal of Applied Physics*, 30 (1959): S120.

Bruchez, M.P. and C.Z. Hotz. *Quantum Dots: Applications in Biology*. Humana Press, Totowa, NJ, 2007.

Campbell, R.J. Biomimetic materials. In *Molecular Biology and Biotechnology: A Comprehensive Desk Reference*, R.A. Meyers. (Ed.). Wiley-VCH, New York, 1995.

Carmeliet, P. and R.K. Jain. Angiogenesis in cancer and other diseases. *Nature*, 407 (2000): 249–257.

Cattaneo, R., T. Miest, E.V. Shashkova, and M.A. Barry. Reprogrammed viruses as cancer therapeutics: Targeted, armed and shielded. *Nature Reviews Microbiology*, 6 (2008): 529–540.

Chasin, M. and R.S. Langer. *Biodegradable Polymers as Drug Delivery Systems*. Taylor & Francis, Inc., Boca Raton, FL, 1990.

Chou, T.C. Theoretical basis, experimental design, and computerized simulation of synergism and antagonism in drug combination studies. *Pharmacological Reviews*, 58 (2006): 621–681.

Cross, S.E., Y.-S. Jin, J. Rao, and J.K. Gimzewski. Nanomechanical analysis of cells from cancer patients. *Nature Nanotechnology*, 2 (2007): 780–783.

Cui, Y., Q. Wei, H. Park, and C.M. Lieber. Nanowire nanosensors for highly sensitive and selective detection of biological and chemical species. *Science*, 293(5533) (2001): 1289–1292.

Deng, W.-G. et al. Synergistic tumor suppression by coexpression of FUS1 and p53 is associated with down-regulation of murine double minute-2 and activation of the apoptotic protease-activating factor 1–dependent apoptotic pathway in human non–small cell lung cancer cells. *Cancer Research*, 67 (2007): 709–717.

Devalapally, H., Z. Duan, M.V. Seiden, and M.M. Amiji. Modulation of drug resistance in ovarian adenocarcinoma by enhancing intracellular ceramide using tamoxifen-loaded biodegradable polymeric nanoparticles. *Clinical Cancer Research*, 14 (2008): 3193–3203.

Everts, M. et al. Covalently linked Au nanoparticles to a viral vector: Potential for combined photothermal and gene cancer therapy. *Nano Letters*, 6(4) (2006): 587–591.

Ferguson, B. and X.-C. Zhang. Materials for terahertz science and technology. *Nature Materials*, 1 (2002): 26–33.

Figg, W.D. and J. Folkman. *Angiogenesis: An Integrative Approach from Science to Medicine.* Springer, New York, 2008.

Folkman, J. Fighting cancer by attacking its blood supply. *Scientific American*, 275(3) (1996): 150–154.

Goodman, T.T., P.L. Olive, and S.H. Pun. Increased nanoparticle penetration in collagenase-treated multicellular spheroids. *International Journal of Nanomedicine*, 2(2) (2007): 265–274.

Gothelf, A., L.M. Mir, and J. Gehl. Electrochemotherapy: Results of cancer treatment using enhanced delivery of bleomycin by electroporation. *Cancer Treatment Reviews*, 29(5) (2003): 371–387.

Grayson, A.C.R. et al. Multi-pulse drug delivery from a resorbable polymeric microchip device. *Nature Materials*, 2 (2003): 767–772.

Grifman, M. et al. Incorporation of tumor-targeting peptides into recombinant adeno-associated virus capsids. *Molecular Therapeutics*, 3(6) (2001): 964–975.

Hanahan, D. and R.A. Weinberg. The hallmarks of cancer. *Cell*, 100(1) (2000): 57–70.

Harold, F.M. *The Way of the Cell: Molecules, Organisms, and the Order of Life.* Oxford University Press, Oxford, NY, 2001.

Hartwell, L., D. Mankoff, A. Paulovich, S. Ramsey, and E. Swisher. Cancer biomarkers: A systems approach. *Nature Biotechnology*, 24 (2006): 905–908.

Hasirci, V., K. Lewandrowski, J.D. Gresser, D.L. Wise, and D.J. Trantolo. Versatility of biodegradable biopolymers: Degradability and an in vivo application. *Journal of Biotechnology*, 86(2) (2001): 135–150.

Hirsch, L.R. et al. Nanoshell-mediated near-infrared thermal therapy of tumors under magnetic resonance guidance. *Proceedings of the National Academy of Sciences USA*, 100(23) (2003): 13549–13554.

Hong, S., P.R. Leroueil, I.J. Majoros, B.G. Orr, J.R. Baker Jr., and Holl M.M. Banaszak. The binding avidity of a nanoparticle-based multivalent targeted drug delivery platform. *Chemical Biology*, 14 (2007): 107–115.

Huwyler, J., D. Wu, and W.M. Pardridge. Brain drug delivery of small molecules using immunoliposomes. *Proceedings of the National Academy of Sciences USA*, 93 (1996): 14164–14169.

Jackson, A.M., J.W. Myerson, and F. Stellacci. Spontaneous assembly of subnanometre ordered domains in the ligand shell of monolayer-protected nanoparticles. *Nature Materials*, 3 (2004): 330–336.

Kano, M. et al. Improvement of cancer-targeting therapy, using nanocarriers for intractable solid tumors by inhibition of TGF-β signaling. *Proceedings of the National Academy of Sciences USA*, 104(9) (2007): 3460–3465.

Keith, C.T., A.A. Borisy, and B.R. Stockwell. Multicomponent therapeutics for networked systems. *Nature Reviews Drug Discovery*, 4 (2005): 1–8.

Khademhosseini, A., R. Langer, J. Borenstein, and J.P. Vacanti. Microscale technologies for tissue engineering and biology. *Proceedings of the National Academy of Sciences USA*, 103(8) (2006): 2480–2487.

Kontermann, R.E. Immunoliposomes for cancer therapy. *Current Opinion in Molecular Therapeutics*, 8(1) (2006): 39–45.

Krishan, A. and P. Arya. Monitoring of cellular resistance to cancer chemotherapy. *Hematology/Oncology Clinics of North America*, 16(2) (2002): 357–372.

Kumar, Challa S.S.R. *Nanomaterials for Medical Diagnosis and Therapy.* Wiley-VCH, Weinheim, Germany, 2007.

Langer, R. and J. Folkman. Sustained release of macromolecules from polymers. In *Polymeric Delivery Systems*, Vol. 68, R.J. Kostelnik. (Ed.). Gordon and Breach Science Publishers, New York, 1978.

Langer, R.A. and D.A. Tirrell. Designing materials for biology and medicine. *Nature*, 428 (2004): 487–492.

LaVan, D.A., T. McGuire, and R. Langer. Small scale systems for in vivo drug delivery. *Nature Biotechnology*, 21 (2003): 1184–1191.

Lee, E.S., D. Kim, Y.S. Youn, K.T. Oh, and Y.H. Bae. A virus-mimetic nanogel vehicle. *Angewandte Chemie International Edition*, 47(13) (2008): 2418–2421.

Liu, D. and R. Sasisekharan. Role of heparan sulfate in cancer. In *Chemistry and Biology of Heparin and Heparan Sulfate*, H.G. Garg, R.J. Linhardt and C.A. Hales. (Eds.). Elsevier, Oxford, U.K., 2005.

Liu, W., H.S. Choi, J.P. Zimmer, E. Tanaka, J.V. Frangioni, and M. Bawendi. Compact cysteine-coated CdSe(ZnCdS) quantum dots for in vivo applications. *Journal of the American Chemical Society*, 129(47) (2007): 14530–14531.

Mann, B.K., A.S. Gobin, A.T. Tsai, R.H. Schmedlen, and J.L. West. Smooth muscle cell growth in photopolymerized hydrogels with cell adhesive and proteolytically degradable domains: Synthetic ECM analogs for tissue engineering. *Biomaterials*, 22 (2001): 3045–3051.

Menikh, A., R. MacColl, C. Mannella, and X.-C. Zhang. Terahertz biosensing technology: Frontiers and progress. *ChemPhysChem*, 3 (2002): 655–658.

Milman, H.A. and E.K. Weisburger. *Handbook of Carcinogen Testing*, Vol. 221. William Andrew Inc., Park Ridge, NJ, 1994.

Murphy, E.A. et al. Nanoparticle-mediated drug delivery to tumor vasculature suppresses metastasis. *Proceedings of the National Academy of Sciences USA*, 105(27) (2008): 9343–9348.

Peer, D., J.M. Karp, S. Hong, O.C. Farokhzad, R. Margalit, and R. Langer. Nanocarriers as an emerging platform for cancer therapy. *Nature Nanotechnology*, 2 (2007): 751–760.

Raman, R., S. Raghuram, G. Venkataraman, J.C. Paulson, and R. Sasisekharan. Glycomics: An integrated systems approach to structure-function relationships of glycans. *Nature Methods*, 2 (2005): 817–824.

Reich, S., C. Thomsen, and J. Maultzsch. *Carbon Nanotubes: Basic Concepts and Physical Properties.* Wiley-VCH, Berlin, Germany, 2004.

Rieth, M. and W. Schommers. *Handbook of Theoretical and Computational Nanotechnology.* American Scientific Publishers, New York, 2006.

Sasisekharan, R., Z. Shriver, G. Venkataraman, and U. Narayanaswamy. Roles of heparan sulfate glycosaminoglycans in cancer. *Nature Reviews Cancer*, 2 (2002): 521–528.

Sengupta, S. and R. Sasisekharan. Exploiting nanotechnology to target cancer. *British Journal of Cancer*, 96 (2007): 1315–1319.

Sengupta, S. et al. Temporal targeting of tumour cells and neovasculature with a nanoscale delivery system. *Nature*, 436 (2005): 568–572.

Shrestha, S. and S.K. Deo. Bioluminescence resonance energy transfer in bioanalysis. In *Photoproteins in Bioanalysis*, S. Daunert and S.K. Deo. (Eds.), pp. 123–128. Wiley-VCH, Weinheim, Germany, 2006.

Shuler, M.L. and H. Xu. Novel cell culture systems: Nano and microtechnology for toxicology. In *Computational Toxicology*, S. Ekins. (Ed.). John Wiley & Sons, Hoboken, NJ, 2007.

Siepmann, J. and A. Gopferich. Mathematical modeling of bioerodible, polymeric drug delivery systems. *Advanced Drug Delivery Reviews*, 48(2) (2001): 229–247.

Simberg, D. et al. Biomimetic amplification of nanoparticle homing to tumors. *Proceedings of the National Academy of Sciences USA*, 104(3) (2007): 932–936.

Timms, J.F. Cancer proteomics: From bench to bedside. *British Journal of Cancer*, 99 (2008): 679.

Tomalia, D.A. and J.M.J. Frechet. *Dendrimers and Other Dendritic Polymers.* John Wiley & Sons, New York, 2003.

Tran, M.A., C.D. Smith, M. Kester, and G.P. Robertson. Combining nanoliposomal ceramide with sorafenib synergistically inhibits melanoma and breast cancer cell survival to decrease tumor development. *Clinical Cancer Research*, 14 (2008): 3571–3581.

Verma, A. et al. Surface-structure-regulated cell-membrane penetration by monolayer-protected nanoparticles. *Nature Materials*, 7 (2008): 588–595.

Visaria, R.K. et al. Enhancement of tumor thermal therapy using gold nanoparticle-assisted tumor necrosis factor-alpha delivery. *Molecular Cancer Therapeutics*, 5(4) (2006): 1014–1020.

von Burkersroda, F., L. Schedl, and A. Gopferich. Why degradable polymers undergo surface erosion or bulk erosion. *Biomaterials*, 23(21) (2002): 4221–4231.

Wagner, E. Programmed drug delivery: Nanosystems for tumor targeting. *Expert Opinion on Biological Therapy*, 7(5) (2007): 587–593.

Wagner, E.K. and M.J. Hewlett. *Basic Virology.* Blackwell Publishing, Malden, MA, 2004.

Weitzman, J. Combing cancer. *Trends in Molecular Medicine*, 7(8) (2001): 337.

West, J.L., R.A. Drezek, and N.J. Halas. Nanotechnology provides new tools for biomedical optics. In *The Biomedical Engineering Handbook*, J.D. Bronzino. (Ed.). CRC Press, Boca Raton, FL, 2006.

Wu, J.T. Identification of risk factors for early neoplasm. In *Cancer Diagnostics*, R.M. Nakamura, W.W. Grody, J.T. Wu and R.B. Nagle. (Eds.). Springer-Verlag, New York, 2004.

Wu, X. et al. Immunofluorescent labeling of cancer marker Her2 and other cellular targets with semiconductor quantum dots. *Nature Nanotechnology*, 21(1) (2003): 41–46.

Yoo, H.S., K.H. Lee, J.E. Oh, and T.G. Park. In vitro and in vivo anti-tumor activities of nanoparticles based on doxorubicin-PLGA conjugates. *Journal of Controlled Release*, 68(3) (2000): 419–431.

Zheng, G., F. Patolsky, Y. Cui, W.U. Wang, and C.M. Lieber. Multiplexed electrical detection of cancer markers with nanowire sensor arrays. *Nature Biotechnology*, 23 (2005): 1294–1301.

Zimmerman, G.R., J. Lehar, and C.T. Keith. Multi-target therapeutics: When the whole is greater than the sum of the parts. *Drug Discovery Today*, 12 (2007): 34–42.

Zur Hausen, H., J.G. Fox, and T.C. Wang. *Infections Causing Human Cancer.* Wiley, John & Sons, Inc., New York, 2006.

Cancer Nanotechnology: Targeting Tumors with Nanoparticles

Erem Bilensoy
Hacettepe University

35.1 Introduction

Cancer is a leading cause of death in the world, with 10 million new diagnosed patients every year ending in 6 million deaths annually, which is 12% of all death causes. The total cost for cancer management in the United States is estimated to be $157 million. By the year 2020, it is estimated that there will be 15 million new cases every year (Brannon-Peppas and Blanchette 2004, Feng et al. 2005). However, it was reported by the World Health Organization (WHO) that the rate of death from cancer has remained unchanged between the years 1950 and 2001 in the United States (Feng et al. 2005). This suggests that although many steps have been taken to improve the efficacy of cancer therapy, a substantial progress has not been reached in the treatment of cancer. Though chemotherapy is successful to some extent, most current anticancer drugs do not greatly differentiate between cancerous and normal cells. Thus, in the process of killing cancer cells, chemotherapeutic agents also damage healthy tissues leading to systemic toxicity and adverse side effects. This is a crucial parameter limiting the maximum allowable dose for chemotherapeutic drugs and reduces patient compliance to a significant extent.

Major challenges are associated with cancer therapy. It is believed that severe side effects and inefficient chemotherapy is mainly caused by formulation factors, pharmacokinetics and toxicity of the drug, drug resistance by cancer cells, and lack of effective oral administration for such drugs. These important challenges in cancer therapy may be briefly summarized as follows:

1. *Formulation factors*: Most anticancer agents that are in use or under clinical trials today are very hydrophobic molecules. Taxane drugs paclitaxel and docetaxel as well as topoisomerase inhibitor camptothecin can be given as examples of such drugs. In order to administer these hydrophobic molecules in injectable form, it is necessary to incorporate significant concentration of co-solvents to solubilize the anticancer agent in water or buffer solution.

A very typical example to this would be Cremophor EL (polyoxyethylated castor oil), which is the co-solvent used in Taxol®, commercial injectable paclitaxel formulation consisting of 6 mg/mL paclitaxel solution in cremophor EL:ethanol mixture (50/50; v/v%). Cremophor EL is associated with severe side effects such as hypersensitivity reactions, nephrotoxicity, hyperlipidemia, abnormal lipoprotein patterns, erythrocyte aggregation, and peripheral neuropathy (Weiss et al. 1990, Lilley and Scott 1993, Gelderblom et al. 2001).

2. *Pharmacokinetics and physicochemical properties of anticancer drugs*: Pharmacokinetics is an important parameter in chemotherapy. It is necessary to expose cancer cells to a sufficiently high concentration of drug for a sufficiently long duration. Current chemotherapy is an intermittent process with periodic injection/iv infusion. Recess is needed between successive treatment periods mostly to allow patient recovery from severe side effects of chemotherapy. These recess periods allow tumor blood vessels to grow rapidly and thus deteriorate the effects of chemotherapy. High peak drug concentration causes severe side effects. However, it is necessary to administer large quantities of drug due to rapid elimination and widespread distribution into target organs and tissues (Parveen and Sahoo 2008).

In general, it is believed that long-term exposure to the drug at modest concentrations would be more effective than a pulsed supply at a high concentration (Feng et al. 2005). In addition, anticancer drugs can be cell-cycle specific or cell-growth-phase specific meaning that they can exert an effective action on cancer cells only during a certain phase of cell growth and proliferation. Therefore, the cell cytokinetics must be carefully considered for effective chemotherapy and dose scheduling is crucial with cell-cycle-specific drugs of high efficacy at the right time to the desired location at a high concentration but safe enough over a sufficiently long period.

Along with pharmacokinetics of anticancer agents, solubility and stability problems of most of these molecules lead to ineffective chemotherapy and reduced patient compliance. Camptothecin, for example, is a potent anticancer agent effective against a wide variety of cancer types in cell culture studies. However, it is clinically ineffective since it is not stable under physiological pH. The active carboxylate form of camptotecin is rapidly hydrolyzed to the inactive lactone form once administered into the body (Takimoto et al. 1998, Hatefi and Amsden 2002).

3. *Drug resistance*: Another serious drawback associated with cancer therapy is drug resistance caused by physiological barriers or cellular mechanisms. This must be solved at the vascular, interstitial, and cellular levels (Links and Brown 1999). Drug transport across tumor microvessels can occur via interendothelial junctions, transendothelial channels, vesicular vacuolar organelles, and fenestrations. It was found that pore cutoff sizes of several tumor models fall within the range of 380–780 nm and certain liposomes were demonstrated to leak out of the porous vessel walls into solid tumors with a pore cutoff size of around 400 nm (Yuan et al. 1995, Hobbs et al. 1998). Unlike most normal tissues, tumor interstitium, which is the dense liquid inner structure of the tumor connecting tumor cells (see Figure 35.1), is characterized by high hydrostatic pressure, which

leads to an outward convective interstitial fluid flow and causes drug resistance (Jain 2001).

Cellular mechanisms such as alteration of enzyme activity or apoptosis (programmed cell death) regulation and transport-based mechanisms such as P-glycoprotein (Pgp)-related multidrug resistance (MDR) also contribute to resistance of tumors to therapeutic agents. Various known physiological barriers to drug delivery such as gastrointestinal (GI) tract or blood–brain barriers (BBB) may have similar mechanisms.

4. *Lack of oral administration/poor oral bioavailability*: In addition to being convenient to patients, oral delivery of anticancer agents may greatly improve their efficacy and reduce side effects. Oral delivery provides sustained exposure to a safe but effective drug concentration to the cancer cells, which may produce better efficacy and fewer side effects than injection/infusion. Unfortunately, most anticancer drugs are not bioavailable upon oral administration. Paclitaxel has an oral bioavailability (rate and extent of absorption of a drug upon oral administration) of only 1%, and exemestane, an antiestrogen anticancer drug, has an oral bioavailability of only 5% (FDA NDA 20753).

The reason for this lack of oral bioavailability is that the drug would be eliminated by the first metabolic process with cytochrome P450 and the efflux pump Pgp (Malingre et al. 2000). Some pharmaceutical companies propose the concurrent use of P450/Pgp suppressors such as Cyclosporin A; however this does not seem feasible since Cyclosporin A would greatly harm the immune system and thus cause medical complications and exert side effects as well as difficulties in formulation. A successful

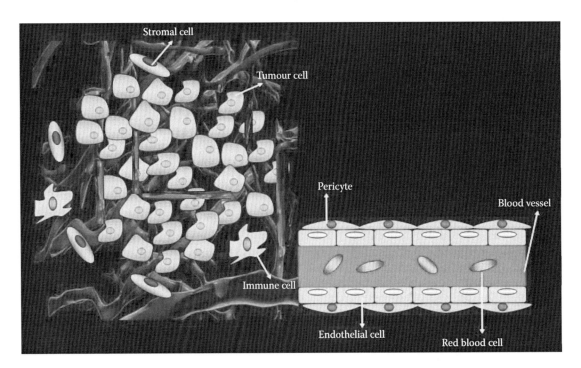

FIGURE 35.1 The structure of a solid tumor and the surrounding vasculature.

oral chemotherapy would allow patients to self-administer the chemotherapy preferably at home and significantly improve patient quality of life.

All these challenges lead to the design and development of nanosized carrier systems capable of extravasation from vascular endothelia in tumor site, meaning that these nanoparticles can leak through the porous blood vessel walls. In this way, nanoparticles provide a modest amount of drug for a prolonged period of time with a controlled release profile for the anticancer agent and hopefully to overcome physiological and cellular barriers for effective chemotherapy.

35.2 Tumor Structure and Physiology: Getting to Know the Battlefield

Tumor is a morbid state characterized by the spontaneous outgrowth of an abnormal mass of cells. The evolution of tumors is random, disorganized, and a condition of numerous mutations. It is a malignant or benign condition that encompasses its own rules of morphogenesis, an immortal state that elucidates different physiologies (Agarwal et al. 2008). Under normal conditions, cells reproduce, grow, divide, multiply, and eventually undergo apoptosis. This maintains proper balance and functioning of the organs. However, tumor cells are subjected to the uncontrolled proliferation of cells, evade apoptosis, and hence develop an abnormal mass of cells that can be life threatening if untreated at an early stage.

Upon undergoing change in the organizational pattern of a gene, the normal cell cycle is disrupted. The gene in question is called proto-oncogene and is thus converted into an oncogene (gene encoding abnormal growth for the cell) and loses its normal growth regulation. As a consequence of mutations, abnormal growth and proliferation of cells begin along with the expression of various surface markers and proteins that facilitate their growth and supplement the cells with necessary nutrients and oxygen at the expense of normal cells. The abnormal signal proteins may induce a faster proliferation of cells. Thus, a gene in a normal cell on being converted to an oncogene proliferates under the stimuli of these agents and leads to the development of tumors (Folkman 2002, Siemann et al. 2006). Pericytes are cellular components of the connective tissue that grows around the basement membrane. This growth of pericytes around tumors provides them with adequate protection against tumor targeting. Pericytes have been known to facilitate tumor growth by promoting cell-to-cell attachment and hence stabilizing the tumor mass (Banfi et al. 2005, Molema 2005).

Most of the ≈10^{13} cells in the human body are within a few cell diameters of a blood vessel. This remarkable feat of organization facilitates the delivery of oxygen and nutrients to the cells that form the tissues of the body. It also enables the delivery of most medicines. However, the homeostatic regulation of tissue and growth of blood vessels tend to break down in solid cancers. Tumor cells often have the potential for more rapid proliferation than the cells that form blood capillaries. So the proliferation of

tumor cells forces vessels apart, reducing vascular density and creating a population of cells distant (>100 μm) from blood vessels, a process that is promoted by a poorly organized blood vessel architecture, irregular blood flow, and the compression of blood and lymphatic vessels by cancer cells (Padera 2004, Minchinton and Tannock 2006), as seen in Figure 35.1. In addition, the disorganized blood vessel network and the absence of functional lymphatics causes increased interstitial fluid pressure (IFP). Finally, the composition and structure of extracellular matrix, the complex group of molecules that exist in tissue outside cells, can slow down the movement of molecules within the tumor (Netti et al. 2000, Heidin et al. 2004).

The vascularization of tumors is heterogeneous, showing dead or bleeding regions as well as regions that are densely vascularized in order to sustain an adequate supply of nutrients and oxygen for rapid tumor growth called angiogenesis (Jain 2001, Brigger et al. 2002). Tumor blood vessels present several abnormalities in comparison to normal physiological vessels, often including a relatively high proportion of proliferating endothelial cells, increasingly curved structures, a deficiency in pericytes, and an aberrant basement membrane formation (Seymour et al. 2002, Baban et al. 1998). The resulting enhanced permeability of tumor vasculature is thought to be regulated by various mediators such as vascular endothelial growth factor (VEGF), bradykinin, nitric oxide, prostaglandins, and matrix metalloproteinases (Maeda 2001). Macromolecular transport pathways across tumor vessels have been shown to occur via open gaps (interendothelial junctions and transendothelial channels), vesicular vacuolar organelles, and fenestrations. However, it remains controversial as to which pathways are predominantly responsible for tumor hyperpermeability and macromolecular transvascular transport. Regardless of the transport mechanism, pore cutoff size of several tumor models has been reported to be ranging between 100 and 780 nm depending on the type of tumor (Haley and Frenkel 2008). This is in contradiction with the tight endothelial junctions of normal vessels that are typically around 5–10 nm in size.

The tumor inner compartment is predominantly composed of a collagen and elastic fiber network. Interdispersed within this cross-linked structure are the interstitial fluid and macromolecular constituents (hyaluronate and proteoglycans) that form a hydrophilic gel (Jain 1987). Tumor interstitial pressure is higher in the tumor center and lower in the periphery favoring decreased drug diffusion to the center of tumors, leading to an outward convective interstitial fluid flow along with the absence of an anatomically well-defined and functioning lymphatic network. Hence, the transport of an anticancer drug in the interstitium will be governed by physiological parameters such as pressure as well as physicochemical properties of the interstitium (composition, structure, charge) and the drug molecule (size, configuration, charge, hydrophobicity) (Jain 1987, Brigger et al. 2002).

As a consequence of poorly organized vasculature in solid tumors, there is limited delivery of oxygen and other nutrients to cells that are distant from functional blood vessels. Poor vascular organization also leads to the buildup of products of metabolism such as lactic acid and carbonic acid, which lower the extracellular pH to ≈6.8–7.0 in contrast to the normal extracellular pH, which is 7.4. Although tumor cells may maintain intracellular pH, the acidic extracellular pH is an important aspect of mutagenesis and crucial for pH-dependent drug delivery to tumors. Figure 35.1 represents a schematic representation of a solid tumor structure with pericytes and immune cells surrounding the tumor mass. The growth process of a tumor, on the other hand, is summarized in Figure 35.2.

Nanoparticles are purposely engineered and constructed systems that are measured in nanometer size. Usually these particles range from a few nanometers to several hundred nanometers depending on their intended use and preparation technique. A variety of materials are used to prepare nanoparticles including natural or synthetic polymers, polysaccharides, lipids, ceramic, and metals. Nanoparticle structures may contain predominantly organic molecules or have inorganic elements as a metal core. Nanoparticulate carriers have a variety of sizes and shapes that include spheres, branched structures, shells, tubes emulsions, and liposomes.

This chapter deals with spherical nanocarriers, namely, nanospheres and nanocapsules, that consists the majority of nanoparticles. Nanoparticles in this sense may be defined as two different types of delivery systems depending on the nature of the system being either matrix-type or membrane-type (reservoir-type) systems. Figure 35.3 depicts nanospheres that are matrix

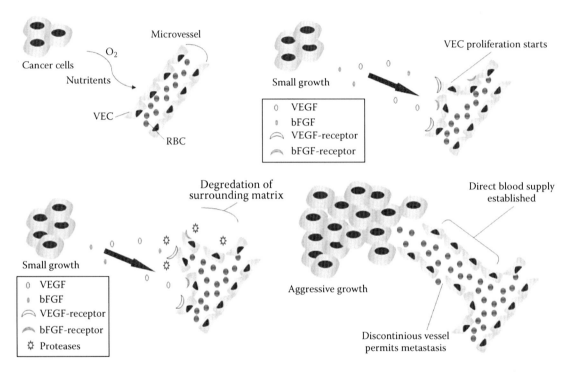

FIGURE 35.2 The growth of a tumor and factors affecting tumor growth. (From Dass, C.R., *J. Drug Target.*, 12, 245, 2004.)

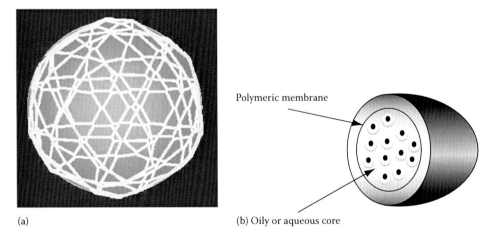

FIGURE 35.3 Schematic representation of spherical nanoparticulate drug carriers; matrix-type nanosphere (a) and membrane-type nanocapsule (b).

systems consisting of a loose network of polymeric bundle in which the drug is dispersed within the matrix and nanocapsules that are vesicular systems in which the drug is confined to a generally oily core surrounded by a thin and continous polymeric film. Therapeutic drugs are incorporated into nanoparticles by entrapment within the matrix, or encapsulation within the core or surface attachment. Nanocapsules schematized in Figure 35.3 are membrane-type systems with an oily core surrounded by a continuous thin polymer film. Nanocapsules are mainly utilized for the encapsulation and controlled release of lipophilic drugs.

35.3 Targeting Strategies

The first important step in improving treatment regimes is the better utilization of the potency of therapeutic agents by more effectively targeting them to tumor tissues. The therapeutic efficacy of drugs can be greatly amplified and their toxicity greatly reduced if high concentrations of anticancer agents could be selectively administered to malignant tissues only. Thus, the need to develop novel cancer therapies and drug delivery strategies that provide specific targeting to tumor cells has been continually at the forefront of medical sciences. Nanotechnology, one of the frontier sciences, can pave the way to overcome the numerous barriers to an efficient and safe drug delivery system. Ideally, drug delivery systems should improve the stability, absorption, and therapeutic action of the drug within the target tissue and permit the long-term release of the drug. Moreover, in addition to reducing the frequency of drug administration and thus improving patient compliance, a novel drug delivery system should offer protection and improve the pharmacokinetics of various drugs. To be successfully used in drug delivery system formulations, materials should be biodegradable, free from leachable impurities, and chemically inert. Polymers should be able to break down into biologically acceptable molecules that can be metabolized and eliminated from the body via normal metabolic pathways. Moreover, the drug delivery system should be able to direct the drug to its target and maintain its concentration at the site for a sufficient period of time for therapeutic action to take effect. Table 35.1 summarizes the ideal characteristics of a targeted nanomedicine.

In chemotherapy, pharmacologically active cancer drugs reach the tumor tissues with poor specificity and dose-limiting

TABLE 35.1 Characteristics of an Ideal Tumor-Targeted Nanomedicine

1. Increase drug localization at tumor through
   a. Passive targeting
   b. Active targeting
2. Decrease drug localization in sensitive, nontarget tissues
3. Ensure minimal drug leakage during transit to target
4. Protect the drug from degradation and from premature clearance
5. Retain the drug at the target site for the desired period of time
6. Facilitate cellular uptake and intracellular trafficking
7. Biocompatible and biodegradable

Source: Lammers, T. et al., *Br. J. Cancer*, 99, 392, 2008.

toxicity, thus resulting in harmful effects to healthy tissues. Targeting cancer cells using nanoparticles loaded with anticancer drugs is a promising tactic that could help overcome these challenges. Drug targeting can be achieved by taking advantage of the distinct pathophysiological features of a tumor tissue (passive targeting) or by targeting the drug carrier using some target-specific ligands also known as active targeting.

35.3.1 Passive Targeting

35.3.1.1 EPR Effect

Twenty years ago, Maeda et al. (1986) found that Evans Blue dye, which binds with plasma albumin, concentrated selectively in tumor tissues following intravenous injection. The same behavior was also noticed with radiolabeled plasma proteins including transferrin (molecular weight MW: 90 kDa (kiloDalton)) and IgG (MW: 160 kDa) whereas smaller proteins such as neocarzinostatin (MW: 12 kDa) and ovomucoid (MW: 29 kDa) did not accumulate in tumors (Figure 35.4). Passive targeting exploits the anatomical differences between normal and tumor tissues. These differences are more heterogeneous with larger vascular density and leaky blood vessel structure unlike the tight endothelium of normal blood vessels. In addition, the extensive production of vascular mediators such as bradykinins, nitric oxide, VEGF, and prostaglandins facilitate extravasation. This, coupled with impaired lymphatic drainage of macromolecules in solid tumors allows enhanced accumulation and retention of high molecular weight drugs in solid tumors. This is popularly known as enhanced permeation and retention (EPR) effect, which allows the extravasation of circulating polymeric nanoparticles within the tumor interstitium and also increases the concentration of the chemotherapeutic agent in the tumor tissue (Greish et al. 2003, Parveen and Sahoo 2008).

Drug carriers in the nanometer size range can extravasate through these gaps into the tumor interstitial space. Because tumors have impaired lymphatic drainage, the carriers concentrate in the tumor, resulting in higher drug concentration in tumor (10-fold or higher) than that can be achieved with the administration of the same dose of free drug. Passive targeting by EPR effect is schematically described in Figure 35.5.

It is now almost certain that polymeric drugs having a molecular weight above the kidney threshold (>40 kDa) accumulate in tumor tissues for prolonged time periods following intravenous injection. Another prerequisite for the EPR effect to take place is that the plasma concentration of the drug must remain high for more than 6 h in mice and rat measured by the area under the time-concentration curve (AUC) (Noguchi et al. 1998, Maeda 2001). The rate of accumulation of the macromolecules and lipids in tumor is conversely proportional to their clearance rate, meaning that the longer the macromolecular drug remains in the circulation, the higher its tumor accumulation will be since only tumor blood vessels are capable of allowing large molecules into the tumor interstitium. Owing to the impaired lymphatic drainage system, its concentration will remain high for a long period fulfilling

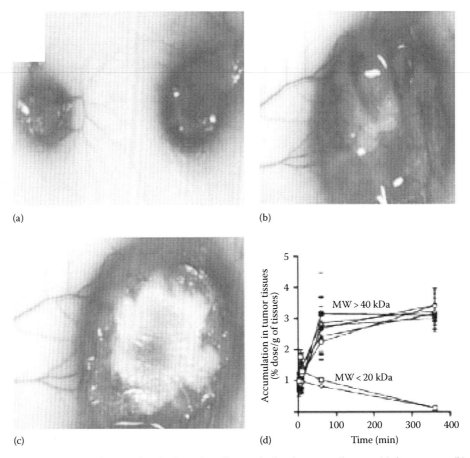

(a)   (b)

(c)   (d)

FIGURE 35.4 Selective accumulation of Evans Blue dye bound to albumin (70 kDa) in a small tumor (a), large tumor (b), and cross section of large tumor (c). The depth of the penetration of the dye was shown. The central region of the tumor is necrotic and avascular and hence does not facilitate the uptake of macromolecules because it is not a growing area (d) shows the accumulation of ^{125}I labeled *N*-(2-hydroxypropyl) methacrylamide) HPMA copolymer in the solid tumor tissues. Note that large macromolecules but not smaller ones manifest progressive accumulation. (Reprinted from Greish, K., *J. Drug Target.*, 15, 457, 2007. With permission.)

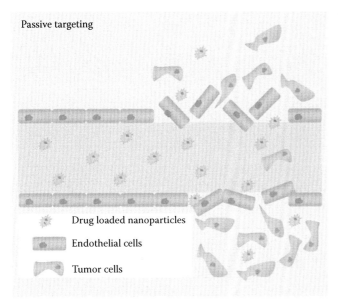

FIGURE 35.5 Passive drug-targeting strategy through enhanced permeation and retention (EPR) effect.

the enhanced permeation and retention targeting strategy. The EPR effect thus does not imply the mere passive delivery of nanosize carriers to tumor tissues but also encompasses the prolonged retention of these drugs for weeks or even months in the tumor tissues ensuring effective and prolonged anticancer effect (Greish 2007). The factors responsible for EPR effect of nanocarriers into solid tumors are seen in Table 35.2.

Passive targeting and accumulation to tumor tissue has been demonstrated using different nanoparticle systems including PLGA, PLA, polyalkylcyanoacrylates, polymethylmethacrylates, poly-epsilon-caprolactone, albumin, gelatin, chitosan, and amphiphilic cyclodextrin (Brigger et al. 2002, Bilensoy, 2008, Bilensoy et al. 2008a,b, Parveen and Sahoo 2008) have been reported with a variety of anticancer agents such as paclitaxel, docetaxel, tamoxifen, doxorubicin, and camptothecin that are associated with bioavailability problems either due to their poor aqueous solubility or poor stability under physiological conditions.

The association of a cytostatic drug to conventional carriers leads to modifications of the drug biodistribution profile, as it is mainly delivered to the mononuclear phagocytic system (MPS) also known as reticuloendothelial system (RES) consisting of

TABLE 35.2   Factors Responsible for the EPR Effect of Macromolecules in Solid Tumors

1. Anatomical factors:

   a. Extensive angiogenesis and high vascular density

   b. Lack of smooth muscle layer, sporadic blood flow, passive dilatation of vessels in the AT-II induced hypertensive state, more leakage

   c. Defective vascular architecture, extensive leakage

   d. Meager lymphatic clearance, enhanced retention of macromolecular drugs, and lipidic particles in the interstitium of tumors

   e. Slow venous return, accumulation of macromolecules in the interstitium

2. Generation of permeability-enhancing factors as follows:

   a. Vascular endothelial growth factor (VEGF/VPF)

   b. Bradykinin/hydroxypropylbradykinin (BK)

   c. Nitric oxide (NO)

   d. Peroxynitrite ($ONOO^-$) (reaction product of superoxide radical and NO)

   e. Matrix metalloproteinases (MMPs), proMMP is activated by $ONOO^-$

   f. Prostaglandins (PGs)

   g. Other proteinases (e.g., kalikrein activated) involved in various protease cascades

   h. Other cytokines (e.g., tumor necrosis factor, IL-2), facilitate EPR effect.

Source: Greish, K., *J. Drug Target.*, 15, 457, 2007.

RES organs of liver, spleen, lungs, and bone marrow. Indeed once in the bloodstream, conventional nanoparticles are rapidly opsonized (coated by proteins) and massively cleared by the fixed macrophages of the RES organs (Brigger et al. 2002). It was reported that both the polymeric composition (type, hydrophobicity, biodegradation profile) of the nanoparticles and the associated drug (molecular weight, charge, localization in the nanospheres: adsorbed or incorporated) have a great influence on the drug distribution pattern in the RES organs (Couvreur et al. 1980). However, the exact underlying mechanism was not fully understood but it was observed that this effect was rapid (0.5 to 3 h) and compatible with endocytosis. This biodistribution can be of benefit for the chemotherapeutic treatment of RES-localized tumors (e.g., hepatocarcinoma or hepatic metastasis arising from digestive tract or gynecological cancers, bronchopulmonary tumors (primitive or metastases) including non–small-cells tumor and small-cells tumor and leukemia).

35.3.1.2 Long-Circulating "Stealth" Nanoparticles

Since the usefulness of conventional (uncoated) nanoparticles is limited by their massive capture by the macrophages of the RES after intravenous administration, other nanoparticulate devices must be considered to target tumors that are not localized in the RES area. Recently, a great deal of work has been devoted to developing the so-called stealth or ghost nanoparticles that can escape from macrophage uptake by surface modification and eventually circulate longer in the bloodstream. These stealth nanoparticles have been shown to be characterized by a prolonged half-life in blood (Gref et al. 1994).

It is known that hydrophobic particles have a greater tendency to be eliminated and larger nanoparticles are more prone to uptake by macrophages. They may be opsonized by serum proteins and then cleared rapidly. On the other hand, carriers possessing hydrophilic surfaces prevent their recognition by the RES. Meanwhile, copolymer micelles having a size of 20–100 nm

are small enough to avoid RES uptake (Kwon and Okano 1996). Hydrophilic modification of the surface is realized through the following approaches:

1. Coating of the nanoparticle surface with hydrophilic surfactants or polymers such as polysorbate 80, poloxamer 188, poloxamer 407, polyethyleneglycol (PEG), polyethyleneoxide (PEO), or poly-L-lysine (PLL)
2. Grafting of the above-mentioned hydrophilic polymers to the main hydrophobic polymer (copolymerization)

Chemical modification of pharmaceutical nanocarriers with certain synthetic polymers, such as PEG, is the approach most frequently used to prolong blood circulation behavior to drug carriers. Hydrophilic polymers have been shown to protect individual molecules and solid particles from interaction with different solutes. The term "steric stabilization" has been introduced to describe the phenomenon of polymer-mediated protection (Torchilin 2007). On the biological level, coating of nanoparticles with PEG sterically hinders interactions of blood components with their surface and reduces the binding of plasma proteins with PEGylated nanoparticles. This approach prevents drug carrier interaction with opsonins (serum proteins responsible of opsonization) and slows the capture by RES. The mechanism by which PEG prevents opsonization include shielding of the surface charge, increased surface hydrophilicity, enhanced repulsive interaction between polymer-coated nanocarriers and blood components, and formation of the polymeric layer over the particle surface, which is impermeable for large molecules of opsonins even at relatively low polymer concentrations.

As a protecting polymer, PEG provides a very attractive combination of properties: excellent solubility in aqueous solutions, high flexibility of its polymer chain, very low toxicity, immunogenicity, and antigenicity meaning that PEG does not induce a strong immune response in the body, lack of accumulation in RES cells, and minimal influence on biological properties of modified pharmaceuticals. It is also important that PEG is not biodegradable and subsequently does not form any toxic metabolites. PEG molecules with a molecular weight below 40 kDa are readily excretable from the body via kidneys. PEGs that are normally used for the modification of drug carriers have a molecular weight from 1 to 20 kDa.

PEGylated polymeric nanoparticles can be prepared based on the block copolymer of PEG and a hydrophobic block such as polylactide-*co*-glycolide (PLGA) (Gref et al. 1994). Using particles with a PLGA-PEG copolymer, one can prepare long-circulating nanoparticles with an insoluble PLGA core and a water-soluble PEG shell covalently linked to the core. Clearance and liver accumulation patterns reveal that the higher the content of PEG blocks, the slower the clearance and better protection from RES uptake. Similarly, coating cyanoacrylate nanoparticles with PEG resulted in their increased maintenance in the circulation allowing their diffusion into even brain tissue (Peracchia et al. 1999, Calvo et al. 2001).

The most significant biological consequence of nanoparticle modification with protecting polymers is a sharp increase in circulation time and decrease in liver accumulation (Torchilin and Trubetskoy 1995). This fact is very important clinically since various long-circulating nanocarriers have been shown to effectively accumulate in many tumors via the EPR effect. From a pharmacokinetic point of view, the association of drugs with any nanocarriers has pronounced effects: delayed drug absorption, restricted drug biodistribution, decreased volume of drug biodistribution, delayed drug clearance, and retarded drug metabolism. The presence of protective polymer on the carrier surface changes these parameters still further. The uptake of nanoparticles by the RES follows Michaelis–Menten kinetics (nonlinear pharmacokinetics). Therefore, the critical point of saturation concentration specifies the limits for doses. Increasing the dose may prevent escape from RES due to saturation of uptake sites but therapeutic indices, side effects, and dose-related drug toxicity need to be taken into consideration (Davis 2000).

35.3.1.3 Localized Delivery

Localized delivery is another approach to targeted delivery in which the therapeutic agent is administered directly to the target site (Labhasetwar 1997, Parveen and Sahoo 2006). This minimizes the systemic toxicity and increases concentration at the site of action. Intratumoral injection is the main administration route in this approach. In a study realized by Sahoo et al. (2004), prostate tumors were treated with direct intratumoral injection of paclitaxel in sustained release biodegradable nanoparticles. This approach was found to be practical and more effective than systemic drug therapy. But this type of targeted drug delivery is only suitable for those sites that are easily accessible. Direct intratumoral delivery of paclitaxel in biodegradable nanoparticles that were conjugated with transferrin ligand demonstrated complete tumor regression in the subcutaneous mice model of prostate cancer (Sahoo et al. 2004). It was also found that the mechanism of higher efficacy of transferrin-conjugated nanoparticles were greater cellular uptake and sustained intracellular retention of the encapsulated drug than that with drug in solution or unconjugated nanoparticles (Sahoo and Labhasetwar 2005).

Unlike water-soluble molecules that are rapidly absorbed though the blood capillary wall and pass into circulation, small particles injected locally infiltrate into the interstitial space around the injection site and are gradually absorbed by the lymphatic capillaries into the lymphatic system (Hawley et al. 1995, Nishioka and Yoshino 2001). For this reason, subcutaneously or locally injected (in the peritumoral region) nanoparticles can be used for lymphatic targeting as a tool for chemotherapy against lymphatic tumors or metastases. For example, aclarubicin adsorbed onto activated carbon particles was tested after subcutaneous injection in mice against a murine model (P388 leukemia cells) of lymph node metastases (Sakakura et al. 1992). The same system was also used in patients after intratumoral and peritumoral injections as a locoregional chemotherapy adjuvant for breast cancer (Hagiwara et al. 1997). In both applications, this carrier system distributed selectively high levels of free aclarubicin to the regional lymphatic system and low levels to the rest of the body (Sakakura et al. 1992, Hagiwara et al. 1997). However, this carrier system is open to criticism as it is not biodegradable and as it is rather large (>100 nm) impeding the drainage from the injection site though aqueous channels (Hawley et al. 1995). Besides, the drug is associated to the particles by adsorption, which leads to a rapid release with possible systemic absorption.

It would seem that PIBCA nanocapsules with an oily core for hydrophobic drugs (Nishikoa et al. 2001) or biodegradable systems coated by adsorption of the surface active agent Poloxamine 904 (Hawley et al. 1995) offer interesting properties for future investigations in this field. Indeed, PIBCA nanocapsules as compared to liposome or emulsion formulations showed a great potential to retain the lipophilic indicator 12-(9-anthroxy) stearic acid (ASA) in the regional lymph nodes for 168 h after intramuscular administration (Nishioka and Yoshino 2001). Poloxamine 904 caused an increased sequestration of the particles in lymph nodes, which would probably reduce systemic absorption of any encapsulated drugs (Hawley et al. 1995).

A study combining intratumoral administration of gadolinium-loaded chitosan nanoparticles (gadopentetic acid-chitosan complex nanoparticles) and neutron capture therapy was performed on the B16F10 melanoma model subcutaneously implanted in mice (Tokumitsu et al. 2000). Results showed an outstanding gadolinium retention in tumor tissue when it was encapsulated with respect to free drug (a highly water soluble compound with rapid elimination). Irradiation of the tumors was performed 8 h after the last intratumoral injection of gadolinium nanoparticles and prevented further tumor growth in the animals treated and increased their life expectancy (Tokumitsu et al. 2000).

Superficial bladder tumors were treated with intravesical injection of nanoparticles prepared from cationic polymers such as gelatin nanoparticles loaded with paclitaxel administered to dogs (Lu et al. 2004). This site is a convenient site to locally administer an anticancer agent as it is a relatively confined space with easy access though intravesical instillation. Thus, tissue concentrations of paclitaxel administered in gelatin nanoparticles were found to be significantly higher than the commercial cremophor formulation (Lu et al. 2004). Bioadhesive nanoparticles were also developed to retain the drug dosage in the bladder for a prolonged period of time and to protect the chemoterapeutic drug mitomycin C from the acidic environment of the bladder using nanoparticles of chitosan, poly-ε-caprolactone coated with chitosan or poly-L-lysine (Sarısözen et al. 2007, Bilensoy et al. 2009). Data proved that cellular uptake of chitosan-coated poly-ε-caprolactone nanoparticles to be significantly higher than other cationic nanoparticles in MB49 mouse bladder carcinoma cell line (Bilensoy et al. 2009) when compared with the normal bladder cell line G/G.

35.3.2 Active Targeting

Active targeting to the tumor can be achieved by molecular recognition of cancer cells either via ligand-receptor, antibody–antigen interactions, or by targeting through aptamers. Active targeting of a therapeutic agent or the carrier is achieved by conjugating the therapeutic agent or the carrier with a tissue or cell-specific ligand, thereby allowing a preferential accumulation of the drug in the tumor tissue, within individual cancer cells, or specific molecules in cancer cells (Lamprecht et al. 2001, Scherer et al. 2002, Parveen and Sahoo 2008). The success of drug targeting depends on the selection of the targeting agent, which should be abundant, have high affinity and specificity of binding to cell surface receptors and should be well suited to chemical modification by conjugation. The receptors and their surface-bound antigens may be expressed uniquely in diseased cells only or may exhibit a differentially higher expression in diseased cells as compared with normal cells. Thus the tumor endothelium and tumor vasculature may provide many targets for cancer therapy, as seen in Figure 35.6.

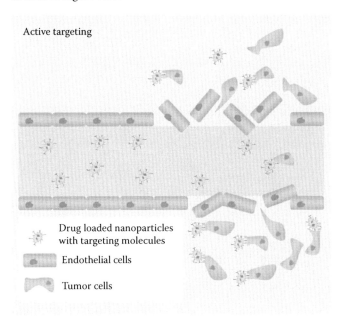

Active targeting

Drug loaded nanoparticles with targeting molecules

Endothelial cells

Tumor cells

FIGURE 35.6 Active targeting of nanoparticles to tumor cells.

In fact, several tumor-targeted nanomedicines are already on the market based mainly on drug–polymer–antigen conjugates, liposomes, micelles, and nanoparticles, as seen in Table 35.3. Antibody-drug conjugates targeted to CD20, CD25, and CD33 that are (over) expressed in non-Hodgkin's lymphoma, T-cell lymphoma, and acute myeloid leukemia, respectively, have been successfully used for delivering radionuclides (Zevalin®), immunotoxins (Ontak®), and antitumor antibiotics (Mylotarg®) more selectively to tumor cells. Parallel to this approach, antibodies, antibody fragments, and peptides have also been used as targeting moieties for drug delivery systems. Clinical evidence in support of this strategy is, however, scarce as has to date only been provided for galactosamine-targeted PHPMA-doxorubicin (PK2) (Seymour et al. 2002) and GAH-treated doxorubicin containing immunoliposomes MCC-465 (Matsumura et al. 2004). For the former, responses were observed for 3 out of 31 patients with liver cancer (with one partial remission lasting for >47 months) and for the latter disease stabilization was detected in 10 out of 18 patients with gastric cancer but no obvious reduction in tumor size was observed. Tumor-activated prodrug therapy uses the approach that a drug conjugated to a tumor-specific molecule will remain inactive until it reaches the tumor (Chari et al. 1998). These systems would ideally be dependent on interactions with cells found specifically on the surface of cancerous cells and not the surface of healthy cells. Most linkers are usually peptidase-cleavable or acid-labile but may not be stable enough in vivo to give desirable clinical outcomes. Limitations also exist due to lower potency of some drugs after being linked to targeting moieties when the targeting portion is not cleaved correctly or at all. Recent research on an adriamycin-conjugated poly(ethylene glycol) linker with enzymatically cleavable peptide sequences (alanyl-valine, alanyl-proline, and glycyl-proline) has shown a greater selectivity to cleavage at tumor cells (Suzawa et al. 2002). One such type of target is monoclonal antibodies, which were first shown to be able to bind specific tumor antigens in 1975 (Kohler and Milstein 1975) but development of these antibodies into tools for cancer treatment took another 20 years. The ideal antigen should be expressed on all tumor cells but not on critical host cells. There should be no mutation or variation and it should be required for cell survival or for a critical cellular function. A number of targeted cancer treatments using antibodies for specific cancer types have been approved by the US Food and Drug Administration (FDA) (Brannon-Peppas and Blanchette 2004). They also have the ability to serve as carriers for drug delivery systems for even more effective and less intrusive cancer therapy. These strategies both exploit the differences between a malignant cell and a normal cell. Some critical features include uncontrolled proliferation, insensitivity to negative growth regulation, angiogenesis, tissue invasion and metastasis, evasion of apoptosis (programmed cell death) and insensitivity to anti-growth signals (Sledge and Miller 2003).

35.3.2.1 Tumor Surface Receptor Targeting

35.3.2.1.1 Lectin–Carbohydrate Interaction

Lectins bind carbohydrate moieties and the lectin–carbohydrate interaction is highly specific, which can be exploited for the

TABLE 35.3 Examples of Clinically Used Tumor-Targeted Nanomedicines

Compound	Brand Name	Indication	Status
Liposomal doxorubicin	Myocet, Caelyx (Doxil)	Breast, ovarian, KS	Approved
Liposomal daunorubicin	Daunoxome,	Kaposi sarcoma	Approved
Liposomal vincristine	Onco-TCS	Non-hodgkin lymphoma	Approved
Liposomal cisplatin	SPI-77	Lung	Phase II
Liposomal lurtotecan	OSI-221	Ovarian	Phase II
Cationic liposomal c-Raf AON	LErafAON	Various	Phase I/II
Cationic liposomal EIA pDNA	PLD-EIA	Breast, ovarian	Phase I/II
Thermosensitive liposomal doxorubicin	ThermoDox	Breast, liver	Phase I
Albumin paclitaxel nanoparticle	Abraxane	Breast	Approved
Albumin methotrexate nanoparticle	MTX-HSA	Kidney	Phase II
Dextran doxorubicin nanoparticle	DOX-OXD	Various	Phase I
PEG-L-asparaginase	Oncaspar	Leukemia	Approved
PEG-IFNα2a/-IFNα2b	PegAsys/PegIntron	Melanoma, leukemia	Phase I/II
PHPMA-doxorubicin	PKI	Breast, lung, colon	Phase II
Galactosamine-targeted PKI	PK2	Liver	Phase I/II
PGA-paclitaxel	Xyotax	Lung, ovarian	Phase III
Paclitaxel-containing polymeric micelles	Genexol-PM	Breast, lung	Phase II
Cisplatin-containing polymeric micelles	Nanoplatin	Various	Phase I
Doxorubicin-containing polymeric micelles	NK911	Various	Phase I
SN38-containing polymeric micelles	LE-SN38	Colon, colorectal	Phase I
^{90}Yttrium-Ibritumab tiuxetan (α-CD20)	Zevalin	Non-hodgkin lymphoma Approved	Approved
DTA-IL2 fusion protein (α-CD25)	Ontak	T-cell lymphoma	Approved
Ozogamycin-gemtuzumab (α-CD33)	Mylotarg	Leukemia	Approved
Doxorubicin-cBR96 (α-CD174)	SGN-15	Lung, prostate, breast	Phase II

Source: Lammers, T. et al., *Br. J. Cancer*, 99, 392, 2008.

development of nanoparticles containing carbohydrate moieties that are directed certain lectins or vice versa (De Meija and Prisearu 2005, Sharon 2007). Several lectins have been found to possess anticancer properties in vitro, in vivo, and in human case studies, they are used as therapeutic agents, preferentially binding to cancer cell membranes or their receptors causing cytotoxicity (cell death), apoptosis, and inhibition of tumor growth. The selection of suitable lectins allows targeting of drugs to specific cells. The high affinity of wheat germ agglutinin (WGA) for the brain endothelia allows drug targeting by binding to cell surface and subsequent internalization without disrupting the barrier properties of brain endothelia (Banks and Kastin 1998, Fischer and Kissel 2001).

In addition, asialogycoprotein receptor expressed on the cell surface of hepatocytes is another example of endogenous lectin that can be exploited for drug targeting by galactose-bearing drug carrier systems (Bies et al. 2004).

35.3.2.1.2 Transferrin Receptor Targeting

Transferrin (Tf) receptors are overexpressed by 2–10-fold in most of the tumor cells unlike normal cells and thus Tf and/or Tf-antibodies may be used for targeting drugs to tumor cells (Vasir and Labhasetwar 2005, Daniels et al. 2006). The merits of PEG-coated biodegradable polycyanoacrylate nanoparticles conjugated with Tf actively targetable nanoparticles loaded with paclitaxel were investigated (Xu et al. 2005). The

distribution profiles of S-180 solid tumor–bearing mice after intravenous administration showed that the tumor accumulation of paclitaxel increased with time and the paclitaxel concentrations in tumor were about 4.8 and 2.1 times higher than those from paclitaxel injection and PEGylated nanoparticles without transferrin at 6 h after intravenous injection. The tumor volume with actively targetable nanoparticles-treated mice was much smaller and in addition the lifespan of tumor-bearing mice was also significantly increased. Thus, their studies showed that PEG-conjugated biodegradable polycyanoacrylate nanoparticles conjugated with Tf could be an effective carrier for paclitaxel delivery.

Recently, the molecular mechanism of greater efficacy of paclitaxel-loaded nanoparticles conjugated with Tf was studied in breast cancer cell line (Sahoo and Labhasetwar 2005). Tf-conjugated nanoparticles demonstrated a greater and sustained antiproliferative activity of the drug in dose- and time-dependent studies when compared to that with drug in solution or unconjugated nanoparticles in MCF-7 and MCF-7/Adr cells. The mechanism of greater antiproliferative activity of the drug with conjugated nanoparticles was determined to be greater cellular uptake and reduced exocytosis compared to that of unconjugated nanoparticles thus leading to higher and sustained intracellular drug levels. The direct intratumoral delivery of paclitaxel-loaded Tf-conjugated nanoparticles were also studied in the subcutaneous mice model of prostate carcinoma and found that those that received a single injection of paclitaxel-loaded Tf-conjugated nanoparticles (single paclitaxel dose: 4 mg/kg) demonstrated complete tumor regression and greater survival rate than those that received either paclixatel-loaded nanoparticles or paclitaxel-cremophor solution (Sahoo et al. 2004).

35.3.2.1.3 Folate Receptor Targeting

The exploitation of folate receptor (FR)–mediated drug delivery has been referred to as a molecular Trojan horse approach whereby drugs attached to folate are shuttled inside a targeted FR-positive cell in a stealth-like fashion. The FR is a highly specific tumor marker that is overexpressed in many human cancers. In addition, this receptor is absent in most normal tissues and thus frequently exploited for drug targeting purposes (Lu and Low 2002, Hilgenbrink and Low 2005, Vasir and Labhasetwar 2005). With the proper design, folate-drug conjugates also display this high affinity property, which enables them to rapidly bind to the FR and become internalized via an endocytic process.

Zhang et al. (2007), designed doxorubicin–loaded, folate-decorated PLGA-vitamin E TPGS nanoparticles for targeted chemotherapy of FR-rich tumors. This study showed that TPGS-FOL cellular uptake was 1.5 and 1.7 times higher in MCF7 and C6 cells, respectively, as compared with nanoparticles with no TPGS-FOL component after 30 min incubation. Furthermore, TPGS-FOL nanoparticles also exhibited a much lower cell viability value than doxorubicin after 24 h incubation. Oyewumi et al. (2004) compared the cellular uptake, biodistribution, and tumor retention of folate-coated and PEG-coated gadolinium (Gd) nanoparticles in tumor-bearing mice. Nanoparticles were characterized in vitro and in vivo studies were carried out in human nasopharyngeal carcinoma tumor-bearing athymic mice. Biodistribution and tumor retention studies were also carried out at predetermined time intervals after injection of the nanoparticles (10 mg/kg). The study demonstrated that though both folate-coated and PEG-coated nanoparticles had comparable tumor accumulation, the cell uptake and tumor retention of folate-coated nanoparticles were significantly enhanced over PEG-coated nanoparticles. Thus, folate ligand coating facilitated tumor cell internalization and retention of gadolinium nanoparticles in the tumor tissue.

Most human tissues lack the folate receptor except the placenta, choroids plexus, lungs, and kidneys. Small, non-immunogenic, stable folate has a high specificity for tumors. Folic acid enters tumor cells either through a carrier protein, a reduced folate carrier, or via receptor-mediated endocytosis facilitated by the folate receptor. Folate-mediated tumor targeting has been used to deliver the following substances to cancer cells such as protein toxins, low molecular weight chemotherapeutic agents, radio imaging agents, MRI contrast agents, radiotherapeutic agents, liposomes containing chemotherapeutics, genes, antisense oligonoucleotides, ribozymes, and immunotherapeutic agents. Several methods were developed to prepare folate-coated nanoparticles utilizing various techniques to amplify the drug delivery.

35.3.2.1.4 LH–RH Receptor Targeting

Many cancer cells overexpress luteinizing hormone-releasing hormone (LH–RH) receptors in their plasma membranes including ovarian, breast, and prostate cancers, though they are detectably expressed in normal visceral organs, providing a target for certain anticancer therapies. Ferric oxide nanoparticles prepared by a reverse micelle colloidal reaction were developed to target LH–RH receptor-bearing cancer cells. On the surface of the nanoparticle shell, LH–RH was coupled to the shell through carbon spacers so as to prevent steric hindrance during the interaction of the targeting agent with its affinity molecule on cells. Following the administration of the nanoparticles to patients and the resultant internalization of the nanoparticles by the tumor cells, the patients were exposed to a DC magnetic field available in standard magnetic resonance imaging equipment. The selective interaction, internalization, and the effect of various conditions on the magnetocytolysis of cells of these nanoparticles were investigated by utilizing LH–RH receptor-expressing cells on oral epithelial carcinoma cells. Data clearly showed that nanoparticles selectively affected specific cell types with a controllable efficiency. The findings revealed that magnetocytolytic activity was effective only in those cells capable of interacting with these nanoparticles and that the nanoparticles likely entered the cells by receptor-mediated endocytosis. Furthermore, results showed that the lytic effect was dependent on the time of exposure to the magnetic field (Praetorius and Mandal 2007).

35.3.2.2 Antigen–Antibody Interaction

In recent years, the overexpressing cancer-specific antigens have become important tools in developing different technologies for cancer therapy. The delivery by targeting cancer-specific antigens concentrates the therapeutics at the cancer site and this helps in diagnostic imaging and treatment of cancers. The advent of monoclonal antibody (mAb) technology in the 1970s and the development of genetically engineered devices in the 1980s along with technological advances in the bulk production of mAbs have led to a number of clinical studies to evaluate the efficacy of cancer-specific mAbs in the targeted delivery of cancer. Antibodies are also effective targeting agents and the discovery of hybridoma technology has heralded an extensive use of antibodies as targeting moieties against specific antigens (proteins expressed on the surface of cells) (Arangoa et al. 2003, Weiner 2006). Some of the ligands that can be exploited for targeted delivery are galactosamine for hepatocytes, melanocyte-stimulating hormone for melanoma, arginine–glycine–aspartate peptide for endothelial tissues, Epstein-Barr virus peptide for lymphocytes, ovarian carcinoma–associated antigen for ovarian carcinoma and Anti-Thy-1, 2 mAb for leukemia and lymphoma (Luo and Prestwich 2002).

The overexpression of receptors and antigens in human cancers also helps in the efficient uptake via receptor-mediated endocytosis. Moreover, due to their transformed nature, tumor cells express new proteins in comparison to normal cells and these markers may be exploited for active targeting. These are known as tumor-associated antigens (TAA) or antibodies and ligands specific to these can be used to target drugs to tumor cells. The encapsulation of TAA in polymer nanoparticles is a promising approach to increase the efficiency of antigen delivery for antitumor vaccines. Solbrig et al. (2007) has optimized a polymer preparation method to deliver both defined tumor-associated proteins and the complex mixtures of tumor antigens present in tumors in which the tumor antigens are encapsulated in a biodegradable 50:50 PLGA copolymer. These antigens retained their antigenicity and functioned better than soluble antigens when tested in vitro assays of <t-cell cytokine formation and in vivo tumor vaccination challenge. The receptor for tumor growth factor-α, endoglin (CD105) is also much favored for tumor imaging and therapy (Fonsatti et al. 2003, Thorpe 2004). Other potential targets include matrix metalloproteinases, integrins, VEGF, platelet-derived growth factors, and angiopoietins and their receptors (Weiner 2006).

Long-circulating immunoliposomes targeted to human epidermal growth factor receptor HER2 were prepared by the conjugation of anti-HER2 mAb fragment to liposome-grafted PEG chains (Kirpotin et al. 2006). These new systems resulted in better and selective cellular internalization into cancer cells in vivo and may provide new opportunities for drug delivery. EGFR was also targeted by another study developing nanocapsules targeted to brain tumors (Tsutsui et al. 2007). The results showed that modified nanocapsules were efficiently delivered to glioma cells but not to normal cells glial cells and they also confirmed that the specific delivery of the nanocapsules to brain tumors in vitro and in vivo.

35.3.2.3 Aptamer-Mediated Targeting

Oligonucleotides (short DNA or RNA oligonucleotide ligands) provide immense promise as a therapeutic agent against cancer as is evident by the rapidly growing class of molecules known as aptamers (Hicke and Stephens 2000, White et al. 2000). Aptamers are DNA or RNA oligonucleotides that are capable of binding to target antigens with high affinity and specificity, analogous to antibodies (Tuerk and Gold 1990). The unique property that labels these as the new generation of targeted drug molecules is their ability to bind to molecular targets such as small molecules, proteins, nucleic acids, and even cells, tissues, and organisms (Burgstaller et al. 2002).

The ideal class of targeting molecules for the delivery of controlled release polymer system should bind with high affinity and specificity to a target antigen to facilitate delivery of drugs. Antibodies offer targeting options but aptamers offer advantages over antibodies as they can be engineered completely in a test tube, are rapidly produced by chemical synthesis, possess desirable storage properties, are easy to characterize and modify, elicit little or no immunogenicity in therapeutic applications, and exhibit a high specificity and affinity for their target antigen. Furthermore due to their small size and similarity with endogenous molecules, aptamers exhibit superior tissue penetration (Hicke and Stephens 2000).

Farokhzad et al. (2006) in a recent study have developed 180 nm docetaxel-encapsulated nanoparticles using biocompatible and biodegradable PLGA-block-PEG copolymer and surface functionalized with the A10 2-fluoropyrimidine RNA aptamers that recognize the extracellular domain of the prostate-specific membrane antigen (PSMA). Their results showed significantly enhanced in vitro cellular toxicity of functionalized nanoparticles as compared with nontargeted nanoparticles that lack the PSMA aptamer. In addition, after a single intratumoral injection of docetaxel–nanoparticle–aptamer bioconjugates, complete tumor reduction was observed in five out of seven xenograft nude mice and 100% survival rate was observed as compared with mice treated with docetaxel-loaded nanoparticles and docetaxel alone, which showed significantly reduced survival rates. Thus, their studies demonstrate the potential utility of nanoparticle–aptamer bioconjugates for a therapeutic application. Very recently, the same group has studied the biodistribution of the aptamer-conjugated PLGA-b-PEG nanoparticles in xenograft mouse model of prostate cancer. The surface functionalization of nanoparticles with the A10 PSMA aptamer significantly enhanced (3.77-fold increase in 24 h, injected dose per gram tissue) the delivery of nanoparticles to tumors versus equivalent nanoparticles lacking the A10 PSMA aptamer.

35.3.3 Stimuli-Sensitive Targeting

The development of stimuli-sensitive nanocarriers is a hot issue in nanomedicine for cancer. The concept is based on the fact that tumors normally have a lower pH value and a higher temperature than normal tissue, and stimuli-sensitive nanocarriers can be built releasing the incorporated drug only when subjected to these special conditions of the tumor. The orientation of the nanoparticles to the tumor site and triggered release of encapsulated drug is realized by a magnetic field or external heat source.

35.3.3.1 pH-Sensitive Targeting

Polymeric components with pH-sensitive (pH-cleavable) bonds are used to produce stimuli-sensitive drug delivery systems that are stable in the circulation or in normal tissues but acquire the ability to degrade and release encapsulated drugs in body areas or cell compartments with lowered pH such as tumors or cell cytoplasm or endosomes (Roux et al. 2002, 2004, Torchilin 2007). Mixed micelles made of pH-sensitive components (polyhistidine and polylactic acid) loaded with adriamycin and targeted with folate residue provided better drug release under lowered pH values and demonstrated better killing of MCF7 cells in vitro (Lee et al. 2003). Such micelles can be prepared from different components (Bae et al. 2005) and loaded with different drugs (Gao et al. 2005) with demonstrated in vivo efficacy. They have also been shown to suppress drug-resistant tumor cells effectively (Lee et al. 2005).

The stimuli sensitivity of PEG coats can also allow for the preparation of multifunctional drug delivery systems with temporarily hidden functions, which, under normal circumstances, are shielded by the protective PEG coat but become exposed after PEG detaches. A nanoparticulate delivery system can be prepared so that it accumulates in the required organ or tissue and then penetrates inside target cells delivering its load (drug or DNA) there. The initial target (tumor) accumulation could be achieved via passive targeting through the EPR effect or by specific ligand-mediated active targeting, whereas the subsequent intracellular delivery could be mediated by certain internalizable ligands (folate, transferrin) or by cell-penetrating peptides. When in the blood, the cell-penetrating function should be temporarily inactivated (sterically shielded) to prevent nonspecific drug delivery into nontarget cells. However, when inside the target, the nanocarrier loses its protective coat, exposes the cell-penetrating function, and provides intracellular delivery. Systems like this require that multiple functions attached to the surface of the nanocarriers work in a certain orchestrated and coordinated way.

For the above system, the following requirements have to be met:

1. The life of the carrier in the circulation should be long enough to fit the EPR effect or targeted delivery requirements meaning that the PEG coating mediated longevity in blood or specific ligand mediating the targeting function should not be lost by the nanocarriers when in circulation.
2. The internalization of the carrier within the target cells should proceed sufficiently fast so as not to allow for carrier degradation and drug loss in the interstitial space (local stimuli-dependent removal of protective function and the exposure of the temporarily hidden second function should proceed fast (Torchilin 2007)).

pH-sensitive polymeric nanoparticles that effectively target the acidic extracellular matrix of the tumor was demonstrated (Sethuraman et al. 2006). Plasmid DNA was complexed to polyethyleneimine (PEI) and further with a pH-sensitive diblock copolymer, poly(methacryloyl sulfadimethoxine) (PSD)-block-PEG, to obtain nanoparticles. The shielding/deshielding of nanoparticles was tested along with cell viability and transfection efficiency at physiological and tumor pH. Size of the nanoparticles were around 300 nm and showed low cytotoxicity and transfection at pH 7.4 due to the shielding of PEI by PSD-b-PEG. This pH-sensitive polymer bound to PEI/DNA complex reduced the interaction of PEI positive charges with cells and reduced the cytotoxicity by 60%. At pH 6.6, nanoparticles demonstrated high cytotoxicity and transfection indicating PSD-b-PEG detachment from nanoparticles and permitting PEI to interact with cells. PS-b-PEG is able to discern the small difference in pH between normal and tumor tissues and was reported to have remarkable potential in drug targeting to tumor areas, as seen in Figure 35.7.

35.3.3.2 Thermotherapy and Magnetic Targeting

Thermotherapy is a method of utilizing hyperthermia directed toward body tissues for the purpose of damaging protein and structures within cancerous cells and in some cases causing tumor cells to directly undergo apoptosis. Healthy cells are capable of surviving exposure to temperatures up to around 46.5°C whereas irreversible damage to diseased cells occurs at temperatures in a range from ~40°C to 46°C. During thermotherapy, surrounding healthy cells are more readily able to dissipate heat and maintain a normal temperature while the targeted tumors

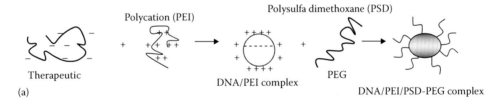

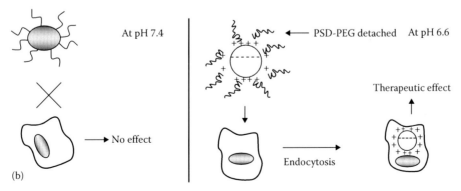

FIGURE 35.7 Targeting based on difference in pH: (a) formation of nanoparticle complex through charge–charge interaction between DNA, cationic polymer (PEI) and PSD-b-PEG; and (b) complex shielded at physiological pH and deshielded at cancer pH. (From Sethuraman, V.A. et al., *Biomacromolecules*, 7, 64, 2006.)

experience difficulty in dissipating heat due to the disorganized and compact vascular structure that is indicative of tumors (Ito et al. 2005). A major technical problem with the application of hyperthermic treatments is the difficulty in heating local tumor region to the intended temperature without damaging normal tissue. Conventional hypothermic systems are designed to heat tissue to ~42.5°C–44°C. However, higher temperatures can kill a greater number of tumor cells and in principle tumor-specific hyperthermia can kill all types of tumors.

Nanoparticles are utilized in various aspects of hyperthermia-based treatments such as serving as the agents of thermotherapy, sensitizers, and for targeting purposes such as utilizing antibody-enhanced targeting efficacy and to reduce hypothermia-associated side effects. Superparamagnetic nanoparticles refer to iron oxide particles or magnetite (Fe_3O_4) particles that are less than 10 nm in diameter (Kelly and Kim 2007). They have been around for years as contrasting agents for magnetic resonance imaging (MRI). Many groups have explored the use of magnetic fields to localize magnetic nanoparticles to target sites, a system known as magnetic drug targeting. As with other nanoparticles, functionalization of these superparamagnetic nanoparticles is exploited so as to permit tumor-specific targeting. Iron oxide nanoparticles can be water-solubilized with hydrophilic polymer coatings such as dextran or PEG. In fact, attaching PEG to nanoparticles in general is a well documented means of sterically preventing opsonization of nanoparticles in the serum and reducing their uptake by the reticuloendothelial system. This effectively enhances biocompatibility and increases circulation time of nanoparticles. Iron oxide nanoparticles can also be made hydrophobic by coating with aliphatic surfactants or liposomes (resulting in magnetoliposomes).

Targeted nanoparticles for the delivery of thermotherapy for cancer treatment with minimal invasion and short duration were developed (Praetorius and Mandal 2007). Prepared ferromagnetic nanoparticles were coated with biocompatible material poly(methacrylic acid-co-methacrylate) using free-radical polymerization. An ionic surfactant (sodium bis-2-ethylhexyl sulfosuccinate) acted to form a stabilizing layer around the magnetic nanoparticles. Antibodies were covalently attached to the coated magnetic nanoparticles. A magnetic field was applied to inductively heat the single domain magnetic nanoparticles attached to the target-specific ligand.

Some researchers have proposed the concept of "intracellular" hyperthermia and have developed submicron magnetic particles for inducing hyperthermia (Jordan et al. 1993, Wada et al. 2001). This concept is based on the principle that under an alternating magnetic field (AMF), a magnetic particle can generate heat by hysteresis loss. Under the influence of an alternating field, superparamagnetic nanoparticles undergo Brownian relaxation in which heat is generated by the rotation of particles in the field. However, concentrations of 0.01%–0.1% iron oxide are necessary to raise the tissue to critical temperatures for thermal ablation and these concentrations are hard to achieve via intravenous administration.

The important properties of magnetic nanoparticles to induce hyperthermia are nontoxicity, biocompatibility, injectability, high-level accumulation in target tumor, and effective absorption of the energy of AMF. It was reported that particle size is a critical factor in obtaining specific absorption rate (SAR) value (Shinkai et al. 1994).

35.4 Future Prospects and Challenges in Tumor-Targeted Nanoparticles

In this chapter, we have primarily restricted ourselves to tumor-targeted nanomedicines designed for the improved delivery of already established low molecular weight chemotherapeutics mainly. Many of the new drugs arising from advances in biotechnology are macromolecules such as proteins and nucleic acids. The clinical development of these challenging and often fragile molecules will likely also profit substantially from the attributes of targeted nanomedicines providing these complex molecules with protection against degradation and elimination and with improved access to target cells and tissues.

Future direction of nanoparticles for cancer therapy seems to be focused on the following issues:

1. The European Science Foundation included in their definition of the discipline of nanomedicine not only the use of nanometer sized materials for the treatment but also for the diagnosis of diseases. In this context, molecular detection of cancer is drawing intense attention and is the subject of ongoing research as diagnostic agents to be used in high-resolution imaging techniques such as MRI and PET (Lammers et al. 2008).

2. The design of systems that are able to respond to external stimuli such as hyperthermia, ultrasound, light, and magnetic fields can be triggered to release their contents during specific time periods or conditions.

3. The targeting of agents other than conventional chemotherapeutics to tumors such as anti-inflammatory agents (e.g., corticosteroids), to inhibit tumor-associated inflammation and siRNA, and to reduce the expression of proteins essential for tumor progression.

4. The development of systems that are able to simultaneously deliver multiple therapeutic agents to tumors such as temporally targeted nanocells, which first release the antiagiogenic agent combrestatin and, subsequently, the chemotherapeutic agent doxorubicin (Sengupta et al. 2005).

5. The translation of the experience gained in oncology into applications improving the treatment of other diseases such as rheumatoid arthritis, Crohn's disease, autoimmune diseases, and infections, which are highly amenable to EPR-mediated drug targeting (Schiffelers et al. 2006).

6. The establishment of treatment regimens in which tumor-targeted nanomedicines are combined with other clinically relevant treatment modalities such as with surgery, with radiotherapy, and with standard chemotherapy.

35.5 Conclusion

Cancer nanotechnology has improved significantly since the last 20 years in terms of therapeutic and diagnostic delivery systems. The wide range of nanocarriers in the forms of nanospheres, nanocapsules, solid lipid nanoparticles, vesicles, micelles, and nanotubes have contributed to the significant progress in this field. The outcome of this research interest has already shown itself in the pharmaceutical market with the first nanoparticulate drug approved by FDA in 2005 for the treatment of breast cancer, namely, Abraxane®. Several other formulations are under trial in Phase I or II stages. Regulatory approach to nanotechnology products are still a challenge to overcome a tripartite approach since nanoparticle fall under the purview of three branches of regulatory agencies such as FDA: drugs, medical devices, and biologics. Therefore, they might be examined by all the three parties. Nanoparticles may also be multifunctional containing diagnostic, therapeutic, and barrier-avoiding agents. With current regulations, it could be expected that regulatory approval will have to be issued for each agent and then for their combination. Nevertheless, nanotechnology will have an important role in realizing the goal of detecting transforming cell populations early by in vivo imaging or ex vivo analysis. It will also allow the appropriate combination of agents to be chosen based on accurate biological information of the tumor, targeting of these agents while avoiding biological barriers to the early cancer lesions to eliminate or contain them without collateral effects on healthy tissue and monitoring treatment effect in real time.

References

Agarwal, A., Asthana, A., Gupta, U., Jain, N.K., Tumor and dendrimers: A review on drug delivery aspects, *J. Pharm. Pharmacol.*, 60, 671–688, 2008.

Arangoa, M.A., Düzgüneş, N., Tros de Harduya, C., Increased receptor mediated gene delivery to the liver by protamine enhanced asialofetuin lipoplexes, *Gene Ther.* 10, 5–14, 2003.

Baban, D., Seymour, L.W., Control of tumor vascular permeability, *Adv. Drug. Deliv. Rev.*, 34, 109–119, 1998.

Bae, Y., Jang, W.D., Nishiyama, N., Multifunctional polymeric micelles with folate mediated cancer cell targeting and ph triggered drug releasing properties for active intracellular drug delivery, *Mol. BioSyst.*, 1, 242–250, 2005.

Banfi, A., Degenfeld, G., von Blau, H.M., Critical role of microenvironment factors in angiogenesis, *Curr. Atheroscler. Rep.*, 7, 227–234, 2005.

Banks, W.A., Kastin, A.J., Characterization of lectin mediated brain uptake of HIV-1 GP120, *J. Neurosci. Res.*, 54, 522–529, 1998.

Bies, C, Lehr, C.M., Woodley, J.F., Lectin-mediated drug targeting: History and applications, *Adv. Drug Deliv. Rev.* 56, 425–435, 2004.

Bilensoy, E., Nanoparticulate delivery systems based on amphiphilic cyclodextrins, *J. Biomed. Nanotechnol.*, 4, 293–303, 2008.

Bilensoy, E., Gürkaynak, O., Doğan, A.L., Hıncal., A.A., Safety and efficacy of amphiphilic ß-cyclodextrin nanoparticles for paclitaxel delivery, *Int. J. Pharm.*, 347, 163170, 2008a.

Bilensoy, E., Gürkaynak, O., Ertan, M., Şen, M., Hıncal, A.A., Development of non-surfactant cyclodextrin nanoparticles loaded with anticancer drug paclitaxel, *J. Pharm. Sci.*, 97(4), 1519–1529, 2008b.

Bilensoy, E., Sarısözen, C., Esendağlı, G, Doğan, A.L., Aktaş, Y., Mungan, A.N., Intravesical cationic nanoparticles of chitosan and polycaprolactone for the delivery of mitomycin c to bladder tumors, *Int. J. Pharm.*, 371(1-2), 170–176, 2009.

Brannon-Peppas, L., Blanchette, J.O., Nanoparticle and targeted systems for cancer therapy, *Adv. Drug Deliv. Rev.*, 56, 1649–1659, 2004.

Brigger, I., Dubernet, C., Couvreur, P., Nanoparticles in cancer therapy and diagnosis, *Adv. Drug Deliv. Rev.*, 54, 631–651, 2002.

Burgstaller, P., Jenne, A., Blind, M., Aptamers and aptazymes: Accelerating small molecule drug discovery, *Curr. Opin. Drug Discov. Devel.*, 5, 690–700, 2002.

Calvo, P., Gouritin, B., Chacun, H., Long circulating PEGylated polycyanoacrylate nanoparticles as new drug carrier for brain delivery, *Pharm. Res.*, 18, 1157–1166, 2001.

Chari, R.V.J., Targeted delivery of chemotherapeutics: Tumor-activated prodrug therapy, *Adv. Drug Deliv. Rev.*, 31, 89–104, 1998.

Couvreur, P., Kante, B., Lenaerts, V., Scailteur, V., Roland, M., Speiser, P., Tissue distribution of antitumor drugs associated with polyalkylcyanoacrylate nanoparticles, *J. Pharm. Sci.*, 69, 199–202, 1980.

Daniels, T.R., Delgado, T., Helguera, G., Penichet, M.L., The transferin receptor part I: Biology and targeting with cytotoxic antibodies for the treatment of cancer, *Clin. Immunol.*, 121, 144–158, 2006.

Davis, S.S., 25 years of drug targeting, *Pharm. J.*, 265, 491–492, 2000.

Dass, C.R., Tumor angiogenesis, vascular biology and enhanced drug delivery, *J. Drug Target.*, 12(5), 245–255, 2004.

De Meija, E.G., Priseau, V.I., Lectin as bioactive plant proteins: A potential cancer treatment, *Crit. Rev. Food Sci., Nutr.*, 45, 425–445, 2005.

Farokhzad, O.C., Cheng, J., Teply, B.A., Sherifi, I., Jon, S., Kantoff, P.W., Richie, J.P., Langer, R., Targeted nanoparticle-aptamer bioconjugates for cancer chemotherapy in vivo, *Proc. Natl. Acad. Sci. USA*, 103, 6315–6320, 2006.

Feng, S.S., Nanoparticles of biodegradable polymers for new-concept chemotherapy, *Expert Rev. Med. Devices*, 1(1), 115–125, 2005.

Fischer, D., Kissel, T., Histochemical characterization of primary capillary endothelial cells from porcine brains using monoclonal antibodies and fluorescein isothiocyanate labeled lectin: Implications for drug delivery, *Eur. J. Pharm. Biopharm.*, 52, 1–11, 2001.

Folkman, J., Incipient angiogenesis, *J Natl. Cancer Inst.*, 92, 94–95, 2002.

Fonsatti, E., Altomonte, M., Arslan, P., Maio, M., Endoglin (CD105): A target for antiangiogenetic cancer therapy, *Curr. Drug Targets*, 4, 291–296, 2003.

Gao, Z.G., Lee, D.H., Kim, D.I., Bae, Y.H., Doxorubicin loaded pH sensitive micelle targeting acidic extracellular pH of human ovarian A2780 tumor in mice, *J. Drug Target.*, 13, 391–397, 2005.

Gelderblom, H., Verweij, J., Nooter, K., Cremophor EL: The drawbacks and advantages of vehicle selection for drug formulation, *Eur. J. Cancer*, 37(13), 1590–1598, 2001.

Gref, R., Minamitake, Y., Peracchia, M.T., Trubetskoy, V., Torchilin, V., Langer, R., Biodegradable long-circulating polymeric nanospheres, *Science*, 263, 1600–1603, 1994.

Greish, K., Enhanced permeability and retention of macromolecular drugs in solid tumors. A royal gate for targeted anticancer nanomedicines, *J. Drug Target.*, 15(7–8), 457–464, 2007.

Greish, K., Fang, J., Inutsuka, T., Nagamitsu, A., Maeda, H., Macromolecular therapeutics: Advantages and prospects with special emphasis on solid tumor targeting, *Clin. Pharmacokinet.*, 42, 1089–1105, 2003.

Hagiwara, A., Takahashi, K., Sawai, K., Sakakura, C., Shirasu, M., Ohgaki, M., Imanashi, T., Yamasaki, J., Takemoto, Y., Kageyama, N., Selective drug delivery to peritumoral region and regional lymphatics by local injection of aclarubicin adsorbed on activated carbon particles in patients with breast cancer—A pilot study, *AntiCancer Drugs*, 8, 666–670, 1997.

Haley, B., Frenkel, E., Nanoparticles for drug delivery in cancer treatment, *Urol. Oncol. Semin. Orig. Invest.*, 26, 57–64, 2008.

Hatefi, A., Amsden, B. Camptothecin delivery methods. *Pharm. Res.* 19, 1389–1399, 2002.

Hawley, A.E., Davis, S.S., Illum, L., Targeting of colloids to lymph nodes: Influence of lymphatic physiology and colloidal characteristics, *Adv. Drug Deliv. Rev.*, 17, 129–148, 1995.

Heidin, C.H., Rubin, K., Pietras, K., Ostman, A., High interstitial fluid pressure-an obstacle in cancer therapy, *Nat. Rev. Cancer*, 4, 806–813, 2004.

Hicke, B.J., Stephens, A.W., Escort aptamers: A delivery service for diagnosis and therapy, *J. Clin. Invest.*, 106, 923–928, 2000.

Hilgenbrink, A.R., Low, P.S., Folate receptor mediated drug targeting: From therapeutics to diagnostics, *J. Pharm. Sci.*, 94, 2135–2146, 2005.

Hobbs, S.K, Monsky, W.L., Yuan, F., Regulation of transport pathways in tumor type and microenvironment, *Proc. Natl. Acad. Sci. USA*, 95, 4607–4612, 1998.

Ito, A., Shinkai, M, Honda, H., Kobayashi, T., Medical application of functionalized magnetic nanoparticles, *J. Biosci. Bioeng.*, 100(1), 1–11, 2005.

Jain, K.K., Delivery of molecular medicine to solid tumors: Lessons from in vivo imaging of gene expression and function, *J. Control. Release*, 74, 2–25, 2001.

Jain, R.K., Transport of macromolecules in the tumor interstitium: A review, *Cancer Res.*, 47, 3039–3051, 1987.

Jordan, A, Wust, P., Fahling, H., John, W., Hinz, A., Felix, R., Inductive heating of ferimagnetic particles and magnetic fluids: Physical evaluation of their potential for hyperthermia, *Int. J. Hyperthermia*, 9, 51–68, 1993.

Kelly, Y., Kim, M.A., Nanotechnology platforms and physiological challenges for cancer therapeutics, *Nanomed. Nanotechnol. Biol. Med.*, 3, 103–110, 2007.

Kirpotin, D.B., Drumond, D.C., Shao, Y, Shalaby, M.R., Hong, K., Nielsen, U.B., Marks, JD., Benz, C.C., Park, W., Antibody targeting of long circulating lipidic nanoparticles does not increase tumor localization but does increase internalization in anal models, *Cancer Res.*, 66, 6732–6740, 2006.

Kohler, G., Milstein, C., Continuous cultures of fuse cells secreting antibody of predefined specificity, *Nature*, 75(suppl 1), 381–394, 1975.

Kwon, G.S., Okano, T., Polymeric micelles as new drug carriers, *Adv. Drug Deliv. Rev.*, 21, 107–116, 1996.

Labhasetwar, V., Nanoparticles for drug delivery, *Pharm. News*, 4, 28–31, 1997.

Lammers, T., Hennink, W.E., Stormi G., Tumour targeted nanomedicines: Principles and practice, *Br. J. Cancer*, 99, 392–397, 2008.

Lamprecht, A., Ubrich, N., Yamamoto, H., Schafer, U., Takeuchi, H., Maincent, P., Kawashima, Y., Lehr, C.M., Biodegradable nanoparticles for targeted drug delivery in treatment of inflammatory bowel disease, *J. Pharmacol. Exp. Ther.*, 299, 775–781, 2001.

Lee, E.S., Na, K., Bae, Y.H, Polymeric micelle for tumor pH and folate mediated targeting, *J. Control. Release*, 91, 103–113, 2003.

Lee, E.S., Na, K., Bae, Y.H, Doxorubcin loaded pH sensitive polymeric micelles for rerfersal of resistant MCF-7 tumor, *J. Control. Release*, 103, 405–418, 2005.

Lilley L.L., Scott, H.B., What you need to know about taxol? *Am. J. Nurs.* 93, 46–50, 1993.

Links, M., Brown, R., Clinical relevance of the molecular mechanisms of resistance to anticancer drugs, *Expert Rev. Mol. Med.*, 1, 1–21, 1999.

Lu, Y., Low, P.S., Folate mediated delivery of macromolecular anticancer therapeutic agents, *Adv. Drug Deliv. Rev.*, 54, 675–693, 2002.

Lu, Z., Yek, T.K., Tsai, M., Au, J.L.S., Wientjes, M.G., Paclitaxel-loaded gelatin nanoparticles for intravesical bladder cancer therapy, *Clin. Cancer Res.*, 10, 7677–7684, 2004.

Luo, Y., Prestwich, G.D., Cancer targeted polymeric drugs, *Curr. Cancer Drug Targets*, 2, 209–226, 2002.

Maeda, H., The enhanced permeability and retention (EPR) effect in tumor vasculature: The key role of tumor-selective macromolecular drug targeting, *Adv. Enzyme Regul.*, 41, 189–207, 2001.

Maeda, H., Matsumura, Y., Oda, T., Cancer selective macromolecular therapeutics : tailoring of antitumor protein drug. In *Protein Tailoring for Food and Medical Uses*, eds. Feeny, R.E., Whitaker, J.R., New York, Marcel Dekker Inc., pp. 353–382, 1986.

Malingre, M.M., Terwogt, J.M.M., Beijnen, J.H., Phase I and pharmacokinetic study of oral paclitaxel, *J. Clin. Oncol.*, 18(12), 2468–2475, 2000.

Matsumura, Y., Gotoh, M., Muro, K., Yamada, Y., Shirao, K., Shimada, Y., Okuwa, M. et al., Phase I and pharmacoknetic study of MCC-465 a doxorubicin (DXR) encapsulated in PEG immunoliposome in patients with metastatic stomach cancer, *Ann. Oncol.*, 15, 517–525, 2004.

Minchinton, A.I., Tannock, I.F., Drug penetration in solid tumors, *Nat. Rev. Cancer*, 6, 583–592, 2006.

Molema, G., Design of vascular endothelium specific drug targeting strategies for the treatment of cancer, *Acta Biochim. Pol.*, 52, 301–310, 2005.

Netti, P.A., Berk, D.A., Swartz, M.A., Grodzinsky, A.J., Jain, R.K., Role of extracellular matrix assembly in interstitial transport in solid tumors, *Cancer Res.*, 60, 2497–2503, 2000.

Nishioka, Y., Yoshino, H., Lymphatic targeting with nanoparticulate system, *Adv. Drug Deliv. Rev.*, 47, 55–64, 2001.

Noguchi, R., Wu, J., Duncan, R., Stroholm, J., Ulbrich, K., Akaike, T., Maeda, H., Early phase of tumor accumulation of macromolecules: A great difference in clearance rate between tumor and normal tissues, *Jpn. J: Cancer Res.*, 89, 307–314, 1998.

Oyewumi, M.O., Yokel, R.A., Jay, M., Coakley, T., Mumper, R.J., Comparison of cell uptake, biodistribution and tumor retention of folate coated and PEG coated gadolinium nanoparticles in tumor bearing mice, *J. Control. Release*, 95, 613–626, 2004.

Padera, T.P, Pathology: Cancer cells compress intratumor vessels, *Nature*, 427, 695, 2004.

Parveen, S., Sahoo, S.K., Nanomedicine: Clinical applications of polyethyleneglycol conjugated proteins and drugs, *Clin. Pharmacokinet.*, 45, 965–988, 2006.

Parveen, S., Sahoo, S.K., Polymeric nanoparticles for cancer therapy, *J. Drug Target.*, 16(2), 108–123, 2008.

Peracchia, M.T., Fattal, E., Desmaele, D., Stealth PEGylated polycyanoacrylate nanoparticles as new drug carrier for brain delivery, *J. Control. Release*, 60, 121–128, 1999.

Praetorius, N.P., Mandal, T.K., Engineered nanoparticles for cancer therapy, *Recent Pat. Drug. Deliv. Formul.*, 1, 37–51, 2007.

Roux, E., Francis, M., Winnik, F.M., Leroux, J.C., Polymer based pH-sensitive carriers as a means to improve the cytoplasmic delivery of drugs, *Int J. Pharm.*, 242, 25–36, 2002.

Roux, E., Passirani, C., Scheffold, S., Serum stable and long circulating PEGylated pH sensitive liposomes, *J Control. Release*, 94, 447–451, 2004.

Sahoo, S.K, Ma, W., Labhasetwar, V, Efficacy of transferrin-conjugated paclixel-loaded nanoparticles in amuriğne model of prostate cancer, *Int. J. Cancer*, 112, 335–340, 2004.

Sahoo, S.K., Labhasetwar, V., Enhanced antiproliferative activity of transferrin-conjugated paclitaxel-loaded nanoparticles is mediated via sustained intracellular drug retention, *Mol. Pharm.*, 2, 373–383, 2005.

Sakakura, C., Takahashi, T., Sawai, S., Ozaki, K., Shirasu, M., Enhancement of therapeutic efficacy of aclarubicin against lymph node metastases using a new dosage form: Aclarubicin adsorbed on activated carbon particles, *AntiCancer Drugs*, 3, 233–236, 1992.

Sarısözen, C., Doğan, A.L., Esendağlı, G., Aktaş, Y., Mungan, N.A., Bilensoy, E., Bioadhesive chitosan nanoparticles for effective delivery of Mitomycin C in superficial bladder cancer therapy. In *Advances in Chitin Science*, Volume X, eds. Şenel, S., Varum, K.M., Şumnu, M.M., Hıncal, A.A., Alp Offset, Ankara, pp. 236–241, 2007.

Scherer, F., Anton, M., Schillinger U., Henke, J., Bergemann, C., Kruger, A., Gansbacher, B., Plank, C., Magnetofection: Enhancing and targeting gene delivery by magnetic force in vitro and in vivo, *Gene Ther.*, 9, 102–109, 2002.

Schiffelers, R.M., Banciu, M., Metselaar, J.M., Storm, G., Therapeutic application of long circulating liposomal glucocorticoids in autoimmune diseases and cancer, *J. Liposome Res.*, 16, 185194, 2006.

Sengupta, S., Eavarone, D., Capila, I., Zhao, G., Watson, N., Kızıltepe, T., Sasikheran, R., Temporal targeting of tumor cells and neovasculature with a nanoscale delivery system, *Nature*, 436, 568–572, 2005.

Sethuraman, V.A., Na, K., Bae, Y.H., pH-responsive sulfonamide/PEI system for tumor specific gene delivery: An in vitro study, *Biomacromolecules*, 7, 64–70, 2006.

Seymour, L.W., Ferry, D.R., Anderson, D., Hesslewod, S., Julyan, P.J., Poyner, R., Doran, J., Young, A.M., Burtles, S., Kerr, D.J., Hepatic drug targeting phase I evaluation of polymer-bound doxorubicin, *J. Clin. Oncol.*, 20, 1668–1676, 2002.

Seymour, L.W., Passive tumor targeting of soluble macromolecules and drug conjugates, *Crit. Rev. Ther. Drug Carrier Syst.*, 9, 7–25, 2001.

Sharon, N., Lectins: Carbohydrate specific reagents and biological recognition molecules, *J. Biol. Chem.*, 282, 2753–2764, 2007.

Shinkai, M., Matsui, M., Kobayashi, T., Heat properties of magnetoipposomes for local hyperthermia. *Jpn. J. Hyperthermic Oncol.*, 10, 168, 1994.

Siemann, D.W., Tumor vasculature: A target for anticancer therapies. In *Vascular Targeted Therapies in Oncology*, John Wiley & Sons Ltd., Chichester, U.K., pp. 1–8, 2006.

Sledge, G., Miller, K., Exploiting the hallmarks of cancer: The future conquest of breast cancer, *Eur. J. Cancer*, 39, 1668–1675, 2003.

Solbrig, C.M., Saucier-Sawyer, J.K., Cody, V., Saltzman, W.M., Hanlon, D.J., Plymer nanoparticles for immunotherapy from encapsulated tumor associated antigens and whole tumor cells, *Mol. Pharm.*, 4, 47–57, 2007.

Suzawa, T., Nagamura, S., Saito, H., Ohta, S., Hanai, N., Kanazawa, J., Okabe, M., Yamasaki, M., Enhanced tumor cell selectivity of adiramycin monoclonal antibody conjugate via a polyethyleneglycol based cleavable linker, *J. Control. Release*, 79, 229–242, 2002.

Takimoto, C.H., Wright, J. Arbuck, S.G. Clinical applications of the camptothecins. *Biochem. Biophys. Acta.*, 1400, 107–119, 1998.

Thorpe, P.E., Vascular targeting agents as cancer therapeutics. *Clin. Cancer Res.*, 10, 415–427, 2004.

Tokumitsu, H., Hiratsuka, J., Sakurai, Y., Kobayashi, T., Ichikawa, H., Fukumori, Y., Gadolinium neutron-capture therapy using novel gadopentetic acid-chitosan complex nanoparticles: In vivo growth suppression of experimental melanoma solid tumor, *Cancer Lett.*, 150, 177–182, 2000.

Torchilin, V.P., Targeted pharmaceutical nanocarriers for cancer therapy and imaging, *AAPS J.*, 9(2), Article 15, E128–E147, 2007.

Torchilin, V.P., Trubetskoy, V.S., Which polymers can make nanoparticulate drug carriers long-circulating? *Adv. Drug Deliv. Rev.*, 16, 141–155, 1995.

Tsutsui, Y, Tomizawa, K., Nagita, M., Michiuie, H., Nishiki, T., Ohmori, I, Seno, M., Matsui, H., Development of bionanocapsules targeting brain tumors, *J. Control. Release*, 122, 159–164, 2007.

Tuerk, C., Gold, L., Systematic evolution of ligands by exponential enrichment: RNA ligands to bacterophage T4 DNA polymerase, *Science*, 249, 505–510, 1990.

Vasir, J.K., Labhasetwar, V., Targeted drug delivery in cancer therapy, *Technol. Cancer Res. Treat.*, 4, 363–374, 2005.

Wada, S., Yue, L. Tazawa, K., Furuta, I., Nagae, H., Takemori, S., Minamimura, T, New local hyperthermia using dextran magnetite complex (DM) for oral cavity: Experimental study in normal hamster tongue, *Oral Dis.* 7, 192–195, 2001.

Weiner, L.M., Fully human therapeutic monoclonal antibodies, *J. Immunother.* 29, 1–9, 2006.

Weiss, R.B., Donehower, R.C., Wiernik, P.H., Ohnuma, T., Gralla, R.J., Trump, D.L., Baker, J.R., VanEcho, D.A., VanHoff, D.D., Leyland-Jones, B., Hypersensitivity reactions with taxol, *J. Clin. Oncol.*, 8, 1263–1268, 1990.

White, R.R., Sullenger, B.A., Rusconi, C.P., Developing aptamers into therapeuytics, *J. Clin. Invest.*, 106, 929–934, 2000.

Xu, Z, Gu, W., Huang, J., Sui, H., Zhou, Z., Yang, Y., Yan, Z., Li, Y., In vitro and in vivo evaluation of actively targetable nanoparticles for paclitaxel delvery, *Int. J. Pharm.*, 288, 361–368, 2005.

Yuan, F., Dellian, M., Fukumura, D., Vascular permeability in a human tumor xenograft: Molecular size dependence and cutoff size, *Cancer Res.*, 55, 3752–3756, 1995.

Zhang, Z., Huey, L.S., Feng, S.S., Folate decorated poly(lactide-co-glycolide)-vitamin E TPGS NPs for targeted drug delivery, *Biomaterials*, 28, 1889–1899, 2007.

Gold Nanoparticles for Plasmonic Photothermal Cancer Therapy

Xiaohua Huang
Georgia Institute of Technology

Ivan H. El-Sayed
University of California at San Francisco

Mostafa A. El-Sayed
Georgia Institute of Technology

36.1 Introduction

Gold nanoparticles (Au NPs) and the myriad of concurrent nanotechnologies under development have engendered a widespread interest in the development of a new generation of nano-biotechnological devices for medical diagnostics and therapy. They are very promising nanostructures for use in oncology as diagnostic optical probes (Boppart et al. 2005, El-Sayed et al. 2005, Huang et al. 2006a, Huang et al. 2007), drug delivery vectors (Paciotti et al. 2004, Paciotti et al. 2006), and photothermal therapy agents (El-Sayed et al. 2006, Huang et al. 2006a,b). They are optically active in multiple imaging domains including fluorescence, absorption, scattering, and Raman spectroscopy. They exhibit intensely enhanced absorption cross sections that are five to six orders of magnitude larger than those of conventional photo-absorbing dyes (Jain et al. 2006) due to the coupling of the electromagnetic wave to the surface plasmon resonance (SPR). This absorbed light energy is able to be nonradiatively converted to heat via a series of photophysical processes (Link et al. 2000). In biologic systems, the dissipation of heat to the surrounding environment can initiate the thermal destruction of the surrounding cells and tissues. Au NPs can be selectively targeted to intracellular targets such as the nucleus, cell cytoplasm, or cell membrane (El-Sayed et al. 2005) and thus enables site selective photothermolysis.

This chapter focuses on the application of photothermal therapy (PTT) for the treatment of cancer using Au NPs, currently one of the most active fields in cancer research. The strongly enhanced absorption property of the nanostructures enables effective laser-based therapy at relatively low energies rendering the therapy method minimally invasive. Compared with photochemical methods, Au NPs have higher photostability and do not photobleach. The current major classes of gold nanostructures for laser-based cancer treatment include gold nanospheres (Pitsillides et al. 2003, Zharov et al. 2003, Hainfeld et al. 2004, Zharov et al. 2004, 2005, El-Sayed et al. 2006, Huang et al. 2006a,b, Khlebtsov et al. 2006), gold nanorods (Huang et al. 2006a,b, Takahashi et al. 2006a,b, Huff et al. 2007, Pissuwan et al. 2007, Tong et al. 2007, Dickerson et al. 2008), gold nanoshells (Hirsch et al. 2003, Loo et al. 2004, O'Neal et al. 2004, Loo et al. 2005, Gobin et al. 2007, Diagaradjane et al. 2008), and gold nanocages (Chen et al. 2005, Hu et al. 2006, Chen et al. 2007, Skrabalak et al. 2007, Melancon et al. 2008). The former two have the advantages of simple synthesis, easy bioconjugation, and ready size tuning by simply varying the amount of reaction regents in the synthesis protocols (Figure 36.1). For spherical Au NPs, changing the quantity of the reducing agent of sodium citrate results in nanoparticles of different sizes from 10 to 200 nm. For gold nanorods made with the seed-mediated growth method, which is the most popular way for nanorod synthesis, altering the silver ion concentration results in nanorods of different aspect ratios (length/width) from 2 to 20.

36.2 Optical Properties of Gold Nanoparticles

As electromagnetic waves pass through matter, energy is lost due to two processes: absorption and scattering. Light absorption results when the photon energy is dissipated due to inelastic processes. Light scattering occurs when the photon energy causes electron oscillations in the matter that emit photons in the form of scattered light either at the same frequency as the incident light (Rayleigh scattering) or at a shifted frequency

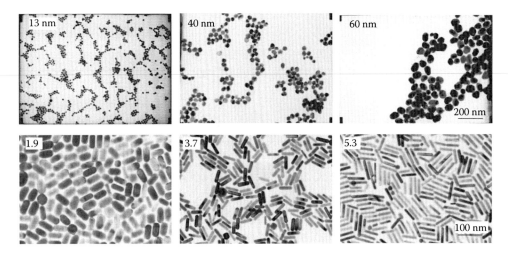

FIGURE 36.1 TEM of gold nanospheres in different sizes (top row) and gold nanorods of different aspect ratios (bottom row).

(Raman scattering). The frequency shift in the Raman scattering corresponds to the energy difference created molecular motion within the matter (molecular bond rotations, stretching, or vibrations). In noble metal nanoparticles, the light absorption, scattering, and total extinction are all strongly enhanced due to the resonance of the coherent collective oscillation of the conduction band with the frequency of the incident light, referred to as the surface plasmon resonance (SPR). The SPR is dependent on factors affecting the electron charge density on the particle surface such as the particle size, shape, structure, and the dielectric properties of the metal and the surrounding medium (Mie 1908, Kerker 1969, Papavassiliou 1980, Bohren and Huffman 1983, Kreibig et al. 1995, Link et al. 1999a,b, Link et al. 2000, Link et al. 2003, Liz-Marzan 2004, Jain et al. 2007). Spherical gold, silver, and copper nanoparticles have a strong SPR band in the visible region, while other metals show broad and weak bands in the UV region. Hollow or core–shell structures show a very large redshift of the SPR wavelength compared with solid structures. Anisotropic nanoparticles such as rods, triangles, and branched structures also exhibit a redshifted SPR band compared with their spherical analogs. The optical properties of spherical and rod shaped Au NPs are discussed in the following sections.

36.2.1 Absorption, Scattering, and Total Extinction of Gold Nanoparticles

36.2.1.1 Gold Nanospheres

In 1908, Gustav Mie first explained the optical properties of symmetric metal nanoparticles by solving the Maxwell equations for an electromagnetic light wave interacting with small nanospheres (Mie 1908). Considering all the multiple oscillations contributing to the absorption and scattering of the incoming light, the solution to Mie's theory leads to the following full expressions for extinction, scattering, and absorption cross sections (Mie theory) for any particle size (Kerker 1969, Papavassiliou 1980, Bohren and Huffman 1983, Kreibig et al. 1995, Link et al. 2000, Hartland 2006):

$$C_{ext} = \frac{2\pi}{|K|^2} \sum_{L=1}^{\infty} (2L+1) R_e(a_L + b_L) \tag{36.1}$$

$$C_{sca} = \frac{2\pi}{|K|^2} \sum_{L=1}^{\infty} (2L+1) \left(|a_L|^2 + |b_L|^2 \right) \tag{36.2}$$

$$C_{abs} = C_{ext} - C_{sca} \tag{36.3}$$

where
k equals $2\pi n_m/\lambda$
n_m is the real refractive index of the surrounding medium
λ is the wavelength of the incident light

L is the summation index of the partial waves with $L = 1$ corresponding to the dipole oscillation and $L = 2$ corresponding to quadrupole oscillation and so on. a_L and b_L are defined as

$$a_L = \frac{\left[(D_L(mx)/m) + (L/x) \right] \psi_L(x) - \psi_{L-1}(x)}{\left[(D_L(mx)/m) + (L/x) \right] \xi_L(x) - \xi_{L-1}(x)} \tag{36.4}$$

$$b_L = \frac{\left[mD_L(mx) + (L/x) \right] \psi_L(x) - \psi_{L-1}(x)}{\left[mD_L(mx) + (L/x) \right] \xi_L(x) - \xi_{L-1}(x)} \tag{36.5}$$

$$D_L(mx) = \frac{d \ln \psi_L(mx)}{d(mx)},$$
$$D_{L-1} = \frac{L/mx - 1}{(D_L + L/mx)} \tag{36.6}$$

$$\psi_{L+1}(x) = \frac{(2L+1)\psi_L(x)}{x - \psi_{L-1}(x)}, \quad \xi_L = \psi_L - i\chi_L \tag{36.7}$$

where

    x is the size parameter equal to kR

    R is the radius of a nanoparticle

    m is the relative refractive index that is defined by $m = n/n_m$

       where n is the complex refractive index of the particle

With the beginning functions of

$$\psi_{-1}(x) = \cos(x), \quad \psi_0(x) = \sin(x),$$
$$\chi_{-1}(x) = \sin(x), \quad \chi_0(x) = \cos(x)$$

(36.8)

the extinction, scattering, and absorption cross sections of nanoparticles of various sizes can thus be attained and their efficiencies are based on $Q_{ext} = C_{ext}/\pi R_{eff}^2$, $Q_{sca} = C_{sca}/\pi R_{eff}^2$, $Q_{abs} = C_{abs}/\pi R_{eff}^2$ where R_{eff} is the effective radius of the particle that is given by $R_{eff} = (3V/4\pi)^{1/3}$. Figure 36.2A through C shows the calculated extinction, scattering, and absorption efficiency of the gold nanospheres of different sizes by Jain and El-Sayed (Jain et al. 2006), which is consistent with the experimental extinction spectra of Au NPs of different sizes in Figure 36.2D (Link et al. 1999a,b).

For nanoparticles much smaller than the wavelength of the light (less than 25 nm in diameter for gold nanoparticles), the dipole oscillation is dominant (consider only $L = 1$, dipole approximation in Equations 36.1 through 36.3) and the extinction cross section (equal to the absorption cross section as the scattering contribution is small) is reduced to the following well-known simple expression (Papavassiliou 1980, Kreibig et al. 1995):

$$C_{ext} = \frac{24\pi^2 R^2 \varepsilon_m^{3/2}}{\lambda} \frac{\varepsilon_i}{(\varepsilon_r + 2\varepsilon_m)^2 + \varepsilon_i^2}$$

(36.9)

where

    ε is the complex dielectric constant of the metal given by $\varepsilon = \varepsilon_r(\omega) + i\varepsilon_i(\omega)$, which is related to the refractive index n of the particle by $\varepsilon = (n + ik)^2$

    ε_m is the dielectric constant of the surrounding medium, which is related to the refractive index of the medium by $\varepsilon_m = n_m^2$

The real part of the dielectric constant of the metal determines the SPR position and the imagery part determines the bandwidth, which is associated with the dephasing of the SPR oscillations by $1/T = \pi c \Gamma$ where T is the total dephasing time, c is the speed of the light, and Γ is the SPR bandwidth. The SPR occurs when $\varepsilon_r(\omega) = -2 \varepsilon_m$.

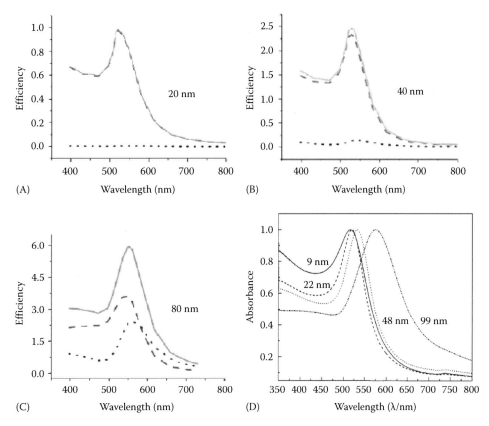

FIGURE 36.2 Calculated absorption (red dashed), scattering (black dotted), and extinction (green solid) spectra of gold nanospheres based on Mie theory $D = 20$ nm for (A), $D = 40$ nm for (B) and $D = 80$ nm for (C). (Reprinted from Jain, P.K. et al., *J. Phys. Chem. B*, 110, 7238, 2006. With permission.) (D) Experimental extinction spectra of gold nanoparticles of different sizes. The surface plasmon resonance of gold nanoparticles induces strong absorption, scattering and extinction at resonant wavelength that is dependent on the size of the nanoparticles. (Reprinted from Link, S. and El-Sayed, M.A., *J. Phys. Chem. B*, 103, 4212, 1999a. With permission.)

Equation 36.9 shows that the extinction cross section is directly related to the particle size as ε is related to the extinction cross section by ε (M^{-1} cm^{-1}) = $10^{-3}N_0C_{ext}$(cm^2)/2.303. However, this does not explain the behavior of particles smaller than 10 nm since the plasmon oscillation in this size region is strongly dampened due to the phase changes resulting from the increased rate of electron-surface collisions compared with larger particles. In this case, the dielectric constant of the metal is both size- and frequency-dependent, and it is formulated according to an interband term and a free conduction electron term (Drude term) in which a damping constant is introduced to describe the lifetime of the electron scattering processes. The plasmon bandwidth is found to be inversely proportional to the radius of the particle, which explains why the SPR of a 5 nm gold sphere is broad and weak and 2 nm gold nanospheres have no SPR. For smaller size nanoparticles, the total light extinction is mainly due to the absorption contributions.

For nanoparticles larger than 25 nm in diameter, the SPR begins to red shift due to higher order electron oscillations resulting from their positions at lower energies. Simulations of the full Mie theory show a minor linear increase of the SPR wavelength and an exponential increase of the extinction cross section on the diameter of the particle. Meanwhile, light scattering significantly intensifies as the particle diameter enlarges (Jain et al. 2006, Jain et al. 2007).

As a result of these properties, smaller nanoparticles offer better light absorption without significant scattering, while larger nanoparticles offer strong light scattering combined with absorption (as shown in Figure 36.1). This data indicates that for photothermal therapy smaller nanoparticles are preferred as photon energy is mainly adsorbed by the nanoparticles and thus efficiently converted to heat to destroy cancer cells and tissue; whereas for imaging applications, larger particles are favored as they show higher scattering efficiency than absorption.

36.2.1.2 Gold Nanorods

Unlike gold nanospheres, gold nanorods exhibit two surface plasmon resonance bands: a weak band in the visible region corresponding to electron oscillations along the short axis, referred to as the transverse band, and a strong band at longer wavelengths corresponding to electron oscillations along the long axis, referred to as the longitudinal band. While the transverse band is insensitive to the size of the nanorods, the longitudinal band is red shifted from the visible to the near infrared region as the aspect ratio (length/width) increases. The optical behavior of gold nanorods can be well understood according to Gan's theory (Gans 1915), which is an extension of Mie's theory. In the dipole approximation, the polarizability of dipole oscillations is given by (Landau et al. 1984)

$$\alpha_i = \frac{4\pi abc}{3}\frac{\varepsilon - \varepsilon_m}{\varepsilon_m + (\varepsilon - \varepsilon_m)n^{(i)}} \quad (36.10)$$

Where $n^{(i)}$ represents the depolarization factors along the three axes a, b, and c of the nanorod. For spheres, $n^{(i)}$ is equal to 1/3.

For ellipsoids with $a > b = c$, and the aspect ratio $R = a/b$, $n^{(i)}$ is defined as

$$n^{(a)} = \frac{1}{R^2-1}\left(\frac{R}{2\sqrt{R^2-1}}\ln\frac{R+\sqrt{R^2-1}}{R-\sqrt{R^2-1}}-1\right) \quad (36.11)$$

$$n^{(b)} = n^{(c)} = \frac{(1-n^{(a)})}{2} \quad (36.12)$$

With the relationship between absorption, scattering cross sections, and polarizability of (Bohren and Huffman 1983)

$$C_{abs} = \frac{k}{3}\text{Im}\left[\alpha_1 + \alpha_2 + \alpha_3\right] \quad (36.13)$$

$$C_{sca} = \frac{k^4}{18\pi}\left[|\alpha_1|^2 + |\alpha_2|^2 + |\alpha_3|^2\right] \quad (36.14)$$

The absorption and scattering cross section are reduced to the following expressions (Qiu et al. 2007):

$$C_{abs} = \frac{2\pi}{3\lambda}\varepsilon_m^{3/2}V\sum_t\frac{\varepsilon_2/(n^{(i)})^2}{\left(\varepsilon_1+\left[(1-n^{(i)})/n^{(i)}\right]\varepsilon_m\right)^2\varepsilon_2^2} \quad (36.15)$$

$$C_{sca} = \frac{8\pi^3}{9\lambda^4}\varepsilon_m^2V^2\sum_i\frac{((\varepsilon_1-\varepsilon_m)^2+\varepsilon_2^2)/(n^{(i)})^2}{\left(\varepsilon_1+\left[(1-n^{(i)})/n^{(i)}\right]\varepsilon_m\right)^2+\varepsilon_2^2} \quad (36.16)$$

And the extinction cross section is the sum of the absorption and scattering cross sections. The resonance occurs at $\varepsilon_1 = -(1 - n^{(i)})\varepsilon_m/n^{(i)}$ where $i = a$ for longitudinal resonance and $i = b, c$ for transverse resonance. Using the linear relationship of the real part of the dielectric constant of gold with light wavelength in the form of $\varepsilon_r(\omega) = 34.66 - 0.07$ when λ is in the range from the 500 to 800 nm region, and the linear fitting between $(1 - n^{(a)})/n^{(a)}$ and R, Link and El-Sayed (Link et al. 1999a,b, Link and El-Sayed 2005) found a linear proportional relationship between the longitudinal SPR absorption maximum and the aspect ratio of nanorods or the dielectric constant of the surrounding medium. These findings are similar to the simulated results of other groups (Yan et al. 2003, Brioude et al. 2005, Spru1nken et al. 2007):

$$\lambda_{max} = (53.71 R - 42.29)\varepsilon_m + 495.14 \quad (36.17)$$

Taking the dielectric constant ε_m of 1.77 for water, the aspect ratio of the nanorods can be derived by measuring the absorption maximum from the following equation:

$$\lambda_{max} = 95 R + 420 \quad (36.18)$$

Tuning the SPR by aspect ratios is unlike the case for spheres, which is only slightly red shifted within the visible spectrum with an increase in the particle size. Increasing the aspect ratio of nanorods produces large redshifts of the SPR absorption maxima from visible to the near infrared (NIR) region. The extinction efficiency is nearly proportional to the aspect ratio in a linear way (Jain et al. 2006). The transverse band is minimally blue shifted with decreasing intensity due to a less longitudinal contribution at longer wavelengths. The longitudinal SPR scattering band is also linearly red shifted with an increase in the aspect ratios while the transverse SPR scattering maximum blue shifts in a nonlinear way of the second-order exponential decay (Zhu et al. 2005).

Gans theory explains the optical properties of the short rods with cylindrical shape as it only considers dipole oscillations. For the optical properties of nanorods with any aspect ratio (and more anisotropic geometry), other models and approximations have been utilized (Purcell and Pennypacker 1973, Ludwig 1991, Kottman et al. 2000, Bruzzone et al. 2003, Barnard and Curtiss 2007). Among these methods, discrete dipole approximation (DDA) has proven to be one of the most powerful electrodynamic and numerical methods for calculating optical properties, especially the scattering problem of targets with any arbitrary geometry (Purcell and Pennypacker 1973, Draine 1988, Draine and Goodman 1993, Draine and Flatau 1994, Draine 2000). In this numerical method, the target particle is viewed as a cubic array of point dipoles with polarization α_i at position r_i and the dipole moment P_i, which is equal to (Lee and El-Sayed 2005)

$$P_i = \alpha_i E_{\text{loc}}(r_i) \qquad (36.19)$$

where E_{loc} is the summation of the incident electric field and the field induced by all other dipoles:

$$E_{\text{loc}}(r_i) = E_{\text{inc},i} + E_{\text{other},i} = E_0 \exp(ik \cdot r_i - i\omega t) - \sum_{j \neq i} A_{ij} \cdot P_j \quad (36.20)$$

$$A_{ij} \cdot P_j = \frac{\exp(ikr_{ij})}{r_{ij}^3} \left\{ k^2 r_{ij} \times (r_{ij} \times p_j) + \frac{(1 - ikr_{ij})}{r_{ij}^2} \right.$$
$$\left. \times \left[r_{ij}^2 P_j - 3r_{ij}(r_{ij} \cdot P_j) \right] \right\}, \quad (j \neq i) \qquad (36.21)$$

Solving Equation 36.21 by an initial guess of the unknown dipole moment P_j, the extinction and absorption cross sections can be derived from the optical theorem (Bohren and Huffman 1983)

$$C_{\text{ext}} = \frac{4\pi k}{|E_0|^2} \sum_{j=1}^{N} \{\text{Im}\} \left(E_{\text{inc},j}^* \cdot P_j \right) \qquad (36.22)$$

$$C_{\text{abs}} = \frac{4\pi k}{|E_0|^2} \sum_{j=1}^{N} \left\{ \{\text{Im}\} \left[P_j (\alpha_j^{-1})^* P_j^* \right] - \frac{2}{3} k^3 |P_j|^2 \right\} \qquad (36.23)$$

DDA provides an easy way to analyze the effects of geometry factors on the SPR wavelength. Results from DDA simulations show that changing the end capping from a prolate ellipsoid to a hemispherical-capped cylinder to a right cylinder results in an SPR redshift of ~50 nm for each (Prescott and Mulvaney 2006). Figure 36.3A through C shows the optical spectra of gold nanorods with different aspect ratios at the same effective radius of 40 nm as calculated by Jain and El-Sayed (Lee and El-Sayed 2005) adopting the DDA code developed by Draine and Flatau (1994). Figure 36.3D is the experimental extinction spectra of gold nanorods of different aspect ratios (Huang et al. 2006a,b). These studies create a gold standard for the choice of gold nanoparticles for biomedical applications. For imaging applications, lager rods are preferred because of their higher scattering efficiency, whereas for photothermal therapy, smaller ones are preferred as light is mainly absorbed by the particles and thus efficiently converted to heat for cell and tissue destruction.

36.2.2 Photothermal Properties

The principle of photothermal therapy for the treatment of cancer is based on the nonradiative photothermal properties of gold nanoparticles. As previously described, electromagnetic waves incident on the nanoparticles induce the electronic surface plasmon resonance oscillation that results in strongly enhanced absorption and scattering orders of a magnitude stronger than the absorption and emission of dye molecules. This strongly absorbed light energy by the nanoparticles is nonradiatively converted into heat quickly via a series of consecutive photophysical processes that have been extensively studied by El-Sayed et al. using ultrafast dynamics (Link et al. 1999a,b, Link et al. 2000, El-Sayed 2001, Mohanmed et al. 2001, Link et al. 2002, Link et al. 2003).

These processes include the following steps: (1) The loss of the phase of the coherently excited electrons on the few femtosecond time scale via electron–electron scattering processes. This electron thermalization results in the formation of a hot electron gas of several thousand kelvins even with laser excitation powers as low as 100 NJ. (2) The loss of electron energy to the phonon bath and the electrons and holes cascade down the energy scale by electron–phonon interactions on the order of 0.5–1 ps. The heat dissipation from the hot electrons to the lattice cause the lattice temperature to rise on the order of a few tens of degrees (Link et al. 2000). (3) The lattice cools by passing its heat to the surrounding medium via phonon–phonon relaxation within ~100 ps. This process will cause the temperature to rise in the species localized around gold nanoparticles. This fast energy conversion and dissipation can be utilized for the efficient heating of local environments by using a selected wavelength of light radiation that overlaps with the nanoparticle SPR absorption band. Furthermore, the size of nanoparticles <200 nm is well matched to biomolecular structures and macromolecules found inside living cells, making them potentially useful for biomedical photothermal and optical diagnostic applications.

In the treatment of cancer, there is great interest in agents that can destroy the tumor cells using novel mechanisms as an adjunct to currently used techniques of surgical removal, radiation

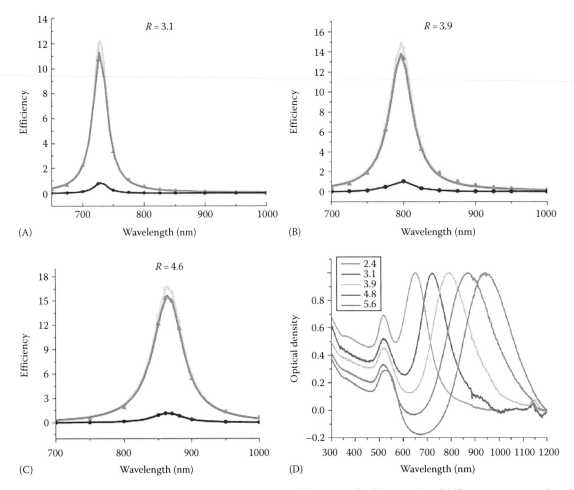

FIGURE 36.3   Calculated absorption (A), scattering (B) and extinction (C) spectra of gold nanorods of different aspect ratios based on DDA method. (Reprinted from Jain, P.K. et al., *J. Phys. Chem. B*, 110, 7238, 2006. With permission.) (D) Experimental extinction spectra of gold nanorods of different aspect ratios. The surface plasmon resonance of gold nanorods induces strong absorption, scattering and extinction at resonant wavelength that is dependent on the aspect ratio of the rods. (Reprinted from Huang, X. et al., *J. Am. Chem. Soc.*, 128, 2115, 2006a. With permission.)

therapy, and chemotherapy. Au NP-based photothermal therapy can damage living cells by rupturing the outer cell membrane or by coagulating and denaturing intracellular proteins and structures. Gold nanospheres and nanorods can be selectively applied to various tumors based on the tumor characteristics. Solid cancers can be located superficially on the skin or deep within various locations of the body. Various techniques of light delivery are available through catheters and fibers that could allow for the delivery of light through some of the body barriers. Gold nanospheres absorb mainly in the visible range where light penetrates minimally through the skin, on the level of a few millimeters. Gold nanorods on the other hand absorb in the infrared region where light can penetrate deeply through some tissues on the order of several centimeters due to the minimal absorption of the light by water or endogenous chromophores in the tissue. This makes NIR-resonant gold nanostructures very useful for clinical therapy applications involving tumors located deep within the body tissue.

The photothermal conversion efficiencies of gold nanoparticles of various sizes are different due to the difference in the absorption efficiency. For gold nanospheres, theoretical studies (Khlebtsov et al. 2006) show that nanoparticles with diameters of 30–40 nm are most preferable, as their normalized absorption is maximal in the visible spectrum region. For nanorods, the most efficient size range is predicted to be between 15 and 70 nm. To achieve optimal photothermal efficiency, the longitudinal absorption maximum must be matched to the NIR laser wavelength.

36.3 Gold Nanoparticles for Plasmonic Photothermal Cancer Therapy

36.3.1 Gold Nanospheres

36.3.1.1 Synthesis

Gold nanospheres are synthesized according to the method described by Frens (1973; Figure 36.4A). Basically, 50 mL of 0.01% $HAuCl_4$ solution (by weight) is heated to boiling with stirring in a 100 mL beaker. A total of 350 µL of 1% (by weight) of trisodium citrate solution is added quickly to the auric acid solution.

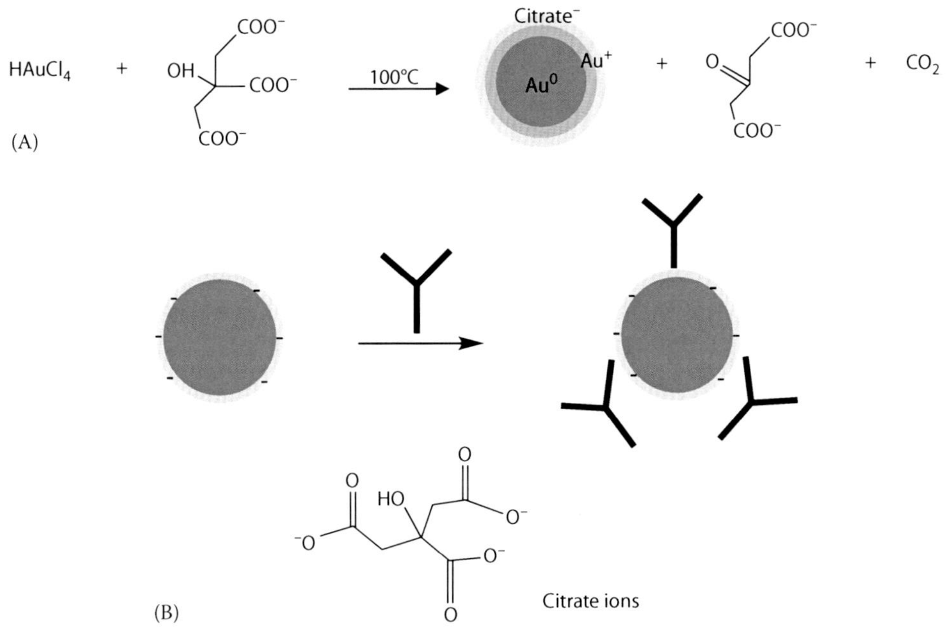

FIGURE 36.4 Synthesis (A) and antibody conjugation (B) of spherical gold nanoparticles.

The solution changes color within a few minutes from yellow to black and then to red. After the color changes, the solution is stirred for 15 min without heating. Nanoparticles prepared in this protocol are generally about 40 nm in diameter with a surface plasmon absorption maximum around 530 nm. The citrate molecules serve as both reducing and capping agents. The citrate-capped Au NPs are negatively charged and thus they are typically stable in solution for months due to the charge repulsion between nanoparticles. The size of gold nanospheres can be adjusted by varying the amount of citrate in the solution. More citrate molecules in the solution cap the gold quicker and result in smaller nanoparticles. Larger nanoparticles are produced when less citrate is used as gold atoms have more time to aggregate to form particles before being capped.

36.3.1.2 Bioconjugation

An exciting aspect of Au NPs is its ability to be conjugated with a range of biomolecules through electrostatic or covalent methods. Targeting strategies have been aimed at the direct delivery of particles through injection, topical application, or intravascular delivery. In the human vascular system, nanoparticles that bypass the immune system can filter into tumor beds through a network of leaky blood vessels. Therefore, the initial strategies of tumor targeting have relied on immunopassivating the nanoparticles with polyethylene glycol (PEG). Other strategies rely on targeting molecules near the tumor or expressed on the tumor surface. Two well studied tumor receptors are the epithelial growth factor receptor (EGFR) and human epithelial growth factor receptor 2 (HER2) that are overexpreseed on the cytoplasmic membrane of many cancers (Reinhard et al. 2007). HER2 is overexpressed in breast cancer while EGFR is an overexpressed in 80% of head and neck upper aerodigestive tract cancers (Harari 2004). Thus, conjugation of these antibodies to

the nanoparticle will enhance the binding of the particles and subsequent uptake into the cells.

Gold is available for conjugation through two mechanisms. One is covalent binding, such as the thiol chemistry, which is a very commonly used technique as gold will readily bind to molecules covalently with Au–S bonds. Thus, PEGylation of gold can be readily achieved with their reactions with thiolated PEG molecules. Antibodies and complex proteins can be conjugated via ethyl-N-(3-dimethylaminopropyl) carbodiimide (EDAC) catalyzed covalent binding between carboxyl or amine groups on the particle surface and amine or carboxyl groups on the proteins.

Another method is a nonspecific interaction such as electrostatic adsorption based on the charge of the particle and targeting ligands, the method which had been used 50 years ago (Hayat 1989). In this method (Figure 36.4B), 40 nm Au NPs are diluted in 20 mM pH 7.4 4-(2-hydroxyethyl)-1-piperazine-ethanesulfonic acid (HEPES) buffer to a final concentration with an optical density of 1.0 at 530 nm. 100 µL 6 mg/mL anti-EGFR monoclonal antibodies are added to a 900 µL 20 mM pH 7.4 HEPES buffer to form a 1 mL diluted antibody solution. Then, a 10 mL gold solution is added drop wisely to the antibody solution while stirring. After stirring for 5 min, the solution is left to react for 20 min. Five hundred microliters of 1% PEG 4000 (polyethylene glycol 4000, PEG 50% W/V) are added to the mixture and the solution is centrifuged at 6000 rpm for 30 min. The anti-EGFR/gold pellet is redispersed in pH 7.4 PBS buffer and stored at 4°C.

At pH = 7.4, the adsorbed antibody on the gold nanoparticle surface still preserves its function (Hayat). The antibodies are bound to negatively charged citrate-capped gold nanospheres by nonspecific interactions. The isoelectric point (PI, zero net charge of the protein) of the antibody is near pH 7. Thus, at pH = 7.4, the antibody has an overall small net negative charge. This suggests that the adsorption of antibody at pH = 7.4 to gold nanoparticles is not purely due to the electrostatic attraction between the negatively charged citrate on the gold surface and the positively charged segment of the antibody. The hydrophobic interaction of the antibody with the hydrophobic three carbon region of the citrates (see the structure in Figure 36.4) may also play an important role.

36.3.1.3 Photothermal Therapy

Earlier studies of light-based cancer therapy is focused on a photochemotherapy technique that employs toxic singlet oxygen or other free radicals produced under light exposure to induce cell or tissue destruction (Wilson and Patterson 1986). Au NPs mediated photothermal therapy is based on the light-to-thermal energy conversion. Initial reports described a process of photothermolysis using Au NPs adhered to the cell membrane in combination with pulsed lasers (Lin et al. 1999, Pitsillides et al. 2003). Lymphocytes incubated with antibody conjugated Au NPs were exposed to laser pulses (20 ns duration) at 565 nm wavelength and cell damage was induced with 100 laser pulses at an energy of 0.5 J/cm² (Pitsillides et al. 2003). Cell death is attributed to the cavitation bubble formation on the cell surface around the nanoparticles. In one other work (Zharov et al. 2003), K562 cancer cells incorporated with Au NPs are killed using a nanosecond Nd-YAG laser at 532 nm. With an energy level of 2–3 J/cm², only one to three laser pulses are necessary to damage a cell containing 10–15 particles of 20 nm in diameter. At a lower energy fluence rate of 0.5 J/cm², 50 pulses and approximately 100 particles are required to produce the same harmful effects on the cells.

Plasmonic photothermal therapy using a continuous wavelength (CW) laser can kill tumor cells based on a hyperthermia mechanism. El-Sayed et al. (2006, Huang et al. 2006a,b) used 40 nm of gold nanoparticle conjugated to anti-EGFR antibodies on cell lines generated from HaCaT normal cells (immortalized human keratinocytes), HOC 313 clone 8 (human oral cancer), and HSC 3 cancer cells (human oral squamous cell carcinoma). In this work, the cells were cultured in DMEM (Dulbecco's Modification of Eagle's Medium) plus 10% FBS (fetal bovine serum) at 37°C under 5% CO_2. The cells were then cleaved by trypsin and replated onto 18 mm glass coverslips in a 12-well tissue culture plate allowing for growth for 3 days. The cover slips were coated with collagen type I in advance for optimum cell growth. The cell monolayer on the cover slips was taken out of the medium from the incubator and rinsed with PBS buffer and then immersed into the anti-EGFR conjugated Au NPs for 30 min at room temperature. The cell was then rinsed with PBS buffer to remove the excess particles and immersed in PBS buffer for the photothermal experiments.

The binding of gold nanoparticles to immortalized benign cells and cancer derived cells was confirmed using dark field light-scattering imaging (Figure 36.5A through C) and surface

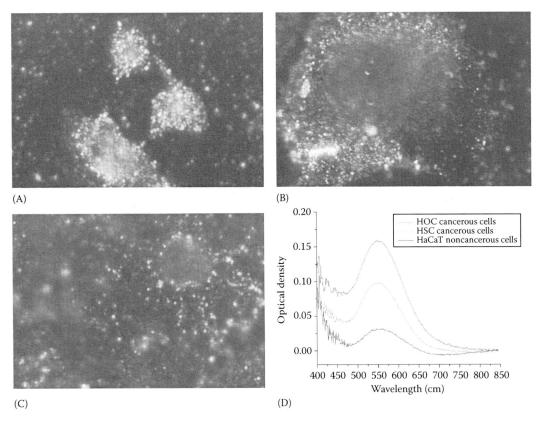

(A)   (B)

(C)   (D)

FIGURE 36.5 Light-scattering images of HSC cancer cells (A), HOC cancer cells (B) and HaCat normal cells (C) after treated with anti-EGFR conjugated gold nanoparticles. (D) is the corresponding absorption spectra averaged from twenty cells for each type of cells. (Reprinted from El-Sayed, I.H. et al., *Cancer Lett.*, 239, 129, 2006. With permission.)

plasmon absorption spectroscopy (Figure 36.5D) from single cells. Anti-EGFR conjugated gold nanoparticles are preferentially and specifically bound to the cancer cells that over-express EGFR by up to 600% more than the benign cells. The benign HaCat cell lines demonstrate fewer nanoparticles with broadening suggesting evidence of particle aggregation resulting from nonspecific binding.

In the *in vitro* photothermal setup (Figure 36.6), the argon ion laser with a wavelength of 514 nm is directed down to the sample by a reflection mirror. The laser beam is focused by a lens (f = 10 cm) to form a 1 mm spot on the sample position where the coverslip with the cells is placed in PBS buffer in a Petri dish. The coverslip is marked with a black marker in a grid on the back side to keep track of each laser irradiation position. The laser irradiates each cell frame sequentially at decreasing energy levels for a constant time of 4 min. After the laser irradiation, the coverslip is taken out of the buffer from the petri dish and stained with a few drops of 0.4% trypan blue solution for 10 min. The trypan blue stain is a test for cell viability. Live cells actively pump it out when placed in buffer solution while dead cells accumulate the dye. Thus, the cells stained blue are the dead cells. After staining, the cells are immersed in a PBS buffer solution in a petrish for bright field imaging under 10× magnification.

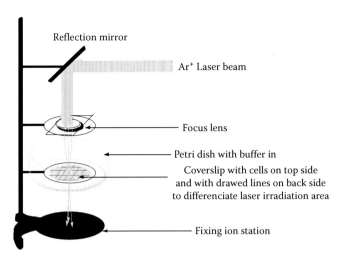

FIGURE 36.6   In vitro photothermal setup. 514 nm Ar⁺ laser is focused and directed to the cells which has been incubated with anti-EGFR antibody conjugated gold nanoparticles before treatment.

Figure 36.7 shows the images of both normal and cancer cells after laser irradiation and trypan blue staining. Cell death is recognized as a blue spot in a circular region that matches the laser spot size. The threshold for HaCat normal cells, HSC cancer cells, and HOC cancer cells is 57, 25, and 19 W/cm², respectively.

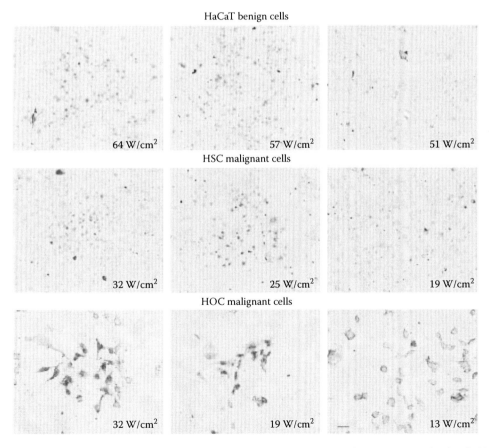

FIGURE 36.7   Selective photothermal therapy of cancer cells using gold nanospheres and Ar⁺ laser at 514 nm. The threshold for HaCat normal cell destruction is 57 W/cm², HSC cancer cells is 25 W/cm² and HOC cancer cells is 19 W/cm². The cancer cells are destructed at lower than half energy than normal cells. (Reprinted from El-Sayed, I.H. et al., *Cancer Lett.*, 239, 129, 2006. With permission.)

The cancer cells are killed at lower than half the energy of normal cells. So the two types of cancer cells can be destroyed at 25 W/cm² while the normal cells survive. This selective photodamage of the cancer cells is clearly attributed to the higher gold loading in cancer cells, as shown in Figure 36.5. Numerical calculations based on the optical density of gold on cells and laser power density shows that the temperature rose to 70°C–80°C in the cells, which correlates well with the temperature required to kill cells when heated directly for the same amount of time (Huang et al. 2006a,b). This method can be extended to other types of cancers since many types of cancer cells overexpress EGF receptors on their cytoplasm membrane. Obviously, other targets exist for each type of cancer that can be exploited for the selective delivery of nanoparticles. However, the use of visible light absorbing nanospheres is limited due to the fact that light in the visible region does not penetrate human epithelium or other organs well.

36.3.2 Gold Nanorods

36.3.2.1 Synthesis

Similar to gold nanospheres activity in the visible region, gold nanorods can be used for plasmonic photothermal therapy (PPTT) in the NIR region. Gold nanorods are synthesized by the seed-mediated growth method developed by the El-Sayed group (Nikoobakht and El-Sayed 2003). This method involves two steps (Figure 36.8A). In the first step, 7.5 mL of 0.2 M CTAB solution is mixed with 2.5 mL 0.001 M HAuCl₄ in a 40 mL beaker. A total of 0.6 mL of ice-cold 0.01 M NaBH₄ is added to the stirred solution and the solution color changes to amber indicating the formation of gold nanospheres of 3–5 nm. Vigorous stirring of the seed solution is continued for 2 min. After stirring, the solution is kept at room temperature.

In the second step, 50 mL of 0.001 M HAuCl₄ is added to 50 mL of 0.2 M CTAB in a 250 mL flat-bottom flask and the solution changes color from yellow (HAuCl₄ dissolved in water) to orange (HAuCl₄ dissolved in CTAB). A total of 2 mL of 0.004 M AgNO₃ solution is added and gently mixed. A total of 700 μL 0.0788 M ascorbic acid is added to reduce the Au³⁺ ions to Au⁺. At this step, the orange gold growth solution becomes colorless due to the Au⁺ formation. To this solution, 120 μL seed solution prepared in the first step is added and the solution is stirred for 10 s followed by still sitting. Usually the color of the solution gradually changes within 30 min indicating nanorod formation. The nanorods growth is completed within 2 h. This protocol produces gold nanorods with an aspect ratio around 3.9 with an absorption maximum around 800 nm, which is optimal for NIR photothermal therapy. By changing the concentration of silver nitrate, the aspect ratios can be tuned. A higher concentration of silver nitrate results in longer rods while lower concentrations produce shorter rods. However, to make nanorods with an aspect ratio over 10, a three step procedure in the absence of silver ions developed by the Murphy group is commonly used (Murphy et al. 2005).

The as-prepared nanorods are capped with CTAB solely on the long axis side in a bilayer structure (Nikoobakht and El-Sayed 2001). The tertiary amino end binds to the gold surface, leaving the hydrophobic carbon chain to interact with that of other CTAB molecules. In this way, the nanorod surface is positively charged due to the amino group of the second CTAB molecules. The growth of gold nanorods is explained by a silver underpotential deposition (UPD) mechanism proposed by Orendorff and Murphy (2006) combing previous UPD, electric-driving-directed, and surfactant preferential binding models. Briefly, AuCl₂⁻ on the CTAB micelles diffuses to CTAB-capped seed spheres by electric field interactions and the sphere symmetry is broken into different facets with preferential binding of CTAB

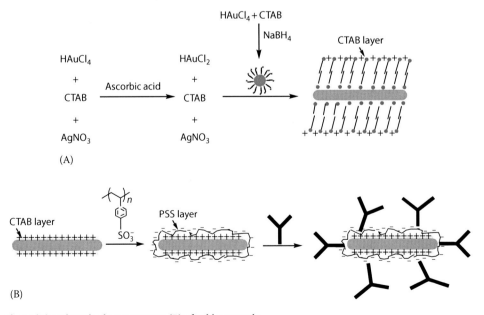

FIGURE 36.8 Synthesis (A) and antibody conjugation (B) of gold nanorods.

onto the {110} facet. Silver ions deposit onto the {110} side facet with a faster rate than they do onto the {100} end facet resulting in the particle growth into rod shape along the [110] direction. The complete deposition of silver ions on to {100} end facet stops the particle growth, which limits the method to make nanorods no longer than 100 nm.

36.3.2.2 Bioconjugation

The nanorods are conjugated to anti-EGFR antibodies via a poly (styrenesulfonate) (PSS) layer (Huang et al. 2006a,b) (Figure 36.8B). The original rods grown by the seed-mediated method are centrifuged at 14,000 rpm twice to get rid of the free CTAB molecules in the solution. A total of 200 µL of 10 mg/mL PSS is added to 10 mL rods with an optical density of 1.5 at 800 nm. The solution is stirred for 15 min. PSS molecules are bound to gold nanorods via electrostatic interactions between the positively charged CTAB on the gold surface and the negatively charged sulfonate ions in the PSS polymer. The extra PSS is separated by centrifuging the rod solution at 8000 rpm for 20 min and the pellet is redispersed in HEPES solution (pH = 7.4). The PSS-capped nanorods are then mixed with antibody solution in HEPES buffer and left to react for 20 min. The solution is centrifuged and redispersed into PBS (pH = 7.4) to form a stock solution with an optical density of around 1.0 at 800 nm. The anti-EGFR/nanorod conjugates are stored at 4°C until used at room temperature.

The antibodies are bound to gold nanorods via nonspecific interactions, similar to the binding of antibodies to gold nanospheres. According to studies by Ai et al. (2002) and Caruso et al. (1997), negatively charged PSS molecules are able to adsorb antibodies at pH = 7.4 to form immunosensing multilayer structures. This suggests that besides the possible electrostatic interaction

between the negatively charged PSS and the positively charged segment of the antibody, the hydrophobic interaction between these two molecules takes roles.

36.3.2.3 Photothermal Therapy

36.3.2.3.1 In Vitro Cancer Therapy

In the *in vitro* experiment, a CW Ti: Sapphire laser at 800 nm is used. This wavelength is located in the NIR window at which human tissue has minimal absorption. It also overlaps efficiently with the longitudinal absorption band of the nanorods. The cell monolayer is immersed into the conjugated nanoparticle solution (OD$_{800\,nm}$ = 1.5) for 30 min, rinsed with PBS buffer, and then exposed to the red laser light at various power densities. The red laser at 800 nm is focused on a 1 mm diameter spot on the sample in a similar setup shown in Figure 36.6. Multiple regions on the slides are exposed to the laser light at different power densities for 4 min each and then stained with 0.4% trypan blue for 10 min to test cell viability (Figure 36.9).

The results show that cell death with the immortalized benign HaCat cells occurs after exposure to the red laser at and above 160 mW (20 W/cm^2; Figure 36.8). Decreasing the laser energy to 120 mW decreases the proportion of blue cells, i.e., the dead cells in the laser spot. The malignant HSC cells suffer photothermal injury at a much lower laser power starting at 80 mW, which corresponds to 10 W/cm^2. This energy threshold is nearly half the value needed to cause the cell death of the nonmalignant HaCaT cells. The HOC malignant cells also undergo photothermal destruction at and above 80 mW while no cell death is observed at a lower power. It can be seen that the two types of malignant cells require less than half the energy required to kill

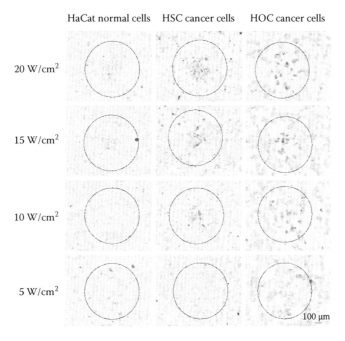

FIGURE 36.9 Selective photothermal therapy of cancer cells using gold nanorods and Ti:+ laser at 514 nm. The threshold for HaCat normal cell destruction is 20 W/cm^2, HSC cancer cells is 10 W/cm^2 and HOC cancer cells is 10 W/cm^2. The cancer cells are destructed at half energy of that required for normal cells. (Reprinted from Huang, X. et al., *J. Am. Chem. Soc.*, 128, 2115, 2006a. With permission.)

the nonmalignant cells. Again, similar to the spherical nanoparticles discussed in Section 36.2.1.3, this is due to the high density of EGFR on the cancer cells and the corresponding higher binding of immunoconjugated gold nanorods that absorb the light and convert it into heat at the cell surface.

The above studies indicate that nanorods conjugated to antibodies can be used as a selective and efficient photothermal agent for cancer cell therapy using a harmless NIR laser. Thus, for further *in vivo* applications, it is expected that the tumor tissue will be selectively destroyed at laser energies that will not harm the surrounding normal tissue due to the higher concentration of nanorods selectively bound to the tumor tissue. The threshold energy to kill the cancer cells using gold nanorods is $10\,W/cm^2$, which is lower than what is needed in the case of the core–shell particles (Hirsch et al. 2003). This difference can result from either a higher absorption cross section, a higher binding affinity constant of the gold nanoparticles to the antibody, or a higher affinity constant of the antibody to the cancer cells.

Similar *in vitro* studies have been conducted by some other groups. Niidome and coworkers (Takahashi et al. 2006a,b) in Japan achieved cell death using phosphatidylcholine-passivated gold nanorods and a pulsed Nd-YAG laser at 1064 nm. Recently, Wei and coworkers at Purdue University (Huff et al. 2007) demonstrated that gold nanorods conjugated to folate ligands can be used for hyperthermic therapy of KB cancer cells with a cw Ti:Sappire laser. Severe blebbing of cell membranes was observed at laser irradiation with a power density as low as $30\,J/cm^2$.

The photothermolysis mechanism is elucidated in more detail by the groups of Wei and Cheng (Tong et al. 2007) using fluorescent staining in *in vitro* studies. Their studies show that the photothermolysis is size- and laser-dependent. The laser energy used to destroy KB cells when the nanoparticles are located on the cytoplasm membrane is 10 times lower than that required when the nanoparticles are internalized inside the cytoplasm. The energy required for a femtosecond laser is 10-fold lower than when a continuous wave laser is used. Based on these results and the staining of cell membrane integrity, cell viability, and actin filaments, they found that cell death is initiated by the disruption of the plasma membrane. The subsequent influx of calcium ions induces membrane blebbing and damage of actin filaments. Obviously, apoptosis is the route of cell destruction by the laser heating of gold nanorods.

36.3.2.3.2 In Vivo Tumor Therapy

For *in vivo* tumor targeting, two targeting strategies are commonly used. One is the use of antibodies specific to biomarkers on the tumor cells; similar routes were used for *in vitro* cellular studies. This method is called active targeting as the nanoparticles bind to tumor cells. Another method is called passive targeting, which takes advantage of the tumor characteristics of leaky vascular and impaired lymphatic drainage (Nie et al. 2007). This allows for the enhanced permeability and retention (EPR) effect of nanoparticles from 5 to 400 nm. The nanoparticles do not bind to cells, but flow into tumor tissue interstice. To avoid uptake by the reticuloendothelial system and maximize blood

circulation time, gold nanorods are conjugated with hydrophilic PEG. mPEG-SH (methoxy-poly(ethylene) glycol, MW5000) is added to the 1 nM gold nanorod solution (calculated using the absorption coefficient of $4.6 \times 10^9\,M^{-1}CM^{-1}$) at a final concentration of 25 μM and allowed for reaction up to one week. The PEGylated nanorods are purified by centrifugation at 8000 rpm for 20 min and re-dispersed in PBS buffer. A dynamic light-scattering measurement shows an increase in size of about 28 nm after PEGylation.

In vivo NIR photothermal therapy using gold nanorods has been demonstrated in a mouse model (Dickerson et al. 2008). Subcutaneous HSC-3 squamous cell carcinoma xenografts are grown on the flank of nude mice to form 3 to 5 mm tumors. 100 μL pegylated gold nanorods ($OD_{\lambda=800} = 120$) are injected into mice via the tail vein or 15 μL pegylated gold nanorods ($OD_{\lambda=800} = 40$) are directly injected into the tumor subcutaneously. Optimal accumulation of the nanorods in the tumor is observed at 24 h after administration by silver staining of the tumor tissue. Nanorod accumulation is semi-quantitatively monitored by NIR transmission imaging (Figure 36.10A) using a portable diode laser with a wavelength of 808 nm. Intensity line-scans of NIR extinction at 24 h post tail vein injection of the nanorods shows extinction approximately three times that observed for control sites. Intensity line-scans of NIR extinction at 2 min after the direct injection are six times higher than that of control sites. This *in vivo* imaging guides the use of different laser energies for the injection method.

Photothermal therapy is conducted using the same laser as that used for NIR imaging. For intravenous injection, an NIR laser with an intensity of 1.7–$1.9\,W/cm^2$ is applied to the tumors 24 h after nanorod injection, and irradiation time is 10 min. For direct administration, tumors are extracorporeally exposed to the laser within 2 min of the injection with an intensity of 0.9–$1.1\,W/cm^2$ for 10 min. In each group, 6–7 mice are studied. Figure 36.10B shows that for direct injection, the average change in tumor volume recorded over a 13 day period decreases over 96% relative to control tumors that are exposed to laser light only. For intravenously treated HSC-3 xenografts, the tumor volume decreases over 74% at day 13. Moreover, resorption of >57% of the directly treated tumors and 25% of the intravenously treated tumors is observed over the monitoring period. In contrast, neither growth suppression nor resorption is observed in any of the control tumors.

The photothermal efficiency is determined by the temperature rise of the tumor tissue. This temperature increase can be simply monitored by a thermocouple during the laser treatment. Figure 36.10C shows that during the laser exposure to the tumors that are intratumorally administered with PEGylated gold nanorods, the temperature increases by 15°C within 2 min and then stabilizes at this temperature for the rest of 8 min. The control tumors show that the temperature rises by 5°C. A similar rise in temperature is observed in the case of tail vein injection (Figure 36.10D). The temperature rises, the NIR image line scans, and the tumor volume changes are in agreement with each other. The lower loading of gold nanoparticles in the tail vein injection compared with the direct injection requires a

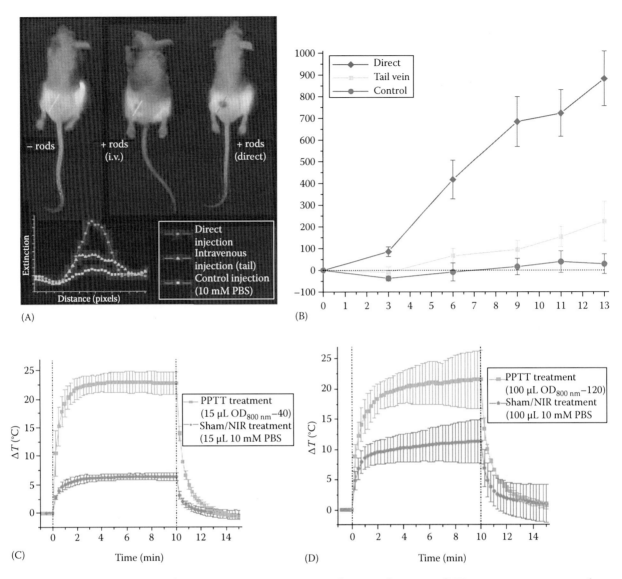

FIGURE 36.10 (A) NIR transmission of mice prior to PPTT treatments and intensity line-scans of NIR extinction at tumor sites for control (—■—), intravenous (—▲—), and direct (—•—) administration of pegylated gold nanorods; (B) Average change in tumor volume for HSC-3 xenografts following near-infrared PPTT treatment by control (—◆—), intravenous , and direct (—•—) injection of pegylated gold nanorods. Error bar is standard error of average value; (C) Temperature rises of tumors during photothermal therapy when the nanoparticles are directly injected into tumors; (D) Temperature rises of tumors during photothermal therapy when the nanoparticles are intravenously injected into mice. (Reprinted from Dickerson, E.B. et al., *Cancer Lett.*, 269, 57, 2008. With permission.)

higher laser power density to reach the same temperature rises for efficient photothermal therapy. This method using gold nanorods also agrees well with that using gold nanoshells (Hirsch et al. 2003).

36.4 Summary

Gold nanoparticles, spheres, and rods exhibit unique optical properties that are under active investigation. A comprehension of the electronic surface plasmon resonance response to electromagnetic radiation provides an explanation for the optical behavior. The surface plasmon resonance enables gold's novel applications as ultrabright and sensitive diagnostic probes using light-scattering techniques and absorption spectroscopy. Furthermore, the absorption plays an additional role in therapy since it provides the basis for thermal activity. Heat production is greater in particles that have higher absorption efficiency (smaller spheres and rods) while scattering is greater in larger particles. Studies using gold nanoparticles conjugated to antibodies demonstrate that the selective labeling of the malignant cells can produce selective photothermal damage. For *in vivo* applications, multiple strategies are available that may allow the selective targeting of gold particles through topical application, direct injection, or intravascular delivery.

References

Ai, H., Fang, M., Jones, S. A., Lvov, Y. M. *Biomacromolecules* 2002, 3, 560.

Barnard, A. S., Curtiss, L. A. *J. Mater. Chem.* 2007, 17, 3315.

Bohren, C. F., Huffman, D. R. *Absorption and Scattering of Light by Small Metal Nanoparticles*, Wiley, New York 1983, 127.

Boppart, S. A., Oldenburg, A. L., Xu, C., Marks, D. L. *J. Biomed. Opt.* 2005, 10, 41208.

Brioude, A., Jiang, X. C., Pileni, M. P. *J. Phys. Chem. B* 2005, 109, 13138.

Bruzzone, S., Arrighini, G. P., Guidotti, C. *Chem. Phys.* 2003, 291, 125.

Caruso, F., Niikura, K., Furlong, D. N., Okahata, Y. *Langmuir* 1997, 13, 3427.

Chen, J., Wiley, B., Li, Z. Y. et al. *Adv. Mater.* 2005, 17, 2255.

Chen, J., Wang, D., Xi, J. et al. *Nano Lett.* 2007, 7, 1318.

Diagaradjane, P., Shetty, A., Wang, J. C. et al. *Nano Lett.* 2008, 8, 1492.

Dickerson, E. B., Dreaden, E. C., Huang, X., Chu, H., Pushpanketh, S., McDonald, J. F., El-Sayed, M. A. *Cancer Lett.* 2008, 269, 57.

Draine, B. T. *Astrophys. J.* 1988, 333, 848.

Draine, B. T. The discrete-dipole approximation for light scattering by irregular targets, in *Light Scattering by Nonspherical Particles: Theory, Measurements, and Geophysical Application* (Eds: Mishchenko, M. I., J. W. Hovenier, L. D. Travis), Academic Press, New York 2000, 131.

Draine, B. T., Goodman, J. J. *Astrophys. J.* 1993, 405, 685.

Draine, B. T., Flatau, P. J. *J. Opt. Soc. Am. A* 1994, 11, 1491.

El-Sayed, M. A. *Acc. Chem. Res.* 2001, 34, 257.

El-Sayed, I. H., Huang, X., El-Sayed, M. A. *Nano Lett.* 2005, 5, 829.

El-Sayed, I. H., Huang, X., El-Sayed, M. A. *Cancer Lett.* 2006, 239, 129.

Frens, G. *Nat. Phys. Sci.* 1973, 241, 20.

Gans, R. *Ann. Phys.* 1915, 47, 270.

Gobin, A. M., Lee, M. H., Halas, N. J. et al. *Nano Lett.* 2007, 7, 1929.

Hainfeld, J. F., Slatkin, D. N, Smilowitz, H. M. *Phys. Med. Biol.* 2004, 49, N309.

Harari, P. M. *Endocr. Relat. Cancer* 2004, 11, 689.

Hartland, G. V. *Annu. Rev. Phys. Chem.* 2006, 57, 403.

Hayat, M. A. *Colloidal Gold: Principles, Methods and Applications*, Vol 1, Academic Press, San Diego, CA 1989.

Hirsch, L. R., Stafford, R. J., Bankson, J. A. *Proc. Natl. Acad. Sci. USA* 2003, 100, 13549.

Hu, M., Petrova, H., Chen, J. et al. *J. Phys. Chem. B* 2006, 110, 1520.

Huang, X., El-Sayed, I. H., El-Sayed, M. A. *J. Am. Chem. Soc.* 2006a, 128, 2115.

Huang, X., Jain, P. K., El-Sayed, I. H., El-Sayed, M. A. *Photochem. Photobiol.* 2006b, 82, 412.

Huang, X., El-Sayed, I. H., Qian, W., El-Sayed, M. A. *Nano Lett.* 2007, 7, 1591.

Huff, T. B., Tong, L, Zhao, Y. et al. *Nanomedicine* 2007, 2, 125.

Jain, P. K., Huang, X., El-Sayed, I. H., El-Sayed, M. A. *Plasmonics* 2007, 2, 107.

Jain, P. K., Lee, K. S., El-Sayed, I. H., El-Sayed, M. A. *J. Phys. Chem. B* 2006, 110, 7238.

Kerker, M. *The Scattering of Light and Other Electromagnetic Radiation*, Academic Press, New York 1969.

Khlebtsov, B., Zharov, V., Melnikov, A., Tuchin, V., Khlebtsov, N. *Nanotechnology* 2006, 17, 5167.

Kottman, J. P., Martin, O. J. F., Smith, D. R., Schultz, S. *Opt. Express* 2000, 6, 213.

Kreibig, U., Vollmer M. *Optical Properties of Metal Clusters*, Springer, Berlin, Germany 1995.

Landau, L. D., Lifshitz, E. M. *Electrodynamics of Continuous Media*, 2nd edn., Pergamon Press, Oxford, U.K. 1984.

Lee, K. S., El-Sayed, M. A. *J. Phys. Chem. B* 2005, 109, 20331.

Lin, C. P., Kelly, M. W., Sibayan, S. A. B, Latina, M. A., Anderson, R. R. *IEEE J. Quant. Elect.* 1999, 5, 963.

Link, S., El-Sayed, M. A. *J. Phys. Chem. B* 1999a, 103, 4212.

Link, S., El-Sayed, M. A. *J. Phys. Chem. B* 1999b, 103, 8410.

Link, S., El-Sayed, M. A. *Int. Rev. Phys. Chem.* 2000, 19, 409.

Link, S., El-Sayed, M. A. *Ann. Rev. Phys. Chem.* 2003, 54, 331.

Link, S., El-Sayed, M. A. *J. Phys. Chem. B* 2005, 109, 10531.

Link, S., Burda, C., Wang, Z. L., El-Sayed, M. A. *J. Chem. Phys.* 1999a, 111, 1255.

Link, S., Mohamed, M. B., El-Sayed, M. A. *J. Phys. Chem. B* 1999b, 103, 3073.

Link, S., Burda, C., Mohamed, M. B., Nikoobakht, B., El-Sayed, M. A. *Phys. Rev. B* 2000, 61, 6086.

Link, S., Furube, A., Mohamed, M. B. et al. *J. Phys. Chem. B* 2002, 106, 945.

Link, S., Hathcock, D. J., Nikoobakht, B., El-Sayed, M. A. *Adv. Mater.* 2003, 15, 5.

Liz-Marzan, L. M. *Mater. Today* 2004, 7, 26.

Loo, C. H., Lin, A., Hirsch, L. R. *Technol. Cancer Res. Treat.* 2004, 3, 33.

Loo, C., Lowery, A., Halas, N. J., West, J. L., Drezek, R. *Nano Lett.* 2005, 5, 709.

Ludwig, A. C. *Comput. Phys. Commun.* 1991, 68, 306.

Melancon, M. P., Lu, W., Yang, Z. *Mol. Cancer Ther.* 2008, 7, 1730.

Mie, G. *Ann. Phys.* 1908, 25, 377.

Mohanmed, M. B., Temer, S. A., Link, S., Braun, M., El-Sayed, M. A. *Chem. Phys. Lett.* 2001, 343, 55.

Murphy, C. J., Sau, T. K., Gole, A. M. *J. Phys. Chem. B* 2005, 109, 13857.

Nie, S., Xing, Y., Kim, G. J., Simons, J. W. *Annu. Rev. Biomed. Eng.* 2007, 9, 257.

Nikoobakht, B., El-Sayed, M. A. *Langmuir* 2001, 17, 6368.

Nikoobakht, B., El-Sayed, M. A. *Chem. Mater.* 2003, 15, 1957.

O'Neal, D. P., Hirsch, L. R., Halas, N. J., Payne, J. D., West, J. L. *Cancer Lett.* 2004, 209, 171.

Orendorff, C. J., Murphy, C. J. *J. Phys. Chem. B* 2006, 110, 3990.

Paciotti, G. F., Kingston, D. G., Tamarkin, L. *Drug Dev. Res.* 2006, 67, 47.

Paciotti, G. F., Myer, L., Weinreich, D. et al. *Drug Deliv.* 2004, 11, 169.

Papavassiliou, G. C. *Prog. Solid State Chem.* 1980, 12, 185.

Pissuwan, D., Valenzuela, S. M., Killingsworth, M. C., Xu, X., Cortie, M. B. *J. Nanoparticle Res.* 2007, 9, 1109.

Pitsillides, C. M., Joe, E. K., Wei, X., Anderson, R. R., Lin, C. P. *Biophys. J.* 2003, 84, 4023.

Prescott, S. W., Mulvaney, P. *J. Appl. Phys.* 2006, 99, 123504.

Purcell, F. M., Pennypacker, C. R. *Astrophys. J.* 1973, 186, 705.

Qiu, L., Larson, T. A., Smith, D. K. et al. *IEEE J. Sel. Top. Quant. Elect.* 2007, 13, 1730.

Reinhard, B., Sheikholeslami, S., Mastroianni, A., Alivisatos, A. P., Liphardt, J. *Proc. Natl. Acad. Sci. USA* 2007, 104, 2667.

Skrabalak, S. E., Chen, J., Au, L. *Adv. Mater.* 2007, 19, 3177.

Sprulnken, D. P., Omi, H., Furukawa, K. *J. Phys. Chem. B* 2007, 111, 14299.

Takahashi, H., Niidome, T., Nariai, A., Niidome, Y., Yamada, S. *Chem. Lett.* 2006a, 35, 500.

Takahashi, H., Niidome, T., Nariai, A., Niidome, Y., Yamada, S. *Nanotechnology* 2006b, 17, 4431.

Tong, L., Zhao, Y., Huff, T. B. et al. *Adv. Mater.* 2007, 19, 3136.

Wilson, B. C., Patterson, M. S. *Phys. Med. Biol.* 1986, 31, 327.

Yan, B., Yang, Y., Wang, Y. *J. Phys. Chem. B* 2003, 107, 9159.

Zharov, V. P., Galitovsky, V., Viegas, M. *Appl. Phys. Lett.* 2003, 83, 4897.

Zharov, V. P., Galitovskaya, E, Viegas, M. *Proc. SPIE* 2004, 5319, 291.

Zharov, V. P., Galitovskaya, E. N, Johnson, C., Kelly, T. *Lasers. Surg. Med.* 2005, 37, 219.

Zhu, J., Huang, L., Zhao, J. et al. *Mater. Sci. Eng. B* 2005, 121, 199.

37

Fullerenes in Photodynamic Therapy of Cancer

Pawel Mroz
Massachusetts General Hospital
Harvard Medical School

Ying-Ying Huang
Massachusetts General Hospital
Harvard Medical School

Tim Wharton
Lynntech Inc.

Michael R. Hamblin
Massachusetts General Hospital
Harvard Medical School
Massachusetts Institute
of Technology

37.1 Introduction

37.1.1 Photodynamic Therapy

Photodynamic therapy (PDT) is a promising new therapeutic procedure for the management of a variety of solid tumors and many nonmalignant diseases. PDT is a two-step procedure that involves the administration of a photosensitizing agent, followed by activation of the drug with the nonthermal light of a specific wavelength (Castano et al., 2004; 2005a,b). PDT generates singlet oxygen and other reactive oxygen species (ROS), which cause oxidative stress and membrane damage in the treated cells and consequently lead to cell death. The anticancer action of PDT is a consequence of a low-to-moderate selective degree of photosensitizer (PS) uptake by proliferative malignant cells, direct cytotoxicity, and a dramatic antivascular action that impairs the blood supply to the area of light exposure (Henderson and Dougherty, 1992). The PDT effectiveness is determined by the oxygen supply and might be decreased in conditions causing tissue hypoxia (Busch, 2006). PDT as a treatment procedure has been accepted by the U.S. Food and Drug Administration for use in endobronchial and endo-esophageal treatment (Dougherty, 2002) and also as a treatment for premalignant and early malignant diseases of the skin (actinic keratoses), bladder, breast, stomach, and oral cavity (Dolmans et al., 2003).

37.1.2 Photosensitizers

A PS is a photosensitive drug that is at the heart of the PDT effect. It is characterized by its ability to absorb the light of a specific wavelength and in turn to transfer the absorbed energy to the oxygen. There are several chemical structures that have been employed as PSs for PDT purposes that will be briefly described in this section.

The characteristics of the ideal PS have been discussed in many reviews (Allison et al., 2004; Detty et al., 2004). They should have low levels of dark toxicity to both humans and experimental animals and a low incidence of toxicity upon administration such as hypotension (decreased blood pressure) or allergic reaction. They should absorb light in the red or far-red wavelengths in order to penetrate the tissue. Absorption bands at shorter wavelengths have less tissue penetration and are more likely to lead to skin photosensitivity (the power in sunlight drops off at $\lambda > 600$ nm). Absorption bands at high wavelengths (>800 nm) mean that the photons will not have sufficient energy for the PS triplet state to transfer energy to the ground state oxygen molecule to excite it to the singlet state. They should have relatively high absorption bands (>20,000–30,000 M^{-1} cm^{-1}) to minimize the dose of PS needed to achieve the desired effect. The synthesis of the PS should be relatively easy and the starting materials should be readily available to make large-scale production feasible. The PS should be a pure compound with a constant composition and a stable shelf-life, and be ideally water soluble or soluble in a harmless aqueous solvent mixture. It should not aggregate unduly in biological environments as this reduces its photochemical efficiency. The pharmacokinetic elimination from the patient should be rapid, i.e., less than one day to avoid the necessity for post-treatment protection from light exposure and prolonged skin photosensitivity. A short interval between injection and illumination is desirable to facilitate outpatient treatment that is both patient-friendly and cost-effective. Pain on treatment is undesirable, as PDT does not usually require anesthesia or heavy sedation. Although high PDT activity is thought to be a good thing, it is possible to have an excessively powerful PS that is unforgiving. With limitations in both PS and light dosimetry, a highly active PS may easily permit treatment overdosage. Currently, it is uncertain whether it is better to have a PS "tailored" to a specific indication and to have families or portfolios of PSs for various diseases or patient types, or to seek one PS that works against most diseases. Lastly, a desirable feature might be to have an inbuilt method of PS dosimetry monitoring that follows a response to treatment by measuring *in vivo* fluorescence and its loss by photobleaching.

The majority of PSs, used both clinically and experimentally, are derived from the tetrapyrrole aromatic nucleus found in many naturally occurring pigments such as heme chlorophyll and bacteriochlorophyll (Figure 37.1). Tetrapyrroles usually have a relatively large absorption band in the region of 400 nm known as the Soret band, and a set of progressively smaller absorption bands as the spectrum moves into the red wavelengths known as the Q-bands. Naturally occurring porphyrins are fully conjugated (nonreduced) tetrapyrroles and vary in the number and type of side groups particularly carboxylic acid groups (uroporphyrin has eight, coproporphyrin has four, and protoporphyrin has two). The longest wavelength band of porphyrim is only at around 630-nm and tends to be very weak (small extinction coefficient). Chlorins are tetrapyrroles with the double bond in one pyrrole ring reduced. This means that the longest wavelength absorption band shifts to the region of 650–690 nm and increases severalfold in height; both these factors are highly desirable for PDT. Bacteriochlorins have two pyrrole rings with reduced double bonds, and this leads to the absorption band shifting even further into the red and increasing further in magnitude. Bacteriochlorins may turn out to be even more effective PSs than

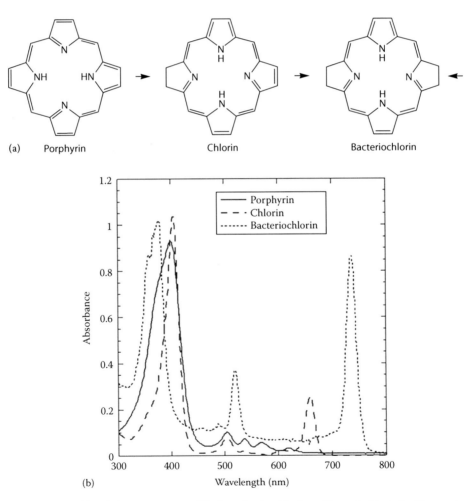

FIGURE 37.1 Chemical structures (a) and absorption spectra (b) of porphyrin, chlorin and bacteriochlorin backbones, Arrows indicate reduced pyrrole rings.

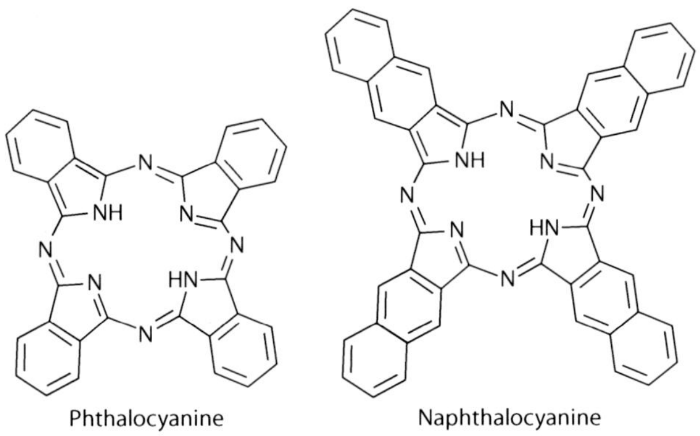

FIGURE 37.2 Chemical structures of phthalocyanine and naphthalocyanine backbones.

chlorins, but with relatively few candidate molecules and some questions about the stability of these molecules upon storage, this remains to be seen. There are a set of classical chemical derivatives generally obtained from naturally occurring porphyrins and chlorins that include such structures as purpurins, pheophorbides, pyropheophorbides, pheophytins, and phorbins some of which have been studied (a few extensively) as PSs for PDT.

A second widely studied structural group of PS is the phthalocyanines (PC), and to a lesser extent, their related cousins the naphthalocyanines (Figure 37.2). Again, their longest absorption band is >650 nm and usually have a respectable magnitude. As can be imagined, the presence of four phenyl groups (or even worse four naphthyl groups) causes solubility and aggregation problems. PCs are frequently prepared with sulfonic acid groups to provide water solubility and with centrally coordinated metal atoms. It was found that the asymmetrically substituted di-sulfonic acids acted as the best PS (compared with mono-, symmetrically di-, tri-, and tetra-substituted sulfonic acids) in both the zinc (Fingar et al., 1993b) and aluminum (Peng et al., 1990) series of PC derivatives.

Another broad class of potential PS includes completely synthetic, nonnaturally occurring, conjugated pyrrolic ring systems. These comprise such structures as texaphyrins (Detty et al., 2004), porphycenes (Szeimies et al., 1996), and sapphyrins (Figure 37.3; Kral et al., 2002). The last class of compounds that has been studied as PS are nontetrapyrrole-derived naturally

occurring or synthetic dyes. Examples of the first group are hypericin (from the plant known as St. Johns wort or *Hypericum perforatum*) (Agostinis et al., 2002) and from the second group are toluidine blue O (Stockert et al., 1996) and Rose Bengal (Figure 37.4; Bottiroli et al., 1997). These compounds have perhaps been more often studied as agents to mediate antimicrobial photoinactivation (Hamblin and Hasan, 2004) rather than as a PS designed to kill mammalian cells for diseases such as cancer.

37.2 Background

37.2.1 Fullerenes as PS

The rapidly growing interest in the medical application of nanotechnology has brought the attention to fullerenes as possible PDT mediators. Pristine C_{60} is highly insoluble in water and biological media and forms only nanoaggregates (Duncan et al., 2008) in aqueous solvents that are not photoactive (Hotze et al., 2008). In a 1991 editorial in *Science*, Culotta and Koshland (1991) described C_{60} (the molecule of the year) as "Buckyballs: A wide open playing field for chemists." Since then, new PSs based on the structures known as functionalized fullerenes that have hydrophilic or amphiphilic side chains or fused ring structures attached to the spherical C_{60} core have been shown to exhibit high efficiency in the production of singlet oxygen, hydroxyl radicals, and superoxide anion and have been proposed as effective PDT mediators in several applications (Figure 37.5). Fullerenes have numerous advantages and disadvantages as PSs for PDT when compared with other more traditional molecular PSs (Mroz et al., 2007b). The advantages are summarized in the following list:

- Fullerenes are particularly effective at the formation of superoxide and oxygen radicals (Type I photochemistry, see below) as well as at the formation of singlet oxygen (Type II photochemistry) compared with tetrapyrroles (Martin and Logsdon, 1987).
- Fullerenes are particularly photostable and manifest little photobleaching compared with tetrapyrroles. Due to the chemical structure of the porphyrin-type PS they are often good reactive substrates for the ROS they produce on illumination (particularly singlet oxygen) and the photoproducts formed upon oxidation of the PS have

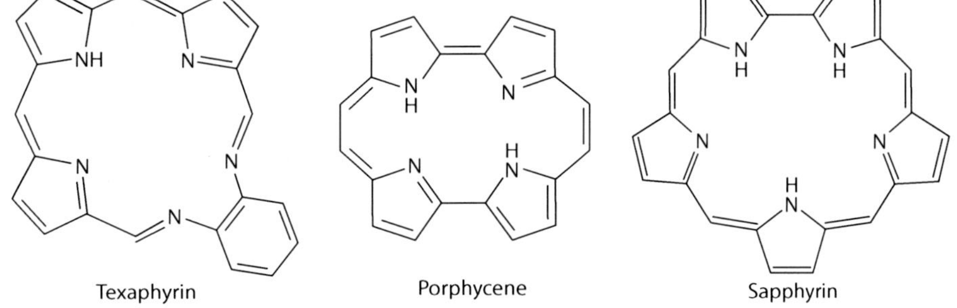

FIGURE 37.3 Chemical structures of some modified tetrapyrrole backbones that have been used in PDT: texaphyrin, porphycene and sapphyrins.

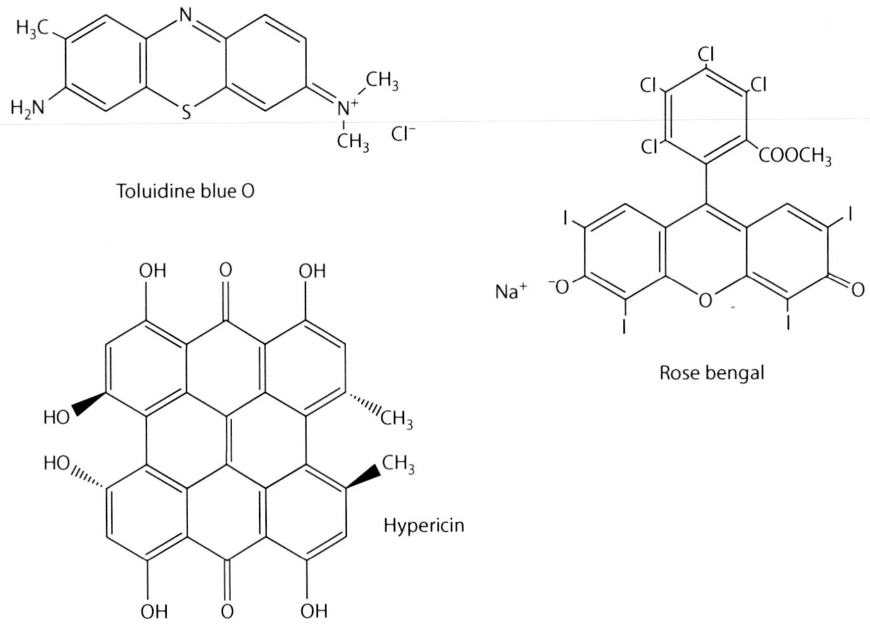

Toluidine blue O

Rose bengal

Hypericin

FIGURE 37.4 Chemical structures of some non-tetrapyrrole dyes that have been used for PDT: toluidine blue, rose Bengal, and hypericin.

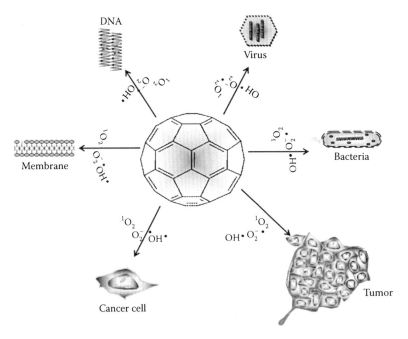

FIGURE 37.5 Examples of potential biomedical applications of fullerene mediated photodynamic therapy.

the chromophore structure disrupted and can no longer act as a PS. By contrast, the carbon cage of the fullerene backbone is less reactive to singlet oxygen in a short timescale and these PSs can therefore carry out many more catalytic cycles, thus dramatically extending the length of time they can respond to illumination.

- Fullerenes can easily be chemically modified for tuning the drug's partition coefficient (Log P for [drug in n-octanol]/[drug in un-ionized H_2O]) and pK_a values for the variation of *in vivo* lipophilicity and the prediction of their distribution in a biological system.

- Light harvesting antenna can be chemically attached on C_{60} to enhance the overall quantum yield and the ROS production yield.

- Molecular self-assembly of fullerene cages into vesicles allows multivalent drug delivery and can produce self-assembled nanoparticles that may have different tissue-targeting properties.

The main disadvantage of fullerenes as PSs lies in the characteristics of their visible absorption spectrum. While tetrapyrrole PSs other than porphyrins (such as chlorins, bacteriochlorins, and phthalocyanines) have been designed to have substantial absorption peaks in the red or far-red regions of the spectrum, the chief absorption of fullerenes is in the blue and green regions. This disadvantage may be dealt with in the following ways:

- One or two red-wavelength absorptive antenna can be chemically attached on a C_{60} to cause ultrafast intramolecular photoinduced energy-transfer processes. In this case, the antenna absorbs the red light photons in high efficiency, then passes the absorbed photon energy to a C_{60} cage within 250 ps that simulates the photoexcitation of fullerenes by red light.
- The attachment of multiple light harvesting antennae on one C_{60} cage can absorb many light photons simultaneously to supply fullerenes with sufficiently high accumulative photoenergy.
- The main application of these PSs may be confined to situations where the target is relatively superficial such as tumors that are superficial in nature, i.e., bladder or skin cancers.
- Two-photon excitation may be used. Femtosecond pulsed lasers at twice the wavelength of the photons to be absorbed (for instance, 800 nm) will have dramatically increased tissue penetration but will have a small spot size that will need to be scanned or rastered over the surface of the tissue to be treated.

37.2.2 Photophysics in PDT

Figure 37.6 graphically illustrates the processes of light absorption and energy transfer that are at the heart of PDT. The ground state PS has two electrons with opposite spins (this is known as a singlet state) in the low energy molecular orbital. Following the absorption of light (photons), one of these electrons is boosted into a high-energy orbital but keeps its spin (first excited singlet state). This is a short-lived (nanoseconds) species and can lose its energy by emitting light (fluorescence) or by internal conversion into heat. The fact that most PSs are fluorescent has led to the development of sensitive assays to quantify the amount of PS in cells or tissues, and allows *in vivo* fluorescence imaging in living animals or patients to measure the pharmacokinetics and distribution of the PS. The excited singlet state PS may also undergo the process known as intersystem crossing whereby the spin of the excited electron inverts to form the relatively long-lived (microseconds to milliseconds) excited triplet state that has electron spins parallel. The long lifetime of the PS triplet state is explained by the fact that the loss of energy by the emission of light (phosphorescence) is a "spin-forbidden" process as the PS would move directly from a triplet to a singlet state.

37.2.3 Photophysics of Fullerenes

The absorption spectra of a typical set of mono-substituted, bis-substituted, and tris-substituted fullerenes are shown in Figure 37.7. The absorption spectrum shows an almost monotonic decay between 300 and 700 nm, and it is noticeable how much difference each succeeding substitution into the fullerene core makes in reducing the visible absorption due to the successive removal of each double bond. When C_{60} is irradiated with visible light, it is excited from the S_0 ground state to a short-lived (~1.3 ns) S_1 excited state (E_S 46.1 kcal/mol). The S_1 state rapidly decays at a rate of 5.0×10^8 s^{-1} and a *triplet* quantum yield (ϕ_T) of 1.0 to a lower lying triplet state $T_1(E_T$ 37.5 kcal/mol) with a long lifetime of 50–100 μs (Equation 37.1). The $S_1 \rightarrow T_1$ decay is formally a spin-forbidden intersystem crossing (*isc*), but is driven by an efficient spin–orbit coupling.

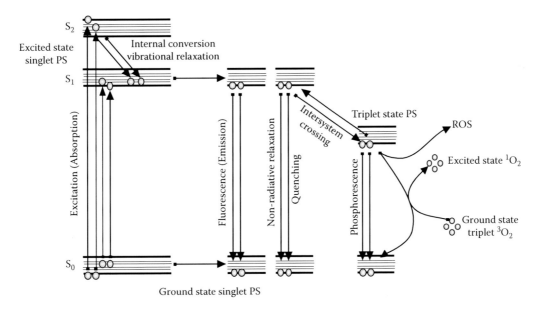

FIGURE 37.6 Jablonsky diagram illustrating absorption of light by PS ground state to form short-lived excited singlet state that can lose energy by fluorescence or internal conversion to heat, or can undergo intersystem crossing to long-lived PS triplet state that can carry out photochemistry.

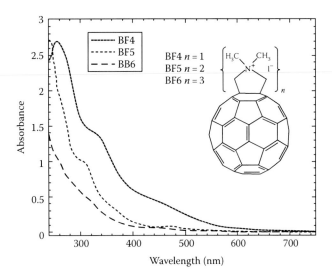

FIGURE 37.7 Absorption spectra of a set of mono-, bis-, and tris-substituted fullerenes.

In the presence of dissolved molecular oxygen ($^3O_2$), which exists as a triplet in its ground state, the fullerene T_1 state is quenched (as a consequence of the quenching, its lifetime is reduced to ~330 ns) to generate singlet oxygen ($^1O_2^*$) by energy transfer at a rate of 2×10^9 M^{-1} s^{-1} (Equation 37.2). The singlet oxygen quantum yield, ϕ_Δ, for this process (at 532 nm excitation) has been reported to be near the theoretical maximum, 1.0 (Arbogast et al., 1991).

$$C_{60} \xrightarrow[\text{vis. light}]{} {}^1C_{60}{}^* \xrightarrow[\text{isc}]{} {}^3C_{60}{}^* \qquad (37.1)$$

$$^{3*}C_{60} + {}^3O_2 \rightarrow C_{60} + {}^1O_2^* \qquad (37.2)$$

37.2.4 Photochemistry in PDT

The PS-excited triplet can undergo three broad kinds of reactions that are usually known as Type I, Type II, and Type III (Figure 37.8). First, in a Type I reaction, the triplet PS can gain an electron from a neighboring reducing agent. In cells, these reducing agents are commonly either nicotinamide adenine dinucleotide (NADH) or nicotinamide adenine dinucleotide (NADPH). The PS is now a radical anion bearing an additional unpaired electron. Alternatively, two triplet PS molecules can react together involving an electron transfer to produce a pair consisting of a radical cation and a radical anion. Radical anions may further react with oxygen with an electron transfer to produce a reactive oxygen species, in particular, superoxide anion. A third possibility is that the PS triplet may directly transfer an electron to ground state oxygen forming the PS radical cation and superoxide. The radical cation may then accept an electron from a reducing agent (e.g., NADH as mentioned above or a protein such as albumin) thus regenerating the PS for further reactions.

In a Type II reaction, the triplet PS can transfer its energy directly to molecular oxygen (itself a triplet in the ground state), to form excited-state singlet oxygen. Both Type I and Type II reactions can occur simultaneously, and the ratio between these processes depends on the type of PS used, the concentrations of the substrate, and oxygen. A less common pathway is known as Type III and here the triplet state PS reacts directly with a biomolecule thus destroying the PS and damaging the biomolecules. The Type II processes are thought to best conserve the PS molecular structure in a photoactive state, and in some circumstances a single PS molecule can generate 10,000 molecules of singlet oxygen. The PS can, in some circumstances, also

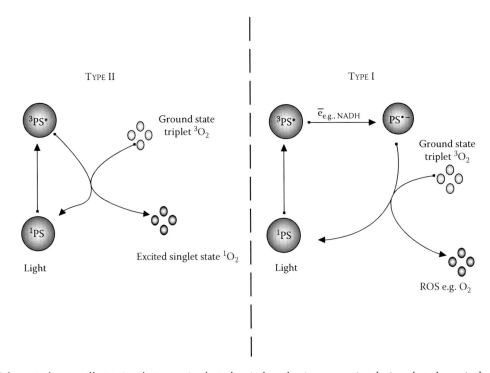

FIGURE 37.8 Schematic diagram illustrating the two main photochemical mechanisms operating during photodynamic therapy.

react with the singlet oxygen it produces in a process known as oxygen-dependent photobleaching.

Singlet oxygen, generated in Type II photochemical reactions, is generally considered as the most important mediator of PDT-induced damage (Weishaupt et al. 1976; Henderson and Miller 1986). Nonetheless, $O_2^{\bullet-}$, H_2O_2, $^\bullet OH$, NO, and other ROS are also detected in cells and tissues exposed to light and photosensitizers (Hariharan et al., 1980). A number of studies provide evidence for the active role of these metabolites in cytotoxic effects induced by PDT (Henderson and Miller 1986; Chekulayeva et al. 2006). Unfortunately, we still lack definitive mechanisms explaining how all these ROS are produced during PDT. ROS other than $^1O_2$ might be produced by Type I photochemical reactions immediately during illumination, especially when O_2 concentration is low (Fuchs and Thiele, 1998). It is possible however, that they can also be formed secondarily, as a result of $^1O_2$ action (Redmond and Kochevar, 2006). Therefore, even if selective $^1O_2$ scavengers abrogate PDT-induced lesions, it is still possible that the damage might have been caused by $^1O_2$-induced secondary ROS.

The presence of $O_2^{\bullet-}$, H_2O_2, and $^\bullet OH$ during PDT has been demonstrated with flow cytometry (using fluorescence probes) (Gilaberte et al. 1997; Kolarova et al. 2007), relatively selective quenchers (Henderson and Miller 1986), and more directly with electron paramagnetic resonance (Viola et al., 1998) and spin trapping methods (He et al. 1998). Fluorescent probes such as 2′,7′-dichlorofluorescein diacetate (H$_2$DCFDA) and its derivatives were used to show that PDT induces $O_2^{\bullet-}$ and H_2O_2 generation (Gilaberte et al. 1997; Lam et al. 2001; Kolarova et al. 2007). One caveat with the use of fluorescent probes is that they are not particularly specific. Although H$_2$DCFDA is insensitive to $^1O_2$, it can react with $^1O_2$-induced peroxyl radicals (Bilski et al., 2002). Moreover, H$_2$DCFDA can be oxidized by peroxynitrite (Myhre et al., 2003).

The damaging effects of superoxide are relatively mild, however, it can cause much more oxidative damage when it reacts with itself to produce hydrogen peroxide and oxygen in the process called "dismutation." The superoxide dismutation can occur spontaneously or can be catalyzed by the enzyme superoxide dismutase (SOD). The resulting product, hydrogen peroxide, similarly to HO$^\bullet$ passes easily through membranes and cannot be kept out of cells. The hydroxyl radical damage is "diffusion rate-limited." This highly reactive radical can add to an organic (carbon-containing) substrate (represented by R below); this could be, for example, a fatty acid that would form a hydroxylated adduct that is itself a radical. The hydroxyl radical can also oxidize the organic substrate by "stealing" or abstracting an electron from it. The resulting oxidized substrate is again itself a radical, and can react with other molecules in a chain reaction. For example, it could react with ground-state oxygen to produce a peroxyl radical (ROO$^\bullet$). The peroxyl radical again is highly reactive, and can react with another organic substrate in a chain reaction. This type of chain reaction is common in the oxidative damage of fatty acids and other lipids and demonstrates why radicals such as the hydroxyl radical can cause so much more damage than one might have expected.

Superoxide is also important in the production of the highly reactive hydroxyl radical (HO$^\bullet$). In this process, superoxide actually acts as a reducing agent, not as an oxidizing agent. This is because superoxide donates one electron to reduce the metal ions (such as ferric iron or Fe^{3+}) that act as the catalyst to convert hydrogen peroxide (H_2O_2) into the hydroxyl radical (HO$^\bullet$). This reaction is called the Fenton reaction and was discovered over 100 years ago. It is important in biological systems because most cells have some level of iron, copper, or other metals, which can catalyze this reaction. The reduced metal (ferrous iron or Fe^{2+}) then catalyzes the breaking of the oxygen–oxygen bond of hydrogen peroxide to produce a hydroxyl radical (HO$^\bullet$) and a hydroxide ion (HO$^-$). Superoxide can react with the hydroxyl radical (HO$^\bullet$) to form singlet oxygen or with nitric oxide (NO$^\bullet$) (also a radical) to produce peroxynitrite (OONO-), another highly reactive oxidizing molecule.

Another potent oxidant associated with PDT is NO; however, its oxidative properties are strongly associated with its potent metabolites such as peroxynitrite or nitrogen dioxide. PDT leads to production in some cells (Gupta et al., 1998), but also depletes it in endothelial cells, a phenomenon that may explain PDT-associated vessel constriction and blood flow stasis (Gilissen et al., 1993). Nitric oxide also has scavenging properties working as a NO chain-breaking antioxidant (Korytowski et al. 2000) and its ability to NO attenuate lipid is 1000-fold more potent than α-tocopherol (O'Donnell et al., 1997). These radical-quenching effects as well as the induction of heme oxygenase (HO-1) expression can in certain circumstances contribute to decreased cell death during PDT (Niziolek et al. 2003, 2006).

These ROS, together with singlet oxygen produced via Type II pathway, are oxidizing agents that can directly react with many biological molecules. Amino acid residues in proteins are important targets that include cysteine, methionine, tyrosine, histidine, and tryptophan (Grune et al., 2001; Midden et al., 1992). Due to their reactivity, these amino acids are the primary target of an oxidative attack on proteins. The reaction mechanisms are rather complex and as a rule lead to a number of final products. Cysteine and methionine are oxidized mainly to sulfoxides, histidine yields a thermally unstable endoperoxide, tryptophan reacts by a complicated mechanism to give *N*-formylkynurenine, and tyrosine can undergo phenolic oxidative coupling. Unsaturated lipids typically undergo ene-type reactions to give lipid hydroperoxides (LOOHs; derived from phospholipids and cholesterol) (Bachowski et al., 1991, 1994; Girotti, 1983, 1985).

DNA can be oxidatively damaged at both the nucleic bases (the individual molecules that make up the genetic code) and at the sugars that link the DNA strands by oxidation of the sugar linkages, or cross-linking of DNA to protein (a form of damage particularly difficult for the cell to repair). Although all cells have some capability of repairing oxidative damage to proteins and DNA, excess damage can cause mutations or cell death. Of the four bases in nucleic acids, guanine is the most susceptible to oxidation by $^1O_2$. The reaction mechanism has been extensively studied in connection with the oxidative cleavage of DNA (Buchko et al., 1995). The first step is a [4 + 2] cycloaddition to the

C-4 and C-8 carbons of the purine ring leading to an unstable endoperoxide (Buchko et al., 1993). The subsequent complicated sequence of reactions and the final products depend on whether the guanine moiety is bound in an oligonucleotide or a double-stranded DNA (Ravanat and Cadet, 1995).

Because of the high reactivity and short half-life of singlet oxygen and hydroxyl radicals, only molecules and structures that are proximal to the area of its production (areas of PS localization) are directly affected by PDT. The half-life of singlet oxygen in biological systems is less than $1\,\mu s$, and, therefore, the radius of the action of singlet oxygen is of the order of $1\,\mu m$ (Moan and Berg, 1991).

37.2.5 Photochemistry of Fullerenes

It is becoming more evident that the ROS produced during PDT with fullerenes are biased towards Type 1 photochemical products (superoxide, hydroxyl radical, lipid hydroperoxides, hydrogen peroxide) when compared with the standard Type 2 ROS, singlet oxygen. It has been known since shortly after the discovery of fullerenes that these compounds will catalyze the formation of ROS after illumination of both pristine and functionalized C_{60} (Foote, 1994). The illumination of fullerenes dissolved in organic solvents in the presence of oxygen leads to the efficient generation of highly reactive singlet oxygen via an energy transfer from the excited triplet state of the fullerene (Arbogast et al., 1991). However, in polar solvents, especially those containing reducing agents (such as NADH at concentrations found in cells), illumination of various fullerenes will generate different reactive oxygen derivatives, such as superoxide anion (Yamakoshi et al., 2003).

Fullerenes with their triply degenerate, low lying lowest unoccupied molecular orbital (LUMO) are excellent electron acceptors capable of accepting as many as six electrons (Koeppe and Sariciftci, 2006). There is some evidence that fullerene excited states (in particular the triplet) are even better electron acceptors than the ground state (Arbogast et al., 1992; Guldi and Prato, 2000). It is thought that the reduced fullerene triplet or radical anion can transfer an electron to molecular oxygen forming the superoxide anion radical $O_2^{-\bullet}$.

Yamakoshi et al. (2003) carried out a photochemical study comparing energy transfer processes (singlet oxygen $^1O_2$) and electron transfer (reduced active oxygen radicals such as superoxide anion radical $O_2^{-\bullet}$), using various scavengers of ROS, physicochemical (electron paramagnetic resonance [EPR] radical trapping and near-infrared [NIR] spectrometry), and chemical methods (nitro blue tetrazolium reaction with superoxide). Whereas $^1O_2$ was generated effectively by photoexcited C_{60} in nonpolar solvents such as benzene and benzonitrile, they found that $O_2^{-\bullet}$ and $OH^\bullet$ were produced instead of $^1O_2$ in polar solvents such as water, especially in the presence of a physiological concentration of reductants including NADH.

Miyata et al. solubilized fullerenes into water with poly-vinylpyrrolidone (PVP) as a detergent (Miyata et al., 2000). Visible-light irradiation of PVP-solubilized C_{60} in water in the presence of NADH as a reductant and molecular oxygen resulted in the formation of $O_2^{-\bullet}$, which was detected by the EPR spin-trapping method. Formation of $O_2^{-\bullet}$ was also evidenced by the direct observation of a characteristic signal of $O_2^{-\bullet}$ by the use of a low-temperature EPR technique at 77 K. On the other hand, no formation of $^1O_2$ was observed by the use of 2,2,6,6-tetramethyl-4-piperidone (TEMP) as a $^1O_2$ trapping agent. No NIR luminescence of $^1O_2$ was observed in the aqueous $C_{60}/PVP/O_2$ system. These results suggest that the photoinduced bioactivities of the PVP-solubilized fullerene are caused not by $^1O_2$, but by reduced oxygen species such as $O_2^{-\bullet}$, which are generated by the electron-transfer reaction of $C_{60}^{-\bullet}$ with molecular oxygen.

We have shown (Mroz et al., 2007a; Yamakoshi et al., 2003) that under biological conditions where mild reducing agents are ubiquitous (and the environment is polar and aqueous), illuminated fullerenes produce superoxide (and possible hydroxyl radical) having switched from a Type II to a Type I mechanism. We used ESR spin traps for superoxide and 1270 nm luminescence measurements for singlet oxygen.

37.3 Presentation of State of the Art

37.3.1 Biocompatibility of Fullerenes

The use of C_{60}-derived materials for biological applications is currently an active area of research and several reviews have been written on the subject (Bosi et al., 2003; Nakamura and Isobe, 2003; Tagmatarchis and Shinohara, 2001). Additionally, some studies on the toxicity of C_{60} and C_{60}-derived compounds have appeared (Chen et al., 1998; Fiorito et al., 2006; Satoh and Takayanagi, 2006). The studies conclude that C_{60} itself is quite nontoxic (Gharbi et al., 2005). For example, PVP solubilized C_{60} showed no noticeable toxic effect in mice after 4 weeks of intraperitoneal administration of 30 mg C_{60} kg^{-1} (Ungurenasu and Airinei, 2000). No genotoxicity was observed at a C_{60} concentration of $0.45\,\mu g/cm^3$ in prokaryotic cells (*E. coli*) and in eukaryotic cells (*Drosophila* somatic wing cells), but at the highest fullerene concentration of $2.24\,\mu g/cm^3$, a slight genotoxic effect was observed in the eukaryotic cells (Zakharenko et al., 1997). No effects were observed in hemolysis tests on sheep red blood cells with a suspension of C_{60} with PVP, even though PVP is known to exhibit cytotoxicity itself (Bosi et al., 2004). Polyalkylsulfonated C_{60} was determined to have an LD_{50} (lethal dose that kills 50%) of 600 mg/kg in female Sprague-Dawley rats (Chen et al., 1998). Another study showed that the fullerenes had no toxic effects at the experimental dose of 2000 mg/kg (also in Sprague-Dawley rats, male) (Mori et al., 2006).

Many organic substances that have aromatic ring systems, such as benzene, are carcinogens because a conjugated carbon ring has the appropriate size and shape to be intercalated into DNA, thus promoting cancer. However, fullerenes appear to be too big to be incorporated into DNA.

Although pristine C_{60} is not soluble in water, it can be suspended as nanoaggregate clusters often termed nC_{60} by dissolution in tetrahydrofuran (THF) followed by the addition to water and the removal of the organic solvent (Fortner et al., 2005;

Zhu et al., 2007). These clusters are spherical clumps of C_{60} between 250 and 350 nm in diameter. Thus, nC_{60} represents a different chemical entity than solutions of C_{60} in which the fullerenes exist as individual molecules. Recently published results suggested that nC_{60} prepared in this manner is moderately toxic to water fleas and juvenile largemouth bass at concentrations in water of around 800 ppb (Oberdorster, 2004). However, upon suggestion that the toxicity could be due to residual THF (Zhu et al., 2006), it was subsequently discovered that the toxicity *was* due largely to THF and not nC_{60}. Additionally, the overwhelming evidence of the essential nontoxicity of C_{60} (**not nC_{60}**) (Isakovic et al., 2006) in previously peer-reviewed articles of C_{60} and many of its derivatives (especially water soluble derivatives) indicates that fullerenes are likely to have little (if any) dark toxicity, especially at the very low concentrations used in PDT.

37.3.2 Tissue Optics in PDT

In PDT, it is important to be able to predict the spatial distribution of light in the target tissue. Light is either scattered or absorbed when it enters tissue and the extent of both processes depends on the tissue type and light wavelength. Tissue optics involve measuring the spatial/temporal distribution and the size distribution of tissue structures and their absorption and scattering properties. This is rather involved because the biological tissue is inhomogeneous and the presence of microscopic inhomogeneities (macromolecules, cell organelles, organized cell structure, interstitial layers, etc.) makes tissue turbid. Multiple scattering within a turbid medium leads to the spreading of a light beam and loss of directionality. Absorption is largely due to endogenous tissue chromophores such as hemoglobin, myoglobin, and cytochromes (Figure 37.9). The complete characterization of light transport in tissue is a formidable task; therefore, heuristic approaches with different levels of approximations have been developed to model it. An effort for modeling light transport also requires accurate values for the optical properties

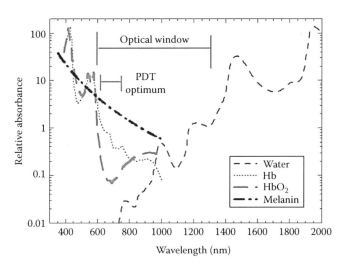

FIGURE 37.9 Tissue optics and the optical window and PDT optimal therapeutic window (600–800 nm) for deep light tissue penetration.

of the tissue. Scattering is generally the most important factor in limiting light penetration into most tissues and is measured by μ_s (reduced scattering coefficient), which for soft tissues is in the range of 100–1000 cm^{-1}. Absorption is usually of less importance and is measured by μ_a (reduced absorption coefficient) with values in the range of 0.1–5 cm^{-1} for most tissue at green and longer wavelengths. The third parameter necessary to define the optical properties of tissue is the anisotropy factor that measures the direction of the scattering of light. It is possible to use mathematical approaches such as the diffusion theory or Monte Carlo modeling to predict how light will travel into the target tissue and the illumination parameters (fluence, fluence rate, wavelength, angle of incidence) may then be adjusted to maximize the light dose. The combination of absorption of the lower wavelength light by the important tissue chromophores (oxy and deoxyhemoglobin and melanin) together with reduced light scattering at longer wavelengths and the occurrence of water absorption at wavelengths greater than 1300 nm has led to the concept of the "optical window" in tissue (see Figure 37.9). In terms of PDT, the average effective penetration depth (intensity reduced to 37%) is about 1–3 mm at 630 nm, the wavelength used for clinical treatment with Photofrin (PF), while penetration is approximately twice that at 700–850 nm (Svaasand, 1984; Wilson et al., 1985). The increased penetration depth of longer wavelength light is a major incentive for the development of PS absorbing at such wavelengths, and a PS of the naphthalocyanine (776 nm) (Mantareva et al., 1997) and bacteriochlorin (780 nm) (Chen et al., 2007) classes fall into this category. The absorption of light by the PS itself can limit tissue light penetration. This phenomenon has been termed "self-shielding" and is particularly pronounced with a PS that absorbs very strongly at the treatment wavelength.

37.3.3 Strategies to Overcome Unfavorable Tissue Optics for Fullerene-Based PDT

37.3.3.1 Two-Photon PDT

Two-photon absorption (TPA) was originally proposed by Maria Göppert-Mayer in 1931 in her doctoral dissertation, but the first experimental verification was provided by Werner Kaiser in 1961 and was facilitated by the recent development of the laser, which was required for excitation because of the intrinsically low intensity of TPA. TPA is a third-order nonlinear optical process and therefore is most efficient at very high intensities. In a centrosymmetric molecule, one- and two-photon allowed transitions are mutually exclusive. In quantum mechanical terms, this difference results from the need to conserve spin. Since photons have a spin of ±1, one-photon absorption requires excitation to involve an electron changing its molecular orbital to one with a spin different by ±1. TPA requires a change of +2, 0, or −2. In nonresonant TPA, two photons combine to bridge an energy gap larger than the energies of each photon individually. If there were an intermediate state in the gap, this could happen via two separate one-photon transitions in a process described as "resonant TPA," "sequential TPA," or "1 + 1 absorption."

In nonresonant TPA, the transition occurs without the presence of the intermediate state. This can be viewed as being due to a "virtual" state created by the interaction of the photons with the molecule.

A long-wavelength near-IR light $\lambda > 780$ nm has relatively low photon energy (>1.5 eV) and is generally too low to activate most PSs by one-photon excitation. Therefore, one-photon absorption (1PA) fails in the phototherapeutic window 780–950 nm, where tissues have maximum transparency to light. Sensitization by simultaneous two-photon absorption (TPA) (Bhawalkar et al., 1997) combines the energy of two identical photons arriving at the PS at the same time, and can provide the energy of a single photon of half the wavelength, which is sufficient to excite the PS to the first excited singlet state. It was not until the 1990s that rational design principles for the construction of two-photon absorbing molecules began to be developed, in response to a need from imaging and data storage technologies, and was aided by the rapid increases in computer power that allowed quantum calculations to be made. The accurate quantum mechanical analysis of two-photon absorbance is orders of magnitude more computationally intensive than that of one-photon absorbance.

The most important features of molecules with strong TPA were found to be a long conjugation system (analogous to a large antenna) and substitution by strong donor and acceptor groups (which can be thought of as inducing nonlinearity in the system and increasing the potential for charge-transfer). Therefore, many push-pull olefins exhibit high TPA transitions, up to several thousand GM (1 GM is defined as 10^{-50} cm^4. s/photon/molecule).

In vitro studies with two-photon activation of Photofrin® showed the killing of vascular endothelial cells, but still required high pulse energy and long illumination times (Karotki et al., 2006). Studies have also reported that two-photon PDT can be used to selectively close small blood vessels in the chicken chorioallantoic membrane (Samkoe et al., 2007; Samkoe and Cramb, 2003). This is accepted as a model of choroidal neovascularization as seen in age-related macular degeneration patients. A recent study (Starkey et al., 2008) demonstrated that two-photon PDT using a specially designed porphyrin construct with a high two-photon cross-section and a femtosecond laser could destroy a tumor when the laser beam had passed through the entire body thickness of the mouse to reach the tumor.

There has been some preliminary work (Chiang et al., 2002; Verma et al., 2005) on design and synthesis of fullerene derivatives that may have an appropriate TPA to allow them to be used in two-photon PDT with NIR light in the 700–800 nm region.

37.3.3.2 Optical Clearing Agents

Obtaining a maximum light penetration depth of >2 mm in opaque tissues such as skin is difficult even with NIR incident light. At visible and NIR wavelengths, optical scattering dominates over absorption and is much more significant in reducing light penetration into biological tissues (Gebhart et al., 2006). Since it is likely that one-photon excitation of fullerenes would use blue or green light, a method to reduce scattering by tissue could be very useful. Recently, a technique known as optical clearing has shown great potential in enhancing the light penetration into tissue (Tuchin et al., 2008). Optical clearing is a method for inducing a transient reduction in optical scattering by biological tissue. Studies have demonstrated the increased light penetration depth using hyperosmotically active chemical agents such as glycerol, propylene glycol, ethylene glycol, dimethyl sulfoxide (DMSO), glucose or dextrose, oleic acid (OA), linoleic acid, etc., applied to the skin or tissue (Khan et al., 2004). Various mechanisms for optical clearing have been proposed but the exact mechanism is still not entirely understood (Hirshburg et al., 2007). It has been inferred that hyperosmotic agents reduce random scattering primarily by better refractive index matching, dehydration, and collagen dissociation.

For the optical clearing of skin tissue by the use of hyperosmotic agents, efficient delivery of the agents to dermis is required. Because skin is such an excellent barrier to topically administered chemical agents, adding enhancers for the effective and reversible enhancement of transdermal penetration (diffusion) will be very helpful. Previous investigations showed that DMSO can be used together with glycerol to synergistically enhance the chemical agent diffusion into the stomach mucosa and skin tissue, due to the carrier effect of DMSO (Jiang et al., 2008). Due to its potential toxicity and the possible side effects, DMSO has been controversial in clinical applications. Therefore, it is necessary to choose other safe hyperosmotic chemical solutions in order to improve the availability of the optical clearing technique for *in vivo* clinical applications (He and Wang, 2004). OA is a good choice because it has been widely used in the field of drug delivery and transdermal penetration (Kogan and Garti, 2006). OA is a mono-unsaturated fatty acid and is generally recognized as being safe. Jiang et al. have shown that when compared with DMSO as an enhancer, OA has a similar clearing effect (Jiang et al., 2008).

37.3.3.3 Covalent Attachment of Light Harvesting Antennae to Fullerenes

Porphyrins and fullerenes are spontaneously attracted to each other. This new supramolecular recognition element can be used to construct discrete host-guest complexes, as well as ordered arrays of interleaved porphyrins and fullerenes. The fullerene porphyrin interaction underlies successful chromatographic separations of fullerenes, and there are promising applications in the areas of porous framework solids and photovoltaic devices (Boyd and Reed, 2005). The structure of a typical fullerene-porphyrin dyad is shown in Figure 37.10. There have been several publications on the synthesis and characterization of the fullerene-porphyrin dyads that may be used as artificial photosynthesis mimics (El-Khouly et al., 2005; Imahori, 2004; Schuster et al., 2004; Vail et al., 2006).

Rancan et al. (2005) used this approach to overcome the necessity to use UV or short-wavelength visible light to photoactivate fullerenes. They synthesized two new fullerene-bis-pyropheophorbide a derivatives: a mono-(FP1) and a hexa-adduct

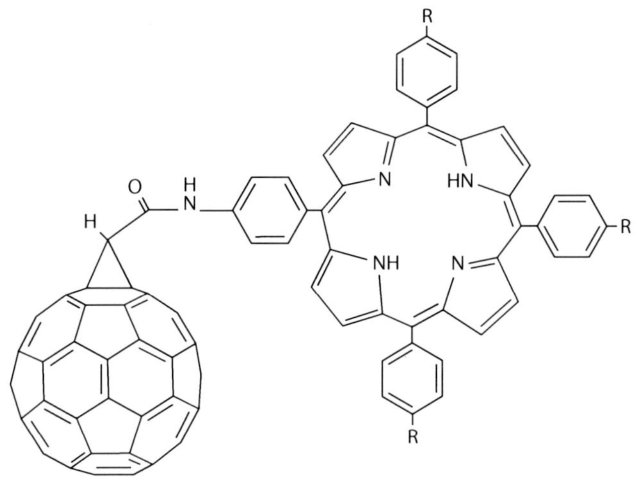

FIGURE 37.10 Chemical structure of a fullerene–porphyrin dyad.

(FHP1). The photophysical characterization of the compounds revealed significantly different parameters related to the number of addends at the fullerene core. In this study, the derivatives were tested with regard to their intracellular uptake and photosensitizing activity towards Jurkat cells in comparison with the free sensitizer, pyropheophorbide a. The C_{60}-hexa-adduct FHP1 had a significant phototoxic activity (58% cell death, after a dose of 400 mJ/cm² of 688 nm light) but the mono-adduct FP1 had a very low phototoxicity and only at higher light doses. Nevertheless, the activity of both adducts was less than that of pure pyropheophorbide a, probably due to the lower cellular uptake of the adducts.

A group from Argentina has also studied the phototoxicity produced by tetrapyrrole–fullerene conjugates. Milanesio et al. (2005) compared PDT with a porphyrin-C_{60} dyad (P-C_{60}) and its metal complex with Zn(II) (ZnP-C_{60}) was compared with 5-(4-acetamidophenyl)-10,15,20-tris(4-methoxyphenyl)porphyrin (P), both in a homogeneous medium containing photo-oxidizable substrates and *in vitro* on the Hep-2 human larynx carcinoma cell line. $^1O_2$ yields (Φ_Δ) were determined using 9,10-dimethylanthracene (DMA). The values of Φ_Δ were strongly dependent on the solvent polarity. Comparable Φ_Δ values were found for dyads and P in toluene, while $^1O_2$ production was significantly diminished for the dyads in dimethylformamide (DMF). In a more polar solvent, the stabilization of charge-transfer state takes place, decreasing the efficiency of the porphyrin triplet-state formation. Also, both dyads photosensitize the decomposition of L-tryptophan in DMF. In a biological medium, no dark cytotoxicity was observed using sensitizer concentrations of ≤1 μM and 24 h of incubation. The uptake of sensitizers into Hep-2 was studied using 1 μM of sensitizer and different times of incubation. Under these conditions, a value of approximately 1.5 nmol/10⁶ cells was found between 4 and 24 h of incubation. The cell survival after irradiation of the cells with visible light was dependent upon the light-exposure level. A higher phototoxic effect was observed for P-C_{60}, which inactivates 80% of cells after 15 min of irradiation. Moreover, both dyads keep a high photoactivity even under argon atmosphere.

In a subsequent paper (Alvarez et al., 2006), they showed the cells died by apoptosis by analysis using Hoechst-33258, toluidine blue staining, terminal deoxynucleotidyl transferase dUTP nick end labeling (TUNEL), and DNA fragmentation. Changes in cell morphology were analyzed using fluorescence microscopy with Hoechst-33258 (*bis*-benzimide) under low oxygen concentration. Under this anaerobic condition, necrotic cellular death predominated over the apoptotic pathway. It was found that P-C_{60} induced apoptosis by a caspase-3-dependent pathway.

37.3.4 Photodynamic Therapy Destroys Tumor Cells *In Vitro*

The direct killing of tumor cells by PDT can involve induction of apoptosis, autophagy, or necrosis.

Apoptosis is a strictly controlled, energy-dependent process of cell death that is catalyzed by the number of hydrolytic enzymes including proteases and nucleases. The activation of these enzymes results in DNA fragmentation and intracellular proteolysis of cellular proteins and organelles. The end result of apoptosis is the apoptotic body that can be readily phagocytosed without triggering any local inflammation.

The induction of apoptosis after PDT has been demonstrated in numerous studies that involved many different cell lines and several structures of PSs (Oleinick et al., 2002). It has been demonstrated that PDT is an efficient apoptosis inducer both in *in vitro* as well as in *in vivo* settings (Agarwal et al., 1991). There are two major pathways that lead to PDT-induced apoptosis that involve either (A) photochemical damage to the mitochondria or (B) signaling from cellular death receptors on the plasma membrane (Almeida et al., 2004).

There are several factors that can influence the mechanism of cell death after PDT that include the total amount of light delivered, the intracellular localization of the photosensitizer (Dellinger, 1996), the incubation time with the PS, and the initial site of binding of the PS (Kessel and Poretz, 2000). It is also important what type of ROS are produced during the PDT process, as they also can differently affect the apoptosis process (Kochevar et al., 2000; Lin et al., 2000).

Autophagy is a process that was first described as a way for cells to recycle the intracellular components during times of starvation. It has also been described as a process that helps cells to dispose of the damaged organelles and replace them with new ones in order to survive (Buytaert et al., 2007). However, if the damage that has occurred is beyond the point of repair, autophagy can be another mechanism of cellular death. Autophagy after PDT has been initially described in the mouse leukemia cell line L1210 as well as in human prostate cancer cells DU145 (Kessel et al., 2006). Subsequently, other cell lines have been described to undergo this process after PDT (Xue et al., 2007), including some cell lines that are defective in apoptosis (Buytaert et al., 2006). It has been reported that the induction of autophagy has been observed whether or not apoptosis was blocked or impaired and as a general early response following PDT (Kessel et al., 2006; Xue et al., 2007).

Necrosis is the least controlled pathway of cell death. It can occur in the course of any significant cell damage that leads to the heavy disruption of the continuity of the cellular membrane or intracellular proteins and organelles. Therefore, factors that promote necrosis during PDT include extra-mitochondrial localization of the PS (e.g., lysosome spill), a high dose of PDT, and glucose starvation (Kiesslich et al., 2005). Also, it has been suggested that the cell genotype may influence the form of cell death following PDT (Wyld et al., 2001). As a result of necrosis, the intracellular contents spill out into the surrounding microenvironment leading to the induction of strong local inflammation.

37.3.5 *In Vitro* PDT with Fullerenes

The first demonstration of phototoxicity in cancer cells mediated by fullerenes was in 1993 when Tokuyama et al. (1993) used carboxylic acid functionalized fullerenes at 6 μM and white light to produce growth inhibition in human HeLa cancer cells. However, these same authors later reported that other carboxylic acid derivatives of C_{60} and C_{70} were completely without any photoactivity as PDT agents at 50 μM (Irie et al., 1996).

Burlaka et al. (2004) used pristine C_{60} at 10 μM with visible light from a mercury lamp to produce some phototoxicity in Ehrlich carcinoma cells or rat thymocytes and used EPR spin-trapping techniques to demonstrate the formation of ROS.

The cytotoxic and photocytotoxic effects of two water-soluble fullerene derivatives, a dendritic C_{60} mono-adduct and the malonic acid C_{60} tris-adduct were tested on Jurkat cells when irradiated with UVA or UVB light (Rancan et al., 2002). The cell death was mainly caused by membrane damage and it was UV dose-dependent.

Tris-malonic acid fullerene was found to be more phototoxic than the dendritic derivative. This result is in contrast to the singlet oxygen quantum yields determined for the two compounds.

Three C_{60} derivatives with two to four malonic acid groups (DMA C_{60}, TMA C_{60}, and QMA C_{60}) were prepared and the phototoxicity of these compounds against HeLa cells was determined by MTT assay ((3-(4,5-Dimethylthiazol-2-yl)-2,5-diphenyltetrazolium bromide-based colorimetric assay in which this tetrazole is reduced by the mitochondria of living cells to purple formazan) and cell cycle analysis (Yang et al., 2002). The relative phototoxicity of these compounds was DMA C_{60} > TMA C_{60} > QMA C_{60}. The hydroxyl radical quencher mannitol (10 mM) was not able to prevent cells from the damage induced by irradiated DMA C_{60}. DMA C_{60}, together with irradiation, was found to decrease the number of cells in the G1 phase of the cell cycle (G(1)) from 63% to 42% and increase cells in the G2 and mitotic phase of the cell cycle (G(2)+M) from 6% to 26%.

Ikeda et al. (2007) used a series of liposomal preparations containing cationic or anionic lipids together with dimyristoylphosphatidylcholine and introduced C_{60} into the lipid bilayer by exchange from cyclodextrin. By adding a phospholipid with an additional fluorochrome, they were able to use fluorescence microscopy to demonstrate uptake of the liposomes by HeLa cells after 24 h incubation. Illumination with 136 J/cm² and 350–500 nm of light gave 85% cell killing in the case of cationic liposomes, and apoptosis was demonstrated.

In our laboratory, we have tested the hypothesis that fullerenes would be able to kill cancer cells by photodynamic therapy *in vitro*. A panel of six functionalized fullerenes were prepared in two groups of three compounds (Figure 37.11). The first series

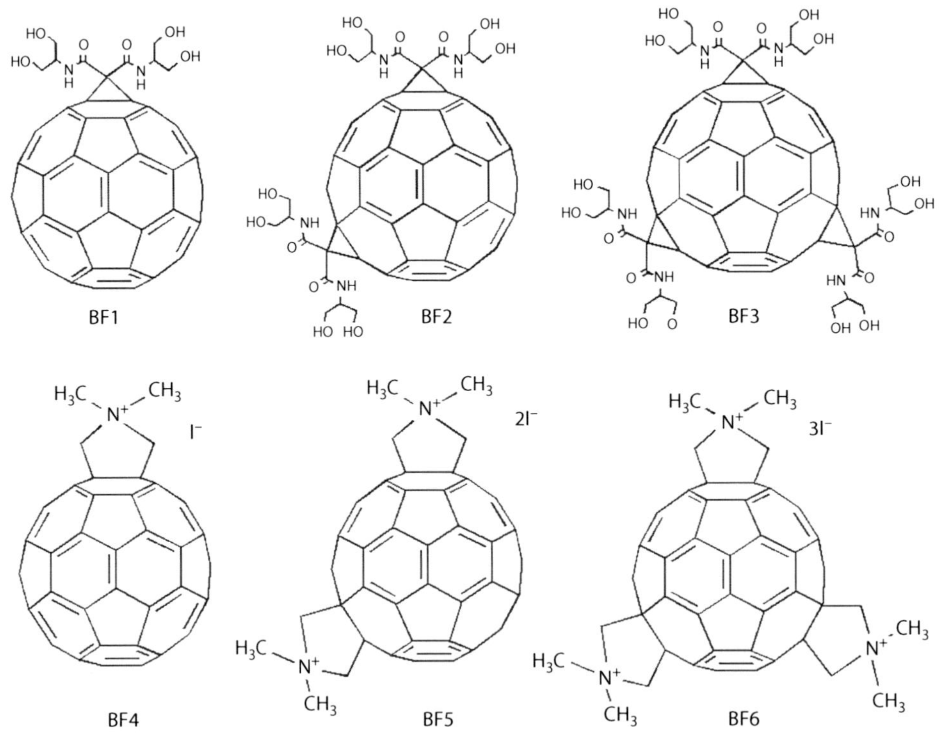

FIGURE 37.11 Chemical structures of six tested functionalized fullerenes (see Mroz et al., 2007a; Tegos et al., 2005).

was constructed with one, two, or three polar di-serinol groups (**BF1–3**) and a second series had one, two, or three quaternary dimethylpyrrolidinium groups (**BF4–6**). We demonstrated that the C_{60} molecule mono-substituted with a single pyrrolidinium group (**BF4**) was a remarkably efficient PS and could mediate the killing of a panel of mouse cancer cells at the low concentration of $2\,\mu M$ with modest exposure to white light. For the first time, we demonstrated that photoactive fullerenes are taken up into cells by measuring the increase in fluorescence of an intracellular probe for the formation of ROS. The complete lack of intrinsic fluorescence of these fullerene derivatives makes traditional confocal fluorescence microscopy experiments and fluorescence extraction from cell pellets impossible to perform. The fact that the fullerenes mediated PDT killing of cells after the medium has been changed and C_{60} has been removed from around the cells certainly implies that the compounds were taken up by cells, but the use of H_2DCFDA provided more direct evidence. The existing literature on the specificity of H_2DCFDA for ROS suggests that the main species that causes fluorescence increase is H_2O_2 and singlet oxygen does not directly cause increased fluorescence, although organic peroxides and peroxyl radicals that could be indirectly formed from singlet oxygen could play a role (Bilski et al., 2002). It was assumed that the superoxide produced from the illuminated fullerene underwent dismutation either catalyzed by superoxide dismutase or spontaneously to produce H_2O_2.

We have also demonstrated the induction of apoptosis by fullerene-PDT in cancer cells at 4–6h after illumination. This is not a surprising finding as there have been many reports of apoptosis occurring after *in vitro* PDT with conventional PS such as Photofrin (He et al., 1994), benzoporphyrin derivative (Granville et al., 1998), and the phthalocyanine, Pc4 (Gupta et al., 1998). The relatively rapid induction of apoptosis after illumination might suggest the fullerenes are localized in subcellular organelles such as mitochondria, as has been previously shown for the aforementioned PS. PS that localize in lysosomes tend to produce apoptosis more slowly than mitochondrial PS, due to the release of lysosomal enzymes that subsequently activate cytoplasmic caspases (Kessel et al., 2000).

The mono-pyrrolidinium substituted fullerene (**BF4**) was the most effective PS out of the six candidates by a considerable margin. The explanation for this observation is probably linked to its relative hydrophobicity as demonstrated by its log *P* value of over 2. It has been established in some structure-function relationship studies that the more hydrophobic a PS is (up to a certain limit when insolubility and aggregation become problems) the more effective it is in producing cell killing (Ben-Dror et al., 2006; Cauchon et al., 2005; Potter et al., 1999). The reason for this relationship is thought to be a combination of a higher cellular uptake and more localization in intracellular membrane organelles such as mitochondria. The single cationic charge possessed by this compound is also likely to play an important role in determining its relative phototoxicity. Many lipophilic monocations have been shown to localize fairly specifically in mitochondria (Ross et al., 2006; Rottenberg, 1984)

and this property has been proposed as a strategy to target drugs to mitochondria (Murphy and Smith, 2007).

37.3.6 PDT with Tumors *In Vivo*

The direct killing effect of PDT on malignant cancer cells that has been studied in detail *in vitro*, as described above, also clearly applies *in vivo*, but in addition, two separate *in vivo* mechanisms leading to PDT-mediated tumor destruction have been described (Figure 37.12).

The first additional mechanism shown to operate during PDT of tumors is the vascular effect in which vascular damage, occurring after the completion of the PDT treatment, contributes to long-term tumor control (Abels, 2004). Microvascular collapse can be readily observed following PDT (Henderson and Fingar, 1987; Star et al., 1986) and can lead to severe and persistent post-PDT tumor hypoxia (Henderson and Fingar, 1987). The mechanisms underlying the vascular effects of PDT differ greatly with different PSs. Photofrin-PDT leads to vessel constriction, macromolecular vessel leakage, leukocyte adhesion, and thrombus formation, all apparently linked to platelet activation and the release of thromboxane (Fingar et al., 1993a). PDT with certain phthalocyanine derivatives primarily causes vascular leakage (Fingar et al., 1993b), and PDT with mono-aspartyl chlorine e6 (MACE) results in blood flow stasis primarily because of platelet aggregation (McMahon et al., 1994). All of these effects may include components related to the damage of the vascular endothelium. PDT may also lead to vessel constriction via the inhibition of the production or release of nitric oxide by the endothelium (Gilissen et al., 1993).

The second additional mechanism that applies when tumors *in vivo* are treated by PDT is the activation of the host immune system (Castano et al., 2006). One of the key experiments used to demonstrate this effect was carried out by Korbelik et al. (1996) who grew the same EMT6 mammary sarcoma tumors in immunocompetent BALB/c mice and immunodeficient nude or scid mice of the same background. PF-PDT led to the initial ablation of the tumors in all mice but long-term cures were obtained only in the immunocompetent mice. An adoptive transfer of bone marrow from BALB/c to SCID mice restored the ability of PDT to produce long-term cures. This established the necessity of a functioning immune system for complete tumor response to PDT. There are now thought to be two aspects to the effect of PDT on the immune response against cancer: (A) the antitumor activity of PDT-induced inflammatory cells and (B) generation of a long-term anti-tumor immune response. These effects can be elicited by phototoxic damage that is not necessarily lethal to all tumor cells and creates an inflammatory stimulus. PDT-induced changes in the plasma membrane and membranes of cellular organelles can prompt a rapid activation of membranous phospholipases (Agarwal et al., 1993) leading to accelerated phospholipid degradation with a massive release of powerful inflammatory mediators (Yamamoto et al., 1991). Other cytokines and growth factors that are potent immunomodulators have been found to be strongly enhanced in PDT-treated mouse tumors (Gollnick et al., 1997; Herman et al., 1996).

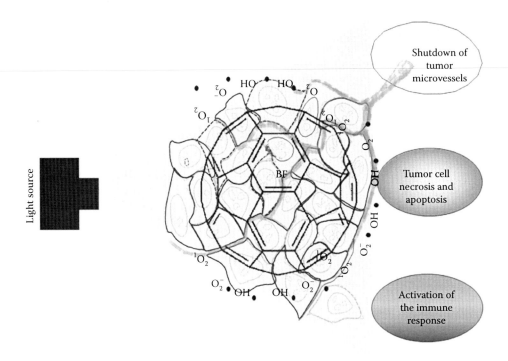

FIGURE 37.12 Three anti-tumor mechanisms that could take place during photodynamic therapy with fullerenes.

37.3.7 *In Vivo* PDT with Fullerenes

Fullerenes should have a photodynamic effect on tumors if the compound is accumulated in the tumor tissue. The first report of fullerenes being used to carry out PDT of actual tumors was by Tabata et al. in 1997. They chemically modified the water-insoluble C_{60} with polyethylene glycol (PEG), not only to make it soluble in water, but also to enlarge its molecular size. When injected intravenously into mice carrying a subcutaneous tumor on the back, the C_{60}-PEG conjugate exhibited higher accumulation and more prolonged retention in the tumor tissue than in normal tissue. The conjugate was excreted without being accumulated in any specific organ. Following intravenous injection of C_{60}-PEG conjugate or Photofrin (a recognized PS) to tumor-bearing mice, coupled with exposure of the tumor site to visible light, the volume increase of the tumor mass was suppressed and the C_{60} conjugate exhibited a stronger suppressive effect than Photofrin. A histological examination revealed that conjugate injection plus light irradiation strongly induced tumor necrosis without any damage to the overlying normal skin. The antitumor effect of the conjugate increased in a fluence and C_{60} dose-dependent manner, and cures were achieved by treatment with a dose of 424 µg/kg at a fluence of 10^7 J/cm².

Liu et al. (2007) conjugated PEG to C_{60} (C_{60}-PEG), and diethylenetriaminepentaacetic acid (DTPA) was subsequently introduced to the terminal group of PEG to prepare C_{60}-PEG-DTPA that was mixed with gadolinium acetate solution to obtain Gd^{3+}-chelated C_{60}-PEG-DTPA-Gd. Following the intravenous injection of C_{60}-PEG-DTPA-Gd into tumor-bearing mice, the PDT anti-tumor effect and MRI of the tumor were evaluated. A similar generation of superoxide upon illumination was observed with or without Gd^{3+} chelation. An intravenous injection of C_{60}-PEG-DTPA-Gd into tumor-bearing mice plus light (400–500 nm, 53.5 J/cm²) showed a significant anti-tumor PDT effect and the effect depended on the timing of light irradiation that correlated with tumor accumulation as detected by the enhanced intensity of the MRI signal.

Chiang and coworkers reported (Chi et al., 1999) a preliminary *in vivo* study of PDT using hydrophilic nanospheres formed from hexa(sulfo-n-butyl)[60]fullerene (FC$_4$S) against tumors in mice. ICR mice bearing sarcoma 180 subcutaneous tumors received either intraperitoneal or intravenous injection of water-soluble FC$_4$S in PBS (5 mg/kg body weight). The tumor site was subsequently irradiated with an argon ion laser beam at a wavelength of 515 or an argon-pumped dye-laser at 633 nm with the beam focused to a diameter of 7–8 mm with the total light dose adjusted to a level of 100 J/cm² in each experiment. Consistently, the inhibition of tumor growth was more effective using the lower wavelength and was better with the 515 nm laser than with the 633-nm laser. Administration of FC$_4$S to mice using the *i.p.* method had a slightly better inhibition effectiveness than the *i.v.* method.

The first fullerene-based clinical treatment of a human patient with rectal adenocarcinoma was attempted by Andrievsky et al. in 2000.

37.4 Summary

This chapter has demonstrated that fullerenes have certain unique features that could make them favorable candidates as PSs for use in the PDT of cancer, and at the same time other unique features that would argue against them as viable PSs for PDT. The most important favorable property is their rather unusual photochemical mechanism. As shown by us and by others, in aqueous solutions and particularly in the presence of reducing agents, these compounds produce a substantial amount of superoxide anion in a Type I photochemical process involving electron transfer from the excited triplet state to molecular oxygen. Although many workers in the PDT field think that the product of the Type II photochemical process, singlet oxygen, is the major cytotoxic species operating in PDT-induced cell killing, there have been reports that Type I mechanisms may be equally effective or even more effective than Type II. This is because hydroxyl radicals are the most reactive and potentially the most cytotoxic of all ROS. It is assumed that hydroxyl radicals are formed from hydrogen peroxide by Fenton chemistry reactions catalyzed by Fe^{2+} or Cu^+ ions, and that the hydrogen peroxide is produced by dismutation of the superoxide anion either by enzyme catalysis or naturally. Another possible mechanism of cytotoxicity is the diffusion-controlled reaction between superoxide and nitric oxide to form the highly toxic species, peroxynitrite. Elevated levels of nitric oxide are present both in cancers and infections thus giving the possibility of additional levels of selectivity for target-specific damage.

37.5 Future Perspectives

At the end of this chapter, we must ask ourselves whether in all reality it is likely that fullerenes will ever be accepted as viable PSs for the PDT of cancer in clinical applications. The chief disadvantage of fullerenes is likely to be their optical absorption properties. The absorption spectrum of fullerenes is highest in the UVA and blue regions of the spectrum where the tissue penetration depth of illumination is shortest due to a combination of light absorption by cellular chromophores and light scattering by cellular structures. However, the molar absorption coefficients of fullerenes are relatively high and the tail of absorption does stretch out into the red regions of the visible spectrum. As noted, there are three approaches that may also help in overcoming the disadvantages of the absorption spectrum: two-photon excitation, optical clearing of tissue, and covalent attachment of light harvesting antennae such as porphyrins. Fullerenes are not the most amenable molecules for drug delivery and choosing appropriate formulations may be difficult. Their pronounced tendency to aggregate in aqueous solutions and therefore in biological media does impose limitations on their photophysical and photochemical activities (Hotze et al., 2008). Nevertheless, in the present days of research progress in nanobiotechnology and in nanomedicine, this problem of aggregation of nanostructures is being increasingly faced and more strategies are being devised to overcome this limitation. Liposomal preparations, micellar preparations, nanoemulsions, and even recombinant lipoproteins may all be a possible delivery vehicle for fullerene PSs (Atkinson and Small, 1986; Koo et al., 2005; Moghimi and Agrawal, 2005; Perkins et al., 2000).

Acknowledgment

This work was supported by the U.S. National Institutes of Health (grants R43CA103268 and R44AI68400 to Lynntech Inc, R01CA/AI838801 and R01AI050875 to MRH).

References

Abels, C. (2004). Targeting of the vascular system of solid tumours by photodynamic therapy (PDT). *Photochem Photobiol Sci*, **3**, 765–771.

Agarwal, M.L., Clay, M.E., Harvey, E.J., Evans, H.H., Antunez, A.R., and Oleinick, N.L. (1991). Photodynamic therapy induces rapid cell death by apoptosis in L5178Y mouse lymphoma cells. *Cancer Res*, **51**, 5993–5996.

Agarwal, M.L., Larkin, H.E., Zaidi, S.I., Mukhtar, H., and Oleinick, N.L. (1993). Phospholipase activation triggers apoptosis in photosensitized mouse lymphoma cells. *Cancer Res*, **53**, 5897–5902.

Agostinis, P., Vantieghem, A., Merlevede, W., and de Witte, P.A. (2002). Hypericin in cancer treatment: More light on the way. *Int J Biochem Cell Biol*, **34**, 221–241.

Allison, R.R., Downie, G.H., Cuenca, R., Hu, X.-H., Childs, C.J., and Sibata, C.H. (2004). Photosensitizers in clinical PDT. *Photodiag Photodyn Ther*, **1**, 27–42.

Almeida, R.D., Manadas, B.J., Carvalho, A.P., and Duarte, C.B. (2004). Intracellular signaling mechanisms in photodynamic therapy. *Biochim Biophys Acta*, **1704**, 59–86.

Alvarez, M.G., Prucca, C., Milanesio, M.E., Durantini, E.N., and Rivarola, V. (2006). Photodynamic activity of a new sensitizer derived from porphyrin-C60 dyad and its biological consequences in a human carcinoma cell line. *Int J Biochem Cell Biol*, **38**, 2092–2101.

Andrievsky, G., Zhmuro, A., Zabobonina, L., and Suchina, E. (2000). First clinical case of treatment of patient (volunteer) with rectal adenocarcinoma by hydrated fullerenes: Natural course of the disease or non-specific anticancer activity? In *Biochemical and Pharmaceutical Aspects of Fullerene Materials*, Vol. Session N9. pp. Poster presentation No. 0377. The Electrochemical Society: Toronto, Canada.

Arbogast, J.W., Darmanyan, A.P., Foote, C.S., Rubin, Y., Diederich, F.N., Alvarez, M.M., Anz, S.J., and Whetten, R.L. (1991). Photophysical properties of C60. *J Phys Chem A Mol Spectrosc Kinet Environ Gen Theory*, **95**, 11–12.

Arbogast, J.W., Foote, C.S., and Kao, M. (1992). Electron-transfer to triplet C-60. *J Am Chem Soc*, **114**, 2277–2279.

Atkinson, D. and Small, D.M. (1986). Recombinant lipoproteins: Implications for structure and assembly of native lipoproteins. *Annu Rev Biophys Biophys Chem*, **15**, 403–456.

Bachowski, G.J., Pintar, T.J., and Girotti, A.W. (1991). Photosensitized lipid peroxidation and enzyme inactivation by membrane-bound merocyanine 540: Reaction mechanisms in the absence and presence of ascorbate. *Photochem Photobiol*, **53**, 481–491.

Bachowski, G.J., Korytowski, W., and Girotti, A.W. (1994). Characterization of lipid hydroperoxides generated by photodynamic treatment of leukemia cells. *Lipids*, **29**, 449–459.

Ben-Dror, S., Bronshtein, I., Wiehe, A., Roder, B., Senge, M.O., and Ehrenberg, B. (2006). On the correlation between hydrophobicity, liposome binding and cellular uptake of porphyrin sensitizers. *Photochem Photobiol*, **82**, 695–701.

Bhawalkar, J.D., Kumar, N.D., Zhao, C.F., and Prasad, P.N. (1997). Two-photon photodynamic therapy. *J Clin Laser Med Surg*, **15**, 201–204.

Bilski, P., Belanger, A.G., and Chignell, C.F. (2002). Photosensitized oxidation of 2′,7′-dichlorofluorescin: Singlet oxygen does not contribute to the formation of fluorescent oxidation product 2′,7′-dichlorofluorescein. *Free Radic Biol Med*, **33**, 938–946.

Bosi, S., Da Ros, T., Spalluto, G., and Prato, M. (2003). Fullerene derivatives: An attractive tool for biological applications. *Eur J Med Chem*, **38**, 913–923.

Bosi, S., Feruglio, L., Da Ros, T., Spalluto, G., Gregoretti, B., Terdoslavich, M., Decorti, G., Passamonti, S., Moro, S., and Prato, M. (2004). Hemolytic effects of water-soluble fullerene derivatives. *J Med Chem*, **47**, 6711–6715.

Bottiroli, G., Croce, A.C., Balzarini, P., Locatelli, D., Baglioni, P., Lo Nostro, P., Monici, M., and Pratesi, R. (1997). Enzyme-assisted cell photosensitization: A proposal for an efficient approach to tumor therapy and diagnosis. The rose Bengal fluorogenic substrate. *Photochem Photobiol*, **66**, 374–383.

Boyd, P.D. and Reed, C.A. (2005). Fullerene-porphyrin constructs. *Acc Chem Res*, **38**, 235–242.

Buchko, G.W., Cadet, J., Ravanat, J.L., and Labataille, P. (1993). Isolation and characterization of a new product produced by ionizing irradiation and type I photosensitization of 2′-deoxyguanosine in oxygen-saturated aqueous solution: (2S)-2,5′-ANHYDRO-1-(2′-deoxy-beta-D-erythro-pentofuranosyl)-5-guanidin ylidene- 2-hydroxy-4-oxoimidazolidine. *Int J Radiat Biol*, **63**, 669–676.

Buchko, G.W., Wagner, J.R., Cadet, J., Raoul, S., and Weinfeld, M. (1995). Methylene blue-mediated photooxidation of 7,8-dihydro-8-oxo-2′-deoxyguanosine. *Biochim Biophys Acta*, **1263**, 17–24.

Burlaka, A.P., Sidorik, Y.P., Prylutska, S.V., Matyshevska, O.P., Golub, O.A., Prylutskyy, Y.I., and Scharff, P. (2004). Catalytic system of the reactive oxygen species on the C60 fullerene basis. *Exp Oncol*, **26**, 326–327.

Busch, T.M. (2006). Local physiological changes during photodynamic therapy. *Lasers Surg Med*, **38**, 494–499.

Buytaert, E., Callewaert, G., Hendrickx, N., Scorrano, L., Hartmann, D., Missiaen, L., Vandenheede, J.R., Heirman, I., Grooten, J., and Agostinis, P. (2006). Role of endoplasmic reticulum depletion and multidomain proapoptotic BAX and BAK proteins in shaping cell death after hypericin-mediated photodynamic therapy. *Faseb J*, **20**, 756–758.

Buytaert, E., Dewaele, M., and Agostinis, P. (2007). Molecular effectors of multiple cell death pathways initiated by photodynamic therapy. *Biochim Biophys Acta*, **1776**, 86–107.

Castano, A.P., Demidova, T.N., and Hamblin, M.R. (2004). Mechanisms in photodynamic therapy: Part one—Photosensitizers, photochemistry and cellular localization. *Photodiagn Photodyn Ther*, **1**, 279–293.

Castano, A.P., Demidova, T.N., and Hamblin, M.R. (2005a). Mechanisms in photodynamic therapy: Part three—Photosensitizer pharmacokinetics, biodistribution, tumor localization and modes of tumor destruction. *Photodiagn Photodyn Ther*, **2**, 91–106.

Castano, A.P., Demidova, T.N., and Hamblin, M.R. (2005b). Mechanisms in photodynamic therapy: Part two—Cellular signalling, cell metabolism and modes of cell death. *Photodiagn Photodyn Ther*, **2**, 1–23.

Castano, A.P., Mroz, P., and Hamblin, M.R. (2006). Photodynamic therapy and anti-tumour immunity. *Nat Rev Cancer*, **6**, 535–545.

Cauchon, N., Tian, H., Langlois, R., La Madeleine, C., Martin, S., Ali, H., Hunting, D., and van Lier, J.E. (2005). Structure-photodynamic activity relationships of substituted zinc trisulfophthalocyanines. *Bioconjug Chem*, **16**, 80–89.

Chekulayeva, L.V., Shevchuk, I.N. et al. (2006). "Hydrogen peroxide, superoxide, and hydroxyl radicals are involved in the phototoxic action of hematoporphyrin derivative against tumor cells." *J Environ Pathol Toxicol Oncol*, **25** (1-2): 51–77.

Chen, H.H., Yu, C., Ueng, T.H., Chen, S., Chen, B.J., Huang, K.J., and Chiang, L.Y. (1998). Acute and subacute toxicity study of water-soluble polyalkylsulfonated C60 in rats. *Toxicol Pathol*, **26**, 143–151.

Chen, Y., Potter, W.R., Missert, J.R., Morgan, J., and Pandey, R.K. (2007). Comparative in vitro and in vivo studies on long-wavelength photosensitizers derived from bacteriopurpurinimide and Bacteriochlorin p6: Fused imide ring enhances the in vivo PDT efficacy. *Bioconjug Chem*, **18**, 1460–1473.

Chi, Y., Canteenwala, T., Chen, H.C., Chen, B.J., Canteenwala, M., and Chiang, L.Y. (1999). Hexa(sulfobutyl)fullerene-induced photodynamic effect on tumors in vivo and toxicity study in rats. *Proc Electrochem Soc*, **99**, 234–249.

Chiang, L.Y., Padmawar, P.A., Canteenwala, T., Tan, L.S., He, G.S., Kannan, R., Vaia, R., Lin, T.C., Zheng, Q., and Prasad, P.N. (2002). Synthesis of C60-diphenylaminofluorene dyad with large 2PA cross-sections and efficient intramolecular two-photon energy transfer. *Chem Commun (Camb)*, 1854–5.

Culotta, L. and Koshland, D.E., Jr. (1991). Buckyballs: Wide Open Playing Field for Chemists. *Science*, **254**, 1706–1709.

Dellinger, M. (1996). Apoptosis or necrosis following Photofrin photosensitization: Influence of the incubation protocol. *Photochem Photobiol*, **64**, 182–187.

Detty, M.R., Gibson, S.L., and Wagner, S.J. (2004). Current clinical and preclinical photosensitizers for use in photodynamic therapy. *J Med Chem*, **47**, 3897–3915.

Dolmans, D.E., Fukumura, D., and Jain, R.K. (2003). Photodynamic therapy for cancer. *Nat Rev Cancer*, **3**, 380–387.

Dougherty, T.J. (2002). An update on photodynamic therapy applications. *J Clin Laser Med Surg*, **20**, 3–7.

Duncan, L.K., Jinschek, J.R., and Vikesland, P.J. (2008). C60 colloid formation in aqueous systems: Effects of preparation method on size, structure, and surface charge. *Environ Sci Technol*, **42**, 173–178.

El-Khouly, M.E., Araki, Y., Ito, O., Gadde, S., McCarty, A.L., Karr, P.A., Zandler, M.E., and D'Souza, F. (2005). Spectral, electrochemical, and photophysical studies of a magnesium porphyrin-fullerene dyad. *Phys Chem Chem Phys*, **7**, 3163–3171.

Fingar, V.H., Siegel, K.A., Wieman, T.J., and Doak, K.W. (1993a). The effects of thromboxane inhibitors on the microvascular and tumor response to photodynamic therapy. *Photochem Photobiol*, **58**, 393–399.

Fingar, V.H., Wieman, T.J., Karavolos, P.S., Doak, K.W., Ouellet, R., and van Lier, J.E. (1993b). The effects of photodynamic therapy using differently substituted zinc phthalocyanines on vessel constriction, vessel leakage and tumor response. *Photochem Photobiol*, **58**, 251–258.

Fiorito, S., Serafino, A., Andreola, F., Togna, A., and Togna, G. (2006). Toxicity and biocompatibility of carbon nanoparticles. *J Nanosci Nanotechnol*, **6**, 591–599.

Foote, C.S. (1994). Photophysical and photochemical properties of fullerenes. *Top Curr Chem*, **169**, 347–363.

Fortner, J.D., Lyon, D.Y., Sayes, C.M., Boyd, A.M., Falkner, J.C., Hotze, E.M., Alemany, L.B., Tao, Y.J., Guo, W., Ausman, K.D., Colvin, V.L., and Hughes, J.B. (2005). C60 in water: Nanocrystal formation and microbial response. *Environ Sci Technol*, **39**, 4307–4316.

Fuchs, J. and Thiele, J. (1998). The role of oxygen in cutaneous photodynamic therapy. *Free Radic Biol Med*, **24**, 835–847.

Gebhart, S.C., Lin, W.C., and Mahadevan-Jansen, A. (2006). In vitro determination of normal and neoplastic human brain tissue optical properties using inverse adding-doubling. *Phys Med Biol*, **51**, 2011–2027.

Gharbi, N., Pressac, M., Hadchouel, M., Szwarc, H., Wilson, S.R., and Moussa, F. (2005). [60]fullerene is a powerful antioxidant in vivo with no acute or subacute toxicity. *Nano Lett*, **5**, 2578–2585.

Gilaberte, Y., Pereboom, D. et al. (1997). "Flow cytometry study of the role of superoxide anion and hydrogen peroxide in cellular photodestruction with 5-aminolevulinic acid-induced protoporphyrin IX." *Photodermatol Photoimmunol Photomed*, **13**(1-2): 43–49.

Gilissen, M.J., van de Merbel-de Wit, L.E., Star, W.M., Koster, J.F., and Sluiter, W. (1993). Effect of photodynamic therapy on the endothelium-dependent relaxation of isolated rat aortas. *Cancer Res*, **53**, 2548–2552.

Girotti, A.W. (1983). Mechanisms of photosensitization. *Photochem Photobiol*, **38**, 745–751.

Girotti, A.W. (1985). Mechanisms of lipid peroxidation. *J Free Radic Biol Med*, **1**, 87–95.

Gollnick, S.O., Liu, X., Owczarczak, B., Musser, D.A., and Henderson, B.W. (1997). Altered expression of interleukin 6 and interleukin 10 as a result of photodynamic therapy in vivo. *Cancer Res*, **57**, 3904–3909.

Granville, D.J., Carthy, C.M., Jiang, H., Shore, G.C., McManus, B.M., and Hunt, D.W. (1998). Rapid cytochrome c release, activation of caspases 3, 6, 7 and 8 followed by Bap31 cleavage in HeLa cells treated with photodynamic therapy. *FEBS Lett*, **437**, 5–10.

Grune, T., Klotz, L.O., Gieche, J., Rudeck, M., and Sies, H. (2001). Protein oxidation and proteolysis by the nonradical oxidants singlet oxygen or peroxynitrite. *Free Radic Biol Med*, **30**, 1243–1253.

Guldi, D.M. and Prato, M. (2000). Excited-state properties of C(60) fullerene derivatives. *Acc Chem Res*, **33**, 695–703.

Gupta, S., Ahmad, N., and Mukhtar, H. (1998). Involvement of nitric oxide during phthalocyanine (Pc4) photodynamic therapy-mediated apoptosis. *Cancer Res*, **58**, 1785–1788.

Hamblin, M.R. and Hasan, T. (2004). Photodynamic therapy: A new antimicrobial approach to infectious disease? *Photochem Photobiol Sci*, **3**, 436–450.

Hariharan, P.V., Courtney, J., and Eleczko, S. (1980). Production of hydroxyl radicals in cell systems exposed to haematoporphyrin and red light. *Int J Radiat Biol Relat Stud Phys Chem Med*, **37**, 691–694.

He, Y.Y. and An, J.Y. et al. (1998). "EPR and spectrophotometric studies on free radicals (O2.-, Cysa-HB.-) and singlet oxygen (1O2) generated by irradiation of cysteamine substituted hypocrellin B." *Int J Radiat Biol*, **74**(5): 647–564.

He, Y. and Wang, R.K. (2004). Dynamic optical clearing effect of tissue impregnated with hyperosmotic agents and studied with optical coherence tomography. *J Biomed Opt*, **9**, 200–206.

He, X.Y., Sikes, R.A., Thomsen, S., Chung, L.W., and Jacques, S.L. (1994). Photodynamic therapy with photofrin II induces programmed cell death in carcinoma cell lines. *Photochem Photobiol*, **59**, 468–473.

Henderson, B.W. and Dougherty, T.J. (1992). How does photodynamic therapy work? *Photochem Photobiol*, **55**, 145–157.

Henderson, B.W. and Fingar, V.H. (1987). Relationship of tumor hypoxia and response to photodynamic treatment in an experimental mouse tumor. *Cancer Res*, **47**, 3110–3114.

Henderson, B.W. and Miller, A.C. (1986). "Effects of scavengers of reactive oxygen and radical species on cell survival following photodynamic treatment in vitro: comparison to ionizing radiation." *Radiat Res*, **108**(2): 196–205.

Herman, S., Kalechman, Y., Gafter, U., Sredni, B., and Malik, Z. (1996). Photofrin II induces cytokine secretion by mouse spleen cells and human peripheral mononuclear cells. *Immunopharmacology*, **31**, 195–204.

Hirshburg, J., Choi, B., Nelson, J.S., and Yeh, A.T. (2007). Correlation between collagen solubility and skin optical clearing using sugars. *Lasers Surg Med*, **39**, 140–144.

Hotze, E.M., Labille, J., Alvarez, P., and Wiesner, M.R. (2008). Mechanisms of photochemistry and reactive oxygen production by fullerene suspensions in water. *Environ Sci Technol*, **42**, 4175–4180.

Ikeda, A., Doi, Y., Nishiguchi, K., Kitamura, K., Hashizume, M., Kikuchi, J., Yogo, K., Ogawa, T., and Takeya, T. (2007). Induction of cell death by photodynamic therapy with water-soluble lipid-membrane-incorporated [60]fullerene. *Org Biomol Chem*, **5**, 1158–1160.

Imahori, H. (2004). Porphyrin-fullerene linked systems as artificial photosynthetic mimics. *Org Biomol Chem*, **2**, 1425–1433.

Irie, K., Nakamura, Y., Ohigashi, H., Tokuyama, H., Yamago, S., and Nakamura, E. (1996). Photocytotoxicity of water-soluble fullerene derivatives. *Biosci Biotechnol Biochem*, **60**, 1359–1361.

Isakovic, A., Markovic, Z., Todorovic-Markovic, B., Nikolic, N., Vranjes-Djuric, S., Mirkovic, M., Dramicanin, M., Harhaji, L., Raicevic, N., Nikolic, Z., and Trajkovic, V. (2006). Distinct cytotoxic mechanisms of pristine versus hydroxylated fullerene. *Toxicol Sci*, **91**, 173–183.

Jiang, J., Boese, M., Turner, P., and Wang, R.K. (2008). Penetration kinetics of dimethyl sulphoxide and glycerol in dynamic optical clearing of porcine skin tissue in vitro studied by Fourier transform infrared spectroscopic imaging. *J Biomed Opt*, **13**, 021105.

Karotki, A., Khurana, M., Lepock, J.R., and Wilson, B.C. (2006). Simultaneous two-photon excitation of photofrin in relation to photodynamic therapy. *Photochem Photobiol*, **82**, 443–452.

Kessel, D. and Poretz, R.D. (2000). Sites of photodamage induced by photodynamic therapy with a chlorin e6 triacetoxymethyl ester (CAME). *Photochem Photobiol*, **71**, 94–96.

Kessel, D., Luo, Y., Mathieu, P., and Reiners, J.J., Jr. (2000). Determinants of the apoptotic response to lysosomal photodamage. *Photochem Photobiol*, **71**, 196–200.

Kessel, D., Vicente, M.G., and Reiners, J.J., Jr. (2006). Initiation of apoptosis and autophagy by photodynamic therapy. *Lasers Surg Med*, **38**, 482–488.

Khan, M.H., Choi, B., Chess, S., Kelly, M.K., McCullough, J., and Nelson, J.S. (2004). Optical clearing of in vivo human skin: Implications for light-based diagnostic imaging and therapeutics. *Lasers Surg Med*, **34**, 83–85.

Kiesslich, T., Plaetzer, K., Oberdanner, C.B., Berlanda, J., Obermair, F.J., and Krammer, B. (2005). Differential effects of glucose deprivation on the cellular sensitivity towards photodynamic treatment-based production of reactive oxygen species and apoptosis-induction. *FEBS Lett*, **579**, 185–190.

Kochevar, I.E., Lynch, M.C., Zhuang, S., and Lambert, C.R. (2000). Singlet oxygen, but not oxidizing radicals, induces apoptosis in HL-60 cells. *Photochem Photobiol*, **72**, 548–553.

Koeppe, R. and Sariciftci, N.S. (2006). Photoinduced charge and energy transfer involving fullerene derivatives. *Photochem Photobiol Sci*, **5**, 1122–1131.

Kogan, A. and Garti, N. (2006). Microemulsions as transdermal drug delivery vehicles. *Adv Colloid Interface Sci*, **123–126**, 369–385.

Kolarova, H., Bajgar, R. et al. (2007). "Comparison of sensitizers by detecting reactive oxygen species after photodynamic reaction in vitro." *Toxicol In Vitro.*

Koo, O.M., Rubinstein, I., and Onyuksel, H. (2005). Role of nanotechnology in targeted drug delivery and imaging: A concise review. *Nanomedicine*, **1**, 193–212.

Korbelik, M., Krosl, G., Krosl, J., and Dougherty, G.J. (1996). The role of host lymphoid populations in the response of mouse EMT6 tumor to photodynamic therapy. *Cancer Res*, **56**, 5647–5652.

Korytowski, W., Zareba, M. et al. (2000). "Nitric oxide inhibition of free radical-mediated cholesterol peroxidation in liposomal membranes." *Biochemistry*, **39**(23): 6918–6928.

Kral, V., Davis, J., Andrievsky, A., Kralova, J., Synytsya, A., Pouckova, P., and Sessler, J.L. (2002). Synthesis and biolocalization of water-soluble sapphyrins. *J Med Chem*, **45**, 1073–1078.

Lam, M., Oleinick, N.L. et al. (2001). "Photodynamic therapy-induced apoptosis in epidermoid carcinoma cells. Reactive oxygen species and mitochondrial inner membrane permeabilization." *J Biol Chem* 276(50): 47379–47386.

Lin, C.P., Lynch, M.C., and Kochevar, I.E. (2000). Reactive oxidizing species produced near the plasma membrane induce apoptosis in bovine aorta endothelial cells. *Exp Cell Res*, **259**, 351–359.

Liu, J., Ohta, S., Sonoda, A., Yamada, M., Yamamoto, M., Nitta, N., Murata, K., and Tabata, Y. (2007). Preparation of PEG-conjugated fullerene containing Gd(3+) ions for photodynamic therapy. *J Control Release*, **117**, 104–110.

Mantareva, V., Shopova, M., Spassova, G., Wohrle, D., Muller, S., Jori, G., and Ricchelli, F. (1997). Si(IV)-methoxyethyleneglycol-naphthalocyanine: Synthesis and pharmacokinetic and photosensitizing properties in different tumour models. *J Photochem Photobiol B*, **40**, 258–262.

Martin, J.P. and Logsdon, N. (1987). Oxygen radicals are generated by dye-mediated intracellular photooxidations: A role for superoxide in photodynamic effects. *Arch Biochem Biophys*, **256**, 39–49.

McMahon, K.S., Wieman, T.J., Moore, P.H., and Fingar, V.H. (1994). Effects of photodynamic therapy using mono-L-aspartyl chlorin e6 on vessel constriction, vessel leakage, and tumor response. *Cancer Res*, **54**, 5374–5379.

Midden, W.R. and Dahl, T.A. (1992). Biological inactivation by singlet oxygen: Distinguishing O2(1 delta g) and O2(1 sigma g+). *Biochim Biophys Acta*, **1117**, 216–222.

Milanesio, M.E., Alvarez, M.G., Rivarola, V., Silber, J.J., and Durantini, E.N. (2005). Porphyrin-fullerene C60 dyads with high ability to form photoinduced charge-separated state as novel sensitizers for photodynamic therapy. *Photochem Photobiol*, **81**, 891–897.

Miyata, N., Yamakoshi, Y., and Nakanishi, I. (2000). Reactive species responsible for biological actions of photoexcited fullerenes. *J Pharmaceut Soc Jpn*, **120**, 1007–1016.

Moan, J. and Berg, K. (1991). The photodegradation of porphyrins in cells can be used to estimate the lifetime of singlet oxygen. *Photochem Photobiol*, **53**, 549–553.

Moghimi, S.M., and Agrawal, A. (2005). Lipid-based nanosystems and complexes in experimental and clinical therapeutics. *Curr Drug Deliv*, **2**, 295.

Mori, T., Takada, H., Ito, S., Matsubayashi, K., Miwa, N., and Sawaguchi, T. (2006). Preclinical studies on safety of fullerene upon acute oral administration and evaluation for no mutagenesis. *Toxicology*, **225**, 48–54.

Mroz, P., Pawlak, A., Satti, M., Lee, H., Wharton, T., Gali, H., Sarna, T., and Hamblin, M.R. (2007a). Functionalized fullerenes mediate photodynamic killing of cancer cells: Type I versus Type II photochemical mechanism. *Free Radic Biol Med*, **43**, 711–719.

Mroz, P., Tegos, G.P., Gali, H., Wharton, T., Sarna, T., and Hamblin, M.R. (2007b). Photodynamic therapy with fullerenes. *Photochem Photobiol Sci*, **6**, 1139–1149.

Murphy, M.P. and Smith, R.A. (2007). Targeting antioxidants to mitochondria by conjugation to lipophilic cations. *Annu Rev Pharmacol Toxicol*, **47**, 629–656.

Myhre, O., Andersen, J.M., Aarnes, H., and Fonnum, F. (2003). Evaluation of the probes 2′,7′-dichlorofluorescin diacetate, luminol, and lucigenin as indicators of reactive species formation. *Biochem Pharmacol*, **65**, 1575–1582.

Nakamura, E. and Isobe, H. (2003). Functionalized fullerenes in water. The first 10 years of their chemistry, biology, and nanoscience. *Acc Chem Res*, **36**, 807–815.

Niziolek, M., Korytowski, W. et al. (2003). "Chain-breaking antioxidant and cytoprotective action of nitric oxide on photodynamically stressed tumor cells." *Photochem Photobiol*, **78**(3): 262–270.

Niziolek, M., Korytowski, W. et al. (2003). "Nitric oxide inhibition of free radical-mediated lipid peroxidation in photodynamically treated membranes and cells." *Free Radic Biol Med*, **34**(8): 997–1005.

Niziolek, M., Korytowski, W. et al. (2006). "Nitric oxide-induced resistance to lethal photooxidative damage in a breast tumor cell line." *Free Radic Biol Med*, **40**(8): 1323–31.

O'Donnell, V.B., Chumley, P.H., Hogg, N., Bloodsworth, A., Darley-Usmar, V.M., and Freeman, B.A. (1997). Nitric oxide inhibition of lipid peroxidation: Kinetics of reaction with lipid peroxyl radicals and comparison with alpha-tocopherol. *Biochemistry*, **36**, 15216–15223.

Oberdorster, E. (2004). Manufactured nanomaterials (fullerenes, C60) induce oxidative stress in the brain of juvenile largemouth bass. *Environ Health Perspect*, **112**, 1058–1062.

Oleinick, N.L., Morris, R.L., and Belichenko, I. (2002). The role of apoptosis in response to photodynamic therapy: What, where, why, and how. *Photochem Photobiol Sci*, **1**, 1–21.

Peng, Q., Moan, J., Nesland, J.M., and Rimington, C. (1990). Aluminum phthalocyanines with asymmetrical lower sulfonation and with symmetrical higher sulfonation: A comparison of localizing and photosensitizing mechanism in human tumor LOX xenografts. *Int J Cancer*, **46**, 719–726.

Perkins, W.R., Ahmad, I., Li, X., Hirsh, D.J., Masters, G.R., Fecko, C.J., Lee, J., Ali, S., Nguyen, J., Schupsky, J., Herbert, C., Janoff, A.S., and Mayhew, E. (2000). Novel therapeutic nano-particles (lipocores): Trapping poorly water soluble compounds. *Int J Pharm*, **200**, 27–39.

Potter, W.R., Henderson, B.W., Bellnier, D.A., Pandey, R.K., Vaughan, L.A., Weishaupt, K.R., and Dougherty, T.J. (1999). Parabolic quantitative structure-activity relationships and photodynamic therapy: Application of a three-compartment model with clearance to the in vivo quantitative structure-activity relationships of a congeneric series of pyropheophorbide derivatives used as photosensitizers for photodynamic therapy. *Photochem Photobiol*, **70**, 781–788.

Rancan, F., Rosan, S., Boehm, F., Cantrell, A., Brellreich, M., Schoenberger, H., Hirsch, A., and Moussa, F. (2002). Cytotoxicity and photocytotoxicity of a dendritic C(60) mono-adduct and a malonic acid C(60) tris-adduct on Jurkat cells. *J Photochem Photobiol B*, **67**, 157–162.

Rancan, F., Helmreich, M., Molich, A., Jux, N., Hirsch, A., Roder, B., Witt, C., and Bohm, F. (2005). Fullerene-pyropheophorbide a complexes as sensitizer for photodynamic therapy: Uptake and photo-induced cytotoxicity on Jurkat cells. *J Photochem Photobiol B*, **80**, 1–7.

Ravanat, J.L. and Cadet, J. (1995). Reaction of singlet oxygen with 2′-deoxyguanosine and DNA. Isolation and characterization of the main oxidation products. *Chem Res Toxicol*, **8**, 379–388.

Redmond, R.W. and Kochevar, I.E. (2006). Spatially resolved cellular responses to singlet oxygen. *Photochem Photobiol*, **82**, 1178–1186.

Ross, M.F., Da Ros, T., Blaikie, F.H., Prime, T.A., Porteous, C.M., Severina, II, Skulachev, V.P., Kjaergaard, H.G., Smith, R.A., and Murphy, M.P. (2006). Accumulation of lipophilic dications by mitochondria and cells. *Biochem J*, **400**, 199–208.

Rottenberg, H. (1984). Membrane potential and surface potential in mitochondria: Uptake and binding of lipophilic cations. *J Membr Biol*, **81**, 127–138.

Samkoe, K.S. and Cramb, D.T. (2003). Application of an ex ovo chicken chorioallantoic membrane model for two-photon excitation photodynamic therapy of age-related macular degeneration. *J Biomed Opt*, **8**, 410–417.

Samkoe, K.S., Clancy, A.A., Karotki, A., Wilson, B.C., and Cramb, D.T. (2007). Complete blood vessel occlusion in the chick chorioallantoic membrane using two-photon excitation photodynamic therapy: Implications for treatment of wet age-related macular degeneration. *J Biomed Opt*, **12**, 034025.

Satoh, M. and Takayanagi, I. (2006). Pharmacological studies on fullerene (C60), a novel carbon allotrope, and its derivatives. *J Pharmacol Sci*, **100**, 513–518.

Schuster, D.I., Cheng, P., Jarowski, P.D., Guldi, D.M., Luo, C., Echegoyen, L., Pyo, S., Holzwarth, A.R., Braslavsky, S.E., Williams, R.M., and Klihm, G. (2004). Design, synthesis, and photophysical studies of a porphyrin-fullerene dyad with parachute topology; charge recombination in the marcus inverted region. *J Am Chem Soc*, **126**, 7257–7270.

Star, W.M., Marijnissen, H.P., van den Berg-Blok, A.E., Versteeg, J.A., Franken, K.A., and Reinhold, H.S. (1986). Destruction of rat mammary tumor and normal tissue microcirculation by hematoporphyrin derivative photoradiation observed in vivo in sandwich observation chambers. *Cancer Res*, **46**, 2532–2540.

Starkey, J.R., Rebane, A.K., Drobizhev, M.A., Meng, F., Gong, A., Elliott, A., McInnerney, K., and Spangler, C.W. (2008). New two-photon activated photodynamic therapy sensitizers induce xenograft tumor regressions after near-IR laser treatment through the body of the host mouse. *Clin Cancer Res*, **14**, 6564–6573.

Stockert, J.C., Juarranz, A., Villanueva, A., and Canete, M. (1996). Photodynamic damage to HeLa cell microtubules induced by thiazine dyes. *Cancer Chemother Pharmacol*, **39**, 167–169.

Svaasand, L.O. (1984). Optical dosimetry for direct and interstitial photoradiation therapy of malignant tumors. *Prog Clin Biol Res*, **170**, 91–114.

Szeimies, R.M., Karrer, S., Abels, C., Steinbach, P., Fickweiler, S., Messmann, H., Baumler, W., and Landthaler, M. (1996). 9-Acetoxy-2,7,12,17-tetrakis-(beta-methoxyethyl)-porphycene (ATMPn), a novel photosensitizer for photodynamic therapy: Uptake kinetics and intracellular localization. *J Photochem Photobiol B*, **34**, 67–72.

Tabata, Y., Murakami, Y., and Ikada, Y. (1997). Photodynamic effect of polyethylene glycol-modified fullerene on tumor. *Jpn J Cancer Res*, **88**, 1108–1116.

Tagmatarchis, N. and Shinohara, H. (2001). Fullerenes in medicinal chemistry and their biological applications. *Mini Rev Med Chem*, **1**, 339–348.

Tegos, G.P., Demidova, T.N., Arcila-Lopez, D., Lee, H., Wharton, T., Gali, H., and Hamblin, M.R. (2005). Cationic fullerenes are effective and selective antimicrobial photosensitizers. *Chem Biol*, **12**, 1127–1135.

Tokuyama, H., Yamago, S., and Nakamura, E. (1993). Photoinduced biochemical activity of fullerene carboxylic acid. *J Am Chem Soc*, **115**, 7918–7919.

Tuchin, V.V., Wang, R.K., and Yeh, A.T. (2008). Optical clearing of tissues and cells. *J Biomed Opt*, **13**, 021101.

Ungurenasu, C. and Airinei, A. (2000). Highly stable C(60)/poly(vinylpyrrolidone) charge-transfer complexes afford new predictions for biological applications of underivatized fullerenes. *J Med Chem*, **43**, 3186–3188.

Vail, S.A., Schuster, D.I., Guldi, D.M., Isosomppi, M., Tkachenko, N., Lemmetyinen, H., Palkar, A., Echegoyen, L., Chen, X., and Zhang, J.Z. (2006). Energy and electron transfer in beta-alkynyl-linked porphyrin-[60]fullerene dyads. *J Phys Chem B*, **110**, 14155–14166.

Verma, S., Hauck, T., El-Khouly, M.E., Padmawar, P.A., Canteenwala, T., Pritzker, K., Ito, O., and Chiang, L.Y. (2005). Self-assembled photoresponsive amphiphilic diphenylaminofluorene-C60 conjugate vesicles in aqueous solution. *Langmuir*, **21**, 3267–3272.

Viola, A., Jeunet, A., Decreau, R., Chanon, M., and Julliard, M. (1998). ESR studies of a series of phthalocyanines. Mechanism of phototoxicity. Comparative quantitation of O2-. using ESR spin-trapping and cytochrome c reduction techniques. *Free Radic Res*, **28**, 517–532.

Weishaupt, K.R., Gomer, C.J. et al. (1976). "Identification of singlet oxygen as the cytotoxic agent in photoinactivation of a murine tumor." *Cancer Res* 36(7 PT 1): 2326–2329.

Wilson, B.C., Jeeves, W.P., and Lowe, D.M. (1985). In vivo and post mortem measurements of the attenuation spectra of light in mammalian tissues. *Photochem Photobiol*, **42**, 153–162.

Wyld, L., Reed, M.W., and Brown, N.J. (2001). Differential cell death response to photodynamic therapy is dependent on dose and cell type. *Br J Cancer*, **84**, 1384–1386.

Xue, L.Y., Chiu, S.M., Azizuddin, K., Joseph, S., and Oleinick, N.L. (2007). The death of human cancer cells following photodynamic therapy: Apoptosis competence is necessary for Bcl-2 protection but not for induction of autophagy. *Photochem Photobiol*, **83**, 1016–1023.

Yamakoshi, Y., Umezawa, N., Ryu, A., Arakane, K., Miyata, N., Goda, Y., Masumizu, T., and Nagano, T. (2003). Active oxygen species generated from photoexcited fullerene (C60) as potential medicines: O2-$^\bullet$ versus 1O2. *J Am Chem Soc*, **125**, 12803–12809.

Yamamoto, N., Homma, S., Sery, T.W., Donoso, L.A., and Hoober, J.K. (1991). Photodynamic immunopotentiation: In vitro activation of macrophages by treatment of mouse peritoneal cells with haematoporphyrin derivative and light. *Eur J Cancer*, **27**, 467–471.

Yang, X.L., Fan, C.H., and Zhu, H.S. (2002). Photo-induced cytotoxicity of malonic acid [C(60)]fullerene derivatives and its mechanism. *Toxicol In Vitro*, **16**, 41–46.

Zakharenko, L.P., Zakharov, I.K., Vasiunina, E.A., Karamysheva, T.V., Danilenko, A.M., and Nikiforov, A.A. (1997). Determination of the genotoxicity of fullerene C60 and fullerol using the method of somatic mosaics on cells of *Drosophila melanogaster* wing and SOS-chromotest. *Genetika*, **33**, 405–409.

Zhu, S., Oberdorster, E., and Haasch, M.L. (2006). Toxicity of an engineered nanoparticle (fullerene, C60) in two aquatic species, Daphnia and fathead minnow. *Mar Environ Res*, **62** Suppl, S5–S9.

Zhu, X., Zhu, L., Li, Y., Duan, Z., Chen, W., and Alvarez, P.J. (2007). Developmental toxicity in zebrafish (*Danio rerio*) embryos after exposure to manufactured nanomaterials: Buckminsterfullerene aggregates (nC60) and fullerol. *Environ Toxicol Chem*, **26**, 976–979.

VIII

Quantum Engines and Nanomotors

Energy Transport and Heat Production in Quantum Engines

Liliana Arrachea
Universidad de Buenos Aires

Michael Moskalets
National Technical University

38.1 Introduction

In 1824, the French physicist Nicolas Léonard Sadi Carnot, better known as Sadi Carnot, in *Reflections on the Motive Power of Fire* settled the principles of the modern theory of thermodynamics by pointing out that motive power (concept later identified as work) is due to the fall of caloric (concept later identified as heat) from a hot to cold body (working substance). These ideas that provided the scientific support for the technological jump based on the steam engine were not well understood at that time. They were actually discovered and further elaborated 30 years later by the German Rudolf Clausius and the British William Thomson (Lord Kelvin). The fundamental principles ruling the operation of the thermal engines were then summarized in the two basic laws of thermodynamics. While the first law simply stresses the conservation of the energy, the second one deals with the subtle distinction between a kind of energy that can be used and another one that is dissipated in a physical process, as well as on the balance between both of them.

In 1851, Lord Kelvin also discovered the Thomson effect, and showed that it was related to other thermoelectric phenomena: the Peltier and Seebeck effects. Unlike Carnot's machines, these effects are related to nonequilibrium processes. However, they also bring about the conceptual distinction between different kinds of energies: one that is transported in some direction due to a voltage or temperature gradient, but being different from the

Joule heating in the sense that the first one is reversible while the latter involves dissipation and, thus, irreversible effects.

Nowadays, we are witnessing a technological trend toward an increasing miniaturization of the electronic components. This is accompanied by a significant activity within communities of the basic sciences, in the search of a better understanding of the behavior of materials and devices with sizes in the range between 1 nm and 10 μm as fundamental pieces of electronic circuits. Paradigmatic examples are the quantum dots fabricated in the interfaces of semiconductor structures where confining gates for electrons and circuits are printed by means of nanolitography within an area of a few μm² (see Figure 38.1). Due to the small scale of these systems, they present some physical features that resemble the molecules. In particular, the landscape of their spectra contains well-defined quantum levels where electrons propagate almost perfectly preserving the phase of their wave functions. However, they are not isolated from the external world but coupled to the substrate, gates, wires, and external fields that induce the transport of electrons. For this reason, they are classified as "open quantum systems" that operate out-of-equilibrium conditions. The "external world," instead, contains pieces that act as macroscopic reservoirs with which the "small quantum systems" exchange particles and energy. Due to the mixed nature of these systems, the concepts of classical electrodynamics and thermodynamics cannot be simply applied to them and theoretical tools that are amenable to capture their

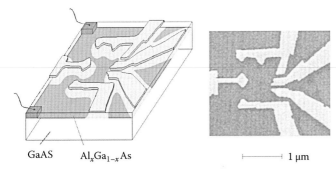

GaAS $Al_xGa_{1-x}As$ |————————————| 1 μm

FIGURE 38.1 Quantum dots fabricated with nanolitography in a semiconductor interface. The two leads going upwards and downwards make the contact between the dot and external reservoirs. The gate at the left can be used to apply a voltage to introduce a rigid shift of the positions of the energy levels of the dot. The two extra leads connected at the right-hand wall of the dot can be used to apply ac potentials at those points. (Courtesy of Charles Marcus, Mesoscopic Lab, Harvard University, Cambridge, MA.)

quantum properties as well as the coupling to their environment are necessary.

In this chapter, we focus on a particular kind of device named "quantum pumps." They have been realized experimentally, precisely, in quantum dots, where ac voltages are applied at their walls, and a current with a dc component is generated in the absence of a stationary voltage difference (see Figure 38.1). From a theoretical point of view, one of the features of these systems is that they can be described in terms of simple Hamiltonians, including macroscopic pieces that represent the wires connected to the quantum dot as well as explicit terms for the time-dependent forces that induce the transport process. The latter feature makes an important difference, in comparison with setups where the transport is induced by means of a stationary voltage difference. This is because in such systems the work of the forces that keep such a bias fixed is not explicitly taken into account in the model Hamiltonian but introduced as a boundary condition.

We can imagine situations in which the macroscopic wires connected at the quantum dot are at different temperatures. Additionally, the time-dependent forces do make work on the system. Thus, we can regard the quantum pump as a microscopic engine where heat can be exchanged between two sources at different temperatures, while work is provided to the system and part of the energy is dissipated. As we will show, such an engine could even operate as a refrigerator, where there is a net heat flow from the reservoir at the lowest temperature to the one at the highest temperature. The goal of this chapter is to introduce theoretical tools for the analysis of the fundamental conservation laws at the microscopic scale, the explicit evaluation of the power developed by the intervening forces, and the distinction between reversible flows and dissipated energy.

This chapter is organized as follows. In Section 38.2, we introduce a simple microscopic model for a quantum pump and we discuss the fundamental principle of the conservation of the charge and the energy in this device. In Section 38.3, we introduce the basics of two well-established many-body techniques

to treat quantum transport in harmonically time-dependent systems. The first one starts from the explicit microscopic Hamiltonian for the system, forces, and environment, which is solved by recourse to nonequilibrium Green's functions. The second one is based on the notion of scattering processes that the electrons experience as they cross through a quantum system under ac driving. Although this chapter is self-contained, we do not include a complete tutorial on these two techniques but we adopt a practical point of view, presenting just the main ideas while we defer the reader to more specific literature on many-body techniques for further details. In Section 38.4, we present explicit expressions for the energy flows in terms of Green's functions and the scattering matrix elements introduced in Section 38.3. We also discuss the nature of the different components contributing to the total energy flow. Section 38.5 is devoted to a summary of the different operating modes that we were able to identify in our quantum engine. Finally, in Section 38.6, we conclude with a discussion of the possible directions to extend these ideas.

38.2 Background

38.2.1 Model

We start by defining explicit Hamiltonians to describe the quantum electronic system as well as its environment. For the sake of simplicity, let us focus on a system like the one sketched in Figure 38.2 where the central system C, on which the time-dependent forces are acting, is placed between two wires: one located at the left (L) and the other at the right (R) of C. We can identify C with the quantum dot of Figure 38.1, with the two ac voltages applied at the two extra leads connected at the wall. The L and R reservoirs of Figure 38.3 correspond to the up and down ones in Figure 38.1.

Along this chapter, we make the following simplifying assumptions on the system: (i) We take into account only the two wires as the external environment for the central system and disregard other effects like, for example, the influence of the phonons of the substrate. Such a simplification is expected to be reasonable, if we concentrate on the behavior at sufficiently low

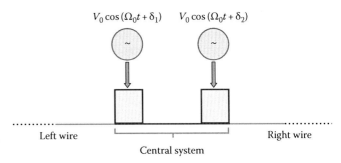

$V_0\cos(\Omega_0 t + \delta_1)$ $V_0\cos(\Omega_0 t + \delta_2)$

Left wire   Central system   Right wire

FIGURE 38.2 Sketch of the setup. The central system is connected to two infinite wires, which play the role of macroscopic reservoirs. In this example, the central system contains a profile of two barriers at which there are ac fields that oscillate with the same amplitude V_0 and frequency Ω_0, and a phase lag $\delta_2 - \delta_1$.

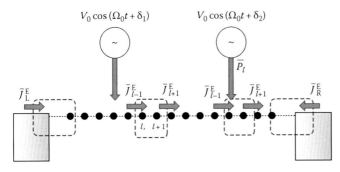

FIGURE 38.3 Analysis of the energy balance in our setup. The dashed boxes enclose the elementary volume of the system where the evolution of the energy is studied. It contains two sites of the underlying lattice. The arrows indicate the direction that we have defined as positive for energy currents along the different pieces of the system as well as the powers done by the fields.

temperatures. (ii) We also assume that the electrons do not experience any kind of many-body interactions, like the Coulomb repulsion between electrons. This assumption is justified in the description of the metallic wires where we can expect an efficient screening, but it is justified within the system C only when the structure is strongly connected to the wires and allows for the screening of the wires to penetrate into it. In this context, the spin degrees of freedom of the electrons behave independent of one another. For this reason, in order to simplify the notation, we do not consider them explicitly. In the case of considering them, we must simply write a factor 2 in front of the final expressions for the currents and densities. The full Hamiltonian reads

$$H(t) = H_{\rm L} + H_{\rm C}(t) + H_{\rm R} + H_{\rm c}, \qquad (38.1)$$

where

$$H_\alpha = \sum_{k\alpha} \varepsilon_{k\alpha} c^\dagger_{k\alpha} c_{k\alpha}, \qquad (38.2)$$

for $\alpha = $ L, R are Hamiltonians of free spinless electrons that represent the wires. We stress that these systems are macroscopic, i.e., they contain a very large number of degrees of freedom and are in thermodynamic equilibrium. This means that they are completely characterized by their density of states

$$\rho_\alpha(\omega) = 2\pi \sum_{k\alpha} \delta(\omega - \varepsilon_{k\alpha}/\hbar), \qquad (38.3)$$

and the Fermi distribution function

$$f_\alpha(\hbar\omega) = \frac{1}{e^{\beta_\alpha(\hbar\omega - \mu_\alpha)} + 1}, \qquad (38.4)$$

with $\beta_\alpha = k_{\rm B} T_\alpha$. The second term, $H_{\rm C}(t)$, describes the quantum structure under consideration as well as the time-dependent gate potentials acting on it. The ensuing Hamiltonian depends on the geometry of the structure as well as on the interactions

that we want to take into account. In the absence of many-body interactions, this system may be described by a single-particle Hamiltonian of the form

$$H_{\rm C}(\mathbf{r},t) = -\frac{\hbar^2 \nabla^2}{2m} + U(\mathbf{r}) + \sum_{l=1}^{M} \delta(\mathbf{r} - \mathbf{R}_l) e V_l(t), \qquad (38.5)$$

m being the mass of an electron, which corresponds to a finite number of time-dependent potentials that we assume to have the simple single-harmonic dependence: $V_l(t) = V_0 \cos(\Omega_0 t + \delta_l)$. The potential $U(\mathbf{r})$ contains the information of the confining walls, barriers, and defects of the structure. For a detailed discussion of the conservation laws and for treating the problem with Green's function formalism, it is convenient to express this Hamiltonian in second quantization. To this end, we must define an appropriate single-particle basis to represent the relevant operators. As the structure under study occupies a reduced region of the space, it is comfortable to work with a single-particle basis that is labeled by spacial coordinates, like that defined by the Wannier functions. It is, thus, useful to work on a discrete lattice containing a finite number (N) of sites and a basis of single-electron states that are localized on the lattice positions. The resulting Hamiltonian in second quantization corresponds to a tight-binding model. For simplicity, we consider this model in one dimension (1D), although this is not an essential assumption:

$$H_{\rm C}(t) = \sum_{l=1}^{N} [\varepsilon_l + e V_l(t)] c^\dagger_l c_l - w \sum_{l=1}^{N-1} (c^\dagger_l c_{l+1} + H \cdot c), \qquad (38.6)$$

where the term with $w = \langle l|(\hbar^2/2m)\partial^2/\partial x^2|l+1\rangle$, $|l\rangle$ being the single-electron basis state localized at the lattice position, "l," describes the kinetic energy of the electrons through jumping processes between nearest-neighbor sites of the underlying lattice. The term with $\varepsilon_l = \langle l|U(x)|l\rangle$ defines a static energy profile for the structure: it contains the information of the existence of barriers and wells. For a system with impurities, this profile can be defined in terms of a random amplitude and this model reduces to the Anderson model. The term with $V_l(t) = \langle l|\sum_{j=1}^{M} \delta(x_l - x_j) V_0 \cos(\Omega_0 t + \delta_j)|l\rangle$ represents the time-dependent gates, being finite at the M pumping centers and vanishing otherwise. Finally, the term $H_{\rm c}$ describes the contacts between the central system and the reservoirs. In our simple 1D model for the central structure, the L lead is connected to the first site, $l = 1$, and the R to the last site, $l = N$, of the central structure. The Hamiltonian reads

$$H_{\rm c} = -w_{\rm cL} \sum_{k{\rm L}} (c^\dagger_{k{\rm L}} c_1 + H \cdot c) - w_{\rm cR} \sum_{k{\rm R}} (c^\dagger_{k{\rm R}} c_N + H \cdot c), \qquad (38.7)$$

which describes hopping processes between the states $k\alpha$ within the wires and the points of C at which the contacts between the two systems are established.

Before closing this subsection, let us mention that, depending on the physical problem under consideration, there may be other time-dependent terms in the Hamiltonian. For systems with ac voltages applied at the L and R wires, we should consider a dispersion relation with a time-dependent component in addition to the static one, $\varepsilon_{k\alpha}^0$. By recourse to a gauge transformation, it can be seen that this type of ac voltages can, equivalently, be included in a time-dependent phase in the contact hopping w_c (see Jauho et al. 1994). Another possible time-dependent term is that originated by an electric field derived from a time-dependent vector potential $\mathbf{A}(t)$ (see Arrachea 2002). That physical situation would take place, for example, when the central system is bended and closed into an annular geometry threaded by a time-dependent magnetic flux. In terms of the Hamiltonian, this introduces a shift: $\mathbf{p} \rightarrow \mathbf{p} - (e/c)\mathbf{A}(t)$ in the momentum of the electrons, c being the velocity of light. In the tight-binding basis this translates into time-dependent phases in the hopping parameter w of $H_C(t)$ along with the periodic boundary condition $N + 1 \equiv 1$. Finally, the study of the coupling to external classical radiation fields is usually treated within the so-called dipolar approximation, which in our second quantization language results in a diagonal voltage profile $V_j(t)$ as in Equation 38.6 with $V_j \sim jV_0$, with V_0 constant and $\delta_j = \delta$, $\forall j$ (see Kohler et al. 2005).

38.2.2 Conservation Laws and Instantaneous Currents

38.2.2.1 Particle Currents and the Conservation of the Charge

A consistent way to define expressions for the electronic currents along the different pieces of the structure is starting from the evolution of the electronic density $n_l = c_l^\dagger c_l$. The variation of n_l is due to the difference between the charge flow exiting and entering the infinitesimal volume that encloses that point

$$-e\frac{\mathrm{d}}{\mathrm{d}t}\langle n_l \rangle = J_l(t) - J_{l-1}(t), \qquad (38.8)$$

where $J_l(t)$ denotes the current exiting the site l toward the neighboring site $l+1$. We denote with a positive sign the flows pointing from left to right. The variation in time of the local charge can be calculated within the Heisenberg picture by recourse to the Eherenfest theorem:

$$e\frac{\mathrm{d}}{\mathrm{d}t}\langle n_l \rangle = -\frac{ie}{\hbar}\langle [H, c_l^\dagger(t)c_l(t)] \rangle. \qquad (38.9)$$

Thus, the explicit evaluation of the above commutator defines an explicit expression for the current. If we consider a site within C, we obtain

$$J_l(t) = \frac{iew}{\hbar}\langle c_l^\dagger(t)c_{l+1}(t) - c_{l+1}^\dagger(t)c_l(t) \rangle. \qquad (38.10)$$

It is easy to verify that the above expression coincides with the mean value of the operator $e\mathbf{v}$, $\mathbf{v}$ being the velocity expressed in second quantization in the basis of localized functions. Similarly, if we consider the contact between C and one of the reservoirs, we get the following expression for the current that exits the reservoir α:

$$J_\alpha(t) = \frac{iew_{c\alpha}}{\hbar}\sum_{k\alpha}\langle c_{k\alpha}^\dagger(t)c_{l\alpha}(t) - c_{l\alpha}^\dagger(t)c_{k\alpha}(t) \rangle, \qquad (38.11)$$

where $l\alpha = 1, N$ for $\alpha = L, R$.

38.2.2.2 Energy Currents, Power, and the Conservation of the Energy

In order to define energy currents, we proceed along a similar line as before. In this case, we analyze the evolution of the energy density at a given elementary volume and write the equation of the conservation of the energy. As our Hamiltonian $H_C(t)$ contains terms involving positions up to nearest neighbors, the smallest volume for analyzing the evolution of the energy density in our 1D lattice is that enclosed by a box confining a bond of nearest-neighbor sites (see Figure 38.3). If we denote by $E_{l,l+1}$ the total energy stored within such a box, the equation for the conservation of the energy reads

$$\frac{\mathrm{d}E_{l,l+1}}{\mathrm{d}t} = J_{l+1}^E(t) - J_l^E(t) + P_l(t) + P_{l+1}(t), \qquad (38.12)$$

where the first two terms denote the difference between outgoing (from $l+1$ to $l+2$) and incoming (from $l-1$ to l) flows, with the same sign convention as in the case of charge flows, while the last two terms denote the power done by the external fields, which are defined as positive when it is provided by the forces. The latter terms vanish if the time-dependent gate potentials are not acting at the points l and $l+1$ enclosed by the box. Our box can also enclose the contact bond between the reservoir α and the central system (see Figure 38.3), in which case we get

$$\frac{\mathrm{d}E_{L,1}}{\mathrm{d}t} = J_1^E(t) - J_L^E(t) + P_1(t),$$

$$\frac{\mathrm{d}E_{N,R}}{\mathrm{d}t} = -J_R^E(t) - J_{N-1}^E(t) + P_N(t), \qquad (38.13)$$

where $J_\alpha^E(t)$ denotes the energy flow that exits the reservoir α. As in Section 38.2.2.1, the explicit expressions for the flows and powers are derived by recourse to the Eherenfest theorem:

$$\frac{\mathrm{d}E_{l,l+1}}{\mathrm{d}t} = -\frac{i}{\hbar}\Big\langle [H, (\varepsilon_l(t)c_l^\dagger(t)c_l(t) + \varepsilon_{l+1}(t)c_{l+1}^\dagger(t)c_{l+1}(t)$$
$$-w\{c_l^\dagger(t)c_{l+1}(t) + c_{l+1}^\dagger(t)c_l(t)\})] \Big\rangle + e\frac{\mathrm{d}V_l(t)}{\mathrm{d}t}\langle c_l^\dagger(t)c_l(t) \rangle$$
$$+e\frac{\mathrm{d}V_{l+1}(t)}{\mathrm{d}t}\langle c_{l+1}^\dagger(t)c_{l+1}(t) \rangle, \qquad (38.14)$$

with $\varepsilon_l(t) = \varepsilon_l^0 + eV_l(t)$, and a similar expression for the volume enclosing the contact bonds between C and the reservoirs. From the evaluation of the previous terms, we obtain the explicit expressions for the energy currents

$$J_{l+1}^E(t) = \frac{iw}{\hbar}\left[w\left\langle c_l^\dagger(t)c_{l+2}(t) - c_{l+2}^\dagger(t)c_l(t)\right\rangle \right.$$
$$\left. - \varepsilon_{l+1}(t)\left\langle c_{l+1}^\dagger(t)c_{l+2}(t) - c_{l+2}^\dagger(t)c_{l+1}(t)\right\rangle \right]; \quad (38.15)$$

$$J_\alpha^E(t) = \frac{iw_{c\alpha}}{\hbar}\sum_{k\alpha}\varepsilon_{k\alpha}\left\langle c_{k\alpha}^\dagger(t)c_{l\alpha}(t) - c_{l\alpha}^\dagger(t)c_{k\alpha}(t)\right\rangle, \quad (38.16)$$

and power developed by the ac voltages

$$P_l(t) = e\frac{dV_l(t)}{dt}\left\langle c_l^\dagger(t)c_l(t)\right\rangle. \quad (38.17)$$

38.2.3 Continuity Equations, DC Charge and Energy Currents and Mean Powers

In the absence of sinks and sources, the average of the charge enclosed by any volume of the sample over one cycle with period $\tau = 2\pi/\Omega_0$ must remain constant, which defines the following continuity equation for the dc charge current (microscopic Kirchoff law):

$$\bar{J}_l = \bar{J}_{l'} = J, \quad (38.18)$$

for arbitrary l, l' along C, where we denote $\bar{A} = 1/\tau\int_0^\tau dtA(t)$.

Analogously, for any volume enclosing lattice points running from $l+1, ..., l'$, we get

$$\bar{J}_l^E = \bar{J}_{l'}^E + \sum_{j=l+1}^{l'}\bar{P}_j, \quad (38.19)$$

where the last term defines the power done by all the voltages enclosed by the volume. The above equation reduces to a "Kirchoff law" for the dc energy current when we enclose a region that is free from the time-dependent voltages, in which case the last term vanishes.

38.2.4 Heat Current

At this point, it is important to mention that the dc energy current $\bar{J}^E$ defined above does not necessarily coincide with the heat current which we will denote J^Q. This is because what is understood as "heat" is usually the energy transferred from one system to another as a consequence of a temperature difference. In order to understand this difference, let us consider that our reservoirs are at temperature $T = 0$, let us place our volume enclosing the contact between one reservoir and the central system and let us assume that we are not enclosing an ac local voltage within it.

Now, let us analyze the following picture based on heuristic arguments that we shall better formalize in subsequent sections. Let us assume that the ac voltages are so weak in amplitude and oscillate with such a low frequency that we can disregard the power they develop. Let us assume that, anyway, they are able to move a small portion of electrons with energies very close to the Fermi energy of the reservoirs μ, that we recall is the same for L and R. This weak motion gives place to currents of particles and energy which, from the definitions (38.11) and (38.16), are approximately related through $J_\alpha^E(t) \sim (\mu/e)J_\alpha(t)$. The same relation holds for L as well as R reservoirs and a small current of particles may translate in a large current of energy, since μ can be large. The dc component of the charge current may be finite and should be positive in one reservoir and negative in the other one, indicating that there is a net flow of charge between L and R or vice versa. The above relation, therefore, tells us that there is a concomitant net flow of energy from a reservoir to the other one, in spite of the fact that we are assuming that both reservoirs are at zero temperature. One could complain that we have not taken into account the power done by the time-dependent fields which would tend to heat the system. However, let us recall that an appropriate choice of μ would allow us an arbitrary large value of the energy flow, against which we can disregard a contribution like (38.17) for weak and slow ac voltages. In summary, the above considerations lead us to conclude that there may exist a net energy flow which cannot be identified as "heat" but has rather a convective nature. In order to better quantify heat, it is thus natural to subtract from the energy flow, the convective component $(\mu/e)J$, i.e.,

$$J_l^Q = \bar{J}_l^E - \frac{\mu}{e}J, \quad J_\alpha^Q = \bar{J}_\alpha^E - \frac{\mu}{e}J. \quad (38.20)$$

Notice that, while the convective term is constant along all the pieces of the system due to the continuity of the charge, the heat and energy currents may have different values due to the contribution of the power done by the external voltages, see Equation 38.19.

To give a more formal definition of the heat flow, J_α^Q, let us consider the particle and the energy balances for a given reservoir α that is kept fixed at both the chemical potential μ_α and the temperature T_α. If the charge current $\bar{J}_\alpha$ and the energy flow $\bar{J}_\alpha^E$ enter the reservoir α, then the number of particles and the energy of a reservoir should change. This, in turn, would change the chemical potential and the temperature of a reservoir. In order to keep μ_α fixed, the electrons have to be removed with the rate $\bar{J}_\alpha/e$ out of the reservoir. Therefore, while keeping μ_α constant, we necessarily remove energy with rate $\mu_\alpha\bar{J}_\alpha/e$. We stress that the convective energy $\mu_\alpha\bar{J}_\alpha/e$ is taken out at equilibrium conditions, therefore, it can be reversibly given back. In general, this energy does not coincide with energy flow $\bar{J}_\alpha^E$ entering the reservoir. To prevent heating of reservoir one needs additionally to take out the energy with rate $J_\alpha^Q = \bar{J}_\alpha^E - \mu_\alpha\bar{J}_\alpha/e$ without taking out particles. Since the reservoir cannot produce work, the only way to remove the remaining energy is to put it in contact with other large body playing the role of a thermostat. The energy exchange between the reservoir α and the thermostat is essentially irreversible. For this reason, we interpret this part of the total energy as "heat" and we identify J_α^Q as the heat flow. If the thermostat would be absent, the temperature of a reservoir would change. As we will show J_α^Q can be directed either to the reservoir from the central system or back. Hence, in the absence of a thermostat, the reservoir can be either heated or cooled. We stress that the energy transported at the rate J_α^Q becomes "heat" only deep inside the macroscopic reservoir, where the electrons scattered by the dynamical central system, are able to equilibrate.

38.3 State of the Art

In this section, we briefly review the many-body techniques to evaluate the currents and powers defined in Section 38.2.

38.3.1 Green's Functions Formalism

38.3.1.1 Expectation Values of Observables and Green's Functions

The expectation value of any one-body observable, $\langle A(t) \rangle = \sum_{l,l'} \langle l'|A(t)|l \rangle \langle c_{l'}^\dagger(t)c_l(t) \rangle$, can be regarded as follows:

$$\langle \hat{A}(t) \rangle = -i \lim_{t' \to t} \sum_{l,l'} \langle l'|A(t)|l \rangle G_{l,l'}^<(t,t'), \qquad (38.21)$$

where

$$G_{l,l'}^<(t,t') = i \langle c_{l'}^\dagger(t')c_l(t) \rangle, \qquad (38.22)$$

a "lesser" Green's function. Our goal, now, is to derive equations for the evolution of this Green's function and strategies to solve them.

38.3.1.2 Brief Review of the Theory of the Nonequilibrium Green's Functions

The formal theory of nonequilibrium Green's function has been developed independently by Kadanoff and Baym (1962), Schwinger (1961), and Keldysh (1964). The structure of that theory is very similar to the one of causal Green's functions at zero temperature (see Mahan 1990), except for the fact that in nonequilibrium situations, the assumption that the state of the system at time $+\infty$ differs just in a phase from the state in $-\infty$ does not hold any longer. The way to overcome this inconvenience is to define the evolution along a special contour C that defines a round trip, first going from $-\infty$ to $+\infty$ and then going back to $-\infty$. As in equilibrium problems, the precise description of that evolution can be accomplished with the help of Wick's theorem and Feynman diagrams and one of the big powers of this technique is the possibility of treating many-body interactions in a systematic way.

We skip here the technical just highlighting the main ideas leading to some useful identities, and we defer the reader to more specialized literature (see Caroli et al. 1971, Mahan 1990, Pastawski 1992, Jauho et al. 1994, Haug and Jauho 1996). Instead of the time-ordering operator used in the equilibrium theory, it is convenient to work with contour-ordered Green's functions:

$$G(1,1') = -i \langle T_C[c(1)c^\dagger(1')] \rangle, \qquad (38.23)$$

where 1, 1' is a schematic notation that labels the electronic degrees of freedom and time in the same index and the operator T_c denotes time-ordering along a contour that begins in $-\infty$ evolves to $+\infty$ (C_+) and then turns back from $+\infty$ to $-\infty$ (C_-). This function corresponds to the casual "time-ordered," the "lesser," the "anti-time-ordered" or "greater" function depending on the position of the two times along the closed time-contour:

$$G(1,1') = G_c(1,1'), \ t_1, t_{1'} \in C_+, \ G(1,1') = G^<(1,1'), \ t_1, \in C_+, t_{1'} \in C_-,$$
$$G(1,1') = G_{\bar{c}}(1,1'), \ t_1, t_{1'} \in C_-, \ G(1,1') = G^<(1,1'), \ t_1, \in C_-, t_{1'} \in C_+,$$
$$(38.24)$$

with

$$G_c(1,1') = -i\Theta(t_1 - t_{1'}) \langle c_1(t_1)c_1^\dagger(t_1') \rangle + i\Theta(t_{1'} - t_1) \langle c_1^\dagger(t_1')c_1(t_1) \rangle,$$
$$G_{\bar{c}}(1,1') = -i\Theta(t_{1'} - t_1) \langle c_1(t_1)c_1^\dagger(t_1') \rangle + i\Theta(t_1 - t_{1'}) \langle c_1^\dagger(t_1')c_1(t_1) \rangle,$$
$$G^<(1,1') = i \langle c_1^\dagger(t_1')c_1(t_1) \rangle, \quad G^<(1,1') = -i \langle c_1(t_1)c_1^\dagger(t_1') \rangle,$$
$$(38.25)$$

which are not independent functions, but satisfy $G_c + G_{\bar{c}} = G^< + G^>$. It is also convenient to define "retarded" and "advanced" functions:

$$G^R(1,1') = \Theta(t_1 - t_{1'})[G^>(1,1') - G^<(1,1')],$$
$$G^A(1,1') = \Theta(t_{1'} - t_1)[G^<(1,1') - G^>(1,1')],$$
$$(38.26)$$

which are also related through $G^R - G^A = G^> - G^<$, $G^A(1, 1') = [G^R(1', 1)]^*$ and $G^<(1, 1') = -[G^>(1', 1)]^*$.

In our simple model of noninteracting electrons, we can split by convenience the Hamiltonian in two parts $H(t) = H_0(t) + H'(t)$, with both terms being of one-body type but H_0 being easily solved. The evolution of this Green's function is given by Dyson's equation, which for a one-body Hamiltonian reads

$$G(1,1') = G_0(1,1') + \hbar^{-1} \int_c d^3x_2 dt_2 G(1,2)H'(2)G_0(2,1'). \quad (38.27)$$

Notice that the above equation actually represents a matricial integral equation, if we distinguish the positions of the times as in (38.25). A convenient tool to derive explicit equations for the different components (38.25) is a theorem due to Langreth, which states given a product of contour-ordered Green's functions of the form:

$$G(t_1, t_{1'}) = \int_c dt_2 G_1(t_1, t_2) G_2(t_2, t_{1'}), \qquad (38.28)$$

then, the following relations hold for the different components:

$$G^{R,A}(t_1, t_{1'}) = \int_{t_{1'}}^{t_1} dt_2 G_1^{R,A}(t_1, t_2) G_2^{R,A}(t_2, t_{1'}), \qquad (38.29)$$

$$G^{<,>}(t_1, t_{1'}) = \int_{-\infty}^{-\infty} dt_2 [G_1^R(t_1, t_2) G_2^{<,>}(t_2, t_{1'}) + G_1^{<,>}(t_1, t_2) G_2^A(t_2, t_{1'})].$$

Therefore, if we want to compute $G^<$, in order to evaluate expectation values of observables, we must also solve two coupled equations for that function and G^R.

38.3.1.3 Green's Functions and Dyson's Equations in Our Problem

Let us first split our Hamiltonian as follows: $H_0(t) = H_L + H_R + H_C(t)$ and $H' = H_c$. Dyson's equation for the retarded function reads

$$G_{j,j'}(t, t') = g_{j,j'}^0(t, t') + \hbar^{-1} \sum_{j1} \int_c dt_1 G_{j,j1}(t, t_1) H_c g_{j1,j'}^0(t_1, t'), \qquad (38.30)$$

where $g_{j,j'}^0(t, t')$ is the contour-ordered Green's function of $H_0(t)$ and j, j' run over all the electronic degrees of freedom of this Hamiltonian. If we write the above equation explicitly for one of indexes in the reservoir and $l \in C$

$$G_{l,k\alpha}(t, t') = -w_c \int_c dt_1 G_{l,l_\alpha}(t, t_1) g_{k\alpha,k\alpha}^0(t_1, t'). \qquad (38.31)$$

For the two indexes $l, l' \in C$

$$G_{l,l'}(t, t') = g_{l,l'}^0(t, t') - w_c \sum_{\alpha = L, R} \int_c dt_1 G_{l,k\alpha}(t, t_1) g_{l\alpha,l'}^0(t_1, t'),$$

$$= g_{l,l'}^0(t, t') + \hbar^{-1} \sum_{\alpha = L, R} \int_{-\infty}^{\infty} dt_1 dt_2 G_{l,l_\alpha}(t, t_1) \Sigma_\alpha(t_2) g_{l\alpha,l'}^0(t_2, t'), \qquad (38.32)$$

where going from the first to the second identity we have substituted (38.31). We have also defined the "self-energy":

$$\Sigma_\alpha(t_1, t_2) = |w_{c\alpha}|^2 \sum_{k\alpha} g_{k\alpha,k\alpha}^0(t_1, t_2). \qquad (38.33)$$

In order to evaluate the currents that we have defined in Section 38.2.3, we need $G_{l,l'}^<(t, t')$ for $l, l' \in C$ and $G_{k\alpha,l\alpha}^{<,>}(t, t')$. Applying the Langreth's rules (38.29) to the above equations, we get (see Arrachea 2002, 2005)

$$G_{l,k\alpha}^{<,>}(t, t') = -w_{c\alpha} \int_{-\infty}^{+\infty} dt_1 [G_{l,l_\alpha}^R(t, t_1) g_{k\alpha,k\alpha}^{0,<,>}(t_1, t') + G_{l,l_\alpha}^{<,>}(t, t_1) g_{k\alpha,k\alpha}^{0,A}(t_1, t')], \qquad (38.34)$$

along the contact and

$$G_{l,l'}^{<,>}(t, t') = \hbar^{-1} \sum_\alpha \int_{-\infty}^{+\infty} dt_1 dt_2 G_{l,l_\alpha}^R(t, t_1) \Sigma_\alpha^{<,>}(t_1, t_2) G_{l\alpha,l'}^A(t_2, t'), \qquad (38.35)$$

for coordinates l, l' belonging to the central system C. The latter equation is obtained after some algebra and after dropping a term that contains $g_{l,l'}^{0,<}(t, t')$ which can be shown to be relevant only in the description of transient behavior (see Jauho et al. 1994). The different components of $G^0(t, t')$ within the reservoirs are straightforwardly evaluated:

$$g_{k\alpha,k\alpha}^{0,R}(t_1, t_2) = -i\Theta(t_1 - t_2) \exp\{-i(\varepsilon_{k\alpha}/\hbar)(t_1 - t_2)\},$$
$$g_{k\alpha,k\alpha}^{0,<,>}(t_1, t_2) = \lambda_\alpha^{<,>}(\varepsilon_{k\alpha}) \exp\{-i(\varepsilon_{k\alpha}/\hbar)(t_1 - t_2)\}, \qquad (38.36)$$

with $\lambda_\alpha^<(\varepsilon_{k\alpha}) = if_\alpha(\varepsilon_{k\alpha})$ and $\lambda_\alpha^>(\varepsilon_{k\alpha}) = -i[1 - f_\alpha(\varepsilon_{k\alpha})]$. With these functions, it is possible to obtain expressions for the different components of Equation 38.33 in terms of the density of states of the reservoir $\rho_\alpha(\omega)$ given in (38.3)

$$\Sigma_\alpha^R(t_1, t_2) = -i\Theta(t_1 - t_2) \int_{-\infty}^{+\infty} \frac{d\omega}{2\pi} e^{-i\omega(t_1 - t_2)} \Gamma_\alpha(\omega),$$

$$\Sigma_\alpha^{<,>}(t_1, t_2) = \int_{-\infty}^{+\infty} \frac{d\omega}{2\pi} e^{-i\omega(t_1 - t_2)} \lambda_\alpha^{<,>}(\hbar\omega) \Gamma_\alpha(\omega), \qquad (38.37)$$

with $\Gamma_\alpha(\omega) = |w_{c\alpha}|^2 \rho_\alpha(\omega)$.

To evaluate (38.34) and (38.35) we still have to calculate the retarded function within the system C (recall that the advanced function can be obtained from $G_{l,l'}^A(t, t') = [G_{l',l}^R(t', t)]^*$). The equation for the retarded function is the retarded component of Equation 38.32 and can be derived by applying Langreth rules on this equation. The result is

$$G_{l,l'}^R(t, t') = g_{l,l'}^{0,R}(t, t') + \hbar^{-1} \sum_\alpha \int_{t'}^t dt_1 dt_2 G_{l,l_\alpha}^R(t, t_1) \Sigma_\alpha^R(t_1, t_2) g_{l\alpha,l'}^{0,R}(t_2, t') \qquad (38.38)$$

with $g_{l,l'}^{0,R}(t, t')$ being the retarded Green's function of the system described only by $H_C(t)$ isolated from the reservoirs. This function is, in turn, still an unknown in our problem. Instead of explicitly evaluating it, we find it more convenient to operate with (38.38) in order to find an equivalent equation for $\hat{G}^R(t, t')$ as follows. We first derive an equation of motion for $g_{l,l'}^{0,R}(t, t')$, starting from the very definition of the retarded function, see Equation 38.26, and by writing the evolution of $c_j(t)$ with $H_C(t)$ in the Heisenberg representation:

$$i\hbar c_j(t) = \left[H_C(t), c_j(t) \right], \qquad (38.39)$$

we get

$$-i\hbar \frac{\partial}{\partial t'} \hat{g}^{0,R}(t,t') - \hat{g}^{0,R}(t,t') \hat{H}_C(t') = \hbar \hat{1} \delta(t-t'), \quad (38.40)$$

where

$\hat{g}^{0,R}(t,t')$ is a $N \times N$ matrix with elements $g_{l,l'}^{0,R}(t,t')$
$\hat{1}$ is the $N \times N$ identity matrix

The above equation means that $\{-i\hbar\partial/\partial t' - \hat{H}_C(t')\} = [\hat{g}^{0,R}]^{-1}$. Therefore, we have not evaluated explicitly the function $\hat{g}^{0,R}(t,t')$ but we have identified an operator which is its inverse. We act with this operator from the right of (38.38) and we consider the following splitting of the central Hamiltonian $H_C(t) = H_C^0 + H_C'(t)$, where $H_C(t)$ collects all the explicit time-dependent terms of $H_C(t)$ and H_C^0, the remaining ones. We get

$$-i\hbar \frac{\partial}{\partial t'} \hat{G}^R(t,t') - \hat{G}^R(t,t') \hat{H}_C^0$$

$$- \int dt_1 \hat{G}^R(t,t_1) \hat{\Sigma}^R(t_1,t') - \hat{G}^R(t,t') H_C'(t') = \hbar \hat{1} \delta(t-t'),$$
$$(38.41)$$

$\Sigma_{l,l'}^R(t,t') = \sum_\alpha \delta_{l,l\alpha} \delta_{l',l\alpha} \Sigma_\alpha^R(t,t')$ being the matrix elements of $\hat{\Sigma}(t,t')$. We define a function $\hat{G}^{0,R}(t,t')$, such that $[\hat{G}^{0,R}]^{-1} = \{-i\hbar\partial/\partial t' - \hat{H}_C^0 - \hat{\Sigma}^R\}$:

$$-i\hbar \frac{\partial}{\partial t'} \hat{G}^{0,R}(t,t') - \hat{G}^{0,R}(t,t') \hat{H}_C^0$$

$$- \int dt_1 \hat{G}^{0,R}(t,t_1) \hat{\Sigma}^R(t_1,t') = \hbar \hat{1} \delta(t-t'). \qquad (38.42)$$

Multiplying (38.38) by the right with $\hat{G}^{0,R}$, we finally find the following equation for the full retarded Green's function

$$\hat{G}^R(t,t') = \hat{G}^{0,R}(t,t') + \hbar^{-1} \int_{t'}^{t} dt_1 \hat{G}^R(t,t_1) H_C'(t_1) \hat{G}^{0,R}(t_1,t'). \quad (38.43)$$

This equation is completely equivalent to (38.38), but has the advantage that the function $\hat{G}^{0,R}$ is an equilibrium Green's function, which evolves according to the stationary terms of the full Hamiltonian H. The above expression has a structure which is particularly adequate for perturbative solutions in the time-dependent part of $H_C(t)$. We shall exploit this property later.

Equation 38.42 can be easily solved by performing the Fourier transform: $\hat{G}^{0,R}(\omega) = \hbar^{-1} \int_0^\tau d\tau e^{-i(\omega+i\eta)\tau} \hat{G}^{0,R}(\tau)$, with $\eta > 0$, since, as we have mentioned before, it corresponds to an equilibrium Green's function that depends on $t - t'$. The result is

$$\hat{G}^{0,R}(\omega) = [\hbar(\omega+i\eta)\hat{1} - \hat{H}_C^0 - \hat{\Sigma}^R(\omega)]^{-1}, \qquad (38.44)$$

and can be explicitly evaluated by simply inverting the above $N \times N$ complex matrix. Substituting in Equation 38.43 and

performing a Fourier transform in $t - t'$, Equation 38.43 results in our specific Hamiltonian (38.6):

$$\hat{G}^R(t,\omega) = \hat{G}^{0,R}(\omega) + \sum_{k=\pm} e^{-ik\Omega_0 t} \hat{G}^R(t,\omega+k\Omega_0) \hat{V}(k) \hat{G}^{0,R}(\omega) \quad (38.45)$$

where the matrix $\hat{V}(1)$ contains elements $V_{l,l'} = \delta_{l,l'} \sum_{j=1}^{M} \delta(x_l - x_j) eV_0 e^{-i\delta_j}$ and $\hat{V}(-1) = [\hat{V}(1)]^*$. The linear set (38.45) has the same structure as the dynamics of a problem in which electrons with a given energy $\hbar\omega$ interact with a potential V emitting or absorbing an energy quantum $\hbar\Omega_0$ and scatter with a final energy $\hbar\omega \pm \hbar\Omega_0$. The solution of (38.45) leads to the complete solution of the problem.

Due to the harmonic dependence on the time t of these equations, the retarded Green's function can be expanded in a Fourier series as follows:

$$\hat{G}^R(t,\omega) = \sum_{n=-\infty}^{+\infty} \hat{\mathcal{G}}(n,\omega) e^{-in\Omega_0 t}. \qquad (38.46)$$

We give the name of *Floquet component* to the functions $\hat{\mathcal{G}}(n,\omega)$, because Equation 38.46 has a similar structure as that proposed by Floquet for the wave functions of time-periodic Hamiltonians. The different components obey the following useful identity

$$\hat{\mathcal{G}}(n,\omega) - \hat{\mathcal{G}}^\dagger(-n,\omega_n) = -i \sum_{n'} \hat{\mathcal{G}}(n+n',\omega_{-n'}) \hat{\Gamma}(\omega_{-n'}) \hat{\mathcal{G}}(n',\omega_{-n'})^\dagger,$$
$$(38.47)$$

where we have introduced the following notation $w_n = w + n\Omega_0$. To prove Equation 38.47, we start from the definition (Equation 38.26) of the retarded Green's function for indexes $l, l' \infty C$. Replacing (38.35) and inserting there the representation (Equation 38.46), we get

$$\hat{G}^R(t,t') = -i\hbar\Theta(t-t') \sum_{k_1,k_2} \int_{-\infty}^{+\infty} \frac{d\omega}{2\pi} e^{-i[\omega(t-t')+\Omega_0(k_1 t - k_2 t')]}$$

$$\times \hat{\mathcal{G}}(k_1,\omega) \hat{\Gamma}(\omega) \hat{\mathcal{G}}^\dagger(k_2,\omega), \qquad (38.48)$$

where $\hat{\Gamma}(\omega)$ contains as matrix element $\Gamma_{l,l'}(\omega) = \delta_{l,l\alpha} \delta_{l',l\alpha} \Gamma_\alpha(\omega)$. Calculating the Fourier transform of this function with respect to $t - t'$ and collecting the nth Fourier coefficient (Equation 38.46) we find

$$\hat{\mathcal{G}}(n,\omega) = \sum_{n'} \int_{-\infty}^{+\infty} \frac{d\omega'}{2\pi} \frac{\hat{\mathcal{G}}(n+n',\omega') \hat{\Gamma}(\omega') \hat{\mathcal{G}}(n',\omega')^\dagger}{\omega - \omega'_{n'} + i\eta}, \quad \eta > 0,$$
$$(38.49)$$

which leads to the identity (Equation 38.47) using

$$\frac{1}{\omega - \omega' + i\eta} = \mathcal{P} \left\{ \frac{1}{\omega - \omega'} \right\} - i\pi\delta(\omega-\omega'). \qquad (38.50)$$

It is important to remark that, for $V_0 = 0$ the identity (Equation 38.47) reduces to the following identity between equilibrium Green's functions:

$$\hat{\rho}(\omega) \equiv \hat{G}^{0.R}(\omega) - \hat{G}^{0.R\dagger}(\omega) = -i\hat{G}^{0.R}(\omega)\hat{\Gamma}(\omega)\hat{G}^{0.R}(\omega)^{\dagger}. \quad (38.51)$$

38.3.1.4 Perturbative Solution of Dyson's Equation

Sometimes, in order to derive analytical expressions, it is convenient to solve the set (38.45) by recourse to a perturbative expansion in $\hat{V}$ (see Arrachea 2005). The solution up to second order in this parameter is obtained by writing (38.45) evaluated at ω_n, for $n = -2, ..., 2$ and back-substituting the equation evaluated at ω_2 into the one evaluated at ω_1, and the latter into the one evaluated at ω, and a similar procedure with $\omega_{-2} \to \omega_{-1}$ and the latter into ω. If we then collect all the coefficients of $e^{-in\Omega_0 t}$ in the resulting expression and recall the representation (Equation 38.46), we obtain

$$\hat{G}^{R}(t,\omega) \sim \sum_{n=-2}^{+2} \mathcal{G}(n,\omega)e^{-in\Omega_0 t}, \quad (38.52)$$

with

$$\hat{\mathcal{G}}(0,\omega) = \hat{G}^{0,R}(\omega) + \sum_{k=\pm 1} \hat{\mathcal{G}}(k,\omega)\hat{V}(-k),$$

$$\hat{\mathcal{G}}(\pm 1,\omega) = \hat{G}^{0,R}(\omega_{\pm 1})\hat{V}(\pm 1)\hat{G}^{0,R}(\omega), \quad (38.53)$$

$$\hat{\mathcal{G}}(\pm 2,\omega) = \hat{G}^{0,R}(\omega_{\pm 2})\hat{V}(\pm 1)\hat{\mathcal{G}}(\pm 1,\omega).$$

The reader can easily extend the procedure to evaluate higher-order terms.

38.3.2 Scattering Matrix Formalism

To calculate the charge and energy flows generated by the driven central system in the wires one can also use the scattering approach. Within this approach, we consider the central system as some scatterer which reflects or transmits electrons incoming from the wires. The electrons coming, for instance, from the left wire can be transmitted to the right wire or can be reflected back to the left wire. To find the current in some wire, we need just to calculate the difference between the number of particles incoming through this wire and the number of particles exiting the central system through the same wire. We do not need to know what happened with an electron inside the central system. We only need to know the quantum-mechanical scattering amplitudes for an electron to be transmitted/reflected through/from the central system. The advantage of the scattering approach is the simplicity and the physical transparency of expressions written in terms of scattering amplitudes. We stress that the scattering approach does not aim to calculate the single particle scattering amplitudes. This approach tells us how to calculate the transport properties of a mesoscopic structure coupled to

wires if the scattering amplitudes are known. To calculate the scattering amplitudes one can use Green's functions method. We will give an explicit expression for the scattering amplitudes in terms of corresponding Green's functions. Actually the combining Green's functions – scattering approach is one of the most powerful and practical approaches for transport phenomena in mesoscopic structures.

38.3.2.1 General Formalism

The scattering approach to transport phenomena in small-phase-coherent samples connected to macroscopic reservoirs was introduced and developed by Landauer and Büttiker (Landauer 1957, 1970, 1975, Büttiker 1990, 1992, 1993).

Within this formalism, we consider electrons only in the one-dimensional wires connecting the central system to macroscopic reservoirs. It is convenient to introduce separate operators $a_\alpha(\varepsilon)$ for incoming and $b_\alpha(\varepsilon)$ for scattered electrons with energy ε.

Then the current $J_\alpha(t)$ flowing into wire α to the central system is the following (Büttiker 1992):

$$J_\alpha(t) = \frac{e}{h} \int_{-\infty}^{\infty} d\varepsilon\, d\varepsilon'\, e^{i\frac{\varepsilon-\varepsilon'}{h}t} \left\{ \left\langle a_\alpha^\dagger(\varepsilon)a_\alpha(\varepsilon') \right\rangle - \left\langle b_\alpha^\dagger(\varepsilon)b_\alpha(\varepsilon') \right\rangle \right\}. \quad (38.54)$$

Here $\langle ... \rangle$ denotes averaging over equilibrium states of reservoirs.

Correspondingly the dc current reads

$$\bar{J}_\alpha = \frac{e}{h} \int_{-\infty}^{\infty} d\varepsilon \left\{ f_\alpha(\varepsilon) - f_\alpha^{(\text{out})}(\varepsilon) \right\}, \quad (38.55)$$

where

$f_\alpha^{(\text{out})}(\varepsilon) = \left\langle b_\alpha^\dagger(\varepsilon)b_\alpha(\varepsilon) \right\rangle$ is the distribution function for electrons exiting the central system through the wire α

$f_\alpha(\varepsilon) = \left\langle a_\alpha^\dagger(\varepsilon)a_\alpha(\varepsilon) \right\rangle$ is the distribution function for electrons incoming through the wire α

This expression tells us that the dc current is the difference per unit time between the number of electrons entering and exiting the system. As the reservoir is at equilibrium, the distribution function for the incoming electrons is the Fermi distribution function. In contrast, the scattered electrons, in general, are nonequilibrium particles. To calculate the distribution function for the scattered electrons, we express the b-operators in terms of a-operators. Since an electron coming from any wire can be scattered into a given wire α, then the operators b_α depend on all the operators for the incoming particles. In the model we consider in this chapter, the number of reservoirs is two, $\beta = L, R$. Therefore, $b_\alpha(\varepsilon) = \sum_{\beta=L,R} S_{\alpha\beta}(\varepsilon)a_\beta(\varepsilon)$, where $S_{\alpha\beta}$ are the scattering amplitudes. These amplitudes are normalized in such a way that their square define corresponding currents (Büttiker 1992). The quantities $S_{\alpha\beta}(\varepsilon)$ can be viewed as the elements of some matrix, which is called the scattering matrix $\hat{S}(\varepsilon)$.

38.3.2.2 Floquet Scattering Matrix

If the scatterer is driven by external forces which are periodic in time with period $\tau = 2\pi/\Omega_0$, then while interacting with such a scatterer an electron can gain or lose some energy quanta $n\hbar\Omega_0$, $n = 0, \pm 1, \dots$. Therefore, in this case, the scattering amplitudes in addition to the two wire indexes become dependent on the two energies, one for the incoming and the other for the outgoing electrons. Such a scattering matrix is called the Floquet scattering matrix $\hat{S}_F$ (see, e.g., Platero and Aguado 2004). Their elements, $S_{F,\alpha\beta}(\varepsilon_n, \varepsilon)$, are related to photon-assisted amplitudes for an electron with energy ε entering the scatterer through the lead β and leaving the scatterer with energy $\varepsilon_n = \varepsilon + n\hbar\Omega_0$ through the lead α. Now the relation between the operators b for outgoing particles and a for incoming particles reads (Moskalets and Büttiker 2002a)

$$b_\alpha(\varepsilon) = \sum_{\beta=L,R} \sum_n S_{F,\alpha\beta}(\varepsilon, \varepsilon_n)\, a_\beta(\varepsilon_n), \qquad (38.56)$$

where the sum over n runs over those n for which $\varepsilon_n > \varepsilon_{0\beta}$, and hence it corresponds to propagating (i.e., current-carrying) states. We denote the Floquet scattering matrix as the submatrix corresponding to transitions between the propagating states only. In the case where $\hbar\Omega_0 \ll \varepsilon$, the sum in Equation 38.56 runs over all the integers: $-\infty < n < \infty$. In what follows, we assume this to be the case. Note that if the scatterer is stationary, the only term that remains nonvanishing is that with $n = 0$, and the Floquet scattering matrix is reduced to the stationary scattering matrix with elements $S_{\alpha\beta}(\varepsilon) = S_{F,\alpha\beta}(\varepsilon, \varepsilon)$.

The conservation of the particle current at each scattering event implies that the Floquet scattering matrix is a unitary matrix (Moskalets and Büttiker 2002a, 2004):

$$\sum_{\alpha=L,R} \sum_{n=-\infty}^{\infty} S_{F,\alpha\beta}^*(\varepsilon_n, \varepsilon)\, S_{F,\alpha\gamma}(\varepsilon_n, \varepsilon_m) = \delta_{m0}\,\delta_{\beta\gamma}, \qquad (38.57)$$

$$\sum_{\beta=L,R} \sum_{n=-\infty}^{\infty} S_{F,\alpha\beta}^*(\varepsilon, \varepsilon_n)\, S_{F,\gamma\beta}(\varepsilon_m, \varepsilon_n) = \delta_{m0}\,\delta_{\alpha\gamma}. \qquad (38.58)$$

Using Equation 38.56, we calculate the distribution function for electrons scattered into wire α:

$$f_\alpha^{(out)}(\varepsilon) = \sum_{\beta=L,R} \sum_{n=-\infty}^{\infty} |S_{F,\alpha\beta}(\varepsilon, \varepsilon_n)|^2 f_\beta(\varepsilon_n). \qquad (38.59)$$

This function is not the Fermi distribution function unless the scatterer is stationary and all the reservoirs have the same chemical potentials and temperatures. This reflects the fact that the particles scattered by the dynamical scatterer (quantum pump) are out of equilibrium.

38.3.2.3 Adiabatic Scattering

If the driving forces change slowly, $\Omega_0 \to 0$, they behave as if they were almost constant for the electrons propagating through the central system. For this reason, the scattering properties of a slowly driven (adiabatic) scatterer are close to those of a stationary one. Nevertheless there is an essential difference: in spite of the slow change of the fields, an electron can still absorb or emit one or several energy quanta $\hbar\Omega_0$ in its travel through the central system. Therefore, although the adiabatic scatterer is characterized by the Floquet scattering matrix dependent on two energies, $\hat{S}_F(\varepsilon_n, \varepsilon)$, it is natural to expect that it could be related to the stationary scattering matrix $\hat{S}(\varepsilon)$ under these conditions.

The stationary scattering matrix $\hat{S}$ depends on the electron energy ε and some properties of the scatterer. To account the latter dependence, we introduce the set of parameters, $\{p_i\}$, $i = 1, \dots, M_p$ and write $\hat{S}(\{p_i\}, \varepsilon)$. Under the action of external periodic forces, the parameters periodically change in time, $p_i(t) = p_i(t + \tau)$. Therefore, the matrix $\hat{S}$ becomes dependent on time, $\hat{S}(t, \varepsilon) \equiv \hat{S}(\{p_i(t)\}, \varepsilon)$, and periodic, $\hat{S}(t, \varepsilon) = \hat{S}(t + \tau, \varepsilon)$. The obtained matrix is called the frozen scattering matrix. This name means that the matrix $\hat{S}(t_0, \varepsilon)$ describes the scattering properties of a stationary scatterer whose parameters coincide with the parameters of a given scatterer frozen at time $t = t_0$. The Fourier coefficient for the frozen matrix

$$\hat{S}_n(\varepsilon) = \int_0^\tau \frac{dt}{\tau} e^{in\Omega_0 t}\, \hat{S}(t, \varepsilon), \qquad (38.60)$$

can be related to the Floquet scattering matrix.

At low driving frequencies, $\Omega_0 \to 0$, one can expand the elements of the Floquet scattering matrix in powers of Ω_0. Up to the first order in Ω_0 we have (Moskalets and Büttiker 2004)

$$\hat{S}_F(\varepsilon_n, \varepsilon) = \hat{S}_n(\varepsilon) + \frac{n\hbar\Omega_0}{2} \frac{\partial \hat{S}_n(\varepsilon)}{\partial \varepsilon} + \hbar\Omega_0 \hat{A}_n(\varepsilon) + \mathcal{O}(\Omega_0^2). \qquad (38.61)$$

Here $\hat{A}_n$ is the Fourier transform for a matrix $\hat{A}(t, \varepsilon)$, which formally encloses corrections that cannot be related to the frozen scattering matrix and has to be calculated independently, see Moskalets and Büttiker (2005) for some examples. Note that in the above equation the frozen scattering matrix and the matrix $\hat{A}$ should be kept as energy-independent within a scale of order $\hbar\Omega_0$.

The unitarity of the Floquet scattering matrix puts some constraint on the matrix $\hat{A}$. Substituting Equation 38.61 into Equation 38.57 and taking into account that the stationary (frozen) scattering matrix is unitary, we get the following relation:

$$\hbar\Omega_0 \left\{ \hat{S}^\dagger \hat{A} + \hat{A}^\dagger \hat{S} \right\} = \frac{i\hbar}{2} \left(\frac{\partial \hat{S}^\dagger}{\partial t} \frac{\partial \hat{S}}{\partial \varepsilon} - \frac{\partial \hat{S}^\dagger}{\partial \varepsilon} \frac{\partial \hat{S}}{\partial t} \right). \qquad (38.62)$$

The advantage of the adiabatic ansatz, Equation 38.61, is that the matrices $\hat{S}$ and $\hat{A}$ depend only on one energy and thus have a much smaller number of elements than the Floquet scattering matrix. In addition, the adiabatic ansatz allows us to draw some conclusions concerning the physical properties of slowly driven systems, in particular, concerning the generated heat flows.

38.3.3 Floquet Scattering Matrix versus Green's Function

There exists a simple relation between the Floquet scattering matrix elements and the Fourier coefficients for Green's function (Arrachea and Moskalets 2006):

$$S_{\mathrm{F},\alpha\beta}(\hbar\omega_m,\hbar\omega_n) = \delta_{\alpha,\beta}\,\delta_{m,n} - i\sqrt{\Gamma_\alpha(\omega_m)\Gamma_\beta(\omega_n)}\,\mathcal{G}_{l\alpha,l\beta}(m-n,\omega_n),$$
$$(38.63)$$

where the Floquet component of the Fourier transformed Green's function $\mathcal{G}(n,\omega)$ was introduced in Equation 38.46. Equation 38.63 is a generalization the periodically driven systems of a formula proposed by Fisher and Lee (1981) for stationary systems. This relation is based on the fact that the unitary property (Equations 38.57 and 38.58) which is fundamental to prove the conservation of the charge within the scattering matrix formalism can be proved from identities between Green's functions, see Equation 38.47 through the relation (38.63). We do not present in this chapter further details on those proofs. Instead, in Section 38.3.4 we explicitly show that both formalisms lead to expressions for the currents through the contacts that are equivalent provided that the above relation holds.

38.3.4 Final Expressions for the DC Currents and Powers

38.3.4.1 Particle Currents and the Conservation of the Charge

We begin with the expression for the dc particle currents within Green's function formalism. In Section 38.3.1.1, we have expressed instantaneous values of observables in terms of lesser Green's functions. Now, we use those expressions to evaluate the dc components of the currents defined in Section 38.2.3. In particular, for the charge currents (38.10) and (38.11), we have

$$\bar{J}_l = \frac{2ew}{\hbar\tau}\int_0^\tau dt\,\mathrm{Re}\!\left[G^<_{l+1,l}(t,t)\right],$$

$$\bar{J}_\alpha = \frac{2ew_{c\alpha}}{\hbar\tau}\int_0^\tau dt\,\mathrm{Re}\!\left[G^<_{l\alpha,k\alpha}(t,t)\right]. \tag{38.64}$$

Using the representation (Equation 38.46) in (38.34) and (38.35) and substituting in the above expressions casts for the charge currents within C:

$$\bar{J}_l = -\frac{2ew}{h}\sum_{\alpha=L,R}\sum_{n=-\infty}^{+\infty}\int_{-\infty}^{+\infty} d\omega\, f_\alpha(\hbar\omega)\Gamma_\alpha(\omega)$$

$$\times \mathrm{Im}\!\left[\mathcal{G}_{l+1,l\alpha}(n,\omega)\mathcal{G}^*_{l,l\alpha}(n,\omega)\right], \tag{38.65}$$

and through the contacts

$$\bar{J}_\alpha = -\frac{2e|w_{c\alpha}|^2}{h}\int_{-\infty}^{+\infty} d\omega\,\mathrm{Re}\Big\{if_\alpha(\hbar\omega)\,\mathcal{G}_{l\alpha,l\alpha}(0,\omega)\rho_\alpha(\omega)$$

$$+\sum_{\beta=L,R}\sum_{n=-\infty}^{+\infty}\sum_{k\alpha} f_\beta(\hbar\omega)\big|\mathcal{G}_{l\alpha,l\beta}(n,\omega)\big|^2\,\Gamma_\beta(\omega)\,g^{0,A}_{k\alpha,k\alpha}(\omega_n)$$

$$= \frac{e}{h}\int_{-\infty}^{+\infty} d\omega\Big\{f_\alpha(\hbar\omega)\Gamma_\alpha(\omega)\,2\,\mathrm{Im}\!\left[\mathcal{G}_{l\alpha,l\alpha}(0,\omega)\right]$$

$$-\sum_{\beta=L,R}\sum_{n=-\infty}^{+\infty} f_\beta(\hbar\omega)\,\Gamma_\alpha(\omega_n)\,|\mathcal{G}_{l\alpha,l\beta}(n,\omega)|^2\,\Gamma_\beta(\omega)\Big\}. \tag{38.66}$$

In the above equations we have used the definitions of the density of states (38.3) and the functions (38.37). Going from the first to the second identity, we have also used the property $\mathrm{Im}\!\left[g^{0,R}_{k\alpha,k\alpha}(\omega)\right] = -\mathrm{Im}\!\left[g^{0,A}_{k\alpha,k\alpha}(\omega)\right] = -\rho_\alpha(\omega)/2$, which can be easily derived by just evaluating the Fourier transforms in (38.36). From the identity (Equation 38.47), this current can also be expressed in the more compact and symmetric form:

$$\bar{J}_\alpha = \frac{e}{h}\sum_{\beta=L,R}\sum_{n=-\infty}^{+\infty}\int_{-\infty}^{+\infty} d\omega[f_\alpha(\hbar\omega_n) - f_\beta(\hbar\omega)]\Gamma_\alpha(\omega_n)$$

$$\times|\mathcal{G}_{l\alpha,l\beta}(n,\omega)|^2\,\Gamma_\beta(\omega). \tag{38.67}$$

Within the Floquet scattering matrix approach, we proceed as follows. Substituting Equation 38.59 into Equation 38.55 we get the current in terms of the Floquet scattering matrix elements:

$$\bar{J}_\alpha = \frac{e}{h}\int_{-\infty}^{+\infty} d\varepsilon\left\{f_\alpha(\varepsilon) - \sum_{\beta=L,R}\sum_{n=-\infty}^{\infty} f_\beta(\varepsilon_n)\,|S_{\mathrm{F},\alpha\beta}(\varepsilon,\varepsilon_n)|^2\right\}. \tag{38.68}$$

An equivalent expression is obtained if we make a shift $\varepsilon_n \to \varepsilon$ (under the integration over energy) and an inversion $n \to -n$ (under the corresponding sum) in the term containing $f_\beta(\varepsilon_n)$. The result is

$$\bar{J}_\alpha = \frac{e}{h}\int_{-\infty}^{+\infty} d\varepsilon\left\{f_\alpha(\varepsilon) - \sum_{\beta=L,R}\sum_{n=-\infty}^{\infty} f_\beta(\varepsilon)\,|S_{\mathrm{F},\alpha\beta}(\varepsilon_n,\varepsilon)|^2\right\}. \tag{38.69}$$

Finally, we can write this equation in an alternative way as follows. We multiply the term $f_\alpha(\varepsilon)$ in Equation 38.69 by the left-hand side of the identity, $\sum_\beta\sum_n|S_{\alpha\beta}(\varepsilon,\varepsilon_n)|^2 = 1$, following from the unitarity condition (Equation 38.58), change $\varepsilon_n \to \varepsilon$ and $n \to -n$ in the resulting expression, and find

$$\bar{J}_\alpha = \frac{e}{h}\sum_{\beta=L,R}\sum_{n=-\infty}^{\infty}\int_{-\infty}^{+\infty} d\varepsilon\left[f_\alpha(\varepsilon_n) - f_\beta(\varepsilon)\right]\big|S_{\mathrm{F},\alpha\beta}(\varepsilon_n,\varepsilon)\big|^2. \tag{38.70}$$

It is important to note that Equation 38.68 coincides with Equation 38.66, while Equation 38.70 coincides with Equation 38.67, if we apply the relation (38.63) between the Floquet scattering matrix and Green's function.

Another feature worthy of being mentioned is the fact that from the expressions (38.67) and (38.70) it can be proved that there is conservation of the charge, which implies

$$\sum_{\alpha=L,R} \bar{J}_\alpha = 0. \tag{38.71}$$

We recall that the $\bar{J}_\alpha$ was defined as the current exiting the reservoir, for this reason current conservation implies that it has different signs at the two reservoirs. A final issue that becomes apparent from Equations 38.67 and 38.70 is the fact that for slow driving, $\Omega_0 \to 0$, only electrons near the Fermi energy $\varepsilon \approx \mu$ will be excited and hence will contribute to the generated current, in agreement with our intuition.

38.3.4.2 Particle Currents within the Adiabatic Approximation

In Section 38.3.2.3, we have introduced an approximation for the low driving limit of the full Floquet scattering matrix that depends on the frozen scattering matrix and a matrix $\hat{A}$. In this section, we present the expression for the current in terms of that approximation.

We have mentioned that the unitary condition imposes a constraint to the matrix $\hat{A}$. Another more specific constraint follows from the conservation of a charge current expressed directly in terms of $\hat{S}$ and $\hat{A}$ matrices. To derive it we calculate the dc pumped current $\bar{J}_\alpha$ up to Ω_0^2 terms for all reservoirs at the same temperature and chemical potential, i.e., $f_\alpha = f_0$, $\forall\alpha$. Since in the adiabatic case under consideration $\Omega_0 \to 0$, then at any finite temperature it is $k_B T \gg \hbar\Omega_0$, and we can expand $f_0(\varepsilon) - f_0(\varepsilon_n) \approx -(\partial f_0/\partial\varepsilon)n\hbar W_0 - (\partial^2 f_0/\partial e^2)(n\hbar\Omega_0)^2/2$. Substituting this expansion and Equation 38.61 into Equation 38.70 and performing the inverse Fourier transformation, we calculate the charge current as a sum of linear (upper index "(1)") and quadratic (upper index "(2)") in driving frequency contributions, $\bar{J}_\alpha = \bar{J}_\alpha^{(1)} + \bar{J}_\alpha^{(2)} + \mathcal{O}(\Omega_0^3)$, with

$$\bar{J}_\alpha^{(1)} = -\frac{e}{2\pi}\int_{-\infty}^{\infty} d\varepsilon \left(-\frac{\partial f_0}{\partial\varepsilon}\right)\int_0^\tau \frac{dt}{\tau}\,\mathrm{Im}\left(\hat{S}(t,\varepsilon)\frac{\partial\hat{S}^\dagger(t,\varepsilon)}{\partial t}\right)_{\alpha\alpha}, \tag{38.72}$$

$$\bar{J}_\alpha^{(2)} = -\frac{e}{2\pi}\int_{-\infty}^{\infty} d\varepsilon \left(-\frac{\partial f_0}{\partial\varepsilon}\right)\int_0^\tau \frac{dt}{\tau}\,\mathrm{Im}\left(2\Omega_0\hat{A}(t,\varepsilon)\frac{\partial\hat{S}^\dagger(t,\varepsilon)}{\partial t}\right)_{\alpha\alpha}. \tag{38.73}$$

The linear behavior of the current as a function of the frequency was calculated by Brouwer (1998) using the scattering approach to low-frequency ac transport in mesoscopic systems developed by Büttiker et al. (1994). The conservation of this current, $\sum_\alpha \bar{J}_\alpha^{(1)} = 0$, was demonstrated by Avron et al. (2004) on the base of the Birman-Krein relation, $d\ln(\det\hat{S}) = -\mathrm{Tr}(\hat{S}d\hat{S}^\dagger)$ (where

$\det(\hat{X})$ and $\mathrm{Tr}(\hat{X})$ are the determinant and the trace of a matrix $\hat{X}$, respectively), applied to the frozen matrix which is unitary.

The conservation of the current up to the second order in frequency, $\sum_\alpha \bar{J}_\alpha^{(2)} = 0$, leads to the constraint for the matrix $\hat{A}$ we are looking for

$$\mathrm{Im}\int_0^\tau \frac{dt}{\tau}\,\mathrm{Tr}\left(\hat{A}\frac{\partial\hat{S}^\dagger}{\partial t}\right) = 0. \tag{38.74}$$

Equations 38.61 and 38.62 show us that the expansion in powers of Ω_0 actually is an expansion in powers of $\hbar\Omega_0/\delta\varepsilon$, where $\delta\varepsilon$ is an energy scale characteristic for the stationary scattering matrix. The energy $\delta\varepsilon$ relates to the inverse time spent by an electron with energy ε inside the scattering region (the dwell time). Therefore, one can say that the adiabatic expansion, Equation 38.61, is valid if the period of external forces is large compared with the dwell time. It is important to stress that this definition of "adiabaticity" is different from that usually used in quantum mechanics one which requires the excitation quantum $\hbar\Omega_0$ to be small compared with the level spacing.

38.3.4.3 Particle Currents within Perturbation Theory

In order to gain physical intuition on the behavior of the dc charge current, let us consider the weak driving regime (low V_0) and let us evaluate (Equation 38.67) $\bar{J}_\alpha$ with the perturbative solution of Green's function we have presented in (38.52). We assume that both reservoirs are at temperature $T_\alpha = 0$. Substituting (38.52) into (38.67), we get

$$\bar{J}_\alpha = \frac{e}{h}\sum_{\beta=L,R}\sum_{k=\pm1}\int_{-\infty}^\infty d\omega\,[f_\alpha(\hbar\omega_k) - f_\beta(\hbar\omega)]\Gamma_\alpha(\omega_k)\left|\mathcal{G}_{l\alpha,l\beta}(k,\omega)\right|^2\Gamma_\beta(\omega). \tag{38.75}$$

In the same spirit as in the adiabatic approximation, let us consider that the driving is slow, i.e., $\Omega_0 \to 0$, and let us expand the integrand of the above equation up to the first order in Ω_0. Replacing the Floquet components evaluated up to second order in perturbation theory (53), we get

$$\bar{J}_\alpha = \frac{2ev_0^2\Omega_0}{h}\sum_{j,j'=1}^M\sum_{\beta=L,R}\Gamma_\alpha(\mu)\Gamma_\beta(\mu)\sin(\delta_j - \delta_{j'})$$
$$\times G_{l\alpha,lj}^{0,R}(\mu)G_{lj,l\beta}^{0,R}(\mu)\left[G_{l\alpha,lj'}^{0,R}(\mu)G_{lj',l\beta}^{0,R}(\mu)\right]^*, \tag{38.76}$$

where j, j' runs over the M pumping potentials. Thus, even without specifying the geometrical details on the structure, which are contained in G^0, Equation 38.76 provides us a valuable piece of information. As a first point, it tells us that at low driving the leading contribution to the dc particle current is $\propto V_0^2\Omega_0$ A second important point is that with local time-dependent potentials, as we are considering in our model, we need at least two of these potentials operating with a phase lag in order to have a nonvanishing value for this lowest-order contribution.

38.3.4.4 Energy and Heat Currents

We can follow a similar procedure as in Section 38.3.4.3 to derive the dc energy and heat currents. In terms of Green's functions we start writing the dc energy currents (38.15) and (38.16) as follows:

$$\overline{J}_l^E = \frac{2w}{\hbar \tau} \int_0^\tau dt \{ \text{Re}[G_{l+2,l}^<(t,t)]w - \text{Re}[G_{l+2,l+1}^<(t,t)]\varepsilon_{l+1}(t) \},$$

$$\overline{J}_\alpha^E = \frac{2w_{c\alpha}}{\hbar \tau} \sum_{k\alpha} \int_0^\tau dt \, \varepsilon_{k\alpha} \text{Re}[G_{l\alpha,k\alpha}^<(t,t)]. \tag{38.77}$$

The energy current within C is

$$\overline{J}_l^E = -\frac{2w}{h} \sum_{\alpha=L,R} \sum_{n=-\infty}^{+\infty} \int_{-\infty}^{+\infty} d\omega f_\alpha(\hbar\omega)\Gamma_\alpha(\omega)\{ w \text{Im}[\mathcal{G}_{l+2,l\alpha}(n,\omega)\mathcal{G}_{l,l\alpha}^*(n,\omega)]$$

$$- \varepsilon_{l+1} \text{Im}[\mathcal{G}_{l+2,l\alpha}(n,\omega)\mathcal{G}_{l+1,l\alpha}^*(n,\omega)] \}, \tag{38.78}$$

where we have assumed that the position $l + 1$ does not coincide with a pumping center, while for the energy current through the contact we get

$$\overline{J}_\alpha^E = \frac{|w_{c\alpha}|^2}{h} \sum_{k\alpha} \int_{-\infty}^\infty d\omega \, \varepsilon_{k\alpha} 2\pi \Big\{ f_\alpha(\hbar\omega)\delta(\omega - \varepsilon_{k\alpha}/\hbar) 2 \text{Im}[\mathcal{G}_{l\alpha,l\alpha}(0,\omega)]$$

$$- \sum_{\beta=L,R} \sum_{n=-\infty}^{+\infty} f_\beta(\hbar\omega)\delta(\omega_n - \varepsilon_{k\alpha}/\hbar)|\mathcal{G}_{l\alpha,l\beta}(n,\omega)|^2 \Gamma_\beta(\omega) \Big\}$$

$$= \frac{\hbar}{2\pi} \int_{-\infty}^\infty d\omega \Big\{ \omega f_\alpha(\hbar\omega)\Gamma_\alpha(\omega) 2 \text{Im}[\mathcal{G}_{l\alpha,l\alpha}(0,\omega)]$$

$$- \sum_{\beta=L,R} \sum_{n=-\infty}^{+\infty} \omega_n f_\beta(\hbar\omega)\Gamma_\alpha(\omega)|\mathcal{G}_{l\alpha,l\beta}(n,\omega)|^2 \Gamma_\beta(\omega) \Big\}, \tag{38.79}$$

which can also be written in the symmetric form

$$\overline{J}_\alpha^E = \frac{\hbar}{2\pi} \sum_{n=-\infty}^{+\infty} \sum_{\beta=L,R} \int_{-\infty}^{+\infty} d\omega \, \omega_n [f_\alpha(\hbar\omega_n) - f_\beta(\hbar\omega)]\Gamma_\alpha(\omega_n)$$

$$\times |\mathcal{G}_{l\alpha,l\beta}(n,\omega)|^2 \Gamma_\beta(\omega). \tag{38.80}$$

We now go back to our heuristic argument introduced in Section 38.2.4 to define the heat current. The above equation shows that for low driving, even for reservoirs at $T = 0$ and very weak driving, such that $\Omega_0 \to 0$, there is a finite energy flow $\overline{J}_\alpha^E \propto \mu \overline{J}_\alpha$, with $\overline{J}_\alpha$ given in Equation 38.67. This energy is transported by the currents from one reservoir to the other one, thus having a convective character and should be subtracted to get a heat flow.

To calculate the heat flow we multiply Equation 38.66 by μ/e and subtract it to Equation 38.79

$$J_\alpha^Q = \frac{\hbar}{2\pi} \int_{-\infty}^{+\infty} d\omega \Big\{ (\omega - \omega_F) f_\alpha(\hbar\omega)\Gamma_\alpha(\omega) 2 \text{Im}[\mathcal{G}_{l\alpha,l\alpha}(0,\omega)]$$

$$- \sum_{\beta=L,R} \sum_{n=-\infty}^{+\infty} (\omega_n - \omega_F) f_\beta(\hbar\omega)\Gamma_\alpha(\omega_n)|\mathcal{G}_{l\alpha,l\beta}(n,\omega)|^2 \Gamma_\beta(\omega) \Big\}, \tag{38.81}$$

where $\hbar\omega_F = \mu$. Equivalently, from (38.80) and (38.67), we can write the heat current flowing through the contact as follows:

$$J_\alpha^Q = \frac{\hbar}{2\pi} \sum_{\beta=L,R} \sum_{n=-\infty}^{+\infty} \int_{-\infty}^{+\infty} d\omega (\omega_n - \omega_F) \big[f_\alpha(\hbar\omega_n) - f_\beta(\hbar\omega) \big]$$

$$\times \Gamma_\alpha(\omega_n)|\mathcal{G}_{l\alpha,l\beta}(n,\omega)|^2 \Gamma_\beta(\omega). \tag{38.82}$$

Within the scattering matrix approach, one can also calculate the heat current J_α^Q by analogy to the charge current, Equation 38.68. As we already mentioned, Equation 38.68 contains the difference between the number of electrons with energy ε entering and leaving the scatterer through the same wire. Each of these electrons has an energy ε. Therefore, to calculate the heat current we multiply the integrand in Equation 38.68 by $(\varepsilon - \mu)$, drop an electron charge e, and get

$$J_\alpha^Q = \frac{1}{h} \int_{-\infty}^{+\infty} d\varepsilon (\varepsilon - \mu) \Big\{ f_\alpha(\varepsilon) - \sum_{\beta=L,R} \sum_{n=-\infty}^\infty f_\beta(\varepsilon_n) |S_{F,\alpha\beta}(\varepsilon,\varepsilon_n)|^2 \Big\}. \tag{38.83}$$

This equation is equivalent to (38.81) through the relation (38.63). Next we make shifts $\varepsilon_n \to \varepsilon$ and $n \to -n$ in the term containing $f_\beta(\varepsilon_n)$ and finally obtain

$$J_\alpha^Q = \frac{1}{h} \int_{-\infty}^{+\infty} d\varepsilon \Big\{ (\varepsilon - \mu) f_\alpha(\varepsilon) - \sum_{\beta=L,R} \sum_{n=-\infty}^\infty (\varepsilon_n - \mu) f_\beta(\varepsilon) |S_{F,\alpha\beta}(\varepsilon_n,\varepsilon)|^2 \Big\}. \tag{38.84}$$

We multiply the term containing $f_\alpha(\varepsilon)$ by the identity $1 = \sum_\beta \sum_n |S_{F,\alpha\beta}(\varepsilon,\varepsilon_n)|^2$. Then we make shifts $\varepsilon_n \to \varepsilon$ and $n \to -n$ in this term and finally get

$$J_\alpha^Q = \frac{1}{h} \sum_{\beta=L,R} \sum_{n=-\infty}^\infty \int_{-\infty}^{+\infty} d\varepsilon (\varepsilon_n - \mu) \big[f_\alpha(\varepsilon_n) - f_\beta(\varepsilon) \big] |S_{F,\alpha\beta}(\varepsilon_n,\varepsilon)|^2, \tag{38.85}$$

which, because of Equation 38.63, is equivalent to Equation 38.82.

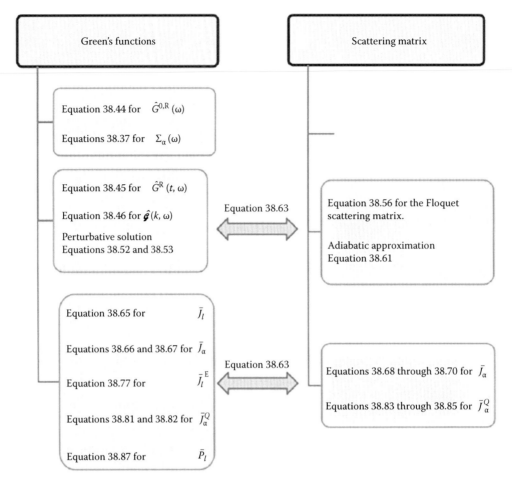

FIGURE 38.4 Diagram summarizing the possible steps to be followed in order to evaluate particle, energy, and heat currents, as well as the power developed by the fields by recourse to the two formalisms presented in this chapter.

38.3.4.5 Mean Power Developed by the Fields

The dc power (38.17) done by the ac fields reads

$$\bar{P}_l = \frac{-i}{\tau} \int_0^\tau dt \frac{d e V_l(t)}{dt} G_{l,l}^<(t,t). \tag{38.86}$$

In terms of the representation (38.46), it results

$$\bar{P}_l = \frac{\hbar\Omega_0 e V_l^0}{2\pi} \sum_{n=-\infty}^{+\infty} \sum_{\alpha=L,R} \sum_{k=\pm 1} \int_{-\infty}^{+\infty} d\omega f_\alpha(\hbar\omega) \Gamma_\alpha(\omega)$$

$$\times \mathrm{Im} \left\{ k e^{-ik\delta_l} \mathcal{G}_{l,l\alpha}(n,\omega) \mathcal{G}_{l,l\alpha}(n+k,\omega)^* \right\}. \tag{38.87}$$

This expression does not have a counterpart in terms of the Floquet scattering matrix. This is because the evaluation of this quantity depends on the microscopic details included explicitly in the Hamiltonian. In fact, notice that the formula (38.63) relates the scattering matrix only with Green's function with the coordinates of the central system, $l\alpha$, $l\beta$, that intervene in

the contacts. For the same reason, we have shown in the previous section equivalent expression within both formalisms only for the currents through the contacts and not for the currents within C. Nevertheless the total power developed by all the fields can be also calculated within the scattering matrix formalism, see Equations 38.88 and 38.97.

38.3.5 Technical Summary

To close this section, we present in Figure 38.4 a diagram with the summary of the procedure to evaluate the different physical quantities we need to discuss the transport behavior of a quantum pump, the alternatives and the possible approximations.

38.4 Results and Critical Discussion

In this section, we apply the concepts and techniques introduced in Section 38.3 to analyze the conservation of the energy and the different mechanisms of heat transport that we can identify in our quantum pump. On the basis of our previous definitions, we can show the existence of three generic effects due to a dynamical scatterer. At any segment of the system, it is possible to verify the conservation laws introduced in Section 38.2.2 by numerically solving the Dyson equation for the retarded Green's functions, evaluating the relevant expectation values of observables following the indications of the diagram of Figure 38.4. In what follows, we present analytical results based on the perturbative solution of Green's function and the adiabatic approximation for the scattering matrix. Without the explicit evaluation of the functions $\hat{G}^{0,R}(\omega)$, which depend only on the geometric statical properties of the system, this procedure allows us to analyze the physical properties of our system within the weak driving regime.

To make the effects clearer, we consider the case when the two reservoirs have, not only the same electrochemical potential $\mu_\alpha = \mu$, but also the same temperature $T_\alpha = T \Rightarrow f_\alpha(\hbar\omega) = f_0(\hbar\omega), \forall\alpha$.

38.4.1 Heating of the Reservoirs by the Quantum Pump

The first effect that takes place in our quantum engine is the heating of the reservoirs (see, e.g., Avron et al. 2001, Moskalets and Büttiker 2002b, Wang and Wang 2002, Avron et al. 2004). Unlike the charge current, $\bar{J}_\alpha$, the sum of heat currents in all the wires, $J_{\text{tot}}^Q = \sum_\alpha J_\alpha^Q$, is nonzero. According to the conservation of the energy expressed in Equation 38.19, the definition of the heat current (38.20) and the conservation of the charge (38.71), it is clear that the total power developed by the fields is equal to the total heat current that enters the reservoirs:

$$\sum_{l=1}^{M} \bar{P}_l = -\sum_{\alpha} J_\alpha^Q. \tag{38.88}$$

For reservoirs at temperature $T = 0$, our intuition suggests us that the total power developed by the fields is fully transformed into heat which flows into the reservoirs (see Figure 38.5). In what follows we analyze the behavior of this flow as a function of the pumping parameters within the low driving regime. To this end, we follow an analogous procedure as in Section 38.3.4.3, and we use perturbation theory to evaluate the powers and heat flows at the contacts.

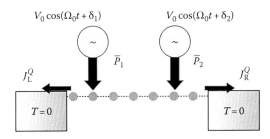

$V_0\cos(\Omega_0 t + \delta_1)$ $V_0\cos(\Omega_0 t + \delta_2)$

$\bar{P}_1$ $\bar{P}_2$

J_L^Q J_R^Q

$T=0$ $T=0$

FIGURE 38.5 Scheme of the working regime of the quantum pump when the two reservoirs are at temperature $T = 0$. All the power developed by the external fields is dissipated in the form of heat that is absorbed by the left and right reservoirs.

38.4.1.1 Heat Current at Weak Driving, $T = 0$

Following exactly the same lines as those presented in the derivation of (38.76), we start from (38.82), and we substitute the perturbative solution of Green's function (38.52) and (38.53) and expand in Taylor series the resulting expression up to the lowest nonvanishing order in Ω_0. The result is

$$J_\alpha^Q \sim \frac{\hbar\Omega_0^2(eV_0)^2}{\pi} \sum_{j,j'=1}^{M}\sum_{\beta=L,R} \cos(\delta_j - \delta_{j'})\Gamma_\alpha(\omega_F)\Gamma_\beta(\omega_F)$$
$$\times G_{l\alpha,lj}^{0,R}(\omega_F) G_{lj,l\beta}^{0,R}(\omega_F)\left[G_{l\alpha,lj'}^{0,R}(\omega_F)G_{lj',l\beta}^{0,R}(\omega_F)\right]^*. \tag{38.89}$$

The total heat flowing through the contacts reads

$$\sum_{\alpha=L,R} J_\alpha^Q = \frac{\hbar\Omega_0^2(eV_0)^2}{\pi}\sum_{j,j'=1}^{M}\sum_{\beta=L,R} \cos(\delta_j - \delta_{j'})\Gamma_\alpha(\omega_F)\Gamma_\beta(\omega_F)$$
$$\times G_{l\alpha,lj}^{0,R}(\omega_F) G_{lj,l\beta}^{0,R}(\omega_F)\left[G_{l\alpha,lj'}^{0,R}(\omega_F)G_{lj',l\beta}^{0,R}(\omega_F)\right]^*$$
$$= \frac{\hbar\Omega_0^2(eV_0)^2}{\pi}\sum_{j,j'=1}^{M} \cos(\delta_j - \delta_{j'})\left|\rho_{lj,lj'}(\omega_F)\right|^2, \tag{38.90}$$

where we have used the identity between equilibrium Green's functions and the definition of the matrix presented in (38.51). Thus, at $T = 0$ and weak driving, there is a net heat flow $\propto V_0^2\Omega_0^2$ into the reservoirs.

38.4.1.2 Mean Power at Weak Driving, $T = 0$

We now follow a similar procedure to evaluate the mean power developed by the jth force. Substituting the perturbative solution (38.52) in (38.87), and keeping terms that contribute at $\mathcal{O}(V_0^2)$, we get

$$\bar{P}_j \sim \frac{\hbar\Omega_0 eV_0}{\pi}\sum_{\alpha=L,R}\sum_{j'=1}^{M}\int_{-\infty}^{+\infty}d\omega\, f_\alpha(\hbar\omega)\,\Gamma_\alpha(\omega)$$
$$\times \text{Im}\left\{e^{i\delta_j}\left[\mathcal{G}_{lj,l\alpha}(0,\omega)\mathcal{G}_{lj,l\alpha}^\star(1,\omega) + \mathcal{G}_{lj,l\alpha}(-1,\omega)\mathcal{G}_{lj,l\alpha}^\star(0,\omega)\right]\right\}. \tag{38.91}$$

Then, replacing (38.53) we derive an equation with several terms which can be collected as follows:

$$\bar{P}_j = \sum_{j'=1}^{M}\left[\lambda_{j,j'}^{(1)}\cos(\delta_j-\delta_{j'}) + \lambda_{j,j'}^{(2)}\sin(\delta_j-\delta_{j'})\right], \tag{38.92}$$

with

$$\lambda_{j,j'}^{(1)} = \frac{\hbar\Omega_0(eV_0)^2}{\pi}\int_{-\infty}^{+\infty}d\omega f_0(\hbar\omega)\,\text{Im}\{\gamma_{j,j'}(\omega)\gamma_{j,j'}^-(\omega)\},$$

$$\lambda_{j,j'}^{(2)} = \frac{\hbar\Omega_0(eV_0)^2}{\pi}\int_{-\infty}^{+\infty}d\omega f_0(\hbar\omega)\,\text{Re}\{\gamma_{j,j'}(\omega)\gamma_{j,j'}^+(\omega)\}, \tag{38.93}$$

being

$$\gamma_{j,j'}(\omega) = \sum_{\alpha=L,R} [G^{0,R}_{lj,l\alpha}(\omega)]^* \Gamma_\alpha(\omega) G^{0,R}_{lj',l\alpha}(\omega) = -i\rho^*_{lj,lj'}(\omega),$$

$$(38.94)$$

$$\gamma^\pm_{j,j'}(\omega) = G^{0,R}_{lj,lj'}(\omega+\Omega_0) \pm G^{0,R}_{lj,lj'}(\omega-\Omega_0).$$

38.4.1.3 Conservation of the Energy

The second term of (38.92) vanishes when we perform a summation over all the fields, since $\lambda^{(2)}_{j,j'}$ is symmetric under a permutation $j \leftarrow j'$ while $\sin(\delta_j - \delta_{j'})$ is antisymmetric under this operation. Thus, the only term contributing to the sum over all the powers is the first one, which for low Ω_0 results in

$$\lambda^{(1)}_{lj,lj'} = \frac{\hbar\Omega_0(eV_0)^2}{\pi} \int_{-\infty}^{+\infty} d\omega$$

$$\times \mathrm{Re}\Big\{ [f_0(\hbar\omega - \hbar\Omega_0)\rho_{lj,lj'}(\omega-\Omega_0)$$

$$- f_0(\hbar\omega+\hbar\Omega_0)\rho_{lj,lj'}(\omega+\Omega_0)][G^{0,R}_{lj,lj'}(\omega)]^*\Big\}$$

$$\sim -\frac{2\hbar\Omega_0^2(eV_0)^2}{\pi}\rho_{lj,lj'}(\omega_F)[G^{0,R}_{lj,lj'}(\omega_F)]^*. \quad (38.95)$$

Performing the sum over j in (38.92) and using $|\rho_{l,l'}(\omega)|^2 = |G_{l,l'}(\omega)|^2 + |G_{l',l}(\omega)|^2 - 2\mathrm{Re}[G_{l,l'}(\omega)G_{l',l}(\omega)]$, we can verify the fundamental law of the conservation of the energy (38.88).

38.4.2 Energy Exchange between External Forces

The evaluation of the coefficient $\lambda^{(2)}_{j,j'}$ at weak driving can be carried out following exactly the same steps as with $\lambda^{(1)}_{j,j'}$. The result is

$$\lambda^{(2)}_{j,j'} \sim -\frac{2\hbar\Omega_0(eV_0)^2}{\pi} \int_{-\infty}^{+\infty} d\omega\, \mathrm{Im}[G^{0,R}_{lj,lj'}(\omega)G^{0,R}_{lj',lj}(\omega)], \quad (38.96)$$

i.e., this contribution is $\propto \Omega_0$, and therefore dominates the behavior of $\bar{P}_j$ at weak driving. Interestingly, this contribution does not exist in a configuration with a single ac field, while it can have different signs at different fields in a configuration with several pumping centers.

Therefore, we present the second general effect taking place in quantum engines: One external force can perform work directly against another external force with a negligible amount of energy being dissipated into the reservoirs. This remarkable mechanism opens the possibility of the coherent energy transfer between pumping centers as indicated in Figure 38.6.

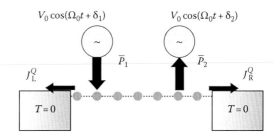

FIGURE 38.6 Scheme of the working regime of the quantum pump when the two reservoirs are at temperature $T = 0$ and low driving: V_0 and Ω_0 small. The dissipated energy flowing into the reservoirs is low, while it is possible that part of the work done by one of the ac fields is coherently transferred to the other one, which receives the ensuing energy.

38.4.3 Directed Heat Transport at Finite Temperature

To show that the dynamical scatterer can induce a directed heat transfer between the reservoirs we, first, calculate the total generated heat $J^Q_{tot} = \sum_\alpha J^Q_\alpha$. Summing up Equation 38.84 over α we find (for $f_\alpha = f_0, \forall \alpha$)

$$J^Q_{tot} = -\frac{\Omega_0}{2\pi} \sum_{\alpha=L,R}\sum_{\beta=L,R}\sum_{n=-\infty}^{\infty}\int_{-\infty}^{\infty} d\varepsilon f_0(\varepsilon) n |S_{F,\alpha\beta}(\varepsilon_n,\varepsilon)|^2. \quad (38.97)$$

The part of the total generated heat that flows into wire α, $J^Q_{tot} = \sum_\alpha J^Q_{\alpha,gen}$, can be defined as follows:

$$J^Q_{\alpha,gen} = -\frac{\Omega_0}{2\pi} \sum_{\beta=L,R}\sum_{n=-\infty}^{\infty}\int_{-\infty}^{+\infty} d\varepsilon f_0(\varepsilon) n |S_{F,\alpha\beta}(\varepsilon_n,\varepsilon)|^2. \quad (38.98)$$

The remaining part of the heat flowing into wire α, $J^Q_{\alpha,pump} = J^Q_\alpha - J^Q_{\alpha,gen}$, is

$$J^Q_{\alpha,pump} = \frac{1}{h} \int_{-\infty}^{+\infty} d\varepsilon(\varepsilon-\mu) f_0(\varepsilon) \left\{ \sum_{\beta=L,R}\sum_{n=-\infty}^{\infty} |S_{F,\alpha\beta}(\varepsilon_n,\varepsilon)|^2 - 1 \right\}.$$

$$(38.99)$$

Using the unitarity condition for the Floquet scattering matrix, Equation 38.57, one can easily show that the part of the heat current $J^Q_{\alpha,pump}$ satisfies the conservation law similar to the one for the charge dc current, Equation 38.71:

$$\sum_{\alpha=L,R} J^Q_{\alpha,pump} = 0. \quad (38.100)$$

This means that $J^Q_{\alpha,pump}$ is transported from one reservoir to another one with the help of a dynamical scatterer. By analogy with the corresponding charge current, we identify this portion of the total heat as a pumped heat (hence the lower index "*pump*"). This is the third general effect we identified in our quantum engine: The dynamical scatterer induces a directed heat transport between the reservoirs (see, e.g., Humphrey et al. 2001. Segal and Nitzan 2006, Arrachea et al. 2007, Rey et al. 2007, Martinez and Hu 2007).

If the pumped heat is, for instance, negative in the L wire, $J_{L,pump}^Q < 0$, then it is necessarily positive in another wire, $J_{R,pump}^Q > 0$. If the absolute value of this heat is larger than the one of the generated component $J_{R,gen}^Q$, then the whole heat flowing into the R wire is positive, i.e., directed from the reservoir to the central system, $J_R^Q = J_{R,gen}^Q + J_{R,pump}^Q > 0$. In this case, the reservoir R will be cooled while L will be heated.

The splitting of J_α^Q into $J_{\alpha,gen}^Q$ and $J_{\alpha,pump}^Q$ helped us to show that J_α^Q can be positive. Strictly speaking, such a splitting is not unique and only the whole heat current J_α^Q has a direct physical meaning. However, at slow driving, one can support such a decomposition of J_α^Q into the generated and the pumped heat by additional physical arguments as follows.

38.4.3.1 Adiabatic Heat Currents

The expansion (38.61) allows us to calculate the heat flow with an accuracy of $\mathcal{O}(\Omega^2)$. To show it explicitly, we rewrite slightly Equation 38.84 (with $f_\alpha = f_0$, $\forall \alpha$). We assume $k_B T \gg \hbar\Omega_0$ and expand the difference of Fermi distribution functions in (38.85) in powers of Ω_0, use Equation 38.61, and find from Equation 38.85 the heat current, $J_\alpha^Q = J_\alpha^{Q,(1)} + J_\alpha^{Q,(2)} + \mathcal{O}(\Omega_0^3)$, where

$$J_\alpha^{Q,(1)} = -\frac{1}{2\pi} \int_{-\infty}^{+\infty} d\varepsilon (\varepsilon - \mu)\left(-\frac{\partial f_0}{\partial \varepsilon}\right) \int_0^\tau \frac{dt}{\tau} \text{Im}\left(\hat{S}(t,\varepsilon)\frac{\partial \hat{S}^\dagger(t,\varepsilon)}{\partial t}\right)_{\alpha\alpha},$$
(38.101)

$$J_\alpha^{Q,(2)} = -\frac{\hbar}{4\pi} \int_{-\infty}^{+\infty} d\varepsilon \left(-\frac{\partial f_0}{\partial \varepsilon}\right) \int_0^\tau \frac{dt}{\tau} \left(\frac{\partial \hat{S}(t,\varepsilon)}{\partial t}\frac{\partial \hat{S}^\dagger(t,\varepsilon)}{\partial t}\right)_{\alpha\alpha}$$
$$-\frac{1}{2\pi} \int_{-\infty}^{+\infty} d\varepsilon (\varepsilon - \mu)\left(-\frac{\partial f_0}{\partial \varepsilon}\right) \int_0^\tau \frac{dt}{\tau} \text{Im}\left(2\Omega_0 \hat{A}(t,\varepsilon)\frac{\partial \hat{S}^\dagger(t,\varepsilon)}{\partial t}\right)_{\alpha\alpha}.$$
(38.102)

Next, we split the heat current into the generated heat and the pumped heat as follows, $J_\alpha^Q = J_{\alpha,gen}^Q + J_{\alpha,pump}^Q$, with

$$J_{\alpha,gen}^Q = -\frac{\hbar}{4\pi} \int_{-\infty}^{+\infty} d\varepsilon \left(-\frac{\partial f_0}{\partial \varepsilon}\right) \int_0^\tau \frac{dt}{\tau} \left(\frac{\partial \hat{S}}{\partial t}\frac{\partial \hat{S}^\dagger}{\partial t}\right)_{\alpha\alpha},$$ (38.103)

$$J_{\alpha,pump}^Q = -\frac{1}{2\pi} \int_{-\infty}^{+\infty} d\varepsilon (\varepsilon - \mu)\left(-\frac{\partial f_0}{\partial \varepsilon}\right)$$
$$\times \int_0^\tau \frac{dt}{\tau} \text{Im}\left(\left[\hat{S} + 2\hbar\Omega_0 \hat{A}\right]\frac{\partial \hat{S}^\dagger}{\partial t}\right)_{\alpha\alpha}.$$ (38.104)

Notice that these equations also remain valid at ultralow temperatures, $k_B T \ll \hbar\Omega_0$, which can be verified by direct calculations taking into account the energy-independence of the matrices

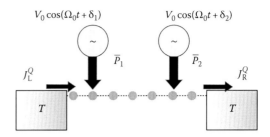

FIGURE 38.7 Scheme of the working regime of the quantum pump when the two reservoirs are at a finite temperature T. There is a net pumping of heat from the one reservoir to the other. The quantum pump, thus works as a refrigerator.

$\hat{S}$ and $\hat{A}$ over a scale of order Ω_0, i.e., over the region of the thermal widening of the edge of the Fermi distribution function.

The above given splitting is justified by the following observations. (i) The quantity $J_{\alpha,gen}^Q$ is negative in each wire α as it should be for the heat generated by the scatterer and flowing into the reservoirs. (ii) At zero temperature, the pumped heat vanishes identically, $J_{\alpha,pump}^Q = 0$, since it is impossible to take heat out of the system kept at zero temperature. To prove the first observation, we show that the integrand in Equation 38.103 is positive. To this end we use the Fourier transformation and get, $1/\tau \int_0^\tau dt (\partial \hat{S}/\partial t\, \partial \hat{S}^\dagger/\partial t)_{\alpha\alpha} = \Omega_0^2 \Sigma_\beta \Sigma_n n^2 |S_{\alpha\beta,n}|^2 > 0$. The second observation follows from the fact that at zero temperature it is $(\varepsilon - \mu)\partial f_0/\partial\varepsilon = 0$, and hence Equation 38.104 vanishes. Note that the conservation of the pumped heat current, $\Sigma_\alpha J_{\alpha,pump}^Q = 0$, directly follows from the conservation of charge currents, Equations 38.72 and 38.73 which implies $1/\tau \int_0^\tau dt\, \text{Im Tr}\,[\hat{S} + 2\hbar\Omega_0 \hat{A}]\partial \hat{S}^\dagger/\partial t = 0$.

From Equation 38.103, it follows that the adiabatic scatterer heats the reservoirs with a rate proportional to $\hbar\Omega_0^2$ (Avron et al. 2001). In contrast, the pumped heat, Equation 38.104, is rather proportional to $k_B T\Omega_0$. At sizable temperatures, $k_B T \gg \hbar\Omega_0$, the amount of pumped heat can exceed the generated heat, $|J_{\alpha,pump}|/J_{\alpha,gen} \sim k_B T/(\hbar\Omega_0) \gg 1$. Therefore, if in the wire α we have $J_{\alpha,pump} > 0$, then the reservoir α will be cooled (see Figure 38.7). This mechanism opens the possibility of using quantum pumps as refrigerators.

38.5 Summary

In this chapter, we have introduced the basic concepts to analyze at the microscopic level the energy transport in quantum systems driven by harmonically time-dependent fields. We have introduced a simple microscopic model for a quantum pump, which consists of a finite structure connected to two macroscopic reservoirs, with ac local fields that oscillate in time with the same frequency and a phase lag. We have analyzed the fundamental conservation laws for the charge and the energy and we have defined the basic concepts to study the transport behavior in these systems: charge currents, energy currents, heat currents, and powers developed by the fields. We have reviewed two complementary techniques to calculate the currents and the powers: the non-equilibrium Green's function formalism for harmonically time-dependent Hamiltonians and the scattering formalism for periodically driven mesoscopic systems. We have shown that the two approaches are equivalent for the evaluation of the charge and heat currents through the contacts between the driven system and the reservoirs. We have also introduced two approximations: the adiabatic approximation to the Floquet scattering matrix and a perturbative solution of Dyson's equations for Green's functions valid within the weak driving regime. Both techniques are important to draw conclusions on general features of the transport behavior without the explicit evaluation of Green's functions or the scattering matrix elements. Such conclusions are, thus, generic and do not depend on the geometrical details of the driven structure. A summary of the technical details, including the main equations and the alternative routes to evaluate them exactly or in an approximate way, is given in a diagram at the end of Section 38.3.

Finally, in Section 38.4, we have applied the concepts and tools we have introduced in the previous sections in order to discuss three important mechanisms of energy transport in quantum pumps. The first one is the fact that the total work done by all the local fields is dissipated in the form of heat that flows to the reservoirs. This effect is rather expected. In any case, we have exploited our theoretical techniques at weak driving to evaluate term-by-term powers and heat currents and explicitly verify the conservation of the energy. To unveil a fundamental law is always a beautiful result in theoretical Physics and an important support for the power of a theoretical tool. In addition we have shown that other two less-expected and subtle transport mechanisms can take place: the coherent transport of energy allowing for regimes where some of the forces make work, while other receive work. This interesting mechanism could be exploited, for instance, to couple two quantum pumps in a combined engine. The final remarkable mechanism is the pumping of heat at finite temperature and weak driving, allowing for the operation of the quantum pump as a refrigerator which extracts heat from a reservoir and injects heat in to the other one.

38.6 Future Perspective

The different operational regimes that we have identified in the quantum pumps have several important outcomes. On the theoretical side, there are several lines to further analyze. A first issue to explore is the role of the geometrical details of the structure, in order to identify the optimal architecture to enhance each mechanism and improve the efficiency of the quantum engine. Another important ingredient is the investigation of the role of many-body interactions. In particular, the electron–electron and the electron–phonon interactions. On the experimental side, it would be very interesting to design an experimental setup to implement these effects. In this sense, it is very promising that quantum refrigeration has been already experimentally explored in mesoscopic structures with superconducting elements under ac driving (Giazzoto et al. 2006).

Acknowledgments

We thank Luis Martin-Moreno for useful discussions and C. Marcus for Figure 38.1. LA acknowledges support from CONICET and UBACyT, Argentina.

References

Arrachea, L. 2002. Current oscillations in a metallic ring threaded by a time-dependent magnetic flux. *Physical Review B* 66: 045315 (11).

Arrachea, L. 2005. A Green-function approach to transport phenomena in quantum pumps. *Physical Review B* 72: 125349 (11).

Arrachea, L. and Moskalets, M. 2006. Relation between scattering matrix and Keldysh formalisms for quantum transport driven by time-periodic fields. *Physical Review B* 74: 245322 (13).

Arrachea, L., Moskalets, M., and Martin-Moreno, L. 2007. Heat production and energy balance in nanoscale engines driven by time-dependent fields. *Physical Review B* 75: 245420 (5).

Avron, J. E., Elgart, A., Graf, G. M., and Sadun, L. 2001. Optimal quantum pumps. *Physical Review Letter* 87: 236601 (4).

Avron, J. E., Elgart, A., Graf, G. M., and Sadun, L. 2004. Transport and dissipation in quantum pumps. *Journal of Statistical Physics* 116: 425–473.

Brouwer, P. W. 1998. Scattering approach to parametric pumping. *Physical Review B* 58: R10135–R10138.

Büttiker, M. 1990. Scattering theory of thermal and excess noise in open conductors. *Physical Review Letter* 65: 2901–2904.

Büttiker, M. 1992. Scattering theory of current and intensity noise correlations in conductors and wave guides. *Physical Review B* 46: 12485–12507.

Büttiker, M. 1993. Capacitance, admittance, and rectification properties of small conductors. *Journal of Physics: Condensed Matter* 5: 9361–9378.

Büttiker, M., Thomas, H., and Prêtre, A. 1994. Current partition in multiprobe conductors in the presence of slowly oscillating external potentials. *Zeitschrift für Physik B Condensed Matter* 94: 133–137.

Caroli, C, Combescot, R., Nozieres, P., and Saint-James, D. 1971. Direct calculation of the tunneling current. *Journal of Physics C: Solid State Physics* 4: 916–929.

Fisher, D. S. and Lee, P. A. 1981. Relation between conductivity and transmission matrix. *Physical Review B* 23: 6851–6854.

Giazotto, F., Heikkila, T. T., Luukanen, A., Savin, A. M., and Pekola, J.P. 2006. Opportunities for mesoscopics in thermometry and refrigeration: Physics and applications. *Rev. Mod. Phys.* 78: 217–274.

Haug, H. and Jauho, A. P. 1996. *Quantum Kinetics in Transport and Optics in Semiconductors*, Springer series in Solid-State Sciences Vol. 123, Springer-Verlag, New York.

Humphrey, T. E., Linke, H., and Newbury, R. 2001. Pumping heat with quantum ratchets. *Physica E* 11: 281–286.

Jauho, A. P., Wingreen, N., and Meir, Y. 1994. Time-dependent transport in interacting and noninteracting resonant-tunneling systems. *Physical Review B* 50: 5528–5544.

Kadanoff, L. P. and Baym, G. 1962. *Quantum Statistical Mechanics*, Benjamin/Cummings Publishing Group, New York.

Keldysh, L. V. 1964. Diagram technique for nonequilibrium processes. *Zhurnal Eksperimentalnoi i Teoreticheskoi Fiziki.* 47: 1515–1527.

Kohler, S., Lehmann, J., and Hänggi, P. 2005. Driven quantum transport on the nanoscale. *Physics Reports* 406: 379–443.

Landauer, R. 1957. Spatial variation of currents and fields due to localized scatterers in metallic conduction. *IBM Journal of Research and Development* 1: 223–231.

Landauer, R. 1970. Electrical resistance of disordered one-dimensional lattices. *Philosophical Magazine* 21: 863–867.

Landauer, R. 1975. Residual resistivity dipoles. *Zeitschrift für Physik B Condensed Matter* 21: 247–254.

Mahan, G. D. 1990. *Many Particle Physics*. New York: Plenum.

Martinez, D. F. and Hu, B. 2007. Operating molecular transistors as heat pumps. arXiv:0709.4660v1.

Moskalets, M. and Büttiker, M. 2002a. Dissipation and noise in adiabatic quantum pumps. *Physical Review B* 66: 035306 (9).

Moskalets, M. and Büttiker, M. 2002b. Floquet scattering theory of quantum pumps. *Physical Review B* 66: 205320 (10).

Moskalets, M. and Büttiker, M. 2004. Adiabatic quantum pump in the presence of external ac voltages. *Physical Review B* 69: 205316 (12).

Moskalets, M. and Büttiker, M. 2005. Magnetic field symmetry of pump currents of adiabatically driven mesoscopic structures. *Physical Review B* 72: 035324 (11).

Pastawski, H. 1992. Classical and quantum transport from generalized Landauer-Büttiker equations. II. Time-dependent resonant tunneling, *Physical Review B* 46: 4053–4070.

Platero, G. and Aguado, R. 2004. Photon-assisted transport in semiconductor nanostructures. *Physics Reports* 395: 1–157.

Rey, M., Strass, M., Kohler, S., Hänggi, P., and Sols, F. 2007. Nonadiabatic electron heat pump. *Physical Review B* 76: 085337 (4).

Schwinger, J. 1961. Brownian motion of a quantum oscillator. *Journal of Mathematical Physics* 2: 407–432.

Segal, D. and Nitzan, A. 2006. Molecular heat pump. *Physical Review E* 73: 026109 (9).

Wang, B. and Wang, J. 2002. Heat current in a parametric quantum pump. *Physical Review B* 66: 125310 (4).

Artificial Chemically Powered Nanomotors

Yu-Guo Tao
University of Toronto

Raymond Kapral
University of Toronto

39.1 Introduction

Self-propelled objects come in all shapes and sizes. While macroscopic motors that provide propulsive force have been the subject of extensive development over centuries and are part of our daily lives, the construction and design of nanoscale molecular motors is a much more recent topic of research. Molecular machines (see, e.g., Ballardini et al. 2001, Balzani et al. 2003, Kinbara and Aida 2005, Kay et al. 2007) have components with nanoscale dimensions and a well-defined structure designed to perform useful functions. When acting as molecular motors, such molecular machines consume energy in some form and covert it into work. In this chapter, we consider the operation of molecular motors that use chemical energy to effect directed motion.

Nature has fabricated a large variety of molecular motors and our very existence depends on the crucial roles they play in biological function (Hess et al. 2004, Mavroidis et al. 2004). The development of experimental techniques that allow one to fabricate and observe structures on the nanoscale has provided the stimulus to construct artificial molecular machines that operate, in some respects, like macroscopic machines. While some functions of molecular machines mimic those of their macroscopic counterparts, the principles of their operation must account for new features. These small machines operate in environments with strong molecular fluctuations and must be able to operate effectively in spite of such strong perturbations. In addition, these machines must be able to overcome the effects of strong frictional forces that their moving components experience (Jülicher et al. 1997, Reimann 2002, Astumian 2007). Biological

systems have evolved in ways that both exploit and mitigate these factors. As a result, nature has provided clues to the design principles that are needed to construct useful synthetic motors and machines.

There is widespread and growing interest in artificial nanoscale and micron-scale motors (Kottas et al. 2005, Shirai et al. 2006, Kay et al. 2007). Work in this area addresses fundamental issues and presents technical challenges. Although a considerable body of work on this topic exists, research is still at an early stage of development. Experimental techniques are being refined, effective molecular design principles are being developed, and mechanisms for self-propulsion are being elucidated. Such fundamental work is a prerequisite for potential applications of nanoscale motors, many of which are still in the realm of science fiction but have the potential to become reality.

This chapter deals with how molecular motors may be modeled and how their operation can be described in theoretical terms. Often the principles of operation of molecular motors are based on macroscopic concepts extended to nanoscales. In more refined approaches, the effects of fluctuations are included by employing stochastic models that represent the environment in which the motors operate by fluctuating random forces. We show how motors and their environments can be described on a mesoscopic scale that accounts for the discrete nature of the solvent molecules and the intermolecular forces that ultimately govern the dynamics.

We begin with a brief overview of biological and synthetic nanomotors, which focuses on the main mechanisms for their operation. This survey also serves to point out some of the distinctive features of molecular motors that distinguish their operation from that of macroscopic motors. Next, we describe the particle-based mesoscopic dynamics that forms the basis for our models of molecular motor operation. In the remaining sections, we discuss specific models of molecular motors, the features that one may extract from simulations of their dynamics, the theoretical description of the origin of self-propulsion, the factors that influence their motion, and the magnitudes of the velocities they can achieve.

39.2 Natural and Synthetic Motors

An examination of the different ways nature has used to effect self-propulsion on molecular and mesoscopic scales reveals certain recurring strategies for generating propulsive forces and coping with strong frictional and fluctuating forces arising from the environment. Synthetic molecular motors are typically much simpler than biological motors but face similar design challenges. Below, we sketch some of the elements that are involved in molecular self-propulsion in both biological and synthetic nanomotors.

39.2.1 Biological Nano- and Micron-Scale Motors

Molecular motors are ubiquitous in biology and biological systems make extensive use of self-propelled motion for essential aspects of their function. Molecular motors such as kinesin, myosin, and ribonucleic acid (RNA) polymerase move along filaments or nucleotide strands and are responsible for the active transport of organelles, vesicles, and other material in the cell, cell division, muscle contraction, etc. A large body of experimental and theoretical work on such motors has identified key elements that are important for the operation of these nanoscale motors (Vale and Milligan 2000, Molloy and Veigel 2003, Yildiz 2006). Chemical energy, often in the form of the conversion of adenosine triphosphate (ATP) to adenosine diphosphate (ADP), is used to drive conformational changes in protein motors. A symmetry-breaking element, for instance, attachment to a microtubule or filament with a definite polarity or an asymmetry in conformational changes in the cycle that returns the motor to its initial conformational state, is responsible for directed motion of the motor. Purcell (1977) presented the simple example of such an asymmetric cyclic sequence of conformational changes shown in Figure 39.1.

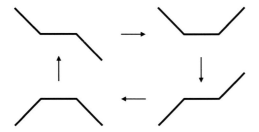

FIGURE 39.1 A sequence of conformational changes that is not the same when carried out in the forward and reverse cyclic directions.

Rotary motors like ATP synthase utilize an electrochemical potential gradient to manufacture ATP or, operating in reverse, consume ATP to generate an electrochemical gradient. The ATP synthase motor is a component of the flagellar drive motor that propels bacteria through solution. Again, the asymmetrical flagellar movement is responsible for directed motion; for instance in *E. coli* (Berg 2003), a micron-scale bacterium, helical flagellar motion with a specific chirality propels the bacterium.

Since protein motors or microorganisms are small nano- or micron-scale objects that move with small velocities in viscous media, the Reynolds numbers that characterize their motion are very small. The Reynolds number $Re = V\ell/\nu$, where V is the speed, ℓ is the linear dimension of the object, and ν is the kinematic viscosity, gauges the relative importance of inertial and viscous forces in the system. The Reynolds number may also be written as the ratio of the Stokes time, $\tau_s = \ell/V$, and the kinematic time, $\tau_\nu = \ell^2/\nu$, $Re = \tau_\nu/\tau_s$. The Stokes time measures the time it takes for a particle to move a distance equal to its length while the kinematic time measures the time it takes for the momentum to diffuse over that distance. For small particle motion in dense fluids, inertial effects are unimportant and the Reynolds number is small. For instance, for an *E. coli* bacterium with a linear dimension of $1\,\mu m$ swimming with a speed of $10\,\mu m/s$ in a medium with kinematic viscosity $10^6\,\mu m^2/s$, the Reynolds number is of the order 10^{-5}. Since such small swimmers operate at low Reynolds number, they cannot make use of the inertia to swim. They must exploit an asymmetry to achieve directed motion (Purcell 1977).

Not all propulsion in biological systems is achieved by chemically induced conformational changes. For example, the cyanobacterium Synechococcus does not possess flagella or other organelles associated with motility but is able to move at speeds of a few $\mu m/min$ on solid surfaces. Strains of this bacterium have been isolated that are able to undergo directed motion in the absence of contact with surfaces. Other gliding bacteria such as *Myxococcus xanthus* also move by gliding over a solid surface. While it is likely that more than one mechanism is responsible for the motion of gliding bacteria, propulsion mechanisms based on surface tension gradients have been proposed to explain the origin of the motion of *M. xanthus*. Such mechanisms are easy to understand. Consider a spherical particle in the interface between two immiscible fluids as shown in Figure 39.2.

The components of the forces parallel to the plane of the interface arising from surface tension will cancel so that there is no net motion. Next, suppose that the (spherical) bacterium excretes surfactant molecules from a site on its surface. This will produce an asymmetric concentration distribution of surfactant molecules that depends on the rate at which surfactant molecules are produced and the effectiveness of diffusion to remove the gradient so generated. Consequently, the surface tension σ will no longer be the same at all points on the phase contact line (see force direction in Figure 39.2) and the net force in the plane, **F**, will not vanish. In the steady state, this force will be balanced by the frictional force acting on the particle, $-\zeta \mathbf{V}$, where ζ is the

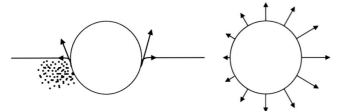

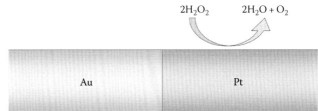

FIGURE 39.2 Schematic diagram showing surface-tension-induced directed motion. (left, side view) A spherical particle in an interface between two immiscible phases. The arrows indicate the forces arising from surface tension and their projections in the plane of the interface. The large particle secrets surfactant molecules (small black dots) from a localized site on its surface. (right, top view) The surfactant molecules lower the surface tension and the asymmetric concentration distribution gives rise to a net directed force that causes the particle to move.

FIGURE 39.3 A gold–platinum nanorod in a hydrogen peroxide solution is propelled by the catalytic decomposition of hydrogen peroxide to oxygen at the Pt end.

friction coefficient and **V** is the particle velocity. Thus, $\mathbf{V} = \mathbf{F}/\zeta$. Usually it is assumed that the surface tension is proportional to the surfactant concentration, which can be determined from the solution of the diffusion equation for a given localized production rate on the surface of the particle. The friction coefficient depends on the object and for a large spherical particle has the Stokes form $\zeta = 6\pi\eta R$ with η as the solvent viscosity. The velocity of the particle propelled by this mechanism is then easily determined.

39.2.2 Synthetic Nanomotors

Synthetic molecular motors with a variety of forms and fabricated from various materials have been constructed (Kay et al. 2007). These molecular motors use chemical, light, or other energy sources to perform directed motion. Like biological motors, some synthetic motors rely on asymmetric molecular motions for propulsion while others have no moving parts and make use of different mechanisms. A few examples will serve to illustrate some of the synthetic motor designs and mechanisms.

A synthetic motor that mimics flagellar swimming motion was constructed by Dreyfus et al. (2005). The motor consists of a cargo region (a red blood cell in this case) and tail constructed from magnetic colloidal particles tethered to each other by double-stranded deoxyribonucleic acid (DNA) biopolymers. The tail orients in an external magnetic field and a perpendicular oscillatory magnetic field is applied to the system to induce motion in the tail. For suitable magnetic field and flexibility of the tail, time reversal symmetry is broken and directed motion is produced. A large number of other types of motors that act as rotors, switches, brakes, and ratchets have also been constructed (Feringa 2007, Balzani et al. 2008, Pollard et al. 2008). Rotaxanes in external electric fields or driven by photochemical stimuli undergo rotational motion (Qu et al. 2005). Chemically driven rotors and motors with other types of actions have also been fabricated (Catchmark et al. 2005, Fletcher et al. 2005, Alvarez-Pérez et al. 2008).

Synthetic motors that do not depend on conformational changes for their operation include striped bimetallic nanorods (Paxton et al. 2004, Fournier-Bidoz et al. 2005) and synthetic catalytic molecules tethered to inactive particles (Vicario et al. 2005). The bimetallic nanorod motors consist of platinum and gold segments (see Figure 39.3). When such rods are immersed in hydrogen peroxide aqueous solution, the rods execute directed motion. The chemical decomposition of hydrogen peroxide to oxygen occurs at the Pt catalytic end. A number of different mechanisms for the motion have been proposed (Ozin et al. 2005, Paxton et al. 2006, Wang et al. 2006, Kovtyukhova 2008, Saidulu and Sebastian 2008). The direction of the movement is toward the platinum ends of the rods. One mechanism relates the motion to an oxygen concentration gradient that occurs from the Pt/Au junction to the end of the gold segment of the nanorod. Oxygen is generated uniformly and selectively on the platinum segment of the rod. As the oxygen concentration decreases, the interfacial tension gradient at the interface increases, causing the nanorod to be propelled in the direction of the platinum end.

39.3 Models of Motor Dynamics

In parallel with these synthetic approaches to the construction of molecular motors, simple theoretical and computational models have been proposed that capture features essential to self-propelled motion (Najafi and Golestanian 2004, Golestanian et al. 2005, Rückner and Kapral 2007). The models for self-propelled motion are often based on continuum deterministic or stochastic descriptions of the solvent in which they move and on macroscopic force-generating mechanisms, such as surface tension gradients, diffusiophoretic effects, etc. Microscopic and mesoscopic simulations of model motors and swimmers have also been carried out (Earl et al. 2007, Pooley et al. 2007, Rapaport 2007, Tao and Kapral 2008).

Simple ratchet models capture some of the essential physics of directed motion. Different types of ratchet models have been devised. The flashing ratchet is one of the most popular and it is easy to see how directed motion is a consequence of an asymmetry in the potential and an externally imposed colored stochastic process or periodic modulation.

A particle moves in an asymmetric periodic potential which is switched on and off according to some protocol (see Figure 39.4).

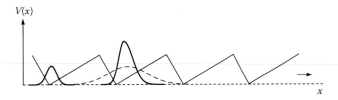

FIGURE 39.4 Schematic diagram showing potential energy functions and position probability densities for a simple flashing ratchet.

The equation of motion for the coordinate x of the particle is assumed to obey overdamped stochastic dynamics described by the Langevin equation,

$$\zeta \frac{dx(t)}{dt} = -\xi(t)\frac{\partial V(x)}{\partial x} + f(t), \qquad (39.1)$$

where

  ζ is the friction coefficient

  $f(t)$ is a Gaussian white noise force with zero mean satisfying the fluctuation-dissipation relation, $\langle f(t_1)f(t_2)\rangle = 2k_B T\zeta\delta(t_1 - t_2)$

  $\xi(t)$ is a random variable that takes the values 0 or 1 sampled from a given distribution

When the system is governed by the asymmetric periodic potential the position probability density is confined to the well regions. When the potential is turned off, the probability density spreads (dashed curve). With the potential on again, the portion of the probability density that lies outside of the maxima will relax into the neighboring well regions in an asymmetric fashion under the overdamped dynamics. This induces a flux of particles to the left. This ratchet model and its variants have been used as simple models of real physical and biological systems.

A very simple model for a one-dimensional swimmer which is driven by asymmetric conformational change has been proposed and studied (Najafi and Golestanian 2004). The model swimmer consists of three spheres linked by rods whose lengths can change between two values. The swimmer undergoes nonreciprocal deformations which break the time-reversal and translational symmetry (see Figure 39.5a). The middle sphere acts as an internal engine which makes the nonreciprocal motion needed to propel the entire system. Initially, the right arm is fixed while the length of the left arm is decreased at a constant relative velocity. Next, the left arm is fixed and the right arm decreases its length at the same constant relative velocity. Following this, the right arm is kept fixed, and the length of the left arm increases. Finally, the left arm is kept fixed and the right arm elongates to its original length. The model device executes these four steps and returns to its original configuration, moving a net distance d to the right. Translational motion occurs as a result of repeated cyclic motion of the motor.

An example of a model reaction-driven nanomotor is shown in Figure 39.5b. An enzyme, modeled as spherical particle with a single active site at a fixed position on its surface, catalyzes

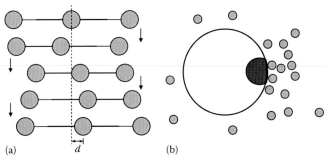

FIGURE 39.5 (a) Complete cycle of the nonreciprocal motion of the one-dimensional swimmer. The model system is displaced to the right by an amount d upon completion of the four-step cycle. (b) A spherical nanoparticle with a single enzymatic site which releases reaction products asymmetrically.

the production of product molecules (Golestanian et al. 2005). This simple molecular motor is driven by phoretic forces on the model enzyme arising from an asymmetric distribution of reaction products. The mechanism is similar to that discussed earlier for models of surface-tension-driven motion of bacteria. The basic elements of such a mechanism will also be employed in the mesoscopic models we discuss below.

39.4 Mesoscopic Dynamics

The environment or solvent plays an important role in the function of molecular motors since fluctuating solvent forces determine many of the motor characteristics on the small scales on which such motors operate. For molecular motors that make use of a sequence of conformational changes for their operation, solvent hydrodynamical effects are important for determining the frictional forces acting on the motor. Even for self-propelled nanoscale objects that do not rely on conformational changes for their propulsion, solvent collective effects influence the motion. Since molecular self-propelled objects swim in highly fluctuating environments, any models for their motion must account for these fluctuations, and theoretical descriptions must explain how such objects can overcome the effects of fluctuations to perform directed motion. Because we are interested in nanoscale self-propelled objects, it is also appropriate to treat the solvent at mesoscopic level. In order to fulfill these requirements, we describe both the molecular motor and its environment at a mesoscopic level. The coupled motion of the motor and its environment is described by a hybrid scheme that combines molecular dynamics (MD) for the motor with mesoscopic multiparticle collision (MPC) dynamics for the solvent (Malevanets and Kapral 1999, Kapral 2008, Gompper et al. 2009). The mesoscale description should be as close as possible to full MD in order to preserve the essential features of the dynamics. We begin with a brief description of MPC dynamics for the solvent, followed by a description of how it may be combined with MD for solute molecules to yield the hybrid model used in the motor simulations presented below.

Molecules move and collide. In MPC dynamics collisions are carried out at mesoscopic level that preserves mass, momentum, and energy conservation (Malevanets and Kapral 1999). More specifically, consider a system of N particles with masses m in a volume V. Particle i has position $\mathbf{r}_i$ and velocity $\mathbf{v}_i$ and the microscopic state of the system is specified by the positions and velocities of all N particles, $\mathbf{x}^N \equiv (\mathbf{r}^N, \mathbf{v}^N) = (\mathbf{r}_1, \mathbf{r}_2, ..., \mathbf{r}_N, \mathbf{v}_1, \mathbf{v}_2, ..., \mathbf{v}_N)$. Particles undergo collisions at discrete time intervals τ and free stream between such collisions. If the position of particle i at time t is $\mathbf{r}_i$, its position at time $t + \tau$ is

$$\mathbf{r}_i^\star = \mathbf{r}_i + \mathbf{v}_i \tau. \tag{39.2}$$

MPC collisions mimic the effects of many real collisions in the system and are carried out as follows: The volume V is divided into N_c cells labeled by cell indices ξ. Each cell is assigned at random a rotation operator $\hat{\omega}_\xi$ chosen from a set Ω of rotation operators. The center of mass velocity of the particles in cell ξ is $V_\xi = N_\xi^{-1} \sum_{i=1}^{N_\xi} \mathbf{v}_i$ where N_ξ is the instantaneous number of particles in the cell. The post-collision velocities of the particles in the cell are then given by

$$\mathbf{v}_i^\star = V_\xi + \hat{\omega}_\xi (\mathbf{v}_i - V_\xi). \tag{39.3}$$

The set of rotation operators can be chosen in various ways and the specific choice will determine the values of the transport properties of the system. Below, we use stochastic rotations about a randomly chosen direction, $\hat{\mathbf{n}}$, by an angle $\alpha = 90°$ (Malevanets and Kapral 1999).

In MPC dynamics, the same rotation operator is applied to each particle in the cell ξ but every cell in the system is assigned a different rotation operator so that collisions in different cells are independent of each other. Not only does MPC dynamics conserve mass, momentum, and energy, but also preserves phase space volumes. These are features shared by full molecular dynamics. The macroscopic transport equations and transport properties of a system obeying MPC dynamics may be determined (Malevanets and Kapral 2000, Ihle and Kroll 2001, Kikuchi et al. 2003) and this provides a link between the mesoscale dynamics and macroscopic behavior. The link between mesoscopic simulation and physical length and time scales has been discussed by Padding and Louis (2006).

In hybrid MPC-MD dynamics, the solute molecules interact with the solvent through an intermolecular potential (Malevanets and Kapral 2000). For a system with N_s solute molecules and N_b solvent particles, let $V_s(\mathbf{r}^{N_s})$ be the intermolecular potential among the N_s solute molecules and $V_{sb}(\mathbf{r}^{N_s}, \mathbf{r}^{N_b})$ the interaction potential between the solute and solvent particles. Solvent–solvent particle intermolecular forces are not needed since these are accounted for by MPC dynamics. The hybrid MD-MPC scheme is obtained by replacing the free streaming step in Equation 39.2 by streaming in the intermolecular potential, $V(\mathbf{r}^{N_s}, \mathbf{r}^{N_b}) = V_s(\mathbf{r}^{N_s}) + V_{sb}(\mathbf{r}^{N_s}, \mathbf{r}^{N_b})$, which is generated by the solution of Newton's equations of motion,

$$\dot{\mathbf{r}}_i = \mathbf{v}_i \quad m_i \dot{\mathbf{v}}_i = -\frac{\partial V}{\partial \mathbf{r}_i} = \mathbf{F}_i. \tag{39.4}$$

Multiparticle collisions are carried out at time intervals τ. This hybrid MD-MPC dynamics satisfies the conservation laws and preserves phase space volumes. As a result, solvent hydrodynamic collective effects and hydrodynamic interactions are properly taken into account. Because the solvent is treated at a particle level, albeit coarse grained, mesoscale fluctuations are also automatically incorporated in the dynamical description. With this mesoscopic dynamics in hand, we may now consider the construction on self-propelled nanomotors.

39.5 Mesoscopic Description of Catalytic Molecular Motors

A class of chemically powered catalytic molecular motors may be constructed using a mesoscopic description of the entire system. These motors operate by mechanisms which are similar to those for self-propelled objects that rely on an inhomogeneous distribution of molecules and force generation by surface tension or phoretic effects. However, our mesoscopic modeling method allows us to discuss the force generation mechanism by directly taking into account the intermolecular forces acting on the motor. As in the reaction-driven self-propelled enzyme model, in these motors there is a site or group that catalyzes a chemical reaction. The product molecules are distributed inhomogeneously in the vicinity of the catalytic site and interact with the motor body. These interactions provide the source of power for the motor. Such motors do not have to carry their own fuel and operate far from equilibrium.

Figure 39.6 shows the type of catalytic molecular motor being considered. The body of the motor consists of a network of beads that are coarse-grained representations of molecular groups. Depending on the nature of the interactions among these beads, this motor body can adopt specific conformations corresponding

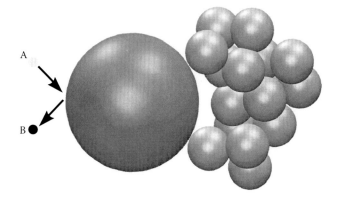

FIGURE 39.6 Structure of a catalytic molecular motor consisting of a catalytic site attached to the body of the motor. The motor body comprises molecular groups which may adopt different configurations depending on the motor design.

to proteins or other macromolecules, nanocolloidal particles or super-molecular structures designed to have specific characteristics. One of the beads is distinguished by being the site that catalyzes chemical reactions. For such an object to function as a motor, we need to specify how chemical reactions occur at the catalytic site and how the various chemical species in the solvent environment interact with the motor.

Suppose there are $N_m + 1$ molecular groups comprising the motor and N_b molecules in the solvent or bath. The $(N_m + 1)$st group is the catalytic site C. The solvent is composed of species $\alpha = A, B, \ldots$ and $N_b = \sum_\alpha N_\alpha$. The total potential energy $V(\mathbf{r}^{N_m+1}, \mathbf{r}^{N_b})$ of the system can be written as the sum of solvent molecule–molecular motor contributions $V_{mb}(r_{ij}^{mb})$ along with the potential energy contributions within the motor itself, $V_{mm}(r_{ij}^{mm})$. We write the total potential energy as the sum of pairwise contributions,

$$V(\mathbf{r}^{N_m+1}, \mathbf{r}^{N_b}) = \sum_{i=1}^{N_b} \sum_{j=1}^{N_m+1} V_{bm}(r_{ij}^{bm}) + \sum_{i=1}^{N_m+1} \sum_{j=1}^{N_m+1} V_{mm}(r_{ij}^{mm}). \quad (39.5)$$

The coordinates are labeled by groups to which they belong and relative coordinates carry two labels. There are no solvent–solvent interactions since these are accounted for by MPC dynamics as discussed above.

The chemical reaction $A + C \rightarrow B + C$ occurs with probability p_R whenever an A molecule encounters (comes within a specified reaction distance from) the catalytic portion of the motor (Tucci and Kapral 2004). While, for simplicity, this reaction is irreversible and thermoneutral and amounts to a change in species identity on encounter, the reaction scheme may be generalized easily to encompass reactions which are not thermoneutral and may involve more complicated reversible reaction mechanisms.

We are interested in the motion of the motor as whole, and so we focus on the velocity of the center of mass of the motor, $\mathbf{V}$. The total force acting on the motor is

$$\begin{aligned}
\mathbf{F} = &-\sum_\alpha \sum_{i=1}^{N_\alpha} \left(\frac{\partial V_{\alpha C}(r_{iC}^{\alpha C})}{\partial \mathbf{r}_C} + \sum_{j=1}^{N_m} \frac{\partial V_{\alpha m}(r_{ij}^{\alpha m})}{\partial \mathbf{r}_j^m} \right) \\
= &-\sum_\alpha \int d\mathbf{r}\, \rho_\alpha(\mathbf{r}; \mathbf{r}^{N_\alpha}) \left(\frac{\partial V_{\alpha C}(|\mathbf{r} - \mathbf{r}_C|)}{\partial \mathbf{r}_C} \right. \\
&\left. + \int d\mathbf{r}_m \frac{\partial V_{\alpha m}(|\mathbf{r} - \mathbf{r}_m|)}{\partial \mathbf{r}_m} \rho_m(\mathbf{r}_m; \mathbf{r}^{N_m}) \right),
\end{aligned} \quad (39.6)$$

where we have explicitly written the contributions from the catalytic site and the body of the motor. In the last line of this equation, we introduced the microscopic densities of solvent and motor body densities: $\rho_\alpha(\mathbf{r}; \mathbf{r}^{N_\alpha}) = \sum_{i=1}^{N_\alpha} \delta(\mathbf{r} - \mathbf{r}_i^\alpha)$ and $\rho_m(\mathbf{r}_m; \mathbf{r}^{N_m}) = \sum_{j=1}^{N_m} \delta(\mathbf{r}_m - \mathbf{r}_i^m)$.

If the solvent contains a large number of A molecules initially, as a result of chemical reactions at the catalytic site a quasi steady state inhomogeneous distribution of B molecules will be formed after a transient period, provided the time is not so long that the total A molecule concentration is significantly depleted. A true far-from-equilibrium steady-state distribution can be established by input and removal of reactants and products from the system. In either case, it is this nonequilibrium steady-state distribution that is of interest. We denote these nonequilibrium averages by $\rho_\alpha(\mathbf{r}) = \langle \rho_\alpha(\mathbf{r}; \mathbf{r}^{N_\alpha}) \rangle$ and $\rho_{\alpha m}(\mathbf{r}, \mathbf{r}_m) = \langle \rho_\alpha(\mathbf{r}; \mathbf{r}^{N_\alpha}) \rho_m(\mathbf{r}_m; \mathbf{r}^{N_m}) \rangle \approx \rho_\alpha(\mathbf{r}) \rho_m(\mathbf{r}_m)$, where we have assumed the two particle bath-motor distribution factors, an approximation which is expected to be valid when the solvent concentration distribution does not strongly perturb the motor body distribution. The nonequilibrium average force on the motor for a fixed set of motor coordinates is then given by

$$\langle \mathbf{F} \rangle = -\sum_\alpha \int d\mathbf{r}\, \rho_\alpha(\mathbf{r}) \left(\frac{\partial V_{\alpha C}(|\mathbf{r} - \mathbf{r}_C|)}{\partial \mathbf{r}_C} + \int d\mathbf{r}_m \frac{\partial V_{\alpha m}(|\mathbf{r} - \mathbf{r}_m|)}{\partial \mathbf{r}_m} \rho_m(\mathbf{r}_m) \right).$$

$$(39.7)$$

At equilibrium the total average force on the motor vanishes so that directed motion is not possible.

In the steady state, the fixed particle force on the motor is balanced by the frictional force that the motor experiences as it moves through the solvent. Neglecting contributions internal to the motor, this force balance may be written as $\boldsymbol{\zeta} \cdot \mathbf{V} = \langle \mathbf{F} \rangle$, where $\boldsymbol{\zeta}$ is the friction tensor (Happel and Brenner 1965, Lee and Kapral 2005). Thus, the motor velocity is given by $\mathbf{V} = \boldsymbol{\zeta}^{-1} \cdot \langle \mathbf{F} \rangle$. This formula contains the elements that are essential for self-propelled directed motion. In particular, in order to compute the velocity, we require a knowledge of the nonequilibrium density fields and the motor friction coefficient, as well as information about the potential energy functions that couple the motor to the solvent. In the next section, we show how the motor dynamics may be simulated and how the motor velocity can be estimated on the basis of the expressions developed here.

39.6 Nanodimer Motors

To illustrate how self-propelled molecular motion can be simulated and the motor velocity can be estimated theoretically, we consider a simple nanodimer motor. In this case, the body of the motor consists of a single noncatalytic sphere $N(N_m = 1)$ linked to a catalytic sphere C by a rigid bond of length R. The solvent comprises a large number $N_b = N_A + N_B$ of point-like A and B molecules with identical masses m. The chemical species in the solvent interact with catalytic and noncatalytic sites through either repulsive or attractive Lennard-Jones (LJ) potentials, which are smoothly truncated to zero so that the potentials have a finite range. The chemical reaction $A \rightarrow B$ takes place with a probability p_R whenever an A molecule approaches to within a reaction distance R_0 of the

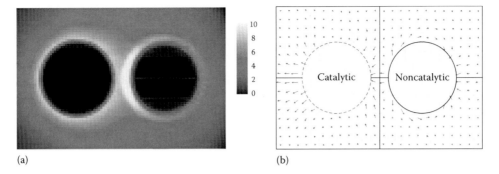

(a) (b)

FIGURE 39.7 (a) The average number density profile of B molecules in the vicinity of the nanodimer, taken from a $24 \times 16 \times 2$ slice parallel to the nanodimer internuclear axis. The simulation was carried out in a 48^3 cubic box. The diameters of both catalytic and noncatalytic spheres are identical, $d_C = d_N = 8.0$. The noncatalytic monomer interacts with solvent B molecules through attractive LJ potentials with $\varepsilon_B = 0.1$. Other parameters are $\varepsilon_A = 1.0$, $T = 1/6$, $R = 10.0$, and $\tau = 0.5$. (b) Solvent molecule velocity field near the dimer spheres.

catalytic sphere. The system is maintained in a steady state by fluxes of reactants and products far from the dimer.*

39.6.1 Simulations of Nanodimer Motion

As is evident from the discussion of the propulsion mechanism, the structure of the nonequilibrium B particle density field in the vicinity of the molecular motor is a quantity of central importance for the existence of directed motion. The A and B concentration fields in the vicinity of the dimer are strongly inhomogeneous. The density gradient of species B is seen more clearly in Figure 39.7a, which shows that the average B density field varies from high to low as one traverses the dimer from the catalytic to noncatalytic ends.

The fluid flow field induced by dimer motion is also an important factor influencing the velocity of the nanodimer. The local solvent velocity field in the vicinity of the dimer is plotted in Figure 39.7b. Collective hydrodynamic effects that emerge from the particle-based MPC dynamics at small scales manifest themselves in the solvent "backflow" seen at the rears of the dimer spheres. The dimer moves in the direction of the catalytic sphere

and collective hydrodynamic effects enhance this motion. Such collective effects have their origin in the coupling between the dimer velocity and the fluid viscous modes. Similar effects have been observed in early molecular dynamics simulations of the velocity correlation function and are responsible for the long-time tails seen in this function (Alder and Wainwright 1967).

Figure 39.8 shows a trajectory of the self-propelled dimer. For this particular trajectory, the component velocity along the x-direction, the direction of the dimer internuclear axis, is much larger than the components in the transverse directions. The average center-of-mass velocity of the dimer projected along the instantaneous internuclear axis is $V_\parallel \equiv \langle \mathbf{V}(t) \cdot \hat{\mathbf{R}}(t) \rangle \approx 0.064$, where the angle brackets denote a time average. The translational D_t and rotational D_r diffusion coefficients can be calculated by computing the mean square displacement and auto correlation function of the unit vector along the internuclear axis, respectively. These transport coefficients are found to be $D_t \approx 3.0$ and $D_r \approx 2.8 \times 10^{-5}$. The orientational relaxation time τ_θ can be determined from D_r and is $\tau_\theta = (2D_r)^{-1} = 1.8 \times 10^4$. Given the value of velocity along the internuclear axis, we see that the dimer will move about 20 times its length on average before

---

* In our simulations, all quantities are reported in dimensionless LJ units based on energy ε, mass m, and distance σ parameters: $r/\sigma \to r$, $t(\varepsilon/m\sigma^2)^{1/2} \to t$ and $k_B T/\varepsilon \to T$. The nanodimer is dissolved in a solvent of A and B molecules within a cubic box of volume V with periodic boundary conditions. The simulation box was then subdivided into $(L_b)^3$ cells in order to perform multiparticle collisions. The rotation angle was fixed at $\alpha = 90°$. We chose an average number density of $n_0 \approx 9.2$ in all simulations; thus, for example, the total number of solvent particles is $\approx 10^6$ in a system with $(48)^3$ collision cells. The masses of both A and B species were taken to be $m = 1$, while the masses of the C and N spheres were adjusted according to their volumes to ensure that the dimer was approximately neutrally buoyant. The MD time step used to integrate Newton's equations of motion with the velocity Verlet algorithm was $\Delta t = 0.01$, while the multiparticle collision time ranged from $\tau = 0.1$ to 1.0. The system temperature varied from $T = 1/12$ to $2/3$. To prevent discontinuous potential changes when the A $\to$ B reaction occurs, the internuclear separation R was fixed by a holonomic constraint in the equation of motion to be larger than the interaction distance. The LJ potential parameter was chosen to be $\varepsilon_A = 1.0$ in most of the simulations, while ε_B varied from 0.1 to 10.0 to change both the speed and direction of the movement of the dimer.

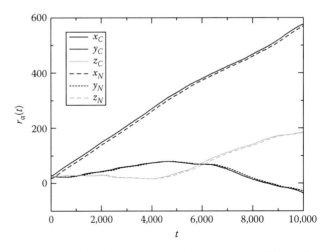

FIGURE 39.8 The three components of the positions of the catalytic (solid) and noncatalytic (dashed) spheres as a function of time, for the same system as in Figure 39.7.

reorientation occurs. Hence, for these dimer parameters there is strong directed motion. Consequently, the directed movement is not masked by Brownian translational and rotational motions.

The dimer velocity depends on factors such as the temperature, solvent viscosity, the dimer bond length R, and the sphere radii, as well as the interactions between the dimer spheres and the solvent. The reduced temperature, $k_BT/\varepsilon \to T$, controls the magnitude of the thermal fluctuations in the system, provides the scale for the dimer–solvent interactions, and enters the reaction rate constant and the friction coefficient. The average velocity along the internuclear axis increases with temperature, but its probability distribution $p(V_\parallel)$ becomes broader, indicating that Brownian thermal motion dominates the directed movement at high temperature. The internuclear separation R affects the nonequilibrium A and B particle density fields in the vicinity of the noncatalytic sphere. Theoretically, $V_\parallel$ is inversely proportional to R^2. Consequently, the two monomers of the nanodimer should be as close as possible to achieve large velocity.

The force on the dimer varies strongly when the interaction potential between the noncatalytic sphere and the solvent B molecules changes from being repulsive to attractive. As one can see from the results in Figure 39.9, the directed motion of the nanodimer dominates Brownian motion for the parameter values used in the simulations. The simulated average velocities are found to be $V_\parallel = 0.056$ and 0.030 for attractive and repulsive LJ potentials with $\varepsilon_B = 0.1$, respectively. In these two examples, "forward"-directed movement of the nanodimers is observed, where the dimer moves in the direction of the catalytic

monomer. Figure 39.9 also plots results for the nanodimer with $\varepsilon_B = 5.0$ (dotted curve). The dimer executes "reverse"-directed motion in contrast to that for $\varepsilon_B = 0.1$. Note that in our model, to construct a backward-moving nanodimer ($\varepsilon_B > \varepsilon_A$) using attractive LJ potentials between the noncatalytic sphere and solvent B molecules, the energy parameter ε must be chosen to be small enough to avoid a large accumulation of solvent B molecules around the nanodimer. However, weak interactions will lead to directed motion with smaller velocities and Brownian thermal fluctuations will again play an important role.

Figure 39.10 shows how $V_\parallel$ changes when the diameters of the catalytic and noncatalytic spheres vary. For fixed d_C, temperature, and energy parameters, increasing the size of the noncatalytic sphere leads to a larger driving force since the B particle gradient over the noncatalytic sphere surface increases. Increasing the size of the noncatalytic monomer in the nanodimer also increases the friction coefficient, but this increase in friction is dominated by the increase in force on the dimer. For a given size of the noncatalytic sphere, reducing the size of the catalytic sphere reduces the chemical reaction rate, so that the nonequilibrium gradient in the A and B particle densities is smaller. Increasing the size of the catalytic sphere also increases the internuclear separation between the two monomers, which tends to reduce the average velocity of the dimer. Thus, the increase of $V_\parallel$ with d_C saturates due to the competition between these two effects and is even seen to decrease slightly.

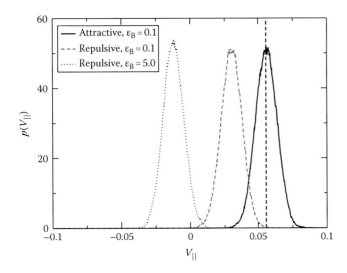

FIGURE 39.9 Probability distribution functions of the center-of-mass velocity of the nanodimer along the internuclear axis. The solid curve corresponds to a system in which B molecules interact with the noncatalytic sphere through attractive LJ potentials with $\varepsilon_B = 0.1$. The dashed and dotted curve plot results for systems where repulsive LJ potentials exist between B solvent molecules and the N sphere, with $\varepsilon_B = 0.1$ and 5.0, respectively. In these examples, the potential parameter $\varepsilon_A = 1.0$. The system temperature is $T = 1/6$, while the diameters of the catalytic and the noncatalytic spheres are $d_C = 4.0$ and $d_N = 8.0$, respectively.

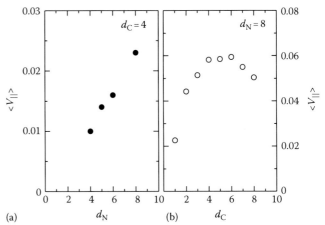

FIGURE 39.10 The average center-of-mass velocity of the nanodimer along the internuclear axis. (a) The diameter of the catalytic sphere is fixed at $d_C = 4.0$, while the noncatalytic sphere diameter d_N varies. In these simulations, repulsive LJ potentials exist between the noncatalytic sphere and both types of solvent molecules with $\varepsilon_A = 1.0$ and $\varepsilon_B = 0.1$, respectively. The internuclear separations are chosen to be as close as possible and consistent with the potential energy cutoff values; the corresponding R values are given in Table 39.1. (b) The diameter of the catalytic sphere d_C varies while the noncatalytic sphere diameter $d_N = 8.0$. The B molecules interact with the noncatalytic sphere through attractive LJ potentials while repulsive LJ potentials exist between the noncatalytic sphere and A molecules. The potential parameters are as same as those used in the simulations shown in Figure 39.7.

39.6.2 Estimate of Nanodimer Velocity

For a nanodimer, Equation 39.5 takes the simple form,

$$V(\mathbf{r}^{N_A}, \mathbf{r}^{N_B}) = \sum_{\alpha=A}^{B} \sum_{i=1}^{N_\alpha} [V_{C\alpha}(r_{i\alpha}) + V_{N\alpha}(r'_{i\alpha})], \qquad (39.8)$$

where
  $\mathbf{r}$ refers to coordinates measured with the catalytic C sphere as the origin
  $\hat{\mathbf{z}}$ is the unit vector along the dimer internuclear bond, pointing in the direction of the N sphere

Also, $\mathbf{r}'$ whose origin is at the center of the sphere N is given by $\mathbf{r}' = \mathbf{r} - R\hat{\mathbf{z}}$. Using this notation, $\mathbf{r}_{i\alpha}$ denotes the vector distance to solvent molecule i of species α and $r_{i\alpha}$ its magnitude, and $r'_{i\alpha} = |\mathbf{r}_{i\alpha} - \mathbf{r}_m| = |\mathbf{r}_{i\alpha} - R\hat{\mathbf{z}}|$. For the nanodimer $\rho_m(\mathbf{r}_m) = \delta(\mathbf{r}_m - \mathbf{r}_N)$ and, using Equation 39.7, the average force on the fixed nanodimer along the dimer internuclear axis is given by

$$\langle \hat{\mathbf{z}} \cdot \mathbf{F} \rangle = -\sum_{\alpha=A}^{B} \int d\mathbf{r}\, \rho_\alpha(\mathbf{r}) \left((\hat{\mathbf{z}} \cdot \hat{\mathbf{r}}) \frac{dV_{C\alpha}(r)}{dr} + (\hat{\mathbf{z}} \cdot \hat{\mathbf{r}}') \frac{dV_{N\alpha}(r')}{dr'} \right). \qquad (39.9)$$

Suppose that the A and B molecules have the same interaction potentials with the catalytic sphere but have different interactions with the noncatalytic sphere. For simplicity, we assume the potentials are short ranged so that the solvent molecules do not simultaneously interact with both spheres in the nanodimer. If the total density is assumed to be uniform space,* we may then write the component of the nanodimer velocity along the bond as

$$V_\parallel = \frac{k_B T}{\zeta} \langle \hat{\mathbf{z}} \cdot \beta \mathbf{F} \rangle = \frac{k_B T}{\zeta} \left[\int d\mathbf{r}' n_A(|\mathbf{r}' + R\hat{\mathbf{z}}|)(\hat{\mathbf{z}} \cdot \hat{\mathbf{r}}') \frac{de^{-\beta[V_{NA}(r')]}}{dr'} \right.$$
$$\left. + \int d\mathbf{r}' n_B(|\mathbf{r}' + R\hat{\mathbf{z}}|)(\hat{\mathbf{z}} \cdot \hat{\mathbf{r}}') \frac{de^{-\beta[V_{NB}(r')]}}{dr'} \right], \qquad (39.10)$$

where we have changed the origin for the integration using the change of variables $\mathbf{r} = \mathbf{r}' + R\hat{\mathbf{z}}$. Once the intermolecular potentials have been specified, the remaining ingredient that needs to be determined is the nonequilibrium density field that is produced by the chemical reactions at the catalytic sphere. When the system is in equilibrium the local density field is

$$\rho_\alpha^{eq}(\mathbf{r}) = \left\langle \rho_\alpha(\mathbf{r}; \mathbf{r}^{N_\alpha}) \right\rangle_{eq} = n_\alpha e^{-\beta[V_{C\alpha}(r) + V_{N\alpha}(r')]}, \qquad (39.11)$$

where the angle brackets $\langle \cdots \rangle_{eq}$ denote an equilibrium canonical average. The equilibrium number density is $n_\alpha = N_\alpha^{eq}/V$. The average force on the nanodimer is zero in equilibrium, and this result shows that the directed nanodimer motion is a nonequilibrium effect that rests on the existence of a nonequilibrium distribution of reactive species in the vicinity of the nanodimer.

By analogy with the equilibrium density field, we may define steady-state density field $n_\alpha(\mathbf{r})$ through the relation $\rho_\alpha(\mathbf{r}) = n_\alpha(\mathbf{r}) e^{-\beta[V_{C\alpha}(r) + V_{N\alpha}(r')]}$. The average force and, thus, the nanodimer velocity can be expressed in terms of this steady-state density field. The steady-state density field can be estimated by solving the diffusion equation with a radiation boundary condition, $4\pi D_A R_0^2 \hat{\mathbf{r}} \cdot (\nabla n_A)(\hat{\mathbf{r}} R_0, t) = k_{f0} n_A(\hat{\mathbf{r}} R_0, t)$, to account for reaction.† The reaction rate constant $k_{f0} = p_R \sigma_C^2 \sqrt{8\pi k_B T/m}$ and characterizes the reactive events that occur within a diffusive boundary layer around the catalytic sphere, and D_A is the diffusion coefficient of the A molecules (Kapral 1981). This approximation neglects perturbations of the density fields arising from the presence of the noncatalytic sphere. Under these conditions, the steady-state solution of the diffusion equation yields,

$$n_A(r) = n_A^0 \left(1 - \frac{k_{f0}}{k_{f0} + k_D} \frac{R_0}{r} \right), \quad n_B(r) = n_A^0 \frac{k_{f0}}{k_{f0} + k_D} \frac{R_0}{r}, \qquad (39.12)$$

where the Smoluchowski rate constant is $k_D = 4\pi R_0 D_A$. (See Rückner and Kapral (2007) and Tao and Kapral (2008) for details.) Substituting Equations 39.12 into Equation 39.10 and performing the angular integration analytically, we obtain

$$V_\parallel = \frac{8\pi}{3} \frac{k_B T}{\zeta} n_A^0 \frac{R_0(2^{1/6}\sigma_N^3)}{R^2} \frac{k_{f0}}{k_{f0} + k_D}$$
$$\times \left[\int_0^{u_c} du\, u e^{-\beta V_{NA}(u)} - \int_0^{u_c} du\, u e^{-\beta V_{NB}(u)} \right], \qquad (39.13)$$

where the reduced cutoff distance $u = r\sigma_N$. The radial integrals depend only on the temperature T and the energy parameter ε and can be easily evaluated numerically.

Simulation results are compared with the theoretical predictions in Table 39.1. The simulation results were obtained by fitting the probability distribution function $p(V_\parallel)$ to a Maxwell–Boltzmann distribution. While not quantitatively accurate, the model is able to predict the magnitudes and trends of the nanodimer velocity. Relatively larger discrepancies are found between the theoretical and simulation results for attractive LJ potentials, which is likely due to errors in the steady-state density field resulting from the neglect of the noncatalytic sphere.

The Reynolds number in our simulations varies from $Re \approx 0.03$ to 1.8, while the Reynolds numbers for most synthetic motors in

---

* The close proximity of the two spheres in the nanodimer will cause deviations from a spherically uniform total density, but we neglect this effect here.

† The reaction distance R_0 is chosen equal to the cutoff distance of the repulsive LJ potential $R_0 = 2^{1/6}\sigma_C$.

TABLE 39.1 Average Velocities of the Center-of-Mass of the Nanodimer Along Its Internuclear Axis

d_N	4.0	5.0	6.0	8.0
R	4.5	5.1	6.0	6.8
$\langle V_\parallel \rangle_{\text{simulation}}$	0.010	0.014	0.016	0.023
$\langle V_\parallel \rangle_{\text{theory}}$	0.018	0.024	0.026	0.038
d_C	1.0	2.0	3.0	4.0
R	6.1	6.6	7.2	7.7
$\langle V_\parallel \rangle_{\text{simulation}}$	0.023	0.044	0.051	0.059
$\langle V_\parallel \rangle_{\text{theory}}$	0.032	0.077	0.108	0.132
d_C	5.0	6.0	7.0	8.0
R	8.3	8.9	9.4	10.0
$\langle V_\parallel \rangle_{\text{simulation}}$	0.058	0.060	0.055	0.051
$\langle V_\parallel \rangle_{\text{theory}}$	0.141	0.144	0.145	0.140

Note: Results in the top and bottom part of the table list internuclear separations R and average velocities $V_\parallel$ for various catalytic or noncatalytic sphere diameters, corresponding to those in Figure 39.10.

aqueous solution range from 10^{-3} to 10^{-4}. Thus, most motors considered in our simulations operate in the small Reynolds number Stokes regime where the viscous forces dominate the inertial forces, similar to the synthetic motors. However, the Reynolds numbers are about two orders of magnitude larger and inertial effects play a small role for the fastest-moving dimers.

Finally, we remark that nanodimer motors that execute both translational and rotary motion have been recently fabricated and studied in the laboratory (Valadares et al. 2009). Simulations of the sort described here can help provide the stimulus for the design of new classes of molecular motors and yield insight into their propulsion mechanisms.

39.6.3 Perspectives

The analysis of a class of simple chemically powered molecular motors showed how intermolecular interactions between the motor and the solvent, in conjunction with the nonequilibrium species density fields in the motor vicinity produced by chemical reactions, provide the ingredients needed for self-propelled motion. Given these general principles, a variety of other translational and rotary motors can be constructed using similar methods.

These chemically powered motors operate in the far-from-equilibrium domain. As a result, a full analysis of their motion must rest on statistical mechanical treatments applicable to this domain. Investigations along these lines will allow one to compute the motor efficiency and understand in more detail the role of the nonequilibrium chemical driving force, the chemical affinity, in the motor motion. Since fluctuations play an essential part in small motor dynamics, the study of nonequilibrium fluctuations in the steady state is central to a full understanding of the dynamics. Thus, studies of chemically powered molecular motors offer opportunities to construct a variety of motors with diverse properties and potentially far-reaching applications.

They can also serve as testing grounds for the most recent theoretical developments in nonequilibrium statistical mechanics.

Acknowledgment

Research supported in part by a grant from the Natural Sciences and Engineering Research Council of Canada.

References

Alder, B. J. and Wainwright, T. E. 1967. Decay of the velocity autocorrelation function. *Phys. Rev. A* 1: 18–21.

Alvarez-Pérez, M., Goldup, S. M., Leigh, D. A., and Slawin, A. M. Z. 2008. A chemically-driven molecular information ratchet. *J. Am. Chem. Soc.* 130: 1836–1838.

Astumian, R. D. 2007. Design principles for Brownian molecular machines: How to swim in molasses and walk in a hurricane. *Phys. Chem. Chem. Phys.* 9: 5067–5083.

Ballardini, R., Balzani, V., Credi, A., Gandolfi, M. T., and Venturi, M. 2001. Artificial molecular-level machines: Which energy to make them work? *Acc. Chem. Res.* 34: 445–455.

Balzani, V., Credi, A., and Venturi, M. 2003. *Molecular Devices and Machines—A Journey into the Nano World*. Weinheim, Germany: Wiley VCH.

Balzani, V., Credi, A., and Venturi, M. 2008. Molecular machines working on surfaces and at interfaces. *Chem. Phys. Chem.* 9: 202–220.

Berg, H. 2003. *E. coli in Motion*. New York: Springer.

Catchmark, J. M., Subramanian, S., and Sen, A. 2005. Directed rotational motion of microscale objects using interfacial tension gradients continually generated via catalytic reactions. *Small* 1: 202–206.

Dreyfus, R., Baudry, J., Roper, M. L., Fermigier, M., Stone, H. A., and Bibette, J. 2005. Microscopic artificial swimmers. *Nature* 437: 862–865.

Earl, D. J., Pooley, C. M., Ryder, J. F., Bredberg, I., and Yeomans, J. M. 2007. Modeling microscopic swimmers at low Reynolds number. *J. Chem. Phys.* 126: 064703.

Feringa, B. L. 2007. The art of building small: From molecular switches to molecular motors. *J. Org. Chem.* 72: 6635–6652.

Fournier-Bidoz, S., Arsenault, A. C., Manners, I., and Ozin, G. A. 2005. Synthetic self-propelled nanorotors. *Chem. Commun.* 441–443.

Fletcher, S. P., Dumur, F., Pollard, M. M., and Feringa, B. L. 2005. A reversible, unidirectional molecular rotary motor driven by chemical energy. *Science* 310: 80–82.

Golestanian, R., Liverpool, T. B., and Ajdari, A. 2005. Propulsion of a molecular machine by asymmetric distribution of reaction products. *Phys. Rev. Lett.* 94: 220801.

Gompper, G., Ihle, T., Kroll, D. M., and Winkler, R. G. 2009. Multiparticle collision dynamics—A particle-based mesoscale simulation approach to the hydrodynamics of complex fluids. *Adv. Polym. Sci.* 221: 1–87.

Happel, J. and Brenner, H. 1965. *Low Reynolds Number Hydrodynamics*. Dordrecht, the Netherlands: Nijhoff.

Hess, H., Bachand, G. D., and Vogel, V. 2004. Powering nanodevices with biomolecular motors. *Chem. Eur. J.* 10: 2110–2116.

Ihle, T. and Kroll, D. M. 2001. Stochastic rotation dynamics: A Galilean-invariant mesoscopic model for fluid flow. *Phys. Rev. E* 63: 020201(R).

Jülicher, F., Ajdari, A., and Prost, J. 1997. Modeling molecular motors. *Rev. Mod. Phys.* 69: 1269–1281.

Kapral, R. 1981. Kinetic theory of chemical reactions in liquids. In *Advances in Chemical Physics Vol. 48*, eds. I. Prigogine and S. A. Rice, pp. 71–181. New York: John Wiley & Sons.

Kapral, R. 2008. Multiparticle collision dynamics: Simulation of complex systems on mesoscales. In *Advances in Chemical Physics Vol. 140*, ed. S. A. Rice, pp. 89–146. New York: John Wiley & Sons.

Kay, E. R., Leigh, D. A., and Zerbetto, F. 2007. Synthetic molecular motors and mechanical machines. *Angew. Chem. Int. Ed.* 46: 72–191.

Kikuchi, N., Pooley, C. M., Ryder, J. F., and Yeomans, J. M. 2003. Transport coefficients of a mesoscopic fluid dynamics model. *J. Chem. Phys.* 119: 6388–6395.

Kinbara, K. and Aida, T. 2005. Toward intelligent molecular machines: Directed motions of biological and artificial molecules and assemblies. *Chem. Rev.* 105: 1377–1400.

Kottas, G. S., Clarke, L. I., Horinek, D., and Michl, J. 2005. Artificial molecular rotors. *Chem. Rev.* 105: 1281–1376.

Kovtyukhova, N. I. 2008. Toward understanding of the propulsion mechanism of rod-shaped nanoparticles that catalyze gas-generating reactions. *J. Phys. Chem. C* 112: 6049–6056.

Lee, S. H. and Kapral, R. 2005. Two-particle friction in a mesoscopic solvent. *J. Chem. Phys.* 122: 214916.

Malevanets, A. and Kapral, R. 1999. Mesoscopic model for solvent dynamics. *J. Chem. Phys.* 110: 8605–8613.

Malevanets, A. and Kapral, R. 2000. Solute molecular dynamics in a mesoscale solvent. *J. Chem. Phys.* 112: 7260–7269.

Mavroidis, C., Dubey, A., and Yarmush, M. L. 2004. Molecular machines. *Annu. Rev. Biomed. Eng.* 6: 363–395.

Molloy, J. E. and Veigel, C. 2003. Myosin motors walk the walk. *Science* 300: 2045–2046.

Najafi, A. and Golestanian, R. 2004. Simple swimmer at low Reynolds number: Three linked spheres. *Phys. Rev. E* 69: 062901.

Ozin, G. A., Manners, I., Fournier-Bidoz, S., and Arsenault, A. 2005. Dream nanomachines. *Adv. Mater.* 17: 3011–3018.

Padding, J. T. and Louis, A. A. 2006. Hydrodynamic interactions and Brownian forces in colloidal suspensions: Coarse-graining over time and length scales. *Phys. Rev. E* 74: 301402.

Paxton, W. F., Kistler, K. C., Olmeda, C. C., Sen, A., St. Angelo, S. K., Cao, Y., Mallouk, T. E., Lammert, P. E., and Crespi, V. H. 2004. Catalytic nanomotors: Autonomous movement of striped nanorods. *J. Am. Chem. Soc.* 126: 13424–13431.

Paxton, W. F., Sundararajan, S., Mallouk, T. E., and Sen, A. 2006. Chemical locomotion. *Angew. Chem. Int. Ed.* 45: 5420–5429.

Pollard, M. M., Meetsma, A., and Feringa, B. L. 2008. A redesign of light-driven rotary molecular motors. *Org. Biomol. Chem.* 6: 507–512.

Pooley, C. M., Alexander, G. P., and Yeomans, J. M. 2007. Hydrodynamic interaction between two swimmers at low Reynolds number. *Phys. Rev. Lett.* 99: 228103.

Purcell, E. M. 1977. Life at low Reynolds number. *Am. J. Phys.* 45: 3–11.

Qu, D.-H., Wang, Q.-C., and Tian, H. 2005. A half adder based on a photochemically driven [2] Rotaxane. *Angew. Chem. Int. Ed.* 44: 5296–5299.

Rapaport, D. C. 2007. Microscale swimming: The molecular dynamics approach. *Phys. Rev. Lett.* 99: 238101.

Reimann, P. 2002. Brownian motors: Noisy transport far from equilibrium *Phys. Rep.* 361: 57–265.

Rückner, G. and Kapral, R. 2007. Chemically powered nanodimers. *Phys. Rev. Lett.* 98: 150603.

Saidulu, N. B. and Sebastian, K. L. 2008. Interfacial tension model for catalytically driven nanorods. *J. Chem. Phys.* 128: 074708.

Shirai, Y., Morin, J.-F., Sasaki, T., Guerrero, J. M., and Tour, J. M. 2006. Recent progress on nanovehicles. *Chem. Soc. Rev.* 35: 1043–1055.

Tao, Y.-G. and Kapral, R. 2008. Design of chemically propelled nanodimer motors. *J. Chem. Phys.* 128: 164518.

Tucci, K. and Kapral, R. 2004. Mesoscopic model for diffusion-influenced reaction dynamics. *J. Chem. Phys.* 120: 8262–8270.

Valadares, L. F., Tao, Y.-G., Zacharia, N., Kitaev, V., Galembeck, F., Kapral, R., and Ozin, G. A. 2009. Self propelled dimers. *Unpublished*.

Vale, R. D. and Milligan, R. A. 2000. The way things move: Looking under the hood of molecular motor proteins. *Science* 288: 88–95.

Vicario, J., Eelkema, R., Browne, W. R., Meetsma, A., La Crois, R. M., and Feringa, B. L. 2005. Catalytic molecular motors: Fuelling autonomous movement by a surface bound synthetic manganese catalase. *Chem. Commun.* 3936–3938.

Wang, Y., Hernandez, R. M., Bartlett D. J., Jr., Bingham, J. M., Kline, T. R., Sen, A., and Mallouk, T. E. 2006. Bipolar electrochemical mechanism for the propulsion of catalytic nanomotors in hydrogen peroxide solutions. *Langmuir* 22: 10451–10456.

Yildiz, A. 2006. How molecular motors move. *Science* 311: 792–793.

40

Nanobatteries

Dale Teeters
The University of Tulsa

Paige L. Johnson
The University of Tulsa

40.1 Introduction

If a battery is simply a device that converts the chemical energy of its own stored components into electrical energy (i.e., produces electrons) via reduction/oxidation reactions, then a nanobattery is the same thing, only smaller. Within this simplified argument, however, lie the questions that make this new technology, the "nanobattery," both so promising and so elusive. How small? What components? Which reactions? How to make it? And most importantly, why? Why develop nanobattery technology?

"How small" is an important definition of terms. Within the scientific literature, the prefix "nano" is used loosely and its specific range within the field of batteries has not been well defined. A definition of "microbatteries," based on power output rather than size, has been utilized for devices with capacities of 200 µAh or less (Julien, 1997). For the purposes of this chapter, a broad definition based on physical dimensions, not power characteristics, will be utilized so that a "nanobattery" will encompass any battery system with one or more components that are substantially submicron in one or more of their dimensions; in other words, incorporating nanostructures that could range from the ones to the hundreds of nanometers in scale.

The battery remains the weak link in the development of new electronic technologies in applications ranging from personal computing and tiny medical devices to electric vehicles and space exploration. Moore's famous prognostication that computing density would double every 2 years has largely been realized. To match such miniaturization, power sources have needed to decrease in size and increase in power as well, yet in spite of generous funding and ample research efforts, annual gains in capacity have averaged only about 6% (Figure 40.1). The lack of progress has hindered the adoption of alternative energies like electrically powered vehicles and solar power, which need battery systems with substantially improved performance characteristics to be truly viable in the marketplace. Individual batteries, moreover, must become smaller to provide the autonomous energy required for ingestable drug delivery systems, smart cards, environmental monitoring dusts, embedded monitors, and remote sensors. Nanobatteries, composed of new materials in novel arrangements and exhibiting enhanced reactivities, have the potential to meet these needs and power the future.

40.2 Battery Basics

If Moore's law had held true for batteries, the power of a heavy-duty car battery would now be available from a coin-sized cell. But unlike computing devices that merely utilize electrons, the batteries that support them must produce those electrons through fundamental chemical reactions that are constrained by the periodic table.

Batteries generate electricity via a set of paired chemical reactions in which electrons are passed from a material with a smaller affinity for electrons to one with a greater affinity. Paired reduction/oxidation (redox) reactions are common, the most familiar example being the rusting of iron in air. A simplified explanation of the complex process of corrosion is that iron gives up its electrons to oxygen and water acts as a conductor of ions, providing the balancing of charges essential for the redox reactions to proceed.

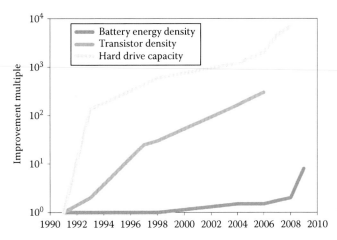

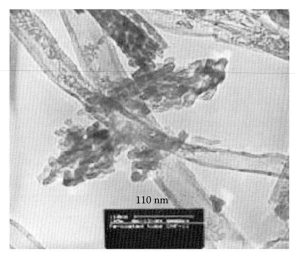

Fe$_2$O$_3$-loaded tubular CNF (1:1)

FIGURE 40.1 An example of the need for better batteries is provided by the lithium-ion systems that have become the energy source of choice for the laptop computer market. The data above show improvement multiples of each technology normalized for values in 1991. Similar gaps exist in other consumer applications. The sharp increase in battery energy density in 2009 is due to a battery system using nanotechnology concepts. Battery energy density. (From Linden, D. and Reddy, T.B. (eds), *Handbook of Batteries*, 3rd edn., McGraw-Hill, New York, 2002, 1–13; Hockenberry, J., *Wired*, 14, 204, 2006; Fehrenbacher, K., Seeo promises a safer lithium battery with higher energy density, earth2tech, http://earth2tech.com/2009/05/07/seeo-promises-a-safer-higher-energy-density-lithium-battery/, 2009.) Transistor density. (From Hockenberry, J., *Wired*, 14, 204, 2006.) Hard drive capacity. (From Five decades of disk drive industry firsts, Disk/trend report, http://www.disktrend.com/5decades2.htm, 2000; Seagate powers next generation of computing with three new hard drives, Seagate Product Press Releases 2008, http://www.seagate.com.)

FIGURE 40.2 Fe$_2$O$_3$ loaded onto carbon nanotubes to make a nanostructured anode for an iron/air battery. Both the type of carbon used and the ratio between the iron and the carbon have been shown to affect the electrochemical performance. (Reprinted from Hang, B.T. et al., *J. Power Sour.*, 150, 261, 2005. With permission.)

$$2Fe_{(s)} \rightarrow 2Fe^+_{(aq)} + 4e^- \quad (40.1)$$

$$O_2 + 4e^- + 4H^+ \rightarrow 2H_2O \quad (40.2)$$

$$Iron + water + air = rust \quad (40.3)$$

$$2Fe + H_2O + O_2 \rightarrow Fe_2O_3 + H_2 \quad (40.4)$$

The difference between this spontaneous redox reaction and those that occur in batteries is largely one of pathway: in a battery, the exchange of electrons is typically passed through a wire to do useful work. An iron/air battery that utilizes the same reactions involved in iron corrosion is one of a family of galvanic technologies that utilize the oxygen in air as the positive electrode. The addition of nanocomposite electrodes to such systems—Fe$_2$O$_3$ loaded onto carbon fibers ranging from 20 to 200 nm in diameter, for example—has resulted in improvements in performance (Figure 40.2; Hang et al., 2006).

All batteries consist of at least one electrochemical "cell." Each cell is made up of three components: the electron producer (smaller electron affinity, negative electrode, anode), the electron absorber (greater electron affinity, positive electrode, cathode), and the electrolyte, which should conduct ions but not electrons and is located between the electrodes to prevent

shorting (Figure 40.3). Each of these materials may be either liquid, plastic, or another solid, and solid-state systems are of increasing importance in nanotechnology research. Multiple cells can be connected into arrays to form a unit that is still referred to as a single "battery"; a concept that makes battery technology scalable. This gives nanobatteries the potential for usefulness beyond single tiny electrochemical cells that could power autonomous microelectromechanical systems (MEMS) and nanoelectromechanical systems (NEMS). They may also be used as connected matrices of a few cells, or at larger ranges up to and including massive parallel arrays of cells forming a single battery that is traditionally sized but more efficient for applications such as electric vehicles and nighttime power delivery for solar systems.

Most battery research is focused on the active ingredients of anode, cathode, and electrolyte. But as a practical matter, these ingredients must be housed in some appropriate container; the anode and cathode must be kept separate, often by a porous physical material; liquid or polymer electrolytes may need to be immobilized; and internal resistances may need to be alleviated by adding conducting materials to semi-conducting electrodes. These architectural considerations are of integral importance to battery designs at any physical level of size, but can be particularly problematic at the nanoscale. Inclusion of additional elements reduces the mass of active materials relative to the mass of the overall battery, thereby limiting the power density that can ultimately be attained. Of particular concern to nanobattery development is the difficulty in making electrical contact with novel nanomaterials. Many nanostructures proposed as new battery components remain of only theoretical interest because this problem has not been overcome.

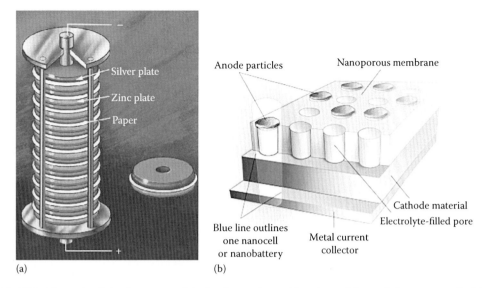

(a)   (b)

FIGURE 40.3 (a) In 1800, Alessandro Volta demonstrated the first battery by stacking plates of zinc and silver separated by blotting paper soaked in brine, which acted as the electrolyte. (b) In 2003, the first nanobattery patent was granted to the University of Tulsa for a system made by stacking graphite and vanadium pentoxide separated by a nanoporous membrane filled with a polymer electrolyte. The Volta cells were connected in series, which increases voltage but not current. The many electrochemical cells of the nanobattery can be connected in parallel, increasing current while maintaining voltage. (From Gillham, O., *Tulsa World*, A13, 2003. With permission.)

40.2.1 Battery Principles—Thermodynamics

The driving force behind the operation of all batteries is the same as for all spontaneous chemical reactions: a favorable change in the free energy of the system as represented by the classic thermodynamic equation:

$$\Delta G = \Delta H - T\Delta S \qquad (40.5)$$

By definition, ΔG, the free energy, represents the net useful energy available from the reaction, which is a function of the enthalpy, ΔH, temperature, and entropy, ΔS, of the system. The equation can be expressed in electrical terms as the net available electrical energy:

$$\Delta G = -nFE^{\circ} \qquad (40.6)$$

where

    n is the number of electrons transferred per mole of reactants
    F is the Faraday constant (~96,485 C/mol or 26.80 Ah), which is equal to the charge of one mole of electrons
    E° is the electromotive force (emf) of the cell reaction in volts

The electromotive force simply quantifies the potential difference between the two electrodes:

$$E^{\circ}_{cell} = E^{\circ}_{anode} - E^{\circ}_{cathode} \qquad (40.7)$$

Each electrode material has a unique potential, based on its inherent chemical affinity for electrons (Figure 40.4). In any arrangement of dissimilar materials in combination with an electrolyte, the material with the lower electron affinity (more negative potential) will tend to give off electrons, or be oxidized,

	Standard Reduction Potential (V)	Electrochemical Equivalents (Ah/g)
Li	−3.01	3.86
Na	−2.71	1.16
Al	−1.66	2.69
Fe	−0.44	0.96
Zn	−0.76 (in acidic electrolyte)	0.82
	−1.25 (in basic electrolyte)	
Pb	−0.13	0.26
PbO_2	1.69	0.224
MnO_2	1.28	0.308
Li_xCoO_2	~2.7	0.137
NiOOH	0.49	0.292

FIGURE 40.4 Standard reduction potentials and electrochemical equivalents for some traditional battery materials.

and the material with the higher electron affinity (more positive potential) will accept them. The simple summation of the individual electrode potentials, then, yields a potential for the entire cell—in the familiar notation of a 1.5, 3, or 6 V battery—also known as the battery's theoretical potential.

Nickel–zinc (NiZn) secondary battery systems, for example, were part of early proposals for MEMS, and are currently marketed as less toxic alternatives to nickel cadmium (NiCd) batteries (Humble et al., 2001). Safety in the use and disposal of batteries remains an important concern in battery design. NiZn systems utilize a metallic anode of zinc rather than cadmium along with a NiOOH cathode in liquid potassium hydroxide (KOH) electrolyte and provide an example of the progress of nanobattery research, both in nanostructuring individual battery components and in assembling a battery architecture on a reduced scale (Figure 40.5).

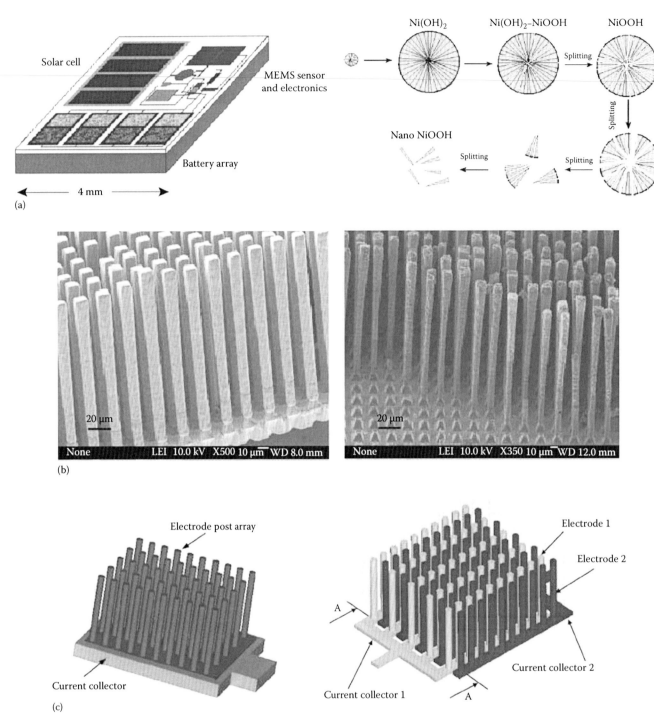

FIGURE 40.5 The application of nanoscale techniques to a NiZn battery system. (a) An early (2001) proposal for a microelectromechanical system (MEMS) utilizing a NiZn cell. The batteries had a 0.02 cm² footprint and a capacity of approx. 0.555 mW-h/cm². (From Humble, P.H. et al., *J. Electrochem. Soc.*, 148, A1357, 2001. With permission.) (b) One approach to improvement is to utilize nanostructured electrode elements in an otherwise traditional battery configuration. Nano-NiOOH obtained by "splitting" spheres of NiOH into rods of 60–150 nm in width has been used in electrodes with super high current densities of 10,000 mA/g and capacities of 276 mA-h/g. (Reprinted from Pan, J. et al., *J. Power Sour.*, 188, 308, 2009. With Permission.) (c) Another approach is to fabricate a new three-dimensional architecture in which there is a high surface area of both electrodes within a small footprint. If a post geometry is used, the electrodes can themselves be placed close together in an interdigitated pattern, reducing cell resistivity. Here, post arrays of both nickel and zinc have been fabricated by electroplating into silicon molds. The nickel posts were coated with nickel hydroxide for battery testing. Gravimetric capacity measurements were approximately the same as those for traditional 2D NiZn systems, however, areal capacity was an order of magnitude greater. (Reprinted from Chamran, F. et al., *J. Microelectromech. Syst.*, 16, 844, 2007. With permission.)

The theoretical potential of the cell is calculated as follows:

$$E^o_{\text{NiZn cell}} = E^o_{\text{Zn}} - E^o_{\text{NiOOH}} = 0.49\,V - (-1.25\,V) = 1.74\,V \quad (40.8)$$

The above calculation assumes standard state conditions of 25°C and 1 atm; for nonstandard state conditions, the voltage is given by the Nernst equation for electrochemical reactions:

$$E = E^\circ + \left(\frac{RT}{nF}\right)\ln\left(\frac{E^o_{\text{anode}}}{E^o_{\text{cathode}}}\right) \quad (40.9)$$

The Nernst equation makes it clear that in general, batteries exhibit better potentials at elevated temperatures. It is essential, however, that some batteries operate in cold environmental conditions, and this problem is also being addressed through nano-engineering. Recent research has shown that nanoscale mixing of graphitic and partially fluorinated domains in Li-CF$_x$ battery materials enhances their performance at low temperatures and may enable cells to function at temperatures of −60°C and below (Whitacre et al., 2007).

The theoretical potential of an electrochemical cell can be measured as well as calculated; the voltage across the terminals of the battery when no current load is being applied is referred to as the open circuit voltage and is usually very close to the theoretical (thermodynamic) voltage. The voltage of a battery measured under actual current withdrawing conditions, or on load, sometimes referred to as the closed-circuit voltage, will always be lower than that measured on an open circuit due to the internal impedance of the cell.

40.2.2 Battery Principles—Capacity and Energy Density

The cell potential does not take into account the amount of material used in the battery, only its nature; a large or a small cell composed of the same active materials would have the same theoretical potential. The parameters of cell capacity and energy, however, incorporate the quantity of material present and are the chief operational parameters for battery evaluation and comparison. A large cell would have a greater capacity, and yield greater energy, than a small cell composed of the same materials.

Experimentally, capacity can be obtained from the discharge curve of the battery (Figure 40.6), and may be stated merely as units of current-time (typically ampere-hours), or in relationship with mass (A-h/g), volume (A-h/cm³), or areal footprint (A-h/cm²). For nanobatteries, the units may be reduced to milli, micro, or nanoamps; and when individual electrodes, rather than the whole cell, are being tested, the data are referred to as the current density.

The theoretical capacity for each individual electrode can be obtained from a rearrangement of Faraday's laws:

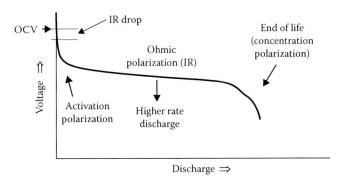

FIGURE 40.6 Discharge curve of voltage versus time at some known discharge current. Capacity can be determined by measuring the time from the beginning of discharge to some defined lower voltage, and multiplying it by the current that was applied. (Reprinted from Winter, M. and Brodd, R.J., *Chem. Rev.*, 104, 4245, 2004. With permission.)

$$I = \frac{gnF}{t\,\text{MW}} \quad (40.10)$$

where
  I is the current flow
  g is equal to the grams of active material
  t is the time of current flow
  MW is the molecular or atomic weight of the active material
  n is the number of electrons involved in the reaction

For ease of calculation, F, MW, the number of electrons involved in the reaction, and a standard time of 1 h are used to generate electrochemical equivalents for anode and cathode materials, simplifying the calculation of the theoretical capacity for the entire battery to a simple addition of terms (Figure 40.4).

As with potentials, experimental capacities lie far below theoretical predictions. Chen et al. have synthesized nanotubes and nanowires of MnO$_2$ using surfactant-assisted hydrothermal techniques, and have synthesized nanorods using potentiostatic and cyclic voltammetric electrodeposition methods (Chen and Cheng 2009). Manganese dioxide (MnO$_2$) is an important and versatile electrode material that can be used as a cathode with either zinc or lithium; coupled with zinc, the theoretical cell capacity would be 0.308 A-h/g + 0.82 A-h/g = 1.128 A-h/g. In a laboratory cell, the nanowires delivered a capacity of 0.267 mA-h/g, which, although still far from the theoretical capacity, was an increase of 28% over the value for bulk MnO$_2$.

In all its forms, the capacity represents the total quantity of electricity that is involved in the battery's redox reactions. The energy delivered by the battery is the product of the voltage and the capacity:

$$E^\circ = VI, \quad \text{usually stated in W-h,} \quad (40.11)$$

and when the mass of the components is considered, the specific energy density is obtained by

$$E = \frac{E^\circ}{(\text{Mass}_{\text{cathode}} + \text{Mass}_{\text{anode}})}, \quad \text{in W-h/g or W-h/kg} \quad (40.12)$$

Or for a complete assembled cell including housing, current collectors, and any structural elements

$$E = \frac{E^\circ}{M_{\text{battery}}} \qquad (40.13)$$

where M is the mass of the entire battery. Energy density is the ultimate quantifier of a battery's performance and the best means for comparing different battery systems. It makes evident the additional advantage of utilizing a light element such as lithium as an electrode material, resulting in a significant reduction of battery mass and thus an increase in energy density. Though energy density is an important parameter for commercial systems, it often exists only as a theoretical projection for nanobattery research, in which laboratory prototypes, rather than complete cells, are being evaluated.

Capacity, and therefore energy output, varies with the discharge rate, time, cell construction, operating temperature, and age/history of the cell. As with potential, even under optimized discharge conditions the actual energy obtained is only 25%–35% of the theoretical energy. Losses in both potential and capacity are due largely to kinetic limitations within the cell, which nanobattery technology is uniquely positioned to address.

40.2.3 Battery Principles—Kinetics

The perfect battery would operate according to its thermodynamic properties by immediately and continuously delivering power at its stated potential (its theoretical or open circuit voltage) until the electrode materials were completely depleted. That no battery does, performance in fact being only on the order of 35% of theoretical maximums, is due largely to limitations on the rate of charge transfer within the cell (Figure 40.7).

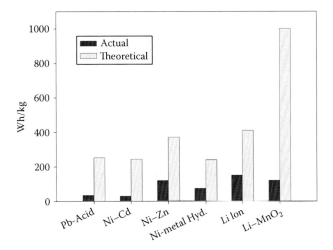

FIGURE 40.7 Actual battery performance is significantly less than thermodynamic predictions, largely due to kinetic limitations that nanobattery innovations are uniquely able to address. (From Linden, D. and Reddy, T.B. (eds), *Handbook of Batteries*, 3rd edn., McGraw-Hill, New York, 2002, pp. 1–13.)

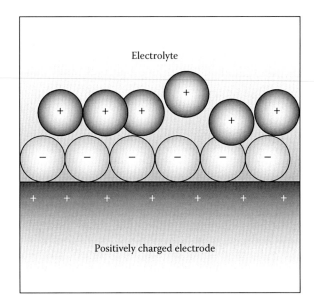

FIGURE 40.8 The electrical double layer forms at the electrode/electrolyte interface, controlling the charge-transfer kinetics within the cell.

All electrochemical reactions take place at the interface between the electrode and the electrolyte, where an electrical double layer forms (Figure 40.8). All ions or atoms that are participating in the battery's redox reactions must pass through this layer. Kinetic limitations to this process result in losses termed activation overpotentials. Additional losses—concentration overpotentials—come from concentration gradients created when active material is used up faster than it can be replaced by the diffusion of new material from the bulk of the electrode. These losses are termed overpotentials because though they result in a reduction of potential available upon discharge, they cause an increase in the potential required to charge the battery. Ohmic losses are proportional to current load and arise from the combination of all resistivities within the cell, including those of the electrodes and electrolytes but also of architectural elements such as terminals and current collectors.

The internal impedance of the cell battery, then, is a combination of the polarization losses at each electrode ($\eta_{\text{electrode}}$) and ohmic losses (IR losses) within all the cell components.

$$V_{\text{discharge}} = V_{\text{theoretical}} - \eta_{\text{anode}} - \eta_{\text{cathode}} - \text{IR} \qquad (40.14)$$

Polarization losses are visible in the discharge curve of the battery, which shows the change in voltage with respect to the time upon battery operation (Figure 40.6).

Nanostructured elements facilitate mass-transport processes and speed diffusion, enabling conditions within the electrode to come closer to equilibrium, or theoretical, conditions. Kinetic overpotentials diminish as increases in the surface area of active materials provide more interfacial area at which redox reactions can occur. Concentration overpotentials are lessened in nanostructured materials because particles with small interior volumes have shorter path lengths by which unreacted material can diffuse to the surface.

Ohmic losses can be addressed by reducing the distance between the electrodes and providing close contact between the electrode and the current collector so that resistivities are minimized. Nanofabrication techniques for achieving this, including the interdigitation of electrode posts (see Figure 40.5c), are an extension of much older battery practices such as the close-interweaving of concentric spirals of lead sheet employed by Gaston Plante in 1859. Plante's battery was likely the first nano-structured battery, albeit inadvertently, for during use its current steadily increased as a layer of lead oxide composed of very small particles built up on the positive plate and the lead of the negative plate became increasingly spongy and porous, building up a nanoscaled architecture that improved the battery's performance by altering its surface area and geometry.

40.3 Nanoengineering of Lithium-Ion Battery Components

Lithium-ion batteries have become the energy storage devices of choice for consumer electronics. They deliver the best combination of high power, fast recharge times, and steady current currently available, and more research dollars are now spent on their investigation than on all other battery chemistries combined (The Economist, 2008). It is not surprising that most of the work on nanobatteries is concerned with this battery technology. However, conventionally scaled lithium systems may be reaching their natural limits. The need to produce more and more power in smaller and smaller devices has led to situations of thermal runaway—in which the battery reactions become self-sustaining and their rate uncontrolled—resulting in some well-publicized fires associated with lithium battery use and storage. These concerns for both safety and performance can be uniquely addressed by nanoengineering.

It is obvious from a theoretical examination of electrode potentials that lithium makes an excellent anode, and that its low mass leads to a high energy density. But lithium's extreme reactivity long prevented any practical use. In the 1980s, researchers documented its intercalation and deintercalation in a graphite electrode, demonstrating its potential for rechargeability, and soon thereafter it was shown that these same intercalation reactions were possible with stable and nonreactive lithium cobalt oxide, leading to the release of the first commercial lithium-ion battery by Sony in 1991 and revolutionizing both consumer electronics and academic battery research (Agarwal, 1982). These same electrodes—graphite and $LiCoO_2$—are still the most common set of active materials in use today and are the subject of a wide range of research efforts dedicated to creating and testing novel nanostructures of these components.

Like other secondary systems, lithium-ion batteries are based on a reversible reaction utilizing not lithium metal itself, but a lithium salt from which metallic lithium ions are generated upon charging:

$$LiCoO_2 \rightarrow Li_{1-x}CoO_2 + xLi^+ + xe^- \qquad (40.15)$$

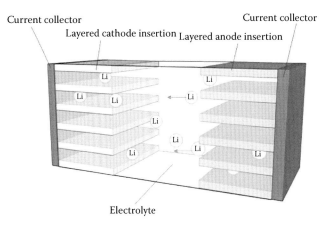

FIGURE 40.9 A standard lithium-ion battery system in which lithium ions "rock" back and forth between intercalation electrodes.

The metallic lithium generated in the anode migrates through a lithium-salt containing electrolyte to the cathode, where it is inserted (intercalated):

$$xLi^+ + xe^- + 6C \rightarrow Li_xC_6 \qquad (40.16)$$

Lithium ions are exchanged back and forth between the two electrodes during charge and discharge, leading the system to be known as a "rocking chair battery" (Figure 40.9). Though the graphite/$LiCoO_2$ redox couple is the most common, a wide variety of other anode/cathode combinations are the subject of research for both macro and nanoscaled systems, including new materials that are only electrochemically active on the nanoscale.

As alluded to in this rudimentary discussion of lithium-ion batteries, nanoengineering efforts are underway to enhance lithium-ion battery performance. Tactics include increasing surface areas and improving surface geometries; constructing path lengths in battery systems that maximize ion and electron diffusion; and new reactions made possible because of the nanoscale dimensions of anodes, cathodes, and electrolyte systems. These efforts involve many of the most important aspects of nanotechnology and as such deserve more detailed discussions.

40.3.1 Surface Area and Geometry Enhancements

Increasing surface area to make more of the electrode material available for reaction is one of the primary goals of battery nanoengineering, and it yields significant gains in capacities. Consider a unit surface area that is flat in nature, for example a current collector on which a nanostructure could be grown or deposited. Two options for such a structure, as shown in Figure 40.10, would be plates of electrode material extending upward from the surface or equally spaced rods of electrode material, where w is the width of the plates or the rod diameter, d is the height of the structures, s is the spacing between the plates or rods, and L is the length of the plates.

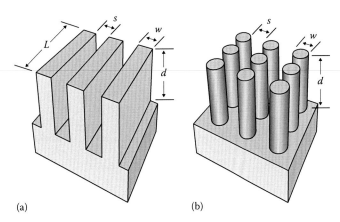

(a)   (b)

FIGURE 40.10 Two potential high aspect ratio geometries for use in nanobattery systems; arrays of (a) plates and (b) posts or rods.

Mathematical relationships can be derived for the ratio of the surface area gain (SAG) for any structure as compared with a flat electrode surface having the same footprint area. Equation 40.17 is an example of such an equation for a structure of plates, as shown in Figure 40.10a, where the complete equations necessary for calculating the ratio of the surface areas are shown followed by the mathematically simplified expression. Equation 40.18 is the simplified expression for rods perpendicular to a surface (Figure 40.10b):

$$\text{SAG}_{\text{Plates}} = \frac{\text{Area with plates}}{\text{Area of flat surface}} = \frac{L(w+s)+2Ld}{L(w+s)} = 1 + \frac{2d}{(w+s)}$$

(40.17)

$$\text{SAG}_{\text{Rods}} = 1 + \frac{\pi d(w/2)^2}{(w+s)^2}$$

(40.18)

For comparison, a surface having plates or rods with values for w, d, and s of 5, 125, and 5 nm, respectively, would increase in surface area over a planar surface by a factor of approximately 26 for both plate and rod structures.

The effect that this increase of surface area would have on the energy capacity of a cathode material for a lithium-ion battery can also be quantified. For example, the cathodic material $LiCoO_2$ has a volumetric energy capacity of 540 mA-h/cm³, which can also be expressed as 54 μA-h/μm·cm², a system of units that allows surface area comparisons to be made more readily. For a 1 μm thick film of $LiCoO_2$ having a square centimeter footprint surface area and as calculated above, a 26-fold increase in surface area due to the rod or plate nanostructures, the new capacity per unit footprint area of 1 cm² of the electrode would be as follows:

$$\text{Capacity} = 54\,\mu\text{A-h} \times 26 \approx 1400\,\mu\text{A-h}$$

(40.19)

The increase in surface area results in a dramatic increase in apparent electrode capacity. Nanorods of $LiCoO_2$ have been

grown with this purpose in mind for lithium batteries (Zhou et al., 2002; Liao et al., 2006; Lu et al., 2008).

Changing the dimensions of the plates or rods changes the surface area gain for the systems. If only the width of the plates or diameters of the rods (w value) are increased, the surface area gain for the rods increases much more quickly than does that for the plates. While the practical matter of mechanical stability and space for the electrolyte must also be considered, this demonstrates the potential to engineer the surface for maximum surface area gain based on the geometry and dimensions chosen. Nanoscale rods (Muraliganth et al., 2008; Zhao et al., 2008; Liu et al., 2009a,b; Subba Reddy et al., 2009; Wang et al., 2009b), plates (Cho et al., 2007; Ma et al., 2008; Seo et al., 2008; Saravanan et al., 2009), pillars (Xiao et al., 2007; Hsu et al., 2008), carbon nanotubes (He et al., 2009; Kawasaki et al., 2009; Lu et al., 2009; Wang et al., 2009a; Yuan et al., 2009), and sponge-like materials with nanopores (Rolison et al., 2009) have been made and investigated with an increase of surface area in mind.

Simple maximization of the surface area is not the only concern, however. One of the difficulties in creating effective nanoarchitectures lies in the conflicting geometric goals for high capacities and low resistance in materials. When utilizing two rod electrode structures facing each other and surrounded by electrolytes (Figure 40.11), the surface area and hence the capacity increases as the height of the rods, d, is increased. But as the dimension d is increased, more electrode material becomes farther from the current collector resulting in a

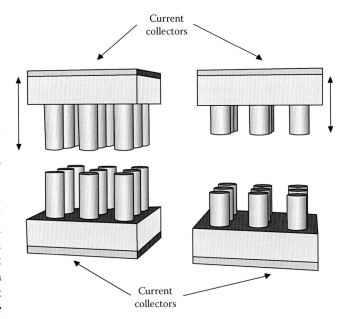

FIGURE 40.11 Two rod structured electrode systems. The length of the rods, d, is greater in the figure on the left than in the figure on the right. While the greater length of the rods will increase surface area and increase the observed capacity, the distance that electrons must flow, indicated by the double arrows, is greater for the larger d value. The greater resistance to electron flow caused by longer travel distances will reduce battery performance resulting in a trade-off in performance characteristics.

farther distance for electrons to travel. Increasing electronic resistance with increasing "*d*" dimensions limits the overall system performance in spite of the capacity gains from greater surface area. In many three-dimensional architectures, therefore, a trade-off between efficient ion diffusion and electron conduction must be made.

40.3.2 Short Path Lengths

One of the most desirable properties of any battery system is a fast rate of charging as well as rapid provision of power (fast discharging). Sluggish discharge in lithium-ion batteries in particular remains a barrier to their use in electric vehicles. Rapid movements of both ions and electrons within the battery's matrix require short path lengths for diffusion. Though the movement of ions in a battery is a complex process, the beneficial effects of short path lengths resulting from nanoscale features can be demonstrated by a simplified but instructive model described by Zhou for lithium-ion systems (Zhou et al., 2005; Jiang et al., 2006).

The ion diffusion coefficient, *D*, can help to understand what path lengths would be necessary for a battery system to achieve a fast charge time. With this constant, having units of cm²/s, the length that a lithium ion could move, *L*, can be calculated by

$$L = \sqrt{D\tau} \qquad (40.20)$$

where τ is the time of the process. A typical diffusion coefficient for an ion in a solid is on the order of 10^{-16} cm²/s (Kavan et al., 2003). Using Equation 40.20 (Jiang et al., 2006), the distance that a lithium ion could diffuse into an electrode material in 10 min would be approximately 2 nm. So for maximum capacity in a structure, as shown in Figure 40.10b, the rods would need to be no more than 4 nm in diameter (2 nm radius). If the rods were any larger than this, the total volume available for lithium intercalation would not be completely utilized in the 10 min time period. This same analysis applies to other geometries such as spherical nanoparticles.

Commercial examples of this nanoscale engineering already exist. A battery developed by A123 Systems using a proprietary nanophosphate electrode system can achieve 80% of its charge capacity in 12 min. While all the technical aspects of this battery are not available, it is known that their battery uses nanoparticles of a phosphate material as the electrodes are reported to be less than 100 nm in diameter (A123 Systems Inc., 2009). The fast charge rate is likely achieved by the ability to insert larger amounts of lithium into the nanoparticles because of the short path length created by their small size.

This leaves room for much nanoengineering, in particular of three-dimensional systems that incorporate high surface areas with short path lengths. One strategy, illustrated in Figure 40.12, involves the direct synthesis of thin films of nanoparticle electrode materials such as NiO and Fe_2O_3 onto a current collecting nickel metal mesh so that the ions and electrons have only a short distance to travel (Hosono et al., 2006a,b). Another approach involves the creation of a hollow nanostructure from which a wire can be grown, as illustrated in Figure 40.13. Individual nanowires acting as current collectors can be grown inside these aligned tubes of electrode material, termed nanobaskets (Johnson and Teeters, 2006), that are 200 nm thick and 700 nm in length. The path length between the nanowire current collectors and the electrolyte on the other side of the tube would be 200 nm. In this case, the path lengths for both mobile species—the lithium ions and the electrons—are relatively short for maximized electrode performance, and intimate contact of the current collector with the electrode has been achieved.

40.3.3 New Electrode Reactions

The electrochemical effects of reducing battery components to the nanodimension are the subject of much experimental research and theoretical modeling. A classic example of how a smaller particle size changes fundamental material properties is the several hundred degree reduction in the melting point of gold as its particle size reaches the nanoscale. This is attributed to a

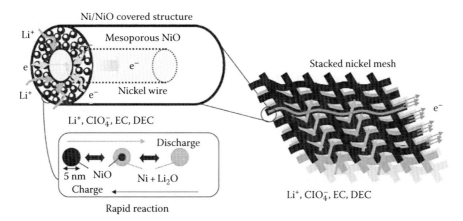

FIGURE 40.12 Nanoparticles such as NiO and Fe_2O_3 surrounding a nickel wire resulting in an electrode system where ions and electrons have only a short distance to travel. The overall capacity is further increased by making a mesh-like structure with the nanoparticle-loaded wires. (Reprinted from Hosono, E. et al., *Electrochem. Commun.*, 8, 284, 2006b. With permission.)

FIGURE 40.13 Copper wires (arrow upper left image) grow from a hollow nanobasket approximately 700 nm high and 200 nm thick (outlined with dashed line), and through the length of a nanoporous membrane (lower right image) to make contact with a bulk current collector on the other side. Such a three-dimensional battery structure addresses the problem of making electrical contact with novel nanostructures, while providing a short diffusion path length to an electrolyte placed on top of the nanobaskets.

change in the excess surface free energy, and can be expressed according to the following equation:

$$T_{\mathrm{m}}(r) = T_{\mathrm{m}}(\infty) - C/r \qquad (40.21)$$

where

$T_{\mathrm{m}}(r)$ is the melting point of a particle of radius r
$T_{\mathrm{m}}(\infty)$ is the melting point of the bulk material
C is a constant incorporating the surface free energies of both the solid and liquid states (Cortie and van Der Lingen, 2002)

The chemical potential, μ, of a system can similarly be defined relative to the radius of a particle:

$$\mu_{\mathrm{particle}} = \mu_{\mathrm{particle}}^{\infty} + 2\frac{\gamma}{r}V \qquad (40.22)$$

where

$\mu_{\mathrm{particle}}^{\infty}$ is the chemical potential for a particle of the bulk material
r is the nanoparticle radius
γ is the area-averaged surface tension (which is proportional to the surface free energy of the solid)
V is the molar volume of the nanoparticle

The second term has been shown to be related to the net available energy of the electrochemical system (see Equation 40.6; Balaya et al., 2006):

$$2\frac{\gamma}{r}V = \Delta G = -nFE^{\circ} \qquad (40.23)$$

So the net available energy increases as the size of the particle decreases. One of the most intriguing aspects of nanotechnology is the potential that materials that are electrochemically inactive on the macro scale may become useful battery materials when nanostructures are employed. Lithium alloy electrodes are one such example and have been proposed for anode use in lithium-ion batteries. The typical lithium-alloying process for lithium alloy electrodes is

$$M + xLi^{+} + xe^{-} \rightleftharpoons Li_{x}M \qquad (40.24)$$

where M can be Sn, Si, Pb, Bi, Sb, Ag, Al, or a multicomponent alloy. Some simple transition metal oxides, sulfides, fluorides, and nitrides are inactive when on the macro scale because they are both poor electronic and ionic conductors. However, upon driving the lithium to react with these transition metal compounds, a metal/LiX nanocomposite is formed with metal nanoparticles embedded in the LiX matrix. The chemistry of this reaction is

$$MX + zLi^{+} + ze^{-} \rightleftharpoons Li_{z}X + M \qquad (40.25)$$

Here, M can be Fe, Co, Ni, Cu, etc. (Poizot et al., 2000) and X is O, S, F, or N (Jiang et al., 2006). The reaction of the transition metal compounds with lithium forms *in situ* metal nanoparticles. This reaction can cycle in the forward and reverse directions making a viable system for the anode in a lithium battery. The ability of these systems to operate as potential battery anodes, i.e., their ability to cycle, is attributed to the enhanced number of atoms on the surface of the nanoparticles relative to the bulk, thus increasing their electrochemical activity (Poizot et al., 2000). The research of Poizot and coworkers showed that capacities of 700 mA-h/g can be obtained for up to 100 charge/discharge cycles in these systems that were once considered electrochemically nonreactive.

Utilizing nanoscale engineering, not only can systems that were thought to be unreactive become useful as electrode materials, but additional storage capacity for lithium ions can also be obtained. A high storage capacity for lithium ions observed in transition metal oxide/metal nanoparticle systems has been attributed to an interfacial charge mechanism whereby lithium ions are stored on the oxide side of the interface and the electrons are localized on the metallic side (Jamnik and Maier, 2003; Zhukovskii et al., 2006). This resulting charge separation increases the capacity of these nanocomposite electrodes beyond that associated with lithium being stored only inside the nanoparticle. This results from the possibility that a complete monolayer of additional lithium ions could be stored at the interface between the metal nanoparticles and the lithium oxide with the electrons being transferred to the metal nanoparticles.

The resulting change in the chemical potential of the system can be evaluated as follows. Lithium ions are at first inserted into the bulk material, which is typical electrode material behavior, and the chemical potential increases accordingly. When the solubility limit of the bulk material for lithium has been reached, new nanoparticle phases and the resulting interfaces are formed. Additional lithium is then stored at the interface of these nanoparticles (Figure 40.14). The chemical potential for this interfacial process is dependent upon the surface area available, i.e., the grain size, and the chemical potential can now be represented by

$$\mu_{Li} = \mu_{Li}(r \to \infty) + 2\frac{\gamma(r)}{r}\nu_{Li} \qquad (40.26)$$

Just as Equation 40.22 describes the change in the nanoscale properties of gold, Equation 40.26 describes the increased ability of these electrode materials to hold lithium. The second term again represents the gain in the chemical potential due to lithium being stored at the interface where r is the nanoparticle size; γ is the area-averaged surface tension, proportional to surface free energy; and ν_{Li} is the molar volume of lithium in the nanoparticle. As the amount of lithium at the nanoparticle interface increases, the surface free energy of the particle increases as well, increasing the chemical potential. Theoretically, charging can continue until the total chemical potential due to lithium inserting into the bulk and interface becomes equal to the chemical potential of pure lithium metal. The additional storage resulting from the interfacial region creates more capacity than was thought

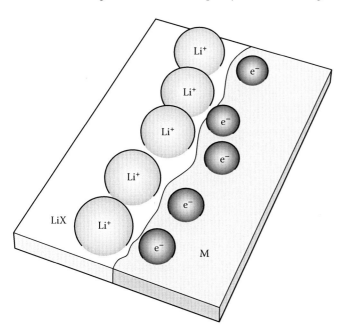

FIGURE 40.14 An illustration of the phenomenon of interfacial charge capacity. The lithium ions reside at the LiX surface while the electrons are at the metal, M, nanoparticle surface.

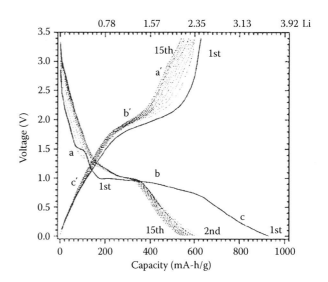

FIGURE 40.15 The electrochemical reaction regions for the Li/TiF$_3$ system. The regions a, b, and c are for lithium insertion while a', b', and c' are for deinsertion. (Reproduced from Li, H. et al., *Adv. Mater.*, 15, 736, 2003. With permission.)

possible for this electrode system. Interestingly, the interfacial storage resembles that of a capacitor and has thus been suggested to be very reversible in nature (Jamnik and Maier, 2003).

Experimentally, this behavior is visible in the voltage profile of the cell, as seen in the analysis of a Li/TiF$_3$ system (Figure 40.15; Li et al., 2003). Lithium insertion occurs in steps a, b, and c and extraction at a', b', and c'. The first step—labeled a—is due to the storage of 0.5 Li to form Li$_{0.5}$TiF$_3$. The next step, b, which demonstrates a clear plateau in the first cycle is attributed to the formation of two phases, nanoparticle in nature, and follows Equation 40.25. For this system, this would be

$$TiF_3 + 3Li^+ + 3e^- \rightleftharpoons 3LiF + Ti \qquad (40.27)$$

Step c is the insertion of lithium at the interface of the nanoparticle. While there is a significant capacity loss in the plateau region after the first cycle, future cycles remain constant with respect to lithium extraction and insertion. This is particularly true of step c, the interfacial storage step. Steps a and b contribute 620 mA-h/g to the lithium storage capacity, while the reversible step c contributes an extra 300 mA-h/g due to interfacial storage. This is an excellent example of the benefits of new reactions due to nanoscale structure.

The electrode material TiO$_2$ provides another example of how nanoscale features can positively affect the availability and accessibility of sites where lithium can be stored. TiO$_2$ exists in three common forms: rutile, anatase, and brookite whose differing crystallographic structures affect their ability to intercalate lithium into the crystal lattice. The rutile form of TiO$_2$ is built of closely packed TiO$_6$ octahedra with few favorable sites for lithium insertion, and so does not readily accept lithium ions, especially at room temperature. When nanoparticles are involved, however, an almost complete loading of the available octahedral sites is seen with a corresponding increase in lithium storage capacity (Jiang et al., 2007, 2008). This is attributed to two factors. First, because of the very short diffusion lengths involved when a nanoparticle is present, the lithium ion can reach the octahedral sites more readily, yet another example of how shortened path lengths improve performance in nanoengineered systems. Second, the diminished particle size increases not just the overall surface area relative to the bulk, but also the proportion of exposed 110 crystallographic planes of TiO_2, which have been shown to be the most favorable path for lithium intercalation (Stashans et al., 1996).

These are but a few examples of the new reactivities that arise due to the exposed surfaces of nanoscale objects, and which can dramatically benefit battery performance. Because the surface of the electrode plays such an important role in its capacity, a carefully crafted surface can allow for a short diffusion length—making the maximum lithium capacity attainable in the shortest time—as well as making more surface area available for a greater total electrode capacity. There is another important advantage to these nanoengineered surfaces: improvements in the stability and therefore the safety of the electrode.

40.3.4 Stable Cycle Performance

Ions take up physical space, and the insertion and deinsertion of ions into an electrode can cause the electrode material to expand and contract. Continuous charging/discharging cycles and the associated changes in volume are an unstable process that can lead to the mechanical failure of the electrode material, and in lithium battery systems can lead to safety hazards including situations of thermal runaway. Nanostructuring can have a stabilizing effect on electrode material, accommodating volumetric changes through more robust geometries.

Some of the most promising anode material types are alloys of metals, such as the lithium alloys of tin and silicon. They can accommodate up to 4.4 atoms of Lithium per tin or silicon atom, according to the following reactions:

$$Li_{4.4}Si \rightarrow Si + 4.4Li^+ + 4.4e^- \qquad (40.28)$$

$$Si + 4.4Li^+ + 4.4e^- \rightarrow Li_{4.4}Si \qquad (40.29)$$

Because of the large amount of lithium that is used to form the alloys, they have high theoretical capacities that range from 2 to 10 times the capacity of graphite, the major component of the anodes in today's lithium-ion batteries. However, problems also exist with the large amount of lithium alloyed to, and then removed from, Sn or Si. Insertion and removal can result in a volume change as large as 200%–300%, an expansion and contraction that a standard electrode configuration cannot withstand (Bruce et al., 2008). This results in an anode that quickly cracks upon charging and discharging and can eventually disintegrate, so even with such desirable high capacities, these materials cannot be used for battery anodes (Yang et al., 1996).

It is evident that a very small volume of anode would better handle the stress of alloying and dealloying than a large volume in which the destruction caused by the volume expansion is magnified by the size of the anode itself. This has been confirmed in the literature (Aricò et al., 2005). The best performance for these alloys can be obtained by using rod structures (see Figures 40.5 and 40.10b), which not only minimize path lengths, but also address volume stability. Such rod configurations can be fabricated by a variety of methods, including etching of the bulk electrode, or by coating copper current collecting rods with an appropriate anode material (Green et al., 2003; Taberna et al., 2006; Hassoun et al., 2007). The nanoscale rod structure can better handle the volume increase and the extra space between the rods provides room to accommodate the volume expansion that does occur. Electrodes with such designs have retained excellent capacities for hundreds of cycles. The ability of nanostructures to address the problem of volume expansion and the resulting electrode instability opens the doors to new battery materials.

40.3.5 Electrolyte Enhancements

Just as electrode materials can be nanoengineered for enhanced performance, electrolytes can also benefit from such treatments, and show similar benefits from improved surface areas and potential new reactions. Though the electrolyte does not factor into equations of potentials and capacities, it is nonetheless essential to battery function, in which electron and ion transport occur separately. The electrolyte must act as both an ionic conductor and an electrical insulator. Electrical neutrality must be maintained for electrons to continue to flow through the circuit; so for each electron that moves, an ion must move as well to balance the change in electrical charge.

The electrolyte has another, physical, function: it keeps the two electrodes from touching so that electrons are not immediately passed between them, resulting in a short circuit. In some applications, the electrolyte itself, in solid or gelled form, may be robust enough to act as its own separator. At other times, porous, nonreactive materials are used. This is especially common in nanoscale battery architectures where the electrolyte is often contained within a porous layer that provides strength while still allowing the free flow of ions. The confinement of electrolytes in nanoporous materials, moreover, can change their properties, enhancing ionic conductivity and benefiting battery performance.

Polymeric electrolyte systems—for reasons of manipulation, fabrication, and adequate stability with electrodes at the electrode/electrolyte interface—are often the phase of choice for nanoscale work and are discussed here. However, many of the same considerations presented apply to rigid solids—such as inorganic solid materials—and polymer electrolyte systems alike.

Solid polymer electrolytes must meet the following criteria to be useful in a reliable battery application.

1. An ionic conductivity approaching 10^{-3} S/cm at room temperature (Gray, 1997). Ionic diffusion through a room temperature polymer will always be lower than the diffusion rates of a liquid system, so obtaining an appreciable ion motion is a concern.

2. An ionic transfer number for the cation—lithium in this discussion—as close to 1 as possible, i.e., $T_{Li} \approx 1$. For battery applications, the electrolyte should conduct ions, but be insulating to electron conduction (the basic definition of a battery electrolyte). The larger the cation transference number, the less concentration polarization will occur during charge/discharge cycles, thus resulting in higher power densities. This is especially relevant in solid polymer electrolytes where $T_{Li} \approx 0.5$ is commonly observed.

3. The polymer electrolyte should be both chemically and electrochemically stable with regards to these unwanted reactions through a broad temperature range. The electrolyte/electrode interface participates in the reactions associated with battery operations, but is also the potential site for many other nondesirable parasitic reactions.

4. The electrolyte should be mechanically stable to promote manufacturing.

Many of these necessary traits can be enhanced by the addition of nanofillers to a solid electrolyte. For instance, when small amounts of nanoparticles are simply dispersed (Croce et al., 1998) in a solid polymer electrolyte such as poly(ethylene oxide) (PEO) complexed with inorganic lithium salts, a nanocomposite polymer electrolyte (NCPE) is formed. The addition of nanoparticles (nanofillers) of inorganic materials such as Al_2O_3, SiO_2, and $LiAlO_2$ to polymer electrolyte materials have been found to increase ion conduction, stabilize the electrode/electrolyte interface, and enhance mechanical properties of the polymer film electrolyte (Wieczorek et al., 1988, 1989; Płocharski et al., 1989; Croce et al., 1998; Appetecchi et al., 2000b; Bronstein et al., 2004).

The ion conduction in many of these systems was found to increase by one or two orders of magnitude, a greatly desired result, but the mechanism by which this increase occurred is not clearly understood. One proposed explanation is that the addition of nanofillers to the polymer electrolyte inhibits the crystallinity of the polymer (Croce et al., 1999). As shown in Figure 40.16, a PEO electrolyte is a semi-crystalline polymer, i.e., a heterogeneous system composed of both amorphous regions where the polymer is disordered and areas where the backbone chains of the polymer are able to fit together in a very orderly fashion making crystalline regions. It has long been felt that ion conduction occurs mostly in the amorphous region of the polymer by a segmental motion of the backbone chain that allows the ion to move through this phase of the polymer. Recently, it has been proposed that the crystalline region is also able to conduct ions (Gadjourova et al., 2001), but conduction through this phase may require an enhanced orientation of the crystalline material that is typically not present in polymer electrolyte systems. In this context, the crystalline regions are considered to significantly block the ion conduction because of their rigid nature.

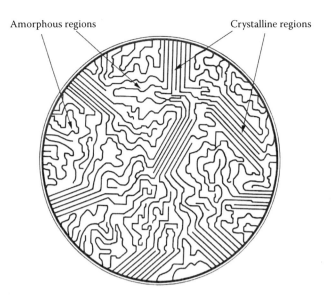

FIGURE 40.16 An illustration of the morphology of a semicrystalline polymer composed of amorphous and crystalline regions.

The nanoparticles present in the polymer matrix may inhibit the ability of the backbone chains to become well ordered and would therefore reduce the amount of crystalline regions in the polymer. The increased amount of amorphous polymer in the NCPE would then result in increased ion conduction.

Other mechanisms for this increased ion conduction have also been proposed. Again, because of the large surface area of these nanofillers, there is much active material available for chemical reactions to occur that would not be accessible on the macroscale. Lewis acid-base reactions are the most common reactions discussed for this interface. By definition, a Lewis acid is a chemical substances that can accept electron pairs from a counterpart substances called a Lewis base. The Lewis base would thus act as an electron-pair donor, forming a new species. These reactions may take place between the new species at the surface of the nanoparticle filler and the dissolved salt in the polymer electrolyte. It has been proposed that these types of reactions, at the nanoparticle/polymer electrolyte interface, provide a pathway consisting of a region of a unique chemical environment by which ions can move more freely (Appetecchi and Passerini, 2000; Aricò et al., 2005).

Another proposed mechanism for increased conduction due to the addition of nanofillers is the formation of a diffuse electrical double layer around the nanoparticle, similar to that described at the interface of the electrode and electrolyte (see Figure 40.8). In this case, a charge on the surface of the nanoparticle is balanced by the appropriately charged ions in the polymer electrolyte. The result is a region close to the nanoparticle surface containing an excess of one charged species with a layer of diffuse charge (and diminishing potential) moving away from the interface. Lewis postulated that if the concentration of the nanoparticles is sufficiently large, the double layers of the particles can overlap, forming a pathway of enhanced conduction (Lewis, 2004).

Nanoparticulate additives to polymer electrolytes can also affect the cation transference number. In many cases, the transference number associated with lithium ions can be increased by interaction with the surface of a solid nanoparticle alone. But the chemistry of the nanoparticle's surface can be tailored to further promote transference. Croce et al. treated ZrO_2 with $(NH_4)_2SO_4$ to make a sulfated-Zirconia (SO_4^{2-}–ZnO_2) surface (Croce et al., 2006b) that allows the lithium ion to "hop" from Lewis acid site to Lewis acid site and results in a lithium ion transference number of $T_{Li} = 0.81$. This is much closer to the desired value of one $T_{Li} = 1$ than the value of $T_{Li} = 0.41$ for the same polymer electrolyte without nanoparticles (Bruce et al., 1988).

The addition of polymer nanoparticles has also been found to stabilize the electrode/electrolyte interface and improve the mechanical properties of the film (Appetecchi et al., 2000a; Bronstein et al., 2004; Croce et al., 2006a). While the exact reason for the stabilization of the interface is still being investigated, several theories have been proposed. Some researchers feel that this effect is due to the ability of the nanoparticles to readily trap impurities on their large and potentially more reactive surface areas, thus removing these compounds from the interface where unwanted reactions can occur. It is also postulated that because of their small particle size, the additional particles can effectively inhibit dendrite growth on electrodes such as pure lithium metal. Nanoparticles have also been shown to enhance the mechanical properties of films, no doubt due to the same factors that enhance the mechanical properties of many composite polymer systems (Park et al., 2009).

Electrolytes can be engineered on the nanoscale by methods other than the addition of nanofillers. Rather than adding nanoparticles, the electrolyte can be confined in nanoscale channels or tubes to enhance ion conduction. Polymer electrolytes have been placed in the nanoscale channels of Al_2O_3 membranes that are used commercially for filtration purposes (Bishop and Teeters, 2009), in TiO_2 nanotubes (Volel et al., 2005), in nanochannels in glass and silicon (Schönhals et al., 2002; Zanotti et al., 2005), and in the nanochannels of polymer membranes (Vorrey and Teeters, 2003). Even the small channels (nanoscale in dimension) formed by the layered structures in montmorillonite and other clays have been used to confine polymer electrolytes (Walls et al., 2003; Wang et al., 2004; Wang and Dong, 2007). Ion conduction enhancement factors as high as two orders of magnitude have been seen in these electrolyte systems. Figure 40.17 is an example of a PEO polymer confined in a nanoporous alumina membrane.

The same explanations proposed earlier for increased ion conduction due to the addition of nanofillers are applicable to the increased ion conduction that results from nanoconfinement. The surface of the nanochannels would be available for chemical reactions such as Lewis acid/Lewis base interactions, and the diffuse double layers can overlap because of the small size of the channels. The confinement in tubes of nanoscale dimensions has also been shown to affect the crystallinity of the polymer electrolyte. Studies of PEO electrolytes confined in nanotubes have shown a reduction in the crystallinity and the

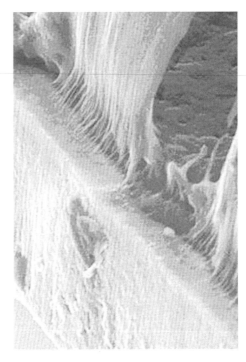

FIGURE 40.17 SEM image of the cross section of a nanoporous alumina membrane. The membrane was filled with PEO electrolyte. As a result of breaking the membrane to form the cross section, the 200 nm diameter nanotubes of PEO polymer can be seen "pulling out" (thread-like structures seen upper left to lower right) of the pores in the membranes where they were originally confined.

melting temperature associated with the crystalline regions in the polymer (Vorrey and Teeters, 2003; Zanotti et al., 2005) or in some cases the complete absence of crystallinity in the confined polymer electrolyte (Volel et al., 2005). Again, the reduction or absence of crystalline regions in the polymer should favor the segmental mobility of the polymer chains correlating to the enhanced conduction.

Additionally, when polymers are nanoconfined, the backbones in the amorphous regions, while still amorphous, can become somewhat oriented or aligned, favoring ion conduction (see Figure 40.16, amorphous regions). Even though the material is still amorphous enough for free segmental motion, the chains may also be partially aligned, allowing the lithium ions to move through the polymer matrix more easily because they do not have to "fight" as many polymer chain entanglements. Studies have shown that polymer electrolytes in nanotubes become stretched (Schönhals et al., 2002), which removes loops and entanglements in the polymer matrix that can hinder the movement of ions. This stretched backbone configuration can thus result in greater ion conduction (Gitelman et al., 2008).

While this discussion has been concerned with solid polymer electrolyte systems, other solid electrolytes such as Li_2ZnI_4 also benefit from the introduction of nanoparticles and the confinement on the nanoscale (Maekawa et al., 2008). Interfacial properties and alignment factors are similarly involved in these materials.

40.4 Nanobattery Strategies and Benefits

The unique advantages of nanostructured materials—increased surface area and interfacial regions that lead to enhanced and/or new reaction processes and increased stability—apply to each of the components of a nanobattery, to electrodes, electrolytes, and current collectors alike. The questions posed in the introduction—How small? What components? Which reactions? How to make it?—have found a wide range of answers within current research efforts and in some commercially available technology, examples of which have already been presented. These may be divided into two main types of approaches, both of which fall under the rubric of "nanobatteries" and require similar considerations in the creation of new electrode and electrolyte systems.

First, and most prevalent, are those efforts that seek to nano-engineer one or more battery components with the goal of assembling a battery that is larger than nanoscaled in its final dimensions but which benefits from internal nanostructures that enhance its performance. Most of the examples given throughout this discussion were of this type; having as their goal the creation of a better anode, cathode, or electrolyte. But what about batteries that are nanoscaled in all their dimensions?

40.4.1 Three-Dimensional Nanobatteries

Although there are a host of research efforts aimed at improving individual nanobattery elements, the approaches to assembling a full nanobattery—one that is nanoscaled in all its dimensions—are relatively few. Yet such three-dimensional nanobatteries are required for energy storage technology to keep pace with the revolution in miniaturization; having potential uses in computer chips, nanosensors, nanostructured medical devices, and all sorts of MEMS and NEMS systems. They are essential elements for the autonomous functioning of these devices.

It appears that the first reported nanobattery was the "nanometer-scale galvanic cell" of Li et al. who, in 1992, used a scanning tunneling microscope to assemble copper and silver pillars in close proximity and found that the copper plating solution acted as an electrolyte, causing the copper to anodically dissolve and plate onto the silver pillars (Li et al., 1992). Though the electrons moved only within the cell, not through external current collectors, this galvanic system, with its largest dimension of 70 nm, represents the first known example of a system in which the anode, cathode, and electrolyte were all on a nanoscale.

Fendler (1999) appears to be one of the first to use self-assembly to construct an entire battery system. He was able to make high-energy density rechargeable lithium-ion batteries by using a layer-by-layer self-assembly of poly(diallyldimethyl-ammonium chloride), graphite oxide nanoplatelets, and polyethylene oxide on indium tin oxide with a lithium wire as a counter electrode. With 10 self-assembled layers, high specific capacities ranging from 1100 to 1200 mA-h/g were achieved. His results represent the second approach to nanobattery systems: fabricating individual electrochemical cells that are fully nanoscaled in all

of their three dimensions and which can be utilized either singly or connected in series or parallel to form large arrays.

Teeters received one of the first patents for a nanobattery system, fabricating structures using the pores of nanoporous alumina membranes as "jackets" for the galvanic cells (Dewan and Teeters, 2003; Vullum and Teeters, 2005; Zhang et al., 2005; Vullum et al., 2006); see Figure 40.3. The pores were filled with a polymer electrolyte and then capped on the ends with various electrode materials resulting in arrays of individual nanobatteries with electrodes of 200 nm and smaller. Individual nanobatteries were charged and discharged by making contact with the electrodes using the tip of an atomic force microscope. While the observed current production was in the nanoamp range, as would be expected for a true nanobattery system, the volumetric capacity was 45 μA-h/μm·cm^2, proving that these batteries were viable for electronic applications.

Prior to this, the initial efforts to create a truly nanoscale battery took the approach of simply making progressively thinner layers of traditional materials in a planar arrangement. Thin-film rechargeable batteries with active layers of 1–10 μm have been of interest since the 1980s. The electrodes are typically formed by RF-magnetron sputtering, and the electrolyte may also be sputtered or be thermally evaporated. Bates and Dudney have made thin-film microbatteries by a deposition technique using a metallic lithium electrode layer with a solid electrolyte (Bates et al., 1993, 1994, 1995a,b). These batteries have lateral dimensions greater than a centimeter and had a capacity of 8.3 μA/cm^2 at an output voltage of approximately 4 V. Salmon et al. developed a microbattery using a Ni/Zn electrode couple with an aqueous KOH electrolyte (Lafollette et al., 1998; Salmon et al., 1998). Once again, fabrication involved a deposition process for the two electrodes with a polymer layer that was later removed to form the electrolyte cavity. The lateral dimensions of thin film batteries continue to be larger than nanoscaled, and a footprint of greater than 1 cm^2 is required to provide appreciable power.

Planar thin-film cells, moreover, are limited by low current and low capacity, the best example to date achieving a reversible capacity of only 0.133 mA-h/cm^2. To address this issue, Peled's group proposed replacing the traditional flat substrates for thin film deposition with perforated ones, demonstrating a 15× increase in battery capacity that correlated closely with geometrical gains calculated using the techniques demonstrated above (Nathan et al., 2005). This battery system, utilizing perforations of 50 μm in size, still lies well above the nanoscale, though further miniaturization is possible (Figure 40.18).

Its use of micro-channels, however, represents one architectural approach to achieving a truly three-dimensional nanobattery—one based on a negative, or open, space. Other open architectures include the nanobasket assembly developed by Johnson et al. (see Figure 40.13; Johnson and Teeters, 2006) and aperiodic sponge-like structures such as aerogels. More common are approaches that first create a solid structure, including the use of posts or meshes as substrates onto which electroactive materials are deposited (see Figure 40.5).

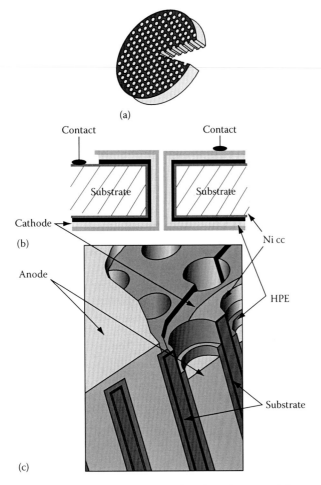

(a)

(b)

(c)

Mesoporous materials have been used extensively as electrode components, and the utilization of their nanostructure to create a truly three-dimensional nanobattery is in the developmental stages. Rolison et al. have extensively studied the physical, electrical, and electrochemical properties of poly (phenylene oxide) (PPO) applied as an ultrathin film of 7–9 nm to MnO_2 ambigels (Figure 40.19; Rolison et al., 2009).

There are also creative new approaches that do not fit neatly into traditional battery strategies, but offer much promise for future energy storage. Biomimetic processes utilize the self-assembly mechanisms already present in natural systems to construct nanoscale battery components. A vesicle-based rechargeable battery, in which synthetic structures mimicking cell membranes were used to encapsulate reductants and oxidants, achieved a maximum operating voltage of 28.9 mV and a discharge rate of 0.8 nA (Figure 40.20) (Stanish et al., 2005).

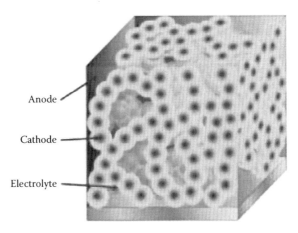

FIGURE 40.18 Schematic design of a three-dimensional battery system (not yet nanoscaled) of thin films on a perforated substrate: (a) microbattery assembly (13 mm diameter); (b) cross-section of single pore (50 μm diameter in assembly; (c) 3-D view of electorde layers within single pore of assembly. (From Nathan, M. et al., *J. Microelectromech. Syst.*, 14, 879, 2005. With permission.)

FIGURE 40.19 Three-dimensional architecture proposed by Rolison in which a sponge-like architecture serves as the cathode. The interior of the sponge could be coated with a nano-thin layer of electrolyte, and the remaining volume filled with anodic material. (Reprinted from Long, J.W. et al., *Chem. Rev.*, 104, 4463, 2004. With permission.)

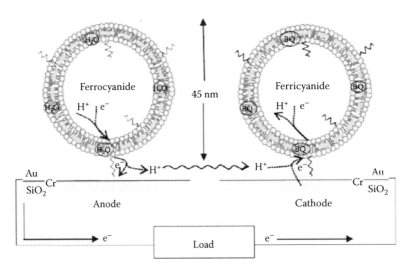

FIGURE 40.20 Schematic (not to scale) of a vesicle-based battery. The polymerized vesicles are tethered to gold current collectors via a disulfide bond. (From Stanish, I. et al., *Adv. Mater.*, 17, 1194, 2005. With permission.)

The U.S. National Center for Biometi Design of Nanoconductors is working toward a biocompatible battery that could be recharged by biological metabolism using as a model called the electric eel, which achieves large voltages and high current densities by stacking electrically polarizable membranes in series. Initially planned for use as part of an artificial retina, the protein-based system could potentially be regenerated by the body's metabolism and could find use in a wide range of implantable devices (Dickerson et al., 2008; Kannan et al., 2009).

M13 viruses genetically engineered to have an affinity for cobalt ions have been used to grow a dense assembly of cobalt oxide nanowires, which were combined with multilayer polyelectrolyte films and used as the negative electrode of a lithium-ion battery system. Charge–discharge data were consistent with that of Co_3O_4 nanoparticles produced by other methods, and the extension of the technique to cathodic materials will allow the construction of a fully self-assembled battery. Virally assembled batteries could be constructed as small transparent films, as large mesh-like architectures, or even as fibers that could be woven into wearable, power-source textiles (Ki et al., 2008).

A different mesh-like, flexible system—cellulose paper—forms the basis of a nanocomposite battery developed by researchers at Rensselaer Polytechnic Institute. Aligned carbon nanotubes embedded into a nanoporous cellulose paper soaked with electrolyte resulted in a printable system with superior mechanical flexibility that can even be cut and trimmed. The battery system, which looks like a piece of matte black paper, can be operated with electrolytes based on bodily fluids, suggesting that the device could be used as a dry-body implant and activated under extreme conditions (Pushparaj et al., 2007).

Sandia National Laboratories is investigating a "nanobattery" that is not an electrochemical cell at all; rather, it is a confined radionuclide source that emits fast electrons deposited in a semiconductor junction and converted to electricity. It offers theoretical energy densities 3–5 orders of magnitude greater than chemical batteries, and potentially extends battery life from hours to years (Crowell, 2007).

A novel nanoengineering approach to the concept of a "reserve battery"—one in which components are kept separated to ensure a long shelf-life—is the nanograss utilized by the company mPhase. Initially developed by scientists at Bell Laboratories, the nanograss is actually lithographically carved silicon filaments several hundred microns high and 300 nm in diameter whose surface chemistry has been altered by the application of a polymeric thin film. This leads to novel behavior in liquid electrolytes, which can sit on the surface of the nanograss without wetting it until an applied jolt of electricity or heat disrupts the surface tension and draws the liquid into the spaces between the grass "blades." This process, known as "electrowetting," starts the operation of the battery. Fine control of this process will allow the battery to be turned on and off at will, or allow the activation of only certain areas of the battery. Because the electrolyte is not in chemical contact with the electrode until this activation occurs, the nanobatteries are expected to have a shelf-life of decades (Lifton and Simon, 2005; Lifton et al., 2008).

40.5 Conclusions

At the time this chapter was written, nanobatteries still remained an emerging but immature technology, though one which is poised to have a significant impact on how energy is stored and delivered. Battery systems whose active materials have been engineered on the nanoscale for improved performance are already on the market, and many more are under investigation. A wide range of approaches to the fabrication of complete batteries on the nanometer scale have been proposed and full systems are in various stages of completion. By delivering more power more quickly from a smaller footprint, nanobatteries can serve as enabling technologies for all manner of advances in other electronic devices from consumer entertainment systems to implantable medical devices and electric vehicles.

Additional Readings

Further information on nanobattery concepts can be found in the following reviews: Nanobatteries (Lowy and Patrut, 2008), nanostructured electrode materials (Long et al., 2004; Jiang et al., 2006; Bisquert, 2008; Bruce et al., 2008; Rolison et al., 2009), 3D architectures (Long et al., 2004; Baggetto et al., 2008; Bruce et al., 2008; Rolison et al., 2009), electrolytes (Tarascon and Armand, 2001; Long et al., 2004; Agrawal and Pandey, 2008; Bruce et al., 2008), nanoelectrochemistry (Bisquert, 2008; Murray, 2008), batteries in general (Tarascon and Armand, 2001; Patil et al., 2008).

References

The Economist (2008) In search of the perfect battery. *Economist*, 386, 22–24.

A123 Systems, Inc. (2009) *A123 Systems Power.* http://www.a123systems.com/technology/power.

Agarwal, R. R. (1982) *Activity and Diffusivity of Lithium Intercalated in Graphite.* Department of Chemical Engineering, Illinois Institute of Technology, Chicago, IL.

Agrawal, R. C. and Pandey, G. P. (2008) Solid polymer electrolytes: Materials designing and all-solid-state battery applications: An overview. *Journal of Physics D: Applied Physics*, 41, 223001.

Appetecchi, G. B. and Passerini, S. (2000) PEO-carbon composite lithium polymer electrolyte. *Electrochimica Acta*, 45, 2139–2145.

Appetecchi, G. B., Croce, F., Persi, L., Ronci, F., and Scrosati, B. (2000a) Transport and interfacial properties of composite polymer electrolytes. *Electrochimica Acta*, 45, 1481–1490.

Appetecchi, G. B., Scaccia, S., and Passerini, S. (2000b) Investigation on the stability of the lithium-polymer electrolyte interface. *Journal of the Electrochemical Society*, 147, 4448–4452.

Aricò, A. S., Bruce, P., Scrosati, B., Tarascon, J. M., and van Schalkwijk, W. (2005) Nanostructured materials for advanced energy conversion and storage devices. *Nature Materials*, 4, 366–377.

Baggetto, L., Niessen, R. A. H., Roozehoom, F., and Notten, P. H. L. (2008) High energy density all-solid-state batteries: A challenging concept towards 3D integration. *Advanced Functional Materials*, 18, 1057–1066.

Balaya, P., Bhattacharyya, A. J., Jamnik, J., Zhukovskii, Y. F., Kotomin, E. A., and Maier, J. (2006) Nano-ionics in the context of lithium batteries. *Journal of Power Sources*, 159, 171–178.

Bates, J. B., Dudney, N. J., Gruzalski, G. R., Zuhr, R. A., Choudhury, A., Luck, C. F., and Robertson, J. D. (1993) Fabrication and characterization of amorphous lithium electrolyte thin films and rechargeable thin-film batteries. *Journal of Power Sources*, 43, 103–110.

Bates, J. B., Gruzalski, G. R., Dudney, N. J., Luck, C. F., and Yu, X. (1994) Rechargeable thin-film lithium batteries. *Solid State Ionics*, 70–71, 619–628.

Bates, J. B., Dudney, N. J., Lubben, D. C., Gruzalski, G. R., Kwak, B. S., Yu, X., and Zuhr, R. A. (1995a) Thin-film rechargeable lithium batteries. *Journal of Power Sources*, 54, 58–62.

Bates, J. B., Lubben, D., and Dudney, N. J. (1995b) Thin film Li-LiMn$_2$O$_4$ batteries. In Anon (Ed.) IEEE *Proceedings of the Annual Battery Conference*. Long Beach, CA.

Bishop, C. and Teeters, D. (2009) Crystallinity and order of poly(ethylene oxide)/lithium triflate complex confined in nanoporous membranes. *Electrochimica Acta*, 54, 4084–4088.

Bisquert, J. (2008) Physical electrochemistry of nanostructured devices. *Physical Chemistry Chemical Physics*, 10, 49–72.

Bronstein, L. M., Karlinsey, R. L., Ritter, K., Joo, C. G., Stein, B., and Zwanziger, J. W. (2004) Design of organic-inorganic solid polymer electrolytes: Synthesis, structure, and properties. *Journal of Materials Chemistry*, 14, 1812–1820.

Bruce, P. G., Evans, J., and Vincent, C. A. (1988) Conductivity and transference number measurements on polymer electrolytes. *Solid State Ionics*, 28–30, 918–922.

Bruce, P. G., Scrosati, B., and Tarascon, J.-M. (2008) Nanomaterials for rechargeable lithium batteries. *Angewandte Chemie International Edition*, 47, 2930–2946.

Chamran, F., Yeh, Y., Min, H. S., Dunn, B., and Kim, C. J. (2007) Fabrication of high-aspect-ratio electrode arrays for three-dimensional microbatteries. *Journal of Microelectromechanical Systems*, 16, 844–852.

Chen, J. and Cheng, F. (2009) Combination of lightweight elements and nanostructured materials for batteries. *Accounts of Chemical Research*, 42, 713–723.

Cho, J., Kim, Y., and Kim, M. G. (2007) Synthesis and characterization of Li[Ni$_{0.41}$Li$_{0.08}$Mn$_{0.51}$]O$_2$ nanoplates for Li battery cathode material. *Journal of Physical Chemistry C*, 111, 3192–3196.

Cortie, M. B. and van Der Lingen, E. (2002) Catalytic gold nanoparticles. In Wuhrer, R., Braach-Maksvytis, V., and Turney, T. (Eds.) *Materials Forum*. Sydney, Australia.

Croce, F., Appetecchi, G. B., Persi, L., and Scrosati, B. (1998) Nanocomposite polymer electrolytes for lithium batteries. *Nature*, 394, 456–458.

Croce, F., Curini, R., Martinelli, A., Persi, L., Ronci, F., Scrosati, B., and Caminiti, R. (1999) Physical and chemical properties of nanocomposite polymer electrolytes. *Journal of Physical Chemistry B*, 103, 10632–10638.

Croce, F., Sacchetti, S., and Scrosati, B. (2006a) Advanced, lithium batteries based on high-performance composite polymer electrolytes. *Journal of Power Sources*, 162, 685–689.

Croce, F., Settimi, L., and Scrosati, B. (2006b) Superacid ZrO$_2$-added, composite polymer electrolytes with improved transport properties. *Electrochemistry Communications*, 8, 364–368.

Crowell, J. (2007) Nuclear nano-batteries: An on-board power supply for MEMS devices. Sandia National Laboratories, Albuquerque, NM, http://www.ca.sandia.gov/8700/projects/content.php?cid=58.

Dewan, C. and Teeters, D. (2003) Vanadia xerogel nanocathodes used in lithium microbatteries. *Journal of Power Sources*, 119–121, 310–315.

Dickerson, M. B., Sandhage, K. H., and Naik, R. R. (2008) Protein- and peptide-directed syntheses of inorganic materials. *Chemical Reviews*, 108, 4935–4978.

Disk/Trend Report (2000) Five decades of disk drive industry firsts, Disk/Trend Report, http://www.disktrend.com/5decades2.htm.

Fehrenbacher, K. (2009) Seeo promises a safer lithium battery with higher energy density, earth2tech, http://earth2tech.com/2009/05/07/seeo-promises-a-safer-higher-energy-density-lithium-battery/.

Fendler, J. H. (1999) Colloid chemical approach to the construction of high energy density rechargeable lithium—ion batteries. *Journal of Dispersion Science and Technology*, 20, 13–25.

Gadjourova, Z., Andreev, Y. G., Tunstall, D. P., and Bruce, P. G. (2001) Ionic conductivity in crystalline polymer electrolytes. *Nature*, 412, 520–523.

Gillham, O. (2003) Micromanaging the future. *Tulsa World*, A13.

Gitelman, L., Israeli, M., Averbuch, A., Nathan, M., Schuss, Z., and Golodnitsky, D. (2008) Polymer geometry and Li$^+$ conduction in poly(ethylene oxide). *Journal of Computational Physics*, 227, 8437–8447.

Gray, F. M. (1997) Polymer electrolytes. *RSC Materials Monographs*, Royal Society of Chemistry, Cambridge, NY.

Green, M., Fielder, E., Scrosati, B., Wachtler, M., and Moreno, J. S. (2003) Structured silicon anodes for lithium battery applications. *Electrochemical and Solid-State Letters*, 6, 75–79.

Hang, B. T., Watanabe, T., Eashira, M., Okada, S., Yamaki, J. I., Hata, S., Yoon, S. H., and Mochida, I. (2005) The electrochemical properties of Fe$_2$O$_3$-loaded carbon electrodes for iron-air battery anodes. *Journal of Power Sources*, 150, 261–271.

Hang, B. T., Watanabe, I., Doi, T., Okada, S., and Yamaki, J. I. (2006) Electrochemical properties of nano-sized Fe$_2$O$_3$-loaded carbon as a lithium battery anode. *Journal of Power Sources*, 161, 1281–1287.

Hassoun, J., Panero, S., Simon, P., Taberna, P. L., and Scrosati, B. (2007) High-rate, long-life Ni-Sn nanostructured electrodes for lithium-ion batteries. *Advanced Materials*, 19, 1632–1635.

He, B. L., Dong, B., Wang, W., and Li, H. L. (2009) Performance of polyaniline/multi-walled carbon nanotubes composites as cathode for rechargeable lithium batteries. *Materials Chemistry and Physics*, 114, 371–375.

Hockenberry, J. (2006) Building a better battery. *Wired*, 14, 204–211.

Hosono, E., Fujihara, S., Honma, I., Ichihara, M., and Zhou, H. (2006a) Fabrication of nano/micro hierarchical Fe_2O_3/Ni micrometer-wire structure and characteristics for high rate Li rechargeable battery. *Journal of the Electrochemical Society*, 153, A1273–A1278.

Hosono, E., Fujihara, S., Honma, I., and Zhou, H. (2006b) The high power and high energy densities Li ion storage device by nanocrystalline and mesoporous Ni/NiO covered structure. *Electrochemistry Communications*, 8, 284–288.

Hsu, C. M., Connor, S. T., Tang, M. X., and Cui, Y. (2008) Wafer-scale silicon nanopillars and nanocones by Langmuir-Blodgett assembly and etching. *Applied Physics Letters*, 93, 133109.

Humble, P. H., Harb, J. N., and Lafollette, R. (2001) Microscopic nickel-zinc batteries for use in autonomous microsystems. *Journal of the Electrochemical Society*, 148, A1357–A1361.

Jamnik, J. and Maier, J. (2003) Nanocrystallinity effects in lithium battery materials. Aspects of nano-ionics. Part IV. *Physical Chemistry Chemical Physics*, 5, 5215–5220.

Jiang, C., Hosono, E., and Zhou, H. (2006) Nanomaterials for lithium ion batteries. *Nano Today*, 1, 28–33.

Jiang, C., Honma, I., Kudo, T., and Zhou, H. (2007) Nanocrystalline rutile TiO_2 electrode for high-capacity and high-rate lithium storage. *Electrochemical and Solid-State Letters*, 10, A127–A129.

Jiang, C., Hosono, E., Ichihara, M., Honma, I., and Zhou, H. (2008) Synthesis of nanocrystalline $Li_4Ti_5O_{12}$ by chemical lithiation of anatase nanocrystals and postannealing. *Journal of the Electrochemical Society*, 155.

Johnson, P. L. and Teeters, D. (2006) Formation and characterization of SnO_2 nanobaskets. *Solid State Ionics*, 177, 2821–2825.

Julien, C. (1997) Solid state batteries. In Gellings, P. J. B. H. J. M. (Ed.) *The CRC Handbook of Solid State Electrochemistry*, CRC Press, Boca Raton, FL.

Kannan, A. M., Renugopalakrishnan, V., Filipek, S., Li, P., Audette, G. F., and Munukutla, L. (2009) Bio-batteries and bio-fuel cells: Leveraging on electronic charge transfer proteins. *Journal of Nanoscience and Nanotechnology*, 9, 1665–1678.

Kavan, L., Procházka, J., Spitler, T. M., Kalbá, M., Zukalová, M., Drezen, T., and Grätzel, M. (2003) Li insertion into $Li_4Ti_5O_{12}$ (spinel). Charge capability vs. particle size in thin-film electrodes. *Journal of the Electrochemical Society*, 150, A1000–A1007.

Kawasaki, S., Iwai, Y., and Hirose, M. (2009) Electrochemical lithium ion storage properties of single-walled carbon nanotubes containing organic molecules. *Carbon*, 47, 1081–1086.

Ki, T. N., Wartena, R., Yoo, P. J., Liau, F. W., Yun, J. L., Chiang, Y. M., Hammond, P. T., and Belcher, A. M. (2008) Stamped microbattery electrodes based on self-assembled M13 viruses. *Proceedings of the National Academy of Sciences of the United States of America*, 105, 17227–17231.

Lafollette, R. M., Salmon, L. G., Barksdale, R. A., Beachem, B., Harb, J. N., Holladay, J. D., Humble, P. H., and Ryan, D. M. (1998) The performance of microscopic batteries developed for MEMS applications. *Proceedings of the 33rd Intersociety Energy Conversion Engineering Conference*, Colorado Springs, CO, IECEC117/1–IECEC117/5.

Lewis, T. J. (2004) Interfaces are the dominant feature of dielectrics at the nanometric level. *IEEE Transactions on Dielectrics and Electrical Insulation*, 11, 739–753.

Li, H., Richter, G., and Maier, J. (2003) Reversible formation and decomposition of LiF clusters using transition metal fluorides as precursors and their application in rechargeable Li batteries. *Advanced Materials*, 15, 736–739.

Li, W., Virtanen, J. A., and Penner, R. M. (1992) A nanometer-scale galvanic cell. *Journal of Physical Chemistry*, 96, 6529–6532.

Liao, C. L., Wu, M. T., Yen, J. H., Leu, I. C., and Fung, K. Z. (2006) Preparation of RF-sputtered lithium cobalt oxide nanorods by using porous anodic alumina (PAA) template. *Journal of Alloys and Compounds*, 414, 302–309.

Lifton, V. A. and Simon, S. (2005) A novel battery architecture based on superhydrophobic nanostructured materials. In Laudon, M. and Romanowicz, B. (Eds.) *2005 NSTI Nanotechnology Conference and Trade Show—NSTI Nanotech 2005 Technical Proceedings*, Anaheim, CA.

Lifton, V. A., Taylor, J. A., Vyas, B., Kolodner, P., Cirelli, R., Basavanhally, N., Papazian, A., Frahm, R., Simon, S., and Krupenkin, T. (2008) Superhydrophobic membranes with electrically controllable permeability and their application to "smart" microbatteries. *Applied Physics Letters*, 93, 043112.

Linden, D. and Reddy, T. B. (eds.) (2002) *Handbook of Batteries*, 3rd edn. McGraw-Hill, New York, pp. 1–13.

Liu, H., Wang, G., Park, J., Wang, J., Liu, H., and Zhang, C. (2009a) Electrochemical performance of α-Fe_2O_3 nanorods as anode material for lithium-ion cells. *Electrochimica Acta*, 54, 1733–1736.

Liu, J., Li, Y., Huang, X., Ding, R., Hu, Y., Jiang, J., and Liao, L. (2009b) Direct growth of SnO_2 nanorod array electrodes for lithium-ion batteries. *Journal of Materials Chemistry*, 19, 1859–1864.

Long, J. W., Dunn, B., Rolison, D. R., and White, H. S. (2004) Three-dimensional battery architectures. *Chemical Reviews*, 104, 4463–4492.

Lowy, D. A. and Patrut, A. (2008) Nanobatteries: Decreasing size power sources for growing technologies. *Recent Patents on Nanotechnology*, 2, 208–219.

Lu, H.-W., Yu, L., Zeng, W., Li, Y.-S., and Fu, Z.-W. (2008) Fabrication and electrochemical properties of three-dimensional structure of $LiCoO_2$ fibers. *Electrochemical and Solid-State Letters*, 11, A140–A144.

Lu, H. W., Li, D., Sun, K., Li, Y. S., and Fu, Z. W. (2009) Carbon nanotube reinforced NiO fibers for rechargeable lithium batteries. *Solid State Sciences*, 11, 982–987.

Ma, M., Tu, J. P., Yuan, Y. F., Wang, X. L., Li, K. F., Mao, F., and Zeng, Z. Y. (2008) Electrochemical performance of ZnO nanoplates as anode materials for Ni/Zn secondary batteries. *Journal of Power Sources*, 179, 395–400.

Maekawa, H., Iwatani, T., Shen, H., Yamamura, T., and Kawamura, J. (2008) Enhanced lithium ion conduction and the size effect on interfacial phase in Li_2ZnI_4-mesoporous alumina composite electrolyte. *Solid State Ionics*, 178, 1637–1641.

Muraliganth, T., Murugan, A. V., and Manthiram, A. (2008) Nanoscale networking of $LiFePO_4$ nanorods synthesized by a microwave-solvothermal route with carbon nanotubes for lithium ion batteries. *Journal of Materials Chemistry*, 18, 5661–5668.

Murray, R. W. (2008) Nanoelectrochemistry: Metal nanoparticles, nanoelectrodes, and nanopores. *Chemical Reviews*, 108, 2688–2720.

Nathan, M., Golodnitsky, D., Yufit, V., Strauss, E., Ripenbein, T., Shechtman, I., Menkin, S., and Peled, E. (2005) Three-dimensional thin-film Li-ion microbatteries for autonomous MEMS. *Journal of Microelectromechanical Systems*, 14, 879–885.

Pan, J., Sun, Y., Wang, Z., Wan, P., Yang, Y., and Fan, M. (2009) Nano-NiOOH prepared by splitting method as super high-speed charge/discharge cathode material for rechargeable alkaline batteries. *Journal of Power Sources*, 188, 308–312.

Park, S. K., Kim, S. H., and Hwang, J. T. (2009) Effect of fumed silica nanoparticles on glass fiber filled thermotropic liquid crystalline polymer composites. *Polymer Composites*, 30, 309–317.

Patil, A., Patil, V., Wook Shin, D., Choi, J. W., Paik, D. S., and Yoon, S. J. (2008) Issue and challenges facing rechargeable thin film lithium batteries. *Materials Research Bulletin*, 43, 1913–1942.

Płocharski, J., Wieczorek, W., Przyłuski, J., and Such, K. (1989) Mixed solid electrolytes based on poly(ethylene oxide). *Applied Physics A Solids and Surfaces*, 49, 55–60.

Poizot, P., Laruelle, S., Grugeon, S., Dupont, L., and Tarascon, J. M. (2000) Nano-sized transition-metal oxides as negative-electrode materials for lithium-ion batteries. *Nature*, 407, 496–499.

Pushparaj, V. L., Shaijumon, M. M., Kumar, A., Murugesan, S., CI, L., Vajtai, R., Linhardt, R. J., Nalamasu, O., and Ajayan, P. M. (2007) Flexible energy storage devices based on nanocomposite paper. *Proceedings of the National Academy of Sciences of the United States of America*, 104, 13574–13577.

Rolison, D. R., Long, J. W., Lytle, J. C., Fischer, A. E., Rhodes, C. P., Mcevoy, T. M., Bourg, M. E., and Lubers, A. M. (2009) Multifunctional 3D nanoarchitectures for energy storage and conversion. *Chemical Society Reviews*, 38, 226–252.

Salmon, L. G., Barksdale, R. A., Beachem, B. R., Harb, J. N., Holladay, J. D., Humble, P. H., Lafollette, R. M., and Ryan, D. M. (1998) Fabrication of rechargeable microbatteries for microelectromechanical system (MEMS) applications. *Proceedings of the 33rd Intersociety Energy Conversion Engineering Conference*, Colorado Springs, CO, IECEC116/1–IECEC116/6.

Saravanan, K., Reddy, M. V., Balaya, P., Gong, H., Chowdari, B. V. R., and Vittal, J. J. (2009) Storage performance of $LiFePO_4$ nanoplates. *Journal of Materials Chemistry*, 19, 605–610.

Schönhals, A., Goering, H., and Schick, C. (2002) Segmental and chain dynamics of polymers: From the bulk to the confined state. *Journal of Non-Crystalline Solids*, 305, 140–149.

Seagate Product Press Releases (2008) Seagate powers next generation of computing with three new hard drives, Seagate Product Press Releases, http://www.seagate.com/ww/v/index.jsp?locale=en-US&name=null&vgnextoid=19549a9dafc0b110VgnVCM100000f5ee0a0aRCRD.

Seo, J. W., Jang, J. T., Park, S. W., Kim, C., Park, B., and Cheon, J. (2008) Two-dimensional SnS_2 nanoplates with extraordinary high discharge capacity for lithium ion batteries. *Advanced Materials*, 20, 4269–4273.

Stanish, I., Lowy, D. A., Hung, C. W., and Singh, A. (2005) Vesicle-based rechargeable batteries. *Advanced Materials*, 17, 1194–1198.

Stashans, A., Lunell, S., Bergstroem, R., Hagfeldt, A., and Lindquist, S.-E. (1996) Theoretical study of lithium intercalation in rutile and anatase. *Physical Review B: Condensed Matter*, 53, 159–70.

Subba Reddy, C. V., Walker, E. H. Jr., Wicker, S. A. Sr., Williams, Q. L., and Kalluru, R. R. (2009) Synthesis of VO_2 (B) nanorods for Li battery application. *Current Applied Physics*, 9, 1195–1198.

Taberna, P. L., Mitra, S., Poizot, P., Simon, P., and Tarascon, J. M. (2006) High rate capabilities Fe_3O_4-based Cu nano-architectured electrodes for lithium-ion battery applications. *Nature Materials*, 5, 567–573.

Tarascon, J. M. and Armand, M. (2001) Issues and challenges facing rechargeable lithium batteries. *Nature*, 414, 359–367.

Volel, M., Armand, M., Gorecki, W., and Saboungi, M. L. (2005) Threading polymer into nanotubes: Evidence of poly(ethylene oxide) inclusion in titanium oxide. *Chemistry of Materials*, 17, 2028–2033.

Vorrey, S. and Teeters, D. (2003) Study of the ion conduction of polymer electrolytes confined in micro and nanopores. *Electrochimica Acta*, 48, 2137–2141.

Vullum, F. and Teeters, D. (2005) Investigation of lithium battery nanoelectrode arrays and their component nanobatteries. *Journal of Power Sources*, 146, 804–808.

Vullum, F., Teeters, D., Nytén, A., and Thomas, J. (2006) Characterization of lithium nanobatteries and lithium battery nanoelectrode arrays that benefit from nanostructure and molecular self-assembly. *Solid State Ionics*, 177, 2833–2838.

Walls, H. J., Riley, M. W., Singhal, R. R., Spontak, R. J., Fedkiw, P. S., and Khan, S. A. (2003) Nanocomposite electrolytes with fumed silica and hectorite clay networks: Passive versus active fillers. *Advanced Functional Materials*, 13, 710–717.

Wang, G., Shen, X., Yao, J., Wexler, D., and Ahn, J. H. (2009a) Hydrothermal synthesis of carbon nanotube/cobalt oxide core-shell one-dimensional nanocomposite and application as an anode material for lithium-ion batteries. *Electrochemistry Communications*, 11, 546–549.

Wang, H., Pan, Q., Cheng, Y., Zhao, J., and Yin, G. (2009b) Evaluation of ZnO nanorod arrays with dandelion-like morphology as negative electrodes for lithium-ion batteries. *Electrochimica Acta*, 54, 2851–2855.

Wang, M. and Dong, S. (2007) Enhanced electrochemical properties of nanocomposite polymer electrolyte based on copolymer with exfoliated clays. *Journal of Power Sources*, 170, 425–432.

Wang, M., Zhao, F., Guo, Z., and Dong, S. (2004) Poly(vinylidene fluoride-hexafluoropropylene)/organo-montmorillonite clays nanocomposite lithium polymer electrolytes. *Electrochimica Acta*, 49, 3595–3602.

Whitacre, J. F., West, W. C., Smart, M. C., Yazami, R., Prakash, G. K. S., Hamwi, A., and Ratnakumar, B. V. (2007) Enhanced low-temperature performance of Li-CFx batteries. *Electrochemical Solid-State Letters*, 10, A166–A170.

Wieczorek, W., Płocharski, J., Przyłuski, J., Głowinkowski, S., and Pajak, Z. (1988) Impedance spectroscopy and phase structure of PEONaI complexes. *Solid State Ionics*, 28–30, 1014–1017.

Wieczorek, W., Such, K., Wyciślik, H., and Płocharski, J. (1989) Modifications of crystalline structure of peo polymer electrolytes with ceramic additives. *Solid State Ionics*, 36, 255–257.

Winter, M. and Brodd, R. J. (2004) What are batteries, fuel cells, and supercapacitors? *Chemical Reviews*, 104, 4245–4269.

Xiao, Z., Feng, C., Chan, P. C. H., and Hsing, I. M. (2007) Formation of silicon nanopores and nanopillars by a maskless deep reactive ion etching process. *TRANSDUCERS and EUROSENSORS '07—4th International Conference on Solid-State Sensors, Actuators and Microsystems*, Lyon, France.

Yang, J., Winter, M., and Besenhard, J. O. (1996) Small particle size multiphase Li-alloy anodes for lithium-ionbatteries. *Solid State Ionics*, 90, 281–287.

Yuan, L., Yuan, H., Qiu, X., Chen, L., and Zhu, W. (2009) Improvement of cycle property of sulfur-coated multi-walled carbon nanotubes composite cathode for lithium/sulfur batteries. *Journal of Power Sources*, 189, 1141–1146.

Zanotti, J. M., Smith, L. J., Price, D. L., and Saboungi, M. L. (2005) Inelastic neutron scattering as a probe of dynamics under confinement. The case of a peo polymer melt. *Annales de Chimie: Science des Materiaux*, 30, 353–364.

Zhang, Z., Dewan, C., Kothari, S., Mitra, S., and Teeters, D. (2005) Carbon nanotube synthesis, characteristics, and micro-battery applications. *Materials Science and Engineering B: Solid-State Materials for Advanced Technology*, 116, 363–368.

Zhao, Z. W., Guo, Z. P., Liu, H. K., and Dou, S. X. (2008) Various carbon metal nanocomposites for lithium ion batteries and direct methanol fuel cells. *ECS Transactions*, 25 edn. Chicago, IL.

Zhou, H., Li, D., Hibino, M., and Honma, I. (2005) A self-ordered, crystalline-glass, mesoporous nanocomposite for use as a lithium-based storage device with both high power and high energy densities. *Angewandte Chemie International Edition*, 44, 797–802.

Zhou, Y., Shen, C., and Li, H. (2002) Synthesis of high-ordered $LiCoO_2$ nanowire arrays by AAO template. *Solid State Ionics*, 146, 81–86.

Zhukovskii, Y. F., Balaya, P., Kotomin Eugene, A., and Maier, J. (2006) Evidence for interfacial-storage anomaly in nanocomposites for lithium batteries from first-principles simulations. *Physical Review Letters*, 96, 058302/1–058302/4.

41

Nanoheaters

Christian Falconi
University of Tor Vergata

and

Italian National Research Council

41.1 Introduction

Heaters are often defined as devices that produce heat or, equivalently, as devices that can heat their surroundings; with such broad definitions, since heat release is associated with all dissipative processes, all practical devices would be heaters (e.g., resistors, tires, personal computers, etc.). As an extreme example, though there is no dissipation in an "ideal" capacitor, "real" capacitors would also be heaters because of their parasitic resistances. It is therefore better to qualitatively define heaters as devices that *efficiently* heat their surroundings; despite the intrinsic vagueness of this definition (how efficiently should a device heat its surroundings to qualify?), it allows to focus on a crucial point: heaters transform an input signal (energy) into heat and, in most cases, it is important that this transformation be as efficient as possible.

Nanotechnology allows the fabrication of nano-sized heaters. At present, the design of nanoheaters is complicated due to insufficient knowledge on nanoscale heat transfer, technological limits, and biocompatibility issues (for *in vivo* applications); nevertheless, the outstanding potential of nanoheaters has already been demonstrated in a few applications. For instance, nanoheaters have been used for high-density memories and imaging (scanning thermal profiler); it has also been shown that "wireless" nanoheaters (e.g., magnetic nanoparticles (Jordan et al., 1996; Jordan et al., 1999; Berry and Curtis, 2003) and gold nanoshells (Loo et al., 2004)) can selectively damage target cells by heating (hyperthermia, or thermal ablation if the temperature is so high that cells are destroyed); "photothermally induced" targeted drug delivery has also been reported (Sershen et al., 2000; Sershen et al., 2001). In future, nanoheaters may become critical components for a wide range of applications (hyperthermia, thermal ablation, targeted drug delivery, imaging, nanofabrication, and data storage, and the analysis of physical, chemical, and biological processes at the nanoscale).

41.2 Background

In Section 41.2.1, we briefly discuss the heat transfer that occurs at nanoscale; in Section 41.2.2, we discuss the equivalent circuits for thermal systems; and in Section 41.2.3, we give the fundamental definitions for nanoheaters.

41.2.1 Heat Transfer at Nanoscale

Heat transfer originates due to temperature differences. Conduction heat transfer is the transport of thermal energy in a given (solid or fluid) medium due to temperature differences. Convection heat transfer occurs when there is a temperature gradient between a surface and a moving fluid on the surface. Radiation heat transfer is the net heat transfer caused by radiation between two surfaces at different temperatures. The theory of heat transfer at macroscale is well established and thoroughly described in several reference textbooks (e.g., Incropera and De Witt, 2002). On the other hand, heat transfer at nanoscale is much more complex. Even from a purely theoretical point of view, heat transfer occurs because of a nonequilibrium. By contrast, temperature is defined as an average quantity for a large number of particles in equilibrium. However, in macroscopic systems, each region of interest (where particles can be considered "almost in equilibrium") comprises such a large number of particles that the definition of a single temperature (i.e., an average quantity) for each region of interest is, in practice, unproblematic. On the contrary, at nanoscale, the regions of interest can be so small that the definition of a single temperature for a given region can be problematic. Beside this "theoretical" issue, in practice, the knowledge on heat transfer at nanoscale is very limited (see Cahill et al., 2003 for an introductory review) and an accurate theoretical estimation of the thermal resistance between a nanoheater and its surroundings is practically impossible. Moreover, even the capabilities of numerical simulations are limited. In fact, in molecular dynamics simulations (Poulikakos et al., 2003) each molecule must be independently considered and, for each molecule, its Newton equation must be solved. With such an approach, clearly, only very small volumes can be investigated. Moreover, beside obvious computational difficulties for solving equations containing an enormous number of variables, molecular dynamics also requires an *a priori* accurate knowledge of all the relevant intermolecular interactions (which is not easy in practical cases) and does not model nonclassical effects.

Despite the complexity of heat transfer at the nanoscale, a qualitative understanding of the major differences between the macroscale and nanoscale heat transfer may provide sufficient insights for the design of effective nanoheaters (e.g., see Falconi et al., 2007a). In particular, the interfaces between different materials become increasingly important with downscaling, so that, at the nanoscale, the thermal boundary resistance often dominates the total thermal resistance. If we consider heat conduction perpendicular to an interface, the reflections of thermal energy waves (i.e., phonon backscattering) limit heat transfer. Equivalently, at the interface between two materials there is a thermal boundary resistance (also referred to as Kapitza resistance), which is inversely proportional to the area of the interface. In practice, at the interface between two materials X and Y, we may consider a (Kapitza) conduction thermal resistance:

$$R_{\text{TH,interface}} = \frac{1}{\kappa_{X,Y} A} \qquad (41.1)$$

where $\kappa_{X,Y}$ is the interface thermal conductivity per unit area that depends on the two materials X and Y. It is sometimes convenient to define the Kapitza length as the crystal length that would have an equivalent thermal resistance. In macroscopic systems, the role of the thermal boundary resistance is generally negligible as this thermal resistance is in series with other thermal resistances that are, generally, much larger (equivalently, all the dimensions are much larger than Kapitza lengths). By contrast, at the nanoscale, the Kapitza resistance can easily dominate the thermal resistance. As an example, in Huxtable et al. (2003), a nanotube was heated by a laser beam and an extremely high Kapitza resistance was derived from the exponential decay of the nanotube temperature. The role of the Kapitza resistance at nanoscale is also discussed in, for instance, Wilson et al. (2002), Costescu et al. (2004), and

Shenogin et al. (2004). If the Kapitza resistance dominates the thermal resistance, though in most cases a quantitative estimation of the thermal resistance R_{TH} between a nanoheater and its surroundings is impossible, it is at least possible to state that the thermal resistance is inversely proportional to the area of the nanoheaters. For instance, in Falconi et al. (2007a), this observation was sufficient to determine the design rules for optimal wireless Joule nanoheaters.

41.2.2 Equivalent Circuits for Thermal Systems

A given thermal system may be translated into an equivalent electric circuit by using the following equivalences:

$$
\begin{aligned}
P &\leftrightarrow I \\
\Delta T &\leftrightarrow \Delta V \\
R_{\text{TH}} &\leftrightarrow R \\
C_{\text{TH}} &\leftrightarrow C
\end{aligned}
\qquad (41.2)
$$

where for each signal or component in the thermal domain (left) there is an associated signal or component in the electrical domain (right).

As an example, let us consider an object that is heated by a power P_0. For simplicity, we assume that all the volume of the object is always at the same temperature. In the thermal domain, the object has its temperature and is separated from the surrounding environment by a thermal resistance. Furthermore, it has its thermal capacitance. Assuming that the environment temperature $T_{\text{environment}}$ is constant, a simple equivalent circuit may be found by defining a node for each volume with a different temperature and using the equivalences (41.2), as shown in Figure 41.1.

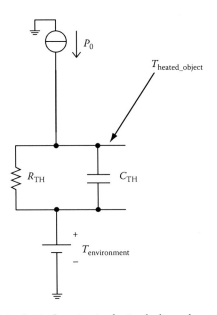

FIGURE 41.1 Equivalent circuit of a simple thermal system.

For a given thermal system, in principle, there is no advantage in the derivation of an equivalent circuit (i.e., the complexity of the thermal system and of its equivalent electric circuit are, obviously, identical). Nevertheless, equivalent circuits of thermal systems are ubiquitously used for important practical reasons. First, properly drawn circuits are very easy to understand. Second, circuit representations are generally well known (not only to electronic engineers). Third, in many cases, the thermal system is interfaced with an electronic system. In such cases, an equivalent electric circuit of the thermal system is the only viable method for simulating the complete (thermal + electronic) system (Falconi et al., 2007b; Falconi and Fratini, 2008). Fourth, once an equivalent electric circuit of the thermal system is obtained, the complete (thermal + electronic) system can easily be analyzed/designed with standard simulators for electronic circuits such as SPICE, thus providing extended simulation capabilities (i.e., using idealized blocks or "analog behavioral models," performing parametric simulations, …). In order to illustrate the advantages of this method, let us consider a temperature-dependent resistor driven by an ideal current source. If the resistance of the resistor linearly depends on the temperature of the resistor, T_{RES},

$$R(T_{RES}) = R_0 \left[1 + \alpha (T_{RES} - T_0) \right] \qquad (41.3)$$

the SPICE electro-thermal model for the temperature-dependent resistor is shown in Figure 41.2. The voltage across the resistor, v_{RES}, is computed as the product of the current through the resistor, I_0, and the temperature-dependent resistance; the thermal inertia is modeled by a thermal resistance R_{TH} and by the thermal capacitance C_{TH}. Clearly, in similar equivalent circuits there is no coherence for the dimensions of the various signals (e.g., the signal labeled with "Power" is, within SPICE, represented by a voltage). We stress that, once a similar "equivalent circuit" of a thermal system is found, all the flexibility and potentialities of standard circuit simulators can be readily applied to the analysis/design of thermal systems. It would, for instance, be very easy to compute all the relevant physical quantities (temperature, resistance, voltage, power, etc.) for various values of the temperature coefficient of the resistor.

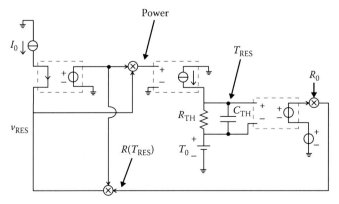

FIGURE 41.2 Equivalent electric circuit of a simple thermal system.

41.2.3 Fundamental Definitions for Nanoheaters

An accurate description of nanoheaters requires a proper nomenclature. In fact, there can be very important differences between traditional heaters and nanoheaters. For instance, an enormous number of nanoheaters will often be used simultaneously, while in traditional systems a single heater is often sufficient. Moreover, for such large number of transducers, calibration may be impractical and other approaches can be more convenient. Here we give the most important definitions for nanoheaters.

41.2.3.1 Signals, Systems, and Nanoheaters

Signals are representations of physical, chemical, or biological quantities that evolve with time (e.g., the voltage across a capacitor, the average temperature of a room, the acceleration of an airplane, etc.). Signals can be classified into six different energy domains (Middelhock et al., 2000): electrical, thermal, mechanical, magnetic, radiant, and chemical.

Systems transform input signals into output signals according to a given transformation

$$T : x(t) \rightarrow y(t) \qquad (41.4)$$

The output signals of nanoheaters are always in the thermal energy domain. In most cases the temperature of the nanoheater itself or the temperature of an object close to the nanoheater (i.e., the temperature of a target cell) or the heating power can be chosen as output signals.

The input signals of nanoheaters can belong to all the energy domains. In order to synthetically classify different types of nanoheaters we will refer to "x activated nanoheaters" where x is relative to the energy domain of the input signal (e.g., a nano-sized resistor can be used as an electrically activated nanoheater).

Transducers are, by definition, systems that convert signals from one energy domain into the signals in a different energy domain. With the exception of thermally activated nanoheaters, all the other nanoheaters are therefore nanotransducers.

Nanoheaters can be wireless or not. For instance gold nanoshells are wireless nanoheaters, while resistive nanoheaters connected to a voltage source by wires are not. Wireless nanoheaters can be especially important for biomedical *in vivo* applications.

A system is time-invariant if and only if, for any given t_0 and for any input signal $x(t)$,

$$T\left[x(t)\right] = y(t) \Rightarrow T\left[x(t - t_0)\right] = y(t - t_0) \qquad (41.5)$$

or, equivalently, if its properties do not change with time.

A system is linear if and only if, for any real numbers c_1 and c_2, and for any input signals $x_1(t)$ and $x_2(t)$,

$$T\left[c_1 x_1(t) + c_2 x_2(t)\right] = c_1 T\left[x_1(t)\right] + c_2 T\left[x_2(t)\right] \qquad (41.6)$$

All real systems are time variant and nonlinear. For instance, even a simple resistor may be regarded as a system that transforms an input signal (current) into an output signal (voltage). Although, we often describe a resistor by means of the time-invariant, linear Ohm's law,

$$v = Ri \qquad (41.7)$$

this description is only accurate within a limited range of the input signal values (beyond those limits significant nonlinear effects will appear). Moreover, for a real resistor, the resistance, R, changes with temperature, contamination, aging, etc. (i.e., the system is time variant). Nevertheless, almost always, real systems are approximated by means of correspondent linear, time invariant systems. In fact, in most cases, this representation is both acceptable (within predefined operative conditions and for the time intervals that are of interest) and extremely convenient, as *time invariant and linear systems* are much easier to be analyzed and designed. However, in the case of nanoheaters, depending on the specific application, a linear, time invariant model can be used or not. As a first issue, in some cases, heaters (macroscopic or nanosized) are inherently nonlinear. As an example, if, for a resistive heater, we consider the current as the input signal and the heating power as the output signal, the system is, clearly, nonlinear. In these cases, a linearized model is only possible within a (limited) range of the operating point (i.e., Taylor first order approximation). As a second issue, in many cases, the properties of (both linear and nonlinear) heaters can strongly change with time. However, though aging can be detrimental and impede proper operations for a sufficient amount of time, modifications with time can even be favorable. For instance, zinc oxide wireless nanoheaters (Falconi et al., 2006b; Falconi et al., 2007a) immersed into the blood, if uncoated, would likely dissolve in relatively short periods of time (Zhou et al., 2006). This, however, could be a very beneficial characteristic from the point of view of safety and low toxicity.

41.2.3.2 Instantaneous Systems: Error, Relative Error, Accuracy, Precision, Sensitivity

A system is instantaneous if, at any given instant, t_0, the output of the system only depends on the input signal at the same instant, so that

$$y_{\text{out}} = f(x_{\text{in}}) \qquad (41.8)$$

Instantaneous systems, by definition, are time invariant (linear or nonlinear) systems. In principle, no real system can be instantaneous. In fact, systems transform input signals into output signals, and, in the real world, signal transformations take time (i.e., are not instantaneous). However, time-invariant systems that are "much faster than their input signals" may be considered as instantaneous (see later for a more quantitative definition). This observation is important as some properties of nanoheaters (e.g., sensitivity) can only be defined under this assumption.

An ideal system transforms an input signal into an output signal according to a desired transformation. However, real systems unavoidably introduce errors. For instantaneous systems, the error may be defined as the difference between the output and the ideal output:

$$e(x_{\text{in}}) = y_{\text{out}}(x_{\text{in}}) - y_{\text{out,ideal}}(x_{\text{in}}) \qquad (41.9)$$

Clearly, since the output signals of nanoheaters are always in the thermal domain, their errors are also in the thermal domain. In general, the error depends on the input; for instance, a system might have a very small error only when the input is within a certain range. For simplicity, we assume that there is no error (or, in practice, a very small error) in the input signal. The ideal output is known without uncertainty and the output may be measured with negligible errors. The accuracy may be qualitatively defined as the capability of the system to produce small errors.

Both the definitions for the error and the accuracy refer to a "single event." However, if many events are considered, the output may be regarded as a random variable $Y_{\text{out}}(x_{\text{in}})$ (since we assume that there is no error on the input signal, the input signal is not a random variable). Under the previously discussed assumptions (no error in the input, known ideal output, and output measured with negligible errors), the average error may be computed:

$$e_{\text{AVG}}(x_{\text{in}}) = E\left[Y_{\text{out}}(x_{\text{in}})\right] - y_{\text{out,ideal}}(x_{\text{in}}) \qquad (41.10)$$

where $E[Y_{\text{out}}(x_{\text{in}})]$ is the mean value of the random variable Y_{out} when the input signal is equal to x_{in}; clearly, the average error e_{AVG} depends on the input. If we consider the accuracy as the capability of the system to produce small average errors, a system may be accurate even if the standard deviation of Y_{out} is very large. However, a similar situation would be unacceptable in many practical cases. For this reason, it is important to specify the precision that is related to the standard deviation of the random variable Y_{out}. In order to intuitively illustrate these concepts, we may consider 10 events and the values reported in Figure 41.3 for both the ideal output and the output corresponding to each event. A system may be accurate and precise (system A); precise, but not accurate (system B); accurate, but not precise (system C, which has a small average error); and neither accurate nor precise (system D). In some cases, nanoheaters must be both accurate and precise; in other cases, it is only important that the average error is low (e.g., when a large number of nanoheaters are attached to the surface of a target cell to be killed by thermal ablation).

In general, the sensitivity of an *instantaneous* system is defined as

$$S = \left.\frac{\partial y_{\text{out}}}{\partial x_{\text{in}}}\right|_Q \qquad (41.11)$$

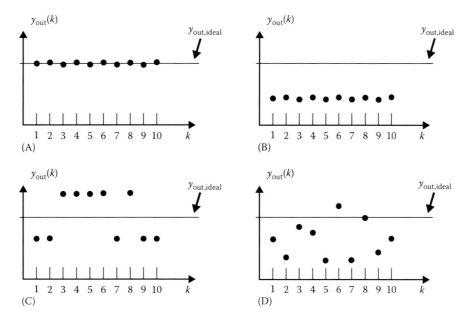

FIGURE 41.3 Schematic representation of systems with different accuracy and precision.

and depends on the operating point Q. We stress that the sensitivity may only be defined for *instantaneous* systems or, equivalently, may only be defined for systems that are much faster than their input signals (i.e., may be considered instantaneous).

If, for a given nanoheater, different signals can be chosen as output signals, different sensitivities will result. For instance, if we again consider a resistive heater (whose equivalent electrothermal SPICE model is shown in Figure 41.2), both the temperature of the resistive heater, T_{RES}, and the heating power can be considered as outputs. Clearly, for consistency, depending on the choice for the output signal, different definitions for the sensitivity must be adopted. In many cases, for wireless nanoheaters, in the absence of a quantitative estimation of the thermal resistance between a nanoheater and its surroundings, the extinction cross-section is a convenient way to express the sensitivity (e.g., see Falconi et al., 2007a).

In general, the error, the accuracy, the precision, the sensitivity, and the resolution depend on the input. However, in many practical cases, these parameters may be approximated by constants for a certain input value's range.

41.2.3.3 Linear Time Invariant Systems: Transfer Functions

If a system is linear and time invariant (or can be approximated by a linear, time invariant model), it can be completely described from the external point of view by its transfer function, which is the Fourier transform of its impulse response. In fact, if the system transfer function is known, for any given input signal $x_{in}(t)$, the output may be found as

$$y_{out}(t) = \Im^{-1}\left\{H(f)X_{in}(f)\right\} \qquad (41.12)$$

where $\Im^{-1}$ denotes the inverse Fourier transform, while $H(f)$ and $X_{in}(f)$ are, respectively, the system transfer function and the Fourier transform of the input signal.

The transfer function of an *instantaneous* linear time invariant system is a constant and is the sensitivity of the system. On the other hand, since instantaneous systems strictly may not exist because of the finite speed of real systems, transfer functions of real systems always depend on frequency. In practical cases, transfer functions may only be approximately constant (e.g., within 3 dB) within a certain range of frequencies that is often called the bandwidth (e.g., 3 dB bandwidth). All real systems have a limited speed and, therefore, have a finite bandwidth (i.e., they may not respond to signals beyond certain frequencies). If the Fourier transforms of all possible input signals is practically zero outside the bandwidth, these systems may then be considered instantaneous. For such systems it is then possible to approximate the transfer function with a sensitivity. As an example, in the equivalent electro-thermal SPICE model of a resistive heater shown in Figure 41.2, both the heating power and the temperature can be considered as the output. In the first case (heating power as the output), the heater is instantaneous (as within the model shown in Figure 41.2, a variation of the input current would instantaneously result in a variation of the heating power). By contrast, in the second case (temperature of the heater as the output) the transfer function has a pole with a time constant equal to $R_{TH}C_{TH}$ (equivalently, the −3 dB bandwidth is $1/[2\pi R_{TH}C_{TH}]$). We mention that the measurement of the thermal time constant of a nanoheater can be useful for measuring the thermal resistance between the nanoheaters and its surroundings (e.g., see Huxtable et al., 2003). Even when the thermal inertia of nanoheaters is considered, by means of their time constants, nanoheaters are generally quite fast. In fact, intuitively, a small object has a small thermal inertia. It is, in fact, possible to demonstrate that down-scaling of all the dimensions reduces the time constant in an approximately linear manner. In fact, at the nanoscale, the Kapitza resistance that is likely to dominate the conduction heat-transfer resistance is inversely proportional to the area of the nanoheater. Moreover, both the thermal resistances associated with the radiation heat transfer and convection heat transfer are inversely proportional to the area of the nanoheater. As a result, the total thermal resistance R_{TH} between the nanoheater and the environment (i.e., the parallel combination of the conduction, radiation, and convection thermal resistances) is also inversely proportional to the area of the nanoheater. By contrast, the thermal capacitance C_{TH} is proportional to the volume of the nanoheater. For this reason, the thermal time constant $R_{TH}C_{TH}$ is proportional to the volume to area ratio and thus is linearly reduced by downscaling. Obviously, this result is only approximate as the coefficients for conduction, convection, and radiation heat transfer, in general, are not constant and can be changed by scaling.

41.2.3.4 Spread

In general, sensitivity, accuracy, thermal resistance, thermal capacitance, and all other properties of nanoheaters unavoidably show variability due to the spread of process parameters, imperfections, aging, variations of the operative conditions, etc. When dealing with large numbers of nominally identical devices, it may be necessary to provide statistical information on the most critical parameters and to specify the range of operative conditions that guarantee a satisfactory behavior. However, in traditional smart systems (Falconi et al., 2007b), a given actuation task is accomplished by a single transducer so that, eventually, a proper calibration can counteract the spread. On the other hand, in many cases (e.g., *in vivo* biomedical applications) a very large number of wireless nanoheaters can be simultaneously used, so that, in practice, calibration may be impossible. As a result, if the "mismatch" among "nominally identical devices" is unacceptable, it may be important to properly select a subset of the fabricated nanodevices or, equivalently, exclude outliers (once a defined selection criterion has been defined). In these cases, post-synthesis separation or selective destruction can represent suitable countermeasures to the spread (Collins et al., 2001; Avouris et al., 2003).

41.2.3.5 Actuation Power/Energy (Densities) and Actuation Time

When a large number of nanoheaters are excited, either the nanodevices are considered all together, or the energy/power densities should be used.

In analogy with actuators, it is convenient to consider the *actuation power/energy* as the input power/energy also for the heaters and the *minimum actuation power/energy* as the minimum actuation power/energy required for action (e.g., for drug delivery or for killing target cells). In the case of wireless nanoheaters, very often densities must be considered (e.g., for gold nanoshells there will be a *minimum actuation power per unit area* that is required for action). The *minimum actuation time* is the time required for enabling action when the *minimum actuation power* (*density*) is applied. However, in some cases a more accurate description is necessary (e.g., in the case of repeated pulses). The *maximum actuation power* (*density*) and the *maximum actuation time* may also be very important for nanostructures' reliability, patient safety, or proper operation (e.g., when wireless nanoheaters are used for hyperthermia, if the input signals become too large, thermal ablation may occur).

41.3 State of the Art

In this section we restrict our attention to the nanoheaters that at present are the most promising, namely, wireless nanoheaters (because of their great potential for medicine), thermally activated nanoheaters, and resistive nanoheaters. Clearly, there are may be other types of nanoheaters as heat release is associated with all dissipative processes.

41.3.1 Wireless Nanoheaters for Thermal Ablation, Hyperthermia, and Targeted Drug Delivery

Nanotechnology has the potential to allow the development of more effective and less invasive methodologies for the diagnosis and treatment of many diseases. In particular, wireless nanoheaters may be useful for hyperthermia, thermal ablation, and targeted drug delivery. Cancer thermal ablation is probably the most investigated application of nanotechnology to medicine. In fact, though hyperthermia and thermal ablation of solid tumors have been applied for many years (with different sources including laser, focused ultrasounds, and microwaves) and even if these techniques are generally considered less invasive than traditional surgery, their use is still limited by side effects on surrounding healthy tissues. Wireless nanoheaters hold promises to solve this "selectivity" issue.

In general, the wireless interaction must be both efficient, so that a limited (i.e., safe) excitation energy/power is sufficient, and selective, so that side effects on healthy tissues are tolerable. Moreover, the nanoheaters must be biosafe and biocompatible (in particular, with reference to toxicity and to the risk of undesired aggregation and obstruction of vessels). In order to obtain a sufficient selectivity the exciting signal may be focused only in the target regions of the body. Such an approach, though useful, is often insufficiently selective, as it does not allow discriminating between the target cells and the surrounding healthy tissues. Better selectivity can be obtained by the so called active targeting, that is, taking advantage of specific interactions (e.g., antibody-antigen, ligand-receptor, or lectin-carbohydrate): after conjugation to proper targeting components, nanoparticles will preferentially accumulate at specific sites, such as tumor, single cancer cells, or intracellular organelles inside cancer cells (Nie et al., 2007). Selectivity can also be "naturally" achieved due to the EPR (enhanced permeability and retention) effect, sometimes referred to as passive targeting. The EPR effect is the cause of the passive extra-vasation of particles with a diameter of less than 400 nm from the blood vessel (permeability), and is due to the leakiness of tumor vessels that contain wide inter-endothelial junctions, an incomplete or absent basement membrane, and large numbers of trans-endothelial channels. The EPR effect may also be enhanced by a dysfunctional lymphatic system that is not able to drain the particles present in the tumor site, thus increasing the retention effect with a resulting increased interstitial fluid pressure. Tumors have abnormal physical structures, such as compromised vasculature, abnormal ECM (Extra Cellular Matrix), highly variable vessel diameters, and highly chaotic vessel organization, which all result in an intermittent blood flow with periodically reversing directions. All these irregularities in the tumor structure, though helpful in nanoparticles' extra-vasation may be an obstacle for them to enter the tumor mass itself or to reach all the cells of the tumor mass.

Wireless nanoheaters can take advantage of many different dissipative mechanisms (e.g., gold nanoshells (O'Neal et al., 2004), magnetic nanoparticles (Johannsen et al., 2007), temperature sensitive liposomes (Mills and Needham, 2004), wireless Joule nanoheaters (Falconi et al., 2007a), and wirelessly-vibrated zinc oxide nanohelixes (Falconi et al., 2006b, 2007a). As an outstanding example, it has already been shown that gold nanoshells (Loo et al., 2004) can selectively damage cancer cells by heating.

Beside thermal ablation and hyperthermia, wireless nanoheaters can also promote a wirelessly controllable, selective, accurate, and noninvasive delivery of drugs (e.g., for the treatment of diabetes (Sershen and West, 2002)). In practice, if a nanoheater is coated by a thermally responsive substance preloaded with a proper drug, the excitation of the nanoheater can trigger the drug release in a selective, controlled, and accurate manner. In principle, drug delivery may use both closed loop and open loop strategies. Closed loop systems should self-regulate the drug delivery in response to changes in the local environment, for instance, by including sensors for measuring physiological parameters. At least in principle, self-regulation at the nanoscale is possible. For instance, in wireless Joule nanoheaters, since the Joule power strongly depends on the ring resistance, if the ring resistance is dominated by the resistance of an outer coating layer, a proper variation of this coating (e.g., due to the interaction with a target cell) may automatically enable or disable the nanoheater (Falconi et al., 2007a). In principle, all wireless nanoheaters may be used for drug delivery modulation. As a successful example, gold nanoshells embedded in a hydrogel with a predetermined phase transition temperature have permitted a wirelessly controlled drug delivery. It must be mentioned that, beside wireless *nanoheaters*, other wireless *nanotransducers* can be used for drug delivery.

Gold nanoshells are nanoparticle beads made of a silica core coated with a thin gold shell (Loo et al., 2005). The absorption/scattering properties of gold nanoshells can be accurately controlled by changing the thickness of both the core and the outer shell; for instance, gold nanoshells with 40 nm core radius and 20 nm shell thickness will mainly scatter the incoming 820 nm radiation, while gold nanoshells with 50 nm core radius and 10 nm shell thickness will mainly absorb a similar radiation

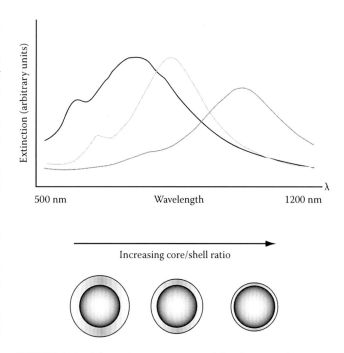

FIGURE 41.4 Schematic representation of the absorption/scattering properties of gold nanoshells with different core/shell ratios. (From Loo, C. et al., *Technol. Cancer Res. Treat.*, 3, 33, 2004; Loo, C. et al., Nano Lett., 5(4), 709, 2005.)

(Loo et al., 2004). Figure 41.4 qualitatively shows the optical resonance of gold nanoshells with different shell thicknesses (i.e., with different core/shell ratios). Gold nanoshells are being investigated for various *in vivo* biomedical applications, such as cancer thermal ablation, cancer imaging, and wirelessly triggered drug delivery. For such applications, the wavelength of the exciting radiation must be chosen in the Near Infra-Red (NIR) spectrum where the optical transmission through tissues is optimal. With the dimensions of 120 nm for the silica core diameter and of 10 nm gold shell layer, a gold nanoshell with an absorption peak at about 800 nm is obtained. Specific targeting to cancer cells can be obtained by active targeting, so that nanoshells will selectively bind to specific cancer cells. After the nanoshells have been attached to a cancer cell, upon wireless activation the nanoshells can be heated up to such high temperatures that the nearby or surrounding cells are killed by thermal ablation, as schematically shown in Figure 41.5. Since tissue chromophores do not significantly absorb light at this wavelength, there is no damage or heating in the healthy tissue surrounding the tumor site to be ablated. The outer shell of gold nanoshells is, in theory, biocompatible as it is made of reduced gold, a noble metal resistant to corrosion, with low toxicity and inert chemical properties. However, as always, the properties of a certain material at the nanoscale may differ from the properties of the same material at the macroscale, so that careful experimental verifications are always important. In order to further improve the biocompatibility of gold nanoshells, "stealthing" polymers like poly(ethylene glycol - PEG) can be grafted to nanoshell surfaces using simple molecular self-assembly techniques (O'Neal et al.,

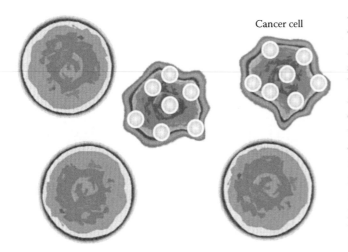

Cancer cell

FIGURE 41.5 Gold nanoshells (or other nanoheaters) can be selectively attached to the surface of cancer cells (shown with irregular shape) by active targeting.

2004). It has already been experimentally shown that intravenously administered nanoshells and NIR treatment may completely eradicate tumors by thermal ablation *in vivo* (O'Neal et al., 2004). Beside nanoshells, many other nanoparticles have been proposed as wireless nanoheaters. For instance, optically activated gold nanoparticles have also been used as wireless nanoheaters, showing that these nanoparticles can heat or even melt a surrounding matrix (Govorov et al., 2006). Wireless nanoheaters have also potential applications in micro-fluidics. For instance, phase-change based microvalves can be activated by a relatively weak laser (Park et al., 2007).

Gold nanoshells have also been used for modulated drug delivery. In particular, gold nanoshells have been embedded in a hydrogel with a predetermined phase transition temperature, so that when the nanoshells are wirelessly heated by a NIR radiation above the transition temperature of the hydrogel the soluble drug withheld in the polymer matrix can be released (Sershen et al., 2000).

When exposed to NIR radiation, *carbon nanotubes* also behave as wireless nanoheaters and may therefore be used for hyperthermia, thermal ablation, and targeted drug delivery. As an example it has been demonstrated that a continuous irradiation with a 808 nm laser at 1.4 W/cm^2, will heat up a 25 mg/L solution of *single walled carbon nanotubes* (SWCN) to 70°C (Kam et al., 2005). As another example, a noninvasive 13.56 MHz RF field has been used for wirelessly heating carbon nanotubes in order to thermally destroy cancer cells (Gannon et al., 2007). In this work three human cancer cell lines were incubated with different concentrations of carbon nanotubes and then exposed to the RF field and, additionally, rabbits were also injected with SWNTs and treated with RF field. After 48 h, all SWNT-treated tumors demonstrated complete necrosis, whereas control tumors that were treated with RF without SWNTs remained completely viable.

Superparamagnetic nanoparticles, that is, iron oxide nanoparticles that are less than 10 nm in diameter have been around for

years as contrast agents for magnetic resonance imaging (MRI). These nanoparticles, after adequate functionalization, permit specific tumor targeting. In order to enhance biocompatibility and to increase blood circulation time, iron oxide particles must be water-solubilized by means of hydrophilic polymer coatings, such as PEG (polyethylene glycol); this coating also prevents opsonization of nanoparticles in the serum and reduces their uptake by the reticuloendothelial system. Iron oxide nanoparticles can also be made hydrophobic by encapsulation in liposomes, instead of PEG, thus resulting in magnetoliposomes. Magnetic nanoparticles can be wirelessly heated by electromagnetic fields. For instance, after a proper electromagnetic excitation, superparamagnetic nanoparticles may undergo Brownian relaxation, thus generating heat due to the rotation of particles in the field (Johannsen et al., 2007). At present, as a major obstacle for their practical use in targeted thermal ablation, the concentration necessary to reach critical temperatures is hard to administer intravenously. For this reason, the challenge is to achieve significant heating with reduced concentrations of nanoparticles (e.g., see Maenosono and Saita, 2006).

Liposomes are vesicles made up of a lipid bilayer that resembles a cell membrane. Their size typically ranges from 90 to 150 nm in diameter and they have excellent biocompatibility properties (after rupture, the liposomes are readily integrated into cells). Liposomes have been used to encapsulate and deliver chemotherapeutics since the 1970s and intensive activities on the liposomes are ongoing, especially, in pharmaceutical research. As there are many different lipids (with different head groups or different fatty acid chain lengths), it is possible to design liposomes with different properties. In particular, temperature sensitive liposomes or pH sensitive liposomes allow the controlled release of their contents in response to, respectively, temperature or pH stimuli. In particular, since temperature sensitive liposomes have a melting temperature slightly above the body temperature, local hyperthermia (microwave, radiofrequency energy, phased array ultrasound, etc.) may easily trigger the release of drugs from circulating and accumulating in the liposomes. Clearly, temperature sensitive liposomes are wireless nanoheaters as drug release is activated only if the liposomes are, somehow, wirelessly heated. Temperature sensitive liposomes circulating in the body at tumor sites can selectively deliver drugs to the tumor (Mills and Needham, 2004). Although the clinical application of temperature-sensitive liposomes for targeted drug delivery in cancer therapy has not yet been realized, liposome nanoparticles are already on the market as chemotherapeutics for ovarian cancer (Doxil, doxorubicin hydrochloride in liposomes) (Kim, 2007).

Zinc oxide nanostructures (Pan et al., 2001; Kong et al., 2004; Gao et al., 2005; Wang and Song 2006) have many ideal properties for *in vivo* wireless applications: their geometries (rings, helixes etc.) permit an efficient wireless interaction and are unlikely to obstruct blood vessels when inserted into the body. Besides, their covalent bonds provide impressive mechanical properties. Moreover, since zinc oxide is a semiconductor, its resistivity may be varied by orders of magnitude with small quantities

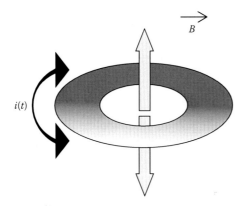

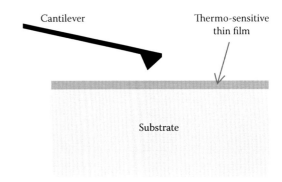

FIGURE 41.6 Schematic representation of a wireless Joule nanoheater. (From Falconi, C. et al., *Sens. Actuators B*, 127, 54, 2007a.)

FIGURE 41.7 Schematic representation of a thermally activated nanoheater (cantilever with nano-tip).

of dopants. Furthermore, zinc oxide is pyroelectric and piezoelectric, thus providing unique nanotransduction mechanisms. Finally, as a striking advantage, ZnO nanowires eventually dissolve into ions that can be completely absorbed by the body and become part of the nutrition (an ideal characteristic for *in vivo* applications). In Zhou et al. (2006), it has been shown that ZnO nanowires can be dissolved by deionized water (pH ≈ 4.5–5.0), ammonia (pH ≈ 7.0–7.1, 8.7–9.0), and NaOH solution (pH ≈ 7.0–7.1, 8.7–9.0). Moreover, ZnO nanowires dissolve completely in less than 12 h when immersed in horse-blood serum diluted with 10% aqueous NaOH (pH ≈ 7.9–8.2). The dissolution time decreases to 6 h when using pure blood serum (pH ≈ 8.5). Due to their piezoelectricity and geometries, ZnO nanostructures offer various opportunities for wireless transduction at the nanoscale. As a first example, wireless Joule nanoheaters, schematically shown in Figure 41.6 (Falconi et al., 2007a), might use zinc-oxide nanorings in order to obtain superior mechanical robustness and gold in order to provide sufficient electrical conductivity (we mention that in these nanoheaters the electromagnetic-to-heat transduction process does not rely on optical properties). As another example, electromagnetic fields may induce mechanical vibrations in piezoelectric nanostructures (nanopumps, nanoscissors, vibration nanoheaters. etc.) that could be useful for thermal treatment or for drug delivery (Falconi et al., 2006b, 2007a). In particular, wirelessly vibrated ZnO nanohelixes could selectively damage the target cells by both heating and inertial cavitation (Falconi et al., 2007a), and then dissolve into nutrition for the body.

41.3.2 Thermally Activated Nanoheaters

Nanosized devices heated through heat transfer from another device in good thermal contact with the nanodevice are termed thermally activated nanoheaters. The auxiliary device can, obviously, be much larger than the nanoheater. In practice a resistive heater or a laser can easily heat a microstructure whose sharp (nano-) tip represents the nanoheater. For instance, Figure 41.7 schematically illustrates a cantilever whose tip may behave as a nanoheater when the cantilever is hot. In all cases, the maximum temperature is dictated by reliability constraints and strongly

depends on the materials (e.g., all-silicon cantilevers can reach much higher temperatures than cantilevers using standard IC-compatible metals for fabricating the heaters).

Atomic force microscopes (Binnig and Quate, 1986) are widely used for sensing and actuating at the nanoscale. However, there are many other approaches for scanning probe microscopy. In particular, in scanning thermal profilers a nano-tip (e.g., the tip of a cantilever with integrated thermal actuators and sensors) is heated and brought in proximity of the surface that is to be investigated (Williams and Wickramasinghe, 1986; Nonnemacher and Wickramasinghe, 1992; Gianchandani and Najafi, 1997; Kim et al., 2007). In these scanning thermal profilers the dimensions of the thermal probe must be extremely small in order to achieve sufficient spatial resolution. Moreover, fast measurements also require a small probe. The most convenient approach for integrating both a temperature sensor and a thermal actuator in a very small volume is, obviously, to use the junction of a thermocouple (Williams and Wickramasinghe, 1986) in order to perform both temperature sensing (Seebeck effect) and thermal actuation (Peltier and Joule effects). Similar scanning thermal profilers have been used for investigating thermal properties at the nanoscale in a wide variety of applications ranging from microelectronics to biology (e.g., measuring thermal properties of a living cell). Scanning thermal profilers can be used in the DC mode and dithered mode. In the dithered mode the tip is oscillated in order to improve the signal to noise ratio (Gianchandani and Najafi, 1997). Another approach is, however, possible if it is not necessary to integrate a temperature sensor together with the nanoheater: a heater can be integrated on a cantilever and heat can be transferred to the sharp tip (nanoheater) by conduction heat transfer. Alternatively, the cantilever, and therefore the tip, can be heated by a laser beam (Hamann et al., 2004). Similar thermally activated nanoheaters have found applications in high-density memory-storage (King et al., 2002; Vettiger et al., 2002; Vettiger and Binnig, 2003; Drechsler et al., 2003; Eleftheriou et al., 2003; Despont et al., 2004; Hamann et al., 2004, 2006; Nam et al., 2007). The so-called "millipede" is an array of thermomechanical probes for storing, reading, and erasing data in very thin polymer films (Vettiger et al., 2002). In the so called thermomechanical writing a heated cantilever tip can form indentations with a thin polymer film in order to write "bits" and can be operated as a thermal impedance sensor in order to read "bits" (King et al., 2002). In Hamann et al. (2004), thermally assisted writing was allowed to achieve higher magnetic storage densities because the tip (nanoheater) heats the magnetic film on the substrate at a temperature close to its Curie temperature and thus lowers its coercivity (allowing to write on otherwise un-writable magnetic materials). In Hamann et al. (2006), phase-change materials that can be rapidly switched from a crystalline to amorphous state and vice versa by the application of heat pulses were used as ultra-high-density memories by using tip-heaters with a diameter less than 5 nm. In Vettiger et al. (2003), thousandths of microcantilevers were integrated in order to increase the speed of data writing/reading. When a cantilever is used, it is possible to combine nanoheating and temperature sensing with piezo-resistive sensing (King et al., 2004; Nam et al., 2007; Lee and King, 2008b). Moreover, the cantilevers can be operated in liquids (Lee and King, 2008a), as required in most biological applications.

Nanoscale manufacturing by means of heated atomic force microscope tips is also being explored, with the potential to become a cheaper alternative to the very high costs of current optical lithography methods (Sheehan et al., 2004; Chiou et al., 2006; Hua et al., 2007; Lee and Oh, 2007). In Hua et al. (2007), local polymer decomposition is used for nanoscale thermal lithography with the time constant for heating and cooling of the tip in the microseconds range. In standard dip pen nanolithography (DPN) (Piner et al., 1999) an AFM-tip coated by proper molecules can write the "ink" (the coating molecules) on a surface. The thermal dip pen nanolithography (Sheehan et al., 2004) uses an AFM tip to deposit an ink which is solid at the environmental temperature and can be melted when the tip is heated (i.e., the AFM tip acts as a controllable nanoheater). With this approach, it has been shown that a solid organic "ink" octadecylphosphonic acid (melting temperature close to 100°C) can be deposited only when the tip of the AFM is heated. In Nelson et al. (2006), thermal dip pen lithography was used for the direct deposition of continuous metal nanostructures. As another interesting nanofabrication-related application of heated cantilevers, we mention that it is possible to grow, by chemical vapor deposition, vertically-aligned carbon nanotubes on a heated cantilever, while the environment is at room temperature (Sunden et al., 2006).

Cantilevers with integrated heaters can also allow new experiments at the nanoscale. For instance, attractive and repulsive thermal forces (i.e., forces originated by temperature gradients in a fluid surrounding a body) were reported (Gotsmann and Dürig, 2005a) and a heated tip was allowed to activate nanowear modes on a polymer surface (Gotsmann and Dürig, 2005b).

41.3.3 Resistive Nanoheaters

Resistive, nanosized devices heated by the flow of an electrical current (Joule effect) are termed resistive nanoheaters. Quasi-1D nanostructures (nanowires, nanobelts, etc.) can then be used as resistive nanoheaters. Since the thermal mass of these

nanostructures is very small, very fast temperature variations can be generated. Moreover, the spatial resolution can be sufficient for investigating temperature-dependent phenomena at the molecular level (Arata et al., 2006a,b; Low et al., 2008). Clearly, an accurate thermal characterization of nanowires is necessary in order to use a nanowire for molecule studies. Moreover, resistive nanoheaters can provide insight for nanoscale heat transfer (e.g., see Lee et al., 2006; Ingvarsson et al., 2007; Nelson and King, 2007). Since traditional methods may not offer sufficient spatial resolution (IR thermal imaging) or sufficiently fast response (Raman spectroscopy and scanning thermal microscopy), other approaches are being explored. For instance it has been proposed (Low et al., 2008) that semiconductor nanocrystals (e.g., CdSe/ZnS nanocrystals) whose fluorescence intensity depends on temperature can be grown or positioned on top of a nanowire in order to monitor its temperature with high spatial resolution and with sufficient speed. Similarly, in (Samson et al., 2008) a fluorescent particle has been glued at the end of a sharp tip in order to investigate heat transfer from a nanoheater.

41.4 Critical Discussion

Nanoscale heat transfer is not yet satisfactorily understood and will likely remain an open problem for many years. Moreover, despite the foreseeable progresses in simulation methodologies and computational tools, significant breakthroughs are needed before accurate numerical predictions of nanoscale heat transfer processes can be used for the design of nanoheaters. Nevertheless, nanoheaters are being intensively studied because of their outstanding potential in a wide variety of applications, including *in vivo* thermal ablation, hyperthermia, and targeted drug delivery; data storage; nanofabrication and lithography; imaging; and investigations of chemical, physical, and biological processes at the nanoscale. As a result, on the one hand, accurate quantitative predictions on the thermal resistances between nanoheaters and their surroundings are, in practice, currently impossible. On the other hand, the theory of nanoscale heat transfer can take great advantage of experiments with nanoheaters (e.g., studies on solid–solid and solid–liquid interfaces at nanoscale can offer insight for nanoscale heat transfer). We also mention that in some cases, despite our limited comprehension of nanoscale heat transfer, it is still possible to determine systematic rules for the design of nanoheaters with optimal efficiency (Falconi et al., 2007a).

The combination of microtechnology and nanotechnology may allow accurate investigations on nanoscale heat transfer and, specifically, the fabrication of tools for the characterization of nanoheaters. In particular, various improvements in scanning thermal microscopy as well as other measurement techniques at the nanoscale (also taking advantage of novel approaches, e.g., Low et al., 2008) hold promises for the design of more and more effective nanoheaters.

In wireless nanoheaters for medical applications, biocompatibility and toxicity are generally fundamental issues. As an important opportunity, zinc oxide nanostructures (Pan et al.,

2001; Kong et al., 2004; Gao et al., 2005; Wang and Song, 2006) are expected to automatically dissolve within a few hours into the nutritive substances of the body (Zhou et al., 2006). Since some of these structures can be used as wireless nanoheaters (Falconi et al., 2007a), wireless nanoheaters that can perform a desired action (hyperthermia, thermal ablation, or targeted drug delivery) and then safely "disappear" can be envisioned.

For thermally activated nanoheaters and for resistive nanoheaters, the precise and accurate control of the temperature of the nanoheater is not necessarily critical (e.g., in data storage). Moreover, in some thermally activated nanoheaters, only the cantilever temperature can be directly monitored. However, in some applications, a more accurate and precise temperature control can be extremely important (Lee et al., 2006), so that accurate thermal calibration and high accuracy, and high precision electronic interfaces for temperature control (Falconi et al., 2006a; Falconi and Fratini, 2008) can be important. As an example, Figure 41.8 shows an electronic interface for temperature control of a resistive heater R_X (which can thermally activate a nanoheater or can be a resistive nanoheater), under the assumption that the temperature coefficient of R_X is positive, as in metal resistors (if the temperature coefficient is negative the input terminals of the comparator must be interchanged). In practice, the feedback loop equates the ratios R_1/R_2 and R_3/R_X (after the system is calibrated, this corresponds to keeping the temperature-dependent resistor at a desired temperature). In fact, if $v_N > v_P$, the output of the comparator is low and, therefore, M_0 is switched on by the flip-flop D, so that a large current flows through R_X, which is then heated, so that its resistance increases (positive temperature coefficient). On the contrary, if $v_N < v_P$, the output of the comparator is high and, therefore, M_0 is switched off by the flip-flop D, so that only a small current can flow through R_X (the auxiliary resistance R_{AUX} is much larger than the on-resistance of the switch M_0). In this circuit, the resistance

R_{AUX} allows both a reliable start-up and a reliable comparison (between v_N and v_P) when M_0 is off. However, it results in a non-zero current through R_X even when M_0 is off and no heating is desired; in order to keep this current small R_{AUX} must be enough large. On the other hand, when M_0 is off, an accurate comparison between the voltages v_N and v_P requires that R_{AUX} be small enough. In practice this trade-off is not an issue if the minimum acceptable temperature can be somewhat above the environmental temperature, as it is often the case.

41.5 Summary

Nanoheaters are nanosized devices that are able to *efficiently* heat their surroundings. Here, we have discussed nanoscale heat transfer and we have given the fundamental definitions for nanoheaters. Afterwards, we have described the state of the art for nanoheaters, focusing on wireless nanoheaters, thermally activated nanoheaters, and resistive nanoheaters. These examples demonstrate the outstanding potential of nanoheaters for a variety of applications, including: hyperthermia, thermal ablation, targeted drug delivery, imaging, nanofabrication, high-density data storage, and nanoscale heat transfer.

41.6 Future Perspective

Nanoscale heat transfer is being intensively investigated because of its great practical importance (e.g., nanotechnology can be the key for reducing the thermal resistance between integrated circuits and their surroundings in order to avoid excessive overheat). Nevertheless, it will not be possible in the near future to accurately predict the thermal resistance between a generic nanoheater and its surroundings (which would be important for design). On the contrary, nanoheaters may be great tools for understanding heat transfer at the nanoscale. Despite this "theoretical" difficulty, possible applications for nanoheaters include *in vivo* medical applications (thermal ablation, hyperthermia, and targeted drug delivery), imaging, data storage, nanofabrication, analysis of temperature-dependent processes at the nanoscale, and more. Such a tremendous potential will attract more and more research efforts in the next years. It is then interesting to consider some of the most important (often cross-disciplinary) challenges for researchers.

The development of novel types of nanoheaters can be very important (e.g., novel wireless nanoheaters with high-extinction cross-section). Moreover, as we mentioned, beside the nanoheaters explicitly described here (i.e., wireless nanoheaters, thermally activated nanoheaters, and resistive nanoheaters), other types of nanoheaters may also be fabricated as heat release is associated with all dissipative processes; as an example, a nanosized device attached to a mechanical actuator can be heated by friction (mechanically activated nanoheater).

For *in vivo* medical applications, biocompatibility is a critical issue. For this reason, nanostructures which can perform desired tasks (e.g., targeted drug delivery) and then automatically dissolve into nutrition for the body may offer a very important

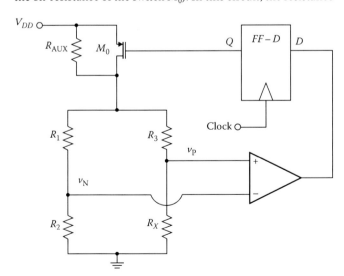

FIGURE 41.8 A simple circuit for temperature control. (From Falconi, C. et al., Temperature and flow velocity control for quartz crystal microbalances, *Proceedings of IEEE ISCAS 2006*, Kos, Greece, 2006a.)

opportunity (e.g., wireless zinc oxide nanotransducers (Falconi et al., 2007a)).

Coatings for targeted drug delivery with high functionality and selectivity must be developed.

The ability to control heat transfer at the nanoscale may enable novel applications not yet identified.

Advances in microfabrication and nanofabrication may allow in further improving the spatial resolution of thermally activated nanoheaters. Tools and methods for the characterization of nanoheaters are needed, with special reference to measuring the temperature with high speed, high accuracy, and high spatial resolution.

Acknowledgments

Arnaldo D'Amico and Giulia Mantini are acknowledged for many useful discussions. Giulia Mantini is also acknowledged for drawing Figures 41.4 and 41.5 as well as for contributing to Section 41.3.1.

References

Arata H. F., Low P., Ishizuka K. et al. 2006a. Temperature distribution measurement on microfabricated thermodevice for single biomolecular observation using fluorescent dye. *Sensors and Actuators B* 117: 339–345.

Arata H. F., Noji H., and Fujita H. 2006b. Motion control of single F1-ATPase rotary biomolecular motor using microfabricated local heating devices. *Applied Physics Letters* 88(083902): 1–3.

Avouris P., Appenzeller J., Martel R., and Wind S. J. 2003. Carbon nanotube electronics. *Proceedings of the IEEE* 91(11): 1772–1784.

Berry C. C. and Curtis A. S. G. 2003. Functionalisation of magnetic nanoparticles for applications in biomedicine. *Journal of Physics D: Applied Physics* 36: R198–R206.

Binnig G. and Quate C. F. 1986. Atomic force microscope. *Physical Review Letters* 56(9): 930–933.

Cahill D. G., Ford W. K., Goodson K. E. et al. 2003. Nanoscale thermal transport. *Journal of Applied Physics* 93(2): 793–818.

Chiou C. H., Chang S. J., Lee G. B. et al. 2006. New fabrication process for monolithic probes with integrated heaters for nanothermal machining. *Japanese Journal of Applied Physics* 45(1A): 208–214.

Collins P. G., Arnold M. S., and Avouris P. 2001. Engineering carbon nanotubes using electrical breakdown. *Science* 292: 706–709.

Costescu R. M., Cahill D. G., Fabreguette F. H. et al. 2004. Ultra-low thermal conductivity in W/Al$_2$O$_3$ nanolaminates. *Science* 303: 989–990.

Despont M., Drechsler U., Yu R. et al. 2004. Wafer scale microdevice transfer/interconnect: Its application in an AFM-based data storage system. *IEEE Journal of Microelectromechanical Systems* 13(6): 895–901.

Drechsler U., Burer N., Despont M. et al. 2003. Cantilevers with nano-heaters for thermomechanical storage application. *Microelectronic Engineering* 67–68: 397–404.

Eleftheriou E., Antonakopoulos T., and Binnig G. K. 2003. Millipede—A MEMS based scanning-probe data-storage system. *IEEE Transactions on Magnetic* 39(2): 938–945.

Falconi C. and Fratini M. 2008. CMOS microsystems temperature control. *Sensors and Actuators B* 129: 59–66.

Falconi C., Zampetti E., Pantalei S. et al. 2006a. Temperature and flow velocity control for quartz crystal microbalances. *Proceedings of IEEE ISCAS 2006*, Kos, Greece.

Falconi C., D'Amico A., and Wang Z. L. 2006b. Wireless nano-actuators and nanosensors for in-vivo biomedical applications. *Proceedings of Eurosensors XX*, Göteborg, Sweden.

Falconi C., D'Amico A., and Wang Z. L. 2007a. Wireless joule nanoheaters. *Sensors and Actuators B* 127: 54–62.

Falconi C., Martinelli E., Di Natale C. et al. 2007b. Electronic interfaces. *Sensors and Actuators, B: Chemical* 121(1): 295–329.

Gannon C. J., Cherukuri P., Yakobson B. I. et al. 2007. Carbon nanotube-enhanced thermal destruction of cancer cells in a noninvasive radiofrequency field. *Cancer* 110(12): 2654–2665.

Gao P. X., Ding Y., Mai W. et al. 2005. Conversion of zinc oxide nanobelts into superlattice-structured nanohelices. *Science* 309: 1700–1704.

Gianchandani Y. B. and Najafi K. 1997. A silicon micromachined scanning thermal profiler with integrated elements for sensing and actuation. *IEEE Transactions on Electron Devices* 44(11): 1857–1868.

Gotsmann B. and Dürig U. 2005a. Experimental observation of attractive and repulsive thermal forces on microcantilevers. *Applied Physics Letters* 87(194102): 1–3.

Gotsmann B. and Dürig U. 2005b. Thermally activated nano-wear modes of a polymer surface induced by a heated tip. *Langmuir* 20: 1495–1500.

Govorov A. O., Zhang W., Skeini T. et al. 2006. Gold nanoparticle ensembles as heaters and actuators: Melting and collective plasmon resonances. *Nanoscale Research Letters* 1: 84–90.

Hamann H. F., Martin Y. C., and Wickramasinghe H. K. 2004. Thermally assisted recording beyond traditional limits. *Applied Physics Letters* 84(2): 810–812.

Hamann H. F., O'Boyle M., Martin Y. C. et al. 2006. Ultra-high-density phase-change storage and memory. *Nature Materials* 5: 383–387.

Hua Y., Saxena S., Henderson C. L. et al. 2007. Nanoscale thermal lithography by local polymer decomposition using a heated atomic force microscope cantilever tip. *Journal of Micro/Nanolithography MEMS MOEMS* 6(2), 023012: 1–6.

Huxtable S., Cahill D. G., Shenogin S. et al. 2003. Interfacial heat flow in carbon nanotube suspensions. *Nature Materials* 2: 731–734.

Incropera F. P. and De Witt D. P. 2002. *Fundamentals of Heat and Mass Transfer*. 5th edition. Wiley, New York.

Ingvarsson S., Klein L. J., Au Y. Y. et al. 2007. Enhanced thermal emission from individual antenna-like nanoheaters. *Optics Express* 15(18): 11249–11254.

Johannsen M., Gneveckow U., Thiesen B. et al. 2007. Thermotherapy of prostate cancer using magnetic nanoparticles: Feasibility, imaging, and three-dimensional temperature distribution. *European Urology* 52(6): 1653–1662.

Jordan A., Wust P., Scholz R. et al. 1996. Cellular uptake of magnetic fluid particles and their effects on human adenocarcinoma cells exposed to AC magnetic fields in vitro. *International Journal of Hyperthermia* 12: 705.

Jordan A., Scholz R., Wust P. et al. 1999. Magnetic fluid hyperthermia (MFH): Cancer treatment with AC magnetic field induced excitation of biocompatible superparamagnetic nanoparticles. *Journal of Magnetism and Magnetic Materials* 201: 413–419.

Kam N. W., O'Connell M., Wisdom J. A. et al. 2005. Carbon nanotubes as multifunctional biological transporters and near-infrared agents for selective cancer cell destruction. *Proceedings of the National Academy of Sciences USA* 102(33): 11600–11605.

Kim K. Y. 2007. Nanotechnology platforms and physiological challenges for cancer therapeutics. *Nanomedicine: Nanotechnology, Biology, and Medicine* 3: 103–110.

Kim K. J., Park K., Lee J. et al. 2007. Nanotopographical imaging using a heated atomic force microscope cantilever probe. *Sensors and Actuators A: Physical* 136(1): 95–103.

King W. P., Kenny T. W., Goodson K. E. et al. 2002. Design of atomic force microscope cantilevers for combined thermomechanical writing and thermal reading in array operation. *IEEE Journal of Microelectromechanical Systems* 11(6): 765–774.

King W. P., Kenny T. W., and Goodson K. E. 2004. Comparison of thermal and piezoresistive sensing approaches for atomic force microscopy topography measurements. *Applied Physics Letters* 85(11): 2086–2088.

Kong X. Y., Ding Y., Yang R. S. et al. 2004. Single-crystal nanorings formed by epitaxial self-coiling of polar-nanobelts. *Science* 303: 1348–1351.

Lee J. and King W. P. 2008a. Liquid operation of silicon microcantilever heaters. *IEEE Sensors Journal* 8(11): 1805–1806.

Lee J. and King W. P. 2008b. Improved all-silicon microcantilever heaters with integrated piezoresistive sensing. *IEEE Journal of Microelectromechanical Systems* 17(2): 432–445.

Lee D. W. and Oh I. K. 2007. Micro/nano-heater integrated cantilevers for micro/nano-lithography applications. *Microelectronic Engineering* 84: 1041–1044.

Lee J., Beechem T., Wrigth T. L. et al. 2006. Electrical, thermal, and mechanical characterization of silicon microcantilever heaters. *IEEE Journal of Microelectromechanical Systems* 15(6): 1644–1655.

Loo C., Lin A., Hirsch L. et al. 2004. Nanoshell-enabled photonics-based imaging and therapy of cancer. *Technology in Cancer Research and Treatment* 3: 33–40.

Loo C., Lowery A., Halas N. et al. 2005. Immunotargeted nanoshells for integrated cancer imaging and therapy. *Nano Letters* 5(4): 709–711.

Low P., Le Pioufle B., Kim B. et al. 2008. Assembly of CdSe/ZnS nanocrystals on microwires and nanowires for temperature sensing. *Sensors and Actuators B* 130: 175–180.

Maenosono S. and Saita S. 2006. Theoretical assessment of FePt nanoparticles as heating elements for magnetic hyperthermia. *IEEE Transactions on Magnetic* 42(6): 1638–1642.

Middelhock S., Audet S. A., and French P. 2000. *Silicon Sensors.* Academic Press, London, U.K.

Mills J. K. and Needham D. 2004. The materials engineering of temperature-sensitive liposomes. *Methods in Enzymology: Liposomes* 387, Part D: 82–113.

Nam H. J., Kim Y. S., Lee C. S. et al. 2007. Silicon nitride cantilever array integrated with silicon heaters and piezoelectric detectors for probe-based data storage. *Sensors and Actuators A* 134: 329–333.

Nelson B. A. and King W. P. 2007. Temperature calibration of heated silicon atomic force microscope cantilevers. *Sensors and Actuators A: Physical* 140(1): 51–59.

Nelson B. A., King W. P., Laracuente A. R. et al. 2006. Direct deposition of continuous metal nanostructures by thermal dippen nanolithography. *Applied Physics Letters* 88(033104): 1–3.

Nie S., Xing Y., Kim G. J. et al. 2007. Nanotechnology applications in cancer. *Annual Review of Biomedical Engineering* 9(12): 1–32.

Nonnemacher M. and Wickramasinghe H. K. 1992. Scanning probe microscopy of thermal conductivity and subsurface properties. *Applied Physics Letters* 61: 168–170.

O'Neal D., Hirsch L., Halas N. et al. 2004. Photo-thermal tumor ablation in mice using near infrared-absorbing nanoparticles. *Cancer Letters* 209(2): 171–176.

Pan Z. W., Dai Z. R., and Wang Z. L. 2001. Nanobelts of semiconducting oxides. *Science* 291: 1947–1949.

Park J. M., Cho Y. K., Lee B. S. et al. 2007. Multifunctional microvalves control by optical illumination on nanoheaters and its application in centrifugal microfluidic devices. *Lab on a Chip* 7: 557–564.

Piner R. D., Zhu J., Xu F., Hong S., and Mirkin C. A. 1999. "Dip-Pen" nanolithography. *Science* 283(29): 661–663.

Poulikakos D., Arcidiacono S., and Maruyama S. 2003. Molecular dynamics: Simulation in nanoscale heat transfer: A review. *Microscale Thermophysical Engineering* 7: 181–206.

Samson B., Aigouy L., Low P. et al. 2008. AC thermal imaging of nanoheaters using a scanning fluorescent probe. *Applied Physics Letters* 92(023101): 1–3.

Sershen S. and West J. 2002. Implantable, polymeric systems for modulated drug delivery. *Advanced Drug Delivery Reviews* 54: 1225–1235.

Sershen S. R., Westcott S. L., Halas N. J. et al. 2000. Temperature-sensitive polymer-nanoshell composites for photothermally modulated drug delivery. *Journal of Biomedical Materials Research* 51: 293–298.

Sershen S. R., West J. L., Westcott S. L. et al. 2001. Nanoshell-polymer composites for photothermally modulated drug delivery. *Proc. of CLEO 2001*, Baltimore, MD.

Sheehan P. E., Whitman L. J., King W. P. et al. 2004. Nanoscale deposition of solid inks via thermal dip pen nanolithography. *Applied Physics Letters* 85(9): 1589–1591.

Shenogin S., Xue L., Ozisik R. et al. 2004. Role of thermal boundary resistance on the heat flow in carbon nanotube composites. *Journal of Applied Physics* 95(12): 8136–8144.

Sunden E. O., Wright T. L., Lee J. et al. 2006. Room-temperature chemical vapor deposition and mass detection on a heated atomic force microscope cantilever. *Applied Physics Letters* 88(033107): 1–3.

Vettiger P. and Binnig G. 2003. The nanodrive project. *Scientific American (International Edition)* 288(1): 34–41.

Vettiger P., Cross G., Despont M. et al. 2002. The "millipede"— Nanotechnology entering data storage. *IEEE Transactions on Nanotechnology* 1(1): 39–55.

Vettiger P., Albrecht T., and Despont M. 2003. Thousands of microcantilevers for highly parallel and ultra-dense data storage. *IEEE IEDM* 03–763 32.1: 1–4.

Wang Z. L. and Song J. 2006. Piezoelectric nanogenerators based on zinc oxide nanowire arrays. *Science* 312: 242.

Williams C. C. and Wickramasinghe H. K. 1986. Scanning thermal progiler. *Microelectronic Engineering* 5: 509–513.

Wilson O. M., Hu X., Cahill D. G. et al. 2002. Colloidal metal particles as probes of nanoscale thermal transport in fluids. *Physical Review B* 66: 224301.

Zhou J., Xu N., and Wang Z. L. 2006. Dissolving behavior and stability of ZnO wires in biofluids: A study on biodegradability and biocompatibility of ZnO nanostructures. *Advanced Materials* 18: 2432–2435.

Nanorobotics

Atomic-Force-Microscopy-Based Nanomanipulation Systems

Cagdas D. Onal
Carnegie Mellon University

Onur Ozcan
Carnegie Mellon University

Metin Sitti
Carnegie Mellon University

42.1 Introduction

We have witnessed an ever-explosive growth of technology dealing with smaller and smaller scales in the past few decades. The quality of working with small scales that attracts more and more researchers to the field is the promise that, one day, it will be possible to build matter from the bottom. The resulting structures will have tuned properties that surpass any material we have ever encountered in the macro domain. Since defects are expected to decrease with decreasing scale, materials might display a different behavior at small scales. For instance, if properties of carbon nanotubes (CNTs) can be somehow translated to larger scales, materials with super-strength and/or superconductivity can be achieved. These materials will be much stronger but much lighter than steel, making them instant ideal candidates for building mechanical structures that need to resist large amounts of stresses, possibly for prolonged periods of time.

A second reason to focus on micro/nanoscale for technological development is the need for compactness. While it may seem as an unnecessary luxury to some, being able to fit more units into the same area is a means of technological propulsion. More transistors mean more processing power of more data being stored on a computer hard drive, which would have taken libraries to store 50 years ago. Having access to more information, and being able to analyze this information faster, technology grows much faster than before, to build devices that can store/process/access more information.

On the other hand, with this rise of small-scale or micro/nanotechnology, handling and manipulating nanoscale objects have become a necessity. All of the possible applications that are mentioned above will need a nanomanipulation system that would enable researchers to assemble complex structures, handle the nano features, and characterize material properties in nanoscale.

Among many nanomanipulation procedures proposed in the literature, a broad taxonomy could be made according to the processes where the first branch is self-assembly. Self-assembly of micro/nanostructures can be achieved by chemical relations between subunits (Castelino et al. 2005), patterned surfaces (Bohringer et al. 1998), carefully tuned electric fields, and even biological carriers. This is a promising method as it offers parallel operation; however, the main problem of self-assembly is that final products are (currently) typically symmetric and the material selection is limited to ones with subunits that automatically assemble or that can be guided with the mentioned methods.

As a different method, a manipulator could interact with the subunits and move them to form an assembly. Many kinds of manipulators are used to realize this task in the literature. The driving force to manipulate subunits into their respective places can be mechanical (AFM probes), electrical (scanning tunneling microscope [STM] probes), magnetic, optical tweezers, chemical, or hybrid (any combination of two or more of the others). Any geometry is achievable with direct interaction with subunits. While mechanical interaction can assemble any structure if the manipulator can touch and move the subunits, other methods have limitations. For example, the electrical manipulators need conductive subunits, and optical tweezers have a lower limit of the size of subunits. Chemical interaction is possible only with specific subunits. Since mechanical approach is not limited with size or material, AFM systems have been widely utilized as a nanomanipulation tool to interact with objects and surfaces, and make changes by mechanical contact manipulation.

AFM systems, invented in 1986, have enabled sub-nanometer resolution three-dimensional (3D) topography images of surfaces (Binnig et al. 1986). The purely mechanical working principle of AFM has enabled not only the imaging of any nanomaterial (conductive, nonconductive, biological, polymer, etc.) in any environment (air, liquid, and vacuum) but also manipulating them using their probe tip by mechanically changing the surface or positioning nanofeatures or particles. AFM provides sub-nanometer resolution 3D topography images by using the interaction forces that act on its compliant probe's sharp tip by touching the surface of the substrate (contact mode) or oscillating just a few nanometers above it (noncontact mode). The most important drawback of AFM as a manipulator is that it is slow due to the serial manipulation of each subunit into position. This problem, however, can be solved using an array of manipulators to operate in parallel, for at least some tasks (Pozidis et al. 2004; Lee and Hong 2005).

AFM-based nanomanipulation implies precise interactions with nanoscale objects. In this context, submicron- or nanometer-scale resolution positioning is imperative. The nature of this interaction can be indentation, cutting, touching, pushing, pulling, or picking and placing nanoentities or nanosurfaces. It is well understood that these interactions and forces among nanoscale objects are very different from objects at larger scales. The reason for these differences stems from the fact that these forces do not scale linearly with the characteristic length of an entity. Inertial (volumetric) forces, such as momentum and weight that are significant at macroscale, become negligible against areal (adhesive) and peripheral (capillary) forces, particularly at micro/nanoscale (Fearing 1995; Sitti 2007). Therefore, experiencing the nanoscale environment is an imperative step for further advancement in nanomanipulation where AFM is advantageous with its force/deflection measurement capability.

The organization of this chapter is as follows: in Section 42.2, we discuss the working principles of AFM; in Section 42.3, we discuss the force measurements, their calibration, possible error sources, and their compensation techniques. We then specify several AFM-based nanomanipulation approaches in Section 42.4. Then, we conclude the chapter in Section 42.5 with some comments on AFM-based nanomanipulation.

42.2 Working Principles of AFM

AFM is initially designed to take high-resolution topographical images of substrates (Binnig et al. 1986). It consists of a microcantilever with a very sharp needle on its free end. As the tip of this needle is brought close to a surface, the cantilever starts to bend down due to interaction forces between the tip and the substrate. This normal deflection of the cantilever is detected with an optical detection system, which includes a laser beam reflected off the cantilever and collected on a position-sensitive photodetector (PSPD). Resulting normal deflection measurement is returned in terms of a voltage signal, namely, the A–B signal ($V_{A–B}$) for historical reasons. Note that, since the cantilever acts as a mirror in this configuration, not the actual deflection,

but the slope of the free end can be detected. While the two are generally related for most applications, this fact should still be taken into consideration when an experiment is designed, especially for manipulation tasks.

The interaction forces are attractive before contact and repulsive during contact. An AFM can operate in both contact and noncontact conditions utilizing static and dynamic modes of force measurement, respectively. In both cases, a servo controller is utilized to keep the interaction forces between the tip and the substrate constant. Imaging is accomplished by enabling this feedback controller and moving the tip over the sample in a raster pattern. The vertical positions of the cantilever base during this motion are collected into an image matrix and displayed as topography images. Note that an AFM is not limited to topographical images as the interaction between the tip and the substrate can stem from many different physical properties of the substrate such as electrostatic, magnetic, and thermal imaging.

Contact mode operation is simpler and generally gives better resolution. In this mode, the cantilever is pressed onto the surface for a given set point force and the measured errors are regulated by the feedback controller. An animation of this operation mode is given in NT-MDTa. One drawback of this method is that it can cause wear on the tip and/or the substrate. Due to this effect, noncontact mode is generally preferred for soft or fragile substrates or specimens not sufficiently immobilized on the surface.

Noncontact mode uses a dynamic measurement technique, where the AFM cantilever is oscillated at a frequency right above its first normal resonant frequency. The amplitude of vibration is detected by the optical measurement system using a phaselock-loop (PLL). As the tip is brought close to the surface, attractive forces cause the resonant frequency to decrease, yielding a smaller vibration amplitude and a different phase. The controller in this mode is set to keep the amplitude (or phase) of the oscillation constant. Interested readers may refer to NT-MDTb for an animation of this operation mode.

Since noncontact mode typically gives worse resolution, a third mode, namely, the tapping mode, is invented. In this mode, the oscillating tip is brought closer to the substrate to let it contact the surface intermittently at each oscillation period. While the working principle of this mode is the same as noncontact mode, the oscillation frequency is selected right below the first normal resonant frequency of the cantilever as the behavior is reversed.

Soon after its invention, researchers realized that AFM is a perfect candidate to interact with and make changes to objects or surfaces on the nanoscale. Since AFM has a purely mechanical working principle, it can be used on a very broad range of materials, providing versatility to AFM-based nanomanipulation. For many applications, having just the normal deflection measurement is not enough and because of this, modern AFMs employ a four-quadrant PSPD to be able to measure the lateral force acting on the tip, as well. Similar to A–B, lateral force microscopy (LFM) signal (V_{LFM}) gives the angle of twist at the end of the cantilever. This measurement translates to a torsion at the cantilever and a lateral force at the AFM tip.

42.3 Force Measurement, Calibration and Error Compensation

In modern AFMs, two force readings can be simultaneously taken as

$$F^* = V_{A-B}\frac{k_n}{s_n} = F_z + 2\frac{L_{tip}}{L}F_y \qquad (42.1)$$

$$F_x = V_{LFM}\frac{k_l}{s_l}, \qquad (42.2)$$

where

- k_n and k_l are the normal and lateral stiffness values of the cantilever, respectively
- s_n and s_l are the normal and lateral sensitivity values of the optical measurement system, respectively, L and L_{tip} are the length of the cantilever and the tip, respectively

In these equations, F^* is a coupled reading of the vertical force (F_z) and the longitudinal force (F_y), while the lateral force (F_x) can be directly measured.

Therefore, to be able to measure the forces acting on the AFM tip, one needs to perform the necessary calibrations on the stiffness and sensitivity values initially. Normal sensitivity calculation is relatively straightforward. s_n can be found by approaching to, elastically indenting, and then retracting from, a hard surface with the AFM probe. The slope of the resulting V_{A-B} vs. z curve in the contact region (as can be seen in Figure 42.2) gives the amount of voltage change per deflection at the cantilever end, since, for a hard surface, the amount of cantilever base motion in the vertical axis is almost entirely equal to the cantilever deflection.

To calculate lateral sensitivity, one typically utilizes friction loops in the lateral direction. These loops are formed by pressing the tip into a hard surface and moving forward and backward in the lateral axis (x direction). The resulting V_{LFM} vs. x curve has a static region where the tip is fixed on the substrate and the cantilever twists due to the lateral friction force, causing an increase in the LFM signal. The slope of this region gives s_l as the amount of voltage change per lateral tip displacement. One needs to be careful in this measurement, since the lateral stiffness of the probe could be very high and close to the lateral stiffness of the substrate, where contact is maintained. If this happens, the lateral deflection of the tip would not be accurately measured leading to a wrong sensitivity value.

There are many methods for the calibration of normal stiffness. Among them, two nondestructive and popular ones can be counted as the thermal noise and Sader's method of stiffness calibration. In the thermal calibration method, the AFM cantilever is left alone in ambient conditions, without any actuation. Due to temperature, air molecules hit the cantilever with random intervals, causing very small oscillations. This external stimulus is truly random, meaning it has a uniform distribution on the frequency domain (i.e., white noise). Using Boltzmann's constant and the energy transferred to the cantilever by the environment, one can calculate the normal stiffness with an error less than 10% (Hutter and Bechhoefer 1993; Butt and Jaschke 1995). The only drawback of this method is the small amplitude of oscillations that need to be measured. Using very sensitive measurement devices such as laser doppler vibrometry, oscillations on the order of tens of picometers can be detected. Even with such a high sensitivity, stiffness values of more than a few newtons per meter cannot be detected with this method.

Sader's method, on the other hand, does not have an upper limit of cantilever stiffness. It, however, needs actuation and the planar dimensions (length and width) of the cantilever. The cantilever is excited in a frequency sweep or a pseudorandom waveform in a given frequency interval. The quality factor and first resonant frequency are extracted from the frequency response. Using these values with the planar cantilever dimensions and the damping coefficient of the fluid medium of oscillation (i.e., air), one can calculate the stiffness of the cantilever in this method (Sader et al. 1999). Similar to the thermal calibration method, the error in the calculated stiffness values is less than 10%.

Calibrating the lateral stiffness is less straightforward. One possibility is using a torsional version of Sader's method. The idea is the same, but with the first torsional mode of the cantilever. Since these modes have much higher resonant frequencies and smaller amplitudes, however, it is more difficult to apply in practice. A novel method called the diamagnetic lateral force calibration is proposed by Li et al. (2006). In this method, a pyrolytic graphite sheet levitated by a strong magnetic field is used as a reference spring to apply a known lateral force on the AFM tip, and by recording the output voltage signals, the force constants can be obtained as a system response. The main advantage of this system is that it gives two coefficients to calculate force values from voltage signals directly and it can also remove the cross-talk effect on just the lateral force measurement. We will discuss the cross-talk issue of AFM force measurements in the next section in detail.

42.3.1 Crosstalk Compensation

A potential problem in AFM applications is the cross-talk effects between the measured force values (Onal et al. 2008). It is simply not acceptable to experience an unreal lateral force while the tip is vertically pressed onto the sample or unreal normal force while one is doing a lateral force measurement due to the cross-talking of signals. Thus, this section focuses on the cross-talk compensation of force measurements in the AFM systems without adding additional complexity to the AFM setup.

To measure and compensate for the cross-talk effect on the two-voltage signals (normal (A–B) and lateral (LFM) deflection signals), we devised a simple strategy. A slope-corrected flat surface would exert a constant normal force for lateral motions and no lateral force for vertical motions if there was no cross-talk effect. Exploiting this fact obviously gives a quantitative measurement of cross-talk between the aforementioned signals.

That is, the observed change in a signal without any physical reason is the result of only the cross-talk and this change can, then, be used to measure and remove this effect. Note that, a significant amount of thermal drift in the normal (z) direction would affect the normal deflection signal causing it not to remain constant. On the other hand, thermal drift effects are more pronounced at the initial stages of experimentation and settles down after a proper waiting time. Hence, we wait for 30 min before starting experiments. Moreover, since drift velocities are reduced (less than 10 nm/min) after the mentioned waiting time, we conduct our experiments in high-substrate velocities in order to further eliminate the thermal drift effect.

The experimental procedure to calibrate cross-talk, thus, uses a flat, smooth, and rigid surface (e.g., freshly cleaved mica or single-crystal silicon). The probe is first approached to this surface, and slope correction is performed using small line scans (about 0.1 μm) since the surface should at least be locally flat in this region of interest. After slope correction, force feedback control (servo) of the AFM is turned off and a lateral friction loop is performed. A typical result of this motion on a single-crystal silicon wafer is given in Figure 42.1. Note that even when the servo is off, the normal deflection of the cantilever should remain the same as no vertical motion is made. Then a vertical approach–retract cycle is performed to the surface with minimal amount of elastic indentation. Resulting voltage curves for normal and lateral deflections for this motion are given in Figure 42.2.

The effect of cross-talk can be generally described by an affine transformation between the measured and actual voltages such as:

$$\begin{pmatrix} V^a_{A-B} \\ V^a_{LFM} \end{pmatrix} = \underbrace{\begin{bmatrix} a_{11} & a_{12} \\ a_{21} & a_{22} \end{bmatrix}}_{\mathbf{A}} \begin{pmatrix} V^m_{A-B} \\ V^m_{LFM} \end{pmatrix}. \tag{42.3}$$

Using the two experimental results, relations between the elements of each row in $\mathbf{A} \in \Re^{2\times2}$ can be found. Explicitly, for a lateral motion on a surface, forward and backward traces of the normal deflection signal should follow the same curve. Therefore, the actual difference between the mean values of V^a_{A-B} should be zero between forward ($\bar{V}^a_{A-B_f}$) and backward ($\bar{V}^a_{A-B_b}$) traces:

$$\Delta V^a_{A-B} = \bar{V}^a_{A-B_f} - \bar{V}^a_{A-B_b} = 0 \tag{42.4}$$

and using (42.3) and (42.4),

$$\Delta V^a_{A-B} = a_{11}\Delta V^m_{A-B} + a_{12}\Delta V^m_{LFM} = 0, \tag{42.5}$$

which gives

$$\frac{a_{11}}{a_{12}} = -\frac{\Delta V^m_{LFM}}{\Delta V^m_{A-B}}, \tag{42.6}$$

for the first experiment that is previously described.

Similarly, the slope of the actual lateral deflection signal, V^a_{LFM}, should again be equal to zero for a vertical approach retract cycle, which means

$$\frac{dV^a_{LFM}}{dz} = a_{21}\frac{dV^m_{A-B}}{dz} + a_{22}\frac{dV^m_{LFM}}{dz} = 0 \tag{42.7}$$

and hence

$$\frac{a_{21}}{a_{22}} = -\frac{dV^m_{LFM}}{dV^m_{A-B}}, \tag{42.8}$$

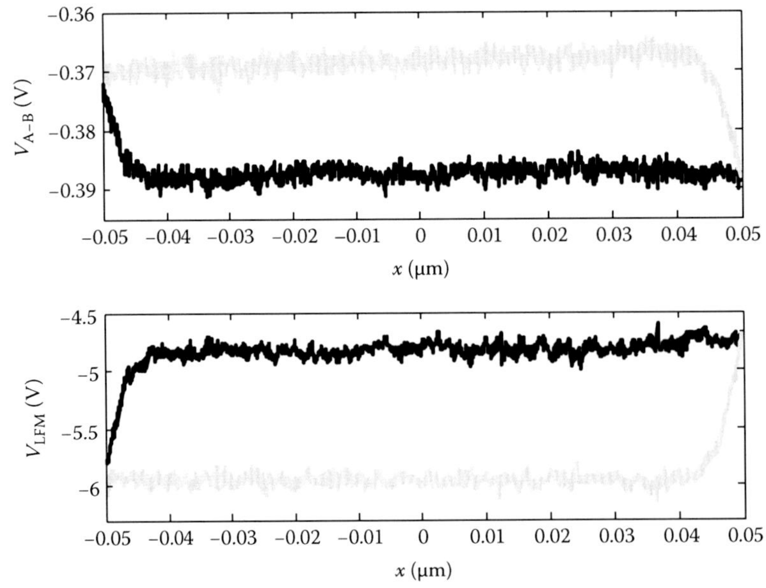

FIGURE 42.1 Normal (A–B) and lateral (LFM) force signals for a forward (black) and backward (gray) line motion in the lateral (x) direction (i.e., friction loop) on a single-crystal silicon wafer. (From Onal, C.D. et al., *Rev. Sci. Instrum.*, 79, 103706, 2008. With permission.)

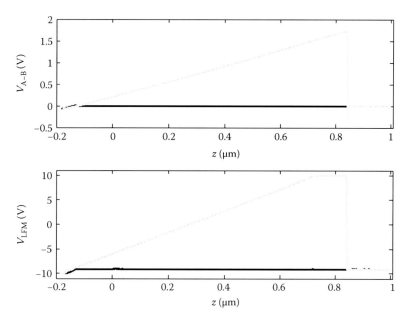

FIGURE 42.2 Normal (A–B) and lateral (LFM) force signals for an approach (black) and retract (gray) cycle in the vertical (z) direction (i.e., force–distance curve) on a single-crystal silicon wafer. Note that cross-talk is so strong that the LFM signal saturates at the retract phase. (From Onal, C.D. et al., *Rev. Sci. Instrum.*, 79, 103706, 2008. With permission.)

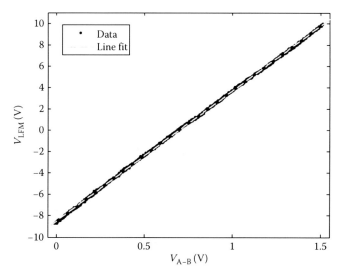

FIGURE 42.3 Lateral (LFM) vs. normal (A–B) deflection signals during the retract phase of the vertical motion before saturation. (From Onal, C.D. et al., *Rev. Sci. Instrum.*, 79, 103706, 2008. With permission.)

in the second experiment. This ratio can be simply found by fitting a line to the V_{LFM} vs. V_{A-B} curve, shown in Figure 42.3, during the retract phase of the vertical motion. In fact, the linearity of this curve can be interpreted as an indication of cross-talk.

Equations 42.6 and 42.8, by themselves, are not enough to calculate **A**. Since vertical and lateral deflection signals are voltage signals, whose sensitivities should be calibrated to achieve the actual forces, without loss of generality, another constraint can be arbitrarily selected as

$$a_{i1}^2 + a_{i2}^2 = 1, \tag{42.9}$$

for $i = 1, 2$. Here, each row of **A** is treated as a vector, whose dot product with the measured voltage vector gives the actual voltage in each axis. This selection ensures that each row vector in the transformation matrix is a unit vector (normalized) and hence do not cause unnecessary scaling of the measured signals.

Using (42.9) with (42.6) and (42.8) allows us to calculate the transformation matrix **A**. Using the calculated transformation to the measured voltage signals, normal and lateral deflection curves, compensated for cross-talk are given in Figures 42.4 and 42.5.

To investigate the error associated with the proposed method, we performed an error propagation analysis for the friction loop data shown in Figure 42.5. Using the standard deviations of the two measured signals as a degree of uncertainty along with Equations 42.6 and 42.7, the propagated amount of uncertainty in the calculated ratio a_{11}/a_{12} comes out to be 11.7% for this particular system configuration and experimental conditions. This value of uncertainty would be further decreased if one uses an AFM with a higher signal-to-noise ratio.

42.3.2 Drift Compensation

Positioning errors in an AFM system are partially caused by hysteresis, creep, and other nonlinearities of the piezo scanning stage. These effects can mostly be reduced by measuring the actual displacement of the scanning unit using position sensors to operate the scanner in a closed-loop. However, position sensors are afflicted with noise, and especially for small scan areas, this often leads to oscillations in the closed-loop control. There exist other approaches that avoid these problems by modeling the piezo-stage characteristics and by applying feedforward strategies (Schitter et al. 2004; Li and Bechhoefer 2007; Mokaberi and Requicha 2007).

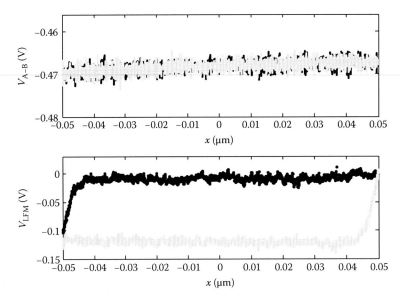

FIGURE 42.4 Normal (A–B) and lateral (LFM) force signals for a forward (black) and backward (gray) line motion in the lateral (x) direction (i.e., friction loop) on a single-crystal silicon wafer after cross-talk compensation. (From Onal, C.D. et al., *Rev. Sci. Instrum.*, 79, 103706, 2008. With permission.)

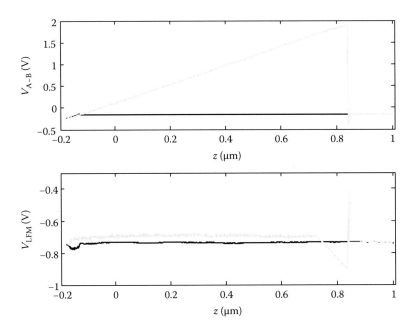

FIGURE 42.5 Normal (A–B) and lateral (LFM) force signals for an approach (black) and retract (gray) cycle in the vertical (z) direction (i.e., force–distance curve) on a single-crystal silicon wafer after cross-talk compensation. Note that the saturation in the measured LFM signal causes an unrealistic change in the corrected signal. (From Onal, C.D. et al., *Rev. Sci. Instrum.*, 79, 103706, 2008. With permission.)

More critical and less straightforward to counteract are spatial uncertainties that are induced by thermal drift. Small variations in temperature cause all AFM components to change slightly in size (due to thermal expansion and contraction), which results in an unknown, time-variant displacement between the AFM tip and the substrate. This motion is generally very slow, but can obviously be detrimental to the success of nanomanipulation in the long run. By operating the AFM under homogeneous environmental conditions, the effect of thermal drift can be reduced, but even in highly temperature-stable conditions thermal drift is still observable and amounts from 0.01 to 0.2 nm/s as reported in Mokaberi et al. (2007). Especially when dealing with feature sizes in the order of a few nanometers, the effect of thermal drift becomes a crucial issue for the success of manipulation (e.g., when pushing a certain object).

The thermal-drift-induced displacement between the AFM probe and the sample surface is not directly observable due to the lack of real-time visual feedback. Therefore, the only possibility to gain information about spatial drift inside the AFM is to use the AFM tip as a sensor itself.

Most commercial AFM softwares provide functions to compensate the effect of thermal drift. However, these functions are usually based on the cross-correlation of successively acquired topography images. This method can obviously only give correct results under the assumption that drift behaves linearly, which has proven not to be the case, especially for inhomogeneous ambient conditions. Additionally, the AFM tip is occupied for minutes during image acquisition and cannot be used simultaneously for manipulation. Hence, this technique is not applicable for nanomanipulation. Other approaches use smaller, local scans to calculate drift, but also these occupy the AFM for a time period, in which manipulation is not possible (Chen et al. 2006).

More sophisticated approaches exist that try to track certain features on the sample (i.e., the center of a nanoparticle (Mokaberi and Requicha 2004, 2006) or the highest point of a nanoobject (Tranvouez et al. 2007)). Even though these techniques are able to measure drift reliably with update rates in the order of seconds, knowledge about the sample (i.e., the shape of a nanoobject) is necessary.

While the first approach is slow but independent of the sample's topography, the latter is faster but requires certain knowledge about the sample surface in terms of a defined feature to track.

A recent approach to the problem of drift estimation is to reduce the drift estimation to a global localization problem as known from mobile robotics (Krohs et al. 2009). To account for the uncertainties that may arise from noisy and faulty sensor data—which is the case for AFM applications—a particle-filter-based state estimation algorithm was implemented to measure the lateral displacement between AFM probe and sample. Even though there also exists thermal drift orthogonal to the sample plane (z direction), this drift can be neglected for manipulation, because the height of the tip in relation to the sample can be obtained by measuring the servo position of the cantilever.

In contrast to existing approaches, this method is fast and does not depend on certain features in the sample's topography. To demonstrate this versatility and robustness, it was evaluated in Krohs et al. (2009) using a highly unstructured surface (a silicon substrate coated with gold) and by performing long-term drift measurements (up to 17 h) with externally triggered temperature changes. Additionally, the algorithm's efficiency was analyzed by performing experiments on a mica substrate with randomly distributed 100 nm gold nanoparticles (Krohs et al. 2009).

The results have shown that drift can be measured reliably over several hours even in very inhomogeneous environments with a very small overhead time of a few seconds per drift estimation iteration. An algorithm to integrate this estimation to manipulation of nanoparticles for drift compensation is also given in Krohs et al. (2009). To demonstrate compensation of drift, a simple experiment is designed on the nanoparticle sample. In this experiment, line scans are taken over the same nanoparticle for 60 min. With the drift compensation turned on, these line scans are more or less identical. The same experiment, with the drift compensation turned off, shows an obvious shift in the line scans in both of the horizontal directions (x and y axes), and the particle is lost after about 10 min.

Moreover, the compensation algorithm is further validated (Krohs et al. 2009) using before and after images of the same $5 \times 5\,\mu m^2$ area over a 60 min time interval, with and without drift compensation. A visible shift without the drift compensation is removed once the compensation is enabled.

42.4 AFM-Based Nanomanipulation

42.4.1 Nanoparticle Manipulation and Assembly

Particle manipulation is the manipulation type where relatively small particles (compared to bulk materials that are often used in other manipulation types) are positioned with high precision using mechanical, electrical, or chemical principles to create more complex structures. Several different kinds of manipulators can be utilized for particle manipulation systems (mechanical or electrostatic gripper, AFM probe tip, etc.) or the positioning can be achieved as a self-assembly process depending on the material or the application.

Particles that are often used in particle manipulation systems have specific geometric shapes (mostly spherical) with usually well-determined dimensions. The particles are either synthesized or grown in a controlled fashion to have uniform shapes and dimensions. Nanoparticles are defined as particles that can be described as above with the biggest dimension of 100 nm. This dimension limit of 100 nm is a commonly accepted limit by several sources; on the other hand, the limit is still subject to change in different sources with different nanoparticle definitions.

Using the definition above, one can also simply define "AFM-based nanoparticle manipulation" as the particle manipulation of nanoparticles: positioning nanoparticles with high precision via mechanical or electrical principles using an AFM probe tip as the manipulator to create more complex structures.

In this context, AFM-based nanoparticle manipulation is significant as a bottom-up approach to manufacturing systems. It is theoretically possible to use nanoparticle manipulation in applications where particles can be precisely positioned-to create miniature sensors, actuators, and man-made materials or to fix protein-or DNA-type biological samples on the surface for enabling characterization. To be more specific, nanoparticle manipulation systems can be utilized for soldering or gluing applications for micro/nanoscale structures or devices. An electrical connection can be established using conductive nanoparticles; on the other hand, polymer nanoparticles can be used as glue droplets for fixing a substrate on a surface at a specific location. Plasmonics is another area that researchers are trying to use nanoparticles as wave-guides or wave-generators, which would require high precision in positioning. Stamps or mask templates can be generated for micro/nano manufacturing purposes; similarly nanoparticle manipulation systems can be used for prototyping for micro/nanoscale devices.

The nanoparticles that are positioned in nanoparticle manipulation systems are smaller than the wavelength of the visible light; therefore, they cannot be seen under optical microscope. Due to this lack of real-time visual feedback, nanoparticle manipulation with AFM stood as a grand challenge for several years.

Several groups worked on AFM-based manipulation, to show its feasibility and acquire more control on the manipulation process. Schaefer et al. published one of the earlier works of this research field, showing that 9–20 nm gold nanoparticles can be manipulated to form clusters using AFM (Schaefer et al. 1995).

The most common approach to the nanoparticle manipulation tasks usually involves a few steps. First the user takes the image of the nanoparticle sample, selects the particle that would be manipulated and a target position for the nanoparticle. The tip is then moved on a line that passes through the center position of the nanoparticle and the target position for the manipulated nanoparticle. In the meantime, servo feedback control is usually turned off or the voltage set point of the signals are set to a lower value in order to decrease the distance between the substrate and the AFM tip, so that particles can be mechanically manipulated, rather than the tip of the AFM probe jumping over the particles. In most of the applications in the literature there is no other control applied on the process other than decreasing the set point or turning the servo feedback off. Therefore, this manipulation procedure is often referred as "blind"- or "push-and-look"-type approach since there is no or little control during manipulation. There are several publications on nanoparticle manipulation (with particles 15–100 nm in diameter) applications with this "blind" pushing technique. These publications can be claimed to be the earlier works that show the feasibility of the idea (Junno et al. 1995; Baur et al. 1997; Ramachandran et al. 1998; Hsieh et al. 2002; Li et al. 2003; Liu et al. 2006b).

Other than the applications above, some groups also worked on improving the "blind" approach for AFM-based nanomanipulation systems. Several efforts on modeling the physics of manipulation interaction are carried. The effects of the tip shape and dimensions on nanomanipulation systems are investigated (Hansen et al. 1998), as well as the effect of the interaction between the substrate and the surface on the manipulation. On conclusions, some substrates interact more with specific surfaces that can cause particles to form assemblies more likely and easily (Resch et al. 1998).

Other than the blind approach, several AFM measurements are being investigated by several groups in order to acquire more control on the manipulation. The cantilever deflection signals, normal and lateral force measurements, or oscillation amplitude during manipulation are among the possible candidates that might enable the researchers to manipulate the nanoparticles

in a more controlled fashion. One of the first attempts on controlled manipulation was to inspect the oscillation amplitude of the AFM cantilever during manipulation operation, which decreases to zero when the tip is in contact with the particle (Martin et al. 1998).

Besides the manipulation systems that are designed to manipulate a single particle in a user-controlled manner, several groups demonstrate automated AFM-based nanomanipulation systems (Li et al. 2005; Mokaberi et al. 2007). The automated manipulation system in Michigan State University (Li et al. 2005) uses the pushing force data to implement a virtual reality system for the user. Results show that the system is working successful enough for coarse positioning of 100 nm particles but there are some mismatches between the actual template and the desired one. The other automated manipulation system (Mokaberi et al. 2007) uses a drift compensator for piezoelectric x–y stages of AFMs (Mokaberi and Requicha 2004), and is able to manipulate 28 particles in a total of 40 min approximate time. The results were impressive for the reliability of the system; on the other hand, success rates of the manipulation attempts were not included in the publication.

Besides these works on reliable nanomanipulation and force-feedback-controlled micromanipulation (Zesch and Fearing 1998), modeling of friction forces in nanoworld (Dedkov 1999), and a combination of these two works with a resulting force model of AFM-based nanomanipulation (Tafazzoli and Sitti 2004) were published and aroused the question whether force-controlled nanomanipulation is possible or not.

In an alternative way to the "blind" approach, to increase the control over the AFM-based nanoparticle manipulation process and to demonstrate the success rate issues in detail, a 2D autonomous nanomanipulation system is designed by Onal et al. (Onal et al. 2009). The AFM tip is utilized to push 100 nm diameter gold nanoparticles on a flat mica substrate covered with Poly-l-lysine (PLL) in 2D. As the most important contribution of this study, control on each individual autonomous manipulation operation is improved using a contact-loss algorithm that continuously tracks the real-time force feedback of the AFM probe, rather than the traditional, blind, push-and-look approach. The contact-loss algorithm dramatically changes the results of the system because if contact loss can be detected during manipulation, it is easier to detect errors in positioning and pushing operation. The performance of the system is investigated through a statistical study. During the performance study, gold nanoparticles are positioned 60 times to target positions in different directions and pushing distances. Eighty-six percent of all the particles can be successfully positioned to the target positions with accuracy below 100 nm. Unsuccessful positioning operations are due to particle sticking to the tip (8%) and particle sticking to the mica substrate (6%). It is suggested that the work can be generalized to different materials and geometries; and does not require the particles to be apart from each other.

Figure 42.6 is a proof of concept that it is even possible to manipulate a particle without disturbing the close neighbor particles using AFM-based nanoparticle manipulation. This is

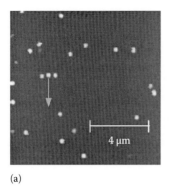

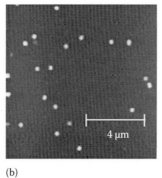

(a)   (b)

FIGURE 42.6 (a) Before and (b) after manipulation of a 100 nm diameter gold nanoparticle indicated by the arrow. It is possible to manipulate a particle without disturbing the close neighboring particles. (From Onal, C. D. et al., *Proc. IEEE ICRA* (in press), 2009. With permission.)

significant because it shows that it is possible to move particles that should be moved however problematic it might be to move. This is crucial to form a given template since the output of the manipulation system should be exactly the template, not a particle more and not a particle less in the template.

Figure 42.7 shows two examples of random templates that are input to an AFM-based manipulation system. Figure 42.7a shows a V-shaped pattern that is formed through 12 particles with 100 nm diameter. Figure 42.7b shows a rectangular pattern of 19 particles with 100 nm diameter. The total manipulation time for both templates is around a few hours for now; however, it is claimed that it can be decreased to approximately half an hour by creating a multiple nanoparticle manipulation where taking images between single particle manipulations would be unnecessary.

Although the Figures 42.6 and 42.7 show that forming templates or complex structures are theoretically possible using AFM-based nanoparticle manipulation, there are certain drawbacks in the nanoparticle manipulation systems. Since AFM is utilized in the systems and it is not possible for any other imaging tool (i.e., optical microscopes, etc.) to sustain visual feedback information to the manipulation system, the reliability, speed, and precision of the nanoparticle manipulation systems are low. "Blind"- or "push-and-look"-type approach is not promising any

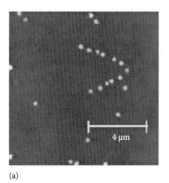

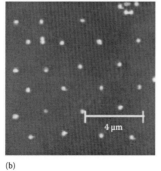

(a)   (b)

FIGURE 42.7 Two resulting sample patterns after a sequence of autonomous single-particle manipulations. (From Onal, C. D. et al., *Proc. IEEE ICRA* (in press), 2009. With permission.)

possible improvements for the performance of the systems. With the current results, even though it is not usually possible to create a nano-featured template with most of the manufacturing techniques like machining or MEMS manufacturing, nanoparticle manipulation systems cannot compete with these techniques that are well established, already cheap, and still rapidly developing. The future of the AFM-based nanoparticle manipulation systems lies in the solutions which can offer high speed, precision, and reliability for nanoparticle manipulation.

42.4.2 Nanoindentation

Indentation typically means making notches, recesses, or sharp depressions on a surface, in a purely mechanical point of view. In a broader perspective, however, small dots of any nature that are marked on a substrate can be interpreted as indentations. Using an AFM for mechanical nanoindentation tasks have become increasingly popular for characterizing surfaces and thin films of many different types of materials.

During mechanical operation, the tip is first brought into contact with the substrate, then pressed into the surface to induce plastic deformation, and finally lifted off the sample, leaving a negative impression of the tip geometry. During this process, the normal deflection of the cantilever can be measured concurrently using the optical lever detection system to determine the amount of load on the substrate. Naturally, a higher load yields a larger indentation depth. Figure 42.8 displays a sequence of AFM images before and after a few nanoindentation experiments at different locations. In these experiments, a polymethyl methacrylate (PMMA) substrate, spincoated on a silicon wafer for a thickness of about 1 μm is used (Onal and Sitti 2009). Spincoating provides a flat film of polymer with RMS roughness on the order of a few nanometers. The tip can be pressed onto the surface for different amounts, with different speeds and different waiting times at maximum penetration to achieve different results, some of which are shown in this figure.

A critical load value that separates elastic and plastic behavior (i.e., surface deforms elastically below the critical load and plastically above it) can be defined for these tasks, for a spherical indenter, as (Fisher-Cribbs 2002)

$$F_c = \left(\frac{R_t}{K}\right)^2 (\pi H)^3, \tag{42.10}$$

where H is the sample hardness. Using a tip radius of 20 nm, and PMMA hardness range of 1.58–3.28 GPa (Liu et al. 2006a), the critical load range for indentation is found to be about 2.76–24.71 μN. The maximum load used in the indentation experiments of Figure 42.8 was calculated to be about 4 μN, which is inside this critical load range.

For plastic nanoindentation, cantilevers of high stiffness (possibly made of stainless steel) and usually with diamond tips are generally utilized. Using this procedure, the Young's modulus and the hardness of the sample material can be determined. While elastic modulus can be deduced from simple elastic force-

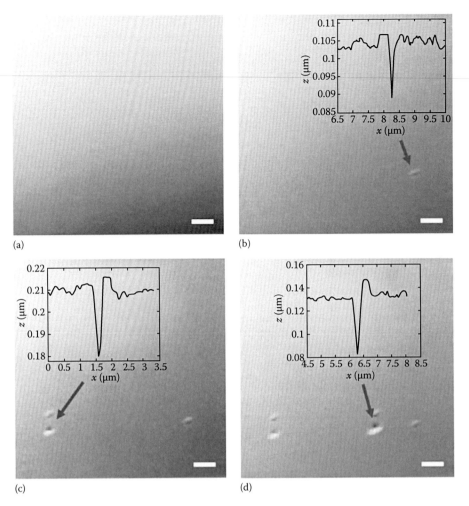

FIGURE 42.8 $10\,\mu m \times 10\,\mu m$ AFM contact mode images of (a) before and (b), (c), (d) after a series of nanoindentation experiments at the same area on a flat PMMA substrate. Scale bars indicate $1\,\mu m$ length and insets provide the depth information of each indentation mark. (From Onal, C. D. and Sitti, M., *Int. J. Rob. Res.*, 28, 484, 2009. With permission.)

distance curves, hardness of a material in the nanoscale can only be calculated by actually deforming the surface plastically.

It has been shown in Bhushan (2004) that material hardness may be increasing in the nanoscale. The power of AFM for this task stems from the fact that it can make an array of such measurements in a very small area with minimal penetration depths for error reduction. Hardness of a substrate can be calculated by measuring the apparent remnant area of indentation after the fact, using preferably noncontact images and the load which caused that depression.

Besides hardness measurements, pure mechanical indentation tasks using AFM are also being utilized for a long time for other mechanical characterization tasks. Other mechanical properties like Young's modulus, relaxation time, or creep properties of a material can also be characterized using mechanical indentation via AFM (Kulkarni and Bhushan 1996). The important part of these experiments is to identify the surface level of the material where the indenter will begin indenting the surface. In some cases, this is implemented using cantilever deflection or servo position (Kulkarni and Bhushan 1996), whereas in some cases

other solutions like electrical contact resistance are tried to be implemented (Bouzehouane et al. 2003).

Another possibility is using resistors to heat a small surface area around the tip and make small indentations in a thermomechanical sense. An application on shape memory polymers is shown in (Yang et al. 2007). The localized heating of a shape memory polymer causes changes in the material properties and the indentation is made after the localized heating. The study is done to investigate the changes in mechanical properties of shape memory polymers due to heating in nanoscale and the possibility of micro-actuation and data storage with shape memory polymers are investigated in the same article.

Other approaches to leave small marks on a surface include electrostatic (Lyuksyutov et al. 2003) and magnetic (Hosaka et al. 1995; Hosaka 2001) methods. In these approaches, the cantilever is brought in close proximity to, or into contact with, the surface and an electric or a magnetic field is applied to locally change sample properties of interest around the tip.

For electrostatic indentation applications, the AFM cantilever tip is brought very close to the surface (on the order of a few nanometers) and a bias voltage of 0–20 V is applied between the tip and the polymer film (Lyuksyutov et al. 2003). It is claimed that compared to mechanical or thermomechanical indentation tasks, electrostatic indentation is faster and more stable which makes it a more preferable choice for data storage systems.

Similarly, the proof of concept for magnetic data storage using an AFM is carried by Hosaka (2001) and Hosaka et al. (1995). Data are written on the magnetic film substrate using voltage pulses. Then the recorded data are read using the MFM mode of an SPM. The advantages of magnetic indentation are suggested as the easy reading of recorded data and the compactness of the storage.

The spatial resolution in AFM-based nanoindentation can be on the order of nanometers, which make it an ideal candidate for high-density data storage systems, exceeding 1 Tb/in^2 limit (Hosaka 2001). High density, however, is not the only requirement for future scanning probe-based data storage systems. The fact that AFM operation is slow compared to modern hard drive heads puts a constraint on the access times that can be achieved with such a system. One possible solution is parallelizing the process, using an array of cantilevers. Extensive research is currently being done in the IBM "millipede" project to this end (Vettiger et al. 2000).

42.4.3 Nanolithography

Lithography is traditionally a method of printing using a flat and smooth plate or stone. In modern technology, patterning photoresists with light at a certain wavelength (photolithography) and using this as a mask to transfer the pattern to an underlying structure is widely employed for microfabrication of small components and circuits. With the ever-increasing demand for miniaturization toward nanoscale, limits of photolithography are constantly being pushed.

Minimum achievable feature size from photolithography is obviously limited by the diffraction limit of light, which is due to its wavelength. One solution researchers pursue is utilizing the wave property of light to create high- and low-intensity regions by the addition or subtraction of light waves from separate sources or, namely, the interference lithography. While interference lithography has the ability to generate structures below the diffraction limit, it usually gives an array of such features like grating patterns, which may not be useful for some applications.

Among other possibilities, a promising method is charged-particle-beam lithography, which includes electron-beam (EBL) and ion-beam lithography. In this method, especially for EBL, very small features can be generated. A high-energy electron-beam is scanned over the substrate in a maskless fashion (using a software mask), to expose the required parts of the resist with very high precision. The cons of this approach are the necessity of a vacuum chamber, quick scattering of electrons in solids, exposure speed due to the serial nature, and high system cost (Madou 2002).

An alternative for nanolithography, however, may be utilizing an AFM tip to pattern surfaces in a serial manner. The nature of this patterning can be mechanical (adding or removing material) or electrical. Since the AFM tip diameter can be on the

order of tens of nanometers, this method has the potential to provide high spatial resolution at a cheap price.

Material removing from the substrate by statically scratching the surface with the tip is the simplest method of AFM-based nanolithography. In this method, the tip is brought into contact with the surface and pressed onto it until plastic deformation. Then, the AFM tip can be moved in arbitrary trajectories and mark the surface. After the surface is patterned, the amount of tip-sample penetration depth is reduced to inside elastic ranges or noncontact imaging is employed to take another scan of the surface. Figure 42.9 displays results of such an experiment. Note that while material is generally removed by the AFM tip leaving grooves on the surface, this removed material is carried with the tip and deposited at positions where a turn is made by the tip on the surface.

Several mechanical lithography examples are tried (Li et al. 2003; Tian et al. 2004; Liu et al. 2006b). The main problem addressed in these applications was the lack of real-time feedback during the use of AFM cantilever tip as an end effector. Using haptic interface (Tian et al. 2004), modeling of forces during manipulation (Li et al. 2003) or adjusting the oscillation set point to control the interaction of tip with surface (Liu et al. 2006b) is offered for mechanical nanolithography using an AFM to prevent applications suffer from uncertainties of nanoscale physics.

Mechanical scratching is a simple and versatile method to pattern a surface on the nanoscale. Nonetheless, it has an important drawback in that it causes tip wear, which limits the repeatability of the procedure. It can also lead to edge irregularities due to the torsion of the cantilever and imaging in contact might create further modifications on the surface.

One possibility to reduce tip wear and accumulation of material at the edges during removal is to use a dynamic scratching technique called dynamic plowing lithography. In this

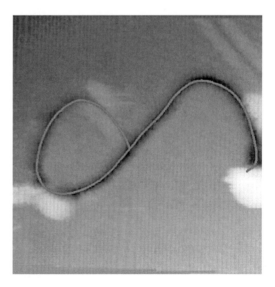

FIGURE 42.9 5 μm × 5 μm AFM contact mode image of a mechanical static nanolithography experiment on a flat PMMA substrate. Blue curve depicts the trajectory of the tip. Note that due to the plowing nature of lithography, material is deposited around the trajectory, especially when a turn is made by the tip.

method, tapping-mode AFM is employed with high oscillation amplitudes to induce only vertical indentations at a very high frequency. The tip is moved over the sample, hammering the substrate down, with less tip wear and hence, better repeatability (NT-MDTc; Klehn and Kunze 1999; Heyde et al. 2001). In Heyde et al. (2001), it is claimed that, no tip wear and absorption of polymer occurred during dynamic plowing lithography technique even after 50 tries, which is a promising result for AFM-based nanolithography in terms of repeatability.

A very popular method to add material on a surface by covering the tip with "ink" and then moving on the substrate to "write" by transferring the "ink" material to the surface is dip-pen lithography (DPN). DPN is a relatively new AFM-based soft-lithography technique where an AFM tip is used to deliver molecules (ink) to a surface via a solvent meniscus, which forms naturally in the ambient atmospheric conditions. As with other AFM-based nanopatterning approaches, it offers high-resolution capabilities for a number of (bio)molecular "inks" on a variety of substrates (metals, semiconductors, and monolayer functionalized surfaces). Nevertheless, due to its original nature of patterning, it certainly has selectivity over materials that can be used (Piner et al. 1999; Lee et al. 2002).

DPN is shown to be a powerful tool for data storage or nanomanufacturing (Piner et al. 1999); on the other hand, it is also shown to be an important tool for characterization purposes. In Lee et al (2002), protein arrays are written and attached on the substrate using DPN, which might allow the researchers to characterize several properties of living organisms like cellular adhesion.

As another nanolithography technique, local anodic oxidation (LAO) is an electrical counterpart for AFM-based nanolithography. This procedure, invented by Dagata (Dagata et al. 1990), is one of the early-developed techniques based on a direct oxidation of the sample by a negative potential applied to the AFM tip with respect to the sample. The oxidation process utilizes the presence of a water-bridge between the tip and the sample under ambient conditions (Garcia et al. 1998). For high local electric fields, water molecules dissolve into H^+ and OH^- ions. OH^- ions get transported toward the positively charged substrate; react with surface atoms to induce oxidation. Oxide layers in forms of lines and dots can be created with nanoscale feature sizes. Current research on this technology generally focuses on reducing the feature size using finite element models or Monte Carlo simulations (Cambel et al. 2006). Its highly material-dependent chemical nature obviously limits the available substrate pool for this method of nanolithography. This fact, however, does not present a huge problem due to its wide possible use in the nanofabrication technology on semiconductors. As a possible application, Held et al (1997) suggest to use the LAO-based nanolithography technique to manufacture self-aligned gate structures for nanotransistors. The smallest structure size that is reported in this article is 30 nm, whereas it is also claimed by other groups that these feature sizes can be as low as 10 nm.

Nanolithography with an AFM tip can yield very precise 2D patterns, but precision comes with a price. As with many AFM-based nanomanipulation applications, it also suffers from the serial nature of the process, which limits the speed and causes the research to generally stay in proof-of-concept demonstrations.

42.4.4 Material Characterization

Characterizing materials for their physical properties is a necessary step to be able to accurately understand and model their behavior. Characterization in the nanoscale has an additional importance, since it is predicted that with a reduction in scale, bulk material properties may change. Every material we encounter in the macroworld has some imperfections. Due to these imperfections, what is perceived becomes an average behavior and does not necessarily reflect the material properties of just a given structure. However, in the nanoscale, these imperfections start to become less pronounced. Moreover, in small scales, dominant factors start to change, generally yielding nonlinear behavior (Bhushan 1999) or even new discoveries (Strukov et al. 2008).

Employing AFM to probe physical properties in the nanoscale is a very popular method due to the aforementioned versatility of this tool. The sharp AFM tip can interact with a very localized region on the surface and make measurements with ease. Applying this knowledge to the scanning principles of AFM operation, one can take "images" of material properties in addition to topography. These additional images can yield friction (friction force microscopy); dielectric constant, or resistance (electrostatic force microscopy); magnetic (magnetic force microscopy); or piezoelectric constant (piezoelectric force microscopy) information of the surface.

Also, some specific experiments can be utilized to measure certain mechanical properties of the interaction between the tip with the substrate. For instance, a vertical force–distance experiment gives a finite adhesive force, from which one can deduce the equivalent work of adhesion using contact mechanics models with a known tip radius. A lateral "friction loop" on the surface gives the sliding friction force of the tip on the surface. Combining this friction force value with the tip radius and some known material properties of the two materials, one can measure the interfacial shear strength, which is the dominant cause of friction in the nanoscale.

Moreover, one can characterize resistance, elastic modulus, or strength of objects lying on the surface, such as nanotubes (Wong et al. 1997; Salvetat et al. 1999), nanoparticles (Sitti 2004), or nanowires (Wu et al. 2005). A recent study on particle friction utilized pushing spherical particles on a flat surface with the AFM tip, while monitoring the forces (Sumer and Sitti 2008). Results of these experiments yielded findings on three different particle motion modes (rolling, sliding, and spinning) and theoretical predictions of particle motion from the force data.

AFM proves to be a useful tool to measure the flexural strength of fibers (or wires) suspended on a trench as well (Wu et al. 2005). In these experiments, the AFM tip is brought to the same height of the fiber and moved laterally to contact the mid-point along the fiber length. Continuing this lateral motion, perpendicular

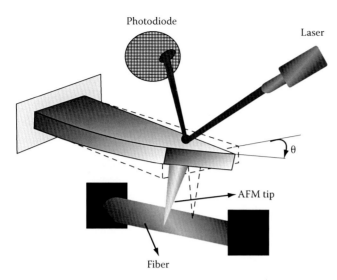

FIGURE 42.10 Schematic of AFM-based lateral pushing and mechanical characterization of fibers: An AFM probe tip is used to push a suspended micro/nano fiber, which is fixed on both ends to the substrate, from the middle in the lateral direction while measuring AMB and LFM signals.

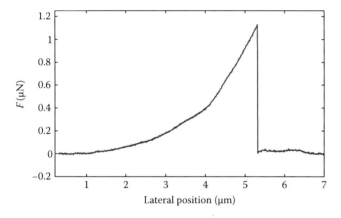

FIGURE 42.11 A sample lateral force curve for a PMMA microfiber breaking experiment.

to the fiber axis with a constant velocity until fracture, one can measure the lateral force and hence the stress value that causes failure. This lateral testing setup is depicted in Figure 42.10 and a sample lateral force curve during a flexural strength test of a PMMA microfiber is given in Figure 42.11.

42.5 Conclusion

With the rise of nanotechnology, handling and manipulating nanoscale objects have become a necessity. Nanotechnology is believed to be offering man-made materials with improved properties like more mechanical strength, miniature and high-sensitivity sensors, miniature actuators, and hybrid systems, characterization of biological samples like proteins or DNA, etc. However, all of these applications require nanomanipulation systems for handling, characterization, or assembly purposes. AFM systems have enabled sub-nanometer resolution

three-dimensional (3D) topography images of surfaces of any nanomaterial (conductive, nonconductive, biological, polymer, etc.) in any environment (air, liquid, and vacuum) due to its pure mechanical working principle. In addition to this, AFM has the ability to change the properties of the surface or the substrate, which makes it even a more powerful tool for nanotechnology. Therefore, as one of the major approaches, AFM systems have also been utilized as a nanomanipulation tool to interact with objects, surfaces, and make changes by mechanical contact manipulation.

In consideration to this, we tried to explain the working principles of AFM in detail. The relation between the cantilever deflection signals and the normal and lateral forces applied on the tip of the cantilever is presented and the calibration methods are suggested. The most common problems of AFM measurements, which are the drift and the cross-talk between the signals, are discussed and possible compensation methods are shown. AFM-based nanomanipulation types are presented in four main headlines as: nanoparticle manipulation, indentation, lithography, and characterization. The techniques that are used in these manipulation types, drawbacks, and possible applications are discussed.

We believe that the drawbacks of AFM-based nanomanipulation should and can be solved in the near future, which might make AFM one of the most powerful tools for nanomanipulation. These future developments in AFM-based nanomanipulation will lead to a great acceleration in the nanotechnology that would first help us to understand the nature of nanoscale physics and then would lead researchers to great innovations in several disciplines like biology, medicine, electronics, or robotics.

References

Baur, C., B. C. Gazen, B. Koel, T. R. Ramachandran, A. A. G. Requicha, and L. Zini 1997. Robotic nanomanipulation with a scanning probe microscope in a networked computing environment. *Journal of Vaccum Science and Technology B*, 15: 1577–1580.

Bhushan, B. 1999. *Handbook of Micro/Nano Tribology*, 2nd edn. CRC Press, Boca Raton, FL.

Bhushan, B. 2004. *Springer Handbook of Nanotechnology*. Springer-Verlag, Heidelberg.

Binnig, G., C. F. Quate, and C. Gerber 1986. Atomic force microscope. Physical Review Letters, 56(9): 930–933.

Bohringer, K.-F., K. Goldberg, M. Cohn, R. Howe, and A. Pisano 1998. Parallel microassembly with electrostatic force fields. In *Proceedings of the IEEE International Conference on Robotics and Automation*, Leuven, Belgium, K. Goldberg, ed., Vol. 2, pp. 1204–1211.

Bouzehouane, K., S. Fusil, M. Bibes, J. Carrey, T. Blon, M. Le Du, P. Seneor, V. Cros, and L. Vila 2003. Nanolithography based on real-time electrically-controlled indentation with an atomic force microscope for nanocontacts elaboration. *Nano Letters*, 3:1599.

Butt, H. J. and M. Jaschke 1995. Calculation of thermal noise in atomic force microscopy. *Nanotechnology*, 6(1):1–7.

Cambel, V., J. Soltys, J. Martaus, and M. Mosko 2006. IV characteristics in structures prepared by tip induced oxidation. *Journal de physique IV France*, 132:171–175.

Castelino, K., S. Satyanarayana, and M. Sitti 2005. Manufacturing of two and three-dimensional micro/nanostructures by integrating optical tweezers with chemical assembly. *Robotica*, 23(4):435–439.

Chen, H., N. Xi, and G. Li 2006. CAD-guided automated nanoassembly using atomic force microscopy-based nanrobotics. *IEEE Transactions on, Automation Science and Engineering*, 3:208–217.

Dagata, J., J. Schneir, H. H. Harary, C. J. Evans, M. T. Postek, and J. Bennett 1990. Modification of hydrogen passivated silicon by a scanning tunneling microscope operating in air. *Applied Physics Letters*, 56(20):2001–2003.

Dedkov, G. V. 1999. Friction on the nanoscale: New physical mechanisms. *Materials Letters*, 38:360–366.

Fearing, R. 1995. Survey of sticking effects for micro parts handling. In *Proceedings IEEE/RSJ International Conference on Intelligent Robots and Systems 95. Human Robot Interaction and Cooperative Robots*, Pittsburgh, PA, Vol. 2, pp. 212–217.

Fisher-Cribbs, A. C. 2002. *Mechanical Engineering Series: Nanoindentation*. Springer-Verlag, Newyork.

Garcia R., M. Calleja, and F. Perez-Murano 1998. Local oxidation of silicon surfaces by dynamic force microscopy: Nanofabrication and water bridge formation. *Applied Physics Letters*, 72(18):2295–2297.

Hansen, L. T., A. Kuhle, A. H. Srensen, J. Bohr, and P. B. Lindelof 1998. A technique for positioning nanoparticles using an atomic force microscope. *Nanotechnology*, 9:337–342.

Held, R., T. Heinzel, P. Studerus, K. Ensslin, and M. Holland 1997. Semiconductor quantum point contact fabricated by lithography with an atomic force microscope. *Applied Physics Letters*, 71:2689.

Heyde, M., K. Rademann, B. Cappella, M. Geuss, H. Sturm, T. Spangenberg, and H. Niehus 2001. Dynamic plowing nanolithography on polymethyl methacrylate using an atomic force microscope. *Review of Scientific Instruments*, 72(1):136–141.

Hosaka, S. 2001. Spm based recording toward ultrahigh density recording with trillion bits/inch2. *IEEE Transactions on Magnetics*, 37(2):855–859.

Hosaka, S., H. Koyanagi, A. Kikukawa, M. Miyamoto, R. Imura, and J. Ushiyama 1995. Fabrication of nanometer-scale structures on insulators and in magnetic materials using a scanning probe microscope. *JVST B-Microelectronics and Nanometer Structures*, 13(3):1307–1311.

Hsieh, S., S. Meltzer, C. Wang, A. Requicha, M. Thompson, and B. Koel 2002. Imaging and manipulation of gold nanorods with an atomic force microscope. *Journal of Physical Chemistry B*, 106(2):231–234.

Hutter, J. L. and J. Bechhoefer 1993. Calibration of atomic-force microscope tips. *Review of Scientific Instruments*, 64(7):1868–1873.

Junno, T., K. Deppert, L. Montelius, and L. Samuelson 1995. Controlled manipulation of nanoparticles with an atomic force microscope. *Applied Physics Letters*, 66(26):3627–3629.

Klehn, B. and U. Kunze 1999. Nanolithography with an atomic force microscope by means of vector-scan controlled dynamic plowing. *Journal of Applied Physics*, 85(7):3897–3903.

Krohs, F., C. D. Onal, M. Sitti, and S. Fatikow 2009. Towards automated nanoassembly with the atomic force microscope: A versatile drift compensation procedure. *ASME Journal of Dynamic Systems, Measurement, and Control*, 131(6):061106-1–8.

Kulkarni, A. V. and B. Bhushan 1996. Nanoscale mechanical property measurements using modified atomic force microscopy. *Thin Solid Films*, 290–291:206–210.

Lee, E. and H. S. Hong 2005. An integrated system of microcantilever arrays with carbon nanotube tips for bio/nano analysis: Design and control. In *Proceeding of the IEEE International Conference on Automation Science and Engineering*, H. S. Hong, ed., pp. 113–117.

Lee, K., S. J. Park, C. A. Mirkin, J. C. Smith, and M. Mrksich 2002. Protein nanoarrays generated by dip-pen nanolithography. *Science*, 295:1702–1705.

Li, Y. and J. Bechhoefer 2007. Feedforward control of a closed-loop piezoelectric translation stage for atomic force microscope. *Review of Scientific Instruments*, 78(1):013702.

Li, G., N. Xi, M. Yu, and W. K. Fung 2003. 3d nanomanipulation using atomic force microscopy. In *Proceedings of the IEEE International Conference on Robotics and Automation ICRA '03*, Taipei, Taiwan, Vol. 3, pp. 3642–3647.

Li, G., N. Xi, H. Chen, C. Pomeroy, and M. Prokos 2005. "Videolized" atomic force microscopy for interactive nanomanipulation and nanoassembly. *IEEE Transactions on Nanotechnology*, 4(5):605–615.

Li, Q., K.-S. Kim, and A. Rydberg 2006. Lateral force calibration of an atomic force microscope with a diamagnetic levitation spring system. *Review of Scientific Instruments*, 77(6):065105.

Liu, C.-K., S. Lee, L.-P. Sung, and T. Nguyen 2006a. Load-displacement relations for nanoindentation of viscoelastic materials. *Journal of Applied Physics*, 100(3):033503.

Liu, Z., Z. Li, G. Wei, Y. Song, L. Wang, and L. Sun 2006b. Manipulation, dissection, and lithography using modified tapping mode atomic force microscope. *Microscopy Research and Technique*, 69(12):998–1004.

Lyuksyutov, S. F., R. A. Vaia, P. B. Paramonov, S. Juhl, L. Waterhouse, R. M. Ralich, G. Sigalovand, and B. Sancaktar 2003. Electrostatic nanolithography in polymers using atomic force microscopy. *Nature Materials*, 2:468–472.

Madou, M. J. 2002. *Fundamentals of Microfabrication. The Science of Miniaturization*, 2nd edn. CRC Press, Boca Raton, FL.

Martin, M., L. Roschier, P. Hakonen, U. Parts, M. Paalanen, B. Schleicher, and E. I. Kauppinen 1998. Manipulation of ag nanoparticles utilizing noncontact atomic force microscopy. *Applied Physics Letters*, 73(11):1505–1507.

Mokaberi, B. and A. Requicha 2004. Towards automatic nanomanipulation: drift compensation in scanning probe microscopes. In *Proceedings of the IEEE International Conference on Robotics and Automation ICRA '04*, New Orleans, LA, Vol. 1, pp. 416–421.

Mokaberi, B. and A. A. G. Requicha 2006. Drift compensation for automatic nanomanipulation with scanning probe microscopes. *IEEE Transactions on Automation Science and Engineering*, (3):199–207.

Mokaberi, B. and A. A. G. Requicha 2007. Compensation of scanner creep and hysteresis for AFM nanomanipulation. *IEEE Transactions on Automation Science and Engineering*, 5(2):197–206.

Mokaberi, B., J. Yun, M. Wang, and A. Requicha 2007. Automated nanomanipulation with atomic force microscopes. In *Proceedings of the IEEE International Conference on Robotics and Automation*, Roma, pp. 1406–1412.

NT-MDTa Afm imaging—constant force mode. http://www.ntmdt. com/spm-notes/view/corrosion-influence-znsse-gaas-laser-structure-spm-mode. Retrieved on April 5th 2009.

NT-MDTb Afm imaging—non-contact mode. http://www.ntmdt. com/spm-principles/view/non-contact-mode. Retrieved on April 5th 2009.

NT-MDTc Afm lithography—dynamic plowing. http://www. ntmdt.com/spm-principles/view/afm-lithography-dynamic-plowing. Retrieved on December 29th 2008.

Onal, C. D. and M. Sitti 2009. A scaled bilateral control system for experimental one-dimensional teleoperated nanomanipulation. *International Journal of Robotics Research*, 28:484–497.

Onal, C. D., B. Sumer, and M. Sitti 2008. Cross-talk compensation in atomic force microscopy. *Review of Scientific Instruments*, 79(10):103706.

Onal, C. D., O. Ozcan, and M. Sitti 2009. Automated 2d nanoparticle manipulation with an atomic force microscope. In *Proceedings of the IEEE International Conference of Robotics and Automation*, Kobe, Japan, Vol. 1, pp. 1814–1819.

Piner, R., J. Zhu, F. Xu, S. Hong, and C. A. Mirkin 1999. Dip-pen nanolithography. *Science*, 283:661–663.

Pozidis, H., W. Haberle, D. Wiesmann, U. Drechsler, M. Despont, T. Albrecht, and E. Eleftheriou 2004. Demonstration of thermomechanical recording at 641 gbit/in/sup 2/. *IEEE Transaction on Magnetics*, 40(4):2531–2536.

Ramachandran, T. R., C. Baur, A. Bugacov, A. Madhukar, and B. E. Koel, A. Requicha, and C. Gazen 1998. Direct and controlled manipulation of nanometer-sized particles using the non-contact atomic force microscope. *Nanotechnology*, 9(3):237–245.

Resch, R., C. Baur, A. Bugacov, B. E. Koel, A. Madhukar, A. Requicha, and P. Will 1998. Building and manipulating 3-d and linked 2-d structures of nanoparticles using scanning force microscopy. *Langmuir*, 14–23:6613–6616.

Sader, J. E., J. W. M. Chon, and P. Mulvaney 1999. Calibration of rectangular atomic force microscope cantilevers. *Review of Scientific Instruments*, 70(10):3967–3969.

Salvetat, J.-P., J.-M. Bonard, N. Thomson, A. Kulik, L. Forró, W. Benoit, and L. Zuppiroli 1999. Mechanical properties of carbon nanotubes. *Applied Physics A: Materials Science & Processing*, 69(3):255–260.

Schaefer, D. M., R. Reifenberger, A. Patil, and R. P. Andres 1995. Fabrication of two-dimensional arrays of nanometer-size clusters with the atomic force microscope. *Applied Physics Letters*, 66(8):1012–1014.

Schitter, G., F. Allgöwer, and A. Stemmer 2004. A new control strategy for high-speed atomic force microscopy. *Nanotechnology*, 15(1):108–114.

Sitti, M. 2004. Atomic force microscope probe based controlled pushing for nanotribological characterization. *IEEE/ASME Transactions on Mechatronics*, 9(2):343–349.

Sitti, M. 2007. Microscale and nanoscale robotics systems [grand challenges of robotics]. *IEEE Robotics & Automation Magazine*, 14(1):53–60.

Strukov, D. B., G. S. Snider, D. R. Stewart, and R. S. Williams 2008. The missing memristor found. *Nature*, 453:80–83.

Sumer, B. and M. Sitti 2008. Rolling and spinning friction characterization of fine particles using lateral force microscopy based contact pushing. *Journal of Adhesion Science and Technology* 22:481–506.

Tafazzoli, A. and M. Sitti 2004. Dynamic modes of nanoparticle motion during nanoprobe-based manipulation. In *Proceedings of 4th IEEE Conference in Nanotechnology*, Munich, Germany.

Tian, X., N. Jiao, L. Liu, Y. Wang, N. Xi, W. Li, and Z. Dong 2004. An afm based nanomanipulation system with 3d nano forces feedback. In *Proceedings of the International Conference on Intelligent Mechatronics and Automation*, Chengdu, China, pp. 18–22.

Tranvouez, E., E. Boer-Duchemin, G. Comtet, and G. Dujardin 2007. Active drift compensation applied to nanorod manipulation with an atomic force microscope. *Review of Scientific Instruments*, 78:115103.

Vettiger, P., M. Despont, U. Drechsler, U. Durig, W. Haberle, M. I. Lutwyche, H. B. Rothuizen, R. Stutz, R. Widmer, and G. K. Binnig 2000. The millipede more than one thousand tips for future afm data storage. *IBM Journal of Research and Development*, 44(3):323–340.

Wong, E. W., P. E. Sheehan, and C. M. Lieber 1997. Nanobeam mechanics: Elasticity, strength, and toughness of nanorods and nanotubes. *Science*, 277(5334):1971–1975.

Wu, B., A. Heidelberg, and J. J. Boland 2005. Mechanical properties of ultrahigh-strength gold nanowires. *Nature Materials Letters*, 4:525–529.

Yang, F., E. Wornyo, K. Gall, and W. King 2007. Nanoscale indent formation in shape memory polymers using a heated probe tip. *Nanotechnology*, 18:285302.

Zesch, W. and R. S. Fearing 1998. Alignment of microparts using force controlled pushing. In *Microrobotics and Micromanipulation Conference*, Boston, MA, Vol. 3519, pp. 148–156.

Nanomanipulation and Nanorobotics with the Atomic Force Microscope

Robert W. Stark
Ludwig-Maximilians-
Universität München

43.1 Introduction

Technologies that control and manipulate matter on the nanoscale will play a key role in the coming decade. We expect nanotechnology to play a key role in medicine, biotechnology, electronics, and in materials research and development. Richard Feynman foresaw the challenges and expectations that come along with the exploration of the "nanocosmos." In his famous talk, "There is plenty of room at the bottom," at the Caltech University in 1959 he addressed the "[…] problem of manipulating and controlling things on a small scale […]" and formulated the fundamental question: "How do we write small?" This is yet to be answered completely.

Methods and technologies for nanolithography and nanopatterning are currently being pushed to their limits to create smaller and smaller structures with highly elaborate functionalities. These technologies include electronic circuits that are shrunk to the nanoscale or nanosensors that are based on a special arrangement of biomolecules. Current research activities also include the spatially well-defined modification of surface properties such as wettability, adhesion, friction, or tailored specific chemical interaction. In the future, the controlled positioning of individual nanoparticles, molecules, and even atoms will enable new chemical reaction paths that are based on a guided self-assembly. These activities include the quest for nanostructured surface functionalization and directed self-assembly of nanoscale objects (Geissler and Xia, 2004). The controlled positioning of particles, clusters, or single molecules allows for the investigation of new chemical processes, tailored molecular interaction, and the direct assembly of new chemical compounds (Heckl, 2004).

On an industrial scale, we observe an evolution of conventional industrial manufacturing processes into nanoscale manufacturing processes. Prominent examples for this transformation are the large-scale production of nanoparticles, the manufacturing of nano-products from nanomaterials, as well as the generation of nanostructures on surfaces. These trends increase the need to control, visualize, and analyze the structures and properties of materials on the nanometer scale in a controlled environment. Science and technology have made tremendous progress in device miniaturization. Various methods have been developed during the recent years that allow for both reading and writing on the smallest scales. Important examples include the scanning electron microscope, the scanning force microscope, electron beam lithography, and extreme UV-lithography, to name just a few technologies. For inspection and patterning, scanning probe methods provide direct access to the nanoworld.

Despite the progress that could be achieved during the last decade, nanomanipulation still presents formidable challenges. One challenge is due to the fact that sensors and effectors have to be tailored to purpose (Requicha, 2003). In the macroscopic world, one can equip a machine with cameras and gripping tools that more or less mimic the functionality of a human hand. Such a straightforward approach does not work on the nanoscale because the objects are much smaller than the wavelength of visible light. Thus, conventional sensors and effectors cannot be used. The imaging problem can easily be solved with an atomic force or an electron microscope. The use of high-resolution microscopes, however, does not solve the challenge of nanogripping. Richard Smalley identified the main challenges as the "fat finger" and "sticky finger" problems (Smalley, 2001; Smalley and

Drexler, 2003), because the manipulation tools are themselves made out of atoms that cannot be made smaller and more precise than the objects they manipulate. In addition to this "fat finger" problem, the nano-object that is being moved may adhere to the effector. Thus, subtle strategies are needed to precisely release the nano-building blocks.

The difficulties of nanogripping can best be illustrated by comparing the forces that occur on the smallest scale. On the nanometer scale, surface and chemical forces are much stronger than the gravitational forces. Gravitational forces are downscaled by the third power of the length scale. This means that, gravitation is 27 orders of magnitude weaker on the nanometer scale as compared to the macroscopic world (meter). The surface area only scales with the second power of the length scale. Thus, the relative strength of surface forces as compared to gravity is increased by nine orders of magnitude. The scaling effects imply that small particles easily adhere to surfaces due to the attractive van der Waals forces but the gravitational forces cannot release them. This makes gripping difficult because the object cannot be dropped after it has been gripped. Similar restrictions are encountered in nano-cutting and fusing.

The dominance of the surface and other short-range forces that are close by defines nanomanipulation as a field of its own right. Currently, there are two important lines of investigation toward nanoscale manipulation: (1) nanopatterning, where strategies are developed that allow us to generate arbitrary nanostructures on the mesoscale or even on the scale of individual atoms; and (2) nanotelerobotics, where the nanoworld is translated into a virtual reality environment. Manipulation of nanobio-objects such as DNA or proteins may even demand a fluidic environment, which on the one hand preserves the molecular functionality and on the other hand prevents unspecific interaction with the container. Thus, surface functionalization and patterning together with the development of microfluidic devices are important prerequisites for the manipulation of biological objects.

In the context of nanomanipulation, a nanorobot is a machine that carries out certain well-defined tasks on the nanoscale. Very similar to a telerobot, the machine establishes an interface between the human operator and a very distant object. The nanoworld, however, is not a planetary system in outer space, but it is a world that is far away from our macro-world in its dimension: nine orders of magnitude have to be bridged from the meter to the nanometer. Thus, we can think of a nanorobot as a (partially autonomous) machine that is equipped with sensors and effectors similar to robots used in industry. Prominent examples for such nanoscale industry robots are based on the atomic force microscope (AFM). The heart of the AFM is a very small tip that is attached to a cantilever. This nanotip can be used to image the surface or to manipulate the sample. This microscope can be used for imaging, for nanolithography, or as a nanomanipulator and nanorobot. For nano-telerobotics, an intuitive human user interface is essential. Such interfaces include haptic interfaces or augmented reality systems

(Guthold et al., 2000; Rubio-Sierra et al., 2003; Li et al., 2004; Vogl et al., 2006). The following discussion will thus focus on nanoscale manipulation with an AFM.

43.2 Atomic Force Microscope as a Nanomanipulation Platform

43.2.1 Overview

Among the family of scanning probe microscopes, the AFM (Binnig et al., 1986) provides the technology platform with a huge pool of methods for quantitative nanoscale measurements of mechanical and electrical surface properties. For imaging, the probing tip (or the sample), are a raster-scanning the specimen while the loading force of the tip is held constant in a feedback operation (Binnig et al., 1986; Hansma et al., 1988). The AFM is usually controlled by a proportional integral (PI) element (Marti et al., 1988). For raster-scanning and z positioning, an actuator with sub-nanometer resolution for all three spatial directions is required. Examples for the actuator design include tripod scanners (Binnig et al., 1986), piezoelectric tube scanners (Binnig et al., 1986; Moheimani, 2008), resonantly vibrating actuators (Humphris et al., 2005), or independent piezoelectric actuators (Ando et al., 2001) for lateral scanning (*X* and *Y* directions) and height regulation (*Z* direction).

The so-called "amplitude modulation" or "tapping mode" (Zhong et al., 1993) operation has become one of the most widely used modes for imaging in ambient conditions. Typically, the cantilever is excited at its fundamental resonance to an oscillation of typically less than 100 nm in amplitude close to the specimen surface. The interaction between the tip and sample limits the oscillatory amplitude. In order to track the surface profile, the feedback circuitry as illustrated in Figure 43.1 keeps the actual oscillatory amplitude constant and thus regulates the relative

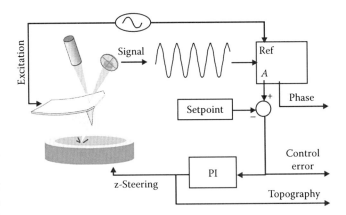

FIGURE 43.1 Feedback regulation in an amplitude modulation AFM. The force sensor is excited at its resonant frequency. The excitation (ref) and the signal are fed into a lock-in amplifier for demodulation. The control error, that is, the difference between the actual amplitude (*A*) and the amplitude set point, is fed into a proportional integral controller (PI). The output of the controller is used to steer the z-actuation of the z-actuator. The control signal is usually referred to as "topography."

tip–sample position during scanning. A mathematical description of the dynamic AFM can be found in Chapter 33 of *Principles and Methods*. The amplitude feedback scheme is in contrast to the "frequency modulation" or "non-contact" mode where the shift of the cantilever resonance is tracked (Albrecht et al., 1991). Frequency modulation (FM) has mostly been used for ultra-high vacuum experiments. Recently, FM feedback also proved valuable for measurements in a liquid environment (Fukuma et al., 2005; Uchihashi et al., 2005; Higgins et al., 2006).

The capabilities of the AFM for mechanical and electrical sensing of material properties reach far beyond imaging and visualization of the specimen. Additional channels can be added to conventional topographic imaging in order to access this "property" contrast. Examples for such a material property contrast include electrostatic (EFM), Kelvin probe (KPFM), or magnetic force microscopy (MFM). Phase imaging (Magonov et al., 1997) in amplitude modulation, that is, the acquisition of the phase lag between the driving force and the cantilever response, has become a widely used method to characterize mechanical and chemical surface properties. The phase signal carries the information of the local energy dissipation (Cleveland et al., 1998) that is determined by the chemical (Noy et al., 1998) and mechanical surface properties (Magonov et al., 1997; Magonov, 2003). Specific chemical information can be obtained by employing functionalized tips (Noy et al., 1998). For example, specific chemical interactions can be measured by recognition imaging (Stroh et al., 2004). Mechanical and chemical properties can also be probed by higher harmonics imaging and related techniques (Hillenbrand et al., 2000; Stark and Heckl, 2000; Sahin et al., 2003; Stark and Heckl, 2003; Preiner et al., 2007). The dynamic nano-indentation of the tip into a polymer surface during dynamic AFM measurements may be used for the characterization of viscoelastic properties (Stark et al., 2002; Sahin et al., 2007). These additional property channels are currently seldom used in a nanomanipulation AFM. In the future, however, it will become more and more important to automatically identify the object of interest. To this end, chemical and mechanical sensing will be important capabilities of a nanomanipulator.

43.2.2 Advanced Control Strategies

A crucial issue in the development of AFM manipulators is the limitation due to the limited imaging speed of conventional AFMs (as of 2008). This section shall illustrate the potential of modern control methods of high-speed positioning for AFM imaging and manipulation. In order to push positioning speed and precision to the limit, a combined approach is needed that includes a careful design of the actuator together with the development of efficient control schemes. In addition to the design and control of the piezo-actuator, the dynamics of the AFM force sensor needs to be addressed in order to achieve maximum control of the tip motion (Sebastian et al., 2007).

Several approaches have been shown to speed up the response of the probing tip. For instance, one can use small cantilevers with very high-resonance frequencies (Viani et al., 1999) or reduce the quality factor of the probe (Mertz et al., 1993; Sulchek et al., 2000). Additionally, every scanning probe system is limited in speed by the dynamic behavior of the actuator. The imaging speed of AFM can be enhanced by increasing the resonance frequencies of the moving parts. This approach requires nanopositioning actuators that are designed as small, light, and stiff as possible (Pohl, 1986; Ando et al., 2001; Humphris et al., 2005; Fantner et al., 2006). By designing small scanning devices, the actuation range is often limited to a relatively narrow area with a width of a few square microns in size (Ando et al., 2001; Humphris et al., 2005). In order to scan larger surface areas, the dynamic response of the actuator has to be fine-tuned. One solution is to add additional high-frequency piezo segments to the probing tip (Lapshin and Obyedkov, 1993; Manalis et al., 1996; Egawa et al., 1998) or to the piezo scanner (Knebel et al., 1997). Due to the limited range of the fast Z-piezo, both actuators are used in a nested feedback loop (Sulchek et al., 1999). The slow actuator operates the long-range action while the small but fast one compensates for small topographical details. Such a solution, however, requires two PI-loops influencing each other. Another approach to improve the positioning performance of AFM is model-based control. In most AFM, the dynamic behavior of the actuator is not compensated, and mechanical oscillations (ringing) may be excited at higher scan-rates due to the higher frequencies introduced by the triangular scanning motion (Croft and Devasia, 1999; Croft et al., 1999). In order to improve the performance, the dynamic behavior of the system has to be compensated. To this end, modern model-based control methods provide a valuable toolbox. Efforts in this direction include open-loop methods (Croft et al., 1999; Schitter and Stemmer, 2004) or model-based feedback controllers to increase the bandwidth of the AFM (Schitter et al., 2001; Salapaka et al., 2002; Schitter et al., 2004; Salapaka and Salapaka, 2008).

The mathematical challenges posed by the demand for high-speed nanopositioning can be tackled by various control engineering methods (Devasia et al., 2007; Moheimani, 2008). Mathematical models of the AFM dynamics are required for all three spatial directions in order to compensate the dynamics of the actuator and thus to improve the control of the AFM. In this approach, it is assumed that there is only negligible cross-talk between the three scan axes. The mathematical models are obtained by system identification procedures that analyze the relation between input and output data (Ljung, 1999). These models describe the system dynamics, that is, the resonances, the zeros in the transfer function, and the high-frequency roll-off. The system inputs are the voltage inputs to the piezo-drivers and the three-dimensional output is the position of the actuator in the X, Y, and Z directions. The mathematical model of the AFM scanner can be used in order to design a new controller with model-based control methods. A model-based controller accounts for the dynamic response of the piezo-scanner and compensates for it. The controller can be designed to allow for the deviations of the mathematical model from the physical plant. Such errors can, for example, be caused by the variations of the resonances due to the varying sample mass or due to a changing piezo-response due to ageing. The robustness of the controller against the variations of the physical plant is essential for the practical application of advanced control strategies in high-speed AFM imaging and AFM manipulation.

Figure 43.2a illustrates a practical implementation of model-based controllers into an existing commercial AFM (Multimode with a Nanoscope IIIa controller, Veeco, Santa Barbara USA) (Schitter et al., 2001; Schitter et al., 2004; Schitter and Stemmer, 2004). The controllers are running in an external digital signal processing (DSP) board. The height controller consists of a model-based feedback (H_∞-feedback) that has been combined with feed-forward (open loop) compensation (H_∞-feedforward). The input for the feed-forward compensation is the previously recorded scan-line that has been delayed (Δt) in order to match the current line. Thus, the model-based closed-loop controller only has to compensate deviations from scan-line to scan-line. Due to the design of the feedback that compensates the actuator dynamics, the control signal cannot be used as a topography signal. Instead, the actual topography has to be calculated from the mathematical scanner model.

In the lateral directions (X and Y), similar concepts can be used. In the actual plant, however, no sensors were available to measure the piezo position. The actuator dynamics could only be measured with additional external sensors such as capacitive sensors or a vibrometer that were used for system identification (Schitter and Stemmer, 2004) but could not be used during scanning. Thus, the scanner dynamics had to be compensated by an open-loop controller that was designed to damp the resonances in the scanning directions.

The AFM images (Figure 43.2b and c), show the distorted height-image obtained with the AFM operating with a well-tuned PI feedback in comparison to data obtained with the model-based controller. Both images of the plasmid DNA were obtained in air at a line-scan rate of 30.5 lines per second. The image of the conventionally steered AFM shows vertical stripes due to the ringing of the piezo-actuator. This ringing was caused by the rapid change between forward and backward motion. The model-based feed-forward controller suppressed ringing and the height imaging could thus be improved.

43.2.3 Nanomanipulator Designs

A simple and straightforward manipulation procedure that can be carried out with most AFM without further modifications,

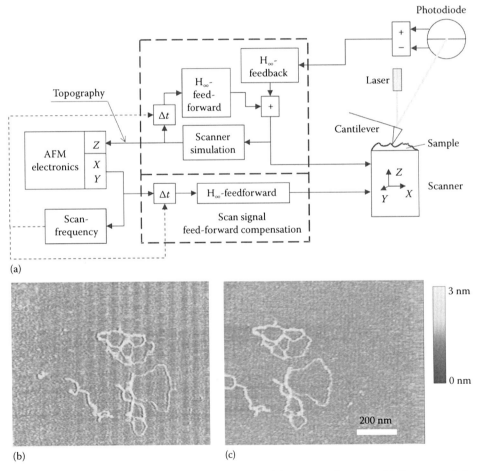

FIGURE 43.2 (a) Scheme of a model-based controller for the compensation of the actuator dynamics. (b) Contact mode image of plasmid DNA in air at a line-scan rate of 30.5 Hz. Ripples due to scanner vibrations distort the image. (c) With the model-based scanner actuation and feedback, the image is smooth even at a scan rate of 30.5 Hz. (Reprinted from Schitter, G. et al., *Ultramicroscopy*, 100, 253, 2004. With permission.)

includes the localization and imaging of the area of interest, followed by a "blind" feed-forward manipulation. Success or failure of the manipulation is determined afterward by reimaging the manipulated area. Such an approach has some drawbacks. First, the probe or sample may change during the manipulation process or the sample drifts out of the defined region. Due to the non-linearity of the tip-sample contact, it will be difficult to predict success or failure of the manipulation. Thus, a time-consuming trial-and-error approach is required. In this situation, elements of virtual reality can assist the human user. For example, haptic interfaces that generate a feedback of the tip-sample forces provide the user with a "feeling" for surface morphology and loading forces during the manipulation process (Guthold et al., 2000; Jobin et al., 2003; Rubio-Sierra et al., 2003; Rubio-Sierra et al., 2005). Entire synthetic worlds facilitate the orientation within the nanoworld by an augmented reality (Li et al., 2004; Luciani et al., 2004; Vogl et al., 2006; Li et al., 2007).

Three types of AFM-based nanomanipulators can be identified:

1. Nanolithography robots. This class includes AFM that are equipped with the hard- and software to execute predefined lithography and manipulation tasks. With these machines, manipulation is usually done in an open-loop fashion, although, there is often a height feedback in order to track the surface profile with the manipulating tip.
2. Nano-telerobots. For example, AFMs with a haptic interface provide an additional feedback to the user. Thus, the user can feel the tip interacting with the sample. The AFM with integrated augmented reality systems provide additional feedback to the user.
3. Autonomous robots. Future developments will include AFM robots that carry out predefined tasks such as arranging nanoparticles to a certain pattern or pulling and analyzing biomolecules (Struckmeier et al., 2008). These robots will be capable of identifying the region of interest, manipulating the object, and controlling success or failure of the manipulation attempt without the user's intervention. Such an approach is, especially, desirable in the case where a large number of highly repetitive manipulation tasks have to be carried out.

For local manipulation in a nanolithography mode, a set of vectors is needed to describe the tip trajectory. The vector path is often created by programs that are part of the AFM software, or by special scripting languages. There are also programs to convert a standard windows metafile into vectors for AFM lithography. The strength and type of manipulation (force assisted or electrical) can be encoded as color in the graphics file (Janusz et al., 2004). Apart from lithography, the vector mode also can be used for a particle manipulation. Single vectors or a series of vectors can be used to exert lateral forces on a nano-object at defined positions. In contrast to the vector-based lithography modes, a point-based lithography can also be used (Jacobs and Stemmer, 1999). Here, a grayscale bitmap serves as a control file. The gray scale encodes the strength of the manipulation. In the nanolithography modes, however, there is no real-time control of the manipulation success. The manipulation process is carried out in a feed-forward manner.

Nano-telerobots provide the experimentalist with additional information on the forces between the tip and the sample, and additional user interfaces are required. A haptic interface can be used in order to transmit forces to the user. Currently, most haptic interfaces are designed as add-ons to commercially available AFMs (Guthold et al., 2000; Jobin et al., 2003; Rubio-Sierra et al., 2003; Rubio-Sierra et al., 2005). Typically, the electronic control unit of the microscope generates the scanning waveform and processes the topographic data during imaging. An additional manipulation system generates the signals that are necessary to take control of the piezoelectric actuators during manipulation.

The scheme of a practical implementation of such a manipulation system is shown in Figure 43.3. In this system, a commercial AFM is equipped with two additional joysticks for steering the tip. The lateral (x and y) motion is steered by a conventional joystick. This joystick is used to control the velocity of the moving tip on the surface. The user is informed about the actual tip position by a marker in a previously acquired AFM image on the computer screen. The AFM image is updated upon a request by the user. A second force-feedback joystick is used to control the loading force of the tip on the sample. With this device, the user can control the vertical velocity of the piezo-actuator. This approach

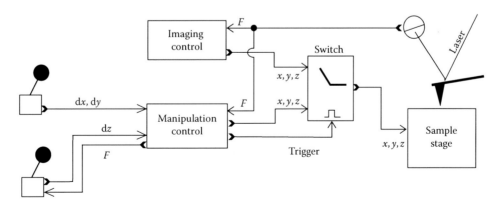

FIGURE 43.3 Simplified scheme of the control circuitry of an AFM equipped with a haptic interface. (Reprinted from Stark, R.W. et al., *Eur. Biophys. J.*, 32, 33, 2003. With permission.)

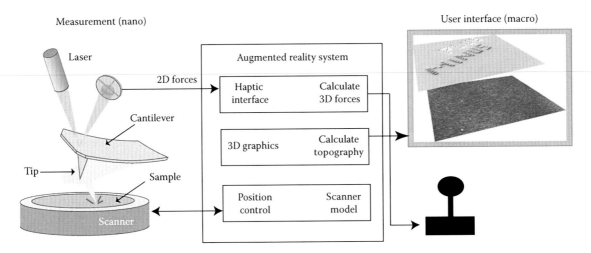

FIGURE 43.4 Concept of an augmented reality system for atomic force microscopy.

implies that the control unit of the AFM (imaging control) generates the waves x, y, and z and processes the scanning information (F) during imaging. The manipulation system converts the input from the positioning joystick (dx, dy) and the force-feedback joystick (dz) into analog signals and generates a corresponding waveform during manipulation (x, y, z). The manipulation computer also sets the return force in the force feedback joystick by D/A conversion of the interactive signal (F) and manages the graphical interface during manipulation. A digital signal processor controls the communication process with the AFM system. An electronic switch allows the user to switch between the imaging and manipulation modes as a response to the logic signal (trigger). This element is designed to allow for a seamless transition between both operation modes and to avoid sudden uncontrolled tip movements due to switching. In the switch element, the scanning waveform is fed into a sample and hold-stage that is controlled by a logic input. After the sample and hold-stage, the scanning waveform is summed to the manipulation waveform. As a force-feedback-joystick, a Wingman Force of Logitech was used. A system of motors and magnetic coils communicates the forces to the user's hand. By using DirectX®, a software interface was developed, which produces a force along the vertical axis of the joystick. The magnitude and direction of the force action can be changed with an acquisition time of 1 ms. In order to use the joystick as the haptic interface between the tip and the user, the force signal is acquired, scaled by the manipulator control system and returned as a force to the user. Thus, the feedback force is proportional to the tip-loading force. The delay between the returned force and the interactive signal is less than 2 ms and gives a realistic feedback to the user. In order to achieve a real-time haptic feedback, however, a bandwidth of about 3 kHz would be required (Jobin et al., 2003).

Although, haptic user interfaces provide feedback about the tip-sample forces, additional visual information can help to improve the nanomanipulation process. In the context of nano-telerobotics, the nanoworld can be considered as a cyberworld (Luciani et al., 2004). Currently, there are no independent means to obtain images with nanometer resolution in real time, for example, by a light or electron microscope. The operator working with a nanomanipulator thus has to rely entirely on a computer created image. This implies a fundamental problem when using an AFM as a nano-telerobot. The tip of the AFM is the sensor and end effector at the same time. The tip, however, can only be used to either image the specimen or to manipulate an object at the same time. During manipulation, the operator thus lacks most of the sensor information apart from force. The augmented reality can help to overcome this limitation. This type of virtual reality combines the measured information about the environment with the simulated data (Figure 43.4). Augmented reality systems generate a composite view by combining real (measured) data with a simulated prediction of the environment. The composite virtual scene on the computer screen helps to enhance the user's perception of the nano-environment. Augmented reality systems for nanomanipulation can, for example, be used to predict the deformation of the sample (Vogl et al., 2006) or the displacement of a nano-object. The deformation can be included in the visual environment and gives a more realistic picture of the manipulation process. Another approach is to locally update the AFM image with the data obtained from the real-time force signal (Li et al., 2004). The augmented reality provides a real-time visual display of the nano-environment. Mathematical models for the manipulation are used to predict the behavior of the manipulated object. The AFM image is then updated locally where changes due to the manipulation process are predicted. This approach makes a video-rate visual feedback feasible (Li et al., 2005), which otherwise cannot be achieved with the current AFM.

43.3 Application Examples

43.3.1 Mask-Less Nanolithography

Surface patterning at the smallest scale is the key for the further development of technological and scientific applications that require rapid prototyping of nanostructures. Currently, optical lithography is the standard technique for the fabrication of structured surfaces. However, the lateral resolution of optical lithography is limited by diffraction. To overcome this limitation, alternative surface patterning methods such as electron beam lithography, masked deposition, scanning probe techniques, or microcontact printing have been developed. Scanning probe microscope–related techniques represent the most versatile approach for rapid prototyping of lateral nanoscale structures (Xie et al., 2006). In the lithography mode of operation, the tip follows a given path while the tip modifies the surface. The surface can be modified mechanically (nanomachining), chemically, or electrically.

Various approaches of mechanical machining exist. For all methods, the loading force of the tip on the surface is a crucial parameter. One can identify mechanical plowing and indentation lithography. In plowing lithography (Jung et al., 1992) the tip machines the surface along a predefined vector trajectory, while in indentation lithography individual points are written by indentation. In plowing lithography, the material can be deformed or removed with high-spatial resolution as can be seen in Figure 43.5. To understand the mechanical machining process one has to keep in mind that the contact area between the tip and the sample amounts only to a few square nanometers. Assuming a typical radius of contact of 10 nm between the tip and the sample, the contact area is only about 300 nm². A loading force of 1 μN thus generates an average contact stress of 3 GPa. At about 10 GPa, even fused silica can be modified because it yields plastically. Thin polymer resist films can be modified much more easily by using such a mechanical machining method. In a subsequent wet chemical etch process the desired nanostructure can be created (Jung et al., 1992). The main parameters that influence the machining process are the applied loading force and the tip speed.

Mechanical AFM nanomachining was demonstrated on a wide variety of materials including metals (Fang and Chang, 2003), semiconductors (Cortes Rosa et al., 1998), polymers (Munz et al., 2003), and biological materials (Stark et al., 1998). Selective removal of superficial layers to study the internal structure of biological specimens has been achieved revealing the internal structure of collagen (Wen and Goh, 2004; Wenger et al., 2008) or a bacterial cell wall (Firtel et al., 2004). Nanomachining of thin polymer resistant films with an AFM is a promising route for the fabrication of nanoscale devices.

For a better control of the machining process, the machining force can be modulated at ultrasonic frequencies in the so-called dynamic plowing lithography (Kunze, 2002; Rubio-Sierra et al., 2006). Figure 43.5a illustrates a set-up where an in-plane acoustic wave is used in order to enhance the machining process. The lateral resolution reached by this acoustical force nanolithography is only limited by the physical size of the AFM tip. Along with the loading force, frequency, and amplitude of the acoustic wave are crucial process parameters (Rubio-Sierra et al., 2006). Figure 43.5b shows a structure that was written into a thin polymer resist layer by this type of acoustic force nanolithography.

The local generation of nanostructures can also be achieved by applying a bias voltage between tip and specimen. In the presence of a water layer, thin oxidic structures can be created by local anodic oxidation on a silicon surface. This bias induces a current through the thin water film-connecting tip and sample. Several methods have been proposed for the local modification of surfaces by the AFM. One powerful patterning strategy is local anodic oxidation as illustrated in Figure 43.6. Electronic devices such as nanowires, metal-oxide field effect transistors (MOSFET), single-electron devices (Irmer et al., 1998), and quantum point contacts can be fabricated (Wouters and Schubert, 2004). The control of the growth direction of organic molecule monolayers with conducting properties is a fundamental issue for the development of molecular electronics devices that can be solved using a silicon oxide template fabricated by local anodic oxidation (García et al., 2004). Figure 43.6b shows a pattern that was generated by the local growth of silicon oxide on the silicon substrate. A water meniscus between the tip and the sample that forms due to capillary condensation acts as an electrolyte and promotes the oxidation.

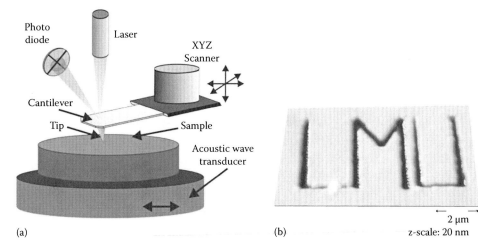

(a) (b)

FIGURE 43.5 (a) Scheme of acoustical force nanolithography. The sample holder consists of an acoustic wave transducer that is used to enhance cantilever flexural vibrations for lithography. (Reprinted from Rubio-Sierra, F.J. et al., *Phys. Stat. Sol. (a)*, 203, 1481, 2006. With permission.) (b) Nanostructures generated by acoustical force nanolithography on a PMMA resist on a silicon substrate.

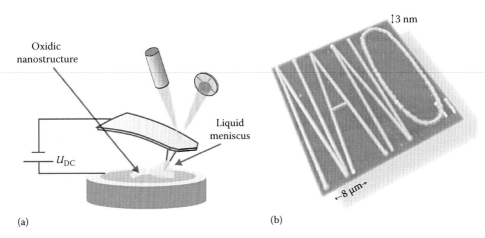

(a)   (b)

FIGURE 43.6 (a) Local anodic oxidation by AFM. A bias voltage U_{DC} is applied between the tip and the sample. The current through the water meniscus between the tip and the sample induces a local oxidation on the substrate. (b) "Nano" written by local anodic oxidation on a silicon surface.

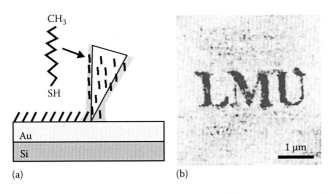

(a)   (b)

FIGURE 43.7 (a) In dip-pen nanolithography, the AFM tip is wetted by a molecular ink that is transferred to the substrate during the writing process. (b) Friction force image of the letters LMU written with alkanethiols on a gold substrate by dip-pen nanolithography.

Chemical surface modification by AFM nanolithography can also be achieved by directly writing molecules onto the surface by dip pen nanolithography (Lee et al., 2002). Figure 43.7 shows a molecular pattern that was generated by the dip-pen nanolithography of 1-hexadecanethiol on a gold substrate. In substitution lithography (nanografting and nanoshaving), self-assembled monolayers are removed by means of mechanical desorption by the AFM tip. The removal is followed by *in situ* replacement with a second component (Krämer et al., 2003; Tinazli et al., 2007).

Nanostructures also can be created by locally modifying a material property. One prominent example is charge writing (Jacobs and Stemmer, 1999; Mesquida et al., 2002). Here, the tip of the AFM is used to locally inject electrical charges (electrons or holes) into an electret. Charge writing in thin electret films is relevant for information storage (Stern et al., 1988; Terris et al., 1989; Terris et al., 1990; Jacobs and Stemmer, 1999). Pre-patterning for a guided self-assembly (Mesquida and Stemmer, 2001) is an example for fabrication and analysis of artificial nanostructures by taking advantage of electrical properties. Figure 43.8 shows a charge pattern created on a polymethylmethacrylate (PMMA) together with the surface topography.

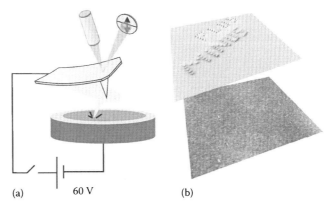

(a)   60 V   (b)

FIGURE 43.8 (a) Charge writing. Electrical charges are injected by a short electrical pulse during imaging. (b) Electrical charges written in a PMMA. Upper image: Surface potential as measured with Kelvin probe microscopy. Lower image: surface topography. (Image size $5\,\mu m \times 5\,\mu m$.)

The nanostructures that can be created by the tip-based various lithography methods can be modified in further processing steps in order to fabricate a functional nanodevice. They may also serve as guiding patterns for molecules used in a nanoelectronic device, nanoparticles, or biomolecules such as proteins for use as future biosensors (Mesquida and Stemmer, 2001; Mesquida and Stemmer, 2002; Mesquida et al., 2006). Tip-based nanostructuring also can be used as a preparation step in order to create the setting for a particle nanomanipulation.

43.3.2 Particle Manipulation

Another important application of AFM nanomanipulation is the arrangement of small particles and atoms on surfaces. For the placement of nanoparticles, the tip is used to push the nanoparticle to the desired location. In order to avoid sticking of the particles to the tip, anti-stiction coatings are used. For imaging, the AFM is operated in a conventional dynamic AFM mode (amplitude or frequency modulation). In order to manipulate the particle the interaction forces between the tip and sample have to be increased. To achieve this goal, several strategies exist (Ramachandran et al., 1997; Requicha, 2003). The simplest approach is to switch the topographic feedback off during manipulation. Thus, the tip follows the preprogrammed trajectory to push the nanoparticle. In the "feedback-off mode," the experimentalist additionally has the option to vary the oscillation amplitude. Without feedback, however, manipulation over large distances becomes difficult to control due to drift. To overcome this limitation, there is also the option to leave the feedback on but to adjust the feedback gains to small values. The gain values are then set to parameters where the AFM still tracks the surface on a larger timescale compensating for drift and slowly varying the topography. The slow feedback does not respond to rapidly varying topographic features thus the tip exerts lateral forces on the nanoparticle.

A more sophisticated approach to particle manipulation is not to vary the feedback parameters but the amplitude set point to increase the forces. In order to control the direction of the particle motion, the particle can be pushed. Direct pushing is, however, difficult to control. In a more advanced approach, the particle is pushed by a tip moving perpendicular to the desired trajectory (Requicha, 2003), as illustrated in Figure 43.9a. The basic idea is that the particle motion in the direction of the tip motion during an up- and a down-scan cancel each other. Only particle movements perpendicular to the tip-motion direction sum up. With this procedure, the motion of the particle can be controlled very precisely.

Particle pushing requires that the oscillation amplitude of the tip be well controlled. In dynamic AFM under ambient conditions, the free amplitude of the tip oscillation far away from the sample is on the order of 20–50 nm. For imaging, the set point is adjusted to about 80%–90% of this value, which ensures gentle imaging conditions. In order to push particles, the amplitude set point is reduced further, ideally to a value smaller than the particle size. Figure 43.9b illustrates the principle. The oscillating tip approaches the nanoparticle. The oscillation amplitude of the tip is reduced due to the interaction with the particle (only the

tip trajectory is drawn) before the feedback can respond. Thus, strong lateral forces are transmitted to the particle pushing it forward.

43.3.3 Optical and Mechanical Nanodissection of DNA

Mechanical nanomanipulation also finds its application in the biological sciences. An application example from the field of cytogenetics shall illustrate the potential of the tip-based nanomanipulation in combination with other manipulation tools: recovery of DNA fragments by chromosomal dissection for cytogenetic studies (Thalhammer et al., 1997a; Thalhammer et al., 1997b; Stark et al., 1998; Schermelleh et al., 1999; Lu et al., 2004; Tsukamoto et al., 2005; Tsukamoto et al., 2006). A first demonstration of micromanipulation of chromosomes by laser beams has already been published in the late 1960s. A pulsed argon laser microbeam was used to manipulate chromosomes (Berns et al., 1969). Later, it was shown that the DNA libraries for painting of chromosomal regions can be generated by microcloning laser-microdissected DNA (Lengauer et al., 1991). Chromosomal DNA can also be collected by manipulation with an ultraviolet (UV) microbeam laser (Schütze et al., 1997; Thalhammer et al., 1997b; Schermelleh et al., 1999; Clement Sengewald et al., 2000; Greulich et al., 2000). The fundamental process of UV-laser dissection is a locally restricted ablative photodecomposition without heating (Srinivasan, 1986; Greulich, 1999; Bäuerle, 2000). Other laser-surgery methods employ infrared or near-infrared lasers (Emmert Buck et al., 1996). Multiphoton induced ablation by 800 nm fs laser pulses is a third viable approach for high-resolution optical dissection (König et al., 2001). The resolution of the light microscope used in manipulation experiments is diffraction limited. The resolution of the AFM, however, is only limited by the diameter of the scanning tip. Complementary to laser dissection the AFM can be used for high-resolution imaging and precise mechanical nanomanipulation. Combined instruments unite the resolution of AFM with the ease of use of the light microscope (Putman et al., 1993; Hillner et al., 1995; Thalhammer et al., 1997b; Stark et al., 2003).

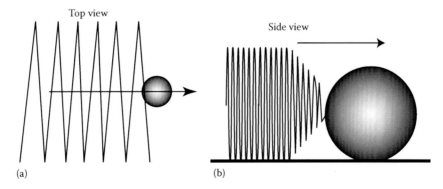

(a) (b)

FIGURE 43.9 Concept of mechanical particle placement by an oscillating AFM tip. (a) Top view: The tip traces the particle in a zigzag motion. Each time the tip hits the particle, the particle is moved forward. (b) Side view: The oscillation trajectory of the tip is sketched by the line. Since the tip has similar dimensions as the particle, the oscillation of the tip is reduced due to the interaction with the tip. This amplitude reduction leads to lateral forces acting on the tip.

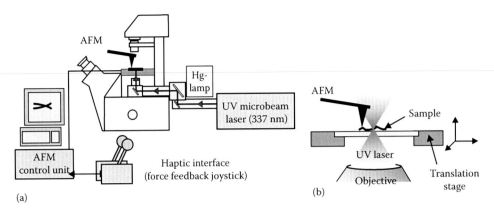

(a)   (b)

FIGURE 43.10 Combined nanomanipulator for life science applications. (a) A multifunctional scanning probe microscope is mounted onto an inverted research microscope. A pulsed UV microbeam laser is coupled into the microscope through the epi-illumination path. A force feedback joystick serves as a simple haptic interface. (b) The scanning stage can be positioned with nanometer resolution in three dimensions with piezo-electric actuators. (From Stark, R.W. et al., *Eur. Biophys. J.*, 32, 33, 2003. With permission.)

Figure 43.10 schematically shows a combined nanomanipulator for applications in life sciences. It integrates a light microscope, a UV-laser microbeam, and an AFM. Optical and mechanical manipulation can be controlled by a haptic interface. The laser—a nitrogen UV-laser ($\lambda = 337.1$ nm)—is coupled into the epi-illumination path. A pulse-width less than 4 ns can be achieved at a pulse energy of 300 µJ. Figure 43.11 shows dissected human metaphase chromosomes. The chromosome to the right was dissected with the UV-laser. A cut width of 380 nm ± 20 nm (full width at half-maximum cut depth) was achieved in cut #UV1. For AFM nanodissection, a medium sized sub-meta-centric chromosome was selected. Different cut depths were realized with the AFM tip. At a loading force of 10 µN only a shallow scratch was generated (cut #2). By increasing the loading force to 20 µN, a 50 nm deep and 170 nm wide cut was achieved (#1). At forces of 40 µN, the chromosome was fully dissected (#3).

In summary, the dissection of genetic material with cut widths close to the diffraction limit of the light microscope were achieved by optical manipulation with a UV-laser as well as by mechanical manipulation with the atomic-force microscope. The results show the resolution capability of UV-laser ablation. With a standard nitrogen laser system, a cut width less than 380 nm is possible in genetic material. The scanning tip of the atomic-force microscope can easily be used for micromanipulation with sub-wavelength resolution. The experiments also demonstrate that a semi-automated nanodissection environment is feasible, where the operator performs the dissection using a haptic interface. A crucial task in a fully automated environment is the identification of the chromosomes from scanning probe data. Although the classification can be done manually for banded chromosomes (Thalhammer et al., 2001), computer-based procedures (Lerner, 1998) will further facilitate chromosomal nanodissection.

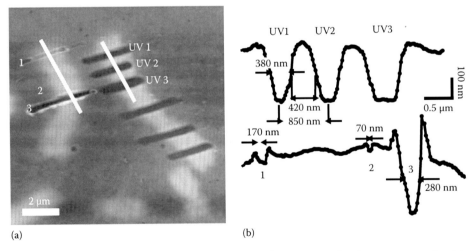

(a)   (b)

FIGURE 43.11 (a) Topographic AFM image of dissected human metaphase chromosomes. The chromosome on the right was dissected by the UV-laser microbeam. The chromosome on the left was dissected by AFM (loading forces: 1, 20 µN; 2 10 µN; B3 40 µN). (b) Cross-sectional analysis. (From Stark, R.W. et al., *Eur. Biophys. J.*, 32, 33, 2003. With permission.)

43.4 Conclusion

Although there are promising approaches for nanohandling and nanomanipulation, the field of nanorobotics is still in an early stage of development. Various research groups work on relevant topics ranging from mechanical and control engineering over surface chemistry to biology. This demonstrates that the development of nanorobots and their application to current research problems also requires a profound knowledge in several areas of science and engineering. Progress in the field of nanorobotics will thus depend on the integration of interdisciplinary knowledge.

From a technological point of view, future developments will lead to massively parallel systems where many sensors and tools can be addressed simultaneously as it is already the case in the data storage system, "Millipede." Certainly, integrated systems will unite the control electronics and the nanoactors together with the nanorobot on a single chip. Together with the chemical and biochemical tools, a lab on a chip system for nanorobotics seems feasible on the long term. Recent progress in the manipulation of individual atoms (Sugimoto et al., 2008) even lets us dream to shape the world atom by atom (Roco, 1999).

References

Albrecht, T.R., P. Grutter, D. Horne, and D. Rugar. 1991. Frequency-modulation detection using high-q cantilevers for enhanced force microscope sensitivity. *J. Appl. Phys.* 69: 668–673.

Ando, T., N. Kodera, E. Takai et al. 2001. A high-speed atomic force microscope for studying biological macromolecules. *Proc. Natl. Acad. Sci. USA.* 98: 12468–12472.

Bäuerle, D. 2000. *Laser Processing and Chemistry.* Springer Verlag, Berlin, Germany.

Berns, M.W., R.S. Olson, and D.E. Rounds. 1969. In vitro production of chromosomal lesions with an argon laser microbeam. *Nature.* 221: 74–75.

Binnig, G., C.F. Quate, and C. Gerber. 1986. Atomic-force microscope. *Phys. Rev. Lett.* 56: 930–933.

Clement Sengewald, A., T. Buchholz, and K. Schütze. 2000. Laser microdissection as a new approach to prefertilization genetic diagnosis. *Pathobiology.* 68: 232–236.

Cleveland, J.P., B. Anczykowski, A.E. Schmid, and V.B. Elings. 1998. Energy dissipation in tapping-mode atomic-force microscopy. *Appl. Phys. Lett.* 72: 2613–2615.

Cortes Rosa, J., M. Wendel, H. Lorenz et al. 1998. Direct patterning of surface quantum wells with an atomic force microscope. *Appl. Phys. Lett.* 73: 2684–2686.

Croft, D. and S. Devasia. 1999. Vibration compensation for high speed scanning tunneling microscopy. *Rev. Sci. Instrum.* 70: 4600–4605.

Croft, D., S. Stilson, and S. Devasia. 1999. Optimal tracking of piezo-based nanopositioners. *Nanotechnology.* 10: 201–208.

Devasia, S., E. Eleftheriou, and S.O.R. Moheimani. 2007. A survey of control issues in nanopositioning. *IEEE Trans. Control Syst. Technol.* 15: 802–823.

Egawa, A., N. Chiba, K. Homma, K. Chinone, and H. Muramatsu. 1998. High-speed scanning by dual feedback control in snom afm. In *5th International Conference on Near Field Optics and Related Techniques (NFO-5)*, Shirahama, Japan, pp. 325–328.

Emmert Buck, M.R., R.F. Bonner, P.D. Smith et al. 1996. Laser capture microdissection. *Science.* 274: 998–1001.

Fang, T.H. and W.J. Chang. 2003. Effects of afm-based nanomachining process on aluminum surface. *J. Phys. Chem. Solids.* 64: 913–918.

Fantner, G.E., G. Schitter, J.H. Kindt et al. 2006. Components for high speed atomic force microscopy. *Ultramicroscopy.* 106: 881–887.

Firtel, M., G. Henderson, and I. Sokolov. 2004. Nanosurgery: Observation of peptidoglycan strands in Lactobacillus helveticus cell walls. *Ultramicroscopy.* 101: 105–109.

Fukuma, T., K. Kobayashi, K. Matsushige, and H. Yamada. 2005. True molecular resolution in liquid by frequency-modulation atomic force microscopy. *Appl. Phys. Lett.* 86: 034103.

García, R., M. Tello, J.F. Moulin, and F. Biscarini. 2004. Size and shape controlled growth of molecular nanostructures on silicon oxide templates. *Nano Lett.* 4: 1115–1119.

Geissler, M. and Y.N. Xia. 2004. Patterning: Principles and some new developments. *Adv. Mater.* 16: 1249–1269.

Greulich, K.O. 1999. *Micromanipulation by Light in Biology and Medicine: The Laser Microbeam and Optical Tweezers.* Birkhäuser, Basel, Switzerland.

Greulich, K.O., G. Pilarczyk, A. Hoffmann et al. 2000. Micromanipulation by laser microbeam and optical tweezers: From plant cells to single molecules. *J. Microsc.* 198: 182–187.

Guthold, M., M.R. Falvo, W.G. Matthews et al. 2000. Controlled manipulation of molecular samples with the nanomanipulator. *IEEE/ASME T. Mech.* 5: 189–198.

Hansma, P.K., V.B. Elings, O. Marti, and C.E. Bracker. 1988. Scanning tunneling microscopy and atomic force microscopy—Application to biology and technology. *Science.* 242: 209–216.

Heckl, W.M. 2004. Molecular self assembly and nanomanipulation—Two key technologies in nanoscience and templating. *Adv. Eng. Mater.* 6: 843–847.

Higgins, M.J., J.E. Sader, and S.P. Jarvis. 2006. Frequency modulation atomic force microscopy reveals individual intermediates associated with each unfolded i27 titin domain. *Biophys. J.* 90: 640–647.

Hillenbrand, R., M. Stark, and R. Guckenberger. 2000. Higher-harmonics generation in tapping-mode atomic-force microscopy: Insights into the tip-sample interaction. *Appl. Phys. Lett.* 76: 3478–3480.

Hillner, P.E., M. Radmacher, and P.K. Hansma. 1995. Combined atomic force and scanning reflection interference contrast microscopy. *Scanning.* 17: 144–147.

Humphris, A.D.L., M.J. Miles, and J.K. Hobbs. 2005. A mechanical microscope: High-speed atomic force microscopy. *Appl. Phys. Lett.* 86: 034106.

Irmer, B., M. Kehrle, H. Lorenz, and J.P. Kotthaus. 1998. Nanolithography by non-contact afm-induced local oxidation: Fabrication of tunnelling barriers suitable for single-electron devices. *Semicond. Sci. Technol.* 13: A79–A82.

Jacobs, H.O. and A. Stemmer. 1999. Measuring and modifying the electric surface potential distribution on a nanometre scale: A powerful tool in science and technology. *Surf. Interf. Anal.* 27: 361–367.

Janusz, L., K. Saveen, S.P. Sunil et al. 2004. Data coding tools for color-coded vector nanolithography. *Rev. Sci. Instrum.* 75: 4646–4650.

Jobin, M., R. Foschia, and A. Kulik. 2003. Nano-scale force feedback manipulator. 12th International Conference on scanning tunneling microscopy/spectroscopy and related techniques (STM'03), Eindhoven (Netherlands). *AIP Conference Proceedings.* 696: 223–226.

Jung, T.A., A. Moser, H.J. Hug et al. 1992. The atomic force microscope used as a powerful tool for machining surfaces. *Ultramicroscopy.* 42: 1446–1451.

Knebel, D., M. Amrein, K. Voigt, and R. Reichelt. 1997. A fast and versatile scan unit for scanning probe microscopy. *Scanning.* 19: 264–268.

König, K., I. Riemann, and W. Fritzsche. 2001. Nanodissection of human chromosomes with near-infrared femtosecond laser pulses. *Opt. Lett.* 26: 819–21.

Krämer, S., R.R. Fuierer, and C.B. Gorman. 2003. Scanning probe lithography using self-assembled monolayers. *Chem. Rev.* 103: 4367–4418.

Kunze, U. 2002. Nanoscale devices fabricated by dynamic ploughing with an atomic force microscope. *Superlattice Microst.* 31: 3–17.

Lapshin, R.V. and O.V. Obyedkov. 1993. Fast-acting piezoactuator and digital feedback loop for scanning tunneling microscopes. *Rev. Sci. Instrum.* 64: 2883–2887.

Lee, K.B., S.J. Park, C.A. Mirkin, J.C. Smith, and M. Mrksich. 2002. Protein nanoarrays generated by dip-pen nanolithography. *Science.* 295: 1702–1705.

Lengauer, C., A. Eckelt, A. Weith et al. 1991. Painting of defined chromosomal regions by in situ suppression hybridization of libraries from laser-microdissected chromosomes. *Cytogenet. Cell Genet.* 56: 27–30.

Lerner, B. 1998. Toward a completely automatic neural-network-based human chromosome analysis. *IEEE Trans. Syst. Man Cyber. Pt B.* 28: 544–552.

Li, G.Y., N. Xi, M.M. Yu, and W.K. Fung. 2004. Development of augmented reality system for afm-based nanomanipulation. *IEEE/ASME T. Mech.* 9: 358–365.

Li, G.Y., N. Xi, H.P. Chen, C. Pomeroy, and M. Prokos. 2005. "Videolized" Atomic force microscopy for interactive nanomanipulation and nanoassembly. *IEEE T. Nanotechnol.* 4: 605–615.

Li, G.Y., L.Q. Liu, N. Xi, and ASME. 2007. Augmented reality enhanced nanomanipulation by atomic force microscopy with local scan. In *ASME International Conference on Manufacturing Science and Engineering*, Atlanta, GA, pp. 643–652.

Ljung, L. 1999. *System Identification—Theory for the User.* PTR Prentice Hall, Upper Saddle River, NJ.

Lu, J.H., H.J. An, H.K. Li et al. 2004. Nanodissection, isolation, and pcr amplification of single DNA molecules. In *Chinese-German Forum on Fundamentals and Technological Perspectives of Nanoscience*, Beijing, China, pp. 1010–1013.

Luciani, A., D. Urma, S. Marliere, and J. Chevrier. 2004. Presence: The sense of believability of inaccessible worlds. *Comput. Graph. UK.* 28: 509–517.

Magonov, S.N. 2003. Visualization of polymer structures with atomic force microscopy. *In Applied Scanning Probe Methods.* B. Bhushan, H. Fuchs, and S. Hosaka, eds. Springer, Heidelberg, Germany.

Magonov, S.N., V. Elings, and M.H. Whangbo. 1997. Phase imaging and stiffness in tapping-mode atomic-force microscopy. *Surf. Sci.* 375: L385–L391.

Manalis, S.R., S.C. Minne, and C.F. Quate. 1996. Atomic force microscopy for high speed imaging using cantilevers with an integrated actuator and sensor. *Appl. Phys. Lett.* 68: 871–873.

Marti, O., S. Gould, and P.K. Hansma. 1988. Control electronics for atomic force microscopy. *Rev. Sci. Instrum.* 59: 836–839.

Mertz, J., O. Marti, and J. Mlynek. 1993. Regulation of a micro-cantilever response by force feedback. *Appl. Phys. Lett.* 62: 2344–2346.

Mesquida, P. and A. Stemmer. 2001. Attaching silica nanoparticles from suspension onto surface charge patterns generated by a conductive atomic force microscope tip. *Adv. Mater.* 13: 1395–1398.

Mesquida, P. and A. Stemmer. 2002. Guiding self-assembly with the tip of an atomic force microscope. *Scanning.* 24: 117–120.

Mesquida, P., H.F. Knapp, and A. Stemmer. 2002. Charge writing on the nanometre scale in a fluorocarbon film. *Surf. Interf. Anal.* 33: 159–162.

Mesquida, P., E.M. Blanco, and R.A. McKendry. 2006. Patterning amyloid peptide fibrils by afm charge writing. *Langmuir.* 22: 9089–9091.

Moheimani, S.O.R. 2008. Invited review article: Accurate and fast nanopositioning with piezoelectric tube scanners: Emerging trends and future challenges. *Rev. Sci. Instrum.* 79: 071101.

Munz, M., B. Cappella, H. Sturm, M. Geuss, and E. Schulz. 2003. Materials contrasts and nanolithography techniques in scanning force microscopy (sfm) and their application to polymers and polymer composites. In *Advances in Polymer Science*, Vol. 164. A. Abe, A.-C. Albertsson, R. Duncan et al., eds. Springer-Verlag, Heidelberg, Germany.

Noy, A., C.H. Sanders, D.V. Vezenov, S.S. Wong, and C.M. Lieber. 1998. Chemically-sensitive imaging in tapping mode by chemical force microscopy: Relationship between phase lag and adhesion. *Langmuir.* 14: 1508–1511.

Pohl, D.W. 1986. Some design criteria in scanning tunneling microscopy. *IBM J. Res. Dev.* 30: 417–427.

Preiner, J., J.L. Tang, V. Pastushenko, and P. Hinterdorfer. 2007. Higher harmonic atomic force microscopy: Imaging of biological membranes in liquid. *Phys. Rev. Lett.* 99: 046102.

Putman, C.A.J., A.M. Van Leeuwen, B.G. De Grooth et al. 1993. Atomic force microscopy combined with confocal laser scanning microscopy: A new look at cells. *Bioimaging.* 1: 63–70.

Ramachandran, T.R., C. Baur, A. Bugacov et al. 1997. Direct and controlled manipulation of nanometer-sized particles using the non-contact atomic force microscope. In *5th Foresight Conference on Molecular Nanotechnology*, Palo Alto, CA. 237–245.

Requicha, A.A.G. 2003. Nanorobots, nems, and nanoassembly. *Proc. IEEE.* 91: 1922–1933.

Roco, M.C. 1999. Nanotechnology: Shaping the world atom by atom. In *Interagency Working Group on Nanoscience Engineering and Technology (IWGN)*, Washington, DC.

Rubio-Sierra, F.J., R.W. Stark, S. Thalhammer, and W.M. Heckl. 2003. Force-feedback joystick as a low-cost haptic interface for an atomic-force-microscopy nanomanipulator. *Appl. Phys. A.* 76: 903–906.

Rubio-Sierra, F.J., W.M. Heckl, and R.W. Stark. 2005. Nanomanipulation by atomic force microscopy. *Adv. Eng. Mater.* 7: 193–196.

Rubio-Sierra, F.J., A. Yurtsever, M. Hennemeyer, W.M. Heckl, and R.W. Stark. 2006. Acoustical force nanolithography of thin polymer films. *Phys. Stat. Sol. (a).* 203: 1481–1486.

Sahin, O., G. Yaralioglu, R. Grow et al. 2003. Harmonic cantilevers for nanomechanical sensing of elastic properties. In *IEEE International Solid State Sensors and Actuators Conference*, Vol. 2, Boston, MA.

Sahin, O., S. Magonov, C. Su, C.F. Quate, and O. Solgaard. 2007. An atomic force microscope tip designed to measure time-varying nanomechanical forces. *Nat. Nanotechnol.* 2: 507–514.

Salapaka, S.M. and M.V. Salapaka. 2008. Scanning probe microscopy. *IEEE Control Syst. Mag.* 28: 65–83.

Salapaka, S., A. Sebastian, J.P. Cleveland, and M.V. Salapaka. 2002. High bandwidth nano-positioner: A robust control approach. *Rev. Sci. Instrum.* 73: 3232–3241.

Schermelleh, L., S. Thalhammer, W. Heckl et al. 1999. Laser micro-dissection and laser pressure catapulting for the generation of chromosome-specific paint probes. *Biotechniques.* 27: 362–367.

Schitter, G. and A. Stemmer. 2004. Identification and open-loop tracking control of a piezoelectric tube scanner for high-speed scanning probe microscopy. *IEEE Trans. Control Syst. Technol.* 12: 449–454.

Schitter, G., P. Menold, H.F. Knapp, F. Allgower, and A. Stemmer. 2001. High performance feedback for fast scanning atomic force microscopes. *Rev. Sci. Instrum.* 72: 3320–3327.

Schitter, G., R.W. Stark, and A. Stemmer. 2004. Fast contact-mode atomic force microscopy on biological specimen by model-based control. *Ultramicroscopy.* 100: 253–257.

Schütze, K., I. Becker, K.F. Becker et al. 1997. Cut out or poke in—The key to the world of single genes: Laser micromanipulation as a valuable tool on the look-out for the origin of disease. *Genet. Anal.* 14: 1–8.

Sebastian, A., A. Gannepalli, and M.V. Salapaka. 2007. A review of the systems approach to the analysis of dynamic-mode atomic force microscopy. *IEEE Trans. Control Syst. Technol.* 15: 952–959.

Smalley, R. 2001. Of chemistry, love and nanobots. *Sci. Am.:* 76–77.

Smalley, R. and E. Drexler. 2003. Point and counterpoint: Nanotechnology. *Chem. Eng. News.* 81: 37–42.

Srinivasan, R. 1986. Ablation of polymers and biological tissue by ultraviolet-lasers. *Science.* 234: 559–65.

Stark, R.W. and W.M. Heckl. 2000. Fourier transformed atomic force microscopy: Tapping mode atomic force microscopy beyond the hookian approximation. *Surf. Sci.* 457: 219–228.

Stark, R.W. and W.M. Heckl. 2003. Higher harmonics imaging in tapping-mode atomic-force microscopy. *Rev. Sci. Instrum.* 74: 5111–5114.

Stark, R.W., S. Thalhammer, J. Wienberg, and W.M. Heckl. 1998. The afm as a tool for chromosomal dissection—The influence of physical parameters. *Appl. Phys. A.* 66: S579–S584.

Stark, M., R.W. Stark, W.M. Heckl, and R. Guckenberger. 2002. Inverting dynamic force microscopy: From signals to time-resolved interaction forces. *Proc. Natl. Acad. Sci. USA.* 99: 8473–8478.

Stark, R.W., F.J. Rubio-Sierra, S. Thalhammer, and W.M. Heckl. 2003. Combined nanomanipulation by atomic force microscopy and uv-laser ablation for chromosomal dissection. *Eur. Biophys. J.* 32: 33–39.

Stern, J.E., B.D. Terris, H.J. Mamin, and D. Rugar. 1988. Deposition and imaging of localized charge on insulator surfaces using a force microscope. *Appl. Phys. Lett.* 53: 2717–2719.

Stroh, C., H. Wang, R. Bash et al. 2004. Single-molecule recognition imaging-microscopy. *Proc. Natl. Acad. Sci. USA.* 101: 12503–12507.

Struckmeier, J., R. Wahl, M. Leuschner et al. 2008. Fully automated single-molecule force spectroscopy for screening applications. *Nanotechnology.* 19: 384020.

Sugimoto, Y., P. Pou, O. Custance et al. 2008. Complex patterning by vertical interchange atom manipulation using atomic force microscopy. *Science.* 322: 413–417.

Sulchek, T., S.C. Minne, J.D. Adams et al. 1999. Dual integrated actuators for extended range high speed atomic force microscopy. *Appl. Phys. Lett.* 75: 1637–1639.

Sulchek, T., R. Hsieh, J.D. Adams et al. 2000. High-speed tapping mode imaging with active q control for atomic force microscopy. *Appl. Phys. Lett.* 76: 1473–1475.

Terris, B.D., J.E. Stern, D. Rugar, and H.J. Mamin. 1989. Contact electrification using force microscopy. *Phys. Rev. Lett.* 63: 2669–2672.

Terris, B.D., J.E. Stern, D. Rugar, and H.J. Mamin. 1990. Localized charge force microscopy. *J. Vac. Sci. Technol. A Vac. Surf. Films.* 8: 374–377.

Thalhammer, S., R.W. Stark, S. Muller, J. Wienberg, and W.M. Heckl. 1997a. The atomic force microscope as a new micro-dissecting tool for the generation of genetic probes. *J. Struct. Biol.* 119: 232–237.

Thalhammer, S., R.W. Stark, K. Schütze, J. Wienberg, and W.M. Heckl. 1997b. Laser microdissection of metaphase chromosomes and characterization by atomic-force microscopy. *J. Biomed. Opt.* 2: 115–119.

Thalhammer, S., U. Koehler, R.W. Stark, and W.M. Heckl. 2001. Gtg banding pattern on human metaphase chromosomes revealed by high resolution atomic-force microscopy. *J Microsc-Oxford.* 202: 464–467.

Tinazli, A., J. Piehler, M. Beuttler, R. Guckenberger, and R. Tampe. 2007. Native protein nanolithography that can write, read and erase. *Nat. Nano.* 2: 220–225.

Tsukamoto, K., S. Kuwazaki, K. Yamamoto et al. 2005. Nanometer-scale dissection of chromosomes by atomic force microscopy combined with heat-denaturing treatment. In *13th International Conference on Scanning Tunneling Microscopy, Spectroscopy and Related Techniques*, Sapporo, Japan, pp. 2337–2340.

Tsukamoto, K., S. Kuwazaki, K. Yamamoto, T. Ohtani, and S. Sugiyama. 2006. Dissection and high-yield recovery of nanometre-scale chromosome fragments using an atomic-force microscope. *Nanotechnology.* 17: 1391–1396.

Uchihashi, T., M. Higgins, Y. Nakayama, J.E. Sader, and S.P. Jarvis. 2005. Quantitative measurement of solvation shells using frequency modulated atomic force microscopy. *Nanotechnology.* 16: S49–S53.

Viani, M.B., T.E. Schaffer, A. Chand et al. 1999. Small cantilevers for force spectroscopy of single molecules. *J. Appl. Phys.* 86: 2258–2262.

Vogl, W., B.K.-L. Ma, and M. Sitti. 2006. Augmented reality user interface for an atomic force microscope-based nanorobotic system. *IEEE Trans. Nanotechnol.* 5: 397–406.

Wen, C.K. and M.C. Goh. 2004. Afm nanodissection reveals internal structural details of single collagen fibrils. *Nano Lett.* 4: 129–132.

Wenger, M.P.E., M.A. Horton, and P. Mesquida. 2008. Nanoscale scraping and dissection of collagen fibrils. *Nanotechnology.* 19: 384006.

Wouters, D. and U.S. Schubert. 2004. Nanolithography and nano-chemistry: Probe-related patterning techniques and chemical modification for nanometer-sized devices. *Angew. Chem. Int. Ed.* 43: 2480–2495.

Xie, X.N., H.J. Chung, C.H. Sow, and A.T.S. Wee. 2006. Nanoscale materials patterning and engineering by atomic force microscopy nanolithography. *Mater. Sci. Eng. R-Rep.* 54: 1–48.

Zhong, Q., D. Inniss, K. Kjoller, and V.B. Elings. 1993. Fractured polymer silica fiber surface studied by tapping-mode atomic-force microscopy. *Surf. Sci.* 290: L688–L692.

Nanorobotic Manipulation

Lixin Dong
Michigan State University

Bradley J. Nelson
Eidgenössische Technische Hochschule Zürich

44.1 Introduction

Nanorobotic manipulation entails the position and/or orientation control of nanometer-scale (dimensions between approximately 1 and 100 nm) objects with nanometer resolution using a robotic manipulator. The legendary physicist Richard Feynman suggested a "weird possibility" for solving "the problem of manipulating and controlling things on a small scale" in his prophetic speech that initiated the idea of nanotechnology (Feynman 1960): "Now comes the interesting question: How do we make such a tiny mechanism? I leave that to you. However, let me suggest one weird possibility. You know, in the atomic energy plants they have materials and machines that they can't handle directly because they have become radioactive. To unscrew nuts and put on bolts and so on, they have a set of master and slave hands, so that by operating a set of levers here, you control the 'hands' there, and can turn them this way and that so you can handle things quite nicely." Such master–slave hands were enabled by the invention of scanning tunneling microscopes (STMs) by Gerd Binnig and Heinrich Rohrer in the early 1980s (Binnig et al. 1982). Their invention has radically changed the way in which we interact with and even regard single atoms and molecules, and it earned them both a Nobel Prize in physics in 1986. The first nanomanipulation experiment was performed by Eigler and Schweizer (1990). They used an STM operating at low temperatures (4 K) and ultrahigh vacuum (UHV) to position individual xenon atoms on a single-crystal nickel surface with atomic precision. The manipulation enabled them to fabricate rudimentary structures of their own design, atom by atom. The result is the famous set of images showing how 35 atoms were moved to form the three-letter logo "IBM," demonstrating

that matter could indeed be maneuvered atom by atom as Feynman envisioned (Feynman 1960). A more generalized form of the STMs called the scanning probe microscopes (SPMs) now allows us to perform "engineering" operations on single molecules, atoms, and bonds, thereby providing a tool that operates at the ultimate limits of fabrication. The SPM enables the exploration of molecular properties on an individual nonstatistical basis, and is itself the primary tool that enabled the field of nanorobotics to emerge.

A nanorobotic manipulation system generally includes nanomanipulators as the positioning device, microscopes as "eyes," various end-effectors including probes and tweezers among others as its "fingers," and types of sensors (force, displacement, tactile, strain, etc.) to facilitate the manipulation and/or to determine the properties of the objects. Key technologies for nanomanipulation include observation, actuation, measurement, system design and fabrication, calibration and control, communication, and human–machine interface.

Nanorobotic manipulation is featured by multi-degrees of freedom and 3D processes, differentiating it from conventional scanning probe techniques. Nanorobotic manipulation was first proposed in the late 1990s as an alternative approach to nanomanipulation using SPMs. In SPMs, the "eyes" and "fingers" are the same probes, and this superposition makes it impossible to see what is happening simultaneously as performing manipulation. Scanning electron microscopes (SEMs) and transmission electron microscopes (TEMs) were introduced as independent "eyes," and more complex manipulators were therefore developed as more agile "fingers." However, the advancement of SPM-based manipulators has blurred the border. The functions of "eyes" and "fingers" of SPMs can be separate in a time-based

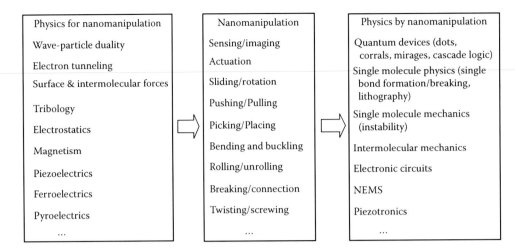

FIGURE 44.1 Physics for and by nanomanipulation.

fashion, and quasi real-time imaging has been achieved. SPMs with large-stroke coarse positioning ability has been developed based on stick–slip motion and they can now be installed inside SEMs and TEMs serving purely as "fingers" with all features of robotic manipulators. So, we use "nanorobotic manipulation" loosely today; both manipulation using scanning probes and inside electron microscopes are referred to as nanorobotic manipulation. The term "nanorobotic manipulation" is used more often to differentiate itself from self-assembly (Whitesides and Grzybowski 2002), directed growth, and other technologies for position and/or orientation control of nanometer-scale objects without involving a manipulator.

The processes of nanorobotic manipulation include sliding/rotation, pushing/pulling, picking/placing, bending/buckling, rolling/unrolling, breaking/connection, twisting/screwing, and so on (Figure 44.1). Nanorobotic manipulation enables a hybrid approach by combining top-down and bottom-up processes for creating nanosystems that can attain a higher functionality because they possess more complex structures. Nanorobotic manipulation expands the lower limit of robotic exploration further into the nanometer scale, and it will provide nanoscale sensors and actuators, structuring and assembly technology for building nanosystems.

Physics has been a strong source for the innovation and insight to nanorobotic manipulation (Figure 44.1). Wave-particle duality serves as the base for electron microscopy; knowledge on electron tunneling, surface and intermolecular forces, tribology, electrostatics, and magnetism enabled the invention of SPMs, and piezoelectric actuators are the most widely used ones in nanorobotic manipulators. Physics is among the fields, which can in turn benefit from the advancement of nanorobotic manipulation (Figure 44.1). Quantum devices, such as quantum dots, corrals, mirages, and cascade logic circuits, have been constructed with a few atoms or molecules using STMs. Some imaginary experiments in the history can now be done physically in a lab. Other examples include single molecule physics (single bond formation/breaking and lithography), single molecule mechanics (instability), intermolecular mechanics, electronic circuits,

nanoelectromechanical systems (NEMS), and piezotronics (Wang 2007).

Nanorobotic manipulation is the essential part of nanorobotics. Nanorobotics (Figure 44.2) is the study of robotics at the nanometer scale, and includes robots that are nanoscale in size, i.e., nanorobots (which have yet to be realized), and large robots capable of manipulating objects that have dimensions in the nanoscale range with nanometer resolution, i.e., nanorobotic manipulators. Knowledge from mesoscopic physics, mesoscopic/supramolecular chemistry, and molecular biology at the nanometer scale converges to form the field. Various disciplines contribute to nanorobotics, including nanomaterial synthesis, nanobiotechnology, and microscopy for imaging and characterization. Topics such as self-assembly, nanorobotic assembly, and hybrid nanomanufacturing approaches for assembling nano building blocks into structures, tools, sensors, and actuators are considered areas of nanorobotic study. A current focus of nanorobotics is on the fabrication of NEMS, which may serve as components for future nanorobots. The main goals of nanorobotics are to provide effective tools for the experimental exploration of the nanoworld, and to push the boundaries of this exploration from a robotics research perspective.

44.2 Scaling to the Nanoworld

When studying nanorobotics, we must first develop an understanding of the physics that underlies interactions at this scale. At the nanoscale, the volume effects associated with inertia, weight, heat capacity, and body forces are dominated by surface effects associated with friction, heat transfer, and adhesion forces (Drexler 1992, Wautelet 2001). For handling an object at this scale, we not only deal with intermolecular physical interactions, but also with much stronger intramolecular/interatomic chemical bonding forces. The intramolecular and intermolecular forces can dominate the surface forces that we see at the microscale. Although the laws of classic Newtonian physics may well suffice to describe changes in behavior down to about 10 nm (Wolf 2004), the change in the magnitude of many important

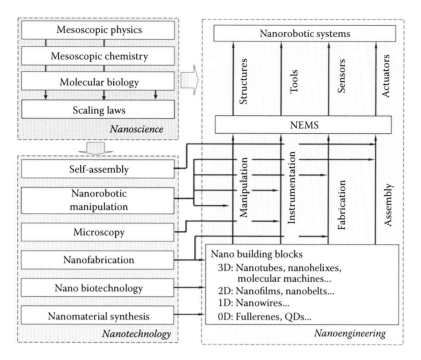

FIGURE 44.2 A roadmap for nanorobotics.

physical properties, such as resonant frequencies, are so great, that completely new applications may appear. Physical effects and chemical reactions can be induced on an individual atomic/molecular basis rather than a statistic one. The most challenging question for nanoroboticists is to understand and exploit these changes in physical behavior that occur at the end of the classical scaling range.

44.2.1 Intermolecular and Interatomic Forces

We begin by referring to some characteristic length L. If we consider the van der Waals force between two molecules with a separation r, then the generalized interaction between molecules is given by the Mie pair potential (Israelachvili 2002)

$$E(r) = -\frac{A}{r^n} + \frac{B}{r^m} \tag{44.1}$$

Note that a repulsive term (positive) as well as an attractive term (negative) is included. A specific case of the Mie potential is the Lennard-Jones potential

$$E(r) = -\frac{A}{r^6} + \frac{B}{r^{12}} \tag{44.2}$$

where A and B are constants, e.g., for solid argon, $A = 8.0 \times 10^{-77} \text{ J m}^6$ and $B = 1.12 \times 10^{-133} \text{ J m}^{12}$ (Jasap 2000). In this potential, the attractive contribution is the van der Waals interaction potential, which varies with the inverse-sixth power of the distance. The repulsive item is sometimes called the repulsive van der Waals potential. The net van der Waals force is given by

$$F_{vdW} = -\frac{dE}{dr} \tag{44.3}$$

If r is scaled as $\sim L$, the attractive force scales as $\sim L^{-7}$, and thus its importance dramatically increases at the nanoscale (Figure 44.3). The repulsive force scales as $\sim L^{-13}$, which is important only at sub-nanometer scales. This provides the fundamentals physics for atomic force microscopy (AFM). For the sphere–halfspace surface pair, we ignored this term and showed that the net van

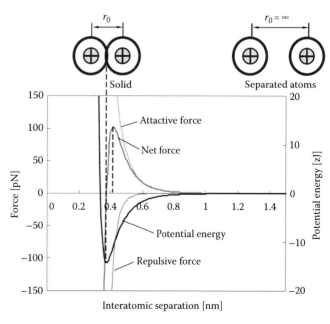

FIGURE 44.3 Intermolecular forces.

der Waals forces scale with separation distance as $\sim L^{-2}$ when keeping the sphere radius unchanged.

The interatomic equilibrium separation r_0 can be solved by setting $F_{\text{vdW}} = 0$. For solid argon, $r_0 = 0.375\,\text{nm}$. The potential energy is at a minimum and corresponds to bond energy $-E_{\text{bond}} = 1.43 \times 10^{-20}\,\text{J}$ or $0.09\,\text{eV}$ (electron volts, $1\,\text{eV} = 1.6022 \times 10^{-19}\,\text{J}$). The maximum value of F_{vdW} is obtained when $d^2E/dr^2 = 0$, or $r = (26B/7A)^{1/6} = 0.416\,\text{nm}$, as $F_{\text{vdW,max}} = 102\,\text{pN}$.

The bond energy of van der Waals-induced dipoles (such as argon solid shown here) is much smaller than electrostatic interaction–based intramolecular ionic bonds (e.g., $3.2\,\text{eV}$ for NaCl rock salt), metallic bonds (e.g., $3.1\,\text{eV}$ for metal Cu) or covalent bonds (e.g., $4\,\text{eV}$ for Si and $7.4\,\text{eV}$ for C (diamond)), which scale as $\sim L^{-2}$.

When understanding the interactions among small nanometer-sized structures, it is important to be aware of the complexity of the forces with which these objects may interact.

44.2.2 Optics

Optics is used for the imaging, characterization, and fabrication of nanorobotic systems. The limitations of optical methods arise from wave optics, in particular diffraction (Wautelet 2001). When a wave of wavelength λ projects onto an element of linear dimension L, the reflected wave diverges. The divergence angle is

$$\beta \approx \lambda / L \qquad (44.4)$$

Hence, $\beta \propto L^{-1}$. In microscopy and lithography, when one irradiates elements with lenses of a fixed numerical aperture ($NA = \mu \sin \beta$, μ is the index of refraction), the resolution is determined by the minimum diameter of the irradiated zone, which is given by the classic Rayleigh criterion

$$L = \frac{2\lambda}{\pi \mu \sin \beta} \qquad (44.5)$$

Hence, the wavelength required to resolve structures scales with λ. For the sake of simplicity, we can approximate $\mu \sin \beta$ by unity and the resolution becomes approximately equal to about half the wavelength of the incoming light. For green light in the middle of the visible spectrum, λ is about $550\,\text{nm}$, so the resolution of a good optical microscope is about $300\,\text{nm}$. This is obviously not small enough for nanoscale objects. If one wishes to attain $L < 100\,\text{nm}$, it becomes necessary to use other radiation sources with shorter wavelengths such as electron beams.

Based on de Broglie's wave-particle duality, we relate particle momentum p to wavelength λ of an electron through Planck's constant h

$$\lambda = \frac{h}{p} \qquad (44.6)$$

If an electron (rest mass: m_0) is accelerated by an electrostatic potential drop eV, the electron wavelength can be described as (Williams and Carter 1996)

$$\lambda = \frac{h}{\sqrt{2m_0 eV \left(1 + (eV/2m_0 c^2)\right)}} \qquad (44.7)$$

If we ignore relativistic effects, we can show that the wavelength of electrons is approximately related to their energy E by

$$\lambda \sim \frac{1.22}{\sqrt{E}} \qquad (44.8)$$

where E is in eV and λ in nm.

So for a $100\,\text{keV}$ electron, we find that $\lambda \sim 4\,\text{pm}$ ($0.004\,\text{nm}$), much smaller than an atomic radius. However, we are nowhere near building microscopes that approach this wavelength limit of resolution, because we cannot make perfect electron lenses, though, many commercially available TEMs have been capable of resolving individual columns of atoms in crystals since the mid-1970s (Williams and Carter 1996). Today, these instruments are routinely used for nanotechnology.

44.2.3 Quantum Effects

When the size of elements decreases to the nanometer scale, quantum effects can become important. In most situations, they arise in electronic properties. Quantum effects must be taken into account as the size of an element L_c approaches the wavelength associated with electrons λ (Equations 44.6 through 44.8). Under these conditions, certain nanoparticles, called quantum dots (QDs), behave as if they were large atoms. Such a system is sometimes referred to as a zero-dimensional (0D) system. Quantum mechanical calculations indicate that the electronic levels are discrete, just as in an atom (and contrary to a solid in which the levels are grouped in energy bands). The spacing of the discrete electronic levels of the quantum dots, E_{QD}, scales as $E_{\text{QD}} \sim L^{-2}$ (Wautelet 2001).

When elements come close together, on the order of nanometers or below, electrons can hop from one element to the other by "tunneling." The tunnel current density, j_{tun}, varies with distance L as

$$j_{\text{tun}} \sim e^{-aL} \qquad (44.9)$$

where a depends on the height of the electronic barrier. This effect is important when L is on the order of a few tenths of a nanometer. This small value determines the rapid decay of the current with L, which also explains the excellent resolution of the STM.

44.3 Imaging at the Nanoscale

Optical microscopes (OMs) have enhanced our knowledge in biology, biomedical research, medical diagnostics and materials science. OMs can magnify objects up to ~ 1000 times but cannot provide a resolution better than $200\,\text{nm}$ due to diffraction limits. This is far beyond the most features of interest at the nanoscale, such as the distance between two atoms in a solid (around 0.2 nm). Recently developed scanning near-field OMs (SNOM) use fiber-optic scattering probes and have achieved spatial resolutions in the 20 nm range. Recently reported far-field approaches for optical microscopy have enabled theoretically unlimited spatial resolution of fluorescent biomolecular complexes. However, imaging tools commonly used in nanorobotics are mainly electron microscopes and SPM, shown schematically in Figure 44.4.

TEMs (Figure 44.4b), like all electron microscopes, use high-energy electrons as a radiation source. Because electrons are much more strongly scattered by a gas than light, optical paths must be evacuated to a pressure better than 10^{-10} Pa. Typical resolution for a TEM can reach the atomic scale down to about 1 Å (0.1 nm).

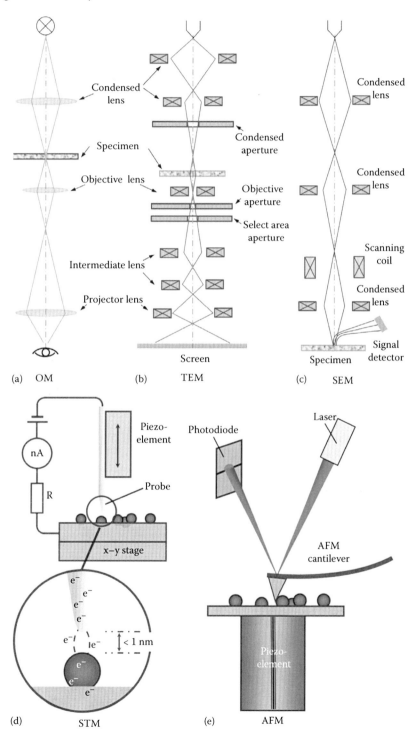

FIGURE 44.4 Imaging at the nanometer scale, (a) optical microscope, (b) transmission electron microscope, (c) scanning electron microscope, (d) scanning tunneling microscopy, and (e) atomic force microscope.

The TEM detects electrons that pass through a given sample, resembling an OM (Figure 44.4a). The electron gun of the TEM operates at high energy levels of between 50 and 1000 kV. In order for proper imaging to take place, the sample must be very thin so that electrons from the beam can pass through the specimen. Electrons that do not pass through the sample cannot be detected, so sample preparation is a critical part of the imaging process. TEMs produce images that are two dimensional in appearance.

The first SEM (Figure 44.4c) became commercially available in 1966, 34 years after the invention of the TEM by Knoll and Ruska. SEMs do not require thin samples and can observe a much larger area of the specimen surface. SEMs have been a valuable resource for viewing samples at a much higher resolution and depth of field than typical OMs. Conventional SEMs can resolve features to the nanometer scale (~1 nm). Unlike conventional OMs, SEMs have a high depth of field, which gives imaged samples a three-dimensional appearance. Early SEMs were limited to viewing conductive samples. However, many of today's SEMs can image nonconductive samples in addition to conductive samples using variable pressure chambers.

Similar to the TEM, the STM (Binnig et al. 1982) can also resolve specimens down to the atomic scale. The scanning probe of the STM is comprised of a noble metal sharpened to an atomic sized tip, which is mounted on a piezoelectrically driven linear stage (Figure 44.4d). The STM makes use of the above-mentioned quantum mechanical effect, tunneling, and has Angstrom-scale resolution (Binnig et al. 1982).

One shortcoming of the STM is that it requires conductive probe tips and samples to work properly. The AFM (Figure 44.4e) was developed in order to view nonconductive samples, giving it a wider applicability than the STM. In addition to imaging nonconductive samples, the AFM can also image samples immersed in liquid, which is useful for biological applications. The AFM has three main modes of operation known as contact mode, non-contact mode, and tapping mode.

Unlike the SEM and TEM, both the STM and the AFM do not require a vacuum environment in order to function. However, a high vacuum is advantageous in order to keep the samples from getting contaminated due to the surrounding environment as well as for controlling humidity. In addition, atomic resolution in air is not possible with an AFM due to humidity. As a result of humidity, a water film is formed and creates capillary forces. This can be solved by operating in a vacuum or completely immersed in a liquid solution.

44.4 Nanorobotic Manipulation Systems

Robotic manipulation at the nanometer scale is a promising technology for handling, structuring, characterizing, and assembling nano building blocks into NEMS.

44.4.1 Strategies

Strategies for nanomanipulation are determined by the necessary environment—air, liquid, or vacuum—which is further determined by the properties and size of the objects. In order to observe objects, STMs can provide sub-angstrom-imaging resolution, whereas AFMs can provide atomic resolutions. Both can obtain 3D surface topology. Because AFMs can be used in an ambient environment, they provide a powerful tool for biomanipulation in a liquid environment. The resolution of SEMs is limited to about 1 nm, whereas field-emission SEMs (FESEM) can achieve higher resolutions. SEM/FESEM can be used for 2D real-time observation for both the objects and the end-effectors of manipulators, and large ultrahigh vacuum (UHV) sample chambers provide enough space to contain a nanorobotica manipulator (NRM) with many degrees of freedom (DOFs) for 3D nanomanipulation. However, the 2D nature of the SEM image makes positioning along the electron-beam direction difficult. High-resolution TEM (HRTEM) can provide atomic resolution. However, the narrow UHV specimen chamber makes it difficult to incorporate large manipulators.

Nanomanipulation processes can be broadly classified into three types: (1) lateral non-contact, (2) lateral contact, and (3) vertical manipulation. Generally, lateral non-contact nanomanipulation is applied for atoms and molecules in UHV with an STM or bio-objects in liquid using optical or magnetic tweezers. Contact nanomanipulation can be used in almost any environment, generally with an AFM, but is difficult for atomic manipulation. Vertical manipulation can be performed by NRMs. Figure 44.5 shows the processes of the three basic strategies.

A lateral non-contact manipulation process is shown in Figure 44.5a. As listed in Table 44.1, motion can be caused by long-range van der Waals forces (attractive) generated by the proximity of the tip to the sample (Eigler and Schweizer 1990, Avouris 1995), by electric field trapping from the voltage bias between the tip and the sample (Whitman et al. 1991), by tunneling current–induced heating or by inelastic tunneling vibration (Avouris 1995). With these methods, nano devices and molecules have been assembled (Lee and Ho 1999). Non-contact manipulation combined with STMs has demonstrated the manipulation of atoms and molecules.

Pushing or pulling nanometer objects on a surface with an AFM is a typical manipulation strategy as shown in Figure 44.5b. Early work demonstrated the effectiveness of this method for the manipulation of nanoparticles (Schaefer et al. 1995). This method has also been shown for nanofabrication (Resch et al. 1998) and biomanipulation. A virtual reality interface may facilitate such manipulation (Falvo et al. 1999, Sitti et al. 2000, Li et al. 2004, Ferreira and Mavroidis 2006). Similar manipulation strategies can yield different results on different objects, e.g., for a nanotube, pushing can induce bending, breaking, rolling, or sliding (Falvo et al. 1999). More examples are listed in Table 44.2.

The pick-and-place task as shown in Figure 44.5c is especially significant for 3D nanomanipulation since its main purpose is to assemble pre-fabricated building blocks into devices. The main difficulty is in achieving sufficient control of the interaction between the tool and the object and between the object and the substrate (Fukuda et al. 2003).

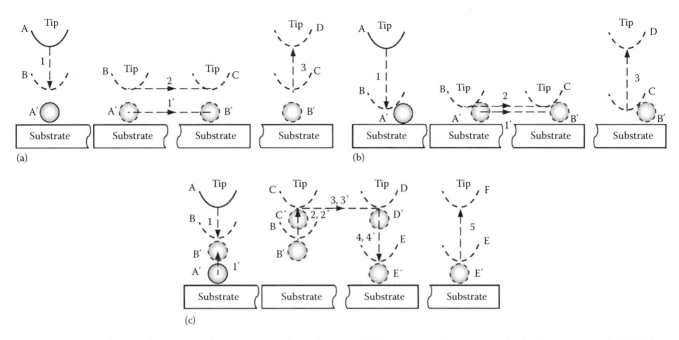

FIGURE 44.5 Fundamental nanomanipulation strategies. In the figure, A, B, C, … represent the positions of end-effector (e.g., a tip); A′, B′, C′, … the positions of objects, 1, 2, 3, … the motions of end-effector; and 1′, 2′, 3′, … the motions of objects. Tweezers can be used in pick-and-place to facilitate the picking up, but are generally not necessarily helpful for placing. (a) Lateral non-contact nanomanipulation (sliding). (b) Lateral contact nanomanipulation (pushing/pulling). (c) Vertical nanomanipulation (picking and placing).

TABLE 44.1 Manipulation Using an STM

Object	Substrate	Temperature	Environment	Strategy of Manipulation	Result/Application	Reference
Xe atom	Ni(110)	4 K	UHV	Van der Waals force	Logo "IBM"	Eigler and Schweizer (1990)
Fe adatom	Cu(111)	4 K	UHV	Van der Waals force	Quantum corrals: wave-particle nature	Crommie et al. (1993)
CO	Cu (111)	(1) 5 K (2) 0.5–40 K	UHV	Van der Waals force for triggering	Molecule cascade logic gates	Heinrich et al. (2002)
Cs	GaAs(110) InSb(110)	RT	UHV	Electric-field-induced electrostatic force	RT bonding dissociation	Whitman et al. (1991)
Si	Si(111)-(7×7)	RT	UHV	Chemical interaction and electrostatic force	Bonding the tip to selected atom	Lyo and Avouris (1991)
$B_{10}H_{14}$	Si(111)	RT	UHV	Tunneling current-induced local heating	Molecule dissociation	Dujardin et al. (1992)
H	H-terminated Si(100)	RT	UHV	Inelastic tunneling caused vibration	Atom/molecule adsorption	Shen et al. (1995)
C_{60}	Step Cu(111)	RT	UHV	Repulsive pushing	Abacus: monoatomic step of molecules	Cuberes et al. (1996)
SWNT, 13 nm long	H-terminated Si(100)	RT	UHV	Repulsive pushing	Splitting two SWNTs, understanding SWNT/Si interaction	Albrecht and Lyding (2007)
CO–Fe, Fe(CO)–CO	Ag(110)	13 K	UHV	Electric-field-induced electrostatic force	Synthesis of molecules	Lee and Ho (1999)

44.4.2 Manipulators

Nanorobotic manipulators are the core components of nanorobotic manipulation systems. The basic requirements for a nanorobotic manipulation system for 3D manipulation include nanoscale positioning resolution, a relative large working space, enough DOFs including rotational ones for 3D positioning and orientation control of the end-effectors, and usually multiple end-effectors for complex operations.

A commercially available nanomanipulator (MM3A™ from Kleindiek) for SEMs is shown in Figure 44.6a. The manipulator has three DOFs, and 5 nm, 3.5 nm, and 0.25 nm resolution in *X*, *Y*, and

TABLE 44.2 Manipulation Using an AFM

Object	Substrate	Temperature	Environment	Strategy of Manipulation	Result/Application	Reference
Au particle	HOPG, WSe$_2$	RT	~500 mTorr, dry N$_2$	Pushing	Nanoclusters assembled	Schaefer et al. (1995)
MoO$_3$ particle	Single crystal MoS$_2$	N/A	N/A	Pushing	Pattern "MoO$_3$," interlocking nanostructures	Sheehan and Lieber (1996)
GaAs particle	GaAs	RT	Air	Pushing	Pattern "nm"	Junno et al. (1995)
Au particle	Mica	RT	Air	Pushing	Pattern "USC"	Baur et al. (1997)
Au particle	Mica	RT	Air	Pushing	3D structures	Resch et al. (1998)
CNTs	Graphite	RT	Air	Bending and buckling	Strength	Falvo et al. (1997)
CNTs	Graphite	RT	Air	Rolling and sliding	Tribology	Falvo et al. (1999)
DNA, fibrin, adeno- and tobacco mosaic virus	Mica	RT	Air	Pushing	Biophysics, mechanical properties	Guthold et al. (2000)
Ag cubes, rods	Polycarbonate	RT	Air	Pushing	CAD-model-based automated manipulation	Chen et al. (2006)

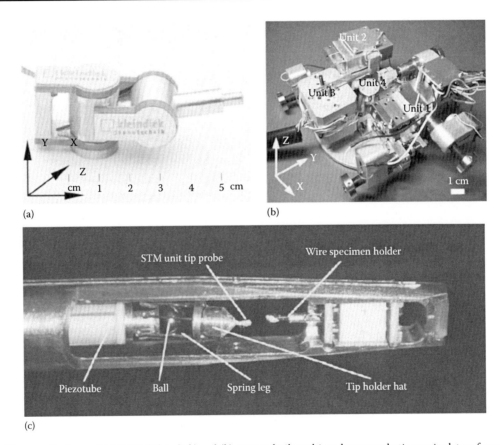

FIGURE 44.6 Nanomanipulators. (a) MM3A (Kleindiek) and (b) custom-built multi-probe nanorobotic manipulators for an SEM. (c) ST1000 STM-TEM holder (Nanofactory Instruments AB).

Z directions at the tip, respectively. Each joint has a piezo-actuator with open-loop control. Kinematic analysis shows that when scanning in the X/Y directions using rotary joints, the additional linear motion in Z direction is very small. For example, when the arm length is 50 mm, the additional motion in the Z direction is only 0.25–1 nm when moving in the X direction for 5–10 μm; these errors can be ignored or compensated with the last prismatic joint, which has a 0.25 nm resolution. Figure 44.6b shows a homemade

nanorobotic manipulation system that has 16 DOFs in total and can be equipped with 3–4 AFM cantilevers as end-effectors for both manipulation and measurement. The positioning resolution is sub-nm over cm range. Such manipulation systems are used not only for nanomanipulation, but also for nanoassembly, nanoinstrumentation, and nanofabrication. Four probe semiconductor measurements are perhaps the most complex manipulation this system can perform, because it is necessary to actuate four probes

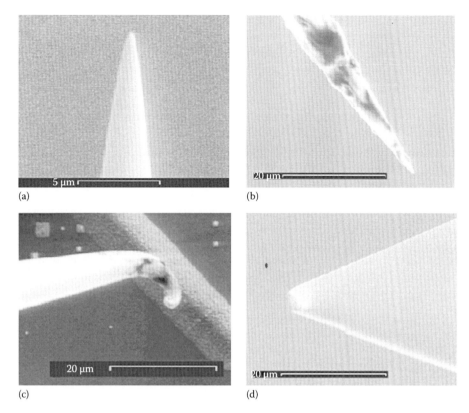

FIGURE 44.7 Tools for nanomanipulation. (a) Sharp tip. (b) Sticky probe. (c) Hook. (d) AFM cantilever.

independently by four manipulators. Figure 44.6c shows an STM built in a TEM holder (Nanofactory Instruments AB, ST-1000) for obtaining high imaging resolution. The STM can serve as a 3-DOF manipulator with sub-nm resolution and over a mm-scale workspace. With the advancement of nanotechnology, one can envision shrinking the size of nanomanipulators and inserting more DOFs inside the limited vacuum chamber of a microscope and, perhaps, achieving the molecular version of manipulators such as those dreamed of by Drexler (1992).

44.4.3 Tools

The standard tool of the manipulator is a commercially available tungsten sharp probe (Picoprobe T-1–10–1 mm (Figure 44.7a) and T-1–10). To facilitate different processes, special tools have been fabricated including a nanohook (Figure 44.7b) that is prepared by controlled "tip-crashing" of a sharp probe onto a substrate, and a "sticky" probe (Figure 44.7c) prepared by tip dipping into a double-sided SEM silver conductive tape (Ted Pella, Inc.). AFM cantilevers (Nanoprobe, NP-S, Figure 44.7d) are used for measuring forces or as electrodes.

44.5 Processes of Nanorobotic Manipulation

Nanorobotic manipulation (NRM) is characterized by multiple DOFs with both position and orientation controls, independently actuated multi-probes, and a real-time observation system. NRM

has proven effective for structuring and characterization nano building blocks and is promising for assembling nanodevices in 3D space (Yu et al. 2000, Dong et al. 2002, Kortschack et al. 2005, Weir et al. 2005, Molhave et al. 2006). It has been applied on various materials such as nanoparticles (0D), nanowires (1D) and nanotubes (1D or 3D), nanobelts and nanofilms (2D), and 3D nanostructures.

44.5.1 Basic Processes

Successful applications of NRM are in the manipulation and characterization of carbon nanotubes (CNTs) and 3D helical structures. The well-defined geometries, exceptional mechanical properties, and extraordinary electric characteristics, among other outstanding physical properties, make CNTs attractive for many potential applications, especially in nanoelectronics, NEMS, and other nanodevices. For NEMS, some of the most important characteristics of nanotubes include their nanometer diameter, large aspect ratio (10–1000), TPa scale Young's modulus, excellent elasticity, ultra-small interlayer friction, excellent capability for field emission, various electric conductivities, high thermal conductivity, high current carrying capability with essentially no heating, sensitivity of conductance to various physical or chemical changes, and charge-induced bond-length change.

Three-dimensional helical structures with nanofeatures, such as carbon nanocoils, helical CNTs (Zhang et al. 1994), and zinc oxide nanobelts (Kong and Wang 2003), have attracted research interest because of their potential applications in NEMS. A new method of creating 3D helical structures with nanometer-scale dimensions has recently been presented (Prinz et al. 2000) and can be fabricated in a controllable way (Zhang et al. 2005). The structures are created through a top-down fabrication process in which a strained nanometer thick heteroepitaxial bilayer curls up to form 3D structures with nanoscale features such as SiGe/Si tubes (Zhang et al. 2008a), Si/Cr rings (Zhang et al. 2008b), SiGe/Si coils (Zhang et al. 2005), InGaAs/GaAs coils (Bell et al. 2006a), Si/Cr spirals (Zhang et al. 2006a), and small-pitch SiGe/Si/Cr coils (Zhang et al. 2006b). Because of their interesting morphology, mechanical, electrical, and electromagnetic properties, potential applications of these nanostructures in NEMS include nanosprings, electromechanical sensors, magnetic field detectors, chemical or biological sensors, generators of magnetic beams, inductors, actuators, and high-performance electromagnetic wave absorbers.

One basic nanomanipulation procedure is to pick up a single tube from nanotube soot (Figure 44.8). This was first demonstrated using dielectrophoresis (Fukuda et al. 2003) through

nanorobotic manipulation. The interaction between a tube and the atomic flat surface of an AFM cantilever tip has been shown to be strong enough for picking up a tube onto the tip (Hafner et al. 2001). By using electron-beam-induced deposition (EBID; Yu et al. 2000), it is possible to pick up and fix a nanotube onto a probe (Dong et al. 2002). For placing a tube being picked up, a weak connection between the tube and the probe is desired.

Bending and buckling a CNT as shown in Figure 44.9 are important for *in situ* property characterization of a nanotube (Dong et al. 2004), which is a simple way to obtain Young's modulus of a nanotube. Stretching is another technique for characterizing a nanostructure. Figure 44.10 shows an example of measuring the spring constant of a helical nanobelt by stretching it.

44.5.2 Special Processes

The construction of NEMS using 3D helical nanostructures involves the assembly of as-fabricated building blocks, which is a significant challenge from a fabrication standpoint. Focusing on

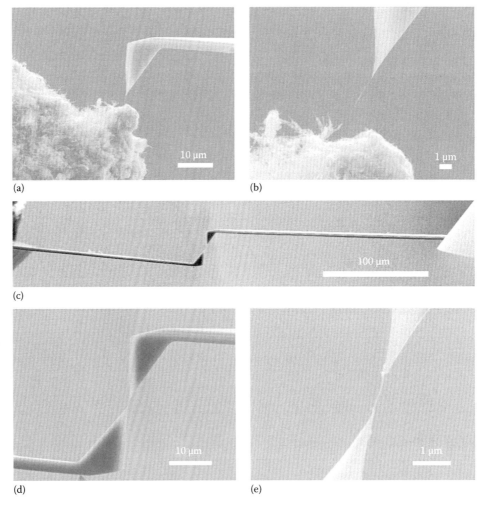

FIGURE 44.8 Nanorobotic manipulation of CNTs. The basic technique is to pick up an individual tube from CNT soot (a) or from an oriented array; (b) shows a free-standing nanotube picked up by van der Waals forces between the probe and the tube, the tube was transferred to another probe (c); (d) and (e) show larger magnification images of attaching the tube to a second probe; comparing (c) and (d), it can be seen that an SEM provides a large field of view at low magnification (c) which can be increased (e) for more precise tasks.

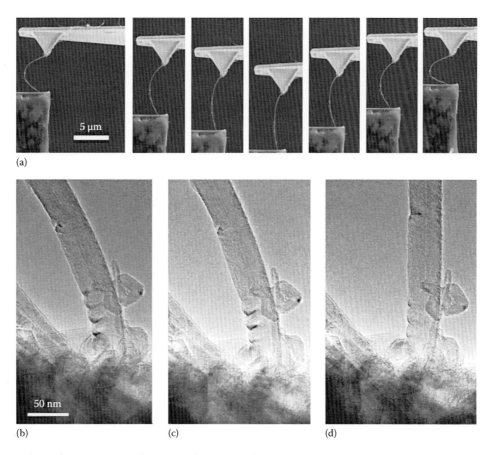

FIGURE 44.9 Nanorobotic characterization of CNTs inside a FESEM (a) and a TEM (b–d). Notice that the TEM can resolve more structural details such as ripples and internal structures such as the inner diameter of a CNT, which is unattainable using a SEM.

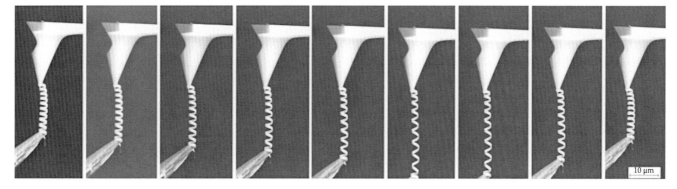

FIGURE 44.10 Nanorobotic characterization of a helical nanobelt. The characterization revealed the spring constant to be 0.003 N/m. Such vertical manipulation remains a challenge for an AFM.

the unique aspects of manipulating 3D helical nanostructures due to their helical geometry, high elasticity, single end fixation, and strong adhesion of the coils to the substrate for wet etching, a series of new processes is presented using the manipulator installed in an SEM. Processes are developed for the manipulation of as-fabricated 3D helical nanostructures. As shown in Figure 44.11, experiments demonstrate that the as-fabricated nanostructures can be released from a chip by picking up with a "sticky" probe from their free ends (Figure 44.11a, tubes), fixed ends (Figure 44.11d, coils), external surfaces (Figure 44.11g,

rings), or internal surfaces (Figure 44.11j, spirals), and bridged between the probe and another probe (Figure 44.11k or an AFM cantilever (Figure 44.11b, e, h), showing a promising approach for robotic assembly of these structures into complex systems. Axial pulling (Figure 44.11f1 through f4)/pushing, radial compressing (Figure 44.5i1 through i5)/releasing, bending/buckling (Figure 44.11c1 through c4), and unrolling (Figure 44.11l1 through l5), and spirals (Figure 44.11n1 through n8, claws) have also been demonstrated for property characterization. The stiffness of the tube, the coil, and the ring has been measured from

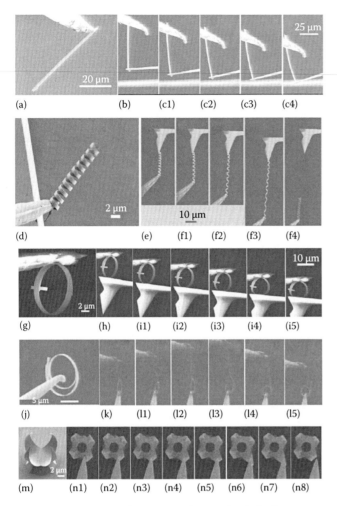

FIGURE 44.11 Nanorobotic manipulation of 3D helical structures. Pick up a tube (a), bridge it between a probe and an AFM cantilever (b), and buckle it (c1–4) for electromechanical property characterization for force measuring. Pick up a small-pitch coil (d), bridge it between a probe and an AFM cantilever (e), and pull it for mechanical property characterization for building a "spring balance" (f1–4). Pick up a ring (External diameter: 12.56 μm. Strip width: 1.2 μm. Number of turns: 2.5. Thickness: Si/Cr 35 nm/10 nm) (g), bridge it between a probe and an AFM cantilever (h), and compress it for mechanical property characterization for understanding its stiffness (i1–5). Pick up a spiral (Si/Cr layer thickness: 35/10 nm) (j), bridge it between a probe and another probe (k), and unroll it for mechanical property characterization for understanding its interlayer interaction (taken from a video clip) (l1–5). Unroll a leaf of claws (m) for mechanical property characterization for understanding its "shape memory" (taken from a video clip) (n1–8).

the SEM images by extracting the AFM tip displacement and the deformation of the structures. The stiffness of the tube, the ring, and the coil springs was estimated to be ~10, 0.137, and 0.003 N/m (calibrated AFM cantilever stiffness: 0.038 N/m), showing a large range for selection. The linear elastic region of the small-pitch coils reaches up to 90%. Unrolling experiments show that these structures have excellent ability on memorizing their original shapes.

The excellent elasticity of nanocoils suggests that they can be used to sense ultra-small forces by monitoring the deformation of the spring as a "spring balance" (Figure 44.11f1 through f4). If working in an SEM, suppose an imaging resolution of 1 nm can be obtained (the best commercially available FESEM can provide such a resolution in an ideal environment), a "spring balance" constructed with the calibrated coil (10 turns, 0.003 N/m) can provide a 3 pN/nm resolution for force measurement. With smaller stripe widths or more turns, nanocoils can potentially provide fN resolution. In the SEM used in these experiments, the available imaging resolution is 10 nm, which provides a 30 pN/10 nm resolution. Figure 44.11f1 through f4 shows a way to use such a coil to measure the adhesive force between a coil and an adhesive silver tape. Comparing the length difference, the extension of the spring can be found and converted to force according to the calibrated spring constant. For Figure 44.11f1 through f3, the relevant forces are determined to be 15.31 ± 0.03 nN, 91.84 ± 0.03 nN (intermediate steps), and 333.67 ± 0.03 nN (maximum holding/releasing force). It can be seen from Figure 44.11f4 that the coil recovered its shape after releasing.

Electrical properties can be characterized by placing a coil between two probes or electrodes (Bell et al. 2006b). An interesting phenomenon found in the measurements is that the SiGe/Si nanocoils with Cr layers can shrink further by passing current through them or by placing a charged probe on them. A 5-turn as-fabricated coil was observed to become an 11-turn coil, showing the possibility of structuring them.

These processes demonstrate the effectiveness of manipulation for the characterization of the 3D helical nanostructures and their assembly for NEMS, which have otherwise been unavailable.

44.5.3 Nanorobotic Assembly

Nanomanipulation is a promising approach for nanoassembly (Fukuda et al. 2003). Key techniques include the control of the position and orientation of the building blocks with nanometer-scale resolution combined with connection techniques.

Random spreading, direct growth, fluidic self-assembly (Rueckes et al. 2000), and dielectrophoretic assembly (Subramanian et al. 2007) have been demonstrated for positioning as-grown nanotubes or other nanostructures on electrodes for the construction of electronic devices or NEMS generally into some type of regular array. Nanorobotic assembly allows for the construction of more complex structures into prototype NEMS. Nanotube intermolecular and intramolecular junctions are basic elements for such assemblies. Although some types of junctions have been synthesized with chemical methods, there is no evidence yet that a self-assembly based approach can provide more complex structures. SPMs have also been used to fabricate junctions, but they are limited to a 2D plane.

In Figure 44.12, we show some examples of the nanorobotic assembly of CNT junctions by emphasizing connection methods. CNT-junctions created using van der Waals forces (a), EBID (b), bonding through mechanochemistry (c), and spot welding via copper encapsulated inside CNTs (d) are shown.

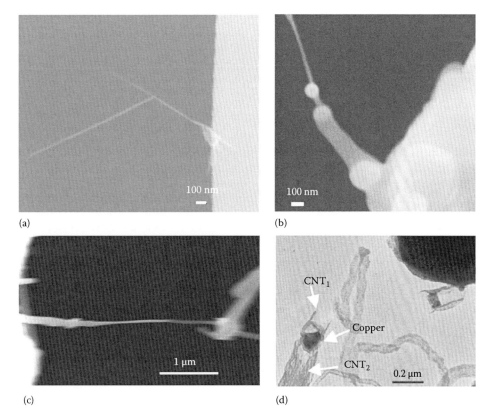

FIGURE 44.12 CNT junctions. (a) CNTs connected with van der Waals force. (b) CNTs joined with EBID. (c) CNTs bonded with mechano-chemical reaction. (d) CNTs welded with copper.

CNT junctions connected with van der Waals forces are currently the most common junctions. Figure 44.12a shows a T-junction connected with van der Waals forces fabricated by positioning the tip of a CNT onto another CNT until they form a bond. The contact quality is determined by measuring the shear connection force.

EBID provides a soldering method to obtain stronger junctions than those connected through van der Waals forces as shown in Figure 44.12a. Hence, if the strength of nanostructures is important, EBID can be applied. Figure 44.12b shows a CNT junction connected through EBID (Dong et al. 2002). The development of conventional EBID has been limited by the expensive electron filament used and low productivity. We have presented a parallel EBID system by using CNTs as emitters because of their excellent field emission properties. As its macro counterpart, EBID works by adding material to obtain stronger connections, but in some cases, the additional material could influence the device function. Therefore, EBID is mainly applied to nanostructures rather than nanomechanisms.

To construct stronger junctions without adding additional materials, mechanochemical nanorobotic assembly is an important strategy. Mechanochemical nanorobotic assembly is based on solid-phase chemical reactions, or mechanosynthesis, which is defined as chemical synthesis controlled by mechanical systems operating with atomic-scale precision, enabling direct positional selection of reaction sites (Drexler 1992). By picking up atoms with dangling bonds rather than stable atoms, it is easier to form primary bonds, which provides a simple but strong connection. Destructive fabrication provides a way to form dangling bonds at the ends of broken tubes. Some of the dangling bonds may close with neighboring atoms, but generally a few bonds will remain reactive. A nanotube with dangling bonds at its end will bind easier to another to form intramolecular junctions. Figure 44.12c shows such a junction (Dong et al. 2003).

EBID involves high-energy electron beams and needs external precursors for getting conductive deposits, which limited its applications. Mechanochemical bonding is promising, but not yet mature. Recently, we developed a nanorobotic spot welding technique (Dong et al. 2007) using copper-filled CNTs for welding CNTs. The solder was encapsulated inside the hollow cores of CNTs during their synthesis, so no external precursors are needed. A bias of just a few volts can induce the migration of the copper, making it a cost-effective approach. Figure 44.12d shows a junction welding using this technique. The quality of the weld is partly determined by the ability to control the mass flow rate from the tube. An ultrahigh precision deposition of 120 ag/s $(1 \text{ ag} = 10^{-18} \text{ g})$ has been realized in our experimental investigation based on electromigration (Figure 44.13).

Nanorobotic manipulation in 3D has opened a new route for nanoassembly. However, nanomanipulation is still performed in a serial manner with master–slave control, certainly not a large-scale production technique. Nevertheless, with advances in the exploration of mesoscopic physics, better control on the material synthesis, more accurate actuators, and effective tools for manipulation, high-speed, parallel, and automatic nanoassembly will be possible.

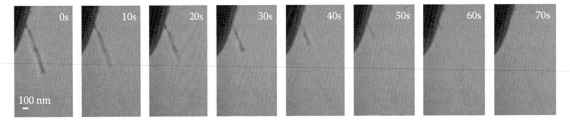

FIGURE 44.13 Attogram precision mass delivery for nanorobotic spot welder.

44.6 NEMS

The next step along the road to fabricating nanorobots is to first fabricate simpler nanoelectromechanical systems. NEMS make it possible to manipulate nanosized objects with nanosized tools, measure mass in femtogram ranges, sense force at pico-Newton scales, and induce GHz motion, among other amazing advancements.

Top-down and bottom-up strategies for manufacturing such nanodevices have been independently investigated by a variety of researchers. Top-down strategies are based on nanofabrication and include technologies such as nano-lithography, nano-imprinting, and chemical etching. Presently, these are 2D fabrication processes with relatively low resolution. Bottom-up strategies are assembly-based techniques. At present, these strategies include techniques such as self-assembly, dip-pen lithography, and directed self-assembly. These techniques can generate regular nano patterns at large scales. With the ability to position and orient nanometer-scale objects, nanorobotic manipulation is an enabling technology for structuring, characterizing, and assembling many types of nanosystems (Fukuda et al. 2003). By combining bottom-up and top-down processes, a hybrid nanorobotic approach based on nanorobotic manipulation provides a third way to fabricate NEMS by structuring as-grown nanomaterials or nanostructures. This new nanomanufacturing technique can be used to create complex 3D nanodevices with such building blocks. Nanomaterial science, bionanotechnology, and nanoelectronics will also benefit from advances in nanorobotic assembly.

The configurations of nanotools, sensors, and actuators based on individual nanotubes that have been experimentally demonstrated are summarized as shown in Figure 44.14. For detecting deep and narrow features on a surface, cantilevered nanotubes (Figure 44.14a; Dong et al. 2002) have been demonstrated as probe tips for AFMs (Dai et al. 1996), STMs, and other types of SPMs. Nanotubes provide ultra-small diameters, ultra-large aspect ratios, and have excellent mechanical properties. Cantilevered nanotubes have also been demonstrated as probes for the measurement of ultra-small physical quantities, such as femto-gram mass (Poncharal et al. 1999), pico-Newton order force sensors, and mass flow sensors (Fukuda et al. 2003) on the basis of their static deflections or change of resonant frequencies detected within an electron microscope. Deflections cannot be measured from micrographs in real time, which limit the application of this kind of sensor. Inter-electrode distance changes

cause emission current variation of a nanotube emitter and may serve as a candidate to replace microscope images. Bridged individual nanotubes (Figure 44.14b; Subramanian et al. 2007) have been the basis for electric characterization. Opened nanotubes (Figure 44.14c; Dong et al. 2006) can serve as an atomic or molecular container, or spot welder (Figures 44.11d and 44.13; Dong et al. 2007).

Controlled exposure of the core of a nanotube (Figure 44.14d) by mechanical breaking or electric breakdown (Cumings and Zettl 2000) is a typical top-down process for fabricating a new family of nanotube devices by taking advantage of the ultra-low interlayer friction. Linear bearings based on telescoping nanotubes have been demonstrated (Cumings and Zettl 2000). A micro actuator with a nanotube as a rotation bearing has been demonstrated (Fennimore et al. 2003). The first demonstration of 3D nanomanipulation of nanotubes took this as an example to show the breaking mechanism of a MWNT, and to measure the tensile strength of CNTs (Yu et al. 2000). A preliminary experiment on a promising nanotube linear motor with field emission current serving as position feedback has been shown with nanorobotic manipulation (Figure 44.14d; Dong et al. 2006). Cantilevered dual nanotubes have been demonstrated as nanotweezers (Kim and Lieber 1999) and nanoscissors (Figure 44.14e; Fukuda et al. 2003) by manual and nanorobotic assembly, respectively.

Based on electric resistance change under different temperatures, nanotube thermal probes (Figure 44.14f) have been demonstrated for measuring the temperature at precise locations. Gas sensors and hot-wire based mass/flow sensors can also be constructed in this configuration rather than a bridged one. The integration of the above-mentioned devices can be realized using the configurations shown in Figure 44.14g and h (Subramanian et al. 2007).

Configurations of NEMS based on 3D helical nanostructures are shown in Figure 44.15. The cantilevered structures shown in Figure 44.15(a, tubes; d, rings; g, coils; and j, spirals) can serve as nanosprings using their elasticity in axial (tubes and coils), radial (rings), and tangential/rotary (spirals) directions. Nanoelectromagnets, chemical sensors nanoinductors, and capacitors involve building blocks bridged between two electrodes (two or four for rings) as shown in Figure 44.15 (b, tubes; e, rings; h, coils; and k, spirals). Electromechanical sensors can use a similar configuration but with one end connected to a moveable electrode as shown in Figure 44.15(c, tubes; f, rings; i, coils; and l, spirals). Mechanical stiffness and electrical conductivity are fundamental

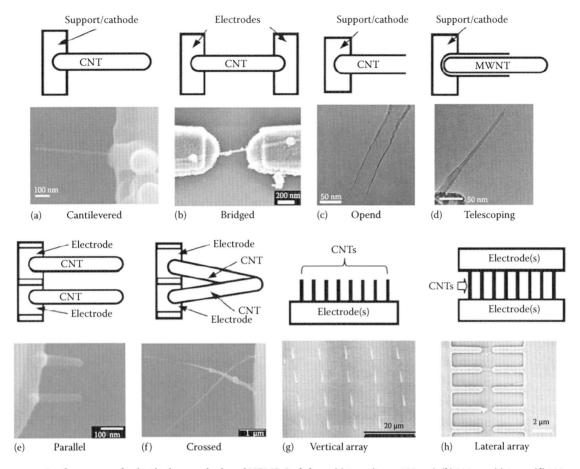

FIGURE 44.14 Configurations of individual nanotube-based NEMS. Scale bars: (a) 1 μm (inset: 100 nm), (b) 200 nm, (c) 1 μm, (d) 100 nm, (e) and (f) 1 μm, (g) 20 μm, and (h) 300 nm. All examples are from the authors' work.

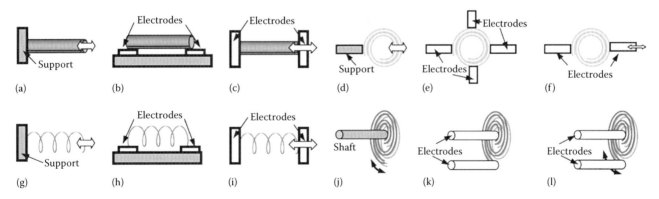

FIGURE 44.15 Configuration of 3D helical nanostructures-based NEMS. (a–c) Tubes. (d–f) Rings. (g–i) Coils. (j–l) Spirals. (a, d, g, j) Cantilevered. (b, e, h, k) Bridged (fixed). (c, f, i, l) Bridged (moveable).

properties for these devices that must be further investigated. Electron microscopy imaging or their intrinsic electromechanical coupling property can serve as readout mechanisms.

44.7 Conclusions

Fundamental knowledge and key technologies for nanorobotic manipulation including scaling laws, imaging, manipulators, tools, processes, and applications have been introduced. Processes of the nanorobotic manipulation of carbon nanotubes and helical nanostructures have been shown including sliding/rotation, pushing/pulling, picking/placing, bending and buckling, rolling/unrolling, breaking/connection, twisting/screwing, and so on. Applications of nanorobotic manipulation in property characterization, structuring, and assembly have been demonstrated. As an enabling technology for nanosystems, nanorobotic manipulation expands the lower limit of robotic exploration further into the nanometer scale. Physics has been a strong source for the innovation and insight to nanorobotic manipulation, and in turn benefits from the advancement of nanorobotic manipulation.

References

Albrecht, P.M. and Lyding, J.W. 2007. Lateral manipulation of single-walled carbon nanotubes on H-passivated Si(100) surfaces with an ultrahigh-vacuum scanning tunneling microscope. *Small* 3: 146–152.

Avouris, P. 1995. Manipulation of matter at the atomic and molecular-levels. *Accounts of Chemical Research* 28: 95–102.

Baur, C., Gazen, B.C., Koel, B. et al. 1997. Robotic nanomanipulation with a scanning probe microscope in a networked computing environment. *Journal of Vacuum Science and Technology B* 15: 1577–1580.

Bell, D.J., Dong, L.X., Nelson, B.J. et al. 2006a. Fabrication and characterization of three-dimensional InGaAs/GaAs nanosprings. *Nano Letters* 6: 725–729.

Bell, D.J., Sun, Y., Zhang, L. et al. 2006b. Three-dimensional nanosprings for electromechanical sensors. *Sensors and Actuators A* 130–131: 54–61.

Binnig, G., Rohrer, H., Gerber, C., and Weibel, E. 1982. Surface studies by scanning tunneling microscopy. *Physical Review Letters* 49: 57–61.

Chen, H.P., Xi, N., and Li, G.Y. 2006. CAD-guided automated nanoassembly using atomic force microscopy-based nanorobotics. *IEEE Transactions on Automation Science and Engineering* 3: 208–217.

Crommie, M.F., Lutz, C.P., and Eigler, D.M. 1993. Confinement of electrons to quantum corrals on a metal surface. *Science* 262: 218–220.

Cuberes, M.T., Schlittler, R.R., and Gimzewski, J.K. 1996. Room-temperature repositioning of individual C-60 molecules at Cu steps: Operation of a molecular counting device. *Applied Physics Letters* 69: 3016–3018.

Cumings, J. and Zettl, A. 2000. Low-friction nanoscale linear bearing realized from multiwall carbon nanotubes. *Science* 289: 602–604.

Dai, H.J., Hafner, J.H., Rinzler, A.G., Colbert, D.T., and Smalley, R.E. 1996. Nanotubes as nanoprobes in scanning probe microscopy. *Nature* 384: 147–150.

Dong, L.X., Arai, F., and Fukuda, T. 2002. Electron-beam-induced deposition with carbon nanotube emitters. *Applied Physics Letters* 81: 1919–1921.

Dong, L.X., Arai, F., and Fukuda, T. 2003. Nanoassembly of carbon nanotubes through mechanochemical nanorobotic manipulations. *Japanese Journal of Applied Physics Part 1* 42: 295–298.

Dong, L.X., Arai, F., and Fukuda, T. 2004. Destructive constructions of nanostructures with carbon nanotubes through nanorobotic manipulation. *IEEE/ASME Transactions on Mechatronics* 9: 350–357.

Dong, L.X., Nelson, B.J., Fukuda, T., and Arai, F. 2006. Towards nanotube linear servomotors. *IEEE Transactions on Automation Science and Engineering* 3: 228–235.

Dong, L.X., Tao, X.Y., Zhang, L., Nelson, B.J., and Zhang, X.B. 2007. Nanorobotic spot welding: Controlled metal deposition with attogram precision from copper-filled carbon nanotubes. *Nano Letters* 7: 58–63.

Drexler, K. 1992. *Nanosystems: Molecular Machinery, Manufacturing and Computation*. New York: Wiley InterScience.

Dujardin, G., Walkup, R.E., and Avouris, P. 1992. Dissociation of individual molecules with electrons from the tip of a scanning tunneling microscope. *Science* 255: 1232–1235.

Eigler, D.M. and Schweizer, E.K. 1990. Positioning single atoms with a scanning tunneling microscope. *Nature* 344: 524–526.

Falvo, M.R., Clary, G.J., Taylor, R.M. et al. 1997. Bending and buckling of carbon nanotubes under large strain. *Nature* 389: 582–584.

Falvo, M.R., Taylor, R.M.I., Helser, A. et al. 1999. Nanometre-scale rolling and sliding of carbon nanotubes. *Nature* 397: 236–238.

Fennimore, A.M., Yuzvinsky, T.D., Han, W.-Q. et al. 2003. Rotational actuators based on carbon nanotubes. *Nature* 424: 408–410.

Ferreira, A. and Mavroidis, C. 2006. Virtual reality and haptics for nanorobotics— A review study. *IEEE Robotics & Automation Magazine* 13: 78–92.

Feynman, R.P. 1960. There's plenty of room at the bottom. *Caltech's Engineering and Science* 23: 22–36.

Fukuda, T., Arai, F., and Dong, L.X. 2003. Assembly of nanodevices with carbon nanotubes through nanorobotic manipulations. *Proceedings of the IEEE* 91: 1803–1818.

Guthold, M., Falvo, M.R., Matthews, W.G. et al. 2000. Controlled manipulation of molecular samples with the nanomanipulator. *IEEE/ASME Transaction on Mechatronics* 5: 189–198.

Hafner, J.H., Cheung, C.L., Oosterkamp, T.H., and Lieber, C.M. 2001. High-yield assembly of individual single-walled carbon nanotube tips for scanning probe microscopies. *Journal of Physical Chemistry B* 105: 743–746.

Heinrich, A.J., Lutz, C.P., Gupta, J.A., and Eigler, D.M. 2002. Molecule cascades. *Science* 298: 1381–1387.

Israelachvili, J.N. 2002. *Intermolecular and Surface Forces*. 2nd ed. London, U.K.: Academic Press.

Jasap, S.O. 2000. *Principles of Electronic Materials and Devices*. 2nd ed. New York: McGraw-Hill Co.

Junno, T., Deppert, K., Montelius, L., and Samuelson, L. 1995. Controlled manipulation of nanoparticles with an atomic-force microscope. *Applied Physics Letters* 66: 3627–3629.

Kim, P. and Lieber, C.M. 1999. Nanotube nanotweezers. *Science* 286: 2148–2150.

Kong, X.Y. and Wang, Z.L. 2003. Spontaneous polarization-induced nanohelixes, nanosprings, and nanorings of piezoelectric nanobelts. *Nano Letters* 3: 1625.

Kortschack, A., Shirinov, A., Truper, T., and Fatikow, S. 2005. Development of mobile versatile nanohandling microbots: Design, driving principles, haptic control. *Robotica* 23: 419–434.

Lee, H.J. and Ho, W. 1999. Single-bond formation and characterization with a scanning tunneling microscope. *Science* 286: 1719–1722.

Li, G.Y., Xi, N., Yu, M.M., and Fung, W.K. 2004. Development of augmented reality system for AFM-based nanomanipulation. *IEEE/ASME Transactions on Mechatronics* 9: 358–365.

Lyo, I.W. and Avouris, P. 1991. Field-induced nanometer-scale to atomic-scale manipulation of silicon surfaces with the STM. *Science* 253: 173–176.

Molhave, K., Wich, T., Kortschack, A., and Boggild, P. 2006. Pick-and-place nanomanipulation using microfabricated grippers. *Nanotechnology* 17: 2434–2441.

Poncharal, P., Wang, Z.L., Ugarte, D., and De Heer, W.A. 1999. Electrostatic deflections and electromechanical resonances of carbon nanotubes. *Science* 283: 1513–1516.

Prinz, V.Y., Seleznev, V.A., Gutakovsky, A.K. et al. 2000. Free-standing and overgrown InGaAs/GaAs nanotubes, nanohelices and their arrays. *Physica E* 6: 828–831.

Resch, R., Baur, C., Bugacov, A. et al. 1998. Building and manipulating 3-D and linked 2-D structures of nanoparticles using scanning force microscopy. *Langmuir* 14: 6613–6616.

Rueckes, T., Kim, K., Joselevich, E. et al. 2000. Carbon nanotube-based non-volatile random access memory for molecular computing science. *Science* 289: 94–97.

Schaefer, D.M., Reifenberger, R., Patil, A., and Andres, R.P. 1995. Fabrication of 2-dimensional arrays of nanometer-size clusters with the atomic force microscope. *Applied Physics Letters* 66: 1012–1014.

Sheehan, P.E. and Lieber, C.M. 1996. Nanomachining, manipulation and fabrication by force microscopy. *Nanotechnology* 7: 236–240.

Shen, T.C., Wang, C., Abeln, G.C. et al. 1995. Atomic-scale desorption through electronic and vibrational-excitation mechanisms. *Science* 268: 1590–1592.

Sitti, M., Horiguchi, S., and Hashimoto, H. 2000. Controlled pushing of nanoparticles: Modeling and experiments. *IEEE/ASME Transaction on Mechatronics* 5: 199–211.

Subramanian, A., Dong, L.X., Tharian, J., Sennhauser, U., and Nelson, B.J. 2007. Batch fabrication of carbon nanotube bearings. *Nanotechnology* 18: 075703.

Wang, Z.L. 2007. The new field of nanopiezotronics. *Materials Today* 10: 20–28.

Wautelet, M. 2001. Scaling laws in the macro-, micro- and nano-worlds. *European Journal of Physics* 22: 601–611.

Weir, N.A., Sierra, D.P., and Jones, J.F. 2005. A review of research in the field of nanorobotics. Sandia Report: SAND2005-6808.

Whitesides, G.M. and Grzybowski, B. 2002. Self-assembly at all scales. *Science* 295: 2418–2421.

Whitman, L.J., Stroscio, J.A., Dragoset, R.A., and Cellota, R.J. 1991. Manipulation of adsorbed atoms and creation of new structures on room-temperature surfaces with a scanning tunneling microscope. *Science* 251: 1206–1210.

Williams, D.B. and Carter, C.B. 1996. *Transmission Electron Microscopy: A Textbook for Material Science.* New York: Plenum Press.

Wolf, E.L. 2004. *Nanophysics and Nanotechnology.* 1st ed. Berlin, Germany: WILEY-VCH.

Yu, M.F., Lourie, O., Dyer, M.J. et al. 2000. Strength and breaking mechanism of multiwalled carbon nanotubes under tensile load. *Science* 287: 637–640.

Zhang, X.B., Bernaerts, D., Tendeloo, G.V. et al. 1994. The texture of catalytically grown coil-shaped carbon nanotubules. *Europhysics Letters* 27: 141–146.

Zhang, L., Deckhardt, E., Weber, A., Schonenberger, C., and Grutzmacher, D. 2005. Controllable fabrication of SiGe/Si and SiGe/Si/Cr helical nanobelts. *Nanotechnology* 16: 655–663.

Zhang, L., Dong, L.X., Bell, D.J. et al. 2006a. Fabrication and characterization of freestanding Si/Cr micro- and nanospirals. *Microelectronic Engineering* 83: 1237–1240.

Zhang, L., Ruh, E., Grützmacher, D. et al. 2006b. Anomalous coiling of SiGe/Si and SiGe/Si/Cr helical nanobelts. *Nano Letters* 6: 1311–1317.

Zhang, L., Dong, L.X., and Nelson, B.J. 2008a. Bending and buckling of rolled-up SiGe/Si microtubes using nanorobotic manipulation. *Applied Physics Letters* 92, 243102.

Zhang, L., Dong, L.X., and Nelson, B.J. 2008b. Ring closure of rolled-up Si/Cr nanoribbons. *Applied Physics Letters* 92, 143110.

45

MRI-Guided Nanorobotic Systems for Drug Delivery

Panagiotis Vartholomeos
*National Technical
University of Athens
Zenon Automation Technologies*

Matthieu Fruchard
University of Orleans

Antoine Ferreira
ENSI Bourges

Constantinos Mavroidis
Northeastern University

45.1 Introduction

This chapter focuses on the state of the art in the emerging field of nanorobotic drug delivery systems guided by magnetic resonance imaging (MRI) devices. Such systems employ the latest nanotechnology breakthroughs and are used to perform optimized drug delivery in the human body, and exhibit substantially increased rates of therapeutic and diagnostic success compared to conventional drug delivery methods. This application lies in the realm of *nanomedicine*, which according to Freitas can be defined "as the process of diagnosing, treating, and preventing disease and traumatic injury, of relieving pain, and of preserving and improving human health, using molecular tools and molecular knowledge of human body." The great potential of this emerging nanomedicine tool stems from the capabilities offered by *nanotechnology*, i.e., the capabilities to conduct engineering activities at a nanometer scale, i.e., at the level of atoms and molecules. The size-related challenge is the ability to measure, manipulate, and assemble matter with features on the scale of 1–100 nm. The tools required to accomplish

such tasks at that scale are the *nanorobots*, which are defined as controllable machines at the nanometer scale that are composed of nanoscale components and algorithmically respond to input forces and information (Ummat et al. 2006). The field of *nanorobotics* studies the design, manufacturing, programming, and control of the nanoscale robots.

There are many differences between macroscale and nanoscale robots. However, they occur mainly in the basic laws that govern their dynamics. Macroscale robots (usually having a volume of several cubic cm or even m) are governed by Newtonian mechanics and are mostly affected by inertial and gravitational forces. If the dimensions of the macroscale robot reduce by three orders of magnitude, it becomes a microrobot. Its motion behavior is also described by Newtonian mechanics, but now it is the surface effects (associated with adhesion forces, heat transfer, and friction) that dominate and govern the motion of a body. As the size is further reduced, down to the nanoworld, Newtonian mechanics still hold and are sufficient to describe the motion of the bodies down to approximately 10 nm (Abbott et al. 2007a, Dong and Nelson 2007). However, the additional three orders of magnitude in size reduction introduce an entire group of effects that have not been encountered in greater sizes. Handling of objects at this scale is strongly affected by intramolecular/inter-atomic chemical bonding forces, which are stronger than surface forces and thus become dominant. The most challenging question in nanorobotics is to obtain an insight of the underlying physics that govern the end of the classical scaling range and to exploit them.

The concept of nanorobotics for medical applications has been inspired by the function of antibodies, i.e., the natural organisms that our body uses to kill or repair diseased cells. Only objects of the size of 100–200 nm can be navigated through the thinnest sections of the vasculature system and would be capable of interacting with cells. Furthermore, if the nanorobots acquire simple functionalities to sense and target specific cells, and to release drug molecules upon reception of a triggering signal, then a radically new approach in medicine would be realizable to conduct curative and reconstructive treatment in the human body at the cellular and molecular level in a controllable manner. Such a groundbreaking advancement would signal a new era in medical diagnosis and treatment of neoplasms, hepatitis, diabetes, and other diseases, where highly controlled and targeted treatment is critical for the survival of the patients. Such nanorobotic devices will hopefully be part of the arsenal of future medical devices and instruments that will (1) perform operations, inspections, and treatments of diseases inside the body and (2) achieve ultra-high accuracy and localization in drug delivery, thus minimizing side effects.

The advent (in the early 1990s) of submicron colloidal systems, made from synthetic or natural polymers (generally termed as nanoparticles) has led to prolonged circulation, controlled drug absorption, improved therapeutic efficiency, and reduced side effects (Langer, 2001). Still this treatment remains a conventional

treatment where the targeting capability at the molecular level is far from being realizable. The idea of MRI-guided nanorobotic systems is to use an MRI to apply to the nanoparticles an external driving force to guide them and retain them at a localized target. The direction and magnitude of the forces applied on the nanoparticles are generated according to a control law, whose feedback—the endovascular position of the nanoparticles—is provided by processing the MRI image. Navigation in combination with specific biomolecules attached on the nanoparticles surface yields improved localization accuracy, increased rates of drug accumulation at the diseased tissues and organs and also controlled duration of therapeutic action. Consequently, a more localized and controlled treatment, be it diagnosis or drug release can be achieved. Developing this system of navigated nanoparticles and allowing the human to intervene in the process for increased targeting accuracy, is a first step in the long, gradual transition from conventional passive endovascular drug circulation to molecular-level active targeted treatment.

Many diverse technologies and disciplines are involved in the development of an MRI-based nanorobotic drug delivery system. Nanorobotics is a field, which calls for collaborative efforts between physicists, chemists, biologists, computer scientists, engineers, and other specialists to work toward this common objective. Figure 45.1 details various scientific domains, which come under the field of nanorobotics (this is just a representative figure and not exhaustive in nature). Currently this field is still developing, but several substantial steps have been taken by great researchers all over the world, who are contributing to this ever challenging and exciting field.

The sections that follow explain the importance of an MRI-based nanorobotic drug delivery system and present its main building blocks. More specifically, Section 45.2 introduces the reader to the main goals of the nanorobotic system, presents

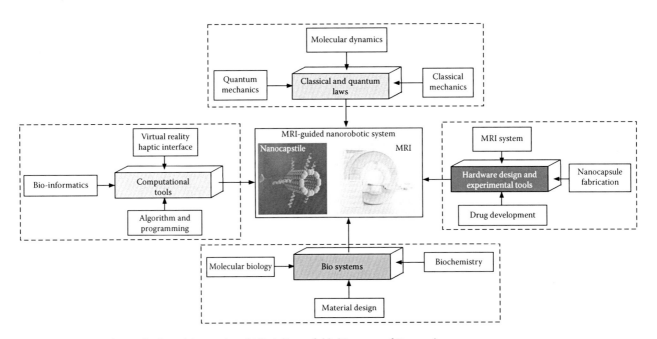

FIGURE 45.1 Nanorobotics for drug delivery: A multidisciplinary field. (Courtesy of Siemens.)

the system architecture, and briefly describes the functionality of each subsystem and how it interfaces to the surrounding subsystems. Section 45.3 provides the essential background knowledge on the principles of systemic drug delivery using nanoparticles–no robotic intervention is assumed at that stage. This piece of information is useful to the reader because it describes the problem of controlled drug delivery and presents the physical process within which the nanorobotic system operates. Section 45.4 provides a description of the nanostructures that are considered suitable candidates as drug-loaded carriers. This section briefly discusses the techniques for drug loading and for developing nanocarriers with targeting capabilities and other critical properties for efficient endovascular circulation. Section 45.5 is devoted to the description of the magnetically guided nanoparticles. These are magnetic nanoparticles that act as nanocapsules themselves or are embedded into other larger in size nanocapsules. Their physical properties are of great importance because they determine the drivability of the system. This section explains the concept of superparamagnetism, why this property is a requisite for most endovascular nanorobotic applications, and mentions the underlying physical properties of superparamagnetic nanostructures. Section 45.6 briefly describes the modeling and computational tools, i.e., the tools required for simulations, conceptualization and characterization of the nanoscale capsules, and their behavior during endovascular navigation. Section 45.7 describes control and navigation techniques for the nanocapsule. This section mainly refers to magnetic steering and the software architecture that is responsible for its implementation. Finally, Section 45.8 concludes this chapter by summarizing the main aspects of the MRI-based nanorobotic drug delivery system, the challenges encountered, their scientific and medical importance and the future perspectives.

45.2 Architecture of MRI-Guided Nanorobotic Systems

45.2.1 General Description

MRI-guided nanorobotic systems aim at diagnosing and treating diseases in cells, organs, or tissues. Their proper function relies on providing engineering and scientific solutions in three technical challenges:

1. *Enhanced diagnostics:* MRI is an advanced imaging system that provides 3D visualization offering the radiologists a detailed three-dimensional view of the tissue or organ of interest. Nanorobotic systems are detectable by MRI without creating artifacts.
2. *In vivo propulsion and navigation:* The MRI system is employed for propulsion and navigation of the nanocapsules. The propulsion of a micro or nanocapsule with magnetic properties in the cardiovasular system is realized through the induction of forces and torques by magnetic gradients and magnetic fields respectively, generated by the MRI coils. The nanocapsules are guided in vivo

to the targeted organs or are accumulated to the tumors' capillary networks. Also, the MRI forces retain the nanocapsules at the target site.

3. *Drug delivery and release:* The magnetic nanocapsules are loaded with drug molecules. Moreover, their surface is chemically processed (coated with polymers and bioconjugated with specific antibodies) so that on one hand they are not detected by the immune system and on the other hand they have increased chances to bind to the receptors of malignant cells. When the nanocapsules are inside the cell or at its vicinity, the release of the drug is triggered. The architecture for implementing the MRI-based nanorobotic system is depicted in Figure 45.2.

Four main *subsystems* can be identified:

- *The Graphical User Interface Module*, which comprises input command prompt, 3D-visualization, and process supervision tools.
- *The Control Module*, which comprises (a) the high-level controller responsible for the nanocapsule navigation tasks and for the generation of the magnetic field gradients and (b) the low-level controller (manufacturer MRI controller) responsible for implementing the actuation commands for the generation of the desired field gradients and for the image acquisition tasks.
- *The Hardware Module*: MRI and the Nanocapsules. This is the controlled hardware, which comprises (a) the MRI hardware and software systems and (b) the nanocapsules that have been injected within the vasculature and are navigated by the field gradients.
- *The Tracking Module*, which comprises the (a) MRI image reconstruction and (b) the image-processing software that estimates the position and accumulation of the nanocapsules within the vasculature, the tissues, and the organs of the human.

45.2.2 Subsystem Description

45.2.2.1 Nanocapsules

This is the most important subsystem of the MRI-based nanorobotic system, and is composed by components (carbon-based structures or molecular elements) that function as actuators, sensors, drug delivery mechanism, and drug release mechanisms. The nanocapsules' drivability, their size, their drug-loading capabilities, and their drug release mechanisms are critical design parameters that largely affect nanocapsules' localization and rates of drug absorption by cancer cells. The determination of the nanocapsule design specifications depends on the MRI system architecture, and on the environment, within which the nanocapsules are found, be it cardiovascular or cell membrane. From a robotics perspective, the components integrated on a nanorobotic capsule can be classified as follows: (1) *Actuation module* such as iron oxide magnetic nanoparticles, which together with the MRI field gradients serve as the propulsion mechanism of the nanocapsule. (2) *Sensor module*,

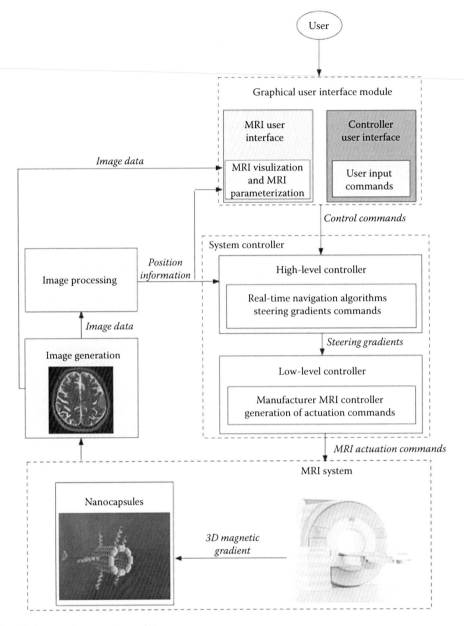

FIGURE 45.2 MRI-guided nanorobotic system architecture.

such as proteins and other biomolecules that respond to external stimuli, to release drug or to detect and bind to the receptors of malignant cells. (3) *Carrier module*, which constitutes the main body of the nanocapsule and is responsible for carrying (by encapsulation or surface conjugation) drug molecules, sensing biomolecules, and polymer decorations that are needed for biocompatibility and sustained circulation of the nanorobotic capsules. These modules, their structure, their physical properties and their design shall be examined in Section 45.4.

45.2.2.2 MRI Propulsion System

The overall concept of the in vivo MRI-tracking system is based on the fact that both tracking and propulsion is possible

by employing the gradient coils of the MRI system. Software-based upgrading of a clinical MRI system is the least expensive approach to convert a platform that is used for imaging to an effective interventional platform. At any instant, only one of the functions can be applied, i.e., either imaging or guidance. However, both can be executed over the same MRI interface. For the implementation of a system capable of tracking and of propelling these particles, the MRI interface has to be shared, and a time-division-multiple-access scheme has to be developed.

The propulsion of the nanocapsules is achieved by generating gradient fields with the gradient coils of the MRI. The exact properties of the gradients needed to propel the nanocapsules have to be simulated and programmed into the MRI system.

Constraints are imposed by the technical limitations of the MRI system (field strength, maximum gradient, slew rate, cooling system) and by the physical properties of the magnetic particles (size, magnetization, etc.), which shall be discussed in more detail in Section 45.5. Typical maximum values of a commercial 3T MRI system are

- Maximum gradient: 45 mT/m
- Slew rate: 200 T/m/s

For example, assuming such values and a duty cycle 50%, the maximum applied field gradient would be 22.5 mT/m.

45.2.2.3 MRI Tracking System

The recognition and tracking of the particles is done by image processing algorithms. Depending on the size of the particles, the observed effect will be either similar to contrast agents or relatively big artifacts and distortion are visible. The algorithms used depend on the observed effect. MRI sequences are employed and tuned to show the desired magnitude of the particle influence. Fast image acquisition and processing is necessary to later enable real-time control of the nanocapsule. The time consumed for imaging purposes should be reduced as much as possible to enable stronger and longer propulsion gradients.

The image-processing software has to run as near as possible to the image generation to avoid communication latencies and delays by the transfer of images which can hinder fast controller reaction. On the other hand, the MRI host computer as well as the MRI workstation are both certified clinical systems, which may not be modified. Therefore, an MRI computer has to be installed and be directly connected to the MRI workstation. This computer would access the built-in interface of the MRI workstation to ensure low latencies. Also, the image-processing software would be executed on this machine. The controller and user interface could be executed on a remote computer. The image processing module is organized in several sub-modules, which usually include

- MRI interface/image acquisition
- Artifact recognition
- Artifact tracking
- Interface to controller

45.2.2.4 Control System

High-level controller, i.e., endovascular navigation can be accomplished by integrating real-time control algorithms with MRI propulsion system and tracking events. The control navigation algorithm is coordinated through the development of proprietary control modules embedded in the clinical MRI system. Optimal navigation performance requires different trade-offs in terms of refresh rate, duty cycle of the propulsion gradients, and repetition time of the tracking sequence. Automatic and stable trajectory tracking require robust controller implementation without modifying the hardware of clinical MRI systems. Different perturbations should be taken into account during the controller design process.

45.3 Nanoparticles as Controlled Drug Delivery Systems

The MRI nanorobotic system aims at enhancing the efficacy of nanoparticles for targeted and controlled drug delivery. Therefore, it is important to introduce the reader to the nanoparticle-controlled drug delivery concept, to its evolution during the past years, to the physical mechanisms that govern delivery and drug release, the challenges, the technology, and to the related terminology. To this end, this section is devoted to provide background on nanoparticle drug delivery systems.

The "magic bullet" concept that was first conceived by Paul Ehrlich in 1891 represents the first early description of the drug-targeting paradigm (Gensini et al. 2006). The concept of controlled drug targeting is to deliver drugs to the right place at the right concentration at the right period of time. Controlled drug targeting in combination with novel drug-carrier technologies are called controlled drug delivery systems (DDSs). It is evident that DDSs can substantially improve the therapeutic and toxicological properties of existing chemotherapies and facilitate the implementation of new ones. By including the drug molecules in technologically optimized drug delivery systems, it is possible to modify *pharmacokinetics* (i.e., the absorption, distribution, metabolism, and excretion of the drug in the organism), and eventually attain improved *biodistribution* (the degree to which a drug becomes available to the target tissue after administration), better therapeutic activity, and minimized intensity of side effects.

45.3.1 Temporal and Distribution Control

The dosage and its administration rate should result in drug temporal and spatial concentrations that are below a toxic drug level and above a level of minimal therapeutic effect. This range of desired drug concentration is called *therapeutic window*. Unfortunately, the body does not handle drugs in a way that maintains the drug time profile or distribution profile within the desired therapeutic window. On one hand, systemic circulation will distribute the drug throughout the body and, consequently, systemic toxicity will affect organs, which are in no need for the drug, and on the other hand, the body always tries to remove foreign substances including drugs according to a *clearance rate*. It is evident that to attain the idealized time profile, a drug should be administered in a controlled way. Two types of control over drug release can be achieved: temporal and distribution control.

In *temporal control*, a drug delivery system aims to deliver the drug over an extended duration or at a specific time during treatment. Controlled release over an extended duration is highly beneficial for drugs that are rapidly metabolized and eliminated from the body after administration. An example of the benefit is shown in Figure 45.3 in which drug concentration at the site of activity within the body is compared after immediate release from three injections administered at 8 hourly intervals and after extended release from a controlled released system.

Drug concentrations may fluctuate widely during the 24 h period when the drug is administered via bolus injection; this means that

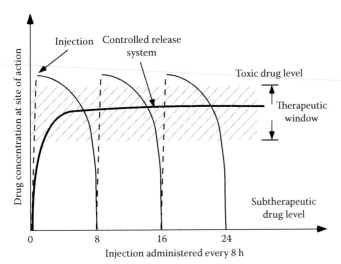

FIGURE 45.3 Drug concentrations at site of therapeutic action after delivery as a conventional injection (thin line) and as a temporal-controlled release system (bold line). (Adapted from Uhrich, K. et al., *Chem. Rev.*, 99, 3181, 1999.)

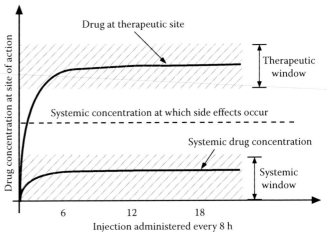

FIGURE 45.4 Drug delivery from an ideal distribution controlled release system. Upper curve: Concentrations at site of therapeutic action. Lower curve: Systemic levels at which side effects occur. (Adapted from Uhrich, K. et al., *Chem. Rev.*, 99, 3181, 1999.)

only for a portion of the treatment period is the drug concentration in the *therapeutic window* (i.e., the drug concentration that produces beneficial effects without harmful side effects) (Uhrich et al. 1999). With the controlled release system, the rate of drug release matches the rate of drug elimination and, therefore, the drug concentration is within the therapeutic window for the vast majority of the 24 h period. Clinically, temporal control can produce a significant improvement in drug therapy. A typical example is the opioid painkiller administered to a patient with terminal cancer: any time the drug concentration is below therapeutic concentration, the patient experiences pain (Uhrich et al. 1999).

In *distribution control*, drug delivery systems aim to target the release of the drug to the precise site of activity within the body. The benefit of this type of control is shown schematically in Figure 45.4, where the drug concentration at the site of activity and the side effect production are compared.

There are two principal situations in which distribution control can be beneficial. The first is when the natural (systemic) distribution causes drug molecules to encounter tissues and cause major side effects that prohibit further treatment. This situation is often the cause of chemotherapy failure when bone marrow cell death prevents the patient from continuing and completing the drug treatment. The second situation is when the natural distribution of the drug prevents drug molecules to reach their molecular site of action. For example, a drug molecule that acts on a receptor in the brain will not be active if, although it is distributed by the patient's blood system, it cannot cross the blood–brain barrier (BBB) (Deboer and Breimer 1994).

45.3.2 Drug-Loaded Nanoparticles

A DDS aims to accomplish both temporal and distribution control. The fundamental idea of a DDS is to associate the drug molecules to a nontoxic and biodegradable carrier such as polymers

micelles, liposomes, gold nanoparticles, and metallic nanoshells. The carrier is designed and processed so that it modifies the pharmacokinetics of the drug molecules in a predefined manner. Drug delivery carriers are usually *nanoparticles*, i.e., have dimensions less than 100–200 nm. The primary reason for nanometer dimensions is that the vessels become thinner and thinner and are finally converted to capillaries that lead to the vicinities of the individual cells and have cross-sectional diameters of at least 2000 nm. For efficient transport through such capillaries, the nanoparticles should be smaller than 300 nm (Gupta and Kompella 2006). Furthermore, in treating the diseases one needs to use the same scale, whether it is correcting a faulty gene, killing a cancer cell, or repairing the cellular metabolism. With the advent of nanoparticles, it is now possible to selectively influence the cellular processes at their scales. Some nanoparticles exhibit unique optical properties, which make them not only suitable for therapeutic use but also for bioimaging. Also, nanoparticles can easily be suspended and form colloids. Finally, nanoparticles dissolve at a much faster rate (and generally yield faster and stronger reactions) because the number of molecules present on a particle surface increases as the particle size decreases, i.e., a much greater percentage of the reactant is exposed to the reaction. A comparison of some biological entities at the nanometer scale is presented in Table 45.1.

In the case of temporal control, the primary goal of DDS nanocarriers is to protect the drug molecules from the surrounding aqueous environment for preprogrammed periods of time. Typical protection mechanisms can involve (1) delaying the dissolution of drug molecules, (2) inhibiting the diffusion of the drug out of the device, or (3) controlling the flow of drug solutions. Carriers employed to delay drug dissolution aim to slow the rate at which drug molecules are exposed to water from the aqueous environment surrounding the drug delivery system. For example, a polymer-coating matrix that dissolves at a slower rate than the drug can achieve this. In diffusion-controlled

TABLE 45.1 Average Dimensions of Various Biological Entities

Carbon Atom	0.1 nm
DNA double helix diameter	3 nm
Ribosome	10 nm
Virus	100 nm
Bacterium	1,000 nm
Red blood cell	5,000 nm
Human hair	50,0000 nm

Source: Gupta, R.B. and Kompella, U.B., *Nanoparticle Technology for Drug Delivery*, Taylor & Francis, New York, 2006.

release, the diffusion of drug molecules is inhibited by the insoluble polymer–matrix that encapsulates the drug. In the case of flow-rate control, water molecules cross the semipermeable membrane (that encapsulates the drug molecules) due to high osmotic gradient. Consequently, the drug molecules dissolve in water and flow through pores at a controlled rate.

45.3.3 Nanoparticle Functionalization

For distribution control, it is required that the carriers be equipped with localization ability in disease sites, i.e., they should target specific malignant cells. In the special case of tumors, two targeting modalities may be defined: *passive* targeting and *active* targeting. Passive targeting is made possible, to a certain extent, by the irregular nature of the tumor it self. It has been determined from pathological, pharmacological, and biochemical studies that tumors possess the following characteristics: hypervascularity, irregular vascular architecture, and absence of effective lymphatic drainage that prevents efficient clearance of macromolecules accumulated in the solid tumor tissues (Matsumura and Maeda 1986, Folkman 1995, Tsukioka et al. 2002). More specifically, tumor blood vessels are distinct from normal vessels in that the endothelial cells in tumors possess wide fenestrations, ranging from 200 nm to 1.2 μm (Hobbs et al. 1998). The large pore sizes allow the passage of nanoparticles into the extravascular spaces and accumulation of nanoparticles inside tumors. These characteristics constitute the basis of the enhanced permeability and retention (EPR) effect (Matsumura and Maeda 1986). Furthermore, macromolecules as well as nanoparticles (30–50 nm) have relatively prolonged plasma half-lives (the time taken by plasma concentration to reduce by 50%) because they are too large to pass through normal vessel walls. Therefore, unless they are trapped by the reticuloendothelial system* (RES), they extravasate into and accumulate within tumor tissues through the EPR effect. Passively targeted nanoparticle agents, such as liposomes, take advantage of both the tumor microphysiology and of their inherent size (Cuenca et al. 2006).

To exploit further this EPR effect and to optimize the targeting capabilities of the drug, active targeting should be employed. This means that the drug carriers should be *functionalized*, i.e., to have their surface chemically modified in order to acquire desired properties. Active targeting is accomplished by attachment of specific molecules on the carrier's surface, thereby enhancing the binding and interactions with antigens or receptors expressed on target cell populations. Vector molecules capable of recognizing tumors include antibodies, lectins, peptides, hormones, folate, and vitamins (Akerman et al. 2002, Paciotti et al. 2004). For example, monoclonal nuclear antibodies can be attached to the drug carriers to promote drug release only to tumor cells and not normal cells (Iakoubov et al. 1995). Interestingly, the high affinity of folic acid for folate receptors provides a unique opportunity to use folic acid as a targeting ligand to target chemotherapeutic agents to cancer cells (Orive et al. 2005). In vitro experiments using folate-tethered liposomes containing calcein or doxorubicin showed a selective drug release in both human cervical cancer HeLa-IU1 cells and human colon cancer Caco-2 cells, because such malignant cells overexpress folate receptors (Zhang et al. 2004). Several groups have reported the use of antibody-conjugated nanoparticles to localize cell surface proteins like c-erbB2 (Wu et al. 2003), epidermal growth factor receptor (EGFR) (Nida et al. 2005), and CA-125. Akermann et al. used quantum dots (see Section 45.4) conjugated to peptides that were specific for either blood or lymphatic vessels to demonstrate specific targeting of vessels (Akerman et al. 2002).

For efficient temporal and distribution control, the drug molecules must have long permanence in the body. Evading *immunogenic reactions* (i.e., bio-recognition and rejection of the nanocarrier by the immune system) is a critical factor for reducing clearance rates and achieving prolonged drug circulation time. This means that critical *phagocytic cells* of the immune system, such as monocytes, macrophages, microglia, Kupffer liver cells, and spleen red pulp macrophages cells, should be circumvented. For acquiring an insight on the immune system refer to Purves et al. (2008). The phagocytic cells migrate through the body—guided by chemoattractant gradients—to encounter, ingest, and degrade the foreign particles or organisms (in this case the drug-loaded nanocarriers). Hepatic and splenic cells can rapidly phagocytize polymeric nanoparticles. The mechanism of phagocyte activation is based on molecules called opsonins, which promote and assist the process of phagocytosis. Prevention of opsonin adsorption can be accomplished via nanocarrier surface modification and has proved to be a successful strategy for sustained circulation. The most well-known surface modifier examined is poly(ethylene glycol) (PEG) (Kingsley et al. 2006).

Moreover, proper functionalization and processing of controlled drug delivery carriers also addresses critical pharmaceutical properties such as solubility, stability, metabolic stability, and cellular permeability. Additional functionalization may include conjugate magnetic nanoparticles for imaging or steering through an MRI system and fluorescent elements for imaging and tracking of the nanocarrier. The idea of a multifunctionalized drug-loaded nanocarrier is schematically

---

* The RES is part of the immune system and critically affects the effectiveness of the DDS. It consists of phagocytic cells, primarily monocytes and macrophages that are accumulated in the spleen and lymph nodes. The Kupffer cells of the liver are also part of the RES.

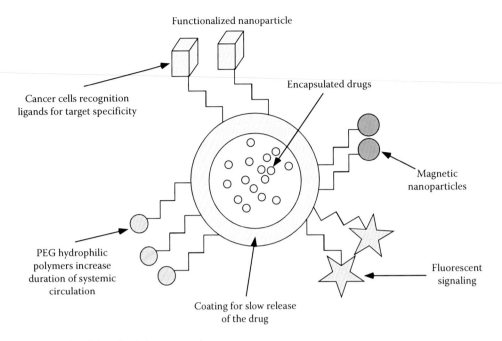

FIGURE 45.5 Multifunctionalized drug-loaded nanoparticle.

TABLE 45.2 Nanoparticle-Based Therapeutics FDA Approved

Product/Brand Name	Nanoparticle Drug Component	FDA Approved Indications	Approval Date
Doxil Caelyx	Pegylated doxorubicin (adriamycin)HC1 liposomes	Metastatic ovarian cancer and AIDS-related Kaposi's sarcoma	1995
Abraxane	Paclitaxel-bound albumin nanoparticles (~130 nm)	Metastatic breast cancer (after combination therapy has failed)	2005
Triglide	Nanocrystalline fenofibrate	Lipid disorders	2005
DepoCyt	Sustained release cytarabine liposomes	Lymphomatous menigitis	1999
Pegasys	Peginterferon alfa-2a	Chronic hepatitis C virus infection	2002
Elestrin	Estradiol gel (0.06%) incorporating calcium phosphate nanoparticles	Treatment of moderate to severe hot flashes in menopausal women	2006
Oncaspar	Pegasparginase	Leukemia	1994

Source: Bawa, R., *Nanotechnol. Law Bus.,* 5(2), 135, 2008.

depicted in Figure 45.5. The functionality of the magnetic nano-particles that appear attached to the coating of the nanoparticle shall be explained in Section 45.5.

45.3.4 Medical Applications and Related Drugs

The potential applications of nanoparticles as therapeutic and diagnostic agents in medicine are vast. There is enormous excitement regarding nanomedicine's potential impact in the fields of diagnostics and therapy. Drug delivery nanoparticles present novel therapeutic opportunities for active agents (drugs or genes) that were previously unsuited to traditional oral or injectable therapeutic formulations, allowing active agents to be delivered efficaciously while minimizing side effects and leading to better patience compliance. Nanoparticle-based therapeutics have enormous potential in addressing the failures of traditional therapeutics that formulated due to factors such as poor solubility or lack of target specificity. Although there are only a few

FDA*-approved nanoparticle-based therapeutics on the market, these formulations are already impacting medicine and promise to alter healthcare (Bawa 2008). Table 45.2 presents FDA-approved drugs that are based on nanoparticle agents. Table 45.3 presents selected nanoparticle-based therapeutics that are in the process of clinical trial (Kingsley et al. 2006, Bawa 2008).

The majority of pharmaceutical research into the utility of nanoparticle drug delivery systems is focused on the area of oncology for the treatment of breast cancer, lung cancer, prostate cancer, endometrial cancer, and cervical cancer. Also, nanoparticle-based drugs have been developed that are responsible for antiangiogenesis (Gupta and Kompella 2006). Furthermore, nanoparticle drug delivery is particularly useful for disorders of the central nervous system (CNS) because some nanoparticles are able to cross the BBB. Moreover, a large number of classes of drugs can benefit from controlled drug delivery systems.

---

* Food and Drug Administration

TABLE 45.3 Nanoparticle-Based Therapeutics in Clinical Trial

Product/Brand Name	Nanoparticle Drug Component	Indications	Approval Status
CALAA-01	Cyclodextrin-containing siRNA delivery nanoparticles	Various cancers	Phase I
INGN-401	Liposome FUS-1	Metastatic, non-small cell lung cancer	Phase I
Aurimmune	Colloidal gold nanoparticles coupled to TNF and PEG-Thiol (~27)	Solid tumors	Phase II
AuroShell	Gold-coated silica nanoparticles (~150)	Solid tumors	Phase I

Source: Bawa, R., *Nanotechnol. Law Bus.*, 5(2), 135, 2008.

These classes include chemotherapeutic drugs (Dang et al. 1994, Walter et al. 1995), immunosuppressants (Katayama et al. 1995), anti-inflamatory agents (Wagenaar and Müller 1994), antibiotics (Schierholz et al. 1997), opioids antagonist (Falk et al. 1997), steroids (Ye and Chien 1996), hormones (Johnson et al. 1996), anesthetics (Maniar et al. 1994), and vaccines (McGee et al. 1994).

45.4 Nanorobotic Carrier Technologies

The nanocarrier is of fundamental importance in a conventional or nanorobotic DDS because it determines the drug pharmacokinetics and cellular penetration. The carrier's physical properties determine to a large extent the efficiency and capabilities of the DDS. Under the term nanocarrier, it is possible to distinguish several reservoirs and technologies. All these capsules differ not only in the structure but also in their biopharmaceutical properties and therapeutic uses. The fabrication protocol of each particle differs considerably and the scale-up could be a challenge for some of these devices. Natural and synthetic polymers including albumin, fibrinogen, alginate, chitosan, and collagen have been used for the fabrication of nanoparticles.

45.4.1 Liposome Nanoparticles

Liposomes are one of the most well-known drug delivery carriers employed in the treatment of cancer. Due to their advantages, liposomal formulations provide a substantial increase in antitumor efficacy compared with free drug or standard chemotherapy regiments. Currently, they are used as drug carriers for administration of several classes of drugs like antiviral, antifungal, antimicrobial, antitubercular, vaccines, and gene therapeutics (Immordino et al. 2006). Liposomes are lipid-based vesicles of nanometer size (typically less than 400 nm). Their membrane is made of phospholipids, which are molecules that have hydrophilic head (attract water) and hydrophobic hydrocarbon tail (repel water). In a phospholipid bilayer (shown in Figure 45.6), the external surface of each layer is composed of phospholipid heads that line up and face the water molecules. Internally to the bilayer, the hydrocarbon tails of one layer faces the hydrocarbon tails of the other layer. These spherical vesicles contain a core of aqueous drug and offer effective protection for the physically entrapped solution

against hydrolysis and enzymatic degradation (Kiparissides and Kammona 2008) (see Figure 45.6). Furthermore, amphiphilic and hydrophobic molecules can be solubilized within the phospholipid layer and in this way liposome can carry both hydrophobic molecules and hydrophilic molecules (see Figure 45.7). A typical

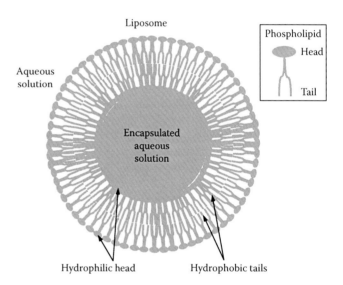

FIGURE 45.6 Liposome formulated by phospholipids.

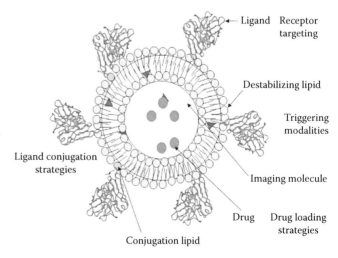

FIGURE 45.7 Functionalized liposome. (Reproduced from Puri, A., http://ccr.cancer.gov/staff/gallery.asp?profileid = 6797. With permission.)

mechanism of delivering the drug molecules to sites of action is the fusion of the lipid bilayer with other bilayers such as the cell membrane, thus delivering into the cell, the liposome contents (drug solution, DNA, etc.), which otherwise would be unable to diffuse through the cell membrane.

The circulation time of liposomes is reduced mainly by the RES and other mechanisms of the immune system (Kingsley et al. 2006) (see Section 45.3). Other factors that limit the therapeutic potential of non-functionalized liposomes are nonspecific uptake and membrane instability (Sahoo and Labhasetwar 2003, Cuenca et al. 2006). To overcome the aforementioned limitations, research on liposome technology has advanced from conventional vesicles, i.e., "first generation liposomes" to "second generation liposomes" with increased bioavailability and pharmacokinetics, in which long-circulating liposomes are obtained by modification of the lipid composition and functionalization of the vesicle surface by various molecules, such as glycolipids, sialic acid, and PEG (Park 2002, Kiparissides and Kammona 2008). Furthermore, liposomes constructed by novel lipid polymers have resulted in significantly increased membrane stability and bioavailability (Torchilin 2005). Also, liposomes that were recently synthesized with self-hydrolyzable lipids may allow for time-controlled release of drugs (Ahmed and Discher 2004). Nonspecific toxicity has been addressed by direct molecular targeting of cancer cells via carrier coating with antibody-mediated or other ligand-mediated interactions (Park 2002, Cuenca et al. 2006).

45.4.2 Polymer Micelles

Polymers micelles have attracted significant attention as nanoscale carriers for the delivery of low-mass drugs, proteins, genes, and imaging agents (Bontha et al. 2006). Polymer micelles are formed by self-assembly of amphiphilic block or graft copolymers in aqueous solutions. They are characterized by a unique core–shell architecture, where hydrophobic blocks are segregated by the aqueous exterior to form an inner core surrounded by a shell of hydrophilic polymer chains. This core–shell architecture is essential for their use as novel functional materials for pharmaceutical applications. The core of the micelles formulates a hollow space that encapsulates water-insoluble drugs or diagnostic agents. Poly(ethylene oxide), PEO, is frequently used as hydrophilic block of micelles-forming copolymers (Kwon and Kataoka 1995). The resulting hydrophilic shell is a brush-like corona (see Figure 45.8) that stabilizes the micelles in aqueous dispersion. This formation affords effective protection of the physically entrapped drug against hydrolysis and enzymatic degradation and allows for prolonged blood circulation times of the polymer micelles (Bontha et al. 2006).

The stability of the micelles can be increased via the synthesis of copolymers with cross-linkable groups. The most commonly used amphiphilic block copolymer is "pluronic," a ternary copolymer of PEG and poly(propylene) (PPO). Upon micellization, PPO forms the core while PEG forms the shell (corona). PEG is preferred as hydrophilic polymer because it is nontoxic,

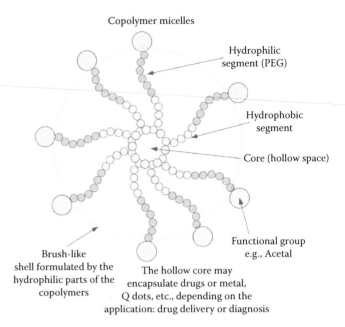

FIGURE 45.8 Copolymer micelle structure.

uncharged, thus preventing undesired electrostatic interactions with plasma proteins and minimizes opsonization (i.e., minimizes the rate of phagocytosis) by effectively reducing micelle uptake by the reticuloendothelial system ("also called stealth micelles"). Similarly to the case of the liposomes, multifunctional micelles can be prepared through conjugation of targeting ligands (e.g., folic acid, RGD peptide, antibodies, RNA aptamer, and carbohydrates like glucose, lactose, etc.) to their shell aiming to induce specific targeting and uptake by the cells. The ligands can be attached either to the block copolymer prior to micellization or to the chemically available sites on the micelle surface. Micelles with sensitivity to external stimuli (e.g., pH, temperature, light, chemical) can also be prepared in order to trigger drug release at the target site (Kiparissides et al. 2006, Sutton et al. 2007).

A very interesting multifunctional drug-loaded antitumor agent called "nanocell," which relies on the concept of polymer micelle has been developed at MIT by Sengupta et al. (2005). This drug-loading method is based on a core–shell architecture, comprising a nanoscale pegylated-phospholipid block copolymer envelope coating a nuclear nanoparticle. Onto this nanoparticle, a chemotherapeutic agent is conjugated and an antiangiogenesis agent is trapped within the lipid envelope. Sengupta et al. proposed that the disruption of this envelope inside a tumor would result in a rapid deployment of the antiangiogenesis agent, leading to vascular collapse and the intra-tumoral trapping of the nanoparticles. The subsequent slow release of the cytotoxic agent from the nanoparticle should then kill the tumor cells.

45.4.3 Quantum Dots

Quantum dots (QDs) are novel semiconductor nanocrystals with broad potential for use in various applications in the research, management, and treatment of cancer (Bruchez et al. 1998, Seydel 2003, Cuenca et al. 2006). These tiny light-emitting particles on the nanometer scale are rapidly emerging as new class of fluorescent probes for biomolecular and cellular imaging. QDs owe their fluorescence emission to electron excitation and exhibit an intrinsic fluorescence emission spectra wavelength between 400 nm and 2000 nm depending on their size and composition (Alper 2005, Cuenca et al. 2006). In comparison with organic dyes and fluorescent proteins, quantum dots possess unique optical and electronic properties that allow them to be tunable to discrete narrow frequencies and also offer them improved signal brightness, resistance against photobleaching, and simultaneous excitation of multiple fluorescent colors (Medintz et al. 2005, Cuenca et al. 2006, Vo-Dinh 2007). These properties are most promising for improving the sensitivity of molecular imaging and quantitative cellular analysis by 1–2 orders of magnitude (Vo-Dinh 2007).

QDs themselves are hydrophobic, but as in the previous nanoparticle cases, they can be solubilized by using amphiphilic polymers that contain both a hydrophobic and hydrophilic segment. The hydrophobic group strongly interacts with tri-n-octylphosphine oxide on the QD's surface, whereas the hydrophilic segments face outward and render the QDs water soluble (Vo-Dinh 2007). For bioconjugation, several types of electrostatic, hydrophobic, and covalent binding have been developed for linking QDs to biomolecules. The resulting conjugates combine the properties of both materials, that is, the spectroscopic characteristic of the QDs and the biomolecular function of the surface-attached entities. Owing to its finite size (comparable to or slightly larger than that of many proteins), a single QD can conjugate several proteins simultaneously, creating a multifunctional nanoparticle–biological hybrid (Medintz 2003). Because of their composition of heavy metals—and previous reports of cytotoxicity—the potential use of quantum dots in humans may be limited (Derfus et al. 2004). Uncoated or non-polymer-protected QDs are unstable when exposed to UV radiation and have been shown to release toxic cadmium (Derfus et al. 2004, Cuenca et al. 2006). Research on the subject of QDs toxicity has indicated that QDs with appropriate stable polymer coating are essentially nontoxic to cells and animals (no effect on cell division or ATP production) (Derfus et al. 2004, Cuenca et al. 2006, Vo-Dinh 2007).

Also results in drug delivery have been reported in (Gao et al. 2004), where the original CdSe QDs were modified by the addition of an impermeable coating of polymer that prevented the leaking of highly toxic cadmium ions from the QD conjugate and provided means to chemically attach tumor-targeting molecules and also allowed drug delivery functionality. In this example, the QDs were conjugated to peptides or antibodies to target human tumor (grown in mice). To prevent tissue damage from the QDs energy emissions, QDs were tuned to radiate in the infrared region. The drug carried by the QDs would be released only when the nanoparticle is targeted by laser light; this helps to control the group of cells that will receive the toxin, thus minimizing the side effects.

45.4.4 Metallic Nanoshells

Metallic nanoshells are colloid nanoparticles coated with a thin metallic layer to form core shell nanoparticles. Two of the most commonly used metals for the synthesis of metallic nanoshells are gold and silver. The core of the nanoshell is composed of polymers such as Teflon or Latex nanosize spheres. Other cores such as alumina, silica, or titanium dioxide can also been employed. The optical properties of nanoshells are different from those of quantum dots. Nanoshells rely on the plasmon-mediated conversion of electrical energy into light (Alper 2005). Similar to quantum dots, nanoshells have the ability to be tunable optically and have emission/absorption properties that range from the UV to the infrared (Alper 2005, Cuenca et al. 2006). Metallic nanoshells are attractive because they offer imaging and potential therapeutic properties similar to those of quantum dots without the possibility for heavy metal toxicity. They have been used in vivo as a contrast agent for imaging with optical coherence tomography and photoacoustic tomography. Nanoparticles of gold half-coated polysteren microspheres containing ferromagnetic material have been developed and have been used as magnetically modulated nanoprobes (Anker et al. 2003).

Metallic nanoshells are most commonly used in medical applications for gene diagmostics, bioimaging, cancer imaging, and intracellular analysis (Anker et al. 2003, Vo-Dinh et al. 2005). Results on drug and gene delivery using metallic nanoshells have also been reported by Halas and coworkers in Sershen et al. (2000). In this case, gold nanoshells whose shell–core ratio allows strong absorption of light in the near infrared (where tissue is relatively transparent) are embedded into a copolymer matrix. Within the copolymer matrix, drug dosage is also entrapped. The copolymer is thermally responsive and exhibits critical solution temperature slightly above body temperature. Illuminating the nanoshells at their corresponding resonance wavelength causes them to transfer heat to the surrounding copolymer matrix gradually increasing its temperature. When the critical solution temperature of the copolymer is exceeded, the matrix collapses and the drugs held within are released.

Metallic nanoshells have been functionalized through surface modification to acquire hydrophilic properties, noncolloidal behavior, molecular recognition and targeting capabilities, and avoidance of immune response. To this end, the gold surfaces of nanoshells are coated with PEG, which through mechanisms described in previous sections reduce protein and cell adsorption on the particles and thus reduce the rate of clearance by the spleen and liver and increase the duration of systemic circulation of in vivo applications (Wang et al. 2004, Loo et al. 2005). The adsorption or grafting of polymers onto particle surface also addresses the problem of aggregation of nanoshells in an aqueous suspension due to their colloidal nature. The repulsive forces and solvation layer of the PEG surface moiety prevent aggregation. For in vivo targeting applications, antibodies are tethered onto the surface of gold nanoshells through covalent linkage. Description of protocols for antibody conjugation and PEG

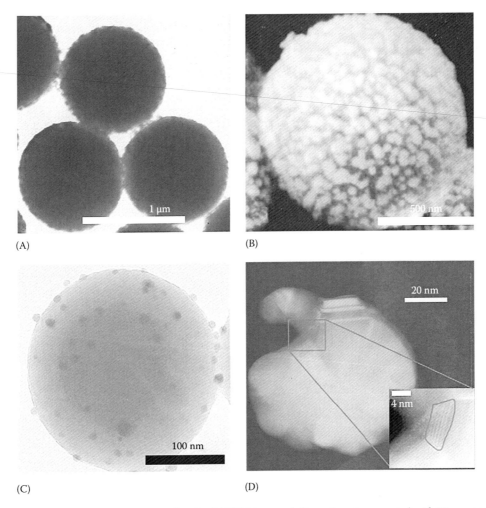

(A)   (B)

(C)   (D)

FIGURE 45.9   Silver nanoshells observed by a SEM and TEM. (A) TEM image of silica sphere 1 μm, coated with 20 nm monocrystalline silver. (B) SEM image of sphere from the same batch. (C) TEM image of 250 nm sphere, with 5–10 nm silver seed decorating the surface. (D) High-resolution TEM of polycrystalline silver grain. (Reproduced from Peterson, M.S.M. et al., *J. Colloid Interface Sci.*, 306, 2007. With permission.)

coating of metallic nanoshells are briefly described in Vo-Dinh (2007). Figure 45.9 demonstrates SEM and TEM images of silver nanoshells.

45.4.5 Gold Nanoparticles

Since 1971, gold nanoparticles, also known as colloidal gold or nanogold, have been used in biological applications (Vo-Dinh 2007). The multifaceted abilities of gold have been investigated in depth right form the electronics industry to medicinal formulations of both modern and traditional practices. Gold nanoparticles usually have size in the range of 10–40 nm, are strong absorbers, photostable, nontoxic, easily conjugated to antibodies or proteins, and have adjustable optical properties (Letfullin et al. 2006). Their optical properties are based on the surface plasmon resonance and have provided excellent detection capabilities for applications such as immunoblotting (i.e., technique for the specific recognition of very small amounts of protein), flow cytometry (i.e., technique for counting, examining, and sorting microscopic particles suspended in a stream of

fluid), and hybridization assays (Vo-Dinh 2007). Gold nanoparticles have been employed to provide photodynamic therapy of malignant tissues. The technique relies on the use of short laser pulses focused on monoclonal antibodies conjugated with light-absorbing microparticles and nanoparticles (Hede and Huilgol 2006). Furthermore, it has been shown that nanogold has 600 times more absorption in cancer cells than normal cells. This property in conjunction with the optical properties of nanogold has been utilized for the detection and imaging of malignant cells. Figure 45.10 demonstrates gold SEM-FEG images of gold nanoparticles (Wostek-Wojciechowska et al. 2004).

The ability of gold nanoparticles to strongly bind with biological molecules has been utilized to target cancer tissues by tagging the nanoparticle with a suitable antibody. This antibody–nanogold conjugate are used as vectors for delivering drug molecules or ionizing radiation in the form of radioactive nanogold (Hainfeld and Powell 2000, Hede and Huilgol 2006). For addressing immunogenicity, surface modification of gold nanoparticles through the PEG spacer is employed. This allows the modified nanoparticles to remain in the systemic circulation for

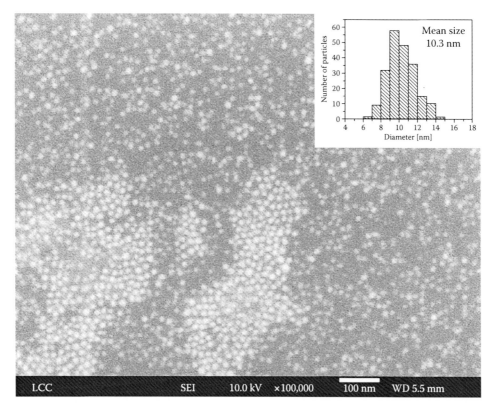

FIGURE 45.10 SEM-FEG images of gold nanoparticles. The mean size of the nanoparticles is demonstrated in the bar chart. (From Wostek-Wojciechowska, D. et al., *Mater. Sci.-Poland*, 22(4), 407, 2004. With permission.)

a prolonged period and provide flexibility to the attached ligand for efficient interaction with its target.

45.4.6 Carbon Nanotubes

One of the most promising candidates for drug delivery applications are carbon nanotubes (CNTs). Their structure can be understood by considering first a perfect graphene sheet. This is a planar layer made of carbon atoms that use three coplanar sp^2 hybrid orbitals to covalently bond to three other carbon atoms to formulate a lattice of coplanar hexagons (Ramsden 2000). These graphene sheets are then rolled up to form a seamless hollow tube. This is the simplest form of carbon nanotube and is called single-wall carbon nanotube (SWNT) (Figure 45.11). These tubes can be capped at the ends by a fullerene-type hemisphere, but are often open. A more complicated carbon nanotube structure

FIGURE 45.11 A TEM image of a SWNT. (From Baughman, R.H. et al., *Science*, 297, 787, 2002. With permission.)

is the multiwall carbon nanotube (MWNT) composed of multiple concentric nanotubes.

These hexagonal networks of carbon atoms exhibit some outstanding characteristics such as exceptional mechanical and thermal properties and extraordinary electrical characteristics. More specifically, they exhibit large aspect ratio (10–1000), terapascal-scale Young's modulus (Yu 2000), excellent elasticity (Walters et al. 1999), ultrasmall interlayer friction (Kis et al. 2006), excellent field-emission properties (Walt et al. 1995), various electric conductivities, and high thermal conductivity (Tombler et al. 2001). Also they exhibit high current carrying capability with essentially no heating and sensitivity of conductance to various physical or chemical changes (Li and Chou 2006). CNTs provide new means and opportunities to improve the distribution and performance of drugs (Liu and Wang 2007). For example, CNTs have relatively large hollow spaces compared to nanoshells, which allows the filling of relatively large amounts of drugs ranging in size from small molecules through to proteins (Mitchel et al. 2002). CNTs have distinct inner and outer surfaces, which can be differentially filled or functionalized allowing for multifunctional carriers (Mitchel et al. 2002). For example, inner fills can be drugs and externally conjugated antibodies can be selected for targeting. CNTs absorb radiation efficiently in the near-infrared region, so they can be used for thermal ablation methods to selectively destroy cancerous cells (Kam et al. 2005). CNTs have open ends, which make the inner surface accessible and subsequent incorporation of molecules within tubes particularly easy (Martin and Kohli 2003). Finally, although no definite conclusions have been drawn on the biocompatibility of CNTs, the first experiment on live animals pursued recently has shown that SWNTs with noncovalently bound PEG can efficiently reach tumor tissues in mice with no apparent toxicity or negative health effects (Liu et al. 2007).

45.4.6.1 Single-Wall Carbon Nanotubes

The diameter of a SWNT is determined by the minimization of the energetic cost to maintain the tubular shape of the nanotube. More specifically, the upper bound on the feasible diameter range is set by theoretical calculations that have shown that the tubular morphology can be maintained up to diameters of ~2.5 nm (Bhushan 2005). For greater diameters, the tubular form is not sustainable because it is energetically more favorable to collapse the single-wall tube into a flattened two-layer ribbon. The lower bound of the diameter range is set by the fact that the shorter the radius of curvature the higher the energetic cost and the stress. A suitable energetic compromise is attained for diameters of ~1.4 nm, the most frequent SWNT diameter encountered regardless of the synthesis technique (at least those based on solid carbon source) when conditions for high SWNT yield are used (Bhushan 2005). No energetic or other physical restrictions are imposed on nanotube length. This depends only on limitations brought by the preparation method and the specific conditions used for the synthesis (thermal gradients, residence time, etc.) (Bhushan 2005). The geometric characteristics described so far have been verified by experimental data. Also, it should be noted that SWNT prepared by the electric arc or the laser vaporization process exhibit similar diameters.

Three different kinds of nanotubes can be defined depending on the orientation of the graphene lattice with respect to the longitudinal axis of the nanotube. The three kinds of nanotubes are (1) the zigzag, where the carbon atoms form a zigzag pattern at the edge of the tube cross section (Figure 45.12a), (2) the armchair where the carbon atoms form an armchair pattern at the edge of the tube cross section (Figure 45.12b), and (3) the chiral, where the hexagonal rings exhibit helicity with respect to the tube axis (Figure 45.12c).

It has been observed that the nanotubes systematically gather into "ropes." This is due to the fact that above and below each hexagon ring electron clouds exist (due to π bonds among electron orbitals perpendicular to the ring) leading to attractive van der Waals forces among the tubes. If SWNTs are isolated, i.e., not allowed to bundle into rope, then each SWNT is considered a single molecule with a unique aspect ratio whose properties are closely influenced by the ways atoms are displayed along the molecule direction. This means that its properties are directly related to the helicity of the carbon rings of nanotube wall and on the radius of curvature.

45.4.6.2 Multi-Wall Carbon Nanotubes

Having described the morphology of the SWNT, it is easy to imagine the MWNTs as a structure in which SWNTs with

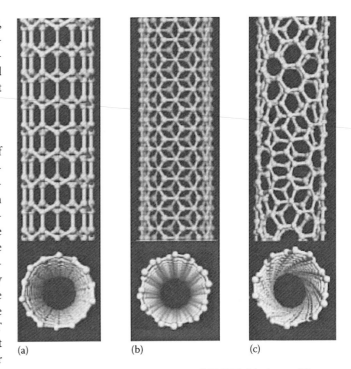

FIGURE 45.12 CAD representation of SWNT: (a) zigzag, (b) armchair, and (c) chiral. Each of them is depicted in a vertical and horizontal arrangement. (From Baughman, R.H. et al., *Science*, 297, 787, 2002. With permission.)

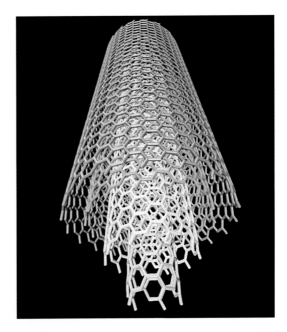

FIGURE 45.13 CAD model of multiwall carbon nanotube. (Courtesy of Alain Rochefort, Nanotechnology-Now.)

regularly increasing diameters are coaxially displayed (see Figure 45.13). The number of walls can be anything starting from two with no upper limit. A characteristic property of MWNTs is that the bond strength depends largely on the direction considered. For example, in-plane direction includes only very strong covalent bonds (~0.142 nm bonds). The direction perpendicular to the plane of the graphene sheets involves only weak van der Waals forces and therefore very loose bonds (~0.34 bond length). In MWNTs, adsorption of molecules (for functionalization purposes) can occur either on or inside the tubes or between aggregated MWNTs.

45.4.6.3 Functionalization of CNTs

As in the case of other nanocarrier candidates, aspects such as functionalization, solubilization as well as dispersion are prerequisites for biomedical applications like cancer targeting using MRI techniques. CNTs functionalization has been accomplished by different methods such as (1) direct covalent functionalization, i.e., direct attachment of functional groups to the surface of the CNT; (2) indirect covalent functionalization, where CNTs are functionalized after they become oxidized by both thermal and chemical oxidation; and (3) non-covalent functionalization of CNTs. This is considered a great alternative to preserve the electronic characteristics of the nanotubes (Geckeler and Rosenberg 2006). The great advantage is the possibility of attaching chemical functionalities to the surface without disrupting the bonding network of the nanotubes. The problem of poor solubility (i.e., they are almost noncompatible with solution chemistry) has been treated in the case of indirect covalent functionalization by the method of sonication, which led not only to disperse products but also to solubilize the metallic residues after thermal oxidation (Geckeler and Rosenberg 2006). A second problem is their tendency to aggregate due to van der Waals forces that are dominant on their surface. Selective cell destruction can be achieved by functionalization of SWNTs with folate moiety, selective internalization of SWNTs inside cells labeled with folate receptor (FR) tumor markers and NIR-triggered cell death, without harming receptor-free normal cells.

45.4.7 Triggered Drug-Release Mechanisms

The use of triggered-release mechanism allows the DDS to become an active participant, where the release of the drug can be activated according to a controlled stimulus as opposed to the case of conventional DDS where the release of the drug is predetermined by the encapsulation material. Several families of molecular assemblies such as the ones described in the previous sections are employed as stimuli-responsive nanocarriers for either passive or active targeting. The composition of each type of carrier can be processed to yield nanocarrier of desired stimuli-responsive property. The benefit of stimuli-responsive nanocarriers is especially important when they are unique to the tissue/organ disease pathology. In this case, the nanocapsule responds specifically to that pathology by releasing drug molecules. Physical properties such as swelling/deswelling, particle disruption, and aggregation change in response to environmental variations. In turn, changing of the value of these properties triggers drug molecules' release. Environmental changes that can trigger the release of drug molecules are pH variations, temperature variations, optical or acoustic stimuli (Kommareddy and Amiji 2005, Shenoy et al. 2005, Torchilin 2007). The most intense research is conducted on pH-based and temperature-based stimuli.

45.4.7.1 pH-Based Release Mechanisms

The pH profile of pathological tissues subject to inflammation, infection, and cancer is significantly different for pH values at healthy tissues (Gerweck and Seetharaman 1996). It is known that the pH at the site of infection, primary tumors, and metastasized tumors is less than the pH of healthy tissue (Ganta et al. 2008). For example, the pH at normal condition is 7.4 and falls to 6.5 after 60 h following onset of inflammatory reaction (Hunt et al. 1986). This behavior can be exploited for the design of stimuli-responsive drug delivery systems. Polymeric nanocarriers that are pH sensitive have been used. For example, the pH-sensitive poly(β-amino ester) (PbAE) constitutes a novel class of biodegradable cationic polymers for the development of site-specific drug delivery systems. In the acidic microenvironment of tumor (pH < 6.5), PbAE undergoes rapid desolution and releases its content (Ganta et al. 2008). Other examples are liposomes, which have also been tailored from pH-sensitive components to achieve sufficient pH sensitivity. The pH-sensitive liposomes are endocytosed in the intact form and fuse with the endovascular membrane as a consequence of the acidic pH inside the endosome and release its active contents into the cytoplasm (Torchilin 2005). Micelles are another example of drug nanocarriers that can be processed to have pH sensitivity and many approaches have been developed to exploit the acidic environment at tumor vicinity to release the micelles drug molecules. The work presented in Bae et al. (2005), describes the development of intracellular pH-sensitive polymeric micelles that can release the anticancer drug, DOX, triggered by the acidic pH at endosomes (pH 5.0–6.0) and lysosomes (pH 4.0–5.0), which resulted in maximization of DOX delivery efficiency to the tumor tissue.

45.4.7.2 Temperature-Based Release Mechanisms

Temperature sensitivity has been extensively investigated to exploit the hyperthermia condition for drug delivery (Meyer et al. 2001). The key concept is that a thermosensitive polymer displays a lower critical solution temperature (LCST) in aqueous solution below which the polymers are water soluble and above which they become water insoluble. The change in the hydration state causes a volume change transition, which has been exploited for drug delivery applications. When thermosensitive polymers are attached on liposome membranes or formulate micelles, their temperature-dependent characteristics endow the nano-vesicles with various temperature-sensitive functionalities. For example, at a certain LCST, polymer chains that formulate a copolymer–micelle turn from hydrophilic into hydrophobic and the micelle collapses. The collapsing of the micelle upon temperature increase may lead to drug release. This way liposomes or micelles acquire control on drug release in a temperature-dependent manner (Kono 2001).

45.5 Magnetically Guided Nanocarriers

45.5.1 Motivation, Past Work, and Related Applications

Magnetically guided NPs have attracted increasing attention by the medical community as potential candidates for (1) diagnosis and visualization; (2) therapeutics, where functionalized and drug-loaded nanocapsules, with embedded magnetic NPs, are guided and retained at a specific location of tissue or organ by means of a magnetic field; and (3) therapeutics where magnetically guided nanoparticles are heated in a magnetic field to produce hyperthermia/ablation of tissue (Arruebo et al. 2007, Misra 2008). Typical applications of magnetically guided NPs are depicted in the diagram of Figure 45.14. As it will be explained shortly, their magnetic properties and their biocompatibility render them tractable and allow their use for in vivo navigation using magnetic field gradients generated externally by MRI coils, magnets, etc. Having such properties, they can be used as nanorobotic sensing and actuation components, especially for in vivo medical applications. In this section, the focus is exclusively on the characterization and the use of magnetic NPs for targeted and controlled drug delivery. These magnetic NPs are employed either as carriers themselves or as embedded nanomagnets into larger carrier structures (such as nanotubes).

Magnetic NPs were initially developed in the late 1980s for diagnostic purposes (Stark et al. 1988). These were superparamagnetic iron oxide contrast agents consisting of 50–100 nm particles that exhibited greater magnetic susceptibility over traditional MR contrast agents such as gadolinium. Also, due to their relatively large size, they had rapid hepatic uptake and thus were useful for hepatic tumor characterization (Suzuki et al. 1996, Cuenca et al. 2006). Subsequently, superparamagnetic iron oxide nanoparticle contrast agents of 5–10 nm size were developed. These had more widespread tissue distribution due to their smaller size, allowing uptake in lymph nodes and bone marrow (Weisseleder et al. 1990). In 1996, the first Phase I clinical trial was carried out by Lubbe et al., in patients whose primary cancer treatments had failed, using magnetic NPs loaded with epirubicin (Lubbe et al. 1996). In more recent efforts, ultrasmall superparamagnetic iron oxide nanoparticles have been used clinically in humans for characterizing lymph node status in patients with breast cancer (Michel et al. 2002), lung cancer, prostate cancer (Nguyen et al. 1999), endometrial cancer, and cervical cancer (Rockall et al. 2005). Numerous start-ups now manufacture magnetic micro- and nanoparticles, which are used in MRI techniques, magnetic fluid hyperthermia, cell sorting and targeting, bio-separation, sensing, enzyme immobilization, immunoassays, and gene transfection and detection systems. Recently, scientists started to conduct research and to employ magnetic NPs conjugated with drug molecules and biomolecules for drug delivery purposes (Jurgons et al. 2006, Arruebo et al. 2007, Cregg et al. 2008, Misra 2008, Balakrishnan et al. 2009).

45.5.2 Synthesis of Magnetic Nanocarriers

Depending on the synthesis process, magnetic NPs can be combined with drug molecules, polymers, and biomolecules to yield different types of nanocarrier pharmaceutical agents. A typical polymer-based approach is to load magnetic nanoparticles (e.g., iron oxides) into the hollow space formulated by the hydrophobic part of copolymer chains (similar processes have been described in Section 45.4.2) and disperse drug molecules among the polymer chains. A characteristic example is given by Jain et al. (2005), where they developed a novel oleic acid (OA)-Pluronic-stabilized

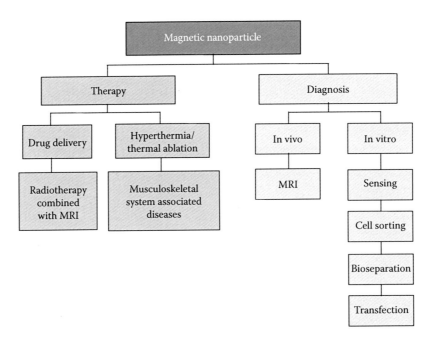

FIGURE 45.14 Applications in medicine of magnetic NPs. (Adapted from Arruebo, M. et al., *Nano Today*, 2(3), 22, 2007.)

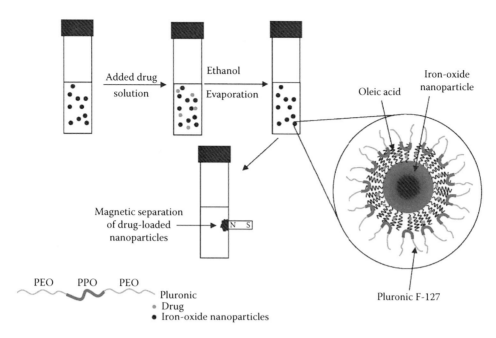

FIGURE 45.15 Schematic representing formulation of iron oxide nanoparticles and the process for drug loading. (From Jain, T. et al., *Mol. Pharm.*, 2(3), 194, 2005. With permission.)

oxide magnetic nanoparticle formulation and characterized it as drug-carrier system for anticancer agents. Their studies demonstrated that the adopted chemical process led the hydrophobic drug to partition into the OA shell and surround the iron oxide core. Also, water dispersity was successfully demonstrated due to the external attachment of Pluronic at the interface of the OA. The process is schematically demonstrated in Figure 45.15.

A different approach is based on functionalization of internally drug-loaded carbon nanotube carriers. The nanotubes can be externally functionalized by attaching chains of magnetic nanoparticles, biomolecules, and polymer parts. Typical steps of MWNT indirect functionalization are treatment with acid

followed by complexation of Fe^{3+}/Fe^{2+} ions and next the introduction of PEGMA-b-AEMA diblock. The strong affinity of the AEMA block toward cations leads to complexation of the polymer with the iron ions instead of micelle formation, and thus their attachment on the MWNTs. Finally, upon addition of a weak base (NH_4OH), iron oxide nanoparticles are formed and are linked on both MWNTs and block copolymers (see Figure 45.16). The latter provide stability to the whole system in water.

A more elaborate study on the subject of functionalization and nanoparticle surface chemistry can be found at Kam et al. (2005), Misra (2008). It is clear, however, that apart from surface chemistry, the key parameters in the pharmacokinetics

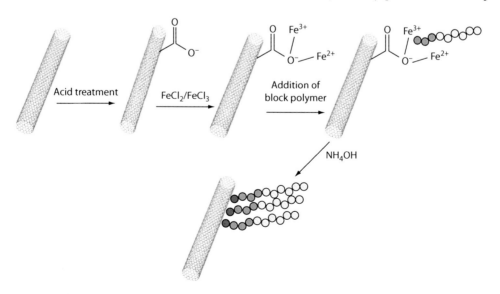

FIGURE 45.16 Preparation of functionalized carbon nanotubes. This example presents functionalization of magnetic nanoparticles and copolymer chains. (Courtesy of Prof. Doumanidis and Prof. Krasia, University of Cyprus, Nicosia Cyprus.)

of magnetic nanocarriers are related to size (magnetic core, hydrodynamic volume, and size distribution), and magnetic properties (magnetic moment, remanence, coercivity) of the magnetic nanoparticles. Furthermore, as it will be shown in the next section, the scaling of magnetic nanoparticles is of major importance for tuning their physical properties and rendering them suitable for control, actuation, and sensing of nanorobotic DDS agents.

45.5.3 Size and Physical Properties of Magnetic NPs

In an MRI-based nanorobotic DDS, the magnetic nanocapsules are guided and/or maintained at a localized target on a tissue by inducing on them appropriate magnetic forces and torques. The magnitude and direction of these forces and torques depend (1) on the external magnetic field gradient (spatial field variation) generated by the MRI coils and (2) on the magnetization properties of the magnetic nanoparticles embedded into the nanocapsule. These physical relations are expressed through the following equations:

$$\vec{F}_m = (\vec{M}V \cdot \nabla)\vec{B} \tag{45.1}$$

$$\vec{\tau}_m = \vec{M}V \times \vec{B} \tag{45.2}$$

where
  $\vec{F}_m$ is the magnetic force (N)
  $\vec{\tau}$ is the magnetic torque (N m)
  $\vec{M}$ is the magnetization of the magnetic nanoparticle (A/m)
  V is the volume of the nanoparticle
  $\vec{B}$ (T) is the magnetic field induced by the MRI coils (where $\vec{B} = \mu_0\vec{H}$ and $\vec{H}$ is the field generated by the MRI coils)

Since the magnetic field gradient decreases with distance to the target, a critical limitation of magnetic drug delivery relates to the strength of the external field $\vec{B}$ that can be applied to obtain the sufficient magnetic gradient $\nabla\vec{B}$ to control the residence time of NPs in the desired area. Other requisites of equal importance for the successful control of the magnetic NPs are high magnetization $\vec{M}$ and large size (large V) of the NPs. It should be noted that magnetic field carriers accumulate not only at the desired site but also throughout the cross section from the external source to the depth marking the effective limit. Therefore, the geometry of the magnetic field is very important and must be taken into account when designing a magnetic targeting process.

On the other hand, a necessary condition for successful circulation of the nanocapsules is the avoidance of magnetic NP agglomerations. Such agglomerations have large size and result in embolization of the vessels. Also, agglomerations having size greater than 300 nm are subject to considerable uptake by parts of the RES system such as the liver and spleen (see Section 45.3). NP agglomerations naturally occur when magnetized NPs are at a close distance to one another and become attracted. Hence, to avoid the permanent formation of agglomerations, the magnetization of NPs should be controlled and when required be diminished. This suggests that the magnetic material be *superparamagnetic*. Superparamagnetism means that once the external magnetic field $\vec{B}$ is removed the nanoparticles loose their magnetization and behave as nonmagnetized particles. Other types of magnetic materials like ferromagnetic or paramagnetic are not suitable, because the former remain magnetized even when the external magnetic filed is removed, and the latter exhibit much lower magnetization compared to superparamagnetic materials. Comparison of magnetization curves of different type of magnetic materials are depicted in Figure 45.17.

The typical process for inducing superparamagnetic properties to ferromagnetic materials is by reducing their size below a critical nanoparticle size. Then, the thermal energy is sufficient to invert the magnetic spin direction responsible for the magnetization of the magnetic nanoparticles. Consequently, in the absence of external magnetic field $\vec{B}$, the spin direction fluctuates

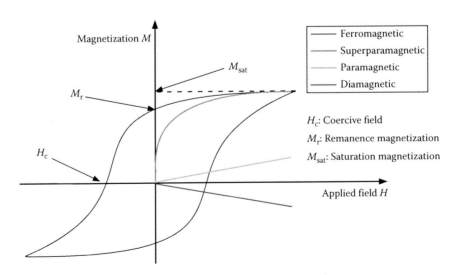

FIGURE 45.17 Magnetization curves for ferromagnetic, paramagnetic, superparamagnetic, and diamagnetic materials.

(due to thermal energy) resulting in a zero net magnetization. However, when an external magnetic field is applied, spins are forced to align in the direction of the external field resulting in nonzero magnetization of the nanoparticle. The critical temperature above which transition occurs from ferromagnetism to superparamagnetism is called blocking temperature T_b and is defined by the following relation:

$$T_b = K_u V / 25k \qquad (45.3)$$

where

K_u is the magnetic anisotropic constant
V is the volume of the magnetic material
k is the Boltzmann's constant

Hence, the smaller the volume, the lower the blocking temperature. In nanoparticles of sufficiently small size, temperature T_b can be 300 K or less. Iron-based NPs become superparamagnetic at sizes <25 nm. A characteristic example is $\gamma - Fe_2O_3$, where nanoparticles of 55 nm exhibit ferromagnetic properties with a coercivity of 4138 A/m at 300 K, but at a size down to 12 nm $\gamma - Fe_2O_3$ nanoparticles show superparamagnetism (Jun et al. 2007). Figure 45.18, demonstrates TEM images and corresponding magnetization loops of iron oxide nanoparticles with size 55 nm (Figure 45.18a and b) and 12 nm (Figure 45.18c and d). As can be observed in plot d, superparamagnetic behavior is characterized by (1) no hysteresis, (2) zero coercive field, and (3) zero remanence magnetization.

As mentioned before, superparamagnetism is a critical property for successful guidance of magnetic nanoparticles and demands that ultra-small nanoparticles be utilized. Therefore, it is important to investigate how fundamental magnetic properties become affected by this extreme scaling. Critical magnetic properties such as susceptibility, coercivity, and magnetic saturation are no longer considered as permanent material characteristics (as in bulk materials). They are strongly affected by size and geometry (Jun et al. 2007). Magnetic coercivity varies as a function of size, as shown in Figure 45.19. Interestingly, just before entering the (zero coercivity) superparamagnetic region, a substantial increase of the NP's coercivity takes place. This phenomenon is due to the transition from larger, multiple domain ferromagnetic structures, to smaller, single-domain superparamagnetic crystals (Jun et al. 2007). In multi-domain nanoparticles, lower reversal magnetic field is required to make magnetization zero because domains cancel each other. On the other hand, magnetic coercivity increases as the size of the single-domain nanoparticle increases. In Figure 45.19, magnetic coercivity is also correlated to the appearance of multiple or single domains.

Saturation magnetization m_s is another physical property of nanoparticles that varies as size scales down. This behavior

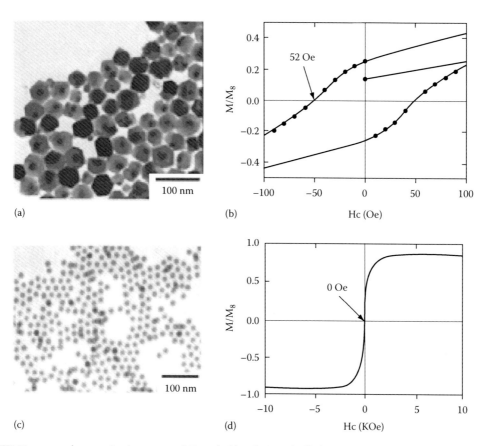

FIGURE 45.18 TEM images and magnetization curves of 55 nm (a, b) and 12 nm (c, d). (From Jun, Y.-W. et al., *Acc. Chem. Res.*, 41(2), 179, 2007. With permission.)

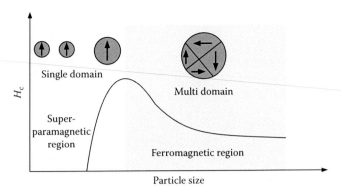

FIGURE 45.19 Variation of coercivity during transition from multi-domain ferromagnetism to single-domain superparamagnetism. (Adapted from Jun, Y.-W. et al., *Acc. Chem. Res.*, 41(2), 179, 2007. With permission.)

is due to the magnetic disordered layers near the surface, i.e., surface canting effects, which become dominant when the ratio of surface over volume is substantially increased. The relation describing magnetic saturation m_s as a function of nanoparticle size is given by

$$m_s = M_s[(r-d)/r]^3 \qquad (45.4)$$

where

  r is the size
  M_s is the saturation magnetization of bulk materials
  d is the thickness of disordered surface layer

Figure 45.20 shows the plot of $m_s^{1/3}$ as a function of r^{-1}.

Apart from magnetic properties like coercivity, susceptibility, and blocking temperature, other important physical properties like spin lifetime τ and anisotropic energy K_u are also affected by size. Furthermore, it is known that shape characteristics like aspect ratio also influence magnetic properties. The interested reader is prompted to refer to Jales (1998), Jun et al. (2007) for a more detailed study.

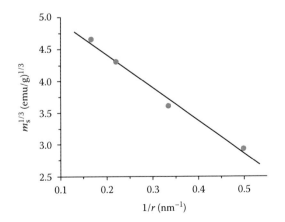

FIGURE 45.20 Plot of $m_s^{1/3}$ as a function of r^{-1}. (From Jun, Y.-W. et al., *Acc. Chem. Res.*, 41(2), 179, 2007. With permission.)

In conclusion, the fundamental magnetic properties are strongly affected by the scaling in the nano range. These scaling relationships can be employed by the nanorobotics engineer to control magnetism from ferromagnetic to superparamagnetic regimes. The challenge in the design of magnetic nanoparticles for nanorobotic DDS becomes apparent; on one hand superparamagnetism is a requisite for successful endovascular navigation and holds only for ultra-small NPs, and on the other hand small size NPs implies a weak magnetization, making it difficult to direct particles and keep them in the proximity of the target while withstanding the drag flow. The drivability of magnetically guided nanoparticles and the limitations imposed by their scale are a subject of current research.

45.6 Nanocapsule Multiphysics Modeling and Computational Tools

The goal of computational nanotechnology is to provide simulation tools for conceptualization, characterization, development, and prototyping of nanoscale capsules for efficient cellular delivery. To aid the designer's understanding of the fundamental properties, the simulation technologies have become also predictive in nature, and many novel concepts and designs have been first proposed based on modeling and simulations. Today, many hundreds if not thousands of entirely different nanovector technology platforms have joined liposomes at this frontier, each with different properties, strengths, and weaknesses. These novel targeted drug delivery systems are mainly tested by *trial and error*. Since this is a very expensive approach, new nanotechnology computational design and prototyping methods are currently developed in order to reduce the risk and uncertainty inherent in the nanocapsule-development process.

Techniques that are currently investigated mainly cover the range of multidisciplines from atomistic mechanics to continuum mechanics (Hamdi and Ferreira 2009, Hamdi, 2009): quantum mechanics, multiparticle simulation, molecular simulation, continuum-based models, stochastic methods, and nanomechanics. The steps in the nanocapsule-targeting of liposomes to cancer cells are shown in Figure 45.21. After the preferential extravasation from hyperfenestrated tumoral vasculature, nanocapsules bind to tumor cells and are endocytosed. We review in this section, the different multiphysical and multiscale models involved in the navigation (phases (1)-(2)), targeting (phase(3)), adhesion (phase(4)), endocytosis (phase (5)), and biodegradation (phase (6)).

45.6.1 Endovascular Navigation

Inspired by the motility or flagella of many bacteria designs, deterministic dynamics models have been proposed where a microrobot capsule is composed of a spiral-type head and an elastic tail (Behkam and Sitti 2006, Li and Zhang 2006) for future controlled drug delivery applications. In order to reach remote areas within the complex cardiovascular pathway, magnetic propulsion (Ritter et al. 1992, McNeil et al. 1995, Ishiyama et al. 2002, Khamesee et al. 2002, 2003, Guo et al. 2005) offers an

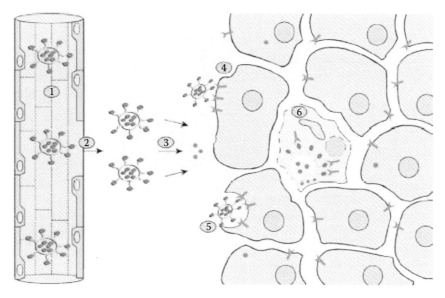

FIGURE 45.21 Schematic diagram illustrating the concept of nanocapsule-targeting of liposomes to cancer cells. Phases (1)–(2): the endovascular navigation of nanocapsules passing through the tumor microvasculature, phase (3): phase (4): the liposomes bind to the folate receptors on the surface of tumor cells *via* the folate ligand, phase (5): liposomes are then internalized by the tumor cells *via* folate receptor–mediated endocytosis. Finally, phase (6) diffuse the drug within the cancer cell. (From Sinek, J.P. et al., *Nanodevices for the Life Sciences*, Vol. 4, Wiley, New York, 2006.)

advantage at such a scale over other proposed actuation methods (Kosa et al. 2005, Yesin et al. 2006) for operation in the human body. Mathieu et al. (2003) mention the difficulty of propelling microdevices in a cardiovascular system due to the relative wide range of vessels' diameters. Thus, conveying the nanocapsule in vessels such as arteries and arterioles depend on the fluids mechanics at nanoscale. In a fluidic environment, and within an external magnetic field, the untethered nanocapsule is subjected to the apparent weight (i.e., the sum of weight and buoyancy forces), magnetic force and torque, and hydrodynamics (see Figure 45.22). The apparent weight is the combined action of weight and buoyancy:

$$\vec{W}_a = V(\rho_b - \rho_f)\vec{g} \qquad (45.5)$$

where
  V is the volume of the bead
  ρ_b and ρ_f are the density of the bead and the density of the fluid, respectively
  $\vec{g}$ the gravitational acceleration

As it was mentioned in the previous section, the magnetic force is related to the magnetic field gradients:

$$\vec{F}_m = \mu_0 V_m (\vec{M}\nabla) \cdot \vec{H} \qquad (45.6)$$

where
  $\mu_0 = 4\pi \times 10^{-7}\,\mathrm{T\,m/A}$ is the permeability of free space
  V_m the volume of the magnetic material in the bead
  $\vec{M}$ the magnetization of the bead
  $\vec{H}$ the external magnetic field
  ∇ being the gradient operator

Magnetic torque is related to the magnetic field and tends to align the magnetization of the bead along the external magnetic field:

$$\vec{T}_m = \mu_0 V_m \vec{M} \times \vec{H} \qquad (45.7)$$

All these forces and torques are volumic ones.

It should be noted that Equations 45.6 and 45.7 are the same as Equations 45.1 and 45.2. They are presented again in this section for reasons of completeness and for clarity.

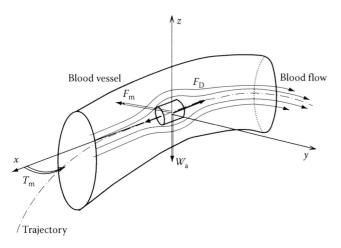

FIGURE 45.22 Kinematics and dynamic motion description of the bead in the horizontal 2-D plane.

Hydrodynamics are modeled by the standard Navier–Stokes equation:

$$\rho_f \left(\frac{\partial \vec{u}_f}{\partial t} + (\vec{u}_f \nabla)\vec{u}_f \right) = \vec{f} - \nabla P + \nabla \cdot \sigma \qquad (45.8)$$

where
$\vec{u}_f$ is the fluid velocity
P the pressure
σ the stress tensor
$\vec{f}$ the body forces acting on the fluid

However, this equation is generically too complex for direct and real-time computations, and the following simplified modeling is usually utilized instead. Under the assumption that the flow is a Newtonian incompressible fluid, hydrodynamic force is referred as drag force:

$$\vec{F}_d = \frac{1}{2}\rho_f u^2 A C_d \frac{\vec{u}}{u} \qquad (45.9)$$

where
$\vec{u} = \vec{u}_b - \vec{u}_f$ is the relative velocity of the bead with respect to the flow, its norm is denoted as u
A is the reference area
C_d the drag coefficient

If the Reynolds number is low, the drag force can be simplified to obtain the Stokes law:

$$\vec{F}_d = 6\pi\mu_f r u \frac{\vec{u}}{u} \qquad (45.10)$$

where
μ_f is the fluid viscosity
r is the Stokes radius

The Stokes' drag given by Equation 45.10 depends on the radius and is linear with respect to velocity. At the opposite, the generic drag force given by Equation 45.9 depends on the area, and is no longer linear because of the square of the velocity. In addition, the drag coefficient is strongly nonlinearly related to the Reynolds number, which is proportional to velocity. Hydrodynamic drag forces are expressed for an infinite extent of fluid, which is not the case in medical applications where the robot is devoted to navigating inside blood vessels, cerebrospinal fluid inside the brain, etc. These boundary effects have an influence on the terminal velocity of the robot in the fluid, which is taken into account using wall effects laws (Goldmann et al. 1967).

At nanoscale, using the Stokes' drag is a good approximation since the Reynold's number is low, which provides a linear model for hydrodynamics, but some other nanophysics effects may dominate in the nanocapsule modeling. Physically based simulation includes kinetics and frictional aspects for object motion with hydrodynamics at low Reynolds number (Re) (Adler 1966,

FIGURE 45.23 Medical complications can arise due to diabetes problems. Nanorobots use sensors to detect glucose levels in bloodstream. (From Cavalcanti, A., *Recent Patents Nanotechnol.*, 1, 1, 2007. With permission.)

Karniadakis and Beskok 2002). At low Re, simplified models have been proposed for simulating the three most promising methods of microrobot swimming—using magnetic fields to rotate helical propellers that mimic bacterial flagella (Sudo et al. 2006), using magnetic fields to oscillate a magnetic head with a rigidly attached elastic tail (Behkam and Sitti 2006), and pulling directly with magnetic field gradients (Mathieu et al. 2006). In the later case, the 3D MRI-based navigation of a ferromagnetic spherical bead in a fluidic *in vitro* environment is considered through simulation. In addition, it is not clear that the fluid is still Newtonian in either of these cases: since lymphocytes and red blood cells are about some micrometers long, their dimensions are similar to the robot's ones. This implies the need to model nanoscale interactions between blood cells and nanocapsules, as shown in Figure 45.23 (Calvacanti et al. 2007).

Some continuum models have been integrated in a virtual reality simulator (Calvacanti and Freitas 2005), where nanorobots moves through a fluid-filled vessel to locate target regions based on random motion and detection of chemical gradients (Hogg 2007). The simulator does allow multiple nanorobots to operate independently and in a cooperative mode (Calvacanti et al. 2008).

45.6.2 Nanocapsule Binding

Once a cell and nanoparticle or liposome have been brought into proximity *via* nonspecific interactions such as Brownian motion and van der Waals forces, receptor–ligand bond formation may occur. Understanding the kinetics of receptor-mediated cell attachment would be of service in the design and optimization of ligand-conjugated nanocapsules (Sinek et al. 2006). An important question is, what are the strength and density of bonds needed to achieve adequate adhesion under various stresses.

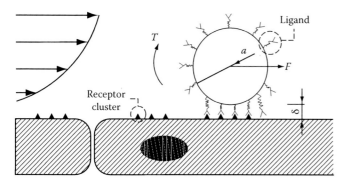

FIGURE 45.24 A spherical nanoparticle adhering to a cell layer under flow. Drawing not to scale, the nanoparticles is 1 or 2 orders of magnitude smaller than the endothelial cells. (Adapted from Decuzzi, P. and M. Ferrari, *Biomaterials*, 29, 337, 2008. With permission.)

A spherical particle is assumed to interact both specifically and nonspecifically with a cell layer under a linear laminar flow. The radius of the particle is a and the shear stress at the wall is μS (Figure 45.24). The particle is decorated with N_l ligands with an uniform surface density ($m_l = N_l/(4\pi a^2)$), whereas the number of receptors N_r with which the particle is specifically interacting is assumed to be fixed and equal to 10, which is generally the characteristic size of a receptor cluster on the cell membrane.

The physiological and biophysical conditions for nanocapsules' firm adhesion have been investigated through deterministic, stochastic, and multiphysics models. Bell's model (Bell 1978) developed a deterministic model for cell attachment in the absence of fluid stress and cell detachment in the presence of fluid stress. The ligand–receptor pairs are described as elastic springs obeying Hook's with a bond spring constant k and an unstressed bond length x. The chemical force exerted on the cell by each bond is then given by $\kappa.(\delta - \xi)$, δ being the stressed bond length. The nonspecific interaction force between the particle and the cell is described through a classical phenomenological relation $\sigma(s + \delta)/(s\delta^2)\exp(-\delta/s)A_c$ (Bell et al. 1984), where σ is the strength and s is the decay length of the nonspecific interactions. Positive values of s are associated with attractive interactions. Ferrari et al. (Decuzzi and Ferrari 2006) defined the area of adhesion on the basis of simple geometrical considerations as

$$\tilde{A}_c = \frac{A_c}{\pi s^2} = \tilde{r}_c^2 = \tilde{a}_2\left[1 - \left(1 - \frac{\tilde{\xi} - \tilde{\xi}_{eq}}{\tilde{a}}\right)^2\right] \quad (45.11)$$

where

$\tilde{r}_c = r_c/s$ is the dimensionless radius of the circular area of interaction

$\tilde{a} = a/s$ is the dimensionless particle radius

This is a reasonable approximation as long as $\tilde{\xi}_{eq}$ is slightly smaller than $\tilde{\xi}$ (small attractive nonspecific forces).

The separation distance δ_{eq} at equilibrium between the particle and the cell substrate can be derived by balancing the specific and nonspecific interaction forces over the area of adhesion.

Therefore, after normalizing the lengths with respect to s $\tilde{\delta} = \delta/s$ and $\tilde{\xi} = \xi/s$, it follows

$$G(\tilde{\delta}_{eq} - \tilde{\xi}) + F\frac{\tilde{\delta} + 1}{\tilde{\delta}^2}\exp(-\tilde{\delta}) = 0 \quad (45.12)$$

with

$$G = \frac{\kappa s^2}{k_B T} \quad (45.13)$$

$$F = \frac{1}{k_B T}\frac{\sigma}{s m_r} \quad (45.14)$$

The force f exerted over the molecular bonds is derived combining the hydrodynamic shear force F, parallel to the vessel wall, and the hydrodynamic torque T, both derived from the work by Goldmann et al. (1967), as described in Piper et al. (1998), to give

$$f = \mu Ss^2\phi(\tilde{\delta}_{eq}, \tilde{a}, \tilde{r}_c) = \mu Ss^2\left[6\pi\left(1 + \frac{\tilde{\delta}_{eq}}{\tilde{a}}\right)F^S + 2\pi\frac{\tilde{a}}{\tilde{r}_c}T^S\right]\cdot\tilde{a}^2 \quad (45.15)$$

where $F^S = 1.668$ and $T^S = 0.944$ for a spherical particle. In Equation 45.15, F is uniformly shared among the ligand–receptor bonds and T is shared uniformly only within the ligand–receptor pairs which are stretched by T. It is important to remark that the above analysis has been developed within the hypothesis that the nanoparticles abruptly adhere to the cell substrate and remain firmly adherent. Recently, Bell's model has been reanalyzed using stochastic mathematical modeling (Gao et al. 2005) and probabilistic modeling (Cozens-Roberts et al. 1990). The authors used a stochastic approach for predicting the adhesion strength of nanoparticles to a cell layer under flow coupled to a mathematical model for the receptor-mediated endocytosis of nanoparticles (Cozens-Roberts et al. 1990). This work allows us to estimate the optimal size and shape of the particles as a function of physiological parameters. These approaches allow "adhesive maps" to be drawn illustrating sites where the adhesive strength of the particle would be larger. For complex structures such as carbon nanotubes (CNTs), molecular dynamics (MD) simulations are preferred to provide the theoretical basis for understanding the crucial factors in targeting and attachment of cell-specific ligands for increased efficiency and selectivity (Decuzzi and Ferrari 2008). As illustration, Figure 45.25 shows the sequence of cell recognition for a computational time $t = 0–8\,ns$. The nanocapsule is coated with specific recognition ligands (peptides, glycoproteins) that bounds to the cell surface in order to perform a site-selective delivery.

45.6.3 Nanocapsule/Cell Phagocytosis

Receptor-mediated endocytosis is one of the most important processes with which therapeutic nanocapsules enter in the infected

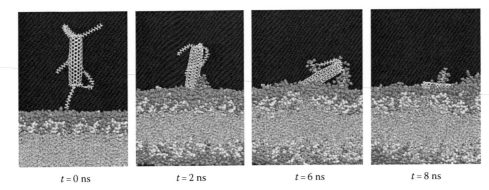

t = 0 ns	*t* = 2 ns	*t* = 6 ns	*t* = 8 ns

FIGURE 45.25 Snapshots of the molecular PEG-peptide and cell recognition for computational time *t* = 0 ns to *t* = 8 ns. (Courtesy of Dr. Mustapha Hamdi, University of Orleans, France.)

cell. Experimental studies on targeted drug delivery into cells have identified particle size as an important factor in cellular uptake of nanomaterials. It has been shown that particles with radii <50 nm exhibit significantly greater uptake compared with particles with radii >50 nm (Desai et al. 1997). Multiphysics models are currently investigated to understand the mechanics of receptor-mediated endocytosis.

Once the contact starts, the receptor density within the contact area is raised to the level of ligand density on the particle surface. Driven by a local reduction in free energy caused by ligand–receptor binding, the receptor in the immediate neighborhood of the adhesion region is drawn to the edge of the contact zone by diffusion, leading to a local depletion of receptors in the vicinity (Figure 45.26a). The diffusive process can be characterized by a nonuniform receptor distribution function $\zeta(s,t)$ (Figure 45.26b).

Receptor-mediated endocytosis analytical models have been proposed. The model, stemming from Gao et al. (2005),

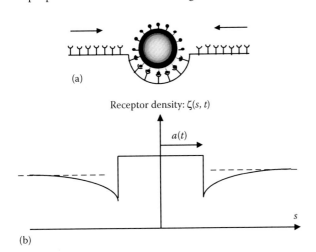

(a)

Receptor density: $\zeta(s, t)$

(b)

FIGURE 45.26 Schematic illustration of the problem. (a) An initially flat membrane containing diffusive receptor molecules wraps around a ligand-coated particle. (b) The receptor density distribution in the membrane becomes nonuniform upon ligand-receptor binding; the receptor density is depleted in the near vicinity of the binding area and induces diffusion of receptors toward the binding site.

assumes that endocytosis is mediated by the formation of stable ligand–receptor bonds and that receptors are collected at the site of adhesion through surface diffusion over the membrane. The explicit expression derived for the dimensionless threshold radius $\tilde{a}_{th} = a/s$ as a function of the ligand–receptor ratio *b* and nonspecific parameter *F* is given as (Decuzzi and Ferrari 2008)

$$\tilde{a}_{th} = H_2\left[\left(C - \frac{1}{2}G(\tilde{\delta}_{eq} - \tilde{\xi})^2 - \frac{F}{\tilde{\delta}_{eq}}\exp(-\tilde{\delta}_{eq})\right) + 1 - \left(\log\beta + \frac{1}{\beta}\right)\right]^{-1/2}$$

(45.16)

where

$$H_2 = \sqrt{2B/\left(s^2 m_1\right)}$$

(45.17)

is a dimensionless parameter related to the membrane bending energy factor *B*, Bk_BT being the bending modulus of the cell membrane, and to the density of ligands available m_1. The parameter *C* is the binding energy factor, Ck_BT being the binding energy, which is related to the surface affinity binding constant K_A^0 through $C = \log[K_A^0 N_A h]$ with N_A the Avogadro number (6.022×10^{23} mol^{-1}) and *h* the effective height of the interfacial volume of solution ($h = 10^{-8}$). The theoretical models give an optimal radius ~27–30 nm for spherical particles. Carbon nanotubes (CNTs) have recently been explored as molecular nanocapsules. Kam et al. (2004) reported that single-walled nanotubes (SWNTs) may enter the cells via endocytosis pathway (length in the range of 100 nm–1 μm and radius in the range of 0.5–2.5 nm). From the analytical models presented in Gao et al. (2005) and Decuzzi and Ferrari (2008), uptake of an isolated CNT is much smaller than the critical radius. The use of fully atomistic computer simulations to probe the interactions of inorganic nanovectors with lipid bilayers is of considerable interest. The accuracies in the atomistic and quantum-mechanical methods have increased to the level whereby simulations have become truly predictive in nature. MD simulations reported by Hamdi and Ferreira (2009)

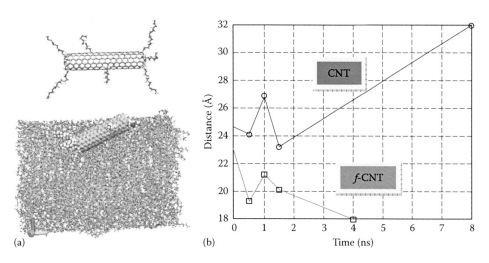

(a)   (b)   Time (ns)

FIGURE 45.27 Functionalized-SWNT (*f*-SWNT)-based nanocapsule spontaneously adsorbs onto the membrane. (a) Organic CNT functionalization with NH_3 peptides. (b) Adsorption distance at $t = 8\,ns$. (Courtesy of Dr. Mustapha Hamdi, University of Orleans, France.)

(Figure 45.27a) show that the endocytosis process can be promoted by certain functionalities on the nanocapsule surface.

When considering amphiphilic CNT, there is mounting evidence that *f*-CNTs are capable of efficient cellular uptake by a mechanism that has not yet been clearly identified (Bianco et al. 2005). However, the nature of the functional group at the CNT surface seems to play a determinant role in the mechanism of interaction with cells. The dynamic simulation studies of Figure 45.27b show clearly that when considering CNTs, the nanovector is not up-taken into the cell and stays at the surface's bilayer with a random thermal fluctuation. On the contrary, the *f*-CNT-based nanovector spontaneously adsorbs onto the membrane, and is gradually embedded by deforming the membrane and inserted into the cell. In a similar way, MD simulations in Lopez et al. (2004) demonstrate that a generic hydrophobic nanotube with hydrophilic functionality at its termini spontaneously inserts into, aligns, and conducts molecules across a lipid bilayer (Figure 45.28). Pore formation occurs in a two-stage process whereby the nanotube is first adsorbed onto and accommodated into the membrane surface with its long axis parallel to the lipid–water interface.

These MD simulations shows that multiphysics and multiscale modeling tools are key components in the nanocapsule's development since the main geometrical, biophysical, and biological parameters govern both cellular binding and uptake events.

45.6.4 Drug Releasing

Once nanocapsules have successfully adsorbed onto the membrane, cellular-level drug kinetics and pharmacodynamics determine modeling concerns. Actually, one major problem is the drug delivery inside the living cell. Even with the phenomenological simplifications (with the Higuchi, power law (Sinek et al. 2006), and Weibull models (Feng and Chien 2003)), cellular-level drug kinetics and transport is highly nonuniform not only because of the inhomogeneous transport of vectors through and extravasation from tumoral vasculature but because of drug gradients due to cellular uptake and metabolism (Sinek et al. 2006). To fully understand the drug-carrier and drug-solvent interaction behaviors and mechanisms, study at the molecular level must be conducted in concert with traditional macroscopic effort.

Ongoing multiscale modeling research makes use of dissipative particle dynamics (DPD) simulations (Maiti et al. 2005) or quantum diffusion dynamics (Panyam and Labhasetwar 2003).

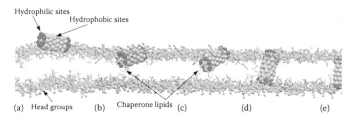

FIGURE 45.28 The functionalized SWNT spontaneously adsorbs onto the membrane. From (a) to (d): Partial immersion subsequently takes place with the SWNT long axis remaining essentially parallel to the membrane plane (b). During this process a few lipids form salt bridges with the hydrophilic termini of the tube. Finally (e), the long axis of the nanotube aligns perpendicular to the bilayer plane allowing the formation of a transmembrane pore, which can now conduct water. (Adapted from Lopez, C.F. et al., *Proc. Natl. Acad. Sci. USA*, 101(13), 4431, 2004. With permission.)

The described nanovector biodegradation computational studies could be used to guide the experimental preparation of drug delivery systems with desired properties.

45.7 Control and Navigation at the Nanoscale

As mentioned in Section 45.2, there are three components necessary for real-time controlled navigation of untethered nanocapsules in the human vasculature, namely: (1) propulsion, (2) tracking, and (3) closed-loop control. These untethered devices have been mainly developed according to three different designs: magnetic bead pulling, biomimetic flagellated robot, and magnetotactic bacteria. Navigation requires observation of the scene in order either to plan the trajectory by off-line mapping, or to correct online the nanocapsule's pose error between the planned and the observed trajectory.

45.7.1 Magnetic Propulsion and Steering

The first propulsion scheme is referred to as magnetic bead: thrust is ensured by pulling the magnetic bead using the magnetic force related to gradients of the external field. This concept was first studied in the 1980s using magnets (Gillies et al. 1994) or superconducting magnets (Quate et al. 1991, Takeda et al. 2007).

The last innovation in this domain has been provided by the Ecole Polytechnique de Montréal, Canada, where the basic idea is the use of magnetic gradients provided by a clinical MRI device to pull the beads (Mathieu et al. 2006, Mathieu and Martel 2007). Such a system combines several advantages: the MRI device provides both fine observation of the scene (thanks to the MRI imager) and actuation of the bead; besides, it makes the implementation easier, since MRI devices are widely implanted in hospitals. This approach is well developed at milli- and microscale, since low-level multiplexed controllers and observers have been developed (Tamaz et al. 2008), and *in vivo* experiments have been leaded on a living animal (Martel et al. 2007) (though the blood flow had been stopped using a balloon catheter).

The main drawback of this approach results from strong limitations on the magnetic gradient amplitude in available MRI devices. As magnetic forces used for propelling are volumetric, whereas the drag force is—at best—dependent on the bead's area, the smaller the bead, the higher the required control forces with respect to hydrodynamic perturbations. Consequently, this approach is well conditioned for beads whose radius is up to a few dozen micrometers with actual MRI devices. At lower scales, it is necessary either to use additional coils to supply higher gradients (Yesin et al. 2006) or to consider other approaches. MRI system upgraded with additional gradient steering coils in order to increase standard MRI gradient amplitudes (100–500 mT) is currently investigated by the authors in Jun et al. (2007).

Another issue is referred to as biomimetic robots using flagella: a red blood cell is fixed to a beating flagella (Dreyfus et al. 2005) imitating the motion of eukaryotic bacterias, or a magnetic bead is attached to a helical nanocoil (Behkam and Sitti 2006) like prokaryotic bacterias. In the first case, the flagella is an elastic rod consisting of magnetic particles, that a periodic transverse magnetic field causes to bend and pivot, inducing a backward motion. In the second case, conversely, the nanocoil is not subjected to any deformation: propulsion is provided by the torque induced by a rotating magnetic field on the bead. In a Stokes flow, swimming is thus obtained through a corkscrew effect in the fluid (Yesin et al. 2005): the higher the rotation frequency of the magnetic field, the higher the thrust. Even if these two swimming methods result from different motions, they are both based on converting mechanical power from the magnetic torque to produce the motion.

Recent results suggest that, under a given size, helical propelling is better than pulling (Abbott et al. 2007b) at low Reynold's number. The comparison is still at the benefit of biomimetic robots as the distance to the magnetic coils increases, which is likely to occur when navigating in the body. However, the assumption of navigating at low Reynold's number can be violated if the rotation frequency becomes too high. Reversing direction can also get complicated because of the rigid rod. Furthermore, this actuation approach is limited in practice due to the difficulty of using it within an MRI device. In fact, additional coils providing a rotational magnetic field in an MRI device is a delicate matter since the precision of an MRI observer relies on a constant magnetic field. This implies the need for an additional imaging system so as to estimate the nanocapsule's pose, otherwise no closed-loop control will be achievable.

Recently, a third approach has been proposed: magnetotactic bacteria. Such bacteria are actuated thanks to embedded or attached ferromagnetic material. The concept had already been studied (Lee et al. 2004) in the field of nanofactory, though not for actuation purpose. Propulsion is supplied directly by the bacteria's flagella, and the magnetic field is used in order to steer the bacteria toward the targeted point (Martel et al. 2009). This technique can also be used in addition to the classic magnetic pulling.

The bacteria are too small to be directly visible by the MRI scanner, however the local magnetic perturbation caused by a swarm of bacteria can be used to locate them, provided that their concentration is sufficient. This raises a crucial issue: the need to control a cloud of robots in order to keep the system observable, which is more difficult than controlling a single robot, since interactions between parts of this multi-agent system have to be modeled. Despite this promising technique, since velocities reached by such systems are about 10 times higher than velocities of biomimetic robots of similar dimensions, it still has to be improved. In fact, some problems are not solved yet (Martel et al. 2006) such as side effects of the fixation of the magnetic material on or in the bacteria, or Joule heating that reduces the bacteria's efficiency and velocity.

45.7.2 Imaging and Tracking in MRI Device

Hydrogen atoms, as the main component of water, are widely spread in the human body, and various tissues can be identified with respect to the hydrogen's density and stability. MRI imaging is based on exciting these atoms and let them relax to equilibrium to obtain a map of their repartition and properties, and thus to get an image of the body tissues in the observed slice. More precisely, clinical MRI imaging can be broken into four major steps. First, the permanent magnet generates a strong and uniform magnetic field in order to align the spin magnetic moment of hydrogen nuclei (protons) along this field. Once polarized, protons are excited by a pulsing RF coil at Larmor frequency to induce a resonance absorption, which modifies the precession motion of the spin. Third, RF excitation is stopped and protons relax to their equilibrium state, producing a free induction decay signal (FID) in the RF coil after some relaxing times. Meanwhile, gradient coils are used first to select the slice to be observed during the polarization step, and second to perform a phase shift/frequency encoding during the readout step, performing a 2D position encoding within the slice. Finally, the slice image is obtained by applying inverse Fourier transforms on the measured FID. Precision of an MRI imaging system is strongly related to the uniformity and strength of the main magnet on the one hand, since they decrease the signal-to-noise ratio. On the other hand, speed and strength of the gradient coils (set by their duty cycle) also affect precision and scanning speed of the imager.

The use of MRI as imaging modality for 3D image guidance of endovascular procedures is of great interest. As stated in Section 45.2, endovascular navigation is rendered possible by integrating propulsion and tracking events using time multiplexing. As stated previously, we have seen how the clinical MRI can be utilized to provide microrobots with locomotion capabilities. The same equipment can be utilized for localization by combining imaging/tracking procedures for precise navigation of an untethered interventional nanorobot. The main drawback

of MRI localization is that the choice of material for fabrication of the microrobot is limited. Ferromagnetic objects cause image artifacts, which are sometimes larger than the object itself to be localized, even though information contained in spatial gradients can overcome this limitation (Sun et al. 2008). Efforts to increase MRI sensitivity have focused on the development of new magnetic core materials (Seo et al. 2006), or in the improvement in nanoparticle size (Jun et al. 2005) or clustering (Lee et al. 2006). An emerging theme in nanoparticle research is to control biological behavior and/or electromagnetic properties by controlling shape. Swarms of nanoparticles were localized in an MRI using contrast agents (Xu et al. 2006), and swarms of magnetotactic bacteria as a single object in Martel et al. (2009). In Park et al. (2008), the authors found that a nanostructure with an elongated assembly of nanoparticles (referred to as nanoworms) influences their efficacy both in vitro and in vivo by enhancing their magnetic relaxivity in MRI. In addition to the trade-off between refresh rate and duty cycle of the propulsion gradients, the repetition time of the tracking sequence is another constraint to be taken into account during imaging and tracking sequences (Tamaz et al. 2008). With insufficient repetition time, MRI signal can be unusable for tracking. Fast and robust image processing algorithms are key issues for future research.

45.7.3 Closed-Loop Control Algorithms

Endovascular navigation through MRI system will be feasible by integrating propulsion and tracking events within the control software that deals with their time sequencing. Optimal navigation performance will require different trade-offs in terms of refresh

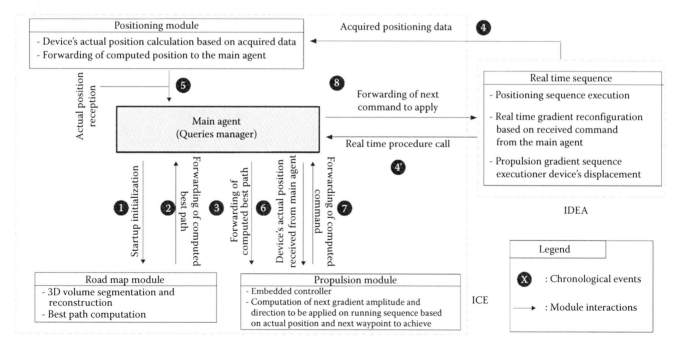

FIGURE 45.29 Basic block diagram representing the software architecture for magnetic resonance propulsion-based applications with the main software modules and communication paths sequentially numbered. (Adapted from Martel et al., *Appl. Phys. Lett.*, 90(14), 105, 2007. With permission.)

rate, duty cycle of the propulsion gradients, and repetition time of the tracking sequence. Recently, the authors in Martel et al. (2007) demonstrated that a ferromagnetic bead was navigated in real time along preplanned trajectories in the carotid artery of living swine at an average velocity of 10 cm/s. The real-time closed-loop control to be executed in the same MRI system is shown in Figure 45.29. The first module known as integrated development environment for applications (IDEA) is the sequence module used to design and run the developed sequences for propulsion and tracking. The other module is the image calculation environment (ICE) and is mainly responsible for the image reconstruction. As the propulsion module used a simple proportional-derivative-integral (PID) controller (Tamaz et al. 2008), the authors reported that navigations instabilities occurred during experiments.

At nanoscale, automatic and stable trajectory control of navigable nanocapsules against environment perturbations (blood flow, drag wall forces) will require advanced controller implementation without modifying the hardware of clinical MRI system due to

- *Magnetic gradients* are used both for observation and control purposes in a time-multiplexed sequence. This sequence design, combined with the MRI overheating avoidance leading to limitations on the MRI duty cycle, tends to increase the disproportional scaling between magnetic forces used for control purpose and perturbation forces (drag forces and net buoyancy forces).
- *Magnetic forces* cannot generically propel the nanocapsule against the blood flow. Therefore, control objectives may be relaxed into steering the nanocapsule within the vessels bifurcations. Predictive controllers including capsule's motion and dynamics with estimation of the versatile flow due to heart pumping are of great interest. Due to nonlinearities of nanocapsule dynamics, advanced control algorithms such as robust Lyapunov-based controllers have to be considered to increase navigation performances.

45.8 Conclusions

This chapter focused on the concept of MRI-guided nanorobotics for conducting diagnostic, curative, and reconstructive treatment in the human body at the cellular and molecular level in a controllable manner. Such nanorobots employ the latest nanotechnology breakthroughs for upgrading the modern conventional drug delivery systems to nanorobotic systems with unique new functionalities and substantially increased rates of therapeutic and diagnostic success. The idea of the MRI-guided nanorobotic system is to use an MRI to apply to the nanoparticles an external driving force to guide them and retain them at a localized target. Navigation in combination with specific biomolecules attached on the nanoparticles surface yields improved localization accuracy, increased rates of drug accumulation at the diseased tissues and organs, and also controlled duration of therapeutic action.

Fully functional, MRI-guided nanorobotic systems are now on the verge of realization. The critical components of the system architecture have reached a sufficient level of technological maturity. Diverse research fields pertinent to nanorobotics and nanomedicine have joined forces to yield fully functionalized, magnetically guided nanoparticles (liposomes, polymer micelles, gold nanoparticles, quantum dots, metallic nanoshells, and carbon nanotubes), which are being mass-produced through standard chemical processes and are readily available in the market. Nanoparticles such as liposomes and polymer micelles have been used extensively by the pharmaceutical industry; they exhibit excellent biocompatibility, they can be loaded with drug molecules, and their properties are well understood. Carbon nanotubes have outstanding physical properties and are considered excellent building blocks for nanorobotic systems. They are on the focus of intense research and already many of the development strategies for functionalizing, processing, and assembling are ready to be applied for the development of fully functionalized nanorobotic devices. Nanocapsules, have been conjugated to polymer/biomolecules that function as sensor modules, and to magnetic nanoparticles embedded into the nanocarrier function as actuator modules as well as sensor modules. From the computational perspective, the state of the art on computational tools and processing power guaranties real-time propulsion and tracking. Simulation tools for conceptualization, characterization, development, and prototyping of nanoscale capsules for efficient cellular delivery have been developed. Dynamic models that predict the behavior of the nanocapsules during critical operations have been proposed. Such models describe endovascular navigation, nanocapsule binding, nanocapsule phagocytosis, and the process of drug release. Several research groups have been experimenting with MRI-based navigation of magnetic particles. This approach has been well developed at milli and microscale, where low-level multiplexed controllers and observers have been designed, and *in vivo* experiments have been conducted on a living animal.

The main drawback of MRI-based navigation stems from the strong limitations on the magnetic gradient amplitude of available MRI devices. As magnetic forces used for propelling are volumetric, whereas the drag force is—at best—dependent on the nanoparticles' area, the smaller the nanoparticle, the higher the required control forces with respect to hydrodynamic perturbations. Consequently, this approach is well conditioned for beads whose radius is up to a few dozen micrometers with actual MRI devices. At the nanoscale, magnetic forces induced by available MRI devices are not sufficient to steer the magnetic nanocapsule within blood vessels. Current research aims to overcome this limitation by upgrading the MRI systems with additional gradient steering coils in order to increase standard MRI gradient amplitudes (100–500 mT).

Acknowledgments

This work was supported by the European Union's 7th FWP (Seventh Framework Programme) and its research area: ICT-2007.3.6 Micro/nanosystems under the project NANOMA: Nano-Actuators and Nano Sensors for Medical Applications.

References

Abbott, J.J., Z. Nagy, F. Beyeler, and B.J. Nelson, 2007a. Robotics in the small, Part I: Microrobotics, *IEEE Robot. Autom. Mag.*, 14(2), 92–103.

Abbott, J.J., K.E. Peyer, M.C. Lagomarsino et al., 2007b. How should microrobots swim? in *International Symposium on Robotics Research*, Hiroshima, Japan.

Adler, J.P., 1966. Chemotaxis in bacteria, *Science*, 153, 708–716.

Ahmed, F. and D. Discher, 2004. Self-porating polymersomes of PEG-PLA and PEG-PCL: Hydrolisis, *J. Control Release*, 96, 37–53.

Akerman, M., W. Chan, P. Laakkonen, S. Bhatia, and E. Ruoslahti, 2002. Nanocrystal targeting in-vivo, *Proc. Natl. Acad. Sci. USA*, 99, 12617–12621.

Alper, J., Shining a light on cancer research, *NCI Alliance for Nanotechnology in Cancer, Monthly Feature*, National Cancer Institute, 2005.

Anker, J.N., C. Behrend, and R. Kopelman, 2003. Magnetically modulated optical nanoprobes, *Appl. Phys. Lett.*, 82, 1102–1104.

Arruebo, M., R. Fernandez-Pacheco, M.R. Ibarra, and J. Santamaria, 2007. Review: Magnetic nanoparticles for drug delivery, *NanoToday*, 2(3), 22–32.

Bae, Y., N. Nishiyama, S. Fukushima et al., 2005. Preparation and biological characterization of polymeric micelle drug carriers with intracellular pH-triggered drug release property: Tumor permeability, controlled subcellular drug distribution, and enhanced in vivo antitumor efficacy, *Bioconjug. Chem.*, 16, 122–130.

Balakrishnan, S., M.J. Bonder, and G.C. Hatjipanayis, 2009. Particle size effect on phase and magnetic properties of polymer-coated magnetic nanoparticles, *J. Magn. Magn. Mater.*, 321, 117–122.

Baughman, R.H., A.A. Zakhidov, and W.A.d. Heer, 2002. Carbon nanotubes—The roote towards applications, *Science*, 297, 787–792.

Bawa, R., 2008. Nanoparticle-based therapeutics in humans: A survey, *Nanotechnol. Law Bus.*, 5(2), 135–155.

Behkam, B. and M. Sitti, 2006. Design methodology for biomimetic propulsion of miniature swimming robots, *ASME J. Dyn. Syst. Meas. Control*, 128, 36–43.

Bell, G.I., 1978. Models for the specific adhesion of cells to cells, *Science*, 200, 618–627.

Bell, G.I., M. Dembo, and P. Bongrand, 1984. Competition between nonspecific repulsion and specific bonding, *Biophys. J.*, 45(6), 1051–1064.

Bhushan, B. (Ed), 2005. *Handbook of Nanotechnology*, Springer-Verlag, Heidelberg, Germany.

Bianco, A., K. Kostarelos, and M. Prato, 2005. Applications of carbon nanotubes in drug delivery, *Curr. Opin. Chem. Biol.*, 9, 674–679.

Bontha, S., A. Kabanov, and T. Brovnich, 2006. Polymer micelles with cross-linked ionic cores for delivery of anticancer drugs, *J. Control. Release*, 114, 163–174.

Bruchez, M.J., M. Moronne, P. Gin, S. Weiss, and A. Alivisatos, 1998. Semiconductor nanocrystals as fluorescent biological labels, *Science*, 281, 2013–2016.

Calvacanti, A. and J.R. Freitas, 2005. Nanorobotics control design: A collective behavior approach for medicine, *IEEE Tans. NanoBioScience*, 4, 133–140.

Calvacanti, A., B. Shirinzadeh, R.A. Freitas, and L.C. Kretly, 2007. Medical nanorobot architecture based on naobioelectronics, *Recent Patents Nanotechnol.*, 1, 1–10.

Calvacanti, A., B. Shririnzadeh, R. Freitas, and T. Hogg, 2008. Nanorobot architecture for medical target identification, *Nanotechnology*, 19, 1–15.

Cozens-Roberts, C., J.A. Quinn, and D.A. Lauffenburger, 1990. Receptor mediated cell attachment and detachment kinetics, *Biophys. J.*, 58, 841–856.

Cregg, P., K. Murphy, and A. Mardinoglu, 2008. Calculations of nanoparticle capture efficiency in magnetic drug targeting, *J. Magn. Magn. Mater.*, 320, 3272–3275.

Cuenca, A.G., H. Jiang, S.N. Hochwald et al., 2006. Emerging implications of nanotechnology on cancer diagnostics and therapeutics, *Cancer*, 107(3), 459–466.

Dang, W., O.M. Colvin, H. Brem, and W.M. Saltzman, 1994. Covalent coupling of methotrexate to dextran enhances the penetration of cytotoxicity into a tissue-like matrix, *Cancer Res.*, 54(7), 1736–1741.

Deboer, A. and D. Breimer, 1994. The blood-brain-barrier—Clinical implications for drug-delivery to the brain, *J. R. Coll. Physicians Lond.*, 28(6), 502–506.

Decuzzi, P. and M. Ferrari., 2006. The adhesive strength of non-spherical particles mediated by specific interactions, *Biomaterials*, 27, 5307–5314.

Decuzzi, P. and M. Ferrari, 2008. Design maps for nanoparticles targeting the diseased microvasculature, *Biomaterials*, 29, 337–384.

Derfus, A., W. Chan, and S. Bjatia, 2004. Proboing the cytotoxicity of semiconductor quantum dots, *Nano Lett.*, 4, 11–18.

Desai, M.P., V. Labhasetwar, E. Walter, R.J. Levy, and G.L. Amidon, 1997. The mechanism of uptake of biodegradable micoparticles in Caco-2 cells is size dependent. *Pharmacy Res.*, 14, 1568–1573.

Dong, L. and B.J. Nelson, 2007. Robotics in the small, *IEEE Robot. Autom. Mag.*, 14(3), 111–121.

Dreyfus, R., J. Beaudry, M.L. Roper et al., 2005. Microscopic artificial swimmers, *Nature*, 437(6), 862–865.

Falk, R., T.W. Randolph, J.D. Meyer, R.M. Kelly, and M.C. Manning, 1997. Controlled release of ionic compounds from poly (image-lactide) microspheres produced by precipitation with a compressed antisolvent, *J. Control. Release*, 44(1), 77–85.

Feng, S.S. and S. Chien, 2003. Chemotherapeutic engineering: Application and further development of chemical engineering principles for chemotherapy of cancer an other diseases, *Chem. Eng. Sci.*, 58, 4087–4114.

Folkman, J., 1995. Angiogenesis in cancer, vascular, rheumatoid and other disease, *Nat. Med.*, 1, 27–31.

Ganta, S., H. Devalapally, A. Shahiwala, and M. Amiji, 2008. A review of stimuli-responsive nanocarriers for drug delivery and gene delivery, *J. Control. Release*, 126, 187–204.

Gao, X., Y. Cui, R.M. Levnson, L.W.K. Chung, and S. Nie, 2004. In vivo cancer targeting and imaging with semiconductor quantum dots, *Nat. Biotechnol.* 22, 969–976.

Gao, H., W. Shi, and L.B. Freund, 2005. Mechanics of receptor-mediated endocytosis, *Proc. Natl. Acad. Sci. USA*, 102(27), 9469–9474.

Geckeler, K.E. and E. Rosenberg (Eds), 2006. *Functional Nanomaterials*, American Scientific Publishers, Valencia, CA.

Gensini, G., A. Conti, and D. Lippi, 2006. The 150th anniversary of the birth of Paul Ehrlich, chemotherapy pioneer, *J. Infect.*, 54(3), 221–224.

Gerweck, L.E. and K. Seetharaman, 1996. Cellular pH gradient in tumor versus normal tissue: Potential exploitation for the treatment of cancer, *Cancer Res.*, 56(6), 1194–1198.

Gillies, G.T., R.C. Ritter, W.C. Broaddus et al., 1994. Magnetic manipulation instrumentation for medical physics research, *Rev. Sci. Instrum.*, 65, 533–562.

Goldmann, A., R. Cox, and H. Brenner, 1967. Slow viscous motion of a sphere parallel to a plane wall. II: Couette flow, *Chem. Eng. Sci.*, 22, 653–660.

Guo, S., J. Sawamoto, and Q. Pan, 2005. A novel type of micro-robot for biomedical appliaction, in *Proceedings of the IEEE/RSJ International Conference on Intelligent Robots and Systems*, pp. 1047–1052.

Gupta, R.B. and U.B. Kompella, 2006. *Nanoparticle Technology for Drug Delivery*, Taylor & Francis, New York.

Hainfeld, J.F. and R.D. Powell, 2000. New frontiers in gold labeling, *J. Histochem. Cytochem.*, 48, 471–480.

Hamid, M., 2009. Computational design and multiscale modeling of a nanoactuator using DNA actuation, Nanotechnol. 20, 485501.

Hamdi, M. and A. Ferreira, 2008. Multiscale design and modeling of protein-based nanomechanisms for nanorobotics, *Int. J. Robot. Res.*, 28(4), 436–449

Hamdi, M. and A. Ferreira, 2009. Virtual reality and multiscale simulation for novel drug delivery nanocapsule designs, *IEEE Nanotechnology Magazine*.

Hede, S. and N. Huilgol, 2006. Nano: The new nemesis of cancer, *J. Cancer Res. Ther.*, 2(4), 186–195.

Hobbs, S., W. Monsky, and F. Yuan, 1998. Regulation of transport pathways in tumour vessels: Role of tumour type and microenvironment, *Proc. Natl. Acad. Sci. USA*, 95, 4607–4612.

Hogg, T., 2007. Coordinating microscopic robots in viscous fluids, *Autonomous Agents and Multi-Agent Syst.*, 14(3), 271–305.

Hunt, C.A., R.D. MacGregor, and R.A. Siegal, 1986. Enginering targeted in vino drug delivery I. The physiological and physicochemical principles governing opportunities & limitations, *Pharm. Res.*, 3, 333–344.

Iakoubov, L., O. Rokhlin, and V. Torchilin, 1995. Antinuclear autoantibodies of the aged reactive against the surface of tumor but not normal cells, *Immunol. Lett.*, 47, 147–149.

Immordino, M.L., F. Dosio, and L. Cattel, 2006. Stealth liposomes: Review of the basic science, rationale, and clinical applications, existing and potential, *Int. J. Nanomed.*, 1(3), 297–315.

Ishiyama, K., M. Sendoh, and K.I. Arai, 2002. Magnetic micromachines for medical applications, *J. Magn. Mater.*, 242–245, 41–46.

Jain, T., M. Morales, S. Sahoo, D.L. Pelecky, and V. Labhasetwar, 2005. Iron oxide nanoparticles for sustained delivery of anticancer agents, *Mol. Pharm.*, 2(3), 194–205.

Jales, D., 1998. *Introduction to Magnetism and Magnetic Materials*, CRC Press, Boca Raton, FL.

Johnson, O.L., J.L. Cleland, H.J. Lee, M. Charnis, and W. Jaworowicz, 1996. A month-long effect from a single injection of microencapsulated human growth hormone, *Nat. Med.*, 2, 795–799.

Jun, Y.W., Y.M. Huh, J.S. Choi et al., 2005. *J. Am. Chem. Soc.*

Jun, Y.-W., J.-W. Seo, and J. Cheon, 2007. Nanoscaling laws of magnetic nanoparticles and their applicabilities in biomedical sciences, *Acc. Chem. Res.*, 41(2), 179–189.

Jurgons, R., C. Seliger, A.H.L. Trahms, S. Odenbach, and C. Alexiou, 2006. Drug loaded magnetic nanoparticles for cancer therapy, *J. Phys.: Condens. Matter*, 18, 2893–2902.

Kam, N.W.S., T.C. Jessop, P.A. Wender, and H.J. Dai, 2004. Nanotube molecular transporters: Internalization of carbon nanotube-protein conjugates into Mammalian cells, *J. Am. Chem. Soc.*, 126(22), 6850–6851.

Kam, N.W.S., M. O'Connell, J.A. Wisdom, and H.J. Dai, 2005. Carbon nanotubes as multifunctional biological transporters and near-infrared agents for selective cancer cell destruction, *Proc. Natl. Acad. Sci. USA*, 102(33), 11600–11605.

Karniadakis, G.E. and A. Beskok, 2002. *Micro Flows: Fundamentals and Simulation*, Springer, New York.

Katayama, N., R. Tanaka, Y. Ohno et al., 1995. Implantable slow release cyclosporin A (CYA) delivery system to thoracic lymph duct, *Int. J. Pharm.*, 115(1), 87–93.

Khamesee, M.B., N. Kato, Y. Nomura, and T. Nakamura, 2002. Design and control of microrobotic system using magnetic levitation, *IEEE/ASME Trans. Mechatronics*, 7(1), 1–14.

Khamesee, M.B., N. Kato, Y. Nomura, and T. Nakamura, 2003. Performance improvement of a magnetically levitated microrobot using an adaptive control, in *Proceedings of the International Conference on MEMS, NANO and Smart Systems*, pp. 332–338.

Kingsley, J.D., H. Dou, J. Morehead et al., 2006. Nanotechnology: A focus on nanoparticles as a drug delivery system, *J. Neuroimmune Pharmacol.*, 1, 340–350.

Kiparissides, C. and O. Kammona, 2008. Nanotechnology advances in controlled drug delivery systems, *Phys. Stat. Sol. (c)*, 5(12), 3828–3833.

Kiparissides, C., A. Alexandridou, K. Kotti, and A. Chaitidou, 2006. Recent advances in novel drug delivery systems, *AZojono J. Nanotechnol. Online*, 2.

Kis, A., K. Jensen, S. Aloni, W. Mickelson, and A. Zettl, 2006. Interlayer forces and ultralow sliding friction in multiwalled carbon nanotubes, *Phys. Rev. Lett.*, 97(2).

Kommareddy, S. and M. Amiji, 2005. Prepeation and evaluation of thiol-modified gelatin nanoparticles for intracellularDNA delivery in response to glutathione, *Bioconjug. Chem.*, 16(6), 1423–1432.

Kono, K., 2001. Thermosensitive polymer-modified liposomes, *Adv. Drug Deliv. Rev.*, 53, 307–319.

Kosa, G., M. Shoham, and M. Zaaroor, 2005. Propulsion of a swimming micromedical robot, in *Proceedings of the 20th International Conference on Robotics and Automation*, Barcelona, Spain, pp. 1327–1331.

Kostarelos, K., L. Lacerda, G. Pastorin et al., 2007. Cellular uptake of functionalized carbon nanotubes is independent of functional group and cellular type, *Nat. Nanotechnol.*, 2(2), 108–113.

Kwon, G.S. and K. Kataoka, 1995. Block copolymer micelles as long-circulating drug vehicles, *Adv. Drug Deliv. Rev.*, 16(2–3), 295–309.

Langer, R., 2001. Perspective: Drug delivery—drugs on target, *Science*, 293, 58–59.

Lee, H., A.M. Purdon, V. Chu, and R.M. Westervelt, 2004. Controlled assembly of magnetic nanoparticles from magnetotactic bacteria using microelectromagnets arrays, *Nano Lett.*, 4(5), 995–998.

Lee, J.H., Y.W. Jun, S.I. Yeon, J.S. Shin, and J. Cheon, 2006. Dual-mode nanoparticle probes for high-performance magnetic resonance and fluorescence imaging of neuroblastoma, *Angew. Chem. Int. Ed.*, 45(48), 8160–8162.

Letfullin, R.R., C. Joenathan, T.F. George, and V.P. Zharov, 2006. Laser-induced explosion of gold nanoparticles: Potential role for nanophotothermolysis of cancer, *Nanomedicine*, 1(4), 473–480.

Li, C.-Y. and T.-W. Chou, 2006. Charge-induced strains in single-walled carbon nanotubes, *Nanotechnology*, 17, 4624–4628.

Li, H. and J.T.M. Zhang, 2006. Dynamics modeling and analysis of a swimming microrobot for controlled drug delivery, in *IEEE International Conference on Robotics and Automation*, IEEE, Orlando, FL, pp. 1768–1773.

Liu, Y.F. and H.F. Wang, 2007. Nanomedicine—Nanotechnology tackles tumours, *Nat. Nanotechnol.*, 2(1), 20–21.

Liu, Z., W.B. Cai, L.N. He et al., 2007. In vivo biodistribution and highly efficient tumour targeting of carbon nanotubes in mice, *Nat. Nanotechnol.*, 2(1), 47–52.

Loo, C., A. Lowery, N. Halas, J. West, and R. Drezek, 2005. Immunotargeted nanoshells for integrated cancer imaging and therapy, *Nano Lett.*, 5, 709–711.

Lopez, C.F., S.O. Nielsen, P.B. Moore, M.L. Klein, and 2004. Understanding natures design for a nanosyringe, *Proc. Natl. Acad. Sci. USA*, 101(13), 4431–4434.

Lubbe, A.S., C. Bergemann, W. Huhnt et al., 1996. Preclinical experiences with magnetic drug targeting: Tolerance and efficacy, *Cancer Res.*, 56, 4694–4701.

Maiti, A., J. Wescott, and G.W. Goldberck, 2005. Mesoscale modelling: Recent developments and applications to nanocomposites, drug delivery and precipitations membranes, *Int. J. Nanotechnol.*, 2(3), 198–214.

Maniar, M., A. Domb, A. Haffer, and J. Shah, 1994. Controlled release of a local anesthetic from fatty acid dimer based polyanhydride, *J. Control. Release*, 30(3), 233–239.

Martel, S., C.C. Tremblay, S. Ngakeng, and G. Langois, 2006. Controlled manipulation and actuation of micro-objects with magnetotactic bacteria, *Appl. Phys. Lett.*, 89(23), 233804–233806.

Martel, S., J.-B. Mathieu, O. Felfoul et al., 2007. Automatic navigation of an untethered device in the artery of a living animal using a conventional clinical magnetic resonance imaging system, *Appl. Phys. Lett.*, 90(14), 105–107.

Martel, S., M. Mohammadi, O. Felfoul, Z. Lu, and P. Pouponneau, 2009. Flagellated magnetotactic bacteria as controlled MRI-trackable propulsion and steering systems for medical nanorobots operating in the human microvasculature, *Int. J. Robot. Res.*, 28(4), 571–582.

Martin, C.R. and P. Kohli, 2003. The emerging field of nanotube biotechnology, *Nat. Rev. Drug Discov.*, 2(1), 47–52.

Mathieu, J.-B. and S. Martel, 2007. Magnetic microparticle steering within the constraints of an MRI system: Proof of concept of a novel targeting approach, *Biomed. Microdevices*, 9, 801–808.

Mathieu, J.-B., S. Martel, L.H. Yahia, G. Soulez, and G. Beaudoin, 2003. MRI systems as a mean of propulsion for a microdevice in blood vessels, in *Proceedings of the IEEE EMBS*, Cancum, Mexico, pp. 3419–3422.

Mathieu, J., G. Beaudoin, and S. Martel, 2006. Method of propulsion of a ferromagnetic core in the cardiovascular system through magnetic gradients generated by an mri system, *IEEE Trans. Biomed. Eng.*, 53(2), 292–299.

Matsumura, Y. and H. Maeda, 1986. A new concept for macromolecular therapeutics in cancer chemotherapy: Mechanism of tumoritropic accumulation of proteins and the antitumor agent SMANCS, *Cancer Res.*, 6, 193–210.

McGee, J.P., S.S. Davis, and D.T. O'Hagan, 1994. The immunogenicity of a model protein entrapped in poly(lactide-co-glycolide) microparticles prepared by a novel phase separation technique, *J. Control. Release*, 31(1), 55–60.

McNeil, R.G., R.C. Ritter, B. Wang et al., 1995. Functional design features and initial performance characteristics of amagnetic-implant guidance system for stereotactic neurosurgery, *IEEE Trans. Biomed. Eng.*, 42(8), 793–801.

Medintz, I.L., 2003. Self-assebled nanoscale biosensors based on quantum dot FRET donors, *Nat. Mater.*, 2, 630–638.

Medintz, I.L., H.T. Uyeda, E.R. Goldman, and H. Mattoussi, 2005. Quantum dot bioconjugates for imaging, labeling and sensing, *Nat. Mater.*, 4, 435–446.

Meyer, D.E., B.C. Shin, G.A. Kong, M.W. Dewhirst, and A. Chilkoti, 2001. Drug targeting unsing thermally responsive polymers and local hyperthermia, *J. Control Release*, 74, 213–224.

Michel, S., T. Keller, and J. Fronhlich et al., 2002 Preoperative breast cancer staging: MR imaging of the axilla with ultrasmall superparamagnetic iron oxide enhancement, *Radiology*, 225, 527–536.

Misra, R.D.K., 2008. Magnetic nanoparticle carrier for targeted drug delivery: Perspective, outlook and design, *Mater. Sci. Technol.*, 24(9), 1011–1019.

Mitchel, D.T., S.B. Lee, L. Trofin et al., 2002. Smart nanotubes for bioseparations and biocatalysis, *J. Am. Chem. Soc.*, 124(40), 11864–11865.

Nanotechnology-Now, http://www.nanotech-now.com/nanotube-buckyball-sites.htm

Nguyen, B., W. Stanford, and B. Thompson et al., 1999. Multicenter clinical trial of ultrasmall superparamagnetic iron oxide in the evaluation of mediastinal lymph nodes in patients with primary lung carcinoma, *J. Magn. Reson. Imaging*, 10, 468–473.

Nida, D., M. Rahman, K. Carlson, R. Richards-Kortum, and M. Follen, 2005. Fluorescent nanocrystals for use in early cervical cancer detection, *Gynecol. Oncol.*, 99(3Suppl 1), S89–S94.

Orive, G., R.M. Hernandez, A.R. Gascon, and J.L. Pedraz, 2005. Micro and nano drug delivery system in cancer therapy, *Cancer Ther.*, 3, 131–138.

Paciotti, G., L. Myer, D. Weinreich, and e. al, 2004. Colloidal gold: A novel nanoparticle vector for tumor directed drug delivery, *Drug Deliv.*, 11, 169–183.

Panyam, J. and V. Labhasetwar, 2003. Biodegradable nanoparticles for drug and gene delivery to cells and tissue, *Adv. Drug Deliv. Rev.*, 55, 329–347.

Park, J.W., 2002. Liposome-based drug delivery in breast cancer treatment, *Breast Cancer Res.*, 4, 95–99.

Park, J.-H., G.v. Maltzahn, L. Zhang et al., 2008. Magnetic iron oxide nanoworms for tumor targeting and imaging, *Adv. Mater.*, 20, 1630–1635.

Piper, J.W., R.A. Swerlick, and C. Zhu, 1998. Determining force dependence of two-dimensional receptor-ligand binding affinity by centrifugation, *Biophys. J.*, 74(1), 492–513.

Puri, A., http://ccr.cancer.gov/staff/gallery.asp?profileid=6797.

Purves, W.K., D. Sadava, G.H. Orians, and H.C. Heller, 2008. *Life—The Science of Biology*, W.H. Freeman, New York.

Quate, E.G., K.G. Wika, M.A. Lawson et al., 1991. Goniometric motion controller for the superconducting coil in amagnetic stereoaxis system, *IEEE Trans. Biomed. Eng.*, 38, 899–905.

Ramsden, E.N., 2000. A-Level Chemistry, Nelson Thornes Ltd, Gloucestershire. UK.

Ritter, R.C., M.S. Grady, I. M.A. Howard, and G.T. Gillies, 1992. Magnetic stereotaxis: Computer-assisted, image-guided remote movement of implants in the brain, *Innov. Technol. Biol. Med.*, 13, 437–449.

Rockall, A., S. Sohaib, and M. Harisinghani et al., 2005 Diagnostic performance of nanoparticle-enhanced magnetic resonance imaging in the diagnosis of lymph node metastases in patients with endometrial and cervical cancer, *J. Clin. Oncol.*, 23, 2813–2821.

Sahoo, S. and V. Labhasetwar, 2003. Nanotech approaches to drug delivery and imaging, *Drug Discov. Today*, 8, 1112–1120.

Schierholz, J., A. Rump, and G. Pulverer, 1997. Ciprofloxacin containing polyurethanes as potential drug delivery systems to prevent foreign-body infections, *Drug Res.*, 47 (I)(1), 70–74.

Sengupta, S., D. Eavarone, I. Capila et al., 2005. Temporal targeting of tumour cells and neovasculature with a nanoscale delivery system, *Nat. Lett.*, 436, 568–572.

Seo, W.S., J.H. Lee, X.M. Sun et al., 2006. FeCo/graphitic-shell nanocrystals as advanced magnetic-resonance-imaging and near-infrared agents, *Nat. Mater.*, 5(12), 971–976.

Sershen, S., S. Westcott, N.J. Halas, and J.L. West, 2000. Temperature-sensitive polymer-nano-shells composites for photthermally modulated drug delivery, *J. Biom. Mat. Res.*, 51, 293–298.

Seydel, C., 2003. Quantum dots get wet, *Science*, 300, 80–81.

Shenoy, D., S. Little, R. Langer, and M. Amiji, 2005. Poly(ethylene oxide)—modified poly(β-amino ester) nanoparticles as a pH-sensitive system for tumor-targeted delivery of hydrophobic drugs: Part 2. In vivo distribution and tumor localization studies, *Pharm. Res.*, 22, 2107–2114.

Sinek, J.P., H.B. Frieboes, B. Sivaraman, S. Sanga, and V. Cristini, 2006. Mathematical and computational modeling: Towards the development and application of nanodevices for drug delivery, In *Nanodevices for the Life Sciences*, Vol. 4, Wiley, New York.

Stark, D., R. Weissleder, G. Elizondo et al., 1988. Superparamagnetic iron oxide: Clinical application as a contrast agent for MR imaging of the liver, *Radiology*, 168, 297–301.

Stryer, L., J. Berg, and J. Tymoczko, 2006. *Biochemistry*, W. H. Freeman, New York.

Sudo, S., S. Segawa, and T. Honda, 2006. Magnetic swimming mechanism in a viscous liquid, *J. Int. Mater. Syst. Struct.*, 17, 729–736.

Sun, C., J.S.H. Lee, and M. Zhang, 2008. Magnetic nanoparticles in MR imaging and drug delivery *Adv. Drug Deliv. Rev.*, 60(11), 1252–1265.

Sutton, D., N. Nasongkla, E. Blanco, and J. Gao, 2007. Functionalized micellar systems for cancer targeted drug delivery, *Pharm. Res.*, 24, 1029.

Suzuki, M., H. Honda, T. Kobayashi et al., 1996. Superparamagnetic iron oxide: Clinical agents using monoclonal antibody-conjugated magnetic particles, *Brain Tumor Pathol.*, 13, 127–132.

Takeda, S.-I., F. Mishima, S. Fujimoto, Y. Izumi, and S. Nishijima, 2007. Development of magnetically targeted drug delivery system using superconducting magnet, *J. Magn. Magn. Mater.*, 311(1), 367–371.

Tamaz, S., R. Gourdeau, A. Chanu, J.-B. Mathieu, and S. Martel, 2008. Real-time MRI-based control of a ferromagnetic core for endovascular navigation, *IEEE Trans. Biomed. Eng.*, 55(7), 1854–1863.

Tombler, T.W., C. Zhou, A. Leo, J. Kong, and H. Dai, 2001. Reversible electromechanical characteristics of carbon nanotubes under local-probe manipulation, *Nature*, 87, 769–771.

Torchilin, V.P., 2005. Recent advances with liposomes as pharmaceutical carriers, *Nat. Rev. Drug Discov.*, 4, 145–160.

Torchilin, V.P., 2007. Targeted pharmaceutical nanocarriers for cancer therapy and imaging, *AAPS J.*, 9, 128–147.

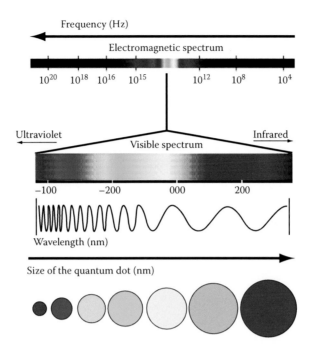

FIGURE 1.4 Size tunable emission of the QD.

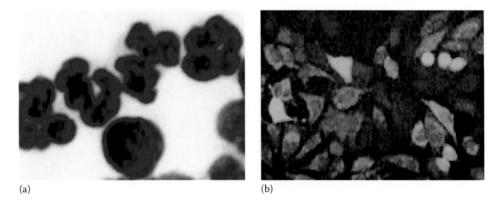

FIGURE 16.4 Intracellular ROS detection. (a) Superoxide anion oxygen free radical (O_2^-) in functional granulocyte leukocytes (Nitro blue tetrasolium method). (b) Hydrogen peroxide (H_2O_2, ROS, but not oxygen free radical) in human uterine cervical carcinoma cell line (HeLa cells) (DCFH-DA fluorescence method).

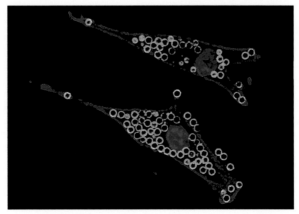

FIGURE 16.12 Metal nanoparticle toxicity: Bone marrow-derived stem cells labeled with large iron oxide particles (2.5 μm) undergo apoptosis (Stained for Annexin V—apoptotic marker). (Courtesy of Dr. Kishore Bhakoo Stem Cell Imaging Group, MRC Clinical Sciences Centre, Faculty of Medicine, Imperial College London, London, U.K., http://sci.csc.mrc.ac.uk.)

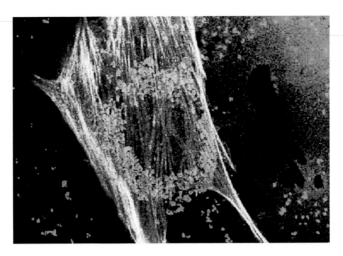

FIGURE 16.13 Bone marrow-derived stem cells labeled with dextran-coated iron oxide nanoparticles conjugated with Tat-FITC. Tat is used as a transfection agent and FITC enables histological verification. (Courtesy of Dr. Kishore Bhakoo Stem Cell Imaging Group, MRC Clinical Sciences Centre, Faculty of Medicine, Imperial College London, London, U.K., http://sci.csc.mrc.ac.uk.)

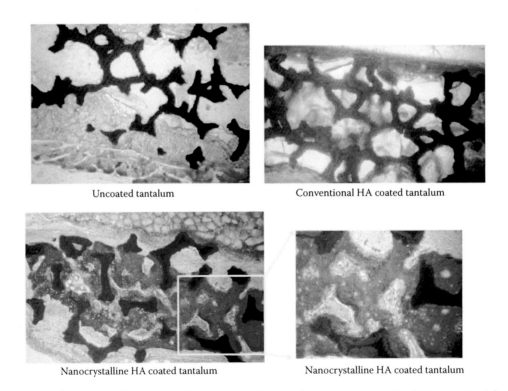

Uncoated tantalum

Conventional HA coated tantalum

Nanocrystalline HA coated tantalum

Nanocrystalline HA coated tantalum

FIGURE 20.3 Histology of rat calvaria after 6 weeks of implantation of uncoated tantalum, conventional HA-coated tantalum and nanocrystalline HA-coated tantalum. Greater amounts of new bone formation occur in the rat calvaria when implanting nanocrystalline HA-coated tantalum than uncoated and conventional HA-coated tantalum. Red represents new bone and blue represents collagen. (Adapted from Sato, M. Nanophase hydroxyapatite coatings for dental and orthopedic applications, PhD thesis, Purdue University, West Lafayette, IN, 2006.)

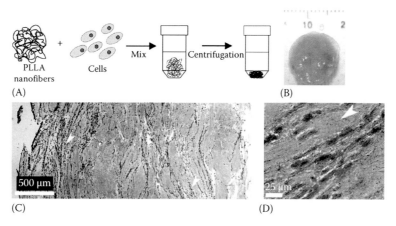

(A) (B)

(C) (D)

FIGURE 20.5 (A) Schematic illustrating an efficient cell seeding method into a cell-nanofiber composite for cartilage tissue engineering applications. (B) Image of a shiny cartilage-like tissue from the cell-nanofiber composite after 42 days of culture. (C) Low-magnification histology showing well-dispersed chondrocyte distribution throughout the nanofiber scaffold after 1 day of cell culture (the cross section). (D) High-magnification histology showing distinct cell populations among the nanofibers. Arrows point to chondrocytes dispersed among nanofibers. (Adapted from Li, W.J. et al., *Tissue Eng. Part A*, 14, 639, 2008.)

| Dtxl-NP-Apt | Dtxl-NP | Dtxl | NP | Saline |

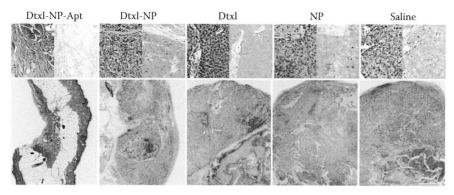

FIGURE 21.5 Effect of targeted bioconjugate nanoparticles. Histological slides demonstrating excised tumors from mice treated with saline, pegylated PLGA nanoparticle without docetaxel, docetaxel alone, docetaxel-encapsulated nanoparticle and docetaxel-encapsulated nanoparticle–aptamer bioconjugates. The docetaxel–nanoparticle–aptamer confirmed absence of residual cancer. All others showed variable PSMA staining showing degrees of tumor viability. (From Farokhzad, O.C. et al., *Proc. Natl. Acad. Sci. USA*, 103(16), 6315, 2006. With permission.)

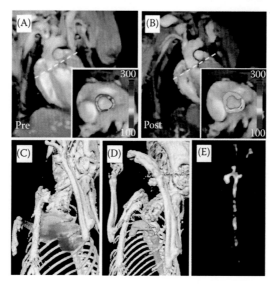

FIGURE 24.4 Trimodal imaging of macrophage expression: T_2-weighted MR imaging of the aortic root of apoE-KO mice (A) pre- and (B) 24 h postinjection with the agent; PET-CT image of (C) an apoE-KO mouse and (D) a wild-type mouse 24 h postinjection. (E) A NIRF image of an excised aorta from an apoE-KO mouse 24 h postinjection. (From Nahrendorf, M. et al., *Circulation*, 117, 379, 2008. With permission.)

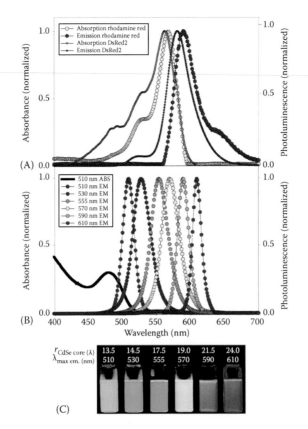

FIGURE 24.6 Comparison of the emission and absorption spectra of (A) Rhodamine red and DsRed2 and (B) quantum dots whose core size varies from 13.5 to 24 Å. A photograph of the quantum dot solutions appears in (C). (Reproduced from Medintz, I.L. et al., *Nat. Mater.*, 4, 435, 2005. With permission.)

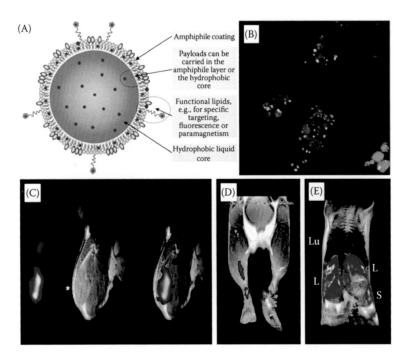

FIGURE 24.8 Emulsions used as contrast agents. (A) Schematic depiction of emulsions. (B) Ahrens et al. used perfluoropolyether emulsions that contained the fluorophore DiI to label dendritic cells. Confocal microscopy indicated a high level of particle uptake, as evidenced by the fluorescence of these cells. (C–E) ^{19}F MRI (red-yellow coloration) combined with conventional MRI (grayscale) allows tracking of these cells *in vivo*. (Adapted from Cormode, D.P. et al., *Arterioscler. Thromb. Vasc. Biol.*, 29, 992, 2009; Ahrens, E.T. et al., *Nat. Biotechnol.*, 23, 983, 2005. With permission.)

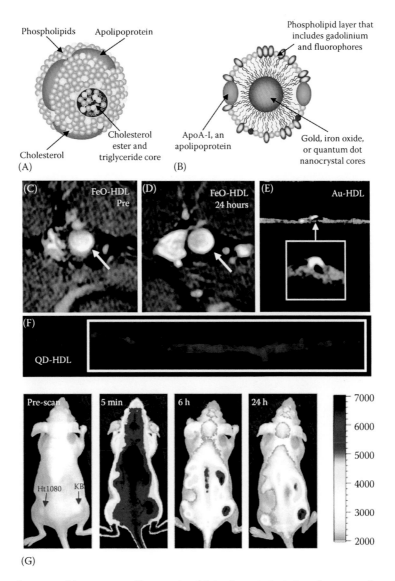

FIGURE 24.9 (A) A schematic depiction of the structure of lipoproteins. (B) A schematic depiction of nanocrystal core HDL. MR images of the aorta of an apoE-KO mouse pre- (C) and 24 h postinjection (D) with FeO-HDL. (E) Micro CT image of the excised aorta of an apoE-KO mouse injected with Au-HDL. (F) Fluorescence image of the aorta of an apoE-KO mouse injected with QD-HDL. (G) *In vivo* fluorescence images of a mouse bearing a KB and a HT1080 tumor on its right and left flank, respectively. The agent became preferentially concentrated in the tumor that overexpressed folate receptors. (Adapted from Cormode, D.P. et al., *Nano Lett.*, 8, 3715, 2008c; Chen, J. et al., *J. Am. Chem. Soc.*, 129, 5798, 2007. With permission.)

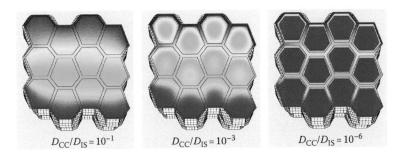

$D_{CC}/D_{IS} = 10^{-1}$ $D_{CC}/D_{IS} = 10^{-3}$ $D_{CC}/D_{IS} = 10^{-6}$

FIGURE 32.3 Color-coded simulations of three diffusing compounds in a 3D stratum corneum model, calculated for different diffusivity ratios D_{CC}/D_{IS} of the compounds between corneocytes (bricks, CC) and in the intercellular space (mortar, IS). (Top view; red indicates high concentration). For nanoparticulate compounds, a strictly intercellular route (right case) will be due. (Courtesy of M. Heisig, IWR, Ruprecht-Karls-Universität Heidelberg, Germany; Feuchter, D. et al., *Comput. Vis. Sci.* 9(2), 117, 2006. With permission.)

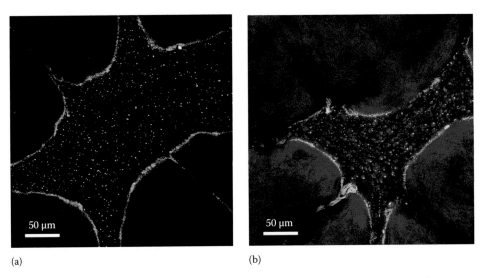

(a)   (b)

FIGURE 32.10 PLGA nanoparticles in human skin furrows 5 h after administration. (a) A two-photon micrograph at $z = 15\,\mu m$ subsurface depth showing superficial keratin fluorescence and clearly resolved single particles. (b) A pseudocolor overlay of two-photon and confocal images at $z = 28\,\mu m$ revealing the release of a drug dummy from the particles and its cutaneous uptake. The two-photon channel (green) shows keratinous layers and single particles, the 488 nm excited confocal channel (blue) addresses the fluorescein-labeled particles solely, and the 543 nm excited confocal channel (red) exclusively shows the dummy compound (Texas Red).

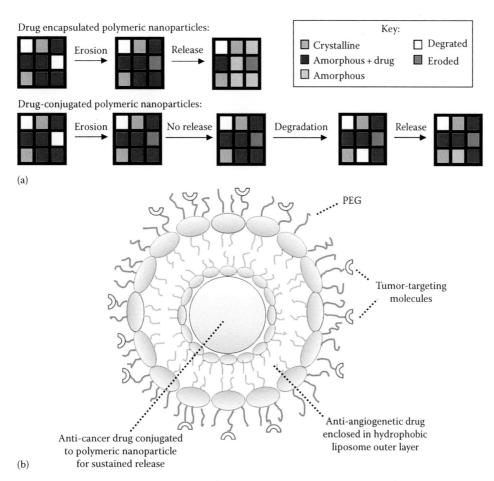

FIGURE 34.5 (a) Mechanistic rationale for sustained release from drug-conjugated polymeric nanoparticles and comparatively faster release from drug-encapsulated polymeric nanoparticles and (b) structure of the nanocell and function of the constituent biomaterials for antiangiogenesis and anticancer combination therapy.

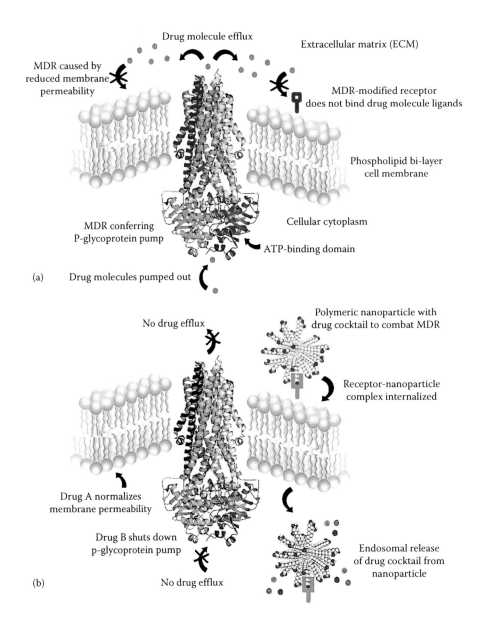

Drug molecule efflux

Extracellular matrix (ECM)

MDR caused by reduced membrane permeability

MDR-modified receptor does not bind drug molecule ligands

Phospholipid bi-layer cell membrane

MDR conferring P-glycoprotein pump

Cellular cytoplasm

ATP-binding domain

(a) **Drug molecules pumped out**

No drug efflux

Polymeric nanoparticle with drug cocktail to combat MDR

Receptor-nanoparticle complex internalized

Drug A normalizes membrane permeability

Drug B shuts down p-glycoprotein pump

Endosomal release of drug cocktail from nanoparticle

(b) **No drug efflux**

FIGURE 34.6 (a) MDR mechanisms; (b) nanoparticle-based delivery for circumvention of MDR.

Tsukioka, Y., Y. Matsumura, T. Hamaguchi et al., 2002. Pharmaceutical and biomedical differences between micellar doxorubicin (NK911) and liposomal doxorubicin (Doxil), *Jpn. J. Cancer Res.*, 93, 1145–1153.

Uhrich, K., S. Cannizaro, R. Langer, and K. Shakeseff, 1999. Polymeric systems for controlled drug release, *Chem. Rev.*, 99, 3181–3198.

Ummat, A., A. Dubey, and C. Mavroidis, 2006. Bio-nanorobotics: A field inspired by nature, In *Biomimetics: Biologically Inspired Technologies*, Y. Bar-Cohen (Ed), CRC Taylor & Francis, Boca Raton, FL.

Vo-Dinh, T. (Ed), 2007. *Nanotechnology in Biology and Medicine*, CRC Press Taylor & Francis, Boca Raton, FL.

Vo-Dinh, T., F. Yan, and M.B. Wabuyele, 2005. Surface-enhanced Raman scattering for medical diagnostics and biological imaging, *J. Raman Spectrosc.*, 36, 640–647.

Wagenaar, B.W. and B.W. Müller, 1994. Piroxicam release from spray dried biodegradable microspheres, *Biomaterials*, 15, 49–54.

Walt, A.d.H., A. Chatelain, and D. Ugarte, 1995. A carbon nanotube field-emission electron source, *Science*, 270, 1179–1180.

Walter, K.A., R.J. Tamargo, A. Olivi, P.C. Burger, and H. Brem, 1995. Intratumoral chemotherapy, *Neurosurgury*, 37(6), 1129–1145.

Walters, D.A., L.M. Ericson, M.J. Casavant et al., 1999. Elastic strain of freely suspended single-wall carbon nanotube ropes, *Appl. Phys. Lett.*, 74(25), 3803–3805.

Wang, Y., X. Xie, X. Wang, et al., 2004. Photoacoustic tomography of a nanoshell contrast agent in the in vivo rat brain, *Technol. Cancer Res. Treat*, 3, 33–40.

Weisseleder, R., G. Elizondo, J. Wittenberg et al., 1990. Ultrasmall superparamagnetic iron oxide: Characteriazation of a new class of contrast agents for MR imaging, *Radiology*, 175, 489–493.

Wostek-Wojciechowska, D., J.K. Jeszka, P. Uznanski et al., 2004. Synthesis of gold nanoparticles in solid state by thermal decomposition of an organometallic precursor, *Mater. Sci.-Poland*, 22(4), 407–413.

Wu, X., H. Liu, and J. Liu, 2003. Immunofluorescent labeling of cancer marker Her2 and other cellular targets with semiconductor quantum dots, *Nat. Biotechnol.*, 21, 41–46.

Xu, T., J.T. Wong, P.M. Shikhaliev et al., 2006. Real-time tumor tracking using implanted positron emission markers: Concept and simulation study, *Med. Phys.*, 33(7), 2598–2609.

Ye, W.-P. and Y.W. Chien, 1996. Dual-controlled drug delivery across biodegradable copolymer. II. Delivery kinetics of levonorgestrel and estradiol from (matrix/matrix) laminate drug delivery system, *J. Control. Release*, 41(3), 259–269.

Yesin, K.B., K. Vollmers, and B.J. Nelson, 2004. Analysis and design of wireless magnetically guided microrobots in body fluids, in *Proceedings of the IEEE International Conference on Robotics and Automation*, New Orleans, LA, pp. 1333–1338.

Yesin, K.B., P. Exner, K. Vollmers, and B.J. Nelson, 2005. Design and control of in vivo magnetic microrobots, in *Proceedings of the 8th International Medical Image Computing and Computer Assisted Intervention (MICCAI)*, pp. 819–826.

Yesin, K.B., K. Vollmers, and B.J. Nelson, 2006. Modeling and control of untethered biomicrorobots in a fluidic environment using electromagnetic fields, *Int. J. Robot. Res.*, 25, 527–536.

Yu, M.F., 2000. Tensile loading of ropes of single wall carbon nanotubes and their mechanical properties, *Phys. Rev. Lett.*, 84(24), 5552–5555.

Zhang, Y., L. Guo, R. Roeske, A. Antony, and H. Jayaram, 2004. Pteroyl-γ-glutamate-cysteine synthesis and its applications in folate receptor-mediated cancer cell targeting using folate-tethered liposomes, *Anal. Biochem.*, 332, 168–177.

46

Medical Micro- and Nanorobots

Sylvain Martel
École Polytechnique de Montréal

46.1 Introduction

Medical microrobots and nanorobots such as the ones capable of navigating in the human blood vessels have been popularly conceptualized as miniature mechanical or electromechanical versions of the more familiar large-scale robots. Indeed, it is not rare to see representations of micro-nanorobots equipped with miniature mechanical tweezers, integrated sonar, and many other components resembling the ones found in larger systems. In reality, due to the complexity, technological limitations, and other technological issues within such scale constraints, the implementation of such small-scale robots and, particularly, nanorobots including the components attached to will mostly rely on other disciplines including but not limited to biology, biochemistry, material sciences, pharmaceutics, nanotechnology, and particularly nanomedicine being the application of nanotechnology to the medical field.

The literature mentions many reasons to justify the use of medical microrobots and nanorobots in the human body: rapid elimination of diseases, repair of biological structures, corrections of genetic defects, augmentation of human capabilities, and many more. But it appears that the most probable medical application in a shorter term for untethered entities that could be considered as precursors or simplified versions of the more complex medical micro- and/or nanorobots, would be tumor targeting based on the direct delivery of therapeutic agents. The use of untethered micro-nanorobots for medical interventions could have major benefits compared to traditional methods by providing less invasive approaches with potential for shorter recovery periods for patients.

Today, engineered nanoparticles are being considered for the delivery of therapeutics to tumor cells. Adding the required functions to such entities with an increase of the embedded complexity to make them more autonomous could lead to true medical nanorobots. Another alternative that is most likely to occur in the shorter term due to technological limitations to embed many of these functions within such space constraints would be to implement some of the complex functions remotely. But first, one must define the essential functions that must be implemented for such entities to be considered as nanorobots capable of performing specific medical tasks. As such, not only traditional engineering practices and techniques, but nanotechnology and, more specifically, nanomedicine will likely play an important role in the implementation of such functions.

In this chapter, a more realistic view of medical interventional microrobots and in particular nanorobots based on actual concepts and technologies available instead of relying on speculations and future visions is described to provide the reader with a more accurate description of what these nanorobots could resemble and do in a not-too-far future. Because of length constraints, only the fundamental principles are explained.

46.2 Medical Microscale Nanorobots versus Microrobots, and Medical Nanorobotics

There exist many definitions for the word "robot." One of them defines a robot as a device that can move and react to sensory input to execute one or more dedicated tasks. As such, a robot is endowed with some type of intelligence or program that runs automatically without human intervention. In turn, the design of a robot will be influenced by the environment in which it operates and the functions that must be integrated to perform some predefined tasks.

Similarly, there exist several definitions of a nanorobot. It can be a relatively large robotic platform capable of precise operations at the nanometer scale, a robot with overall dimensions in the nanometer scale, or a larger robot relying on nanometer-scale components to be able to perform one or more given tasks. The first definition does not apply to medical nanorobots operating in the human body whereas the implementation of a nanorobot that would match the second definition would not only be extremely difficult if feasible to implement, but its small overall dimension would not be justified for most applications being performed inside the human body.

For instance, the vascular network of an adult includes nearly 100,000 km of blood vessels, providing the highest accessibility to the different regions inside the human body. For medical robots traveling in the human blood vessels, the overall dimensions (except in some regions such as the blood–brain barrier) of each nanorobot do not need to be much smaller than approximately 2 μm across, being approximately half the diameter of the smallest human capillaries. Hence, nanorobots with overall dimensions of a few micrometers and defined here as microscale nanorobots are likely to be among the most viable untethered robots for operations in the human microvasculature. Although there is no official consensus, as for nanotechnology where the definition generally includes components of less than 100 nm, microscale nanorobots could refer to nanorobots with overall dimensions being less than 100 μm. From 100 μm to less than 1 mm, such nanorobots would typically be referred to as submillimeter nanorobots.

Therefore, medical microscale nanorobots can be defined as robots with overall dimensions in the micrometer scale and typically less than 100 μm that are designed to operate in particular environments inside the human body while being built with nanometer-scale components to enable one or more functions necessary to accomplish one or more medically oriented tasks. As such, a robot with overall dimensions in the micrometer range but not relying on one or more properties of nanometer-scale components to embed one or more functions essential to accomplish its task, would typically be referred to as a microrobot instead of a microscale nanorobot.

Furthermore, there exist two major trends in the development of microscale nanorobots: synthetic and hybrid. A synthetic approach aims at developing nanorobots with synthetic parts only, whereas a hybrid nanorobot aims at developing nanorobots with a mix of biological and synthetic components. Each approach would offer advantages as well as disadvantages and be more appropriate depending on the application and the working environment.

As for robotics and nanorobotics, medical nanorobotics can be defined as the science and technology of medical nanorobots, and their design, fabrication, and application. Unlike robotics that has connections mainly to electronics, mechanics, and software, nanorobotics also often includes many others such as, but not limited to, biology, biochemistry, nanotechnology, and nanomedicine.

Since actual technologies are not advanced enough to implement fully autonomous nanorobots, medical nanorobotics must provide platforms capable of compensating for such a lack of autonomy. As such, as for traditional robotics, besides non-real-time functions, such medical nanorobotic platforms would typically provide support for the three essential real-time functions, namely, actuation, positioning, and control.

46.3 Actuation

Although there may be many forms of actuation for microscale nanorobots, one essential form of actuation for such robots would be propulsion. However, one of the main challenges for medical nanorobotics is that it must deal with extremely small robots that must, in most cases, be untethered. The absence of a tether poses serious constraints on power and, particularly, on selecting an appropriate method of propulsion for microscale robots designed to operate in the human body. Propulsion needs steering to enable such microscale nanorobots to be propelled toward a desired direction, and this will be discussed later. Presently, it appears that there are two main propulsion methods suitable for operations in the human body: one method relies on magnetism while the other uses flagellated bacteria as propellers.

46.3.1 Magnetic Propulsion

The possibility of using magnetic propulsion for navigating untethered nanorobots in the human blood vessels have been demonstrated experimentally and reported for the first time in Martel et al., (2007). During the experiment, a part of the project MR-SUB (magnetic resonance submarine), an untethered ferromagnetic bead was propelled and controlled automatically using the magnetic gradients of a clinical magnetic resonance imaging (MRI) system at an average velocity of 10 cm/s along a preplanned trajectory in the carotid artery of a living pig—an animal model close to human. This milestone stimulated many research groups of the international nanorobotics community to pursue R&D in medical nanorobotics, which would most probably result in new significant achievements in the relatively near future.

Faraday's law of magnetic induction states that when a material is placed within a magnetic field, the magnetic forces of the material's electron will be affected. Hence, embedding the right material inside microscale nanorobots operating in a magnetic field could provide a suitable method of propulsion in various regions inside the human body. But the type of material used must be selected adequately with particular attention to the atomic and molecular structure of the material, including the magnetic moments associated with the atoms, the latter being affected by the electron orbital motion, the change in orbital motions caused by an external field, and the spin of the electrons. The choice here is limited since in most atoms, electrons occur in pairs spinning in opposite directions, causing their magnetic fields to cancel each others. On the other hand, materials with unpaired electrons will be more suitable for actuation since they will have a net magnetic field enabling them to react more or in other words, be more susceptible to an external field.

Diamagnetic metals such as most elements in the periodic table including copper, silver, and gold, for instance, have a very weak and negative susceptibility to magnetic fields. Paramagnetic metals such as magnesium, molybdenum, lithium, and tantalum have also small but positive susceptibility to an external magnetic field. On the other hand, ferromagnetic materials such as iron, nickel, and cobalt have unpaired electrons and as such, they have a large and positive susceptibility to an external field exhibiting a strong attraction to magnetic fields while being able to retain their magnetic properties after the external field is removed. Ferromagnetic materials get their magnetic properties not only because their atoms carry a magnetic moment but because the material is made up of nanometer-scaled regions known as magnetic domains. In each magnetic domain, all of the magnetic dipoles are coupled together in a preferential direction. When the material is not magnetized, the magnetic domains are nearly randomly organized leading to a zero net magnetic field for the whole part. By magnetically saturating the bulk material in a strong magnetic field to a level known as saturation magnetization, more propulsion force can be induced. Increasing the homogeneous magnetic field beyond what is required to reach the specific saturation magnetization level of the material, will not contribute to an increase of the propulsion force. As depicted in Equation 46.1, the magnetic force F_{mag} in Newton (N) acting on a magnetized particle is proportional to its magnetization M expressed in amperes per meter (A/m) and to the gradient of the magnetic field B in tesla (T). The magnetization of the magnetic material M is a function of the ambient magnetic field. Nevertheless, it reaches a plateau value called saturation magnetization (M_{sat}) when this ambient magnetic field is high enough, which is typically the case in the tunnel of a clinical MRI system. The last remark is important as MRI can also provide a valuable imaging modality for tracking magnetic microscale nanorobots in regions in the human body where line of sight is not possible:

$$\vec{F}_{mag} = \left(\sum_{n=1}^{N} V_n \right) (\vec{M}.\nabla)\vec{B} \tag{46.1}$$

In each domain, the atom moment parallel to the magnetic force points to different directions since each domain in the bulk material may not be all aligned properly—a lower propulsion force will be obtained. Hence, instead of embedding a relatively large piece in each microscale nanorobot, one solution is to embed N single-domain (SD) particles within a volume V that must be in the nanometer range (typically up to a few tens of nanometers) where SD property exists and being mechanically coupled as to form a volume equivalent to the initial larger piece. By doing so, these particles a cluster of nano-propulsion engines (Equation 46.1), will show superparamagnetic behaviors because of their small size. Superparamagnetic particles do not have remanent magnetization, which may make it less difficult to integrate

them in microscale nanorobots made of polymers or some other types of materials by avoiding the problem of clustering, since the formation of clusters could result in a loss of single magnetic domain structures.

Not only superparamagnetic particles can be guided, but they also exhibit heating properties when submitted to appropriate alternating magnetic fields, and this may prove to be another interesting avenue for the implementation of actuation mechanisms for microscale nanorobots. Indeed, smaller size particles require lower magnetic field to show the same loss of power. Hence, when integrated in appropriate materials (such as some types of hydrogels), the heat generated can be used to shrink or expand (depending upon the properties of the material) particular parts or the entire microscale nanorobot. In the context of medical interventions in the vascular systems, for instance, this feature could be used for several purposes such as controlled embolization, controlled drug release, etc. For therapeutic purpose, these same SD magnetic nanoparticles (MNP) can be used for local hyperthermia where the temperature is elevated a few degrees locally in order to enhance therapeutic efficacy.

46.3.2 Bacterial Propulsion

Magnetic propulsion when applied for humans becomes extremely difficult due to technological constraints. This is due in great part to the larger inner diameter of the coils responsible for generating the magnetic field that must accommodate the human body. This in turn, makes it very difficult to sustain an adequate propulsion cycle through larger electrical current circulating in the coils without overheating the system. Although, magnetic propulsion based on the induction of force on microscale nanorobots could be efficient to larger microvascular vessels, it would most likely be much less efficient or appropriate for traveling in the smaller microvasculature where blood vessel diameters can be as small as 4 μm.

As such, a propulsion mechanism that does not require external power may be desirable when operating in the human microvasculature. To move or actuate untethered microscale nanorobots without the need for external propulsion–dedicated hardware, artificial molecular machines are needed (Drexler, 1992) but their conception for useful tasks in the microvasculature are beyond present technological possibilities. As such, efficient nanomotors such as the flagellated propulsion mechanisms embedded in several species of bacteria have inspired many researchers trying to mimic them using modern engineering development methods. One of the best examples of this trend for potential medical applications is an artificial flagellum in the form of a nanocoil that has been propelled using a rotating magnetic field (Bell et al., 2007)—an approach that still relies on external hardware for propulsion.

To overcome the technological limits associated with magnetic-based propulsion, another approach is to harness existing flagellated nanomotors (Martel et al., 2006) which includes the use and integration of bacteria and more specifically magnetotactic bacteria (MTB) (Blakemore, 1975) with their molecular motors as a means of propulsion for microscale nanorobots. MTB are particularly interesting for medical microscale nanorobots because their swimming direction can be controlled inside the human body as discussed later. The molecular motor as found in flagellated bacteria measures less than 300 nm across. It has a flagellum attached to a rotor that acts as a propeller capable of full rotations. This rotary engine composed of proteins is powered by a flow of protons. The shape of the flagellum consists of a 20 nm-thick hollow tube with a helical shape with a sharp bend outside and next to the outer membrane. As for the macroscale counterpart, it has a shaft that passes through protein rings in the cell's membrane that act as bearings. Counterclockwise rotations of a polar flagellum will thrust the bacterium forward and reversing it will propel the cell backward. MTB of type MC-1 is of special interest for microscale nanorobots since their swimming speeds are much faster than most flagellated bacteria.

The terminal velocity, v_T, of a single bacterium can then be estimated (discounting the effect of the trailing flagellum (or flagella bundles) and assuming a constant rotational speed of the flagellum or a constant protons flux in the molecular motor) from Stokes' equation as

$$v_T = \frac{F_T}{3\pi\eta d} \tag{46.2}$$

When flagellated bacteria of type MC-1 are subjected to specific experimental and cultivation conditions, $4.7 \times 10^{-12} \geq F_T \geq 4.0 \times 10^{-12}$ N with $\eta = 1.0$ mPa·s (milli-pascal second) in water at 20°C represent the thrust force and the viscosity of the medium respectively, while $d \approx 2 \times 10^{-6}$ m (meter) represents the diameter of the cell when unloaded (notice that the MC-1 cell is round in shape). It is important to keep in mind that the diameter of the entire moving body could increase when loaded or when being part of a hybrid microscale nanorobot of larger dimensions than the bacterial cell itself. Although several bacteria could be attached to a larger nanorobot to increase the thrust force, for operations in the smaller diameter capillaries, a single bacterium per microscale nanorobot is likely to be used considering that the diameter of a single cell is approximately half the diameter of the smallest capillaries. Beyond this diameter, the bacterial nanorobot would experience an increase of the drag force acting against its motion due to wall effect as explained in Francis (1933) and Fidleris and Whitmore (1961).

As depicted in Equation 46.2, for a particular flagellated bacterium acting as a propulsion system, the terminal velocity of the microscale nanorobots assuming that the overall diameter remain approximately the same as the bacterium will be influenced by the fluid viscosity. When operating in blood, one should keep in mind that the bacterium itself is approximately half the diameter of a red blood cell and as

such, the viscosity of plasma (approximated by water) instead of whole blood as it would be the case for larger scale robots, is considered. As such, it should be noted that these microscale nanorobots being propelled by bacteria with a terminal velocity computed from Equation 46.2, would operate in a low Reynolds number ($Re < 1$) regime. Reynolds number is used to characterize different flow regimes, such as laminar (low Re), transitional, or turbulent flow (high Re). In the laminar flow regime, viscous forces become dominant and are characterized by smooth and constant fluid motion unlike the turbulent flow regime which is dominated by inertial forces tending to produce random eddies and vortices, to name but a couple types of fluctuations.

Equation 46.2 does not account for factors related to living entities. For instance, from prior experiments, it was shown that the average swimming speed v_{B37} from a large group of MC-1 MTB (without a preselection process) from the same culture decreases gradually over time t (expressed in minutes up to the maximum lifespan or displacement time of the MTB) in human blood at an internal body temperature of 37°C from an initial average swimming speed v_{MTB} (typically being approximately 200 μm/s although much higher velocities have been obtained with different culture conditions) prior to being in contact with blood, according to

$$v_{B37} = 0.09\, t^2 - 8.10\, t + v_{MTB} \tag{46.3}$$

Although more experiments need to be performed to fully validate Equation 46.3, the fact is that unlike a synthetic version, such bioactuators by being nonpathogenic will have a limited lifespan in the human body with a continuous decrease of the velocity. Still, MTB-tagged nanorobots will most likely be more efficient than magnetically propelled nanorobots of similar dimensions during a limited but often sufficient amount of time (approximately 40 min) to conduct a target operation when transiting through anarchic arteriocapillar networks stimulated by tumoral angiogenesis with capillaries located near the targeted tumoral lesion as small as 4–5 μm in diameter. As such, they should operate only in smaller capillaries where the blood flow is lowered unless the higher blood velocities in larger vessels is reduced sufficiently through techniques such as embolization or by the use of a balloon catheter or similar approaches which in all cases may prove to be difficult in parts of the vascular network.

46.4 Steering or Directional Control

Steering or directional control of medical nanorobots being propelled magnetically or with flagellated bacteria or other means is essential, particularly for target medical interventions. As for propulsion, magnetism is very well suited for steering purpose inside the human body. As such, the following sections are dedicated to magnetic steering for magnetically propelled and MTB-tagged nanorobots, respectively.

46.4.1 Magnetic Steering for Magnetic Microscale Nanorobots

Magnetic steering of magnetic entities such as MNP consists essentially in a direction propulsion force with a magnitude computed using Equation 46.1. Ideally, 3D directional control is suitable for efficient navigation in the human body such as in the vascular network. As such, an orthogonal coil configuration capable of inducing sufficient propulsion force in any direction becomes highly desirable. Although a custom coil configuration could be implemented, MRI systems as first proposed and validated *in vivo* in Martel et al. (2007) seem to be a serious alternative for the implementation of magnetically steerable microscale nanorobots dedicated to target interventions in the human body and, particularly, in the vascular network where line of sight is not possible.

A clinical MRI system consists of an imaging superconducting electromagnet providing a typical homogeneous field known as the B_0 field of 1.5 or 3 T in more recent models and sufficient to achieve the magnetization saturation level of the magnetic material embedded in the micrometer-scale nanorobots allowing them to achieve maximum propulsion/steering forces (see Equation 46.1). The system also includes gradient coils implemented in an orthogonal configuration which produce linear magnetic gradients in any directions in the B_0 field. Such linear magnetic field gradients are used for image slice selection in the MR image volume. As such, linear magnetic field gradients in any directions within such 3D volume can be generated, allowing for 3D directional propulsion. Because propulsion gradients can interfere with MR imaging, imaging gradients and propulsion/steering gradients are applied successively in a time-multiplexed fashion.

The magnitude of conventional clinical MRI systems allows for a maximum gradient of 40 mT/m. From Equation 46.1, one can see that this will limit the usage to microscale nanorobots with a larger effective volume of embedded magnetic material in order to achieve sufficient propulsion/steering force for effective navigation in the vascular network, limiting navigation in larger blood vessels. Although it would be possible for such microscale nanorobots to operate beyond catheterization, hence offering a real advantage for some types of interventions compared to existing modern medical instruments, the gradient amplitude would not be sufficient to operate in smaller diameter vessels such as the arterioles where the diameters may vary between ~50 and 150 μm. As such, special propulsion and imaging coils configurations capable of higher propulsion/steering gradients while retaining MR imaging capability is required. Nonetheless, due to various factors including technological constraints such as cooling limitation, the maximum propulsion/steering gradient amplitude that can be generated for effective navigation in any direction in humans is presently estimated at ~400–500 mT/m.

From Equation 46.1 and considering the blood flow velocities in different vessels of different diameters, one can see that as the effective volume of magnetic material decreases with a diminution of the overall size of the microscale nanorobot to enable it to travel in smaller diameter vessels, the induced magnetic force decays such that with smaller overall dimensions, propulsion would rely less on magnetic force and more on the blood flow velocity. In this case, the force induced by the magnetic gradients must be sufficient to steer such microscale nanorobots in the right directions at vessels bifurcations. This is still difficult for a single microscale nanorobot considering the relatively high blood velocity in smaller diameter vessels and the distances between vessels bifurcations. Embolization if possible would not be suitable here since the blood flow would eliminate the source of propulsion, and controlling the blood flow may prove to be extremely difficult and would most likely not be a reliable technique in the shorter term, if the approach would be possible.

One approach to increase the steering force is to increase the effective volume of magnetic material (Equation 46.1) with an aggregation or cluster of microscale nanorobots. Higher magnetophoretic velocity can be achieved with larger clusters but if they are too large, they could cause unplanned embolization, unintentionally blocking smaller diameter vessels. The interactive forces acting on microscale nanorobots in the same aggregation is a complex topic beyond the objectives of this chapter since several parameters (e.g., use of surfactant, colloidal parameters, ionic content, viscosity, magnetic properties of the microscale nanorobots, their overall dimensions, volume fraction, amplitude of the magnetic field, etc.) can be taken into account for tuning the behaviors and characteristics of the aggregation. A loosely coupled cluster will more likely break apart and is more likely to deform when flowing through a smaller diameter vessel, hence reducing the risk of clogging. Nonetheless, as the diameter of the vessel decreases, the size of the cluster is also decreased and as such, would become less effective in smaller diameter vessels in the microvasculature.

46.4.2 Magnetotaxis-Based Steering for Bacterial Microscale Nanorobots

Magnetotaxis is the ability of MTB to sense a magnetic field and to coordinate their directional motion in response to such a magnetic field. Magnetotaxis is, therefore, an interesting avenue to steer or control the directional motion of the bacteria inside the human body compared to chemotaxis used by most species, or by other means such as phototaxis.

The cell of MTB can transport iron ions from their surrounding medium into magnetosome membrane vesicles to form a saturated solution in which a short time later will synthesize in the MC-1 cell, magnetite (Fe_4O_3) crystals with dimensions of a few tens of nanometers. In the cell, these SD superparamagnetic nanoparticles separated by a membrane and referred to as magnetosomes form inside the cell a magnetosome chain oriented in the axis of propulsion of the bacterium. The size and linear arrangement of the magnetosomes within the MTB significantly affect magnetotaxis. This chain of magnetosomes acts similar to a compass needle. This is depicted in Figure 46.1.

Hence, the flagellated nanomotors combined with the nanometer-sized magnetosomes of a single magnetotactic bacterium can be used as effective integrated propulsion and

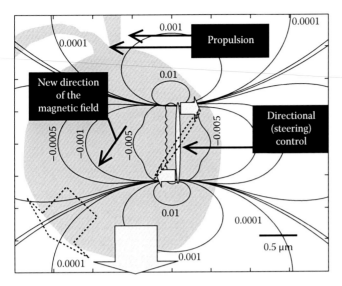

FIGURE 46.1 Image of one MC-1 magnetotactic bacterium showing the two flagella bundles used for propulsion. Directional (steering) control is achieved by inducing a torque on aligned nanoparticles (magnetosomes) synthesized in the cell itself. The new position of the magnetosomes and the new direction of motion are shown as dotted shapes.

steering systems for medical nanorobots designed for targeting locations only accessible through the smallest capillaries in humans. This control can be done simultaneously on a whole swarm or aggregations of these bacteria, as depicted in Figure 46.2a.

Figure 46.2b shows a simple example demonstrating the accuracy of such steering technique for bacterial microscale nanorobots. In this particular example, a single flagellated bacterium is pushing a 3 μm bead representing the synthetic part of a hybrid nanorobot. The propulsion and steering have been done along predetermined paths using computer control.

46.5 Positioning and Tracking

Being able to position and track such medical nanorobots inside the human body is essential for controlling them, particularly for applications involving direct targeting. Many scientific papers have proposed solutions to propel such miniature robots in the vascular network without considerations on other important yet essential aspects for practical applications in the human body and being able to track them is one of them.

The easiest method is to rely on optical microscopy techniques. But since this method needs direct line of sight, it is a very restrictive approach considering that most locations in the human body and particularly in the vascular network cannot be observed with a microscope. Therefore, tracking such nanorobots must rely on an appropriate medical imaging modality and as such, MRI proves to be powerful in this respect.

As for the nanoparticles used for propulsion in synthetic nanorobots propelled by the gradients generated by an upgraded MRI platform as discussed earlier in this chapter, the magnetosomes

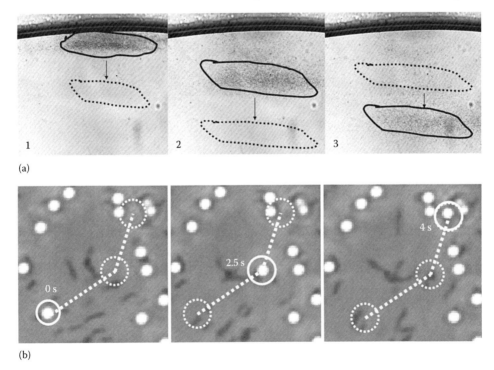

FIGURE 46.2 (a) Directional control of a swarm of flagellated bacteria. (b) Computer steering (directional) control of a single flagellated bacterium pushing a 3 μm bead along a preprogrammed path. The image shows that the beads without bacteria attached remain at the same locations.

embedded in each MC-1 MTB, as shown in Figure 46.1, are Fe_3O_4 single magnetic domain crystals of a few tens of nanometers in size. Similar to nanoparticles used for MRI contrast agents, magnetosomes cause a local distortion of the magnetic field inside the bore of a clinical MRI system. The local magnetic field distortion from each magnetosome can be approximated at a point P of coordinate r *(x, y, z)* by that of a magnetic dipole, as described by

$$\vec{B}'(P) = \frac{\mu_0}{4\pi}\left(3\frac{(\vec{m}\cdot\vec{r})\vec{r}}{r^5} - \frac{\vec{m}}{r^3}\right) \tag{46.4}$$

where $\mu_0 = 4\pi 10^{-7}$ H/m is the permeability of free space. For a uniformly magnetized object, the dipolar magnetic moment $(A\cdot m^2)$ is given by

$$\vec{m} = \frac{4}{3}\pi a^3 \vec{M}_{SAT} \tag{46.5}$$

where
M_{sat} is the saturation magnetization of the object
a its radius (m) considering a spherical shape

A numerical simulation of a single bacterium magnetic field was plotted over an electron microscopy image in Figure 46.1. The simulation results assumed 11 aligned magnetosomes each with a diameter of 70 nm. As shown, the magnetosome chain significantly disturbs the local magnetic field.

This magnetic field perturbation is relatively very significant near the bacterium as observed where values greater than 100 ppm (parts per million) have been computed. This value decreases further away from the bacterium. This is quite significant for medical nanorobotics considering that an accepted homogeneity level of modern MRI clinical scanner is approximately 5 ppm over a 50 cm diameter spherical volume at 1.5 T. These results suggest that MRI could potentially be used to track MTB-tagged nanorobots in 3D inside the human body providing a superior imaging modality compared to existing medical imaging methods.

46.6 Conclusion

The real challenge in the development of medical micro-nanorobots designed to operate in the human blood vessels goes behind the aspect of propulsion alone. To provide the basic functionalities and to integrate them successfully into workable nanorobotic platforms within known technological and physiological constraints, principles of nanophysics would need to be considered and applied correctly. In this chapter, although only fundamental principles were briefly explained, the importance of nanophysics in the development of medical nanorobots has been demonstrated.

References

Bell, D.J., Lentenegger, S., Hammar, K.M., Dong, L.X., and Nelson, D.J. 2007. Flagella like propulsion for microrobots using a nanocoil and a rotating electromagnetic field. In *Proc. of the IEEE Int. Conf. on Robotics and Automation (ICRA)*, pp. 1128–1133.

Blakemore, R.P. 1975. Magnetotactic bacteria. *Science*, 190, 377–379.

Drexler, K.E. 1992. *Nanosystems: Molecular Machinery, Manufacturing, and Computation.* John Wiley & Sons, New York.

Fidleris, V. and Whitmore, R.L. 1961. Experimental determination of the wall effect for spheres falling axially in cylinder vessels. *Br. J. Appl. Phys.*, 12, 490–494.

Francis, A.W. 1933. Wall effect in falling ball method for viscosity. *Physics*, 4, 403–406.

Martel, S., Tremblay, C., Ngakeng, S., and Langlois, G. 2006. Controlled manipulation and actuation of micro-objects with magnetotactic bacteria. *Appl. Phys. Lett.*, 89, 233804–233806.

Martel, S., Mathieu, J.-B., Felfoul, O., Chanu, A., Aboussouan, É., Tamaz, S., Pouponneau, P. Beaudoin, G., Soulez, G., Yahia, L'H., and Mankiewicz, M., 2007. Automatic navigation of an untethered device in the artery of a living animal using a conventional clinical magnetic resonance imaging system, *Appl. Phys. Lett.*, 90, 114105.

47

Nanohandling Robot Cells

Sergej Fatikow
Carl von Ossietzky Universität Oldenburg

Thomas Wich
Carl von Ossietzky Universität Oldenburg

Christian Dahmen
Carl von Ossietzky Universität Oldenburg

Daniel Jasper
Carl von Ossietzky Universität Oldenburg

Christian Stolle
Carl von Ossietzky Universität Oldenburg

Volkmar Eichhorn
Carl von Ossietzky Universität Oldenburg

Saskia Hagemann
Carl von Ossietzky Universität Oldenburg

Michael Weigel-Jech
Carl von Ossietzky Universität Oldenburg

47.1 Automated Nanohandling: Overview and Trends

47.1.1 Introduction

The handling of microscale and nanoscale objects is an important application field in robotic technology. It is often referred to as nanohandling, keeping in mind the range of aspired positioning accuracy for the manipulation of microscale and nanoscale objects of different natures. The nanohandling of objects may include their finding, grasping, moving, tracking, releasing, positioning, pushing, pulling, cutting, bending, twisting, etc. Additionally, different characterization methods such as indenting or scratching on the nanoscale, measurement of different features of the object, probe positioning with nanometer accuracy, structuring or shaping of nanostructures, and generally all kinds of changes to matter at the nanolevel could also be defined as nanohandling in the broadest sense. This chapter addresses several approaches that can be automated with the help of nanohandling robots. As in the field of "classical" industrial robotics, where humans leave hard, unacceptable work to robots, robots with nanohandling capabilities can help humans to handle extremely small objects with very high accuracy. The size of these robots also plays an important role in many applications. Highly miniaturized robots, often referred to as microrobots, are able to operate in constricted work spaces, e.g., under a light microscope or in the vacuum chamber of a scanning electron microscope (SEM). In particular, microsystem technology (MST) and nanotechnology require this kind of robot since humans lack capabilities in sensing, precision, and direct manipulation at those scales. Automated nanohandling by microrobots will have a great impact in both these technologies.

The development of nanohandling robot systems is a big technological challenge for the robotics research community. Advanced actuator and sensor technologies that are suitable for nanohandling have to be investigated and implemented. Another crucial issue is the development of real-time robot control methods that meet the demands of automated nanomanipulation. The state of the art for nanohandling control approaches includes teleoperated and semiautonomous control strategies. The reader will find a good review of the current work on these approaches in the article by Sitti (2003). Here, the operator controls the nanohandling robot directly or sends task commands to the nanorobot controller using vision, force, or tactile feedback to control the nanohandling process. The further development of MST and nanotechnology, however, requires automated control approaches, so that the robots can accomplish work without any user intervention by using feedback information from different sensors. The automated approach is very challenging, especially due to the difficulty in getting available real-time nanoscale visual feedback and the lack of advanced control strategies able to deal with changing and uncertain physical parameters and disturbances. This chapter presents several promising solutions and applications of the robot-based nanohandling.

47.1.2 Trends in Nanohandling

The following three approaches are being pursued by the majority of the nanohandling labs in the robotic research society, and they seem to be most promising and versatile for future developments in this field:

- Top-down approach utilizing serial nanohandling by microrobot systems. The main goal is the miniaturization of robots, manipulators, and their tools as well as the adaptation of the robotic technology (sensing, actuating, control, automation) to the demands of MST and nanotechnology. This approach is the major topic of this chapter.
- Bottom-up approach or self-assembly utilizing parallel nanohandling by the autonomous organization of micro- and nano-objects into patterns or structures without human intervention.
- The use of a scanning probe microscope (SPM) as a nanohandling robot. In this approach, the (functionalized) tip of an atomic force microscope (AFM) probe or of a scanning tunneling microscope (STM) probe acts as a robot end-effector affecting the position or the shape of a nano-scale part.

Several other approaches such as the use of optical tweezers or electrophoresis might also be adapted for automated nanohandling. They are primarily used for the manipulation of fragile biological samples because of the low grasping forces lying in the pN range. The latter is clearly one of the limitations of these noncontact methods, which are not covered by this chapter. The interested reader can get some inputs from e.g., Holmlin et al. (2000), Sinclair et al. (2004), Yu et al. (2004), Yamamoto et al. (1996), and Christofanelli et al. (2002).

Self-assembly is an approach for parallel nano- and microfabrication, which draws its inspiration from strategies used by nature for the development of complex functional structures. Self-assembly can be seen as the spontaneous formation of higher ordered structures from basic units. Recent technological advances have led to the development of novel "bottom-up" self-assembly strategies capable of creating ordered structures with a wide variety of tunable properties.

Generally, the self-assembly process involves recognition and making connections to the other parts of the system. For this reason, each part has to be equipped with a mechanism supporting its process of self-assembly, i.e., the ability to recognize (self-assembly programming mechanism) and connect (self-assembly binding/driving force) to the proper adjacent part or template. Additionally, an external agitation mechanism is often needed to drive the system to the correct self-assembly. This approach is being increasingly exploited to assemble systems at the micro and nanoscale (Parviz et al. 2003, Morris et al. 2005).

The goal on self-assembly in the microscale is usually the exact planar positioning of parts on a substrate (2D self-assembly) or the creation of 3D-shaped microstructures that cannot be fabricated by existing micromachining methods. To guide the self-assembly, e.g., gravitational, magnetic, or capillary forces can be utilized. A typical application of gravity includes the agitation of parts to make them move on the substrate surface until they "find" suitable binding sites—particularly shaped recesses in the substrate—and get stuck in them. Self-assembly using capillary forces is performed by the exploitation of the hydrophobic and/or hydrophilic features of substrate and microparts, which can be modulated in different ways to improve the controllability and selectivity. Typically, a large number of parts to be self-assembled are put into a fluid on the substrate surface. The parts are attracted by the corresponding binding sites and spontaneously build an ordered structure on the substrate. The main advantage of electrostatic forces is the ability to dynamically control the self-assembly process by modulating the force. In comparison with 2D self-assembly, the 3D approaches are just at the very beginning of their active investigation.

Especially at the nanoscale, when the assembly process deals with a large number of parts, the ability to efficiently manipulate single parts gradually diminishes with the decreasing size of the parts, and the need for a parallel manipulation method arises. Typically, the self-assembly of nano-objects such as nanocrystals, nanowires, or carbon nanotubes (CNT) exploits biologically inspired interaction paradigms such as shape complementarity, van der Waals forces, hydrogen bonding, hydrophobic interactions, or electrostatic forces. A good, and maybe the best-known, example of self-assembled nanostructures are the so-called self-assembled monolayers (SAMs), which are built from organic molecules that chemically bind to a substrate and form an ordered lattice. SAMs can be used for the modulation of surface-dependent phenomena, which is of interest for different applications of nanotechnology, especially for nanoelectronics and nano-optics. Also, 3D self-assembled nanostructures are possible (Fritz et al. 2000) e.g., utilizing a molecular recognition process for binding complementary DNA strands.

The self-assembly of nanowires and CNTs has recently attracted significant attention. The reason is to pursue many promising applications both in nanoelectronics and nano-optics as well as in nano-micro interface technologies. The assembly of nanowires and CNTs is a challenging task due to their shape anisotropy that makes their proper integration into a device difficult. Electric fields between the electrodes on a substrate are widely used in dealing with this task and to trigger the self-alignment of rod-shaped nano-objects (Nagahara et al. 2002, Kamat et al. 2004). The above-mentioned SAM approach is another option for self-assembling CNTs, which is based on the fabrication of binding sites through SAM patterning (Rao et al. 2003).

Self-assembly can be combined with the robot-based handling, leading to hybrid approaches that use the advantages of both serial and parallel technologies. This combination might be a promising solution for different applications in order to achieve higher complexity or productivity. For example, a major European research project that started in 2006 aims at combining ultra-precision robots with innovative self-assembly technologies. The goal is to develop a new versatile 3D automated production system with a positioning accuracy of at least 100 nm for complex microscale products (Hydromel 2006). The combination of serial robot-based handling and parallel self-assembly has not yet been achieved at the industrial scale, and the project team is going to prove the viability of this new production concept.

Self-assembly is a powerful approach that has the potential to radically change the automated fabrication of microscale and nanoscale devices as it enables the parallel handling in a very selective and efficient way. This research field attracts a rapidly increasing number of research groups from multiple disciplines. However, despite promising results achieved up to now, this technology still remains on the level of basic research. One of the most critical challenges in the development of future devices through self-assembly is the limited availability of suitable integration tools that enable automatic localization and integration of parts into the system, especially when the number of sites is very large. Another challenge is the increasing complexity of parts due to the necessary fabrication steps for the implementation of binding features. The study of defects in self-assembled systems and the introduction of fault-tolerant approaches, like in biological systems, will also play a prominent role in transferring self-assembly from research laboratories to device manufacturing. To sum up, the ability to make a complete device by only using self-assembly steps and to become one of the key assembly approaches for the products of MST and nanotechnology remains to be seen.

47.1.2.1 SPM as a Nanohandling Robot

SPMs can deliver high-resolution images of a wide class of hard and soft samples, which are used e.g., for materials and surface sciences, bioscience research, or nanotechnology. Additionally, these devices can be used to interact with nanoscale parts in a controllable way, which results in a change of their position or their shape. The latter approach, the use of an AFM acting as a nanohandling robot, has been actively investigated in the past (Ramachandran et al. 1998, Chen et al. 2006, Mokaberi et al. 2007).

The ultimate goal of this approach is to automatically assemble nanoscale parts in nanosystems in ambient conditions, aiming at the rapid prototyping for nanodevices. The part first has to be localized on the substrate by an imaging scan performed in dynamic mode. In the second step, the AFM tip is brought to the immediate vicinity of the part and is moved afterward—staying in dynamic mode without AFM feedback in the z-direction—to the center of the part toward a predetermined location. As a result, the part is pushed in a "blind" feed-forward way by repulsive forces. The re-imaging of the area of interest afterward reveals the results of the manipulation, which are often not satisfying and require frequent experiments by trial and error. The current research aims at developing a high-level AFM control system to perform predictable nanohandling operations, which might open the door to high-throughput automated nanomanipulation processes (Makaliwe and Requicha 2001, Chen et al. 2006). The whole variety of operational modes of SFM (Meyer et al. 2004) has not been fully investigated with regard to nanomanipulation.

The SPM tip can also be used to modify surfaces with nanometer resolution or to change the object shape, e.g., by scratching, indenting, cutting, dissecting, etc. (Heckl 1997, Schimmel et al. 1999, Villarroya et al. 2004). A destructive interaction between the tip and the sample is usually an unwanted effect while imaging. However, for nanomachining purposes, the SPM tip can be exploited as a nanohandling tool such as nanoscalpel or nanoindenter. Nanoscratching is implemented by moving an AFM tip on a surface and applying a high load force to the tip. This technique can be used among others for mask-free lithography on the nanoscale level. Biological specimens can also be handled in this way. The chromosomal microdissection by AFM was used e.g., for isolating DNA (Lü et al. 2004). It was possible to extract a DNA chromosome by one AFM linescan and pick it up by the AFM tip through hydrophilic attraction.

The mask-free nanolithography mentioned above can also be implemented by anodic oxidation or by the so-called dip-pen nanolithography (DPN). AFM is used as a "writing" device capable of drawing lines with the width of a few tens of nanometers. To perform nanostructuring by anodic oxidation, a thin metal layer is deposited on the substrate surface and a voltage is applied between the metal and the conductive AFM tip. Since the metal surface is moistened in an ambient atmosphere, an electrolytic process is triggered by the voltage, resulting in a tiny metal oxide dot on the surface. By using proper process control, these dots can form a sophisticated nanopattern on the substrate surface. Another way of writing on the nanoscale is DPN, which works in a manner analogous to that of a dipped pen. The AFM tip is coated with a chemical reagent ("ink") that is to be deposited. The molecules of the ink are transported from the AFM tip to a target substrate using capillary forces, through a solvent meniscus forming between the tip and the substrate under ambient conditions. This simple method of directly depositing molecules onto a substrate has recently become an attractive tool for nanoscientists, especially because of its versatility as it enables molecular deposition of virtually any material (hard and soft) on any substrate. However, ink/substrate combinations must be chosen carefully so that the ink does not agglomerate or diffuse. Additionally, the ink molecules have to be able to anchor themselves to their deposition location (molecular "glue"). These challenges are the subject of the current research activities.

The main problem of AFM-based nanohandling concerning automation is the lack of real-time visual feedback. The same AFM tip cannot be simultaneously used for both imaging and handling, so that the results of nanohandling have to be frequently visualized by an AFM scan to verify the performance. This procedure makes the nanohandling process inefficient, unsuitable for high throughput, and includes uncertainties due to being "blind" during the manipulation steps.

One of the ways to overcome this problem is to model the nanopart behavior, including all the relevant interactions between the tip, part, and substrate. Having a valid model, it is possible to mathematically simulate the behavior of the nano-objects during manipulation and to calculate the expected position of the part in real-time. This approach will enable nanomanipulation in open-loop mode without real-time visual feedback. Such a model is the basis of the so-called augmented reality systems transferring the real workspace into virtual reality and delivering calculated position feedback during manipulation (Li et al. 2005). This approach, however, requires exact knowledge of nanomanipulation phenomena, which is not available in the current state of the nanosciences. Due to the lack of understanding of what exactly is going on during nanomanipulation, the usability of augmented reality systems for automated handling is currently limited with regard to reliability and reproducibility.

Another challenge of the AFM-based nanohandling arises when accuracies in the sub-nanometer range are required. Most commercial AFM devices cannot offer a reliable position feedback at this level, and spatial uncertainty in AFM—because of the thermal drift of AFM components, creep and hysteresis of piezo actuators, and other variant effects and nonlinearities—cannot be taken care of in a direct way. Some solutions to this problem have been addressed in the articles by Mokaberi and Requicha (2006) and Schitter et al. (2004).

A combination of the AFM with other imaging techniques that supply independent visual feedback from the work scene during nanomanipulation by the AFM tip seems to be most promising for automation. For positioning accuracies down to about 0.5 μm, the manipulation can be monitored by an optical microscope. This approach is e.g., frequently used in bioscience research to provide multimodal imaging capabilities for yielding extensive information on biomolecules and biological processes. For applications on the nanoscale, SPM–SEM hybrid systems are currently attracting rising interest. An SPM head is integrated into the vacuum chamber of an SEM so that both microscopy methods are used in a complementary fashion to analyze the sample properties, building a sophisticated nanocharacterization device (Troyton et al. 1998, Joachimsthaler et al. 2003). To exploit such a system for nanohandling, the SEM is used as a sensor for real-time visual feedback during nanomanipulation or modification of the sample surface by the AFM. Important issues to be addressed are the synchronization of both microscopes and proper system engineering enabling the AFM tip to act in SEM's field of view.

The latter concept builds a bridge to the main topic of this chapter, which is introduced in the next section. Indeed, if we exclude the visualization feature of AFM and just think of the nanopositioning capability of the AFM scanner, then we are left with a nanohandling robot carrying a tiny cantilever with a (functionalized) nanoscale tip as an end-effector—a three degrees-of-freedom (DoF) nanohandling robot operating in the vacuum chamber of the SEM and using SEM images for vision feedback to automatically control the end-effector position in real-time.

47.1.3 Automated Microrobot-Based Nanohandling

The following two concepts are currently being followed to carry out micro and nanomanipulation:

- Purely manual manipulation is a common practice e.g., in medicine and biological research. Even in industry, such tasks are often carried out by specially trained technicians. However, with progressive part miniaturization and the positioning tolerances going down to the nanoscale, the capabilities of the human hand have their natural limitations.

- The application of teleoperated manipulation systems, which transforms the user's hand motions into the finer motions of the system manipulators. Here, special effort is devoted to the development of methods that allow the transmission of feedback information in a user-friendly form from the work scene (images, forces, noises). The user interface can include a haptic device, providing tactile information that helps the user to operate in a more intuitive way. The system might also use an augmented reality environment that is based on the mathematical modeling of the application-relevant phenomena. However, the fundamental problems of the resolution of the fine motion and of speed as well as of repeatability remain, since the motion of the tool is a direct imitation of the user's hand.

The focus of the recent research is, however, on the use of automated nanohandling stations containing miniaturized nanohandling robots that exploit direct drives typically implemented by using piezoelectric, electrostatic, or thermal microactuators. The flexibility of such a microrobot can be enhanced by dividing the actuator system into a coarse positioning module (often being referred to as a mobile platform) and a fine positioning module or nanomanipulator carrying an application-specific end-effector. The user instructions are given through a graphical user interface to the station's control system that generates corresponding commands for the robot actuators. The degree of abstraction of the user instructions is determined by the capabilities of the control system. Different aspects of this approach are discussed later in this chapter, along with promising applications in MST, nanotechnology, biotechnology, and material science.

Microrobotics for handling microscale and nanoscale parts has been established as a self-contained research field for nearly 15 years (Fearing 1992, Fujita 1993, Morishita and Hatamura 1993,

Aoyama et al. 1995, Codourey et al. 1995, Fatikow et al. 1995, Hatamura et al. 1995, Magnussen et al. 1995, Arai and Tanikawa 1997, Hesselbach et al. 1997, Menciassi et al. 1997, Rembold and Fatikow 1997, Sitti and Hashimoto 2000, Ferreira et al. 2001, Bourjault and Chaillet 2002, Fukuda et al. 2003, Mazerolle et al. 2004, Brufau et al. 2005, Driesend et al. 2005b, Martel 2005). In recent years, a trend toward the microrobot-based automation of nanohandling processes emerged (Fatikow and Rembold 1997, Fatikow et al. 1997, Fatikow et al. 1998, Fahlbusch et al. 1999, Kasaya et al. 1999, Bleuler et al. 2000, Fatikow 1996, 2000, Fatikow et al. 2000, Schmoeckel et al. 2000, Tanikawa et al. 2001, Fatikow et al. 2002, Yang et al. 2003, Hülsen et al. 2004, Clevy et al. 2005, Wich et al. 2005, Fatikow et al. 2007b). Process feedback, i.e., the transmission of information from the nano-world to the macroworld to facilitate the control of the handling process, has emerged as the most crucial aspect of nanohandling automation. With the current technology, it is rather difficult to obtain reliable force information while handling micro-scale and especially nano-scale parts. For this reason, vision feedback is often the only way to control a nanohandling process. The capability of an optical microscope rapidly decreases with the parts being scaled down to the nanoscale level. Scanning near-field optical microscopy (SNOM) may, however, be exploited for nanoscale manipulations in an ambient environment (Fukuda et al. 2003). The vacuum chamber of a SEM is, however, for many applications the best place for a nanohandling robot. It provides an ample workspace, very high resolution up to 1 nm, and a large depth of field. Quite a few research groups have been

investigating different aspects of nanohandling in SEM, e.g., Aoyama et al. 1995, Hatamura et al. 1995, Kasaya et al. 1999, Schmoeckel et al. 2000, Fatikow et al. 2002, Mazerolle et al. 2004, Misaki et al. 2004, and Nakajima et al. 2004. However, real-time visual feedback from changing work scenes in the SEM containing moving nanorobots is still a challenging issue.

Figure 47.1 presents a generic concept of an automated micro-robot-based nanohandling station (AMNS).

The microrobots are usually driven by piezoactuators that enable positioning resolutions down to sub-nm ranges, which is the main precondition for the development of an AMNS. The travel range is comparatively large with several tens of millimeters for stationary microrobots and with almost no limitation for mobile microrobot platforms (Kortschack and Fatikow 2004). The mobile robots of the station have a nanomanipulator integrated in their platform, which makes them capable both of moving over longer distances and of manipulating with nanometer accuracy. The former leads to more flexibility, as the robots can be deployed everywhere inside the SEM vacuum chamber. Stationary robots are, on the other hand, easier to control, which makes them more suitable for high-throughput automation. Depending on the application, different combinations of both robot types are to be implemented. Various tools can be attached to the nanomanipulator and exchanged according to the task to be accomplished. The already mentioned use of a (functionalized) AFM probe with its extremely sharp tip (10–20 nm) as a robot tool is currently one of the most actively pursued approaches (Sitti and Hashimoto 2000, Fukuda et al. 2003, Mircea et al. 2007).

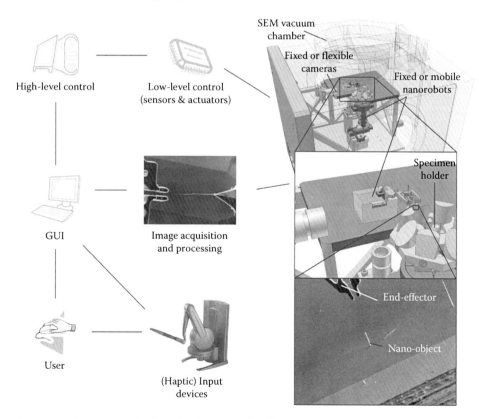

FIGURE 47.1 Generic concept of an automated microrobot-based nanohandling station.

The robots usually have to be tethered and get driving voltages for the actuators over the vacuum-sealed mechanical robot interface integrated into the SEM wall.

The sensor system of the AMNS includes the SEM, video cameras, force sensors, as well as—if available—position sensors integrated into the robot axes. The SEM delivers near-field sensor information for the fine positioning of the robot tool, and video cameras provide necessary far-field information for the coarse positioning of the robot. They typically are integrated into the wall of the SEM's vacuum chamber. Additionally, a mobile microrobot can be operated as a "cameraman," carrying a miniature video microscope (Jasper et al. 2007). Such a mobile microscope can deliver images from virtually any point of view in the work space of the nanohandling robots. The combination of a mobile video microscope and SEM with an integrated video camera can contribute to the station's flexibility and versatility, providing a smooth transition between the different magnifications during a nanohandling process. The sensor data are sent to the station's control system for real-time signal processing. Its task is to calculate the positions of the robots and their tools as well as the positions of the parts to be handled or other objects of interest. The calculated positions serve as input data for the closed-loop robot control.

Teleoperation is often the first step on the way to automation, as it helps the user to learn more about the nanohandling task to be implemented. For this reason, even though the AMNS is designed for automated nanohandling, a user interface for the teleoperation of the robot system is often a relevant component of the AMNS. The user can influence the handling process by a haptic interface and/or by a graphical user interface (GUI). This interface may include an augmented reality system for simple nanohandling tasks. A good overview of the teleoperation techniques and applications is given in the article by Ferreira and Mavroidis (2006).

The positioning accuracy of the microrobots during automated nanohandling is affected by several factors, so that a powerful robot control system is required. In the low-level control system, driving voltages for the robot actuators are calculated in real-time, which keeps the robot and its end-effector on the desired path. The high-level control system is responsible among others for path planning, error handling, and the time-saving parallel execution of tasks. Both user interfaces, GUI and haptic interface, are supported by the high-level control system as well. An advanced control system architecture tailored for nanohandling automation in the SEM was suggested in the article by Stolle and Fatikow (2008).

The AMNS concept has been implemented in different application fields in which (semi-) automated nanomanipulation is required (Trüper et al. 2004, Sievers and Fatikow 2006, Fatikow et al. 2007b, Jähnisch and Fatikow 2007, Luttermann et al. 2007, Mircea et al. 2007, Wich et al. 2008, Eichhorn et al. 2008, Fatikow et al. 2008a,b, Fatikow 2008, Krohs et al. 2008). The relevant aspects of these developments will be introduced in the following sections.

47.1.4 Structure of the Chapter

The following is a brief summary of the topics covered in the following sections.

Section 47.2 introduces the fundamental considerations for automated robot-based nanoassembly. The main challenge is the control of the adhesive surface forces during the assembly process. The single sub-processes—joining, separation, transport, etc.—are explained and a brief overview over a known setup for (semi-) automated handling on the micro and nanoscale is given. Additionally, the necessary tools and methods for the successful automation of assembly processes on the nanoscale are described.

Section 47.3 covers the vision feedback in nanohandling robot stations. The focus is on the use of the SEM in combination with real-time image processing algorithms, which is used as the near-field sensor for the automation of nanohandling tasks. The specific properties of an SEM as image sensor are explained, and the requirements for real-time SEM image processing methods are outlined. The integration of an SEM into an image processing system enables real-time image access and electron beam control. The main part of the chapter is devoted to the description, implementation, and validation of real-time tracking algorithms for SEM images, using cross-correlation and active contours with region-based minimization. Additionally, methods for the extraction of depth-information from SEM images are discussed.

Section 47.4 deals with the control issues in automated nanohandling. An AMNS usually consists of several nanorobots that need to be controlled. Due to the working principles and characteristics of the employed actuators, the key problems differ from macroscale robotics. There are major differences on all layers from the electrical actuation to open- and closed-loop control. Additionally, due to the required nanometer-precision, the world model required for effective automization can only be a partial model and needs to be adaptive. Several approaches are presented that tackle the arising issues.

Section 47.5 describes the nanorobotic characterization of CNT starting with a motivation of the research activities on this novel material. Fabrication techniques and possible applications for CNTs are given. The physical basics needed for the mechanical characterization of CNTs are introduced and a short state of the art in this area is given. An implementation example of a nanorobotic handling cell setup is presented. A mechanical characterization technique based on piezoresistive AFM probes is developed and experimental results are discussed with respect to theoretical models. Finally, the Young's modulus of multi-wall carbon nanotubes (MWCNTs) is calculated.

Section 47.6 deals with the characterization and manipulation of biological objects by methods used with nano and micro-robotic systems. An overview of the methods that can be used is given. Also, the current state-of-the-art applications with these methods are presented. So, the use of AFM, cell injection methods, SEM, environmental scanning electron microscope (ESEM), dielectrophoresis, optical tweezers, and nanotweezers as tools and methods for the handling of biological objects are described, including a critical look at the advantages and disadvantages of these methods. This section finishes with an overview of the nanorobotic systems the "Division Microbotics and Control Engineering" (AMiR) has developed to handle and characterize biological objects.

47.2 Automated Robot-Based Nano-Assembly

This section describes the challenges for assembly on the nanoscale from a technological and process-oriented point of view. With respect to automation, the critical process steps are identified and tools and methods are introduced to allow for efficient handling, manipulation, and assembly of objects on the nanoscale.

47.2.1 Assembly Processes on the Nanoscale

Assembly processes on the nanoscale are commonly differentiated from micro or macroscale assembly by the size of the objects to be assembled. Thus, the term nano-assembly is used when the object size is below one micrometer (Böhringer et al. 1999). The interaction between tool and object during assembly is, however, determined by the smallest dimension of the part (Wich and Hülsen 2008), which again determines the necessary actuator accuracy (Fatikow 2008). Therefore, assembly processes where one object dimension—referred to as the critical size—is in the sub-μm range are referred to as nano-assembly processes.

When assembly processes are scaled down from the macro to the nanoscale, i.e., the critical object size is reduced from the millimeter to the nanometer range, the surface-to-volume ratio increases dramatically. Consequently, the influence of adhesive surface forces increases compared with gravitation. The components of adhesive surface forces are van-der-Waals, electrostatic, and capillary forces. The effect of these dominant surface forces on the nanoscale is a strong adhesive bond between the objects, the underlying substrate, and the tools used for handling and manipulation. Hence, objects stick to the tool and can hardly be released. This is one of the major challenges for assembly processes on the nanoscale, resulting in a completely new approach for assembly process planning and development.

Assembly processes on the macroscale include the sub-processes handling, adjustment, joining, and inspection. During handling, the object is transported and oriented between two locations; an essential part is the picking and placing of the object. Before joining objects, they have to be adjusted to each other. Joining is used to assemble objects by utilizing form, material, or force closure. Usually, the joining process is followed by the sub-process inspection, in order to verify the assembly.

On the nanoscale, handling the sub-process is very complex. When picking objects, specialized separation techniques have to be utilized as the objects adhere due to the adhesive surface forces to the substrate, i.e., the force between the tool and the object has to exceed the force between the object and the substrate. Typical

examples for breaking the adhesive bond between the object and the substrate are rolling off (Saito et al. 2001), pulling, or peeling off (Mølhave et al. 2006), as well as particle beam induced etching and sputtering, respectively (Wich and Fatikow 2008).

When placing the object, the force between the substrate and the object has to exceed the force between the tool and the object. Again, the above mentioned specialized techniques have to be applied to break the adhesive bond between the tool and the object. Thus, handling and assembly processes on the nanoscale correspond to each other; both involve the sub-processes separation, transport, joining, adjustment, and inspection.

47.2.2 Survey on Nano-Assembly Stations

Nano-assembly stations consist of the assembly system, a sensor network, and the control infrastructure. The assembly system controls the material flow through separating, moving, and joining the objects. The whole process is surveyed by the sensor network with the magnifying vision sensor as the main component. The information flow, i.e., the control of the assembly process is realized with the control framework.

The main components of the assembly system are the tools used for pick-and-place operations on the objects, actuators for positioning and orienting the tool to the objects, and additional components for joining and separation sub-processes, respectively. For the handling of objects, two types of tools are commonly used. Due to the strong adhesive forces, simple tools such as finely etched metal tips can be utilized for the picking and placing of objects. These metal tips—commonly etched from tungsten or platinum-iridium wires—are easy to produce (Campbell et al. 1998). The production of two-jaw grippers for the handling of nanoscale objects requires sophisticated MEMS technologies. However, these tools allow for better control during gripping. Both types of tools require specialized handling methods, which will be described in Section 47.2.3.

For positioning the tool and the object relative to each other, a wide variety of actuators can be used. The time-variant behavior of the positioning system and other perturbations can be compensated for, therefore, generally, closed-loop positioning systems are preferred. As a rule of thumb, the position accuracy or the minimum step resolution of open-loop actuators, respectively, should be one tenth of the critical object size. Commonly applied actuators utilize the piezo-based stick–slip principle (Bergander et al. 2000), allowing for small steps in the (sub-) nanometer range at a traverse path in the centimeter range. These actuators can be implemented as linear and rotational actuators as well as mobile microrobots (Kortschack and Fatikow 2004). Piezo stack actuators, whose travel is transformed with flexible hinges to prevent play, are also applied commonly. Due to their restricted travel in the 100 μm range, they are preferably applied for the end-effector's movement.

The main component of the sensor network is the magnifying vision sensor. For the observation of the nano-assembly process, typically SEMs or SPMs are used. In exceptional cases, light optical microscopes can also be used for assembly of nanoscopic

TABLE 47.1 Overview of Published Micro/Nanoassembly Stations with an at Least Semi-Automated Operation Mode

Publication	Position Sensors	Objects with (Critical) Dimensions	Operation	Mode
Codourey et al. (1997)*	Light optical microscope	Diamond crystals, 100 µm	Pick & place	Semi-automated
Yang et al. (2001)*	Integrated in handling robot	Metal sheets, 0.1 mm	Pick & place	Automated
Thompson and Fearing (2001)*	Linear axes with integrated position sensors	Metal blocks, 100 µm	Pick & place	Automated
AMiR(2006)	Light optical microscope with CCD-cameras	Glass spheres, diameter 30 µm	Pick & place	Automated
Fatikow et al. (2007b)	SEM, CCD-cameras	TEM-lamellae	Pick & place	Semi-automated
ZuNaMi-project (unpublished)	SEM, linear axes with integrated position sensors	Carbon nanotubes (CNT), diameter 200 nm	Assembly	Automated

Notes: The publications marked with an asterisk (*) report on assembly processes but refer more precisely to pick & place operations.

objects with a high aspect ratio, i.e., silicon nanowires (Reynolds et al. 2008). The resolution of the SEM is in the nanometer range, whereas SPMs can achieve atomic resolution. Both instruments are thus well suited for the observation of the nano-assembly process. However, the image acquisition time of SPMs is generally longer compared with SEMs. This criterion is important for the automation of nano-assembly processes where the positions of objects and tools are derived by means of object recognition from the vision sensor images, referred to as "visual servoing" (Vikramaditya and Nelson. 1997, Fatikow et al. 2000, Sievers 2006). Further information about object recognition and tracking can be found in Section 47.3. The SEM's advantage over the SPM, however, is its ability to image objects not only on the nanoscale but also at a very low magnification, which simplifies navigation on the substrate. A more detailed discussion of the advantages and disadvantages of different microscope types with respect to assembly, handling, and manipulation on the nanoscale can be found in the article by Wich and Hülsen (2008).

For the control of automated nano-assembly processes, a complex infrastructure is necessary. The main components are the sensor server, which acquires and pre-processes sensor data, e.g., from the vision system; the low-level controller, which is used for positioning the objects and tools; and the high-level controller. The latter is the central control unit for sequencing the single sub-processes and their procedures during assembly.

In Table 47.1, a brief overview on (semi-) automated assembly stations for micro or nanoscale parts is given.

47.2.3 Tools and Methods for (Automated) Nano-Assembly

The automation of nano-assembly processes generally aims at increasing the key process figures, which are throughput and reliability. For optimization of these parameters, it is necessary to analyze the process chain and to identify the critical procedures within it.

Automated nano-assembly processes consist of a first sub-process, in which the objects to be assembled are indexed, and the assembly process itself, where new products are generated (cp. Figure 47.2). These two steps are performed utilizing a magnifying vision sensor, e.g., the SEM. Additionally, these steps can be completed by pre- or post-processing sub-processes, e.g., mounting the single objects to an adaptor or releasing the products from the adaptor. The indexing is performed to determine the poses of objects and can additionally be used for checking quality parameters on them. This enables adaptive assembly processes. On the nanoscale, the indexing and the assembly processes consist of very distinct methods needed for localization, handling, joining and separation of objects, and for contact detection between the objects and tools. These methods

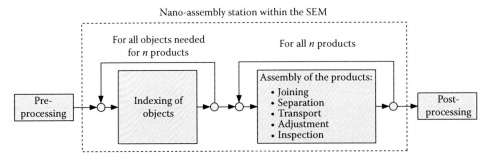

FIGURE 47.2 Automated nano-assembly processes consist of a step for indexing all objects needed to assemble *n* products and the assembly step itself. Additional pre- and post-processing steps complete the process chain.

and techniques have a major impact on the key figures of the automated assembly-process, either because of poor reliability, long duration, or because of their repeated executions in the process chain. In the next paragraphs, tools and methods that are important for the successful automation of the nano-assembly processes will be discussed.

Localization of objects—objects to be assembled are usually mounted on adapters, allowing the operator to handle the nanoscale objects. These adapters have macroscopic dimensions, i.e., general precision engineering tolerances apply. However, these tolerances are huge compared with the critical dimensions of the objects. Thus, the single positions of the objects are unknown and have to be measured by means of a localization step. Starting at a low magnification, a known characteristic feature on the adapter or the substrate is localized by means of object recognition and then centered in the middle of the vision sensor image. Afterward, the vision sensor zooms in to the next magnification step and the next characteristic feature or the object itself is localized. The above-mentioned steps form the so-called zoom and center (ZAC) procedure (Wich and Hülsen 2008), which is the key element for the localization of objects. This procedure can be adopted to allow for the recognition of similar but not identical objects by means of an optimized object recognition algorithm (Wich et al. 2008). Measurements on different objects in the SEM showed that generally 2–3 ZAC steps are sufficient to zoom from the lowest magnification to a magnification where objects with critical dimensions in the sub-µm range are recognizable. The ZAC procedure is applied when the initial poses of the single objects have to be recorded during the localization procedure, but also several times in each assembly cycle.

Contact/distance detection—the SEM, which is widely used as a vision sensor in nano-assembly, only generates 2D-image information. During the adjustment sub-process, the object and the tool have to be approached and brought into contact with each other in all three dimensions in order to enable the following sub-process, e.g., joining. However, due to the high depth-of-focus of the SEM (Reimer 1998), the out-of-image-plane distance between the tool and the object is hardly detectable. Hence, specialized contact/distance detection methods have been developed to allow for adaptive automation on the nanoscale. Table 47.2 gives an overview of the

well-established methods, which can also be applied for light optical microscopes.

The depth-from-focus (DfF) method performs a focus sweep during which two regions of interest (ROI) within the images are evaluated for sharpness (Estana et al. 2004, Fatikow et al. 2007a), e.g., one ROI images the tool, the other one images the object to be picked up. The resulting measurement curves for the sharpness (arbitrary units) over focus distance show distinct peaks, where the tool or the object within the ROI is imaged sharply. From the focus distance between the two peaks, the out-of-image-plane distance between the tool and object can be derived.

For 3D-image reconstruction, the acquisition of two images under slightly tilted angles is necessary. In the case of an SEM, this can be achieved through electron-beam tilting by means of a separate beam deflection coil (Jähnisch 2008a) or by tilting the specimen under the electron beam. The former method is preferred because of the higher image acquisition speed. Subsequent image processing allows for the generation of 3D-images of the scene. From this data set, the position and orientation of the objects and tools relative to each other can be derived.

Contact between the objects can be detected through measuring the contact force between them. This force leads to a deformation, which can be measured by means of object recognition if the object is deformed in the image plane. This method has been proved to work reliably, especially when objects such as CNTs with a high aspect ratio are deformed in the field of view (Wich et al. 2006b).

Force sensors can be used to detect contact between objects. However, with respect to force measurements on the nanoscale, very sensitive force measurement methods are necessary for detecting forces in the nano- and micro-Newton range. A very reliable method is the touchdown technique, where the tools or objects are mounted onto the sensor (Fatikow et al. 2007a). The sensor including the mounted tools/objects is then forced to oscillate at the systems resonance frequency. Contact between the tool/object and another object results in a phase-shift between excitatory and excited oscillation. The phase shift is used as a measure for the contact force.

Handling strategies of small objects—due to the dominant adhesive surface forces, objects on the nanoscale tend to stick

TABLE 47.2 Overview of Commonly Used Techniques for Contact or Distance Detection between Objects or Tools and Objects

Method	Measurement Direction Related to Image Acq. Plane	Measurement Value	Object Requirements	Additional Sensors Needed
Depth-from-Focus (DfF)	Out-of-plane	Distance	Textured objects	—
3D-image acquisition	Out-of-plane	Distance	Textured objects	Beam shift/tilt for SEMs
Optical deformation analysis	In-plane	Contact force	Recognizable deformation on imaged scale	—
Tactile sensors	Independent	Contact (force)	—	Tactile sensor

to the substrate surface when picked up and to stick to the tool when released. Consequently, specialized techniques are necessary to allow for defined pick-and-place operations.

When simple tools, e.g., fine metal tips, are used as a tool, the object can stick to the tool through adhesion forces. With respect to automation, a defined gripping force between the tool and the object is necessary. This can be achieved by material closure, i.e., the object is bonded to the tool by means of an electron beam or focused ion beam induced deposition (EBiD, FIBiD). For placing the object, it is bonded to the substrate and the connection between the tool and the object is released, e.g., by means of electron beam or ion beam induced etching or sputtering (Wich and Fatikow 2008). Alternatively, more specialized guided movements between the tool and the object can be used to overcome the adhesive forces (Saito et al. 2001, Mølhave et al. 2006). These techniques, however, are hard to automate.

The application of two-jaw grippers allows for a dynamic adaptation of the gripping forces during the handling sub-process. However, this technique faces two major challenges: First, the jaw cross-section has to be adapted to the object to be handled, resulting in a low stiffness of the jaws. Hence, the transmittable gripping force is low. To overcome this, grippers with optimized topology are in development (Sardan et al. 2008). Second, the object sticks to the gripper jaws during placing. Again, the object can be bonded to the substrate using the above-mentioned techniques (Mølhave et al. 2006, Wich et al. 2006a). The gripper jaws have to be optimized for the reduced contact area (Arai et al. 1998) or specialized placing techniques have to be used. With respect to assembly automation, the placing of objects utilizing a bonding technology is sensible, as this is commonly equivalent to the "joining" sub-process.

Bonding and separation methods—in the SEM and the focused ion beam microscope (FIB), the electron or ion beam can be used to trigger chemical reactions on the substrate in the presence of precursor molecules. These precursor molecules dissociate under electron impact into nonvolatile and volatile fragments. The nonvolatile fragments form a deposition suitable for bonding objects to each other. Depending on the precursor material, the mechanical and electrical properties of the deposits can be widely adjusted. A detailed overview of the charged particle beam induced processes is given in the article by Utke et al. (2008). Typical precursors used for nano-assembly purposes are tungsten-hexacarbonyle ($W(CO)_6$) for the deposition of hard, brittle, and electrically conducting mechanical interconnects and di-cobalt-octacarbonyle ($Co_2(CO)_8$) for the deposition of soft and flexible mechanical interconnects (Wich and Fatikow 2008). Specialized precursors such as xenon-difluorine form reactive fragments under electron impact and can thus be used for etching purposes. The electron/ion beam triggered deposition and etching processes are suitable for the joining and releasing of objects in nano-assembly processes. The application of etching processes is described in the article by Wich and Fatikow (2008).

47.3 Vision Feedback for Automated Nanohandling

On the nanoscale, object behavior is not as predictable as on the macroscale. The increasing influence of different factors such as parasitic forces leads to effects that can disturb and hinder the handling processes. Also, the sheer small dimensions of the objects and distances have been shown to be problematic. One reason is e.g., that the thermal expansion of the parts of the setup such as the actuators, which are carrying manipulation tools, may exceed the dimensions of the objects to be handled. This means that the object cannot be precisely located using only actuator information. The automation of nanohandling processes, therefore, requires additional sensor feedback to ensure its success. In order to increase knowledge about the setup and the objects, image processing algorithms are applied on images acquired from imaging sources suitable for these dimensions. After the introduction of some basic image processing terminology in Section 47.3.1, Section 47.3.2 discusses the mainly used imaging source, the SEM, and its limitations and properties. The applied techniques of object tracking are presented in Sections 47.3.3 through 47.3.5, after which the problem of missing depth information and possible depth estimation methods are evaluated in Section 47.3.6.

47.3.1 Image Processing Basics

Images are represented as matrices containing values representing the brightness at the position of the entry in the matrix:

$$I(x, y) = g_{xy}$$

Operations can be executed on all separate values

$$I_2(x, y) = f\big(I_1(x, y)\big)$$

which can be used for pixel-wise operations such as thresholding.

Operations may also include, e.g., the image gradient

$$\nabla = \begin{pmatrix} \dfrac{\delta}{\delta x} \\ \dfrac{\delta}{\delta y} \end{pmatrix}$$

resulting in a gradient image

$$\nabla I(x, y) = \begin{pmatrix} \dfrac{\delta I(x, y)}{\delta x} \\ \dfrac{\delta I(x, y)}{\delta y} \end{pmatrix}.$$

47.3.2 SEM as Imaging Sensor

As the observed dimensions are smaller than the resolution of optical microscopes, the scanning electron microscope is used. It delivers high resolution and high magnification images combined with high flexibility and a fast possible scanning speed. Though it has these advantages, it also has disadvantages and problems to overcome. Some of the problems result from the scanning nature of the SEM, and the fact that pixels contained in an image differ in their acquisition time:

$$I_{SEM}(x, y) = I_{Scene}(x(t), y(t))$$

The main problems for the image processing and automation are the following:

- Increased level of noise: one problem of SEM images is that, if they are acquired at high speed, they contain an increased level of additive noise due to the low duration of scanning for each pixel (see Figure 47.3):

$$I_{real}(x(t), y(t)) = I_{ideal}(x(t), y(t)) + \eta(t)$$

- Distortions due to movement: this additional problem also results from the scanning nature of the SEM. It occurs when the object movement is in the same order of magnitude as the scanning time:

$$\left(\frac{\delta x_{object}}{\delta t} \geq \frac{\delta x_{scan}}{\delta t} \right) \vee \left(\frac{\delta y_{object}}{\delta t} \geq \frac{\delta y_{scan}}{\delta t} \right)$$

- Brightness variation: charge accumulation may lead to a change in object appearance:

$$I_{SEM}(x, y) = f(x, y, Q(x, y, t))$$

- Missing depth information: The high depth of field is in some cases a feature of the SEM, but for the automation, the missing depth information introduces additional problems.

These problems have to be taken into account by image processing algorithms analyzing SEM images in automation setups.

47.3.3 Object Tracking

Object tracking is used to continuously extract the position of objects in the nanohandling scene. Tracking algorithms use data about the previous pose of the object to determine the current pose, as depicted in Figure 47.4. This decreases the complexity of the problem of locating the object and enables the usage of different approaches for localization.

Many different algorithms for object tracking exist, but not all are suitable for the specific requirements of SEM imaging. Two approaches that turned out to be robust enough for use in nanohandling automation setups will be described in the following sections: template matching using cross-correlation and region-based active contours.

47.3.4 Template Matching Using Cross-Correlation

A simple yet powerful approach for tracking objects in SEM images is template matching using cross-correlation. If an input image matrix *I* contains the brightness data of the SEM image and a template image, *T*, exists for the object of interest, the cross correlation of both matrices gives the correlation matrix

$$C = I * T$$

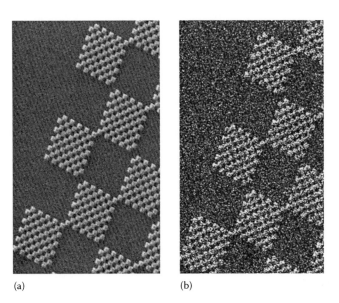

(a)   (b)

FIGURE 47.3   SEM image using (a) slow and (b) fast scanning.

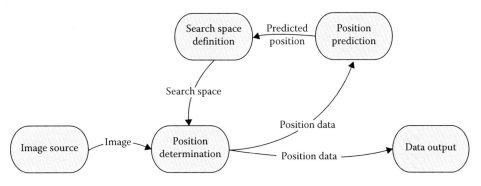

FIGURE 47.4   Basic principle of object tracking.

with

$$C(x_p, y_p) = \sum_{x=0}^{x_t} \sum_{y=0}^{y_t} I(x_p + x, y_p + y) \cdot T(x, y)$$

The correlation matrix contains a measure for the similarity of the template image with each similarly sized part of the input image. If the object of interest is included in the input image, it can be concluded that the object is at the position determined by the maximal value in the correlation matrix.

The position of the maximal value may be determined to fractions of a pixel by applying a center of gravity calculation to a region R surrounding the maximal value determined by thresholding. The coordinate calculations are

$$x_{\text{subpixel}} = \frac{\displaystyle\sum_{(x,y)\in R} x \cdot (C(x,y) - t)}{\displaystyle\sum_{(x,y)\in R} (C(x,y) - t)}$$

and

$$y_{\text{subpixel}} = \frac{\displaystyle\sum_{(x,y)\in R} y \cdot (C(x,y) - t)}{\displaystyle\sum_{(x,y)\in R} (C(x,y) - t)}$$

If the image data contains additive noise, the correlation matrix theoretically is not altered, because the template image and the additive noise are uncorrelated. Practically, the effect of a decreasing signal to noise ratio will be a decreasing ratio of correlation matrix peak and correlation matrix values. An example of tracking using cross-correlation can be seen in Figure 47.5.

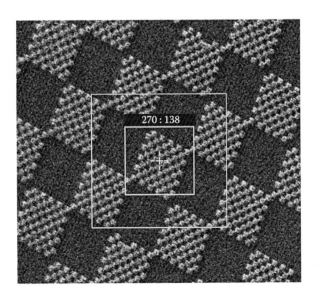

FIGURE 47.5 Chessboard pattern tracked with cross correlation.

47.3.5 Region-Based Active Contours

Another useful approach for object tracking is the use of active contours, or snakes (Blake and Isard 1998).

Snakes can be represented as curves in 2D space:

$$s \rightarrow v(s) = \begin{pmatrix} x(s) \\ y(s) \end{pmatrix}$$

The active contour has an energy function consisting of an internal and an external part. The original formulation of the contour energy function is given according to Goudail (2004) as

$$E(C) = E_{\text{int}}(C) + E_{\text{ext}}(C),$$

with

$$E_{\text{int}}(C) = \int_0^1 \left[\alpha \cdot \left|\frac{\delta v}{\delta s}\right|^2 + \beta \cdot \left|\frac{\delta^2 v}{\delta s^2}\right|^2 \right] ds$$

and

$$E_{\text{ext}}(C) = -\int_0^1 \gamma \left|\nabla I(v)\right|^2 ds.$$

Because of the additive noise in SEM images, the traditional formulation of snakes using edge-based methods is not robust enough. Therefore, region-based energy functions are used. The external part of the energy function is dependent on the image data, it is chosen in dependency on statistical image properties. An external energy function that, according to Sievers (2007), is suitable for SEM images is

$$E_{\text{ext}} = N_a f\left(\frac{1}{N_a}\sum_{(x,y)\in a} I(x,y)\right) + N_b f\left(\frac{1}{N_b}\sum_{(x,y)\in b} I(x,y)\right)$$

with

$$f(z) = z \cdot \ln(z)$$

and a, b being the two regions into which the snake segments the image.

Depending on the desired shape properties of the snake, different internal energy functions have to be chosen. The original formulation has two factors influencing the weight, length, and straightness of the contour. For certain object shapes, the internal energy function has to be replaced by a different function. An example scene for active contour tracking can be seen in Figure 47.6.

47.3.6 Depth Estimation

In order to deliver the z-component of object coordinates, different approaches can be taken. The SEM as such does not deliver information about this component.

FIGURE 47.6 Chessboard pattern tracked with region-based active contours.

The two approaches that are suitable and used for this are as follows:

- Displacement-based methods
- Focus-based methods

Displacement-based methods either rely on the displacement of different visible features relative to each other due to perspective projection in the SEM or on the displacement of the same features in corresponding images with a different viewing angle.

In the first case, an algorithm fitting a rigid body model to the image can extract depth information by analyzing these projection displacements (Kratochvil et al. 2007).

In the second case, an electron beam deflecting lens (see Jähnisch 2008b and Jähnisch 2007) is used to obtain two images from slightly different angles. The corresponding features have to be determined and from the displacement, the z-position of the feature can be calculated. Difficulties come up when the

scanning speed is increased because correspondency analysis is getting more difficult in noisy images.

Focus-based methods can also be used to obtain depth estimates. Though the SEM has a high depth of view, defocusing is still evident in the images. If the current working distance is known and can be set, DfF methods are possible (see Eichhorn et al. 2008).

The basic principle of the DfF method is to obtain a number of images with known working distances, in most cases with equidistant steps (see Figure 47.7). The region or object of interest has to be located and the sharpness has to be analyzed over the whole image sequence. The result is a curve similar to the one depicted in Figure 47.8. By determining the image with the maximal object sharpness, the distance of the object to the electron gun is determined. It is the corresponding working distance for the image. Due to the fact that the object is often moving in the image during focus sweeps, the approach can be improved by incorporating tracking information to increase robustness (see Dahmen 2008).

47.3.7 Conclusion

The algorithms presented have already been used in the automation of nanohandling processes. The SEM is the image sensor of choice to deliver information about objects at the nanoscale, and most of the inherent difficulties can be solved by using the algorithms presented.

47.4 Control Issues in Automated Nanohandling

The automation itself is a major challenge of an AMNS and is only possible if several tools and a reliable interface to the employed nanorobotic systems are provided. The control of the individual robots is a necessary link between the physical system and automation sequences. This section describes the special control issues that arise for microrobotic and nanorobotic systems and differ significantly from macroscale robotics.

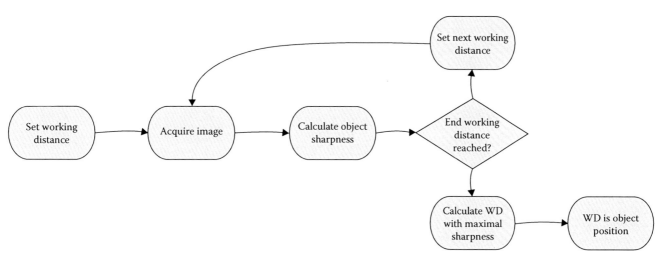

FIGURE 47.7 Flow diagram of the DfF algorithm.

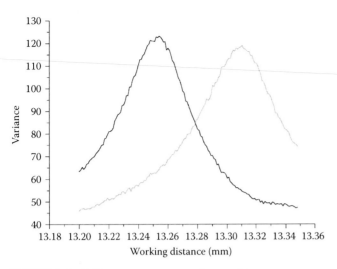

FIGURE 47.8 DfG measurement curves for two objects.

TABLE 47.3   Examples of Different Actuator Topologies

	Single DoF	Multiple DoF
Scanning	Electrostatic comb drive	Steward platform
Step-wise	Linear actuator	Mobile robot

There are several differences on all layers ranging from low-level electrical actuation to high-level closed-loop control.

This section is divided into four parts. First, the special properties and problems of nanorobotic systems are outlined in Section 47.4.1. Then, the control issues in low-level electrical actuation as well as the open-loop control are described in Section 47.4.2. Based on these considerations, Section 47.4.3 outlines the problems when implementing closed-loop microrobot control. Section 47.4.4 then describes issues when implementing a world model for nanoscale handling, which is necessary for effective automation.

47.4.1 Nanorobotic Systems

Nanorobotic systems usually consist of centimeter-sized robots that can move with nanometer-resolution. Thereby, they are able to perform nanohandling operations, i.e., manipulate or assemble parts with at least one dimension on the nanometer scale (see Section 47.2). However, the robot needs to move several millimeters or even centimeters in order to perform an approach to the nano-object and then needs to operate with nanometer accuracy. Thus, about six orders of magnitude need to be covered by a single system, which is a big difference compared with macroscale robotics.

There are basically two different actuator types: scanning and step-wise actuators. The most common scanning actuators are based on piezoceramic elements or on electrostatic actuation. They do not perform steps, i.e., a specific control value (voltage) applied to the actuator leads to the same actuator position when hysteresis effects are neglected. While these actuators can move very precisely and are comparatively easy to model, they have a very limited range, usually below 100 μm. This is solved by the step-wise actuators, the most common of which are piezoceramic actuators using the stick–slip driving principle (Breguet et al. 2000). These actuators can cover large distances by performing many small and precise steps (Table 47.3).

Both actuator types can either be designed to have a single isolated degree of freedom (DoF), i.e., linear or rotary actuators, or to integrate multiple degrees of freedom in a single actuator, e.g., steward platform or mobile robot (Jasper and Edeler 2008). Section 47.4.3 gives examples for each of these actuator scenarios. Each topology involves different problems for control and automation.

On the micro and nanoscale, several error sources significantly affect the control and automation approaches. First, there is always a non-negligible amount of drift in a system (Mokaberi and Requicha 2006). This can be thermal drift due to the expansion of the employed materials, sensor drift, e.g., because of the charging effects in the SEM, or creep effects in the actuators or mountings. Secondly, due to the many orders of magnitude that are involved, it is impossible to create perfectly independent degrees of freedom. If for example, two linear actuators for *x* and *y* are combined orthogonally, there will always be a slight misalignment leading to relevant displacements, i.e., displacements on the nanometer scale.

In order to integrate nanorobotic systems into automation architectures, they need to be controlled. The control of a nanorobot can be subdivided into three tiers: physical tier, open-loop control, and closed-loop control. The physical tier consists of the supply voltage, signal generation, and amplification hardware necessary to drive the actuators electrically. The open-loop control then tries to select signal parameters in order to generate a desired movement without sensor feedback. In order to do so, an inverted model of the actuator is used, which either can be derived from design considerations or obtained through characterization measurements (Hülsen 2007). Based on the open-loop control, the closed-loop control incorporates sensor information in order to create a reliable and repeatable movement. Each of the tiers differs from macroscale robotics and, therefore, different issues need to be solved.

47.4.2 Electrical Actuation and Open-Loop Control

The physical tier creates electric signals for the different actuators. This layer differs greatly from macroscale robotics and needs special consideration, because the driving principles and employed actuators are entirely different. For both piezo-based and electrostatic actuators, high voltages in the range of 50–200 V are required (Fatikow 2000). These voltages need to be controlled very precisely. Therefore, digital to analog conversion circuitry is used to generate a specific voltage signal, which is then amplified by linear high-voltage amplifiers.

Especially for stick–slip actuators, signal generation is a difficult task because the required high slew rate in combination with the capacitive behavior of piezo actuators make a stable amplifier circuit difficult to achieve. Furthermore, to minimize vibrations, the signals need to be generated with very precise timing and the transition between different actuation parameters needs to be smooth. With respect to the physical tier, the most complex actuators are stick–slip actuators with multiple degrees of freedom such as mobile platforms. Here, multiple channels need to be supplied with high-voltage sawtooth-shaped signals that are synchronous, i.e., have an identical frequency and phase (Jasper and Edeler 2008).

The task of an open-loop controller is the selection of specific actuation parameters in order to create a desired movement. This is usually achieved by deriving a model of the actuator's behavior and inverting this model (Hülsen 2007). However, actuators for micro and nanoscale handling are hard to model. The main problem sources are hysteresis and drift as well as environmental influences. Additionally, the exact dynamic motion behavior, e.g., of a stick–slip drive, is not fully understood with all the influences of vibrations and exact friction modeling (Breguet et al. 2000). Thus, an actuator cannot be modeled from design parameters, and characterization sequences are required to approximate the model of actuators (Jasper and Edeler 2008).

Figure 47.9 shows the results of such a characterization sequence. A stick–slip actuator was actuated with sawtooth-shaped signals of different amplitudes leading to different step sizes. There is no movement for amplitudes below $50\,V_{pp}$ and for higher voltages, the step size increases linearly.

With the obtained results, a model of the actuator's behavior, in this case the step size s dependent on the amplitude a, can be approximated using the following equation:

$$s = \begin{cases} (a - 50\,V_{pp}) \cdot 0.5\,nm/V_{pp} & a > 50\,V_{pp} \\ 0 & a \le 50\,V_{pp}. \end{cases}$$

An inversion of this model leads to

$$a = s \cdot 2(V_{pp}/nm) + 50\,V_{pp},$$

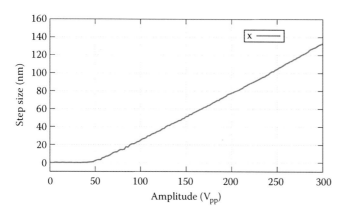

FIGURE 47.9 Applying different stick–slip signal amplitudes to an actuator.

which can be used to derive the necessary actuation parameter to create a certain movement. If the behavior of the actuator is less predictable or strongly time-variant, i.e., the approximated model changes during the actual operation, adaptive methods need to be used (Hülsen 2007).

47.4.3 Closed-Loop Control

The closed-loop control has the task of moving the robot to a precise position based on the open-loop control and with respect to sensor values. There are several differences between nanoscale and macroscale closed-loop control. Macroscale control mostly solves the issues of the system's dynamic behavior, e.g., inertia. On the micro and nanoscale, these effects become negligible. If e.g., a linear stick–slip actuator moves a total mass m of 100 g and can create a force F of approximately 1 N, its maximum acceleration a is

$$a = F/m = 10\left[m/s^2\right].$$

However, during the handling operation, it only needs to move within a range of $50\,\mu m$, and thus the time required to perform any movement is

$$t = \sqrt{2s/a} \approx 0.003\,[s].$$

Also, for this actuator, the maximum velocity v of the actuator is $10\,mm/s$, which would be reached in

$$t = \frac{v}{a} = 0.0001\,[s].$$

The values are not identical for all actuators, but are usually within one order of magnitude to these values. For virtually all actuators, the dynamic effects can thus be neglected for closed-loop control. However, different challenges arise due to the nature of the sensors as well as the required precision.

Due to the drift effects described in Section 47.4.1, a sensor local to a certain actuator cannot generate the required precision. Thus, a global sensor that can measure the relative position of, e.g., tool and specimen is required. For nanohandling, high resolution microscopes such as SEM and AFM in combination with image processing algorithms are the only suitable global sensor. Using the SEM with object recognition and tracking algorithms, however, leads to different control issues. First, there is a significant amount of time required for measuring the robots position. This time is partly required for the line-wise scanning of the microscope and partly required by the image processing algorithm (Sievers 2006). Thus, there is a delay, i.e., dead-time, between the robot's movement and the change in sensory data. This dead-time makes an efficient and stable control loop difficult. Second, due to the sequential, line-wise scanning, the amount of dead-time depends on the position of the robot in the microscope's image. If it is located close to the top of the image, it will be scanned earlier and the dead-time will be longer because the position will be calculated after the whole image is scanned and processed. Third, also due to the line-wise scanning, the images get distorted if the robot moves during the image acquisition.

To solve the problems mentioned above, a special closed-loop control scheme can be derived taking the negligible inertia into account (Jasper and Edeler 2008). The approach relies on either an effective open-loop control or local sensors in each actuator leading to a known upper bound for a relative position error. This means that if the robot is supposed to cover a certain distance d, there is a known maximal proportional position error, i.e., ±10%. This assumption can be made for most of the currently employed actuators. The closed-loop control in such a system is mostly limited by the slow sensor updates of the global sensors, e.g., the SEM. Therefore, it should be synchronized with the sensor and do as much movement as possible with each sensor update before allowing the system some time to settle in order to obtain another reliable sensor reading. Thus, with each sensor update, the robot is moved as far as possible without risking an overshoot with the known upper error limit. If for example, the limit is 10%, then the robot will cover 90% of the remaining distance with each sensor update. Using this approach, the remaining distance decreases exponentially and the robot reaches its target position with only a few sensor updates.

47.4.4 Partial, Time-Variant World Model

Besides the challenges in control of nanorobotic systems, there are also many challenges for high-level control. The basic task of a high-level control system is the coordination of sensors and actuators (i.e., low-level control programs). This includes the quality control of the SEM images with respect to brightness, contrast, and to keep the objects of interest in the focus plane of the SEM. At this control tier, the control of the actuators consists of high-level control commands, which, e.g., include a set of goal positions on a trajectory. On top of these basic tasks, the automation control subsystem is part of the high-level control system. This subsystem is responsible for the flow control and fault handling in the nanoassembly task. An example for this kind of task is the assembly of scanning tunneling microscope (STM) super tips (Wich et al. 2008). For this task, a CNT is attached to the tip of an STM for better depth resolution. The third field of responsibility is path planning and collision avoidance. In most cases, the output of the automation subsystem are goal positions of the end-effector (tool center point). Given these goal positions, a trajectory needs to be planned that avoids collisions of the robot in its work space. Possible obstacles are e.g., other robot systems or the surface of a wafer, which carries assembly parts such as CNTs.

Many of the problems such as trajectory planning and collision avoidance are well studied problems in the macrorobotic world. However, most of the techniques cannot be directly applied to the nanorobotic world. Unlike in macrorobotic systems, not every robot joint has its own built-in position and force sensor. The small joint sizes often prevent the integration of sensors into a device. Robots such as mobile platforms, which

can move freely on a plane without any guidance, do not have built-in sensors at all. Therefore, visual sensors such as cameras and the SEM are used as external sensors for position tracking (see Section 47.3). For movements with high magnifications, the SEM is often the best choice because of its high resolution. However, this choice comes with some drawbacks. At high magnifications, the field of view is rather small—about $30 \times 20\,\mu m$ at a magnification factor of 4000. Only the position of the end-effector tip and the nanoparts can be monitored during their relative movement. Therefore, only end-effector collisions can be detected, while collisions of the robot and an obstacle outside the field of view stay hidden. Another drawback is the missing third dimension. The 2D SEM images make it hard to extract 3D data that, however, is required for the alignment of parts (e.g., for nanoassembly). Therefore, it is quite challenging to bring a CNT into contact with an STM tip without damaging either of them. There are several techniques for extracting 3D data from the SEM. The z-position can be calculated out of two images for two different perspectives. They can be generated by electron beam deflection or sample stage tilting (Jähnisch 2008a,b). A different technique includes the distance measurement by comparing the focal planes of two objects, which is called depth from focus (see Section 47.3). However, the resolution and accuracy of these methods are quite low (e.g., ±5 μm for depth from focus) compared with those of the actuators. Most methods only measure the relative distance between two objects and not the absolute position.

The lack of sufficient sensor information and the very restricted field of view of the SEM lead to the fact that a global world coordinate system is hard to be defined. In fact, most of the nanorobotic systems are only capable of relative path planning inside the view of the SEM. A problem that arises due to the lack of global coordinates is repeatedly finding work pieces on a wafer substrate. This problem is tackled by taught-in positions of the robots, which correspond to working positions and work piece positions. However, a loss of calibration or drift effects as described in Section 47.4.1 might lead to a time consuming manual search process.

A possible solution to the problem of a missing world coordinate system can be provided by a world model built into the high-level control system. This world model needs to be capable to bridge all orders of magnitude from a centimeter scale down to a nanometer scale, with a sufficient level of detail. All centimeter scale objects inside the work space such as nanorobots, cameras, and sample stages need to be part of this model. This can be achieved by including the technical drawings of the computer aided design of an AMNS into the model. Another group of objects that need to be addressed are the end-effectors and working parts such as grippers and CNTs. These objects are at the micro and nanometer scale. The challenge of including these objects is two-fold. First, the objects of interest are at the same size as dust particles, which might also be attached to the surface of the sample. Due to the tremendous number of particles, it is impossible to include all of them into the world model. Therefore, the world model is a partial model of the work space. It needs to include all work pieces required as well as all obstacles that might collide with the end-effectors during automation. Second, drift effects such as thermal drift or electrical charging (see Section 47.4.1) lead to uncertainty in the exact position of the parts, if they are not in the current field of view. Therefore, the world model is a time variant, which introduces the need for continuous updates as well as special considerations in path planning.

The world model needs to be initialized before the automation starts. This includes two steps: calibration and indexing. During calibration, the position of the camera and SEM image planes as well as the distortion effects of the optical lens system needs to be determined. The determination of a suitable transformation matrix can be performed with the aid of calibration samples. These samples consist of continuous patterns, e.g., a chess board. For SEM calibration, the calibration pattern needs to be analyzed at different magnification levels and working distances. Another important step is the calibration of the robots, end-effectors, and the sample stage, which might largely vary from the CAD model data. This is due to the fact that while at the macroscale manufacturing tolerances of several micrometers are acceptable, at the nanometer scale, this might lead to an error of several thousand percent. The kinematic equations for the robots also need to be adapted. Due to the long lever and tolerances of the axes (see Section 47.4.1), the movement directions are always coupled. An x axis movement always leads to a small y and z movement. This is true even for Cartesian robot systems. This can be compensated for by moving the end-effector to fixed target positions at low magnifications, measuring the displacement, and taking it into account in the kinematic equations. Note that this requires a well calibrated sensor system.

After calibrating the world model, an indexing step is required. The goal of indexing is to determine the position of all objects of interest at all magnification levels required for the automation. These objects include the end-effectors relative position to the robot, obstacles that might lead to collisions, as well as the localization of work pieces such as CNTs. The localization mainly utilizes pattern recognition techniques (see Section 47.3). This step might include a full inspection of the specimen (e.g., wafer) at the target magnification. In many cases, background knowledge about the specimen (e.g., the periodic alignment pattern of grown CNTs) can be utilized to speed up the localization process. The object positions are stored inside the world model for later use in automation sequences.

Once the world model is initialized, it can be used for path and automation planning. The global coordinate system of the model enables the application of macrorobotic algorithms. It enables collision-free movements of the actuators without visual feedback.

This can speed up automation cycles tremendously since fewer visual inspection steps are required. However, especially the position of objects at the nanoscale might change due to drift effects. The sensor accuracy can be modeled as an expansion of the surface of the objects in all three dimensions. This expansion is representing the volume, which may not be intersected by trajectories without additional sensor feedback. In order to compensate for the drift effects, the path planner needs to utilize the age of a position and drift models of the materials. The drift can also be represented as surface expansion. However, this expansion is time variant. It may be necessary to start a local search for an object, if it cannot be found in the field of view. This may include switching to a lower magnification as well as a systematic search in the surrounding area of the expected position. Once the object is found again, the world model needs to be updated with the new position of the object.

The world model's graphical representation can be used for another application. It can assist a human operator to assess the z distance of two objects and see the robots' positions even though only the end-effector is visible in the SEM image. For well-known structures, which are quite common in MEMS, the top view of a scene in the SEM image can be replenished by the z information contained in the model. A human operator requires several trial-and-error cycles to compensate for the missing z-position during the approach. Therefore, a free movable and rotatable view on the handling scene can save lots of time of a human operator.

47.4.5 Conclusion

On the control level of an AMNS, several issues arise that are not comparable to macroscale robotics. However, this chapter describes several solutions on all levels of microrobot control. With an effective control system and a suitable world model, the automation of micro and nanoscale handling can be achieved.

47.5 Application I: Characterization of Carbon Nanotubes

CNTs are one of the most promising materials that have been discovered in recent years (Iijima 1991). CNTs show unique physical properties and therefore have a high application potential in various products (Baughman et al. 2002, Robertson 2004). Especially their mechanical and electrical properties exceed conventional materials such as steel or copper. CNTs can be used to develop novel CNT-based composite materials because their tensile strength and Young's modulus is a multiple compared with steel (Thostenson et al. 2001, Harris 2004). The electrical current density is about 1000 times higher compared with copper so that CNTs are the perfect material for novel interconnects in microchip fabrication (Kreupl et al. 2002, Patel-Predd 2008). Several production techniques have been investigated in the past. The three most widely spread techniques are production by arc-discharge (Ebbesen and Ajayan 1992), by laser ablation (Guo et al. 1995), and by chemical vapor deposition (CVD) (Teo et al. 2001). The big advantage of CVD is that CNTs can be produced and grown at defined positions and with controllable geometries. Finding the correct growth parameters that lead to desired CNTs with well known mechanical and electrical properties is one of the central issues. Vice versa, novel methods and strategies have to be developed allowing for reliable characterization of CNTs grown by CVD. Nanorobotic systems provide the perfect basis for this task since they realize positioning in the nanometer range and can be integrated into the vacuum chamber of SEMs for real-time visual feedback. Furthermore, the nanorobotic approach enables the assembly of prototypic CNT-based devices.

In this section, we present an example for the implementation of an AMNS that can be used for the reliable mechanical characterization of CNTs. Section 47.5.1 gives a short state of the art concerning the mechanical characterization of CNTs. The physical basics, needed for the characterization, are presented in Section 47.5.2. The setup of the AMNS is described in Section 47.5.3. The experimental results of mechanical CNT characterization are discussed in Section 47.5.4 and a conclusion is given in Section 47.5.5.

47.5.1 State of the Art: Mechanical Characterization of CNTs

A CNT can be seen as a 2D sheet of carbon atoms that is wrapped up to a cylinder. The strong chemical sp^2-bonding of graphite is the reason for the extraordinary properties of CNTs. In principle, CNTs exist in two configurations: single-wall carbon nanotubes (SWCNTs) and MWCNTs. Over the past few years, several experimental measurements have been carried out worldwide to mechanically and electrically characterize CNTs. One of the interesting mechanical parameters that can be calculated is the Young's modulus E. There is a wide range of reported values that will be summarized in this section.

For SWCNTs, the Young's modulus given in different publications varies between 0.9 and 1.7 TPa. Most of the authors report values of about 1 TPa (Pantano et al. 2004). For MWCNTs, the Young's modulus seems to be even higher because of the additional van-der-Walls forces between individual shells. The range of reported values for the Young's modulus of MWCNTs is between hundreds of GPa (Krüger 2007) and 4.15 TPa (Treacy et al. 1996). Most of the publications give a Young's modulus of MWCNTs between hundreds of GPa (Wang et al. 2001, Krüger 2007) and 1.3 TPa (Han 2005, Nakajima et al. 2006).

47.5.2 Physical Basics for Mechanical CNT Characterization

In this section, the fundamentals needed for the characterization of CNTs are introduced. The mechanical characterization and the calculation of Young's modulus are realized by performing bending experiments of individual MWCNTs. A previous TEM analysis of the present MWCNTs has shown that the multi-wall structure is covered by a conical layer of amorphous carbon (Sardan et al. 2008). For this reason, a theoretical model has to be developed in order to describe the conical geometry of the CNTs correctly.

The elastic curve for a conical structure is given by the differential equation

$$E \frac{\partial^2}{\partial x^2} \left(I(x) \frac{\partial y^2(x)}{\partial x^2} \right) = w(x)$$

where
  $I(x)$ is the geometrical moment of inertia and dependent on the place x of the CNT
  $w(x)$ is the distributed load
  $y(x)$ is the deflection of the CNT

The geometrical moment of inertia for the conical geometry of the MWCNTs is given by

$$I(x) = \frac{\pi}{4} \left(\left(\frac{(R_0 - R_1)}{L} \right) x + R_1 \right)^4$$

with a base radius of R_0, a radius at the point of deflection R_1, and the length of the CNT L (compare with Figure 47.10).

The solution for the above differential equation is

$$y(x) = -\frac{2FL^2(L-x)^2(2LR_1 + 3R_0 x - 2R_1 x)}{3E\pi R_0^3 \left(LR_1 + R_0 x - R_1 x \right)^2}$$

with the boundary conditions $y(L) = y'(L) = 0$ for the fixed support and $y''(0) = 0$ as well as $y'''(0) = F/EI$ for the flexible tip of the CNT where the force F is acting. The maximum deflection will appear at the tip of the CNT where $x = 0$. Therefore, the solution given above can be rewritten as

$$y(0) = \frac{4FL^3}{3E\pi R_0^3 R_1}.$$

For a cylindrical CNT, the solution would be

$$y(0) = \frac{4FL^3}{3E\pi r}$$

with radius r. The elastic curves for a conical and a cylindrical CNT with the same length and base radius are plotted in Figure 47.11. The radius of the conical CNT reduces to the tip by a quarter of the base radius. These theoretical elastic curves will be compared with the experimental results in Section 47.5.4.

The Young's modulus E will then be calculated based on the most accurate theoretical model. Vice versa, the influence of the conical layer can be determined. If the experimental elastic curve of the CNTs is similar to the cylindrical theoretical model, the influence of the conical layer is negligible and the mechanical stiffness of the MWCNT structure is dominant. In case the

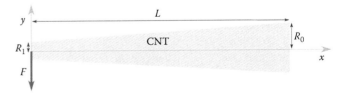

FIGURE 47.10 Geometric model of the conical CNT with base radius R_0, the radius at the point of deflection R_1, the length of the CNT L and the acting force F.

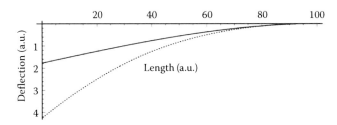

FIGURE 47.11 Graphical representation of theoretical elastic curves for a conical (dashed line) and cylindrical (line) CNT with same base radius, length and acting force. The radius of the conical CNT is reducing to the tip by a quarter of the base radius.

conical model is more suitable for the experimental data, the conical layer of amorphous carbon has a dominating influence.

47.5.3 Setup of the Automated Microrobot-Based Nanohandling Station

In this section, an AMNS is described that is capable of performing the required nanomanipulation tasks for mechanical characterization of CNTs and that can easily be integrated into the vacuum chamber of an SEM for getting real-time visual feedback.

A technical drawing of the developed AMNS is shown in Figure 47.12. The whole setup is mounted to the door (1) of the vacuum chamber of a LEO 1450 SEM. All parts of the setup are aligned to the SEM's region of interest. The secondary electron detector (8) of the SEM was taken into particular consideration to avoid shadowing effects. Special vacuum cable feedthroughs (9) connect the control electronics to the nanorobotic devices inside the SEM's chamber.

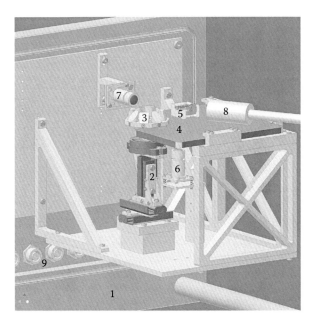

FIGURE 47.12 CAD-model of the AMNS: SEM chamber door (1), stage (2), sample holder (3), glass plate for mobile microrobots (4), mobile microrobot (5), camera to track the mobile microrobots (6), camera to supervise the coarse approach (7), SEM secondary electron detector (8), and vacuum cable feedthroughs (9).

The CNT samples are located on the nanorobotic stage (2) that provides four degrees of freedom and is used for the positioning of up to eight sample holders (3), bringing the desired sample into the SEM's focus and into the working range of the mobile platform. The nanorobotic stage consists of three linear positioners (x, y, and z) and a rotatory axis (φ). All axes are manufactured by SmarAct GmbH, Oldenburg, Germany and provide internal optical sensors allowing for closed-loop actuation with 50 nm resolution.

In addition, a mobile microrobot (5) is actuated on a glass surface (4) realizing a flexible task organization. The concept of using mobile microrobots provides multiple degrees of freedom and long travel ranges with minimum space requirements. Different mobile platforms (Driesen et al. 2005a, Edeler et al. 2008) can be used for carrying manipulation tools such as AFM probes or electrothermal microgrippers (Andersen et al. 2009). The position of the mobile microrobot can be tracked from below by using a CCD camera (6) that detects the light of LEDs mounted to the bottom of the mobile microrobot. Thus, the mobile microrobot and the tool, respectively, can be moved into the ROI of the SEM or can be parked at a secure position. Additionally, a CCD camera (7) provides a side view of the scenario in order to monitor the distance between the end-effector and the sample during coarse positioning.

47.5.4 Experimental Results: Mechanical Characterization of CNTs

The mechanical characterization is realized by performing bending experiments of individual MWCNTs (compare with Figure 47.13). Self-sensing piezoresistive AFM probes are used to deflect the CNTs and to measure the acting force on the CNT during deflection.

To evaluate the experimental elastic curve of a MWCNT against the theoretical models given in Section 47.5.2, the elastic curve of a MWCNT is extracted out of the SEM image and plotted in Figure 47.14. It can be seen that the experimental elastic curves correspond exactly to the conical theoretical model meaning that the conical layer of amorphous carbon has to be considered. For this reason, Young's modulus of the given MWCNTs is calculated based on the conical model. Therefore, Young's modulus is given by

$$E = -\frac{4FL^3}{3y(x)\pi R_0^3 R_1}.$$

In order to determine Young's modulus, the absolute force is needed and thus the piezoresistive AFM probe has to be calibrated.

The complete calibration sequence is described in the article by Eichhorn et al. (2007). The simultaneous measurement of force and deflection during CNT bending allows for acquiring characteristic force–displacement curves, thus calculating the CNT's

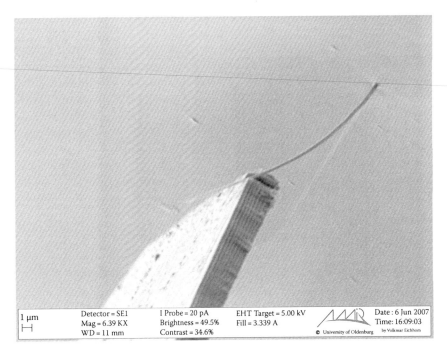

FIGURE 47.13 SEM image showing the typical bending of an individual MWCNT by using a piezoresistive AFM probe.

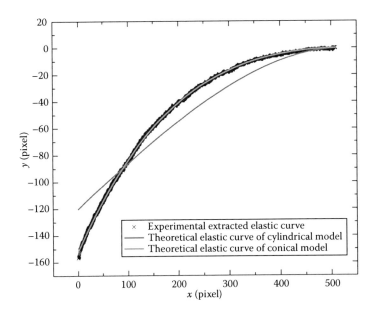

FIGURE 47.14 Comparison of extracted experimental elastic curve and theoretical elastic curves for conical and cylindrical CNT geometry.

spring constant. By determination of the CNT's geometrical properties using SEM images, it is possible to calculate the corresponding Young's modulus. For systematic mechanical CNT characterization, several measurements have been performed at different positions of one MWCNT. Figure 47.15 shows the resulting Young's modulus depending on the distance between the CNT base and the point of interaction. The measured values vary slightly but lead to an average Young's modulus of 132 ± 15 GPa, which is in good agreement with values that are reported in different publications.

47.5.5 Conclusion

The presented AMNS and cantilever-based characterization technique allows for the nondestructive mechanical characterization of individual MWCNTs by using piezoresistive AFM probes. A theoretical model for conical CNTs has been developed and proven with experimental elastic curves that have been extracted from SEM images of deflected MWCNTs. Furthermore, the Young's modulus has been determined. The AMNS can also be used to realize the pick-and-place manipulation of individual

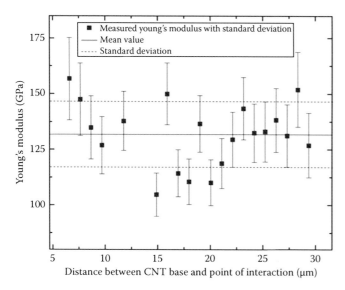

FIGURE 47.15 Graphical representation of calculated Young's modulus for different points of CNT interaction.

CNTs for the assembly of prototypic CNT-based components (Sardan et al. 2008). The control architecture and SEM image processing software presented in Sections 47.3 and 47.4 will allow for the automation of nanorobotic CNT characterization and handling.

47.6 Application II: Handling of Biological Objects

Current research areas in molecular and cell biology, medicine, and process sensor technology (sensor systems to monitor various processes in industry and science) often require advances in the nano-engineering technologies to look for molecular phenomena with the highest possible resolution in order to enlighten the "black-box", which still shades most of metabolic reactions. The aim of this work, starting from current medical problems, is to propose novel studies especially in molecular microbiology, cellular microbiology, and virology with direct applications in medicine and in the industry, closing the bottom-up cycle. Thanks to the advances in micro and nanofabrication and micro and nanorobotics over the last few decades, robotic systems that offer the possibility to characterize and manipulate biological objects can be developed today. Using such systems, new studies of single cell phenomena with nanometer resolution (e.g., studying the local mechanical and electrical properties of single bacteria for characterization and evaluation of the resistance to antibiotics or to the spread of infections) up to complete cell compounds (e.g., studying the mechanical properties of bacterial biofilms) are possible and deepen the appreciation of the processes at the nanoscale. According to this, the following section will give a short overview of the methods of handling of biological objects, which can be used for the design of micro and nanorobotic systems. So, the use of AFM-based approaches and cell injection approaches will be described as well as the use of SEM and ESEM for biohandling. Information about contactless

methods for the handling of biological objects that can be integrated into micro and nanorobotic systems (e.g., dielectrophoresis, optical tweezers) will be given. A short overview of the work at AMiR in the area of handling biological objects is given at the end of this section.

47.6.1 Handling and Characterization of Biological Objects Using an Atomic Force Microscope

Since the development of the AFM by Binnig, Quate, and Gerber (Binnig et al. 1986), the potential of the AFM for the characterization and manipulation of biological objects and materials has quickly been perceived. The main advantages of the AFM are its high resolution (down to the sub-nanometer range) and the possibility to work under physiological conditions as well as under vacuum conditions or in a liquid environment, depending on the examined object. Furthermore, in addition to the visualization possibilities, an AFM can measure different properties (mechanical and electrical) and can manipulate biological objects. For example, one of the greatest advantages in the field of biomaterial characterization and manipulation is the possibility to examine the samples in a liquid environment under nearly lifelike conditions (Lal and John 1994, Castillo et al. 2009). The AFM also enables the monitoring of the forces exerted during manipulation to provide important information about the interactions between cantilever tip, biological object, and surface.

Currently, the AFM fulfills two tasks—the imaging and the manipulation of the samples. Therefore, the first step is always an imaging step, followed by a manipulation step with the same cantilever. These two steps have to be repeated until the manipulation is done. Unfortunately, with this method, it is not possible to get a real-time view of the manipulation process because the AFM-based imaging process currently needs several minutes depending on the quality of the picture. Such control schemes can be described as "look and move" schemes and require the use of an SEM for the observation of the manipulated object and the AFM probe under real-time (Sitti 2007). For a manipulation, the samples have to be immobilized on the substrates via a reversible method such as electrostatic interactions or low covalent bindings. However, compared with other methods and instruments, the use of AFM-based manipulation and characterization opens the widest field of applications in this area of research. Thus, a high number of different applications concerning the imaging, characterization, and manipulation of biological objects can be found. In the case of the manipulation of biological samples with the AFM, a pushing, cutting, pulling, touching, indenting, and structuring of a broad range of materials (biological and nonbiological) is possible (Figure 47.16).

Among the applications are the manipulation as well as the structural and mechanical characterization of various viruses and virus capsids (Kuznetsov et al. 2008); the force mapping of the elastic properties of synaptic vesicles (Laney et al. 1997); the mechanical, structural, and electrical characterization

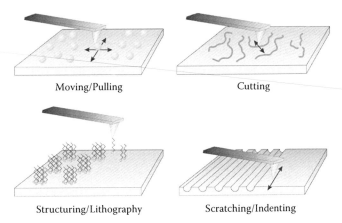

Moving/Pulling   Cutting

Structuring/Lithography   Scratching/Indenting

FIGURE 47.16 Schematically view of manipulation tasks done by AFM-based robotic systems as described by Sitti (2007).

of DNA molecules (Pablo et al. 2000, Seidel et al. 2004); the manipulation of single-stranded DNA to form simple nano-electric circuits (Kufer et al. 2008); the characterization of the mechanical properties of lactate oxidase (Parra et al. 2007) and other proteins; the dissection of human chromosomes (Stark et al. 2003, Rubio-Sierra et al. 2005); the mechanical properties (pretransition and progressive softening depending on mechanical stress) of surface immobilized antibodies (Afrin et al. 2005); the characterization of the mechanical properties of microtubules (Schaap et al. 2006); as well as the physicochemical and mechanical properties of bacteria such as *Pseudomonas putida* by using single force spectroscopy (Abu-Lail and Camesano 2003).

Independent of these wide application ranges and the suitability for the imaging, characterization, and manipulation of biological objects, some disadvantages of the AFM-based handling of biological object exist. First of all, the separation of the imaging and manipulation, followed by the restriction to small working areas of about $100 \times 100\,\mu m^2$ (e.g., JPK Nanowizard, Nanotec AFM) and the limited working speed efficiency (Rubio-Sierra et al. 2005) leads to a difficult handling of this method. However, the principle of the method, imaging, characterization, and manipulation by using a small cantilever is important for usage in micro and nanorobotic systems. By using piezoresistive cantilevers, swarms of mobile platforms are also possible (Brufau et al. 2005), which will lead to a highly automated and paralleled work. With such systems, the limitations concerning the working time, efficiency, and the small working areas can be circumvented. New methods to build high speed AFMs are also being developed and published (Ando et al. 2008).

The AFM can also be used for combined robotic systems to suppress the disadvantages by using additional video microscopy systems and by adding noncontact ablation by ultraviolet (UV) microbeam laser for large scale manipulation of biological objects such as chromosomes (Rubio-Sierra et al. 2005). However, AFM-based robotic systems are also used for the imaging, characterization, and handling of all possible materials besides biological objects.

47.6.2 Handling of Biological Objects Using Microgrippers and Nanotweezers

Another method for the handling of biological objects are microgrippers and nanotweezers, which use small and variable forces to pick and place cells (Kim et al. 2008) or other biological objects. The main advantages of such grippers are a better control over the applied forces and a well defined mechanical grip effect on the object. Thus, they are more effective in gripping experiments than cantilever-based structures of AFM- and STM-based microrobotic systems. A microgripper also gives the possibility of making direct electrical and mechanical measurements on the grabbed objects if the gripper jaws are conductive or can provide a force feedback signal (Mølhave et al. 2006). Most of these grippers work by using electrostatic, thermal, or piezoelectric effects to control the gripping forces (Kim et al. 2004, Kim et al. 2008). In the case of the thermal gripper, the effect of heat-induced material expansion is used. In the case of the piezoelectric gripper, the effect is used so that piezoelectric ceramics can be expanded and contracted by the use of specific electric fields. Another major advantage of these grippers is the use as a tool for mobile platforms and micro- and nanorobotic systems with a combination of linear and rotary positioners (e.g., linear axes of Smaract GmbH) for fine positioning.

Currently, there are some groups and vendors that sell various types of microgrippers and nanotweezers for micro- and nanorobotic systems. Peter Bøggild's group at the Technical University of Denmark has developed silicon microgrippers, nanotweezers, and nanostructured nonstick coatings to avoid adhesive effects during nanomanipulation tasks (Mølhave et al. 2006, Carlson et al. 2007, Sardan et al. 2007). Also, nanotube nanotweezers for the manipulation of GaAs nanowires and polystyrene nanoclusters exist (Kim and Lieber 1999), which are manufactured by Kim and Lieber. Similar nanotweezers have also been used for the successful manipulation of nanowires (Bøggild et al. 2001) and DNA (Hashiguchi et al. 2003). Other companies that produce types of microgrippers and nanotweezers are SmarAct GmbH, Nascatec, and Zyvex Instruments. Some of these grippers can operate at UHV conditions as well as in a standard atmosphere. There are differences in the opening widths and the accuracy of the grippers and tweezers, so that it is possible to choose the best gripper with respect to the biological objects that shall be manipulated.

Some of these biological objects, where microgrippers and nanotweezers were used for manipulation tasks, are cancer cells, e.g., the aggressive HeLa cells (Chronis and Lee 2005, Beyeler et al. 2007), porcine aortic valve interstitial cells (Kim et al. 2008), and a wide spectrum of other biological cells (Solano and Wood 2007, Kim et al. 2008).

Unfortunately, the working principles of such microgrippers and nanotweezers are also the main disadvantage for use with biological objects. The thermal gradients or electric fields to open the grippers can destroy some of the biological samples. To overcome these problems, the development of mechanically actuated grippers has shown an alternative approach, but there is still a risk of damaging and contaminating the biological samples (Blideran et al. 2006a,b).

47.6.3 Manipulation of Biological Objects Using Cell Injection Methods

To understand the fundamental elements of biological systems, the need to analyze individual cells is of great importance. Even in the identification of genes, in gene therapy, and in the bacterial synthesis of specific DNA sequences, the full knowledge of the reactions and mechanisms of single cells has to be examined. One approach is the infiltration of known DNA sequences into single cells by using single-cell injection. For this, a gripper, an optical trap, or a micropipette is used to hold the cell while another micropipette performs an injection process. Currently, this cell injection is conducted manually and needs well-trained scientists. However, even with fully trained scientists, the success rate is extraordinary small. One reason for this is the dependence of the injection on the injection speed and trajectory.

To overcome the problems of a manual cell injection and to suppress the contamination and destruction of cells, automated solutions are necessary. Furthermore, automated cell injection can be highly reproducible with precision and control of the pipette motion (Sun and Nelson 2002). By using automated cell injection, the risk of contamination can be decreased and the reproducibility can be increased leading to significantly increased success rates of the cell injection process.

For example, this is shown by the work of Sun and Nelson (2002). The authors describe a completely automated robotic cell injection system to conduct autonomous pronuclei DNA injection of mouse embryos with a success rate of 100%.

47.6.4 Manipulation of Biological Objects Using SEM and ESEM

Besides the possibility of manipulating and imaging nonbiological nanoscopic objects by using a SEM or TEM, special systems have been developed to manipulate and characterize biological objects, e.g., manipulation of collagen fibers (Layton et al. 2005). One can see the necessary ultra-high vacuum and low temperatures are a critical problem for use in case of the biological samples (Fahlbusch et al. 2005). Also, the preparation of the biological samples is more complicated and a little bit harsh (biological objects need to be fixed, dehydrated, critical point dried, and coated). With an ESEM, the imaging and manipulation of biological samples is possible without special treatments (Tai and Tang 2001). Some difficulties when using ESEM in wet samples have been reported (Mestres et al. 2007, Stokes 2003), so that it seems necessary to use an SEM in addition to the ESEM to provide a real image of the sample and resolve finer structures in detail (Muscariello et al. 2005).

47.6.5 Manipulation and Handling of Biological Objects Using Contactless Methods

According to various reasons, efforts in science are directed toward the miniaturization of existing processes. One field is the lab-on-a-chip system, where microfluidic channels only require small volumes of samples and buffers for the chemical and biochemical reactions. Due to the decreasing size of these systems, the problem of handling these samples occurs. To solve this problem as well as avoid mechanical damage when handling biological objects, contactless methods can be used. For the contactless handling of biological objects, the use of optical and magnetic tweezers (by using magnet beads that are connected to biomolecules), electrophoresis (e.g., patch clamp methods), dielectrophoresis (DEP), and acoustic methods (e.g., ultrasonic probes) are possible.

DEP—among the many manipulation techniques, the electric field-based approach is well suited for miniaturization because of the relative ease of microscale generation and structuring of an electric field on microchips (Andersson and van den Berg 2004). DEP can be described as a phenomenon in which a force is exerted on a dielectric particle subjected to an asymmetric electric field. According to the point that all particles exhibit dielectrophoretic activity in the presence of electric fields, the resulting dielectric force does not require the charging of the particle. However, the strength of the force strongly depends on the properties of the surrounding medium, the electrical properties of the surrounding electric field, and on the electrical properties of the particles. This leads to a highly selective manipulation of particles by using an electric field of a particular frequency so that this method can be used for the manipulation, transport, separation, and sorting of different types of biological objects. It appears that in literature about DEP on biological structures, currently the treatment and handling of cells is examined more frequently than other biological objects such as DNA. The described applications start with the cell sorting of living and dead cells (Tai et al. 2007), the sorting of healthy and cancer cells (Yang et al. 1999, An et al. 2008), and the sorting of other different cell types (Wang et al. 2000, Yang et al. 2000). Few applications for the use of DEP with DNA are described (Regtmeier et al. 2007, Tuukkanen et al. 2007).

Compared with other contactless methods for the manipulation of biological objects, the DEP technique also offers the possibility to be easily included into microrobotic and nanorobotic systems to handle biological objects fully automatedly.

Optical trapping by using optical tweezers—since the demonstration of the use of strongly focused laser beams to manipulate various objects at the micro and nanoscale by Arthur Ashkin (Ashkin 1970, 1992), the contactless measurement of forces and the contactless handling of biological objects have been revolutionized. The working principle of optical tweezers is shown in Figure 47.17 and can be described as a sharp focused laser beam that interacts with a micro or nanoparticle. Optical traps use the radiation pressure of the light (photons). The most familiar form is the scattering force (F_{scatt}), which describes the force due to the light that is proportional to the light intensity and acts in the direction of the propagation of light.

$$F_{scatt} = n_m \frac{\langle s \rangle \sigma}{c} \quad \text{with}$$

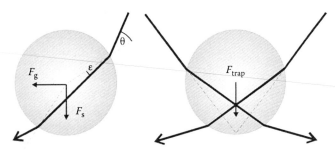

FIGURE 47.17 Schematic view of the interactions between a particle and a laser beam and the resulting trap forces.

$$\sigma = \frac{8}{3}\pi(k_L r)^4 r^2 \left(\frac{m^2 - 1}{m^2 + 2}\right)^2.$$

$\langle S \rangle$ represents the POYNTING-Vector

r is the radius of the particle

n_m is the refractive index of the surrounding media

m is the relative refractive index between the media and the particle

k_L is the wave number of the light

Optical tweezers owe their trapping to the gradient force (F_{grad}), which is instead proportional to the spatial gradient in light intensity and acts in the direction of the gradient.

$$F_{grad} = \frac{\alpha}{2}\nabla\langle E^2 \rangle \quad \text{with}$$

$$\alpha = n_m^2 r^2 \left(\frac{m^2 - 1}{m^2 + 2}\right).$$

In this case, α represents the polarizability of the trapped microparticle.

The gradient force used by optical tweezers arises from fluctuating electric dipoles that are induced when light passes through transparent objects (Svoboda and Block 1994). When a dielectric sphere is placed in a light gradient, the sum of all the rays passing through it generates an imbalance in force. This force pushes the sphere toward the brighter region of the light. So a focus functions as a trap because the strong light gradients in its neighborhood points toward the center. With these conditions, trapping is stable if $F_{grad} > F_{scatt}$. Only with microscope objects with a high numerical aperture are such optical traps possible in practice.

With these optical tweezers, it was possible to get basic information about the interactions of the laser beam and small nanoparticles of silicon and various organic polymers (Wright et al. 1993) as well as about the resulting forces and the modeling of these interactions (Barton and Alexander 1989, Wright et al. 1990). Experiments to determine the speed of bacteria in various flow schemes (Svoboda and Block 1994), to manipulate chromosomes and sperm cells (Wright et al. 1990), to manipulate cell organelles (Ruiz et al. 2003), and to determine the deformation effects on erythrocytes and fibroblasts (Guck et al. 2001) were performed. With the optical tweezers, it was also possible

to characterize DNA in order to provide information about the mechanical, elastic, and entropic properties (Svoboda and Block 1994, Baumann et al. 1997, Yu 2003). Also, the direct manipulation of actin filaments has been demonstrated by using optical tweezers (Arai et al. 1999).

47.6.6 Handling of Biomaterials in AMiR

This section provides a short overview of the activities of the "Division Microbotics and Control Engineering" (AMiR) in the field of handling and characterization of biological objects. Using experiences during the design, construction, and control of microrobotic systems, AMiR developed nano and microrobotic systems for the handling and characterization of biological objects. The aims of AMiR cover the development of novel nanohandling robot stations for the characterization and handling of biological objects by AFM probes (Fatikow et al. 2006a,b), the development of systems for an automated characterization, and the manipulation of biological cells (Figure 47.18; Hagemann et al. 2007, Krohs et al. 2007) up to the development of a nanorobotic station for the manipulation and characterization of nanoscale biological objects (Weigel-Jech et al. 2008) such as DNA and single cell organelles. With these systems, e.g., the characterization of biofilms, the construction of small nanoelectronic circuits of DNA and CNTs and the characterization, direct control, and manipulation of various single cell phenomena shall be enabled.

47.6.7 Conclusion and Outlook

In this chapter, some methods for the robot-based characterization, manipulation, and handling of biological objects were presented. Today, it is possible to use AFM-based methods, microgrippers and nanotweezers, the SEM/ESEM, dielectrophoresis, cell injection methods, as well as contactless methods within automated robotic systems. Principally, there are more methods to characterize, manipulate, and handle biological objects on the nanoscale. However, most of these methods are

FIGURE 47.18 Biohandling station for cell characterization and cell handling as presented by Hagemann et al. (2007).

currently not useable for robot-based applications or miniaturization. With advantages in nanotechnology and microsystem technology, this will change over the next years, so that almost all possible methods of handling nanoscale biological objects can be used for robotic systems.

It has also been shown that the development of such nanorobotic systems to characterize and handle nanoscale biological objects is part of the work of AMiR. Over the next years, it will be possible to build a new modular nanostation for almost all tasks in the research area of bionanotechnology. The first steps to reach this ambitious goal have been done and the next steps will follow.

References

Abu-Lail, N. I. and Camesano, T. A. 2003. Polysaccharide properties probed with atomic force microscopy. *Journal of Microscopy* 212: 217–238.

Afrin, R., Alam, M. T., and Ikai, A. 2005. Pretransition and progressive softening of bovine carbonic anhydrase II as probed by single molecule atomic force microscopy. *Protein Science* 14: 1447–1457.

An, J., Lee, J., Kim, Y., Kim, B., and Lee, S. 2008. Analysis of cell separation efficiency in dielectrophoresis-activated cell sorter. In *Third International IEEE Conference on Nano/Micro Engineered Molecular Systems*, Hainan Island, China, pp. 965–969.

Andersson, H. and van den Berg, A. 2004. Microtechnologies and nanotechnologies for single-cell analysis. *Current Opinion in Biotechnology* 15: 44–49.

Andersen, K. N., Petersen, D. H., Carlson, K., Molhave, K., Sardan, O., Horsewell, A., Eichhorn, V., Fatikow, S., and Bøggild, P. 2009. Multimodal electrothermal silicon microgrippers for nanotube manipulation. *IEEE Transactions on Nanotechnology*, 8(1): 76–85.

Ando, T., Uchihashi, T., Kodera, N. et al. 2008. High-speed AFM and nano-visualization of biomolecular processes. *Pflügers Archiv—European Journal of Physiology* 456: 211–225.

Aoyama, H., Iwata, F., and Sasaki, A. 1995. Desktop flexible manufacturing system by movable miniature robots. In *International Conference on Robotics and Automation*, Nagoya, Japan, pp. 660–665.

Arai, T. and Tanikawa, T. 1997. Micro manipulation using two-finger hand. In *IEEE/RSJ International Conference on Intelligent Robots and Systems, Proceedings of the International Workshop on Working in the Micro- and Nano-Worlds: Systems to Enable the Manipulation and Machining of Micro-Objects*, Grenoble, France, pp. 12–19.

Arai, F., Andou, D., Nonoda, Y., Fukuda, T., Iwata, H., and Itoigawa, K. 1998. Integrated microendeffector for micromanipulation. *Transactions on Mechatronics, IEEE/ASME* 3(1): 17–23.

Arai, Y., Yasuda, R., Akashi, K. et al. 1999. Tying a molecular knot with optical tweezers. *Letters to Nature* 399: 446–448.

Ashkin, A. 1970. Acceleration and trapping of particles by radiation pressure. *Physical Review Letters* 24: 156–159.

Ashkin, A. 1992. Forces of a single-beam gradient laser trap on a dielectric sphere in the ray optics regime. *Biophysical Journal* 61: 569–582.

Barton, J. P. and Alexander, D. R. 1989. Fifth order corrected electromagnetic field components for a fundamental Gaussian beam. *Journal Applied Physics* 66: 2800–2802.

Baughman, R. H., Zakhidov, A. A., and de Heer, W. A. 2002. Carbon nanotubes—the route toward applications. *Science* 297: 787–792.

Baumann, C. G., Smith, S. B., Bloomfield, V. A., and Bustamante, C. 1997. Ionic effects on the elasticity of single DNA molecules. *The Proceedings of the National Academy of Sciences USA* 94: 6185–6190.

Bergander, A., Schmitt, C., Clavel, R., Bleuler, H., Breguet, J.-M., and Crez, R. P. 2000. Piezoactuators for motion control from centimeter to nanometer. In *Proceedings of the 2000 IEEE/RS/ International Conference on Intelligent Robots and Systems*, Takamatsu, Japan, pp. 492–497.

Beyeler, F., Neild, A., Oberti, S. et al. 2007. Monolithically fabricated microgripper with integrated force sensor for manipulating microobjects and biological cells aligned in an ultrasonic field. *Journal of Microelectromechanical Systems* 16: 7–15.

Binnig, G., Quate, C. F., and Gerber, C. 1986. Atomic force microscope. *Physical Review Letters* 56: 930–933.

Blake, A. and Isard, M. 1998. *Active Contours*. London, U.K.: Springer.

Bleuler, H., Clavel, R., Breguet, J.-M., Langen, H., and Pernette, E. 2000. Issues in precision motion control and microhandling. In *International Conference on Robotics & Automation*, San Francisco, CA.

Blideran, M. M., Bertsche, G., Henschel, W., and Kern, D. P. 2006a. A mechanically actuated silicon microgripper for handling micro- and nanoparticles. *Microelectronic Engineering* 83: 1382–1385.

Blideran, M. M., Fleischer, M., Henschel, W. et al. 2006b. Characterization and operation of a mechanically actuated silicon microgripper. *The Journal of Vacuum Science and Technology B* 24: 3239–3243.

Bøggild, P., Hansen, T. M., Tanasa, C., and Grey, F. 2001. Fabrication and actuation of customized nanotweezers with a 25 nm gap. *Nanotechnology* 12: 331–335.

Böhringer, K. F., Fearing, R. S., and Goldberg, K. Y. 1999. Microassembly. In *Handbook of Industrial Robotics*, ed. S. Y. Nof, 2nd edn. New York: John Wiley & Sons, pp. 1045–1066.

Bourjault, A. and Chaillet, N. 2002. *La Microrobotique*. Paris, France: Hermes.

Breguet, J.-M., Perez, R., Bergander, A., Schmitt, C., Clavel, R., and Bleuler, H. 2000. Piezoactuators for motion control from centimeter to nanometer. In *International Conference on Intelligent Robots and Systems*.

Brufau, J., Puig-Vidal, M., López-Sánchez, J. et al. 2005. Micron: Small autonomous robot for cell manipulation applications. In *Proceedings of the International Conference on Robotics and Automation*, Barcelona, Spain, pp. 856–861.

Campbell, S. A., Walsh, F. C., Kerfriden, S., Nahlt, A. H., and Smiths, J. R. 1998. The electrochemical etching of tungsten STM tips. *Electrochimica Acta* 43(12–13): 1939–1944.

Carlson, K., Andersen, K. N., Eichhorn, V. et al. 2007. A carbon nanofiber scanning probe assembled using an electrothermal microgripper. *Nanotechnology* 18: 345501–345508.

Castillo, J., Dimaki, M., and Svendsen, W. E. 2009. Manipulation of biological samples using micro and nano techniques. *Integrative Biology* 1: 30–42.

Chen, H., Xi, N., and Li, G. 2006. CAD-guided automated nano-assembly using atomic force microscopy-based nanorobotics. *IEEE Transactions on Automation Science & Engineering* 3: 208–217.

Christofanelli, M., De Gasperis, G., Zhang, L., Hung, M.-C., Gascoyne, P. R. C., and Hortobagyi, G. N. 2002. Automated electrorotation to reveal dielectric variations related to HER2/neu overexpression in MCF-7 sublines. *Clinical Cancer Research* 8: 615–619.

Chronis, N. and Lee, L. P. 2005. Electrothermally activated SU-8 microgripper for single cell manipulation in solution. *Journal of Microelectromechanical Systems* 14: 857–863.

Clevy, C., Hubert, A., Agnus, J., and Chaillet, N. 2005. A micro-manipulation cell including a toll changer. *Journal of Micromechanics and Microengineering* 15: 292–301.

Codourey, A., Zesch, W., and Büchi, R. 1995. A robot system for automated handling in mirco-world. In *IEEE/RSJ International Conference on Intelligent Robots and Systems*, Pittsburgh, PA, pp. 185–190.

Codourey, A., Rodriguez, M., and Pappas, I. 1997. A task-oriented teleoperation system for assembly in the microworld. In *Proceedings of the 8th International Conference on Advanced Robotics (ICAR 1997)*, Monterey, CA.

Dahmen, C. 2008. Focus-based depth estimation in the SEM. In *SPIE International Symposium on Optomechatronic Technologies*, San Diego, CA.

Division Microrobotics and Control Engineering (AMiR), University of Oldenburg. 2006. Setup for automated handling of micro glass spheres, *Presented at the Hannover Fair International*.

Driesen, W., Varidel, T., Mazerolle, S., Bergander, A., and Breguet, J.-M. 2005a. Flexible micro manipulation platform based on tethered cm-sized mobile micro robots. In *Proceedings of IEEE International Conference on Robotics and Biomimetics (ROBIO)*, Macau, Hong Kong, pp. 145–150.

Driesen, W., Varidel, T., Regnier, S., and Breguet, J.-M. 2005b. Micro manipulation by adhesion with two collaborating mobile microrobots. *Journal of Micromechanics and Microengineering* 15: 259–267.

Ebbesen, T. W. and Ajayan, P. M. 1992. Large-scale synthesis of carbon nanotubes. *Nature* 358: 220–222.

Edeler, C., Jasper, D., and Fatikow, S. 2008. Development, control and evaluation of a mobile platform for microrobots. In *Conference Proceedings of International Federation of Automatic Control*, Seoul, Korea, pp. 12739–12744.

Eichhorn, V., Carlson, K., Andersen, K. N., Fatikow, S., and Bøggild, P. 2007. Nanorobotic manipulation setup for pick-and-place handling and nondestructive characterization of carbon nanotubes. In *Proceedings of IEEE/RSJ International Conference on Intelligent Robots and Systems* (IROS), San Diego, CA, pp. 291–296.

Eichhorn, V., Fatikow, S., Wich, Th. et al. 2008. Depth-detection methods for microgripper based CNT manipulation in a scanning electron microscope. *Journal of Micro-Nano Mechatronics*, 4(1): 27–36. Springer.

Estana, R., Seyfried, J., Schmoeckel, F., Thiel, M., Buerkle, A., and Woern, H. 2004. Exploring the micro- and nanoworld with cubic centimeter-sized autonomous microrobots. *Industrial Robot* 31(2): 159–178.

Fahlbusch, St., Buerkle, A., and Fatikow, S. 1999. Sensor system of a microrobot-based micromanipulation desktop-station. In *International Conference on CAD/CAM, Robotics and Factories of the Future*, Campinas, Brazil, Vol. 2, pp. 1–6.

Fahlbusch, S., Mazerolle, S., Breguet, J. M. et al. 2005. Nanomanipulation in a scanning electron microscope. *Journal of Materials Processing Technology* 167: 371–382.

Fatikow, S. 1996. An automated micromanipulation desktop-station based on mobile piezoelectric microrobots. In *SPIE International Symposium on Intelligent Systems & Advanced Manufacturing 2906: Microrobotics: Components and Applications*, Boston, MA, pp. 66–77.

Fatikow, S. 2000. *Microrobotics and Microassembly*. Stuttgart Leipzig, Germany: Teubner.

Fatikow, S. (Ed.) 2008. *Automated Nanohandling by Microrobots*. ed. S. Fatikow. Springer Series in Advanced Manufacturing. London, U.K.: Springer.

Fatikow, S. and Rembold, U. 1997. *Microsystem Technology and Microrobotics*. Berlin/Heidelberg/New York: Springer.

Fatikow, S., Magnussen, B., and Rembold, U. 1995. A piezoelectric mobile robot for handling of microobjects. In *International Symposium on Microsystems, Intelligent Materials and Robots*, Sendai, Japan, pp. 189–192.

Fatikow, S., Rembold, U., and Wörn, H. 1997. Design and control of flexible microrobots for an automated microassembly desktop-station. In *SPIE International Symposium on Intelligent Systems & Advanced Manufacturing IS02: Micro-robotics and Microsystem Fabrication*, Pittsburgh, PA, pp. 66–77.

Fatikow, S., Munassypov, R., and Rembold, U. 1998. Assembly planning and plan decomposition in an automated micro-robot-based microassembly desktop station. *Journal of Intelligent Manufacturing* 9: 73–92.

Fatikow, S., Seyfried, J., Fahlbusch, St., Buerkle, A., and Schmoeckel, F. 2000. A flexible microrobot-based microassembly station. *Journal of Intelligent and Robotic Systems*, 27: 135–169. Dordrecht, the Netherlands: Kluwer Academic Publishers.

Fatikow, S., Fahlbusch, St., Garnica, St. et al. 2002. Development of a versatile nanohandling station in a scanning electron microscope. In *Third International Workshop on Microfactories, Minneapolis*, Minneapolis, MN, pp. 93–96.

Fatikow, S., Eichhorn, V., Hagemann, S., and Hülsen, H. 2006a. AFM probe-based nanohandling robot station for the characterization of CNTs and biological cells. In *Fifth International Workshop on Microfactories (IWMF)*, Besancon, France.

Fatikow, S., Kray, S., Eichhorn, V., and Tautz, S. 2006b. Development of a nanohandling robot station for nanocharacterization by an AFM probe. In *IEEE Mediterranean Conference on Control and Automation (MED)*, Ancona, Italy.

Fatikow, S., Eichhorn, V., Wich, T., Sievers, T., Hänßler, O., and Norstrom-Andersen, K. 2007a. Depth-detection methods for CNT manipulation and characterization in a scanning electron microscope. In *Proceedings of the IEEE International Conference on Mechatronics and Automation (ICMA 2007)*, Harbin, Heilongjiang, China, pp. 45–50.

Fatikow, S., Wich, Th., Hülsen, H., Sievers, T., and Jähnisch, M. 2007b. Microrobot system for automatic nanohandling inside a scanning electron microscope. *IEEE-ASME Transactions on Mechatronics* 12: 244–252.

Fatikow, S., Wich, Th., Mircea, I. et al. 2008a. Automatic nanohandling station inside a scanning electron microscope. *Proceedings of the Institution of Mechanical Engineers, Part B: Journal of Engineering Manufacture, PEP* 222: 117–128.

Fatikow, S., Eichhorn, V., Stolle, Ch., Sievers, T., and Jähnisch, M. 2008b. Development and control of a versatile nanohandling robot cell. *IFAC Journal on Mechatronics* 18: 370–380.

Fearing, R. S. 1992. A miniature mobile platform on an air bearing. In *Third International Symposium on Micro Machine and Human Science*, Nagoya, Japan, pp. 111–127.

Ferreira, A. and Mavroidis, C. 2006. Virtual reality and haptics for nanorobotics. *IEEE Robotics & Automation Magazine* 13: 78–92.

Ferreira, A., Fontaine, J.-G., and Hirai, S. 2001. Virtual reality-guided microassembly desktop workstation. In *Fifth Japan-France Congress on Mechatronics*, Besançon, France, pp. 454–460.

Fritz, J., Baller, M. K., Lang, H. P. et al. 2000. Translating biomolecular recognition into nanomechanics. *Science* 288: 316–318.

Fujita, H. 1993. Group work of microactuators. In *International IARP-Workshop on Micromachine Technologies and Systems*, Tokyo, Japan, pp. 24–31.

Fukuda, T., Arai, F., and Dong, L. 2003. Assembly of nanodevices with carbon nanotubes through nanorobotic manipulations. *Proceedings of the IEEE* 91: 1803–1818.

Goudail, F. 2004. *Statistical Image Processing Techniques for Noisy Images*. New York: Kluwer Academic/ Plenum Publishers.

Guck, J., Ananthakrishnan, R., Mahmood, H., Moon, T. J., Cunningham, C. C., and Kaes, J. 2001. The optical stretcher: A novel laser tool to micromanipulate cells. *Biophysical Journal* 81: 767–784.

Guo, T., Nikolaev, P., Thess, A., Colbert, D. T., and Smalley, R. E. 1995. Catalytic growth of single-walled nanotubes by laser vaporization. *Chemical Physics Letters* 243: 49–54.

Hagemann, S., Krohs, F., and Fatikow, S. 2007. Automated characterization and manipulation of biological cells by a nanohandling robot station. In *Nanotech Northern Europe (NTNE)*, Helsinki, Finland.

Han, J. 2005. Structures and properties of carbon nanotubes. In *Carbon Nanotubes Science and Applications*, ed. M. Meyyappan. Boca Raton, FL: CRC Press, pp. 1–24.

Harris, P. J. F. 2004. Carbon nanotube composites. *International Materials Reviews* 49(1): 31–43.

Hashiguchi, G., Goda, T., Hosogi, M. et al. 2003. DNA manipulation and retrieval from an aqueous solution with micromachined nanotweezers. *Analytical Chemistry* 75: 4347–4350.

Hatamura, Y., Nakao, M., and Sato, T. 1995. Construction of nano manufacturing world. *Microsystem Technologies* 1: 155–162.

Heckl, W. M. 1997. *Visualization and Nanomanipulation of Molecules in the Scanning Tunnelling Microscope. Pioneering Ideas for the Physical and Chemical Sciences*. New York: Plenum Press.

Hesselbach, J., Pittschellis, R., and Thoben, R. 1997. Robots and grippers for micro assembly. In *Ninth International Precision Engineering Seminar*, Braunschweig, Germany, pp. 375–378.

Holmlin, R. E., Schiavoni, M., Chen, C. Y., Smith, S. P., Prentiss, M. G., and Whitesides, G. M. 2000. Light-driven microfabrication: Assembly of multicomponent, three-dimensional structures by using optical tweezers. *Angewandte Chemie-International Edition* 39: 3503–3506.

Hülsen, H. 2007. Self-organising locally interpolating maps in control engineering. PhD thesis, University of Oldenburg, Germany.

Hülsen, H., Trüper, T., Kortschack, A., Jähnisch, M., and Fatikow, S. 2004. Control system for the automatic handling of biological cells with mobile microrobots. In *American Control Conference*, Boston, MA, pp. 3986–3991.

Hydromel 2006. Hydromel—Integrated Project under the Sixth Framework Programme of the European Community (2002–2006), < NMP-2-CT-2006-026622 >.

Iijima, S. 1991. Helical microtubules of graphitic carbon. *Nature* 354: 56–58.

Jähnisch, M. 2008a. 3D imaging systems for SEM. In *Automated Nanohandling by Microrobots*, ed. S. Fatikow. Springer Series in Advanced Manufacturing. London, U.K.: Springer, pp. 129–165.

Jähnisch, M. 2008b. 3D-Bildsystem für die Nanohandhabung im Rasterelektronenmikroskop. PhD thesis, University of Oldenburg, Germany.

Jähnisch, M. and Fatikow, S. 2007. 3D vision feedback for nanohandling monitoring in a scanning electron microscope. *International Journal of Optomechatronics*, 1(1): 4–26. *Taylor & Francis*.

Jasper, D. and Edeler, C. 2008. Characterization, optimization and control of a mobile platform. In *International Workshop on Microfactories, IWMF 2008*, Evanston, IL.

Jasper, D., Dahmen, Ch., and Fatikow, S. 2007. CameraMan—robot cell with flexible vision feedback for automated nanohandling inside SEM. In *Third IEEE Conference on Automation Science and Engineering*, Scottsdale, AZ, pp. 51–56.

Joachimsthaler, J., Heiderhoff, R., and Balk, L. J. 2003. A universal scanning-probe-microscope-based hybrid system. *Measurement Science and Technology* 14: 87–96.

Kamat, P. V., Thomas, K. G., Barazzouk, S., Girishkumar, G., Vinodgopal, K., and Meisel, D. 2004. Self-assembled linear bundles of single wall carbon nanotubes and their alignment and deposition as a film in a dc field. *Journal of the American Chemical Society* 126: 10757–10762.

Kasaya, T., Miyazaki, H., Saito, S., and Sato, T. 1999. Micro object handling under SEM by vision-based automatic control. In *International Conference on Robotics and Automation*, Detroit, MI, pp. 2736–2743.

Kim, P. and Lieber, C. M. 1999. Nanotube nanotweezers. *Science* 286: 2148–2150.

Kim, D.-H., Kim, B., and Kang, H. 2004. Development of a piezoelectric polymer-based sensorized microgripper for microassembly and micromanipulation. *Microsystem Technologies* 10: 275–280.

Kim, K., Liu, X., Zhang, Y., and Sun, Y. 2008. Nanonewton force-controlled manipulation of biological cells using a monolithic MEMS microgripper with two-axis force feedback. *Journal of Micromechanics and Microengineering* 18: 55013–55021.

Kortschack, A. and Fatikow, S. 2004. Development of a mobile nanohandling robot. *Journal of Micromechatronics* 2(3): 249–269.

Kratochvil, B. E., Dong, L. X., and Nelson, B. J. 2007. Real-time rigid-body visual tracking in a scanning electron microscope. In *Proceedings of the 7th IEEE Conference on Nanotechnology (IEEE-NANO2007)*, Hong Kong, China.

Kreupl, F., Graham, A. P., Duesberg, G. S., Steinhögl, W., Liebau, M., Unger, E., and Hönlein, W. 2002. Carbon nanotubes in interconnect applications. *Microelectronic Engineering* 64: 399–408.

Krohs, F., Hagemann, S., and Fatikow, S. 2007. Automated cell characterization by a nanohandling robot station. In *IEEE Mediterranean Conference on Control and Automation (MED)*, Athens, Greece.

Krohs, F., Luttermann, T., Stolle, Ch., Fatikow, S., Brousseau, E., and Dimov, St. 2008. Towards automation in AFM based nanomanipulation and electron beam induced deposition for microstructuring. In *Multi-Material Micro Manufacture*, eds. S. Dimov and W. Menz. Cardiff, U.K.: Whittles Publishing, pp. 118–123.

Krüger, A. 2007. *Neue Kohlenstoffmaterialien*. Wiesbaden, Germany: Teubner Verlag.

Kufer, S. K., Puchner, E. M., Gumpp, H., Liedl, T., and Gaub, H. E. 2008. Single-molecule cut-and-paste surface assembly. *Science* 319: 594–596.

Kuznetsov, Y., Gershon, P. D., and McPherson, A. 2008. Atomic force microscopy investigation of vaccinia virus structure. *Journal of Virology* 82: 7551–7566.

Lal, R. and John, S. A. 1994. Biological applications of atomic force microscopy. *American Journal of Physiology—Cell Physiology* 266: c1–c21.

Laney, D. E., Garcia, R. A., Parsons, S. M., and Hansma, H. G. 1997. Changes in the elastic properties of cholinergic synaptic vesicles as measured by atomic force microscopy. *Biophysical Journal* 72: 806–813.

Layton, B. E., Sullivan, S. M., Palermo, J. J., Buzby, G. J., Gupta, R., and Stallcup III, R. E. 2005. Nanomanipulation and aggregation limitations of self-assembling structural proteins. *Microelectronics Journal* 36: 644–649.

Li, G. Y., Xi, N., Chen, H. P., Pomeroy, C., and Prokos, M. 2005. Videolized AFM for interactive nanomanipulation and nano-assembly. *IEEE Transactions on Nanotechnology* 4: 605–615.

Lü, J., Li, H., An, H. et al. 2004. Positioning isolation and biochemical analysis of single DNA molecules based on nanomanipulation and single-molecule PCR. *Journal of the American Chemical Society* 126: 11136–11137.

Luttermann, T., Wich, Th., Stolle, Ch., and Fatikow S. 2007. Development of an automated desktop station for EBiD-based nano-assembly. In *Second International Conference on Micro-Manufacturing*, Greenville, SC, pp. 284–288.

Magnussen, B., Fatikow, S., and Rembold, U. 1995. Actuation in microsystems: Problem field overview and practical example of the piezoelectric robot for handling of microobjects. In *INRIA/IEEE Conference on Emerging Technologies and Factory Automation*, Paris, France, Vol. 3, pp. 21–27.

Makaliwe, J. H. and Requicha, A. A. G. 2001. Automatic planning of nanoparticle assembly tasks. In *IEEE International Symposium on Assembly and Task Planning*, Fukuoka, Japan, pp. 288–293.

Martel S. 2005. Fundamental principles and issues of high-speed piezoactuated three-legged motion for miniature robots designed for nanometer-scale operations. *International Journal of Robotics Research* 24: 575–588.

Mazerolle, S., Rabe, R., Fahlbusch, S., Michler, J., and Breguet, J.-M. 2004. High precision robotics system for scanning electron microscopes. *Proceedings of the IWMF* 1: 17–22.

Menciassi, A., Carozza, M. C., Ristori, C., Tiezzi, G., and Dario, P. 1997. A workstation for manipulation of micro objects. In *IEEE International Conference on Advanced Robotics*, Monterey, CA, pp. 253–258.

Mestres, P., Pütz, N., and Laue M. 2007. Consequences of tilting of biological specimens in wet mode ESEM imaging. *Microscopy and Microanalysis* 13: 244–245.

Meyer, E., Jarvis, S. P., and Spencer, N. D. 2004. Scanning probe microscopy in materials science. *MRS Bulletin* 29: 443–445.

Mircea, J., Fatikow, S., and Sill, A. 2007. Microrobot-based nano-indentation of an epoxy-based electrically conductive adhesive. In *IEEE NANO-Conference*, Hong Kong, pp. 719–722.

Misaki, D., Kayano, S., Wakikaido, Y., Fuchiwaki, O., and Aoyama, H. 2004. Precise automatic guiding and positioning of microrobots with a fine tool for microscopic operations. In *IEEE/RSJ International Conference on Intelligent Robots and Systems*, Sendai, Japan, pp. 218–223.

Mokaberi, B. and Requicha, A. A. G. 2006. Drift compensation for automatic nanomanipulation with scanning probe microscopes. *IEEE Transactions on Automation Science and Engineering* 3: 199–207.

Mokaberi, B., Yun, J., Wang, M., and Requicha, A. A. G. 2007. Automated nanomanipulation with atomic force microscopes. In *Proceedings IEEE International Conference on Robotics & Automation*, Rome, Italy.

Mølhave, K., Wich, T., Kortschack, A., and Bøggild, P. 2006. Pick-and-place nanomanipulation using microfabricated grippers. *Nanotechnology* 17: 2434–2441.

Morishita, H. and Hatamura, Y. 1993. Development of ultra precise manipulator system for future nanotechnology. In *International IARP Workshop on Micro Robotics and Systems*, Karlsruhe, Germany, pp. 34–42.

Morris, C. J., Stauth, S. A., and Parviz, B. A. 2005. Self-assembly for microscale and nanoscale packaging: Steps toward self-packaging. *IEEE Transactions on Advanced Packaging* 28: 600–611.

Muscariello, L., Rosso, F., Marino, G. et al. 2005. A critical overview of ESEM applications in the biological field. *Journal of Cellular Physiology* 205: 328–334.

Nagahara, L. A., Amlani, I., Lewenstein, J., and Tsui, R. K. 2002. Directed placement of suspended carbon nanotubes for nanometer-scale assembly. *Applied Physics Letters* 80: 3826–3829.

Nakajima, M., Arai, F., Dong, L., Nagai, M., and Fukuda, T. 2004. Hybrid nanorobotic manipulation system inside scanning electron microscope and transmission electron microscope. In *IEEE/RSJ International Conference on Intelligent Robots and Systems*, Sendai, Japan, pp. 589–594.

Nakajima, M., Arai, F., and Fukuda, T. 2006. In situ measurement of Young's modulus of carbon nanotubes inside a TEM through a hybrid nanorobotic manipulation system. *IEEE Transactions on Nanotechnology* 5(3): 243–248.

Pablo, P. J. de, Moreno-Herrero, F., Colchero, J. et al. 2000. Absence of dc-conductivity in λ-DNA. *Physical Review Letters* 85: 4992–4995.

Pantano, A., Parks, D. M., and Boyce, M. C. 2004. Mechanics of deformation of single and multiwalled carbon nanotubes. *Journal of the Mechanics and Physics of Solids* 52: 789–821.

Parra, A., Casero, E., Lorenzo, E., Pariente, F., and Vázquez, L. 2007. Nanomechanical properties of globular proteins: Lactate oxidase. *Langmuir* 23: 2747–2754.

Parviz, B. A., Ryan, D., and Whitesides, G. M. 2003. Using self-assembly for the fabrication of nano-scale electronic and photonic devices. *IEEE Transactions on Advanced Packaging* 26: 233–241.

Patel-Predd, P. 2008. Update: Carbon-nanotube wiring gets real. *Spectrum IEEE* 45: 14.

Ramachandran, T. R., Baur, C., Bugacov, A. et al. 1998. Direct and controlled manipulation of nanometer-sized particles using the non-contact atomic force microscope. *Nanotechnology* 9: 237–245.

Rao, S. G., Huang, L., Setyawan, W., and Hong, S. 2003. Nanotube electronics: Large-scale assembly of carbon nanotubes. *Nature* 425: 36–37.

Regtmeier, J., Duong, T. T., Eichhorn, R., Anselmetti, D., and Ros, A. 2007. Dielectrophoretic manipulation of DNA: Separation and polarizability. *Analytical Chemistry* 79: 3925–3932.

Reimer, L. 1998. *Scanning Electron Microscopy—Physics of Image Formation and Microanalysis*, 2nd edn, Springer Series in Optical Sciences, Vol. 45. New York: Springer.

Rembold, U. and Fatikow, S. 1997. Autonomous microrobots. *Journal of Intelligent and Robotic Systems* 19: 375–391.

Reynolds, K., Komulainen, J., Kivijakola, J. et al. 2008. Probe based manipulation and assembly of nanowires into organized mesostructures. *Nanotechnology* 19: 485301.

Robertson, J. 2004. Realistic applications for CNTs. *Materials Today* 7: 46–52.

Rubio-Sierra, J., Heckl, W., and Stark, R. W. 2005. Nanomanipulation by atomic force microscopy. *Advanced Engineering Materials* 7: 193–196.

Ruiz, I., Wang, P., Schaffer, C., and Kleinfeld, D. 2003. Optical trapping and ablation. Neurophysics laboratory final report PHYS 173/BGGN 266 Lab. University of California, San Diego, CA.

Saito, S., Miyazaki, H. T., Sato, T., Takahashi, K., and Onzawa, T. 2001. Dynamics of micro-object operation considering the adhesive effect under an SEM. *Microrobotics and Microassembly III* 4568(1): 12–23.

Sardan, O., Alaca, B. E., Yalcinkaya, A. D., Bøggild, P., Tang, P. T., and Hansen, O. 2007. Microgrippers: A case study for batch-compatible integration of MEMS with nanostructures. *Nanotechnology* 18: 375501–375512.

Sardan, O., Eichhorn, V., Petersen, D. H., Fatikow, S., Sigmund, O., and Bøggild, P. 2008. Rapid prototyping of nanotube-based devices using topology-optimized microgrippers. *Nanotechnology* 19: 495503.

Schaap, I., Carrasco, C., de Pablo, P. J., MacKintosh, F. C., and Schmidt, C. F. 2006. Elastic response, buckling, and instability of microtubules under radial indentation. *Biophysical Journal* 91: 1521–1531.

Schimmel, Th., von Blanckenhagen, P., and Schommers, W. 1999. Nanometer-scale structuring by application of scanning probe microscopes and self-organisation processes. *Applied Physics A* 68: 263.

Schitter, G., Stark, R. W., and Stemmer A. 2004. Fast contact-mode atomic force microscopy on biological specimen by model-based control. *Ultramicroscopy* 100: 253–257.

Schmoeckel, F., Fahlbusch, St., Seyfried, J., Buerkle, A., and Fatikow, S. 2000. Development of a microrobot-based micromanipulation cell in an SEM. In *SPIE International Symposium on Intelligent Systems & Advanced Manufacturing: Conference on Microrobotics and Microassembly*, Boston, MA, pp. 129–140.

Seidel, R., Colombi Ciacchi, L., Weigel, M., Pompe, W., and Mertig M. 2004. Synthesis of platinum cluster chains on DNA templates: Conditions for a template-controlled cluster growth. *Journal of Physical Chemistry B* 108: 10801–10811.

Sievers, T. 2006. Global sensor feedback for automatic nanohandling inside a scanning electron microscope. In *Proceedings of IPROMS NoE Virtual International Conference on Intelligent Production Machines and Systems*, pp. 289–294.

Sievers, T. 2007. Echtzeit-Objektverfolgung im Rasterelektronenmikroskop. PhD thesis, University of Oldenburg, Germany.

Sievers, T. and Fatikow, S. 2006. Real-time object tracking for the robot-based nanohandling in a scanning electron microscope. *Journal of Micromechatronics* 3(3): 267–284.

Sinclair, G., Jordan, P., Laczik, J., Courtial, J., and Padgett, M. 2004. Semi-automated 3-dimensional assembly of multiple objects using holographic optical tweezers. *SPIE's Optical Trapping and Optical Micromanipulation* 5514: 137–142.

Sitti, M. 2003. Teleoperated and automatic nanomanipulation systems using atomic force microscope probes. *IEEE Conference on Decision and Control*, Maui, Hawaii.

Sitti, M. 2007. Microscale and nanoscale robotics systems [Grand Challenges of Robotics]. *IEEE Robotics Automation Magazine* 14: 53–60.

Sitti, M. and Hashimoto, H. 2000. Two-dimensional fine particle positioning under optical microscope using a piezoresistive cantilever as a manipulator. *Journal of Micromechatronics* 1: 25–48.

Solano, B. and Wood, D. 2007. Design and testing of a polymeric microgripper for cell manipulation. *Microelectronic Engineering* 84: 1219–1222.

Stark, R. W., Rubio-Sierra, J., Thalhammer, S., and Heckl, W. 2003. Combined nanomanipulation by atomic force microscopy and UV-laser ablation for chromosomal dissection. *European Biophysics Journal* 32: 33–39.

Stokes, D. 2003. Low vacuum & ESEM imaging of biological specimens. *Microscopy and Microanalysis* 9: 190–191.

Stolle, Ch. and Fatikow, S. 2008. Towards automated nanohandling in a scanning electron microscope. In *Proceedings of the 6th IEEE Conference on Industrial Informatics*, Daejeon, Korea, pp. 160–165.

Sun, Y. and Nelson, B. J. 2002. Biological cell injection using an autonomous microRobotic system. *The International Journal of Robotics Research* 21: 861–868.

Svoboda, K. and Block, S. M. 1994. Biological applications of optical forces. *Annual Reviews Biophysical Biomolecular Structures* 23: 247–285.

Tai, S. S. W. and Tang, X. M. 2001. Manipulating biological samples for environmental scanning electron microscopy observation. *Scanning* 23: 267–272.

Tai, C.-H., Hsiung, S.-K., Chen, C.-Y., Tsai, M.-L., and Lee, G.-B. 2007. Automatic microfluidic platform for cell separation and nucleus collection. *Biomedical Microdevices* 9: 533–545.

Tanikawa, T., Kawai, M., Koyachi, N. et al. 2001. Force control system for autonomous micromanipulation. In *International Conference on Robotics and Automation*, Seoul, South Korea, pp. 610–615.

Teo, K. B. K., Chhowalla, M., Amaratunga, G. A. J., Milne, W. I., Hasko, D. G., Pirio, G., Legagneux, P., Wyczisk, F., and Pribat, D. 2001. Uniform patterned growth of carbon nanotubes without surface carbon. *Applied Physics Letters* 79(10): 1534–1536.

Thompson, J. A. and Fearing, R. S. 2001. Automating microassembly with ortho-tweezers and force sensing. In *Proceedings of the 2001 IEEE/RSJ International Conference on Intelligent Robots and Systems*, Wailea, HI.

Thostenson, E. T., Ren, Z., and Chou, T.-W. 2001. Advances in the science and technology of carbon nanotubes and their composites: A review. *Composites Science and Technologies* 61: 1899–1912.

Treacy, M., Ebbesen, T., and Gibson, J. 1996. Exceptionally high Young's modulus observed for individual carbon nanotubes. *Nature* 381: 678–681.

Troyton, M., Lei, H. N., Wang, Z., and Shang, G. 1998. A scanning force microscope combined with a scanning electron microscope for multidimensional data analysis. *Scanning Microscopy* 12: 139–148.

Trüper, T., Kortschack, A., Jähnisch, M., Hülsen, H., and Fatikow, S. 2004. Transporting cells with mobile microrobots. *IEE Proceedings—Nanobiotechnology* 151: 145–150.

Tuukkanen, S., Kuzyk, A., Toppari, J. J. et al. 2007. Trapping of 27 bp-8 kbp DNA and immobilization of thiol-modified DNA using dielectrophoresis. *Nanotechnology* 18: 295204–295214.

Utke, I., Hoffmann, P., and Melngailis, J. 2008. Gas-assisted focused electron beam and ion beam processing and fabrication. *Journal of Vacuum Science and Technology. B*, 26(4): 1197–1276.

Vikramaditya, B. and Nelson, B. J. 1997. Visually guided microassembly using optical microscopes and active vision techniques. In *Proceeding of the 1997 IEEE International Conference on Robotics and Automation*, Albequerque, NM.

Villarroya, M., Perez-Murano, F., Martin, C. et al. 2004. AFM lithography for the definition of nanometre scale gaps: Application to the fabrication of a cantilever-based sensor with electrochemical current detection. *Nanotechnology* 15: 771–776.

Wang, X.-B., Yang, J., Huang, Y., Vykoukal, J., Becker, F. F., and Gascoyne, P. R. C. 2000. Cell separation by dielectrophoretic field-flow-fractionation. *Analytical Chemistry* 72: 832–839.

Wang, Z., Poncharal, R. G. P., de Heer, W., Dai, Z., and Pan, Z. 2001. Mechanical and electrostatic properties of carbon nanotubes and nanowires. *Materials Science and Engineering* 16: 3–10.

Weigel-Jech, M., Hagemann, S., and Fatikow, S. 2008. Development of a nanostation for manipulation and characterization of biomaterials to support sensor development in bioNanotechnology. In *Proceedings of the International Conference on Nanosensors for Industrial Applications*, Vienna, Austria.

Wich, T. and Fatikow, S. 2008. Electron beam induced processing for nanomanufacturing. In *Proceedings of the 3rd International Conference on Micromanufacturing (ICOMM)*, Pittsburgh, PA, pp. 92–97.

Wich, T. and Hülsen, H. 2008. Robot-based automated nanohandling. In *Automated Nanohandling by Microrobots*, ed. S. Fatikow, Springer Series in Advanced Manufacturing. London, U.K.: Springer, pp. 24–56.

Wich, Th., Sievers, T., Jähnisch, M., Hülsen, H., and Fatikow, S. 2005. Nanohandling automation within a scanning electron microscope. In *IEEE International Symposium on Industrial Electronics*, Dubrovnik, Croatia, pp. 1073–1078.

Wich, T., Sievers, T., and Fatikow, S. 2006a. Assembly inside a scanning electron microscope using electron beam induced deposition. In *Proceedings of the IEEE/RSJ International Conference on Intelligent Robots and Systems*, Beijing, China.

Wich, Th., Sievers, T., and Fatikow, S. 2006b. Assembly inside a scanning electron microscope using electron beam induced deposition. In *IEEE International Conference on Intelligent Robots and Systems*, Beijing, China, pp. 294–299.

Wich, T., Stolle, C., Frick, O., and Fatikow, S. 2008. Automated nano-assembly in the SEM I: Challenges in setting up a warehouse. In *Proceedings of the 17th International Federation of Automatic Control World Congress (IFAC)*, Seoul, South Korea.

Wright, W. H., Sonek, G. J., Tadir, Y., and Berns, M. W. 1990. Laser trapping in cell biology. *IEEE Journal of Quantum Electronics* 26: 2148–2157.

Wright, W. H., Sonek, G. J., and Berns, M. W. 1993. Radiation trapping forces on microspheres with optical tweezers. *Applied Physics Letters* 63: 715–717.

Yamamoto, K., Akita, S., and Nakayama, Y. 1996. Orientation of carbon nanotubes using electrophoresis. *Japanese Journal of Applied Physics* 35: L917–L918.

Yang, J., Huang, Y., Wang, X.-B., Becker, F. F., and Gascoyne, P. R. C. 1999. Cell separation on microfabricated electrodes using dielectrophoretic/gravitational field-flow fractionation. *Analytical Chemistry* 71: 911–918.

Yang, J., Huang, Y., Wang, X.-B., Becker, F. F., and Gascoyne, P. R. C. 2000. Differential analysis of human leukocytes by dielectrophoretic field-flow-fractionation. *Biophysical Journal* 78: 2680–2689.

Yang, G., Gaines, J. A., and Nelson, B. J. 2001. A flexible experimental workcell for efficient and reliable wafer-level 3D micro-assembly. In *Proceedings of the IEEE International Conference on Robotics and Automation (ICRA)*, Seoul, South Korea.

Yang, G., Gaines, J. A., and Nelson, B. J. 2003. A supervisory wafer-level 3D micro-assembly system for hybrid MEMS fabrication. *Journal of Intelligent and Robotic Systems* 37: 43–68.

Yu, Y. 2003. Introduction to probing DNA with optical tweezers. *Introduction to Biophysics—Term Paper*.

Yu, T., Cheong, F.-C., and Sow, C.-H. 2004. The manipulation and assembly of CuO nanorods with line optical tweezers. *Nanotechnology* 15: 1732–1736.

Index